HANDBOOK OF

GRID GENERATION

HANDBOOK OF GRID GENERATION

Edited by

Joe F. Thompson
Bharat K. Soni
Nigel P. Weatherill

CRC Press
Boca Raton London New York Washington, D.C.

Acquiring Editor: B. Stern
Project Editor: Sylvia Wood
Marketing Manager: J. Stark
Cover design: Dawn Boyd
Manufacturing Manager: Carol Slatter

Library of Congress Cataloging-in-Publication Data

Thompson, Joe F.
Handbook of Grid Generation / Joe F. Thompson, Bharat Soni, Nigel Weatherill, editors.
p. cm.
Includes bibliographical references and index.
ISBN 0-8493-2687-7 (alk. paper)
1. Numerical grid generation (Numerical analysis) I. Thompson, Joe F. II. Soni, B.K. III. Weatherill, N.P.
QA377.H3183 1998
519.4--dc21
98-34260
CIP

International Standard Book Number 0-8493-2687-7
Library of Congress Card Number 98-34260
Printed in the United States of America 1 2 3 4 5 6 7 8 9 0
Printed on acid-free paper

Foreword

Grid (mesh) generation is, of course, only a means to an end: a necessary tool in the computational simulation of physical field phenomena and processes. (The terms *grid* and *mesh* are used interchangeably, with identical meaning, throughout this handbook.)

And grid generation is, unfortunately from a technology standpoint, still something of an art, as well as a science. Mathematics provides the essential foundation for moving the grid generation process from a user-intensive craft to an automated system. But there is both art and science in the design of the mathematics for — not of — grid generation systems, since there are no inherent laws (equations) of grid generation to be discovered. The grid generation process is not unique; rather it must be designed. There are, however, criteria of optimization that can serve to guide this design.

The grid generation process has matured now to the point where the use of developed codes — freeware and commercial — is generally to be recommended over the construction of grid generation codes by end users doing computational field simulation. Some understanding of the process of grid generation — and its underlying principles, mathematics, and technology — is important, however, for informed and effective use of these developed systems. And there are always extensions and enhancements to be made to meet new occasions, especially in coupling the grid with the solution process thereon.

This handbook is designed to provide essential grid generation technology for practice, with sufficient detail and development for general understanding by the informed practitioner. Complete details for the grid generation specialist are left to the sources cited. A basic introduction to the fundamental concepts and approaches is provided by Chapter l, which covers the state of practice in the entire field in a very broad sweep. An even more basic introduction for those with little familiarity with the subject is given by the Preface that precedes this first chapter. Appendixes provide information on a number of available grid generation codes, both commercial and freeware, and give some representative and illustrative grid configurations.

The grid generation process in general proceeds from first defining the boundary geometry as discussed in Part III. Points are distributed on the curves that form the edges of boundary sections. A surface grid is then generated on the boundary surface, and finally, a volume grid is generated in the field. Chapter 13, although directed at structured grids, gives a general overview of the entire grid generation process and the fundamental choices and considerations involved from the standpoint of the user. Chapter 2, though also largely directed at structured grids, covers essential mathematical elements from tensor analysis and differential geometry relevant to the entire subject, particularly the aspects of curve and surfaces.

The other chapters of this handbook cover the various aspects of grid generation in some detail, but still from the standpoint of practice, with citations of relevant sources for more complete discussion of the underlying technology. The chapters are grouped into four parts: structured grids, unstructured grids, surface definition, and adaptation/quality. An introduction to each part provides a road map through the material of the chapters.

A source of fundamentals on structured grid generation is the 1985 textbook of Thompson, Warsi, and Mastin, now out of print but accessible on the Web at www.erc.msstate.edu. A recent comprehensive text of both structured and unstructured grids is that of Carey 1997 from Taylor and Francis publishers.

The first step in generating a grid is, of course, to acquire and input the boundary data. This boundary data may be in the form of output from a CAD system, or may simply be sets of boundary points acquired from drawings. CAD boundary data are generally in the form of some parametric description of boundary curves and surfaces, typically consisting of multiple segments for which assembly and some adjustments may be required. Point boundary data may be in the form of 1D arrays of points describing boundary curves and 2D arrays for boundary surfaces, or could be an unorganized cloud of points on a surface. In the latter case, conversion to some surface tessellation or parametric description is required. These initial steps of boundary definition are common in general to both structured and unstructured grid generation. And, unfortunately, considerable human intervention may be necessary in this setup phase of the process.

The setup of the boundary definition from the CAD approach is discussed in general in Chapter 13, while details of application, together with procedures for boundary curve and surface parametric representations, are covered in Part III. There is then the fundamental choice of whether to use a structured or unstructured grid. Structured grids are covered in Part I, and unstructured grids are covered in Part II.

The next step with either type of grid is the generation of the corresponding type of grid on the boundary surfaces — preceded, of course, by a distribution on points on the curves that form the edges of these surfaces. This surface grid generation is covered in Chapters 9 and 19 for structured and unstructured grids, respectively.

Finally, the quality of the grid, with relation to the accuracy of the numerical solution being done on the grid, and the adaptation of the grid to improve that accuracy are covered in Part IV.

Grid generation is still under active research and development, particularly in regard to automation, adaptation, and hybrid combinations. This handbook is therefore necessarily a snapshot in time, especially in these areas, but much of the material has matured now, and this collection should be of enduring value as a source and reference.

Bharat K. Soni
Joe F. Thompson
Nigel P. Weatherill

Starkville, MS, and Swansea, Wales, UK

Contributors

Michael J. Aftosmis
NASA Ames Research Center
Moffett Field, CA

Timothy J. Baker
Princeton University
Princeton, NJ

Mark W. Beall
Rensselaer Polytechnic Institute
Troy, NY

Marsha J. Berger
Courant Institute
New York University

William M. Chan
MCAT, Inc. at NASA Ames Research Center
Moffett Field

Zheming Cheng
Program Development Corporation
White Plains, NY

Hugues L. de Cougny
Rensselaer Polytechnic Institute
Troy, NY

Luís Eça
Technical University of Lisbon
Lisbon, Portugal

Peter R. Eiseman
Program Development Corporation
White Plains, NY

Austin L. Evans
NASA Lewis Research Center
Cleveland, OH

Gerald Farin
Arizona State University
Tempe, AZ

David R. Ferguson
The Boeing Company
Seattle, WA

Luca Formaggia
Ecole Polytechnique Federale de Lausanne
Lausanne, Switzerland

Timothy Gatzke
The Boeing Company
St. Louis, MO

Paul-Louis George
INRIA
Le Chesnay Cedex, France

Bernd Hammann
University of California at Davis
Davis, CA

O. Hassan
University of Wales Swansea
Swansea, UK

Jochem Häuser
CLE Salzgitter Bad
Salzgitter, Germany

Frédéric Hecht
INRIA
Le Chesnay Cedex, France

Sergey A. Ivanenko
Computer Center of the Russian Academy of Sciences
Moscow, Russia

Olivier-Pierre Jacquotte
Research Directorate (DRET)
Paris, France

Brian A. Jean
U.S. Army Corps of Engineers Waterways Experiment Station
Vicksburg, MS

Yannis Kallinderis
University of Texas
Austin, TX

O.B. Khairullina
Urals Branch of the Russian Academy of Sciences
Ekaterinburg, Russia

Ahmed Khamayseh
Los Alamos National Laboratory
Los Alamos, NM

Andrew Kuprat
Los Alamos National Laboratory
Los Alamos, NM

Kelly R. Laflin
North Carolina State University
Raleigh, NC

Kunwoo Lee
Seoul National University
Seoul, Korea

David L. Marcum
Mississippi State University
Starkville, MS

C. Wayne Mastin
Nichols Research Corporation
Vicksburg, MS

D. Scott McRae
North Carolina State University
Raleigh, NC

Robert L. Meakin
Army Aeroflightdynamics Directorate (AMCOM)
Moffett Field, CA

John E. Melton
NASA Ames Research Center
Moffett Field, CA

David P. Miller
NASA Lewis Research Center
Cleveland, OH

K. Morgan
University of Wales Swansea
Swansea, UK

Robert M. O'Bara
Rensselaer Polytechnic Institute
Troy, NY

Sangkun Park
Information Technology R&D Center
Seoul, Korea

J. Peraire
Massachusetts Institute of Technology
Cambridge, MA

J. Peiró
Imperial College
London, UK

E. J. Probert
University of Wales Swansea
Swansea, UK

Anshuman Razdan
Arizona State University
Tempe, AZ

Robert Schneiders
MAGMA Giessereitechnologie GmbH
Aachen, Germany

Jonathon A. Shaw
Aircraft Research Association
Bedford, U.K.

A.F. Sidorov
Urals Branch of the Russian Academy of Sciences
Ekaterinburg, Russia

Mark S. Shephard
Rensselaer Polytechnic Institute
Troy, NY

Robert E. Smith
NASA Langley Research Center
Hampton, VA

Bharat K. Soni
Mississippi State University
Starkville, MS

Stefan P. Spekreijse
National Aerospace Laboratory (NLR)
Emmeloord, The Netherlands

Joe F. Thompson
Mississippi State University
Starkville, MS

O.V. Ushakova
Urals Branch of the Russian Academy of Sciences
Ekaterinburg, Russia

Zahir U.A. Warsi
Mississippi State University
Starkville, MS

Nigel P. Weatherill
University of Wales Swansea
Swansea, UK

Yang Xia
CLE Salzgitter Bad
Germany

Tzu-Yi Yu
Chaoyang University of Technology
Wufeng, Taiwan

Paul A. Zegeling
University of Utrecht
Utrecht, The Netherlands

Acknowledgments

Grid (mesh) generation is truly a worldwide active research area of computation science, and this handbook is the work of individual authors from around the world. It has been a distinct pleasure, and an opportunity for professional enhancement, to work with these dedicated researchers in the course of the preparation of this book over the past two years. The material comes from universities, industry, and government laboratories in 10 countries in North America, Europe, and Asia. And we three are from three different countries of origin, though we have collaborated for years.

The attention to quality that has been the norm in the authoring of these many chapters has made our editing responsibility a straightforward process. These chapters should serve well to present the current state of the art in grid generation to practitioners, researchers, and students.

The assembly and editing of the material for this handbook from all over the world via the Internet has been a rewarding experience in its own right, and speaks well for the potential for worldwide collaborative efforts in research.

Our thanks go to Mississippi State University and the University of Wales Swansea for the encouragement and support of our efforts to produce this handbook. Specifically at Mississippi State, the work of Roger Smith in administering the electronic communication is to be noted, as are the efforts of Alisha Davis, who handled the word processing.

Bob Stern of CRC Press has been great to work with and appreciation is due to him for recognizing the need for this handbook and for his editorial guidance and assistance throughout its preparation. His efforts, and those of Sylvia Wood, Suzanne Lassandro and Dawn Mesa, also at CRC, have made this a pleasant process.

We naturally are especially grateful for the support of our wives, Purnima, Emilie, Barbara, and our families in this and all our efforts. And finally, Mississippi and Wales — two great places to live and work.

Bharat K. Soni
Joe F. Thompson
Nigel P. Weatherill
Author/Editors

Preface: An Elementary Introduction

Joe F. Thompson, Bharat K. Soni, and Nigel P. Weatherill

This first section is an elementary introduction provided for those with little familiarity with grid (mesh) generation in order to establish a base from which the technical development of the chapters in this handbook can proceed. (The terms *grid* and *mesh* are used interchangeably throughout with identical meaning.) The intent is not to introduce numerical solution procedures, but rather to introduce the idea of using numerically generated grid (mesh) systems as a foundation of such solutions.

P-1 Discretizations

The numerical solution of partial differential equations (PDEs) requires first the discretization of the equations to replace the continuous differential equations with a system of simultaneous algebraic difference equations. There are several bifurcations on the way to the construction of the solution process, the first of which concerns whether to represent the differential equations at discrete points or over discrete cells. The discretization is accomplished by covering the solution field with discrete points that can, of course, be connected in various manners to form a network of discrete cells. The choice lies in whether to represent the differential equations at the discrete points or over the discrete cells.

P-1.1 Point Discretization

In the former case (called *finite difference*), the derivatives in the PDEs are represented at the points by algebraic difference expressions obtained by performing Taylor series expansions of the solution variables at several neighbors of the point of evaluation. This amounts to taking the solution to be represented by polynomials between the points. This can be unrealistic if the solution varies too strongly between the points. One remedy is, of course, to use more points so that the spacing between points is reduced. This, however, can be expensive, since there will then be more points at which the equations must be evaluated.

This is exacerbated if the points are equally spaced and strong variations in the solution occur over scattered regions of the field, since numerous points will be wasted in regions of small variation. An alternative, of course, is to make the points unequally spaced.

P-1.2 Cell Discretization

The other possibility of this first bifurcation is to return the PDEs to their more fundamental integral form and then to represent the integrals over discrete cells. Here there is yet another bifurcation — whether to represent the solution variables over the cell in terms of selected functions and then to integrate

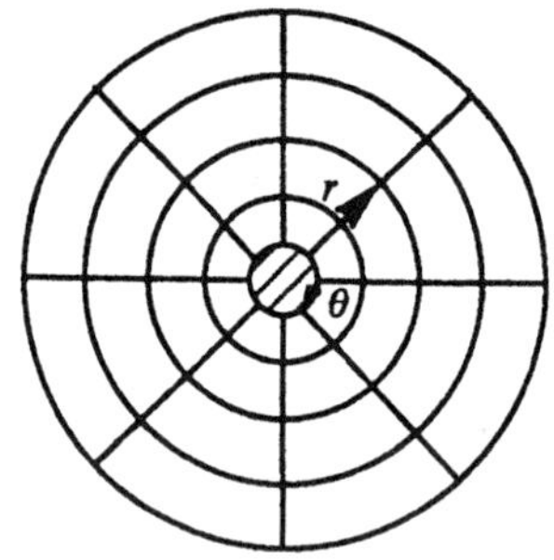

FIGURE 1

these functions analytically over the volume (*finite element*), or to balance the fluxes through the cell sides (*finite volume*).

The finite element approach itself comes in two basic forms: the *variational*, where the PDEs are replaced by a more fundamental integral variational principle (from which they arise through the calculus of variations), or the *weighted residual* (Galerkin) approach, in which the PDEs are multiplied by certain functions and then integrated over the cell.

In the finite volume approach, the solution variables are considered to be constant within a cell, and the fluxes through the cell sides (which separate discontinuous solution values) are best calculated with a procedure that represents the dissolution of such a discontinuity during the time step (Riemann solver).

P-2 Curvilinear (Structured) Grids

The finite difference approach, using the discrete points, is associated historically with rectangular Cartesian grids, since such a regular lattice structure provides easy identification of neighboring points to be used in the representation of derivatives, while the finite element approach has always been, by the nature of its construction on discrete cells of general shape, considered well suited for irregular regions, since a network of such cells can be made to fill any arbitrarily shaped region and each cell is an entity unto itself, the representation being on a cell, not across cells.

P-2.1 Boundary-Fitted Grids

The finite difference method is not, however, limited to rectangular grids and has long been applied on other readily available analytical coordinate systems (cylindrical, spherical, elliptical, etc.) that still form a regular lattice. albeit curvilinear, that allows easy identification of neighboring points. These special curvilinear coordinate systems are all orthogonal, as are the rectangular Cartesian systems, and they also can exactly cover special regions (e.g., cylindrical coordinates covering the annular region between two concentric circles) in the same way that a Cartesian grid fills a rectangular region. The *cardinal feature* in each case is that some coordinate line is coincident with each portion of the boundary.

In fact, these curvilinear systems can be considered to be *logically rectangular*, and from a programming standpoint are no different, conceptually, from the Cartesian system. Thus, for example, the cylindrical grid in Figure 1, where the radial coordinate r varies from r_1 on the inner boundary to r_2 on the outer and the azimuthal coordinate θ varies from 0 to 2π, can be diagrammed logically as shown in Figure 2.

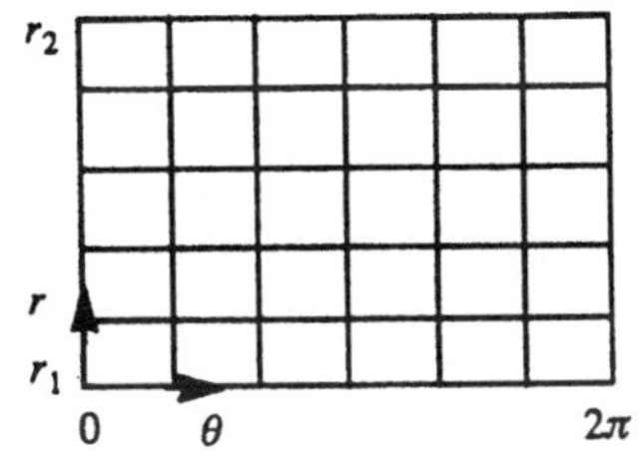

FIGURE 2

The continuity of the azimuthal coordinate can be represented by defining extra "phantom" columns to the left of 0 and to the right of 2π and setting values on each phantom column equal to those on the corresponding "real" columns inside of 2π and 0, respectively. This latter, logically rectangular, view of the cylindrical grid is the one used in programming anyway, and without being told of the cylindrical configuration, a programmer would not realize any difference here from programming in Cartesian coordinates — there would simply be a different set of equations to be programmed on what is still a logically rectangular grid, e.g., the Laplacian on a Cartesian grid (with $\xi = x$ and $\eta = y$),

$$\nabla^2 f = f_{\xi\xi} + f_{\eta\eta}$$

becomes (with $\xi = \theta$ and $\eta = r$)

$$\nabla^2 f = \frac{f_{\xi\xi}}{\eta^2} + f_{\eta\eta} + \frac{f_\eta}{\eta}$$

on a cylindrical grid. The key point here is that in the logical (i.e., programming) sense there is really no essential difference between Cartesian grids and the cylindrical systems: both can be programmed as nested loops; the equations simply are different.

Another key point is that the cylindrical grid *fits the boundary* of a cylindrical region just as the Cartesian grid fits the boundary of a rectangular region. This allows boundary conditions to be represented in the same manner in each case also (see Figure 3). By contrast, the use of a Cartesian grid on a cylindrical region requires a stair-stepped boundary and necessitates interpolation in the application of boundary conditions (Figure 4) — the proverbial square peg in a round hole.

P-2.2 Block Structure (The Sponge Analogy)

The best way to visualize the correspondence of a curvilinear grid in the physical field with a logically rectangular grid in the computational field is through the *sponge analogy.* Consider a rectangular sponge within which an equally spaced Cartesian grid has been drawn. Now wrap the sponge around a circular cylinder and connect the two ends of the sponge together. Clearly the original Cartesian grid in the sponge now has become a curvilinear grid fitted to the cylinder. But the rectangular logical form of the grid lattice is still preserved, and a programmer could still operate in the logically underformed sponge in constructing the loop and the difference expressions, simply having been given different equations to program. The correspondence of "phantom" points just outside one of the connected faces of the sponge

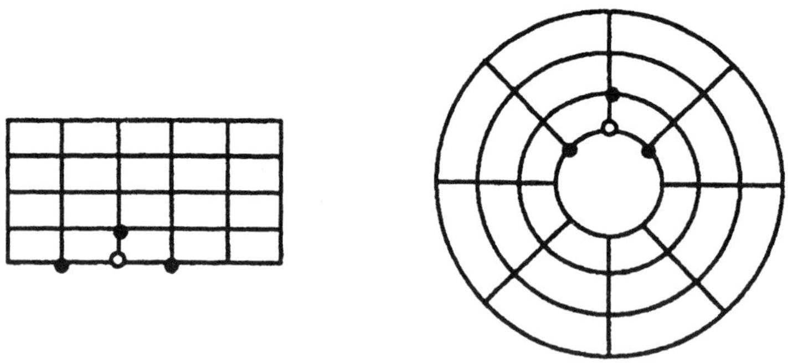

FIGURE 3

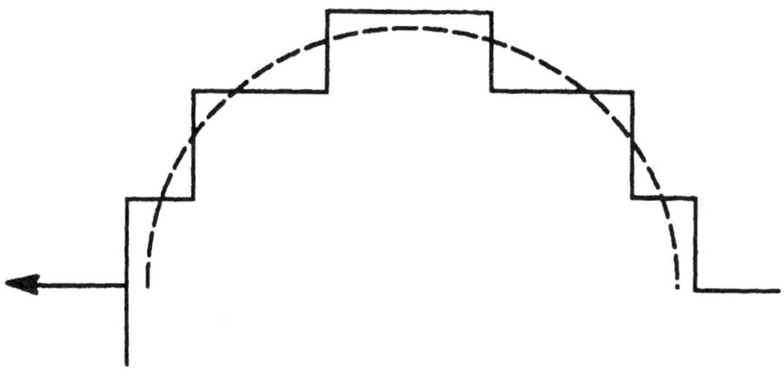

FIGURE 4

with "real" points just inside the face to which it is connected is clear — this is simply the correspondence of 370° with 10° in a cylindrical system.

Such a sponge could just as well be around a cylinder of noncircular cross section, regardless of the cross-sectional shape. To carry the analogy further, the sponge could, in principle, be wrapped around a body of any shape, or could be expanded and compressed to fill any region (e.g., expanding to fill a sphere), again producing a curvilinear grid filling the region and having the same correspondence to a logically rectangular grid (Figure 5). The programmer need not know, in fact, what has been done to the sponge. It is also clear from this analogy that the sponge could deform in time; i.e., the curvilinear grid could move in physical space, while the logically rectangular grid could still be considered fixed in computational space (image the sponge filling a beating heart). Again, the programmer need not be told that the boundaries are moving, but simply again be given a different set of equations that will include a transformation of the time derivatives as well.

It is not hard to see, however, that for some boundary shapes the sponge may have to be so greatly deformed that the curvilinear grid will be so highly skewed and twisted that it is not usable in a numerical solution. The solution to this problem is to use not one, but rather a group of sponges to fill the physical field. Each sponge has its own logically rectangular grid that deforms to a curvilinear grid when the sponge is put in place in the field. Each sponge now abuts with adjacent sponges, and the correspondence

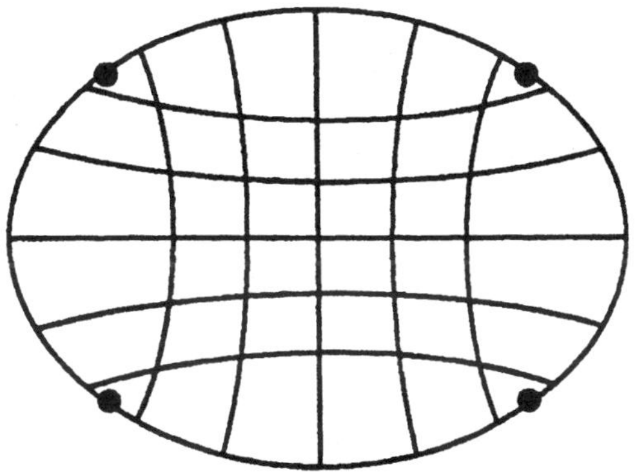

FIGURE 5

across an interface is analogous to that across the two connected faces of the single sponge in the cylindrical case above — here it is simply that the "phantom" points just outside one sponge correspond to "real" points just inside a different sponge.

Block-structured grid codes are based on this multiple-sponge analogy, with the physical field being filled with a group of grid blocks with correspondence of grid lines, and in fact complete continuity, across the interfaces between blocks. This approach has been carried to a high degree of application in the aerospace industry (cf. Chapter 13), with complete aircraft configurations being treated with a hundred or so blocks. Current grid generation systems seek to make the setup of this block structure both graphical and easy for the user. The ultimate goal is to automate the process (cf. Chapter 10).

2.3 Grid Generation Approaches

With these obvious advantages of specialized curvilinear coordinate systems fitted to the boundaries of cylindrical, spherical, elliptical, and other analytic regions, it has been natural to use grids based on these systems for finite difference solutions on such regions. In the late 1960s the visual analogy between potential solutions (electrostatic potential, potential flow, etc.) that are solutions of Laplace's equation, $\nabla^2\phi = 0$, and curvilinear grids led to the idea of generating grid lines in arbitrary regions as the solution of Laplace's equation. Thus, whereas potential flow is described in terms of a stream function ψ and a velocity potential ϕ that are orthogonal and satisfy $\nabla^2\psi = 0$, $\nabla^2\phi = 0$ (Figure 6), a curvilinear grid could be generated by solving the system $\nabla^2\xi = 0$, $\nabla^2\eta = 0$ with η a constant on the upper and lower boundaries in the above region, while ξ is constant on the left and right boundaries (Figure 7).

Here again, for purposes of programming, the grid forms a logically rectangular lattice (Figure 8). The problem of generating a curvilinear grid to fit any arbitrary region thus becomes a *boundary value problem* — the generation of interior values for the curvilinear coordinates from specified values on the boundary of the region (cf. Chapter 4). In order to set this up, note that we have for the boundary value problem the generation of interior values of the curvilinear coordinates ξ and η from specified constant values on opposing boundaries (Figure 9).

Clearly ξ and η must vary monotonically and over the same range over the boundaries on which they are not specified, else the grid would overlap on itself. Thus, on the lower and upper boundaries, ξ here must vary monotonically from ξ_1 on the left to ξ_2 on the right. Similarly, on the left and right boundaries,

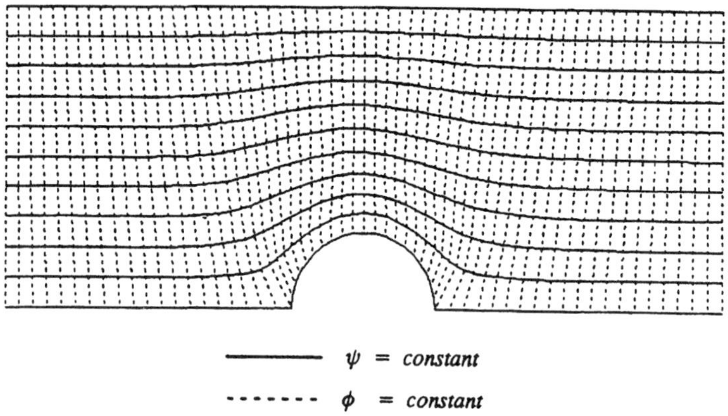

FIGURE 6

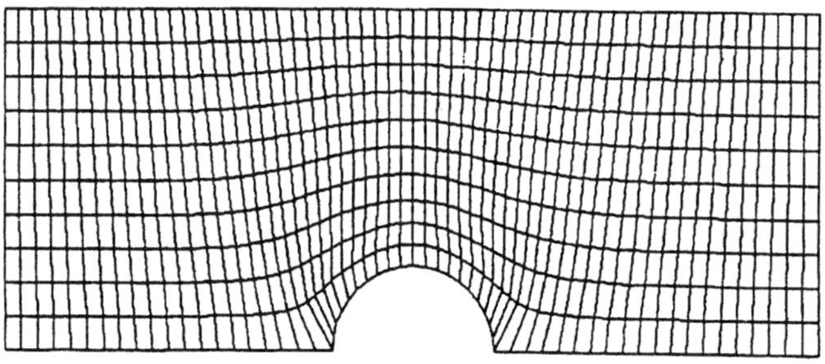

FIGURE 7

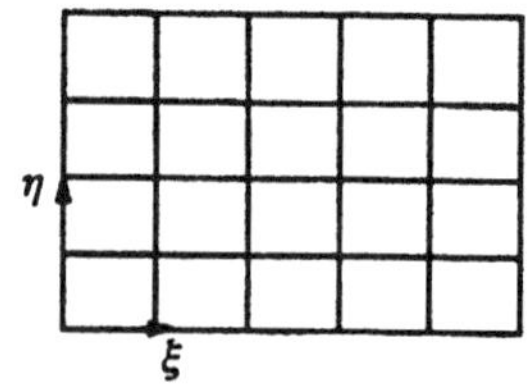

FIGURE 8

η must vary monotonically from η_1 at the bottom to η_2 at the top. The next question is what this variation should be. This is, in fact, up to the user. Ultimately, the discrete grid will be constructed by plotting lines of constant ξ and lines of constant η at equal intervals of each, with the size of the interval determined by the number of grid lines desired. Thus, if there are to be 10 grid lines running from side to side between the top and bottom of the region, 10 points would be selected on the left and right sides — with their locations being up to the user. Once these points are located, η can be said to assume, at the 10 points on each side, 10 values at equal intervals between its top and bottom values, η_1 and η_2. With this specification on the sides, the curvilinear coordinate η is thus specified on the entire boundary of

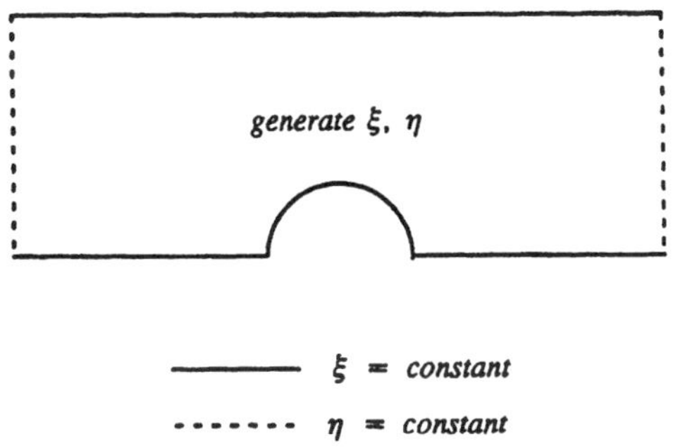

FIGURE 9

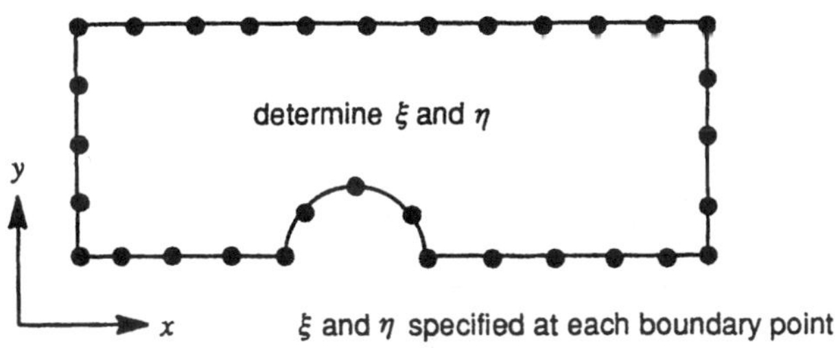

FIGURE 10

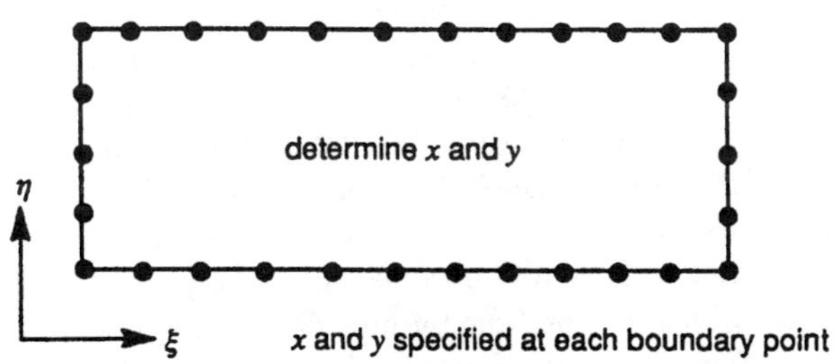

FIGURE 11

the region, and its interior values can be determined as a boundary value problem. A similar specification of ξ on the bottom and top boundaries by placing points on these boundaries sets up the determination of ξ in the interior from its boundary values. Now the problem can be considered a boundary value problem in the physical field for the curvilinear coordinates ξ and η (Figure 10) or can be considered a boundary value problem in the logical field for the Cartesian coordinates, x and y (Figure 11).

Note that the boundary points are by nature equally spaced on the boundary of the logical field regardless of the distribution on the boundaries of the physical field. Continuing the potential analogy, the curvilinear grid can be generated by solving the system $\nabla^2\xi = 0$, $\nabla^2\eta = 0$, in the first case, or by solving the transformation of these equations (transformation relations are covered in Chapter 2), in the

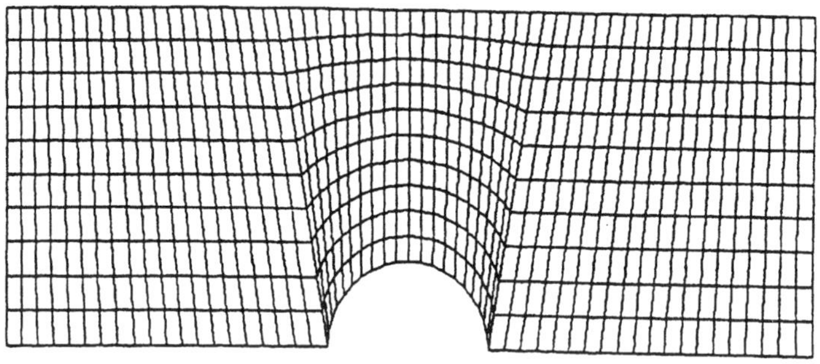

FIGURE 12

$$\alpha x_{\xi\xi} - 2\beta x_{\xi\eta} + \gamma x_{\eta\eta} = 0$$
$$\alpha y_{\xi\xi} - 2\beta y_{\xi\eta} + \gamma y_{\eta\eta} = 0$$
$$\alpha = x_\eta^2 + y_\eta^2$$
$$\gamma = x_\xi^2 + y_\xi^2$$
$$\beta = x_\xi x_\eta + y_\xi y_\eta$$

second case. Although the equation set is longer in the second case, the solution region is rectangular, and the differencing can be done on a uniformly spaced rectangular grid. This is, therefore, the preferred approach. Note that the placing of points in any desired distribution on the boundary of the physical region, where x and y are the independent variables, amounts to setting (x,y) values at equally spaced points on the rectangular boundary of the logical field, where ξ and η are the independent variables. This is the case regardless of the shape of the physical boundary.

This boundary value problem for the curvilinear grid can be generalized beyond the analogy with potential solutions, and in fact is in no way tied to the Laplace equation. The simplest approach is to generate the interior values by interpolation from the boundary values — a process called *algebraic* grid generation (cf. Chapter 3). There are several variants of this process. Thus for the region considered above, a grid could be generated by interpolating linearly between corresponding points on the top and bottom boundaries (Figure 12). Note that the point distributions on the side boundaries have no effect here. Alternatively, the interpolation could be between pairs of points on the side boundaries (Figure 13). The second case is, however, obviously unusable since the grid overlaps the boundary. Here the lack of influence from the points on the bottom boundary is disastrous.

Another alternative is *transfinite interpolation* in which the interpolation is done in one (either) direction as above, but then the resulting error on the two sides not involved is interpolated in the other direction and subtracted from the first result. This procedure includes effects from all of the boundary and consequently matches the point distribution that is set on the entire boundary. This is the preferred approach, and it provides a framework for placing any one-dimensional interpolation into a multiple-dimensional form. It is possible to include any type of interpolation, such as cubic, which gives orthogonality at the boundaries, in the transfinite interpolation format.

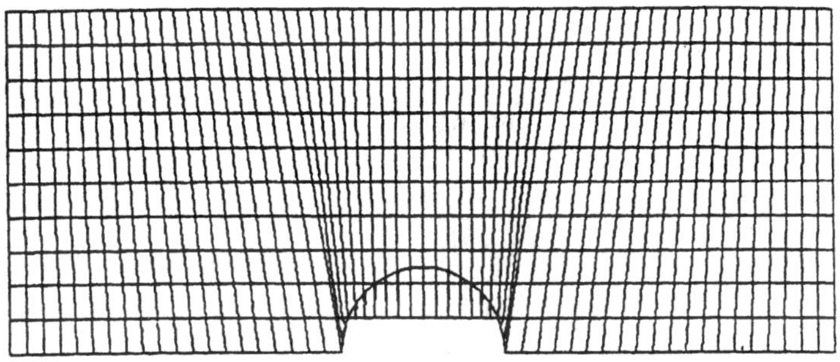

FIGURE 13

It is still possible in some cases for the grid to overlap the boundaries with transfinite interpolation, and there is no control over the skewness of the grid. This gives incentive to now return to the grids generated from solving the Laplace equation.

The Laplace equation is, by its very nature, a smoother, tending to average values at points with those at neighboring points. It can be shown from the calculus of variations, in fact, that grids generated from the Laplace equation are the smoothest possible. There arises, however, the need to concentrate coordinate lines in certain areas of anticipated strong solution variation, such as near solid walls in viscous flow. This can be accomplished by departing from the Laplace equation and *designing* a partial differential equation system for grid generation: designing because, unlike physics, there are no laws governing grid generation waiting to be discovered.

The first approach to this, historically, was the obvious: simply replace the Laplace equation with Poisson equations $\nabla^2\xi = P, \nabla^2\eta = Q$ and leave the *control functions* on the right-hand sides to be specified by the user (with appeal to Urania, the muse of science, for guidance). This does in fact work but the approach has evolved over the years, guided both by logical intuition and the calculus of variations, to use a similar set of equations but with a somewhat different right-hand side. Also, the user has been relieved of the responsibility for specifying the control functions, which are now generally evaluated automatically by the code from the boundary point distributions that are set by the user (cf. Chapter 4). These functions may also be adjusted by the code to achieve orthogonality at the boundary and/or to reduce the grid skewness or otherwise improve the grid quality (cf. Chapter 6).

Algebraic grid generation, based on transfinite interpolation, is typically used to provide an initial solution to start an iterative solution of the partial differential equation for this *elliptic* grid generation system that provides a smoother grid, but with selective concentration of lines, and is less likely to result in overlapping of the boundary.

This elliptic grid generation has an analogy to stretching a membrane attached to the boundaries (cf. Chapter 33) Grid lines inscribed on the underformed membrane move in space as the membrane is selectively stretched, but the region between the boundaries is always covered by the grid. Another form of grid generation from partial differential equations has an analogy with the waves emanating from a stone tossed into a pool This *hyperbolic* grid generation uses a set of hyperbolic equations, rather than the Poisson equation, to grow an orthogonal grid outward from a boundary (cf. Chapter 5). This approach is, in fact, faster than the elliptic grid generation, since no iterative solution is involved, but it is not possible to fit a specified outer boundary. Hyperbolic grid generation is thus limited in its use to open regions. As with the elliptic system, it is possible to control the spacing of the grid lines, and the orthogonality helps prevent skewness.

The control of grid line spacing can be extended to dynamically couple the grid generation system with the physical solution to be performed on the grid in order to resolve developing gradients in the solution wherever such variations appear in the field (cf. Chapter 34 and 35). With such *adaptive* grids, certain solution variables, such as pressure or temperature, are made to feed back to the control functions in the grid generations system to adjust the grid before the next cycle of the physical solution algorithm on the grid.

P-2.4 Variations

Structured grids today are typically generated and applied in the block-structured form described above with the multiple-sponge analogy. A variation is the *chimera* (from the monster of Greek mythology, composed of disparate parts) approach in which separate grids are generated about various boundary components, e.g., bodies in the field, and these separate grids are simply overlaid on a background grid and perhaps on each other in a hierarchy (cf. Chapter 11). The physical solution process on this composite grid proceeds with values being transferred between grids by interpolation. This approach has a number of advantages: (1) simplicity in grid generation since the various grids are generated separately, (2) bodies can be added to, or taken out of, the field easily, (3) bodies can carry their grids when moving relative to the background (think of simulating the kicking of a field goal with the ball and its grid tumbling end over end), (4) the separate grids can be used selectively to concentrate points in regions of developing gradients that may be in motion. The disadvantages are the complexity of setup (but this is being attacked in new code development) and the necessity for the interpolation between grids.

Another approach of interest is the *hybrid* combination with separate structured grids over the various boundaries, connected by unstructured grids (cf. Chapter 23). There is great incentive to use structured grids over boundaries in viscous flow simulation because the boundary layer requires very small spacing out from the wall, resulting either in very long skewed triangular cells or a prohibitively and unnecessarily large number of small cells when unstructured grids are used. This hybrid approach is less well developed but can be expected to receive more attention.

P-2.5 Transformation

The use of numerically generated nonorthogonal curvilinear grids in the numerical solution of PDEs is not, in principle, any more difficult than using Cartesian grids: the differencing and solution techniques can be the same; there are simply more terms in the equations. For instance, the first derivative f_x could be represented in difference form on a Cartesian grid as

$$\left(f_x\right)_{ij} = \frac{f_{i+1,j} - f_{i-1,j}}{2\Delta x}$$

or if the spacing is not uniform, though the grid is still rectangular, by

$$\left(f_x\right)_{ij} = \frac{f_{i+1,j} - f_{i-1,j}}{x_{i+1,j} - x_{i-1,j}}$$

To use a curvilinear grid, this derivative is transformed so that the curvilinear coordinate (ξ,η) rather than the Cartesian coordinate x,y, are the independent coordinates. Thus

$$f_x = \frac{\left(x_\xi f_\eta - x_\eta f_\xi\right)}{J}$$

where $J = x_\xi\, y_\eta - x_\eta\, y_\xi$ is the Jacobian of the transformation and represents the cell volume. This then could be written in a difference form, taking $\Delta\xi$ and $\Delta\eta$ to be unity without loss of generality, using

$$\left(f_\xi\right)_{ij} = \frac{1}{2}\left(f_{i+1,j} - f_{i-1,j}\right)$$

$$\left(f_\eta\right)_{ij} = \frac{1}{2}\left(f_{i,j+1} - f_{i,j-1}\right)$$

with analogous expressions for x_ξ, x_η, y_ξ, y_η.

Movement of the grid, either to follow moving boundaries or to dynamically adapt to developing solution gradients, is not really a complication, since the time derivative can also be transformed as

$$\left(f_t\right)_{\mathbf{r}} = \left(f_t\right)_\xi - \left(f_x\dot{\mathbf{x}} + f_y\dot{\mathbf{y}}\right)$$

where the time derivative on the left is taken at a fixed position in space, i.e., is the time derivative appearing in the PDEs while the one on the right is that seen by a particular grid point moving with a speed $(\dot{\mathbf{x}},\dot{\mathbf{y}})$. The spatial derivatives (f_x, f_y) are transformed as was discussed above. There is no need to interpolate solution values from the grid at one time step to the displaced grid at the next time step, since that transfer is accomplished by the grid speed terms $(\dot{\mathbf{x}},\dot{\mathbf{y}})$ in the above transformation relation.

The straightforwardness of the use of curvilinear grids is further illustrated by the appearance of the generic convection–diffusion equations;

$$f_t + \nabla\cdot(\mathbf{u}f) + \nabla\cdot(\nu\nabla f) + S = 0$$

where $\mathbf{u}$ is the velocity, ν is a diffusion coefficient, and S is a source term, after transformation:

$$A_t + \sum_{i=1}^{3}\left(U^i + \nu\nabla^2\xi^i\right)A_{\xi^i} + \sum_{i=1}^{3}\sum_{j=1}^{3} g^{ij}\left(\nu A_{\xi^j}\right)_{\xi^i} + A\sum_{i=1}^{3}\mathbf{a}^i\cdot\mathbf{u}_{\xi^i} + S = 0$$

where now the time derivative is understood to be that seen by a certain (moving) grid point. Here the elements of the contravariant metric tensor g^{ij} are given by

$$g^{ij} = \mathbf{a}^i\cdot\mathbf{a}^j$$

where the $\mathbf{a}^i$ are the contravariant base vectors (which are simply normals to the cell sides):

$$\mathbf{a}^i = \left(\mathbf{a}_j \times \mathbf{a}_k\right) / \sqrt{g} \quad (i, j, k \text{ cyclic})$$

with the $\mathbf{a}_i$ the covariant base vectors (tangents to the coordinate lines):

$$\mathbf{a}_i = \mathbf{r}_{\xi i}$$

where $\mathbf{r}$ is the Cartesian coordinate of a grid point, and $\sqrt{g}$ is the Jacobian of the transformation (the cell volume):

$$\sqrt{g} = \mathbf{a}_1 \cdot \left(\mathbf{a}_2 \times \mathbf{a}_3\right)$$

Also, the contravariant velocity (normal to the cell sides) is

$$U^i = \mathbf{a} \cdot (\mathbf{u} - \mathbf{r})$$

where $\mathbf{u}$ is the fluid velocity and $\mathbf{r}$ is the velocity of the moving grid. For comparison, the Cartesian grid formulation is

$$A_t + \sum_{i=1}^{3} u_i A_{x_i} + \sum_{i=1}^{3} \sum_{j=1}^{3} \delta_{ij} \left(\nu A_{x_j}\right)_{x_i} + A \sum_{i=1}^{3} \left(u_i\right)_{x_i} + S = 0$$

The formulation has thus been complicated by the curvilinear grid only in the sense that the coefficient u_i has been replaced by the coefficient $U^i + \nu(\nabla^2 \xi^i)$, and the Kronecker delta in the double summation has been replaced by g^{ij} (thus expanding that summation from three terms to nine terms), and through the insertion of variable coefficients in the last summation. When it is considered that the transformed equation is to be solved on a fixed rectangular field with a uniform square grid, while the original equation would have to be solved on a field with moving curved boundaries, the advantages of using the curvilinear systems are clear.

These advantages are further evidenced by consideration of boundary conditions. In general, boundary conditions for the example being treated would be of the form

$$\alpha A + \beta \mathbf{n} \cdot (u \nabla A) = \gamma$$

where $\mathbf{n}$ is the unit normal to the boundary and α, β, and γ are specified. These conditions transform to

$$\alpha A + \beta \frac{\nu}{\sqrt{g^{ii}}} \sum_{j=1}^{3} g^{ij} A_{\xi j} = \gamma$$

for a boundary on which ξ^i is constant. For comparison, the original boundary conditions can be written in the form

$$\alpha A + \beta v \sum_{i=1}^{3} n_j A_{x_j} = \gamma$$

The transformed boundary conditions thus have the same form as the original conditions, but with the coefficient n_j replaced by $g^{ij}/\sqrt{g^{ii}}$. The important simplification is the fact that the boundary to which the transformed conditions are applied is fixed and flat (coincident with a curvilinear coordinate surface). This permits a discrete representation of the derivative A_{ξ^j} along the transformed boundary without the need for interpolation. By contrast, the derivative A_{x^j} in the original conditions cannot be discretized along the physical boundary without interpolation since the boundary is curved and may be in motion.

Although the transformed equation clearly contains more terms, the differencing is the same as on a rectangular grid, i.e., it is done on the logically rectangular computational lattice, and the solution field is logically rectangular. Note that it is not necessary to discover and implement a transformation for each new boundary shape — rather the above formulation applies for all, simply with different values of (x, y, z) at the grid points.

The transformed PDE can also be expressed in conservative form as

$$\left(\sqrt{g}A\right)_t + \sum_{i=1}^{3}\left[\sqrt{g}\left(U^i A + v\sum_{j=1}^{3} g^{ij} A_{\xi^j}\right)\right]_{\xi^i} + \sqrt{g}S = 0$$

for use in the finite volume approach. For more information on transformations, see Chapter 2.

P-3 Unstructured Grids

P-3.1 Connectivities and Data Structures

The basic difference between structured and unstructured grids lies in the form of the data structure which most appropriately describes the grid. A structured grid of quadrilaterals consists of a set of coordinates and connectivities that naturally map into elements of a matrix. Neighboring points in a mesh in the physical space are the neighboring elements in the mesh matrix (Figure 14).

Thus, for example, a two-dimensional array $x(i,j)$ can be used to store the x-coordinates of points in a 2D grid. The index i can be chosen to describe the position of points in one direction, while j describes the position of points in the other direction. Hence, in this way, the indices i and j represent the two families of curvilinear lines. These ideas naturally extend to three dimensions.

For an unstructured mesh the points cannot be represented in such a manner and additional information has to be provided. For any particular point, the connection with other points must be defined explicitly in the connectivity matrix (Figure 15).

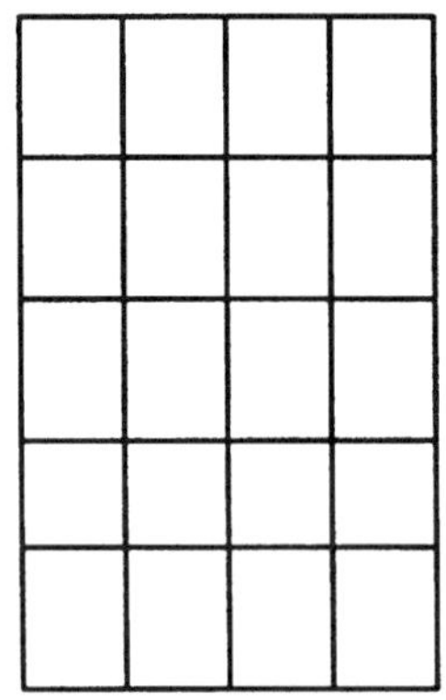

FIGURE 14

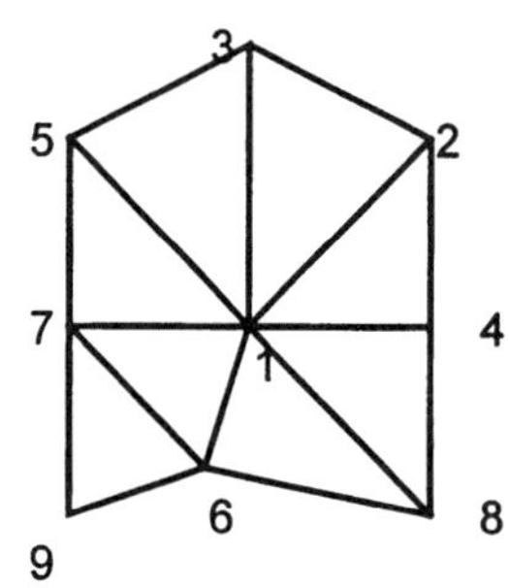

Element,	Nodes
1	1, 5, 7
2	1, 4, 2
3	1, 2, 3
4	1, 3, 5
5	1, 7, 6
6	1, 6, 8
7	1, 8, 4
8	7, 9, 6

FIGURE 15

A typical form of data format for an unstructured grid in two dimensions is

Number of Points,
Number of Elements

$$x_1, y_1$$
$$x_2, y_2$$
$$x_3, y_3$$
$$\ldots$$
$$n_1, n_2, n_3$$
$$n_4, n_5, n_6$$
$$n_7, n_8, n_9$$
$$\ldots$$

where (x_1, y_1) are the coordinates of point i, and n_i, $1=1,N$ are the point numbers with, for example, the triad (n_1, n_2, n_3) forming a triangle.

Other forms of connectivity matrices are equally valid, for example, connections can be based upon edges. The real advantage of the unstructured mesh is, however, because the points and connectivities do not possess any global structure. It is possible, therefore, to add and delete nodes and elements as the geometry requires or, in a flow adaptivity scheme, as flow gradients or errors evolve. Hence the unstructured approach is ideally suited for the discretization of complicated geometrical domains and complex flowfield features. However, the lack of any global directional features in an unstructured grid makes the application of line sweep solution algorithms more difficult to apply than on structured grids.

P-3.2 Grid Generation Approaches

In contrast to the generation of structured grids, algorithms to construct unstructured grids are frequently based upon geometrical ideas. There are now many techniques available, many of which are described within this Handbook. For this elementary overview it is not appropriate to discuss details but to comment on general procedures.

P-3.2.1 Triangle and Tetrahedra Creation by Delaunay Triangulation

The Delaunay approach to unstructured grid generation is now popular. The basic concepts go back as far as Dirichlet, who in a paper in 1850 discussed the basic geometrical concept. Dirichlet proposed a method whereby a given domain could be systematically decomposed into a set of packed convex polygons. Given two points in the plane, P and Q, the perpendicular bisector of the line joining the two points subdivides the plane into two regions, V and W. The region V is closer to P than it is to Q. Extending these ideas, it is clear that for a given set of points in the plane, the regions V_i are territories that can be assigned to each point so that V_i represents the space closer to P_i than to any other point in the set. This geometrical construction of tiles is known as the Dirichlet tessellation. This tessellation of a closed domain results in a set of non-overlapping convex polygons, called Voronoï regions, covering the entire domain.

From this description, it is apparent that in two dimensions, the territorial boundary that forms a side of a Voronoï polygon must be midway between the two points it separates and is thus a segment of the perpendicular bisector of the line joining these two points. If all point pairs that have some segment of a boundary in common are joined by straight lines, the result is a triangulation of the convex hull of the set of points P_i. This triangulation is known as the Delaunay triangulation.

Equivalent constructions can be defined in higher dimensions. In three dimensions, the territorial boundary that forms a face of a Voronoï polyhedron is equidistant between the two points it separates. If all point pairs that have a common face in the Voronoï construction are connected, then a set of tetrahedra is formed that covers the convex hull of the data points.

For the number of points which may be required in grid for computational analysis, it might appear that the above procedure would be difficult and computationally expensive to construct. However, there are several algorithms that can form the construction in a very efficient manner. These are discussed at length in Chapters 1, 16 and 20. The approach is very flexible in that it can automatically create grids with the minimum of user interaction for arbitrary geometries.

P-3.2.2 Triangle and Tetrahedra Creation by the Advancing Front Method

A grid generation technique based on the simultaneous point generation and connection is the *advancing front method.* Unlike the Delaunay approach, advancing front methods are not based on any geometrical criteria. They encompass the logical procedure of starting with a boundary grid of edges, in two dimensions, triangular faces, in three dimensions, and creating a point and constructing an element. Slowing the initial boundary *advances* into the domain until the domain is filled with elements. The placing of points within the domain is, like the Delaunay approach, controlled by a combination of a background mesh and sources that provides the required data to ensure adequate resolution of the domain. The algorithms that generate grids in this way are based on fast geometrical search routines. Details are to be found in Chapter 17.

It is possible to combine techniques from both the Delaunay and the Advancing Front methods to produce effective grid generation procedures – a sort of combination that tries to utilize the advantages of both approaches. Chapter 18 discusses one such approach.

The Delaunay triangulation produces elements that are isotropic in nature. Although the Advancing Front method can produce elements with stretching, it cannot produce high quality meshes with stretching factors applicable to some problems, such as high Reynolds number viscous flows. Hence, it is necessary to augment the standard procedures outlined above. In general, this is done by introducing a mapping that ensures that regular isotropic grids can be generated but once mapped back to the physical space are distorted in a well defined manner to give appropriate element stretching. Such a method is described in detail in Chapter 20.

P-3.2.3 Unstructured Grids of Quadrilaterals and Hexahedra

The preference of some developers for quadrilateral or hexahedral element based unstructured meshes has resulted in effort devoted to the generation of such meshes. In two dimensions, it is possible to modify the Advancing Front algorithm to construct quadrilaterals, although the additional complexity in extending this approach to three dimensions has not yet been overcome for practical geometries. An alternative approach that has seen some success is that of "paving." This approach relies upon iteratively layering or paving rows of elements in the interior of a region. As rows overlap or coincide they are carefully connected together. It is fair to conclude that almost without exception the methods for the construction of unstructured hexahedral based grids are heuristic in nature, requiring considerable effort to include the many possible geometrical occurrences. Chapter 21 discusses in detail aspects of this kind of grid generation.

P-3.2.4 Surface Mesh Generation

The generation of unstructured grids on surfaces is, in itself, one of the most difficult and yet important aspects of mesh generation in three dimensions. The surface mesh influences the field mesh close to the boundary. Surface meshes have the same requirement for smoothness and continuity as the field meshes for which they act as boundary conditions, but in addition, they are required to conform to the geometry surfaces, including lines of intersection and must accurately resolve regions of high curvature.

The approach usually taken to generate grids on surfaces is to represent the geometry in parametric coordinates. A parametric representation of a surface is straightforward to construct and provides a description of a surface in terms of two parametric coordinates. This is of particular importance, since the generation of a mesh on a surface then involves using grid generation techniques developed for two space dimensions. A full description of these procedures is given in Chapter 19.

P-3.3 Grid Adaptation Techniques

To resolve features of a solution field accurately it is, in general, necessary to introduce grid adaptivity techniques. Adaptivity is based on the equidistribution of errors principle, namely,

$$w_i ds_i = \text{constant}$$

where w_i is the error or activity indicator at node i and ds_i is the local grid point spacing at node i.

Central to adaptivity techniques and the satisfaction of this equidistribution principle is to define an appropriate indicator w_i. Adaptivity criteria are based on an assessment of the error in the solution of the governing equations or are constructed to detect features of the field. These estimators are intimately connected to the analysis equations to be solved. For example, some of the main features of a solution of the Euler equations can be shock waves, stagnation points and vortices, and any indicator should accurately identify these flow characteristics. However, for the Navier-Stokes equations, it is important not only to refine the mesh in order to capture these features but, in addition, to adequately resolve viscous dominated phenomena such as the boundary layers. Hence, it seems likely that, certainly in the near future, adaptivity criteria will be a combination of measures, each dependent on some aspects of the flow and, in turn, on the flow equations.

There is also an extensive choice of criteria based on error analysis. Such measures include, a comparison of computational stencils of different orders of magnitude, comparison of the same derivatives on different meshes, e.g., Richardson extrapolation, and resort to classical error estimation theory. No generally applicable theory exists for errors associated with hyperbolic equations, hence, to date combinations of rather *ad hoc* methods have been used.

Once an adaptivity criterion has been established, the equidistribution principle is achieved through a variety of methods, including point enrichment, point derefinement, node movement and remeshing, or combinations of these. For more information on grid adaption techniques, see Chapter 35.

P-3.3.1 Grid Refinement

Grid refinement, or *h*-refinement, involves the addition of points into regions where adaptation is required. Such a procedure clearly provides additional resolution at the expense of increasing the number of points in the computation.

Grid refinement on unstructured grids is readily implemented. The addition of a point or points involves a local reconnection of the elements, and the resulting grid has the same form as the initial grid. Hence, the same solver can be used on the enriched grid as was used on the initial grid.

It is important that the adaptivity criteria resolve both the discontinuous features of the solution (i.e. shock waves, contacts) and the smooth features as the number of grid points are increased. A desirable feature of any adaptive method to ensure convergence is that the local cell size goes to zero in the limit of an infinite number of mesh points.

Grid refinement on a structured or multiblock grids is not so straightforward. The addition of points will, in general, break the regular array of points. The resulting distributed grid points no longer naturally fit into the elements of an array. Furthermore, some points will not "conform" to the grid in that they have a different number of connections to other points. Hence grid refinement on structured grids requires a modification to the basic data structure and also the existence of so-called non-conforming nodes requires modifications to the solver. Clearly, point enrichment on structured grids is not as natural a process as the method applied on unstructured grids and hence is not so widely employed. Work has been undertaken to implement point enrichment on structured grids and the results demonstrate the benefits to be gained from the additional effort in modifications to the data structure and the solve.

P-3.3.2 Grid Movement

Grid movement satisfies the equidistribution principle through the migration of points from regions of low activity into regions of high activity. The number of nodes in this case remains fixed. Traditionally, algorithms to move points involve some optimization principle. Typically, expressions for smoothness, orthogonality and weighting according to the analysis field or errors are constructed and then an optimization is performed such that movement can be driven by a weight function, but not at the expense of loss of smoothness and orthogonality. Such methods are in general, applicable to both structured and unstructured grids.

An alternative approach is to use a weighted Laplacian function. Such a formulation is often used to smooth grids, and of course the formal version of the formulation is used as the elliptic grid generator presented earlier.

P-3.3.3 Combinations of Node Movement, Point Enrichment and Derefinement

An optimum approach to adaptation is to combine node movement and point enrichment with derefinement. These procedures should be implemented in a dynamic way, i.e., applied at regular intervals within the simulation. Such an approach also provides the possibility of using movement and enrichment to independently capture different features of the analysis.

P-3.3.4 Grid Remeshing

One method of adaptation which, to date, has been primarily used on unstructured grids, is adaptive remeshing. As already indicated, unstructured meshes can be generated using the concept of a background mesh. For an initial mesh, this is usually some very coarse triangulation that covers the domain and on which the spatial distribution is consistent with the given geometry. For adaptive remeshing, the solution achieved on an initial mesh is used to define the local point spacing on the background mesh which was itself the initial mesh used for the simulation. The mesh is regenerated using the new point spacing on the background mesh. Such an approach can result in a second adapted mesh that contains fewer points than that contained in the initial mesh. However, there is the overhead of regeneration of the mesh which in three dimensions can be considerable. Nevertheless, impressive demonstrations of its use have been published.

Contents

PART II Unstructured Grids

PART III Surface Definition

PART IV Adaptation and Quality

I

Block-Structured Grids

Joe F. Thompson

Introduction to Structured Grids

The grid generation process, in general, proceeds from first defining the boundary geometry as discussed in Part III. Then points are distributed on the curves that form the edges of boundary sections. A surface grid is then generated on the boundary surface, and finally a volume grid is generated in the field. Chapter 13 gives a general overview of the entire grid generation process and the fundamental choices and considerations involved from the standpoint of the user.

The underlying essential mathematics of structured grid generation, including essential concepts from differential geometry and tensor analysis, is collected in Chapter 2. The mathematical constructs explained in this chapter are utilized throughout the chapters of this handbook.

The distribution of points on boundary curves (edges of boundary surfaces) is commonly done through several distribution functions as described in Section 3.6 of Chapter 3. (The mathematics of curves is covered in Section 2.3 of Chapter 2.) These functions have been adopted over time as providing point distributions that comply with certain constraints that must be applied in order to control error that can be introduced into the solution by the grid if the spacing changes too rapidly, as discussed in Chapter 32 of Part IV.

Structured grids can be generated algebraically or as the solution of PDEs. Algebraic grid generation is simply some form of interpolation from boundary points — the variants just use different kinds of interpolation. The most fundamental and versatile form — and now commonly incorporated in grid generation codes — is TFI (transfinite interpolation), which is introduced in Section 1.3.5 of Chapter 1 and described in Chapter 3. The basic equations of TFI are given in Section 3.4 of Chapter 3, and the specific equations for application with and without orthogonality at the boundaries are given in Section 3.5.

Algebraic grid generation based on TFI is the fastest procedure for structured grids, and is also commonly used to generate an initial grid in generation systems based on PDEs. Grids generated algebraically can, however, have some problems with smoothness and may overlap strongly convex portions of boundaries.

Generation systems based on PDEs can produce smoother grids with fewer problems with boundary overlap. Such generation systems are therefore often used to smooth algebraic grids.

Since grid generation is essentially a boundary-value problem, grids can be generated from point distributions on boundaries by solving elliptic PDEs in the field. The smoothness properties and extremum principles inherent in some such PDE systems can serve to produce smooth grids without boundary overlap. The PDE solution is generally one by iteration, and therefore elliptic grid generation is not as fast as algebraic grid generation.

The elliptic PDEs for grid generation are not unique, of course, but must be designed. This design has converged over the years to the elliptic system given in Section 1.3.3 of Chapter 1, which forms the basis for most grid generation codes today. This formulation incorporates control functions that are determined from the boundary point distribution to control the grid line spacing and orientation in the field to be compatible with that on the boundary. Procedures for the determination of these control functions in grid codes have evolved in time to the forms noted in this section of Chapter 1, which can accomplish boundary orthogonality through iterative adjustment during the generation process. A more recent and general formulation, with a sounder basis for evaluation of the control functions, is given here in Chapter 4: for 2D in Section 4.2 and for 3D in Section 4.4. This iterative solution of the elliptic system is often done by SOR, but a Picard iteration is given in Section 4.2.2 of Chapter 4, and a conjugate gradient solution is given in Section 12.10.4 of Chapter 13, in connection with parallel implementation.

The generation of a grid on a boundary surface is a necessary prelude to the generation of a volume grid, and this is generally done by representing the boundary surface parametrically by NURBS or another spline formulation, and then generating the grid in parameter space either algebraically or using PDEs. This is perfectly analogous to 2D grid generation except that surface curvature terms appear in the PDEs. With the generation system operating in parameter space, the resulting grid is guaranteed to lie on the boundary surface. The parametric representation of the boundary surface is covered in Chapter 29, utilizing the underlying curve and surface constructs given in Chapter 28. Other aspects of surface generation are covered in the other chapters in Part III, and the mathematical foundations are given in Section 2.4 and in Section 2.5.2 of Chapter 2.

Algebraic surface grid generation is simply the application of TFI to generate values of the surface parameters on the surface from the values set on the edges of the boundary surface by the grid point distribution on those edges, as covered in Section 9.2 of Chapter 9. Elliptic surface grid generation operates with the PDEs formulated in terms of the surface parameters, and surface curvature terms appearing in the PDEs (see Section 2.5.2 of Chapter 2). A commonly applied procedure is given in Section 9.3 of Chapter 9, and a more recent and general procedure is given in Section 4.3 of Chapter 4. Hyperbolic surface grid generation is covered in Section 5.3 of Chapter 5.

It is generally advantageous, in view of such things as boundary layer phenomena and turbulence models, to have the grid orthogonal to boundaries even though orthogonality is not imposed in the field. This is commonly done through iterative adjustment of the control functions as described in Chapter 6: in Section 6.2 for 2D grids, Section 6.3 for surface grids, and Section 6.4 for volume grids. Another procedure in 2D, also using the control functions, is given in Section 4.2 of Chapter 4.

An alternative approach to grid generation via PDEs is to use a hyperbolic generation system rather than an elliptic. Elliptic equations admit boundary conditions, i.e., grid point distributions, on all boundaries of a region. Hyperbolic systems, however, can take boundary conditions only on a portion of the boundary. Therefore, while elliptic grid generation systems can produce a grid in the entire volume from point distributions of the entire boundary, hyperbolic systems generate the grid by marching outward from a portion of the boundary. Hyperbolic grid generation systems therefore cannot be used to generate a grid in the entirety of a volume defined by a complete boundary. Chapter 5 covers hyperbolic grid generation in volumes in Section 6.2 and on surfaces in Section 6.3.

Structured grids are not generally made orthogonal, although orthogonality at boundaries is often incorporated, as has been noted above. In fact, 3D orthogonality is not, in general, possible without imposing certain conditions on the grids on the boundary surfaces. And even in 2D, orthogonality imposes severe restrictions on the grid distribution. Transformed PDEs, however, take a much simpler form on orthogonal grids, providing some incentive for their use when feasible — with relatively simple boundary configurations and physical problems without strong localized gradients. Chapter 7 covers orthogonal grid generation systems.

As has been noted, PDEs for grid generation are designed, not discovered. Considerable research has gone into this topic, leading to generally standard elliptic (Chapter 4) and hyperbolic (Chapter 5) grid generation systems. The underlying theory of harmonic mappings provides a framework for the development of elliptic grid generation systems, and this topic is treated in some depth in Chapter 9. This theoretical base also leads to the formulation of adaptive grid systems, also covered in this chapter. Adaptive grids are most fundamentally formulated from variational principles, and this is covered in Chapter 36 of Part IV. Adaptive grids and grid quality are covered in the chapters of Part IV.

A strong and versatile alternative to block-structured grids is the overset grid approach (originally called chimera, after the composite monster of Greek mythology). With this approach, individual structured grids are generated around separate boundary components, e.g., bodies, and these separate grids simply overlap each other in some hierarchy. Data is transferred between overlapping grids by interpolation. The overset grid approach is covered here in Chapter 11. The grid generation involved is typically done by hyperbolic generation systems, described in Chapter 5.

The mathematics and technology of structured grid generation have matured now so that the techniques covered in Part I can be expected to be of enduring utility. The block structure is versatile, and serves as the foundation for efficient solutions because of its inherently simple data structure. Construction of the block configuration by hand, even with graphically interactive tools, is very labor intensive, however, as noted in Chapter 13. Automation of the block structure, rather than graphical interaction, is the goal, and this is an area of active research and development (Section 21.2 of Chapter 21 is relevant here). A very promising recent approach is included in Chapter 11. Finally, operation on parallel processors is essential now, and the block structure provides a natural means of domain decomposition, as covered in Sections 12.8–12.10 of Chapter 12.

The operation of the block structure is discussed in Sections 12.2–12.6 of Chapter 12. Chapter 12 also covers a script-based meta-language approach to structured grid generation in Section 12.7. Although most available grid generation systems have departed from the script-based approach in favor of graphical interaction, the script-based approach has definite advantages in design cycles.

1

Fundamental Concepts and Approaches

Joe F. Thompson

Nigel P. Weatherill

1.1 Introduction

This introductory chapter uses fluid mechanics as an example field problem for reference; the applicability of the concepts discussed is, however, not in any way limited to this area.

Fluid mechanics is described by nonlinear equations, which cannot generally be solved analytically, but which have been solved using various approximate methods including expansion and perturbation methods, sundry particle and vortex tracing methods, collocation and integral methods, and finite difference, finite volume, and finite element methods. Generally the finite difference, finite volume, and finite element discretization methods have been the most successful, but to use them it is necessary to discretize the field using a grid (mesh). (The terms *grid* and *mesh* are used interchangeably throughout with identical meaning.) The mesh can be structured or unstructured, but it must be generated under some of the various constraints described below, which can often be difficult to satisfy completely. In fact, at present it can take orders of magnitude more person-hours to construct the grid than it does to construct and analyze the physical solution on the grid. This is especially true now that solution codes of wide applicability are becoming available. Computational fluid dynamics (CFD) is a prime example, and grid generation has been cited repeatedly as a major pacing item (cf. Thompson [1996]). The same is true for other areas of computational field simulation.

The proceedings of the several international conferences on grid generation (Thompson [1982], Hauser and Taylor [1986], Sengupta, et al. [1988], Arcilla, et al. [1991], Eiseman, et al. [1994], Soni et al. [1996]) as well as those of the NASA conferences (Smith [1980], Smith [1992], Choo [1995]) provide numerous illustrations of application to CFD and some other fields.

A recent comprehensive text is Carey [1997].

0-8493-2687-7/99/$0.00+$.50

1.2 Mesh Generation Considerations

The generated mesh must be sufficiently dense that the numerical approximation is an accurate one, but it cannot be so dense that the solution is impractical to obtain. Generally, the grid spacing should be smoothly and sufficiently refined to resolve changes in the gradients of the solution. If the grid is also boundary-conforming and curvilinear, the application of boundary conditions is simplified. Boundary-conforming curvilinear grids may also allow the use of various approximate equations such as boundary-layer equations. The grid should also be constructed with computational efficiency in mind. The accuracy of a numerical approximation can also be impaired, if a grid changes discontinuously or is too skewed. Various computers often require well-organized data, and memory requirements can grow to impractical limits unless the data is organized well. Finally, the choice of a grid should not lead to overly complex computer codes.

A mesh is a set of points distributed over a calculation field for a numerical solution of a set of partial differential equations (PDEs). This set may be structured, e.g., formed by the intersections of curvilinear coordinate surfaces, or unstructured, i.e., with no relation to coordinate directions. In the first case the points form quadrilateral cells in 2D, or hexahedral cells in 3D (with nonplanar sides). The unstructured mesh generally consists of triangles and tetrahedra in 2D and 3D, respectively, in its most basic form, but may be made of hexahedra or elements of any shape in general.

The structured grid can be generated algebraically by interpolation from boundaries, e.g., transfinite interpolation, or by solving a set of partial differential equations in the region. An entire subject, complete with textbook (Thompson, Warsi, and Mastin [1985], now on the Web at www.erc.msstate.edu), has developed around the generation of structured grids having the fundamental characteristic that some curvilinear coordinate surface is coincident with each boundary segment, i.e., boundary conforming. A later text is Knupp and Steinberg [1993]. Castillo [1991] provides a compilation of mathematical aspects as well. Structured grid generation is also covered in the recent text of Carey [1997]. Several earlier surveys of the field are still useful for basic understanding (Thompson, Warsi, and Mastin [1982], Thompson [1984], Thompson [1985], Eiseman [1985]).

Structured boundary-conforming meshes have been widely applied in computational fluid dynamics. Basically, the algebraic generation systems (Chapter 3) are faster, but the grids generated from partial differential equations are generally smoother. The hyperbolic generation systems (Chapter 5) are faster than the elliptic systems, but are more limited in the geometries that can be treated. The elliptic systems (Chapter 4) are the most generally applicable with complicated boundary configurations, but transfinite interpolation is also effective in a composite grid framework.

The generation of unstructured meshes can be done by tessellation of a point distribution that could be random but is more likely to have been produced by some ordered procedure. This tessellation is not unique, and involves some type of nearest-neighbor search, such as the Delaunay triangulation (Chapter 16). Other approaches are the advancing front procedure (Chapter 17) and the finite octree method (Chapter 15). The recent text of Carey [1997] covers unstructured grid generation as well as structured grid generation. An earlier text on unstructured grids is George [1991].

General configurations can conceivably be treated with either type of mesh, and hybrid combinations (Chapter 23) are also possible, using individual structured meshes near boundaries, with these subregions being connected by an unstructured mesh. Still another approach is overlaid grids (chimera) (Chapter 11), in which separate boundary-conforming structured grids are generated for each component of a complex configuration, and data is communicated between the various component grids by interpolation.

Of particular importance is the development of dynamically adaptive meshes (Chapters 33–36) coupled with the physical solution. In this mode the mesh is locally refined by the selective addition of points, and/or is moved to concentrate points, in order to resolve developing gradients in the physical solution on the mesh. Both of these approaches have seen considerable development and show much promise in particular areas.

Implementation of solution algorithms on structured, unstructured, and overlaid grids places differing requirements on the algorithms. Various conflicts arise between the grid and solution procedures in regard to requirements and ease of operation. In particular, unstructured grids require a much more complex solution data structure, but are more easily generated and adapted. Structured grids provide a more natural representation of normal derivative boundary conditions and allow more straightforward approximations based on prevailing directions, e.g., parallel or normal to a boundary or flow direction. The structure also leads to a much simpler data set construction, and allows the use of directional time splitting and flux representations. On the other hand, unstructured grids can be much more readily imagined for complicated boundary configurations.

1.3 Structured Grids

1.3.1 Composite Grids

Structured grid generation had its roots in the U.S. in the work of Winslow and Crowley at Lawrence Livermore National Lab in the late 1960s (Winslow [1967]), and in Russia from Godunov and Prokopov [1967] at about the same time. (There is also that enigmatic Biblical reference to the "four corners of the earth," once thought to proclaim a flat earth but now seen to be a prophesy of structured grid generation.) Another very fundamental component was the work of Bill Gordon at Drexel on transfinite interpolation for the automotive industry, introduced to the emerging grid generation community at the grid conference in Nashville in 1982 (Gordon and Thiel [1982]).

1.3.1.1 Terminology

The use of composite grids has been the key to the treatment of general 3D configurations with structured grids. Here in general, *composite* refers to the fact that the physical region is divided into subregions, within each of which a structured grid is generated. These subgrids may be patched together at common interfaces, may be overlaid, or may be connected by an unstructured grid. Considerable confusion has arisen in regard to terminology for composite grids, making it difficult to immediately classify papers on the subject.

Composite grids in which the subgrids share common interfaces are referred to as *block, patched, embedded,* or *zonal* grids in the literature. The use of the first two of these terms is fairly consistent with this type of grid (*patched* comes from the common interfaces, *block* from the logically rectangular structure), but the last two are sometimes also applied to overlaid grids. *Overlaid* (overset) grids are often called *chimera* grids after the composite monster of Greek mythology. Unfortunately, the common interface grids can also be said to overlap, since they typically use surrounding layers of points to achieve continuity. Embedded grids can be almost anything, and the term is probably best avoided. The use of *zonal* comes mostly from CFD applications where the suggestion of applying different solution equations sets in different flow regions is made. Perhaps *block* or *patched* would be best for the common interface grids, *chimera* for the overlaid (avoiding *overlaid*) grids, and *hybrid* for the structured–unstructured combinations.

1.3.1.2 Forms

With this terminology adopted, the block (or patched) grids may be completely continuous at the interfaces, have slope or line continuity, or be discontinuous (sharing a common interface but not common points thereon). (*Block* seems to cover all of these possibilities, but *patched* is being stretched a bit in the latter case.) Complete continuity is achieved through a surrounding layer of *(image, phantom)* points at which values are kept equal to those at corresponding *object* points inside an adjacent block. This requires a data indexing procedure to link the blocks across the interfaces. With complete continuity, the interface is not fixed (not even in shape), but is determined in the course of the solution. This type of interface necessitates an elliptic generation system. Slope continuity requires that the grid generation procedure incorporate some control over the intersection angle at boundaries (usually, but not necessarily, orthogonality), as can be done through Hermite interpolation in algebraic generation systems or through iterative adjustment of the control functions in elliptic systems. In this case the points on the interface are fixed, and the subgrids are generated independently, except for the use of the common interface points and a common (presumably orthogonal) angle of intersection with the interface. The PDE coding construction is greatly simplified with either complete or slope continuity, since then no algorithm modifications are necessary at the interfaces.

The *chimera* (overlaid) grids are composed of completely independent component grids that overlap a background grid, other component grids and/or other component boundary elements, creating *holes* in the component grids. This requires flagging procedures to locate grid points that lie out of the field of computation, but such holes can be handled even in tridiagonal solvers by placing ones at the corresponding positions on the matrix diagonal and all zeros off the diagonal. These overlaid grids also require interpolation to transfer data between grids, and that subject is the principal focus of effort in regard to the use of this type of composite grid.

The *hybrid* structured–unstructured grids avoid this interpolation by replacing the overlaid region with an unstructured grid connecting logically rectangular structured component grids. This can require modification of solution codes, however.

1.3.2 Block-Structured Grids

Block-structured grids opened the door to real-world CFD in the late 1980s, and many real applications are still based on these grids (see Chapters 12 and 13). The idea appears in the proceedings of the grid conference in Nashville in 1982 (Rubbert and Lee [1982]), but it was Weatherill and Forsey [1984], and Miki and Takagi [1984] that attracted attention to the block-structured approach. Today's structured grid codes are based on this approach. Although the grid is logically rectangular within each block, the blocks fit together in an unstructured manner. Block-structured generation systems that maintain complete continuity across block interfaces allow difference representations to be applied on the block interfaces as in the rest of the field. Complete continuity across block interfaces in the field is accomplished by treating the interface in the manner of a branch cut, with correspondence between "phantom" points outside one block with "real" points inside the adjacent block.

The curvilinear grid system can be constructed simply by setting values in a rectangular array of position vectors,

$$\mathbf{r}_{ijk}(i=1,2,\ldots,I \quad j=1,2,\ldots,J \quad k=1,2,\ldots,K)$$

and identifying the indices *i, j, k* with the three curvilinear coordinates. The position vector $\mathbf{r}$ is a three-vector giving the values of the *x, y,* and *z* Cartesian coordinates of a grid point. Since all increments in the curvilinear coordinates cancel out of the transformation relations for derivative operators, there is no loss of generality in defining the discretization to be on integer values of these coordinates.

Fundamental to this curvilinear coordinate system is the coincidence of some coordinate surface with each segment of the boundary of the physical region, in the same manner that surfaces of constant radius coincide with the inner and outer boundary segments of the region between two concentric spheres filled with a polar coordinate system. This is accomplished by placing a two-dimensional array of points on a physical boundary segment and entering these values into the array $\mathbf{r}_{ijk}$ of position vectors, with one index constant, e.g., in , $\mathbf{r}_{ijk}$ with *i* from 1 to *I* and *j* from 1 to *J*. The curvilinear coordinate *k* is thus constant on this physical boundary segment. With values set on the sides of the rectangular array of position vectors in this manner, the generation of the grid is accomplished by determining the values in its interior, e.g., by interpolation or a PDE solution. The set of values $\mathbf{r}_{ijk}$ then forms the nodes of a curvilinear coordinate system filling the physical region. A physical region bounded by six generally curved sides can thus be considered to have been transformed to a logically rectangular computational region on which the curvilinear coordinates are the independent variables.

Although in principle it is possible to establish a correspondence between any physical region and a single empty logically rectangular block for general three-dimensional configurations, the resulting grid is likely to be much too skewed and irregular to be usable when the boundary geometry is complicated. A better approach with complicated physical boundaries is to segment the physical region into contiguous subregions, each bounded by six curved sides (four in 2D) — each of which transforms to a logically rectangular block in the computational region. Each subregion has its own curvilinear coordinate system irrespective of that in the adjacent subregions (see Figure 13.5).

This then allows both the grid generation and numerical solutions on the grid to be constructed to operate in a logically rectangular computational region, regardless of the shape or complexity of the full physical region. The full region is treated by performing the solution operations in all of the logically rectangular computational blocks. With the block-structured framework, partial differential equation solution procedures written to operate on logically rectangular regions can be incorporated into a code for general configurations in a straightforward manner, since the code only needs to treat a rectangular block. The entire physical field then can be treated in a loop over all the blocks. Transformation relations for partial differential equations are covered in Chapter 2 and in Thompson, Warsi, and Mastin [1985], on the Web. Discretization error related to the grid is covered in Chapter 32. The evaluation and control of grid quality (Chapter 33) is an ongoing area of active research.

The generally curved surfaces bounding the subregions in the physical region form internal interfaces across which information must be transferred, i.e., from the sides of one logically rectangular computational block being paired with another on the same, or different, block, since both correspond to the same physical surface. Grid lines at the interfaces may meet with complete continuity, with or without slope continuity, or may not meet at all.

Complete continuity of grid lines across the interface requires that the interface be treated as a branch cut on which the generation system is solved just as it is in the interior of blocks. The interface locations are then not fixed, but are determined by the grid generation system. This is most easily handled in coding by providing an extra layer of points surrounding each block. Here the grid points on an interface of one block are coincident in physical space with those on another interface of the same or another block, and also the grid points on the surrounding layer outside the first interface are coincident with those just inside the other interface, and vice versa. This coincidence can be maintained during the course of an iterative solution of an elliptic generation system by setting the values on the surrounding layers equal to those at the corresponding interior points after each iteration. All the blocks are thus iterated to convergence together, so that the entire composite grid is generated at once. The same procedure is followed by PDE solution codes on the block-structured grid.

The construction of codes for complicated regions is greatly simplified by the block structure since, with the use of the surrounding layer of points on each block, a PDE code is only required basically to operate on a logically rectangular computational region. The necessary correspondence of points on the surrounding layers (image points) with interior points (object points) is set up by the grid code and made available to the PDE code.

1.3.3 Elliptic Systems

Elliptic grid generation is treated in detail in Chapter 4. This section provides an overview of the technology as applied in the EAGLE system (Thompson [1987]), as an example of the technology applied in several current grid generation codes.

1.3.3.1 Generation System

An elliptic grid generation system used in many codes is

$$\sum_{m=1}^{3}\sum_{n=1}^{3} g^{mn}\,\mathbf{r}_{\xi^m\xi^n} + \sum_{n=1}^{3} g^{nn} P_n \mathbf{r}_{\xi^n} = 0 \tag{1.1}$$

where the g^{mn} are the elements of the contravariant metric tensor,

$$g^{mn} = \nabla\xi^m \cdot \nabla\xi^n$$

These elements are more conveniently expressed for computation in terms of the elements of the covariant metric tensor, g_{mn},

$$g_{mn} = \mathbf{r}_{\xi^m} \cdot \mathbf{r}_{\xi^n}$$

which can be calculated directly. Thus

$$g^{mn} = \left(g_{ik}\, g_{j\ell} - g_{i\ell}\, g_{jk}\right)/g$$
$$(m,i,j)\text{ cyclic},\quad (n,k,\ell)\text{ cyclic}$$

where g, the square of the Jacobian, is given by

$$g = \det\left|g_{mn}\right| = \mathbf{r}_{\xi^1} \cdot \left(\mathbf{r}_{\xi^2} \times \mathbf{r}_{\xi^2}\right)$$

In these relations, $\mathbf{r}$ is the Cartesian position vector of a grid point ($\mathbf{r} = \mathbf{i}x + \mathbf{j}y + \mathbf{k}z$) and the ξ^i ($i = 1,2,3$) are the three curvilinear coordinates. The P_n ($n = 1,2,3$) are the control functions that serve to control the spacing and orientation of the grid lines in the field.

The first and second coordinate derivatives are normally calculated using second-order central differences. One-sided differences dependent on the sign of the control function P_n (backward for $P_n < 0$ and forward for $P_n > 0$) are useful to enhance convergence with very strong control functions. The control functions are evaluated either directly from the initial algebraic grid and then smoothed, or by interpolation from the boundary-point distributions.

1.3.3.2 Control Functions

The now-standard procedure in block-structured systems is to first generate surface grids on block faces — both boundary and in-field block interfaces — from point distributions placed on the face edges by distribution functions. Then volume grids are generated within the blocks. In both this surface and volume grid generation, the first step is normally TFI, to be followed by elliptic generation with control functions interpolated into the field in accordance with boundary point distribution and surface curvature.

The three components of the elliptic grid generation system, Eq. 1.1, provide a set of three equations that can be solved simultaneously at each point for the three control functions, P_n ($n = 1,2,3$), with the derivatives here represented by central differences. This produces control functions that will reproduce the algebraic grid from the elliptic system in a single iteration, of course. Thus, evaluation of the control functions in this manner would be of trivial interest except that these control functions can be smoothed before being used in the elliptic generation system. This smoothing is done by replacing the control function at each point with the average of the four neighbors in the two curvilinear directions (one in 2D) other than that of the function. Thus P_i is smoothed in the ξ^j and ξ^k directions, where i, j, k are cyclic. No smoothing is done in the direction of the function because to do so would smooth the spacing distribution.

An algebraic grid is generated by transfinite interpolation (Chapter 3) from the boundary point distribution, to serve as the starting solution for the iterative solution of the elliptic system. With the boundary point distribution set from the hyperbolic sine or tangent functions, which have been shown to give reduced truncation error (Chapters 3 and 32), this algebraic grid has a good spacing distribution but may have slope breaks propagated from corners into the field. The use of smoothed control functions evaluated from the algebraic grid produces a smooth grid that retains essentially the spacing of the algebraic grid.

The elliptic generation system can be solved by point SOR iteration using a field of locally optimum acceleration parameters. These optimum parameters make the solution robust and capable of convergence with strong control functions.

Control functions can also be evaluated on the boundaries using the specified boundary point distribution in the generation system, with certain necessary assumptions (orthogonality at the boundary) to eliminate some terms, and then can be interpolated from the boundaries in this manner. More general regions can, however, be treated by interpolating elements of the control functions separately. Thus control functions on a line on which ξ^n varies can be expressed as

$$P_n = A_n = + \frac{S_n}{\rho_n} \tag{1.2}$$

where A_n is the logarithmic derivative of the arc length, S_n is the arc length spacing, and P_n the radius of curvature of the surface on which ξ^n is constant.

The arc length spacing, S_n, and the arc length contribution, A_n, to the control function can be interpolated into the interior of the block from the four sides on which they are known by two-dimensional interpolation. The radius of curvature, ρ_n, however is interpolated one-dimensionally between the two surfaces on which it is evaluated. The control function is finally formed by adding the arc length spacing divided by the radius of curvature to the arc length contribution according to Eq. 1.2. This procedure allows very general regions with widely varying boundary curvature to be treated. A more general construction of the control functions is given in Chapter 4.

1.3.3.3 Boundary Orthogonality

The standard approach used to achieve orthogonality and specified off-boundary spacing on boundaries has been the iterative adjustment of control functions in elliptic generation systems, first introduced by Sorenson in the GRAPE code in the 1980s (Sorenson [1989]). Various modifications of this basic concept have been introduced in several codes, and the general approach is now common (see Chapter 6).

A second-order elliptic generation system allows either the point locations on the boundary or the coordinate line slope at the boundary to be specified, but not both. It is possible, however, to iteratively adjust the control functions in the generation system until not only a specified line slope but also the spacing of the first coordinate surface off the boundary is achieved, with the point locations on the boundary specified. These relations can be applied on each successive coordinate surface off the boundary, with the off-surface spacing determined by a hyperbolic sine distribution from the spacing specified at the boundary. The control function increments are attenuated away from the boundary, and contributions are accumulated from all orthogonal boundary sections. Since the iterative adjustment of the control functions is a feedback loop, it is necessary to limit the acceleration parameters for stability. This allows the basic control function structure evaluated from the algebraic grid, or from the boundary-point distributions, to be retained, and thus relieves the iterative process from the need to establish this basic form of the control functions. The extent of the orthogonality into the field can also be controlled. This orthogonality feature is also applicable on specified grid surfaces within the field, allowing grid surfaces in the field to be kept fixed while retaining continuity of slope of the grid lines crossing the surface. This is quite useful in controlling the skewness of grid lines in some troublesome areas.

Alternatively, boundary orthogonality can be achieved through Neumann boundary conditions, which allow the boundary points to move over a surface spline. The boundary point locations by Newton iteration on the spline to be at the foot of normals to the adjacent field points. This is done as follows: The Neumann point on the section currently closest to the field point **R** is first located. This is done by sweeping the section in ever expanding squares centered on the Neumann point. (These squares are actually limited by the section edges, of course, and hence, may become rectangles.) Next the quadrant about this closest point above which the field point lies is determined by comparing the dot products of the distance vector (from the closest point to the field point) with the tangent vectors ($\mathbf{r}_\xi$, $\mathbf{r}_\eta$) to the two grid lines on the section. The quadrant is identified by the signs of these two dot products. The Neumann boundary point in question, **r**, is then moved to the foot of the normal from the adjacent field point to the surface. This position is found as the solution of the nonlinear system

$$\mathbf{r}_\xi \cdot (\mathbf{R}-\mathbf{r})=0, \quad \mathbf{r}_\eta \cdot (\mathbf{R}-\mathbf{r})=0 \tag{1.3}$$

by Newton iteration. The location of the closest current boundary point and the examination of dot products described above has determined the surface cell, i.e., the quadrant, on which this solution lies so that the iteration can be confined to a single cell.

Provision can also be made for extrapolated zero-curvature boundary conditions and for mirror-image reflective boundary conditions on symmetry planes.

1.3.3.4 Surface Grids

In the case of a curved surface, the surface is splined and the surface grid is generated in terms of surface parametric coordinates. The generation of a grid on a general surface (Chapter 9) is a two-dimensional grid problem in its own right, which can also be done either by interpolation or a PDE solution. In general, this is a 2D boundary value problem on a curved surface, i.e., the determination of the locations of points on the surface from specified distributions of points on the four edges of the surface. This is best approached through the use of surface parametric coordinates, whereby the surface is first defined by a 2D array of points, $\mathbf{r}_{mn}$, e.g., a set of cross sections. The surface is then splined, and the spline coordinates (u,v; surface parametric coordinates) are then made the dependent variables for the interpolation or PDE generation system. The generation of the surface grid can then be accomplished by first specifying the boundary points in the array $\mathbf{r}_{ij}$ on the four edges of the surface grid; converting these Cartesian coordinate values to spline coordinate values (u_{ij}, v_{ij}) on the edges; then determining the interior values in the arrays u_{ij} and v_{ij} from the edge values by interpolation or PDE solution; and finally converting these spline values to Cartesian coordinates $\mathbf{r}_{ij}$ (see Figure 9.1).

1.3.4 Hyperbolic System

Elliptic generation systems operate throughout the entirety of a region, while hyperbolic systems move outward from boundaries. An alternate approach to boundary orthogonality and spacing is to incorporate a hyperbolic generation system near the boundary, transitioning to an elliptic system in the far field.

It is also possible to base a grid generation system on hyperbolic partial differential equations, rather than elliptic equations (Chapter 5). In this case the grid is generated by numerically solving a hyperbolic system, marching in the direction of one curvilinear coordinate between two boundary curves in two dimensions, or between two boundary surfaces in three dimensions. The hyperbolic system, however, allows only one boundary to be specified, and is therefore of interest only for use in calculation on physically unbounded regions where the precise location of a computational outer boundary is not important. The hyperbolic grid generation system has the advantage of being generally faster than elliptic generation systems but, as just noted, is applicable only to certain configurations. Hyperbolic generation systems can be used to generate orthogonal grids.

In two dimensions the condition of orthogonality is simply $g_{12} = 0$. If either the cell area $\sqrt{g}$ or the cell diagonal length (squared), $g_{11} + g_{22}$, is a specified function of the curvilinear coordinates, i.e.,

$$\sqrt{g}=F(\xi,\eta) \quad \text{or} \quad g_{11}+g_{22}=F(\xi,\eta)$$

then the system consisting of $g_{12} = 0$ and either of the two equations just above is hyperbolic. Since the system is hyperbolic, a noniterative marching solution can be constructed proceeding in one coordinate direction away from a specified boundary.

The cell volume distribution in the field can be controlled by the specified *control function* F. One form of this specification is as follows: Let points be distributed on a circle having a perimeter equal to that of the specified boundary at the same arc length distribution as on that boundary. Then specify a radial distribution of concentric circles about this circle according to some distribution function, e.g., the hyperbolic tangent. Then use the volume distribution from this unequally spaced cylindrical coordinate system as F. The specification of the cell volume prevents the coordinate system from overlapping even off a concave boundary. In this case the line spacing will expand rapidly away from the boundary in order to keep the cell volume from vanishing. Although this prevents overlap, the rapid expansion that occurs can lead to problems with truncation error in some cases. This approach is extendable to 3D with the coordinate lines emanating from the boundary being orthogonal to the other two coordinates, but the latter two lines not being orthogonal. There apparently is no system, hyperbolic or elliptic, that will give complete orthogonality in 3D in general.

This hyperbolic grid generation system is faster than the elliptic generation systems by one or two orders of magnitude, the computational time required being equivalent to about that for one iteration in a solution of the elliptic system. The specification of the cell volume distribution avoids the grid line overlapping that otherwise can occur with concave boundaries in a method involving projection away from a boundary. The grid may, however, be somewhat distorted when concave boundaries are involved. The cell volume specification also allows control of the grid line spacing, but again concave boundaries may cause the intended spacing to occur in the wrong coordinate direction, since it is only the volume, and not the spacing in the two separate coordinate directions, that is controlled. As has been noted, the grid is constructed to be orthogonal.

The hyperbolic generation system is not as general as the elliptic systems, however, since the entire boundary of the region cannot be specified. Boundary slope discontinuities are propagated into the field, so that the metric elements will be discontinuous along coordinate lines emanating from boundary slope discontinuities. Finally, since hyperbolic partial differential equations can have shock-like solutions in some circumstances, it is possible for very unsuitable grids to result with some specifications of boundary point and cell volume distributions. This is in contrast with the elliptic generation system, which tends to emphasize smoothness because of the nature of elliptic partial differential equations.

1.3.5 Algebraic System

Transfinite interpolation (TFI) has become the standard for algebraic grid generation systems, and is now incorporated in most large codes. TFI can accomplish interpolation from any combination of faces, edges, and corners — with boundary orthogonality and with blending functions interpolated from boundary point distributions.

Algebraic grid generation is treated in detail in Chapter 3. This section provides an overview of the technology as applied in the EAGLE system, for example (Thompson [1988]).

1.3.5.1 Transfinite Interpolation

An algebraic three-dimensional generation system based on transfinite interpolation (using either Lagrange or Hermite interpolation) generates an initial solution to start the iterative solution of the elliptic generation system. The interpolation, in general complete transfinite interpolation from all boundaries, can be restricted to any combination of directions or lesser degrees of interpolation, and the form (Lagrange, Hermite, or incomplete Hermite) can be different in different directions or in different blocks. The blending functions can be linear or, more appropriately, based on interpolated arc length from the boundary point distributions. (This arc length is interpolated by 2D transfinite interpolation from four sides of the block.)

Hermite interpolation, based on cubic blending functions, allows orthogonality at the boundary. Incomplete Hermite uses quadratic functions and hence can give orthogonality atone of two opposing boundaries, while Lagrange, with its linear functions, does not give orthogonality.

The transfinite interpolation is done by the appropriate combination of 1D projectors, F_i, for the type of interpolation specified. (Each projector is simply the 1D interpolation in the direction indicated.) For interpolation from all sides of the section, if all three directions are indicated and the section is a volume, this interpolation is from all six sides, and the combination of projectors is the Boolean sum of the three projectors:

$$F_1 + F_2 + F_3 - F_1F_2 - F_2F_3 - F_3F_1 + F_1F_2F_3$$

With interpolation in only the two directions j and k, or if the section is a surface on which ξ^i is constant, the combination is the Boolean sum of F_j and F_k:

$$F_j + F_k - F_j F_k \quad (i, j, k)\,\text{cyclic}$$

With interpolation in only a single direction i, or if the section is a line on which ξ^i varies, the interpolation is between the two sides on which ξ^i is constant using only the single projector F_i.

With interpolation from the edges of the section, with all three directions indicated and the section a volume, the interpolation is from all 12 edges using the Boolean combination

$$F_1 F_2 + F_2 F_3 + F_3 F_1 - 2F_1 F_2 F_3$$

Interpolation from the eight corners of the section is done using $F_1\ F_2\ F_3$. There are also other possible combinations.

Blocks can be divided into subblocks for the purpose of generation of the algebraic grid and the control functions. Point distributions on the sides of the subblocks can either be specified or automatically generated by transfinite interpolation from the edge of the side. This allows additional control over the grid in general configurations, and is particularly useful in cases where point distributions need to be specified in the interior of a block or to prevent grid overlap in highly curved regions. This also allows points in the interior of the field to be excluded if desired, e.g., to represent holes in the field.

1.3.6 Adaptive Grid Schemes

Adaptive grid systems are treated in detail in Chapters 33 and 34. This section provides an overview of the technology as applied in the EAGLE system as an example, (Tu and Thompson [1991], Kim and Thompson [1990]).

Dynamically adaptive grids continually adapt to follow developing gradients in the physical solution. This adaptation can reduce the oscillations associated with inadequate resolution of large gradients, allowing sharper shocks and better representation of boundary layers. Another advantageous feature is the fact that in the viscous regions where real diffusion effects must not be swamped, the numerical dissipation from upwind biasing is reduced by the adaptation. Dynamic adaptation is at the frontier of numerical grid generation and may well prove to be one of its most important aspects, along with the treatment of real three-dimensional configurations through the composite grid structure.

There are three basic strategies that may be employed in dynamically adaptive grids coupled with the partial differential equations of the physical problems. (Combinations are also possible, of course.)

1.3.6.1 Redistribution of a Fixed Number of Points

In this approach, points are moved from regions of relatively small error or solution gradient to regions of large error or gradient. As long as the redistribution of points does not seriously deplete the number of points in other regions of possible significant gradients, this is a viable approach. The increase in spacing that must occur somewhere is not of practical consequence if it occurs in regions of small error or gradient, even though in a formal mathematical sense the global approximation is not improved. The redistribution approach has the advantage of not increasing the computer time and storage during the solution, and of being straightforward in coding and data structure. The disadvantages are the possible deleterious depletion of points in certain regions and the possibility of the grid's becoming too skewed.

1.3.6.2 Local Refinement of a Fixed Set of Points

In this approach, points are added (or removed) locally in a fixed point structure in regions of relatively large error or solution gradient. Here there is, of course, no depletion of points in other regions, and therefore no formal increase of error occurs. Since the error is locally reduced in the area of refinement, the global error does formally decrease. The practical advantage of this approach is that the original point structure is preserved. The disadvantages are that the computer time and storage increase with the refinement, and that the coding and data structure are difficult, especially for implicit flow solvers.

1.3.6.3 Local Increases in Algorithm Order

In this approach, the solution method is changed locally to a higher-order approximation in regions of relatively large error or solution gradient without changing the point distribution. This again increases the formal global accuracy elsewhere. The advantage is that the point distribution is not changed at all. The disadvantage is the great complexity of implementation in implicit flow solvers.

1.3.6.4 Formulations

Adaptive redistribution of points traces its roots to the principle of equidistribution of error, by which a point distribution is set so as to make the product of the spacing and a weight function constant over all the points:

$$w \, \nabla x = \text{constant}$$

With the point distribution defined by a function $x(\xi)$, where ξ varies by a unit increment between points, the equidistribution principle can also be expressed as

$$w \, x_\xi = \text{constant} \tag{1.4}$$

This one-dimensional equation can be applied in each direction in an alternating fashion, but a direct extension to multiple dimensions can also be made in either of two ways as follows.

From the calculus of variations, the above equation can be shown to be the Euler variational equation for the function $x(\xi)$, which minimizes the integral

$$I = \int w(x) x_\xi^{\,2} dx$$

Generalizing this, a competitive enhancement of grid smoothness, orthogonality, and concentration can be accomplished by representing each of these features by integral measures over the grid and minimizing a weighted average of the three.

The second approach is to note the correspondence between Eq. 1.4 and the one-dimensional form of the following commonly used elliptic grid generation system, Eq. 1.1. Here the *control functions, P_n,* serve to control the grid line spacing and orientation. The 1D form of Eq. 1.1 is

$$x_{\xi\xi} + P x_\xi = 0 \tag{1.5}$$

Differentiation of Eq. 1.4 yields

$$w x_{\xi\xi} + w_\xi \, x_\xi = 0 \tag{1.6}$$

Then, from Eq. 1.5 and 1.6,

$$\frac{x_{\xi\xi}}{x_\xi} = -P = -\frac{w_\xi}{w}$$

from which the control function can be taken as

$$P = \frac{w_\xi}{w} \tag{1.7}$$

It is logical then to represent the control functions in 3D as

$$P_n = \frac{w_{\xi^n}}{w}, n = 1,2,3 \tag{1.8}$$

This can be generalized to 3D as

$$P_i = \sum_j \frac{g_{ij}}{g^{ii}} \frac{(w_i)_{\xi^j}}{w_i} \tag{1.9}$$

which, in fact, does arise from a variational form (Warsi and Thompson [1990]). An example of the use of adaptive grids is shown in Figure 34.9.

1.4 Unstructured Grids

Unstructured grid generation has its roots in the finite element world of structures modeling. The real introduction to the CFD community came in the 1980s primarily from Baker, Weatherill, and Lohner. Unstructured grids have inherent simplicity of construction in that, by definition, no structure is required. Also it is not inherently necessary to communicate the actual topology of the configuration to the grid generator. Although largely synonymous with tetrahedral grids, unstructured grids may alternatively be composed of hexahedral cells (without directional structure). The term might strictly encompass any combination of cell shapes, but in the grid generation literature combinations of regions with structure (e.g., structured or prismatic grids near body surfaces) with regions without structure are generally called hybrid grids. For that matter, block-structured grids are unstructured in the large.

Traditionally, unstructured grids have been used with the finite element method. There is, therefore, an extensive literature that covers techniques to generate unstructured grids (cf. Carey [1997], George [1991], Thacker [1980]).

In this introductory chapter, it is not possible to present, in detail, all the different techniques. Instead, emphasis here will be given to one particular approach, based upon the Delaunay triangulation, which provides a powerful unstructured grid generation method. This will be used to illustrate the flexibility and characteristics of unstructured grid methods when applied to complicated geometries in two and three dimensions and in grid adaptation. Brief details of other methods will be given.

1.4.1 The Delaunay Triangulation

Structured grid generation methods place an emphasis on creating the position of points. The subsequent connections between points are defined automatically given the (i, j, k) ordering. Such ordering does not exist in unstructured grids and hence connections between points, in addition to the position of points, have to be defined by an unstructured grid method.

Grid generation based on the Delaunay triangulation (Chapter 16) uses a particularly simple criterion for connecting points to form conforming, nonintersecting elements. This geometrical construction has been known for many years, but only relatively recently has it been used for grid generation for computational

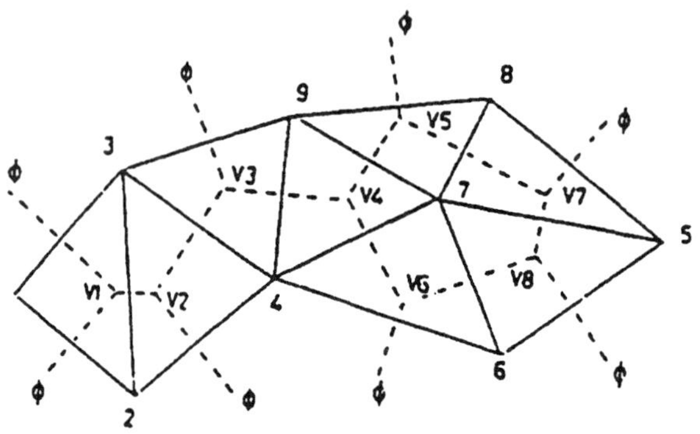

FIGURE 1.1 The Delaunay triangulation (solid line), and Voronoï regions (hashed line).

fluid dynamics. The geometrical criterion provides a mechanism for connecting points. The task of point generation must be considered independently. Hence, grid generation by Delaunay triangulation involves the two distinct problems of point connection and point creation.

1.4.1.1 Delaunay–Voronoï Geometrical Construction

Dirichlet [1850] first proposed a method whereby a given domain, in arbitrary space, could be systematically decomposed into a set of packed convex regions. For a given set of points (P_i), the space is subdivided into regions (V_i), in such a way that the region (V_i) is the space closer to P_i than to any other point. This geometrical construction of tiles is known as the Dirichlet tessellation. This tessellation of a closed domain results in a set of non-overlapping convex regions called Voronoï regions (Voronoï [1908]) that cover the entire domain. More formally, if a set of points is denoted by (P_i), then the Voronoï regions (V_i) can be defined as

$$(V_i) = \left\{P : \left\|p - P_i\right\| < \left\|p - P_j\right\|\right\}, \forall j \neq i \tag{1.10}$$

i.e., the Voronoï regions (V_i) are the set of points P that are closer to P_i than to any other point. The sum of all points p forms a Voronoï region.

From this definition, it is clear that, in two dimensions, the territorial boundary that forms a side of a Voronoï polygon must be midway between the two points that it separates and is thus a segment of the perpendicular bisector of the line joining these two points. If all point pairs that have some segment of boundary in common are joined by straight lines, the result is a triangulation within the convex hull of the set of points (P_i). This triangulation is known as the Delaunay triangulation (Delaunay [1934]). An example of this geometrical construction is given in Figure 1.1.

The construction is also valid in three dimensions. Territorial boundaries are faces that form Voronoï polyhedra and are equidistant between point pairs. If points with a common face are connected, then a set of tetrahedra is formed that covers the convex hull of points.

The Delaunay triangulation possesses some interesting properties. One such property is the in-circle criterion, which states that the circumcircles of the triangles $T(P_i)$ contain no points of the set (P_i). This applies in arbitrary dimensions and is the property used to construct an algorithm for the triangulation. As a consequence of the in-circle criterion, in two dimensions, the triangulation $T(P_i)$ also satisfied the maximum–minimum criterion, which states that if the diagonal of any strictly convex quadrilateral is replaced by the opposite one, the minimum of the six internal angles will not decrease. This is a particularly attractive property, since it ensures that the triangulation maximizes the angle regularity of the triangles, and in this way is analogous to the smoothness property of grids generated by elliptic partial differential equations.

Voronoï Vertex	Delaunay Triangle	Neighbor Voronoï Vertex
1	1 2 3	2 ϕ ϕ
2	2 3 4	1 3 ϕ
3	3 4 9	2 4 ϕ
4	4 7 9	3 5 6
5	7 8 9	4 7 ϕ
6	4 6 7	4 8 ϕ
7	5 8 7	5 8 ϕ
8	5 7 6	6 7 ϕ

FIGURE 1.2 The data structure for the Voronoï diagram and Delaunay triangulation shown in Figure 1.1.

The structure of the Voronoï diagram and Delaunay triangulation can be described by constructing two lists for each Voronoï vertex: a list of the points that define a triangle for a given vertex of the Voronoï construction (so-called forming points), and a free data structure containing the neighboring Voronoï vertices to a given Voronoï vertex. As an example, Figure 1.2 contains the vertex structure for the construction shown in Figure 1.1.

This data structure naturally extends to applications in three dimensions, where each Voronoï vertex has four forming points (tetrahedra of the Delaunay triangulation) and four neighboring Voronoï vertices.

1.4.1.2 Algorithm to Construct the Delaunay Triangulation

There are several algorithms used to construct the Delaunay triangulation. One approach, which is flexible in that it readily applies to two and three dimensions, is due to Bowyer [1981]. Each point is introduced into an existing Delaunay satisfying structure, which is locally broken and then reconstructed to form a new Delaunay-satisfying construction. In the presentation here the terms in italics indicate the interpretation for three dimensions.

Algorithm I

Step 1

Define the convex hull within which all points will lie. It is appropriate to specify four points (*eight points*) together with the associated Voronoï diagram structure.

Step 2

Introduce a new point anywhere within the convex hull.

Step 3

Determine all vertices of the Voronoï diagram to be deleted. A point that lies within a circle (*sphere*) centered at a vertex of the Voronoï diagram and passes through its three (*four*) forming points results in the deletion of that vertex. This follows from the "in-circle" criterion.

Step 4

Find the forming points of all the deleted Voronoï vertices. These are the contiguous points to the new point.

Step 5

Determine the neighboring Voronoï vertices to the deleted vertices that have not themselves been deleted. These data provide the necessary information to enable valid combinations of the contiguous points to be constructed.

Step 6

Determine the forming points of the new Voronoï vertices. The forming points of the new vertices must include the new point together with the two (*three*) points that are contiguous to the new point and form an edge (*face*) of a neighbor triangle (*tetrahedron*). These are the possible combinations obtained from Step 5.

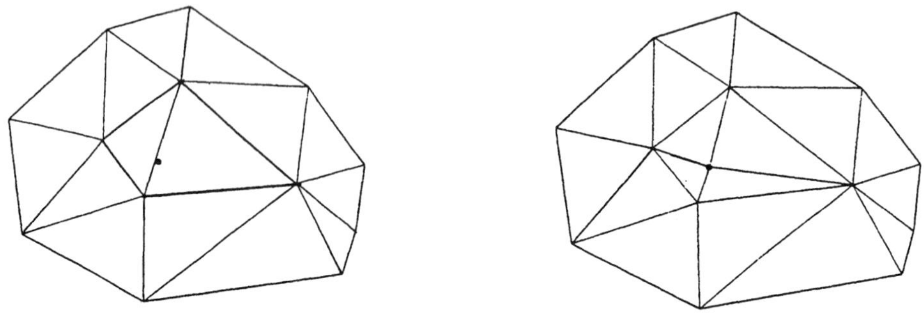

FIGURE 1.3 The addition of a new point results in deletion of some triangles and the construction of new ones.

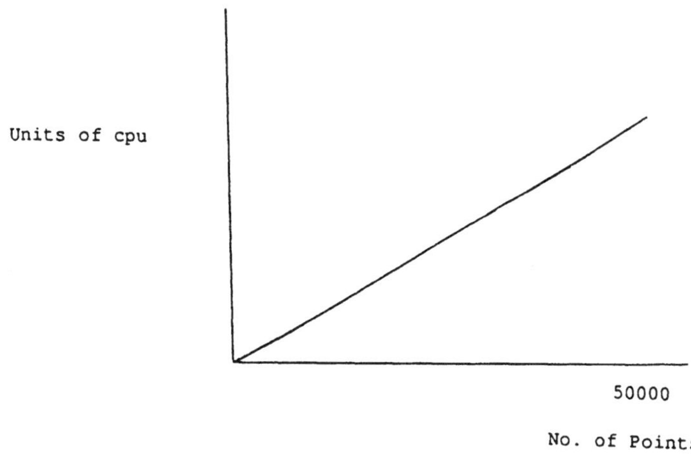

FIGURE 1.4 CPU time v. number of connected points.

Step 7

Determine the neighboring Voronoï vertices to the new Voronoï vertices. Following Step 6, the forming points of all new vertices have been computed. For each new vertex, perform a search through the forming points of the neighboring vertices, as found in Step 5, to identify common pairs (*triads*) of forming points. When a common combination occurs, then the two (*three*) associated vertices are neighbors of the Voronoï diagram.

Step 8

Reorder the Voronoï diagram structure, overwriting the entries of the deleted vertices.

Step 9

Repeat Steps 2–8 for the next point.

Figure 1.3 indicates that for a given point, the local region of influence is detected, i.e., the triangles associated with circles which contain the point. These triangles are deleted, and the new point connected to the nodes which form the enclosing polygon. This new construction is Delaunay satisfying.

The algorithm described here can be used to connect an arbitrary set of points that lie within a convex hull. The efficiency with which this can be achieved depends upon the use of appropriate data structures. The tree structure of neighbor vertices, indicated in Figure 1.2, is central to the implementation. To illustrate performances, Figure 1.4 shows a plot of CPU time against a number of elements generated in 3D on a workstation.

The algorithm described provides an important basis for an unstructured grid method. To illustrate its use and to demonstrate an additional problem, consider the problem of generating a boundary

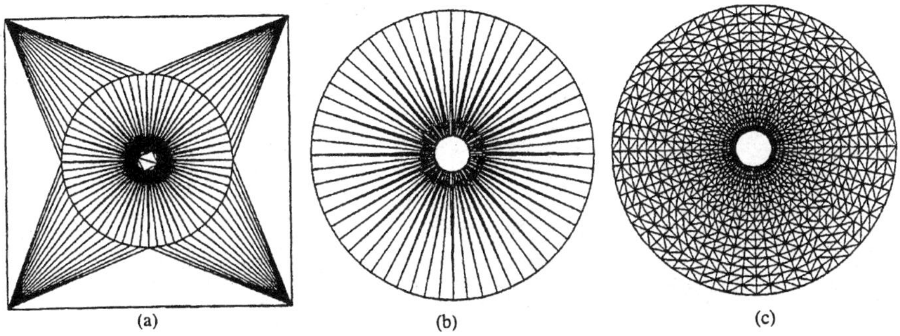

FIGURE 1.5 Delaunay triangulation of points on two circles. (a) Delaunay construction including the convex hull points. (b) Delaunay construction after the removal of the convex hull points. (c) Delaunay construction with points from a polar grid.

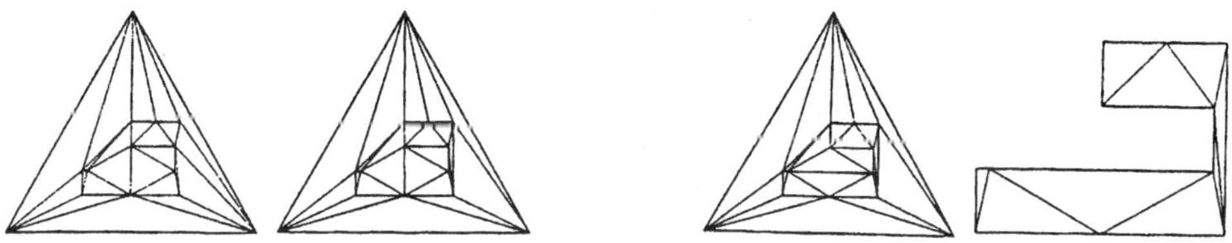

FIGURE 1.6 The boundary is completed by swapping edges in the Delaunay triangulation.

conforming grid within a multiply connected domain defined between two concentric circles. The circles are defined by a set of discrete points. Following the algorithm outlined, these points can be contained within an appropriate hull and then connected together. The result is shown in Figure 1.5a. It is clear that a set of valid triangles has been derived that covers the region of the hull. Two issues are immediately raised. First, to derive a triangulation in the specified region, triangles outside this region must be deleted. Second, if the triangles are to provide a boundary conforming grid it is necessary that edges in the Delaunay triangulation form the given geometrical boundaries of the domain. Unfortunately, given a set of points which define prespecified boundaries there is no guarantee that the resulting Delaunay triangulation will contain the edges which define the domain boundaries. Such a case is also true in three dimensions, where boundary faces must be included in the tetrahedra of the Delaunay triangulation for the resulting grid to be boundary conforming. It is necessary, therefore, to check the integrity of boundaries, and if found not to be complete, appropriate steps must be taken.

Prespecified boundary connectivities can be reconstructed by combinations of edge swapping to recover boundaries in two dimensions is given in Figure 1.6. The given boundary edges are recovered through edge swapping. In 3D, this problem is more severe and requires careful attention.

Once the boundary is complete, it is a simple task to delete triangles exterior to the region of interest. Deletion of unwanted triangles in Figure 1.5a leads to the triangulation shown in Figure 1.5b.

Figure 1.5b represents a valid triangulation of the points that define the two concentric circles. However, the triangles span the entire region and are clearly inappropriate for any form of analysis. Hence, it is necessary to address the problem of point creation.

1.4.2 Point Creation

1.4.2.1 Points Created by an Independent Generation Technique

Points for connection by the Delaunay algorithm could be derived by a method external to the triangulation routine. For example, in the case of the two circles, a polar grid could be generated and the set of points then connected together to form the grid. Such a triangulation with polar grid points is overly

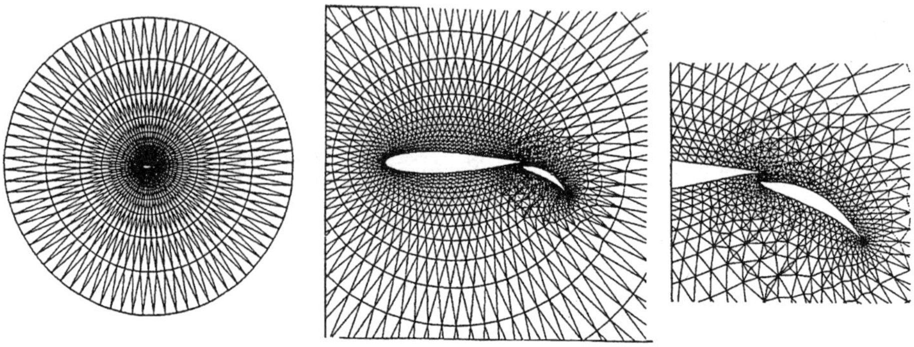

FIGURE 1.7 Unstructured grid with points generated from a structured method.

FIGURE 1.8 Delaunay triangulation of a regular set of points superimposed over the domain.

complicated. However, for more realistic domains, which may be more geometrically complex, the approach can prove to be effective. Taking an example from aerospace engineering, Figure 1.7 shows a grid in which two structured grids have been independently generated around the two components and the total set of points connected together to form the unstructured mesh. For more general geometries, alternative, more flexible point creation routines are required.

1.4.2.2 Points Created by Grid Superposition and Successive Subdivision

It is possible to extend the use of an independent grid generation technique to include grid superposition and successive subdivision. The basic idea is to superimpose a regular grid over the domain. The regular grid can be generated using a quadtree or octree data structure that allows point density in the regular grid to be consistent with point spacing at the boundary. An example of this approach is shown in Figure 1.8. In general, this approach results in good spatial discretization in the interior of the domain, although in the vicinity of boundaries the grid quality can be poor.

1.4.2.3 Point Creation Driven by the Boundary Point Distribution

For grid generation purposes, the domain is defined by points on the geometrical boundaries. It will be assumed that this point distribution reflects appropriate geometrical features, such as variation in boundary curvature and gradient. Ideally, any method for automatic point generation should ensure that the boundary point distribution is extended into the domain in a smooth manner. A procedure that has proved successful in creating a smooth point distribution consistent with boundary point spacing and that naturally extends to three dimensions is as follows.

Algorithm II

Step 1

Compute the point distribution function dp_i for each boundary point $\mathbf{r}_i = (x, y)$, where it is assumed that points i+1 and i are contiguous:

$$dp_i = 0.5\left[\sqrt{(\mathbf{r}_{i+1} - \mathbf{r}_i)^2} + \sqrt{(\mathbf{r}_i - \mathbf{r}_{i-1})^2}\right]$$

Step 2

Generate the Delaunay triangulation of the boundary points.

Step 3

Initialize $j = 0$.

Step 4

For each triangle T_m within the domain, perform the following:

a. Define a point at the centroid of the triangle T_m, with nodes n_1, n_2, n_3:

$$\mathbf{P}_c = \frac{1}{3}(\mathbf{r}_{n1} + \mathbf{r}_{n2} + \mathbf{r}_{n3})$$

b. Derive the point distribution function dp_c by interpolating the point distribution function from the nodes n_1, n_2, n_3:

$$dp_c = \frac{1}{3}(dp_{n1} + dp_{n2} + dp_{n3})$$

c. If

$$|\mathbf{P}_c - \mathbf{r}_{nk}| < \alpha\, dp_c \quad k = 1,2,3$$

then reject point P_c; next triangle. If

$$|\mathbf{P}_c - \mathbf{r}_{nk}| > \alpha\, dp_c \quad k = 1,2,3$$

then, if

$$|P_j - P_c| > \beta\, dp_c \quad j = 1,\ldots,N$$

accept point $\mathbf{P}_c$ and add to list $\mathbf{P}_j$, $j = 1, N$. If

$$|\mathbf{P}_j - \mathbf{P}_c| < \beta\, dp_c \quad j = 1,\ldots,N$$

then reject point $\mathbf{P}_c$; next triangle

Step 5

If $j = 0$, go to Step 7.

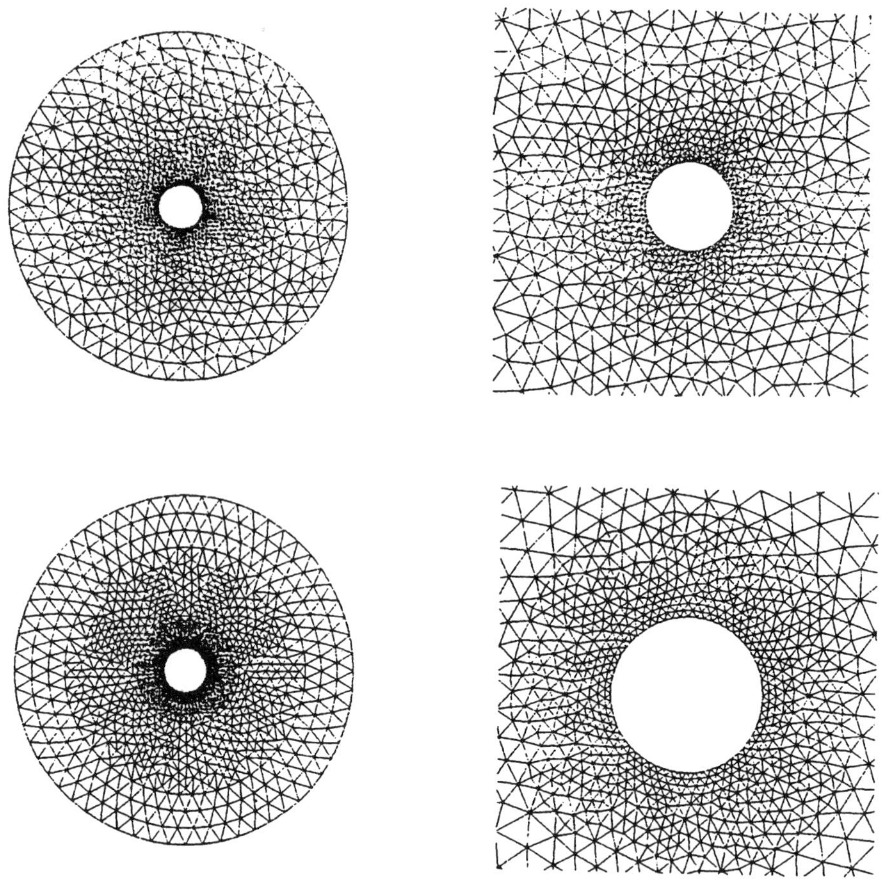

(a)

(b)

FIGURE 1.9 Grid point creation and distribution controlled by the boundary point distribution.

Step 6

Perform Delaunay triangulation of the derived points $\mathbf{P}_j$, $j = 1, N$. Go to Step 3.

Step 7

Smooth the grid.

It proves beneficial to smooth the position of the grid points using a Laplacian filter. The coefficient α controls the grid point density, while β has an influence on the regularity of the triangulation. Figure 1.9 shows two triangulations produced using this point creation algorithm. In Figure 1.9a, $\alpha = 1$, $\beta = 10$, while in Figure 1.9b $\alpha = 1$, $\beta = 0.02$. The effect of β is clearly evident. A more realistic example is given in Figure 1.10, where a grid is shown for a finite element stress analysis of a tooth.

The algorithm outlined is equally applicable in three dimensions. Figure 1.11 shows an unstructured tetrahedral grid around an airplane. Appropriate point clustering has been achieved close to the plane. A flow solution has been computed on this mesh as a further demonstration of the applicability of the approach.

The procedure outlined creates points consistent with the point distribution on the boundaries. However, in many problems information is known about features within the domain that require a suitable spatial discretization. It proves possible to modify the above algorithm to take such effects into account. Two techniques can be readily implemented. The first utilizes the idea of point and line sources, while the second uses the concept of a background mesh.

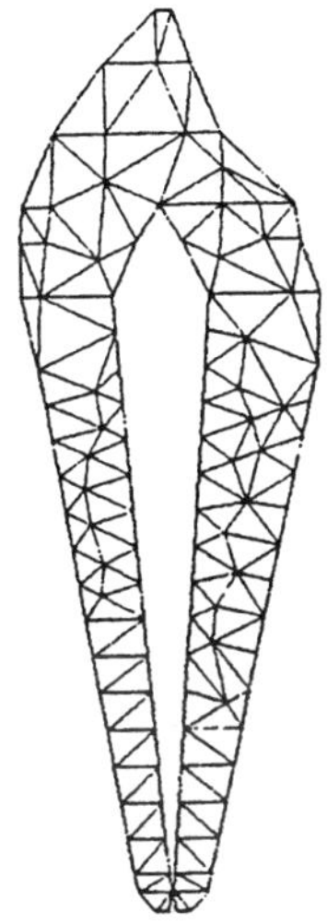

FIGURE 1.10 Unstructured grid used for finite element stress analysis.

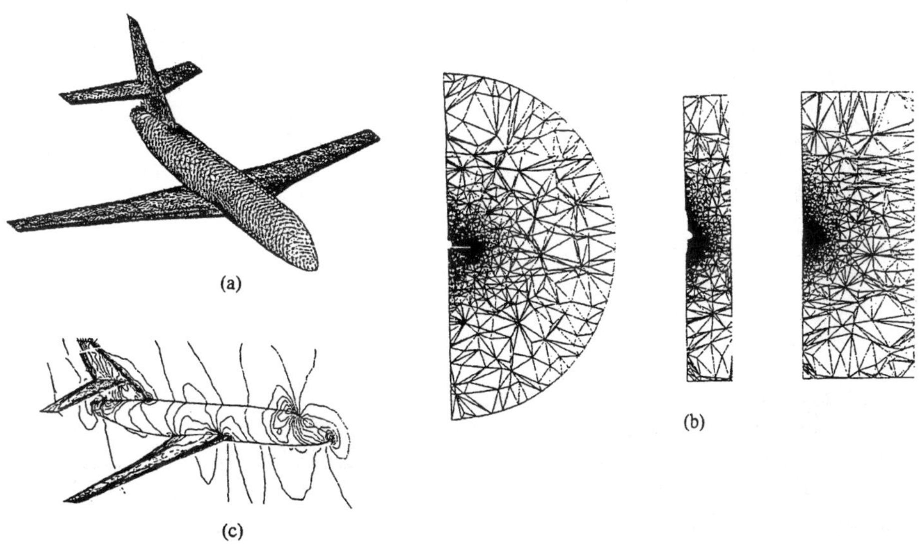

FIGURE 1.11 Automatic point creation in three dimensions driven by boundary point distribution. (a) Surface grid. (b) Cuts through the field. (c) Solution of inviscid flow.

1.4.2.4 Point Creation Controlled by Point and Line Sources

In somewhat of an analogous way to point sources used as control functions with elliptic PDEs, it is possible to define line and point sources to provide grid control for unstructured meshes. Local grid point spacing can be defined as

$$dp_i = \min\left\{A_j e^{B_j|\mathbf{R}_j - \mathbf{r}_i|}\right\},\ j = 1,\ldots \text{number of sources} \tag{1.11}$$

where A_j, B_j, and $\mathbf{R}_j$ (j = 1, number of sources) are user-controlled amplification and decay parameters and the position of sources, respectively. Grid point creation is then performed as outlined, but in Step 4b, the appropriate point distribution function at the centroid is determined by Eq. 1.10. In practice, the substitution of Eq. 1.10 for Step 4b is trivial.

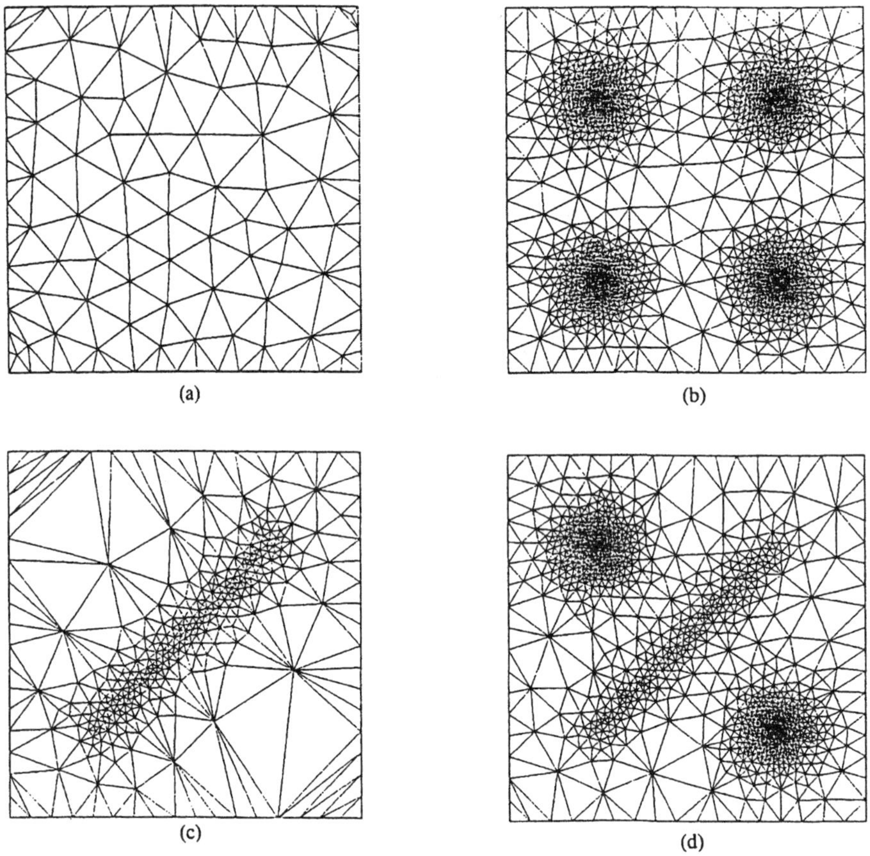

FIGURE 1.12 Point density controlled through point and line sources.

Examples of the use of point sources are given in Figure 1.12a. Figure 1.12a shows the mesh controlled through the boundary point distribution, while in Figure 1.12b a point source has been specified. It is clear that grid spacing is controlled by the source position and associated parameters. Line sources can also be introduced. For simplicity, line sources are treated as a series of point sources. In this way Eq. 1.10 is also applicable. An example of grid control by a line source is given in Figure 1.12c.

Combinations of line and point sources can also be used, such as the example shown in Figure 1.12d. It should be noted that the user-specified information to implement the sources is minimal.

It is clear that the sources provide a mechanism for clustering points. To ensure an adequate point spacing in regions not influenced by the sources, it is appropriate to use a combination of point spacing derived from the sources and the boundary point distribution. In practice, this can be implemented by defining the local length parameter to be

$$dp_i = \min\left\{ A_j e^{B_j|\mathbf{R}_j - \mathbf{r}_i|}, dp_{\text{boundary}} \right\}, j = 1, \ldots \text{number of sources}$$

where dp_{boundary} is the point spacing parameter defined from the boundary point distribution and derived using Algorithm II. An example of the use of this approach is shown in Figure 1.13, for ocean modeling of the North Atlantic. An unstructured grid generated from the boundary point spacing is shown in Figure 1.13a, while in Figure 1.13b, a single line source has been appropriately positioned to ensure a higher point resolution to capture the Gulf Stream.

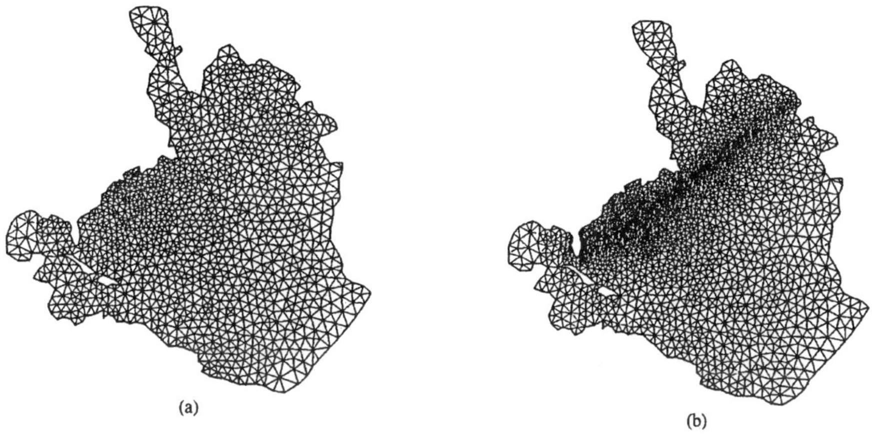

FIGURE 1.13 Unstructured grid for the North Atlantic. (a) Points created from boundary point spacing. (b) Points created from boundary point spacing and line source.

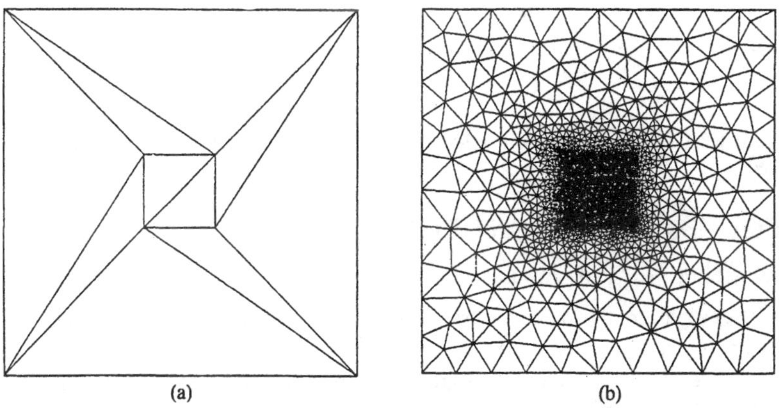

FIGURE 1.14 Grid control using a background mesh.

1.4.2.5 Point Creation Controlled by a Background Mesh

An alternative way to control grid point spacing is by defining a background mesh on which the local point spacing is defined. To implement this approach in the proposed algorithm the point spacing in Step 4b is derived from the background mesh.

As an example, consider the rectangular domain in Figure 1.14a. A background mesh is defined of 10 elements at whose nodes a point spacing is specified. If small spacing is defined at interior nodes and larger spacing on the boundary, then the use of the automatic point creation algorithm results in the mesh shown in Figure 1.14b. Modification to the topology of the background mesh or the point spacing function at nodes allows complete control of the unstructured grid density. In practice, the background mesh is a previous mesh used for analysis in which a measure of the analysis is converted, by an appropriate transformation, to spacings in the physical domain. An example of this is given in the section on adaptation techniques.

The above ideas provide some examples of the way unstructured grids can be generated. There is considerable flexibility to introduce points where required. Algorithms to construct such grids are not over complicated to program and are fast and efficient provided, as already emphasized, appropriate care is taken with respect to data structures.

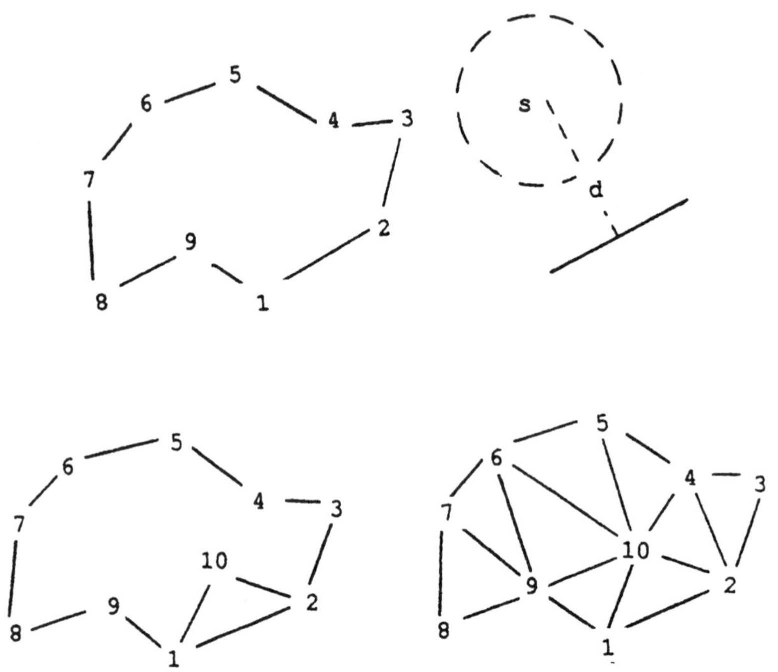

FIGURE 1.15 Test case example indicating the steps in the advancing front technique.

1.4.3 Other Unstructured Grid Techniques

1.4.3.1 Advancing Front Methods

Another class of unstructured grid generators is based upon the idea of an advancing front (Chapter 17). Such methods construct a mesh of a domain from its boundary information. The method is applicable in both two and three dimensions, where triangles and tetrahedra are generated, respectively.

The basic ideas of the method are best illustrated in two dimensions. Consider a region bounded by points (P_i) and edges ($P_m P_n$). The edges are called the front. A test case example is shown in Figure 1.15. To construct a grid in the domain, perform the following:

1. Choose an edge on the front, say P_1 P_2.
2. Construct the perpendicular bisector of P_1 P_2 and create a point s a distance d into the domain.
3. Create a circle, center s, of radius r.
4. Determine any points which lie within this circle (a_i) and order them in distance from the center s.
5. Form triangles ($P_1 P_2$ a_i) and accept the first triangle that satisfies the following conditions:
   a. Edges ($P_1 a_i$) and ($P_2 a_i$) do not intersect any other edges.
   b. Triangle ($P_1 P_2$ a_i) satisfies an appropriate quality indicator. (Such indicators are based upon regularity of interior angles, etc.)
6. If a_i is a new point, add to the list of points. Add any new edges to the front and delete the old edge ($P_1 P_2$).

This procedure is repeated until the front is empty, i.e., there are not edges left in the front.

In the above algorithm, the grid point density can be controlled by the length d_1, i.e., the distance away from the mid-point of the current edge on the front. This length parameter can be obtained using a background mesh, or a distribution of line or point sources. The effects are the same as those indicated for the Delaunay algorithm.

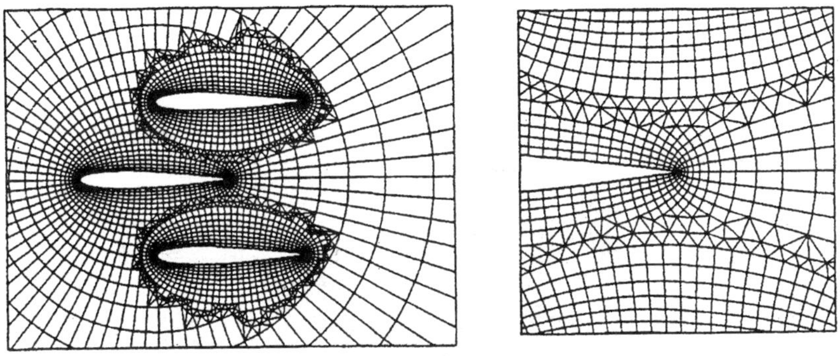

FIGURE 1.16 Hybrid grid for multiple airfoils.

In principle, this basic procedure holds for applications in three dimensions, where the front consists of a set of triangular faces which bound the domain. In practice, the implementation of this basic procedure requires an efficient data structure to ensure realistic computational times.

It is worth commenting that the advancing front technique can be used to also generate grids with elements aligned in given directions. This is achieved by introducing a directional parameter, in addition to a length parameter d. In this way, instead of constructing a line perpendicular to the edge on the front, a line inclined in the specified direction can be generated. Again the directional parameters can be specified in the background mesh.

1.4.3.2 Quadtree and Octree

Grid generation based upon quadtree (2D) and octree (3D) have recently been introduced (Chapter 15). Such methods begin with a point definition of the boundaries. Superimposed over the domain is a sparse regular grid that is subdivided so that at boundaries the cell size is consistent with the boundary point spacing. The data points and cells are contained in the quadtree or octree data structure. The grid is made to be boundary conforming by appropriate cutting and reconnecting to form triangles and tetrahedra.

1.4.3.3 Hybrid Grids

To achieve an optimum compromise between regularity and flexibility, it is possible to combine grid types in the form of hybrid or structured–unstructured grids (Chapter 23). Figure 1.16 shows three airfoils where each is locally discretized using a structured grid that is connected using an unstructured grid. The idea can be also applied in three dimensions. Figure 1.17 shows a surface grid for a fuselage, wing, pylon, and nacelle, where the pylon and nacelle components have been incorporated into a structured grid using locally unstructured grid. Such grid generation techniques require analysis modules that can utilize mixed element types.

1.4.4 Unstructured Grid Generation on Surfaces

In most engineering applications it is not possible to define a surface in a closed form mathematical expression. In general, a given surface is defined as a discrete set of points, that map to a regular array. In such cases, it is possible to define the surface in terms of two parametric coordinates (u, v) where each pair maps to a point on the surface. With this description of a surface it is possible to construct grids on surfaces by utilizing 2D techniques applied in the parametric coordinates (Chapter 19). The final grid on the surface is then obtained by mapping to the physical space. The point connections remain fixed through the transformation.

Complicated surfaces can be defined by using more than one set of rectangular point sets. Figure 1.18 shows a grid in parametric coordinates, which when mapped to physical space becomes an unstructured grid on the surface of a wing. The grid at the tip of the wing is treated separately from the grid on upper and lower surfaces of the wing.

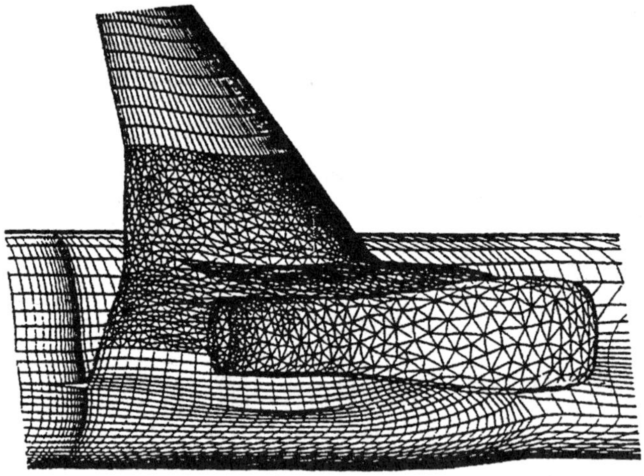

FIGURE 1.17 Hybrid grid for a wing, fuselage, pylon, and nacelle.

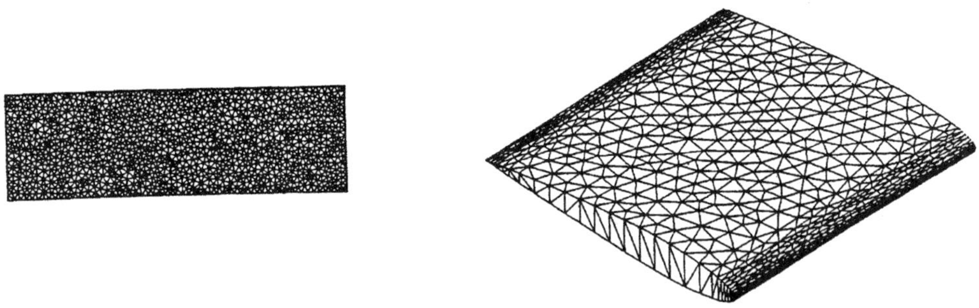

FIGURE 1.18 Grid in parametric space converted to unstructured surface grid in physical space.

1.4.5 Adaptation on Unstructured Grids

The basic principles of adaptation have been given above in Section 1.3.6. Here comments will be restricted to grid adaptation on unstructured grids (Chapter 35).

1.4.5.1 Point Enrichment

Local point enrichment can be achieved on any grid type. It is an effective way to ensure greater resolution of the domain in critical regions. It is most naturally employed on unstructured grids where, on the addition of a point and the subsequent connections, the data format of the grid does not change. Hence, solution modules do not require any modifications. This is not true for structured grids where the addition of a point breaks the (i, j, k) data format and hanging or nonconforming nodes are created.

Points are added to the domain in regions where some measure of error or solution activity is large. Dependent upon the problem, a suitable indicator f is chosen. From a computation on a grid, the indicator f is known at all points. It is then possible to construct operators that indicate where additional points should be added. For example, point enrichment could be driven by detecting changes in f along edges. If

$$\frac{|\phi_a - \phi_b|}{|\phi_i - \phi_j|_{\max}} > \text{tolerance,} \tag{1.12}$$

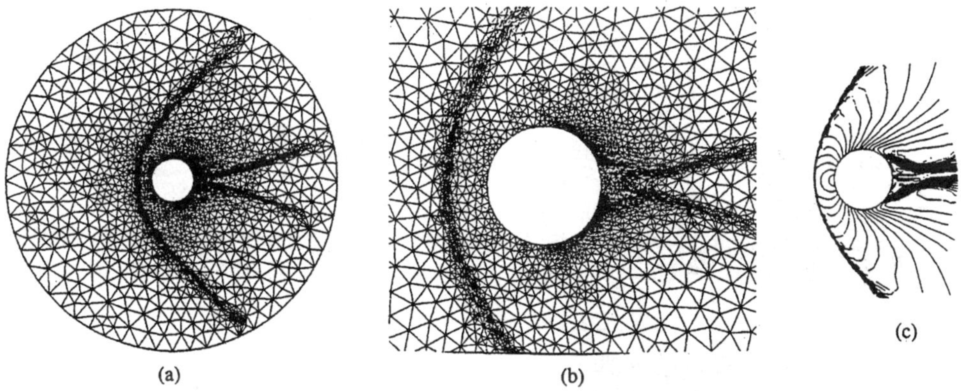

FIGURE 1.19 Mach 3 flow around a cylinder showing point enrichment. Flowfield contours of density also shown.

then add a point along the edge *a* to *b*. Connections to the new node can then be made. Similar expressions can be constructed for triangles, tetrahedra, etc. For some class of problems, more sophisticated error indicators can be used. These can be applied to give a solution which can have a prespecified bound on the errors. In some regions of the domain it may be possible to delete nodes. This process can be driven by criteria similar to the one for enrichment.

Examples of point enrichment and derefinement are given in Figure 1.19 and Figure 1.20a. The first example illustrates the use of point enrichment, driven by gradients of density, on an unstructured grid, for a simulation of Mach 3 flow around a cylinder. Contours of density for the flowfield are shown in Figure 1.19c. It is clear that points have been added where gradients in density are high. Figure 1.20a shows an enriched structured multiblock grid. As indicated earlier, once points are added to such a grid, the data format has to be modified. To provide flexibility for grid point enrichment on such grids, the data format has been converted to quadtree. In the example shown the point addition was driven by gradients in density. Contours of density are shown in Figure 1.20d, again confirming that the point enrichment has occurred in the relevant regions.

1.4.5.2 Node Movement

Node enrichment successfully enhances the resolution of an analysis. However, it can become computationally expensive and provides a diminishing return on successive enrichments. After ensuring that there is sufficient mesh point resolution, node movement can provide the required mechanism to achieve high resolution at a negligible cost. Many techniques have been explored to move points. One that is particularly simple, is applicable to all grid types, and is effective, is based on a weighted Laplacian formulation. A typical form is the following:

$$\mathbf{r}_0^{n+1} = \mathbf{r}_0^n + \omega \frac{\sum_{i=1}^{M} C_{io}\left(\mathbf{r}_i^n - \mathbf{r}_0^n\right)}{\sum_{i=1}^{M} C_{io}} \tag{1.13}$$

where $\mathbf{r} = (x, y)$, $\mathbf{r}_o^{n+1}$ is the position of node *o* at relaxation level $n + 1$, C_{io} is the adaptive weight function between nodes *i* and *o*, and ω is the relaxation parameter. The summation is taken over all edges connecting points *o* and *i*, where it is taken that there are *M* surrounding nodes. The weight C_{io} can be taken as a measure of activity, and a typical form is

$$C_{io} = \alpha_1 + \alpha_2 \frac{\left|\phi_i - \phi_o\right|}{\left|\phi_i + \phi_o\right|} \quad i = 1, \ldots, M \tag{1.14}$$

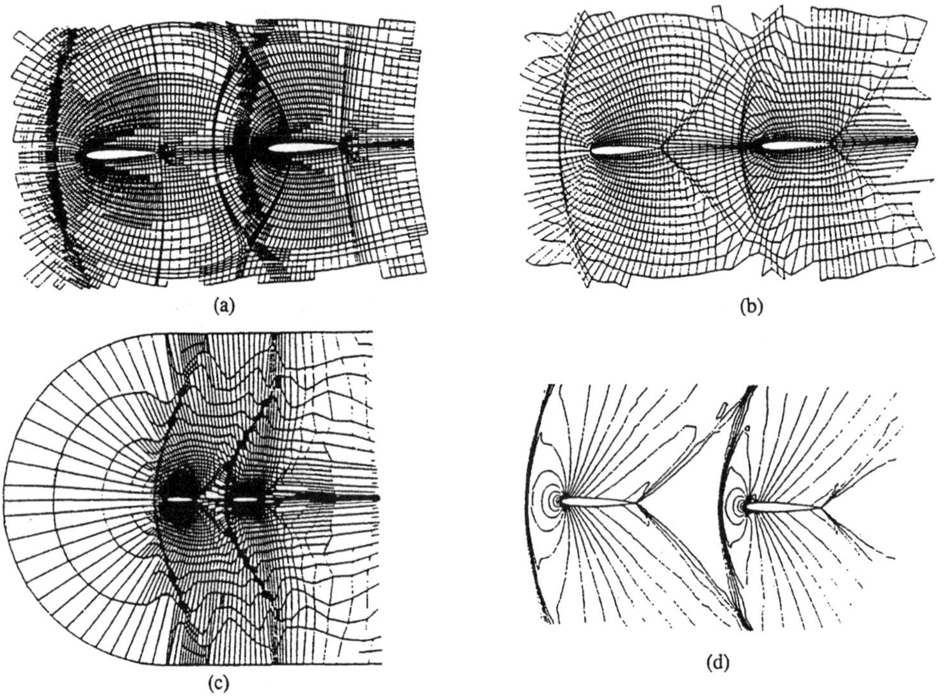

FIGURE 1.20 Adaption on a multiblock grid; (a) point enrichment, (b) node movement, (c) point enrichment, derefinement and node movement, (d) contours of density.

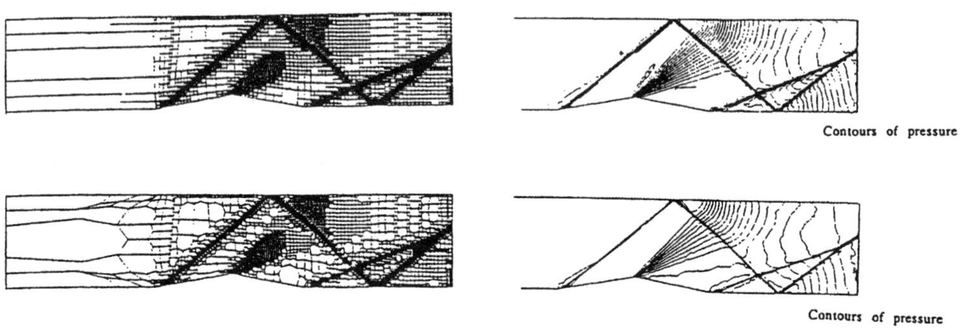

FIGURE 1.21 Mach 2 flow in a channel showing refined/derefined grid with flow contours of density.

where ϕ is the adaptive parameter, α_1 and α_2 are constants.

Figure 1.20b shows a multiblock structured grid that has been adapted using node movement driven by density gradients. As was the case for point enrichment, comparison with the contours of density confirms the point movement has been into regions with high gradients.

For extensive numerical studies it appears that the use of point enrichment, derefinement, and movement should be closely coupled. Sequences of these adaptive procedures give the optimum results, as judged by solution accuracy and computational efficiency. Examples of computations where these adaptive mechanisms have been cycled are given in Figure 1.20c and Figure 1.21. In Figure 1.21, the contours of density on a refined and derefined grid can be compared with those obtained after the grid points have been moved. A clear improvement in shock capturing is evident.

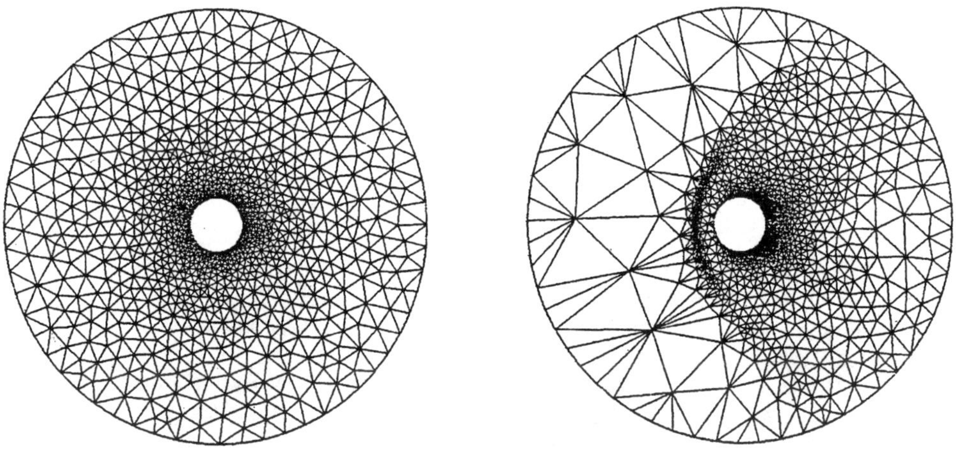

FIGURE 1.22 Mach 3 flow over a cylinder showing remeshing.

1.4.5.3 Remeshing

The concept of grid point generation driven by the spacing specified on a background mesh can be utilized for adaptation. In this case, the result of an analysis can be used to construct spacings, which are then assigned to the mesh, which in turn are used in the background mesh. There are several ways of performing the transformation between results and local length scales, but typically they take the form of

$$dp_i^{\text{new}} = dp_i^{\text{old}} \cdot \frac{\phi_{\text{average}}}{\phi_i} \tag{1.15}$$

where dp_i^{new} and dp_i^{old} are the new and old point distributions, ϕ_i is the adaptive indicator, ϕ_{average} is the indicator averaged throughout the domain. An example of remeshing, where the initial mesh is used as a background mesh and pressure was used as the adaptive indicator, is given in Figure 1.22. It is seen that local point clustering has occurred in the vicinity of the bow shock wave, but not in the region rear of the cylinder, which might be expected from the contours shown in Figure 1.19c. This illustrates a key area in grid adaption, in that, although there are steep gradients in density rear of the cylinder, there are no such gradients in pressure. Hence, the adaption process for remeshing, driven in this example by pressure, does not detect the features in density rear of the cylinder. As yet, for flow problems, there is no universal indicator and hence the selection of the parameter has to be made on a case-by-case basis.

1.4.6 Summary

Unstructured grids provide considerable flexibility for the discretization of complex geometries and grid adaptation. In these sections a brief outline has been given on such techniques. Particular details have been given on the use of the Delaunay triangulation. It should be emphasized that, although the majority of applications have been drawn from aerospace engineering, the ideas and principles discussed are equally applicable to other fields.

References

1. **Arcilla, A. S., Hauser, J., Eiseman, P. R., and Thompson, J. F.**, (Eds.), *Numerical Grid Generation in Computational Fluid Dynamics and Related Fields, Proceedings of the 3rd International Conference*, North Holland, 1991.
2. **Bowyer, A.**, Computing Dirichlet Tessellations, *The Computer Journal*, Vol. 24, pp. 162–166, 1981.

3. **Carey, G. F.**, *Computational Grids: Generation, Adaptation, and Solution Strategies*, Taylor & Francis, 1997.
4. **Castillo, J. E.**, (Ed.), *Mathematical Aspects of Numerical Grid Generation*, SIAM Press, 1991.
5. **Choo, Y.**, (Ed.), *Proceedings of the Surface Modeling, Grid Generation and Related Issues in Computational Fluid Dynamics Workshop*, NASA Conference Publication 3291, NASA Lewis Research Center, Cleveland, OH, May 1995, p. 359.
6. **Delaunay, B.**, Sur la sphere vide, Bulletin of Academic Science URSS, Class. Science National, 1934, pp. 793–800.
7. **Dirichlet, G. L.**, Uber die Reduction der positiven Quadratischen formen mit drei Underestimmten Ganzen Zahlen, *Z. Reine Angew. Mathematics*, Vol. 40, No. 3, pp. 209–227, 1850.
8. **Eiseman, P. R.**, Grid generation for fluid mechanics computations, *Annual Review of Fluid Mechanics*, Vol. 17, 1985.
9. **Soni, B. K., Thompson, J.F., Eiseman, P.R., and Hauser, J.**, (Eds.), *Numerical Grid Generation in Computational Field Simulation. Proceedings of the 5th International Conference*, MSU Publisher, Mississippi State, MS, U.S., April 1996.
10. **Eiseman, P. R., Hauser, J., Thompson, J. F., and Weatherill, N. P.**, (Eds.), *Numerical Grid Generation in Computational Field Simulation and Related Fields, Proceedings of the 4th International Grid Conference*, Pineridge Press, Swansea Wales, U.K., 1994.
11. **George, P. L.**, Automatic Mesh Generation, Wiley Publications, 1991.
12. **Godunov, S. K. and Propokov, G. P.**, On the computation of conformal transformations and the construction of difference meshes, *Zh. Vychisl. Mat. Mat. Fiz.*, Vol. 7, p. 209, 1967.
13. **Gordon, W. J. and Thiel, L. C.**, Transfinite mappings and their application to grid generation, *Numerical Grid Generation*, Thompson, J. F., (Ed.), North Holland, 1982.
14. **Hauser, J. and Taylor, C.**, (Ed.), *Numerical Grid Generation in Computational Fluid Dynamics, Proceedings of the 1st International Conference*, Pineridge Press, 1986.
15. **Kim, J. K. and Thompson, J. F.**, Three-dimensional solution-adaptive grid generation on a composite block configuration, *AIAA Journal*, Vol. 28, p. 420, 1990.
16. **Knupp, P. and Steinberg, S.**, *Fundamentals of Grid Generation*, CRC Press, Boca Raton, FL, 1993.
17. **Miki, K. and Takagi, T.**, A domain decomposition and overlapping method for the generation of three-dimensional boundary-fitted coordinate systems, *Journal of Computational Physics*, Vol. 53, p. 319, 1984.
18. **Rubbert, P. E. and Lee, K. D.**, Patched coordinate systems, *Numerical Grid Generation*, Thompson, J.F., (Ed.), North-Holland, 1982.
19. **Sengupta, S., Hauser, J., Eiseman, P. R., and Thompson, J. F.**, (Eds.), *Numerical Grid Generation in Computational Fluid Dynamics 1988, Proceedings of the 2nd International Conference,* Pineridge Press, 1988.
20. **Smith, R. E.**, (Ed.), *Numerical Grid Generation Techniques*, NASA Conference Publication 2166, NASA Langley Research Center, 1980.
21. **Smith, R. E.**, (Ed.), *Proceedings of the Software Systems for Surface Modeling and Grid Generation Workshop*, NASA Conference Publication 3143, NASA Langley Research Center, Hampton, VA, 1992, p. 161.
22. **Sorenson, R. L.**, The 3DGRAPE book: theory, users' manual, examples, NASA TM-10224, 1989.
23. **Thacker,** *Int. J. Numer. Meth. Eng.*, 15, p. 1335, 1980.
24. **Thompson, J. F.**, (Ed.), *Numerical Grid Generation*, North Holland, 1982. (Also published as Vol. 10 and 11 of *Applied Mathematics and Computation*, 1982.)
25. **Thompson, J. F.**, Grid generation techniques in computational fluid dynamics, *AIAA Journal*, Vol. 22, p. 1505, 1984.
26. **Thompson, J. F., Warsi, Z. U. A., and Mastin, C. W.**, *Numerical Grid Generation: Foundations and Applications*. North Holland, 1985.
27. **Thompson, J. F.**, A survey of dynamically adaptive grids in the numerical solution of partial differential equations, *Applied Numerical Mathematics*, Vol. 1, p. 3, 1985.

28. **Thompson, J. F.**, A general three-dimensional elliptic grid generation system on a composite block structure, *Computer Methods in Applied Mechanics and Engineering*, Vol. 64, p 377, 1987.
29. **Thompson, J. F.**, A composite grid generation code for 3D regions —the EAGLE code, *AIAA Journal*, Vol. 26, p. 915, 1988,.
30. **Thompson, J. F.**, A reflection on grid generation in the '90s: trends, needs and influences, *Numerical Grid Generation in Computational Field Simulation*. Soni, B. K., Thompson, J. F., Hauser, J., Eiseman, P. R., (Eds.), *Proceedings of the 5th International Conference*, MSU Publisher, Mississippi State, MS, U.S., April 1996, p. 1029.
31. **Thompson, J. F., Warsi, Z. U. A., Mastin, C. W.**, Boundary-fitted coordinate systems for numerical solution of partial differential equations — a review, *J. of Computational Physics*, Vol. 47, p. 1, 1982.
32. **Tu, Y. and Thompson, J. F.**, Three-dimensional solution-adaptive grid generation on composite configurations, *AIAA Journal*, Vol. 29, p. 2025, 1991.
33. **Voronoï, G.**, Nouvelles applications des parametres continus a la theorie des formes quadratiques. recherches sur les parallelloedres primitifs, *Journal Reine Angew. Mathematics*, Vol. 134, 1908.
34. **Warsi, Z. U. A. and Thompson, J. F.**, Application of variational methods in the fixed and adaptive grid generation, *Computers in Mathematics with Applications*, Vol. 19, p. 31–41, 1990.
35. **Weatherill, N. P. and Forsey, C. R.**, Grid generation and flow calculations for complex aircraft geometries using a multi-block scheme, AIAA-84-1665, *AIAA 17th Fluid Dynamics, Plasma Dynamics, and Laser Conference*, Snowmass, CO, 1984.
36. **Winslow, A. M.**, Equipotential zoning of two-dimensional meshes, *J. of Computational Physics*, Vol. 1, p. 149, 1966.
37. **Winslow, A. M.**, Numerical solution of the quasilinear Poisson equation in a nonuniform triangle mesh, *Journal of Computational Physics*, Vol. 135, pp. 128–138, 1997; reprinted from November 1966, Vol. 1, Number 2, pp. 149–172.

2

Mathematics of Space and Surface Grid Generation

Zahir U. A. Warsi

2.1 Introduction

The purpose of this chapter is to provide a comprehensive mathematical background for the development of a set of differential equations that are geometry-oriented and are generally applicable for obtaining curvilinear coordinates or grids in intrinsically curved surfaces. To achieve this aim it is imperative to consider some geometrical results on curvilinear coordinates in the embedding space. The geometrical results are usually a consequence of some differential operations in the embedding space which also lead toward the theory of curves. The embedding space for non-relativistic problems is Euclidean or flat. Sections 2.2, 2.3, and 2.4 contain some basic results that are more fully explained in the books by Struik [1], Kreyszig [2], Willmore [3], Eisenhart [4], Aris [5], and McConnell [6] among others, and in a monograph by Warsi [7]. In the course of development of the subject in this chapter, some elementary tools and results of tensor analysis have helped to provide concise results

0-8493-2687-7/99/$0.00+$.50

with full generality.* This chapter mainly focuses on one aspect of grid generation, which is the method of elliptic partial differential equations. It has been shown that the developed equations automatically satisfy some important results of the theory of surfaces. From this we conclude that the developed equations should be preferred to any other arbitrarily chosen set of equations to generate coordinates or grids in a surface. Another important outcome of these model equations is that the "fundamental theorem of surface theory" can be re-stated in a computationally realizable form. In other words, the proposed model equations can also be used to generate a surface if appropriate metric data** has been specified. Thus the proposed model equations have dual use, viz., generating the coordinate lines in a given surface, or generating a surface based on the metric data. Further, because of the elliptic nature of the equations, the generated grid lines will be smooth.

The idea of coordinate generation by using the elliptic partial differential equations in a plane is essentially due to Winslow [8]. However, if one stretches backward from Winslow to trace the foundations of the theory of coordinate generation by elliptic partial differential equations, then it is not possible to escape from the conclusion that the seed work was done by Allen [9], though in a different context. Later Chu [10] and Thompson, et al. [11] have used Winslow's model for applications. In [11] extensive work was done to choose the coordinate control functions for application to a variety of problems. The application of the methods developed in [11] to extremely difficult problems involving geometries encountered in aeronautical engineering made the method of grid generation an important tool in CFD. Many years of work by a number of researchers and workers was published in a book [12]. Other books have followed in recent times ([13, 14]).

In an attempt to generalize the Winslow model of numerical coordinate generation, and further, to provide a mathematical foundation to the model equations, Warsi [15–18] has used the formulae of Gauss to arrive at the model equations as discussed in the cited references and in this article. These model equations are applicable for coordinate generation on generally curved surfaces with the coordinate generators (the control functions) appearing in them in a natural way. As noted earlier the same equations can also be used to generate a surface. For a plane these model equations reduce to those given in [8–11]. Some authors have also developed the surface coordinate generation model by using variational methods [19–21].

2.2 A Résumé of Differential Operations in Curvilinear Coordinates

For a presentation of a connected account of the theory of numerical coordinate mapping, it is imperative to review some basic concepts and formulae pertaining to the differential operations in curvilinear coordinates. As noted in the introduction, the formulae obtained by using simple tensor operations expose themselves effectively and in their full generality. Thus we use the symbol x^i, $i = 1, 2, \ldots, n$ to represent a curvilinear coordinate system in either a Euclidean or non-Euclidean n-space. In a Euclidean 3-space, denoted by E^3, one can introduce a rectangular Cartesian coordinate system x_k, $k = 1, 2, 3$, or $x_1 = x$, $x_2 = y$, $x_3 = z$, and the corresponding unit vectors i_k, $k = 1, 2, 3$, or $\underset{\sim}{i}_1 = \underset{\sim}{i}$, $\underset{\sim}{i}_2 = \underset{\sim}{j}$, $\underset{\sim}{i}_3 = \underset{\sim}{k}$. The position vector $\underset{\sim}{r}$ is

$$\underset{\sim}{r} = \underset{\sim}{i}_1 x_1 + \underset{\sim}{i}_2 x_2 + \underset{\sim}{i}_3 x_3 = x_k \underset{\sim}{i}_k \tag{2.1a}$$

$$= \underset{\sim}{i} x + \underset{\sim}{j} y + \underset{\sim}{k} z \tag{2.1b}$$

*For those readers who have not used tensor calculus in their works, the material presented here is, nevertheless, useful if the tensor quantities are viewed as abbreviations. For example, a Christoffel symbol is nothing but an abbreviated name of an algebraic sum of the first partial derivatives of the metric coefficients.

**Metric data means the first and second fundamental coefficients. Refer to Section 2.4

(In general the repeated indices, when one is a subscript and the other a superscript, will imply summation over the range of index values. Exceptions to this rule will sometimes occur when the background system is rectangular Cartesian, as in Eq. 2.1a where both repeated indices are subscripts.) By introducing a general coordinate system x^i, $i = 1, 2, 3$, in E^3 and assuming the functions

$$x_i = f_i\left(x^1, x^2, x^3\right), \quad i = 1,2,3 \tag{2.2a}$$

to be continuously differentiable and which are also invertible, i.e.,

$$x^i = \Phi^i\left(x_1, x_2, x_3\right), \quad i = 1,2,3 \tag{2.2b}$$

we form the covariant base vectors

$$\underset{\sim}{a}_j = \frac{\partial \underset{\sim}{r}}{\partial x^j}, \quad j = 1,2,3 \tag{2.3a}$$

where $\underset{\sim}{a}_j$ is tangent to the coordinate curve x^j. A system of reciprocal base vectors $\underset{\sim}{a}^i$ are formed that satisfy the equations

$$\underset{\sim}{a}^i \cdot \underset{\sim}{a}_j = \delta^i_j \tag{2.3b}$$

where

$$\begin{aligned} \delta^i_j &= 0 \quad \textit{if } i \neq j \\ &= 1 \quad \textit{if } i = j \end{aligned}$$

is the Kronecker symbol. (In a purely rectangular Cartesian setting it is a common practice to use δ_{ij} as the Kronecker symbol.) Since the coordinates x^j are independent among themselves, the simple result

$$\frac{\partial x^i}{\partial x^j} = \delta^i_j$$

leads one to the formula

$$\underset{\sim}{a}^i = grad\ x^i \tag{2.3c}$$

where

$$grad = \underset{\sim}{\nabla} = \frac{\partial(\ \ )}{\partial x_m} \underset{\sim}{i}_m \tag{2.4}$$

is the gradient operator.

2.2.1 Representations in Terms of $\underset{\sim}{a}_i$ and $\underset{\sim}{a}^i$

All quantities that follow certain transformation of coordinate rules are called tensors. Tensors of various orders (ranks) can either be formed or appear naturally. In particular, scalars and vectors are tensors of order zero and one, respectively. A vector $\underset{\sim}{u}$ can be represented in either of the following forms:

$$\underset{\sim}{u} = u^i \underset{\sim}{a}_i \tag{2.5a}$$

$$= u_i \underset{\sim}{a}^i \tag{2.5b}$$

$$= U\underset{\sim}{i} + V\underset{\sim}{j} + W\underset{\sim}{k} \tag{2.5c}$$

In Eqs. 2.5a, 2.5b u^i and u_i are the contravariant and covariant components of $\underset{\sim}{u}$, respectively. In the same fashion a tensor $\tilde{T}$ of second order can be represented in any one of the following forms:

$$\tilde{T} = T^{ij} \underset{\sim}{a}_i \underset{\sim}{a}_j \tag{2.6a}$$

$$= T_{ij} \underset{\sim}{a}^i \underset{\sim}{a}^j \tag{2.6b}$$

$$= T^i{}_j \underset{\sim}{a}_i \underset{\sim}{a}^j \tag{2.6c}$$

$$= T_i{}^j \underset{\sim}{a}^i \underset{\sim}{a}_j \tag{2.6d}$$

Here T^{ij} are the contravariant components and T_{ij} are the covariant components of $\tilde{T}$. In Eqs. 2.6c, 2.6d the components are of the mixed type. Further $\underset{\sim}{a}_i \underset{\sim}{a}_j$ is the dyadic product of the vectors $\underset{\sim}{a}_i$ and $\underset{\sim}{a}_j$. A unit tensor $\tilde{I}$ has units on the main diagonal and zeros elsewhere. Thus using either Eq. 2.6c or Eq. 2.6d we have

$$\tilde{I} = \delta^i_j \underset{\sim}{a}_i \underset{\sim}{a}^j = \delta^j_i \underset{\sim}{a}^i \underset{\sim}{a}_j$$

In short,

$$\tilde{I} = \underset{\sim}{a}_i \underset{\sim}{a}^i = \underset{\sim}{a}^i \underset{\sim}{a}_i \tag{2.7}$$

The transpose of the tensor $\tilde{T}$ is denoted as $\tilde{T}^T$, and has the representation

$$\tilde{T}^T = T^{ji} \underset{\sim}{a}_i \underset{\sim}{a}_j = T^{ij} \underset{\sim}{a}_j \underset{\sim}{a}_i \tag{2.8}$$

and similarly with the other representations. A tensor is symmetric if

$$\tilde{T}^T = \tilde{T} \tag{2.9a}$$

and skew-symmetric if

$$\tilde{T}^T = -\tilde{T} \tag{2.9b}$$

Vectors and tensors in the rectangular Cartesian system can be written in a straightforward manner using summation on repeated subscripts, e.g., [22].

2.2.2 Differential Operations

Let the position vector $\underset{\sim}{r}$ be expressed in terms of the curvilinear coordinates x^i. The first differential $d\underset{\sim}{r}$ is then

$$d\underset{\sim}{r} = \frac{\partial \underset{\sim}{r}}{\partial x^i} dx^i$$

Using Eq. 2.3a,

$$d\underset{\sim}{r} = \underset{\sim}{a}_i \, dx^i \tag{2.10}$$

On comparison with Eq. 2.5a we note that dx^i are the contravariant components of the differential displacement vector $d\underset{\sim}{r}$. It must, however, be noted that x^i are not the contravariant components of any vector.

Let ϕ (x^1, x^2, x^3) be a scalar point function. Then its first differential is

$$d\phi = \frac{\partial \phi}{\partial x^i} dx^i \tag{2.11a}$$

From Eq. 2.10, using Eq. 2.3b we have

$$dx^i = \underset{\sim}{a}^i \cdot d\underset{\sim}{r} \tag{2.11b}$$

which when used in Eq. 2.11a yields

$$d\phi = (\nabla \phi) \cdot d\underset{\sim}{r} \tag{2.11c}$$

where

$$\nabla \phi = \frac{\partial \phi}{\partial x^i} \underset{\sim}{a}^i \tag{2.11d}$$

is the gradient of ϕ, and is a vector.

Let $\underset{\sim}{u}$ be a vector function of position; then its first differential is

$$d\underset{\sim}{u} = \frac{\partial \underset{\sim}{u}}{\partial x^i} dx^i$$

Using Eq. 2.11b, we have

$$d\underset{\sim}{u} = \left(\frac{\partial \underset{\sim}{u}}{\partial x^i} \underset{\sim}{a}^i \right) \cdot d\underset{\sim}{r}$$

We shall use the definition of the gradient of a vector as

$$grad\ \underset{\sim}{u} = \frac{\partial \underset{\sim}{u}}{\partial x^i} \underset{\sim}{a}^i \tag{2.12}$$

so that

$$d\underset{\sim}{u} = (grad\ \underset{\sim}{u}) \cdot d\underset{\sim}{r} \tag{2.13}$$

The divergence of a vector field $\underset{\sim}{u}$ is obtained by adding the diagonal terms of the tensor grad $\underset{\sim}{u}$, which in vector operational form is

$$div\ \underset{\sim}{u} = \frac{\partial \underset{\sim}{u}}{\partial x^i} \cdot \underset{\sim}{a}^i \tag{2.14}$$

Taking a lead from Eq. 2.14, the divergence of a tensor is

$$div\ \tilde{T} = \frac{\partial \tilde{T}}{\partial x^i} \cdot \underset{\sim}{a}^i \tag{2.15}$$

To complete this discussion, the curl of a vector field $\underset{\sim}{u}$ is defined as

$$curl\ \underset{\sim}{u} = \underset{\sim}{a}^i \times \frac{\partial \underset{\sim}{u}}{\partial x^i}$$

2.2.3 Metric Tensor and the Line Element

In E^3 we introduce a system of curvilinear coordinates x^i. The differential displacement vector is then given by Eq. 2.10 and the length element ds is given by

$$ds^2 = d\underset{\sim}{r} \cdot d\underset{\sim}{r} = (\underset{\sim}{a}_i \cdot \underset{\sim}{a}_j) dx^i dx^j$$

Writing

$$g_{ij} = \underset{\sim}{a}_i \cdot \underset{\sim}{a}_j \tag{2.16}$$

we obtain

$$ds^2 = g_{ij} dx^i dx^j \tag{2.17}$$

The coefficient g_{ij} are the covariant components of the metric tensor. Though Eq. 2.17 has been obtained for a Euclidean space, it is applicable to both the Euclidean and non-Euclidean spaces. In fact, Eq. 2.17 forms the one and the only postulate of Riemannian geometry. Obviously, g_{ij} are symmetric components, i.e.,

$$g_{ij} = g_{ji}$$

and the determinant of the matrix formed by g_{ij} is

$$g = det\left(g_{ij}\right) \tag{2.18}$$

which is strictly positive for E^3. The contravariant components of the metric tensor are

$$g^{ij} = \underset{\sim}{a}^i \cdot \underset{\sim}{a}^j \tag{2.19}$$

which are easily obtained in terms of g_{ij} as

$$g^{ij} = \left(g_{rp}g_{lt} - g_{rt}g_{lp}\right)/g \tag{2.20}$$

where the groups (i, r, l) and (j, p, t) separately assume values in the cyclic permutations of 1, 2, 3, in this order. Introducing the following subdeterminants,

$$\begin{aligned}
G_1 &= g_{22}g_{33} - \left(g_{23}\right)^2 \\
G_2 &= g_{11}g_{33} - \left(g_{13}\right)^2 \\
G_3 &= g_{11}g_{22} - \left(g_{12}\right)^2 \\
G_4 &= g_{13}g_{23} - g_{12}g_{33} \\
G_5 &= g_{12}g_{23} - g_{13}g_{22} \\
G_6 &= g_{12}g_{13} - g_{11}g_{23}
\end{aligned} \tag{2.21}$$

we have on using Eq. 2.20:

$$\begin{aligned}
g^{11} &= G_1/g, \quad g^{22} = G_2/g, \quad g^{33} = G_3/g \\
g^{12} &= G_4/g, \quad g^{13} = G_5/g, \quad g^{23} = G_6/g
\end{aligned} \tag{2.22}$$

As is shown in the cited references, e.g. [2, 7],

$$g_{jl}g^{kl} = \delta_j^k \tag{2.23}$$

$$\underset{\sim}{a}_i = g_{ik}\,\underset{\sim}{a}^k \tag{2.24a}$$

$$\underset{\sim}{a}^j = g^{jk}\,\underset{\sim}{a}_k \tag{2.24b}$$

$$\underset{\sim}{a}_1 \cdot \left(\underset{\sim}{a}_2 \times \underset{\sim}{a}_3\right) = \underset{\sim}{a}_2 \cdot \left(\underset{\sim}{a}_3 \times \underset{\sim}{a}_1\right) = \underset{\sim}{a}_3 \cdot \left(\underset{\sim}{a}_1 \times \underset{\sim}{a}_2\right) = \sqrt{g} \tag{2.25a}$$

Writing $x^1 = \xi$, $x^2 = \eta$, $x^3 = \zeta$, and denoting a partial derivative by a variable subscript, one of the expanded forms of $\sqrt{g}$ is

$$\left(y_\xi z_\eta - y_\eta z_\xi\right)x_\zeta + \left(x_\eta z_\xi - x_\xi z_\eta\right)y_\zeta + \left(x_\xi y_\eta - x_\eta y_\xi\right)z_\zeta = \sqrt{g} \tag{2.25b}$$

Using Eq. 2.21, we also have

$$\begin{aligned} g &= G_1 g_{11} + G_4 g_{12} + G_5 g_{13} \\ &= G_2 g_{22} + G_4 g_{12} + G_6 g_{23} \\ &= G_3 g_{33} + G_5 g_{13} + G_6 g_{23} \end{aligned} \tag{2.25c}$$

Other representations of the base vectors are

$$\underset{\sim}{a}^i = \frac{1}{2\sqrt{g}}\left(\underset{\sim}{a}_j \times \underset{\sim}{a}_k\right)e^{ijk} \tag{2.26a}$$

$$\underset{\sim}{a}_i = \frac{\sqrt{g}}{2}\left(\underset{\sim}{a}^j \times \underset{\sim}{a}^k\right)e_{ijk} \tag{2.26b}$$

where e^{ijk} and e_{ijk} are the permutation symbols. In terms of the metric tensor, the unit tensor defined in Eq. 2.7 is

$$\tilde{I} = g_{ij}\, \underset{\sim}{a}^i\, \underset{\sim}{a}^j \tag{2.27a}$$

$$= g^{ij}\, \underset{\sim}{a}_i\, \underset{\sim}{a}_j \tag{2.27b}$$

$$= \delta^i_j\, \underset{\sim}{a}_i\, \underset{\sim}{a}^j \tag{2.27c}$$

Using Eq. 2.24 in Eq. 2.5, we have

$$u^i = g^{ik} u_k \tag{2.28a}$$

and

$$u_i = g_{ik} u^k \tag{2.28b}$$

2.2.4 Differentiation of the Base Vectors

The main aim is to express the partial derivatives of the base vectors in terms of the base vectors. First, from the definition of the covariant base vectors, Eq. 2.3a, it is readily obvious that

$$\frac{\partial \underset{\sim}{a}_i}{\partial x^j} = \frac{\partial \underset{\sim}{a}_j}{\partial x^i} \tag{2.29}$$

Using this result and the simple derivations given in [7] we have the following results:

$$\frac{\partial \underset{\sim}{a}_i}{\partial x^j} = [ij, k]\underset{\sim}{a}^k \tag{2.30a}$$

$$= \Gamma_{ij}^k \underset{\sim}{a}_k \tag{2.30b}$$

where the abbreviations

$$[ij,k] = \frac{1}{2}\left(\frac{\partial g_{ik}}{\partial x^j} + \frac{\partial g_{jk}}{\partial x^i} - \frac{\partial g_{ij}}{\partial x^k}\right) \tag{2.31a}$$

and

$$\Gamma_{ij}^k = g^{sk}[ij,s] \tag{2.31b}$$

are called the Christoffel symbols of the first and second kind, respectively. Note that $[ij, k] = [ji, k]$ and $\Gamma_{ij}^k = \Gamma_{ji}^k$. Eq. 2.30b can also be stated as

$$\frac{\partial^2 \underset{\sim}{r}}{\partial x^i \partial x^j} = \Gamma_{ij}^k \underset{\sim}{a}_k \tag{2.32}$$

To obtain the partial derivatives of the contravariant base vectors $\underset{\sim}{a}^i$, we differentiate Eq. 2.3b with respect to any coordinate, say x^k, and use the previous results to obtain

$$\frac{\partial \underset{\sim}{a}^i}{\partial x^k} = -\Gamma_{jk}^i \underset{\sim}{a}^j \tag{2.33}$$

Taking the dot product of Eq. 2.33 with $\underset{\sim}{a}^k$ and using the definition in Eq. 2.3c, we readily get

$$\nabla^2 x^i = -g^{jk}\Gamma_{jk}^i \tag{2.34}$$

where

$$\nabla^2 = \frac{\partial^2}{\partial x_m \partial x_m}$$

is the Laplacian operator and x_m are the Cartesian coordinates.

2.2.5 Covariant and Intrinsic Derivatives

When one takes the partial derivative of a vector in its entity form, i.e.,

$$\frac{\partial \underset{\sim}{u}}{\partial x^k} = \frac{\partial}{\partial x^k}\left(u^i \underset{\sim}{a}_i\right)$$

and uses Eq. 2.30b, the result is

$$\frac{\partial \underset{\sim}{u}}{\partial x^k} = u^i_{;k} \underset{\sim}{a}_i \tag{2.35a}$$

where

$$u^i_{;k} = \frac{\partial u^i}{\partial x^k} + \Gamma^i_{jk} u^j \tag{2.35b}$$

is called the covariant derivative of a contravariant component. A semicolon before an index implies covariant differentiation. Similarly,

$$\frac{\partial \underset{\sim}{u}}{\partial x^k} = \frac{\partial}{\partial x^k}\left(u_i \underset{\sim}{a}^i\right)$$

and then on using Eq. 2.33, one gets

$$\frac{\partial \underset{\sim}{u}}{\partial x^k} = u_{i;k} \underset{\sim}{a}^i \tag{2.36a}$$

where

$$u_{i;k} = \frac{\partial u_i}{\partial x^k} - \Gamma^j_{ik} u_j \tag{2.36b}$$

is called the covariant derivative of a covariant component. The idea of covariant differentiation can be extended to tensors of any order. Refer to [5] and [22] for some explicit formulae for a second-order tensor. In particular it can be shown that the covariant derivatives of the metric tensor components are zero. That is

$$g_{ij;k} = 0, \quad g^{ij}_{;k} = 0$$

These two equations yield explicit formulae for the partial derivatives of the covariant and contravariant metric components, which are

$$\frac{\partial g_{ij}}{\partial x^k} = \Gamma^r_{ik} g_{rj} + \Gamma^r_{jk} g_{ri} \tag{2.37a}$$

and

$$\frac{\partial g^{ij}}{\partial x^k} = -\Gamma^i_{rk} g^{rj} - \Gamma^j_{rk} g^{ri} \tag{2.37b}$$

Let G^{rm} be the cofactor of g_{rm} in the determinant g. Then

$$g\delta^r_p = g_{pm} G^{rm}$$

and

$$G^{rm} = gg^{rm}$$

Thus

$$\frac{\partial g}{\partial x^j} = gg^{rm} \frac{\partial g_{rm}}{\partial x^j} \tag{2.38}$$

Using Eq. 2.37a in Eq. 2.38, one readily obtains

$$\Gamma^r_{rj} = \frac{1}{2g} \frac{\partial g}{\partial x^j} \tag{2.39a}$$

$$= \frac{\partial}{\partial x^j}\left(\ln \sqrt{g}\right) \tag{2.39b}$$

Using Eqs. 2.3b, 2.30b, and 2.39a in Eq. 2.14, the formula for the divergence of a vector $\underset{\sim}{u}$ becomes

$$div\ \underset{\sim}{u} = \frac{1}{\sqrt{g}} \frac{\partial}{\partial x^i}\left(\sqrt{g} u^i\right) \tag{2.40}$$

Similarly, the formulae for the divergence of a tensor can be developed.

Let the curvilinear coordinates x^i be functions of a single parameter t, i.e.,

$$x^i = x^i(t), \quad t_0 \le t \le t_1$$

Then $\underset{\sim}{u}$ becomes a function of t, i.e.,

$$\underset{\sim}{u}\left(x^i\right) = \underset{\sim}{u}\left(x^i(t)\right)$$

and the total derivative of $\underset{\sim}{u}$ with t is

$$\frac{d\underset{\sim}{u}}{dt} = \frac{d}{dt}\left(u^i \underset{\sim}{a}_i\right)$$

$$= \frac{du^i}{dt} = \underset{\sim}{a}_i + u^i \frac{d\underset{\sim}{a}_i}{dt}$$

Using the chain rule of partial differentiation and the definition of the covariant derivative, one obtains

$$\frac{du}{dt} = \left(\frac{\partial u^i}{\partial t} + u^i_{;j} \frac{dx^j}{dt}\right)\underset{\sim}{a}_i$$

The intrinsic derivative of u^i is defined as

$$\frac{\delta u^i}{\delta t} = \frac{\partial u^i}{\partial t} + u^i_{;j} \frac{dx^j}{dt} \tag{2.41}$$

and then

$$\frac{d\underset{\sim}{u}}{dt} = \frac{\delta u^i}{\delta t}\underset{\sim}{a}_i \tag{2.42}$$

2.2.6 Laplacian of a Scalar

Let ϕ (x^1, x^2, x^3) be a scalar. The Laplacian of ϕ is defined as

$$\nabla^2\phi = div\left(grad\ \phi\right) \tag{2.43}$$

From Eq. 2.11d, the components $\frac{\partial\phi}{\partial x^i}$ are the covariant components of the vector grad ϕ. According to Eq. 2.28a, the contravariant components are

$$g^{ij}\frac{\partial\phi}{\partial x^i}$$

Thus using Eq. 2.40,

$$\nabla^2\phi = \frac{1}{\sqrt{g}}\frac{\partial}{\partial x^j}\left(\sqrt{g}g^{ij}\frac{\partial\phi}{\partial x^i}\right) \tag{2.44}$$

which is one of the form for the Laplacian. Another form can be obtained by opening the differentiation on right-hand side and using Eqs. 2.37b and 2.39a, or else using Eq. 2.11d in Eq. 2.14 and using the preceding developed formulae. In either case, we get

$$\nabla^2\phi = g^{ij}\left(\frac{\partial^2\phi}{\partial x^i\partial x^j} - \Gamma^k_{ij}\frac{\partial\phi}{\partial x^k}\right) \tag{2.45a}$$

or, by using Eq. 2.34,

$$\nabla^2\phi = g^{ij}\frac{\partial^2\phi}{\partial x^i\partial x^j} + \left(\nabla^2 x^k\right)\frac{\partial\phi}{\partial x^k} \tag{2.45b}$$

Note that if $\phi = x^r$, a curvilinear coordinate, then from Eq. 2.45a,

$$\nabla^2 x^r = -g^{ij}\Gamma^r_{ij}$$

which is Eq. 2.34.

2.3 Theory of Curves

Practically all standard texts on differential geometry describe the theory of curves in formal details [1–6]. This section is intended to supplement the textual material in later sections for reference.

In E^3 using the rectangular Cartesian coordinates x_m, $m = 1, 2, 3$, the position vector at a point on the curve is stated as a function of an arbitrary parameter t as

$$\underset{\sim}{r}(t) = x_m(t)\,\underset{\sim}{i}_m, \quad t_o \le t \le t_1$$

The main assumption here is that at least one derivative,

$$\dot{x}_m = \frac{dx_m}{dt}, \quad m = 1,2,3$$

is different from zero. A simple example of the parametric equation of a curve is that of a straight line, which is

$$\underset{\sim}{r}(t) = \underset{\sim}{a} + \underset{\sim}{b}t$$

where $\underset{\sim}{a}$ and $\underset{\sim}{b}$ are constant vectors with the components of $\underset{\sim}{b}$ being proportional to the direction cosines of the line.

On a curve the arc length from a point P_0 of parameter t_0 to a point P of parameter t can be obtained by using Eq. 2.17 in Cartesian coordinates. Thus,

$$\begin{aligned} ds^2 &= d\underset{\sim}{r} \cdot d\underset{\sim}{r} \\ &= \dot{\underset{\sim}{r}} \cdot \dot{\underset{\sim}{r}}(dt)^2 \end{aligned} \tag{2.46}$$

so that

$$s(t) \int_{t_0}^{t} \sqrt{\dot{\underset{\sim}{r}} \cdot \dot{\underset{\sim}{r}}} dt$$

If instead of t one takes the arc length as a parameter, then from Eq. 2.46

$$\underset{\sim}{t} \cdot \underset{\sim}{t} = 1 \tag{2.47a}$$

where

$$\underset{\sim}{t} = \frac{d\underset{\sim}{r}}{ds} \tag{2.47b}$$

From Eq. 2.47a it is obvious that $\underset{\sim}{t}(s)$ is a unit vector tangent to the curve. Further,

$$\dot{\underset{\sim}{r}} = \underset{\sim}{t}\frac{ds}{dt} \tag{2.47c}$$

is also a tangent vector. Differentiating Eq. 2.47a, we get

$$\underset{\sim}{t} \cdot \frac{d\underset{\sim}{t}}{ds} = 0$$

Writing

$$\hat{\underset{\sim}{k}} = \frac{d\underset{\sim}{t}}{ds} \tag{2.48}$$

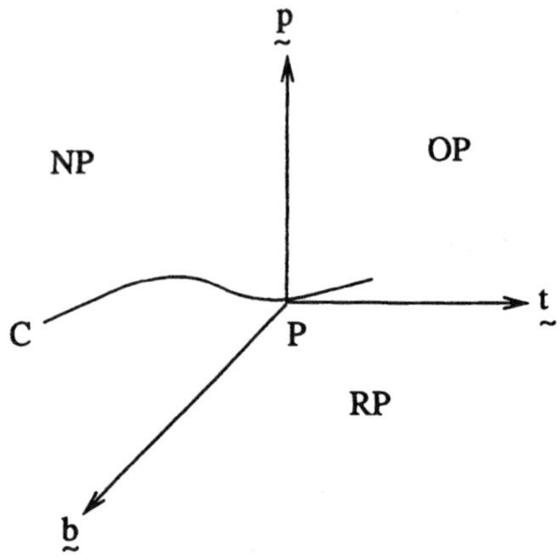

FIGURE 2.1 Right-handed triad $\underset{\sim}{t}$, $\underset{\sim}{p}$, $\underset{\sim}{b}$, of unit vectors at P on a space curve C. OP = osculating plane; NP = normal plane; RP = rectifying plane.

we note that the vectors $\underset{\sim}{t}$ and $\hat{\underset{\sim}{k}}$ are orthogonal. The vector $\hat{\underset{\sim}{k}}$ is the curvature vector because it expresses the rate of change of the unit tangent vector as one follows the curve. Now forming the unit vector

$$\underset{\sim}{p} = \hat{\underset{\sim}{k}} \big/ k \tag{2.49a}$$

where

$$k = \left| \hat{\underset{\sim}{k}} \right| \tag{2.49b}$$

is the curvature of the curve at a point. The unit vector $\underset{\sim}{p}$ is called the principal normal vector. The plane containing $\underset{\sim}{t}$ and $\underset{\sim}{p}$ is called the osculating plane.

Another vector $\underset{\sim}{b}$ is now formed as

$$\underset{\sim}{b} = \underset{\sim}{t} \times \underset{\sim}{p} \tag{2.50}$$

The triad of vectors $\underset{\sim}{t}$, $\underset{\sim}{p}$, $\underset{\sim}{b}$, in this order, form a right-hand system of unit vectors at a point of the curve. Besides the osculating plane the two other planes, termed the normal plane and the rectifying plane, are shown in Figure 2.1.

The vector $\underset{\sim}{b}$ is called the binormal vector and is associated with the torsion of the space curve. Based on simple arguments, e.g. [7], we can obtain the famous formulae of Frenet, or of Serret– Frenet, which are

$$\frac{d\underset{\sim}{t}}{ds} = k\underset{\sim}{p} \tag{2.51a}$$

$$\frac{d\underset{\sim}{p}}{ds} = -k\underset{\sim}{t} + \tau\underset{\sim}{b} \tag{2.51b}$$

$$\frac{d\underset{\sim}{b}}{ds} = -\tau\underset{\sim}{p} \tag{2.51c}$$

The scalar τ is called the torsion of a curve at a point and it is zero for plane curves.

Eqs. 2.51 are fundamental to the theory of curves. In fact, the fundamental theorem for space curves is stated as follows. "If s > 0 is the arc length along a curve and the functions $k(s)$ and $\tau(s)$ are single-valued and prescribed functions of s, then the solution of Eqs. 2.51 yields a space curve which is unique except for its position in space." For prescribed $k(s)$ and $\tau(s)$ Eqs. 2.51 can be solved in analytical forms for some very small number of cases. Eqs. 2.51 form a set of nine scalar equations, and if the initial conditions at some $s = s_0$ are prescribed for $\underset{\sim}{t}$ and $\underset{\sim}{p}$ (initial condition for $\underset{\sim}{b}$ can then be obtained from Eq. 2.50), then according to the theory of existence of the ordinary differential equations, the set of nine equations can be solved by any standard numerical method, such as the Runge–Kutta method. If k and τ are prescribed in terms of some other parameter t, then the same program can be slightly altered by prescribing ds/dt and replacing $k(s)$ by $k(t)$, etc., in the program.

2.3.1 A Collection of Usable Formulae for Curves

The formulae of curvature and torsion in terms of the arc length s for a curve $\underset{\sim}{r}(s)$ are as follows:

$$k(s) = \left(\frac{d^2\underset{\sim}{r}}{ds^2} \cdot \frac{d^2\underset{\sim}{r}}{ds^2} \right)^{1/2} \tag{2.52a}$$

$$\tau(s) = \rho^2 \frac{d\underset{\sim}{r}}{ds} \cdot \left(\frac{d^2\underset{\sim}{r}}{ds^2} \times \frac{d^3\underset{\sim}{r}}{ds^3} \right) \tag{2.52b}$$

where

$$\rho(s) = \frac{1}{k(s)}$$

is the radius of curvature. If the curve is expressed in terms of a parameter t as $\underset{\sim}{r}(t)$, then denoting differentiation with t by a dot, we have

$$k(t) = \left[\left(\dot{\underset{\sim}{r}} \cdot \dot{\underset{\sim}{r}} \right) \left(\ddot{\underset{\sim}{r}} \cdot \ddot{\underset{\sim}{r}} \right) - \left(\dot{\underset{\sim}{r}} \cdot \ddot{\underset{\sim}{r}} \right)^2 \right]^{1/2} \Big/ \left(\dot{\underset{\sim}{r}} \cdot \dot{\underset{\sim}{r}} \right)^{3/2} \tag{2.53a}$$

$$\tau(t) = \rho^2 \dot{\underset{\sim}{r}} \cdot \left(\ddot{\underset{\sim}{r}} \times \dddot{\underset{\sim}{r}} \right) \Big/ \left(\dot{\underset{\sim}{r}} \times \dot{\underset{\sim}{r}} \right)^3 \tag{2.53b}$$

Let a space curve be defined as the intersection of the two surfaces $f(x, y, z) = 0$ and $g(x, y, z) = 0$. Then the unit tangent vector of the curve is given by [1]

$$\underset{\sim}{t} = \left(\underset{\sim}{i} J_1 + \underset{\sim}{j} J_2 + \underset{\sim}{k} J_3 \right) \Big/ \left(J_1^2 + J_2^2 + J_3^2 \right)^{1/2} \tag{2.54}$$

where

$$J_1 = f_y g_z - f_z g_y, \quad J_2 = f_z g_x - f_x g_z, \quad J_3 = f_x g_y - f_y g_x$$

and a variable subscript denotes a partial derivative.

2.4 Geometrical Elements of the Surface Theory

The theory of surfaces embedded in E^3 was developed with all its essential aspects in the 19th century. Almost all of the useful concepts and formulae presently used in engineering and applied sciences were developed by Gauss, Monge, Darboux, Beltrami, and Christoffel, just to name a few. For a detailed discussion of the topics discussed in this section, the reader is referred to Refs. [1–3].

In the theory of surfaces embedded in E^3 we can either use the rectangular Cartesian coordinates x_m or some general coordinates x^i. For the sake of generality, let us first use a general system of coordinates x^i. A surface is then defined parametrically by the use of two parameters $u^\alpha = (u^1, u^2)$ as

$$x^i = x^i\left(u^1, u^2\right), \quad i = 1,2,3 \tag{2.55a}$$

The functions x^i defined in Eq. 2.55a are continuously differentiable with respect to the parameters u^1 and u^2, and the matrix

$$\left[\frac{\partial x^i}{\partial u^\alpha}\right]$$

is of rank two, i.e., at least one square subdeterminant is not zero. From Eq. 2.55a,

$$dx^i = \frac{\partial x^i}{\partial u^\alpha} du^\alpha \tag{2.55b}$$

where the Greek indices assume values 1 and 2. Also, the displacement vector $d\underset{\sim}{r}$, which belongs both to the surface and the embedding space E^3, can be represented either as

$$d\underset{\sim}{r} = \frac{\partial \underset{\sim}{r}}{\partial x^i} dx^i = \underset{\sim}{a}_i\, dx^i \tag{2.55c}$$

or as

$$d\underset{\sim}{r} = \frac{\partial \underset{\sim}{r}}{\partial u^\alpha} du^\alpha \tag{2.55d}$$

The element of length

$$ds^2 = d\underset{\sim}{r} \cdot d\underset{\sim}{r}$$

from Eq. 2.17, or alternatively from Eq. 2.55c by using Eq. 2.55b, can be stated as

$$ds^2 = a_{\alpha\beta} du^\alpha du^\beta \tag{2.56}$$

where

$$a_{\alpha\beta} = g_{ij} \frac{\partial x^i}{\partial u^\alpha} \frac{\partial x^j}{\partial u^\beta} \tag{2.57}$$

Obviously $a_{\alpha\beta}$ are symmetric. Since the embedding space is Euclidian, one can also use the rectangular Cartesian coordinates x_m in place of the curvilinear coordinates x^i. In such a case $g_{ij} = \delta_{ij}$, and from Eq. 2.57,

$$a_{\alpha\beta} = \frac{\partial x_m}{\partial u^\alpha}\frac{\partial x_m}{\partial u^\beta} = \frac{\partial \underset{\sim}{r}}{\partial u^\alpha} \cdot \frac{\partial \underset{\sim}{r}}{\partial u^\beta}$$

From here onward we shall return to the previous symbolism and use $g_{\alpha\beta}$ in place of $a_{\alpha\beta}$ so that

$$g_{\alpha\beta} = \frac{\partial \underset{\sim}{r}}{\partial u^\alpha} \cdot \frac{\partial \underset{\sim}{r}}{\partial u^\beta} \tag{2.58}$$

and Eq. 2.56 is written as

$$ds^2 = g_{\alpha\beta} du^\alpha du^\beta \tag{2.59}$$

which gives an elemental arc on a surface of parameters/coordinates u^1, u^2. The metric, Eq. 2.59, for an element of length in the surface is called the "first fundamental form." For the purpose of having expanded formulae we write $x_1 = x$, $x_2 = y$, $x_3 = z$; $u^1 = \xi$, $u^2 = \eta$ and then from Eq. 2.58:

$$g_{11} = x_\xi^2 + y_\xi^2 + z_\xi^2 \tag{2.60a}$$

$$g_{12} = x_\xi x_\eta + y_\xi y_\eta + z_\xi z_\eta \tag{2.60b}$$

$$g_{22} = x_\eta^2 + y_\eta^2 + z_\eta^2 \tag{2.60c}$$

$$G_3 = g_{11}g_{22} - (g_{12})^2 \tag{2.60d}$$

where a variable subscript implies a partial derivative. Further, similar to Eq. 2.23 we have

$$g_{\alpha\beta} g^{\alpha\gamma} = \delta_\beta^\gamma \tag{2.61a}$$

so that

$$g^{11} = g_{22}/G_3, \quad g^{12} = g^{21} = -g_{12}/G_3, \quad g^{22} = g_{11}/G_3 \tag{2.61b}$$

The vectors

$$\underset{\sim}{a}_\alpha = \frac{\partial \underset{\sim}{r}}{\partial u^\alpha}, \quad \alpha = 1,2 \tag{2.62}$$

are the covariant surface base vectors and they form a tangent vector field. The angle θ between the coordinate lines $\xi = u^1$ and $\eta = u^2$ at a point in the surface is obviously given by

$$\begin{aligned} cos\theta &= \underset{\sim}{a}_1 \cdot \underset{\sim}{a}_2 \Big/ \left|\underset{\sim}{a}_1\right|\left|\underset{\sim}{a}_2\right| \\ &= g_{12}\Big/\sqrt{g_{11}g_{22}} \end{aligned} \tag{2.63a}$$

and

$$|\underset{\sim}{a}_1 \times \underset{\sim}{a}_2|^2 = g_{11} g_{22} \sin^2 \theta$$
$$= g_{11} g_{22} (1 - \cos^2 \theta)$$

Thus, using Eq. 2.63a, we have

$$\begin{aligned} |\underset{\sim}{a}_1 \times \underset{\sim}{a}_2|^2 &= g_{11} g_{22} - (g_{12})^2 \\ &= G_3 \end{aligned} \tag{2.63b}$$

Coordinates in the surface at a point are orthogonal if $g_{12} = 0$ at that point.

The surface base vectors in Eq. 2.62 define the unit normal vector $\underset{\sim}{n}$ at each point of the surface through the equation

$$\underset{\sim}{n} = (\underset{\sim}{a}_1 \times \underset{\sim}{a}_2) / |\underset{\sim}{a}_1 \times \underset{\sim}{a}_2|$$

Thus

$$\underset{\sim}{n} = \frac{1}{\sqrt{G_3}} (\underset{\sim}{a}_1 \times \underset{\sim}{a}_2) \tag{2.64}$$

The rectangular Cartesian components of $\underset{\sim}{n}$ denoted by X, Y, Z are

$$X = J_1 / \sqrt{G_3}, \quad Y = J_2 / \sqrt{G_3}, \quad Z = J_3 / \sqrt{G_3} \tag{2.65}$$

where

$$J_1 = y_\xi z_\eta - y_\eta z_\xi, \quad J_2 = x_\eta z_\xi - x_\xi z_\eta, \quad J_3 = x_\xi y_\eta - x_\eta y_\xi$$

2.4.1 The Surface Christoffel Symbols

The surface Christoffel symbols can be formed by the same technique as noted in Section 2.2, independent of any other consideration. For clarity in the analysis to follow, we shall denote the surface Christoffel symbols of the second kind by $\Upsilon^{\sigma}_{\alpha\beta}$. The formula is

$$\Upsilon^{\sigma}_{\alpha\beta} = g^{\sigma\delta} [\alpha\beta, \delta] \tag{2.66}$$

where

$$[\alpha\beta, \delta] = \frac{1}{2} \left(\frac{\partial g_{\beta\delta}}{\partial u^\alpha} + \frac{\partial g_{\alpha\beta}}{\partial u^\beta} - \frac{\partial g_{\alpha\beta}}{\partial u^\delta} \right) \tag{2.67}$$

and $[\alpha\beta, \delta]$ are the surface Christoffel symbols of the first kind. The technique mentioned above can concisely be stated as follows:

Obviously (similar to Eq. 2.29),

$$\frac{\partial \underset{\sim}{a}_\alpha}{\partial u^\beta} = \frac{\partial \underset{\sim}{a}_\beta}{\partial u^\alpha} \tag{2.68a}$$

Next

$$\frac{\partial}{\partial u^\gamma}\left(\underset{\sim}{a}_\alpha \cdot \underset{\sim}{a}_\beta\right) = \frac{\partial g_{\alpha\beta}}{\partial u^\gamma} \tag{2.68b}$$

$$\frac{\partial}{\partial u^\beta}\left(\underset{\sim}{a}_\alpha \cdot \underset{\sim}{a}_\gamma\right) = \frac{\partial g_{\alpha\gamma}}{\partial u^\beta} \tag{2.68c}$$

$$\frac{\partial}{\partial u^\alpha}\left(\underset{\sim}{a}_\beta \cdot \underset{\sim}{a}_\gamma\right) = \frac{\partial g_{\beta\gamma}}{\partial u^\alpha} \tag{2.68d}$$

Adding Eq. 2.68c and Eq. 2.68d and subtracting Eq. 2.68b while using Eq. 2.68a, one obtains

$$\frac{\partial \underset{\sim}{a}_\alpha}{\partial u^\beta} \cdot \underset{\sim}{a}^\theta = \Upsilon^\theta_{\alpha\beta} \tag{2.69}$$

where $\underset{\sim}{a}^\theta$ are the contravariant surface base vectors satisfying

$$\underset{\sim}{a}^\alpha \cdot \underset{\sim}{a}_\beta = \delta^\alpha_\beta \tag{2.70a}$$

and

$$\underset{\sim}{a}^\theta = g^{\theta\alpha}\, \underset{\sim}{a}_\alpha \ \text{ etc.} \tag{2.70b}$$

As a caution, one must not hurriedly conclude an equation similar to Eq. 2.30b from Eq. 2.69. It must also be mentioned here that according to Eq. 2.70a, $\underset{\sim}{a}^1$ is orthogonal to $\underset{\sim}{a}_2$ and $\underset{\sim}{a}^2$ is orthogonal to $\underset{\sim}{a}_1$, but still $\underset{\sim}{a}^1$ and $\underset{\sim}{a}^2$ lie in the tangent plane to the surface.

2.4.2 Normal Curvature and the Second Fundamental Form

A plane containing the unit tangent vector $\underset{\sim}{t}$ and the unit surface normal vector $\underset{\sim}{n}$ at a point *P* of the surface cuts the surface in different curves when rotated about $\underset{\sim}{n}$ as an axis. We refer to Figure 2.2, where the vectors $\underset{\sim}{t}$, $\underset{\sim}{n}$, the curvature vector $\hat{\underset{\sim}{k}}$, and another unit vector $\underset{\sim}{e}$ in the tangent plane are shown.

Each curve obtained by rotating the $\underset{\sim}{t}-\underset{\sim}{n}$ plane is called a normal section of the surface at *P*. Since these curves belong both to the surface and also the embedding space, a study of the curvature properties of these curves also reveals the curvature and torsion properties of the surface itself.

We decompose the vector $\hat{\underset{\sim}{k}}$ at *P* of *C*, defined by Eq. 2.48, as

$$\hat{\underset{\sim}{k}} = \underset{\sim}{k}_n + \underset{\sim}{k}_g \tag{2.71}$$

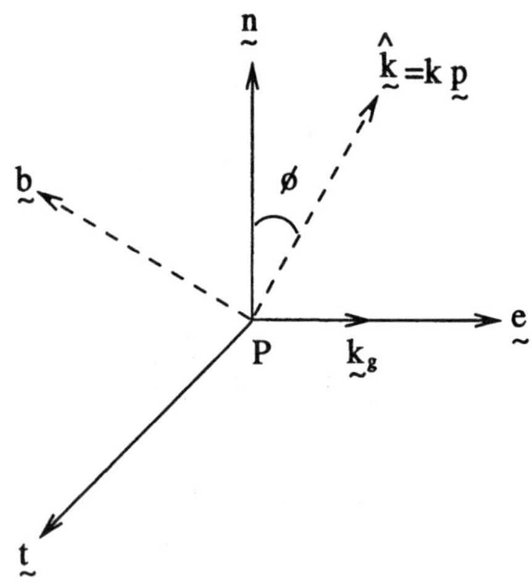

FIGURE 2.2 Right-handed triad $\underset{\sim}{t},\underset{\sim}{e},\underset{\sim}{n}$ of unit vectors at P on a surface. The vectors $\underset{\sim}{p}$ and $\underset{\sim}{b}$ are perpendicular to $\underset{\sim}{t}$ and lie in the $\underset{\sim}{e}-\underset{\sim}{n}$ plane.

where the vector $\underset{\sim}{k}_n$, is normal to the surface, and the vector $\underset{\sim}{k}_g$ is tangent to the surface as shown in Figure 2.2. The vector $\underset{\sim}{k}$ is called the normal curvature vector at the point, and it is directed either toward or against the direction of the surface normal $\underset{\sim}{n}$. Thus

$$\underset{\sim}{k}_n = \underset{\sim}{n} k_n \tag{2.72}$$

where k_n is the normal curvature of the normal section of the surface, and is an algebraic number. To find a formula for k_n we consider the equation

$$\underset{\sim}{n} \cdot \underset{\sim}{t} = 0$$

and differentiate it with respect to s, which yields

$$k_n = \frac{-d\underset{\sim}{n} \cdot d\underset{\sim}{r}}{(ds)^2} \tag{2.73}$$

Next we differentiate

$$\underset{\sim}{n} \cdot \underset{\sim}{a}_\beta = 0$$

with respect to u^α and have

$$\frac{\partial \underset{\sim}{n}}{\partial u^\alpha} \cdot \underset{\sim}{a}_\beta = -\underset{\sim}{n} \cdot \frac{\partial^2 \underset{\sim}{r}}{\partial u^\alpha \partial u^\beta}$$

Further

$$d\underset{\sim}{n} = \frac{\partial \underset{\sim}{n}}{\partial u^\alpha} du^\alpha, \quad d\underset{\sim}{r} = \underset{\sim}{a}_\beta du^\beta$$

Eq. 2.73 yields

$$k_n = \left(\underset{\sim}{n} \cdot \frac{\partial^2 \underset{\sim}{r}}{\partial u^\alpha \partial u^\beta} \right) \frac{du^\alpha du^\beta}{(ds)^2} \tag{2.74}$$

A set of new coefficients $b_{\alpha\beta}$ are now defined as

$$b_{\alpha\beta} = \underset{\sim}{n} \cdot \frac{\partial^2 \underset{\sim}{r}}{\partial u^\alpha \partial u^\beta} \tag{2.75a}$$

$$= -\frac{\partial \underset{\sim}{n}}{\partial u^\alpha} \cdot \underset{\sim}{a}_\beta \tag{2.75b}$$

$$= \frac{1}{\sqrt{G_3}} \underset{\sim}{a}_1 \cdot \left(\underset{\sim}{a}_2 \times \frac{\partial^2 \underset{\sim}{r}}{\partial u^\alpha \partial u^\beta} \right) \tag{2.75c}$$

Thus Eq. 2.74, beside having the form given in Eq. 2.73, can also be stated as

$$k_n = \frac{b_{\alpha\beta} du^\alpha du^\beta}{(ds)^2} \tag{2.76a}$$

$$= \frac{b_{\alpha\beta} du^\alpha du^\beta}{g_{\mu\nu} du^\mu du^\nu} \tag{2.76b}$$

It is easy to see from Eq. 2.76a that

$$\left(k_n d\underset{\sim}{r} + d\underset{\sim}{n} \right) \cdot d\underset{\sim}{r} = 0$$

But $d\underset{\sim}{r}$ is arbitrary, so that

$$k_n d\underset{\sim}{r} + d\underset{\sim}{n} = 0 \tag{2.77}$$

which is due to Rodrigues [1].

The form

$$b_{\alpha\beta} du^\alpha du^\beta$$

is called the "second fundamental form," and $b_{\alpha\beta}$ the coefficients of the second fundamental form. In expanded form, writing $\xi = u^1$, $\eta = u^2$, we have

$$b_{11} = Xx_{\xi\xi} + Yy_{\xi\xi} + Zz_{\xi\xi} \tag{2.78a}$$

$$b_{12} = Xx_{\xi\eta} + Yy_{\xi\eta} + Zz_{\xi\eta} = b_{21} \tag{2.78b}$$

$$b_{22} = Xx_{\eta\eta} + Yy_{\eta\eta} + Zz_{\eta\eta} \tag{2.78c}$$

$$b = b_{11}b_{22} - (b_{12})^2 \tag{2.78d}$$

Returning to the consideration of k_n we note that from Eqs. 2.71 and 2.72

$$\underset{\sim}{n} \cdot \hat{\underset{\sim}{k}} = k_n$$

which on using Eq. 2.51a gives

$$k_n = k\cos\phi \tag{2.79}$$

where

$$\underset{\sim}{p} \cdot \underset{\sim}{n} = \cos\phi$$

and k is the curvature of the curve C. Introducing the radius of curvatures

$$\rho = 1/k, \quad \rho_n = 1/k_n$$

we get from Eq. 2.79

$$\rho = \rho_n \cos\phi \tag{2.80}$$

which is due to Meusnier [1].

2.4.3 Principal Normal Curvatures

Let us introduce the directions

$$l = \frac{d\xi}{ds}, \quad m = \frac{d\eta}{ds}$$

then Eq. 2.76a takes the form

$$k_n = b_{11}l^2 + 2b_{12}lm + b_{22}m^2 \tag{2.81}$$

If only the direction

$$\lambda = \frac{d\eta}{d\xi}$$

is introduced, then

$$k_n = \frac{b_{11} + 2\lambda b_{12} + \lambda^2 b_{22}}{g_{11} + 2\lambda g_{12} + \lambda^2 g_{22}} \tag{2.82}$$

With the coefficients $g_{\alpha\beta}$ and $b_{\alpha\beta}$ as constants at a point, the quantity k_n is a function of λ. The extremum values of k_n are obtained by

$$\frac{dk_n}{d\lambda} = 0$$

and the roots of this equation determine those directions for which the normal curvatures k_n assumes extreme values. These extreme values are called the principal normal curvatures at P of the surface, which we shall denote by k_I and k_{II}. The corresponding directions λ are called the principal directions. Following the details given in [2], we obtain the following important equations for the sum and product of the principal curvatures:

$$k_I + k_{II} = b_{\alpha\beta} g^{\alpha\beta} \tag{2.83}$$

$$k_I k_{II} = \frac{b}{G_3} \tag{2.84}$$

where G_3 and b have been defined in Eqs. 2.60d and 2.78d, respectively. Here a few definitions are in order.

(i) *Lines of curvature:*
The line of curvature is a curve in a surface whose curvature at any point is either k_I or k_{II}. The tangent to the line of curvature falls in the principal direction. The equations for the determination of the lines of curvature are obtained by differentiating Eq. 2.82 with respect to λ and setting the result equal to zero. Thus

$$A\left(\frac{d\eta}{d\xi}\right)^2 + B\frac{d\eta}{d\xi} + C = 0 \tag{2.85}$$

where

$$A = b_{22}g_{12} - b_{12}g_{22}$$
$$B = b_{22}g_{11} - b_{11}g_{22}$$
$$C = b_{12}g_{11} - b_{11}g_{12}$$

Note that Eq. 2.85 is equivalent to two first-order ordinary differential equations, and their solutions define two families of curves in a surface which are the lines of curvature. Further, these curves are orthogonal. It is obvious from Eq. 2.85 that if $A = 0$, then $d\xi = 0$, and if $C = 0$ then $d\eta = 0$. Thus the curves $\xi =$ const. and $\eta =$ const. are the lines of curvature if $A = 0$ and $C = 0$.

In an actual computation if the coefficients of the first and second fundamental forms are known throughout the surface as functions of ξ and η, and further the initial point ξ_0, η_0 is prescribed, then the curves of curvature can be obtained by a numerical method, e.g., the Runge–Kutta method. If the curves ξ and η are themselves the curves of curvature, then as discussed above in these coordinates $g_{12} = 0$ and $b_{12} = 0$, and from Eq. 2.82,

$$k_n = b_{11}\left(\frac{d\xi}{ds}\right)^2 + b_{22}\left(\frac{d\eta}{ds}\right)^2 \tag{2.86}$$

a formula due to Euler. The normal curvatures are then

$$k_I = \frac{b_{11}}{g_{11}} \quad for \quad \eta = const. \ (\xi - curve)$$
$$k_{II} = \frac{b_{22}}{g_{22}} \quad for \quad \xi = const. \ (\eta - curve).$$

(ii) *Asymptotic directions:*
Points on a surface where $k_n = 0$ give two directions, which from Eq. 2.82 are

$$\frac{d\eta}{d\xi} = \frac{-b_{12} \pm \sqrt{(b_{12})^2 - b_{11}b_{22}}}{b_{22}}$$

(iii) *Results for a surface of the form $z = f(x, y)$:*
When the equation of a surface is given in the form $z = f(x, y)$, then it is convenient to take

$$x = \xi, \quad y = \eta, \quad z = f(\xi, \eta)$$

Then

$$\begin{aligned}
&\underset{\sim}{r} = \underset{\sim}{i}\xi + \underset{\sim}{j}\eta + \underset{\sim}{k} f(\xi, \eta) \\
&\underset{\sim}{a}_1 = \underset{\sim}{i} + \underset{\sim}{k} f_x \\
&\underset{\sim}{a}_2 = \underset{\sim}{j} + \underset{\sim}{k} f_y \\
&g_{11} = 1 + f_x^2, \quad g_{12} = f_x f_y, \quad g_{22} = 1 + f_y^2 \\
&G_3 = 1 + f_x^2 + f_y^2 \\
&\underset{\sim}{n} = \left(-\underset{\sim}{i} f_x - \underset{\sim}{j} f_y + \underset{\sim}{k}\right)\sqrt{G_3} \\
&dA = \sqrt{G_3}\,dxdy, \quad element \ of \ area \\
&b_{11} = f_{xx}/\sqrt{G_3}, \quad b_{12} = f_{xy}/\sqrt{G_3}, \quad b_{22} = f_{yy}/\sqrt{G_3}
\end{aligned} \tag{2.87}$$

As an example, for a monkey saddle

$$z = y^3 - 3yx^2$$

for which all the geometrical elements can be computed from Eq. 2.87.

(iv) *Results for a body of revolution:*
Let a curve $z = f(x)$ in the plane $y = 0$ be rotated about the z-axis. The surface of revolution so generated has the parametric representation

$$x = \xi \cos\eta, \quad y = \xi \sin\eta, \quad z = f(\xi)$$

where $\xi > 0$ and $\frac{df}{d\xi} = f'$ is bounded. For this case,

$$\underset{\sim}{a}_i = \underset{\sim}{i} \cos\eta + \underset{\sim}{j} \sin\eta + k\, f'$$

$$\underset{\sim}{a}_2 = \xi\left(-\underset{\sim}{i} \sin\eta + \underset{\sim}{j} \cos\eta\right)$$

$$g_{11} = 1 + f'^2, \quad g_{12} = 0, \quad g_{22} = \xi^2, \quad G_3 = \xi^2\left(1 + f^2\right)$$

$$\underset{\sim}{n} = \frac{1}{\sqrt{1 + f'^2}}, \quad \left(\underset{\sim}{i}\, f' \cos\eta + \underset{\sim}{j}\, f' \sin\eta - \underset{\sim}{k}\right)$$

$$b_{11} = \frac{f'f''}{\sqrt{1 + f'^2}}, \quad b_{12} = 0, \quad b_{22} = \frac{\xi f'}{\sqrt{1 + f'^2}}$$

Also referring to Eq. 2.93,

$$\Upsilon^1_{11} = \frac{f''}{1 + f^2}, \quad \Upsilon^1_{22} = -\frac{\xi}{1 + f^2}, \quad \Upsilon^2_{12} = \frac{1}{\xi}$$

and all other Christoffel symbols are zero.

As a particular case, for a cone

$$\xi = r \sin\alpha, \quad f(\xi) r \cos\alpha$$

where r is the radial distance from the origin (apex of the cone) to a point on the cone's surface, and α is the angle made by r with the z-axis. Then

$$x = r \sin\alpha \cos\eta, \quad y = r \sin\alpha \sin\eta, \quad z = r \cos\alpha$$

which yields the equation of a cone:

$$x^2 + y^2 = z^2 \tan^2\alpha$$

2.4.4 Mean and Gaussian Curvatures

The mean curvature K_m of a surface at a point is defined as

$$K_m = \tfrac{1}{2}\left(k_I + k_{II}\right) \tag{2.88a}$$

while the Gaussian or total curvature at a point is defined as

$$K = k_I k_{II} \tag{2.88b}$$

Surfaces for which $K_m = 0$ are called "minimal" surfaces, while surfaces for which $K = 0$ are called developable surfaces. The manner in which k_I and k_{II} have been obtained and the Gaussian curvature K has been formed suggests that K is an extrinsic property. In fact, K is an intrinsic property of a surface, that is, it depends only on the first fundamental form and on the derivatives of its coefficients [1, 2, 7].

2.4.5 Derivatives of the Surface Normal; Formulae of Weingarten

From the simple identity

$$\underset{\sim}{n} \cdot \underset{\sim}{n} = 1$$

one obtains by differentiation the following two equations:

$$\underset{\sim}{n} \cdot \frac{\partial \underset{\sim}{n}}{\partial u^{\alpha}} = 0, \quad \alpha = 1,2$$

These two equations suggest that $\frac{\partial \underset{\sim}{n}}{\partial u^{\alpha}}$, $\alpha = 1, 2$, lie in the tangent plane to the surface. Thus

$$\frac{\partial \underset{\sim}{n}}{\partial u^1} = P \underset{\sim}{a}_1 + Q \underset{\sim}{a}_2$$

$$\frac{\partial \underset{\sim}{n}}{\partial u^2} = R \underset{\sim}{a}_1 + S \underset{\sim}{a}_2$$

To find the coefficients P, Q, R, S, we differentiate $\underset{\sim}{n} \cdot \underset{\sim}{a}_1 = 0$ with respect to the u^2 and $\underset{\sim}{n} \cdot \underset{\sim}{a}_2 = 0$ with respect to u^1. The solution of the four scalar equations yields [7],

$$\frac{\partial \underset{\sim}{n}}{\partial u^{\alpha}} = -b_{\alpha\beta} g^{\beta\gamma} \underset{\sim}{a}_{\gamma}, \quad \alpha = 1,2 \tag{2.89}$$

Eq. 2.89 were obtained by Weingarten [2, 7], and provide the formulae for the partial derivatives of the surface normal vector with respect to the surface coordinates.

2.4.6 Formulae of Gauss

In E^3 the vectors $\underset{\sim}{a}_1$, $\underset{\sim}{a}_2$, $\underset{\sim}{n}$ form a system of independent vectors. It should therefore be possible to express the first partial derivatives of a base vector in terms of the base vectors themselves. Based on the preceding developments, the logical outcome is to have

$$\begin{aligned} \frac{\partial \underset{\sim}{a}_{\alpha}}{\partial u^{\beta}} &= \frac{\partial^2 \underset{\sim}{r}}{\partial u^{\alpha} \partial u^{\beta}} \\ &= \Gamma^{\gamma}_{\alpha\beta} \underset{\sim}{a}_{\gamma} + \underset{\sim}{n} b_{\alpha\beta} \end{aligned} \tag{2.90}$$

As a check we note that the dot products of Eq. 2.90 with $\underset{\sim}{a}^{\theta}$ and $\underset{\sim}{n}$ yield Eqs. 2.69 and 2.75a, respectively. Eq. 2.90 provides the formulae of Gauss for the second derivatives $\partial^2 \underset{\sim}{r} / \partial u^{\alpha} \partial u^{\beta}$.

The coefficients of the second fundamental form $b_{\alpha\beta}$ for a surface have already been defined in Eq. 2.75a. One can obtain a new formula for them by considering the Gauss' formulae, Eq. 2.90, and the space Christoffel symbols as stated in Eq. 2.32. In E^3 consider a surface defined by $x^3 =$ const., and let $x^1 = u^1$ and $x^2 = u^2$. Then from Eq. 2.32,

$$\frac{\partial^2 \underset{\sim}{r}}{\partial u^\alpha \partial u^\beta} = \Gamma^1_{\alpha\beta}\underset{\sim}{a}_1 + \Gamma^2_{\alpha\beta}\underset{\sim}{a}_2 + \Gamma^3_{\alpha\beta}\underset{\sim}{a}_3; \quad x^3 = const.$$

Since both $\underset{\sim}{a}_1$ and $\underset{\sim}{a}_2$ have been evaluated at x^3 = const., taking the dot product with the unit surface normal vector $\underset{\sim}{n}$, one gets

$$\underset{\sim}{n} \cdot \frac{\partial^2 \underset{\sim}{r}}{\partial u^\alpha \partial u^\beta} = \Gamma^3_{\alpha\beta}\left(\underset{\sim}{n} \cdot \underset{\sim}{a}_3\right)$$

Writing

$$\underset{\sim}{n} \cdot \underset{\sim}{a}_3 = \lambda \tag{2.91a}$$

and comparing with Eq. 2.90, one obtains

$$b_{\alpha\beta} = \left(\lambda \Gamma^3_{\alpha\beta}\right)_{x^3=const.} \tag{2.91b}$$

which can also be used to find the coefficients $b_{\alpha\beta}$, [16]. Thus the formulae of Gauss can also be stated as

$$\frac{\partial^2 \underset{\sim}{r}}{\partial u^\alpha \partial u^\beta} = \Upsilon^\nu_{\alpha\beta}\underset{\sim}{a}_\gamma + \underset{\sim}{n}\left(\Gamma^3_{\alpha\beta}\lambda\right)_{x^3=const.} \tag{2.92}$$

From Eq. 2.66, the expanded form of the surface Christoffel symbols for the surface $x^3 = const.$ and with $u^1 = \xi$, $u^2 = \eta$ are as follows:

$$\begin{aligned}
\Upsilon^1_{11} &= \left[g_{22}\frac{\partial g_{11}}{\partial \xi} + g_{12}\left(\frac{\partial g_{11}}{\partial \eta} - 2\frac{\partial g_{12}}{\partial \xi}\right)\right] \Big/ 2G_3 \\
\Upsilon^2_{22} &= \left[g_{11}\frac{\partial g_{22}}{\partial \eta} + g_{12}\left(\frac{\partial g_{22}}{\partial \xi} - 2\frac{\partial g_{12}}{\partial \eta}\right)\right] \Big/ 2G_3 \\
\Upsilon^1_{22} &= \left[g_{22}\left(2\frac{\partial g_{12}}{\partial \eta} - \frac{\partial g_{22}}{\partial \xi}\right) - g_{12}\frac{\partial g_{22}}{\partial \eta}\right] \Big/ 2G_3 \\
\Upsilon^2_{11} &= \left[g_{11}\left(2\frac{\partial g_{12}}{\partial \xi} - \frac{\partial g_{11}}{\partial \eta}\right) - g_{12}\frac{\partial g_{11}}{\partial \xi}\right] \Big/ 2G_3 \\
\Upsilon^1_{12} = \Upsilon^1_{21} &= \left(g_{22}\frac{\partial g_{11}}{\partial \eta} - g_{12}\frac{\partial g_{22}}{\partial \xi}\right) \Big/ 2G_3 \\
\Upsilon^2_{12} = \Upsilon^2_{21} &= \left(g_{11}\frac{\partial g_{22}}{\partial \xi} - g_{12}\frac{\partial g_{11}}{\partial \eta}\right) \Big/ 2G_3 \\
\Upsilon^1_{11} + \Upsilon^2_{12} &= \frac{1}{2G_3}\frac{\partial G_3}{\partial \xi} \\
\Upsilon^1_{12} + \Upsilon^2_{22} &= \frac{1}{2G_3}\frac{\partial G_3}{\partial \eta} \\
G_3 &= g_{11}g_{22} - \left(g_{12}\right)^2
\end{aligned} \tag{2.93}$$

2.4.7 Gauss–Codazzi Equations

Consider the identity

$$\frac{\partial}{\partial u^{\gamma}}\left(\frac{\partial^2 \underset{\sim}{r}}{\partial u^{\alpha}\partial u^{\beta}}\right)=\frac{\partial}{\partial u^{\beta}}\left(\frac{\partial^2 \underset{\sim}{r}}{\partial u^{\alpha}\partial u^{\beta}}\right)$$

for any choice of α, β, and γ. Using Eq. 2.90 and then Eq. 2.89, one obtains

$$R_{\mu\alpha\gamma\beta}-\left(b_{\alpha\beta}b_{\gamma\mu}-b_{\alpha\gamma}b_{\beta\mu}\right)=0 \tag{2.94}$$

and

$$\frac{\partial b_{\alpha\beta}}{\partial u^{\gamma}}-\frac{\partial b_{\alpha\gamma}}{\partial u^{\beta}}+\Upsilon^{\delta}_{\alpha\beta}b_{\delta\gamma}-\Upsilon^{\lambda}_{\alpha\gamma}b_{\beta\lambda}=0 \tag{2.95}$$

where $R_{\mu\alpha\gamma\beta}$ is the two-dimensional Riemann curvature tensor, given as

$$R_{\mu\alpha\gamma\beta}=g_{\mu\delta}\left(\frac{\partial \Upsilon^{\delta}_{\alpha\beta}}{\partial u^{\gamma}}-\frac{\partial \Upsilon^{\delta}_{\alpha\gamma}}{\partial u^{\beta}}+\Upsilon^{\sigma}_{\alpha\beta}\Upsilon^{\delta}_{\gamma\sigma}-\Upsilon^{\lambda}_{\alpha\gamma}\Upsilon^{\delta}_{\beta\lambda}\right) \tag{2.96}$$

Eq. 2.94 is called the equation of Gauss and is exhibited here in tensor form. In two dimensions, only four components are non-zero. That is

$$R_{1212}=R_{2121}=b$$

and

$$R_{2112}=R_{1221}=-b$$

where

$$b=b_{11}b_{22}-\left(b_{12}\right)^2$$

The Gaussian curvature K is given by

$$K=R_{1212}/G_3 \tag{2.97}$$

On the other hand, Eq. 2.95 yields two equations: one for $\alpha = 1, \beta = 1, \gamma = 2$ and the other for $\alpha = 2, \beta = 2, \gamma = 1$. The resulting two equations are called the Codazzi or Codazzi–Mainardi equations.

2.4.8 Second-Order Differential Operator of Beltrami

First of all, it is of interest to note that Eqs. 2.35b and 2.36b for the covariant derivative and Eqs. 2.37a, 2.37b and 2.39 are all valid in any space including E^3, and are equally applicable to a surface that is nothing but a two-dimensional non-Euclidean space. Thus the above-noted formulae for a surface are as follows:

$$
\begin{aligned}
u^{\alpha}_{,\beta} &= \frac{\partial u^{\alpha}}{\partial u^{\beta}} + \Upsilon^{\alpha}_{\gamma\beta} u^{\gamma} \\
u_{\alpha,\beta} &= \frac{\partial u_{\alpha}}{\partial u^{\beta}} - \Upsilon^{\gamma}_{\alpha\beta} u_{\gamma} \\
\frac{\partial g_{\alpha\beta}}{\partial u^{\gamma}} &= \Upsilon^{\delta}_{\alpha\gamma} g_{\delta\beta} + \Upsilon^{\delta}_{\beta\gamma} g_{\delta\alpha} \\
\frac{\partial g_{\alpha\beta}}{\partial u^{\gamma}} &= -\Upsilon^{\alpha}_{\delta\gamma} g^{\delta\beta} - \Upsilon^{\beta}_{\delta\gamma} g^{\delta\alpha} \\
\Upsilon^{\delta}_{\delta\beta} &= \frac{1}{2G_3} \frac{\partial G_3}{\partial u^{\beta}} \\
&= \frac{\partial}{\partial u^{\beta}} \left(\ln \sqrt{G_3} \right)
\end{aligned}
\tag{2.98}
$$

The second-order differential operator of Beltrami when applied to a function ϕ yields [2]

$$
\Delta_2 \phi = \frac{1}{\sqrt{G_3}} \frac{\partial}{\partial u^{\alpha}} \left(\sqrt{G_3} g^{\alpha\beta} \frac{\partial \phi}{\partial u^{\beta}} \right) \tag{2.99}
$$

Suppose $\phi = u^{\delta}$, a surface coordinate, then

$$
\Delta_2 u^{\delta} = \frac{1}{\sqrt{G_3}} \frac{\partial}{\partial u^{\alpha}} \left(\sqrt{G_3} g^{\alpha\delta} \right) \tag{2.100}
$$

Using the formulae given in Eq. 2.98, we get

$$
\Delta_2 u^{\delta} = -\Upsilon^{\delta}_{\alpha\beta} g^{\alpha\beta} \tag{2.101a}
$$

Note the exact similarity between Eqs. 2.44 and 2.100, and between Eqs. 2.34 and 2.101a. Using the formulae given in Eqs. 2.98, Eq. 2.99 becomes

$$
\Delta_2 \phi = g_{\alpha\beta} \left(\frac{\partial^2 \phi}{\partial u^{\alpha} \partial u^{\beta}} - \Upsilon^{\gamma}_{\alpha\beta} \frac{\partial \phi}{\partial u^{\gamma}} \right) \tag{2.101b}
$$

or, by using Eq. 2.101a,

$$
\Delta_2 \phi = g^{\alpha\beta} \frac{\partial^2 \phi}{\partial u^{\alpha} \partial u^{\beta}} + \left(\Delta_2 u^{\delta} \right) \frac{\partial \phi}{\partial u^{\delta}} \tag{2.101c}
$$

2.4.9 Geodesic Curves in a Surface

The geodesic curves in a surface are defined in two ways [1]:

(i) Geodesics are curves in a surface that have zero geodesic curvature.
(ii) Geodesic curves are lines of shortest distance between points on a surface.

In the first definition, we must first obtain the formula for the geodesic curvature. Referring to Eq. 2.71 and Figure 2.2, we write the curvature vector of a curve C as

$$\hat{\underset{\sim}{k}} = \underset{\sim}{n} k_n + \underset{\sim}{e} k_g \tag{2.102}$$

where the unit vector $\underset{\sim}{e}$ lies in the tangent plane to the surface. Refer to Figure 2.2. Note that

$$\underset{\sim}{e} = \underset{\sim}{n} \times \underset{\sim}{t}$$

and

$$\begin{aligned} k_g &= \underset{\sim}{e} \cdot \hat{\underset{\sim}{k}} \\ &= \underset{\sim}{e} \cdot \frac{d\underset{\sim}{t}}{ds} \\ &= \left(\underset{\sim}{n} \times \underset{\sim}{t}\right) \cdot \frac{d\underset{\sim}{t}}{ds} \\ &= \left(\underset{\sim}{t} \times \frac{d\underset{\sim}{t}}{ds}\right) \cdot \underset{\sim}{n} \end{aligned} \tag{2.103a}$$

Further

$$\frac{d\underset{\sim}{t}}{ds} = \frac{\partial \underset{\sim}{a}_\alpha}{\partial u^\beta} \frac{du^\alpha}{ds} \frac{du^\beta}{ds} + \underset{\sim}{a}_\alpha \frac{d^2 u^\alpha}{ds^2} \tag{2.103b}$$

Using the formulae of Gauss, Eq. 2.90, in Eq. 2.103b, putting the result in Eq. 2.103a, and writing $u^1 = \xi$, $u^2 = \eta$, we get after some simplification

$$\begin{aligned} k_g \Big/ \sqrt{G_3} = {} & \Upsilon^2_{11}\left(\frac{d\xi}{ds}\right)^3 - \Upsilon^1_{22}\left(\frac{d\eta}{ds}\right)^3 + \left(2\Upsilon^2_{12} - \Upsilon^1_{11}\right)\left(\frac{d\xi}{ds}\right)^2 \frac{d\eta}{ds} \\ & - \left(2\Upsilon^1_{12} - \Upsilon^2_{22}\right)\left(\frac{d\eta}{ds}\right)^2 \frac{d\xi}{ds} + \frac{d\xi}{ds}\frac{d^2\eta}{ds^2} - \frac{d\eta}{ds}\frac{d^2\xi}{ds^2} \end{aligned} \tag{2.104}$$

Eq. 2.104 is the formula for the geodesic curvature of a curve C in the surface with reference to the surface coordinates ξ, η. Here s is the arc length along the curve C. From Eq. 2.104, the geodesic curvature of the coordinate curve η or ξ = const. is

$$\left(k_g\right)_{\xi=const.} = -\sqrt{G_3}\,\Upsilon^1_{22} / g_{22}^{3/2} \tag{2.105a}$$

and the geodesic curvature of the coordinate curve ξ or $\eta = const.$ is

$$\left(k_g\right)_{\eta=const.} = \sqrt{G_3}\,\Upsilon_{11}^2 / g_{11}^{3/2} \tag{2.105b}$$

Obviously if the η-curve is a geodesic then $\Upsilon_{22}^1 = 0$, while if the ξ-curve is a geodesic then $\Upsilon_{11}^2 = 0$. The differential equation for the geodesic curve is obtained from Eq. 2.104 by putting $k_g = 0$. For brevity, writing

$$\xi' = \frac{d\xi}{ds}, \quad \eta' = \frac{d\eta}{ds}$$

and using

$$\xi'\eta'' - \eta'\xi'' = \xi'^2 \frac{d}{ds}\left(\frac{\eta'}{\xi'}\right) = \xi'^2 \frac{d}{ds}\left(\frac{d\eta}{d\xi}\right)$$

we get

$$\frac{d^2\eta}{d\xi^2} - \Upsilon_{22}^1\left(\frac{d\eta}{d\xi}\right)^3 - \left(2\Upsilon_{12}^1 - \Upsilon_{22}^2\right)\left(\frac{d\eta}{d\xi}\right)^2 + \left(2\Upsilon_{12}^2 - \Upsilon_{11}^1\right)\frac{d\eta}{d\xi} + \Upsilon_{11}^2 = 0 \tag{2.106}$$

By solving Eq. 2.106 under the initial conditions

$$\eta(\xi_0), \ (point); \ \ and \ \left(\frac{d\eta}{d\xi}\right)_{\xi=\xi_0}, \ \ (direction)$$

a unique geodesic can be obtained. According to [3], a geodesic can be found to pass through any given point and have any given direction at that point. If the Christoffel symbols are known for all points of a surface in terms of the surface coordinates ξ, η, then a numerical method, e.g., the Runge–Kutta method, can be used to solve Eq. 2.106.

In E^3 a straight line is the shortest distance between two points. A generalization of this concept to Riemannian or non-Euclidean spaces can be accomplished by using the integral of Eq. 2.46 and applying the Euler–Langrange equations. The end result (refer to [2]) is that the intrinsic derivative (Eq. 2.41) applied to the contravariant components of the unit tangent vector $\underset{\sim}{t}$ with the parameter t replaced by the arc length s is zero. That is,

$$\frac{\delta}{\delta s}\left(\frac{du^\gamma}{ds}\right) = 0$$

which yields

$$\frac{d^2u^\alpha}{ds^2} + \Upsilon_{\beta\gamma}^\alpha \frac{du^\beta}{ds}\frac{du^\gamma}{ds} = 0, \quad \alpha = 1,2 \tag{2.107}$$

The two second-order ordinary differential equations from Eq. 2.107 can be solved simultaneously to yield the geodesic curves $u^1 = u^1(s)$, $u^2 = u^2(s)$ by specifying the initial conditions. Alternatively, writing $u^1 = \xi$, $u^2 = \eta$ and

$$\frac{d\eta}{d\xi} = \frac{d\eta}{ds} \cdot \frac{ds}{d\xi} = \frac{\eta'}{\xi'}$$

$$\frac{d^2\eta}{d\xi^2} = \frac{\eta''}{\xi'^2} - \frac{\xi''\eta'}{\xi'^3}$$

and using the two equations from Eq. 2.107, one obtains Eq. 2.106.

2.4.10 Geodesic Torsion

The torsion of the geodesic of a surface is called the geodesic torsion and is denoted by τ_g. Before we proceed further, it is important to note that the basic triads of vectors for space curves is $(\underset{\sim}{t}, \underset{\sim}{p}, \underset{\sim}{b})$ and for the surface curves is $(\underset{\sim}{t}, \underset{\sim}{e}, \underset{\sim}{n})$. It can be proved (refer to [2]) that for a surface geodesic the unit normal $\underset{\sim}{n}$ to a surface at a point is equal to the principal normal $\underset{\sim}{p}$ of the surface geodesic at the same point, i.e., $\underset{\sim}{p} = \underset{\sim}{n}$. Thus from Eq. 2.50,

$$\underset{\sim}{b} = \underset{\sim}{t} \times \underset{\sim}{n}$$

and from Eq. 2.51c,

$$\frac{d\underset{\sim}{b}}{ds} = -\tau_g \underset{\sim}{n}$$

Thus

$$\frac{d\underset{\sim}{t}}{ds} \times \underset{\sim}{n} + \underset{\sim}{t} \times \frac{d\underset{\sim}{n}}{ds} = -\tau_g \underset{\sim}{n}$$

The first term is zero, since $\hat{\underset{\sim}{k}}$ is parallel to $\underset{\sim}{n}$, and we obtain

$$\tau_g = \underset{\sim}{n} \cdot \left(\frac{d\underset{\sim}{n}}{ds} \times \underset{\sim}{t} \right) \tag{2.108}$$

To establish a relation between the torsion τ of a curve C lying on a surface and the torsion of the geodesic τ_g which touches C at the point P, we consider Eq. 2.102 and write it as

$$k\underset{\sim}{p} = \underset{\sim}{n}k_n + \underset{\sim}{e}k_g$$

where k is the curvature of the curve C and k_g is the geodesic (tangential) curvature of the surface at P. Further, using the relation

$$k_g = k \sin\phi$$

from Figure 2.2 and Eq. 2.79, we get

$$\underset{\sim}{p} = \underset{\sim}{n}\ \cos\phi + \underset{\sim}{e}\ \sin\phi \tag{2.109}$$

On differentiating Eq. 2.109 with respect to s, using Eq. 2.51b, and taking the dot product with $\underset{\sim}{n}$, we obtain

$$\tau\left(\underset{\sim}{b}\cdot\underset{\sim}{n}\right) = \underset{\sim}{n}\cdot\frac{d\underset{\sim}{e}}{ds}\sin\phi - \frac{d\phi}{ds}\sin\phi$$

Differentiating

$$\underset{\sim}{e} = \underset{\sim}{n}\times\underset{\sim}{t}$$

using $\underset{\sim}{b}\cdot\underset{\sim}{n} = \sin\phi \neq 0$, and Eq. 2.108, we get

$$\tau_g = \tau + \frac{d\phi}{ds} \tag{2.110}$$

2.5 Elliptic Equations for Grid Generation

In this section we shall develop the elliptic equations for grid generation, or numerical coordinate mapping, in both the Euclidean and non-Euclidean spaces. The mathematical apparatus to achieve this aim has already been developed in Sections 2.2 through 2.4. In this regard the following two important points should be noted.

(i) Depending on the number of space dimensions, one has to choose a set of "grid or coordinate generators," which form a sort of constraints on the variables of computational or logical space.
(ii) The resulting grid generation equations should be obtained in a form in which the computational space variables appear as the independent variables rather than the dependent variables.

2.5.1 Elliptic Grid Equations in Flat Spaces

First by setting $\phi = \underset{\sim}{r}$ in Eq. 2.45a and noting that $\underset{\sim}{r} = \underset{\sim}{i}_m x_m$ so that its Laplacian is zero, we have

$$g^{ij}\left(\frac{\partial^2 \underset{\sim}{r}}{\partial x^i \partial x^j} - \Gamma^k_{ij}\frac{\partial \underset{\sim}{r}}{\partial x^k}\right) = 0$$

Using Eq. 2.34, we get

$$g^{ij}\frac{\partial^2 \underset{\sim}{r}}{\partial x^i \partial x^j} + \left(\nabla^2 x^k\right)\frac{\partial \underset{\sim}{r}}{\partial x^k} = 0 \tag{2.111}$$

If we now take the grid generators as a set of Poisson equations, i.e.,

$$\nabla^2 x^k = P^k \tag{2.112}$$

where P^k are arbitrary functions of the coordinates x^i, then from the identity shown as Eq. 2.111 a deterministic set of equations is obtained, which is

$$D\underset{\sim}{r} + gP^k \frac{\partial \underset{\sim}{r}}{\partial x^k} = 0 \tag{2.113}$$

where D is a second-order differential operator defined as

$$D = gg^{ij} \frac{\partial^2}{\partial x^i \partial x^j}$$

Writing $\underset{\sim}{r} = \underset{\sim}{i}_m x_m$, where $x_m(x^1, x^2, x^3)$ with $m = 1, 2, 3$, one can readily write three coupled quasilinear partial differential equations for x_1, x_2, x_3 from Eq. 2.113. Writing $x^1 = \xi$, $x^2 = \eta$, $x^3 = \zeta$, denoting a partial derivative by a variable subscript, and using Eq. 2.22, the operator D is written as

$$D = G_1\partial_{\xi\xi} + G_2\partial_{\eta\eta} + G_3\partial_{\zeta\zeta} + 2G_4\partial_{\xi\eta} + 2G_5\partial_{\xi\zeta} + 2G_6\partial_{\eta\zeta} \tag{2.114a}$$

In two dimensions there is no dependence on z and $g_{33} = 1$, so that

$$D = g_{22}\partial_{\xi\xi} - 2g_{12}\partial_{\xi\eta} + g_{11}\partial_{\eta\eta} \tag{2.114b}$$

and the two equations for $x_1 = x$, $x_2 = y$, from Eq. 2.113 are

$$g_{22}x_{\xi\xi} - 2g_{12}x_{\xi\eta} + g_{11}y_{\eta\eta} + g\left(P^1 x_\xi + P^2 x_\eta\right) = 0 \tag{2.115a}$$

$$g_{22}y_{\xi\xi} - 2g_{12}y_{\xi\eta} + g_{11}y_{\eta\eta} + g\left(P^1 y_\xi + P^2 y_\eta\right) = 0 \tag{2.115b}$$

A more general choice for P^k is to take it as [15–17]

$$P^k = g^{ij} P^k_{ij} \tag{2.116}$$

where $P^k_{ij} = P^k_{ji}$ are arbitrary functions. As an example, with this choice the P^1 and P^2 appearing in Eqs. 2.115 become

$$P^k = \left(g_{22}P^k_{11} - 2g_{12}P^k_{12} + g_{11}P^k_{22}\right)/g, \quad k = 1,2 \tag{2.117}$$

Note that the g appearing in Eqs. 2.115 and 2.117 is

$$g = g_{11}g_{22} - \left(g_{12}\right)^2$$

With the choice of Eq. 2.116, Eq. 2.113 becomes

$$D\underset{\sim}{r} + g g^{ij} P_{ij}^{k} \frac{\partial \underset{\sim}{r}}{\partial x^{k}} = 0 \tag{2.118}$$

Either Eq. 2.113 or Eq. 2.118 forms the basic coordinate generation equations of the elliptic type in Euclidean spaces. For engineering and applied sciences, usually the Euclidean spaces of two (E^2) or three (E^3) dimensions are needed. In all cases these equations are quasilinear and are solved numerically under the Dirichlet or mixed Dirichlet and Neumann boundary conditions. Note that both Eqs. 2.113 and 2.118 are elliptic partial differential equations in which the independent variables are x^i or ξ, η, ζ, and the dependent variables are the rectangular Cartesian coordinates $\underset{\sim}{r} = (x_m) = (x, y, z)$.

2.5.1.1 Coordinate Transformation

Let $\bar{x}^i$ be another coordinate system such that

$$\bar{x}^i = f^i\left(x^1, x^2, x^3\right), \quad i = 1,2,3$$

A transformation from one coordinate system to another is said to be admissible if the transformation Jacobian $J \neq 0$, where

$$J = det\left(\frac{\partial \bar{x}^i}{\partial x^j}\right) \tag{2.119a}$$

Under the condition $J \neq 0$, the inverse transformation

$$x^i = \phi^i\left(\bar{x}^1, \bar{x}^2, \bar{x}^3\right)$$

exists and

$$\bar{J} = \det\left(\frac{\partial x^i}{\partial \bar{x}^j}\right) \tag{2.119b}$$

where $\bar{J} \neq 0$.

The theory of coordinate transformation plays two key roles in grid generation. First, if the coordinates $\bar{x}^i$ are considered, then Eq. 2.118 takes the form

$$\bar{D}\underset{\sim}{r} + \bar{g}\,\bar{g}^{ij} \bar{P}_{ij}^{k} \frac{\partial \underset{\sim}{r}}{\partial \bar{x}^{k}} = 0 \tag{2.120}$$

How are the control system function P_{ij}^k and $\bar{P}_{ij}^k$ related? An answer to this question may provide a significant advancement towards the problem of adaptivity. For details on the relationships between P_{ij}^k and $\bar{P}_{ij}^k$ refer to [15] and [23]. Second, the consideration of coordinate transformation leads one to the generating equations in which the dependent variables are not the rectangular Cartesian coordinates. For example, in some problems the dependent variables may be cylindrical coordinates.

Before proceeding on the second topic it will be helpful to summarize some basic transformation formulae. Refer to [2, 7], etc.,

$$\bar{g}^{pn} = g^{ij} \frac{\partial \bar{x}^p}{\partial x^i} \frac{\partial \bar{x}^n}{\partial x^j} \tag{2.121a}$$

$$\bar{g}_{pn} = g_{ij} \frac{\partial x^i}{\partial \bar{x}^p} \frac{\partial x^j}{\partial \bar{x}^n} \tag{2.121b}$$

$$\bar{\Gamma}^p_{kn} = \Gamma^s_{ij} \frac{\partial \bar{x}^p}{\partial x^s} \frac{\partial x^i}{\partial \bar{x}^k} \frac{\partial x^j}{\partial \bar{x}^n} + \frac{\partial^2 x^j}{\partial \bar{x}^k \partial \bar{x}^n} \frac{\partial \bar{x}^p}{\partial x^j} \tag{2.121c}$$

$$\frac{\partial^2 \bar{x}^p}{\partial x^k \partial x^n} = \Gamma^s_{kn} \frac{\partial \bar{x}^p}{\partial x^s} - \bar{\Gamma}^p_{rt} \frac{\partial \bar{x}^r}{\partial x^k} \frac{\partial \bar{x}^t}{\partial x^n} \tag{2.121d}$$

Using Eq. 2.121c in Eq. 2.121d, we get

$$\frac{\partial^2 \bar{x}^p}{\partial x^k \partial x^n} = -\frac{\partial^2 x^j}{\partial \bar{x}^r \partial \bar{x}^t} \frac{\partial \bar{x}^p}{\partial x^j} \frac{\partial \bar{x}^r}{\partial x^k} \frac{\partial \bar{x}^t}{\partial x^n} \tag{2.121e}$$

Inner multiplication yields

$$\frac{\partial^2 x^s}{\partial \bar{x}^m \partial \bar{x}^n} = -\frac{\partial^2 \bar{x}^p}{\partial x^k \partial x^r} \frac{\partial x^s}{\partial \bar{x}^p} \frac{\partial x^k}{\partial \bar{x}^m} \frac{\partial x^r}{\partial \bar{x}^n} \tag{2.121f}$$

Eq. 2.121e, 2.121f provide the formulae for the second derivatives. The first partial derivatives of x^i with respect to $\bar{x}^j$ are given by

$$\frac{\partial x^i}{\partial \bar{x}^j} = \frac{C^i_j}{J} \tag{2.121g}$$

$$C^i_j = \frac{\partial \bar{x}^r}{\partial x^s} \frac{\partial \bar{x}^k}{\partial x^n} - \frac{\partial \bar{x}^r}{\partial x^n} \frac{\partial \bar{x}^k}{\partial x^s} \tag{2.121h}$$

where (i, s, n) and (j, r, k) are cyclic permutations of (1, 2, 3), and J is defined by Eq. 2.119a.

According to Eq. 2.34 the Laplacian of the coordinates $\bar{x}^s$ is

$$\nabla^2 \bar{x}^s = -\bar{g}^{ij} \bar{\Gamma}^s_{ij} \tag{2.122}$$

and

$$\begin{aligned} \nabla^2 x^k &= -g^{ij} \Gamma^s_{ij} \\ &= g^{ij} P^k_{ij} = P^k \end{aligned} \tag{2.123}$$

Thus writing $\phi = \bar{x}^s$ in Eq. 2.45b and using Eqs. 2.122 and 2.123, we get

$$g^{ij} = \frac{\partial^2 \bar{x}^s}{\partial x^i \partial x^j} + P^k \frac{\partial \bar{x}^s}{\partial x^k} = \bar{g}^{ij} \bar{\Gamma}^s_{ij} \tag{2.124}$$

Writing

$$g^{ij} = \bar{g}^{mn} \frac{\partial x^i}{\partial \bar{x}^m} \frac{\partial x^j}{\partial \bar{x}^n}$$

in Eq. 2.124 and using Eq. 2.121g, we get

$$C^i_m C^j_n \bar{g}^{mn} \frac{\partial^2 \bar{x}^s}{\partial x^i \partial x^j} + J^2 P^k \frac{\partial \bar{x}^s}{\partial x^k} = -J^2 \bar{g}^{ij} \overline{\Gamma}^s_{ij} \tag{2.125}$$

For prescribed functions P^k, the set of Eq. 2.125 generates the $\bar{x}^s$ coordinates as functions of x^i coordinates. Here $\bar{x}^s$ can be either rectangular Cartesian or any other coordinate system, e.g., cylindrical. Note that if $\bar{x}^s$ are rectangular Cartesian coordinates, then

$$C^i_m C^j_n \bar{g}^{mn} = \sum_{m=1}^{3} C^i_m C^j_m$$

and

$$\overline{\Gamma}^s_{ij} = 0$$

so that Eq. 2.125 becomes Eq. 2.113.

2.5.1.2 Non-Steady Coordinates

There are many situations in which the curvilinear coordinates are changing with time. This occurs mostly in problems where the coordinates move in an attempt to produce an adaptive solution. For a review of the time-dependent coordinates the reader is referred to [22]. For our present purposes we consider one possible grid generator to obtain time-dependent coordinates.

Basically a time-dependent coordinate system x^i is stated as

$$x^i = x^i(\underset{\sim}{r}, t), \quad i = 1,2,3 \tag{2.126a}$$

$$\tau = t \tag{2.126b}$$

and its inverse as

$$\underset{\sim}{r} = \underset{\sim}{r}(x^i, \tau) \tag{2.127a}$$

$$t = \tau \tag{2.127b}$$

From [22], we have the result

$$\frac{\partial \underset{\sim}{r}}{\partial \tau} = -\frac{\partial \underset{\sim}{r}}{\partial x^k} \frac{\partial x^k}{\partial t} \tag{2.128}$$

Suppose for time-dependent coordinates we change the grid generator, Eq. 2.113, to the form

$$\nabla^2 x^k = P^k + \phi \frac{\partial x^k}{\partial t} \tag{2.129}$$

where $\phi = \phi(x^k)$. One may choose $\phi = c/g$, or, $\phi = c$, where c is a constant. Substitution of Eq. 2.129 in Eq. 2.111 with $\phi = c$ and using Eq. 2.128 yields

$$g \frac{\partial \underset{\sim}{r}}{\partial \sigma} = D\underset{\sim}{r} + gP^k \frac{\partial \underset{\sim}{r}}{\partial x^k} \tag{2.130}$$

where $\sigma = \tau / c$ and the operator D is same as used in Eq. 2.113. Eq. 2.130 is parabolic in σ and may be used to proceed in stepwise fashion from some initial time. It must, however, be noted that the success of the grid generator, Eq. 2.129, depends upon a proper choice of the control functions P^k or P^k_{ij} if the form of Eq. 2.116 is used. The proper choice of the control functions depends on the physical problem. Much work in this area remains to be done.

2.5.1.3 Nonelliptic Grid Generation

Besides the elliptic grid generation methodology as discussed in the preceding subsections, which gives the smoothest grid lines, many authors have used the parabolic and hyperbolic equation methodologies. In the hyperbolic grid generation as developed in [24] the grid generators are formed of the following three equations:

$$g_{13} = 0, \quad g_{23} = 0, \quad \sqrt{g} = \Delta V \tag{2.131}$$

where ΔV is a prescribed cell volume. One may take a certain distribution of x^1 and x^2 at the surface x^3 = const. and march along the x^3 direction. Efficient numerical schemes can be used if Eq. 2.131 are combined as a set of simultaneous first-order equations. It must, however, be noted that Eqs. 2.131 are not invariant to a coordinate transformation.

2.5.2 Elliptic Grid Equations in Curved Surfaces

The basic formulation of the elliptic grid generation equations for a curved surface, forming a two-dimensional Riemannian space, is available in [15–18], and [25]. Here we summarize the salient features of the equations with the intent of establishing the fact that the proposed equations are not the result of any sort of simplifying assumptions. (In this regard, readers are referred to [26].) Further, every coordinate system in a surface must satisfy the proposed equations irrespective of the method used to obtain them.

We consider a curved surface embedded in E^3 and use the formulae of Gauss as given in Equation 2.90. Inner multiplication of Equation 2.90 by $g^{\alpha\beta}$ while using Eqs. 2.83 and 2.101a results in having

$$g^{\alpha\beta} \frac{\partial^2 \underset{\sim}{r}}{\partial u^\alpha \partial u^\beta} + \left(\Delta_2 u^\delta\right) \frac{\partial \underset{\sim}{r}}{\partial u^\delta} = \underset{\sim}{n}\left(k_I + k_{II}\right) \tag{2.132}$$

From Eq. 2.101c we note that by setting $\phi = \underset{\sim}{r}$, the left-hand side of Eq. 2.132 can be written as $\Delta_2 \underset{\sim}{r}$. Thus

$$\Delta_2 \underset{\sim}{r} = \underset{\sim}{n}\left(k_I + k_{II}\right) \tag{2.133}$$

where in both Eqs. 2.132 and 2.133 $\underset{\sim}{n}$ is the surface unit normal vector. Also by using Eq. 2.99 we have

$$\Delta_2 \underset{\sim}{r} = \frac{1}{\sqrt{G_3}} \frac{\partial}{\partial u^\alpha} \left(\sqrt{G_3} g^{\alpha\beta} \frac{\partial \underset{\sim}{r}}{\partial u^\beta} \right) \tag{2.134}$$

We will return to Eqs. 2.133 and 2.134 subsequently. First, in Eq. 2.132 writing $x^1 = \xi$, $x^2 = \eta$, and

$$\Delta_2 \xi = -g^{\alpha\beta} \Upsilon^1_{\alpha\beta} = P \tag{2.135a}$$

$$\Delta_2 \eta = -g^{\alpha\beta} \Upsilon^2_{\alpha\beta} = Q \tag{2.135b}$$

while using the operator D defined as

$$\begin{aligned} D &= G_3 g^{\alpha\beta} \frac{\partial^2}{\partial u^\alpha \partial u^\beta} \\ &= g_{22} \partial_{\xi\xi} - 2 g_{12} \partial_{\xi\eta} + g_{11} \partial_{\eta\eta} \end{aligned} \tag{2.136}$$

we get

$$D \underset{\sim}{r} + G_3 \left(P \underset{\sim}{r}_\xi + Q \underset{\sim}{r}_\eta \right) = \underset{\sim}{n} R \tag{2.137}$$

where

$$R = G_3 \left(k_I + k_{II} \right) = g_{22} b_{11} - 2 g_{12} b_{12} + g_{11} b_{22} \tag{2.138}$$

Eq. 2.137 is a deterministic equation for grid generation if the control functions P and Q, which are the Beltramians of ξ and η, respectively, given in Eq. 2.135, are prescribed. The three scalar equations from Eq. 2.137 are

$$Dx + G_3 \left(P x_\xi + Q x_\eta \right) = XR \tag{2.139a}$$

$$Dy + G_3 \left(P y_\xi + Q y_\eta \right) = YR \tag{2.139b}$$

$$Dz + G_3 \left(P z_\xi + Q z_\eta \right) = ZR \tag{2.139c}$$

where $\underset{\sim}{n} = (X, Y, Z)$.

For prescribed P and Q, which may be chosen as zero, the set of elliptic equations stated in Eq. 2.139 form a model for surface coordinate generation. Looking back we note that the basis of these equations are the formulae of Gauss. To check whether the same equations can be obtained by using the formulae of Weingarten stated in Eq. 2.89 we proceed from Eq. 2.134. First we use the easily verifiable identity

$$g^{\alpha\beta} \underset{\sim}{a}_\beta = \varepsilon^{\alpha\delta} \underset{\sim}{a}_\delta \times \underset{\sim}{n}$$

in Eq. 2.134. Here

$$\varepsilon^{11}=0,\quad \varepsilon^{12}=1\sqrt{G_3},\quad \varepsilon^{21}=-1\sqrt{G_3},\quad \varepsilon^{22}=0$$

and as before

$$\underset{\sim}{a}_\beta=\frac{\partial \underset{\sim}{r}}{\partial u^\beta}\quad etc.$$

Thus

$$\Delta_2\underset{\sim}{r}=\frac{1}{\sqrt{G_3}}\frac{\partial}{\partial u^\alpha}\left(\sqrt{G_3}\varepsilon^{\alpha\delta}\underset{\sim}{a}_\delta\times\underset{\sim}{n}\right)$$

Opening the differentiation and using Eq. 2.89 along with the definition of given in Eq.2. 64, we obtain

$$\begin{aligned}\Delta_2\underset{\sim}{r}&=\underset{\sim}{n}\,b_{\alpha\beta}g^{\alpha\beta}\\&=\underset{\sim}{n}\left(k_I+k_{II}\right)\end{aligned}$$

which is precisely Eq. 2.133 or Eq. 2.132. From this analysis we conclude that the proposed set of equations, i.e., Eq. 2.132, satisfies both the formulae of Gauss and Weingarten. In summary, we may state the following:

(i) The solution of the proposed equations automatically satisfies the formulae of Gauss and Weingarten.
(ii) When the curved surface degenerates to a plane z = const., then the proposed equations reduce to the elliptic coordinate generation equation given as Eq. 2.115. In this situation the Beltrami operator reduces to the Laplace operator, i.e.,

$$\Delta_2\xi=\nabla^2\xi,\quad \Delta_2\eta=\nabla^2\eta$$

The key term in the solution of Eq. 2.139 is the term $k_I + k_{II}$ appearing on the right-hand side. For a given surface if this term can be expressed as a function of x, y, z, then there is no difficulty in solving the system of equations. Suppose the equation of the surface is given as $F(x, y, z) = 0$, then from [17],

$$\begin{aligned}k_I+k_{II}&\left[\left(F_y^2+F_z^2\right)\left(2F_xF_zF_{xz}-F_z^2F_{xx}-F_x^2F_{zz}\right)\right.\\&+2F_xF_y\left(F_z^2F_{xy}+F_xF_yF_{zz}-F_yF_zF_{xz}-F_xF_zF_{yz}\right)\\&\left.+\left(F_x^2+F_z^2\right)\left(2F_yF_zF_{yz}-F_z^2F_{yy}-F_y^2F_{zz}\right)\right]\Big/P^3F_z^2,\quad F_z\neq 0\end{aligned}\tag{2.140}$$

where

$$P^2=F_x^2+F_y^2+F_z^2$$

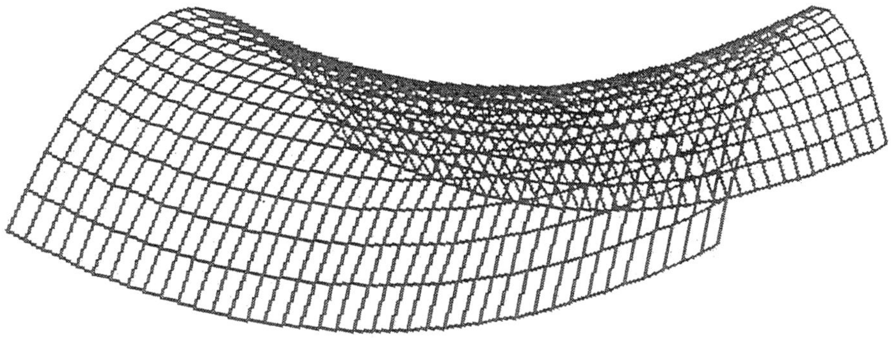

FIGURE 2.3 A demonstrative example of the solution of Eq. 2.137 for a hyperbolic paraboloidal shell.

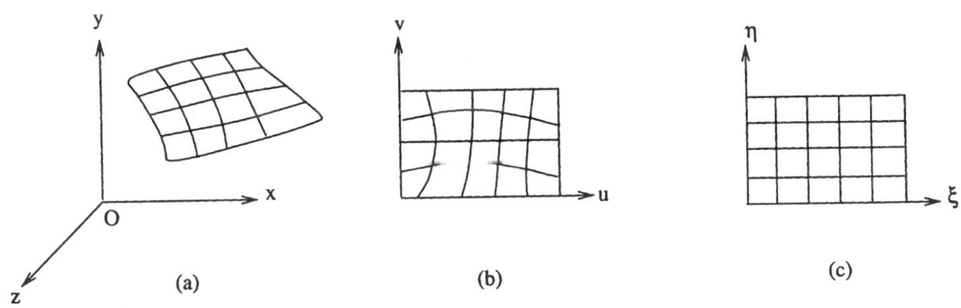

FIGURE 2.4 Transformation from the physical space (a) to the parametric space (b) to the logical space (c).

If $F_z = 0$, then a cyclic interchange of the subscripts will yield a formula in which F_z does not appear in the denominator. Thus we see that the whole problem of coordinate generation in a surface through Eq. 2.139 depends on the availability of the surface equation $F(x, y, z) = 0$. Numerical solutions of Eq. 2.139 have been carried out for various body shapes, including the fuselage of an airplane [25]. Here the function $F(x, y, z) = 0$ was obtained by a least square fit on the available data. As an example, Figure 2.3 shows the distribution of coordinate curves on a hyperbolic paraboloidal shell.

To alleviate the problem of fitting the function $F(x, y, z) = 0$, another set of equations can be obtained from Eq. 2.139. The basic philosophy here is to introduce an intermediate transformation (u,v) between E^3 and (ξ,η), as shown in Figure 2.4.

Let u and v be the parametric curves in a surface in which the curvilinear coordinates ξ and η are to be generated. Introducing

$$\bar{g}_{11} = \underset{\sim}{r}_u \cdot \underset{\sim}{r}_u, \quad \bar{g}_{12} = \underset{\sim}{r}_u \cdot \underset{\sim}{r}_v, \quad \bar{g}_{22} = \underset{\sim}{r}_v \cdot \underset{\sim}{r}_v$$

$$\bar{G}_3 = \bar{g}_{11}\bar{g}_{22} - \left(\bar{g}_{12}\right)^2, \quad J_3 = u_\xi v_\eta - u_\eta v_\xi$$

then from the expressions such as

$$\underset{\sim}{r}_\xi = \underset{\sim}{r}_u u_\xi + \underset{\sim}{r}_v v_\xi, \quad \underset{\sim}{r}_\eta = \underset{\sim}{r}_u u_\eta + \underset{\sim}{r}_v v_\eta \quad etc.$$

and simple algebraic manipulations, Eq. 2.137 yields the following two equations.

$$au_{\xi\xi} - 2bu_{\xi\eta} + cu_{\eta\eta} + J_3^2\left(Pu_\xi + Qu_\eta\right) = J_3^2\bar{\Delta}_2 u \tag{2.141a}$$

$$av_{\xi\xi} - 2bv_{\xi\eta} + cv_{\eta\eta} + J_3^2\left(Pv_\xi + Qv_\eta\right) = J_3^2\overline{\Delta}_2 v \tag{2.141b}$$

where

$$a = g_{22}/\overline{G}_3, \quad b = g_{12}/\overline{G}_3, \quad c = g_{11}/\overline{G}_3$$

and

$$\overline{\Delta}_2 u = \frac{1}{\sqrt{\overline{G}_3}}\left[\frac{\partial}{\partial u}\left(\frac{\overline{g}_{22}}{\sqrt{\overline{G}_3}}\right) - \frac{\partial}{\partial v}\left(\frac{\overline{g}_{12}}{\sqrt{\overline{G}_3}}\right)\right]$$

$$\overline{\Delta}_2 v = \frac{1}{\sqrt{\overline{G}_3}}\left[\frac{\partial}{\partial v}\left(\frac{\overline{g}_{11}}{\sqrt{\overline{G}_3}}\right) - \frac{\partial}{\partial u}\left(\frac{\overline{g}_{12}}{\sqrt{\overline{G}_3}}\right)\right]$$

Eqs. 2.141 were also obtained independently in [27] and recently in [28] by using the Beltrami equations of quasiconformal mapping. Nevertheless, the simple conclusion remains that Eqs. 2.141 are a direct outcome of Eq. 2.137.

2.5.2.1 Transformation of the Surface Coordinates

Let $\bar{u}^\alpha = f^\alpha(u^1, u^2)$ be an admissible coordinate transformation in a surface. It is a matter of direct verification that

$$\underline{n}\left(u^1, u^2\right) = \overline{\underline{n}}\left(\bar{u}^1, \bar{u}^2\right), \quad \textit{invariant}$$

and

$$k_I + k_{II} = \bar{k}_I + \bar{k}_{II}, \quad \textit{invariant}$$

Using these and other derivative transformations, it can be shown that Eq. 2.132 transforms to

$$\bar{g}^{\alpha\beta}\frac{\partial^2 \underline{r}}{\partial\bar{u}^\alpha\partial\bar{u}^\beta} + \left(\overline{\Delta}_2\bar{u}^\delta\right)\frac{\partial\underline{r}}{\partial\bar{u}^\delta} = \overline{\underline{n}}\left(\bar{k}_I + \bar{k}_{II}\right) \tag{2.142}$$

where

$$\overline{\Delta}_2\bar{u}^\delta = -\bar{g}^{\alpha\beta}\overline{\Upsilon}^\delta_{\alpha\beta} = \bar{g}^{\alpha\beta}\overline{P}^\delta_{\alpha\beta}$$

Similarly

$$\Delta_2\underline{r} = \overline{\Delta}_2\underline{r}$$

The above analysis shows that Eq. 2.132 is form-invariant to coordinate transformation. The same result was obtained previously with regard to Eq. 2.118. How are the control functions $P^\delta_{\alpha\beta}$ and $\overline{P}^\delta_{\alpha\beta}$ related? An answer to this question is similar to the one addressed in [23] and is given in [17, Appendix A]. If initially a harmonic coordinate system is chosen [29], then a recursive relation gives the subsequent surface coordinate control functions.

2.5.2.2 The Fundamental Theorem of Surface Theory

The fundamental theorem of surface theory proves the existence of a surface if the coefficients of the first and the second fundamental forms satisfy certain conditions. Referring to [1] the statement of the theorem is as follows: "If $g_{\alpha\beta}$ and $b_{\alpha\beta}$ are given functions of u^δ, sufficiently differentiable, which satisfy the Gauss and Codazzi equations as given in Eqs. 2.94 and 2.95, respectively and $G_3 \neq 0$, then there exists a surface that is uniquely determined except for its position in space." The demonstration of this theorem consists in showing that the formulae of Gauss and Weingarten as given in Eqs. 2.90 and 2.89 respectively have to be solved under proper conditions. It may be noted that Eqs. 2.89 and 2.90 are 5 vector equations that yield 15 scalar equations, and the proper conditions are

$$\underset{\sim}{n} \cdot \underset{\sim}{n} = 1, \quad \underset{\sim}{a}_\alpha \cdot \underset{\sim}{n} = 0, \quad \alpha = 1,2$$

$$\underset{\sim}{a}_\alpha \cdot \underset{\sim}{a}_\beta = g_{\alpha\beta}, \quad \alpha, \beta = 1,2$$

$$\underset{\sim}{n} \cdot \frac{\partial^2 \underset{\sim}{r}}{\partial u^\alpha \partial u^\beta} = b_{\alpha\beta}, \quad \alpha, \beta = 1,2$$

The above statement poses an elaborate scheme and is quite involved for practical computations if one wants to generate a surface based on a knowledge of $g_{\alpha\beta}$ and $b_{\alpha\beta}$. A restatement of the fundamental theorem of surface theory is now possible because Eq. 2.132 already satisfies Eqs. 2.89 and 2.90. Thus, a restatement of the theorem is as follows: "If the coefficient $g_{\alpha\beta}$ and $b_{\alpha\beta}$ of the first and second fundamental forms have been given that satisfy the Gauss and Codazzi equations (Eqs. 2.94, 2.95), then a surface can be generated by solving only one vector equation (Eq. 2.132) to within an arbitrary position in space." This theorem has been checked numerically for a number of cases [30].

2.5.2.3 Time-Dependent Surface Coordinates

If in a given surface the coordinates are time-dependent, then we take the "grid generator" similar to Eq. 2.129 with $\phi = c$ as

$$\Delta_2 u^\delta = g^{\alpha\beta} P^\delta_{\alpha\beta}, + c \frac{\partial u^\delta}{\partial t}, \quad \delta = 1,2 \tag{2.143}$$

Realizing that the surface is defined by x^3 = const., the resulting surface grid generation equation becomes

$$G_3 \frac{\partial \underset{\sim}{r}}{\partial \sigma} = D\underset{\sim}{r} + \left(P\underset{\sim}{r}_\xi + Q\underset{\sim}{r}_\eta \right) - \underset{\sim}{n} R \tag{2.144}$$

where $\sigma = \tau / c$ and all other quantities are similar to those given in Eq. 2.137. The choice $\phi = c/G_3$ has been used to generate the surface coordinates in a fixed surface by parametric stepping and using a spectral technique [31].

2.5.2.4 Coordinate Generation Equations in a Hypersurface

In the course of an effort to extend the fundamental basis of Eq. 2.132 we have considered an extension of the embedding space E^3 to a Riemannian-4 (M^4) space. In M^4 let the local coordinates be x^i, $i = 1, \ldots, 4$ and let S be an immersed hypersurface of local coordinates ξ^α, $\alpha = 1, \ldots, 3$. In the ensuing analysis, a comma preceding an index denotes a partial derivative. From

$$dx^i = x^i_{,\alpha} d\xi^\alpha$$

we note that $x^i_{,\alpha}$ are the tangent vectors. Here, and in what follows, a comma preceding an index will denote a partial derivative while a semicolon will denote a covariant derivative. Further g_{ij} and $a_{\alpha\beta}$ are the covariant metric tenors and Γ^i_{jk} and $\Upsilon^\alpha_{\beta\gamma}$ are the Christoffel symbols in M^4 and S, respectively. The metric coefficients are related as

$$a_{\alpha\beta} = g_{ij} x^i_{,\alpha} x^j_{,\beta} \tag{2.145a}$$

$$a^{\alpha\beta} = g^{mn} \xi^\alpha_{,m} \xi^\beta_{,n} \tag{2.145b}$$

Let a^i_α be a contravariant vector in M^4 and a covariant vector in S, then from [2], the covariant derivative of $a^i_{,\alpha}$ in S is given by

$$a^i_{\alpha;\beta} = x^i_{\alpha,\beta} + a^r_\alpha \Gamma^i_{rk} x^k_{,\beta} - a^i_\gamma \Upsilon^\gamma_{\alpha\beta} \tag{2.145c}$$

Replacing a^i_α by x^i_α in Eq. 2.145c, we get

$$x^i_{,\alpha;\beta} = x^i_{,\alpha\beta} + \Gamma^i_{rk} x^r_{,\alpha} x^k_{,\beta} - \Upsilon^\gamma_{\alpha\beta} x^i_{,\gamma} \tag{2.146}$$

From [2], the formulae of Gauss in a Riemannian manifold are

$$x^i_{,\alpha;\beta} = b_{\alpha\beta} n^i \tag{2.147}$$

and the formula of Weingarten is

$$n^k_{,\beta} = -b_{\alpha\beta} a^{\alpha\gamma} x^k_{,\gamma} - \Gamma^k_{rp} x^r_{,\beta} n^p$$

where n^i are the components of the normal to S in M^4 and $b_{\alpha\beta}$ is the covariant tensor of the second fundamental form. Using Eq. 2.147 in Eq. 2.146 and taking the inner multiplication of every term with $a^{\alpha\beta}$, we get

$$a^{\alpha\beta} x^i_{,\alpha\beta} + \left(\Delta_2 \xi^\gamma\right) x^i_{,\gamma} = -g^{rk} \Gamma^i_{rk} + Pn^i \tag{2.148}$$

where

$$\Delta_2 = a^{\alpha\beta}\left(\frac{\partial^2}{\partial\xi^\alpha \partial\xi^\beta} - \Upsilon^\gamma_{\alpha\beta} \frac{\partial}{\partial\xi^\gamma}\right)$$

and

$$P = a^{\alpha\beta} b_{\alpha\beta}$$

Eq. 2.148 is a generalization of Eq. 2.132 for a Riemannian hypersurface [32, 33]. The main difference is the appearance of the space Christoffel symbols, which vanish when M^4 becomes E^3.

2.6 Concluding Remarks

1. If Dirichlet data is prescribed on the bounding curves of a given surface, then the three scalar equations from Eq. 2.132 can be used to generate coordinates in the surface. The distribution of these coordinates can be controlled by assigning suitable functions P and Q.
2. If the coefficients of the first and the second fundamental forms have been given as functions of some surface coordinates, then the surface suitable to these coefficients can be generated by solving the three scalar equations from Eq. 2.132. In this case, $\Delta_2 u^\delta$ is expressed in terms of the given $g_{\alpha\beta}$, and $k_I + k_{II}$ is expressed in terms of $g^{\alpha\beta} b_{\alpha\beta}$.
3. For a recent account of the use of elliptic equations in grid generation with algebraic parametric transformations, refer to [34].

References

1. Struik, D.J., *Lectures on Classical Differential Geometry.* Addison-Wesley Press, 1950.
2. Kreyszig, E., *Introduction to Differential Geometry and Riemannian Geometry.* University of Toronto Press, Mathematical Exposition No. 16, 1968.
3. Willmore, T.J., *An Introduction to Differential Geometry.* Oxford University Press, 1959.
4. Eisenhart, L.P., *Riemannian Geometry.* Princeton University Press, 1926.
5. Aris, R., *Vectors, Tensors, and the Basic Equations of Fluid Mechanics.* Prentice Hall, Englewood Cliffs, NJ, 1962.
6. McConnell, A.J., *Application of the Absolute Differential Calculus.* Blackie, London, 1931.
7. Warsi, Z.U.A., Tensors and differential geometry applied to analytic and numerical coordinate generation, MSSU-EIRS-81-1, *Engineering and Industrial Research Station,* Mississippi State University, 1981.
8. Winslow, A.M., Numerical solution of the quasi-linear poisson equation in a non-uniform triangular mesh, *J. Computational Phys.* 1967, 1, pp 149–172.
9. Allen, D.N. de. G., Relaxation methods applied to conformal transformations, *Quart. J. Mech. Appl. Math.* 1962, 15, pp 35–42.
10. Chu, W-H., Development of a general finite difference approximation for a general domain, part i: machine transformation, *J. Computational Phys.* 1971, 8, pp 392–408.
11. Thompson, J.F., Thames, F.C., and Mastin, C.W., Automatic numerical generation of body-fitted curvilinear coordinate system for field containing any number of arbitrary two-dimensional bodies, *J. Computational Phys.* 1974, 15, pp 299–319.
12. Thompson, J.F., Warsi, Z.U.A., and Mastin, C.W., *Numerical Grid Generation: Foundations and Applications.* North-Holland, Elsevier, New York, 1985.
13. Knupp, P. and Steinberg, S., *Fundamentals of Grid Generation.* CRC Press, Boca Raton, FL, 1993.
14. George, P.L., *Automatic Mesh Generation: Application to Finite Element Methods.* Wiley, NY, 1991.
15. Warsi, Z.U.A., Basic differential models for coordinate generation, *Numerical Grid Generation.* Thompson J.F. (Ed.), Elsevier Science, 1982, pp 41–77.
16. Warsi, Z.U.A., A note on the mathematical formulation of the problem of numerical coordinate generation, *Quart. Applied Math.* 1983, 41, pp 221–236.
17. Warsi, Z.U.A., Numerical grid generation in arbitrary surfaces through a second-order differential-geometric model, *J. Computational Phys.* 1986, 64, pp 82–96.
18. Warsi, Z.U.A., Theoretical foundation of the equations for the generation of surface coordinates, *AIAA J.* 1990, 28, pp 1140–1142.
19. Castillo, J.E., The discrete grid generation method on curves and surfaces, *Numerical Grid Generation in Computation Fluid Dynamics and Related Fields.* Arcilla, A.S. et al. (Eds.), Elsevier Science, 1991, pp 915–924.
20. Saltzman, J. and Brackbill, J.U., Application and generalization of variational methods for generating adaptive grids, *Numerical Grid Generation.* Thompson, J.F. (Ed.), North-Holland, 1982, pp 865–878.

21. Warsi, Z.U.A. and Thompson, J.F., Application of variational methods in the fixed and adaptive grid generation, *Computer and Mathematics with Applications.* 1990, 19, pp 31–41.
22. Warsi, Z.U.A., *Fluid Dynamics: Theoretical and Computational Approaches.* CRC Press, Boca Raton, FL, 1993.
23. Warsi, Z.U.A., A Synopsis of elliptic PDE models for grid generation, *Appl. Math. and Computation.* 1987, 21, pp 295–311.
24. Steger, J.L. and Rizk, Y.M., *Generation of Three-Dimensional Body-Fitted Coordinates Using Hyperbolic Partial Differential Equations.* NASA TM 86753, 1985.
25. Warsi, Z.U.A. and Tiarn, W.N., Numerical grid generation through second-order differential-geometric models, *IMACS, Numerical Mathematics and Applications.* Vichnevetsky, R. and Vigners, J. (Eds.), Elsevier Science, 1986, pp 199–203.
26. Thomas, P.D., Construction of composite three-dimensional grids from subregion grid generated by elliptic systems, AIAA Paper No. 83-1905, 1983.
27. Garon, A. and Camerero, R., Generation of surface-fitted coordinate grids, *Advances in Grid Generation.* Ghia, K.N. and Ghia, U. (Eds.), ASME, FED-5, 1983, pp 117–122.
28. Khamayesh, A., Ph.D. Dissertation, Mississippi State University, May 1994.
29. Dvinsky, A.S., Adaptive grid generation from harmonic maps on Riemannian manifolds, *J. Computational Phys.* 1991, 95, pp 450–476.
30. Beddhu, M., private communication, 1994.
31. Koomullil, G.P. and Warsi, Z.U.A., Numerical mapping of arbitrary domains using spectral methods, *J. Computational Phys.* 1993, 104, pp 251–260.
32. Sritharan, S.S. and Smith, P.W., Theory of harmonic grid generation, *Complex Variables.* 1988, 10, pp 359–369.
33. Warsi, Z.U.A., Fundamental Theorem Of The Surface Theory And Its Extension To Riemannian manifolds of general relativity, *GANITA.* 1995, 46, pp 119–129.
34. Spekreijse, S.P., Elliptic grid generation based on Laplace equations and algebraic transformations, *J. Computational Phys.* 1995, 118, pp 38–61.

3

Transfinite Interpolation (TFI) Generation Systems

Robert E. Smith

3.1 Introduction

This chapter describes an algebraic grid generation produced called transfinite interpolation (TFI). It is the most widely used algebraic grid generation procedure and has many possible variations. It is the most often-used procedure to start a structured grid generation project.

The advantage of using TFI is that it is an interpolation procedure that can generate grids conforming to specified boundaries. Grid spacing is under direct control. TFI is easily programmed and is very computationally efficient.

Before discussing TFI, a background on grid requirements and the concepts of computational and physical domains is presented. The general formulation of TFI is described as a Boolean sum and as a recursion formula. Practical TFI for linear, Lagrangian, and Hermite cubic interpolation is described. Methods for controlling grid point clustering in the application of TFI are discussed. Finally, a practical TFI recipe to conform an existing grid to new specified boundaries is described.

3.2 Grid Requirements

Grids provide mathematical support for the numerical solution of governing field equations in a continuum domain. The physics is expressed as a system of differential or integral equations subject to initial and boundary conditions. A numerical solution is obtained by superimposing a grid onto the continuum

0-8493-2687-7/99/$0.00+$.50

domain, discretizing the governing equations relative to the grid, and applying a numerical solution algorithm to the discrete approximation of the governing equations. The result is an evaluation of the solution at the grid points. Two key ingredients necessary for obtaining an accurate and efficient solution are (1) the numerical solution algorithm, and (2) the grid.

A grid generation technique should be as efficient as possible to achieve the desired characteristics. However, the importance of a particular characteristic or combination of characteristics can outweigh alone in determining which grid generation technique is applied to a particular problem.

The most efficient grid generation techniques are algebraic and are based on the application of interpolation formulas. Algebraic grid generation techniques relate a computational domain, which is a rectangular parallelepiped (a square in two dimensions and a box in three dimensions), to an arbitrarily shaped physical domain with corresponding sides. The computational domain is a mathematical abstraction. The physical domain is the bounded continuum domain where a numerical solution to a system of governing equations of motion is desired.

A side in the computational domain can map into a line or a point in the physical domain, in which case a singularity occurs in the mapping. Singularities can pose problems for the computation of numerical solutions when the governing equations are expressed in differential form. However, grid singularities usually do not cause problems when the governing equations are expressed in integral form.

A single block (square or box in the computational domain and deformed block in the physical domain) is not usually sufficient to fit to boundaries of a complex solution domain. Therefore, the complex domains must be divided into subdomains and multiple blocks used to cover the subdomains. Depending on the solution technique used to solve the governing equations, the grid points at the boundaries of adjoining blocks must be contiguous.

TFI is a multivariate interpolation procedure. When TFI is applied for algebraic grid generation, a physical grid is constrained to lie on or within specified boundaries. TFI is a Boolean sum of univariate interpolations in each of the computational coordinates. Virtually any univariate interpolation (linear, quadratic, spline, etc.) can be applied in a coordinate direction. Therefore, there are a limitless number of possible variations of TFI that can be created by using different combinations and forms of the univariate interpolations. Often for a particular application, a high order and more sophisticated interpolation is used in one coordinate direction, which we will call the primary coordinate direction, and a low-order interpolation, such as linear interpolation, is used in the remaining coordinate directions.

3.3 Transformations and Grids

Algebraic grid generation techniques are transformations from a rectangular computational domain to an arbitrarily shaped physical domain. This is shown schematically in Figure 3.1 and as a general equation

$$\mathbf{X}(\xi,\eta,\zeta)=\begin{bmatrix} x(\xi,\eta,\zeta) \\ y(\xi,\eta,\zeta) \\ z(\xi,\eta,\zeta) \end{bmatrix} \tag{3.1}$$
$$0\le\xi\le 1 \quad 0\le\eta\le 1 \quad \text{and} \quad 0\le\zeta\le 1$$

A discrete subset of the vector-valued function $\mathbf{X}$, (ξ_I, η_J, ζ_K) is a structured grid for

$$0\le\xi_I=\frac{I-1}{\hat{I}-1}\le 1 \quad 0\le\eta_J=\frac{J-1}{\hat{J}-1}\le 1 \quad 0\le\zeta_K=\frac{K-1}{\hat{K}-1}\le 1$$

where

$$I=1,2,3\cdots,\hat{I} \quad J=1,2,3\cdots,\hat{J} \quad K=1,2,3,\cdots,\hat{K}$$

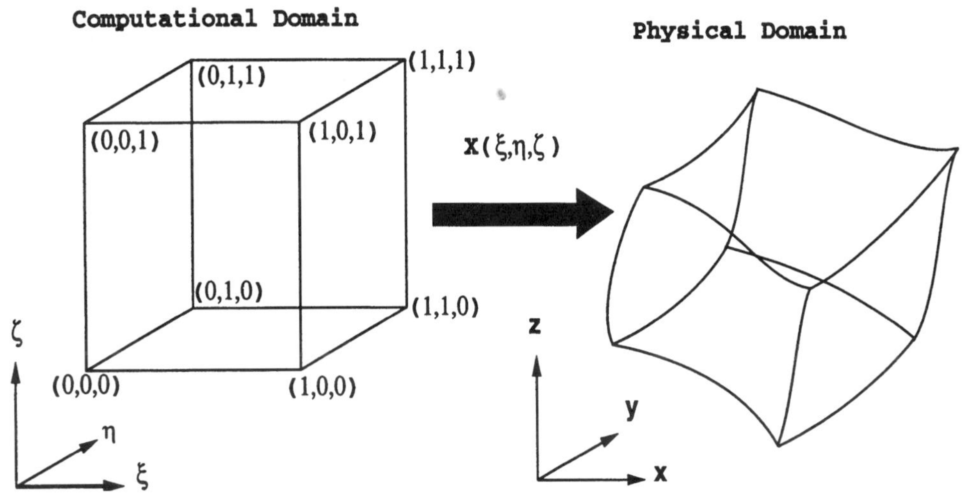

FIGURE 3.1 Transformation between computational and physical domains.

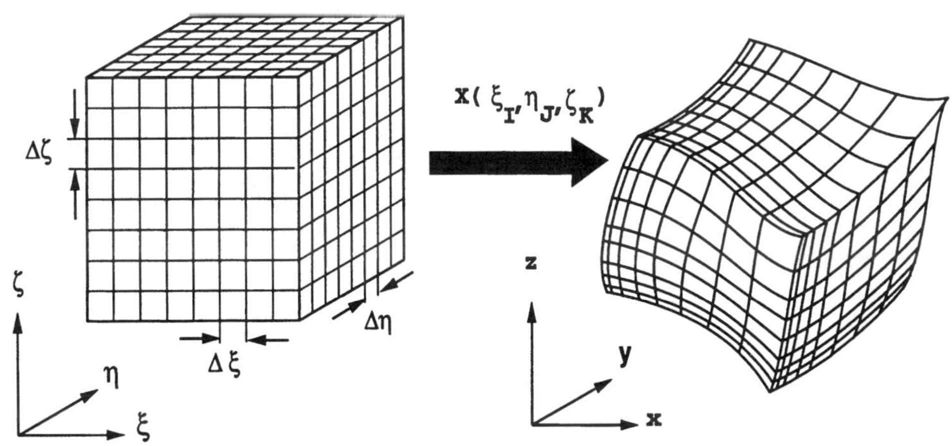

FIGURE 3.2 Grids in computational and physical domains.

The relationships between the indices *I*, *J*, and *K* and the computational coordinates (ξ,η,ζ) uniformly discretize the computational domain and imply a relationship between discrete neighboring points. The transformation to the physical domain produces the actual grid points, and the relationship of neighboring grid points is invariant under the transformation (Figure 3.2). A grid created in this manner is called a structured grid. TFI provides a single framework creating the function $\mathbf{X}(\xi,\eta,\zeta)$.

3.4 Transfinite Interpolation (TFI)

Transfinite interpolation (TFI) was first described by William Gordon in 1973 [1]. TFI has the advantage of providing complete conformity to boundaries in the physical domain. In the early 1980s, Lars Eriksson described TFI for application to grid generation for computational fluid dynamics (CFD) [2,3,4]. Variants of TFI have since been described many times [5,6,7].

3.4.1 Boolean Sum Formulation

The essence of TFI is the specification of univariate interpolations in each of the computational coordinate directions, forming the tensor products of the interpolations, and finally the Boolean sum. The univariate interpolation functions are a linear combination of known (user-specified) information in the physical domain (positions and derivatives) for given values of the computational coordinate and coefficients that are blending functions whose independent variable is the computational coordinate. The general expressions of the univariate interpolations for three dimensions are

$$
\begin{aligned}
\mathbf{U}(\xi,\eta,\zeta) &= \sum_{i=1}^{L}\sum_{n=0}^{P}\alpha_i^n(\xi)\frac{\partial^n \mathbf{X}(\xi_i,\eta,\zeta)}{\partial \xi^n} \\
\mathbf{V}(\xi,\eta,\zeta) &= \sum_{j=1}^{M}\sum_{m=0}^{Q}\beta_j^m(\eta)\frac{\partial^m \mathbf{X}(\xi,\eta_j,\zeta)}{\partial \eta^m} \\
\mathbf{W}(\xi,\eta,\zeta) &= \sum_{k=1}^{N}\sum_{l=0}^{R}\gamma_k^\ell(\zeta)\frac{\partial^\ell \mathbf{X}(\xi,\eta,\zeta_k)}{\partial \zeta^\ell}
\end{aligned}
\tag{3.2}
$$

Conditions on the blending functions are

$$
\frac{\partial^{\bar n}\alpha_i^n(\xi_{\bar i})}{\partial \xi^{\bar n}} = \delta_{i\bar i}\delta_{n\bar n} \qquad \frac{\partial^{\bar m}\beta_j^m(\eta_{\bar j})}{\partial \eta^{\bar m}} = \delta_{j\bar j}\delta_{m\bar m} \qquad \frac{\partial^{\bar \ell}\gamma_k^\ell(\zeta_{\bar k})}{\partial \zeta^{\bar \ell}} = \delta_{k\bar k}\delta_{\ell\bar \ell}
\tag{3.3}
$$

$$
\bar i = 1,2,\cdots,L \quad \bar j = 1,2,\cdots,M \quad \bar k = 1,2,\cdots,N
$$

$$
\bar n = 0,1,\cdots,P \quad \bar m = 0,1,\cdots,Q \quad \bar \ell = 0,1,\cdots,R
$$

The tensor products are

$$
\begin{aligned}
\mathbf{UW} = \mathbf{WU} &= \sum_{i=1}^{L}\sum_{k=1}^{N}\sum_{\ell=0}^{R}\sum_{n=0}^{P}\alpha_i^n(\xi)\gamma_k^\ell(\zeta)\frac{\partial^{\ell n}\mathbf{X}(\xi_i,\eta,\zeta_k)}{\partial\zeta^\ell\partial\xi^n} \\
\mathbf{UV} = \mathbf{VU} &= \sum_{i=1}^{L}\sum_{j=1}^{M}\sum_{m=0}^{Q}\sum_{n=0}^{P}\alpha_i^n(\xi)\beta_j^m(\eta)\frac{\partial^{nm}\mathbf{X}(\xi_i,\eta_j,\zeta)}{\partial\eta^m\partial\xi^n} \\
\mathbf{VW} = \mathbf{WV} &= \sum_{j=1}^{M}\sum_{k=1}^{N}\sum_{m=0}^{Q}\sum_{\ell=0}^{R}\beta_j^m(\xi)\gamma_k^\ell(\zeta)\frac{\partial^{\ell m}\mathbf{X}(\xi,\eta_j,\zeta_k)}{\partial\zeta^\ell\partial\eta^m} \\
\mathbf{UVW} &= \sum_{i=1}^{L}\sum_{j=1}^{M}\sum_{k=1}^{N}\sum_{\ell=0}^{R}\sum_{m=0}^{Q}\sum_{n=0}^{P}\alpha_i^n(\xi)\beta_j^m(\eta)\gamma_k^\ell(\zeta)\frac{\partial^{\ell mn}\mathbf{X}(\xi_i,\eta_j,\zeta_k)}{\partial\zeta^\ell\partial\eta^m\partial\xi^n}
\end{aligned}
\tag{3.4}
$$

The commutability in the above tensor products is assumed in most practical situations, but in general, it is not guaranteed. It is dependent upon the commutability of the mixed partial derivatives.

The Boolean sum of the three interpolations is

$$
\mathbf{X}(\xi,\eta,\zeta) = \mathbf{U}\oplus\mathbf{V}\oplus\mathbf{W} = \mathbf{U}+\mathbf{V}+\mathbf{W}-\mathbf{UV}-\mathbf{UW}-\mathbf{VW}+\mathbf{UVW}
\tag{3.5}
$$

3.4.2 Recursion Formulation

The application of TFI as a Boolean sum of univariate interpolations in the computational coordinate directions implies that each of the terms in the sum be evaluated and then the sum is evaluated.

Alternately, TFI can be expressed as a three-step recursion formula. The first step is to express the univariate interpolation in one coordinate direction

$$\mathbf{X}_1(\xi,\eta,\zeta)=\sum_{i=1}^{L}\sum_{n=0}^{P}\alpha_i^n(\xi)\frac{\partial^n \mathbf{X}(\xi_i,\eta,\zeta)}{\partial \xi^n} \tag{3.6}$$

The second and third steps use the preceding step. That is

$$\mathbf{X}_2(\xi,\eta,\zeta)=\mathbf{X}_1(\xi,\eta,\zeta)+\sum_{j=1}^{M}\sum_{m=0}^{Q}\beta_j^m(\eta)\left[\frac{\partial^m \mathbf{X}(\xi,\eta_j,\zeta)}{\partial \eta^m}-\frac{\partial^m \mathbf{X}_1(\xi,\eta_j,\zeta)}{\partial \eta^m}\right] \tag{3.7}$$

$$\mathbf{X}(\xi,\eta,\zeta)=\mathbf{X}_2(\xi,\eta,\zeta)+\sum_{k=1}^{N}\sum_{\ell=0}^{R}\gamma_k^\ell(\zeta)\left[\frac{\partial^\ell \mathbf{X}(\xi,\eta,\zeta_k)}{\partial \zeta^\ell}-\frac{\partial^\ell \mathbf{X}_2(\xi,\eta,\zeta_k)}{\partial \zeta^\ell}\right] \tag{3.8}$$

3.4.3 Blending Function Conditions

In the above equations, $\alpha_i^n(\xi)$, $\beta_j^m(\eta)$, and $\gamma_k^\ell(\zeta)$ are blending functions subject to δ function conditions. The defining parameters $\frac{\partial^{\ell mn}X(\xi_i,\eta_j,\zeta_k)}{\partial\zeta^\ell\partial\eta^m\partial\xi^n}$ in the equations are positions and partial derivatives in the physical domain and are user-specified. In this definition, the implicit assumption is that coordinate curves are to be interpolated along with their derivatives. This occurs through a network of intersecting surfaces and derivatives that must be specified.

3.5 Practical Application of TFI

In the practical process of generating grids, it is necessary to minimize, or at least keep to a manageable level, the amount of input geometry data (position and derivatives along curves or surfaces). At the same time, it is necessary to maintain a high degree of control, particularly near boundary surfaces for which there may be high gradients in the solution of the governing equations.

3.5.1 Linear TFI

The simplest application of TFI is to use linear interpolation functions for all coordinate directions and specify the positional data on the six bounding surfaces (Figure 3.3). $P = Q = R = 0$ and $L = M = N = 2$ in Eq. 3.2.

The linear blending functions that satisfy the δ function conditions in Eq. 3.3 are

$$\begin{aligned}
\alpha_1^0(\xi)&=1-\xi\\
\alpha_2^0(\xi)&=\xi\\
\beta_1^0(\eta)&=1-\eta\\
\beta_2^0(\eta)&=\eta\\
\gamma_1^0(\zeta)&=1-\zeta\\
\gamma_2^0(\zeta)&=\zeta
\end{aligned}$$

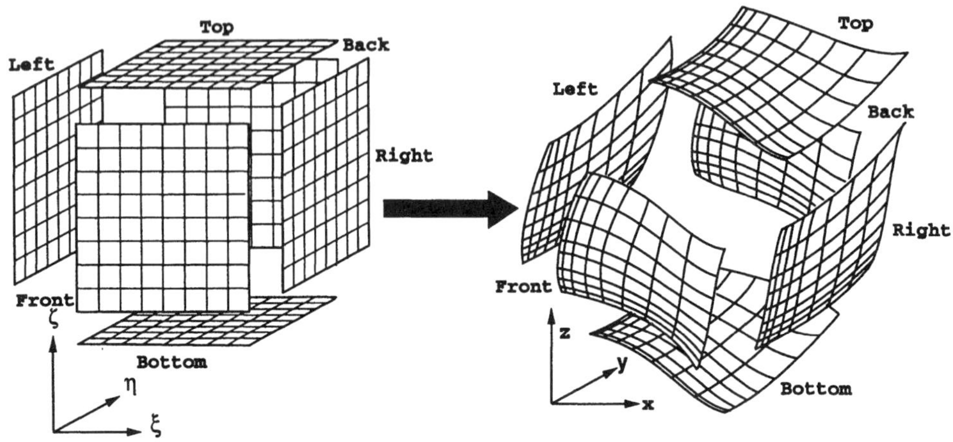

FIGURE 3.3 Boundary surfaces for linear TFI.

The univariate interpolations and tensor products are

$$
\begin{aligned}
\mathbf{U}(\xi_I,\eta_J,\zeta_K) &= (1-\xi_I)\mathbf{X}(0,\eta_J,\zeta_K)+\xi_I\,\mathbf{X}(1,\eta_J,\zeta_K)\\
\mathbf{V}(\xi_I,\eta_J,\zeta_K) &= (1-\eta_J)\mathbf{X}(\xi_I,0,\zeta_K)+\eta_I\,\mathbf{X}(\xi_I,1,\zeta_K)\\
\mathbf{W}(\xi_I,\eta_J,\zeta_K) &= (1-\zeta_K)\mathbf{X}(\xi_I,\eta_J,0)+\zeta_I\,\mathbf{X}(\xi_I,\eta_J,1)\\
\mathbf{UW}(\xi_I,\eta_J,\zeta_K) &= (1-\xi_I)(1-\zeta_K)\mathbf{X}(0,\eta_J,0)+(1-\xi_I\zeta_K)\mathbf{X}(0,\eta_J,1)\\
&+\ \xi_I(1-\zeta_K)\mathbf{X}(1,\eta_J,0)+\xi_I\zeta_K\,\mathbf{X}(1,\eta_J,1)\\
\mathbf{UV}(\xi_I,\eta_J,\zeta_K) &= (1-\xi_I)(1-\eta_J)\mathbf{X}(0,0,\zeta_K)+(1-\xi_I)\eta_J\mathbf{X}(0,1,\zeta_K)\\
&+\ \xi_I(1-\eta_J)\mathbf{X}(1,0,\zeta_K)+\xi_I\eta_J\mathbf{X}(1,1,\zeta_K)\\
\mathbf{VW}(\xi_I,\eta_J,\zeta_K) &= (1-\eta_J)(1-\zeta_K)\mathbf{X}(\xi_I,0,0)+(1-\eta_J)\zeta_K\mathbf{X}(\xi_I,1,0)\\
&+\ \eta_J(1-\zeta_K)\mathbf{X}(\xi_I,0,1)+\eta_J\zeta_K\,\mathbf{X}(\xi_I,1,1)\\
\mathbf{UVW}(\xi_I,\eta_J,\zeta_K) &= (1-\xi_I)(1-\eta_J)(1-\zeta_K)\mathbf{X}(0,0,0)+(1-\xi_I)(1-\eta_J)\zeta_K\mathbf{X}(0,0,1)\\
&+\ (1-\xi_I)\eta_J(1-\zeta_K)\mathbf{X}(0,1,0)+(1-\xi_I)\eta_J\zeta_K\,\mathbf{X}(0,1,1)\\
&+\ \xi_I(1-\eta_J)(1-\zeta_K)\,\mathbf{X}(1,0,0)+\xi_I(1-\eta_J)\zeta_K\,\mathbf{X}(1,0,1)\\
&+\ \xi_I\eta_J(1-\zeta_K)\mathbf{X}(1,1,0)+\xi_I\eta_J\zeta_K\,\mathbf{X}(1,1,1)
\end{aligned}
$$

The expression for a TFI grid ($I = 1, \ldots, \hat{I}, \quad J = 1, \ldots, \hat{J}, \quad K = 1, \ldots, \hat{K}$) with linear interpolation functions (Eq. 3.5) is

$$
\begin{aligned}
&\mathbf{X}(\xi_I,\eta_J,\zeta_K)=\mathbf{U}(\xi_I,\eta_J,\zeta_K)+\mathbf{V}(\xi_I,\eta_J,\zeta_K)+\mathbf{W}(\xi_I,\eta_J,\zeta_K)-\mathbf{UW}(\xi_I,\eta_J,\zeta_K)\\
&-\mathbf{UV}(\xi_I,\eta_J,\zeta_K)-\mathbf{VW}(\xi_I,\eta_J,\zeta_K)+\mathbf{UVW}(\xi_I,\eta_J,\zeta_K)
\end{aligned}
\tag{3.9}
$$

3.5.2 Lagrangian TFI

When additional surfaces corresponding to the interior of the computational box can be provided (see Figure 3.4 for the case of two interior surfaces that would correspond to cubic Lagrangian interpolation),

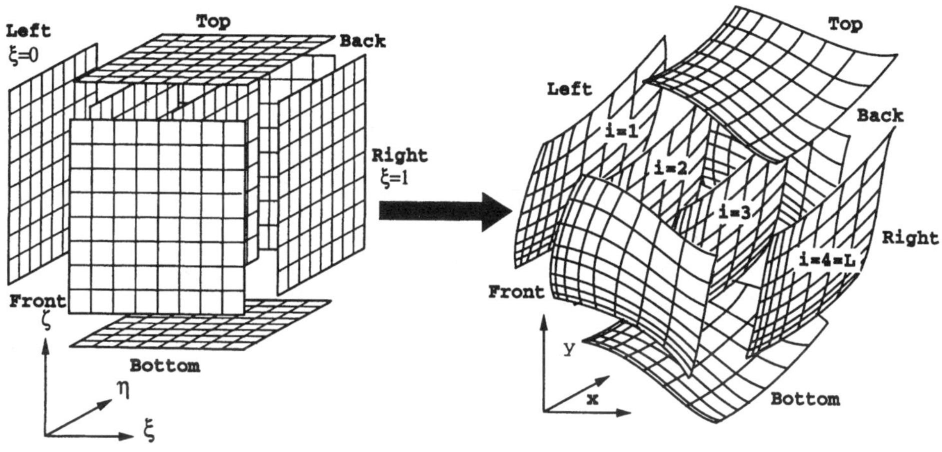

FIGURE 3.4 Transfinite interpolation with Lagrangian blending functions.

a general formula for the blending functions can be used. The formula for a computational coordinate, for instance, the ξ coordinate is

$$\alpha_i^0(\xi) = \frac{\Pi_{\substack{\bar{i}\neq i\\ \bar{i}=1}}^{L}\left(\xi - \xi_{\bar{i}}\right)}{\Pi_{\substack{\bar{i}\neq i\\ \bar{i}=1}}^{L}\left(\xi_i - \xi_{\bar{i}}\right)} \tag{3.10}$$

The univariate interpolation function in the ξ computational coordinate direction is

$$\mathbf{U}(\xi,\eta,\zeta) = \sum_{i=1}^{L} \alpha_i^0(\xi)\frac{\partial^0 \mathbf{X}(\xi,\eta,\zeta)}{\partial \xi^0} = \sum_{i=1}^{L} \alpha_i^0(\xi)\mathbf{X}(\xi,\eta,\zeta) \tag{3.11}$$

The Lagrange blending function allows a polynomial interpolation of degree $L - 1$ through L points and satisfies the cardinal condition $\alpha_i^0(\xi_{\bar{i}}) = \delta_{i\bar{i}}$. It is not recommended that high-degree Lagrangian blending functions be used for grid generation because of the large quantity of geometric data that must be supplied and the potential excessive movement in the interpolation. Using $L = 2$ results in the linear interpolation above being a special case of Lagrangian interpolation.

3.5.3 Hermite Cubic TFI

Often in grid generation, the outward derivative at one or more sides of the physical domain corresponding to sites of the computational domain can be specified. It is then feasible to use Hermite blending functions in the coordinate direction in which derivative information can be specified. For example, if ξ is the coordinate direction, the univariate Hermite interpolation ($L = 2$, $P = 1$) corresponding to Eq. 3.2 is

$$\begin{aligned}\mathbf{U}(\xi,\eta,\zeta) &= \sum_{i=1}^{2}\sum_{n=0}^{1} \alpha_i^n(\xi)\frac{\partial^n \mathbf{X}(\xi_i,\eta,\zeta)}{\partial \xi^n} = \\ &\alpha_1^0(\xi)\mathbf{X}(\xi_1,\eta,\zeta) + \alpha_1^1(\xi)\frac{\partial \mathbf{X}(\xi_1,\eta,\zeta)}{\partial \xi} + \alpha_2^0(\xi)\mathbf{X}(\xi_2,\eta,\zeta) + \alpha_2^1(\xi)\frac{\partial \mathbf{X}(\xi_2,\eta,\zeta)}{\partial \xi}\end{aligned} \tag{3.12}$$

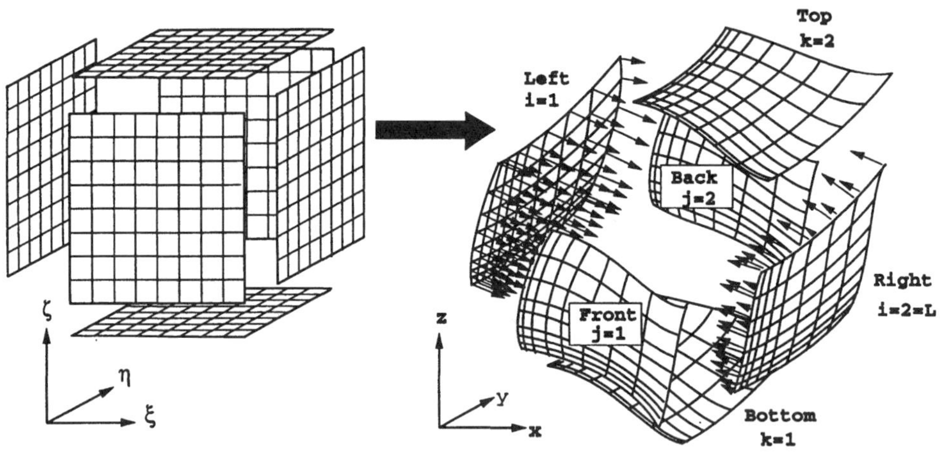

FIGURE 3.5 Transfinite interpolation with Hermite cubic blending functions.

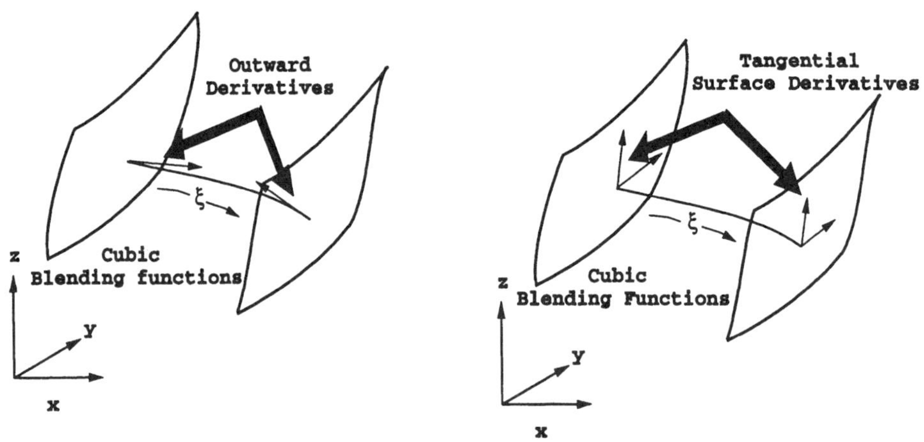

FIGURE 3.6 Outward derivatives obtained from cross-product of surface derivatives.

where

$$\begin{aligned}
\alpha_1^0(\xi) &= 2\xi^3 - 3\xi^2 + 1 \\
\alpha_1^1(\xi) &= \xi^3 - 2\xi^2 + \xi \\
\alpha_2^0(\xi) &= -2\xi^3 + 3\xi^2 \\
\alpha_2^1(\xi) &= \xi^3 - \xi^2
\end{aligned}$$

The outward derivatives in the ξ coordinate direction can be specified by the cross-product of the tangential surface derivatives in the η and ζ coordinate directions at $\xi = 0$ and $\xi = 1$. This effectively creates the trajectories of grid curves that are orthogonal to the surfaces $\mathbf{X}(\xi_1, \eta, \zeta)$ and $\mathbf{X}(\xi_2, \eta, \zeta)$. That is,

$$\frac{\partial \mathbf{X}(\xi_1,\eta,\zeta)}{\partial \xi} = \left[\frac{\partial \mathbf{X}(\xi_1,\eta,\zeta)}{\partial \eta} \times \frac{\partial \mathbf{X}(\xi_1,\eta,\zeta)}{\partial \zeta}\right]\psi_1(\eta,\zeta) \tag{3.13}$$

and

$$\frac{\partial \mathbf{X}(\xi_2,\eta,\zeta)}{\partial \xi} = \left[\frac{\partial \mathbf{X}(\xi_2,\eta,\zeta)}{\partial \eta} \times \frac{\partial \mathbf{X}(\xi_2,\eta,\zeta)}{\partial \zeta} \right] \psi_2(\eta,\zeta) \tag{3.14}$$

The scalar functions $\psi_1(\eta,\zeta)$ and $\psi_2(\eta,\zeta)$ are magnitudes of the outward derivatives in the ξ direction at $\mathbf{X}(\xi_1\eta,\zeta)$ and $\mathbf{X}(\xi_2\eta,\zeta)$. The derivative magnitude parameters can be constants or surface functions. Increasing the magnitudes of the derivatives extends the orthogonality effect further into the physical domain between the two opposing surfaces. However, the magnitudes can be excessively large, resulting in the interpolations equation being multivalued. This is manifested by grid crossover and is remedied by lowering the magnitudes. Note that when the interpolations in the η and ζ directions are applied, the orthogonality effect achieved with the above application of Hermite interpolation in the ξ direction can be altered.

3.6 Grid Spacing Control

TFI transforms a rectangular computational domain to a physical domain with irregular boundaries. A uniform grid in the computational domain is obtained by partitioning each computational coordinate into equal increments. With the transformation, the discrete points in the computational domain map into irregular spaced points in the physical domain creating a physical grid. The spacing between points in the physical domain is controlled by the blending functions $\alpha_i^n(\xi)$, $\beta_j^m(\eta)$ and $\gamma_k^\ell(\zeta)$. Blending functions that produce the desired shape of a grid (i.e., relative orientation between points) may not produce the desired spacing between points. In order to create grids with desired grid concentrations, additional information must be provided. One approach is to design or modify the blending functions to exactly produce the desired concentrations. Another approach, which is effective and practical, is to define an intermediate control domain between the computational domain and the physical domain.

An intermediate control domain is defined to be a rectangular domain where each intermediate coordinate is related to the computational coordinates by

$$u = f(\xi,\eta,\zeta) \qquad v = g(\xi,\eta,\zeta) \qquad w = h(\xi,\eta,\zeta) \tag{3.15}$$

Under the application of these functions, uniformly spaced grid points in the computational domain map to nonuniformly spaced grid points in the control domain enclosed by the unit cube (Figure 3.7). The intermediate coordinates u, v, and w must be single-valued functions of $f(\xi,\eta,\zeta)$, $g(\xi,\eta,\zeta)$, and $h(\xi,\eta,\zeta)$, respectively.

The blending functions are redefined with the intermediate coordinates as the independent variables. That is $\alpha_i^n(u)$, $\beta_j^m(v)$ and $\gamma_k^\ell(w)$.

There are many practical considerations to be exercised at this point. The overall TFI formulation will shape a grid to fit the six boundary surfaces. Control functions that manipulate the grid point spacing are applied. These functions can be simple and be applied universally, or they can be complex and blend from one form to another, transversing from one boundary to an opposite boundary. It may be desirable for a control function to cause concentration of grid points at the extremes of the computational coordinate or somewhere in between. A low slope in a control function leads to grid concentration and high slope leads to grid dispersion. Several control functions are described.

3.6.1 Single-Exponential Function

A useful function that maps an independent variable, ρ, $0 \le \rho \le 1$, to a monotonically increasing dependent variable, r, $0 \le r \le 1$, is

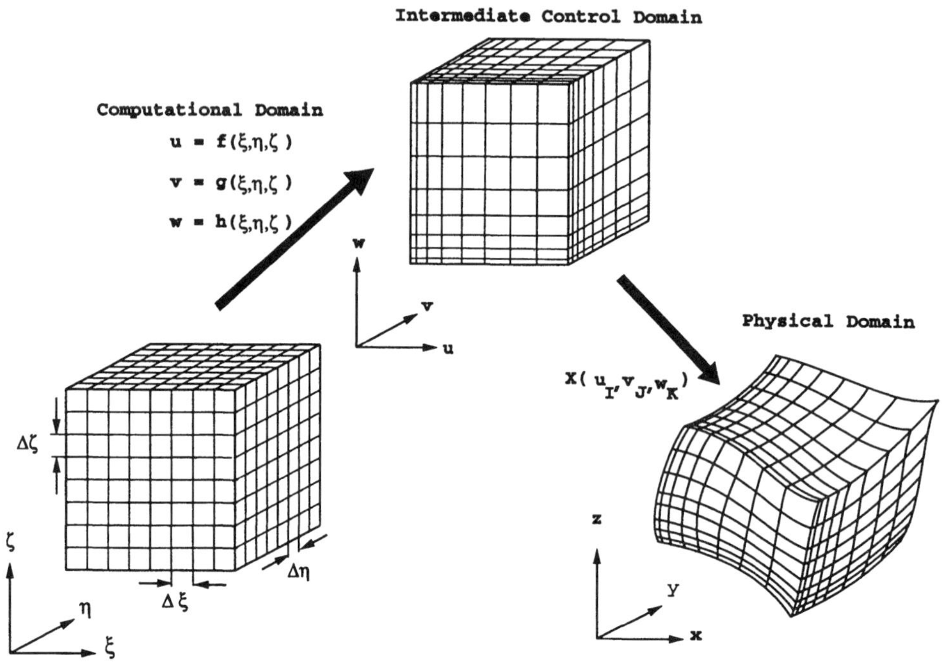

FIGURE 3.7 Intermediate control domain.

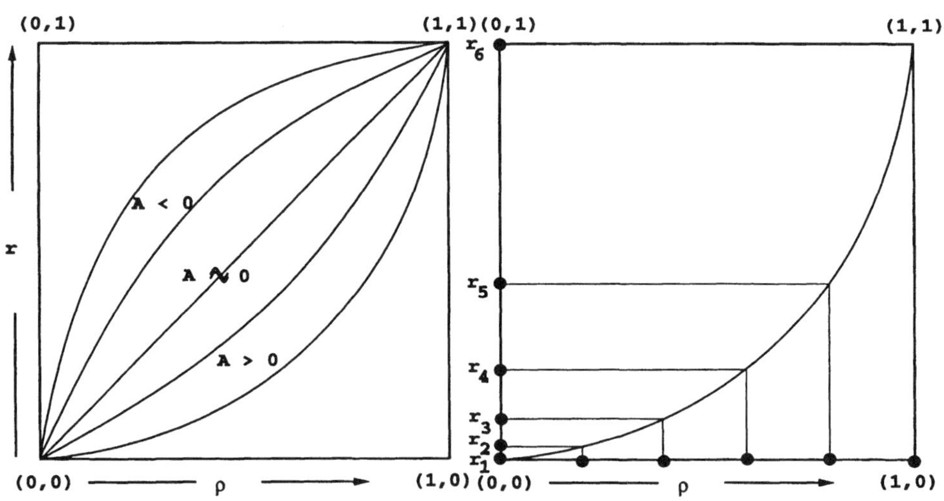

FIGURE 3.8 Single-exponential control function example.

$$r = \frac{e^{A\rho} - 1}{e^{A} - 1} \tag{3.16}$$

where ρ is assumed to be a computational coordinate and r is assumed to be an intermediate variable.

The sign and magnitude of the parameter A specifies whether the lowest slope is near (0, 0) or (1, 1) and the magnitude of the slope (Figure 3.8). For $A = 0$ the single exponential function is singular and is not useful for producing an exact straight line between (0, 0) and (1, 1). This would correspond to a uniform clustering of the dependent variable. However, a magnitude of $|A| = .0001$ will produce a very near straight line. A uniform discrete spacing of the independent variable evaluation of the control function produces concentration or dispersion in the discrete values of the dependent variable. Often the r_2 value ($r_1 = 0$) at $\Delta\rho$ or the $r_{\hat{N}-1}$ value ($r_{\hat{N}} = 1$) at $1 - \Delta\rho$ is specified, and the value of A that causes the function to pass through the point $(\Delta\rho, r_2)$ or $(1 - \Delta\rho, r_{\hat{N}-1})$ is determined with a Newton–Raphson iteration. This creates a control function that specifies the spacing between the first and second grid point or the next to last and last grid point in a coordinate direction. $\hat{N}$ is the index for the last grid point.

3.6.2 Double-Exponential Function

Another function that maps an independent variable, $\rho, 0 \le \rho \le 1,$ to a monotonically increasing dependent variable $r, 0 \le r \le 1,$ and provides more flexibility than the single exponential is

$$
\begin{aligned}
&r = A_1 \frac{e^{\frac{A_2}{A_3}\rho} - 1}{e^{A_2} - 1} \\
&0 \le \rho \le A_3 \qquad 0 \le r \le A_1 \\
&r = A_1 + (1 - A_1)\frac{e^{A_4 \frac{\rho - A_3}{1 - A_3}} - 1}{e^{A_4} - 1} \\
&A_3 \le \rho \le 1 \qquad A_1 \le r \le 1 \\
&A_4 \quad \text{chosen} \ni \frac{Dr(A_3)}{D\rho} \subset C^1
\end{aligned}
\tag{3.17}
$$

The user-specified parameters in Eq. 3.17 are A_1, A_2, and A_3. The parameter A_4 is computed. A_3 and A_1 are the abscissa and ordinate of a point inside the unit square through which the function will pass. A_2 and A_4 are exponential parameters for the two segments. The derivative condition at the joining of the two exponential functions is satisfied by applying a Newton–Raphson iteration that adjusts the value of the parameter A_4. The double exponential control function provides added spacing control as compared to the single exponential function for concentrations near (0, 0) or (1, 1). Also, the double-exponential function allows a grid concentration in the interior or the domain (Figure 3.9). The concept of the double-exponential function can be extended to an arbitrary number of segments, but it is recommended to keep the number of segments small.

3.6.3 Hyperbolic Tangent and Sine Control Functions

Two other single-segment control functions that are used for grid clustering are the hyperbolic tangent (tanh) control function and the hyperbolic sine (sinh) control function. They are

$$
r = 1 + \frac{\tanh\left(B(\rho - 1)\right)}{\tanh B}
\tag{3.18}
$$

$$
\begin{aligned}
&r = 1 + \frac{\sinh\left(C(1 - \rho)\right)}{\sinh C} \\
&0 \le \rho \le 1 \qquad 0 \le r \le 1
\end{aligned}
\tag{3.19}
$$

where the parameters B and C govern the control functions and their derivatives. The hyperbolic tangent function in many references is a preferred control function for clustering grid points in a boundary-layer for computational fluid dynamics applications.

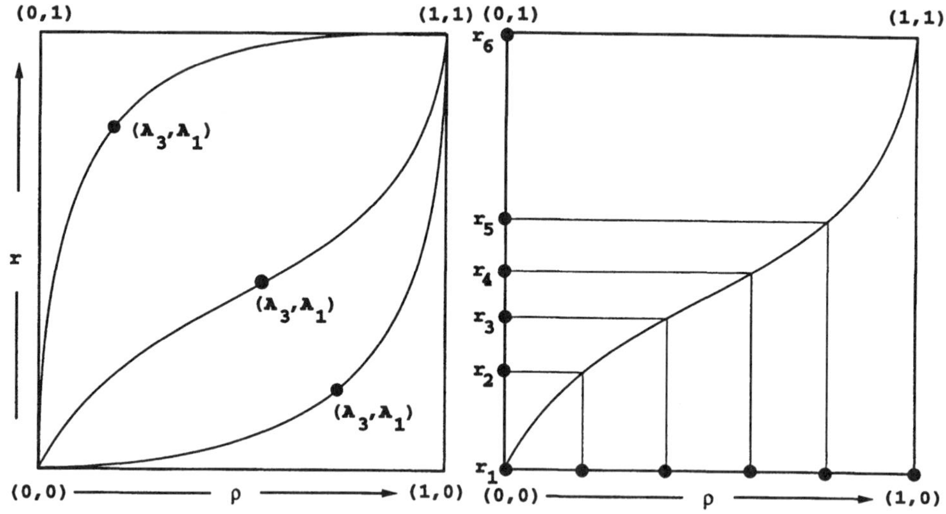

FIGURE 3.9 Double-exponential control function example.

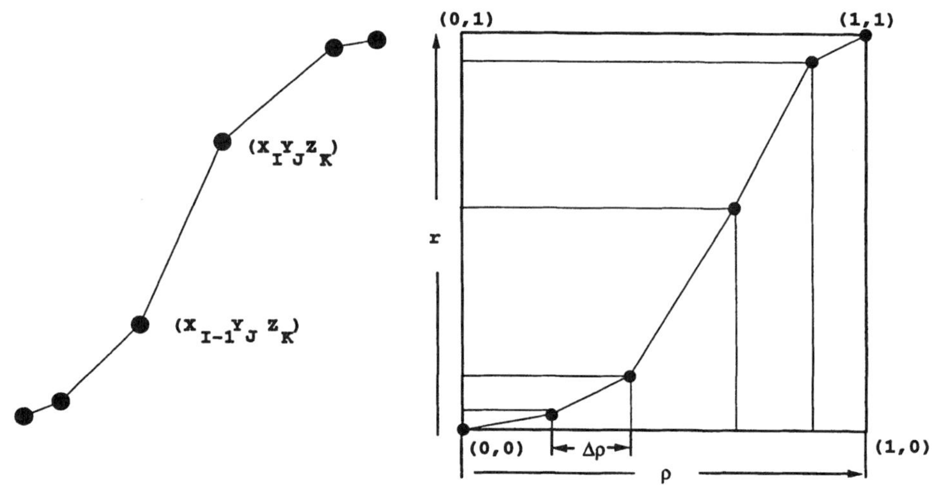

FIGURE 3.10 Arclength control function example.

3.6.4 Arclength Control Functions

Very often an existing sequence of grid points along a coordinate curve, for instance, along a boundary curve, is known (Figure 3.10). It is desirable to use the sequence of points to create a control function. This can be done by normalizing the indices of the points to create the independent variable and computing the normalized accumulated chord lengths along the sequence of points to create the dependent variable. This process approximates the normalized arclength along the curve. A sequence of points is $\{x_{I,J,K}, y_{I,J,K}, z_{I,J,K}, I = 1, 2, \ldots \hat{N}\}$ and J and K are fixed, the formulae for the independent variable $\rho, 0 \le \rho \le 1,$ and the dependent variable $r, 0 \le r \le 1,$ are

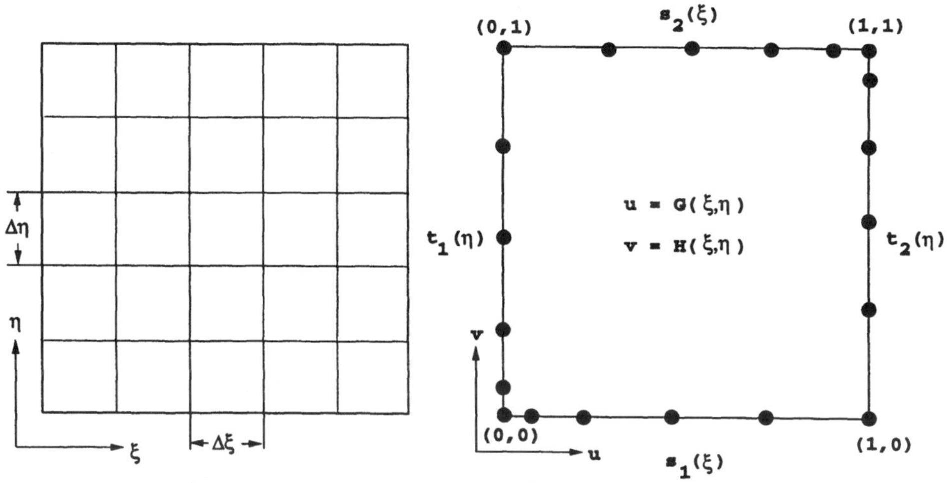

FIGURE 3.11 Boundary-blended control function example.

$$
\begin{aligned}
\rho_I &= \frac{I-1}{\hat{N}-1} \\
s_I &= \sqrt{\left(x_{I,J,K}-x_{I-1,J,K}\right)^2+\left(y_{I,J,K}-y_{1,J,K}\right)^2+\left(z_{I,J,K}-z_{I-1,J,K}\right)^2}+s_{I-1} \\
r_I &= \frac{s_I}{s_{\hat{N}}}
\end{aligned}
\tag{3.20}
$$

Note that if the number of grid points to be used in the grid generation formula (i.e., TFI) is $\hat{N}$, there is no need to compute the independent variable ρ_I. If, however, the number of grid points in the coordinate direction is different from $\hat{N}$, then the dependent variable r_I must be interpolated from the normalized approximate arclength evaluation, and the independent variable values ρ_I are necessary.

3.6.5 Boundary-Blended Control Functions

One of the practical problems that occurs in grid generation is the need to have different control functions specified along each edge of the intermediate domain and compute blended values of the intermediate variables interior to the domain. Soni [8] has proposed a blending formula for arclength control functions along the boundary edges that is very useful. This formula also is applicable for other control functions defined along the edges (Figure 3.11). A two-dimensional description of this type of blending is shown. Let $s_1(\xi)$, $0 \le s_1(\xi) \le 1$, and $s_2(\xi)$, $0 \le s_2(\xi) \le 1$, be control functions along the edges spanning between $t_1(\eta = 0)$, $t_2(\eta = 0)$ and $t_1(\eta = 1)$, $t_2(\eta = 1)$. Let $t_1(\eta)$, $0 \le \eta \le 1, 0 \le t_1(\eta) \le 1$ and $t_2(\eta)$, $0 \le \eta \le 1, 0 \le t_2(\eta) \le 1$, be control functions along the edges spanning between $s_1(\xi = 0)$, $s_2(\xi = 0)$ and $s_1(\xi = 1)$, $s_2(\xi = 1)$. The blended values of intermediate control variables are

$$
\begin{aligned}
u &= \frac{\left(1-t_1(\eta)\right)s_1(\xi)+t_1(\eta)s_2(\xi)}{1-\left(s_2(\xi)-s_1(\xi)\right)\left(t_2(\eta)-t_1(\eta)\right)} \\
v &= \frac{\left(1-s_1(\xi)\right)t_1(\eta)+s_1(\xi)t_2(\eta)}{1-\left(t_2(\eta)-t_1(\eta)\right)\left(s_2(\xi)-s_1(\xi)\right)}
\end{aligned}
\tag{3.21}
$$

3.7 Conforming an Existing Grid to New Boundaries

TFI is normally used to generate a grid given three pairs of defined opposing boundaries. A variation of TFI can also be used to adjust an existing grid to three new pairs of opposing boundaries. This TFI variation can be stated in the following way. Note that ξ_I, η_J, and ζ_K are replaced with the indices I, J, and K.

Given a grid $\hat{\mathbf{X}}(I, J, K)$, $I = 1, 2 \ldots \hat{I}$, $J = 1, 2, \ldots \hat{J}$, $K = 1, 2, \ldots \hat{K}$ and boundary surface grids $\mathbf{X}(1, J, K)$, $\mathbf{X}(\hat{I}, J, K)$, $\mathbf{X}(I, 1, K)$, $\mathbf{X}(I, \hat{J}, K)$, $\mathbf{X}(I, J, 1)$, and $\mathbf{X}(I, J, \hat{K})$, an adjusted grid $\mathbf{X}(I, J, K)$, can be produced by

$$\mathbf{X}_1(I,J,K) = \hat{\mathbf{X}}(I,J,K)$$
$$+\alpha_1^0(\xi)\left[\mathbf{X}(1,J,K) - \hat{\mathbf{X}}(1,J,K)\right] + \alpha_2^0(\xi)\left[\mathbf{X}(\hat{I},J,K) - \hat{\mathbf{X}}(\hat{I},J,K)\right]$$
$$\mathbf{X}_2(I,J,K) = \mathbf{X}_1(I,J,K)$$
$$+\beta_1^0(\eta)\left[\mathbf{X}(I,1,K) - \mathbf{X}_1(I,1,K)\right] + \beta_2^0(\eta)\left[\mathbf{X}(I,\hat{J},K) - \mathbf{X}_1(I,\hat{J},K)\right]$$
$$\mathbf{X}(I,J,K) = \mathbf{X}_2(I,J,K)$$
$$+\gamma_1^0(\zeta)\left[\mathbf{X}(I,J,1) - \mathbf{X}_2(I,J,1)\right] + \gamma_2^0(\xi)\left[\mathbf{X}(I,J,\hat{K}) - \mathbf{X}_2(I,J,\hat{K})\right]$$

$$\begin{aligned}
\alpha_1^0(\xi) &= 1 - u_1(\xi) \\
\alpha_2^0(\xi) &= u_2(\xi) \\
\beta_1^0(\eta) &= 1 - v_1(\eta) \\
\beta_2^0(\eta) &= v_2(\eta) \\
\gamma_1^0(\zeta) &= 1 - w_1(\zeta) \\
\gamma_2^0(\zeta) &= w_2(\zeta) \\
u_1(\xi) &= \frac{e^{C_1\xi} - 1}{e^{C_1} - 1} \\
u_2(\xi) &= \frac{e^{C_2\xi} - 1}{e^{C_2} - 1} \\
v_1(\eta) &= \frac{e^{C_3\eta} - 1}{e^{C_3} - 1} \\
v_2(\eta) &= \frac{e^{C_4\eta} - 1}{e^{C_4} - 1} \\
w_1(\zeta) &= \frac{e^{C_5\zeta} - 1}{e^{C_5} - 1} \\
w_2(\zeta) &= \frac{e^{C_6\zeta} - 1}{e^{C_6} - 1}
\end{aligned}$$

where the constants C_1, C_2, $\ldots C_6$ specify how far into the original grid the effect of the six boundary surfaces is carried.

3.8 Summary

TFI generates grids that conform to specified boundaries. The recipe is a Boolean sum of univariate interpolations, and it is also expressed as a recursion formula. Since any univariate interpolation subject to δ conditions can be applied in a coordinate direction, there are an infinite number of variations of TFI. However, low-order univariate interpolation functions are the most practical. Lagrangian and Hermite cubic formulae have been presented.

Grid spacing control can be best achieved by creating intermediate variables to be used in the interpolation functions. The intermediate variables are computed with control functions whose independent variables are computational coordinates and have adjustable parameters affecting spacing. Several examples of practical control functions have been presented.

A variation of TFI to conform an existing grid to new specified boundaries has also been represented. This minor variation is highly useful in a practical grid generation environment.

References

1. **Gordon, W.N. and Hall, C.A.**, Construction of curvilinear coordinate systems and application to mesh generation, *International J. Num. Methods in Eng.*, Vol. 7, pp. 461–477, 1973.
2. **Eriksson, L.-E.**, Three-dimensional spline-generated coordinate transformations for grids around wing-body configurations, *Numerical Grid Generation Techniques*, NASA CP 2166, 1980.
3. **Eriksson, L.-E.**, Generation of boundary conforming grids around wing-body configurations using transfinite interpolation, *AIAA J.*, Vol. 20, pp. 1313–1320, 1982.
4. **Eriksson, L.-E.**, *Transfinite Mesh Generation and Computer-Aided Analysis of Mesh Effects*, Ph.D. Dissertation, University of Uppsala, Sweden, 1984.
5. **Smith, R.E. and Wiese, M.R.**, *Interactive Algebraic Grid Generation*, NASA TP 2533, 1986.
6. **Eiseman, P.R. and Smith, R.E.**, Applications of algebraic grid generation, *AGARD Specialist Meeting Applications of Mesh Generation to Complex 3-D Configurations*, 1989.
7. **Samareh-Abolhassani, J., Sadrehaghighi, I., Smith, R.E., and Tiwari, S.N.**, Applications of Lagrangian blending functions for grid generation around airplane geometries, *J. Aircraft*, 27(10), pp. 873–877, 1990.
8. **Soni, B.K.**, Two- and three-dimensional grid generation for internal flow applications, AIAA Paper 85-1526, 1985.

4

Elliptic Generation Systems

Stefan P. Spekreijse

4.1 Introduction

Since the pioneering work of Thompson on elliptic grid generation, it is known that systems of elliptic second-order partial differential equations produce the best possible grids in the sense of smoothness and grid point distribution. The grid generation systems of elliptic quasi-linear second-order partial differential equations are so-called *Poisson systems* with control functions to be specified. The secret of each "good" elliptic grid is the method to compute the control functions [3].

Originally Thompson and Warsi introduced the Poisson systems by considering a curvilinear coordinate system that satisfies a system of Laplace equations and is transformed to another coordinate system [30,35]. Then this new coordinate system satisfies a system of Poisson equations with control functions completely specified by the transformation between the two coordinate systems. However, Thompson did not advocate to use this approach for grid generation. Instead he proposed to use the Poisson system with control functions specified directly rather than through a transformation [30]. Since then, the general approach is to compute the control functions at the boundary and to interpolate them from the boundaries into the field [5,29]. The standard approach used to achieve grid orthogonality and specified cell height on boundaries has been the *iterative* adjustment of the control functions in the Poisson systems (Chapter 6), first introduced by Sorenson of NASA Ames in the GRAPE code in the 1980s [24]. Various modifications of this basic concept have been introduced in several codes, and the general approach is now common [23,5,29]. Although successful, it appears that the method is not easy to apply in practice [14]. Even today, new modifications are proposed to improve the grid quality and to overcome numerical difficulties in solving the Poisson grid generation equations [23,16,12].

In this chapter we describe a useful alternative approach to specify the control functions. It is based on Thompson's and Warsi's original idea to define the control functions by a transformation. The transformation, which we call a *grid control map*, is a differentiable one-to-one mapping from computational space to parameter space. The independent variables of the parameter space are harmonic functions in physical space. The map from physical space to parameter space is called the *harmonic map*

0-8493-2687-7/99/$0.00+$.50

(Chapter 8). The composition of the grid control map and the inverse of the harmonic map obeys the familiar Poisson systems with control functions completely defined by the grid control map. The construction of appropriate grid control maps such that the corresponding grid in physical space has desired properties is the main issue of this chapter.

One of the main advantages of this approach is that the method is *noniterative.* If an appropriate grid control map has been constructed, then the corresponding grid control functions of the Poisson system are computed and their values remain unchanged during the solution of the Poisson system. Picard iteration appears to be a simple and robust method to solve the Poisson system with fixed control functions.

Another advantage is that the construction of an appropriate grid control map can be considered as a numerical implementation of the constructive proof for the existence of the desired grid in physical space. If the grid control map is one-to-one, then the composition of the grid control map and the inverse of the harmonic maps exist so that the solution of the Poisson system is well-defined.

This chapter is organized as follows. Section 4.2 concerns the two-dimensional case. Although published earlier [25], the 2D Poisson system together with the expressions to compute the control functions from the grid control map are given for completeness. The solution of the Poisson system by Picard iteration is shortly described. Section 4.2.3 describes methods to construct appropriate grid control maps. Boundary orthogonality is obtained by applying Dirichlet–Neumann boundary conditions for the harmonic map and by applying cubic Hermite interpolation in parameter space. In that case, the harmonic map is *quasi-conformal.* This observation leads to the construction of appropriate grid control maps such that the solution of the Poisson system generates an *orthogonal* grid in physical space with boundary grid points fixed on two adjacent edges but moved along the other two opposite edges (see Chapter 7). This result is similar to that reported by Kang and Leal [13], although they used the Ryskin–Leal grid generation equations [19] instead of the Poisson grid generation equations. Section 4.2.4 shows generated grids in physical space for well-defined geometries so that the reader is able to recompute the grids (by the methods presented in this chapter or by his/her own favorite methods for comparison). The corresponding constructed grid control maps are shown as grids in parameter space.

Section 4.3 briefly describes how the same methods to construct appropriate grid control maps for 2D grids can also be used for grid generation on surfaces in 3D physical space (see Chapter 9). It is shown that surface grid generation on *minimal surfaces* (soap films) is in fact the same as 2D grid generation. Conceptually, the same methods can also be used for parametrically defined surfaces, although the numerical implementation is completely different.

The extension to volume grid generation is described in Section 4.4. The construction of appropriate grid control maps for 3D domains is less well developed than for 2D domains. However, a method to construct a grid control map has been proposed which works surprisingly well for many applications.

The now-standard procedure in multi-block structured grid generation codes is to first generate surface grids on block faces, both boundary and interior block interfaces, from grid point distributions placed on the face edges by distribution functions. Then volume grids are generated within the blocks. For this reason, the elliptic grid generation methods described in this chapter assume fixed position of the prescribed boundary grid points.

4.2 Two-Dimensional Grid Generation

4.2.1 Harmonic Maps, Grid Control Maps and Poisson Systems

Consider a simply connected bounded domain D in two-dimensional space with Cartesian coordinates $\vec{x}(x, y)^T$. Suppose that D is bounded by four edges E_1, E_2, E_3, E_4. Let (E_1, E_2) and let (E_3, E_4) be the two pairs of opposite edges as shown in Figure 4.1.

A *harmonic map* is defined as a differentiable one-to-one map from D onto a unit square such that

1. The boundary of D is mapped onto the boundary of the unit square,
2. The vertices of D are mapped, in the proper sequence, onto the corners of the unit square,
3. The two components of the map are harmonic functions in the interior of D.

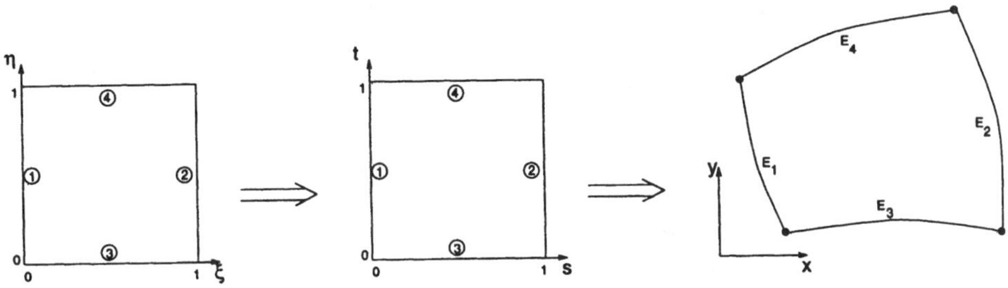

FIGURE 4.1 Composite map from computational (ξ,η) space to a domain D in Cartesian (x,y) space.

Let $\vec{s}: D \mapsto P$ be a harmonic map where the parameter space P is the unit square in a two-dimensional space with Cartesian coordinates $\vec{s} = (s, t)^T$. Assume that

- $s \equiv 0$ at edge E_1 and $s \equiv 1$ at edge E_2,
- $t \equiv 0$ at edge E_3 and $t \equiv 1$ at edge E_4.

The problem of generating an appropriate grid in the physical domain D can be effectively reduced to a simpler problem of generating an appropriate grid in the parameter space P, which can after that be mapped into D, by using the inverse of the harmonic map $\vec{x}: P \mapsto D$.

Define the computational space C as the unit square in a two-dimensional space with Cartesian coordinates $\vec{\xi} = (\xi, \eta)^T$. A *grid control map* $\vec{s}: C \mapsto P$ is defined as a differentiable one-to-one map from C onto P and maps a uniform grid in C to a nonuniform (in general) grid in P. Assume that

- $s(0, \eta) \equiv 0$ and $s(1, \eta) \equiv 1$,
- $t(\xi, 0) \equiv 0$ and $t(\xi, 1) \equiv 1$.

Then the computational coordinates also fulfill

- $\xi \equiv 0$ at edge E_1 and $\xi \equiv 1$ at edge E_2,
- $\eta \equiv 0$ at edge E_3 and $\eta \equiv 1$ at edge E_4.

The composition of a grid control map $\vec{s}: C \mapsto P$ and the inverse of the harmonic map $\vec{x}: P \mapsto D$ define a map $\vec{x}: C \mapsto D$ which transforms a uniform grid in C to a nonuniform (in general) grid in D. The composite map obeys a quasi-linear system of elliptic partial differential equations, known as the *Poisson grid generation equations*, with control functions completely defined by the grid control map. The secret of each "good" elliptic grid generation method is the method of computing appropriate control functions, which is thus equivalent to constructing appropriate grid control maps.

We will now derive the quasi-linear system of elliptic partial differential equations which the composite mapping $\vec{x} = \vec{x}(\vec{s}(\vec{\xi}))$ has to fulfill. Suppose that the harmonic map and the grid control map are defined so that the composite map exists. Introduce the two covariant base vectors (see Chapter 2)

$$\vec{a}_1 = \frac{\partial \vec{x}}{\partial \xi} = \vec{x}_\xi, \quad \vec{a}_2 = \frac{\partial \vec{x}}{\partial \eta} = \vec{x}_\eta \tag{4.1}$$

and define the covariant metric tensor components as the inner product of the covariant base vectors

$$\vec{a}_{i,j} = \left(\vec{a}_i, \vec{a}_j\right) \quad i = \{1,2\} \quad j = \{1,2\} \tag{4.2}$$

The two contravariant base vectors $\vec{a}^1 = \nabla\xi = (\xi_x, \xi_y)^T$ and $\vec{a}^2 = \nabla\eta = (\eta_x, \eta_y)^T$ obey

$$\left(\vec{a}^i, \vec{a}_j\right) = \delta^i_j \quad i = \{1,2\} \quad j = \{1,2\} \tag{4.3}$$

with δ_j^i the Kronecker symbol. Define the contravariant metric tensor components

$$a^{ij} = \left(\vec{a}^i, \vec{a}^j\right) \quad i = \{1,2\} \quad j = \{1,2\} \tag{4.4}$$

so that

$$\begin{pmatrix} a_{11} & a_{12} \\ a_{12} & a_{22} \end{pmatrix} \begin{pmatrix} a^{11} & a^{12} \\ a^{12} & a^{22} \end{pmatrix} = \begin{pmatrix} 1 & 0 \\ 0 & 1 \end{pmatrix} \tag{4.5}$$

and

$$\begin{aligned} \vec{a}^1 &= a^{11}\vec{a}_1 + a^{12}\vec{a}_2 & \vec{a}^2 &= a^{12}\vec{a}_1 + a^{22}\vec{a}_2 \\ \vec{a}_1 &= a_{11}\vec{a}^1 + a_{12}\vec{a}^2 & \vec{a}_2 &= a_{12}\vec{a}^1 + a_{22}\vec{a}^2 \end{aligned} \tag{4.6}$$

Introduce the determinant J^2 of the covariant metric tensor: $J^2 = a_{11}a_{22} - a^2_{12}$.

Now consider an arbitrary function $\phi = \phi(\xi, \eta)$. Then ϕ is also defined in domain D, and the Laplacian of ϕ is expressed as

$$\Delta\phi = \phi_{xx} + \phi_{yy} = \frac{1}{J}\left\{\left(Ja^{11}\phi_\xi + Ja^{12}\phi_\eta\right)_\xi + \left(Ja^{12}\phi_\xi + Ja^{22}\phi_\eta\right)_\eta\right\} \tag{4.7}$$

which may be found in Chapter 2 and in every textbook on tensor analysis and differential geometry (for example, see [15]). Take as special cases respectively $\phi \equiv \xi$ and $\phi \equiv \eta$. Then Eq. 4.7 yields

$$\Delta\xi = \frac{1}{J}\left\{\left(Ja^{11}\right)_\xi + \left(Ja^{12}\right)_\eta\right\} \quad \Delta\eta = \frac{1}{J}\left\{\left(Ja^{12}\right)_\xi + \left(Ja^{22}\right)_\eta\right\} \tag{4.8}$$

Thus the Laplacian of ϕ can also be expressed as

$$\Delta\phi = a^{11}\phi_{\xi\xi} + 2a^{12}\phi_{\xi\eta} + a^{22}\phi_{\eta\eta} + \Delta\xi\phi_\xi + \Delta\eta\phi_\eta \tag{4.9}$$

Substitution of respectively $\phi \equiv s$ and $\phi \equiv t$ in this equation yields

$$\Delta s = a^{11}s_{\xi\xi} + 2a^{12}s_{\xi\eta} + a^{22}s_{\eta\eta} + \Delta\xi s_\xi + \Delta\eta s_\eta \tag{4.10}$$

$$\Delta t = a^{11}t_{\xi\xi} + 2a^{12}t_{\xi\eta} + a^{22}t_{\eta\eta} + \Delta\xi t_\xi + \Delta\eta t_\eta \tag{4.11}$$

Using these equations and the property that s and t are harmonic in domain D, thus $\Delta s = 0$ and $\Delta t = 0$, we find the following expressions for the Laplacian of ξ and η:

$$\begin{pmatrix} \Delta\xi \\ \Delta\eta \end{pmatrix} = a^{11}\vec{P}_{11} + 2a^{12}\vec{P}_{12} + a^{22}\vec{P}_{22} \tag{4.12}$$

where

$$\vec{P}_{11} = -T^{-1}\begin{pmatrix} s_{\xi\xi} \\ t_{\xi\xi} \end{pmatrix} \quad \vec{P}_{12} = -T^{-1}\begin{pmatrix} s_{\xi\eta} \\ t_{\xi\eta} \end{pmatrix} \quad \vec{P}_{22} = -T^{-1}\begin{pmatrix} s_{\eta\eta} \\ t_{\eta\eta} \end{pmatrix} \tag{4.13}$$

and the matrix T is defined as

$$T = \begin{pmatrix} s_\xi & s_\eta \\ t_\xi & t_\eta \end{pmatrix} \tag{4.14}$$

The six coefficients of the vectors $\vec{P}_{11} = (P^1_{11}, P^2_{11})^T$, $\vec{P}_{12} = (P^1_{12}, P^2_{12})^T$ and $\vec{P}_{22} = (P^1_{22}, P^2_{22})^T$ are the so-called *control functions.* The six control functions are completely defined and easily computed for a given grid control map $\vec{s} = \vec{s}(\vec{\xi})$. Different and less useful expressions of these control functions can also be found in [30,35].

Finally, substitution of $\phi \equiv \vec{x}$ in Eq. 4.9 yields

$$\Delta\vec{x} = a^{11}\vec{x}_{\xi\xi} + 2a^{12}\vec{x}_{\xi\eta} + a^{22}\vec{x}_{\eta\eta} + \Delta\xi\vec{x}_\xi + \Delta\eta\vec{x}_\eta \tag{4.15}$$

Substituting Eq. 4.12 into this equation and using the fact that $\Delta\vec{x} \equiv 0$, we arrive at the familiar Poisson grid generation system:

$$\begin{aligned} a^{11}\vec{x}_{\xi\xi} + 2a^{12}\vec{x}_{\xi\eta} + a^{22}\vec{x}_{\eta\eta} &+ \left(a^{11}P^1_{11} + 2a^{12}P^1_{12} + a^{22}P^1_{22}\right)\vec{x}_\xi \\ &+ \left(a^{11}P^2_{11} + 2a^{12}P^2_{12} + a^{22}P^2_{22}\right)\vec{x}_\eta = 0 \end{aligned} \tag{4.16}$$

Using Eqs. 4.2, and 4.5 we find the following well-known expressions for the contravariant metric tensor components:

$$J^2a^{11} = a_{22} = \left(\vec{x}_\eta, \vec{x}_\eta\right) \quad J^2a^{12} = -a_{12} = -\left(\vec{x}_\xi, \vec{x}_\eta\right) \quad J^2a^{22} = a_{11} = \left(\vec{x}_\xi, \vec{x}_\xi\right) \tag{4.17}$$

Thus the Poisson grid generation system defined by Eq. 4.16 can be simplified by multiplication with J^2. Then we obtain:

$$\begin{aligned} a_{22}\vec{x}_{\xi\xi} - 2a_{12}\vec{x}_{\xi\eta} + a_{11}\vec{x}_{\eta\eta} &+ \left(a_{22}P^1_{11} - 2a_{12}P^1_{12} + a_{11}P^1_{22}\right)\vec{x}_\xi \\ &+ \left(a_{22}P^2_{11} - 2a_{12}P^2_{12} + a_{11}P^2_{22}\right)\vec{x}_\eta = 0 \end{aligned} \tag{4.18}$$

This equation, together with the expressions for the control functions P^k_{ij} given by Eq. 4.13, is the two-dimensional grid generation system. For a given grid control map, so that the six control functions in Eq. 4.18 are given functions of ξ and η, boundary conforming grids in the interior of domain D are computed by solving this quasi-linear system of elliptic partial differential equations with prescribed boundary grid points as Dirichlet boundary conditions. The discretization and solution method of this Poisson system is discussed in the next section. The construction of appropriate grid control maps such that the corresponding grid in physical space has desired properties is discussed in the remaining sections.

4.2.2 Discretization and Solution Method

Consider a uniform rectangular grid of $(N + 1) \times (M + 1)$ points in computational space C defined as

$$\xi_{ij} = \xi_i = i/N \quad \eta_{i,j} = \eta_j = j/M \quad i = 0\ldots N \quad j = 0\ldots M \tag{4.19}$$

Assume that $\vec{x}_{i,j}$ is prescribed on the boundary of this grid and consider the computation of $\vec{x}_{i,j}$ in the interior of the computational grid based on the solution of the Poisson system defined by Eq. 4.18.

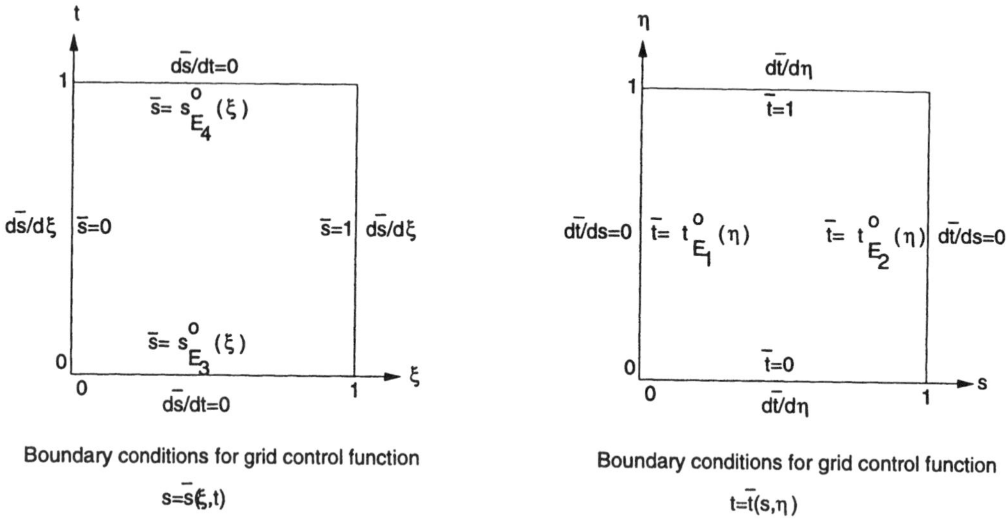

FIGURE 4.2 Boundary conditions for both control of orthogonality and first grid cell height.

Assume that a grid control map $\vec{s}$: $C \mapsto P$ has been constructed. Thus the values s_{ij} and t_{ij} are known at each grid point. At each interior grid point $(i, j) \in (1 \cdots N-1, 1 \cdots M-1)$, the six control functions P^1_{11}, P^2_{11}, P^1_{12}, P^2_{12}, P^1_{22}, P^2_{22} defined by Eq. 4.13 are now easily computed using central differences for the discretization of $s_{\xi\xi}$, $s_{\xi\eta}$, $s_{\eta\eta}$, s_ξ, s_η and $t_{\xi\xi}$, $t_{\xi\eta}$, $t_{\eta\eta}$, t_ξ, t_η.

The iterative solution process of the nonlinear elliptic Poisson grid generation system defined by Eq. 4.18 can be simply obtained by Picard iteration. Rewrite the Poisson system as

$$P\vec{x}_{\xi\xi} - 2Q\vec{x}_{\xi\eta} + R\vec{x}_{\eta\eta} + S\vec{x}_\xi + T\vec{x}_\eta = 0 \tag{4.20}$$

with

$$\begin{aligned} &P = (\vec{x}_\eta, \vec{x}_\eta) \quad Q = (\vec{x}_\xi, \vec{x}_\eta) \quad R = (\vec{x}_\xi, \vec{x}_\xi) \\ &S = PP^1_{11} - 2QP^1_{12} + RP^1_{22} \\ &T = PP^2_{11} - 2QP^2_{12} + RP^2_{22} \end{aligned} \tag{4.21}$$

The iterative solution by Picard iteration can be written as

$$P^{k-1}\vec{x}^k_{\xi\xi} - 2Q^{k-1}\vec{x}^k_{\xi\eta} + R^{k-1}\vec{x}^k_{\eta\eta} + S^{k-1}\vec{x}^k_\xi + T^{k-1}\vec{x}^k_\eta = 0 \tag{4.22}$$

where k is the Picard index and

$$\begin{aligned} &P^{k-1} = (\vec{x}^{k-1}_\eta, \vec{x}^{k-1}_\eta) \quad Q^{k-1} = (\vec{x}^{k-1}_\xi, \vec{x}^{k-1}_\eta) \quad R^{k-1} \approx (\vec{x}^{k-1}_\xi, \vec{x}^{k-1}_\xi) \\ &S^{k-1} = P^{k-1}P^1_{11} - 2Q^{k-1}P^1_{12} + R^{k-1}P^1_{22} \\ &T^{k-1} = P^{k-1}P^2_{11} - 2Q^{k-1}P^2_{12} + R^{k-1}P^2_{22} \end{aligned} \tag{4.23}$$

Thus, a current approximate solution

$$\vec{x}^{k-1} = \left\{\vec{x}^{k-1}_{ij}, i = 0 \ldots N, j = 0 \ldots M\right\} \tag{4.24}$$

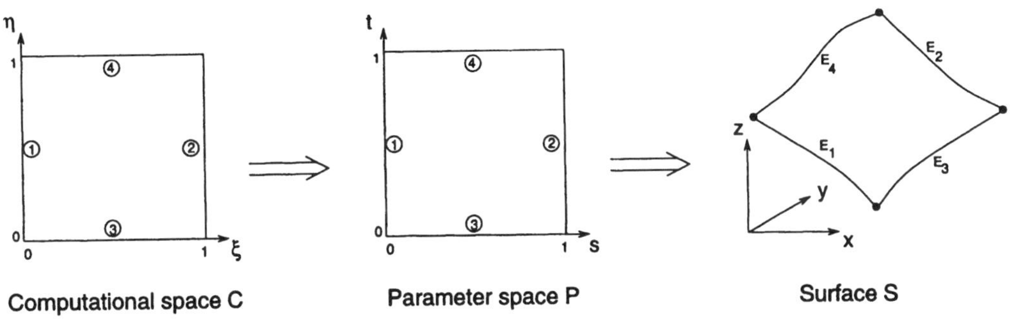

FIGURE 4.3 Composite map from computational (ξ, η) space to a surface S in Cartesian (x, y, z) space.

is improved by the following steps:

- Compute at interior grid points the coefficients $P^{k-1}, Q^{k-1}, R^{k-1}, S^{k-1}, T^{k-1}$ by applying central differences for the discretization of $\vec{x}_{\xi}^{k-1}$ and $\vec{x}_{\eta}^{k-1}$. Note that the six control functions remain unchanged during the iterative procedure.
- Discretize at interior grid points $\vec{x}_{\xi\xi}^k, \vec{x}_{\xi\eta}^k, \vec{x}_{\eta\eta}^k, \vec{x}_{\xi}^k, \vec{x}_{\eta}^k$ using central differences.
- After the discretization of $\vec{x}_{\xi\xi}^k, \vec{x}_{\xi\eta}^k, \vec{x}_{\eta\eta}^k, \vec{x}_{\xi}^k, \vec{x}_{\eta}^k$ we arrive at a linear system of equations for the unknowns $\vec{x}_{ij}^k\; i = 1 \cdots N-1, j = \cdots M-1$. At each interior grid point we have a nine-point stencil. Boundary grid points are prescribed and remain unchanged. This linear system can be solved by a black-box multigrid solver. Such a multigrid solver is called twice to compute the two components x_{ij}^k and y_{ij}^k of $\vec{x}_{ij}^k$. The solution of the linear system provides a better approximate solution $\vec{x}^k$.

The following algorithm describes the computation of an interior grid in domain D with prescribed boundary grid points and a given grid control map.

Algorithm 1. Grid Generation.

1. Compute the six control functions from the grid control map.
2. Compute an initial grid in the interior of domain D by a simple algebraic grid generation method (see Chapter 3). The quality of the initial grid is unimportant, and severe grid folding is allowed. The initial grid is used as starting solution for the Picard iteration process. The final grid will be independent of the initial grid.
3. Solve the quasi-linear Poisson grid generation equations iteratively by Picard iteration. The fixed position of the boundary grid points define Dirichlet boundary conditions. In general, a sufficiently converged grid is obtained in about 10 Picard iterations. The residual is then typically decreased by a factor 1000.

4.2.3 Construction of Grid Control Maps

4.2.3.1 Laplace Grids

The simplest grid control map is the identity map $\vec{s} = \vec{\xi}$. The six control functions are identical zero and the Poisson grid generation system defined by Eq. 4.18 simplifies to $a_{22}\vec{x}_{\xi\xi} - 2a_{12}\vec{x}_{\xi\eta} + a_{11}\vec{x}_{\eta\eta} = 0$, which is equivalent with $\Delta\xi = 0$ and $\Delta\eta = 0$, according to Eq. 4.12.

Grids based on this equation are the so-called Laplace (or Harmonic) grids, which were first introduced by Winslow [34]. The inherent smoothness of the Laplace operator makes the grid evenly spaced in the interior. Therefore, the quality of a Laplace grid will be acceptable only as long as the boundary grid points are evenly spaced along the edges.

This is illustrated in Figure 4.5 and Figure 4.6 where a region about a NACA0012 airfoil is subdivided into four domains. The domains have common edges, and more or less evenly spaced boundary grid

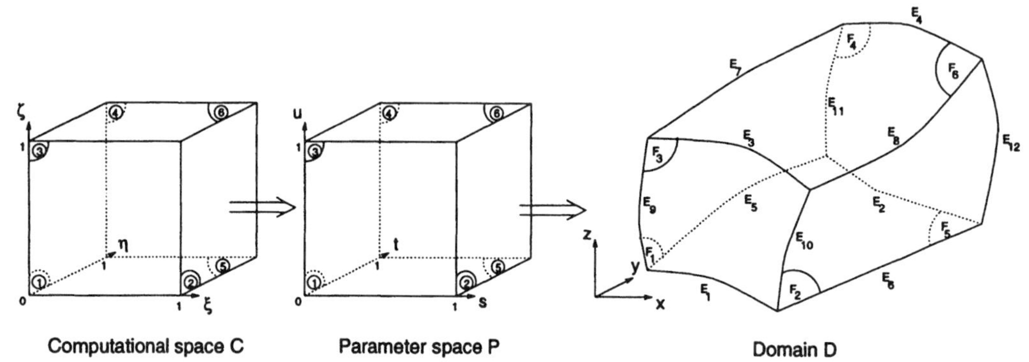

FIGURE 4.4 Composite mapping from computational (ξ, η, ζ) space to a domain D in Cartesian (x, y, z) space.

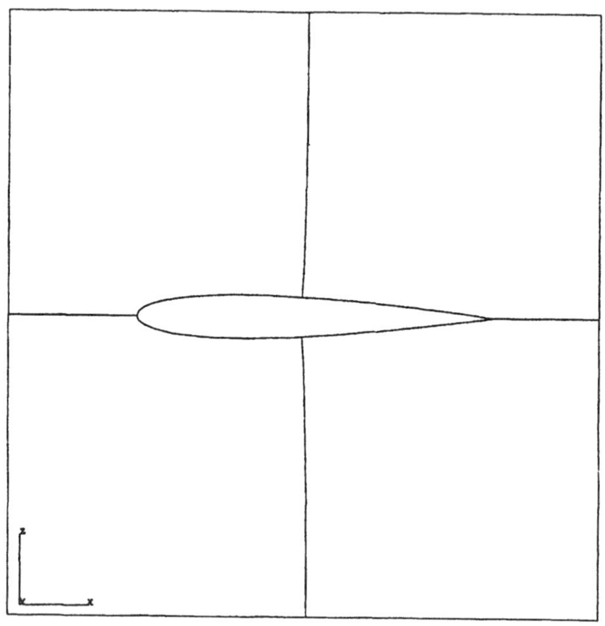

FIGURE 4.5 Domain boundaries near NACA0012 airfoil. The location of grid points on the domain boundaries is prescribed and fixed.

points are prescribed. Figure 4.6 shows Laplace grids in each domain. The result is not bad for this O-type Euler mesh. (Only smooth grids are required for the solution of the Euler equations for nonviscous flow, where strong gradients near boundaries do not occur.) Laplace grids provide no control about the angle distribution between internal grid lines and the boundary. This causes slope discontinuity of the grid lines across internal domain boundaries, as shown in Figure 4.6.

The situation is completely different for Navier–Stokes type of meshes where the grid must contain a boundary layer grid. Highly stretched grids are required for solutions of the Navier–Stokes equations for viscous flow, where large gradients occur near boundaries. Figure 4.9 shows a region about a RAE2822 airfoil also subdivided into four domains. The boundary grid point distribution is highly dense near the leading and trailing edge of the airfoil. Figure 4.10 shows the Laplace grids in the four domains. These grids are unacceptable because the inherent smoothness of the Laplace operator causes evenly spaced grids so that the interior grid contains no boundary layer at all. Therefore, Laplace grids are in general unusable in most practice.

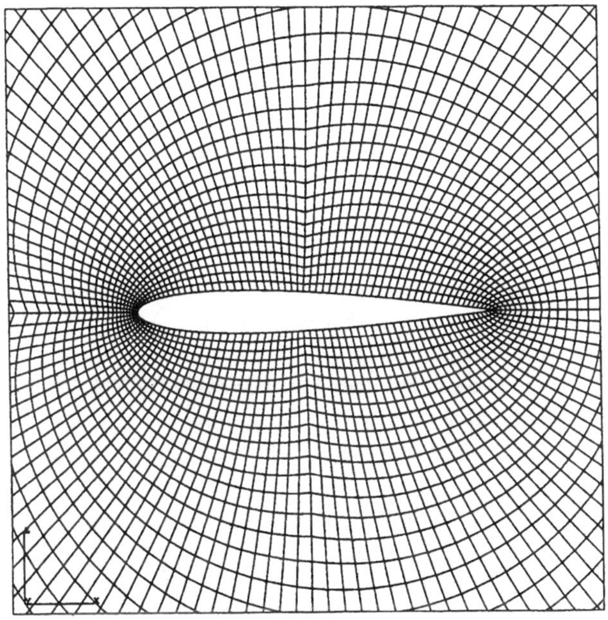

FIGURE 4.6 Laplace grid. Grid control map is the identity map.

4.2.3.2 Arc Length Based Grids

Consider domain D as shown in Figure 4.1. Assume that the boundary grid points are prescribed at the four edges of D. A boundary-conforming grid in the interior of domain D with an interior grid point distribution which is a good reflection of the prescribed boundary grid point distribution can be obtained by constructing a grid control map based on normalized arc length. In order to construct such a grid control, we define

- $s \equiv 0$ at edge E_1 and $s \equiv 1$ at edge E_2,
- s is the normalized arc length along edges E_3 and E_4,
- $t \equiv 0$ at edge E_3 and $t \equiv 1$ at edge E_4,
- t is the normalized arc length along edges E_1 and E_2.

For example, this means that along edge E_3 we define $s(u) = \int_o^u \|\vec{x}_u\| du \Big/ \int_0^1 \|\vec{x}_u\| du$ where $\vec{x}: u \in [0, 1] \mapsto (x, y) \in R^2$ is a parametrization of edge E_3 in the right direction. Thus $\vec{s} : \partial D \mapsto \partial P$ is defined by these requirements. The two Laplace equations $\Delta s = 0$ and $\Delta t = 0$, together with the above-specified Dirichlet boundary conditions, define the harmonic map $\vec{s}: D \mapsto P$. Note that this map depends only on the shape of domain D and is independent of the prescribed boundary grid point distribution.

The boundary grid points are prescribed at the four edges of D. Thus $\vec{x}: \partial C \mapsto \partial D$ is prescribed. Because $\vec{x}: \partial C \mapsto \partial D$ is prescribed and $\vec{s}: \partial D \mapsto \partial P$ is defined as described above, it follows that $\vec{s} : \partial C \mapsto \partial P$ is also defined.

From the preceding requirements it follows that

$$s(0,\eta) = 0 \quad s(1,\eta) = 1 \quad s(\xi,0) = s^a_{E_3}(\xi) \quad s(\xi,1) = s^a_{E_4}(\xi) \tag{4.25}$$

where the functions $s^a_{E_3}$, $s^a_{E_4}$ are monotonically increasing, and

$$t(\xi,0) = 0 \quad t(\xi,1) = 1 \quad t(0,\eta) = t^a_{E_1}(\eta) \quad t(1,\eta) = t^a_{E_2}(\eta) \tag{4.26}$$

where the functions $t^a_{E_1}$, $t^a_{E_2}$ are also monotonically increasing. The superscript a is used to indicate that these functions measure the normalized arc length at the boundary grid points.

The grid control map $\vec{s}$: $C \mapsto P$ is now defined by the following two algebraic equations:

$$s = s^a_{E_3}(\xi)(1-t) + s^a_{E_4}(\xi)t \tag{4.27}$$

$$t = t^a_{E_1}(\eta)(1-s) + t^a_{E_2}(\eta)s \tag{4.28}$$

Eq. 4.27 implies that a coordinate line ξ = const. is mapped to the parameter space P as a straight line: s is a linear function of t, and Eq. 4.28 implies that a grid line η = const. is also mapped to P as a straight line: t is a linear function of s. For given values of ξ and η, the corresponding s and t values are found as the intersection point of the two straight lines. It can be easily verified that the grid control map is a differentiable and one-to-one because of the positiveness of the Jacobian: $s_\xi t_\eta - s_\eta t_\xi > 0$.

The discrete computation of the grid control map is straightforward. For a grid of $(N+1) \times (M+1)$ points, the distance between succeeding grid points at the boundary are computed as

$$\bar{d}_{0,j} = \left\| \vec{x}_{0,j} - \vec{x}_{0,j-1} \right\| \quad \bar{d}_{N,j} = \left\| \vec{x}_{N,j} - \vec{x}_{N,j-1} \right\| \quad j = 1 \ldots M \tag{4.29}$$

$$\bar{d}_{i,0} = \left\| \vec{x}_{i,0} - \vec{x}_{i-1,0} \right\| \quad \bar{d}_{i,M} = \left\| \vec{x}_{i,M} - \vec{x}_{i-1,M} \right\| \quad i = 1 \ldots N \tag{4.30}$$

Define the length of edges E_1, E_2 E_3, E_4 by

$$L_{E_1} = \sum_{j=1}^{M} \bar{d}_{0,j} \quad L_{E_2} = \sum_{j=1}^{M} \bar{d}_{N,j} \quad L_{E_3} = \sum_{i=1}^{N} \bar{d}_{i,0} \quad L_{E_4} = \sum_{i=1}^{N} \bar{d}_{i,M} \tag{4.31}$$

and the normalized distances as

$$d_{o,j} = \bar{d}_{o,j} / L_{E_1} \quad d_{N,j} = \bar{d}_{N,j} / L_{E_2} \quad j = 1 \ldots M \tag{4.32}$$

$$d_{i,0} = \bar{d}_{i,0} / L_{E_3} \quad d_{i,M} = \bar{d}_{i,M} / L_{E_4} \quad i = 1 \ldots N. \tag{4.33}$$

The discrete components $s_{i,j}$ and $t_{i,j}$ of the grid control map are computed at the boundary by

$$s_{o,j} = 0 \quad s_{N,j} = 1 \quad j = 0 \ldots M \tag{4.34}$$

$$t_{i,0} = 0 \quad t_{i,M} = 1 \quad i = 0 \ldots N \tag{4.35}$$

and

$$s_{i,0} = s_{i-1,0} + d_{i,0} \quad s_{i,M} = s_{i-1,M} + d_{i,M} \quad i = 1 \ldots N \tag{4.36}$$

$$t_{o,j} = t_{0,j-1} + d_{0,j} \quad t_{N,j} = t_{N,j-1} + d_{N,j} \quad j = 1 \ldots M \tag{4.37}$$

The interior values are defined according to Eq. 4.27 and Eq. 4.28 and are thus found by solving simultaneously the two linear algebraic equations,

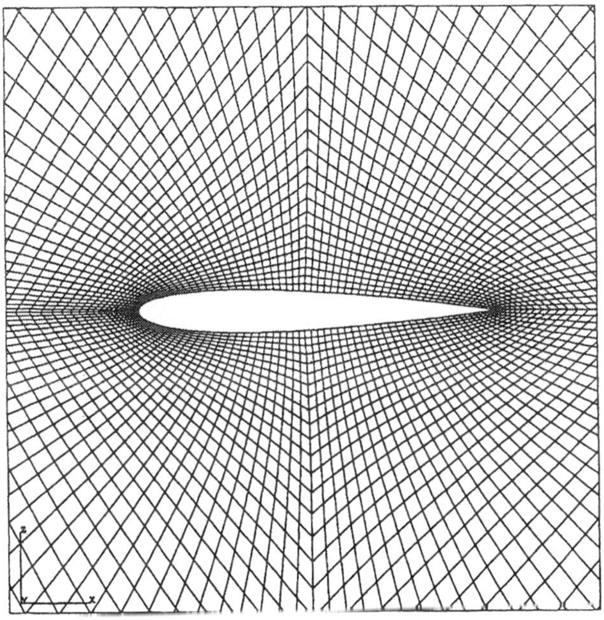

FIGURE 4.7 Arc length-based grid.

$$s_{i,j} = s_{i,0}\left(1 - t_{i,j}\right) + s_{i,M}\, t_{i,j} \tag{4.38}$$

$$t_{i,j} = t_{0,j}\left(1 - s_{i,j}\right) + t_{N,j}\, s_{i,j} \tag{4.39}$$

for each pair $(i, j) \in (1 \cdots N - 1, 1 \cdots M - 1)$.

The next algorithm summarizes the computation of arc length-based grid in the interior of D.

Algorithm 2. Arc length-based grids

1. Compute the four edge functions $t^a_{E_1}$, $t^a_{E_2}$, $s^a_{E_3}$ and $s^a_{E_4}$ from the boundary grid point distribution.
2. Compute the grid control map according to Eq. 4.27 and Eq. 4.28.
3. Compute the corresponding interior grid in D as described in Algorithm 1.

Illustrations of boundary conforming grids obtained with this grid control map are shown in Figure 4.7 and Figure 4.11. As opposed to Laplace grids, the interior grid point distribution is always a good reflection of the prescribed boundary grid point distribution. Grid folding hardly ever occurs, because both the grid control map and the harmonic map are one-to-one. When grid folding occurs, then it must be caused by discretization errors [18]. Hence, grid folding will always disappear when the grid is sufficiently refined.

A shortcoming of this grid control map is that there is no control about the angle distribution between interior grid lines and the boundary edges of the domain. It is often desired that the interior grid lines are orthogonal at the boundary edges. For example, viscous flow simulations often require orthogonality of the grid in a boundary layer. This can be achieved with a grid control map as constructed below.

4.2.3.3 Grid Orthogonality at the Boundary

Consider domain D with prescribed boundary grid points. Suppose that it is desired to generate a boundary-conforming grid in the interior of D which is orthogonal at all four edges of domain D. This can be achieved by imposing Dirichlet–Neumann boundary conditions for the harmonic map:

- $s \equiv 0$ at edge E_1 and $s \equiv 1$ at edge E_2,
- $\partial s/\partial n$ along edges E_3 and E_4, where n is the outward normal direction,
- $t \equiv 0$ at edge E_3 and $t \equiv 1$ at edge E_4,
- $\partial t/\partial n$ along edges E_1 and E_2, where n is the outward normal direction.

The two Laplace equations $\Delta s = 0$ and $\Delta t = 0$, together with the above specified boundary conditions, define the harmonic map $\vec{s}$: $D \mapsto P$. Again this map depends only on the shape of domain D and is independent of the prescribed boundary grid point distribution.

The Neumann boundary conditions $\partial s/\partial n = 0$ along edges E_3 and E_4 imply that a parameter line $s =$ const. in P will be mapped into domain D by the inverse of the harmonic map as a curve which is orthogonal at those edges. Similarly, a parameter line $t =$ const. in P will be mapped as a curve in D which is orthogonal at edge E_1 and edge E_2. These properties can be used to construct a grid control map such that the interior grid in D will be orthogonal at the boundary.

The boundary grid points are prescribed at the four edges of D. Thus $\vec{x}$: $\partial C \mapsto \partial D$ is prescribed. Because $\vec{x}: \partial C \mapsto \partial D$ is prescribed and $\vec{s}: \partial D \mapsto \partial P$ is also defined, it follows that $\vec{s}: \partial C \mapsto \partial P$ is also defined.

From the preceding requirements it follows that

$$s(0,\eta)=0 \quad s(1,\eta)=1 \quad s(\xi,0)=s^0_{E_3}(\xi) \quad s(\xi,1)=s^0_{E_4}(\xi) \tag{4.40}$$

where the functions $s^0_{E_3}$, $s^0_{E_4}$ are monotonically increasing, and

$$t(\xi,0)=0 \quad t(\xi,1)=1 \quad t(0,\eta)=t^0_{E_1}(\eta) \quad t(1,\eta)=t^0_{E_2}(\eta) \tag{4.41}$$

where the functions $t^0_{E_1}$, $t^0_{E_2}$ are also monotonically increasing. The superscript 0 is used to indicate that these functions are constructed in a way to obtain grid orthogonality at the boundary.

The grid control map $\vec{s}$: $C \mapsto P$ is now defined by

$$s = s^0_{E_3}(\xi)\,H_0(t) + s^0_{E_4}(\xi)\,H_1(t) \tag{4.42}$$

$$t = t^0_{E_1}(\eta)\,H_0(s) + t^0_{E_2}(\eta)\,H_1(s) \tag{4.43}$$

where H_0 and H_1 are cubic Hermite interpolation functions defined as

$$H_0(s)=(1+2s)(1-s)^2 \quad H_1(s)=(3-2s)s^2 \quad 0 \le s \le 1 \tag{4.44}$$

Note that $H_0(0) = 1$, $H_0'(0) = 0$, $H_0(1) = 0$, $H_0'(1) = 0$ and $H_1(0) = 0$, $H_1'(0) = 0$, $H_1(1) = 1$, $H_1'(1) = 0$. It follows from Eq. 4.42 that a coordinate line $\xi =$ const. in C is mapped to parameter space P as a cubic curve (with t as dependent variable) which is orthogonal at both edge E_3 and edge E_4 in P. Such a curve in parameter space P will thus be mapped by the inverse of the harmonic map $\vec{x}$: $P \mapsto D$ as a curve which is orthogonal at both edge E_3 and edge E_4 in D. Similar observations can be made for coordinate lines $\eta =$ const. Thus the grid will be orthogonal at all four edges in domain D.

Grid orthogonality at boundaries may introduce grid folding. Fortunately, grid folding will not easily arise. From Eq. 4.42 it follows that two different coordinate lines $\xi = \xi_1$, $\xi = \xi_2$, $\xi_1 \neq \xi_2$ are mapped to parameter space P as two disjunct cubic curves which are orthogonal at both edge E_3 and edge E_4 in P. This is due to the fact that $s^0_{E_3}(\xi)$ and $s^0_{E_4}(\xi)$ are monotonically increasing functions. The same holds for different coordinate lines $\eta = \eta_1$, $\eta = \eta_2$, $\eta_1 \neq \eta_2$. For given values of ξ and η, the corresponding s and t values are found as intersection point of two cubic curves. However, such two cubic curves may have more than one intersection point. In that case, grid folding will occur. However, in practice we hardly ever encounter grid folding due to orthogonalization of the grid at the boundary.

We have described a method to obtain an orthogonal grid at all four edges of domain D. In practice, orthogonality of the grid is often only desired at less than four edges. Suppose for example that it is only desired to have an orthogonal grid at edge E_3. Then take $t_{E_1}(\eta) = t^0_{E_1}(\eta)$, $t_{E_2}(\eta) = t^0_{E_2}(\eta)$, $s_{E_4}(\xi) = s^0_{E_4}(\xi)$ and $s_{E_3}(\xi) = s^0_{E_3}(\xi)$. Furthermore, the grid control map $\vec{s}: C \mapsto P$ is such that a coordinate line η = const. is mapped to P as a straight line and a coordinate line ξ = const. is mapped to P as a parabolic curve (with t as dependent variable) which is only orthogonal at edge E_3 in P. For given values of ξ and η, the corresponding s and t values are then found as intersection point of a straight line and a parabolic curve.

The discrete computation of the grid control map is more complicated when grid orthogonality is required. We have seen that for a grid control map based on normalized arc length, the functions $t^0_{E_1}$, $t^0_{E_2}$, $s^0_{E_3}$ and $s^0_{E_4}$ can be directly computed from the prescribed boundary grid points only. However, when grid orthogonality is required, the functions $t^0_{E_1}$, $t^0_{E_2}$, $s^0_{E_3}$ and $s^0_{E_4}$ can only be found by solving the Laplace equations $\Delta s = 0$ and $\Delta t = 0$ supplied with the above mentioned Dirichlet–Neumann boundary conditions. The solution of the Laplace equations $\Delta s = 0$ and $\Delta t = 0$ supplied with the boundary conditions requires an initial folding-free grid in the interior of domain D. Therefore, an orthogonal grid at the boundary is in general obtained in three steps:

Algorithm 3. Grid orthogonality at boundary

1. Compute an initial boundary conforming grid in the interior of D without grid folding. Such a grid can be computed using the grid control map based on normalized arc length as described in Algorithm 2.
2. Solve on this mesh $\Delta s = 0$ and $\Delta t = 0$ supplied with the above specified Dirichlet–Neumann boundary conditions. A solution method is described in [19]. The solution at the boundary defines the edge functions $t^0_{E_1}$, $t^0_{E_2}$, $s^0_{E_3}$ and $s^0_{E_4}$.
3. Compute the grid control map according to Eq. 4.42 and Eq. 4.43.
4. Compute the corresponding interior grid in D as described in Algorithm 1.

Illustrations of boundary conforming grids obtained with this grid control map are shown in Figure 4.8 and Figure 4.19. The common interior boundary edges of the four domains can hardly be recognized any more because of the excellent grid orthogonality at these edges. The grid spacing of the interior grid is also good in both cases. For more information on grid orthogonality at the boundary, see Chapter 6.

In the next section we will prove that the harmonic map $\vec{s}: D \mapsto P$ supplied with Dirichlet–Neumann boundary conditions is *quasi-conformal.* This observation leads to the construction of appropriate grid control maps such that the corresponding grid is orthogonal, not only at the boundary but also in the interior of D.

4.2.3.4 Orthogonal Grids

There is a famous theorem in conformal mapping theory which states that each simply connected domain D can be mapped conformally to a rectangle R in such a way that the vertices of domain D are mapped, in the proper sequence, onto the corners of the rectangle [8,11]. The ratio of the length of two adjacent sides of the rectangle is called the conformal module M, which is a characteristic and fundamental property of each domain.

Let $\vec{u}: D \mapsto R$ be the conformal map where R is the rectangle $[0, 1] \times [0, M]$ in a two-dimensional space with Cartesian coordinates $\vec{u} = (u, v)^T$. The components of the conformal map obey the Cauchy–Riemann relations:

$$\begin{pmatrix} u_x \\ u_y \end{pmatrix} = \begin{pmatrix} v_y \\ -v_x \end{pmatrix} \tag{4.45}$$

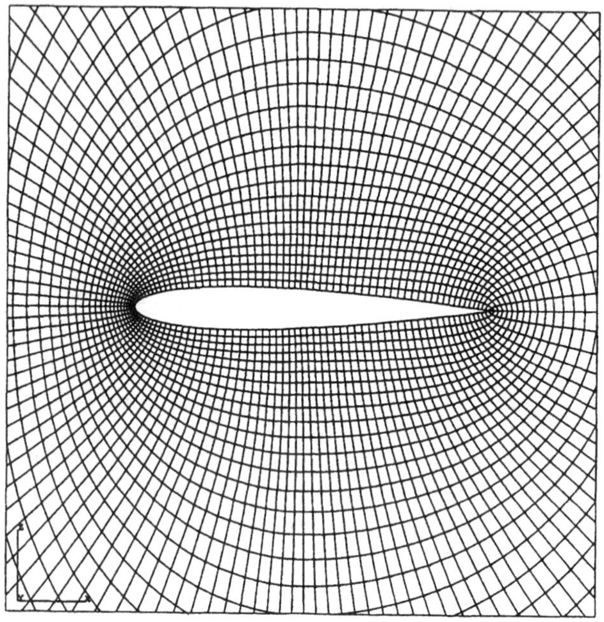

FIGURE 4.8 Grid with boundary orthogonality. Boundary orthogonality makes the grid smooth across internal domain boundaries.

Hence $\Delta u = 0$ and $\Delta v = 0$ in the interior of domain D. Furthermore, we may assume that the map $\vec{u}$: $D \mapsto R$ obeys

- $u \equiv 0$ at edge E_1 and $u \equiv 1$ at edge E_2,
- $v \equiv 0$ at edge E_3 and $v \equiv M$ at edge E_4.

From these boundary conditions and using the Cauchy–Riemann relations we can also conclude that

- $\partial u/\partial n = 0$ along edges E_3 and E_4, where n is the outward normal direction,
- $\partial v/\partial n = 0$ along edges E_1 and E_2, where n is the outward normal direction.

Thus the conformal map $\vec{u}$: $D \mapsto R$ is harmonic and obeys the same set of Dirichlet–Neumann boundary conditions as the harmonic map $\vec{s}$: $D \mapsto P$. Therefore the two maps are related to each other according to

$$s = u \qquad t = \frac{v}{M} \tag{4.46}$$

This means that the harmonic map is quasi-conformal and obeys

$$\begin{pmatrix} s_x \\ s_y \end{pmatrix} = M \begin{pmatrix} t_y \\ -t_x \end{pmatrix} \tag{4.47}$$

Thus the two contravariant vectors are orthogonal but have different lengths. It is not difficult to show, using the relations between covariant and contravariant vectors given by Eq. 4.6, that the covariant vectors fulfill

$$\begin{pmatrix} x_s \\ y_s \end{pmatrix} = \frac{1}{M} \begin{pmatrix} y_t \\ -x_t \end{pmatrix} \tag{4.48}$$

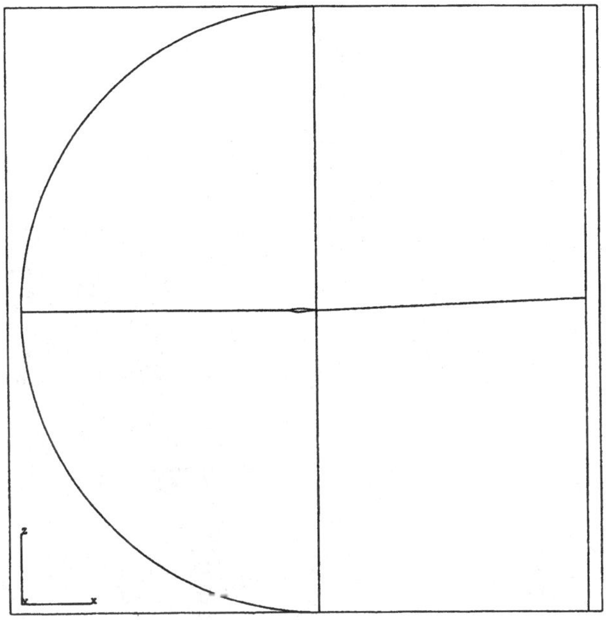

FIGURE 4.9 Region about RAE2822 airfoil subdivided into four domains.

so that the inverse mapping obeys

$$M^2 \vec{x}_{ss} + \vec{x}_{tt} = 0 \tag{4.49}$$

which is the well-known partial differential equation for quasi-conformal maps [14, page 96]. It can also be easily verified that the conformal module can be computed from

$$M = \int_{E_2} \frac{\partial s}{\partial n} d\sigma \tag{4.50}$$

where n is the outward normal direction and σ a line element along edge E_2 in D [11].

Conformal maps are angle preserving. The inverse of the conformal map $\vec{u}$: $D \mapsto R$ is also conformal and maps an orthogonal grid in the rectangle R to an orthogonal grid in D. Therefore, an algorithm to compute an orthogonal grid in the interior of D with a prescribed boundary grid point distribution at all four edges may consist of the following steps:

1. Compute an initial boundary conforming grid in the interior of D without grid folding. This can be achieved using the grid control map based on normalized arc length.
2. Solve on this mesh $\Delta s = 0$ and $\Delta t = 0$ supplied with Dirichlet–Neumann boundary conditions. Compute the edge functions $t^0_{E_1}$, $t^0_{E_2}$, $s^0_{E_3}$, and $s^0_{E_4}$ and the conformal module M according to Eq. 4.50.
3. Map the edge functions in P to the rectangle R, using Eq. 4.46, and compute an orthogonal boundary conforming grid in R.
4. Map the orthogonal grid in R to P, again using Eq. 4.46. This grid in P defines a grid control map that will create an orthogonal grid in the interior of D.

Thus, a difficult problem of generating an orthogonal grid in a domain D can be effectively reduced to a simpler problem of generating an orthogonal grid in the rectangle R. Unfortunately, there is no simple algorithm available to generate an orthogonal grid in the interior of a rectangle

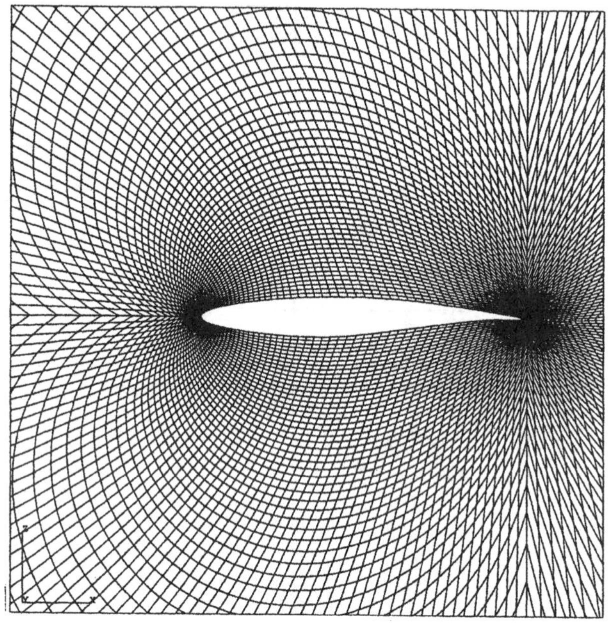

FIGURE 4.10 Laplace grid near airfoil. Grid control map is the identity map.

with prescribed boundary grid points at all four sides. The question of an existence proof for this problem still remains unanswered [17]. Numerical experiments indicate that even for a rectangle it is probably not possible to generate an orthogonal grid for all kinds of boundary grid point distributions [9].

However, if the boundary grid points have fixed positions on two adjacent edges of domain D but are allowed to move along the boundary of the other two edges, then a simple algorithm does exist to generate an orthogonal grid in D. This result is similar to that reported by Kang and Leal [13], although they used the Ryskin–Leal grid generation equations [19] instead of the Poisson grid generation equations. For example, suppose that the boundary grid points are fixed at edges E_1 and E_3 and are allowed to move along edges E_2 and E_4. Then the algorithm becomes the following.

Algorithm 4. Grid orthogonality

1. Compute an initial boundary conforming grid in the interior of D without grid folding. Such a grid can be computed using the grid control map based on normalized arc length as described in Algorithm 2.
2. Solve on this mesh $\Delta s = 0$ and $\Delta t = 0$ supplied with Dirichlet–Neumann boundary conditions and compute the edge functions $t^0_{E_1}$, $t^0_{E_2}$, $s^0_{E_3}$, and $s^0_{E_4}$.
3. The initial position of the boundary grid points at edge E_2 corresponds with the edge function $t^0_{E_2}$. Move the boundary grid points along edge E_2 in such a way that the new position corresponds with $t^0_{E_1}$. This is simply a matter of interpolation. The points along edge E_4 should be moved such that their new position corresponds with $s^0_{E_3}$.
4. Define the grid control map as $s(\xi,\eta) = s^0_{E_3}(\xi)$ and $t(\xi,\eta) = t^0_{E_1}(\eta)$.
5. Compute the corresponding orthogonal grid in D as described in Algorithm 1.

The grid in parameter space P is a simple nonuniform rectangular mesh. Such a mesh also corresponds to a nonuniform rectangular grid in the rectangle R so that the corresponding grid in D will indeed be orthogonal.

An illustration of this algorithm is shown in Figure 4.13, which consists of two grids in a channel with a circular arc. The lower part shows a grid obtained with Algorithm 3. The grid points are prescribed and their position is fixed while grid orthogonality is obtained at all four edges. The upper part shows

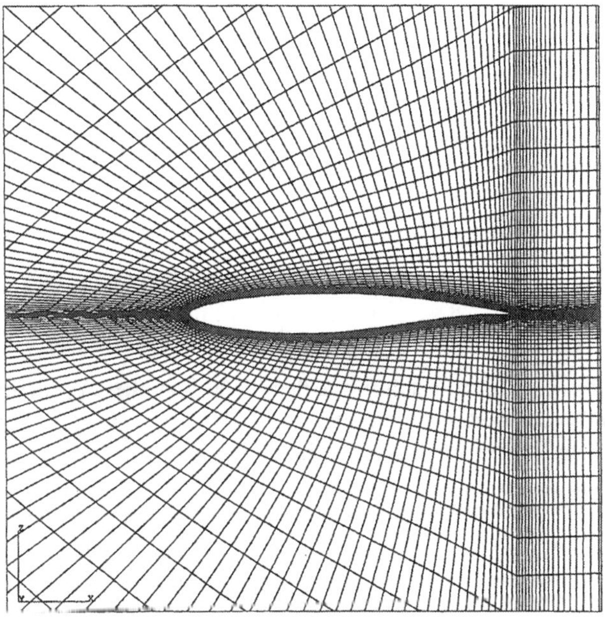

FIGURE 4.11 Arc length-based grid.

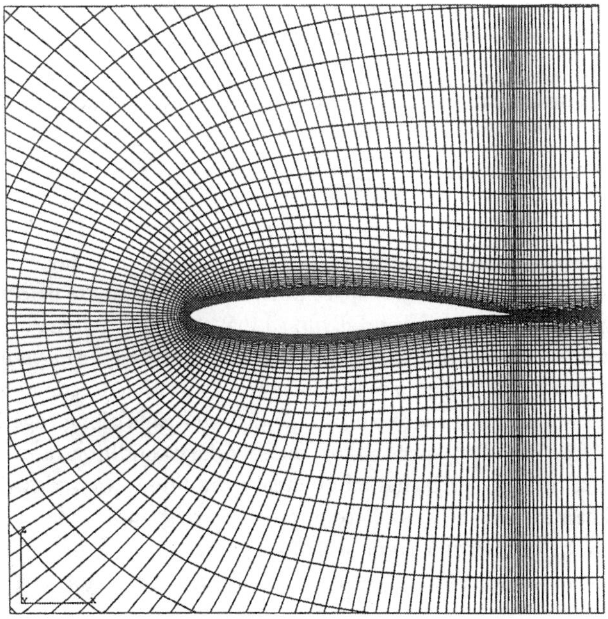

FIGURE 4.12 Grid with boundary orthogonality.

an orthogonal grid obtained by Algorithm 4. The figure clearly demonstrates how the boundary grid points have to move in order to obtain an orthogonal grid. For more information on orthogonal grids, see Chapter 7.

4.2.3.5 Complete Grid Control at the Boundary

In Section 4.2.3.3 we described the construction of a grid control map such that grid orthogonality is obtained at the boundary of D. However, the method provides no precise control of the height of the first grid cells along the boundary. In general, the cell height distributions of the first grid cell along the boundary in D is fairly good, as illustrated in Figure 4.8 and Figure 4.12. However, there are applications, especially in grid boundary layers for viscous flows, where not only grid orthogonality but also grid spacing should be precisely controlled. For example, it may be required that the first grid cell height is constant in the complete grid boundary layer, in spite of convex or concave parts of the boundary shape.

In order to have precise control about both grid orthogonality and grid cell height, we have to consider more general grid control maps. Both the grid control map based on normalized arc length, defined by Eq. 4.27 and Eq. 4.28, and the one based on Dirichlet–Neumann boundary conditions, defined by Eq. 4.42 and Eq. 4.43, have the form

$$s = \bar{s}(\xi,t) \quad t = \bar{t}(s,\eta) \tag{4.51}$$

Grid control maps of this type have the advantage that the two families of grid lines are independent: a grid line ξ = const. in C is mapped to parameter space P as a curve defined by $s = \bar{s}(\xi,t)$, which will be mapped by the inverse of a harmonic map to a curve in domain D. For given values of ξ and η, the corresponding grid point in P is found as the intersection point of the two curves $s = \bar{s}(\xi,t)$, $t = \bar{t}(s,\eta)$. When the boundary grid point distribution is changed in one set of opposite edges and remains unchanged in the other set, then one family of grid lines remains unchanged in both P and D.

Suppose that grid orthogonality and first-cell height specification are required at all four edges. Then the boundary conditions for the grid control map defined by Eq. 4.51 are shown in Figure 4.11. The boundary condition $\partial\bar{s}/\partial t = 0$ at E_3 and E_4 in (ξ, t)-space is needed for grid orthogonality at E_3 and E_4 in D. The values of $\partial\bar{s}/\partial\xi$ at E_1 and E_2 in (ξ, t)-space control the cell height of the first grid cells at E_1 and E_2 in D. Similarly, the boundary condition $\partial\bar{t}/\partial s = 0$ at E_1 and E_2 in (s, η)-space is needed for grid orthogonality at E_1 and E_2 in D. The values of $\partial\bar{t}/\partial\eta$ at E_3 and E_4 in (s, η)-space control the cell height of the first grid cells at E_3 and E_4 in D.

The algorithm for complete control of both grid orthogonality and cell height along the four edges becomes the following.

Algorithm 5. Complete grid control at boundary

1. Use Algorithm 3 to compute an initial boundary conforming grid in the interior of D which is orthogonal at the boundary. The corresponding grid control map is based on Eq. 4.42 and Eq. 4.43.
2. Compute $\partial\bar{s}/\partial\xi$ at E_1 and E_2 in (ξ, t)-space from Eq. 4.42. Compute $\partial\bar{t}/\partial\eta$ at E_3 and E_4 in (s, η)-space from Eq. 4.43. Adapt $\partial\bar{s}/\partial\xi$ and $\partial\bar{t}/\partial\eta$ so that the grid in domain D gets the desired grid cell height distribution along the corresponding edges. Note that the harmonic map and its inverse depend only on the shape of domain D. Therefore it is possible to compute how a change, in for example $\partial\bar{s}/\partial\xi$ at E_1 in (ξ, t)-space will change the cell height along edge E_1 in D.
3. Compute $s = \bar{s}(\xi, t)$ in (ξ, t)-space so that all boundary conditions are satisfied. Also compute $t = \bar{t}(s, \eta)$ in (s, η)-space such that all boundary conditions are satisfied. Compute the corresponding grid control map $\bar{s} : C \mapsto P$ for given values of ξ and η. The corresponding grid point in P is found as the intersection point of the two curves $s = \bar{s}(\xi, t)$, $t = \bar{t}(s, \eta)$.
4. Compute the corresponding interior grid in D as described in Algorithm 1.

The question remains how to compute $s = \bar{s}(\xi, t)$ and $t = \bar{t}(s, \eta)$ such that all boundary conditions are fulfilled. The boundary data $\bar{s}(0, t)$, $\bar{s}(1,t)$, $\bar{s}(\xi,0)$, $\bar{s}(\xi,1)$ and $\partial\bar{s}/\partial\xi\,(0,t)$, $\partial\bar{s}/\partial\xi\,(1,t)$, $\partial\bar{s}/\partial t\,(\xi,0)$, $\partial\bar{s}/\partial t\,(\xi,1)$, can be interpolated by using a bicubically blended Coon's patch [10,36]. However, the use of such an algebraic interpolation method has a severe shortcoming because twist vectors have to be specified at the four corners.In general, the tangent boundary conditions $\partial\bar{s}/\partial\xi$, $\partial\bar{s}/\partial t$, are conflicting at a corner when the two edges of domain D are not orthogonal at the corresponding vertex. In that case, the twist vector is not well-defined at the corner. Because of the conflicting tangent boundary conditions at the corners, we prefer to apply an elliptic partial differential equation to interpolate the boundary data. A fourth-order elliptic operator is needed to satisfy all boundary conditions. Therefore, the biharmonic equations

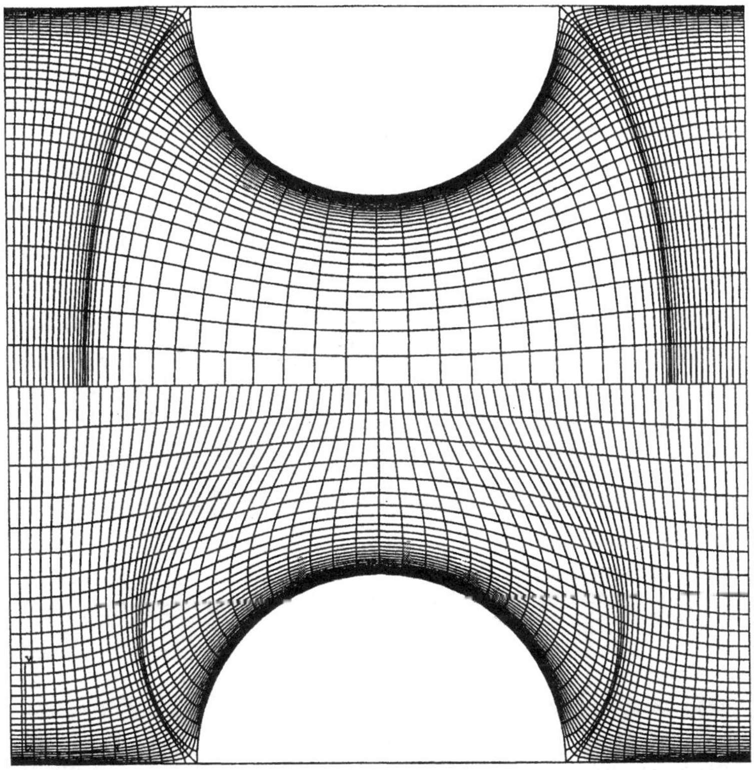

FIGURE 4.13 Orthogonal grid generation by boundary grid point movement along an edge. The grid in the lower part is orthogonal only at the boundary. The grid in the upper part is also orthogonal in the interior.

$$\Delta\Delta\bar{s} = 0 \tag{4.52}$$

where $\Delta = \partial^2/\partial\xi^2 + \partial^2/\partial t^2$, and

$$\Delta\Delta\bar{t} = 0 \tag{4.53}$$

where $\Delta = \partial^2/\partial s^2 + \partial^2/\partial\eta^2$ is a proper choice. The advantage of the use of the biharmonic equation to interpolate the boundary data is that the solution is always a smooth function even when the tangent boundary conditions are conflicting at the corners. A disadvantage is that the biharmonic operator does not fulfill a maximum principle. When there is a grid boundary layer along for example edge E_1 in D then the monotonic boundary functions $s^0_{E_3}(\xi)$ and $s^0_{E_4}(\xi)$ have very small values in a large part of the interval $0 << \xi << 1$. In that case, the solution of the biharmonic equation may have small negative values in the interior, which is of course unacceptable. This problem is solved by applying a change in variables. In fact, we solve $\Delta\Delta f = 0$ where $f\!: \bar{s} : \in [0, 1] \mapsto [0, 1]$ is a monotonic function which maps a unit interval onto a unit interval. The boundary conditions for $\bar{s}$ are transferred to corresponding boundary conditions for f. After solving $\Delta\Delta f = 0$, we find f values at interior grid points and the corresponding $\bar{s}$ values are found using f^{-1}. In practice, we define $f\!: \bar{s} \in [0, 1] \mapsto [0, 1]$ so that $f(\frac{1}{2}(s^0_{E_3}(\xi) + s^0_{E_4}(\xi)) \equiv \xi$. A similar change in variable is used for the grid control function $t = \bar{t}(s, \eta)$.

The biharmonic equations are solved by the black-box biharmonic solver BIHAR [3], which is available on the electronic mathematical NETLIB library.

Algorithm 5 describes complete boundary control for both grid orthogonality and grid spacing. It is also possible to have only grid spacing control without boundary grid orthogonality. In that case,

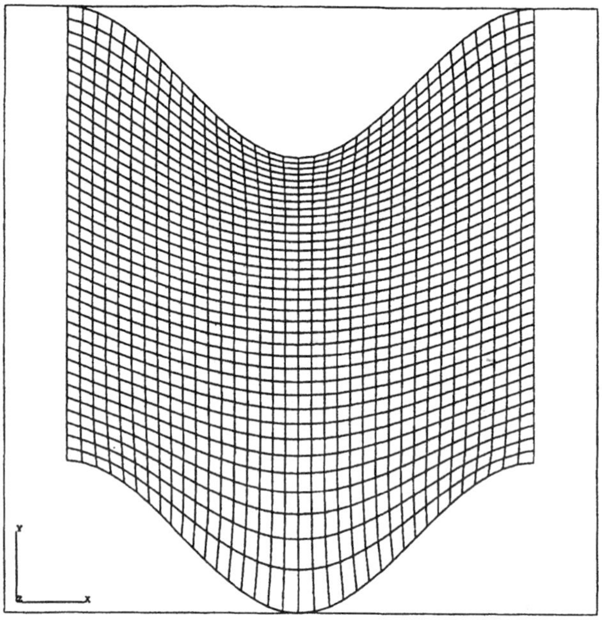

FIGURE 4.14 Laplace grid.

Algorithm 2 must be used instead of Algorithm 3 in the first step of Algorithm 5. An illustration of the result of grid spacing control is shown in Figure 4.14 through Figure 4.17. The same test case was also used by Eiseman [28]. The upper side of the domain is convex; the lower side is concave. The boundary grid points are prescribed and evenly distributed. Figure 4.14 shows a Laplace grid with the typical behavior near the convex and concave parts of the boundary. Figure 4.15 shows the grid with mesh spacing control at the upper and lower side. Clearly, the cell height becomes constant at both the convex and concave sides. Figure 4.16 shows the grid with grid orthogonality only at the convex and concave sides and Figure 4.17 shows the grid with combined control of both mesh spacing and grid orthogonality at the convex and concave sides.

4.2.4 Best Practices

In this section we show how the previously discussed algorithms work in practice. The chosen examples mainly concern simple well-defined geometries so that the reader is able to recompute the generated grids. In all cases, the boundary grid points are predefined and their location is fixed.

Example 1. Triangular domain

This example illustrates Algorithm 3 to obtain grid orthogonality at the boundary. Figure 4.19 shows the grid obtained with Algorithm 2. The corresponding grid control map, based on Eq. 4.27 and Eq. 4.28, is shown in Figure 4.18 as a grid in parameter space P. Notice that the grid lines are straight in P. Figure 4.21 shows the grid in parameter space obtained by solving $\Delta s = 0$ and $\Delta t = 0$ on the grid shown in Figure 4.19 supplied with Neumann boundary conditions on the two bottom edges of the triangle. It should be noticed that although this grid control map is completely different from the grid control map shown in Figure 4.18, the corresponding grid in the interior of the triangle will still be the same. Figure 4.22 shows the new grid control map based on Eq. 4.42, 4.43. Thus the position of the boundary grid points is the same in both Figure 4.22 and Figure 4.21. Notice that the grid is orthogonal at the left and bottom edge of P. These two edges in P correspond with the two bottom edges of the triangle. The corresponding grid is shown in Figure 4.23. The grid is clearly orthogonal at the two bottom edges of the triangle. Figure 4.23 shows the nice behavior of the grid near the 0-type singularity.

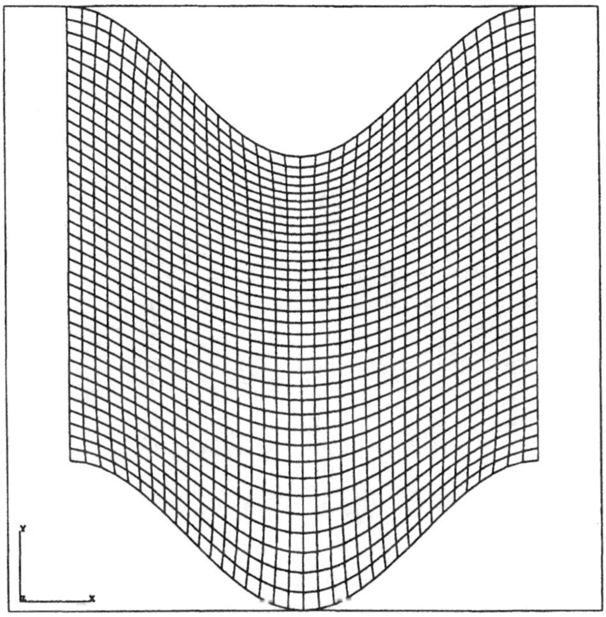

FIGURE 4.15 Grid with cell height control at upper and lower side.

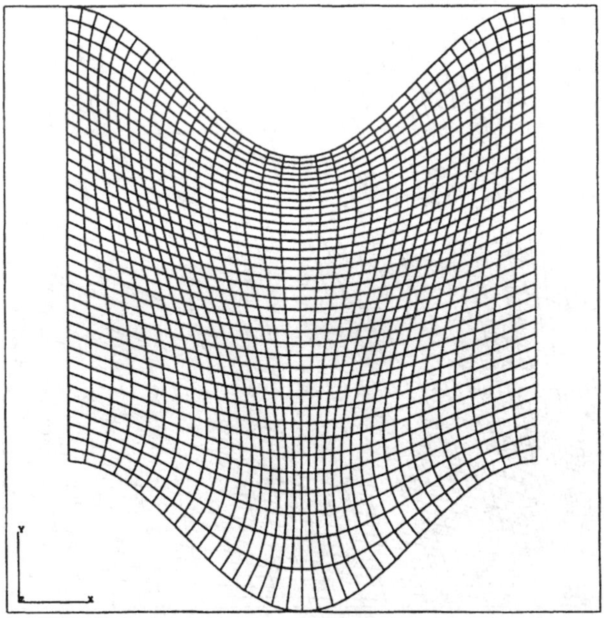

FIGURE 4.16 Grid with boundary orthogonality at upper and lower side.

Example 2. Circular domain

This example illustrates Algorithm 5 for complete grid control at the boundary. The prescribed boundary grid points are evenly spaced as shown in Figure 4.26. The grid in parameter space P, based on Eq. 4.27 and Eq. 4.28, is shown in Figure 4.25 and is thus uniform so that the corresponding grid in Figure 4.26 is a Laplace grid. Figure 4.27 shows the grid in parameter space obtained by solving $\Delta s = 0$ and $\Delta t = 0$ supplied with Neumann boundary conditions at all four sides. Figure 4.28 shows the new

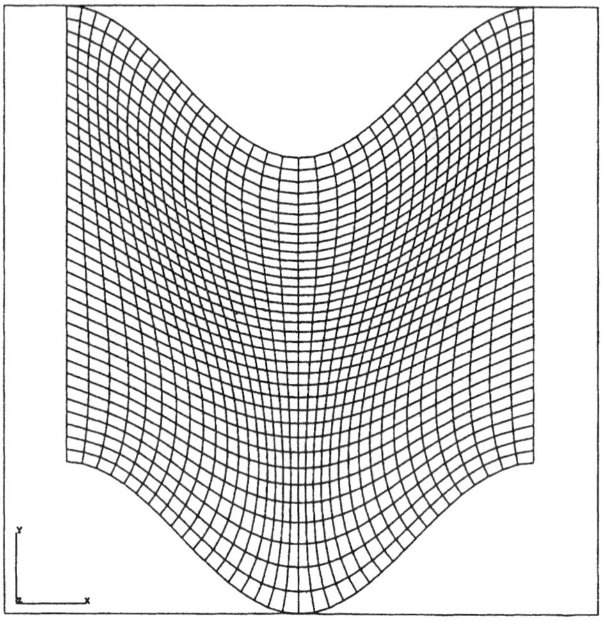

FIGURE 4.17 Grid with both cell height control and boundary orthogonality at upper and lower side.

FIGURE 4.18 Initial grid in parameter space based on normalized arc length.

grid control map based on Eq. 4.42 and Eq. 4.43. This grid in parameter space is no longer uniform but remains rectangular because of the symmetry in both geometry and boundary grid. The corresponding grid in physical space, shown in Figure 4.29, is thus orthogonal as explained in Section 4.2.3.4. Notice the bad mesh spacing along the boundary of this orthogonal grid. The adapted grid in parameter space to obtain also a good mesh spacing is shown in Figure 4.30. This adapted grid is obtained by the method described in Section 4.2.3.5. Figure 4.31 shows the corresponding grid in physical space and demonstrates the successful combination of boundary grid orthogonality and good mesh spacing.

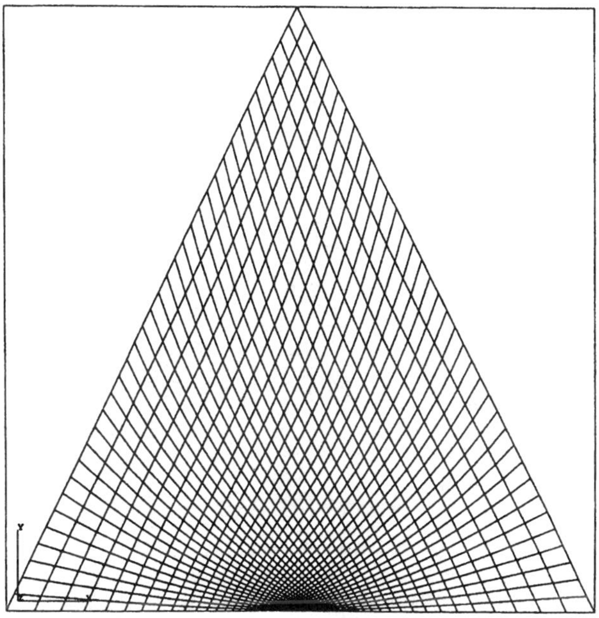

FIGURE 4.19 Corresponding grid in physical space.

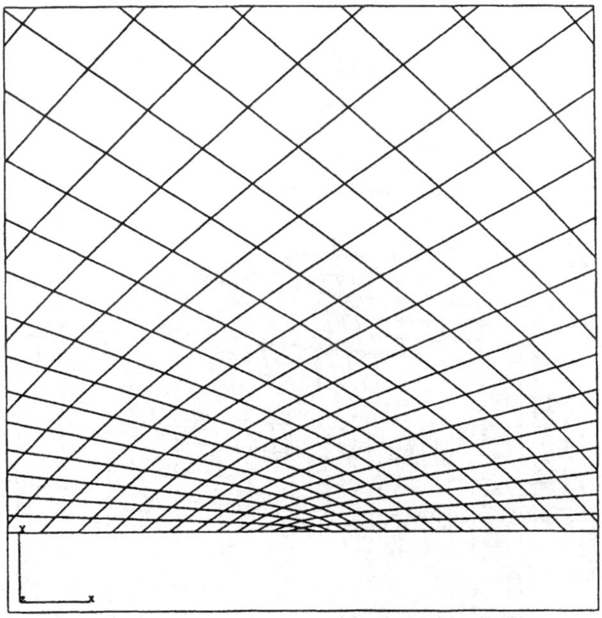

FIGURE 4.20 Blow up near O-type singularity.

Example 3. Domain bounded by semicircles on the four sides of the unit square

This geometry is also used by Duraiswami and Prosperetti [8] and Eça [9]. The prescribed boundary grid points are no longer evenly spaced but dense near the four corners of the domain. Figure 4.32 shows the grid in parameter space based on Eq. 4.27 and Eq. 4.28. Figure 4.33 shows the corresponding grid in physical space. Figure 4.34 shows the grid in parameter space obtained by solving $\Delta s = 0$ and $\Delta t = 0$ supplied with Neumann boundary conditions at all four sides. Figure 4.35 shows the new grid control map based on Eq. 4.42 and Eq. 4.43. This grid in parameter space is rectangular because of the symmetry

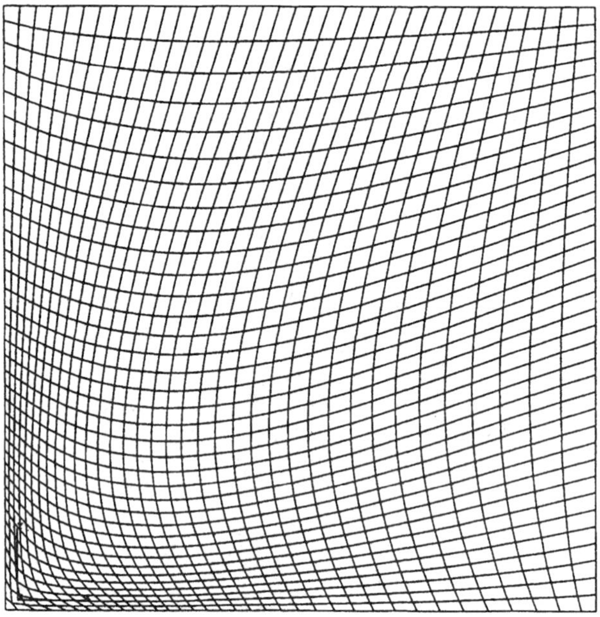

FIGURE 4.21 Grid in parameter space obtained by solving Laplace equations with Neumann boundary conditions at the two bottom edges of the triangle.

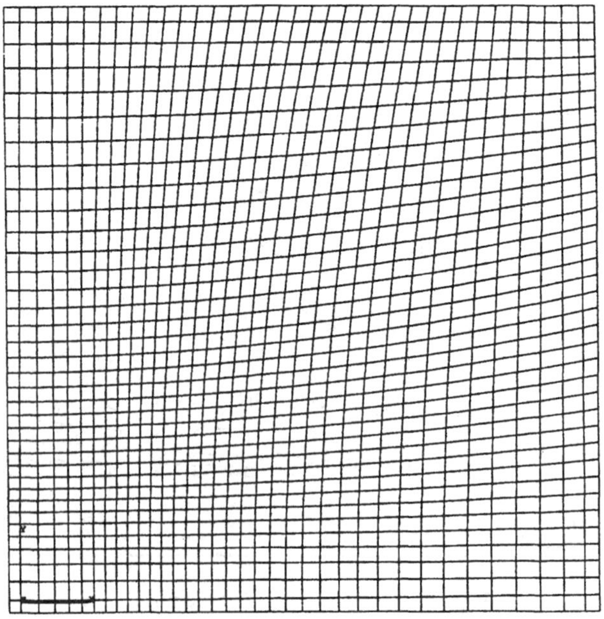

FIGURE 4.22 New grid in parameter space for boundary orthogonality. Position of boundary grid points are the same as in Figure 4.21.

in both geometry and boundary grid. The corresponding grid in physical space, shown in Figure 4.36, is thus orthogonal as explained in Section 4.2.3.4. The adapted grid in parameter space to obtain also a good mesh spacing is shown in Figure 4.37 and Figure 4.38 shows the result in physical space.

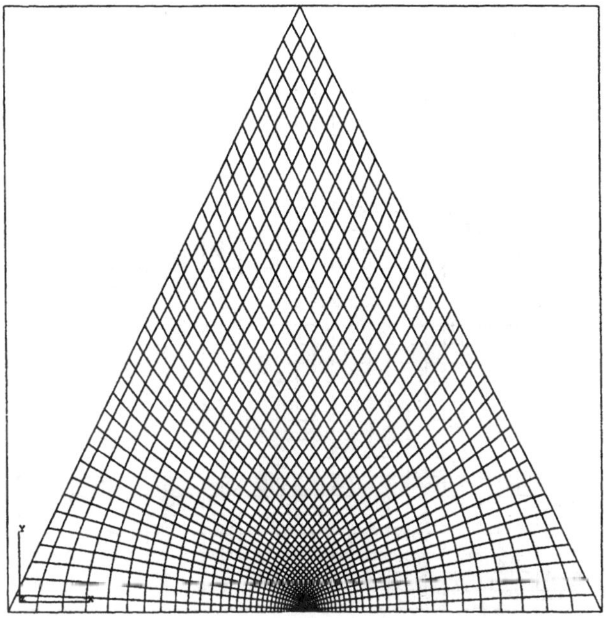

FIGURE 4.23 Corresponding grid in physical space.

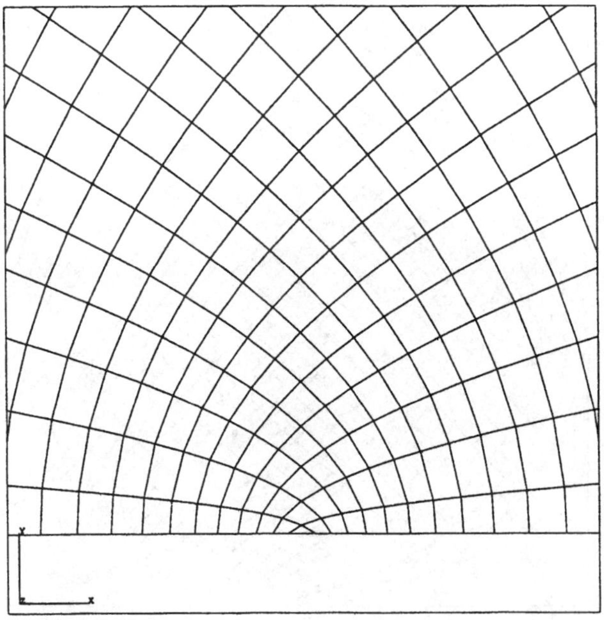

FIGURE 4.24 Blow up near O-type singularity.

Example 4. Degenerated domains

Two degenerated domains are considered: a lune bounded by the curves $y = x(1 - x)$ and $y = -x(1 - x^2)$ and a trilateral. The lune has two degenerated edges, the trilateral only one. Both geometries are also used by Duraiswami and Prosperetti [8] and Eça [9].

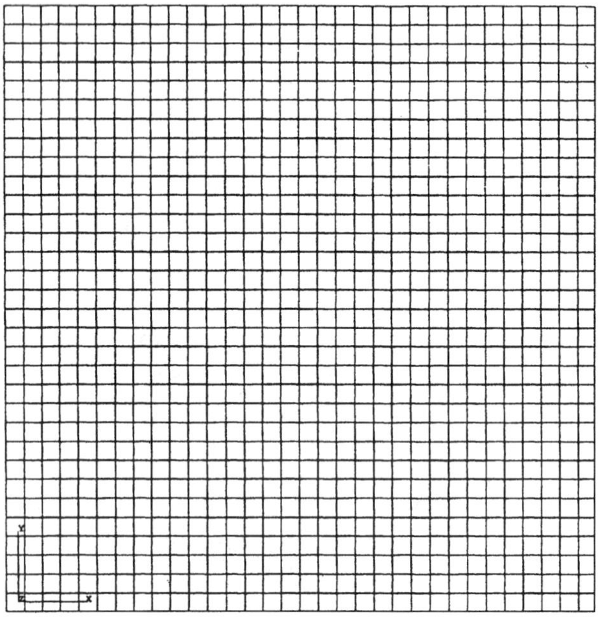

FIGURE 4.25 Initial uniform grid in parameter space based on normalized arc length.

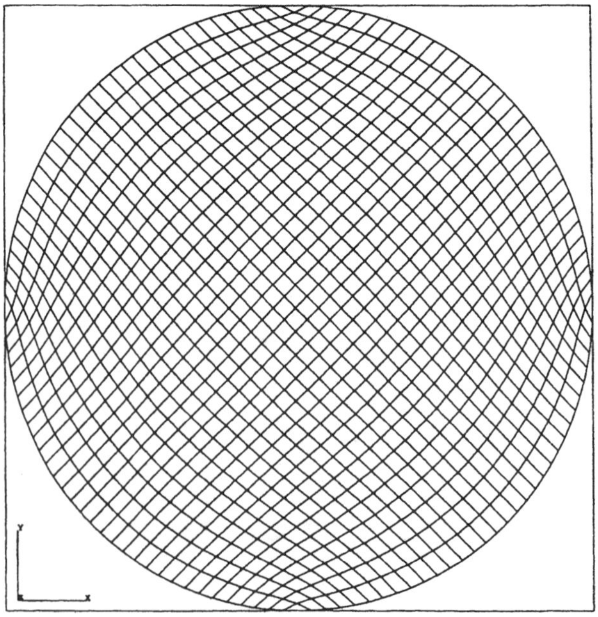

FIGURE 4.26 Corresponding Laplace grid in physical space.

In case of the lune, an evenly spaced boundary grid point distribution is used so that the grid in parameter space based on Eq. 4.27 and Eq. 4.28 is uniform and the corresponding grid in physical space is harmonic. See Figure 4.39 and Figure 4.40. Figure 4.41 shows the grid in parameter space obtained by solving $\Delta s = 0$ and $\Delta t = 0$ supplied with Neumann boundary conditions at the two nondegenerated edges. Notice the large change in the position of the boundary grid points in parameter space compared to the initial uniform grid. Figure 4.42 shows the new grid control map based on Eq. 4.42 and Eq. 4.43. This

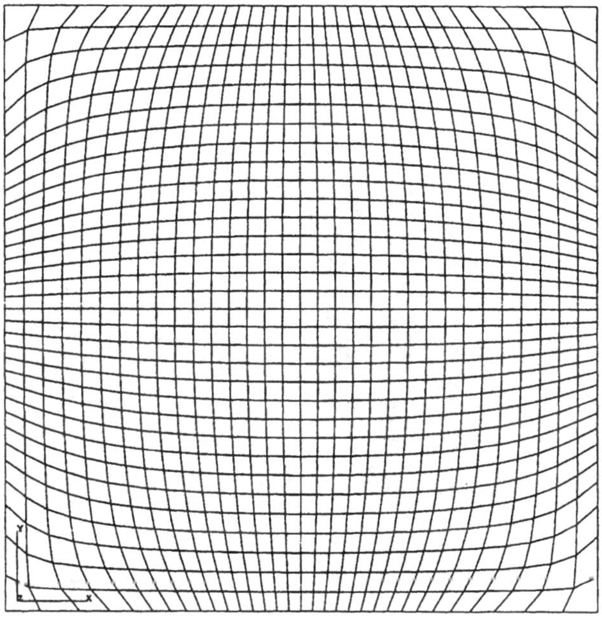

FIGURE 4.27 Grid in parameter space obtained by solving the Laplace equations with Neumann boundary conditions at all four sides.

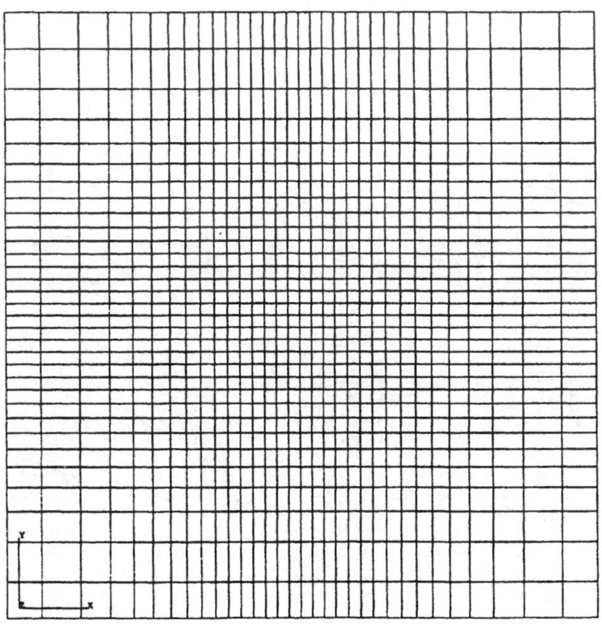

FIGURE 4.28 New grid in parameter space for boundary orthogonality at all four sides. Position of boundary points is the same as in Figure 4.27.

grid in parameter space is almost rectangular. The corresponding grid in physical space, shown in Figure 4.43, is therefore almost orthogonal.

For the trilateral, we show only the final grid in parameter space, obtained by Algorithm 5, and the corresponding grid in physical space. See Figure 4.44 and Figure 4.45.

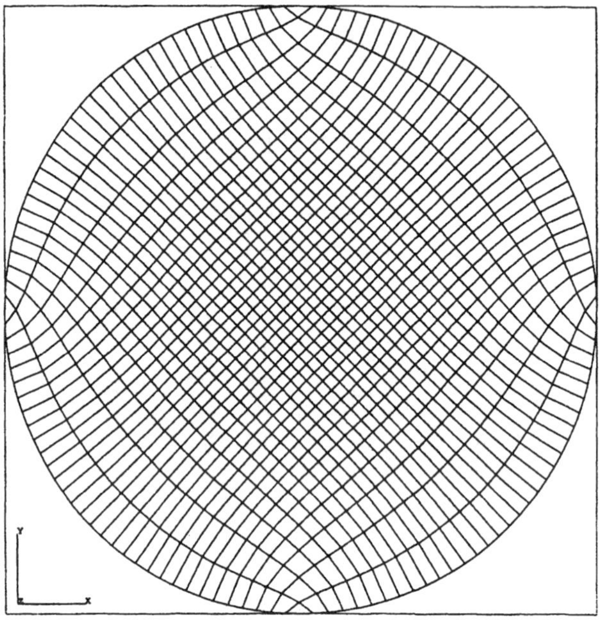

FIGURE 4.29 Corresponding grid in physical space. Interior grid is also orthogonal.

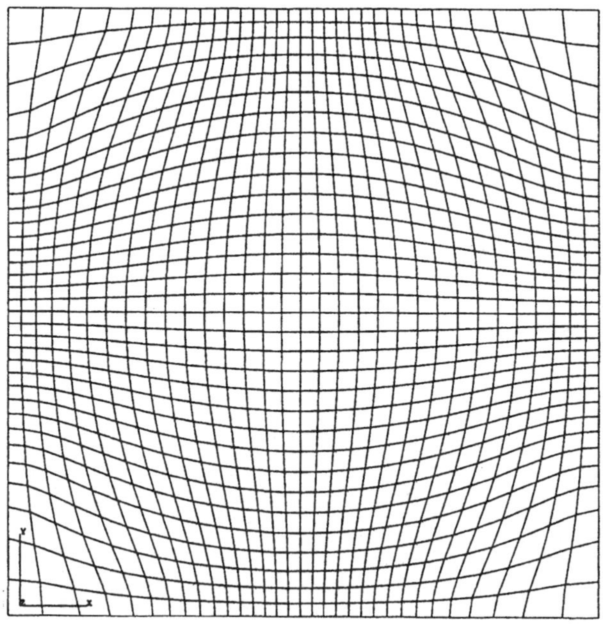

FIGURE 4.30 Adapted grid in parameter space for complete boundary control.

Example 5. Navier–Stokes grid around a complex artificial boundary

This example is used to demonstrate the robustness of the proposed algorithms. Figure 4.46 shows the grid in parameter space based on Eq. 4.27 and Eq. 4.28, and Figure 4.47 shows the corresponding *C*-type Navier–Stokes grid in physical space. Figure 4.49 shows the grid in parameter space obtained by solving $\Delta s = 0$ and $\Delta t = 0$ with Neumann boundary conditions at the lower boundary of the domain (three edges). Figure 4.50 shows the new grid in parameter space based on Eq. 4.42 and Eq. 4.43. The grid is orthogonal at the left, right, and lower side of the parameter space. The corresponding grid in physical space is shown in Figure 4.51 and Figure 4.52.

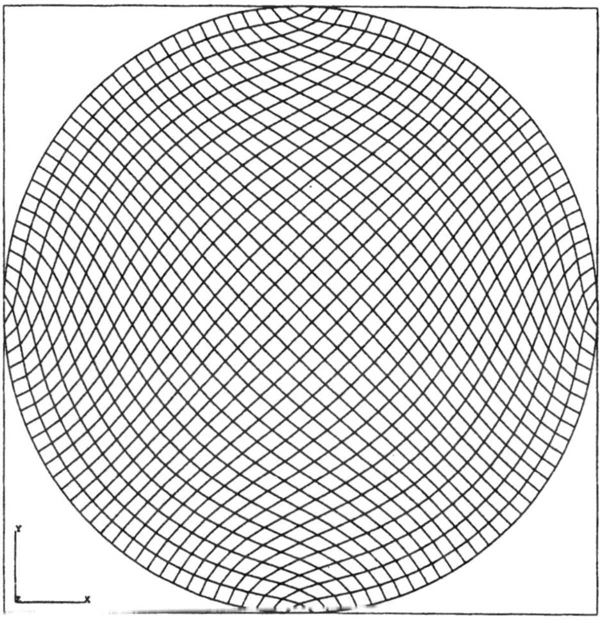

FIGURE 4.31 Corresponding grid in physical space.

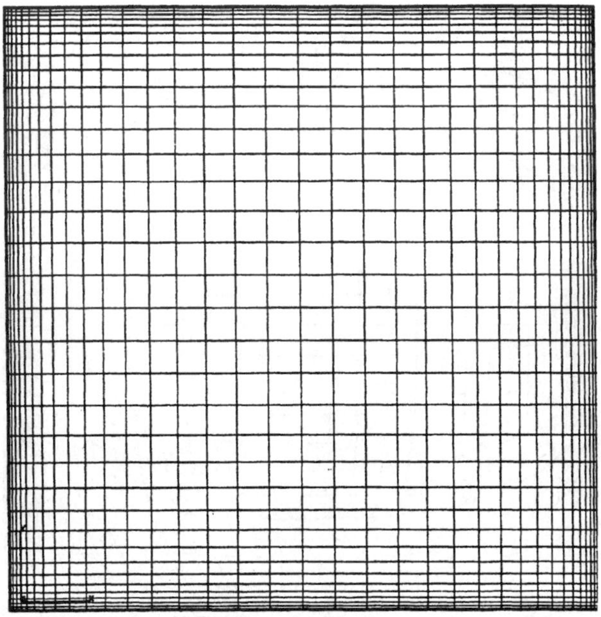

FIGURE 4.32 Initial grid in parameter space based on normalized arc length.

4.3 Surface Grid Generation

The concepts of harmonic maps and grid control maps as used for grid generation in 2D domains can also be used for grid generation on surfaces in 3D.

Consider a surface S bounded by four edges E_1, E_2, E_3, E_4. Let (E_1, E_2) and (E_3, E_4) be the two pairs of opposite edges as shown in Figure 4.3.

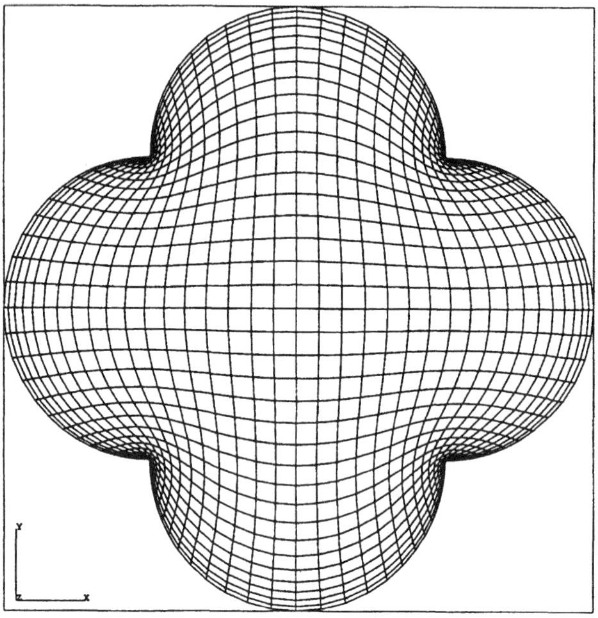

FIGURE 4.33 Corresponding grid in physical space.

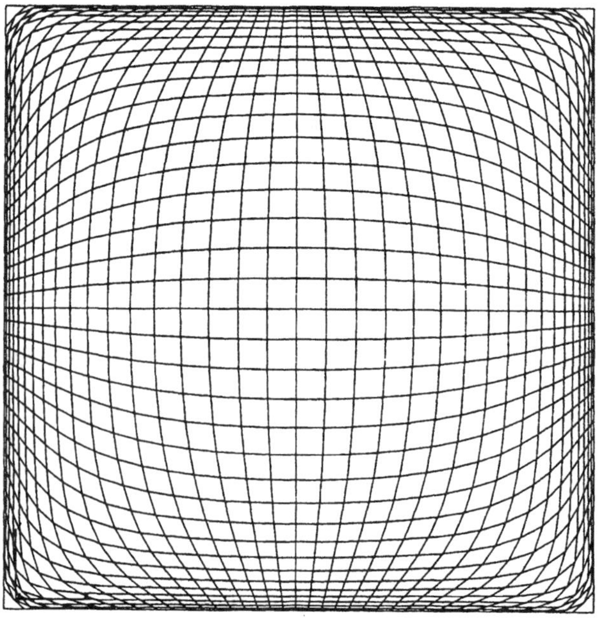

FIGURE 4.34 Grid in parameter space obtained by solving the Laplace equations with Neumann boundary conditions at all four sides.

A *harmonic map* is defined as a differentiable one-to-one map from S onto a unit square such that

1. The boundary of S is mapped onto the boundary of the unit square,
2. The vertices of S are mapped, in the proper sequence, onto the corners of the unit square,
3. The two components of the map are harmonic functions on S. This means that the two components obey the Laplace–Beltrami equations for surfaces (see Part II of Section 2.5 in Chapter 2).

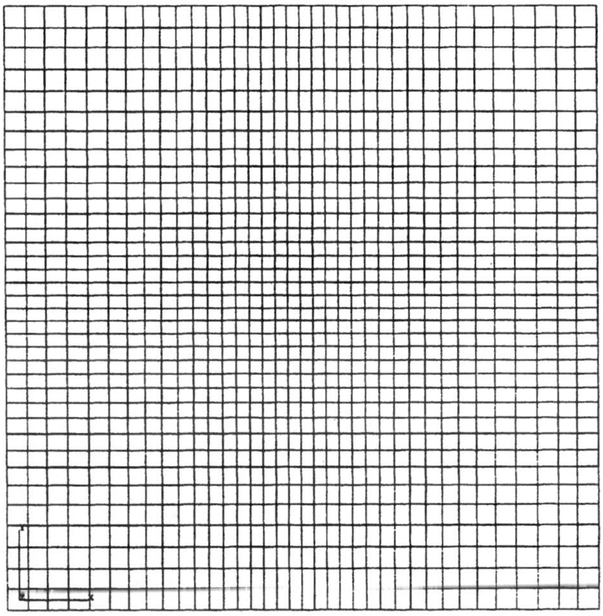

FIGURE 4.35 New grid in parameter space for boundary orthogonality. Position of boundary grid points are the same as in Figure 4.34

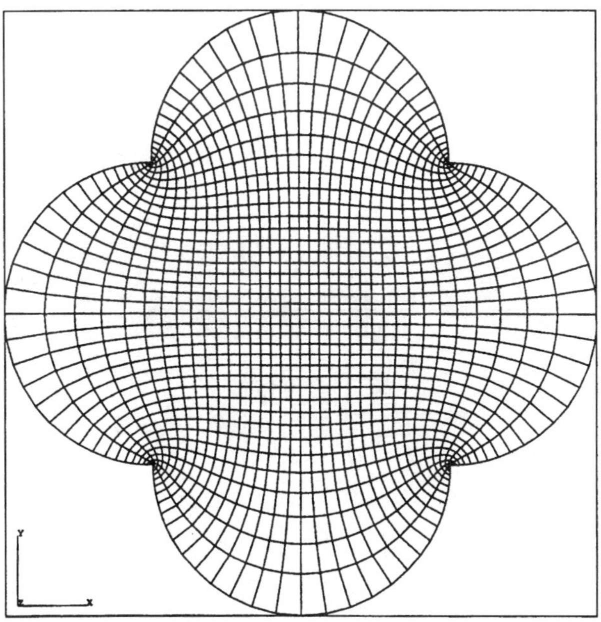

FIGURE 4.36 Corresponding grid in physical space. Interior grid is also orthogonal.

Let $\vec{s}$: $S \mapsto P$ be a harmonic map where the parameter space P is the unit square in a two-dimensional space with Cartesian coordinates $\vec{s} = (s, t)^T$. Thus $\Delta s = 0$ and $\Delta t = 0$ where Δ is the Laplace–Beltrami operator for surfaces [15].

The problem of generating an appropriate grid on surface S can be effectively reduced to a simpler problem of generating an appropriate grid in the parameter space P, which can after that be mapped on S, by using the inverse of the harmonic map $\vec{x}$: $P \mapsto S$.

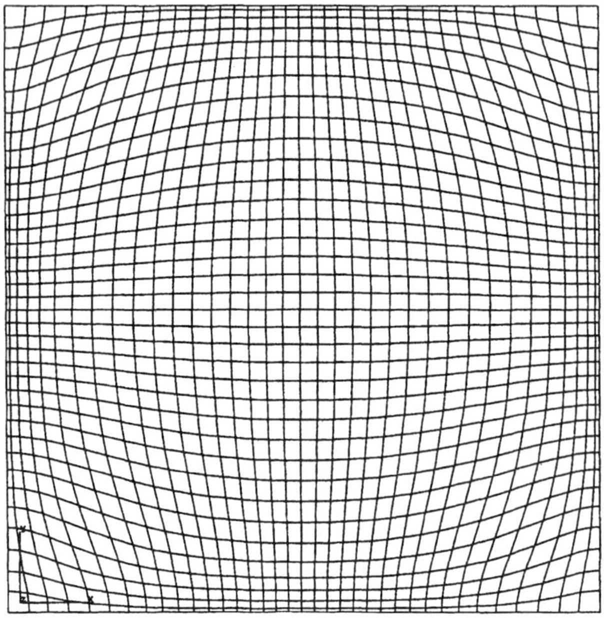

FIGURE 4.37 Adapted grid in parameter space for complete boundary control.

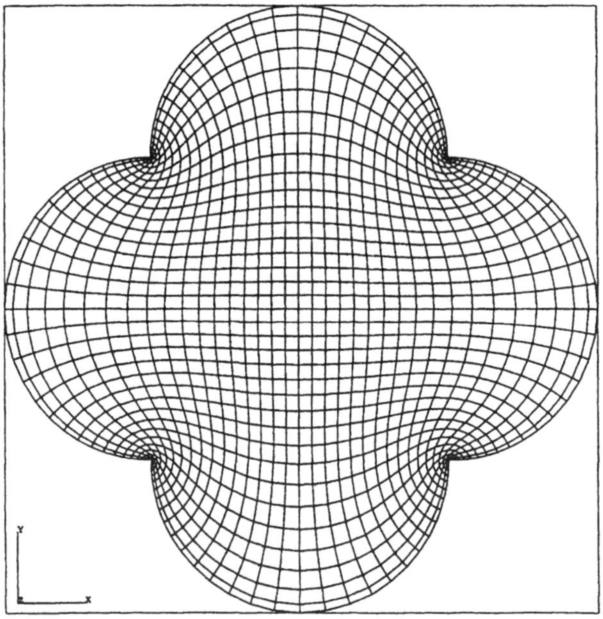

FIGURE 4.38 Corresponding grid in physical space.

Define the computational space C as the unit square in a two-dimensional space with Cartesian coordinates $\vec{\xi} = (\xi, \eta)^T$ A *grid control map* $\vec{s}$: $C \mapsto P$ is defined as a differentiable one-to-one map from C onto P and maps a uniform grid in C to a, in general, nonuniform grid in P.

The composition of a grid control map $\vec{s}$: $C \mapsto P$ and the inverse of the harmonic map $\vec{x}$: $P \mapsto S$ defines a map $\vec{x}$: $C \mapsto S$ which transforms a uniform grid in C to a, in general, nonuniform grid on surface S. The same ideas as used for 2D domains can be applied to construct appropriate grid control maps such that the corresponding surface grid has desired properties.

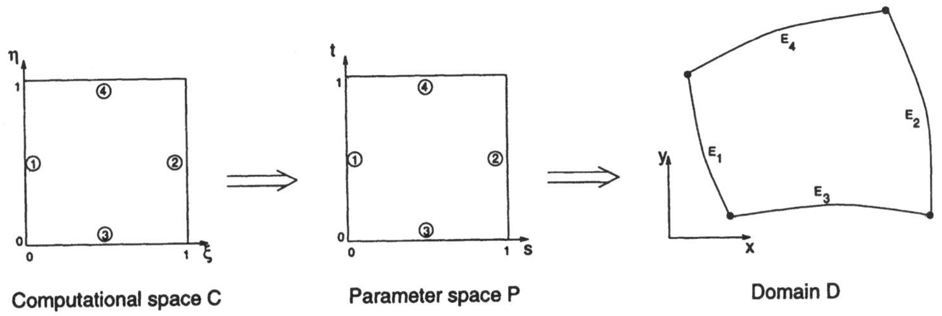

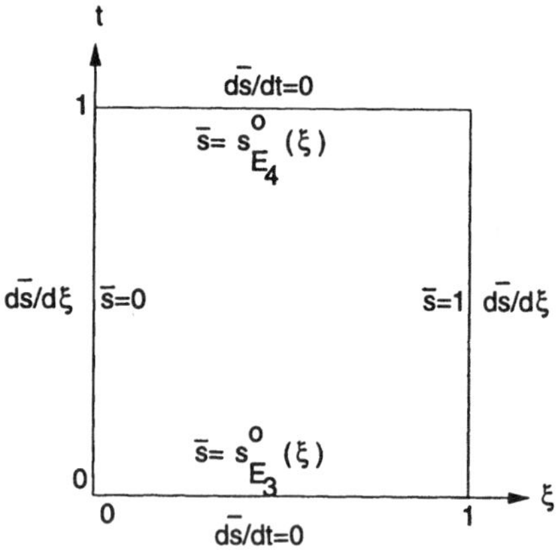

Boundary conditions for grid control function

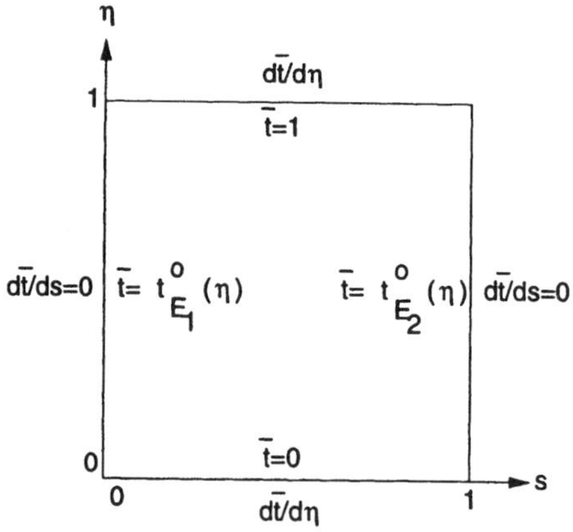

Boundary conditions for grid control function

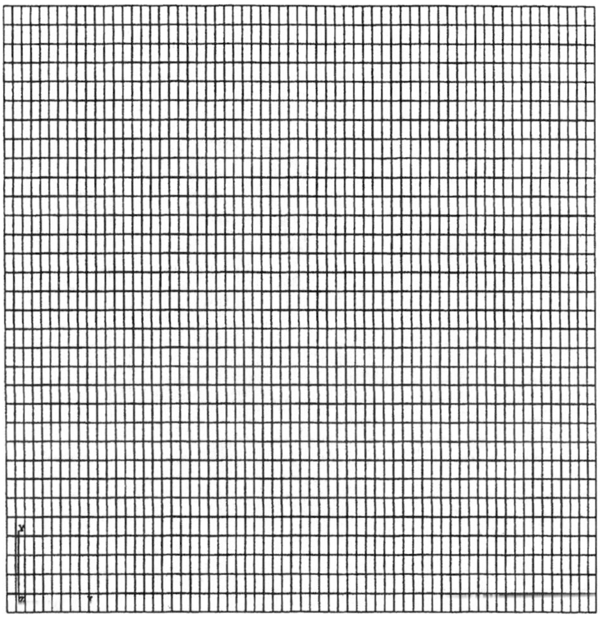

FIGURE 4.39 Initial uniform grid in parameter space based on normalized arc length.

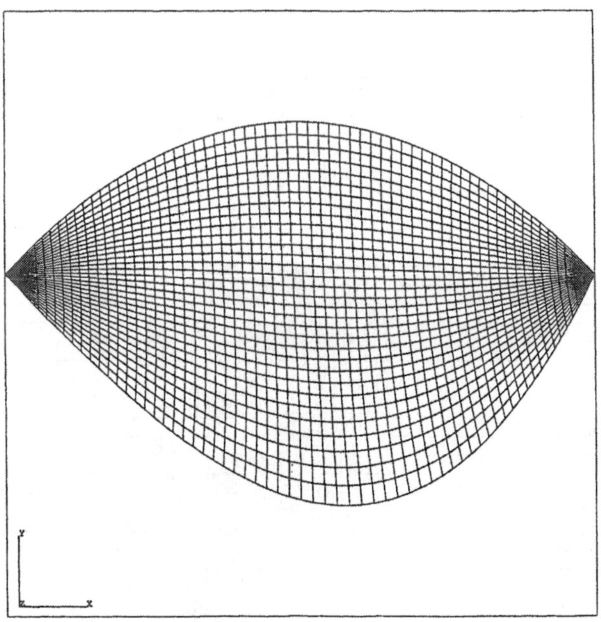

FIGURE 4.40 Corresponding Laplace grid in physical space.

For example, assume that the boundary grid points are prescribed on surface *S* and suppose that it is desired to construct a boundary conforming grid on *S* which is orthogonal at all four edges. Then the same Neumann boundary conditions as used in Section 4.2.3.3. must be used to define the harmonic map. Furthermore, the grid control map must be defined by Eq. 4.42 and Eq. 4.43. Then the composite map defines a boundary conforming grid on *S* that is orthogonal at all four edges.

However, the numerical implementation of these ideas is different from the 2D case because the composite map no longer fulfills a simple Poisson system as defined by Eq. 4.18. There is an exception,

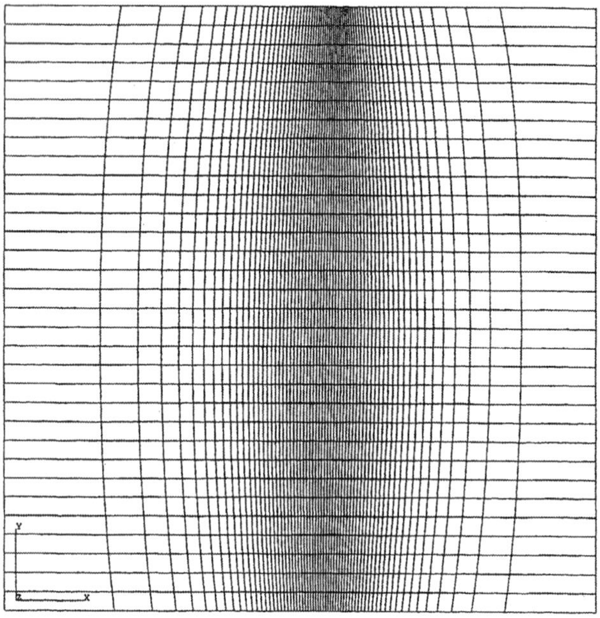

FIGURE 4.41 Grid in parameter space obtained by solving Laplace equations with Neumann boundary conditions at the two nondegenerated edges.

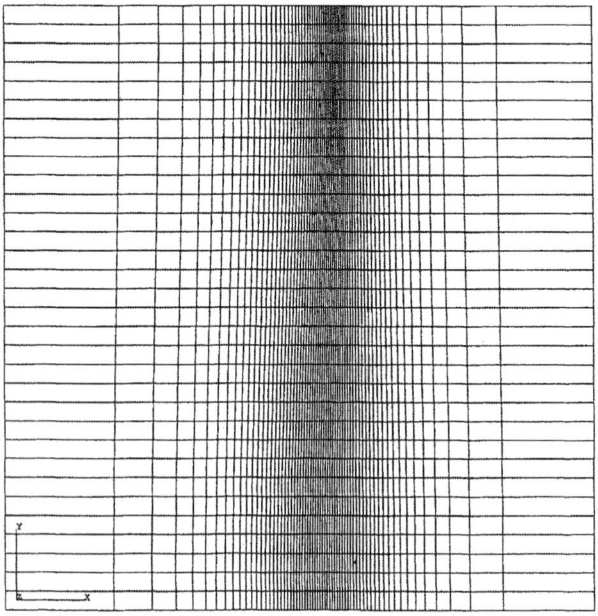

FIGURE 4.42 New grid in parameter space for boundary orthogonality at the two nondegenerated edges. Position of boundary grid points are the same as in Figure 4.41.

namely when S is a *minimal surface.* A minimal surface has zero mean curvature, and its shape is a soap film bounded by its four edges. There is a famous theorem in differential geometry which states that the Laplace–Beltrami operator applied on the position vector of an arbitrary surface S obeys

$$\Delta\vec{x} = 2H\vec{n} \tag{4.54}$$

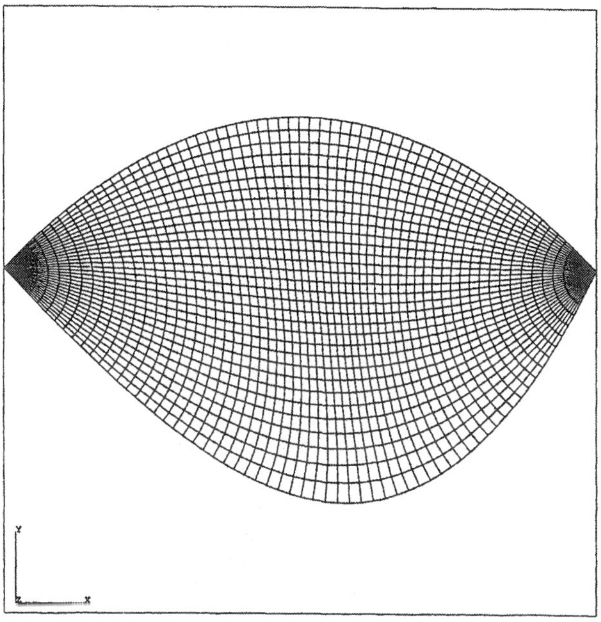

FIGURE 4.43 Corresponding grid in physical space.

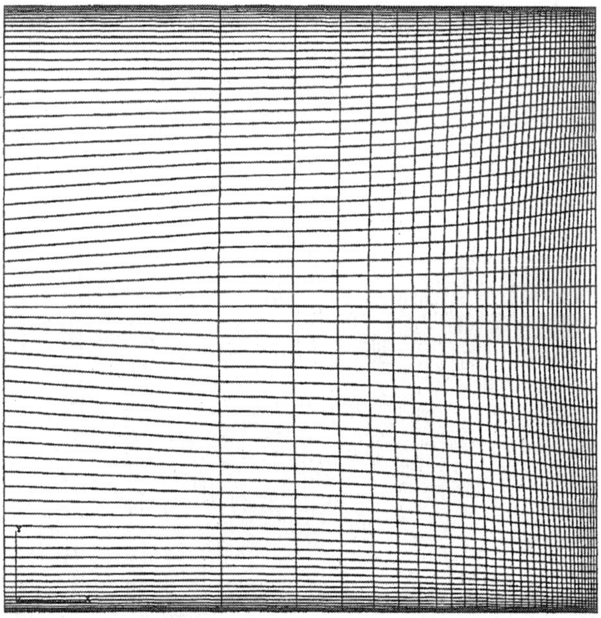

FIGURE 4.44 Constructed grid in parameter space for both grid orthogonality and mesh spacing control at the boundary of a trilateral.

where $\vec{n}$ is the unit vector normal to the surface and H is the mean curvature. (See Part II of Section 2.5 in Chapter 2, or Theorem 1 in Dierkes, et al. [7]). The requirement of zero mean curvature implies

$$\Delta \vec{x} = 0 \tag{4.55}$$

Thus for minimal surfaces we also have $\Delta s = 0$, $\Delta t = 0$ and $\Delta \vec{x} = 0$. Following the same derivation as in Section 4.2.1 for 2D domains, we find that the composite map obeys the same Poisson system given

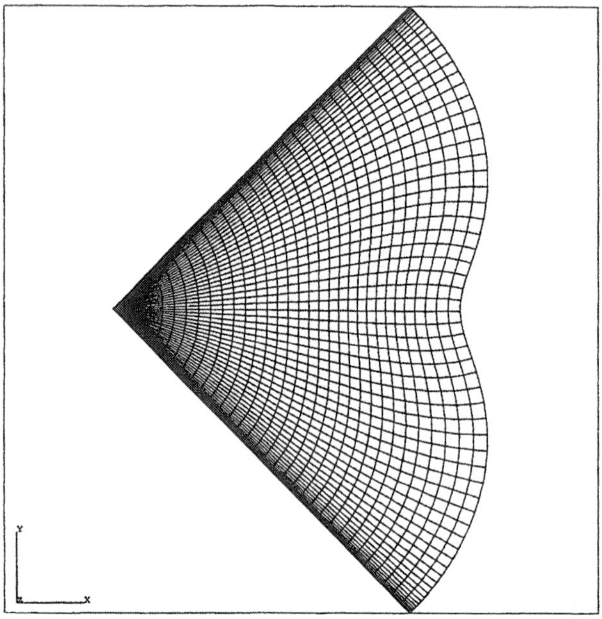

FIGURE 4.45 Corresponding grid in trilateral.

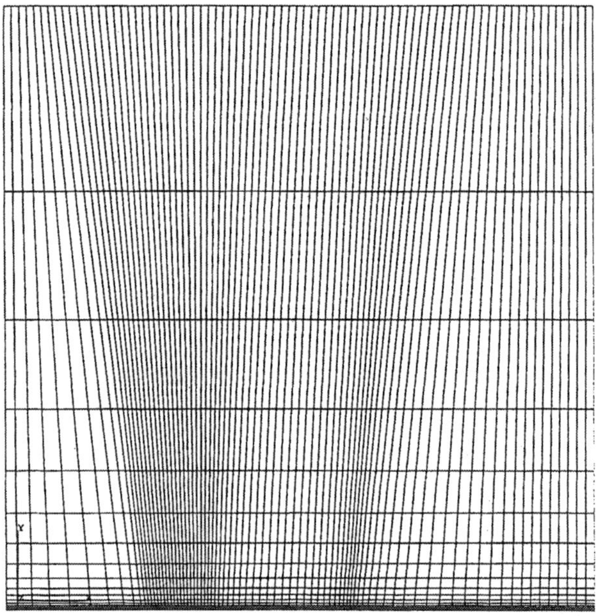

FIGURE 4.46 Initial grid in parameter space map based on normalized arc length.

by Eq. 4.18 (for more details see [25]). Thus an interior grid point distribution on a minimal surface is found by solving Eq. 4.18 with the prescribed boundary grid points as Dirichlet boundary conditions. The only difference compared with the two-dimensional case is that now $\vec{x} = (x, y, z)^T$ instead of $\vec{x} = (x, y)^T$. The same ideas to construct appropriate grid control maps and their corresponding grids in 2D domains can also be directly applied to minimal surfaces. In fact, all previously discussed 2D examples are generated as minimal surface grids where the four boundary edges are lying in a plane in three-dimensional space.

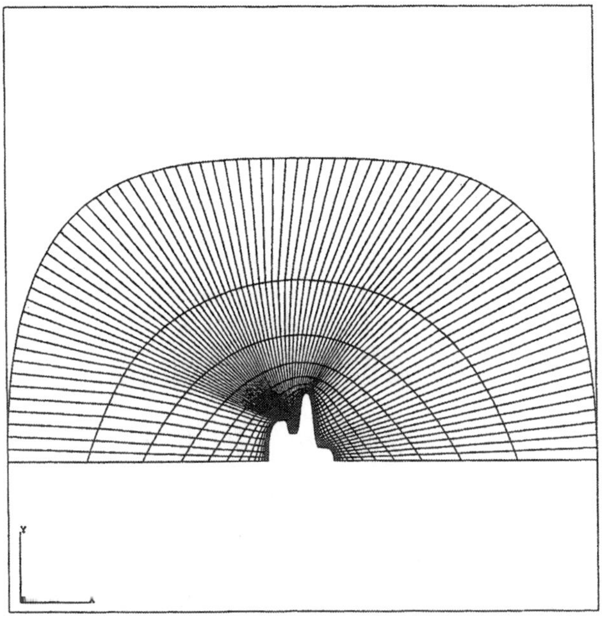

FIGURE 4.47 Corresponding grid in physical space.

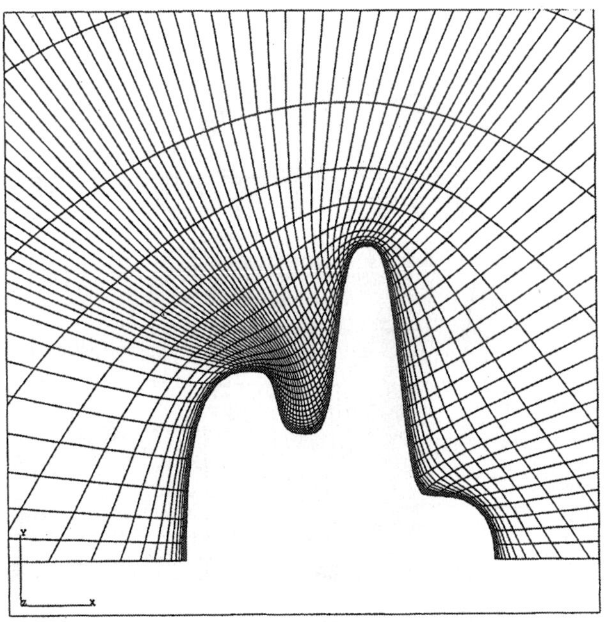

FIGURE 4.48 Blow up.

Examples of characteristic minimal surface grids are shown in Figures 4.53–4.57. Figure 4.53 is a so-called square Scherk surface [7]. Figure 4.54 shows what happens when the boundary edges of the Scherk surface are replaced by semicircular arcs. Figure 4.55 and Figure 4.56 show the change in the shape of the minimal surface when these semicircular arcs are bent together. Boundary orthogonality is imposed at all four sides for all these three cases. Because of the symmetry in both geometry and boundary grid point distribution, the generated surface grids are not only orthogonal at the boundary but also in the interior. Finally, Figure 4.57 is Schwarz's *P*-surface [7], which is in fact constructed as a collection of connected minimal surfaces.

FIGURE 4.49 Solution of Laplace equations with Neumann boundary conditions at the three bottom edges of the domain.

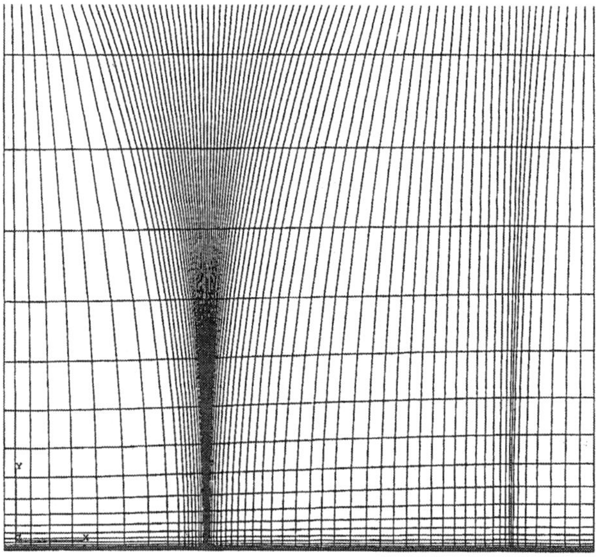

FIGURE 4.50 New grid in parameter space for boundary orthogonality at the three bottom edges of the domain. Position of the boundary grid points is the same as in Figure 4.49.

In general, surface S is not a minimal surface but a parametrically defined surface with a prescribed geometrical shape given by a map $\vec{x}$: $Q \mapsto S$ where Q is some parameter space defined as a unit square in 2D. In order to construct, for example, a boundary conforming grid on S which is orthogonal at all four edges, we solve on an initial surface grid on S the Laplace–Beltrami equations with the same Neumann boundary conditions as used in Section 4.2.3.3. The solution can be written as a map $\vec{s}$: $Q \mapsto P$. The appropriate grid control map, defined by Eq. 4.42 and Eq. 4.43, defines a nonuniform grid in P. The corresponding grid in Q can then be found by using the inverse map $\vec{s}^{-1}$: $P \mapsto Q$. This is done numerically

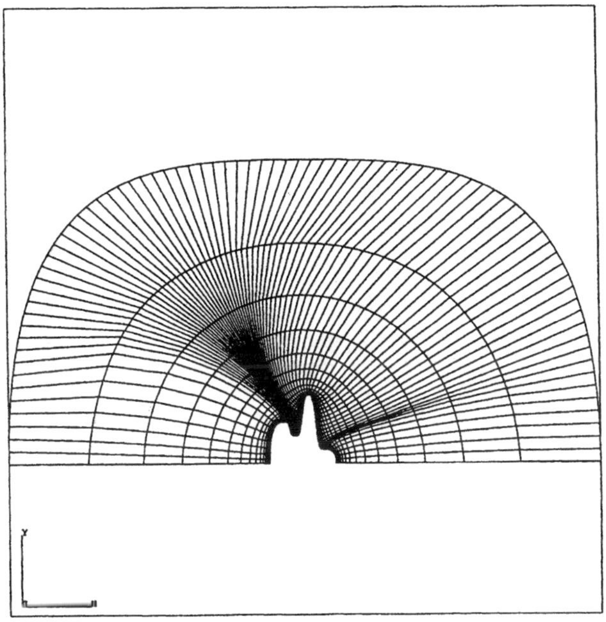

FIGURE 4.51 Corresponding grid in physical space.

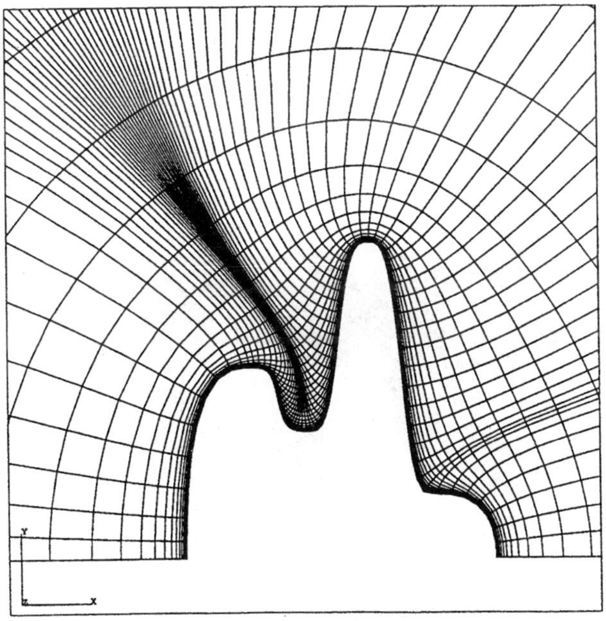

FIGURE 4.52 Blow up.

in a way described in [25]. Once the corresponding grid in Q is found, then the corresponding surface grid on S is computed using the parametrization $\vec{x}$: $Q \mapsto S$. This new surface grid on S differs from the initial surface grid S. The complete process should be repeated until the surface grid on S (and the corresponding grids in parameter space P and Q) do not change anymore. In practice, only a few (Eq. 4.2–4.5) iterations appear to be sufficient. After convergence, the final surface grid not only isorthogonal at the boundary but is also independent of the parametrization and depends only on the shape of the surface and the position of the boundary grid points.

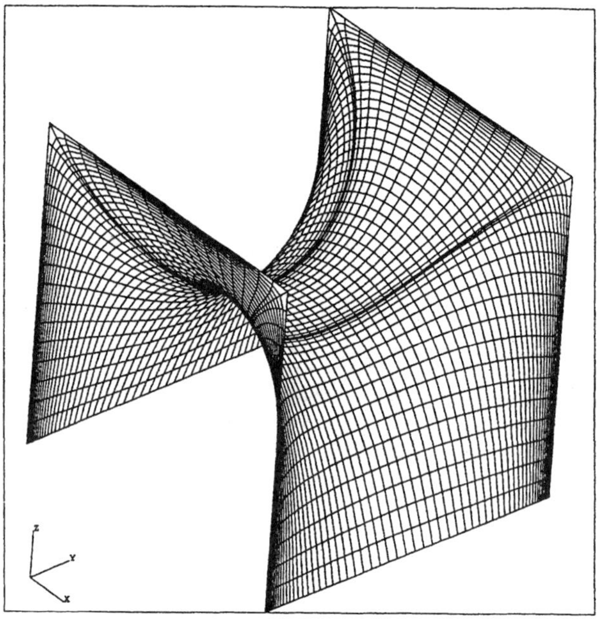

FIGURE 4.53 Minimal surface grid (Scherk surface). Surface grid is orthogonal.

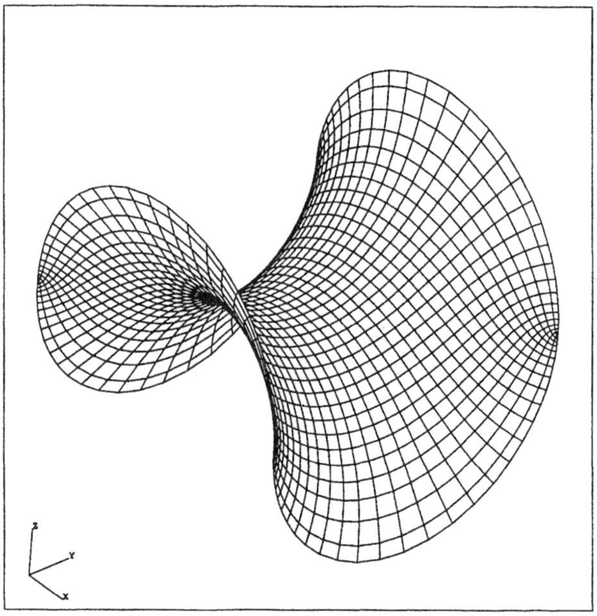

FIGURE 4.54 Minimal surface grid bounded by four orthogonal circular arcs. Surface grid is orthogonal.

4.4 Volume Grid Generation

Consider a simply connected bounded domain D in three-dimensional space with Cartesian coordinates $\vec{x} = (x, y, z)^T$. Suppose that D is bounded by six faces $F_1, F_2, F_3, F_4, F_5, F_6$. Let (F_1, F_2), (F_3, F_4), and (F_5, F_6) be the three pairs of opposite faces. Furthermore, consider the 12 edges $\{E_i, i = 1 \cdots 12\}$ and assume that these edges are related to the six faces as shown in Figure 4.4.

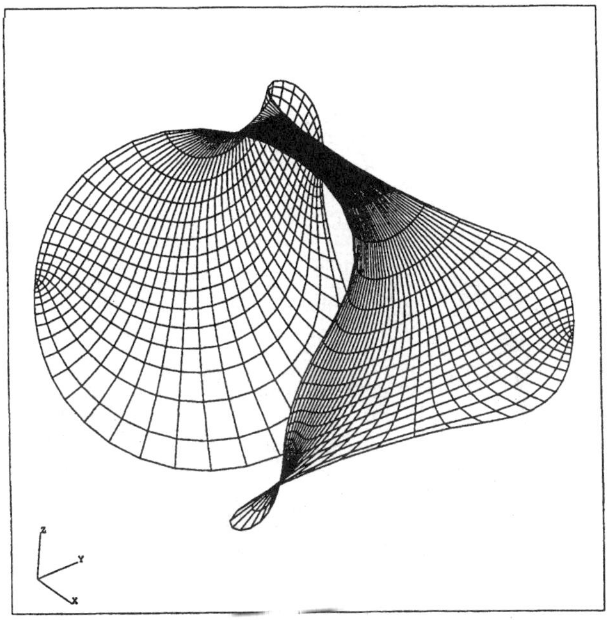

FIGURE 4.55 Change in shape by bending opposite circular arcs together.

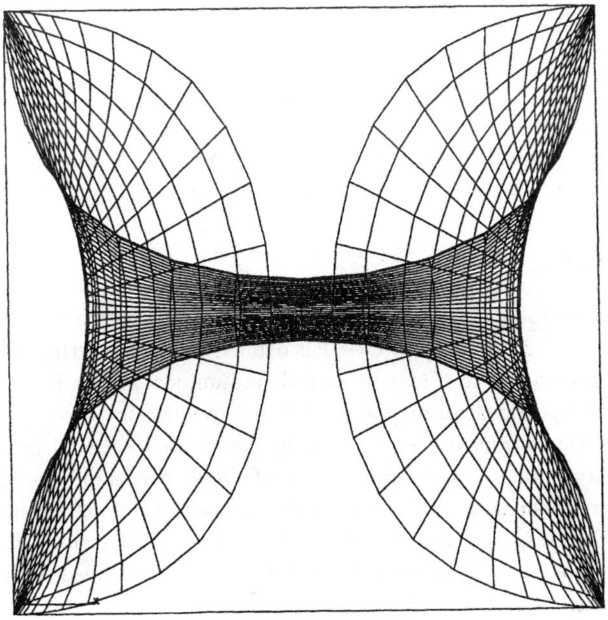

FIGURE 4.56 Projection on xy-plane.

In 3D, a harmonic map is defined as a differentiable one-to-one map from D onto a unit cube such that

1. The boundary of D is mapped onto the boundary of the unit cube,
2. The vertices, edges, and faces of D are mapped onto the corresponding vertices, edges, and faces of the unit cube,
3. The three components of the map are harmonic functions in the interior of D.

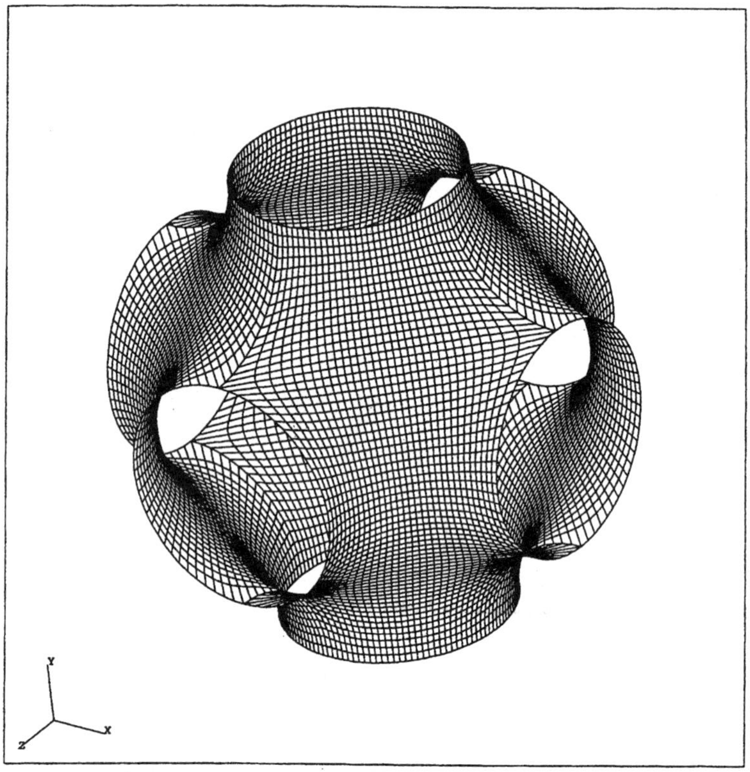

FIGURE 4.57 Schwarz's *P*-minimal surface.

Let $\vec{s}: D \mapsto P$ be a harmonic map where the parameter space P is the unit cube in a three-dimensional space with Cartesian coordinates $\vec{s} = (s, t, u)^T$. Inside D the components obey

$$\Delta s = s_{xx} + s_{yy} + s_{zz} = 0 \quad \Delta t = t_{xx} + t_{yy} + t_{zz} = 0 \quad \Delta u = u_{xx} + u_{yy} + u_{zz} = 0 \tag{4.56}$$

Define the computational space C as the unit cube in a three-dimensional space with Cartesian coordinates $\vec{\xi} = (\xi, \eta, \zeta)^T$. A *grid control map* $\vec{s}: C \mapsto P$ is defined as a differentiable one-to-one map from C onto P and maps a uniform grid in C to a, in general, nonuniform grid in P.

The composition of a grid control map $\vec{s}: C \mapsto P$ and the inverse of the harmonic map $\vec{x}: P \mapsto D$ defines a map $\vec{x}: C \mapsto D$ that transforms a uniform grid in C to a, in general, nonuniform grid in D. As in 2D, the composite map obeys a quasi-linear system of elliptic partial differential equations, known as the *Poisson grid generation equations,* with control functions completely defined by the grid control map.

The derivation of the Poisson grid generation equations can be done along the same lines as for the 2D case. Suppose that the harmonic map and grid control map are defined so that the composite map exists. Introduce the three covariant base vectors

$$\vec{a}_1 = \vec{x}_\xi \quad \vec{a}_2 = \vec{x}_\eta \quad \vec{a}_3 = \vec{x}_\zeta \tag{4.57}$$

and the covariant metric tensor components

$$a_{ij} = (\vec{a}_i, \vec{a}_j) \quad i = \{1,2,3\} \quad j = \{1,2,3\} \tag{4.58}$$

The three contravariant base vectors $\vec{a}^1 = \nabla\xi = (\xi_x, \xi_y, \xi_z)^T$, $\vec{a}^2 = \nabla\eta = (\eta_x, \eta_y, \eta_z)^T$, and $\vec{a}^3 = \nabla\zeta = (\zeta_x, \zeta_y, \zeta_z)^T$ obey

$$\left(\vec{a}^i, \vec{a}_j\right) = \delta^i_j \quad i = \{1,2,3\} \quad j = \{1,2,3\} \tag{4.59}$$

Define the contravariant metric tensor components

$$a^{ij} = \left(\vec{a}^i, \vec{a}^j\right) \quad i = \{1,2,3\} \quad j = \{1,2,3\} \tag{4.60}$$

so that

$$\begin{pmatrix} a_{11} & a_{12} & a_{13} \\ a_{12} & a_{22} & a_{23} \\ a_{13} & a_{23} & a_{33} \end{pmatrix} \begin{pmatrix} a^{11} & a^{12} & a^{13} \\ a^{12} & a^{22} & a^{23} \\ a^{13} & a^{23} & a^{33} \end{pmatrix} = \begin{pmatrix} 1 & 0 & 0 \\ 0 & 1 & 0 \\ 0 & 0 & 1 \end{pmatrix} \tag{4.61}$$

Define J^2 as the determinant of the covariant metric tensor.

Consider an arbitrary function $\phi = \phi(\xi, \eta, \zeta)$. Then ϕ is also defined in domain D and the Laplacian of ϕ can be expressed as

$$\Delta\phi = \frac{1}{J}\left\{\left(Ja^{11}\,\phi_\xi + Ja^{12}\,\phi_\eta + Ja^{13}\,\phi_\zeta\right)_\xi + \left(Ja^{12}\,\phi_\xi + Ja^{22}\,\phi_\eta + Ja^{23}\,\phi_\zeta\right)_\eta + \left(Ja^{13}\,\phi_\xi + Ja^{23}\,\phi_\eta + Ja^{33}\,\phi_\zeta\right)_\zeta\right\} \tag{4.62}$$

As in the two-dimensional case, substitution of $\phi \equiv \xi$, $\phi \equiv \eta$ and $\phi \equiv \zeta$ into this equation yields expressions for $\Delta\xi$, $\Delta\eta$, and $\Delta\zeta$. Combining these expressions with Eq. 4.62 gives

$$\Delta\phi = a^{11}\,\phi_{\xi\xi} + 2a^{12}\,\phi_{\xi\eta} + 2a^{13}\,\phi_{\xi\zeta} + a^{22}\,\phi_{\eta\eta} + 2a^{23}\,\phi_{\eta\zeta} + a^{33}\,\phi_{\zeta\zeta} + \Delta\xi\phi_\xi + \Delta\eta\phi_\eta + \Delta\zeta\phi_\zeta \tag{4.63}$$

Substitute $\phi = (s, t, u)^T$ in Eq. 4.63 and use the property that s, t, and u are harmonic in domain D, i.e., $\Delta s = 0$, $\Delta t = 0$, and $\Delta u = 0$. Then the following expressions for the Laplacian of ξ, η, and ζ are found:

$$\begin{pmatrix} \Delta\xi \\ \Delta\eta \\ \Delta\zeta \end{pmatrix} = a^{11}\,\vec{P}_{11} + 2a^{12}\,\vec{P}_{12} + 2a^{13}\,\vec{P}_{13} + a^{22}\,\vec{P}_{22} + 2a^{23}\,\vec{P}_{23} + a^{33}\,\vec{P}_{33} \tag{4.64}$$

where

$$\begin{aligned} \vec{P}_{11} &= -T^{-1}\begin{pmatrix} s_{\xi\xi} \\ t_{\xi\xi} \\ u_{\xi\xi} \end{pmatrix} & \vec{P}_{12} &= -T^{-1}\begin{pmatrix} s_{\xi\eta} \\ t_{\xi\eta} \\ u_{\xi\eta} \end{pmatrix} & \vec{P}_{13} &= -T^{-1}\begin{pmatrix} s_{\xi\zeta} \\ t_{\xi\zeta} \\ u_{\xi\zeta} \end{pmatrix} \\ \vec{P}_{22} &= -T^{-1}\begin{pmatrix} s_{\eta\eta} \\ t_{\eta\eta} \\ u_{\eta\eta} \end{pmatrix} & \vec{P}_{23} &= -T^{-1}\begin{pmatrix} s_{\eta\zeta} \\ t_{\eta\zeta} \\ u_{\eta\zeta} \end{pmatrix} & \vec{P}_{33} &= -T^{-1}\begin{pmatrix} s_{\zeta\zeta} \\ t_{\zeta\zeta} \\ u_{\zeta\zeta} \end{pmatrix} \end{aligned} \tag{4.65}$$

and the matrix T is defined as

$$T = \begin{pmatrix} s_\xi & s_\eta & s_\zeta \\ t_\xi & t_\eta & t_\zeta \\ u_\xi & u_\eta & u_\zeta \end{pmatrix} \tag{4.66}$$

The 18 coefficients of the six vectors $\vec{P}_{11}, \vec{P}_{12}, \vec{P}_{13}, \vec{P}_{22}, \vec{P}_{23}, \vec{P}_{33}$ are so-called control functions. Thus the 18 control functions are completely defined and easily computed for a given grid control map $\vec{s} = \vec{s}(\vec{\xi})$.

Finally, substituting $\phi \equiv \vec{x}$ in Eq. 4.63 and using the fact that $\Delta\vec{x} \equiv 0$, we arrive at the following equation:

$$a^{11}\,\vec{x}_{\xi\xi} + 2a^{12}\,\vec{x}_{\xi\eta} + 2a^{13}\,\vec{x}_{\xi\zeta} + a^{22}\,\vec{x}_{\eta\eta} + 2a^{23}\,\vec{x}_{\eta\zeta} + a^{33}\,\vec{x}_{\zeta\zeta} + \Delta\xi\vec{x}_\xi + \Delta\eta\vec{x}_\eta + \Delta\zeta\vec{x}_\zeta = 0 \tag{4.67}$$

The final form of the Poisson grid generation system can now be derived from this equation by substitution of Eq. 4.64, by multiplication with J^2, and by expressing the contravariant tensor components in the covariant tensor components according to Eq. 4.61. The result can be written as

$$\begin{aligned} & a^{11}\,\vec{x}_{\xi\xi} + 2a^{12}\,\vec{x}_{\xi\eta} + 2a^{13}\,\vec{x}_{\xi\zeta} + a^{22}\,\vec{x}_{\eta\eta} + 2a^{23}\,\vec{x}_{\eta\zeta} + a^{33}\,\vec{x}_{\zeta\zeta} \\ + & \left(a^{11}\,P^1_{11} + 2a^{12}\,P^1_{12} + 2a^{13}\,P^1_{13} + a^{22}\,P^1_{22} + 2a^{23}\,P^1_{23} + a^{33}\,P^1_{33}\right)\vec{x}_\xi \\ + & \left(a^{11}\,P^2_{11} + 2a^{12}\,P^2_{12} + 2a^{13}\,P^2_{13} + a^{22}\,P^2_{22} + 2a^{23}\,P^2_{23} + a^{33}\,P^2_{33}\right)\vec{x}_\eta \\ + & \left(a^{11}\,P^3_{11} + 2a^{12}\,P^3_{12} + 2a^{13}\,P^3_{13} + a^{22}\,P^3_{22} + 2a^{23}\,P^3_{23} + a^{33}\,P^3_{33}\right)\vec{x}_\zeta = 0 \end{aligned} \tag{4.68}$$

with

$$\begin{aligned} a^{11} &= a_{22}\,a_{33} - a^2_{23} \quad & a^{12} &= a_{13}\,a_{23} - a_{12}\,a_{33} \quad & a^{13} &= a_{12}\,a_{23} - a_{13}\,a_{22} \\ a^{22} &= a_{11}\,a_{33} - a^2_{13} \quad & a^{23} &= a_{13}\,a_{12} - a_{11}\,a_{23} \quad & a^{33} &= a_{11}\,a_{22} - a^2_{12} \end{aligned} \tag{4.69}$$

and

$$\begin{aligned} a_{11} &= \left(\vec{x}_\xi, \vec{x}_\xi\right) \quad & a_{12} &= \left(\vec{x}_\xi, \vec{x}_\eta\right) \quad & a_{13} &= \left(\vec{x}_\xi, \vec{x}_\zeta\right) \\ a_{22} &= \left(\vec{x}_\eta, \vec{x}_\eta\right) \quad & a_{23} &= \left(\vec{x}_\eta, \vec{x}_\zeta\right) \quad & a_{33} &= \left(\vec{x}_\zeta, \vec{x}_\zeta\right) \end{aligned} \tag{4.70}$$

This equation, together with the expressions for the control functions P^k_{ij} given by Eq. 4.65, forms the 3D grid generation system. For a given grid control map, so that the 18 control functions in Eq. 4.68 are given functions of (ξ, η, ζ), boundary conforming grids in the interior of domain D are computed by solving this quasi-linear system of elliptic partial differential equations with prescribed boundary grid points as Dirichlet boundary conditions.

The construction of appropriate grid control maps for 3D domains is less well developed than for 2D domains. In [25], a grid control map has been proposed which works surprisingly well for many applications. The grid control map is the 3D extension of the 2D grid control map defined by Eq. 4.27 and Eq. 4.28. The map $\vec{s}$: $C \mapsto P$ is defined by

$$s = s_{E_1}(\xi)(1-t)(1-u) + s_{E_2}(\xi)t(1-u) + s_{E_3}(\xi)(1-t)u + s_{E_4}(\xi)tu \tag{4.71}$$

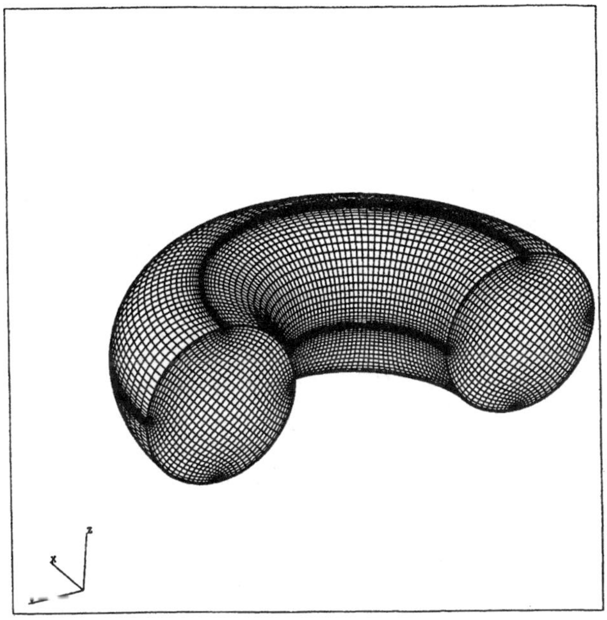

FIGURE 4.58 Boundary surface grid of a semi-torus.

$$t = t_{E_5}(\eta)(1-s)(1-u)+t_{E_6}(\eta)s(1-u)+t_{E_7}(\eta)(1-s)u+t_{E_8}(\eta)su \tag{4.72}$$

$$u = u_{E_9}(\zeta)(1-s)(1-t)+u_{E_{10}}(\zeta)s(1-t)+u_{E_{11}}(\zeta)(1-s)t+u_{E_{12}}(\zeta)st \tag{4.73}$$

where the twelve edge functions $s_{E_1}, \ldots, u_{E_{12}}$ measure the normalized arc length along the corresponding twelve edges of domain D (see Figure 4.4).

Equation 4.71 implies that a grid plane ξ = const. is mapped to the parameter space P as a bilinear surface: s is a bilinear function of t and u. Similarly, Eq. 4.72 and Eq. 4.73 imply that grid planes η = const. and ζ = const. are also mapped to the parameter space P as bilinear surfaces. For a given computational coordinate (ξ, η, ζ) the corresponding (s, t, u) value is found as the intersection point of three bilinear surfaces. Newton iteration is used to compute the intersection points. It can be easily verified that two bilinear surfaces corresponding to two different ξ-values will never intersect in parameter space P. The same is true for two different η or ζ values. This observation indicates that the grid control map is a differentiable one-to-one mapping.

An illustration of a volume grid computed by solving Eq. 4.68, with the grid control map defined by Eq. 4.71–4.73, is shown in Figures 4.58–4.61. The domain is a semi-torus. The prescribed boundary grid points on the surface of the semi-torus are shown in Figure 4.58. Figure 4.59 shows the surface grid on the two exterior circular grid planes. Figure 4.60 shows the computed interior grid depicted on some internal circular planes. Figure 4.61 shows the computed interior grid on the circular plane exactly halfway inside the torus. The mesh spacing of the interior grid is excellent despite the concave boundary. The angles between the interior grid lines and the boundary surface are reasonable but no longer orthogonal. This is not surprising, because the grid control map provides no control about the angle distribution between interior grid lines and the boundary of the domain.

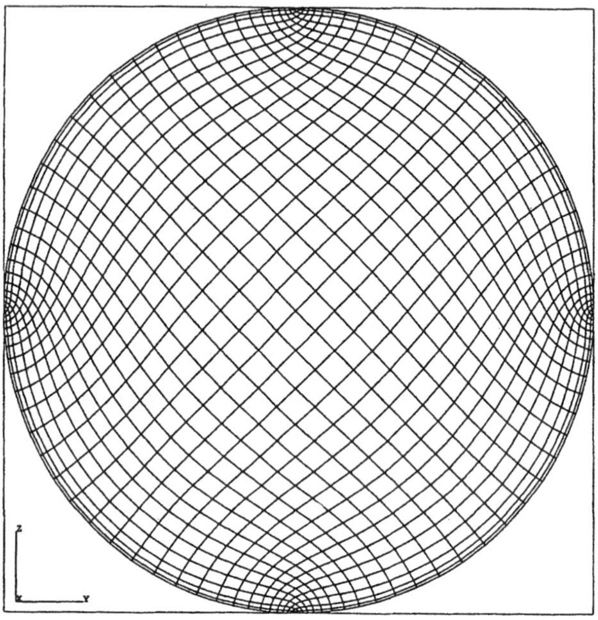

FIGURE 4.59 Surface grid on the two exterior circular planes.

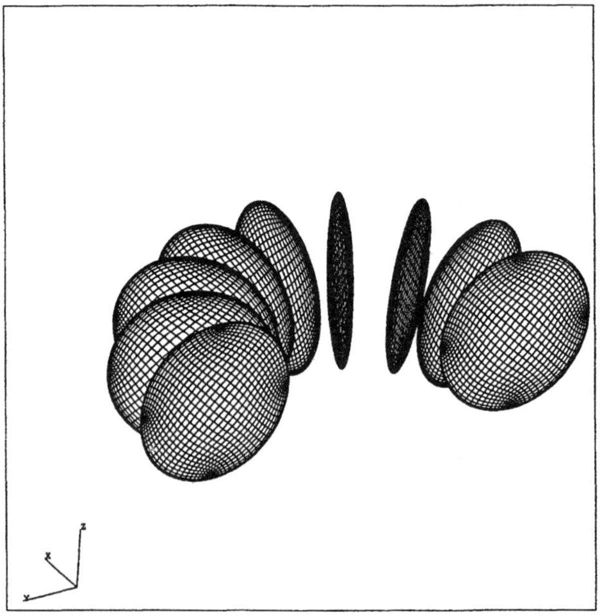

FIGURE 4.60 Interior grid planes inside the torus.

4.5 Research Issues and Summary

The grid generation systems of elliptic quasi-linear second-order partial differential equations are the familiar so-called Poisson systems with control functions to be specified. In this chapter, a Poisson system is considered as a system of partial differential equations that the composition of a grid control map and the inverse of a harmonic map has to obey. The control functions in the Poisson system are then completely defined by the grid control map. Boundary conforming grids in physical space are computed by solving the Poisson system with control functions specified by a grid control map.

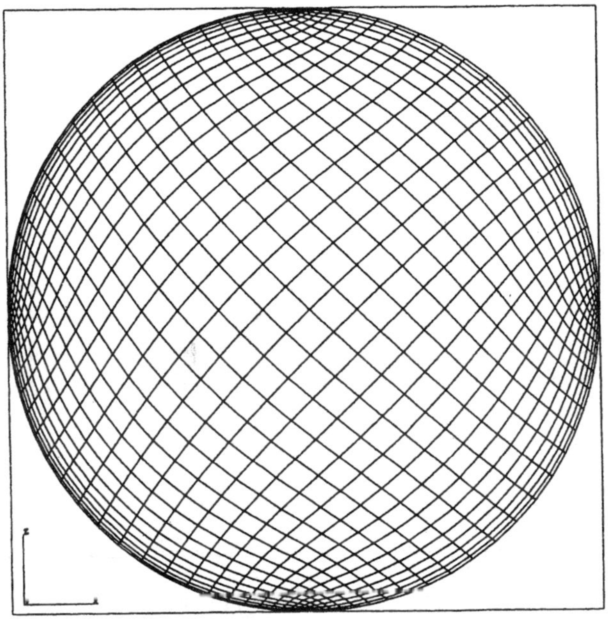

FIGURE 4.61 Interior grid inside the torus on a circular plane halfway between the two exterior circular planes.

One of the main advantages of this approach is that the method is noniterative. If an appropriate grid control map has been constructed, then the corresponding grid control functions of the Poisson system are computed and their values remain unchanged during the solution of the Poisson system. Another advantage is that the construction of an appropriate grid control map can be considered as a numerical implementation of the constructive proof for the existence of the desired grid in physical space. If the grid control map is one-to-one, then the composition of the grid control map and the inverse of the harmonic maps exist so that the solution of the Poisson system is well-defined.

In two dimensions, boundary orthogonality is obtained by applying Dirichlet–Neumann boundary conditions for the harmonic map. In that case, the harmonic map is quasi-conformal. This property shows the relation with orthogonal grid generation.

The use of harmonic maps and grid control maps for surface grid generation is also shortly described. The two-dimensional Poisson systems can be directly extended to surface grid generation on minimal surfaces (soap films). The extension to volume grid generation is also given.

The construction of appropriate grid control maps such that the corresponding grid in physical space has desired properties is the main issue of this chapter. The chosen examples concern mainly simple well-defined geometries so that the reader is able to recompute the grids. However, the elliptic grid generation methods described in this chapter have been implemented in ENGRID, NLR's multi-block grid generation code [26,27,2], and are nowadays used on a routine basis to construct Euler or Navier–Stokes grids in blocks and block-faces with complex geometrical shapes. The construction of appropriate grid control maps for 3D domains is less well developed than for 2D domains and surfaces. Further investigation is expected in this direction.

4.6 Further Information

The book of Thompson, Warsi, and Mastin [30] is still the best introduction to elliptic grid generation systems. A more recent book is Carey [4]. Also the book of Knupp and Steinberg [14] is a valuable source about the fundamentals of structured grid generation and related topics, like tensor analysis and differential geometry. The book of Kreyszig [15] and Dierkes, et.al. [7] are excellent textbooks about differential geometry and tensor analysis.

The proceedings of the grid generation conferences [1,22,33], the VKI lecture series about grid generation [31,32], and the NASA conference publications [5,6] contain a lot of useful information about the application of elliptic grid generation systems, often embedded in multiblock grid generation systems.

The *Journal of Computational Physics* provides many good more or less fundamental articles about elliptic grid generation systems.

References

1. **Arcilla, A. S.**, (Ed.) et al., Numerical grid generation in computational fluid dynamics and related fields, *Proceedings of the 3rd International Conference*, Barcelona, Spain, North-Holland, 1991.
2. **Boerstoel, J. W., Kassies, A., Kok, J. C., and Spekreijse, S. P.**, ENFLOW — a full-functionality system of CFD codes for industrial Euler/Navier–Stokes flow computations, *Proceedings of 2nd International Symposium on Aeronautical Science and Technology*, IASTTI'96, Jakarta, Indonesia, 1996.
3. **Bjorstad, P. E.**, Numerical solution of the biharmonic equation, Ph.D. thesis, Stanford University, 1980.
4. **Carey, G. F.**, *Computational Grids*, Taylor & Francis, 1997.
5. **Chawner, J. R. and Steinbrenner, J. P.**, Automatic Structured Grid Generation using GRIDGEN, *Surface Modeling, Grid Generation, and Related Issues in Computational Fluid Dynamic (CFD) Solutions*, Choo, Y. K., (Ed.), NASA-CP-3291, 1995, pp. 463–476.
6. **Choo, Y. K.**, (Ed.), *Surface Modeling, Grid Generation, and Related Issues in Computational Fluid Dynamic (CFD) Solutions*, NASA-CP-329 1, Proceedings of a workshop held at NASA Lewis, Cleveland, Ohio, 1995.
7. **Dierkes, U., Hildebrandt, S., Kuster, A., and Wohlrab, O.**, Minimal surfaces I, *Grundlehren der Mathematischen Wissenschaffen* 295, Springer Verlag, Berlin, 1991.
8. **Duraiswami, R. and Prosperetti, A.**, Orthogonal mapping in two dimensions, *J.Comput. Phys.*, 98, pp. 254–268, 1992.
9. **Eça, L.**, 2D orthogonal grid generation with boundary point distribution control, *J. Comput. Phys.*, 125, pp. 440–453, 1996.
10. **Farin, G.**, *Curves and Surfaces for Computer Aided Geometric Design — A Practical Guide*. Academic Press, San Diego, 1990.
11. **Henrici, P.**, *Applied and Computational Complex Analysis*, Vol. 3, Wiley, New York, 1986.
12. **Hsu, K. and Lee, S.L.**, A numerical technique for two-dimensional grid generation with grid control at all of the boundaries, *J.Comput. Phys.*, 96, pp. 451–469, 1991.
13. **Kang, I.S. and Leal, L.G.**, Orthogonal grid generation in a 2D domain via the boundary integral technique, *J. Comput. Phys.*, 102, pp. 78–87, 1992.
14. **Knupp, P. and Steinberg, S.**, *Fundamentals of Grid Generation*, CRC Press, Boca Raton, FL, 1993.
15. **Kreyszig, E.**, *Differential Geometry*, Dover, New York, 1991.
16. **Mastin, C. W.**, Multilevel elliptic smoothing of large three-dimensional grids, *Surface Modeling, Grid Generation, and Related Issues in Computational Fluid Dynamic (CFD) Solutions*, Choo, Y. K, (Ed.), NASA-CP-3291, pp. 689–696, 1995.
17. **Oh, H. J. and Kang, I. S.**, A non-iterative scheme for orthogonal grid generation with control function and specified boundary correspondence on three sides, *J. Comput. Phys.*, 112, pp. 138–148, 1994.
18. **Roache, P. J. and Steinberg, S.**, A new approach to grid generation using a variational formulation, *AIAA 7th Computational Fluid Dynamics Conference*, AIAA paper 85-1527, pp. 360–370, 1985.
19. **Ryskin, G. and Leal, L. G.**, Orthogonal mapping, *J. Comput. Phys.*, 50, pp. 71–100, 1983.
20. **Smith, R. E.**, (Ed.), Software systems for surface modeling and grid generation, NASA-CP-3143, Proceedings of a workshop held at NASA Langley, Hampton, VA, 1992.
21. **Sonar, T.**, Grid generation using elliptic partial differential equations, *DFVLR Forschungsbericht* 89–15, 1989.

22. **Soni, B. K.**, (Ed.), et al., Numerical grid generation in computational field simulations, *Proceedings of the 5th International Conference*, Mississippi State University, NSF Engineering Research Center for Computational Field Simulation, 1996.
23. **Sorenson, R. L. and Alter, S. J.**, 3D GRAPE/AL: The Ames/Langley technology update, *Surface Modeling, Grid Generation, and Related Issues in Computational Fluid Dynamic (CFD) Solutions*, Choo, Y. K., (Ed.), NASA-CP-3291, pp 447–462, 1995.
24. **Sorenson, R. L. and Steger, J. L.**, Numerical Generation of Two-Dimensional Grids by Use of Poisson Equations with Grid Control, *Numerical Grid Generation Techniques*, Smith, R. E., (Ed.), NASA-CP-2166, pp. 449–461, 1980.
25. **Spekreijse, S. P.**, Elliptic grid generation based on Laplace equations and algebraic transformations, *J. Comput. Phys.*, 118, pp. 38–61, 1995.
26. **Spekreijse, S. P. and Boerstoel, J. W.**, Multiblock grid generation, Part 1: Elliptic grid generation methods for structured grids, *Computational Fluid Dynamics*, VKI-Lecture-Series 1996-06, Deconinck, H., (Ed.), Von Karman Institute for Fluid Dynamics, pp. 1–39, 1996.
27. **Spekreijse, S. P. and Boerstoel, J. W.**, Multiblock Grid Generation, Part 2: Multiblock Aspects, *Computational Fluid Dynamics*, VKI-Lecture-Series 1996-06, Deconinck, H., (Ed.), Von Karman Institute for Fluid Dynamics, pp. 1–48, 1996.
28. **Takahashi, S. and Eiseman P. R.**, Adaptive grid movement with respect to boundary curvature, *Numerical Grid Generation in Computational Fluid Dynamics and Related Fields*, 4th International Grid Conference, Weatherill N. P., et al. (Eds.), Pineridge Press Limited, Swansea, Wales (UK),1994, p. 563.
29. **Thompson, J. E.**, A composite grid generation code for general 3D regions — the EAGLE code, *AIAA Journal*, 1988, 26, Vol.3, p 271.
30. **Thompson, J. F., Warsi, Z.U.A., and Mastin, C.W.**, *Numerical Grid Generation: Foundations and Applications*, Elsevier, New York, 1985.
31. **Weatherill, N. P.**, (Ed.), Grid generation, VKI-Lecture-Series 1994-02, Von Karman Institute for Fluid Dynamics, 1994.
32. **Weatherill, N. P.**, (Ed.), Numerical grid generation, VKI-Lecture-Series 1990-06, Von Karman Institute for Fluid Dynamics, 1990.
33. **Weatherill, N. P.**, (Ed.), et al., Numerical Grid Generation in Computational Fluid Dynamics and Related Fields, *Proceedings of the 4th International Conference*, Swansea, Wales, Pineridge Press, 1994.
34. **Winslow, A.**, Numerical solution of the quasilinear poisson equations in a nonuniform triangle mesh, *J. Comput. Phys.*, 2, pp. 149–172, 1967.
35. **Warsi, Z. U. A.**, Basic differential models for coordinate generation, *Proceedings Numerical Grid Generation*. Thompson, J. F. (Ed.), North-Holland, Amsterdam, pp. 41–77, 1982.
36. **Yamaguchi, F.**, *Curves and Surfaces in Computer Aided Geometric Design*, Springer Verlag, Berlin, 1988.

5

Hyperbolic Methods for Surface and Field Grid Generation

William M. Chan

5.1 Introduction

Two of the most widely used classes of methods for structured grid generation are algebraic methods and partial differential equation (PDE) methods. The PDE methods can be classified into three types: elliptic, parabolic, and hyperbolic. This chapter will focus on the use of hyperbolic partial differential equation methods for structured surface grid generation and field grid generation.

In hyperbolic grid generation, a mesh is generated by propagating in the normal direction from a known level of points to a new level of points, starting from a given initial state. For two-dimensional (2D) field grid generation and for surface grid generation, the initial state is a curve. For three-dimensional (3D) field grid generation, the initial state is a surface. The governing equations are typically derived from grid angle and grid cell size constraints. Local linearization of these equations allows a mesh to be generated by marching from a known state to the next. The total marching distance from the initial state and the marching step sizes at each level can be prescribed based on requirements of the specific application.

When generating 2D field grids or surface grids using algebraic interpolation or elliptic methods, grid points on all four boundaries of a nonperiodic mesh have to be specified prior to the generation of the interior points. Thus, exact control of the mesh boundaries is inherent with such methods. When using

0-8493-2687-7/99/$0.00+$.50

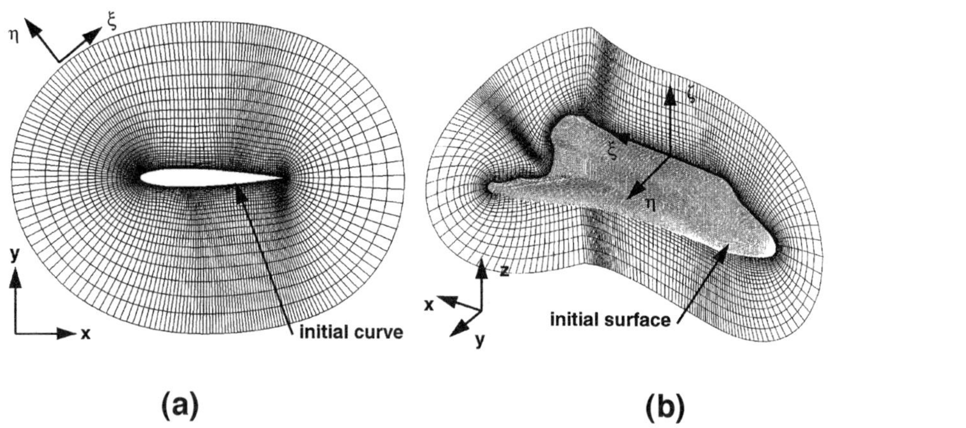

FIGURE 5.1 (a) Example of hyperbolic field grid in two dimensions (airfoil O-grid). The marching direction is given by η. (b) Example of hyperbolic field grid in three dimensions (simplified Space Shuttle Orbiter). The marching direction is given by ζ.

hyperbolic methods, only the initial state can be exactly prescribed as one of the boundaries of the mesh. Exact specification of the side and outer boundaries is not possible with a one sweep marching scheme but limited control is achievable. When exact control of all the mesh boundaries is not needed, less work is required using hyperbolic methods since only one boundary has to be prescribed instead of four. The reduction in effort becomes more significant in 3D field grid generation where only the initial surface needs to be prescribed instead of the six boundary surfaces required for a nonperiodic grid using algebraic interpolation or elliptic methods. Excellent orthogonality and grid clustering characteristics are inherently provided by hyperbolic methods. Since a marching scheme is used, the grid generation time can be one to two orders of magnitude faster than typical elliptic methods.

In a structured grid approach for solving field simulation problems for complex configurations, the complex domain is typically decomposed into a number of simpler subdomains. A grid is generated for each subdomain, and communications between subdomains are managed by a domain connectivity program. The two main methods for domain decomposition are the patched grid approach [Rai, 1986]; [Thompson, 1988] and the Chimera overset grid approach [Steger, Dougherty, and Benek, 1983]. In the patched grid approach, neighboring grids are required to abut each other. Since exact specification of all the grid boundaries is needed, algebraic and elliptic methods are best suited for generating grids for this scheme. In the overset grid approach, neighboring grids are allowed to overlap with each other. This freedom is particularly well suited to hyperbolic grid generation methods. Thus, hyperbolically generated grids are heavily used in most overset grid computations on complex geometries (see Section 5.5 for a sample list of applications; also see Chapter 11).

Field grid generation in 2D using hyperbolic equations was introduced by Starius [1977] and Steger and Chaussee [1980]. A 2D field grid in the Cartesian x-y plane is generated by marching from an initial curve in the plane (see Figure 5.1a). Related work in two dimensions includes that by Kinsey and Barth [1984] for implicitness enhancements, Cordova and Barth [1988] for non-orthogonal grid control, Klopfer [1988] for adaptive grid applications, and Jeng, Shu and Lin [1995] for internal flow problems. Exact prescription of the side boundaries can be achieved by performing elliptic iterations at the end of each step [Cordova, 1991]. Extension of the hyperbolic grid generation scheme to 3D was presented in [Steger and Rizk, 1985]. A 3D volume grid is generated by marching from an initial surface (see Figure 5.1b). Enhancements to the robustness of the basic field grid generation scheme were developed by Chan and Steger [1992]. Hybrid schemes formed by mixing hyperbolic with elliptic and parabolic equations have been used by Nakamura [1987], Steger [1989a], Takanashi and Takemoto [1993].

Surface grid generation using hyperbolic equations was introduced by Steger [1989b]. A surface grid is generated by marching from an initial curve that lies on a reference surface (see Figure 5.2). The basic

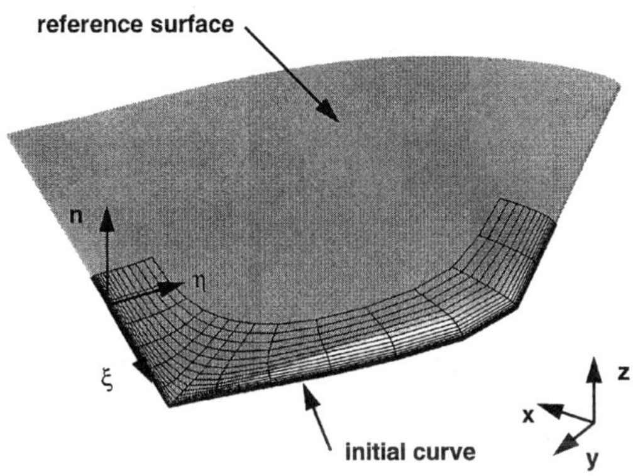

FIGURE 5.2 Example of hyperbolic surface grid (cap grid over wing tip region). The local initial curve direction, local marching direction and local surface normal are indicated by ξ and η, and **n**, respectively.

scheme allowed only a single rectangular array of quadrilateral cells (single panel network) to be the reference surface. In practical situations, a complex surface geometry is typically described by numerous surface patches where each patch may be a nonuniform rational B-spline (NURBS) surface or some other geometric entity (cf. Part III). For hyperbolic surface grid generation to be applicable in such cases, the scheme has to be able to grow surface grids that can span over multiple patches. A first step toward this goal was made by extending the basic scheme to generate surface grids that can span over multiple panel networks [Chan and Buning, 1995]. Each panel network is used to model a surface patch where the point distribution on the panel network can be made as fine as necessary from the original patch definition.

5.2 Hyperbolic Field Grid Generation

In hyperbolic field grid generation, a field grid is generated in two or three dimensions by marching from a specified initial state. A new grid level is produced by linearizing about the current known level and solving the governing equations. In two dimensions (2D), the initial state is a curve on the Cartesian *x-y* plane. In three dimensions (3D), the initial state is a surface in three-dimensional space. For practical applications, the initial state is typically chosen to coincide with the configuration body surface to produce body-fitted grids.

5.2.1 Governing Equations for Hyperbolic Field Grid Generation

The governing equations presented below are derived from grid orthogonality and cell size constraints. By demanding that the marching direction be orthogonal to the current known state, an orthogonality relation can be derived for 2D, and two orthogonality relations can be derived for 3D. The system of equations can be closed by a cell area/volume constraint where the local cell areas/volumes are user-specified. A convenient method for specifying cell areas/volumes is described in Section 5.2.3. Other formulations of the governing equations have been used. For example, locally nonorthogonal grids in 2D can be generated by the introduction of an angle source term [Cordova and Barth, 1988]; an arc length constraint can be used instead of the cell area constraint in 2D [Steger and Chaussee, 1980].

In 2D, consider generalized coordinates $\xi(x,y)$ and $\eta(x,y)$. The 2D field grid generation equations can be written as

$$x_\xi x_\eta + y_\xi y_\eta = 0 \tag{5.1a}$$

$$x_\xi y_\eta - y_\xi x_\eta = \Delta A \tag{5.1b}$$

where ΔA is the user-specified local cell area. The initial state is chosen to be at the first η = const. curve.

In 3D, consider generalized coordinates $\xi(x, y, z)$, $\eta(x, y, z)$, and $\zeta(x, y, z)$ corresponding to grid indices j, k, and l, respectively. The 3D field grid generation equations can be written as

$$\vec{r}_\xi \cdot \vec{r}_\zeta = x_\xi x_\zeta + y_\xi y_\zeta + z_\xi z_\zeta = 0 \tag{5.2a}$$

$$\vec{r}_\eta \cdot \vec{r}_\zeta = x_\eta x_\zeta + y_\eta y_\zeta + z_\eta z_\zeta = 0 \tag{5.2b}$$

$$\vec{r}_\zeta \cdot \left(\vec{r}_\xi \times \vec{r}_\eta\right) = x_\xi y_\eta z_\zeta + x_\zeta y_\xi z_\eta + x_\eta y_\zeta z_\xi - x_\xi y_\zeta z_\eta - x_\eta y_\xi z_\zeta - x_\zeta y_\eta z_\xi = \Delta V \tag{5.2c}$$

where $\vec{r} = (x, y, z)^T$ and ΔV is the user-specified local cell volume. The initial state is chosen to be at the first ζ = const. surface.

5.2.2 Numerical Solution of Hyperbolic Field Grid Generation Equations

Local linearlization of Eq. 5.1a,b and Eq. 5.2a,b,c results in systems of grid generation equations in 2D and 3D, respectively. Such systems have been shown to be hyperbolic for marching in η in 2D and for marching in ζ in 3D (see [Steger, 1991]). The system of grid generation equations is solved with a noniterative implicit finite difference scheme.

Second-order central differencing is used in the ξ direction in 2D and in the ξ and η directions in 3D. In these directions, appropriate boundary conditions have to be employed (see Section 5.2.4), and smoothing terms have to be added to provide numerical stability (see Section 5.2.5). A first-order implicit scheme is used in the marching direction. An unconditionally stable implicit scheme has the advantage that the marching step size can be selected based only on considerations of grid accuracy. At each marching step, linearization is performed about the previous marching step. More details of the numerical scheme are now presented.

Local linearization of Eq. 5.1 about a known state results in the system of grid generation equations

$$A_0 \vec{r}_\xi + B_0 \vec{r}_\eta = \vec{f} \tag{5.3}$$

where the subscript 0 denotes evaluation at the known state 0, and

$$A = \begin{bmatrix} x_\eta & y_\eta \\ y_\eta & -x_\eta \end{bmatrix} \quad B = \begin{bmatrix} x_\xi & y_\xi \\ -y_\xi & x_\xi \end{bmatrix} \quad \vec{f} = \begin{bmatrix} 0 \\ \Delta A + \Delta A_0 \end{bmatrix} \tag{5.4}$$

The matrix B_0^{-1} exists if $x_\xi^2 + y_\xi^2 \neq 0$. Moreover, $B_0^{-1}A_0$ is a symmetric matrix. Hence, the system in Eq. 5.3 is hyperbolic for marching in η. The numerical solution of the 2D equations follows closely that of the 3D equations. Only the details for the 3D equations are given below.

Local linearization of Eq. 5.2 about a known state 0 results in the system of grid generation equations

$$A_0 \vec{r}_\xi + B_0 \vec{r}_\eta + C_0 \vec{r}_\zeta = \vec{e} \tag{5.5}$$

where

$$A = \begin{bmatrix} x_\zeta & y_\zeta & z_\zeta \\ 0 & 0 & 0 \\ \left(y_\eta z_\zeta - y_\zeta z_\eta\right) & \left(x_\zeta z_\eta - x_\eta z_\zeta\right) & \left(x_\eta y_\zeta - x_\zeta y_\eta\right) \end{bmatrix} \tag{5.6a}$$

$$B=\begin{bmatrix} 0 & 0 & 0 \\ x_\zeta & y_\zeta & z_\zeta \\ \left(y_\zeta z_\xi - y_\xi z_\zeta\right) & \left(x_\xi z_\zeta - x_\zeta z_\xi\right) & \left(x_\zeta y_\xi - x_\xi y_\zeta\right) \end{bmatrix} \tag{5.6b}$$

$$C=\begin{bmatrix} x_\xi & y_\xi & z_\xi \\ x_\eta & y_\eta & z_\eta \\ \left(y_\xi z_\eta - y_\eta z_\xi\right) & \left(x_\eta z_\xi - x_\xi z_\eta\right) & \left(x_\xi y_\eta - x_\eta y_\xi\right) \end{bmatrix} \tag{5.6c}$$

and $\vec{e} = (0, 0, \Delta\dot{V} + 2\Delta V_0)^T$. The matrix C_0^{-1} exists unless $(\Delta V_0) \to 0$. Moreover, $C_0^{-1} A_0$ and $C_0^{-1} B_0$ are symmetric matrices and hence the system of equations is hyperbolic for marching in ζ.

Eq. 5.5 is solved numerically by a noniterative implicit marching scheme in ζ. Additional smoothing and implicitness are attained by differencing $\nabla_\zeta \vec{r} = \vec{F}$ as $\vec{r}_{l+1} - \vec{r}_l = (1 + \theta)\vec{F}_{l+1} - \theta\vec{F}_l$, where values of the implicitness factor θ range between 0 and 4 [Kinsey and Barth, 1984]. After approximate factorization and addition of numerical smoothing terms, the equations to be solved can be written as

$$\begin{aligned} &\left[I+\left(1+\theta_\eta\right)C_l^{-1} B_l \delta_\eta - \varepsilon_{i\eta}(\Delta\nabla)_\eta\right]\left[I+\left(1+\theta_\xi\right)C_l^{-1} A_l \delta_\xi - \varepsilon_{i\xi}(\Delta\nabla)_\xi\right]\left(\vec{r}_{l+1}-\vec{r}_l\right) \\ &= C_l^{-1}\vec{g}_{l+1} - \left[\varepsilon_{e\xi}(\Delta\nabla)_\xi + \varepsilon_{e\eta}(\Delta\nabla)_\eta\right]\vec{r}_l \end{aligned} \tag{5.7}$$

with

$$\delta_\xi \vec{r}_j = \frac{\vec{r}_{j+1} - \vec{r}_{j-1}}{2} \qquad \delta_\eta \vec{r}_k = \frac{\vec{r}_{k+1} - \vec{r}_{k-1}}{2} \tag{5.8a}$$

$$(\Delta\nabla)_\xi \vec{r}_j = \vec{r}_{j+1} - 2\vec{r}_j + \vec{r}_{j-1} \qquad (\Delta\nabla)_\eta \vec{r}_k = \vec{r}_{k+1} - 2\vec{r}_k + \vec{r}_{k-1} \tag{5.8b}$$

where $\vec{g}_{l+1} = (0, 0, \Delta\dot{V}_{l+1})^T$. I is the identity matrix; θ_ξ, θ_η are the implicitness factors in ξ and η, respectively; $\varepsilon_{e\xi}$, $\varepsilon_{e\eta}$ are second-order explicit smoothing coefficients in ξ and η, respectively; $\varepsilon_i \approx 2\varepsilon_e$ in ξ and η; and the subscript l indicates the grid level in the marching direction. The smoothing coefficients can be chosen to be constants or made to vary spatially depending on local geometric demands. Proper choice of spatially varying smoothing coefficients can significantly enhance the robustness of the scheme in cases involving complex geometry (see Section 5.2.5 for more details). Only the indices that change are shown in Eq. 5.8, i.e., $\vec{r}_j \equiv \vec{r}_{j,k,l}$.

The coefficient matrices A_l, B_l and C_l contain derivatives in ξ, η, and ζ. The ξ- and η-derivatives are computed by central differencing, while the ζ-derivatives are obtained from Eq. 5.2 as a linear combination of ξ- and η-derivatives as follows:

$$\begin{pmatrix} x_\zeta \\ y_\zeta \\ z_\zeta \end{pmatrix} = \frac{\Delta V}{Det(C)}\begin{pmatrix} y_\xi z_\eta - y_\eta z_\xi \\ x_\eta z_\xi - x_\xi z_\eta \\ x_\xi y_\eta - x_\eta y_\xi \end{pmatrix} = C^{-1}\vec{g} \tag{5.9}$$

with $Det(C) = (y_\xi z_\eta - y_\eta z_\xi)^2 + (x_\eta z_\xi - x_\xi z_\eta)^2 + (x_\xi y_\eta - x_\eta y_\xi)^2$. In regions with sudden grid spacing jumps in the ξ or η direction, a more robust method for computing the ζ derivatives in described in [Chan and Steger, 1992].

Extra robustness at sharp convex corners can be achieved by demanding that the marching increment $\Delta\vec{r}_l = \vec{r}_{l+1} - \vec{r}_l$ at the corner point be the average of the marching increments of its neighbors. This can be achieved by solving the following averaging equation at the corner point:

$$\Delta\vec{r}_{j,k} = \frac{1}{2}\left(\mu_\xi + \mu_\eta\right)\Delta\vec{r}_{j,k} \tag{5.10}$$

where

$$\mu_\xi \Delta\vec{r}_{j,k} = \frac{1}{2}\left(\Delta\vec{r}_{j+1,k} + \Delta\vec{r}_{j-1,k}\right) \qquad \mu_\eta \Delta\vec{r}_{j,k} = \frac{1}{2}\left(\Delta\vec{r}_{j,k+1} + \Delta\vec{r}_{j,k-1}\right) \tag{5.11}$$

Approximate factorization of Eq. 5.10 gives

$$\left(I - \frac{1}{2}\mu_\xi\right)\left(I - \frac{1}{2}\mu_\eta\right)\Delta\vec{r} = 0 \tag{5.12}$$

which has the same form as the block tridiagonal matrix factors of the hyperbolic equations. A switch is made to solve Eq. 5.12 if a sharp convex corner exists in either the ξ or η direction. For example, the switch can be made if the external angle of the corner is greater than 240°. The averaging equation works particularly well if the surface grid spacings on the two sides of the corner are equal.

5.2.3 Specification of Cell Sizes

The local cell sizes (ΔA in 2D and ΔV in 3D) have to be specified at each point on each grid level as the mesh is marched from the initial state. There is no unique method for prescribing the cell size but a convenient method is described below. Other schemes for cell size specification are described in [Steger and Chaussee, 1980] and [Steger and Rizk, 1985].

For both 2D and 3D, a one-dimensional (1D) stretching function in the marching direction is specified by the user. The stretching function provides the step sizes to be used at each grid level in the marching direction. In 2D, ΔA is computed by the product of the local arc length and the marching step size at the current level. In 3D, ΔV is computed by the product of the local cell area and the marching step size at the current level. The step size specification via the stretching function provides good grid clustering control near the initial state which is usually chosen to be at a solid surface for fluid flow computations. Such clustering control is particularly important in viscous calculations.

The stretching function used is typically geometric or hyperbolic tangent, although any arbitrary stretching function can be employed (cf. Chapter 32). For both geometric and hyperbolic tangent stretching, the total marching distance and the number of points to be used in the marching direction have to be specified. The geometric stretching allows the prescription of grid spacing at one end of the domain only (usually at the initial state). The hyperbolic tangent stretching allows grid spacing at one or both ends of the domain to be specified. This is convenient when it is necessary to control the outer boundary grid spacing. Such a situation frequently arises in overlapping grid systems where it is desirable to have comparable grid spacings at the boundaries of neighboring grids.

For typical applications, the same initial/final spacings and marching distance are applied at every grid point on the initial state. However, certain applications require the use of different initial/final spacings and marching distances at different points on the initial state (see three-element airfoil example in Section 5.5). A convenient method for specifying the variable grid spacings and marching distances is to prescribe these parameters at key control points on the initial state, and then use interpolation to provide their values at the remaining points.

Grid smoothness can be enhanced to a certain degree by performing smoothing steps on the prescribed cell sizes. This has the effect of making the cell sizes more uniform which is typically a desirable

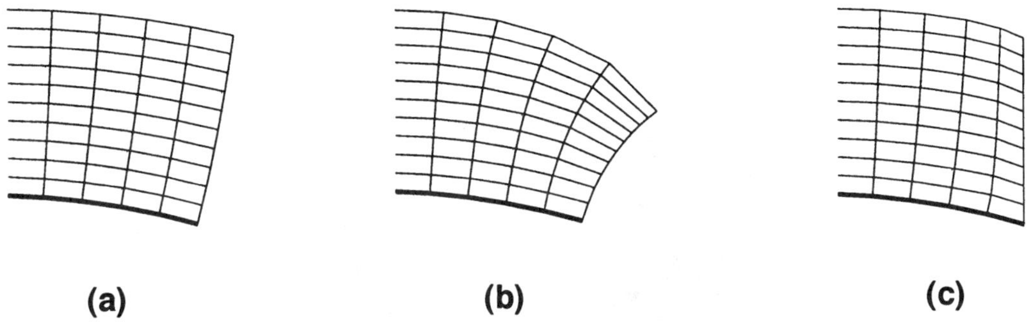

FIGURE 5.3 A field grid slice with (a) a free floating boundary, (b) a free floating boundary with splay, and (c) a constant plane boundary (▬ initial curve).

characteristic in the far field regions of a grid. For example, a smoothed cell volume $\Delta\bar{V}_{j,k,l}$ in 3D can be computed as

$$\Delta\bar{V}_{j,k,l} = (1-\nu_a)\Delta V_{j,k,l} + \frac{\nu_a}{4}\left(\Delta V_{j+1,k,l} + \Delta V_{j-1,k,l} + \Delta V_{j,k+1,l} + \Delta V_{j,k-1,l}\right) \tag{5.13}$$

where this is applied one or more times under each marching step. A typical value of ν_a that has been employed is 0.16.

5.2.4 Boundary Conditions

Numerical boundary conditions have to be supplied in ξ for 2D cases, and in ξ and η for 3D cases. The boundary conditions used are dictated by the topology of the specified initial state or by the desired boundary behavior of the grid being generated. For example, a periodic initial curve in 2D demands the use of a periodic boundary condition in ξ. For a nonperiodic initial curve in 2D, the user has several choices in influencing the behavior of the grid side boundaries emanating from the two endpoints of the initial curve. The boundaries can be allowed to float freely, splay outward, or made to be at a constant Cartesian plane station (see Figure 5.3).

An implicit boundary scheme is used to implement the above boundary conditions. A periodic solver is used to invert the left-hand-side factor in Eq. 5.7 that corresponds to a periodic direction. A mixed zeroth- and first-order extrapolation scheme is used for the free-floating and splay conditions. For example, the dependent variable $\Delta\vec{r} = (\Delta x, \Delta y, \Delta z)^T = \vec{r}_{l+1} - \vec{r}_l$ at the $j = 1$ boundary can be made to satisfy

$$(\Delta\vec{r})_{j=1} = (\Delta\vec{r})_{j=2} + \varepsilon_x\left[(\Delta\vec{r})_{j=2} - (\Delta\vec{r})_{j=3}\right] \tag{5.14}$$

where $0 \le \varepsilon_x \le 1$ is the extrapolation factor. The appropriate elements at the end points of the block tridiagonal matrix on the left-hand side of Eq. 5.7 are modified by Eq. 5.14. A free-floating condition is achieved by setting $\varepsilon_x = 0$. Increasing ε_x from zero has the effect of splaying the boundary of the field grid away from the grid interior. A constant plane condition in x, y, or z can be imposed by simply setting the appropriate component of $\Delta\vec{r}$ to zero. For example, a constant ξ plane condition at the $j = 1$ boundary is set by imposing $(\Delta x, \Delta y, \Delta z)^T_{j=1} = (0, \Delta y, \Delta z)^T_{j=2}$.

In 3D, more complicated topologies are possible with a surface as the initial state. The surface may be

1. Nonperiodic in both ξ and η directions,
2. Periodic in one direction and nonperiodic in the other direction (cylinder topology),
3. Periodic in both directions (torus topology).

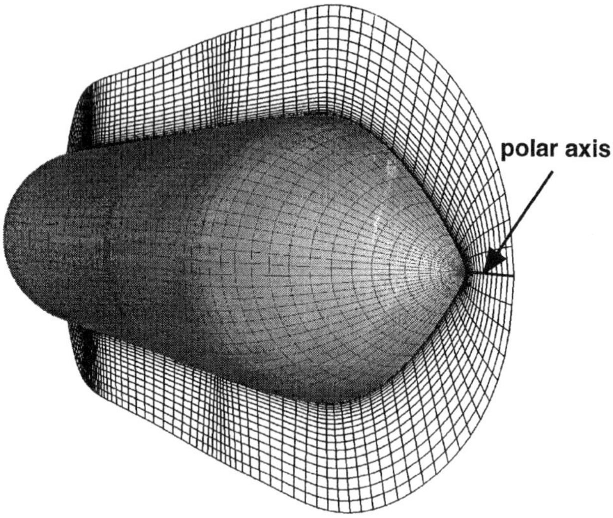

FIGURE 5.4 Surface grid with a singular axis point and slices of volume grid with a polar axis emanating from the singular axis point.

At a nonperiodic boundary, the same nonperiodic boundary schemes may be applied as in the 2D case (see Figure 5.3). Furthermore, singularities may be present at a surface grid boundary such as a singular axis point or a collapsed edge. Special numerical boundary treatment is needed at these boundaries. A singular axis point is a surface grid boundary where all the points are coincident. The volume grid contains a polar axis emanating from the axis point on the surface grid (see Figure 5.4). A collapsed edge condition is sometimes applied at a wing tip under a C-mesh or O-mesh topology. The C-type or O-type grid lines on the wing surface grid collapse to zero thickness at the wing tip to form a collapsed edge. Figure 5.5 shows a collapsed edge case for a C-mesh of a wing. The slice of the volume grid emanating from the collapsed edge forms a singular sheet ($k = kmax$ slice in Figure 5.5). Further illustration of different boundary conditions in 3D are shown in [Chan, Chiu, and Buning, 1993].

5.2.5 Grid Smoothing Mechanisms

There are three mechanisms through which smoothing is supplied to a grid generated with the scheme described above. All three mechanisms can be controlled by the user. The first is through the implicitness factors θ_ξ and θ_η in Eq. 5.7. Values of these parameters in the range 1–4 are mildly effective in preventing crossing of grid lines in concave corner regions in the ξ and η directions, respectively. The second smoothing mechanism is introduced by the number of times the specified cell areas/volumes are smoothed as described in Section 5.2.3. This has the effect of spreading clustered grid lines apart so that the cells sizes tend towards a uniform distribution as the number of smoothing steps is increased. The strongest and the most important is the third smoothing mechanism governed by the second-order smoothing coefficients in Eq. 5.7. These are discussed in more detail below.

The second-order smoothing applied in Eq. 5.7 serves to provide numerical dissipation needed for the central differencing scheme. A direct effect of this smoothing term is the enhancement of grid smoothness, but at the same time, a reduction in grid orthogonality also occurs. For a complex geometry, it is clear that different regions of the field grid require different amounts of added numerical smoothing. A low amount of smoothing is desired in regions where grid orthogonality should dominate. This is typically needed in regions near the body surface and in low curvature regions of the geometry. In concave regions of the surface, a high amount of smoothing is needed to prevent grid lines from crossing. A spatially variable dissipation coefficient based on the above attributes was designed and shown to work well for

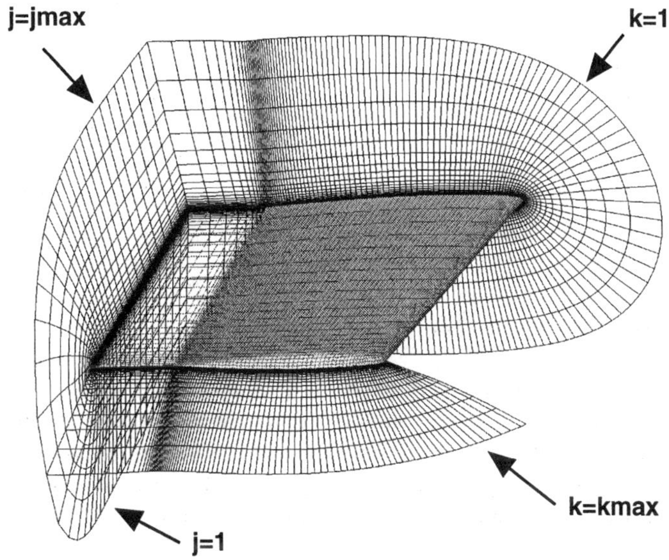

FIGURE 5.5 Surface grid wand slices of volume grid near a collapsed edge for a C-mesh topology. The surface grid has *jmax* by *kmax* points in the *j* and *k* directions with the collapsed edge at *k*=*kmax*.

a wide variety of cases [Chan and Steger, 1992]. Essential highlights of the dissipation model are discussed below. The original reference can be consulted for further details.

Let D_e be the explicit second-order dissipation added to the right-hand side of Eq. 5.7 given by

$$D_e = -\left[\varepsilon_{e\xi}(\Delta\nabla)_\xi + \varepsilon_{e\eta}(\Delta\nabla)_\eta\right]\vec{r}_l \tag{5.15}$$

The coefficients $\varepsilon_{e\xi}$ and $\varepsilon_{e\eta}$ are designed to depend on five quantities as follows:

$$\varepsilon_{e\xi} = \varepsilon_c N_\xi S_l d_\xi a_\xi \qquad \varepsilon_{e\eta} = \varepsilon_c N_\eta S_l d_\eta a_\eta \tag{5.16}$$

The only user-adjustable parameter is ε_c All other quantities in Eq. 5.16 are automatically computed by the scheme.

1. ε_c is a user-supplied constant of $O(1)$. A default of 0.5 can be used but the level of smoothing in difficult cases can be raised by changing ε_c.
2. Scaling with the local mesh spacing is provided through N_ξ and N_η, which are approximations to the matrix norms $\|C^{-1}A\|$ and $\|C^{-1}B\|$, respectively, given by

$$N_\xi = \sqrt{\frac{x_\zeta^2 + y_\zeta^2 + z_\zeta^2}{x_\xi^2 + y_\xi^2 + z_\xi^2}} \qquad N_\eta = \sqrt{\frac{x_\zeta^2 + y_\zeta^2 + z_\zeta^2}{x_\eta^2 + y_\eta^2 + z_\eta^2}} \tag{5.17}$$

3. The scaling function S_l is used to control the level of smoothing as a function of normal distance from the body surface. It is designed to have a value close to zero near the body surface where grid orthogonality is desired, and to gradually increase to a value of one at the outer boundary.
4. The grid convergence sensor functions d_ξ and d_η are used to locally increase the smoothing where grid line convergence is detected in the ξ and η directions, respectively. The d_ξ function is made to depend on the ratio of the average distances between grid points in the ξ direction at

level $(l-1)$ to that at level l. This ratio is high in concave regions where grid lines are converging and hence more dissipation is provided here. It is of order one or smaller in flat or convex regions where less dissipation is needed. A limiter is used to prevent the value of the d_ξ function from becoming too low in convex regions. The d_η function behaves similarly in the η direction.

5. The grid angle functions a_ξ and a_η are used to locally increase the smoothing at severe concave corner points in the ξ and η directions, respectively. Both a_ξ and a_η are designed to have the value of one except at a severe concave corner point. Extra smoothing is added only at the concave corner point as opposed to the entire concave region as supplied by d_ξ or d_η. Grids for concave angles down to 5° have been obtained with this scheme.

5.3 Hyperbolic Surface Grid Generation

In hyperbolic surface grid generation, a surface grid is generated by marching from a specified initial curve on a given surface geometry (reference surface). As in hyperbolic field grid generation, a new grid level is produced by linearizing about the current known level and solving the governing equations. After each marching step, the new set of points are projected on to the reference surface prior to the next marching step. The scheme described below is independent of the form of the reference surface (see Section 5.3.3).

5.3.1 Governing Equations for Hyperbolic Surface Grid Generation

Consider generalized coordinates $\xi(x, y, z)$ and $\eta(x, y, z)$ and let $\hat{n} = (\hat{n}_1, \hat{n}_2, \hat{n}_3)^T$ be the local unit surface normal. An orthogonality relation is derived by demanding that the local marching direction η be orthogonal to the local curve direction ξ of the current known state. A cell area constraint and the surface tangency of the marching direction are used to close the system by providing the remaining two equations. The governing equations can be written as

$$\vec{r}_\xi \cdot \vec{r}_\eta = x_\xi x_\eta + y_\xi y_\eta + z_\xi z_\eta = 0 \tag{5.18a}$$

$$\hat{n} \cdot \left(\vec{r}_\xi \times \vec{r}_\eta\right) = \hat{n}_1\left(y_\xi z_\eta - z_\xi y_\eta\right) + \hat{n}_2\left(z_\xi x_\eta - x_\xi z_\eta\right) + \hat{n}_3\left(x_\xi y_\eta - y_\xi x_\eta\right) = \Delta S, \tag{5.18b}$$

$$\hat{n} \cdot \vec{r}_\eta = \hat{n}_1 x_\eta + \hat{n}_2 y_\eta + \hat{n}_3 z_\eta = 0, \tag{5.18c}$$

where $\vec{r} = (x, y, z)^T$ and ΔS is a user-specified surface mesh cell area. This can be prescribed using a similar method as that for ΔA described in Section 5.2.3.

5.3.2 Numerical Solution of Hyperbolic Surface Grid Generation Equations

Local linearization of Eq. 5.18 about a known state 0 results in a system of grid generation equations

$$A_0 \vec{r}_\xi + B_0 \vec{r}_\eta = \vec{f}, \tag{5.19}$$

with

$$A = \begin{bmatrix} x_\eta & y_\eta & z_\eta \\ \hat{n}_3 y_\eta - \hat{n}_2 z_\eta & \hat{n}_1 z_\eta - \hat{n}_3 x_\eta & \hat{n}_2 x_\eta - \hat{n}_1 y_\eta \\ 0 & 0 & 0 \end{bmatrix} \tag{5.20a}$$

$$B = \begin{bmatrix} x_\xi & y_\xi & z_\xi \\ -\hat{n}_3 y_\xi + \hat{n}_2 z_\xi & -\hat{n}_1 z_\xi + \hat{n}_3 x_\xi & -\hat{n}_2 x_\xi + \hat{n}_1 y_\xi \\ \hat{n}_1 & \hat{n}_2 & \hat{n}_3 \end{bmatrix} \tag{5.20b}$$

$$\vec{f} = \begin{bmatrix} 0 \\ \Delta S + \Delta S_0 \\ 0 \end{bmatrix} \tag{5.20c}$$

The matrix B_0^{-1} exists unless the arc length in ξ is zero. Moreover, $B_0^{-1} A_0$ is symmetric and the system of equations is hyperbolic for marching in η (see [Steger 1989b, 1991] for more details). A local unit vector in the marching direction η can be obtained by the cross product of the local unit surface normal $\hat{n}$ with a local unit vector in the ξ direction.

Eq. 5.19 is solved numerically by a non-iterative implicit marching scheme in η, similar to the scheme employed for solving the field grid generation equations described in Section 5.2.2. The nearby known state 0 is taken from the previous marching step. Central differencing with explicit and implicit second-order smoothing is employed in ξ while a two-point backward implicit differencing is employed in η. The numerical scheme can be written as

$$\left[I + (1+\theta) B_k^{-1} A_k \delta_\xi - \varepsilon_i (\Delta\nabla)_\xi\right](\vec{r}_{k+1} - \vec{r}_k) = B_k^{-1} \vec{g}_{k+1} - \left[\varepsilon_e (\Delta\nabla)_\xi\right]\vec{r}_k \tag{5.21}$$

where

$$\delta_\xi \vec{r}_j = \frac{\vec{r}_{j+1} - \vec{r}_{j-1}}{2} \qquad (\Delta\nabla)_\xi \vec{r}_j = \vec{r}_{j+1} - 2\vec{r}_j + \vec{r}_{j-1} \tag{5.22}$$

and $\vec{g}_{k+1} = (0, \Delta S_{k+1}, 0)^T$. I is the identity matrix, j, k are the grid indices in ξ and η, respectively, θ is the implicitness factor as introduced for Eq. 5.7, ε_e and ε_i are the explicit and implicit smoothing coefficients, respectively, with $\varepsilon_i \approx 2\varepsilon_e$; These can be chosen to vary spatially as described in Section 5.2.5. Only the indices that change are shown in Eq. 5.21 and Eq. 5.22, i.e., $\vec{r}_{j+1} \equiv \vec{r}_{j+1,k}$, etc.

The elements of A contain derivatives in η. These derivatives can be expressed in terms of derivatives in ξ using Eq. 5.18 and are computed as

$$\begin{pmatrix} x_\eta \\ y_\eta \\ z_\eta \end{pmatrix} = B^{-1}\vec{g} = \frac{1}{\beta}\begin{bmatrix} x_\xi - \hat{n}_1 w & \hat{n}_2 z_\xi - \hat{n}_3 y_\xi & \hat{n}_1 s_\xi^2 - x_\xi w \\ y_\xi - \hat{n}_2 w & \hat{n}_3 x_\xi - \hat{n}_1 z_\xi & \hat{n}_2 s_\xi^2 - y_\xi w \\ z_\xi - \hat{n}_3 w & \hat{n}_1 y_\xi - \hat{n}_2 x_\xi & \hat{n}_3 s_\xi^2 - z_\xi w \end{bmatrix}\vec{g} \tag{5.23}$$

where

$$w = \hat{n} \cdot \vec{r}_\xi = \hat{n}_1 x_\xi + \hat{n}_2 y_\xi + \hat{n}_3 z_\xi \tag{5.24a}$$

$$s_\xi^2 = \vec{r}_\xi \cdot \vec{r}_\xi = x_\xi^2 + y_\xi^2 + z_\xi^2 \tag{5.24b}$$

$$\beta = Det(B) = s_\xi^2 - w^2 \tag{5.24c}$$

5.3.3 Communications with the Reference Surface

At the beginning of each marching step, the local unit surface normal at each point on the current known state have to be computed. These normals are needed in the matrices for marching the grid generation equations. After each marching step, the new points have to be projected back onto the reference surface. For a high-quality grid, the local step size should be small relative to the local curvature of the surface. This ensures that the distances moved by the grid points due to projection are small, which would in turn guarantee that the final grid spacings in the marching direction are close to the step sizes originally specified.

Each hyperbolic marching step is performed independently from the surface normal evaluation before the step and the point projection after the step. This implies that different representations of the reference surface can be easily substituted if routines are provided to

1. Compute the surface normal at a given point on the reference surface,
2. Project a given point onto the reference surface.

A scheme is outlined below for the above two steps for a reference surface consisting of a collection of multiple panel networks. Each panel network contains a rectangular array of points. The surface patch represented by these points is typically approximated by a set of bilinear quadrilaterals with vertices located at the points.

Surface normals on a panel network can be computed as follows. The surface normal of a quadrilateral is given by the cross-product of its diagonals. The surface normal at a vertex point on the panel network is then computed as the average of the surface normals of the quadrilaterals that share the vertex. For a given point on the panel network, bilinear interpolation of the normals at the vertices of the quadrilateral on which the point lies is used to obtain the normal at the point.

A stencil walk method can be used to project a given point onto the multiple panel networks. First, Cartesian bounding boxes of each panel network are employed to determine the set of panel networks that may contain the point. Next, a quadrilateral from the set of candidate panel networks that is closest to the given point is taken to be the starting location of the stencil walk. On seeking bilinear interpolation coefficients of the given point on the chosen quadrilateral, the results either indicate the point is inside the quadrilateral, or the next quadrilateral in the appropriate direction should be tried. When the stencil walk hits a boundary of the panel network, the walk continues on to the neighboring network if there is one. In practical situations, small gaps may exist between neighboring panel networks. A tolerance parameter may be used to extrapolate the boundaries of each panel network to cover the gaps for projection purposes.

The stencil walk continues until the point converges inside a quadrilateral. For points close to the reference surface, the stencil walk typically converges very quickly. However, if the point is far away from the reference surface (e.g., as a result of taking too large a marching step relative to the curvature of the surface), convergence may not occur or the point may converge to an erroneous location. For further information, see Chapter 29.

5.4 General Guidelines for High-Quality Grid Generation

Hyperbolic grid generation requires the specification of an initial state from which a field or surface grid is generated. The grid point distribution on the initial state directly affects the quality of the hyperbolic grid that can be produced. Two important areas of concern are described below.

5.4.1 Grid Stretching

In a given direction, let the grid spacings on each side of an interior point be Δs_1 and Δs_2. The grid stretching ratio R at an interior point in the given direction is defined to be

$$R = \max(\Delta s_1, \Delta s_2)/\min(\Delta s_1, \Delta s_2) \tag{5.25}$$

TABLE 5.1 Approximate Speeds of Hyperbolic Field Generator on a Variety of Platforms

Platform	Approximate Speed (number of points generated per CPU second)
CRAY C-90	220,000
SGI R10000 175 MHz	28,000
SGI R4400 250 MHz	20,000
HP 9000/755 99 MHz	16,000
Pentium PC 90 MHz	1,400

In order to limit truncation error, a stretching ratio of about 1.3 should not be exceeded in any direction on the initial state and in the marching direction, i.e., large and sudden jumps in the grid spacings in any direction should be avoided.

5.4.2 Point Distribution Near Corners

Proper grid point placement near convex and concave corners on the initial state can significantly enhance the quality of the resulting hyperbolic grid. The grid spacings on each side of a convex or concave corner should be equal. Moreover, grid points should be clustered toward a convex corner with sharper corners requiring more clustering. These grid properties are desirable for producing smooth grids and are also essential for satisfactorily resolving the flow around the corner. On the other hand, grid points should *not* be clustered into a concave corner on the initial state. A uniform or declustered grid spacing at the concave corner can significantly reduce the tendency for grid lines to converge as the grid is marched out from the corner.

5.5 Applications

In field grid generation, hyperbolic methods are usually used to produce body-fitted grids, i.e., the initial states are chosen to lie on the body surface of the configuration. Such methods have been frequently employed in single grid computations where the outer boundary of the grid lies in the far field. These methods have been equally successful in producing multiple body-fitted grids in complex configurations using the overset grid approach. In such applications, individual grids are typically generated independently of each other and the outer boundaries are not too far from the body surface. The freedom of allowing neighboring grids to overlap makes hyperbolic grids particularly well suited for this gridding approach.

Typical speeds of a hyperbolic field grid generator for a 3D problem on a number of computing platforms are given in Table 5.1. The speed is given by the number of grid points generated per CPU second, e.g., a 3D field with 220,000 points requires about 1 CPU second to generate on the Cray C-90. Typical speed of a hyperbolic surface grid generator is about 20,000 points per CPU second on a SGI R10000 machine.

Sample grids from several overset grid configurations are presented in the subsections below. Other interesting applications not shown here include the F-18 Aircraft [Rizk and Gee, 1992], a joined-wing configuration [Wai, Herling, and Muilenburg, 1994], the RAH-66 Comanche helicopter [Duque and Dimanlig, 1994, 1995], and various marine applications [Dinavahi and Korpus, 1996].

5.5.1 Applications Using 2D Hyperbolic Field Grids

5.5.1.1 Three-Element Airfoil

The first example on the use of 2D hyperbolic field grid generation is a three-element airfoil configuration consisting of five grids shown in Figures 5.6a–5.6d ([Rogers, 1994]). Hyperbolic grids are generated

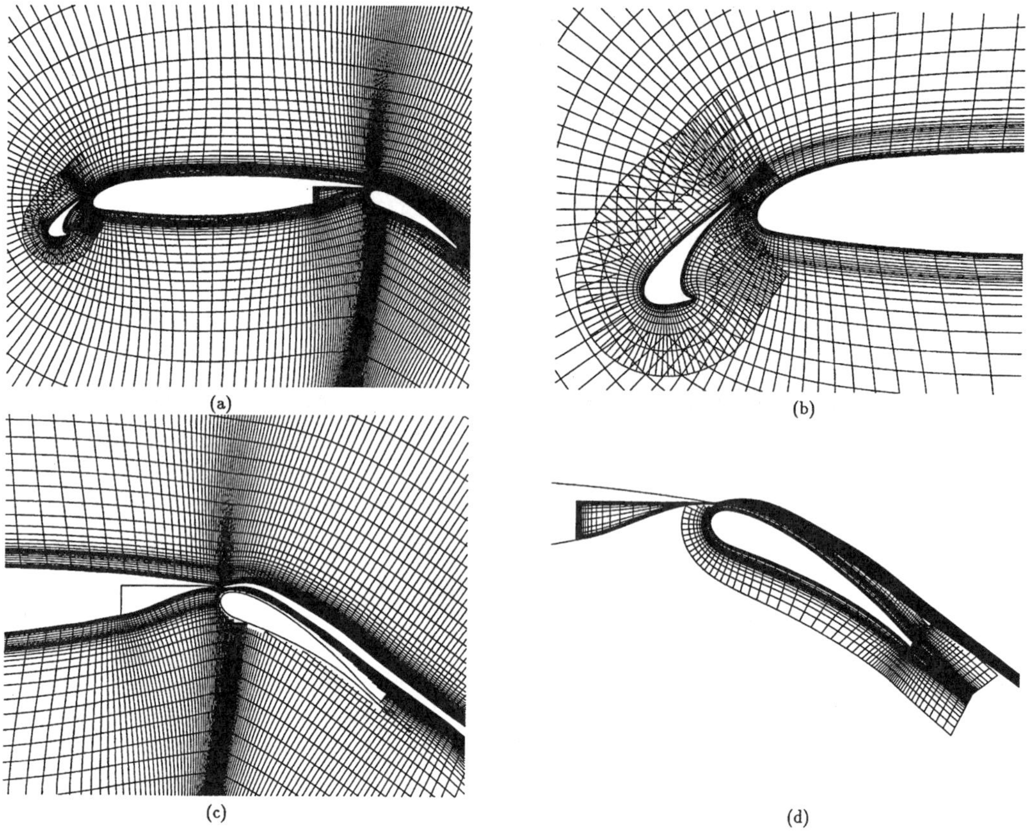

FIGURE 5.6 Field grids for three-element airfoil. Only every two points are shown in the normal direction. Field points that lie in the interior of a neighboring element have been blanked by a domain connectivity program. (a) Overview. (b) Close-up view of the slat region. (c) Field grid of main element in the flap region. (d) Field grid of flap, flap wake, and cove/wake grid of main element in the flap region.

independently around the slat, main element, and flap. In order to properly resolve the shear layers in the wake regions for this configuration, two specially tailored algebraic grids are needed. One is used downstream of the finite thickness trailing edge region of the flap and the other is used in the cove and wake regions of the main element.

A nonuniform stretching function in the normal direction (see Section 5.2.3) is used to specify variable marching step sizes to accomplish two effects in this configuration:

1. *The fanned wake in the slat.* In standard C-mesh topologies, a uniform viscous wall spacing is used along the wake cut and along the body surface. If such a grid spacing is used for the slat grid, the downstream boundary in the wake would contain viscous spacing. However, the grid spacing is much coarser in the region of the main element field grid which overlaps the slat grid downstream boundary. Such drastic differences in grid resolution between neighboring grids at the grid boundaries can be highly undesirable for intergrid communication. Flow features from the fine grid may not be resolvable by the coarse grid. Moreover, interpolation of information from the coarse grid onto the fine grid may contaminate the fine grid solution. In the slat grid shown (Figure 5.6b), the wall spacing is kept constant along the body surface but is increased with distance downstream from the trailing edge along the wake cut. The declustered spacing at the downstream wake boundary now provides better quality communication with the main element grid.
2. *The clustered regions in the field grid of the main element.* In multi-element airfoil configurations, the flow in the wake of an element has to be sufficiently resolved by the field grid associated with

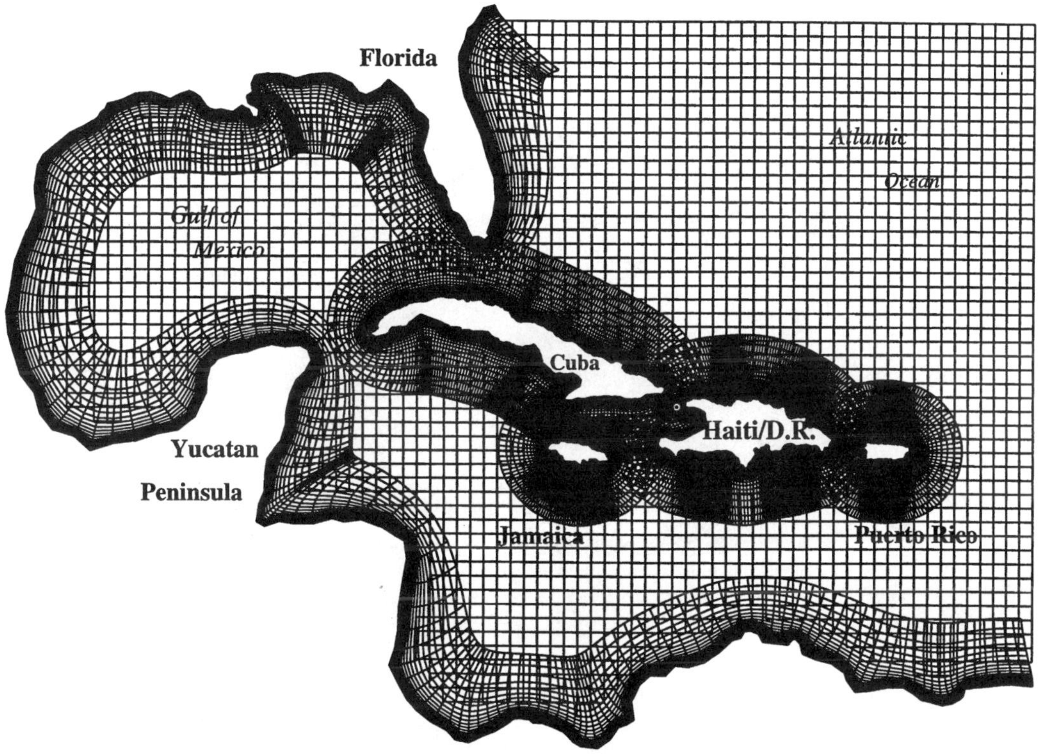

FIGURE 5.7 Hyperbolic field grids around the Greater Antilles Islands and the Gulf of Mexico. The body-fitted grids are embedded in a uniform Cartesian background grid.

the element downstream. For example, the slat grid wake passes into the main element field grid. A special stretching function is used in the normal direction for the main element to achieve a tight normal spacing in the vicinity of the wake of the slat (see Figure 5.6a, 5.6b). A similar grid clustering is installed in the flap field grid in the vicinity of the main element wake (see Figure 5.6d).

5.5.1.2 Greater Antilles Islands and Gulf of Mexico

The second example of the use of 2D hyperbolic field grids is taken from grids around the Greater Antilles Islands and the Gulf of Mexico in geophysical simulations [Barnette and Ober, 1995]. Body-fitted grids are generated using hyperbolic methods around the coastlines of the islands and the gulf (see Figure 5.7). Each grid is grown to a distance not too far from the initial state. The set of curvilinear grids is embedded in a uniform background Cartesian mesh. This approach makes generation of the body-fitted grids much easier than when one of the body-fitted grids is also made to serve as a background grid by growing to a large distance from the body surface. Moreover, the use of a uniform Cartesian mesh in the background has the desirable advantage of providing a uniform resolution in the space between the different body-fitted grids.

5.5.2 Applications Using 3D Hyperbolic Field Grids

5.5.2.1 SOFIA Telescope

SOFIA stands for Stratospheric Observatory for Infrared Astronomy. A telescope is placed in an open cavity in a Boeing 747 aircraft for airborne astronomical observations. Part of the structure that houses the SOFIA telescope is shown in Figure 5.8. The lower structure is the truss base, which contains the primary and tertiary mirrors. A ring-like structure called the truss yoke is situated above the truss base. Only half of the truss yoke grid is shown so that the entire truss base grid is visible. Flowfield computations on the SOFIA configuration have been performed by Atwood and Van Dalsem [1993], and Srinivasan

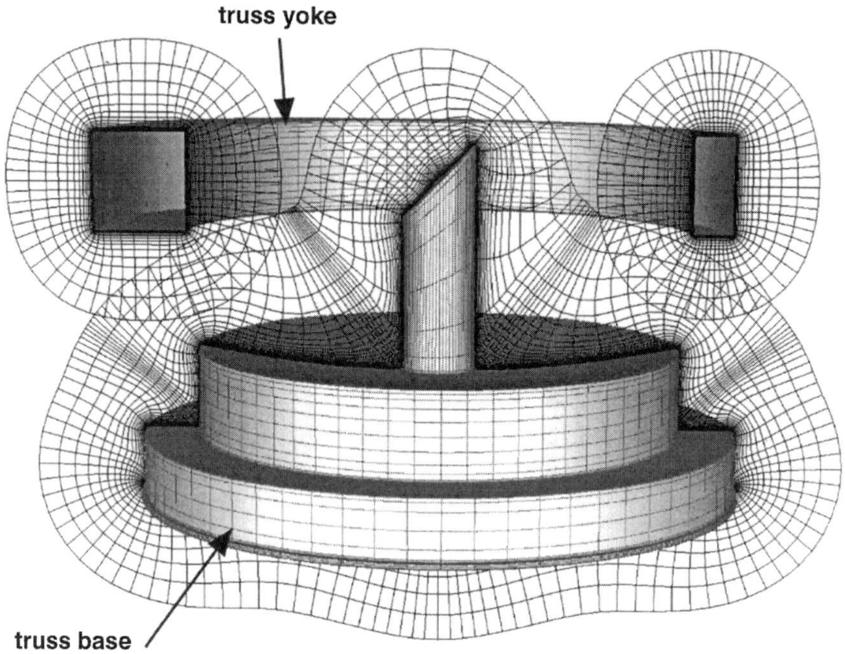

FIGURE 5.8 Surface grids and slices of volume grids for SOFIA truss base and truss yoke.

and Klotz [1997]. Surface grids for most of the SOFIA configuration were generated using GRIDGEN [Chawner and Steinbrenner, 1995], while most of the body-fitted volume grids were generated using hyperbolic methods with HYPGEN [Chan, Chiu, and Buning, 1993].

5.5.2.2 Apache Helicopter

The tail section of the Apache helicopter is shown in Figure 5.9. Surface grids were generated using elliptic methods with ICEMCFD [Wulf and Akdag, 1995] and GRIDGEN. Body-fitted volume grids were generated using hyperbolic methods with HYPGEN. The overlapping volume grids in the tail section are embedded in a background Cartesian grid for the tail alone to provide a uniform grid resolution in the off-body region of the tail. Then, the entire vehicle (fuselage and tail) is embedded in a larger background Cartesian mesh.

5.5.2.3 Space Shuttle Launch Vehicle

The Space Shuttle Launch Vehicle configuration consists of the Orbiter, External Tank (ET), Solid Rocket Boosters (SRB), and the various attach hardware between the main components. A high-fidelity grid system consisting of 111 grids and approximately 16 million points was constructed using overset grids and a number of flow computations were performed [Pearce, et al., 1993]; [Gomez and Ma, 1994]; [Slotnick, Kandula, and Buning, 1994]. Surface grids were primarily generated directly on the CAD data using algebraic interpolation and elliptic methods with ICEMCFD. All the body-fitted volume grids were generated using hyperbolic methods with HYPGEN except for the volume grids in the elevon gaps where algebraic interpolation was used to provide control of multiple grid boundaries. A sample of the volume grids for some of the components are shown in Figures 5.10a, 5.10b, and 5.10c.

5.5.3 Applications Using Hyperbolic Surface Grids

In practical applications, the initial states of hyperbolic surface grids are typically chosen to be control curves of the surface geometry. These control curves can be one of the following types:

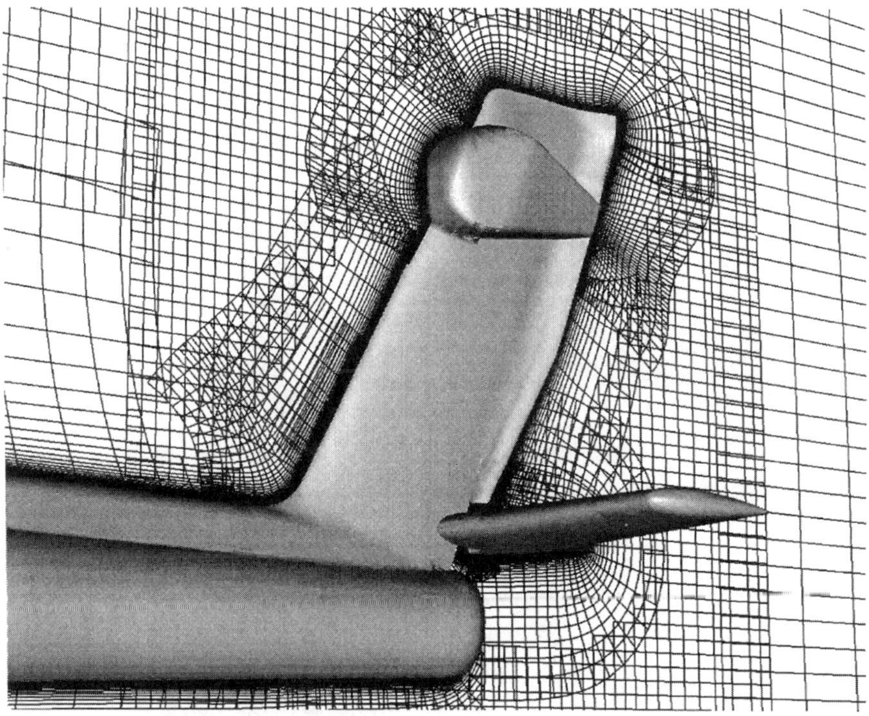

FIGURE 5.9 Body-fitted volume grids and background Cartesian grids for the tail region of the Apache helicopter.

1. Intersection curve between surface components, e.g., an intersection curve between a wing and a fuselage.
2. Curve along a surface discontinuity.
3. Curve along a high curvature contour, e.g., along the leading edge and tip of a wing.
4. Curve along a surface domain boundary.
5. Special curve at which clustering is needed for nongeometrical reasons.

The possibility of covering a complex surface geometry with overlapping surface grids was suggested by Steger [1991]. Such surface grids may be conveniently generated using hyperbolic methods and algebraic methods. For a grid that is bounded by just one control curve, a hyperbolic or algebraic marching scheme is the most convenient method for generating the grid. Variable marching distances and step sizes for different points on the initial curve are frequently used to ensure sufficient overlap between neighboring grids. For a grid bounded by two or more control curves, algebraic interpolation methods are more appropriate.

There is currently only one software package that the author is aware of for performing hyperbolic surface grid generation — a code called SURGRD developed at NASA Ames Research Center [Chan and Buning, 1995]. The code has hyperbolic and algebraic marching options for surface grid generation on a reference surface consisting of multiple panel networks. Surface descriptions derived from CAD data (e.g., NURBS surfaces) are usually converted to a high fidelity multiple panel network description prior to using SURGRD. This data translation is typically performed via some other grid generation package such as GRIDGEN [Chawner and Steinbrenner, 1995], ICEMCFD [Wulf and Akdag, 1995], or NGP [Gaither, et al., 1995]. The initial curves needed for hyperbolic surface grid generation can also be generated from these packages or selected directly from curve subsets of the multiple panel network description.

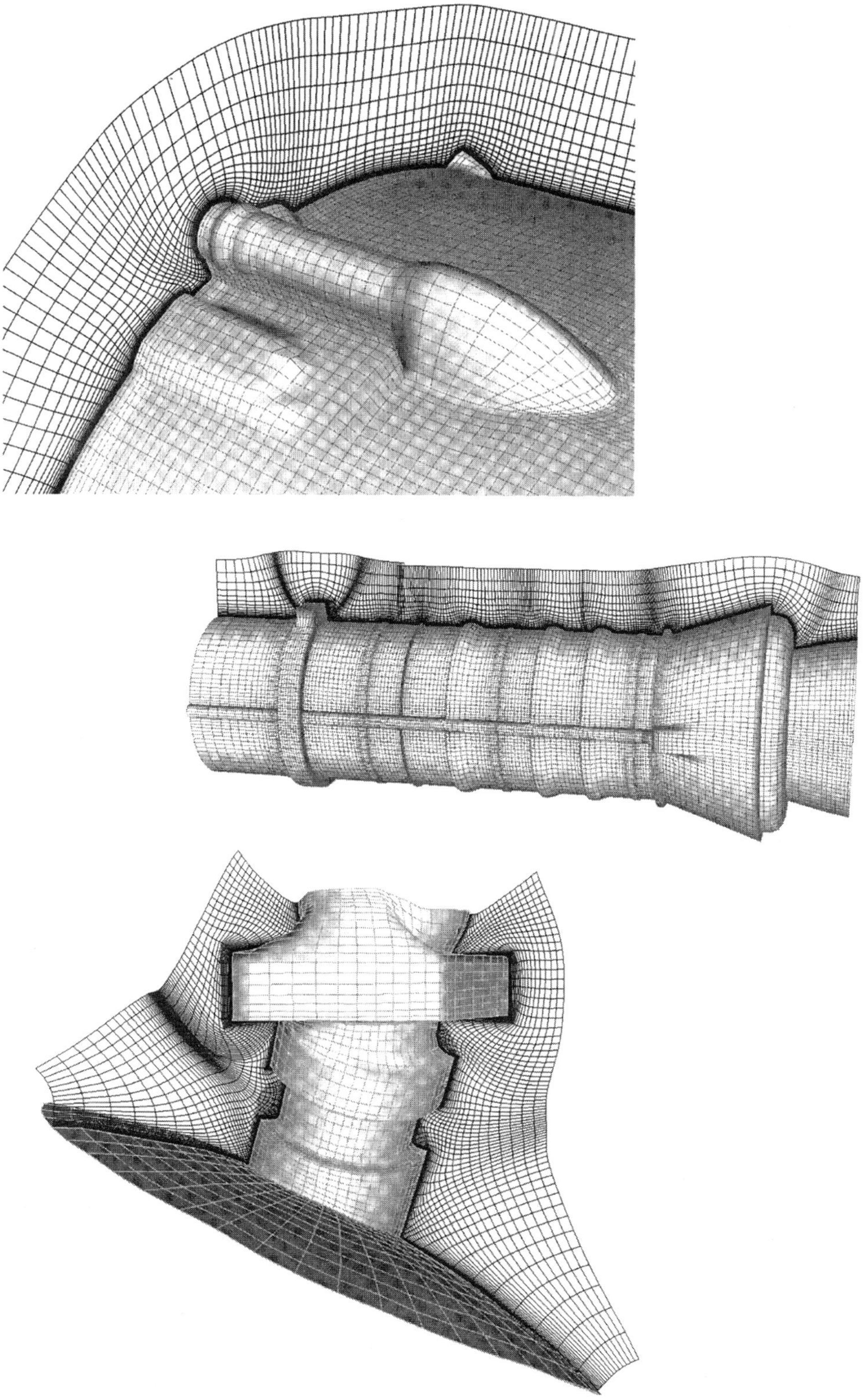

(a)

(b)

(c)

FIGURE 5.10 A sample of surface grids and slices of volume grids from the Space Shuttle Launch Vehicle configuration. (a) Forward top region of External Tank. (b) Back half of Solid Rocket Booster. (c) Liquid hydrogen feedline.

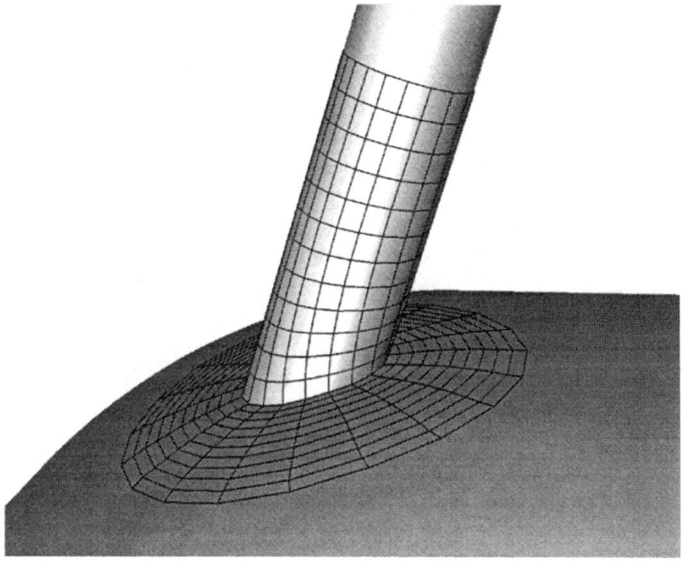

FIGURE 5.11 Collar surface grid for pipe/curved wall intersection region. The intersection curve is used as the intial curve for hyperbolic marching.

5.5.3.1 Collar Grid

When applying overset grid methods on two intersecting geometric components, the surface and volume grids for the two components are usually generated independently of each other. A third grid called a collar grid is typically used in the intersection region to resolve the local geometry of both components [Parks, et al., 1991]. One of the first practical applications of hyperbolic surface grid generation was on producing collar surface grids. The intersection curve between the two components is used as the initial curve for the hyperbolic marching scheme. Surface grids are then generated onto the two components by marching out from both sides of the initial curve. The two resulting surface grids are concatenated to form the collar surface grid. Figure 5.11 shows a collar surface grid for the junction between a pipe and a curved wall.

5.5.3.2 Pylon

Figure 5.12 shows a set of overlapping surface grids generated using hyperbolic methods for a pylon. The surface definition consists of multiple patches converted from IGES format to panel network format. Initial curves are selected from intersection curves and surface discontinuity curves of the geometry. Highly skewed grids are avoided by using more grids and allowing them to overlap.

5.5.3.3 V-22 Tiltrotor

Flow computations on the V-22 Tiltrotor configuration were performed by Meakin [1993]. The surface geometry was described by 22 panel networks. Since a hyperbolic surface grid generator was not available at the time, all the surface grids for the simulation were generated using algebraic methods with the GRIDGEN package. Figure 5.13 shows surface grids that were more recently generated using the SURGRD code as a demonstration of hyperbolic surface grid generation capability. All the grids were produced using hyperbolic methods except for the following two grids where algebraic marching was used: the wing portion of the wing/fuselage collar grid and the wing portion of the wing/nacelle collar grid.

5.5.3.4 X-CRV Crew Return Vehicle

The X-CRV Crew Return Vehicle is presented as a further example that features one of the first production computations on complex configurations that uses hyperbolic surface grids extensively (see Figures 5.14a–d). Surface and volume grid generation and flow computations were performed by Gomez

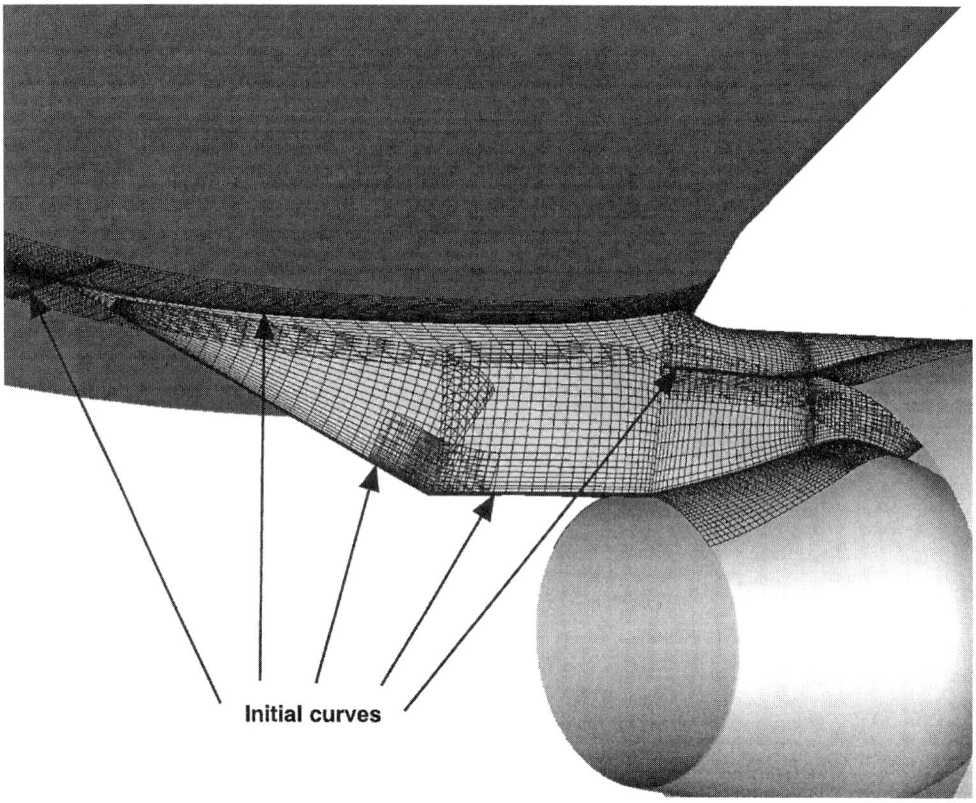

FIGURE 5.12 Surface grids for pylon region of subsonic transport. Initial curves are indicated by thick lines.

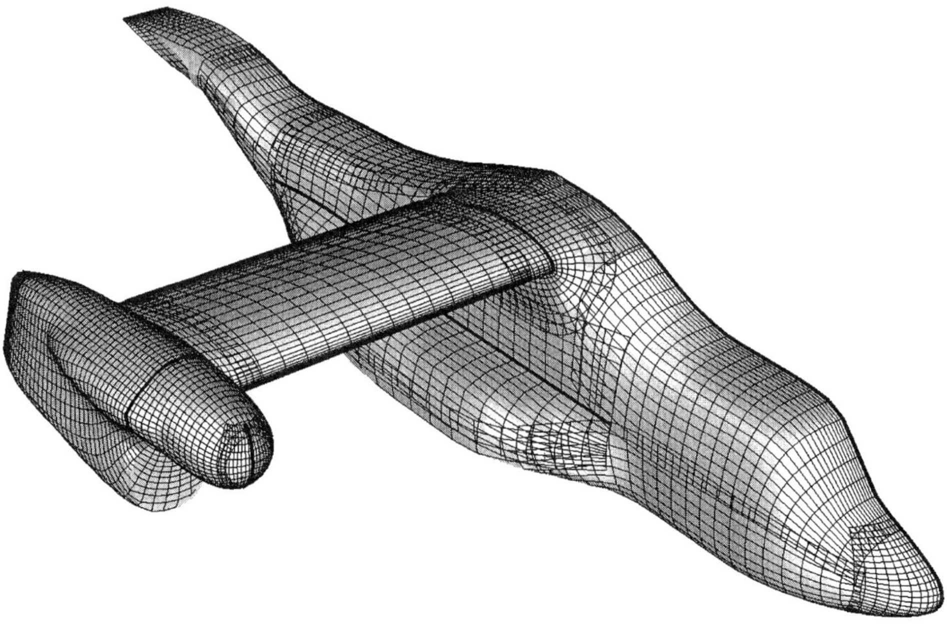

FIGURE 5.13 Surface grids generated using hyperbolic methods for the V-22 Tiltrotor fuselage, wing, and nacelle. Initial curves are indicated by thick lines.

and Greathouse [1996]. The surface geometry consists of 62 trimmed NURBS surfaces. These were converted to a multiple panel network format using NGP. Selection of initial curves and distribution of grid points were accomplished using GRIDGEN. Most of the surface grids were generated using hyperbolic methods with SURGRD except for several grids in the flap and rudder area where GRIDGEN was employed to produce surface grids via algebraic interpolation. The symmetric configuration contains 20 surface grids while the full configuration contains 33 surface grids. Body-fitted volume grids were generated using hyperbolic methods with HYPGEN.

The approximate time spent by the user on each grid generation step is given below. About half a day was spent on cleaning up the surface geometry. Selection of the initial curves required about one hour while generation of the hyperbolic surface grids took about half an hour. Most of the time was spent on adjusting the variable marching distances of each grid to ensure sufficient overlap between neighboring grids. The hyperbolic volume grids were produced in about half an hour. Surface grids were generated on an SGI Power Onyx and volume grids were generated on a Cray J-90.

5.6 Summary and Research Issues

Hyperbolic grid generation methods and a sample of working applications have been presented. The scheme requires the solution of a set of nonlinear hyperbolic partial differential equations and can be formulated for 2D and 3D field grid generation as well as surface grid generation. Orthogonal or nearly orthogonal grids can be generated by a fast marching method. Grid clustering near a boundary is naturally accomplished by specifying the cell sizes via a 1D stretching function. A variety of grid topologies can be produced with different boundary conditions. Robustness is achieved by the use of spatially varying smoothing coefficients and proper treatment of convex corners.

The exact specification of side and outer boundaries is not allowed in a one-sweep hyperbolic marching scheme. This restriction makes hyperbolic grid generation methods unsuitable for the patched grid approach for computations on complex configurations. However, hyperbolic grid generation methods are particularly well suited for the overset grid approach where neighboring grids are permitted to overlap. Numerous overset grid applications have successfully used hyperbolic methods for field grid generation.

Structured grid generation on complex configurations has typically been a highly time-consuming step for the user. As the geometric configurations of interest become more and more complex, there is a growing demand to automate the grid generation process. It is unclear whether a totally "black box" grid generation method can be devised using overset grids where absolutely no input is required from the user. However, the grid generation procedure can be divided into substeps where some of the substeps can be automated. For the substeps that are difficult to automate, schemes can be developed to reduce the human effort needed. The resulting process will require some user interaction but may still be acceptable and quite fast for many applications.

For reasons already discussed in this chapter, hyperbolic grid generation methods will most likely play a key role in the future of automating the overset grid generation process. Some potential research issues are highlighted below. Although hyperbolic field grid generation is currently fairly robust, some adjustments of the smoothing parameters (see Section 5.2.5) are still needed for very complex cases. Since the scheme is fast, these iterations typically do not take much time. However, the smoothing mechanisms may still be improved to a point where no adjustments are necessary. Further research in mixed hyperbolic–elliptic methods in 3D would allow specifications of the side and outer boundaries as needed in certain special situations.

In order for hyperbolic surface grid generation to be more practical and convenient, the software has to be developed to grow surface grids directly onto different types of surface definitions, e.g., NURBS surfaces and triangulated surfaces. More convenient methods for specifying the initial curves will have to be devised. As the emphasis shifts toward automation, less user input will be expected. This implies the following:

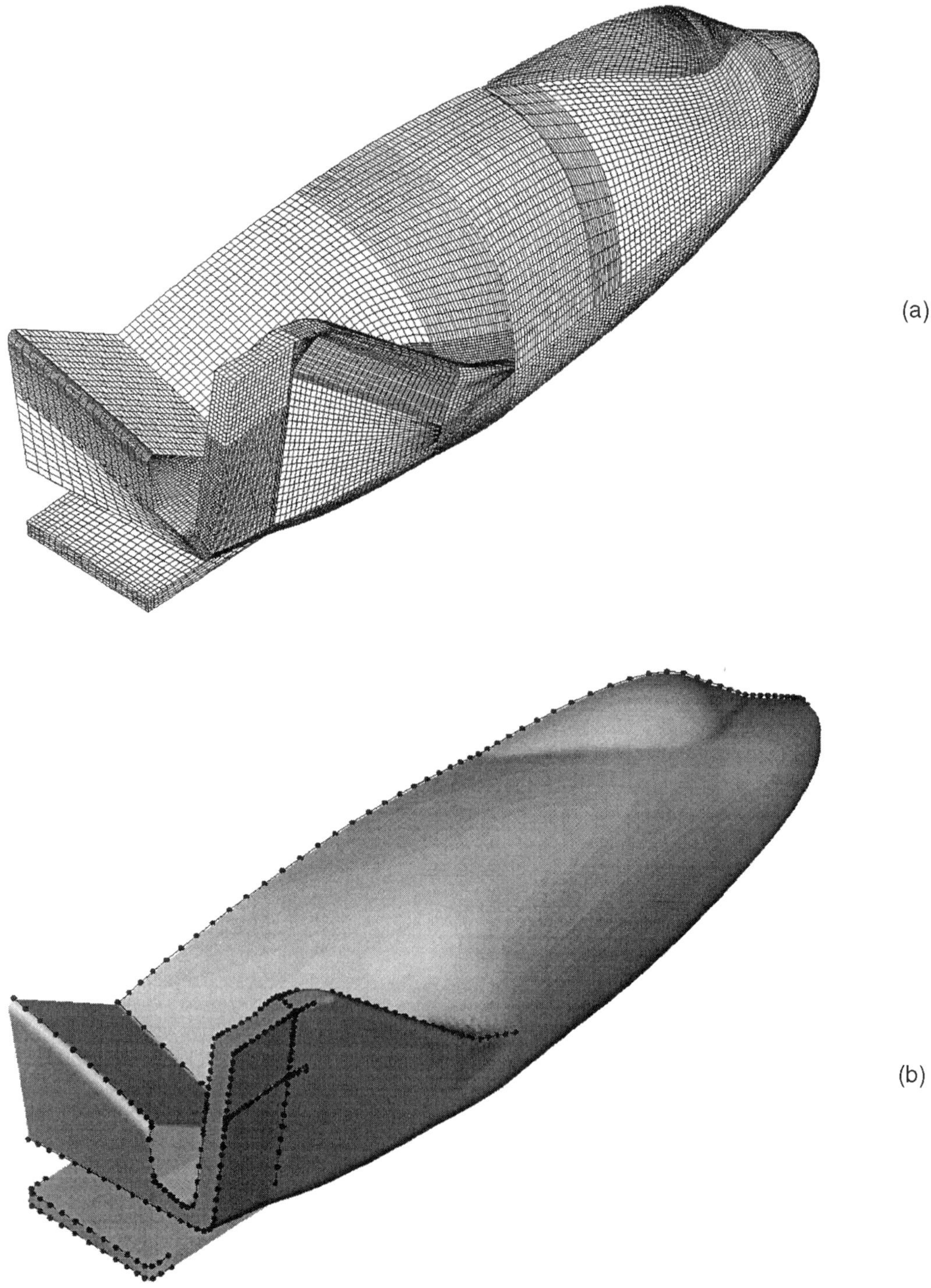

FIGURE 5.14 X-CRV Crew Return Vehicle. (a) Surface definition. (b) Initial curves for hyperbolic surface grid generation. Points on curves are indicated by black dots.

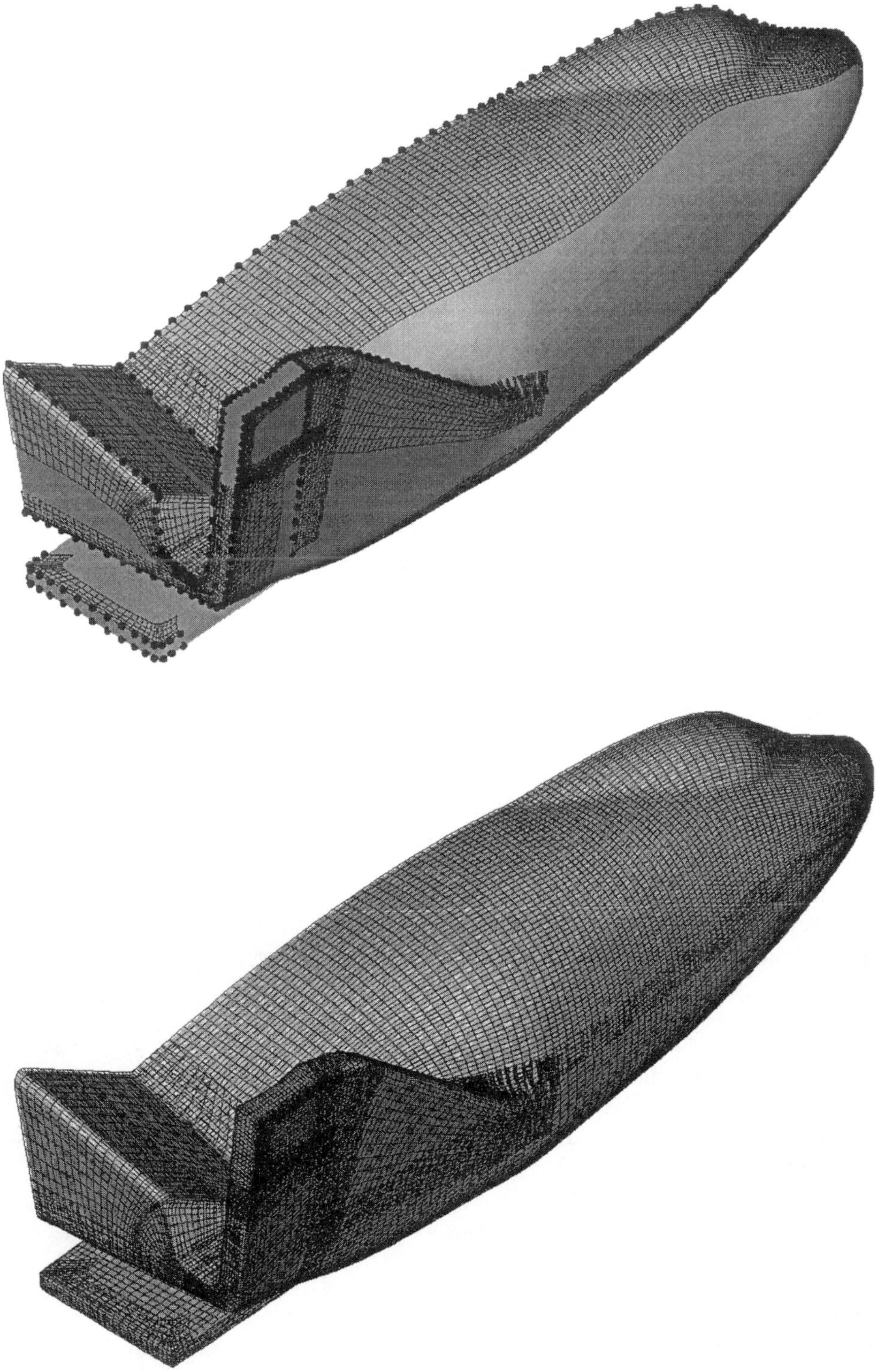

(c)

(d)

FIGURE 5.14 (continued) X-CRV Crew Return Vehicle. (c) Partially completed surface grids generated by hyperbolic methods. (d) Final hyperbolic surface grids.

1. Complicated surface topologies requiring significant user interaction will be avoided. Simple surface grid topologies will be favored, which typically results in an increase in the number of grids.
2. User-specified spatially varying marching distances and step sizes to provide proper overlap between neighboring grids will be employed less often. Surface grids may be marched hyperbolically to a constant distance from their initial curves and the gaps between the surface grids may be filled by algebraic grids via an automatic scheme [Chan and Meakin, 1997].
3. The resulting set of relatively simple overlapping surface grids may be radiated out into the field using hyperbolic methods. These body-fitted field grids may be grown to some constant distance away from the body surface and embedded in layers of background Cartesian meshes of decreasing resolutions with distance from the body [Meakin, 1995].

Acknowledgments

The author would like to thank the following for providing the image files for some of the examples shown in this chapter: Dr. Daniel Barnette for the Greater Antilles Islands and the Gulf of Mexico, Mr. Jim Greathouse for the X-CRV, and Dr. Earl Duque for the Apache helicopter.

References

1. **Atwood, C. A. and Van Dalsem, W.R.**, Flowfield simulation about the stratospheric observatory for infrared astronomy, *J. of Aircraft*, 30, 5, pp. 719–727, 1993.
2. **Barnette, D. W. and Ober, C. C.**, Progress report on a method for parallelizing the overset grid approach," *Proceedings of the 6th International Symposium on Computational Fluid Dynamics*, Lake Tahoe, NV, 1995.
3. **Chan, W. M. and Buning, P. G.**, Surface grid generation methods for overset grids, *Computers and Fluids*, 24, 5, pp. 509–522, 1995.
4. **Chan, W. M. and Meakin, R. L.**, Advances towards automatic surface domain decomposition and grid generation for overset grids, *Proceedings of the 13th AIAA Computational Fluid Dynamics Conference*, 1997, AIAA Paper 97-1979, Snowmass Village, CO.
5. **Chan, W. M. and Steger, J. L.**, Enhancements of a three-dimensional hyperbolic grid generation scheme, *Appl. Math. and Comput.*, 51, pp. 181–205, 1992.
6. **Chan, W. M., Chiu, I. T., and Buning, P. G.**, *User's Manual for the HYPGEN Hyperbolic Grid Generator and the HGUI Graphical User Interface*, NASA TM 108791, 1993.
7. **Chawner, J. R. and Steinbrenner, J. P.**, Automatic structured grid generation using GRIDGEN (Some Restrictions Apply), *Proceedings of NASA Workshop on Surface Modeling, Grid Generation, and Related Issues in Computational Fluid Dynamics (CFD) Solutions*, NASA CP 3291, 1995.
8. **Cordova, J. Q.**, Advances in hyperbolic grid generation, *Proceedings of the 4th International Symposium on Computational Fluid Dynamics*, Davis, CA, *U.S.A.*, Volume 1, pp. 246–251, 1991.
9. **Cordova, J. Q. and Barth, T. J.**, Grid generation for general 2D regions using hyperbolic equations, AIAA Paper 88-0520, 1988.
10. **Dinavahi, S. P. G. and Korpus, R. A.**, Overset Grid Methods in Ship Flow Problems, Unpublished results, Science Applications International Corp., 1996.
11. **Duque, E. P. N. and Dimanlig, A. C. B.**, Navier-Stokes simulation of the RH-66 comanche helicopter, *Proceedings of the 1994 American Helicopter Society Aeromechanics Specialist Meeting*, San Francisco, CA, 1994.
12. **Duque, E. P. N., Berry, J. D., Budge, A. M., and Dimanlig, A. C. B.**, A comparison of computed and experimental flowfields of the RAH-66 helicopter, *Proceedings of the 1995 American Helicopter Society Aeromechanics Specialist Meeting*, Fairfield County, CT, 1995.
13. **Gaither, A., Gaither, K., Jean, B., Remotigue, M., Whitmire, J., Soni, B., Thompson, J., Dannenhoffer, J., and Weatherill, N.**, The National Grid Project: a system overview, *Proceedings of NASA Workshop on Surface Modeling, Grid Generation, and Related Issues in Computational Fluid Dynamics (CFD) Solutions*, NASA CP 3291, 1995.

14. **Gomez, R. J. and Greathouse, J. S.**, Manned spacecraft overset grid applications, Unpublished results, NASA Johnson Space Center, 1996.
15. **Gomez, R. J. and Ma, E. C.**, Validation of a large scale chimera grid system for the space shuttle launch vehicle, *Proceedings of the 12th AIAA Applied Aerodynamics Conference*, AIAA Paper 94-1859, Colorado Springs, CO, 1994.
16. **Jeng, Y. N., Shu, Y. L. and Lin, W. W.**, Grid generation for internal flow problems by methods using hyperbolic equations. *Numer. Heat Transf. Part B* 27, pp. 43–61, 1995.
17. **Kinsey, D. W. and Barth, T. J.**, Description of a hyperbolic grid generation procedure for arbitrary two-dimensional bodies, AFWAL TM 84-191-FIMM, 1984.
18. **Klopfer, G. H.**, Solution adaptive meshes with a hyperbolic grid generator, *Proceedings of the Second International Conference on Numerical Grid Generation in Computational Fluid Dynamics*, Miami, FL, pp. 443–453, 1988.
19. **Meakin, R. L.**, Moving body overset grid methods for complete aircraft tiltrotor simulations, *Proceedings of the 11th AIAA Computational Fluid Dynamics Conference*, AIAA Paper 93-3350, Orlando, FL, 1993.
20. **Meakin, R. L.**, An efficient means of adaptive refinement within systems of overset grids, *Proceedings of the 12th AIAA Computational Fluid Dynamics Conference*, AIAA Paper 95-1722, San Diego, CA, 1995.
21. **Nakamura, S.**, Noninterative three dimensional grid generation using a parabolic-hyperbolic hybrid scheme, AIAA Paper 87-0277, 1987.
22. **Parks, S. J., Buning, P. G., Steger, J. L., and Chan, W. M.**, Collar grids for intersecting geometric components within the chimera overlapped grid scheme, *Proceedings of the 10th AIAA Computational Fluid Dynamics Conference*, AIAA Paper 91-1587, Honolulu, HI, 1991.
23. **Pearce, D. G., Stanley, S. A., Martin, F. W., Gomez, R. J., Le Beau, G. J., Buning, P. G., Chan, W. M., Chiu, I. T., Wulf, A., and Akdag, V.**, Development of a large scale chimera grid system for the space shuttle launch vehicle, AIAA Paper 93-0533, 1993.
24. **Rai, M. M.**, A conservative treatment of zonal boundaries for Euler equation calculations, *J. Comput. Phys.* 62, pp. 472–503, 1986.
25. **Rizk, Y.M. and Gee, K.**, Unsteady simulation of viscous flowfield around F-18 aircraft at large incidence, *J. of Aircraft*, 29, 6, pp. 986–992, 1992.
26. **Rogers, S.E.**, Progress in high-lift aerodynamic calculations, *J. of Aircraft*, 31, 6, pp. 1244–1251, 1994.
27. **Slotnick, J. P., Kandula, M., and Buning, P. G.**, Navier-Stokes simulation of the space shuttle launch vehicle flight transonic flowfield using a large scale chimera grid system, *Proceedings of the 12th AIAA Applied Aerodynamics Conference*, AIAA Paper 94-1860, Colorado Springs, CO, 1994.
28. **Srinivasan, G. R. and Klotz, S. P.**, Features of cavity flow and acoustics of the stratospheric observatory for infrared astronomy, *Proceedings of the ASME Fluids Engineering Conference*, Vancouver, British Columbia, Canada, June 1997.
29. **Starius, G.**, Constructing Orthogonal Curvilinear Meshes by Solving Initial Value Problems, *Numerische Mathematik* 28, pp. 25–48, 1977.
30. **Steger, J. L.**, Generation of three-dimensional body-fitted grids by solving hyperbolic partial differential equations, NASA TM 101069, 1989a.
31. **Steger, J. L.**, Notes on surface grid generation using hyperbolic partial differential equations, Internal Report TM CFD/UCD 89-101, Department of Mechanical, Aeronautical and Materials Engineering, University of California, Davis, 1989b.
32. **Steger, J. L.**, Grid generation with hyperbolic partial differential equations for application to complex configurations, *Numerical Grid Generation in Computational Fluid Dynamics and Related Fields*, Ascilla, A. S., Hauser, J., Eiseman P. R., Thompson, J. F., (Ed.), Elsevier Science, B.V., North-Holland, 1991.
33. **Steger, J. L. and Chaussee, D. S.**, Generation of body-fitted coordinates using hyperbolic partial differential equations, *SIAM J., Sci. Stat. Comput.*, 1, pp. 431–437, 1980.

34. **Steger, J. L., Dougherty, F. C., and Benek, J. A.,** A chimera grid scheme, *Advances in Grid Generation,* Ghia K.N. and Ghia, U., (Ed.), ASME FED, Vol. 5, 1983.
35. **Steger, J. L. and Rizk, Y. M.,** Generation of three-dimensional body-fitted coordinates using hyperbolic partial differential equations, NASA TM 86753, 1985.
36. **Takanashi, S. and Takemoto, M.,** Block-structured grid for parallel computing, *Proceedings of the 5th International Symposium on Computational Fluid Dynamics,* Sendai, Japan, Vol. 3, pp. 181–186, 1993.
37. **Thompson, J. F.,** A composite grid generation code for general 3D regions — the Eagle code, *AIAA J.,* 26, 3, pp. 271–272, 1988.
38. **Wai, J., Herling, W. W., and Muilenburg, D. A.,** Analysis of a joined-wing configuration, 32nd Aerospace Sciences Meeting & Exhibit, AIAA Paper 94-0657, Reno, NV, 1994.
39. **Wulf, A. and Akdag, V.,** Tuned grid generation with ICEM CFD, *Proceedings of NASA Workshop on Surface Modeling, Grid Generation, and Related Issues in Computational Fluid Dynamics (CFD) Solutions,* NASA CP 3291, 1995.

Further Information

In addition to the references given above, further information on applications of hyperbolic grid generation methods can be found in Chapter 11.

6

Boundary Orthogonality in Elliptic Grid Generation

Ahmed Khamayseh
Andrew Kuprat
C. Wayne Mastin

6.1 Introduction

Experience in the field of computational simulation has shown that grid quality in terms of smoothness and orthogonality affects the accuracy of numerical solutions. It has been pointed out by Thompson et al. [8] that skewness increases the truncation error in numerical differentiation. Especially critical in many applications is orthogonality or near-orthogonality of a computational grid near the boundaries of the grid. If the boundary does not correspond to a physical boundary in the simulation, orthogonality can still be important to ensure a smooth transition of grid lines between the grid and the adjacent grid presumed to be across the nonphysical boundary. If the grid boundary corresponds to a physical boundary, then orthogonality may be necessary near the boundary to reduce truncation errors occurring in the simulation of boundary layer phenomena, such as will be present in a Navier–Stokes simulation. In this case, fine spacing near the boundary may also be necessary to accurately resolve the boundary phenomena.

In elliptic grid generation, an initial grid (assumed to be algebraically computed using transfinite interpolation of specified boundary data) is relaxed iteratively to satisfy a quasi-linear elliptic system of partial differential equations (PDEs). The most popular method, the Thompson, Thames, Mastin (TTM) method, incorporates user-specifiable *control functions* in the system of PDEs. If the control functions are not used (i.e., set to zero), then the grid produced will be smoother than the initial grid, and grid folding (possibly present in the initial grid) may be alleviated. However, nonuse of control functions in general leads to nonorthogonality and loss of grid point spacing near the boundaries.

Imposition of boundary orthogonality can be effected in two different ways. In *Neumann orthogonality*, no control functions are used, but boundary grid points are allowed to slide along the boundaries until boundary orthogonality is achieved and the elliptic system has iterated to convergence. This method, which is taken up in this chapter, is appropriate for nonphysical (internal) grid boundaries, since grid spacing present in the initial boundary distribution is usually not maintained. Previous methods for

0-8493-2687-7/99/$0.00+$.50

implementing Neumann orthogonality have relied on a Newton iteration method to locate the orthogonal projection of an adjacent interior grid point onto the boundary. The Neumann orthogonality method presented here uses a Taylor series to move boundary points to achieve approximate orthogonality. Thus, there is no need for inner iterations to compute boundary grid point positions.

In *Dirichlet orthogonality*, also taken up in this chapter, control functions (called *orthogonal control functions*) are used to enforce orthogonality near the boundary while the initial boundary grid point distribution is not disturbed. Early papers using this approach were written by Sorenson [3] and Thomas and Middlecoff [6]. In Sorenson's approach, the control functions are assumed to be of a particular exponential form. Orthogonality and a specified spacing of the first grid line off the boundary are achieved by updating the control functions during iterations of the elliptic system. Thompson [7] presents a similar technique for updating the orthogonal control functions. This technique evaluates the control functions on the boundary and interpolates for interior values. A user-specified grid spacing normal to the boundary is required.

The technique of Spekreijse [5] automatically constructs control functions solely from the specified boundary data without explicit user-specification of grid spacing normal to the boundary. Through construction of an intermediate parametric domain by arclength interpolation of the specified boundary point distribution, the technique ensures accurate transmission of the boundary point distribution throughout the final orthogonal grid. Applications to planar and surface grids are given in [5].

In this chapter, we present a technique similar to [7] for updating of orthogonal control functions during elliptic iteration. However, our technique does not require explicit specification of grid spacing normal to the boundary but, as in [5], employs an interpolation of boundary values to supply the necessary information. However, unlike [5], this interpolation is not constructed in an auxiliary parametric domain, but is simply the initial algebraic grid constructed using transfinite interpolation. Although this grid is very likely skewed at the boundary, the first interior coordinate surface is assumed to be correctly positioned in relation to the boundary, which is enough to give us the required normal spacing information for iterative calculation of the control functions. Ghost points, exterior to the boundary, are constructed from the interior coordinate surface, leading to potentially smoother grids, since central differencing can now be employed at the boundary in the direction normal to the boundary. Since our technique does not employ the auxiliary parametric domain of [5], theory and implementation are simpler. The implementation of this technique for the case of volume grids is straightforward, and indeed we present an example.

We mention here that Soni [2] presents another method of constructing an orthogonal grid by deriving spacing information from the initial algebraic grid. However, unlike our method which uses ghost points at the boundary, this method does not emphasize capture of grid spacing information at the boundary. Instead, the algebraic grid influences the grid spacing of the elliptic grid in a uniform way throughout the domain. With no special treatment of spacing at the boundary, considerable changes in normal grid spacing can occur during the course of elliptic iteration. This may be unacceptable in applications where the most numerically challenging physics occurs at the boundaries.

In Section 6.2, we present Neumann and Dirichlet orthogonality as applied to planar grid generation. We also present a control function blending technique that allows for preservation of interior grid point spacing in addition to preservation of boundary grid point spacing. In Section 6.3, we present analogous techniques for construction of orthogonal surface grids, and in Section 6.4, we present the analogous techniques for volume grids. To demonstrate these techniques, examples are presented in these sections. We present our conclusions in Section 6.5.

6.2 Boundary Orthogonality for Planar Grids

We assume an initial mapping $\mathbf{x}(\xi,\eta) = (x(\xi,\eta), y(\xi,\eta))$ from computational space $[0, m] \times [0, n]$ to the bounded physical domain $\Omega \subset I\!R^2$. Here m, n are positive integers and *grid lines* are the lines $\xi = i$ or $\eta = j$, with $0 \le i \le m$ or $0 \le j \le n$ being integers. The initial mapping $\mathbf{x}(\xi,\eta)$ is usually obtained using algebraic grid generation methods such as linear transfinite interpolation.

Given the initial mapping, a general method for constructing curvilinear structured grids is based on partial differential equations (see Thompson et al. [8]). The coordinate functions $x(\xi,\eta)$ and $y(\xi,\eta)$ are iteratively relaxed until they become solutions of the following quasi-linear elliptic system:

$$g_{22}\left(\mathbf{x}_{\xi\xi}+P\mathbf{x}_{\xi}\right)-2g_{12}\mathbf{x}_{\xi\eta}+g_{11}\left(\mathbf{x}_{\eta\eta}+Q\mathbf{x}_{\eta}\right)=0 \tag{6.1}$$

where

$$g_{11}=\mathbf{x}_{\xi}\cdot\mathbf{x}_{\xi}=x_{\xi}^{2}+y_{\xi}^{2},$$
$$g_{12}=\mathbf{x}_{\xi}\cdot\mathbf{x}_{\eta}=x_{\xi}x_{\eta}+y_{\xi}y_{\eta},$$
$$g_{22}=\mathbf{x}_{\eta}\cdot\mathbf{x}_{\eta}=x_{\eta}^{2}+y_{\eta}^{2}.$$

The control functions P and Q control the distribution of grid points. Using $P = Q = 0$ tends to generate a grid with a uniform spacing. Often, there is a need to concentrate points in a certain area of the grid such as along particular boundary segments — in this case, it is necessary to derive appropriate values for the control functions.

To complete the mathematical specification of system Eq. 6.1, boundary conditions at the four boundaries must be given. (These are the $\xi = 0$, $\xi = m$, $\eta = 0$, and $\eta = n$ or "left," "right," "bottom," and "top" boundaries.) We assume the orthogonality condition

$$\mathbf{x}_{\xi}\cdot\mathbf{x}_{\eta}=0, \qquad \text{on } \xi=0,m, \quad \text{and } \eta=0,n. \tag{6.2}$$

We assume that the initial algebraic grid neither satisfies Eq. 6.1 nor Eq. 6.2. Nevertheless, the initial grid may possess grid point density information that should be present in the final grid. If the algebraic grid possesses important grid density information, such as concentration of grid points in the vicinity of certain boundaries, then it is necessary to invoke "Dirichlet orthogonality" wherein we use the freedom of specifying the control functions P, Q in such a fashion as to allow satisfaction of Eq. 6.1, Eq. 6.2 without changing the initial boundary point distribution at all, and without greatly changing the interior grid point distribution. If, however, the algebraic grid does not possess relevant grid density information (such as may be the case when the grid is an "interior block" that does not border any physical boundary), we attempt to solve Eq. 6.1, Eq. 6.2 using the simplest assumption $P = Q = 0$. Since we are not using the degrees of freedom afforded by specifying the control functions, we are forced to allow the boundary points to "slide to allow satisfaction of Eq. 6.1, Eq. 6.2. This is "Neumann orthogonality." The composite case of having some boundaries treated using Dirichlet orthogonality, some treated using Neumann orthogonality, and some boundaries left untreated will be clear after our treatment of the pure Neumann and Dirichlet cases.

6.2.1 Neumann Orthogonality

As is typical, let us assume that the boundary segments are given to be parametric curves (e.g., B-splines). If we set the control functions P, Q to zero, then it will be necessary to slide the boundary nodes along the parametric curves in order to satisfy Eq. 6.1, Eq. 6.2. A standard discretization of our system is central differencing in the ξ and η directions. The system is then applied to the interior nodes to solve for $\mathbf{x}_{i,j} = (x_{i,j}, y_{i,j})$ using an iterative method.

With regard to the implementation of boundary conditions, suppose along the boundary segments $\xi = 0$ and $\xi = m$ the variables x and y can be expressed in terms of a parameter u as $x = x(u)$ and $y = y(u)$. For the $\xi = 0$ and $\xi = m$ boundaries, let $(\mathbf{x}_{\eta})_{i,j}$ denote the central difference $(1/2(\mathbf{x}_{i,j+1} - \mathbf{x}_{i,j-1}))$ along the boundaries ($i = 0$ or $i = m$). Using one-sided differencing for $\mathbf{x}_{\xi}$, Eq. 6.2 is discretized as

$$\left(\mathbf{x}_{i+1,j}-\mathbf{x}_{i,j}\right)\cdot\left(\mathbf{x}_{\eta}\right)_{0,j}=0, \quad \text{along } \xi=i=0 \tag{6.3}$$

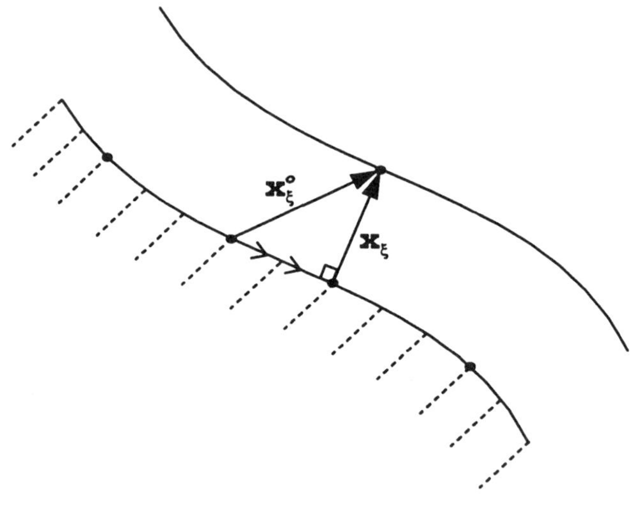

FIGURE 6.1 Change in x_ξ when boundary point is repositioned in Neumann orthogonality.

$$\left(\mathbf{x}_{i,j}-\mathbf{x}_{i-1,j}\right)\cdot\left(\mathbf{x}_\eta\right)_{m,j}=0,\quad \text{along } \xi=i=m. \tag{6.4}$$

Solution of Eq. 6.3 or Eq. 6.4 for $\mathbf{x}_{i,j} = (x_{i,j}, y_{i,j})$ in effect causes the sliding of $\mathbf{x}_{i,j}$ along the boundary so that the grid segment between $\mathbf{x}_{i,j}$ and its neighbor on the first interior coordinate curve ($\xi = 1$ or $\xi = m - 1$) is orthogonal to the boundary curve. (See Figure 6.1.)

To solve for $\mathbf{x}_{i,j}$ the old parameter value u_0 is used to solve for the new u to compute the new $\mathbf{x}_{i,j}$. Using the Taylor expansion of $\mathbf{x}(u)$ about u_0 to give

$$\mathbf{x}_{i,j}=\mathbf{x}(u)\approx\mathbf{x}(u_0)+\mathbf{x}_u(u_0)(u-u_0), \tag{6.5}$$

substituting Eq. 6.5 in Eq. 6.3 implies that

$$u=u_0+\frac{\left(\mathbf{x}_\eta\right)_{0,j}\cdot\left(\mathbf{x}_{1,j}-\mathbf{x}(u_0)\right)}{\left(\mathbf{x}_\eta\right)_{0,j}\cdot\mathbf{x}_u(u_0)} \tag{6.6}$$

to give $\mathbf{x}_{i,j} = \mathbf{x}(u)$ along the boundary $\xi = 0$. Whereas, substituting Eq. 6.5 in Eq. 6.4 implies that

$$u=u_0+\frac{\left(\mathbf{x}_\eta\right)_{m,j}\cdot\left(\mathbf{x}_{m-1,j}-\mathbf{x}(u_0)\right)}{\left(\mathbf{x}_\eta\right)_{m,j}\cdot\mathbf{x}_u(u_0)} \tag{6.7}$$

to give $\mathbf{x}_{i,j} = \mathbf{x}(u)$ along the boundary $\xi = m$.

Consider next the case where the boundaries are $\eta = 0$ and $\eta = n$. Orthogonality Eq. 6.2 with central differencing in the ξ direction and one-sided differencing in the η direction implies

$$u=u_0+\frac{\left(\mathbf{x}_\xi\right)_{i,0}\cdot\left(\mathbf{x}_{i,1}-\mathbf{x}(u_0)\right)}{\left(\mathbf{x}_\xi\right)_{i,0}\cdot\mathbf{x}_u(u_0)}, \tag{6.8}$$

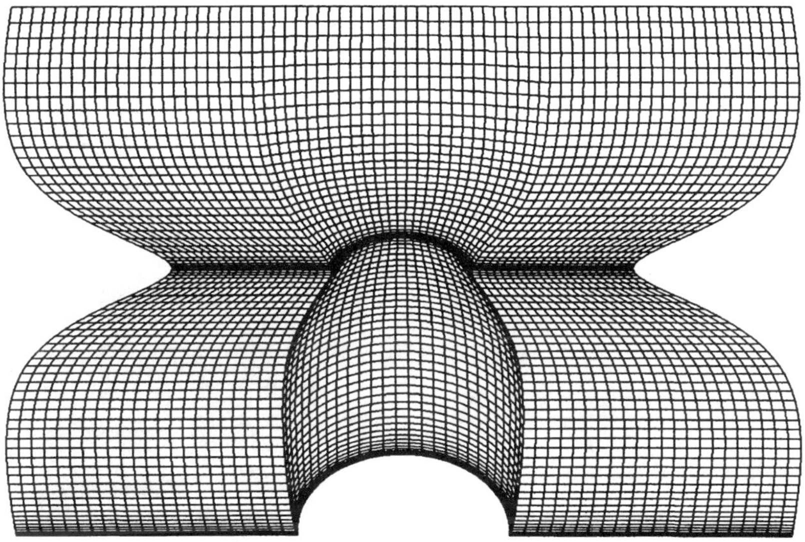

FIGURE 6.2 An algebraic planar grid on a bicubic geometry.

which gives $\mathbf{x}_{i,j} = \mathbf{x}(u)$ along the boundary $\eta = 0$, and

$$u = u_0 + \frac{\left(\mathbf{x}_\xi\right)_{i,n} \cdot \left(\mathbf{x}_{i,n-1} - \mathbf{x}(u_0)\right)}{\left(\mathbf{x}_\xi\right)_{i,n} \cdot \mathbf{x}_u(u_0)} \tag{6.9}$$

to give $\mathbf{x}_{i,j} = \mathbf{x}(u)$ along the boundary $\eta = n$.

These boundary condition equations are to be evaluated for each cycle in the course of the iterative procedure. Note that a periodic boundary condition is used in the case of doubly connected regions. Also note that during the relaxation process, "guards" must be used to prevent a given boundary point from overtaking its neighbors when sliding along the boundaries. Indeed, near obtuse corners, there is a tendency for grid points to try to slide along the boundary curves past the corners in order to satisfy the orthogonality condition. An appropriate guard would be to limit movement of each grid point so that its distance from its two boundary-curve neighbors is reduced by at most 50% on a given iteration, down to a user-specified minimum length δ in physical space.

As an application of Neumann orthogonality, consider Figure 6.2, which is an initial algebraic planar grid on a bicubic geometry. The mesh is highly nonorthogonal at certain points along the boundaries, and it possesses an undesirable concentration of points in the interior of the grid. In fact, there is folding of the algebraic grid in this central region.

Figure 6.3 shows an elliptically smoothed grid using Neumann orthogonality. The grid is clearly seen to be smooth, boundary-orthogonal, and no longer folds in the interior. For certain applications, this grid may be entirely acceptable. However, if the bottom boundary of the grid corresponded to a physical boundary, then the results of Figure 6.3 might be deemed unacceptable. This is because, although orthogonality has been established, grid point distribution (both along the boundary and normal to the boundary) has been significantly altered. In this case, the Dirichlet orthogonality technique will have to be employed.

6.2.2 Dirichlet Orthogonality

The above discussion shows how orthogonality can be imposed without use of control functions, by sliding grid points along the boundary. Orthogonality can also be imposed by adjusting the control

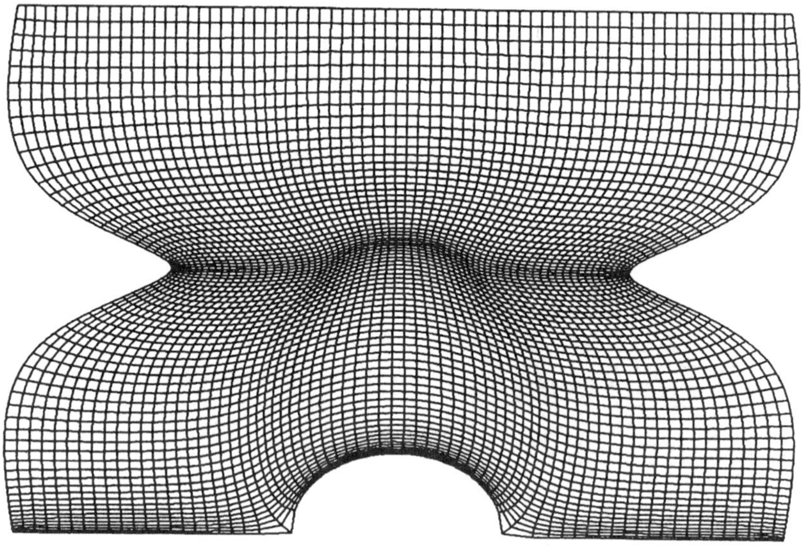

FIGURE 6.3 An elliptic planar grid on a bicubic geometry with Neumann orthogonality.

functions near the boundary and keeping the boundary points fixed. This approach was originally developed by Sorenson [3] for imposing boundary orthogonality in two dimensions. Sorenson [4] and Thompson [7] have extended this approach to three dimensions. However, as mentioned in the introduction, our approach does not require user specification of grid spacing normal to the boundary. Instead, our technique automatically derives normal grid spacing data from the initial algebraic grid.

Assuming boundary orthogonality Eq. 6.2, substitution of the inner product of $\mathbf{x}_\xi$ and $\mathbf{x}_\eta$ into Eq. 6.1 yields the following two equations for the control functions on the boundaries:

$$P = -\frac{\mathbf{x}_\xi \cdot \mathbf{x}_{\xi\xi}}{g_{11}} - \frac{\mathbf{x}_\xi \cdot \mathbf{x}_{\eta\eta}}{g_{22}}$$
$$Q = -\frac{\mathbf{x}_\eta \cdot \mathbf{x}_{\eta\eta}}{g_{22}} - \frac{\mathbf{x}_\eta \cdot \mathbf{x}_{\xi\xi}}{g_{11}} \tag{6.10}$$

These control functions are called the *orthogonal control functions* because they were derived using orthogonality considerations. They are evaluated at the boundaries and interpolated to the interior using linear transfinite interpolation. These functions need to be updated at every iteration during solution of the elliptic system.

We now go into detail on how we evaluate the quantities necessary in order to compute P and Q on the boundary using Eq. 6.10. Suppose we are at the "left" boundary $\xi = 0$, but not at the corners ($\eta \neq 0$ and $\eta \neq n$). The derivatives $\mathbf{x}_\eta$, $\mathbf{x}_{\eta\eta}$ and the spacing $g_{22} = \|\mathbf{x}_\eta\|^2$ are determined using centered difference formulas from the boundary point distribution and do not change. However, the g_{11}, $\mathbf{x}_\xi$, and $\mathbf{x}_{\xi\xi}$ terms are not determined by the boundary distribution. Additional information amounting to the desired grid spacing normal to the boundary must be supplied.

A convenient way to infer the normal boundary spacing from the initial algebraic grid is to assume that the position of the first interior grid line off the boundary is correct. Indeed, near the boundary, it is usually the case that all that is desired of the elliptic iteration is for it to swing the intersecting grid lines so that they intersect the boundaries orthogonally, *without changing the positions* of the grid lines parallel to the boundary. This is shown graphically in Figure 6.4, where we see a grid point, from the first interior grid line, swung along the grid line to the position where orthogonality is established. The

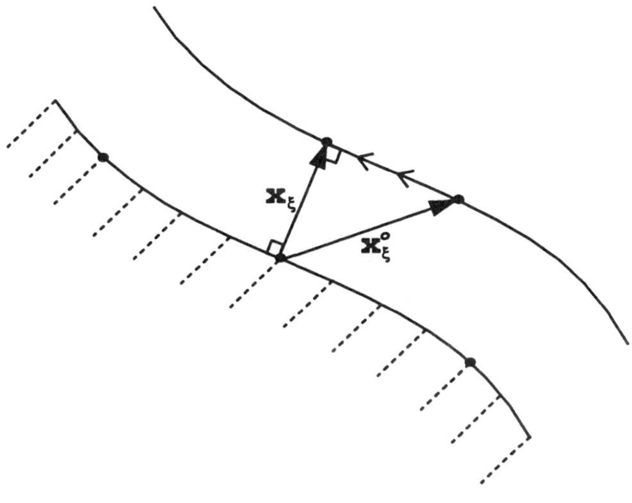

FIGURE 6.4 Projection of interior algebraic grid point to orthogonal position.

effect of forcing all the grid points to swing over in this fashion would thus be to establish boundary orthogonality, but still leave the algebraic interior grid line unchanged. The similarity of Figure 6.1 and Figure 6.4 seems to indicate that this process is analogous to, and hence just as "natural" as, the process of sliding the boundary points in the Neumann orthogonality approach with zero control functions.

Unfortunately, this preceding approach entails the *direct specification* of the positions of the first interior layer of grid points off the boundary. This is not permissible for a couple of reasons. First, since they are adjacent to *two* different boundaries, the points $\mathbf{x}_{1,1}$, $\mathbf{x}_{m-1,1}$, $\mathbf{x}_{1,n-1}$, and $\mathbf{x}_{m-1,n-1}$ have contradictory definitions for their placement. Second, and more importantly, the direct specification of the first layer of interior boundary points together with the elliptic solution for the positions of the deeper interior grid points can lead to an undesirable "kinky" transition between the directly placed points and the elliptically solved-for points. (This "kinkiness" is due to the fact that a perfectly smooth boundary-orthogonal grid will probably exhibit some small degree of nonorthogonality as soon as one leaves the boundary — even as close to the boundary as the first interior line. Hence, forcing the grid points on the first interior line to be *exactly* orthogonal to the boundary cannot lead to the smoothest possible boundary-orthogonal grid.)

Nevertheless, our "natural" approach for deriving grid spacing information from the algebraic grid can be modified in a simple way, as depicted in Figure 6.5. Here, the orthogonally-placed interior point is reflected an equal distance across the boundary curve to form a "ghost point." Repeatedly done, this procedure in effect forms an "exterior curve" of ghost points that is the reflection of the first (algebraic) grid line across the boundary curve. The ghost points are computed at the beginning of the iteration and do not change. They are employed in the calculation of the normal second derivative $\mathbf{x}_{\xi\xi}$ at the boundary and the normal spacing $\sqrt{g_{11}}$ off the boundary; the fixedness of the ghost points assures that the normal spacing is not lost during the course of iteration, as it sometimes is in the Neumann orthogonality approach. Conversely, all of the interior grid points are free to change throughout the course of the iteration, and so smoothness of the grid is not compromised.

More precisely, again at the "left" $\xi = 0$ boundary, let $(\mathbf{x}_\eta)_{0,j}$ denote the centrally differenced derivative $1/2(\mathbf{x}_{0,j+1} - \mathbf{x}_{0,j-1})$. Let $(\mathbf{x}_\xi^o)_{0,j}$ denote the one-sided derivative $\mathbf{x}_{1,j} - \mathbf{x}_{0,j}$ evaluated on the initial algebraic grid. Then condition Eq. 6.2 implies that if $\mathbf{a}$ is the unit vector normal to the boundary, then

$$\mathbf{a} \equiv \frac{\mathbf{x}_\xi}{\|\mathbf{x}_\xi\|} = \frac{y_\eta, -x_\eta}{\sqrt{x_\eta^2 + y_\eta^2}} = \frac{(y_\eta, -x_\eta)}{\sqrt{g_{22}}},$$

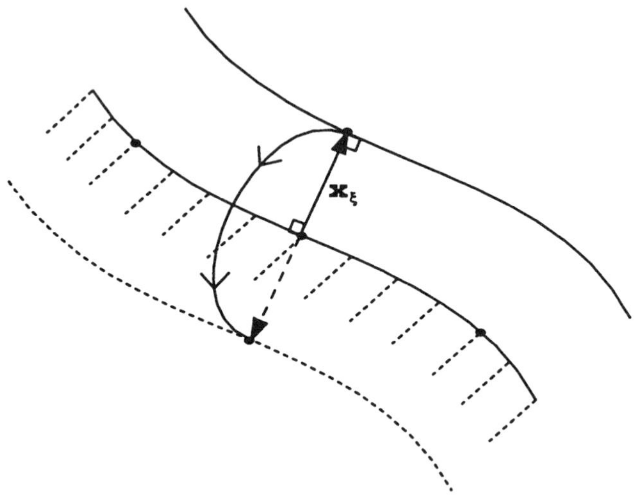

FIGURE 6.5 Reflection of orthogonalized interior grid point to form external ghost point.

Now the condition from Figure 6.4 is

$$\mathbf{x}_\xi = \mathbf{P}_\mathbf{a}\left(\mathbf{x}_\xi^0\right), \tag{6.11}$$

where $\mathbf{P}_\mathbf{a} = \mathbf{a}\mathbf{a}^T$ is the orthogonal projection onto the one-dimensional subspace spanned by the unit vector **a**. Thus we obtain

$$\mathbf{x}_\xi = \mathbf{a}\left(\mathbf{a}\cdot\mathbf{x}_\xi^0\right) = \frac{\left(y_\eta, -x_\eta\right)}{g_{22}}\left(y_\eta x_\xi^0 - x_\eta y_\xi^0\right). \tag{6.12}$$

Finally, the reflection operation of Figure 6.5 implies that the fixed ghost point location should be given by

$$\mathbf{x}_{-1,j} = \mathbf{x}_{0,j} - \left(\mathbf{x}_\xi\right)_{0,j}.$$

This can also be viewed as a first-order Taylor expansion involving the orthogonal derivative $(\mathbf{x}_\xi)_{0,j}$:

$$\mathbf{x}_{-1,j} = \mathbf{x}_{0,j} + \Delta\xi\left(\mathbf{x}_\xi\right)_{0,j},$$

with $\Delta\xi = -1$. The orthogonal derivative $(\mathbf{x}_\xi)_{0,j}$ is computed in Eq. 6.12 using only data from the boundary and the algebraic grid. Now in Eq. 6.10, the control function evaluation at the boundary, the second derivative $\mathbf{x}_{\xi\xi}$ is computed using a centered difference approximation involving a ghost point, a boundary point, and an iteratively updated interior point. The metric coefficient g_{11} describing spacing normal to the boundary is computed using Eq. 6.12 and is given by

$$\left(g_{11}\right)_{0,j} = \left(\mathbf{x}_\xi\right)_{0,j}\cdot\left(\mathbf{x}_\xi\right)_{0,j}.$$

Finally, note that the value for $(\mathbf{x}_\xi)_{0,j}$ used in Eq. 6.10 is not the fixed value given by Eq. 6.12, but is the iteratively updated one-sided difference formula given by

$$\left(\mathbf{x}_\xi\right)_{0,j} = \mathbf{x}_{i,j} - \mathbf{x}_{0,j}.$$

Evaluation of quantities at the $\xi = m$ boundary is similar. Note, however, that the ghost point locations are given by

$$\mathbf{x}_{m+1,j} = \mathbf{x}_{m,j} + \left(\mathbf{x}_\xi\right)_{m,j},$$

where $(\mathbf{x}_\xi)_{m,j}$ is evaluated in Eq. 6.12, which is also valid for this boundary.

On the "bottom" and "top" boundaries $\eta = 0$ and $\eta = n$, it is now the derivatives x_η, $\mathrm{x}_{\eta\eta}$, and the spacing g_{11} that are evaluated using the fixed boundary data using central differences. Using similar reasoning to the "left" and "right" boundary case, we obtain that for the "bottom" boundary the ghost point location is fixed to be

$$\mathbf{x}_{i,-1} = \mathbf{x}_{i,0} - \left(\mathbf{x}_\eta\right)_{i,0},$$

where we use

$$\mathbf{x}_\eta = \frac{\left(-y_\xi, x_\xi\right)}{g_{11}}\left(-y_\xi x_\eta^0 + x_\xi y_\eta^0\right). \tag{6.13}$$

Here, $(-y_\eta, x_\eta)$, g_{11} is evaluated using central differencing of the boundary data, and (x_η^o, y_η^o) represents a one-sided derivative $\mathbf{x}_{i,1} - \mathbf{x}_{i,0}$ evaluated on the initial algebraic grid. The metric coefficient $(g_{22})_{i,0} = (\mathbf{x}_\eta)_{i,0} \cdot (\mathbf{x}_\eta)_{i,0}$ is now computed using Eq. 6.13, and $\mathbf{x}_{\eta\eta}$ is computed using a ghost point, a boundary point, and an iteratively updated interior point. The value of $(\mathbf{x}_\eta)_{i,0}$ used in Eq. 6.10 is not the fixed value given in Eq. 6.13, but is the iteratively updated one-sided difference formula given by

$$\left(\mathbf{x}_\eta\right)_{i,0} = \mathbf{x}_{i,1} - \mathbf{x}_{i,0}.$$

Finally, the "upper" $\eta = n$ boundary is similar, and we note that the ghost-point locations are given by

$$\mathbf{x}_{i,n+1} = \mathbf{x}_{i,n} + \left(\mathbf{x}_\eta\right)_{i,n},$$

with $(\mathbf{x}_\eta)_{i,n}$, evaluated using Eq. 6.13.

Quantities for the four corner points, $\mathbf{x}_{0,0}$, $\mathbf{x}_{m,0}$, $\mathbf{x}_{0,n}$ and $\mathbf{x}_{m,n}$, are computed somewhat differently in that no orthogonality considerations or ghost points are used. Indeed, the values $\mathbf{x}_\xi$, $\mathbf{x}_{\xi\xi}$, $\mathbf{x}_\eta$, $\mathbf{x}_{\eta\eta}$, g_{11}, g_{22} are all evaluated once using one-sided difference formulas that use the specified boundary values and do not change during the course of iteration. We forego imposition of orthogonality at the corners, because at the corners *conformality* is more important than *orthogonality.* In other words, orthogonality at the corners should be sacrificed in order to ensure that the resulting grid does not spill over the physical boundaries in the neighborhood of the corners. For the case of highly obtuse or highly acute corners, it may in fact be necessary to relax orthogonality in the regions that are within *several* grid lines of the corners. One way to do this is to construct ghost points near the corners with the orthogonal projection operation Eq. 6.11 omitted (i.e., constructed by simple extrapolation), and to use a blend of these ghost points and the ghost points derived using the orthogonality assumption.

To further ensure that the elliptic system iterations do not cause grid folding near the boundaries, "guards" may be employed, similar to those mentioned in the previous section on Neumann orthogonality. In practice, however, we have found these to be unnecessary for Dirichlet orthogonality.

6.2.2.1 Blending of Orthogonal and Initial Control Functions

The orthogonal control functions in the interior of the grid are interpolated from the boundaries using linear transfinite interpolation and updated during the iterative solution of the elliptic system. If the initial algebraic grid is to be used only to infer correct spacing at the boundaries, then it is sufficient to use these orthogonal control functions in the elliptic iteration. However, note that the orthogonal control functions do not incorporate information from the algebraic grid beyond the first interior grid line. Thus if it is desired to maintain the entire initial interior point distribution, then at each iteration the orthogonal control functions must be smoothly blended with control functions that represent the grid density information in the whole algebraic grid. These latter control functions we refer to as "initial control function," and their computation is now described.

The elliptic system Eq. 6.1 can be solved simultaneously at each point of the algebraic grid for the two functions P and Q by solving the following linear system:

$$\begin{pmatrix} g_{22}\,x_\xi & g_{11}\,x_\eta \\ g_{22}\,y_\xi & g_{11}\,y_\eta \end{pmatrix} \begin{pmatrix} P \\ Q \end{pmatrix} = \begin{pmatrix} R_1 \\ R_2 \end{pmatrix} \tag{6.14}$$

where

$$R_1 = 2g_{12}x_{\xi\eta} - g_{22}x_{\xi\xi} - g_{11}x_{\eta\eta} \quad \text{and}$$
$$R_2 = 2g_{12}y_{\xi\eta} - g_{22}y_{\xi\xi} - g_{11}y_{\eta\eta}.$$

The derivatives here are represented by central differences, except at the boundaries where one-sided difference formulas must be used. This produces control functions that will reproduce the algebraic grid from the elliptic system solution in a single iteration. Thus, evaluation of the control functions in this manner would be of trivial interest except when these control functions are smoothed before being used in the elliptic generation system. This smoothing is done by replacing the control function at each point with the average of the nearest neighbors along one or more coordinate lines. However, we note that the P control function controls spacing in the ξ-direction and the Q control function controls spacing in the η-direction. Since it is desired that grid spacing normal to the boundaries be preserved between the initial algebraic grid and the elliptically smoothed grid, we cannot allow smoothing of the P control function along ξ-coordinate lines or smoothing of the Q control function along η-coordinate lines. This leaves us with the following smoothing iteration where smoothing takes place only along allowed coordinate lines:

$$\begin{aligned} P_{i,j} &= \frac{1}{2}\left(P_{i,j+1} + P_{i,j-1}\right) \\ Q_{i,j} &= \frac{1}{2}\left(Q_{i+1,j} + Q_{i-1,j}\right). \end{aligned} \tag{6.15}$$

Smoothing of control functions is done for a small number of iterations.

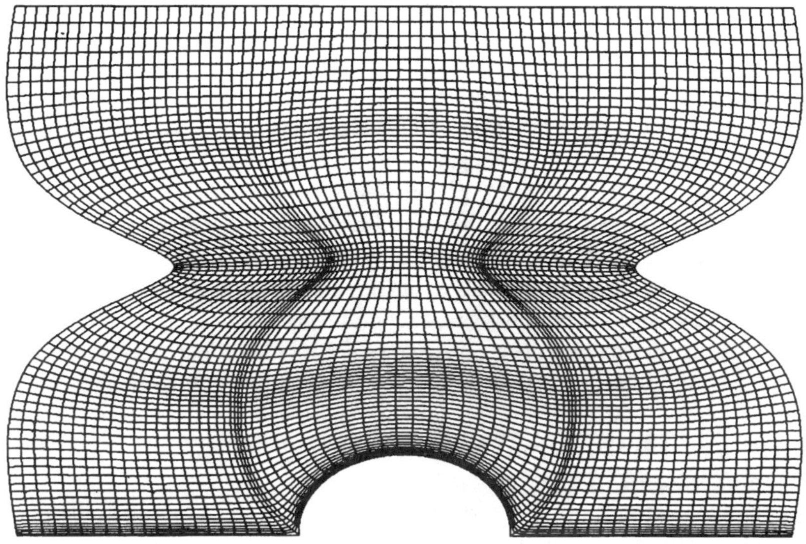

FIGURE 6.6 An elliptic planar grid on a bicubic geometry with Dirichlet orthogonality.

Finally, by blending the smoothed initial control functions together with orthogonal control functions, we will produce control functions that will result in preservation of grid density information throughout the grid, along with boundary orthogonality. An appropriate blending function for this purpose is

$$b_{i,j} = e^{-\frac{1}{\delta}\frac{i}{m}\frac{j}{n}\frac{m-i}{m}\frac{n-j}{n}},$$

where δ is some positive number chosen such that the exponential decays smoothly from unity on the boundary to nearly zero in the interior. δ can be considered to be the characteristic length of the decay of the blending function in the (ξ,η) domain. So, for example, if $\delta = .05$, the orthogonal control functions heavily influence a region consisting of 5% of grid lines which are nearest to each boundary. Now the new blended values of the control functions are computed as follows:

$$\begin{aligned} P(i,j) &= b_{i,j}P_O(i,j) + \left(1-b_{i,j}\right)P_I(i,j) \\ Q(i,j) &= b_{i,j}Q_O(i,j) + \left(1-b_{i,j}\right)Q_I(i,j) \end{aligned} \tag{6.16}$$

where P_O and Q_O are the orthogonal control functions from Eq. 6.10. P_I and Q_I are the smoothed initial control functions computed using Eqs. 6.14 and 6.15.

As an application of Dirichlet orthogonality, in Figure 6.6 we show the results of smoothing the algebraic grid of Figure 6.2 using orthogonal control functions only. Like the grid produced using Neumann orthogonality, the grid is smooth, boundary-orthogonal, and no longer folds in the interior. However, unlike the grid of Figure 6.3, we see that the grid of Figure 6.6 *preserves* the grid point density information of the algebraic grid at the boundaries. The effect of smoothing near the boundaries has been essentially to slide nodes along the coordinate lines parallel to the boundaries, without affecting the spacing between the coordinate lines normal to the boundary.

We note that if the user for some reason wished to preserve the *interior* clustering of grid points in the algebraic grid, then the above scheme given for blending initial control functions with orthogonal control functions would have to be slightly modified. This is because the fact that the algebraic grid is actually folded in the interior makes the evaluation of the initial control functions using Eq. 6.14 ill-defined. This is easily remedied by evaluating the initial control functions using Eq. 6.14 at the boundaries only using one-sided derivatives, and then defining them over the whole mesh using transfinite interpolation. Since there is no folding of the algebraic grid at the boundaries, this is well-defined. (The interpolated initial control functions will reflect the grid density information in the interior of the initial grid, because the interior grid point distribution of the initial grid was computed using the same process — transfinite interpolation of boundary data.) Then we proceed as above, smoothing the initial control functions and blending them with the orthogonal control functions.

Finally we note that if the algebraic initial grid possesses folding *at the boundary*, then using data from the algebraic grid to evaluate either the initial control functions or the orthogonal control functions at the boundary will not work. In this case, one could reject the algebraic grid entirely and manually specify grid density information at the boundary. This would however defeat the purpose of our approach, which is to simplify the grid generation process by reading grid density information off of the algebraic grid. Instead, we suggest that in this case the geometry be subdivided into patches sufficiently small so that the algebraic initial grids on these patches do not possess grid folding at the boundaries.

6.3 Boundary Orthogonality for Surface Grids

Now we turn our attention to applying the same principles of the previous section to the case of surface grids. Our surface is assumed to be defined as a mapping $\mathbf{x}(u,v)$: $\mathbb{R}^2 \to \mathbb{R}^3$. The (u,v) space is the *parametric space*, which we conveniently take to be $[0,1] \times [0,1]$. The parametric variables are themselves taken to be functions of the computational variables ξ, η, which live in the usual $[0, m] \times [0, n]$ domain. Thus

$$\mathbf{x} = (x,y,z) = (x(u,v), y(u,v), z(u,v)) \text{ and } (u,v) = (u(\xi,\eta), v(\xi,\eta)) \tag{6.17}$$

The mapping $\mathbf{x}(u,v)$ and its derivatives $\mathbf{x}_u$, $\mathbf{x}_v$, etc., are assumed to be known and evaluatable at reasonable cost. It is the aim of surface grid generation to provide a "good" mapping $(u(\xi,\eta), v(\xi,\eta))$ so that the composite mapping $\mathbf{x}(\mathbf{u}(\xi,\eta), \mathbf{v}(\xi,\eta))$ has desirable features, such as boundary orthogonality and an acceptable distribution of grid points.

A general method for constructing curvilinear structured surface grids is based on partial differential equations (see Khamayseh and Mastin [1], Warsi [9], and Chapter 9). The parametric variables u and v are solutions of the following quasi-linear elliptic system:

$$g_{22}(u_{\xi\xi} + Pu_\xi) - 2g_{12}u_{\xi\eta} + g_{11}(u_{\eta\eta} + Qu_\eta) = J^2\Delta_2 u \quad \text{and} \tag{6.18}$$

$$g_{22}(v_{\xi\xi} + Pv_\xi) - 2g_{12}v_{\xi\eta} + g_{11}(v_{\eta\eta} + Qv_\eta) = J^2\Delta_2 v, \tag{6.19}$$

where

$$\begin{aligned}
g_{11} &= \bar{g}_{11}u_\xi^2 + 2\bar{g}_{12}u_\xi v_\xi + \bar{g}_{22}v_\xi^2,\\
g_{12} &= \bar{g}_{11}u_\xi u_\eta + \bar{g}_{12}(u_\xi v_\eta + u_\eta v_\xi) + \bar{g}_{22}v_\xi v_\eta,\\
g_{22} &= \bar{g}_{11}u_\eta^2 + 2\bar{g}_{12}u_\eta v_\eta + \bar{g}_{22}v_\eta^2,\\
\Delta_2 u &= \bar{J}\left[\frac{\partial}{\partial u}\left(\frac{\bar{g}_{22}}{\bar{J}}\right) - \frac{\partial}{\partial v}\left(\frac{\bar{g}_{12}}{\bar{J}}\right)\right],\\
\Delta_2 v &= \bar{J}\left[\frac{\partial}{\partial v}\left(\frac{\bar{g}_{11}}{\bar{J}}\right) - \frac{\partial}{\partial u}\left(\frac{\bar{g}_{12}}{\bar{J}}\right)\right],\\
\bar{g}_{11} &= \mathbf{x}_u \cdot \mathbf{x}_u, \bar{g}_{12} = \mathbf{x}_u \cdot \mathbf{x}_v, \bar{g}_{22} = \mathbf{x}_v \cdot \mathbf{x}_v,\\
\bar{J} &= \sqrt{\bar{g}_{11}\bar{g}_{22} - \bar{g}_{12}^2}, \quad J = u_\xi v_\eta - u_\eta v_\xi, \quad \text{and}\\
\mathbf{x} &= \mathbf{x}(u,v), \ 0 \le u,v \le 1.
\end{aligned} \tag{6.20}$$

Note that if $x \equiv u$, $y \equiv v$, $z \equiv 0$, then $\bar{g}_{11} = 1$, $\bar{g}_{12} = 0$, $\bar{g}_{22} = 1$, $\bar{J} = 1$, and $\Delta_2 u = \Delta_2 v = 0$, making Eqs. 6.18–6.20 identical to the homogeneous elliptic system for two-dimensional grid generation Eq. 6.1 presented in the previous section.

As in the previous section, the control functions P and Q can be set to zero, and Neumann orthogonality can be imposed by sliding points along the "left," right," "bottom," and "top" boundaries. These four boundaries are respectively $(0, v(0,\eta))$, $(1,v(m,\eta))$, $(u(\xi,0),0)$, $(u(\xi,n), 1)$ in parametric space, which are mapped to the boundaries $\mathbf{x}(0,v)$, $\mathbf{x}(1,v)$, $\mathbf{x}(u,0)$, and $\mathbf{x}(u,1)$ in physical space. Of course orthogonality must be established in physical space. As before, if there is a need to respect the grid point concentration in the initial algebraic grid, we implement Dirichlet orthogonality, deriving appropriate values for P and Q.

6.3.1 Neumann Orthogonality

We require the condition of orthogonality in physical space:

$$\mathbf{x}_\xi \cdot \mathbf{x}_\eta = 0, \quad \text{on } \xi = 0, m, \quad \text{and } \eta = 0, n. \tag{6.21}$$

Symbolically this is identical to Eq. 6.2, but here we understand that $\mathbf{x}$ is a composite function Eq. 6.17 which takes on values in $I\!R^3$. Expanding Eq. 6.21 using the chain rule yields the equation

$$\bar{g}_{11}\, u_\xi u_\eta + \bar{g}_{22}\, v_\xi v_\eta + \bar{g}_{12}\left(u_\xi v_\eta + u_\eta v_\xi\right) = 0.$$

This orthogonality condition is used to formulate derivative boundary conditions for the elliptic system. If the "left" and "right" boundary curves $u = 0$ and $u = 1$ are considered, we have $u_\eta = 0$ and the orthogonality condition reduces to

$$\bar{g}_{22}\, v_\xi + \bar{g}_{12}\, u_\xi = 0. \tag{6.22}$$

Similarly, along the "bottom" and "top" curves $v = 0$ and $v = 1$, $v_\xi = 0$ and orthogonality is imposed by

$$\bar{g}_{11}\, u_\eta + \bar{g}_{12}\, v_\eta = 0. \tag{6.23}$$

When solving the elliptic system, Eq. 6.22 determines the values of v on the boundary segments $u = 0$ and $u = 1$, and Eq. 6.23 determines the values of u on the boundary segments $v = 0$ and $v = 1$.

To implement this numerically, we use forward differencing on the boundaries $u = 0$ and $v = 0$ and backward differencing on the boundaries $u = 1$ and $v = 1$ to compute the new values for $u_{i,j}$ and $v_{i,j}$:

$$\begin{aligned} v_{0,j} &= \frac{\bar{g}_{12}}{\bar{g}_{22}}\left(u_{1,j} - u_{0,j}\right) + v_{1,j} \\ & \qquad\qquad\qquad\qquad 0 < j < n \\ v_{m,j} &= -\frac{\bar{g}_{12}}{\bar{g}_{22}}\left(u_{m,j} - u_{m-1,j}\right) + v_{m-1,j} \\ u_{i,0} &= \frac{\bar{g}_{12}}{\bar{g}_{11}}\left(v_{i,1} - v_{i,0}\right) + u_{i,1} \\ & \qquad\qquad\qquad\qquad 0 < i < m \\ u_{i,n} &= -\frac{\bar{g}_{12}}{\bar{g}_{11}}\left(v_{i,n} - v_{i,n-1}\right) + u_{i,n-1}. \end{aligned}$$

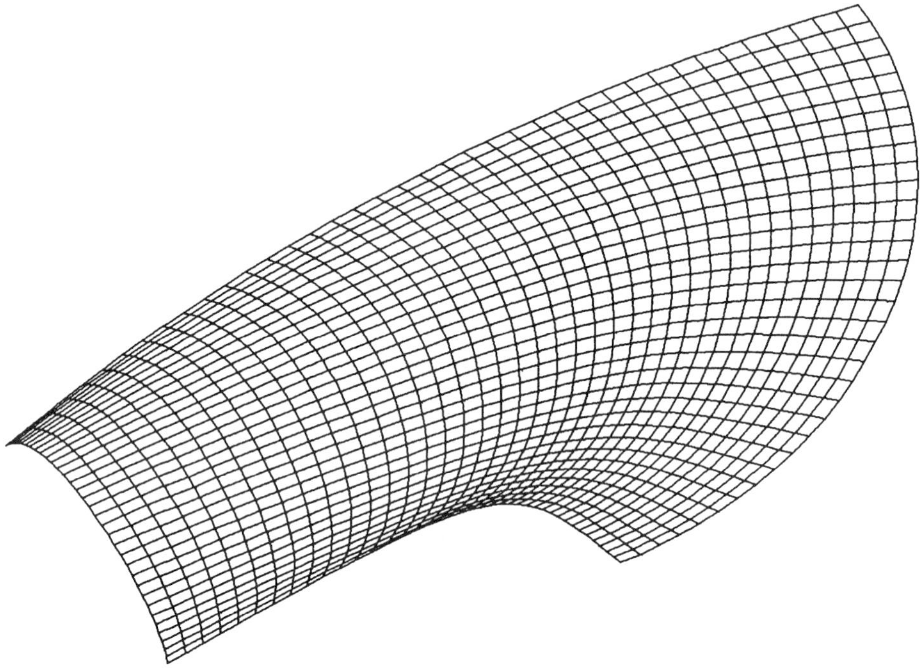

FIGURE 6.7 An algebraic surface grid on a bicubic geometry.

Since the boundary points are permitted to float with the solution as a means to achieve orthogonality (Figure 6.3), the values of $\bar{g}_{ij}$ must, of course, be reevaluated after each cycle using the definition of the geometry $\mathbf{x}(u,v)$. Also, as in the last section, "guards" must be used to prevent a given boundary point from overtaking its neighbors when sliding along the boundaries.

Figure 6.7 shows an initial algebraic grid on a bicubic surface geometry. The grid was obtained using linear transfinite interpolation and is the starting iterate for our elliptic smoothing. Clearly, the initial grid is not orthogonal at the boundaries where orthogonality is often desired, especially for Navier–Stokes computation.

Figure 6.8 shows the elliptically smoothed surface grid on the same geometry. Neumann orthogonality was applied to allow the boundary points to float so that the grid is orthogonal on the boundary. Significant changes in boundary grid spacing occur near some of the corners.

6.3.2 Dirichlet Orthogonality

For the case of Dirichlet orthogonality for surface grids, we essentially follow the same technique as that used in Section 6.2.2. Expressions for the control function P and Q are derived at the boundary using the assumption of orthogonality, and then to facilitate evaluation of these expressions, ghost points are placed orthogonally off the boundary with normal spacing derived from the initial grid (Figure 6.5).

We rewrite the elliptic system Eqs. 6.18–6.19 in vector form:

$$g_{22}\left(\mathbf{u}_{\xi\xi}+P\mathbf{u}_{\xi}\right)-2g_{12}\,\mathbf{u}_{\xi\eta}+g_{11}\left(\mathbf{u}_{\eta\eta}+Q\mathbf{u}_{\eta}\right)=J^2\Delta_2\mathbf{u}, \tag{6.24}$$

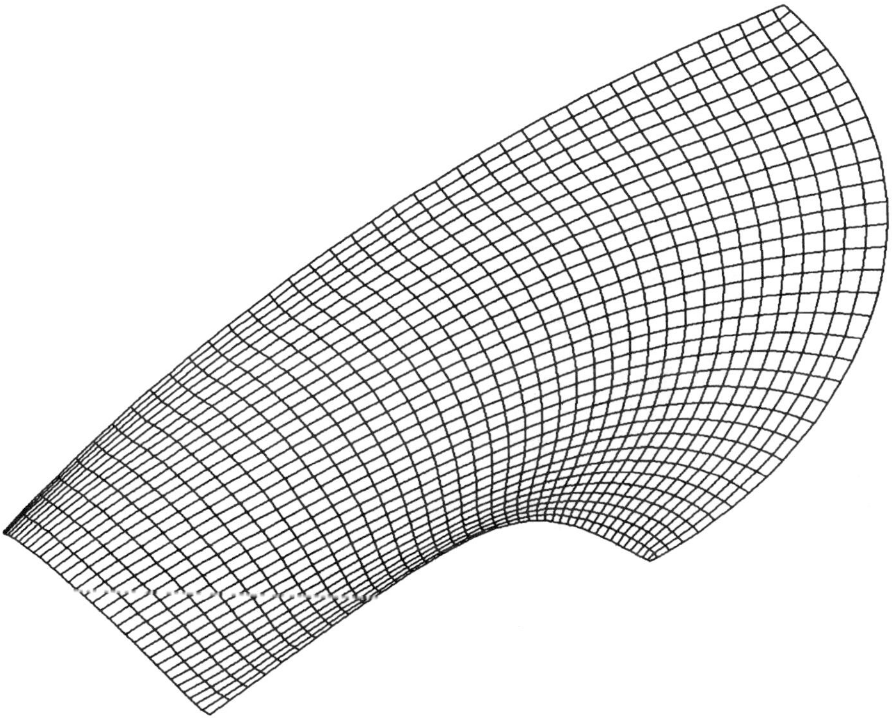

FIGURE 6.8 An elliptic surface grid on a bicubic geometry with Neumann orthogonality.

where $\mathbf{u} = (u,v)$. For $\mathbf{u}_1 = (u_1, v_1)$ and $\mathbf{u}_2 = (u_2, v_2)$, define

$$\mathbf{u}_1 \circ \mathbf{u}_2 = \bar{g}_{11} u_1 u_2 + \bar{g}_{12}(u_1 v_2 + u_2 v_1) + \bar{g}_{22} v_1 v_2 .$$

Note that $\mathbf{u}_1 \circ \mathbf{u}_2 \equiv \mathbf{u}_1^T \ \bar{G} \ \mathbf{u}_2$, which is the inner product in parametric space induced by the metric tensor $\bar{G} = \begin{pmatrix} \bar{g}_{11} & \bar{g}_{12} \\ \bar{g}_{12} & \bar{g}_{22} \end{pmatrix}$. Orthogonality in this inner product is equivalent to orthogonality in physical space.

Suppose that the grid lines are orthogonal, i.e., $\mathbf{x}_\xi \cdot \mathbf{x}_\eta = \mathbf{u}_\xi \circ \mathbf{u}_\eta$ vanishes. Applying $\circ \mathbf{u}_\xi$ to Eq. 6.24 yields

$$g_{22}\left(\mathbf{u}_{\xi\xi} \circ \mathbf{u}_\xi + P\mathbf{u}_\xi \circ \mathbf{u}_\xi\right) + g_{11}\,\mathbf{u}_{\eta\eta} \circ \mathbf{u}_\xi = J^2 \Delta_2\, \mathbf{u} \circ \mathbf{u}_\xi .$$

In the same manner, applying $\circ \mathbf{u}_\xi$ to Eq. 6.24 yields the following equation for the second control function on the boundaries:

$$g_{22}\mathbf{u}_{\xi\xi} \circ \mathbf{u}_\eta + g_{11}\left(\mathbf{u}_{\eta\eta} \circ \mathbf{u}_\eta + Q\mathbf{u}_\eta \circ \mathbf{u}_\eta\right) = J^2 \Delta_2\, \mathbf{u} \circ \mathbf{u}_\eta .$$

The values of P and Q can be determined from the complete expansion of the above equations as follows:

$$P = J^2 \frac{\bar{g}_{11}\Delta_2 u\, u_\xi + \bar{g}_{22}\Delta_2 v\, v_\xi + \bar{g}_{12}\left(\Delta_2 u\, v_\xi + \Delta_2 v\, u_\xi\right)}{g_{11}g_{22}} - \frac{\bar{g}_{11}u_{\eta\eta}u_\xi + \bar{g}_{22}v_{\eta\eta}v_\xi + \bar{g}_{12}\left(u_{\eta\eta}v_\xi + v_{\eta\eta}u_\xi\right)}{g_{22}} - \frac{\bar{g}_{11}u_{\xi\xi}u_\xi + \bar{g}_{22}v_{\xi\xi}v_\xi + \bar{g}_{12}\left(u_{\xi\xi}v_\xi + v_{\xi\xi}u_\xi\right)}{g_{11}}$$

$$Q = J^2 \frac{\bar{g}_{11}\Delta_2 u\, u_\eta + \bar{g}_{22}\Delta_2 v\, v_\eta + \bar{g}_{12}\left(\Delta_2 u\, v_\eta + \Delta_2 v\, u_\eta\right)}{g_{11}g_{22}} - \frac{\bar{g}_{11}u_{\eta\eta}u_\eta + \bar{g}_{22}v_{\eta\eta}v_\eta + \bar{g}_{12}\left(u_{\eta\eta}v_\eta + v_{\eta\eta}u_\eta\right)}{g_{22}} - \frac{\bar{g}_{11}u_{\xi\xi}u_\eta + \bar{g}_{22}v_{\xi\xi}v_\eta + \bar{g}_{12}\left(u_{\xi\xi}v_\eta + v_{\xi\xi}u_\eta\right)}{g_{11}} \quad (6.25)$$

As in the previous section, these control functions derived using orthogonality considerations are called *orthogonal control functions,* are interpolated to the interior using linear transfinite interpolation, and are updated at every iteration during solution of the elliptic system.

We now go into some detail about the exact way these control functions are evaluated at the boundary. The terms $\bar{g}_{11}$, $\bar{g}_{12}$, $\bar{g}_{22}$, $\Delta_2 u$, $\Delta_2 v$ are evaluated at the boundary from the geometry definition $\mathbf{x}(\mathbf{u})$ and do not change during the course of iteration.

At non-corner points on the "left" $u = 0$ and "right" $u = 1$ boundaries, as in Section 6.2.2 we have that the derivatives $\mathbf{u}_\eta$, $\mathbf{u}_{\eta\eta}$ and the spacing $g_{22} = \|\mathbf{x}_\eta\|^2$ are determined using centered difference formulas from the boundary point distribution and do not change. The normal derivative $\mathbf{u}_\xi$ off the boundary is computed using one-sided difference formulas that involve one boundary point and the adjacent interior point. Dependence on the interior point implies that this value must be updated during the course of iteration. Also updated during the course of iteration is $\mathbf{u}_{\xi\xi}$, which is computed using a centered difference formula involving an interior point, a boundary point, and a *ghost point* $\mathbf{u}_{-1,j}$ or $\mathbf{u}_{m+1,j}$ off the boundary. The ghost point value is derived once at the beginning of iteration by doing an analysis of the correct grid spacing off the boundary and by imposing physical orthogonality.

We now derive the location of the ghost points at the "left" $u = 0$ boundary. Similar to Section 6.2.2, let $(\mathbf{u}_\eta)_{0,j}$ denote the centrally differenced derivative $1/2(\mathbf{u}_{0,j+1} - \mathbf{u}_{0,j-1})$ and let $(\mathbf{u}_\xi^o)_{0,j}$ denote the initial one-sided derivative $\mathbf{u}_{1,j}^o - \mathbf{u}_{0,j}$, where $\mathbf{u}_{1,j}^o \equiv \mathbf{u}_{1,j}$ on the initial algebraic grid, and $\mathbf{u}_{0,j}$ is the unchanging boundary value.

Now to define $\mathbf{u}_\xi$, used in the definition of ghost points and grid spacing off the boundary, we again make the assumption of Figure 6.4 that in physical space $\mathbf{x}_\xi$ is the projection of $\mathbf{x}_\xi^o$ $(= \mathbf{x}_u u_\xi^o + \mathbf{x}_v v_\xi^o)$ onto the direction $\mathbf{a} \equiv \frac{x_\xi}{\|x_\xi\|}$ physically orthogonal to the boundary. This is equivalent to Eq. 6.11 or, in terms of the grid spacing off the boundary, this is equivalent to

$$\left\|\mathbf{x}_\xi\right\| = \mathbf{x}_\xi^0 \cdot \frac{\mathbf{x}_\xi}{\left\|\mathbf{x}_\xi\right\|}. \quad (6.26)$$

Combining Eq. 6.26 with the parametric space orthogonality condition Eq. 6.22, we obtain

$$\mathbf{u}_\xi = \left(u_\xi^0, -\frac{\overline{g}_{12}}{\overline{g}_{22}} u_\xi^0 \right). \tag{6.27}$$

The grid point locations are then defined by the reflection operation in physical space shown in Figure 6.5 or equivalently, the first order Taylor expansion in parametric space involving the orthogonal boundary derivative:

$$\begin{aligned}\mathbf{u}_{-1,j} &= \mathbf{u}_{0,j} + \Delta\xi\left(u_\xi\right)_{o,j} \\ &= \mathbf{u}_{0,j} - \left(u_\xi\right)_{0,j}\end{aligned}$$

This leads to ghost point locations at the left boundary given by

$$\begin{aligned}u_{-1,j} &= u_{0,j} - \left(u_\xi\right)_{0,j} \\ &= u_{0,j} - \left(u_\xi^0\right)_{0,j} \\ &= u_{0,j} - \left(u_{1,j}^0 - u_{0,j}\right) \\ &= -u_{1,j}^0\end{aligned}$$

and

$$\begin{aligned}v_{-1,j} &= v_{0,j} - \left(v_\xi\right)_{0,j} \\ &= v_{0,j} + \frac{\overline{g}_{12}}{\overline{g}_{22}}\left(u_\xi^0\right)_{0,j} \\ &= v_{0,j} + \frac{\overline{g}_{12}}{\overline{g}_{22}}\left(u_{1,j}^0 - u_{0,j}\right) \\ &= v_{0,j} + \frac{\overline{g}_{12}}{\overline{g}_{22}} u_{1,j}^0 .\end{aligned}$$

The last quantity required for computation of the control functions at the $u = 0$ boundary using Eq. 6.25 is the grid spacing orthogonal to the boundary $g_{11} = \|\mathbf{x}_\xi\|^2$ orthogonal to the boundary. We have that

$$g_{11} = \overline{g}_{11} u_\xi^2 + 2\overline{g}_{12} u_\xi v_\xi + \overline{g}_{22} v_\xi^2 .$$

Substituting Eq. 6.27 into this formula, we easily obtain

$$g_{11} = \frac{\overline{g}}{\overline{g}_{22}}\left(u_\xi^0\right)^2, \tag{6.28}$$

where $\overline{g} \equiv \overline{g}_{11}\overline{g}_{22} - \overline{g}_{12}^2$. Since the boundary points are fixed, this quantity is constant at each boundary point throughout the iteration.

Computation of the control functions at the $u = 1$ boundary is done in the same way as that for the $u = 0$ boundary. We note that Eq. 6.27 is still valid, and using the first-order Taylor expansion

$$\begin{aligned} \mathbf{u}_{m+1,j} &= \mathbf{u}_{m,j} + \Delta\xi\left(u_\xi\right)_{m,j} \\ &= \mathbf{u}_{m,j} + \left(u_\xi\right)_{m,j}, \end{aligned}$$

the ghost point locations are given by

$$\begin{aligned} u_{m+1,j} &= u_{m,j} + \left(u_\xi\right)_{m,j} \\ &= u_{m,j} + \left(u_\xi^0\right)_{m,j} \\ &= u_{m,j} + \left(u_{m,j} - u_{m-1,j}^0\right) \\ &= 2 - u_{m-1,j}^0 \end{aligned}$$

and

$$\begin{aligned} v_{m+1,j} &= v_{m,j} + \left(v_\xi\right)_{m,j} \\ &= v_{m,j} - \frac{\overline{g}_{12}}{\overline{g}_{22}}\left(u_\xi^0\right)_{m,j} \\ &= v_{m,j} - \frac{\overline{g}_{12}}{\overline{g}_{22}}\left(u_{m,j} - u_{m-1,j}^0\right) \\ &= v_{m,j} - \frac{\overline{g}_{12}}{\overline{g}_{22}}\left(1 - u_{m-1,j}^0\right). \end{aligned}$$

Also note that the expression for grid spacing off the "left" boundary Eq. 6.28 is still valid for the "right" boundary.

For the non-corner "bottom" and "top" boundaries, we have that $\mathbf{u}_\xi$, $\mathbf{u}_{\xi\xi}$, $g_{11} = \|\mathbf{x}_\xi\|^2$ are computed once using centered difference formulas, $\mathbf{u}_\eta$ is computed repeatedly using a one-sided difference formula, and $\mathbf{u}_{\eta\eta}$ is computed repeatedly using a centered difference formula involving a ghost point value $\mathbf{u}_{i,-1}$ or $\mathbf{u}_{i,n+1}$ that is computed once using grid spacing and physical orthogonality considerations. In fact, analogous to the orthogonal boundary derivative Eq. 6.27 which is valid for the "left" and "right" boundary, we can derive with similar reasoning that for the "bottom" and "top" boundaries we should have

$$\mathbf{u}_\eta = \left(-\frac{\overline{g}_{12}}{\overline{g}_{11}} v_\eta^0, v_\eta^0\right),$$

where v_η^0 is a one-sided difference computed using the initial algebraic grid. This corresponds to the orthogonal projection in physical space shown in Figure 6.5. By similar reasoning as that used for the "left" and "right" boundaries, this leads to fictitious boundary point locations

$$\begin{aligned} u_{i,-1} &= u_{i,0} + \frac{\overline{g}_{12}}{\overline{g}_{11}} v_{i,1}^0 \\ v_{i,-1} &= -v_{i,1}^0 \end{aligned}$$

on the "bottom" boundary, and

$$u_{i,n+1} = u_{i,n} - \frac{\overline{g}_{12}}{\overline{g}_{11}}\left(1 - v^0_{i,n-1}\right)$$

$$v_{i,n+1} = 2 - v^0_{i,n-1}$$

for the "top" boundary. Similar to Eq. 6.28, the grid spacing off the "bottom" and "top" boundaries is given by

$$g_{22} = \frac{\overline{g}}{\overline{g}_{11}}\left(v^0_\eta\right)^2$$

Using the same rationale as used in Section 6.2.2, quantities for the four corner points,

$$(0,0),\quad (m,0),\quad (0,n),\quad (m,n),$$

are computed without orthogonality considerations or ghost points. The values $\mathbf{u}_\xi$, $\mathbf{u}_{\xi\xi}$, $\mathbf{u}_\eta$, $\mathbf{u}_{\eta\eta}$, g_{11}, g_{22} are all evaluated once using one-sided difference formulas using the specified boundary values and do not change during the course of iteration.

If blending of orthogonal and initial control functions is desired to maintain the initial interior point distribution, we follow the same program followed in Section 6.2.2, which is to compute the initial control functions that would reproduce the algebraic grid, smooth them, and then blend them with orthogonal control functions using Eq. 6.16. However, now the blending is done in the parametric domain, so that the blending function is given by

$$b_{i,j} = e^{-\frac{1}{\delta}u_{i,j}v_{i,j}\left(1-u_{i,j}\right)\left(1-v_{i,j}\right)},\quad 0 \le u_{i,j},\ v_{i,j} \le 1$$

and δ can be considered to be the characteristic length of the decay of the blending function in the (u,v)-parametric domain.

Figure 6.9 exhibits an elliptically smoothed orthogonal grid on the surface geometry depicted in Figure 6.7. The elliptic grid was generated using control functions computed from an initial algebraic grid that had been blended with orthogonal control functions computed on the boundaries using Eq. 6.25. We see that initial spacing is preserved throughout the grid, and the grid near the boundaries is almost perfectly orthogonal.

6.4 Boundary Orthogonality for Volume Grids

The elliptic system of partial differential equations for generating curvilinear coordinates in volumes is given by (see Chapter 4 and Thompson [7])

$$\sum_{m=1}^{3}\sum_{n=1}^{3} g^{mn}\mathbf{x}_{\xi^m\xi^n} + \sum_{n=1}^{3} g^{nn}P_n\mathbf{x}_{\xi^n} = 0 \tag{6.29}$$

where ξ^i, $i = 1, 2, 3$ are the curvilinear coordinates and $\mathbf{x} = (x_1, x_2, x_3)$ is the vector of physical coordinates. The construction of a three-dimensional grid on a given geometry in physical space (x_1, x_2, x_3) may be viewed as construction of a mapping $\mathbf{x}(\xi)$ to physical space from a convenient computational space (ξ^1, ξ^2, ξ^3), which we take to be the brick $[\xi^1_{min}, \xi^1_{max}] \times [\xi^2_{min}, \xi^2_{max}] \times [\xi^3_{min}, \xi^3_{max}]$.

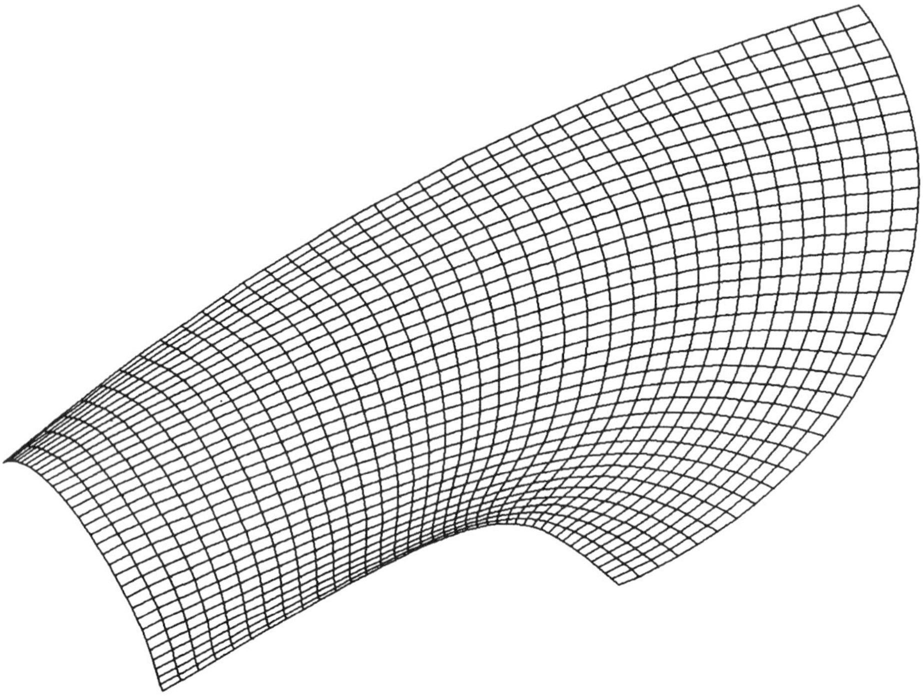

FIGURE 6.9 An elliptic surface grid on a bicubic geometry with Dirichlet orthogonality.

The P_n are the three control functions that serve to control the spacing and the orientation of the grid lines in the field. The elements of the contravariant metric tensor g^{mn} and the elements of the covariant metric tensor g_{mn} are expressed by

$$g^{mn} = \nabla \xi^m \cdot \nabla \xi^n$$
$$g_{mn} = \mathbf{x}_{\xi^m} \cdot \mathbf{x}_{\xi^n}.$$

Moreover, the contravariant and covariant metrics are matrix inverses of each other and are related as

$$g^{mn} = \left(g_{ik} g_{jl} - g_{il} g_{jk}\right) / g, \quad (m,i,j),(n,k,l) \quad \text{cyclic}$$

where g, the square of the Jacobian of the mapping $\mathbf{x}(\xi)$, is given by

$$g = \det\left[g_{mn}\right] = \left(\mathbf{x}_{\xi^1} \cdot \left(\mathbf{x}_{\xi^2} \times \mathbf{x}_{\xi^3}\right)\right)^2.$$

The elliptic generation system in Eq. 6.29 is the one used in smoothing volume grids. The first step in solving the system in Eq. 6.29 is to generate grids on the six surfaces bounding the physical subregion. Then the initial algebraic volume grid is generated between six faces using transfinite interpolation. The initial grid is considered to be the initial solution to the elliptic system Eq. 6.29 and the faces of the grid provide boundary conditions for (x_1, x_2, x_3).

The concept of volume orthogonality proceeds in the same spirit as the surface case.

6.4.1 Neumann Orthogonality

The first technique of achieving boundary orthogonality requires moving the physical coordinates on the surface (face) $S_{\xi^l_{min}}$ (or $S_{\xi^l_{max}}$) so that the orthogonality conditions

$$\begin{aligned}\mathbf{x}_{\xi^l}\cdot\mathbf{x}_{\xi^m}&=0\\ \mathbf{x}_{\xi^l}\cdot\mathbf{x}_{\xi^n}&=0\end{aligned}\tag{6.30}$$

are satisfied with (l, m, n) cyclic. Assume for the moment that our objective is to move the node $\mathbf{x}_{i,j,k}$ on the surface $S_{\xi^l_{min}}$ represented parametrically by $\mathbf{x}(u_0, v_0)$ to a new location $\mathbf{x}(u,v)$ on the surface. To determine the position of the new node $\mathbf{x}$ we need to solve for u and v. Denoting the node off the surface by $\mathbf{x}^{\perp}$ using one-sided differencing, we can write

$$\mathbf{x}_{\xi^l}\approx\mathbf{x}^{\perp}-\mathbf{x}\quad\text{on}\quad S_{\xi^l_{min}}.$$

Thus, the orthogonality conditions in Eq. 6.30 are expressed as

$$\begin{aligned}\left(\mathbf{x}^{\perp}-\mathbf{x}\right)\cdot\mathbf{x}_{\xi^m}&=0\\ \left(\mathbf{x}^{\perp}-\mathbf{x}\right)\cdot\mathbf{x}_{\xi^n}&=0.\end{aligned}\tag{6.31}$$

Taylor expansion of $\mathbf{x}(u,v)$ about (u_0, v_0) gives

$$\mathbf{x}(u,v)\approx\mathbf{x}^{o}+\mathbf{x}_u^{o}\left(u-u_0\right)+\mathbf{x}_v^{o}\left(v-v_0\right),\tag{6.32}$$

where $\mathbf{x}^o = \mathbf{x}(u_0, v_0)$, $\mathbf{x}_u^o = \mathbf{x}_u(u_0, v_0)$, and $\mathbf{x}_v^o = \mathbf{x}_v(u_0, v_0)$. Substituting Eq. 6.32 in the system Eq. 6.31 yields

$$\begin{aligned}\left(\mathbf{x}_u^{o}\cdot\mathbf{x}_{\xi^m}\right)\left(u-u_0\right)+\left(\mathbf{x}_v^{o}\cdot\mathbf{x}_{\xi^m}\right)\left(v-v_0\right)&=\left(\mathbf{x}^{\perp}-\mathbf{x}^{o}\right)\cdot\mathbf{x}_{\xi^m}\\ \left(\mathbf{x}_u^{o}\cdot\mathbf{x}_{\xi^n}\right)\left(u-u_0\right)+\left(\mathbf{x}_v^{o}\cdot\mathbf{x}_{\xi^n}\right)\left(v-v_0\right)&=\left(\mathbf{x}^{\perp}-\mathbf{x}^{o}\right)\cdot\mathbf{x}_{\xi^n}.\end{aligned}\tag{6.33}$$

Using the chain rule of differentiation on $\mathbf{x}_\xi$ and $\mathbf{x}_\eta$

$$\begin{aligned}\mathbf{x}_{\xi^m}&=\mathbf{x}_u^{o}u_{\xi^m}+\mathbf{x}_v^{o}v_{\xi^m}\\ \mathbf{x}_{\xi^n}&=\mathbf{x}_u^{o}u_{\xi^n}+\mathbf{x}_v^{o}v_{\xi^n}\end{aligned}$$

and substituting in Eq. 6.33, we obtain the linear system

$$\mathbf{A}\mathbf{w}=\mathbf{b}$$

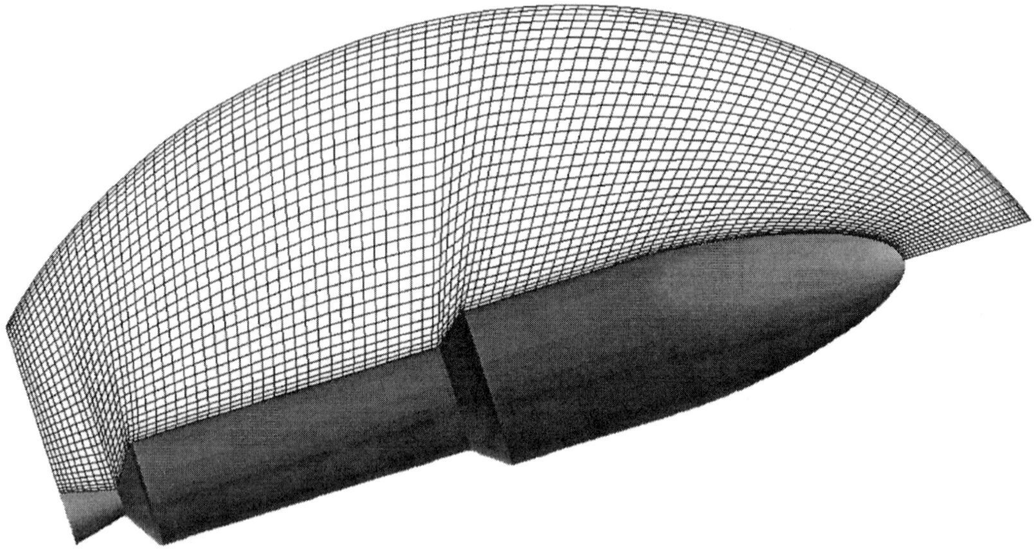

FIGURE 6.10 A cross section of an algebraic volume grid exterior to a booster.

where

$$\mathbf{A} = \begin{pmatrix} (\mathbf{x}_u^o \cdot \mathbf{x}_u^o) u_{\xi^m} + (\mathbf{x}_u^o \cdot \mathbf{x}_v^o) v_{\xi^m} & (\mathbf{x}_u^o \cdot \mathbf{x}_v^o) u_{\xi^m} + (\mathbf{x}_v^o \cdot \mathbf{x}_v^o) v_{\xi^m} \\ (\mathbf{x}_u^o \cdot \mathbf{x}_u^o) u_{\xi^n} + (\mathbf{x}_u^o \cdot \mathbf{x}_v^o) v_{\xi^n} & (\mathbf{x}_u^o \cdot \mathbf{x}_u^o) u_{\xi^n} + (\mathbf{x}_u^o \cdot \mathbf{x}_v^o) v_{\xi^n} \end{pmatrix}$$

$$\mathbf{w} = \begin{pmatrix} w_1 \\ w_2 \end{pmatrix} = \begin{pmatrix} u - u_0 \\ v - v_0 \end{pmatrix}$$

$$\mathbf{b} = \begin{pmatrix} (\mathbf{x}^\perp - \mathbf{x}^o) \cdot \mathbf{x}_u^o u_{\xi^m} & (\mathbf{x}^\perp - \mathbf{x}^o) \cdot \mathbf{x}_v^o v_{\xi^m} \\ (\mathbf{x}^\perp - \mathbf{x}^o) \cdot \mathbf{x}_u^o u_{\xi^n} & (\mathbf{x}^\perp - \mathbf{x}^o) \cdot \mathbf{x}_v^o v_{\xi^n} \end{pmatrix}.$$

Solving the above system for w_1 and w_2, we then compute $u = u_0 + w_1$ and $v = v_0 + w_2$. Finally, we compute new coordinates $\mathbf{x}(u,v)$ to get the location of the grid point on the surface $S_{\xi^1_{min}}$.

Figure 6.10 shows the cross section of an algebraic volume grid on a booster geometry. Clearly the grid is highly nonorthogonal at various points on the booster surface.

Figure 6.11 shows the same grid after elliptic smoothing with imposed Neumann orthogonality. The grid points successfully moved along the booster surface to achieve orthogonality, but with the unfortunate side effect of some degradation of the initial boundary node distribution.

6.4.2 Dirichlet Orthogonality

As in the case of planar or surface grids, an alternative way of constructing orthogonal volume coordinates is to keep the surface nodes fixed and to allow the interior values in the array $\mathbf{x}_{i,j,k}$ to move. This type of orthogonality can be enforced using the control functions P_1, P_2, and P_3 computed on the surfaces.

An iterative solution procedure for the determination of the three control functions for the general three-dimensional case was initially developed by Sorenson [4]. Expressions for the control functions on a coordinate surface on which ξ^l is constant can be obtained from the two coordinate lines lying on the surface, i.e., the lines on which ξ^m and ξ^n vary, (l,m,n) being cyclic. The development presented here follows that of Thompson [7].

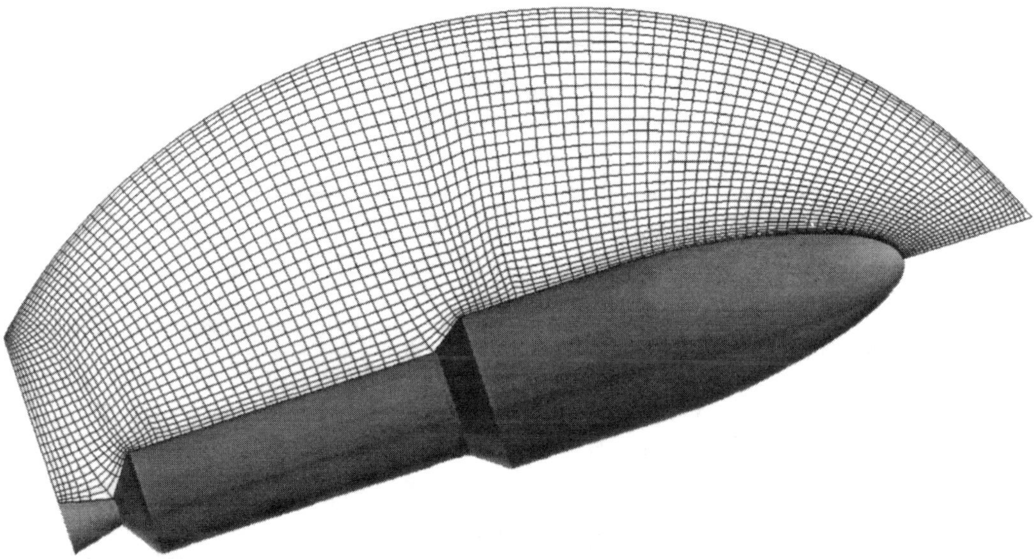

FIGURE 6.11 A cross section of an elliptically smoothed volume grid exterior to a booster with imposed Neumann orthogonality at the surface.

The inner product of $\mathbf{x}_{\xi^l}$, $\mathbf{x}_{\xi^m}$, and $\mathbf{x}_{\xi^n}$ with Eq. 6.29 and using the orthogonality condition Eq. 6.30 yields the following three equations for P_l, P_m, and P_n on the surfaces ξ^l = const.

$$
\begin{aligned}
P_l = &-\frac{1}{g_{ll}}\mathbf{x}_{\xi^l}\cdot\mathbf{x}_{\xi^l\xi^l} \\
&-\frac{\mathbf{x}_{\xi^l}}{g_{mm}g_{nn}-g_{mn}^2}\cdot\left(g_{nn}\mathbf{x}_{\xi^m\xi^m}+g_{mm}\mathbf{x}_{\xi^n\xi^n}-2g_{mn}\mathbf{x}_{\xi^m\xi^n}\right)
\end{aligned}
\tag{6.34}
$$

$$
\begin{aligned}
P_m = &-\frac{1}{g_{ll}}\left(\mathbf{x}_{\xi^m}-\frac{g_{mn}}{g_{nn}}\mathbf{x}_{\xi^n}\right)\cdot\mathbf{x}_{\xi^l\xi^l} \\
&-\left(\frac{\mathbf{x}_{\xi^m}-\frac{g_{mn}}{g_{nn}}\mathbf{x}_{\xi^n}}{g_{mm}g_{nn}-g_{mn}^2}\right)\cdot\left(g_{nn}\mathbf{x}_{\xi^m\xi^m}+g_{mm}\mathbf{x}_{\xi^n\xi^n}-2g_{mn}\mathbf{x}_{\xi^m\xi^n}\right)
\end{aligned}
\tag{6.35}
$$

$$
\begin{aligned}
P_n = &-\frac{1}{g_{ll}}\left(\mathbf{x}_{\xi^n}-\frac{g_{mn}}{g_{mm}}\mathbf{x}_{\xi^m}\right)\cdot\mathbf{x}_{\xi^l\xi^l} \\
&-\left(\frac{\mathbf{x}_{\xi^n}-\frac{g_{mn}}{g_{mm}}\mathbf{x}_{\xi^m}}{g_{mm}g_{nn}-g_{mn}^2}\right)\cdot\left(g_{nn}\mathbf{x}_{\xi^m\xi^m}+g_{mm}\mathbf{x}_{\xi^n\xi^n}-2g_{mn}\mathbf{x}_{\xi^m\xi^n}\right).
\end{aligned}
\tag{6.36}
$$

Proceeding as in the planar case, we construct ghost points for the evaluation of $\mathbf{x}_{\xi^l\xi^l}$. At the $\xi = \xi_{min}$ boundary, we define the unit vector orthogonal to the boundary,

$$\mathbf{a} \equiv \frac{\mathbf{x}_{\xi^l}}{\left\|\mathbf{x}_{\xi^l}\right\|} = \frac{\mathbf{x}_{\xi^m} \times \mathbf{x}_{\xi^n}}{\left\|\mathbf{x}_{\xi^m} \times \mathbf{x}_{\xi^n}\right\|}.$$

The fixed derivative orthogonal to the boundary is then defined by

$$\mathbf{x}_{\xi^l} = \mathbf{P}_{\mathbf{a}}\left(\mathbf{x}^o_{\xi^l}\right),$$

where $\mathbf{x}^o_{\xi^l}$ is the one-sided derivative obtained from the initial algebraic grid, and $\mathbf{P}_{\mathbf{a}} = \mathbf{a}\mathbf{a}^{\mathrm{T}}$ is the orthogonal projection onto the one-dimensional subspace by the unit vector **a**. Thus we obtain

$$\begin{aligned} \mathbf{x}_{\xi^l} &= \mathbf{a}\left(\mathbf{a} \cdot \mathbf{x}^o_{\xi^l}\right) \\ &= \frac{\mathbf{x}_{\xi^m} \times \mathbf{x}_{\xi^n}}{\left\|\mathbf{x}_{\xi^m} \times \mathbf{x}_{\xi^n}\right\|^2}\left(\mathbf{x}_{\xi^m} \times \mathbf{x}_{\xi^n} \cdot \mathbf{x}^o_{\xi^l}\right). \end{aligned} \tag{6.37}$$

So, for the ξ^l_{min} surface (i.e., $i = 0$), our ghost point locations would be given by

$$\mathbf{x}_{-1,j,k} = \mathbf{x}_{0,j,k} - \left(\mathbf{x}_{\xi^l}\right)_{0,j,k},$$

where $(\mathbf{x}_{\xi^l})_{0,j,k}$ was computed using Eq. 6.37 and is fixed, since it depends only on fixed boundary data and data from the initial grid. For the ξ^l_{max} surface (i.e., $i = m$), our ghost point locations would be given by

$$\mathbf{x}_{m+1,j,k} = \mathbf{x}_{m,j,k} + \left(\mathbf{x}_{\xi^l}\right)_{m,j,k},$$

again using the fixed orthogonal derivative Eq. 6.37. The ghost points for the ξ^2_{min}, ξ^2_{max}, ξ^3_{min}, and ξ^3_{max} surfaces are similarly computed.

Note that for $\mathbf{x}_{\xi^l}$ computed by Eq. 6.37, we have that

$$\mathbf{x}_{\xi^m} \times \mathbf{x}_{\xi^n} \cdot \mathbf{x}_{\xi^l} = \mathbf{x}_{\xi^m} \times \mathbf{x}_{\xi^n} \cdot \mathbf{x}^o_{\xi^l}.$$

This means that the ghost points will form cells with the same volume as the first layer of cells in the algebraic grid. This is expected because, as in Figure 6.5 for the planar case, the ghost points have been constructed to form a surface that is the reflection of the first interior coordinate surface, and so cell volume must be conserved. Of course, the ghost points will form cells which are orthogonal to the boundary, while the first layer of cells from the algebraic grid are probably not.

Now, similar to the planar case, the $\mathbf{x}_{\xi^l\xi^l}$ terms in Eqs. 6.34–6.36 are computed using a ghost point, a boundary point, and an iteratively updated interior point, while $g_{ll} = \|\mathbf{x}_{\xi^l}\|^2$ computed using Eq. 6.37 and is fixed for the whole iteration. The $\mathbf{x}_{\xi^l}$ terms appearing in Eq. 6.34 are evaluated using one-sided differencing involving a boundary point and an iteratively updated interior point. The remaining terms in Eqs. 6.34–6.36 are computed using central differencing on the fixed boundary data. At the 8 corners and the 12 edges, the terms in Eqs. 6.34–6.36 are evaluated using all one-sided differences (for the corners) or a combination of one-sided and central differences (for the edges). As in the planar case, no orthogonality information is incorporated into the calculation of the orthogonal control functions at these

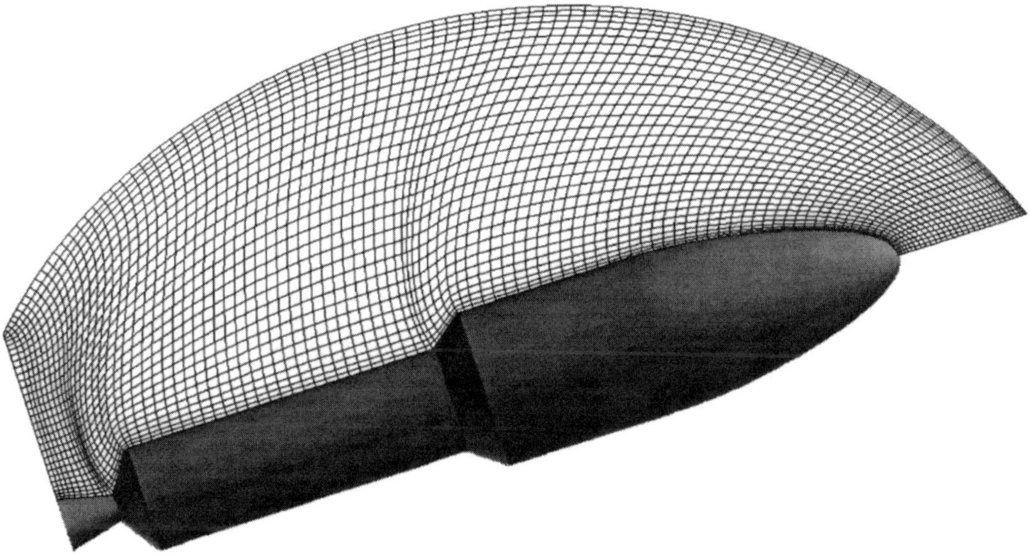

FIGURE 6.12 A cross section of an elliptically smoothed volume grid exterior to a booster with imposed Dirichlet orthogonality at the surface.

points that are at the boundaries of the boundary surfaces. Finally, the orthogonal control functions computed using Eqs. 6.34–6.36 are interpolated to the interior by linear transfinite interpolation.

If blending of orthogonal and initial control functions is desired to maintain the initial interior point distribution, we follow the same program followed in Section 6.2.2, which is to compute the initial control functions that would reproduce the algebraic grid, smooth them, and then blend them with orthogonal control functions using Eq. 6.16. However, now the blending is done on a brick rather than on a rectangle, and so the blending function is given by

$$b_{i,j,k} = e^{-\frac{1}{\delta}u_{i,j,k}v_{i,j,k}w_{i,j,k}\left(1-u_{i,j,k}\right)\left(1-v_{i,j,k}\right)\left(1-w_{i,j,k}\right)}$$

where

$$u_{i,j,k} = \frac{i-\xi^1_{min}}{\xi^1_{max}-\xi^1_{min}}$$

$$v_{i,j,k} = \frac{j-\xi^2_{min}}{\xi^2_{max}-\xi^2_{min}}$$

$$w_{i,j,k} = \frac{k-\xi^3_{min}}{\xi^3_{max}-\xi^3_{min}}.$$

As in the planar case, δ is some positive number that can be considered to be the characteristic length of the decay of the blending function in the computational domain.

In Figure 6.12 we show the cross section of the grid of Figure 6.10 after elliptic smoothing using Dirichlet orthogonality. Clearly the grid is orthogonal at the surface, and the effect of smoothing has been to slide nodes along the coordinate surfaces parallel to the boundary, without affecting the spacing of the coordinate surfaces normal to the boundary.

6.5 Summary

A comprehensive development has been presented for the implementation of boundary orthogonality in elliptic grid generation for planar domains, surfaces, and volumes. For each of these three cases, two techniques have been presented. One technique, Neumann orthogonality, involves sliding points along the boundaries to establish orthogonality. Our implementation of the other technique, Dirichlet orthogonality, involves sliding points along the first interior coordinate surface of the initial grid and then reflecting them across the boundary to form the ghost points which will be used in the computation of the orthogonal control functions in the elliptic system. The former technique is appropriate for interior boundaries between different grid patches, while the latter technique is appropriate for physical boundaries where grid point density must be preserved under elliptic iteration. These techniques can be applied at all or selected boundaries.

In the case of Dirichlet orthogonality, orthogonal control functions can be blended with initial control functions if preservation of interior grid point distribution is desired. These orthogonality techniques have proven to be reliable and efficient in the construction of planar, surface, and volume grids.

References

1. Khamayseh, A. and Mastin, C W., Computational conformal mapping for surface grid generation, *J. Comput. Phys.* 1996, 123, pp 394–401.
2. Soni, B.K., Elliptic grid generation system: control functions revisited-I, *Appl. Math. Comput.* 1993, 59, pp 151–163.
3. Sorenson, R.L., A computer program to generate two-dimensional grids about airfoils and other shapes by the use of Poisson's equations, NASA TM 81198. NASA Ames Research Center, 1980.
4. Sorenson, R.L., Three-dimensional elliptic grid generation about fighter aircraft for zonal finite difference computations, AIAA-86-0429. AIAA 24th Aerospace Science Conference, Reno, NV, 1986.
5. Spekreijse, S.P., Elliptic grid generation based on laplace equations and algebraic transformations, *J. Comput. Phys.* 1995, 118, pp 38–61.
6. Thomas, P.D. and Middlecoff, J.F., Direct control of the grid point distribution in meshes generated by elliptic equations, *AIAA J.* 1980, 18, pp 652–656.
7. Thompson, J.F., A general three-dimensional elliptic grid generation system on a composite block structure, *Comp. Meth. Appl. Mech. and Eng.* 1987, 64, pp 377–411.
8. Thompson, J.F., Warsi, Z.U.A., and Mastin, C.W., *Numerical Grid Generation: Foundations and Applications.* North-Holland, New York, 1985.
9. Warsi, Z.U.A., Numerical grid generation in arbitrary surfaces through a second-order differential geometric model, *J. Comput. Phys.* 1986, 64, pp 82–96.

7

Orthogonal Generating Systems

Luís Eça

7.1 Introduction

The generation of orthogonal grids is still one of the great challenges of grid generation. An orthogonal grid offers significant advantages in the solution of systems of partial differential equations:

- The transformation of partial differential equations produces the smallest number of additional terms.
- In general, the accuracy of the numerical differencing techniques is the highest in orthogonal grids.
- The boundary conditions on rigid boundaries can be enforced in the simplest possible way.
- The implementation of turbulence models, which often require information along perpendicular directions, is simplified.

However, for a three-dimensional complex geometry, a fully orthogonal grid may not exist. In fact, as noted in [1], the coordinate lines on the bounding surfaces of an orthogonal three-dimensional grid must follow lines in the direction of the maximum or minimum curvature of the surface. Therefore, this chapter will be limited to orthogonal generating systems for planes and curved surfaces.

In an orthogonal grid, all the off-diagonal components of the metric tensor are equal to zero. This strong restriction on the grid construction is often in conflict with the possibility to have direct control of the grid line spacing. Conformal mapping is a well-known technique (see for example [2]) for orthogonal grid generation in two dimensions, which enforces all the grid cells to have the same aspect ratio.* Therefore, conformal mapping has no control of the grid line spacing. Although some successful applications of conformal mapping are still reported, for example [3], this chapter is mainly dedicated to orthogonal generating systems that allow control of the grid line spacing.

As reported in [4] and [5], there are basically two types of orthogonal generating systems:

- Trajectory methods, which generate an orthogonal grid from an existing nonorthogonal grid.
- Field methods, which are based on the solution of a system of partial differential equations.

*Conformal mapping preserves the grid cell aspect ratio. In grid generation, the standard procedure is to adopt a uniform computational domain, which implies that in physical space all the grid cells have the same aspect ratio.

0-8493-2687-7/99/$0.00+$.50

In the first approach, the grid is constructed from a known nonorthogonal grid, where one set of coordinate lines is retained. In general, these methods use a marching process to recalculate the grid node distribution along the retained set of grid lines in such a way that the intersection between the new grid lines and the retained set of grid lines is orthogonal. The grid line spacing is determined by the retained set of coordinate lines of the nonorthogonal grid and by the grid node distribution on the boundary where the new set of grid lines starts. This type of methods allows the specification of the grid node distribution on three of the four boundaries of the domains. Several of these types of methods are discussed in references [1] and [4]. The main difficulties reported are the dependency of the orthogonal system on the nonorthogonal original grid and the requirement that in singly connected regions, the components of the boundary must be orthogonal; otherwise, the orthogonal trajectories may leave the physical domain.

In the field approach, the grid is generated by the solution of a system of partial differential equations. Two types of generating systems have been used to generate orthogonal grids: elliptic systems and hyperbolic systems. Hyperbolic systems, which have some resemblances with the orthogonal trajectories methods, require that one of the boundaries must be left completely free. The solution is obtained by a marching procedure that starts from a known boundary and proceeds toward the free boundary. Hyperbolic generating systems are discussed in Chapter 5 of this book.

This chapter will focus on orthogonal generating systems based on elliptic systems of partial differential equations, which require the knowledge of the boundary shape of all the domain. The control of the grid line spacing may be exercised by the specification of the boundary node distribution or by the specification of the grid cells aspect ratio. Elliptic systems of equations offer a wide range of possibilities for the generation of orthogonal grids. Unfortunately, there are only proofs of the existence and uniqueness of such orthogonal mappings for a restricted number of conditions [6]. Nevertheless, the numerical solution of elliptic systems of partial differential equations shows that it is possible to obtain orthogonal grids for a wide range of practical domains, with some control of the grid line spacing.

7.2 Generating Systems

In an orthogonal grid, all the off-diagonal components of the metric tensor are identical to zero, which means that

$$g_{ij} = \vec{a}_i \cdot \vec{a}_j = \frac{\partial x}{\partial \xi^i}\frac{\partial x}{\partial \xi^j} + \frac{\partial y}{\partial \xi^i}\frac{\partial y}{\partial \xi^j} + \frac{\partial z}{\partial \xi^i}\frac{\partial z}{\partial \xi^j} = 0 \text{ with } i \neq j \tag{7.1}$$

where g_{ij} are the components of the covariant metric tensor, $\vec{a}_i$ are the covariant base vectors, (x, y, z) are the coordinates in the physical domain, and $(\xi^1, \xi^2, \xi^3) \equiv (\xi, \eta, \zeta)$ are the coordinates of the transformed plane.

It is also known, [4] and [5], that any orthogonal grid has to satisfy the following system of partial differential equations:

$$\frac{\partial}{\partial \xi}\left(\frac{h_\eta h_\zeta}{h_\xi}\frac{\partial x^i}{\partial \xi}\right) + \frac{\partial}{\partial \eta}\left(\frac{h_\xi h_\zeta}{h_\eta}\frac{\partial x^i}{\partial \eta}\right) + \frac{\partial}{\partial \zeta}\left(\frac{h_\xi h_\eta}{h_\zeta}\frac{\partial x^i}{\partial \zeta}\right) = 0 \tag{7.2}$$

where $(x^1, x^2, x^3) \equiv (x, y, z)$ and h_{ξ^i} are the scale factors defined by:

$$h_{\xi^i} = \sqrt{g_{ii}} = \sqrt{\left(\frac{\partial x}{\partial \xi^i}\right)^2 + \left(\frac{\partial y}{\partial \xi^i}\right)^2 + \left(\frac{\partial z}{\partial \xi^i}\right)^2} \tag{7.3}$$

7.2.1 Two-Dimensional Regions

In an orthogonal two-dimensional grid, Eq. 7.1 reduces to

$$g_{12} = \frac{\partial x}{\partial \xi}\frac{\partial x}{\partial \eta} + \frac{\partial y}{\partial \xi}\frac{\partial y}{\partial \eta} = 0 \tag{7.4}$$

The ratio between the grid cell area in the physical and transformed domains is given by the Jacobian, $\sqrt{g}$, of the transformation:

$$\sqrt{g} = \frac{\partial x}{\partial \xi}\frac{\partial y}{\partial \eta} - \frac{\partial x}{\partial \eta}\frac{\partial y}{\partial \xi} = \sqrt{g_{11}\, g_{22}} = h_\xi h_\eta \tag{7.5}$$

From the orthogonality condition, Eq. 7.4, and the definition of the Jacobian in a 2D orthogonal grid, Eq. 7.5, it is easy to see that a 2D orthogonal grid must also satisfy the Beltrami equations

$$\begin{aligned} f\frac{\partial x}{\partial \xi} &= \frac{\partial y}{\partial \eta} \\ f\frac{\partial y}{\partial \xi} &= -\frac{\partial x}{\partial \eta} \end{aligned} \tag{7.6}$$

where f is the so-called distortion function, which defines the grid cell aspect ratio

$$f = \frac{h_\eta}{h_\xi} = \sqrt{\frac{\left(\frac{\partial x}{\partial \eta}\right)^2 + \left(\frac{\partial y}{\partial \eta}\right)^2}{\left(\frac{\partial x}{\partial \xi}\right)^2 + \left(\frac{\partial y}{\partial \xi}\right)^2}} \tag{7.7}$$

The equality of the second-order cross-derivatives of x and y and the Beltrami equations imply that

$$\begin{aligned} \frac{\partial}{\partial \xi}\left(f\frac{\partial x}{\partial \xi}\right) + \frac{\partial}{\partial \eta}\left(\frac{1}{f}\frac{\partial x}{\partial \eta}\right) &= 0 \\ \frac{\partial}{\partial \xi}\left(f\frac{\partial y}{\partial \xi}\right) + \frac{\partial}{\partial \eta}\left(\frac{1}{f}\frac{\partial y}{\partial \eta}\right) &= 0 \end{aligned} \tag{7.8}$$

Eq. 7.8 are no more than the two-dimensional form of Eq. 7.2. If f is known, Eq. 7.8 are a set of linear elliptic partial differential equations. Otherwise, Eq. 7.8 becomes nonlinear, which implies that its solution must be iterative. The two equations are coupled through the specification of the boundary conditions or through the distortion function determination, if f is assumed to be unknown.

It is interesting to note that Eq. 7.8 multiplied by the Jacobian of the transformation, $\sqrt{g}$, may be rewritten as

$$\begin{aligned} h_\eta^2\left(\frac{\partial^2 x}{\partial \xi^2} + P\frac{\partial x}{\partial \xi}\right) + h_\xi^2\left(\frac{\partial^2 x}{\partial \eta^2} + Q\frac{\partial x}{\partial \eta}\right) &= 0 \\ h_\eta^2\left(\frac{\partial^2 y}{\partial \xi^2} + P\frac{\partial y}{\partial \xi}\right) + h_\xi^2\left(\frac{\partial^2 y}{\partial \eta^2} + Q\frac{\partial y}{\partial \eta}\right) &= 0 \end{aligned} \tag{7.9}$$

with

$$P = \frac{1}{f}\frac{\partial f}{\partial \xi} = -\frac{\frac{\partial x}{\partial \xi}\frac{\partial^2 x}{\partial \xi^2}+\frac{\partial y}{\partial \xi}\frac{\partial^2 y}{\partial \xi^2}}{h_\xi^2} - \frac{\frac{\partial x}{\partial \xi}\frac{\partial^2 x}{\partial \eta^2}+\frac{\partial y}{\partial \xi}\frac{\partial^2 y}{\partial \eta^2}}{h_\eta^2}$$

$$Q = f\frac{\partial}{\partial \eta}\left(\frac{1}{f}\right) = -\frac{\frac{\partial x}{\partial \eta}\frac{\partial^2 x}{\partial \xi^2}+\frac{\partial y}{\partial \eta}\frac{\partial^2 y}{\partial \xi^2}}{h_\xi^2} - \frac{\frac{\partial x}{\partial \eta}\frac{\partial^2 x}{\partial \eta^2}+\frac{\partial y}{\partial \eta}\frac{\partial^2 y}{\partial \eta^2}}{h_\eta^2} \tag{7.10}$$

Equations 7.9 are the well-known elliptic generating system proposed by Thompson et al., [5], and the control functions P and Q, given by Eq. 7.10, are the control functions calculated iteratively at the boundaries with the GRAPE approach, [7], to obtain orthogonality at the boundaries (cf. Chapter 6). Although this result shows that Eq. 7.9 may also be used as an orthogonal generation system, for orthogonal grid generation it is better* to adopt Eq. 7.8 as the generating system.

7.2.1.1 Distortion Function and Boundary Conditions

The specification of the distortion function and of the boundary conditions in Eq. 7.8 are closely related. In a closed domain, two types of boundary conditions may occur:

- The coordinates of the boundary grid nodes are prescribed, which corresponds to Dirichlet boundary conditions.
- The shape of the boundary line is prescribed and the orthogonality condition Eq. 7.4 is satisfied, which leads to a Neumann–Dirichlet boundary condition.

The distortion function may be seen as a known function or as an unknown that has to be determined by the simultaneous solution of Eq. 7.8 and Eq. 7.7. If f is a known function, then Neumann–Dirichlet boundary conditions must be applied to ensure that the grid is orthogonal. The specification of x, y, and f at a boundary makes the problem overdetermined and will not guarantee that the orthogonality condition is satisfied. On the other hand, if f is assumed to be an unknown quantity to be determined in the solution procedure by Eq. 7.6 or Eq. 7.7, then the boundary grid coordinates should be prescribed.

Unfortunately, it is only possible to prove that Eq. 7.8 has a unique solution [6] when f is given by an equation of the type

$$f(\xi,\eta) = M\Pi(\xi)\Theta(\eta) \tag{7.11}$$

where M is the conformal module of the physical domain, which guarantees that the four corners of the physical domain are mapped into the four corners of the transformed domain. The conformal module, M, is an intrinsic property of any quadrilateral domain which depends only on the boundary lines that define the domain. M may be calculated *a priori*, as in [6], or it may be calculated iteratively as suggested by Arina in [8] using

$$M^2 = \frac{\iint \frac{h_\eta}{h_\xi}\,\mathrm{d}\xi\,\mathrm{d}\eta}{\iint \frac{h_\xi}{h_\eta}\,\mathrm{d}\xi\,\mathrm{d}\eta} \tag{7.12}$$

*See Section 7.3.1

If f is constant, and therefore equal to M, the grid is quasi-conformal,* which means that all the grid cells have the same aspect ratio.

The functions $\Pi(\xi)$ and $\Theta(\eta)$ represent one-dimensional stretching functions. Eq. 7.11 may be rewritten in an alternative way, where the one-dimensional stretching functions are determined iteratively from a prescribed boundary point distribution on two adjacent boundaries, ξ_0 and η_0:

$$f(\xi,\eta) = \frac{f(\xi_0,\eta)\, f(\xi,\eta_0)}{f(\xi_0,\eta_0)} \tag{7.13}$$

There is no analytical proof that the system of partial differential Eq. 7.8 has a unique solution, or even a solution, if f is not prescribed by a function of type Eq. 7.11, which is equivalent to specifying the boundary point distribution in two boundaries. However, it is possible to solve numerically the system of Eq. 7.8 with different approaches. Other forms of distortion functions may be used when Neumann–Dirichlet boundary conditions are applied on all the boundaries. It is also possible to generate orthogonal grids with the boundary point distribution prescribed on all the boundaries, if f is determined iteratively as a part of the solution.

For complete boundary point correspondence, two different techniques have been attempted:

- The distortion function is calculated at the boundaries from its definition equation, and the field values are obtained from the boundary values by algebraic interpolation or by the solution of a partial differential equation.
- The distortion function is calculated from its definition equation in the whole field.

The first approach, which was introduced by Ryskin and Leal [9], allows control of the grid line spacing from the boundary point distribution and from the definition of the field values of f. However, this method is strongly dependent on the geometry of the physical domain and, in general, it is only able to produce nearly orthogonal grids [10]. More promising results can be obtained with the second approach, as reported in [11, 12, 13].

7.2.1.2 Orthogonality Parameters

The off-diagonal metric terms of an orthogonal grid are equal to zero. In general, these terms are not calculated analytically. Therefore, in numerical solutions, it is important to quantify the orthogonality of a given grid. Usually, the deviation from orthogonality $\left|\frac{\pi}{2} - \theta\right|$, where θ is given by

$$\cos(\theta) = \frac{g_{12}}{h_\xi\, h_\eta} \tag{7.14}$$

is used to quantify the grid orthogonality. Another parameter which may also be used to quantify the grid orthogonality is the mean quadratic error of the Beltrami Eq. 7.6, which can be defined as

$$\phi_b = \iint \left[\frac{1}{\sqrt{g}} \left(\frac{1}{f} \left(f \frac{\partial x}{\partial \xi} - \frac{\partial y}{\partial \eta} \right)^2 + f \left(\frac{1}{f} \frac{\partial x}{\partial \eta} + \frac{\partial y}{\partial \xi} \right)^2 \right) \right] d\xi\, d\eta \tag{7.15}$$

7.2.2 Curved Surfaces

On a curved surface, the orthogonality condition Eq. 7.1 reduces to

*$M = 1$ corresponds to a conformal mapping, where Eq. 7.6 become the Cauchy–Riemann equations.

$$g_{12} = \frac{\partial x}{\partial \xi}\frac{\partial x}{\partial \eta} + \frac{\partial y}{\partial \xi}\frac{\partial y}{\partial \eta} + \frac{\partial z}{\partial \xi}\frac{\partial z}{\partial \eta} = 0 \tag{7.16}$$

In a curved surface there are only two independent variables, which means that any curved surface may be described by a parametric representation with independent coordinates (u, v):

$$\begin{aligned} x &= X(u,v) \\ y &= Y(u,v) \\ z &= Z(u,v) \end{aligned} \tag{7.17}$$

As described in [14], an orthogonal grid must satisfy the following relations:

$$\begin{aligned} f\frac{\partial u}{\partial \xi} &= \frac{a_{12}}{\sqrt{a}}\frac{\partial u}{\partial \eta} + \frac{a_{22}}{\sqrt{a}}\frac{\partial v}{\partial \eta} & \frac{1}{f}\frac{\partial u}{\partial \eta} &= -\frac{a_{12}}{\sqrt{a}}\frac{\partial u}{\partial \xi} - \frac{a_{22}}{\sqrt{a}}\frac{\partial v}{\partial \xi} \\ f\frac{\partial v}{\partial \xi} &= -\frac{a_{11}}{\sqrt{a}}\frac{\partial u}{\partial \eta} - \frac{a_{12}}{\sqrt{a}}\frac{\partial v}{\partial \eta} & \frac{1}{f}\frac{\partial v}{\partial \eta} &= \frac{a_{11}}{\sqrt{a}}\frac{\partial u}{\partial \xi} + \frac{a_{12}}{\sqrt{a}}\frac{\partial v}{\partial \xi} \end{aligned} \tag{7.18}$$

where a_{ij} are the components of the metric tensor of the transformation between the physical domain, (x, y, z), and the parametric space (u, v):

$$\begin{aligned} a_{11} &= \left(\frac{\partial x}{\partial u}\right)^2 + \left(\frac{\partial y}{\partial u}\right)^2 + \left(\frac{\partial z}{\partial u}\right)^2 \\ a_{22} &= \left(\frac{\partial x}{\partial v}\right)^2 + \left(\frac{\partial y}{\partial v}\right)^2 + \left(\frac{\partial z}{\partial v}\right)^2 \\ a_{12} &= \frac{\partial x}{\partial u}\frac{\partial x}{\partial v} + \frac{\partial y}{\partial u}\frac{\partial y}{\partial v} + \frac{\partial z}{\partial u}\frac{\partial z}{\partial v} \end{aligned} \tag{7.19}$$

$$\sqrt{a} = \sqrt{a_{11}a_{22} - a_{12}^2} \tag{7.20}$$

As in the two-dimensional regions, f defines the grid cell aspect ratio, which in this case is defined by

$$f = \frac{h_n}{h_\xi} = \frac{\sqrt{\left(\frac{\partial x}{\partial \eta}\right)^2 + \left(\frac{\partial y}{\partial \eta}\right)^2 + \left(\frac{\partial z}{\partial \eta}\right)^2}}{\sqrt{\left(\frac{\partial x}{\partial \xi}\right)^2 + \left(\frac{\partial y}{\partial \xi}\right)^2 + \left(\frac{\partial z}{\partial \xi}\right)^2}} \tag{7.21}$$

Adding Eq. 7.18 differentiated with respect to ξ and η [14] it is possible to obtain the following elliptic system of partial differential equations:

$$\begin{aligned} \frac{\partial}{\partial \xi}\left(f\frac{\partial u}{\partial \xi}\right) + \frac{\partial}{\partial \eta}\left(\frac{1}{f}\frac{\partial u}{\partial \eta}\right) &= \frac{\partial u}{\partial \eta}\frac{\partial}{\partial \xi}\left(\frac{a_{12}}{\sqrt{a}}\right) + \frac{\partial v}{\partial \eta}\frac{\partial}{\partial \xi}\left(\frac{a_{22}}{\sqrt{a}}\right) - \frac{\partial u}{\partial \xi}\frac{\partial}{\partial \eta}\left(\frac{a_{12}}{\sqrt{a}}\right) - \frac{\partial v}{\partial \xi}\frac{\partial}{\partial \eta}\left(\frac{a_{22}}{\sqrt{a}}\right) \\ \frac{\partial}{\partial \xi}\left(f\frac{\partial v}{\partial \xi}\right) + \frac{\partial}{\partial \eta}\left(\frac{1}{f}\frac{\partial v}{\partial \eta}\right) &= \frac{\partial u}{\partial \xi}\frac{\partial}{\partial \eta}\left(\frac{a_{11}}{\sqrt{a}}\right) + \frac{\partial v}{\partial \xi}\frac{\partial}{\partial \eta}\left(\frac{a_{12}}{\sqrt{a}}\right) - \frac{\partial u}{\partial \eta}\frac{\partial}{\partial \xi}\left(\frac{a_{11}}{\sqrt{a}}\right) - \frac{\partial v}{\partial \eta}\frac{\partial}{\partial \xi}\left(\frac{a_{12}}{\sqrt{a}}\right) \end{aligned} \tag{7.22}$$

Eq. 7.22 is a coupled system of partial differential equations which, in general, are non-linear. Eq. 7.22 will become linear if f is assumed to be known and if the derivatives of the components of the a_{ij} metric tensor are independent of u and v.

In the generating system defined by Eq. 7.22, the coefficients of the left-hand-side terms are functions of the transformation between physical domain, (x, y, z), and computational domain, (ξ, η), and the coefficients on the right-hand-side terms are functions of the transformation between the physical domain and the parametric space, (u, v). In [15] it is shown that it is possible to derive a generating system, which does not include explicitly the transformation between physical domain and computational domain, which is based on the orthogonality condition Eq. 7.16 written for the parametric coordinates:

$$\frac{\partial u}{\partial \xi}\frac{\partial u}{\partial \eta}+\frac{\partial v}{\partial \xi}\frac{\partial v}{\partial \eta}+H=0 \tag{7.23}$$

where

$$H=\frac{1}{a_{11}+a_{22}}\left[a_{12}\left(\frac{\partial u}{\partial \xi}\frac{\partial v}{\partial \eta}+\frac{\partial u}{\partial \eta}\frac{\partial v}{\partial \xi}\right)-a_{11}\frac{\partial v}{\partial \xi}\frac{\partial v}{\partial \eta}-a_{22}\frac{\partial u}{\partial \xi}\frac{\partial u}{\partial \eta}\right] \tag{7.24}$$

Eq. 7.23 is written in a form similar to the off-diagonal component of the covariant metric tensor of a 2D coordinate transformation. Therefore, with an algebraic manipulation equivalent to the one which enables the derivation of the Beltrami equations in a 2D orthogonal transformation [4] it is possible to obtain the following equations:

$$\begin{aligned}
\frac{\partial u}{\partial \xi}&=\frac{H}{\sqrt{b}}\frac{\partial v}{\partial \xi}+\frac{b_{11}}{\sqrt{b}}\frac{\partial v}{\partial \eta} &\qquad \frac{\partial u}{\partial \eta}&=-\frac{b_{22}}{\sqrt{b}}\frac{\partial v}{\partial \xi}-\frac{H}{\sqrt{b}}\frac{\partial v}{\partial \eta}\\
\frac{\partial v}{\partial \xi}&=-\frac{H}{\sqrt{b}}\frac{\partial u}{\partial \xi}-\frac{b_{11}}{\sqrt{b}}\frac{\partial u}{\partial \eta} &\qquad \frac{\partial v}{\partial \eta}&=\frac{b_{22}}{\sqrt{b}}\frac{\partial u}{\partial \xi}+\frac{H}{\sqrt{b}}\frac{\partial u}{\partial \eta}
\end{aligned} \tag{7.25}$$

where b_{ij} stands for the component of the covariant metric tensor of the 2D coordinate transformation between parametric space and computational domain:

$$\begin{aligned}
b_{11}&=\left(\frac{\partial u}{\partial \xi}\right)^2+\left(\frac{\partial v}{\partial \xi}\right)^2\\
b_{22}&=\left(\frac{\partial u}{\partial \eta}\right)^2+\left(\frac{\partial v}{\partial \eta}\right)^2\\
\sqrt{b}&=\frac{\partial u}{\partial \xi}\frac{\partial v}{\partial \eta}-\frac{\partial u}{\partial \eta}\frac{\partial v}{\partial \xi}
\end{aligned} \tag{7.26}$$

From the equality of the cross-derivatives of the parametric coordinates, u and v with respect to ξ and η and Eq. 7.25, it is possible to construct the following generating system:

$$\begin{aligned}
\frac{\partial}{\partial \xi}\left(\frac{b_{22}}{\sqrt{b}}\frac{\partial u}{\partial \xi}\right)+\frac{\partial}{\partial \eta}\left(\frac{b_{11}}{\sqrt{b}}\frac{\partial u}{\partial \eta}\right)+\frac{\partial}{\partial \xi}\left(\frac{H}{\sqrt{b}}\frac{\partial u}{\partial \eta}\right)+\frac{\partial}{\partial \eta}\left(\frac{H}{\sqrt{b}}\frac{\partial u}{\partial \xi}\right)&=0\\
\frac{\partial}{\partial \xi}\left(\frac{b_{22}}{\sqrt{b}}\frac{\partial v}{\partial \xi}\right)+\frac{\partial}{\partial \eta}\left(\frac{b_{11}}{\sqrt{b}}\frac{\partial v}{\partial \eta}\right)+\frac{\partial}{\partial \xi}\left(\frac{H}{\sqrt{b}}\frac{\partial v}{\partial \eta}\right)+\frac{\partial}{\partial \eta}\left(\frac{H}{\sqrt{b}}\frac{\partial v}{\partial \xi}\right)&=0
\end{aligned} \tag{7.27}$$

Eq. 7.27 is a nonlinear set of partial differential equations that relate the parametric coordinates (u, v) to the computational domain coordinates (ξ, η). This system of Eq. 7.27 was suggested by Niederdrenk [16] as an alternative to the system proposed in [15], which is based on an equivalent form of Eq. 7.25 that led to a coupled system of equations.

The generating system Eq. 7.27 does not include the distortion function f explicitly. Therefore, when the distortion function f is assumed to be known, it is better to adopt the generating system defined by Eq. 7.22. On the other hand, if f is assumed to be an unknown, then the numerical solution of Eq. 7.27 is the simplest.

7.2.2.1 Distortion Function and Boundary Conditions

With the introduction of the parametric space (u, v), grid generation on a curved surface reduces to a two-dimensional transformation between the parametric space and the computational domain, (ξ, η). Therefore, in general, the specification of the distortion function f and of the boundary conditions is similar to what occurs in a two-dimensional region, which is described in Section 7.2.1.1.

As in the two-dimensional regions, the boundary nodes must be allowed to move along the boundaries when the distortion function is specified, and f should be calculated iteratively when the coordinates of the boundary nodes are fixed.

As shown by Arina [14], Eq. 7.22 reduces to a two-dimensional plane mapping when (u, v) are isothermic or conformal coordinates, [17], for which the right-hand side of Eq. 7.22 is zero. Therefore, the analytical proofs of existence and uniqueness of orthogonal mappings on curved surfaces are equivalent to the ones existing for two-dimensional plane regions [14]. The definition of f on a curved surface should also follow Eq. 7.11, Eq. 7.12, and Eq. 7.13, where the conformal nodule of the curved surface, M, also guarantees that the four corners of the physical domain are transformed into the four corners of the computational domain. As in the two-dimensional case, although there is no proof of existence and uniqueness of the solution, it is possible to solve numerically Eq. 7.22 with different types of distortion functions or with Dirichlet boundary conditions in more than two boundaries.

For complete boundary point correspondence, it is better to solve Eq. 7.27 where the distortion function is not calculated explicitly. In this case, the metric coefficients of the transformation between parametric space and computational domain are calculated iteratively.

Both generating systems Eq. 7.22 and Eq. 7.27, require the calculation of the covariant metric tensor components of the transformation between the physical and parametric domains, a_{ij}. In general, the best results are obtained when all the derivatives are discretized with the computational domain variables, ξ and η, as the independent variables. Therefore, the derivatives of x, y, and z with respect to u and v are obtained from

$$\begin{aligned}\frac{\partial x^i}{\partial u} &= \frac{\partial x^i}{\partial \xi}\frac{\partial \xi}{\partial u} + \frac{\partial x^i}{\partial \eta}\frac{\partial \eta}{\partial u}\\ \frac{\partial x^i}{\partial v} &= \frac{\partial x^i}{\partial \xi}\frac{\partial \xi}{\partial v} + \frac{\partial x^i}{\partial \eta}\frac{\partial \eta}{\partial v}\end{aligned} \tag{7.28}$$

with

$$\frac{\partial \xi}{\partial u} = \frac{1}{\sqrt{b}}\frac{\partial v}{\partial \eta} \qquad \frac{\partial \xi}{\partial v} = -\frac{1}{\sqrt{b}}\frac{\partial u}{\partial \eta} \qquad \frac{\partial \eta}{\partial u} = -\frac{1}{\sqrt{b}}\frac{\partial v}{\partial \xi} \qquad \frac{\partial \eta}{\partial v} = \frac{1}{\sqrt{b}}\frac{\partial u}{\partial \xi} \tag{7.29}$$

7.2.2.2 Orthogonality Parameters

On a curved surface, the deviation from orthogonality can be calculated in the same way as in a two-dimensional region, $\left|\frac{\pi}{2} - \theta\right|$, with θ give by Eq. 7.14.

On a curved surface, the relations between the first derivatives of the parametric coordinates, u and v, with respect to ξ and η may be written in several ways, Eq. 7.18 or Eq. 7.25. These equations may be seen as generalized forms of the Beltrami equations in a two-dimensional mapping. The definition of a mean quadratic error for these equations is not unique. However, the closest form to Eq. 7.6 is given by Eq. 7.18, which lead to a mean quadratic error, ϕ_c, given by

$$\phi_c = \iint \left[\frac{1}{\sqrt{b}}\left(g_1(\xi,\eta)+g_2(\xi,\eta)\right)\right] d\xi\, d\eta \tag{7.30}$$

where

$$\begin{aligned} g_1(\xi,\eta) &= \left| f\frac{\partial u}{\partial \xi} - \frac{a_{12}}{\sqrt{a}}\frac{\partial u}{\partial \eta} - \frac{a_{22}}{\sqrt{a}}\frac{\partial v}{\partial \eta}\right| \left|\frac{1}{f}\frac{\partial v}{\partial \eta} - \frac{a_{11}}{\sqrt{a}}\frac{\partial u}{\partial \xi} - \frac{a_{12}}{\sqrt{a}}\frac{\partial v}{\partial \xi}\right| \\ g_2(\xi,\eta) &= \left| f\frac{\partial v}{\partial \xi} + \frac{a_{11}}{\sqrt{a}}\frac{\partial u}{\partial \eta} + \frac{a_{12}}{\sqrt{a}}\frac{\partial v}{\partial \eta}\right| \left|\frac{1}{f}\frac{\partial u}{\partial \eta} + \frac{a_{12}}{\sqrt{a}}\frac{\partial u}{\partial \xi} + \frac{a_{22}}{\sqrt{a}}\frac{\partial v}{\partial \xi}\right| \end{aligned} \tag{7.31}$$

7.3 Numerical Solutions

The generation of orthogonal grids on planes and curved surfaces with systems of partial differential equations is a nonlinear problem. In general, the nonlinearity is introduced by an unknown value of the distortion function *f*, which can be simply the conformal module of the domain, *M*. Even in the case where the distortion function is known, the orthogonality condition and the specified boundary shape will lead to a nonlinear equation at the boundary. Although there are methods to estimate *a priori* the unknown quantities, when *f* is defined by a product of two one-dimensional stretching functions [6,18], the following iterative algorithm may be applied to the generation of an orthogonal grid with a system of elliptic partial differential equations:

1. Construct an initial approximation for the grid. In general, linear transfinite interpolation provides an acceptable initial guess.
2. Calculate the metric coefficients that appear as coefficients of the generating system.
3. Solve the elliptic system of partial differential equations with fixed coefficients and the appropriate boundary conditions.
4. Go back to Step 2 if the convergence criteria are not satisfied.

7.3.1 Discretized Equations

There are several discretization techniques that can be applied to elliptic systems of partial differential equations. The advantages and drawbacks of the different discretization techniques are not discussed in this chapter. Although some of the basic ideas may be extended to other discretization techniques, the present discussion will be restricted to finite-difference discretizations. For the sake of simplicity, the discretization of the generating system of equations is exemplified for the *x* equation of a two-dimensional orthogonal mapping, Eq. 7.8. The integration of the *x* equation in a typical control volume with the unknowns collocated at the center of the control volume, as shown in Figure 7.1, leads to

$$f_{i+\frac{1}{2},j}\left(\frac{\partial x}{\partial \xi}\right)_{i+\frac{1}{2},j} - f_{i-\frac{1}{2},j}\left(\frac{\partial x}{\partial \xi}\right)_{i-\frac{1}{2},j} + \frac{1}{f_{i,j+\frac{1}{2}}}\left(\frac{\partial x}{\partial \eta}\right)_{i,j+\frac{1}{2}} - \frac{1}{f_{i,j-\frac{1}{2}}}\left(\frac{\partial x}{\partial \eta}\right)_{i,j-\frac{1}{2}} = 0 \tag{7.32}$$

The discretization of the first-order derivatives of *x* with central differencing schemes produces the following pentadiagonal system of algebraic equations:

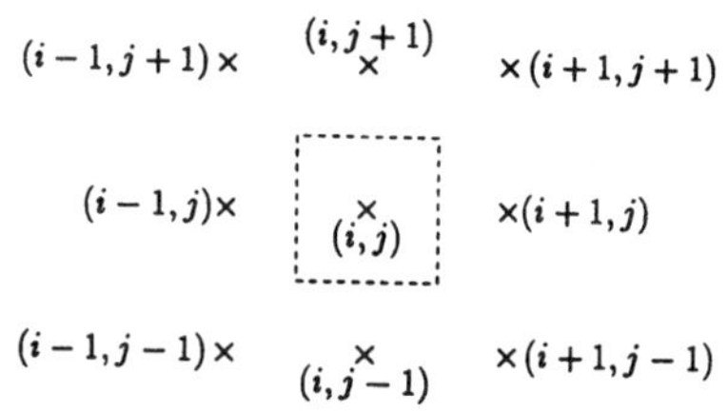

FIGURE 7.1 Typical control volume used in the discretization.

$$f_{i+\frac{1}{2},j}\,x_{i+1,j} + f_{i-\frac{1}{2},j}\,x_{i-1,j} + \frac{1}{f_{i,j+\frac{1}{2}}}x_{i,j+1} + \frac{1}{f_{i,j-\frac{1}{2}}}x_{i,j-1} - F_{i,j}\,x_{i,j} = 0 \tag{7.33}$$

where

$$F_{i,j} = f_{i+\frac{1}{2},j} + f_{i-\frac{1}{2},j} + \frac{1}{f_{i,j+\frac{1}{2}}} + \frac{1}{f_{i,j-\frac{1}{2}}} \tag{7.34}$$

In each iteration of the solution procedure, Eq. 7.33 represent a linear algebraic system of equations, which, for example, can be easily solved with a *successive line over-relaxation* method.

If the distortion function is an unknown quantity, its value at the boundaries of the control volume can be calculated using central differencing schemes in Eq. 7.7, where the (x, y) coordinates at the corners of the control volume are interpolated from the four surrounding nodes.

$$\begin{aligned}
f_{i+\frac{1}{2},j} &\cong \frac{\sqrt{\left(x_{i+1,j+1}+x_{i,j+1}-x_{i+1,j-1}-x_{i,j-1}\right)^2+\left(y_{i+1,j+1}+y_{i,j+1}-y_{i+1,j-1}-y_{i,j-1}\right)^2}}{2\sqrt{\left(x_{i+1,j}-x_{i,j}\right)^2+\left(y_{i+1,j}-y_{i,j}\right)^2}}\\
f_{i-\frac{1}{2},j} &\cong \frac{\sqrt{\left(x_{i-1,j+1}+x_{i,j+1}-x_{i-1,j-1}-x_{i,j-1}\right)^2+\left(y_{i-1,j+1}+y_{i,j+1}-y_{i-1,j-1}-y_{i,j-1}\right)^2}}{2\sqrt{\left(x_{i,j}-x_{i-1,j}\right)^2+\left(y_{i,j}-y_{i-1,j}\right)^2}}\\
f_{i,j+\frac{1}{2}} &\cong \frac{2\sqrt{\left(x_{i,j+1}-x_{i,j}\right)^2+\left(y_{i,j+1}-y_{i,j}\right)^2}}{\sqrt{\left(x_{i+1,j}+x_{i+1,j+1}-x_{i-1,j}-x_{i-1,j+1}\right)^2+\left(y_{i+1,j}+y_{i+1,j+1}-y_{i-1,j}-y_{i-1,j+1}\right)^2}}\\
f_{i,j-\frac{1}{2}} &\cong \frac{2\sqrt{\left(x_{i,j}-x_{i,j-1}\right)^2+\left(y_{i,j}-y_{i,j-1}\right)^2}}{\sqrt{\left(x_{i+1,j}+x_{i+1,j-1}-x_{i-1,j}-x_{i-1,j-1}\right)^2+\left(y_{i+1,j}+y_{i+1,j-1}-y_{i-1,j}-y_{i-1,j-1}\right)^2}}
\end{aligned} \tag{7.35}$$

The accuracy of the calculation may be strongly affected by the determination of f at the faces of the control volume, or if Eq. 7.9 is adopted as the generating system of a two-dimensional orthogonal grid. The numerical errors that can be introduced by the discretization of the generating system are illustrated with a simple example. Consider a two-dimensional orthogonal mapping between two square domains. The computational domain has square grid cells defined by $\Delta\xi = \Delta\eta = 1$. In the physical domain, a

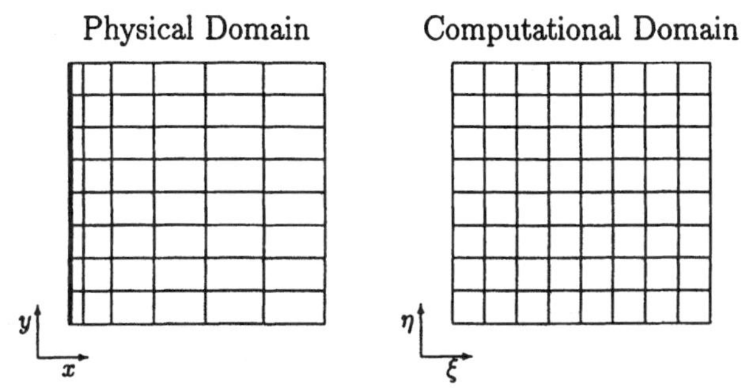

FIGURE 7.2 Two-dimensional orthogonal mapping with one-dimensional stretching applied in the x direction.

one-dimensional stretching function is applied in such a way that Δy is constant and $\Delta x_i = \sqrt{\Delta x_{i-1}}$, where $\Delta x_i = x_i - x_{i-1}$. The two regions are illustrated in Figure 7.2. In this mapping, the distortion function f and the x coordinate are independent of η, and so Eq. 7.33 reduces to

$$f_{i+\frac{1}{2}}\Delta x_{i+1} - f_{i-\frac{1}{2},j}\Delta x_i = 0 \tag{7.36}$$

Eq. 7.36 is numerically satisfied if $f_{i+\frac{1}{2}}$ and $f_{i-\frac{1}{2}}$ are calculated by Eq. 7.35. However, if the distortion function at the boundaries of the control volume is calculated from the mean of f at the two surrounding grid nodes, Eq. 7.36 is not satisfied numerically, which means that the discretized equations indicate that the grid is not orthogonal!

In the present example, it is easy to see that the application of central differencing schemes to Eq. 7.7 at a grid node produces

$$f_i \cong \frac{2\Delta y}{\Delta x_{i+1}}\frac{1}{1+\Delta x_{i+1}} = \frac{2\Delta y}{\Delta x_i}\frac{\sqrt{\Delta x_i}}{1+\sqrt{\Delta x_i}} = \frac{2\Delta y}{\Delta x_i}\frac{\Delta x_{i+1}}{1+\Delta x_{i+1}}. \tag{7.37}$$

The mean values of f at the faces of the control volume are

$$\begin{aligned} f_{i+\frac{1}{2}} &\approx \frac{\Delta y}{\Delta x_{i+1}}\left(\frac{1}{1+\Delta x_{i+1}} + \frac{\sqrt{\Delta x_{i+1}}}{1+\sqrt{\Delta x_{i+1}}}\right) \\ f_{i-\frac{1}{2}} &\approx \frac{\Delta y}{\Delta x_i}\left(\frac{1}{1+\Delta x_{i+1}^2} + \frac{\Delta x_{i+1}}{1+\Delta x_{i+1}}\right) \end{aligned} \tag{7.38}$$

The substitution of Eq. 7.38 in Eq. 7.36 shows that with this approach, the discretized equations are not satisfied in an orthogonal grid!*

A similar problem occurs with the generating system defined by Eq. 7.9, where the second-order derivatives have been expanded into two terms. Therefore, for the numerical generation of orthogonal grids, Eq. 7.8 is written in a more suitable form than Eq. 7.9.

---

*This result is in agreement with one of the first remarks made by Joe Thompson in the first Lecture Series on Grid Generation held at the von Kármám Institute in 1990: "Do not average metric coefficients! It is better to interpolate grid coordinates and to calculate the metric coefficients from the interpolated coordinates."

7.3.2 Boundary Conditions

The generation of orthogonal grids on plane and curved surfaces may include two types of boundary conditions:

- Dirichlet boundary conditions. The coordinates of the grid nodes are specified.
- Neumann–Dirichlet boundary conditions. The orthogonality condition is directly satisfied at the boundary, and the grid nodes lie on a specified boundary shape.

The numerical application of Dirichlet boundary conditions is straightforward. However, in general, the Neumann–Dirichlet boundary conditions lead to a nonlinear equation at the boundary. In general, it is easier to uncouple the solution of the system of algebraic equations that determines the coordinates of the interior grid nodes from the application of the Neumann–Dirichlet boundary conditions. Therefore, the linear algebraic system of equations obtained from the discretization of the generating system of partial differential equations is usually solved with Dirichlet boundary conditions. The orthogonality condition at the boundary is enforced *a posteriori*. However, if an iterative solver is adopted for the solution of the algebraic system of equations, the orthogonality condition at the boundary can be enforced after each iteration of the solver.

The easiest way to implement the orthogonality condition at a boundary is to represent the boundary line in a parametric form. For example, in a ξ boundary, the derivatives of the grid coordinates with respect to ξ are obtained from the parametric representation of the boundary line. Using backward or forward differencing schemes for the derivatives in the η direction, the orthogonality condition becomes a nonlinear equation with ξ as the independent variable. This nonlinear equation may be solved by Newton iteration.

7.3.3 Convergence Criteria

Any iterative solution procedure requires a convergence criterion to determine when to stop the iterative process. The maximum difference between grid coordinates of consecutive iterations, ϕ_x, can be used to define the convergence criterion.

$$\phi_x^n = \max\left(\left|x^n - x^{n-1}\right|, \left|y^n - y^{n-1}\right|\right), \tag{7.39}$$

where the superscript ^{n} refers to the iteration number. In surface grid generation (x, y) are substituted by the parametric coordinates (u, v).

When f is calculated iteratively as part of the solution, it is also necessary to specify a convergence criterion for the determination of the distortion function. The examples presented in [13] show that it is difficult to specify a convergence criterion based on the maximum relative difference between the distortion function of consecutive sweeps,

$$\phi_f^n = \max\left(\left|\frac{f^n - f^{n-1}}{f^n}\right|\right). \tag{7.40}$$

However, the same results suggest that the difference between ϕ_f of consecutive iterations,

$$\psi^n = \phi_f^{n-1} - \phi_f^n, \tag{7.41}$$

may be used as the convergence criterion of the determination of the distortion function. The application of this convergence criterion based on ψ^n, allows the use of Neumann–Dirichlet boundary conditions when f is obtained from its definition equation, which, as shown in [13], leads to an unstable calculation if no convergence criteria is applied in the iterative determination of f.

7.3.4 Two-Dimensional Regions

The generation of two-dimensional orthogonal grids with systems of elliptic partial different equations is exemplified for three types of domains: nonsymmetric, symmetric, and domains which do not have orthogonal boundary lines.

In all these examples, the maximum and mean deviations from orthogonality, MDO and ADO, are calculated with the coordinate derivatives discretized by central differencing schemes. The mean quadratic error of the Beltrami equations, ϕ_b defined by Eq. 7.15, is calculated assuming that the integrand is constant in each control volume.

The convergence criterion applied in these examples is $\phi_x \leq 1.0 \times 10^{-6}$. The convergence criterion of the iterative calculation of the distortion function is assumed to be $\psi_f \leq 1.0 \leq 10^{-5}$ in more than two iterations. The boundary lines are represented by cubic splines, based on the initial boundary point distribution. The initial grids are generated with linear transfinite inter-polation.

7.3.4.1 Nonsymmetric Domains

The different possibilities of orthogonal grid generation in a nonsymmetric region, with and without control of the grid line spacing, are illustrated in a very popular test case of orthogonal grid generation. The physical domain is defined by $0 \leq x \leq \frac{1}{2} + \frac{1}{3}\cos(\pi y)$ and $0 \leq y \leq 1$. The following options are considered:

1. Quasi conformal mapping. Neumann–Dirichlet boundary conditions on all the boundaries and $f = M$.
2. Grid note distribution fixed on two boundaries and f given by the product of two one-dimensional stretching functions, Eq. 7.13.
3. Neumann–Dirichlet boundary conditions on all the boundaries and f given by the sum of linear and sine functions.
4. f obtained from Eq. 7.7 and grid note distribution fixed on three or four boundaries.

The first two options correspond to situations for which there is an analytical proof of the existence and uniqueness of the solution. Although there is no proof that the solutions are unique for the remaining two options, the numerical solutions illustrate the versatility of the elliptic system of partial differential Eq. 7.8 in the generation of two-dimensional orthogonal mappings.

Figure 7.3 presents the quasi-conformal grid and two grids where the one-dimensional stretching functions are iteratively determined from a fixed boundary node distribution on two boundaries. In both cases, Figures 7.3b and 7.3c, the boundary point distribution is prescribed on the boundary $x = \frac{1}{2} + \frac{1}{3}\cos(\pi y)$. In grid 7.3b, Dirichlet boundary conditions are also applied at the boundary $y = 1$, whereas, in grid 7.3c, an equidistant grid node distribution is prescribed on boundary $y = 0$. The quasi-conformal grid illustrates the lack of control of the grid line spacing of this technique, which, in this case, is caused by the boundary curvature. In grids 7.3b and 7.3c, the control of the grid spacing is determined by the two boundaries with fixed boundary nodes.

The control of the grid line spacing can be achieved through the definition of the distortion function. As an example of such control, Figure 7.4 includes two grids where f is given by the sum of linear and sine functions of ξ and η. The definition of f in these examples is not included in the general class of distortion functions defined by Eq. 7.11. Therefore, there is no analytical proof of the existence of such mapping. Nevertheless, the numerical results show that, in practice, it is possible to adopt more general distortion functions to obtain a different grid line spacing. In the previous examples, the control of the grid line spacing is determined by the specification of the distortion function. In some cases, it may be useful to control the grid line spacing from the boundary point distribution. Figure 7.5 presents three grids where f is calculated iteratively from its definition equation. In grid 7.5a, the boundary nodes are prescribed on all the boundaries. In general, it is difficult to guess a boundary point distribution that produces a smooth orthogonal grid. In this example, there is a region where the grid line spacing tends to zero, which, in most cases, is unacceptable for numerical purposes. However, if the grid nodes are allowed to move in one of the boundaries, the grid becomes smooth, as illustrated in Figures 7.5b and 7.5c.

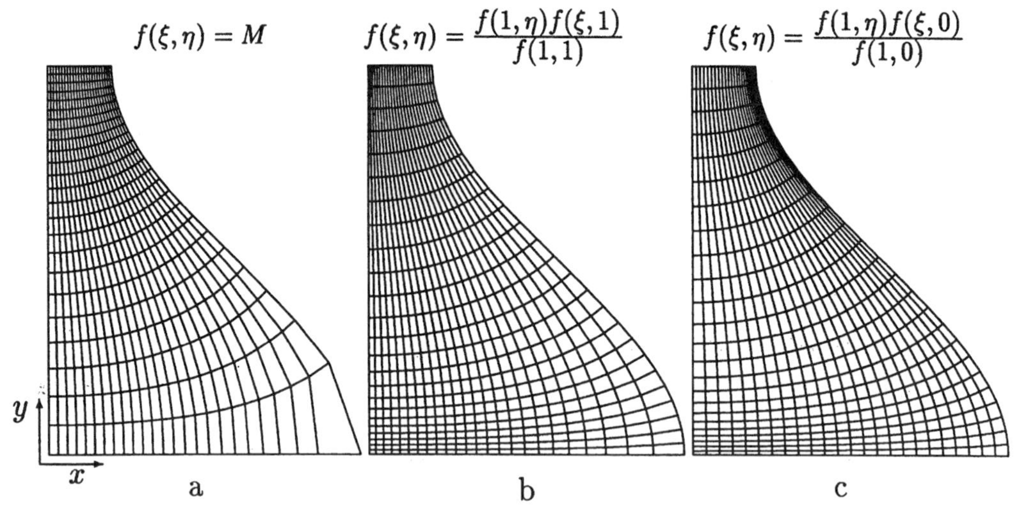

FIGURE 7.3 25 × 25 orthogonal grids in a nonsymmetric region. Distortion function equal to constant, $f = M$, quasi-conformal mapping, and f given as the product of two one-dimensional stretching functions.

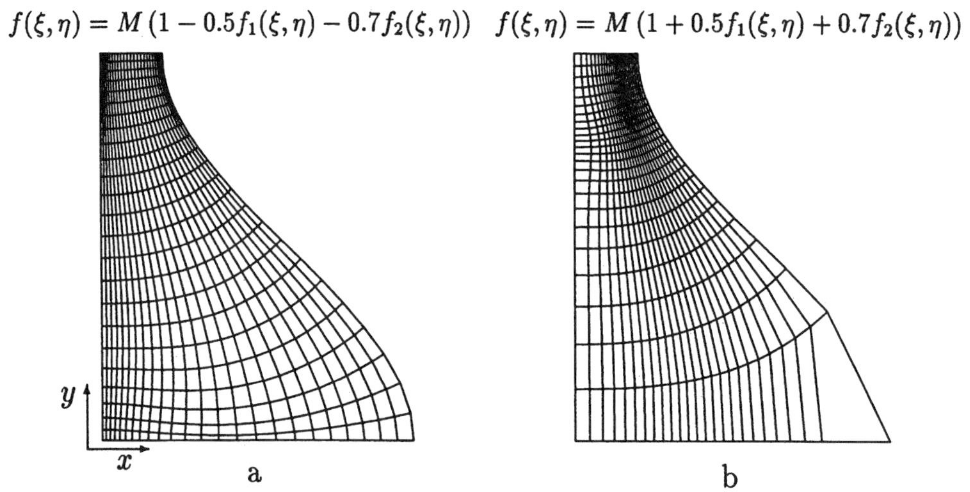

FIGURE 7.4 25 × 25 orthogonal grids in a nonsymmetrical region. Distortion function equal to the sum of linear and sine functions, $f_1(\xi,\eta) = \xi - \eta$ and $f_2(\xi,\eta) = \sin(\pi\xi) - \sin(\pi\eta)$.

Table 7.1 includes the orthogonality parameters, maximum deviation from orthogonality (MDO), mean deviation from orthogonality (ADO), and the mean quadratic error of the Beltrami Eq. 7.15, of the 25 × 25 grids plotted in Figures 7.3, 7.4, and 7.5. The large values of MDO of the grids 7.3a and 7.4b are related to the lack of resolution at the lower right corner, whereas the large value of MDO of the grid 7.5a is originated by the distortion imposed by the orthogonality condition and the fixed boundary point distribution at the upper boundary. All the grids exhibit small values of ADO and ϕ_b.

Figure 7.6 presents the variation of the orthogonality parameters with the number of grid nodes per direction, i.e., the effect of the discretization truncation error in the grid orthogonality. There are two different patterns in the variation of the orthogonality parameters, MDO, ADO, and ϕ_b, with the number of grid nodes per direction. As expected, in the mappings calculated with the distortion function equal to constant or given by the product of two one-dimensional stretching functions, grids 7.3, 7.3a, and 7.3b, the orthogonality parameters tend to zero with the increase in the number of grid nodes. The same

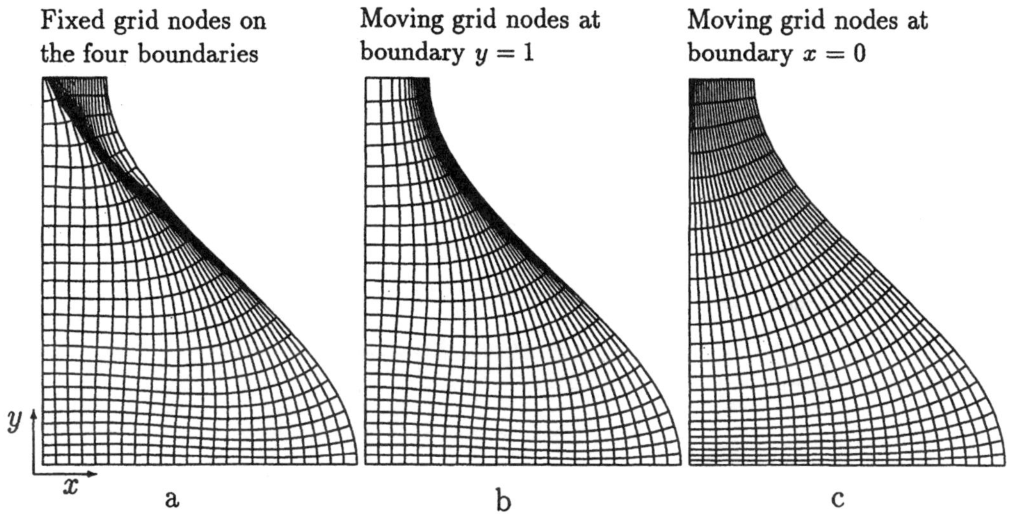

FIGURE 7.5 25 × 25 orthogonal grids in a nonsymmetric region. Distortion function obtained from the definition equation.

TABLE 7.1 Orthogonality Parameters of Two-Dimensional Orthogonal Mappings in a Nonsymmetric Region (25 × 25 grids)

Distortion Function	Boundary Conditions	MDO (degrees)	ADO (degrees)	$\phi_b \times 10^3$	Figure
Constant = M	Neumman–Dirichlet on the four boundaries	14.16	0.63	0.61	7.3a
Two one-dimensional stretching functions	Dirichlet on two boundaries	4.25	1.40	1.11	7.3b
Two one-dimensional stretching functions	Dirichlet on two boundaries	1.72	0.66	0.22	7.3c
Linear and sine functions	Neumann–Dirichlet on the four boundaries	4.38	0.46	0.17	7.4a
Linear and sine functions	Neumann–Dirichlet on the four boundaries	22.76	1.43	2.49	7.4b
Definition Equation 7.7	Dirichlet on the four boundaries	6.79	0.82	0.56	7.5a
Definition Equation 7.7	Dirichlet on three boundaries	1.19	0.21	0.03	7.5b
Definition Equation 7.7	Dirichlet on three boundaries	0.41	0.07	0.002	7.5c

behavior is obtained when f is calculated by the definition equation and the coordinates of the boundary nodes are fixed in three boundaries, 7.5a and 7.5b. However, in the mappings calculated with f defined by a sum of linear and sine functions, grids 7.4a and 7.4b, and in the mapping with complete boundary point correspondence and f determined by the definition equation, grid 7.5a, the orthogonality parameters become almost independent of the number of grid nodes per direction. This result suggests that the conditions of grids 7.4a, 7.4b and 7.5a correspond only to a nearly orthogonal mapping.

7.3.4.2 Symmetric Domains

In many cases of practical importance, the geometry exhibits one or more axes of symmetry. If the boundary point distribution is also symmetric, it is possible that there is more than one orthogonal mapping that satisfies the prescribed boundary point distribution. In fact, the grid orthogonality is completely independent of the grid node distribution along the symmetry line. Therefore, if different orthogonal mappings are generated in half-domain with fixed boundary point distributions, which only differ in the grid coordinates along the symmetry line, there is more than one orthogonal mapping for the full domain. This means that in these types of mappings, the distortion function should be specified to determine the mapping and, therefore, the boundary conditions should allow the grid notes to move

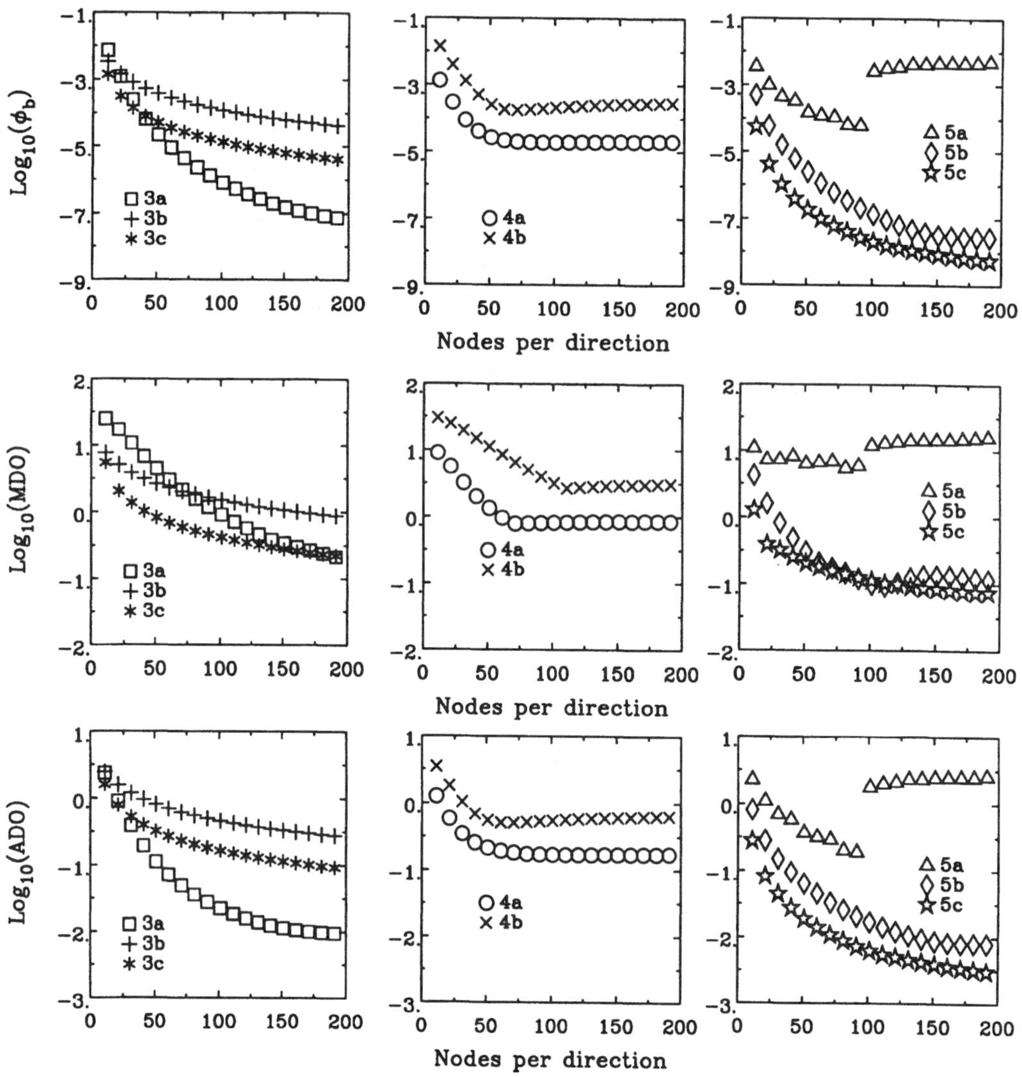

FIGURE 7.6 Variation of the orthogonality parameters with the number of grid nodes per direction. Orthogonal mappings in a symmetric region.

along the boundary. However, as shown in [13], it is possible to specify the grid nodes in all the boundaries and to determine f from the definition Eq. 7.7. Although the solution of the problem may not be unique, it is possible to generate numerically a grid which may be useful for practical purposes.

A widely used geometry has been selected to illustrate the results of orthogonal mappings in symmetric regions with f determined by its definition Eq. 7.7. It is a concave region limited by the lines $x = 0$, $x = 1$, $y = 0$, and $y = \frac{3}{4} + \frac{1}{4}\sin(\pi(\frac{1}{2} - 2x))$.

Figure 7.7 presents three 25 × 25 grids generated with fixed boundary point distributions on all the boundaries. The corresponding orthogonality parameters are given in Table 7.2. The three grids have the same boundary point distribution on the top boundary, but very different grid node distributions along the remaining three boundaries. The orthogonality parameters of the three grids confirm the ability to generate orthogonal grids with a complete boundary point correspondence.

The variation of MDO, ADO and ϕ_b with the number of grid nodes per direction is illustrated in Figure 7.8. The orthogonality parameters tend to zero with the increase in the number of grid nodes per

TABLE 7.2 Orthogonality Parameters of Two-Dimensional Orthogonal Mappings in a Symmetric Region (25 × 25 grids)

Distortion Function	Boundary Conditions	MDO (degrees)	ADO (degrees)	$\phi_b \times 10^3$	Figure
Definition Equation 7.7	Dirichlet on the four boundaries	1.67	0.30	0.06	7.7a
Definition Equation 7.7	Dirichlet on the four boundaries	2.20	0.36	0.09	7.7b
Definition Equation 7.7	Dirichlet on the four boundaries	4.03	0.49	0.17	7.7c

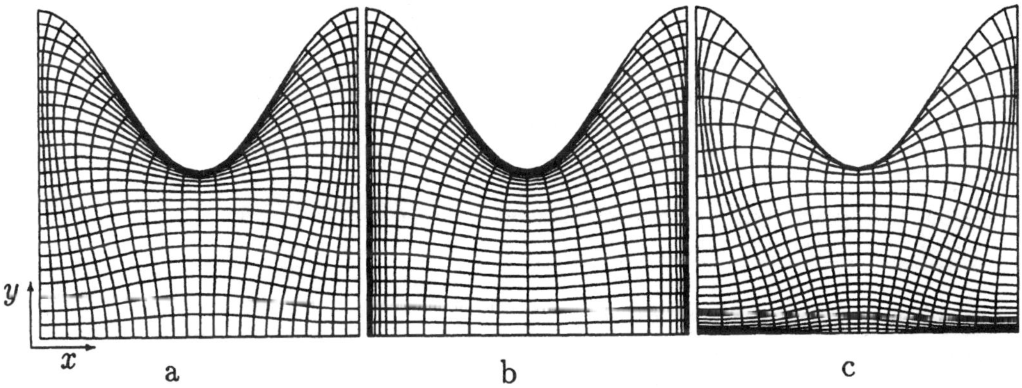

FIGURE 7.7 25 × 25 orthogonal grids in a symmetric region. Boundary nodes fixed on the four boundaries. Distortion function obtained from the definition equation.

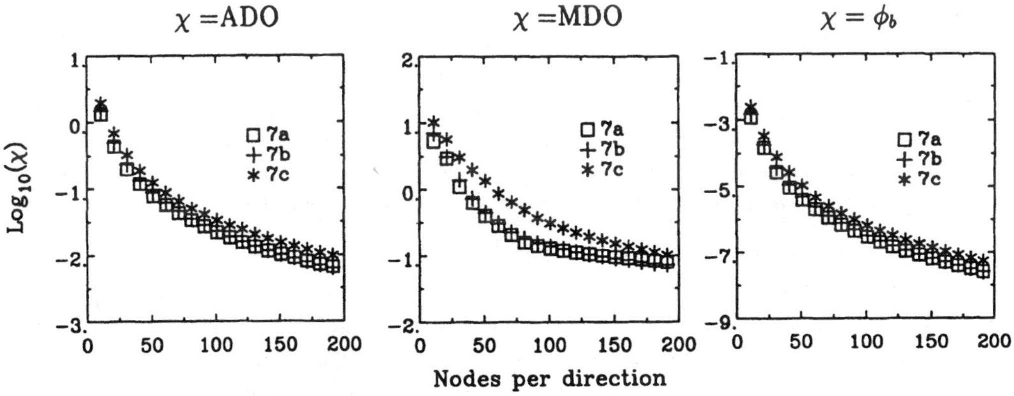

FIGURE 7.8 Variation of the orthogonality parameters with the number of grid nodes per direction. Orthogonal mappings in a symmetric region.

direction in the three cases, which are orthogonal mappings with complete boundary point correspondence. This result is not obtained in a nonsymmetric region, as illustrated in Figure 7.6. However, in a symmetric region, the symmetry line corresponds to a boundary with moving grid nodes, which means that this result is in agreement with the one obtained in the grids of the previous example, where the same behavior of the orthogonality parameters is obtained for a grid with fixed grid nodes on three boundaries.

7.3.4.3 Domains with Nonorthogonal Boundaries

The grid topology and/or the geometry of the domain may imply boundary lines which are not orthogonal. At the corners of the domain where the boundary lines are not perpendicular, the orthogonal

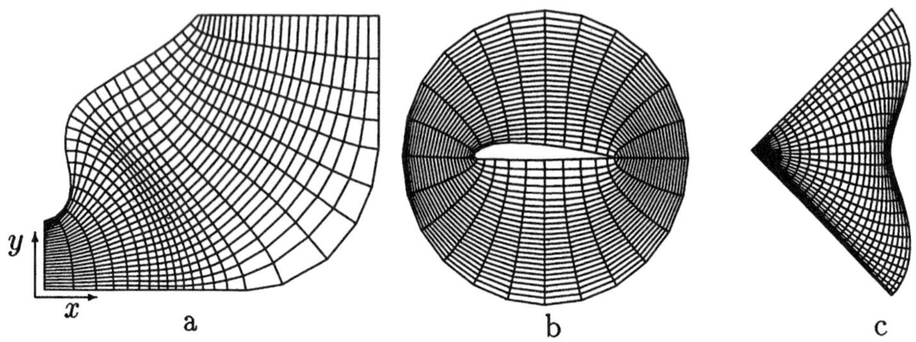

FIGURE 7.9 25 × 25 orthogonal grids in domains with nonorthogonal boundaries. Boundary nodes fixed on the four boundaries. Distortion function obtained from the definition equation.

TABLE 7.3 Orthogonality Parameters of Two-Dimensional Orthogonal Mappings in a Domains with Nonorthogonal Boundaries (25 × 25 grids)

Distortion Function	Boundary Conditions	MDO (degrees)	ADO (degrees)	$\phi_b \times 10^3$	Figure
Definition Equation 7.7	Dirichlet on the four boundaries	4.47	0.51	0.18	7.9a
Definition Equation 7.7	Dirichlet on the four boundaries	6.95	0.39	0.14	7.9b
Definition Equation 7.7	Dirichlet on the four boundaries	1.42	0.24	0.04	7.9c

mapping becomes singular, which means that the Beltrami equations are not satisfied and so the elliptic generating system 7.8 cannot be applied. When Dirichlet boundary conditions are applied, it is not necessary to solve any differential equation at the boundary. Therefore, grid singularities can be handled very easily when f is determined iteratively from its definition equation. Grid singularities can also be dealt with when the distortion function is prescribed, as in quasi- conformal mapping. Examples of distortion functions appropriate to domains with grid singularities are given in [6].

To illustrate the possibilities of the elliptic generating system in geometries with nonorthogonal boundaries, three geometries with different types of singularities are considered:

- A typical cross-section of a ship stern, where the intersection of the ship surface with the waterline is not orthogonal.
- An O-grid for a NACA 2412 airfoil, where the grid lines angle at the trailing edge is close to π.
- A trilateral region, limited by the lines $y = x$, $y = -x$, and the line defined by $x = r\cos\theta$, $y = r\sin\theta$, with $r(\theta) = 1.0 - 0.15(1.0 - \sin\theta)$. In this case, one of the sides of the computational domain is transformed into a single point in the physical domain.

In these examples, Dirichlet boundary conditions are applied on all the boundaries, which means that f is determined iteratively from the definition equation.

Figure 7.9 presents 25 × 25 orthogonal grids in the three domains, and the correspondent orthogonality parameters are given in Table 7.3. With the chosen boundary point distribution, the grid 7.9c is not symmetric. The orthogonality parameters of these grids are very similar to the ones obtained without grid singularities. The influence of the number of grid nodes per direction in the orthogonality parameters is illustrated in Figure 7.10. The three parameters tend to a constant value, which is the behavior obtained in a domain without grid singularities and fixed grid nodes in all the boundaries. It is also possible to consider mappings with moving grid nodes along the boundaries. However, the implementation of Neumann–Dirichlet boundary conditions in the vicinity of grid singularities may be troublesome.

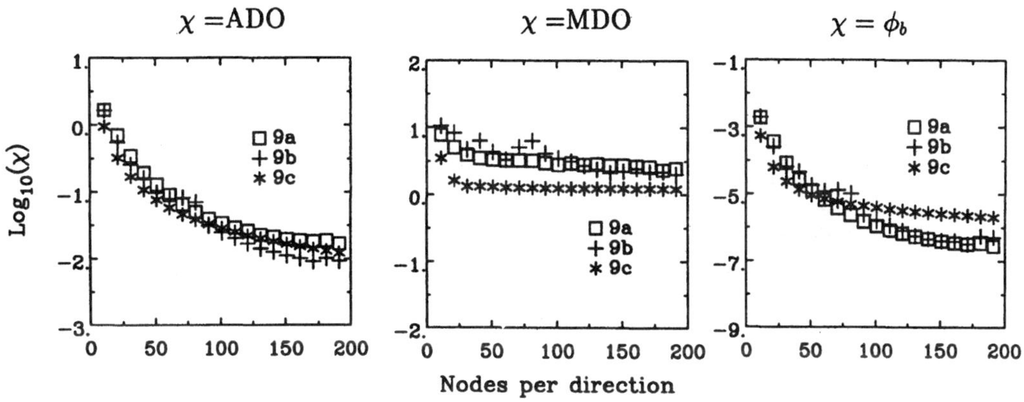

FIGURE 7.10 Variation of the orthogonality parameters with the number of grid nodes per direction. Orthogonal mappings in regions with nonorthogonal boundaries.

7.3.5 Curved Surfaces

The generation of orthogonal grids on curved surfaces has the same possibilities as two-dimensional orthogonal mappings in plane regions, which have been described in the previous section. On curved surfaces, the grid coordinates are determined in a parametric space (u, v), where u and v are obtained from a mapping between parametric space and computational domain, which ensures that the mapping between physical space, (x, y, z), and computational domain, (ξ, η), is orthogonal. As in the two-dimensional case, three types of domains are considered: nonsymmetric, symmetric and domains with non-orthogonal boundaries. In the present examples, Eq. 7.22 are solved when the distortion function is prescribed, whereas Eq. 7.28 are adopted when the distortion function is assumed to be unknown.

The orthogonality parameters, MDO, ADO, and ϕ_c are calculated with the coordinate derivatives discretized by central differencing schemes. ϕ_c, defined by Eq. 7.32, is calculated assuming that the integrand is constant in each control volume. The initial grid is obtained with linear transfinite interpolation in the parametric space. In many practical problems, the surfaces do not have an analytical representation and some type of interpolation is required. In these examples, the surface geometry is represented by a cubic spline interpolation based on a fixed number of nodes. All the coordinate derivatives are discretized in the computational domain, which means that the derivatives of x, y, and z with respect to u and v are obtained from Eq. 7.30.

The convergence criterion applied in these examples is $\phi_x \leq 1.0 \times 10^{-5}$. The convergence criterion of the iterative calculation of the coefficients of Eq. 7.27 is assumed to be $\psi_f \leq 1.0 \times 10^{-4}$ in more than two iterations.

7.3.5.1 Nonsymmetric Domains

On a curved surface it is possible to generate orthogonal grids with a prescribed distortion function, f, without control of the boundary point distribution, or with a specified boundary point distribution and an unknown f. The first case is equivalent to a specified boundary point distribution on two adjacent sides of the domain, when f is given by the product of two one-dimensional stretching functions. The following options are considered:

1. Quasi-conformal mapping. Neumann–Dirichlet boundary conditions on all the boundaries and $f = M$.
2. Boundary point distribution fixed on two boundaries and f given by the product of two one-dimensional stretching functions, Eq. 7.13.
3. f obtained from Eq. 7.21 and boundary point distribution fixed on three or four boundaries.

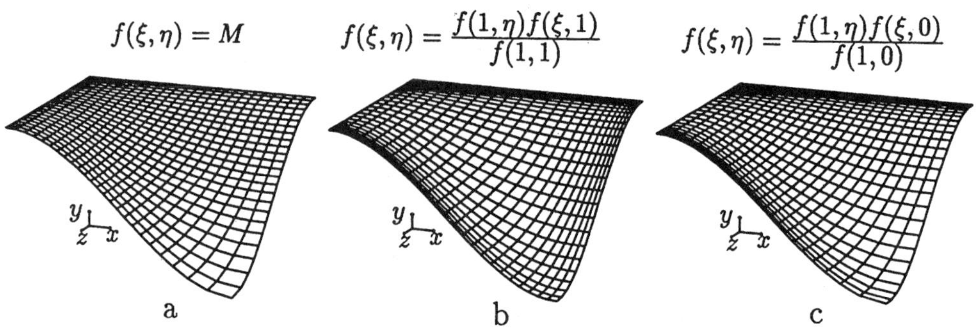

FIGURE 7.11 25 × 25 orthogonal grids in a nonsymmetric curved surface.

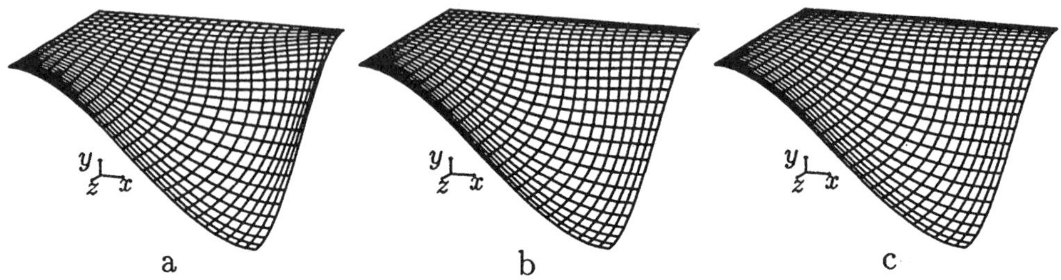

FIGURE 7.12 25 × 25 orthogonal grids in a nonsymmetric curved surface.

TABLE 7.4 Orthogonality Parameters of Mappings in a Nonsymmetric Curved Surface (25 × 25 grids)

Distortion Function	Boundary Conditions	MDO (degrees)	ADO (degrees)	$\phi_c \times 10^4$	Figure
Constant = M	Neumann–Dirichlet on the four boundaries	2.06	0.09	0.14	7.11a
Two one-dimensional stretching functions	Dirichlet on two boundaries	0.90	0.22	0.30	7.11b
Two one-dimensional stretching functions	Dirichlet on two boundaries	0.82	0.15	0.23	7.11c
Definition equation 7.21	Dirichlet on the four boundaries	1.42	0.63	1.68	7.12a
Definition equation 7.21	Dirichlet on three boundaries	0.36	0.11	0.07	7.12b
Definition equation 7.21	Dirichlet on three boundaries	0.38	0.10	0.06	7.12c

In these options, only the first two have analytical proofs of the existence and uniqueness of the solution. However, as in the two-dimensional case, it is possible to obtain numerical solutions for the remaining option and, therefore, to increase the possibilities of control of the grid line spacing. The results of these mappings are illustrated for a surface defined by $0 \le x \le 1$, $y = 1.0 - 0.5(x^2(3-2x)(1.0 - \sin\pi(\frac{1}{2} - z))$ and $0 \le z \le 1$. In this case, the parametric coordinates (u, v) are defined in the (x, z) plane.

Figure 7.11 presents 25 × 25 grids correspondent to the first two options, which are calculated from the solution of Eq. (7.22). In the grids 7.11b and 7.11c, the boundary point distribution is prescribed on the boundary $x = 1$. In grid 7.11b, Dirichlet boundary conditions are also applied at the boundary $z = 1$, whereas, in grid 7.11c, an equidistant grid node distribution is prescribed on boundary $z = 0$. The 25 × 25 grids plotted in Figure 7.12 were obtained with the generating system (7.27). Grid 7.12a is a mapping with complete boundary point correspondence and grids 7.12b and 7.12c include moving grid nodes on one of the boundaries. Table 7.4 presents the orthogonality parameters of the grids plotted in Figures 7.11 and 7.12. The values of MDO, ADO, and ϕ_c of these grids confirm the ability to generate orthogonal grids on curved surfaces with different types of control of the grid line spacing.

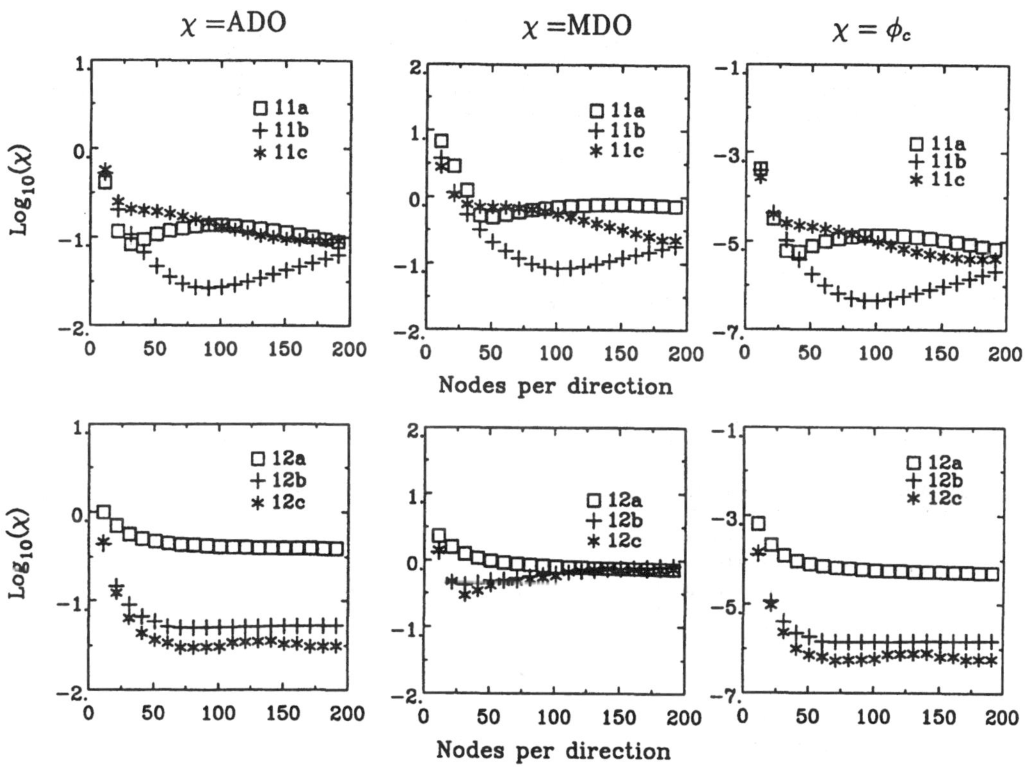

FIGURE 7.13 Variation of the orthogonality parameters with the number of grid nodes per direction. Orthogonal mappings in a nonsymmetric curved surface.

Figure 7.13 illustrates the influence of the number of grid nodes per direction on the orthogonality parameters of the mappings plotted in Figures 7.11 and 7.12. Although different values are obtained for each mapping, all the curves exhibit the tendency to converge to a constant value. This result may be unexpected for the mappings of Figure 7.11. However, it is important to note that the proof of existence and uniqueness of an orthogonal mapping on a curved surface, given in [14], is based on the use of isothermic parametric coordinates, for which the problem reduces to a two-dimensional orthogonal mapping between (u, v) and (ξ, η) . In the present example, u and v are not isothermic coordinates, which means that the right-hand side of Eq. 7.22 does not vanish and so the mapping between parametric space and computational domain is not orthogonal. Therefore, the proof presented in [14] is not applicable to the present example.

7.3.5.2 Symmetric Domains

If the surface exhibits an axis of symmetry and the boundary node distribution is also symmetric, the grid orthogonality becomes independent of the boundary point distribution along the symmetry line. Therefore, the grid cell aspect ratio should be specified. However, as in the two-dimensional case, it is possible to generate orthogonal grids on a symmetric domain assuming that the grid cell aspect ratio is unknown, as shown in [15].

The generation of orthogonal grids on symmetric curved surfaces is illustrated on a surface defined by $0 \le x \le 1,\ 0 \le z \le \frac{3}{4} + \frac{1}{4}\sin(\pi(\frac{1}{2} - 2x))$ and $y = 1.0 - \frac{1}{4}(1.0 - \sin(\pi(\frac{1}{2} - 2x))\ (1{,}0 - \sin(\pi(\frac{1}{2} - z))$, with the u and v parametric coordinates defined in the (x, z) plane.

Figure 7.14 presents three 25 × 25 grids calculated with complete boundary point correspondence, which are obtained from the solution of the system of Eq. 7.27. The orthogonality parameters of these grids are given in Table 7.5 and the influence of the number of grid nodes in the orthogonality parameters

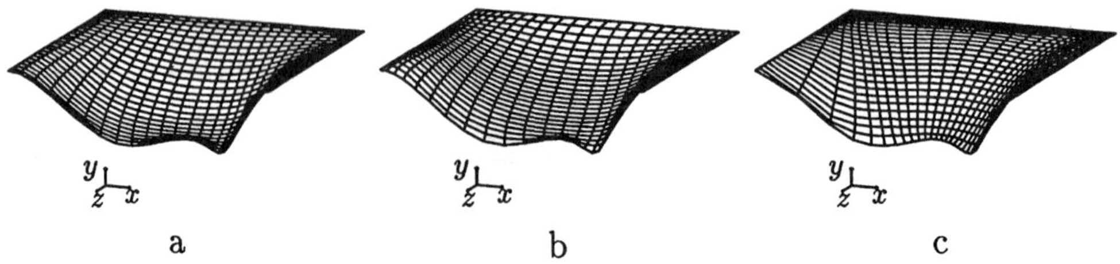

FIGURE 7.14 25 × 25 orthogonal grids in a symmetric curved surface. Distortion function obtained from the definition equation. Complete boundary point correspondence.

TABLE 7.5 Orthogonality Parameters of Mappings in a Symmetric Curved Surface (25 × 25 grids)

Distortion Function	Boundary Conditions	MDO (degrees)	ADO (degrees)	$\phi_c \times 10^4$	Figure
Definition Equation 7.21	Dirichlet on the four boundaries	1.56	0.29	0.65	7.14a
Definition Equation 7.21	Dirichlet on the four boundaries	1.34	0.31	0.68	7.14b
Definition Equation 7.21	Dirichlet on the four boundaries	3.56	0.56	3.22	7.14c

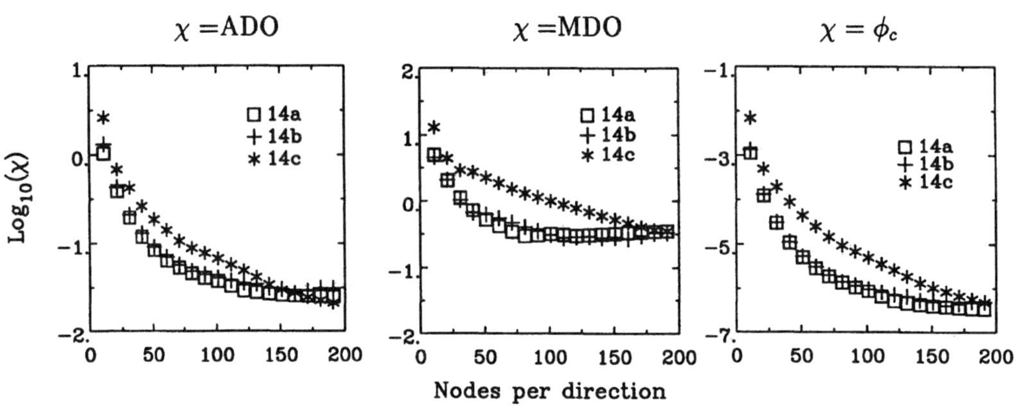

FIGURE 7.15 Variation of the orthogonality parameters with the number of grid nodes per direction. Orthogonal mappings in a symmetric curved surface.

is illustrated in Figure 7.15. The results confirm the possibility to control the boundary point distribution of an orthogonal mapping on a curved surface, even in a symmetric domain. As in the nonsymmetric region, the values of MDO, ADO, and ϕ_c tend to a constant value with the increase in the number of grid nodes per direction. These constant values are almost independent of the specified boundary point distribution.

7.3.5.3 Domains with Nonorthogonal Boundaries

In many practical problems, a surface may exhibit boundary lines that are not orthogonal. At these locations, an orthogonal mapping becomes singular. However, when Dirichlet boundary conditions are applied, it is not necessary to solve any equation at the boundary. Therefore, non-orthogonal boundaries can be handled easily when Dirichlet boundary conditions are applied.

The ability to generate orthogonal grids on curved surfaces with non-orthogonal boundaries is illustrated on two different geometries: the nose and cockpit of a fighter aircraft and a wing of elliptical planform

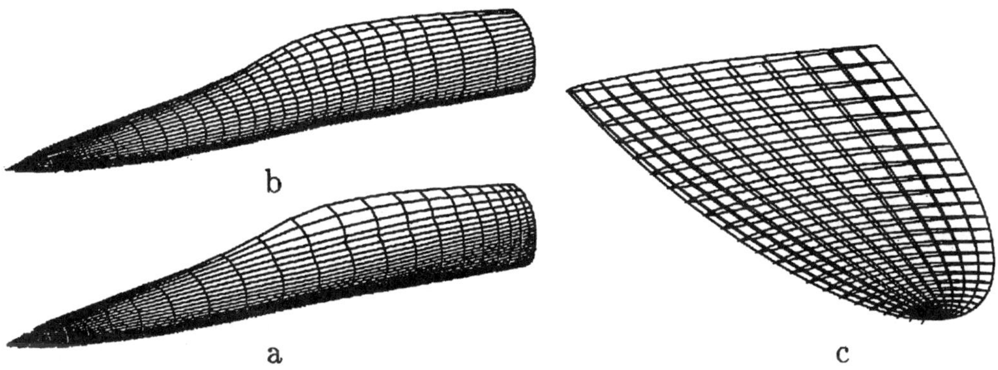

FIGURE 7.16 25 × 25 orthogonal grids in curved surfaces with grid singularities. Distortion function obtained from the definition equation. Complete boundary point correspondence.

TABLE 7.6 Orthogonality Parameters of Orthogonal Mappings in Curved Surfaces with Nonorthogonal Boundaries (25 × 25 grids)

Distortion Function	Boundary Conditions	MDO (degrees)	ADO (degrees)	$\phi_c \times 10^4$	Figure
Definition Equation 7.21	Dirichlet on the four boundaries	4.22	1.26	8.19	7.16a
Definition Equation 7.21	Dirichlet on the four boundaries	2.14	0.97	4.12	7.16b
Definition Equation 7.21	Dirichlet on the four boundaries	1.79	0.13	0.24	7.16c

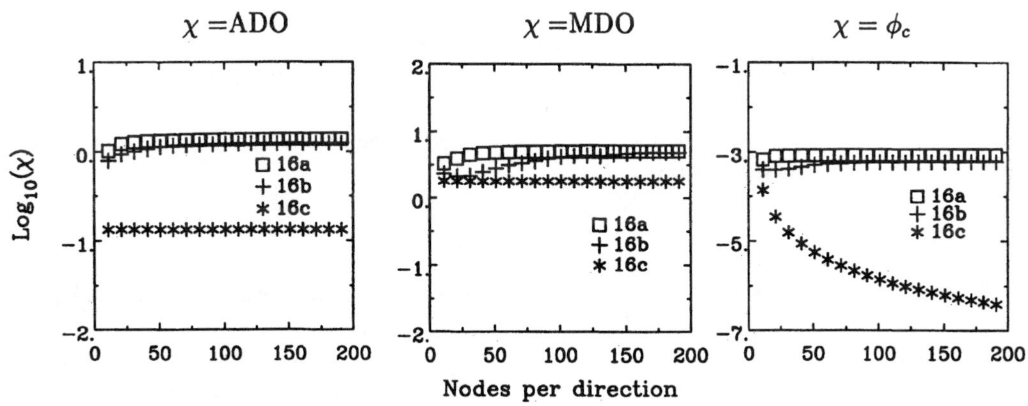

FIGURE 7.17 Variation of the orthogonality parameters with the number of grid nodes per direction. Orthogonal mappings in curved surfaces with nonorthogonal boundaries.

with a NACA 4412 airfoil section. In both cases, one side of the parametric domain is transformed into a single point of the physical space. The present examples are restricted to mappings with Dirichlet boundary conditions on all the boundaries and, therefore, the generating system defined by Eq. 7.27.

Three 25 × 25 grids are plotted in Figure 7.16 and the respective orthogonality parameters are given in Table 7.6. Figure 7.17 presents the influence of the number of grid nodes per direction on MDO, ADO, and ϕ_c. The results are equivalent to the ones obtained for curved surfaces without grid singularities. As in the two-dimensional mappings, the application of Neumann–Dirichlet boundary conditions with finite-difference discretizations in the vicinity of grid singularities may be troublesome.

7.4 Summary

This chapter presents an overview of orthogonal generating systems based on the solution of elliptic partial differential equations. In three-dimensional geometries, it is impossible to generate fully orthogonal grids in most of the cases, which implies that the main research effort in orthogonal grid generation is concentrated in two-dimensional regions and curved surfaces.

The use of generating systems based on elliptic systems of partial differential equations allows the control of the grid line spacing from the definition of the grid cell aspect ratio or through the specification of the boundary point distribution. However, the number of situations for which there is a theoretical proof of the existence and uniqueness of an orthogonal mapping is rather small. In two-dimensional regions, it is possible to obtain such a proof when the grid cell aspect ratio is defined as the product of two one-dimensional stretching functions, which is equivalent to the specification of the boundary point distribution on two adjacent boundaries. This proof can be extended to curved surfaces when isothermic parametric coordinates are adopted to describe the surface.

For practical purposes, it is possible to generate numerically orthogonal grids on two-dimensional regions with more general distributions of the grid cell aspect ratio or with the boundary point distribution fixed in more than two boundaries. In the latter case, the grid cell aspect ratio is determined iteratively as part of the solution. Although there is no theoretical proof that these mappings yield well-posed problems, the numerical solutions obtained in nonsymmetric and symmetric domains show that there are several possibilities to control the grid line spacing in orthogonal mappings. However, in mappings with complete boundary point correspondence, the orthogonality restriction may produce an interior grid line spacing which is unacceptable for numerical purposes. In general, in these cases, the use of moving grid nodes along one of the boundaries is sufficient to obtain a smooth interior grid line spacing.

The numerical results also show that it is possible to generate orthogonal grids on curved surfaces adopting nonisothermic parametric coordinates. Overall, the properties of such mappings are similar to the ones of two-dimensional mappings. However, the numerical results suggest that complete orthogonality will only be achieved with isothermic parametric coordinates.

Elliptic generating systems are also able to handle orthogonal mappings that include grid singularities at the boundary. The ability to specify Dirichlet boundary conditions at the boundaries allows the generation of orthogonal grids in domains with nonorthogonal boundaries with the same approach used in domains without grid singularities.

References

1. **Eiseman, P. R.**, Orthogonal grid generation, *Numerical Grid Generation,* Thompson, Joe F., (Ed.), Elsevier Science, pp. 193–233, 1982.
2. **Henrici, P.**, *Applied and Computational Complex Analysis,* Vol. III, John Wiley & Sons, NY, 1986.
3. **Moretti, G.**, Orthogonal grids around difficult bodies, *AIAA J.*, 30(4) pp. 933–938,1992.
4. **Thompson, J. R., Warsi, Z. U. A., and Mastin, C. W.**, Boundary-fitted coordinate systems for numerical solution of partial differential equations — a review, *J. Comput. Phys.*, 47(1), pp. 1–108, 1982.
5. **Thompson, J. F., Warsi, Z. U. A., and Mastin, C. W.**, *Numerical Grid Generation — Foundations and Applications,* Elsevier Science, 1985.
6. **Duraiswami, R. and Prosperetti, A.**, Orthogonal mapping in two dimensions, *J. Comput. Phys.*, 98, pp. 254–268, 1992.
7. **Sorensen, R. L.**, Grid generation by elliptic partial differential equations for tri-element augmentor-wing airfoil, *Numerical Grid Generation,* Thompson, Joe F., (Ed.), Elsevier Science, pp. 193–233, 1982.
8. **Arina, R.**, Orthogonal grids with adaptive control, *Proceedings of the 1st International Conference on Numerical Grid Generation in CFD,* Hauser, J. and Taylor, C., (Ed.), Pineridge Press, pp. 113–124, 1986.

9. **Ryskin, G. and Leal, L. G.**, Orthogonal Mapping, *J. Comput. Phys.*, 50, pp. 71–100, 1983.
10. **Chikhliwala, E. D. and Yortsos, Y. C.**, Application of orthogonal mapping to some two-dimensional domains, *J. Comput. Phys.*, 57, pp. 391–402, 1985,.
11. **Albert, M. R.**, Orthogonal Curvilinear Coordinate Generation for Internal Flows, *Proceedings of the 2nd International Conference on Numerical Grid Generation in CFD*, Hauser, J. and Taylor, C., (Ed.), Pineridge Press, pp. 113–124, 1988.
12. **Allievi, A. and Calisal, S.M.**, Application of Bubnov-Galerkin formulation to orthogonal grid generation, *J. Comput. Phys.*, 98, pp. 163–173, 1992.
13. **Eça, L.**, 2D orthogonal grid generation with boundary point distribution control, *J. Comput. Phys.* 125, pp. 440–453, 1996.
14. **Arina, R.**, Adaptive orthogonal surface coordinates, *Proceedings of the 2nd International Conference on Numerical Grid Generation in CFD*, Hauser, J. and Taylor, C., (Ed.),Pineridge Press, pp. 351–359, 1988.
15. **Eça, L.**, Orthogonal grid generation with systems of partial differential equations, *Proceedings of the 5th International Conference on Numerical Grid Generation in Computational Field Simulations*, Soni, B. K., Thompson, J. F., Hauser, J. and Eiseman, P., (Ed.), Mississippi State University, 1996, pp. 25–36.
16. **Niederdrenk, P.**, private communication, 1996.
17. **Doubrovine, B., Novikov, S., and Fomenko, A.**, Géométrie contemporaine — méthodes et applications — géométrie des surfaces, des groupes de transformations et des champs, (French translation), *Éditions MIR*, Moscow, 1982.
18. **Kang, I. S. and Leal, L. G.**, Orthogonal Grid Generation in a 2D Domain via the Boundary Integral Technique, *J. Comput. Phys.*, 102, pp. 77–87, 1992.

Further Information

The proceedings of the Conferences in Numerical Grid Generation in Computational Field Simulations include several papers dedicated to orthogonal grid generation. The conferences have been held since 1986.

The monthly *Journal of Computational Physics* has reported most of the advances in orthogonal grid generation in the last few years.

8

Harmonic Mappings

Sergey A. Ivanenko

0-8493-2687-7/99/$0.00+$.50

8.1 Introduction

Methods of grid generation based on the theory of harmonic maps are presented in this chapter. Algorithms for structured and unstructured adaptive grids in two-dimensional and three-dimensional cases as well as for grids on surfaces are described in detail. All methods are based on grid nodes movement (r-refinement).

Two fundamental problems in grid generation are considered in the present chapter.

The first problem is to find conditions for discrete mappings to the nondegenerate. The condition of convexity of all the grid cells in two dimensions is assumed as a discrete analog of the Jacobian positiveness. It guarantees the grid to be nondegenerate. Indeed, if all grid cells are convex, then all grid nodes do not leave a domain, and such a grid does not contain self-intersecting cells. In the three-dimensional case, a more complicated analog of Jacobian positiveness is presented.

The second problem is to develop a suitable theoretical framework for grid generation. The theory of harmonic maps has been chosen as a basis for this purpose. The problem of constructing harmonic coordinates on the surface of the graph of control functions is formulated. Harmonic coordinates are constructed from harmonic mapping of the surface onto a parametric square (or cube in the three-dimensional case). The projection of these coordinates onto a physical region produces an adaptive–harmonic grid [Liseikin, 1991, 1993; Ivanenko, 1993, 1995]. The application of such monitoring surfaces was also considered by Dwyer, et al. [1982], Eiseman [1987], and Spekreijse, et al. [1996].

Two methods are used for numerical solution. The first one is based on the finite-difference approximation of Euler equations. The second method is based on a direct minimization of the discrete analog of the harmonic functional.

The variational approach has been extended to the case of irregular meshes [Ivanenko, 1995b]. The main principle can be formulated as follows. Recall that harmonic coordinates are generated by the global harmonic mapping of the physical domain or the surface of control function onto a parametric square. The result will be a regular grid. Irregular (unstructured) grids can be considered as a set of local coordinates, different for each cell or element. Hence, each cell, for example a quadrilateral, can be harmonically mapped onto the same auxiliary unit square. The total irregular grid with fixed connections can be computed by minimizing the sum of harmonic functionals, written for each grid cell. This will be a smoothing and adaption stage in the method of irregular grid generation. For triangular grids, each triangle should be mapped harmonically onto an equilateral triangle and so on.

A very important property of variational approaches is that the functionals are approximated in such a way that all their discrete analogues have infinite barrier on the boundary of the set of nondegenerate grids. The resulting algorithms assure generation of nondegenerate grids according to developed discrete conditions of the Jacobian positiveness. Consequently, the theory of harmonic maps, applied to grid generation, can be assumed as a general framework for the development of fully automated algorithms. Moreover, as on the continuous level, the theory of harmonic maps provides construction of nondegenerate curvilinear coordinates; on the discrete level, the developed application of this theory guarantees generation of nondegenerate grids in arbitrary domains.

8.2 Nondegenerate Planar Grids

Two types of grids/meshes are used in computations: regular (structured) and irregular (unstructured). Regular grids contain only regular nodes, or nodes whose neighbors are known only from the indexation. A typical example is a curvilinear grid constructed by a mapping of a parametric square onto a physical domain. Grid nodes are enumerated with double indices in the two-dimensional and by triple indices in the three-dimensional case. This is not the case of irregular meshes. For such a mesh, neighbors of nodes must be specified. In spite of the fact that the set of regular grids is a reduction of the set of irregular meshes, we will start with the consideration of regular grids. The condition of the Jacobian positiveness is considered as the condition for a regular grid to be nondegenerate. An irregular mesh can

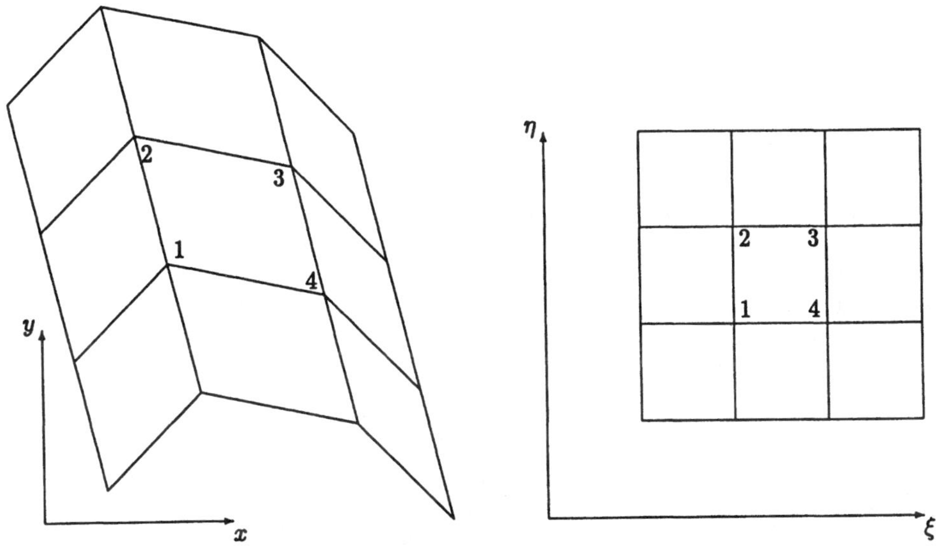

FIGURE 8.1 Correspondence of nodes numbers for a mapping of the square cell 2+1/2, 2+1/2 in the plane ξ, η onto a corresponding quadrilateral cell in the plane x, y.

be assumed as a set of local coordinates, so the condition of the Jacobian positiveness can be used also to define discrete conditions for an irregular mesh to be nondegenerate.

8.2.1 Two-Dimensional Regular Grids

The problem of grid generation in two dimensions will be considered in the following formulation. In a simply connected domain Ω on the plane x, y a grid

$$(x,y)_{i,j} \qquad i=1,\ldots,i^* \qquad j=1,\ldots,j^* \tag{8.1}$$

must be constructed with given coordinates of boundary nodes

$$(x.y)_{i1} \qquad (x,y)_{i,j^*} \qquad (x,y)_{1j} \qquad (x,y)_{i^*j} \tag{8.2}$$

The problem can be treated as a discrete analog of the problem of finding functions $x(\xi, \eta)$ and $y(\xi, \eta)$, ensuring one-to-one mapping of the parametric square

$$0<\xi<1 \qquad 0<\eta<1 \tag{8.3}$$

onto a domain Ω (see Figure 8.1) with a given transformation of the square boundary onto the boundary of Ω, associated with the boundary conditions Eq. 8.2, i.e., on each side of the parametric square the following eight functions are specified:

$$x(\xi,0)=x_{down}(\xi) \quad x(\xi,1)=x_{up}(\xi) \quad x(0,\eta)=x_{left}(\eta) \quad x(1,\eta)=x_{right}(\eta)$$
$$y(\xi,0)=y_{down}(\xi) \quad y(\xi,1)=y_{up}(\xi) \quad y(0,\eta)=y_{left}(\eta) \quad y(1,\eta)=y_{right}(\eta)$$

Instead of the parametric square Eq. 8.3 on the plane ξ, η the parametric rectangle is often introduced to simplify the computational formulas

$$1 < \xi < i^* \qquad 1 < \eta < j^* \tag{8.4}$$

associated with the square grid (ξ_i, η_j) on the plane ξ, η such that

$$\xi_i = i \qquad \eta_j = j \qquad i = 1,\ldots,i^* \quad j = 1,\ldots,j^*$$

In the paper by Bobilev, Ivanenko, and Ismailov [1996], the following theorem has been proven:

THEOREM 1. If a smooth mapping of one domain onto another with a one-to-one mapping between boundaries possesses a positive Jacobian not only inside a domain but also on its boundary, then such a mapping will be one-to-one.

Hence, the curvilinear coordinate system constructed in a domain Ω will be nondegenerate if the Jacobian of the mapping $x(\xi, \eta)$, $y(\xi, \eta)$ is positive:

$$J = x_\eta y_\xi - x_\eta y_\xi > 0 \qquad 0 \le \xi \le 1 \qquad 0 \le \eta \le 1 \tag{8.5}$$

Thus, the problem of constructing curvilinear coordinates in a domain Ω can be formulated as the problem of finding of smooth mapping of a parametric square onto a domain Ω that satisfies the condition of the Jacobian positiveness Eq. 8.5. The mapping between boundaries must be one-to-one, which can be easily provided from the condition of monotonic variations of ξ and η along the appropriate parts of the boundary of a domain Ω.

Consequently, in the discrete case for the grid (Eq. 8.1) a discrete analog of the Jacobian positiveness must be also applied.

8.2.2 Discrete Analog of the Jacobian Positiveness

The condition of grid cell convexity was introduced by Ivanenko and Charakhch'yan [1988] as a discrete analog of the Jacobian positiveness. The mapping $x(\xi, \eta)$, $y(\xi, \eta)$ was approximated by quadrilateral finite elements.

Let the coordinates $(x, y)_{ij}$ of grid nodes be given. To construct the mapping $x^h(\xi, \eta)$, $y^h(\xi, \eta)$ of the parametric rectangle Eq. 8.4 onto the domain Ω such that $x^h(i, j) = x_{i,j}$ and $y^h(i, j) = y_{ij}$ we use quadrilateral isoparametric finite elements [Strang and Fix, 1973]. The square cell numbered $i + 1/2$, $j + 1/2$ on the plane ξ, η is mapped onto the quadrilateral cell on the plane x, y, formed by nodes with coordinates

$$(x,y)_{i,j} \qquad (x,y)_{i,j+1} \qquad (x,y)_{i+1,j+1} \qquad (x,y)_{i+1,j}$$

The cell vertices are numbered from 1 to 4 in the clockwise direction as is shown in Figure 8.1. The node (i, j) corresponds to the vertex 1, node $(i, j + 1)$ to vertex 2 and so on. Each vertex is associated with a triangle: vertex 1 with Δ_{412}, vertex 2 with Δ_{123} and so on. The doubled area J_k, $k = 1, 2, 3, 4$, of these triangles is introduced as follows:

$$J_1 = (x_4 - x_1)(y_2 - y_1) - (y_4 - y_1)(x_2 - x_1) =$$
$$(x_{i+1,j} - x_{i,j})(y_{i,j+1} - y_{i,j}) - (y_{i+1,j} - y_{i,j})(x_{i,j+1} - x_{i,j})$$

In the first expression the vertex indices are used and in the second the corresponding node indices are used. Functions x^h, y^h for $i \le \xi \le i + 1$, $j \le \eta \le j + 1$ are represented in the form

$$\begin{aligned} x^h(\xi,\eta) &= x_1 + (x_4 - x_1)(\xi - i) + (x_2 - x_1)(\eta - j) + (x_3 - x_4 - x_2 + x_1)(\xi - i)(\eta - j) \\ y^h(\xi,\eta) &= y_1 + (y_4 - y_1)(\xi - i) + (y_2 - y_1)(\eta - j) + (y_3 - y_4 - y_2 + y_1)(\xi - i)(\eta - j) \end{aligned} \tag{8.6}$$

Each side of the square is linearly transformed onto the appropriate side of the quadrilateral. Consequently, the global transformation x^h, y^h is continuous on the cell boundaries. To check the one-to-one property of the transformation Eq. 8.6, we write out the expression for the Jacobian

$$J^h = x^h_\xi \, y^h_\eta - x^h_\eta \, y^h_\xi = \det\begin{pmatrix} x_4 - x_1 + A(\eta - j) & x_2 - x_1 + A(\xi - i) \\ y_4 - y_1 + B(\eta - j) & y_2 - y_1 + B(\xi - i) \end{pmatrix}$$

where $A = x_3 - x_4 - x_2 + x_1$, $B = y_3 - y_4 - y_2 + y_1$. Jacobian is linear, not bilinear, since the coefficient before $\xi\eta$ in this determinant is equal to zero. Consequently, if $J^h > 0$ in all corners of the square, it does not vanish inside this square. In the corner 1 ($\xi = i$, $\eta = j$) of the cell $i + 1/2, j + 1/2$ the Jacobian

$$J^h(i,j) = (x_4 - x_1)(y_2 - y_1) - (y_4 - y_1)(x_2 - x_1)$$

i.e., $J^h(i, j) = J_1$ is the doubled area of triangle Δ_{412}, introduced above.

From this it follows that the condition of the Jacobian positiveness for the mapping $x^h(\xi, \eta)$, $y^h(\xi, \eta)$

$$x^h_\xi \, y^h_\eta - x^h_\eta \, y^h_\xi > 0 \qquad 1 \le \xi \le i^* \qquad 1 \le \eta \le j^*$$

is equivalent to the system of inequalities

$$[J_k]_{i+1/2, j+1/2} > 0 \quad k = 1,2,3,4 \qquad i = 1,\ldots,i^* - 1 \qquad j = 1,\ldots,j^* - 1 \tag{8.7}$$

where $J_k = (x_{k-1} - x_k)(y_{k+1} - y_k) - (y_{k-1} - y_k)(x_{k+1} - x_k)$, and in expressions for J_k one should put $k - 1 = 4$ if $k = 1$, and $k + 1 = 1$ if $k = 4$.

If conditions Eq. 8.7 are satisfied, then all grid cells are convex quadrilaterals. Hence, if the mapping $x(\xi, \eta)$, $y(\xi, \eta)$ is approximated by piecewise-bilinear functions, then the one-to-one condition is equivalent to the condition of convexity of all grid cells Eq. 8.7. Such grids were called convex grids [Ivanenko and Charakhch'yan, 1988], and only convex grids can be used in the finite element method with conforming quadrilateral elements.

The set of grids satisfying inequalities Eq. 8.7 is called a convex grid set and denoted by D. This set belongs to the Euclidean space R^N, where $N = 2(i^* - 2)(j^* - 2)$ is the total number of degrees of freedom of the grid equal to double the number of its internal nodes. In this space D is an open bounded set. Its boundary ∂D is the set if grids for which at least one of the inequalities Eq. 8.7 becomes an equality.

8.2.3 Irregular Two-Dimensional Meshes

In the employment of irregular meshes we must define the correspondence between local (for each element) and global nodes numeration. In Figure 8.2 the simplest example of an irregular mesh is shown. Element numbers are shown in circles. The local numeration is shown only for the element 1. The global numeration is shown with a bold font.

The function of $COR(N, k)$ is introduced to define a correspondence between local and global node numbers:

$$COR(N,k) = n \qquad n = 1,\ldots,N_n \qquad N = 1,\ldots N_e \qquad k = 1,2,3,4$$

where n is a global node number, N_n is a total number of mesh nodes, N is an element number, N_e is a number of elements, k is a local node number in the element. This function is implemented in the computer program as a function for a regular grid and as an array for an irregular mesh. For example,

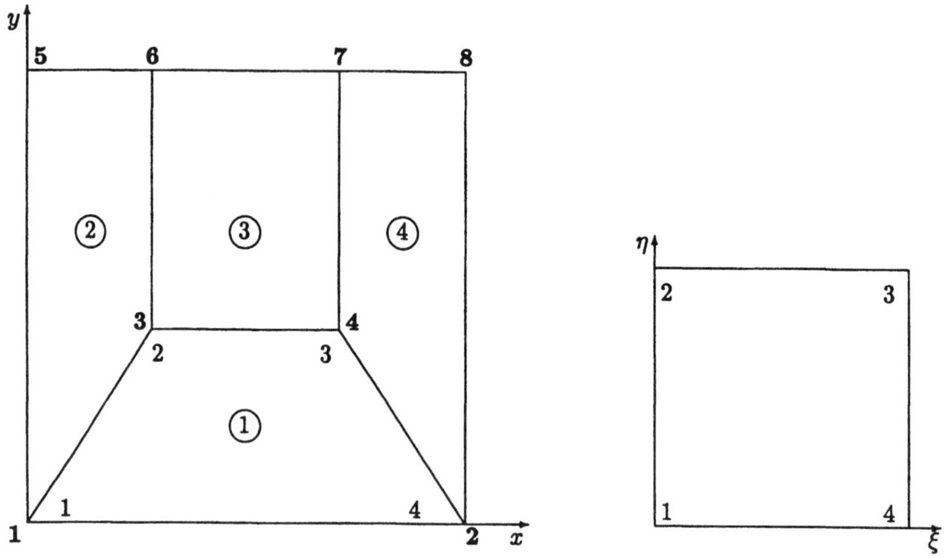

FIGURE 8.2 Correspondence of nodes numbers for a mapping of the unit square in the plane ξ, η onto the quadrilateral cell 1 of irregular mesh in the plane x, y.

for the irregular mesh shown in Figure 8.2 the correspondence between local and global numerations is defined as follows:

$$COR(1,1)=1 \quad COR(1,2)=3 \quad COR(1,3)=4 \quad COR(1,4)=2$$

For irregular meshes the array *COR* is filled up during the mesh construction, for example, by a front method. It is often necessary to use other correspondence functions, for example, when we must define numbers of two elements from the number of their common edge or to define the neighbor numbers for a given node. The choice of these functions depends on the type of elements used and on the solver peculiarities. We will consider below only the simplest data structure, defined by *COR(N, k)*, which is enough for our purposes.

For regular grids we can use the function with the same name instead of the array *COR*. It is convenient to use one-dimensional numeration instead of double indices. For node numbers of a regular grid, introduced above in Eq. 8.1, we have

$$N(i,j)=i+(j-1)(i^*-1) \quad i=1,\dots,i^*-1 \quad j=1,\dots,j^*-1$$
$$n(i,j)=i+(j-1)i^* \quad i=1,\dots,i^* \quad j=1,\dots,j^*$$

where $n(i, j)$ corresponds to the node i, j, and $N(i, j)$ corresponds to the cell number $i + 1/2$, $j + 1/2$. Then the correspondence function is defined as follows:

$$COR(N(i,j),1)=n(i,j) \quad COR(N(i,j),2)=n(i,j+1)$$
$$COR(N(i,j),3)=n(i+1,j+1) \quad COR(N(i,j),4)=n(i+1,j)$$

Now we consider conditions for the mesh node coordinates to assure a mesh to be nondegenerate. Note, that in the case of a regular grid instead of the mapping $x(\xi, \eta)$, $y(\xi, \eta)$ of the parametric rectangle Eq. 8.4 onto a domain Ω, a bilinear mapping of the same unit square onto each quadrilateral cell can be considered. All argumentation in Section 8.2.1 will be true in this case, since the Jacobian of the mapping $x^h(\xi, \eta)$, $y^h(\xi, \eta)$ is not changed if the square cell is shifted in the plane ξ, η. Hence, for each cell of irregular mesh a bilinear mapping of the unit square on the plane ξ, η onto this cell can be introduced (see Figure 8.2). The condition of the Jacobian positiveness can be written as follows:

$$[J_k]_N > 0 \qquad k = 1,2,3,4 \qquad N = 1,\ldots,N_e \tag{8.8}$$

where $J_k = (x_{k-1} - x_k)(y_{k+1} - y_k) - (y_{k-1} - y_k)(x_{k+1} - x_k)$ is the area of the triangle, written in local numeration. Consequently, all the mesh cells satisfying inequalities Eq. 8.8 will be convex quadrilaterals.

As in the case of regular grids, irregular meshes, satisfying inequalities Eq. 8.8 will be called convex meshes.

As in the previous subsection the set of meshes, satisfying inequalities Eq. 8.8 is called a convex mesh set and denoted by D. This set belongs to the Euclidean space $R^{N_{in}}$, where N_{in} is the total number of degrees of freedom of the mesh equal to double the number of its internal nodes. In this space D is an open bounded set. Its boundary ∂D is the set of meshes for which at least one of the inequalities Eq. 8.8 becomes an equality.

8.3 Planar Harmonic Grid Generation

Experience has shown the efficiency and the reliability of the method based on harmonic mapping, proposed by Winslow [1966]. This is consistent with the theoretical foundation of the method, since the theory guarantees that the generated curvilinear coordinate system is nondegenerate. This property follows from the general result on existence and uniqueness of the one-to-one harmonic mapping of an arbitrary domain onto a parametric square.

Development of the method suggested by Godunov and Prokopov [1972] is based on the use of such additional parameters that there was no loss of the one-to-one property. This approach was introduced to control the grid spacing (adaption). Further developments of this approach were presented by Thompson, et al. [1985].

The system of two Laplace equations is used for constructing harmonic mapping. The natural way to extend this method is to use more common elliptic equations with right-hand sides. However, in the general case it is not clear how to obtain conditions on control parameters under which the generation of a nondegenerate curvilinear coordinate system (regular grid) is guaranteed.

8.3.1 Problem Formulation

The simplest and the most investigated elliptic equation is Laplace equation. That is why the system

$$x_{\xi\xi} + x_{\eta\eta} = 0 \qquad y_{\xi\xi} + y_{\eta\eta} = 0$$

or its direct extensions may be considered for grid generation.

However, these equations cannot guarantee the generation of a nondegenerate grid. A simple example was constructed by Prokopov [1993]. Let us consider the transformation

$$x(\xi,\eta) = \frac{1}{2}(\xi^2 - \eta^2) - \frac{2}{3}\xi, \quad y(\xi,\eta) = \xi\eta + \frac{1}{2}\xi - \frac{1}{3}\eta$$

defined on the unit square $0 < \xi < 1$, $0 < \eta < 1$.

Obviously, this transformation satisfies Laplace equations and the Jacobian

$$J(\xi,\eta) = x_\xi y_\eta - x_\eta y_\xi = \left(\xi - \frac{2}{3}\right)\left(\xi - \frac{1}{3}\right) + \left(\eta + \frac{1}{2}\right)\eta$$

Since $J(\xi, 0) = (\xi - {}^2/_3)(\xi - {}^1/_3) < 0$ on the interval $\eta = 0, {}^1/_3 < \xi < {}^2/_3$, the transformation is folded near the image of the lower part of the square boundary. The example is interesting because the image of the square has a very simple form so the transform degeneration and the grid folding seems absolutely unexpected.

The method of grid generation guaranteeing the one-to-one mapping on the continuous level was proposed by Winslow [1966]. Two families of grid lines are constructed as contours of functions $\xi(x, y)$, $\eta(x, y)$ satisfying two Laplace equations

$$\xi_{xx} + \xi_{yy} = 0 \quad \eta_{xx} + \eta_{yy} = 0 \tag{8.9}$$

with Dirichlet boundary conditions associated with the one-to-one mapping of the boundary of parametric square Eq. 8.4 onto the boundary of domain.

After transforming to independent variables ξ, η, these equations take the form

$$\alpha x_{\xi\xi} - 2\beta x_{\xi\eta} + \gamma x_{\eta\eta} = 0 \qquad \alpha y_{\xi\xi} - 2\beta y_{\xi\eta} + \gamma y_{\eta\eta} = 0 \tag{8.10}$$

where $\alpha = x_\eta^2 + y_\eta^2$, $\beta = x_\xi x_\eta + y_\xi y_\eta$, $\gamma = x_\xi^2 + y_\xi^2$.

The standard approximation of Eq. 8.10 with centered differences for the first-order derivatives was used by Winslow [1966] and Godunov and Prokopov [1972]. Computational formulas for the extension of the method to the case of adaptive planar grids will be described in detail in the next section.

8.3.2 Variational Method for Irregular Planar Mesh Smoothing

The process of irregular mesh generation usually contains two stages. The meshes produced at the first stage by automated techniques often exhibit large variations of mesh cells. The smoothing techniques are used then to form better shaped cells and yield more accurate analyses. Various approaches have been developed, but the most promising is, in our opinion, an approach based on harmonic mappings. For regular grids such algorithms were proposed by Yanenko, et al. [1977], Brackbill and Saltzman [1982], and Ivanenko and Charakhch'yan [1988]. In this section we will consider extension of the method presented in papers by Ivanenko and Charakch'yan [1988] and Ivanenko [1988], guaranteeing the convexity of all the grid cells to the case of irregular meshes.

The Dirichlet (harmonic) functional was considered by Brackbill and Saltzman [1982]:

$$I = \int \frac{x_\xi^2 + y_\xi^2 + x_\eta^2 + y_\eta^2}{J} \, d\xi \, d\eta \tag{8.11}$$

The minimum of this functional is attained on the harmonic mapping of a domain Ω onto a parametric square. This functional and its generalizations have been used in many papers for regular grid generation.

The problem of irregular mesh smoothing or relaxation is formulated as follows. Let the coordinates of irregular mesh be given:

$$(x, y)_n \qquad n = 1, \ldots, N_n \tag{8.12}$$

The mesh is formed by quadrilateral elements, i.e., the array $COR(N, k)$ is also defined. The problem is to find new coordinates of the mesh nodes, minimizing the sum of the functional Eq. 8.11 values, computed for a mapping of the unit square onto an each cell of a mesh.

It is clear that for a regular grid, this formulation reduces to a discrete analog of the problem to construct harmonic coordinates ξ and η in a domain Ω. Now we will consider the approximation of the functional Eq. 8.11.

The present algorithm is based on a particular approximation of the functional Eq. 8.11 whereby the minimum ensures all mesh cells to be convex quadrilaterals and guarantees no folding for the mesh. In its implementation the peculiarity of vanishing the Jacobian when the one-to-one property is lost can be used explicitly.

The mapping $x(\xi, \eta)$, $y(\xi, \eta)$ is approximated by functions $x^h(\xi, \eta)$, $y^h(\xi, \eta)$ introduced above. Substituting these expressions into Eq. 8.11 and replacing integrals over the square cell by the quadrature formulas with nodes coinciding with the square corners on the plane ξ, η, the following discrete analog can be obtained:

$$I^h = \sum_{N=1}^{N_e} \sum_{k=1}^{4} \frac{1}{4} \left[F_k\right]_N \tag{8.13}$$

where F_k is the integrand evaluated in the kth grid node

$$F_k = \left[\left(x_{k+1} - x_k\right)^2 + \left(x_k - x_{k-1}\right)^2 + \left(y_{k+1} - y_k\right)^2 + \left(y_k - y_{k-1}\right)^2\right] J_k^{-1}$$

and J_k is the doubled area of triangle introduced above.

Note that the approximation Eq. 8. 13 of the functional Eq. 8.11 can be obtained as follows. The square cell on the plane ξ, η is divided into two triangles first by the diagonal 13, and then by 24. The mapping of the square onto a quadrilateral cell in the plane x, y is approximated by a function which is linear in each triangle. Denote this function as before $x^h(\xi, \eta)$, $y^h(\xi, \eta)$. All derivatives in the integrand of Eq. 8.11 are easy to compute, for example, for one of two triangles obtained by splitting the quadrilateral cell with the diagonal 13 we have

$$x_\xi^h = x_3 - x_2 \qquad y_\xi^h = y_3 - y_2 \qquad x_\eta^h = x_2 - x_1 \qquad y_\eta^h = y_2 - y_1$$

$$J^h + \left(x_1 - x_2\right)\left(y_3 - y_2\right) - \left(y_1 - y_2\right)\left(x_3 - x_2\right)$$

The integral Eq. 8.11 over the quadrilateral cell in the plane ξ, η is approximated by half of the sum of values of this integral, computed for piecewise-linear approximations on triangles, obtained for the first and the second splittings. The result is the approximation Eq. 8.13.

The function I^h has the following property, which can be formulated as a theorem:

THEOREM 2. The function I^h has an infinite barrier at the boundary of the set of convex meshes, i.e., if at least one of the quantities J_k tends to zero for some cell while remaining positive, then $I^h \to +\infty$.

Proof. In fact, suppose that $J_k \to 0$ in Eq. 8.13 for some cell, but I^h does not tend to $+\infty$. Then the numerator in Eq. 8.13 must also tend to zero, i.e., the lengths of two sides of the cell tend to zero. Consequently, the areas of all triangles that contain these sides must also tend to zero. Repeating the argument as many times as necessary, we conclude that the lengths of the sides of all grid cells, including those at the boundary of the domain, must tend to zero, i.e., the mesh compresses into a point, which is impossible.

Thus, if the set D is not empty, the system of algebraic equations

$$R_x = \frac{\partial I^h}{\partial x_n} = 0 \qquad R_y = \frac{\partial I^h}{\partial y_n} = 0$$

has at least one solution that is a convex mesh. To find it, one must first find a certain initial convex mesh, and then use a method of unconstrained minimization. Since the function Eq. 8.13 has a infinite barrier on the boundary of the set of convex meshes, each step of the method can be chosen so that the mesh always remains convex.

We first consider a method of minimizing the function assuming that the initial convex mesh has been found. Suppose the mesh at the lth step of the iterations is determined. We use the quasi-Newtonian procedure when the $(l+1)$-th step is accomplished by solving two linear equations for each interior node:

$$\begin{aligned} \tau R_x + \frac{\partial R_x}{\partial x_n}\left(x_n^{l+1} - x_n^l\right) + \frac{\partial R_x}{\partial y_n}\left(y_n^{l+1} - y_n^l\right) = 0 \\ \tau R_y + \frac{\partial R_y}{\partial x_n}\left(x_n^{l+1} - x_n^l\right) + \frac{\partial R_y}{\partial y_n}\left(y_n^{l+1} - y_n^l\right) = 0 \end{aligned} \tag{8.14}$$

From this follows

$$\begin{aligned} x_n^{l+1} = x_n^l - \tau\left(R_x \frac{\partial R_y}{\partial y_n} - R_y \frac{\partial R_x}{\partial y_n}\right)\left(\frac{\partial R_x}{\partial x_n}\frac{\partial R_y}{\partial y_n} - \frac{\partial R_y}{\partial x_n}\frac{\partial R_x}{\partial y_n}\right)^{-1} \\ y_n^{l+1} = y_n^l - \tau\left(R_y \frac{\partial R_x}{\partial x_n} - R_x \frac{\partial R_y}{\partial x_n}\right)\left(\frac{\partial R_x}{\partial x_n}\frac{\partial R_y}{\partial y_n} - \frac{\partial R_y}{\partial x_n}\frac{\partial R_x}{\partial y_n}\right)^{-1} \end{aligned} \tag{8.15}$$

where τ is the iteration parameter, that is chosen so that the mesh remains convex. For this purpose, after each step the conditions Eq. 8.8 are checked and if they are not satisfied, this parameter is multiplied by 0.5. Note that Eq. 8.15 is not the Newton–Raphson iteration process, because not all the second derivatives are taken into account. The rate of convergence is low by comparison. At the same time, the Newton–Raphson method gives a much more complex system of linear equations.

Each of the derivatives in Eq. 8.15 is the sum of a proper number of terms, in accordance with the number of triangles containing the given node as a vertex. For example, for the irregular mesh shown in Figure 8.2, the number of such triangles for the node 3 is equal to 9. Rather than write out such cumbersome expressions, we consider the first and second derivatives of the terms in Eq. 8.15. Arrays storing the derivatives are first cleared, and then all mesh triangles are scanned and the appropriate derivatives are added to the relevant elements of the arrays. The use of formulas Eq. 8.15 for the boundary node (if its position on the boundary is not fixed) should be completed by the projection of this node onto the boundary.

If the initial mesh is not convex, the computational formulas should be modified so that the initial grid need not belong to the set of convex meshes [Ivanenko, 1988]. To achieve this, the quantities J_k appearing in the expressions for R_x, R_y and in their derivatives are replaced with new quantities $\tilde{J}_k$:

$$\tilde{J}_k = \begin{cases} J_k & \text{if } J_k > \varepsilon \\ \varepsilon & \text{if } J_k \le \varepsilon \end{cases}$$

where $\varepsilon > 0$ is some sufficiently small quantity.

It is important to choose an optimal value of ε so that the convex mesh is constructed as fast as possible. The method used for specifying the value of ε is based on the computation of the absolute value of the average area of triangles with negative areas:

$$\varepsilon = \max\left[\alpha S_{neg}/\left(N_{neg} + 0.01\right), \varepsilon_1\right]$$

where S_{neg} is double the absolute value of the total area of triangles with negative areas, and N_{neg} is the number of these triangles. The quantity $\varepsilon_1 > 0$ sets a lower bound on ε to avoid very large values appearing in computations. The coefficient α is chosen experimentally and is in the range $0.3 \le \alpha \le 0.7$.

Computational formulas for the direct extension of the method to the case of adaptive planar grids will be described in detail below.

8.4 Harmonic Maps Between Surfaces. Derivation of Governing Equations

8.4.1 Introductory Remarks

Recall that for grid generation in a domain Ω the auxiliary problem of constructing a harmonic mapping of this domain onto the parametric square is involved. A mapping of the domain boundary onto the square boundary is given. Laplace equations for unknown functions ξ and η are "inverted" into the equations for the functions x and y, Eq. 8.10, which are then solved numerically, as described in Section 8.3.1. On the other hand, the problem can be stated as a variational minimization of the functional Eq. 8.12 dependent on the unknown functions $x(\xi, \eta)$ and $y(\xi, \eta)$. The variational approach is convenient for the method extension to the case of surfaces. To achieve this, the problem of finding the harmonic mapping of the surface onto the parametric square is formulated. The one-to-one mapping between boundaries should be specified.

In the following subsection a more common problem of constructing harmonic maps between manifolds is considered. The emphasis is placed on the formulation of the conditions, providing the one-to-one mapping.

8.4.2 Theory of Harmonic Maps

First we present some common definitions from the survey by Eells and Lemaire [1988]. Let M and N be two n-dimensional manifolds (surfaces) with metrics g and h, defined in local coordinates u^i and ξ^α, $i, \alpha = 1, \ldots, n$. The energy density of a map $\xi(u)$: $(M, g) \to (N, h)$ is called the function $e(\xi)$: $M \to R(\geq 0)$, defined in local coordinates as follows:

$$e(\xi)(u) = g^{ij}(u)\frac{\partial \xi^\alpha(u)}{\partial u^i}\frac{\partial \xi^\beta(u)}{\partial u^j}h_{\alpha\beta}(\xi(u)) \tag{8.16}$$

where the standard summation convention is assumed, g_{ij} and h_{ij} are the elements of metric tensors G and H manifolds M and N, and g^{ij} is the inverse metric:

$$g^{ij}g_{jk} = \delta^i_k = \begin{cases} 1 & if\ i = k \\ 0 & if\ i \neq k \end{cases}$$

This means if g_{ij} are the elements of matrix G, then g^{ij} are the elements of the inverse matrix G^{-1}.

The generalization of Dirichlet functional for the mapping $\xi(u)$ is called the energy of the mapping and is defined as follows:

$$E(\xi) = \int_M e(\xi)(u)dM, \text{ where } dM = \sqrt{\det(G)}\, du^1 \ldots du^n \tag{8.17}$$

A smooth map $\xi(u)$: $(M, g) \to (N, h)$ is called harmonic if it is an external of the energy functional E.

The Euler equations, whose solution minimizes the energy [Eells and Lemaire, 1988] contain Christoffel symbols. The simplified solution form of these equations will be presented below.

The fundamental result on the sufficient conditions of existence and uniqueness of harmonic maps, proved by Hamilton [1975] and Shoen and Yau [1978], can be formulated as the theorem.

THEOREM 3. Let the smooth one-to-one map $\phi : M \to N$ exist that is also one-to-one between boundaries ∂M and ∂N. The curvature of the manifold N is nonpositive, and its boundary ∂N is convex. Then there exists a unique harmonic map $\varphi : M \to N$, such that φ is homotopy equivalent to ϕ and $\varphi(\partial M) = \phi(\partial M)$.

Here we consider the case when N has a simple shape, for example, it is a unit cube in the Euclidean space. The conditions of the theorem (nonpositive curvature and convex boundary) are obviously satisfied in this case. Consequently, the theory of harmonic maps includes the theoretic foundation of the method, proposed by Winslow [1966].

So, consider when M is a n-dimensional manifold, N is a unit cube in R^n: $0 < \xi^i < 1$, $i = 1, \ldots, n$. The Euclidean metric in R^n is $h_{\alpha\beta} = \delta_{\alpha\beta}$. If the local coordinates u^i and ξ^α are the same, then Eq. 8.16 can be simplified to give

$$e(\xi) = g^{ij}\frac{\partial\xi^\alpha}{\partial\xi^i}\frac{\partial\xi^\beta}{\partial\xi^j}\delta_{\alpha\beta} = g^{ij}\,\delta_{ij} = g^{ii} = Tr\left(G^{-1}\right)$$

Hence, the energy functional Eq. 8.17 will be

$$E(\xi) = \int_M g^{ii}(\xi)dM = \int_0^1 \ldots \int_0^1 Tr\left(G^{-1}\right)\sqrt{\det(G)}\, d\xi^1 \ldots d\xi^n \tag{8.18}$$

The Euler equations for the functional Eq. 8.18 can be also simplified, and we can avoid the appearance of Christoffel symbols in these equations. Now we will derive these equations following Liseikin [1991].

8.4.3 Derivation of Governing Equations

We denote by S^{rn} a n-dimensional in R^{n+k} with a local coordinate system

$$\left(u^i, \ldots, u^n\right) = u \in S^n \subset R^n$$

The surface is defined by a nondegenerate transform

$$r(u): S^n \to S^{rn} \qquad r = \left(r^1, \ldots, r^{n+k}\right) \tag{8.19}$$

The new parameterization of the surface S^{rn} is defined by a mapping of a unit cube $Q^n : \{0 < \xi^i < 1, i = 1, \ldots, n\}$ in R^n onto a surface S^{rn}:

$$r(u(\xi)): Q^n \to S^{rn} \qquad \xi = \left(\xi^1, \ldots, \xi^n\right) \in Q^n \tag{8.20}$$

which is the composition of $r(u)$ and some nondegenerate transform

$$u(\xi): Q^n \to S^n \tag{8.21}$$

The problem of finding a new parameterization of the surface is stated as the problem of construction at this transformation $u(\xi)$. The mapping of $r(u(\xi))$ defines on a surface S^{rn} a new coordinate system $(\xi^1 \ldots \xi^n) = \xi$, which generates a local metric tensor

$$G^{r\xi} = \left\{g_{ij}^{r\xi}\right\} \qquad i, j = 1, 2 \ldots, n$$

whose elements are scalar products of the vectors $r_i = \partial r/\partial \xi^i$ and $r_j = \partial r/\partial \xi^j$:

$$g_{ij}^{r\xi} = r_i r_j = \sum_{m=1}^{n+k} \frac{\partial r^m}{\partial \xi^i} \frac{\partial r^m}{\partial \xi^j}$$

The elements of the metric tensor defined by the transformation $r(u)$ are given by

$$G^{ru} = \left\{ g_{ij}^{ru} \right\} \qquad i, j = 1, 2 \ldots, n$$

These elements are the scalar products of the vectors $\partial r/\partial u^i$ and $\partial r/\partial u^j$:

$$g_{ij}^{ru} = \sum_{m=1}^{n+k} \frac{\partial r^m}{\partial u^i} \frac{\partial r^m}{\partial u^j}$$

Consider the contravariant metric tensors whose elements form the symmetric matrices $G_{\xi r}$ and G_{ur}, inverse to the matrices $G^{r\xi}$ and G^{ru}:

$$g_{\xi r}^{ij} = (-1)^{i+j} d_{\xi r}^{ji} / \det\left(G^{r\xi}\right) \qquad g_{ur}^{ij} = (-1)^{i+j} d_{ur}^{ji} / \det\left(G^{ru}\right)$$

where $d_{\xi r}^{ji}$ and d_{ur}^{ji} are the determinants of cofactors of the elements $g_{ij}^{r\xi}$ and g_{ij}^{ru} in the matrices $G^{r\xi}$ and G^{ru} correspondingly.

Let us prove the following relation:

$$g_{\xi r}^{ij} = \sum_{m,l}^{n} g_{ur}^{ml} \frac{\partial \xi^i}{\partial u^m} \frac{\partial \xi^j}{\partial u^l} \tag{8.22}$$

Indeed, substituting in the following identity the right-hand side of Eq. 8.22 instead of $g_{\xi r}^{lp}$ we obtain

$$\delta_i^p = g_{il}^{r\xi} g_{\xi r}^{lp} = \frac{\partial r^\alpha}{\partial \xi^i} \frac{\partial r^\alpha}{\partial \xi^l} g_{\xi r}^{lp} = \frac{\partial r^\alpha}{\partial u^t} \frac{\partial r^\alpha}{\partial u^h} \frac{\partial u^t}{\partial \xi^i} \frac{\partial u^h}{\partial \xi^l} g_{\xi r}^{lp} = g_{th}^{ru} g_{ur}^{mj} \frac{\partial u^t}{\partial \xi^i} \frac{\partial u^h}{\partial \xi^l} \frac{\partial \xi^l}{\partial u^m} \frac{\partial \xi^p}{\partial u^j} =$$
$$g_{th}^{ru} g_{ur}^{mj} \delta_m^h \frac{\partial u^t}{\partial \xi^i} \frac{\partial \xi^p}{\partial u^j} = g_{th}^{ru} g_{ur}^{hj} \frac{\partial u^t}{\partial \xi^i} \frac{\partial \xi^p}{\partial u^j} = \delta_t^j \frac{\partial u^t}{\partial \xi^i} \frac{\partial \xi^p}{\partial u^j} = \delta_i^p$$

The summation is performed on repeated indices, here $\alpha = 1, 2, \ldots, n + k$; $i, j, l, p, t, h, m = 1, \ldots, n$. Now taking Eq. 8.22 into account, the functional Eq. 8.17 takes the form

$$I = \int_{S^{rn}} g_{\xi r}^{ii} \, dS^{rn} = \int_{S^{rn}} \sum_{i,m,l}^{n} g_{ur}^{ml} \frac{\partial \xi^i}{\partial u^m} \frac{\partial \xi^i}{\partial u^l} \, dS^{rn} \tag{8.23}$$

In the derivation of the Euler equations the integration domain in Eq. 8.23 will be replaced by S^{rn}, and the surface element is transformed as follows:

$$dS^{rn} = \sqrt{\det\left(G^{ru}\right)} dS^n = \sqrt{\det\left(G^{ru}\right)} du^1 \ldots du^n$$

Consequently, the functional Eq. 8.23 can be written as

$$I=\int_{S^n}\sqrt{\det(G^{ru})}\left(\sum_{i,m,l}^{n} g_{ur}^{ml}\frac{\partial\xi^i}{\partial u^m}\frac{\partial\xi^i}{\partial u^l}\right)du^1\ldots du^n \tag{8.24}$$

The quantities $\sqrt{\det(G^{ru})}$ and g_{ur}^{ml} in the functional Eq. 8.24 are independent on the functions $\xi^i(u)$ and their derivatives, and hence remain unchanged when $\xi(u)$ is varied. Therefore the Euler equations for the functions $\xi^i(u)$, minimizing Eq. 8.24 are of the form

$$L(\xi^i)=\sum_{m=1}^{n}\frac{\partial}{\partial u^m}\left(\sqrt{\det(G^{ru})}\sum_{l=1}^{n} g_{ur}^{ml}\frac{\partial\xi^i}{\partial u^l}\right)=0 \qquad i=1,\ldots,n \tag{8.25}$$

The equations which each component $u^i(\xi)$ of the function $u(\xi)$ satisfies can be derived from Eq. 8.25. To achieve this, *i*th equation of the system Eq. 8.25 is multiplied by $\partial u^j/\partial\xi^i$ and summed over *i*. As a result, we have

$$\sum_{i=1}^{n}L(\xi^i)\frac{\partial u^j}{\partial\xi^i}=\sum_{i,m=1}^{n}\frac{\partial}{\partial u^m}\left(\sum_{p=1}^{n}\sqrt{\det(G^{ru})}\,g_{ur}^{mp}\frac{\partial\xi^i}{\partial u^p}\right)\frac{\partial u^j}{\partial\xi^i}=$$

$$\sum_{m=1}^{n}\frac{\partial}{\partial u^m}\left(\sum_{i,p=1}^{n}\sqrt{\det(G^{ru})}\,g_{ur}^{mp}\frac{\partial\xi^i}{\partial u^p}\frac{\partial u^j}{\partial\xi^i}\right)-\sum_{i,m,p,t=1}^{n}\sqrt{\det(G^{ru})}\,g_{ur}^{mp}\frac{\partial\xi^i}{\partial u^p}\frac{\partial\xi^t}{\partial u^m}\frac{\partial^2u^j}{\partial\xi^i\partial\xi^t}=0$$

Here $j = 1, \ldots, n$.

Now, multiplying each equation on $1/\sqrt{\det(G^{ru})}$ and taking into account Eq. 8.22 and the relation

$$\sum_{i=1}^{n}\frac{\partial\xi^i}{\partial u^p}\frac{\partial u^j}{\partial\xi^i}=\delta_p^j$$

we finally obtain

$$g_{\xi r}^{it}\frac{\partial^2u^j}{\partial\xi^i\partial\xi^t}=\frac{1}{\sqrt{\det(G^{ru})}}\sum_{m=1}^{n}\frac{\partial}{\partial u^m}\left(\sqrt{\det(G^{ru})}\,g_{ur}^{mj}\right) \qquad j=1,\ldots,n \tag{8.26}$$

This is a quasilinear system of elliptic equations that is a direct extension of the system Eq. 8.10. It will be the basis of the algorithms for structured two-dimensional adaptive grids, grids on surfaces and three-dimensional grids. For derivation of governing equations in all these cases, we need only to express the contravariant components $g_{\xi r}^{ij}$ and g_{ur}^{ij} as functions on the covariant components $g_{ij}^{\xi r}$ and g_{ij}^{ur} and substitute the associate expressions into Eq. 8.26 for $n = 2$ and $n = 3$.

8.5 Two-Dimensional Adaptive-Harmonic Structured Grids

8.5.1 Derivation of Equations

Let Ω be a two-dimensional domain in R^2, and let in a Euclidean space R^3, the surface S^{r2} is given as $z = f(x, y)$. We introduce new notations

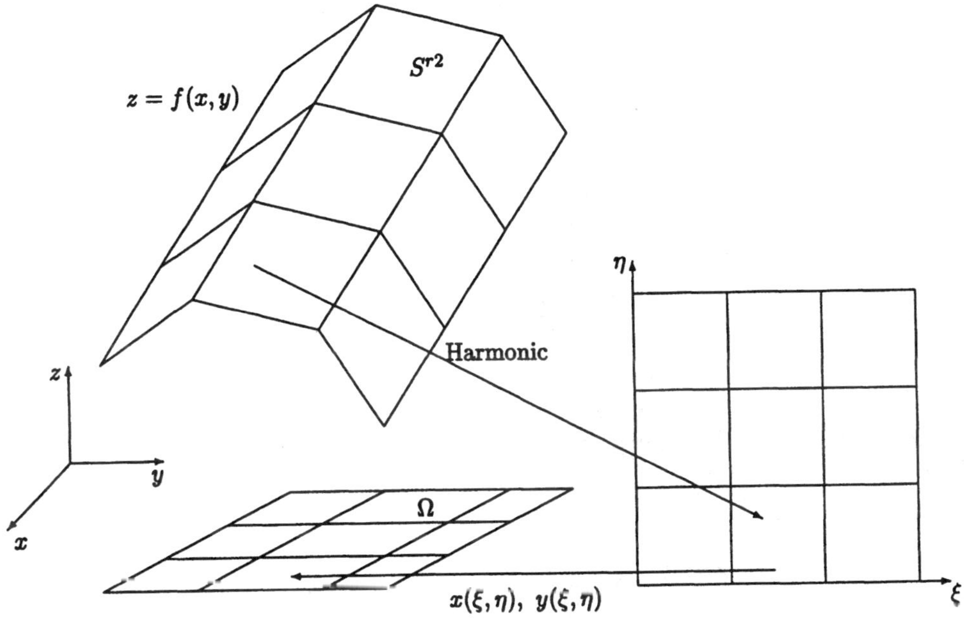

FIGURE 8.3 Harmonic coordinates on the surface of the graph of the graph of a function $z = f(x, y)$.

$$r = (r^1, r^2, r^3) = (x, y, z) = (x, y, f(x, y)) \in S^{r2} \subset R^3 \qquad u = (u^1, u^2) = (x, y) \in \Omega \subset R^2$$

$$\xi = (\xi^1, \xi^2) = (\xi, \eta) \in Q^2 \subset R^2 \qquad r_\xi = (x_\xi, y_\xi, z_\xi),\ r_\eta = (x_\eta, y_\eta, z_\eta)$$

The problem formulation is the following. Suppose we are given a simply connected domain Ω with a smooth boundary in the x, y plane. Consider the surface $z = f(x, y)$ of the graph of the function $f \in C^2(\Omega)$. It is required to find a mapping of the parametric square Q^2 onto the domain Ω under a given mapping between boundaries such that the mapping of the surface onto the parametric square be harmonic (see Figure 8.3). Thus, the problem is to minimize the Dirichlet functional, written for a surface

$$I = \int \left(g^{11}_{\xi r} + g^{22}_{\xi r} \right) dS^{r2} \tag{8.27}$$

where $g^{11}_{\xi r}$ $g^{12}_{\xi r}$ $g^{22}_{\xi r}$ are the elements of the contravariant metric tensor $G_{\xi r}$ dependent on the elements of the covariant metric tensor $G^{r\xi}$ as follows:

$$g^{11}_{\xi r} = g^{r\xi}_{22} / \det(G^{r\xi}) \qquad g^{22}_{\xi r} = g^{r\xi}_{11} / \det(G^{r\xi}) \qquad g^{12}_{\xi r} = g^{21}_{\xi r} = -g^{r\xi}_{12} / \det(G^{r\xi})$$

where

$$g^{r\xi}_{11} = r_\xi^2 = x_\xi^2 + y_\xi^2 + z_\xi^2 \qquad g^{r\xi}_{12} = g^{r\xi}_{21} = (r_\xi \cdot r_\eta) = x_\xi x_\eta + y_\xi y_\eta + z_\xi z_\eta \qquad g^{r\xi}_{22} = r_\eta^2 = x_\eta^2 + y_\eta^2 + z_\eta^2$$
$$\det(G^{r\xi}) = g^{r\xi}_{11} g^{r\xi}_{22} - (g^{r\xi}_{12})^2 \qquad z_\xi = f_x x_\xi + f_y y_\xi, \quad z_\eta = f_x x_\eta + f_y y_\eta \tag{8.28}$$

Inverting dependent and independent variables in Eq. 8.27 and taking in account

$$dS^{r2} = \sqrt{g_{11}^{r\xi} g_{22}^{r\xi} - (g_{12}^{r\xi})^2}\; d\xi \; d\eta$$

we obtain

$$I = \int \frac{g_{11}^{r\xi} + g_{22}^{r\xi}}{\sqrt{g_{11}^{r\xi} g_{22}^{r\xi} - (g_{12}^{r\xi})^2}}\, d\xi \, d\eta \tag{8.29}$$

Euler equation for the functional Eq. 8.29 follow from Eq. 8.26 for $n = 2$, $k = 1$. We need only to compute the elements of the covariant metric tensor $G^{r\xi}$ and contravariant metric tensor $G_{\xi r}$ of the transform $r(u) = r(x, y) : \Omega \rightarrow S^{r2}$:

$$r = (x, y, f(x,y)) \qquad r_x = (1, 0, f_x) \qquad r_y = (0, 1, f_y)$$
$$g_{11}^{ru} = r_x^2 = 1 + f_x^2 \qquad g_{12}^{ru} = g_{21}^{ru} = r_x \cdot r_y = f_x f_y \qquad g_{22}^{ru} = r_y^2 = 1 + f_y^2$$
$$\det(G^{ru}) = g_{11}^{ru} g_{22}^{ru} - (g_{12}^{ru})^2 = 1 + f_x^2 + f_y^2 \qquad \det(G^{r\xi}) = \det(G^{ru})(x_\xi y_\eta - x_\eta y_\xi)^2$$
$$g_{ur}^{11} = g_{22}^{ru} / \det(G^{ru}) = (1 + f_y^2)/(1 + f_x^2 + f_y^2)$$
$$g_{ur}^{12} = -g_{21}^{ru} / \det(G^{ru}) = -f_x f_y / (1 + f_x^2 + f_y^2)$$
$$g_{ur}^{22} = g_y^{ru} / (1 + f_x^2 + f_y^2) \qquad g_{ur}^{22} = g^{rT}$$

Substituting these expressions into Eq. 8.26, we obtain equations, written in a form convenient for practical use:

$$\begin{aligned} L(x) &= \alpha x_{\xi\xi} - 2\beta x_{\xi\eta} + \gamma x_{\eta\eta} - J^2 D\left[\frac{\partial}{\partial x}\frac{1 + f_y^2}{D} - \frac{\partial}{\partial y}\frac{f_x f_y}{D}\right] = 0 \\ L(y) &= \alpha y_{\xi\xi} - 2\beta y_{\xi\eta} + \gamma y_{\eta\eta} - J^2 D\left[\frac{\partial}{\partial x}\frac{-f_x f_y}{D} + \frac{\partial}{\partial y}\frac{1 + f_x^2}{D}\right] = 0 \end{aligned} \tag{8.30}$$

where

$$D = \sqrt{1 + f_x^2 + f_y^2}, \quad J = x_\eta y_\xi, \quad \alpha = x_\eta^2 + y_\eta^2 + f_\eta^2, \quad \beta = x_\xi x_\eta + y_\xi y_\eta + f_\xi f_\eta, \quad \gamma = x_\xi^2 + y_\xi^2 + f_\xi^2.$$

8.5.2 Numerical Implementation

Eq. 8.30 are approximated on the square grid with the unit size Eq. 8.4, introduced above with the simplest difference relations

$$x_\xi \approx \left[x_\xi\right]_{ij} = 0.5\left(x_{i+1,j} - x_{i-1,j}\right) \quad x_\eta \approx \left[x_\eta\right]_{ij} = 0.5\left(x_{i,j+1} - x_{i,j-1}\right)$$
$$y_\xi \approx \left[y_\xi\right]_{ij} = 0.5\left(y_{i+1,j} - y_{i-1,j}\right) \quad y_\eta \approx \left[y_\eta\right]_{ij} = 0.5\left(y_{i,j+1} - y_{i,j-1}\right)$$
$$f_\xi \approx \left[f_\xi\right]_{ij} = 0.5\left(f_{i+1,j} - f_{i-1,j}\right) \quad f_\eta \approx \left[f_\eta\right]_{ij} = 0.5\left(f_{i,j+1} - f_{i,j-1}\right)$$
$$x_{\xi\xi} \approx \left[x_{\xi\xi}\right]_{ij} = x_{i+1,j} - 2x_{ij} + x_{i-1,j}$$
$$x_{\xi\eta} \approx \left[x_{\xi\eta}\right]_{ij} = 0.25\left(x_{i+1,j+1} - x_{i+1,j-1} - x_{i-1,j+1} + x_{i-1,j-1}\right) \tag{8.31}$$

$$x_{\eta\eta} \approx \left[x_{\eta\eta}\right]_{ij} = x_{i,j+1} - 2x_{i,j} + x_{i,j-1} \quad y_{\xi\xi} \approx \left[y_{\xi\xi}\right]_{ij} = y_{i+1,j} - 2y_{ij} + y_{i-1,j}$$
$$y_{\xi\eta} \approx \left[y_{\xi\eta}\right]_{ij} = 0.25\left(y_{i+1,j+1} - y_{i+1,j-1} - y_{i-1,j+1} + y_{i-1,j-1}\right)$$
$$y_{\eta\eta} \approx \left[y_{\eta\eta}\right]_{ij} = y_{i,j+1} - 2y_{ij} + y_{i,j-1}$$
$$\alpha \approx \left[x_\eta\right]_{ij}^2 + \left[y_\eta\right]_{ij}^2 + \left[f_\eta\right]_{ij}^2 \quad \beta \approx \left[x_\xi\right]_{ij}\left[x_\eta\right]_{ij} + \left[y_\xi\right]_{ij}\left[y_\eta\right]_{ij} + \left[f_\xi\right]_{ij}\left[f_\eta\right]_{ij} \quad \gamma \approx \left[x_\xi\right]_{ij}^2 + \left[y_\xi\right]_{ij}^2 + \left[f_\xi\right]_{ij}^2$$

Substitute these expressions into Eq. 8.30 and denote the difference approximations of $L(x)$ and $L(y)$ as $[L(x)]_{ij}$ and $[L(y)]_{ij}$ correspondingly. Suppose that the coordinates of grid nodes $(x, y)_{ij}$ at the lth step of iterations are determined. Then the $(l + 1)$-th step is accomplished as follows:

$$x_{ij}^{l+1} = x_{ij}^l + \tau\frac{\left[L(x)\right]_{ij}}{2[\alpha]_{ij} + 2[\gamma]_{ij}} \quad y_{ij}^{l+1} = y_{ij}^l + \tau\frac{\left[L(y)\right]_{ij}}{2[\alpha]_{ij} + 2[\gamma]_{ij}} \tag{8.32}$$

The expressions in square brackets denote the corresponding approximations of expressions in the grid node (i, j) at the lth iteration step. The value of iteration parameter τ is chosen in limits $0 < \tau < 1$, usually $\tau = 0.5$.

Derivatives $[f_x]_{ij}$ and $[f_y]_{ij}$ in the ijth grid node are evaluated with the centered differences

$$\left[f_x\right]_{ij} = \frac{(f_{i+1,j} - f_{i-1,j})(y_{i,j+1} - y_{i,j-1}) - (f_{i,j+1} - f_{i,j-1})(y_{i+1,j} - y_{i-1,j})}{(x_{i+1,j} - x_{i-1,j})(y_{i,j+1} - y_{i,j-1}) - (x_{i,j+1} - x_{i,j-1})(y_{i+1,j} - y_{i-1,j})}$$
$$\left[f_y\right]_{ij} = -\frac{(f_{i+1,j} - f_{i-1,j})(x_{i,j+1} - x_{i,j-1}) - (f_{i,j+1} - f_{i,j-1})(x_{i+1,j} - x_{i-1,j})}{(x_{i+1,j} - x_{i-1,j})(y_{i,j+1} - y_{i,j-1}) - (x_{i,j+1} - x_{i,j-1})(y_{i+1,j} - y_{i-1,j})}$$

These formulas must be modified for the boundary nodes. Indices, "leaving" the computational domain must be replaced by the nearest boundary indices. For example, if $j = 1$, then $(i, j - 1)$ must be replaced by (i, j).

Note that if $[f_\xi]_{ij} = 0$ and $[f_\eta]_{ij} = 0$, then $[f_x]_{ij} = 0$ and $[f_y]_{ij} = 0$ and the method Eq. 8.32 reduces to the Winslow method, described briefly in Section 8.3.1.

The adaptive-harmonic grid generation algorithm is formulated as follows:

1. Compute the values of the control function at each grid node. The result is f_{ij}.
2. Evaluate derivatives $(f_x)_{ij}$ and $(f_y)_{ij}$ and other expressions in Eq. 8.32 using the above formulas.
3. Make one iteration step and compute new values of x_{ij} and y_{ij}.
4. Repeat, starting with Step 1 to convergency.

The resulting algorithm can be used in the numerical solution of the partial differential equations. In this case, at the first step of the algorithm the values f_{ij} in each grid node are taken from the finite-difference or finite element solution of the host equations.

Note that for control of the number of grid nodes in the layers of high gradients, it is convenient to use Cf instead of $f(x, y)$. The larger the coefficient C, the greater the number of nodes in the layer of high gradients of the function f.

8.6 Two-Dimensional Adaptive-Harmonic Irregular Meshes

8.6.1 Problem Formulation

In notations of Section 8.5.1 the problem is formulated as follows. Suppose we are given a simply connected domain Ω with a smooth boundary in the x, y plane. Consider the surface $z = f(x, y)$ of the graph of the function $f \in C^1(\Omega)$. It is required to find a mapping of the parametric square Q^2 onto a domain Ω under a given mapping between boundaries such that the mapping of the surface onto the parametric square be harmonic (see Figure 8.3). Thus, the problem is to minimize the harmonic functional Eq. 8.27.

Substituting expressions Eq. 8.28 for z_ξ and z_η into Eq. 8.29, we obtain the functional from the paper by Ivanenko [1993] to define adaptive-harmonic grid, clustered in regions of high gradients of the function $f(x, y)$:

$$I = \int \frac{\left(x_\xi^2 + x_\eta^2\right)\left(1 + f_x^2\right) + \left(y_\xi^2 + y_\eta^2\right)\left(1 + f_y^2\right) + 2 f_x f_y (x_\xi y_\xi + x_\eta y_\eta)}{\left(x_\xi y_\eta - x_\eta y_\xi\right)\sqrt{1 + f_x^2 + f_y^2}} \, d\xi \, d\eta \tag{8.33}$$

The problem of irregular mesh smoothing and adaption is formulated as follows. Let the coordinates of irregular mesh be given. The mesh is formed by quadrilateral elements, i.e., the array $COR(N, k)$ is also defined. The problem is to find new coordinates of the mesh nodes, minimizing the sum of the functional Eq. 8.33 values, computed for a mapping of the unit square onto each cell of a mesh (see Figure 8.3).

8.6.2 Approximation of the Functional

The functional Eq. 8.33 possesses the same properties as the functional Eq. 8.11, and it can be also approximated in such a way that its minimum is attained on a grid/mesh of convex quadrilaterals:

$$I^h = \sum_{N=1}^{N_e} \sum_{k=1}^{4} \frac{1}{4} \left[F_k\right]_N \tag{8.34}$$

where

$$F_k = \frac{D_1\left[1 + (f_x)_k^2\right] + D_2\left[1 + (f_y)_k^2\right] + 2D_3 (f_x)_k (f_y)_k}{J_k\left[1 + (f_x)_k^2 + (f_y)_k^2\right]^{1/2}}$$

$$D_1 = \left(x_{k-1} - x_k\right)^2 + \left(x_{k+1} - x_k\right)^2 \qquad D_2 = \left(y_{k-1} - y_k\right)^2 + \left(y_{k+1} - y_k\right)^2$$

$$D_3 = \left(x_{k-1} - x_k\right)\left(y_{k-1} - y_k\right) + \left(x_{k+1} - x_k\right)\left(y_{k+1} - y_k\right)$$

$$J_k = \left(x_{k-1} - x_k\right)\left(y_{k+1} - y_k\right) - \left(x_{k+1} - x_k\right)\left(y_{k-1} - y_k\right)$$

Here $(f_x)_k$ and $(f_y)_k$ are the values of derivatives, computed in the node number k of the cell number N.

If the set of convex meshes D is not empty, the system of algebraic equations

$$R_x = \frac{\partial I^h}{\partial x_n} = 0 \qquad R_y = \frac{\partial I^h}{\partial y_n} = 0$$

has at least one solution which is a convex mesh. To find it, one must first find a certain initial convex mesh, and then use some method of unconstrained minimization of the function I^h. Since this function has an infinite barrier on the boundary of the set D, each step of the method can be chosen so that the mesh always remains convex.

8.6.3 Minimization of the Functional

Suppose the mesh at the lth step of the iterations is determined. We use the quasi-Newtonian procedure when the $(l+1)$-th step is accomplished as follows:

$$\begin{aligned} x_n^{l+1} &= x_n^l - \tau\left(R_x \frac{\partial R_y}{\partial y_n} - R_y \frac{\partial R_x}{\partial y_n}\right)\left(\frac{\partial R_x}{\partial x_n}\frac{\partial R_y}{\partial y_n} - \frac{\partial R_y}{\partial x_n}\frac{\partial R_x}{\partial y_n}\right)^{-1} \\ y_n^{l+1} &= y_n^l - \tau\left(R_y \frac{\partial R_x}{\partial x_n} - R_x \frac{\partial R_y}{\partial x_n}\right)\left(\frac{\partial R_x}{\partial x_n}\frac{\partial R_y}{\partial y_n} - \frac{\partial R_y}{\partial x_n}\frac{\partial R_x}{\partial y_n}\right)^{-1} \end{aligned} \tag{8.35}$$

where τ is the iteration parameter, which is chosen so that the mesh remains convex. For this purpose after each step conditions Eq. 8.8 are checked and if they are not satisfied, this parameter is multiplied by 0.5. Then conditions Eq. 8.8 are checked for the grid, computed with a new value of τ and if they are not satisfied, this parameter is multiplied by 0.25 and so on.

The adaptive-harmonic algorithm for r–refinement is formulated as follows:

1. Generate an initial mesh by the use of a marching method.
2. Compute the values of the control function f_n at each mesh node.
3. Evaluate derivatives $(f_x)_n$ and $(f_y)_n$ and other expressions in Eq. 8.35.
4. Make one iteration step and compute new values of x_n and y_n.
5. Repeat starting with Step 2 to convergency.

Computational formulas for $[f_x]_n$ and $[f_y]_n$ will be presented below.

8.6.4 Derivation of Computational Formulas

Note that the approximation Eq. 8.34 of the functional Eq. 8.33 can be obtained as it was done for the functional Eq. 8.11 in Section 8.3.2. The square cell on the plane ξ, η is divided into two triangles first by the diagonal 13, and then by 24. The mapping of the square onto a quadrilateral cell in the plane x, y is approximated by two functions which are linear in each triangle. Denote this functions as before $x^h(\xi, \eta)$, $y^h(\xi, \eta)$. All derivatives in the integrand of Eq. 8.33 is easy to compute, as it was done in Section 8.3.2. Then the integral Eq. 8.33 over the square cell in the plane ξ, η is approximated by a half of the sum of values of this integral, computed for piecewise linear approximations on triangles, obtained for the first and the second splittings. The result is the approximation Eq. 8.34.

Four triangles, introduced above are considered for the quadrilateral cell number N. Each of these triangles corresponds to a corner with the number k and gives a proper contribution to the functional and also to the values of its derivatives. Since the integrand in Eq. 8.33 does not depend on the rotation of the coordinate system ξ, η, then all the computational formulas will be the same for all triangles. We enumerate nodes of triangle corresponding to the corner with the local number k from 1 to 3 as follows:

node 1 corresponds to the local node number $k-1$ of the cell N,
node 2 corresponds to the local node number k of the cell N,
node 3 corresponds to the local node number $k+1$ of the cell N.

Then in the new numeration the expression for F_k will be

$$F=\frac{D_1\left[1+(f_x)_k^2\right]+D_2\left[1+(f_y)_k^2\right]+2D_3(f_x)_k(f_y)_k}{J_2\left[1+(f_x)_k^2+(f_y)_k^2\right]^{1/2}} \tag{8.36}$$

where

$$D_1=(x_1-x_2)^2+(x_3-x_2)^2 \qquad D_2=(y_1-y_2)^2+(y_3-y_2)^2$$
$$D_3=(x_1-x_2)(y_1-y_2)+(x_3-x_2)(y_3-y_2)$$
$$J_2=(x_1-x_2)(y_3-y_2)-(x_3-x_2)(y_1-y_2)$$

We introduce notations

$$U=\frac{D_1\left[1+(f_x)_k^2\right]+D_2\left[1+(f_y)_k^2\right]+2D_3\,(f_x)_k\,(f_y)_k}{\left[1+(f_x)_k^2+(f_y)_k^2\right]^{1/2}}$$
$$V=(x_1-x_2)(y_3-y_2)-(x_3-x_2)(y_1-y_2)$$

We use formulas for the derivatives of the relation of two functions. Differentiating, we obtain

$$F=\frac{U}{V}$$
$$F_x=\frac{U_x-FV_x}{V} \qquad F_y=\frac{U_y-FV_y}{V} \qquad F_{xx}=\frac{U_{xx}-2F_xV_x-FV_{xx}}{V} \tag{8.37}$$
$$F_{yy}=\frac{U_{yy}-2F_yV_y-FV_{yy}}{V} \qquad F_{xy}=F_{yx}=\frac{U_{xy}-F_xV_y-F_yV_x-FV_{xy}}{V}$$

For the triangle vertex with the number 1, we should substitute appropriate expressions instead of U and V, U_x and V_x and so on into Eq. 8.37 and replace x and y by x_1 and y_1.

For the vertex 1 we have

$$V_x=y_3-y_2, \quad V_y=x_2-x_3$$
$$V_{xx}=0, \quad V_{xy}=0, \quad V_{yy}=0$$

$$U_x=2\frac{\left[1+(f_x)_k^2\right](x_1-x_2)+(f_x)_k(f_y)_k(y_1-y_2)}{\left[1+(f_x)_k^2+(f_y)_k^2\right]^{1/2}} \qquad U_y=2\frac{\left[1+(f_y)_k^2\right](y_1-y_2)+(f_x)_k(f_y)_k(x_1-x_2)}{\left[1+(f_x)_k^2+(f_y)_k^2\right]^{1/2}}$$

$$U_{xx}=2\frac{1+(f_x)_k^2}{\left[1+(f_x)_k^2+(f_y)_k^2\right]^{1/2}} \qquad U_{xy}=2\frac{(f_x)_k(f_y)_k}{\left[1+(f_x)_k^2+(f_y)_k^2\right]^{1/2}} \qquad U_{yy}=2\frac{1+(f_y)_k^2}{\left[1+(f_x)_k^2+(f_y)_k^2\right]^{1/2}}$$

For the vertex 2 we have

$$V_x = y_1 - y_3 \qquad V_y = x_3 - x_1$$

$$V_{xx} = 0 \qquad V_{xy} = 0 \qquad V_{yy} = 0$$

$$U_x = 2\frac{\left[1+(f_x)_k^2\right](2x_2 - x_1 - x_3) + (f_x)_k (f_y)_k (2y_2 - y_1 - y_3)}{\left[1+(f_x)_k^2 + (f_y)_k^2\right]^{1/2}}$$

$$U_y = 2\frac{\left[1+(f_y)_k^2\right](2y_2 - y_1 - y_3) + (f_x)_k (f_y)_k (2x_2 - x_1 - x_3)}{\left[1+(f_x)_k^2 + (f_y)_k^2\right]^{1/2}}$$

$$U_{xx} = 4\frac{1+(f_x)_k^2}{\left[1+(f_x)_k^2 + (f_y)_k^2\right]^{1/2}} \qquad U_{xy} = 4\frac{(f_x)_k (f_y)_k}{\left[1+(f_x)_k^2 + (f_y)_k^2\right]^{1/2}} \qquad U_{yy} = 2\frac{1+(f_y)_k^2}{\left[1+(f_x)_k^2 + (f_y)_k^2\right]^{1/2}}$$

For the vertex 3 we have

$$V_x = y_2 - y_1 \qquad V_y = x_1 - x_2$$

$$V_{xx} = 0 \qquad V_{xy} = 0 \qquad V_{yy} = 0$$

$$U_x = 2\frac{\left[1+(f_x)_k^2\right](x_3 - x_2) + (f_x)_k (f_y)_k (y_3 - y_2)}{\left[1+(f_x)_k^2 + (f_y)_k^2\right]^{1/2}}$$

$$U_y = 2\frac{\left[1+(f_y)_k^2\right](y_3 - y_2) + (f_x)_k (f_y)_k (x_3 - x_2)}{\left[1+(f_x)_k^2 + (f_y)_k^2\right]^{1/2}}$$

$$U_{xx} = 2\frac{1+(f_x)_k^2}{\left[1+(f_x)_k^2 + (f_y)_k^2\right]^{1/2}} \qquad U_{xy} = 2\frac{(f_x)_k (f_y)_k}{\left[1+(f_x)_k^2 + (f_y)_k^2\right]^{1/2}} \qquad U_{yy} = 2\frac{1+(f_y)_k^2}{\left[1+(f_x)_k^2 + (f_y)_k^2\right]^{1/2}}$$

Computations are performed as follows. Let F and its derivatives on x_1 and y_1 be computed with the use of formulas Eq. 8.37 for the cell number N and triangle number k. Then the computed values are added to the appropriate array elements

$$I^h += F \qquad [R_x]_n += F_x \qquad [R_y]_n += F_y$$

$$[R_{xx}]_n += F_{xx} \qquad [R_{xy}]_n += F_{xy} \qquad [R_{yy}]_n += F_{yy}$$

where $n = COR(N, k - 1)$.

Similarly for the vertex 2, the correspondence between local and global number is $n = COR(N, k)$. Similarly for the vertex 3, the correspondence between local and global number is $n = COR(N, k + 1)$.

Derivatives $[f_x]_n$ and $[f_y]_n$ are computed as follows. All triangles of the mesh are scanned and for the triangle number k of the cell number N the following values are computed:

$$f_x = (f_1 - f_2)(y_3 - y_2) - (f_3 - f_2)(y_1 - y_2)$$
$$f_y = (x_1 - x_2)(f_3 - f_2) - (x_3 - x_2)(f_1 - f_2)$$
$$J_2 = (x_1 - x_2)(y_3 - y_2) - (x_3 - x_2)(y_1 - y_2)$$

where f_1, f_2, and f_3 are values of the function f at vertices of the triangle, numbered 1, 2, and 3, corresponding to local numbers of corners of a quadrilateral cell $k - 1$, k and $k + 1$. Computed values are added to corresponding array elements (which were first cleared):

$$[f_x]_n += f_x \qquad [f_y]_n += f_y \qquad [J]_n += J_2 \qquad n = COR(N,k)$$

New values of derivatives are computed as follows:

$$[f_x]_n /= [J]_n \qquad [f_y]_n /= [J]_n$$

Here, according to C-language notations, $a+ = b$ means that the new value of a becomes equal to $a + b$, and $a/ = b$ means that the new value of a becomes equal to a/b.

So, the iteration method for irregular mesh relaxation and adaption is described in detail.

8.7 Adaptive-Harmonic Structured Surface Grid Generation

8.7.1 Derivation of Equations

Introduce the following notations:

$$r = (r^1, r^2, r^3, r^4) = (x, y, z, f) \in S^{r2} \subset R^4$$
$$u = (u^1, u^2) = (u, v) \in Q^2 \subset R^2 \qquad \xi = (\xi^1, \xi^2) = (\xi, \eta) \in Q^2 \subset R^2$$
$$r_\xi = (x_\xi, y_\xi, z_\xi, f_\xi) \quad r_\eta = (x_\eta, y_\eta, z_\eta, f_\eta) \quad r_u = (x_u, y_u, z_u, f_u) \quad r_v = (x_v, y_v, z_v, f_v)$$

Thus, consider a two-dimensional surface in a four-dimensional space, defined as $x = x(u,v)$, $y = y(u, v)$, $z = z(u,v)$, $f = f(u,v)$. Let functions $\xi = \xi(u,v)$, $\eta = \eta(u,v)$ are used to define a new parameterization of a surface.

The problem of construction the adaptive-harmonic grid on a surface is stated as the problem of finding the new parameterization $u = u(\xi,\eta)$, $v = v(\xi,\eta)$, minimizing the functional Eq. 8.24, specified for this surface.

The result of minimization will be a new parameterization $u = u(\xi,\eta)$, $v(\xi,\eta)$, defining the adaptive-harmonic grid on a surface. Difficulties encountered in this problem are concerned with nonunique solutions of its discrete analog, in spite of the result from the harmonic map theory that the continuous problem has a unique solution [Steinberg and Roache, 1990].

Metric tensor elements g_{ij}^{ru} are defined

$$g_{11}^{ru} = x_u^2 + y_u^2 + z_u^2 + f_u^2 \qquad g_{12}^{ru} = x_u x_v + y_u y_v + z_u z_v + f_u f_v \qquad g_{22}^{ru} = x_v^2 + y_v^2 + z_v^2 + f_v^2$$

We write out the Euler equations in the case of adaption. These equations follow from Eq. 8.26 if $n = 2$, $k = 2$:

$$L(u) = \alpha u_{\xi\xi} - 2\beta u_{\xi\eta} + \gamma u_{\eta\eta} - J^2 D\left[\frac{\partial}{\partial u}\frac{g_{22}^{ru}}{D} - \frac{\partial}{\partial v}\frac{g_{12}^{ru}}{D}\right] = 0$$
$$L(v) = \alpha v_{\xi\xi} - 2\beta v_{\xi\eta} + \gamma v_{\eta\eta} - J^2 D\left[\frac{\partial}{\partial u}\frac{-g_{12}^{ru}}{D} + \frac{\partial}{\partial v}\frac{g_{11}^{ru}}{D}\right] = 0 \tag{8.38}$$

where

$$D = \sqrt{g_{11}^{ru} g_{22}^{ru} - \left(g_{12}^{ru}\right)^2} \qquad J = u_\xi v_\eta - u_\eta v_\xi \qquad \alpha = g_{22}^{r\xi} D^2 J^2 = x_\eta^2 + y_\eta^2 + z_\eta^2 + f_\eta^2$$
$$\beta = g_{12}^{r\xi} D^2 J^2 = x_\xi x_\eta + y_\xi y_\eta + z_\xi z_\eta + f_\xi f_\eta \qquad \gamma = g_{11}^{r\xi} D^2 J^2 = x_\xi^2 + y_\xi^2 + z_\xi^2 + f_\xi^2$$

8.7.2 Numerical Implementation

The method Eq. 8.32 is used for the numerical solution of Eq. 8.38, where x and y are replaced by u and v and $[L(u)]_{ij}$ and $[L(v)]_{ij}$ are the approximations of Eq. 8.38 at the grid node ij. All derivatives on u and v are computed with the use of formulas similar to formulas from Section 8.5.2:

$$[f_u]_{ij} = \frac{\left(f_{i+1,j} - f_{i-1,j}\right)\left(v_{i,j+1} - v_{i,j-1}\right) - \left(f_{i,j+1} - f_{i,j-1}\right)\left(v_{i+1,j} - v_{i-1,j}\right)}{\left(u_{i+1,j} - u_{i-1,j}\right)\left(v_{i,j+1} - v_{i,j-1}\right) - \left(u_{i,j+1} - u_{i,j-1}\right)\left(v_{i+1,j} - v_{i-1,j}\right)}$$
$$[f_v]_{ij} = -\frac{\left(f_{i+1,j} - f_{i-1,j}\right)\left(u_{i,j+1} - u_{i,j-1}\right) - \left(f_{i,j+1} - f_{i,j-1}\right)\left(u_{i+1,j} - u_{i-1,j}\right)}{\left(u_{i+1,j} - u_{i-1,j}\right)\left(v_{i,j+1} - v_{i,j-1}\right) - \left(u_{i,j+1} - u_{i,j-1}\right)\left(v_{i+1,j} - v_{i-1,j}\right)}$$

The adaptive-harmonic surface grid generation algorithm is formulated as follows:

1. Generate a quasi-uniform harmonic surface grid using the same algorithm as for adaption, but $f = 0$.
2. Compute the values of the control function at each grid node. The result is f_{ij}.
3. Evaluate derivatives $(f_u)_{ij}$ and $(f_v)_{ij}$ and other expressions in Eq. 8.38 using the above formulas.
4. Make one iteration step and compute new values of u_{ij} and v_{ij}.
5. Repeat starting with Step 2 to convergency.

The resulting algorithm is simple in implementation but can demand a special procedure for the choice of the parameter τ to achieve the numerical stability.

8.8 Irregular Surface Meshes

8.8.1 Problem Formulation

In notations of the previous section, consider a two-dimensional surface in a four-dimensional space, defined as $x = x(u, v)$, $y = y(u, v)$, $z = z(u, v)$, $f = f(u, v)$. Let functions $\xi = \xi(u, v)$, $\eta = \eta(u, v)$ are used to define a new parameterization of a surface.

The problem of construction of the adaptive-harmonic grid on a surface is stated as the problem of finding the new parameterization $u = u(\xi, \eta)$, $v = v(\xi, \eta)$ minimizing the functional

$$I = \int \frac{g_{11}^{ru}\left(u_\xi^2 + u_\eta^2\right) + 2g_{12}^{ru}\left(u_\xi v_\xi + u_\eta v_\eta\right) + g_{22}^{ru}\left(v_\xi^2 + v_\eta^2\right)}{\sqrt{g_{11}^{ru} g_{22}^{ru} - \left(g_{12}^{ru}\right)^2}\left(u_\xi v_\eta - u_\eta v_\xi\right)}\, d\xi\, d\eta \tag{8.39}$$

where

$$g_{11}^{ru} = x_u^2 + y_u^2 + z_u^2 + f_u^2 \quad g_{12}^{ru} = x_u x_v + y_u y_v + z_u z_v + f_u f_v \quad g_{22}^{ru} = x_v^2 + y_v^2 + z_v^2 + f_v^2 \tag{8.40}$$

The result of minimization will be a new parameterization $u = u(\xi, \eta)$, $v = v(\xi, \eta)$.

Now we can formulate the problem of irregular surface mesh smoothing and adaption. Let coordinates of an irregular mesh in the plane u, v be given:

$$(u,v)_n \qquad n = 1,\ldots,N_n$$

The mesh is formed by quadrilateral elements, i.e., the array $COR(N, k)$ is also defined. Functions $x = x(u, v)$, $y = y(u, v)$, $z = z(u, v)$ and $f = f(u, v)$ are assumed to be specified, for example, can be computed by analytic formulas.

The problem is to find new coordinates of the mesh nodes, minimizing the sum of the functional Eq. 8.39 values, computed for a mapping of the unit square in the plane ξ, η onto each cell of a mesh in the plane x, y.

8.8.2 Approximation of the Functional

Note that if in the functional Eq. 8.33 we replace expressions for $1 + (f_x)^2$ by g_{11}^{ru}, $f_x f_y$ by g_{12}^{ru}, and $1 + (f_y)^2$ by g_{22}^{ru}, we obtain the functional Eq. 8.39. Hence, the last one possesses all the properties of the functional Eq. 8.33 and also can be approximated in such a way that the minimum of its discrete analog is attained on a nondegenerate grid of convex quadrilaterals on the plane u, v. The algorithm from the Section 8.5 can be used for its approximation and minimization:

$$I^h = \sum_{N=1}^{N_e} \sum_{k=1}^{4} \frac{1}{4} \left[F_k\right]_N \tag{8.41}$$

where

$$F_k = \frac{D_1 g_{11}^{ru} + D_2 g_{22}^{ru} + 2D_3 g_{12}^{ru}}{J_k \sqrt{g_{11}^{ru} g_{22}^{ru} - \left(g_{12}^{ru}\right)^2}}$$

$$D_1 = \left(u_{k-1} - u_k\right)^2 + \left(u_{k+1} - u_k\right)^2 \qquad D_2 = \left(v_{k-1} - v_k\right)^2 + \left(v_{k+1} - v_k\right)^2$$

$$D_3 = \left(u_{k-1} - u_k\right)\left(v_{k-1} - v_k\right) + \left(u_{k+1} - u_k\right)\left(v_{k+1} - v_k\right)$$

$$J_k = \left(u_{k-1} - u_k\right)\left(v_{k+1} - v_k\right) - \left(u_{k+1} - u_k\right)\left(v_{k-1} - v_k\right)$$

Here the values g_{ij}^{ru} are computed at the node number k of the cell number N.

If the set D of convex meshes on the plane u, v is not empty, the system of algebraic equations

$$R_u = \frac{\partial I^h}{\partial u_n} = 0 \qquad R_v = \frac{\partial I^h}{\partial v_n} = 0$$

has at least one solution that is a convex mesh. To find it, one must first find a certain initial convex mesh, and then use some method of unconstrained minimization of the function I^h. Since this function has an infinite barrier on the boundary of the set of convex meshes, each step of the method can be chosen so that the mesh always remains convex.

8.8.3 Minimization of the Functional

Suppose the mesh at the lth step of the iterations is determined. We use the quasi-Newtonian procedure when the (l+1)-th step is accomplished by solving two linear equations for each interior node:

$$\tau R_u + \frac{\partial R_u}{\partial u_n}\left(u_n^{l+1} - u_n^l\right) + \frac{\partial R_u}{\partial v_n}\left(v_n^{l+1} - v_n^l\right) = 0$$
$$\tau R_v + \frac{\partial R_v}{\partial u_n}\left(u_n^{l+1} - u_n^l\right) + \frac{\partial R_v}{\partial v_n}\left(v_n^{l+1} - v_n^l\right) = 0 \qquad (8.42)$$

where τ is the iteration parameter, which is chosen so that the mesh remains convex. For this purpose after each step the conditions of grid convexity on the plane u, v are checked and if they are not satisfied, this parameter is multiplied by 0.5.

The adaptive-harmonic algorithm for the mesh smoothing and adaption on a surface is formulated as follows:

1. Generate an initial mesh with the use of a marching method.
2. Compute new values x_n, y_n, z_n and f_n at each mesh node.
3. Evaluate derivatives $[x_u]_n$ and $[x_v]_n$, $[y_u]_n$ and $[y_v]_n$, $[z_u]_n$ and $[z_v]_n$, $[f_u]_n$ and $[f_v]_n$ used in Eq. 8.42.
4. Make an iteration step and compute new values of u_n and v_n.
5. Repeat starting with Step 2 to convergency.

Computational formulas for $[f_u]_n$ and $[f_v]_n$ can be obtained as described in Section 8.6.4.

8.8.4 Derivation of Computational Formulas

Recall that if in the functional Eq. 8.33 we replace expressions for $1 + (f_x)^2$ by g_{11}^{ru}, $f_x f_y$ by g_{12}^{ru}, and $1 + (f_y)^2$ by g_{22}^{ru}, we obtain the functional Eq. 8.39. From this follows that for derivation of computational formulas for surface meshes we need only to perform these replacements in computational formulas for adaptive planar meshes, described in Section 8.6.4.

8.9 Three-Dimensional Regular Grids

8.9.1 Derivation of Equations

We will derive equations at once for the case of adaptation. Introduce notations

$$r = \left(r^1, r^2, r^3, r^4\right) = (x, y, z, f) \in S^{r3} \subset R^4$$
$$u = \left(u^1, u^2, u^3\right) = (x, y, z) \in \Omega \subset R^3 \qquad \xi = \left(\xi^1, \xi^2, \xi^3\right) = (\xi, \eta, \mu) \in Q^3 \subset R^3$$
$$r_\xi = \left(x_\xi, y_\xi, z_\xi, f_\xi\right) \qquad r_\eta = \left(x_\eta, y_\eta, z_\eta, f_\eta\right) \qquad r_\mu = \left(x_\mu, y_\mu, z_\mu, f_\mu\right),$$
$$r_x = \left(1, 0, 0, f_x\right) \qquad r_y = \left(0, 1, 0, f_y\right) \qquad r_z = \left(0, 0, 1, f_z\right).$$

The functional Eq. 8.24 in the three-dimensional case has the form

$$I = \int \left(g_{\xi r}^{11} + g_{\xi r}^{22} + g_{\xi r}^{33}\right) dS^{r3} \qquad (8.43)$$

where dS^{r3} is the element of the surface S^{r3}.

The functional Eq. 8.43 can be used for constructing harmonic coordinates on the surface of the graph of control function dependent on three variables. Projection of these coordinates onto a physical domain gives an adaptive-harmonic grid, clustered in regions of high gradients of adapted function $f(x, y, z)$.

The Euler equations of the functional Eq. 8.43 follow from Eq. 8.26 for $n = 3$, $k = 1$. We need only to compute the elements of the covariant metric tensor G^{ru} and contravariant tensor G_{ur} of the transform $r(u) = r(x, y, z) : \Omega \to S^{r3}$:

$$g_{11}^{ru} = r_x^2 = 1 + f_x^2 \quad g_{22}^{ru} = r_y^2 = 1 + f_y^2 \quad g_{33}^{ru} = r_z^2 = 1 + f_z^2,$$

$$g_{12}^{ru} = g_{21}^{ru} = r_x \cdot r_y = f_x f_y \quad g_{13}^{ru} = g_{31}^{ru} = r_x \cdot r_z = f_x f_z \quad g_{23}^{ru} = g_{32}^{ru} = r_y \cdot r_z = f_y f_z$$

$$\det(G^{ru}) = g_{11}^{ru}\left[g_{22}^{ru}g_{33}^{ru} - \left(g_{23}^{ru}\right)^2\right] - g_{12}^{ru}\left(g_{12}^{ru}g_{33}^{ru} - g_{13}^{ru}g_{23}^{ru}\right) + g_{13}^{ru}\left(g_{12}^{ru}g_{23}^{ru} - g_{22}^{ru}g_{13}^{ru}\right) =$$

$$\left(1 + f_x^2\right)\left(1 + f_y^2 + f_z^2\right) - f_x^2 f_y^2 - f_x^2 f_z^2 = 1 + f_x^2 + f_y^2 + f_z^2$$

$$\det(G^{r\xi}) = g_{11}^{r\xi}\left[g_{22}^{r\xi}g_{33}^{r\xi} - \left(g_{23}^{r\xi}\right)^2\right] - g_{12}^{r\xi}\left(g_{12}^{r\xi}g_{33}^{r\xi} - g_{13}^{r\xi}g_{23}^{r\xi}\right) + g_{13}^{r\xi}\left(g_{12}^{r\xi}g_{23}^{r\xi} - g_{22}^{r\xi}g_{13}^{r\xi}\right) =$$

$$\det(G^{ru})\left[x_\xi\left(y_\eta z_\mu - y_\mu z_\eta\right) - y_\xi\left(x_\eta z_\mu - x_\mu z_\eta\right) + z_\xi\left(x_\xi y_\eta - x_\eta y_\xi\right)\right]^2$$

$$g_{\xi r}^{11} = \left[g_{22}^{r\xi}g_{33}^{r\xi} - \left(g_{23}^{r\xi}\right)^2\right]\Big/\det(G^{r\xi}) \qquad g_{\xi r}^{12} = -\left(g_{12}^{r\xi}g_{33}^{r\xi} - g_{13}^{r\xi}g_{23}^{r\xi}\right)\Big/\det(G^{r\xi})$$

$$g_{\xi r}^{13} = \left(g_{12}^{r\xi}g_{23}^{r\xi} - g_{13}^{r\xi}g_{22}^{r\xi}\right)\Big/\det(G^{r\xi}) \qquad g_{\xi r}^{22} = \left[g_{11}^{r\xi}g_{33}^{r\xi} - \left(g_{12}^{r\xi}\right)^2\right]\Big/\det(G^{r\xi})$$

$$g_{\xi r}^{23} = -\left(g_{11}^{r\xi}g_{23}^{r\xi} - g_{13}^{r\xi}g_{12}^{r\xi}\right)\Big/\det(G^{r\xi}) \qquad g_{\xi r}^{33} = \left[g_{11}^{r\xi}g_{22}^{r\xi} - \left(g_{12}^{r\xi}\right)^2\right]\Big/\det(G^{r\xi})$$

$$g_{ur}^{11} = \left(1 + f_y^2 + f_z^2\right)\Big/\left(1 + f_x^2 + f_y^2 + f_z^2\right) \qquad g_{ur}^{12} = -f_x f_y\Big/\left(1 + f_x^2 + f_y^2 + f_z^2\right)$$

$$g_{ur}^{13} = -f_x f_z\Big/\left(1 + f_x^2 + f_y^2 + f_z^2\right) \qquad g_{ur}^{22} = \left(1 + f_x^2 + f_z^2\right)\Big/\left(1 + f_x^2 + f_y^2 + f_z^2\right)$$

$$g_{ur}^{23} = -f_y f_z\Big/\left(1 + f_x^2 + f_y^2 + f_z^2\right) \qquad g_{ur}^{33} = \left(1 + f_x^2 + f_y^2\right)\Big/\left(1 + f_x^2 + f_y^2 + f_z^2\right)$$

Substituting these expressions into Eq. 8.26, we obtain equations convenient for practical use:

$$\begin{aligned}
L(x) &= g_{\xi r}^{11} x_{\xi\xi} + 2g_{\xi r}^{12} x_{\xi\eta} + 2g_{\xi r}^{13} x_{\xi\mu} + g_{\xi r}^{22} x_{\eta\eta} + 2g_{\xi r}^{23} x_{\eta\mu} + g_{\xi r}^{33} x_{\mu\mu} - \\
&\frac{1}{D}\left[\frac{\partial}{\partial x}\frac{1 + f_y^2 + f_z^2}{D} + \frac{\partial}{\partial y}\frac{-f_x f_y}{D} + \frac{\partial}{\partial z}\frac{-f_x f_z}{D}\right] = 0 \\
L(y) &= g_{\xi r}^{11} y_{\xi\xi} + 2g_{\xi r}^{12} y_{\xi\eta} + 2g_{\xi r}^{13} y_{\xi\mu} + g_{\xi r}^{22} y_{\eta\eta} + 2g_{\xi r}^{23} y_{\eta\mu} + g_{\xi r}^{33} y_{\mu\mu} - \\
&\frac{1}{D}\left[\frac{\partial}{\partial x}\frac{-f_x f_y}{D} + \frac{\partial}{\partial y}\frac{1 + f_x^2 + f_z^2}{D} + \frac{\partial}{\partial z}\frac{-f_y f_z}{D}\right] = 0 \\
L(z) &= g_{\xi r}^{11} z_{\xi\xi} + 2g_{\xi r}^{12} z_{\xi\eta} + 2g_{\xi r}^{13} z_{\xi\mu} + g_{\xi r}^{22} z_{\eta\eta} + 2g_{\xi r}^{23} z_{\eta\mu} + g_{\xi r}^{33} z_{\mu\mu} - \\
&\frac{1}{D}\left[\frac{\partial}{\partial x}\frac{-f_x f_z}{D} + \frac{\partial}{\partial y}\frac{-f_y f_z}{D} + \frac{\partial}{\partial z}\frac{1 + f_x^2 + f_y^2}{D}\right] = 0
\end{aligned} \tag{8.44}$$

where

$$D = \sqrt{1 + f_x^2 + f_y^2 + f_z^2}$$

8.9.2 Numerical Implementation

The problem of grid generation in three dimensions will be considered in the following formulation. In a simply connected domain Ω in the space x, y, z a grid

$$(x,y,z)_{ijm} \quad i=1,\ldots,i^* \quad j=1,\ldots,j^* \quad m=1,\ldots,m^*$$

must be constructed with given coordinates of boundary nodes

$$(x,y,z)_{ij1} \quad (x,y,z)_{ijm^*} \quad (x,y,z)_{i1m} \quad (x,y,z)_{ij^*m} \quad (x,y,z)_{1jm} \quad (x,y,z)_{i^*jm}$$

Instead of the parametric cube the following parametric domain can be introduced to simplify the computational formulas:

$$1<\xi<i^* \quad 1<\eta<j^* \quad 1<\mu<m^*$$

associated with the cube grid (ξ_i, η_j, μ_m) such that

$$\xi_i=\iota, \quad \eta_j=j \quad \mu_m-m \quad i=1,\ldots,i^* \quad j=1,\ldots,j^* \quad m=1,\ldots,m^*$$

Eq. 8.44 are approximated on this grid with the use of simplest finite-difference relations for derivatives on ξ, η, μ. For example, derivatives of $f(\xi, \eta, \mu)$ are approximated as

$$f_\xi \approx \left[f_\xi\right]_{ijm}=\frac{1}{2}\left(f_{i+1,j,m}-f_{i-1,j,m}\right) \quad f_\eta \approx \left[f_\eta\right]_{ijm}=\frac{1}{2}\left(f_{i,j+1,m}-f_{i,j-1,m}\right)$$

$$f_\mu \approx \left[f_\mu\right]_{ijm}=\frac{1}{2}\left(f_{i,j,m+1}-f_{i,j,m-1}\right) \quad f_{\xi\xi} \approx \left[f_{\xi\xi}\right]_{ijm}=f_{i+1,j,m}-2f_{ijm}+f_{i-1,j,m}$$

$$f_{\xi\eta} \approx \left[f_{\xi\eta}\right]_{ijm}=\frac{1}{4}\left(f_{i+1,j+1,m}-f_{i-1,j+1,m}-f_{i+1,j-1,m}+f_{i-1,j-1,m}\right)$$

$$f_{\xi\mu} \approx \left[f_{\xi\mu}\right]_{ijm}=\frac{1}{4}\left(f_{i+1,j,m+1}-f_{i+1,j,m-1}-f_{i-1,j,m+1}+f_{i-1,j,m-1}\right)$$

$$f_{\eta\eta} \approx \left[f_{\eta\eta}\right]_{ijm}=f_{i,j+1,m}-2f_{ijm}+f_{i,j-1,m}$$

$$f_{\eta\mu} \approx \left[f_{\eta\mu}\right]_{ijm}=\frac{1}{4}\left(f_{i,j+1,m+1}-f_{i,j+1,m-1}-f_{i,j-1,m+1}+f_{i,j-1,m-1}\right)$$

$$f_{\mu\mu} \approx \left[f_{\mu\mu}\right]_{ijm}=f_{i,j,m+1}-2f_{ijm}+f_{i,j,m-1}$$

The method similar to Eq. 8.32 is used for the numerical solution of the resulting finite-difference equations:

$$x_{ijm}^{l+1}=x_{ijm}^{l}+\tau\frac{\left[L(x)\right]_{ijm}}{2\left[g_{\xi r}^{11}\right]_{ijm}+2\left[g_{\xi r}^{22}\right]_{ijm}+2\left[g_{\xi r}^{33}\right]_{ijm}}$$

$$y_{ijm}^{l+1}=y_{ijm}^{l}+\tau\frac{\left[L(y)\right]_{ijm}}{2\left[g_{\xi r}^{11}\right]_{ijm}+2\left[g_{\xi r}^{22}\right]_{ijm}+2\left[g_{\xi r}^{33}\right]_{ijm}} \tag{8.45}$$

$$z_{ijm}^{l+1}=z_{ijm}^{l}+\tau\frac{\left[L(z)\right]_{ijm}}{2\left[g_{\xi r}^{11}\right]_{ijm}+2\left[g_{\xi r}^{22}\right]_{ijm}+2\left[g_{\xi r}^{33}\right]_{ijm}}$$

Consider formulas for the transformation of derivatives in the three-dimensional case:

$$x_\xi f_x + y_\xi f_y + z_\xi f_z = f_\xi \quad x_\eta f_x + y_\eta f_y + z_\eta f_z = f_\eta \quad x_\mu f_x + y_\mu f_y + z_\mu f_z = f_\mu$$

From this follows

$$f_x = f_\xi \left(y_\eta z_\mu - y_\mu z_\eta\right)/J - f_\eta \left(y_\xi z_\mu - y_\mu z_\xi\right)/J + f_\mu \left(y_\xi z_\eta - y_\eta z_\xi\right)/J$$

$$f_y = -f_\xi \left(x_\eta z_\mu - x_\mu z_\eta\right)/J + f_\eta \left(x_\xi z_\mu - x_\mu z_\xi\right)/J - f_\mu \left(x_\xi z_\eta - x_\eta z_\xi\right)/J$$

$$f_z = f_\xi \left(x_\eta y_\mu - x_\mu y_\eta\right)/J - f_\eta \left(x_\xi y_\mu - x_\mu y_\xi\right)/J + f_\mu \left(x_\xi y_\eta - x_\eta y_\xi\right)/J$$

where $J = x_\xi(y_\eta z_\mu - y_\mu z_\eta) - x_\eta(y_\xi z_\mu - y_\mu z_\xi) + x_\mu(y_\xi z_\eta - y_\eta z_\xi)$.

Approximating all derivatives in these expressions with the use of the above formulas, we obtain the approximation of derivatives $[f_x]_{ijm}$, $[f_y]_{ijm}$, and $[f_z]_{ijm}$, used in Eq.8.45.

The adaptive-harmonic grid generation algorithm is formulated as follows:

1. Generate a quasi-uniform grid using the same algorithm as for adaption, but $f = 0$.
2. Compute the values of the control function f_{ijm} at each grid node.
3. Evaluate derivatives $[f_x]_{ijm}$, $[f_y]_{ijm}$, and $[f_z]_{ijm}$ and substitute them into Eq. 8.45.
4. Make on iteration step and compute new values of x_{ijm}, y_{ijm}, and z_{ijm}.
5. Repeat starting with Step 2 to convergency.

The resulting algorithm is simple in implementation and can be used for meshing the three-dimensional domains until the increased complexity of domain or boundary layers produce the appearance of self-intersecting cells. Then the special algorithm should be employed, based on a variational formulation and guaranteeing nondegenerate grid generation.

8.10 Three-Dimensional Irregular Meshes

8.10.1 Discrete Analog of the Jacobian Positiveness

The three-dimensional case is much more complicated than the two-dimensional case, because simple conditions of the Jacobian positiveness cannot be obtained for the trilinear mapping of the unit cube onto a hexahedral cell. The notation of convexity also cannot be used, since faces of a hexahedron are not plane. This is why the approach developed for two-dimensional meshes in Section 8.2 cannot be directly extended to the three-dimensional case.

Nevertheless, the discrete analog of the Jacobian positiveness for the mapping of the unit cube onto a hexahedral cell can be obtained. We use the decomposition of the parametric cube to tetrahedra, which are mapped onto the corresponding tetrahedra of the decomposed hexahedral cell. The mapping of each tetrahedra is one-to-one. This approach is analogous to the approach used in 2D case for approximation of the functional Eq. 8.11 in such a way that it has an infinite barrier at the boundary of the set of nondengenerate meshes. Recall that in Section 8.3.2 the quadrilateral cell is decomposed to two triangles first by the one diagonal and then by the other. In the first and second decompositions the mapping is approximated by the functions which are linear in each triangle. All the conditions of the Jacobian positiveness for each of such mappings coincide with the condition for all the mesh cells to be convex quadrilaterals.

Consider a unit cube in the three-dimensional space ξ, η, μ, shown in Figure 8.4. We divide it into two prisms by the plane 1584. Then we devide the prism shown in Figure 8.4 into three tetrahedra drawing the diagonals 14, 25, 58, 45, and 46. Obtained tetrahedra denote as T^ξ_{5124}, T^ξ_{5684} and T^ξ_{5624}. Note

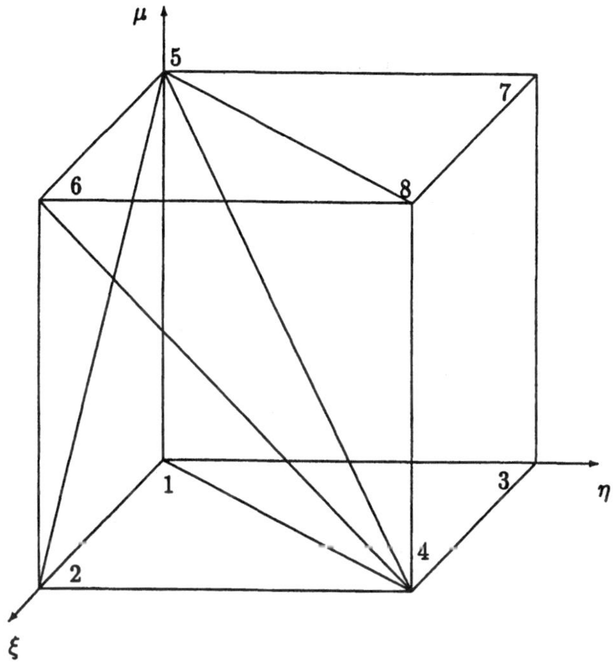

FIGURE 8.4 Vertex numeration and decomposition of the cube to tetrahedrons.

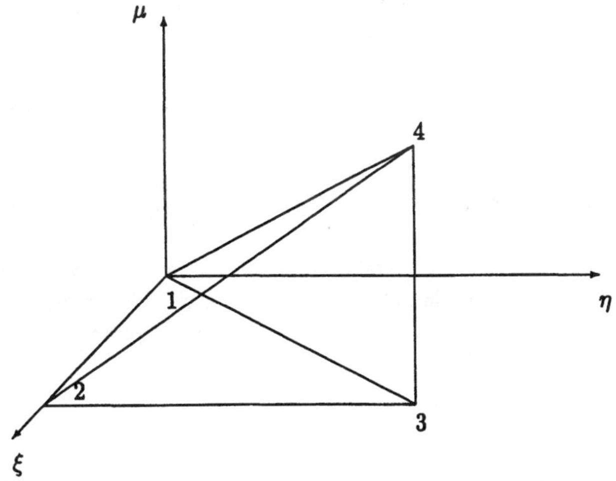

FIGURE 8.5 Vertex numeration for the base tetrahedron.

that all these tetrahedra are equal to each other (with rotation and reflection taken into account) and one of the edges of the cube corresponds to each of them. For example, tetrahedron T^{ξ}_{5124} can be referred to the edge 12. Only one extra tetrahedrons is referred to this edge, namely T^{ξ}_{3126}. What is the difference between tetrahedra T^{ξ}_{5124} and T^{ξ}_{3126}? The answer is that each of them corresponds to a proper type of coordinate system, right-hand or left-hand. It is easy to compute the total number of such tetrahedra. It is equal to double the number of the cube edges, i.e., 24. For the unit cube the volume of one tetrahedron is equal to $^1/_6$, and the total volume of all such tetrahedra is equal to 4.

Consider the base tetrahedron shown in Figure 8.5. Vertices are enumerated from 1 to 4 as shown in Figure 8.5. Each vertex corresponds to a radius-vector r_1, r_2, r_3, or r_4 in the space x, y, z. All these vectors define tetrahedron in the space x, y, z. We introduce the base vectors

$$e_1 = r_2 - r_1, \quad e_2 = r_3 - r_2, \quad e_3 = r_4 - r_3.$$

Note that the coordinate system e_1, e_2, e_3 is a right-hand system, which is easy to see from the orientation of the base tetrahedron in Figure 8.5. Hence, the volume of the "right" tetrahedron is equal to

$$J_{T\,right} = (e_1 \times e_2) \cdot e_3$$

At the same time, the volume of the "left" tetrahedron is equal to

$$J_{T\,left} = -(e_1 \times e_2) \cdot e_3$$

Now, in analogy with the two-dimensional case, the condition for the mesh to be nondegenerate for the three-dimensional hexahedral mesh can be expressed as follows:

$$\left[\left(J_{T\,left}\right)_m\right]_N > 0 \quad \left[\left(J_{T\,right}\right)_m\right]_N > 0 \quad m = 1, \ldots, 12; \quad N = 1, \ldots, N_e \tag{8.46}$$

where $(J_{T\,left})_m$ is a volume of the tetrahedron corresponding to the edge number m and defining the left-hand coordinate system, $(J_{T\,right})_m$ is a volume of the tetrahedron corresponding to the edge number m and defining the right-hand coordinate system (each cube has 12 edges), N is the cell number, N_e is the total number of cells. Conditions Eq. 8.46 define the discrete analog of the Jacobian positiveness in the three-dimensional case. Meshes satisfying inequalities Eq. 8.46 we will call nondegenerate hexahedral meshes.

As in the two-dimensional case, we should introduce the function $COR(N,k)$ to define a correspondence between local and global node numbers:

$$n = COR(N,k) \quad n = 1, \ldots, N_n \quad N = 1, \ldots, N_e \quad K = 1, \ldots, 8$$

where n is a global node number, N_n is a total number of mesh nodes, N is an element number, N_e is a number of elements, k is a local node number in the element. This function is implemented in the computer program as a function for a regular grid and as an array for an irregular mesh.

8.10.2 Problem Formulation

Let adapted function $f(x, y, z)$ define a three-dimensional surface in the four-dimensional space. In notations of the previous section, the functional Eq. 8.24 can be written as follows:

$$I = \int \frac{g_{11}^{r\xi} g_{22}^{r\xi} - \left(g_{12}^{r\xi}\right)^2 + g_{11}^{r\xi} g_{33}^{r\xi} - \left(g_{13}^{r\xi}\right)^2 + g_{22}^{r\xi} g_{33}^{r\xi} - \left(g_{23}^{r\xi}\right)^2}{\sqrt{g_{11}^{r\xi}\left[g_{22}^{r\xi} g_{33}^{r\xi} - \left(g_{23}^{r\xi}\right)^2\right] - g_{12}^{r\xi}\left(g_{12}^{r\xi} g_{33}^{r\xi} - g_{13}^{r\xi} g_{23}^{r\xi}\right) + g_{13}^{r\xi}\left(g_{12}^{r\xi} g_{23}^{r\xi} - g_{22}^{r\xi} g_{13}^{r\xi}\right)}} \, d\xi d\eta d\mu \tag{8.47}$$

where

$$g_{11}^{r\xi} = r_\xi^2 \quad g_{22}^{r\xi} = r_\eta^2 \quad g_{33}^{r\xi} = r_\mu^2 \quad g_{12}^{r\xi} = g_{21}^{r\xi} = \left(r_\xi \cdot r_\eta\right) \quad g_{13}^{r\xi} = g_{31}^{r\xi} = \left(r_\xi \cdot r_\mu\right) \quad g_{23}^{r\xi} = g_{32}^{r\xi} = \left(r_\eta \cdot r_\mu\right)$$

here

$$f_\xi = f_x x_\xi + f_y y_\xi + f_z z_\xi \quad f_\eta = f_x x_\eta + f_y y_\eta + f_z z_\eta \quad f_\mu = f_x x_\mu + f_y y_\mu + f_z z_\mu$$

The functional Eq. 8.47 can be used for constructing harmonic coordinates on the surface of the graph of control function dependent on three variables. Projection of these coordinates onto a physical domain gives an adaptive-harmonic grid, clustered in regions of high gradients of adapted function $f(x, y, z)$.

The problem of irregular three-dimensional mesh smoothing and adaption is formulated as follows. Let the coordinates of irregular mesh be given:

$$(x, y, z)_n \quad n = 1, \ldots, N_n \tag{8.48}$$

The mesh is formed by hexahedral elements, i.e., the array *COR*(*N*, *k*) is also defined. The problem is to find new coordinates of the mesh nodes, minimizing the sum of the functional Eq. 8.47 values, computed for a mapping of the unit cube onto each cell of a mesh.

8.10.3 Approximation of the Functional

First consider the case, where $f(x, y, z) = 0$. The functional Eq. 8.47 in this case can be written in a more simple form:

$$I = \int \frac{\left(r_\xi \times r_\eta\right)^2 + \left(r_\xi \times r_\mu\right)^2 + \left(r_\eta \times r_\mu\right)^2}{\left(r_\xi \times r_\eta\right) \cdot r_\mu} d\xi d\eta d\mu \tag{8.49}$$

where $\times$ is a vector product, and $\cdot$ is a scalar product,

$$r_\xi = \left(x_\xi, y_\xi, z_\xi\right) \quad r_\eta = \left(x_\eta, y_\eta, z_\eta\right) \quad r_\mu = \left(x_\mu, y_\mu, z_\mu\right)$$

Let the linear transform $x^h(\xi, \eta, \mu)$, $y^h(\xi, \eta, \mu)$, $z^h(\xi, \eta, \mu)$ map the base tetrahedron T_{1234}^{ξ} in the space ξ, η, μ onto a tetrahedron T_{1234} in the space x, y, z. The value of the functional with the linear functions $x^h(\xi, \eta, \mu)$, $y^h(\xi, \eta, \mu)$ and $z^h(\xi, \eta, \mu)$ can be computed precisely. Consequently, the approximation of this functional can be written as

$$I^h = \sum_{N=1}^{N_e} \sum_{m=1}^{12} \frac{1}{24} \left[\left(F_m\right)_{left} + \left(F_m\right)_{right} \right]_N \tag{8.50}$$

where

$$\left(F_m\right)_{left} = \frac{\left(r_\xi^h \times r_\eta^h\right)^2 + \left(r_\xi^h \times r_\mu^h\right)^2 + \left(r_\eta^h \times r_\mu^h\right)^2}{\left(J_m\right)_{left}} \quad \left(F_m\right)_{right} = \frac{\left(r_\xi^h \times r_\eta^h\right)^2 + \left(r_\xi^h \times r_\mu^h\right)^2 + \left(r_\eta^h \times r_\mu^h\right)^2}{\left(J_m\right)_{right}}$$

$$\left(J_m\right)_{left} = -\left(r_\xi^h \times r_\eta^h\right) \cdot r_\mu^h \quad \left(J_m\right)_{right} = \left(r_\xi^h \times r_\eta^h\right) \cdot r_\mu^h$$

Consider one term in Eq. 8.50, for example, $(F_m)_{left}$, and suppose that the Jacobian $(J_m)_{left}$ tends to zero, remaining positive. For I^h not to tend to infinity in this situation it is necessary that the numerator in $(F_m)_{left}$ must also tend to zero. From the form of the numerator it follows that vectors $e_1 = r_2 - r_1$, $e_2 = r_3$

$-r_2$ and $e_3 = r_4 - r_3$ are parallel, hence all points r_1, r_2, r_3, and r_4 lie on a straight line. Consequently, the volumes of all tetrahedra that contain corresponding faces must also tend to zero, including the tetrahedron defined by the edge 34 and containing the edge 23. Repeating the argument as many times as necessary, we conclude that all mesh nodes, including those at the boundary of the domain, must lie on a straight line, which is impossible.

From this follows that the function I^h has an infinite barrier at the boundary of nondegenerate three-dimensional hexahedral meshes, satisfying inequalities Eq. 8.46. Hence, if this set is not empty, the system of algebraic equations

$$R_x = \frac{\partial I^h}{\partial x_n} = 0 \quad R_y = \frac{\partial I^h}{\partial y_n} = 0 \quad R_z = \frac{\partial I^h}{\partial z_n} = 0$$

has at least one solution which is a nondegenerate mesh. To find it, one must first find a certain initial nondegenerate mesh, and then use some method of unconstrained minimization of the function I^h. Since this function has an infinite barrier on the boundary of the set of nondegenerate meshes, each step of the method can be chosen so that the mesh always satisfies inequalities (Eq. 8.46).

For adaptive mesh generation with the employment of the functional Eq. 8.47, we use the same approach: consider T tetrahedra, described above. Then the mapping of the base tetrahedron onto each of these tetrahedra is approximated by linear functions, with assumption that f is also approximated by a linear function defined by its values in tetrahedron vertices. Then the integrand in Eq. 8.47 will be equal to constant. Note that the integrand in Eq. 8.47 differs from Eq. 8.49: the first is an invariant for the orthogonal transformations of the base tetrahedron. This means that we do not need to use two terms in the approximation of Eq. 8.47 corresponding to right-hand and left-hand coordinate systems. The value of this functional depends only on the numeration of nodes of the base tetrahedron, not on its type.

8.10.4 Minimization of the Functional

Suppose the mesh at the lth step of the iterations is determined. We use the quasi-Newtonian procedure when the $(l+1)$-th step is accomplished by solving two linear equations for each interior node:

$$\begin{aligned}
&\tau R_x + \frac{\partial R_x}{\partial x_n}\left(x_n^{l+1} - x_n^l\right) + \frac{\partial R_x}{\partial y_n}\left(y_n^{l+1} - y_n^l\right) + \frac{\partial R_x}{\partial z_n}\left(z_n^{l+1} - z_n^l\right) = 0 \\
&\tau R_y + \frac{\partial R_y}{\partial x_n}\left(x_n^{l+1} - x_n^l\right) + \frac{\partial R_y}{\partial y_n}\left(y_n^{l+1} - y_n^l\right) + \frac{\partial R_y}{\partial z_n}\left(z_n^{l+1} - z_n^l\right) = 0 \\
&\tau R_z + \frac{\partial R_z}{\partial x_n}\left(x_n^{l+1} - x_n^l\right) + \frac{\partial R_z}{\partial y_n}\left(y_n^{l+1} - y_n^l\right) + \frac{\partial R_z}{\partial z_n}\left(z_n^{l+1} - z_n^l\right) = 0
\end{aligned} \tag{8.51}$$

where τ is the iteration parameter, which is chosen so that the mesh remains nondegenerate. For this purpose after each step the conditions Eq. 8.46 are checked and if they are not satisfied, this parameter is multiplied by 0.5.

The adaptive-harmonic algorithm for the three-dimensional mesh is formulated as follows:

1. Generate initial mesh with the use of a marching method.
2. Compute new values f_n at each mesh node.
3. Make one iteration step Eq. 8.51 and compute new values of x_n, y_n, and z_n.
4. Repeat starting with Step 2 to convergency.

Note, that the algorithm contains computational formulas for $[f_x]_n$, $[f_y]_n$ and $[f_z]_n$ which will be presented below.

8.10.5 Derivation of Computational Formulas

We will obtain computational formulas in the case of adaption, i.e., we will approximate the functional Eq. 8.47. The used approach is similar to the method of approximation to the functional described in Section 8.3.

Consider the linear transform $x^h(\xi, \eta, \mu)$, $y^h(\xi, \eta, \mu)$, $z^h(\xi, \eta, \mu)$ of the base tetrahedron shown in Figure 8.5 onto one of tetrahedra of the cell decomposition. Function f will be approximated by the linear function $f^h(\xi, \eta, \mu)$. Derivatives of these functions can be easily computed taking into account the numeration of the vertices of the base tetrahedron:

$$r_\xi^h = \left(x_\xi^h, y_\xi^h, z_\xi^h, f_\xi^h\right) = r_2 - r_1 = \left(x_2 - x_1, y_2 - y_1, z_2 - z_1, f_2 - f_1\right)$$
$$r_\eta^h = \left(x_\eta^h, y_\eta^h, z_\eta^h, f_\eta^h\right) = r_3 - r_2 = \left(x_3 - x_2, y_3 - y_2, z_3 - z_2, f_3 - f_2\right)$$
$$r_\mu^h = \left(x_\mu^h, y_\mu^h, z_\mu^h, f_\mu^h\right) = r_4 - r_3 = \left(x_4 - x_3, y_4 - y_3, z_4 - z_3, f_4 - f_3\right)$$

From this follows

$$g_{ij} = \left(r_{i+1} - r_i\right) \cdot \left(r_{j+1} - r_j\right)$$

i.e.,

$$\begin{gathered} g_{11} = \left(r_2 - r_1\right)^2 \quad g_{22} = \left(r_3 - r_2\right)^2 \quad g_{33} = \left(r_4 - r_3\right)^2 \\ g_{12} = g_{21} = \left(\left(r_3 - r_2\right) \cdot \left(r_2 - r_1\right)\right) \\ g_{13} = g_{31} = \left(\left(r_4 - r_3\right) \cdot \left(r_2 - r_1\right)\right) \\ g_{23} = g_{32} = \left(\left(r_4 - r_3\right) \cdot \left(r_3 - r_2\right)\right) \end{gathered} \tag{8.52}$$

Substituting these expressions into the integrand of Eq. 8.47 we obtain

$$F = \frac{U}{V}$$

where

$$U = g_{11}g_{22} - \left(g_{12}\right)^2 + g_{11}g_{33} - \left(g_{13}\right)^2 + g_{22}g_{33} - \left(g_{23}\right)^2 \tag{8.53}$$

$$V = \sqrt{g_{11}\left[g_{22}g_{33} - \left(g_{23}\right)^2\right] - g_{12}\left(g_{12}g_{33} - g_{13}g_{23}\right) + g_{13}\left(g_{12}g_{23} - g_{22}g_{13}\right)} \tag{8.54}$$

We use formulas for differentiating the relation of two functions. After differentiating we obtain

$$\begin{gathered} F_x = \frac{U_x - FV_x}{V} \quad F_y = \frac{U_y - FV_y}{V} \quad F_z = \frac{U_z - FV_z}{V} \\ F_{xx} = \frac{U_{xx} - 2F_x V_x - FV_{xx}}{V} \quad F_{yy} = \frac{U_{yy} - 2F_y V_y - FV_{yy}}{V} \quad F_{zz} = \frac{U_{zz} - 2F_z V_z - FV_{zz}}{V} \\ F_{xy} = F_{yx} = \frac{U_{xy} - F_x V_y - F_y V_x - FV_{xy}}{V} \quad F_{xz} = F_{zx} = \frac{U_{xz} - F_x V_z - F_z V_x - FV_{xz}}{V} \\ F_{yz} = F_{zy} = \frac{U_{yz} - F_z V_y - F_y V_z - FV_{yz}}{V} \end{gathered} \tag{8.55}$$

For the vertex 1 of the tetrahedron we should substitute the expressions Eq. 8.52, Eq. 8.53, and Eq. 8.54 into Eq. 8.55, and also replace x, y and z by x_1, y_1 and z_1 in the resulting formulas.

For the vertex 2 x, y, and z in Eq. 8.55 are replaced by x_2, y_2, and z_2.
For the vertex 3 x, y, and z in Eq. 8.55 are replaced by x_3, y_3, and z_3.
For the vertex 4 x, y, and z in Eq. 8.55 are replaced by x_4, y_4, and z_4.

In computing the derivatives of f_i on x_j, y_j, and z_j, $i = 1, \ldots, 4$, $j = 1, \ldots, 4$, we use the formulas for the transformation of derivatives in the three-dimensional space:

$$x_\xi f_x + y_\xi f_y + z_\xi f_z = f_\xi \quad x_\eta f_x + y_\eta f_y + z_\eta f_z = f_\eta \quad x_\mu f_x + y_\mu f_y + z_\mu f_z = f_\mu$$

From this follows

$$\begin{aligned} f_x &= f_\xi\left(y_\eta z_\mu - y_\mu z_\eta\right)/J - f_\eta\left(y_\xi z_\mu - y_\mu z_\xi\right)/J + f_\mu\left(y_\xi z_\eta - y_\eta z_\xi\right)/J \\ f_y &= -f_\xi\left(x_\eta z_\mu - x_\mu z_\eta\right)/J + f_\eta\left(x_\xi z_\mu - x_\mu z_\xi\right)/J - f_\mu\left(x_\xi z_\eta - x_\eta z_\xi\right)/J \\ f_z &= f_\xi\left(x_\eta y_\mu - x_\mu y_\eta\right)/J - f_\eta\left(x_\xi y_\mu - x_\mu y_\xi\right)/J + f_\mu\left(x_\xi y_\eta - x_\eta y_\xi\right)/J \end{aligned} \tag{8.56}$$

where

$$J = x_\xi\left(y_\eta z_\mu - y_\mu z_\eta\right) - x_\eta\left(y_\xi z_\mu - y_\mu z_\xi\right) + x_\mu\left(y_\xi z_\eta - y_\eta z_\xi\right)$$

Note that the derivatives on x, y, and z are independent on which system of coordinates, right-hand or left-hand is used in Eq. 8.56. Substituting the expressions for the derivatives of x^h, y^h and z^h on ξ, η, μ into Eq. 8.56, we obtain formulas for the derivatives f_x^h, f_y^h, and f_z^h. We use the following formulas in computations:

$$\frac{\partial f_i}{\partial x_j} = \begin{cases} f_x^h & \text{if } i = j \\ 0 & \text{if } i \neq j \end{cases} \quad \frac{\partial f_i}{\partial y_j} = \begin{cases} f_y^h & \text{if } i = j \\ 0 & \text{if } i \neq j \end{cases} \quad \frac{\partial f_i}{\partial z_j} = \begin{cases} f_z^h & \text{if } i = j \\ 0 & \text{if } i \neq j \end{cases}$$

Computations are performed as follows. Let F and its derivatives on x_1, y_1 and z_1 in the numeration of the base tetrahedron be computed with the use of formulas Eq. 8.55 for the cell number N and the local node number k. Then the computed values are added to the appropriate array elements (which were first cleared):

$$\begin{aligned} &I^h += F \quad \left[R_x\right]_n += F_x \quad \left[R_y\right]_n += F_y \quad \left[R_z\right]_n += F_z \\ &\left[R_{xx}\right]_n += F_{xx} \quad \left[R_{yy}\right]_n += F_{yy} \quad \left[R_{zz}\right]_n += F_{zz} \\ &\left[R_{xy}\right]_n += F_{xy} \quad \left[R_{xz}\right]_n += F_{xz} \quad \left[R_{yz}\right]_n += F_{yz} \end{aligned} \tag{8.57}$$

where $n = COR(N, k_1)$. Here, $a+ = b$ means that the new value of a becomes equal to $a + b$.

Similarly for the vertex 2, the correspondence between local and global number is $n = COR(N, k_2)$.

Similarly for the vertex 3, the correspondence between local and global number is $n = COR(N, k_3)$.

Similarly for the vertex 4, the correspondence between local and global number is $n = COR$ (N, k_4).

So, the iteration method for irregular three-dimensional mesh relaxation and adaption is described in detail.

8.11 Results of Test Computations

8.11.1 Comparison Between the Winslow Method and the Variational Approach

Comparison between the variational algorithm described in Section 8.3 and the Winslow method was presented in the paper by Ivanenko and Charakhch'yan [1988]. We will describe here results of computations shown in Figure 8.6. In Figure 8.6 the regular grids 10×10, 19×19 and 37×37, generated for backward facing step by the Winslow method (Figures 8.6a, 8.6c, 8.6e) and by the variational barrier methods (Figs. 8.6b, 8.6d, 8.6f) are shown. The choice of this example is concerned with the discussion about the applicability of the Winslow method. There is an opinion that this method can generate quite satisfactory grids if the number of grid nodes is sufficiently large, despite the fact that in many cases this method generates grids with self-intersecting cells. Indeed, if the number of grid nodes tends to infinity, the limit will be a continuous mapping which is one-to-one. Such a mapping can be used then for the replacement of independent variables (Jacobian is positive inside a domain). This is not the case of a discrete mapping (a grid). If the Jacobian is negative on the boundary, then the Winslow method might generate grids with degenerate cells near the boundary for any number of grid nodes. As shown in the presented example, the form of degenerate cell near the internal corner is worse with increasing the number of nodes (the Winslow method, Figures 8.6a, 8.6c, 8.6e). At the same time, the variational method generates satisfactory (convex) grids for any number of grid nodes (Figures 8.6b, 8.6d, 8.6f).

The geometric sense of the smoothing procedure defined by harmonic functional is that the shape of each cell tends to be a square. From this follows constraints on the application of the variational method for irregular meshes. In fact, satisfactory mesh with square cells might not exist for the given mesh structure. It is clear that if the square cell is used as initial, the variational method will not change it (the Winslow finite-difference method will not change it, too). If the initial mesh has the form shown in Figure 8.7a we obtain the irregular smoothed mesh shown in Figure 8.7b after 700 iterations.

The grid quality was estimated with the following parameters: J_{min} is the minimum of the areas of all triangles, scaled by the maximum area, *Aspect* is the maximum ratio of edge lengths in quadrilateral, and *Skew* is the minimum cells angle in degrees. For meshes in Figure 8.7 the minimum area decreases from 0.13 to 0.0002, the maximum ratio increases from 10 to 10.3 and minimum angle decreases from 13.9 to 11.7. But the mesh in Figure 8.7b looks more smooth than the mesh in Figure 8.7a. This means that all these quality parameters do not estimate the mesh quality properly. Note that the mesh after smoothing looks like several cobwebs and is extremely nonuniform. This example shows that in some cases the variational method can be unsatisfactory for smoothing of irregular meshes, for example, if refinement is used for several blocks with regular grid structure in each as shown in Figure 8.7a.

8.11.2 Comparison Between the Finite-Difference Method for Two-Dimensional Adaptive-Harmonic Meshes and the Variational Approach

Methods for adaptive mesh generation are illustrated by the following example of control function [Ivanenko, 1993]. The square domain $0 < x < 1$, $0 < y < 1$ is considered. The cubic curve

$$y_0(x) = 25(x - 0.5)(x - 0.75)(x - 0.25) + 0.5$$

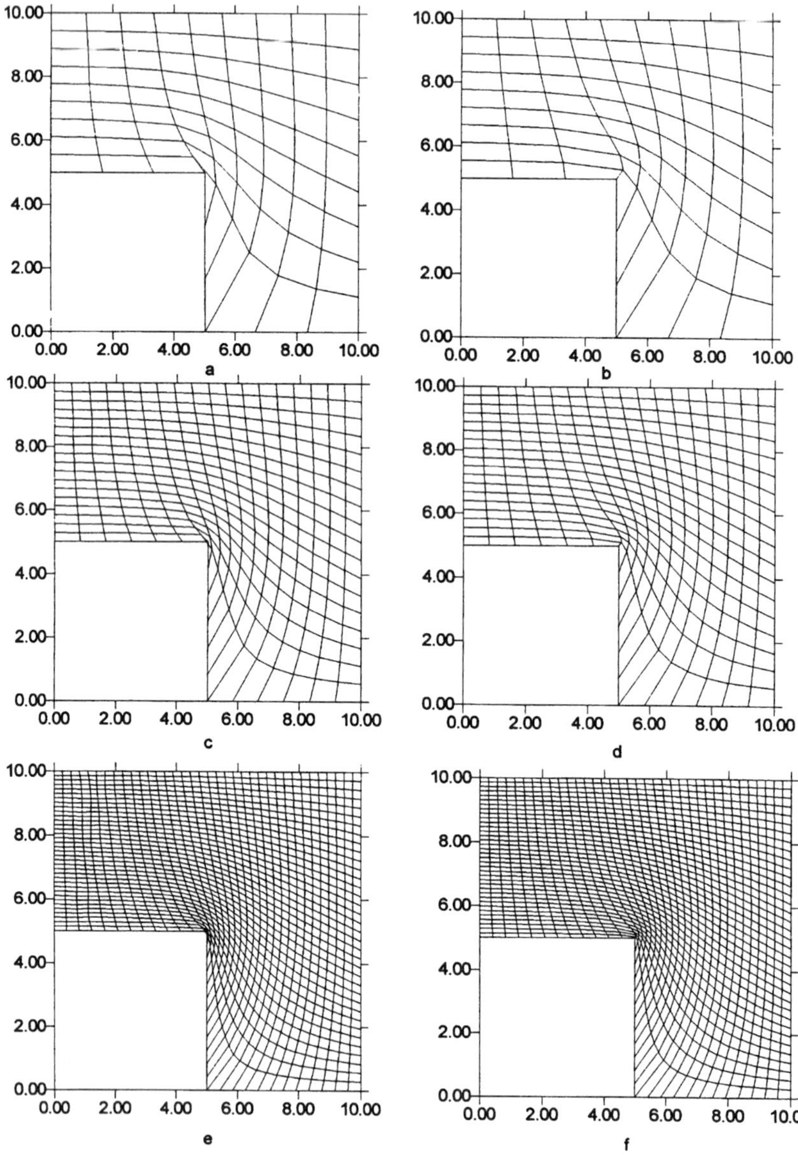

FIGURE 8.6 Regular grid 10 × 10, 19 × 19, and 37 × 37 generated by the Winslow method (a,c,e), and by the variational barrier method (b, d, f).

determines the form of a layer of high gradients. For a given point *x, y* the function *f*(*x, y*) is calculated as follows:

$$f = \begin{cases} 1 & \text{if} \quad y \geq y_0 + \delta \\ 0.5(y - y_0 + \delta)/\delta & \text{if} \quad y_0 + \delta \geq y \geq y_0 - \delta \\ 0 & \text{if} \quad y \leq y_0 - \delta \end{cases}$$

Here

$$\delta = \delta_0 \left[1 + \left(\frac{\partial y_0}{\partial x} \right)^2 \right]^{1/2}$$

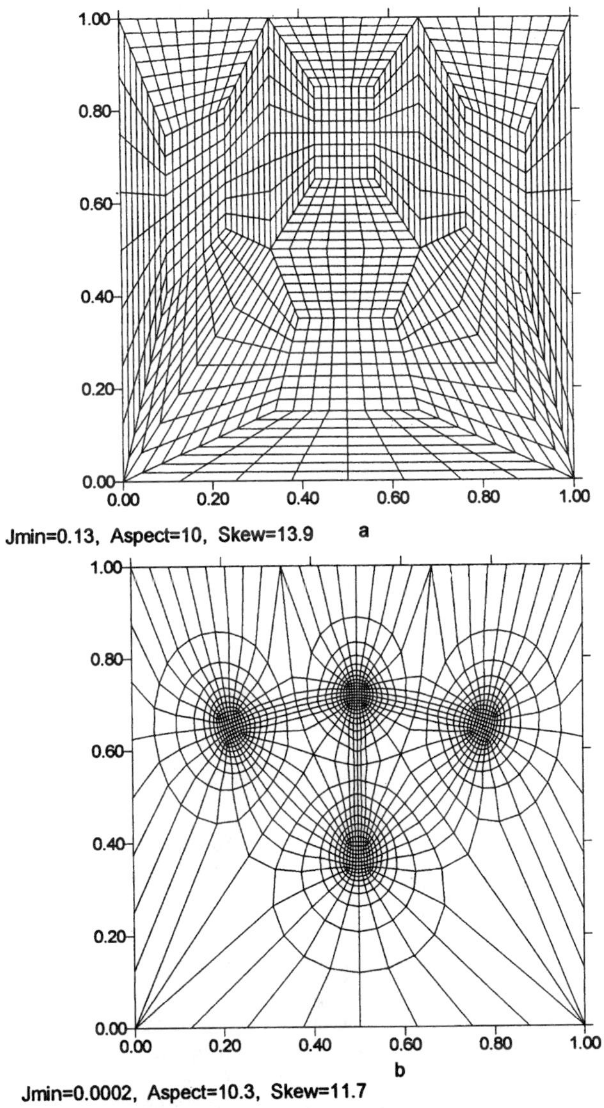

FIGURE 8.7 Smoothing of irregular mesh; (a) initial mesh, (b) smoothed mesh.

The value of δ is chosen so that the width of the layer will be about $2\delta_0$ everywhere along the curve. In all test computation this value was chosen to be $\delta_0 = 0.02$.

An additional control parameter C is introduced to control the number of mesh nodes inside the boundary or internal layers. The function $Cf(x, y)$ is used in computational formulas instead of $f(x, y)$. Increasing the value C, more mesh nodes will be in the layer of high gradients. This value is chosen in the range from 0.1 to 0.5. A number of points in a layer is approximately $C/(C + 1)$, i.e., if $C = 0.5$ one third part of points will be in a layer of high gradients.

The grid, generated by the finite-difference method with $C = 0.2$ slightly differs from the grid generated by variational method with the same value of parameter C. But with the value of parameter $C = 0.5$, the satisfactory grid cannot be generated by the finite-difference method (Figure 8.8a). The grid generated for this value of parameter by the variational method is shown in Figure 8.8b. All grid cells are convex.

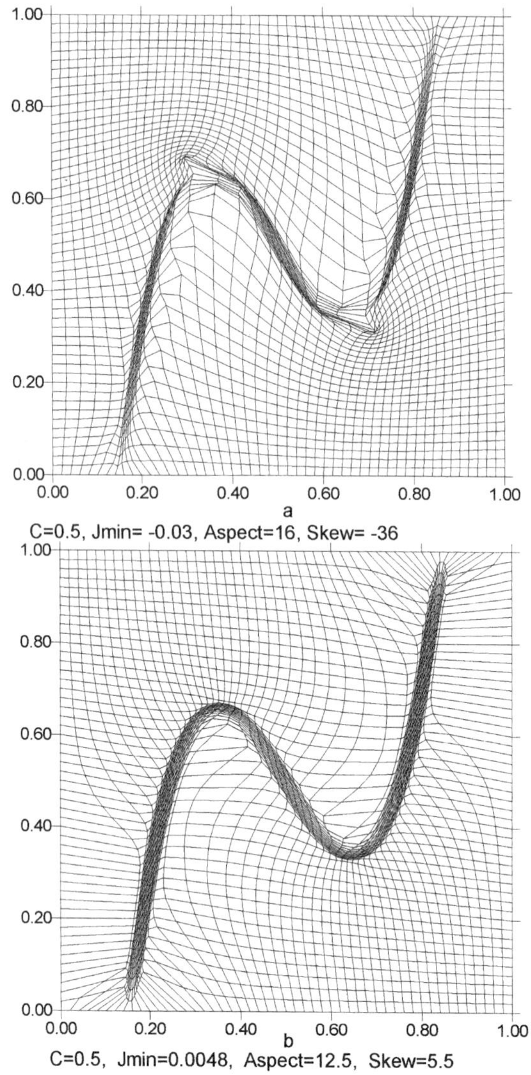

FIGURE 8.8 Adaptive-harmonic grids; (a) generated by the finite-difference method, (b) generated by the variational method.

8.11.3 Comparison Between the Finite-Difference Method for Adaptive-Harmonic Grid Generation on Surfaces and the Variational Approach

The comparison of the finite-difference method for grid generation on surfaces with the variational method was performed on an example of a surface defined parametrically:

"Monkeys saddle"

$$x = u, \quad y = v \quad z = 8(v-0.3)^3 - 24(u-0.5)^2(v-0.5) \quad 0 < u < 1 \quad 0 < v < 1$$

Methods for adaptive mesh generation on surfaces are illustrated on the example of control function, defined in previous subsection with u and v replaced by x and y. An additional control parameter C is also introduced to control the number of mesh nodes inside the boundary or internal layer. If $C < 0.4$, the finite-difference method generates quite satisfactory grids on the surface. But if $C = 0.5$, the finite-difference method generates

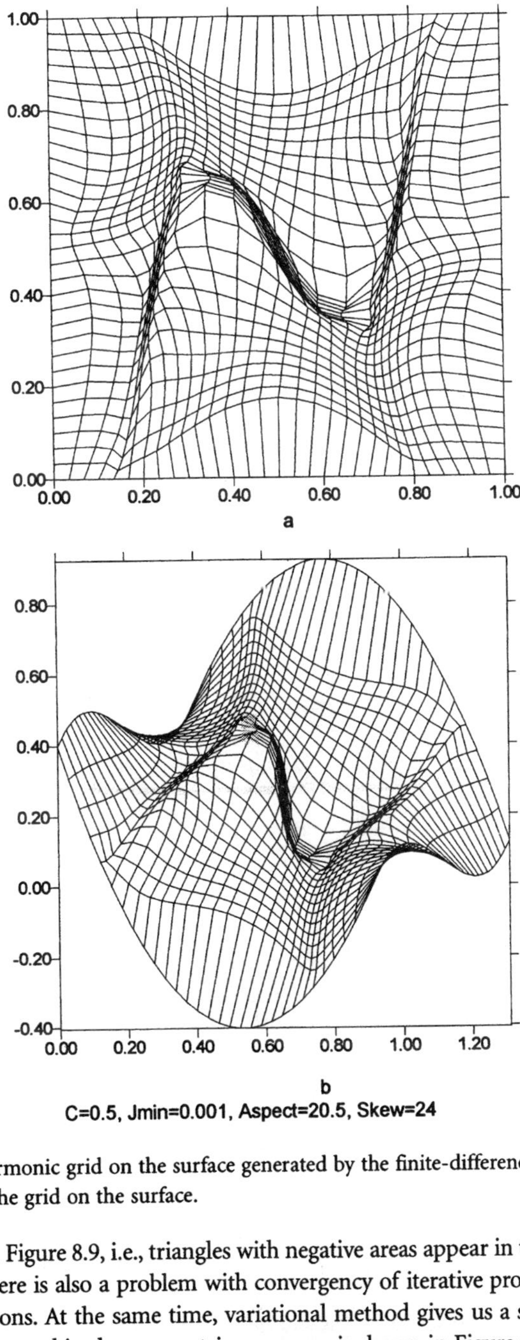

C=0.5, Jmin=0.001, Aspect=20.5, Skew=24

FIGURE 8.9 Adaptive-harmonic grid on the surface generated by the finite-difference method; (a) the grid in the parametric space *u, v*, (b) the grid on the surface.

degenerate grid shown in Figure 8.9, i.e., triangles with negative areas appear in the parametric space *u, v*, as shown in Figure 8.9a. There is also a problem with convergency of iterative process. Such meshes are often unsuitable for computations. At the same time, variational method gives us a satisfactory mesh, shown in Figure 8.10. The grid generated in the parametric space *u, v* is shown in Figure 8.10a.

8.11.4 Comparison Between the Finite-Difference Method for Adaptive-Harmonic Three-Dimensional Meshes and the Variational Approach

The comparison between variational and finite-difference methods was performed with the grid quality estimated by the following parameters: J_{min} is the minimum of the tetrahedra volumes, scaled by the maximum volume, *Aspect* is the maximum ratio of lengths of adjacent edges, and *Skew* is the minimal angle between edges in degrees.

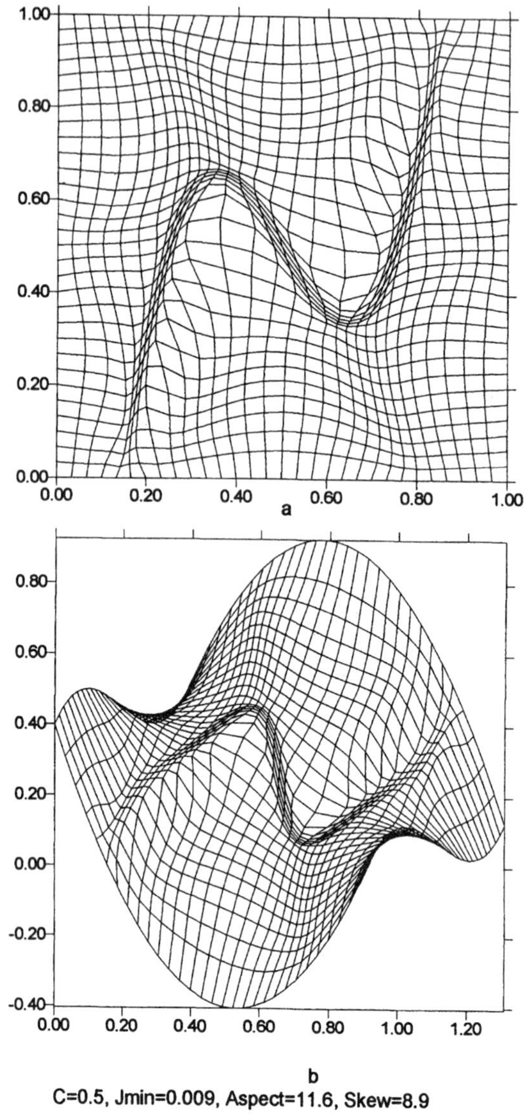

FIGURE 8.10 Adaptive-harmonic grid on the surface generated by the variational barrier method. (a) The grid in the parametric space u, v, (b) the grid on the surface.

Methods for adaptive mesh generation are illustrated using the same example of the control function dependent only on two variables x and y. An additional control parameter C is introduced to control the number of mesh nodes inside the boundary or internal layer.

The domain is a cube with a pedestal in the middle of the down face.

An adaptive grid generated in the domain by the finite-difference method with $C = 0.2$ is shown in Figure 8.11. Values of quality parameters are shown in the figure. The projection of the mesh surface $\mu = 3$ onto the plane $z = 0$ is shown in Figure 8.11a. The section of the mesh in Figure 8.11c shows the presence of degenerate cells ($J_{min} = -0.3$). At the same time, the mesh shown in Figure 8.12 generated for the same domain with the same parameter C by the variational method does not contain degenerate cells ($J_{min} = 0.02$).

Note that the control function is two-dimensional, but the generated adaptive grids are substantially three-dimensional. Moreover, variational method generates are more fitted to control function mesh. The same results can be obtained for irregular mesh smoothing and adaption.

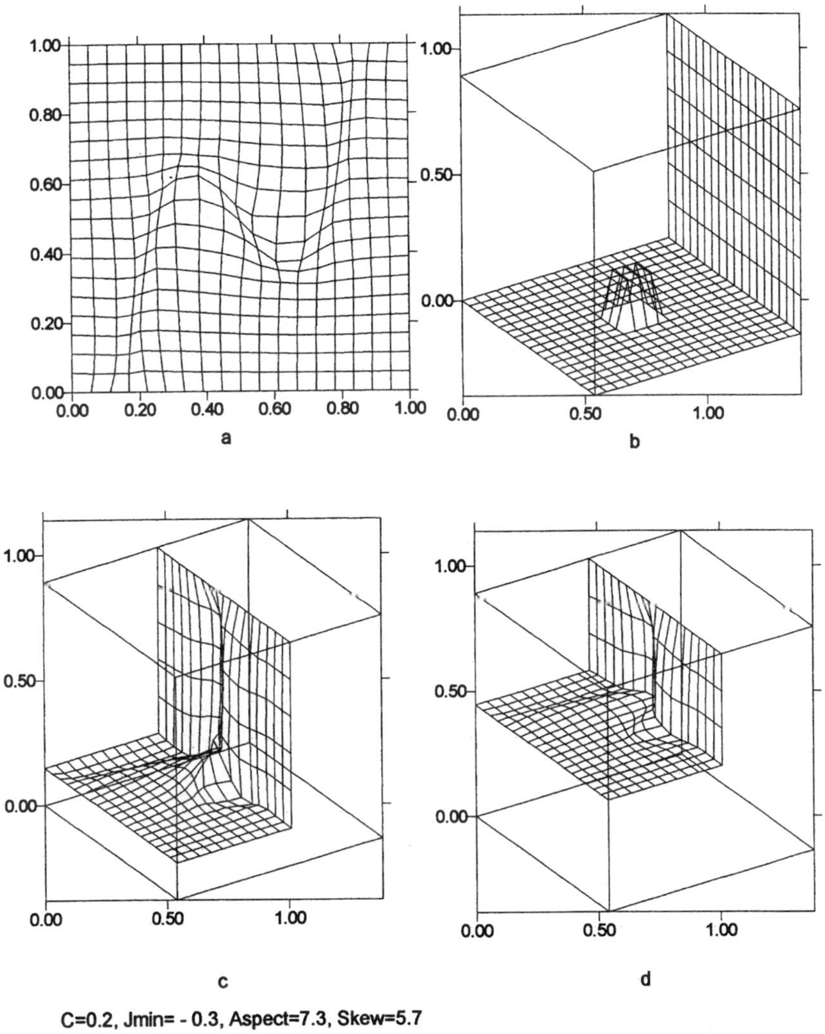

FIGURE 8.11 Adaptive-harmonic three-dimensional grid 19 × 19 × 7 generated by the finite-difference method; (a) projection of the coordinate surface $\mu = 3$ onto the x, y plane, (b) coordinate surfaces $\mu = 1$ and $\eta = 19$, (c) coordinate surfaces $\mu = 2$ and $\eta = 11$, (d) coordinate surfaces $\mu = 4$ and $\eta = 11$.

8.12 Conclusions

Algorithms for adaptive regular and irregular mesh generation in two and three dimensions as well as for surfaces are considered in the present chapter. The approach is based on the theory of harmonic maps. Formulated algorithms can be used for grid/mesh generation with strong clustering of mesh nodes and assure generation of nondegenerate meshes. The main conclusion is the following. The meshes produced by irregular mesh smoothing and adaption are better for more regular meshes.

The variational algorithm for three-dimensional meshes appear to be cumbersome. At the same time it is approximately 10 times more expensive than the finite-difference method for regular grids.

These investigations have been stimulated by the need in fully automatic numerical solvers for the complex problems of mathematical physics. This means that the human intervention into the solution process, especially into adaptive grid generation, should be minimized. Modern methods do not always satisfy these conditions, so the development of new fully automatic grid generation algorithms is of great importance today.

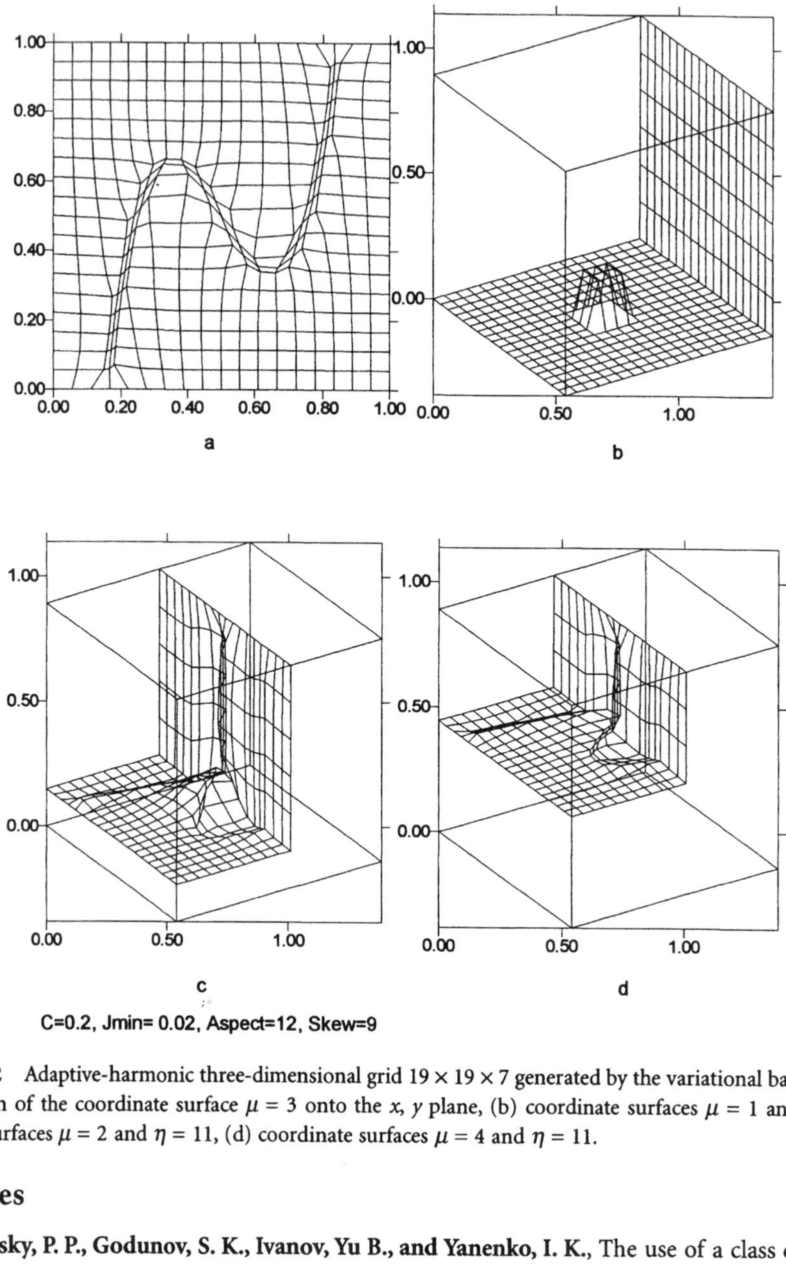

FIGURE 8.12 Adaptive-harmonic three-dimensional grid $19 \times 19 \times 7$ generated by the variational barrier method; (a) projection of the coordinate surface $\mu = 3$ onto the x, y plane, (b) coordinate surfaces $\mu = 1$ and $\eta = 19$, (c) coordinate surfaces $\mu = 2$ and $\eta = 11$, (d) coordinate surfaces $\mu = 4$ and $\eta = 11$.

References

1. **Belinsky, P. P., Godunov, S. K., Ivanov, Yu B., and Yanenko, I. K.,** The use of a class of quasiconformal mappings to construct difference nets in domains with curvilinear boundaries, *USSR Comput. Maths. Math. Phys.,* 15(6), pp. 133–139, 1975.
2. **Bobilev, N. A., Ivanenko, S. A., and Ismailov, I. G.,** Some remarks on homeomorphysms, *Russian Mathematical Notes,* Vol. 60(4), pp. 593–596, 1996.
3. **Brackbill, J. U.,** An adaptive grid with directional control, *J. Comp. Phys.,* 108(1), pp. 38–50, 1993.
4. **Brackbill, J. U. and Saltzman, J. S.,** Adaptive zoning for singular problems in two dimensions, *J. Comput. Phys.,* Vol. 46(3), pp. 342–368, 1982.
5. **Dvinsky, A. S.,** Adaptive grid generation from harmonic maps on Riemanian manifolds, *J. Comp. Phys.,* 95(3), pp. 450–476, 1991.
6. **Dwyer, H. A., Smooke, M.D., and Kee, R.J.,** Adaptive gridding for finite difference solution to heat and mass transfer problems, *Appl. Math. and Comput.,* 10/11, pp. 339–356, 1982.

7. **Eells, J. E. and Lemaire, L.**, Another report on harmonic maps, *Bulletin of the London Mathematical Society*, 20(86), pp. 387–524, 1988.
8. **Eells, J. E. and Sampson, J. H.**, Harmonic mappings of Riemannian manifolds, *Amer. J. Math.*, 86(1), pp. 109–160, 1964.
9. **Eiseman, P. R.**, Adaptive grid generation, *Comput. Methods in Appl. Mech. and Engineering*, 64, pp. 321–376, 1987.
10. **Godunov, S. K. and Prokopov, G.P.**, The use of moving meshes in gas-dynamics calculations, *USSR Comput. Maths. Math. Phys.*, 12(2), pp. 182–191, 1972.
11. **Godunov, S. K., Zabrodin, A.V., Ivanov, M. Ya, Prokopov, G P., and Kraiko, A.N.**, *Numerical Solution of Multidimensional Problems of Gas Dynamics*, Nauka, Moscow (in Russian), 1976.
12. **Hamilton, R.**, Harmonic maps of manifolds with boundary, *Lecture Notes in Math.*, 471, pp. 165–172, 1975.
13. **Ivanenko, S. A.**, Generation of non-degenerate meshes, *USSR Comput. Maths. Math. Phys.*, 28(5), pp. 141–146, 1988.
14. **Ivanenko, S. A.**, Adaptive grids and grids on surfaces, *Comput. Maths. Math. Phys.*, 33(9), pp. 1179–1193, 1993.
15. **Ivanenko, S. A.**, Adaptive curvilinear grids in the finite element method, *Comput. Maths. Math. Phys.*, 35(9), pp. 1071–1087, 1995a.
16. **Ivanenko, S. A.**, Adaptive-harmonic grid generation and its application for numerical solution of the problems with boundary and interior layers, *Comput. Maths. Math. Phys.*, 35(10), pp. 1203–1220, 1995b.
17. **Ivanenko, S. A. and Charakhch'yan, A.A.**, Curvilinear grids of convex quadrilaterals, *USSR Comput. Maths. Math. Phys.* 1988, 28(2), pp. 126–133
18. **Liseikin, V. D.**, Construction of structured grids on n-dimensional surfaces, *USSR Comput. Maths. Math. Phys.* 1991, 31(11), pp. 1670–1683.
19. **Liseikin, V. D.**, On some interpretations of a smoothness functional used in constructing regular and adaptive grids, *Russ. J. Numer. Anal. Modelling*, 8(6), pp. 507–518, 1993.
20. **Prokopov, G. P.**, About the comparative analysis of algorithms and programs for regular two-dimensional grid generation, (in Russian) *Topics of Nuclear Science and Technology. Ser. Mathematical Modelling of Physical Processes*, Issue 1, pp. 7–12, 1993.
21. **Spekreijse, S. P., Hagmeijer, R., Boerstoel, J. M.**, Adaptive grid generation by using Laplace-Beltrami operator on a monitoring surface, In *Proceedings of the 5th International Conference on Numerical Grid Generation in Computational Field Simulations*, April 1–5, 1996, Mississippi State University, pp. 137–146.
22. **Steinberg, S. and Roache, P.**, Anomalies in grid generation in curves, *J. Comput. Phys.*, 91, pp. 255–277, 1990.
23. **Strang, G. and Fix, G. J.**, *An Analysis of the Finite Element Method*, Prentice-Hall, Englewood Cliffs, NJ, 1973.
24. **Thompson, J. F., Warsi, Z. U. A., and Mastin, C. W.**, *Numerical Grid Generation*, North-Holland, NY, 1985.
25. **Winslow, A. M.**, Numerical solution of quasilinear Poisson equation in nonuniform triangle mesh, *J. Comput. Phys.*, 1(2), pp. 149–172, 1966.
26. **Yanenko, N. N., Danaev, N. T., and Liseikin, V. D.**, *On a Variational Method for Generating Grids*, (in Russian) pp. 157–163, 1977.

9

Surface Grid Generation Systems

Ahmed Khamayseh

Andrew Kuprat

9.1 Introduction

Structured surface grid generation entails the generation of a curvilinear coordinate grid on a surface. It may be necessary to generate such a grid in order to perform a two-dimensional numerical simulation of a physical process involving the surface. Alternately, surface grid generation may represent a stage in the generation of a volume grid, which itself would be used in a three-dimensional numerical simulation involving the volume or volumes bounded by the surface. We mention here that *unstructured* surface mesh generation (wherein the surface is usually decomposed into a collection of triangles but no obvious curvilinear coordinate system exists) is covered in Chapter 19. Unstructured surface meshes are arguably easier to construct and have found wide application in numerical simulation as well.

Grid quality is a critical area for many numerical simulation problems. The distribution of the grid points and the geometric properties of the grid such as skewness, smoothness, and cell aspect ratios have a major impact on the accuracy of the simulation. The solution of a system of partial differential equations can be greatly simplified by a well-constructed grid. It is also true that a grid which is not well suited to the problem can lead to an unsatisfactory result. In some applications, improper choice of grid point locations can lead to an apparent instability or lack of convergence. This chapter will cover techniques for the generation of structured surface meshes of sufficient quality for use in physical simulations.

Before a grid can be generated, the surface geometry itself must be created, usually by one of two methods. In the first method, the object to be simulated has a shape that can be calculated from a mathematical formula, such as a sphere. There are a wide variety of shapes in this class, including airfoils, missile geometries, and sometimes even complete wings. These types of shapes are very easy to define, and lead to an efficient grid generation process, with high-quality resulting grids.

The second manner in which surface geometries are specified involves representation of the initial geometry as a computer-aided design (CAD) surface, where CAD systems typically represent the surfaces of a certain geometry with a set of structured points or patches. The CAD surface is then typically converted to a nonuniform rational B-splines (NURBS) surface representation (cf. Part III).

0-8493-2687-7/99/$0.00+$.50

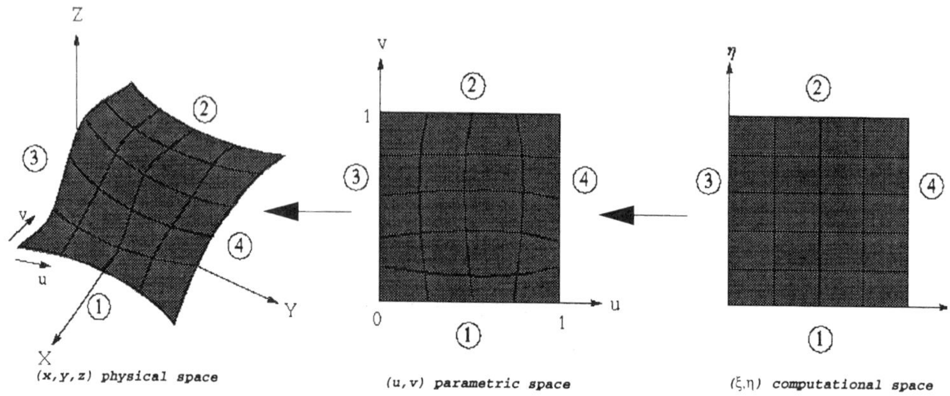

FIGURE 9.1 Mapping from computational ("ξ,η") space to physical ("x,y,z") space via parametric ("u,v") space.

In any event, we presume that the surface geometry is available as a *parametrically defined surface* such as a *quadric surface, Bezier surface, B-spline surface,* or *NURBS surface.* We thus presume the existence of a surface geometry definition in the form of a mapping $(x(u,v), y(u,v), z(u,v))$ from a *parametric* (u,v) *domain* to a *physical* (x,y,z) *domain.* This mapping is assumed differentiable, and we assume that the mapping and its derivatives can be quickly evaluated. We compactly denote this mapping as $\mathbf{x}(\mathbf{u})$, where $\mathbf{x} = (x,y,z)$, and $\mathbf{u} = (u,v)$.

In structured surface grid generation, the actual grid generation process is the generation of a mapping from the discrete rectangular *computational* (ξ,η) *domain* to the parametric (u,v) domain, which results in the composite map $\mathbf{x}(\xi,\eta) = (x(\xi,\eta), y(\xi,\eta), z(\xi,\eta))$ (see Figure 9.1).

As seen in the figure, the physical space is a subset of IR^3; the parametric space is a subset of IR^2, which is taken to be the $[0,1] \times [0,1]$ unit square. Technically speaking, the computational space is a discrete rectangular set of points (ξ,η), $\xi \in \{0,1,\ldots, m\}$, $\eta \in \{0,1,\ldots,n\}$. However, in order for us to be able to apply the powerful machinery of differentiable mappings between spaces, we extend the computational space to be a continuum, so that it is the rectangle $[0,m] \times [0,n]$. This is what is depicted in Figure 9.1. (Note: In this chapter the coordinates of a point in computational space are sometimes denoted by (ξ,η), and other times (i,j). The (i,j) notation is usually used in algorithms where i,j take on only integer values, while the (ξ,η) notation is usually used in mathematical derivations where ξ,η can take on continuum values.)

With regard to the composite map $\mathbf{x}(\xi,\eta)$ *or* the mapping $\mathbf{u}(\xi,\eta)$, we define *grid lines* to be lines of constant ξ or η, *grid points* to be points where ξ,η are integers, and *grid cells* to be the quadrilaterals formed between grid lines. It will always be clear if by *grid lines, grid points,* or *grid cells* we are referring to objects on the gridded surface or to objects in the parametric domain.

The surface geometry $\mathbf{x}(\mathbf{u})$ may contain some singularities (e.g., the mapping of a line to a point in a certain parameterization of a cone). We require that the composite map $\mathbf{x}(\xi,\eta) = \mathbf{x}(\mathbf{u}) \circ \mathbf{u}(\xi,\eta)$ not contain any additional degeneracies. This leads to the requirement that $\mathbf{u}(\xi,\eta)$ be one-to-one and onto. If a $\mathbf{u}(\xi,\eta)$ mapping is generated which is not one-to-one and onto, quite often the problem will be detected as a visible "folding" of grid lines when the gridded surface is viewed using computer graphics.

That $\mathbf{u}(\xi,\eta)$ should be an isomorphism is a "bare bones" requirement. It is usually also required that the $\mathbf{u}(\xi,\eta)$ map be constructed such that the composite map $\mathbf{x}(\xi,\eta)$ have the following properties in the interest of reducing errors occurring in numerical simulations that use the grid:

1. Grid lines should be smooth to provide continuous transformation derivatives.
2. Grid points should be closely spaced in the physical domain where large numerical errors are expected.
3. Grid cells should have areas that vary smoothly across the surface.
4. Excessive grid skewness (nonorthogonal intersection of grid lines) should be avoided, since it sometimes increases truncation errors.

In order to generate surface grids with the above requirements, two approaches, *algebraic* and *elliptic*, have been most popularly embraced in the mesh generation community. This chapter covers these two techniques in some detail, presenting practical algorithms as well as theoretical development. Both these methods complement each other and both are typically used in a complete grid generation system.

Algebraic mesh generation proceeds in stages. The grid is first constructed on the boundary curves, and a surface grid is then constructed by algebraic interpolation between the boundary curves. In fact, one could then continue further by constructing an interpolated volume grid between bounding surface grids. This process can itself be a complete method for the generation of meshes. Indeed, a certain interpolation method that we describe — cubic Hermite interpolation — can be used to generate surface meshes that possess boundary orthogonality required in certain numerical simulations. Usually, however, the simplest form of algebraic mesh generation — linear transfinite interpolation — is used to produce a valid "initial" mesh that can then be smoothed by another method to satisfy possible requirements on grid line orthogonality or grid point distribution.

Elliptic mesh generation is the natural complement to the above process. An initial grid, usually produced by algebraic methods, is smoothed by iteratively solving a system of partial differential equations that relate the physical (x,y,z) and computational (ξ,η) variables. Desired orthogonality properties and desired grid point distributions in the physical domain are effected by imposing appropriate boundary conditions and/or source terms in the elliptic system of equations. An alternative technique for smoothing initial grids to produce desired properties are the variational methods in Brackbill and Saltzman [5], Castillo [6], and Saltzman [18]. They will not be covered in this chapter.

Related surveys on algebraic methods and the use of transfinite interpolation in grid generation can be found in Abolhassani and Stewart [1], Chawner and Anderson [7], Smith [19], and Soni [20]. For surveys on elliptic methods in grid generation, we refer the reader to Khamayseh and Mastin [12], Sorenson [21], Spekreijse [22], Thomas and Middlecoff [26], Thompson et al. [27], Thompson [29], Warsi [30], and Winslow [33]. For further study on the foundations and fundamentals of grid generation, we refer to Knupp and Steinberg [13] and Thompson et al. [28].

Finally, we refer the reader to other related chapters in this book; these are Chapter 3 on TFI generation systems, Chapter 4 on elliptic generation systems, Chapter 6 on boundary orthogonality, Chapter 7 on orthogonal generation systems, and Part III on surface generation. Although we cite individual papers throughout this chapter, in most cases referral to these chapters will suffice.

9.2 Algebraic Surface Grid Generation

Algebraic surface grid generation involves (1) distribution of grid points along the boundary curves and (2) bidirectional interpolation usually called *transfinite interpolation* (TFI), which defines the remaining points, while simultaneously matching all four boundary curves (cf. Chapter 3). Step (2) can be done by unidirectional interpolation between boundaries, but this is not as reliable or popular an approach. The transfinite interpolation will incorporate the specified spacing at the boundaries, and possibly orthogonality conditions as well. Grid orthogonality at the boundaries, wherein the grid intersects the boundaries as close as possible to a 90° angle, can be crucial in certain numerical applications.

Since interpolation is fundamentally projection from boundaries, problems can arise in configurations in which the line of sight to boundaries in the parametric plane is not present. In this case, the user must break the surface into a sufficient number of subsurface patches to alleviate the problem. In the following, we assume that we are to generate a grid on a reasonably well-behaved subsurface patch.

9.2.1 Distribution of Grid Points on the Boundary Curves

The methodology of constructing an $(m + 1) \times (n + 1)$ algebraic grid on a physical surface starts with the specification of the boundary distribution along the physical boundaries of the surface. This is equivalent to specifying the distribution of the four boundary curves in the parametric domain:

$$\{u(\xi,0),\quad u(\xi,n),\quad v(0,\eta),\quad v(m,\eta)\,|\,0\le\xi\le m\quad 0\le\eta\le n\}$$

Without loss of generality, let us generate the points on the "lower" boundary curve $\{u(\xi,0)|0 \le \xi \le m\}$. This curve in parametric space corresponds to the curve $\{\mathbf{x}(u,0)|0 \le u \le 1\}$ in physical space. The treatment of the other three ("upper," "left," and "right") boundary curves will be similar. For convenience, we suppress the constant second arguments of **x** and *u*, so that we have

$$u(\xi) \equiv u(\xi,0)$$
$$x(u) \equiv x(u,0)$$

and our task is to find $\{u(\xi)|0\le\xi\le m\}$ so that $\{\mathbf{x}(u(\xi))|0\le\xi\le m\}$ is a "good" parameterization of the boundary curve $\mathbf{x}(\mathbf{u})$.

The task of finding $u(\xi)$ is of course equivalent to finding $\xi(u)$. Now let us define $\rho(u)\equiv\frac{d\xi}{du}$. Then

$$\xi(u)=\int_0^u \rho(w)dw$$

We see that finding ρ is equivalent to obtaining ξ. However, ρ is readily seen to be the desired *grid point density*, which can be dictated in a straightforward manner from physical considerations.

Indeed, physical considerations may guide the user to desire

1. Equal arc length spacing wherein points are spaced at equal distances in physical space. In this case, grid point density should be proportional to the *rate of change of arc length*. That is, $\rho \propto \|\mathbf{x}'\|$.
2. Curvature-weighted arc length spacing, wherein points are connected in areas of large curvature. In this case, we have

$$\rho \propto \kappa(u)\|\mathbf{x}'\|$$

where $\kappa(u)$ is the curvature of the boundary curve $\mathbf{x}(u)$ at u.

3. Grid attraction to an *attractor point* u^* in parametric space corresponding to a point $\mathbf{x}^* = \mathbf{x}(u^*)$ in physical space. A typical case is $u^* = 0$ or $u^* = 1$, when one has interesting physical phenomena (such as a Navier–Stokes boundary layer) at one end of the boundary curve. Or perhaps we might have $0 < u^* < 1$, with a point in the interior of the curve being of interest. In either case, a good choice for ρ is

$$\rho_u^*(u) \propto \frac{1}{\sqrt{\left(\kappa\left(u-u^*\right)\right)^2+1}}$$

where κ is a *strength factor* that determines the degree of attraction to u^*.

9.2.1.1 Hybrid Grid Density Functions

In practice, the user will likely desire a hybrid grid density function that is a linear combination of several other grid density functions. Assume we have grid density functions ρ_i, each normalized so that $\int_0^1 \rho_i\, du = \xi(1)-\xi(0)=m$. Then if we have positive constants λ_i such than $\Sigma\lambda_i = 1$, we have that $\rho = \Sigma\, \lambda_i\, \rho_i$ is a grid density function with suitable normalization. This hybrid density function will attempt to move grid points into regions where any one of the functions ρ_i desires grid points. Thus one could distribute grid points based on the hybrid criteria of arc length, curvature, and attraction to a set $\{u_i^*\}$ of distinct points.

This hybrid approach is the most useful, since it can accommodate many different situations that arise in practice. In this section, we will present an algorithm for grid point distribution along boundary curves based on a hybrid grid density function. The general principle of the algorithm is that (1) we construct $\rho(u)$ on a relatively fine grid of points $\tilde{u}_i = \frac{i}{m}, 0 \le i \le M,$ where M is 5–10 times m, (2) the *grid function* $\xi(u) = \int_0^u \rho(w)dw$ is evaluated by integrating ρ on the fine grid, and (3) the curve points $u(\xi)$ are generated in the parametric space of the curve by inverting the grid function $\xi(u)$. *Note*: Without computing $\xi(u)$ on a finer grid than that desired for $u(\xi)$, step (3) would be prone to inaccuracy, possibly leading to an unacceptable grid distribution.

Before we present the algorithm, we touch on a few technical points.

1. The grid density function for arc length is given by

$$\rho_s(u) = \frac{m\|\mathbf{x}'(u)\|}{\int_0^1 \|\mathbf{x}(w)\|\, dw}$$

Here $\frac{m}{\int_0^1 \|\mathbf{x}(w)\| dw}$ is the normalization required so that $\int_0^1 \rho_s\,(u)du = m$. If $u = \tilde{u}_i$ and $du = \tilde{u}_i - \tilde{u}_{i-1}$, we use the approximation.

$$\|\mathbf{x}'(\tilde{u}_i)\| du \approx \|\, \mathbf{x}(\tilde{u}_i) - \mathbf{x}(\tilde{u}_{i-1})\|$$

2. The grid density function for curvature-weighted arc length is

$$\rho_\kappa(u) = \frac{m\kappa(u)\|\mathbf{x}'(u)\|}{\int_0^1 \kappa(w)\|\mathbf{x}'(w)\| dw}$$

By definition $\kappa(u) = \frac{d\theta}{ds}$ where $d\theta$ is the angular change in the direction of the tangent of the curve during a small traversal of arc length ds along the curve. Thus

$$\kappa(u)\|\mathbf{x}'(u)\| = \frac{d\theta}{ds}\frac{ds}{du} = \frac{d\theta}{du}$$

If $u = \tilde{u}_i$ we use the approximation

$$\kappa(\tilde{u}_i)\|\mathbf{x}'(\tilde{u}_i)\| du \approx \theta_i - \theta_{i-1} = \|t_i - t_{i-1}\|$$

where $t_i \equiv \frac{\mathbf{x}'(\tilde{u}_i)}{\|\mathbf{x}'(\tilde{u}_i)\|}$ is the unit tangent vector to the physical curve at $\tilde{u}_i$ If the total integrated curvature $\int_0^1 \kappa(u)\|\mathbf{x}'(u)\| du \approx \sum_{i=1}^{M} \|\mathbf{t}_i - \mathbf{t}_{i-1}\|$ is less than some minimal angular tolerance (say $\varepsilon_\kappa =$ .01 radian), then we remove curvature weighted arc length as a criterion for grid point distribution and replace it with a simple arc length criterion. We do this to avoid distributing points based on a quantity which is essentially absent, which can lead to a nonsmooth distribution.

3. The grid density function for attraction (with strength k) to a point u^* is given by

$$\rho_{u^*}(u) = m\frac{1}{\sqrt{\left(k\left(u-u^*\right)\right)^2+1}} \bigg/ \int_0^1 \frac{1}{\left(k\left(w-u^*\right)\right)^2+1}dw$$

$$\int_0^u \rho_{u^*}(w)dw = m\frac{\operatorname{arcsinh}\left(k\left(u-u^*\right)\right)+\operatorname{arcsinh}\left(ku^*\right)}{\operatorname{arcsinh}\left(k\left(1-u^*\right)\right)+\operatorname{arcsinh}\left(ku^*\right)} \tag{9.1}$$

If $u^* = 0$, we have

$$\int_0^u \rho_0(w)dw = m\frac{\operatorname{arcsinh}(ku)}{\operatorname{arcsinh}(k)}$$

This leads to a grid distribution of the form

$$u(\xi) = \frac{\sinh\left(\frac{\alpha\xi}{m}\right)}{\sinh\alpha}$$

It has been noted that the smoothness of this distribution in the vicinity of $u^* = 0$ results in smaller truncation errors in finite difference discretizations than "exponential" distributions that approach the point of attraction in a more severe fashion, see Chapter 32 and Thompson et al. [28].

Algorithm 2.1 *Hybrid Curve Point Distribution Algorithm*

Assume physical curve $\mathbf{x}(u)$, $0 \le u \le 1$. Given weights $\lambda_s, \lambda_\kappa$, points $\{u_i^* \mid 0 \le u_i^* \le 1, 1 \le i \le p\}$, weights $\{\lambda_i \mid 1 \le i \le p\}$ and strengths $\{k_i \mid k_i \ge 0, 1 \le i \le p\}$ with $\lambda_s + \lambda_\kappa + \Sigma_{i=1}^p \lambda_i = 1$, we create a distribution of $m + 1$ points $u_0, u_1, \ldots, u_m$ that are simultaneously attracted to each of the points in $\{u_i^*\}$, placed in regions of high curvature, and placed to avoid large gaps in arc length. User also specifies a parametric grid size $M \le m$ and minimum integrated curvature tolerance ε_κ. (We suggest $M = 5m$ and $\varepsilon_\kappa = .01$)

1. Initialize grid function ξ to zero.

Do $i = 1,\ldots,M$

$\xi_i \leftarrow 0$

2. Compute arc lengths. Rescale so that maximum scaled arc length is m. Add to ξ, weighted by λ_s.

$s_0 \leftarrow 0$

Do $i = 1,\ldots,M$

$$s_i \leftarrow s_{i-1} + \left\|\mathbf{x}\left(\frac{i}{M}\right) - \mathbf{x}\left(\frac{i-1}{M}\right)\right\|$$

Do $i = 1,\ldots,M$

$$s_i \leftarrow m\frac{s_i}{s_M}$$

$$\xi_i \leftarrow \xi_i + \lambda_s s_i$$

3. Compute curvature weighted arc lengths on fine grid. Check if curve has nontrivial amount of curvature. If so, normalize to m, and add into ξ, weighted by λ_κ. Otherwise, use arc length instead.

$$
\begin{aligned}
&\text{Do } i = 0,\ldots M\\
&\qquad \mathbf{t}(i) \leftarrow \mathbf{x}'\left(\frac{i}{M}\right) \Big/ \left\| \mathbf{x}'\left(\frac{i}{M}\right) \right\|\\
&\qquad \theta_0 \leftarrow 0\\
&\text{Do } i = 1,\ldots M\\
&\qquad \theta_i \leftarrow \theta_{i-1} + \left\| \mathbf{t}(i) - \mathbf{t}(i-1) \right\|\\
&\text{If } \left(\theta_M \ge \varepsilon_\kappa\right) \text{ then}\\
&\quad \text{Do } i = 1,\ldots M\\
&\qquad \theta_i \leftarrow m\frac{\theta_i}{\theta_M}\\
&\qquad \xi_i \leftarrow \xi_i + \lambda_\kappa \theta_i\\
&\text{Else}\\
&\quad \text{Do } i = 1,\ldots M\\
&\qquad \xi_i \leftarrow \xi_i + \lambda_\kappa s_i
\end{aligned}
$$

4. Add in contributions to grid function due to attractor points.

$$
\begin{aligned}
&\text{Do } j = 1,\ldots,p\\
&\quad \text{Do } i = 1,\ldots,M\\
&\qquad \xi_i \leftarrow \xi_i + \lambda_j m \frac{\operatorname{arcsinh}\left(k_j\left(\frac{i}{M} - u_j^*\right)\right) + \operatorname{arcsinh}\left(k_j u_j^*\right)}{\operatorname{arcsinh}\left(k_j\left(1 - u_j^*\right)\right) + \operatorname{arcsinh}\left(k_j u_j^*\right)}
\end{aligned}
$$

5. Obtain point distribution by inverting grid function.

$$
\begin{aligned}
&\xi_M \leftarrow m \ \text{(Force final grid function value to be exactly } m,\text{)}\\
&u_0 \leftarrow 0\\
&j \leftarrow 1\\
&\text{Do } i = 1,\ldots,M\\
&\quad \text{Do while}\left(j \le \xi_i\right)\\
&\qquad u_j \leftarrow \frac{i}{M} - \frac{i}{M}\frac{\xi_i - j}{\xi_i - \xi_{i-1}} \ \left(\text{Obtain } \frac{i-1}{M} < u_j \le \frac{i}{M} \text{ using linear interpolation.}\right)\\
&\qquad j \leftarrow j + 1
\end{aligned}
$$

9.2.1.2 Determination of Weights $\lambda_s, \lambda_\kappa, \lambda_1$ and Strengths k_i

When using the boundary point distribution algorithm, one must choose weights $\lambda_s, \lambda_\kappa, \lambda_i$ and strengths k_i. As a rough guide, we find it is sufficient to set the weights for each desired criterion to be equal (and to add to 1). So for example, if we desire distribution on *arc length* and *two attractor points*, we would set $\lambda_s = \lambda_1 = \lambda_2 = \frac{1}{3}$. (In this case, we would set $\lambda_\kappa = 0$.)

As far as setting the strengths k_i on the attractor points u_i^*, one needs to consider the degree of concentration required by the particular application. We consider the case of a single attractor point $u^* = u_1^*$. From Eq. 9.1 we have that

$$p(u) = m\frac{k/\sqrt{\left(k\left(u-u^*\right)\right)^2+1}}{\operatorname{arcsinh}\left(k\left(1-u^*\right)\right)+\operatorname{arcsinh}\left(ku^*\right)}$$

So

$$\rho\left(u^*\right) = m\frac{k}{\operatorname{arcsinh}\left(k\left(1-u^*\right)\right)+\operatorname{arcsinh}\left(ku^*\right)} \geq m\frac{k}{2\operatorname{arcsinh}\left(\frac{k}{2}\right)} \tag{9.2}$$

Thus, for example, setting $k = 100$ would give us $\rho(u^*) \geq 10m$, which means that the grid lines are packed in the neighborhood of u^* at a density in excess of 10 times of the average grid density $\rho_{ave} = m$. Now suppose that the user is required to construct a grid with a specified value of $\rho(u^*)/m$ – that is, a specified excess grid density at the attractor u^*. As a rough guide, we recommend trying the heuristic

$$k = 15\,\frac{\rho\left(u^*\right)}{m} \tag{9.3}$$

and adjusting it as needed. Although one could solve the nonlinear Eq. 9.2 for k exactly, the presence of other criteria (such as arc length, curvature, or other attractor points) muddles the analysis, so that one in practice tries Eq. 9.3 and adjusts k as necessary.

If one desires a certain grid spacing Δx in the region near $\mathbf{x}^* = \mathbf{x}(u^*)$, we note that

$$\Delta x = \left\|\mathbf{x}_\xi\right\|\Big|_{u=u^*} = \left\|\mathbf{x}'\cdot u_\xi\right\|\Big|_{u=u^*} = \frac{\left\|\mathbf{x}'\left(u^*\right)\right\|}{\rho\left(u^*\right)}$$

Using Eq. 9.3, we conclude that

$$k = 15\frac{\left\|\mathbf{x}'\left(u^*\right)\right\|}{m\Delta x}$$

is a rough estimate for the strength k required to obtain a grid with the desired spacing Δx near the attractor $\mathbf{x}^* = \mathbf{x}(u^*)$ on the physical curve $\mathbf{x}(u(\xi))$.

9.2.2 Interpolation of Grid Points Between Boundary Curves

The second step in algebraic grid generation involves interpolation from the boundary curve distributions onto the interior of the surface. This is equivalent to finding the interior points in parametric space:

$$\left\{u(\xi,\eta),\quad v(\xi,\eta)\,\middle|\,0<\xi<m\quad 0<\eta<n\right\}$$

given that we know the boundary distributions in parametric space:

$$\{u(\xi,0),\quad u(\xi,n),\quad v(0,\eta),\quad v(m,\eta) | 0 \le \xi \le m \quad 0 \le \eta \le n\}$$

The technique for accomplishing this is called *transfinite interpolation* (Chapter 3), which generates an interpolated grid while matching all four boundaries at all points. When performing interpolation calculations, it is mathematically convenient to rescale the domain (ξ,η) space to be the unit square. We thus define

$$s(\xi,\eta) \equiv \xi/m \quad t(\xi,\eta) \equiv \eta/n$$

and our task is made equivalent to finding

$$\{u(s,t),\quad v(s,t) | 0 < s < 1 \quad 0 < t < 1\}$$

given that we know the boundary curves

$$\{u(s,0),\quad u(s,1),\quad v(0,t),\quad v(1,t) | 0 \le s \le 1 \quad 0 \le t \le 1 \tag{9.4}$$

As always, "*i,j*" will denote coordinates in computational space. So, for example $(\mathbf{u}_s)_{i,j}$ means du/ds evaluated at $\xi = i, \eta = j$, or equivalently at $s = s_i \equiv \frac{i}{m}, t = t_j \equiv \frac{j}{n}$.

Transfinite interpolation involves the sum of unidirectional interpolation in both the "*s*" and "*t*" directions, *minus* a tensor product interpolation that ensures the simultaneous matching of all four boundaries. Symbolically, this is written as

$$\mathbf{u}_{i,j} = \mathbf{u}^s_{i,j} + \mathbf{u}^t_{i,j} - \mathbf{u}^{st}_{i,j} \tag{9.5}$$

Here $\mathbf{u}^s_{i,j}$ is obtained by interpolation in *s* between the $\mathbf{u}_{0,j}$ and $\mathbf{u}_{m,j}$ and $\mathbf{u}^t_{i,j}$ is obtained by interpolation in *t* between $\mathbf{u}_{i,0}$ and $\mathbf{u}_{i,n}$. $\mathbf{u}^{st}_{i,j}$ is obtained by the composite operation of (1) interpolation in *t* between the four corners $\mathbf{u}_{0,0}, \mathbf{u}_{0,n}, \mathbf{u}_{m,0}, \mathbf{u}_{m,n}$ to produce *interpolated* $\mathbf{u}_{0,j}, \mathbf{u}_{m,j}$ values, and (2) interpolation in *s* between the *interpolated* $\mathbf{u}_{0,j}, \mathbf{u}_{m,j}$ values. (*Note:* It will be seen in the expressions that follow that the order of *s*- and *t*-interpolation in the evaluation of $\mathbf{u}^{st}_{i,j}$ could be interchanged with no change in the result.) In this section, we give explicit formula for two kinds of transfinite interpolation schemes corresponding to two different choices for the underlying unidirectional interpolation scheme.

Our first set of transfinite interpolation formulas assume that the underlying unidirectional interpolation scheme is simply *linear interpolation.* The formulas for this kind of interpolation are given by

$$\begin{aligned}
\mathbf{u}^s_{i,j} &= \begin{bmatrix} 1-s_i \\ s_i \end{bmatrix}^T \begin{bmatrix} \mathbf{u}_{0,j} \\ \mathbf{u}_{m,j} \end{bmatrix} \\
\mathbf{u}^t_{i,j} &= \begin{bmatrix} \mathbf{u}_{i,0} \\ \mathbf{u}_{i,n} \end{bmatrix}^T \begin{bmatrix} 1-t_j \\ t_j \end{bmatrix} \\
\mathbf{u}^{st}_{i,j} &= \begin{bmatrix} 1-s_i \\ s_i \end{bmatrix}^T \begin{bmatrix} \mathbf{u}_{0,0} & \mathbf{u}_{0,n} \\ \mathbf{u}_{m,0} & \mathbf{u}_{m,n} \end{bmatrix} \begin{bmatrix} 1-t_j \\ t_j \end{bmatrix}
\end{aligned} \tag{9.6}$$

The (u,v) values computed by the above formula may produce a surface grid suitable for many applications. However, it is possible that the grid might be unsuitable due to *nonorthogonality* of the grid lines. In this case, the grid is still suitable as a starting grid for elliptic smoothing iterations which can impose orthogonality of the grid lines at the boundaries.

Alternately, if the surface grid generated using Eq. 9.6 is unacceptable due to nonorthogonality at the boundaries, one may rectify the problem by using *Hermite cubic* transfinite interpolation. The formulas for this kind of interpolation allow the direct specification of derivatives at the boundaries, which means that orthogonality can be imposed.

Cubic Hermite transfinite interpolation is given by Eq. 9.5, where now

$$
\begin{aligned}
\mathbf{u}^{s}_{i,j} &= \begin{bmatrix} H_0^0(s_i) \\ H_0^1(s_i) \\ H_1^1(s_i) \\ H_1^0(s_i) \end{bmatrix}^T \begin{bmatrix} \mathbf{u}_{0,j} \\ (\mathbf{u}_s)_{0,j} \\ (\mathbf{u}_s)_{m,j} \\ \mathbf{u}_{m,j} \end{bmatrix} \\
\mathbf{u}^{t}_{i,j} &= \begin{bmatrix} \mathbf{u}_{i,0} \\ (\mathbf{u}_t)_{i,0} \\ (\mathbf{u}_t)_{i,n} \\ \mathbf{u}_{i,n} \end{bmatrix}^T \begin{bmatrix} H_0^0(t_j) \\ H_0^1(t_j) \\ H_1^1(t_j) \\ H_1^0(t_j) \end{bmatrix} \\
\mathbf{u}^{st}_{i,j} &= \begin{bmatrix} H_0^0(s_i) \\ H_0^1(s_i) \\ H_1^1(s_i) \\ H_1^0(s_i) \end{bmatrix}^T \begin{bmatrix} \mathbf{u}_{0,0} & (\mathbf{u}_t)_{0,0} & (\mathbf{u}_t)_{0,n} & \mathbf{u}_{0,n} \\ (\mathbf{u}_s)_{0,0} & (\mathbf{u}_{st})_{0,0} & (\mathbf{u}_{st})_{0,n} & (\mathbf{u}_s)_{0,n} \\ (\mathbf{u}_s)_{m,0} & (\mathbf{u}_{st})_{m,0} & (\mathbf{u}_{st})_{m,n} & (\mathbf{u}_s)_{m,n} \\ \mathbf{u}_{m,0} & (\mathbf{u}_t)_{m,0} & (\mathbf{u}_t)_{m,n} & \mathbf{u}_{m,n} \end{bmatrix} \begin{bmatrix} H_0^0(t_j) \\ H_0^1(t_j) \\ H_1^1(t_j) \\ H_1^0(t_j) \end{bmatrix}
\end{aligned}
\tag{9.7}
$$

Here

$$
\begin{aligned}
H_0^0\,(t) &= (t-1)^2\,(2t+1) \\
H_0^1\,(t) &= t^2(3-2t) \\
H_1^1\,(t) &= (t-1)^2 t \\
H_1^0\,(t) &= (t-1)t^2
\end{aligned}
$$

which obey the conditions

$$
\frac{d^{\alpha}\, H_{\beta'}^{\alpha'}}{dt^{\alpha}}(\beta t) = \delta_{\beta,\beta'}^{\alpha,\alpha'} \qquad \alpha, \alpha', \beta, \beta' \in \{0,1\}
$$

Note: The above expressions for $\mathbf{u}^s_{i,j}$, $\mathbf{u}^t_{i,j}$, $\mathbf{u}^{st}_{i,j}$ can be also used in the context of *surface generation,* rather than grid generation. In other words, by viewing $\mathbf{u}(s,t)$ as a mapping from parametric space to physical space, one could use these expressions to generate a surface patch that matches the specified physical boundary curves. This type of surface patch is known as a Coons patch, see Part III and Farin [9] and Yamaguchi [34].

The above formulas are not complete until we can supply the normal derivatives $\mathbf{u}_s$ at the left and right boundaries and $\mathbf{u}_t$ at the bottom and top boundaries. We also need the "twists" $\mathbf{u}_{st}$ at the four corners. It turns out that the assumption of orthogonality of grid lines at the boundaries in the physical domain will allow us to supply the normal derivatives in the parametric domain. The twists will then be chosen to be consistent with these normal derivatives.

We now consider computation of the normal derivative $\mathbf{u}_s$ at the left and right boundaries, and $\mathbf{u}_t$ at the top and bottom boundaries. The computation of these derivatives is equivalent to the computation of $\mathbf{u}_\xi$ and $\mathbf{u}_\eta$, since $\mathbf{u}_s = m\mathbf{u}_\xi$ and $\mathbf{u}_t = n\mathbf{u}_\eta$. To compute $\mathbf{u}_\xi$, at the left and right boundaries in parametric space, we first assume boundary orthogonality in physical space. That is, we assume

$$\mathbf{x}_\xi \cdot \mathbf{x}_\eta = 0$$

Thus,

$$\left(\mathbf{x}_u u_\xi + \mathbf{x}_v v_\xi\right)\cdot\left(\mathbf{x}_u u_\eta + \mathbf{x}_v v_\eta\right) = 0$$

Now on these boundaries we know that $u_\eta = 0$. We also know that $v_\eta \neq 0$ because the density of grid points on the boundaries is finite everywhere. Using this, we easily derive

$$\left(\mathbf{x}_u \cdot \mathbf{x}_v\right)u_\xi + \left(\mathbf{x}_v \cdot \mathbf{x}_v\right)v_\xi = 0$$

Denoting the *metric tensor components* by $\bar{g}_{11} = \mathbf{x}_u \cdot \mathbf{x}_u$, $\bar{g}_{12} = \mathbf{x}_u \cdot \mathbf{x}_v$ $\bar{g}_{22} = \mathbf{x}_v \cdot \mathbf{x}_v$, this is equivalently written as

$$\bar{g}_{12}\, u_\xi + \bar{g}_{22}\, v_\xi = 0 \tag{9.8}$$

This determines the normal derivatives $\mathbf{u}_\xi$ to within a constant. To determine the magnitudes of the derivatives, we need to add one more piece of data, which is the spacing off of the boundary:

$$\left\|\mathbf{x}_\xi\right\| = \sqrt{\bar{g}_{11}\, u_\xi^2 + 2\bar{g}_{12}\, u_\xi\, v_\xi + \bar{g}_{22}\, v_\xi^2}$$

We have found that a good spacing is obtained from linear transfinite interpolation as follows. We compute Eq. 9.5 using Eq. 9.6, denoting the normal derivatives computed at the boundary by

$$\mathbf{x}_\xi^0 = \mathbf{x}_u\, u_\xi^0 + \mathbf{x}_v\, v_\xi^0$$

Here, for the left boundary, $(u_\xi^0)_{0,j} = u_{1,j} - u_{0,j}$, where $u_{1,j}$ was computed by Eq. 9.5 and Eq. 9.6. For the right boundary $(u_\xi^0)_{m,j} = u_{m,j} - u_{m-1,j}$ where again $u_{m-1,j}$ was computed by Eq. 9.5 and Eq. 9.6. Now we specify that the new grid spacing $\|\mathbf{x}_\xi\|$ should be equal to $\mathbf{x}_\xi^0$ projected onto the orthogonal direction $\mathbf{x}_\xi / \|\mathbf{x}_\xi\|$ off the boundary. The idea is that the correct positions of the interior grid points in our final grid will be obtained by having the interior grid points of the linear TFI grid slide along the first interior grid line until they are in orthogonal position (see Figure 9.2). This condition is

$$\left\|\mathbf{x}_\xi\right\| = \mathbf{x}_\xi^0 \cdot \frac{\mathbf{x}_\xi}{\left\|\mathbf{x}_\xi\right\|} \tag{9.9}$$

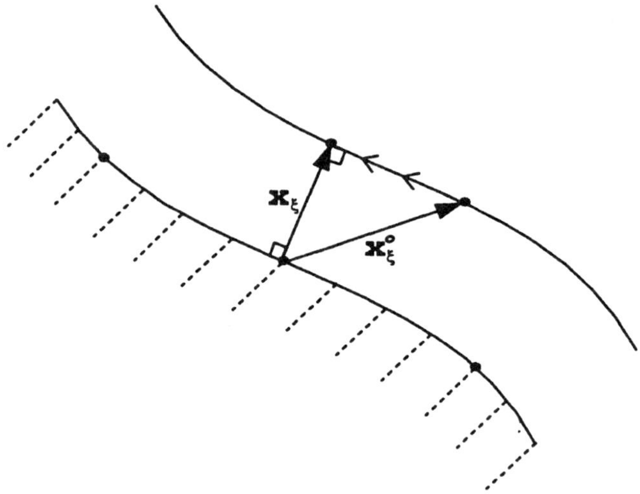

FIGURE 9.2 Derivation of grid spacing off boundary from linear TFI.

Solving both Eq. 9.8 and Eq. 9.9, we obtain

$$\mathbf{u}_\xi = \left(u^0_\xi, \quad -\frac{\bar{g}_{12}}{\bar{g}_{22}} u^0_\xi \right)$$

Using similar reasoning at the "bottom" and "top" boundaries, we obtain

$$\mathbf{u}_\eta = \left(-\frac{\bar{g}_{12}}{\bar{g}_{11}} v^0_\eta, \quad v^0_\eta \right)$$

Here, for the bottom boundary, $(u^\circ_\eta)_{i,0} = u_{i,1} - u_{i,0}$, where $u_{i,1}$ was computed by Eq. 9.5 and Eq. 9.6. For the top boundary, $(u^\circ_\xi)_{i,n} = u_{i,n} - u_{i,n-1}$ where again $u_{i,n-1}$ was computed by Eq. 9.5 and Eq. 9.6. Thus, the desired normal derivatives are given by

$$\begin{aligned} \mathbf{u}_s &= m\left(u^0_\xi, -\frac{\bar{g}_{12}}{\bar{g}_{22}} u^0_\xi \right) \\ \mathbf{u}_t &= n\left(-\frac{\bar{g}_{12}}{\bar{g}_{11}} v^0_\eta, v^0_\eta \right) \end{aligned} \tag{9.10}$$

Thus it appears that we can use substitution of Eq. 9.10 into Eq. 9.7 to obtain algebraic surface grids with perfectly orthogonal grid lines at the boundary. Unfortunately, our normal derivatives will in general not satisfy the following compatibility conditions:

$$\begin{aligned} &\lim_{t\to\beta} \mathbf{u}_s(\alpha,t) = \mathbf{u}_s(\alpha,\beta) \\ & \qquad\qquad\qquad\qquad \alpha,\beta \in \{0,1\} \\ &\lim_{s\to\alpha} \mathbf{u}_t(s,\beta) = \mathbf{u}_t(\alpha,\beta). \end{aligned} \tag{9.11}$$

This is because the right-hand side values are determined by the boundary data Eq. 9.4, while the left-hand side values are determined by the orthogonality conditions Eq. 9.10, and these can be very easily inconsistent.

Since Eq. 9.11 is violated, it is necessary to relax the orthogonality conditions in some vicinity of the corners. Although elliptic methods in the next section allow this vicinity to be quite small, algebraic methods are quite fragile, and so it is in practice best to impose exact orthogonality Eq. 9.10 only at the midpoint positions

$$(s,t)=\left(0,\frac{1}{2}\right),\ \left(1,\frac{1}{2}\right),\ \left(\frac{1}{2},0\right),\ \left(\frac{1}{2},1\right)$$

Normal derivatives between the midpoints and the corners are then computed using cubic Hermite interpolation. Thus for the derivatives along the "left" and the "right" boundaries,

$$\mathbf{u}_s(\alpha,t)=\begin{cases} H_1^0(2t)\mathbf{u}_s\left(\alpha,\frac{1}{2}\right)+H_0^0(2t)\mathbf{u}_s(\alpha,0) & 0<t<\frac{1}{2} \\ H_1^0(2-2t)\mathbf{u}_s\left(\alpha,\frac{1}{2}\right)+H_0^0(2-2t)\mathbf{u}_s(\alpha,1) & \frac{1}{2}<t<1 \end{cases} \qquad \alpha\in\{0,1\} \tag{9.12a}$$

Similarly, for the "top" and "bottom" boundaries, we have

$$\mathbf{u}_t(s,\beta)=\begin{cases} H_1^0(2s)\mathbf{u}_t\left(\frac{1}{2},\beta\right)+H_0^0(2s)\mathbf{u}_t(0,\beta) & 0<s<\frac{1}{2} \\ H_1^0(2-2s)\mathbf{u}_t\left(\frac{1}{2},\beta\right)+H_0^0(2-2s)\mathbf{u}_t(1,\beta) & \frac{1}{2}<s<1 \end{cases} \qquad \beta\in\{0,1\} \tag{9.12b}$$

Note: In case one or more of the four boundaries does not require orthogonality (e.g., the boundary is an internal boundary dividing two subsurface patches), we can use a Hermite interpolation scheme similar to Eq. 9.12 to interpolate *all* the derivatives on the curve. So for example, for the bottom curve, a purely interpolated (nonorthogonal) derivative would be

$$\mathbf{u}_t(s,0)=H_1^0(s)\mathbf{u}_t(1,0)+H_0^0(s)\mathbf{u}_t(0,0) \qquad 0<s<1$$

Violation of consistency conditions also causes problems for the twists $\mathbf{u}_{st}$, see Farin [9]. In general, neither the orthogonal derivatives Eq. 9.10 nor the interpolated derivatives Eq. 9.12 will satisfy

$$\lim_{s\to\alpha}\frac{\mathbf{u}_t(s,\beta)-\mathbf{u}_t(\alpha,\beta)}{s-\alpha}=\lim_{t\to\beta}\frac{\mathbf{u}_s(\alpha,t)-\mathbf{u}_s(\alpha,\beta)}{t-\beta} \qquad \alpha,\beta\in\{0,1\} \tag{9.13}$$

This means that the twists $\mathbf{u}_{st}(\alpha,\beta)$ are not necessarily well-defined. (Indeed, if Eq. 9.11 is also false, the one or both sides of Eq. 9.13 may be infinite!)

A practical resolution of this is to compute the twists $\mathbf{u}_{st}(\alpha,\beta)$ using a finite difference formula with a sufficiently large finite difference increment to "blur" the inconsistencies. For the twist $\mathbf{u}_{st}(0,0)$, such a formula is suggested by Figure 9.3. Here

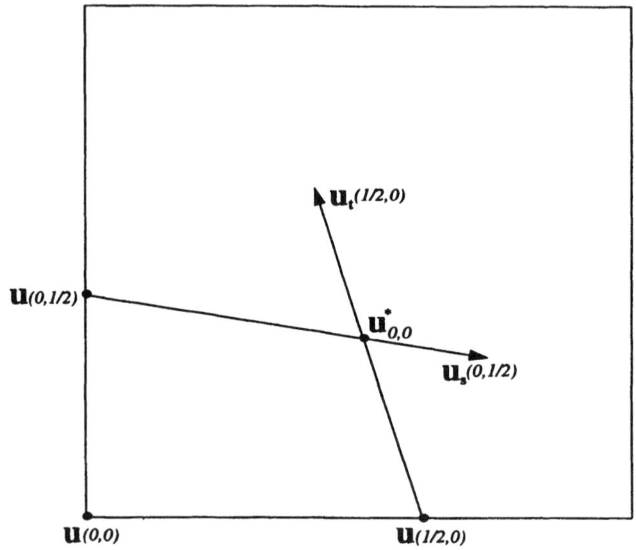

FIGURE 9.3 Heuristic scheme for computing a reasonable twist $\mathbf{u}_{st}(0,0)$.

$$\mathbf{u}_{st}(0,0) \approx \frac{\dfrac{\mathbf{u}^*_{0,0} - \mathbf{u}\left(\frac{1}{2},0\right)}{\frac{1}{2}} - \dfrac{\mathbf{u}\left(0,\frac{1}{2}\right) - \mathbf{u}(0,0)}{\frac{1}{2}}}{\frac{1}{2}} = 4\left(\mathbf{u}^*_{0,0} - \mathbf{u}\left(\frac{1}{2},0\right) - \mathbf{u}\left(0,\frac{1}{2}\right) + \mathbf{u}(0,0)\right)$$

where $\mathbf{u}^*_{0,0}$ is the intersection point between (1) the line with direction $\mathbf{u}_t(^1/_2,0)$ passing through $\mathbf{u}(^1/_2,0)$ and (2) the line with direction $\mathbf{u}_s(0,^1/_2)$ passing through $\mathbf{u}(0,^1/_2)$.

For the general twist $\mathbf{u}_{st}(\alpha,\beta)$, we thus use

$$\mathbf{u}_{st}(\alpha,\beta) = 4\left(\mathbf{u}^*_{\alpha,\beta} - \mathbf{u}\left(\frac{1}{2},\beta\right) - \mathbf{u}\left(\alpha,\frac{1}{2}\right) + \mathbf{u}(\alpha,\beta)\right)$$

where $\mathbf{u}^*_{\alpha,\beta}$ is the intersection point between (1) the line with direction $\mathbf{u}_t(^1/_2,\beta)$ passing through $\mathbf{u}(^1/_2,\beta)$ and (2) the line with direction, $\mathbf{u}_s(\alpha,^1/_2)$ passing through $\mathbf{u}(\alpha,^1/_2)$.

9.2.3 NURBS Surface Grid Generation Examples

After generation of a grid $\{\mathbf{u}_{ij} \mid 0 \le i \le m, 0 \le j \le n\}$ in parametric space, the actual grid in physical space is simply $\{\mathbf{x}(\mathbf{u}_{ij}) \mid 0 \le i \le m, 0 \le j \le n\}$. The examples in this chapter all utilize a NURBS surface representation (cf. Chapter 30):

$$\mathbf{x}(u,v) = \frac{\sum_{j=0}^{n}\sum_{i=0}^{m} \omega_{i,j}\, \mathbf{d}_{i,j}\, N_i^k(u) N_j^l(v)}{\sum_{j=0}^{n}\sum_{i=0}^{m} \omega_{i,j}\, N_i^k(u) N_j^l(v)}$$

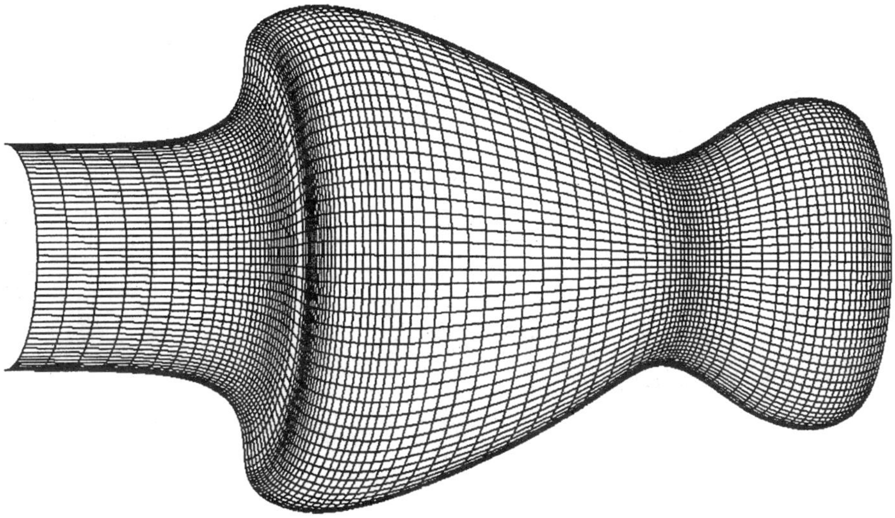

FIGURE 9.4 Linear TFI surface grid with boundary point distribution based on arc length and curvature.

defined by

- Two orders k and l,
- Control points $\mathbf{d}_{i,j} = (x_{i,j}, y_{i,j}, z_{i,j})$, $i = 0, \ldots, m, j = 0, \ldots, n$,
- Real weights $\omega_{i,j}, i = 0, \ldots, m, j = 0, \ldots, n$,
- A set of real u – knots, $\{u_0, \ldots, u_{m+k} \mid u_i \leq u_{i+1}, i = 0, \ldots, (m+k-1)\}$,
- A set of real v – knots, $\{v_0, \ldots, v_{n+l} \mid v_j \leq v_{j+1}, j = 0, \ldots, (n+l-1)\}$,
- B-spline basis functions $N_i^k(u)$, $u \in [u_i, u_{i+k}]$, $i = 0, \ldots, m$,
- B-spline basis functions $N_j^l(v)$, $v \in [v_j, v_{j+1}]$, $j = 0, \ldots, n$, and
- Surface segments $\mathbf{x}_{i,j}(u,v)$, $u \in [u_i, u_{i+1}]$, $i = (k-1), \ldots, m$, $v \in [v_j, v_{j+1}]$, $j = (l-1), \ldots, n$.

The advantage of using a NURBS-based geometry definition is the ability to represent both standard analytic shapes (e.g., conics, quadrics, surfaces of revolution, etc.) and free-form curves and surfaces. Therefore, both analytic and free-form shapes are represented precisely, and a unified database can store both. Another potential advantage of using NURBS is the fact that positional as well as derivative information of surfaces can be evaluated analytically. For the use of NURBS in grid generation we refer to Khamayseh and Hamann [11]. For a detailed discussion of B-spline and NURBS curves and surfaces we refer the reader to Bartels et al. [3], de Boor [8], Farin [9], and Piegl [16]. We also refer the reader to Part III on CAGD techniques for surface grids.

In our first example (Figure 9.4), we use linear TFI to generate a surface grid on a portion of a surface of revolution. The boundary point distribution on these curves was generated by using Algorithm 2.1 with $\lambda_s = \lambda_\kappa = 1/2$. That is, the points are distributed equally according to both arc length and curvature considerations. The effect of curvature distribution is clearly seen: boundary grid points are clustered in areas of high curvature. The fact that arc length is still considered to some degree is seen in the fact that a nonzero density of grid points is still distributed where curvature is small or absent. The linear TFI uniformly propagates these boundary distributions into the interior of the grid.

In the next example (Figure 9.5), we again use linear TFI to generate a grid on a similar surface of revolution. However, in addition to distribution on arc length and curvature, we instruct Algorithm 2.1

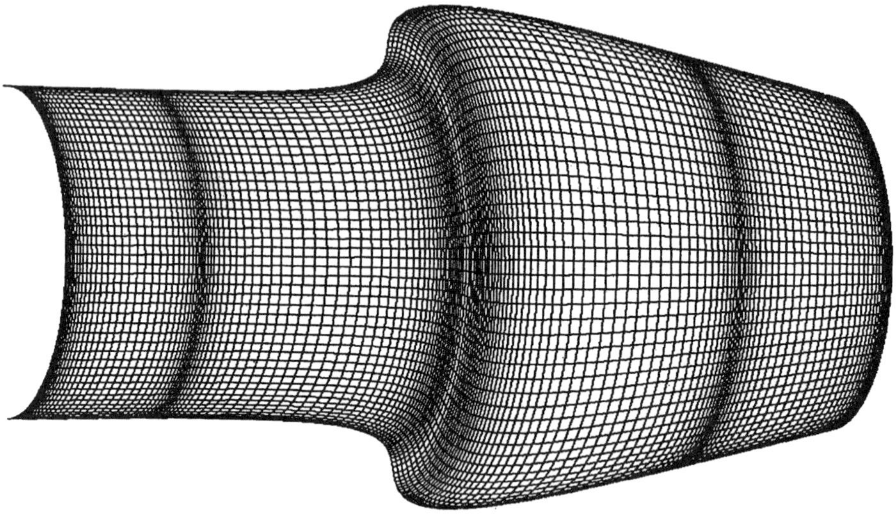

FIGURE 9.5 Linear TFI surface grid with boundary point distribution based on arc length, curvature, and four attractor points.

to heed the influence of four attractor points on the "top" and "bottom" boundary curves. These attractors can be seen to be at both endpoints and at two interior points. (The concentration of the grid at the center, however, is not due to any attractor, but is due to the physical curvature of the surface.) The parameters used for the top and bottom curves were $\lambda_s = \lambda_\kappa = \lambda_1 = \lambda_2 = \lambda_3 = \lambda_4 = 1/6, k_1 = k_2 = k_3 = k_4 = 120$, and $u_1^* = 0, u_2^* = .1, u_3^* = .9$, and $u_4^* = 1$. By Eq. 9.3, $k_i = 120$ implies that the grid should be packed in the neighborhood of the attractors u_i^* at a density approximately 8 times higher than the average grid line density. This is consistent with the appearance of Figure 9.5.

The final example of algebraic surface grid generation in this section (Figure 9.6) uses cubic Hermite TFI in conjunction with uniform arc length boundary point distribution. Orthogonality at the boundaries is clearly visible on this surface. However, boundary orthogonality can easily cause cubic Hermite grids to "fold" in the interior on more challenging geometries. In practice, a more robust approach to enforcing boundary orthogonality is to generate an initial linear TFI grid and then use it as a starting grid for the elliptic grid generation system described in the next section.

9.3 Elliptic Surface Grid Generation

Elliptic grid generation is a technique of smoothing an initial (usually algebraic) mesh to improve grid quality. Grid improvement may involve forcing grid line orthogonality, forcing smooth grading of cell sizes, etc. What makes elliptic grid generation challenging is that grid smoothing must always ensure that the resulting grid points stay on the surface. With this constraint in mind, the efficiency approach of constructing a smooth grid is to work in the parametric space rather than on the physical surface. However, there are some disadvantages associated with this approach. The differential equations become more complicated and contain two sets of derivatives, the derivatives of the physical variables with respect to the parametric variables $(x_u, x_v, y_u, y_v, z_u, z_v, x_{uu}, x_{uv}, x_{vv}, \ldots)$ and the derivatives of the parametric variables with respect to the computational variables $(u_\xi, u_\eta, v_\xi, v_\eta, u_{\xi\xi}, u_{\xi\eta}, u_{\eta\eta}, \ldots)$.

The elliptic system may preserve the original distribution of grid points or redistribute points based upon the choice of the *control functions* that are commonly used in adaptive grid generation. The control functions are evaluated either directly from the initial algebraic grid, or by interpolation from the boundary point distributions and then smoothed. Orthogonality of the grid may be imposed along certain boundary components of the physical region. Boundary orthogonality can be achieved through Neumann boundary conditions, which allow the boundary points to float along the boundary of the

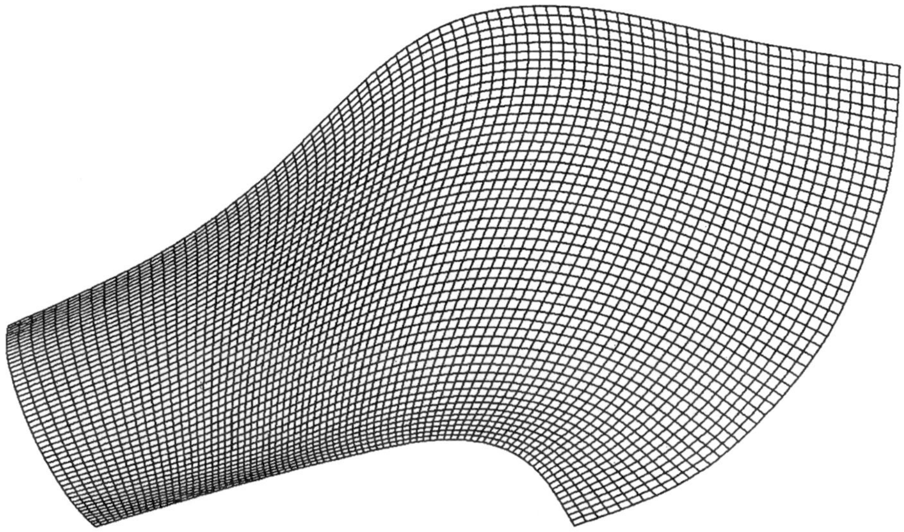

FIGURE 9.6 Cubic Hermite TFI surface grid with uniform arc length boundary point distribution.

surface. Alternatively, the control functions can be determined to provide orthogonality at boundaries with specified normal spacing.

The use of elliptic models to generate curvilinear coordinates is quite popular, see Chapter 4 and Thompson et al. [28]. Since elliptic partial differential equations determine a function in terms of its values on the entire closed boundary of a region, such a system can be used to generate the interior values of a surface grid from the values on the sides. An important property is the inherent smoothness in the solutions of elliptic systems. As a consequent of smoothing, slope discontinuities on the boundaries are not propagated into the field.

Early progress on the generation of surface grids using elliptic methods was made by Takagi et al. [25], Warsi [31], and Whitney and Thomas [32]. The elliptic grid generation system and the surface equations obtained by Warsi [30, 31] were based on the *fundamental theory of surfaces* from differential geometry, which says that for any surface, the surface coordinates must satisfy the formulas of Gauss and Weingarten, see Struik [24]. On the other hand, the same generation system was derived by Takagi et al. [25] and Whitney and Thomas [32] based on Poisson's differential equation in three dimensions.

Distinct from the previous two approaches to deriving a system of elliptic equations for grid generation, there is the approach based on conformal mappings. Mastin [15], Thompson et al. [27], and Winslow [33] developed elliptic generation systems based on conformal transformations between the physical and computational regions. Planar two-dimensional smooth orthogonal boundary-fitted grids were produced using these techniques. These methods have been extended by Khamayseh and Mastin [12] to develop the analogous elliptic grid generation system for surfaces.

In this section we present the derivation of the standard elliptic surface grid generation system using the theory of conformal mappings. First, conformal mapping of smooth surfaces onto rectangular regions is utilized to derive a first-order system of partial differential equations analogous to Beltrami's system for quasi-conformal mapping of planar regions. Second, the usual elliptic generation system for three-dimensional surfaces, including *source terms,* is formulated based on Beltrami's system and quasi-conformal mapping. We conclude this section with a detailed description of how the elliptic grid generation system is implemented numerically.

9.3.1 Conformal Mapping on Surfaces

A surface grid generated by the conformal mapping of a rectangle onto the surface is orthogonal and has a constant aspect ratio. These two conditions can be expressed mathematically as the system of equations

$$\begin{aligned} \mathbf{x}_\xi \cdot \mathbf{x}_\eta &= 0 \\ F\left\|\mathbf{x}_\xi\right\| &= \left\|\mathbf{x}_\eta\right\| \end{aligned} \tag{9.14}$$

where F is the grid aspect ratio. These two equations can be rewritten as

$$\begin{aligned} & x_\xi x_\eta + y_\xi y_\eta + z_\xi z_\eta = 0 \\ & F^2\left(x_\xi^2 + y_\xi^2 + z_\xi^2\right) = x_\eta^2 + y_\eta^2 + z_\eta^2 \end{aligned}$$

Using the chain rule for differentiation, the physical derivatives are expanded as

$$\begin{aligned} \mathbf{x}_\xi &= \mathbf{x}_u u_\xi + \mathbf{x}_v v_\xi \\ \mathbf{x}_\eta &= \mathbf{x}_u u_\eta + \mathbf{x}_v v_\eta \\ \mathbf{x}_{\xi\xi} &= \mathbf{x}_{uu} u_\xi^2 + 2\mathbf{x}_{uv} u_\xi v_\xi + \mathbf{x}_{vv} v_\xi^2 + \mathbf{x}_u u_{\xi\xi} + \mathbf{x}_v v_{\xi\xi} \\ \mathbf{x}_{\xi\eta} &= \mathbf{x}_{uu} u_\xi u_\eta + \mathbf{x}_{uv}\left(u_\xi v_\eta + u_\eta v_\xi\right) + \mathbf{x}_{vv} v_\xi v_\eta + \mathbf{x}_u u_{\xi\eta} + \mathbf{x}_v v_{\xi\eta} \quad \text{and} \\ \mathbf{x}_{\eta\eta} &= \mathbf{x}_{uu} u_\eta^2 + 2\mathbf{x}_{uv} u_\eta v_\eta + \mathbf{x}_{vv} v_\eta^2 + \mathbf{x}_u u_{\eta\eta} + \mathbf{x}_v v_{\eta\eta} \end{aligned}$$

Thus, the above system is equivalent to

$$\begin{aligned} & x_u^2 u_\xi u_\eta + x_u x_v\left(u_\xi v_\eta + u_\eta v_\xi\right) + x_v^2 v_\xi v_\eta \\ & + y_u^2 u_\xi u_\eta + y_u y_v\left(u_\xi v_\eta + u_\eta v_\xi\right) + y_v^2 v_\xi v_\eta \\ & + z_u^2 u_\xi u_\eta + z_u z_v\left(u_\xi v_\eta + u_\eta v_\xi\right) + z_v^2 v_\xi v_\eta = 0 \end{aligned}$$

and

$$\begin{aligned} & F^2(x_u^2 u_\xi^2 + 2x_u x_v u_\xi v_\xi + x_v^2 v_\xi^2 \\ & + y_u^2 u_\xi^2 + 2y_u y_v u_\xi v_\xi + y_v^2 v_\xi^2 \\ & + z_u^2 u_\xi^2 + 2z_u z_v u_\xi v_\xi + z_v^2 v_\xi^2) \\ & = x_u^2 u_\eta^2 + 2x_u x_v u_\eta v_\eta + x_v^2 v_\eta^2 \\ & + y_u^2 u_\eta^2 + 2y_u y_v u_\eta v_\eta + y_v^2 v_\eta^2 \\ & + z_u^2 u_\eta^2 + 2z_u z_v u_\eta v_\eta + z_v^2 v_\eta^2. \end{aligned}$$

The above equations are combined to give the complex equation

$$\begin{aligned} & \left(x_u^2 + y_u^2 + z_u^2\right)\left(Fu_\xi + iu_\eta\right)^2 \\ & + 2\left(x_u x_v + y_u y_v + z_u z_v\right)\left(Fu_\xi + iu_\eta\right)\left(Fv_\xi + iv_\eta\right) \\ & + \left(x_v^2 + y_v^2 + z_v^2\right)\left(Fv_\xi + iv_\eta\right)^2 = 0 \end{aligned}$$

This equation can be put into a compact form:

$$\bar{g}_{11} Z^2 + 2\bar{g}_{12} ZW + \bar{g}_{22} W^2 = 0 \tag{9.15}$$

where

$$\begin{aligned}
\bar{g}_{11} &= \mathbf{x}_u \cdot \mathbf{x}_u = x_u^2 + y_u^2 + z_u^2 \\
\bar{g}_{12} &= \mathbf{x}_u \cdot \mathbf{x}_v = x_u x_v + y_u y_v + z_u z_v \\
\bar{g}_{22} &= \mathbf{x}_v \cdot \mathbf{x}_v = x_v^2 + y_v^2 + z_v^2 \\
Z &= Fu_\xi + iu_\eta \quad \text{and} \quad W = Fv_\xi + iv_\eta
\end{aligned} \tag{9.16}$$

Solving the quadratic Eq. 9.15 either for Z or W, say Z, we have

$$Z = \frac{-\bar{g}_{12} \pm \sqrt{\bar{g}_{12}^2 - \bar{g}_{11}\bar{g}_{22}}}{\bar{g}_{11}} W$$

or in terms of u and v,

$$Fu_\xi + iu_\eta = \frac{-\bar{g}_{12} \pm i\bar{J}}{\bar{g}_{11}} \left(Fv_\xi + iv_\eta \right)$$

where $\bar{J} = \sqrt{\bar{g}}$ and $\bar{g} = \bar{g}_{11}\bar{g}_{22} - \bar{g}_{12}^2$ is the Jacobian of the mapping from the parametric space to the surface. We equate the real and the imaginary parts of the above equation to obtain

$$\begin{aligned}
Fu_\xi &= -\frac{\bar{g}_{12}}{\bar{g}_{11}} Fv_\xi \pm \frac{\bar{J}}{\bar{g}_{11}} v_\eta \\
u_\eta &= \pm \frac{\bar{J}}{\bar{g}_{11}} Fv_\xi - \frac{\bar{g}_{12}}{\bar{g}_{11}} v_\eta
\end{aligned}$$

The above system of equations can be expressed in the form of a first-order elliptic system:

$$\begin{aligned}
Fu_\xi &= av_\eta - bu_\eta \\
Fv_\xi &= bv_\eta - cu_\eta
\end{aligned} \tag{9.17}$$

where

$$\begin{aligned}
a &= -\frac{\bar{g}_{22}}{\pm\bar{J}} \\
b &= \frac{\bar{g}_{12}}{\pm\bar{J}} \quad \text{and} \\
c &= -\frac{\bar{g}_{11}}{\pm\bar{J}}
\end{aligned}$$

Note that $ac - b^2 = 1$ which is sufficient for ellipticity. The sign $\pm$ needs to be chosen such that the Jacobian

$$J = u_\xi v_\eta - u_\eta v_\xi > 0$$

We have that by definition $\bar{g}_{11} \geq 0$ and $\bar{g}_{22} \geq 0$, so choosing the negative sign will make $a \geq 0$ and $c \geq 0$. From the system Eq. 9.17, we see that $FJ = F(u_\xi v_\eta - u_\eta v_\xi) = av_\eta^2 - 2bu_\eta v_\eta + cu_\eta^2$. Noting $b^2 = ac - 1$ implies that $b = \sqrt{ac-1} < \sqrt{ac}$, we have that $FJ > av_\eta^2 - 2\sqrt{ac}u_\eta v_\eta + cu^2{}_\eta = (\sqrt{a}v_\eta - \sqrt{c}u_\eta)^2 \geq 0$ and hence $J > 0$.

9.3.2 Formulation of the Elliptic Generator

In this subsection, we will see that conformal mappings on surfaces produce an elliptic system equivalent to that produced by quasi-conformal mappings of planar regions. The inhomogeneous form of this system will be our elliptic grid generator for surfaces.

A quasi-conformal mapping is a homeomorphism:

$$\psi(u,v) = \xi(u,v) + i\eta(u,v)$$

that maps the (u,v) space onto (ξ,η) space so that the real and the imaginary parts of ψ satisfy Beltrami's system of equations:

$$\begin{aligned} M\eta_v &= p\xi_u + q\xi_v \\ -M\eta_u &= q\xi_u + r\xi_v \end{aligned} \tag{9.18}$$

where p, q, and r are functions of u and v with $p,r > 0$ and satisfy the equation $pr - q^2 = 1$. The quasi-conformal quantity M is invariant and often referred to as the module or the aspect ratio of the region of consideration. For further study of the theory and application of quasi-conformal mappings, we refer to Ahlfors [2] and Renelt [17].

It is the system Eq. 9.18 that forms the basis of general elliptic grid generation for the planar two-dimensional case, see Mastin and Thompson [14]. An earlier approach was proposed by Belinskii et al. [4] and Godunov and Prokopov [10] to handle the problem of quasi-conformal mappings to construct curvilinear grids.

In fact, Eq. 9.18 forms the basis of a general elliptic grid generator for surfaces as well. We first invert the system Eq. 9.17 so that the computational variables (ξ,η) are the dependent variables and the parametric variables (u,v) become the independent variables.

Assume that ξ and η are twice continuously differentiable and the Jacobian of the inverse transformation $J = u_\xi v_\eta - u_\eta v_\xi$ is nonvanishing in the region under consideration. Then the metrics $(u_\xi, u_\eta, v_\xi, v_\eta)$ and $(\xi_u, \xi_v, \eta_u, \eta_v)$ are uniquely related by the following:

$$\begin{aligned} \xi_u &= \frac{v_\eta}{J} \qquad \xi_v = -\frac{u_\eta}{J} \quad \text{and} \\ \eta_u &= -\frac{v_\xi}{J} \qquad \eta_v = \frac{u_\xi}{J} \end{aligned} \tag{9.19}$$

Using these quantities the system Eq. 9.17 so that the parametric variables become the independent variables, the system can be expressed either in the form

$$\begin{aligned} F\eta_v &= a\xi_u + b\xi_v \\ -F\eta_u &= b\xi_u + c\xi_v \end{aligned} \tag{9.20}$$

or

$$\begin{aligned} \xi_u &= F(c\eta_v + b\eta_u) \\ -\xi_v &= F(a\eta_u + b\eta_v) \end{aligned} \tag{9.21}$$

These first-order elliptic systems (which represent conformal mapping of a parametric surface onto a square) are thus in the form of Beltrami's system of equations for quasi-conformal mapping of planar regions.

The elliptic system of equations actually used for surface grid generation is a straight-forward generalization of the above systems. Indeed these systems are equivalent to the following uncoupled second-order elliptic system:

$$a\xi_{uu} + 2b\xi_{uv} + c\xi_{vv} + (a_u + b_v)\xi_u + (b_u + c_v)\xi_u = 0$$
$$a\eta_{uu} + 2b\eta_{uv} + c\eta_{vv} + (a_u + b_v)\eta_u + (b_u + c_v)\eta_v = 0$$

This implies ξ and η are solutions of the following second-order linear elliptic system with $\Phi = \Psi = 0$:

$$\begin{aligned} \bar{g}_{22}\xi_{uu} - 2\bar{g}_{12}\xi_{uv} + \bar{g}_{11}\xi_{vv} + (\Delta_2 u)\xi_u + (\Delta_2 v)\xi_v &= \Phi \\ \bar{g}_{22}\eta_{uu} - 2\bar{g}_{12}\eta_{uv} + \bar{g}_{11}\eta_{vv} + (\Delta_2 u)\eta_u + (\Delta_2 v)\eta_v &= \psi \end{aligned} \tag{9.22}$$

where $\Delta_2 u$ and $\Delta_2 v$ are defined by

$$\Delta_2 u = \bar{J}(a_u + b_v) = \bar{J}\left[\frac{\partial}{\partial u}\left(\frac{\bar{g}_{22}}{\bar{J}}\right) - \frac{\partial}{\partial v}\left(\frac{\bar{g}_{12}}{\bar{J}}\right)\right]$$
$$\Delta_2 v = \bar{J}(b_u + c_v) = \bar{J}\left[\frac{\partial}{\partial v}\left(\frac{\bar{g}_{11}}{\bar{J}}\right) - \frac{\partial}{\partial u}\left(\frac{\bar{g}_{12}}{\bar{J}}\right)\right]$$

It is this system which forms the basis of the elliptic methods for generating surface grids. The source terms (or control functions), Φ and Ψ, are added to allow control over the distribution of grid points on the surface. In the computation of a surface grid, the points in the computational space are given and the points in the parametric space must be computed. Therefore, in an implementation of a numerical grid generation scheme, it is convenient to interchange variables again so that the computational variable ξ and η are the independent variables. Introducing Eq. 9.19 in Eq. 9.22, the transformation is reduced to the following system of equations:

$$\begin{aligned} -Au_\xi + Bv_\xi &= \psi \\ Au_\eta - Bv_\eta &= \Phi \end{aligned} \tag{9.23}$$

where

$$A = \frac{g_{22}\,v_{\xi\xi} - 2g_{12}\,v_{\xi\eta} + g_{11}\,v_{\eta\eta}}{\bar{J}J^3} - \frac{\Delta_2 v}{\bar{J}J}$$
$$B = \frac{g_{22}\,u_{\xi\xi} - 2g_{12}\,u_{\xi\eta} + g_{11}\,u_{\eta\eta}}{\bar{J}J^3} - \frac{\Delta_2 u}{\bar{J}J}$$
$$g_{11} = \mathbf{x}_\xi \cdot \mathbf{x}_\xi = \bar{g}_{11}\,u_\xi^2 + 2\bar{g}_{12}\,u_\xi\,v_\xi + \bar{g}_{22}\,v_\xi^2$$
$$g_{12} = \mathbf{x}_\xi \cdot \mathbf{x}_\eta = \bar{g}_{11}\,u_\xi\,u_\eta + \bar{g}_{12}(u_\xi\,v_\eta + u_\eta\,v_\xi) + \bar{g}_{22}\,v_\xi\,v_\eta \quad \text{and}$$
$$g_{22} = \mathbf{x}_\eta \cdot \mathbf{x}_\eta = \bar{g}_{11}\,u_\eta^2 + 2\bar{g}_{12}\,u_\eta\,v_\eta + \bar{g}_{22}\,v_\eta^2$$

Solving the system Eq. 9.23 for A and B, we have

$$A = -\frac{1}{J}\left[\Phi v_\xi + \psi v_\eta\right] \quad \text{and}$$
$$\mathrm{B} = -\frac{1}{\mathrm{J}}\left[\Phi u_\xi + \psi u_\eta\right]$$

From the above equations, we see that u and v are solutions of the following quasi-linear elliptic system:

$$g_{22}\left(u_{\xi\xi} + Pu_\xi\right) - 2g_{12}\, u_{\xi\eta} + g_{11}\left(u_{\eta\eta} + Qu_\eta\right) = J^2\Delta_2 u \tag{9.24a}$$

$$g_{22}\left(v_{\xi\xi} + Pv_\xi\right) - 2g_{12}\, v_{\xi\eta} + g_{11}\left(v_{\eta\eta} + Qv_\eta\right) = J^2\Delta_2 v \tag{9.24b}$$

where

$$P = \frac{\bar{J}J^2}{g_{22}}\Phi \quad \text{and}$$
$$Q = \frac{\bar{J}J^2}{g_{11}}\psi$$

We thus have completed our derivation of the standard elliptic generation system Eq. 9.24 from the conformal mapping conditions for surfaces Eq. 9.14. This system is solved for the parametric functions $u(\xi,\eta)$ and $v(\xi,\eta)$ at the grid points using the techniques of the next section.

Note that if $x \equiv u,\ y \equiv v,\ z \equiv 0$, then $\bar{g}_{11} = 1, \bar{g}_{12} = 0, \bar{g}_{22} = 1, \bar{J} = 1$, and $\Delta_2 u = \Delta_2 v = 0$, making the generation system identical to the well-known homogeneous elliptic system for planar grid generation presented in Thompson et al. [28].

9.3.3 Numerical Implementation

In this subsection, we deal with the numerical discretization and implementation of the elliptic generation system derived in this chapter. We first examine the basic concept of finite difference approximation, and the derivation of the difference schemes for the elliptic equations. Later we present the effect and the methodology of computing control functions in elliptic surface grid generation.

We begin our discussion of finite difference schemes for the elliptic generation system Eq. 9.24. The basic idea of finite difference schemes is to replace derivatives by finite differences. As before, $u_{i,j}$ denotes $u(\xi,\eta)$ evaluated at the $\xi = i$, $\eta = j$ grid point, and similarly for $v_{i,j}$. The first derivatives are computed using difference approximations of the form

$$\frac{\partial u}{\partial \xi}(\xi,\eta) \approx \frac{u_{i+1,j} - u_{i,j}}{\Delta\xi}$$
$$\frac{\partial u}{\partial \xi}(\xi,\eta) \approx \frac{u_{i,j} - u_{i-1,j}}{\Delta\xi}$$
$$\frac{\partial u}{\partial \xi}(\xi,\eta) \approx \frac{u_{i+1,j} - u_{i-1,j}}{2(\Delta\xi)}$$

where $\Delta\xi$ is the computational grid spacing in the ξ-direction. The above discretizations are known as forward, backward, and central differences, respectively. The second derivatives are approximated with central difference expressions of the form

$$\frac{\partial^2 u}{\partial \xi^2}(\xi,\eta) \approx \frac{u_{i+1,j} - 2u_{i,j} + u_{i-1,j}}{(\Delta\xi)^2}$$

and expressions of the form

$$\frac{\partial^2 u}{\partial \xi \partial \eta}(\xi,\eta) \approx \frac{u_{i+1,j+1} - u_{i-1,j+1} - u_{i+1,j-1} + u_{i-1,j-1}}{4(\Delta\xi)(\Delta\eta)}$$

for the mixed partial derivatives. Now we apply central difference discretization to approximate the solution of the elliptic system Eq. 9.24 for $u_{i,j}$ and $v_{i,j}$. Knowing that $\Delta\xi = \Delta\eta = 1$, we obtain the following finite difference schemes:

$$\begin{aligned} &g_{22}\left(u_{i+1,j} - 2u_{i,j} + u_{i-1,j}\right) + \frac{g_{22}P}{2}\left(u_{i+1,j} - u_{i-1,j}\right) + \\ &g_{11}\left(u_{i,j+1} - 2u_{i,j} + u_{i,j-1}\right) + \frac{g_{11}Q}{2}\left(u_{i,j+1} - u_{i,j-1}\right) = \\ &\frac{g_{12}}{2}\left(u_{i+1,j+1} - u_{i-1,j+1} - u_{i+1,j-1} + u_{i-1,j-1}\right) + J^2\Delta_2 u \end{aligned} \tag{9.25a}$$

$$\begin{aligned} &g_{22}\left(v_{i+1,j} - 2v_{i,j} + v_{i-1,j}\right) + \frac{g_{22}P}{2}\left(v_{i+1,j} - v_{i-1,j}\right) + \\ &g_{11}\left(v_{i,j+1} - 2v_{i,j} + v_{i,j-1}\right) + \frac{g_{11}Q}{2}\left(v_{i,j+1} - v_{i,j-1}\right) = \\ &\frac{g_{12}}{2}\left(v_{i+1,j+1} - v_{i-1,j+1} - v_{i+1,j-1} + v_{i-1,j-1}\right) + J^2\Delta_2 v \end{aligned} \tag{9.25b}$$

The quantities $g_{i,j}$, J, $\Delta_2 u$, and $\Delta_2 v$ in the difference equations involve two types of approximations. The derivative of the parametric variables with respect to the computational variables are approximated using finite difference approximation, whereas the derivative terms of the physical variables with respect to the parametric variables are computed analytically from the surface definition **x**(**u**). For ease of notation, quantities with subscripts omitted are assumed evaluated at (i,j), so that for example $g_{11} = (g_{11})_{i,j} = g_{11}(u_{i,j}, v_{i,j})$.

As a convenience, we present the expanded forms of $\Delta_2 u$ and $\Delta_2 v$, which must be evaluated in the numerical scheme

$$\Delta_2 u = \frac{1}{\bar{J}}\left\{\bar{J}\left[(\bar{g}_{22})_u - (\bar{g}_{12})_v\right] - \left[\bar{g}_{22}(\bar{J})_u - \bar{g}_{12}(\bar{J})_v\right]\right\}$$

$$\Delta_2 v = \frac{1}{\bar{J}}\left\{\bar{J}\left[(\bar{g}_{11})_v - (\bar{g}_{12})_u\right] - \left[\bar{g}_{11}(\bar{J})_v - \bar{g}_{12}(\bar{J})_u\right]\right\}$$

with

$$(\bar{J})_u = \frac{1}{2\bar{J}}\left[\bar{g}_{11}(\bar{g}_{22})_u + \bar{g}_{22}(\bar{g}_{11})_u - 2\bar{g}_{12}(\bar{g}_{12})_u\right]$$

$$(\bar{J})_v = \frac{1}{2\bar{J}}\left[\bar{g}_{11}(\bar{g}_{22})_v + \bar{g}_{22}(\bar{g}_{11})_v - 2\bar{g}_{12}(\bar{g}_{12})_v\right]$$

$$\left(\bar{g}_{11}\right)_u = 2\mathbf{x}_u \cdot \mathbf{x}_{uu} ,$$
$$\left(\bar{g}_{11}\right)_v = 2\mathbf{x}_u \cdot \mathbf{x}_{uv} ,$$
$$\left(\bar{g}_{22}\right)_u = 2\mathbf{x}_v \cdot \mathbf{x}_{uv} ,$$
$$\left(\bar{g}_{22}\right)_u = 2\mathbf{x}_v \cdot \mathbf{x}_{vv} ,$$
$$\left(\bar{g}_{12}\right)_u = \mathbf{x}_u \cdot \mathbf{x}_{uv} + \mathbf{x}_v \cdot \mathbf{x}_{uu}, \quad \text{and}$$
$$\left(\bar{g}_{12}\right)_v = \mathbf{x}_u \cdot \mathbf{x}_{vv} + \mathbf{x}_v \cdot \mathbf{x}_{uv} .$$

Now we consider the iterative method known as successive overrelaxation (SOR) to solve the elliptic generation system 9.24. This method is relatively easy to implement and requires little extra computer storage when we use the Gauss–Seidel methodology of *immediate replacement* of the "old" values by the "new" values at each iteration. For these reasons, this technique is very widely used in the numerical solution of elliptic equations.

Solving for $u_{i,j}$ from Eq. 9.25a, and for $v_{i,j}$ from Eq. 9.25b, we have

$$\begin{aligned} u_{i,j} = \frac{1}{4\left(g_{11}+g_{22}\right)} \{ & 2g_{22}\left(u_{i+1,j}+u_{i-1,j}\right) + g_{22}P\left(u_{i+1,j}-u_{i-1,j}\right) + \\ & 2g_{11}\left(u_{i,j+1}+u_{i,j-1}\right) + g_{11}Q\left(u_{i,j+1}-u_{i,j-1}\right) - \\ & 2g_{12}\left(u_{i+1,j+1}-u_{i-1,j+1}-u_{i+1,j-1}+u_{i-1,j-1}\right) - 2J^2\Delta_2 u \} \end{aligned} \tag{9.26a}$$

$$\begin{aligned} v_{i,j} = \frac{1}{4\left(g_{11}+g_{22}\right)} \{ & 2g_{22}\left(v_{i+1,j}+v_{i-1,j}\right) + g_{22}P\left(v_{i+1,j}-v_{i-1,j}\right) + \\ & 2g_{11}\left(v_{i,j+1}+v_{i,j-1}\right) + g_{11}Q\left(v_{i,j+1}-v_{i,j-1}\right) - \\ & 2g_{12}\left(v_{i+1,j+1}-v_{i-1,j+1}-v_{i+1,j-1}+v_{i-1,j-1}\right) - 2J^2\Delta_2 v \} \end{aligned} \tag{9.26b}$$

To update the solution through an iterative method, SOR is used so that the values of the parametric coordinates given by Eq. 9.26 are taken as intermediate values, and the acceleration process yields the new values at the current iteration as

$$\mathbf{u}_{i,j}^{k+1} = \omega_{i,j}\,\mathbf{u}_{i,j}^{k+1} + \left(1-\omega_{i,j}\right)\mathbf{u}_{i,j}^{k}$$

where $\omega_{i,j}$ is the acceleration parameter. It is well known, see Strikwerda [23], that for *linear systems* a necessary condition for convergence is that the acceleration parameter $\omega_{i,j}$ should satisfy

$$0 < \omega_{i,j} < 2 \tag{9.27}$$

However, Eq. 9.27 does not in general imply convergence for linear systems, or our system Eq. 9.26 which is usually nonlinear. In practice, we have found that for most geometries, the choice of $\omega_{i,j} = 1$ leads to convergence. This is the usual Gauss–Seidel relaxation scheme. For certain highly curved geometries, the system is *highly* nonlinear, and underrelaxation (choosing $0 < \omega_{i,j} < 1$) may be required to ensure convergence. In practice, we have never used overrelaxation ($1 < \omega_{i,j} < 2$) for the solution of Eq. 9.26.

9.3.4 Control Function Computation

For the elliptic generation system, the source terms or control functions P and Q are used to control the specified distribution of grid points on the surface. In the computation of the elliptic surface grid, the control functions are evaluated once and then used in the iterative technique to update the grid. The control functions must be selected so that the grid has the required distribution of grid points on the surface.

In the absence of control function, i.e., $P = Q = 0$, the generation system tends to produce the smoothest possible uniform grid, with a tendency of grid lines to concentrate over convex boundary regions and to spread out over concave regions.

The elliptic system Eq. 9.24 can be solved simultaneously at each point of the algebraic grid for the two functions P and Q by solving the following linear system:

$$\begin{pmatrix} g_{22}u_\xi & g_{11}u_\eta \\ g_{22}v_\xi & g_{11}v_\eta \end{pmatrix} \begin{pmatrix} P \\ Q \end{pmatrix} = \begin{pmatrix} R_1 \\ R_2 \end{pmatrix} \tag{9.28}$$

where

$$R_1 = J^2\Delta_2 u + 2g_{12}u_{\xi\eta} - g_{22}u_{\xi\xi} - g_{11}u_{\eta\eta} \quad \text{and}$$
$$R_2 = J^2\Delta_2 v + 2g_{12}v_{\xi\eta} - g_{22}v_{\xi\xi} - g_{11}v_{\eta\eta}$$

The derivatives here are represented by central differences, except at the boundaries where one-sided difference formulas must be used. This produces control functions that will reproduce the algebraic grid from the elliptic system solution in a single iteration. Thus, evaluation of the control functions in this manner would be of trivial interest except when these control functions are smoothed before being used in the elliptic generation system. This smoothing is done by replacing the control function at each point with the average of the nearest neighbors along one or more coordinate lines. However, we note that the P control function controls spacing in the ξ-direction and the Q control function controls spacing in the η-direction. Since it is usually desired that grid spacing normal to the boundaries be preserved between the initial algebraic grid and the elliptically smoothed grid, it is advisable to not allow smoothing of the P control function along ξ-coordinate lines or smoothing of the Q control function along η-coordinate lines. This leaves us with the following smoothing iteration where smoothing takes place only along allowed coordinate lines:

$$P_{i,j} = \frac{1}{2}\left(P_{i,j+1} + P_{i,j-1}\right)$$
$$Q_{i,j} = \frac{1}{2}\left(Q_{i+1,j} + Q_{i-1,j}\right)$$

Smoothing of control functions is done for a small number of iterations.

The effect of using smoothed initial control functions is that the final elliptic grid is smoother as well as more orthogonal than the initial grid, while essentially maintaining the overall distribution of grid points.

As presented up to this point, the elliptic smoothing scheme with nonzero control functions is well-defined only if the Jacobian of the transformation from computational to parametric variables *for the initial grid* is non-vanishing. If, for example, the initial grid was produced by linear TFI and contains "folded" grid lines, the system Eq. 9.28 for generating control functions $P_{i,j}$, $Q_{i,j}$ will in fact be singular. If the "folding" of initial grid lines occurs at the boundary, this is a fatal flaw and the surface patch must

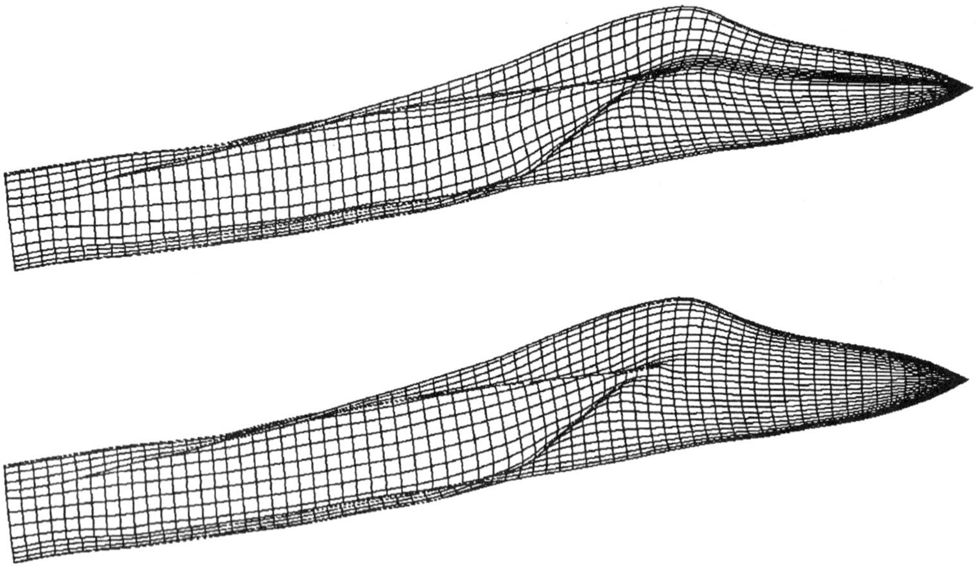

FIGURE 9.7 Aircraft geometry — algebraic grid (top) and elliptic grid (bottom).

be divided into sufficiently small subsurface patches for which we can generate nonfolded initial meshes in the vicinities of the patch boundaries. If, however, the initial mesh has valid Jacobians at the boundaries, with folding restricted to the interior, then the surface patch need not be subdivided. In this case, control functions can be computed at the boundaries from Eq. 9.28 using one-sided derivatives, and then linear transfinite interpolation (discussed in Section 9.2.2) can be used to define the control functions in the interior of the grid.

Figure 9.7 shows the effect of elliptic smoothing (with zero control functions) applied to an aircraft geometry. The initial algebraic mesh computed using linear TFI with uniform arc length distribution clearly exhibits kinked grid lines in front of the aircraft engine inlet, as well as a nonuniform distribution of grid points in this region. These grid defects could conceivably lead to unacceptable artifacts in a Navier–Stokes flow computation involving the grid. The elliptically smoothed grid has created orthogonality of grid lines and uniformity of grid point distribution. Of course the *shape* of the gridded surface has not been affected whatsoever, since all smoothing is done in the parametric domain.

We close this section by noting that our derivation of the elliptic grid generation equations from the conformal mapping conditions for surface Eq. 9.14 did not take boundary conditions into account. A consequence of this is that even with zero control functions ($P = Q = 0$) the elliptic generator Eq. 9.24 may produce nonorthogonal grids in the vicinity of the surface boundaries, especially if a highly nonuniform grid point distribution is specified on the boundary curves. Grid orthogonality at the boundaries is often necessary for accuracy of numerical simulations.

In this book, Chapter 6 covers in detail two techniques for achieving grid orthogonality at the boundaries. The first technique allows the grid points to move along the boundary. This technique involves derivative boundary conditions for the elliptic grid generation equations and is referred to as Neumann orthogonality. The second technique leaves the boundary points fixed, but modifies the elliptic equations through the control functions to achieve orthogonality and a specified grid spacing off the boundary. This technique is referred to as Dirichlet orthogonality, since the boundary conditions for the elliptic system are of Dirichlet type.

9.4 Summary and Research Issues

Algebraic and elliptic techniques for the efficient construction of high-quality structured surface grids have been presented in this chapter. We have seen that surface grids are first generated using algebraic methods, and usually improved by applying elliptic smoothing iterations.

Algebraic techniques start with the distribution of points along the boundary curves of the surface. For this, we have presented a sophisticated algorithm which takes into account arc length, curvature, and attraction to an arbitrary set of "attractor" points.

Linear and cubic Hermite transfinite interpolation methods are presented for algebraic surface grid generation. We have described the simplest and most widely used algebraic grid generator — linear transfinite interpolation. This method is usually sufficient for producing the initial grids required by elliptic methods. We have also presented a detailed algorithmic description of cubic Hermite transfinite interpolation, which is an algebraic method capable of imposing boundary orthogonality — a common requirement for the success of numerical simulations. However, in practice cubic Hermite TFI is not very robust and might force the user to subdivide the surface into an excessive number of subsurface patches in order to achieve the desired result.

A complete development of elliptic surface grid generation with control functions has been presented. Our development follows from the properties of conformal mappings of surfaces. Elliptic smoothing is a robust method of enforcing desired grid properties such as orthogonality and smoothness of grid lines. Elliptic smoothing is especially useful when the surface is "poorly parameterized" and the algebraic interpolation of parametric values does not give a satisfactory grid. Since this situation arises frequently when surfaces are defined by CAD packages, the capability to smooth and improve surface grids is essential in any state-of-the-art grid generation code.

The techniques covered in this chapter have been incorporated into several grid generation packages that have the capability of producing high-quality surface grids on complex design geometries. Nevertheless, research issues still exist.

The iterative solution of the nonlinear elliptic system Eq. 9.24 is considerably more expensive than the analogous system for planar two-dimensional grids. This is because of the presence of the geometry-dependent terms $\Delta_2 u, \Delta_2 v, g_{11}, g_{12}, g_{22}$ which must be reevaluated every iteration. These terms require evaluation of the geometry definition $\mathbf{x}(\mathbf{u})$, which can be relatively expensive. Thus very large surface meshes may require a nontrivial amount of computer time to smooth elliptically. Multigrid or *ad hoc* grid sequencing methods are a promising avenue of research addressing this problem.

Much more daunting than any amount of computer time required to generate a mesh is the much larger amount of "people time" required to "block" complex surface geometries. "Blocking" of a complex surface is the task of decomposing a surface into an adequate set of subsurface patches. Subsurface patches must for the most part be four-sided, although some degeneracies are allowed. Moreover, it is better (for good performance of the algebraic and elliptic techniques covered in this chapter) if the subpatch boundaries are aligned in a natural way with the distinctive geometrical features of the overall surface. This process thus represents an area of expensive human intervention and is usually the most time-consuming component of the grid generation process. "Autoblocking" — the automation of the blocking task — is thus a hot area of research. For a description of progress in this area, see Chapter 10.

Finally, we mention that *adaptive* surface grid generation is very much an open problem. Given a computational *field* (such as temperature, pressure, etc.) defined over a surface grid, it may be desired to concentrate grid lines in areas where the field has a large gradient or second derivative. This problem has been addressed in *planar* two-dimensional grid generation by modifying the control functions in the elliptic grid generation system to force adaptation of grid lines to the field being simulated. Analogous modification of control functions for surface grid generation has not been undertaken to our knowledge. We note that the rewards of adaptive grid generation are potentially large, especially in time-dependent simulations where it is desirable to have a dense region of grid lines track moving solution features.

References

1. **Abolhassani, J.S. and Stewart, J.E.**, Surface grid generation in parameter space, *J. Comput. Phys.*, 113, pp. 112–121, 1994.
2. **Ahlfors, L.V.**, *Lectures on Quasiconformal Mappings*, Van Nostrand, New York, 1996.
3. **Bartels, R.H., Beatty, J. C., and Barsky, B.A.**, *An Introduction to Splines for Use in Computer Graphics and Geometric Modeling*, Morgan Kaufmann, Los Altos, CA, 1987.
4. **Belinskii, P.P., Godunov, S.K., and Yanenko, I.K.**, The use of a class of quasiconformal mappings to construct difference nets in domains with curvilinear boundaries, *USSR Comp. Math. Math. Phys.*, 15, pp 133–144, 1975.
5. **Brackbill, J.U. and Saltzman, J.S.**, Adaptive zoning for singular problems in two dimensions, *J. Comput. Phys.*, 46, pp 342–368, 1982.
6. **Castillo, J.E.**, Discrete variations grid generation, *In Mathematical Aspects of Numerical Grid Generation*, (Ed.), Castillo, J.E., SIAM, Philadelphia, pp 35–58, 1991.
7. **Chawner, J.R. and Anderson, D.A.**, Development of an algebraic grid generation method with orthogonality and clustering control, in *Numerical Grid Generation in Computational Fluid Dynamics*, (Ed.), Arcilla, A.S., Häuser, J., Eiseman, P.R., Thompson, J.F., North-Holland, NY, pp 107–117, 1991.
8. **de Boor, C.**, *A Practical Guide to Splines*. Springer-Verlag, NY, 1978.
9. **Farin, G.**, *Curves for Surfaces for Computer Aided Geometric Design*, 3rd Edition, Academic Press, Boston, 1993.
10. **Godunov, S.K. and Prokopov, G.P.**, On the computational of conformal transformations and the construction of difference meshes, *USSR Comp. Math. Math. Phys.*, 7, pp. 89–124, 1967.
11. **Khamayseh, A. and Hamann, B.**, Elliptic grid generation using NURBS surfaces, *Comput. Aid. Geom. Des.*, 13, pp. 369–386, 1996.
12. **Khamayseh, A. and Mastin, C.W.**, Computational conformal mapping for surface grid generation, *J. Comput. Phys.*, 123, pp. 394–401, 1996.
13. **Knupp, P. and Steinberg, S.**, *Fundamentals of Grid Generation*, CRC Press, Boca Raton, FL, 1993.
14. **Mastin, C.W. and Thompson, J.F.**, Quasiconformal mappings and grid generation, *SIAM J. Sci. Stat. Comput.*, 5, pp. 305–310, 1984.
15. **Mastin, C.W.**, Elliptic grid generation and conformal mapping, in *Mathematical Aspects of Numerical Grid Generation*, Castillo, J.E., (Ed.), SIAM, Philadelphia, pp. 9–17, 1991.
16. **Piegl, L. and Tiller, W.**, *The NURBS Book*, Springer-Verlag, Berlin, Germany, 1995.
17. **Renelt, H.**, *Elliptic Systems and Quasi-Conformal Mappings*. Wiley, NY, 1988.
18. **Saltzman, J.S.**, Variations methods for generating meshes on surfaces in three dimensions, *J. Comput. Phys.*, 63, pp. 1–19, 1986.
19. **Smith, R.E.**, Algebraic Grid Generation, in *Numerical Grid Generation*, Thompson, J.F., (Ed.), North-Holland, NY, pp. 137–170, 1982.
20. **Soni, B.K.**, Two and three dimensional grid generation for internal flow applications of computational fluid dynamics, AIAA-85-1526, *AIAA 7th Computational Fluid Dynamics Conference*, Cincinnati, OH, 1985.
21. **Sorenson, R.L.**, Three dimensional elliptic grid generation about fighter aircraft for zonal finite difference computations, AIAA-86-0429. *AIAA 24th Aerospace Science Conference*, Reno, NV, 1986.
22. **Spekreijse, S.P.**, Elliptic grid generation based on laplace equations and algebraic transformations, *J. Comput. Phys.*, 118, pp. 38–61, 1995.
23. **Strikwerda, J.C.**, *Finite Difference Schemes and Partial Differential Equations*, Wadsworth & Brooks/Cole, Pacific Grove, CA, 1989.
24. **Struik, D.J.**, *Lectures on Classical Differential Geometry*, Dover, NY, 1988.
25. **Takagi, T., Miki, K., Chen, B.C., and Sha, W.T.**, Numerical generation of boundary-fitted curvilinear coordinate systems for arbitrarily curved surfaces, *J. Comput. Phys.*, 58, pp. 67–79, 1985.

26. **Thomas, P.D. and Middlecoff, J.F.**, Direct control of the grid point distribution in meshes generated by elliptic equations, *AIAA J.*, 18, pp. 652–656, 1980.
27. **Thompson, J.F., Thames, F.C., and Mastin, C.W.**, automatic numerical generation of body-fitted curvilinear coordinate system for field containing any number of arbitrary two dimensional bodies, *J. Comput. Phys.*, 15, pp. 299–319, 1974.
28. **Thompson, J.F., Warsi, Z.U.A., and Mastin, C.W.**, *Numerical Grid Generation: Foundations and Applications.* North-Holland, NY, 1985.
29. **Thompson, J.F.**, A general three-dimensional elliptic grid generation system on a composite block structure, *Comp. Meth. Appl. Mech. and Eng.*, 64, pp. 377–411, 1987.
30. **Warsi, Z.U.A.**, Numerical grid generation in arbitrary surfaces through a second-order differential geometric model, *J. Comput. Phys.*, 64, pp. 82–96, 1986.
31. **Warsi, Z.U.A.**, Theoretical foundation of the equations for the generation of surface coordinates, *AIAA J.*, 28, pp. 1140–1142, 1990.
32. **Whitney, A.K. and Thomas, P.D.**, Construction of grids on curved surfaces described by generalized coordinates through the use of an elliptic system, in *Advances in Grid Generation*, Ghia, K.N. and Ghia, U., (Ed.), ASME Conference, Houston, TX, pp. 173–179, 1983.
33. **Winslow, A.M.**, Numerical solution of the quasilinear poisson equations in a nonuniform triangle mesh, *J. Comput. Phys.*, 2, pp. 149–172, 1967.
34. **Yamaguchi, F.**, *Curves and Surfaces in Computer Aided Geometric Design*, Springer–Verlag, NY, 1988.

10

A New Approach to Automated Multiblock Decomposition for Grid Generation: A Hypercube++ Approach

Sangkun Park

Kunwoo Lee

10.1 Introduction

A wide variety of grids may be desired in various applications depending on the solution technique employed. The typical types of grids used in the field of computational fluid dynamics (CFD) are block structured [1–8], unstructured [9–12], overset [13–15], hybrid [16–18], and Cartesian grids [19]. Among them, the block-structured grid method is the most established (see Chapter 13). These grids tend to be computationally efficient, and high aspect ratio cells that are necessary for efficiently resolving viscous layers can be easily generated. But, in general, it takes too much time to generate the associated grids due to the lack of the automated techniques for block decomposition.

All the methods including the block-structured approach for grid generation have their own advantages and have been used with satisfactory results. However, a critical obstacle to be overcome for the effective use of such approaches is the automatic decomposition of the spatial domain. The multiblock decomposition of a flow domain is the first and the most important step in the generation of the grids for computational flow simulations, and is considered as the most labor intensive task in any CFD application. Soni et al. [20] pointed out that it can take a significantly longer labor time to generate a computational grid than to execute the flow field simulation code on the grid or to analyze the results. Similarly, Vatsa

0-8493-2687-7/99/$0.00+$.50

et al. [21] also noted the biggest bottleneck in the grid generation process is the domain decomposition and asserted that efforts should be focused on automating or simplifying the domain decomposition process.

Allwright [4] has devised various rules and strategies from the experience gained in graphical block decomposition. These rules are being progressively implemented in his automated method, which generates a wire-frame schematic to represent the grid topology when a simple block representation of the configuration to be modeled is given. Shaw and Weatherill [5] also proposed a similar approach. They used a Cartesian H-type block structure globally and C- or O-type topology was locally embedded around certain components. Stewart [6] has developed the search rules for driving directional probing from the boundary for an appropriate block decomposition, in analogy with balloons inflating to obtain a coarse approximation to the outer boundary of a region. Dannenhoffer [7] suggested an abstraction concept of the geometry to capture the basic topology. In his scheme, the grid topology is specified by placing blocking objects on the background grid, and then a set of transformations [8] is used to generate a suitable assembly of grid blocks. This approach is now being developed for three-dimensional cases.

In general, the multiblock structure is, to a large extent, capable of filling up topologically complex flow domains in an efficient way. This multiblock approach also allows different flow models in different blocks and different grid refinement strategies for different blocks. Furthermore, it may be expected that this multiblock approach naturally leads to parallel executions of calculations per block on different computing resources if blocks are constrained to satisfy a supplementary constraint; the block's dimensionality has to be consistent with a suitable load balancing.

This chapter presents a new algorithm for an automatic multiblock decomposition. The main idea proposed in this chapter is inspired by the hypercube introduced by Allwright and the abstraction concept by Dannenhoffer. All procedures related to this algorithm are automatically performed with some defaults or can be customized using any user-specified parameter values for a special purpose. Thus, this algorithm would enable any grid generation system to simply and efficiently construct both a block topology and its geometry for general geometries in a systematic fashion.

10.2 Underlying Principles

The basic idea behind an automatic domain decomposition into multiblocks suggested in this chapter is to carry out the decomposition not in a complex space in which the curved or complicated geometries exist, but in a simple space in which the transformed simple shapes appear. This transformation is accomplished by introducing a nonuniform rational B-spline volume that maps a physical domain onto a parameter domain. Then, all the geometric operations related to the multiblock domain decomposition are carried out in the parametric space. These procedures include the hypercube++ generation and hypercube++ merging algorithms to be described later. Grid generation or grid refinement can also be implemented in the parameter space in an effective way. Once the grids are generated in the parameter space, the grids in the physical space are derived by remapping, which is basically evaluating the NURBS volume.

The basic idea described above can be illustrated as shown in Figure 10.1. The hypercube++ generation algorithm allows a real curved body and its surroundings to transform into simple brick-shaped elements, and the hypercube++ merging algorithm allows the production of a sum of the brick-shaped elements when a space surrounding multiple bodies is considered, and is similar to the Boolean sum used in solid modeling systems. Each brick-shaped element in the hypercube++ structure is mapped onto the corresponding physical space by the NURBS volume such that the face of a brick element adjacent to the internal body is transformed into the curved surface of the corresponding physical body.

10.2.1 NURBS Volume

Nonuniform rational B-splines, commonly called NURBS (see Chapter 30), have become very popular in curve and surface description, and in the representation, design, and data exchange of geometric information in many applications, especially in numerical grid generation. [22]

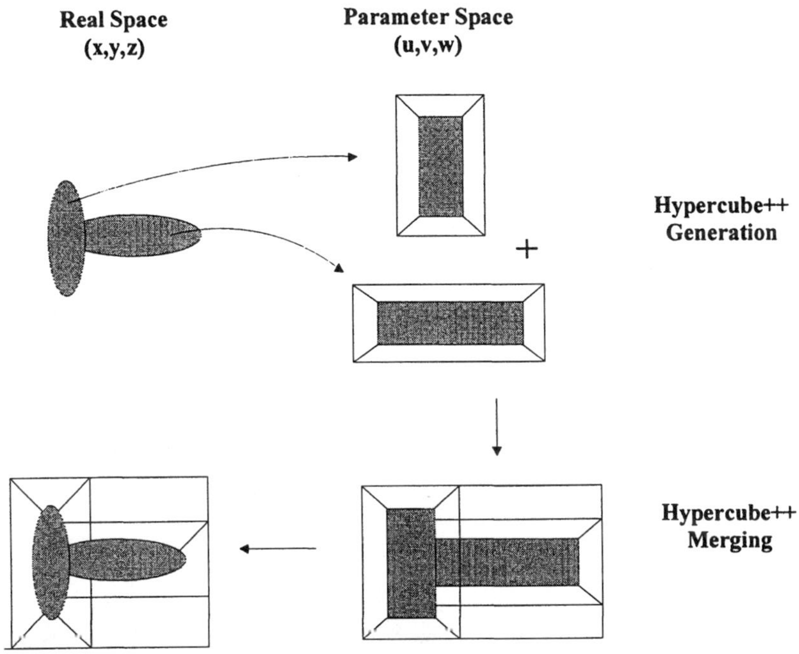

FIGURE 10.1 Global steps of the suggested multiblock decomposition and its related algorithms.

While in the past, the computer-aided geometric design (CAGD) has been mostly concerned with curves and surfaces, more recently there has been an increasing interest in higher-dimensional multivariate objects such as volumes and hypersurfaces in $\mathbf{R}^n$, $n > 3$. Almost all of the methods developed for surfaces in the CAGD literature can be generalized to higher-dimensional objects. A typical example is a tensor product Bezier volume, B-spline volume, or their generalized form, NURBS volume. As noted earlier [23, 24], the NURBS volume is an extension of the well-known NURBS surface, in the same manner that the NURBS surface is an extension of the NURBS curve.

A NURBS volume of order ku in the u direction, kv in the v direction, and kw in the w direction is a trivariate vector-valued piecewise rational function of the form

$$\mathbf{B}(u,v,w) = \frac{\Omega(u,v,w)}{h(u,v,w)} = \frac{\sum_{i=0}^{nu}\sum_{j=0}^{nv}\sum_{k=0}^{nw} h_{ijk}\mathbf{B}_{ijk}N_i^{ku}(u)N_j^{kv}(v)N_k^{kw}(w)}{\sum_{i=0}^{nu}\sum_{j=0}^{nv}\sum_{k=0}^{nw} h_{ijk}N_i^{ku}(u)N_j^{kv}(v)N_k^{kw}(w)} \tag{10.1}$$

The $\{\mathbf{B}_{ijk}\}$ form a tridirectional control net, the $\{h_{ijk}\}$ are the weights, and the $\{N_i^{ku}(u)\}$, $\{N_j^{kv}(v)\}$, and $\{N_k^{kw}(w)\}$ are the nonrational B-spline basis functions defined on the knot vectors

$$\mathbf{U} = \{u_i\}_{i=0}^{nu+ku} = \{\bar{u}_0,\cdots,\bar{u}_{ku-1},\bar{u}_{ku},\cdots\cdots,\bar{u}_{nu},\bar{u}_{nu+1},\cdots,\bar{u}_{nu+ku}\}$$
$$\text{where } \bar{u}_0 = \cdots = \bar{u}_{ku-1} \text{ and } \bar{u}_{nu+1} = \cdots = \bar{u}_{nu+ku},$$
$$\mathbf{V} = \{v_j\}_{j=0}^{nv+kv} = \{\bar{v}_0,\cdots,\bar{v}_{kv-1},\bar{v}_{kv},\cdots\cdots,\bar{v}_{nv},\bar{v}_{nv+1},\cdots,\bar{v}_{nv+kv}\}$$
$$\text{where } \bar{v}_0 = \cdots = \bar{v}_{kv-1} \text{ and } \bar{v}_{nv+1} = \cdots = \bar{v}_{nv+kv},$$
$$\mathbf{W} = \{w_k\}_{k=0}^{nw+kw} = \{\bar{w}_0,\cdots,\bar{w}_{kw-1},\bar{w}_{kw},\cdots\cdots,\bar{w}_{nw},\bar{w}_{nw+1},\cdots,\bar{w}_{nw+kw}\}$$
$$\text{where } \bar{w}_0 = \cdots = \bar{w}_{kw-1} \text{ and } \bar{w}_{nw+1} = \cdots = \bar{w}_{nw+kw}.$$

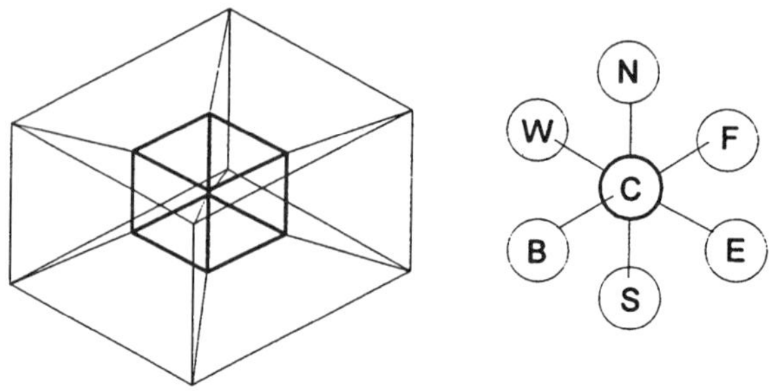

FIGURE 10.2 Hypercube structure.

Notice that this parametric representation maps a cube in the parameter space onto a three-dimensional space. The domain of the mapping, which is sometimes referred to as parametric space, has axes *u*, *v*, and *w*, and the range, which is called model space, has the usual *x*, *y*, and *z* axes.

10.2.2 Hypercube++ Structure

10.2.2.1 Hypercube and Its Limitations

As shown in Figure 10.2, the hypercube structure introduced by Allwright and his colleagues [15] is useful for a multiblock decomposition of a region around a simple convex body by wrapping around the body. In this wrap-around strategy, a convex-shaped body is located in the central region and the other six regions are placed around the body.

Thus, a hypercube is composed of seven blocks, called *east*, *west*, *south*, *north*, *front*, *back*, and *center* block, as shown in Figure 10.2. This naming convention naturally defines the relative position of the seven blocks. In addition to this elementary structure, degenerate structures can also be considered. They are referred to as seven basic hypercubes [18], which are shown in Figure 10.3. The combination of these basic hypercubes can lead to better geometric flexibility.

However, a basic hypercube has a limitation in representing more general configurations. It is basically impossible to represent a region surrounding body surfaces by any one of the basic hypercubes in such cases that the body shape is not convex or there are multiple bodies. Therefore, we need an enhanced hypercube structure to solve two such problems. For this purpose, a hierarchical hypercube++ structure is proposed in this chapter.

10.2.2.2 Hypercube++ Structure

The hypercube++ structure, which is a hierarchical extension of the hypercube, represents the parent/child relations between the related hypercubes with the relative positions (e.g., east, west, etc.) of the blocks in each hypercube, and thus provides all the topological information between decomposed blocks.

The hypercube++ structure allows such topological structures, as shown in Figure 10.4. These examples demonstrate the capabilities of the hypercube++ structure.

In the hypercube++ structure, a hypercube structure can be located in one of the blocks of the parent hypercube, as is shown in Figure 10.4a, where the west and the east block of the parent hypercube located in the center has a pointer to its child hypercube located to the left and the right, respectively. And also the center block can be degenerated into one face so that only two blocks exist in the hypercube as shown in Figure 10.4b, where only the back and the front block can be found. These enhanced structures make it possible to have any number of hypercubes stand in a line as shown in Figure 10.4c, or in a combined way as in Figure 10.4d. In Figure 10.4, the hypercube++ data representation of each example is shown at the right-hand side. The circles in the figure mean the blocks, and their terminal nodes represent the true blocks having the geometric definition, i.e., a NURBS volume. The blocks corresponding to the nonterminal nodes have no geometric meaning, but are introduced to represent the hierarchy between the hypercubes. In the figure, the hierarchical parent/child relation is displayed with an arrow.

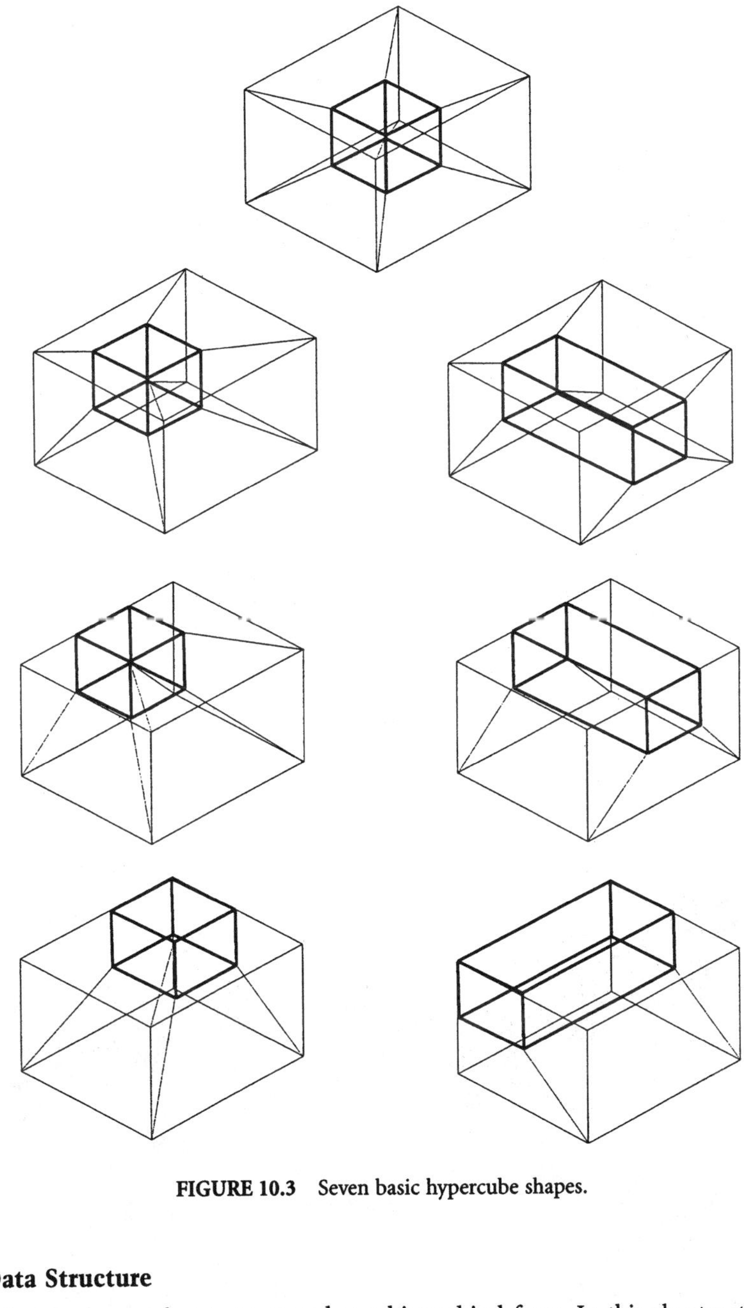

FIGURE 10.3 Seven basic hypercube shapes.

10.2.2.3 Data Structure

As noted earlier, the hypercube++ structure has a hierarchical form. In this chapter, the hierarchy is implemented by the combination of the *Hycu* and *Blk* data structure written in the *C* language shown in Figure 10.5. The *Hycu* data structure is composed of seven blocks, *blk* [*7*], and also has a pointer to its *parent* block. The *Blk* data structure has pointers to its *parent* and *child* hypercube for a hierarchical structure, and *bspvol* to point the corresponding NURBS volume. Also, it has *grid* or *mesh* pointer for creating or modifying grid points or mesh elements.

By using some operators or procedures for adding a child or parent hypercube to the hierarchical structure of a given hypercube++, the hypercube++ structure can be grown up to represent a multiblock decomposition of any complex configuration.

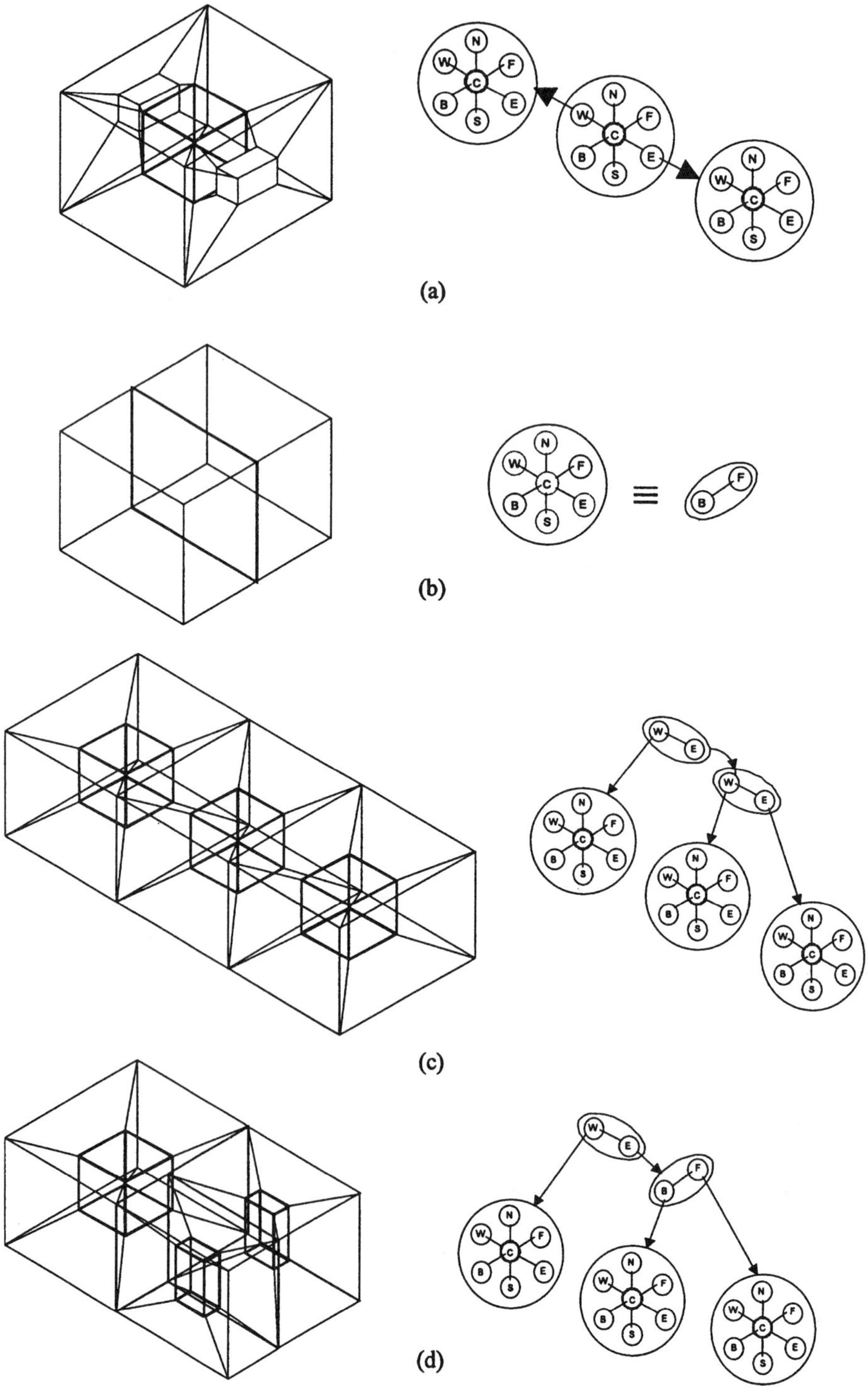

FIGURE 10.4 Examples of hypercube++ structure.

```
  typedef struct Hycu {          /* Hypercube Data Structure */
 int      situation;            /* situation flag, i.e., 1 if east-west */
 Blk      *parent;              /* pointer to parent block */
 Blk      *blk[7];              /* pointers to child block */
}  Hycu;

  typedef struct Blk {           /* Block Data Structure */
 int         stand;             /* position taken up, i.e., 1 if east */
 int          env;              /* flag to whether if contact to a body */
 int          map;              /* mapping function */
 Hycu     *parent;              /* pointer to parent hypercube++ */
 Hycu      *child;              /* pointer to child hypercube++ */
 BSPVOL *bspvol;                /* pointer to NURBS volume */
 BGrid     *grid;               /* pointer to grid structure */
 BMesh    *mesh;                /* pointer to mesh structure */
  }  Blk;
```

FIGURE 10.5 Hypercube++ data structure.

10.3 Best Practices

10.3.1 Hypercube++ Generation

For a given component, the region around the component is represented by one of seven basic hypercubes once a user or a system specifies all feature surfaces of the component in the given configuration. The hypercube++ generation algorithm can be summarized by the following:

Step 1: For a given component, input the boundary surfaces, as shown in Figure 10.6a.

Step 2: Generate an inner box that minimally encloses the input surfaces and an outer box that wraps around the inner box. The size of the outer box is calculated from a characteristic length in the flow condition, e.g., the thickness of a boundary layer, or determined by a user's input. See Figure 10.6b.

Step 3: Generate a NURBS volume of which the size is the same as the outer box. In this chapter, the volume is called the local mapping volume.

Step 4: Increase the number of control points of the mapping volume by knot insertion. The knots inserted into the volume are the parametric values of maximal, minimal, and center point of the inner box: three knots are inserted along each parametric direction. In general, the knots are inserted to increase the geometric flexibility in shape control. See Figure 10.6c.

Step 5: Move the control points of the mapping volume, which are located on the boundary faces of the inner box, onto the input boundary surfaces. The new position of each control point is obtained such that the distance between the control point and the new position is the minimum distance from the initial control point to the boundary surfaces. See Figures 10.6d and 10.6h. Now we can notice that splitting the volume at the inserted knots results into the approximate shape of the input surfaces as shown in Figure 10.6e.

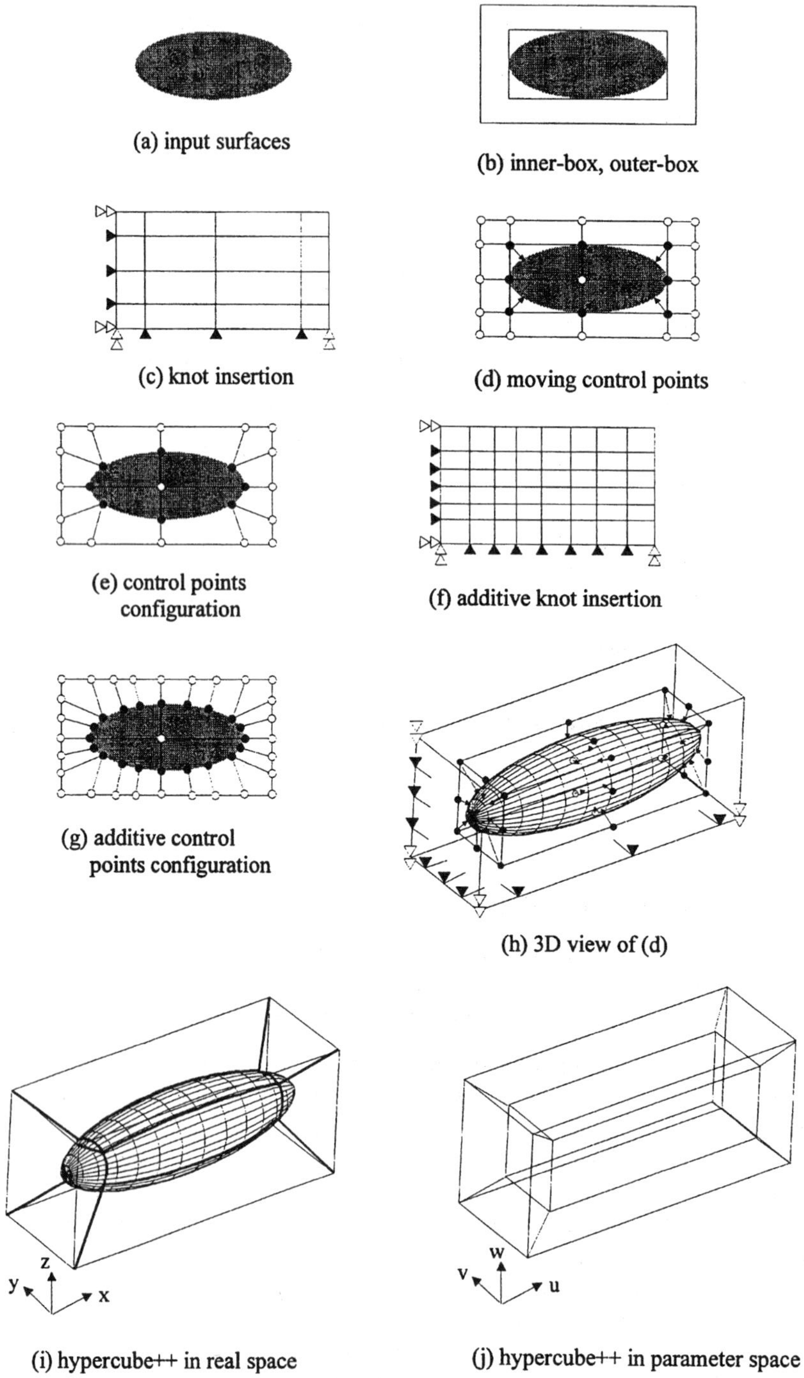

FIGURE 10.6 Hypercube++ generation.

Step 6: Add more control points by inserting knots at appropriate points such that they are uniformly distributed on the inner box as shown in Figure 10.6f. Then translate the new control points onto the input boundary surfaces as in Step 5. See Figure 10.6g. These steps are necessary to approximate the inner shape of the mapping volume more closely. After moving the control points onto the input boundary surfaces, we can see that the curved boundary surfaces are

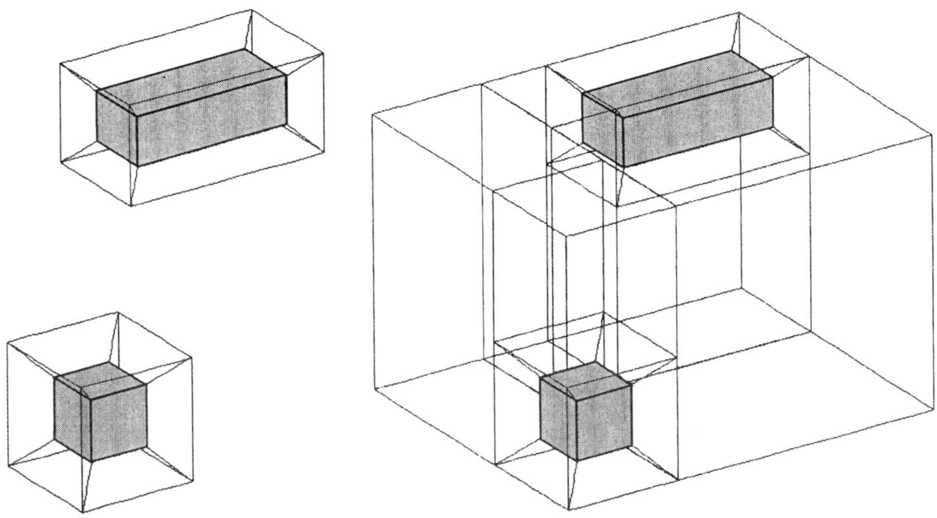

(a) two initial hypercube++'s (b) a merged hypercube++

FIGURE 10.7 Hypercube++ merging for a separate case.

transformed into the planes in the parametric domain. That is, the curved object in real space (x, y, z) is transformed into the box-like shape in parameter space (u,v,w), and the region around the curved object shown in Figure 10.6i is simplified into the parametric region bounded by the inner box and the outer box as shown in Figure 10.6j.

Step 7: Generate a hypercube structure in the parametric domain of the mapping volume. That is, the inner box is located in the center block of the hypercube and the other blocks are created by connecting the vertices of the inner box to the corresponding vertices of the outer box in the parametric domain. The surrounding blocks except the center have their different NURBS volumes as their geometric objects, which are called as block volumes in this chapter. The center block does not need to have a NURBS volume because the grid will not be generated in the center block, i.e. inside the object.

10.3.2 Hypercube++ Merging

The hypercube++ merging algorithm permits that two basic hypercubes be merged into one hypercube++ structure in a hierarchical form, or the merged hypercube++'s are also combined into a single hypercube++. In this way, an arbitrary number of the hypercube++'s are merged into one complex hypercube++, regardless of whether two hypercube++'s are overlapping or not. The relative position between two hypercube++'s can be classified into three cases: a separate, a contained, and an overlapped condition.

The necessary steps in the hypercube++ merging algorithm for the three cases mentioned above are outlined as follows:

Step 1: Check the relative position between two given hypercube++'s. The possible situations are: "separate" as shown in Figure 10.7, "contained" as in Figure 10.8, and "overlapped" as in Figure 10.9. We will briefly describe how these situations are handled as below. Note that the center blocks are colored dark in the figures.

Step 2: For a separate case, the *outer-merge* algorithm allows a new merged hypercube++ to include two given ones as a child in its new hierarchical structure as shown in Figure 10.7.

Step 3: For a contained case, the *inner-merge* algorithm allows a larger hypercube++ to include a smaller hypercube++ in its new hierarchical structure as shown in Figure 10.8.

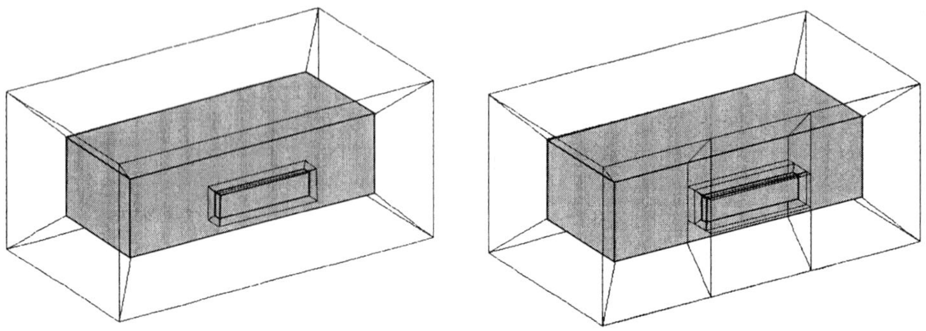

(a) two initial hypercube++'s (b) a merged hypercube++

FIGURE 10.8 Hypercube++ merging for a contained case.

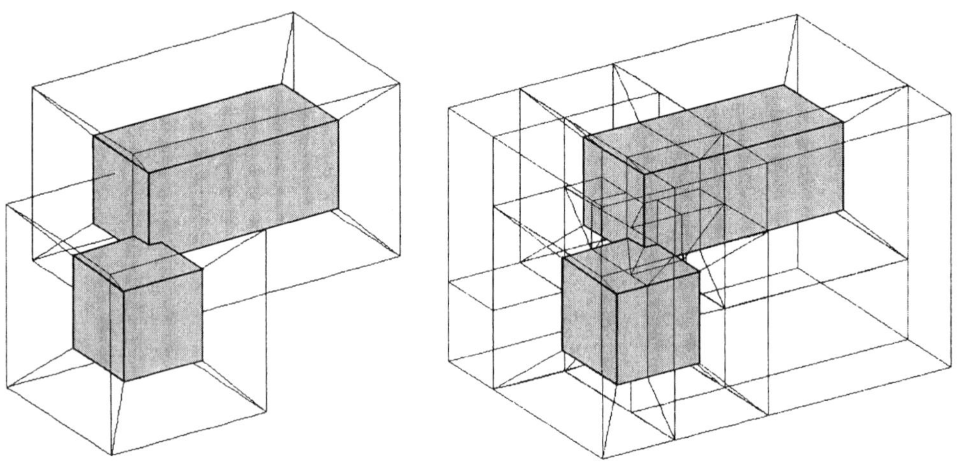

(a) two initial hypercube++'s (b) a merged hypercube++

FIGURE 10.9 Hypercube++ merging for an overlapped case.

Step 4: For an overlapped case, the *hypercut* algorithm allows one of two hypercube++'s to be cut by all the infinite cutting planes which are obtained from the outer boundary faces of the box which minimally encloses the other, resulting in maximal six pieces which also have a hypercube++ structure. Next, the hypercube++ that originated the cutting planes, called a cutting hypercube++, is merged with one of the cut pieces located inside by using the *inner-merge* algorithm. Finally, the result is also merged with the cut pieces located outside the cutting hypercube++ by using the *outer-merge* algorithm. The above merging processes are executed by calling the *overlap-merge* algorithm. Two initial hypercube++'s and their merged hypercube++ are shown in Figure 10.9.

To implement the three algorithms described above, two operators, i.e., *hycucut* and *hypercut* (A,B,m) algorithm, need to be developed. The *hycucut* operator cuts a single hypercube++ by a given cutting plane, and creates two cut hypercube++'s as shown in Figure 10.10.

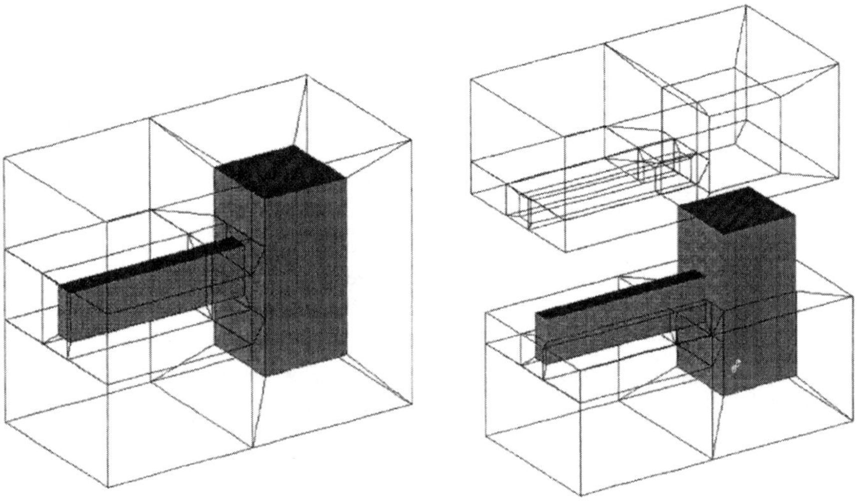

(a) an initial hypercube++ (b) two resulting hypercube++'s

FIGURE 10.10 *Hycucut* operator.

- **hycucut** operator

Input: a hypercube++, a cutting plane
Output: two hypercube++'s
Procedure: Step 1 ~ 2
Step 1: Generate two hypercube++'s copied from the given hypercube++.
Step 2: For each hypercube++, geometrically, cut all the block volumes that can be cut by the cutting plane. Topologically, remove the unnecessary blocks that do not exist in the half-space selected, where the half-space is one of the two regions separated by the cutting plane. See Figure 10.10.

The *hypercut* (*A,B,m*) operator is an elementary mechanism for the cutting process between *A* and *B* where *A* and *B*, respectively, are a hypercube++ or a single block. The algorithm is briefly described as follows:

- **hypercut (*A,B,m*)** operator

Input: A, B, and *m*, {*A,B*} can have the following forms: {*H,H*}, {*H,b*}, {*b,H*}, and {*b,b*} where *H* = hypercube++ and b = block, and the *m* indicates a merging option, no action if *m* is equal to 0, and perform a merging process if *m* is 1.
Output: separated hypercube++'s or their combined hypercube++
Procedure: Step 1 ~ 5
(*Cutting process*) Step 1 ~ Step 3
Step 1: With the *hycucut* algorithm, *B* is cut by the cutting planes which are generated by infinitely extending the boundary planes of the box which minimally encloses *A*. Here, the cutting planes are orthogonal to the maximal length direction of *B*. If not cut, continue to cut with the boundary planes orthogonal to the next maximal-length direction of *B*. See Figures 10.11a and 10.11b.
Step 2: Among the hypercube++'s or the blocks that are cut from *B*, find one which overlaps *A*. If not found, then the cutting process is terminated. Otherwise, the selected one becomes *B* to be used in Step 1. See Figure 10.11b.

Step 3: Repeat Step 1 until B does not overlap A. See Figure 10.11c.
(Note that this cutting process has the purpose of minimizing the total area of the boundary faces of each cut volume, which is desirable for the parallel computing in that the load on the processors is balanced and communications among processors are minimized.)

(*Merging process*) Step 4 ~ Step 5

Step 4: Check the merging option, m. If m is 0, then this merging process is skipped, and return the cut hypercube++'s as outputs. See Figure 10.11c.

Step 5: Otherwise, the cut hypercube++'s are merged into a single combined hypercube++ of which a hierarchical structure is built in a reverse sequence of the cutting process, and return the combined hypercube++ as an output. See Figure 10.11d.

With an appropriate choice of A, B, and m in the *hypercut* (A,B,m) operator explained above, the *outer-merge*, the *inner-merge*, and the *overlap-merge* algorithms can be easily implemented as shown below.

- **outer-merge algorithm**

Input:two hypercube++'s, H_1 and H_2

Output: a single merged hypercube++

Procedure: Step 1 ~ 4

Step 1: Generate a block which encloses two given hypercube++'s minimally. See Figures 10.12a and 10.12b.

Step 2: Generate a new hypercube++ by cutting the block in Step 1 into three blocks, b_1, b_2, and the middle block such that H_1 and H_2 are located in b_1 and b_2, respectively. See Figure 10.12c.

Step 3: Execute the *hypercut* (A,B,m) algorithm where $A = H_1$, $B = b_1$, and $m = 1$. See Figure 10.12d.

Step 4: Execute the *hypercut* (A,B,m) algorithm where $A = H_2$, $B = b_2$, and $m = 1$. See Figure 10.12d.

(Note that the new hypercube++ includes two given hypercube++'s in its hierarchical structure.)

- **inner-merge algorithm**

Input: two hypercube++'s, H_1 (contains H_2) and H_2 (inside H_1)

Output: a single merged hypercube++

Procedure: Step 1 ~ 4

Step 1: For each center block of H_1 where a real body is located, perform the following Step 2 and 3.

Step 2: Execute the *hypercut* (A,B,m) algorithm where $A = b_c$ (= center block), $B = H_2$, and $m = 0$. See Figure 10.13a.

Step 3: Kill the cut hypercube++ inside b_c and combine the remainders into a single hypercube++ (= H_2 again) of which a hierarchical structure is built in a reverse sequence of the cutting process in Step 2. See Figure 10.13b.

Step 4: Finally, execute the *hypercut* (A,B,m) algorithm where $A = H_2$, $B = H_1$, and $m = 1$. See Figure 10.13c. (Note that H_2 is absorbed into H_1 while H_1 and H_2 are cut by each other.)

- **overlap-merge algorithm**

Input: two hypercube++'s, H_1 (supplies the cutting planes) and H_2 (is cut)

Output: a single merged hypercube++

Procedure: Step 1 ~ 3

Step 1: Execute the *hypercut* (A,B,m) algorithm where $A = H_1$, $B = H_2$, and $m = 0$. See Figures 10.14a, 10.14b and 10.14c.

Step 2: Execute the *inner-merge* algorithm with H_1 and the cut piece located inside H_1. See Figure 10.14c.

Step 3: Execute the *outer-merge* algorithm with the merged result in Step 2 and the cut pieces located outside H_1 in a reverse sequence of the cutting process in Step 1. See Figure 10.14d.

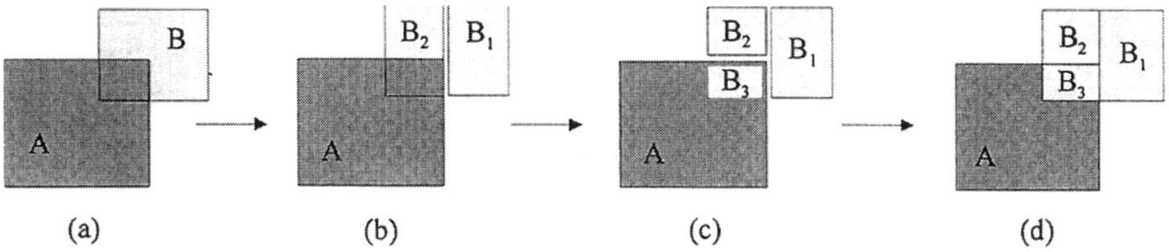

FIGURE 10.11 *Hypercut* (*A*,*B*,*m*) operator.

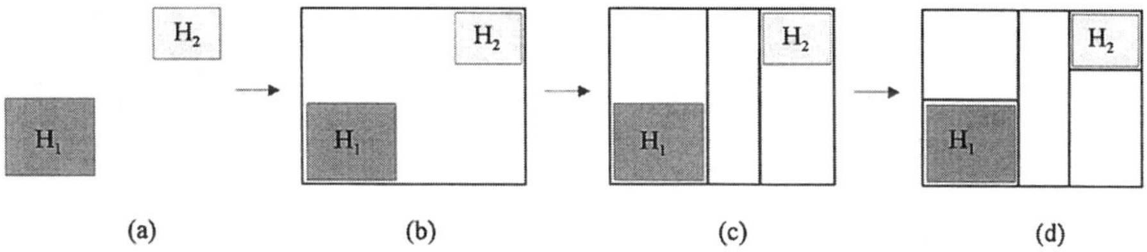

FIGURE 10.12 *Outer-merge* algorithm.

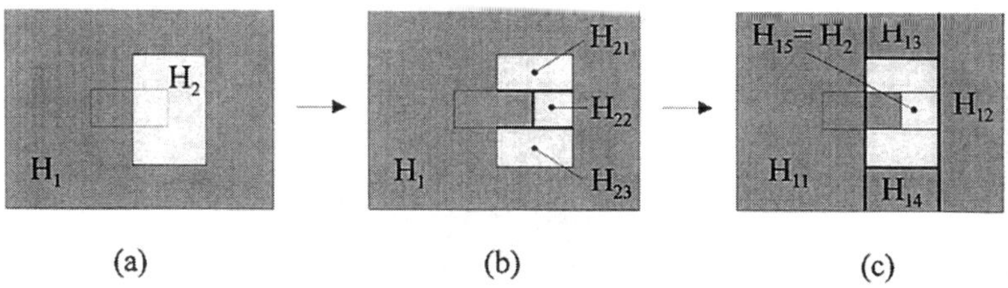

FIGURE 10.13 *Inner-merge* algorithm.

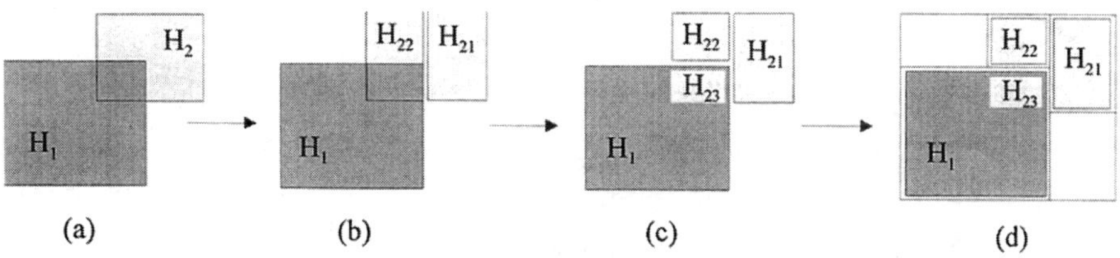

FIGURE 10.14 *Overlap-merge* algorithm.

Figure 10.15 illustrates a hierarchical structure of the merged hypercube++ shown in Figure 10.7b. This example aids understanding of a merged hierarchical structure caused by the hypercube++ merging algorithm. Figure 10.15a shows the physical shape of the hypercube++ at each hierarchical level while Figure 10.5b shows its corresponding schematic data representation of the topological information. Note that the final blocks decomposed by the suggested hypercube++ approach are colored dark in Figure 10.15b.

10.3.3 Main Features of Hypercube++ Approach

The hypercube++ approach has many features or advantages over current graphics-based approaches that rely on high-speed graphics to allow expert users to interactively design the block topology and generate the block geometry with the trial-and-error process. The main features of this new approach are summarized as follows:

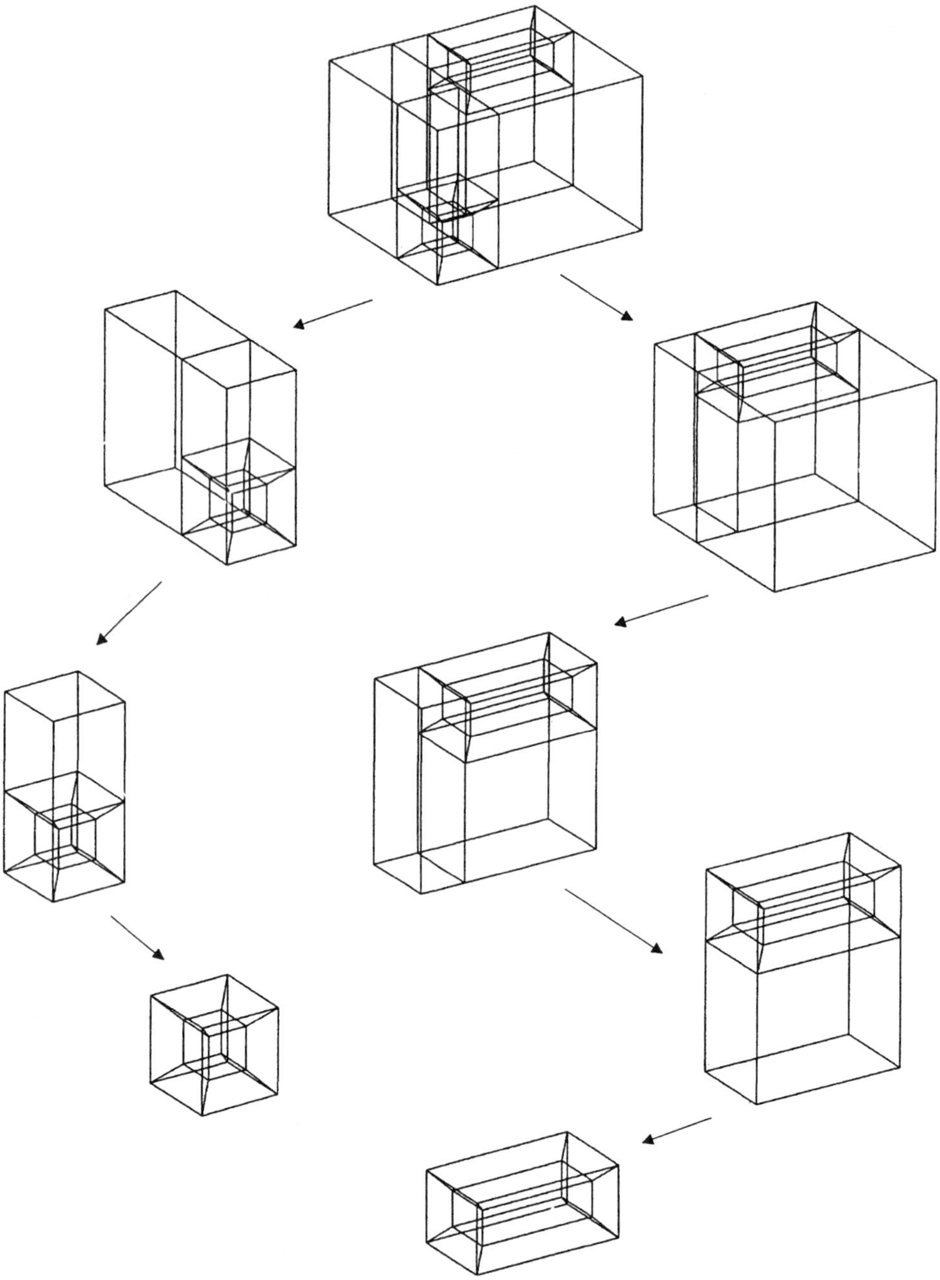

FIGURE 10.15a A hierarchical structure of the merged hypercube++ shown in Figure 10.7b. An example for illustrating the hierarchical structure of the hypercube++ merged by the hypercube++ merging algorithm.

- A multiblock decomposition is derived in about an order-of-magnitude less time than is typically required by traditional techniques and in an automatic manner.
- It is easy to search the neighboring blocks of a specific block by a simple evaluation of the hypercube++ structure. The neighboring information is necessary for the generation of contiguous grids, especially for the communication of the flow data between the blocks when solving the flow problems.

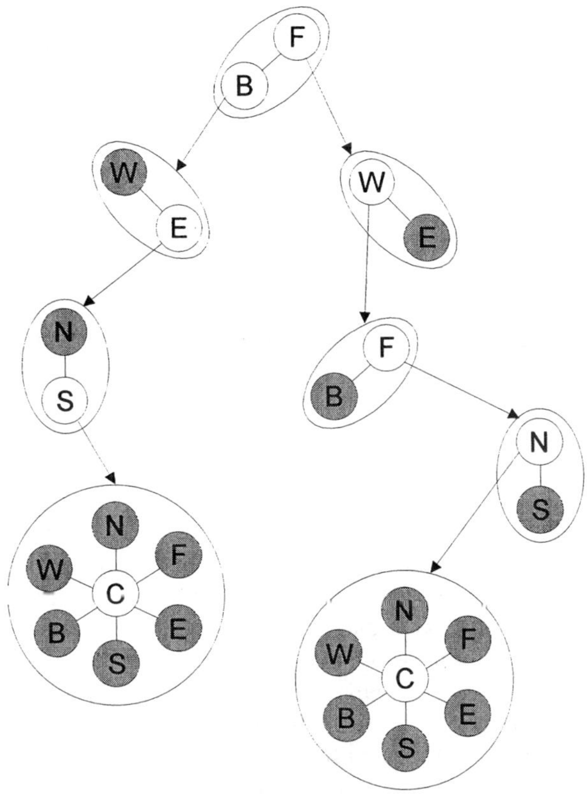

FIGURE 10.15b A schematic data representation of Figure 10.11a.

- It is simple to find the blocks which are in contact with the body surfaces. As is well known, the region near the body surfaces is very important in the flow computations, especially in the boundary layer flow. A higher resolution and orthogonality of grids are commonly required in a boundary layer.
- The change of the shape of any geometry can be confined locally. This local property supported by the NURBS volume makes it possible to automatically modify the blocks in compliance to any change of the body surfaces in a given configuration without intensive computations that are needed in traditional techniques for the redistribution of the grids already generated.
- It is not necessary to completely reconstruct the multiblock decomposition for any changed configuration when a new component is added to a given configuration. In the current systems based on the graphics-oriented approach, a complete multiblock reconstruction is needed to accommodate the new component. However, the hypercube++ merging algorithm allows the local region near a new component to be assembled into the global region around a given configuration without any reconstruction.
- It is independent of the number of bodies and their relative positions in a given configuration, and thus is applicable to any complex configuration.
- It is independent of the grid generator to be used together, and thus is immediately applicable to many current systems. Note that any type of grid generator, i.e., structured, unstructured, or hybrid approach, requires a domain decomposition as the preliminary step to resolve any three-dimensional complex configuration. Therefore, any type of grid can be generated for each decomposed block, so resulting in the creation of any grids to be desired.
- It is possible to define some templates for widely used topologies and configurations. That is, some hypercube++ structures can be reserved as templates for their reuse.

10.3.4 Applications

Three different examples have been selected to demonstrate the applicability of the present approach. These examples illustrate decomposed multiblocks and structured initial grids. The initial grids have been generated in a simple way that all grid points of each block are generated in the parameter space and then transformed into the real space by the mapping function of each block. Even though the initial grids generated in these examples have a structured type, it is possible to generate any type of grids with an appropriate grid generator, since all topological information can be derived from the hypercube++ structure generated, and all the geometric information can be calculated from the NURBS volume corresponding to each block.

Figure 10.16 shows an example of an impeller configuration. The hypercube++ generation algorithm is applied to the blade surfaces of each impeller, resulting into the creation of 12 basic hypercubes, and then the hypercube++'s for all the blades are merged into a single hypercube++ by the hypercube++ merging algorithm. Figure 10.16a shows the multiblock architecture of the impeller, which is made of 140 blocks, and Figure 10.16b shows the block-structured grids, which globally have the grid dimensions of $50 \times 16 \times 240$ in the respective (i,j,k) directions.

The second example shown in Figure 10.17 is a complex airplane configuration consisting of the fuselage, the main wing, the nacelle, the pylon, the tail, and the tail wing as the shape components. The hypercube++ generation algorithm is applied to each shape component resulting in the six basic hypercubes, and then, as in the impeller case, all generated hypercubes are merged into a single one by the hypercube++ merging algorithm. It takes about 3 minutes to generate the hypercube++ structure for the airplane on a 10 MIPS engineering workstation. Figure 10.17a shows the multiblock architecture of the airplane, which has 157 blocks, and Figure 10.17b shows the block-structured grids, which globally have the grid dimensions of $80 \times 30 \times 50$ in the three coordinate directions, i, j, k, respectively. Figure 10.17c gives another view of the wing-nacelle configuration in detail.

The final example shown in Figure 10.18 is a building complex that consists of 43 buildings. To each building, the hypercube++ generation algorithm is applied into the creation of 43 basic hypercubes, and then all generated hypercubes are merged into a single one by the hypercube++ merging algorithm as in the two cases above. Figure 10.18a shows the multiblock architecture of the building complex composed of 304 blocks, and Figure 10.18b shows the block-structured grids, which globally have the grid dimensions of $70 \times 50 \times 10$ in the three coordinate directions, i, j, k, respectively.

10.4 Research Issues and Summary

A new method for an automatic multiblock decomposition of a field around any number of complex geometries has been proposed. This method is based on hypercube++ data structure to represent the hierarchical relationship between various types of hypercubes, while the geometry of the hypercube is represented by nonuniform rational B-splines (NURBS) volume, which maps the physical space of a hypercube onto the parameter space.

The generation of grid topology based on the hypercube++ structure consists of two main steps: (1) the hypercube++ generation step, which is applied to the region around a single shape element, e.g., a wing in an airplane, to generate an appropriate hypercube, and (2) the hypercube++ merging step, which merges simple hypercubes or the ones merged already into a single but more complex hypercube++ to represent the regions around the shape composed of several shape elements.

This approach has been demonstrated with some examples to show that it allows a user to construct a multiblock decomposition in a matter of minutes for any three-dimensional configurations in an automatic manner.

The multiblock approach proposed in this chapter currently has two problems. First, the number of the resulting blocks may be too big in certain cases. A scheme to reduce the number of the blocks needs to be developed and inserted into the hypercube++ merging algorithm. One way to solve this problem

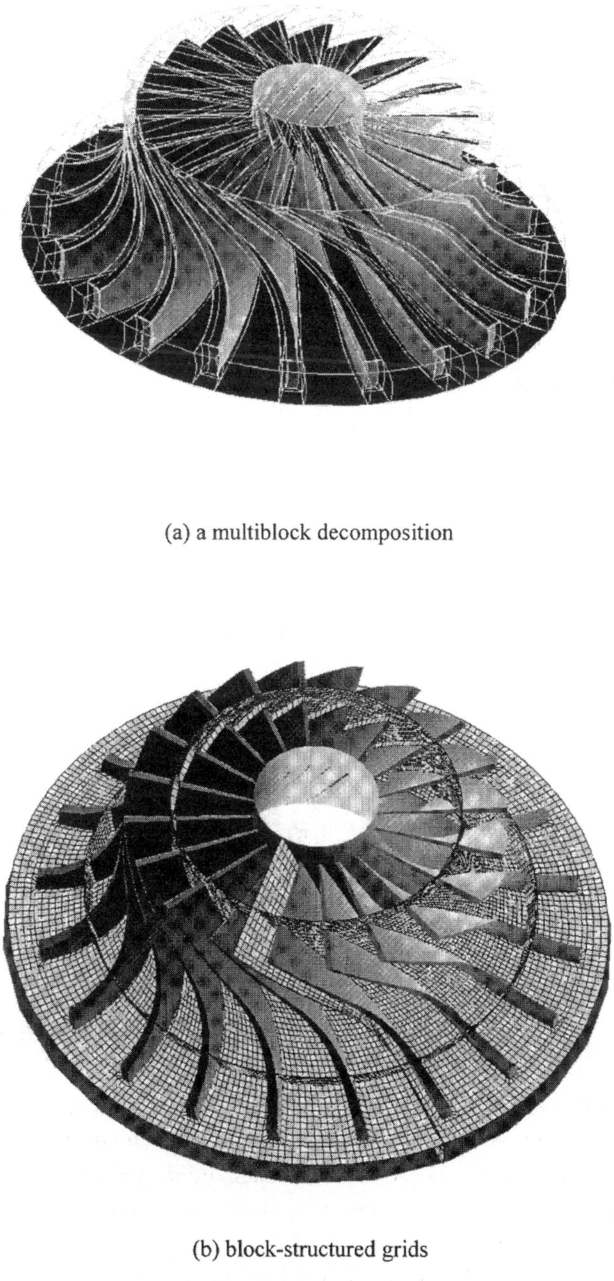

(a) a multiblock decomposition

(b) block-structured grids

FIGURE 10.16 Application of the hypercube++ approach to an impeller configuration.

would be to impose the size constraint to the hypercube in the hypercube++ generation algorithm. The appropriate size limit on the hypercube++ will not allow the blocks to be cut unnecessarily in the hypercube++ merging algorithm. Second, the current approach cannot generate the hypercube for strongly nonconvex shape elements without dividing them into a set of convex shape elements. A method to generate a well-structured hypercube is desired to deal with a strongly nonconvex shape element. In some cases, the given configuration may have strong nonconvex shape elements as its component. This problem may be resolved by introducing the technique of FFD [25, 26].

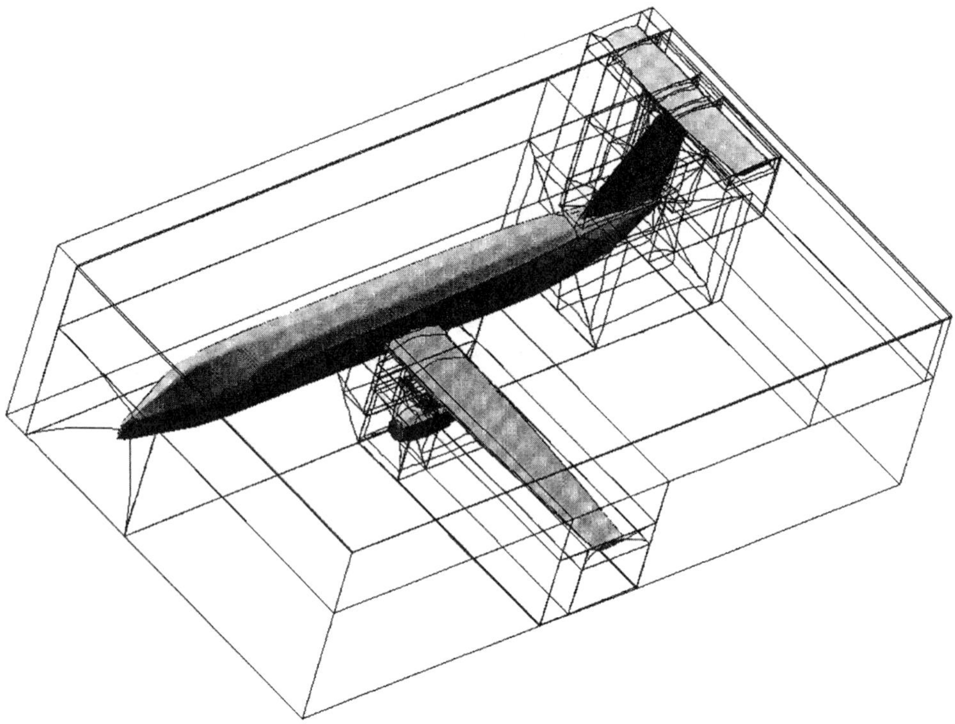

(a) a multiblock decomposition

FIGURE 10.17 Application of the hypercube++ approach to an airplane configuration.

Further Information

A number of Internet sites have World Wide Web home pages displaying grid- or mesh-related topics. The following is just a sample. Other sites containing the electronic information related to the computational fluid dynamics can be found from the following lists.

- http://www-users.informatik.rwth-aachen.de/~roberts/meshgeneration.html (Information on people, research groups, literature, conferences, software, open positions, and related topics)
- http://www.ce.cmu.edu/NetworkZ/sowen/www/mesh.html (A good overview of the current literature available on the subject of mesh generation; conferences, symposiums, selected topics, authors, and other resources)
- http://www.erc.msstate.edu/thrusts/grid/ (Grid technology overview: Historical perspective and state-of-the-art, and accomplishments and significant events in research)
- http://www.erc.msstate.edu/thrusts/grid/cagi/content.html (Introduction to a CAGI system, which can either read the standard IGES format or generate grids from NURBS definition)
- http://www.erc.msstate.edu/education/gumb/html/index.html (Tutorial on a modular multiblock structured grid generation system derived from the structured grid system embedded within the NGP system)
- http://www.tfd.chalmers.se/CFD_Online/ (An overview of the vast resources available on the Internet for people working in CFD)

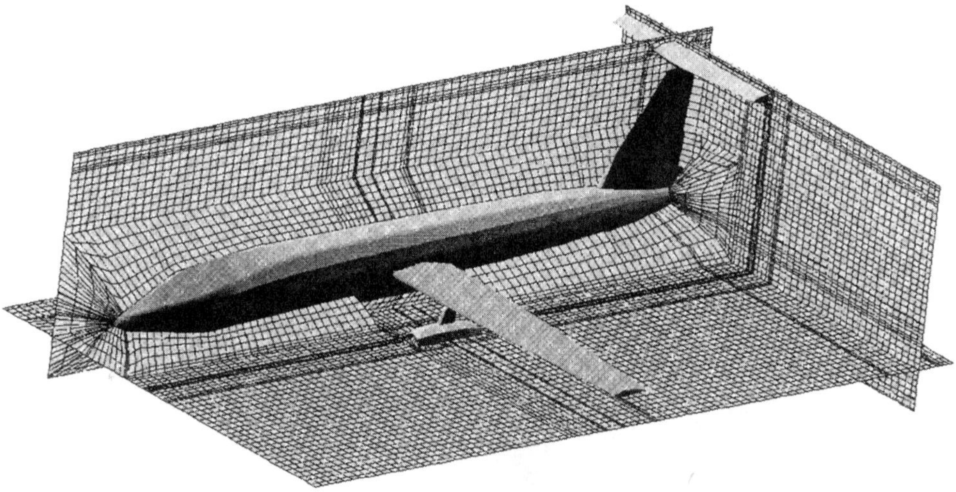

(b) block-structured grids

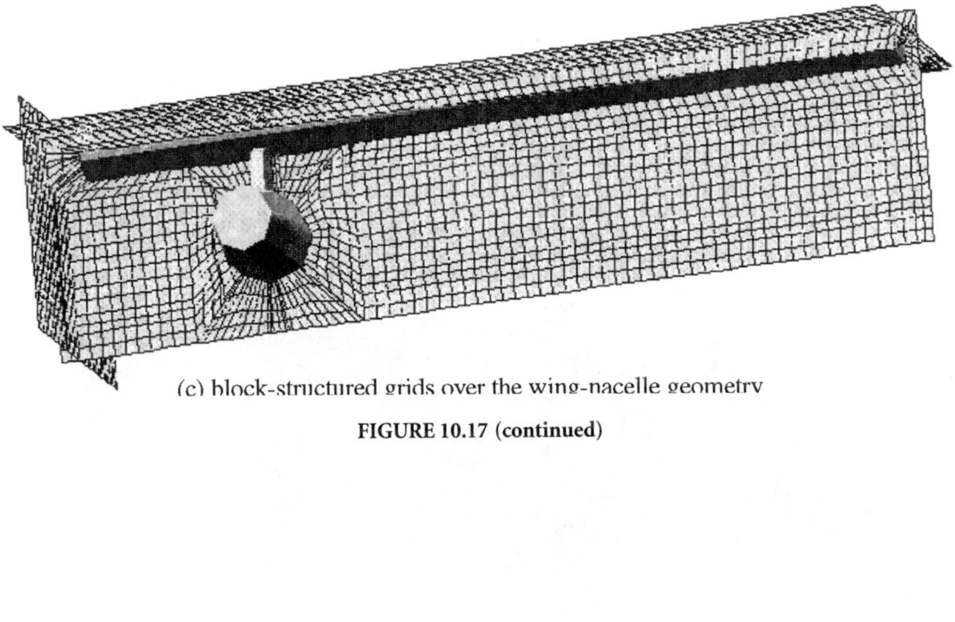

(c) block-structured grids over the wing-nacelle geometry

FIGURE 10.17 (continued)

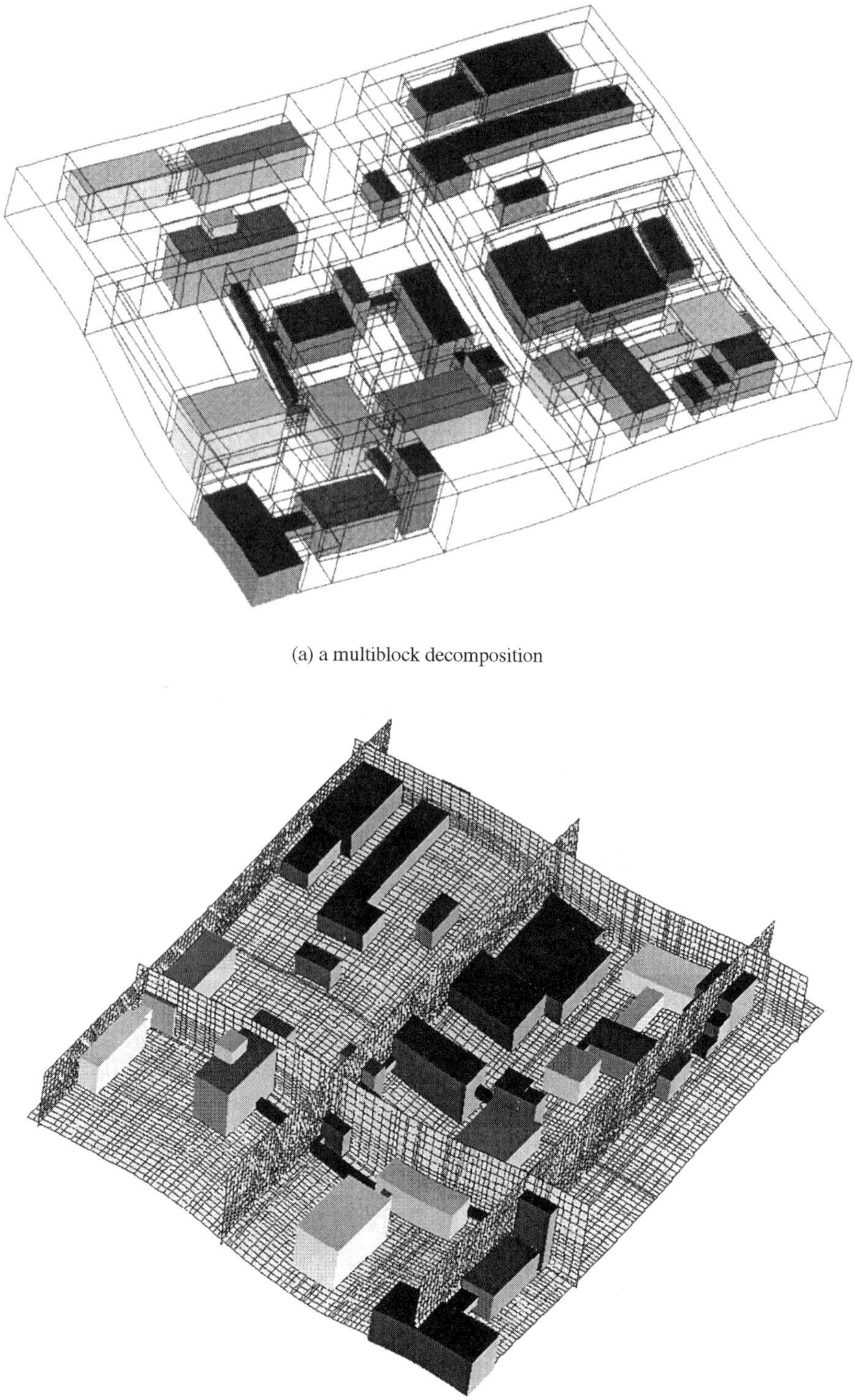

(a) a multiblock decomposition

(b) block-structured grids

FIGURE 10.18 Application of the hypercube++ approach to a building complex configuration.

References

1. Weatherill, N.P. and Forsey, C.R., Grid generation and flow calculations for complex aircraft geometries using a multi-block scheme, AIAA Paper 84-1665. 1984.
2. Arabshahi, A. and Whitfield, D.L., A multi-block approach to solving the three-dimensional unsteady Euler equations about a wing-pylon-store configuration, AIAA Paper 89-3401.1989.
3. Sorenson, R.L. and McCann, K.M., A method for interactive specification of multiple-block topologies, AIAA Paper 91-0147. 1991.
4. Allwright, S.E., Techniques in multiblock domain decomposition and surface grid generation, *Numerical Grid Generation in Computational Fluid Mechanics '88*. Sengupta, S., Thompson, J.F., Eiseman, P.R., and Hauser, J., (Eds.), Pineridge Press, Miami, FL, 1988, pp 559–568.
5. Shaw, J.A. and Weatherill, N.P., Automatic topology generation for multiblock grids, *Applied Mathematics and Computation.* 1992, 53, pp 355–388.
6. Stewart, M.E.M., Domain-decomposition algorithm applied to multielement airfoil grids, *AIAA J.*, 1992, 30.
7. Dannenhoffer, J.F., A new method for creating grid abstractions for complex configurations, AIAA Paper 93-0428. 1993.
8. Dannenhoffer, J.F., A Block-structuring technique for general geometries, AIAA Paper 91-0145. 1991.
9. Blake, K.R. and Spragle, G.S., Unstructured 3D Delaunay mesh generation applied to planes, trains and automobiles, AIAA Paper 93-0673. 1993.
10. Baker, T.J., Prospects and expectations for unstructured methods, *Proceedings of the Surface Modeling, Grid Generation and Related Issues in Computational Fluid Dynamics Workshop*, NASA Conference Publication 3291, NASA Lewis Research Center, Cleveland, OH, May 1995.
11. Marcum, D.L. and Weatherill, N.P., Unstructured grid generation using iterative point insertion and local reconnection, *AIAA J.* 1995, 33, pp 1619–1625.
12. Lohner, R. and Parikh, P., Generation of three-dimensional unstructured grids by the advancing-front method, AIAA Paper 88-0515. 1988.
13. Meakin, R.L., Grid related issues for static and dynamic geometry problems using systems of overset structured grids, *Proceedings of the Surface Modeling, Grid Generation and Related Issues in Computational Fluid Dynamics Workshop*, NASA Conference Publication 3291, NASA Lewis Research Center, Cleveland, OH, May 1995.
14. Wang, Z.J. and Yang, H.Q., A unified conservative zonal interface treatment for arbitrarily patched and overlapped grids, AIAA Paper 94-0320. 1994.
15. Kao, K.H., Liou, M.S., and Chow, C.Y., Grid Adaptation using chimera composite overlapping meshes, *AIAA J.* 1994, 32, pp 942–949.
16. Kallinderis, Y., Khawaja, A., and McMorris, H., Hybrid prismatic/tetrahedral grid generation for viscous flows around complex geometries, *AIAA J.* 1996, 34, pp 291–298.
17. Parthasarathy, V. and Kallinderis, Y., Adaptive prismatic-tetrahedral grid refinement and redistribution for viscous flows, *AIAA J.* 1996, 34, pp 707–716.
18. Steinbrenner, J.P. and Noack, R.W., Three-dimensional hybrid grid generation using advancing front techniques, *Proceedings of the Surface Modeling Grid Generation and Related Issues in Computational Fluid Dynamics Workshop*, NASA Conference Publication 3291, NASA Lewis Research Center, Cleveland, OH, May 1995.
19. Aftosmis, M.J., Melton, J.E., and Berger, M.J., Adaptation and surface modeling for Cartesian mesh methods, *AIAA-95-1725-CP.* 12th AIAA Computational Fluid Dynamics Conference, San Diego, CA, June 1995.
20. Soni, B.K., Huddleston, D.H., Arabshahi, A., and Vu, B., A study of CFD algorithms applied to complete aircraft configurations, *AIAA Paper 93-0784.* 1993.

21. Vatsa, V.N., Sanetrick, M.D., Parlette, E.B., Block-structured grids for complex aerodynamic configurations, *Proceedings of the Surface Modeling Grid Generation and Related Issues in Computational Fluid Dynamics Workshop*, NASA Conference Publication 3291, NASA Lewis Research Center, Cleveland, OH, May 1995.
22. Yu, T.-Y., Soni, B.K., and Shih, M.H., CAGI : Computer Aided Grid Interface," AIAA Paper 95-0243. 1995.
23. Casale, M.S. and Stanton, E.L., An overview of analytic solid modeling, *IEEE Computer Graphics and Applications.* 1985, 5, pp 45–56.
24. Lasser, D., Bernstein-Bezier representation of volumes, *Computer Aided Geometric Design.* 1985, 2, pp 145–149.
25. Barr, A.H., Global and local deformations of solid primitives, *Computer Graphics.* 1984, 18, pp 21–30.
26. Coquillart, S., Extended free-form deformation: a sculpturing tool for 3D geometric modeling, *Computer Graphics.* 1990, 24, pp 187–196.

11

Composite Overset Structured Grids

Robert L. Meakin

11.1 Introduction

The use of composite overset structured grids is an effective means of dealing with a wide variety of flow problems that spans virtually all engineering disciplines. Numerous examples involving steady and unsteady three-dimensional viscous flow for aerospace applications exist in the literature. The literature also chronicles a host of applications of the approach in areas as diverse as biomedical fluid mechanics and meteorology. Many factors provide incentive for adopting the approach. A geometrically complex problem can be reduced to a set of simple components. Arbitrary relative motion between components of multiple-body configurations is accomplished by allowing grid components to move with six degrees of freedom in response to applied and dynamic loads. Limited memory resources can be accommodated by problem decomposition into appropriately sized components. Scalability on parallel compute platforms can be realized through problem decomposition into components (or groups of components) of approximately equal size.

In many ways, a composite overset grid approach is similar to the so-called patched, or block-structured approach (see Chapter 13). However, even though differences between the approaches may appear slight (i.e., one requires neighboring grid components to overlap and the other does not), they are in fact substantial. In an overset approach, grid components are not required to align with neighboring components in any special way. Accordingly, the approach offers an additional degree of flexibility that is not available with patched grids. Steger [1992] observed that an overset grid approach assumes "… characteristics of an unstructured grid finite element scheme that uses large powerful elements in which each element is itself discretized." Indeed, the approach should enjoy many of the grid generation freedoms commonly associated with unstructured grids, while retaining, on a component-wise basis, all of the computational advantages that are inherent to structured data.

The maturation process for overset grid generation tools is ongoing. Historically, application scientists and engineers have used grid generation software designed for patched grids to generate required overset

0-8493-2687-7/99/$0.00+$.50

grid components. Since available software has not been designed with overlapping grids in mind, problem components are typically gridded independently in a sequential fashion. Given the level of geometric and physical complexity that is often required for flow simulation, this practice places a heavy burden on the analyst in terms of time and expertise required to generate needed grids. Fortunately, grid generation schemes that exploit the flexibility inherent to an overset approach are active areas of research [Petersson, 1995; Chan and Meakin, 1997]. Efficient and highly automated methods of overset grid generation and domain decomposition should be available in the near future.

The present chapter is divided into three main sections covering the topics of domain decomposition, domain connectivity, and research issues. These sections are followed by brief sections that define terms, references, and sources for more detailed information on subjects related to overset grids. Terms peculiar to overset grid nomenclature appear in italic at their first occurrence, and are defined in Section 11.5.

For the purposes of this chapter, the starting point for grid generation is assumed to be a trimmed "water-tight" definition of problem surface geometry in a suitable format (e.g., NURBS, or panel networks). Note that the subjects of surface and volume grid generation are covered in other chapters of the handbook (Chapter 9 and 4, respectively) and will be referred to only indirectly in the present chapter. Chapter 5 on hyperbolic grid generation should be of particular interest to anyone seeking more information about the overset grid approach.

11.2 Domain Decomposition

This section covers domain decomposition issues for composite overset structured grids. Included in the discussion are surface geometry decomposition, volume decomposition, and issues peculiar to multiple-body applications.

11.2.1 Surface Geometry Decomposition

All real objects can be viewed as composites of discontinuities (point and line) and simple surfaces. A finite cone, for example, has both point and line discontinuities. Surface geometry entities not associated with point or line discontinuities are simple surfaces. The task of surface geometry decomposition is to partition given problem definitions into sets of surface areas that can readily be converted into overlapping surface grid components. It is worth noting that surface geometry decomposition problems do not have unique solutions. A number of trivial shapes can be represented very well with a single surface (e.g., a sphere, a rectangular flat plate, etc.). However, even simple shapes can be decomposed into component parts and represented with an infinite variety of sets of component surface areas. The present objective is simply to define a convenient set of surface areas to form the basis for surface grid generation. In this chapter, the term *seam* is used to denote surface areas that are associated with either point or line discontinuities in a geometry definition. The term *block* is used to denote simple surface areas. Hence, the task of surface geometry decomposition can be restated as one of partitioning a given problem geometry into a *quilt** of overlapping seams and blocks (see Figure 11.1).

Once a surface definition has been decomposed into seams and blocks, generation of a corresponding number of overlapping surface grids is a conceptually simple task. Most of the basic algorithms needed to develop fully automated surface grid generation software currently exist. Algebraic and elliptic surface grid generation techniques, appropriate for simple surfaces, have long been available (see Chapter 9 of this handbook). The idea for hyperbolic surface grid generation (Chapter 5) was put forward more recently [Steger, 1989], and has since been generalized [Chan and Buning, 1995].

*Quilt nomenclature has been adopted here to describe surface geometry decomposition issues unique to composite overset structured grids. The patches of material stitched together in "patchwork quilts" are commonly know as "blocks." Hence, in this analogy, seam and block surface components correspond to quilt stitches and square quilt patches, respectively.

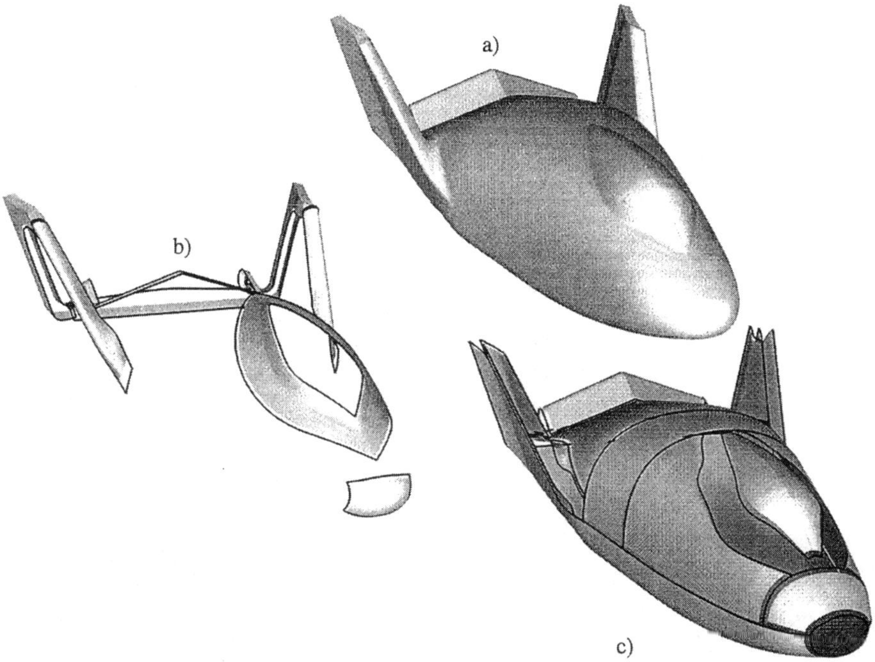

FIGURE 11.1 Surface geometry decomposition into a quilt of seams and blocks. (a) X-38 surface geometry definition, (b) seams over control lines and line discontinuities, (c) blocks over simple surfaces.

11.2.1.1 Seam Topologies

Point discontinuities can exist as a natural feature of an object, such as the tip of a cone. Such situations may dictate the use of a *tip* topology for the surface area in the immediate vicinity of the discontinuity. A tip topology is defined by placing a grid point coincident with the discontinuity and marching away from the point an acceptable distance on the surface. A tip decomposition preserves the point discontinuity in the corresponding surface grid. In addition, a volume grid generated from the surface will have a polar axis that extends from the discontinuity. The existence of an axis generally implies that the flow solver will be required to implement special boundary conditions along the axis. Typically, this means that the flow solution along the axis will be derived from an averaging process involving the nearby off-axis solution. If the point discontinuity is mild, as in a wide-angle cone, it may be acceptable to ignore the discontinuity and use a block topology in the vicinity of the point.

Figure 11.2 indicates two surfaces that could be decomposed with a tip topology. The tip of a generic finned-store is shown in Figure 11.2a, and an aircraft fuselage nose tip is shown in Figure 11.2b. The figure contrasts narrow and wide-angle tips, and illustrates how a wide-angle tip can be appropriately represented as a block (i.e., a simple surface) rather than as a seam.

Surface intersections on an object result in line discontinuities, such as at the junction between an aircraft fuselage and wing. An object can also have line discontinuities as a result of "forced contouring," such as exterior mold lines on an automobile, or crease lines that result from plastic deformation of an object due to stress, or fold lines as on the edges of a box. All line discontinuities that are germane to the flow analysis problem at hand must be faithfully represented in the surface grid system. A seam topology can be defined in the vicinity of a line discontinuity by marching in both line-normal directions away from the line an acceptable distance on the surface, resulting in a quadrilateral patch. In this way, a seam topology can be used as the basis for surface grids that are aligned with the discontinuity and accurately represent the surface shape. Figure 11.3 indicates three example seams aligned with surface line discontinuities. Seam components are indicated in the figure for a fuselage crease-line, rotor-blade trailing edge line, and fin-store intersection line. In some of the Chimera literature*, seam topologies like that shown in Figure 11.3c are referred to as *collars* [Parks et al., 1991].

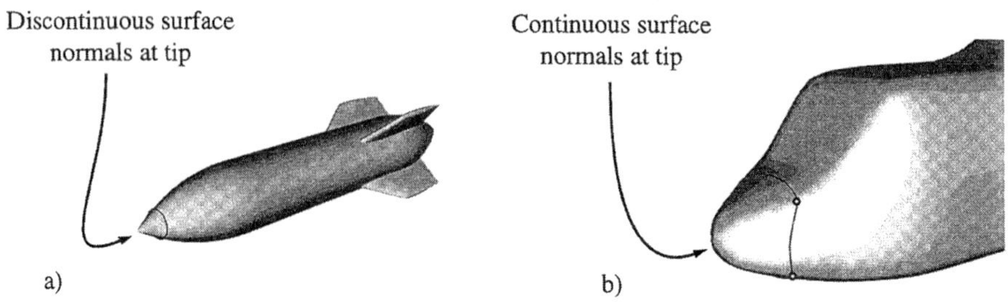

FIGURE 11.2 Seam topologies. (a) Sharp nose of a store decomposed into a surface tip. The radial boundary of the seam is indicated by a thin black line, (b) blunt nose of a fuselage decomposed into a quadrilateral surface area. Seam boundaries are indicated by thin black line segments. Dots indicate seam boundary corners.

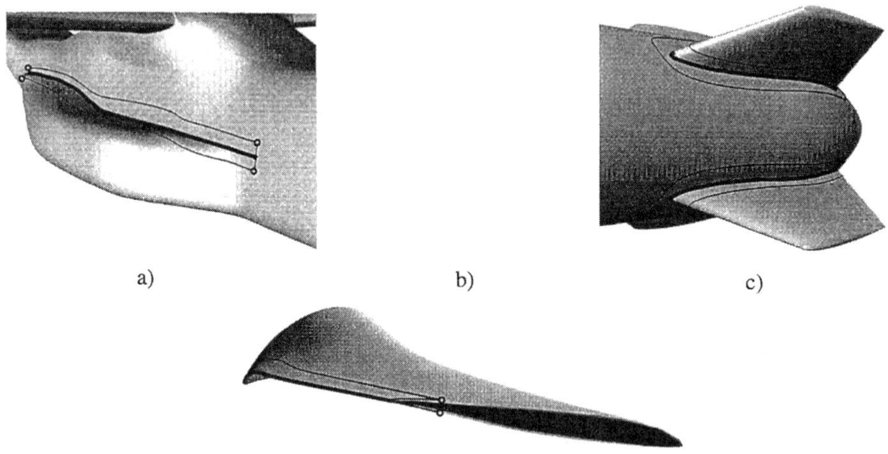

FIGURE 11.3 Surface geometry decomposition into seams over line discontinuities. Discontinuities are indicated by thick black lines. Seam boundaries are indicated by thin black lines. Dots indicate seam boundary corners. (a) V-22 fuselage/sponson crease, (b) rotor-blade trailing edge, (c) fin-store intersections.

In addition to actual line discontinuities in an object surface, it is sometimes desirable to align grid lines on a surface for other reasons. For example, even though the leading edge of a wing generally has a smooth radius of curvature, and is not a surface discontinuity, accurate flow simulations require a high degree of geometric fidelity of this aspect of a wing surface definition. This is easily done by identifying the wing leading edge as a surface control line, and decomposing the wing surface with a seam topology in the vicinity of the leading edge (see Figure 11.4a). Other examples of seam topologies are shown in Figures 11.1 and 11.4b. Figure 11.1 shows a possible surface geometry decomposition of the X-38 (crew return vehicle). Specifically, Figure 11.1b shows seam components at the vehicle nose, around the canopy, and over the rims of the twin vertical tails. Additional seam topologies are also indicated in the figure (less visible) for various components of the aft portion of the vehicle. Figure 11.4b shows a seam component over the tip of a rotor blade, which avoids the special boundary conditions required by "slit" topologies commonly used as wing and blade tip endings. Seams like this can provide a higher degree of geometric fidelity to the grid system employed than is realizable by collapsing a wing, or blade-tip, into a slit.

*A "Chimera" is a mythological creature made up of incongruent parts of other beasts. Steger appropriately coined the term "Chimera overset grids" to indicate a powerful way to apply structured grid solution techniques to geometrically complex multibody configurations.

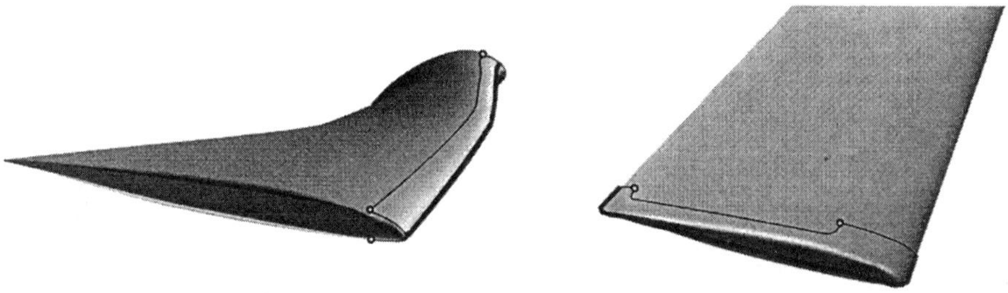

a) b)

FIGURE 11.4 Surface geometry decomposition into seams over control lines. (a) Rotor-blade leading edge control line, (b) outboard blade-tip ending control line. Control lines are indicated by thick black lines. Seam topology boundaries are indicated by thin black lines. Dots indicate seam boundary corners.

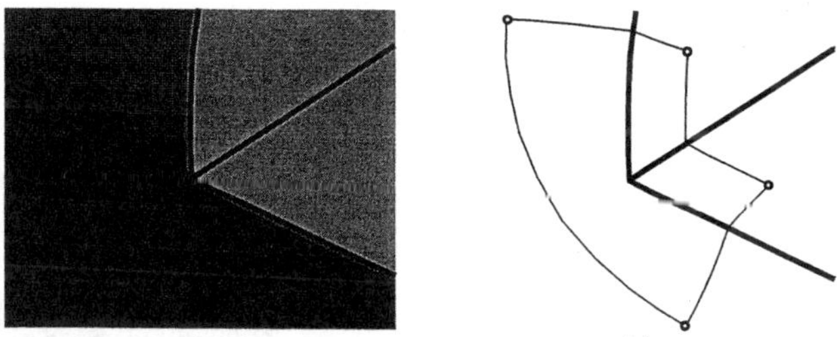

a) b)

FIGURE 11.5 Surface geometry decomposition in the vicinity of a point discontinuity. (a) Intersection of three line discontinuities, (b) seam topology over the point of intersection. Line discontinuities are indicated by thick black lines. Seam topology boundaries are indicated by thin black lines. Dots indicate seam boundary corners.

A final topology that deserves mention here is one for point discontinuities that result from the intersection of three surfaces, such as exist at the corners of a box. This type of discontinuity defines the point of intersection of three line discontinuities. In most cases of this type, the appropriate decomposition is a seam topology like that which is used for simple line discontinuities. The situation is illustrated in Figure 11.5 for a component of the X-38. For this type of decomposition, two of the three intersecting lines are concatenated into one coordinate line. The seam is then defined by marching in both line-normal directions away from the concatenated line an acceptable distance on the surface, while constraining one of the line-normal seam lines to be co-linear with the third line discontinuity. If the three angles of intersection of a corner point are all narrow, then the corner will approximate a cone and a tip topology can be used instead.

11.2.1.2 Block Topologies

Blocks are simple surface areas, or areas that contain mild discontinuities that can effectively be ignored (as in the case shown in Figure 11.2b). Typically the surface area of a given geometry definition can be decomposed primarily into such blocks. For example, Figure 11.1c shows the blocks corresponding to one possible decomposition of the X-38. Block boundaries are always quadrilateral and represent the simplest basis from which structured surface grid components can be generated.

11.2.2 Volume Geometry Decomposition

A convenient way to think of volume decomposition is to categorize the physical domain of a problem into near-body and off-body regions. The near-body portion of a domain is defined to include all seams and blocks required to describe problem surface geometry and the volume of space extending a short distance away from the respective surfaces. The off-body portion of a domain is defined to be the domain

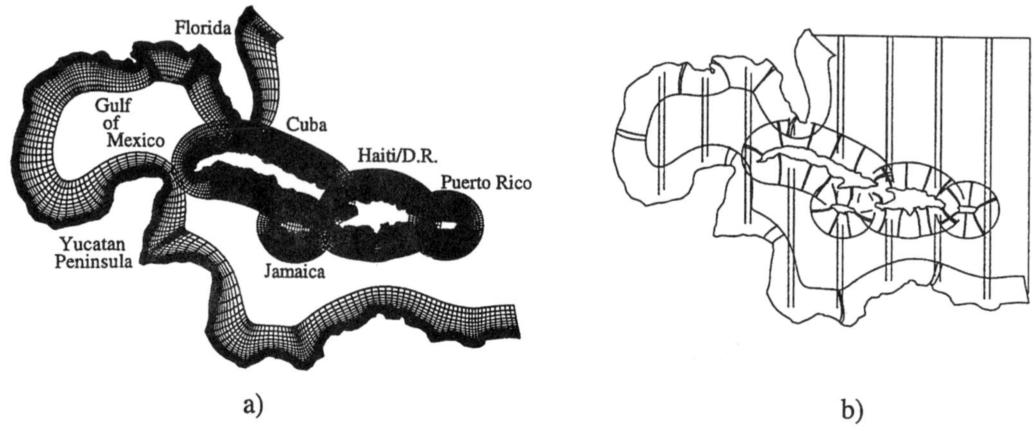

FIGURE 11.6 Overset grid components for unsteady simulation of basin-scale oceanic flows. (a) Body-fitted grid components for the Gulf of Mexico and the Greater Antilles Islands, (b) background Cartesian grid component boundaries.

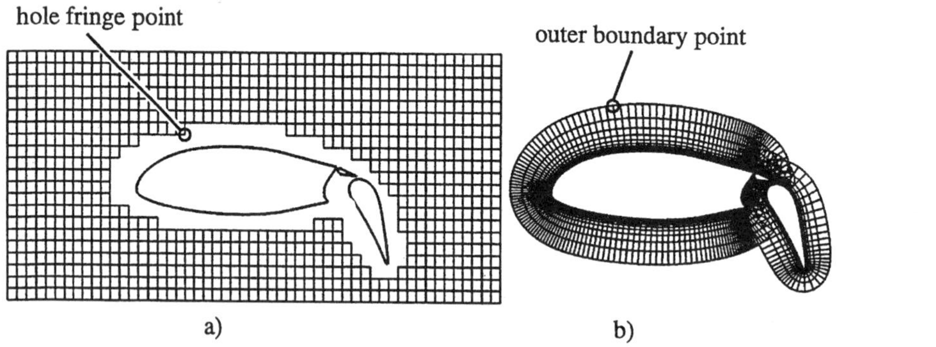

FIGURE 11. 7 Grid components for a flapped-wing configuration. (a) Background Cartesian grid with Chimera hole caused by the wing, (b) body-fitted grid components about the wing and double-slotted flap.

volume not covered (except that required for minimum overlap) by the near-body volumes. The aspect of a *Chimera* overset grid approach that trivializes off-body grid generation is the fact that off-body volume components can overlap the near-body domain by an arbitrary amount. Hence, the off-body domain can be filled using any convenient set of topologies. Typically, Cartesian systems are used for this purpose (e.g., see Figures 11.6b and 11.7a). Hyperbolic grid generation schemes can efficiently generate high quality near-body grids radiating from appropriate quilts of overlapping surface grid components. Generation of off-body Cartesian volume grids (Chapter 22) is trivial for this application.

Although the idea of solving differential equations on overlapping domains is very old [Schwarz, 1869], the idea did not blossom into a practical analysis tool until relatively recent times. Steger et al. [1983] introduced the Chimera method of domain decomposition to treat geometrically complex multiple-body configurations using composite over-set structured grids. In the approach, curvilinear body-fitted structured grids are generated about body components and overset onto systems of topologically simple background grid components. Solutions to the governing flow equations are then obtained by solving the requisite systems of difference equations according to some prescribed sequence on all grid components. Physical boundary conditions are enforced as usual (e.g., no-slip conditions at solid surfaces), while intergrid boundary conditions are supplied from solutions in the overlap regions of neighboring grid components. The solution procedure is repeated iteratively to facilitate free transfer of information between all grids, and to drive the overall solution to convergence. Intergrid boundary conditions are typically updated explicitly.

FIGURE 11.8 Selected surface grid components for a tiltrotor and flapped-wing configuration. Rotor blade grids move relative to stationary wing, nacelle, and background grid components.

Examples of the Chimera method of domain decomposition are illustrated in Figures 11.6 through 11.9 (see Chapter 5 of this handbook for more examples). Figure 11.6 indicates a set of overlapping grids for unsteady simulation of basin-scale oceanic flows in the Gulf of Mexico [Barnette and Ober, 1994]. Body-fitted grids are used to discretize the Gulf coastline and the Greater Antilles Islands. The body-fitted grids are overset onto a system of Cartesian grids that cover the rest of the oceanic region enclosed within the Gulf Coast solution domain. In the figure, the outlines of nine Cartesian grid components are indicated. However, the number of off-body Cartesian grids used is arbitrary. The body-fitted island grids of Figure 11.6 are topologically similar to the body-fitted grids used to discretize the flapped-wing illustrated in Figures 11.7 and 11.8. Figure 11.8 is illustrative of the capacity of an overset grid approach to accommodate relative motion between problem components. The grid components shown in Figure 11.8 are for a tiltrotor and flapped-wing configuration [Meakin, 1995]. Grids associated with the rotor-blades move relative to stationary wing and background grid components. Figure 11.9 shows a detail of some of the overlapping surface grid components of the integrated space shuttle vehicle [Gomez and Ma, 1994]. The figure indicates the degree of geometric complexity and fidelity that has been realized with the approach.

The novel contribution of Chimera to the overall approach of structured grid based domain decomposition is the allowance for *holes* within grid components. For example, the rotor-blade grids shown in Figure 11.8 cut through several other grid components during the course of a simulation. Likewise, the flapped-wing grids cut holes in background Cartesian grid systems. A detail of this is shown in Figure 11.7, where a hole is cut in a background Cartesian grid by the flapped-wing. As indicated in the figure, a Chimera domain decomposition gives rise to two types of intergrid boundary points: hole *fringe points* and grid component *outer boundary points.*

It is a relatively simple matter to adapt any viable structured grid flow solver to function within the framework of Chimera overset grids. For example, the implicit approximately factored algorithm of Warming and Beam [1978] for the thin-layer Navier–Stokes equations

$$\partial_\tau \hat{Q} + \partial_\xi \hat{E} + \partial_\eta \hat{F} + \partial_\zeta \hat{G} = Re^{-1} \partial_\zeta \hat{S} \tag{11.1}$$

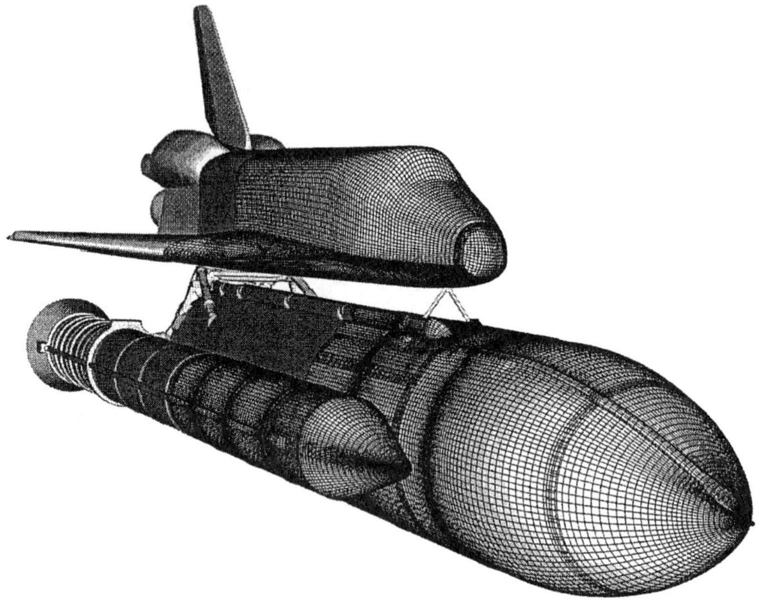

FIGURE 11.9 Selected surface grid components from a composite overset grid discretization of the integrated space shuttle vehicle.

is easily expressed in Chimera form as

$$\left[I + i_b \Delta t \delta_\xi \hat{A}^n\right]\left[I + i_b \Delta t \delta_\eta \hat{B}^n\right]\left[I + i_b \Delta t \delta_\zeta \hat{C}^n - i_b \Delta t Re^{-1} \delta_\zeta J^{-1} \hat{M}^n J\right]\Delta \hat{Q}^n = -i_b \Delta t\left(\delta_\xi \hat{E}^n + \delta_\eta \hat{F}^n + \delta_\zeta \hat{G}^n - Re^{-1} \delta_\zeta \hat{S}^n\right) \tag{11.2}$$

The single and overset grid versions of the algorithm are identical except for the variable i_b, which accommodates the possibility of having arbitrary holes in the grid. The array i_b has values of either 0 (for hole points), or 1 (for conventional *field points*). Accordingly, points inside a hole are not updated (i.e., $\Delta Q = 0$) and the solution values on intergrid boundary points are supplied via interpolation from corresponding solutions in the overlap region of neighboring grid systems. By using the i_b array, it is not necessary to provide special branching logic to avoid hole points, and all vector and parallel properties of the basic algorithm remain unchanged [Steger et al., 1983].

11.2.3 Chimera Hole-Cutting

Definition of the Chimera i_b array is an important step in the realization of the several advantages of a general composite overset structured grid approach. The i_b array accommodates the possibility of arbitrary holes in grid components, and thereby, allows efficient execution of the structured grid flow solver being used. The only nontrivial task associated with the definition of i_b is to determine whether points in a given grid component lie inside specified hole-cutting surfaces. A number of procedures are available to make this determination. Consider point P and a surface S defined by a group of surface grid components taken from an existing set of volume grids (see Figure 11.10).

11.2.3.1 Surface Normal Vector Test

Let $\boldsymbol{r}_P$ be the position vector of point P, $\boldsymbol{r}_{ij}$, the position vectors* of discrete points on S, and $\boldsymbol{n}_{ij}$ the surface normal vectors at points $\boldsymbol{r}_{ij}$. Point P is outside of surface S if any of the dot products $(\boldsymbol{r}_P - \boldsymbol{r}_{ij}) \cdot \boldsymbol{n}_{ij} > 0$.

*Note that the use of "ij" here is to denote grid indices, rather than tensor rank.

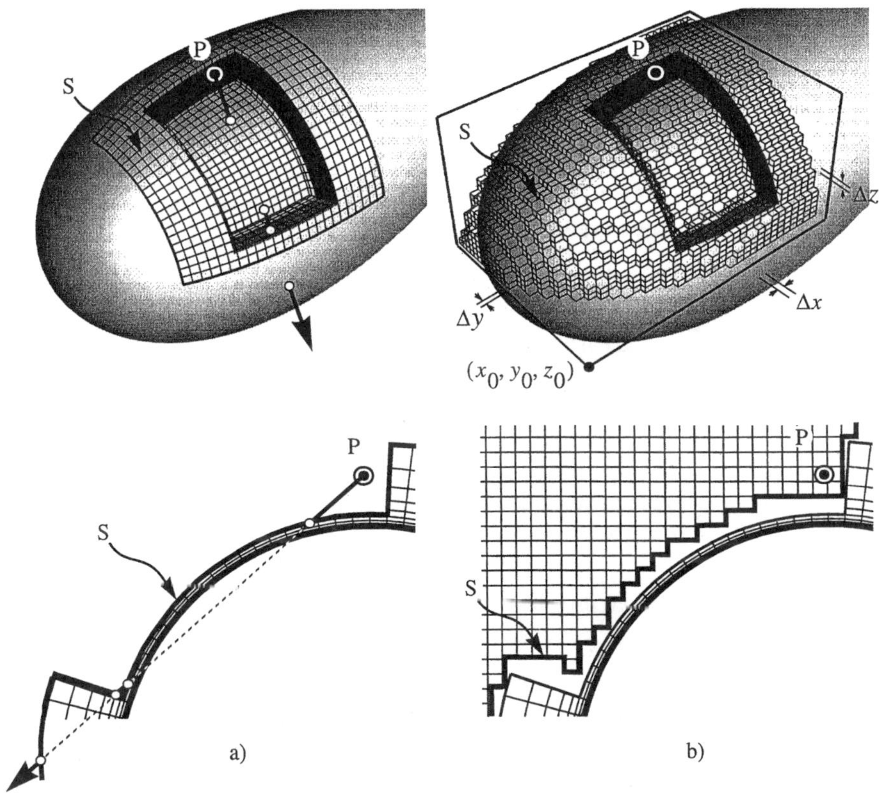

FIGURE 11.10 Chimera hole cutting. Given cutting surface S, determine the hole/field status of point P. (a) Hole cutting surface defined by collection of surfaces from existing grid components. (b) Approximate hole cutting surface represented with a uniform Cartesian "hole-map."

The cost of this operation is proportional to the number of points in the grid component being tested and the number of points used to define the hole-cutting surfaces. Typically, the cost of the test is reduced by trading dot product computations for computations to determine the Euclidean distance between point couples. Hence, the hole-cutter surface point nearest to point P is first identified. Then, only the dot product between P and the nearest hole-cutter surface point needs to be computed.

The surface normal vector test has one significant failure mode. Hole-cutting surfaces, viewed from the outside, must be convex. Even if hole-cutting surfaces are constructed from multiple surface grid components, the composite surface must be closed and convex. Hole-cutting surfaces that have concavities must be broken into multiple closed convex surfaces.

11.2.3.2 Vector Intersection Test

The number of intersections between an arbitrary ray extended from a point P and any closed hole-cutting surface can be used as the basis of a robust and unambiguous inside/outside test. If a ray intersects the closed surface an odd number of times, then the point is inside. If the ray intersects the surface zero or an even number of times, the point is outside. The test is illustrated in Figure 11.10a with an arbitrary ray drawn from a test point in the proximity of S. Results of the test are independent of the direction in which rays are extended from the test points, and do not require that the hole-cutting surfaces be convex.

If a ray extended from a test point intersects the hole-cutting surface at an edge of a face that is coplanar with the ray, the test will fail. However, the failure is easily overcome by redefining the ray in a random direction away from the offending face. Implementation of this test is more complicated than for the surface normal vector test. Still, the test is practical and may provide a more robust mechanism for hole determination.

11.2.3.3 Uniform Cartesian Test

An idea suggested by Steger [1992] offers an efficient means of hole point determination that may prove to be the basis of future fully automatic domain connectivity algorithms. A closed surface S can be enclosed within a uniform Cartesian grid, as indicated in Figure 11.10b. Points in the grid can be marked as being inside or outside of S very easily [Chiu and Meakin, 1995]. A uniform Cartesian grid so marked becomes an approximate representation of S and is referred to as a *hole-map*. The proximity of any point P with respect to surface S can be determined by consulting the hole-map of S.

Given the coordinates of P, the corresponding bounding hole-map element can be computed directly as

$$j=\frac{(x_P-x_0)}{\Delta x}+1,\quad k=\frac{(y_P-y_0)}{\Delta y}+1,\quad l=\frac{(z_P-z_0)}{\Delta z} \tag{11.3}$$

where x_0, y_0, z_0 are the coordinates of the hole-map origin, and Δx, Δy, Δz are the hole-map spacings. If the eight vertices of hole-map element j, k, l are all marked as a hole, then P is inside the hole-cutting surface. If the eight vertices are all unmarked, P is outside the surface. However, if the eight vertices are of mixed type (marked and unmarked), P is near a hole-cutting plane and a simple radius test, or the vector intersection test can be used to determine the actual status of P.

11.2.4 Identification of Intergrid Boundary Points

The solution of field equations on overlapping systems of grids requires numerical boundary conditions to be supplied at all intergrid boundary points. Given a definition of the i_b array, it is very easy to identify the intergrid boundary points that exist in all components of an overset system of grids. Points on grid component outer boundaries that are not physical boundaries (e.g., no-slip surfaces), conventional numerical boundaries (e.g., planes of symmetry, inflow/outflow, etc.), or Chimera hole-points, are intergrid boundary points. In addition, field points bordered by one or more hole-points are also intergrid boundary points (i.e., *fringe-points*).

Accordingly, a list of all intergrid boundary points can be made by inspecting the i_b array on a grid-by-grid basis. The list must include the grid component identity, and the j, k, l and x, y, z coordinates of each intergrid boundary point in the system of grids. Such a list completely defines the domain connectivity needs associated with a given overset system of grids and specified hole-cutter definitions.

11.3 Domain Connectivity

The costs of the advantages inherent to an overset grid approach are reflected in the need to establish domain connectivity. Domain connectivity is the communication of dependent variable information between grid components. Transmission of this information occurs through the intergrid boundary points of a problem. Specifically, values of the dependent variables are defined on intergrid boundary points by interpolation from the interior of overlapping neighboring grid systems. Accordingly, the domain connectivity solution for a given system of overlapping grids is the identity of a suitable *donor* element for each intergrid boundary point in the system. The present section describes basic algorithms for establishing domain connectivity among general systems of overlapping structured grids.

General implementations of the method must allow for grid components posed in curvilinear coordinate systems. This fact makes the task of establishing domain connectivity nontrivial. The position of points within all grid components is defined relative to a fixed reference frame. Data structure is realized on a component-wise basis due to the fact that grid points are distributed along curvilinear coordinate lines. However, the coordinate systems of the respective grid components are mutually independent. Hence, there is no direct correspondence between the computational space of one grid component and that of any other component in the system. The task of establishing domain connectivity can be stated for a single intergrid boundary point as follows. Given an intergrid boundary point P, identify a grid

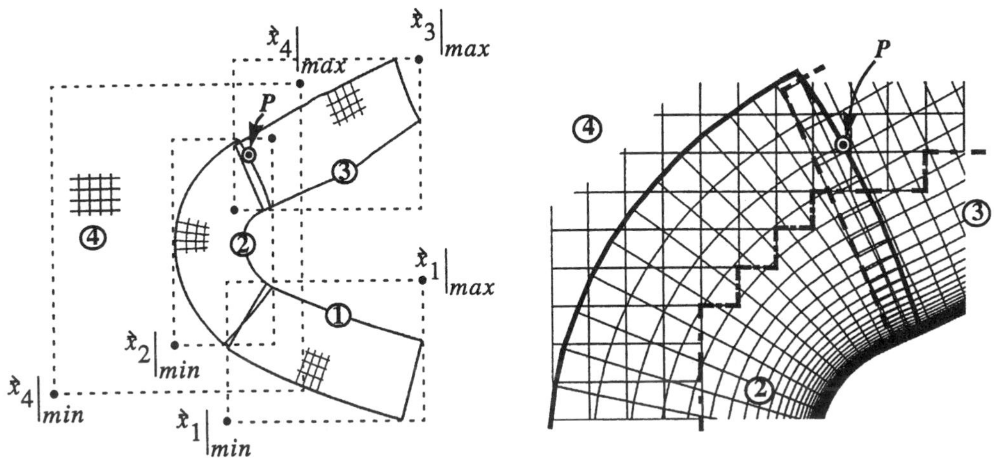

a) Overlapping grid components. b) Detail of intergrid boundaries.

FIGURE 11.11 Donor grid identification using simple min/max bounding boxes. (a) Three overlapping grid components overset onto a background Cartesian grid. Bounding-boxes are indicated by light dashed lines. Black dots are used to indicate bounding box diagonal end points. (b) Detail of intergrid boundaries. Point *P* is bounded by an element from component 3 and from component 4.

component that can satisfy the domain connectivity needs of *P*, and the position of *P* within the computational space of the donating component. The following sections of this chapter describe alternative methods of establishing domain connectivity for a single intergrid boundary point. Of course, to establish domain connectivity for an entire overset system of grids, any such method would need to be applied to all of the intergrid boundary points in the system.

11.3.1 Donor Grid Identification

Typically, only one donor element will exist for a given intergrid boundary point. Indeed, an individual intergrid boundary point can be bounded by only one element from any one overlapping grid component. However, it is not uncommon for some intergrid boundary points to be overlapped by more than one neighboring grid component, leading to the possibility of multiple donor elements for such points. The situation is illustrated in Figure 11.11, where intergrid boundary point *P* is overlapped by an element from two different grid components. Identification of the grid which provides interpolation information for point *P* depends on which donor element provides the best match. A discussion of "best" donor is given shortly. However, first consider a method for identifying all grid components that might contain a donor element for point *P*.

The extreme values, $(x_{min}, y_{min}, z_{min})$ and $(x_{max}, y_{max}, z_{max})$, of the reference frame coordinates of any grid component define the diagonal end-points of a rectilinear box that encompasses the entire component (e.g., see Figure 11.11a). Even for an overset system of grids that contains numerous grid components, the information required to define all diagonal end-points is minimal (*viz.*, $6 \times N$, where N is the number of grids). A necessary condition for donor grid identification is that *P* be bounded by the diagonal end-points of the grid component in question, i.e.,

$$x_{min} < x_P < x_{max}, \quad y_{min} < y_P < y_{max}, \quad z_{min} < z_P < z_{max} \tag{11.4}$$

If the grid component is Cartesian, then Eq. 11.4 is sufficient proof that the component contains a donor element for *P*. However, in general, overset grid discretizations are comprised of (at least some) non-Cartesian grid components. Therefore, in general, Eq. 11.4 is only an indicator of donor potential. For example, Figure 11.11a illustrates three overlapping grid components overset onto a background Cartesian grid component. The bounding-box diagonals readily indicate donor potential between the four components shown. The present discussion considers identification of the donor grid and element for a single intergrid boundary point. In practice, this information is sought for groups of intergrid boundary points at a time. In this sense, the information available through Eq. 11.4 also provides a simple mechanism for identifying all intergrid boundary points that may have a donor in a given grid component.

For intergrid boundary points that are bounded by an element from more than one neighboring grid component, a choice must be made as to which element will be allowed to provide the needed donor information. Current domain connectivity algorithms employ only a rudimentary set of rules for determining the acceptability of available donor elements. The most fundamental rule is that none of the vertices which define a donor element can be hole points. Values of the dependent variables are not defined at hole points. Hence, acceptance of donor elements with any number of hole-point vertices would corrupt the transfer of dependent variable information to the receiving intergrid boundary point. Typically, the first donor element identified that satisfies the rudimentary set of donor acceptability rules is used when more than one bounding element exists for a given intergrid boundary point.

The accuracy of dependent variable information transfer is maximum when the geometric properties of donor and recipient elements are comparable. The relative volume size, orientation, and aspect ratios between donor and recipient elements govern sources of numerical error in the intergrid communication process. Of course, the magnitude of numerical error is proportional to the gradients of the dependent variables being communicated between the grids. Hence, if the dependent variables are represented smoothly in donor and recipient grids, then the error will be small. Indeed, formal solution accuracy can be maintained on overlapped systems of grids using simple interpolation [Meakin, 1994]. Hence, given a robust method of adaption to guarantee smooth variation of dependent flow variables in computational space, the existing rudimental rules of donor element acceptability should be sufficient.

There are two reasons why this is not the case in practice, and that donor acceptability rule definition constitutes a valid area of research. First, very few flow solvers that accommodate an overset grid approach also have an adaptive capability. Therefore, resolution of gradients of the dependent variables cannot be ensured in most overset grid solvers currently available. In addition, the magnitude of interpolation error for resulting applications can only be estimated after the fact. Second, donor acceptability rules based on geometric measures of goodness will accept only the best available donor element in instances where multiple donors exist. Probably much more significantly, rules based on geometric measures of goodness can form the basis of an iterative procedure to obtain the best realizable domain connectivity solution from a given system of grids. Maximization of domain connectivity solution quality will even contribute to error reduction when coupled with solution adaption.

11.3.2 Donor Element Identification

Once it has been determined that a given grid component may contain a suitable donor for an intergrid boundary point P, the task remains to identify the actual element that bounds P and evaluate its acceptability. Some of the acceptability issues have been discussed in the previous section. In the present section, methods of donor element identification are given. By far the simplest method of donor element identification is an exhaustive search. Such a scheme would involve testing all possible elements within a grid component to determine if point P is inside, or not. Although simple inside/outside tests can be devised, the cost of an exhaustive search is prohibitive for all but highly idealized problems. Fortunately, a variety of alternatives to an exhaustive search exist. Since all search procedures require an inside/outside test, a particularly useful test is described below. Then, some of the search algorithms in common use within available domain connectivity codes are discussed.

11.3.2.1 Inside/Outside Test

Let $\mathbf{x}(\mathbf{s}_P)$ define the reference frame coordinates of intergrid boundary point P as a function of the computational space coordinates $\mathbf{s}$ of a candidate interpolation donor element. Values of $\mathbf{x}$ and $\mathbf{s}$ are known for the eight vertices of the candidate donor element. The value of $\mathbf{x}$ is also known for point P.

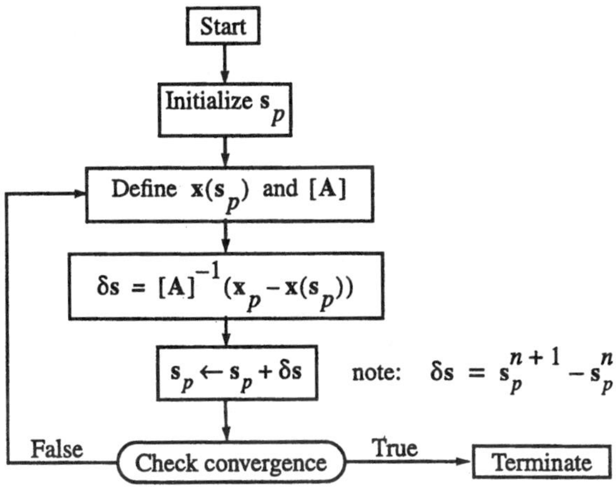

FIGURE 11.12 Iteration to solve Eq. 11.5 for the computational space position of a point relative to the origin of a candidate interpolation donor element.

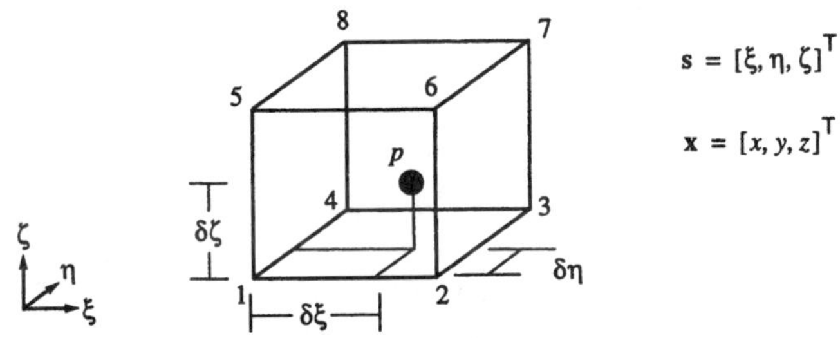

FIGURE 11.13 The computational space of a candidate interpolation donor element for point P.

We seek values of **s** for point P, $\mathbf{s}_P$. If P is inside the element, values of $\mathbf{s}_P$ will be bounded between 0 and 1. A quadratically convergent iterative scheme for $\mathbf{s}_P$ can be constructed from Eq. 11.5 and is outlined in Figure 11.12.

$$\mathbf{x}_P = \mathbf{x}(\mathbf{s}_P) + [\mathbf{A}]\delta\mathbf{s} \tag{11.5}$$

If the solution to Eq. 11.5 produces values of $\mathbf{s}_P < 0$, or $\mathbf{s}_P > 1$, P is outside the candidate donor element. The expressions used to define $\mathbf{x}(\mathbf{s}_P)$ and [**A**] depend on the specific interpolation scheme used to define the variation **s** of within the donor element. A definition of $\mathbf{x}(\mathbf{s}_P)$ is given by Eq. 11.6 below, assuming the use of trilinear interpolation and the element notation indicated in Figure 11.13.

$$\begin{aligned}\mathbf{x}(\mathbf{s}_P) = \mathbf{x}_1 &+ (-\mathbf{x}_1 + \mathbf{x}_2)\xi + (-\mathbf{x}_1 + \mathbf{x}_4)\eta + (-\mathbf{x}_1 + \mathbf{x}_5)\zeta \\ &+ (\mathbf{x}_1 - \mathbf{x}_2 + \mathbf{x}_3 - \mathbf{x}_4)\xi\eta \\ &+ (\mathbf{x}_1 - \mathbf{x}_2 - \mathbf{x}_5 + \mathbf{x}_6)\xi\zeta \\ &+ (\mathbf{x}_1 - \mathbf{x}_4 - \mathbf{x}_5 + \mathbf{x}_8)\eta\zeta \\ &+ (-\mathbf{x}_1 + \mathbf{x}_2 - \mathbf{x}_3 + \mathbf{x}_4 + \mathbf{x}_5 - \mathbf{x}_6 + \mathbf{x}_7 - \mathbf{x}_8)\xi\eta\zeta\end{aligned} \tag{11.6}$$

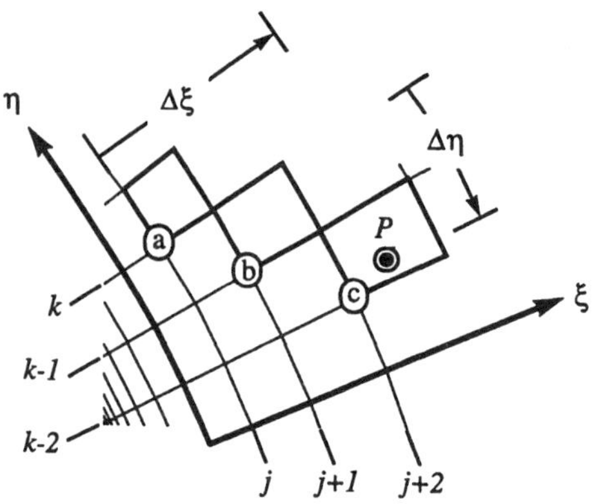

FIGURE 11.14 Gradient search of a given grid component for the element that bounds point P. Search initiated in element "a" and terminated in bounding element "c."

The gradient matrix [**A**] is simply

$$[\mathbf{A}]=\begin{bmatrix} \frac{\partial}{\partial \xi}x(s_p) & \frac{\partial}{\partial \eta}x(s_p) & \frac{\partial}{\partial \zeta}x(s_p) \\ \frac{\partial}{\partial \xi}y(s_p) & \frac{\partial}{\partial \eta}y(s_p) & \frac{\partial}{\partial \zeta}y(s_p) \\ \frac{\partial}{\partial \xi}z(s_p) & \frac{\partial}{\partial \eta}z(s_p) & \frac{\partial}{\partial \zeta}z(s_p) \end{bmatrix} \tag{11.7}$$

where the elements of [**A**] are the corresponding derivatives of Eq. 11.6.

Although other methods exist to determine whether, or not, a point is inside a given element, the iteration defined by Eq. 11.5 is certainly adequate. Equation 11.5 is an expression of the method of steepest descents, and can be used to drive a gradient search procedure* for the bounding element of *P*.

11.3.2.2 Gradient Search

The use of a gradient search procedure to find an element within a structured grid component that bounds a given intergrid boundary point can be very effective [Benek et al., 1986]. A search for the element that bounds point *P* can be initiated from an arbitrary element in the donating grid component. However, typical implementations of the method often begin by evaluating the euclidean distance between *P* and points on the grid component outer boundary. The grid component outer boundary point nearest to *P* defines a convenient element from which to initiate the search. If this element fails, at least the actual donor is nearby and, hopefully, only a few steps of the search procedure will be required to find it.

In any case, the element identified as the search starting point is considered as a candidate donor element for point *P*, and Eq. 11.5 is solved for the local coordinate increments $\mathbf{s}_P$ from the candidate donor element origin. If the vector components of $\mathbf{s}_P$ are bounded between 0 and 1, then *P* is inside the element and the search is complete. If any of the components of $\mathbf{s}_P$ are outside these bounds, the search must be continued. However, the direction (i.e., gradient), in computational space, to the element that bounds *P* is indicated by $\mathbf{s}_P$. For example, consider the situation indicated in Figure 11.14. If Eq. 11.5 is solved for the element "a" and point *P* indicated, the vector components (2D example) of $\mathbf{s}_P$ would be $[\xi > 1), (\eta < 0)]^T$. Accordingly, the gradient to the actual bounding element points in the $+j$, $-k$ direction in computational space. Further, the correct donor to the example indicated in Figure 11.14 would be identified on the second application of Eq. 11.5 from element "a."

*A gradient search method is commonly referred to as "stencil-walking" in the Chimera overset grid literature.

General implementations of the method must take into account certain grid topological situations that can obscure the path (in computational space) to needed donor elements. A periodic plane, for example, poses a minor difficulty. The same is true for "slit" topologies (e.g., wake cuts). The search procedure will step in computational space toward the element that bounds *P* from the side of the periodic (or slit) plane where the search is initiated. The search will terminate on either the actual bounding element, or on the element nearest *P* on the "wrong" side of the periodic (or slit) plane. A grid about a thin body may also pose a similar type of problem. If a search is initiated for the element that bounds *P* near the actual bounding element in physical space, but far from it in computational space (i.e., because it's on the opposite side of the body), the search may fail. In this case, the search procedure would step in computational space toward the actual bounding element, but terminate on the thin body surface nearest the actual bounding element, but on the "wrong" side of the body. Pathological search situations that arise because of topological issues can easily be remedied by allowing for restart locations within the grid component.

General search algorithms must also allow for degenerate elements. For example, the donor element for a given intergrid boundary point may reside in the volume grid generated from a surface seam component that has a "tip" topology. Accordingly, the volume grid will have a polar axis that emanates from the point discontinuity at the body surface, and all grid elements associated with the axis will have a collapsed edge. Axis elements are prismatic, rather than hexahedral. This type of element can be acceptable. Therefore, a robust domain connectivity algorithm must detect candidate donor element degeneracies and allow the gradient search procedure to continue when encountered.

11.3.2.3 Spatial Partitioning

A viable alternative to an exhaustive search approach to donor identification is spatial partitioning. There are numerous methods of this type. Applied to the domain connectivity problem, the methods involve partitioning the physical space of a grid component into rectilinear volumes of space, and establishing a correspondence between partitions and the grid points they contain. Then, the task of domain connectivity is to identify the partition that bounds a given intergrid boundary point, and apply an exhaustive search within the partition to find the actual bounding element. The methods differ primarily in the allowable levels and mechanisms of partitioning.

The simplest spatial partitioning approach is known as the "bucket" method, and allows only one level of partitioning. Applied to the domain connectivity problem, the approach partitions the domain of a grid component into a three-dimensional array of rectilinear buckets. Then, the grid points are sorted into the resulting buckets. In order to find the grid component element that bounds intergrid boundary point *P*, the bucket that contains *P* is first identified. If the data structure used to define the buckets is Cartesian, identification of the bucket that bounds *P* is trivial, otherwise this step could become computationally significant for large problems. Given the identity of the bucket that bounds *P*, an exhaustive search of the grid points contained in the bucket is conducted to find the actual bounding element.

It is possible that none of the points inside the bucket that bounds *P* define the origin of the grid element that bounds *P*. In fact, since the possibility of empty buckets exists, it is possible that the bucket that bounds *P* is empty. Accordingly, if the search of the bounding bucket fails, neighboring buckets must be searched until the actual bounding element is identified. The bucket method is an improvement over an exhaustive search, but is limited by costs associated with nonuniform distribution of grid points among existing buckets. Large numbers of empty buckets may exist, requiring cost to identify the bucket which contains the donor element. Other buckets may have a large number of points, requiring substantial computational effort to do an exhaustive search of individual buckets.

Multilevel partitioning methods exist that remedy many of the shortcomings of the simple bucket method. A "split tree" (binary alternating direction tree) is one example. In this approach, the physical space of a grid component is split into two partitions at each level. The partitioning occurs alternately along planes of constant *x*, *y*, and *z*. Ideally, positioning of the planes is done such that the grid points are divided equally between the two partitions that exist at any level. In this approach, the empty bucket problem does not exist, and grid points are more evenly distributed among partitions. However, the highest level partition (tree branch) that bounds a given intergrid boundary point must be identified. This requires a recursive search of the tree-structure of the partitioning. Once the high-level bounding partition has been identified, an exhaustive search of the points inside the partition is required to identify the actual element that bounds *P*.

As with the bucket method, it is possible that the origin of the bounding element of *P* does not lie inside the partition that bounds *P*. For such cases, neighboring partitions on the same branch that bounds *P* must be searched until the bounding element is found.

11.3.2.4 Combined Spatial Partitioning and Gradient Search

Many overset grid applications involve very large geometrically complex three-dimensional domains. Discretization of such problems can involve numerous grid components and millions of grid points. Problems of this magnitude demand an efficient and robust domain connectivity algorithm. A way to enhance the computational efficiency of the search aspects of domain connectivity is to combine the strengths of spatial partitioning with a gradient search procedure. Such an approach has been introduced to provide domain connectivity for problems involving relative motion between component parts [Meakin, 1991].

In the implementation just noted, a set of rectilinear partitions, or "buckets," are used to completely enclose the physical space of a curvilinear grid component, the partitions forming a stair-step enclosure around the grid component boundaries. An example of such a partitioning is illustrated in Figures 11.15a and 11.15b. Each partition encloses a unique portion of the grid component. The computational space of the grid component bounded by a given partition is mapped to a uniform Cartesian grid defined within the partition. Discretization of the several partitions into separate uniform Cartesian grids is a second level of partitioning of the domain, but one that retains data structure to facilitate search by truncation, rather than by exhaustion.

Partitioning and mapping the computational space of a grid component is a one-time expense, even for moving body problems (assuming rigid-body motion). The object of the mapping is to define values of the grid component curvilinear coordinates (j, k, l) at each uniform Cartesian grid point within the partitions. This can be done with a single sweep through a grid component. For example, consider the curvilinear grid indicated in Figure 11.15c. All uniform Cartesian points in a partition will be bounded at least once by the bounding box of a curvilinear element from the grid component. Figure 11.15c indicates three curvilinear element bounding boxes associated with points j,k, $j,k+1$, and $j,k+2$, respectively. The values of j, k, l assigned to a bounding-box are the coordinates of the curvilinear element origin bounded by the box. Then, all partition uniform Cartesian points that are also enclosed within the bounding box are defined to have the same j, k, l values as the box. Accordingly, in Figure 11.15c, curvilinear grid indices j, k (2D) are mapped to the uniform Cartesian points marked with solid squares during the k-sweep indicated. The curvilinear grid indices mapped to the two upper solid squares are then overwritten by $j,k+1$. Also, $j,k+1$ is mapped to the uniform Cartesian points marked with solid diamonds. This procedure is continued through the k-sweep indicated, and on through the curvilinear grid. A single pass through the curvilinear grid is sufficient to define j, k, l (3D) at all uniform Cartesian points in all partitions associated with the grid component. This is true even though spacing of the uniform Cartesian grids may be small or large relative to the curvilinear grid. In Figure 11.15c, the uniform Cartesian elements are slightly smaller than the curvilinear grid. In contrast, the uniform Cartesian elements are large relative to the curvilinear grid shown in Figure 11.15d. Figure 11.15e indicates the mapping of the "k" coordinate direction to the partitions of the curvilinear grid indicated in Figure 11.15a (shades of gray are proportional to values of k).

Given this type of spatial partitioning, a very good approximation to the grid component element that bounds a given intergrid boundary point *P* can be identified directly. First, the partition that contains *P* is identified by evaluating Eq. 11.4 for each partition. Then, the uniform Cartesian element (of the bounding partition) that bounds *P* is computed directly using Eq. 11.3. In many cases, these two steps correctly identify the element that bounds *P*. If the element identified does not bound *P*, Eq. 11.5 can be used to drive a gradient search for the correct bounding element. Since the actual bounding element is

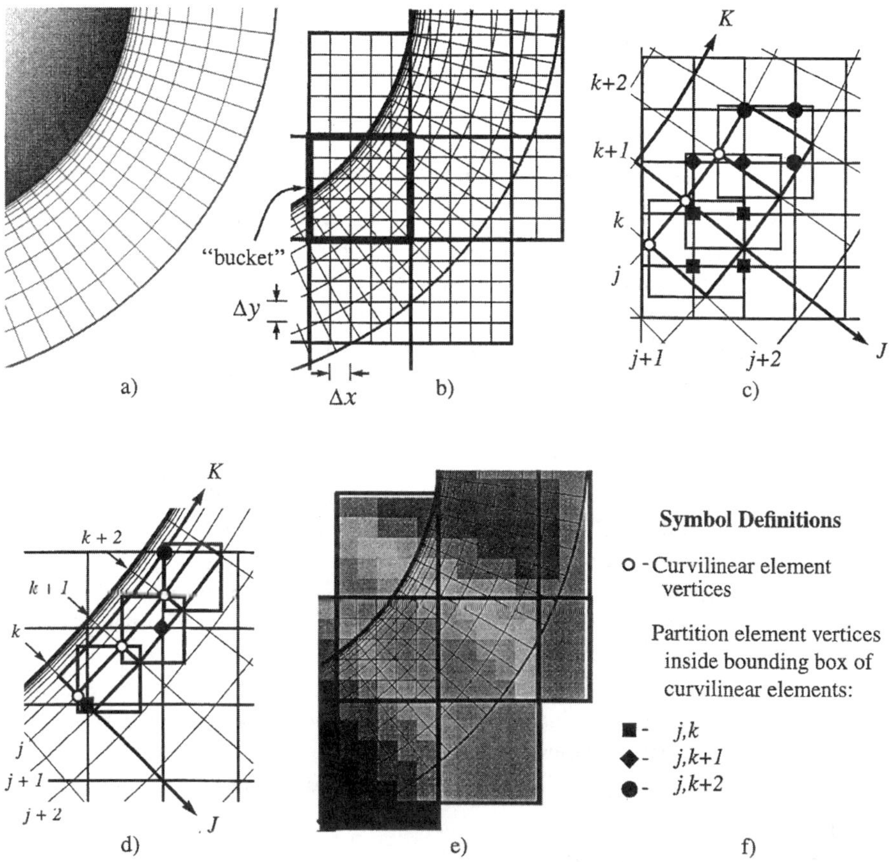

FIGURE 11.15 Simple spatial partitioning of curvilinear grid component. (a) Curvilinear grid, (b) partition boundaries, or "buckets," (c) *j*, *k* space of curvilinear grid mapped to uniform Cartesian points within a partition, (d) partition uniform Cartesian grid that is coarse relative to the curvilinear grid component being mapped, (e) *k*-coordinate of curvilinear grid component mapped onto uniform Cartesian grids of several partitions, (f) definition of symbols used in (c) and (d).

guaranteed to be near the element identified by the spatial partitioning, only a few steps (at most) of the gradient search routine should be required.

11.4 Research Issues

The preceding sections have described basic concepts in domain decomposition and domain connectivity required for implementation of a composite overset structured grid approach. It has been noted that the approach has been used advantageously on a wide variety of applications of practical importance. The approach is indeed very powerful. However, it is still a maturing technology. The following paragraphs provide a brief statement of areas that require further development to realize the full advantages inherent to the approach.

11.4.1 Surface Geometry Decomposition

Software tools are needed to simplify the task of surface geometry decomposition. Automation of many aspects of this task are possible (see Chapter 29). This aspect of grid generation is the most fundamental part of grid generation and affects all subsequent processes of analysis from grid generation to solution of PDEs on the final system of grids. Moreover, this aspect of grid generation currently relies most heavily on user expertise, but has the least amount of software available to assist in the task. Some research is being carried out in this area by Petersson [1995], and Chan and Meakin [1997]. Continued effort in the area is needed.

11.4.2 Surface and Volume Grid Generation

Research in the areas of surface and volume grid generation for overset grid applications is much less pressing than for surface geometry decomposition. This is so because a number of excellent software tools already exist for performing these tasks automatically (see Chapter 5 of this handbook). The main problem here is that grid generation software designed for overset grids exist as stand-alone entities. A user must be familiar with many codes and input requirements to use the software. A software engineering effort to combine existing overset grid generation tools into a stand-alone and easy-to-use package is needed.

11.4.3 Adaptive Refinement

A criticism sometimes leveled against structured grid approaches is that adaptive refinement cannot be done, or is very difficult to do within a structured grid framework. This is simply not true. Adaptive refinement schemes have been developed and applied within structured grid codes for many years. The first adaptive mesh refinement (AMR) schemes began to appear in the international literature in the early 1980s [Berger and Oliger, 1984]. Many advances have been realized since then. Perhaps a more accurate criticism would be to note that structured-grid adaptive refinement applications involving geometrically complex configurations have been very limited. Adaptive refinement needs to be demonstrated for applications of practical importance using overset structured grids. In general, all methods of adaptive refinement require further research to improve generality and robustness. The area of error estimation and feature detection is independent of discretization methodology, and requires further investigation.

11.4.4 Domain Connectivity

Software exists to establish domain connectivity among systems of overset structured grids [Benek et al., 1986; Brown et al., 1989; Meakin, 1991; Suhs and Tramel, 1991; Maple and Belk, 1994]. Existing domain connectivity software is very close to providing the degree of automation required for this task. Software advances in the areas of surface geometry decomposition and volume grid generation will eliminate many of the overset grid related problems that currently do not become apparent until domain connectivity is attempted. Still, existing domain connectivity software is deficient in some respects. Requirements for user specified hole-cutting surfaces need to be eliminated. For problems of practical importance, hole-cutter shape specification is a tedious task that is prone to human error. Given a set of volume grids and corresponding topological and boundary condition information, fully automatic, high-quality domain connectivity solutions should be realizable. Advances in methods to create Chimera holes and the establishment of robust definitions of geometric measures of donor element goodness are basic to the realization of fully automatic domain connectivity software.

Even with fully automatic domain connectivity, improved computational efficiency in the areas of Chimera hole-cutting and donor element identification will probably always be desirable. This is especially true for unsteady moving body applications that require domain connectivity to be established at every time-step.

Defining Terms

Block: simple surface area in a geometry definition that can be covered with a quadrilateral patch (see Figure 11.1c).

Chimera: a type of domain decomposition that allows arbitrary *holes* in overlapping grid components (see Figure 11.7).

Collar: grid component generated from a *seam* about the junction of two surfaces, such as the junction between an aircraft wing and fuselage (see Figure 11.3c).

Donor element: the element of a grid component used to supply values of the dependent variables (typically by interpolation) to an intergrid boundary point (see Figure 11.14).

Field points: points in a grid component where values of the dependent variables are determined by numerical solution to the governing set of equations to be solved on the grid system.

Fringe points: points in a grid component that define the border between conventional *field points* and Chimera *hole points* (see Figure 11.7).

Hole-map: an approximate representation of a Chimera hole-cutting surface (see Figure 11.15e).

Hole points: points in a grid component for which values of the dependent variables will not be updated or defined (see Figure 11.7).

Outer boundary points: points on the exterior surfaces of a grid component that are not flow boundaries or *hole points* (see Figure 11.7).

Quilt: surface geometry decomposition that results in a set of overlapping *seams* and *blocks* (see Figure 11.1).

Seam: surface areas that are associated with point or line discontinuities, or control lines, in a geometry definition (see Figure 11.1b).

Tip: surface topology for an area associated with a point discontinuity in the geometry definition (see Figure 11.2a).

Acknowledgment

A chapter on composite overset structured grids, such as presented here, must include an acknowledgment of the seminal role of the late Professor Joseph L. Steger to this area of computational mechanics. Recently, the Third Symposium on Overset Composite Grid and Solution Technology was held at the Los Alamos National Laboratory. The impact of Steger's "Chimera" method of domain decomposition was clearly apparent. Applications ranging from biological issues regarding the mechanisms of food particle entrapment inside the oral cavities of vertebrate suspension feeding fish, to the aerodynamic performance of atmospheric reentry vehicles were also presented. Simulations of blast wave propagation to consider safety regulations for launch facilities located near populated regions, studies of the acoustic noise levels of high-speed trains passing through tunnels, and simulations of the aeroacoustic performance of rotary wing aircraft were also presented. Demonstrations of analysis capability that relate to many other aspects of our society were also given. Truly, Professor Steger's influence has been great.

Further Information

Many domain connectivity issues are actually problems in computational geometry, which has a large literature of its own. The text by O'Rourke [1994] is very good. Melton's Ph.D. thesis [1996] also describes a number of algorithms that are particularly relevant to domain connectivity. A complete discussion of spatial partitioning methods is given in the book by Samet [1990]. *Computational Fluid Dynamics Review 1995* includes a review article on "The Chimera Method of Simulation for Unsteady Three-Dimensional Viscous Flow" [Meakin, 1995a] and has a substantial set of references that point to basic research being carried out in a number of areas related to composite overset structured grids. Henshaw [1996] recently published a review paper on automatic grid generation that devotes a section to overlapping grid generation.

References

1. Barnette, D. and Ober, C., Progress report on high-performance high resolution simulations of coastal and basin-scale ocean circulation *Proceedings of the 2nd Overset Composite Grid Sol. Tech. Symp.*, Fort Walton Beach, FL, 1994.
2. Benek, J., Steger, J., Dougherty, F., and Buning, P., Chimera: A grid-embedding technique *AEDC-TR-85-64*, 1986.

3. Berger, M. and Oliger, J., Adaptive mesh refinement for hyperbolic partial differential equations *J. Comput. Phys.* 1984, 53: 484–512.
4. Brown, D., Chesshire, G., Henshaw, W., and Kreiss, O., On composite overlapping grids *7th Int. Conf. Finite Element Methods in Flow Probs.,* Huntsville, AL, 1989.
5. Chan, W. and Buning, P., Surface grid generation methods for overset grids *Computers and Fluids.* 24, (5): 509–522.
6. Chan, W. and Meakin, R., Advances towards automatic surface domain decomposition and grid generation for overset grids(submitted for publication) In *Proceedings of the 13th AIAA CFD Conf.*, Snowmass, CO, 1997.
7. Chiu, I.T. and Meakin, R., On automating domain connectivity for overset grids AIAA Paper 95-054, *33rd Aero. Sci. Mtg.* Reno, NV, 1995.
8. Gomez, R. and Ma, E., Validation of a large-scale chimera grid system for the Space Shuttle Launch Vehicle In *Proceedings of the 12th AIAA Applied Aero. Conf.* 1994, Paper 94-1859-CP., pp 445–455.
9. Henshaw, W., Automatic grid generation *Acta Numerica.* 1996, pp 121–148.
10. Maple, R. and Belk, D., Automated set up of blocked, patched, and embedded grids in the beggar flow solver *Numerical Grid Generation in Computational Fluid Dynamics and Related Fields,* Weatherill, N.P., et al., (Ed.), Pine Ridge Press, 1994, pp 305–314.
11. Meakin, R., A new method for establishing intergrid communication among systems of overset grids *Proceedings of the 10th AIAA CFD Conf.,* Paper 91-1586-CP, 1991, pp 662–671.
12. Meakin, R., On the spatial and temporal accuracy of overset grid methods for moving body problems *Proceedings of the 12th AIAA Appl. Aero. Conf.*, 1994, Paper 94-1925-CP, pp 857–871.
13. Meakin, R., The chimera method of simulation for unsteady three-dimensional viscous flow *Computational Fluid Dynamics Review.* Hafez, M. and Oshima, K., (Eds.), John Wiley & Sons, Chichester, England, 1995, pp 70–86.
14. Meakin, R., Unsteady simulation of the viscous flow about a V-22 rotor and wing in hover *Proceedings of the AIAA Atmos. Flght. Mech. Conf.*, 1995, Paper 95-3463-CP, pp 332–344.
15. Melton, J., Automated three-dimensional Cartesian grid generation and Euler flow solutions for arbitrary geometries Ph.D. thesis, University of California, Davis, 1996.
16. O'Rourke, J., *Computational Geometry in C.* Cambridge University Press, 1994.
17. Parks, S., Buning, P., Steger, J., and Chan, W., Collar grids for intersecting geometric components within the chimera Overlapped Grid Scheme *Proceedings of the 10th AIAA CFD Conf.* Paper 91-1587-CP, 1991, pp 672–682.
18. Petersson, N.A., A new algorithm for generating overlapping grids *CAM-report 95-31,* (submitted to *SIAM J. Sci. Comp.*)UCLA, 1995.
19. Samet, H., *The Design and Analysis of Spatial Data Structures.* Addison-Wesley, Reading, MA, 1990.
20. Schwarz, H.A., Ueber einige Abbildungsaufgaben *J. Reine Angew. Math.* 1869, 70, pp 105–120.
21. Steger, J., Notes on surface grid generation using hyperbolic partial differential equations. (unpublished report), Dept. Mech., Aero. & Mat. Eng., University of California, Davis, 1989.
22. Steger, J., Notes on composite overset grid schemes — chimera. (unpublished report), Dept. Mech., Aero. and Mat. Eng., University of California, Davis, 1992.
23. Steger, J., Dougherty, F.C., and Benek, J., 1983. A chimera grid scheme *Advances in Grid Generation,* Ghia K.N. and Ghia, U., (Eds.), ASME FED, 1983, Vol 5., pp 59–69.
24. Suhs, N. and Tramel, R., *PEGSUS 4.0 User's Manual.* AEDC-TR-91-8, 1991.
25. Warming, R. and Beam, R., On the construction and application of implicit factored schemes for conservation laws *SIAM-AMS Proc.* 1978, 11: 85–129.

12

Parallel Multiblock Structured Grids

Jochem Häuser

Peter Eiseman

Yang Xia

Zheming Cheng

12.1 Overview

In this overview the lesson learned from constructing 3D multiblock grids for complex geometries are presented, along with a description of their interaction with fluid dynamics codes used in parallel computing. A brief discussion of the remaining challenging problems is given, followed by an outlook of what can be achieved within the next two or three years in the field of parallel computing in aerospace combined with advanced grid generation.

The overall objective of this chapter is to provide parallelization concepts independent of the underlying hardware — regardless whether parallel or sequential — that are applicable to the most complex topologies and flow physics. At the same time, the solver must remain efficient and effective. An additional requirement is that once a grid is generated, the flow solver should run immediately without any further human interaction.

The field of CFD (computational fluid dynamics) is rapidly changing and is becoming increasingly sophisticated: grids define highly complex geometries, and flows are solved involving very different length and time scales. The solution of the Navier–Stokes equations must be performed on parallel systems, both for reasons of overall computing power and cost effectiveness.

Complex geometries can either by gridded by completely unstructured grids or by structured multiblock grids. In the past, unstructured grid methods almost exclusively used tetrahedral elements. As has been

0-8493-2687-7/99/$0.00+$.50

shown recently this approach has severe disadvantages with regard to program complexity, computing time, and solution accuracy as compared to hexahedral finite volume grids [Venkatakrishnan, 1994]. Multiblock grids that are unstructured on the block level but structured within a block provide the geometrical flexibility and retain the computational efficiency of finite difference methods. Consequently, this technique has been implemented in the majority of the flow solvers.

In order to have the flow solution independent of the block topology, grids must be slope continuous. This causes a certain memory overhead: if n is the number of internal points in each direction for a given block, this overhead is the factor $(n + 2)^3/n^3$, where an overlap of two rows or columns has been assumed. The overhead is mainly caused by geometrical complexity, i.e., generating a block topology that aligns the flow with the grid as much as possible requires a much more sophisticated topology.

Since grid topology is determined by both the geometry and the flow physics, blocks are disparate in size, and hence load balancing is achieved by mapping a group of blocks to a single processor. The message passing algorithm — we will specify our parallelization strategy in more detail in Section 12.8 — must be able to efficiently handle the communication between blocks that reside on the same processor, meaning that only a copy operation is needed. For message passing, only standard library packages are used, namely Parallel Virtual Machine (PVM) and Message Passing Interface (MPI). Communication is restricted to a handful of functions that are encapsulated, thus providing full portability. A serial machine is treated as a one-processor parallel machine without message passing. Available parallelism (the maximum number of processors that can be used for a given problem) is determined by the number of points in the grid: a tool is available to split large blocks, if necessary. Grids generated employ NASA's standard Plot3D format.

In particular, a novel numerical solution strategy has been developed to solve the 3D N–S equations for arbitrary complex multiblock grids in conjunction with complex physics on parallel or sequential computer systems.

In general, numerical methods are of second order in space. A set of two ghost cells in each direction exists for each block, and parallelization is simply introduced as a new type of boundary condition. Message passing is used for updating ghost cells, so that each block is completely independent of its neighbors. Since blocks are of different size, several blocks are mapped onto a single processor to achieve almost always a perfect static load balancing. This implementation enables the code to run on any kind of distributed memory system, workstation cluster, massively parallel system as well as on shared memory systems and in sequential mode.

A comprehensive discussion of the prevailing concepts and experiences with respect to load balancing, scaling, and communication is presented in this article. Extensive computations employing multiblock grids have been performed to investigate the convergence behavior of the Newton-CG (conjugate gradient) scheme on parallel systems, and to measure the communication bandwidth on workstation clusters and on large parallel systems.

There can be no doubt that the future of scientific and technical computing is parallel. The challenging tasks to be tackled in the near future are those of **numerical scaling** and of **dynamic load balancing.** Numerical scaling means that the computational work increases with $O(N)$ or at most $O(N\log N)$ where N denotes problem size. This is normally not the case. A simple example, the inversion of a matrix with N elements, needs an order of $O(N^3)$ floating point operations. For instance, increasing the problem size from 100,000 points to 10 million points would increase the corresponding computing time a million times. Obviously, no parallel architecture could keep pace with this computational demand.

Therefore, one of the most challenging tasks is the development of algorithms that scale numerically. The so-called **tangled web** approach (see Section 12.10), based on the idea of adaptive coupling between grid points during the course of a computation, is an important technique that might have the potential to achieve this objective. It should be clear that in order to obtain numerical scaling, this tangled web approach will not result in an algorithm that is scalable and parallel. This is due to the high load imbalance that may be caused during the computation, based on the dynamic behavior of the algorithm. Thus, this approach is inherently dynamic and therefore needs dynamic load balancing. Only if these two requirements are satisfied can **very large scale applications (VLSA)** be computed in CFD.

From the lessons learned so far it can be confidently predicted that the techniques are available both for numerical scaling and dynamic load balancing for these VLSA. It remains to implement the basic ideas in such a way that routine computations for the complex physics and the complex geometries that characterize today's aerospace design can be performed at the increased level of accuracy demanded by the CFD applications of the future.

12.2 Multiblock Grid Generation and Parallelization

Structured grids use curvilinear coordinates to produce a body-fitted mesh. This has the advantage that boundaries can be exactly described and hence boundary conditions can be accurately modeled. In early grid generation it was attempted to always map the physical solution domain to a single rectangle or a single box in the computational domain. For multiply connected solution domains, branch cuts had to be introduced, a procedure well known in complex function theory and analytic mapping of 2D domains, e.g., for the Joukowski airfoil. However, it became soon obvious that certain grid line configurations could not be obtained. If one considers, for example, the 2D flow past an infinitely long cylinder with a small enough Reynold number, it would be advantageous if the grid line distribution would be similar to the streamline pattern. A 2D grid around a circle which is mapped on a single rectangle necessarily has O-type topology, unless additional slits (double-valued line or surface) or slabs (blocks that are cut out of the solution domain) are introduced. However in this case, the main advantage of the structured approach, namely that one has to deal logically only with a rectangle or a box, that is, the code needs only two or three "*for*" loops (C language), no longer holds. The conclusion is that this structuredness is too rigid, and some degree of unstructuredness has to be introduced. From differential geometry the concept of an atlas consisting of a number of charts is known. A set of charts covers the atlas where charts may be overlapping. Each chart is mapped onto a single rectangle. In addition, now the connectivity of the charts has to be determined. This directly leads to the multiblock concept, which provides the necessary geometrical flexibility and the computational efficiency of the finite volume or finite difference techniques used in most CFD codes.

For a vehicle like the Space Shuttle a variety of grids can be constructed. One can start with a simple monoblock topology that wraps around the vehicle in an O-type fashion. This always leads to a singular line, which normally occurs in the nose region. This line needs special treatment in the flow solution. It has been observed that the convergence rate is reduced; however, special numerical schemes have been devised to alleviate this problem. Furthermore, a singularity invariably leads to a clustering of grid points in an area where they are not needed. Hence, computing time may be increased substantially. In addition, with a monoblock mesh, gridline topology is fixed and additional requirements with regard to grid uniformity and orthogonality cannot be matched. However, with a multiblock mesh, a grid has to be smooth across block boundaries.

Since multiblock grids are unstructured at the block level, information about block connectivity is needed along with six faces of each block. For reasons of topological flexibility it is mandatory that each block has its own local coordinate system. Hence blocks are rotated with respect to each other. Slope continuity of grid lines across neighboring block boundaries, for instance as shown in Figures 12.1 and 12.2, is achieved by overlapping edges (2D) or faces (3D). For grid generation an overlap of exactly one row or one column is necessary (see Figure 12.8). A flow solver that retains second-order accuracy, however, needs an overlap of two rows or columns.

The solution domain is subdivided into a set of blocks or segments (in the following the words block and segment are used interchangeably). The multiblock concept, used as a domain decomposition approach, allows the direct parallelization of both the grid generation and the flow codes on massively parallel systems. Employment of the overlap feature directly leads to the message passing concept, i.e., the exchange of faces between neighboring blocks.

Each (curvilinear) block in the physical plane is mapped onto a Cartesian block in the computational plane (see Figure 12.3). The actual solution domain on which the governing physical equations are solved is therefore a set of connected, regular blocks in the computational plane. However, this does not mean

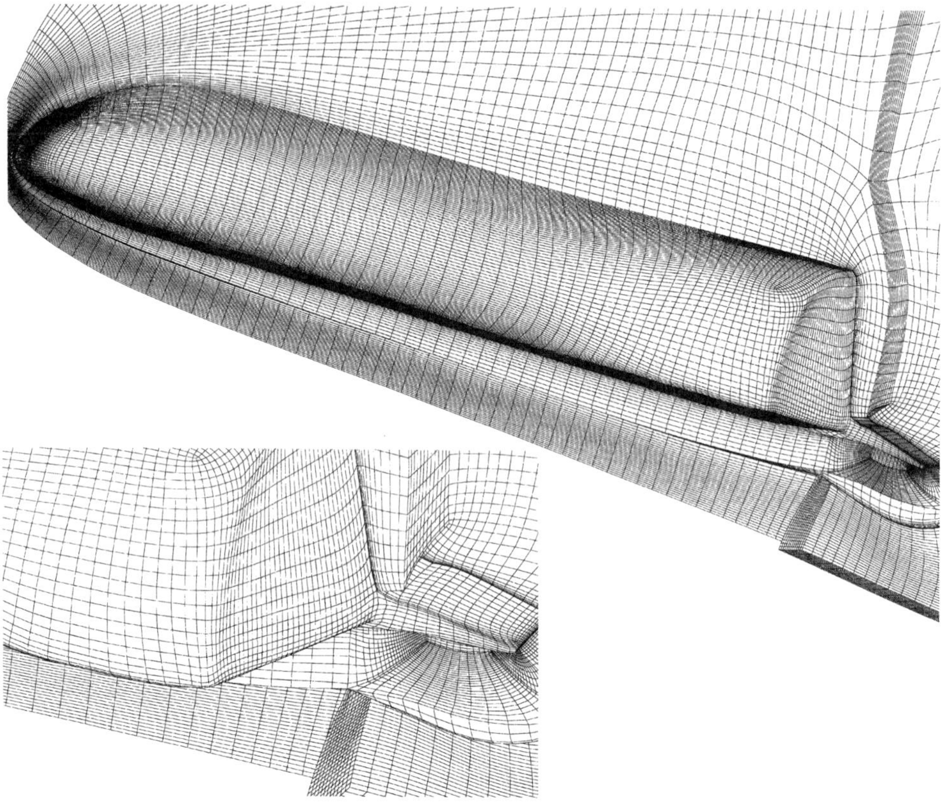

FIGURE 12.1 Halis Space Shuttle grid with local three dimensional clustering around the body flap. This clustering leads to high resolution at the body flap, but prevents the extension of this resolution into the farfield, thus substantially reducing the number of grid points. The grid is generated fully automatically, once the basic wireframe topology has been given. This procedure leads to a more complex topology and to blocks that are of different size. Vector computers that need long vector lengths will perform poorly on this topology. On the other hand, parallel machines in combination with the parallelization strategy described in this article, will give a high parallel efficiency.

that the solution domain in the computational plane has a regular structure, rather it may look fragmented. Therefore, an important point is that the parallelization approach must not rely on a nearest neighbor relation of the blocks. Hence, communication among blocks follows a random pattern. A parallel architecture based on nearest neighbor communication, e.g., for lattice gauge problems, will not perform well for complex aerodynamic applications, simply because of communication overhead, caused by random communication. However, as we will see in Section 12.9, communication time is not a problem for implicit VFD codes, but load balancing is a crucial issue.

The grid point distribution within each block is generated by the solution of a set of three Poisson equations, one for each coordinate direction, in combination with transfinite interpolation and grid quality optimization (cf. Chapter 4). In this context, a grid point is denoted as a boundary point if it lies on one of the six faces of a block in the computational plane. However, one has to discern between physical boundary points on fixed surfaces and matching boundary points on overlap surfaces of neighboring blocks. The positions of the latter ones are not known *a priori* but are determined in the solution process. Parallelization, therefore, in a multiblock environment simply means the introduction of a new boundary condition for block interfaces. Even on a serial machine, block updating has to be performed. The only difference is that a copy from and to memory can be used, while on a parallel system block updating is performed via message passing (PVM or MPI). The logic is entirely the same, except for the additional packing and sending of messages.

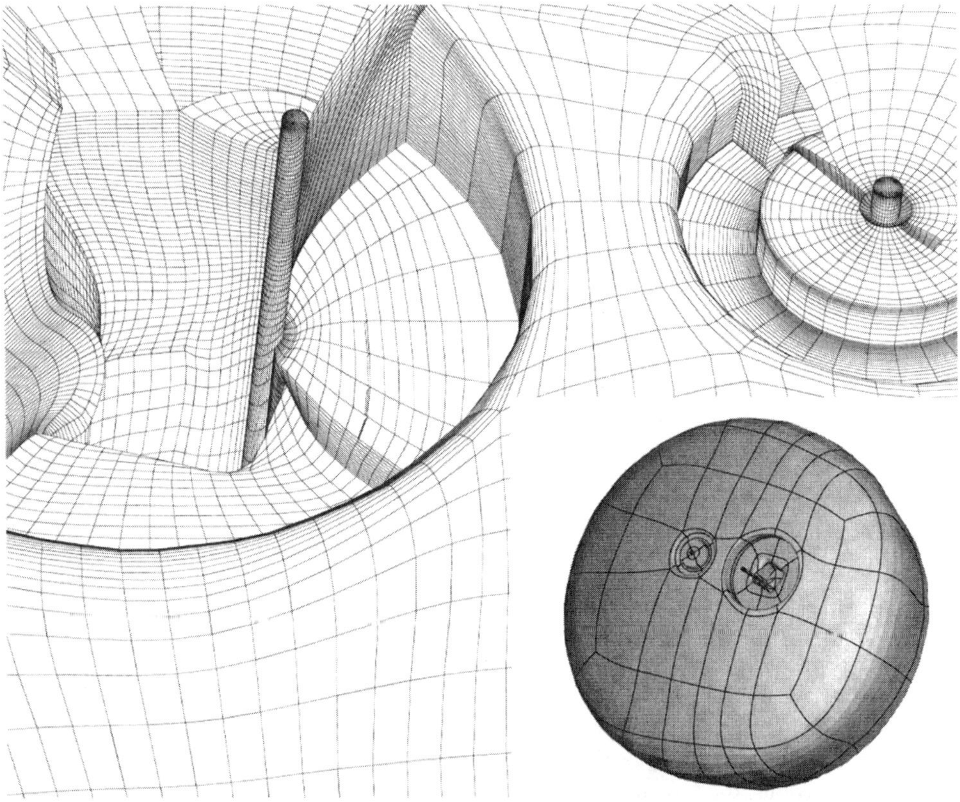

FIGURE 12.2 Cassini–Huygens Space Probe grid. The probe will enter Titan's atmosphere in 2004 to measure its composition. On the windward side several instruments are shown, leading to microaerodynamics phenomena [Bruce, 1995]. The grid comprises a complex topology, consisting of 462 blocks that are of different size.

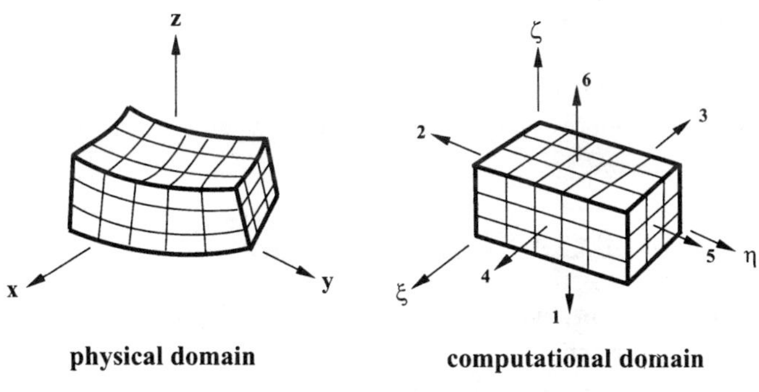

FIGURE 12.3 Mapping of a block from solution domain to computational domain. Arrows indicate orientation of faces, which are numbered in the following way: 1 bottom, 2 left, 3 back, 4 front, 5 right, 6 top. The rule is that plane $\zeta = 1$ corresponds to 1, plane $\eta = 1$ to 2, and plane $\xi = 1$ to 3.

12.3 Computational Aspects of Multiblock Grids

As has been discussed previously, boundary-fitted grids have to have coordinate lines, i.e., they cannot be completely unstructured. In CFD in general, and in high speed flows in particular, many situations are encountered for which the flow in the vicinity of the body is aligned with the surface, i.e., there is a prevailing flow direction. This is especially true in the case of hypersonic flow because of the high kinetic energy. The use of a structured grid, allows the alignment of the grid, resulting in locally 1D flow. Hence,

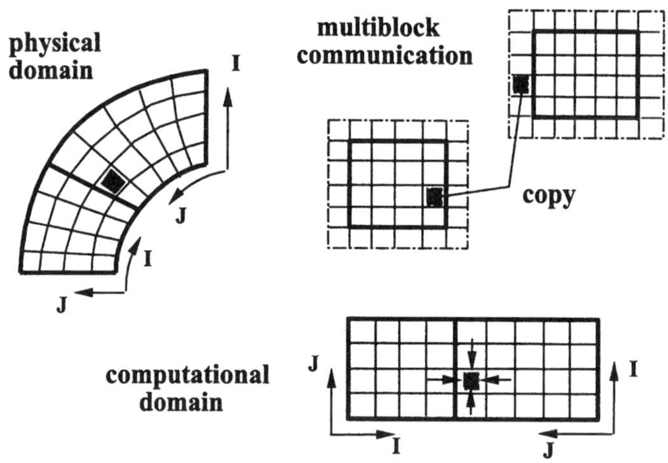

FIGURE 12.4 Multiblock grids are constructed using an overlap of one row or column. The information from an internal cell of the neighboring block is transferred via message passing (or memory copying) in the overlap cell of the current block.

numerical diffusion can be reduced, i.e., better accuracy can be achieved. In the present approach, a solution domain may be covered by hundreds or thousands of blocks. Second, structured grids can be made orthogonal at boundaries and almost orthogonal within the solution domain, facilitating the implementation of boundary conditions and also increasing numerical accuracy. This will be of advantage when turbulence models are employed using an almost orthogonal mesh. In the solution of the Navier–Stokes equations, the boundary layer must be resolved. This demands that the grid is closely wrapped around the body to describe the physics of the boundary layer (some 32 layers are used in general for structured grids). Here some type of structured grid is indispensable. For instance, the resolution of the boundary layer leads to large anisotropies in the length scales of the directions along and off the body. Since the time-step size in an explicit scheme is governed by the smallest length scale or, in the case of reacting flow, by the magnitude of the chemical production terms, extremely small time steps will be necessary. This behavior is not demanded by accuracy considerations but to retain the stability of the scheme. Thus, implicit schemes will be of advantage. In order to invert the implicit operator, a structured grid produces a regular matrix, and thus makes it easier to use a sophisticated implicit scheme.

12.4 Description of the Standard Cube

A formal description of block connectivity is needed to perform the block updating, i.e., to do the message passing. To this end, grid information is subdivided into topology and geometry data. The following format is used for both the grid generator and the flow solver, using the same topology description. All computations are done for a standard cube in the computational plane as shown in Figure 12.5. The coordinate directions in the computational plane are denoted by ξ, η, and ζ, and block dimensions are given by I, J, and K, respectively.

In the computational plane, each cube has its own right-handed coordinate system (ξ, η, ζ), where the ξ direction goes from back to front, the η direction from left to right, and the ζ direction from bottom to top, see Figure 12.5. The coordinate values are given by proper grid point indices i, j, k in the ξ, η, ζ directions, respectively. That means that values range from 1 to I in the ξ direction, from 1 to J in the η direction, and from 1 to K in the ζ direction. Each grid point represents an integer coordinate value in the computational plane.

A simple notation of planes within a block can be defined by specifying the normal vector along with the proper coordinate value in the specified direction. For example, face 2 can be uniquely defined by

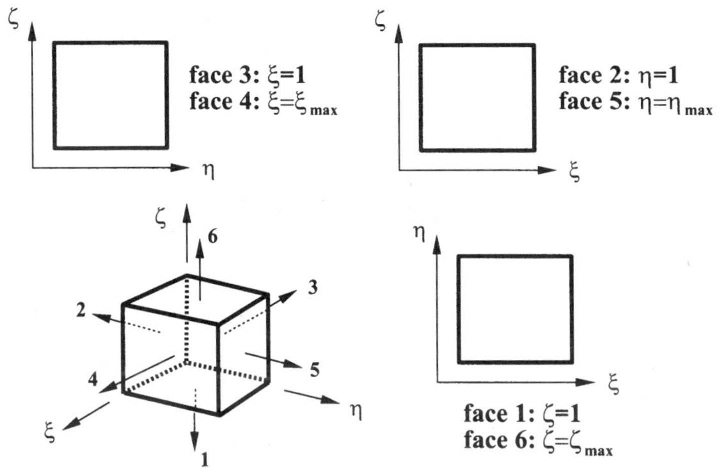

FIGURE 12.5

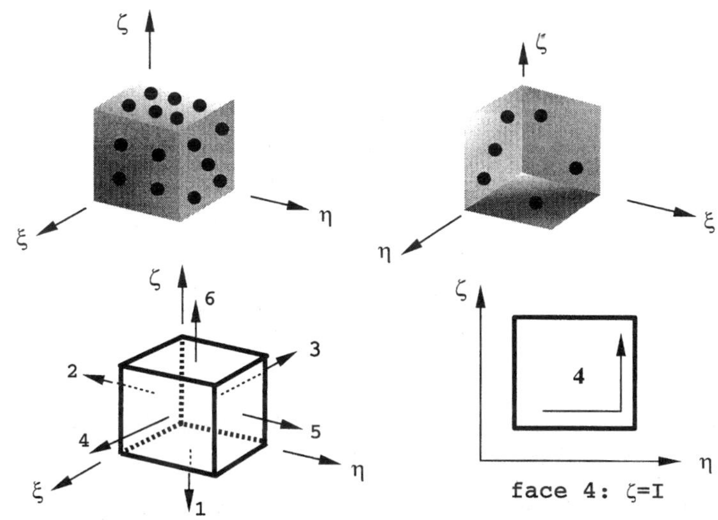

FIGURE 12.6 Orientation of faces. Coordinates ξ, η, ζ are numbered 1, 2, 3 where coordinates with lower numbers are stored first.

describing it as a $J(\eta)$ plane with a j value 1, i.e., by the pair (J, 1) where the first value is the direction of the normal vector and the second value is the plane index. Thus, face 4 is defined by the pair (I, J). This notation is also required in the visualization module.

Grid points are stored in such a way that the I direction is treated first, followed by the J and K directions, respectively. This implies that K planes are stored in sequence. In the following the matching of blocks is outlined. First, it is shown how the orientation of the face of a block is determined. Second, rules are given how to describe the matching of faces between neighboring blocks. This means the determination of the proper orientation values between the two neighboring faces.

To determine the orientation of a face, arrows are drawn in the direction of increasing coordinate values. The rule is that the lower-valued coordinate varies first, and thereby the orientation is uniquely determined. The orientation of faces between neighboring blocks is determined as follows, see Figure 12.7. Suppose blocks 1 and 2 are oriented as shown. Each block has its own coordinate system (right-handed). For example, orientation of block 2 is obtained by a rotation of π of block 1 about the ζ-axis — rotations

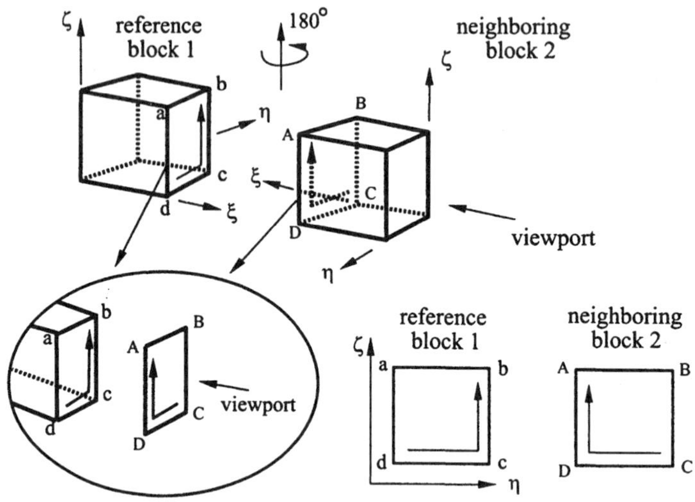

FIGURE 12.7 Determination of orientation of faces between neighboring blocks as seen from block 1 (reference block). The face of the reference block is always oriented as shown and then the corresponding orientation of the neighboring face is determined (see Figure 12.9).

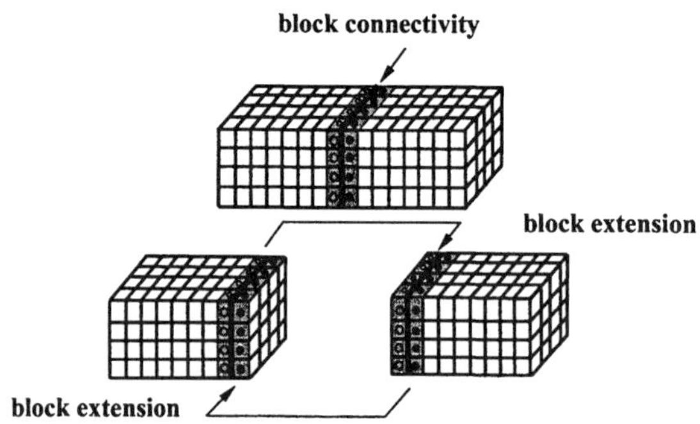

FIGURE 12.8 The figure shows the overlap of two neighboring blocks. For the flow solver, an overlap of two rows or columns is needed. The algorithm is not straightforward, because of the handling of diagonal points.

are positive in a counterclockwise sense. Thus face 4 of block 1 (used as the reference block) and face 4 of block 2 are matching with the orientations as shown, determined from the rules shown in Figure 12.9. All cases in group 1 can be obtained by rotating a face about an angle of 0, $1/2\pi$, π, or $3/2\pi$. This is also valid for elements in group 2. The code automatically recognizes when the orientation between two faces needs to be mirrored. Thus cases 1 and 7 in Figure 12.9 are obtained by rotating case 1 by $\pi/2$. Here, the rotations are denoted by integers 0, 1, 2, and 3, respectively.

12.5 Topology File Format for Multiblock Grids

To illustrate the topology description for the multiblock concept, a simple six-block grid for a diamond shape is shown in Figure 12.10. Only control file information for this grid is presented. The meaning of the control information is explained below. Since the example is 2D, the first line of this file starts with **\cntrl2d**. In 3D, this line has the form **\cntrl3d**. The corresponding coordinate values have been omitted, since they only describe the outer boundaries in pointwise form.

After the control line, object specific information is expected. The object specification is valid until the next control line is encountered or if the end of the current input file is reached. Control lines that cannot be identified are converted to the internal object type error.

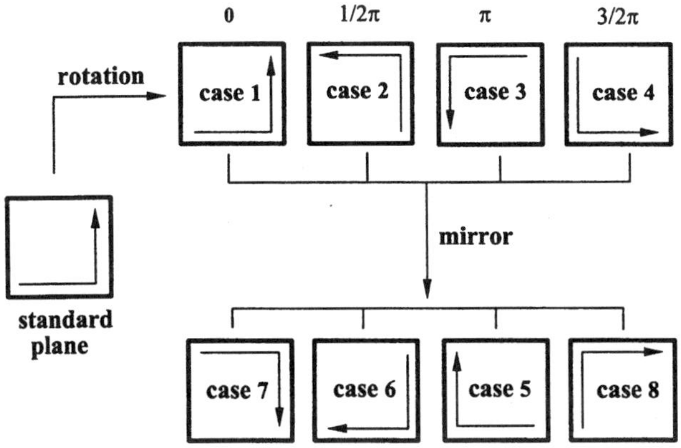

FIGURE 12.9 The 8 possible orientations of neighboring faces are shown. Cases 1 to 4 are obtained by successive rotations e.g., 0, $\frac{1}{2}\pi$, π, and $\frac{3}{2}\pi$. The same situation holds for cases 5 to 8 upon being mirrored.

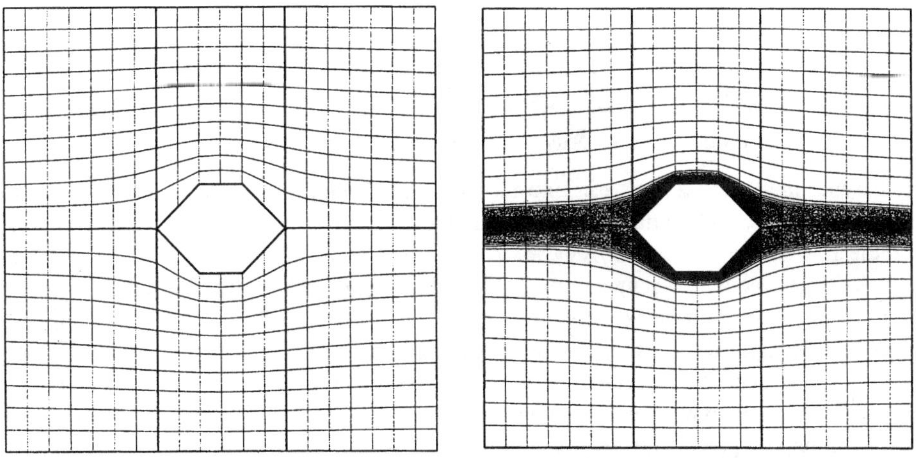

FIGURE 12.10 A six-block grid for a diamond-shaped body. This type of grid line configuration cannot be obtained by a monoblock grid. Grid lines can be clustered to match the flow physics, e.g., resolving a boundary layer. The topology information of this grid is shown in Table 12.1.

The control information in 2D has the form

\cntrl12d
nos
I J
*s*1 *st nb ns*
*s*2 *st nb ns*
*s*3 *st nb ns*
*s*4 *st nb ns*

where *nos* is the block number, and *I*, *J* are the number of grid points in the respective directions. The next four lines describe the four edges (or sides) of a block. *s*1 to *s*4 denote the side number where 1 is east, 2 north, 3 west, and 4 south. *st* is the side-type. 0 means fixed side, 1 is a fixed side used to compute the initial algebraic grid. A side of type 2 is a matching side (overlap). In this case, the corresponding values for *nb* and *ns* have to be given where *nb* is the number of the neighboring block and *ns* the number of the matching side of this block. If *st* is 0 or 1, these values should be set to zero. The edge control information can be in any order. The only restriction is that the same order must be used when boundary data are read. A similar format is used for the control information in 3D:

\cntrl3d
nos
I J K
s1 st nb ns nr
s2 st nb ns nr
s3 st nb ns nr
s4 st nb ns nr
s5 st nb ns nr
s6 st nb ns nr

Again, *nos* denotes the block number, and *I*, *J*, and *K* are the dimensions in *x*, *y*, and *z*-direction, respectively. Each block has six faces, so for each face there is one line with face-specific information. *s*1 to *s*6 are the face numbers as used for the standard block, see Figure 12.5, *st* is the face type, where a 1 denotes a face used for initialization (interpolated initial grid). In addition, to specify the neighboring block *nb* and the neighboring face *ns*, the rotation value *nr* is necessary.

- *s*1..*s*6: [1,6] → face number
- *st:* [0,3] → face type
- *nb:* [1,*N*] (*N* is total number of blocks) → neighboring block number
- *ns:* [1,6] → neighboring face of block *nb*
- *nr:* [0,3] → rotation needed to orient current face to neighboring face

Once the coordinates of a grid have been computed, the topology file as described above is constructed automatically from the grid points. While in the six-block example the command file could be set up by the user, the grid for the Cassini–Huygens Space Probe (see Figure 12.2), with its detailed microaerodynamics description, required a fully automatic algorithm. It would be too cumbersome for the user to find out the orientation of the blocks. Moreover, the generic aircraft (see Figure 12.15 later in this chapter), comprises 2200 blocks. All these tools are provided to the engineer in the context of the PAW (Parallel Aerodynamics Workbench) environment that serves as a basis from the conversion of CAD data to the realtime visualization of computed flow data by automating as much as possible the intermediate stages or grid generation and parallel flow computation.

12.6 Local Grid Clustering Using Clamp Technique

In the following we briefly describe a technology to obtain a high local resolution without extending this resolution into the far field where it is not needed. We thus substantially reduce the total number of grid points. This local clustering, however, changes block topology and leads to blocks of widely different size [Häuser et al., 1996]. Thus, it has a direct impact on the parallelization strategy, because size and number of blocks cannot be controlled. Therefore the parallelization strategy has to be adjusted to cope with this kind of sophisticated topology.

It is well known that along fixed walls a large number of grid lines is required in order to capture the boundary layer. In the remaining solution domain this requirement would cause a waste of grid points and reduce the convergence speed of flow solver. It is therefore mandatory to localize the grid line distribution. To this end, the so-called clamp clip technique was developed. Its principle is to build a closed block system connected to the physical boundary. The number of grid lines can be controlled within the block. Using clamp clips, grid lines are closed in clamp blocks. The local grid refinement can be achieved without influencing the far field grid. This is demonstrated in the lower part of Figure 12.11 and in Figure 12.1.

TABLE 12.1 Control Information for the Six-Block Diamond Grid

```
\cntrl2d
1              4
8  11          7  11
1  2  3  3     1  2  6  3
2  3  2  4     2  0  0  0
3  1  0  0     3  2  2  1
4  1  0  0     4  1  0  0
2              5
8  11          8  11
1  3  4  3     1  0  0  0
2  0  0  0     2  2  6  4
3  1  0  0     3  3  3  1
4  2  1  2     4  1  0  0
3              6
7  11          8  11
1  2  5  3     1  0  0  0
2  0  0  0     2  0  0  0
3  3  1  1     3  3  4  1
4  1  0  0     4  3  5  2
                 \file diamond.lin
```

Note: This command file is also used by the parallel flow solver. File **diamond.lin** contains the actual coordinates values.

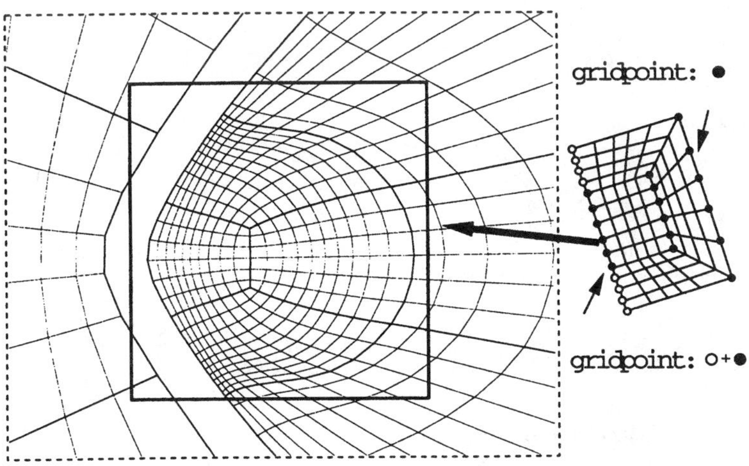

FIGURE 12.11 Clamp technique to localize grid line distribution. This figure shows the principle of a clamp. The real power of this technique is demonstrated in the Space Shuttle grid (see Figure 12.1).

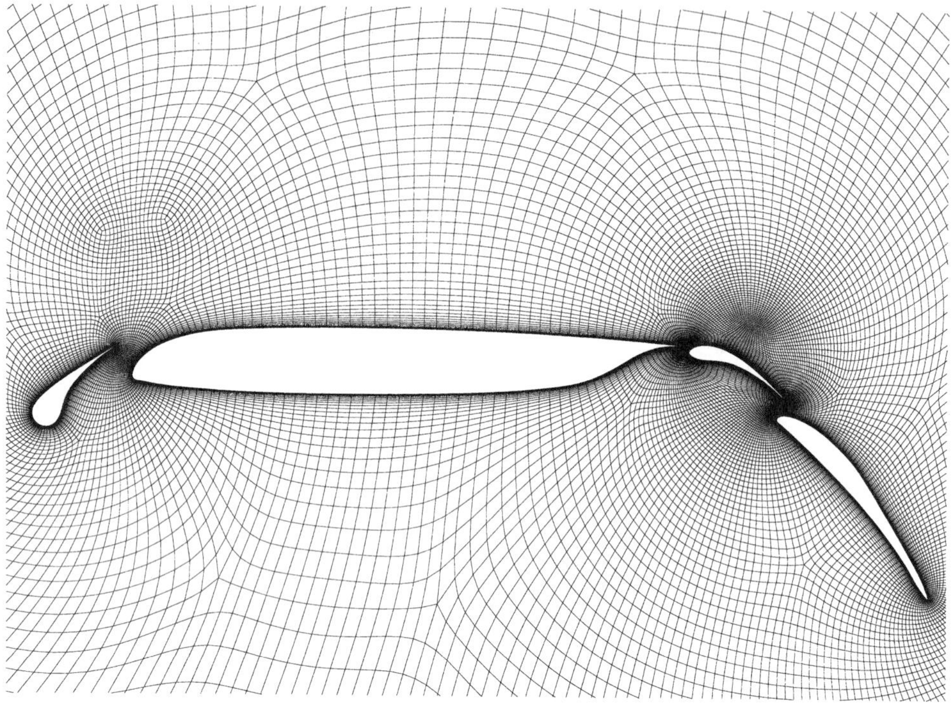

FIGURE 12.12 Navier–Stokes grid for a four-element airfoil, comprising 79 blocks. The first layer of grid points off the airfoil contour is spaced on the order of 10^{-6} based on chord length. Only the Euler grid is generated by the grid generator, the enrichment in the boundary layer is generated within a few seconds by the clustering module.

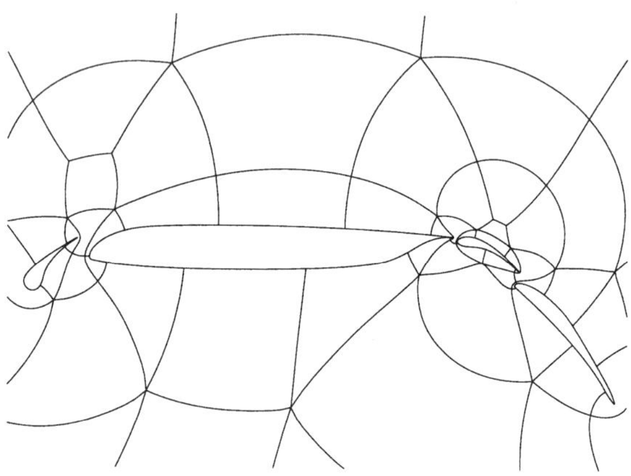

FIGURE 12.13 The figure shows the block structure of the four-element airfoil.

12.7 A Grid Generation Meta Language

12.7.1 Topology Input Language

With the parallel computers of today substantially more complex fluid flow problems can be tackled. The simulation of complete aircraft configurations, complex turbine geometries, or flows including combustion is now being computed in industry. Consequently, geometries of high complexity are now of interest as well as very large meshes, for instance, computations of up to 30 million grid points have

FIGURE 12.14 This picture shows the grid of the generic aircraft including flaps in a wind tunnel; however, the topology is exactly the same as for a real aircraft.

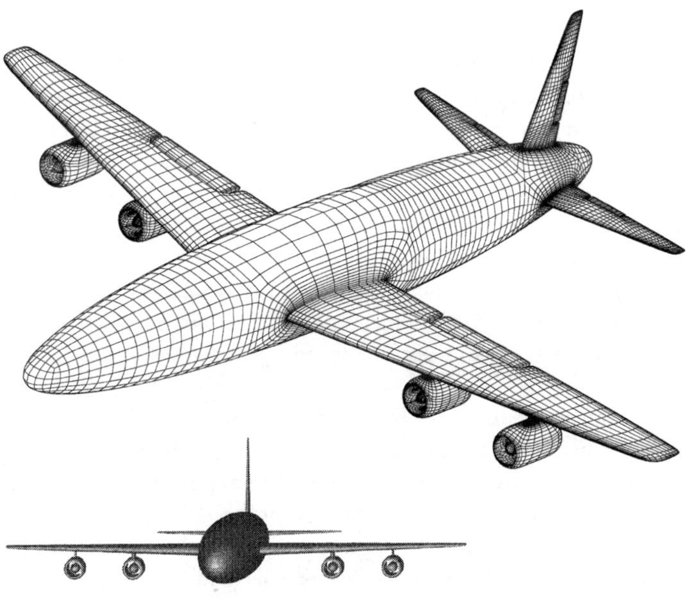

FIGURE 12.15 By modifying only a few lines of the **TIL** code, a four-engine generic aircraft is generated. The original two-engine grid was used as a starting grid.

been performed. Clearly, both grid generation codes and flow solvers have to be capable of handling this new class of application.

Conventional grid generation techniques derived from CAD systems that interactively work on the CAD data to generate the surface grid and then the volume grid are not useful for these large and complex meshes. The user has to perform tens of thousands of mouse clicks with no or little reusability of his/her input. Moreover, a separation of topology and geometry is not possible. An aircraft, for example, has a certain topology, but different geometry data may be used to describe different aircraft. Topology definition consumes a certain amount of work, since it strongly influences the resulting grid line configuration. Once the topology has been described, it can be reused for a whole class of applications. One step further would be the definition of objects that can be translated, rotated, and multiplied. These features could be used to build an application-specific data base that can be employed by the design engineer to quickly generate the grid needed.

In the following a methodology that comes close to this ideal situation is briefly described. A complete description can be found in [Eiseman, et al., 1996]. To this end, a completely different grid generation approach will be presented. A compiler-type grid generation language has been built, based on the ANSI-C syntax that allows the construction of objects. The user provides a (small) input file that describes the so-called **TIL** (Topology Input Language) code to build the wireframe model, see below, and specifies filenames used for geometry description of the configuration to be gridded. There is also the possibility to interactively construct this topology file, using the so-called AZ-Manager [Eiseman et al., 1996] package that works as a "topology generation engine."

For the description of the surface of a vehicle, a variety of surface definitions can be used. The surface can be described as a set of patches (quadrilaterals) or can be given in triangular form. These surface definitions are the interface to the grid generator. In general, a preprocessor is used that accepts surface definitions following the *NASA IGES CFD* [NASA, 1994] standard and converts all surfaces into triangular surfaces. That is, internally only triangular surfaces are used. In addition, the code allows the definition of analytic surfaces that are built in or can be described by the user
in a C function type syntax. The user does not have to input any surface grids, that is, surface and volume grids are generated fully automatically. This approach has the major advantage that it is reusable, portable, modular, and based on the object-oriented approach. Highly complex grids can be built in a step-by-step fashion from the bottom up, generating a hierarchy of increasingly complex grid objects.

For example, the grid around an engine could be an object (also referred to as component). Since an aircraft or spacecraft generally has more than one engine located at different positions beneath its wing, the basic engine object would have to be duplicated and positioned accordingly. In addition, the language is hierarchical, allowing the construction of objects composed of other objects where, in turn, these objects may be composed of more basic objects, etc. In this way, a library can be built for different technical areas, e.g., a turbomachinery library, an aircraft library, or a library for automotive vehicles. The **TIL** has been devised with these features in mind. It denotes a major deviation from the current interactive blocking approach and offers substantial advantages in handling both the complexity of the grids that can be generated and the human effort needed to obtain a high quality grid. No claims are made that **TIL** is the only (or the best) implementation of the concepts discussed, but it is believed that it is a major step toward a new level of performance in grid generation, in particular when used for parallel computing.

The versatility and relative ease of use — the effort is comparable with mastering LaTex, but the user need not write **TIL** code, because a **TIL** program can be generated by the interactive tool AZ-Manager, a procedure similar to the generation of applets using a Java applet builder — will be demonstrated by presenting TIL code for the six-block Cassini–Huygens probe. All examples presented in this chapter demonstrate both the versatility of the approach and the high quality of the grids generated.

In the following we present the **TIL** code to generate a 3D grid for the Cassini-Huygens space probe. Cassini-Huygens is a joint NASA-ESA project launched in 1997. After a flight time of seven years, the planet Saturn will be reached, and the Huygens probe will separate from
the Cassini orbiter and fly on to Titan, Saturn's largest moon. Titan is the only moon in the solar system possessing an atmosphere (mainly nitrogen). During the two-hour descent, measurements of the composition of the atmosphere will be performed by several sensors located at the windward side of the space probe. In order to ensure that laser sensors will function properly, no dust particles must be convected to any of the lens surfaces. Therefore, extensive numerical simulations have been performed investigating this problem.

TABLE 12.2 TIL Code for Six-Block Huygens Space Probe

```
SET GRIDDEN 47

COMPONENT Huygens()

BEGIN

s 1 -tube "HuygensProfil.lin";

s 2 -ellip(0.00015625 0.00015625 0.00015625) -o;

INPUT 1 (650 0 0 0 650 0 0 0 650)*face(sIN (1),cIN (-4),(-4),cOUT (1..4));

INPUT 2 (0 0 -4525)*(4525 0 0 0 4525 0 0 0 4525)*face(sIN (2),cIN (1:1..4),(-4),cOUT

(1..4));

INPUT 3 (0 0 325)*(650 0 0 0 650 0 0 0 650)*face(sIN (1),cIN (1:1..4),(-4),cOUT (1..4));

INPUT 4 (0 0 4525)*(4525 0 0 0 4525 0 0 0 4525)*face(sIN (2),cIN (3:1..4),(2:1..4),cOUT

(1..4));

END

COMPONENT face(sIN s, cIN c[1..4], C[1..4])

BEGIN

c 1 -1 - 1 + 0 -s s -L c:1 C:1;

c 2  1 - 1 + 0 -s s -L c:2 C:2 1;

c 3  1 + 1 + 0 -s s -L c:3 C:3 2;

c 4 -1 + 1 + 0 -s s -L c:4 C:4 1 3;

END
```

Note: The topology of this grid is explained in Figure 12.16.

In order to compute the microaerodynamics caused by the sensors, the proper grid has to be generated. A sequence of grids of increasing geometrical complexity has been generated. The simplest version, comprising six blocks, does not contain the small sensors that are on the windward side of the probe. With increasing complexity the number of blocks increases as well. The final grid, modeling the sensors, comprises 462 blocks. However, it is important to note that each of the more complex grids was generated by modifying the **TIL** code of its predecessor.

The general approach for constructing the Cassini-Huygens grids of increasing complexity is to first produce an initial mesh for the plain space probe without any instruments. Thus the first topology is a grid that corresponds to a *box in a box*, shown in Figure 12.16. The refinement of the grid is achieved by adding other elements designed as different objects. This topology describes the spherical far field and the body. The final grid is depicted in Figure 12.2 and Figure 12.17. This grid has a box-in-box structure: the outer box illustrates the far field and the interior one is the Huygens body. It should be noted that AZ-Manager was employed to automatically produce the **TIL** code from graphical user input [Ref.: "AZ-Manager"].

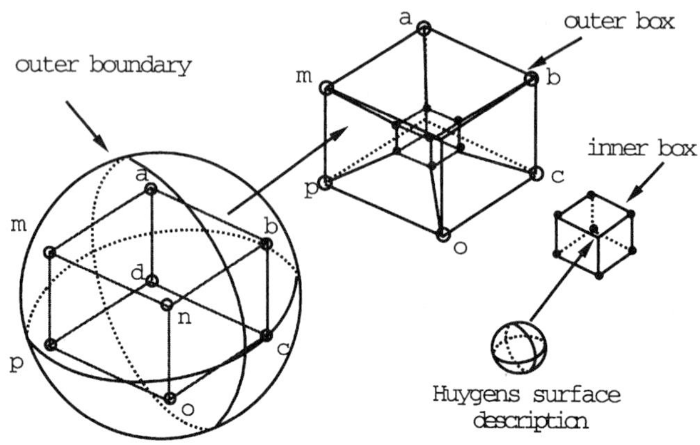

FIGURE 12.16 Topological design for the Huygens space probe grid. In this design all sensors are ignored. The topology is that of a 4D hypercube. The wireframe model consists of 16 vertices (corners). Vertices are placed interactively close to the surface (automatic projection onto the surface is performed) to which they are assigned. The grid comprises 6 blocks.

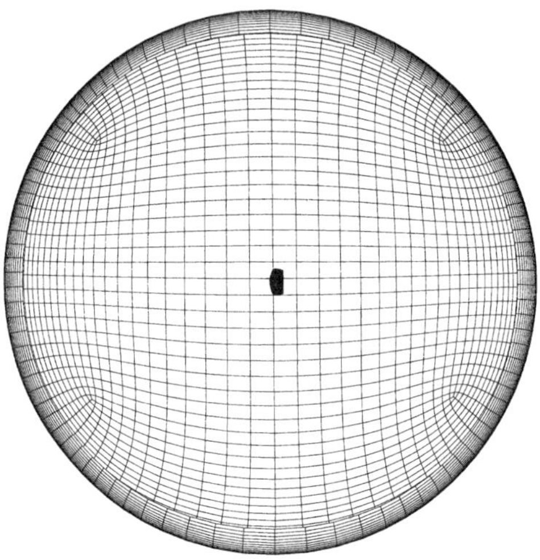

FIGURE 12.17 The 462-block grid for the Cassini–Huygens Space Probe launched in 1997 to fly to Saturn's moon Titan and to measure the composition of Titan's atmosphere after a flight time of seven years. This grid is bounded by a large spherical far field, in which the Huygens space probe is embedded. The ratio of the far field radius and the Huygens radius is about 20.

12.8 Parallelization Strategy for Complex Geometry

There are basically three ways of parallelizing a code. First, a simple and straightforward approach is to **parallelize the do-loops** in the code. Many so-called automatic parallelizers analyze do-loops and suggest a parallelization strategy based on this analysis. This concept, however, is not scalable to hundreds or thousands of processors, and results in a very limited speedup. A second approach therefore is to **parallelize the numerical solution process** for these equations. For example, if a matrix-vector multiplication occurs, this multiplication could be distributed on the various processors and performed in parallel. Again, scalability to a large number of processors cannot be obtained. Moreover, this technique would work only for large regular matrices. If a discretized problem were represented by a large number of smaller matrices (often the case in practice, e.g., multiblock grids), parallelization would be impossible.

The third approach adopts a simple idea and is denoted as **domain decomposition,** sometimes also refereed to as **grid partitioning.**

The solution domain is subdivided into a set of subdomains that exchange information to update each other during the solution process. The numerical solution takes place within each domain and is thus independent of the other domains. The solution space can be the actual space–time continuum or it can be an abstract space. For the computer simulation, this space is discretized and thus is described by a set of points. Domain decomposition is the most general and versatile approach. It also leads to the best parallel efficiency, since the number of points per subdomain (or block) can be freely varied as well as the number of subdomains per processor. A large number of codes in science and engineering use finite elements, finite differences, or finite volumes on either unstructured or structured grids. The process of parallelizing this kind of problem is to domain decompose the physical solution domain. Software [Williams et al., 1996] is available to efficiently perform this process both for unstructured and structured grids. Applying this strategy results in a fully portable code, and allows the user to switch over to new parallel hardware as soon as it becomes available.

There is, however, an important aspect in parallelization, namely the geometrical complexity of the solution domain. In the following, a brief discussion on geometrical complexity and how it affects parallelization is given. If the solution domain comprises a large rectangle or box, domain decomposition is relatively straightforward. For instance, the rectangle can be decomposed into a set of parallel stripes, and a box can be partitioned into a set of planes. This leads to a one-dimensional communication scheme where messages are sent to left and right neighbors only.

However, more realistic simulations in science and engineering demand a completely different behavior. For example, the calculation past an entire aircraft configuration leads to a partitioning of the solution domain that results in a large number of subdomains of widely different size, i.e., the number of grid points of the various blocks differ considerably. As a consequence, it is unrealistic to assume that a solution domain can be partitioned into a number of equally sized subdomains. In addition, it is also unrealistic to assume a nearest-neighbor communication. On the contrary, the set of subdomains is unordered (unstructured) on the subdomain level, leading to random communication among subdomains. In other words, the communication distance cannot be limited to nearest neighbors, but any distance on the processor topology is possible (processor topology describes how the processors are connected, for instance in a 2D mesh, in a torus or in a hypercube etc.). Hence, the efficiency of the parallel algorithm must not depend on nearest-neighbor communication. Therefore, the parallelization of solution domains of complex geometry requires a more complex communication pattern to ensure a load-balanced application. It also demands more sophisticated message passing among neighboring blocks, which may reside on the same, on a neighboring, or on a distant processor. The basic parallelization concept for this kind of problem is the introduction of a new type of boundary condition, namely the interdomain boundary condition that is updated in the solution process by neighboring subdomains via message passing. Parallelization then is simply achieved by the **introduction of a new type of boundary condition.** Thus, parallelization of a large class of complex problems has been logically reduced to the well-known problem of specifying boundary conditions.

12.8.1 Message Passing for Multiblock Grids

The following message-passing strategy has been found useful in the implementation of a parallel multiblock code. The parallelization of I/O can be very different with respect to the programming models (SPMD, Single Program Multiple Data, host-node — not recommended) and I/O modes (host-only I/O, node local I/O, fast parallel I/O hardware, etc.) supported by the parallel machines. The differences can also be hidden in the interface library.

Portability: Encapsulation of message passing routines helps to reduce the effort of porting a parallel application to different message passing environments.

Source code: Encapsulation allows the use of one source code both for sequential and parallel machines.

Maintenance and further development: Encapsulation keeps message-passing routines local. Thus, software maintenance and further development will be facilitated.

Common message-passing subset: Portability can be highly increased by restricting oneself to use only operations included in the common subset for implementing the interface routines.

- Since each processor of the parallel machine takes one or more blocks, there may not be enough blocks to run the problem on parallel machines. There are tools to automatically split the blocks to allow the utilization of more processors.
- In general, blocks are of very different sizes, so that the blocks must be distributed to the processors to produce a good load balance. There are tools to solve this bin-packing problem by a simple algorithm that takes virtually no time.

An extremely simple message passing model is implemented, consisting of only *send* and *receive.* The simplicity of this model implies easy portability.

For an elementary Laplace solver on a square grid, each grid point takes the average of its four neighbors, requiring 5 flops, and communicates 1 floating-point number for each gridpoint on the boundary. For a more sophisticated elliptic solver, needing 75 flops per internal grid point, grid coordinates have to be exchanged across boundaries. Our flow solver, *ParNSS* [Williams et al., 1996], in contrast, does a great deal of calculation per grid point, while the amount of communication is still rather small.

Thus we may expect any implicit flow solver to be highly efficient in terms of communication. When the complexity of the physics increases, with turbulence models and chemistry, we expect the efficiency to get even better. This is why a flow solver is a viable parallel program even when running on a workstation cluster with slow communication (Ethernet).

12.8.2 Parallel Machines and Computational Fluid Dynamics

For the kinds of applications that we are considering, we have identified four major issues concerning parallelism, whether on workstation clusters or parallel machines.

Load balancing. As discussed above, the number of blocks in the grid must be equal to or larger than the number of processors. We wish to distribute the blocks to processors to produce an almost equal number of grid points per processor; this is equivalent to minimizing the maximum number of grid points per processor. We have used the following simple algorithm to achieve this.

The largest unassigned block is assigned to the processor with the smallest total number of grid points already assigned to it, then the next largest block, and so on until all blocks have been assigned.

Given the distribution of blocks to processors, there is a maximum achievable parallel efficiency, since the calculation proceeds at the pace of the slowest processor, i.e., the one with the maximum number of grid points to process. This peak efficiency is the ratio of the average to the maximum of the number of grid points per processor, which directly proceeds from the standard definition of parallel efficiency.

Convergence. For convergence acceleration a block-implicit solution scheme is used, so that with a monoblock grid, the solution process is completely implicit, and when blocks are small, distant points become decoupled. Increasing the number of processors means that the number of blocks must increase, which in turn may affect the convergence properties of the solver. It should be noted that any physical fluid has a finite information propagation speed, so that a fully implicit scheme is neither necessary nor desirable.

Performance. It is important to establish the maximum achievable performance of the code on the current generation of supercomputers. Results from the Intel Paragon machine and SGI Power Challenge are presented.

Scalability. Parallel processing is only useful for large problems. For a flow solver, we wish to determine how many processors may be effectively utilized for a given problem size, since we may not always run extremely large problems.

TABLE 12.3 Intel Paragon Computations for a 192-Block Halis Grid

Node	Wall Time	Corrected Time (per step)	Iterations	Speedup	Maximum Efficiency
128	606	14.46	35	3.647	0.835
64	631	27.59	21	1.932	0.918
32	648	53.28	12	1.000	0.986

Note: Distributing 192 blocks of different size onto 128 processors leads to a certain load imbalance, hence speedup is somewhat reduced.

12.9 Parallel Efficiency for Multiblock Codes

It is often stated the scientific programs have some percentage of serial computational work, s, that limit the speedup, S, of parallel machines to an asymptotic value of $1/s$, according to Amdahl's law where $s + p = 1$ (normalized) and n is the number of processors:

$$S = \frac{s+p}{s+p/n} = \frac{1}{s+p/n} \tag{12.1}$$

This law is based on the question, given the computation time on the serial computer, how long does it take on the parallel system? However, the question can also be posed in another way: Let s', p' be the serial and parallel time spent on the parallel system, then $s' + p'n$ is the time spent on a uniprocessor system. This gives an alternative to Amdahl's law and results in the speedup which is more relevant in practice:

$$S = \frac{s'+p'n}{s'+p'} = n-(n-1)s' \tag{12.2}$$

It should be noted that domain decomposition does not demand the parallelization of the solution algorithm but is based on the partitioning of the solution domain; i.e., the same algorithm on different data is executed. In that respect, the serial s or s' can be set to 0 for domain decomposition and both formulas give the same result. The important factor is the ratio r_{CT} (see below), which is a measure for the communication overhead. In general, if the solution algorithm is parallelized, Amdahl's law gives a severe limitation of the speedup, since for $n \to \infty$, S equals $1/s$. If, for example, s is 2% and n is 1000, the highest possible speedup from Amdahl's law is 50. However, this law does not account for the fact that s and p are functions of n. As described below, the number of processors, the processor speed, and the memory are not independent variables which simply means, if we connect more and faster processors, a larger memory is needed, leading to a larger problem size and thus reducing the serial part. Therefore speedup increases. If s' equals 2% and $n = 1024$, the scaled sized law will give a speedup of 980, which actually has been achieved in practice. However, one has to keep in mind that s and s' are different variables. If s' denoted the serial part on a parallel processor in floating point operations, it is not correct to set $s = s'n$, since the solution algorithms on the uniprocessor and parallel system are different in general.

For practical applications the type of parallel systems should be selected by the problem that has to be solved. For example, for routine applications to compute the flow around a spacecraft on 10^7 grid points, needing around 10^{14} floating point operations, computation time should be some 15 minutes. Systems of 1000 processors can be handled, so each processor has to perform about 10^{11} computations, and therefore a power (sustained!) of 100 MFlops per processor is needed. Assuming that 200 words, 8 bytes/word, are needed per grid point, the total memory amounts to 16 GB: that means 16 MB of private memory for each processor, resulting in 22 grid points in each coordinate direction. The total amount of processing time per block consists of computation and communication time:

TABLE 12.4 Convergence Behavior for 2D NACA 0012 Airfoil Speedup as Function of Number of Blocks

Block	Grid Points Per Block	Iteration	Computing Time	Speedup
2	2400	253	52519	1.00
32	1560	305	33930	1.55
120	435	317	22577	2.326
256	213	333	19274	2.725
480	119	349	17752	2.958
1024	61	380	18012	2.916

Note: This table clearly demonstrates that a fully coupled implicit solution scheme is not optimal.

$$t_p = N^3 * 10000 * t_c + 6N^2 * 10 * 8 * t_T \qquad (12.3)$$

where we assumed that 10,000 floating point operations per grid point are needed, and 10 variables of 8 byte length per boundary point have to be communicated. Variables t_c, t_T are the time per floating point operation and the transfer time per byte, respectively. For a crude estimate, we omit the set-up time for a message. Using a bus speed of 100 MB/s, we find for the ratio of computation time and communication time.

$$r_{CT} := \frac{N^3 * 10000 * 100}{6N^2 * 10 * 8 * 100} \approx 20N \qquad (12.4)$$

That is, for $N = 22$, communication time per block is less than 0.25% of the computation time. In that respect, implicit schemes should be favored, because the amount of computation per time step is much larger than for an explicit one.

In order to achieve the high computational power per node a MIMD (multiple instruction multiple data) architecture should be chosen; that means that the system has a parallel architecture. It should be noted that the condition $r_{CT} >> 1$ is not sufficient. If the computation speed of the single processor is small, e.g., 0.1 MFlops, this will lead to a large speedup, which would be misleading because the high value for r_{CT} only results from low processor performance.

12.10 Parallel Solution Strategy: A Tangled Web

12.10.1 Parallel Numerical Strategy

In this section a brief overview of the parallel strategy for the solution of large systems of linear equations as may be obtained from the discretization of the Navier-Stokes equations or elliptic grid generation equations is presented. The "tangled web" of geometry, grid and flow solver is discussed. The solution strategy is multifaceted, with

- **space strategy:** halving the grid spacing, termed grid sequencing,
- **linear solver strategy:** domain decomposition conjugate gradient–GMRES,
- **time-stepping strategy:** explicit/implicit Newton schemes for the N–S equations.

First we distinguish space discretization from time discretization. In case of the elliptic equations, we only have to solve the linear system once. The Navier–Stokes equations require the solution of a system of linear equations at each time step. If we are interested in a steady-state solution, a Newton–Raphson scheme is used. In addition, there is a sequence of grids, each with 8 times as many points as the last, and we loop through these from coarsest to finest, interpolating the final solution on one grid as the initial solution on the next finer grid. At the same time coarsening is used to compress the eigenvalue spectrum.

On each grid, the spatial discretization produces a set of ordinary differential equations: $d\mathbf{U}/dt = f(\mathbf{U})$, and we assume the existence of a steady-state $\mathbf{U}^*$ such that $f(\mathbf{U}^*) = 0$. We approach $\mathbf{U}^*$ by a sequence of explicit or implicit steps, repeatedly transforming an initial state $\mathbf{U}^0$ to a final state $\mathbf{U}$.

12.10.2 Time Stepping Procedure

For the Navier–Stokes equations, the following time stepping approach is used.

The explicit step is the two-stage Runge–Kutta:

$$\mathbf{U}^{n+1} = \mathbf{U}^n + f\left(\mathbf{U}^n + f'\left(\mathbf{U}^n\right)\Delta t/2\right)\Delta t \tag{12.5}$$

The implicit time step is a backward Euler:

$$\mathbf{U}^{n+1} = \mathbf{U}^n + f\left(\mathbf{U}^{n+1}\right)\Delta t \tag{12.6}$$

Third, we have the final step, getting to the steady-state directly via Newton, which can also be thought of as an implicit step with infinite Δt:

$$\text{solve} \quad \mathbf{f}(\mathbf{U}) = 0 \tag{12.7}$$

There is also a weaker version of the implicit step, which we might call the linearized implicit step, that is actually just the first Newton iteration of the fully nonlinear implicit step:

$$\mathbf{U}^{n+1} = \mathbf{U}^n + \left[1 - df/d\mathbf{U}\Delta t\right]^{-1} f\left(\mathbf{U}^n\right)\Delta t \tag{12.8}$$

The most time-consuming part in the solution process is the inversion of the matrix of the linear system of equations. Especially for fluid flow problems, we believe conjugate gradient (CG) techniques to be more robust than multigrid techniques, and therefore the resulting linear system is solved by the CG–GMRES method.

12.10.3 Parallel Solution Strategy

We use the inexpensive explicit step, as long as there is sufficient change in the solution, $\Delta\mathbf{U}$. When $\|\Delta\mathbf{U}\|/\|\mathbf{U}\|$ is too small, we begin to use the implicit step. Also, with each block the implicit solution scheme, the so-called dynamic GMRES, might exhibit a different behavior, that is a *Krylov* basis of different size may be used, eventually requiring dynamic load balancing.

12.10.4 Solving Systems of Linear Equations: The CG Technique

The conjugated gradient (CG) technique is a powerful method for systems of linear equations and therefore is used in many solvers. Its derivatives for nonpositive and nonsymmetric matrices (as obtained from discretizing the governing equations on irregular domains), for instance, GMRES (see Section 12.10.5), has a direct impact on the parallel efficiency of a computation. Its Krylov space dimension and hence the numerical load per grid point, varies during the computation depending on the physics. For instance, very high grid aspect ratios, a shock moving through the solution domain, or the development of a shear layer may have dramatic effects on the computational load within a block. Therefore, this kind of algorithm requires dynamic load balancing to ensure a perfect load balanced application.

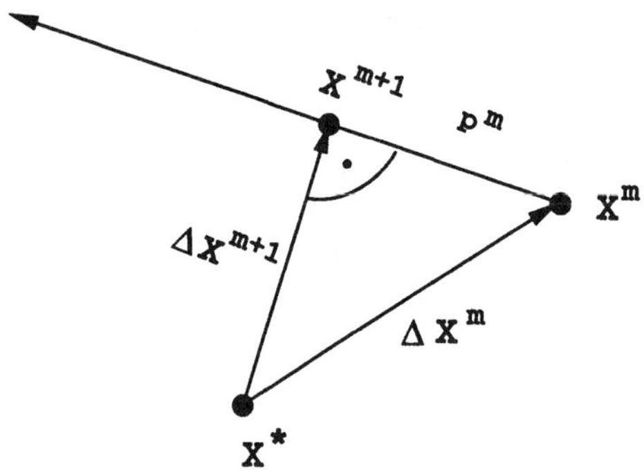

FIGURE 12.18 Geometrical interpretation of CG method. Let $\mathbf{x}^*$ denote the exact (unknown) solution, $\mathbf{x}^m$ an approximate solution, and $\Delta\mathbf{x}^m$ the distance from the exact solution. Given any search direction $\mathbf{p}^m$, except for $\mathbf{p}^m$ orthogonal to $\Delta\mathbf{x}^m$, it is straightforward to see that the minimal distance from $\mathbf{p}^m$ to $\mathbf{x}^*$ is found by constructing $\Delta\mathbf{x}^{m+1}$ perpendicular to $\mathbf{p}^m$.

In the following we give a brief description of the conjugate gradient method, explaining the geometric ideas on which the method is based. We assume that there is a system of linear equations derived from the grid generation equations or an implicit step of the N–S equations, together with an initial solution vector. This initial vector may be obtained by an explicit scheme, or simply may be the flow field from the previous step. It should be noted that the solution of this linear system is mathematically equivalent to minimizing a quadratic function. The linear system is written as

$$M\ \Delta U = R \Leftrightarrow \mathbf{Ax} = \mathbf{b} \tag{12.9}$$

using the initial solution vector x^0. The corresponding quadratic function is

$$f(x) = \frac{1}{2}\mathbf{x}^T\mathbf{Ax} - \mathbf{x}^T\mathbf{b} \tag{12.10}$$

where gradient $\nabla f = \mathbf{Ax} - \mathbf{b}$. For the solution of the Navier–Stokes equations, $\mathbf{x}^0$ is obtained from the most recent time steps, that is $\mathbf{x}^0 := \mathbf{U}^n - \mathbf{U}^{n-1}$ where index n denotes the number of explicit steps that have been performed. In the conjugate gradient method, a succession of one-dimensional search directions $\mathbf{p}^m$ is employed, i.e., the search is done along a straight line in the solution space — how these directions are constructed is of no concern at the moment — and a parameter α_m is computed such that function $f(\mathbf{x}^m - \alpha_m\mathbf{p}^m)$ along the $\mathbf{p}^m$ direction is minimized. Setting $\mathbf{x}^{m+1}$ equal to $\mathbf{x}^m - \alpha_m\mathbf{p}^m$, the new search direction is then to be found. In two dimensions, the contours $f =$ const. form a set of concentric ellipses, see Figure 12.19, whose common center is the minimum of f. The conjugate gradient method has the major advantage that only short recurrences are needed, that is, the new solution vector depends only on the previous one and the search direction. In other words, storage requirements are low. The number of iterations needed to achieve a prescribed accuracy is proportional to the square root of the condition number of the matrix, which is defined as the ratio of the largest to the smallest eigenvalue. Note that for second-order elliptic problems, the condition number increases by a factor of four when the grid-spacing is halved.

It is clear from Figure 12.18 that the norm of the error vector $\mathbf{x}^{m+1}$ is smallest being orthogonal to the search direction $\mathbf{p}^m$.

$$\left(\mathbf{x}^m - \mathbf{x}^n\right)\cdot\mathbf{p}^m = 0 \tag{12.11}$$

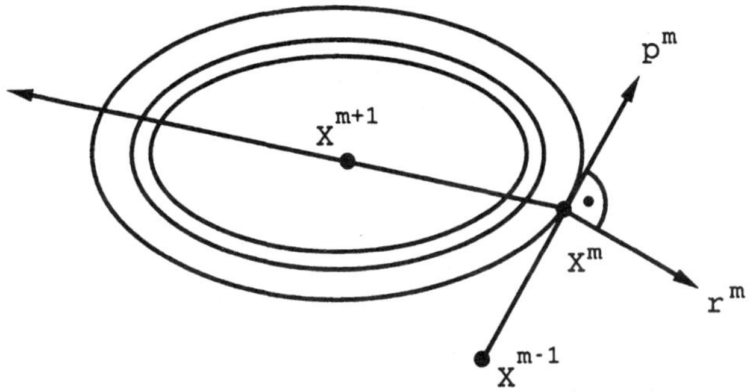

FIGURE 12.19 Geometrical interpretation of conjugate gradient method: since $\mathbf{r}^m$ is perpendicular to $\mathbf{p}^m$, a plane is spanned by these two vectors. The residual $\mathbf{r}^m$ is the gradient of the quadratic form $f(\mathbf{x})$ and thus perpendicular to the tangent of $f(\mathbf{x})$ = const. = $f(\mathbf{x}^m)$ at $\mathbf{x}^m$, because $\mathbf{x}^m$ is a minimal point. The next search direction pm+1 must therefore go through the midpoint of the ellipse, which is the projection of $f(\mathbf{x})$ onto this plane. The midpoint is the optimal point, i.e. gives the lowest residual in this plane. It is straightforward to show that $\mathbf{p}^{m+1}$ must satisfy $(\mathbf{p}^{m+1}, A\mathbf{p}^m) = 0$, simply because we are dealing with an ellipse. Moreover, $\mathbf{p}^{m+1}$ must be a linear combination of $\mathbf{r}^m$ and $\mathbf{p}^m$, and thus can be expressed as $\mathbf{p}^{m+1} = \mathbf{r}^m + \beta_k \mathbf{p}^m$.

From this **first orthogonality condition,** α_m can be directly computed. Figure 12.18 shows a right-angled triangle, and it directly follows (*Euclidean norm*) that the sequence of error vectors is **strictly monotonic decreasing.** In other words, *if the linear system derived from the Navier–Stokes equations,* **A x = b,** *has a unique solution, convergence is guaranteed, if N linear independent search vectors* $\mathbf{p}^m$ *are used.* This, however, is not of practical relevance, because in the solution of the Navier–Stokes equations there may be millions of variables, and only a few hundred or thousand iterations are acceptable to reach the steady state.

Since the exact change in the solution is not known, in practical computations the residual is used that is defined as

$$\mathbf{r}^m := \mathbf{b} - \mathbf{A}\mathbf{x}^m \tag{12.12}$$

Minimizing the quadratic function $f(\mathbf{x}^m - \alpha_m \mathbf{p}^m)$ along search direction $\mathbf{p}^m$ and using the expression for the residual directly gives

$$\alpha_m = \frac{(\mathbf{r}^m, \mathbf{p}^m)_A}{(\mathbf{p}^m, \mathbf{p}^m)_A} \tag{12.13}$$

In addition, it is required that $f(\mathbf{x}^m - \alpha_m \mathbf{p}^m)$ also be the minimum over the vector space spanned by all previous search directions $\mathbf{p}^0, \mathbf{p}^1, \ldots, \mathbf{p}^{m-1}$, because we must not destroy the minimal property when a new search direction is added. Hence the search directions are chosen to be A orthogonal, denoted as the **second orthogonality condition** defining the scalar product $(\mathbf{p}^k, \mathbf{p}^m)_A := (\mathbf{p}^k, A\mathbf{p}^m) = 0$ for $k \neq m$.

In determining the direction vectors, $\mathbf{p}^m$, a natural condition is that if a minimum in direction $\mathbf{p}^m$ is computed, the minimization already performed in the previous search directions, $\mathbf{p}^0, \mathbf{p}^1, \ldots, \mathbf{p}^{m-1}$ must not be affected. This is clearly the case if $\mathbf{p}^m$ is orthogonal to all previous basis vectors, because then $\mathbf{p}^m$ has no components in these directions and thus the minimum of f with respect over the subspace of $\mathbf{p}^0$, $\mathbf{p}^1, \ldots, \mathbf{p}^{m-1}$ is not changed by adding $\mathbf{p}^m$.

The original conjugate gradient method, however, has a requirement that matrix A by symmetric and positive definite (i.e., the quadratic form $\mathbf{x}^T \mathbf{A}\,\mathbf{x} > 0$). Clearly, matrix **A** of Eq. 12.9 does not possess these features. Therefore, an extension of the conjugate gradient method, termed *Dynamic GMRES* is employed that is described next.

12.10.5 Basic Description of GMRES

We have seen that the Navier–Stokes equations can be reduced to a system of linear equations, Eq. 12.9. Since a problem may comprise several million variables, an efficient method is needed to invert the matrix on the LHS. The system resulting from the Navier–Stokes equations is linear but neither positive definite nor symmetric, the term ($\mathbf{p}^m$, $\mathbf{Ap}^m$) is not guaranteed to be positive, and the search vectors are not mutually orthogonal. Therefore the conjugate gradient technique cannot be used directly. The extension of the conjugate gradient technique is termed the *generalized minimum residual* (GMRES) method [Saad, 1996]. It should be remembered that $\mathbf{p}^{m+1} = \mathbf{r}^m + \alpha^m \mathbf{p}^m$ and that the α^m are determined such that the second orthogonality condition holds, but this is no longer possible for the nonsymmetric case. However, this feature is mandatory to generate a basis of the solution space. Hence, this basis must be explicitly constructed. GMRES minimizes the norm of the residual in a subspace spanned by the set of vectors $\mathbf{r}^0$, $\mathbf{Ar}^0$, $\mathbf{A}^2\mathbf{r}^0$, ..., $\mathbf{A}^{m-1}\mathbf{r}^0$, where vector $\mathbf{r}^0$ is the initial residual, and the mth approximation to the solution is chosen from this space. The above-mentioned subspace, a Krylov space, is made orthogonal by the well-known Gram–Schmidt procedure, known as an Arnoldi process when applied to a Krylov subspace. When a new vector is added to the space (multiplying by A), it is projected onto all other basis vectors and made orthogonal with the others. Normalizing it and storing its norm in entry $h_{m,m-1}$, a matrix H_m is formed with nonzero entries on and above the main diagonal as well as in the subdiagonal. Inserting the equation for x_m into the residual equation, and after performing some modifications, a linear system of equations for the unknown coefficients $\gamma^l{}_m$ involving matrix H_m is obtained. H_m is called an upper Hessenberg matrix. To annihilate the subdiagonal elements, a 2D rotation (Givens rotation) is performed for each column of H_m until $h_{m,m-1} = 0$. A Givens rotation is a simple 2×2 rotation matrix. An upper triangular matrix R_m remains, which can be solved by back substitution.

It is important to note that the successful solution of the parallel flow equations can only be performed by a Triad numerical solution procedure. Numerical Triad is the concept of using **grid generation, domain decomposition,** and the **numerical solution scheme** itself. Each of the three Triad elements has its own unique contribution in the numerical solution process. However, in the past, these topics were considered mainly separately and their close interrelationship has not been fully recognized. In fact, it is not clear which of the three topics will have the major contribution to the accurate and efficient solution of the flow equations. While it is generally accepted that grid quality has an influence on the overall accuracy of the solution, the solution dynamic adaptation process leads to an intimate coupling of numerical scheme and adaptation process, i.e., the solution scheme is modified by this coupling as well as the grid generation process. When domain decomposition is used, it may produce a large number of independent blocks (or subdomains). Within each subdomain a block-implicit solution technique is used, leading to a decoupling of grid points. Each domain can be considered to be completely independent of its neighboring domains, **parallelism** simply being achieved by introducing **a new boundary condition,** denoted as inter-block or inter-domain boundary condition. Updating these boundary points is done by message passing. It should be noted that exactly the same approach is used when the code is run in serial mode, except that no messages have to be sent to other processors. Instead, the updating is performed by simply linking the receive buffer of a block to its corresponding neighboring send buffer. Hence, **parallelizing a multiblock code demands neither rewriting the code nor changing its structure.**

A major question arises in how the **decomposition process affects the convergence rate** of the implicit scheme. First, it should be noted that the N–S equations are not elliptic, unless the time derivative is omitted and inertia terms are neglected (Stokes equations). This only occurs in the boundary layer when a steady state has been reached or has almost been reached. However, in this case the Newton method will converge quadratically, since the initial solution is close to the final solution. The update process via boundaries therefore should be sufficient. In all other cases, the N–S equations can be considered hyperbolic. Hence, a full coupling of all points in the solution domain would be unphysical, because of the finite propagation speed, and is therefore not desired and not needed. To retain second-order accuracy across block (domain) boundaries, an overlap of two points in each coordinate direction has to be implemented. This guarantees the numerical solution is independent of

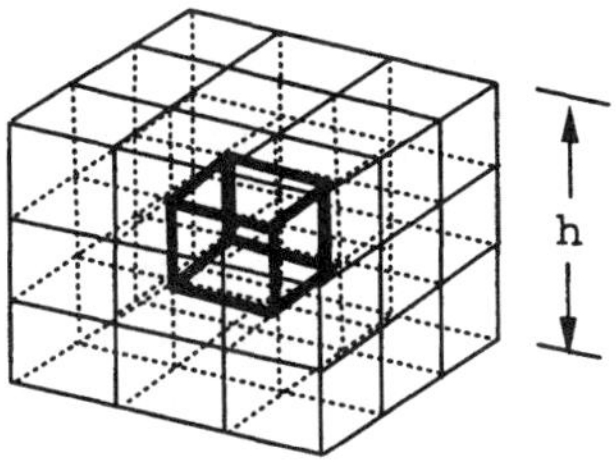

FIGURE 12.20 Flow variables are needed along the diagonals to compute mixed second derivatives for viscous terms. A total of 26 messages would be needed to update values along diagonals. This would lead to an unacceptable large number of messages. Instead, only block faces are updated (maximal six messages), and values along diagonals are approximated by a finite difference stencil.

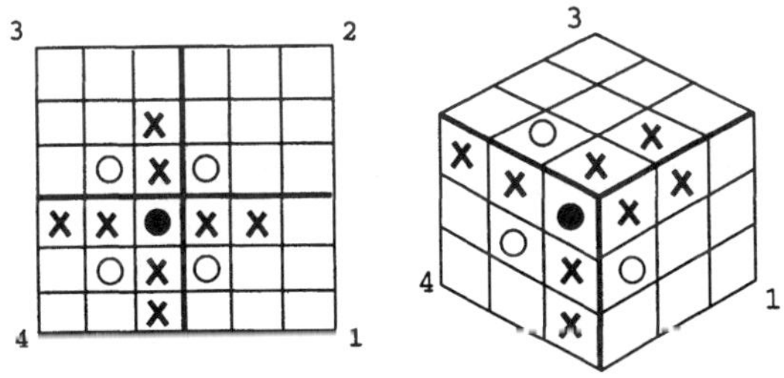

FIGURE 12.21 The figure shows the computational stencil. Points marked by a cross are used for inviscid flux computation. Diagonal points (circles) are needed to compute the mixed derivatives in the viscous fluxes. Care has to be taken when a face vanishes and 3 lines intersect.

block topology. The only restriction comes from the computation of flow variables along the diagonals on a face of a block (see Figure 12.20), needed to compute the mixed derivatives in viscous terms.

It would be uneconomical to send these diagonal values by message passing. Imagine a set of 27 cubes with edge length $h/3$ assembled into a large cube of edge length h. The small cube in the middle is surrounded by 26 blocks that share a face, an edge, or a point with it. Thus, 26 messages would have to be sent (instead of 6 for updating the faces) to fully update the boundaries of this block. Instead, the missing information is constructed by finite difference formulas that have the same order of truncation error, but may have larger error coefficients.

To continue the discussion of convergence speed it should be remembered that for steady-state computations implicit techniques converge faster than fully explicit schemes. The former are generally more computationally efficient, in particular for meshes with large variations in grid spacing. However, since a full coupling is not required by the physics, decomposing the solution domain should result in a convergence speed-up, since the inversion of a set of small matrices is faster than the inversion of the single large matrix, although boundary values are dynamically updated. On the other hand, if the decomposition leads to a block size of one point per block, the scheme is fully explicit and hence computationally less efficient than the fully implicit scheme. Therefore, an optimal decomposition topology must exist that most likely depends on the flow physics and the type of implicit solution process. So far, no theory has been developed.

Second, domain decomposition may have a direct influence on the convergence speed of the numerical scheme. In this chapter, the basis of the numerical solution technique is the Newton method, combined with a conjugate gradient technique for convergence acceleration within a Newton iteration. In the preconditioning process used for the conjugate gradient technique, domain decomposition may be used to decrease the condition number (ratio of largest to smallest eigenvalues) of the matrix forming the left-hand side, derived from the discretized N–S equations. In other words, the eigenvalue spectrum may be compressed, because the resulting matrices are smaller. Having smaller matrices the condition number should not increase; using physical reasoning it is concluded that in general the condition number should decrease.

From these remarks, it should be evident that only a combination of grid generation scheme, numerical solution procedure, and domain decomposition approach will result in an effective, general numerical solution strategy for the parallel N–S equations on complex geometries. Because of their mutual interaction these approaches must not be separated. Thus, the concept of numerical solution procedure is much more general than devising a single numerical scheme for discretizing the N-S equations. Only the implementation of this interconnectedness in a parallel solver will lead to the optimal design tool.

12.11 Future Work in Parallel Grid Generation and CFD

Since neither vector nor parallel computing is of interest to the scientist or engineer who has to compute an application, a simple but general rule is that scalar architectures requiring the smallest number of processors to provide a certain computing power should be favored. As experience shows, it is the input and output that becomes cumbersome when a large massively parallel system is used. The paradigm of having each processor read its own file and write its own file starts to tax the file system greatly. This is because there is a single disk controller converting file names into disk track locations, and this constitutes a sequential bottleneck. It is better to have all the processors opening a single large file, and each reading and writing large records from that file whose size is a power of two number of bytes. For instance, this I/O approach has been implemented for the Intel version of *ParNSS* running on several hundred processors.

One of the most challenging tasks is the development of algorithms that scale numerically. The so-called **Tangled Web** approach, see Section 12.10, based on the idea of a varying coupling strength among grid points during the solution process, will be one of the most important novel techniques that might have the potential to achieve this objective.

Acknowledgment

We are grateful to our colleagues Jean Muylaert and Martin Spel from ESTEC, Noordwijk, The Netherlands for many stimulating discussions. This work was partly funded under EFRE Contract 95.016 of the European Union and the Ministry of Science and Culture of the State of Lower Saxony, Germany.

References

1. **Bruce, A.**, et al., JPL sets acoustic checks of cassini test model, *Aviation Week and Space Technology*, 143(9), pp. 60–62, 1995.
2. **Eiseman, P.**, et al., GridPro/AZ3000, User's guide and reference manual, *PDC*, 300 Hamilton Ave, Suite 409, White Plains, N, 10601, pp. 112, 1996.
3. **Häuser, J., et al.**, Euler and N–S grid generation for halis configuration with body flap," *Proceedings of the 5th International Conference on Numerical Grid Generation in Computational Field Simulation*, Mississippi State University, pp. 887–900, 1996.
4. NASA Reference Publication 1338, NASA geometry data exchange specification for CFD, (NASA IGES), Ames Research Center, 1994.
5. **Saad, Y.**, *Iterative Methods For Sparse Linear Systems*, PWS Publishing, 1996.
6. **Venkatakrishnan, V.**, Parallel implicit unstructured grid Euler solvers, *AIAA Journal*, Vol. 32, 10, 1994.
7. **Williams, R.**, Strategies for approaching parallel and numerical scalability in CFD codes, *Parallel CFD*, Elsevier, North-Holland 1996.

II

Unstructured Grids

Nigel P. Weatherill

Introduction to Unstructured Grids

The fundamental difference between a structured and an unstructured grid is the ordering of the nodes to form the elements or cells within the grid. If the nodes can be ordered into a regular array (i,j,k), with the assumption that the nodes (i,j,k), $(i,j,k+1)$, etc., are neighbors, then the grid is described as *structured.* If the nodes cannot be arranged in such a form the grid is *unstructured.* Hence, an unstructured grid must include, as part of its definition, the connection between nodes that form the mesh. Clearly, any method that generates an unstructured mesh must include a procedure for providing the explicit definition of the connections between nodes to form elements, in addition to the coordinates of the nodes themselves.

In common with the procedure of generating structured grids, the process of constructing unstructured grids begins with a geometrical definition of the domain to be meshed (see Part III). Such a definition will be in the form of NURBs curves and surfaces, or an equivalent, such as splines and bicubic splines. Most unstructured grid methods then build a grid based on a hierarchical approach that involves generating grids on boundary curves, boundary surfaces, and finally a volume grid. The shape of the elements generated in unstructured grids can vary; traditionally, triangles on surfaces and tetrahedra in the volume have been used; however, quadrilaterals and hexahedra are favored in some applications (Chapter 21). The requirement to define a connectivity matrix between nodes may appear an unnecessary burden when compared with structured grids, but such a requirement provides the flexibility to generate a mesh of any element type and more recently grids of mixed (hybrid) element type (Chapters 23 and 25).

By their very nature, the irregular ordering of the connections between nodes within an unstructured mesh places great emphasis on techniques that enable searches to be made through the grid in a fast and efficient manner. Hence, data structures play a major role both in the generation of unstructured meshes and in the subsequent use of such grids with solution algorithms. Techniques to generate unstructured grids are, in most cases, based on relatively straightforward concepts. However, the practical implementation of some of these methods within a computer code is a major challenge. Hence, it is appropriate, prior to any discussion, to introduce an in-depth discussion on data structures (Chapter 14). Basic data structures are described, including linear lists and tree structures. These techniques are then applied to multidimensional search algorithms, with some details given on the alternating digital tree that has proven to be so effective in many unstructured grid generation algorithms, in particular, the advancing front method.

With all the research activity devoted to automatic grid generation, there are now many techniques for the construction of unstructured grids. However, three approaches are very widely used. They can be broadly described as tree-based methods, such as octree (Chapter 15), point insertion methods based on Delaunay triangulation (Chapters 16, 18, 20, and 26), and advancing front methods (Chapter 17). Chapter 15 describes the method whereby a domain is broken down into elements using a recursive subdivision based on a spatial tree structure. Such approaches can be thought of as starting with a cube that encloses the geometry of the domain on which a grid has already been generated. The initial cube is subdivided into eight cubes. Hence, from one cube there are eight branches, which is the beginning of a tree data structure. After subdivision, a check is performed to determine if the length scale of one of the cubes is consistent, i.e., is of the same order, as the local length scale of the grid on the boundary that is enclosed by the cube. If there is no consistency, then the cube is further subdivided; if there is consistency then no further subdivision is required. When no further cubes need to be subdivided, then the final step requires the subdivided grid to be connected to the boundary surface mesh. This approach, which can clearly admit directional refinement, makes full use of tree data structures and is often referred to as quad-tree in two dimensions, and octree in three dimensions.

Many unstructured grid generation methods are based on Delaunay–Voronoï methods (Chapters 16, 18, and 20). These geometrical constructions have been known for many years, with a paper by Dirichlet appearing in 1850. The basic concept of the Delaunay triangulation is simple and elegant. Given a set of nodes, the Voronoï diagram subdivides the space into tessellations, in which each tile is the space closer to a particular node that any other node. Clearly, the boundaries of the Voronoï diagram represent the perpendicular bisectors between adjacent nodes. If nodes are connected that have a common boundary of the Voronoï diagram, then a triangulation of the nodes is formed. In two dimensions the triangulation is a set of triangles, in three dimensions the triangulation consists of tetrahedra. The Delaunay triangulation has some interesting properties, and the so-called in-circle criterion, in which no node is contained within a circle (in two dimensions) or sphere (in three dimensions) passing through the nodes that form the element, can be used to construct the triangulation in an efficient manner. Delaunay–Voronoï methods provide a mechanism for connecting nodes; they do not provide a method for creating nodes. Hence, it is necessary to consider methods for the automatic creation of nodes. Such methods are based on the iterative refinement of the initial triangulation formed when the boundary nodes are connected using a Delaunay triangulation. A variety of methods have been investigated, including simply adding nodes at centroids of elements, along element edges, or, more generally, using Steiner points. Following the generation of interior nodes, a major issue with Delaunay–Voronoï methods is to ensure that the elements of the mesh conform to the boundary of the domain. In general, this will not be the case everywhere within the grid, and hence, steps must be taken to ensure boundary integrity. This issue can be addressed by introducing what is termed a constrained Delaunay triangulation or using postprocessing methods which, through element face and edge swapping, recover the boundary mesh within the global unstructured mesh.

The advancing front method (Chapter 17) takes a boundary descretization and creates elements within the domain, advancing in from the boundary until the entire domain is filled with elements. Given an initial front, which in two dimensions is the set of edges forming the boundary discretization, and in three dimensions is the set of triangular faces of the surface mesh, a node is created from which a valid element is made. Clearly, in forming a new element, it is essential that the edges of the element do not intersect any existing elements and that the element quality is satisfactory. Such checks highlight the need for effective data structures.

Common to all unstructured grid methods is the requirement to control the grid point spacing (Chapters 16, 17, 18, 20, 35). The construction of grids usually involves a subdivision of the boundary geometry into a surface grid followed by the volume grid generation. Hence, if a grid is to have consistent point spacing both on boundaries and within the domain, it is essential the grid point density is specified before the boundary grid generation. The grid point density is commonly controlled by a background mesh. This can be a very coarse grid that covers the domain and at each node of the background mesh the grid point density is specified. Hence, in the grid generation, the required point spacing at any position in the domain is interpolated from the background mesh spacing. In practice, for relatively simple geometries, it is possible to define the background grid automatically and then allow the user to set the spacing at each node of the mesh. However, this becomes more problematic for more complicated geometrical shapes and the method has been supplemented with the use of grid sources. A grid source is defined in terms of a position in a mesh where the required grid point spacing is specified, together with the region over which the source should influence the grid (Chapter 35). Such an approach does not require the user to construct a coarse background mesh and hence is more time-efficient. A source can be defined to be effective as a point, line, surface, or even a volume.

The basic mechanics of grid point control can be readily extended to enable grid adaptation to solution data. Chapter 35 presents an in-depth discussion on adaptation techniques based around the use of a background mesh and sources, together with the more conventional techniques of point enrichment (h-refinement) and point movement (r-refinement).

Most grid generation techniques require the surface of a domain to be meshed prior to the generation of the volume grid. Hence, surface grid generation is an essential and important step in the unstructured grid procedure. In Chapter 19, details are presented of how high-quality surface meshes of triangles can be generated on geometrical support surfaces. This step in the grid generation procedure links grid generation techniques with geometrical representation, and it is essential that a good understanding of surface modeling is acquired (see Part III). In most methods, surface grids are generated in the parametric space, which can be interpreted as two dimensional with additional information that represents surface curvature. Hence, standard grid generation techniques can be used to construct surface grids.

The Voronoï–Delaunay method does naturally allow for the construction of highly stretched or non-isotropic elements. The advancing front method does allow elements to be created that are stretched and aligned in prespecified directions, although the degree of stretching of elements that can be formed is limited. Hence, there is a major interest in unstructured grid methods that can form elements that are aligned in specified directions and have arbitrary aspect ratios. A typical application for such meshes is in the simulation of high Reynolds number flow fields where the efficient resolution of boundary layers is required. In Chapter 20, the generation of nonisotropic grids is discussed within the framework of the Voronoï–Delaunay approach. In the approach described, the unstructured grid is generated in a mapped space using the notion of a metric to distort regular elements into nonisotropic elements within the computational domain.

As computational methods advance and mature, there is both a requirement to attempt simulations with larger meshes and to use new parallel processing computer hardware. Both these requirements place an additional burden on grid generation technology. To meet these challenges, it is necessary to consider the generation of grids in parallel. Chapter 24 introduces some of the issues involved in generating unstructured grids in parallel.

It is clear from the contents of this handbook and a review of the literature that there are now many different approaches to the generation of grids. An obvious question is "Which is the best approach?"In this handbook we have not addressed this issue; we are content to present descriptions of key techniques and leave the reader to decide which is the most appropriate approach for any given application or problem. In fact, in the grid generation community, there is a realization that there is no such thing as the best grid generation approach — it is problem-dependent. However, there are now emerging grid generation packages that provide a user with the capability to generate structured grids and unstructured grids, and provide an ability to generate grids that are combinations of structured and unstructured grids or so-called hybrid grids. Chapter 23 presents an in-depth discussion for the motivation of hybrid grids and furthermore describes a system that provides a capability to generate grids that are totally structured (multiblock) to hybrid to totally unstructured (see also Chapter 25).

The application of unstructured grids to realistic problems requires techniques described in several chapters in the handbook to be used and integrated. To provide an overview of the complete procedure, Chapter 26 provides illustrated examples of the use of unstructured grids. In particular, a real example is taken and details provided on how an unstructured grid was generated starting from the initial geometry specified as point strings through to the final unstructured tetrahedral grid and solution using a finite element algorithm.

In the Foreword, it was emphasized that grid generation is only a means to an end. Once the spatial discretization, that is the grid, has been generated, attention can focus on developing the solution algorithm for the particular equation or set of equations. Chapter 26 provides some introductory material that describes how mathematical operations can be performed on unstructured grids. Some elementary concepts relating to the finite element method are described.

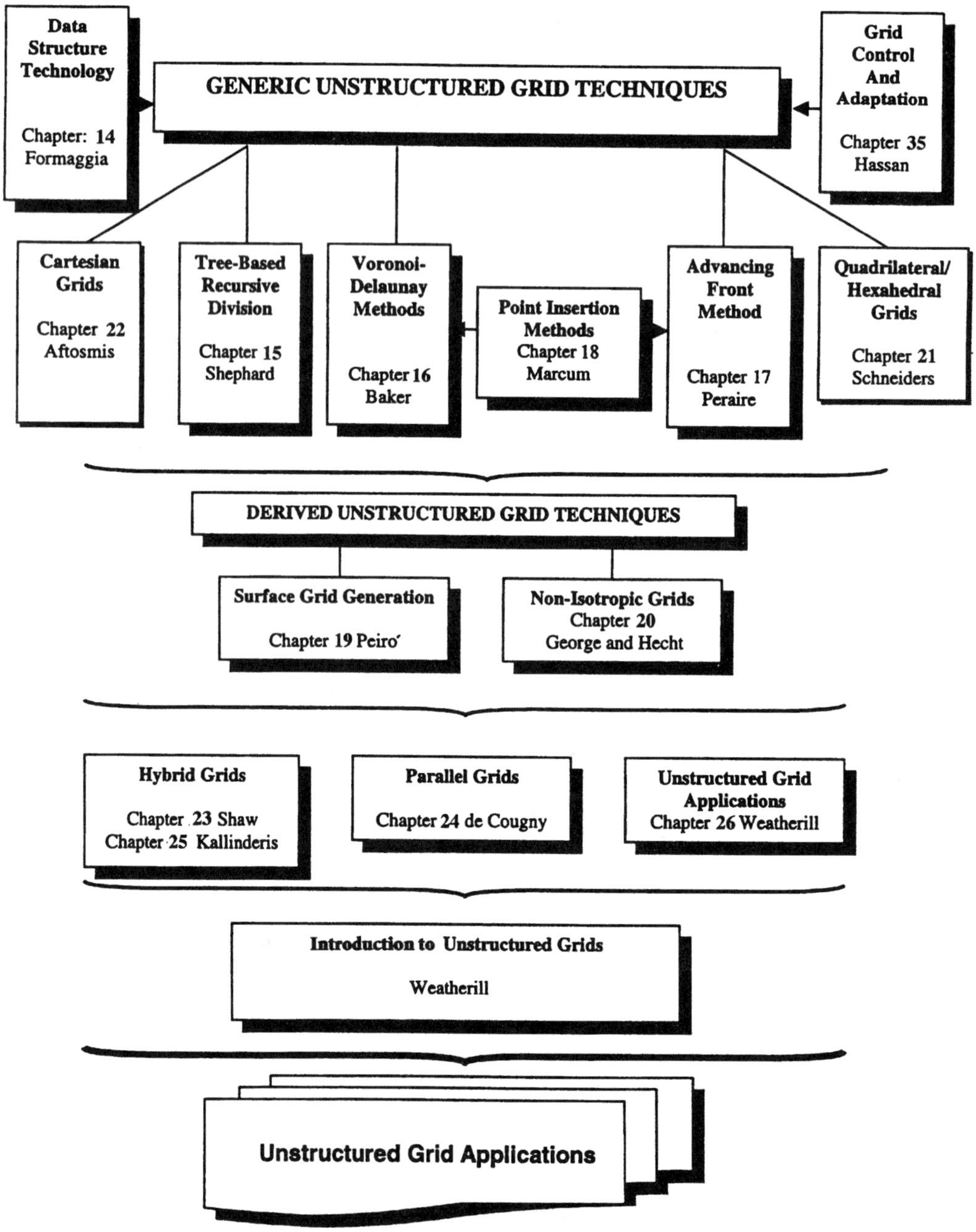

Diagrammatic Layout of the Chapters in Part II – Unstructured Mesh Generation

13

Block-Structured Applications

Timothy Gatzke

13.1 Introduction

The goal of computational fluid dynamics (CFD), and computational field simulation in general, is to provide answers to engineering problems using computational methods to simulate fluid physics. CFD has demonstrated the capability to predict trends for configuration modifications and parametric design studies. Its most valuable contribution today may be in allowing detailed understanding of the flowfield to determine causes of specific phenomena. Surface pressure data is routinely accepted subject to the limitations of the solution algorithms used. Careful application of CFD can provide reasonably accurate increments between configurations. A great deal of care (and validation) is required to get absolute quantities, such as drag, skin friction, or surface heat transfer, on full vehicles.

Grid generation is a necessary step in the process, and includes the bulk of the setup time for the problem. The grid generated will impact many aspects of the study. The rate of stretching in the grid, and the grid resolution in regions of curvature and/or high flowfield gradients will affect the quality of the results. The number of grid points will dictate the CPU requirements and the computational and calendar time for the study. A rough rule of thumb is that the CPU time for a flow solution is proportional to the number of grid points raised to the 3/2 power. The complexity of the grid will drive the personnel costs.

Engineers would look forward to grid generation if it were a low-stress and straightforward task that could be performed in a morning and success were guaranteed. Someday that may be the case, but for now, grid generation is often challenging, and usually very time-consuming. However, as can be seen in other chapters of this handbook, grid generation methods have come a long way. The simple geometries of a few years ago have been replaced by very complex configurations, such as fighter aircraft with stores, the underhood of an automobile, and human respiratory and circulatory systems.

0-8493-2687-7/99/$0.00+$.50

The goal of this chapter will be to focus on the implementation of the technology rather than the development of the technology. This chapter will take a broad view of "applications" and discuss three categories of applications: (1) application of grid generation tools to generate grids for engineering studies, (2) general application of CFD to engineering problems, and (3) application of grid generation technology in the development of grid generation codes. For each of these categories, a key element is managing three types of risk: cost, schedule, and technical. Can the project be completed at a cost that is competitive with other approaches? Can the product or results be obtained in time to meet the end user's needs? And when complete, will the results satisfy expectations? An understanding of the process and key issues should help control technical, budget, and schedule risk.

13.2 Guidelines for Generating Grids

This section looks at issues that arise in the grid generation environment for production CFD applications. Creating a grid for a specific application is highly dependent on several factors. These include details and features of the configuration to be analyzed, the grid generation and flow solver codes to be used, and the type of grids being generated, i.e., structured, overset, or unstructured. Grid generation is also subject to management issues, such as schedule, budget, and resource allocation.

An important requirement for confidence in any CFD study is a validation of the approach. This validation should include use of the same grid approach on a configuration of similar complexity, with similar run conditions on the same solver (since different solvers can have very different grid requirements), and comparison with trusted experimental results. Similar smaller-scale validations should also be performed for novel grid generation techniques used to control grid resolution, and to compare the effectiveness of different gridding approaches.

One important consideration for production CFD usage is the availability of sufficient resources to meet the study goals. Setting aside for now schedule issues, the issues affecting grid generation boil down to disk space to store the grid (and eventually the related flow solution) and memory to compute the solution (or to smooth the grid). Disk space limitations can cause skimping on the number of points in the grid. Splitting into more blocks does not always reduce the total number of points, although it may allow portions of the grid to be generated separately and combined on another system with more memory. Memory limitations can often be addressed by splitting the problem into more blocks, presuming a multiblock approach. This may mean a little more overhead in the solution process and a little more work generating the grids, but tools can be developed to automate the splitting or combining of blocks (Dannenhoffer, 1995) which set up the block-to-block connectivity information for the user as well. It is suggested that skimping on grid points be avoided as much as possible, as it endangers both the accuracy and convergence of the solution. Often a better approach is to simplify the geometry being used in the study to produce a smaller grid.

Similarly, schedule constraints often lead to trying to do more with fewer points, to reduce the time required for both grid generation and flow solution. Schedules rarely include enough time to perform a study as it ought to be done. In this case, it is better to eliminate unnecessary detail rather than crudely model more complex detail. It is important to examine the goals of the study closely, and rescope the problem as necessary to fit within the required schedule. Then a suitable grid may be generated to meet these new goals. If it is clear that the schedule requirements and study goals cannot be met in a manner that benefits the end customer, it may be best to abandon the study early, rather than expending significant resources and ultimately failing to obtain results in time to be useful.

The block-structured grid generation process can be divided into several discrete steps, as shown in Figure 13.1, although some of these steps may be interchanged or overlap. The boundary condition and connectivity step may or may not be considered part of the grid generation process depending on whether the boundary conditions are viewed as associated with the geometry and grid, or with the solution algorithm. We will take the view in this chapter that the boundary conditions and block connectivity are associated with the grid as part of the process of providing a complete model of the flowfield around the vehicle.

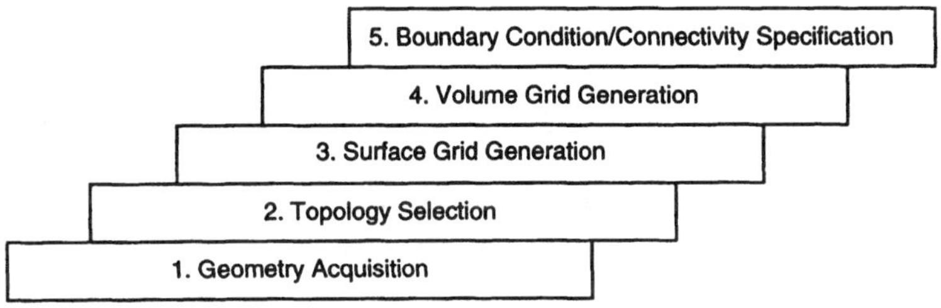

FIGURE 13.1 Steps in the gird generation process.

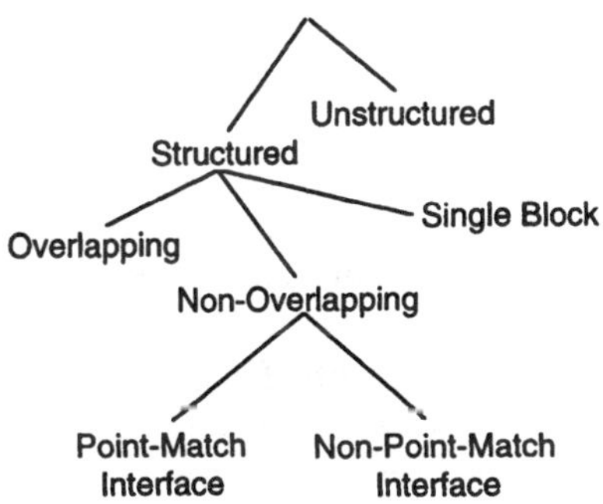

FIGURE 13.2 Method selection decision tree.

13.2.1 Basic Decisions

Before actually starting the grid generation process, several decisions need to be made. The selection of a grid method such as structured or overset, a grid code, and a solution code will ultimately define much of the process. The grid code may determine the specific order in which steps are performed, or leave the order to the discretion of the user. The process around which the tools are built, along with the expertise of the user, will determine how long the process will take. The flow- solver selection will affect the choice of the grid code, and can have a large impact on how the grid is generated. Rarely is the flow solver chosen based on the fact that it might make the grid generation task easier. The most common reasons for choosing a particular flow solver are availability and familiarity with it, and confidence in its accuracy for the type of application.

Figure 13.2 illustrates several basic decisions that quickly narrow the list of candidate flow solvers, based on the fundamental needs of the application. The first is often made unconsciously: should a structured or unstructured approach be used? At times this decision is made based on the prevailing approach used by the user's organization rather than the strengths and weaknesses of these approaches for a specific application. Similarly, the decision to use an overlapping or non-overlapping method is also not always made based on technical merit, but rather the availability and familiarity with a given code. If a nonoverlapping approach is chosen, a third decision that falls into the same category is whether to use a code with point-match interfaces or arbitrary non-point-match interfaces. As indicated previously, many of these decisions may be predetermined by factors beyond the user's control. But for the user without a large commitment to any one code, or with the flexibility to choose from a variety of codes, the benefits and limitations of different approaches and their impact on the grid generation process will be presented.

The primary use for unstructured grid methods currently is Euler analyses, although many efforts at unstructured viscous analysis may someday gain their share of production applications. The primary benefit is a reduction in grid generation time. Unstructured methods also offer hope to exploit adaptive solution strategies. Structured grid methods predominate in Navier–Stokes analyses and applications using existing grids that need little or no modification. Overset grids offer simplified generation of grid points at the expense of more complex generation of holes and connectivity between overlapping blocks. Overset grid quality issues related to appropriate resolution in the overlapping regions are more complex due to the 3D nature of the grid-to-grid interface. The use of point-match block interfaces for nonoverlapping grids simplifies the passing of information between blocks in the flow solver, at the expense of a much more restrictive grid generation problem. These restrictions generally result in more grid points or require innovative strategies to control grid resolution. These innovative strategies require more effort to keep an appropriate resolution in critical areas, without large numbers of points propagating where they are not needed. Some of these techniques will be illustrated later in this chapter. One alternate decision that can avoid the block interface issue is to use a single block grid. For some applications this may be optimum, but most complex configurations quickly eliminate this option from consideration, due to difficulties obtaining suitable resolution on the block boundaries and still being able to generate an acceptable grid on the interior.

13.2.2 Preparation for Grid Generation

A fundamental requirement for grid generation is a geometric definition of the configuration, be it a simple 2D airfoil, a complex fighter aircraft, the underhood area of an automobile, or the passages of a human heart. This geometry often is contained in a CAD (computer-aided design) model. Other possible sources of the information are drawings (only useful for very simple shapes), a series of cross sections, arrays of discrete points defining "patches" of the geometry, clouds of points triangulated to form a stereo lithography (STL) model of the geometry, or sometimes even select sections of an existing grid of the model. More detail on geometry modeling is discussed in Part III.

One factor that can affect generation of grids on a geometry definition is the inherent skewness of the geometry definition. This can occur in a point definition geometry if the cross-section cuts or defining curves have kinks or discontinuities in spacing distributions. For analytic surfaces, this skewness can be embedded in the geometry parametrization. For both analytic and point definition surfaces, this skewness can affect the ability to obtain a smooth grid. The user generally has more control over a point definition geometry, and can break the surface or redistribute to get a more uniform distribution. Analytic surfaces with unusual parameterizations require careful selection of operations that are not sensitive to the parametrization of the surface.

Another issue related to geometry that is often overlooked is verification that the geometry definition actually corresponds to the desired configuration of interest. For simple studies this may not be a problem, especially if only one geometry definition is available. However, in the design environment, configurations can be changing rapidly. As a matter of fact, it may not be possible to complete the grid for a given configuration before it is obsolete. In this environment, documentation of the specific details of the geometry definition that went into a particular grid is critical. (Record-keeping is often the first casualty of the hectic environment where it is most critical.)

Once the user has the right geometry, how faithful must the grid be to the geometry definition? Some grid generation codes ensure that the grids generated lie precisely on the geometry definition, while others leave it to the user to verify the integrity of the grid. For example, when redistributing points on surfaces that have been manipulated, is the new grid on the original geometry definition, or on an approximation resulting from subsequent manipulation? Will the grid automatically be projected back onto the original surfaces, or is that up to the user? Many times the user needs to modify the geometry to correct deficiencies, and may not want the grid to lay on the original geometry. An important point to remember is that by its nature, CFD is a discretization of the problem to be solved. A geometry definition is suitable if it provides the appropriate level of detail, and sufficient fidelity to the "real" geometry to make solution effects due to geometry inaccuracy small compared to the overall accuracy of the analysis. Obviously, the higher the accuracy required, the more important geometry fidelity becomes.

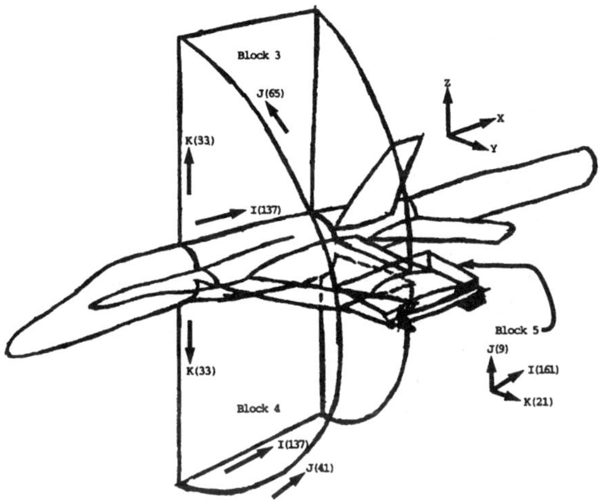

FIGURE 13.3 Sketch of a block layout for a point-match grid about a fighter aircraft with wing tip missile.

It is highly recommended that the user sketch out the block layout prior to generating the grid. Drawing the topology in this manner provides several benefits. First it verifies that the topology is possible. Many times verbal discussions of topology, especially with less experienced users, lead to misunderstandings. When the user is asked to draw a picture, it is easy to point out good features and problem areas, and when trying to put their ideas on paper, users will often realize on their own the flaw in the mental visualization. The information on the layout, such as number of points and preliminary distributions, can aid the user when sitting at the tube generating the grid or preparing an input file for the solution code. Another benefit is as a visual aid for communicating about the study with others. This is especially important if more than one individual will be working to generate the grid, or run the flow solver. Finally, this picture will contain information that will be used for postprocessing the solution and communicating the results to the customer. An example of a block layout for selected blocks of a 17-block grid about a fighter aircraft configuration is shown in Figure 13.3. Notice the inclusion of the number of points and the direction of the indices. As the grid is generated, the user may wish to include additional data such as the spacings used.

13.2.2.1 Level of Detail

Another key decision that often arises is how much detail should be included for a particular study. For example, when looking for forces on the radome of a fighter aircraft, the aft part of the vehicle is not needed, and the actual break will depend on the accuracy that is needed. On the other hand, afterbody drag effects will be highly dependent on at least gross effects of the forward elements of the configuration. Does an antenna sticking out on the lower side of the fuselage need to be modeled when predicting cruise drag on a commercial transport? In modeling a high-lift system, do the struts that support the wing slats and flaps need to be modeled or is modeling the gaps enough? When optimizing a wing, do the gaps between the flaps need to be modeled or can they be blended together?

In the past, many of these decisions were made for us by the limitations of our solution algorithms and grid generation tools. Now the reasons stem from time and schedule constraints, limited computing resources, and a practical decision as to what is really needed to get an answer to the design question being asked. A few guidelines are presented below.

If it is worth modeling a particular feature, use sufficient resolution. Conversely, if you cannot model a feature in enough detail, why include it at all? This is not to say that features cannot be modeled at one level of detail for gross effects, and at a finer level of detail for more accurate analysis. For example, the study of the control effectiveness of a horizontal stabilator might require only a coarse modeling of

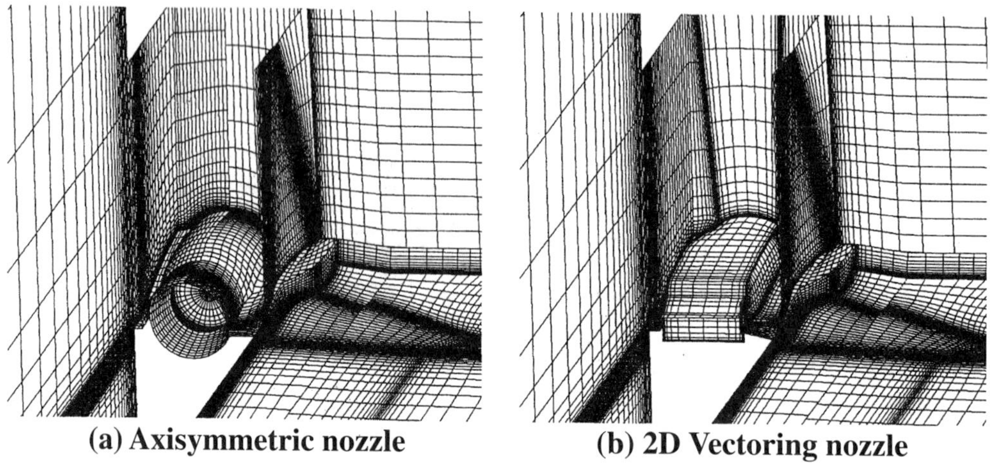

FIGURE 13.4 Multi-block approach promotes efficient grid generation for local geometry changes. (From Gatzke, T.D., et al., "MACGS: A Zonal Grid Generation System for Complex Aero-Propulsion Configurations," 1991, copyright McDonnell Douglas Corporation, with permission.)

the wing to get the downwash effects, while a highly resolved wing grid would be necessary for computation of absolute drag numbers. In a study with a wing, fuselage, and stabilator, it would be possible to model the wing with a grid so coarse that the gross effects would be so inaccurate that they would render the results meaningless. This is more likely to be an issue for small-scale features such as gaps between components, or local protuberances on the geometry, where in their absence, the grid would have had low resolution. Modeling them well increases grid points, and therefore disk requirements and solution time. The temptation is to not increase grid resolution, or increase it only slightly, and let the coarse grid fall on the feature where it will. A danger is that this seemingly innocuous treatment will cause the solution to behave badly. Possibly sharp turning of the grid will make the solution unstable, and much more time may be spent trying to keep it running, and often, eventually regridding and starting the run over.

13.2.3 Getting Started

Once the appropriate configuration and geometry source have been determined, the user is faced with a decision requiring careful thought. Should the grid be generated from scratch, or is there an existing grid that can be used as a starting point? Views vary on whether starting from an existing grid can really save a significant amount of time.

The advantages of starting from an existing grid are that potentially, much of the grid may not need to be changed. If perturbations of the surfaces are limited, the same topology and distributions may be used eliminating a time-consuming part of the process. Figure 13.4 illustrates major changes to a fighter aircraft grid that were simplified using a multi-block approach. The configuration included a full fuselage and part of the wing in addition to the detailed aft-end geometry. The baseline grid with axisymmetric nozzle contained just over 800,000 points in 22 blocks. The 2D nozzle grid was generated by replacing about 6 blocks in the baseline grid with 19 blocks. This change required one third of the time that it took to generate the baseline grid from scratch.

There can also be several disadvantages of using an existing grid. The original grid was developed for a purpose that may not support the new goals. The distributions used in the original grid and even the grid topology could be the cause of problems when computing a flow solution. If the grid is to be used with a different flow solver, that flow solver may have different grid requirements. Changes from Euler to Navier–Stokes analysis or vice-versa can make the original grid less desirable as a starting point. The time it takes to investigate a complicated grid must be considered along with any required changes. After investigating the grid, it may require such extensive rework that the user will generate a new grid from scratch anyway.

There are many times when an existing grid can be used as a starting point. An organized approach to using existing grids will be presented later. However, careful consideration up front, and a willingness to start from scratch when necessary, can avoid spending a lot of time heading down the wrong path.

13.2.4 Generating the Grid

The goal of any application is to get a reasonable answer to a particular problem. The difficulty of a study depends on how accurate the answer must be to be "reasonable." Besides using capable tools, the quality of a solution is based on getting suitable resolution where it is required by the physics of the problem. The efficiency of the solution will be driven by minimizing unnecessary grid points.

How does one determine resolution requirements? Knowing ahead of time what flow features to expect would be very useful here. While we don't know the solution before we start, we can make some educated guesses. These may be based on the user's understanding of fluid flow. This can be very difficult for even the expert fluid dynamicist when it comes to complex configurations, and this experience takes a long time to acquire. Therefore, it is important to rely on other experts as the user travels up the learning curve. Another place to look for understanding is other studies that have been run for similar cases. This can also include analyses using lower-order methods, such as panel codes, to get rough estimates of the flowfield. While the features may not occur in the same place, there is a distinct probability that some of them will occur. Another place to look is experimental data. While this does not often precede the computational analysis, CFD studies are sometimes run to give a better understanding of a feature observed in test, and if the data is available, use it.

13.2.4.1 Controlling Grid Resolution

There are several ways to control grid resolution. Some of these are inherent in a particular grid generation approach. Some are "tricks" that improve control of grid resolution over brute force methods that require more grid points.

Structured grids that require one-to-one point-match at block interfaces are one of the most restrictive approaches from the grid generation standpoint. Points added to resolve a feature in one block propagate to adjacent blocks that may not require the additional points. The primary means to control this is through the grid topology.

In the 2D multiblock example shown in Figure 13.5, a technique is used to get a C-topology grid around an airfoil while satisfying a restriction that the upstream and downstream boundaries have the same number of points and spacing distribution. The C-topology close to the airfoil provides better resolution of the leading edge, and allows conversion to a viscous grid by increasing the number of points only in the normal direction of this block.

When gridding a surface which is conceptually triangular, i.e., one edge is singular, a common technique involves introducing an "artificial" corner. In the leading edge extension (LEX) surface grid shown in Figure 13.6a, the downstream section of the outer edge of the LEX belongs to the streamwise family of grid lines while the upstream section of the edge belongs to the spanwise family of grid lines. This introduces the artificial corner along the outer edge of the LEX and avoids a singularity at the upstream corner. However, this technique does introduce some skewness at the corner.

In another common approach, the streamwise family of grid lines on the LEX coalesce to form a singular line on the fuselage. Variations of this approach have the grid lines fan back out forward of the leading edge of the LEX, or instead of coming to a true singular point, they may come to a near singular point before fanning out, as shown in Figures 13.6b and 13.6c, respectively. These two methods also eliminate the singular line, but at the expense of the stretching rate (in the transitions from normal to the singular or near singular point) instead of the increased skewness that is often inherent in the artificial corner method. Suitability of any of these methods depends on the ability of the solution algorithm to accurately compute on such grids with embedded singularities.

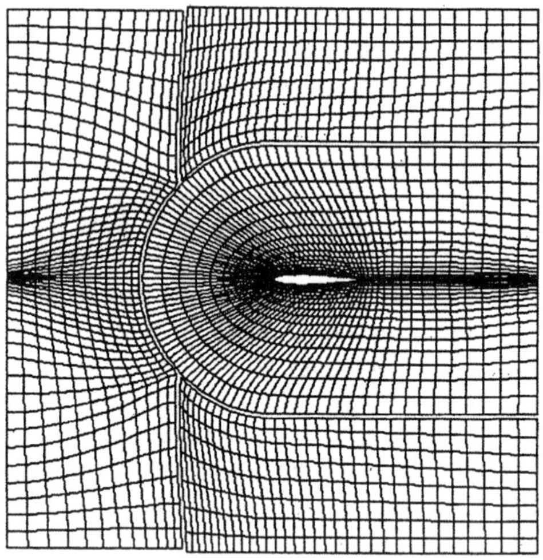

FIGURE 13.5 Multi-block approach allows good resolution of the airfoil while satisfying constraints on the edges of the grid.

There are many ways to generate any grid. For a grid in a duct, the most obvious approach is a polar grid, as shown in Figure 13.7a. This type of grid makes it easy to cluster grid points to the wall surface, but it may be difficult to use this topology if the block is to connect to a block without a singularity. Another approach is to use a rectangular grid topology as shown in Figure 13.7b. Note the use of four artificial corners that continue down the length of the duct. A disadvantage of this approach is that clustering points toward the wall involves adding points normal to four faces of the grid block, as shown in Figure 13.7c.

Other topologies can be implemented to replace singularities where desired. An example of a multi-block approach that avoids the singular axis down the middle of a duct is shown in Figure 13.7d. Note that this approach does not involve artificial corners, but it does increase the number of blocks. This topology does not lend itself to connecting the end of the duct with other blocks if a point-match scenario is being used.

One of the main benefits of block-structured grids that do not require point-to-point matching at boundaries is the ability to provide more resolution in the block adjacent to the vehicle and lesser resolution in adjacent blocks. Without the point-match restriction, finer resolutions do not need to propagate to adjacent blocks. The user must make sure that resolution in each block is sufficient for flow features that may occur. Proximity to a vehicle surface alone is not enough to determine the level of resolution. For example, a wake behind a wing will have a shear layer downstream from the wing that requires adequate resolution in this downstream region.

A word of caution: it is best to limit the variation in grid spacing across a block interface, especially in the vicinity of strong flowfield gradients. In simple flows, grids with poor orthogonality and stretching can produce acceptable results. However, if large flow gradients are present at the block interface, severe spacing mismatch can introduce convergence problems.

13.2.4.2 Overset Grid Methods

As with non-point-match grids, overset grids also offer more flexibility in distributing points. Increased resolution can be added where desired by simply overlapping the region with a finer resolution block. To a large degree, the key concept for overset grids is really boundary condition and block-to-block interface specification. Instead of all boundaries being faces of the block, now some of the boundaries are defined by the edge of a hole within a grid block, as shown in Figure 13.8. At any overset boundary, the solution values must be interpolated from some other block of the grid.

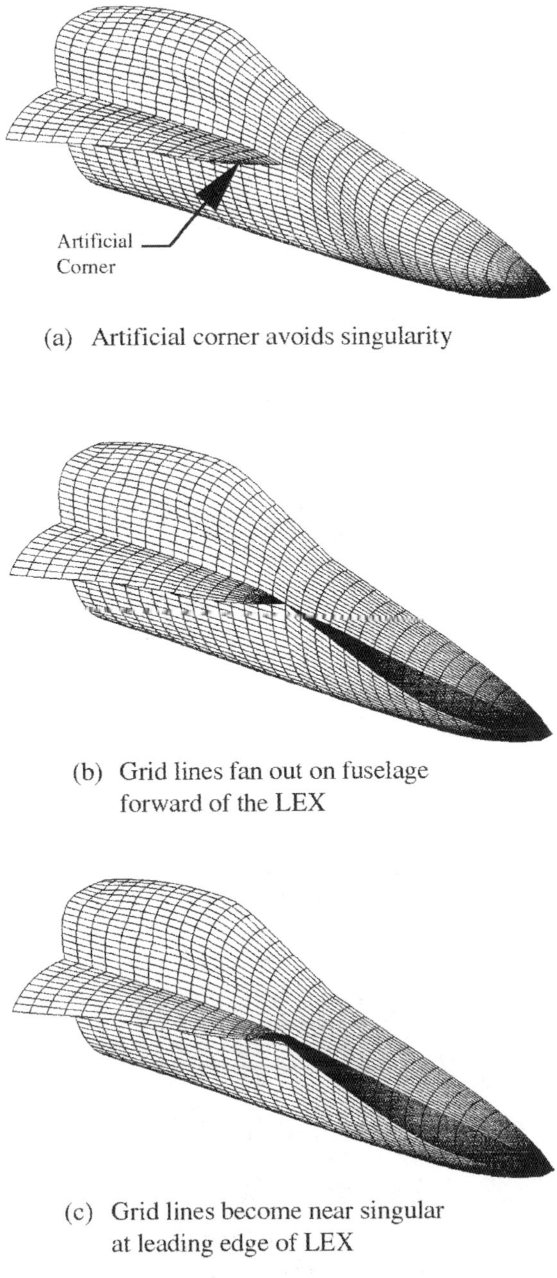

FIGURE 13.6 Gridding techniques for singularities.

From a production grid generation standpoint, controlling the resolution of the background and overlapping grid blocks, particularly in the region of overlap, is of primary importance. Generating an independent grid sounds easy, but in reality the overlapping grid layout is often dependent on flow features present in the background grid. For example, with an airfoil and flap, the flap is so close to the airfoil that the flap grid must be able to resolve features such as shocks or wakes from the airfoil. The resolution in the region of overlap should be comparable to avoid smearing of gradients. In 3D, the resolution of the background grid and the overlapping grids may vary drastically throughout the region of overlap.

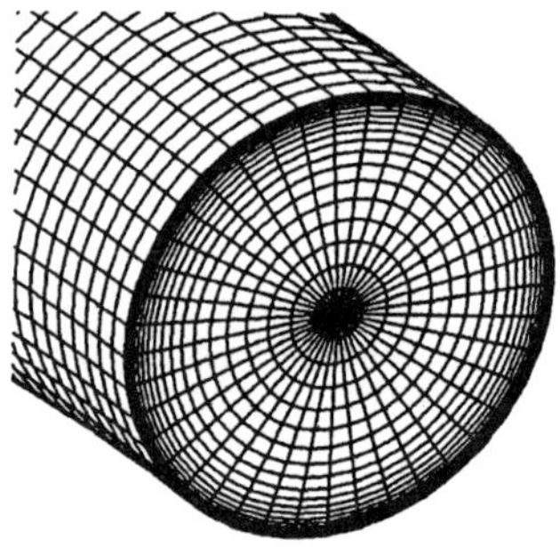

(a) Polar grid

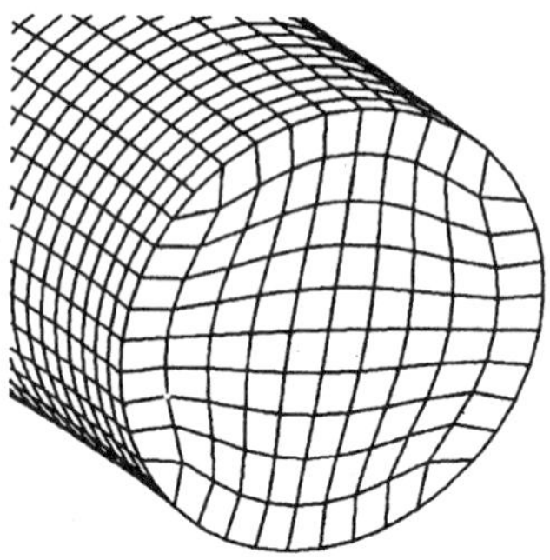

(b) Rectangular grid

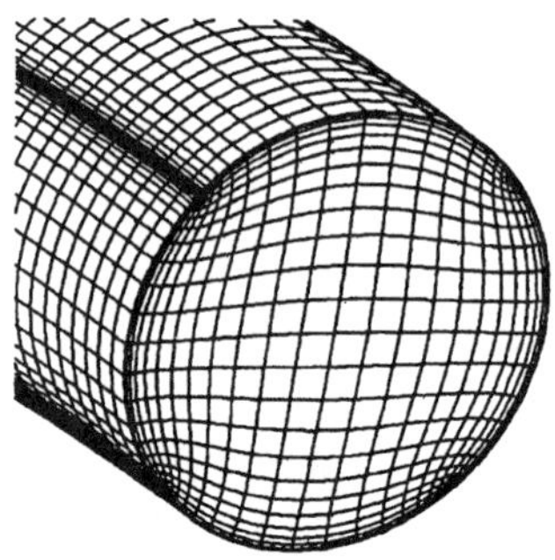

(c) Rectangular grid with clustering toward wall

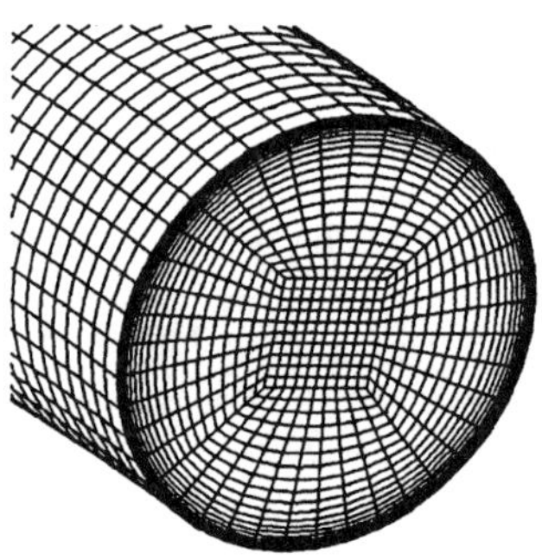

(d) Multiblock grid

FIGURE 13.7 Techniques for avoiding a singularity in a duct grid.

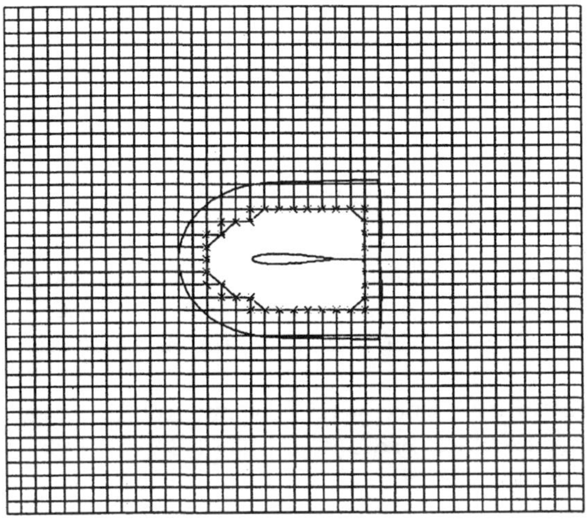

FIGURE 13.8 Boundary defined by the edge of a hole in an overset grid.

The nature of overset grids often causes overlapping regions to occur in critical regions of the flowfield, such as in the junction between the wing and the fuselage. Collar grids, which are used to provide suitable surface resolution where independent overlapping component grids come together, can help this problem. But the overlap region between the collar grid and the wing or fuselage grid still goes all the way to the surface of the vehicle where large gradients may be found. It takes care to make sure that the blocks which will contain features such as shocks, vortices, and wakes have appropriate resolution, especially if these features cross overlapping boundaries.

Because the overlapping issues are more complex due to their 3D nature, it is important to use grid quality assessment tools for overset grids. These tools should check for smooth hole region boundaries, a sufficient amount of overlap for adjacent regions, and comparable resolution in the overlapping region.

13.2.4.3 Spacing Normal to a Wall

When determining the grid spacing normal to a wall for a viscous analysis, there are several factors that influence the decision. The normal spacing is a function of the flow condition at which the analysis will be run and also a function of the length scale of the geometry. For a wing, the reference length is usually taken to be the root chord. For a blended body, or a duct, the length would be the total length of the geometry. It is also a factor of the flow solver parameters such as turbulence model, and the sensitivity of the algorithms to wall spacing. The goal is to get enough resolution in the boundary layer to adequately define the boundary layer profile, and get reasonably accurate turbulence effects (depending on study goals), without slowing down convergence excessively due to tight grid spacing. One method of assessing this spacing is through calculation of a quantity called "y^+". For practical applications, the reference length is used in the y^+ calculation and a fixed spacing is usually applied at the wall, even though the thickness of the boundary layer grows as flow moves downstream. This means that a good distribution needs plenty of points at the reference location, so that the distribution still has some points in the boundary layer upstream where the boundary layer is thinner.

The quantity y^+ is the first grid spacing increment normal to the wall, measured in units of the Law of the Wall. It is based on flat plate boundary layers. An appropriate equation is

$$\Delta y_{physical} = \frac{Ly^+}{R_{e,L}} \left(\frac{\nu_{wall}}{\nu_\infty} \right) \sqrt{\frac{\rho_{wall}}{\rho_\infty} \frac{2}{C_f}} \tag{13.1}$$

where L is the length scale used in $R_{e,L}$ (L could be chord, diameter, body length, or any other dimension); R_e is the Reynold's number, ρ is the density, ν is the kinematic viscosity, and C_f is the skin friction coefficient. The subscript "wall" denotes values at the wall, and the subscript "∞" denotes freestream values. If a better estimate is not available, a suitable value of C_f is 0.002. The flat plate relationship for C_f also is useful:

$$C_f = 0.025 \quad \left[R_{e,L}\right]^{\left(\frac{1}{7}\right)} \tag{13.2}$$

Except for hypersonic applications, the engineer can generally assume $(\nu_{wall}/\nu_{\infty})$ and $(\rho_{wall}/\rho_{\infty})$ are 1.0. However, these are functions of pressure and temperature: if the pressure and temperature (especially temperature) differ strongly (say 50%) from the freestream in critical regions, then those differences must be recognized.

The following guidelines for y^+ are based on the flow solver NASTD (Bush, 1988) used at McDonnell Douglas Corp. For NASTD, the recommended y^+ is 1–3 for the Spalart-Allmaras turbulence model, 3–5 for the Baldwin–Barth turbulence model, and less than 1 for two-equation turbulence models. For hypersonic analyses where wall heat transfer rates or adiabatic wall temperature is to be predicted, y^+ should be in the range of 0.1 to 0.5 (based on experience with hypersonic aerothermal predictions). The preferred y^+ for other flow solution algorithms and turbulence models would be determined from appropriate benchmarking and validation studies.

13.2.4.4 Typical Distributions

How does the user determine what is a good distribution? There are really two parts embedded in that question; "How many points are required?" and "How should the points be spaced?" These questions are interdependent, since a poor spacing scheme will require more points than an "optimum" scheme. First, consider point spacing. For the distribution normal to the surface of the configuration, the most common distribution is some form of a hyperbolic tangent or hyperbolic sine distribution (cf. Chapter 3, Section 3.6, and Chapter 32.) Thompson, et al., (1985) discuss advantages and disadvantages of these and other distribution functions. For distributions along the surface of the geometry, the choice of distributions is more open; however, most cases can be handled using primarily hyperbolic tangent and equal arc distributions, as well as the ability to match an existing distribution from some other part of the grid. The driving issues for surface grids are resolution of geometric features, such as curvature and smooth spacing transitions.

Once the grid spacing normal to the wall for viscous analyses is set as discussed above, the required number of points can be found by setting a maximum stretching rate. Of course, this is a starting value that may need to be adjusted to resolve additional features of the flowfield. Along the surface of a geometry, the number of points is based upon resolving geometric features such as curvature, discontinuities, etc., combined with limits on stretching rate. Typical distributions for certain geometric features are presented in Table 13.1. These guidelines were compiled from a survey of several "expert" users. Additional variation can be expected for different solution algorithms that may require finer resolution or tolerate coarser resolution to achieve comparable results.

While these distribution guidelines were developed from aerodynamic studies using a particular solution algorithm, some of the information may be extended to the general case. When extrapolating these guidelines to other applications, the normal spacings are generally applicable. If the spacing is being generated from the surface to the far field, the larger number of points is preferable. If the distribution is for a block that has a much shorter normal distance, the smaller number of points may be adequate. For inviscid analyses, the normal distribution may generally be on the order of magnitude of the streamwise spacing, but usually the normal spacing is smaller than the streamwise spacing. But other features such as curvature or the presence of other components may increase the needed number of points in any location. Add points to resolve expected gradient regions. If possible, limit the ratio of adjacent cell sizes to about 1.2 (preferably smaller for most distributions).

TABLE 13.1 Typical Grid Point Distributions

Wing:	Chordwise	41–100 Points
		Low end for gross effects
		High end for detailed pressure/lift/drag
		Hyperbolic tangent distribution
		Leading edge spacing
		Enough to define radius of curvature
		Not more than defining geometry
		0.1% of chord
		20 points in first 5% of chord
		Trailing edge spacing
		1 to 10 times leading edge spacing
		10 points in last 5% of chord
	Spanwise	21–33 Points (Euler)
		41–57 Points (viscous cluster at root)
		Root spacing 20% of largest (Euler)
		$y^+ = 1$ to 5 (viscous)
		Tip spacing 10% of largest
	Normal	33–41 Points (Euler)
		41–65 Points (viscous)
		Spacing at wall
		Match leading edge cell size (Euler)
		0.2% of chord (Euler)
		$y^+ = 1$ to 5 (viscous)
		0.002% of chord (viscous)

13.2.5 Checking Quality

Once the surface grids are complete, it is useful to generate a shaded view of the surface grids. This often highlights cusps or dimples in the grid surface, as well as highly skewed grid areas. The shading of the grid can be compared with the shading of the geometry to quickly look for variances. This cannot replace numerical grid quality checks, but it does often pick up discrepancies that may otherwise remain undetected until much further into the analysis. The last place you want this kind of problem to show up is when you are showing surface contours of the final solution to your customer or boss.

Grid quality is discussed in more detail in Chapters 33 and 34, but it cannot be overemphasized how important quality is. Just about everyone uses negative volume checks or Jacobian checks. Most codes also have other quality assessment available. USE THEM! Yes, it takes more time, and rarely do you have a lot of extra time in the typical CFD engineering environment. But checking quality will save time in the long run. Some of the checks most useful for finding grid flaws are checks for stretching (from one cell to the next cell), discontinuity or turning angle (to look for kinks or corners in the grid), and spacing (to verify appropriate spacing at wall boundaries). Other checks that can often guide refinement of the grid include orthogonality and aspect ratio. Just as important as grid quality checks are automated checking of boundary conditions and connectivity between blocks. Checks of this sort are especially important for emerging technologies such as overset grids where 3D visualization is more difficult.

Tools that compare the surface grid to the analytical geometry definition can increase confidence that the grid generation process has remained faithful to the original geometry. They are very valuable when a grid is obtained from another source, and it is not known which specific configuration was used to generate the grid. The grid can be compared with a variety of components to determine which most closely matches the grid. There is a limit to the usefulness however, since deviation is often due to decisions to exclude geometric details, or differences between the geometry as designed, and the geometry as gridded with features such as deflected flaps, specific nozzle settings, etc.

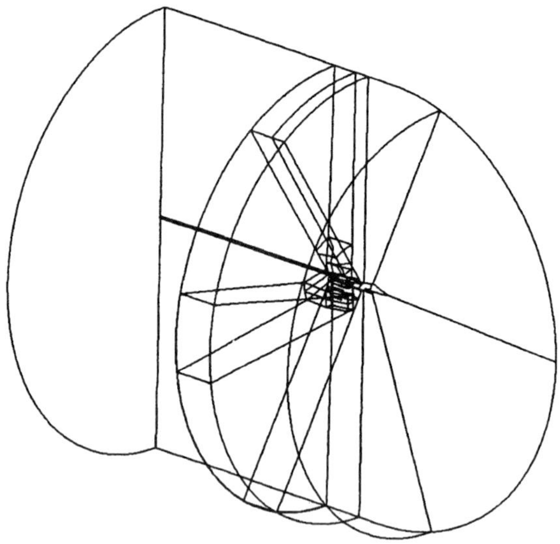

FIGURE 13.9 Block layout for a non-point-match grid of the NEWPLANE configuration.

13.2.6 Grid Generation Example

As an example, we will look briefly at two grid approaches for a fictitious fighter called "NEWPLANE." NEWPLANE is a geometry developed to evaluate geometry and gridding tools. Generated within a CAD system, it uses variety of surface types, and includes intentional gaps, overlap, and mismatch to fully evaluate the capabilities of the grid generation system. The first grid generated uses non-point-match interfaces, and consists of 18 blocks and over 1.3 million points. The block layout for this grid is shown in Figure 13.9. Figure 13.10 shows the surface grids and select grid planes through the wing and vertical tail. Note the "C" topology around each of these components. This grid required about one and a half weeks to generate.

A second grid was generated using point-match interfaces. It contains 167 blocks and about 960,000 points. Because of the point-match feature, many of these blocks could be combined to end up with far fewer blocks. In comparing this grid to the non-point-match grid, the point-match grid has much less resolution in several areas, including chordwise on the wing and normal to wall surfaces. Figures 13.11 and 13.12 illustrate the overall block topology and symmetry plane, respectively. Note in the symmetry plane view, the block downstream of the tail, which is "D" shaped. Wrapping the block in this manner allows two opposite faces of this block to connect with the upstream blocks. Another face of the block that does not contain as many points will connect with the downstream zones. This technique keeps the fine resolution near the body from propagating to the outer blocks where high resolution is not required.

13.2.7 Summary

Because of the variety of grid codes available, it is impossible to assess the effectiveness of each of these codes. However, we will lay out a generic process and try to estimate a level of effort assuming a "state of the art" code. The times quoted here are meant to be engineering estimates for production use, which take into account real-world issues. These issues include the fact that (1) not every user has the same level of experience and ability, (2) the user may find it difficult to sit at a tube doing grid generation at peak efficiency for 8 hours a day for study after study, and (3) tasks almost always take longer than people estimate.

Let's define a few configurations to give a rough estimate of times. The first configuration that we will define is a simple wing-body, where the wing is clean, and the fuselage is not overly complex, and each component is defined by no more than two surfaces. For an Euler study, grid generation for this case

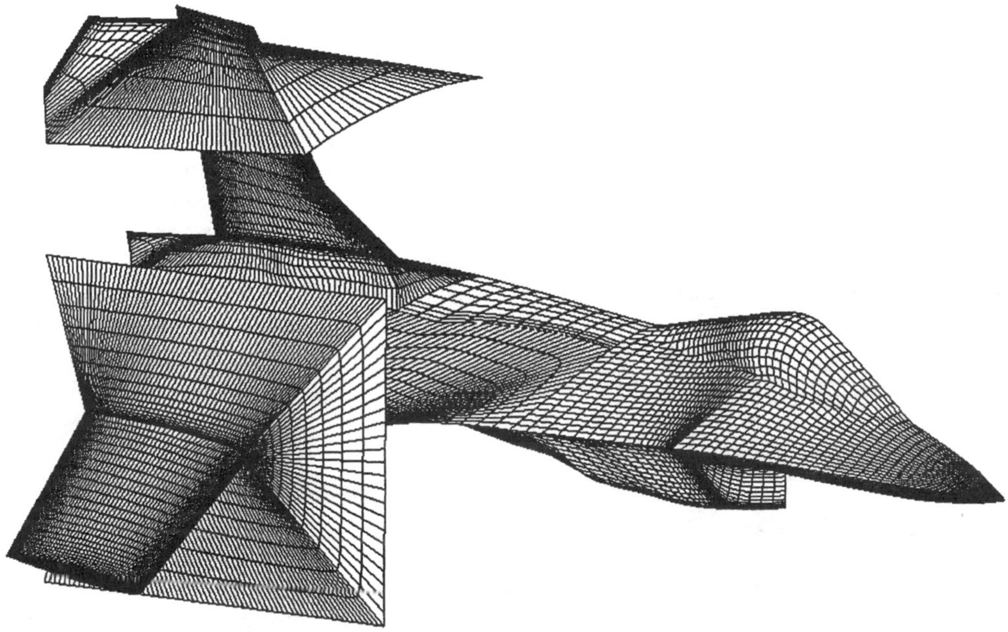

FIGURE 13.10 Surface grids and select grid planes for the non-point-match NEWPLANE grid.

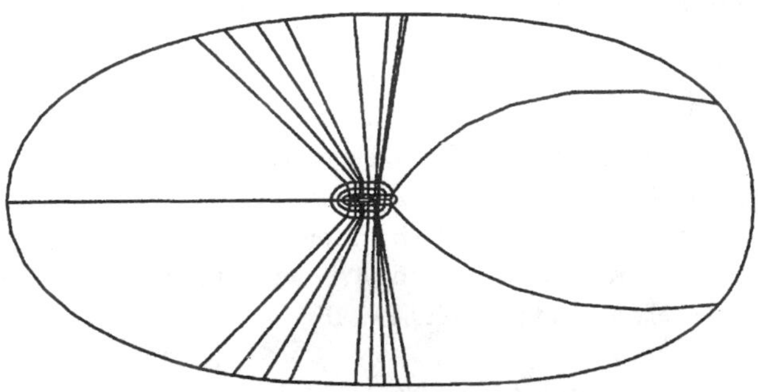

FIGURE 13.11 Symmetry plane block layout for a point-match grid of the NEWPLANE configuration.

should not take more than about 3 days with current tools, and tools tailored to this narrow class of problem may operate in a matter of hours.

For a second case, let's consider a fighter configuration with wing including deployed flaps and slats, fuselage, inlets, nozzles, pylons, and stores. The CAD model for this configuration will contain hundreds to thousands of individual surfaces. It may also contain many additional details besides the external aero surfaces that will need to be sorted out. For a Navier–Stokes analysis using a nonoverlapping structured grid, it would likely take about 6 weeks to generate the grid. This time would include several days just to figure out and verify what is in the CAD model, a few days to determine a suitable topology for the grid, 2 to 3 weeks generating the grids on the vehicle surface, with attention to number of grid points and spacing distributions to get suitable grid resolution in key areas. This would be followed by the generation of the remaining faces of the blocks and then generation of the interior grid. As mentioned earlier, quality checks must be performed and problems corrected. And finally, specification of boundary conditions and block-to-block connectivity associated with the grid is performed. These estimates are highly dependent on the code, the application, and the skill level of the user.

13.2.7.1 When Is It Time To Change Codes?

When is it worth switching to a new code? How important is experience and familiarity? The developer of the new code (or the salesman) will be very optimistic about the ease of switching. The user who will have to learn the new code may have a different view. First, is there a compelling reason to replace the current code, i.e., things that it cannot do, far too slow, too hard for new users to learn, etc.? Second, does the new code have all of the capabilities required, or is the problem just going to shift to some other aspect of the grid generation process? If the user gets to a point and then is stuck, the code is worthless no matter how quickly the impasse was reached. Third, what is the cost to purchase/license the new code and provide training? These first three reasons are just common sense. A fourth consideration that is underestimated is how does the new code fit into the organization's process? Typically, there are a lot of related tools that are developed over a period of time, as a particular code is used. It may be format conversion tools, tailoring for the flow solver or post processors, or ways to get geometry ready for the gridding process. The cost to replace the function of these tools needs to be taken into account. If the new code already includes most of these capabilities, this impact will be small. Otherwise, building a new process to go around a new code can be more expensive than the code itself.

13.3 CFD Application Study Guidelines

A primary motivation for grid generation is to facilitate CFD application studies. These studies can be managed in several ways. One individual may be tasked with the full process of generating the grid, running the flow solver, postprocessing the data, and extracting the engineering information to satisfy the study goals. Or, one individual may generate the grid, another run the solver, and yet another postprocess the data. For a large study, there may be multiple persons performing each step. The more complex the study, the more involved management usually is, but there are a few common points that apply to all.

For any study, it is important to *agree on the goals* up front. Putting this in writing minimizes misunderstandings later. The most common format is a memo that should include the goals of the study, a description of the configuration to be modeled, the approach being taken, a schedule and estimate, as well as assumptions, technical, schedule, and budget risks, and resource requirements. The required accuracy should also be stated, and should be the basis for determining the appropriate level of detail to be included in the analysis configuration. The schedule should be realistic, and all parties should be aware of firm deadlines, possible delays, and the resulting risk.

13.3.1 Managing Large CFD Studies

Large CFD studies can pose special challenges. By large, we mean studies that have large and/or complex grids, studies that require a large number of solutions with different flow conditions and/or grid variations, and/or studies that utilize more than two persons. A hardware issue that arises is finding enough disk space and CPU availability to perform the study. These studies either involve large files (~30 to 500 megabytes each), or a large number of files, or both. This necessitates documentation and a clear file naming convention, so that data can be easily located at a future date. It also often means regular backing up of files, and archiving data or moving it to some other location as soon as a run is finished. Similarly, input parameters and history files need to be correlated with the grids and solutions to be able to extract the real meaning of the results.

The large CFD study will have schedules measured in weeks and months rather than days. It is impractical to plan for an individual to spend 100% of his time on a task of this duration and continuously maintain peak productivity. Depending on the user's expertise, it may be difficult at the beginning of a study to sit at a terminal doing grid generation for eight hours straight each day. Also, there is a tendency when estimating a task to look at how long it would take the most efficient person to perform the task. However, the actual user may not match this productivity, so estimates should take into account a "nominal" efficiency.

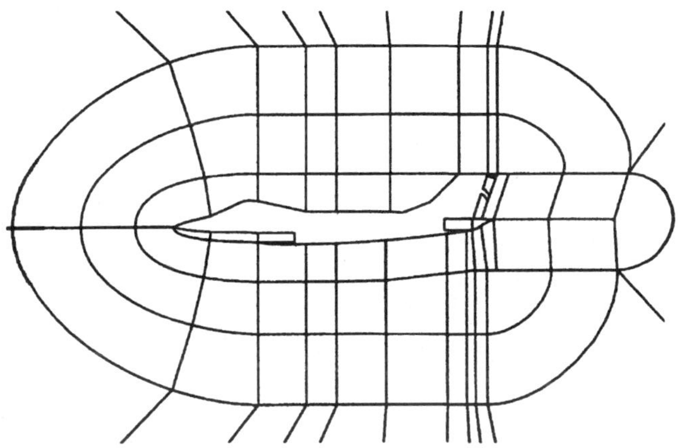

FIGURE 13.12 "D" shaped block behind tail illustrates a method for reducing grid points in the far field.

Large CFD studies also lend themselves more readily to parallel computing. If this is done on a block-by block basis, this information may influence the grid generation effort with respect to the number and relative size of blocks. Attention must be paid to the sizes and number of blocks to aid in load balancing among the available processors. It must also be remembered that the speed of all processors may not be equal. If solutions are computed in a distributed (workstation) parallel environment, there is the additional need to track multiple machines, and the increased vulnerability to network or individual workstation problems.

There can be an infinite number of possibilities for file naming conventions. An important consideration is embedding as much information into the name as possible to distinguish one solution from another. Written notes can be separated from the files themselves, and self documentation within the files may be the only way to answer questions that arise. This is especially important in light of the fact that large studies are more likely to be performed by more than one individual, through team efforts or as a result of personnel turnover/reassignment. Self-documenting files avoid many of these issues.

As a body of studies is performed, it is essential to develop a method of cataloging the configurations analyzed, grids generated, and solutions obtained, along with the miscellaneous input and post-processing files that accompany the solution. Then as the need arises to find old data or extract additional information, or to reuse or modify a grid, the data is easily available to avoid starting from scratch.

13.3.2 Modular "Master Grid" Approach

Another method that has been used to reduce cycle time is a "master grid" approach. The object of this approach is to generate one grid that can be used for any study. This means that it must have sufficient resolution for any study. This approach leads to a very dense grid. This is a modular approach, with new blocks being generated for each variation in the configuration, retaining the blocks in regions where the configuration does not change. This approach has been used at McDonnell Douglas Corp. for analyses of the F-15 Eagle fighter. The grid produced has in excess of 6.2 million grid points in approximately 104 blocks for a half model utilizing symmetry. Up to 12.4 million points have been used for asymmetric cases.

In order to create this master grid that can be used for a wide range of studies, additional resolution is built in when the grid is first created, rather than modeling less important features coarsely. This adds a little more time up front, with the expectation of saving time on grid generation for future studies. For actual analyses, the blocks in regions of high interest are run as is, while blocks away from the region of interest can be coarsened by solving on a subset (such as every other point) in one or more directions to reduce run times. As additional studies are performed, a library of grid blocks is accumulated that can be plugged in for various studies.

13.3.3 Communication

Due to the length of time that a study can take, it is important to have good communication with the customer. This starts with the written statement of goals, and should continue with regular progress reporting, either verbal or written. There are also several points at which specific concurrence should be obtained. The first is when geometry is available prior to gridding. Presumably, the customer is familiar with the configuration, and can pass judgment on its suitability for the study goals, and give guidance on problem areas in the geometry model. The customer should view the final grid to verify its fidelity to the configuration and the impact of simplifications. The customer should review the first solutions as data becomes available, to guide postprocessing requirements and to guide changes in direction of the study as necessary. The goal is not to perform CFD studies, but to solve engineering problems. Many times this will lead to replanning during the course of the study, based on initial data.

Communication is also very important among those working on a large study. There are several ways in which the labor can be divided. Several people may each work on a different variation of the configuration while collaborating on common areas. Or, the grid task may be divided into regions, such as fore and aft, inboard and outboard. This requires careful selection of breakpoints and coordination at these interfaces. In this latter case, best results are obtained if the individuals are co-located so that communication can be continuous (this is really a benefit for any study).

In small CFD application studies, one person may generate the grid and also run the flow solver and process the data. In a larger study, these tasks are often split among several individuals. If several individuals are generating grids, it is beneficial if the person who will run the flow solver is also responsible for "assembling" the grid. This refers to the process of combining blocks generated separately into a complete grid file, and setting boundary conditions and block connectivity. This gives the person who will perform the solution, more intimate knowledge of the grid that is useful for looking for problems in the solution process, and setting up an efficient postprocessing method.

13.4 Grid Code Development Guidelines

Several decisions must be made before embarking on writing a grid generation code. A key issue is whether to write a new code, purchase a commercially available code, or utilize public domain/free software. There are a number of grid generation codes that are products of significant development efforts and incorporate a wide range of technology. Another decision is whether to develop a code for a specific application, or write a general code encompassing a variety of grid generation approaches (LaBozzetta, et al., 1994). Other decisions include interactive or batch, type of graphical user interface, platforms to be supported, grid methodology, etc. Many of these decisions will fall out naturally from need that the code is to address.

While many decisions are a matter of preference, the following discussion highlights some features that should be included in any modern system.

13.4.1 Development Approach

One of the most important features for efficient development is modularity. By this we mean isolating I/O, graphical display, and machine-dependent features from the computational "guts" of the computational method. This offers several benefits including ease of maintenance. If care is taken in the development, a library of grid algorithms can be developed and then the developer need only remember the interface to the routine. Since the "program" includes calls to the library rather than these algorithms themselves, the code is much cleaner and easier to understand.

Another benefit is reusability. The same modular functions, or library, can be called from a general purpose interactive graphical system, a specialized batch program for a common class of problem, or an optimization routine which might be automatically perturbing the geometry. This can speed development of new applications and specialized analysis tools.

Isolating machine-dependent operations, which may include platform-dependent I/O formats and graphics drivers, enhances portability across platforms. The library of machine-dependent operations can be created for each platform of interest, and the rest of the code may be the same for all systems.

The selection of input, output, and intermediate file formats affects compatibility with other grid tools, flow solvers, and postprocessors, and thus the efficiency of the whole process. An approach such as the McDonnell Douglas "common file format," or the CGNS (computational grid Navier–Stokes) effort supported by NASA and Boeing, defines a flexible file format. This approach consists of two fundamental parts, the software routines to read and write data "variables" into the file, and a standard naming convention that allows programs to access data by name rather than requiring knowledge of the structure of the file. Using a binary direct access format allows rapid reading and writing of files in a compact format. A powerful feature of the software is its ability to read from and write to binary files that have been created on other types of machines. All translation is handled by library routines transparent to the user and the application calling the routine. This provides maximum portability of both code and data files.

Even though grid generation has come a long way, keeping up with all of the literature on grid generation would be a full-time job. Without time to read everything that is written, it is ludicrous to expect to include every grid technology. Yet, it is tempting to make a list of all the proven and or promising technologies and try to combine them into the ultimate system. But such a system would (1) be confusing to learn, and (2) probably never get done. So common sense dictates the following guidelines.

The real goal of a system should be a *seamless integration of tools to go start to finish*, without being a strain on the user's endurance, patience, or blood pressure. This does not mean the process will be flawless for every case, but should always have a reasonable approach to work around problems. The process should be natural so that the user will easily understand the organization and quickly move up the learning curve. When features are added, they should enhance the process, not just the raw technology of the program. Candidates are those features that a user would say "If only the program could …" as they are using it.

The "state of the art" in CFD requires that grid generation algorithms operate using double precision and that the resulting grids also be stored that way. Figure 13.13 compares a grid stored in double precision with the same grid stored in single precision. The initial grid was generated in double precision, and all storage was in a binary file. This grid could represent a polar grid around a body where the body grid becomes singular. The spacing normal to the wall is 0.0005, and the singular axis is not located at the origin. (Shifting to the origin might be precluded by the existence of multiple singular axes.) It should be apparent that calculation of such a grid in single precision would be troublesome and have little value.

13.4.2 Geometry Issues

Geometric data can come from a variety of sources, including computer aided design (CAD) systems, IGES (initial graphics exchange specification) files from various sources, point definition surfaces (which may come from existing grids), and stereo lithography (STL) models. From an applications perspective, the user has limited control over the source of geometry. The user may be able to request the data in a particular format if it is an option of the software being used to create the geometry, but may still require additional processing to get it into the format acceptable to the grid generation system. From a code development standpoint, the code must first meet the most common requirements of the targeted end user, and from there the more flexible the better.

Linking the grid generation tools directly to a particular CAD system could eliminate the need to convert CAD data to some format outside the CAD system, such as IGES. This is an aggressive approach which faces several technical challenges. Most CAD system internal data structures are proprietary, which means the developer is limited to "hooks" to the data provided by the CAD system. Often, using these hooks means being limited to a machine that is running the CAD software, and requires a license. CAD systems can contain a large number of surface types, which can significantly increase development costs unless tools exist to make these different types transparent to the application. CAD system software is

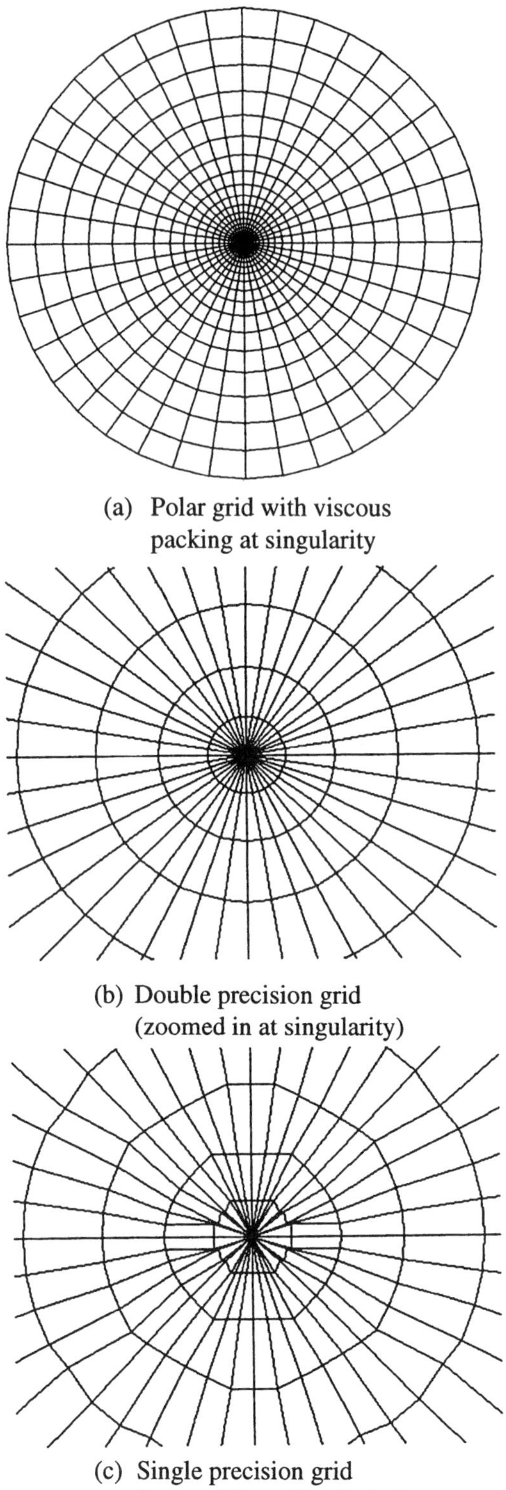

(a) Polar grid with viscous packing at singularity

(b) Double precision grid (zoomed in at singularity)

(c) Single precision grid (zoomed in at singularity)

FIGURE 13.13 Limitations of single precision grid generation.

upgraded frequently, making it a moving target, which requires more maintenance time. Also the CAD model may contain large numbers of surfaces or patches, and lots of unneeded detail that makes gridding more difficult. Maintenance is difficult, as the CAD vendor may change internal data formats on future releases, and be reluctant to provide timely and detailed discussion of these formats. In spite of these issues, this approach has been used successfully for several systems (Gatzke and Melson, 1995).

IGES is a standard for the exchange of geometry data. As with CAD systems, IGES has a large number of entity types. However, not all CAD systems and design tools support all of the standard IGES types. In addition to the standard entity types, IGES permits creation of user-defined entity types. These will not be portable among systems unless both systems know how to interpret the user-defined entity type. To simplify working with IGES files for computational analysis, two standards have been proposed for subsets of the IGES entity types: these are the NASA IGES standard and the NASA IGES NURBS Only (NINO) standard. These standards are discussed in detail in Chapter 31. These standards simplify development of the grid generation system, because only a small subset of the IGES entities need be supported. Unfortunately, designers often do not restrict themselves to these limited subsets, so tools are required to convert or approximate the actual geometry with entity types available in the subset, primarily NURBS (non-uniform rational B-spline) surfaces. Currently, this seems to be the most popular approach (Steinbrenner and Chawner, 1995)(Gaither et al., 1995).

If one can expect the majority of geometry data to subscribe to the NASA IGES or NINO specifications, or be readily translatable into these formats, then the reduced subset of IGES will simplify development of the grid generation tools. Additional details on geometry modeling are given in Part III.

Some geometries may not be available in an analytical form (CAD or IGES). If the geometry exists only on paper, in which case it is generally simpler, CAD tools can be built into the grid generation system. These tools are also important for modifying and repairing geometry models, unless the user will rely on a designer to change the model whenever required (not very realistic in most organizations). But unless the grid generation system is actually built within a CAD system (Akdag and Wulf, 1992), these tools will not be expected to be a full-blown CAD system. Again, the minimum reasonable set of CAD tools that enhance the process as it is envisioned in its ideal state is a good guideline. The CAD tools can be augmented as needs arise.

Many designs and CAD systems utilize trimmed surfaces. These surfaces combine an analytic shape of the surface with information about how to limit the surface to a subset of the "shape." These bounds for the surface are referred to as trimming curves. The developer must decide whether trimmed surfaces will be supported and if so, how to implement this support. It support for trimmed surfaces is not required, the development effort will be much simpler.

13.4.3 Attention to Detail

A final thought on grid generation code development. The code developer needs to pay attention to details —The Little Things. This might seem a bit odd, and puts us on the edge of a fine line between practicality and extravagance. Are we going to make things work, or fine tune the details? Yes, there are certain things that the program needs or it is worthless, and these have to get done. There are codes that have the necessary ingredients, but that are not used because they are considered tedious. The easiest way for the developer to know about these needs is to also be a user of the code, as well as getting as much feedback from other user's as possible. Soon there will be a mile-long "scroll" of desired features, and it is some of the seemingly minor features from this list that may have the greatest effect in making the code a success. Attention to detail can make or break a code.

An example for a graphical program is the quality of the zoom capability. The program may generate superb grids, but users need to be able to visually inspect their grids. If the grid is for viscous analyses, the user will want to zoom in to see the spacing at the wall. All graphical programs have zoom capabilities, but not all give the user the control to easily zoom in on the wall spacing at a particular spot. In some codes, you can get just so close (usually not enough for a tightly packed grid) and then your "eye" passes through the grid. Others make it awkward to keep the feature of interest centered in the view when zooming extremely close. If you're going to do it, do it right! Users often decide in the first 15 minutes that they don't like a code because of the interface or other problems right off the bat, and they rarely come back if they have other alternatives.

There are many forms that documentation can take: user manuals, on-line help, HTML documents on the Internet. These are all important, but should not be counted on to overcome shortcomings in the interface or intuitiveness of the process. More important than the type of documentation is the quality and commitment to keep it up to date. The type of documentation should promote easy updates as new features are added.

A geometry issue that comes down to the philosophy of the grid generation approach, rather than the details of the geometry model, is the handling of intersections. If two surfaces intersect to form the corner of a block, how will the intersection be defined and to what tolerance will it be computed? If the grids on the two adjoining faces have large spacing at the corner, as shown in Figure 13.14a, the tolerance can be very loose. However, if the clustering toward the corner is tight for one or both surfaces, as shown in Figure 13.14b, the tolerance can be very important. The zoomed in view of Figure 13.14c shows the problem that can be buried in the boundary layer. Moving the edge of either surface will cause a major kink in that surface grid. Even if points on the adjacent grid faces stay on their respective surfaces, if the tolerance for the intersection is larger than the spacing normal to the corner, a discontinuity or jump in the spacing can occur. So care must be taken if the exact intersection curve cannot be found.

A related issue is edge preservation. It is not always enough to make sure that the grid lies on the original surfaces. It may also be critical that the edge grid lies on a specific defining curve. As is the case with intersections, this defining curve should be an exact curve. This avoids problems that can arise when an approximate curve is used to generate surface grids which, when projected onto other surfaces, may not be compatible with their common edge definition.

Many of the modern codes, such as GRIDGEN (Steinbrenner and Chawner, 1995), MACGS (LaBozzetta et al., 1994), and NGP (Gaither, et al., 1995) have built-in boundary condition specification as part of the grid generation process. This was rare several years ago. However, it makes quite a bit of sense if looked at from a generic boundary condition perspective. By that, we mean boundary conditions that are not defined in a particular format for a specific code, but represent a fundamental property of the grid being generated. Examples of such boundary conditions are physical solid wall surface of a vehicle, interface between two blocks, area through which mass is entering the flowfield (jet/nozzle), freestream (far field) boundary, etc. This information can be associated with surfaces of the grid without knowing the code that the flow solver expects to see, or indeed, even without knowing what flow solver will be used with this grid.

13.5 Research Issues and Summary

Mainstream grid generation methods are fairly mature in their technical capabilities, but still do not meet the speed requirements that industry would like to see. Research areas that would be of great value are automation of the process for real-world problems, and more validation of grid and CFD techniques. As more and more new methods are developed, a great deal of management and technical support is needed to take these methods to the production environment.

In spite of the limitations, grid generation and CFD methods are being used successfully for a wide variety of applications. Many users move up the learning curve rapidly. The rate at which workstations capabilities are growing is phenomenal. However, when taking these factors into account, it is also essential to include the fact that the "typical" problem size is also growing very rapidly.

Increased problem size, computer resources, and user experience base lead to the bottom line for CFD applications. There are more new customers for CFD applications everyday, and for most of those applications, grid generation turn-around time is a limiting factor. Whoever can solve that problem will provide a great service to the CFD engineer.

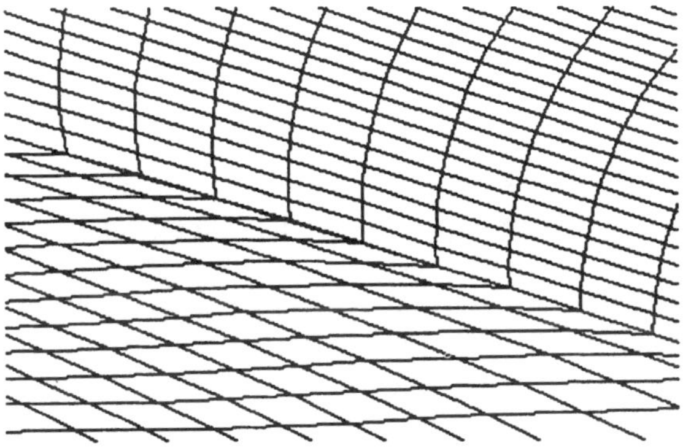

(a) Coarse grids on intersecting surfaces provide minimal difficulty.

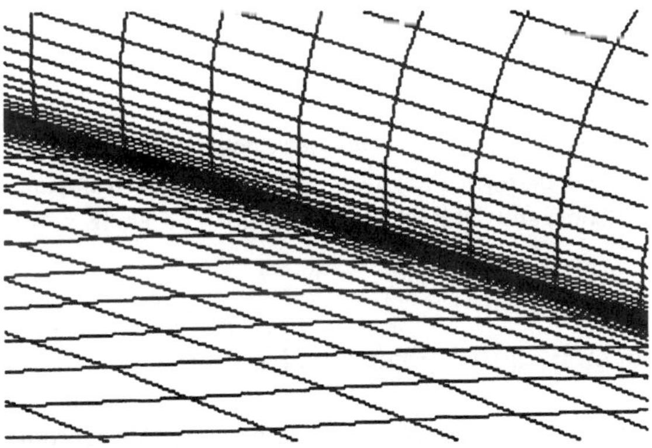

(b) Tightly clustered grids are more challenging

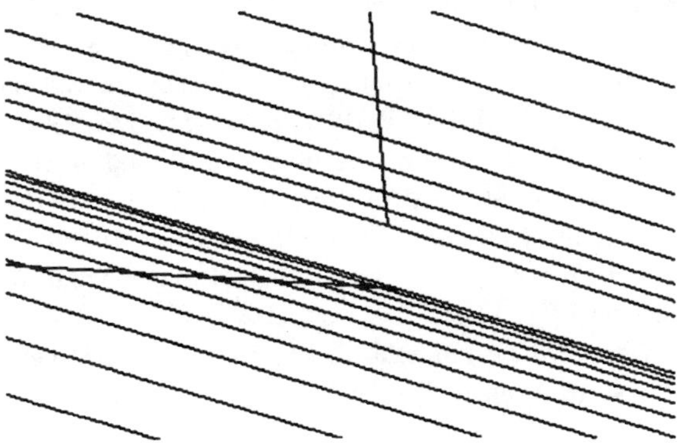

(c) Mismatches can occur deep within the boundary layer.

FIGURE 13.14 Accuracy of intersections and edge definitions is critical for viscous grids.

13.6 Defining Terms

Grid: the set of all the blocks that model a configuration for analysis

Block (or zone): a three-dimensional array of grid points modeling a region or subregion that will be used in computation of a solution.

Topology: the layout, orientation, and/or organization of one or more blocks that form a grid.

Point-match: a grid in which each point on a block face that interfaces with an adjacent block face will have an identical point on the adjacent block face.

Non-point-match: a grid in which the points on a block face that interfaces with an adjacent block face will not in general have an identical point on the adjacent block face, but will rather be arbitrarily located within some cell on the block face.

Inviscid (Euler) **analysis**: analysis in which viscous terms are neglected. The resulting non-zero velocity at the surface, and lack of a boundary layer velocity profile, reduces the required grid resolution at the wall in the normal direction.

Viscous (Navier–Stokes) **analysis**: analysis that includes viscous terms. The zero velocity condition at the wall and the resulting boundary layer velocity profile require fine resolution normal to the wall for proper modeling.

y^+: a measure of the spacing from the wall to the first grid point off the wall measured in units of the *Law of the Wall.*

Further Information

For more information, readers are encouraged to check papers describing CFD studies of the type in which they are interested, and perform their own systematic demonstration and validation for their specific grid methods, grid code, and solution code, or contact others using these same codes.

References

1. **Akdag, V. and Wulf, A.**, Integrated geometry and grid generation system for complex configurations, NASA CP 3143, pp. 161–171, April 1992.
2. **Bush, R. H.**, A three dimensional zonal Navier–Stokes code for subsonic through transonic propulsion flowfields, AIAA Paper No. 88-2830, July 1988.
3. **Dannenhoffer, J. F.**, A technique for optimizing grid blocks, NASA CP 3291, pp. 751–762, May 1995.
4. **Gaither, A., Gaither, K., Jean, B., Remotigue, M., Whitmire, J., Soni, B., and Thompson, J.**, The National Grid Project: a system overview, NASA CP 3291, pp. 423–446, May 1995.
5. **Gatzke, T. D. and Melson, T. G.**, Generating grids directly on cad database surfaces using a parametric evaluator approach, NASA CP 3291, pp. 505–515, May 1995.
6. **LaBozzetta, W. F., Gatzke, T. G., Ellison, S., Finfrock, G. P., and Fisher, M. S.**, MACGS - toward the complete grid generation system, AIAA Paper No. 94–1923, June 1994.
7. **Panton, R. L.**, *Incompressible Flow.* 1st ed., Wiley Interscience, NY, 1984.
8. **Steinbrenner, J. P. and Chawner, J. R.**, The GRIDGEN user manual: version 10, available from Pointwise, Inc., Jan 1995.
9. **Thompson, J. F., Warsi, Z. U. A., Mastin, C. W.**, *Numerical Grid Generation Foundations and Applications*, 1st ed., North–Holland, NY, 1985.

14

Data Structures for Unstructured Mesh Generation

Luca Formaggia

14.1 Introduction

The term *data structures,* or information structures, signifies the structural relationships between the data items used by a computer program. An algorithm needs to perform a variety of operations on the data stored in computer memory and disk; consequently, the way the data is organized may greatly influence the overall code efficiency.

For example, in mesh generation there is often the necessity of answering queries of the following kind: give the list of mesh sides connected to a given node, or find all the mesh nodes laying inside a certain portion of physical space, for instance, a sphere in 3D. The latter is an example of a range search operation, and an inefficient organization of the node coordinate data will cause the looping over all mesh nodes to arrive at the answer. The time for this search operation would then be proportional to the number of nodes n, and this situation is usually referred to by saying that the algorithm is of order n, or more simply $O(n)$. We will see later in this chapter that a better data organization may reduce the number of operations for that type of query to $O(\log_2 n)$, with considerable CPU time savings when n is large.

The final decision to use a certain organization of data structure may depend on many factors; the most relevant are the type of operations we wish to perform on the data and the amount of computer memory available. Moreover, the best data organization for a certain type of operation, for instance searching if an item is present in a table, is not necessarily the most efficient one for other operations, such as deleting that item from the table. As a consequence, the final choice is often a compromise. The fact that an efficient data organization strongly depends on the kind of problem at hand is probably the major reason that a large number of information structures are described in the literature. In this chapter, we will describe only a few of them: the ones that, in the author's opinion, are most relevant to unstructured mesh generation. The reader interested in a more ample surveys may consult specialized

0-8493-2687-7/99/$0.00+$.50

texts, among which we mention [10, 2] for a general introduction to data structures and related algorithms, and [11, 20, 17] for a more specific illustration of range searching and data structures relevant to computational geometry.

It is a commonly held opinion that writing sophisticated data structures is made simpler by adopting programming languages that allow for recursion, dynamic memory allocation, pointer and structure data types. This is probably true, and languages such as C/C++ are surely among the best candidates for the purpose. However, all the data structures presented in this work may be (and indeed they have been) implemented in Fortran, and a Fortran implementation is often more cumbersome but normally not less efficient than the best C implementation. I am not advocating the use of Fortran for this type of problem — quite the contrary — but I wish to make the point that also Fortran programs may well benefit from the use of appropriate information structures.

This chapter is addressed to people with a mathematical or engineering background, and only a limited knowledge of computer science, who would like to understand how a more effective use of data structures may help them in developing or improving a mesh generation/adaption algorithms. Readers with a strong background in computer science will find this chapter rather trivial, apart from possibly the last section on multidimensional searching.

14.2 Some Basic Data Structures

In this section we review some basic data structures. It is outside the scope of this work to give detailed algorithmic descriptions and analysis. We have preferred to provide the reader with an overview of some of the information structures that may be profitably used in mesh generation/adaption procedures, together with some practical examples, rather than to delve into theoretical results.

First, some nomenclature will be given. A *record* **R** is a set of one or more consecutive memory locations where the basic pieces of information are kept, in separate *fields*. Many authors use the term *node* instead of record. We have chosen the latter to avoid possible confusion with mesh nodes. The *location* &**R** of record **R** is the pointer to the position where the record is stored, while $*p$ indicates the record whose location is p. In the algorithm descriptions, I will use a C-like syntax, for example i++ is equivalent to $i = i+1$. Moreover, with the expression A.b we will indicate the attribute b, which can be either a variable or a function, associated to item A. For instance **R**.f may indicate the field f of the record **R**.

14.2.1 Linear Lists

Quoting from Knuth [10] "A *linear list* is a set of records whose structural properties essentially involve only the linear (one-dimensional) relative position of the records." In a list with n records, the record positions can be put in a one-to-one correspondence with the set of the first n integer numbers, so that we may speak of the ith element in the list (with $1 < i < n$), which we will indicate with $L.i$. The type of operations we normally want to perform on linear lists are listed in the following.

Record Access **RA**(k). This operation allows the retrieval of the content of a record at position k.

Record Insert **RI**(**r**,k). After this operation the list has grown by one record and the inserted record will be at position k. The record previously at position k will be at $k + 1$ and the relative position of the other records remains unchanged.

Record Delete **RD**(k). The kth record is eliminated, all the other records remain in the same relative position.

14.2.1.1 Stacks and Queues

Linear lists where insertion and deletion are made only "at the end of the list" are quite common, so they have been given the special name of *deques*, or double-ended queues. Two special types of deques are of particular importance: stacks and queues.

With *stack*, or *LIFO* list, it is indicated a linear list where insertion, deletion, and accesses are made only at one end. For example, a list where the operations allowed are **RA**(n), **RI**(n+1), and **RD**(n), i.e., all the operations made on the last list position, is a stack. The insert operation is often called a "push," while the combination of **RA**(n) and **RD**(n) is referred to as a "pop" operation.

In a *queue*, also called *FIFO* list, the elements are inserted at one end, and accessed and deleted at the other end. For instance, a linear list where only **RD**(1), **RA**(1), and **RI**(n+1) operations are allowed is a queue.

The stack is a very common data structure. It occurs every time we wish to "accumulate items" one by one and then retrieve them in the inverted order. For instance, when in a triangulation process we are searching the nodes that lie inside a sphere, every time a new node is found it may be pushed onto a stack. At the end of the search, we may "pop" the nodes from the stack one by one. We have so far identified a linear list by its properties and the set of operations that may be performed on it. Now, we will investigate how a linear list could be actually implemented, looking in some detail at the implementations based on sequential and linked allocation.

14.2.1.2 Sequential Allocation

The method of *sequential allocation* is probably the most natural way of keeping a linear list. It consists in storing the records one after the other in computer memory, so that there is a linear mapping between the position of the record in the list and the memory location where that record is actually stored. With sequential allocation, direct addressing is, therefore, straightforward. A sequentially allocated list broadly corresponds to the ARRAY data structure, present in all high-level computer languages. In the following, we use the C convention that the first element in array A is $A[0]$.

As an example, let us consider how to implement a stack using sequential allocation. One possibility is to store the stack S in a structure formed by two integers. $S.max$ and $S.n$ indicating the maximum and the actual number of records on the stack, respectively, and an array $S.A[max]$ containing the records. Unless we know beforehand that the program will never try to store more than $S.max$ records on the stack, we need to consider the possibility of stack overflowing. When such condition occurs, we could simply set an error indicator and exit the push function. A more sophisticated approach would consider the possibility of increasing the stack size. In that case, we will probably store an additional variable s_{grow} indicating how much the stack should increase if overflow occurs. In that situation, we could then allocate memory for an array sized $S.max + s_{grow}$, adjourn $S.max$ to the new value, and move the old array on the new memory location. We must remember to verify that there are enough computer resources available for the new array. If not, we have a "hard" overflow and we can only exit the function with an error condition.

We have just considered the possibility of letting the stack grow dynamically. What about shrinking it when there is a lot of unused space in S? We should first decide on a strategy, in order to avoid growing and shrinking the stack too often, since these may be costly operations. For instance, we could shrink only when $S.max - S.n > 1.5s_{grow}$. The value of s_{grow} may itself be a result of a compromise between memory requirement and efficiency. A too small value could mean performing too many memory allocation/deallocations and array copying operations. Too large a value will imply a waste of memory resources. We will not continue this discussion further. We wanted only to show how, even when dealing with a very simple information structure such as stack, there are subtle details that could be important for certain applications.

The sequential implementation just described may be readily modified to be used also for a general double-ended queue Q. Figure 14.1 shows how this may be done. We use an array Q.$A[max]$, plus the integer quantities Q.n, Q.max, Q.$start$, and Q.end, respectively, indicating the actual and maximum number of records in the deque and the position of the initial and final record in the array. In Table 14.1, we illustrate the algorithms for the four basic operations, **RI**(1), **RI**(n+1), **RD**(1), and **RD**(n). As a matter of fact, Q.n is not strictly necessary, yet it makes the algorithms simpler. When an overflow occurs we may decide to grow the structure by a given amount, and the same considerations previously made for stacks apply here.

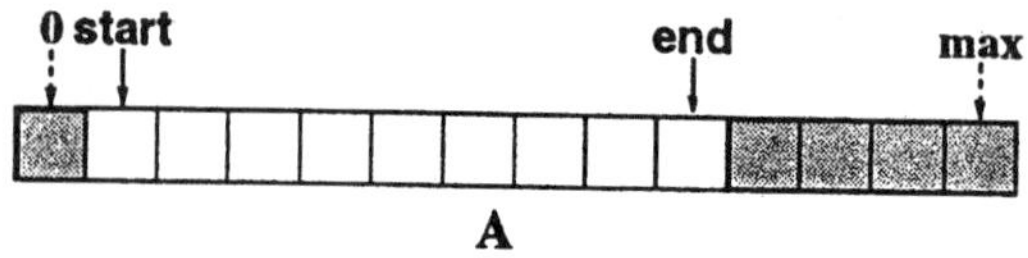

FIGURE 14.1 How to implement a double-ended queue (deque) using sequential allocation.

TABLE 14.1 An Example of Algorithms for Inserting and Removing Records from the Sequentially Allocated Deque Illustrated in Figure 14.1.

Deque::RI (**R**,1)	Deque::RI (**R**,n+1)
1. $(n > max) \mapsto$ OVERFLOW; 2. $start = (start + max - 1) \bmod max$; 3. A [start] = **R**; 4. $n++$.	1. $(n > max) \mapsto$ OVERFLOW; 2. $end = (end + 1) \bmod max$; 3. $A\,[end]$ = **R**; 4. $n++$.
Deque::RD (1)	Deque::RD(n)
1. $n = 0 \mapsto$ UNDERFLOW; 2. $start = (start + 1) \bmod max$; 3. $n--$.	1. $n = 0 \mapsto$ UNDERFLOW; 2. $end = (end + max - 1)$ mod max; 3. $n--$.

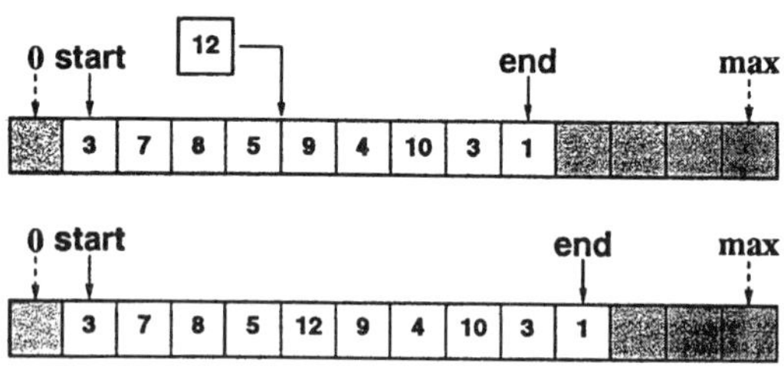

FIGURE 14.2 Adding a record in a sequentially stored list.

14.2.1.3 Linked Allocation

What happens if we have to add a record at a random location inside a sequentially stored list? Figure 14.2 graphically shows that we should move "in place" a slice of the array. This procedure requires, in general, $O(n)$ operations and therefore it should be avoided. In order to increase the flexibility of a linear list by allowing an efficient implementation of random record insertion and deletion, we need to change the way the structure is implemented. This may be done by adding to each record the link to the next record in the list. For instance, we could add to **R** a field **R**.*next*, containing the location of the successive record. A list which uses this type of layout is called a *linked list*, or, more precisely a *singly linked list*. There are, in fact, many types of linked lists. If we have also the link to the previous record **R**.*prev*, we have a *doubly linked list* that permits enhanced flexibility, as it allows one to sequentially traverse the list in both directions and to perform record insertions in $O(1)$ operations. Figure 14.3 illustrates an example of a singly and of a doubly linked list.

A disadvantage of linked allocation is that direct addressing operations are costly, since they require traversing the list until the correct position is reached. Moreover, a linked list uses more memory space per record than the corresponding sequential lists. However, in many practical applications direct addressing is not really needed. Furthermore, with a linked list it is normally easier to manage the memory requirements dynamically and to organize some sharing of resources among different lists.

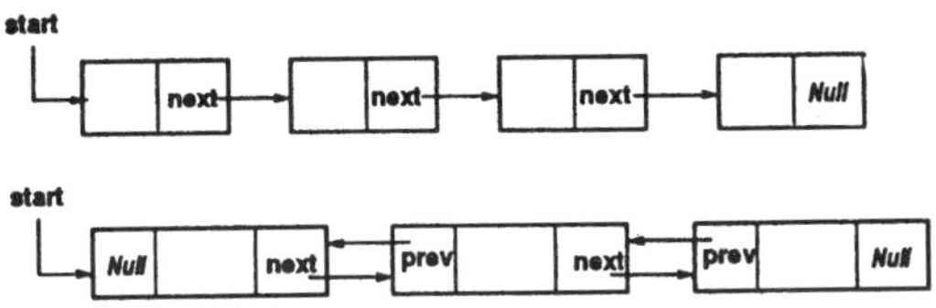

FIGURE 14.3 An illustration of a singly and of a doubly linked list.

TABLE 14.2 Algorithmic Implementation of the Addition and Deletion of a Record from a Doubly Linked Circular List

Dcllist::RI (**R**, **Q**). Insert record **R** in list, after record Q	Dcllist::RD (**R**). Delete record **R** from list
1. p = **Q**.*next*;	1. r = **R**.*prev*;
2. **R**.*next* = p;	2. p = **P**.*next*;
3. (*p).*prev* = &R;	3. (*r).*next* = p;
4. **Q**.*next* = &R;	4. (*p).*prev* = r.
5. **R**.*prev* = &Q.	

It is often convenient to use a variant of the linked list, called a *circular linked list*. In a circular (singly or doubly) linked list every record has a successor and a predecessor and the basic addition/deletion operation has a simpler implementation. There is also usually a special record called *header* that contains the link to the first record in the list, and it is pointed to by the last one. Table (14.2) shows a possible algorithm for the implementation of the basic addition/deletion operations on a circular doubly linked list *L*. The memory location for a new record could be dynamically allocated from the operating system, where we would also free the ones deleted from the list. However, this type of memory management could be not efficient if we expect to have frequent insertions and deletions, as the operations of allocating and deallocating dynamic memory have a computational overhead. Moreover, it cannot be implemented with programming languages that do not support dynamic memory management. It is then often preferable to keep an auxiliary list, called *list of available space* (*LAS*), or free list, which acts as a pool where records could be dumped and retrieved. At start-up the *LAS* will contain all the initial memory resources available for the linked list(s). The *LAS* is used as a stack and is often singly linked. Here, for sake of simplicity, we assume that also the *LAS* is stored as a doubly linked circular list. Figure 14.4 shows graphically an example of a doubly linked circular list and the corresponding *LAS*, plus the operation required for the addition of a record. In the implementation shown in the table we have two attributes associated with a list *L*, namely *L.head*, and *L.n*, which gives the location of the header and the number of records currently stored in the list, respectively. Consequently, *LAS.n* indicates the number of free records currently available for the linked list(s). In Table (14.3) we illustrate the use of the *LAS* for the insert and delete operation. We have indicated with **R**.*cont* the field where the actual data associated with **R** is kept.

It remains to decide what to do when an overflow occurs. Letting the list grow dynamically is easy: we need to allocate memory for a certain number of records and join them to the *LAS*. The details are left to the reader. If we want to shrink a linked list we can always eliminate some records from the *LAS* by releasing them to the operating system. Again, we should take into account that many fine grain allocations/deallocations could cause a considerable computational overhead, and a compromise should be found between memory usage and efficiency. We have mentioned the possibility that the list of available storage could be shared among many lists. The only limitation is that the records in all those lists should be of equal size. Linked lists may be implemented in languages, such as Fortran, that do not provide pointer data type. Pointers would be substituted by array indices, and both the linked list and the *LAS* could be stored on the same array. The interested reader may consult [2] for some implementation details.

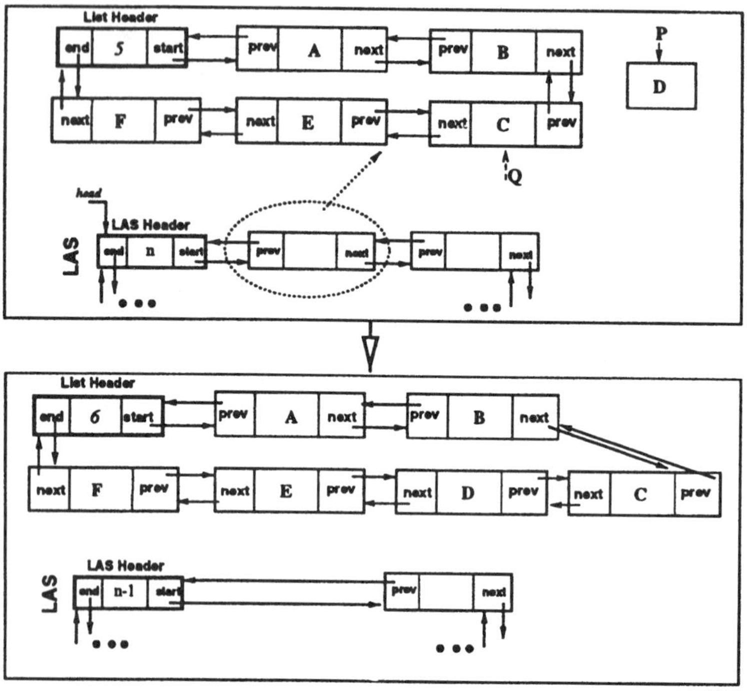

FIGURE 14.4 An example of a doubly linked circular list and of the associate list of available storage. The operations involved in the addition of record D after position Q are graphically illustrated.

TABLE 14.3 Record Addition and Deletion from a Doubly Linked Circular List, Using a List of Available Space for Record Memory Management

Insert data x in list L in a record placed after record **Q**	Delete **R** from list L
1. *LAS*.$n = 0 \rightarrow$ OVERFLOW;	1. $n = 0 \rightarrow$ UNDERFLOW;
2. p = (*LAS.head*).*next*;	2. *RD*(**R**);
3. **R** = *p;	3. $n--$;
4. *LAS*.RD(**R**);	4. **Q** = **LAS.head*;
5. *LAS*.$n--$;	5. *LAS*.RI(**R**,**Q**);
6. **R**.*cont* = x;	6. *LAS*.n++.
7. RI(**R**,**Q**)	
8. n++.	

14.2.2 A Simple Hash Table

It may be noted that searching was not present among the set of operations to be performed on a linear list. This is because linear lists are not well suited for this type of application. We will now introduce a data structure used in unstructured grid generation and grid adaption procedures and that is better designed for simple search queries.

Let's first state in a general form the problem we wish to address. Let us assume that we need to keep in a table H some records that are uniquely identified by a set of keys $K = \{k_1, k_2, \ldots, k_l\}$ and let us indicate with **R**.k_i the ith key associated to **R**, respectively. The type of operations we want to perform on H are as follows:

1. Search if a record with given keys is present in the table;
2. Add a new entry to the table;
3. Delete an entry from the table.

A possible implementation that may allow efficiently solving the problem is to consider one of the keys as the *principal key.* Without loss of generality, we assume that the first key k_1 is the principal key, and in the following it will be simply referred to as k. Let U be a set of keys and $k \in U$ a generic key of the set. We now build a function $h(k)$, called *hashing function,** $h(k)$: $U \to \{0, \ldots, m-1\}$, that assigns to each key of U an integer number between 0 and $m-1$. We have various ways of building the hashing table H, depending whether h is a one-to-one mapping or not. However, before proceeding further, let us consider a practical example.

Assume that we want to keep track of the triangular faces of a 3D tetrahedral mesh, when the mesh layout is constantly changing, for example during a mesh generation or adaption process. A face **F** could be identified by its node numbering $\{i_1, i_2, i_3\}$, and to make the identification unique we could impose that

$$k \equiv k_1 = \min\ (i_1, i_2, i_3),\ k_3 = \max\ (i_1, i_2, i_3),\ k_2 = \{i_1, i_2, i_3\} - \{k_1, k_2\}$$

Since k is an integer number, and we expect that the k's will be almost uniformly distributed, a simple and effective choice for $h(k)$ is the identity function $h(k) = k$. A *hash table H* may then be formed by an array $H.A[m]$, where m is the maximum node numbering allowed in the mesh. The array will be addressed directly using the key k. Each table entry $H.A[k]$ will either contain a *null* pointer, indicating that the corresponding key is not present in the table, or a pointer to the beginning of a *linked list* whose records contain the remaining keys for each face with principal key k, plus possible ancillary information. If we use doubly linked lists, each entry in the table may store two pointers, pointing to the head and the tail of the corresponding list, respectively. In practice, each array element acts as a header for the list. Figure 14.5 illustrates this information structure, where, for simplicity a 2D mesh has been considered. If we use doubly linked list, add and delete operations are $O(1)$ while simple search is a $O(1 + n/m)$ operation, where n indicates the number of records actually stored in the table. Since in many practical situations, such as the one in the example, the average value of n/m is relatively small ($\approx$ 6 for a 2D mesh), the searching is quite efficient. As for memory usage, if we assume that no ancillary data is stored, we need approximately $2mP + n_{max}[2P + (l-1)I]$ memory locations, where P and I are the storage required for a pointer and an integer value, respectively, while l is the number of keys (3 in our case), and n_{max} is the maximum number of records that could be stored at the same time in the linked lists. In this case, n_{max} is at most the maximum number of faces in the mesh. All chained lists have records of the same size, therefore a common *LAS* is normally used to store the available records. Some memory savings may be obtained by storing the first record of each chained list directly on the corresponding array element, at the expense of a more complicated bookkeeping.

The structure just presented is an example of a hash table with collision resolved by chaining. The term *collision* means the event caused by two or more records that hash to the same slot in the table. Here, the event is resolved by storing all records with the same hash function in a linked list. This is not the only possibility, however, and many other hashing techniques are present in the literature, whose description is here omitted. In the previous example we have assumed that we know the maximum number of different keys. What can be done if we do not know this beforehand, or if we would like to save some memory by having a smaller table? We need to use a different hash function. There are many choices for hash functions that may be found in the literature. However, for the type of problems just described, the division method, i.e.,

$$h(k) = k \bmod m$$

*In general, h may be a function of all keys, i.e., $h = h(K)$. For sake of simplicity, we neglect the general case.

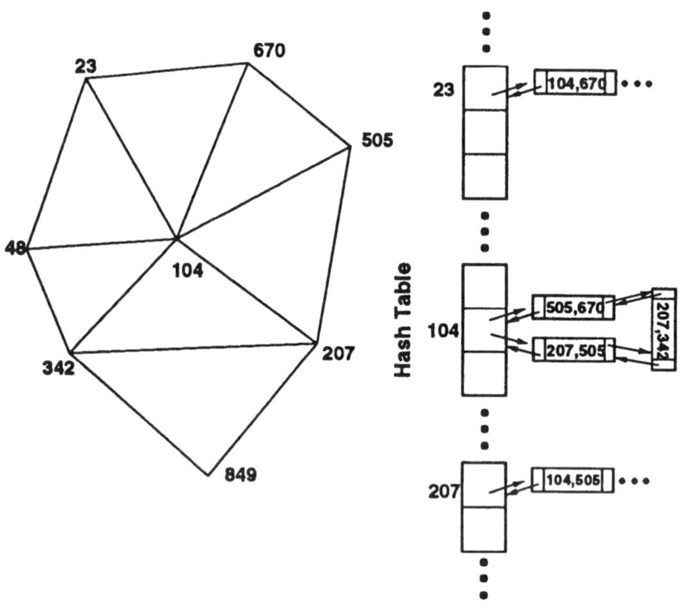

FIGURE 14.5 A simple hash table to keep track of triangular faces.

is simple and effective. Going back to our example, if we choose $m = 103$, then faces $\{104, 505, 670\}$ and $\{342, 207, 849\}$ have the same hash value $h = 1$, even if their principal key is different (104 and 207, respectively). In order to distinguish them, we need to store also the principal key in the chained linked list records, changing the memory requirement to approximately $2m_{max} P + n_{max} [2P + lI]$. Comparing with the previous expression, it is clear that this method is convenient when $n_{max} < < m_{max}$. In which particular situations would a hash table like the one presented in the example be useful? Let us assume that somebody has given you a tetrahedral grid, without any indication of the boundary faces. How do you find the boundary faces? You may exploit the fact that each mesh face, apart from the ones at the boundary, belongs to two tetrahedra, set up a hash table H of the type just described, and run the following algorithm.

1. Loop over the elements e of the mesh
   1.1. Loop over element faces
      1.1.1. Compute the keys K for the face and the principal key k
      1.1.2. Search K in H
         - If K is present then delete the corresponding record
         - Otherwise add to H the record containing the face keys
2. Traverse all items in the hash table and push them onto stack F

The stack F will now obtain all boundary faces. A similar structure may be used also to dynamically store the list of nodes surrounding each mesh node, or the list of all mesh sides and many other grid data. We have found this hash table structure very useful and rather easy to program.

The implementation just described is useful in a dynamic setting, when add and delete operations are required. In a static problem, when the grid is not changing, we may devise more compact representations based on sequential storage and direct addressing. Again, let's consider a practical problem, such as storing a table with the nodes surrounding each given mesh node, when the mesh, formed by n nodes and n_e elements, is not changing. We use a structure K with the following attributes:

$K.n = n$, number of entries in the table;
$K.IA[n + 1]$, the array containing the pointer to array JA;
$K.JA[3n_e]$, the array containing the list of nodes.

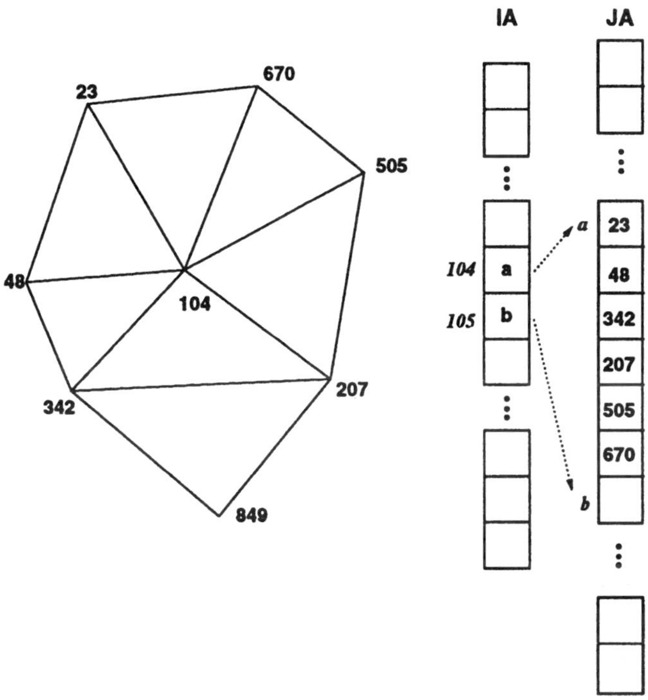

FIGURE 14.6 The structure used for searching the point surrounding each mesh node in the static case.

Figure 14.6 graphically shows how the structure works. The indices $\{IA[i], \ldots, IA[i+1]-1\}$ are used to directly address the entries in array *JA* that contain the numbering of the nodes surrounding node i (here, we have assumed that the smallest node number is 0). The use of the structure is straightforward. The problem remains of how to build it in the first place. A possible technique consists of a two-sweep algorithm. We assume that we have the list of mesh sides.* In the first sweep we loop over the sides and we count the number of nodes surrounding each node, preparing the layout for the second pass:

1. For ($i = 0, i \leq n; i++$) $IA[i] = 0$
2. *Begin sweep 1:* loop over the mesh sides i_1, i_2
   2.1. For $i \in \{i_1, i_2\}$) $IA[i]$ ++
3. For ($i = 1, i \leq n, i$++) $IA[i] += IA[i-1]$
4. For ($i = n-1, i \geq 1, i--$) $IA[i] = IA[i-1]$
5. $IA[0] = 0$
6. *Begin sweep 2:* loop over the mesh side i_1, i_2
   6.1. For ($i \in \{i_1, i_2\}$)
      6.1.1. $JA[IA[i]] = i_1 + i_2 - i$
      6.1.2. $IA[i]$ ++
7. For ($i = n, i \geq 1, i--$) $IA[i] = IA[i-1]$
8. $IA[0] = 0$

It is worth mentioning that this structure is also the basis of the *compressed sparse row format,* used to efficiently store sparse matrices.

*The algorithm that works using the element connectivity list, i.e., the list of the nodes on each element, is only a bit more complicated, and it is left to the reader.

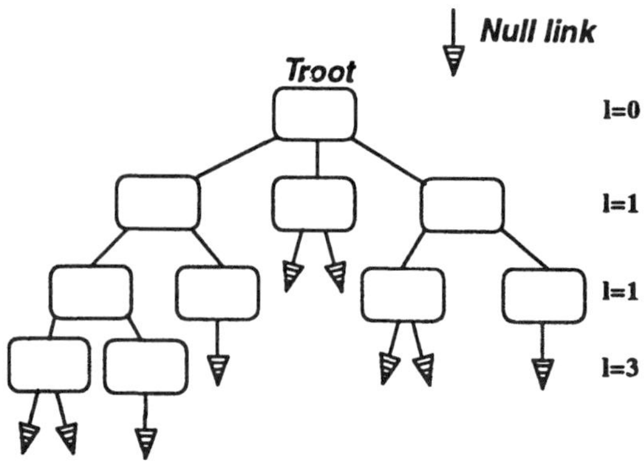

FIGURE 14.7 An example of the representation of a generic tree.

14.3 Tree Structures

The data structures seen so far are not optimal when the following operations are to be made on the data:

1. Simple search: search if a record is present in the structure;
2. Range search: find all data which are within a certain range of values;
3. Tracking the maximum (or minimum) value: return the record which has the maximum value of a given field.

Those queries are normally better answered with the adoption of a tree-type structure. Before explaining why this is so, let us give some nomenclature. Formally, a *tree* T is a finite set whose elements are called nodes (here we cannot avoid the ambiguity with mesh nodes), such that

1. There is a special node called *root*, which will be normally indicated by *T*.**root**;
2. The other nodes may be partitioned into n disjoint sets, $\{T_1,\ldots, T_n\}$, each of which is itself a tree, called *subtrees* of the root.

If the order of the subtrees is important, then the tree is called an *ordered tree.* The one just given is a recursive definition, and the simplest tree has only one node. In that case the node is called a *leaf node.*

A tree takes its name by the way it is usually represented. Actually, the most used representation, such as the one shown in Figure 14.7, is an upside-down tree, with the root at the top. The tree root is linked to the root of its subtrees, and the branching continues until we reach a leaf node. Sometimes *null links* are also indicated, as in Figure 14.7. In all graphical representations of a tree contained in the remaining part of this work, we have omitted the null links.

The number of subtrees of a node is called the *degree* of the node. A leaf has degree 0. A tree whose nodes have at most degree n is termed an *N-ary tree.* The root of a tree T is said to be at *level* $l = 0$ with respect to T and to be the *parent* of the roots of its subtree, which are called its *children* and are at level $l = 1$, and so on. If a node is at level k it takes at least k steps to reach it starting from the tree root, and the highest level reached by the nodes of given tree is called the *height* of that tree. A tree is called *complete* if it has the minimum height possible for the given number of nodes.

An important case is that of *binary trees.* In a binary tree each node is linked to at most two children, normally called the left and right child, respectively. Binary trees are important because many information structures are based on them and also because all trees may in fact be represented by means of a binary tree [10]. Tree structure is applied to grid generation in Chapter 15.

TABLE 14.4 Inorder and Preorder Traversal of a Binary Tree

inorder_walk (T)	preorder_walk (T)
1. *T*.**root** = NIL ↦ RETURN;	1. *T*.**root** = NIL ↦ RETURN;
2. inorder_walk (T.*left*);	2. VISIT (*T*.**root**);
3. VISIT (*T*.**root**);	3. preorder_walk (T.*left*);
4. inorder_walk (T.*right*);	4. preorder_walk (T,*right*);

Note: VISIT indicated whatever operation we wish to do on the tree node, for example printing its content.

14.3.1 Binary Trees

In a binary tree each node may be considered as a record N whose two fields N.*left* and N.*right* contain a pointer to the left and right subtree root, respectively. Therefore, a binary tree may be thought of as a different layout of the same record used in a doubly linked list. Indeed, all considerations about using a *LAS* for records management apply straight away to binary trees. While there is an unequivocal meaning to the term "traversing a linear list," i.e., examine the records in their list ordering, this is no longer true for a tree structure. So, if we want to list all the nodes in a binary tree, we must first decide which path to follow. There are basically three ways of traversing a binary tree, depending whether the root is visited before traversing the subtrees (preorder traversal), between the traversing of the left and the right subtree (inorder, or symmetric, traversal), or after both subtrees have been traversed (postorder traversal). Table 14.4 shows two possible recursive algorithms for inorder and preorder traversal, respectively, the extension to postorder being obvious. We have used recursion, since it allows to write concise algorithms. We should warn, however, that recursion causes some computational overhead. Therefore, nonrecursive algorithms should be preferred when speed is an issue.

14.3.1.1 How to Implement Binary Trees

We have already seen that a binary tree implementation is similar to that of a doubly linked list. Each node is represented by a record which has two special fields containing pointers that may be either null or pointing to the node subtrees. If both pointers are null, the node is a leaf node.

However, if the tree is a complete binary tree, a more compact representation could be adopted, which uses an array and no pointers, but rather integer operations on the array addresses. We will use a data organization of this type in the next section dedicated to a special complete binary tree: the heap. A more general discussion may be found in [10].

14.3.2 Heaps

Often, there is the necessity to keep track of the record in a certain set that contains the maximum (or minimum) value of a key. For example, in a 2D mesh generation procedure based on the advancing front method [14] we need to keep track of the front side with the minimum length, while the front is changing. An information structure that answers this type of query is a *priority queue,* and a particular data organization which could be used for this purpose is the *heap.* A heap is normally used for sorting purposes and indeed the *heap-sort* algorithm exploits the properties of a heap to sort a set of N keys in $O(N\log_2 N)$ time, with no need of additional storage. We will illustrate how a heap may also be useful as a priority queue.

We will indicate in the following with > the *ordering relation.* A heap is formally defined as a binary tree with the following characteristics: If k is the key associated with a heap node, and k_l and k_r are the keys associated to the not-empty left and right subtree root, respectively, the following relation holds:

$$k >= k_l \text{ and } k >= k_r$$

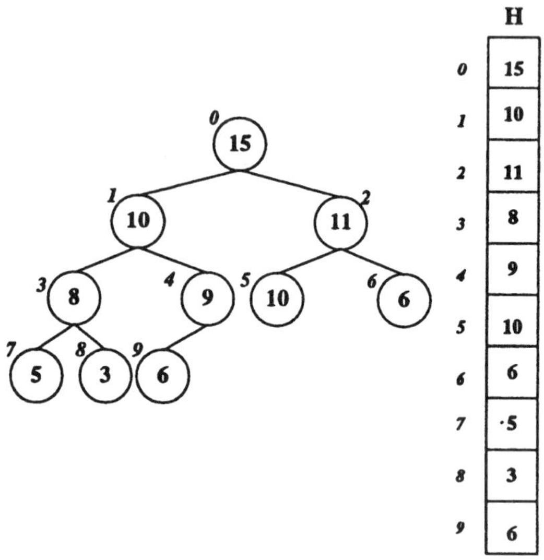

FIGURE 14.8 Tree representation of a heap and an example of its sequential allocation.

As a consequence, the key associated to each node is not "smaller" than the key of any node of its subtrees, and the heap root is associated to the "largest" key in the set. We have placed in quotes the words "largest" and "smaller" because the ordering relation > may in fact be arbitrary (as long as it satisfies the definition of an ordering relation), and it does not necessarily correspond to the usual meaning of "greater than." An interesting feature of the heap is that, by employing the correct insertion and addition algorithms, the heap can be kept complete, and the addition, deletion, and simple query operations are, even in the worst case, of $O(\log_2 n)$, while accessing the "largest" node is clearly $O(1)$.

A heap T may be stored using sequential allocation. We will indicate by $T.n$ and $T.max$ the current and maximum number of records stored in T, respectively, while $T.H[max]$ is the array that will hold the records. The left and right subtrees of the heap node stored at $H[i]$ are rooted at $H[2i+1]$ and $H[2i+2]$, respectively, as illustrated in Figure 14.8. Therefore, by using the integer division operation, the parent of the node stored in $H[j]$ is $H[(j-1)/2]$. The heap definition may then be rewritten as

$$H[j].key <= H[(j-1)/2].key \ (0 < j < n) \tag{14.1}$$

When inserting a new node, we provisionally place it in the next available position in the array and we then climb up the heap until the appropriate location is found. Deletion could be done employing a top-down procedure, as shown in Figure 14.9. We consider the heap rooted at the node to be deleted, and we recursively move the "greatest" subtree root to the parent location, until we reach a leaf where we move the node stored on the last array location. Finally, a bottom-up procedure analogous to that used for node insertion is performed.*

Since the number of operations is clearly proportional to the height of the tree, we can deduce that, even in the worst case, insertion and deletion are $O(\log_2 n)$. Simple and range searches could be easily implemented with a heap as well. However, a heap is not optimal for operations of this type.

*The terms top-down and bottom-up refer to the way a tree is normally drawn. So, by climbing up a tree we reach the root!

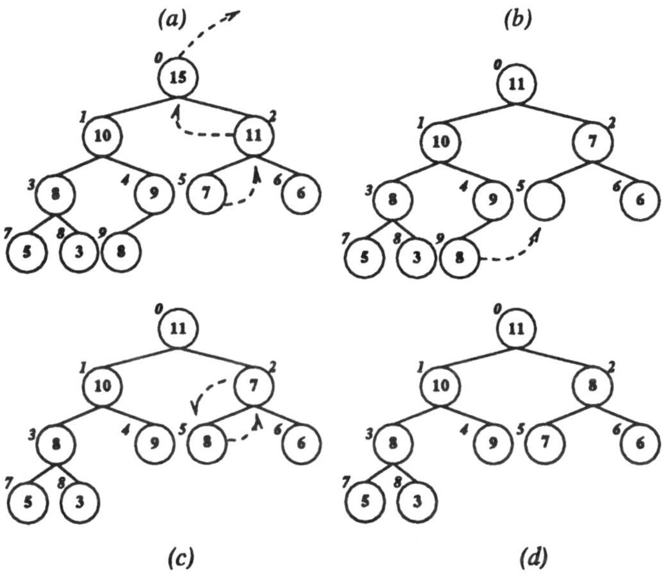

FIGURE 14.9 Deletion of the root of a heap. (a) The "highest" subtree root is recursively "promoted" until we reach a leaf. (b) The last node is placed in the empty leaf and (c) it is sifted-up to the right place by a succession of exchanges with its parent, until the final position (d) is reached.

14.3.3 Binary Search Tree

Better techniques for simple and range searching make use of a *binary search tree.* Indicating again with k, k_l, and k_r the keys associated with a node, its left and right subtree roots, respectively, a binary search tree is an oriented binary tree where the following expression is satisfied:

$$k_l <= k \text{ and } k_r > k. \tag{14.2}$$

As before, > indicates an ordering relation. It should be noted that we must disambiguate the case of equal keys, so that the comparison may be used to discriminate the records that would follow the left branch from the ones that would go to the right. Inorder traversal of a binary search tree returns the records in "ascending" order.

The simple search operation is obvious. We recursively compare the given key with the one stored in the root, and we choose the right or left branch according to the comparison, until we reach either the desired record or a leaf node. In the latter case, the search ends unsuccessfully. In the worst case, the number of operations for a simple search is proportional to the height of the tree. For a complete tree the search is then $O(\log_2 n)$. However, the shape of a binary tree depends on the order in which the records are inserted and, in the worst case, (which, for example, happens when a set of ordered records is inserted) the tree degenerates and the search becomes $O(n)$. Fortunately, if the keys are inserted in random order, it may be proved the search is, on average, still $O(\log_2 n)$ [10].

Node addition is trivial and follows the same lines of the simple search algorithm. We continue the procedure until we reach a leaf node, of which the newly inserted node will become a left or right child, according to the value of the key comparison. Node deletion is only slightly more complicated, unless the deleted node has fewer than two children. In that case the deletion is indeed straightforward if the node is a leaf, while if it has a single child, we can slice it out by connecting its parent with its child.

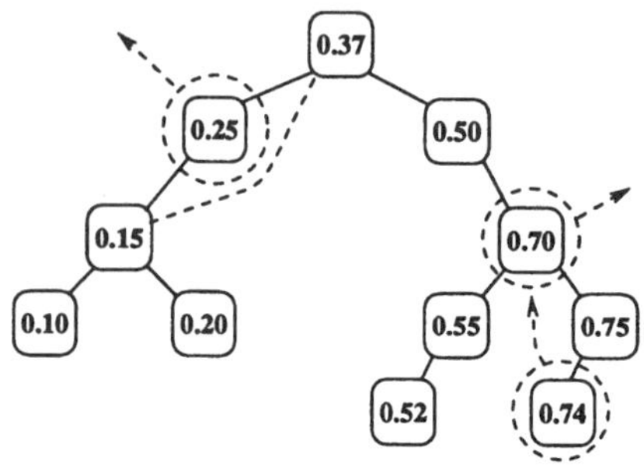

FIGURE 14.10 An example of binary search tree and a graphical illustration of the operations necessary to delete a "single parent" node (0.25) and a node with two children (0.70). The tree has been created inserting the keys in the following order: 0.37, 0.25, 0.50, 0.10, 0.70, 0.15, 0.20, 0.55, 0.75, 0.52, 0.74.

In the case that the deleted node has two children, we have to find its successor **S** in the inorder traversal of the tree, which has necessarily at most one child. We then slice **S** out of the tree and put it at the deleted node location. The resulting structure is still a binary search tree, Figure 14.10 illustrates the procedure. It can be proved that both insert and delete operations are $O(\log_2 n)$ for complete binary search trees. Other algorithmic details may be found, for example, in [2].

A binary search tree may be kept almost complete by various techniques, for example, by adopting the *red-black tree* data structure [7] or the AVL tree, whose description may be found in [22].

14.3.4 Digital Trees

In a binary search tree the discrimination between left and right branching is made by a comparison between keys. What happens if we make the comparison with fixed values instead? Let us suppose that the keys are floating point numbers within the range $[A, B)$. We will say that a tree node N is associated to the interval $[a, b)$ if all keys stored on the tree rooted at N fall within that range. Then, we can put $[a, b) = [A, B)$ for the root of tree T, and we may recursively build the intervals associated to the subtrees as follows: given a tree associated to the interval $[a, b)$, the left and right root subtrees will be associated to the intervals $[a, r)$ and $[r, b)$, respectively, where $r \in [a, b)$ is a discriminating value, which is usually taken as $r = (a + b)/2$. This data structure is called *digital search tree.* Adding a node to a digital binary tree is simple, and resembles the algorithms used for binary search trees. We start from the root and follow the left or right path according to the result of the test $k > r$? The difference with binary search trees is that the discriminant r now has a value that does not depend on the record currently stored at the node, but on the node position in the tree structure. We want again to point out that the result of the discriminating test must be unique: that is why the intervals are open at one end. In this way, the case $k = r$ will not be ambiguous, by leading to follow the path on the right.

Deleting a node is even more trivial, since *every* node of a digital search tree can be placed at the root position! Therefore, when deleting a node **N**, we just have to substitute it with a convenient leaf node of the subtree rooted at **N** (of course, we do not move the nodes, we just reset the links). For example, if **N** is not a leaf node (in which case we could just release it), we would traverse in postorder the subtree rooted at **N**, as the first "visited" node will certainly be a leaf, and substitute **N** with it. In order to keep a better balanced tree in a highly dynamic setting, when many insertion and deletion operations are expected, we could keep at each node **N** a link to a leaf node of the tree rooted at **N** with the highest level l. That node will be used to substitute **N** when it has to be deleted. With this technique the algorithms for insertion/deletion are just a little more involved, and it could be useful to store at each node **N** also the link **N**.*parent* to the parent node. With a digital tree we cannot slice out of the tree a single-child node as we have seen in binary trees, since this operation will cause a change of node level in the subtree rooted at the sliced-out location, and the discriminant value r, which is function of the node level, will

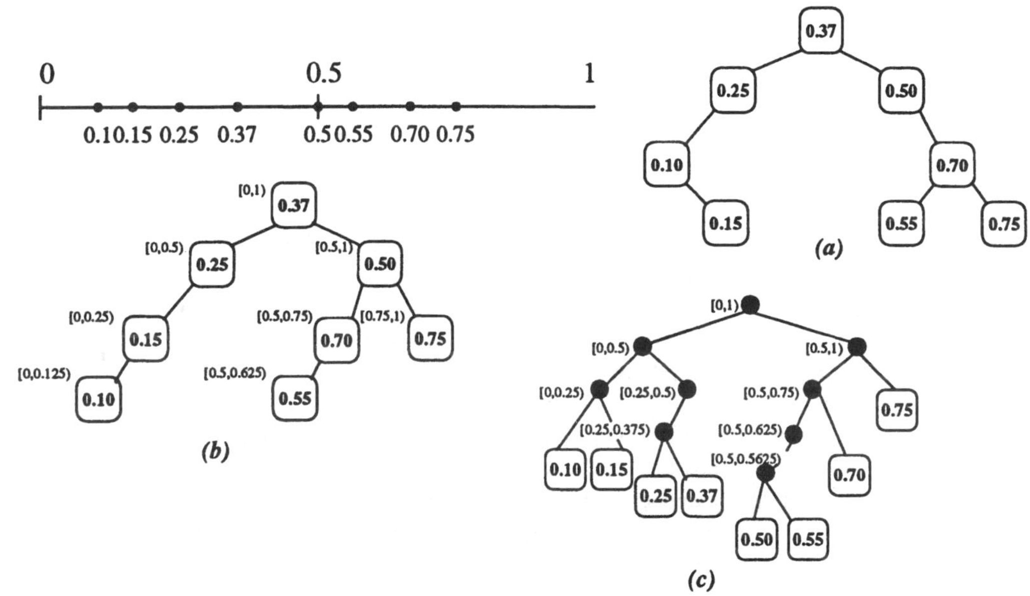

FIGURE 14.11 An example of a binary search tree (a), a digital search tree (b), and binary trie structure (c), for a given set of data.

change. As a consequence the structure resulting from this operation will, in general, not be a digital search tree anymore.

The name of this data structure derives from the fact that it is in principle possible to transform a key into an ordered set of binary digits d_0, d_1, …, d_m so that, at tree level l, the decision to follow the right or left path could be made by examining the lth digit. In particular, for the example just shown, the decision could be made by considering the lth significant digit of the binary representation of the number $(k - a)/(b - a)$, and following the left path if $d_l = 0$.

Directly related to the digital search tree is the trie structure [11], which differs from the digital search tree mainly because the actual data is only stored at the leaf nodes. The trie shape is completely independent from the order in which the nodes are inserted. The shape of a digital tree, instead, still depends on the insertion order. On the other hand, a trie uses up more memory, and the algorithms for adding and deleting nodes are a little more complex. Contrary to a binary search tree, both structures require the knowledge of the maximum and minimum value allowed for the key.

Figure 14.11 shows an example of the search structures seen so far for a given set of keys $k \in [0, 1)$. For the digital and binary trie we have put beside each node the indication of the associated interval.

14.4 Multidimensional Search

In mesh generation procedures there is often the necessity of finding the set of nodes or elements lying within a specified range, for instance, finding the points that are inside a given sphere. There also arises the need to solve geometric intersection problems, such as finding the triangular faces that may intersect a tetrahedron.

Needless to say, these are fundamental problems in computational geometry (cf. Part III of this handbook) and there has been a great deal of research work in the last years aimed at devising optimal data structures for this purpose. There is not, however, a definite answer. Therefore, we will again concentrate only on those structures suited for mesh generation procedures, where we usually have dynamic data and where memory occupancy is a critical issue.

14.4.1 Searching Point Data

Given a set of points P in R^d, where d is either 2 or 3, we will consider the following queries:

- *Point search:* is the point P present in the P?
- *Range search:* that are the points of P that lie inside a given interval $I \subset R^d$?

In addition to those operations, we may want to be able to efficiently add and delete nodes to the set. For the case $d = 1$, it was shown in the previous section that a binary search tree could efficiently answer these queries. It would be natural to ask whether it can be used also for multidimensional searches. Unfortunately, binary search is based on the existence of an ordering relation between the stored data, and there is normally no way of determining an ordering relation between multidimensional points. In fact, we will see that in principle an ordering relation may be found, for instance using a technique called bit interleaving, but in practice this procedure is not feasible, as it would require costly operations, both in terms of computation and memory.

The most popular procedures for handling multidimensional searches are either based on hierarchical data structure or on grid methods [20]. We will illustrate some of the former, and in particular data structures based either on binary trees quadtrees, or octrees. For sake of simplicity, we will consider only a Cartesian coordinate system and the two-dimensional case, the extension to 3D being obvious.

14.4.2 Quadtrees

The *quadtree* is "4-ary" tree whose construction is based on the recursive decomposition of the Cartesian plane. Its three-dimensional counterpart is the *octree*. There are basically two types of quadtrees, depending whether the space decomposition is driven by the stored point data (point-based quadtrees) or it is determined *a priori* (region-based quadtrees). Broadly speaking, this subdivision is analogous to the one existing between a binary and a digital search tree. We will in the following indicate with **B** the *domain bounding box* defined as the smallest interval in R^2 enclosing the portion of space where all points in P will lie. Normally, it can be determined because we usually know beforehand the extension of the domain that has to meshed. For sake of simplicity, we will often assume in the following that **B** is unitary, that is $\mathbf{B} \equiv [0, 1) \times [0, 1)$. There is no loss of generality when using this assumption, as an affine transformation can always be found that maps our point data set into a domain enclosed by the unitary interval.

14.4.2.1 Region-Based Quadtrees

A region-based quadtree is based on the recursive partitioning of **B** into four equally sized parts, along lines parallel to the coordinate axis. We can associate to each quadtree node **N** an interval $\mathbf{N}.I = [a, b) \times [a, b)$ where all the points stored in the tree rooted at **N** will lie. Each node **N** has four links, often denoted by SW, SE, NW, NE, that point to the root of its subtrees, which have associated the intervals obtained by the partitioning. Point data are usually stored only at leaf nodes, though it is also possible to create variants where point data can be stored on any node. Figure (14.12) illustrates an example of a region quadtree. The particular implementation shown is usually called PR-quadtree [20,19].

Point searching is done by starting from the root and recursively following the path to the subtree root whose associated interval encloses the point, until we reach a leaf. Then the comparison is made between the given node and the one stored in the leaf. Range searching could be performed by examining only the points stored in the subtrees whose associated interval has a non-empty intersection with the given range. Details for point addition/deletion procedures may be found in the cited reference. The shape of the quadtree here presented, and consequently both search algorithm efficiency and memory requirement, is independent of the point data insertion order, but it depends on the current set of points stored. If the points are clustered, as often happens in mesh generation, this quadtree can use a great deal of memory because of many empty nodes. Compression techniques have been developed to overcome this problem: details may be found in [15]. In unstructured mesh generation the region quadtree, and the octree in 3D, is often used not just for search purposes [12], but also as a region decomposition tool (see also Chapter 22). To illustrate the idea behind this, let us consider the example shown in Figure 14.13,

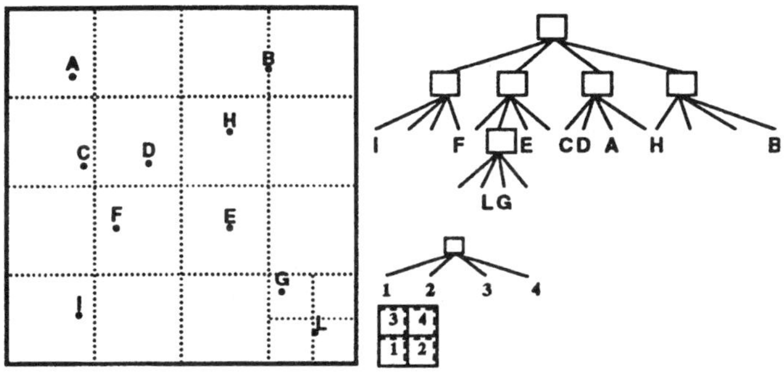

FIGURE 14.12 An example of a region-based quadtree.

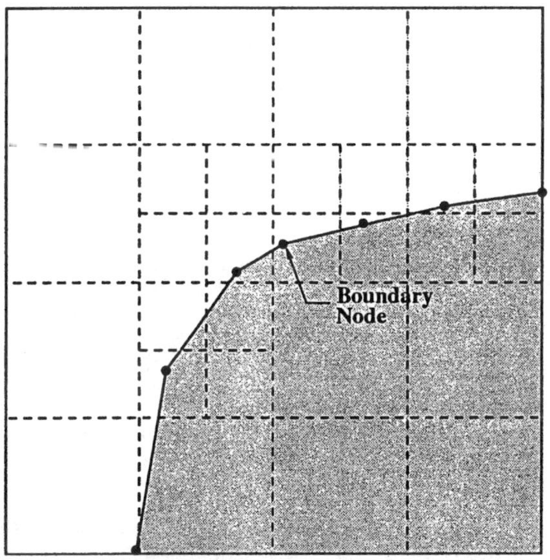

FIGURE 14.13 Domain partitioning by a region quadtree. The quadtree contains the boundary points. The partitions associated with the quadtree nodes are shown with dotted lines.

where the line at the border with the shaded area represents a portion of the domain boundary. A region quadtree of the boundary nodes has been built, and we are showing the hierarchy of partitions associated to the tree nodes. It is evident that the size of the partitions is related to the distance between boundary points, and that the partitioning is finer near the boundary. Therefore, structures of this type may be used as the basis for algorithms for the generation of a mesh inside the domain, in various ways. For instance, a grid may be generated by appropriately splitting the quad/octree partitions into triangle/tetrahedra [23]. Alternatively, the structure may be used to create points to be triangulated by a Delaunay type procedure [21] (cf. Chapter 16). Finally, it can be adopted for defining the mesh spacing distribution function in an advancing front type mesh generation algorithm [9] (cf. Chapter 17).

14.4.2.2 Point-Based Quadtrees

A point quadtree is a type of multidimensional extension of the binary search tree. Here the branching is determined by the stored point, as shown in Figure 14.14. It has a more compact representation than the region quadtree, since point data is stored also at non-leaf nodes. However, the point quadtree shape strongly depends on the order in which the data is inserted, and node deletion is rather complex. Therefore, it is not well suited for a dynamic setting. However, for a static situation, where the points

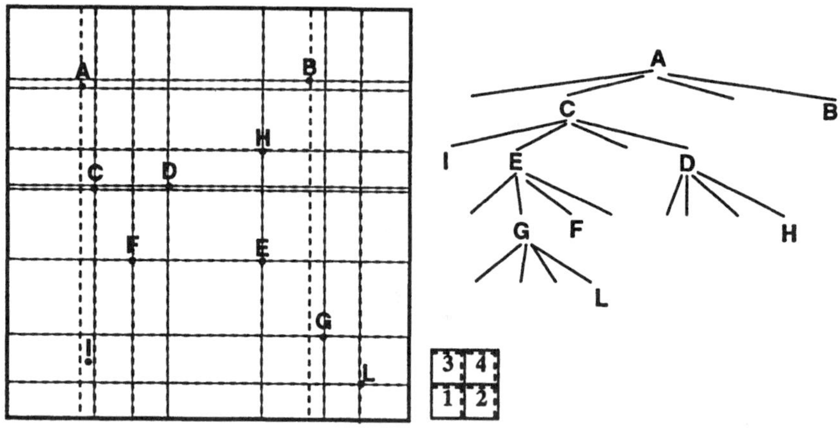

FIGURE 14.14 An example of a point-based quadtree. Nodes have been inserted in lexicographic order.

are known *a priori*, a simple technique has been devised [4] to generate optimized point quadtree, and this fact makes this structure very interesting for static situations, since simple search operations become $O(\log_4 n)$ n being the total number of points stored. It should be mentioned that a procedure that allows for dynamic quadtree optimization has also been devised [16]. Its description is beyond the scope of this chapter.

14.4.3 Binary Trees for Multidimensional Search

A shortcoming of quad/octrees is that they are rather costly in terms of the memory required. It is possible, however, to use a binary tree also for a multidimensional search. Indeed many of the ideas illustrated for the one-dimensional case may be extended to more dimensions if we allow the discriminating function r at each tree level to alternate between the coordinates. We may use the same technique adopted for the binary search tree, but we now discriminate according to the x coordinate at even-level nodes, while nodes at odd level are used to discriminate according to the y coordinate. We have now a two-dimensional search tree denoted *k-d tree*, which may be also considered as the binary tree counterpart of a point-based quadtree. Figure (14.15) shows an example of a k-d tree. As the node where point D is stored by a y-discriminator, its left subtree contains only points P that satisfy $P.y < D.y$. K-d trees are a valid alternative to point-based quad/octrees. According to Samet [20]: "we can characterize the k-d tree as a superior serial data structure and the point quadtree as a superior parallel data structure." However, they also share the same defects. Their shape strongly depends on the node insertion order, unless special techniques are adopted [3], and node deletion is a quite complex operation also for k-d trees.

An alternative that has encountered great success in unstructured mesh generation is the *alternating digital tree* (ADT) [1], which is a digital tree where, as in a k-d tree, the discrimination is alternated between the coordinates. It differs from a k-d tree because here the discrimination is made against fixed space locations. To each node **N** we associate an interval in $[x_1, x_2) \times [y_1, y_2)$, and all point data in the subtree rooted at **N** will lie in that interval. The tree root is associated to the bounding interval **B**. If a node is an x-discriminator, its left and right child will be associated to the intervals $[x_1, r) \times [y_1, y_2)$ and $[r, x_2) \times [y_1, y_2)$, respectively, with $r = (x_1 + x_2)/2$. A y-discriminating node will act in a similar fashion by subdividing the interval along the y axis. Figure (14.15) illustrates an example of an ADT tree.

The algorithms for node addition and deletion are analogous to the ones shown for the one-dimensional digital tree. Simple searching is $O(\log^2 n)$ if the tree is complete. Unfortunately, the tree shape is not independent of the order of node insertion, even if, in general, ADT trees are better balanced than their k-d counterpart. For static data, while a special insertion order has been devised to get a balanced k-d tree, no similar techniques are currently available for an ADT. Therefore, we may claim that ADTs are better than k-d trees for dynamic data, while for a static situation, a k-d tree with optimal point

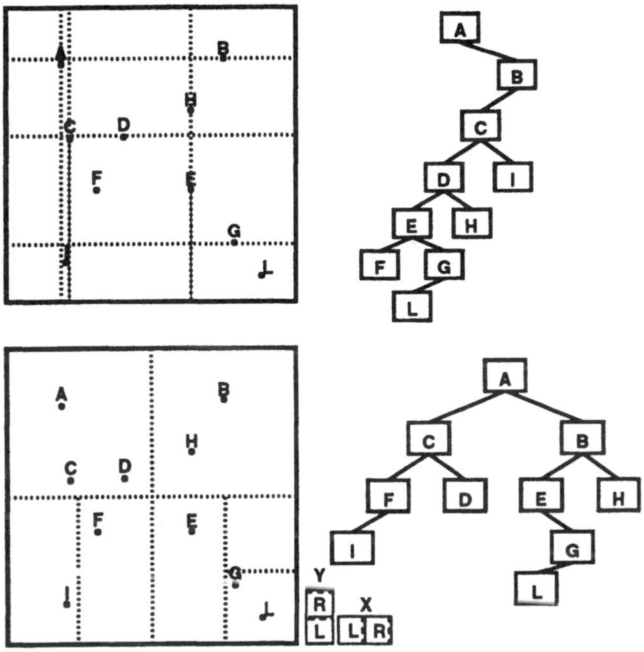

FIGURE 14.15 An example of a two-dimensional k-d tree (top) and an ADT (bottom), on the same set of data. Nodes at even levels discriminate the x coordinate, while odd nodes discriminate the y coordinate. In the k-d tree the discrimination is made against the data stored at the node, while in the ADT structure we use fixed spatial locations. Nodes have been inserted in lexicographic order in both cases.

insertion order is more efficient. Range searches in an ADT are made by traversing the subtrees associated with intervals which intersect the given range.

A region decomposition based structure similar to ADT, where the data points are stored only at leaf nodes is the *bintree* [20]: it has not been considered here because of its higher memory requirement compared with ADT.

14.4.3.1 Bit Interleaving

For sake of completeness, we mention how, at least theoretically, a binary search tree may be used also for multi-dimensional searching using a technique called *bit interleaving*. Let us assume that **B** is unitary. Then, given a point P = (x, y) we may consider the binary representation of its coordinates, which we will indicate as $x_0, x_1, x_2, \ldots, x_d$ and $y_0, y_1, y_2, \ldots, y_d$. We may now build a code by interleaving the binary digits, obtaining $x_0, y_0, \ldots, x_d, y_d$ and define an ordering relation by treating the code as the binary representation of a number. The code is unique for each point in **B**, and we can use it as a key for the construction of a binary search tree. This technique, however, is not practical because it would require storing at each node a code that has a number of significant digits twice as large as the one required for the normal representation of a float (three times as large for 3D cases!). It may be noted, however, that the ADT may indeed be interpreted as a digital tree where the discrimination between left and right branching at level l is made on the base of the lth digit of the code built by a bit interleaving procedure (without actually constructing the code!).

14.4.4 Intersection Problems

Geometry intersection problems frequently arise in mesh generation procedures (see also Chapter 29). In a front advancing algorithm, for instance, we have to test whether or not a new triangle intersects the current front faces. General geometrical intersection problems may be simplified by adopting a two-step

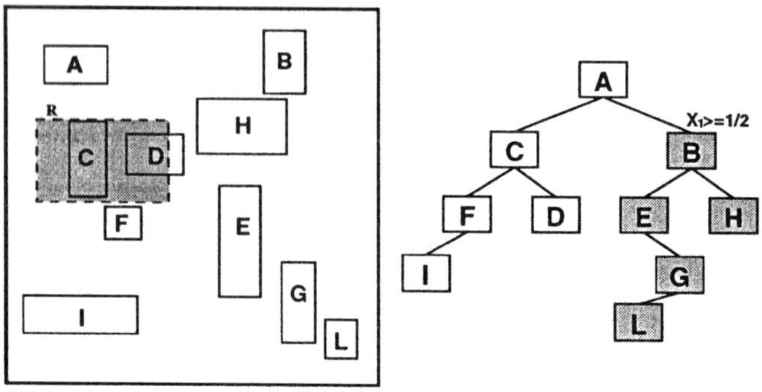

FIGURE 14.16 Intersection problem solved by means of an ADT structure. The subtree rooted at node B (shaded nodes) does not need to be examined, since all subtree nodes correspond to rectangles that lie in the half-hyperspace $x_1 \geq 1/2$, which cannot intersect **R**, since $\mathbf{R}.x_2 \leq 1/2$.

procedure. In the first step we associate with each geometrical entity of interest G its smallest enclosing interval $I^G \equiv [x_1^G, x_2^G] \times [y_1^G, y_2^G]$, and we then build specialized data structure which enables one to efficiently solve the following problem.

> Given a set I of intervals (rectangles in 2D or hexahedra in 3D), find the subset $H \subset I$ of all elements of I which intersect a given arbitrary interval.

In the second phase, the actual geometrical intersection test will be made, restricted only to those geometrical entities associated to the elements of H.

Data structures that enable solving efficiently this type of problem may be subdivided into two categories: the ones that represent an interval in R^n as a point in R^{n^2}, and those that directly store the intervals. An example of the latter technique is the R-tree [20], which has been recently exploited in a visualization procedure for three-dimensional unstructured grid for storing sub-volumes so that they can be quickly retrieved from disk [13]. We will here concentrate on the first technique: i.e., how to represent an interval as a point living in a greater dimensional space.

14.4.4.1 Representing an Interval as a Point

Let us consider the 2D case, where intervals are rectangles with edges parallel to the coordinate axis. A rectangle $R \equiv [x_1, x_2] \times [y_1, y_2]$ may be represented by a point $\mathrm{P} \in R^4$. There are different representations possible, two of which are listed in the following:

1. $\mathrm{P} \equiv [x_1, y_1, x_2, y_2]$
2. $\mathrm{P} \equiv [xc, yc, dx, dy]$ where $xc = (x_1 + x_2)/2$, $dx = (x_2 - xc)$, ...

In the following we will consider the first representation. Once the rectangle has been converted into a point, we can adopt either a k-d or an ADT data structure both for searching and geometrical intersection problems. If we use an ADT tree, the problem of finding the possible intersections can be solved by traversing the tree in preorder, excluding those subtrees whose associated interval in R^4 cannot intersect the given rectangle. Figure 14.16 shows a simple example of this technique.

14.5 Final Remarks

We have given an overview of some of the information structures that may be successfully adopted within mesh generation schemes. This survey is, of course, not exhaustive. Many ingenious data structures have been devised by people working on grid generation and related fields in order to solve particular problems in the most effective way. We will in this final section, just mention a few of the efforts in this direction that we have not had the possibility to describe in detail.

The *List* [10] structure (with uppercase L!) has been adopted [18] to control a hierarchy of grids for multigrid computations. An edge-based structure [8] has been devised for storing mesh topology data, which should be more efficient for Delaunay mesh generation algorithms. Doubly linked circular lists are used for the implementation of a grid topology model that allows an efficient automatic block detection for multiblock structured mesh generation procedures [5, 6].

Many other examples could be made. We hope that the reader now has and idea of how appropriate data structures may help in devising efficient grid generation procedures.

References

1. **Bonet, J. and Peraire, J.**, An alternating digital tree (ADT) algorithm for 3D geometric searching and intersection problems, *Int. J. Num. Meths. Eng.*, 31, pp. 1–17, 1991.
2. **Cormen, T. H., Leiserson, C. E., and Rivest, R. L.**, *Introduction to Algorithms,* The MIT Electrical Engineering and Computer Science Series, McGraw-Hill, 1990.
3. **Fiedman, J. H., Bentley, J. L., and Finkel, R. A.**, An algorithm for finding best matches in logarithmic expected time, *ACM Transactions on Mathematical Software,* 3(3), pp. 209–226, September 1977.
4. **Finkel, R. A. and Bentley, J. L.**, Quad trees: a data structure for retrieval on composite keys, *Acta Inform.* 1974, 4: pp 1–9.
5. **Gaither, A.**, A topology model for numerical grid generation, Weatherill, N., Eiseman, P. R., Hauser, J., Thompson, J. F., (Eds.), *Proceedings of the 4th International Conference on Numerical Grid Generation and Related Fields,* Swansea, Pineridge Press, 1994.
6. **Gaither, A.**, An efficient block detection algorithm for structured grid generation, Soni, B. K., Thompson, J. F., Hauser, J., Eiseman, P. R., (Eds.), *Numerical Grid Generation in Computational Field Simulations,* Vol. 1, 1996.
7. **Guibas, L. J. and Sedgewick, R.**, A diochromatatic framework for balanced trees, *Proceedings of the 19th Annual Symposium on Foundations of Computer Science,* IEEE Computer Society, 1978, pp. 8–21.
8. **Guibas, L. J. and Stolfi, J.**, Primitives for the manipulation of general subdivisions and the computation of Voronoï diagrams, *ACM Transaction on Graphics,* April 1985, 4(2).
9. **Kallinderis, Y.**, Prismatic/tetrahedral grid generation for complex geometries, *Computational Fluid Dynamics,* Lecture Series 1996-06. von Karman Institute for Fluid Dynamics, Belgium, March 1996.
10. **Knuth, D. E.**, *The Art of Computer Programming.* Vol. 1, *Fundamental Algorithms of Addison-Wesley Series in Computer Science and Information Processing.* Addison–Wesley, 2nd ed., 1973.
11. **Knuth, D. E.**, *The Art of Computer Programming,* Vol. 3, *Sorting and Searching of Addison–Wesley Series in Computer Science and Information Processing.* Addison-Wesley, 1973.
12. **Lohner, R.**, Generation of three dimensional unstructured grids by advancing front method, AIAA Paper 88-0515, 1988.
13. **Ma, K. L., Leutenegger, S., and Mavriplis, D.**, Interactive exploration of large 3D unstructured-grid data, Technical Report 96-63, ICASE, 1996.
14. **Morgan, K., Peraire, J., and Peirò, J.**, Unstructured grid methods for compressible flows, *Special Course on Unstructured Grid Methods for Advection Dominated Flows,* 1992, AGARD-R-787.
15. **Ohsawa, Y. and Sakauchi, M.**, The BD-tree, a new n-dimensional data structure with highly efficient dynamic characteristics, Mason, R.E.A., (Ed.), *Information Processing 83,* North-Holland, Amsterdam, 1983, pp. 539–544.
16. **Overmars, M. H. and van Leeuwev, J.**, Dynamic multi-dimensional data structures based on quad- and k-d trees, *Acta Informatica,* 17(3), pp. 267–285, 1982.
17. **Preparata F. P. and Shamos, M. I.**, *Computational Geometry: An Introduction,* Springer–Verlag, 1985.
18. **Rivara, M. C.**, Design and data structures of fully adaptive, multigrid, finite-element software, *ACM Trans. Math. Soft.,* 10, 1984.

19. **Samet, H.**, The quadtree and related hierarchical data structures, *Computing Surveys*, 1984, 16, pp. 188–260.
20. **Samet, H.**, *The Design and Analysis of Spatial Data Structures*, Addison–Wesley, 1990.
21. **Schroeder, W. J. and Shephard, M. S.**, A combined octree/Delaunay method for fully automatic 3D mesh generation, *Int. Journal on Numerical Methods in Eng.*, 29, pp. 37–55, 1990.
22. **Wirth, N.**, *Algorithm + Data Structures = Programs*, Prentice-Hall, Englewood Cliffs, NJ, 1976.
23. **Yerry, M. A. and Shephard, M. S.**, A modified quadtree approach to finite element mesh generation, *IEEE Computer Graphics and Applications*, 3, pp. 39–46, January/February 1983.

15

Automatic Grid Generation Using Spatially Based Trees

Mark S. Shephard

Hugues L. de Cougny

Robert M. O'Bara

Mark W. Beall

15.1 Introduction

This chapter examines the use of spatially based trees defined by recursive subdivision methods in the automatic generation of numerical analysis grids. The application of recursive subdivision over a spatial domain begins with a regular shape that is subdivided, in some regular manner, into a number of similarly shaped pieces, to be referred to as tree cells. The subdivision process is recursively applied until the smallest individual cells satisfy a given criteria. This subdivision process leads naturally to the definition of a spatially based tree structure where the root node of the tree corresponds to the starting regular shape, and the nodes of the tree defined by its recursive subdivision correspond to a specific portion of the spatial domain. The terminal nodes represent the smallest cell defined for that portion of the domain.

Recursive subdivision provides a natural means to decompose a geometric domain into a set of terminal cells that can be related to the grids or elements used in a numerical analysis. The associated tree structure provides an effective means for supporting various operations common to grid generation and numerical analysis, including determining the cell covering a particular location in space and determining neighbors. If the shape of the geometric domain of the analysis corresponded directly to the regular shape of the root node, the process of automatic grid generation using recursive subdivision would be trivial. Since

0-8493-2687-7/99/$0.00+$.50

the geometric domain of the analysis typically has a complex shape, specific consideration must be given to the interaction of the cells of the tree and the geometric domain of the analysis. Alternative methods for determining and representing those interactions have been devised for use in automatic grid generation. The method selected strongly influences all aspects of the grid generation process.

Determining the interactions of the cells of the tree with the analysis geometry and the decomposition of the cells into elements represents the most complex aspect of automatic grid generation using spatially based trees. In those cases where the tree cells are directly allowed to represent whatever portion of the analysis geometry included within them, the grid generation process is straightforward. The only technical issues relate to indicating the appropriate information to the analysis procedure for those cells containing some portion of the boundary of the domain on their interior. In those cases, where the elements defined in the tree cells have to conform to the geometry, the creation of elements in cells containing portions of the boundary of the domain is far more complex. In the worst case, the element creation procedures used in those cells represent complete automatic mesh generation procedures.

Section 15.2 outlines spatial subdivision techniques and associated trees that have been used in automatic mesh generation. Section 15.3 describes the basic issues that must be addressed in the use of spatial subdivision in automatic grid generation. Section 15.4 presents the techniques used in conjunction with automatic grid generation to construct the spatially based tree. Section 15.5 discusses the issues and approaches used to create elements within the cells of the tree. Finally, Section 15.6 indicates procedures that can be applied to improve the mesh after the basic mesh has been constructed.

15.2 Recursive Domain Subdivision to Define Spatially Based Trees

The application of recursive subdivision of a domain into subdomains, and the definition of an associated tree structure, has a long history (see Samet for a review of the area [3]) in a number of application areas including computer graphics, image processing, and computational geometry [9,10] (grid generation can be considered a computational geometry application). There are a variety of means in which the domains can be subdivided and the associated trees defined. For purposes of this discussion, emphasis will be placed on the quadtree structures for two-dimensional domains, and octree structures for three-dimensional domains, which have been most commonly used in grid generation (see also Section 3 of Chapter 14).

Considering the two-dimensional case, the first step in the generation of a quadtree for a given object is the definition of a rectangular-piped, typically a square, which covers the domain of the object. The rectangle is then subdivided into the four quadrants defined by bisecting each of the sides of the rectangle. Each quadrant is then examined to determine if it is to be subdivided based on given subdivision criteria. If they are to be subdivided, the process of creating the four quadrants for that rectangle is repeated. The process continues until the subdivision criteria are satisfied throughout the domain.

The process naturally defines a tree structure where the nodes in the tree correspond to rectangles at a particular point in the process. The tree is referred to as a quadtree, since four children are defined each time a node is subdivided an additional level in the tree. The original rectangle that encloses the object defines the root of the tree. The four quadrants defined by the subdivision of the root define the next level. These quadrants are each tested against the subdivision criteria. If they pass the criteria, they are marked as terminal quadrants. Any quadrant that does not pass the criteria is subdivided into its four quadrants, which form level two of the tree. This process is continued until all quadrants satisfy the given criteria, or the maximum tree level is reached. An octree for a three-dimensional object is defined in the same way, with the only difference being that the rectangular hexahedron, typically a cube, is subdivided into its eight octants such that each parent node in the domain has eight children.

A common cell (quadrant or octant) subdivision criteria used by many applications of spatial quadtrees and octrees is to refine the cell if it contains any of the boundary of the object, that is if the cell is neither fully within a single material region or exterior to the model. Figure 15.1 demonstrates the generation

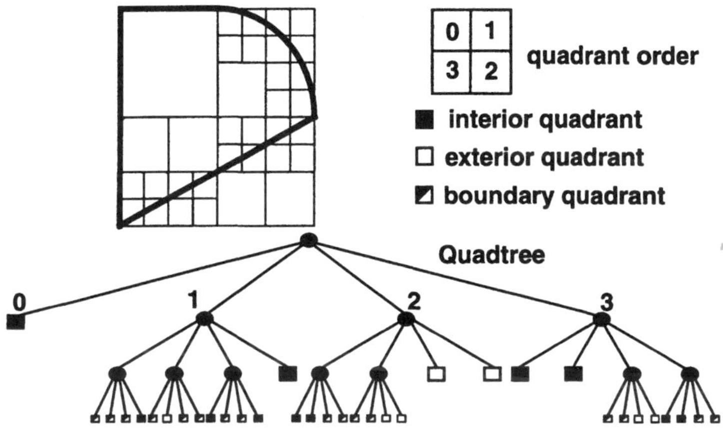

FIGURE 15.1 Quadtree example.

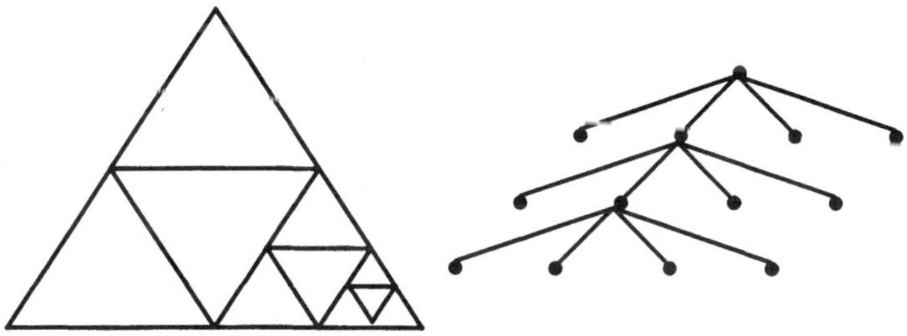

FIGURE 15.2 Quadtree defined by the subdivision of a triangle.

of a four-level quadtree for a simple two-dimensional domain bounded by three line segments and a circular arc. The object and tree quadrants resulting from three levels of subdivision of quadrants containing portions of the boundary are shown in the upper portion of Figure 15.1. The bottom portion of the figure shows the resulting tree structure. By the subdivision criteria used in this example, each parent quadrant contains a portion of the boundary and is neither fully inside or outside the object. The terminal quadrants are marked are either interior, exterior, or boundary depending on their relationship to the geometric domain.

There are alternative spatial decompositions and associated storage structures. One possibility is to consider the recursive subdivision of cells with alternative shapes. For example, in two dimensions, the root cell could be a single equilateral triangle and its four children defined by the bisection of the three sides forming four similar triangles, as shown in Figure 15.2. The extension of this procedure to three-dimensional simplices is not straightforward since the subdivision of a tetrahedron does not yield a set of similar tetrahedra (a regular tetrahedron does not close pack).

An alternative possibility to construct a spatially based structure is to consider anisotropic refinement of cells in which cells are only bisected in selected directions. Such subdivision processes do require the introduction of alternative structures for their definition. One example is a switching function representation [27,28] in which subdivision of cells can be limited to whichever coordinate direction is desired. Figure 15.3 shows the application of a switching function representation to the simple two-dimensional domain used earlier.

15.3 Quadtrees and Octrees for Automatic Mesh Generation

Octree and quadtree structures have been used to support the development of two- and three-dimensional mesh generators for a number of years [1,2,3,11,12,16–18,20,22,24,26,31,32]. Although each of the

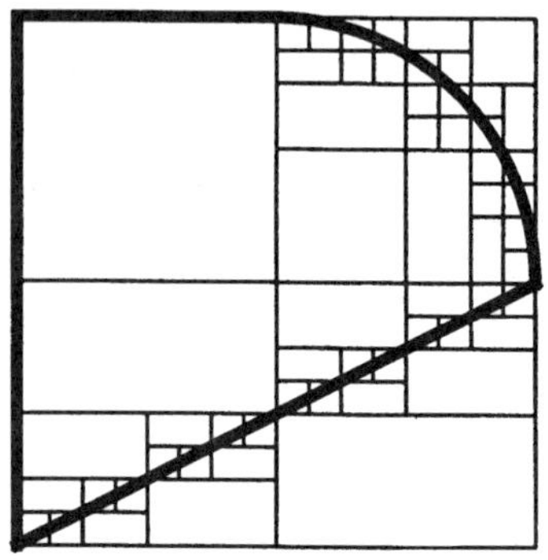

FIGURE 15.3 Switching function representation of a two-dimensional example.

quadtree- and octree-based mesh generators are different, there are specific basic aspects common to all the procedures: the mesh generation process is implemented as a two-step discretization process. The quadtree or octree is generated in the first step. The tree is then used to localize many of the element-generation processes, which constitute the second step. Those cells (quadrants or octants) containing portions of the object's boundary receive specific consideration to deal with the boundary of the object. The corners of the cells are used as nodes in the mesh. In specific procedures, additional nodes are defined by the interaction of the boundary of the object being meshed with the cells' boundaries. The mesh gradation is controlled by varying the level of the cells within the tree through the domain occupied by the object.

The specific algorithmic steps used within a quadtree- or octree-based mesh generator depend strongly on the assumptions made with respect to the representation of the boundary of the model, and on the form of interaction between the boundary of the model and a tree cell that is represented. Before discussing the alternative tree construction and element creation algorithms, these basic options for the representation of the model boundary and its interaction with the tree are discussed.

In general, the geometric domains to be meshed are curvilinear models defined within a geometric modeling system. The tree-based mesh generation procedures can attempt to interact directly with this curvilinear geometry, or require a polygonal approximation. The use of a polygonal approximation greatly simplifies the determination of the interactions of the geometric model with the tree cells. The polygonal approximation may be constructed through a process which is independent of the mesh generation process, or it may be the boundary triangulation that defines the surface mesh. These two polygonal forms are typically handled differently.

Factors that enter into the selection of the approach to account for the interactions of the model boundary with the boundary of the tree cells include (1) level of geometric approximation desired, (2) sensitivity of the element creation procedures to small features created by the model and cell boundary interactions, (3) importance of maintaining spatial associativity of the resulting tree cells. Figure 15.4 demonstrates three basic options (columns) for representing the interactions of the tree and model boundary (top row) and the potential influence on the resulting mesh (bottom row). The first option (left column) employs exact interactions of the model and tree as defined by the intersections of the model and cell boundaries. This option maintains the spatial associativity of the tree cells*, and does not introduce any geometric approximations. However, under the normal assumption in mesh generation that the trees cells are on the order of the size of the desired elements, this approach has the disadvantage of producing disproportionally small and distorted elements (see mesh in lower-left corner of Figure 15.4) when the model and cell interactions leave small portions of a cell in the model.

*Spatial associativity of the tree cells is maintained when the cells remain undistorted. When spatial associativity is maintained, the appropriate tree cell can be obtained by traversing down from the root using the coordinates of a point.

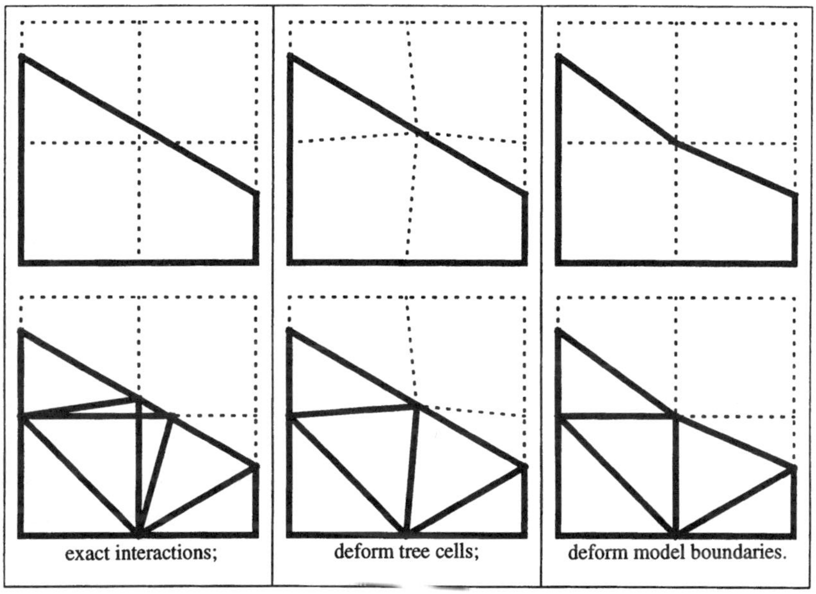

FIGURE 15.4 Options for the interactions of the model boundary with the boundary of the tree cells.

The other two options eliminate the influence on the mesh of these small portions of the cells by either distorting the necessary cells (center column, Figure 15.4), or distorting the geometry of the appropriate model entities (right column, Figure 15.4). Both approaches require the development of specific logic to determine when and how to perform the needed distortion. If tree cells are deformed (center column, Figure 15.4) they no longer can employ operations that rely on spatial associativity, while the deformation of the model can introduce undesirable geometric approximations into the process.

There are a variety of options available to create elements once the tree has been defined and the cells qualified with respect to the model. The tree cells that are interior to the domain of the object are typically meshed quickly employing procedures that take specific advantage of the simplicity of the cell's topology and shape, and use knowledge of the tree structure to determine the influence of neighboring cells on the mesh within the cell of interest.

Meshing of the cells that contain portions of the boundary of the object is a more complex process with the details being strongly influenced by the representation of the boundary of the model being used and the method used to represent the tree's boundary cells (see Section 15.5).

The input information required to generate the tree structure for use in mesh generation is the geometric model and information on the size of elements desired throughout the domain. For purposes of this discussion, no specific representation of the geometric domain is assumed. Instead, it is assumed that it exists and there is support for the interrogations of that representation needed to obtain the information required for the various operations performed during the tree construction and meshing processes. This approach allows a more uniform presentation of the tree building and element creation procedures, and provides a generalized method to link the meshing procedures to the domain geometry in a consistent manner [21,23].

In this discussion it is explicitly assumed that the size of the terminal cells throughout the domain of the geometric model is on the order of the element sizes required. Therefore, the information on desired element sizes will define the sizes of the terminal cells in the tree. Any spatially based mesh control functions can be easily represented using such an approach.

TABLE 15.1 Topological Entities for the Three Models

Model	Geometric	Octree	Mesh
Regions	G_i^3	O_i^3	M_i^3
Faces	G_i^2	O_i^2	M_i^2
Edges	G_i^1	O_i^1	M_i^1
Vertices	G_i^0	O_i^0	M_i^0

15.4 Tree Construction for Automatic Mesh Generation

15.4.1 Preliminaries

It is convenient to view the process of octree- and quadtree-based mesh generation as one of discretizing the geometric model into a model defined by the cells of the tree, and then discretizing the cells of the tree into the mesh model. Both of these steps require interactions with the geometric model. Irrespective of the algorithmic details used to carry out these steps, the key issue in ensuring the resulting mesh is valid is understanding the relationship of the mesh to the geometric model [19,23]. At the most basic level, the relationships between the models can be described in terms of the association of the topological entities defining the boundaries of the various model entities. In three dimensions the primary topological entities are regions, faces, edges and vertices which will be denoted for the geometric, octree, and mesh models, as indicated in Table 15.1.

To support the mesh generation requirements for the entire range of engineering analyses, the models must be non-manifold models [8,29], in which the entities and their adjacencies, in terms of which entities bound each other, are defined for general combinations of regions, faces, edges, and vertices.

The association of topological entities of the mesh with respect to the geometric model is referred to as classification, in which the mesh topological entities are classified with respect to the geometric model topological entities upon which they lie.

Definition: Classification *— The unique association of mesh topological entities of dimension* d_i, $M_i^{d_i}$ *to the topological entity of the geometric model of dimension* d_j, $G_j^{d_j}$ *where* $d_i \leq d_j$, *on which it lies is termed classification and is denoted* $M_i^{d_i} \sqsubset G_j^{d_j}$ *where the classification symbol,* $\sqsubset$, *indicates that the left-hand entity, or set, is classified on the right-hand entity.*

In specific implementations it is possible to employ the classification of the mesh entities against octree entities. Octree entities can cover portions of more than one model entity; therefore the use of classification of octant entities against model is not possible for all octant entities. However, understanding the relationship of the octant to the model is important to track during the tree and mesh construction processes. One device used to aid in the process of understanding the relationship of the closure* of the octant, $\overline{O}_j^3$, with respect to the geometric model is to assign each octant a type. The four octant types indicate if $\overline{O}_j^3$ is inside the geometric model region, $T(O_j^3) = in(G_i^3)$ where G_i^3 is the model region the octant and all its bounding entities are classified within, outside the domain, $T(O_j^3) = out$, contains a portion of the model boundary, $T(O_j^3) = bdry$, or its status is not yet determined $T(O_j^3) = unk$.

*The closure of an octant includes the octant, its 6 faces, 12 edges, and 8 vertices. Although it possibly can define an octant's relationship either with respect to the entity or its closure, the specific choice made influences the details of the various algorithms that carry out algorithmic steps based on the octant status.

15.4.2 Mesh Control and Octant Sizes

Since the edges of the terminal octants will become the edges of the elements in the grid, the size of the octants is dictated by the mesh control information applied. For a given root octant, the size of a terminal octant is controlled by its level in the octree; therefore, the sizes of the elements are controlled by specifying octant root size and levels throughout the object being meshed. Since the octree is, at least initially, spatially addressable, any mesh control function that can indicate the element size in a particular location in space can be used.

Although general functions to define element sizes as a function of position have application, alternative methods to specify mesh control tend to be easier to use. For *a priori* mesh size specification, users of automatic mesh generators find it advantageous to associate mesh size parameters with the topological entities of the model. For example, to indicate the maximum element size associated with an edge, vertex, face, or region. Users also like to be able to control the mesh size based on the local curvature of the model faces. *A posteriori* mesh size specification as defined by an adaptive procedure, which typically associates a desired element size with elements in the mesh of the previous steps. In an octree mesh generator, there is some advantage to associating this information directly with the octree octants to define the level variation.

In most octree mesh generators, the final octant size at a location is equal to or less than that indicated by the mesh control parameters. The octant size at a location can be forced to be less than requested by the mesh control parameters when the octant is subdivided to satisfy the commonly applied one-level difference rule. The one-level difference rule [31] (also known as the 2:1 rule [10]) is commonly used in octree-based meshing procedures to control mesh gradations and element aspect ratios. This rule forces octants that share an edge to have no more than a one-level difference. (This forces the maximum difference for octants that share only a corner to two levels.)

15.4.3 Definition of Octree

The first step in the construction of the octree is to define the size and position of the root octant, O_1^3, typically referred to as the universe. The object must be contained within the closure of the universe. If the domain has a polygonal representation, the minimum limits of the root can be easily defined in terms of the extreme coordinate components of the model vertices. However, if the model is curved, the extreme coordinate values have to be determined using more complex algorithms, which typically have some known degree of approximation error. In these cases, the conservative approach is to expand the coordinates defining the universe by some amount greater than the possible approximation error to ensure $\overline{O_1^3} \cup \overline{G} = \overline{G}$, where $\overline{G}$ is the closure of the model. Note that $T(O_1^3) = bdry$.

A number of alternative approaches have been proposed to decompose the root octant into the final octants that will be meshed [10,12,18,22]. Most of these rely on a recursive subdivision of a given parent octant into its children until the children are of the desired size as defined by the local mesh control information. Given a function that indicates the smallest element size desired within an octant, it is a simple process to examine the size of the current octant, and to subdivide it if it has not yet been refined to a sufficient level. The more critical issues of octant refinement are associated with determining, and representing, the interactions of octants with the portion of the geometric domain that are fully or partly contained within it, particularly in the case when these operations are performed directly with respect to the solid model representation.

The minimum information requirement during octant refinement is the octant type for each child. Since this understanding is gained by qualifying the interactions of the children octants with the model entities that interacted with the child's parent, the process of octree creation focuses on the most effective means to determine these interactions. In the case when the mesh control parameters are associated with the model's topological entities, determining which model entities interact with the octant is central to determining if a given octant is to be subdivided further. Octants can also be forced to subdivide simply due to the complexity of the portion of the geometric model within them because of limitations of specific octant triangulation procedures used to handle that level of complexity.

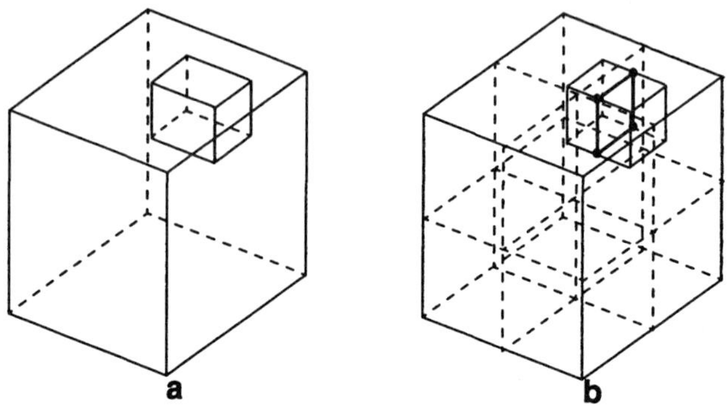

FIGURE 15.5 Octant subdivision and determination of model/octant interactions.

If an octant to be subdivided is inside, $S(O^3_j) = in(G^i_3)$, or outside, $S(O^3_j) = out$, each of the eight children receives the same octant type. The octants that contain portions of the model boundary, $S(O^3_j) = bdry$, or possibly contain portions of the model boundary, $S(O^3_j) = unk$, require execution of geometric operations to determine which of those entities are associated with each child octant, so that the octant's type can be properly set and the proper model entities associated with the children octants. In general the determination of the interactions of the model entities with an octant requires performance of intersections of octant boundary entities with model boundary entities, as well as operations to determine when model entities are entirely contained within an octant. Since these intersection operations can dominate the cost of an octree meshing process, their effective execution to determine the octant type and the specific intersection information needed for further tree refinement and later creation of elements is critical. The reader is referred to Kela [10] for details of a complete and effective procedure for this process.

As a demonstration of the type of operations that would be performed when the full set of interactions between the octant and model entities are desired, consider Figure 15.5a, which shows an octant with a rectangular prism in the upper rear portion of the octant. The model vertices, edges, and faces of this simple model are entirely inside the root octant. Key to determining the relationship of the model with the eight children created by subdivision of the root octant is determining the interactions of the three bisection planes shown in Figure 15.5b. The basic intersection operations performed to determine these interactions are the intersection of the model edges fully or partly within the parent octant with the three planes, and the intersections of the edges of the planes and the edges defined by the intersections of the three planes, with the model faces contained fully or partly in the original octant. In the particular example shown in Figure 15.5a, the result of these operations determines four intersections of the edges of the model with one bisection of the planes. The resulting intersection vertices, shown as darkened vertices in Figure 15.5b, are used in an edge and loop building algorithm to create the darkened edges that complete the qualification of the model information in the children octants and are used as edges and vertices in the finite element mesh. The result of subdividing the original boundary octant yields six children octants that are outside, and two that are boundary octants. Note that only performing intersections with the bisection planes is not sufficient to properly qualify the children octants in all cases. Information on portions of model entities associated with the original octant that do not interact with the bisection planes has to be transferred to the appropriate children. Information on the interactions of the model entities with octant entities, and model entity bounding boxes, allows this information to be determined quickly in most cases. When the results of these operations are inconclusive, more costly geometric operations are required [10].

In some octree-based mesh generators, the interaction that can be represented between the octant and model is more limited. For example, a procedure may allow interaction which can be adequately approximated by the diagonals between octant corners with only one model face cut per octant. If the model complexity at the requested octant level is too great to be properly approximated in the prescribed manner, the octant must be subdivided further until the number and complexity of model entities within the octant can be represented. This process does introduce refinement past what was requested. In addition, it is always possible to devise situations, particularly on nonmanifold models, where the topological complexity at the boundary of a particular model entity is such that no level of refinement will allow a topologically correct approximation of the situation when there are preset limits on the model topological complexity allowed within an octant.

15.4.4 Information Stored in the Tree

As the octree representation for a geometric domain is constructed, information about the interactions of the geometric model and the octants is associated with the octant in preparation for the creation of the elements in the next step. The amount of information stored is a strong function of the type of model/octree interaction information used to create the elements inside the octants.

Once an octant has been given the type *outside*, no additional information need be stored with it. In the case of octants inside a model region, the basic information stored with the octant is a pointer to the model region it is inside of, and information on the local element sizes, or at least the means to obtain that information through the region pointer.

Boundary octants carry additional information which aids in qualifying the interactions of the octant with the boundary of the domain. The specific model information stored is a function of what is needed to control octant subdivision and by the element creation procedures. In the simplest of cases where the analysis procedure will use the entire octant geometry and only account for a volume fraction correction, the information can be limited to a knowledge of the model boundary entities interacting with the octant, as is sufficient to calculate the volume fraction and control further octant subdivision.

Since there are no *a priori* limits on the number of model entities interacting with an octant, general octree mesh generation algorithms employ a more complete representation of the interactions of the model and the octant. The approach used to do this employs a localized boundary representation consisting of the entities defined by the intersection of the model and octant entities. As octants are subdivided, the octant level topological information is updated to indicate the information that is associated with the children octants and the new entities created by the intersection of the model entities with the new octant entities.

As a more explicit example of the information that may be stored in an octant [22], consider the boundary octant shown in Figure 15.6, where most of the octant is interior to a model region and one corner is exterior to the domain due to a reentrant corner in the geometric model. Since the octant level information stored will be used to drive the octant meshing process, the specific entities defined at the octant level will consist of mesh vertices, mesh edges, and octant level loops which are classified against the original model. Figure 15.6 shows the visible mesh vertices and mesh edges for our example.

Visible mesh vertices M_i^0 through M_6^0 are classified on octant vertices, $M_i^0 \sqsubset O_j^0$ and interior to a model region, $M_i^0 \sqsubset G_k^3$. M_8^0, M_{10}^0, and M_{12}^0 are classified on octant edges, $M_i^0 \sqsubset O_k^1$, and model faces, $M_i^0 \sqsubset G_k^2$. M_7^0, M_9^0 and M_{11}^0 are classified on octant faces, $M_i^0 \sqsubset O_k^2$, and model edges, $M_i^0 \sqsubset G_k^1$. M_{13}^0 is classified in the octant interior, $M_{13}^0 \sqsubset O_k^3$, and a model vertex $M_{13}^0 \sqsubset G_k^0$. The one invisible mesh vertex is classified on an octant vertex, $M_i^0 \sqsubset O_j^0$, and interior to a model region, $M_i^0 \sqsubset G_k^3$.

Visible mesh edges M_1^1 through M_6^1 are classified on octant edges, which they span, and interior to a model region $M_i^1 \sqsubset G_k^3$. M_7^1 through M_9^1 are classified on the octant edges, which they partly span, and interior to a model region $M_i^1 \sqsubset G_k^3$. M_{10}^1 through M_{15}^1 are classified on octant faces, and on model face $M_i^1 \sqsubset G_k^2$. M_{16}^1 through M_{18}^1 are classified in an octant region, and on model edge $M_i^1 \sqsubset G_k^1$. There are three invisible mesh edges which are classified on octant edges, which they span, and interior to a model region $M_i^1 \sqsubset G_k^3$.

There are six visible loops of mesh edges in the example octant. The mesh edge loops $M_4^1 - M_5^1 - M_9^1 - M_{14}^1 - M_{15}^1 - M_5^1$, $M_1^1 - M_7^1 - M_{10}^1 - M_{11}^1 - M_8^1 - M_2^1$ and $M_3^1 - M_8^1 - M_{12}^1 - M_{13} - M_9 - M_6$ are classified on octant faces, $M_i^0 \sqsubset O_k^2$, with four of the edges interior to a model region $M_i^1 \sqsubset G_k^3$, and two on model faces. The mesh edge loops $M_{14}^1 - M_{13}^1 - M_{18}^1 - M_{16}^1$, $M_{11}^1 - M_{17}^1 - M_{18}^1 - M_{12}^1$ and

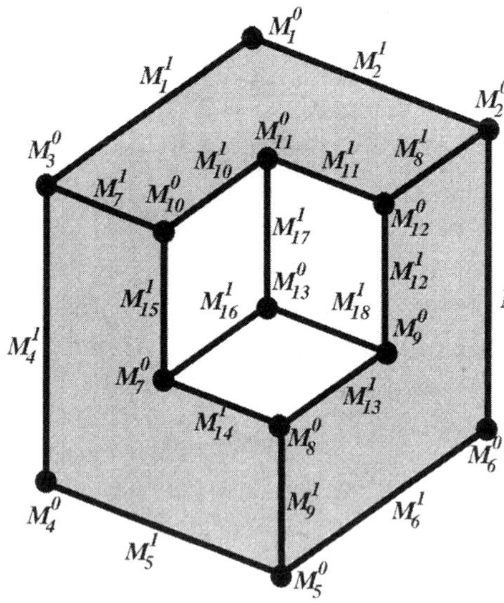

FIGURE 15.6 Information stored at the octant level.

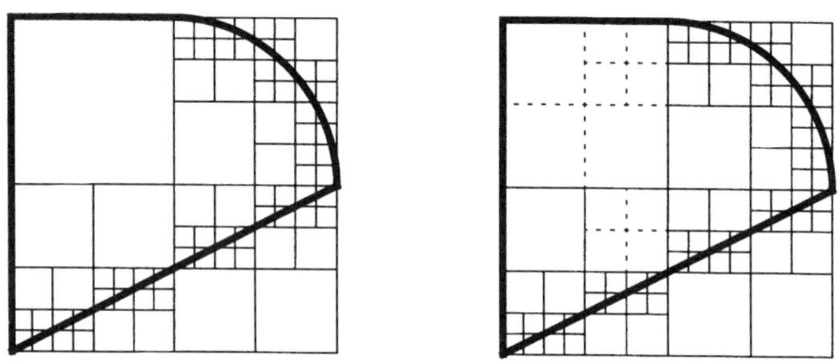

FIGURE 15.7 Quadtree example before (left) and after (right) one-level difference enforcement.

$M_{10}^1 - M_{15}^1 - M_{16}^1 - M_{17}^1$ are classified in the octant interior, $M_{13}^0 \sqsubset O_k^3$, and on model faces, $M_i^1 \sqsubset G_k^2$. The three invisible loops each have four mesh edges that correspond to the four octant edges that bound the octant face. They are classified interior to a model region $M_i^1 \sqsubset G_k^3$.

As a last step before generating the mesh within the octants, most octree-based mesh generators will enforce a one-level difference between octants sharing edges and neighbors. This process helps control element gradations and shapes, and makes the meshing of interior octants easier. Figure 15.7 demonstrates this for a two-dimensional quadtree case. The left image shows a tree before the application of a one-level difference operation, while the right image shows the tree with the additional quadrant refinements (dashed lines) required for one-level difference between edge neighbors. The determination of the tree cells needing refinement is easily determined using tree traversal [31]. It should be noted that when this process forces boundary cells to be refined, the process of determining the appropriate boundary interaction must be carried out with respect to the refined cells.

15.5 Mesh Generation Within the Tree Cells

15.5.1 Meshing Interior Cells

It is common to take specific advantage of the simple geometric shape of the interior cells when creating the elements within those cells. In some cases, the interior octants are treated as individual hexahedral elements. If the tree level through the domain is uniform, the use of one hexahedron per interior octant is possible without further consideration. In the case where there are level differences between neighboring octants, it becomes necessary to account for the fact that the faces of neighboring hexahedra across level differences will not be conforming. For example, in the case of a one-level difference, the one face of the hexahedron will be covered by four quadrilateral faces of the lower level neighbors. These situations can be addressed by the imposition of appropriate multipoint constraint equations. The tree structure can be effectively used to determine the neighboring information needed to construct these constraints.

It is possible to construct conforming meshes that will account for the level differences when tetrahedral elements are used. Again, the tree structure is used to determine the required neighboring information. In some implementations, template structures have been devised to mesh most or all of the internal octants. The simple six-pyramid procedure [31] is easy to implement, but yields more than the desired number of elements in the cases of level differences. More elaborate schemes that maintain the minimum number of elements are possible [18].

Template procedures for interior octants which produce conforming Delaunay meshes have also been developed [18]. By using a slightly reduced circumsphere concept the Delaunay triangulation for an octant, which has all eight vertices on the same circumsphere, becomes uniquely defined by the order in which points are inserted during octant Delaunay point insertion. Combining the ability to control the triangulation, by the order of point insertion, coupled with the knowledge of the octants neighbors available from the tree structure, allows the automatic construction of octant template codes for the interior octants. These procedures can account for neighbors with a level difference. Note that interior octants neighboring boundary octants with non-corner mesh vertices near the interior octants will require the overriding of the template defining the interior octant triangulations to regain a globally Delaunay triangulation. The triangulation process in this case must consider information from neighboring octants.

15.5.2 Meshing Boundary Cells

The process of meshing the boundary cells is a strong function of the level of geometric complexity supported by the mesh generator. In cases where there is only a limited amount of geometric complexity allowed per octant, simple templates are possible. When there is no specific limitation on the level of geometric complexity allowed within the octant, the process of meshing the boundary octant requires all the functionality of an automatic mesh generator applied to the local region [12,19,22].

To demonstrate the issues and options associated with meshing boundary octants, the basics of four different approaches will be considered for the creation of elements in the boundary octants. The first two create tetrahedral elements assuming that the surface has not been pre-triangulated. The first of these approaches applies an element removal procedure starting from a basic octant level boundary representation as outlined in the previous section. The second approach develops a Delaunay triangulation based on the mesh vertices of the octant level boundary representation, which is then followed by an assurance algorithm that insures the resulting surface triangulation is topologically compatible and geometrically similar. Since the first two procedures operate strictly accounting for the intersections of the model and octant boundary entities, they are susceptible to the small, poorly shaped elements caused by boundary octants nicking the model boundary.

The third procedure creates tetrahedral elements from a given surface triangulation using an element removal procedure. The last boundary octant meshing procedure considers the creation of hexahedral elements to fill the region between the interior octants and the model boundary. These two procedures create the elements in the regions between the model boundary and interior octants without strict adherence to the boundary octant's boundary. Therefore, they are not susceptible to the creation of poorly shaped elements caused by the boundary octants nicking the model boundary.

15.5.2.1 Element Removal to Mesh Boundary Octants

One approach to generate meshes in the boundary octants is to apply a general set of element removal operations to the local octant boundary representation developed during the octree creation process. In this approach the only interaction with neighboring octants which must be taken into account is to copy the surface triangulations of any common neighboring interior or boundary octant's faces that already have been triangulated.

The most general procedure for the creation of elements in the boundary octants is to apply the three-element removal operators of vertex removal, edge removal, and face removal [22,30], working from the boundary representation defined in terms of octant face loops. These removal operators are capable of creating the surface triangulation on octant face loops that have not yet been triangulated, while matching existing triangulation for those that have been previously triangulated. Preference is given to the application of the vertex removal and edge removal operations since they do not create any new mesh vertices. However, situations can arise where face removal must be applied.

To demonstrate the application of element removal on the boundary octant, the process of meshing the boundary octant of Figure 15.6 with the octant faces already triangulated (Figure 15.8, upper-left image) is considered. The first three tetrahedral elements are created by the removal of mesh vertices M_{10}^0, M_8^0, and M_{12}^0. The upper-right image of Figure 15.8 shows the octant after the three vertex removals. The next three elements are created by edge removal of edges $M_7^0 - M_{11}^0$, $M_{11}^0 - M_9^0$, and $M_9^0 - M_7^0$. The lower-left image of Figure 15.8 shows the octant after the application of the three edge removals. The next six elements are created by the application of three edge removals and three vertex removals. For example, the three edge removals could be $M_7^0 - M_3^0$, $M_5^0 - M_9^0$, and $M_2^0 - M_{11}^0$. This process creates edges $M_4^0 - M_{13}^0$, $M_6^0 - M_{13}^0$, and $M_1^0 - M_{13}^0$, thus allowing the application of vertex removal at vertices M_7^0, M_9^0, and M_{11}^0. The lower-right image of Figure 15.8 shows the octant after the removal of these six elements. The last six elements are created by one edge removal and five vertex removals. For example, if edge $M_4^0 - M_5^0$ is removed, edge $M_0^0 - M_{13}^0$ is created, thus allowing vertex removal at vertex M_5^0. The last four vertex removals are then applied to vertices M_6^0, M_2^0, M_1^0, and M_3^0 in order.

15.5.2.2 Delaunay Point Insertion to Mesh Boundary Octants

An alternative approach to meshing boundary octants has been used in an octree-Delaunay mesh generation procedure [18]. In this procedure each complete boundary octant is first meshed without consideration of the model boundary, using the same procedure that produces compatible triangulations for the interior octants. Assuming that the surface has not already been pre-triangulated, the remaining steps in meshing the boundary octant in this procedure include the following:

1. Insert the mesh vertices necessary to account for the interaction of the model boundary with the octant.
2. Perform topological compatibility and geometrically similarities of the octant level mesh edges and faces classified in the model's boundary to ensure a valid geometric triangulation of the octant [19,23].
3. Eliminate all tetrahedra exterior to the model.

The vertices inserted in the first step are defined by (1) model vertices within the octant, (2) the intersection of model edges with the octant faces, and (3) the intersection of the octant edges with the model faces. The creation of a globally Delaunay triangulation as these points are inserted requires consideration of the triangulation of, at a minimum, those octants the mesh vertex being inserted bounds. In addition, when the mesh vertices are close to other octants, their triangulation may also need to be considered during the vertex insertion process. Specific methods to know which octants must be considered have been developed [18].

A generalized topological compatibility and geometric similarity algorithm [19,23] must be applied after the points have been inserted. In some cases it is not possible with the given set of points to recover a valid geometric triangulation which satisfies the Delaunay empty circumsphere requirement. In these cases, additional points can be generated using octree subdivision or specific point insertion processes [14,18]. After a valid boundary triangulation has been constructed, it is a simple task to complete the boundary octant triangulation process by deleting those elements outside the domain of the object.

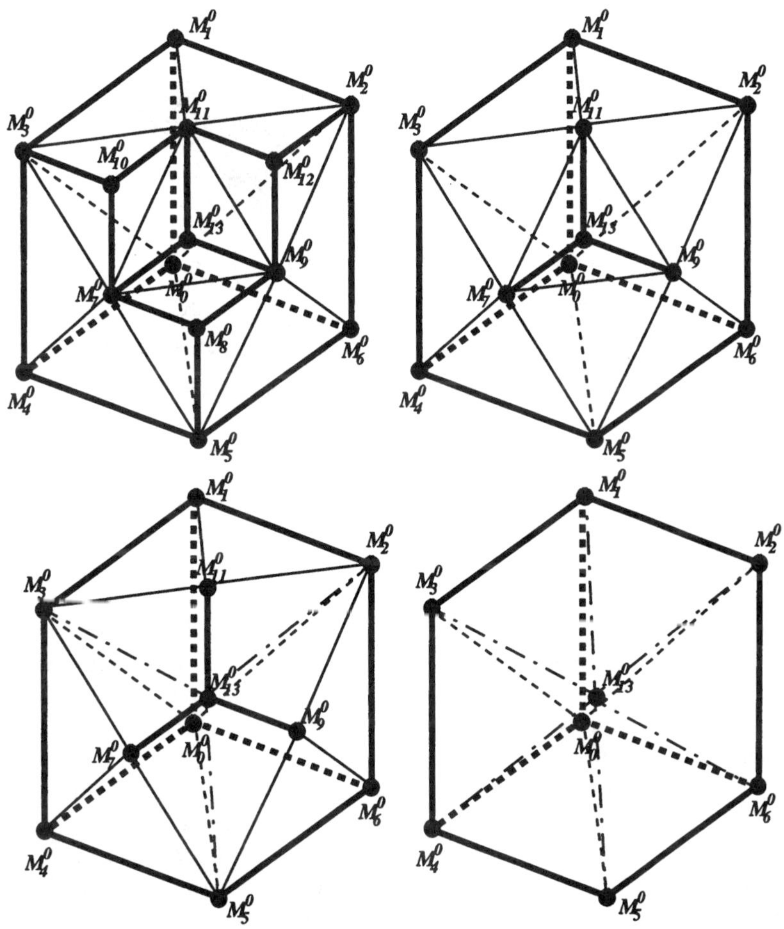

FIGURE 15.8 Mesh generation in a boundary octant by element removal.

15.5.2.3 Element Removal from a Pre-Triangulated Surface to Interior Octants

In this octree mesh generator the tetrahedral mesh is created from a pre-triangulated surface mesh [6]. The octree for this procedure is created such that the octants containing the surface triangles are sized to have edge lengths equivalent to that of the edges of the surface triangulation that is partly or completely interior to them. The interior octants are created such that they satisfy the one-level difference rule. To avoid the poorly shaped elements caused by close interaction of surface triangles and interior octants, the additional cell type of boundary-like interior cells is introduced. These are interior cells that are closer than some fraction of the surface triangle edge length to the surface triangulation. Using one half an edge length of the near-by surface triangle as the distance criterion works well for this purpose.

Figure 15.9 demonstrates the application of this process to a simple two-dimensional domain. The left image shows the set of domain boundary segments and the quadrants generated based on them. The image shows the boundary quadrants that contain portions of the boundary segments, the interior quadrants that are more than half an edge length from the boundary segments, and the boundary-like interior quadrants that are interior octants within one half an edge length of the boundary segments. The interior cells are meshed using templates and are indicated by the shaded triangles in the right image of Figure 15.9.

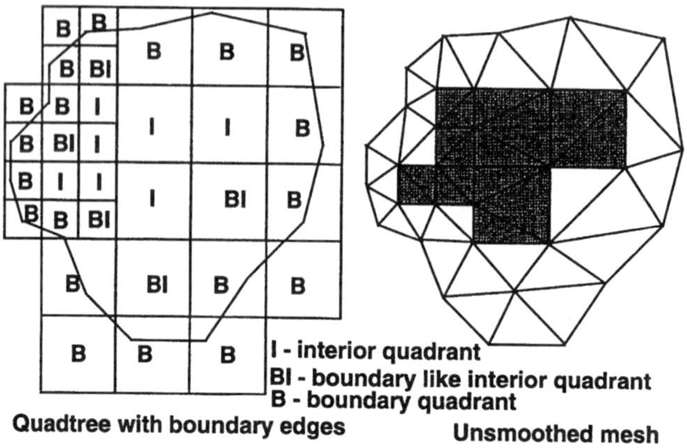

FIGURE 15.9 Mesh generation given a discretized boundary

After the interior octants are meshed, the remaining portion of the domain to be meshed is that region lying between the outer faces of the meshed interior octants and the surface triangulation. This region is meshed employing element removal operations in a manner similar to that used in the current advancing front mesh generators.

The description of the procedure to mesh the remaining portion of the domain focuses on the mesh faces defining the surface triangulation and those on the exterior of the interior octants. Since the completion of the meshing process requires connecting tetrahedra to one or, in multiregion problems, up to two sides of these faces, these faces are referred to as partly connected faces. The mesh generation process is complete when all mesh faces are fully connected.

The element creation process to connect partly-connected faces is not constrained to following cell boundaries. It is guided solely by the creation of elements to fill the region between partly connected faces. This is depicted in the right image of Figure 15.9 by the unshaded triangles created during this step. The tree structure is used during this process to efficiently locate neighbor information. Each partly connected face is associated with one or more octants, thus allowing the tree neighbor-finding procedure to be used to locate neighboring partly connected faces that a current face can be connected to.

Given a partly connected mesh face, the face removal consists of connecting it to a mesh vertex of a nearby partly connected face. Since the volume to be meshed consists of the region between the given surface triangulation and the interior octree, the vertex used is usually an existing one. In some situations it is desirable to create a new vertex. The choice of this vertex must be such that the created element is of good quality and its creation does not lead to poor (in terms of shape) subsequent face removals in that neighborhood.

Early element removal procedures had some difficulty in the process of determining the vertex to connect to, in that the criteria used emphasized the element being created with little consideration for the situation remaining for subsequent face removals. Consideration of the influence on subsequent face removals is a difficult process since one does not know about them until they arise. One possible solution is to make sure that any element creation does not make new mesh entities too close (relatively) to existing mesh entities. This process requires an exhaustive set of geometric checks against mesh entities in the neighborhood. Although it is possible to develop the appropriate set of checks, it is in general an expensive process since the number and complexity of checks required is quite high even when efficient procedures are used to provide a proper set of candidate mesh entities to consider. An alternative method is to use a more efficient criterion that indirectly accounts for the various situations that can arise. The Delaunay circumsphere criteria does provide a quality mesh when given a well-distributed set of points that avoids the creation of flat elements. The use of Delaunay criteria in general element removal mesh generation procedures has been shown to be an effective means to control this process.

One procedure [6] combines the use of the octree, Delaunay meshing criteria, and more exhaustive checks when a local Delaunay solution is not available. Starting with the mesh vertices closest to the face to be removed, the Delaunay circumsphere test is performed. Since the Delaunay mesh for a given set of points is unique to within the degeneracy of more than four points on the circumsphere, the first vertex which satisfies this criteria is used to create the element. If there are degeneracies, consideration must be given to the other points on the circumsphere to ensure a proper selection is made. If none of the candidate vertices satisfy the Delaunay criteria for that face, a more exhaustive checking procedure is undertaken which explicitly considers the shapes of the element created as well as shapes of future elements dictated by other nearby connections.

Face removals are performed in waves. A new partly connected mesh face resulting from some face removal is not processed until all other partly connected mesh faces existing at the beginning of that wave are processed. Also, partly connected mesh faces resulting from the meshing of interior terminal octants are never processed for face removals so long as other partly connected faces exist. This gives priority to partly connected mesh faces coming from the model boundary. The process of removing partly connected mesh faces ends when there are no more partly connected mesh faces.

15.5.2.4 Hexahedral Element Creation from the Interior Octants to the Model Boundary

A technique for the generation of hexahedral elements for the boundary octants has also been proposed [17]. The implementation discussed here requires the use of a uniform tree level throughout the domain. Octants more than one-half element length away from the model boundary are defined as interior octants and meshed with a single hexahedra. The region from those interior octants to the model surface is then meshed using a projection method.

The basic idea is to define one or more projection lines from each of the vertices on the outer surface of the interior octree to the outer surface of the model. The square faces on the outer surface of the interior octants and the projection lines are used to define hexahedra. The number and direction of projectors defined from a vertex on the surface of the interior octree depends on which of the eight octants the vertex bounds are interior octants.

In the case when the exterior of the object is a single smooth closed face, it is reasonably straightforward to use the default projectors and directions to define a set of hexahedra in the volume between the interior octants and boundary. The existence of model edges and vertices will force decisions to be made to alter the direction used for the projectors so that those edges and vertices are properly represented. In cases where they can be represented using the default numbers of projectors, it is possible to define elements that are topologically hexahedral in the volume from the interior octree to model surface. The problem that arises is that often some number of those hexahedra have unacceptable shapes in that some element angles are in the invalid range. Specific subdivision techniques can be used to produce a valid element at the cost of local increases in the number of elements. In addition to the geometric complexities introduced by the model edges and vertices, there can be configurations of edge and vertex interaction that will not always produce topological hexahedral polyhedra using the default projection edges.

15.6 Mesh Finalization Processes

Some analysis procedures take specific advantage of the regular shape, square in two dimensions and cube in three dimensions, of the tree cells. In other cases, particularly when confirming meshes of triangular and tetrahedral are generated, the analysis calculation does not require that the elements stay strictly aligned with the cell boundaries. In these cases, it is possible to apply procedures to improve the shapes and gradation of elements.

The most commonly, and easily, applied operation for such element shape improvements is to reposition node point locations. Although node point repositioning can lead to substantial improvements in element shape measures, there are many cases where the constraints of the neighboring elements are such that the element shape remains poor. The inclusion of various local mesh modifications can yield more dramatic improvements in the mesh. In addition, the application of a full set of mesh modification operators can be used to eliminate the adverse effects of the small elements caused when the model boundary and octant boundaries are close.

15.6.1 Node Point Repositioning

The application of node point repositioning within a quadtree or octree mesh generator can follow the normal process of iteratively repositioning one node at a time based on a specific repositioning criteria using the mesh connectivity information. It is also possible to use information based on the tree structure to define alternative connectivities.

Although any of the standard criteria for node point positioning can be applied, it is advisable to only apply criteria which ensures that the elements will remain valid. The application of Laplacian smoothing often yields good element shape improvements, but should only be applied in a constrained manner such that a node is allowed to move only if the shape of the worst shape element connected to it improves.

One approach that has worked well in an octree mesh generator is to employ a combination of two smoothing operators. The first operator applied is a constrained Laplacian operator where the standard average of all connected nodes is used as the target for the node point. If this location is found to yield improvement in the shape of the worst-shaped element connected to the node, the node is moved to that position. If that location does not yield improvement in the worst-shaped element, locations on the line from the current position to the centroid are checked and the first that yields improvement is selected.

The overall result of the Laplacian smoothing step is general improvement in the overall quality of the elements and reasonable improvements in the mesh gradation. However, a small number of the most poorly shaped elements are not improved, since the direction of motion defined by the centroid only degrades the shape of this element.

The second smoothing operator focuses its attention on the small number of poorly shaped elements. Several approaches that specifically focus on improving the shape of the worst shaped element connected to a node are possible. One reasonably efficient means to this is to employ a line search approach where the direction of motion is selected to ensure improvement in the shape of the worst element connected to the vertex. Given a node and the worst shaped element connected to it, it is possible to determine the position of the node which will optimize the shape of that element [5].

Moving the node all the way to the optimum position of the initially worst-shaped element can degenerate the shapes of other elements connected to the node being moved. Therefore, care must be taken to move the node to a location which does improve the shape of the current worst shape element, but limits the degradation of any other connected element such that its current shape and the shape of the starting connected worst-shaped-element are equivalent. Using the vector from the original model position to the optimum location for the worst shape element, it can be shown that the worst-shaped of any of the connected elements defines a precise function with a unique minimum that can be effectively determined using an efficient golden section search procedure [5].

A important aspect of node point smoothing in quadtree and octree mesh generators is to apply smoothing to all nodes classified on the boundary of the domain, as well as interior nodes near the model boundary. The procedures to smooth nodes classified on nodal edges and faces must consider the resulting shapes of all connected elements, while being constrained to stay on the model edge or face the node is classified on.

15.6.2 Elimination of Poorly Sized and Shaped Elements Caused by Interactions of the Object Boundary and the Tree

One undesirable feature of octree based mesh generators is the disproportionately small and poorly shaped elements that can arise when the boundary of the model comes close to the boundary of an octant. As indicated in Figure 15.4, for the two-dimensional case, it is possible to deform the octree cells so the model and octree boundaries yield elements of the desired size. In the two-dimensional case it is a reasonably straightforward process to perform the quadrant distortion as the model boundary quadrant interactions are determined.

The complexity of the possible model boundary/octant boundaries in the three-dimensional case makes the immediate distortion of the octants as the interactions are determined much more difficult. An alternative approach to meet the same goal is to carry out the octree creation and mesh generator process without octant distortion and to then apply an appropriate set of mesh modification operators to eliminate the appropriate entities.

To successfully meet the goal of eliminating the adverse effects of the small features, the set of operators must include deletion and splitting operators in addition to the swapping operator commonly used to improve the element shapes. As a specific example of the usefulness of a deletion operator, consider the two-dimensional mesh shown in the lower part of Figure 15.4a. A collapse operator that eliminates the short edge on the boundary of the model yields the mesh shown in Figure 15.4b.

15.6.3 Three-Dimensional Mesh Modifications to Improve Mesh Quality

The set of generalized three-dimensional mesh modification operators includes

1. Edge and face swaps [4,7].
2. Edge, face, and region split operators [4].
3. Edge collapse operators [4,25].

The edge collapse operator is a key tool in the elimination of disproportionally small and poorly shaped elements caused by close model/octant boundary interactions. The majority of these situations are characterized by the existence of one or more mesh edges that are substantially shorter than that dictated by the local mesh control information. Once detected, edge collapse operations can be applied to eliminate these and their influence on the mesh. Although the majority of these edges can be collapsed, there are situations for curved geometric domains where the direct application of an edge collapse would yield an invalid mesh in that vicinity. In these cases the application of various swapping and splitting operations will produce a mesh configuration where an edge collapse can be applied, or the local mesh quality measures are improved.

The application of mesh modification operators with the goal of locally improving the quality of the elements must first decide the mesh quality measure to be concerned with. Since the application of the general mesh modification operators is reasonably expensive, these are often applied in an attempt to improve the quality of only elements that have shapes worse than some specific limit. In the case when mesh modifications are applied after disproportionately small mesh edges have been applied, reducing the maximum dihedral angles below some upper limit works well.

15.6.4 A Couple of Examples

The first example (Figure 15.10) demonstrates the influence of the mesh finalization procedures and the alignment of the octree with respect to the model. One concern often expressed in the literature about the use of octree mesh generation techniques is the influence of the alignment of the octree has with respect to the mesh generated for a model. In general, the alignment of the octree does affect the final mesh generated. However, the degree and importance of the influence depends strongly on details of how the mesh is generated within the octree and the level of mesh finalization used. In this example, the mesh was generated with the octree aligned with the vertical axis of the model (Figures 15.10a and 15.10c) and tilted at 30° to the vertical axis (Figures 15.10b and 15.10d) of the model. In the top images (Figures 15.10a and 15.10b) the mesh was generated strictly following the interactions of the model and octree cells. Although the meshes generated look very different, and one may feel the aligned mesh (Figure 15.10a) is superior to the one tilted at 30° (Figure 15.10b): both meshes lack acceptable control of mesh quality in that the maximum dihedral angle of the worst element in the mesh in both cases is greater than 179.9°. In contrast, the application of a full set of mesh finalization operations which

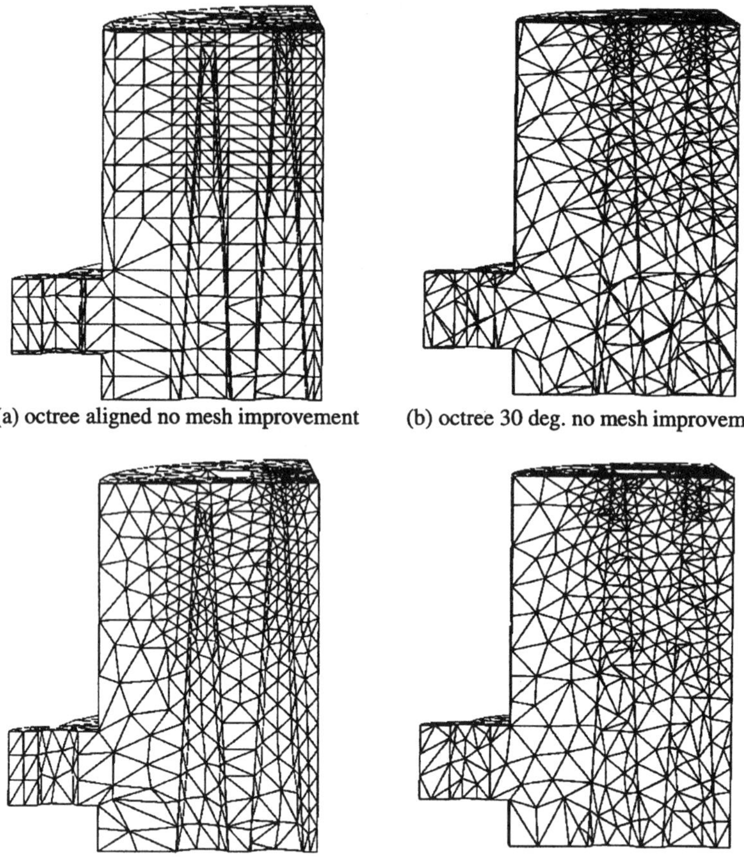

(a) octree aligned no mesh improvement (b) octree 30 deg. no mesh improvement

(c) octree aligned mesh improvement (d) octree 30 deg. mesh improvement

FIGURE 15.10 An example problem with different octree alignments, with and without mesh finalization applied.

reposition nodes, and modify the mesh to eliminate small segments caused by nicks and elements with large dihedral angles (Figures 15.10c and 15.10d), yields meshes that do not show any clear indication of the influence of the alignment of the octree with respect to the model. The quality of the element shapes produced is also similar, with the maximum dihedral angles of any elements in the meshes for the two cases being the same to three significant figures (145°) in this particular example.

The ability of the mesh finalization procedures to produce the control mesh entity shape quality is a strong function of the set of mesh finalization tools employed. If only the smoothing procedures described in Section 15.6.1 are applied to the meshes in Figures 15.10a and 15.10b, the resulting meshes would look much the same as those in Figures 15.10c and 15.10d. However, the quality of the worst-shaped elements would not be the same as when the full set of mesh finalization procedures are applied. In this example, smoothing would reduce the largest dihedral angle down to 169° and increase the smallest dihedral angle to only 1.23°. Including the mesh modifications to eliminate the small nicks does not eliminate the largest dihedral angle, but does increase the minimum angle to 6.7°. Finally, inclusion of the mesh modification operations to decrease the dihedral angle of elements with too large of a dihedral angle reduces the maximum dihedral angle to 145° and increases the smallest dihedral angle in the mesh to 11°.

Octree-based mesh generators easily support the application of any spatially based isotropic mesh size control functions. Figure 15.11 demonstrates the application of mesh controls associated with the entities of the geometric model. Figure 15.11a shows a uniform mesh in which the element sizes requested were the same throughout the model. In the mesh shown in Figure 15.11b, the element size requested for two

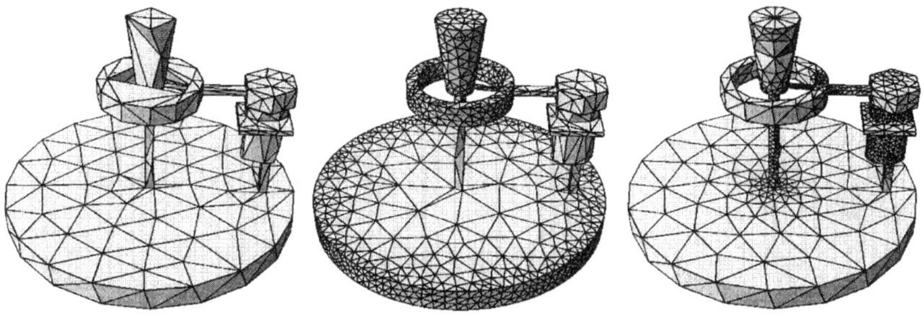

(a) uniform mesh (b) three faces refined (c) curvature refinement

FIGURE 15.11 Application of mesh size controls.

cylindrical and conical faces were set to be smaller than the rest of the model faces (Figure 15.11b). The mesh shown in Figure 15.11c employs a procedure that sets the element sizes associated with the model faces based on the local curvature of the face.

15.7 Closing Remarks

Spatially based tree structures, primarily octrees in three dimensions and quadtrees in two dimensions, have been employed in the development of automatic mesh generation algorithms. An investigation of the literature shows that the alternative procedures developed vary greatly in how the tree is employed in the process of element creation. In all cases, the mesh generation procedures must deal with the same basic issues as other mesh generators. That is, they must employ criteria and procedures to create the elements, and they must ensure that the resulting mesh is a valid representation of the domain being meshed.

All octree and quadtree mesh generators employ the tree structure in the efficient determination of neighbor information. Most procedures also take advantage of the regular shape of interior cells to efficiently create elements in those cells.

Some procedures create meshes which strictly adhere to the boundaries of the undeformed tree cells. In most cases these procedures are taking explicit advantage of this during the analysis process. When such procedures are combined with a general set of mesh finalization procedures, it is possible to eliminate the adverse influence of the shape and alignment of the tree cells on the final mesh.

References

1. **Baehmann, P. L. and Shephard, M. S.**, Adaptive multiple level h-refinement in automated finite element analysis, *Eng. with Computers,* 5(3/4), pp. 235–247, 1989.
2. **Baehmann, P. L., Witchen, S. L., Shephard, M. S., Grice, K.R., and Yerry, M.A.**, Robust, geometrically based, automatic two-dimensional mesh generation, *Int. J. Num. Meth. Eng.*, 1987, 24(6), pp. 1043–1078.
3. **Buratynski, E. K.**, A fully automatic three-dimensional mesh generator for complex geometries, *Int. J. Num. Meth. Eng.*, 30, pp. 931–952, 1990.
4. **de Cougny, H. L. and Shephard, M. S.**, Parallel mesh adaptation by local mesh modification, Scientific Computation Research Center, Report 21-1995, Rensselaer Polytechnic Institute, Troy, NY, 12180-3590, 1995.
5. **de Cougny, H. L., Shephard, M. S. and Georges M.K.**, Explicit node point smoothing within the finite octree mesh generator, Scientific Computation Research Center, Report 10-1990, Rensselaer Polytechnic Institute, Troy, NY, 12180-3590, 1990.
6. **de Cougny, H. L., Shephard, M.S., and Ozturan, C.**, Parallel three-dimensional mesh generation on distributed memory MIMD, *Eng. with Computers*, 12(2), pp. 94–106, 1996.

7. **de l'Isle, E. B. and George, P. L.**, Optimization of tetrahedral meshes, INRIA, Domaine de Voluceau, Rocquencourt, 1993, BP 105, 78153, Le Chesnay Cedex, France.
8. **Gursoz, E. L., Choi, Y., and Prinz, F. B.**, Vertex-based representation of non-manifold boundaries, *Geometric Modeling Product Engineering*, (Eds.) Wozny, M. J., Turner, J. U., Priess, K., North Holland, pp. 107–130, 1990.
9. **Jackins, C. L. and Tanimoto, S. L.**, Octrees and their use in the representation of three-dimensional objects, *Comput. Graphics and Image Processing*, 14, pp. 249–270, 1980.
10. **Kela, A.**, Hierarchical octree approximations for boundary representation-based geometric models, *Computer Aided Design*, 21, pp. 355–362, 1989.
11. **Kela, A., Perucchio, R., and Voelcker, H. B.**, Toward automatic finite element analysis, *Comput. Mech. Eng.*, pp. 57–71, July, 1986.
12. **Perucchio, R., Saxena, M., Kela, A.**, Automatic mesh generation from solid models based on recursive spatial decompositions. *Int. J. Num. Meth. Eng.*, 28, pp. 2469–2501, 1989.
13. **Samet, H.**, *Applications of Spatial Data Structures*, Addison-Wesley, Reading, MA, 1990.
14. **Sapidis, N. and Perucchio, R.**, Combining recursive spatial decompositions and domain delaunay triangulation for meshing arbitrary shaped curved solid models, *Comp. Meth. Appl. Mech. and Eng.*, 108, pp. 281–302, 1993.
15. **Saxena, M. and Perucchio, R.**, Parallel FEM algorithms based on recursive spatial decompositions - {i}.automatic mesh generation, *Computers and Structures*, 45, pp. 817–831, 1992.
16. **Saxena, M., Finnigan, P.M., Graichen, C.M., Hathaway, A.F., and Parthasarathy, V.N.**, Octree-Based Automatic Mesh Generation for Non-Manifold Domains, *Engineering with Computers*, 11, pp. 1–14, 1995.
17. **Schneiders, R. and Bunten, R.**, Automatic mesh generation of hexahedral finite element meshes, *Computer Aided Geometric Design*, 12, pp. 693–707, 1995.
18. **Schroeder, W.J. and Shephard, M. S.**, A combined octree/Delaunay method for fully automatic 3-D mesh generation, *Int. J. Num. Meth. Eng.*, 29, pp. 37–55, 1990.
19. **Schroeder, W. J. and Shephard, M. S.**, On rigorous conditions for automatically generated finite element meshes, *Product Modeling for Computer-Aided Design and Manufacturing*, Turner, J. U., Pegna, J., Wozny, M. J., (Ed.), North-Holland, Amsterdam, 1991, pp. 267–281.
20. **Shephard, M.S.**, Update to: Approaches to the automatic generation and control of finite element meshes, *Applied Mechanics Reviews*, 49(10-2), S5–S14, 1996.
21. **Shephard, M. S. and Finnigan, P. M.**, Toward automatic model generation, *State-of-the-Art Surveys on Computational Mechanics*, 3, A.K. Noor, A.K., Oden, J.T., (Eds.), ASME, pp. 335–366, 1989.
22. **Shephard, M. S. and Georges, M. K.**, Automatic three-dimensional mesh generation by the finite octree technique, *Int. J. Num. Meth. Eng.*, 32, pp. 709–739, 1991.
23. **Shephard, M. S. and Georges, M. K.**, Reliability of automatic 3-D mesh generation, *Comp. Meth. Appl. Mech. and Eng.*, 101, pp. 443–462, 1992.
24. **Shephard, M. S., Baehmann, P. L., and Grice, K. R.**, The versatility of automatic mesh generators based on tree structures and advanced geometric constructs, *Comm. Appl. Num. Meths.*, 4, pp. 145–155, 1988.
25. **Shephard, M. S., Flaherty, J. E., de Cougny, H. L., Bottasso, C. L., and Ozturan, C.**, Parallel automatic mesh generation and adaptive mesh control, *Solving Large Scale Problems in Mechanics: Parallel and Distributed Computer Applications*, Papadrakakis, M., (Ed.), John Wiley & Sons, Chichester, U.K., 1997, pp. 459–493.
26. **Shephard, M. S., Yerry, M. A. , and Baehmann, P. L.**, Automatic mesh generation allowing for efficient *a priori* and *a posteriori* mesh refinements, *Comp. Meth. Appl. Mech. and Eng.*, 55, pp. 161–180, 1986.
27. **Shpitalni, M.**, Switching function based representation — an alternative to quadtree encoding for manufacturing systems, *Annals of the CIRP*, 34, pp. 163–167, 1985.
28. **Shpitalni, M., Bar-Yoseph, P., and Krimberg, Y.**, Finite element mesh generation via switching function representation, *Finite Elements in Analysis and Design*, 5(2), pp. 119–130, 1989.

29. **Weiler, K. J.**, The radial-edge structure: A topological representation for non-manifold geometric boundary representations, *Geometric Modeling for CAD Applications*, Wozny, M.J., McLaughlin, H. W., Encarnacao, J. L., (Eds.), North Holland, pp. 3–36, 1988.
30. **Wordenweber, B.**, Automatic mesh generation, *Computer-Aided Design*, 16(5), pp. 285-291, 1984.
31. **Yerry, M. A. and Shephard, M. S.**, Automatic three-dimensional mesh generation by the modified-octree technique, *Int. J. Num. Meth. Eng.*, pp. 1965–1990, 1984.
32. **Yerry, M. A. and Shephard, M. S.**, Finite element mesh generation based on a modified-quadtree approach, *IEEE Computer Graphics and Applications,* 3(1), pp 36–46, 1983.

16

Delaunay–Voronoï Methods

Timothy J. Baker

16.1 Introduction

It is a remarkable fact that seemingly simple concepts can often lead to whole new fields of research and find extensive applications in many diverse areas. This phenomenon is well illustrated by the Voronoï diagram [Voronoï, 1908] and its dual the Delaunay triangulation [Delaunay, 1934]. Though formulated early in the twentieth century long before the rise of scientific computing, these fundamental geometric ideas have recently found a wealth of applications ranging from interpolation of data [Farin, 1990], to medical imaging [Boissonnat, 1988], computer animation and grid generation [Cavendish, et al., 1985; Shenton and Cendes, 1985; Baker, 1987; George, et al., 1988; Perronnet, 1988; Schroeder and Shephard, 1988; Weatherill and Hassen, 1992; Sharov and Nekahashi, 1996].

Each application has its own specific requirements that lead to interesting and often difficult questions. For example, medical imaging usually requires the detection and representation of biological tissues and features (i.e., complicated surfaces embedded in 3D space). The input data provided by the imaging device is often in the form of a cloud of points. A precise representation of the surface geometry is usually not required, but a faithful rendering of the topology certainly is. The requirements for computer animation are somewhat similar, although there is often a greater need for correct rendering of surface geometry, especially sharp corners and cusp-like features. In addition, the constraints of computer memory and processing speed put a premium on efficient data management. Thus it is preferable to choose a set of points and surface triangles with small cardinality to represent a given object. Consequently, one would like to know what is the best surface representation for a given number of points and triangles.

Grid generation places a great emphasis on achieving a good representation of the surface geometry. This in turn requires a close link between the CAD definition and grid generator and a stringent need to ensure that not only the surface points but also the edges and faces of the surface grid lie on the true surface. At the same time, it would be highly desirable to automate the grid generation process and allow the user to proceed directly from CAD definition to surface and volume grid, and then finite element solution, without any user intervention.

0-8493-2687-7/99/$0.00+$.50

Despite the large number of grid generation papers whose titles contain the adjective "automatic" (the author is himself guilty of this hyperbole), truly automatic grid generation still remains an elusive goal. Although fully automated grid generators have been created for very specific problems (e.g., a structured grid around a fuselage and two lifting surfaces [Baker, 1991a], grid generation for arbitrary domains is still insufficiently robust to qualify as completely automatic. Despite this cautionary note, it is fair to say that tetrahedral grid generation, and in particular Delaunay-based methods, are at a highly advanced stage and tantalizingly close to achieving the ultimate goal of black box grid generation.

The most difficult issue is the preservation of surface integrity. Since the Delaunay triangulation of a set of points does not necessarily have edges and faces that coincide with the desired boundary surface, some extra algorithm or procedures must be imposed to ensure this property. One early method [Baker, 1987; Baker, 1989] allowed the boundary surface to be defined by a cloud of points arranged as a series of cuts or space curves (see also [Boissonnat, 1988]). An algorithm to generate a layer of additional points just offset from the boundary surface [Baker, 1991b] provided the means to create a Delaunay triangulation whose edges and faces almost always lay on the boundary surface. In the few instances that this technique failed to preserve boundary integrity, the defining cloud of points could be modified until the desired end was achieved. The disadvantage of this approach is the lack of any direct control over boundary surface integrity; varying degrees of user interaction are required depending on the complexity of the domain being triangulated.

An alternative approach that does, in principle, lead to full automation is based on the idea of modifying the Delaunay triangulation by a series of edge/face swaps until boundary surface integrity is achieved. First proposed by George et al. [1988] (see [George, 1988] for a more detailed description), this idea has been pursued by several others [Weatherill and Hassan, 1992; Sharov and Nakahashi, 1996] and a number of grid generators that exploit this technique are now available. One advantage of this approach is the opportunity to treat surface grid generation and volume grid generation as independent operations. Thus surface grid generation can be closely coupled to the CAD systems, allowing the user to create a good quality surface grid that conforms to the true boundary. The volume grid generator, as a separate module, then creates a grid of tetrahedra that conform to the prescribed surface grid.

For planar domains the situation is very satisfactory. In this case, the boundary surface is given by a prescribed set of points and edges, and a grid of triangles is required that conforms to the boundary edges. Given a pair of triangles with a common edge that form a convex quadrilateral, one can replace the common edge by connecting the other pair of points instead. Using this technique of diagonal swapping, it is known [Guibas and Stolfi, 1985] that any planar triangulation of a fixed point set can be converted into the Delaunay triangulation. Moreover, it is possible to convert the Delaunay triangulation of the set of boundary points into a triangulation whose edges match the prescribed boundary edges (the so-called constrained Delaunay triangulation [Lee and Lin, 1986]). Selective insertion of points inside the domain will then lead to a planar triangulation whose triangles meet certain guaranteed quality measures [Ruppert, 1992; Chew, 1993; Baker 1994].

In three dimensions the theory is far less developed. The main difficulties are the following: (1) there exist configurations of boundary points and faces for which no conforming grid of tetrahedra exists unless extra points are inserted, (2) although 3D analogs of diagonal swapping exist, it does not appear possible to convert an arbitrary 3D triangulation into the corresponding Delaunay triangulation, (3) the presence of slivers, formed by four coplanar points, can arise and indeed will often arise when efforts are made to create a constrained Delaunay triangulation that conforms with a prescribed boundary.

In practice, it is possible to generate a constrained Delaunay triangulation in 3D provided the prescribed surface triangulation is sufficiently nice, and what distinguishes a good tetrahedral grid generator from one that is not so good lies in how nice the surface triangulation has to be for the grid generator to create a valid grid of boundary conforming tetrahedra. For example, if the boundary surface is extracted from the Delaunay triangulation of the boundary surface points, then there should clearly be no difficulty in creating a conforming grid of tetrahedra, since this is precisely the Delaunay triangulation to which

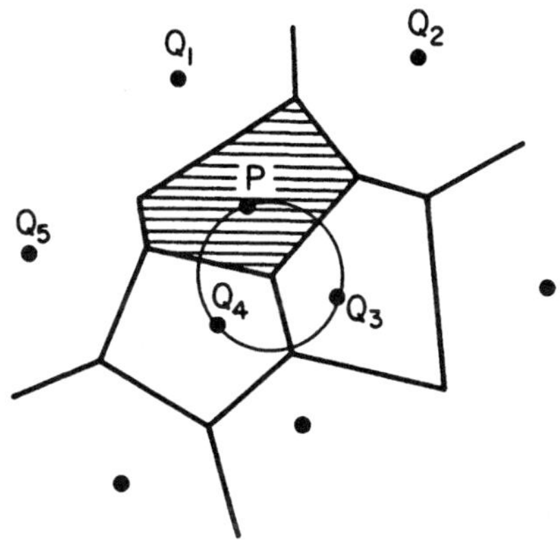

FIGURE 16.1 Voronoï diagram of a planar set of points.

the boundary surface corresponds. If, as is usually the case, the surface triangulation is close to but not completely Delaunay, then a combination of edge/face swaps and point insertions will establish a constrained Delaunay triangulation that does conform to the boundary. For boundary triangulations which deviate greatly from the Delaunay state, it will be difficult and perhaps impossible to construct a conforming set of tetrahedra. Surface triangulation is addressed in Chapter 19.

Once the initial boundary-conforming set of tetrahedra has been established, a final grid can be created by selectively adding points into the domain in order to produce a set of good quality tetrahedra whose size varies gradually, leading to a grid suitable for accurate computation by a finite element method. Because of the appearance of slivers, it is also necessary to apply a grid optimization procedure to remove these singular tetrahedra. The following sections provide more detail about the current procedures and outline those areas that are actively being researched.

16.2 Underlying Principles

16.2.1 Voronoï Diagram and Delaunay Triangulation

The Delaunay triangulation [Delaunay, 1934] of a set of points and the dual geometric construct, the Voronoï diagram [Voronoï, 1908], are extremely fertile concepts that have been the subject of considerable theoretical investigation and have found numerous practical applications. The Voronoï diagram marks off the region of space that lies closer to each point than the other points. This is illustrated for the planar case in Figure 16.1. The solid lines make up the Voronoï diagram, forming a tessellation of the space surrounding the points. Each Voronoï tile (e.g., the hatched area around point *P*) consists of the region of the plane that is closer to that point than any other. The edges of the Voronoï diagram are formed from the perpendicular bisectors of the lines connecting neighboring points (e.g., points *P*, Q_3, Q_4 in Figure 16.1) and, hence, each vertex is the circumcenter of the triangle formed by three points. This determines a unique triangulation known as the Delaunay triangulation and is such that the circumcircle through each triangle contains no points other than its forming points.

These concepts generalize to higher dimensions. In particular, the Delaunay triangulation of three-space is the unique triangulation such that the circumsphere through each tetrahedron contains no points other than its forming points. In two dimensions, this circle criterion can be shown [Sibson, 1978] to be equivalent to the equiangular property that selects the triangulation that maximizes the minimum of the six angles in any pair of two triangles that make up a convex quadrilateral. No equivalent characterization appears to be known in three dimensions, but the circle criterion can still be regarded as selecting a good triangulation for the given set of points.

16.2.2 Bowyer–Watson Algorithm

A particularly straightforward method for generating the Delaunay triangulation is the Bowyer–Watson algorithm [Bowyer, 1981; Watson, 1981], which can readily be applied to any number of dimensions. It is an incremental algorithm that directly exploits the circle criterion of the Delaunay triangulation as follows.

Let T_n be the Delaunay triangulation of the set of n points, $V_n = \{P_i \mid i = 1, \ldots, n\}$. For any simplex $S \in T_n$, let R_s be the circumradius and Q_s the circumcenter. Now introduce a new point P_{n+1} inside the convex hull of V_n, and define $B = \{S \mid S \in T_n, d(P_{n+1}, Q_s) < R_s\}$ where $d(P, Q)$ is the Euclidean distance between points P and Q. Now B is non-empty, since P_{n+1} is inside the convex hull of V_n and hence inside some simplex $S' \in T_n$, from which it follows that $S' \in B$. The region C formed when B is removed from T is simply connected, contains P_{n+1} (since P_{n+1} is inside $S' \in B$), and P_{n+1} is visible from all points on the boundary of C. It is therefore possible to generate a triangulation of the set of points $V_{n+1} = V_n \cup \{P_{n+1}\}$ by connecting P_{n+1} to all points on the boundary of C. Furthermore, this triangulation is precisely the Delaunay triangulation T_{n+1}.

Proofs that the cavity C is simply connected, that P_{n+1} is visible from the cavity boundary and that the new triangulation is also Delaunay can be found in [Baker, 1987; Baker, 1989].

The implementation of the Bowyer–Watson algorithm in three dimensions starts with a supertetrahedron, or supercube partitioned into five tetrahedra, which contains all the other points. The remaining points, which comprise the grid to be triangulated, are introduced one at a time and the Bowyer–Watson algorithm is applied to create the Delaunay triangulation after each point insertion.

It is necessary to maintain two lists, each of length four, for each tetrahedron in the existing structure. One list holds the forming points of the tetrahedron, the other holds the addresses of the four neighboring tetrahedra that have a common face. The second list, which provides information about the contiguities between the tetrahedra, is not strictly needed for the implementation of the algorithm. However, it allows all tetrahedra belonging to a cavity to be found by means of a tree search, once one such tetrahedron has been found. Without the contiguity information the algorithm would be hopelessly inefficient. It is also convenient to store the radius of the circumsphere and the coordinates of the circumcenter for each tetrahedron.

The remaining step in the Bowyer–Watson algorithm is the requirement to update the data structure. Tetrahedra belonging to the set B are deleted from the lists, and new tetrahedra, obtained by connecting the new point to all triangular faces of the cavity boundary, are added. Finally, it is necessary to determine the contiguities that exist among the new tetrahedra, and also between the new tetrahedra and the old tetrahedra that have faces on the cavity boundary.

The only floating point operations required in this algorithm occur in the Delaunay test for each tetrahedron that is examined when searching for those tetrahedra that make up the cavity. Owing to the finite precision arithmetic that is used, the Delaunay test will make an ambiguous decision if the new point falls on the circumsphere of a tetrahedron [Baker, 1992]. High-precision arithmetic must therefore be used. Moreover, it is particularly important when forming the set of B of cavity tetrahedra to exclude from B any tetrahedron whose circumsphere does not strictly contain the new point. This can be achieved by introducing a tolerance $\mu > 0$ and including in B only those tetrahedra S for which $d(P, Q) < R_s - \mu$, where μ is chosen sufficiently large to ensure strict inclusion.

When a new point is inserted, a search is made through the list of tetrahedra to find the first tetrahedron that fails the Delaunay test. The remaining tetrahedra that make up the cavity can be found by a tree search. After these tetrahedra have been removed, the points on the boundary of the cavity are connected to the new point P and the new tetrahedra thus formed are added to the data structure.

The time required to triangulate N points will be given by

$$T = \sum_{k}^{N} (T_k + T'_k)$$

Here, T_k is the time taken to search for the first cavity tetrahedron that arises from the introduction of the kth point into the triangulation of $k - 1$ points. T_k' is the time taken to find all remaining tetrahedra in the cavity and construct the new triangulation. The time T_k' will be proportional to the number of tetrahedra in the cavity. If the points are inserted in a widely distributed manner corresponding to a coarse sprinkling followed by a finer distribution [Baker, 1987], the cavity size, and hence time T_k', should be roughly independent of k. The majority of points are field points that are introduced selectively (e.g., at the circumcenter of the tetrahedron having maximum volume). Under fairly mild conditions on the current state of the triangulation, the time T_k' can therefore be regarded as $O(1)$.

Thus, the time complexity of the algorithm is dominated by the search time T_k. In general, the list of tetrahedra will be randomly ordered and, in the worst case, T_k will be $O(k)$, leading to an overall time complexity for the triangulation that is $O(N^2)$.

It is therefore necessary to introduce a data structure that allows an efficient search for the first tetrahedron that fails the Delaunay test irrespective of the point ordering. To achieve this, one can exploit an octree structure to store the points that have previously been inserted [Baker, 1989]. The octree data structure is used to find the point nearest to a newly introduced point. With each previously introduced point, one associates a tetrahedron that has this point as a vertex. The search for the first tetrahedron in the Delaunay cavity thus starts with the tetrahedron associated with the point nearest to the new point, and proceeds to examine all neighboring tetrahedra that have this nearest point as a vertex. In this way, it is possible to find the first cavity tetrahedron in a time T_k, that is $O(\log k)$. It follows that the overall time complexity of the algorithm is $O(N \log N)$. Other data structures have been proposed that also achieve a fast search time [Bonet and Peraire, 1991].

16.2.3 Tanemura–Ogawa–Ogita Algorithm

An alternative algorithm [Tanemura, et al., 1983] for generating the Delaunay triangulation can be described for the planar case as follows.

Given a set of points V_n and a Delaunay edge e, we can construct a circle C_i through the endpoints of e and any one of the remaining points P_i. One of these circles, say C_1, will be empty and thus defines the Delaunay triangle T containing point P_1 and e as the edge opposite P_1. For a constrained Delaunay triangulation with respect to a fixed edge e, we require only the segment of the circle on the same side of e as the candidate point P_i to be empty. The triangle T contains two edges other than e. If either of these edges is not among the list of edges already generated, it is added to the list. Any internal edge (i.e., nonboundary edge) that is associated with only one triangle is considered an active edge on which a new Delaunay triangle should be constructed by the above procedure. The algorithm stops, and the triangulation is complete, when every boundary edge corresponds to the side of one triangle and every internal edge forms the common side of precisely two triangles. This approach is the basis of an algorithm first proposed by Tanemura et al. [1983] and subsequently exploited by Merriam [1991] and Mavriplis [1993].

In the planar case, the Tanemura–Ogawa–Ogita algorithm is well suited to the task of generating the constrained Delaunay triangulation with respect to a prescribed set of boundary edges. After establishing a triangulation consisting of the boundary points and conforming to the boundary edges, it is then possible to use the Bowyer–Watson algorithm to selectively add points until an acceptable grid has been created. Various point placement strategies including circumcenter point insertion [Weatherill and Hassan, 1992] and the Voronoï segment method [Rebay, 1993] have been proposed and analyzed [Chew, 1993; Baker 1994].

In 3D the possible nonexistence of a constrained Delaunay triangulation that will conform to a prescribed surface triangulation severely limits the usefulness of the Tanemura–Ogawa–Ogita algorithm. In this case, the preferred approach appears to be based on the Bowyer–Watson algorithm followed by a series of edge/face swaps to establish any edges and faces of the prescribed boundary surface [George, et al., 1988; Weatherill and Hassan, 1992; Sharov and Nakahashi, 1996].

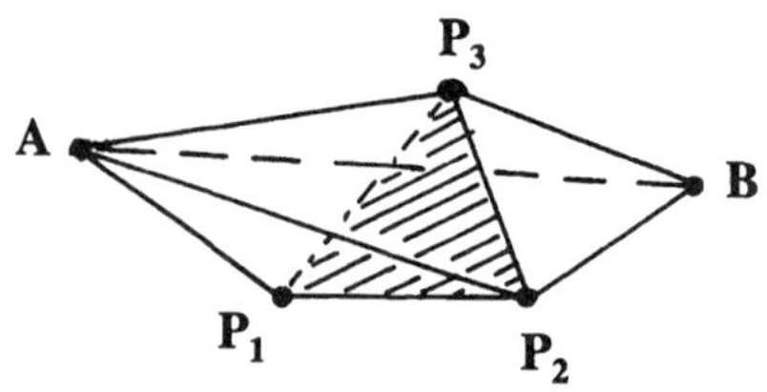

FIGURE 16.2 Three tetrahedra with common edge AB, or two tetrahedra with common face $P_1P_2P_3$.

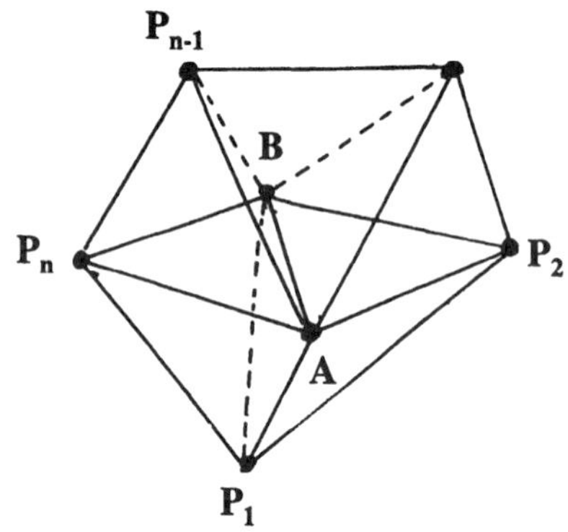

FIGURE 16.3 Several tetrahedra surrounding edge AB.

16.2.4 Edge/Face Swapping

The simplest swappable combination in 3D occurs when three tetrahedra share a common edge. In Figure 16.2, the three tetrahedra ABP_1P_2, ABP_2P_3, and ABP_3P_1 together with the common edge AB can be replaced by the two tetrahedra $AP_1P_2P_3$ and $BP_1P_2P_3$ together with the common face $P_1P_2P_3$. Provided that the ensemble of tetrahedra is convex (i.e., edge AB intersects face $P_1P_2P_3$), then either combination can exist without affecting the remaining tetrahedra. In the general case, when n tetrahedra share a common edge (see Figure 16.3), a transformation that replaces the n tetrahedra by $2(n-2)$ tetrahedra can be found provided the ensemble of tetrahedra is convex. In order to determine the new set of tetrahedra it is necessary to cover the interior of the polygon $\{P_1, \ldots, P_n\}$ by triangular facets. For $n \geq 4$ the set of new tetrahedra is not uniquely defined [Briére de L'isle and George, 1993].

The utility of these swapping operations lies in the opportunity to establish any prescribed boundary edges and faces that do not exist in the volume grid. Suppose, for example, that a given edge AB lies on the prescribed boundary but does not exist in the volume mesh. After identifying the face $P_1P_2P_3$ in the volume mesh that intersects the line segment AB, one can apply the reverse of the edge/face swap illustrated in Figure 16.2. It is possible, however, for the line segment AB to lie in or very close to one of the faces AP_1P_2, AP_2P_3, AP_3P_1. For example, suppose that AB lies in face AP_3P_1 and thus intersects edge P_1P_3. In this singular case it is necessary to identify the ring of tetrahedra incident to edge P_1P_3 and use an edge/face swap that removes edge P_1P_3 and inserts edge AB assuming, of course, that points A and B are both vertices associated with the tetrahedral ring. If this is not the case, or if the ring of tetrahedra is not convex, then the line segment AB cannot be established as an edge of the volume grid. In practice, an edge that was not established initially can often appear in the volume grid or be established by swapping procedures after further missing boundary edges have been inserted. If some missing boundary edges stubbornly remain, then one can resort to the insertion of extra grid points, either inside the domain [George, et al., 1988] or perhaps on the boundary surface [Weatherill and Hassan, 1992; Sharov and Nakahashi, 1996] at the midpoint of the missing edge.

After all boundary edges and possible additions of new boundary points and edges have been established, it is necessary to ensure that all boundary faces are contained in the volume grid. In practice, a volume grid that contains all boundary edges will at worst be missing only a handful of boundary faces. Suppose, for example, that a missing face has vertices $P_1P_2P_3$. Then (see Figure 16.2) it is necessary to identify the edge AB that intersects the missing face and carry out the edge/face swap that removes edge AB and establishes face $P_1P_2P_3$.

It is, however, possible for more than one edge of the volume grid to intersect the missing face. In this case, it is again necessary to add an extra grid point either in the domain or on the boundary surface at say, the barycenter of the missing face. The reader is referred to the literature [George, et al.,1988; Weatherill and Hassan, 1992; Sharov and Nakahashi, 1996] for a more detailed discussion of the various ways in which edge/face swaps and grid point insertion can be applied to establish the prescribed boundary surface.

16.2.5 Grid Optimization

The objective of grid optimization is to achieve a volume grid with a smooth grading in size of tetrahedra and good tetrahedral quality as measured by some criterion such as dihedral angle or ratio of circumradius to in-radius. Perhaps the most pressing requirement is the identification and removal of slivers. These tetrahedra are formed by four co-planar or nearly co-planar points and hence will have a volume that is extremely small [Cavendish, et al., 1985]. Although it is possible to monitor the formation of such tetrahedra during the grid generation process, any attempt to prevent their formation at this stage will usually sabotage efforts to establish the boundary surface or lead to a final volume grid whose overall quality is in fact worse. It appears best to apply grid optimization as a post processing operation on the final grid [Briére de L'isle and George, 1993]. Sliver-like tetrahedra can be found by searching for edges which have adjacent incident faces with a dihedral angle close to 180°. It follows that at least one of the incident tetrahedra is a sliver, and an edge/face swap that removes this edge will also remove the sliver. If the ring of tetradedra incident to the edge is nonconvex, then this approach fails. In practice, it is usual to apply the edge/swap procedure to remove as many slivers as possible and then smooth the grid (i.e., adjust the positions of the nonboundary grid points). This two-step process can be iterated until a grid of acceptable quality has been obtained. A popular smoothing technique is Laplacian smoothing, although care must be taken to ensure that no grid points cross any faces, thus leading to an invalid grid. Another technique is based on moving each grid point until all incident edges have nearly equal length. Still other techniques have been based on linear programming. Smoothing changes the positions of the nonboundary grid points but leaves the topology intact. Edge/face swaps leave the grid positions fixed but change the topology. It is therefore reasonable to expect that an iterative process alternating between these two procedures should lead to an improved grid.

16.2.6 Constrained Delaunay Triangulation

The key procedure that lies at the heart of the Bowyer–Watson algorithm is the determination of the cavity tetrahedra whose circumspheres contain the new point P. For a Delaunay triangulation, the cavity is simply connected and the point P is visible from all faces of the cavity. When the triangulation is no longer Delaunay but constrained by the presence of fixed faces that arise, for example, when one or more of the interior cavity faces is a boundary surface face, then the visibility issue needs to be reexamined. Since some of the faces of the restricted cavity need not be visible from P, it is necessary to find the tetrahedron that contains P and then examine neighboring tetrahedra by means of a tree search. Clearly P lies within the circumsphere of the tetrahedron that contains P and every face of this tetrahedron is visible from P. Each face of this tetrahedron is examined to determine whether it is a fixed (i.e., protected) face that must not be removed. If it is not a protected face, then the tree search proceeds to the adjacent tetrahedron on the other side of this face. If P lies within the circumsphere of the new tetrahedron and P is visible from its other three faces, then this tetrahedron is added to the list of cavity tetrahedra. After each of the neighboring faces and tetraheda have been examined in this way, the process is repeated for each of the tetradehra that has been newly admitted to the cavity list.

Starting from the original set B of tetrahedra whose circumspheres contain the point P we arrive at a reduced set $B_1 \subset B$ whose faces found by the tree search were judged visible from the point P. Since the tree search examines the tetrahedra in a particular sequence, it is possible that there exist one or more tetrahedra in the subset B_1 whose faces are not visible from P when the visibility test is reapplied to the reduced set B_1. The tree search and visibility test must therefore be repeated for set B_1 to create a new subset $B_2 \subset B_1$. If the sets B_2 and B_1 are identical, then all faces of B_1 are visible from P. The points on the boundary of the restricted cavity C_1, formed when B_1 is removed from the triangulation T, can then be connected to point P to form a valid retriangulation. A detailed discussion of these issues has been given by Wright and Jack [1994].

Let $\mathbf{r}_i = (x_i, y_i, z_i)$ be the coordinate vector of the ith vertex of a tetrahedron where $i = 1, \ldots, 4$ and let $\mathbf{r}_p = (x_p, y_p, z_p)$ be the coordinate vector of the point P. The face opposite vertex 4 is visible with respect to point P if point P lies on the same side of this face as vertex 4. An alternative statement is that point P and vertex 4 must lie in the same half space formed by the plane containing vertices 1, 2, and 3. The visibility test thus amounts to checking whether the volume of the tetrahedron formed by the points r_1, r_2, r_3, and r_4, has the same sign as the volume of the tetrahedron formed by the points r_1, r_2, r_3, and r_p. In other words, the sign of the determinant

$$\begin{vmatrix} 1 & 1 & 1 & 1 \\ x_1 & x_2 & x_3 & x_4 \\ y_1 & y_2 & y_3 & y_4 \\ z_1 & z_2 & z_3 & z_4 \end{vmatrix}$$

must be compared with the sign of the determinant formed by replacing x_4, y_4, and z_4 with x_p, y_p, and z_p. The validity of the retriangulation therefore rests on the accuracy of the determinant evaluation and hence on the precision of the computer arithmetic that is used. Difficulties arise when one or both determinants are extremely close to zero leading to uncertainty as to whether the correct sign has been computed. Various schemes using variable precision arithmetic [Shewchuk, 1996] and also integer arithmetic have been proposed to handle this situation. An interesting development by Edelsbrunner and Mücke [1990] could lead to a systematic handling of these situations.

16.3 Research Issues

At the present stage of knowledge, it is fair to say that planar triangulations are well understood and that they enjoy a number of properties that do not apparently extend to three dimensions. The existence in the planar case of a constrained Delaunay triangulation that conforms to any set of prescribed edges makes it possible to construct a grid of triangles for any two-dimensional region whether simply connected or multiply connected. The refinement of the planar grid by insertion of points inside the region can be shown to generate isotropic grids of high quality. The issue of generating anisotropic grids that are designed to contain high aspect ratio triangles, aligned with particular features in a finite element solution, is less well developed but is currently an area of active research.

The most obvious difference between three-dimensional triangulations and the planar counterpart is the existence in 3D of valid boundary surface triangulations for which no space filling, conforming set of tetrahedra exists. This precludes the possibility of generating a constrained Delaunay triangulation (in fact, any triangulation) containing only the boundary points and conforming with the boundary surface for every possible boundary surface configuration. It should always be possible, although the author is not aware of a proof, to achieve a conforming triangulation if extra points (so-called Steiner points) are inserted inside the domain. To ensure that the final grid has good quality tetrahedra, the inserted points should not be too close to the boundary surface. In order to guarantee a grid for any boundary surface configuration one therefore needs to know when and where any extra points should be inserted inside the domain.

One potential research area of great importance is the classification of those boundary surface triangulations for which no conforming volume triangulation exists. Given a boundary surface triangulation one can either (1) create a space-filling, boundary-conforming grid of tetrahedra, (2) be unable to create any conforming tetrahedra, or (3) build a number of tetrahedra inside the domain until no further tetrahedra can be introduced because of the boundary constraint. Case (3) can be viewed as a situation in which tetrahedra are created as one would with an advancing front method until the front, interpreted as a boundary surface triangulation, falls into category (2). It seems likely that the number of triangles in any boundary surface in category (2) should be relatively small, and that the number of distinct cases that fall into category (2) may therefore be a finite and perhaps not too large number. If this is the case, then it may be possible to classify the different cases in category (2) and provide an algorithm for adding a point, or points, inside the domain to create a conforming grid. A satisfactory answer to this question would solve completely the problem of automatic grid generation for tetrahedral grids.

A less ambitious question that may prove easier to tackle is to ask how far one can proceed with swapping techniques to change a non-boundary conforming grid of tetrahedra into a grid that does match a given boundary surface triangulation. Is it possible to quantify the extent to which a given boundary surface triangulation fails to be Delaunay and is it possible to relate this characteristic to whether a Delaunay volume grid can be converted by a given set of swapping techniques into a boundary conforming grid? These are all rather difficult questions and although some important work has been done in this area [Joe, 1991], it may be a long time before they are answered in any reasonably satisfactory way.

Another important area of research relates to the quality of the final tetrahedral grid that is obtained after the initial grid has been refined by successive insertion of points inside the domain. As pointed out earlier, slivers will almost invariably appear and these extremely small volume tetrahedra wreak havoc on most finite element methods. Edge/face swaps can be applied as described earlier to remove slivers and generally improve grid quality. This approach works well most of the time but there are situations where slivers persist in the optimized grid (the edge/face swap may fail because the tetrahedral ensemble is non-convex for example). The application of smoothing may move the grid into a configuration in which a further application of edge/face swaps will remove all slivers, but there is no guarantee that this will always be the case. Any insight into this problem would be very useful, and a recipe for optimizing the grid, with a guarantee of removing all slivers, would have a profound impact on grid generation.

A related question concerns the different optimization criteria and whether these will lead to a global optimum or whether they might generate optimization schemes that get stuck in local optima. There are numerous ways of defining the quality of a tetrahedron including minimum and maximum dihedral angle, ratio of circum-radius to in-radius and several other criteria that have been reported in the literature. Some interesting work has already been carried out to establish which criteria are associated with optimization problems that have global optima and continuing research in this area will undoubtedly lead to better methods for grid optimization. A similar though somewhat different question is how should a grid be optimized to achieve the most accurate finite element solution for a given problem? This will obviously depend on the partial differential equation being solved as well as the finite element discretization being used.

Even partial answers to these questions will go a long way to making existing tetrahedral grid generation more efficient and more reliable. With the much deeper knowledge that will eventually be gained, we can look forward one day to achieving truly automatic three-dimensional grid generation.

References

1. Baker, T.J., Three-dimensional mesh generation by triangulation of arbitrary point sets, *AIAA 8th Computational Fluid Dynamics Conference*, AIAA Paper 87-1124-CP, Hawaii, June 1987.
2. Baker, T.J. Automatic mesh generation for complex three-dimensional regions using a constrained Delaunay triangulation," *Engineering with Computers.* 1989, 5, p 161.
3. Baker, T.J., Single block mesh generation for a fuselage plus two lifting surfaces, *Proc. 3rd International Conference on Numerical Grid Generation in Computational Fluid Dynamics.* Arcilla, A.S., (Ed.), Elsevier Science Publishers B.V., North-Holland, 1991a, p 261.

4. Baker, T.J., Shape reconstruction and volume meshing for complex solids, *Int. J. Num. Meth. Eng.*, 1991b, 32, p 665.
5. Baker, T.J., Tetrahedral mesh generation by a constrained Delaunay triangulation, *Artificial Intelligence, Expert Systems and Symbolic Computing.* Houstis, E.N. and Rice, J.R., (Eds.), Elsevier Science, Publishers B.V., North-Holland, 1992.
6. Baker, T.J., Triangulations, mesh generation and point placement strategies, *Frontiers of Computational Fluid Dynamics,* Caughey, D.A. and Hafez, M.M., (Eds.), John Wiley and Sons, 1994, p 101.
7. Boissonnat, J.D., Shape reconstruction from planar cross sections, *Computer Vision, Graphics and Image Processing,* 1988, 4, p 1.
8. Bonet, J. and Peraire, J., An Alternating Digital Tree (ADT) algorithm for geometric searching and intersection problems, *Int. J. Num. Meth. Eng.,* 1991, 31, p 1.
9. Bowyer, A., Computing Dirichlet tessellations, *Comput. J.,* 1981, 24, p 162.
10. Brière de L'isle, E. and George, P.L., Optimisation de maillages tridimensionnels, *INRIA Report.* (2046), 1993.
11. Cavendish, J.C., Field, D.A., and Frey, W.H., An approach to automatic three-dimensional finite element mesh generation, *Meth. Eng.,* 1985, 21, p 329.
11. Chew, P., Guaranteed quality mesh generation for curved surfaces, *Proc. 9th Symp. On Comp. Geom.* ACM Press, 1993, p 274.
12. Delaunay, B., Sur la sphère vide, *Bull. Acad. Science USSR VII: Class Sci. Mat. Nat.,* 1934, 6, p 793.
13. Edelsbrunner, H. and Mücke, E.P. Simulation of simplicity: a technique to cope with degenerate cases in geometric algorithms, *ACM Trans. Graphics,* 1990, 9, p 66.
14. Farin, G., Surfaces over Dirichlet tessellations. *Computer Aided Geometric Design,* 1990, 7, p 281.
15. George, P.L., Hecht. F., and Saltel, E., Tétraedrisation automatique et respect de la frontière, *INRIA Report.* 835, 1988.
16. George, P.L., Hecht, F., and Saltel, E., Constraint of the boundary and automatic mesh generation, *Proc. 2nd International Conference on Numerical Grid Generation in Computational Fluid Mechanics,* Sengupta, S., (Ed.), Pineridge Press, 1988, p 589.
17. Guibas, L. and Stolfi, J., Primitives for the manipulation of general subdivisions and the computation of Voronoï diagrams, *ACM Trans. Graphics.* 1985, 4, p 74.
18. Joe, B., Construction of three-dimensional Delaunay triangulations using local transformations, *Computer Aided Geometric Design,* 1991, 8, p 123.
19. Lee, D. and Lin, A., Generalized Delaunay Triangulation for planar graphs, *Discrete Comp. Geom.* 1986, 1, p 201.
20. Mavriplis, D., An advancing front Delaunay triangulation algorithm designed for robustness, AIAA Paper 93-0671, 1993.
21. Merriam, M., An efficient advancing front algorithm for Delaunay triangulation, AIAA Paper 91-0792, 1991.
22. Perronnet, A., Un algorithme de tètraèdrisation d'un objet multi-matériaux ou de l'extérieur d'un objet, *Numerical Analysis Laboratory Report.* (R88005). University Pierre et Marie Curie, 1988.
23. Rebay, S., Efficient unstructured mesh generation by means of delaunay triangulation and Bowyer–Watson algorithm, *J. Comp. Physics,* 1993, 106, p 125.
24. Ruppert, J., Results on Triangulation and high quality mesh generation, Ph.D. thesis. University of California at Berkeley, 1932.
25. Schroeder, W.J. and Shephard, M.S., Geometry-based fully automatic mesh generation and the Delaunay triangulation, *Int. J. Num. Meth. Eng.,* 1988, 26, p 2503.
26. Sharov, D. and Nakahashi, K., A boundary recovery algorithm for Delaunay tetrahedral meshing, *Proc. 5th International Conference on Numerical Grid Generation in Computational Field Simulations,* Soni, B.K. and Thompson, J.F., (Eds.), NSF Engineering Research Center for Computational Field Simulation, 1996, p 229.
27. Shenton, D.N. and Cendes, Z.J., Three-dimensional finite element mesh generation using Delaunay tessellation, *IEEE Trans. Magnetics,* 1985, MAG-21, p 2535.

28. Shewchuk, J.R., Adaptive precision floating-point arithmetic and fast robust geometric predicates, *Computer Science Report.* Carnegie Mellon University, 1996, CMU-CS-96-140.
29. Sibson, R., Locally equiangular triangulations, *Comput. J.*, 1978, 21, p 243.
30. Tanemura, M., Ogawa, T., and Ogita, N., A new algorithm for three-dimensional Voronoï tessellation, *J. Comp. Physics*, 1983, 51, p 191.
31. Voronoï, G., Nouvelles applications des paramètres continues à la théorie des formes quadratiques, dieuxieme memoire: researches sur les parallelloedres primitif, *J. Reine Angew. Math.*, 1908, 134, p 198.
32. Watson, D., Computing the *n*-dimensional Delaunay tessellation with application to Voronoï polytopes, *Comput. J.*, 1981, 24, p 167.
33. Weatherill, N.P. and Hassan, O., Efficient three-dimensional grid generation using the Delaunay triangulation, *Proc. First European Computational Fluid Dynamics Conference*, Brussels, 1992.
34. Weatherill, N.P., Hassan, O., and Marcum, D.L., Calculation of steady compressible flowfields with a finite element method, 1993, AIAA Paper, 93-0341.
35. Wright, J.P. and Jack, A.G., Aspects of three-dimensional constrained Delaunay meshing, *Int. J. Num. Meth. Eng.*, 1994, 37, p 1841.

17

Advancing Front Grid Generation

J. Peraire, J. Peiró

K. Morgan

17.1 Introduction

The advancing front technique (AFT) for the generation of unstructured triangular meshes was first formulated by George [14], but this original publication did not receive significant immediate attention. It seems that the first reference to this work appeared in an appendix of the book by Thomasset [32].

The first journal publication of the method was that of Lo [19], where the AFT was used to produce a triangulation by linking a set of points, which had been generated beforehand in a Cartesian fashion.

The algorithm was modified by Peraire et al. [25], using a new formulation in which elements and points were simultaneously generated. The method also allowed for the generation of high aspect ratio triangles and, more importantly, grid control was introduced through the specification of a spatial variation of the desired element size and shape. This facility was later used for adaptive computations in computational fluid dynamics.

The methodology was subsequently extended to three dimensions (3D) in [21,26,20,15,16]. The use of the AFT for 3D adaption in compressible flows is described in [28]. Recent implementations of the AFT that improve the generation times and/or the point placement/selection strategies have been reported [13,17,23,22,12]. In addition, the method has also been modified to produce a procedure for the generation of unstructured meshes of quadrilaterals in [34, 4] and of hexahedrals in [5].

0-8493-2687-7/99/$0.00+$.50

17.2 Mesh Generation Requirements

A computational domain of complex geometrical shape may be discretized in terms of an unstructured mesh of tetrahedra. This is an alternative to the approach based upon the use of the multi-block method of grid generation in which the domain is initially subdivided into an unstructured assembly of hexahedral blocks and a structured hexahedral mesh is employed within each block (see, for example, Chapter 13 and [1,31,33]). The unstructured mesh approach is attractive, as it offers the possibility of automating this procedure so that mesh generation times can be significantly reduced. In an unstructured mesh, the number of points and elements that are neighbors to an interior point will vary through the domain. This lack of regularity in the mesh means that the use of an unstructured mesh solution algorithm generally involves an additional cost, in terms of computer time and memory, when compared with its structured mesh counterparts. On the other hand, the unstructured mesh approach offers, as a counter-balance, a greater versatility and geometrical flexibility to the mesh generating process.

To take full advantage of these characteristics, the mesh generation procedure will be required to comply with the following requirements:

- The algorithm should be able to handle arbitrary geometries in a fully automatic manner and with minimum user intervention.
- The input data should be reduced to a computerized geometric representation of the domain to be discretized.
- The approach followed should provide control over the spatial variation of element size and shape through the domain.
- Adaptive methods should be incorporated into the process, with the objective of producing the most accurate approximation of the solution for a given number of points.

The algorithmic procedure for the generation of elements and nodes to be described in the following is a three-dimensional extension of the AFT method originally proposed in [25]. This method has been implemented in the FELISA system [24].

17.3 Geometry Modeling

The boundary of the computational domain has to be represented in a convenient mathematical form before the solution procedure can begin. As the objective is that the discretization of a domain of arbitrary geometrical complexity should be accomplished in an automatic manner; the method adopted to achieve this mathematical description ought to possess the greatest possible generality. In addition, the computer implementation of this description must provide the means for automatically computing any geometrical quantity relevant to the generation procedure. The area of solid modeling provides [29] the most general up-to-date set of methods for the computational representation and analysis of general shapes matching the above requirements.

In this section we give a brief description of the geometry modeling strategy that is employed. More sophisticated representations that ease the task of performing quick geometry modifications, could also be used [11].

17.3.1 Description of the Computational Domain

In the case of a planar two-dimensional analysis, the boundary of the computational domain is represented by closed loops of orientated composite cubic spline curves (cf. Chapter 27) [11]. For simply connected domains these boundary curves are oriented in a counter-clockwise sense while for multiply connected regions the exterior boundary curves are given a counterclockwise orientation and all the interior boundary curves are oriented in a clockwise sense (Figure 17.1).

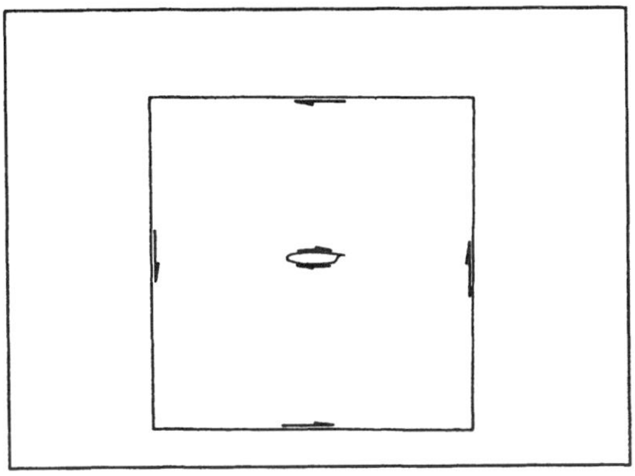

FIGURE 17.1 Boundary orientation for a two-dimensional domain.

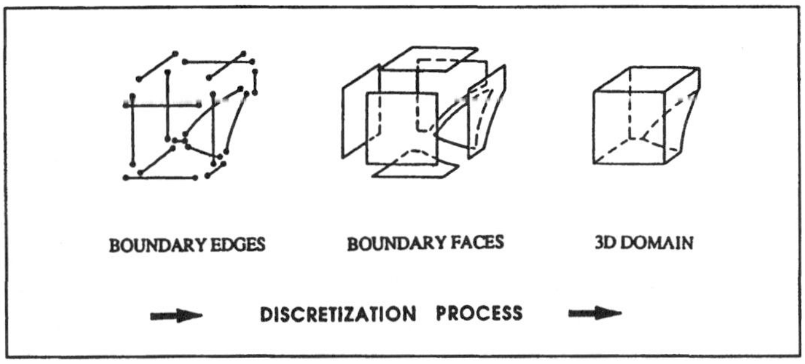

FIGURE 17.2 Decomposition of the boundary of a three-dimensional domain into its surface and curve components.

In three dimensions, following solid modeling ideas, the domain to be discretized is viewed as a region in R^3, which is bounded by a general polyhedron whose vertices are points on curved surfaces which intersect along curves. The edges of the polyhedron are segments on these intersection curves. In our notation, the portions of these curves and surfaces needed to define the boundary of the three-dimensional domain of interest are called *curve segments* and *surface regions*, respectively. A surface region is represented as a region — a patch — on a surface delimited by curve segments. Each curve segment is common to two surface regions and is a segment of the intersection curve between their respective support surfaces. Figure 17.2 shows the decomposition of the boundary of a three-dimensional domain into its surface and curve components. The approximate representation of the curves and surfaces where curve segments and surface regions is accomplished by means of composite curves and surfaces (Chapter 29 and [11]). These are called the curve and surface components.

In addition, boundary curves and surfaces are oriented (see Figure 17.3). This is important during the generation process as it defines the location of the region that is to be discretized. The orientation of a boundary surface is defined by the direction of the inward normal. The orientation of the boundary curves is defined with respect to the boundary surfaces that contain them. Each boundary curve will be common to two boundary surfaces and will have opposite orientations with respect to each of them.

17.3.2 Curve and Surface Representation

The problem of curve and surface representation is not considered here, as it has been described in detail in Part III of this Handbook.

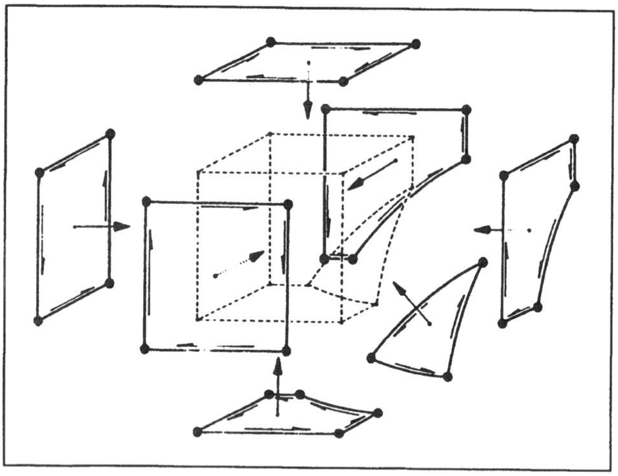

FIGURE 17.3 Orientation of the boundary components in three dimensions.

17.3.3 The Advancing Front Technique

The algorithmic procedure adopted for mesh generation is based upon the method originally proposed in [25] for two dimensions and then extended to three dimensions in [26, 27]. The approach is regarded as a generalization of the advancing front technique [14, 19] with the distinctive feature that elements, i.e., triangles or tetrahedra, and points are generated simultaneously. This enables the generation of elements of variable size and stretching and differs from the approach followed in tetrahedral generators based upon Delaunay concepts [2, 10], which generally connect mesh points that have already been distributed in space (Chapter 16).

The generation problem consists of subdividing an arbitrarily complex domain into a consistent assembly of elements. The consistency of the generated mesh is guaranteed if the generated elements cover the entire domain and the intersection between elements occurs only on common points, sides or triangular faces in the three dimensional case. The final mesh is constructed in what may be defined as a bottom-up manner. This means that the process starts by discretizing each boundary curve in turn. Nodes are placed on the boundary curve components and then contiguous nodes are joined with straight line segments. In the later stages of the generation process, these segments will become sides of triangular faces. The length of these segments must therefore, be consistent with the desired local distribution of mesh size. This operation is repeated for each boundary curve in turn.

The next stage consists of generating planar faces. For each two-dimensional region or surface to be discretized, all the sides produced when discretizing its boundary curves are assembled into the so-called initial front. The relative orientation of the curve components with respect to the surface must be taken into account in order to give the correct orientation to the sides in the initial front. This front is used to generated a triangular mesh on the surface. The size and shape of the generated triangles must be consistent with the local desired size and shape. These triangles will become faces of the tetrahedra to be generated later.

For the generation of tetrahedra the advancing front procedure is taken one step further. The front is now made up of the triangular faces that are available to form a tetrahedron. The initial front is obtained by assembling the triangulation of the boundary surfaces. Nodes and elements will be simultaneously created. When forming a new tetrahedron, the three nodes belonging to a triangular face from the front are connected either to an existing node or to a new node. After generating a tetrahedron, the front is updated. The generation procedure is completed when the number of triangles in the front is zero.

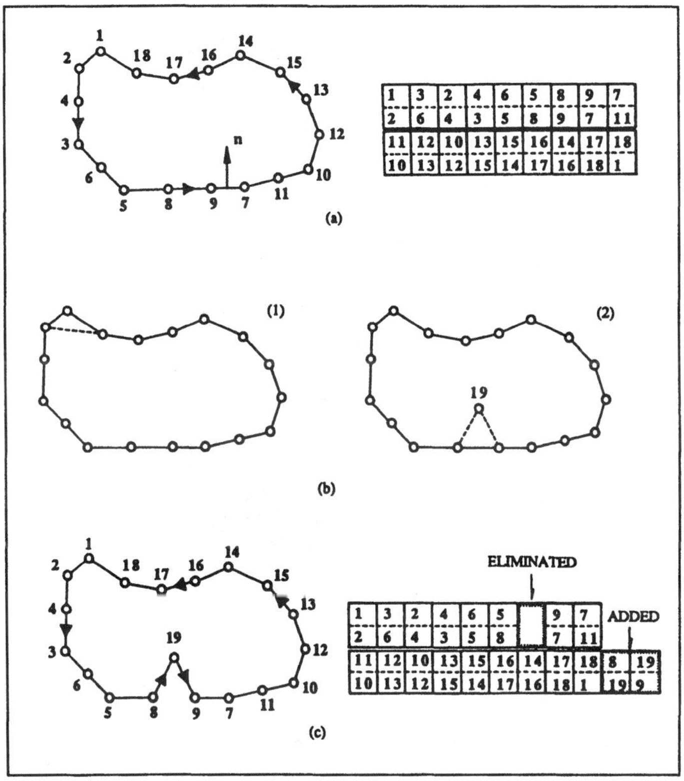

FIGURE 17.4 The front updating procedure in two dimensions. (a) The initial generation front. (b) Creation of a new element: (1) no new point is created; (2) the new point 19 is created. (c) The updating of the front for the case (b) (2).

17.3.4 Front Updating

The triangle generation algorithm utilizes the concept of a generation front. The front is a dynamic data structure that changes continuously during the generation process. At the start of the process the front consists of the sequence of straight line segments that connect consecutive boundary nodes. At any given time, the front contains the set of all the sides which are currently available to form a triangular face.

Any straight line segment that is available to form an element side is termed active, whereas any segment no longer active is removed from the front. During the generation process an active side is selected from the front and a triangular element is generated. This may involve creating a new node or simply connecting to an existing one. After the triangle has been generated, the front is updated. This updating process is illustrated in Figure 17.4.

Thus while the domain boundary will remain unchanged, the generation front changes continuously and needs to be updated whenever a new element is formed. The generation proceeds until the front is empty. Figure 17.5 illustrates the idea of the advancing front technique for a circular planar domain by showing the initial front and the form of the mesh at various stages during the generation process.

17.3.5 Characterization of the Mesh: Mesh Parameters

The geometrical characteristics of a general mesh are locally defined in terms of certain mesh parameters. If N is the number of dimensions (two or three) then the parameters used are a set of N mutually orthogonal directions α_i; $i = 1, \ldots, N$, and N associated element sizes δ_i; $i = 1, \ldots, N$ (see Figure 17.6). Thus, at a certain point, if all N element sizes are equal, the mesh in the vicinity of that point will consist

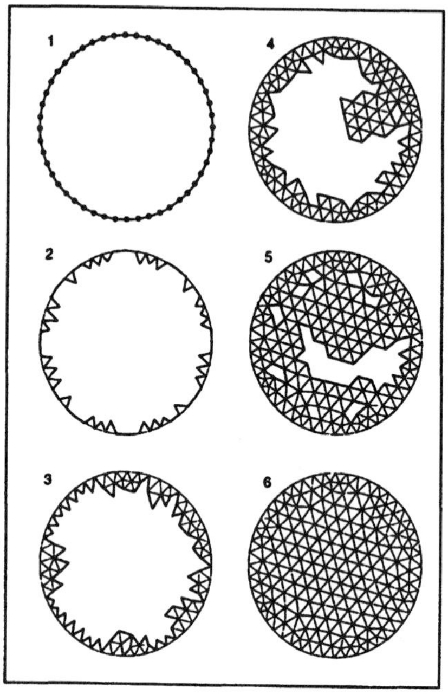

FIGURE 17.5 The advancing front technique showing different stages during the triangulation process.

of approximately equilateral elements. To aid the mesh generation procedure, a transformation **T** which is a function of α_i and δ_i is defined. This transformation is represented by a symmetric $N \times N$ matrix and maps the physical space onto a space in which elements, in the neighborhood of the point being considered, will be approximately equilateral with unit average size. This new space will be referred to as the normalized space. For a general mesh this transformation will be a function of position. The transformation **T** is the result of superimposing N scaling operations with factors $\frac{1}{\delta_i}$ in each α_i direction. Thus

$$\mathbf{T}(\alpha_i, \delta_i) = \sum_{i=1}^{N} \frac{1}{\delta_i} \alpha_i \otimes \alpha_i \tag{17.1}$$

where $\otimes$ denotes the tensor product of two vectors. The effect of this transformation in two dimensions is illustrated in Figure 17.7 for the case of constant mesh parameters throughout the domain.

17.3.6 Mesh Control

The inclusion of adequate mesh control is a key ingredient in ensuring the generation of a mesh of the desired form. Control over the characteristics is obtained by the specification of a spatial distribution of mesh parameters. This is accomplished by means of the *background mesh* supplemented by a *distribution of sources.*

17.3.7 Background Mesh

The background mesh is used for interpolation purposes only and is made up of triangles in two dimensions and tetrahedra in three dimensions. Values of α_i and δ_i are defined at the nodes of the background mesh. The background mesh employed must cover the region to be discretized (see Figure 17.8). In the generation of an initial mesh for the analysis of a particular problem, the background mesh will usually consist of a small number of elements. The generation of the background mesh can

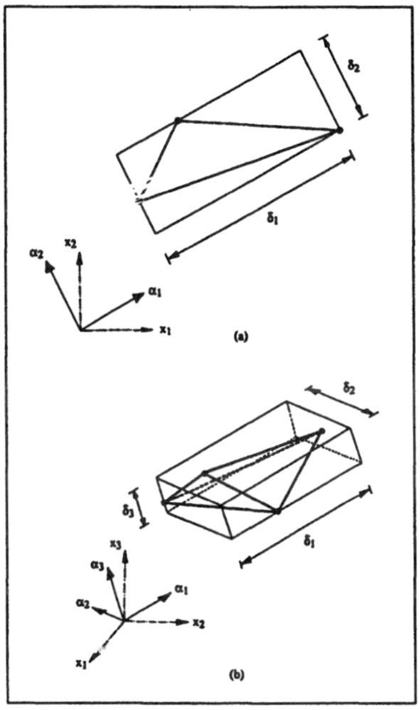

FIGURE 17.6 Characterization of the mesh: (a) the mesh parameters in two dimensions, (b) the mesh parameters in three dimensions.

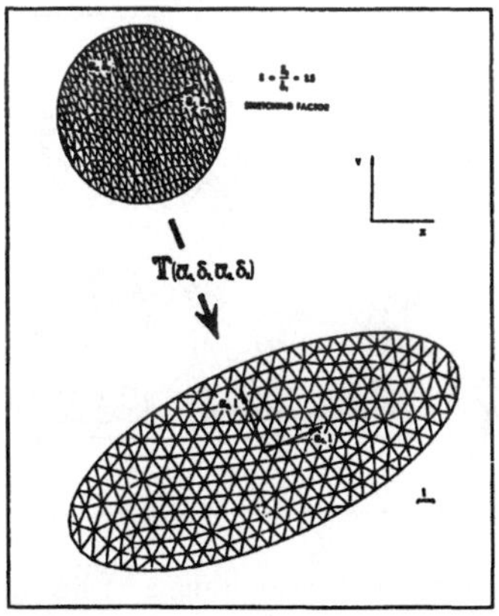

FIGURE 17.7 The effect of the transformation T for a constant distribution of the mesh parameters.

in this case be accomplished without resorting to sophisticated procedures, e.g., a background mesh consisting of a single element can be used to impose the requirement of a linearly varying or a constant spacing and stretching through the computational domain.

The effect of prescribing a variable mesh spacing and stretching is illustrated in Figure 17.9 for a problem involving a rectangular domain, using a background mesh consisting of two triangular elements.

17.3.8 Distribution of Sources

For complex geometries, the manual definition of a background mesh can become a very tiresome task. The use of a distribution of sources eases the problem of ensuring the desired specification of the mesh parameters in specific regions in the computational domain, such as the leading and trailing edges of wings. In this approach, an isotropic* spatial distribution of element sizes is specified as a function of

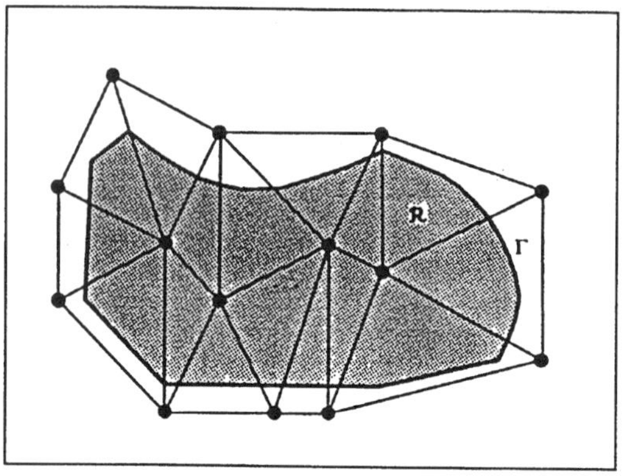

FIGURE 17.8 The background mesh for the specification of a spatial distribution of mesh parameters.

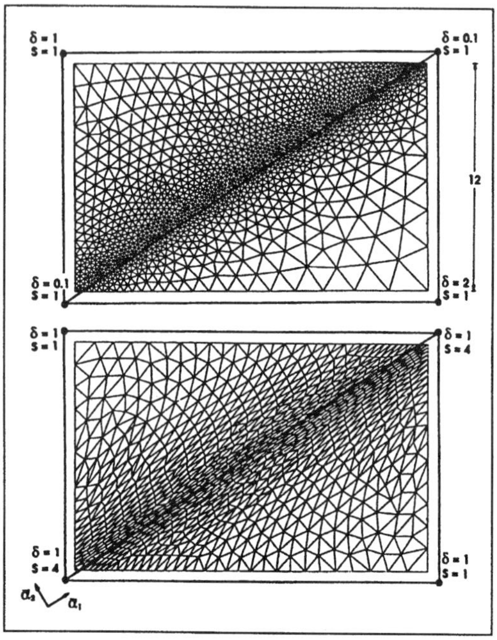

FIGURE 17.9 Mesh generated for a rectangular domain using a background mesh consisting of two elements to illustrate the effect of variable mesh spacing and stretching.

the distance x from the point of interest to a "source." The source may take the form of a point, a line segment or a triangle. The form adopted for the function is

$$\delta(x) = \begin{cases} \delta_1 & \text{if } x \leq x_c \\ \delta_1 e^{\left|\frac{x-x_c}{D-x_c}\right| \log 2} & \text{if } x \geq x_c \end{cases} \tag{17.2}$$

This function is local in character and allows for a rapid increase in element size, thus ensuring that the number of generated elements around the source can be kept within reasonable bounds. The quantities

*The spacing at a point is the same in all directions.

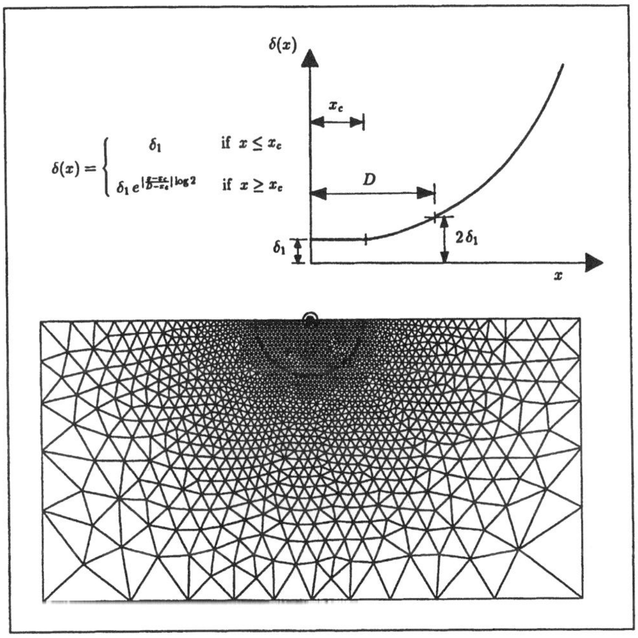

FIGURE 17.10 Mesh generated for a rectangular domain using a point source.

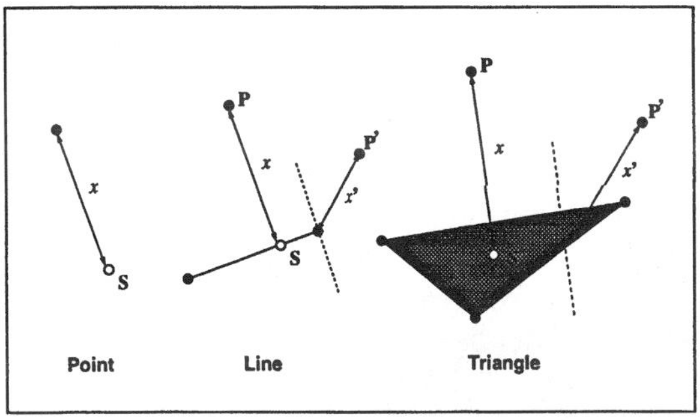

FIGURE 17.11 Point, line, and triangle sources.

δ_1, D, and x_c denote user-specified values that can be altered to control the form of $\delta(x)$. An example of a mesh produced by such a function is shown in Figure 17.10.

For line and triangle sources the spacing δ at a point **P** is defined in a similar manner. We choose the closest point **S** in the line or triangle to the point **P** — see Figure 17.11 — as a point source. The distance x is now the distance between the points **P** and **S** and the quantities δ_1, D, and x_c at point **S** are linearly interpolated from the nodal values at the points defining the line or triangle sources.

The spacing at a point is computed for the background mesh and for each of the user-defined point, line and triangle sources. The final spacing is computed as the minimum of all of them.

17.3.9 Calculation of the Transformation T

The generation process is always performed in a normalized space. The transformation **T**, given by Eq. 17.1, is repeatedly used to transform regions in the physical space into regions in the normalized space. In this way the process is greatly simplified, as the desired size for a side, triangle, or tetrahedron in this space is always unity. After the element has been generated, the coordinates of the newly created point, if any, are transformed back to the physical space, using the inverse transformation.

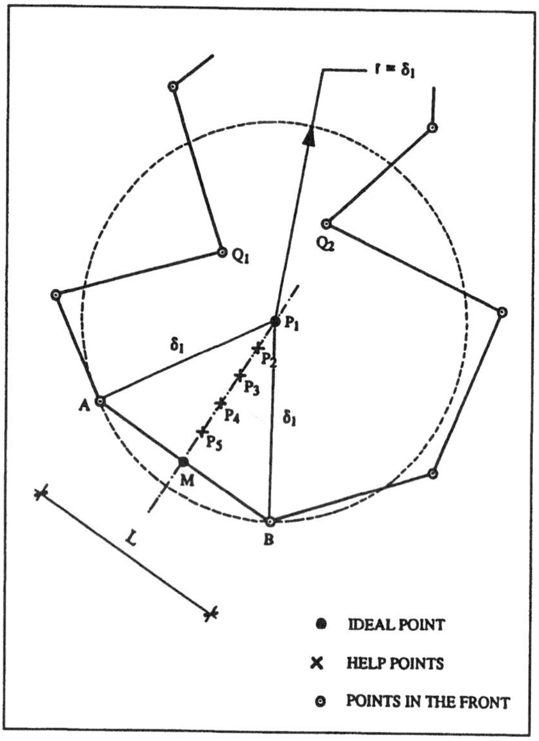

FIGURE 17.12 The generation of a new triangle.

At any point of the computational domain the transformation **T** is computed as follows. First, the element of the background mesh that contains the point is found and the transformation $\mathbf{T}^b$ is computed by linearly interpolating its components from the element nodal values. The stretching directions α_i^b and corresponding spacings δ_i^b; $i = 1, 2, 3$ are obtained from the eigenvalues and eigenvectors of the matrix $\mathbf{T}^b$. The spacings δ_i^b are then modified to account for the distribution of sources. The new spacings δ_i^* at the point are computed as the smallest of the spacings defined by all the sources and the current spacing δ_i^b. Finally, the transformation **T** is obtained by substituting the values of α_i^b and δ_i^* in the formula Eq. 17.1.

17.3.10 Curve Discretization

The discretization of the boundary curve compounds is achieved by positioning nodes along the curve according to a spacing dictated by the local value of the mesh parameters. Consecutive points are joined by straight lines to form sides. The manner by which this can be accomplished has been described in detail in Chapter 16.

17.3.11 Triangle Generation in Two-Dimensional Domains

The triangle generation algorithm utilizes the concept of a generation front. At the start of the process the front of the sequence of straight-line segments that connect consecutive boundary nodes. During the generation process, any straight-line segment that is available to form an element side is termed active, whereas any segment that is no longer active is removed from the front. Thus, while the domain boundary will remain unchanged, the generation front changes continuously and needs to be updated whenever a new element is formed along the steps described in Section 17.3.4.

In the process of generating a new triangle the following steps are involved (Figure 17.12):

1. Select a side *AB* of the front to be used as a base for the triangle to be generated. Here, the criterion is to choose the shortest side. This is especially advantageous when generating irregular meshes.
2. Interpolate from the background mesh the transformation **T** at the center of the side *M* and apply it to the nodes in the front that are relevant to the triangulation. In our implementation we define the relevant points to be all those that lie inside the circle of center *M* and radius three times the length of the side being considered. Let $\hat{A}$, $\hat{B}$ and $\hat{M}$ denote the positions in the normalized space of the points *A, B,* and *M,* respectively.
3. Determine, in the normalized space, the ideal position $\hat{P}_1$ for the vertex of the triangular element. The point $\hat{P}_1$ is located on the line perpendicular to the side that passes through the point $\hat{M}$ and at a distance δ_1 from the points $\hat{A}$ and $\hat{B}$. The direction in which $\hat{P}_1$ is generated is determined by the orientation of the side. The value δ_1 is chosen according to the criterion

$$\delta_1 = \begin{cases} 1.00 & \textit{if } 0.55\times L < 1.00 < 2.00\times L \\ 0.55\times L & \textit{if } 0.55\times L < 1.00 \\ 2.00\times L & \textit{if } 1.00 > 2.00\times L \end{cases} \tag{17.3}$$

   where L is the distance between points $\hat{A}$ and $\hat{B}$. Only in situations where the side *AB* happens to have characteristics very different from those specified by the background mesh will the value of δ_1 be different from unity. However, the above inequalities must be taken into account to ensure geometrical compatibility. Expression (3) is pure empirical, and different inequalities could be devised to serve the same purpose.
4. Select other possible candidates for the vertex and order them in a list. Two types of points are considered viz. (a) all the nodes $\hat{Q}^1$, $\hat{Q}^2$, … in the current generation front that are, in the normalized space, interior to a circle with center $\hat{P}^1$ and radius $r = \delta_1$, and (b) the set of points $\hat{P}^1, \ldots, \hat{P}^5$ generated along the height $\hat{P}^1\hat{M}$. For each point $\hat{Q}^i$, construct the circle, with center $\hat{C}_Q^i$ on the line defined by points $\hat{P}^1$ and $\hat{M}$, and that passes through the points $\hat{Q}^i$, $\hat{A}$ and $\hat{B}^1$. The position of the centers, $\hat{C}_Q^i$, of these circles, on the line $\hat{P}^1\hat{M}$, defines an ordering of the $\hat{Q}^i$ points. A list is created that contains all the $\hat{Q}^i$ points in which the point with the furthest center from $\hat{P}^1$ in the direction $\hat{P}^1\hat{M}$ appears at the head of list. The points $\hat{P}^1, \ldots, \hat{P}^5$ are added at the end of this list.
5. Select the best connecting point. This is the first point in the order list which gives a consistent triangle. Consistency is guaranteed by ensuring that none of the newly created sides intersects with any of the existing sides in the front.
6. Finally, if a new node is created, its coordinates in the physical space are obtained by using the inverse transformation $\mathbf{T}^{-1}$.
7. Store the new triangle and update the front by adding/removing the relevant sides.

This mesh generation procedure is schematically presented in the diagram shown in Figure 17.13.

17.3.12 Mesh Quality Enhancement

In order to enhance the quality of the generated mesh, two post-processing procedures are applied. These procedures, which are local in nature, do not alter the total number of points or elements in the mesh.

- *Diagonal swapping:* This changes the connectivities among nodes in the mesh without altering their position. This process requires a loop over all the element sides, excluding those sides on the boundary. For each side *AB* (Figure 17.14), common to the triangles *ABC* and *ADB,* one considers the possibility of swapping *AB* by *CD,* thus replacing the two triangles *ABC* and *ADB* by the triangles *ADC* and *BCD.* The swapping is performed if a prescribed regularity criterion is better satisfied by the new configuration than by the existing one. In our implementation, the swapping operation is performed if the minimum angle occurring in the new configuration is larger than that in the original configuration.

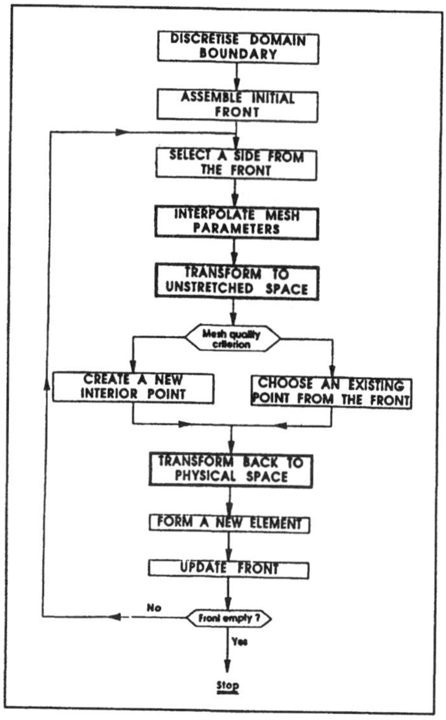

FIGURE 17.13 Flow chart for mesh generation using the advancing front technique. Double lined boxes are only required if the effect of variable mesh size and stretching are to be included.

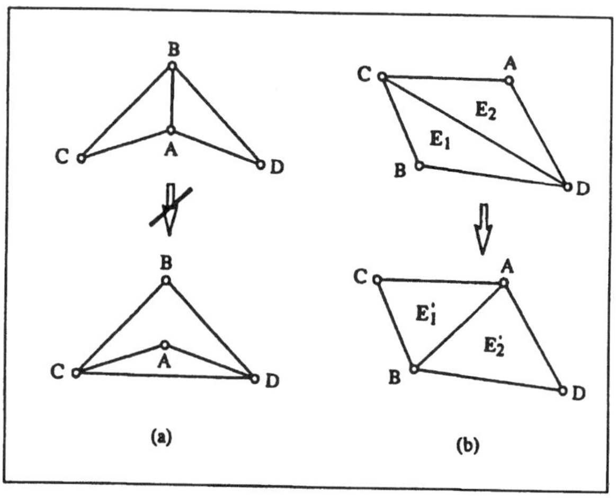

FIGURE 17.14 The diagonal swapping procedure: (a) nonadmissible, (b) admissible.

- *Mesh smoothing*: This alters the positions of the interior nodes without changing the topology of the mesh. The element sides are considered as springs and the stiffness of a spring is assumed to be proportional to its length. The nodes are moved until the spring system is in equilibrium. The equilibrium positions are found by iteration. Each iteration amounts to performing a loop over the interior points and moving their coordinates to coincide with those of the centroid of the neighboring points. Usually three to five iterations are performed.

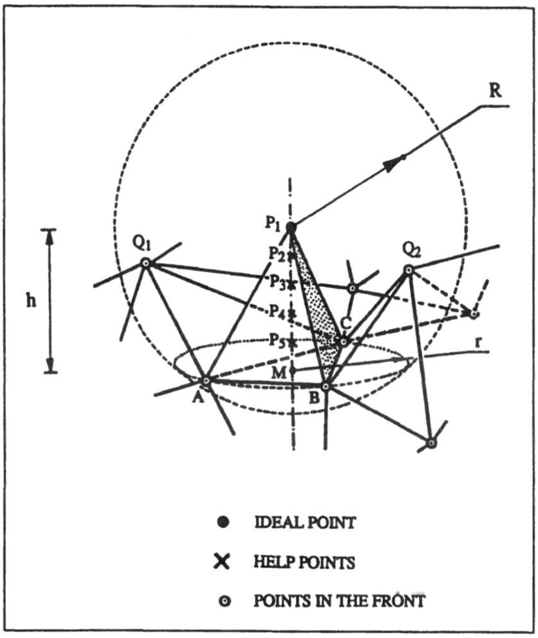

FIGURE 17.15 The generation of a tetrahedral element.

The combined application of these two post-processing algorithms is found to be very effective in improving the smoothness and regularity of the generated mesh.

17.3.13 Surface Discretization

The method followed for the triangulation of the surface components is an extension of the mesh generation procedure for planar domains described above. The discretisation of each surface component is accomplished by generating a two-dimensional mesh of triangles in the parametric plane (u_1, u_2) and then using the mapping $\mathbf{r}(u_1, u_2)$ defined in Section 17.3. This mapping establishes a one-to-one correspondence between the surface component and a region on the parametric plane (u_1, u_2). Thus, a consistent triangular mesh in the parametric plane will be transformed, by the mapping $\mathbf{r}(u_1, u_2)$, into a valid triangulation of the surface component. The construction of the triangular mesh in the parameter plane (u_1, u_2) using the two dimensional mesh generator, requires the determination of an appropriate spatial distribution of the two dimensional mesh parameters. This problem has been addressed in detail in Chapter 19.

17.3.14 Generation of Tetrahedra

The starting point for the discretization of the three-dimensional domain into tetrahedra is the formation of the initial generation front. The initial front is the set of oriented triangles that constitutes the discretized boundary of the domain and is formed by assembling the discretized boundary surface components. The order in which the nodes of these triangles are given defines the orientation, which is the same as that of the corresponding boundary surface component. The algorithm for generating tetrahedra is analogous to that described above for the generation of triangles (see Figure 17.13). However, in the three-dimensional case the range of possible options at each stage is much wider and the number of geometrical operations involved increases considerably. Thus, the ability of the method to produce a mesh, and the efficiency of its implementation, relies heavily upon the type of strategy selected. The generation of a generic tetrahedral element involves the following steps (Figure 17.15):

1. Select a triangular face *ABC* from the front to be a base for the tetrahedron to be generated. In principle, any face could be chosen, but we have found it to be advantageous in practice to consider the smallest faces first. For this purpose, the size of the face is defined in terms of the size of its shortest height.
2. Interpolate from the background mesh the transformation **T** at the centroid of the face *M* and apply it to the nodes in the front that are relevant to the triangulation. In our implementation, we define the relevant points to be those which lie inside the sphere of center *M* with radius equal to three times the value of the maximum dimension of the face being considered. Let $\hat{A}$, $\hat{B}$, $\hat{C}$ and $\hat{M}$ denote the positions in the normalized space of the points *A, B, C,* and *M,* respectively.
3. Determine, in the transformed space, the ideal position $\hat{P}^1$ for the vertex of the tetrahedral element. The point $\hat{P}^1$ lies on the line that passes through the point $\hat{M}$ and is perpendicular to the face. The direction in which $\hat{P}^1$ is generated is determined by the orientation of the face. The location of $\hat{P}^1$ is computed so that the average length of the three newly created sides, which join point $\hat{P}^1$ with points $\hat{A}$, $\hat{B}$, and $\hat{C}$, is unity. For faces whose size in the parametric plane is very different from unity, this step may have to be modified, as in Eq. 17.3, to ensure geometrical compatibility. However, such cases rarely occur in practice. Let δ_1 be the maximum of the distances between point $\hat{P}^1$ and points $\hat{A}$, $\hat{B}$, and $\hat{C}$.
4. Select other possible candidates for the vertex and order them in a list. Two types of points are considered viz. (a) all the nodes $\hat{Q}^1$, $\hat{Q}^2$, ... in the current generation front which are, in the normalized space, interior to a sphere with center $\hat{M}$ and radius $r = \delta_1$, and (b) a new set of points $\hat{P}^1, \ldots, \hat{P}^5$ generated along the height $\hat{P}^1\hat{M}$. Consider the set of points $\hat{A}$, $\hat{B}$, and $\hat{C}$ and denote by $\hat{D}$ the member of this set that is furthest away from $\hat{M}$. For each point $\hat{Q}^i$, construct the sphere with center $\hat{C}_Q^i$ on the line defined by points $\hat{P}^1$ and $\hat{M}$ and which passes though points $\hat{Q}^i$ and $\hat{D}$. The position of the centers $\hat{C}_Q^i$ of these spheres on the line $\hat{P}^1\hat{M}$ defines an ordering of the $\hat{Q}^1$ points in which the point with the furthest center from $\hat{P}^1$ in the direction $\hat{P}^1\hat{M}$ appears at the head of list. The points $\hat{P}^1, \ldots, \hat{P}^5$ are added at the end of this list.
5. Select the best connecting point. This is the first point in the ordered list that gives a consistent tetrahedron. Consistency is guaranteed by ensuring that none of the newly created sides intersects with any of the existing faces in the front, and that none of the existing sides in the front intersect with any of the newly created faces.
6. If a new node is created, its coordinates in the physical space are obtained by using the inverse transformation $\mathbf{T}^{-1}$.
7. Store the new tetrahedron and update the front by adding/removing the necessary triangles.

17.3.15 Mesh Quality Assessment

Any discussion of mesh quality should be intimately related to the form of the solution we are trying to represent on that mesh. Two factors need to be considered here:

1. Determination of the characteristics of the optimal mesh for the problem at hand. This introduces the concept of adaptivity and this aspect is considered elsewhere.
2. Assessment on how well the generated mesh meets the requirements specified by the mesh parameters. This assessment can be made by examining the generated mesh and determining the statistical distribution of certain indicators. For example, in Figure 17.16 we have chosen as indicators the number of elements around a side, the magnitude of the element dihedral angles, and the length of the side. These indicators are compared with optimal values, i.e., those of a regular tetrahedron that has the exact dimensions specified by the mesh parameters.

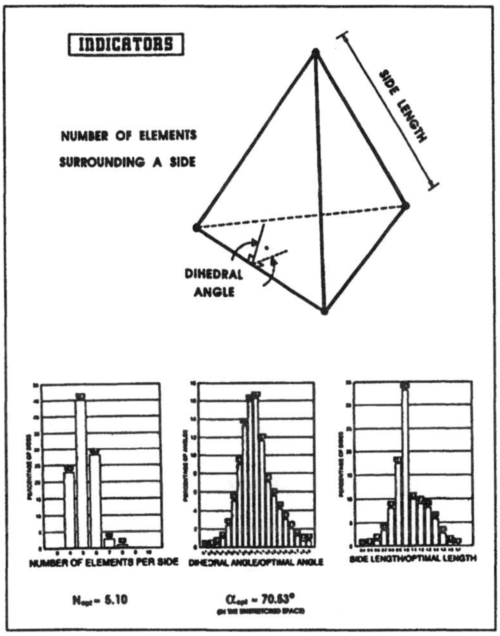

FIGURE 17.16 Mesh quality statistics.

17.4 Data Structures

From Section 17.2 it is apparent that a successful implementation of the mesh generation algorithm will require the use of data structures that enable certain sorting and searching operations to be performed efficiently. For instance, the generation front will require a data structure that allows for the efficient insertion/deletion of sides/faces and that also allows for the efficient identification of the sides/faces that intersect with a prescribed region in space.

The problem of determining the members of a set of n points that lie inside a prescribed subregion of an N-dimensional space is known as geometric searching. Several algorithms have been proposed [3, 30, 9] that solve this problem, or equivalent problems, with a computational expense proportional to $\log(n)$. The problem complexity increases considerably when, instead of considering points, one deals with finite size objects such line segments, triangles, or tetrahedra. A common problem encountered here, namely geometric intersection, consists of finding the objects that overlap a certain subregion of the space being considered. The algorithm adopted here for solving these problems in three dimensions is based on the use of the alternating digital tree [7].

The alternating digital tree (ADT) algorithm allows for the efficient solution of the geometric searching and intersection problem. It naturally offers the possibility of inserting and removing points and optimally searching for the points contained inside a given region. The ADT algorithm is an extension of the so-called digital binary tree search technique, which is exhaustively used in [18] for one-dimensional problems. It is applicable to any number of dimensions, and allows any geometrical object in an N-dimensional space to be treated as a point in a $2N$-dimensional space.

The following sections describe the ADT algorithm, and the associated data structures employed, for the efficient solution of the geometric searching problem.

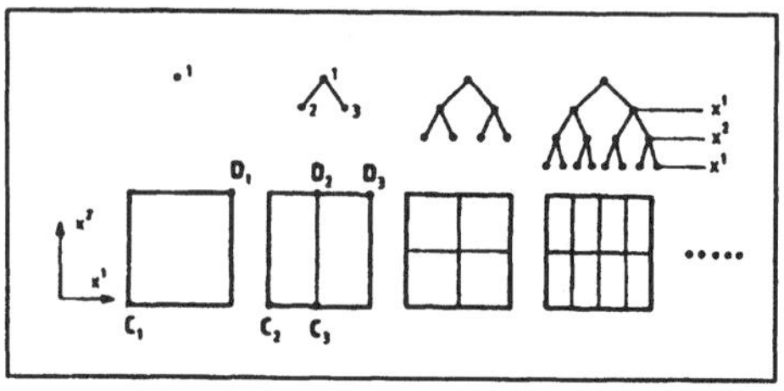

FIGURE 17.17 The relation between a binary tree and a bisection process.

17.4.1 The Alternating Digital Tree

Binary trees provide the basis for several searching algorithms, including the one to be presented here. A detailed exposition of binary tree structures can be found in Chapter 14 and references therein.

Consider a set of n points in a N-dimensional space (R^N) and assume for simplicity that the coordinate values of their position vectors $\mathbf{x}_1, \mathbf{x}_2, \ldots, \mathbf{x}_n$, after adequate scaling, vary within the interval [0, 1). The aim of geometric searching algorithms is to select from this set those points that lie inside a given subregion of the space. To facilitate their representation, only rectangular — or "hyper-rectangular" — regions will be considered, thereby allowing their definition in terms of the scaled coordinates of the lower and upper vertices as $(\mathbf{a}, \mathbf{b})$.

Comparing the coordinates of each point k with the vertex coordinates of a given subregion, to check whether the condition $a^i \le x^i_k \le b^i$ is satisfied for $i = 1, 2, \ldots, N$, would render the cost of the searching operation proportional to the number of points n. This computational expense, however, can be substantially reduced by storing the points in a binary tree, in such a way that the structure of the tree reflects the positions of the points in space. There exist several well-known algorithms that will accomplish this effect for one dimensional problems; the most popular are the binary search tree and digital tree methods [18]. Binary search trees have been extended to N-dimensional problems in [6], but the resulting tree structure, known as N-d trees, do not allow for the efficient deletion of nodes. The algorithm presented here is a natural extension of the one- dimensional digital tree algorithm and overcomes the difficulties encountered in N-d trees.

Broadly speaking, an alternating digital tree can be defined as a binary tree in which a set of n points are stored according to certain geometrical criteria. These criteria are based on the similarities arising between the hierarchical and parental structure of a binary tree and a recursive bisection process: each node in the tree has two sons, likewise a bisection process divides a given region into two smaller subregions. Consequently, it is possible to establish an association between tree nodes and subregions of the unit hypercube as follows: the root represents the unit hypercube itself; this region is now bisected across the x^1 axis and the region for which $0 \le x^1 < 0.5$ is assigned to the left son and the region for which $0.5 \le x^1 < 1$ is assigned to the right son; at each of these nodes the process is repeated across the x^2 direction as shown in Figure 17.17. In a two-dimensional space this process can be repeated indefinitely by choosing x^1 and x^2 directions in alternating order; similarly, in a general N-dimensional space, the process can be continued by choosing directions $x^1, x^2, \ldots, x^N$ in cyclic order.

Generally, if a node k at the hierarchy level m — the root being level 0 — represents a region $(\mathbf{c}_k, \mathbf{d}_k)$, the subregions associated to its left and right sons, (c^l_k, d^l_k) and (c^r_k, d^r_k) result from the bisection of $(\mathbf{c}_k, \mathbf{d}_k)$ by a plane normal to the jth coordinate axis, where j is shown cyclically from the N space directions as:

$$j = 1 + \text{mod}(m, N) \tag{17.4}$$

and mod(m, N) denotes the remainder of the quotient of m over N. Hence (c^l_k, d^l_k) and (c^r_k, d^r_k) are obtained as

$$c^i_{kl} = c^i_k,\ d^i_{kl} = d^i_k \quad \text{for } i \ne j \quad \text{and} \quad c^j_{kl} = c^j_k,\ d^j_{kl} = \frac{1}{2}\left(c^j_k + d^j_k\right) \tag{17.5}$$

$$c^i_{kr} = c^i_k,\ d^i_{kr} = d^i_k \quad \text{for } i \neq j \quad \text{and} \quad c^j_{kl} = \frac{1}{2}\left(c^j_k + d^j_k\right),\ d^j_{kl} = d^j_k \tag{17.6}$$

This correlation between nodes and subdivisions of the unit hypercube allows an ADT to be further defined by imposing that each point in the tree should lie inside the region corresponding to the node where it is stored. Consequently, if node k of an ADT structure contains a point with coordinates $\mathbf{x}_k$, the following condition must be satisfied:

$$c^i_k \leq x^i_k < d^i_k \quad \text{for} \quad i = 1,2,\ldots,N \tag{17.7}$$

17.4.2 Geometric Searching

Consider now a set of points stored in an ADT structure. The fact that Eq. 17.7 is satisfied by every point provides the key to the efficient solution of a geometric searching problem. To illustrate this, note first that the recursive structure of the bisection process described above implies that the region related to a given node k contains all the subregions related to notes descending from k; consequently, all points stored in these nodes must also lie inside the region represented by node k. For instance, all points in the ADT structure are stored in nodes descended from the root and, clearly, all of them lie inside the unit hypercube — the region associated with the root. Analogously, the complete set of points stored in any subtree is inside the region represented by the root of the subtree.

This feature can be effectively used to reduce the cost of a geometric searching process by checking, at any node k, the intersection between the searching range $(\mathbf{a}, \mathbf{b})$ and the region represented by node k, namely $(\mathbf{c}_k, \mathbf{d}_k)$. If these two regions fail to overlap, then the complete set of points stored in the subtree rooted at k can be disregarded from the search, thus avoiding the need to examine the coordinates of every single point.

Consequently, a systematic procedure to select the points that lie inside a given searching range $(\mathbf{a}, \mathbf{b})$ can be derived from the traversal algorithm previously presented. Now the generic operation "visit the root" can be reinterpreted as checking whether the point stored in the root falls inside the searching range. Additionally, the left and right subtrees need to be traversed only if the regions associated with their respective root nodes intersect with the range.

17.4.3 Geometric Intersection

Geometrical intersection problems can be found in many applications; for instance, a common problem that may emerge in contact algorithms [8], hidden-line removal applications, or in the advancing front mesh generation algorithm presented in Section 17.2, is to determine from a set of three-noded triangular elements those which intersect with a given line segment. Similar problems, involving other geometrical objects, are encountered in a wide range of geometrical applications. In general, a geometric intersection problem consists of finding from a set of geometrical objects those which intersect with a given object. If every one-to-one intersection is investigated, the solution of these problems can become very expensive, especially when complex objects such as curves or surfaces are involved. Fortunately, many of these one-to-one intersections can be quickly discarded by means of a simple comparison between the coordinate limits of every given pair of objects. For instance, a triangle with x-coordinate varying from 0.5 and 0.7 cannot intersect with a segment with x-coordinate ranging from 0.1 to 0.3. Generally, the intersection between two objects in the N-dimensional Euclidean space, requires each of the N pairs of coordinate ranges to overlap. Consider for instance the intersection problem between triangular facets and a target straight line segment in R^3; then, if $(\mathbf{x}_{k,min}, \mathbf{x}_{k,max})$ are the coordinate limits of element k and $(\mathbf{x}_{0,min}, \mathbf{x}_{0,max})$ are the lower and upper limits of the target segment (see Figure 17.18), an important step toward the solution of a geometric intersection problem is to select those that satisfy the inequality

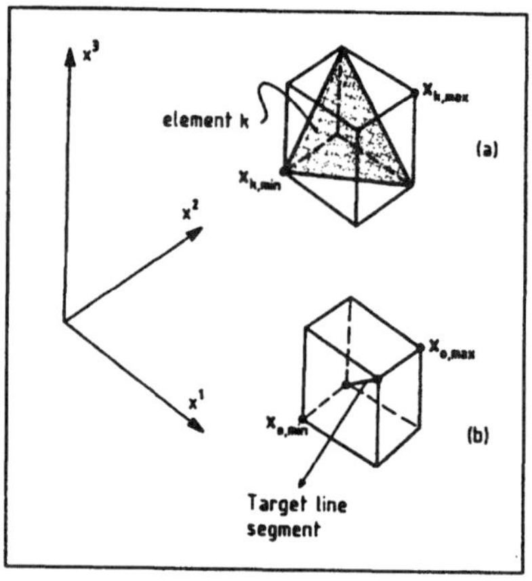

FIGURE 17.18 The definition of coordinate limits for triangular elements and straight line segments.

$$\begin{aligned} x^i_{k,min} &\le x^i_{0,max} \\ & \qquad \text{for} \quad i = 1,2,\ldots,N \\ x^i_{k,max} &\le x^i_{0,min} \end{aligned} \tag{17.8}$$

The cost of checking Eq. 17.8 for every element grows proportionally to n, and for very numerous sets may become prohibitive. This cost, however, can be substantially reduced by using a simple device whereby the process of selecting those elements that satisfy Eq. 17.8 can be interpreted as a geometric searching problem. Additionally, since the number of elements that satisfy Eq. 17.8 will normally be much smaller than n, the cost of determining which of these intersects with the target segment becomes affordable.

In order to interpret Eq. 17.8 as a geometric searching problem, it is first convenient to assume that all the elements to be considered lie inside a unit hypercube — a requirement that can be easily satisfied through adequate scaling of the coordinate values. Consequently, Eq. 17.8 can be rewritten as

$$\begin{aligned} 0 \le\ & x^1_{k,min} \le x^1_{0,max} \\ & \vdots \\ 0 \le\ & x^N_{k,min} \le x^N_{0,max} \\ x^1_{0,min} \le\ & x^1_{k,max} \le 1 \\ & \vdots \\ x^N_{0,min} \le\ & x^N_{k,max} \le 1 \end{aligned} \tag{17.9}$$

Consider now a given object k in R^N with coordinate limits $\mathbf{x}_{k,min}$, and $\mathbf{x}_{k,max}$; combining this two sets of coordinate values, it is possible to view an object k in R^N as a point in R^{2N} with coordinates x^i_k for i = 1, 2, …, $2N$ defined as (see Figure 17.19):

$$\mathbf{x}_k = \left[x^1_{k,min} \ldots,\ x^N_{k,min},\ x^1_{k,max}, \ldots,\ x^N_{k,max}\right]^T \tag{17.10}$$

Using this representation of a given object k, Eq. 17.9 becomes simply:

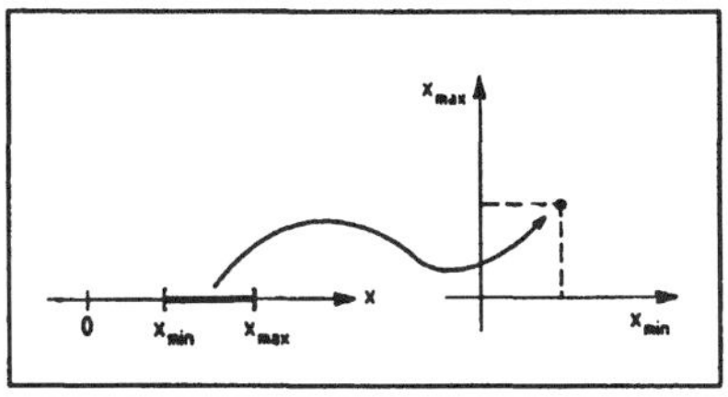

FIGURE 17.19 The representation of a region in R^1 as a point in R^2.

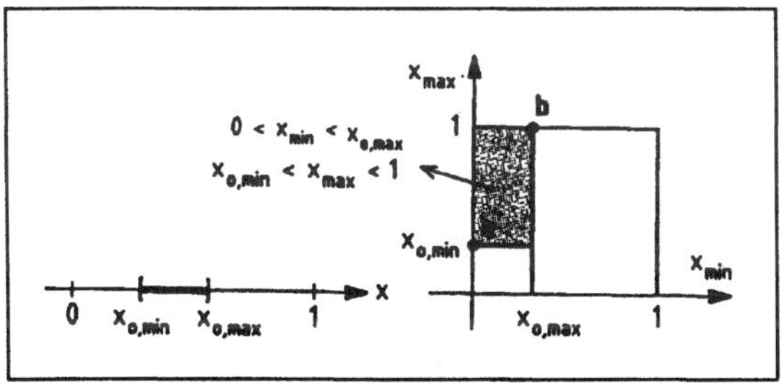

FIGURE 17.20 The intersection problem in R^1 as a searching problem in R^2.

$$a^i \leq x_k^i \leq b^1 \quad for \quad i = 1,2,\ldots,2N \tag{17.11}$$

where **a** and **b** can be interpreted as the lower and upper vertices of a "hyper-rectangular" region in R^{2N} and, recalling Eq. 17.9, their components can be obtained in terms of the coordinate limits of the target object (see Figure 17.20) as

$$\begin{aligned} \mathbf{a} &= \left[0,\ldots,0,x_{0,max}^1,\ldots,x_{o,max}^N\right]^T \\ \mathbf{b} &= \left[x_{0,min}^1,\ldots,x_{0,min}^N,1,\ldots,1\right]^T \end{aligned} \tag{17.12}$$

Consequently, the problem of finding which objects in R^N satisfy Eq. 17.8 becomes equivalent to a geometric searching problem in R^{2N}, i.e., obtaining the points $\mathbf{x}_k$ that lie inside the region limited by **a** and **b**. Once this subgroup of elements has been selected, the intersection of each one of them with the target object must be checked to complete the solution of the geometric intersection problem.

17.4.4 The Use of the ADT for Mesh Generation

The advancing front algorithm described in Section 17.2 requires frequent use of operations such as searching, for the points inside a certain region of the space, and determining intersections between geometrical objects — in this case sides and faces. The complexity of the problem is increased by the fact that the set of faces forming the generation front changes continuously as new faces need to be inserted and deleted during the process. Clearly, for meshes consisting of a large number of elements the cost of performing this operations can be very important.

A successful implementation of the above algorithms has been accomplished by making extensive use of the ADT data structure. For instance, the algorithm for tetrahedra generation employs two tree structures; one for the faces in the front and the other for the sides defined by the intersection between each pair of faces in the front. This combination allows a high degree of flexibility so that the operations

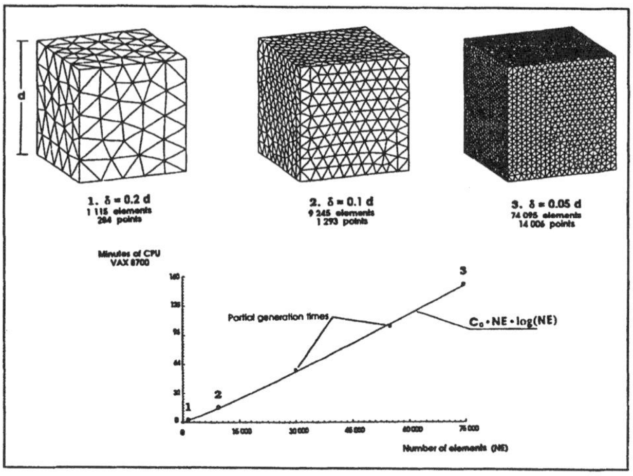

FIGURE 17.21 Mesh generation CPU times.

of insertion, deletion, geometric searching, and geometric intersection can be performed optimally. The overall computational performance of the algorithm is demonstrated by generating tetrahedral meshes, using the above method, for a unit cube (see Figure 17.21). Different numbers of elements have been obtained by varying the mesh size. In Figure 17.21 the computer time required on a VAX 8700 machine has been plotted against the number *NE* of elements generated. It can be observed that a typical $NE \times \log(NE)$ behavior is attained. Using this approach, meshes containing up to one million elements have been generated and no degradation in the performance has been detected.

17.5 Conclusions

A detailed description of the basics of a mesh generation procedure, based upon advancing front concepts, has been presented. Although no meshes for three-dimensional computational domains have been included, there are numerous examples in the literature of the power of the approach when it is applied to the problem of discretizing three dimensional domains of general complex shape [26, 17]. Recent implementations [22] have been shown to be extremely robust and achieve a high level of computational efficiency.

References

1. **Allwright, S.**, Multiblock topology specification and grid generation for complete aircraft configurations, *Applications of Mesh Generation to Complex 3D Configurations,* AGARD Conference Proceedings, No. 464, pp. 11.1–11.11, 1990.
2. **Baker, T. J.**, Unstructured mesh generation by a generalized Delaunay algorithm, *Applications of Mesh Generation to Complex 3D Configurations,* AGARD Conference Proceedings 1990, No. 464, pp. 20.1–20.10.
3. **Bentley J. L. and Friedman, J.H.**, Data structures for range searching, *Computing Surveys,* 11, No 4, 1979.
4. **Blacker T. D. and Sthepenson, M.B.**, Paving: a new approach to automated quadrilateral mesh generation, *Int. J. Num. Meth. Eng.,* 32, pp. 811–847, 1991.
5. **Blacker T. D. and Meyers, R.J.**, Seams and wedges in plastering: a 3d hexahedral mesh generation algorithm, *Eng. Computers,* 9, pp. 83–93, 1993.
6. **Bentley, J. L.**, Multidimensional binary search trees used for associative searching, *Comm. ACM.* 18, No 1, 1975.

7. **Bonet J. and Peraire, J.**, An alternating digital tree (adt) algorithm for geometric searching and intersection problems, *Int. J. Num. Meth. Eng.*, 31, pp. 1–17, 1990.
8. **Bonet, J.**, Finite element analysis of thin sheet superplastic forming process, Ph.D. Thesis, University of Wales, C/PhD/128/89, 1989.
9. **Boris, J.**, A vectorised algorithm for determining the nearest neighbours, *J. Comp. Phys.*, 66, pp. 1–20, 1986.
10. **Cavendish, J. C., Field, D. A., and Frey, W. H.**, An approach to automatic three dimensional finite element mesh generation, *Int. J. Num. Meth. Eng.*, 21, pp 329–348, 1985.
11. **Faux, I. D. and Pratt, M. J.**, *Computational Geometry for Design and Manufacture*, Ellis Horwood, Chichester, 1981.
12. **Formaggia, L.**, An unstructured mesh generation algorithm for three-dimensional aircraft configurations, *Numerical Grid Generation in CFD and Related Fields.* (Ed.) Sanchez-Arcilla, A., et al., North-Holland, pp. 249–259, 1991.
13. **Frykestig, J.**, Advancing front mesh generation techniques with application to the finite element method, Dept. of Structural Mechanics Publication 94, 10, Chalmers University of Technology, Göteborg, Sweden, 1994.
14. **George, A. J.**, Computer implementation of the finite element method, Ph.D. Thesis, Stanford University, STAN–CS–71–208, 1971.
15. **Golgolab, A.**, Mailleur 3D automatique pour des géométries complexes, INRIA Research Report No 1004, March 1989.
16. **Huet, F.**, Generation de maillage automatique dans des configurations tridimensionnelles complexes. Utilisation d'une Methode de Front, *Applications of Mesh Generation to Complex 3D Configurations*, AGARD Conference Proceedings, No. 464, pp 17.1–17.12, 1990.
17. **Jin H. and Tanner, R.I.**, Generation of unstructured tetrahedral meshes by advancing front technique, *Int. J. Num. Meth. Eng.*, 36, pp 1805–1823, 1993.
18. **Knuth, D.**, *The Art Of Computer Programming — Sorting And Searching*, Vol. 3, Addison-Wesley, 1973.
19. **Lo, S. H.**, A new mesh generation scheme for arbitrary planar domains, *Int. J. Num. Meth. Eng.*, 21, pp. 1403–1426, 1985.
20. **Lo, S. H.**, Volume discretization into tetrahedra — II. 3D triangulation by advancing front approach, *Comp. Struct.*, 39, No 5, pp. 501–511, 1991.
21. **Löhner R. and Parikh, P.**, Generation of three-dimensional unstructured grids by the advancing-front method, AIAA Paper AIAA-88-0515, 1988.
22. **Löhner, R.**, Extensions and improvements of the advancing front grid generation technique, *Comm. Num. Meth. Eng.*, 12, pp 683–702, 1996.
23. **Möller P. and Hansbo, P.**, On advancing front mesh generation in three dimensions, *Int. J. Num. Meth. Eng.*, 38, pp. 3551–3569, 1995.
24. **Peiró, J., Peraire J., and Morgan, K.**, **FELISA** system reference manual. Part **I:** basic theory, Civil Eng. Dept. Report, CR/821/94, University of Wales, Swansea, U.K., 1994. (More information about the **FELISA** system is available at http://ab00.larc.nasa.gov/~kbibb/felisa.html.)
25. **Peraire, J., Vahdati, M., Morgan, K., and Zienkiewicz, O.C.**, Adaptive remeshing for compressible flow computations, *J. Comp. Phys.*, 72, pp. 449–466, 1987.
26. **Peraire, J. Peiró, J., Formaggia, L., Morgan, K., and Zienkiewicz, O.C.**, Finite element Euler computations in three dimensions, *Int. J. Num. Meth. Eng.*, 26, pp. 2135–2159, 1988.
27. **Peraire, J., Morgan, K., and Peiró, J.**, Unstructured finite element mesh generation and adaptive procedures for CFD, *Applications of Mesh Generation to Complex 3D Configurations*, AGARD Conference Proceedings, No. 464, pp 18.1–18.12, 1990.
28. **Peraire, J., Peiró, J., and Morgan, K.**, Adaptive remeshing for three-dimensional compressible flow computations, *J. Comp. Phys.*, 103, pp. 269–285, 1992.
29. **Requicha, A. A. G. and Voelcher, H. B.**, Solid modeling: a historical summary and contemporary assessment, *IEEE Computer Graphics and Applications*, 3, 2, pp. 9–24, 1982.

30. **Shamos M. I. and Hoey, D.**, Geometric intersection problems, *17th Annual Symposium on Foundations of Computer Science,* IEEE, 1976.
31. **Thompson, J. F., Warsi, Z. U. A., and Mastin, C. W.**, *Numerical Grid Generation — Foundations and Application,* North-Holland, 1985.
32. **Thomasset, F.**, *Implementation of Finite Element Methods for Navier–Stokes Equations,* Springer Series in Comp. Physics, 1981.
33. **Weatherill, N. P.**, Mesh generation in computational fluid dynamics, von Karman Institute for Fluid Dynamics Lecture Series 1989-04, Brussels, 1989.
34. **Zhu, J. Z., Zienkiewicz, O. C., Hinton, E., and Wu, J.**, A new approach to the development of automatic quadrilateral mesh generation, *Int. J. Num. Meth. Eng.*, 32, pp. 894–866, 1991.

18

Unstructured Grid Generation Using Automatic Point Insertion and Local Reconnection

David L. Marcum

18.1 Introduction

Unstructured grid generation procedures for triangular and tetrahedral elements have typically been based on either an octree [Shepard and Georges, 1991], advancing-front [Lohner and Parikh, 1988; Peraire et al., 1988], or Delaunay [Baker, 1987; George et al., 1990; Holmes and Snyder, 1988; Weatherill, 1985] approach. Efficiency is the primary advantage of the octree approach (see Chapter 15). The advancing-front approach (see Chapter 17) offers advantages of high-quality elements and integrity of the boundary. And, the Delaunay approach (see Chapter 16) offers advantages of efficiency and a sound mathematical basis. None of these procedures offers combined advantages of efficiency, quality, robustness, and sound mathematics. Recent research has focused on improving these methods and combining them in order to provide improved overall characteristics. Methods using a combined approach with advancing-front-type point placement and a Delaunay connectivity have been developed for triangular elements [Mavriplis, 1993; Muller et al., 1993; Rebay, 1993]. These methods can produce grids with quality similar to that of traditional advancing-front methods along with the robustness and sound mathematics of a Delaunay approach. However, efficiency has not been substantially improved.

0-8493-2687-7/99/$0.00+$.50

Alternative approaches have been developed using automatic point insertion and connectivity optimization. In this type of approach, point placement and connectivity schemes can be devised that are independent processes. For connectivity optimization, variations of the edge-swapping or local-reconnection algorithm of Lawson [1986] can be used. In this scheme, the grid is repetitively reconnected to locally satisfy a desired criterion. A Delaunay triangulation can be obtained using an in-circle criterion. Barth [1995] has implemented this approach with a Delaunay criterion and circumcenter point placement. However, alternative local reconnection criteria are desired for optimal grid quality. This is especially true in three dimensions, where a Delaunay satisfied grid typically contains many "sliver" elements (which have four, nearly coplanar points). Lawson's method can be used with alternative criteria which should not produce slivers. Unfortunately, in three dimensions, most criteria quickly converge to optimum local states which are far from the desired global optimum.

Marcum and Weatherill [1995] developed a very efficient local reconnection procedure using advancing-front point placement and a combined Delaunay/min–max (minimize the maximum angle) type local-reconnection criterion for generation of triangular or tetrahedral element grids. It is often referred to as the advancing-front/local–reconnection or AFLR method. This procedure differs substantially from the previously cited methods in that the combined Delaunay/min–max reconnection criterion is the only criteria developed to date that allows effective optimization of a three-dimensional tetrahedral element connectivity; it makes effective use of the existing grid as a search data structure, and point insertion is performed using direct subdivision. This methodology has also been extended for generation of high-aspect-ratio elements, right-angle elements, and solution-adapted grids [Marcum, 1995a; 1995b; 1996a; Marcum and Gaither, 1997]. High-quality grids have been generated about geometrically complex configurations in two and three dimensions for a variety of applications using this method. The combined Delaunay criterion can be used effectively with optimization criteria other than min–max. Various point placement strategies and connectivity optimization criteria have been implemented and compared within this procedure. Results verify that, for isotropic grid generation, advancing-front point placement with a combined Delaunay/min–max connectivity criterion consistently produces the highest element quality in an efficient manner [Marcum, 1995c]. Fully compatible edge and surface grid generation components using this procedure have also been developed [Marcum, 1996b].

In this chapter, an overview of the AFLR method for planar, surface, and volume grid generation is presented. Several application examples are presented demonstrating the capabilities, consistency, efficiency, and quality of this approach. In addition, a discussion on best practices using this methodology is presented.

18.2 Unstructured Grid Generation Procedure

The AFLR triangular/tetrahedral grid generation procedure used in the present work is a combination of automatic point creation, advancing type ideal point placement, and connectivity optimization schemes. A valid grid is maintained throughout the grid generation process. This provides a framework for implementing efficient local search operations using a simple data structure. It also provides a means for smoothly distributing the desired point spacing in the field using a point distribution function. This function is propagated through the field by interpolation from the boundary point spacing or by specified growth normal to the boundaries. Points are generated using either advancing-front-type point placement for isotropic elements, advancing-point-type point placement for isotropic right angle elements, or advancing-normal type point placement for high-aspect-ratio elements. The connectivity for new points is initially obtained by direct subdivision of the elements that contain them. Connectivity is then optimized by local reconnection with a min–max type (minimize the maximum angle) type criterion. The overall procedure is applied repetitively until a complete field grid is obtained.

The basic steps in the procedure are briefly outlined below. More complete details and results are presented in Marcum and Weatherhill [1995] and Marcum [1995c].

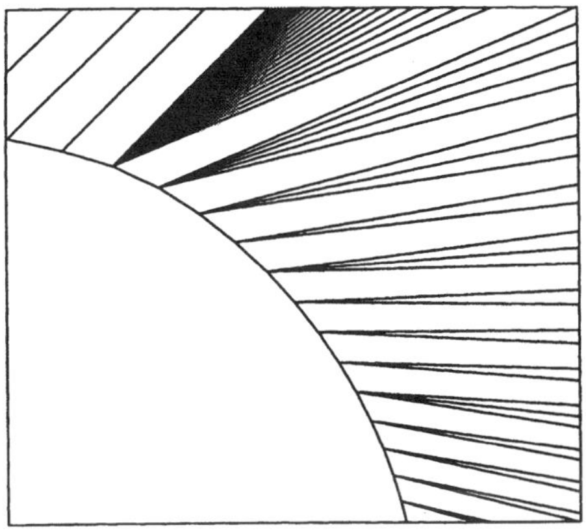

(a)

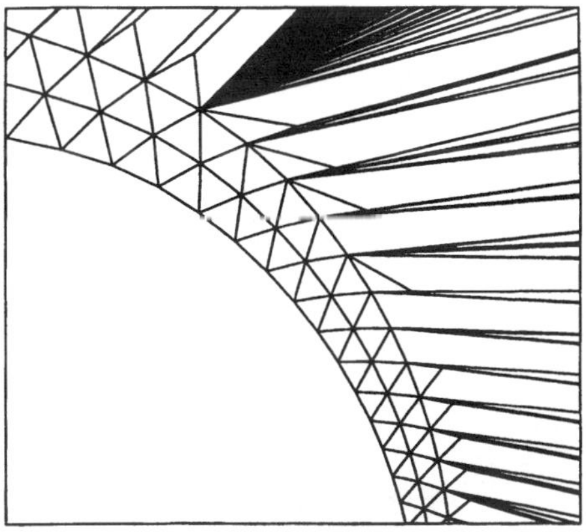

(b)

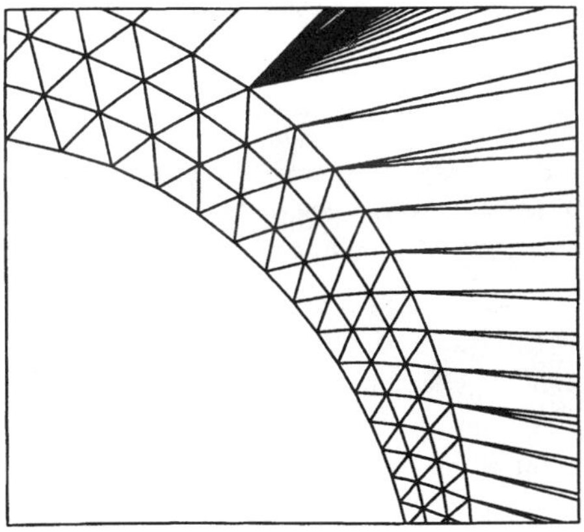

(c)

FIGURE 18.1 Unstructured grid generation process. (a) Initial triangulation, (b) triangulation after direct point insertion on third grid generation iteration, (c) triangulation after local reconnection on third grid generation iteration.

1. Specify point spacing on the boundary surface.
2. Generate a boundary surface grid.
3. Generate a valid initial triangulation of the boundary surface points only and recover all boundary surfaces. An example initial triangulation is shown in Figure 18.1a.
4. Assign a point distribution function to each boundary point based on the local point spacing. Also, optionally assign geometric growth rates normal to the boundary surface.

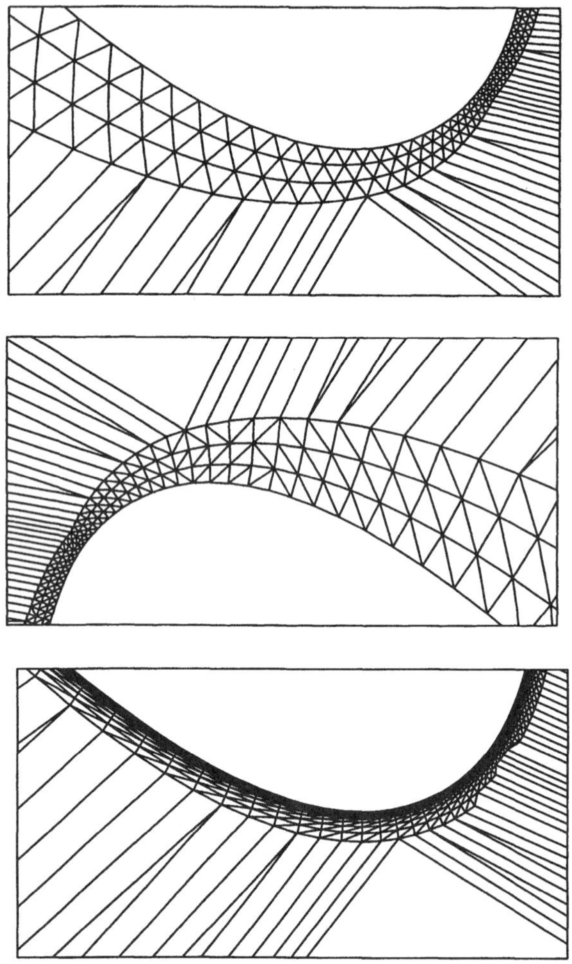

(a)

(b)

(c)

FIGURE 18.2 Different point placement strategies. (a) Advancing-front point placement for isotropic equiangular elements, (b) advancing-point point placement for isotropic right-angle elements, (c) advancing-normal point placement for high-aspect-ratio right-angle elements.

5. For isotropic elements, generate points using advancing-front-type point placement. Points are generated by advancing from the edge/face that satisfies the point distribution function of elements that only satisfy the point distribution function on one edge/face. An example triangulation generated using advancing-front point placement is shown in Figure 18.2a.
6. For right angle elements, generate points using advancing-point-type point placement. Points are generated by advancing as in step 5, except two points are created by advancing along edge/face normals from the two/three points of the satisfied edge/face. An example triangulation generated using advancing-point point placement is shown in Figure 18.2b.
7. For high-aspect-ratio elements, generate points using advancing-normal-type point placement. Points are generated one layer at a time from the boundaries by advancing along normals dependent upon the boundary surface geometry. An example triangulation generated using advancing-normal point placement is shown in Figure 18.2c. A key aspect of the present implementation is the use of multiple normals. At points where the boundary surface is discontinuous, multiple normals are assigned to produce optimal grid quality. An example high-aspect-ratio element grid with multiple normals is shown in Figure 18.3.

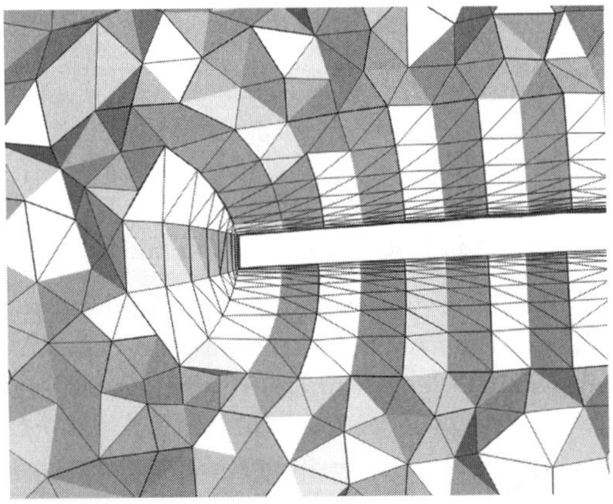

FIGURE 18.3 Tetrahedral field cut for high-aspect-ratio elemnt grid with multiple surface normals.

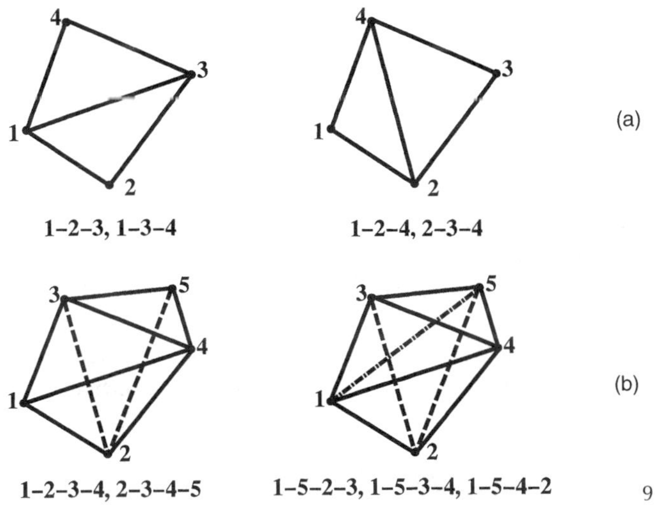

FIGURE 18.4 Possible triangulations for reconnectable element pairs. (a) Four reconnectable points in two dimensions, (b) five reconnectable points in three dimensions.

8. Interpolate the point distribution function for new points from the containing elements. If geometric growth is specified, then the distribution function is determined from an approximate distance to the nearest boundary and the specified geometric growth from that boundary.
9. Reject new points that are too close to other new points.
10. Insert the accepted new points by directly subdividing the elements which contain them. A triangulation after direct insertion is shown in Figure 18.1b.
11. Optimize the connectivity using local reconnection. For each element pair, compare the reconnection criterion for all possible connectivities and reconnect using the most optimal one. Possible triangulations in two and three dimensions are shown in Figures 18.4a and 18.4b, respectively. Repeat this local reconnection process until no elements are reconnected. In three dimensions a combined Delaunay/min–max type criterion is used [Marcum and Weatherill, 1995; Marcum, 1995a]. In this process, a Delaunay criterion is used initially and then the min–max criterion is applied. This improves the overall grid quality substantially and overcomes most of the problems associated with optimum local states that prohibit a global optimum from being obtained. Triangulations before and after local reconnection are shown in Figures 18.1b and 18.1c.

12. Repeat the point generation and local reconnection process, steps 5 through 11, until no new points are generated.
13. Optionally combine elements to form quadrilaterals in two dimensions or hexahedral, prism, or pyramid elements in three dimensions. Elements are combined by advancing from boundary surfaces and selecting combinations based on alignment and quality.
14. Smooth the coordinates of the field grid.
15. Optimize the connectivity using the local reconnection process (step 11).

The procedure described in the above steps allows complete control over the type and quality of grid to be generated with minimal user interaction. In generating the boundary surface grid, user input is required to specify point spacings at selected control points. Further control over the spacing of the field points can be obtained using specified geometric growth, fixed field points, embedded boundary surfaces or adaptation sources [Marcum, 1995b; 1996a]. Once a boundary surface grid is generated no further user input or adjustment of parameters is required other than selecting desired options such as type of point placement or geometric growth.

With advancing-normal-type point placement for high-aspect-ratio elements, the procedure described above does produce sliver elements in three dimensions. These elements are generated only in regions of high-aspect-ratio elements with a very structured alignment. Elimination of these elements with local reconnection is not feasible. There may be no nearby optimization path which produces a better connectivity. The problem is inherently due to the very structured nature of the grid in these regions. Only a limited set of possible triangulations, that do not contain sliver elements, exists for a set of tetrahedra aligned in prismatic groups. A modified process is used for three- dimensional cases. In the present approach, the element connectivity is generated along with new points in high-aspect-ratio regions. Local reconnection is not used to determine the connectivity in these regions. Instead, the connectivity is directly determined as each new point is generated. This produces a very structured connectivity and allows the tetrahedral elements to be easily combined into structured type elements. Typically, the majority of the tetrahedral elements within the high-aspect-ratio region can be combined into six-node pentahedrons (prisms). The outer layer of this region may have some five-node pentahedrons (pyramids) to match the outer tetrahedral elements. In all cases, the pentahedral elements have strict node, edge, and face matching to each other and to neighboring tetrahedral elements.

18.3 Two-Dimensional Application Examples

Selected application examples are presented here to demonstrate the capabilities of the present procedure for generation of two-dimensional unstructured grids. A summary of grid quality and required CPU time for the primary examples is presented in Table 18.1.

Grid quality distributions and statistics are presented for each example. Element angle is used as the grid quality measure. The complete set of grid quality data consists of the three corner angles for all triangles. Maximum and standard deviation values are presented along with distribution plots in 5° increments. The results for the examples presented are representative of those obtained for a variety of configurations. Typically, for an isotropic grid, the maximum element angle is 120° or less, the standard deviation is 7° or less, and 99.5% or more of the elements have angles between 30° and 90°. The minimum angle is usually dictated by the geometry. Standard deviation is not applicable for grids with high-aspect-ratio elements, as there should be peak distributions at a small angle, 60°, and 90°. Also, the minimum angle typically depends upon the maximum aspect ratio with high-aspect-ratio elements.

CPU time required on a laptop PC, desktop PC, and workstation is presented for each example. Computer routines for the two-dimensional grid generator are written in Fortran. All floating-point calculations are performed using 64 bit precision with 8 byte data. The CPU times reported include all I/O and generation of grid quality data. A discretized boundary edge grid file is the input. The output includes a grid coordinate and connectivity file and a quality data file. The efficiency of the overall procedure is such that generation of a typical grid requires only seconds on any current PC or workstation. All of the cases presented can be generated on a PC with at least 16 MB of RAM.

TABLE 18.1 Summary of Grid Quality and CPU Requirements for Two-Dimensional Example Cases

			CPU Time (sec)		
2D Case	Max. Angle (deg)	Std. Dev. Angle (deg)	Pentium 120 Toshiba Tecra 500 128 MB Solaris, g77	Pentium Pro 200 Gateway 2000 G6-200 128 MG Solaris, g77	Ultra SPARC II 300 Sun Ultra 2 512 MB Solaris, f77 single processor
Wake-adapted Multi-element airfoil; 140,609 triangles	127	n/a	40	16	9.5
Mediterranean; 213,323 triangles	118	7.2	61	27	13

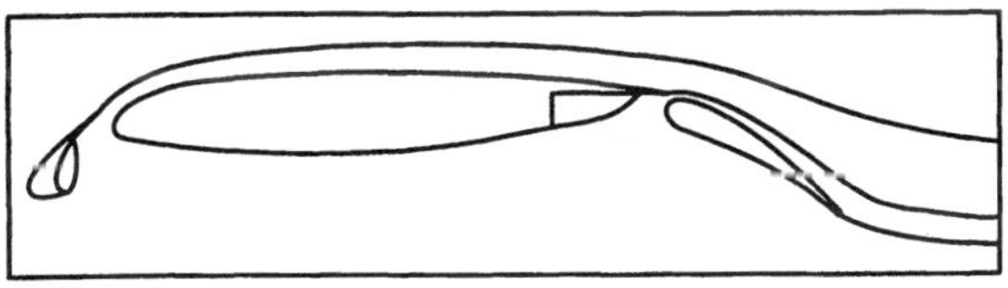

FIGURE 18.5 Boundary edges for multi-element airfoil with multiple wakes.

User input required to generate a complete grid is minimal and includes specifying the point spacing at selected control points on boundary curves and selection of options such as growth from boundary curves or generation of high-aspect-ratio elements. There are no user adjustable parameters that need to be changed from case to case. Specification of point spacings is minimized by automatic reduction of the boundary point spacing in regions where the spacing is greater than the distance between nearby boundaries. The present code is very robust and thoroughly tested. It does not fail to produce a valid grid, given a set of boundary curves that are valid and have a reasonable discretization.

18.3.1 Multi-Element Airfoil

A grid was generated for CFD analysis of the multielement airfoil configuration shown in Figure 18.5. In multielement airfoil configurations, viscous effects and the interaction of multiple wakes can impact overall performance. For optimum solution accuracy, a viscous grid with solution-adapted wakes was used. An initial grid without adaptation was generated and a viscous solution was obtained. Selected streamlines were tracked in the wake regions. These streamlines were then discretized and treated as embedded boundary edges for generation of aligned high-aspect-ratio elements in wake regions. Aligned high-aspect-ratio elements produce optimal resolution and grid quality. The boundary edges and embedded wakes are shown in Figure 18.5. The final solution-adapted grid contains 71,032 points and 140,609 elements and is shown Figure 18.6. Grid quality distribution for this grid is shown in Figure 18.7. Element angle distribution and maximum angle verify that the grid is of very high quality. The maximum element angle is generated within the high-aspect-ratio element region adjacent to one of the corners of the blunt trailing edges. Required CPU time is listed in Table 18.1. Details of solution-adaptation for viscous flow fields are presented in Marcum, [1996a].

18.3.2 Mediterranean Sea

A grid was generated for the geometrically complex coastline of the Mediterranean Sea. The boundary edges for the computational domain are shown in Figure 18.8. The initial boundary curve discretization

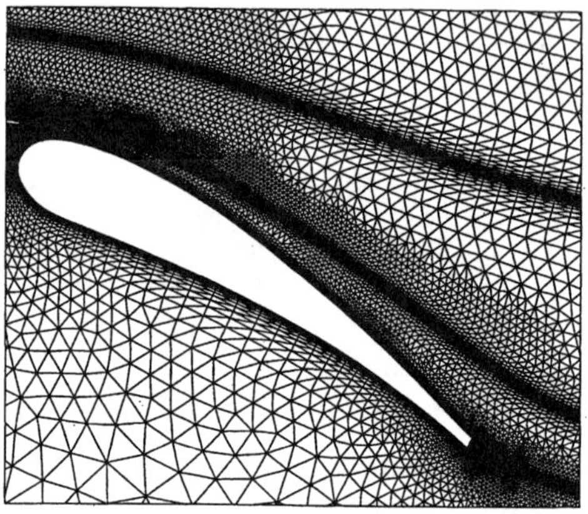

FIGURE 18.6 Final solution-adapted grid for multi-element airfoil with multiple wakes.

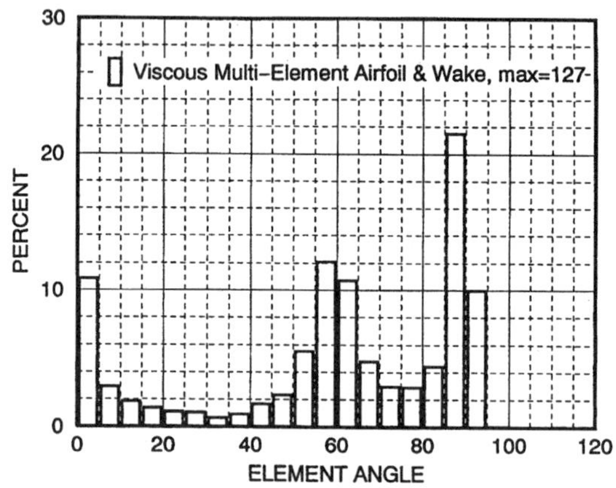

FIGURE 18.7 Grid quality distribution for multi-element airfoil grid.

FIGURE 18.8 Boundary edges of Mediterranean Sea grid.

was nearly uniform. Automatic point spacing reduction was used to reduce the point spacing near points of high curvature and in regions where boundaries are close to one another. Views of the grid near the Aegean Sea and Sea of Crete are shown in Figures 18.9a and 18.9b, respectively. The grid generated contains 111,612 points and 213,323 elements. Point distribution function growth was used to increase the element size away from the coastline. Element size varies smoothly within the grid. Grid quality distribution for the grid is shown in Figure 18.10. Element angle distribution, maximum value, and standard deviation verify that the grid is of very high quality. Required CPU time is listed in Table 18.1.

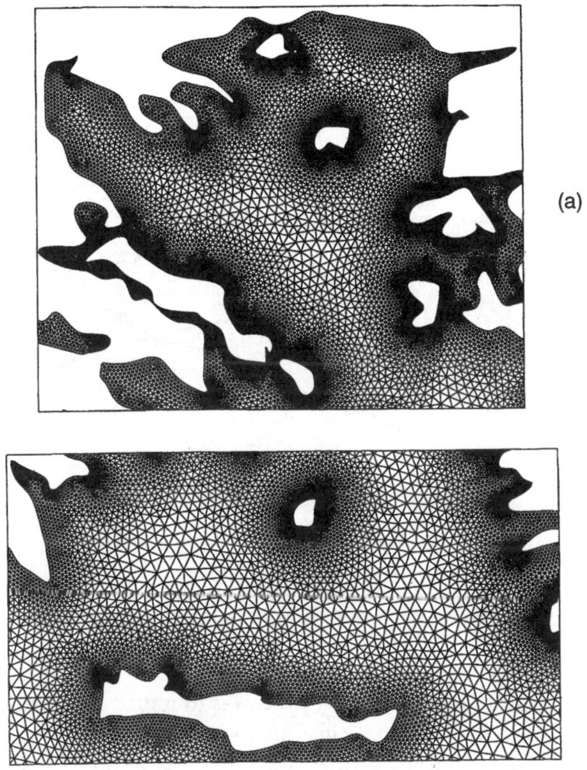

FIGURE 18.9 Mediterranean Sea grid generated using point distribution function growth. (a) Grid near Aegean Sea, (b) grid near Sea of Crete.

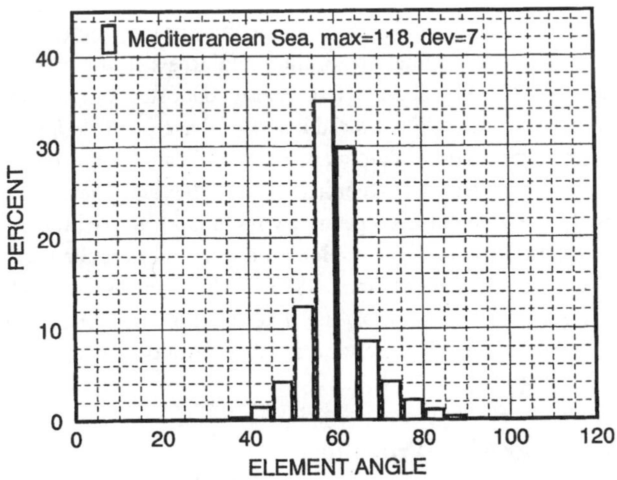

FIGURE 18.10 Grid quality distributions for Mediterranean Sea grid.

18.4 Three-Dimensional Surface Grid Generation

For grid generation with the present methodology, the grid point distribution is automatically propagated from specified control points to edge grids, from edge to surface grids, and finally from surface grids to

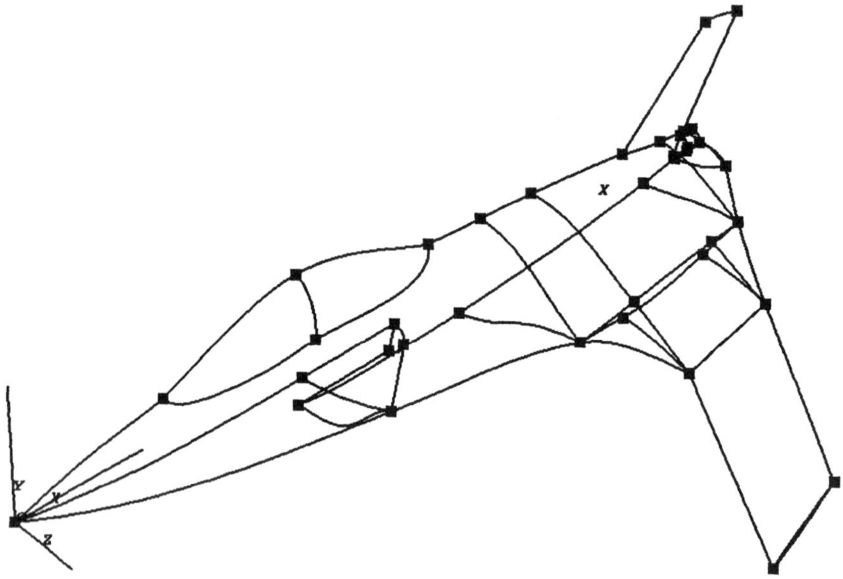

FIGURE 18.11 Surface patches, edges, and corner points for fighter geometry definition.

the volume grid. Surface patches, edges, and corner points for a fighter geometry definition are shown in Figure 18.11. The first step in the grid generation process is to initially set the desired point spacing to a global value at all edge end points. Point spacings are then set to different values at desired control points on edges in specific regions requiring further resolution. For example, endpoints along leading edges and trailing edges would typically be set to a very fine point spacing. Point spacings can be set anywhere along an edge. A point in the middle of a wing section would typically be set to a larger point spacing than at the leading or trailing edges. As control point spacings are set, a discretized edge grid is created for each edge. Specification of desired control point spacings is typically the only user input required in the overall grid generation process.

A CAD geometry system is used to define and evaluate the surface geometry. Edge and surface grid generation requires use of geometry evaluation routines and access to the geometry database. Surface topology is extracted from the CAD database and a separate data structure is used for grid generation [Gaither, 1994; 1997]. The grid generation procedures used have been designed to isolate geometry evaluation access. All access to geometry evaluation routines and data base is outside the grid generation routines. This approach produces a very clean interface between the grid generation and geometry system. It also makes it very easy to use different CAD geometry systems with very little modification. The edge grid generation and subsequent surface grid generation procedures are described in the following sections. Additional information can be found in [Marcum, 1996b]. The only CAD-related routine required for the present edge and surface grid generation is one that determines physical space coordinates, x,y,z, given mapped space coordinates, u,v. This routine is labeled routine xyz_from_uv in the following sections.

18.4.1 Edge Grid Generation Procedure

Edge grids are created using a one-dimensional version of the standard grid generation procedure. This ensures that point distribution and growth rates are fully compatible for optimal final grid quality. For each edge or segment the point spacing is specified at both ends, as shown in Figure 18.12a. Edge grid generation is then used to produce the point distribution shown in Figure 18.12b. The basic steps in the edge grid procedure are outlined as follows.

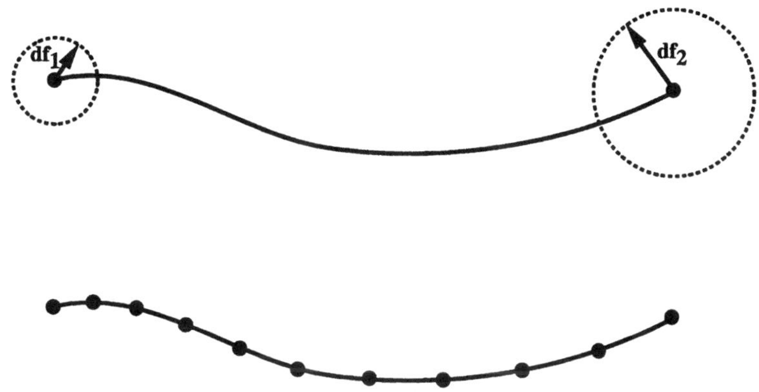

FIGURE 18.12 Specified point spacings and final point distribution for a surface edge.

1. Create an interpolation table for the mapped space coordinates versus arc length using the geometry evaluation routine xyz_from_uv.
2. Advance from each end of the edge segment in arc-length space and create two new points. The point spacings for these points are interpolated from the exposed interior endpoints of the edge.
3. Reject a new point if it is too close to the other new point.
4. Repeat the edge grid point generation process steps 2 and 3 until no new points are created.
5. Smooth the arc-length coordinates of the edge grid.
6. Interpolate for mapped space coordinates, u,v, at the generated arc-length coordinates.
7. Obtain the physical space coordinates, x,y,z, at the interpolated mapped space coordinates, u,v, using the geometry evaluation routine, xyz_from_uv.

The edge grid generation routine consists of steps 2 through 6 above. All generation parameters for details such as interpolation, limiting, rejection, and smoothing are identical to those used in the standard planar and volume grid generation procedures.

18.4.2 Surface Grid Generation Procedure

Given a geometry definition that uses a surface mapping, one can generate a surface grid in either mapped or physical space. With a mapped space approximation (MSA), a standard two-dimensional grid generator can be used as is. The advantage of this approach is efficiency. However, for realistic configurations, the mappings are often distorted in physical space and an MSA approach produces a poor-quality surface grid. The two-dimensional grid generation procedure can be modified to generate near optimal grids on a surface using a physical space approximation (PSA). In this approach, an approximate surface definition is used within the surface grid generation procedure to determine point placement, such that ideal surface triangles are created in physical space. The approximate surface definition provides an efficient means of iterating on the surface and allows the CAD geometry system to be decoupled from the grid generation procedure. A valid grid in both mapped and physical space is maintained throughout the procedure and all searching is done in mapped space. Local reconnection is performed in both mapped and physical space. The physical space reconnection cannot be used for elements that are considerably larger than the desired element size. These elements exist early in the process and can be composed of highly curved edges. The physical space reconnection does not account for edge curvature. However, in mapped space these edges are not curved and mapped space reconnection can be used. The PSA procedure produces an output grid in mapped space which corresponds to an approximately optimal surface grid when mapped back to the actual surface definition. The basic steps in the overall procedure are listed below.

1. Generate a surface grid entirely in mapped space using the standard two-dimensional procedure. This grid will be used to define the physical space approximation. Any triangulation of the surface that adequately resolves the geometry can be used for the physical space approximation.
2. For the grid from step 1 above, obtain the physical space coordinates, *x,y,z,* at the mapped space coordinates, *u,v,* using the geometry evaluation routine xyz_from_uv.
3. Generate a valid initial triangulation of the edge points only and recover all discrete edges.
4. Assign a point distribution function to each edge point based on the local physical point spacing.
5. Generate points using advancing-front point placement by advancing from satisfied edges. These points are generated to obtain approximately optimal elements in physical space. Iteration and interpolation of physical space coordinates from mapped space coordinates are required. The grid from steps 1 and 2 is used as a locally linear approximation to the surface definition.
6. Interpolate the point distribution function for new points from the containing elements.
7. Reject new points that are too close to other new points in physical space.
8. For each accepted new point, search in mapped space for the containing element and directly inset the point.
9. Optimize the connectivity using local reconnection in mapped space.
10. Optimize the connectivity using local reconnection in physical space. Only elements that are close to satisfying the distribution function are allowed to be reconnected.
11. Repeat the point generation and local-reconnection process, steps 5 through 10, until no new points are generated.
12. Smooth the mapped space coordinates of the surface grid using physical space edge length weighting. This is equivalent to smoothing directly in physical space.
13. Interpolate for the smoothed physical space coordinates using the grid from steps 1 and 2.
14. Optimize the connectivity using physical space local reconnection.
15. Obtain the "true" physical space coordinates, *x,y,z,* on the surface at the generated mapped space coordinates, *u,v,* using the geometry evaluation routine, xyz_from_uv.

The PSA surface grid generation routine consists of steps 3 through 14 above. All generation parameters for details such as interpolation, limiting, rejection, and smoothing are identical to those used in the standard planar and volume grid generation procedures. For both the edge and surface grid generation procedures, the final physical space grid is located on the actual surface defined by the geometry data base. The approximate physical space surface grid is used only within the grid generation procedures.

18.5 Three-Dimensional Surface Grid Generation Application Examples

Two selected application examples are presented here to demonstrate the capabilities of the present procedure for generation of unstructured surface grids. Grid quality distributions and statistics are presented for each example. Element angle is used as the grid quality measure. The complete set of grid quality data consists of the three corner angles for all surface triangles. Maximum and standard deviation values are presented along with distribution plots in 5° increments. The results for the examples presented are representative of those obtained for a variety of surfaces. Typically, the resulting grid quality is the same as that expected for the two-dimensional grid generator. Required CPU times are about three times that required of the the two-dimensional grid generator.

18.5.1 Generic Shell

The first case is a generic shell that was derived from a circular surface patch with a circular hole that is distorted in physical space. Surface grids were generated for this case with two different procedures. One was generated with the mapped space approximation (MSA) approach. With MSA the standard two-dimensional grid generator is used in mapped space. The other grid was generated using the physical

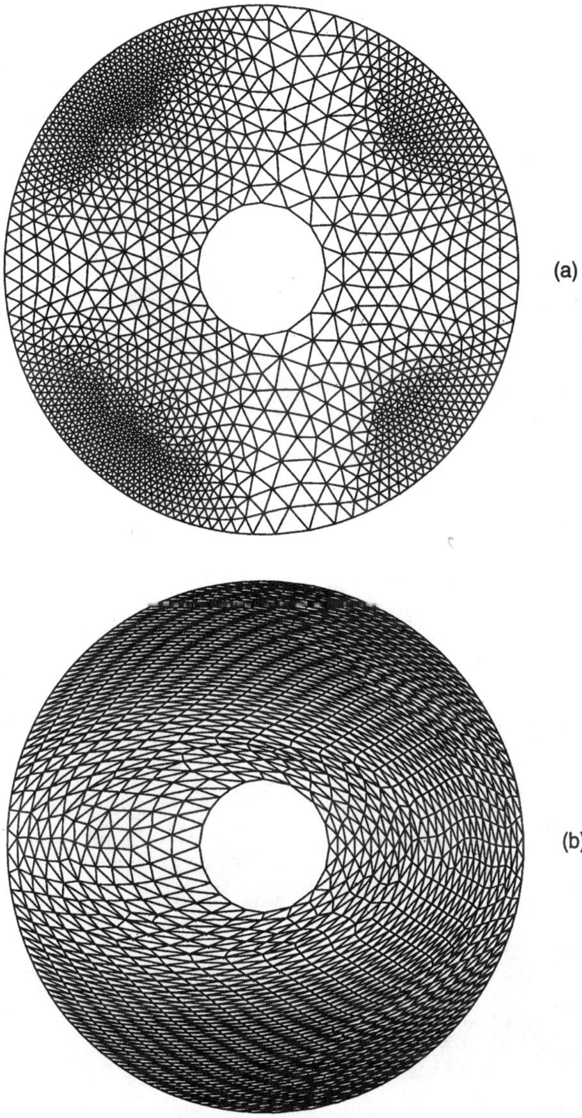

FIGURE 18.13 Surface grids in mapped space for generic shell. (a) MSA grid, (b) PSA grid.

space approximation (PSA) approach. Both grids are shown in Figures 18.13a and 18.13b in mapped space. The MSA grid is optimal in mapped space while the PSA grid is not. The grids in physical are shown in Figures 18.14a and 18.14b. The MSA grid contains distorted elements in physical space while the PSA grid is of very high quality. Grid quality distributions in physical space for these grids are shown in Figure 18.15. Element angle distribution, maximum value, and standard deviation verify that the PSA surface grid is of very high quality and the MSA surface grid is not.

18.5.2 Hawaiian Islands

A surface grid was generated on the geometrically complex ocean bottom around the Hawaiian Islands. For this case, a usable grid can only be obtained using some form of physical space grid generation. The surface grid generated using the PSA approach is shown in Figure 18.16. A nearly uniform point spacing was specified and, as shown in Figure 18.16, a nearly uniform grid is generated. Grid quality distribution

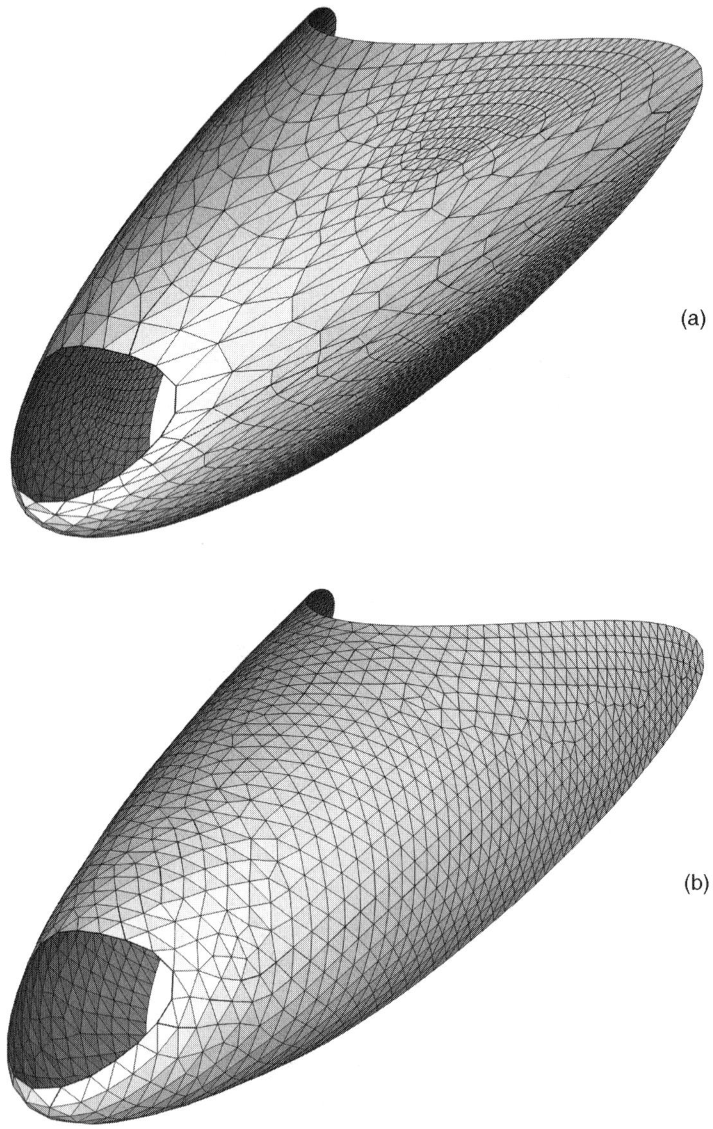

FIGURE 18.14 Surface grids in physical space for generic shell. (a) MSA grid, (b) PSA grid.

in physical space for this case are shown in Figure 18.17. Element angle distribution, maximum value, and standard deviation verify that the PSA surface grid is of very high quality.

18.6 Surface and Volume Grid Generation Best Practice

The AFLR procedures previously described are very automated and require minimal user interaction. User input can affect the usefulness and quality of the grid. Optimum quality can usually be approached by reducing the element size. Unfortunately, a grid of optimal quality may often require an excessive number of elements. Obtaining a solution with such a grid may require a prohibitive level of CPU effort. The task for the user is to obtain a grid that offers the best compromise. An ideal grid is often not the most optimal one from just a quality perspective. Instead, an ideal grid is one within the size limits dictated by CPU resources or time for the solution process, resolves the primary geometric features or

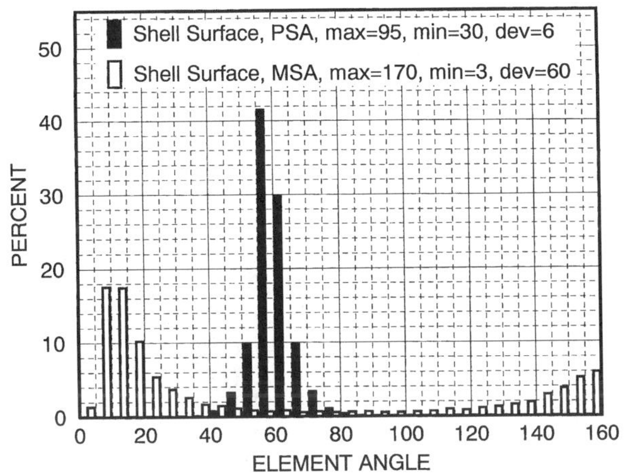

FIGURE 18.15 Grid quality distributions for generic shell surface grids.

FIGURE 18.16 PSA surface grid for ocean bottom near Hawaiian Islands.

those of interest for the given analysis, and has quality at a level that will not impact the solver performance or accuracy.

Problem size limits are usually well defined for a given problem. For grid resolution requirements, there is typically at least a consensus on acceptable levels of resolution for a given method of analysis and class of configurations. Requirements for grid quality are not often as well established. Significant differences in how quality affects solver performance and accuracy can exist between solution algorithms of a similar class. Very low quality elements, however, are always detrimental to the solution process. The impact of low-quality elements on solver accuracy can be very localized and is not usually the critical issue. Solver performance, e.g., convergence rate, can be significantly reduced due to presence of even just a few low-quality elements. Other aspects of the solution process can also be impacted by a low-quality grid. For example, a low-quality element can create difficulties in cases where there may be grid deformation during the solution process.

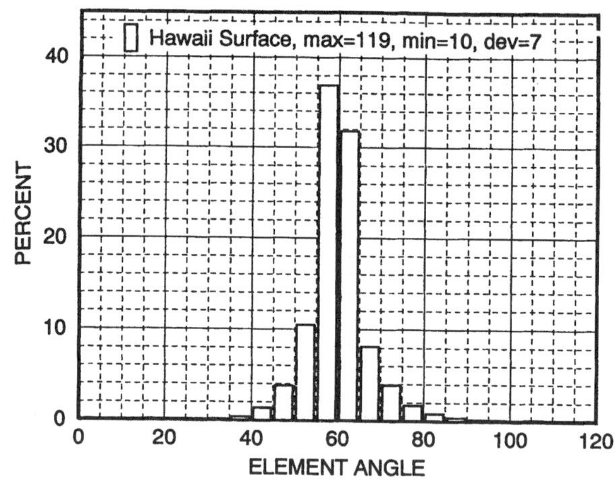

FIGURE 18.17 Grid quality distributions for Hawaiian Island surface grid.

The quality of a tetrahedral element may be defined in many ways. At the extremes, grid quality is well defined. A very flat or sliver element with four nearly coplanar points is always considered a very low-quality element. An ideal element, in isotropic cases, is one that approaches a tetrahedron with equal length sides and equal dihedral angles. However, this definition is not appropriate for high-aspect-ratio elements. In this case, an ideal high-aspect-ratio element contains one perfectly structured and aligned corner with right angles. Element quality can be quantified by a variety of measures. Among those, dihedral angle offers distinct advantages. Element dihedral angle is advocated in this chapter as it is directly related to the solution algorithm performance and accuracy and it is fairly universal. Barth [1991] demonstrates how the dihedral angle contributes to the diagonal term in the solution matrix of a Laplacian or Hessian. This applies to the solution of many equations, especially in CFD analysis. Large dihedral element angles produce a significant negative contribution to the diagonal terms. Angles approaching 180° will degrade the performance of the solver. Another aspect of using the dihedral element angle is that it applies to both isotropic and high-aspect-ratio elements. A large angle in either case is a low-quality element. Quality for a given surface or volume grid can be evaluated by inspecting worst-case and overall measures. Worst-case quality can be quantified by the maximum angle for all of the grid elements. Overall quality, for isotropic cases, can be quantified by the standard deviation in the angle. In the case of high-aspect-ratio elements, there are multiple peak values and a single deviation is not appropriate. Inspection of the distribution near expected peak values of 0°, 70°, and 90° can verify the overall quality. The minimum angle peak in this case is dictated by the maximum aspect ratio.

Several other measures of grid quality have been proposed (see Chapter 33). Many of these can be obtained as ratios of element properties. The following element quality measures are of this type.

$$Q_l = \sqrt{24}\, R_i / L_{\max} \tag{18.1}$$

$$Q_r = 3 R_i / R_c \tag{18.2}$$

$$Q_v = (9/8)\sqrt{3}\, V / R_c^3 \tag{18.3}$$

Q_l is a length ratio based measure, Q_r is a radius ratio based measure, Q_v is a volume ratio based measure, L_{max} is the maximum edge length, R_e is the circumsphere radius, R_i is the inscribed sphere radius, and V is the volume. The constants in these equations are chosen such that a quality measure value of one is an ideal isotropic element and a value of zero is a perfectly flat element with four coplanar points. These measures are only appropriate for isotropic type elements. Perfectly aligned and

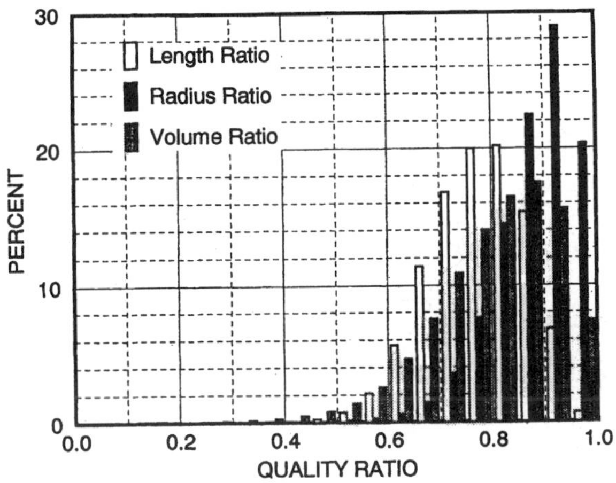

FIGURE 18.18 NASA space shuttle orbiter volume grid quality ratios.

structured like high-aspect-ratio elements are identified as being of low quality (quality measure value near zero). Even a grid generated to be isotropic may contain some high-aspect-ratio elements, if the surface grid contains any high-aspect-ratio triangles. In most cases, these elements pose no problem for the solver if they are not skewed. The measures defined above do not distinguish between skewed and high-aspect-ratio elements. Skewed elements with large dihedral angles are identified as low-quality elements. However, a high-aspect-ratio element with a maximum dihedral angle of 90° is also identified as being of low quality. Another characteristic of quality ratio measures is that they all are very sensitive to deviations from ideal. For example, a perfect isotropic right-angle element has values of $Q_l = Q_r = 0.732$ and $Q_v = 0.5$ which are relatively far from the equiangular ideal of $Q_l = Q_r = Q_v = 1$. True "ideal" elements cannot be generated for most geometries of interest. Ideal elements are arranged in groups of five surrounding an edge and cannot match up to a flat surface or even a typical curved surface. Also, ideal elements cannot exist if the element size varies.

Typical distributions in 0.05 increments of the quality ratios given by Eq. 18.1, Eq. 18.2, and Eq. 18.3 are shown in Figure 18.18. These distributions are for an isotropic type grid about a geometrically complex NASA Space Shuttle Orbiter geometry (presented in the next section on three-dimensional application examples). The high quality of the volume grid is reflected in the clustering of the distributions at high values of the quality ratios. The peaks at an ideal value less than one represent real limitations on quality, which are independent of methodology. For each quality ratio, the maximum ideal value is always one and the minimum value is usually dictated by the geometry. Typical average values are $\overline{Q_l} > 0.75$, $\overline{Q_r} > 0.85$, and $\overline{Q_v} > 0.75$. Typical limits on quality ratio distributions are 99.99% of elements have $Q_l > 0.3$, $Q_r > 0.4$, and $Q_v > 0.1$. Also, 99.5% of elements have $Q_l > 0.5$, $Q_r > 0.6$, and $Q_v > 0.35$. For comparison, 99.99% of element dihedral angles are less than 135° and 99.5% are less than 120°.

As previously mentioned, the user of a grid generation procedure can impact the final grid quality. With the procedure described in this article, volume element size and distribution is determined from the boundary. A low-quality surface grid will produce low-quality volume elements near the surface. In most cases, a high-quality surface grid will produce a high-quality volume grid. Low-quality surface elements are usually the result of inappropriate edge spacing. With fast surface grid regeneration and simple point spacing specification, optimizing the surface quality is a quick process. An example of a surface mesh with a low quality triangle, which can be corrected by point spacing specification, is shown in Figure 18.19a. In this case, the surface patch has close edges that cannot be eliminated. In Figure 18.19a, the initial choice of a uniform spacing at the edge end-points produces a single low-quality triangle. Specifying a single point spacing at the middle of the edge near the close edges eliminates the low-quality element, as shown in Figure 18.19b. Alternatively, the spacing near the close edges can be reduced to produce a more "perfect" grid, at the expense of an increased number of elements, as shown in Figure 18.19c.

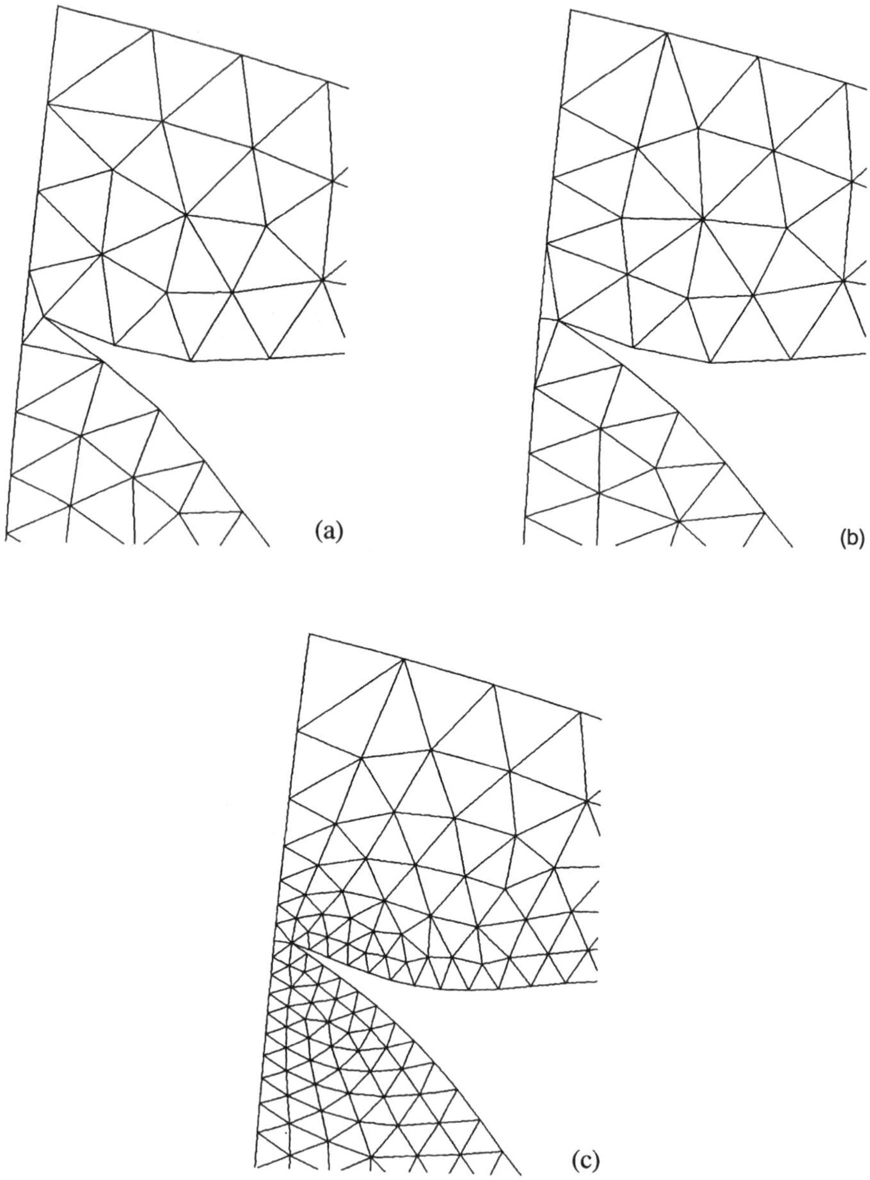

FIGURE 18.19 Surface grid problem due to close edges. (a) Surface grid patch with distorted surface element, (b) surface grid patch improved by applying a point spacing near problem edge, (c) surface grid patch improved by applying a reduced point spacing near problem edge.

Surface definition can also impact surface grid quality. This type of problem is usually due to a surface patch with a width that is smaller than the desired element size. An example is shown in Figure 18.20a. The original surface definition contains four patches, each with a minimum width less than the element spacing, as shown in Figure 18.20a. The resulting surface grid contains elements with edges that are shorter than the desired element size, as shown in Figure 18.20b. Combining the four patches into one surface patch improves the quality, as shown in Figure 18.20c. Spacings between the nearby edges could be reduced for further improvement.

Other conditions can affect volume quality even if the surface grid is of high-quality. An example is shown in Figure 18.21. In this case, there are two nearby surfaces with large differences in element size.

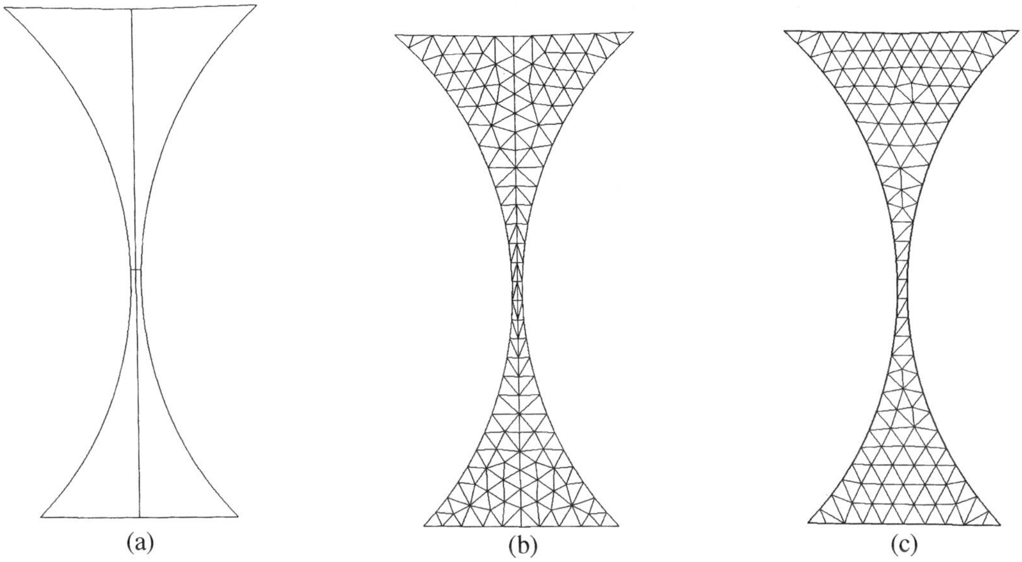

FIGURE 18.20 Surface grid problem due to multiple surface definitions. (a) Four original surface definition patches, (b) surface grid with four surface definition patches, (c) surface grid with one combined surface definition patch.

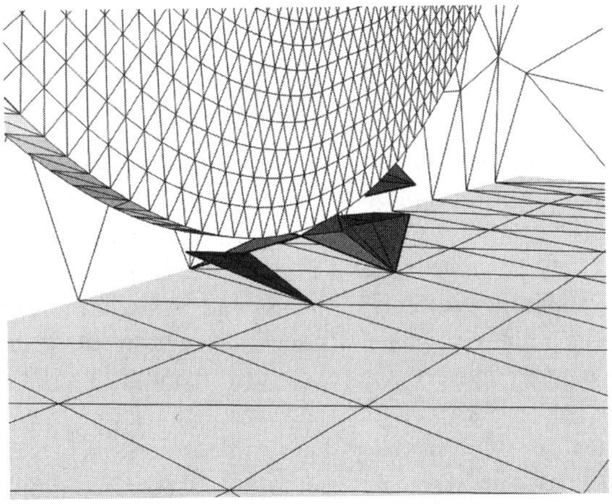

FIGURE 18.21 Distorted tetrahedral elements between surface grids that are close and have large differences in surface element size.

This results in distorted volume elements between the surfaces, as shown in Figure 18.21. These elements can be eliminated by increasing the spacing on the surface that has the smaller elements and/or decreasing the spacing on the surfaces which have the larger elements. From a solution algorithm, perspective, the spacings should probably be reduced. The region between the two objects cannot be resolved by the solver without additional grid points.

Usability of the volume grid must also be considered along with quality. A high-quality surface grid with desired geometric resolution may produce too many volume elements for efficient analysis. Often, high resolution is only required near the surfaces. Geometric growth can be used in this case to produce a volume grid with substantially fewer elements. With growth, element size is constant very close to the surface and grows geometrically away from the surface. An example with growth is presented in the next section.

TABLE 18.2 Summary of Grid Quality and CPU Requirements for Three-dimensional Example Cases

3D Case	Max. Angle (deg)	Std. Dev. Angle (deg)	CPU Time (sec) Pentium 120 Toshiba Tecra 500 128 MB Solaris, gcc	Pentium Pro 200 Gateway 2000 G6-200 128 MG Solaris, gcc	Ultra SPARC II 300 Sun Ultra 2 512 MB Solaris, cc single processor
Pump; 123,439 tetrahedra	154	19	4.3	2.1	0.9
SUV interior; 527,563 tetrahedra	156	17	21	9.5	4.4
Space shuttle orbiter; 3,026,562 tetrahedra	155	17	n/a	n/a	44
Launch vehicle; 2,107,774 tetrahedra	160	n/a	34	16	5.2
Destroyer hull; 4,268,192 tetrahedra	163	n/a	69	34	15

18.7 Three-Dimensional Application Examples

Selected application examples are presented here to demonstrate the capabilities of the present procedure for generation of three-dimensional unstructured grids. All surface grids were generated using the previously described PSA surface grid generation procedure. A summary of grid quality and required CPU time for the primary examples is presented in Table 18.2.

Grid quality distributions and statistics are presented for each example. Element angle is used as the grid quality measure. The complete set of grid quality data consists of the six dihedral angles for all tetrahedra. Maximum and standard deviation values along with distribution plots in 5° increments are presented for both the surface and volume grids. The results for the examples presented are representative of those obtained for a variety of configurations. Typically, for an isotropic grid, the maximum element angle is 160° or less, the standard deviation is 17° or less, and 99.5% or more of the elements have angles between 30° and 120°. The minimum angle is usually dictated by the geometry. Standard deviation typically increases when geometric growth is used to increase the field point spacing.

CPU time required on a laptop PC, desktop PC, and workstation is presented for each primary example. Computer routines for the three-dimensional grid generator are written in C with dynamic memory that is automatically reallocated based upon actual requirements. All floating-point calculations are performed using 64 bit precision with 8 byte data. The CPU times reported include all I/O and generation of grid quality data. A boundary surface grid file is the input. The output includes a grid coordinate and connectivity file and a quality data file. The efficiency of the overall procedure is such that generation of a typical grid requires only minutes on many current PCs or workstations. Generation of a typical surface grid requires only seconds. Memory required is about 100 bytes per isotropic element generated. For grids with high-aspect-ratio elements, the memory requirements are considerably less.

User input required to generate a complete grid is minimal and includes specifying the point spacing at selected control points on the boundary curves for surface grid generation. Selection of options such as growth from boundaries is the only required user input for volume grid generation. There are no user adjustable parameters that need to be changed from case to case. The present code is very robust and thoroughly tested. It does not fail to produce a valid volume grid, given a set of boundary surface triangulations that are valid and have a reasonable discretization. Currently, the PSA surface and AFLR volume generation routines are used in the SolidMesh grid generation system [Gaither, 1997] for research and education at the MSU ERC. All of the example cases presented in this section were generated using

FIGURE 18.22 Pump cover surface grid.

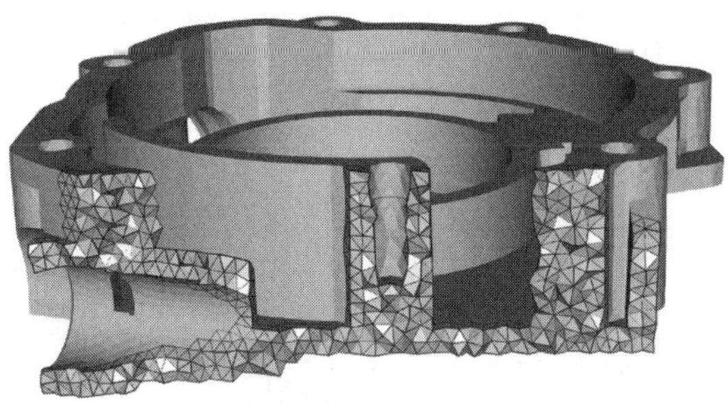

FIGURE 18.23 Tetrahedral field cut for pump cover grid.

SolidMesh. Also, the AFLR volume generation routines are used in the HyperMesh® finite element pre- and post-processing commercial code from Altair Computing, Inc.

18.7.1 Pump Cover

A grid suitable for structural analysis was generated for a pump cover. The surface grid contains 40,534 boundary faces and is shown in Figure 18.22. Distribution of grid points within the volume grid can be visualized using a tetrahedral field cut, which displays the exposed surfaces of tetrahedron that intersect a given plane, as shown in Figure 18.23. Element size is uniform within the volume grid. The complete volume grid contains 30,897 points and 123,439 elements. High-order tetrahedrons can be obtained by adding midpoints on the element edges. Midpoints on the surface must be evaluated using the geometry definition. Grid quality distributions for the surface and volume grids are shown in Figures 18.24 and 18.25, respectively. Element angle distributions, maximum values, and standard deviations verify that the surface and volume grids are of very high quality. The standard deviation is higher than typical, as there are several areas where there are only one or two rows of elements between surfaces. This limits the overall quality that can be obtained. Required CPU time is listed in Table 18.2.

18.7.2 SUV Interior

A grid was generated for interior airflow and thermal management analysis of a sport utility vehicle (SUV). Exterior and interior surfaces are shown in Figures 18.26a and 18.26b, respectively. The surface grid contains 69,744 boundary faces. A tetrahedral field cut near the drivers seat is shown in Figure 18.27.

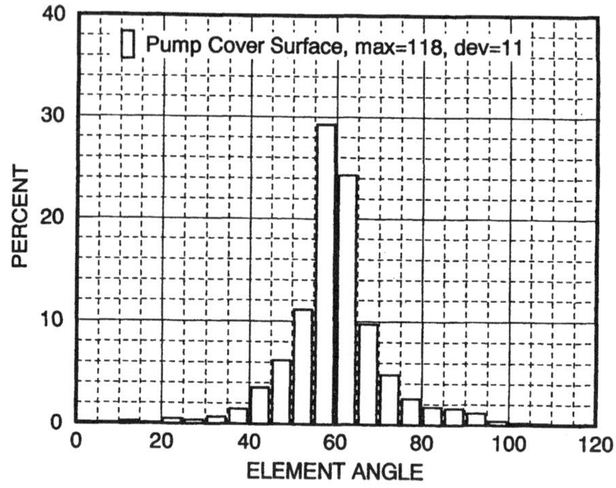

FIGURE 18.24 Pump cover surface grid quality.

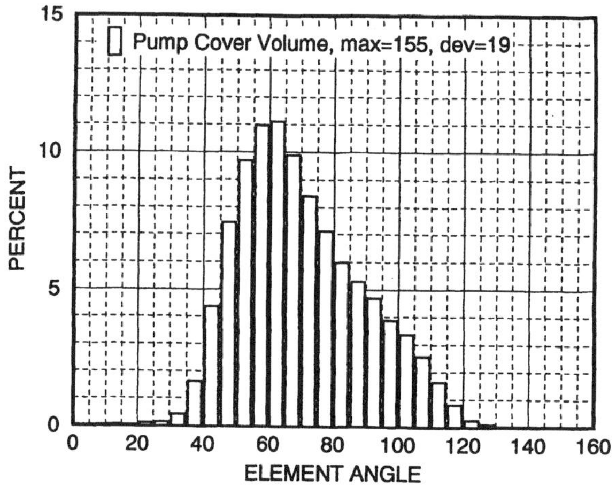

FIGURE 18.25 Pump cover volume grid quality

Point distribution function growth was used to automatically increase element size within the interior. Element size grows smoothly away from the surfaces, as shown in Figure 18.27. The complete volume grid contains 106,095 points and 527,563 elements. Without growth, the volume grid contains approximately twice as many points and elements. For this case, a growth rate of 1.2 was used. The growth rate can be increased to further decrease the number of elements. However, the quality begins to degrade with high growth rates. Quality degradation is typically not a significant factor for growth rates of 1.5 or less. Grid quality distributions for the surface and volume grids are shown in Figures 18.28 and 18.29, respectively. Element angle distributions, maximum values, and standard deviations verify that the surface and volume grids are of very high quality. Required CPU time is listed in Table 18.2.

18.7.3 NASA Space Shuttle Orbiter

A grid suitable for inviscid CFD analysis was generated for the NASA Space Shuttle Orbiter. This case demonstrates the level of geometric complexity that can be handled routinely using the present methodology. Geometry clean-up and preparation required approximately 3 days to complete. However, geometry work is highly dependent on the state of the starting geometry definition. Total time for geometry preparation can range from none to a couple of weeks. Surface and volume grid generation

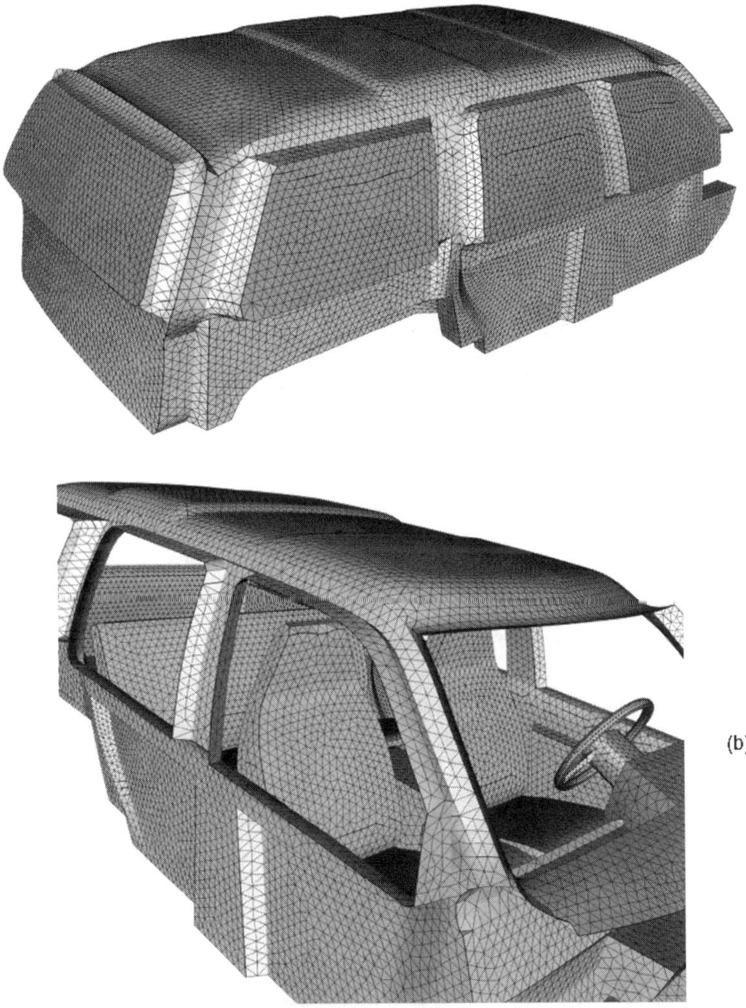

FIGURE 18.26 SUV surface grid. (a) Exterior surfaces, (b) windows removed to show interior surfaces.

related work required approximately 4 hours. This time included modifications for grid quality optimization and resolution changes based upon preliminary CFD solutions. The surface grid on the orbiter surface is shown in Figure 18.30. The total surface grid contains 150,206 boundary faces. A tetrahedral field cut is shown in Figure 18.31. Element size varies smoothly in the field. The complete volume grid contains 547,741 points and 3,026,562 elements. Grid quality distributions for the surface and volume grids are shown in Figures 18.32 and 18.33, respectively. Element angle distributions, maximum values, and standard deviations verify that the surface and volume grids are of very high quality. Required CPU time is listed in Table 18.2. CPU times are not available for the PCs tested as they each are configured with 128 MB of RAM and this case requires about 300 MB of RAM.

18.7.4 Launch Vehicle

A grid suitable for high Reynolds number viscous CFD analysis was generated for a generic launch vehicle. The surface grid on the launch vehicle surface is shown in Figure 18.34. The total surface grid contains 47,392 boundary faces. A tetrahedral field cut is shown in Figure 18.35. Element size varies smoothly in the field, and there is a smooth transition between high-aspect-ratio and isotropic element regions. Also, in areas where there are small distances between surfaces, the merging high-aspect-ratio regions transition

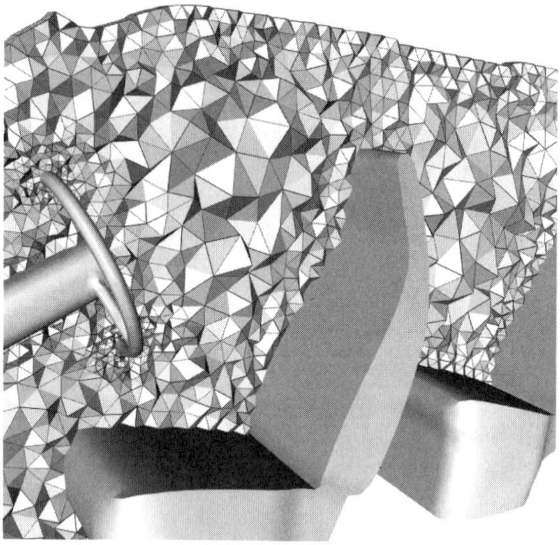

FIGURE 18.27 Tetrahedral field cut for SUV grid.

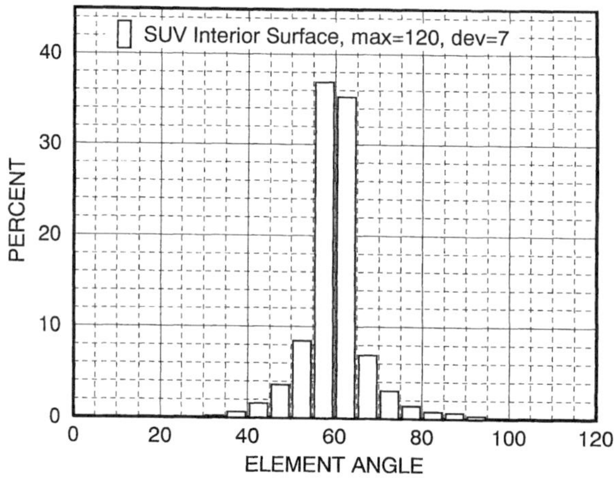

FIGURE 18.28 SUV surface grid quality.

(locally) to isotropic generation. If these regions advance too close, without transition, the element quality can be substantially degraded. The complete volume grid contains 363,664 points and 2,107,774 elements. Most of the tetrahedral elements in the high-aspect-ratio regions can be combined into pentahedral elements for improved solver efficiency. With element combination, the complete volume grid contains 461,241 tetrahedrons, 4,757 five-node pentahedrons (pyramids), and 545,673 six-node pentahedrons (prisms). Grid quality distributions for the surface and volume grids are shown in Figures 18.36 and 18.37, respectively. Element angle distributions and maximum values verify that the surface and volume grids are of very high quality. The distribution peaks are at the expected values of near 0°, 70°, and 90°. Required CPU time is listed in Table 18.2. The CPU times listed for this case reflect the fact that generation of high-aspect-ratio elements requires considerably less time than generation of isotropic elements. For the PCs tested, the very last process, which merges the isotropic and high-aspect-ratio regions, was unable to finish. This process requires about 160 MB of RAM and the PCs are configured with 128 MB of RAM. However, the CPU times shown in Table 18.2 are valid for the PCs, as this process and writing of the output grid file requires a small fraction (approximately 6%) of the total time and the times shown have been adjusted up to account for the work not done.

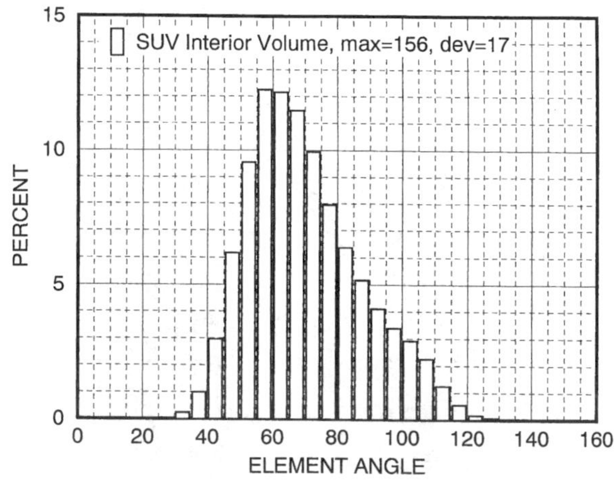

FIGURE 18.29 SUV volume grid quality.

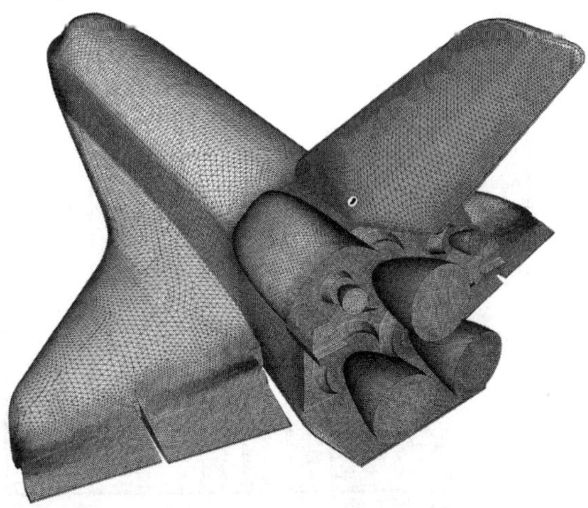

FIGURE 18.30 NASA space shuttle orbiter surface grid.

18.7.5 Destroyer Hull

A grid suitable for high Reynolds Number viscous CFD analysis was generated for the Navy model 5415 destroyer hull. Multiple views of the surface grid on the water-line, hull, and propeller surfaces are shown in Figures 18.38a, 18.38b and 18.38c. The total surface grid contains 86,026 boundary faces. A tetrahedral field cut is shown in Figure 18.39. Element size varies smoothly in the field and there is a smooth transition between high-aspect-ratio and isotropic element regions. The complete volume grid contains 734,330 points and 4,268,192 elements. Most of the tetrahedral elements in the high-aspect-ratio regions can be combined into pentahedral elements for improved solver efficiency. With element combination, the complete volume grid contains 822,604 tetrahedrons, 9,398 five-node pentahedrons (pyramids), and 1,142,264 six-node pentahedrons (prisms). Grid quality distributions for the surface and volume grids are shown in Figures 18.40 and 18.41, respectively. Element angle distributions and maximum values verify that the surface and volume grids are of very high quality. The distribution peaks are at the expected values of near 0°, 70°, and 90°. Required CPU time is listed in Table 18.2. The CPU times listed for this case reflect the fact that generation of high-aspect-ratio elements requires considerably less time than generation of isotropic elements. For the PCs tested, the very last process, which merges the isotropic

FIGURE 18.31 Symmetry plane surface grid and tetrahedral field cut for NASA space shuttle orbiter grid.

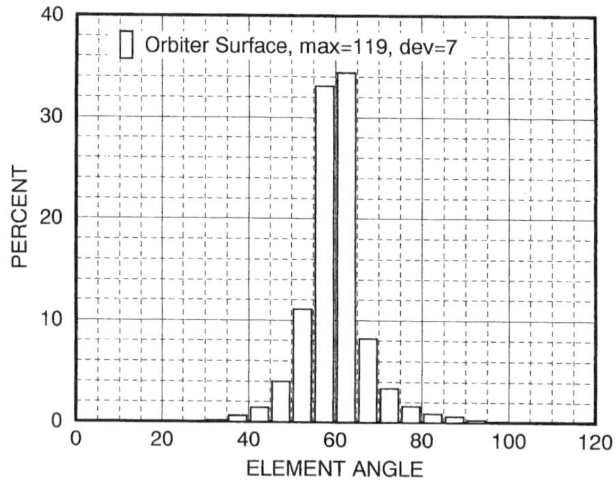

FIGURE 18.32 NASA space shuttle orbiter surface grid quality.

and high-aspect-ratio regions, was unable to finish. This process requires about 320 MB of RAM and the PCs are configured with 128 MB of RAM. However, the CPU times shown in Table 18.2 are valid for the PCs as this process and writing of the output grid file requires a small fraction (approximately 3%) of the total time and the times shown have been adjusted up to account for the work not done.

18.8 Summary

Methods for generation of unstructured planar, surface, and volume grids using the AFLR procedure have been presented. This procedure is based on an automatic point insertion scheme with local-reconnection connectivity optimization. Results for a variety of configurations have been presented. The results demonstrate that the procedure consistently produces grids of very high quality. Efficiency is such that standard PCs or workstations can be used to generate three-dimensional unstructured grids for complex configurations. The combined quality and efficiency of the AFLR procedure represents the current state of the art in unstructured tetrahedral grid generation.

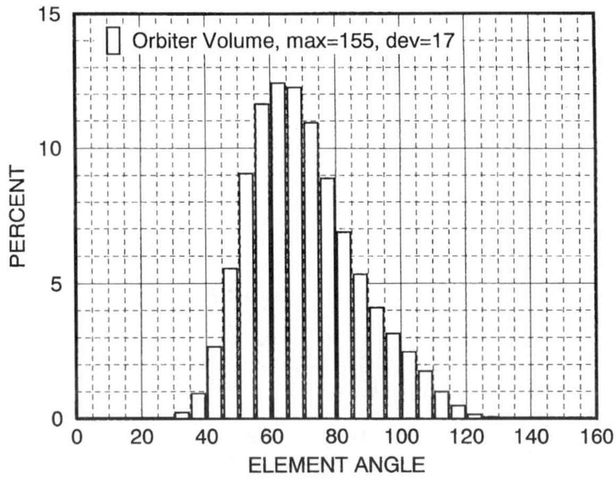

FIGURE 18.33 NASA space shuttle orbiter volume grid quality.

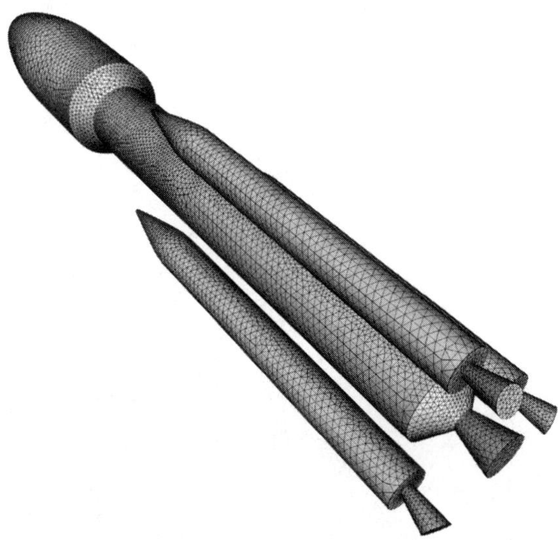

FIGURE 18.34 Surface grid for launch vehicle.

Acknowledgments

The author would like to acknowledge the efforts of Adam Gaither at the MSU ERC for preparing the CAD geometry definitions, generating the surface grids, and integrating, within SolidMesh, the software used to produce the results presented in this article. The author would also like to acknowledge support for this work from the Air Force Office of Scientific Research, Dr. Leonidas Sakell, Program Manager, Ford Motor Company, University Research Program, Dr. Thomas P. Gielda, Technical Monitor, Boeing Space Systems Division, Dan L. Pavish, Technical Monitor, National Science Foundation, ERC Program, Dr. George K. Lea, Program Director. In addition, the author would like to acknowledge Dr. Thomas Gielda of Ford Motor Company for providing the SUV interior geometry, Reynaldo Gomez of NASA Johnson Space Center for providing the Space Shuttle Orbiter geometry, Dr. Jim Johnson of General Motors Corporation for providing the pump cover geometry, and Dr. Edwin Rood of the Office of Naval Research for providing the destroyer model 5415 hull geometry.

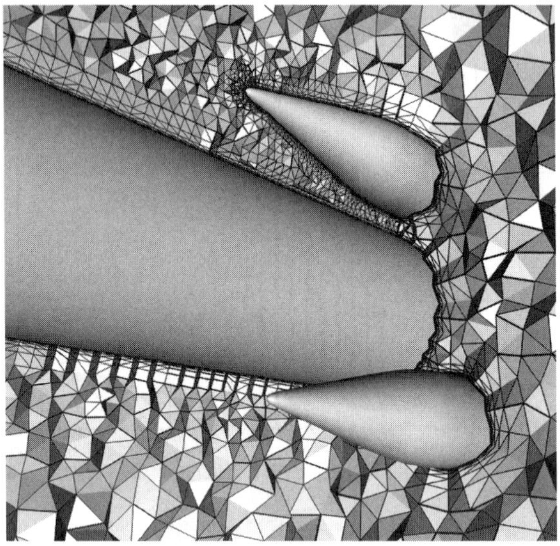

FIGURE 18.35 Tetrahedral field cuts for launch vehicle grid.

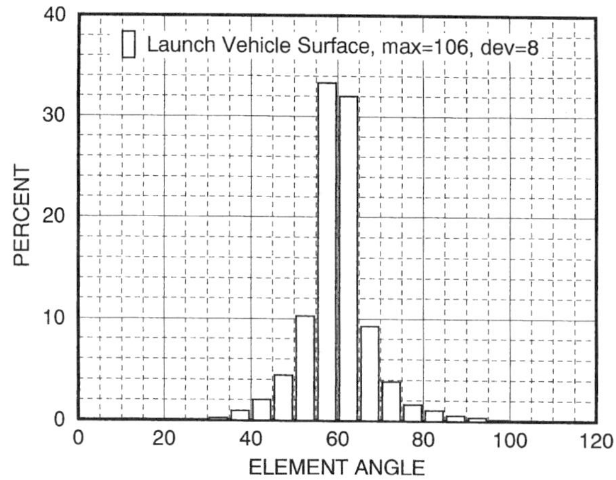

FIGURE 18.36 Launch vehicle surface grid quality.

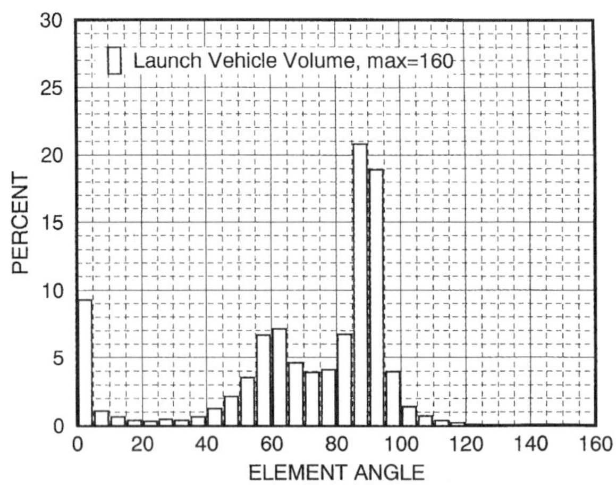

FIGURE 18.37 Launch vehicle volume grid quality.

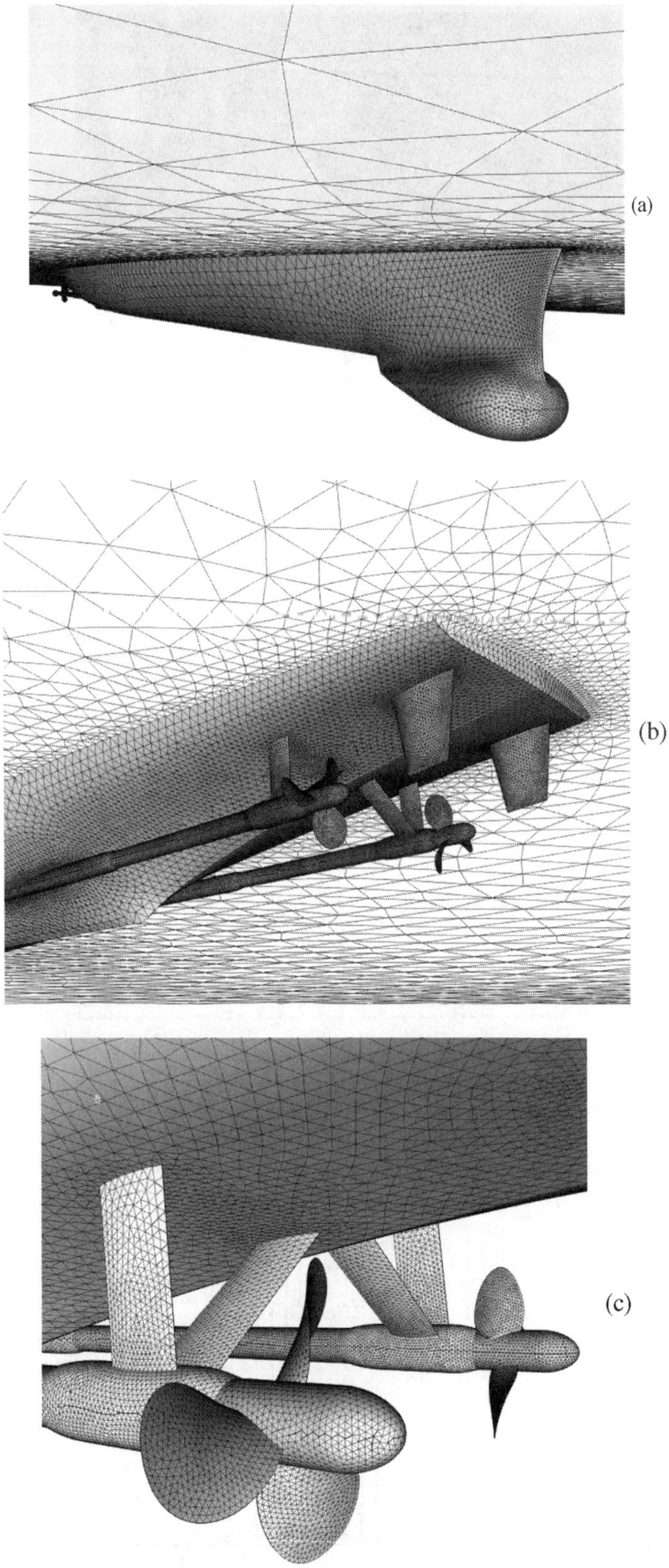

FIGURE 18.38 Destroyer hull surface grid. (a) Complete hull and water-line surfaces, (b) hull and propellers, (c) propellers.

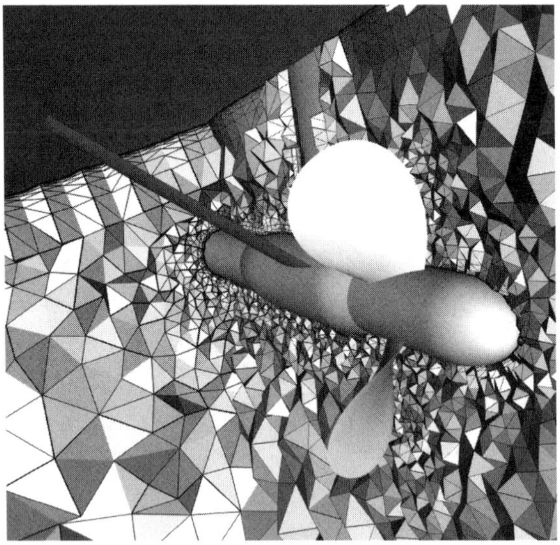

FIGURE 18.39 Tetrahedral field cut for destroyer hull grid.

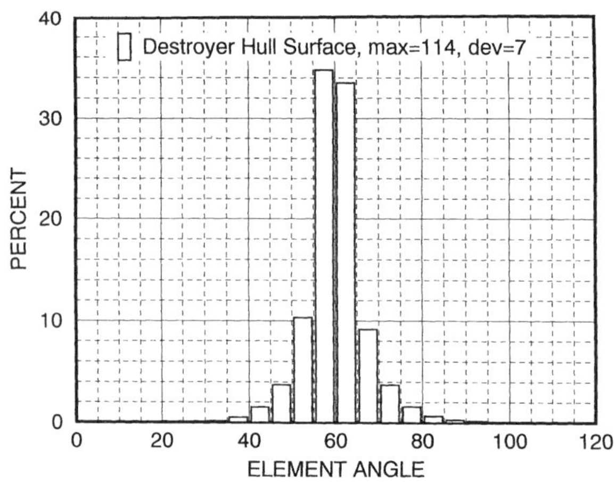

FIGURE 18.40 Destroyer hull surface grid quality.

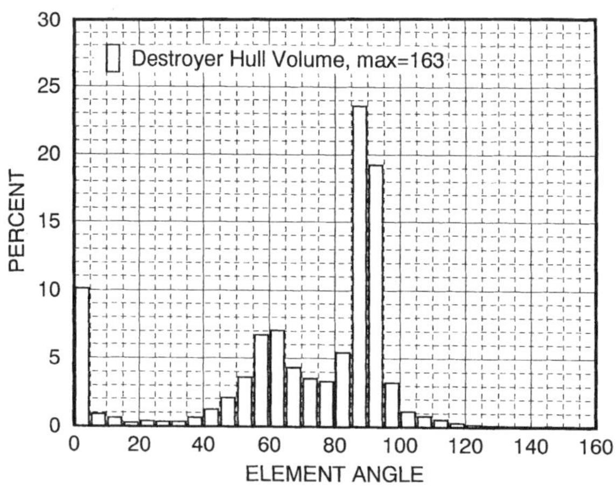

FIGURE 18.41 Destroyer hull volume grid quality.

References

1. **Baker, T. J.**, Three-dimensional mesh generation by triangulation of arbitrary point sets, AIAA Paper 87-1124, 1987.
2. **Barth, T. J.**, Steiner triangulation for isotropic and stretched elements, AIAA Paper 95-0213, 1995.
3. **Barth, T. J.**, Numerical aspects of computing viscous high Reynolds number flows on unstructured meshes, AIAA Paper 91-0721, 1991.
4. **Gaither, J. A.**, A solid modelling topology data structure for general grid generation, MS Thesis, Mississippi State University, 1997.
5. **Gaither, J. A.**, A topology model for numerical grid generation, *Proceedings of the Fourth International Conference on Numerical Grid Generation in Computational Fluid Dynamics*, Weatherill, N. P., Eiseman, P. R., Hauser, J., Thompson, J. F., (Ed.), Pineridge Press Ltd, 1994.
6. **George, P. L., Hecht, F., and Saltel, E.**, Fully automatic mesh generator for 3D domains of any shape, *Impact of Computing in Science and Engineering*, 2, p. 187, 1990.
7. **Holmes, D. G. and Snyder, D.D.**, The generation of unstructured meshes using Delaunay triangulation, *Proceedings of the Second International Conference on Numerical Grid Generation in Computational Fluid Dynamics*, Sengupta, S., Hauser, J., Eiseman, P. R., Thompson, J. F., (Ed.), Pineridge Press Ltd., 1988.
8. **Lawson, C. L.**, Properties of n-dimensional triangulations, *Computer Aided Geometric Design*, 3, p. 231, 1986.
9. **Lohner, R. and Parikh, P.**, Three-dimensional grid generation by the advancing-front method, *International Journal of Numerical Methods in Fluids*, 8, p. 1135, 1988.
10. **Marcum, D. L.**, Generation of unstructured grids for viscous flow applications, AIAA Paper 95-0212, 1995.
11. **Marcum, D. L.**, Generation of high-quality unstructured grids for computational field simulation, *6th International Symposium on Computational Fluid Dynamics*, Lake Tahoe, NV, 1995.
12. **Marcum, D. L.**, Adaptive Unstructured Grid Generation for Viscous Flow Applications, *AIAA Journal*, 1996, 34, p. 2440.
13. **Marcum, D. L.**, Control of Point Placement and Connectivity in Unstructured Grid Generation Procedures, *IX International Conference on Finite Elements in Fluids*, Venice, Italy, 1995.
14. **Marcum, D. L.**, Unstructured Grid Generation Components for Complete Systems, *5th International Conference on Grid Generation in Computational Fluid Simulations*, Starkville, MS, 1996.
15. **Marcum, D. L. and Gaither, K.P.**, Solution adaptive unstructured grid generation using pseudo-pattern recognition techniques, AIAA Paper 97-1869, 1997.
16. **Marcum, D. L. and Weatherill, N.P.**, Unstructured grid generation using iterative point insertion and local reconnection, *AIAA Journal*, 33, p. 1619, 1995.
17. **Mavriplis, D. J.**, An advancing front delaunay triangulation algorithm designed for robustness, AIAA Paper 93-0671, 1993.
18. **Muller, J. D., Roe, P. L., and Deconinck, H.**, A frontal approach for internal node generation in delaunay triangulations, *International Journal of Numerical Methods in Fluids*, 17, p. 256, 1993.
19. **Peraire, J., Peiro, J., Formaggia, L., Morgan, K., and Zienkiewicz, O. C.**, Finite element Euler computations in three-dimensions, *International Journal of Numerical Methods in Engineering*, 26, p. 2135, 1988.
20. **Rebay, S.**, Efficient unstructured mesh generation by means of Delaunay triangulation and Bowyer–Watson algorithm, *Journal of Computational Physics*, 106, p. 125, 1993.
21. **Shepard, M. S. and Georges, M. K.**, Automatic three-dimensional mesh generation by the finite octree technique, *International Journal of Numerical Methods in Engineering*, 32, p. 709, 1991.
22. **Weatherill, N. P.**, A method for generation of unstructured grids using dirichlet tessellations, *MAE Report No. 1715*, Princeton University, 1985.

19

Surface Grid Generation

J. Peiró

19.1 Introduction

The triangular surface grid generation procedure to be described in this chapter has been developed with the primary intention of employing it as the first step of 3D tetrahedral grid generation methods such as the Delaunay or the advancing front (AFT) techniques described in Chapter 16–18. However, the approach here discussed will be of more general interest, and applications to others areas such as, for instance, finite element analysis of shells, graphical display of surfaces, and the calculation of surface intersections in CAD systems, to name but a few, can also be envisaged.

The construction of a surface grid consists of approximating the surface by a set of planar triangular facets. In the rest of this chapter we will consider boundary-fitted grids only, i.e., the vertices of the triangulation lie on the surface. The discretization of a surface (or a part of it) into a general body conforming grid consists of positioning points on the surface, which will constitute the nodes of the grid, and defining the links to be established between a node and its neighbors. Therefore, any surface generation method requires an analytical definition of the surface that permits locating grid nodes on it, and a criterion for positioning the grid nodes on the surface and defining their connectivities according to a spatial distribution of the size and shape of the grid elements.

In current engineering practice, most of the geometrical data required in design is generated, stored and manipulated using CAD systems [5]. Applications such as weather forecast modeling or medical imaging, on the other hand, require the generation and handling of discrete data This type of data can either be suitably transformed into a format compatible with that of a CAD system or be dealt directly with in discrete form. The later approach is outside the scope of this chapter and the interested reader is referred to [8] for a discussion of appropriate grid generation techniques. In what follows we will assume that the required geometrical data is available in the form of CAD parametric curves and surfaces represented by spline composite curves and tensor-product surfaces, such as Ferguson, Bezier, or NURBS [10] (see Part III of this Handbook).

0-8493-2687-7/99/$0.00+$.50

Although a surface is topologically a two-dimensional region, the location of the grid nodes will be three-dimensional. This allows for two possible strategies to be employed in the generation of triangular surface grids. One can either generate grid nodes and connectivities directly in 3D or take advantage of the 2D character of the surface and reduce the surface grid generation to a 2D problem. Both strategies have their advantages and disadvantages.

The generation of triangulations directly on the surface presents several difficulties. The advancing front technique can be easily extended to deal with surfaces. However, determining the validity of a new triangle in 3D by verifying whether it intersects with the sides in the generation front is not a trivial task. A triangle and a side might not intersect in space, but they can cross and still produce an invalid triangle. The main problem associated with Delaunay-based methods is the absence, for surfaces of variable curvature, of circumcircle, and circumsphere criteria equivalent to those available for 2D and 3D grid generation, respectively.

On the other hand, if a definition of the surface as a mapping from a 2D region and IR^3 exists, this can be used to generate a grid in the 2D region which, at a later stage, will be transformed onto the surface. Nevertheless, existing 2D mesh generation methods will require considerable enhancements to deal with the added difficulty of controlling the size and shape of the elements to be generated in the 2D region since these grid characteristics will depend on the surface mapping employed.

In the approach adopted here, the use of geometrical definitions of surfaces in the form tensor-product spline surfaces leads to a parametrization that defines the region of the surface to be discretized as a mapping between a 2D region in a parameter plane and IR^3. The grid on the surface is obtained as the image of a triangulation of the region in the 2D region. The spatial distribution of grid size and shape in the parameter plane is defined in such a way that, after applying the mapping, the image grid on the surface presents the geometrical characteristics required by the user. These are specified by means of a 3D grid control function. The triangular grid is generated using a modified 2D AFT that accounts for the rapid variation of the grid characteristics in the parameter plane that the surface parametrization might induce.

19.2 Surface modeling

In the following, the domain to be discretized, termed here *computational domain,* will be viewed as a three-dimensional object that will be described by means of the surfaces that enclose it. This is known as a *boundary representation* (B-Rep) of the domain [5, 10]. This is the internal solid representation used by the majority of commercial and research solid modelers.

In a boundary representation, the computational domain is the region interior to a boundary surface. This surface can be considered as a generalized polyhedron that is the union of a set of *faces,* bound by *edges,* which in turn are bound by *vertices.* The faces lie on surfaces, the edges lie on curves, and the vertices are endpoints of the edges. An illustration of the notation utilized here is depicted in Figure 19.1. Therefore, a B-Rep model requires the storage of two types of data: geometrical and topological.

The geometrical data consist of the basic parameters defining the shape of the surfaces and curves, and the point coordinates of the vertices. The topological data are concerned with the adjacency relations between the different components of the boundary surface: vertices, edges, and faces. Finally, a convention of orientations designates on which side of a face to find the computational domain. It will be seen later that, by restricting the domain and the faces forming its boundary to be connected[†] regions, an orientation compatible with the geometric definition can be obtained automatically.

[†]A region is said to be connected if any two points in the interior of the region can be joined by a continuous curve whose points are all interior to the region.

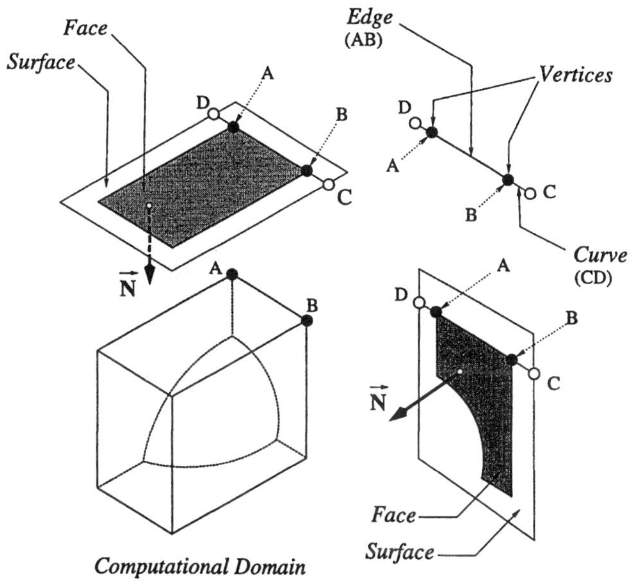

FIGURE 19.1 B-Rep of the boundary of the computational domain showing the orientation of the faces and the notation employed.

19.2.1 Geometrical Definition

The B-Rep of the domain provides a description of the computational domain in terms of a set of oriented faces. The generation of a boundary-fitted grid for this domain will require an analytical definition of the surfaces on which the faces are defined and their intersection curves. This mathematical representation should permit us to perform operations such as, for example, locating a point in space and calculating lengths and tangent vectors of curves as well as normal vectors and areas of surfaces.

19.2.1.1 Curves

Although curves are represented in the B-Rep model of the computational domain as the intersection of two surfaces, the use of such approach for grid generation is not recommended, since it results in an implicit representation of the intersection curve. This curve is given as the solution of a system of two nonlinear equations representing each of the intersecting surfaces (usually high-order polynomials). This means that some of the most common operations required in grid generation such as positioning a point on the curve, calculating the length of the curve, etc., will involve an iterative procedure for the solution of such system.

A more straightforward approach that eases the process of discretization is to adopt a parametric representation of the curve that accurately approximates the true intersection. This curve is computed once during a preprocessing stage. A method commonly employed is to locate a set of ordered points along the surface intersection through which a spline curve is later interpolated. The distribution of points should be such that the distance between the interpolated curve and the true surface intersection, using an appropriate norm, is within the accepted bounds of accuracy. This is a procedure which is readily available in most of the state-of-the-art systems for CAD.

Adopting the CAD representation of spline curves, e.g., Ferguson, Bezier, or NURBS, as described in Part III, curves are given by a parametric representation such as

$$\mathbf{r}(u) = \begin{Bmatrix} x_1(u) \\ x_2(u) \\ x_3(u) \end{Bmatrix} \qquad 0 \le u \le U \tag{19.1}$$

Here, and in the following, **r** will be denote the position vector of a point with respect to a Cartesian frame of reference (x_1, x_2, x_3). The tangent vector **t** to the curve, at a point with parametric coordinate u, is given by

$$\mathbf{t}(u) = \frac{d\mathbf{r}}{du} \tag{19.2}$$

19.2.1.2 Surfaces

Tensor products of splines are the most common form of CAD surface representation. Such surfaces can be described by a parametric representation such as

$$\mathbf{r}(u_1, u_2) = \begin{Bmatrix} x_1(u_1, u_2) \\ x_2(u_1, u_2) \\ x_3(u_1, u_2) \end{Bmatrix} \qquad 0 \le u_1 \le U_1; \quad 0 \le u_2 \le U_2 \tag{19.3}$$

The normal vector **n** to the surface, at a point of parametric coordinates (u_1, u_2), is given by

$$\mathbf{n}(u_1, u_2) = \frac{\partial \mathbf{r}}{\partial u_1} \times \frac{\partial \mathbf{r}}{\partial u_2} \tag{19.4}$$

where × denotes vector product.

Eq. 19.3 defines the surface as a mapping between a 2D rectangular region on a parameter plane (u_1, u_2) and IR^3. Such a parametric representation is provided by the majority of surface representation systems used in CAD.

For grid generation purposes, we will require that the mapping defining the surface is bijective almost everywhere and that a normal to the surface can be defined, and is continuous, for all the interior points. Singular points, i.e., those where the normal is not defined such as, for instance, the apex of a cone or the pole of a sphere, are allowed to appear only on the boundary.

19.2.2 Topological Description

The B-Rep model provides a hierarchical definition of the computational domain as the 3D region interior to a boundary partitioned into a set of faces. A face is a region on a surface delimited by an oriented set of edges.† Finally, an edge is the segment on a curve bound by two vertices.

The topological data required by the model is the definition of the boundary of a region at a certain level of the hierarchical model: domain, face, and edge, in terms of a list of regions in the next lower level: faces, edges, and vertices, respectively.

Vertices are points common to three or more faces and are represented by their 3D Cartesian coordinates. An edge is defined by the parametric curve on which it lies and the two end vertices. This representation admits the definition of several nonoverlapping edges on the same curve. If the computation domain is assumed to be connected, then an edge will be common to two faces only. A face is defined by the surface on which it lies and a set of edges forming its boundary. Again, several non-overlapping faces can be defined on the same surface.

†A face is sometimes referred to as a "trimmed" surface.

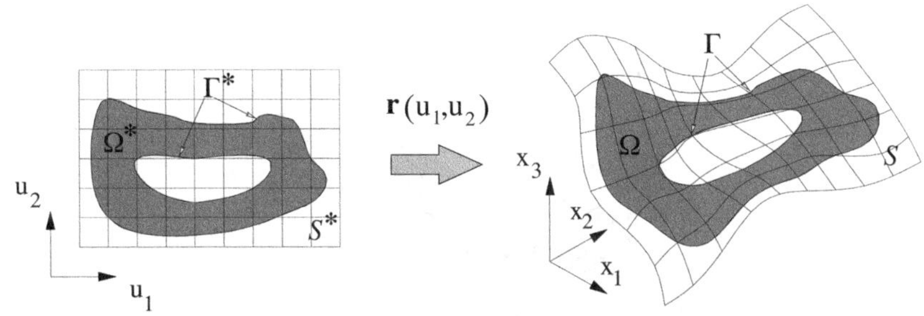

FIGURE 19.2 Definition of a face on a surface as a mapping.

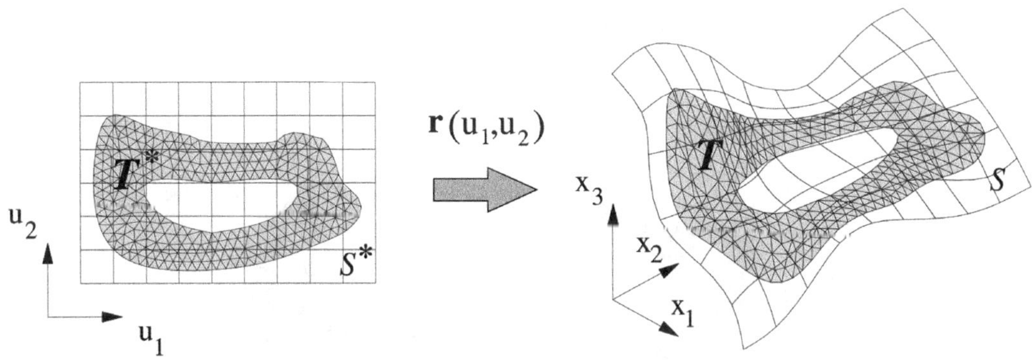

FIGURE 19.3 Mapping of a triangular grid T^* in the parameter plane onto the surface.

19.3 Surface Discretization

The representation of a surface S given by Eq. 19.3 allows us to define a face as a region Ω on the surface with boundary Γ, which is the image, by the mapping (Eq. 19.3), of a region Ω^* in the parameter plane (u_1, u_2). This region is delimited by a boundary Γ^* which is the preimage in the parameter plane of the boundary of the face Γ. The notation used here is illustrated in Figure 19.2.

If the mapping representing the surface is bijective, i.e., the normal to the surface does not vanish and is continuous, for all the points interior to the face, then such a mapping will transform a valid triangulation T^* in the parameter plane into a valid surface triangulation T (Figure 19.3).[†] This suggests the idea of generating a grid in the parameter plane that will later be mapped onto the surface to produce an appropriate surface discretization. This is accomplished by ensuring that the size and shape of the triangles generated in the parameter plane are such that when mapped onto the surface the size and shape of the resulting triangles comply with those specified by a suitably defined *grid control function.*

19.3.1 Grid Control Function

The inclusion of adequate grid control is a key ingredient in ensuring the generation of a grid of suitable characteristics for the performance of a numerical simulation. In this approach, the shape and size of the elements in the grid are assumed to be a function of the position, and they are locally defined in terms of a set of *mesh parameters.* Here the mesh parameters used are a set of three mutually orthogonal directions α_i; $i = 1, 2, 3$, and three associated element sizes, or spacings, δ_i ; $i = 1, 2, 3$ (see Figure 19.4). Thus, at a certain point, if all three element sizes are equal, the grid in the vicinity of that point will consist of approximately equilateral elements.

[†]Strictly speaking, it suffices that the mapping be bijective at the nodes of the triangulation only.

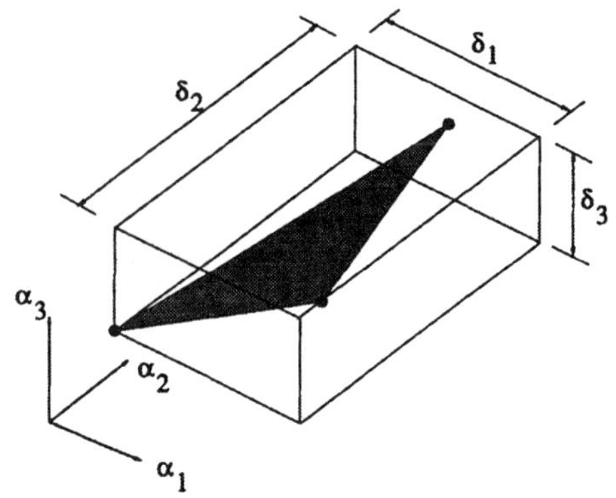

FIGURE 19.4 Mesh parameters.

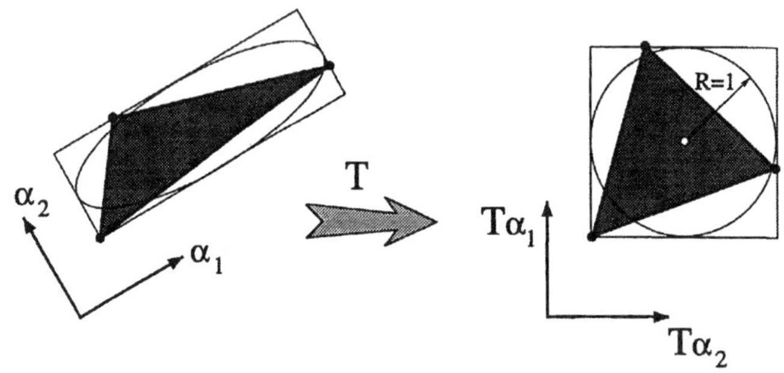

FIGURE 19.5 The effect of the local mapping **T** (2D).

The grid control is accomplished by defining a function which represents the characteristics of an element in the neighborhood of a point. This function is represented by means of a linear transformation that locally maps the 3D space onto a space where elements, in the neighborhood of the point being considered, will be approximately equilateral with unit average size. This new space will be referred to as the *normalized space.* For a general grid, this transformation will be a function of position. The mapping, denoted by **T**, is represented by a symmetric 3 × 3 matrix; it is a function of the mesh parameters α_i and δ_i at that position, and can be expressed as $\mathbf{r} = (x_1, x_2, x_3)$.

$$\mathbf{T}(\mathbf{r}) = \sum_{j=1}^{3} \frac{1}{\delta_j(\mathbf{r})} \alpha_j(\mathbf{r}) \otimes \alpha_j(\mathbf{r}) \qquad (19.5)$$

where ⊗ denotes the tensor product of two vectors. The effect of this transformation in two dimensions is illustrated in Figure 19.5.

The spatial variation of the mesh parameters (or equivalently **T**) is obtained through the definition of their values at a set of discrete positions and a procedure for interpolation at intermediate points. The most commonly used methods for the definition of the grid control function are the *background grid* and the distribution of *sources* [12]. In the first method, the mesh parameters are obtained via linear interpolation from a grid of tetrahedra in which each node is assigned a set of grid parameters. In the second method, the mesh parameters at a point are given as a user-defined function of the distance from the point to the reference sources.

19.3.2 Grid Quality

It is possible to impose restrictions on the grid spacing to ensure that some measure of grid quality is satisfied. The method proposed here tries to avoid rapid spatial changes in grid spacing, since they usually cause problems to the grid generation procedure and might lead to the creation of badly distorted grids.

A simple argument in one dimension provides us with a criterion for ensuring a smooth spatial variation of elements sizes. Consider two adjacent element of sizes (or spacings) δ_1 and δ_2, for which we would like to impose that the size should not change more than a certain fraction, K, of their average size. This condition can be written as

$$|\delta_2 - \delta_1| \leq K \frac{\delta_2 + \delta_1}{2} \tag{19.6}$$

A continuous analogue of Eq. 19.6 is given by

$$\left|\frac{d\delta(x)}{dx}\right| \leq K \tag{19.7}$$

This can be easily extended to the multi-dimensional case by simply imposing that

$$|\nabla\delta \cdot \mathbf{S}| \leq K \tag{19.8}$$

where **S** denotes a 3D unit vector.

19.4 Triangulation of Surfaces

The surface grid generation method proposed here is based on the idea that, using a tensor-product representation of a surface, a face can be obtained as a mapping between a region on a 2D parameter plane and 3D. If the mapping is not singular, i.e., the normal to the surface is non-zero and finite, at interior points on the face, then a valid triangulation in the parameter plane will subsequently transform onto a valid triangulation on the surface. The characteristics of the triangular grid in the parameter plane should be calculated so as to guarantee that the distribution of element size and shape in the transformed surface grid approximately complies with the user-specified 3D grid control function. The following sections describe how this is achieved in practice.

19.4.1 Grid Generation Procedure

The grid generation proceeds in a bottom-up fashion. Edges on the curves are discretized first into straight sides. Triangular grids are independently generated in each of the faces on the surfaces forming the boundary of the computational domain. A set of previously generated sides forms the initial generation front on the surface.

The procedural steps are the following:

1. Read the geometrical definition and a suitable distribution of mesh parameters.
2. Discretize the edges.
   a. Calculate the local coordinate u of the points defining the end of the edges.
   b. Position points along the edge according to the grid control function.
3. Discretize the faces.
   a. Calculate the local coordinates (u_1, u_2) in the parameter plane of the points generated in the previous step that belong to the edges in the boundary of the face.
   b. Form the initial front in the parameter plane and orientate the surface in a form compatible with its normal, as defined by its parametrization (Eq. 19.3) and according to Eq. 19.4.

   c. Generate an appropriate triangulation in the parameter plane using a suitably modified 2D advancing front technique.
   d. Perform grid enhancement techniques in the parameter plane to achieve a better representation of the surface curvature and to improve the quality of the surface grid.
   e. Map the resulting grid onto the surface definition.
4. Orientate the discretized boundary.

19.4.2 Computation of the Local Coordinates of the Edge Endpoints

An edge is a region on a curve delimited by two endpoints. These endpoints are vertices of the boundary of the computational domain. However, since the curve is only an approximation to the true intersection, the vertices will not, in general, lie exactly on the curve. For this reason, the delimiting points of the edge are taken to be the points on the curve which are the closest to the vertices. The distance between the vertex and the closest point on the curve has to be smaller than a certain threshold distance D_t. Its value is utilized to determine whether two points are coincident and it should be either known from the geometrical tolerance used in the creation of the CAD data or, if this is not available, calculated from the machine roundoff error.

The problem of finding the parametric coordinate of a vertex can be formulated as a point projection problem, i.e., given a vertex $\mathbf{r}^*$, find the parametric coordinate u of the point $\mathbf{r}(u)$ on the curve such that

$$D = \|\mathbf{r}(u) - \mathbf{r}^*\| = \min \tag{19.9}$$

The solution to the above equation is obtained by means of a standard iterative procedure for function minimization [4]. An initial bracketing of the minimum in Eq. 19.9 is given by a triplet of parametric coordinates $u_{(1)} \le u_{(2)} \le u_{(3)}$. The interval end values are taken to be those corresponding to the endpoints of the curve, $u_{(1)} = 0$ and $u_{(3)} = U$, and the third value, $u_{(2)}$, is obtained as follows. The curve is first divided into a few straight segments, then the segment closest to the point is found and, finally, $u_{(2)}$ is taken to be the average value of the parametric coordinates of the endpoints of the closest segment. Once the initial bracketing is done, the bracket is contracted, using a combination of sectioning by golden section search and parabolic interpolation, until the position corresponding to a minimum of the distance, $D = D_{\min}$, is found. If the geometrical data is correctly defined, the value of this distance should not be larger than the threshold distance ($D_{\min} \le D_t$).

19.4.3 Curve Discretization

This procedure consists of dividing the edge into straight sides. The sides should be such that their length is approximately compliant with the spacing specified by the grid control function. Here we will consider two approaches which are equivalent in the hypothetical case that we could define a continuous grid control function and that the length integrations involved could be carried out exactly. The first method is based in the placement of points along the curve according to a *distribution function* which is reminiscent of those employed in PDE based grid generation methods [2]. In the second approach, the linear mapping $\mathbf{T}$ is used to transform the curve to a new space where the grid spacing is uniform.

19.4.3.1 Discretization Using a Distribution Function

The discretization of the edges in the surface definition is achieved by positioning nodes along the curve according to a certain function $\delta(s)$, the grid spacing, which represents the size of the sides to be generated along the curve. The parameter s denotes the arc length of the curve which, for a curve represented in parametric form as $\mathbf{r}(u)$, $0 \le u \le U$, is given by

$$s(u) = \int_0^{s(u)} ds = \int_0^u \left\| \frac{d\mathbf{r}(t)}{dt} \right\| dt \tag{19.10}$$

where $\|\mathbf{a}\|$ denotes the Euclidean norm of the vector $\mathbf{a}$. In what follows, the edge is taken to be the region on the curve given by the parametric interval $0 \leq U_1 \leq u \leq U_2 \leq U$, where U_1 and U_2 are the parametric coordinates of the points on the curve which are the closest to the vertices representing the endpoints as computed by the procedure described in Section 19.4.2. The distribution of spacing along the edge, $\delta(s)$, is calculated using the information about the spatial distribution of mesh parameters provided by the grid control function described in Section 19.3.1. Consecutive points generated in the discretization procedure will then by joined by means of straight lines to form sides. The procedure employed here to determine the position and number of nodes to be created on the edges is based on the definition of an appropriate node distribution function.

Consider an interval of length ds at a point $\mathbf{r}(u)$ corresponding to an associated arc length s and assume that the interval is small enough so that the spacing $\delta(s)$ can be taken to be approximately constant. Under these assumptions, the number of subdivisions dA_e of the interval will be

$$dA_e = \frac{ds}{\delta(s)} \tag{19.11}$$

The distribution function will be obtained through the integration of Eq. 19.11. To achieve this, the definition of the spacing function $\delta(s)$ along the curve is required first. Here, this is accomplished by generating a set of uniformly spaced *sampling points* $\mathbf{r}(u_i)$; $i = 1, \ldots, m$ along the curve. A safe choice for the distance between sampling points is the minimum user specified element size but, often, considerably larger values can be used. The position of the sampling points, i.e., the value of u_i, is computed by numerically solving the equation

$$s_i = s(u_i) = L_1 + \frac{i-1}{m-1}(L_2 - L_1) = \int_0^{u_i} \left\| \frac{d\mathbf{r}}{du} \right\| du; \qquad i = 1, \ldots, m \tag{19.12}$$

where $L_1 = s(U_1)$ and $L_2 = s(U_2)$ denote the arc length values corresponding to the endpoints of the edge.

For all the sampling points $\mathbf{r}(u_j)$; $j = 1, \ldots, m$, the mesh parameters are obtained by interpolation from the grid control function and the spacing δ_j associated to a sampling point is computed as

$$\delta_j^c = \left\| \mathbf{T}_j \mathbf{t}_j \right\|^{-1} \tag{19.13}$$

where $\mathbf{T}_j$ is the value of the auxiliary transformation at the sampling point given by formula 19.5 and $\mathbf{t}_j$ represents the tangent to the curve at that point, $\frac{d\mathbf{r}}{du}(u_j)$. Then, a piecewise linear distribution of spacings $\delta(s)$ along the edge is obtained from the values δ_j^c computed in Eq. 19.13 and may be written as

$$\delta(s) = \sum_{i=1}^{m} \delta_i^c N_i(s) \tag{19.14}$$

where $N_i(s)$ represents the linear finite element shape function

$$N_i(s_j) = \begin{Bmatrix} 0 & \text{if} & i \neq j \\ 1 & \text{if} & i = j \end{Bmatrix} \tag{19.15}$$

The positions s_k, $k = 1, \ldots, N_e - 1$ of the internal nodes to be created are the solutions of the equation

$$\phi(s_k) = \frac{N_e}{A_e} \int_{L_1}^{s_k} \frac{1}{\delta(s)} ds = k; \qquad k = 1, \ldots, N_e - 1 \tag{19.16}$$

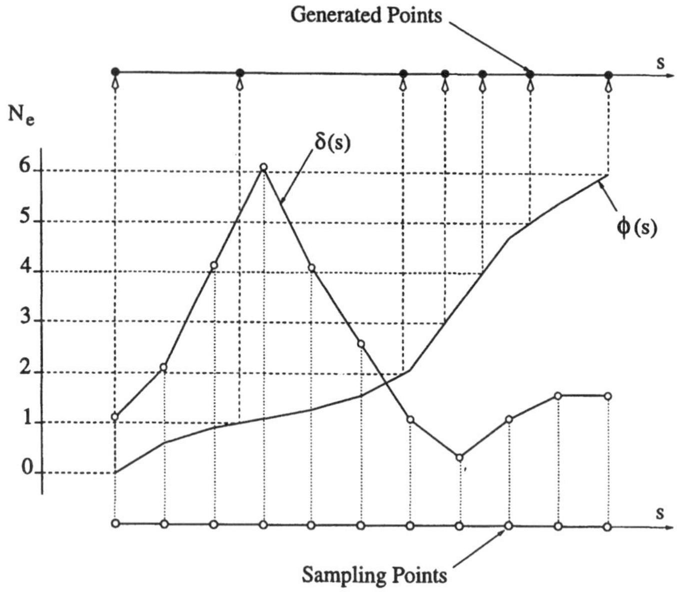

FIGURE 19.6 Curve discretization by means of a distribution function.

$\phi(s)$ is commonly referred to as the distribution function and N_e denotes the number of sides generated on the curve. Its value is chosen to be the nearest integer value to A_e, which is computed by integrating expression 19.10 as

$$A_e = \int_{L_1}^{L_2} \frac{1}{\delta(s)}\, ds \tag{19.17}$$

The positioning of the points along the curve using the discretization procedure described above is illustrated in Figure 19.6.

The integrals in Eqs. 19.16 and 19.17 can be computed explicitly since the $\delta(s)$ is taken to be the piecewise linear function Eq. 19.14. The solution of Eq. 19.16 is obtained via the Newton's iteration

$$s_k^{(i+1)} = s_k^{(i)} - \frac{\delta\left(s_k^{(i)}\right)}{A_e}\left\{\phi\left(s_k^{(i)}\right) - k\right\} \tag{19.18}$$

where $s_k^{(i)}$ denotes the value of the arch length s_k at iteration i and the initial value for the iteration is taken to be $s_k^{(0)} = s_{k-1}$.

19.4.3.2 Discretization Using the Mapping T

Here the placement of points along the edge is based on a transformation of the curve to a normalized space where the spacing along the curve is uniform. In order to determine the position and number of nodes to be created on each edge, the following steps are followed:

1. Subdivide recursively each edge into smaller curves until their length is smaller than a certain prescribed value, i.e., define a set of sampling points $\mathbf{r}_j = \mathbf{r}(u_j)$; $j = 1, \ldots, n$ as described previously. When subdividing an edge, the position and tangent vectors corresponding to these new points, $\mathbf{t}_j$ can be readily found directly from the original definition of the curve.
2. For each data point $\mathbf{r}_j$; $j = 1, \ldots, n$ obtain from the grid control function the coefficients of the transformation $\mathbf{T}_j$ and transform the position and tangent vectors according to $\hat{\mathbf{r}}_j = \mathbf{T}_j\,\mathbf{r}_j$ and $\hat{\mathbf{t}}_j = \mathbf{T}_j\,\mathbf{t}_j$. The new position and tangent vectors, $\hat{\mathbf{r}}_j$ and $\hat{\mathbf{t}}_j$; $j = 1, \ldots, n$, define a spline curve that can

be interpreted as the image of the original edge in the normalized space. It must be noted that, because of the approximate nature of this procedure, the new curve will in general have discontinuities of curvature, even if the curvature of the original curve varies continuously.

3. Compute the length of the edge in the normalized space, subdivide it into segments of approximately unit length, and calculate the parametric coordinate of each newly created point. This information is then used to determine the coordinates of the new nodes in the physical space, using the parametric representation of the curve.

19.4.4 Computation of Coordinates in the Parameter Plane

A face is defined as a region on a surface delimited by a set of edges. These have been discretized in the previous step of the generation process, and the assembly of the discretized edges forms the boundary of the triangular grid to be generated. However, in the approach adopted here, the AFT generation of the triangular grid will take place on the parameter plane. Consequently, to form the initial generation front in the parameter plane, the (u_1, u_2) parametric coordinates of the nodes generated on the edges need to be computed. Since the mapping $\mathbf{r}(u_1, u_2)$ cannot, in general, be inverted analytically, the coordinates (u_1, u_2) of such points are found numerically by means of an iterative procedure.

The curve where the edge is defined is only an approximation to the intersection curve of the surfaces to which the two adjacent faces belong to. As a result of this, the nodes generated on the edges are not exactly on the surface. The distance between these nodes and the surface depends on the accuracy used to approximate the true intersection between the surfaces by a spline curve.

In this formulation, the parametric coordinates (u_1, u_2) of a node, denoted by $\mathbf{r}^*$, are taken to be those of the point $\mathbf{r}(u_1, u_2) = (x_1, x_2, x_3)$ in the surface closest to $\mathbf{r}^*$. This can be formulated as the minimization problem of finding the parametric coordinates (u_1, u_2) for which

$$\left\|\mathbf{r}^* - \mathbf{r}\left(u_1, u_2\right)\right\| = \min \tag{19.19}$$

It should be pointed out that the discretization of the edges if performed directly in the 3D space and not in the parameter plane in order to ensure compatibility of nodal coordinates between contiguous faces.

The non-linear Eq. 19.19 is solved by means of an iterative procedure that involves the following steps:

1. The distance $\|\mathbf{r} - \mathbf{r}^*\|$ is calculated for all the singular points on the surface boundary. If for one of them, this value is smaller than the threshold distance D_t used to determine whether two points coincide, then its parametric coordinates are the sought solution.
2. If the answer is not found among the singular points, the search for the minimum continues on the boundary. The minimization is performed using the 1D procedure described in the Section 19.4.2. The iteration stops if a point $\mathbf{r}$ is found that verifies $\|\mathbf{r} - \mathbf{r}^*\| < D_t$.
3. Finally, we look for the minimum in the interior of the region. The closest point found on the boundary is used as the initial guess for a conjugate gradient method with line minimization [4]. This method is very efficient but might fail in certain circumstances, e.g., for interior points in the vicinity of a singular point. In such cases, a more robust, but also more expensive, "brute force" approach is used. This method starts with an initial uniform subdivision of the parameter plane into rectangular regions along coordinate lines. Amongst these rectangles, the closest to the target point is selected for further subdivision into four. The distance between the centroid of the rectangle and the target point is used for this purpose. This procedure is repeated until a point $\mathbf{r}$ is found which verifies the convergence criteria, i.e., $\|\mathbf{r} - \mathbf{r}^*\| < D_t$.

19.4.5 Orientation of Initial Front

A simple procedure for automatically orientating the initial front can be devised if we assume that the region in the parameter plane representing a face on the surface is a connected region. For this type of

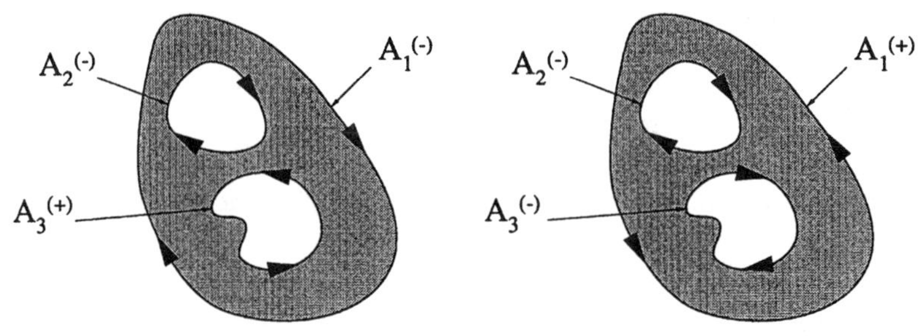

(a) Initial orientation (b) Final orientation

FIGURE 19.7 Automatic boundary orientation in the parameter plane.

regions, the boundary of the face is formed by one or more closed non-self-intersecting loops of edges. The edges in a loop join other edges at their end vertices and a vertex is always shared by two edges in the face. Under these assumptions, the loops of edges can be identified and their points ordered so as to assign a unique orientation to the closed curve. There are two possible orientations for a curve that can be determined as follows. The area of a region in the parameter plane (u_1, u_2) delimited by a closed curve C can be expressed, using Green's theorem, as the absolute value of the line integral

$$A = \frac{1}{2}\oint_C - u_2 du_1 + u_1 du_2 \tag{19.20}$$

The sign of A is used to characterize the orientation of the curve.

The initial front representing the discretized boundary of the region in the parameter plane is formed by one or more loops of discretized edges. This provides a piecewise linear representation of the loop as a set of straight segments that permits a simple numerical evaluation of the integral Eq. 19.20. For a connected region, the loop representing the outer boundary will have the largest area in absolute value. The final orientation of the loops defining the boundary of the region is selected so the area of the exterior loop is positive and the area of the interior loops, if any, is negative according to Eq. 19.20. This is depicted in Figure 19.7.

19.4.6 Grid Generation in the Parameter Plane

The definition of the surface where a face lies as a mapping permits the surface grid generation to be performed in the parameter plane by a suitably modified 2D grid generation method. The grid generation method employed here is a modification on the 2D AFT, which is briefly summarized in Section 19.4.6.1. This procedure requires the definition of a suitable distribution of mesh parameters in the parameter plane such that, when the triangular grid generated there is transformed onto the surface, the resulting surface triangulation approximately complies with the grid characteristics specified by 3D grid control function. The utilization of a bijective mapping permits to establish a correspondence between the 3D mesh parameters on the surface and the 2D mesh parameters in the parameter plane. This is described in Section 19.4.6.2.

19.4.6.1 The Modified 2D AFT

The modified AFT follows these algorithmic steps:

1. Select a side from the current generation front. The sides of the front are ordered, using a heap structure, according to their length in 3D. The side selected is the shortest side, which is located at the root of the binary tree representing the heap.
2. Determine the position of the "ideal" point to form a triangle. The position of the ideal point should be such that the size and shape of the resulting surface triangle complies with those specified

by the 3D grid control function. A detailed description of the nonlinear iterative procedure employed to achieve this is given in Section 19.4.7.

3. Generate a list of alternative locations and select a list of possible candidates among the nodes in the generation front.
4. Go through the list of candidate nodes (which are organized in a heap structure according to a measure of quality in 3D) and select the best among those producing a compatible triangle, i.e., one that does not intersect with the current generation front. This compatibility condition is verified in the 2D parameter plane, thus avoiding the problem of crossing if checked directly on the surface.
5. Update the generation front and repeat the process if there are sides left in the front.

19.4.6.2 Grid Characteristics in the Parameter Plane

The discretization of each face is accomplished by generating a two-dimensional grid of triangles in the parametric plane (u_1, u_2) and then transforming it onto the surface using the mapping $\mathbf{r}(u_1, u_2)$ defined in Eq. 19.3. This mapping establishes a one-to-one correspondence between the face and a region on the parametric plane (u_1, u_2) (Figure 19.2). Thus, a consistent triangular grid in the parametric plane will be transformed, by the mapping $\mathbf{r}(u_1, u_2)$, into a valid triangulation of the surface component. The construction of the triangular grid in the parameter plane (u_1, u_2) using the two-dimensional grid generator, requires the determination of an appropriate spatial distribution of the two-dimensional mesh parameters. These consist of a set of two mutually orthogonal directions α_i^*; $i = 1, 2$, and two associated element sizes δ_i^*; $= 1, 2$.

The two-dimensional mesh parameters in the (u_1, u_2) plane can be evaluated from the spatial distribution of the three-dimensional mesh parameters and the metric tensor that locally represents the deformation characteristics of the mapping. To illustrate this process, consider a point P^*, in the parametric plane of coordinates (u_1^*, u_2^*), where the values of the mesh parameters α_i^*, δ_i^*; $i = 1, 2$ are to be computed. Its image on the surface will be the point P given by the position vector $\mathbf{r}(u_1^*, u_2^*)$. The transformation between the physical space and the normalized space at this point $\mathbf{T}$ can be obtained from the grid control function. A new mapping can now be defined at the point P between the parametric plane (u_1, u_2) and the normalized space as

$$\mathbf{R}(u_1, u_2) = \mathbf{Tr}(u_1, u_2) \tag{19.21}$$

A curve in the parametric plane passing through point P^* and with unit tangent vector $\beta = (\beta_1, \beta_2)$ at this point, is transformed by the above mapping into a curve in the normalized space passing through the point of coordinates $\mathbf{R}(u_1, u_2)$. The arc length parameters ds^* and ds, along the original and transformed curves, respectively, are related by the expression 19.14.

$$ds^2 = \left\{ \sum_{i,j=1}^{2} \frac{\partial \mathbf{R}}{\partial u_i} \cdot \frac{\partial \mathbf{R}}{\partial u_j} \beta_i \beta_j \right\} ds^{*2} \tag{19.22}$$

Assuming that this relation between the arc length parameters also holds for the spacings, we can compute the spacing δ_β at the point P^* and along the direction β in the parameter plane as

$$\frac{1}{\delta_\beta} = \sqrt{\sum_{i,j=1}^{2} \frac{\partial \mathbf{R}}{\partial u_i} \cdot \frac{\partial \mathbf{R}}{\partial u_j} \beta_i \beta_j} \tag{19.23}$$

The two-dimensional mesh parameters α_i^*, δ_i^*; $i = 1, 2$ are determined from the direction in which δ_β attains an extremum. This reduces to finding the eigenvalues and eigenvectors of a symmetric 2×2 matrix.

19.4.6.3 Influence of the Surface Parametrization

It is clearly apparent from the discussion in the previous section that there is a direct relation between the mesh parameters in the parameter plane required to produce a surface grid compliant with the 3D grid control function and the parametrization of the surface. The main problem associated with this is that the parametrization of a surface is not unique. A region in the parameter plane can be transformed into a surface using an unlimited number of parametrizations. However, the choice of parametrization will influence the performance and accuracy of the grid generation method, since different parametrizations will induce different degrees of distortion between the parameter plane and the surface. An example of this is illustrated in Figure 19.8 in which a uniform triangular grid for a square region, defined on a planar surface, is obtained by using three different parametrizations. Figures 19.8(a), 19.8(c) and 19.8(e) show the triangulation and a set of 5 × 5 coordinate lines $u_1 = 0, 1, 2, 3, 4$ and $u_2 = 0, 1, 2, 3, 4$ in the parameter plane. Figures 19.8(b), 19.8(d) and 19.8(f) show their respective images on the 3D surface.

The first parametrization, shown in Figures 19.8(a) and 19.8(b), preserves the length and area ratios; the mapping does not introduce any distortion and the grids on the surface and the parameter plane are alike. The second parametrization of the surface mapping, Figures 19.8(c) and 19.8(d), maintains the length ratio along u_1 but introduces distortion in the u_2-direction. The pre-image of the 3D square region is no longer a square since its sides are not straight lines due to the deformation induced by the mapping. To account for this deformation, stretched triangles must be generated in the parameter plane in order to produce a uniform triangulation in 3D. This certainly makes the task of generating a suitable grid in the parameter plane more difficult. The third mapping, Figures 19.8(c) and 19.8(d), introduces distortion in both directions. Stretched elements are required for this case too, but now the variation of the mesh parameters through the parameter plane is more rapid than before, which further increases the difficulties associated with generating the grid in the parameter plane. A slight deterioration of the grid quality is readily noticeable in Figure 19.8(d).

As a consequence of the additional deformation introduced by the surface parametrization, large variations of the mesh parameters in a relatively small neighborhood of a point in the parameter plane might occur.† The best parametrization, from the point of view of grid generation, is the one that uses parametric coordinates based on arc length. This results in a surface mapping that produces a small distortion between the parameter plane and the surface. However, such a parametrization is not easy to obtain in practice. Therefore, provisions should always be made to account for mapping-induced distortion in the grid generation procedure.

The method originally proposed in [11] assumed that the values of the mesh parameters at the midpoint of the side selected for the generation of a triangle were approximately constant in the neighborhood of the side. In the presence of rapid local changes in the mesh parameters, the quality of the surface grids deteriorates; therefore, it had to be modified to account for the rapid variation of mesh parameters. An improved nonlinear iteration procedure is used here to determine the position of the grid nodes on the surface. This is described in detail in the following section.

19.4.7 Finding the Location of the Ideal Point

Following the notation displayed in Figure 19.9, let us consider a side AB in the generation front to be used to generate a new triangle in the surface grid. A candidate location, the so-called "ideal point" P, is sought as the vertex of a triangle with base AB that complies with the size and shape characteristics prescribed by the 3D grid control function. The location of the ideal point is obtained as follows.

The matrix $\mathbf{T}$ is calculated at the midpoint M of the side AB using the values of the 3D mesh parameters given by the user-specified grid control function. It is assumed that the local mapping, represented by $\mathbf{T}$, can be taken to be constant in the (3D) neighborhood of the side. This assumption is reasonably correct if criteria of grid quality such those described in Section 19.3.2 are enforced on the grid control

†See [15] for a more detailed exposition of this problem.

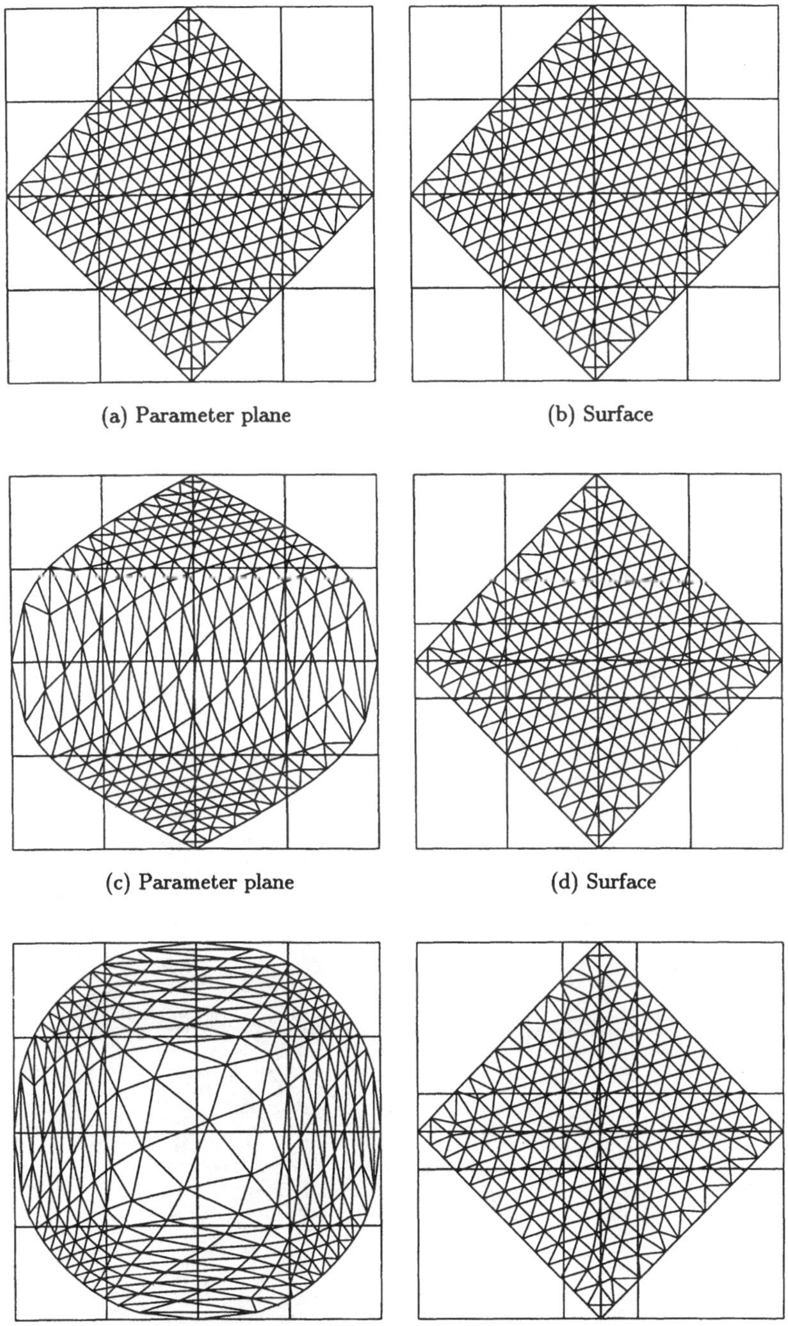

(a) Parameter plane (b) Surface

(c) Parameter plane (d) Surface

(e) Parameter plane (f) Surface

FIGURE 19.8 Influence of the surface parametrization. The network of lines on the surface represents the set of coordinate curves $u_1 = 0, 1, 2, 3, 4$ and $u_2 = 0, 1, 2, 3, 4$.

function. It must be stressed that this is not necessarily true for the triangle in the parameter plane since, as discussed in Section 19.4.6.3, the surface mapping might introduce rapid variations of the 2D mesh parameters.

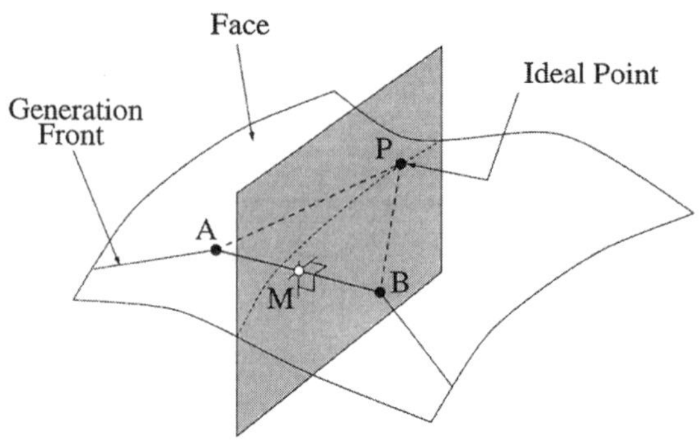

FIGURE 19.9 Location of the ideal point.

The location of the ideal point P is calculated by first transforming, using **T**, the coordinates of the relevant points in the triangle to a 3D normalized space. Then its parametric coordinates (u_1, u_2) are determined by requesting that its position $\mathbf{r}(u_1, u_2)$ in the normalized space satisfies

$$\{\mathbf{r}(u_1, u_2) - \mathbf{r}_M\} \cdot \{\mathbf{r}_B - \mathbf{r}_A\} = 0 \tag{19.24}$$

$$\left\|\mathbf{r}(u_1, u_2) - r_A\right\|^2 = 1 \tag{19.25}$$

where $\mathbf{r}_A$, $\mathbf{r}_B$ and $\mathbf{r}_M$ denote the positions in the normalized space of the points A, B, and M, respectively.

The system of Eqs. 19.24 and 19.25 is nonlinear. Its solution is achieved by iteration using Newton's method. The iterative procedure can be written in abbreviated matrix form as

$$\Delta\mathbf{u}^{(k)} = \mathbf{u}^{(k+1)} - \mathbf{u}^{(k)} = -\,\mathbf{J}^{-1}\left[\mathbf{u}^{(k)}\right]\mathbf{f}\left[\mathbf{u}^{(k)}\right] \tag{19.26}$$

with

$$\Delta\mathbf{u}^{(k)} = \begin{bmatrix} u_1^{(k+1)} \\ u_2^{(k+1)} \end{bmatrix} - \begin{bmatrix} u_1^{(k)} \\ u_2^{(k)} \end{bmatrix}; \qquad \mathbf{f}\left[\mathbf{u}^{(k)}\right] = \begin{bmatrix} \{\mathbf{r}^{(k)} - \mathbf{r}_M\} \cdot \{\mathbf{r}_B - \mathbf{r}_A\} \\ \left\|\mathbf{r}^{(k)} - \mathbf{r}_A\right\|^2 - 1 \end{bmatrix} \tag{19.27}$$

and

$$\mathbf{J}\left[\mathbf{u}^{(k)}\right] = \begin{bmatrix} \dfrac{\partial \mathbf{r}^{(k)}}{\partial u_1} \cdot \{\mathbf{r}_B - \mathbf{r}_A\} & \dfrac{\partial \mathbf{r}^{(k)}}{\partial u_2} \cdot \{\mathbf{r}_B - \mathbf{r}_A\} \\ 2\dfrac{\partial \mathbf{r}^{(k)}}{\partial u_1} \cdot \{\mathbf{r}^{(k)} - \mathbf{r}_A\} & 2\dfrac{\partial \mathbf{r}^{(k)}}{\partial u_2} \cdot \{\mathbf{r}^{(k)} - \mathbf{r}_A\} \end{bmatrix} \tag{19.28}$$

where the index (k) denotes the value of the corresponding variable at the kth iteration of the Newton procedure.

The convergence of this iterative method depends on the choice of initial guess $\mathbf{u}^{(0)}$. If the surface mapping does not introduce severe distortions, an initial guess of the location of the ideal point calculated using the values of the 2D mesh parameters from expression 19.23 usually leads to convergence of the Newton method. However, in general, it is not always possible to avoid or reduce the deformation induced by the mapping and, therefore, an alternative method for handling such situations is required.

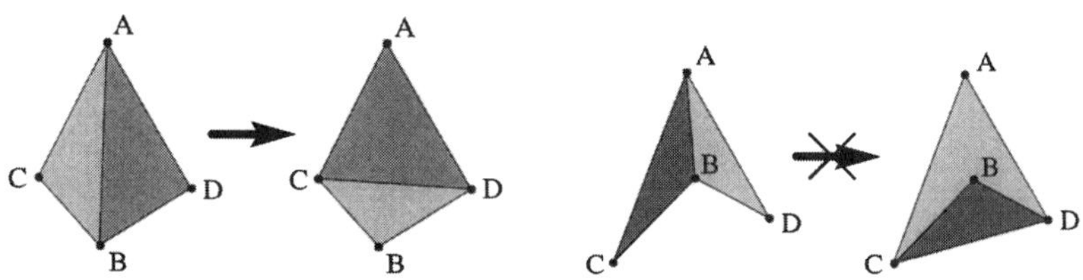

FIGURE 19.10 The diagonal swapping procedure; (a) admissible, (b) inadmissible.

The approach adopted here is to improve upon this initial guess, if the nonlinear iteration fails to converge, by means of a "brute force" approach based on selective recursive subdivision. In the event of convergence failure of the Newton method, a conservative estimate of the maximum ratio between the length of a side in the parameter plane and their image on the surface is calculated first. This ratio is used to determine the size of a rectangular region in the parameter plane that is to be attached to the front side and will contain the location of the ideal point. The selection of the new initial guess for the Newton iteration is based on a quadtree recursive subdivision. The rectangular region is first divided into four rectangles that and the new guess for the position of the ideal point is the center of the rectangle that best approximates the requirements Eqs. 19.24 and 19.25. If the Newton iteration fails to converge, the previously chosen rectangle is further subdivided into four to produce a new initial guess for another iteration. The procedure is repeated until convergence of the Newton iteration is achieved.

19.4.8 Surface Grid Enhancement Techniques

The triangular grid generated on the face in the previous step may contain some badly distorted triangles, especially if the mapping-induced distortions are large. In order to enhance the quality of the generated grid, two post-processing are applied: *diagonal swapping* and *grid smoothing*. These procedures are local in nature and do not alter the total number of nodes and elements in the grid. A description of the implementation of these two methods follows.

19.4.8.1 Diagonal Swapping

This procedure modifies the grid connectivity without altering the positions of the nodes. This process requires a loop over all the element sides, excluding those sides on the boundary. Following the notation of Figure 19.10, for each internal side *AB* common to two triangles *ACB* and *ABD*, one considers the possibility of swapping *AB* by *CD*, thus replacing the triangles *ACB* and *ABD* by the triangles *ACD* and *BDC*, as shown in Figure 19.10(a). This operation is admissible only if the region bound by the rectangle *ACBD* is convex. If it is not, the swapping procedure will result in an incompatible grid connectivity as depicted in Figure 19.10(b).

When the alternative configuration is admissible, the swapping operation is performed if a user-defined quality criterion is better satisfied by the new configurations than by the existing one. In the present implementation, three grid quality criteria for swapping are used: optimal node connectivity, maximizing the minimum angle, and accurate representation of surface curvature.

The optimal node connectivity is represented by the ideal number of sides joining at an internal node. This number is taken to be six for an internal node, which is the number of sides at a node for a grid of equilateral triangles. For a boundary node, the ideal number of connectivities depends on the boundary geometry. The difference between the actual and the ideal number of connectivities, the defect value, is computed for each of the four nodes in the current configuration. The swapping is performed if the new configuration reduces the sum of nodal defect values.

The criteria of maximizing the minimum angle requires to perform an admissible swapping if the minimum of the angles between adjacent sides of the surface triangles in the new configuration is larger than that in the original configuration.'

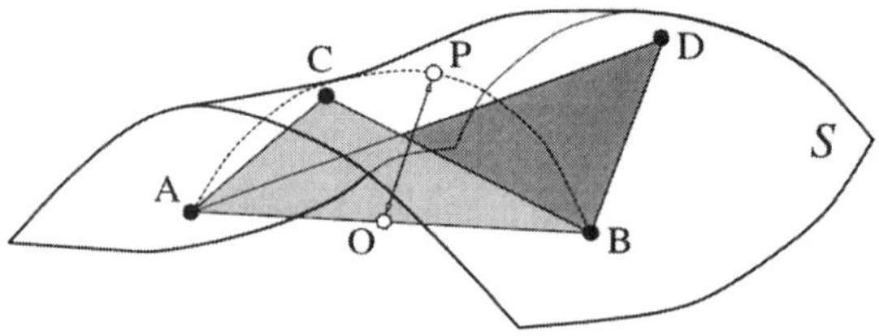

FIGURE 19.11 Accounting for surface curvature in the diagonal swapping procedure.

The final criterion is based on improving the representation of the curvature of the surface. Following the notation of Figure 19.11, *A*, *B*, *C*, and *D* are nodes of the triangular grid and are located on the surfaces *S*. *O* is the midpoint of side *AB* and *P* is the image of the midpoint of the side in the parameter plane. The length *OP* provides a measure of the accuracy of the approximation of the surface by triangles. In this case the surface will be better represented by using the triangles *ACD* and *BDC*. The swapping procedure aims at reducing the distance *OP;* here the swapping is performed only if the distance in the current configuration is three to four times larger than that of the new configuration.

In practice, the strategy employed consists of performing two of three loops of side swapping according to the first two criteria and then it concludes with an optimal loop over the internal sides to improve on the representation of the surface curvature.

19.4.8.2 Grid Smoothing

This method modifies the positions of the interior nodes without changing the connectivity of the grid. The element sides are considered as springs. The stiffness of a spring is assumed to be proportional to its length in 3D. The nodes are removed until the spring system is in equilibrium. The equilibrium positions are found by relaxation. Each step of this iterative procedure amounts to performing a loop over the interior nodes in which each node is move independently. In order to move a node *I*, only the sides that connect with the node are considered to be active springs, and the rest of the nodes $J = 1, \ldots, N_I$ connected with *I* by active sides are taken to be fixed. Denoting the coordinates in the parameter plane by the vector $\mathbf{u} = (u_1, u_2)$, the node *I* is then moved to an equilibrium position $\mathbf{u}_I$ which is the solution of

$$\mathbf{f}(\mathbf{u}_I) = \sum_{J=1}^{N_1} \omega_{IJ} \frac{\mathbf{u}_J - \mathbf{u}_I}{\|\mathbf{u}_J - \mathbf{u}_I\|} = 0 \tag{19.29}$$

ω_{IJ} represents the spring stiffness, which is taken to be proportional to the difference between the 3D length of the side and the length δ_{IJ} along the side *IJ* as specified by the 3D grid control function, i.e.,

$$\omega_{IJ}(\mathbf{u}_I) = \|\mathbf{r}_J - \mathbf{r}(\mathbf{u}_I)\| - \delta_{IJ} \tag{19.30}$$

The new position of the node *I* is approximately calculated by using one step of a Newton method for the solution of Eq. 19.29 starting from an initial guess $\mathbf{u}_0$. Here $\mathbf{u}_0$ is taken to be the centroid of the surrounding nodes

$$\mathbf{u}_0 = \frac{1}{N_I} \sum_{J=1}^{N_I} \mathbf{u}_J \tag{19.31}$$

and the new position $\mathbf{u}_J$ is given by

$$\mathbf{u}_J = \mathbf{u}_0 - \left[\frac{\partial \mathbf{f}}{\partial \mathbf{u}_I}(\mathbf{u}_0)\right]^{-1} \mathbf{f}(\mathbf{u}_0) \tag{19.32}$$

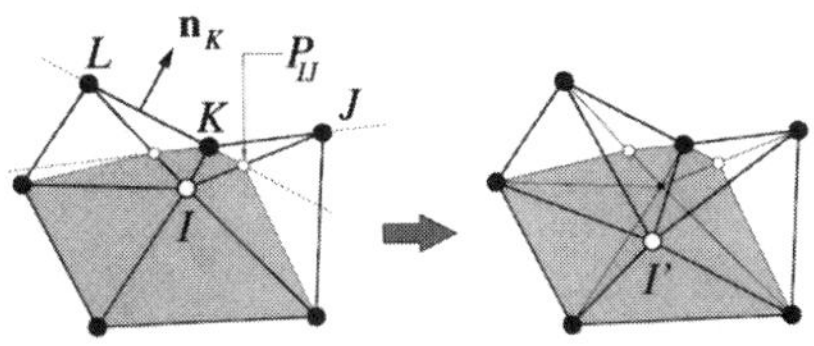

FIGURE 19.12 Mesh smoothing: The node *I* is moved to the equilibrium position *I′* within the shaded area.

The procedure is repeated for all the interior nodes. Usually two to four loops over the nodes are performed to enhance the grid.

This procedure works well if the region formed by the triangles surrounding the node is convex. If it is not, following the method suggested in [6], the motion of the point is restricted to the interior of a convex region, represented by the shaded area in Figure 19.12. This area is defined by a new set of vertices P_{IJ}, on the sides *IJ* surrounding point *I*, which are obtained as follows.

The coordinates of a point along the side *IJ* can be expressed as

$$\mathbf{u} = \mathbf{u}_I + \lambda(\mathbf{u}_J - \mathbf{u}_I) \qquad \text{with} \qquad 0 \le \lambda \le 1 \tag{19.33}$$

The intersection between the straight lines along the sides *IJ* and *KL* will correspond to a value $\lambda = \lambda_K$ in Eq. 19.33 with

$$\lambda_K = -\frac{\mathbf{u}_I \cdot \mathbf{n}_K}{(\mathbf{u}_J - \mathbf{u}_I) \cdot \mathbf{n}_K} \qquad \text{with} \qquad 0 \le \lambda_K \le 1 \tag{19.34}$$

where $\mathbf{n}_K$ denotes the normal to side *KL*. Finally, the position of the vertex P_{IJ} is represented by λ_P given by

$$\lambda_P = \min\left(\lambda_1, \ldots, \lambda_{N_I}\right) \tag{1935}$$

When the region defined by the triangular elements surrounding node *I* is nonconvex, the vertices P_{IJ} determined in this fashion are used instead of the original nodes $J = 1, \ldots, N_I$ in the smoothing procedure previously described.

The combined application of these two post-processing techniques is found to be very effective in improving the smoothness and regularity of the triangular grid generated on the surface.

19.5 Orientation of the Assembled Surface

Following arguments similar to those presented in Section 19.4.5, by assuming that the 3D computational domain is connected, the faces forming its boundary can be automatically oriented.

Here the discretized faces generated in the previous step of the generation procedure can be joined together to form closed surfaces by using the available information on the common edges. Since these surfaces are assumed to be connected, an edge is always common to two faces. This also permits to assign a consistent orientation to the set of faces forming the closed surface. This orientation is given by the direction of its normal. Using the Gauss' Theorem, the volume of the region interior to a closed surface *S* can be expressed as the absolute value of the surface integral

$$V = \frac{1}{3}\oint_S \mathbf{r} \cdot \mathbf{n}\, dS \tag{19.36}$$

where **r** is the position of a point on the surface and **n** denotes the unit normal to the surface. The sign of V characterizes the orientation of the surface, i.e., a positive value of V indicates that the adopted normal **n** is the outer normal. The integral 19.36 can be computed numerically given a triangulation of the face.

The closed surface that gives the maximum volume in absolute value is taken to be the outer boundary and is assigned an orientation compatible with a positive value of V according to Eq. 19.36. The other surfaces, if any, are assigned an orientation such that the value of V is negative.

The imposition of the (not very severe) restriction that the computational domain and the boundary faces forming its boundary should be connected provides a simple method for their automatic orientation. This greatly reduces the amount of information about the topology of the computational domain that the user has to provide to the grid generation code.

Further Information

A presentation of the discretization of surfaces using the advancing front directly on the surface and a discussion of the problems associated with verifying the validity of a new element directly in 3D space by means of an auxiliary projection for triangular and quadrilateral grids are given in [8] and [3], respectively.

An alternative method for the discretization of curves in which the grid control function defines a variable metric tensor **M** along the curve is presented in [7]. Using the notation employed in this chapter, the metric tensor can be written as $\mathbf{M} = \mathbf{T}^t\mathbf{T}$, where **T** is given by Eq.19.5.

A discussion of the generation of grids on surfaces that are piecewise continuous approximations of discrete data and hence are not defined via a single mapping from a parameter plane is given in [8].

Surface grid generation in the parameter plane using the Delaunay approach requires the introduction of a modified circumcircle criterion or the use of an auxiliary transformation to account for grid stretching. Examples of such approaches have been proposed in [9, 1].

References

1. Borouchaki, H. and George, P.L., Maillage de surfaces paramétriques. partie I: Aspects Théoriques, INRIA Research Report No. 2928, July 1996.
2. Casey, G.F. and Dinh, H.T. Grading functions and mesh redistribution, *SIAM J. Num. Anal.,* 1985, 22, No. 3, pp. 1028–1040.
3. Cass, R.J., Benzley, S.E., Meyers, R.J., and Blacker, T.D., Generalized 3-D paving: an automated quadrilateral surface mesh generation algorithm, *Int. J. Num. Meth. Eng.,* 1996, 39, pp 1475–1489.
4. Fletcher, R., *Practical Methods of Optimization,* John Wiley, New York, 1987.
5. Hoffmann, C.M., *Geometric and Solid Modeling,* Morgan Kaufmann, San Mateo, CA, 1989.
6. Formaggia, L. and Rapetti, F., MeSh2D (Unstructured mesh generator in 2D) Algorithm overview and description, CRS4 Technical Report COMPMECH-96/1, February, 1996.
7. Laug, P., Borouchaki, H., and George, P.L., Maillage de courbes gouverné par une carte de métriques, INRIA Research Report No. 2818, March, 1996.
8. Löhner, R., Regridding surface triangulations, *J. Comp. Phys.* 1996, 126, pp 1–10.
9. Mavriplis, D.J., Unstructured mesh generation and adaptivity, ICASE report No. 95-26, April, 1995.
10. Mortenson, M.E., *Geometric Modeling,* John Wiley, New York, 1985.
11. Peiró, J., Peraire, J., and Morgan, K., The generation of triangular meshes on surfaces, Creasy, C. and Craggs, C., (Eds.), *Applied Surface Modelling,* Ellis Horwood, 1989, Chapter 3, pp 25–33.
12. Peiró, J., Peraire, J., and Morgan, K., FELISA system reference manual, part I: basic theory, Civil Eng. Dept. report, CR/821/94, University of Wales, Swansea, U.K., 1994.
13. Peiró, J., Peraire, J., and Morgan, K., Adaptive remeshing for three-dimensional compressible flow computations, *J. Comp. Phys.* 1992, 103, pp 269–285.
14. Stoker, J.J., *Differential Geometry,* Wiley Interscience, New York, 1969.
15. Samareh-Abolhassani, J. and Stewart, J.E., Surface grid generation in parameter space, *J. Comp. Phys.,* 1994, 113, pp 112–121.

20

Nonisotropic Grids

Paul Louis George

Frédéric Hecht

20.1 Introduction

Nonisotropic or anisotropic grids or meshes have a wide range of applications in engineering. An important domain in which such grids can be beneficial is the numerical simulation of certain PDE systems by the finite element method.

Local mesh adaptation, and specifically anisotropic adaptation, is a useful technique to improve the accuracy of the numerical solution, see for example [Peraire et al., 1987], [Lohner, 1989], [Lo, 1991], [Mavriplis, 1994], and [Weatherill et al., 1994]. It is a way of capture rapid variations of the solution with a reasonable number of degrees of freedom. Isotropic adaptation allows a mesh to be obtained that has a variable density in some regions, while anisotropic adaptation leads to an ability to capture directional features requested by the physical problem.

0-8493-2687-7/99/$0.00+$.50

Coupling regularization methods and local mesh refinement together is a possible solution to create adapted meshes. At first, an initial mesh of the domain is constructed using any mesh generation method, then the solution is computed. Owing to a pertinent choice of a criterion (gradient, *a posteriori* error estimate, etc.), the regions of the domain requiring some level of adaptation are emphasized. Then, a new mesh is created, that is better suited, and the process is repeated until the convergence is achieved.

Irrespective of the space dimension (practically, in dimension three), the refinement procedures are well known (see for example, [Berger and Jameson, 1985], [Bristeau and Periaux, 1986] and [Lee and Lo, 1992]); however, the derefinement procedures are rather difficult to implement. Thus, a global method is proposed in place of a method based on local modifications. This global method relies on the adequate use of a fully automatic mesh generation algorithm governed by a criterion (or a set of criteria) in an iterative process. A mesh is reconstructed at each iteration step according to a function of the solution resulting of the previous iteration.

In general, the adaptation criterion indicates the element sizes that are required. It can also specify the desired sizes in a general metric rather than in the classical metric, see for instance [Peraire et al., 1987] or [Vallet, 1992], thus making the treatment of anisotropic cases possible.

This chapter aims at discussing such an approach. We are primarily interested in a Delaunay-type method; hence only unstructured meshes will be considered. This chapter will show how to extend this well-known method to the case where anisotropic meshes are expected. To clarify the discussion, the different steps involved in a classical Delaunay mesh generation algorithm are first recalled (Section 20.2). In Section 20.3, the classical scheme is extended to the adapted or anisotropic mesh generation context. The main features of such a scheme are given and further details are given in the following sections. The notions of a metric and length are both introduced in Section 20.4, in a Riemannian space. The Delaunay method is extended to this context (Section 20.5) revealing, in particular, that the proposed extension results from the flexibility of the classical method. Field point construction is discussed in Section 20.6, while optimization procedures are developed in Section 20.7. In Section 20.8, a solution is proposed for the construction of the metric
used to govern the mesh generation algorithm and in Section 20.9 the use of previous materials to define a loop of adaptation is discussed. Application examples are provided in Section 20.10, including computational fluid dynamics computations and an application of anisotropic mesh generation for parametric surface meshing. To conclude this chapter, several extensions to three dimensions will be briefly mentioned.

20.2 The Classical Delaunay Mesh Generation Method

This section recalls a series of well-known issues regarding the classical Delaunay mesh generation method (see also Chapter 16). For sake of simplicity, the bidimensional case is assumed. Usually, the Delaunay mesh generation procedure enables the mesh construction of a domain from the sole data of a boundary discretization of this domain. Thus, the mesh of a (closed) domain in R^2 is governed by geometrical considerations only (the boundary discretization), irrespective of the physics of the given problem. Resulting mesh sizes, as well as element densities or anisotropic features, are solely related to the given discretization, this data being the only available information. Our objective is to recall this classical scheme, as the method has already been described in numerous references, using the set of edges forming the boundary discretization as data.

20.2.1 Scheme of a Classical Mesh Generator

Let Ω be a domain and let $F(\Omega)$ be the set of edges discretizing its boundary. These edges, the so-called *constrained edges,* have a set of points, denoted as $S(\Omega)$ as endpoints.

In order to obtain a mesh of Ω, an *empty mesh* is first constructed, which is (according to [George, 1991]) an "empty" mesh in the sense that its vertices are (at least in two dimensions) the sole members of $S(\Omega)$. Then, a new mesh is constructed by adding the field points inside the empty mesh, this new mesh being then optimized so as to complete the final mesh of Ω. The field points are defined so that all internal mesh edges have an acceptable length. This *classical mesh generation algorithm* includes two main steps.

1. Creation of the empty mesh of Ω.
   - Creation of a rectangle enclosing Ω and construction of a mesh of this box,
   - Insertion of the points of $S(\Omega)$ into the previous mesh,
   - Regeneration of $F(\Omega)$, the constrained edges, in the current mesh in order to define a mesh of Ω
2. Creation of the mesh of Ω
   - Initialization by means of the previous mesh.
   - Field points creation (loop)
     - Generation of points along the internal edges of the current mesh using a length criterion,
     - Insertion of these points,
     - Iteration if the current mesh has been modified.
   - Optimization of the resulting mesh.

Some of these steps will be valid without significant changes in classical situations (say, isotropic case) as well as in anisotropic situations, thus they will be only briefly mentioned. However, a few steps involved in the mesh generation algorithm will be widely affected in the case of anisotropic requirements. These changes mainly concern the way the field points are computed and the way a mesh point is inserted (the so-called Watson algorithm). These two aspects will be detailed in the classical approach to clarify the proposed extension in the anisotropic context.

20.2.2 Boundary Mesh Creation

At first the construction of the "empty" mesh by means of the Watson algorithm is explained. To this end, this basic tool, later referred to as the *Delaunay kernel*, is recalled. This kernel is an incremental process allowing the insertion of a point in a given (Delaunay) triangulation. The generation of the constrained edges of $F(\Omega)$ in order to obtain the empty mesh is then recalled. Afterwards, the field points creation and insertion procedures are described.

20.2.2.1 The Delaunay Kernel

The Delaunay kernel is a procedure resulting in the insertion of an internal point in a (Delaunay) triangulation. This procedure is primarily based on a *proximity criterion*. The latter, based on length evaluation, appears to be well suited in respect of the envisaged extension.

The Delaunay kernel in any dimension, using the classical Euclidean metric, has been proposed by several authors, including Bowyer [1981], Watson [1981] and Hermeline [1982]. In two dimensions this algorithm leads, cf. Figure 20.1, to replace the set of triangles whose open circumdiscs include the point under consideration (i.e., the *cavity*) by the *ball* composed of the triangles formed by joining this point to the edges that constitute the boundary of the current cavity. The fundamental idea is, on the one hand, that the cavity is a *star-shaped* polygon with respect to the point considered, and on the other hand, that the mesh of the complement of this polygon is not affected. Formally speaking, the Delaunay kernel can be written as

$$T^{n+1} = T^n - C(P) + B(P) \tag{20.1}$$

where $C(P)$ is the cavity associated with point P, $B(P)$ is the corresponding ball and T^n denotes the mesh resulting of the insertion of the first n points. The cavity is constructed using the proximity criterion, which can be written as

$$\{K, K \in T, \quad \text{such that } P \in Disc(K)\} \tag{20.2}$$

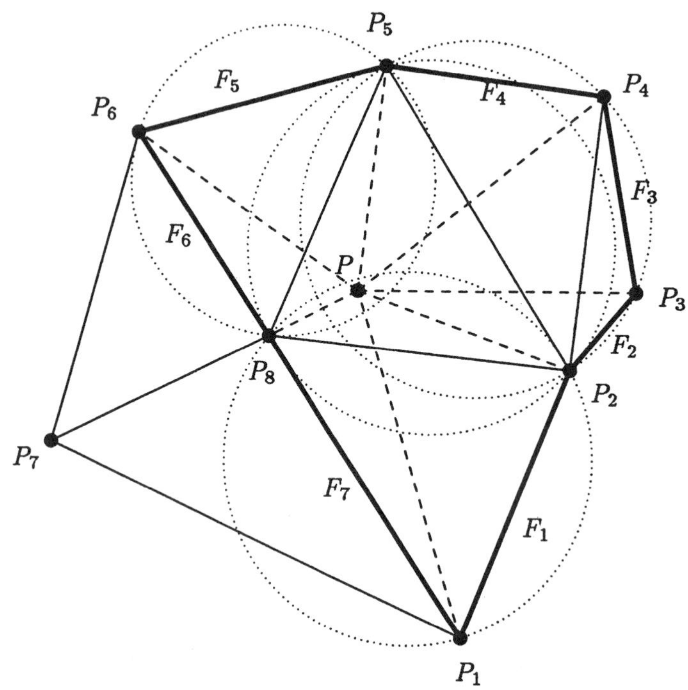

FIGURE 20.1 Insertion of point P, cavity and ball.

where $Disc(K)$ is the open circumdisc with respect to element K.

Numerous implementations of this algorithm have been developed by several authors including, for recent papers only [Borouchaki et al., 1995] and [Borouchaki, George and Lo, 1996].

20.2.2.2 Meshing the Enclosing Rectangle

The introduction of a rectangle enclosing the domain allows us to be in a convex situation and guarantees that all the boundary points are enclosed within this box. Thus, the above Delaunay kernel can be easily applied. The box, defined using four extra points, is covered with a two-triangle mesh. The points of $S(\Omega)$ are inserted in this mesh using the Delaunay kernel. Notice that the resulting mesh is a mesh of the box rather than a mesh of the domain Ω. To obtain a mesh of Ω, the edges of $F(\Omega)$ are recreated in the current mesh using local modifications (basically, diagonal swappings). At the time the edges of $F(\Omega)$ have been regenerated in the mesh, it is possible to classify the elements in this mesh with respect to Ω. The internal elements are specifically tagged, while the elements outside Ω are marked distinctively. Nevertheless, these exterior elements are not removed at this time, in order to preserve a convex environment and simplify the further procedures.

20.2.3 Creating the Mesh of a Domain Ω

At this point, a mesh of Ω is given. In principle, this mesh does not contain any point interior to Ω. Hence, to obtain the desired mesh, field points are created according to an iterative procedure. To start, the current mesh is initialized by the "empty" mesh. At each iteration, the current internal mesh edges are analyzed and internal points are

- Constructed along the edges so that, on the one hand, the so-created subdivisions are of *ideal length* and on the other hand, a point is not too close to an already existing point, and
- Inserted in the current mesh via the Delaunay kernel (specifically a *constrained variation* of it).

This process is repeated as long as the current mesh is modified.

To complete this algorithm we will have to define the concept of a constrained Delaunay kernel and we have to discuss the notion of an ideal distance between two points in case the desired element sizes are specified.

The *constrained Delaunay kernel* is a variation of the classical kernel that maintains the boundary integrity during the point insertion process.

Let (P, Q) be a pair of points, let h_P (resp. h_Q) be the desired size at point P (resp. Q) and let $h(t)$ be a monotonous continuous function that indicates the size variations along the segment $[P, Q]$, such that $h(0) = h_P$ and $h(1) = h_Q$. The length $l(P, Q)$ of the segment $[P, Q]$ is ideal with respect to $h(t)$ if and only if (cf. [Laug et al., 1996]).

$$l(P,Q) = \left(\int_0^1 \frac{1}{h(t)} dt \right)^{-1}. \tag{20.3}$$

The function $h(t)$ is a size interpolant inside the domain. The desired size at a boundary point is considered as the average of the lengths of the edges sharing this point. The internal point size is defined via the function $h(t)$ associated with the supporting edge of this point.

A variation of this procedure regarding the creation of the internal points consists in processing the edges according to their lengths. In this way, the most significant edges are first processed.

To close the description of the classical mesh generation process, we still have to mention that the mesh resulting from the internal points insertion is optimized. The optimization is based on diagonal swapping and point relocation procedures. These tools are driven by the element qualities and will be described in a general context, in Section 20.7.

20.3 Scheme of an Anisotropic Mesh Generator

We assume now that the desired mesh must have anisotropic features and we consider that the anisotropy is defined by means of a specified metric. More precisely, a *metric* is specified for which the desired size is 1 (cf. [Vallet, 1992]). In this context, the mesh generator shall provide an acceptable mesh with respect to this metric. To summarize, this leads to

- Generalizing the notion of a desired size, which can vary in two different directions, and
- Normalizing the classical ideal length to unity, with respect to the considered metric.

From a practical point of view, the metric is known as a discrete function and by interpolating the metric everywhere it is not specified, a *Riemannian structure* is obtained on the domain. Associated with the so-defined metric, the domain is called a *control space.* A mesh is *satisfactory* if all its elements are equilateral with respect to this control space. Therefore, the problem is to extend the classical method (as introduced in the previous section) to permit the construction of an (almost) satisfactory mesh, with respect to the control space. Consequently, meshing the domain includes two main stages:

1. The mesh of the boundary of Ω and
2. The mesh of Ω using the boundary mesh as a support.

These two steps are governed by the control space.

Before discussing the extension of the previously described classical tools to the anisotropic case, it may be noticed that, at this time, the mesh of the boundary of the domain is supposed to conform to the control space. Thus, the anisotropic mesh of the domain can be obtained using an extension of the classical Delaunay kernel and a generalization of the previous internal point creation procedure. To summarize, both aspects lead us to define properly the lengths with respect to the control space.

20.3.1 The Mesh of the Domain

As mentioned before, the mesh of the domain according to the control space, can be obtained by

- Generalizing the Delaunay kernel to a Riemannian space (Section 20.5),
- Replacing the ideal length by the *unity measure* in the control space during the field point creation (Section 20.6),
- Extending the triangle quality notion to a Riemannian space (Section 20.7).

Consequently, the scheme of an *anisotropic mesh generation method* governed by a control space, the mesh of the boundary being supplied, can be summarized as

- Creation of the empty mesh resulting from the insertion of the boundary points and then regeneration of the boundary edges.
- Generation and insertion (loop) of the field points.
  - Computation of the edge lengths of the current mesh,
  - Subdivision of the edges whose lengths exceed the "unity" in the control space,
  - Insertion of these points in the current mesh,
  - Iteration if any modification arises.
- Optimization.

20.4 Fundamental Definitions

Several fundamental definitions are provided in this section, before returning to our purpose.

20.4.1 Metric at a Point

The metric or metric tensor at a point X of the domain Ω is the specification at X of a definite positive matrix

$$M(X) = \begin{pmatrix} a_X & b_X \\ b_X & c_X \end{pmatrix} \tag{20.4}$$

such that $a_X > 0$, $c_X > 0$ and $a_X c_X - b^2_X > 0$.

The metric field $(M(X))_{X\in\Omega}$ induces a Riemannian structure on Ω. The latter, along with this structure, is denoted by $(\Omega, (M(X))_{X\in\Omega})$. If, for all points X in the domain, the metrics are identical, then the Riemannian structure is nothing other than a Euclidean structure; Ω applied with this structure is then denoted by $(\Omega, M(X))$ or simply by (Ω, M), X being an arbitrary point of Ω.

20.4.2 Length of a Segment

In the Riemannian space defined by $(\Omega, (M(X))_{X\in\Omega})$, the length L of a curve Γ of IR^n, parametrized by $\gamma(t)_{t=0..1}$, is

$$L = \int_0^1 \sqrt{{}^t\nabla_\gamma(t) M(\gamma(t)) \nabla_\gamma(t)}\, dt \tag{20.5}$$

consequently, the length of a segment $[P, Q] = (P + t\overrightarrow{PQ})_{0\le t\le 1}$ in Ω is given by

$$l(P,Q) = \int_0^1 \sqrt{{}^t\overrightarrow{PQ} M\left(P + t\overrightarrow{PQ}\right) \overrightarrow{PQ}}\, dt \tag{20.6}$$

where $\overrightarrow{PQ}$ is a vector of origin P and extremity Q. With $\overrightarrow{PQ} = \binom{u_1}{u_2}$ and $M(P + t\overrightarrow{PQ}) = \begin{pmatrix} a(t) b(t) \\ b(t) c(t) \end{pmatrix}$, then

$$l(P,Q) = \int_0^1 \sqrt{a(t)u_1^2 + 2b(t)u_1u_2 + c(t)u_2^2}\, dt \tag{20.7}$$

Notice that in the case of an Euclidean space defined by (Ω, M), with $M(X) = \begin{pmatrix} a & b \\ b & c \end{pmatrix}$, one has

$$l(P,Q) = \sqrt{au_1^2 + 2bu_1u_2 + cu_2^2}. \tag{20.8}$$

20.5 The Anisotropic Delaunay Kernel

The key idea of the Delaunay kernel is the adequate definition of the cavity associated with the point considered (cf. relations 20.1 and 20.2). The sole component requisite, the evaluation of the cavity, is based on length computations, thus, this sole operation shall be extended to the Riemannian space context.

In the classical situation, the cavity can be evaluated from a *base*, i.e., the set of triangles enclosing the point to be inserted, enriched by adjacency with the triangles whose open circumdiscs enclose this point. According to this algorithm, the cavity is necessarily connected. The circumdiscs are evaluated in the usual Euclidean metric. This proximity criterion shall now be extended to the Riemannian context.

Let $K = [P_1, P_2, P_3]$ be a triangle in the current mesh and P be the point to be inserted. The problem we face is to decide if a triangle K belongs to the cavity associated with P. It could be observed that in this case, the edges $([P, P_i])_{1 \le i \le 3}$ will be part of the ball associated with P, and therefore will be automatically formed.

The problem is then to find a suitable proximity criterion that enables us to construct the cavity. Finding a general solution of this problem is very difficult, as will be seen, because the metric can vary widely from one point to another. The easiest solution consists in replacing the Riemannian space by an Euclidean space whose metric is that at point P. This is a quite natural choice (cf. Section 20.5.2). A second solution results from the above remarks and leads to take into account two Euclidean spaces, one of them associated with the point P and the other one associated with the vertex of element K not yet in the cavity (cf. Section 20.5.3). A third solution consists in using the four points of the configuration, the points P, P_1, P_2, and P_3, (cf. Section 20.5.4). Before debating the different options, we define locally, Section 20.5.1, the proximity criterion by means of a measure related to a metric, the so-called *Delaunay measure*.

20.5.1 The Delaunay Measure

Let Z be a point of Ω. Considering the Euclidean space $(\Omega, M(Z))$, we denote by l^Z the distance between two points of Ω in this space. The circumdisc associated with a triangle K, whose center is denoted O^Z, is defined in this space by

$$\left(l^Z(O^Z, X)\right)^2 = {}^t\overrightarrow{OX}M(Z)\overrightarrow{OX} = k^2 \tag{20.9}$$

where $X \in R^2$ and k is a real value such that the disc is circumscribed to triangle K. Hence, the center O^Z is the solution of the linear system

$$\begin{cases} l^Z(O^Z, P_1) = l^Z(O^Z, P_2) \\ l^Z(O^Z, P_1) = l^Z(O^Z, P_3) \end{cases} \tag{20.10}$$

and k is precisely $l^Z(O^Z, P_1)$. The circumdisc of triangle K encloses the point P, if and only if

$$l^Z(O^Z,P) < l^Z(O^Z,P_1), \tag{20.11}$$

and, in this case, the Delaunay criterion associated with the pair (P, K) is said to be violated according to the metric at point Z. By normalizing to one the above inequality, a dimensionless measure is obtained, defined by

$$\alpha_Z(P,K) = \frac{l^Z(O^Z,P)}{l^Z(O^Z,P_1)}. \tag{20.12}$$

The violation of the Delaunay criterion associated with the pair (P, K) in the metric at Z means that $\alpha_Z(P, K) < 1$. The coefficient $\alpha_Z(P, K)$ is named the Delaunay measure of the triple (P, K, Z) [George and Borouchaki, 1997].

20.5.2 Approach Using Only One Metric

According to this notion of measure, we consider the case where the Delaunay criterion depends only upon the metric at one point P, the point to be inserted. This approach is the easiest one. The triangle K belongs to the cavity if

$$\alpha_P(P,K) < 1. \tag{20.13}$$

It is obvious to check that the so-defined cavity remains star-shaped with respect to P. This is a consequence of the fact that the given circumdiscs are convex and that the cavity is constructed by adjacency using an edge that separates the discs into two disconnected parts. Consequently, a valid solution results from this choice, although this solution is a coarse approximation, as pointed out by numerical experiences. Actually, a Riemannian space is locally approached by only one Euclidean space.

20.5.3 Approach Using Two Metrics

A more precise analysis of the process used to construct the cavity shows that at least one triangle exists in the cavity, that is adjacent to K. Let f be the common edge and let P_j be the vertex of K such that $K = [f, P_j]$. Then, following the way the edges of the ball associated with P are constructed, it could be seen that, if K belongs to the cavity, then the edge $[P, P_j]$ will be formed. So, it is quite natural to consider the metric at the point P along with that at the point P_j, during the evaluations of the proximity criterion. Hence, the triangle K belongs to the cavity if

$$\alpha_P(P,K) + \alpha_{P_j}(P,K) < 2. \tag{20.14}$$

Similarly, it is easy to check that the cavity constructed in this way is star-shaped with respect to P. Thus, a valid solution is obtained, which is a better approximation, as the violation of the Delaunay criterion is evaluated using two metrics, that at the point P and that at the vertex of K previously defined.

20.5.4 Approach Using Four Metrics

This approach leads to the best approximation. Four metrics are considered, that at the point P and those at the three vertices of the triangle K considered. In this case, the triangle K is valid for the Delaunay criterion if

$$\alpha_P(P,K) + \alpha_{P_1}(P,K) + \alpha_{P_2}(P,K) + \alpha_{P_3}(P,K) < 4. \tag{20.15}$$

Similarly to the case of two metrics, this solution is valid. Nevertheless, numerical experiments indicate that there is no significant difference with the previous approximation, except a bigger cost in terms of CPU.

Remark: From the pure mathematical point of view, this problem is not well posed, as in Riemannian geometry the side of a triangle is a geodesic curve and the triangles are generally not straight-sided.

20.6 The Field Points Definition

The field point creation is a crucial step in the mesh generation process. The aim is to create the appropriate number of points and to locate them properly. The field points must be defined so that the resulting mesh edges are of a unit length in the control space.

As for the classical method, an iterative procedure is proposed. At each iteration, field points are constructed along the current mesh edges, the current mesh being initialized by the empty mesh. Hence, at the iteration i, the mesh edges of the iteration $i-1$ are considered as supporting edges along which the internal points will be created. The points are created so as to divide the edges into *unit segments*. Each point constructed is stored and will be inserted if it is not too close to an already existing point. The generation, *filtration* as well as the insertion of the internal points are governed by the control space. The latter is a Riemannian space allowing the computation of edge lengths and distances between two points. The control space is discussed in Section 20.6.1 and the different procedures involved in the field points definition are described in Sections 20.6.2 and 20.6.3.

20.6.1 The Control Space

The control space is constructed from a *background mesh* and a *metric map* defined at each mesh vertex. To obtain the first governed mesh, the background mesh is set to the classical mesh. Within an adaptation loop (Section 20.9), the background mesh at stage j is the mesh at stage $j-1$. The Riemannian structure of the control space is explicitly defined at each mesh vertex and implicitly known at any other location in the domain. Indeed, if X is an interior point

- Either X is a vertex of the background mesh and the matrix at X exists in the metric map,
- Or X is enclosed in a triangle and the metric at X is defined using an interpolation based on the triangle vertices.

The metric map is a finite set of positive definite 2×2 matrices. These matrices define locally the size as well as the desired element shapes. To define the above interpolation, let us recall first the geometrical meaning of the metrics.

20.6.1.1 Geometrical Interpretation of the Metrics

In the *isotropic* case, the metric is defined by λI_2, where I_2 is the 2×2 identity matrix and λ is a strictly positive number. Let h be the desired element size in any direction, then the metric can be interpreted in terms of h. In the Euclidean space supplied with this metric, the unit circle centered at the origin is a circle of radius h in the space supplied with the usual metric; this circle is defined by

$$\phi(X) = x_1^2/h^2 + x_2^2/h^2 = 1 \tag{20.16}$$

where X corresponds to the point (x_1, x_2). Consequently, h and λ are such that

$$\lambda = \frac{1}{h^2}. \tag{20.17}$$

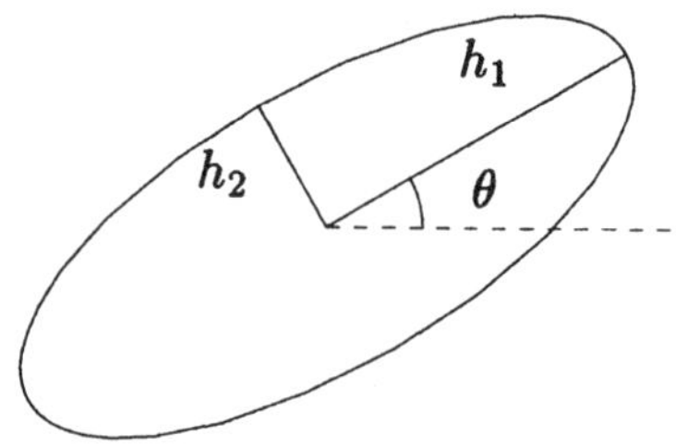

FIGURE 20.2 Unit circle, anisotropic case.

In the *anisotropic* case, the metric is defined by a symmetric positive definite 2×2 matrix

$$M = \begin{pmatrix} a & b \\ b & c \end{pmatrix}, \tag{20.18}$$

and the unit circle, centered at the origin, is the ellipse of equation

$$\phi(X) = ax_1^2 + 2bx_1x_2 + cx_2^2, \tag{20.19}$$

in the usual Euclidean space. Obviously, in the base associated with the principal axis of this ellipse, ϕ can be replaced by

$$\phi'(Y) = y_1^2/h_1^2 + y_2^2/h_2^2 = 1; \tag{20.20}$$

where h_1 and h_2 are the desired sizes along the principal axis of the ellipse. Figure 20.2 illustrates the unit circle associated with a metric where $h_1 = 2.5$, $h_2 = 1$, and $\theta = \pi/6$, θ being the angle of anisotropy.

20.6.1.2 Interpolation on a Segment

The question that arises is how to interpolate a metric on a segment from the metrics of its endpoints or how to perform the interpolation (in terms of size), by means of a *monotonous* and *continuous* function, from one ellipse, say M_1, to another ellipse, say M_2.

In the isotropic case, the solution is obvious. Actually, if the first metric is defined by λI_2 and the second by μI_2, the desired size specification with respect to the first metric is $h_1 = 1/\sqrt{\lambda}$ and $h_2 = 1/\sqrt{\mu}$ for the second. Hence, the interpolation function for an *arithmetic progression* (in terms of size), is defined by

$$M(t) = \frac{1}{\left(h_1 + t(h_2 - h_1)\right)^2} I_2 \quad 0 \le t \le 1, \tag{20.21}$$

with $M(0) = M_1$ and $M(1) = M_2$. In the anisotropic situation, several solutions are possible. They will be discussed hereafter.

20.6.1.2.1 Interpolation According to a Matrix Exponentiation

According to the isotropic case, where a metric is written as $M = h^{-2} I_2$, one can observe that the variations of h are "equivalent" to the variations of $M^{-1/2}$. Thus,

$$M(t) = \left((1-t)M_1^{-1/2} + tM_2^{-1/2}\right)^{-2} \quad 0 \le t \le 1. \tag{20.22}$$

The computation of $M^{-1/2}$ requires the evaluation of the eigenvalues of M. To avoid this evaluation, it is possible to consider an interpolation such that

$$M(t) = \left((1-t)M_1^{-1} + tM_2^{-1}\right)-1 \quad 0 \le t \le 1, \tag{20.23}$$

and observe that this formulation promotes the smallest sizes.

Both interpolation schemes are well defined because, if M is a metric then tM^α is a metric, where $t > 0$ and α are two reals, and if M_1 and M_2 are two metrics then $M_1 + M_2$ is a metric.

These schemes are not fully satisfactory as variations in terms of hs are not explicitly controlled. The following solution allows us to control h according to two directions.

20.6.1.2.2 Simultaneous Matrix Reduction

We now consider a "better" interpolation scheme, the simultaneous reduction of two metrics, the latter being two quadratic forms. The simultaneous reduction of two forms results in a base where the two forms are defined by two diagonal matrices. Let M_1 and M_2 by the two metrics. Let us introduce the matrix $N = M_1^{-1} M_2$. This matrix is M_1-symmetric, so it can be diagonalized in R^2. Let (v_1, v_2) be the eigenvectors of N, they define a base in R^2, and

$${}^t v_1 M_1 v_2 = {}^t v_1 M_2 v_2 = 0. \tag{20.24}$$

Let $X = x_1 v_1 + x_2 v_2$ be a real vector in the base $(v_1 v_2)$; if $(\lambda_i = {}^t v_i M_1 v_i)_{1 \le i \le 2}$ and $(\mu_i = {}^t v_i M_2 v_i)_{1 \le i \le 2}$ then, by definition, for all i, $1 \le i \le 2$, $\lambda_i > 0$, $\mu_i > 0$, and

$${}^t X M_1 X = \lambda_1 x_1^2 + \lambda_2 x_2^2 \quad \text{and} \quad {}^t X M_2 X = \mu_1 x_1^2 + \mu_2 x_2^2. \tag{20.25}$$

Let us define $(h_{1,i} = 1/\sqrt{\lambda_i})_{1 \le i \le 2}$ and $(h_{2,i} = 1/\sqrt{\mu_i})_{1 \le i \le 2}$. The value $h_{1,i}$ (resp. $h_{2,i}$) is precisely the unit length in the metric M_1 (resp. M_2), with respect to the axis v_i. The metric interpolation between M_1 and M_2 is defined by

$$M(t) = {}^t P^{-1} \begin{pmatrix} 1/h_1^2(t) & 0 \\ 0 & 1/h_2^2(t) \end{pmatrix} P^{-1} \quad 0 \le t \le 1, \tag{20.26}$$

where P is the matrix whose columns are (v_1, v_2) and $(h_1(t), h_2(t))$ are monotonous continuous functions such that $h_i(0) = h_{1,i}$ and $h_i(1) = h_{2,i}$ for $1 \le i \le 2$. In practice, one can consider two kinds of interpolation functions:

- $h_i(t) = h_{1,i} + t(h_{2,i} - h_{1,i})$ (arithmetic progression),
- $h_i(t) = h_{1,i}(h_{2,i}/h_{1,i})^t$ (geometric progression).

It could be observed that this interpolation is controlled for the axes (v_1, v_2) solely. To illustrate this process, Figure 20.3 depicts the examples of two initial metrics and the related interpolated metrics in the case of an arithmetic progression.

20.6.1.3 Interpolation over a Triangle

To interpolate over a triangle, we simply have to extend the interpolation scheme suitable for a segment. Let X be a point in the triangle $K = [P_1, P_2, P_3]$ and $(\alpha_i)_{1 \le i \le 3}$ be the barycentric coordinates of X in K. Then, for the $M^{-1/2}$ interpolation scheme, we have

$$M(X) = \left(\sum_{i=1}^{3} \alpha_i M(P_i)^{-\frac{1}{2}} \right)^{-2} \tag{20.27}$$

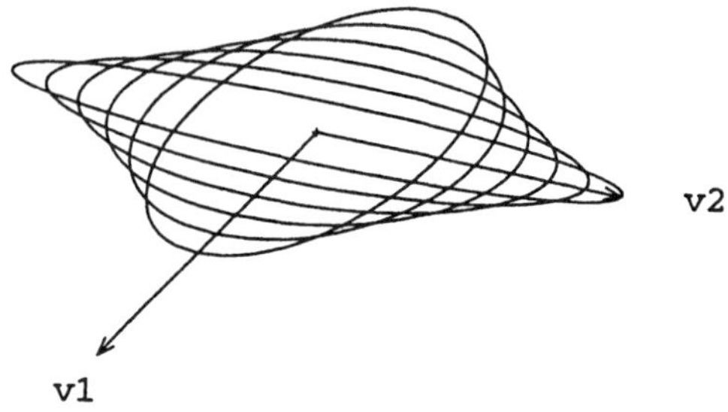

FIGURE 20.3 Arithmetic interpolation.

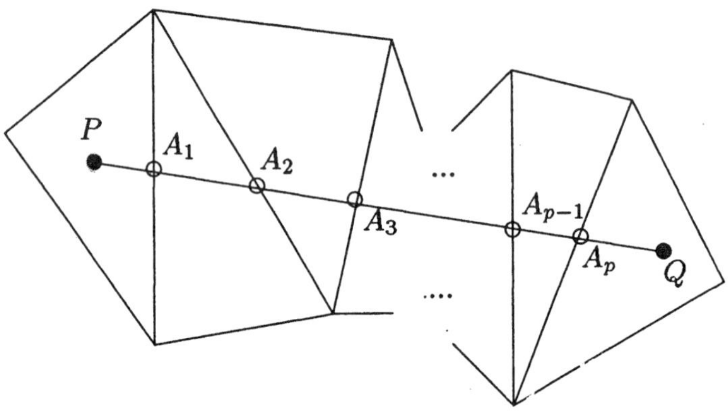

FIGURE 20.4 Edge length.

The interpolation scheme using the simultaneous reduction of matrices is not associative. To overcome this drawback, we consider a global ordering of the point numbers. The vertices of K are ordered and the scheme is applied accordingly. Assuming that the vertices of K are such that $P_1 < P_2 < P_3$, where < stands for the above ordering, then two reals α and β and a point P_3^* exist, such that

$$P_3^* = (1-\alpha)P_1 + \alpha P_2 \quad and \quad X = (1-\beta)P_3^* + \beta P_3; \tag{20.28}$$

the scheme is applied at first on the segment $[P_1, P_2]$ to interpolate the metric at P_3^* and afterwards on the segment $[P_3^*, P_3]$ to interpolate the metric at X.

20.6.2 Computation of the Edge Length

Each edge of the current mesh is embedded in the background mesh (in practice, this mesh is the current mesh and its complement in the given bounding box). The visited edge is then subdivided into several segments defined by the intersections of this edge with the edges of the background mesh elements (Figure 20.4). This process is valid, as the rectangle enclosing the background mesh is also a bounding box of the current mesh, such that every segment is included in a triangle of this background mesh.

The set of intersection points, the A_i's on the figure, forms the *discrete specification* of the metrics needed to analyze the current edge. Using this specification, the edge length can be evaluated. Let $[P, Q]$ be an edge of the current mesh, let $(A_j)_{1 \le j \le p}$ be the intersections of this edge with the background triangles and let $(t_j)_{1 \le j \le p}$ be such that $A_j = P + t_j \overrightarrow{PQ}$, with $A_0 = P$ and $A_{p+1} = Q$. Then the edge length of $[P, Q]$ is

$$l(P,Q) = \sum_{j=0}^{p} l(A_j, A_{j+1}), \tag{20.29}$$

and the length of each segment $[A_j, A_{j+1}]$ is evaluated by considering a metric interpolation on $[A_j, A_{j+1}]$ (cf. Section 20.6.1.2).

The metrics at the points P and Q are known. Actually, the set points includes the boundary points (for which the metric is well defined) and some previously created points whose metric was fixed at the time they were created.

It is now possible to propose a numerical method for computing the length of each segment in the above subdivision. In the isotropic case, the length of a segment can be computed exactly from the metrics at its endpoints, using the interpolation scheme on the segment [Laug et al., 1996]. In the anisotropic case, the length of the segment $[A_j, A_{j+1}]$ is given by Eq. 20.6. To compute this integral form, an approximate scheme is used. Let l_a be the approximation solution. Then

- Let $L = \dfrac{\sqrt{{}^t\overrightarrow{A_jA_{j+1}}M(A_j)\overrightarrow{A_jA_{j+1}}} + \sqrt{{}^t\overrightarrow{A_jA_{j+1}}M(A_{j+1})\overrightarrow{A_jA_{j+1}}}}{2}$;
- If $L < L_0$ ($L_0 < 1$) then $l_a(A_j, A_{j+1}) = L$; or, if M is the midpoint of segment $[A_j, A_{j+1}]$, then $l_a(A_j, A_{j+1}) = l_a(A_j, M) + l_a(M, A_{j+1})$.

This process is recursive. The resulting value is satisfactory if the approximate values, i.e., the l_a's, are smaller than a given value L_0 (in practice, $L_0 = 0.5$ seems adequate). This process subdivides the segment into subsegments whose length is smaller than L_0. As a consequence, the proposed method provides a series of points $(S_i^j)_{1\le i\le r_j}$ on the segment $[A_j, A_{j+1}]$, such that

$$\begin{cases} l\left(A_j, S_1^j\right) < L_0 \\ l\left(S_i^j, S_{i+1}^j\right) < L_0 \quad 1 < i < r_j \,. \\ l\left(S_{r_j}, A_{j+1}\right) < L_0 \end{cases} \tag{20.30}$$

20.6.3 Field Point Creation

The edge lengths are computed and points are created along these edges so as to subdivide them into subsegments of unit length. The latter represents the goal to achieve in order to create a (almost) satisfactory mesh in the control space.

The subsegments whose length is smaller than L_0 resulting from the edge length analysis are now used to define a subdivision into unit length segments. According to Figure 20.4, we have

$$[P,Q] = \bigcup_{j=0}^{p} \left[A_j, A_{j+1}\right]; \tag{20.31}$$

and for each segment $[A_j, A_{j+1}]$, the subdivision $(S_i^j)_{0\le i\le r_j+1}$ is known such that $S_0^j = A_j$, $S_{r_j+1}^j = A_{j+1}$ and $l(S_i^j, S_{i+1}^j) < L_0$ for $0 \le i \le r_j$. Then

$$l(P,Q) = \sum_{j=0}^{p} \sum_{i=0}^{r_j} l\left(S_i^j, S_{i+1}^j\right). \tag{20.32}$$

The method relies upon the definition of m such that $m \le l(P, Q) < m + 1$. The edge $[P, Q]$ will be splitted in m or $m + 1$ segments if

$$\frac{m}{l(P,Q)} > \frac{l(P,Q)}{m+1} \quad \text{or} \quad \frac{m}{l(P,Q)} < \frac{l(P,Q)}{m+1} \tag{20.33}$$

holds. To clarify this choice, let us assume that m is selected. The edge must be divided into m segments whose length is $\delta = l(P, Q)/m$. Let $(C_k = P + k\delta)_{1 \le k < m}$ be the subdivision points. For a given k, j_α and i_β exists, such that

$$l\left(P, S_{i_\beta}^{j_\alpha}\right) \le k\delta < l\left(P, S_{i_\beta+1}^{j_\alpha}\right); \tag{20.34}$$

thus, on $C_k \in [S_{i\beta}^{j\alpha}, S_{i\beta+1}^{j\alpha}]$, and

$$C_k = S_{i_\beta}^{j_\alpha} + \frac{k\delta - l\left(P, S_{i_\beta}^{j_\alpha}\right)}{l\left(S_{i_\beta}^{j_\alpha}, S_{i_\beta+1}^{j_\alpha}\right)}\left(S_{i_\beta+1}^{j_\alpha} - S_{i_\beta}^{j_\alpha}\right). \tag{20.35}$$

As the point C_k belongs to the segment $[A_{j_\alpha}, A_{j_\alpha+1}]$, the metric at C_k is well-defined using an interpolation on this segment. It can be observed that the value δ is bounded by the values $\delta_{min} = 1/\sqrt{2}$ and $\delta_{max} = \sqrt{2}$, which are two tolerance thresholds relative to the desired unit value.

20.6.4 Filtration of the Field Points

At the times the field points have been created along all edges, a filtration process is employed to discard those points that are too close to the others. The threshold value used is the above value δ_{min}. This step is strictly required because the point generation process is local to every edge. To this end, a control grid is introduced consisting of regular cells. The points already retained are stored within a cell, and a point P will be retained or discarded if the enclosing cell (or the neighboring cells, at an appropriate distance) already contains a point Q, such that $l^P(P, Q) < \delta_{min}$ and $l^Q(P, Q) < \delta_{min}$ is satisfied or not.

20.6.5 Insertion of the Field Points

The set of points retained after filtration is inserted in the current mesh using the extended Delaunay kernel in its constrained version.

20.7 Optimization

To improve the resulting mesh, two procedures can be used, the diagonal swapping and the internal points relocation operators. The target is to achieve equilateral (or close to equilateral) triangles with respect to the control space. The optimization procedure consists in successively applying the diagonal swapping operator, then moving the points, these two steps being then repeated.

20.7.1 Element Quality

Let $K = [P_1, P_2, P_3]$ be a triangle. In the usual Euclidean space, a possible definition of its quality is, according to [Lo, 1991],

$$Q(K) = \alpha \frac{\left|Det\left(\overrightarrow{P_1P_2}, \overrightarrow{P_1P_3}\right)\right|}{\sum_{1 \le j < k \le 3} \left\|\overrightarrow{P_jP_k}\right\|^2}, \tag{20.36}$$

where $Det(\overrightarrow{P_1P_2}, \overrightarrow{P_1P_3})$ is the determinant of the matrix whose columns are $\overrightarrow{P_1P_2}$ and $\overrightarrow{P_1P_3}$. Det represents twice the surface of the triangle K, while $\left\|\overrightarrow{P_jP_k}\right\|$ is the length of edge $[P_j, P_k]$ of K and $\alpha = 2\sqrt{3}$ is a normalization factor such that the quality of an equilateral triangle is 1. Accordingly, $0 \le Q(K)$

≤ 1 and a nice-shaped triangle quality is close to 1, while an ill-shaped triangle quality is close to 0. In a Riemannian space, the quality of a triangle K can be defined as

$$\Gamma(K) = \min_{1 \leq i < k \leq 3} Q^i(K), \tag{20.37}$$

where $Q^i(K)$ is the triangle quality in the Euclidean space associated with vertex P_i of K, and a simple calculus gives

$$Q^i(K) = \alpha \frac{\left|\sqrt{Det(M_i)}Det\left(\overrightarrow{P_1P_2}, \overrightarrow{P_1P_3}\right)\right|}{\sum_{1 \leq j < k \leq 3} {}^t\overrightarrow{P_jP_k} M_i \overrightarrow{P_jP_k}}. \tag{20.38}$$

with $M^i = M(P_i)$.

20.7.2 Diagonal Swapping

Diagonal swapping is a way to improve the mesh quality using a topological modification. This tool allows edges to be removed, if possible. Let f be a mesh edge. We term the *shell* of f, the set of triangles sharing f. The quality of a shell is that of its worst element. The diagonal swapping operator is then applied if the resulting mesh quality improves, as compared to that of the initial shell. Each edge f is associated with a ratio g_f representing the quality improvement factor after the diagonal swapping is applied to f.

In view of optimizing the mesh quality, diagonal swapping is applied iteratively depending on the variation of g_f. Initially, the ratio of improvement is set to a value $\omega > 1$ (in practice, a value $\omega = 2$ is advised), then the coefficient ω is decreased to 1. According to this procedure, the most significant diagonal swapping are done first.

20.7.3 Point Relocation

Let P be an internal mesh point and (K_i) be the ball of P (the set of elements having P as vertex). The point relocation process consists in moving P to improve the quality of the ball (i.e., that of its worst element). Two procedures have been developed, the first one leading to unit edge lengths, the other one leading to optimal elements (in terms of shape).

20.7.3.1 Relocation with Unit Length

Let (P_i) be the vertices of (K_i) other than P. Each point P_i is associated with an optimal point P_i^* such that

$$\overrightarrow{P_iP_i^*} = \overrightarrow{P_iP} / l(P_i, P), \tag{20.39}$$

for which $l(P_i, P_i^*) = 1$ holds. The process consists in moving the point P step by step toward the centroid Q of the points P_i^*, if the quality of the set (K_i) is improved. This process [Briere de l'Isle and George, 1995] leads to establishing unit length for the edges sharing P.

20.7.3.2 Relocation with Optimal Shape

Let (f_i) be the edges opposite to vertex P in the triangles (K_i)'s ($K_i = [P, f_i]$). The optimal point P_i^* is associated with each edge f_i, such that the triangle $K_i^* = [P_i^*, f_i]$ enjoys the best possible quality $\Gamma(K_i^*)$.

Let Q be the centroid of the points P_i^*, then the point P is moved step by step towards the point Q, as the quality variation is controlled.

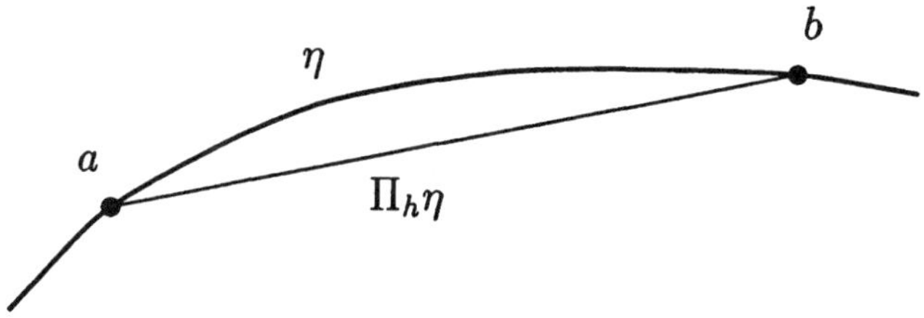

FIGURE 20.5 Piecewise linear interpolation in 1D.

This process leads to optimally shaped triangles. To obtain the point P_i^*, one can possibly consider the centroid of the optimal points associated with f_i, each of them being evaluated in the metric specified at the vertices of the triangle K_i. To clarify this approach, let us consider the edge $f_i = [P_i, P_{i+1}]$, $(K_i = [P, f_i])$ and let us compute the optimal point P_i^* related to f_i, with respect to an Euclidean structure associated with a given metric $M = \left(\begin{smallmatrix} a & b \\ b & c \end{smallmatrix}\right)$. The point P_i^* lies in the same half-plane as P, with respect to f_i and is defined so that $K_i^* = [P_i^*, f_i]$ is an equilateral triangle in the Euclidean structure related to M. If P is the matrix mapping the canonical base into the base of the eigenvectors of M, and Λ is the diagonal matrix formed by the eigenvectors of M, the optimal point P_i^* is defined by

$$P_i^* = P_i + P\Lambda^{-\frac{1}{2}} R(\pi/3) \Lambda^{\frac{1}{2}} P^{-1} \overrightarrow{P_i P_{i+1}}; \tag{20.40}$$

as $M = P\Lambda P^{-1}$, and as $R(\theta)$ is a rotation matrix of angle θ, one has

$$P_i^* = P_i + M^{-\frac{1}{2}} R(\pi/3) M^{\frac{1}{2}} \overrightarrow{P_i P_{i+1}}, \tag{20.41}$$

or

$$P_i^* = P_i + \frac{1}{2d} \begin{pmatrix} d-b & -c \\ a & d+b \end{pmatrix} \overrightarrow{P_i P_{i+1}}, \tag{20.42}$$

where $d = \sqrt{(ac - b^2)/3}$.

20.8 Metric Construction

We now discuss the construction of the *metric tensor* M in order to satisfy an adaptation criterion. Suppose that we have only one unknown, denoted by η. We are trying to determine the metric tensor in order to equilibrate the interpolation error for the piecewise linear continuous finite element. The error equilibration idea is natural if we want to minimize the number of unknowns for solving the given problem with a given error (**don't put too many grid points to get a too- small error in some place**).

We assume that an initial solution has been computed for a given mesh and we denote by $\Pi_h \eta$ the piecewise linear interpolation of η supposed to be regular enough.

In one dimension, the interpolation error can be defined by $\epsilon_\infty = |\eta - \Pi_h \eta|_\infty$.* On a segment $[a, b]$ of the 1D mesh (see Figure 20.5), we have $(\Pi_h \eta)(a) = \eta(a)$ and $(\Pi_h \eta)(b) = \eta(b)$, so by using Taylor expansion, for all $x \in]a, b[$, we have

$$(\eta - \Pi_h \eta)(x) = (x-a)\eta'(a) + \frac{(x-a)^2}{2}\eta''(a) - (x-a)(\Pi_h \eta)'(a) + O((x-a)^3), \tag{20.43}$$

*$|.|_\infty$ is the L^∞ norm.

By construction of Π_h,

$$(\Pi_h\eta)'(a) = \frac{(\Pi_h\eta)(b) - (\Pi_h\eta)(a)}{b-a} = \frac{\eta(b) - \eta(a)}{b-a}, \tag{20.44}$$

and with again the Taylor expansion to evaluate $\eta(b)$, we get

$$(\Pi_h\eta)'(a) = \eta'(a) + \frac{(b-a)}{2}\eta''(a) + O(b-a)^2. \tag{20.45}$$

From Eq. 20.43 and Eq. 20.45 we obtain

$$\eta(x) - (\Pi_h\eta)(x) = \frac{(x-a)^2}{2}\eta''(a) - \frac{(x-a)(b-a)}{2}\eta''(a) + O(b-a)^3, \tag{20.46}$$

therefore,

$$\eta(x) - (\Pi_h\eta)(x) = \frac{(x-a)(x-b)}{2}\eta''(a) + O(b-a)^3. \tag{20.47}$$

But, we have also

$$\max_{x\in[a,b]} |(x-a)(x-b)| = (a-b)^2/4; \tag{20.48}$$

so the interpolation error ϵ_∞ on a segment $[ab]$ is

$$\varepsilon_\infty = \frac{(b-a)^2}{8}\eta''(a) + 0(b-a)^3. \tag{20.49}$$

In a two-dimensional space, the interpolation error is related to the Hessian matrix of η (see [d'Azevedo and Simpson, 1989] and [d'Azevedo and Simpson, 1991] for the proof)

$$\varepsilon = |\eta - \Pi_h\eta| \le c_0 h^2 |H(\eta)|, \tag{20.50}$$

where $|.|$ is the $H^1(\Omega)$ norm, or the $L^\infty(\Omega)$ norm and where the Hessian matrix is defined by

$$H(\eta) = \begin{pmatrix} \dfrac{\partial^2\eta}{\partial x_1^2} & \dfrac{\partial^2\eta}{\partial x_1 \partial x_2} \\ \dfrac{\partial^2\eta}{\partial x_1 \partial x_2} & \dfrac{\partial^2\eta}{\partial x_2^2} \end{pmatrix}. \tag{20.51}$$

Let us define the absolute value of a symmetric matrix by

$$|H| \overset{def}{=} R \begin{pmatrix} |\lambda_1| & 0 \\ 0 & |\lambda_2| \end{pmatrix} R^{-1} \tag{20.52}$$

where R is the unit matrix (i.e., ${}^tR = R^{-1}$), which diagonalizes the symmetric matrix H, and let λ_1, λ_2 be the eigenvalues of H such that

$$H = R\begin{pmatrix} \lambda_1 & 0 \\ 0 & \lambda_2 \end{pmatrix}R^{-1}. \tag{20.53}$$

The error on a mesh edge a_i can be computed as

$$\varepsilon_i \approx c_0\left|{}^t a_i H a_i\right|, \tag{20.54}$$

where c_0 is the constant of the relation 20.50 or $c_0 = 1/8$ if we consider Eq. 20.49. In order to minimize the number of vertices, we have to equilibrate this error. So the error ϵ_i must be close to a given constant ε_0, a given threshold.

In the previous section, we have introduced several tools to construct a unit mesh with respect to a metric M. Consequently, the length of the segment a_i in the metric $|H|$ is $\sqrt{\left|{}^t a_i H a_i\right|} \approx \sqrt{e_0/c_0}$. To achieve a unit mesh size, we simply use the metric tensor M as defined

$$M \overset{def}{=} \frac{c_0}{\varepsilon_0}|H| \tag{20.55}$$

Remark: M is a dimensionless matrix.

For other interpolation, such as a quadratic triangle, we can compute an interpolation error $\epsilon(x, d)$ in all the direction d, and the problem is now find, for all X, the biggest $M(x)$ such that

$${}^t dM(x)d \le \varepsilon(x,d) \quad \forall d \in \mathrm{IR}^2. \tag{20.56}$$

20.8.1 Computation of the Hessian

The second derivatives are the fundamental key point in the metric definition, in case of a piecewise linear finite element (i.e., the second derivatives are equal to zero in each triangle). Therefore, a weak formulation (by means of the Green's formula) has to be used to compute the Hessians as

$$\int_\Omega H_{i,j}.v_h = -\int_\Omega \frac{\partial \eta_h}{\partial x_i}\frac{\partial v_h}{\partial x_j} + \int_{\partial\Omega} \frac{\partial \eta_h}{\partial x_i}\cdot v_h \tag{20.57}$$

where v_h is the classical P^1 test function and $H = (H_{ij})_{i=1,2,j=1,2}$ and η_h is the numerical approximate of η (remark $\eta_h \neq \Pi_h\eta$, but we assume η_h and $\Pi_h\eta$ are close enough).

To solve the liner problem Eq. 20.57, we use a mass lumping technique so as to obtain a diagonal problem, and the discrete Hessian H_{ij}^k at a vertex k is thus computed by

$$H_{ij}^k = \frac{\left(-\int_\Omega \frac{\partial \eta_h}{\partial x_i}\frac{\partial v_h^k}{\partial x_j} + \int_{\partial\Omega}\frac{\partial \eta_h}{\partial x_i}\cdot v_h^k\right)}{\int_\Omega v_h^k} \tag{20.58}$$

where v_h^k is the piecewise linear finite element hat function associated with the vertex k.

20.8.2 Remark on Metric Computation

In the metric definition, we have to introduce the maximum and minimum mesh edge lengths in order to avoid unrealistic metrics. This is not really a restriction as we have usually a pretty good idea of what these quantities should be. More precisely, the eigenvalues of the metric defined in Eq. 20.53 are founded as follows:

$$\tilde{\lambda}_{1,2} = min\left(max\left(|\lambda_{1,2}|, \frac{1}{h_{max}^2} \right), \frac{1}{h_{min}^2} \right),$$

h_{min} and h_{max} being the minimal and maximal allowable edge lengths in the mesh.

20.8.3 Metric Associated with Classical Norms

All the results given in this section are obtainable by a compilation of the results in [Castro and Diaz, 1996] and [d'Azevedo and Simpson, 1991]

It is then possible to change the norm in Eq. 20.50 so as to compute the error $\epsilon = |\eta - \Pi_h\eta|$. To this end, we introduce a new class of metrics defined by

$$M = \frac{c_0^p}{\varepsilon_0^p} |A|^{\frac{1}{2}} \quad |H|^p \quad |A|^{\frac{1}{2}} \tag{20.59}$$

where the exponent p of a matrix is defined by, according to the notation of Eq. 20.52,

$$|H|^p \stackrel{def}{=} R \begin{pmatrix} |\lambda_1|^p & 0 \\ 0 & |\lambda_2|^p \end{pmatrix} R^{-1}. \tag{20.60}$$

The given number p and the given symmetric definite positive matrix A can be defined for the different classical norms as follows:

- For the L^∞ norm, $p = 1$, $A = Id_2$, and the error is

$$\epsilon = \|f - \Pi_h(f)\|_\infty, \tag{20.61}$$

- For the H^1 norm, $p = 1$, $A = Id_2$, and the error is

$$\epsilon = \|\nabla(f - \Pi_h(f))\|_{L^2}, \tag{20.62}$$

- For the energy norm, $p = 2$ and the error is

$$\epsilon = \left\| \sqrt{\nabla(f - \Pi_h(f)) A \nabla(f - \Pi_h(f))} \right\|_\infty, \tag{20.63}$$

- For the L^2 norm, $p = {}^1/_2$, $A = Id_2$, and the error is

$$\varepsilon = \|f - \Pi_h(f)\|_{L^2}. \tag{20.64}$$

20.8.4 Metric with Relative Error

In the previous computation we have used a global error. But the cutoff definition of the error becomes a problem when the magnitude of the variation solution is greater than 10^3. The global error is not sufficient, and we have to use a local relative error. The relative error ϵ_r is defined by

$$\epsilon_r = \left\| \frac{\eta - \Pi_h(\eta)_X}{max(|\eta|, \text{cutoff})} \right\| \tag{20.65}$$

where cutoff is a positive number that allows to avoid a division by zero.

The metric tensor M related to the relative error ϵ_r is

$$M_r = \frac{c_0^p \ |A|^{\frac{1}{2}} \ \ |H|^p \ \ |A|^{\frac{1}{2}}}{\varepsilon_0^p \ max\left(\text{cutoff}, |\eta|^p\right)} \tag{20.66}$$

Remark: This is a dimensionless error.

20.8.5 Metric Intersection

In the case where several metric maps are specified (for instance for multicriteria problems), we propose a method that enables us to merge these maps so as to retrieve the one metric case and therefore to define the control space. The problem is to define a metric at a point that is consistent with two or more initially specified metrics.

Let us consider the unit circles (ellipses) associated with two initial metrics. The sought solution is the metric associated with the intersection of these two circles. In general, this intersection is not an ellipse, so we retain the largest ellipse included in it as the solution. The latter defines a metric, namely the *intersection metric.*

To obtain this intersection metric, we use the simultaneous reduction applied to the initial metrics. If M_1 and M_2 denote these two metrics, the two corresponding circles can be written in the base associated with the simultaneous reduction of the matrices M_1 and M_2 as

$$^t X M_1 X = \lambda_1 x_1^2 + \lambda_2 x_2^2 = 1 \quad \text{and} \quad ^t X M_2 X = \mu_1 x_1^2 + \mu_2 x_2^2 = 1. \tag{20.67}$$

The intersection metric $M_1 \cap M_2$ is then defined by

$$M_1 \cap M_2 = {}^t P^{-1} \begin{pmatrix} \max(\lambda_1, \mu_1) & 0 \\ 0 & \max(\lambda_2, \mu_2) \end{pmatrix} P^{-1}, \tag{20.68}$$

where P is the matrix that allows to transform the canonical base to the base associated with the reduction. On Figure 20.6 the intersection metric of two given metrics can be seen.

If more than two metrics are involved, the above scheme is applied recursively,

$$M_1 \cap \cdots \cap M_q = \left(\ldots\left(\left(M_1 \cap M_2\right) \cap M_3\right) \cap \ldots\right) \cap M_q. \tag{20.69}$$

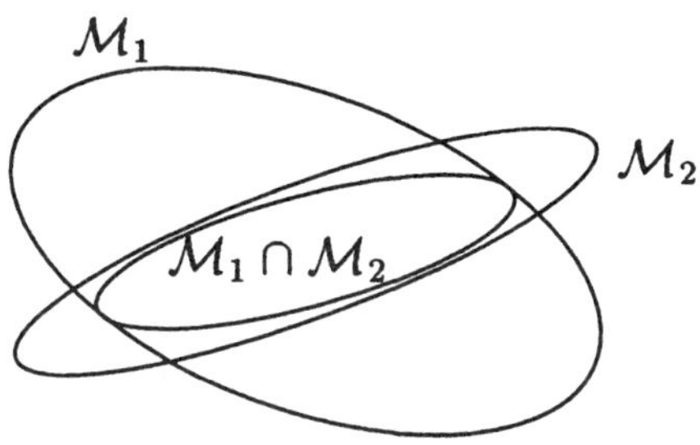

FIGURE 20.6 Ellipse of the intersection of two metrics.

20.9 Loop of Adaptation

The aforementioned framework can be easily extended to construct a loop of adaptation. To this end, the control space at iteration j is defined by the mesh at iteration $j-1$ and a metric map is specified at every mesh vertex. In this general situation, it is also necessary to create the mesh of the boundary at each step with respect to the control space. To this end, the geometry of the boundary is strictly required. We assume that this geometry is known by T_{geom} a mesh constructed so as to permit the creation of a suitable mathematical support, denoted as $Supp_{geom}$, which can be easily handled. In so doing, we avoid being too closely coupled with a CAD system. With this background, the scheme of a *loop of adaptation* is given as

- Input of T_{geom}, the mesh serving as definition for the geometry and construction of the support $Supp_{geom}$.
- Construction of the initial boundary discretization F_0 according to a size map H_0 given on $Supp_{geom}$.
- Initial mesh T_0 using F_0 and H_0 as data.
- Adaptation loop (starting at $j = 1$).
  - Input of a metric map H_j on T_{j-1}.
  - Discretization F_j of the support $Supp_{geom}$ governed by the control space (T_{j-1}, H_j).
  - Mesh adaptation T_j using F_j and the control space (T_{j-1}, H_j).
  - Iteration $j = j + 1$, if required.

The diagram associated with an adaptation loop is summarized in Figure 20.7.

The above procedure is repeated until an almost satisfactory mesh T_j is obtained, with respect to (T_j, H_{j+1}). In other words, the edges in T_j tend to have a length close to one, in the control space associated with the current metric.

20.10 Application Examples

The CFD examples depicted in this section are the result of the NSC2KE solver. We describe two configurations of compressible inviscid and viscous flows. In all cases, the initial meshes have been generated using the EMC2 software [Hecht and Saltel, 1990], and the adaption loop can be done using the Bamg software [Hecht, 1997].

We compare the normalized residual evolution of the adapted computation with a direct computation for which we started on the last adapted mesh from an uniform solution using the same time integration procedure. The residual is based on the norm of the right-hand side of the equations and not on the time derivative terms. In this way, the time step size does not influence the convergence history. This gives an idea of the cost of similar computations on a large uniform grid. In fact, for the same resolution we would need many more grid points in an uniform mesh, but this just enforces our conclusions. For both cases, the convergence is accelerated by the adaptation technique. In terms of CPU, as in the

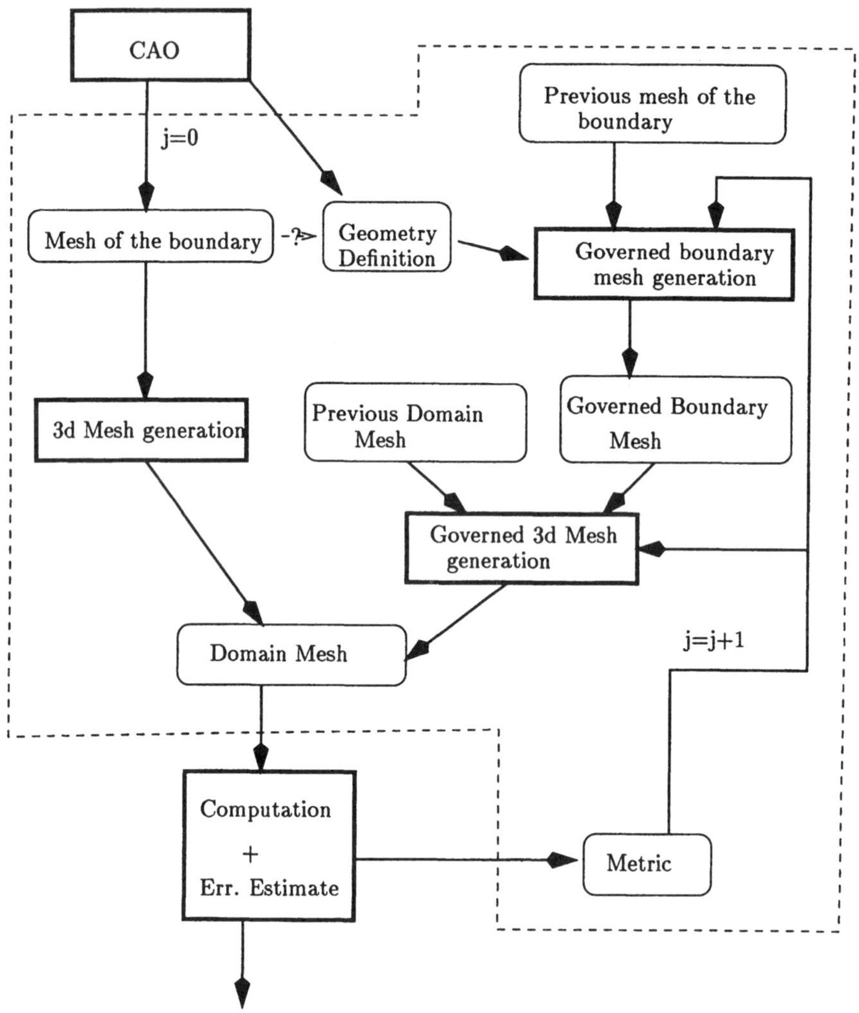

FIGURE 20.7 Adaptation diagram.

adaptation loop, most of the work is done at the coarser levels, the computation cost is reduced by at least a factor of 20.

The stop criterion for the adaptation loop has been the independence of the results, with respect of the mesh (especially regarding the wall coefficients).

These techniques have been validated on several other configurations, such as multielement airfoils. These computations can be found in [Castro, Diaz, et al., 1995].

20.10.1 Navier–Stokes Solver

We use the NSC2KE fluid solver for the computations; details can be found in [Mohammadi, 1994]. A finite-volume-Galerkin formulation of Navier–Stokes equations in conservation form has been considered. A four stage Runge–Kutta scheme is used for time integration. The Roe Riemann solver [Roe, 1981] has been used for the Euler part together with a MUSCL type reconstruction and Van Albada limiters for second-order accuracy. P^1 finite element has been used for the viscous part of the operation. A Steger–Warming [Steger and Warming, 1982] flux splitting has been used at the inflow and outflow boundaries, while nonpenetration or slip boundary conditions have been applied to solid walls depending on the flow nature. Turbulent modeling is done using the classical $k - \varepsilon$ [Launder and Spalding, 1972] model with special wall-laws, enabling the computations for separated and unsteady flows [Mohammadi and Pironneau, 1994]; [Hecht and Mohammadi, 1997].

20.10.2 Flow Over a Backward Step

This is the classical backward-step (ratio 2 between the step and the channel heights) at $Re_{\infty/H} = 44{,}000$ and inflow Mach number of 0.1. The parameter δ is set to $0.005H$ which corresponds to $y^+ \sim 20$. For mesh adaptation the parameters h_{min}, h_{max}, and h_n are respectively $0.002H$, H, and $0.002H$. We can see that these parameters are quite easy to choose for a given configuration.

We compare the mesh obtained using the global and relative estimations presented above for the same interpolation error ($c\epsilon = 10^{-2}$).

20.10.3 Transonic Turbulent Flow Over a RAE2822

This is a transonic flow at inflow Mach number of 0.734 and 2.79 degrees of incidence. The chord-based Reynolds number is $Re_C = 6.5 \times 10^6$. Experimental data are available for the pressure and friction coefficients distribution. One difficulty here is to correctly predict the shock position. The aim here is to show the impact of the ingredients described above regarding mesh generation algorithms for boundary layers. The parameters h_{min}, h_{max}, and h_n are respectively $0.01C$, $3.C$, and $0.0002C$. The interpolation error is $\epsilon = 5 \times 10^{-3}$.

20.11 Application to Surface Meshing

We like to give a different application of the anisotropic mesh generation method. To this end, we consider the problem of parametric surface meshing.

Let Ω be a domain of R^2 and σ be a smooth function, then the surface Σ defined by the following parametrization σ

$$\sigma : \Omega \rightarrow R^3, (u,v) \mapsto \sigma(u,v)$$

can be meshed in Ω by means of an anisotropic two-dimensional mesh generation method, the resulting mesh being then mapped in R^3 by means of σ.

As the purpose is to obtain an accurate approximation of Σ, the mesh generation process must be governed so as to provide this issue. The question is then to construct an adequate metric map in Ω, the domain where the construction is made, to obtain, after mapping the resulting mesh onto Σ, an accurate surface mesh.

The key-idea is to use the intrinsic properties of the surface Σ to define a metric map in R^3, referred to as M_3, enabling us to construct the needed metric map, M_2 in R^2. In this way, the scheme of the mesh generation process is what follows

$$\begin{array}{c} \text{governed mesh} \quad \left(\Sigma, M_3(P)_{P\in\Sigma}\right) \\ \Updownarrow \\ \text{governed mesh} \quad \left(\Omega, M_2(X)_{X\in\Omega}\right) \end{array},$$

where $X \in \Omega$ and $P \in \Sigma$ are related by $P = \sigma(X)$.

Assuming that M_3 is given, $M_2(X)$ is the metric induced by $M_3(P)$ on the tangent plane of the surface Σ at P. If $\Pi(P)$ denotes the matrix transforming the canonical base of R^3 in the local base at P, the metric $M_2(X)$ is then defined by the two first rows and the two first lines of the matrix

$$^t\Pi(P)M_3(P)\Pi(P). \tag{20.70}$$

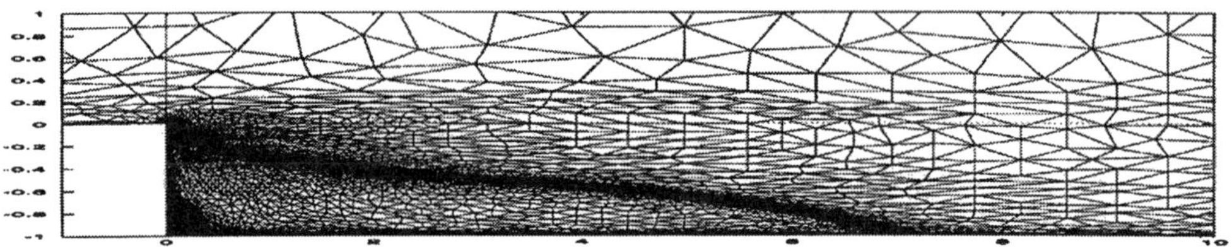

FIGURE 20.8 Partial view of the mesh obtained with the relative criterion. The main and secondary recirculations are correctly identified.

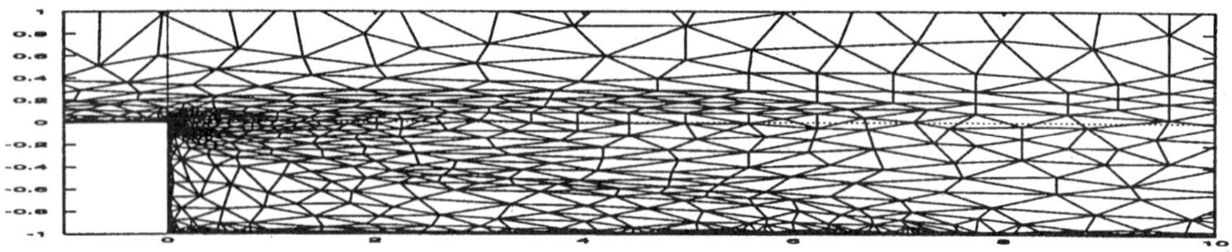

FIGURE 20.9 Partial view of the mesh obtained with the global criterion. The main recirculation is weakly detected and the secondary one has not been captured.

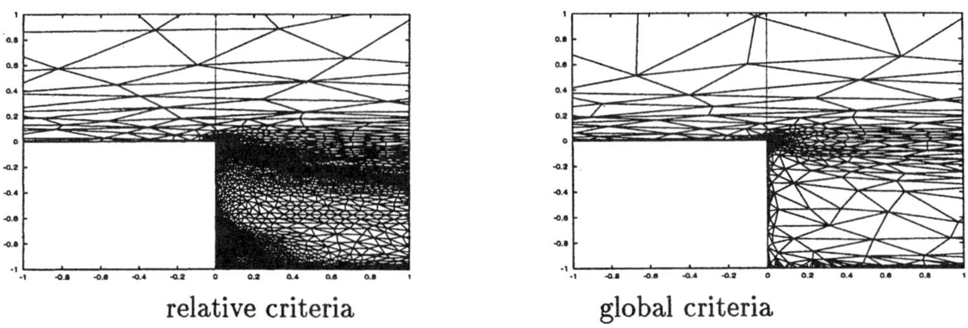

FIGURE 20.10 Partial view of the mesh obtained, main and secondary recirculation are correctly identified with the relative criteria and the secondary one has not been captured with the global error.

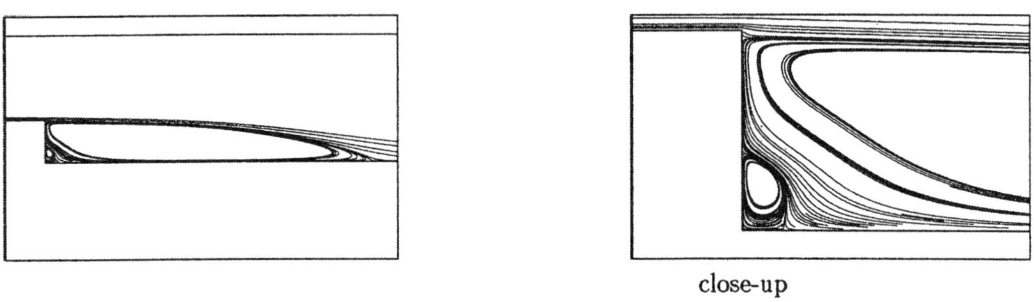

FIGURE 20.11 Backward facing step: particle tracking for the computation using the relative error estimation.

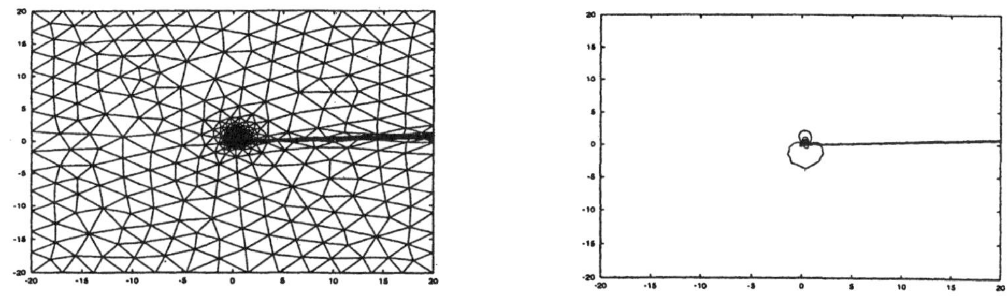

FIGURE 20.12 RAE 2822: Adapted mesh with about 7000 nodes and Iso-Mach contours.

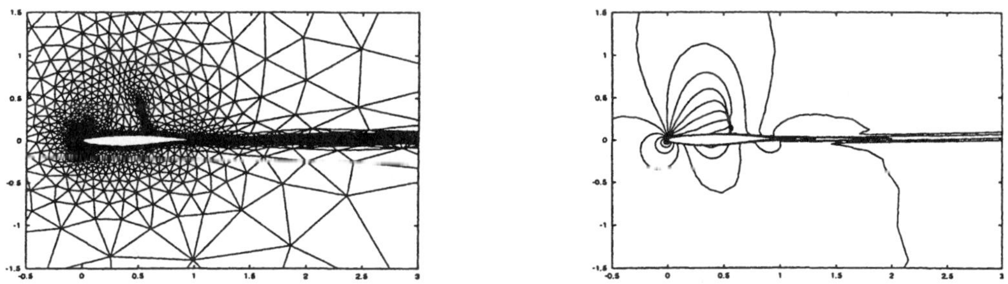

FIGURE 20.13 RAE 2822: Adapted mesh and Iso-Mach contours (partial view).

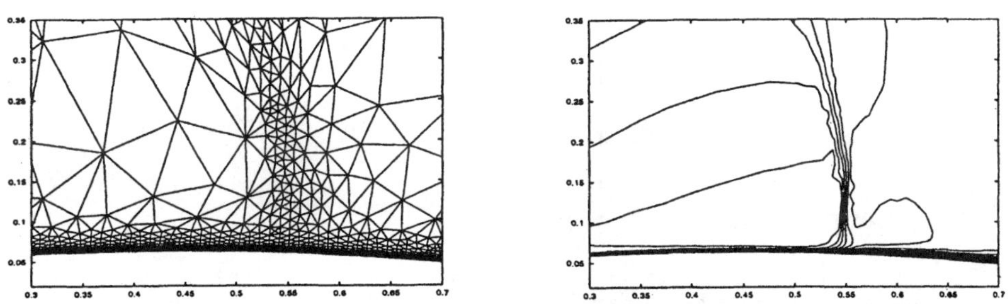

FIGURE 20.14 RAE 2822: Zoom over the region of shock-boundary layer interaction.

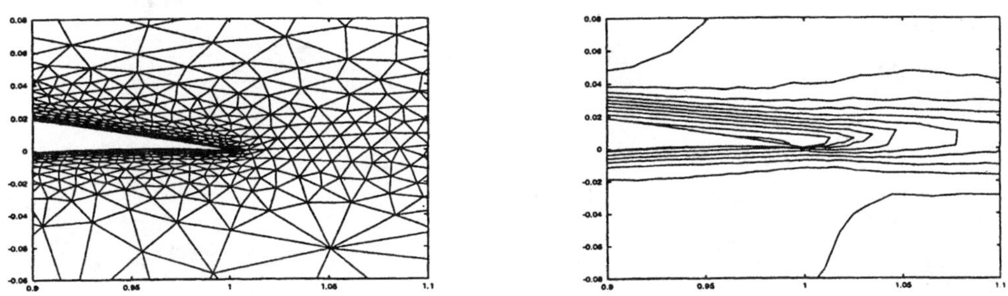

FIGURE 20.15 RAE 2822: Zoom around the trailing edge.

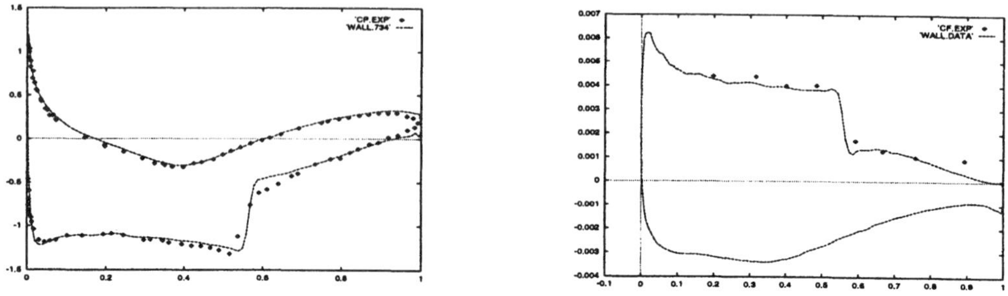

FIGURE 20.16 RAE 2822: Pressure and friction coefficients distribution.

Then, different types of approximate meshes can be obtained as a function of the metric $M_3(P)$. In this respect, we can specify isotropic meshes with constant or variable size as well as pure anisotropic meshes. For instance, a metric of the type

$$M_3(P,h)=\begin{pmatrix}\frac{1}{h^2} & 0 & 0\\ 0 & \frac{1}{h^2} & 0\\ 0 & 0 & \frac{1}{h^2}\end{pmatrix} \tag{20.71}$$

leads to specify an isotropic mesh where h is the expected size at P (if h is independent of P, a uniform mesh will be obtained), while

$${}^tB(P)\begin{pmatrix}\frac{1}{h_1^2(P)} & 0 & 0\\ 0 & \frac{1}{h_2^2(P)} & 0\\ 0 & 0 & \frac{1}{h_3^2(P)}\end{pmatrix}B(P) \tag{20.72}$$

leads to specify an anisotropic mesh of sizes h_1, h_2 and h_3 in the base vector directions of $B(P)$ at P.

Two specific metric can be constructed in this way, referred to as $M_3(P, \rho)$ and $M_3(P, \rho_1, \rho_2)$ where ρ_1 and ρ_2 are the principal radii of curvature of the surface at point P while ρ is the minimal radius of curvature of Σ at point P.

More precisely, if ρ, in fact $\rho(P)$, is the smaller of the principal radii of curvature ρ_1 and ρ_2 then the metric map

$$M_3(P,\rho)=\begin{pmatrix}\frac{1}{\rho^2} & 0 & 0\\ 0 & \frac{1}{\rho^2} & 0\\ 0 & 0 & \frac{1}{\rho^2}\end{pmatrix} \tag{20.73}$$

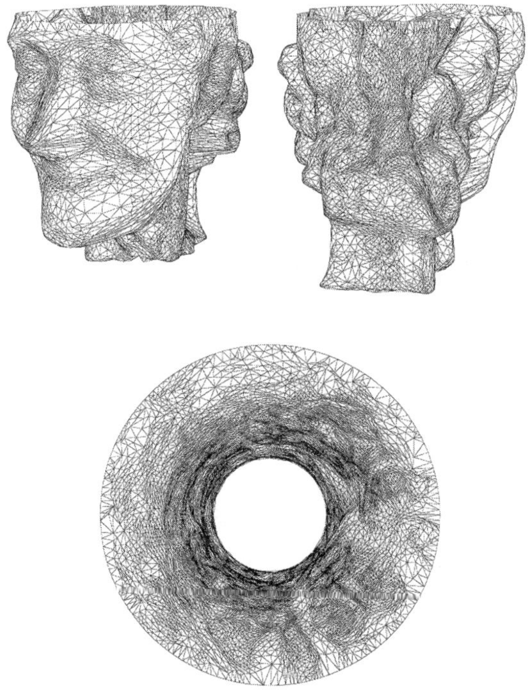

FIGURE 20.17 An isotropic geometric mesh of a cylinder using as domain Ω a ring.

is called the *isotropic map related to the minimal radius of curvature* and, according to [Borouchaki and George, 1996], this map enables us to obtain an isotropic mesh with a second-order approximation of the surface Σ. Similarly, assuming $\rho_1 < \rho_2$ (where ρ_1 and ρ_2 are functions of P), the metric map

$$M_3(P,\rho_1,\rho_2) = {}^tB_p(P)\begin{pmatrix} \frac{1}{\rho_1^2} & 0 & 0 \\ 0 & \frac{1}{\rho_2'^2} & 0 \\ 0 & 0 & \lambda \end{pmatrix} B_p(P) \tag{20.74}$$

where λ is an arbitrary scalar value is called the *anisotropic map related to the principal radii of curvature,* and this map allows to obtain an anisotropic mesh with a second order approximate of the surface.

To illustrate the surface meshing application, we given Figure 20.17 where the surface Σ is the head section depicted on top while the parametric domain Ω is a ring. The anisotropic mesh generation process is governed by a metric map of the form $M_3(P, \rho_1, \rho_2)$ so as to obtain a second order approximate of the surface. The resulting mesh includes 20,374 triangles and 10,190 vertices.

20.12 Concluding Remarks

In this paper, the main lines of a classical Delaunay-type mesh generation algorithm have been recalled. Special attention has been paid to the incremental insertion point procedure, the so-called Delaunay kernel, and to the field point construction using the supporting edges.

We have proposed a scheme for a Delaunay generator in the anisotropic case, or more generally an adaptation problem, based on the classical scheme as the usual metric has been replaced by a Riemannian structure. The Delaunay kernel as well as the field point creation procedure have been rewritten. To this end, the notion of a control space has been introduced, leading to a new definition of the lengths involved in the different steps of the method.

Application examples for two-dimensional CFD computations and parametric surfaces have been presented to illustrate the features of the governed mesh generation algorithm.

The extension of the proposed method to three dimensions shall not induce any major difficulty. The process used to define the field points along the edges is still valid. The Delaunay kernel can be formally extended, while verifying that the star-shaped property of the cavity still holds. The delicate aspects expected concern the proper definition of the set $Supp_{geom}$, geometric support of the domain and also the ability to remesh the surface of the domain according to a specified map.

Acknowledgment

We are greatly indebted to Houman Borouchaki (currently at Université Technologique de Troyes) as well as to Bijan Mohammadi (from INRIA) for helping us in this work.

References

1. Berger, M.J. and Jameson, A., 1985. Automatic adaptive grid refinement for Euler equations, *AIAA J.* 1985, 23,4, pp 561–568.
2. Borouchaki, H. and George, P.L., Maillage de Surfaces parametriques. partie i; aspects théoriques, *Rapport de Recherche. 1996, INRIA,* 2928.
3. Borouchaki, H., George, P.L., and Lo, S.H., Optimal Delaunay point insertion, *Int. Jour. Num. Meth. Eng.* 1996, 39, 20, pp 3407–3438.
4. Borouchaki, H., George, P. L., and Mohammadi, B., Delaunay mesh generation governed by metric specifications. part 2: application examples, *Finite Elements in Analysis and Design,* 1996, 25(1–2), pp 85–109.
5. Borouchaki, H., George, P. L., Hecht, F., Laug, P., and Saltel, E., Delaunay mesh generation governed by metric specifications. part 1: algorithms, *Finite Elements in Analysis and Design* 1996, 25(1–2), pp 61–83.
6. Borouchaki, H., George, P. L., Hecht, F., and Saltel, E. Reasonably efficient Delaunay based mesh generator in 3 dimensions, *4th International Meshing Roundtable,* Albuquerque, NM, 1995, pp 3–14.
7. Bowyer, A., Computing dirichlet tessellations, *Comput. J.* 1981, 24, pp 162–166.
8. Briere de l'Isle, E. and George, P.L., 1995. Optimization of tetrahedral meshes, *IMA Volumes in Mathematics and its Applications.* Babuska, I., Henshaw, W.D., Oliger, J.E., Flaherty, J.E., Hopcroft, J.E., and Tezduyar, T., (Eds.), Springer-Verlag, 1995, Vol. 75, p. 97–128.
9. Bristeau, M.O. and Periaux, J., Finite element methods for the calculation of compressible viscous flows using self-adaptive refinement, in *VKI Lecture Notes on CFD,* 1986.
10. Castro Díaz, M.J., Mesh refinement over surfaces, *INRIA, Rocquencourt,* 1994, RR-2462.
11. Castro Díaz, M.J., Generación y adaptatión anisotrópa de mallados de elementos finitos para la resolutión numérica de e.d.p. aplicaciones, (Tesis doctoral), Universidad de Málaga,1996.
12. Castro Díaz, M.J. and Hecht, F. Anisotropic surface mesh generation, *INRIA Research Report, Rocquencourt.* 1995, RR-2672.
13. Castro Díaz, M. J., Hecht, F., and Mohammadi, B., New progress in anisotropic mesh adaption for inviscid and viscous flow simulations, *INRIA Research Report, Rocquencourt.* 1995, RR-2671.
14. Cherfils, C. and Hermeline, F., Diagonal Swap procedures and characterizations of 2D-Delaunay triangulations, *M² AN.* 1990, 24,5, pp 613–625.
15. d'Azevedo, E.F. and Simpson, R.B., On optimal interpolation triangle incidences, *SIAM's Journal of Scientific and Statistical Computing, 1989,* 6, pp1063–1075.
16. d'Azevedo, E.F. and Simpson, R.B.,On Optimal triangular-meshs for minimizing the gradient error, *Numerische Mathematik,* 1991, 59, 4, pp 321–348.

17. George, P.L., *Automatic Mesh Generation. Applications to Finite Element Methods.* Wiley, 1991.
18. George, P.L. and Borouchaki, H., *Triangulation de Delaunay et Maillage. Applications aux Éléments Finis,* Hermès, Paris, 1997.
19. George, P.L., Hecht, F., and Saltel, E., Automatic mesh generator with specified boundary, *Comp. Meth. in Appl. Mech. and Eng.* 1991, 92, pp 269–288.
20. Hecht, F., Bidimensional anisotropic mesh generator, Technical report, *INRIA, Rocquencourt.* 1997 Source code: http://ftp.inria.fr/INRIA/Projects/Gamma/bamg.tar.gz
21. Hecht, F. and Mohammadi, B., Mesh adaption by metric control for multi-scale phenomena and turbulence, *AIAA Paper* 97-0859, 1997.
22. Hecht, F. and Saltel, E., EMC[2] un logiciel d'édition de maillages et de contours bidimensionnels, Technical report *INRIA, Rocquencourt,* 1990, RT-0118.
Source code: http://ftp.inria.fr/INRIA/Projects/Gamma./emc2.tar.gz
23. Hermeline, F., Triangulation automatique d'un polyèdre en dimension *N. R.A.I.R.O. Analyse Num.* 1982, 16, 3, pp 211–242.
24. Laug, P. and Borouchaki, H., The BL2D mesh generator: beginner's guide, user's and programmer's manual, *R.T. INRIA.* 0194, 1996.
Source code: http://ftp.inria.ft/INRIA/Projects/Gamma/bl2d.tar.Z
25. Laug, P., Borouchaki, H., and George, P. L., Maillage de courbes gouverné par une carte de métriques, *R. R. INRIA.* 2818, 1996.
26. Launder, B.E. and Spalding, D.B., Mathematical models of turbulence, *Academic Press.* 1972, 40, pp 263–293.
27. Lawson, C.L., Properties of n-dimensional triangulations, *Comput. Aided Geom. Design.* 1986, 3, pp 231–246.
28. Lee, C.K. and Lo, S.H., An automatic adaptive refinement finite element procedure for 2D Elastostatic analysis, *Int. Jour. Num. Meth. Eng.* 1992, 35, pp 1967–1989.
29. Lo, S.H., Automatic mesh generation and adaptation by using contours, *Int. Jour. Num. Meth. Eng.* 1991, 31, pp 689–707.
30. Lohner, R., Adaptive remeshing for transient problems, *Comp. Meth. in Appl. Mech. and Eng.* 1989, 75, pp 195–214.
31. Mavriplis, D.J., Adaptive mesh generation for viscous flows using Delaunay triangulation, *Jour. of Comput. Phys.*1990, 90, 2, pp 271–291.
32. Mohammadi, B., CFD with NSC2KE: A user guide, Technical report, *INRIA, Rocquencourt.* RT-0164, 1994. Source code: http://frp.inria.fr/INRIA/Projects/Gamma/NSC2KE.tar.gz
33. Mohammadi, B. and Pironneau, O., *Analysis of the K-Epsilon Turbulence Model.* J. Wiley and Masson Pub, 1994.
34. Peraire, J., Vahdati, M., Morgan, K., and Zienkiewicz, O.C., Adaptive remeshing for compressible flow computations, *J. Comput. Phys.* 1987, 72, pp 449–466.
35. Preparata, F.P. and Shamos, M.I., *Computational geometry, an introduction.* Springer-Verlag, 1985.
36. Roe, P. L. 1981. Approximate Riemann solvers, parameters vectors and difference schemes, *JCP.* 43.
37. Steger, J. and Warming, R.F., Flux vector splitting for the inviscid gas dynamic with applications to finite-difference methods, *Journal Comp. Phys.* 1982, 40, pp 263–293.
38. Vallet, M.G., Génération de maillages éléments finis anisotropes et adaptatifs, *Thèse d'Université.* Paris 6, 1992.
39. Watson, D.F., Computing the n-dimensional Delaunay tessellation with application to Voronoï polytopes, *Comput. J.* 1981, 24, pp 167–172.
40. Weatherill, N.P., Marchant, M.J., Hassan, O., and Marcum, D.L., Grid adaptation using a distribution of sources applied to inviscid compressible flow simulations, *Int. J. Num. Meth. Eng.* 1994, 19, pp 739–764.

21
Quadrilateral and Hexahedral Element Meshes

Robert Schneiders

21.1 Introduction

This chapter explains techniques for the generation of quadrilateral and hexahedral element meshes. Since structured meshes are discussed in detail in other parts of this volume, we focus on the generation of unstructured meshes, with special attention paid to the 3D case.

Quadrilateral or hexahedral element meshes are the meshes of choice for many applications, a fact that can be explained empirically more easily than mathematically. An example of a numerical experiment is presented by Benzley [1995], who uses tetrahedral and hexahedral element meshes for bending and torsional analysis of a simple bar, fixed at one end. If elastic material is assumed, second-order tetrahedral elements and first-order hexahedral elements both give good results (first-order tetrahedral elements perform worse). In the case of elastic–plastic material, a hexahedral element mesh is significantly better.

A mathematical argument in favor of the hexahedral element is that the volume defined by one element must be represented by at least five tetrahedra. The construction of the system matrix is thus computationally more expensive, in particular if higher order elements are used. Unstructured hex meshes are often used in computational fluid dynamics, where one tries to fill most of the computational domain with a structured grid, allowing irregular nodes but in regions of complicated shape, and for the simulation of processes with plastic deformation, e.g., metal forming processes.

In contrast to the favorable numerical quality of quadrilateral and hexahedral element meshes, mesh generation is a very difficult task. A hexahedral element mesh is a very "stiff" structure from a geometrical point of view, a fact that is illustrated by the following observation: Consider a structured grid and a new node that must be inserted by using local modifications only (Figure 21.1). While this can be done in 2D, in the three-dimensional case it is no longer possible! Thus, it is not possible to generate a hexahedral element mesh by point insertion methods, a technique that has proven very powerful for the generation of tetrahedral element meshes (Delaunay–type algorithms, Chapter 16).

Many algorithms for the generation of tetrahedral element meshes are advancing front methods (Chapter 17), where a volume is meshed starting from a discretization of its surface and building the volume mesh layer by layer. It is very difficult to use this idea for hex meshing, even for very simple

0-8493-2687-7/99/$0.00+$.50

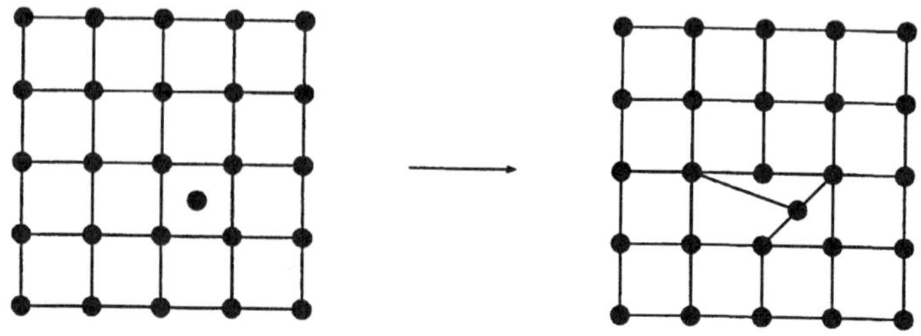

FIGURE 21.1 Inserting a point into a structured quadrilateral element mesh.

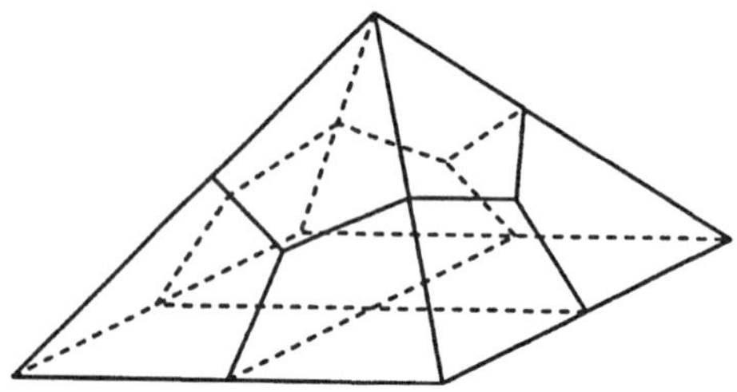

FIGURE 21.2 Surface mesh for a pyramid.

structures! Figure 21.2 shows a pyramid whose basic square has been split into four and whose triangles have been split into three quadrilateral faces each. It has been shown that a hexahedral element mesh whose surface matches the given surface mesh exactly exists [Mitchell 1996], but all known solutions have degenerated or zero-volume elements.

The failure of point-insertion and advancing-front type algorithms severely limits the number of approaches to deal with the hex meshing problem. Most proposed algorithms can be classified either as block-decomposition or superposition methods. The situation is better for the generation of quadrilateral element meshes.

In the remainder of the chapter, we will explain the basic techniques for quadrilateral and hexahedral element mesh generation, with special attention paid to the three-dimensional case. Much of the research work has been presented in the *Numerical Grid Generation in Computational Fluid Dynamics* and in the *Mesh Generation Roundtable and Conference* conference series, and detailed information can be found in the proceedings. The proceedings of the latter are available online at the Meshing Research Corner [Owen 1996], a large database with literature on mesh generation maintained at Carnegie Mellon University by S. Owen.

21.2 Block-Decomposition Methods

In the early years of the finite element method, hexahedral element meshes were the meshes of choice. The geometries considered at that time were not very complex (beams, plates), and a hexahedral element mesh could be generated with less effort than a tetrahedral mesh (graphics workstations were not available at that time). Meshes were generated by using mapped meshing methods: A mesh defined on the unit cube is transformed onto the desired geometry Ω with the help of a mapping $F : [0, 1]^3 \to \Omega$. This method can generate structured grids for cube-like geometries (Figure 21.3).

The mapping F can be specified explicitly (isoparametric or conformal mapping) or implicitly (solution of an elliptic or hyperbolic partial differential equation). The problem of finding a suitable mapping F has been the object of major research efforts in recent years, and an overview is given elsewhere in this handbook. A summary of the results can be found in the books of Thompson [1985] and Knupp [1995].

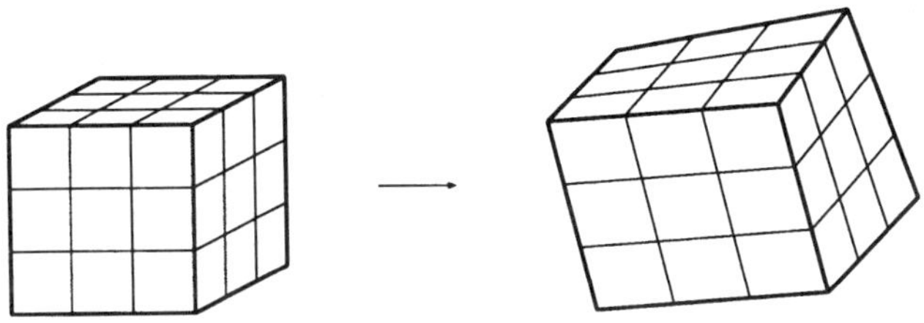

FIGURE 21.3 Mapped meshing.

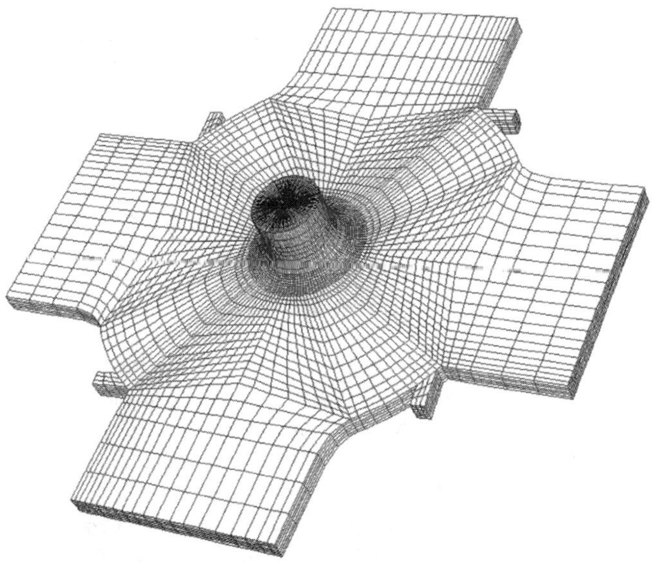

FIGURE 21.4 Multiblock-structured mesh.

If the geometry to be meshed is too complicated or has reentrant edges, meshes generated by mapped meshing methods usually have poorly shaped elements and cannot be used for numerical simulations. In this case, a preprocessing step is required: The geometry is interactively partitioned into blocks that are meshed separately (the meshes at joint interfaces must match, a problem considered in [Tam and Armstrong 1993] and [Hannemann 1995]). These multiblock-type methods are state of the art in university and industrial codes (see Chapter 13). Figure 21.4 shows an example mesh that was generated with Fluent Inc.'s GEOMESH* preprocessor.

In principle, most geometries can be meshed in this way. However, there is a limitation in practice: The construction of the multiblock decomposition, which must be done interactively by the engineer. For complex geometries, e.g., a flow field around an airplane or a complicated casting geometry, this task can take weeks or even months to complete. This severely prolongs the simulation turnaround time and limits the acceptance of numerical simulations (a recent study suggests that in order to obtain a 24-hour simulation turnaround time, the time spent for mesh generation has to be cut to at most one hour).

One way to deal with that problem is to develop solvers based on unstructured tetrahedral element meshes. In the 1980s, powerful automatic tetrahedral element meshers were developed for that purpose (they are described elsewhere in this volume).

The first attempt to develop a truly automated hex mesher was undertaken by the finite element modeling group at Queens University in Belfast (C. Armstrong). Their strategy is to automate the block

*GEOMESH is a trademark of Fluent Inc.

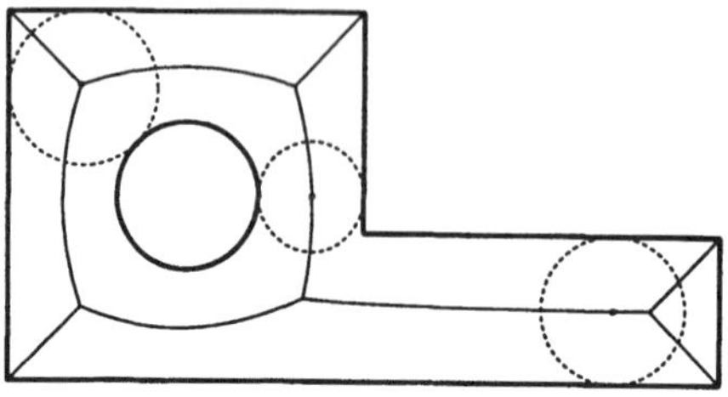
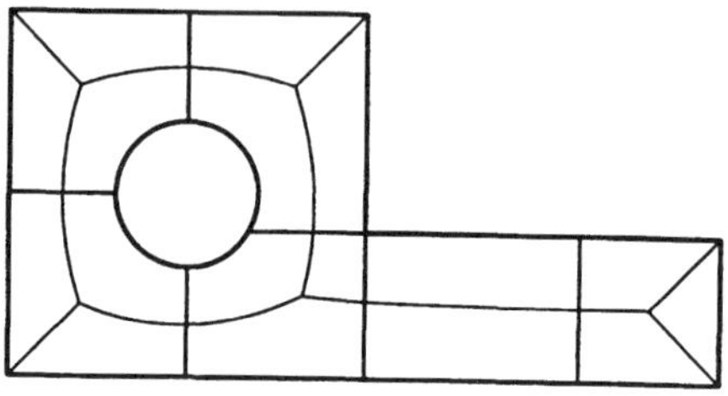

FIGURE 21.5 Medical axis and domain decomposition.

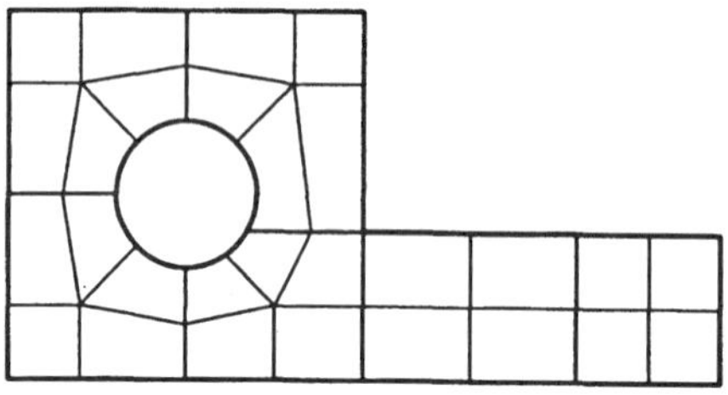
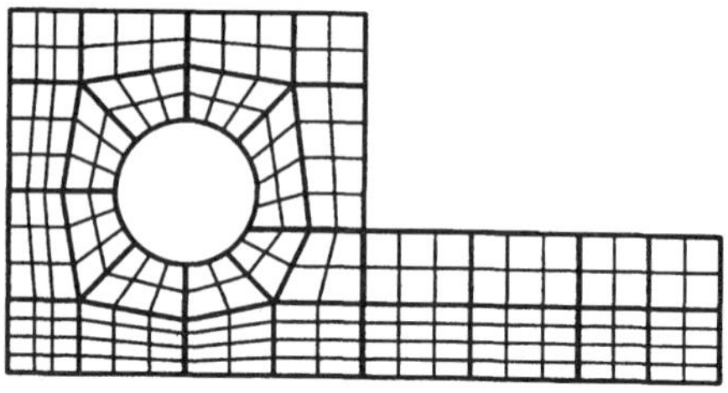

FIGURE 21.6 Multiblock-decomposition and resulting mesh.

decomposition process. The starting point is the derivation of a simplified geometrical representation of the geometry, the medial axis in 2D and the medial surface in 3D. In the following we will explain the idea (see [Price, Armstrong, Sabin 1995] and [Price and Armstrong 1997] for the details).

We start with a discussion of the 2D algorithm. Consider a domain A for which we want to find a partition into subdomains A_i. We define the *medial axis* or skeleton of A as follows: for each point $P \in A$, the touching circle $U_r(P)$ is the largest circle around P that is fully contained in A. The medial axis $M(A)$ is the set of all points P whose touching circles touch the boundary δA of A more than once.

The medial axis consists of nodes and edges and can be viewed as a graph. An example is shown in Figure 21.5: two circles touch the boundary of A exactly twice; the respective midpoints fall on edges of the medial axis. A third circle has three points common with δA, the midpoint is a branch point (node) of the medial axis. The medial axis is a unique description of A: A is the union of all touching circles $U_r(P)$, $P \in M(A)$.

The medial axis is a representation of the topology of the domain and can thus serve as a starting point for a block decomposition (Figures 21.5 and 21.6). For each node of $M(A)$ a subdomain is defined, its boundary consisting of the bisectors of the adjacent edges and parts of δA (a modified procedure is used if nonconvex parts of δA come into play [Price, Armstrong, Sabin 1995]). The resulting decomposition of A consists of n–polygons, $n \geq 3$, whose interior angle are smaller than 180°. A polygon is then split up by using the midpoint subdivision technique [Tam and Armstrong 1993], [Blacker and Stephenson 1991]: Its centroid is connected to the midpoints of its edges; the resulting tesselation consists of convex quadrilaterals. Figure 21.6 shows the multiblock decomposition and the resulting mesh, which can be generated by applying mapped meshing to the faces.

It remains to explain how to construct the medial axis. This is done by using a Delaunay technique (Figure 21.7a): The boundary δA of the domain A is approximated by a polygon p, and the constrained Delaunay triangulation (CDT) of p is computed. One gets an approximation to the medial axis by connection of the circumcircles of the Delaunay triangulation (the approximation is a subset of the Voronoï diagram of p).

By refining the discretization p of δA and applying this procedure, one gets a series of approximations that converges to the medial axis (Figure 21.7b). Consider a triangle of the CDT to p: part of its circumcircle overlaps the complement of A. The overlap for the circumcircle of the respective triangle

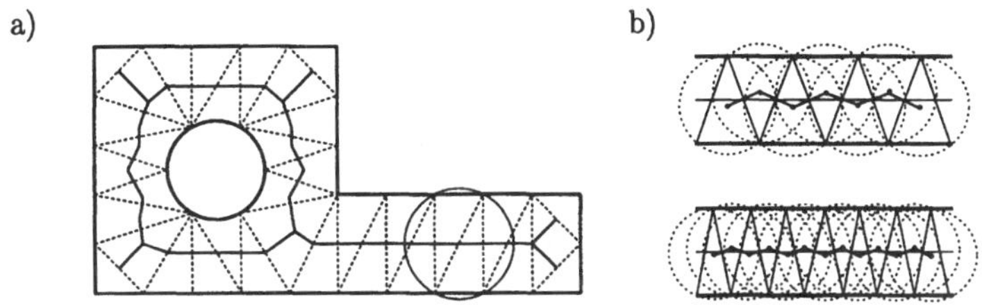

FIGURE 21.7 Approximating the medial axis.

of the refined polygon's CDT is significantly smaller. If the edge lengths of p tend to zero, the circumcircles converge to circles contained in A that touch δA at least twice. Their midpoints belong to the medial axis.

In three dimensions, the automization of the multiblock decompositions is found by using the medial surface. The medial surface is a straightforward generalization of the medial axis and is defined as follows: Consider a point P in the object A and let $U_r(P)$ the maximum sphere centered in P that is contained in A. The medial surface is defined as the set of all points P for which $U_r(P)$ touches the object boundary δA more than once. P lies on

- A face of the medial surface, if $U_r(P)$ touches δA twice.
- An edge of the medial surface, if $U_r(P)$ touches δA three times.
- A node of the medial surface, if $U_r(P)$ touches δA four times or more.

The medial surface is a simplified description of the object (again, A is the union of the touching spheres $U_r(P)$ for all points P on the medial surface). The medial surface preserves the topology information and can therefore be used for finding the multiblock decomposition.

Armstrong's algorithm for hexahedral element mesh generation follows the line of the 2D algorithm (Figure 21.8). The first step is the construction of the medial surface with the help of a constrained Delaunay triangulation (Shewchuk [1998] shows how to construct a surface triangulation for which a constrained Delaunay triangulation exists). The medial surface is then used to decompose the object into simple subvolumes. This is the crucial step of the algorithm, and it is much more complex than in the two-dimensional case. A number of different cases must be considered, especially if nonconvex edges are involved; they will not be discussed here, the interested reader is referred to [Price and Armstrong 1997] for the details.

Armstrong identifies 13 polyhedra an object is decomposed to (Figure 21.9 shows a selection). These meshable primitives have convex edges, and each node is adjacent to exactly three edges. The midpoint subdivision technique [Tan and Armstrong 1993] can therefore be used to decompose the object into hexahedra: the midpoints of the edges are connected to the midpoints of the faces (Figure 21.10). Then both the edge and face midpoints are connected to the center of the object, and the resulting decomposition consists of valid hexahedral elements.

Figure 21.11 shows a mesh generated for a geometry with a nonconvex edge. The example highlights the strength of the method: the mesh is well aligned to the geometry, it is a "nice" mesh — an engineer would try to create a mesh like this with an interactive tool.

The medial surface technique tries to emulate the multiblock decomposition done by the engineer "by hand." This leads to the generation of quality meshes, but there are some inherent problems: namely, it does not answer the question whether a good block decomposition exists, which may not be the case if the geometry to be meshed has small features. Another problem is that the medial surface is an unstable entity. Small changes in the object can cause big changes in the medial surface and the generated mesh.

Nevertheless, the medial surface is extremely useful for engineering analysis: It can be used for geometry idealization and small feature removal, which simplifies the medial surface, enhances the stability of the algorithm and leads to better block decompositions. The method delivers relatively coarse meshes that are well aligned to the geometry, a highly desirable property especially in computational mechanics. It

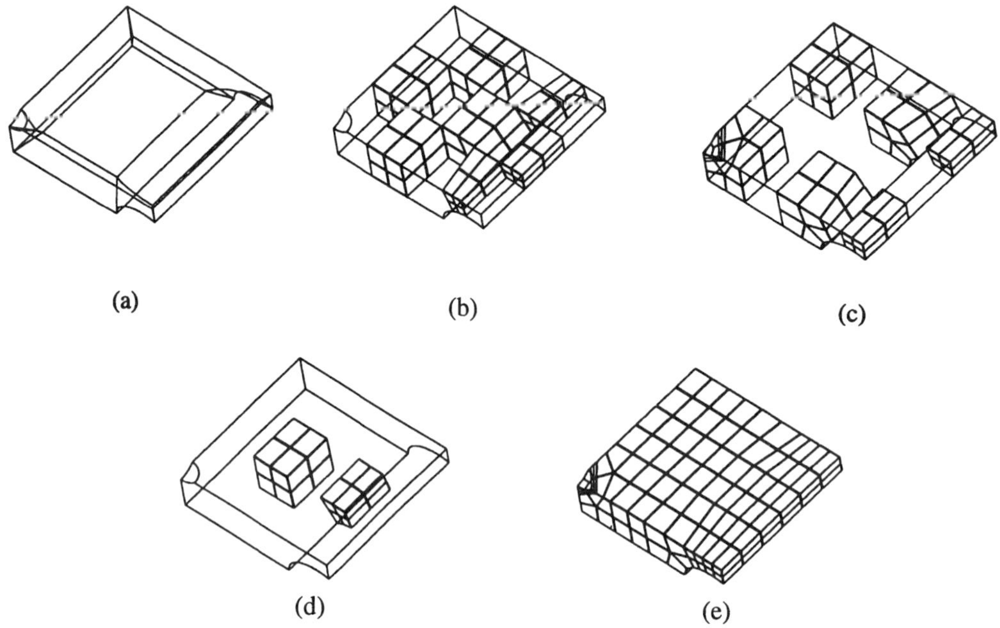

FIGURE 21.8 Medial-surface algorithm for the generation of hexahedral element meshes. (a) medial surface, (b) edge primitives, (c) vertex primitives, (d) face primitives, and (e) final mesh.

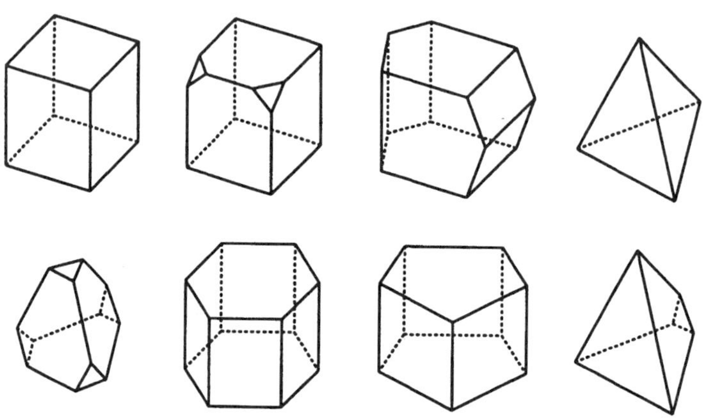

FIGURE 21.9 Meshable primitives (selection).

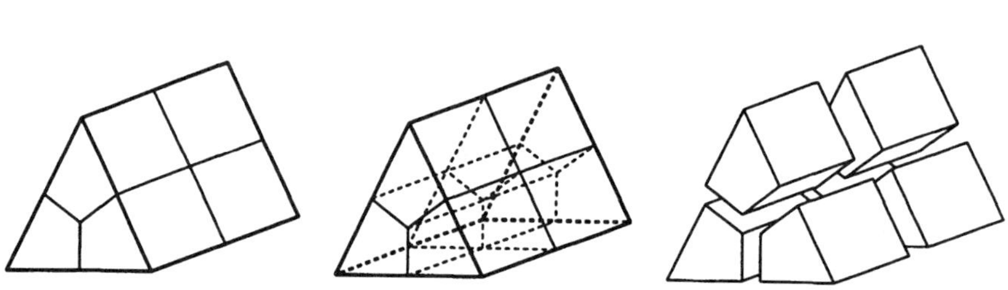

FIGURE 21.10 Volume decomposition by midpoint subdivision.

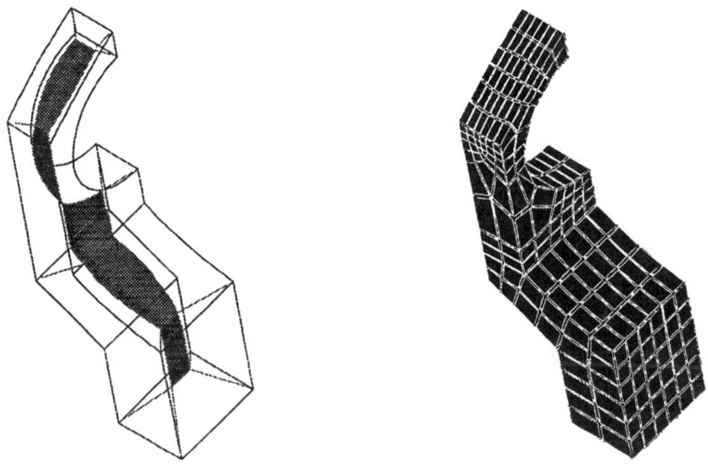

FIGURE 21.11 Medial surface and mesh for a mechanical part.

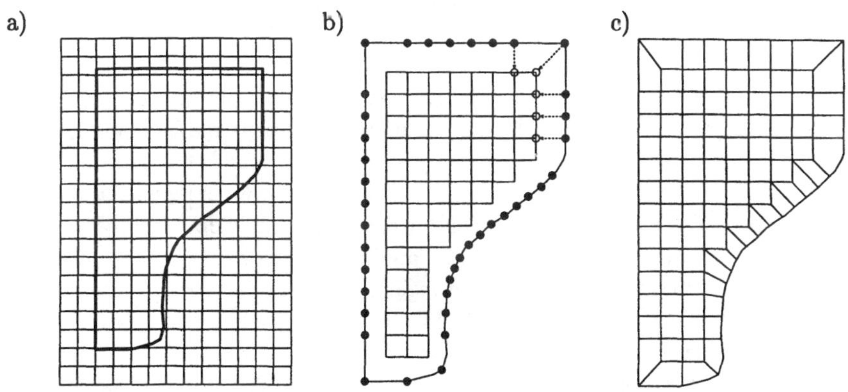

FIGURE 21.12 2D grid-based algorithm.

is natural that an approach to high-quality mesh generation leads to a very complex algorithm, but the problems are likely to be solved.

Two other hex meshing algorithms based on the medial surface are known in the literature. Holmes [1995] uses the medial surface concept to develop meshing templates for simple subvolumes. Chen [Turkkiyah 1995] generates a quadrilateral element mesh on the medial surface which is then extended to the volume.

21.3 Superposition Methods

The acronym *superposition methods* refers to a class of meshing algorithms that use the same basic strategy. All these algorithms start with a mesh that can be more or less easily generated and covers a sufficiently large domain around the object, which is then adapted to the object boundary. The approach is very pragmatic, but the resulting algorithms are very robust, and there are several promising variants.

Since we have actively participated in this research, we will concentrate on a description of our own work, the grid-based algorithm [Schneiders 1996a]. Figure 21.12 shows the 2D variant: A sufficiently large region around the object is covered by a structured grid. The cell size h of the grid can be chosen arbitrarily, but should be smaller than the smallest feature of the object. It remains to adapt the grid to the object boundary — the most difficult part of the algorithm.

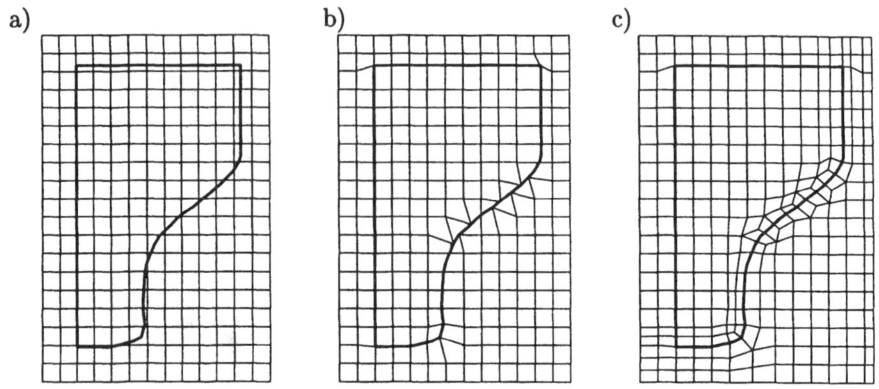

FIGURE 21.13 Grid-based algorithm — boundary adaption by projection technique.

According to [Schneiders 1996a], all elements outside the object or too close to the object boundary are removed from the mesh, with the remaining cells defining the *initial mesh* (Figure 21.12a, note that the distance between the initial mesh and the boundary is approximately h). The region between the object boundary and the initial mesh is then meshed with the isomorphism technique. The boundary of the initial mesh is a polygon, and for each polygon node, a node on the object boundary is defined (Figure 21.12b). Care must be taken that characteristic points of the object boundary are matched in this step, a problem that is not too difficult to solve in 2D. By connecting polygon nodes to their respective nodes on the objective boundary, one gets a quadrilateral element mesh in the boundary region (Figure 21.12c).

The "principal axis" of the mesh depends on the structure of the initial mesh, and in the grid- based algorithm the element layers are parallel to one of the coordinate axis. Consequently, the resulting mesh (Figure 21.12) has a regular structure in the object interior and near boundaries that are parallel to the coordinate axis; irregular nodes can be found in regions close to other parts of the boundary. This is typical for a grid-based algorithm, but can be avoided by choosing a different type of initial mesh.

The only input parameter for the grid-based algorithm is the cell size h. In case of failure, it is therefore possible to restart the algorithm with a different choice of h, a fact that greatly enhances the robustness of the algorithm.

Another way to adapt the initial mesh to the boundary, the projection method, was proposed in [Taghavi 1994] and [Ives 1995]. The starting point is the construction of a structured grid that covers the object (Figure 21.13a), but in contrast to the grid-based algorithm, all cells remain in place. Mesh nodes are moved onto the characteristic points of the object and then onto the object edges, so that the object boundary is fully covered by mesh edges (Figure 21.13b). Degenerate elements may be constructed in this step, but disappear after buffer layers have been inserted at the object boundary (Figure 21.13c, the mesh is then optimized by Laplacian smoothing).

The projection method allows the meshing objects with internal faces; the resulting meshes are similar to those generated with the isomorphism techniques, although there tend to be high aspect ratio elements at smaller features of the object. In contrast to the isomorphism technique, the mesh is adapted to the object boundary before inserting the buffer layer.

Superposition methods can be used for the 3D case. The idea of the grid-based algorithm is shown for a simple geometry, a pyramid (1 quadrilateral, 4 triangular faces, Figure 21.14). The whole domain is covered with a structured uniform grid with cell size h. In order to adapt the grid to the boundary, all cells outside the object that intersect the object boundary or are closer than $0.5 \cdot h$ to the boundary are removed from the grid. The remaining set of cells is called the initial mesh (Figure 21.14a).

The isomorphism technique [Schneiders 1996a] is used to adapt the initial mesh to the boundary, a step that poses many more problems in 3D than in 2D. The technique is based on the observation that the boundary of the initial mesh is an unstructured mesh M of quadrilateral elements in 3D. An

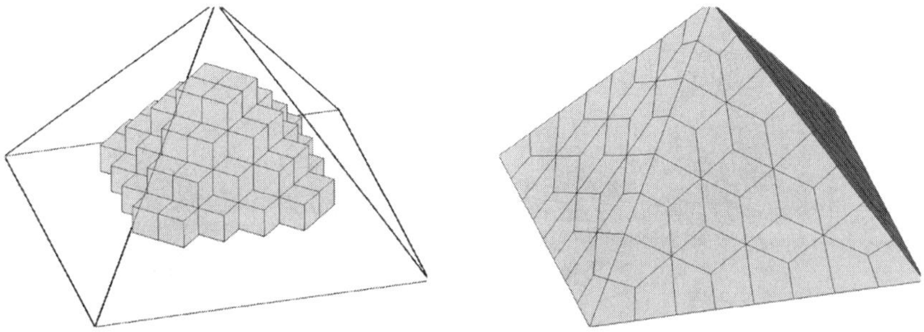

FIGURE 21.14 Initial mesh (a) and isomorphic mesh (b) on the boundary.

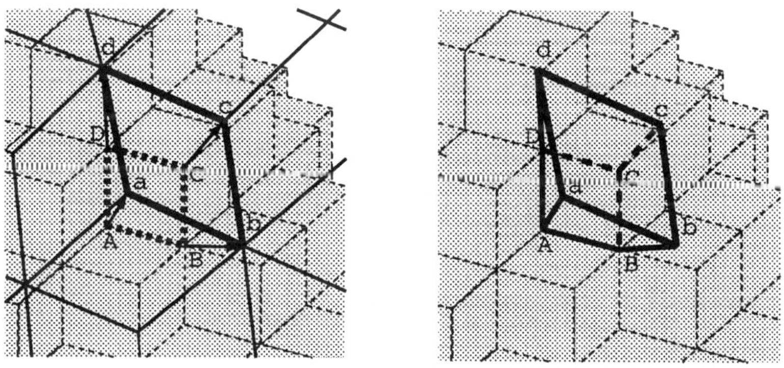

FIGURE 21.15 Construction of hexahedral elements in the boundary region.

isomorphic mesh M' is generated on the boundary. For each node of $v \in M$ a node $v' \in M'$ is defined on the object boundary, and for each edge $(v, w) \in M$ an edge $(v', w') \in M'$ is defined. It follows that for each quadrilateral $f \in M$ of the initial mesh's surface there is exactly one face $f' \in M'$ on the object boundary. Figure 21.14b shows the isomorphic mesh for the initial mesh of Figure 21.14b.

Figure 21.15 shows the situation in detail: The quadrilateral face $(A, B, C, D) \in M$ corresponds to the face $(a, b, c, d) \in M'$. The nodes A, B, C, D, a, b, c, d define a hexahedral element in the boundary region! This step can be carried out for all pairs of faces, and the boundary region can be meshed with hexahedral elements in this way.

The crucial step in the algorithm is the generation of a good quality mesh M' on the object boundary. All object edges must be matched by a sequence of mesh edges, and the shapes of the faces $f' \in M'$ must be nondegenerate. If the surface mesh does not meet these requirements, the resulting volume mesh does not represent the volume well or has degenerate elements. Fulfilling this requirement is a nontrivial task; also, the implementation becomes a problem (codes based on superposition techniques usually have more than 100,000 lines of code). We will not describe the process in detail, but some important steps will be discussed for the example shown in Figures 21.16–21.21.

Figure 21.16a shows the initial mesh for another geometry that does not look very complicated but nevertheless is difficult to mesh. The first step of the algorithm is to define the coordinates of the nodes of the isomorphic mesh. Therefore, normals are defined for the nodes on the surface of the initial mesh by averaging the normals N_f of the n adjacent faces f (cf. Figure 21.16b):

$$N_v = \frac{1}{n} \sum_{f \text{ adj. } v} N_f$$

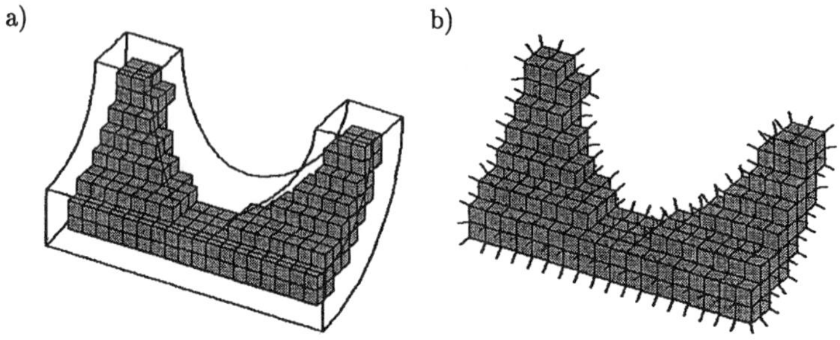

FIGURE 21.16 Initial mesh (a) and normals (b).

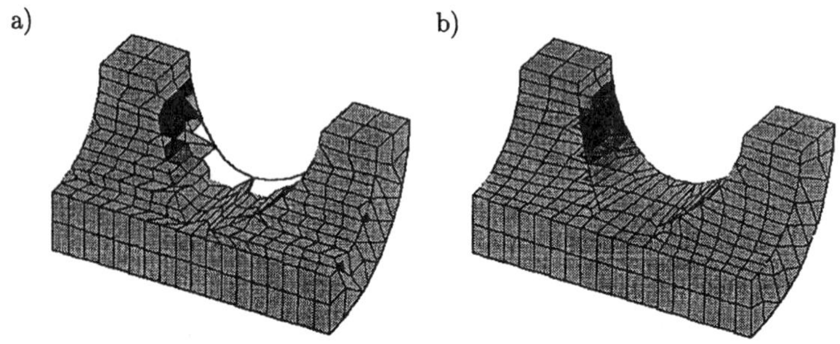

FIGURE 21.17 Isomorphic surface mesh.

For each point $v \in M$, the position of the corresponding point $v' \in M'$ is calculated as the intersection of the normal N_v with the object boundary. The point v' is then projected onto

- A characteristic vertex P of the object, if $dist(v', P) \leq 0.1 \cdot h$.
- A characteristic edge E of the object, if $dist(v', E) \leq 0.1 \cdot h$.

In case of projection, a flag is set for the respective node to indicate the entity it has been fixed to. Figure 21.17a shows that the quality of the generated surface mesh is unsatisfactory but that at least some of the characteristic vertices and edges of the object are covered by mesh nodes and edges.

For the generation of hexahedral elements in the boundary region, the topology of the surface mash M' must not be changed, but we are free to modify the location of the nodes in spaces. This allows the optimization of the surface mesh by moving the nodes v' to appropriate positions (Figure 21-17b shows that the quality of the surface mesh can be improved significantly). A Laplacian smoothing is applied to the nodes of the surface mesh. The new position x_i^{new} of a node v' is calculated as the average of the midpoints S_k of the N adjacent faces.

$$x_i^{new} = \frac{1}{N} \sum_{k=1}^{N} S_k$$

The following rules are applied in the optimization phase:

- After a correction step, the nodes are reprojected onto the object boundary.
- Nodes that are fixed to a characteristic vertex of the object are not considered.
- Nodes that are fixed to a characteristic edge are reprojected onto that edge.
- Nodes that are fixed to a characteristic edge but whose neighbors are not fixed are released from that edge.

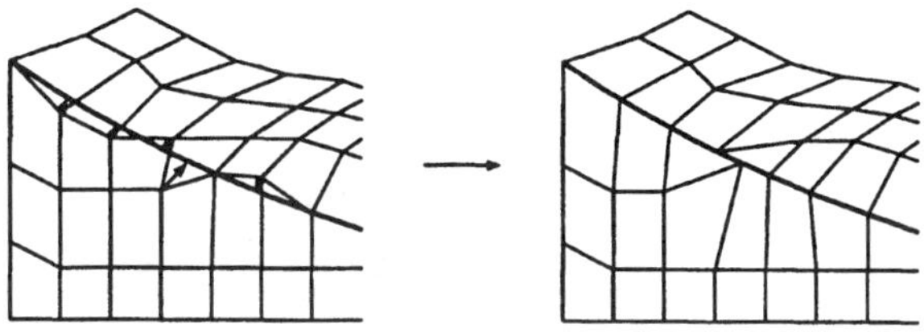

FIGURE 21.18 Edge repair.

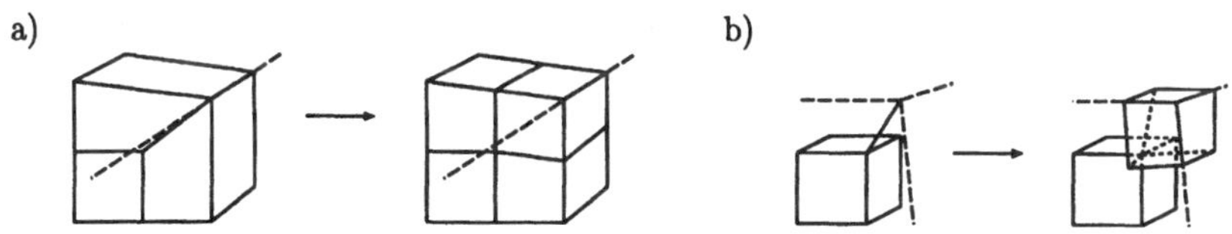

FIGURE 21.19 Inserting additional elements at sharp convex edges.

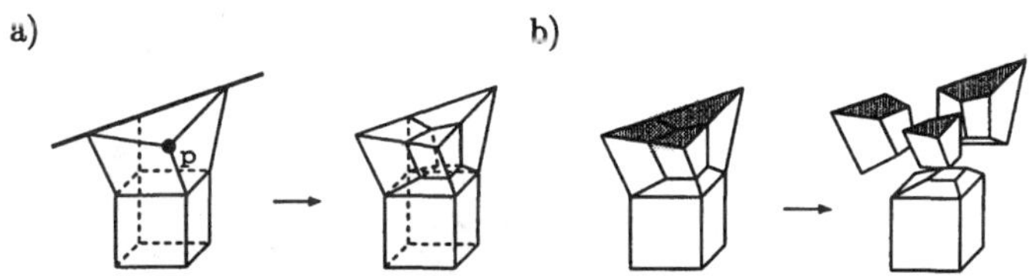

FIGURE 21.20 Splitting degenerated elements.

In the next step, the object vertices and edges are covered by mesh nodes and edges:

- Each object vertex is assigned the closest mesh node.
- Edge capturing: Starting from a vertex, mesh nodes are projected onto an object edge (Figure 21.18).
- The smoothing procedure is reapplied.

Figure 21.17b shows that the surface mesh accurately represents the object geometry and that the overall mesh quality has been improved. Nevertheless, degenerate faces can result from the edge capturing process if three nodes of a face are fixed to the same characteristic edge. This cannot be avoided if the object edges are not aligned to the "principal axes" of the mesh (cf. Figure 21.18). There are two ways to deal with the problem.

First, the boundary region is filled with a hexahedral element mesh. Due to the meshing procedure, there are two rows of elements adjacent to a convex edge (Figure 21.19a). If the solid angle alongside the edge is sufficiently smaller than 180°, the mesh quality can be improved by inserting an additional row of elements, followed by a local resmoothing. At object vertices where three convex edges meet, one additional element is inserted.

Figure 21.21a shows the resulting mesh after the application of the optimization step (note that many degeneracies have been removed). The remaining degenerate elements are removed by a splitting procedure.

Figure 21.20 shows the situation: Three points of a face have been fixed to a characteristic edge; the node P is "free." This face is split up into three quadrilaterals in a way that the flat angle is removed (Figure 21.20b). The adjacent element can be split in a similar way into four hexahedral elements. In order to maintain the conformity of the mesh, the neighbor elements must be split up also; it is, however, important that only neighbor elements adjacent to P must be refined — the initial mesh remains unchanged.

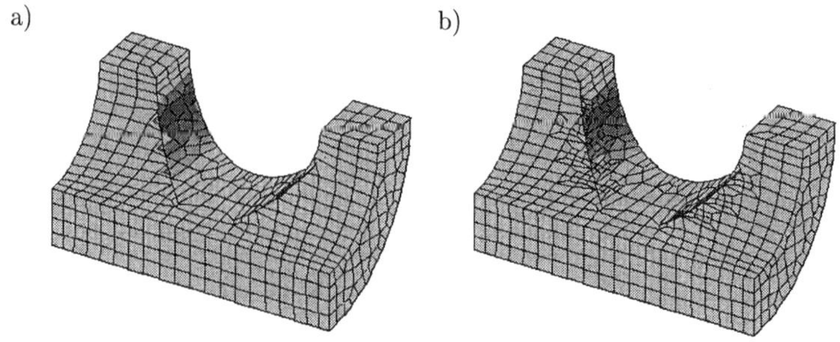

FIGURE 21.21 Removing degenerated elements.

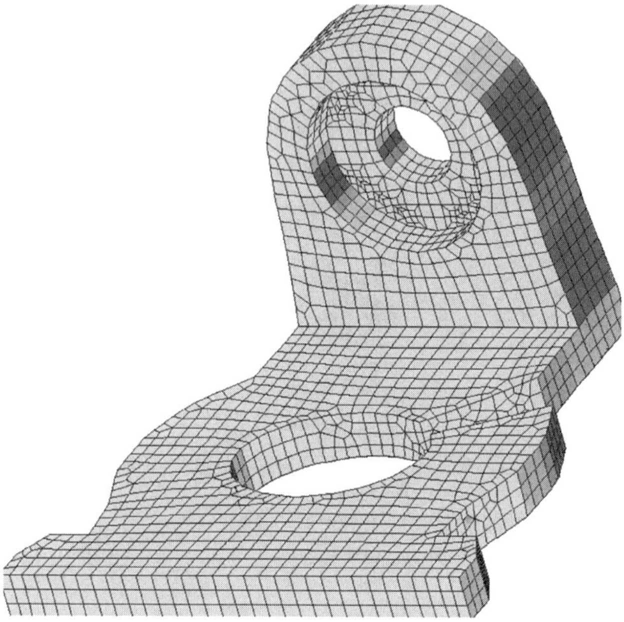

FIGURE 21.22 Mesh for a mechanical part.

Figure 21.21b shows the resulting mesh. Note that the surface mesh is no longer isomorphic to the initial mesh (Figure 21.16a) since removing the degenerated elements has had an effect on the topology (the mesh in Figure 21.17b is isomorphic to the initial mesh). The mesh has a regular structure at faces and edges that are parallel to one of the coordinate axes. The mesh is unstructured at edges whose adjacent edges include a "flat" angle and where degenerate elements had to be removed by the splitting operation. Figure 21.22 shows another mesh for a mechanical part.

The grid-based algorithm is only one out of many possible mesh generators that use the superposition principle. Figure 21.23 shows an examples where a nonuniform initial mesh has been generated. One can then apply the isomorphism (or projection) technique to adapt the mesh to the object boundary.

A weak point of the grid-based method is the fact that the elements are nearly equal sized. This can cause problems, since the element size h must be chosen according to the smallest feature of the object — a mesh with an unacceptable number of elements may result. The natural way to overcome this drawback is to choose an octree-based structure as an initial mesh, which would allow the adaption of the element size to the geometry. In the following we will explain the basic ideas and the problems that must be solved in this approach (see [Schneiders 1996b] for the details).

For reasons that will become clear later, we choose a special kind of octree structure (cf. Chapter 14). The root octant (a box that contains the object to be meshed) is subdivided into 27 octants (children).

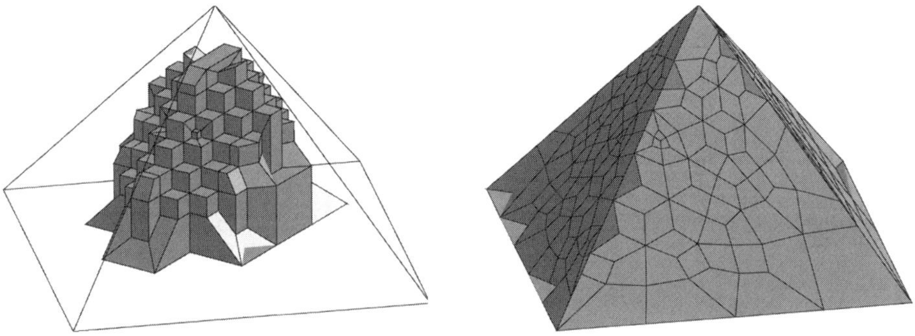

FIGURE 21.23 Octree-based initial mesh and isomorphic surface mesh.

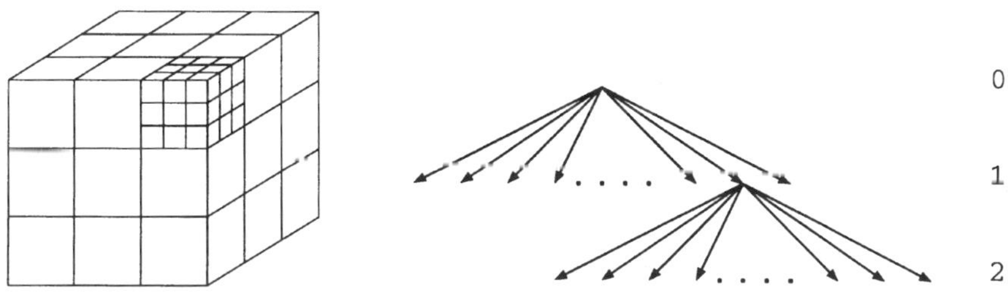

FIGURE 21.24 Octree decomposition.

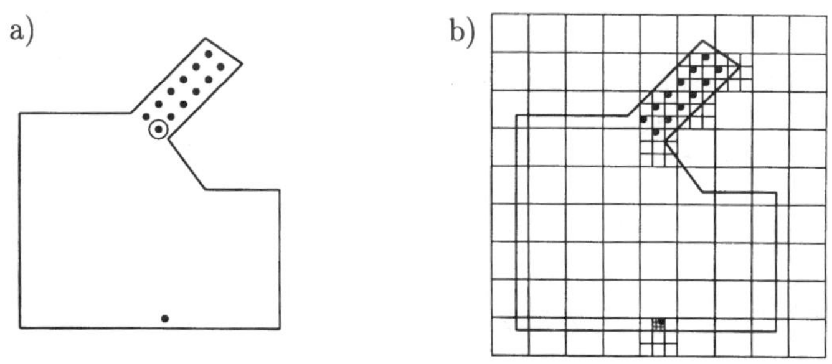

FIGURE 21.25 Mesh density information and initial quadtree.

These octants can be split up recursively until the mesh has the desired level of resolution. Figure 21.24 shows an example where one suboctant has been split. The example also shows that each octant can be assigned a level in a natural way:

- The root octant is assigned level 0.
- If an octant of level l is split, its children get level $l + 1$.

The octree structure has hanging nodes that have to be removed — one has to find the "conforming hull." This is the difficult problem to be solved in octree-based meshing, and it is equivalent to the refinement problem for hexahedral element meshes. For ease of understanding, we will treat the 2D case first.

Figure 21.25 shows the object to be meshed. The mesh density is represented by tupels (p, h), which means that the element size at the point p should not exceed h (although there are better ways to represent mesh density, this method has been chosen for ease of explanation). These points can be set according to the object geometry or deliberately, for example to get a dense mesh in an area where a point load is applied.

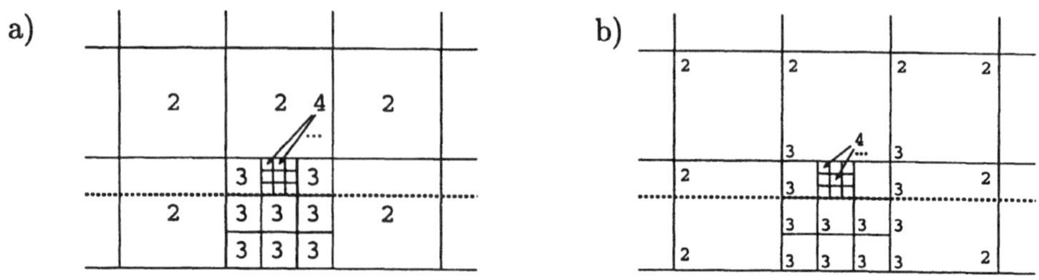

FIGURE 21.26 Quadrant and node levels.

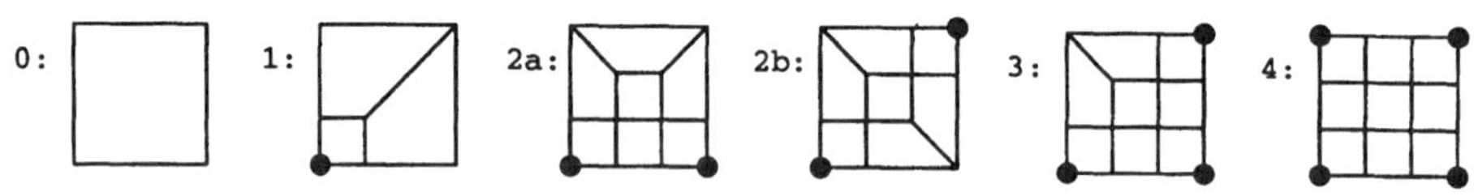

FIGURE 21.27 2D templates.

Starting from a box that contains the object to be meshed, the following procedure generates the quadtree:

```
procedure refine_quadrant (quadrant)

begin
  if the quadrant contains a point p whose associated edge length
  is smaller than the quadrant size then
    split up the quadrant into 9 (3D: 27) quadrants;
    for all new quadrants q_i
      refine_quadrant (q_i);

end;

refine_quadrant (root_quadrant);
```

Figure 21.26a shows a part of the quadtree and the quadrant levels. There are quadrants with hanging nodes at one or more edges if the level of a neighboring quadrant is different. These quadrants must be split up in order to get a conforming mesh.

First, the level information is transferred to the nodes of the quadtree: A node v is assigned the maximum level of its adjacent quadrants (Figure 21.26b):

$$l(v) = \max\{l(q) \mid v \text{ is a node of } q\}$$

The hanging nodes are removed by inserting appropriate templates from the list shown in Figure 21.27. The insertion is done successively for the quadrants with level 0, 1, 2, … . The nodal subdivision levels help in finding the correct template. Consider an arbitrary quadrant with level $l(q)$: the nodes v of q with $l(v) > l(q)$ are marked (Figure 21.28a, $l = 2$). The configuration of the marked nodes uniquely determines the template that must be inserted into q. Figure 21.28a shows the result after all quadrants with level 2 have been processed (the templates 1 and 2a were used).

The newly generated nodes and faces are assigned the level $l(q) + 1$. The procedure is then repeated until no hanging nodes are left (Figure 21.28b):

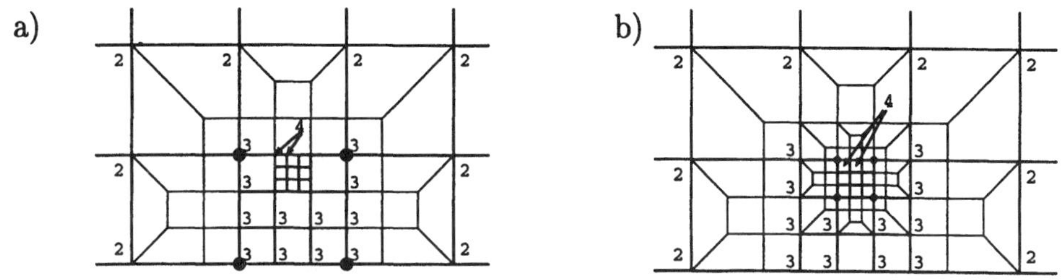

FIGURE 21.28 Construction of the conforming closure.

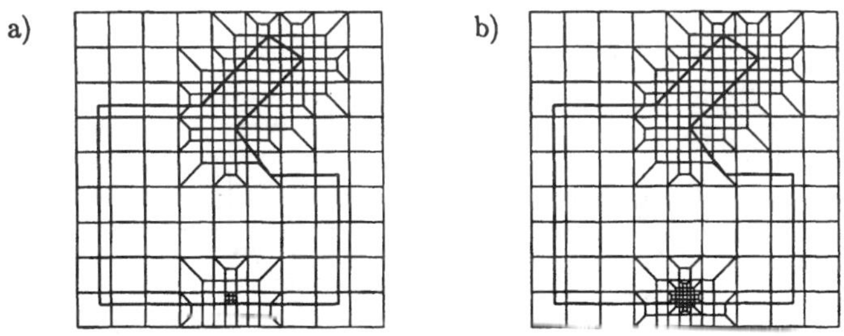

FIGURE 21.29 Generation of the conforming closure.

```
procedure conforming_closure

for l = 0 to maximum_level
  for all quadrants q with level l
    mark nodes v with level (v) > l;
    insert appropriate template;
    set new levels;
```

The choice of templates guarantees that the process results in a conforming mesh. An edge is

- Split into three, if both nodes are marked.
- Split into two, if one of its nodes is marked.
- Not split, if no node is marked.

Only those elements with a perfect shape may be split up recursively, and it can be shown that the minimum angle in the mesh does not depend on the refinement level [Schneiders 1996c]. Figure 21.29 shows the situation after applying the conforming closure to level 3 and level 4 quadrants.

Boundary fitting of the mesh can be done by using either the projection or the isomorphism technique; a short review of the latter one will be given here (see [Schneiders 1996b] for the details). A subset of the conforming quadrilateral element mesh is selected as the initial mesh (Figure 21.30a). This is not as straightforward as for the grid-based algorithm: care must be taken that the distance of each boundary edge *e* to the object boundary roughly equals the edge length (if this condition is not respected, elements with unacceptable aspect ratios may be generated).

One can then construct normals for the boundary nodes of the initial mesh, generate mesh nodes on the object boundary and construct elements in the boundary region (Figure 21.30b). The mesh is then optimized, in a manner similar to grid-based mesh generation.

The 3D algorithm follows the same line. For ease of explanation, we choose as an example a block where we want a very fine mesh at one location on the boundary (Figure 21.31). First, a three-level octree is constructed. Octant and node levels are then computed as in the 2D algorithm.

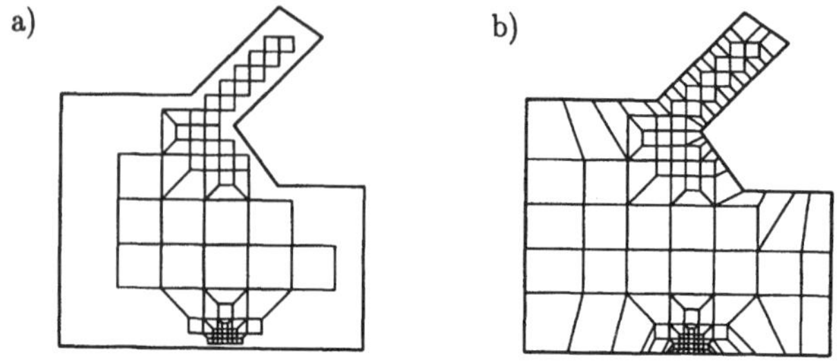

FIGURE 21.30 Adapting the mesh to the boundary.

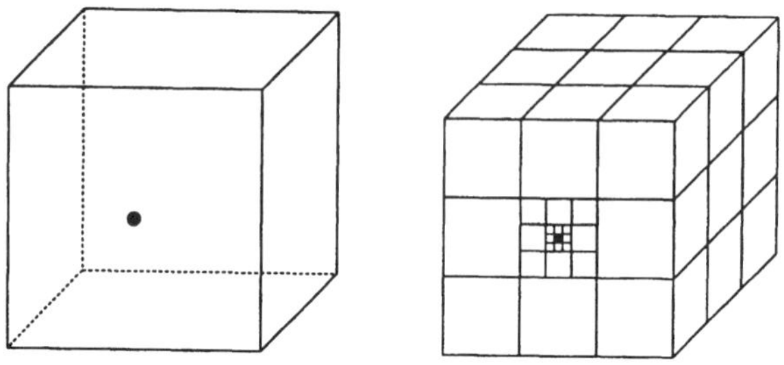

FIGURE 21.31 Three-dimensional refinement requirement.

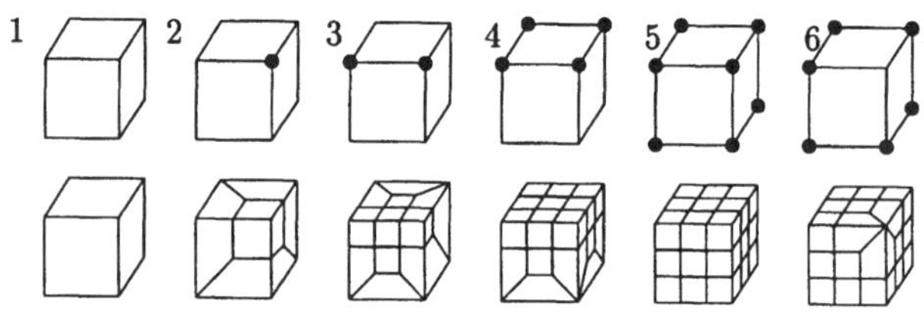

FIGURE 21.32 Selected 3D-templates.

As in 2D, the problem to be solved is the construction of the conforming hull. This is done by inserting appropriate templates into the octree structure. In 3D, a total of 22 templates are needed; Figure 21.32 shows a selection. The templates are constructed by applying the 2D templates (Figure 21.27) to the octant faces — this guarantees that the process results in a conforming mesh. In this way, the problem of how to find the conforming hull is reduced to finding volume meshes for these templates.

In the example in Figure 21.31, the templates 1, 2, 3, and 4 from the list are needed (this set forms an important subset, templates for "convex" refinement specifications). The solution for template 2 is similar to the splitting operation in Figure 21.20; template 3 is more complex. Template 4 may look confusing at first glance, but is easier to understand if its construction is done in two steps. A sweep with face template 2a is used in one direction, the three newly generated hexahedra at the face to be refined are split in the same way but in the opposite direction. Note that the new elements at the marked nodes have perfect shape, so that they can be refined further without reducing the smallest angle in the mesh. Both the isomorphism and projection techniques can be used to fit the mesh to the object boundary [Schneiders 1996b].

Unfortunately, the proposed method does not work in every case. Template 6 in Figure 21.32 is the weak point: it has a total of 55 quadrilateral faces on the surface. According to [Mitchell 1996], a hexahedral element mesh has an even number of boundary faces, so a mesh that "fits into" template 6 cannot exist.

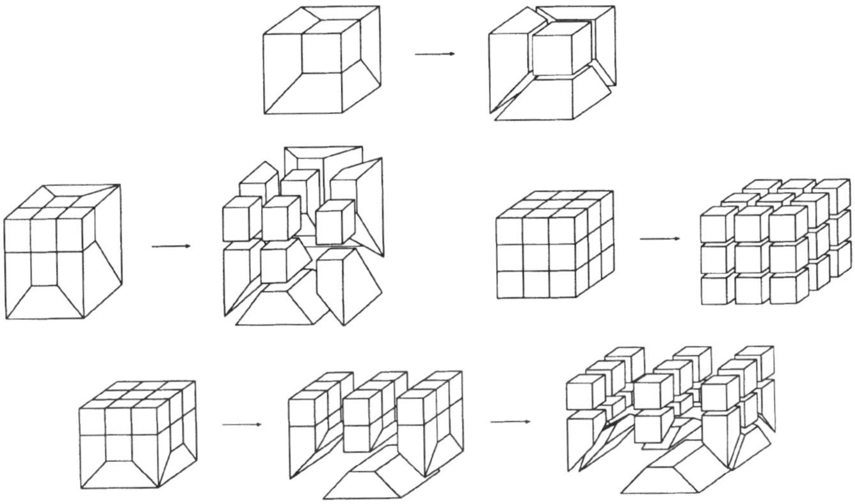

FIGURE 21.33 3D templates — details.

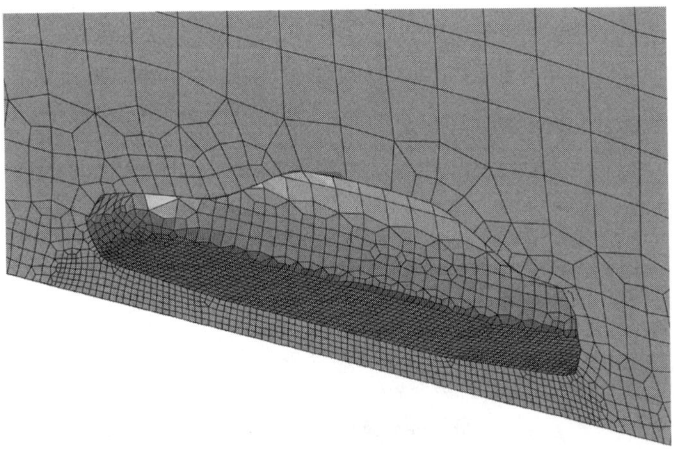

FIGURE 21.34 Hexagedral element mesh for the simulation of flow around a car.

The algorithm in the form presented here can only be applied to a limited set of problems, "convex" refinement specifications. In practice, even the limited set of templates is useful, if the region where a fine mesh is needed is relatively small. Further, there exist two workarounds for the problem, level propagation and buffer layer insertion [Schneiders 1996b].

If one accepts hanging nodes in the mesh, finding the confinement hull is not necessary. This removes one obstacle, but makes boundary adaptation more difficult. Algorithms of this type were developed by Smith [1996], who uses the isomorphism technique for body fitting, and by Tchon [1997] who uses the projection method. The algorithms are implemented in Fluent Inc.'s Gambit and NUMECA's IGG/Hexa preprocessor.

Octree-based meshing without hanging nodes, based on the standard octree structure, is complicated by the fact that the transitioning cannot be localized as in the case of the 1-27-octree. This problem is treated in [Schneiders 1998]. The paper also presents a new approach to deal with the conforming hull problem. Figure 21.34 shows part of a mesh that has been generated for the simulation of flow around a car.

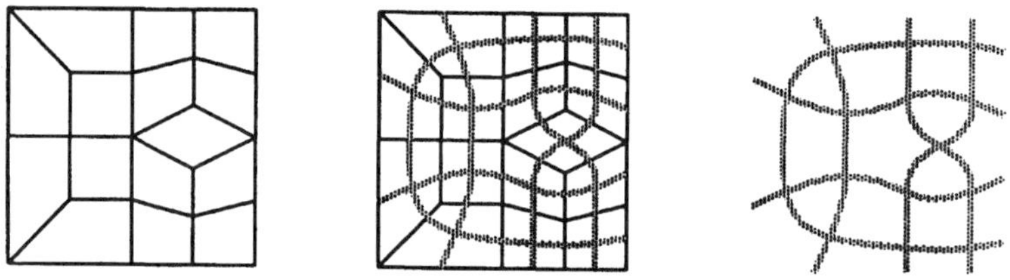

FIGURE 21.35 Quadrilateral mesh and spatial twist continuum.

The grid- and octree-based algorithms presented here prove that the superposition principle is an algorithmic tool to successfully deal with the hex meshing problem. They are, however, not the only methods of choice; combinations with the other methods outlined in this chapter seem promising. Further research may reveal the full potential of superposition methods.

21.4 The Spatial Twist Continuum

The techniques presented so far can also be used for the generation of tetrahedral element meshes. In contrast to that, the spatial twist continuum is a unique concept for quadrilateral and hexahedral element mesh generation.

The results presented here were mainly achieved by the CUBIT team, a joint research group at Sandia National Laboratories and Brigham Young University that has been working on quadrilateral and hexahedral element meshing since the beginning of the 1990s. The group is working on algorithms that generate a mesh starting from discretization of the object surface into quadrilaterals. As part of their research, the paving [Blacker and Stephenson 1991] and plastering [Blacker 1993] advancing-front type mesh generators have been developed. These algorithms are described in section 21.5; here we will describe other results.

Given an unstructured quadrilateral element mesh $M = (V, E, F)$, the *spatial twist continuum* (STC) [Murdock et al. 1997] $M' = (V', E', F')$ is defined as follows:

- For each face $f \in F$, the midpoint v' is a node of V'.
- For each edge $e \in E$,we define an edge $e' = (v'_1, v'_2) \in E'$ where v'_1 and v'_2 are the midpoints of the two quadrilaterals that share e.

For each node $v \in V$, a face $f' \in F'$ is defined by the midpoints of the adjacent quadrilaterals. The STC is the combinatorial dual [Preparata and Shamos 1985] of the quadrilateral mesh.

Figure 21.35 shows a quadrilateral mesh and the corresponding STC (gray lines). The edges of the STC are displayed not as straight lines but as curves. This allows the recognition of chords, a very important structure: one can start at a node, follow an edge e_1 to the next node, then choose the edge e_2 "straight ahead" that is not adjacent to e_1, continue to the next edge e_3 and so on. The sequence e_1, e_2, ... forms a chord (displayed as a smooth curve in Figure 21.35). Chords can be closed or open curves and can have self-intersections, and a chord corresponds to a row of quadrilaterals in the mesh.

By definition, an STC corresponds to a quadrilateral element mesh. Thus, in order to generate a quadrilateral mesh, one can just construct an STC by arranging a set of chords, and then generate the mesh by constructing the dual. The problem with this strategy is that elements with unacceptable shape can be constructed. Figure 21.36 gives an example: two chords intersect twice, and the quadrilaterals corresponding to the intersection points have two edges in common. One can overcome this problem by adding a chord to the STC. This corresponds to the insertion of an additional row of elements into the mesh, and the degeneracy is resolved.

A quadrilateral mesh that respects a given boundary discretization can be constructed by first inserting chords that connect the boundary segments, then adding chords to resolve the degeneracies.

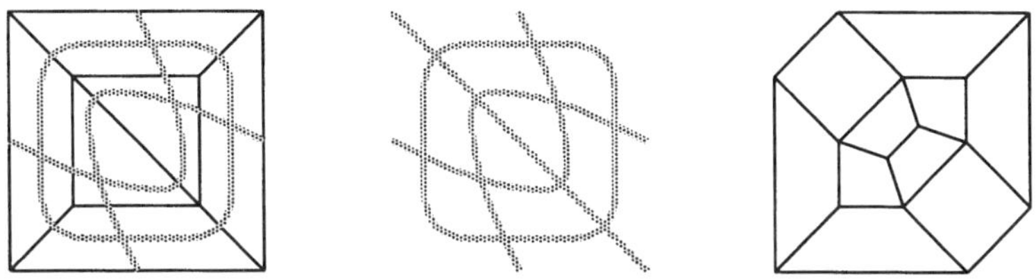

FIGURE 21.36 Removing a degeneracy from a quadrilateral mesh.

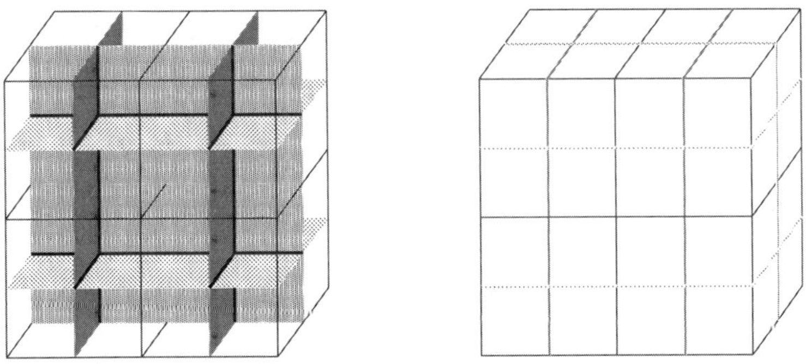

FIGURE 21.37 STC for a mesh of four hexahedra and corresponding surface STC.

The STC is a very good construct to analyze and improve mesh generation algorithms, and the idea can also be used in 3D. This was noticed by the CUBIT team and led to important theoretical results [Mitchell 1996] and to the whisker weaving algorithm [Tautges 1996], which will be described in the following.

As in 2D, the STC is the combinatorial dual of a hexahedral element mesh. Figure 21.37 shows an example: The midpoints of the hexahedra are the nodes of the STC, each pair of adjacent hexahedra gives an edge, and the set of hexahedra that have an edge in common defines a face of the STC. As in 2D, one can identify chords that correspond to rows of elements; in the example of Figure 21.37 all chords start and end at the mesh boundary, but there may also be cyclic chords in an STC. The faces of the STC can be combined to a sheet that corresponds to a layer of elements. A chord is defined by the intersection of two sheets or by a self-intersecting sheet. Vertices of the STC that correspond to hexahedra are defined by the intersection of three chords or three sheets (or less in case of self-intersections).

Basically, the STC is a set of intersecting sheets, and "dualizing" the STC gives a hexahedral mesh. From this, it is clear that it is difficult to apply a local change to a hexahedral mesh, since that is equivalent to a modification of the STC. The only operations allowed for an STC are the insertion or deletion of a sheet, and both will likely have a global effect.

These ideas can be applied to the construction of a hexahedral mesh from a surface discretization. Given a hexahedral element mesh, the STC of its surface mesh matches the intersection of the hexahedral mesh's STC with the surface: An intersection of a sheet with the surface is a chord of the surface STC. These intersections are called *loops*. Figure 21.38 shows an example surface mesh and the corresponding four loops. The generation of a hexahedral mesh is thus an inverse problem and can be solved as follows:

- Generate the surface STC.
- For each loop, construct a sheet whose intersection with the surface matches that loop.
- Add sheets in the interior to remove degeneracies.
- Dualize to get the hexahedral mesh.

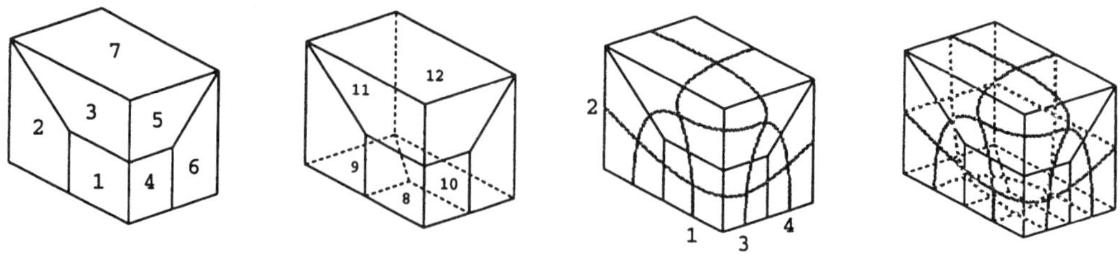

FIGURE 21.38 STC of a surface mesh.

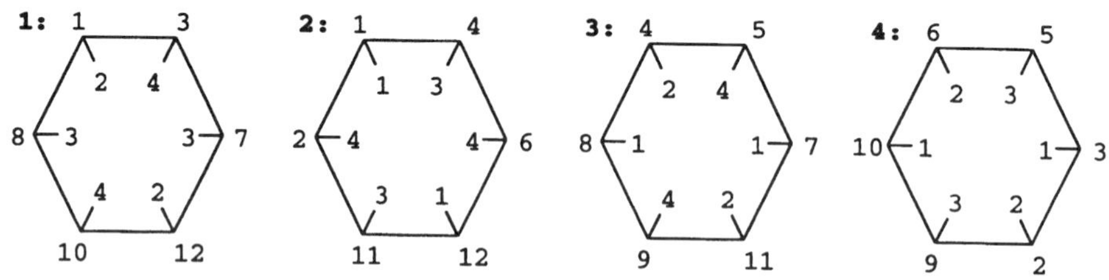

FIGURE 21.39 Initial set of whisker sheets.

In [Mitchell 1996], it is proved that, given a surface discretization with an even number of elements, a hexahedral element exists. Mitchell shows that an STC that respects all constraints can be generated by inserting sheets into the original STC. His proof is, however, nonconstructive, since he does not give an algorithm for the construction of the first STC. This is done by the whiskerweaving algorithm [Tautges 1996], which is described in the following.

The first step of the algorithm is the initialization of *whisker sheet diagrams.* A whisker sheet corresponds to a sheet of the STC to be constructed, so there is one whisker sheet for each loop. Figure 21.39 shows the whisker sheets for the loops of Figure 21.38. The vertices of a sheet correspond the faces that the loop intersects, and are labeled outside by the face numbers. Since the faces also correspond to the intersection of two loops, the vertices are labeled inside with the number of the intersecting loop (whisker sheet).

The next step in whisker weaving is the formation of a hex by crossing three chords on three sheets. Two sheets correspond to two chords on the third sheet, and it is required that the chords start at adjacent faces. The chords are pairwise crossed and define three vertices which correspond to the same STC vertex (hexahedron).

In the example of Figure 21.39 the sheets 1, 2, and 3 have been selected, Figure 21.40a shows the result. By duality, this step is equivalent to the construction of a hexahedron at the faces 1, 4, and 8 (Figure 21.41a).

Next the sheets 2, 3, and 4 that correspond to the chords starting at the faces 2, 9, and 11 are selected. The result is shown in Figure 21.40b and is equivalent to the construction of another hexahedron (Figure 21.41b). Obviously, glueing the hexahedra is the next step (Figure 21.40c), which is equivalent to joining the chords 2 and 3 in the sheets 2 and 3 (Figure 21.40c).

In the following the chords corresponding to the faces 3, 5, 7 and 6, 10, 12 are joined. The gluing operation completes the construction of the whisker sheet (Figure 21.42). Having dualized the STC, one gets the mesh shown in Figure 21.38d.

The dualizing process does not always result in a valid hex mesh. Hexahedra with more than two faces in common may be present in the mesh or invalid elements may be constructed if their base faces are nearly coplanar. Mitchell [1996] identifies 7 constraints an STC must fulfil in order to guarantee that dualizing results in a valid hex mesh. This is done by inserting additional sheets into the STC, see [Mitchell 1996] for the details.

It proved to be very difficult to derive a stable version of the whisker weaving algorithm. If the surface STC has self-intersections, it may be nearly impossible (the STC corresponding to the surface in

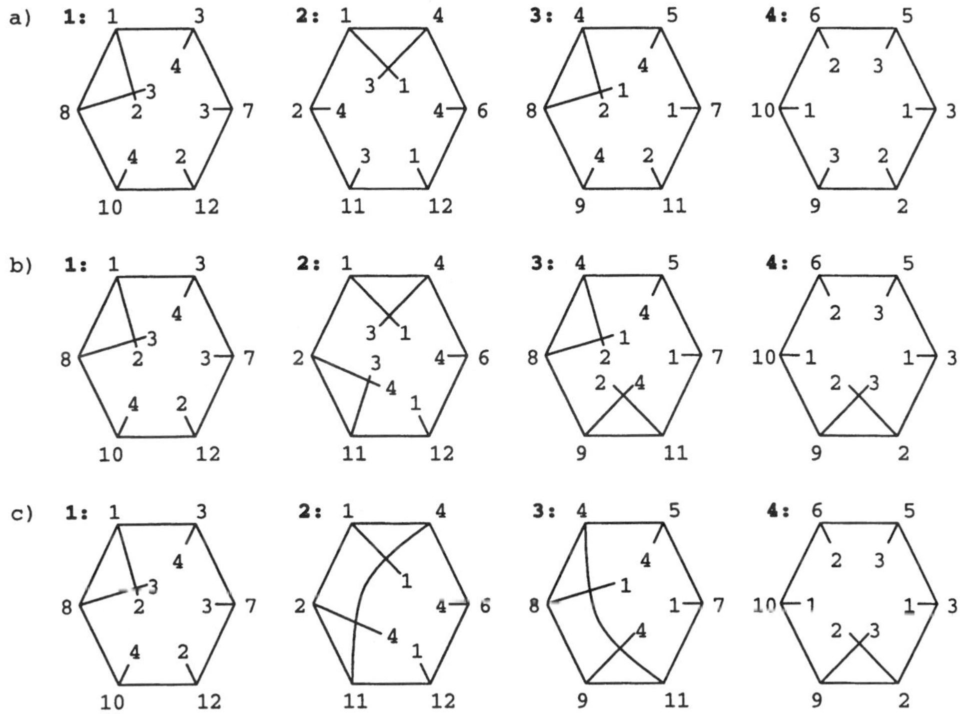

FIGURE 21.40 Whisker weaving: example.

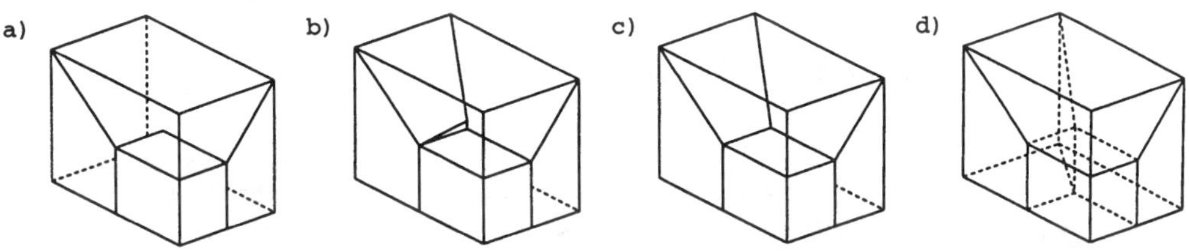

FIGURE 21.41 Dualized view of whisker weaving.

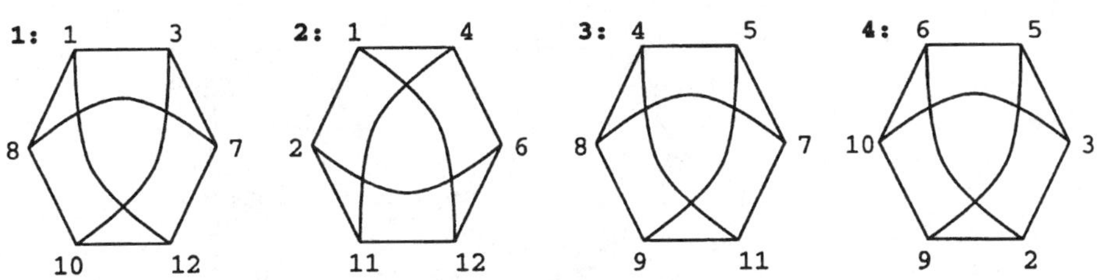

FIGURE 21.42 Resulting whisker sheets.

Figure 21.2 consists of two loops, one of them with 8 self-interactions — whisker weaving done "by hand" is very difficult). This is probably due to the fact that the algorithm is quite indeterministic and relies more on topological than on geometrical information.

Compared to whisker weaving, block-decomposition and superposition methods are easier to realize, since they are not constrained by a given surface discretization. Algorithmic complexity is the price one

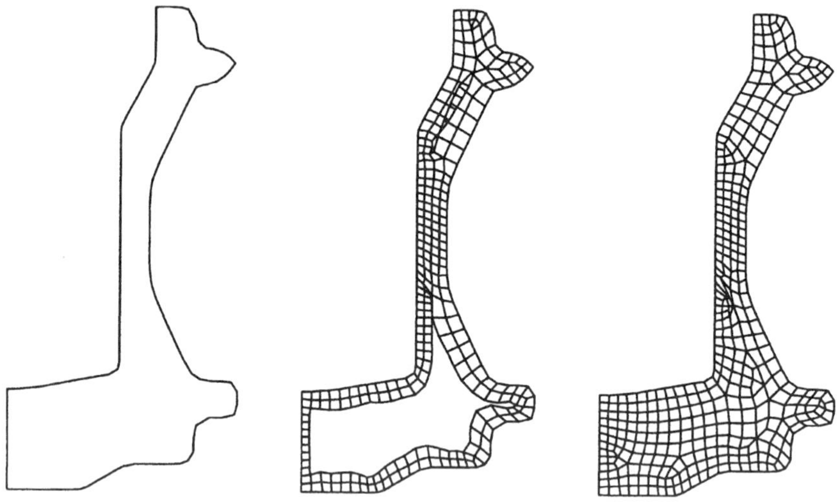

FIGURE 21.43 Paving algorithm.

has to pay for the potential benefit of the whisker weaving algorithm, and it is still a subject of research. Nevertheless, the concepts presented in this chapter give much insight into the nature of hexahedral mesh generation, and the techniques are useful in enhancing block-decomposition of superposition type algorithms (the templates in Figure 21.32 were constructed by using the STC concept).

21.5 Other Approaches

Advancing-front type methods are very popular for the generation of tetrahedral element meshes. First, a mesh of triangles is generated for the surface, then a volume mesh is generated layer by layer. This allows the control of mesh quality near the boundary; internal faces can be represented in the mesh, and the method can be parallelized.

An advancing-front type algorithm for the generation of quadrilateral element meshes was proposed in [Blacker and Stephenson 1991]. Figure 21.43 gives an idea of how it works: starting from a boundary discretization, the interior is "paved" with quadrilaterals layer by layer. If the layers overlap, a "seaming" procedure is invoked, and the procedure is repeated until the remaining cavities have been filled.

The paving algorithm is probably the best mesh generator for quadrilateral element meshes. It generates meshes of high-quality elements with an acceptable number of elements. Irregular nodes (interior nodes with other than four elements adjacent) are more likely to be found in the interior than close the boundary. The algorithm is very complex and not easy to implement.

An attempt to develop a three-dimensional version, the plastering algorithm, has been made in the CUBIT project. Starting from a quadrilateral discretization of the object boundary, layers of hexahedral elements are generated in the volume (Figure 21.44). The number of different cases to be considered is far greater than in the 2D case. Degenerated elements (wedges) are removed by propagating them through the mesh [Blacker 1993].

Unfortunately, it turned out to be impossible to develop a robust algorithm. The problems arise when two fronts intersect. In the 2D case, one can glue ("seam") the fronts if the change in element size is not too large. This is not sufficient in 3D, since these surface meshes must be isomorphic in order to seam them. This condition is unlikely to hold for practical problems. Another problem is that the cavities cannot be meshed in every case. Figure 21.2 shows an example (the cavity can be reduced to an octahedron, a nonmeshable object in the sense that a valid hex mesh that matches the octahedron surface has not yet been found).

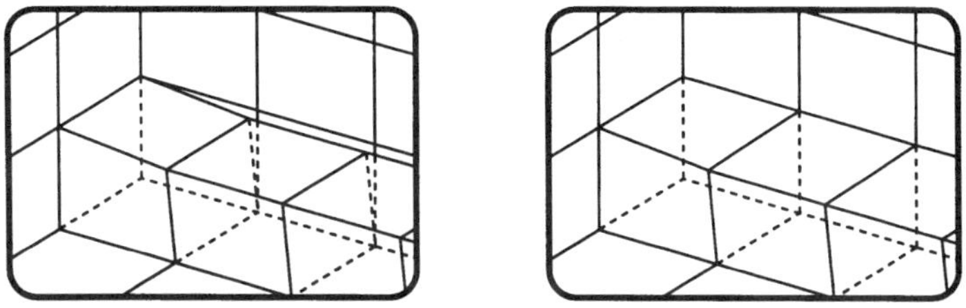

FIGURE 21.44 Plastering algorithm.

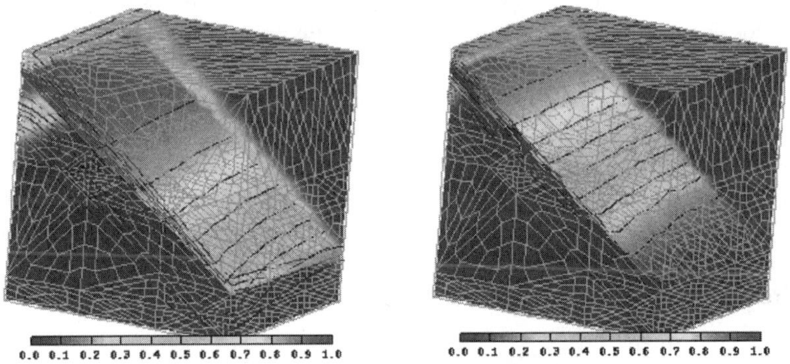

FIGURE 21.45 Mesh for a geological structure with internal boundaries.

It turns out that generating a hex mesh from a surface discretization is hard to realize if the decisions are made purely on local information. So the original idea was rejected, and global information was incorporated using the concept of the dual described in section 20.4.

An algorithm for the generation of hexahedral element meshes for very complicated domains (geological structures with internal boundaries) was proposed by Taniguchi [1996]. His approach is similar to Armstrong's algorithm in that he decomposes the domain into simple subvolumes (tetrahedra, pentrahedra, etc.) that are then meshed separately. The method is based on Delaunay triangulation, and therefore can be applied for arbitrary convex domains that consist of a set of convex subdomains that are surrounded by fracture planes. Figure 21.45 shows a mesh generated for the simulation of groundwater flow; for simulations like this it is very important that the boundaries between different layers of material are present in the mesh.

A similar method for hexahedral element meshing of mechanical parts was proposed by Sakurai [Shih and Sakurai 1996] (volume decomposition method). Also notable is the work of Shang-Sheng Liu [Liu 1996]; he tries to integrate the mesh generation into a solid modeling environment, an approach that is attractive particularly for mechanical engineering CAD systems.

So far we have concentrated on meshing strategies that can be applied both in two and three dimensions. There are, however, strategies for quadrilateral element mesh generation that cannot be extended to the 3D case. Two of these shall be discussed briefly.

The block decomposition approach used by Armstrong poses far fewer problems in 2D. Whereas in 3D one must take care to generate subvolumes that can be split up into hexahedra, this is not really a problem in 2D, since every polygon with an even number of edges can be meshed with quadrilateral elements. So, there is much more room for finding a good partitioning strategy. An algorithm of this type is describe by Nowottny [1997] (Figure 21.46). First the holes of the polygon to be meshed are removed by connecting them to the outer boundary. Then appropriate cuts are inserted until sufficiently small subregions have been generated. These are then meshed directly.

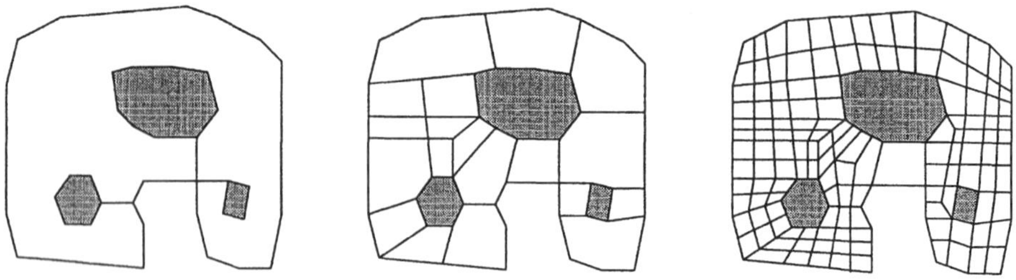

FIGURE 21.46 Quadrilateral mesh generation by geometrically optimized domain decomposition.

This strategy works in 2D since a sufficiently large polygon can always be split into two meshable subpolygons. This does not hold in 3D (Figure 21.2), and thus an extension of this strategy seems unlikely to be realized.

Another approach for the generation of quadrilateral element meshes was proposed in [Shimada 1994]: first one generates a triangular mesh with an even number of elements; pairs of triangles are then combined to quadrilaterals until no triangles remain.

This approach is very elegant, especially since it allows the use of the work done on triangulation algorithms. Obtaining graded meshes or meshes for geometries with internal boundaries is especially straightforward using this approach. Unfortunately, it cannot be used for 3D, since combining tetrahedra into hexahedra is not possible except for tet meshes with a very regular structure.

21.6 Software and Online Information

In the followings a selection of quadrilateral or hexahedral element mesh generators is given. We restrict ourselves to unstructured mesh generation. Information on multiblock-based systems can be found elsewhere in this handbook or on the web page "Mesh Generation and Grid Generation on the Web" [Schneiders 1996d], which has links to all programs described here.

HEXAR (Cray Research, http://www.cray.com/products/applications/directory/codes /HEXAR.html)
A grid-based mesh generator, available for Cray's parallel machines.

FAM4 (FEGS Ltd, http://www.fegs.co.uk/FAM_products.html)
Medial-surface based mesh generator, an implementation of C. Armstrongs ideas.

KUBRIX (Simulation Works, Inc. http://www.siw.com, http://kubrix.com)
This mesh generator uses a fuzzy logic-based block decomposition method.

Houdini (Algor Inc., http://www.algor.com/houdini/homepage.htm)
This mesh generator uses an advancing front technique for the generation of tetrahedral and hexahedral element meshes.

FAME (AVL-LIST GmbH, http://www.avl.co.at/html/11.htm)
Preprocessor for the FIRE cfd code, comes with a grid-based mesh generator.

PEP (http://www.rwth-aachen.de/ibf/)
A preprocessor for the simulation of metal forming processes, with 2D and 3D grid-based mesh generators [Schneiders 1996a] and an implementation of the paving algorithm [Blacker and Stephenson 1991].

CUBIT (http://sass577.endo.sandia.gov/SEACAS/CUBIT_sw/Cubit.html)
A tool for the generator of hexahedral element meshes: whisker weaving and mapped meshing.

ICEM CFD Hexa (http://www.icemcfd.com/hexa.html)
A automatic hex mesher, part of a powerful preprocessor for cfd applications.

IGG/Hexa (http://stro5.vub.ac.be/pub/numeca/numeca.html)
Numeca's octree-based hex mesher: Body-fitted meshes with hanging nodes.

Cooper Tool/Gambit (http://www.fluent.com)
Fluent's semiautomatic hex mesher and the octree-based automatic unstructured hex mesher (hanging nodes are allowed).

Many groups who are active in the field have online information (cf. [Schneiders 1996d]):

http://web.cs.ualberta.ca/~barry/
Barry Joe, University of Alberta: Author of the GEOMPACK tet mesher, is now working on algorithms for the generation of hex meshes.

http://caor.ensmp.fr/Francais/Personnel/Mounoury/Francais/Introduction.html
Valéry Mounoury, CAOR (Paris), uses semantical analysis of volumes as a starting point for mesh generation.

http://www-users.informatik.rwth-aachen.de/~roberts/index.html
Grid-based and octree-based hex meshing at the Techical University of Aachen.

http://www.rwth-aachen.de/ibf/
Automatic remeshing for the simulation of metal forming processes.

http://www.unibw-hamburg.de/MWEB/ikf/fft/home_e.html
The Institute for Production Technology at the University of Hamburg develops a grid-based mesh generator for the preprocessing of the simulation of metal-forming processes.

http://daimler.me.metu.edu.tr/users/tekkaya/
A. Tekkaya, Middle East Technical University (Ankara), uses a grid-based mesh generator for the simulation of metal-forming processes.

http://www.inria.fr/Equipes/GAMMA-eng.html
The GAMMA project at INRIA (France, director: Paul-Louis George) has an outstanding record in algorithms for tet meshing and is also considering the hex meshing problem.

http://sog1.me.qub.ac.uk/femgroup.html
Web server for the finite element group at Queens University in Belfast (C. Armstrong, medial-surface tools).

http://www.et.byu.edu/~cubit/
CUBIT is a joint project of SANDIA National Laboratories and Brigham Young University, sponsored by the Department of Energy and an industrial consortium, working on advancing-front and whisker-weaving methods.

http://smartcad.me.wisc.edu/~shang-sh/homepage.html
Shang-Sheng Liu, University of Wisconsin, works on hex meshing in a solid modeling environment.

http://swhite.me.washington.edu/~cdam/PEOPLE/HAO/hao.html
Information on Hao Chen's thesis work at the University of Washington, medial-surface-based algorithm.

http://www.lance.colostate.edu/~hiroshi/mesh.html
A description of the advancing-layer mesh generator developed at Colorado State University.

Acknowledgment

This work benefitted from the support of the following people: C. Armstrong, T. Taniguchi, and D. Nowottny contributed some of the figures. M. Schneider helped in translating the text. The author wishes to thank them for their help.

References

1. Benzley, S. E., Perry, E., Merkley, K., Clark, B., and Sjaardema, G. A comparison of all hexagonal and all tetrahedral finite element meshes for elastic and elastic-plastic analysis, *Proc. 4th International Meshing Roundtable*, Sandia National Laboratories, Albuquerque, NM, pp 179–192, 1995.
2. Blacker, T.D. and Stephenson, M.B., Paving: a new approach to automated quadrilateral mesh generation, *Int. J. Num. Meth. Eng.* 32, pp 811–847, 1991.
3. Blacker, T.D. and Meyers, R.J., Seams and wedges in plastering: a 3D hexahedral mesh generation algorithm, *Engineering with Computers*, 9, pp 83–93, 1993.
4. Brodersen, O., Hepperle, M., Ronzheimer, A., Rossow, C.-C., and Schöning, B., The parametric grid generation system megacads, *Proc. 5th Int. Conf. on Numerical Grid Generation Computational Field Simulations.* Soni, B.K., Thompson, J.F., Häuser, J., Eiseman, P., (Eds.), NSF, Mississippi, pp 353–362, 1996.
5. George, P.L., *Automatic Mesh Generation: Applications to Finite Element Methods*, John Wiley & Sons, 1991.
6. Holmes, D., Generalized method of decomposing solid geometry into hexahedron finite elements, *Proc. 4th International Meshing Roundtable*, Sandia National Laboratories, pp 141–152, 1995.
7. Ives, D., geometric grid generation. surface modeling, grid generation, and related issues in computational fluid dynamic (CFD)Solutions, *Proc. NASA-Conference*, Cleveland, OH, NASA CP-3291, 1995.
8. Knupp, P. and Steinberg, S., *Fundamentals of Grid Generation.* CRC Press, Boca Raton, FL.
9. Liu, S.-S. and Gadh, R., Basic logical bulk shapes (blobs) for finite element hexahedral mesh generation, *Proceedings 5th International Meshing Roundtable*, 1996.
10. Mitchell, S.A., A characterization of the quadrilateral meshes of a surface which admit a compatible hexahedral mesh of the enclosed volume, *Proceedings STACS '96.* Grenoble, 1996.
11. Möhring, R., Müller-Hannemann, M., and Weihe, K. Using network flows for surface modeling, *Proceedings of the Sixth Annual ACM-SIAM Symposium on Discrete Algorithms*, pp 350–359, 1995.
12. Murdock, P., Benzley, S.E., Blacker, T.D., and Mitchel, S.A., The spatial twist continuum: a connectivity based bethod for representing and constructing all-hexahedral finite element meshes, *Finite Elements in Analysis and Design*, 28, pp 137–149, 1997.
13. Nowottny, D., Quadrilateral mesh generation via geometrically optimized domain decomposition, *Proc. 6th International Meshing Roundtable*, Park City, UT, pp 309–320, 1997.
14. Owen, S., Meshing research corner, Literaturdatenbank, URL, 1996. http://www.ce.cmu.edu/~sowen/mesh.html
15. Preparata and Shamos, *Computational Geometry: An Introduction.* Springer Verlag, NY, pp 24–26, 1985.
16. Price, M.A., Armstrong, C.G., and Sabin, M.A., Hexahedral mesh generation by medial axis subdivision: I. Solids with convex edges, *Int. J. Num. Meth. Eng.*, 38, pp 3335–3359, 1995.
17. Price, M.A. and Armstrong, C.G., Hexahedral mesh generation by medial axis subdivision: ii. solids with flat and concave edges, *Int. J. Num. Meth. Eng*, 40, pp 111–136, 1997.
18. Sabin, M., Criteria for comparison of automatic mesh generation methods, *Adv. Eng. Softw*, 13, pp 220–225., 1991.
19. Schneiders, R., A Grid-based algorithm for the generation of hexahedral element meshes, *Engineering with Computers*,12, pp 168–177, 1996a.

20. Schneiders, R., Schindler, R., and Weiler, F., Octree-based generation of hexahedral element meshes, *Proc. 5th International Meshing Roundtable*, Sandia National Laboratories, pp 205–216, 1996b.
21. Schneiders, R., Refining quadrilateral and hexahedral element meshes, *Proc. 5th International Conference on Numerical Grid Generation in Computational Field Simulations*, pp 699–708, 1996c.
22. Schneiders, R., Mesh generation and grid generation on the Web, http://www-users.informatik.rwth-aachen.de/~roberts/meshgeneration.html., 1996d.
23. Schneiders, R., Octree-based hexahedral mesh generation, to appear in *Journal of Computational Geometry and Applications*, special issue on mesh generation, 1998.
24. Shewchuk, J.R., A condition guaranteeing the existence of higher-dimensional constrained Delaunay triangulations, submitted to the Fourteenth Annual Symposium on Computational Geometry, 1998.
25. Shih, B.-Y. and Sakurai, H., Automated hexahedral mesh generation by swept volume decomposition and recomposition, *Proc. 5th Int. Meshing Roundtable.* 1996.
26. Shimada, K. and Itoh, T., Automated conversion of 2D triangular meshes into quadrilateral meshes, *Proc. Int. Conf. on Computational Engineering Science*, 1994.
27. Smith, R.J. and Leschziner, M.A., A novel approach to engineering computations for complex aerodynamic flows, *Proc. 5th Int.Conf. on Numerical Grid Generation in Computational Field Simulations*, pp 709–716, 1996.
28. Taghavi, R., Automatic, parallel and fault tolerant mesh generation from CAD on Cray Research Supercomputers, *Proc. CUG Conf.* Tours, France, 1994.
29. Tam, T.K.H. and Armstrong, C.G., Finite element mesh control by integer preprogramming, *Int. J. Num. Meth. Eng.*, 36, pp 2581–2605, 1993.
30. Taniguchi, T., New concept of hexahedral mesh generation for arbitrary 3D domain — block degeneration method, *Proc. 5th Int. Conf. on Numerical Grid Generation in Computational Field Simulations*, pp. 671–678, 1996.
31. Tautges, T.J. and Mitchell, S., Progress report on the whisker weaving all-hexahedral meshing algorithm, *Proc. 5th Int. Conf. on Numerical Grid Generation in Computational Field Simulations*, pp 659–670, 1996.
32. Tchon, K.-F., Hirsch, C., and Schneiders, R., Octree-based hexahedral mesh generation for viscous flow simulations, *Proc. 13th AIAA Computational Fluid Dynamics Conference*, Snowmass, CO, 1997.
33. Thompson, J.F., Warsi, Z.U.A., and Mastin, C.W., *Numerical Grid Generation: Foundations and Applications.* North-Holland, 1985.
34. Turkkiyah, G.M., Ganter, M.A., Storti, D.W., and Chen, H., Skeleton-based hexahedral finite element mesh generation of general 3D solids, *Proc. 4th Int. Meshing Roundtable*, Sandia National Laboratories, late addition, 1995.

22

Adaptive Cartesian Mesh Generation

Michael J. Aftosmis

Marsha J. Berger

John E. Melton

22.1 Introduction

The last decade has witnessed a resurgence of interest in Cartesian mesh methods for CFD. In contrast to body-fitted structured or unstructured methods, Cartesian grids are inherently *non-body-fitted*; i.e., the volume mesh structure is independent of the surface discretization and topology. This characteristic promotes extensive automation, dramatically eases the burden of surface preparation, and greatly simplifies the reanalysis processes when the topology of a configuration changes. By taking advantage of these important characteristics, well-designed Cartesian approaches virtually eliminate the difficulty of grid generation for complex configurations. Typically, meshes with millions of cells can be generated in minutes on moderately powerful workstations [1, 2].

As the name suggests, Cartesian non-body-fitted grids use a regular, underlying, Cartesian grid. Solid objects are carved out from the interior of the mesh, leaving a set of irregularly shaped cells along the surface boundary. Since most of the volume mesh is completely regular, highly efficient and accurate finite volume flow solvers can be used. All the overhead for the geometric complexity is at the boundary, where the Cartesian cells are cut by the body. This boundary overhead is only two-dimensional, with typically 10–15% of the cells intersecting the body. Fundamentally, Cartesian approaches exchange the case-specific problem of generating a body-fitted mesh for the more general problem of intersecting hexahedral cells with a solid geometry. Fortunately, the geometry and mathematics of this problem have been thoroughly studied, and robust algorithms are available in the literature of computational geometry and computer graphics [25,53,38,41].

Although Cartesian grid methods date back to the 1970s, it was only with the advent of adaptive mesh refinement (AMR) that their use became practical [11]. Without some provision for grid refinement,

0-8493-2687-7/99/$0.00+$.50

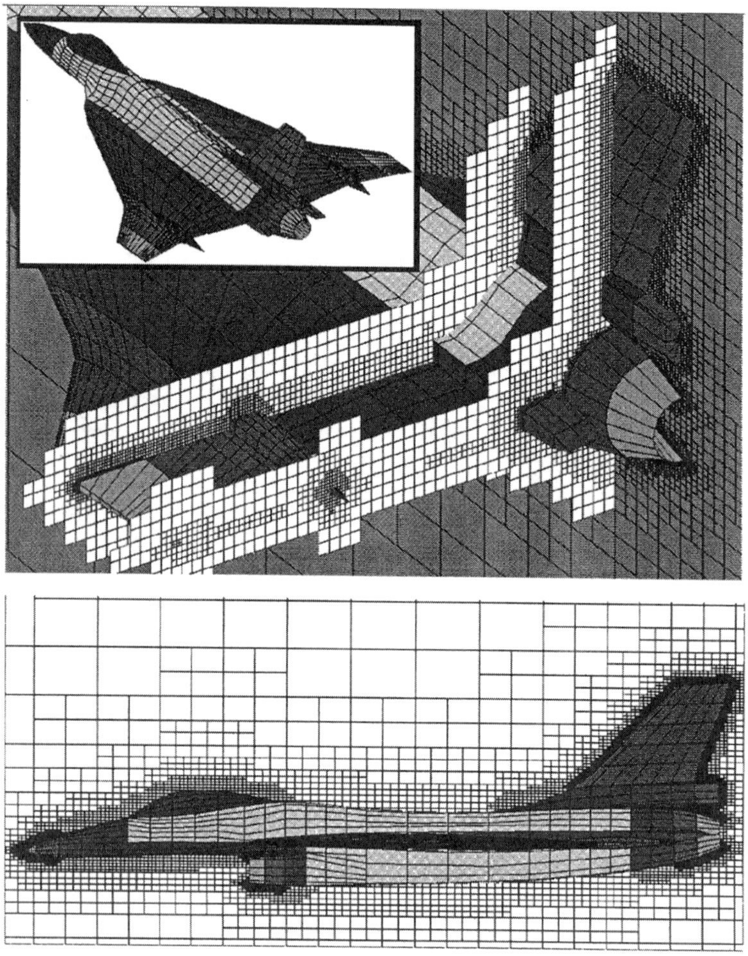

FIGURE 22.1 Cartesian grid for an F16Xl.

Cartesian grids would lack the ability to efficiently resolve fluid and geometry features of various sizes and scales. This resolution is readily incorporated into structured meshes via grid point clustering. Many algorithms for automatic Cartesian grid refinement have, however, been developed in the last decade, largely alleviating this shortcoming. Figure 22.1 illustrates a typical grid with refinement for discretizing the flow around the General Dynamics F16XL.

Early work with Cartesian grids used a staircased representation of the boundary. In contrast, modern Cartesian grids allow planar surface approximations at walls, and some even retain subcell descriptions of the boundary within the body-intersected cells. Obviously, this additional complexity places a greater burden on the flow solver, and recent research has focussed on developing numerical methods to accurately integrate along the surface boundaries of a Cartesian grid [3, 8, 9, 19, 26, 27]. The most serious current drawback of Cartesian grids is that their use is restricted to inviscid or low Reynolds number flows [28, 20]. An area of active research is their coupling to prismatic grids (see [11, 30, 36, 54, 50]) or other methods for incorporating boundary layer zoning into the Cartesian grid framework [20, 13].

A fairly extensive literature on the flow solvers developed for Cartesian grids with embedded adaptation is now available. This chapter therefore focuses on efficient approaches for Cartesian mesh generation. Section 22.2 contains an overview of Cartesian grids, including the geometric information needed by our finite volume flow solver, and a brief discussion of data structures. Most important are the surface geometry requirements for the volume mesh generator. Section 22.3 presents the details of the volume mesh generation, including the geometric adaptation criteria and the treatment of the cut cells. Section 22.4 contains a variety of examples of both Cartesian meshes and flow solutions. Section 22.5 includes a discussion of remaining research issues including approaches for viscous flow. For more thorough discussions of Cartesian mesh topics, see references [1, 33] or the alternative approaches documented in [15, 22, 43].

22.2 Overview of Cartesian Grids

22.2.1 Geometric Requirements of Cartesian Finite Volume Flow Solvers

Cartesian grids pose some unique challenges to the design of efficient finite volume schemes, accurate surface boundary conditions, and associated data structures. While most of the cells in the volume mesh may be regular, cells at the boundary between refinement levels and cells that intersect the surface may have irregular neighbor connections and computational stencils. Nevertheless, a cell-centered finite-volume scheme is easily implemented as a summation of flux contributions from each of a cell's faces:

$$\frac{\partial}{\partial t}\int q dV + \sum_{faces} \bar{f} \cdot \hat{n} dS = 0 \tag{22.1}$$

where the flux, $\bar{f}$, is computed using the normal vector $\hat{n}$ and surface area dS associated with each face. For a simple first-order scheme, the contributions from the flow faces require the face area vector. More accurate approaches require the positions of the face and volume centroids. This level of geometric information is sufficient to support a linear reconstruction of the solution to the face centroid and a second-order midpoint rule for the flux quadrature.

Since the cells of a Cartesian grid can intersect the surface geometry in a completely arbitrary way, general strategies for imposing the surface boundary conditions and computing the flux contributions from the surface faces must be devised. For inviscid flow simulations about solid objects, the surface pressure, normal direction, and area must be available to form the flux contribution from the solid face. Decisions about the surface representation within each mesh cell must therefore be made. Frequently, schemes utilize the average surface normal and surface area within each cut cell. Applying the divergence theorem to cell C and its closed boundary ∂C yields:

$$\int_C (\nabla \bullet \mathbf{F}) dV = \oint_{\partial C} (\mathbf{F} \cdot \hat{n}) dS$$

Substituting the vector function $\mathbf{F} = (1, 0, 0)$ yields an expression for n_x, the x-component of the surface vector within cell C:

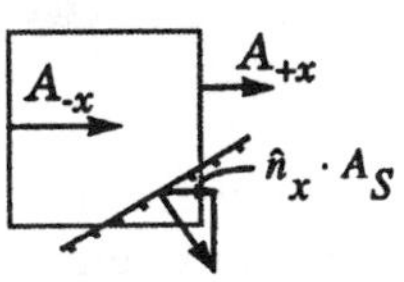

$$\int_{bodySurface} n_x (dS) = A_{-x} - A_{+x} = \hat{n}_x \cdot A_{Surface} \tag{22.2}$$

A_{-x} and A_{+x} are the exposed areas of the cell's x-normal faces. This approach for determining the components of the average surface normal is consistent with the use of a zeroth-order (constant) extrapolation of the pressure to the surface. Improved accuracy requires at least a linear extrapolation of the pressure to the surface. Thus, volume centroids of the cut-cells and area centroids and normals of the individual surface facets within each cut-cell are required. Borrowing the terminology from Harten, we refer to this additional geometric data as *subcell information* [29]. Although the accuracy improvement that this provides is still being quantified, [9] the mesh generation algorithm described in this chapter is designed to extract this maximal level of geometric detail. The surface flux contributions are incorporated into the summation of Eq. 22.1 in a straightforward manner.

The final piece of geometric information required for an accurate flow solution is provided by an algorithm for recognition and treatment of "split-cells," i.e., those Cartesian cells that are divided into two or more disjoint regions by thin pieces of surface geometry. Without an accurate treatment of split cells, the effective chord of a thin wing may be reduced up to 15% due to an inadequate resolution of thin leading and trailing edges [33, 30]. Successive grid refinements could be used to resolve thin pieces of geometry. This approach, however, quickly becomes prohibitively expensive in three dimensions [34]. Although it complicates the mesh generation, it is far more economical to recognize split-cells during the grid generation process and subdivide a cell into its distinct and separate flow regions.

22.2.2 Data Structures

Successful algorithms for Cartesian grid generation can be implemented using a variety of data structures. There are three predominant types usually encountered in the literature. The obvious first choice, suggested by the nested hierarchical nature of the grid itself, is to use an octree in 3D or quadtree in 2D [22, 19, 44, 40, 14] (See also Chapter 14). The connectivity of the tree also provides the information needed in a multigrid method. Although local refinement is easy to implement with this data structure, drawbacks to tree approaches include the difficulties of vectorizing (on vector architectures) and minimizing bandwidth to preserve locality (on cache-based machines). To avoid the tree traversal overhead, a mapping of leaf nodes to some other data structure is often used [45].

A second alternative is the use of block structured Cartesian meshes, typically associated with the adaptive mesh refinement (AMR) approach [10, 7, 11, 39, 43]. In this approach, cells at a given level of refinement are organized into rectangular grid patches, usually containing on the order of hundreds to thousands of cells per patch. This blocking process necessarily flags for inclusion some cells that do not need refinement. However, this overhead is typically less than 30% of the flagged cells in time dependent simulations. When the refinement stems from geometry alone, this number approaches 50% [6]. Nevertheless, the use of a structured array with prescribed connectivity permits an entire grid patch to be stored very compactly in approximately 20 words of memory. Offsetting this advantage is the fact that efficient schemes for patch-to-patch communication are relatively complex to program.

The third alternative, and the one adopted throughout this chapter, is to use an unstructured data structure where the connectivity is explicitly stored with the mesh. The simplifications of using Cartesian grids lead to an extremely compact data structure. We use a face-based data structure, where the mesh is described by a list of cell faces that point to the Cartesian cells on either side. Adjacent cells at different levels of refinement (which can differ by at most one level) are incorporated into this structure by having the refined faces point to their respective finer cells on one side, and the *same* coarse cell on the other side. Despite the unstructured framework for this approach, the Cartesian nature of the hexahedra permit cell and face structures in the volume mesh to be stored with approximately 9 words per cell. This number increases to an average of 15 words per cell when including storage for the geometry and cut-cell information [2].

22.2.3 Surface Geometry

Three-dimensional geometries can be specified in a variety of formats. Examples include proprietary CAD formats, trimmed NURBS, stereolithography formats, networks of grid patches, and others (see Part III). The mesh generation process begins by assembling the surface descriptions of each component into a configuration. Separate watertight triangulations of wings, fuselages, ailerons, and other components are then created and positioned relative to each other. The individual component triangulations need not be constrained to the intersection curves between components, and neighboring components are not required to have commensurate length scales. Once created, the components can be easily translated and/or rotated as necessary to quickly create new configurations. Adopting this component-based approach greatly alleviates the CAD burden for studies of multiple component configurations.

Overlapped components can create internal (unexposed) geometry, which greatly complicates surface operations in the volume mesh generator. In comparison to field cells, cells that intersect the surface geometry are much more expensive to generate, and when the geometry is in fact internal to another component, this expense is wasted. This inefficiency can be eliminated by preprocessing the component geometry to extract the wetted (exposed) surface of the entire configuration as follows. Taking as input the union of component descriptions, the triangulations are intersected against each other to produce a triangulation containing only the wetted surface of the configuration. The original component triangulations are therefore free to overlap in an arbitrary way, while the mesh generator ultimately receives only a triangulation of the wetted surface.

While conceptually straightforward, the efficient implementation of such an intersection algorithm is delicate. The algorithm must be designed to perform a series of computational geometry operations:

1. Intersect the triangles from different components.
2. Retriangulate the intersected triangles, keeping the intersection line segments as constraints in the new triangulation.
3. Discard those triangles that are inside of other components.

These steps will be discussed in detail in the following sections.

A major criterion for the design of the preprocessor is the robust treatment of geometric degeneracies. In three dimensions, the vast majority of coding effort can be consumed by the special case handling required for perhaps less than 1% of the intersections [24,16]. The following presentation initially assumes that no degeneracies arise. This restriction is lifted in later sections, where a consistent algorithmic approach for treating degeneracies is discussed. The approach is automatic, and does not require special case coding.

22.2.3.1 Triangle Intersections

The intersection of possibly hundreds of thousands of surface triangles requires an efficient algorithm for finding lists of candidate intersecting triangles. While a variety of spatial data structures return this list in log N time, where N is the number of surface triangles, a particularly attractive structure is the alternating digital tree (ADT)[12] (see also section 14.4.3 of Chapter 14). Although the ADT requires $O(N \log N)$ time to initially insert the triangles into the tree, the approach compares very favorably to brute force algorithms which can take $O(N)$ time to find all the intersecting triangles for each cell. The size of the tree can be minimized by using simple bounding box checks on each component in the region of possible intersection to screen the triangles as they are inserted. If there is no possibility of a triangle in one component intersecting any other component, it is not inserted into the tree.

For robustness, the intersection of two triangles is computed in two steps. First, the topological connectivity is determined using geometric primitives and robust arithmetic. This step treats the input triangles as "exact." Once the logical connectivity has been established, the actual location of the intersection points of the two triangles is computed using (unreliable) floating-point arithmetic. For example, due to the limited precision of floating-point math, a constructed intersection point may actually lie slightly outside a triangle's interior. To avoid robustness problems arising from such circumstances, these situations are resolved using the robustly computed logical connectivity.

Triangle–triangle intersection is easily reduced to computing the intersection of line segments and triangles. One characterization is as follows:

1. Two edges of one triangle must cross the plane of the other.
2. There must be a total of two edges (of the available six) that pierce within the boundaries of the triangles.

Both of these tests can be recast as the evaluation of the signed volume of a tetrahedron, where the points p, q, r, s are vertices of triangles in R^3. The volume of a tetrahedron is

$$6V\left(T_{p,q,r,s}\right)=\begin{bmatrix} p_0 & p_1 & p_2 & 1 \\ q_0 & q_1 & q_2 & 1 \\ r_0 & r_1 & r_2 & 1 \\ s_0 & s_1 & s_2 & 1 \end{bmatrix} \tag{22.3}$$

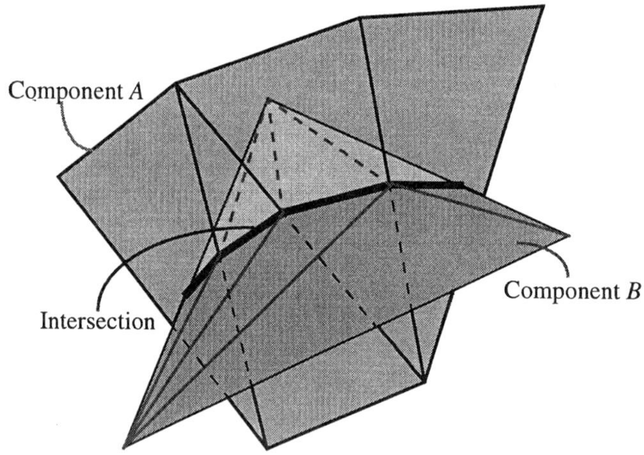

FIGURE 22.2 Constrained retriangulation of an intersected triangle divides it into regions that are completely interior and exterior to the flow.

For example, let triangle $T1$ have vertices {0,1,2} and let (a,b) be an edge of triangle $T2$. The edge intersects the plane of $T1$ if $V(T_{0,1,2,a})$ has a different sign than $V(T_{0,1,2,b})$. The edge intersects in the *interior* of $T1$ if $V(T_{a,1,2,b})$, $V(T_{a,0,1,b})$ and $V(T_{a,2,0,b})$ all have the same sign. Thus at most five determinant evaluations are done for each of the six triangle edges. Note that the only information needed from the evaluation of the determinant is its sign.

Using the adaptive floating point precision package of [47], for example, this determinant can be computed reliably and quickly, even for degenerate cases where the determinant evaluates to exactly zero (indicating a degeneracy, see Section 22.2.3.4). Most of the time, the computation of the sign of the determinant can be done using ordinary floating-point arithmetic. This sign is valid, provided that it is larger than an error bound which is computed using knowledge of the properties guaranteed by the IEEE floating-point arithmetic standard [48,1]. Only if the error bound exceeds the computed value of the determinant does a more accurate evaluation need to done using an adaptive-precision floating-point library (see [41 or 47]). After the *existence* of an intersection is robustly established, the algorithm uses the usual floating-point arithmetic to construct the actual location of the intersection point.

22.2.3.2 Constrained Retriangulation

The result of the preceding intersection step is a list of line segments linked to each intersected triangle. These segments divide the intersecting triangles into polygonal regions that are either completely inside or outside the body. In order to remove the portions of the triangles that are interior, we first triangulate the polygonal regions, treating the original intersection segments as constraints, and then discard those triangles lying inside the body. In an effort to maintain well-behaved triangles, we use a constrained Delaunay triangulation algorithm to maximize the minimum angles produced [55] (see Chapter 16). References [17, 21, 49] give two different approaches for generating a constrained triangulation. Figure 22.2 shows two polygonal regions decomposed into sets of triangles. The constraints from the original component intersections are highlighted.

22.2.3.3 Inside/Outside

The final step in the intersection process is the classification of the resulting set of triangles into those that are either internal to the geometry or exposed and on the wetted surface of the configuration. The algorithm for inside/outside classification is also used during volume mesh generation and is presented in Section 22.3.2.3.

22.2.3.4 Automatic Treatment of Degeneracies

The preceding discussion assumed that the input geometry was free from degenerate data. In other words, the determinants in Section 22.2.3.1 always evaluate to a non-zero number. However, degeneracies are common in input geometry, and most of the complication in the grid generation arises from such cases [16]. For example, if four input points are exactly co-planar, the determinant in Eq. 22.3 will return exactly zero. For an algorithm to be robust, such degeneracies must be resolved in a consistent manner[58, 57].

One approach toward uniform treatment of degenerate geometry is offered by simulation of simplicity which is a method of virtual displacements [24]. The idea is that all data points p_k can be thought of as being perturbed by ε_k, where ε is large enough to break all degeneracies but small enough to not perturb the general data. As long as all determinant evaluations use this same perturbation, this tie-breaking algorithm consistently resolves degeneracies by reporting the sign of the perturbed determinant as positive or negative. By basing it on the global index of a node, the perturbation is consistent across all points in the geometry.

The tie-breaking is implemented as follows. When evaluating determinants, if $det(T) = 0$, the more complicated determinant $det(T + E)$ is evaluated, where E is a perturbation matrix given by

$$(E)_{i,j} = \varepsilon_{i,j} = \varepsilon^{2^{i\delta - j}}, \qquad 1 < j < d, \quad \delta \geq d \tag{22.4}$$

and i denotes the index of the point, $i \in \{0, \ldots, (|V| - 1)\}$, and d is the spatial dimension ($d = 3$ for triangles in R^3).

Eq. 22.3 is an asymptotic expansion of the determinant in powers of an infinitesimal parameter ε. Note that the perturbations $\varepsilon_{i,j}$ are virtual; the geometric data itself is never altered. The first non-zero term in the asymptotic expansion of the determinant gives the sign of the determinant. As a simple two dimensional example, let T and E be the 2 x 2 matrices

$$T = \begin{bmatrix} a_0 & a_1 \\ b_0 & b_1 \end{bmatrix}, \quad E = \begin{bmatrix} \varepsilon^{1/2} & \varepsilon^{1/4} \\ \varepsilon^{2} & \varepsilon^{1} \end{bmatrix} \tag{22.5}$$

Then

$$det(T + E) = det(T) + (-b_0)\varepsilon^{1/4} + (b_1)\varepsilon^{1/2} + (a_0)\varepsilon + \varepsilon^{3/2} + (-a_1)\varepsilon^2 + (-1)\varepsilon^{9/4} \tag{22.6}$$

The fifth term in the expansion has a coefficient of 1. Thus, if each of the first four terms evaluate to 0, the sign of the result would be taken to be positive. In three dimensions there are 15 possible terms in the expansion that generalizes Eq. 22.6 before a constant term is reached (the sign of which conclusively establishes the sign of the original determinant). In practice, rarely are more than two or three terms evaluated before a non-zero coefficient is found. The virtual perturbation computations can be easily incorporated into the low-level subroutine which evaluates the determinant in Eq. 22.3.

22.3 Cartesian Volume Mesh Generation

22.3.1 Overview

Cartesian mesh generation is ostensibly a simple task, where the only complications stem from the presence of body-cut cells and refinement boundaries. Since these occur as lower-dimensional features, the vast majority of the cells in the final mesh are regular, non-body-intersecting, Cartesian hexahedra. Since generation of uniform Cartesian cells is extremely fast, the performance of the overall algorithm

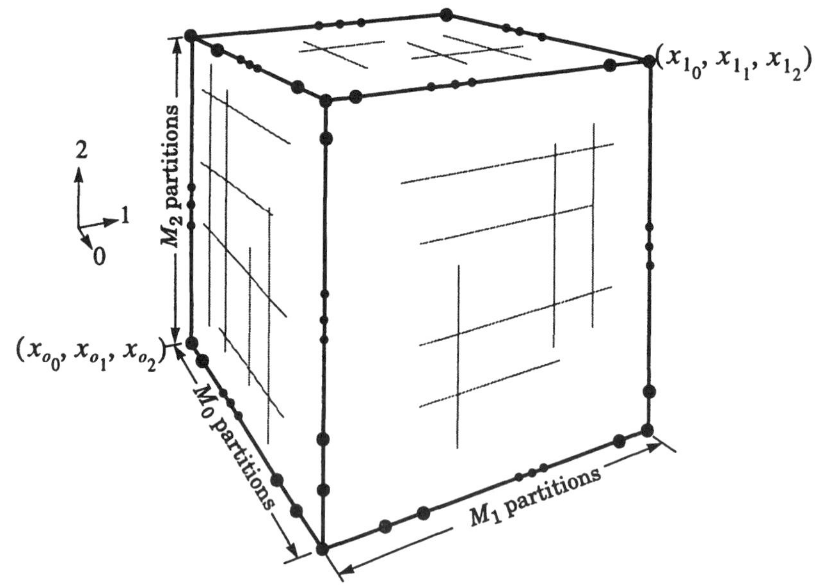

FIGURE 22.3 Cartesian mesh with M_j total divisions in each direction discretizing the region from $\bar{x}_0$ or $\bar{x}_1$.

depends directly on the treatment of cut-cells and the scheme used for adaptive refinement. The following discussions place special emphasis on the performance of the algorithms for generating large numbers (10^6–10^8) of cells. Wherever possible, the methods seek to maintain linear or logarithmic time complexity so that the expense of generating the volume mesh is not dominated by poor algorithmic performance on lower-dimensional collections of cells.

This section begins by highlighting the important algorithms for volume mesh generation, including the detection of cells which lie within solid portions of the geometry and the geometric criteria for cell division. Note that in addition to geometric refinement, flow-field refinement is possible, and in fact, essential. Finally, a variety of algorithms are presented which permit very rapid computation of the geometric information necessary to describe the body-cut cells themselves.

22.3.2 Volume Mesh Generation

The mesh generation process begins with an initial coarse mesh (or even a single cell) covering the domain of interest. This mesh is then repeatedly subdivided to resolve the boundary of the geometry. After each refinement, cells which lie completely inside the body are removed from the mesh. Only when the generation of the volume mesh is complete does the algorithm compute the details of the cut-cell intersections with the surface geometry. Adopting this strategy decouples operations within the body-cut cells from the volume mesh generation process.

22.3.2.1 Initial Mesh Specification and Integer Coordinates

Figure 22.3 shows an example of a coordinate aligned Cartesian mesh defined by its minimum and maximum coordinates $\bar{x}_o$ and $\bar{x}_1$. This region is subdivided with M_j possible coordinates in each dimension, $j = \{0, 1, 2\}$. Thus, each node in the mesh may be specified exactly by the integer vector, $\vec{i}$, and the Cartesian coordinates, $\bar{x}_i$, of any allowable location in this mesh are reconstructed when needed from

$$x_{i_j} = x_{0_j} + \frac{i_j}{M_j}\left(x_{1_j} - x_{0_j}\right) \tag{22.7}$$

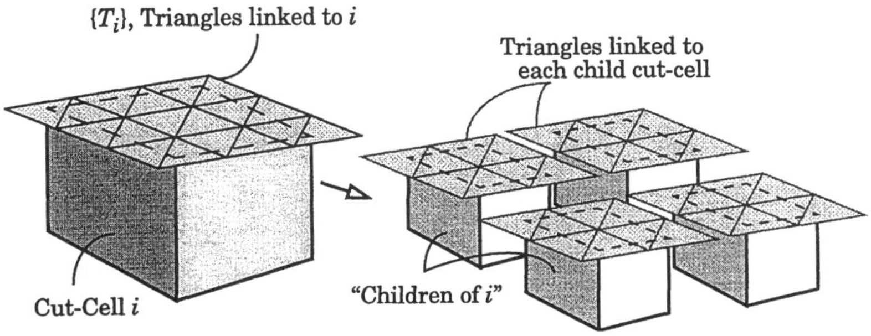

FIGURE 22.4 List of triangles associated with children of a cut-cell may be obtained using ADT, or by exhaustively searching over the parent cell's triangle list.

The use of integer coordinates makes it possible to unambiguously compare vertex locations and leads to compact storage schemes. These properties make integer numbering schemes particularly attractive for the construction of Cartesian meshes. Appendix 1 of this chapter provides details of one such integer numbering scheme which is amenable to adaptively refined Cartesian meshes. This scheme is extremely compact and provides all geometric information and cell-to-vertex pointers with only 96 bits per cell.

22.3.2.2 Efficient Spatial Searches

Assume that the intersection algorithm of Section 22.3 returns a set of triangles $\{T\}$ that describe the wetted surface of the configuration. If the N_T surface triangles in $\{T\}$ are inserted into a efficient spatial data structure such as an ADT, then locating the subset $\{T_i\}$ of triangles actually intersected by the ith Cartesian cell will have complexity proportional to $\log(N_T)$. When a cell is subdivided, a child cell inherits the triangle list of its parent. As the mesh subdivision continues, the triangle lists connected to a surface intersecting ("cut") Cartesian cell will get shorter by approximately a factor of 4 with each successive subdivision. Figure 22.4 illustrates the passing of a parent cell's triangle list to its children.

This observation implies that there is a machine dependent crossover beyond which it becomes faster to simply perform an exhaustive search over a parent cell's triangle list rather than perform an ADT lookup to get a list of intersection candidates for cell i. This is easy to envision, since all of the triangles that are linked to a child cut-cell must have originally been members of the parent cell's triangle list. If a parent cell intersects only a very small number of triangles, then there is no reason to perform a full intersection check using the ADT. The crossover point is primarily determined by the number of elements in N_T and the processor's data cache size.

22.3.2.3 Inside/Outside Determination

A body-intersecting parent cell may find that some of its children cells lie completely inside the body. These cells must be identified and removed from the mesh. Determination of a cell's status as "flow" or "solid" is a specific application of the *point-in-polyhedron* problem that is frequently encountered in computational geometry.

Figure 22.5 illustrates two common containment tests for a cell q and a simply connected polygon P. On the left side of the sketch, the *winding number* [25] is computed by completely traversing the closed boundary P from the perspective of an observer located on cell q, and keeping a running total of the signed angles between successive polygonal edges. As shown in the left of the sketch, if $q \notin P$ then the positive angles are erased by the negative contributions, and the total angular turn is identically zero. If, however, $q \in P$, then the winding number is 2π.

The alternative to computing the winding number is to use a *ray-casting* approach based on the Jordan Curve Theorem. As indicated in the right sketch of Figure 22.5, one casts a ray, r, from q and simply counts the number of intersections of r with ∂P. If the point lies outside, $q \notin P$, this number is even; if the point is contained, $q \in P$, the intersection count is odd.

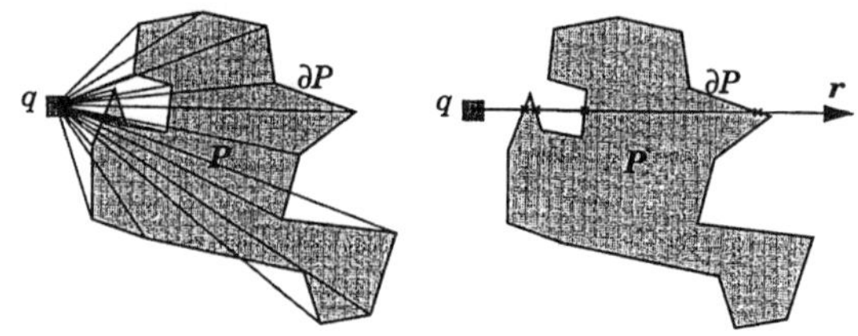

FIGURE 22.5 Illustration of point-in-polygon testing using the (left) winding number and (right) "ray-casting" approaches for determining if q is inside or outside of P.

While both approaches are conceptually straightforward, they are considerably different computationally. Computation of the winding number involves floating-point computation of many small angles, each of which is prone to round-off error. The running sum will make these errors cumulative, increasing the likelihood of robustness pitfalls. In addition, the method answers the topological question "inside or outside?" with a floating-point comparison. By contrast, the ray-casting algorithm poses the inside/outside question in topological terms (i.e., "Does it cross?").

The ray-casting approach fits well within the search and intersection framework developed earlier. Let point q lie on any nonintersected cell in the domain. Then assume r is cast along a coordinate axis ($+x$ for example) and truncated just outside the $+x$ face of the bounding-box for the entire configuration. This ray may then be represented by a line segment from the test point (q_0, q_1, q_2) to $(\lceil \partial\Omega \rceil_x + \varepsilon, q_1, q_2)$ and the problem reduces to that of finding a list of intersection candidates for the segment–triangle intersection algorithm as in Section 22.2.3.1. The tree returns the list of intersection candidates while the signed volume in Eq. 22.3 checks for intersections. Counting the number of such intersections determines a cell's status as inside or outside. Using a spatial data structure like an ADT to return the list of intersection candidates for r makes it possible to identify this list in a time proportional to log (N_T). In addition, computing intersections between the Cartesian cells and surface triangulation via signed volume computations opens the possibility of utilizing exact arithmetic and generalized tie-breaking algorithms from Section 22.2.3.4 to address issues of robustness.

22.3.2.4 Neighborhood Traversal

The ray casting operation in the preceding section takes log (N_T) time. However, it is common to have to perform the in/out test on potentially large lists of Cartesian cells. A painting algorithm makes it possible to avoid casting as many rays as there are cells. Such an algorithm traverses a topologically connected set of cells while passing the status ("flow/solid") of one cell to other cells in its neighborhood. Some details of such an algorithm are presented in [1], where it is demonstrated that mesh traversal may be accomplished with a linear time bound. These techniques make it possible to cast only as many rays as there are topologically disjoint regions of cells.

22.3.3 Cell Subdivision and Mesh Adaptation

Cell subdivision may be triggered by either geometric or flow field requirements. While a variety of sources document various strategies for solution adaptive refinement (see for example [3]), no discrete solution is available during the initial mesh generation. This section therefore focuses on geometry-based adaptation strategies.

All surface intersecting Cartesian cells in the domain are initially automatically refined a specified number of times $(R_{min})_j$. Typically this level is set to be four divisions less than the maximum allowable number of divisions $(R_{max})_j$ in each direction. Anytime a cut-cell is tagged for division, the refinement must be propagated several (usually 3–5) layers into the mesh using a "buffering" algorithm that operates by sweeps over the faces of the cells. Buffering is required to maintain mesh smoothness and avoid corruption of the difference stencil in the immediate vicinity of the body.

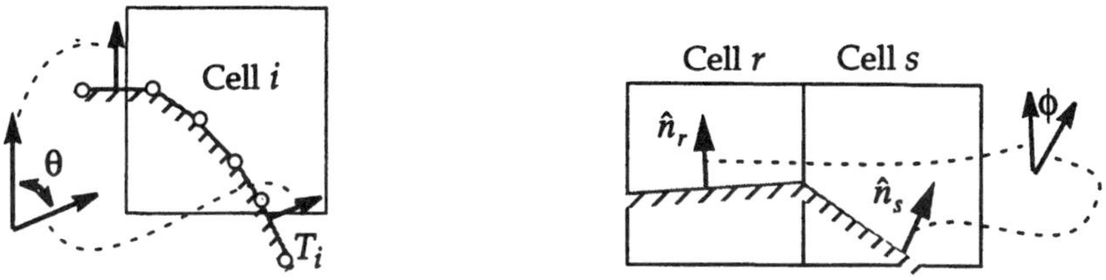

FIGURE 22.6 (a) Measurement of the maximum angular variation within cut-cell i, (b) measurement of the angular variation between adjacent cut-cells.

Further refinement is based upon a curvature detection strategy similar to that originally presented in [33]. This is a two-pass strategy which first detects angular variation of the surface normal, $\hat{n}$, within each cut cell and then examines the average surface normal behavior between adjacent cut cells.

Taking k as a running index to sweep over the set of triangles $\{T_i\}$, let V_j represent the jth component of the vector subtraction between the maximum and minimum components of the normal vectors in each Cartesian direction:

$$V_j = \max_k(n_j) - \min_k(n_j) \qquad \forall k \in \{T_i\}. \tag{22.8}$$

The direction cosines of $\bar{V}$ then provide a measure of the angular variation of the surface normal within cell i.

$$\cos(\theta_{j_i}) = \frac{V_j}{|\bar{V}|} \tag{22.9}$$

Similarly, $(\phi_j)_{r,s}$ measures the jth component of the angular variation of the surface normals between any two adjacent cut cells r and s. With $\hat{n}_i$ denoting the *average* unit normal vector within any cut cell i, the components of $\bar{\phi}_{r,s}$ are

$$\cos(\phi_j)_{r,s} = \frac{|n_{j_r} - n_{j_s}|}{|\hat{n}_r - \hat{n}_s|}. \tag{22.10}$$

If θ_j or ϕ_j in any cell exceeds a preset angle threshold, the offending cell is tagged for subdivision in direction j. Figures 22.6a and 22.6b illustrate the construction of ϕ and θ in two dimensions.

Obviously, by varying these thresholds, one may control the number of cut-cells that are tagged for geometric refinement. When both thresholds are identically 0°, all the cut cells will be tagged for refinement, and when they are 180° only those at sharp cusps will be tagged. Reference [1] presents an exploration of the sensitivity to variation of these parameters for angles ranging from 0° to 179° on several example configurations. In practice, both of these thresholds are generally set at 20°.

22.3.4 Body Intersecting Cells

In three dimensions, the surface triangulation will cut arbitrarily through the body intersecting Cartesian cells. The resulting intersections can therefore be quite complex. We can begin to understand the details of such an intersection by considering the generic cut-cell illustrated in Figure 22.7. The abstraction shown in the sketch presents a single cut-cell, c, which is linked to a set $\{T_c\}$ of four triangles ($T_0 - T_3$) that compose the small swatch of the configuration's surface triangulation intersected by the cell. Since both the Cartesian cell and the triangles are convex, the intersection of each triangle with the cell produces

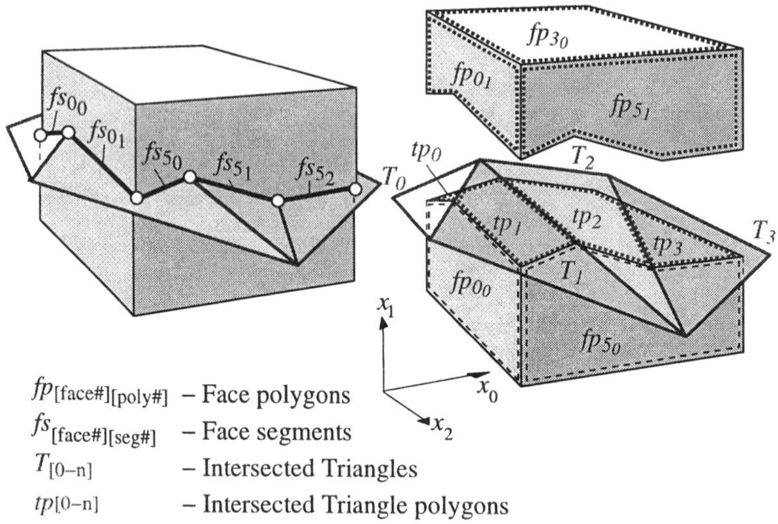

FIGURE 22.7 Anatomy of an abstract cut-cell.

a convex polygon referred to as a *triangle-polygon, tp*. Edges of the triangle-polygons are formed by the clipped edges of the triangles themselves, and the *face-segments, fs*, that result from the intersection of the triangles with the faces of the Cartesian cell. On the Cartesian cells themselves, these segments lead to *face-polygons, fp*, which consist of edges from the Cartesian cell and the face segments from the triangle-face intersection. Note that triangle-polygons are always convex, while face-polygons may not be (e.g., face-polygons $fp_{0,1}$, $fp_{5,0}$, and $fp_{5,1}$ in Figure 22.7).

Clearly, these intersections may become very complex. It is easy to envision the pathological case where an entire configuration intersects only one or two Cartesian cells, creating tens of thousands of triangle polygons. Thus, an efficient implementation is of paramount importance. Many of the algorithms for efficiently constructing this geometry rely on techniques from the literature on computer graphics and are highly specialized for use with coordinate aligned regions [18, 51]. In principle, similar methods could be adopted for non-Cartesian hexahedra, or even other cell types; however, speed and simplicity would be compromised. Since rapid cut-cell intersection is an important part of Cartesian mesh generation, we present a few central operations in detail.

22.3.4.1 Rapid Intersection with Coordinate Aligned Regions

Figure 22.8 shows a two-dimensional Cartesian cell c that covers the region $[\bar{c}, \bar{d}]$. The points (p, q,...,v) are assumed to be vertices of c's candidate triangle list T_c. Each vertex is assigned an "outcode" associated with its location with respect to cell c. This code is really an array of flags which has a "low" and a "high" bit for each coordinate direction, $[\mathrm{lo}_0, \mathrm{hi}_0, \ldots, \mathrm{lo}_{d-1}, \mathrm{hi}_{d-1}]$. Since the region is coordinate aligned, a single inequality must be evaluated to set each bit in the outcode of the vertices. Points inside the region, $[\bar{c}, \bar{d}]$, have no bits set in their outcode.

Using the operators & and | to denote bitwise applications of the "and" and "or" Boolean primitives, candidate edges (like $\overline{rs}$) can be trivially rejected as not intersecting cell c if:

$$\texttt{outcode}_r \,\&\, \texttt{outcode}_s \neq 0 \tag{22.11}$$

This reflects the fact that the outcodes of both r and s will have their low x bit set, thus neither point can be inside the region. Similarly, since ($\texttt{outcode}_t$ | $\texttt{outcode}_v$) = 0, the segment $\overline{tv}$ must be completely contained by the region $[\bar{c}, \bar{d}]$ in Figure 22.8.

If all the edges of a triangle, like Δ_{tuv}, cannot be trivially rejected, then there is a possibility that it intersects the 0000 region. Such a polygon can be tested against the face-planes of the region by

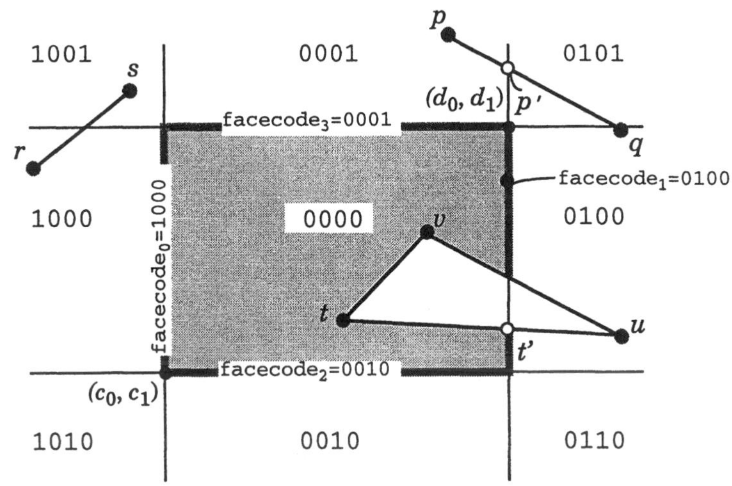

FIGURE 22.8 `outcode` and `facecode` setup of coordinate aligned region $[\bar{c}, \bar{d}]$ in two dimensions.

constructing a logical bounding box (using a bitwise "or") and testing against each facecode of the region. In Figure 22.8, testing

$$\texttt{facecode}_j \,\&\left(\texttt{outcode}_t|\texttt{outcode}_u|\texttt{outcode}_v\right) \quad \forall j \in \{0,1,2,\ldots,2d-1\} \tag{22.12}$$

produces a non-zero result only for the `0100` face. In Eq. 22.12, the logical bounding box of Δ_{tuv} is constructed by taking the bitwise "or" of the outcodes of its vertices.

Once a constructed intersection point, such as p' or t', is computed, it can be classified and tested for containment on the boundary of $[\bar{c}, \bar{d}]$ by an examination of its outcode. However, since these points lie degenerately on the `01XX` boundary, the contents of this bit may not be trustworthy. For this reason, we mask out the questionable bit before examining the contents of these outcodes. Applying "not" in a bitwise manner yields

$$\begin{aligned} &(\texttt{outcode}_{p'} \,\&\, (\neg\texttt{facecode}_1)) = 0 \quad \text{while} \\ &(\texttt{outcode}_{t'} \,\&\, (\neg\texttt{facecode}_1)) \neq 0 \end{aligned} \tag{22.13}$$

which indicates that t' is on the face, while p' is not.

There are clearly many alternative approaches for implementing the types of simple queries that this section describes. However, an efficient implementation of these operations is central to the success of a Cartesian mesh code. The bitwise operations and comparisons detailed in the proceeding paragraphs generally execute in a single machine instruction making this a particularly attractive approach. Further discussion of the use of outcodes may be found in [18].

22.3.4.2 Polygon Clipping

With the fast spatial comparison operators in the previous section outlined, we are ready to construct the triangle-polygons and face-segments that describe the surface within the Cartesian cell. The triangle-polygons ($tp_0 - tp_4$) in Figure 22.7 are the regions of the triangles that lie *within* the cut-cells. Thus, extraction of the triangle-polygons is properly thought of as a *clipping* operation performed on each triangle.

The term "clipping" refers to a process where one object acts as a "window" and we compute the parts of a second object visible through this window [25]. Numerous algorithms have been proposed for the clipping of an object against a rectangular or cubical window [32,37]. In this section we apply an algorithm

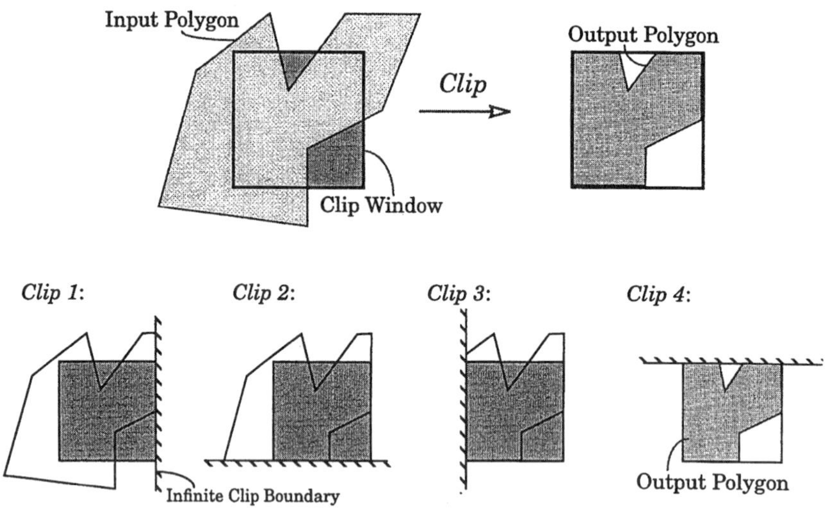

FIGURE 22.9 Illustration of divide-and-conquer strategy of Sutherland–Hodgman polygon clipping.The problem is recast as a series of simpler problems in which a polygon is clipped against a succession of infinite edges.

due to Sutherland and Hodgman for clipping against any convex window [51]. While slightly more general than is absolutely necessary, this algorithm has the attractive property that the output polygon is kept as an ordered list of vertices.

The asymptotic complexity of this clipping algorithm is $O(pq)$, where p is the degree of the clip window and q is the degree of the clipped object. While this time bound is formally quadratic, p for a 3D Cartesian cell is only 6, and the fast intersection checks of the previous section permit very effective filtering of trivial cases.

The Sutherland–Hodgman algorithm adopts a divide-and-conquer strategy that views the entire clipping operation as a sequence of identical, simpler problems. In this case the process of clipping one polygon against another is transformed into a sequence of clips against an infinite edge. Figure 22.9 illustrates the process for an arbitrary polygon clipped against a rectangular window. The input polygon is clipped against infinite edges constructed by extending the boundaries of the clip window.

The algorithm is conveniently implemented as two nested loops. The outer loop sweeps over the clip-border (cell faces in 3D), while the inner is over the edges of the polygon. In our application to the intersected triangles, the initial input polygon is the triangle T, and the clip-window is the cut Cartesian cell. Implementation of the algorithm requires testing of the input triangle's edges against the clip region, so it is useful to combine this algorithm with the outcode flags discussed in the previous section.

Figure 22.10 illustrates the clipping problem (in 2D) for generating the triangle-polygons shown in the view of an abstract cut-cell in Figure 22.7. In Figure 22.10, the triangle T is formed by the set of directed edges, $\overline{v_1 v_0}$, $\overline{v_2 v_1}$, and $\overline{v_0 v_2}$, and the clipped polygon, tp, is a quadrilateral.

As the edges of the input polygon are processed by each clip-boundary the output polygon is formed according to a set of four rules. For each directed edge in the input polygon we denote the vertex at the origin of the edge as "`orig`" and the vertex of the destination as "`dest`." "IN" implies that the test vertex is on the same side of the clip-boundary as the clip-window. We may test for this by examining the `outcode` of each vertex, and comparing to the `facecode` of the current-clip boundary. A test vertex is "IN" if its `outcode` does not have the bit associated with the `facecode` of the clip-boundary set, while "OUT" implies that this bit is set. Using the bitwise operators from the previous section,

$$\begin{aligned} &\text{if}\left(\texttt{facecode}(\text{clip-boundary})\,\&\,\texttt{outcode}(\text{vertex}) = 0\right) \text{ then IN} \\ &\text{if}\left(\texttt{facecode}(\text{clip-boundary})\,\&\,\texttt{outcode}(\text{vertex}) \neq 0\right) \text{ then OUT} \end{aligned} \tag{22.14}$$

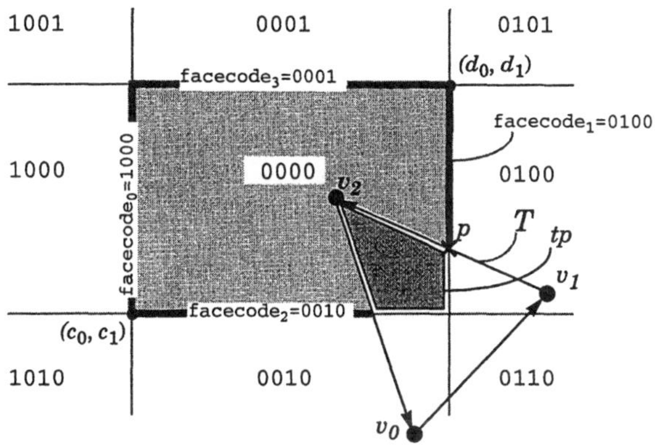

FIGURE 22.10 Setup for clipping a candidate triangle *T*, against a coordinate aligned region and extracting the clipped triangle, *tp*.

TABLE 22.1 Rules for Sutherland–Hodgmen Polygon Clipping

Case	Origin	Destination	Action
SH.1	IN	IN	Add `dest` to the output polygon.
SH.2	IN	OUT	Add intersection of edge and clip-boundary to the output polygon.
SH.3	OUT	OUT	Do nothing.
SH.4	OUT	IN	Add both intersection and `dest` to output polygon.

With these definitions, the output polygon is constructed by traversing around the perimeter of the input polygon and applying the following rules to each edge. Table 22.1 summarizes the actions of the Sutherland–Hodgman algorithm.

Notice that both **SH.2** and **SH.4** describe cases where the edge of the input polygon crosses the clip-boundary. In both of these cases, we must add the point of intersection of the edge with the clip-boundary to the output polygon. This point may be almost trivially constructed since the clip-boundary is coordinate aligned. For the example in Figure 22.10, the constructor for point *p*, which is the intersection of edge $\overline{v_2 v_1}$ with the right side of the clip-boundary, reduces to

$$\vec{p} = \vec{v}_1 + \alpha(\vec{v}_2 - \vec{v}_1) \tag{22.15}$$

where α is simply the distance fraction in the horizontal coordinate of the clip boundary between vertices v_l and v_2.

Returning to the cut-cell shown in Figure 22.7, we note that the face-segments are the edges of the triangle-polygons (just created) that result from a clip. The face-polygons are formed by simply connecting loops of cut-cell edges with these face-segments. Thus, all the necessary elements of the cut-cell have been constructed.

Since the Sutherland–Hodgman algorithm was originally developed for window clipping in computer graphics, both hardware and software versions of it are available on many platforms. Thus, on platforms with advanced graphics hardware, it is frequently possible to make direct calls to the hardware clipping routines to perform the polygon clipping discussed in the preceding paragraphs. Such hardware implementations typically execute tens to hundreds of times faster than software implementations. Similarly, many of the fast bitwise comparisons in the previous section are often available as hardware routines.

Figure 22.11 shows an example of the intersection between the body-cut Cartesian cells and the surface triangulation of a high wing transport configuration. In this case approximately 500,000 cells in the Cartesian mesh intersected the surface triangulation. The figure shows a view of the port side of the

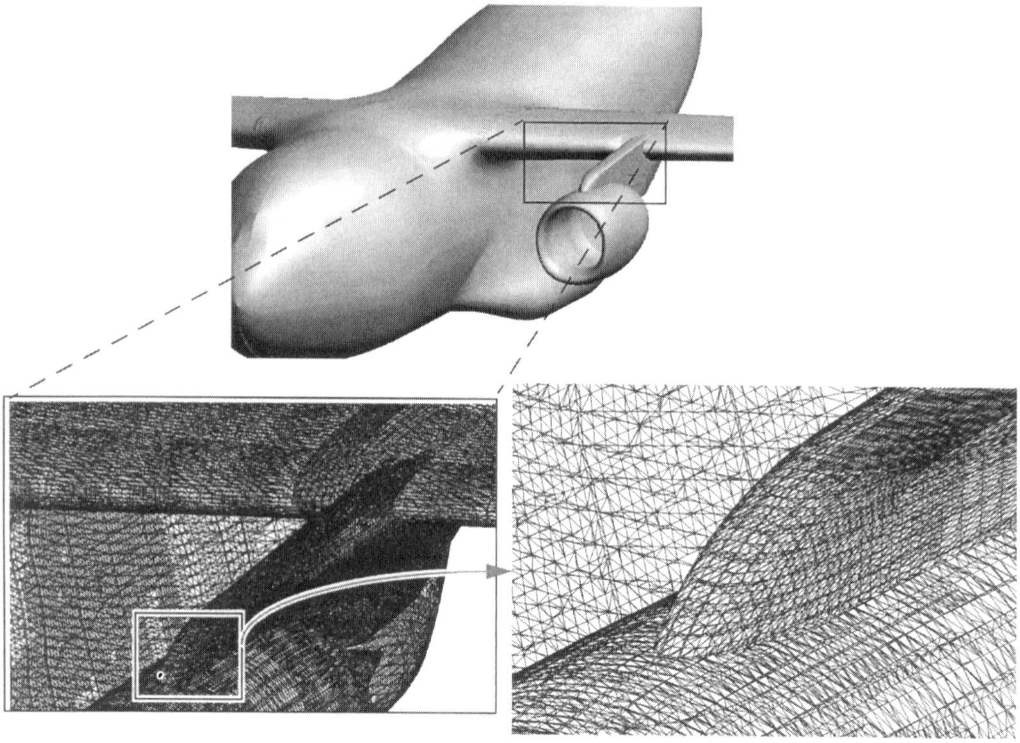

FIGURE 22.11 Triangle-polygons on surface of high wing transport configuration resulting from intersection of body-cut Cartesian cells with surface triangulation.

aircraft and two zoom-boxes with successive enlargements of the triangle-polygons resulting from the intersection. In this example, the triangle-polygons themselves have been triangulated before plotting. This example contained about 2.9M cells in the full Cartesian mesh.

22.4 Examples

22.4.1 Steady State Simulations

Cartesian grids generated automatically about complex geometries are, of course, only useful if those same grids are suitable for engineering analysis. In this section, numerous examples of complex grids and their associated steady and unsteady flow field solutions are discussed in order to demonstrate that non-body-fitted Cartesian methods are indeed suitable for a variety of demanding applications.

22.4.1.1 ONERA M6

The flow field about the ONERA M6 wing was computed at the standard test conditions of Mach 0.84 and $\alpha = 3.06°$ [4, 46]. The cells in the original mesh were subdivided up to nine times, resulting in a total of 1.2 million cells. The left frame in Figure 22.12 shows an isometric view of this final mesh, including the symmetry plane and portions of the mesh at three outboard stations, while the frame at the right contains the corresponding surface and flow field isobars. Figure 22.13 compares computed pressure distributions for this wing at five locations along the span with experimental data [46].

As is typical of other high-resolution Euler computations for this case, these solutions overpredict the strength of the main shock, but in general, the pressure distributions compare well with those presented by other researchers. Additional information about these computations is presented in [33]. The lift and drag coefficients for this case were 0.275 and 0.0128, respectively.

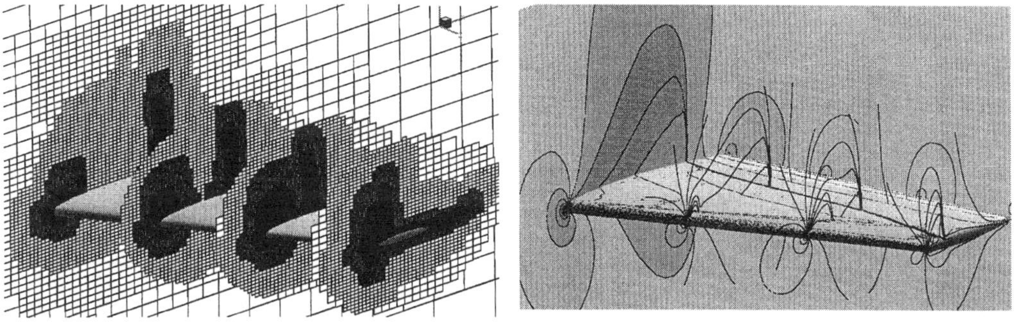

FIGURE 22.12 Adapted mesh and computed isobars for inviscid flow over an ONERA M6 wing at Mach 0.84 and $\alpha = 3.06°$.

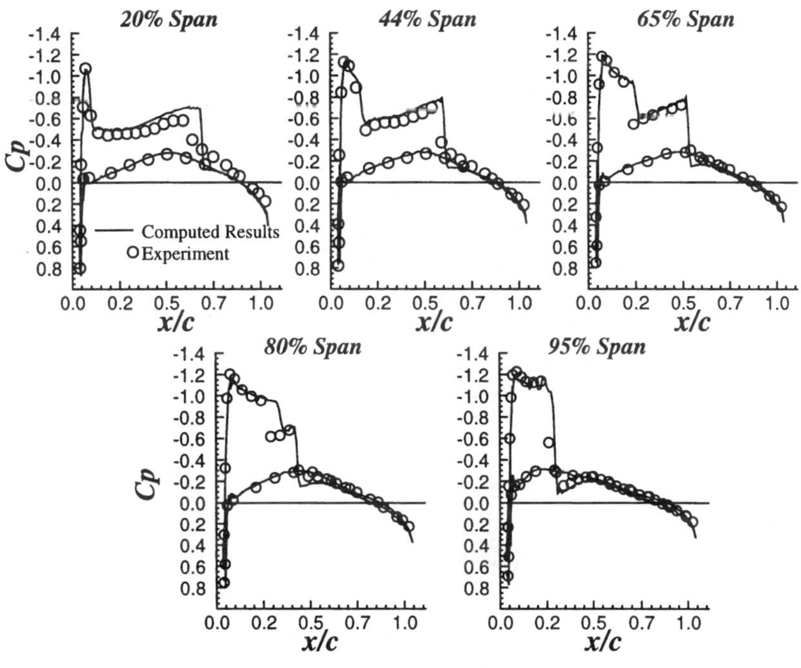

FIGURE 22.13 C_p vs. x/c at $2y/b$ = 0.2, 0.4, 0.65, 0.8, and 0.95.

22.4.1.2 Examples with Complex Geometry

The next four examples of Cartesian grids and steady-state simulations illustrate the geometric complexity that is now routinely simulated with Cartesian methods. Designers, project engineers, and other non-CFD-experts must often repeatedly analyze realistic configurations such as these in order to improve aerodynamic performance. The level of automation attainable with Cartesian approaches makes them particularly attractive for time-critical applications.

Figure 22.14 shows a Cartesian mesh with 5.81 M cells discretizing the space around a McDonnell Douglas Apache attack helicopter. The configuration is composed of 320,000 triangles describing 85 separate components, including armaments, wing stores, night-vision equipment, and avionics packages. The surrounding flow field mesh was generated in 320 seconds on a moderately powerful engineering workstation (MIPS 195 Mhz R10000 CPU). The only user inputs to the mesh program were the dimensions of the bounding box of the outer domain, a clustering parameter that controls the refinement on

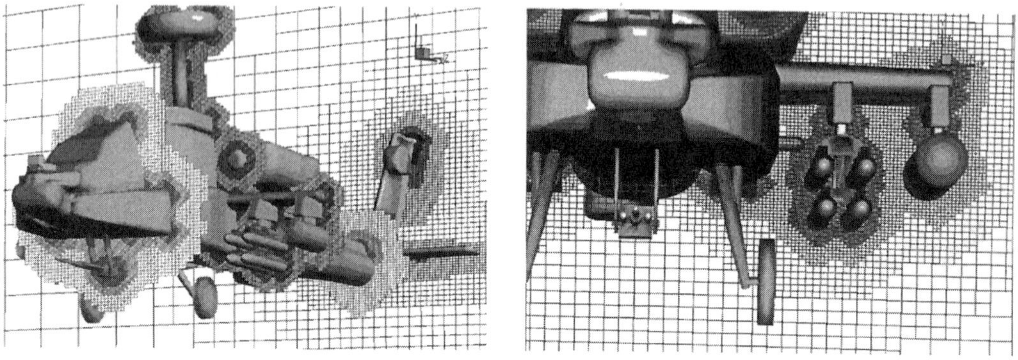

FIGURE 22.14 Left: Cartesian mesh for attack helicopter configuration with 5.81 M cells. Right: Close-up of mesh through left wing and stores.

FIGURE 22.15 Isobars resulting from inviscid flow analysis of attack helicopter configuration computed on mesh with 1.2 M cells.

the surface, and a target number of cells in the final mesh. Figure 22.15 displays the computed isobars on this same configuration on a coarser mesh of approximately 1.2 M cells.

Figure 22.16 shows two views of a mesh generated after positioning three F-15 aircraft in formation with the Apache helicopter. The helicopter is offset from the axis of the lead fighter to emphasize the asymmetry of the mesh. Each fighter has flow-through inlets and is described by 13 individual component triangulations and 201,000 triangles. After surface preprocessing, the entire four-aircraft configuration contained 121 components described with 683,000 triangles. The lower frame in Figure 22.16 shows portions of three cutting planes through the mesh and geometry, while the upper frame shows one cutting plane at the tail of the rear two aircraft, and another just under the helicopter geometry. The final mesh includes 5.61 M cells, and required a maximum of 365 Mb to compute. Mesh generation time was approximately 6 minutes and 30 seconds on a workstation with a MIPS 195 Mhz R10000 CPU.

22.4.1.3 Transport Aircraft with High-Lift System Deployed

Figure 22.17 shows the mesh and flow field about a high-wing transport (HWT) aircraft with its high-lift devices deployed in a landing configuration. The aircraft was composed of 18 components and a total of 700,000 triangles. This solution contained approximately 1.7 million cells and had ten levels of cell refinement. Flowfield adaptation was triggered by a simple criterion formed from the undivided first difference of density. At a low subsonic Mach number and a moderate angle of attack, this indicator

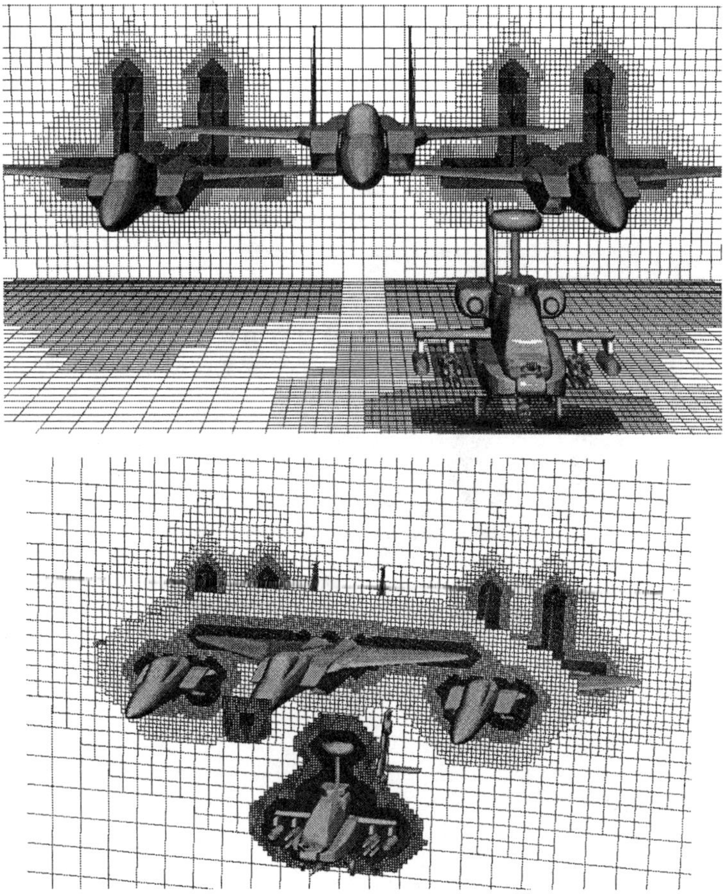

FIGURE 22.16 Cutting planes through mesh of multiple aircraft configuration with 5.61 M cells and 683,000 triangles in the triangulation of the wetted surface.

targets refinement of the suction peaks on the leading edge slat and main element, as well as the inviscid jet through the flap system. Despite the fact that this simulation is inviscid, the sharp outboard corner of the flap has correctly spawned a flap vortex, which is evidenced by the twisting stream ribbon in the figure. Additional information about the solution can be found in [3].

22.5 Research Issues

22.5.1 Moving Geometry

Developments in several directions would greatly extend the applicability of Cartesian grid methods. The most obvious extension is to applications involving moving and/or deforming geometry. A very successful first step in this direction was demonstrated in two space dimensions in [5]. The sequence in Figure 22.18 shows a jet-powered projectile in a quiescent stream that penetrates a deformable shell structure. A simple fracture model was used in calculating the deformation of the shell.

22.5.2 NURBS Surface Definitions

The mesh generation method presented in this chapter requires component surface triangulations as input geometry. Basing the method on simplicial geometry such as this has many advantages, since the

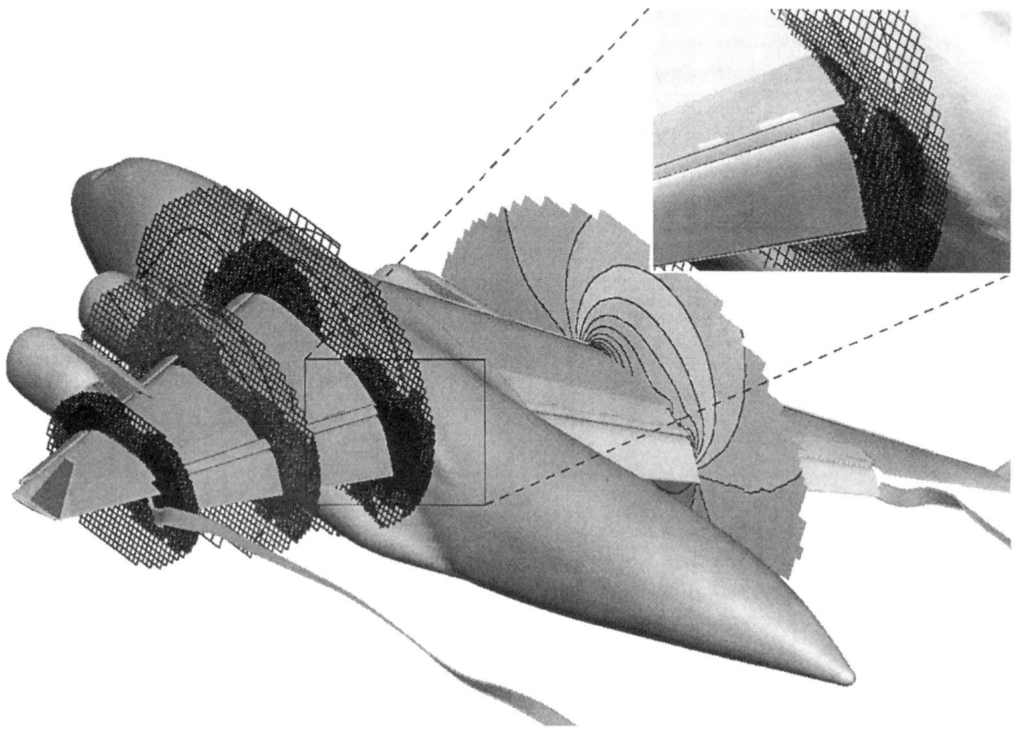

FIGURE 22.17 HWT example with high-lift system deployed. The mesh contains 1.65 M cells at 10 levels of refinement. The mesh is presented by cutting planes at 3 spanwise locations, and the cutting plane on the starboard wing is flooded by isobars of the discrete solution.

input geometry is known explicitly to a specified level of precision. Extending the methodology to accept alternative descriptions of the input geometry would further simplify and improve the analysis process. For example, it would be convenient and expedient to work with a geometry format native to current CAD/CAM systems, such as the NURBS description of the geometry [23] (see Part III). This approach was investigated in [35]; however, the need to compute non-linear intersections of splines and Cartesian hexahedra at each step made the procedure extremely expensive. The NURBS representation of a geometry can be extremely flexible, and an ability to work directly from it would eliminate any errors due to the surface faceting inherent in triangulations.

22.5.3 Viscous Applications

Finally, the ability to capture boundary layers with a nonisotropic refinement strategy will be necessary for this method to be applicable to high Reynolds number viscous flows. A very interesting but not entirely successful first attempt at combining Cartesian data structures with variable boundary layer zoning is presented in [20]; however, the mesh was too irregular to accurately compute the viscous terms using simple stencils. Other approaches currently under investigation use either integral boundary layer models or hybrid grids (see Chapter 23) that combine a near-body fitted grid and a background Cartesian grid [56, 54]. Although this latter approach only needs a small region around the body to have a viscous grid, this severely compromises the automation of the Cartesian approach since it effectively couples the surface discretization with part of the volume mesh. Developments in these directions will have a great impact in extending the usefulness of Cartesian grids.

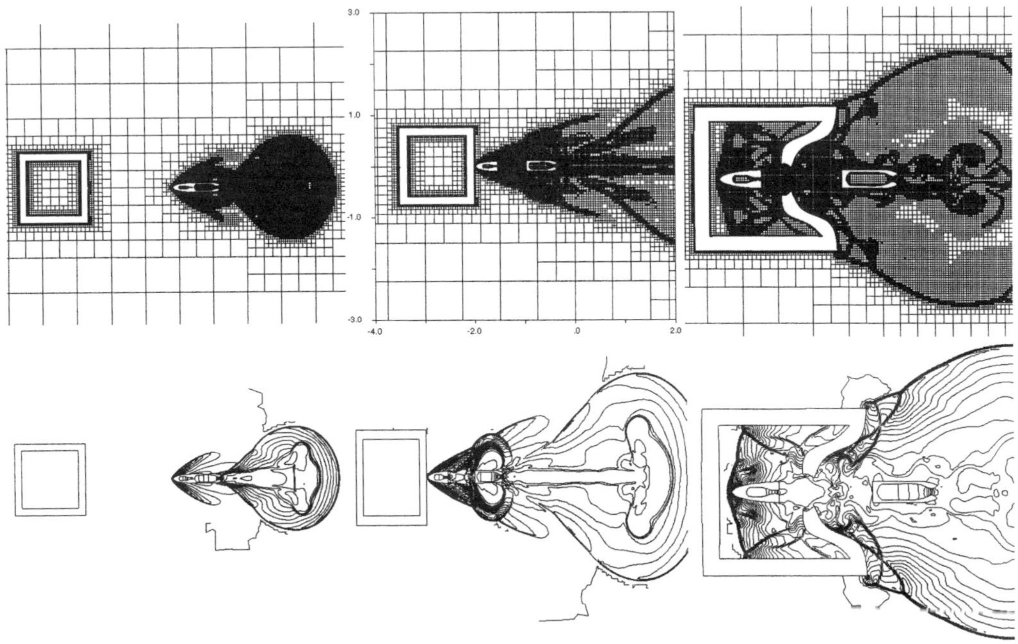

FIGURE 22.18 Density contours and adapted quadtree grids showing a time history of a projectile penetration problem. (Powell, K., von Karman Institute for Fluid Dynamics, Lecture Series 1994-05, Rhode-Saint-Genèse, Belgium, March 1996. With permission.)

22.6 Summary

The adaptive Cartesian mesh approach demonstrates great potential for dramatically accelerating the routine inviscid analyses of complex configurations. Many of the advantages of Cartesian grids arise from the independence of the surface description from the flow field discretization and the resultant ease and speed with which grids can be generated. Incorporating a component-based Cartesian approach also streamlines the surface definition process. New configurations can be quickly assembled from libraries of existing components, and individual components can be easily repositioned using simple transformations. Additionally, conventional inviscid finite volume flow solver schemes can be straightforwardly modified and implemented on Cartesian grids.

Although many of the geometric algorithms described in this chapter have their roots in the fields of computer graphics and computational geometry, they are well-suited for robust Cartesian grid generation. With appropriate attention to algorithmic complexity and careful programming, the resulting codes can be designed to run extremely efficiently on current workstations. By taking full advantage of the natural simplicity of Cartesian grids, a fast, automated, robust, and low-memory grid generation scheme can be developed.

Appendix 1: Integer Numbering of Adaptive Cartesian Meshes

Figure 22.A.1 shows a model of the jth direction of a Cartesian mesh covering the region $[\bar{x}_0, \bar{x}_1]$. As shown in the sketch, specifying the domain with $\bar{x}_0$ and $\bar{x}_1$ and the initial partitioning by N_j uniquely identifies a set of possible Cartesian cell locations in this region. Each additional refinement increases the maximum integer coordinate by a factor of $2(N_j - 1)$. This relationship suggests a natural mapping to a system of integer coordinates. If one defines a maximum number of permissible cell divisions in this

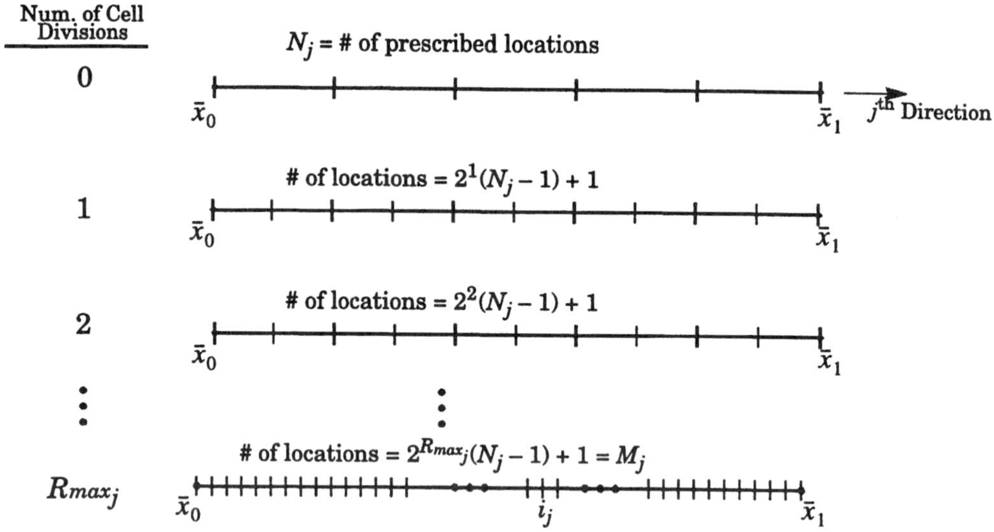

FIGURE 22.A.1 Specification of integer coordinate locations for a coordinate direction with N_j prescribed boundaries.

direction, $Rmax_j$, then any point in such a mesh can be uniquely located by its integer coordinates (i_0, i_1, i_2). Allocating m bits of memory to store each integer i_j, the upper bound on the permissible total number of vertices in each coordinate direction becomes 2^m.

Figure 22.A.1 demonstrates that on a mesh with N_j prescribed nodes, performing R_j cell refinements in each direction will produce a mesh with a maximum integer coordinate of $2^{R_j}(N_j - 1) + 1$ which must be resolvable in m bits.

$$2^{R_j}\left(N_j - 1\right) + 1 \le 2^m \tag{22.A.1}$$

Thus, the maximum number of cell subdivisions that can be addressed by a set of m-bit integer coordinates is

$$(Rmax)_j = \left\lfloor \log_2\left(2^m - 1\right) - \log_2\left(N_j - 1\right) \right\rfloor \tag{22.A.2}$$

where the floor "$\lfloor \ \rfloor$" indicates rounding down to the next lower integer. Substituting back into Eq. 22.A.1 gives the total number of vertices we can address in each coordinate direction using m-bit integers and with N_j prescribed nodes in the direction.

$$M_j = 2^{Rmax_j}\left(N_j - 1\right) + 1 \tag{22.A.3}$$

Thus, the floor in Eq. 22.A.2 ensures that M_j can never exceed 2^m. The mesh in Figure 22.A.3 is an illustration of this numbering scheme in three dimensions.

The examples in this chapter use up to $m = 21$ bits per direction, which provides over 2.1×10^6 addressible locations in each coordinate direction. This choice has the advantage that all three indices may then be packed into a single 64-bit integer for storage*.

*This is a choice of convenience. All three integer coordinates may, of course, be sorted separately, permitting $2^{64} - 1 = 1.84 \times 10^{19}$ addressible locations using 64-bit integers.

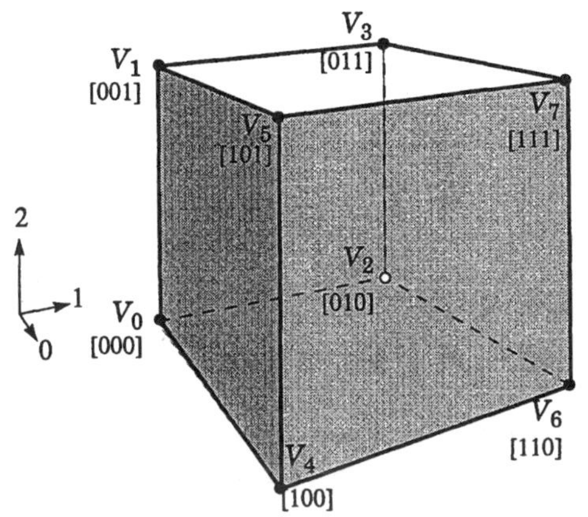

FIGURE 22.A.2 Vertex numbering within a cell. Square brackets [-] indicate crystal directions.

Cell-to-Node Pointers

Figure 22.A.2 gives an example of the vertex numbering within an individual Cartesian cell. This system has been adopted by analogy to the study of crystalline structures specialized for cubic lattices [52]. Within this framework, the cell vertices are numbered with a boolean index of 0 (low) or 1 (high) in each direction. Following this ordering, Figure 22.A.2 shows the crystal direction of each vertex in square brackets (with no commas). Reinterpreting this 3-bit pattern as an integer yields a unique numbering scheme (from 0 to 7) for each vertex on the cell.

For any cell *i*, $\overline{V}_0$ is the integer position vector $(V_{0_0}, V_{0_1}, V_{0_2})$ of its vertex nearest to the x_0 corner of the domain. Knowing the number of times that cell *i* has been divided in each direction, R_j, one may express its other 7 vertices directly.

$$\begin{aligned}
\overline{V}_1 &= \overline{V}_0 + (0, \; 0, \; 2^{Rmax_2 - R_2}) \\
\overline{V}_2 &= \overline{V}_0 + (0, \; 2^{Rmax_1 - R_1}, \; 0) \\
\overline{V}_3 &= \overline{V}_0 + (0, \; 2^{Rmax_1 - R_1}, \; 2^{Rmax_2 - R_2}) \\
\overline{V}_4 &= \overline{V}_0 + (2^{Rmax_0 - R_0}, \; 0, \; 0) \\
\overline{V}_5 &= \overline{V}_0 + (2^{Rmax_0 - R_0}, \; 0, \; 2^{Rmax_2 - R_2}) \\
\overline{V}_6 &= \overline{V}_0 + (2^{Rmax_0 - R_0}, \; 2^{Rmax_1 - R_1}, \; 0) \\
\overline{V}_7 &= \overline{V}_0 + (2^{Rmax_0 - R_0}, \; 2^{Rmax_1 - R_1}, \; 2^{Rmax_2 - R_2})
\end{aligned} \tag{22.A.4}$$

Since the powers of two in this expression are simply a left shift of the bitwise representation of the integer subtraction $Rmax_j - R_j$, vertices $\overline{V}_1$ through $\overline{V}_7$ can be computed from $\overline{V}_0$ and R_j at very low cost. In addition, the total number of refinements in each direction will be a (relatively) small integer, thus it is possible to pack all three components of $\overline{R}$ into a single 32-bit word.

Acknowledgment

This work was supported in part by NASA Ames Research center, by DOE Grants DE-FG02-88ER25053 and DE-FG02-92ER25139, and by AFOSR grant F49620-97-0322. Thanks also to RIACS, whose support of Dr. M. Berger is gratefully acknowledged.

References

1. Aftosmis, M.J., Solution adaptive cartesian grid methods for aerodynamic flows with complex geometries, von Karman Institute for Fluid Dynamics, Lecture Series 1997-02, Rhode-Saint-Genèse, Belgium, Mar. 3-7, 1997.
2. Aftosmis, M.J., Berger, M.J., and Melton, J.E., Robust and efficient Cartesian mesh generation for component-based geometry, *AIAA Paper 97-0196*, Jan. 1997.
3. Aftosmis, M.J., Melton, J.E., and Berger, M.J., Adaptation and surface modeling for Cartesian mesh methods, *AIAA Paper 95-1725-CP*, June 1995.
4. AGARD Fluid dynamics panel, test cases for inviscid flow field methods, *AGARD Advisory Report AR-211.* May 1985.
5. Bayyuk, S., *Euler Flows with Arbitrary Geometries and Moving Boundaries.* Ph.D thesis, Dept. of Aero. and Mech. Eng., University of Michigan, 1996.
6. Berger M.J., Aftosmis, M.J., and Melton, J.E., Accuracy, adaptive methods and complex geometry, *Proc. 1st AFOSR Conf. on Dynam. Mot. in CFD.* Rutgers, NJ, 1996.
7. Berger M.J. and Colella, P., Local adaptive mesh refinement for shock hydrodynamics. *J. Comp. Physics.* 1989, 82, pp 64–84.
8. Berger, M. and LeVeque, R., Stable boundary conditions for Cartesian grid calculations, *ICASE Report No. 90-37*, 1990.
9. Berger, M. and Melton, J.E., An accuracy test of a cartesian grid method for steady flow in complex geometries, *Proc. 8th Int. Conf. Hyp. Problems,* (also RIACS Report 95-02) Uppsala, Stonybrook, NY, June 1995.
10. Berger, M.J. and LeVeque, R., Cartesian meshes and adaptive mesh refinement for hyperbolic partial differential equations, *Proc. 3rd Int. Conf. Hyp. Problems,* Uppsala, Sweden, 1990.
11. Berger, M.J. and Oliger, J., Adaptive mesh refinement for hyperbolic partial differential equations, *J. Comp. Physics,* 1984, 53, pp 482–512.
12. Bonet, J. and Peraire, J., An alternating digital tree (ADT) algorithm for geometric searching and intersection problems, *Int. J. Num. Meth. Eng.*, 1991, 31, pp 1–17.
13. Chan, W.M. and Meakin, R.L., Advances towards automatic surface domain decomposition and grid generation for overset grids, *Proc. of the AIAA 13th Comp. Fluid Dyn. Conf., AIAA Paper 97-1979*, Snowmass, Colorado, June 1997.
14. Charlton. E.F. and Powell, K.G., An octree solution to conservation-laws over arbitrary regions (OSCAR), *AIAA Paper 97-0198*, Jan. 1997.
15. Charlton. E.F., An octree solution to conservation-laws over arbitrary regions (OSCAR) with applications to aircraft aerodynamics, Ph.D. thesis, Dept. of Aero. and Astro. Eng., Univ. of Michigan, 1997.
16. Chazelle, B., et al., *Application Challenges to Computational Geometry: CG Impact Task Force Report.* TR-521-96. Princeton Univ., April 1996.
17. Chew, L.P., Constrained Delaunay triangulations, *Algorithmica,* 1989, 4, pp 97–108.
18. Cohen, E., Some mathematical tools for a modeler's workbench, *IEEE Comp. Graph. and App.* Oct. 1983, 3, p 7.
19. Coirier, W.J. and Powell, K.G., An accuracy assessment of Cartesian-mesh approaches for the euler equations, *AIAA Paper 93-3335-CP*, July 1993.
20. Coirier, W.J., An adaptively refined, Cartesian, cell-based scheme for the Euler equations, NASA TM-106754, Oct., 1994. also Ph.D. thesis, Dept. of Aero. and Astro. Eng., Univ. of Mich., 1994.
21. De Floriani, L. and Puppo, E., An on-line algorithm for constrained Delaunay triangulation, *CVGIP: Graphical Models and Image Proc.* 1992, 54, 3, pp 290–300.
22. De Zeeuw, D. and Powell, K., An adaptively refined Cartesian mesh solver for the Euler equations, *AIAA Paper 91-1542*, 1991.

23. *DT_NURBS Spline Geometry Subprogram Library Theory Document*, version 3.3. USN Surface Warfare Center/Carderock Div. David Taylor Model Basin, Bethesda MD. CARDEROCKDIV-94/000, Dec. 1996.
24. Edelsbrunner, H. and Mücke, E.P., Simulation of simplicity: a technique to cope with degenerate cases in geometric algorithms. *ACM Transactions on Graphics*, Jan. 1990, 9, 1, pp 66-104.
25. Foley, J., van Dam, A., Feiner, S., and Hughes, J., *Computer Graphics: Principles and Practice*, ISBN 0-201-84840-6, Addison-Wesley, Reading, MA, 1995.
26. Forrer, H., *Boundary Treatment for a Cartesian Grid Method*, Seminar für Angewandte Mathmatic, ETH Zürich, ETH Research Report No. 96-04, 1996.
27. Forrer, H., *Second Order Accurate Boundary Treatment for Cartesian Grid Methods*, Seminar für Angewandte Mathmatic, ETH Zürich, ETH Research Report No. 96-13, 1996.
28. Gooch, C.F., Solution of the Navier–Stokes equations on locally refined Cartesian meshes, Ph.D. dissertation, Dept. of Aero. Astro. Stanford Univ., Dec. 1993.
29. Harten, A., ENO schemes with subcell resolution, *ICASE Report 87-56*, Aug. 1987.
30. Karman, S.L., Jr., SPLITFLOW: A 3D Unstructured Cartesian/prismatic grid CFD code for complex geometries, *AIAA 95-0343*, Jan. 1995.
31. Keener, E.R., Pressure-distribution measurements on a transonic low-aspect ratio wing, NASA TM-86683, 1985.
32. Liang, Y. and Barsky, B.A., An analysis and algorithm for polygon clipping, *Comm. ACM*, 1983, 26, 3, pp 868–877.
33. Melton, J.E., *Automated Three-Dimensional Cartesian Grid Generation and Euler Flow Solutions for Arbitrary Geometries*, Ph.D. thesis, Univ. California Davis, 1996.
34. Melton, J.E., Berger, M.J., Aftosmis, M.J., and Wong, M.D., 3D applications of a Cartesian grid Euler method, *AIAA Paper 95-0853*, Jan. 1995.
35. Melton, J.E., Enomoto, F.Y., and Berger, M.J., 3D Automatic Cartesian grid generation for Euler flows, *AIAA Paper -93-3386-CP*, July 1993.
36. Melton, J.E., Pandya, S., and Steger, J., 3-D Euler solutions using unstructured Cartesian and prismatic grids, *AIAA Paper 93-0331*, July 1993.
37. Newman, W.M. and Sproull, R.F., *Principles of Interactive Computer Graphics*, 2nd Ed. McGraw-Hill, NY, 1979.
38. O'Rourke, J., *Computational Geometry in C.*, Cambridge Univ. Press, NY, 1993.
39. Pember, R.B., Bell, J.B., Colella, P., Crutchfield, W.Y., and Welcome, M.L., An adaptive Cartesian grid method for unsteady compresible flow in irregular regions, *J. Comp. Phy.* 1995, 120, pp 278–304.
40. Powell, K., *Solution of the Euler and Magnetohydrodynamic Equations on Solution-Adaptive Cartesian Grids*, von Karman Institute for Fluid Dynamics, Lecture Series 1994-05, Rhode-Saint-Genèse, Belgium, Mar. 1996.
41. Preparata, F.P. and Shamos, M.I., *Computational Geometry: An Introduction*, Springer–Verlag, 1985.
42. Priest, D.M., Algorithms for arbitrary precision floating point arithmetic, *10th Symp. on Computer Arithmetic*, IEEE Comp. Soc. Press, 1991, pp 132-143.
43. Quirk, J., An Alternative to unstructured grids for computing gas dynamic flows around arbitrarily complex two dimensional bodies, *ICASE Report 92-7*, 1992.
44. Finkel, R.A. and Bentley, J.L., *Quad trees: a data structure for retrieval on composite keys.* Acta Informatica, 1974, 4,1, pp 1–9.
45. Samet, H., *The Design and Analysis of Spatial Data Structures.* Addison-Wesley Series on Computer Science and Information Processing. Addison–Wesley, 1990.
46. Schmitt, V. and Charpin, F., Pressure distributions on the ONERA-M6-Wing at transonic mach numbers, *Experimental Data Base for Computer Program Assessment*, AGARD Advisory Report AR-138, 1979.

47. Shewchuk, J.R., Robust Adaptive Floating-point Geometric Predicates, *Proceedings of the Twelfth Annual Symposium on Computational Geometry*, ACM, May 1996, pp 141–150.
48. Shewchuk, J.R., Adaptive precision floating-point arithmetic and fast robust geometric predicates. CMU-CS-96-140, School of Computer Science, Carnegie Mellon Univ., 1996.
49. Sloan S.W., A fast algorithm for generating constrained Delaunay triangulations, *Computers and Structures*, Pergammon Press Ltd., 1993, 47, 3, pp 441–450.
50. Stern, L.G., *An Explicitly Conservative Method for Time-Accurate Solution of Hyperbolic Partial Differential Equations on Embedded Chimera Grids*, Ph.D. thesis, Univ. of Wash, 1996.
51. Sutherland, I.E. and Hodgman, G.W., Reentrant polygon clipping, *Comm ACM,* 1974, 17,1, pp 32–42.
52. Van Vlack, L.H., *Elements of Material Science and Engineering*, Addison-Wesley, 1980.
53. Voorhies, D., *Graphics Gems II: Triangle-Cube Intersections.* Academic Press, 1992.
54. Wang, Z.J., Przekwas, A., and Hufford, G., Adaptive Cartesian/adaptive prism grid generation for complex geometry, *AIAA Paper 97-0860,* Jan. 1997.
55. Watson, D.F., Computing the n-dimensional Delaunay Tessellation with application to Voronoï polytopes, *Computer J.* 1981, 24, 2, pp 167–171.
56. Welterlen, T.J. and Karman, S.L., Jr., Rapid assessment of F-16 store trajectories using unstructured CFD, *AIAA 95-0354*, Jan. 1995.
57. Yap, C. and Dubé, T., The exact computation paradigm, *Computing in Euclidean Geometry*, 2nd Ed. Du, D.-Z. and Hwang, F.K., (Eds.), World Scientific Press, 1995, pp. 452-492.
58. Yap, C-.K., Geometric consistency theorem for a symbolic perturbation scheme, *J. Comp. Sys. Sci.*, 1990, 40, 1, pp 2–18.

23

Hybrid Grids

Jonathon A. Shaw

23.1 Introduction

Recent years have witnessed much conjecture over the relative merits of the various methodologies that have emerged as candidates for providing a robust, effective, high-quality mesh generation capability for gridding complex three-dimensional domains. These methods are generally classified into one of two categories, namely *structured* or *unstructured* approaches, with strong advocates of each still existing amongst both the method development and user communities. Promoters of structured schemes highlight the efficiency and accuracy that is attained through the employment of regularly arranged hexahedral volumes. Supporters of unstructured schemes emphasize the geometric flexibility and suitability for adaptation inherent to the use of irregularly connected tetrahedral volumes.

This Handbook will serve to further the debate on the absolute superiority of one of these approaches over the other without, one suspects, enabling a definitive conclusion to be reached.

However, a review of this handbook in conjunction with the proceedings of the now firmly established series of conferences devoted to numerical grid generation indicates that there *is* an underlying trend within the field of grid generation. This trend is toward an increasing cross-fertilization of ideas and techniques between the two camps. Practitioners of the unstructured approach are having to use *directional* information to achieve elements of suitable quality near boundaries, while structured grid generators are devising increasingly *irregular* schemes to attain appropriate geometric flexibility.

The limit of this trend is to replace the sole use of one mesh type by the use of combined meshes composed of both structured and unstructured grids — hybrid grids. This combination of grid types not only allows the benefits of structured and unstructured grids to be attained simultaneously, but also allows high grid quality to be achieved throughout the domain due to the appropriate use of each element type.

In this chapter, the prime interest is the generation of grids containing more than one element type. This will be termed *hybrid grid generation.* However, reference will also be made to the generation of single element type meshes where it is felt that the work particularly demonstrates the movement of ideas between the two main fields of mesh generation. This will be termed *hybrid grid technology.*

0-8493-2687-7/99/$0.00+$.50

This chapter has three main sections and a summary. Section 23.2 is devoted to "*Underlying Principles*" and contains a general description of both work in the field of hybrid grid generation and the use of hybrid grid technology. It begins by tracing the roots of hybrid grid generation back to two quite distinct sources. The move to hybrid grid generation/technology from the purely unstructured approach is then reviewed, followed by observations on the progression of the structured community to hybrid grids. The section ends with a discussion on the potential savings in execution times and memory requirements that can be made through using hybrid grids instead of solely unstructured grids.

In Section 23.3, entitled *"Best Practices,"* the discussion becomes more focused around the author's own experience in generating hybrid grids. This is because, in spite of the very real potential benefits that are to be gained through the use of hybrid grids, there is at present a dearth of evidence that other capabilities exist that are able to form *general* three-dimensional hybrid meshes. The section begins with a brief overview of the evolution of a mesh generation system that can be used to form either solely structured (hexahedral cells), semistructured (prismatic cells), unstructured (tetrahedral cells) or a hybrid combination of any of these grid types. Attention is then focused briefly on the key elements of the capabilities that are used to form the different types of elements, with the details left either to references or study of other chapters in the book. The very important area of interfacing the different grid types is then covered, and this is followed by a discussion on data structures for describing a hybrid grid. Finally, three examples of hybrid grids for aero- and hydrodynamic applications are presented along with a description of the main considerations that have been borne in mind while forming these grids.

Section 23.4 covers some of the open research issues that will need to be addressed within the field of hybrid grid generation for the approach to realize its potential. The discussion also focuses on some of the practical implications that lie behind the adoption of a hybrid grid strategy, which possibly indicate why there are currently so few general, three-dimensional hybrid capabilities.

23.2 Underlying Principles

23.2.1 Historical Review

As with many other ideas, the origins of the concept of hybrid grid generation can be traced back to two unrelated workers, namely Nakahashi from Japan and Weatherill from the U.K.

Nakahashi advocated the use of hybrid grids in conjunction with a zonal finite difference (FD) and finite element (FE) flow solution methodology [Nakahashi and Obayashi, 1987a, b]. An implicit finite difference method was applied on structured grids to viscous flow modeling near geometric surfaces. The remaining regions were modeled by an explicit, node-based finite element solution of the Euler equations formulated on unstructured grids. Communication between the FD and FE zones was achieved by allowing the grids to overlap by one cell, with the grids sharing common nodes in these regions. Hence, information required at the zonal boundary of one region could be taken from the interior of the adjacent grid.

The observation that the approach combined both the computational efficiency of the FD method and the geometric flexibility of the FE method was central to Nakahashi's promotion of the use of hybrid grids.

In his early work, Nakahashi does not present sufficient detail about the techniques used to generate the grids for it to be possible to judge the generality of the mesh generation tools he used. Nevertheless, the fact that he was able to demonstrate that three-dimensional zonal flow solutions could be achieved on hybrid grids composed of tetrahedra and hexahedra is indeed worthy of note.

Weatherill proposed the use of hybrid grids by considering the apparent advantages and disadvantages of both the structured and unstructured approaches [Weatherill, 1988a] (at this time, he was well placed to give a pragmatic view on both approaches, having been involved in pioneering work in both block-structured [Weatherill and Forsey, 1985] and unstructured [Jameson, Baker, and Weatherill, 1986] mesh generation for complete aircraft.) He observed that the structured grid approach provides high-quality meshes at a relatively low cost and, because of inherent directional qualities, also provides an ideal environment for accurate and efficient flow algorithm techniques. However, structured meshes can be somewhat restrictive when applied to complex geometries and do not readily admit mesh point enrichment. In contrast, the unstructured mesh generation techniques have almost total flexibility for complex shapes and readily accept mesh enrichment. These advantages are counterbalanced, however, by their relatively high computational costs and lack of directional properties.

The observation that lies at the heart of Weatherill's proposition of hybrid grids is that the real advantages of one approach are the disadvantages of the other. The combination of the approaches is an attempt to capitalize on the merits of both approaches. This was demonstrated in two dimensionals by embedding unstructured regions of triangular grid in a background structured quadrilateral grid to

1. Form grids for multielement aerofoils.
2. Perform mesh adaptation to the flow over an aerofoil.
3. Improve mesh quality locally.

In this work, the structured regions were formed using the block-structured approach and the unstructured regions were created using the Delaunay connectivity algorithm [Weatherill, 1988b].

In contrast to Nakahashi, Weatherill [1988a] developed a single finite-volume flow algorithm for use with hybrid grids, as an extension of the scheme of Jameson, Baker and Weatherill [1986]. In this cell-vertex scheme, the control volume for a node was viewed as being the sum of elements containing the node, thereby creating overlapping control volumes. Hence, the flux balancing for nodes at the interface of the two mesh types was achieved by operating over both triangular and quadrilateral elements.

23.2.2 The Trend from Unstructured to Hybrid Grids

The unstructured grid approach, based primarily around the Delaunay [Weatherill, 1988b] (see Chapter 16) and moving-front (advancing front) [Morgan, Peraire and Peiro, 1992] (see Chapter 17) algorithms, has been shown to provide a highly effective basis for simulating inviscid flows over complex configurations, particularly when coupled with solution adaptive point enrichment and removal algorithms. However, considerable obstacles have been encountered in attempting to extend these algorithms to the generation of the highly compressed tetrahedra that are necessary for the efficient computation of viscous flows. These difficulties arise principally because both techniques use the properties of a sphere to determine the suitability of point connectivities, which works very well for the generation of isotropic grids, but not for the highly anisotropic grids required to allow shear layers to be resolved.

These problems have motivated workers [Pirzadeh, 1992; Kallinderis, 1996; Marchant and Weatherill, 1994] to investigate employing *structured* grid generation techniques locally to march triangulations of the geometric surfaces a distance sufficient to cover the expected extent of the shear layer (see Chapter 25). The conventional unstructured mesh generation techniques are then employed to yield a triangulation of the remainder of the domain. This approach allows most of the flexibility of the unstructured approach to be maintained through the use of triangles to cover the surface of the geometry, while also enabling the required point density close to solid surfaces to be achieved.

In some cases, the semistructured layers of prismatic elements that are formed by this surface inflation approach are retained for the flow simulation for reasons of efficiency [Kallinderis, 1996]. In others, each prism is subdivided into three tetrahedra [Marchant and Weatherill, 1994] to avoid the need to have a flow algorithm that operates over more than one element type. Whichever the case, it is possible to make use of the structured nature of the grid normal to the surface to enhance the sophistication of the subsequent modeling, as demonstrated by Weatherill, et al., [1987] in their use of locally structured triangular grids for multielement aerofoil flows.

This approach has met with a considerable degree of success. However, it is prone either to lack geometric flexibility, require excessive user intervention, or produce grids whose quality is not sufficient to support an accurate flow simulation. These limitations are observed in junction regions at discontinuities in surface slope and where a geometry has a high degree of surface curvature in one direction only. Combinations of these features exacerbate matters. Furthermore, the polar-like topology of the semistructured region of grid is such that the point distribution normal to the wake center-line is not of sufficient density for the wake to be adequately resolved.

An alternative approach to the generation of unstructured grids for viscous flows, which uses hybrid grid technology, centers on the use of *directional* refinement, as proposed in two dimensions by Barth [1994] and extended to three dimensions by Peraire and Morgan [1996]. Initially, an isotropic grid is formed, which is subsequently enriched until each point in the grid satisfies user specified stretching distributions that have been defined for curves and surfaces within the domain.

The scheme appears to offer significant potential savings in that, in addition to the stretching of the grid normal to geometric and wake surfaces, it allows anisotropic surface triangulations to be established in regions where the surface curvature is only high in one direction. However, it is not clear how well point density and element quality can be controlled in junction regions and at the edge of the refined regions. Furthermore, there remains the question of how well viscous flows can be simulated on highly stretched tetrahedral elements. The analysis of Baker [1996] adds significantly to this particular debate.

23.2.3 The Trend from Structured to Hybrid Grids

Within the class of mesh generation schemes that have been proposed to extend the application of structured grids to geometrically complex domains, the block-structured [Weatherill and Forsey, 1985] (see Chapter 13) and overlying [Benek, Steger, and Dougherty, 1983] (see Chapter 11) approaches have met with most success. However, neither approach has yet matured sufficiently for novel configurations to be treated accurately in a *routine* manner.

The block-structured approach has the potential to be the ultimate demonstration of hybrid grid technology. Within each block the grid is formed of regularly arranged hexahedra that can be generated by either of the established structured methods, namely the solution of elliptic partial differential equations or transfinite interpolation. The blocks have an irregular connectivity, however, which for all but the simplest of domains is not amenable to efficient manual specification, as discussed in Shaw and Weatherill [1992]. This motivates a requirement to be able to decompose a domain automatically into a suitable block structure, which can be cast as the need to generate a coarse unstructured grid of hexahedra.

Schonfeld and Weinerfelt [1991] proposed a scheme for this in two dimensions based on the use of the moving front technique to form quadrilateral cells. However, while the scheme was demonstrated for multielement aerofoil configurations, the block structures created did not form the most natural topology for each component, which is a key feature in the successful application of block-structured grids. The objective of forming *effective* block structures, which can be readily *controlled*, remains an open problem which if ever realized may be so irregular as to negate most of the advantages of structured grids. The semiautomatic approaches of Shaw and Weatherill [1992], Eiseman, Cheng and Hauser [1994], and Dannenhoffer [1996] represent the most advanced solutions to the problem to date (see Chapter 10).

The proposition that there is a limited range of problems that can be efficiently resolved using the block-structured approach has led Shaw, et al., [1991] to discuss situations where the use of hybrid grids would be favored. This is discussed further in Section 23.3.

Overlying grids do not have some of the restrictions of block-structured grids. However, the time taken to establish the meshes can be significant because of the need to ensure that sudden changes in mesh size are not encountered in overlapping regions. The *FAME* (feature associated mesh embedding) scheme of Albone [1988; 1992], which adopts a unified treatment to both geometric and flow features, appears to overcome this particular problem. For each feature (whose topology is either a corner, line, or surface), the approach forms individual meshes that are ordered hierarchically for the flow modeling based on the degrees of constraint possessed by the feature. An octree grid, formed by the repetitive subdivision of Cartesian hexahedral cells into eight, is then used to cover the remainder of the domain, with the refinement driven by the mesh spacing of the feature associated grids. This use of very many overlapping regular grids, coupled to the employment in the background of multiple levels of unstructured, embedded hexahedra, appears very flexible. However, it suffers along with other overlying methods with *conservation* and *nonuniqueness* problems when transferring solutions between meshes.

Increasingly, overlying mesh generators are migrating toward the use of hybrid grids. Liou and Kao [1994] demonstrate an approach in two dimensions whereby an initial set of regular, overlying grids is formed. The quadrilateral cells in the regions of overlap are then identified and removed from the grids, leaving a void which is subsequently filled with triangles. The approach adopted allows much of the technology developed for overlapping grids to be retained while overcoming the problems of conservation.

Noack, Steinbrenner, and Bishop [1996] have pursued a similar approach in three dimensions. In their work, the background mesh is an octree grid. Structured body-conforming meshes are formed adjacent to solid surfaces, and tetrahedra are used to fill the void between the background and local sets of hexahedra. In contrast to the methodology described in Section 23.3, the triangular faces of the tetrahedra are allowed to abut the quadrilateral faces of the hexahedra directly. The flow algorithm that operates on these meshes is described in Bishop and Noack [1995].

23.2.4 Potential Computational Benefits of Using Hybrid Meshes

In this section, observations are made on the computational benefits that structured grids offer in comparison with unstructured grids. If the majority of a hybrid mesh is composed of structured grid, then it is apparent that these benefits will also extend to the hybrid environment.

Shaw, et al., [1993] undertook a study to model inviscid flow over a wing-foreplane-fuselage configuration with both a solely block-structured grid and with a hybrid grid, with the unstructured region containing the foreplane. While this study cannot be viewed as totally rigorous, their findings were that to achieve a similar accuracy in the flow simulation, the surface mesh density of the unstructured grid had to be nearly an order of magnitude *more* dense than the structured grid. This was due primarily to the isotropic nature of the surface mesh, which meant that to resolve the streamwise curvature of the surface, the spanwise density of the mesh needed to be very high.

Also, it is apparent that independent of the strategy that is pursued to create a mesh for modeling shear layer flow, the required point density normal to the surface will be the same. These factors lead to the conclusion that in the viscous region of the flow domain the point density in an unstructured grid will be approximately ten times greater than in a structured grid. Furthermore, the rate at which time marching can be performed on the unstructured grid will be about one half (for a cell-vertex scheme) of that on a structured grid of the same point density. In addition to this, the amount of work done per time step will be greater due to the larger number of faces and edges in the unstructured grid, and there will also be increased processing time due to the amount of indirect addressing that needs to be done.

This short discussion suggests conservatively that to achieve the same level of convergence and accuracy, computations of viscous flow on an unstructured grid will be more than *20 times* as expensive as on a structured grid.

Turning to memory requirements, the findings reported in Shaw, Peace and Weatherill [1994] indicate that the storage requirements per point for a flow solution on an unstructured grid are about four times those of a structured grid. Even if the unstructured grid is coarser in the farfield, the storage of the unstructured grid, with its greater total number of points, would typically be *40 times* greater than for a structured grid.

There are thus very clear incentives to use structured grids whenever possible. For hybrid grids the implication is that the extent of unstructured grid employed should be as minimal as possible.

23.3 Best Practices

There is general agreement on what is needed in a mesh. It must conform to boundaries, contain points that are distributed effectively, be defined in a manner amenable to efficient computations, and have connections between points that form elements whose geometric properties satisfy certain criteria. The relative importance attached to these requirements will depend upon the problem being addressed.

In this chapter, the application is the modeling of the high Reynolds number turbulent flows associated with the complex bodies studied by aero- and hydrodynamicists, for which reviews of the demands on a mesh are given by Albone [1988], Albone and Swift [1996] and Patis and Bull [1996]. The task of simulating the flows over these geometries presents the severest of challenges to the traditional use of a single element type in a mesh and represents the principal industrial need that has motivated the current interest in hybrid grid generation/technology.

An example of this, which will be the focus of the remainder of this chapter, is the work of the Research Group at ARA in hybrid grid generation that has led to the development of the *SAUNA* (Structured and Unstructured Numerical Analysis) CFD system.

23.3.1 Mesh Generation Techniques Employed in the *SAUNA* System

In this section the basic mesh generation techniques that have been developed for the *SAUNA* system, which has been used to generate the grids described later, are reviewed. The system is capable of forming either solely block-structured, semistructured, or unstructured grids. In addition, it is capable of forming a hybrid combination of any of these mesh types. Hence, the same system can be used to form meshes efficiently for problems as diverse as the steady, viscous flow over a civil aircraft or the unsteady inviscid flow over a store released from a carriage bay.

23.3.1.1 Overview of Development

The initial approach to grid generation pursued within *SAUNA* was coined "Multi-block" [Weatherill and Forsey, 1985]. It centered around the formation of a global grid through the patching together of many structured, nonoverlapping grid systems, each of which covered a region that was topologically equivalent to a cuboidal block. This block-structured approach was applied to increasingly complex problems through the 1980s with a considerable degree of success [Shaw, Georgala and Weatherill, 1988]. However, as more and more complex configurations were attempted, so an appreciation of the limits to the range of problems the approach can handle developed.

Following a study of the hybrid approach in two dimensions [Weatherill, 1988a], work began on the initial development of a three-dimensional hybrid capability [Shaw, Peace and Weatherill, 1994]. The objectives were to explore ideas and gain an appreciation of the major issues that would need to be addressed to create a CFD system based on the hybrid philosophy.

The full development of a hybrid capability then began in earnest. The grid generation strategy for inviscid flow modeling centered around the use of hexahedral volumes combined into blocks, wherever they are readily attained, with pockets of tetrahedral grid embedded as appropriate to model local regions of high geometric complexity [Shaw, Georgala, Peace and Childs, 1991].

In the extension of this hybrid approach to the creation of grids for viscous flow modeling, the use of prismatic grid regions has been addressed, this additional grid type fitting in naturally to the hybrid grid framework [Chappell, Shaw, and Leatham, 1996; Peace, Chappell, and Shaw, 1996]. For geometric regions that are sufficiently complex to require an unstructured surface grid, the structured extension of the grid away from the surface allows layers of semistructured prismatic elements to be created. The regular nature of the grid normal to the surface is seen as being preferable to a fully unstructured approach in terms of both accuracy and efficiency. However, to achieve high mesh quality in junction regions, the approach requires to be augmented by a capability to create local block-structured regions between two intersecting surfaces from which prisms are grown. This avoids the need to generate prismatic elements in the regions highlighted as being difficult in Section 23.2.2.

A natural hierarchy of mesh elements for viscous flows can be drawn from the discussion to date and indeed this is the order in which the elements are created:

1. Block-structured hexahedral grid.
2. Semistructured prismatic grid.
3. Unstructured tetrahedral grid.

The generation of these grids is now considered in turn.

23.3.1.2 Structured Grid Generation

The multi-block approach [Weatherill and Forsey, 1985] is employed for the generation of structured grids. The domain is decomposed into an assemblage of topologically cuboidal blocks, each of which possesses its own curvilinear coordinate system. Grid lines are constrained to pass between block interfaces with continuity of position, slope and curvature. The technique allows the embedding of appropriate mesh structures local to components. The connectivity arrangement of blocks, known as the block topology, is determined via a semiautomatic approach, based on an input schematic representation of the configuration [Shaw and Weatherill, 1992].

23.3.1.2.1 Surface Grid Generation.

The generation of the surface grids is accomplished via the solution of elliptic PDEs [Thompson, Thames, and Mastin, 1974], with the initial boundary point distribution established automatically using an algorithm that is sensitive to local grid topology and geometry [Shaw and Weatherill, 1992] (see Chapter 9). If the default grids are found to be of insufficient quality, a graphics-based module is employed to modify boundary point distributions and add constraints to the mesh. The meshes are subsequently regenerated until satisfactory quality is achieved.

23.3.1.2.2 Field Grid Generation

The field mesh for inviscid flows is also generated by solution of elliptic PDEs with the source terms calculated using the method proposed by Thomas and Middlecoff [1980] (see Chapter 4). Algebraic techniques are employed to enrich the mesh for viscous flow modeling to allow exact control of the first cell height away from the surface (see Chapter 3). A capability to regenerate the mesh automatically in response to a perturbation of the geometry allows the system to be embedded within a design optimization strategy [Lovell and Doherty, 1994].

Mesh adaptation to either viscous or inviscid flow phenomena is performed using the LPE method of Catherall [1996]. This involves the numerical solution of equations for node positions that are formed as a linear combination of an inverted Laplace equation, an inverted Poisson equation, and an equidistribution equation. The Laplace term promotes smoothness and orthogonality, the Poisson term enables the retention of favorable features of the initial mesh, and the equidistribution term controls the redistribution of nodes according to a measure of solution activity. Mesh adaptation is covered in Part IV of this Handbook.

Prior to performing a flow simulation, the grid is decomposed into microblocks containing four cells in each coordinate direction. This micro-block structure is then recombined into macro-blocks based on either the requirement to distribute the grid effectively over a number of processors or to allow long loops to be achieved on vector machines. This recombination capability is also used in the generation of hybrid grids to redefine the grid into blocks when part of the initial block-structured grid has been removed to be replaced by tetrahedra and/or prisms.

23.3.1.3 Semistructured Grid Generation

The technique employed [Chappell, 1996] for generating prismatic elements is a marching method, and as such starts from a defined surface and propagates outwards to an outerboundary, the exact shape or location of which cannot be predetermined. The prismatic grid is built up one layer at a time. At each stage, the positions of points in the next layer are determined as a function of the current outer grid surface, which will initially be the input unstructured surface grid.

The generation of a prismatic layer can be separated into two distinct processes: the evaluation of *normal vectors* and the determination of *marching distances* along these vectors.

23.3.1.3.1 Evaluation of Normal Vectors

The first stage of the prismatic grid generation process is the determination of marching direction vectors at all points on the unstructured surface. This is achieved by evaluating the normals to all surface triangles and sending contributions to the forming nodes weighted according to the angle subtended at the node. All nodal vectors are then normalized to unit magnitude.

This yields an approximately normal marching vector for every point on the current grid surface. If these vectors are used in this form, however, the normal grid lines will converge from concave surface regions, leading to grid crossover. This undesirable feature can be overcome by an iterative smoothing of the vectors using a Laplacian filter, with the amount required being surface-topology- dependent. The trade-off is a reduction in grid orthogonality.

23.3.1.3.2 Marching Distances along Normal Vectors

In the development of prismatic grid generation methods, workers have given much attention to the determination of appropriate marching distances along each grid line. If the initial surface features any concave regions, then the maximum distance away from the body to which the grid can extend will be limited, unless some form of marching distance variation is employed.

The goal of marching distance variation within a layer is to compensate for regions of high concave and convex curvature, increasing marching distances in the former case and reducing them in the latter. The overall effect is that the grid tends toward a spherical effect as it moves away from the geometric surface. Several approaches to this problem have been investigated with a spring analogy approach found to be the most successful [Chappell, 1996].

By treating the normal vectors connecting a point to its neighbors as springs and summing their effects, an overall "spring force" vector for the point can be calculated. The scalar product of this vector with the nodal normal vector gives a measure of the local surface curvature. In convex regions, where the net effect of the adjoining points will act in opposition to the marching direction, a negative measure will be returned, and vice-versa in concave regions. This measure can form the basis of a marching distance modification function which, with appropriate use of unit vectors, is independent of the distance between a node and its neighbors. The modification function is subject to two constraints. The first checks that the value lies within an appropriate range, the second ensures stability as the grid propagates radially. An average distance for the layer is calculated based on user-defined parameters, which is then multiplied by the modification function to give the nodal marching distance.

23.3.1.4 Unstructured Grid Generation

The optimal properties of the Delaunay connection algorithm, and efficient algorithms that exist for its implementation, led to its adoption within the *SAUNA* system for forming the regions of tetrahedral grid [Childs, et al., 1992; Childs and Shaw, 1993]. The mesh generation is performed in two stages: surface grids, followed by volume grids. For the former, the generation of grids that are independent of the geometry definition has been a particular focus of effort. For the latter, the problem of boundary integrity requires careful attention. See Chapters 19 and 16 for a discussion of unstructured surface and volume grid generation.

23.3.1.4.1 Unstructured Surface Grid Generation

Separate meshes are formed for each surface of the configuration and for the boundary of the domain. For each, boundary point distributions are defined in a graphics-based working environment, with boundary lines delimited into segments to facilitate precise control over distributions. These point distributions can be augmented by fixed internal lines either to exercise precise control of the local grid or ensure that a feature (i.e., a slope discontinuity) is resolved accurately.

To be consistent with the creation of a high quality Delaunay field mesh, it is required that the surface meshes consist of triangles that are approximately equilateral in physical space. To this end a pseudo-Delaunay surface triangulation procedure has been developed [Childs and Shaw, 1993], which is coupled to a grid point location algorithm. Control of grid density in regions of high surface curvature is assured through the solution of an optimization problem based on determining a desired edge length distribution. Each surface grid is generated independently, and they are then unified to form the bounding grid for the field grid.

23.3.1.4.2 Boundary Integrity

The Delaunay approach is beset by its inability to ensure that the resulting triangulation conforms to the boundaries of the flow domain–boundary integrity. Therefore, if the scheme is to be applied routinely, the basic methodology must be supplemented by a procedure that overcomes this limitation.

To this end, an automatic boundary integrity algorithm has been developed that consists of local modifications to the datum bounding surface grids so that they more closely match the Delaunay triangulation of the boundary points. Such modifications are limited only by topological considerations and the need to keep a faithful geometric description of curved surfaces. The procedure is an iterative one that is deemed to have converged when all edges and faces of the boundary triangulation are contained in the tetrahedrization. The full implications of this approach to boundary integrity are discussed later, in the section covering the interfacing of different grid types.

23.3.1.4.3 Unstructured Field Grid Generation

The three-dimensional grid is determined automatically from the bounding surface grids with grid points positioned according to boundary grid density, curvature and desired rate of change of grid density. The procedure commences with the creation of an initial octree model of the flow domain (see Chapters 14 and 15). Each octant is subdivided as necessary until the density of the terminal octants cutting boundary surfaces is comparable with that of the boundary grid. Further levels of refinement of the octree are then performed based on surface curvature. Finally, the octree is graded so that adjacent octants do not differ by more than one level. Grid points are then located within the empty octants that lie interior to the unstructured domain and connected together to form a coarse tetrahedrization of the domain. This grid is used as the basis for solving a coupled set of PDEs which yield a desired edge length in the field.

A denser set of points is then formed by selective addition of suitable points to the Delaunay grid, via an automatic edge refinement procedure, until the optimal edge lengths for the tetrahedra are attained. Throughout, it is found essential to employ the generalized Delaunay algorithm wherein the grid is allowed to become non-Delaunay, due to boundary influences, but only if grid quality is enhanced.

Mesh smoothing techniques, coupled with point addition and removal algorithms are used to regenerate the grid in response to a change to the shape of the boundary of the domain. This technology can be used to achieve meshes rapidly either as a result of a design modification or in response to the motion of a body, as in a store release [Leatham, 1996].

23.3.2 Interfacing Different Grid Types

The interfacing of the different elements of a hybrid grid represents a major component in the development of a hybrid grid generation system, which must be performed in an automatic manner. In this section, the interfacing of block-structured, unstructured, and semistructured grids is discussed.

23.3.2.1 Interfacing Structured and Semistructured Grids

At the interface of block-structured and prismatic grid regions, the quadrilateral faces of the elements must abut. This means that all points on the interface will be fixed points to which the prismatic grid generator must conform as the layers are formed.

To make the transition from block-structured to prismatic grid as smooth as possible, the vectors resulting from the fixed boundary points are used in the smoothing process for the normals in the prismatic region [Chappell, 1996]. This has the effect of preventing any sharp changes of direction near the interface.

To obtain a representative marching distance for the prismatic grid, a Laplacian equation is solved for each layer, with the multi-block mesh spacing providing the necessary boundary data.

23.3.2.2 Interfacing Structured and Unstructured Grids

Clearly, some form of special treatment is required at the interface between regions of structured hexahedral grid and unstructured tetrahedral grid. One strategy could be to allow a number of tetrahedral faces to abut the face of a hexahedra. However, while this would simplify the grid generation process, a significant burden would be placed on the flow solver, which would not only have to perform well on different types of elements but would also have to be insensitive to hanging nodes, edges, and faces.

Alternatively, an additional element, the pyramid can be used. For if the quadrilateral base of this element adjoins a hexahedron, the remaining triangular faces can abut to this tetrahedra, thereby maintaining a one-to-one connectivity of all faces within the mesh. This is the approach that has been followed.

However, due to the point addition and edge swapping techniques adopted to ensure that the Delaunay algorithm conforms to the boundary of the domain, the interface of the pyramid elements is augmented locally by a buffer "layer" of tetrahedra, prior to the generation of the unstructured grid. These tetrahedra are formed in two stages, the first of which protects the faces, and the second the edges of the pyramids from the unstructured grid generator. The pyramids and initial layer of tetrahedra are both formed in an automatic manner [Shaw, et al., 1991].

Following the generation of the unstructured region of grid, the initial layer of tetrahedra at the interface needs to be adjusted as a result of the steps taken to ensure boundary integrity. An automatic module has been developed to accomplish this task in response to knowledge of the edges that have been swapped on, and nodes that have been added to, the boundary of the unstructured domain.

23.3.2.3 Interfacing Semistructured and Unstructured Grids

There are three principal factors governing the ideal extent of the prismatic region, the first two of which place a lower limit and the third an upper limit on the extent of the prismatic region:

1. The grid should extend to a distance where viscous effects become negligible.
2. The cell aspect ratio (height/average side length) should be as close to unity as possible to promote a smooth transition to the tetrahedral region.
3. The quality of the triangulation on the outer layer should be as good as possible in order to achieve a good quality tetrahedral mesh.

The concept of a buffer layer is also used to interface prisms and tetrahedra [Chappell, Shaw, and Leatham, 1996]. In this case the buffer is not needed to ensure compatibility of element faces but rather to eliminate the need to modify the prismatic grid after the generation of the unstructured grid. The prismatic region can be insulated from the effects of the procedure followed to achieve boundary integrity by breaking down the outer layer of prismatic cells, with each prism becoming three tetrahedra. This operation must be performed in such a way that the diagonal introduced by splitting a quadrilateral face matches for both prisms abutting that face. An iterative algorithm for achieving this type of decomposition of the outer-most layer of semistructured grid was originally proposed by Lohner [1993]. The set of face splits derived from this are used to determine the initial make-up of the tetrahedral buffer layer.

Following the generation of the unstructured grid, the same procedures that are used to modify the definition of the tetrahedra in the structured/unstructured interface region are used to modify the tetrahedra in the semistructured/unstructured interface.

On completion of the generation of the unstructured grid, all grid types are passed to a separate module that forms the complete data structure describing the grid.

23.3.3 Data Structures for Describing Hybrid Grids

The data structure that describes the hybrid grid to the flow solver is central to the success of the approach. The description that has been adopted for a cell-vertex scheme is detailed in Peace and Shaw [1992].

The nodes are all uniquely numbered, with all nodes at which a given boundary condition is applied stored contiguously. Nodes that either lie inside each block or solely within the unstructured or semistructured field grids are also stored contiguously.

Connectivity matrices are used to describe the joining of faces of tetrahedra to the triangular faces of either other tetrahedra, or prisms, or pyramids. Similarly, for the prisms, the unstructured surface grids are stored in edge-based connectivity matrices, with surface node-based pointers used to define the nodes lying along the lines of structure in the grid. The block-structured region is stored in a block-based structured manner for points that do not lie on block faces. A pointer system, based on the faces of each block, is used to access nodes that lie on block faces; these nodes might be either part of more than one block or be part of other elements. All edges in the unstructured grid and on the boundaries of both blocks and regions of semistructured grid are stored also.

In Section 23.3.1, the main emphasis of the discussion is on the techniques that have been developed to position nodes for each of the grid types. It is worth nothing that a significant part of the total work undertaken to develop the individual mesh generation modules has focused on creating and communicating data that allows the data structure described above for the complete hybrid grid to be defined automatically.

23.3.4 Examples of Hybrid Meshes

The creation of grids for three different aerodynamic and hydrodynamic configurations is discussed in order to highlight what are considered to be *"best practices"* in the generation of hybrid grids. The configurations are chosen because they demonstrate the three possible types of mesh that can be formed with the *SAUNA* grid generation system. The examples begin with the combination of block-structured and unstructured grids, followed by the use of semistructured and unstructured grids and end with an example which utilizes all mesh types.

Due to the commercial sensitivity of some of the configurations shown, there are no results presented in this section from flow calculations performed on these meshes. However, details of the flow algorithm that operates on these meshes can be found in Peace and Shaw [1992], with results from both inviscid and viscous flow calculations given in Shaw, et al., [1993], Shaw, et al., [1994b], Peace, et al., [1994], and Peace, et al., [1996]. Further discussions on the generation of hybrid meshes can be found in these references and in Shaw, et al., [1994b].

23.3.4.1 Creation of a Block-Structured/Unstructured Grid for a Civil Aircraft*

To illustrate the creation of a block-structured/unstructured grid, a civil wing-fuselage-pylon-nacelle configuration has been chosen. While block-structured grids have been formed for this type of layout, significant time was taken to establish these meshes, which are of a questionable quality around the pylon. Furthermore, apparently minor changes to the pylon geometry can lead to a major requirement to modify the local block topology.

However, if the configuration is considered without the pylon, then the remaining components, both individually and collectively, are well-suited to the generation of a block-structured grid. The highly three-dimensional pylon, with its complex shaping and intersection with both the wing and nacelle surface, is readily modeled by an unstructured grid. The complete configuration can therefore be addressed efficiently by the hybrid approach without having to incur the overhead associated with the completely unstructured approach.

The creation of the hybrid grid commenced with the decomposition of the domain around the wing-fuselage and nacelle into blocks. A polar topology was embedded around the fuselage, with a spherical polar topology chosen to model the nose region. A "C" topology conformed to the wing leading-edge geometry. Finally, a polar topology internal and external to the nacelle, with a "C" topology around the intake lip, yielded a total of 642 blocks.

The next stage in the creation of the mesh was to identify the extent of the structured grid that should be removed and replaced by unstructured grid. In this case, the pylon geometry was introduced into the structured grid and all micro-blocks that either contained the pylon or lay within a user specified distance of the pylon were removed. The remainder of the structured grid was combined into 34 macro-blocks and the initial structured/unstructured interface formed as depicted in Figure 23.1.

An unstructured grid was formed on the pylon and part of the wing and nacelle surfaces. In conjunction with the inner triangulation of the structured/unstructured interface this formed the boundary data for the generation of the unstructured field mesh. The meshes were then fused together as shown in Figure 23.2.

*Grid generated by A. Shires, DERA, Bedford, UK and C.M. Newbold, DERA, Farnborough, UK.

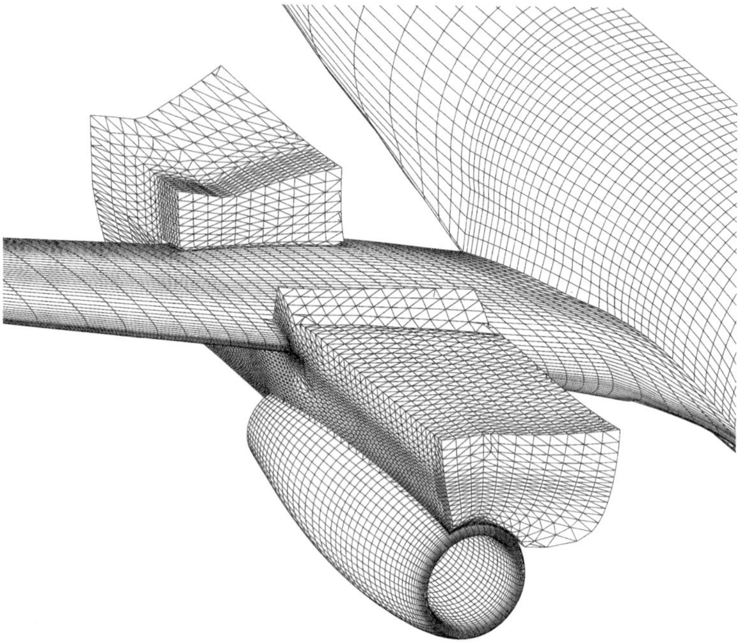

FIGURE 23.1 Boundary for unstructured part of hybrid grid.

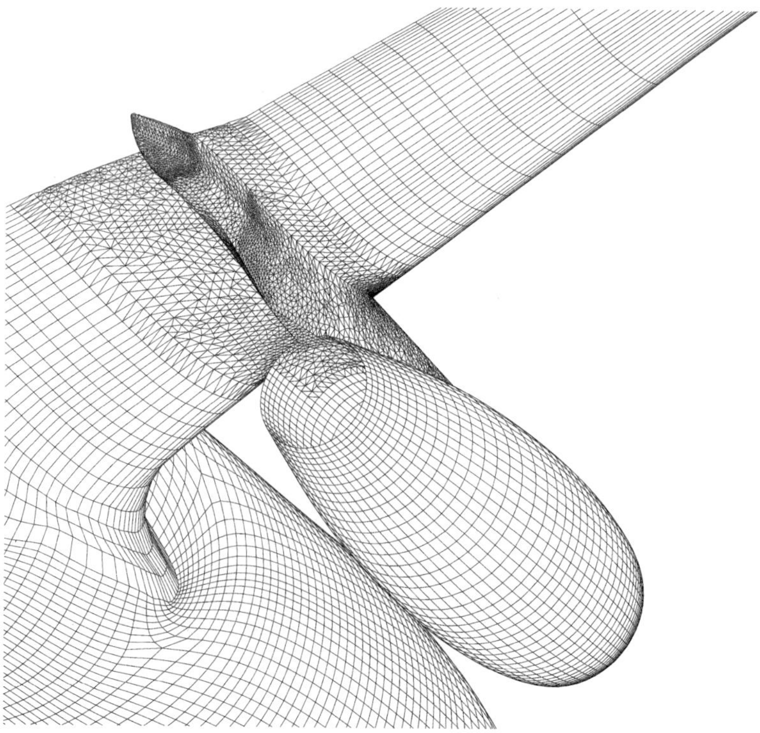

FIGURE 23.2 Hybrid structured/unstructured grid configuration 1.

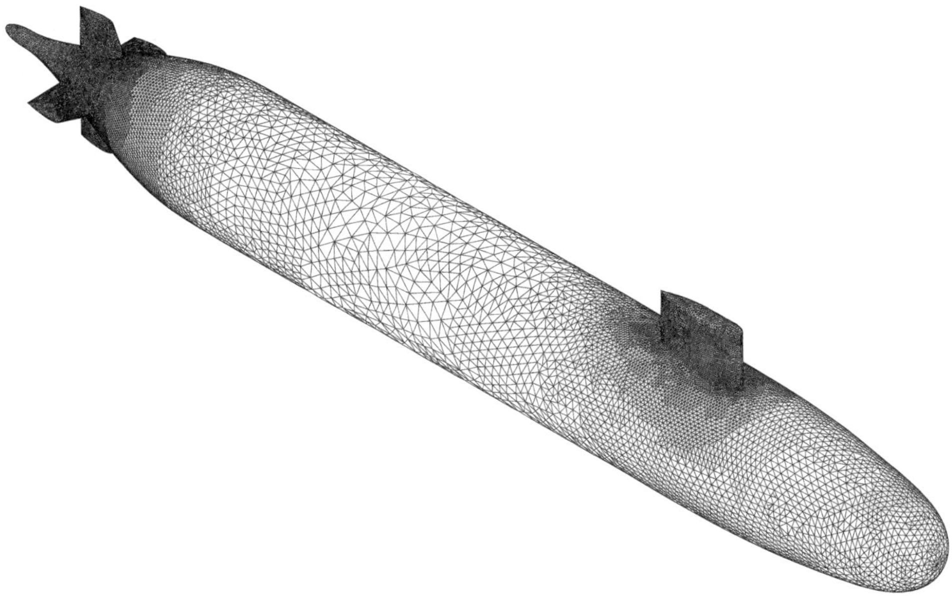

FIGURE 23.3 Surface triangulation for fully appended submarine.

23.3.4.2 Creation of a Semistructured/Unstructured Grid for a Submarine*

To demonstrate the use of semistructured/unstructured grids, a fully appended submarine is chosen. The configuration features several regions of high geometric difficulty for the generation of prismatic grids, i.e., where the trailing-edges of the control surfaces intersect the hull, and a large range of length scales.

Unstructured grids were generated on all surfaces, as shown in Figure 23.3, and the surface inflation technique employed to generate the semistructured grid away from the submarine. In practice, several attempts had to be made to achieve a valid extent of prismatic grid, with user input parameters that control the amount of smoothing and stretching of the grid adjusted to achieve the desired result.

Unstructured surface grids were then formed for the farfield boundary and the remaining extent of the symmetry plane that had not yet been covered. These were used in conjunction with the outer triangulation of the semistructured/unstructured buffer to provide boundary data for the generation of the unstructured field grid.

Part of the unified hybrid mesh is shown in Figure 23.4. Note in particular how the marching distance of the grid adjusts to the size of the local surface triangulation to achieve a smooth transition of cell sizes at the prismatic/tetrahedral interface.

While the example is an impressive demonstration of prismatic mesh generation, it does illustrate some of the weaknesses of a strategy based on the sole use of prisms and tetrahedra that were alluded to in Section 23.2.2.

23.3.4.3 Creation of a General Hybrid Grid for a Store Below a Research Aircraft**

The final configuration examined is a wing/fuselage/foreplane research model below the wing of which is a finned store. To model the full trajectory of the store as it is released from the aircraft, possibly pitching, yawing, and rolling, is beyond the efficient application of block-structured grids.

* Grid generated by J.A. Chappell, ARA, Bedford, UK.

**Grid generated by J.A. Shaw and J.A. Chappell, ARA, Bedford, UK.

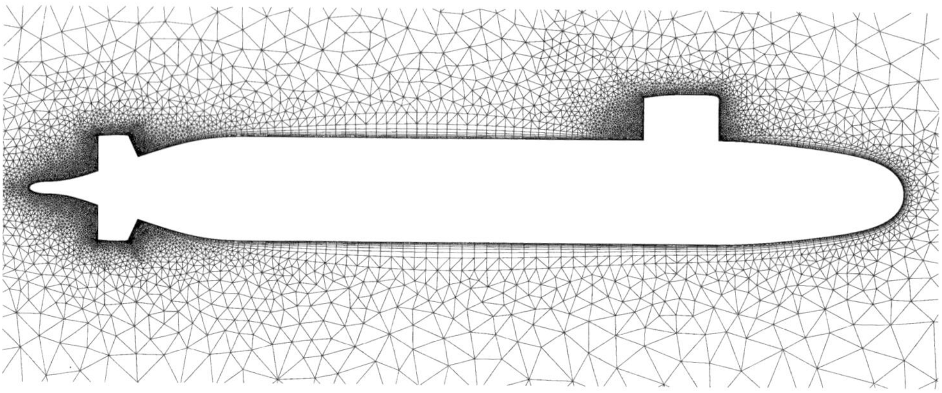

FIGURE 23.4 Hybrid semistructured/unstructured grid for configuration 2.

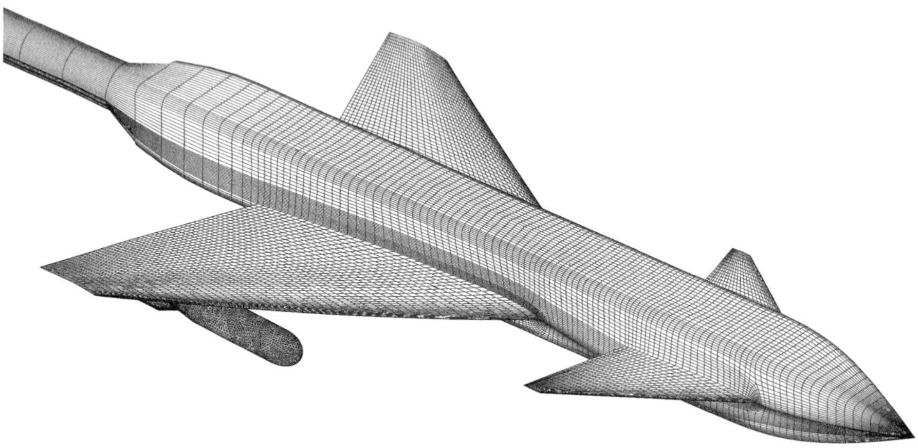

FIGURE 23.5 Hybrid block-structured/semistructured/unstructured grid for configuration 3.

However, the parent aircraft is amenable to the generation of a block-structured grid, which was readily attained. A region of this grid below the wing was then removed and the block-structured/unstructured interface constructed.

Layers of prismatic grid were grown from a surface triangulation of the store and fins. The field mesh was completed by forming tetrahedral grid in the region between the block-structured/unstructured and semistructured/unstructured buffers.

Figures 23.5 and 23.6 illustrate the full hybrid surface grid and a section through the field grid, respectively. The case amply demonstrates the building block route to forming efficient, high-quality grids for configurations of great complexity that is possible with hybrid grids.

23.4 Research Issues and Summary

It is generally accepted that the techniques for generating meshes are well established and that the main challenge lies in developing highly usable systems around these techniques. This is particularly the case for hybrid grids.

For hybrid grids to become acceptable in the user environment, clear indication of where a given mesh type should be used for the problem of interest needs to be available. A user's own knowledge base, coupled with good training, on-line support, and documentation will go some way toward meeting this objective. However, what is and is not a difficult region of geometry to mesh with a structured grid is

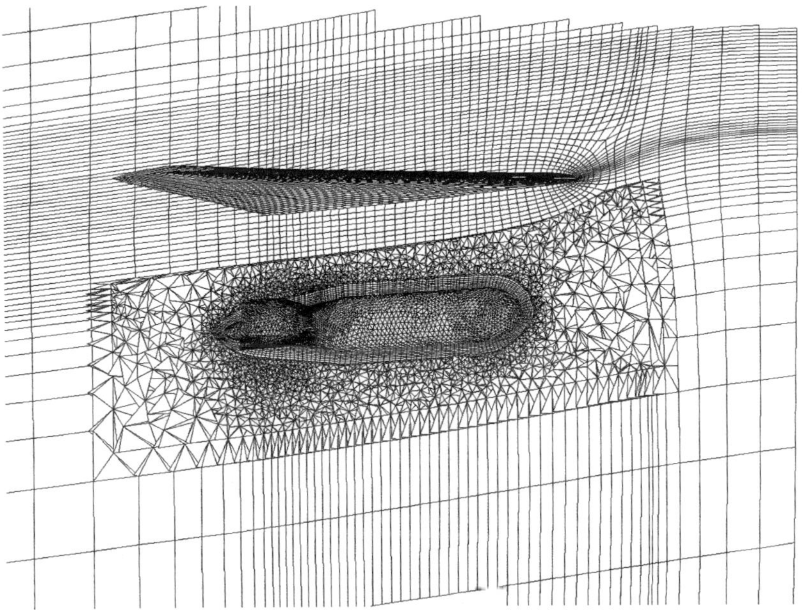

FIGURE 23.6 Slice through hybrid field grid for configuration 3.

not always readily appreciated, even for experienced practitioners. Some form of artificial intelligence that interrogates the local geometric properties of the boundaries of a domain would appear to be required, but the level of sophistication that would be needed should not be underestimated.

It is apparent that the simulation of aerodynamics and hydrodynamic flows will be performed increasingly on parallel processors. For effective computations to be achieved on these platforms, the algorithms used to decompose the domain need to be capable of providing a good load balance across all processors. This becomes an increasingly significant issue in a hybrid grid where the topology of the structured regions imposes significant constraints on the decomposition and the different elements require different processing times per time step.

It was expected initially that significant problems may be encountered in the flow simulations at the interface between the different element types. To date this has not been observed, which may be testament to the care taken to join the grids together. However, it would be naive to suggest that this region of mesh, which inevitably contains significant changes in element size, could not lead to difficulties. Further validation of the flow solution in these regions is needed.

Furthermore, each of the tools within the hybrid system must be of a similarly high quality and easy to use since the number of modules that need to be executed to produce the complete grid is inevitably significant compared to single element systems.

The inevitable impact of this is the expense and long term commitment to the philosophy that is required to develop a usable capability. When many groups have already invested heavily in either structured or unstructured grid technology, the decision to move to hybrid grids is not taken easily.

While numerous papers are now appearing on the approach in two dimensions, the evidence of work in three dimensions is sparse. The formation of strategic alignments between major industrial companies and/or government bodies, which allow specialists in the two main fields of grid generation to collaborate, could arguably have the greatest impact on changing this situation.

Hybrid grid generation offers the potential of combining the advantages of structured and unstructured grids, enabling high quality, efficient meshes to be formed for a wide range of problems. The meshes will inevitably take longer to form and require greater expertise then totally unstructured grids. However, the potential efficiency and modeling gains that hybrid grids offer are such that the total elapsed time and cost to achieve the end result the engineer needs justifies this required investment.

Acknowledgments

This work has been undertaken with the support of the Procurement Executive, United Kingdom Ministry of Defence. I am grateful to my past and present colleagues at ARA who have contributed their ideas and effort to the work described here and am indebted to those who have so willingly provided the figures for the examples discussed.

Finally, I would like to dedicate this article to Dr. David Catherall, who has acted as technical monitor for the work described here throughout its development and who is shortly to retire from full-time work at DERA Farnborough, U.K. Dave's consistent support over many years for the work described has been, and still is, greatly appreciated.

References

1. Albone, C.M., An approach to geometric and flow complexity using Feature-Associated Mesh Embedding (FAME): Strategy and First Results, *Numerical Methods for Fluid Dynamics III.* Morton K.W. and Baines, M.J., (Eds.), Clarendon Press, Oxford, U.K., 1988, p 215–235.
2. Albone, C.M., Embedded meshes of controllable quality synthesized from elementary geometric features, *AIAA Paper 92-0663.* 30th AIAA Aerospace Sciences Meeting, Reno, NV, 1992.
3. Albone, C.M. and Swift, V.J., Resolution of high Reynolds number flow features using dynamically-overlying meshes, *Numerical Grid Generation in Computational Field Simulations.* 1996, Soni, B.K., Thompson, J.F., Hauser, J., Eiseman, P.R., (Eds.) Mississippi State University, MS, pp 855–864.
4. Baker, T.J., Discretization of the Navier–Stokes equations and mesh induced errors, *Numerical Grid Generation in Computational Field Simulations.* Soni, B.K., Thompson, J.F., Hauser, J. and Eiseman, P.R., (Eds.), Mississippi State University, MS, 1996, pp 209–218.
5. Barth, T., Aspects of unstructured grids and finite volume solvers for the Euler and Navier–Stokes equations, *VKI Lecture Series on Computational Fluid Dynamics,* 1994, 05.
6. Bishop, D.G. and Noack, R.W., An implicit flow solver with upwind differencing for three-dimensional hybrid grids, *AIAA Paper 95-1707,* 12th AIAA Computational Fluid Dynamics Conference, 1995.
7. Benek, J.A., Steger, J.L., Dougherty, F.C., A flexible grid embedding technique with applications to the Euler equations, *AIAA Paper 83-1944,* Danvers, MA, 1983.
8. Catherall, D., 1996. Adaptivity via mesh movement with three-dimensional block-structured grids, *Numerical Grid Generation in Computational Field Simulations.* Soni, B.K., Thompson, J.F., Hauser, J. and Eiseman, P.R., (Eds.), Mississippi State University, MS, 1996, pp 57–66.
9. Chappell, J.A., Private communication, 1996.
10. Chappell, J.A., Shaw, J.A., Leatham, M., 1996. The generation of hybrid grids incorporating prismatic regions for viscous flow calculations, *Numerical Grid Generation in Computational Field Simulations.* Soni, B.K., Thompson, J.F., Hauser, J. and Eiseman, P.R., (Eds.), Mississippi State University, MS, 1996, pp 537–546.
11. Childs, P.N. and Shaw, J.A., Generation and analysis of hybrid structured /unstructured grids, *Numerical Methods for Fluid Dynamics IV,* Baines, M.J. and Morton, K.W., (Eds.), Clarendon Press, Oxford, U.K., 1992, pp 499–507.
12. Childs, P.N., Shaw, J.A., Peace, A.J., Georgala, J.M., SAUNA: A system for grid generation and flow simulation using hybrid structured/unstructured grids, *Computational Fluid Dynamics '92.* Hirsch, Ch., Periaux, J., Kordulla, W., (Eds.), Elsevier, Amsterdam, Holland, 1992, pp 875–882.
13. Dannenhoffer III, J.F., Automatic generation of block structures — progress and challenges, *Numerical Grid Generation in Computational Field Simulations.* Soni, B.K., Thompson, J.F., Hauser, J. and Eiseman, P.R., (Eds.), Mississippi State University, MS, 1996, pp 403–412.
14. Eiseman, P.R., Cheng, Z., Hauser, J., 1994. Applications of multi-block grid generators with automatic zoning, *Numerical Grid Generation in Computational Fluid Dynamics and Related Fields.* Weatherill, N.P., Eiseman, P.R., Hauser, J., Thompson, J.F., (Eds.), Pineridge Press, Swansea, U.K., 1994, pp 123–134.

15. Jameson, A., Baker, T.J., and Weatherill, N.P., Calculation of Inviscid transonic flow over a complete aircraft, *AIAA Paper 86-0103.* 24th AIAA Aerospace Sciences Meeting, Reno, NV, 1986.
16. Kallinderis, Y., Discretisation of complex 3D flow domains with adaptive hybrid grids, *Numerical Grid Generation in Computational Field Simulations.* Soni, B.K., Thompson, J.F., Hauser, J. and Eiseman, P.R., (Eds.), Mississippi State University, MS, 1996, pp 505–515.
17. Leatham, M., Private communication, 1996.
18. Liou, M.S. and Kao, K.-H., Progress in grid generation: from chimera to DRAGON grids, *Frontiers of CFD.* Caughey, D.A. and Hafez, M.M., (Eds.), John Wiley, Chichester, England, 1994, pp 385-412.
19. Lohner, R., Matching Semi-Structured and Unstructured Grids for Navier–Stokes Calculations, *AIAA Paper 93-3348.* 1993.
20. Lovell, D.A. and Doherty, J.J., Aerodynamic design of aerofoils and wings using a constrained optimisation method, *Proc. of 19th ICAS Congress, Paper-94-2.1.2,* 1994.
21. Marchant, M.J. and Weatherill, N.P., Unstructured grid generation for viscous flow simulations, *Numerical Grid Generation in Computational Fluid Dynamics and Related Fields.* Weatherill, N.P., Eiseman, P.R., Hauser, J., Thompson, J.F., (Eds.), Pineridge Press, Swansea, U.K., 1994, pp 151–162.
22. Morgan, K., Peraire, J., Peiro, J., Unstructured mesh methods for compressible flows, *AGARD Report 787. Special Course on Unstructured Grid Methods for Advection Dominated Flows,* 1992., 5.1–5.39.
23. Nakahashi, K. and Obayashi, S., FDM - FEM Zonal method for viscous flow computations over multiple bodies, *AIAA Paper 87-0604.* 25th AIAA Aerospace Sciences Meeting, Reno, NV, 1987a.
24. Nakahashi, K. and Obayashi, S., Viscous flow computations using a composite grid, *AIAA Paper 87-1128,* 1987b.
25. Noack, R.W., Steinbrenner, J.P., Bishop, D.G., A three dimensional hybrid grid generation technique with application to bodies in relative motion, *Numerical Grid Generation in Computational Field Simulations.* Soni, B.K., Thompson, J.F., Hauser, J. and Eiseman, P.R., (Eds.), Mississippi State University, MS, 1996, pp 547–556.
26. Patis, C.C. and Bull, P.W., 1996. Generation of grids for viscous flows around hydrodynamic vehicles, *Numerical Grid Generation in Computational Field Simulations.* Soni, B.K., Thompson, J.F., Hauser, J. and Eiseman, P.R., (Eds.), Mississippi State University, MS, 1996, pp 825–834.
27. Peace, A.J. and Shaw, J.A., The modeling of aerodynamic flows by solution of the Euler equations on mixed polyhedral grids, *Int. J. Num. Meth. Eng.* 1992, 35, pp 2003–2029.
28. Peace, A.J., May, N.E., Pocock, M.F., Shaw, J.A., Inviscid and viscous flow modeling of complex aircraft configurations using the CFD simulation system SAUNA, *Proc. of 19th ICAS Congress, Paper-94-2.6.3,* 1994.
29. Peace, A.J., Chappell, J.A., Shaw, J.A., Turbulent flow calculations for complex aircraft geopmetries using prismatic grid regions in the SAUNA CFD code. *Proc. of 20th ICAS Congress,* Paper 96-1.4.2, 1996.
30. Peraire, J. and Morgan, K., Viscous Unstructured mesh generation using directional refinement, *Numerical Grid Generation in Computational Field Simulations.* Soni, B.K., Thompson, J.F., Hauser, J. and Eiseman, P.R., (Eds.), Mississippi State University, MS, 1996, pp 1151–1164.
31. Pirzadeh, S., Viscous unstructured three-dimensional grids by the advancing layers method, *AIAA Paper 94-0417.* 32nd AIAA Aerospace Sciences Meeting, Reno, NV, 1994.
32. Schonfeld, T. and Weinerfelt, P., The automatic generation of quadrilateral multi-block grids by the advancing front technique, *Numerical Grid Generation in Computational Fluid Dynamics and Related Fields.* Arcilla, A.S., Hauser, J., Eiseman, P.R., Thompson, J.F., (Eds.), Elsevier Science Publishers B.V. (North-Holland), Amsterdam, Holland, 1991, pp 743–754.
33. Shaw, J.A. and Weatherill, N.P., Automatic topology generation for multi-block grids, *App. Maths Computation,* 1992, 52, pp 355–388.
34. Shaw, J.A., Georgala, J.M., Weatherill, N.P., The construction of component adaptive grids for aerodynamic geometries, *Numerical Grid Generation in CFD '88.* Sengupta, S., Hauser, J., Eiseman, P.R., J.F.Thompson, J.F., (Eds.), Pineridge Press, Swansea, U.K., 1988, pp 383–394.

35. Shaw, J.A., Peace, A.J., Weatherill, N.P., A three-dimensional hybrid structured–unstructured method : motivation, basic approach and initial results, *Computational Aeronautical Fluid Dynamics.* Fezoui, L., Hunt, J.C.R., Periaux, J., (Eds.), Clarendon Press, Oxford, U.K., 1994, pp 157–201.
36. Shaw, J.A., Georgala, J.M., May, N.E., Pocock, M.F., Application of three dimensional hybrid structured/unstructured grids to land, sea and air vehicles, *Numerical Grid Generation in Computational Fluid Dynamics and Related Fields.* Weatherill, N.P., Eiseman, P.R., Hauser, J., Thompson, J.F., (Eds.), Pineridge Press, Swansea, U.K., 1994a, p151–162.
37. Shaw, J.A., Georgala, J.M., Peace, A.J., Childs, P.N., The construction, application and interpretation of three-dimensional hybrid meshes, *Numerical Grid Generation in Computational Fluid Dynamics and Related Fields.* Arcilla, A.S., Hauser, J., Eiseman, P.R., Thompson, J.F., (Eds.), Elsevier Science Publishers B.V. (North-Holland), Amsterdam, Holland, 1991, pp 887–898.
37. Shaw, J.A., Peace, A.J., Georgala, J.M., Childs, P.N., Validation and evaluation of the advanced aeronautical CFD system SAUNA – A method developer's view, *Recent Developments and Applications in Aeronautical CFD.* Paper 3. Royal Aeronautical Society, London, U.K., 1993.
38. Shaw, J.A., Peace, A.J., May, N.E., Pocock, M.F., Verification of the CFD Simulation system SAUNA for complex aircraft configurations, *AIAA Paper 94-0393.* 32nd AIAA Aerospace Sciences Meeting, Reno, NV, 1994.
39. Thomas, P.D. and Middlecoff, J.F., Direct control of grid point distributions in meshes generated by elliptic equations, *AIAA J. 18:652-656,* 1980.
40. Thompson, J.F., Thames, F.C., Mastin, W., Automatic Numerical generation of body fitted curvilinear co-ordinate systems for field containing arbitrary two-dimensional bodies, *J. Comp. Phys.* 1994, 15, pp 299–319.
41. Weatherill, N.P., On the Combination of structured–unstructured meshes, *Numerical Grid Generation in CFD '88.* Sengupta, S., Hauser, J., Eiseman, P.R., J.F.Thompson, J.F., (Eds.), Pineridge Press, Swansea, U.K., 1988a, pp 729–739.
42. Weatherill, N.P., A method for generating irregular computational grids in multiply connected planar domains, *Int. J. Num. Methods Fluids,* 1988b, 8, pp 181–197.
43. Weatherill, N.P. and Forsey, C.R., Grid generation and flow calculations for aircraft geometries, *J. of Aircraft.* 1985, 22,10, pp 855–860.
44. Weatherill, N.P., Johnston, L.J., Peace, A.J., Shaw, J.A., A method for the solution of the Reynolds-averaged Navier–Stokes equations on triangular grids, In *Proc. of the 7th GAMM Conf. on Numerical Methods in Fluid Dynamics.* Louvain-La-Neuve, Belgium, 1987.

24

Parallel Unstructured Grid Generation

Hugues L. de Cougny

Mark S. Shepherd

24.1 Introduction

Scalable parallel computers have enabled researchers to apply finite element and finite volume analysis techniques to larger and larger problems. As problem sizes have grown into millions of grid points, the task of meshing models on a serial machine has become a bottleneck for two reasons: (1) it will take too much time to generate meshes, and (2) meshes will not fit in the memory of a single machine.

Parallel mesh generation is difficult, because it requires the ability to decompose the domain to be meshed into subdomains that can be handed out to processors. This is referred to as partitioning. Partitioning in the context of parallel mesh generation is hard because it has to be done with an input that is either a geometric model or a surface mesh. This means one is trying to partition a 3D domain having only the knowledge of its boundary, at least initially. In contrast, it is much easier to partition a 3D mesh, which is what finite element or finite volume parallel solvers typically do. Proper evaluation of the work load is also a challenge in parallel mesh generation. It is problematic to accurately predict the number of elements to be generated in a given subdomain, or how much computation per element will be required. This leads to difficulties in maintaining good load balance at all times.

There are two types of commercially viable parallel architectures: (1) distributed memory, and (2) shared memory [11]. Distributed memory machines are such that each node has its own local memory. They are often associated with message passing libraries, such as MPI [1]. With a message passing library, the programmer is explicitly responsible for communicating data across processors if needed. With a shared memory machine, there is a global address space that each node can read and/or write to. To gain full efficiency, and reduce communication (at the machine level) to a minimum, on today's shared memory computers, the programmer may have to arrange the data in a specific form depending on how the problem is partitioned. Also, high-level programming languages, such as FORTRAN 90 [12], may not be well-suited for parallel mesh generation because of the lack of a static structure to the problem. In the following, focus is given to the distributed memory parallel architecture. It is assumed parallelism is driven by a message passing library, and in particular, MPI [1].

0-8493-2687-7/99/$0.00+$.50

The next two sections discuss the requirements that any parallel mesh generator should fulfill and how parallel mesh generators can be classified into three separate classes. The following three sections describe parallel mesh generation techniques presented in the literature using this classification. The last section will conclude this chapter with remark and comments.

24.2 Requirements for Parallel Mesh Generation

The ideal parallel mesh generator should be

1. Scalable with respect to time and memory
2. Efficient in a parallel sense
3. Stable

A process is considered "time" scalable if the running time increases slowly with the number of processors, assuming the ratio of problem size to number of processors stays constant. As an example, a process with a complexity of $O((n/n_p)\log(n_p))$, where n related to the problem size and n_p is the number of processors, is scalable since the $\log(n_p)$ term increases slowly with n_p. The same concept applies to "memory" scalability. The memory requirements on a single processor should increase slowly as the problem size increases with the number of processors at the same rate. Scalability is an absolute requirement. If the parallel procedure is not scalable, there will be a limit, sooner or later, on how big a problem can be. Parallel efficiency refers to how well the parallel procedure makes use of the computing resources that are available [11]. Idling processors should be avoided as much as possible. Parallel efficiency is usually related to how well the work load is balanced across the available processors (load balancing). Parallel efficiency should not be confused with "sequential" efficiency, which relates to "sequential" algorithms and has nothing to do with parallelism. In the following, efficiency will refer to parallel deficiency unless noted otherwise. Parallel efficiency is not an absolute requirement but is very desirable. Note that a parallel procedure can be scalable but inefficient, and vice-versa. Stability is with respect to the quality of the produced triangulations. If the quality degrades as the number of processors increases, the parallel mesh generator is not stable.

24.3 Classification of Parallel Mesh Generators

Parallel unstructured mesh generators presented to date all employ the concept of domain partitioning. Figure 24.1 shows a partitioned domain in 2D as well as the associated partition graph obtained by connecting neighboring subdomains with a graph edge. Typically, a processor will be given the task to mesh a subdomain. What differentiates the various approaches is how they treat the interfaces between subdomains. In this paper, three classes of parallel unstructured mesh generators are considered:

1. Those that mesh interfaces as they mesh the subdomains
2. Those that premesh the interfaces
3. Those that postmesh the interfaces

The first class of parallel mesh generators refers to those that neither premesh nor postmesh interfaces. Interfaces are meshed at the same time as subdomains. In the second class, objects are partitioned in such a way that subdomain meshing requires no communication. This is possible by meshing interfaces before the subdomains. In the third class subdomains are meshed, and interfaces are left out for later processing.

24.4 Meshing Interfaces Along with subdomains

The parallel implementations of the Bowyer–Watson algorithm [2, 21] (see Chapter 16) by Chrisochoides and Sukup [4] and Okusanya and Peraire [13] are examples of meshing interfaces at the same time as

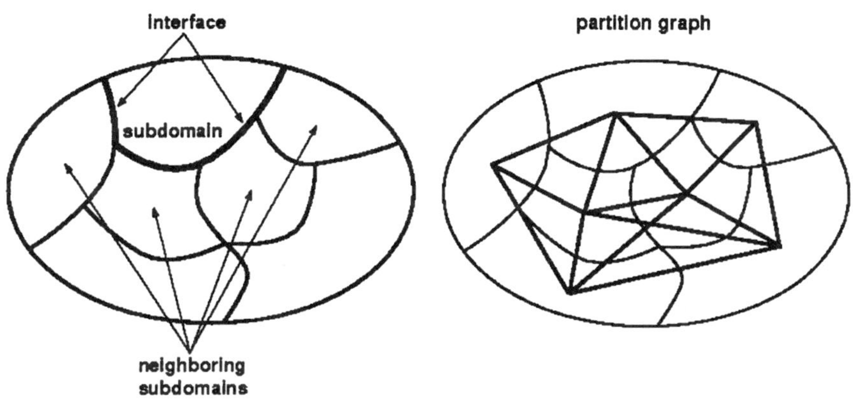

FIGURE 24.1 Example of partition in 2D.

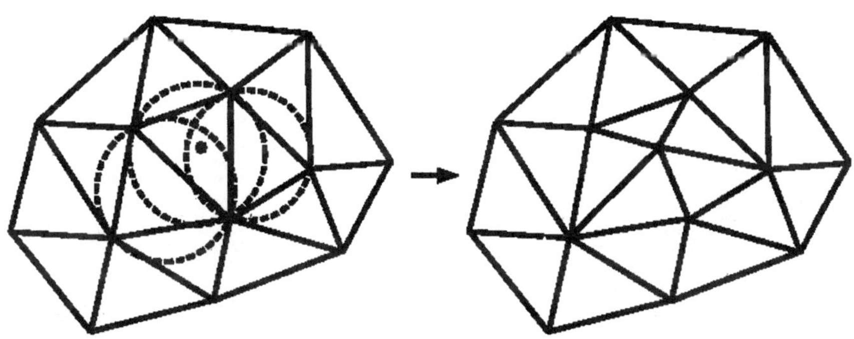

FIGURE 24.2 Delaunay insertion in 2D.

subdomains. The input is a distributed initial mesh that is boundary conforming. It should be noted that this initial mesh could potentially be obtained using the same algorithm, assuming a parallel boundary recovery procedure is available [9].

Assuming element sizes have been prescribed across the domain, any mesh edge in the triangulation that is too long is refined by inserting one or more vertices along the edge using the Bowyer–Watson algorithm [10]. In practice, imposed sizes are stored in a secondary structure such as a background grid or a tree (see Chapter 14 and 15). It should be noted that if the number of grid cells or octants is proportional to the size of the input, the grid or tree has to be distributed to ensure "memory" scalability.

Given a point to insert, the Bowyer–Watson algorithm proceeds as follows (in two dimensions):

1. Find one mesh face that contains the new point.
2. From that mesh face, find all mesh faces whose circumcircles contain the new point using mesh adjacency.
3. Delete the mesh faces (this creates a cavity).
4. Connect the boundary edges of the cavity to the new point.

A graphical description is given in Figure 24.2. If the mesh is distributed, the insertion of a new point on a given processor may not be possible if the cavity extends to neighboring processors due to the mesh being distributed.

The parallel Bowyer–Watson algorithm as described by Chrisochoides and Sukup [4] operates by looping over the following inner loop:

```
for each point to insert do
  get triangle that contains it
  perform task:
    expand cavity
    if cavity cannot be obtained then
      add uncompleted task to blocking-queue
      send a request to neighboring processor(s)
        for needed triangles
    else then
      delete cavity
      connect cavity's boundary to point
    endif
  poll for pending requests
  put received requests (if any) in ready-queue
  move any task from the blocking-queue
    which has been serviced (by a neighboring processor) to
    the ready-queue
  while ready-queue not empty do
    perform task
    if task can be serviced on processor then
      notify requesting processor that
        task has been serviced
    endif
  endwhile
endfor
```

The *blocking-queue* contains tasks that are suspended due to missing information residing on other processors. The *ready-queue* contains tasks that can be performed on a processor. Tasks can switch from the *ready-queue* to the *blocking-queue*, and vice-versa. The complete procedure is actually an outer loop that adds to the inner loop the processing of the *ready-queue* and a check for termination. The outer loop is needed since some processors may still have points to insert while others are done.

This procedure has been implemented using Active Messages [3] on the IBM-SP2. From Chang et al. [3], "Active Messages is a low-latency communication mechanism that minimizes overheads and allows communication and computation to be overlapped in multiprocessors." With Active Messages, a processor must poll for pending messages. If the poll is a hit, the message is received. Polling induces negligible overhead (at least on the IBM-SP2).

This procedure is scalable since a processor usually needs to communicate with its neighbors when inserting a point close to the partition boundary. This is usually true if the partitions are initially, and remain, "bulky." A "bulky" partition is such that the ratio of surface to volume is high.

For this procedure to work well, and therefore have a chance to be efficient, communication must overlap computation well. Beside this communication/computation overlapping issue, the efficiency of the above procedure depends upon how well the computation load is distributed. It is difficult to evaluate how much work is needed to "refine" a subdomain, or more exactly, how many vertices a processor will have to insert. It is assumed that the work required to insert a point is, on average, constant. Note that the number of vertices to be inserted on a subdomain is proportional to the number of elements that will be generated. A rough estimate of the number of elements that need to be generated on a given subdomain can be obtained after building a secondary structure such as a quadtree (in 2D) from imposed sizes that, for example, satisfies the maximum 2:1 level of difference rule. This tree construction is similar to the one described in [5]. The number of interior and boundary terminal quadrants (in 2D) provides a rough estimate of the number of elements that will be generated on the subdomain. Here load balancing is more difficult, since work on a given processor may be induced by a neighboring processor. This typically happens when points are inserted near partition interfaces. It is assumed that, for a given processor, the work induced by neighboring processors should average out the work the processor itself has relayed to neighboring processors. This means that points to be inserted near interfaces should be evenly distributed among processors. Although very different from interface postmeshing, discussed later, this raises the same basic issue of how to partition the interface for proper load balance. Another issue related to efficiency is how much time is spent updating the various mesh data structures as neighboring processors answer to sent requests. The updating procedures must be very fast, typically as fast as the deletion and creation procedures used in the course of the Bowyer–Watson algorithm.

This parallel Bowyer–Watson algorithm is stable with respect to triangulation quality as the number of processors increases since (1) the Delaunay triangulation is unique for a random input, and (2) no interior artificial boundaries are introduced (see Section 24.5 for when this happens).

24.5 Premeshing Interfaces

This class has been further subdivided into three subclasses depending on how the partitioning into subdomains is performed:

1. Partitioning of an initial coarse mesh
2. Partitioning of a background tree
3. Direct partitioning (prepartitioning) of the input surface mesh

24.4.1 Initial Coarse Mesh Partitioning

A commonly used approach [7, 22] consists of the following:

1. Generate a coarse initial mesh.
2. Partition that coarse mesh into n_p subdomains.
3. Refine interface edges of coarse mesh to proper sizes.
4. Distribute the subdomains to the n_p processors.
5. Mesh subdomains.
6. Optimize locally the submeshes.
7. Optimize globally the assembled mesh.

Figure 24.3 gives a graphical description of the procedure.

Initial mesh generation, partitioning, and global optimization are performed on one processor (host), and are therefore not scalable. Subdomain mesh generation and local mesh optimization phases are performed in parallel. These steps are scalable.

The partitioning of the coarse mesh should be such that the subsequent parallel subdomain mesh generation phase is load-balanced. This is a difficult task. The best one can do is to define heuristics to estimate the number of elements the mesher will generate on a given subdomain. If partitioning is done well, then it is expected that the speed-up will be nearly perfect for the subdomain meshing generation.

It is important to keep in mind that the quality of the meshes generated should not degrade as the number of processors increases. This is a concern since this form of partitioning produces "artificial" boundaries. A constrained Delaunay mesher is usually likely to be minimally influenced by these artificial boundaries as long as they are not too close to "natural" boundaries. On the other hand, an advancing front method is likely to create triangulations that degrade is artificial boundaries multiply. This is due to the nature of the advancing front method which has, in general, a tendency to create poor elements as fronts collide [7]. To alleviate this problem, it is necessary in this case to optimize the mesh.

24.4.2 Tree Partitioning

Saxena and Perucchio [14] suggest a tree decomposition of the geometric model to drive the parallel meshing process. Here the input is a geometric model, not a boundary mesh. In 3D, the terminal octants

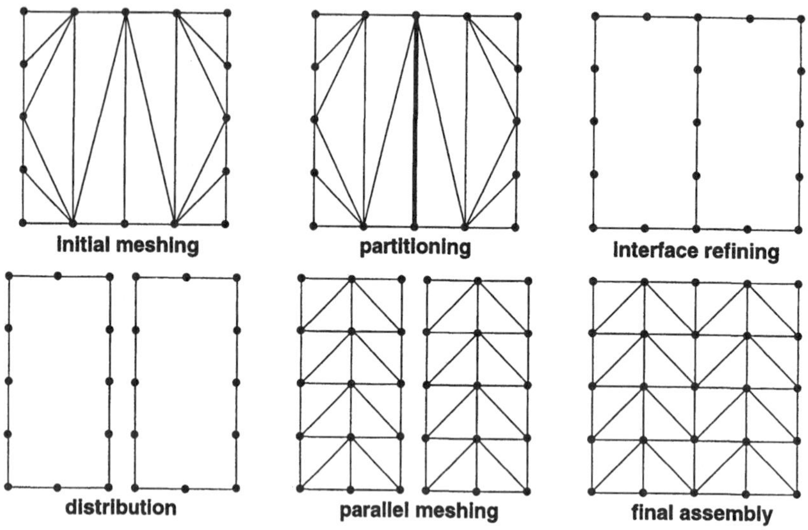

FIGURE 24.3 Meshing of subdomains coming from the partitioning of an initial coarse mesh.

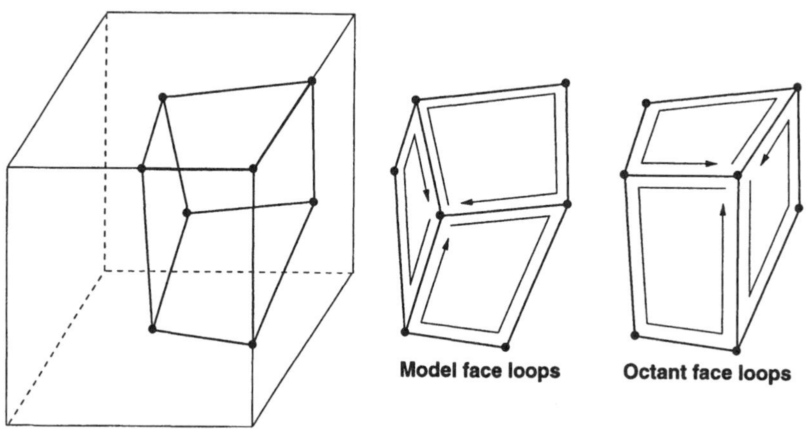

FIGURE 24.4 Example of model face and octant face loops.

are such that their sizes correspond to the sizes imposed by the meshing attributes. Interior terminal octants are meshed using meshing templates. Terminal octants that interact with the domain's boundary (boundary octants) are intersected with the model and then meshed using either a meshing template or an element extraction technique. The interaction between a boundary terminal octant and the model results in the creation of model face loops and interior (to the model) octant face loops [16]. Figure 24.4 shows model face loops and octant face loops for an octant that interacts with three model faces joining at a model vertex.

The set of interior and boundary octants is partitioned among the available processors. The process of intersecting boundary terminal octants with the model and meshing the terminal octants is performed in parallel and without communication. Since an octant face can be shared by several processors (two if the tree is uniform) and meshes on interfaces have to match, care must be taken when meshing octant face loops. The Delaunay triangulation is very attractive here since it is unique, assuming vertices are not in a degenerate situation (four vertices forming a rectangle). Because octant faces are rectangles, it is likely a loop on an octant face has degeneracies. By inserting loop vertices in a given order, the uniqueness of the Delaunay triangulation can again be guaranteed [15]. Note that the meshing of model face loops does not require any such consideration since model face loops cannot be on interfaces. Once octant

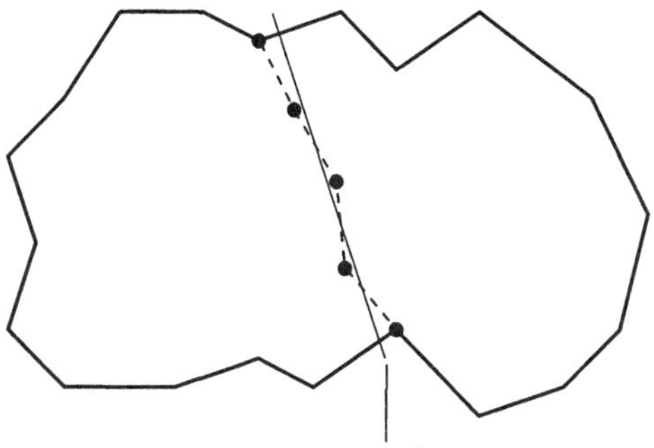

FIGURE 24.5 A separator and its associated triangulation in 2D.

face loops and model face loops have been meshed, interiors of octants are meshed with meshing templates or element extraction techniques.

Octree generation and partitioning are performed sequentially, and are therefore not scalable. It should be noted that a parallel scalable procedure to perform both at the same time is described later in the present chapter. The subdomain meshing procedure is performed in parallel and is scalable.

The performance of the parallel steps of the meshing procedure depends upon how well the partitioner can anticipate how much work will be spent meshing an octant. It is easy to figure this out for an interior octant. It is, however, difficult to estimate how much work will be spent on meshing a boundary octant since one does not know *a priori* how complex the interaction with the model will be.

Stability of the meshing procedure (with respect to triangulation quality) is not an issue here, since identical meshes are created irrespective of the number of processors.

24.4.3 Prepartitioning

Galtier and George [8] prepartition a surface mesh by triangulating appropriately placed separators. A separator cuts a domain into two parts. Given a surface mesh and a separator (say, a plane), the triangulation of the separator is such that

1. It separates, without modification, the initial surface mesh into two subsurface meshes.
2. Sizes of mesh entities on the separator are consistent with imposed sizes.

The separator is not triangulated in the usual sense. The geometry of the separator is used to guide the meshing of the domain, defined by the input surface mesh, in the vicinity of the separator. The triangulation associated with the separator is made of triangles. In other words, a separator and its associated triangulation have the same dimension. Figure 24.5 shows a line separator and its associated triangulation (dashed line segments) when the input is a 2D polygonal mesh (solid line segments). How separators are actually meshed is explained next after a short discussion of the properties of the "projective" Delaunay concept. (Delaunay mesh generation is covered in Chapter 16.)

The technique used to mesh the separator is based on a rather new concept, referred to as "projective" Delaunay. In classic Delaunay, given a set of vertices in 3D, the Voronoï domain at a vertex is defined as the locus of points that are closer to that vertex than to any other vertex in the set. Any two vertices whose Voronoï domains share a side are connected by an edge in the associated Delaunay triangulation. With projective Delaunay onto a surface, given a set of vertices in 3D space, the Voronoï domain at a vertex is defined as the locus of points on the surface that are closer to that vertex than to any other vertex in the set. This defines a Voronoï diagram on the surface. This Voronoï diagram on the surface defines a projective Delaunay triangulation in 3D space. The Voronoï diagram is constructed on the surface and the resulting projective Delaunay triangulation is built in 3D space by connecting vertices whose Voronoï domains on the surface are adjacent. The term "projective" is misleading in this context, since there is actually no projection involved here. Figure 24.6 shows a simple example of "projective"

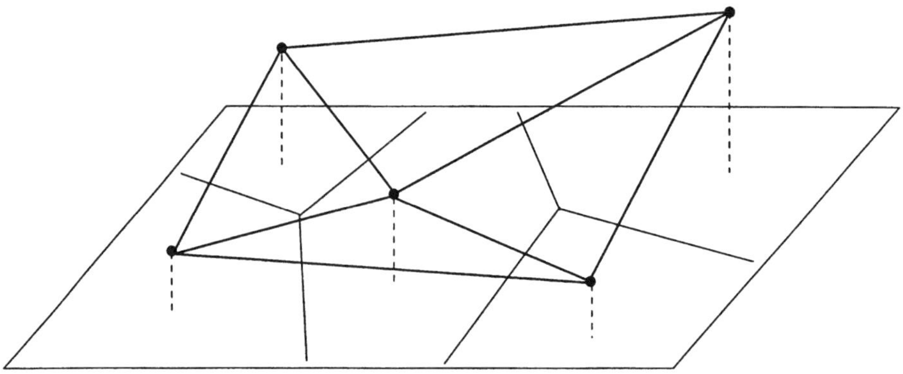

FIGURE 24.6 Example of projective Delaunay triangulation with respect to a plane.

Delaunay triangulation when the surface is a plane. Given a set of vertices in 3D space and a separator, only a subset of these vertices are involved in the projective Delaunay triangulation of the separator. This means the meshing of the separator is local to the separator.

Given an input surface mesh and a separator, the projective Delaunay triangulation of the separator is obtained as follows:

1. Define the poly-line boundary of what will be the triangulation associated with the separator by intersecting edges of the input surface mesh with the separator.
2. Build the "projective" Delaunay triangulation of the separator using only vertices from the input surface mesh.
3. Recognize the poly-line boundary.
4. Delete any mesh face that is outside the poly-line boundary.
5. Insert additional vertices on edges that are too long according to meshing attributes.

Figure 24.7 shows the meshing of a planar separator on a cube. The bottom left picture corresponds to step 2. The bottom right picture corresponds to step 5. Note that only one additional vertex has been inserted. This process is similar to the building of constrained Delaunay triangulations using insertion [9]. If the poly-boundary is not part of the projective Delaunay triangulation, there is no attempt in trying to recover the boundary. If boundary edges are missing, the meshing of the separator is aborted, and an alternate separator is considered. Mesh entities resulting from the meshing of two different separators cannot intersect each other because (1) the two projective Delaunay triangulations are part of the Delaunay triangulation of the set of vertices appearing on these two triangulations, and (2) the Delaunay triangulation is unique. It is assumed there are not Delaunay degenerate situations, that is, more than four vertices on a sphere. However, a mesh entity on a separator can possibly intersect another mesh entity on the surface mesh. If this happens, the meshing of the separate is again aborted. In the context of prepartitioning, if a separator cannot be meshed, a nearby separator can be considered in its place. This means that, even if a specific separator cannot be meshed, the prepartitioner may still succeed. Nevertheless, due to the possible failure of meshing a given separator, there is no real guarantee the prepartitioner will always succeed at placing the separators where they were meant to be. The cost of meshing a separator depends upon the number of generated mesh faces.

Two different techniques for prepartitioning are considered [8]:

1. Cuts along a single direction.
2. Recursive cuts.

In both methods, the separator surfaces are planes. It should be noted that separators do not have to be planes. A separator plane is always perpendicular to the cutting direction. It can be chosen so that it separates any domain into two subdomains with nearly equal number of surface mesh faces. Given the

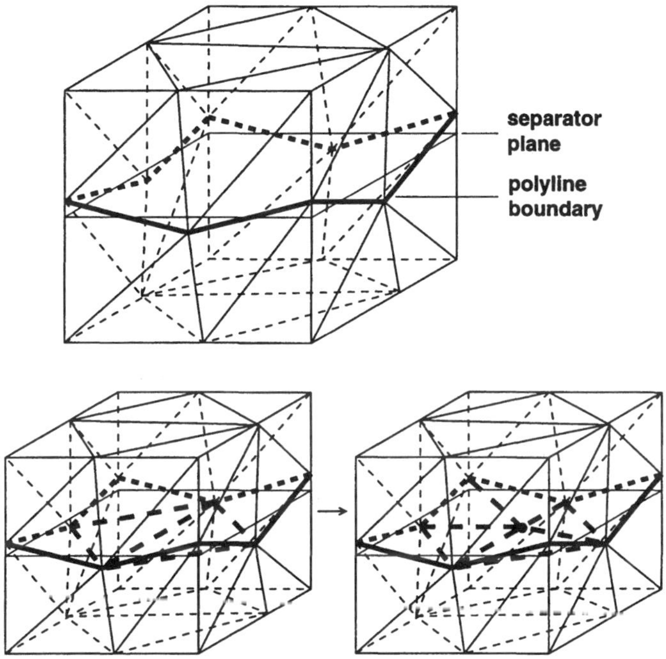

FIGURE 24.7 Meshing of a separator plane on a cube.

surface mesh of a model, the principal axis of minimum inertia that generates the lowest rotation momentum is a good choice for a cutting direction. In the first method, all cuts are made along that axis. In the second method, the first cut is along that axis. This defines two subdomains that can themselves be cut independently along their respective principal axes of inertia, thus, forming four subdomains. Recursive cutting continues until the desired number of subdomains has been obtained. *A priori*, the desired number of subdomains is equal to the number of processors.

Efficient parallelization of the first method requires separators to be sufficiently distant from each other so that there is no interference when meshing them. Each processor, except one, is responsible for the meshing of a separator. Note that this requirement may conflict with proper load balance during the volume mesh generation phase. The cost of meshing the separators in parallel depends on the maximum number (among all processors) of generated mesh faces. This does not scale since, as the number of surface mesh faces increases, the number of mesh faces to generate on a separator increases (at the same rate in the case of uniform sizing) irrespective of how many processors are used.

The second method can be easily parallelized with the *divide and conquer* paradigm [11]. At each recursive cut, one half of the problem goes to one half of the processors. The cost of meshing the first separator (corresponding to the first cut) depends upon the number of mesh faces generated. This again does not scale if only one processor is involved in meshing the separator. Scalability can be achieved only if the meshing of an individual separator can be performed in parallel. This was not discussed [8].

Once prepartitioning is complete, subdomain meshing is performed in parallel and without any communication using constrained Delaunay triangulation in 3D [9]. Note that prepartitioning implicitly distributes the input meshed needed by the volume mesher on each processor. Subdomain meshing is clearly scalable.

The quality of generated triangulations may degrade as the number of processors increases. This is due to the fact that artificial boundaries are created that can potentially be close to original boundaries. Because a Delaunay method is used here, the creation of artificial boundaries far enough away from original boundaries will not cause degradation. It is possible to check for closeness of mesh entities as the separators are being meshed and take appropriate decisions. Checking for closeness is, however, expensive and will lower the performance of prepartitioning in terms of speed.

24.6 Postmeshing Interfaces

This technique has been used first by Shostko and Löhner [18]. Given an input surface mesh, a background grid is built serially on processor 0 (host). The role of the background grid is twofold:

1. To keep track of desired element size information in space.
2. To control the parallel execution.

The task parallel paradigm drives the parallel mesh generator. Tasks are handed out by a dedicated host processor. Other processors are referred to as nodes. The host processor is responsible for:

1. Building a background grid.
2. Partitioning the background grid in at least n_p subdomains.
3. Handing out a background grid subdomain along with the front faces it contains to the next available node.

The individual n_p nodes do the following:

1. Mesh background grid subdomains given by the host.
2. Send back to the host the front faces that could not be processed.

On a given node, the advancing front method is used to mesh the subdomain defined by the background grid elements. To prevent overlapping of submeshes coming from different nodes, a mesh region will not be created if it crosses the subdomain's boundary. More precisely, it will not even be created if it is too close. Assuming the distance between interfaces is always large enough, stability of the parallel mesh generation procedure with respect to triangulation quality degradation will be maintained.

Subdomains should be such that the rate of success of the advancing front method is as high as possible. This rate of success can be defined as the ratio of front faces for which mesh regions could be created to the total number of mesh faces processed. If this ratio gets too low, nodes spend most of their effort determining that they cannot create mesh regions. Partitioning should define "bulky" partitions, that is, partitions with a low surface-to-volume ratio. Note that, when computing this ratio, only the surface shared by two processors should be considered. A "greedy" algorithm [6] that looks at element adjacency to build subdomains is used here. It should be stressed that this partitioning is performed on one processor. This is fine with respect to scalability as long as the size of the background grid is constant, that is, is not a function of the size of the input mesh. If the size of the background grid depends upon the input surface mesh, partitioning must be performed in parallel for that step to be scalable. Concerning "memory" scalability, if the size of the background grid (number of elements) is of the order of the size of the input surface mesh, the background grid should be distributed.

The host-nodes paradigm also poses a problem for scalability. This can be easily seen when the host is handing out subdomains to available nodes for meshing. Assume that the host initially holds n input mesh faces and that it is handing out n/n_p mesh faces to each processor in turn. The cost of this operation is $O(n)$, which is not scalable. The host-nodes paradigm poses a problem any time the host has to communicate with a nonconstant number of processors at the same time.

After the nodes have created the mesh regions within their respective subdomains, the space in between the meshed subdomains remains to be meshed. The "skeleton" of this empty space is made up of the interfaces between the subdomains. In 3D problems, there are three types of interfaces:

1. Faces.
2. Edges.
3. Vertices.

Figure 24.8 shows the interfaces after subdomain meshing on a simple 2D example with four processors. In this particular example, there are four edge interfaces and one vertex interface.

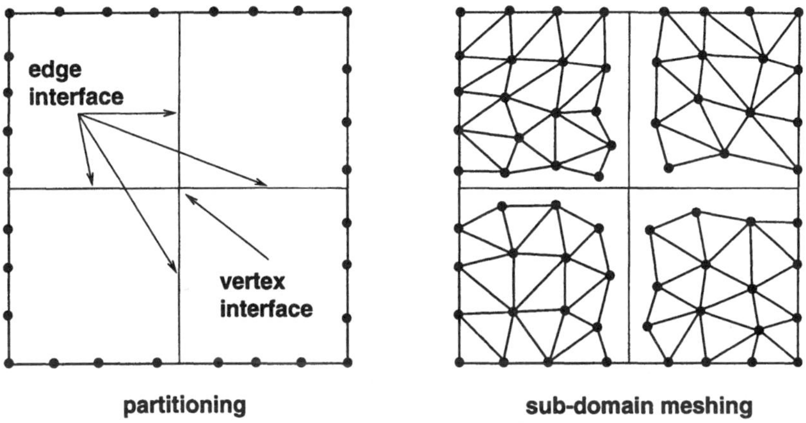

FIGURE 24.8 Interfaces in 2D.

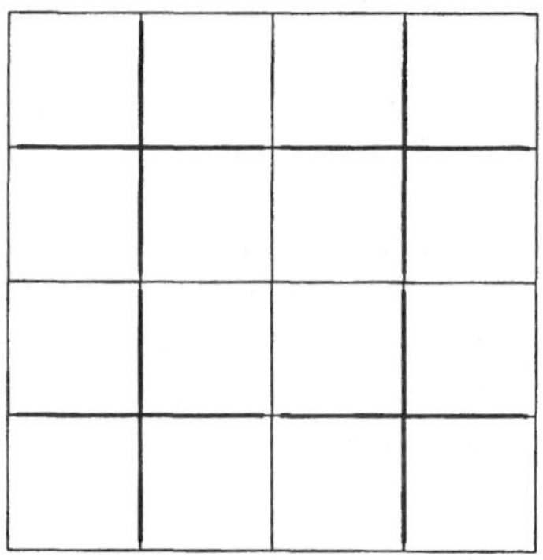

FIGURE 24.9 Interface meshing in 2D using "base" methodology.

The "base" methodology for interface meshing is as follows:

```
for each corner interface do
  if no adjacent subdomain is marked as being used then
    mark adjacent subdomains as being used
    hand out data to next available node (for meshing)
  endif
endfor
```

Implementation-wise, the node is given (by the host node) the background grid elements and the active faces of the subdomains adjacent to the corner interface being considered. This means the node will mesh all interfaces coming to that corner. The "base" methodology is repeated until there are no more interface vertices remaining. Figure 24.9 explains the "base" methodology on a 2D example. What is shown are 16 subdomains belonging to 16 processors. The thicker lines represent interfaces handed out, for meshing, to nodes. To be more precise, each + sign is given to a node for meshing.

The cost of one iteration of the "base" methodology is equal (in 3D) to the maximum number of face interfaces coming to a vertex interface times the maximum size of a face interface. The maximum number of iterations is equal to the maximum number of face interfaces coming to a vertex interface. As the

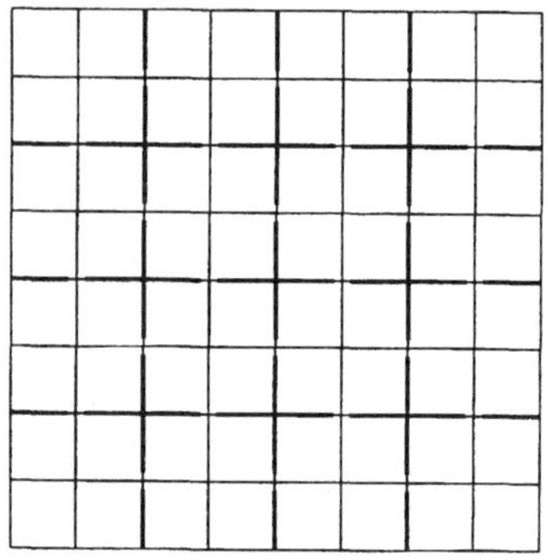

FIGURE 24.10 Subdomain refinement to improve interface meshing in "improved" methodology.

problem size increases proportionally with the number of processors, one can assume these quantities remain nearly constant. This means that the procedure used for meshing the interfaces is scalable. It is, however, not efficient with respect to parallelism. Considering Figure 24.9, only four processors out of the 16 available can work at meshing the interfaces at the same time. Better efficiency can be obtained by further subdividing the subdomains once subdomain meshing is complete. Figure 24.10 shows the effect of subdomain refinement on interface meshing. Note that the initial subdomains were the same as in Figure 24.9, meaning the vertex interfaces are at the same locations. Comparing with Figure 24.9 where only four processors could be work at the same time, here, all processors can work at the same time. This "improved" methodology leads to better load balance and, therefore, better parallel efficiency.

The parallel mesh generator presented by de Cougny, et al. [5] also uses an advancing front method to mesh the volume in between a surface mesh and template-meshed interior octants. Given a distributed surface mesh, the latest version of this procedure builds a distributed octree in parallel [20]. The octree is such that

1. The root octant fully encloses the input mesh.
2. The size of any terminal octant is comparable to the size of the input mesh entities it contains or will contain.
3. There is no more than one level of difference between octant edge neighbors.

The purpose of the tree is to

1. Enable data localization during volume meshing
2. Have a quickly defined spatial structure that can be partitioned
3. Use fast octant meshing techniques on interior terminal octants that are more than one element deep from the surface mesh

The input for tree building is a distributed array of points in 3D space associated with a tree level. It is referred to as the (*point, level*) array. This array is built by considering, for each mesh vertex on the input surface mesh, the average length of the connected mesh edges transformed into a tree level. Any length d can be transformed into a tree level by applying the formula, $level = \log_2(D/d)$ where D is the dimension of the root octant. Tree building is decomposed into two steps:

1. Local root building
2. Subtree subdivision

The process for local root building is as follows:

```
1. Initialize processor set to all processors
2. Create global root octant on each processor in processor set
```

```
3. Initialize control of global root octant to processor set
while all processor sets not of cardinality 1 do
   only consider terminal octants under processor set's control
   4. subdivide (once) any terminal octant
      that needs to be subdivided
   5. assign terminal octants to processors
      according to "load ratio rule"
   6. split processor set into subsets
endwhile
   7. Delete trees on all processors but 0
   8. Migrate terminal octants according to their owning processors
```

The concept of processor sets is an attractive feature of MPI [1]. It enables subdivision of a set of processors into subsets that can run in parallel independently of each other. A processor set can be subdivided into subsets, and each subset can be further subdivided as many times as desired.

In step 1, the processor set is initialized to the complete set of processors. Its cardinality is n_p. The goal of the procedure is to recursively split the processor sets until all processor sets contain a single processor. When a processor set has control of a terminal octant, it means that (1) the terminal octant exists on all processors in the set, and (2) only that set can make decisions regarding whether or not to refine it. In steps 2 and 3, the initial processor set is given control of the root octant. Critical to the effectiveness of the tree-building procedure is what happens within the *while* loop (steps 4, 5, and 6). Consider a processor set controlling a set of terminal octants. As a reminder, the set of terminal octants exist on all processors in the set at this time. The (*point, level*) array known to the processor set contains only entries relevant to the set of terminal octants the processor set is responsible for. If necessary, this array can be evenly distributed among the processors in the set for load balance using a simple data migration scheme. In step 4, the decision to refine, or not refine, at terminal octant is made after (1) having each processor in the set examined its (*point, level*) array keeping track of the maximum level, and (2) having communicated its maximum level with all processors in the set. If the global maximum level is more than the octant's level, the terminal octant is refined once at this point. The reason why a terminal octant is refined only once is because the tree must be as shallow as possible for subtree building, described next. The shallower the tree at that point, the more efficient the complete tree building procedure will be, since subtree building requires no communication. Once the terminal octants under the processor set's control have been processed for refinement (new ones are ignored), they are assigned to processors within the processor set according to a "load ratio rule" (step 5) and the current processor set is split into subsets (step 6).

The "load ratio rule" attempts to make sure that processor subsets will carry a load, measured by the number of points within the volumes of the terminal octants they will be in charge of, that is close to the load average times the number of processors in the subset. Considering Figure 24.11, if the number of processors in the set is three, then, octants 0 and 1 are assigned to processor 0, octants 2 and 3 are assigned to processor 1 and 2. The processor set is split into two subsets: one containing processor 0 and the other one containing processors 1 and 2. The (*point, level*) array is redistributed so that (1) each entry ends up on a processor in the subset that is in charge of the terminal octant that contains the point, and (2) processors in subsets hold the same number of entries. This guarantees locality of data and an even distribution of the (*point, level*) array. Each subset resulting from the split is now considered a processor set in the next iteration of the *while* loop. This process continues until all current processor sets contain a single processor. The rest of the procedure can then be run without using the concept of processor sets, since they have all been reduced to single processors. In Steps 7 and 8, the current tree is actually distributed by deleting it everywhere except on processor 0 and migrating terminal octants based upon which processors have control of them. This procedure builds a distributed "partial" tree where each terminal octant can be seen as a local root of a constructed subtree. Figure 24.12 shows a distributed "partial" tree in 2D. Each terminal octant exists only on a single processor. Details about the data structure

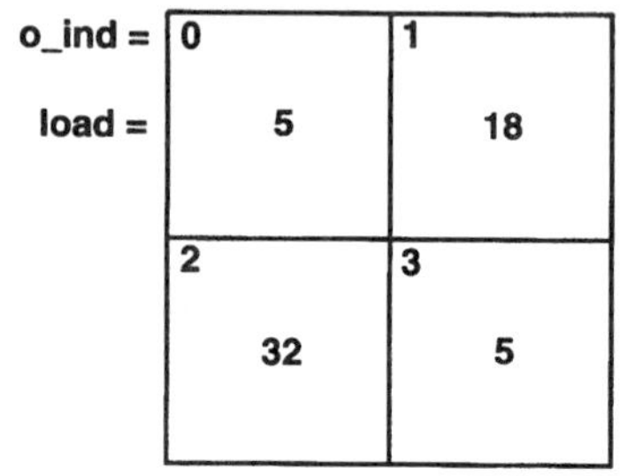

FIGURE 24.11 Example for the "load ratio rule."

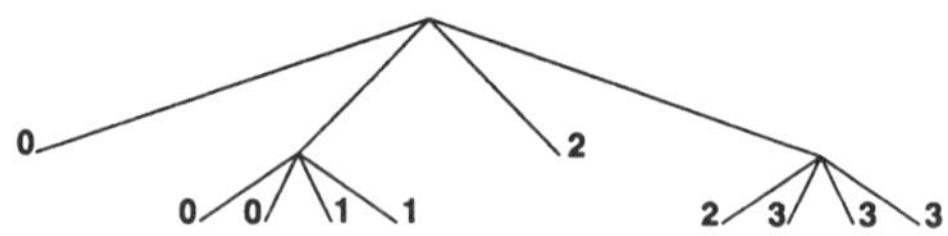

FIGURE 24.12 "Partial" distributed tree.

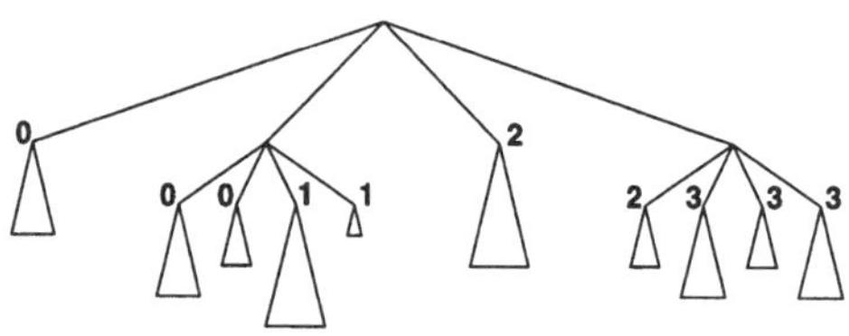

FIGURE 24.13 Complete distributed tree.

used for this distributed tree can be found in [20]. Subtree building, described below, is implicitly load balanced with respect to the (*point, level*) array, that is, each processor will have approximately the same number of points to insert into its subtree(s). This is due to the "load ratio rule" used in the above procedure for local root building.

The process of subtree subdivision is as follows:

```
Each processor is responsible for 1 or more local roots
for each (point, level) do
  get octant(s) the point is in
  if octant_level < level then
    refine octant recursively to level
  endif
enddo
```

Each processor builds subtrees rooted at the local roots. These subtrees only exist on the processors they have been built in. Here terminal octants can be recursively subdivided until the desired level is reached. This procedure requires no communication. Figure 24.13 shows the complete tree structure after subtree building in 2D.

The cost for local root building is $O((n/n_p)\log(n_p))$. The n/n_p term represents how many (*point, level*) each processor is holding. The $\log(n_p)$ factor reflects the number of iterations in the *while* loop. The cost for subtree subdivision is $O((n/n_p)\log(n/n_p))$. The n/n_p term indicates how many (*point, level*) each processor has to insert into its subtree(s). The $\log(n/n_p)$ term is for tree traversals of the subtree(s). The

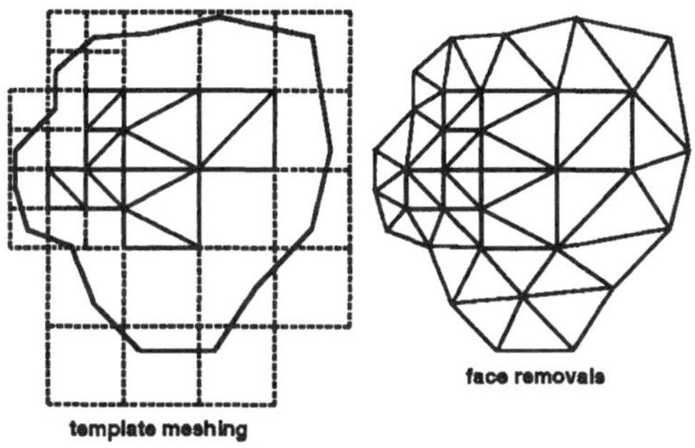

FIGURE 24.14 Template meshing and face removals.

total cost for tree building is dominated by the subtree subdivision cost. Tree building is therefore a scalable process.

Terminal octants are classified according to their interactions with the input surface mesh. If a terminal octant has mesh entities from the input mesh within its volume, it is classified boundary. Once all boundary terminal octants have been recognized, any unclassified terminal octant is classified either interior to a model region or outside. Interior terminal octants at least one element length away from the input surface mesh are meshed using meshing templates. Interior terminal octants are partitioned using a parallel recursive inertial bisection procedure, described below, to ensure load balance while meshing templates. The way templates have been designed is such that triangulations on octant faces shared by two processors are guaranteed to implicitly conform.

Once interior terminal octants have been meshed, the domain between the input surface mesh and the meshed octants is meshed in parallel by applying face removals. A face removal is the basic operation in the advancing front method which, given a front face, creates a mesh origin. Figure 24.14 graphically shows the two meshing steps of the procedure on a 2D example.

It is important to explain how face removals are performed in parallel in order to understand the full parallel face removal procedure. A front face is not removed if the tree neighborhood from which target vertices are drawn is not fully present on processor. This implicitly guarantees this parallel mesh generator is stable with respect to triangulation quality degradation. Face removals can be applied until there is no front face that can be removed. At that point, the tree is repartitioned. The process of applying face removals and repartitioning the tree continues until the front is empty. The parallel face removal procedure is as follows:

```
where there are boundary terminal octants then
  repartition the boundary terminal octants
  apply face removals
  reclassify boundary terminal octants which have no more
    front faces in their volumes as complete
endwhile
```

Tree repartitioning is performed by a parallel recursive inertial bisection procedure [17] based upon the *divide and conquer* paradigm [11]. The input to the parallel recursive inertial bisection procedure is a set of distributed boundary terminal octants. All processors participate in the first bisection along the axis of minimum momentum. Half the processors are given the task to further bisect the terminal octants before the median. The other half is responsible for the terminal octants after the median. The median is such that it separates the set of boundary terminal octants into two nearly equal parts. Bisection continues until the number of partitions is equal to the number of processors. Terminal octants are migrated according to their destinations. When a terminal octant is migrated, the front faces in its volume are migrated to the same destination. If a front face is connected on one side to a mesh region, the mesh region is migrated. The cost of repartitioning n entities is $O((n/n_p)\log(n/n_p)\log(n_p))$. This is a scalable process. The $(n/n_p)\log(n/n_p)$ term comes from sorting the entities along the axis of minimum momentum. The $\log(n_p)$ factor represents how many times recursive bisection needs to be applied. Details about the implementation of this parallel repartitioning scheme can be found in [17]. This repartitioning method has been chosen because (1) it is relatively easy to parallelize, (2) it generates relatively good partitions [19], and (3) it is multipurpose in the sense that it can be used for other applications than parallel mesh generation. Note that other parallel repartitioners could be used. After a face removal step, boundary terminal octants that have been "filled up" with mesh regions are reclassified as complete and do not participate in the next tree repartitioning.

The number of available processors for meshing is reduced when the rate of success of the face removal step drops considerably. This rate of success is defined as the ratio of successfully removed faces to tried faces. To study the scalability of the face removal meshing loop, assume that, at each step, the number of processors is reduced by half. Without loss of generality, the initial number of processors is assumed to be a power of two. The proposed face removal meshing loop can only be scalable if the number of octants to repartition is reduced by half at each iteration. Reducing the problem by half at each iteration cannot be guaranteed in theory. Although test case results have shown promising speed-ups for up to 32 processors with removal rates greater than one half for most steps, scalability of the described face removal meshing loop is questionable in a theoretical sense.

Scalability can, however, be ensured by explicitly meshing the interface resulting from subdomain meshing. subdomain meshing is the combination of the first partitioning and face removal step. The following procedure to mesh interfaces is similar to the one described by Shostko and Löhner [18]. The main difference resides in the fact that here the host-nodes paradigm is not used. Decision making concerning the repartitioning of interfaces and the actual migration of data is performed in parallel. Interface meshing is hierarchical, that is, face, edge, and vertex interfaces are considered in turn. Also, here, "very" fine-grain parallelism coming from the tree is used to improve parallel efficiency. Since this procedure can be used by other parallel mesh generators, it is discussed without considering the use of template meshing for interior terminal octants. Template meshing on interior terminal octants reduces the sizes of face and edge interfaces, which makes the procedure, described next, more efficient.

A face interface can be meshed by migrating to one processor boundary terminal octants that are closer to that face interface than any other face interface. After the face removal step, each processor assigns its remaining boundary terminal octants, that is, those that have not been "filled up," to its bounding interfaces based on distance consideration. Within a subdomain, any remaining boundary terminal octant is assigned to the closest bounding interface. Figure 24.15 shows the assignment (to interfaces) of boundary terminal octants resulting from subdomain meshing. In this case, the subdomain of interest is bounded by three edge interfaces denoted as 0, 1, and 2. Assignment of boundary terminal octants to face interfaces is performed in parallel by all processors. Figure 24.15 only shows what happens on one processor. The idea is to have each face interface meshed by a single processor by migrating to that processor (unless already there) all boundary terminal octants associated with the face interface. In practice, to avoid unnecessary migration, a processor adjacent to the face interface will be chosen to mesh it. Since the initial partitioning is "bulky" and terminal octants are similar in sizes to the front mesh faces they contain, *a priori* all face interfaces can be meshed within the same step without interference except at edge and vertex interfaces, which is expected. The work needed to mesh a face interface can be accurately estimated by counting the boundary terminal octants that have been associated with it. This means that good load balance during face interface meshing is possible. The cost for face interface meshing is equal to the maximum number of face interfaces a subdomain has, times the maximum number of elements to be generated on a face interface. As remarked previously, this leads to a scalable procedure. Edge interface meshing uses the exact same methodology and is not described here. Vertex interfaces can efficiently be meshed independently of each other since they have become small and bounded subdomains. Note that the tree is not needed anymore when meshing the vertex interfaces.

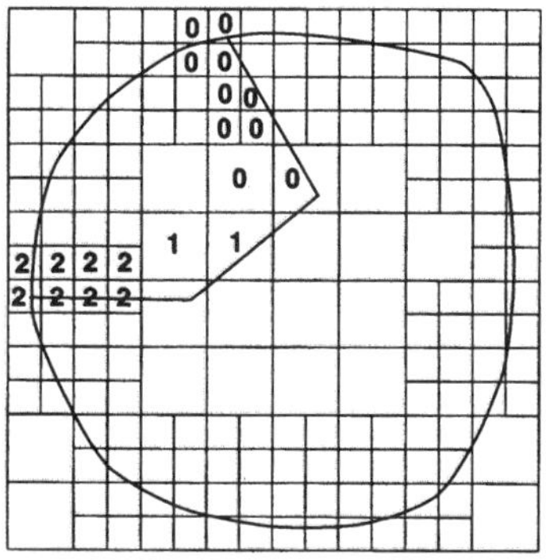

FIGURE 24.15 Assignment of boundary terminal octants to interfaces.

24.7 Conclusion

This chapter has reviewed parallel unstructured mesh generation procedures with respect to: (1) scalability, (2) parallel efficiency, and (3) stability relative to triangulation quality. Scalability appears to be a requirement since it guarantees that, as more processors are used, bigger problems can be solved in reasonable clock time. Parallel mesh generation is difficult because there is no real structure than can "perfectly" drive a parallel algorithm. The final structure appears only upon completion with the generated 3D mesh. Parallel mesh generation is also tedious because it usually involves several processes and data structures that all need to be "time" and "memory" scalable, respectively. The parallel mesh generation field is still very young, which means that the algorithms presented in this chapter are probably evolving very fast and completely new algorithms are being written. Because the development of efficient scalable parallel techniques takes much more time than their sequential counterparts, it may take a while before parallel mesh generation comes to a state of maturity.

References

1. Message passing interface forum MPI: a message-passing interface standard, *International Journal of Supercomputer Applications and High-Performance Computing.* 8(3/4), 1994. Special issue on MPI.
2. Bowyer, A., Computing Dirichlet tessellations, *The Computer Journal.* 1981, 24, 2, pp 162–166.
3. Chang, C.-C., Czajkowski, G., von Eicken, T., design and performance of active messages on the IBM SP-2, Technical report, Cornell University, 1996.
4. Chrisochoides, N. and Sukup, F., Task parallel implementation of the Bowyer–Watson algorithm, *5h Int. Conf. on Numerical Grid Generation in Computational Field Simulations.* Mississippi State University, 1996, pp 773–782.
5. de Cougny, H.L., Shephard, M.S., Özturan, C., Parallel three-dimensional mesh generation on distributed memory MIMD computers, *Engineering with Computers.* 1996, 12, pp 94–106.
6. Farhat, C. and Lesoinne, M., Automatic partitioning of unstructured meshes for the parallel solution of problems in computational mechanics. *Int. J. for Numerical Methods in Engineering.* 1993, 36, pp 745–764.
7. Gaither, A., Marcum, D., Reese, D., A paradigm for parallel unstructured grid generation, *5th Int. Conf. on Numerical Grid Generation in Computational Field Simulations.* Mississippi State University, 1996, pp 731–740.
8. Galtier, J. and George, P.-L., Prepartitioning as a way to mesh subdomains in parallel, *5th Int. Meshing Roundtable.* 1996, Pittsburgh, PA, pp 107–121.

9. George, P. L., Hecht, F., Saltel, E., Automatic mesh generator with specified boundary, *Computer Methods in Applied Mechanics and Engineering.* 92, pp 269–288.
10. George, P. L. and Hermeline, F., Delaunay's mesh of a convex polyhedron in dimension d. Application to arbitrary polyhedra, *Inte. J. for Numerical Methods in Engineering.* 1992, 33, pp 975–995.
11. Jájá, J., *An Introduction to Parallel Algorithms.* Addison-Wesley, 1992.
12. Metcalf, M. and Reid, J., *Fortran 90/95 Explained.* Oxford University Press, 1995.
13. Okusanya, T. and Peraire, J., Parallel unstructured mesh generation, *5th Int. Conf. on Numerical Grid Generation in Computational Field Simulations.* Mississippi State University, 1996, pp 719–729.
14. Saxena, M. and Perucchio, R., Parallel FEM algorithms based on recursive spatial decomposition. I. automatic mesh generation, *Computers & Structures.* 1992, 45(5–6), pp 817–831.
15. Schroeder, W.J. and Shephard, M.S., A combined octree/Delaunay method for fully automatic 3-D Mesh Generation, *Int. J. for Numerical Methods in Engineering.* 1990, 29, pp 37–55.
16. Shephard, M.S. and Georges, M.K., Automatic three-dimensional mesh generation by the finite octree technique, *Int. J. for Numerical Methods in Engineering.* 1991, 32(4), pp 709–749.
17. Shephard, M.S., Flaherty, J.E., de Cougny, H.L., Özturan, C., Bottasso, C.L., Beall, M.W., Parallel automated adaptive procedures for unstructured meshes, *Special Course on Parallel Computing in CFD.* AGARD, 1995, number R-807, pp 6.1–76.49.
18 Shostko, A. and Löhner, R., Three-dimensional parallel unstructured grid generation, *Int. J. for Numerical Methods in Engineering.* 1995, 38, pp 905–925.
19. Simon, H.D. and Teng, S.-H., How good is recursive bisection, technical report, NASA Ames Research Center, 1993.
20. Simone, M., de Cougny, H.L., Shephard, M., Tools and Techniques for parallel grid generation, *5th Int. Conf. on Numerical Grid Generation in Computational Field Simulations.* Mississippi State University, 1996, Vol. II, pp 1165–1174.
21. Watson, D., Computing the n-dimensional Delaunay tessellation with applications to Voronoï polytopes, *The Computer Journal.* 1981, 24(2), pp 167–172.
22. Wu, P. and Houstis, E.N., Parallel adaptive mesh generation and decomposition, *Engineering with Computers.* 1996, 12, pp 155–167.

25

Hybrid Grids and Their Applications

Yannis Kallinderis

25.1 Introduction

There is an ever-increasing demand to perform flow simulations that incorporate the complete details of geometry as well as sophisticated field physics. The success of numerical flow simulators depends to a great extent on the computational grid that is employed. As a consequence, grid generation has become a task of primary importance. Books and surveys on grid generation include [1–6]. Structured meshes consisting of blocks of hexahedra and unstructured grids consisting of tetrahedra have been the traditional means of discretizing 3D flow domains [2, 3].

Hybrid grids usually consist of prisms and tetrahedra in 3D, and correspondingly quadrilaterals and triangles in 2D. Layers of prisms are employed to resolve boundary layers and wakes, while tetrahedra cover the rest of the domain. Hybrid meshes are intended to provide flexibility by combining essential features of the two broad types of meshes, namely the structured and the unstructured grids [7–15]. Hybrid meshes consisting of triangles and quadrilaterals have been employed in two dimensions in [16–26]. Other hybrid mesh techniques involve generating a mesh made up of tetrahedral and prismatic elements and then destructuring the prisms to form tetrahedra [27, 28]. Adaptation and load balancing for parallel computation of hybrid grids have been presented in [29, 30].

There are a number of issues to be addressed when dealing with turbulent flow simulations involving complex geometries. These considerations include: (1) the different orientation of the viscous flow features, (2) the disparate length scales that need to be resolved within the same domain, (3) the requirements of the Navier–Stokes solvers, (4) the grid generation time, (5) the required user expertise, as well as (6) the university of application of the grid generator.

The main features that are encountered in flow fields include boundary layers, wakes, shock waves, and vortices. These features have different orientations that make generation of a single grid that conforms to them very difficult. In addition, the mesh has to follow the boundaries of the computational domain. A *hybrid* grid that combines elements of different orientation appears to be much more flexible in

0-8493-2687-7/99/$0.00+$.50

conforming to the flow features. The prisms are assigned the task of capturing the features that are following the body surface, while the tetrahedra are used for the features that are away (e.g., shocks and vortices).

The different spatial scales encountered in viscous flows vary by orders of magnitude from each other. These scales are imposed by the flow features and the geometry. The laminar sublayer requires placement of grid points at distances away from the wall of the order of 10^{-6} times the scale of the geometry, while the points at the farfield may be at a distance of order 1 from one another. Shock waves and vortices have very different scales as well. Furthermore, the details of the geometry frequently impose scales on the grid generator. The gaps between the main wing and the flap and the tip clearances in turbomachinery geometries are typical examples of small scales.

The issue becomes even more complex when taking into account the directionality of the different scales. The small scale required in the boundary layers is in the direction normal to the surface, while much larger sizes of the mesh are sufficient in the lateral directions. Similar directionality also exists in wakes and shock waves. This directionality leads to the issue of generating high aspect ratio grid cells. Generation of thin prismatic grids for the boundary layers and wakes has the advantages of being feasible and fast, and also results in a smaller number of elements compared to tetrahedra. On the other hand, the isotropic nature of tetrahedra appears to be appropriate for the vortices and other regions of the domain where the flow is changing equally in all directions.

Navier–Stokes solvers place strict requirements on the mesh. Accuracy and stability of the numerical methods depend crucially on the local resolution and the uniformity of the grid. Smooth transition of element sizes at the prism/tetrahedra interface is important for accuracy and robustness of Navier–Stokes numerical methods [8, 9]. Furthermore, computing resources in terms of CPU time and memory storage are dictated by the number of grid elements. These facts place several requirements on mesh generation.

Employment of the thin semistructured prismatic elements in the regions of shear layers results in sufficient accuracy with significantly reduced computing resources compared to all-tetrahedral meshes. The flow field on the body surface usually contains regions of strong flow directionality such as the leading and trailing edges of a wing. Generation of anisotropic surface grid elements results in significant savings in the number of elements without sacrificing accuracy.

Minimum user expertise and universal application are also primary considerations placed on grid generators. A generation method must use a relatively small number of control parameters whose effects are obvious even to an inexperienced user. It is highly desired that a grid generation method be applicable to a great variety of geometries without modification. Furthermore, the setup time to apply the generator should be kept to a minimum.

25.2 Underlying Principles

The hybrid grid generator consists of two major parts: (1) the prisms generator, which is an algebraic, marching-type technique, and (2) the tetrahedra generator which is an advancing front type of method (see Chapter 19). Details of the two techniques can be found in [6, 8, 9].

25.2.1 The Structured Marching Method for Prisms

An unstructured triangular grid is employed as the starting surface to generate a prismatic mesh. This grid, covering the body surface, is marched away from the body in distinct steps, resulting in generation of semistructured prismatic layers in the marching direction (Figure 25.1).

The process can be visualized as a gradual inflation of the body's volume. A major issue with marching methods is the avoidance of crossing of the grid lines. There are three main aspects of the algebraic grid generation process: (1) determination of the directions along which the nodes will march (marching vectors), (2) determination of the distance by which the nodes will march along the marching vectors, and (3) smoothing operations on positioning of the nodes on the new layer.

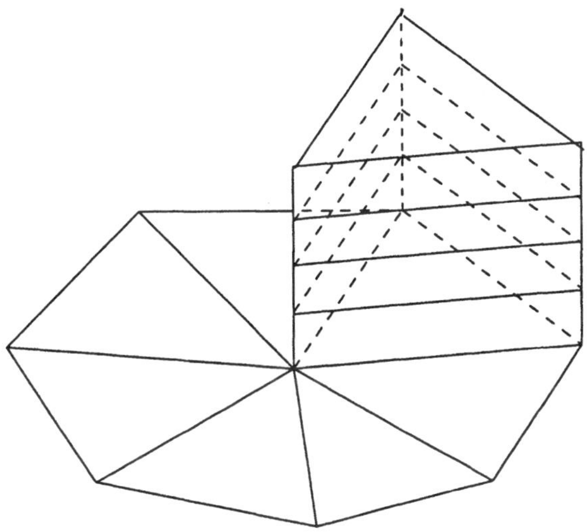

FIGURE 25.1 Generation of a prismatic grid from a triangular boundary surface.

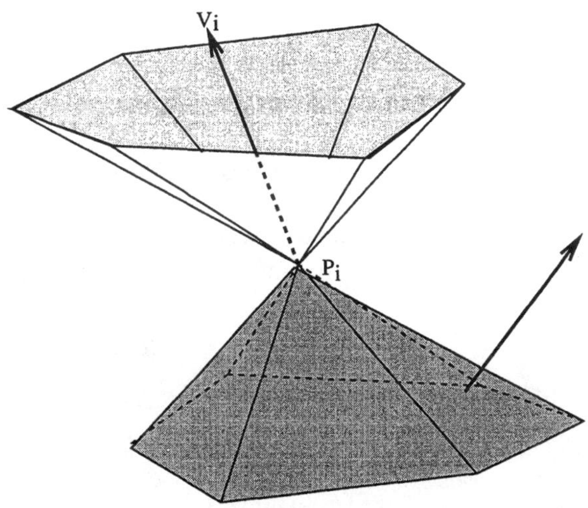

FIGURE 25.2 Example of the manifold of point P_i and its corresponding visibility region.

25.2.1.1 Determination of the Marching Vectors

Each node on the marching surface is advanced along a marching vector. The marching direction is based on the *node-manifold,* which consists of the group of faces sharing the node to be marched. The primary criterion to be satisfied when marching is that the new node should be visible from all the faces on the manifold (*visibility condition*) [7]. An example of a manifold and its corresponding visibility region is shown in Figure 25.2. The dark-shaded region is the manifold of node P_i, and the polyhedral cone above the node is the visibility region. The vector, V_i, is one possible node-normal satisfying the visibility constraint.

The node-normal vector lies on the bisection plane of the two faces on the manifold that form the wedge with the smallest angle. This process has yielded consistently valid normal vectors at the nodes by constructing the vector most normal to the most acute face planes. Essentially, it does this by maximizing the minimum angle between the node-normal and all the surrounding face normals. This vector is then used as the marching direction for the nodes on the surface to form the new layer. A more detailed description of the marching procedure can be found in [7].

25.2.1.2 Marching Step Size

Determination of marching distances is based on the characteristic angle β_{ave} of the manifold of each node to be marched. This angle is computed using the average dot product between the pairs of faces forming the manifold. The marching distance is a linear function of β_{ave}. It yields relatively large marching steps in the concave regions, and small steps in the convex areas of the marching surface. Specifically, the distance Δn is:

$$\Delta n = (1+\alpha)\Delta n_{ave}, \tag{25.1}$$

where Δn_{ave} is the averaging marching step for the layer, and α is a linear function of the manifold angle β_{ave}. The sign of α is positive for concave regions and negative for convex regions.

The average marching step for each layer (j), Δn_{ave} is computed based on a user-specified initial marching step Δn_o on the body surface and a stretching factor st, as follows:

$$\Delta n^j_{ave} = \Delta n_o \times st^{(j-1)}. \tag{25.2}$$

25.2.1.3 Smoothing Steps

The initial marching vectors are the normal vectors. However, this may not provide a valid grid, since overlapping may occur, especially in concave regions of the grid surface with closely spaced nodes. To prevent overlapping, the directions of the marching vectors must be altered. Altering of the directions should not end abruptly in the local neighborhood of the nodes involved, since this may cause overlapping in nearby regions. A gradual reduction of the magnitude of the change in the vector direction is accomplished via a number of weighted Laplacian type smoothing operations over the marching vectors of all nodes. Typically, ten smoothing passes are performed. These smoothing steps rotate each original marching vector based on the normal vectors of its surrounding manifold nodes as follows:

$$\vec{V}_i = \omega \vec{V}_i' + (1-\omega)\frac{1}{\Sigma_j 1/d_{ij}}\sum_j (1/d_{ij})\vec{V}_j, \tag{25.3}$$

where $\vec{V}_i'$ and $\vec{V}_i$ are the initial and final marching vectors of node i, while $\vec{V}_j$ are the marching vectors of the surrounding nodes j that belong to the manifold of node i. The weighting factor ω is a function of the manifold characteristic angle β_{ave}. It has small values in concave regions, and relatively large ones in convex areas. The averaging of the marching vectors of the neighboring nodes is distance-weighted with d_{ij} denoting the distance between nodes i and j.

A similar procedure is employed for the smoothing of the marching steps Δn to eliminate abrupt changes in cell sizes.

25.2.1.4 Constraints Imposed to Enhance Quality

Typical Navier–Stokes integration methods impose restrictions on the spacing of the points along the marching lines and on the smoothness of these lines. In other words, the prismatic grid should not be excessively stretched or skewed. Constraints are imposed on the lateral and normal distribution of marching step sizes and the deviation of the direction of the marching vectors from one layer to the next.

The lateral distribution of cell sizes are constrained so that any node on the current marching surface cannot have a step size (Δn_i) that is very different from the size (Δn_j) corresponding to any of its surrounding nodes. Specifically,

$$0.5 \times \Delta n_j < \Delta n_i < 2.0 \times \Delta n_j, \tag{25.4}$$

The constraints on the step size variation along each marching line are applied in a similar manner. A node on the prismatic layer j cannot have a step size that is smaller than that on the previous layer $(j-1)$. Also, it cannot exceed the size of the previous step by more than a factor of st_{max} (usually set to 1.3). Specifically,

$$\Delta n_{j-1} < \Delta n_j < st_{max} \times \Delta n_{j-1}. \tag{25.5}$$

Another constraint limits the deviation between two consecutive marching vectors $\vec{V}_{j-1}$ and $\vec{V}_j$ to be less than a specified angle (typically 30°).

The above-mentioned constraints reduce "kinks" in the marching vector directions as well as abrupt changes in step sizes, thus providing a smooth mesh suitable for viscous flow computations. Since the visibility criterion is the ultimate test for the validity of the mesh, this criterion is the final constraint imposed on the grid.

25.2.1.5 Automatic Adjustment of the Prism Layer Thickness

Treatment of narrow gaps and cavities in regions such as wing–engine configurations and in between different bodies in multiply connected domains has been a major concern for structured and semistructured mesh generators. The structured nature of prisms prohibits filling such complex geometries without overlapping layers if special measures are not taken. A method has been developed that adjusts the marching step of the prism layers for the treatment of such gaps [8]. The technique allows entirely automatic generation of single-block, nonoverlapping prismatic meshes. Two key features of the method are no user interaction and universality of its application to different geometries. The nodes in the vicinity of a cavity are detected by a special algorithm. The marching distances of these flagged nodes are reduced so that the mesh does not overlap. This may result in prismatic meshes of significantly varying local thickness. Smooth variation of the thickness is attained via lateral smoothing of the size of the marching steps.

The local thickness of the prism layer in the cavity or gap region is reduced to avoid overlapping prism layers. This is done by recomputing the initial marching distance Δn_o for all the flagged nodes according to the following equation:

$$\Delta n_o = \frac{C_1 \times d_G}{\Sigma_j st^{(j-1)}}, \tag{25.6}$$

where d_G denotes the gap distance computed by a special gap-detection algorithm, C_1 is a user- specified constant controlling the extent of reduction (usually chosen to be 0.25), st is the stretching factor, and j is the prism layer index. Thus, the total thickness of the prism layers in the vicinity of the gap is approximately $C_1 \times d_G$, with slight variations depending on the local curvature of the marching surface. The exact step size for every node on each layer is then determined by Eqs. 25.1 and 25.2.

In order to avoid abrupt changes in the thickness of the prism layers due to the local receding, the *unflagged* nodes in the neighborhood of the cavity are also receded to a certain extent. This extent gradually reduces to zero as the nodes get farther away from the cavity or gap.

25.2.2 The Octree-Advancing Front Methods for Tetrahedra

A combined octree-advancing front method is used to generate the unstructured grid [9]. Advancing front type methods require specification by the user of the distribution of three parameters over the entire domain to be gridded. These field functions are (1) are node spacing, (2) the grid stretching, and (3) the direction of the stretching. Using the octree-advancing front method, these parameters do not need to be specified. Instead, they are determined via an automatically generated octree.

The octree is constructed via a *divide-and-conquer* process, which starts with a *master* hexahedron that contains the body. This hexahedron is recursively subdivided into eight smaller hexahedra called *octants*.

Any octant that intersects the body is a *boundary octant* and is subdivided further (inward refinement). The subdivision of a boundary octant ceases when its size matches the local length scale of the geometry. The choice of the local length scale depends on the particular application of the octree. The length scale can be chosen to be local prism thickness, edge length, or curvature. This flexibility allows the same octree creation technique to be used for many different unstructured applications.

Then, the hexahedral grid is further refined in a *balancing process* (outward refinement) to prevent neighboring octants whose depth differs by more than one. Outward refinement is performed to ensure that the final octree varies smoothly in size away from the original surface. The sole criterion for outward refinement is a depth difference greater than one between the octant itself and any of its neighbors. The outward refinement continues until no octants meet the refinement criterion. Typically, five sweeps are performed to produce a balanced octree. The octree data structure is similar to earlier data structures used for search operations during the grid generation process [31] (see Section 14.4.2.1 of Chapter 14).

Two important features of the octree-advancing front method are its capability to match disparate length scales and its geometry independence. The octree is able to insure a smooth size transition over the large range of length scales which are present in a "viscous" mesh. The octree is also able to be used for many different types of geometries with minimal user interaction.

25.2.2.1 Length Scales

Octree refinement is terminated when the size of a boundary octant is the same size as the local length scale of the geometry. This local length scale depends on the application. Three different applications are considered, namely, surface mesh generation, tetrahedral mesh generation for hybrid grids, and all tetrahedral mesh generation.

For surface mesh generation, the local length scale is determined by the local curvature of the geometry. This length scale is small in areas where the curvature is large, i.e., the trailing edge of a wing, and large where the geometry is flat. The distance between surfaces is another length scale used for surface mesh generation. The local length scale is proportional to this distance. This allows for automatic clustering in regions where surfaces are in close proximity.

For hybrid prismatic/tetrahedral mesh generation, the local length scale is simply the local thickness of the last prismatic layer. This will ensure that the size of the tetrahedra in the direction normal to the outer prismatic surface is the same as the height of the neighboring prisms. This smooth transition in size from the prisms to the tetrahedra is important for accuracy of the numerical method. Finally, for an all tetrahedral mesh, the local length scale is the local edge length of the original triangulated surface. The octree-advancing front method can also be used to create meshes for inviscid simulations. Given an initial surface triangulation, the octree is refined until the boundary octants match the size of the local surface triangulation.

Figure 25.3 shows plane cuts of the octree for two different geometries. The first case corresponds to the High Speed Civil Transport (HSCT) aircraft, while the second to a two-element wing. A plane cut of the prismatic part of the hybrid mesh is also shown. The size of the octants intersecting the outermost prismatic surface matches the thickness of the last prismatic layer, even in the region of the engine where the thickness of the prisms is several orders of magnitude smaller than their thickness away from the engine. The same observations apply to the second case of the two-element wing.

25.2.2.2 Octree Guides Advancing Front Mesh Generation

The advancing front volume grid generation starts from the surface of the body or the outermost prismatic surface for the case of a hybrid grid. The triangular faces of this surface form the initial front list. A face from this list is chosen to start the tetrahedra generation. Then, a list of points is created that consists of a new node, as well as of "nearby" existing points of the front. One of these points is chosen to connect to the vertices of the face. Following the choice of the point, a new tetrahedron is formed. The list of the faces, edges, and points of the front is updated by adding and/or removing elements [32]. The method requires a data structure that allows for efficient addition/removal of faces, edges, and points, as well as for fast identification of faces and edges that intersect a certain region. The alternating digital tree (ADT) algorithm is employed for these tasks [33] (see Section 14.25.4.3 of Chapter 14).

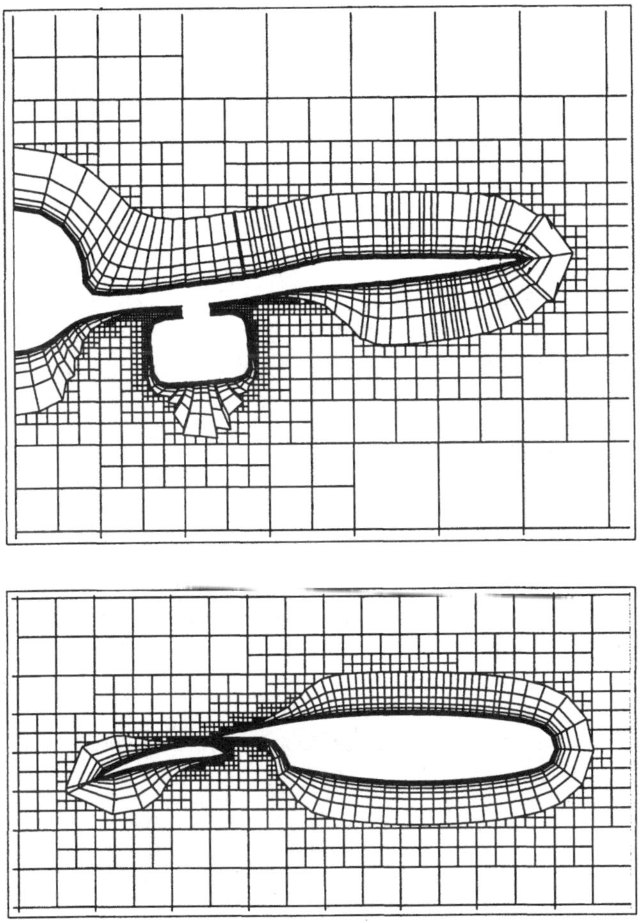

FIGURE 25.3 Plane cuts of octree meshes. The top figure shows a plane cut of an octree mesh for the HSCT aircraft. The bottom figure shows a plane cut of an octree mesh for the partial-flap high-lift wing at mid span of the flapped region. Both figures show how the octant sizes match the local thickness of the final prismatic layer. Every third layer of the prismatic mesh is shown for clarity of the figure.

The tetrahedra that are generated using this octree method grow in size as the front advances away from the original surface. Their size, the rate of increase of their size, as well as the direction of the increase, are all given from the octree. The octants are progressively larger with distance away from the body. Their sizes determine the characteristic size of the tetrahedra that are generated in their vicinity. This method is flexible and can be used to generate tetrahedra around different types of geometry.

The surface mesh generation proceeds in the same manner as the tetrahedral mesh generation, except that surface triangles are generated from an initial front made up of edges [32]. The surface geometry is treated as a patchwork of CAD panels (see Chapter 19 and Part III). An interface is required between the CAD representation and the surface grid generator. The interior of each panel is filled with triangles using the same octree for each panel to insure smooth size transitions across panel boundaries. New triangles are generated using either already existing points, or new points generated on the surface using information from the octree. The octree allows for a smooth transition in size on the surface from areas where the triangles are small (i.e., trailing edge) to areas where the triangles are larger.

The advancing front method creates a new element by connecting each face or edge of the current front to either a new or an existing node. This new point is found by using a characteristic distance δ calculated from the size of the local octant to which the face of the front belongs. Specifically,

$$\delta = \alpha st^{(l_t - l)}, \tag{25.7}$$

where α is a scaling factor, *st* is the stretching parameter, l_t is the total number of octant levels, and l is the level of the local octant. The value of *st* controls the rate of growth of the mesh. The lower the value of *st*, the less the mesh increases in size away from the body. A typical value of the stretching parameter *st* is 1.8. The level l of the local octant is the number of subdivisions of the master octant required to get to the size of the local octant.

For hybrid mesh generation, smooth transition in size from the prisms to the tetrahedra is important for accuracy of the numerical methods. The value of the scaling factor α is calculated so that the initial marching size (δ) of the tetrahedra equals the local thickness of the outermost prismatic layer.

For surface mesh generation, α can be varied to generate different meshes using the *same* octree. Higher values of α result in coarser meshes, while lower values of α yield finer meshes. Both the coarse and fine meshes will have similar local variation of the sizes of the surface triangles.

25.2.2.3 Anisotropic Surface Meshes

The octree-advancing front method can also create anisotropic surface meshes. Anisotropic meshes are useful in reducing the number of triangular faces needed to capture all the flow features in a simulation. Allowing high aspect ratio triangles aligned with geometry and flow features in regions that exhibit strong directionality enables a substantial savings in number of both surface and volume grid elements. A user only needs to specify the following: (1) a line segment that defines the direction of the stretching of the mesh, (2) the aspect ratio (*AR*) of the triangles desired along that line segment, and (3) the area of influence (d_{max}) of the line segment. Examples of such line segments include the leading edges, trailing edges and engine inlets. The method for generating anisotropic meshes starts with the size, δ_{oct}, given by the octree and augments it with the perpendicular distance, *d* from the user-specified line segment. The local mesh size is now characterized by three sizes, δ_1, δ_2 and δ_3 given by

$$\delta_1 = c \times \delta_{oct} \qquad \delta_2 = \delta_{oct} \qquad \delta_3 = \delta_{oct}, \tag{25.8}$$

with

$$c = \frac{AR - 1}{d_{max}} d + AR, \tag{25.9}$$

and δ_1 is the size of the mesh in the direction of the line segment, while δ_2 and δ_3 are the sizes of the mesh in directions perpendicular to the line segment and perpendicular to each other. The method is flexible and robust using multiple line segments at different locations and directions to define directionality on different parts of the surface. Furthermore, it provides a smooth transition between regions of different directionality.

Figure 25.4 shows both an isotropic surface mesh for the M6 wing and an anisotropic mesh created with line segments extending over the entire leading and trailing edges with the same aspect ratios as the previous mesh. The isotropic mesh has 39,290 faces while the anisotropic mesh has 6333 faces while maintaining the same chord-wise point density obtained from the same octree. These meshes show the 6.2:1 reduction in the number of generated faces when an anisotropic method is used. This reduction in faces leads to a substantial reduction in the number of elements of the corresponding volume mesh.

25.2.2.4 Automatic Partial Remeshing

Grids generated using an advancing front type scheme can contain regions of low quality within the mesh domain. These low-quality regions must be altered before the mesh can be used with a flow solver. A method for improving low quality regions has been developed [9]. This method removes low quality regions from the mesh and fills the resulting cavities using the same advancing front generator on the new front defined by the surface of these holes.

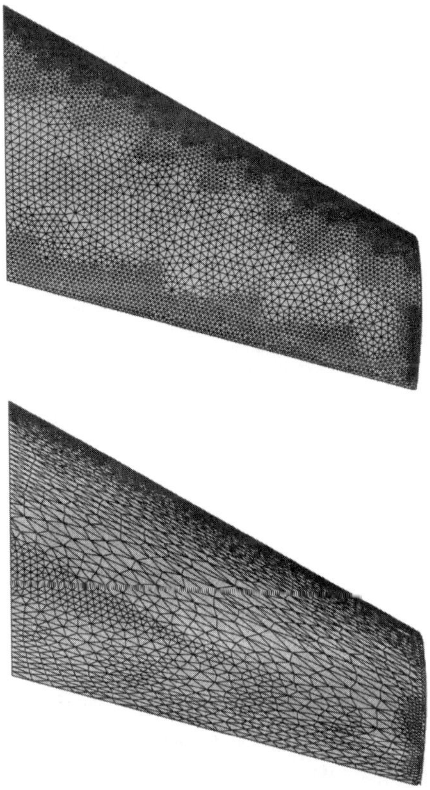

FIGURE 25.4 Significant savings in number of triangles are realized due to the use of leading- and trailing-edge line segments for the ONERA M6 wing. The top mesh is an isotropic mesh with 39,290 faces. The bottom mesh is an anisotropic mesh with only 6,333 faces. Note that even though the isotropic mesh has six times the number of faces, the anisotropic mesh has the same chordwise point distribution.

In order to properly define the low quality regions of the mesh, the quality of a given region must be quantified. There are several measures of mesh quality. One such indicator that has been used is the volume ratio of the two tetrahedra sharing each face, $R = \text{Vol}_{max}/\text{Vol}_{min}$. Large values of R indicate a very stretched mesh. If $R = 1$, the mesh is locally uniform.

Once the low quality regions of the mesh have been located using the quality measure R, these regions must be removed from the mesh. For each face with a value of R greater than a user-specified value, R_{sp}, a cavity is opened around the low quality region by removing tetrahedra. The radius of the opened cavity is dependent on the local length scale of the mesh.

After cavities have been formed around each of the low quality regions of the mesh, the exposed triangular faces inside the cavities are put together to form a new initial front. Then, the advancing front generator refills the cavities with better quality tetrahedra. This process of cavity definition and cavity remeshing is repeated until a specific level of quality is reached.

The entire process of cavity definition and remeshing is performed automatically with no user intervention. The remeshing process is efficient and typically takes a quarter of the time that the initial tetrahedral generation requires.

25.3 Best Practices

This section presents applications of hybrid grids that include both external and internal geometries. The cases are chosen in order to demonstrates the suitability of the hybrid grids for complex geometries, as

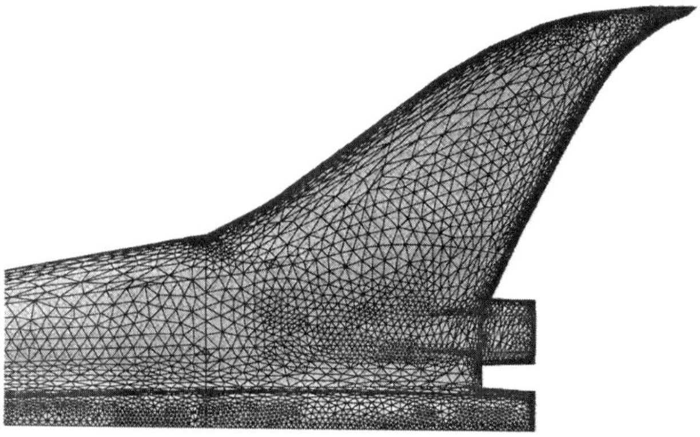

FIGURE 25.5 Anisotropic surface mesh for the HSCT with engines. The figure shows the anisotropic regions near the leading and trailing edge of the wing. The mesh has 30,189 faces, while a similarly spaced isotropic mesh would have 60,583 faces.

well as the robustness and generality of the generator to yield meshes for very different topologies. The specific cases are: (1) an aircraft configuration, (2) an adapted hybrid mesh, (3) resolution of multiple wakes past a wing, (4) a deformable hybrid grid in two dimensions, (5) a turbomachinery blade with tip clearance, and (6) a burner.

25.3.1 High Speed Civil Transport (HSCT) Aircraft

The High Speed Civil Transport (HSCT) aircraft is a next-generation aircraft being designed to travel at supersonic speeds. It has a double-delta wing configuration emerging from the nose. The cavity between the engine and the wing presents a challenge in grid generation.

An example of a locally directional surface mesh for the aircraft is illustrated in Figure 25.5. It is observed that the method generates a reduced number of points on the wing in the spanwise direction while maintaining a large number of nodes in the chordwise direction. A strongly directional mesh has been generated primarily in the leading and trailing edge regions of the wing.

A view of the hybrid mesh is shown in Figure 25.6. The third is shown on two surfaces that are perpendicular to each other. The first is the symmetry plane with the quadrilateral faces corresponding to the prisms and the triangular faces corresponding to the tetrahedra. The second surface is a field cut intersecting the fuselage and engine. A field cut is a cut through the discretized grid showing all the cells that intersect the plane cut, thus emphasizing the 3D nature of the grid. The prisms are assigned the task of capturing the features that are following the aircraft surface, while the tetrahedra are used for the features that are away (e.g., shocks and vortices). Figure 25.7 illustrates the widely varying length scales of the hybrid grid. The field cut shows portion of the fuselage, the wing, as well as the engine. Note the varying thickness of the prismatic layer which is dictated not only by the thickness of the boundary layer, but also by the size of the cavity between the engine and the wing. The tetrahedral part of the mesh is very dense in the cavity area in order to match the sizes of the local prisms and becomes isotropic away from the cavity.

25.3.2 Adapted Hybrid Mesh

The case of an adaptively embedded hybrid mesh is presented next for the same HSCT aircraft geometry. Turbulent flow is simulated with Mach number (M_∞) equal to 3, angle of attack (α) equal to 5° and Reynolds number (Re) equal to 6.3×10^6. The grid is locally embedded according to the magnitude of flow gradients [29]. Figure 25.8 shows a plane cut of the adapted hybrid grid employed for simulation of turbulent supersonic flow around the aircraft. A view of the solution via entropy contours on the initial coarse grid and the corresponding locally refined hybrid mesh is shown. The right hand side of

FIGURE 25.6 View of the hybrid mesh around the HSCT aircraft with engines on two different planes that are perpendicular to each other. The first plane is that of the symmetry while the second is a field cut intersecting the fuselage and engine.

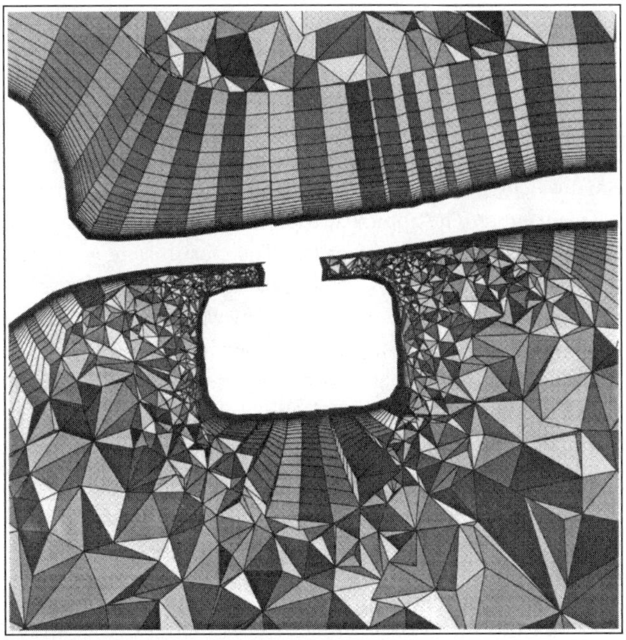

FIGURE 25.7 Close up of the hybrid grid for the HSCT aircraft around the engine cavity. The tetrahedral mesh is very dense here compared with other regions so as to match the thin local prism cell sizes.

FIGURE 25.8 View of the solution (entropy contours) on the coarse grid and the corresponding adapted grid for the HSCT configuration. The right-hand side of the figure shows the initial mesh superimposed with entropy contours. The adapted hybrid grid (left side) has been refined in the vicinity of the vortex, and near the wing/fuselage junction. Case of turbulent flow with $M_\infty = 3$, $\alpha = 5°$ and $Re = 6.3 \times 10^6$.

the figure illustrates the initial grid superimposed with entropy contours of the solution. Two are the main flow features here. The boundary layer conforms to the surfaces of the fuselage and wing, while the vortex has a totally independent orientation. The prismatic mesh used follows the shape of the boundary layer, while the tetrahedral grid appears to be more appropriate for the vortex. Furthermore, local refinement has been applied by the adaptation algorithm in the region of the vortex.

25.3.3 Resolution of Multiple Wakes

The ability of the prismatic elements to capture multiple wakes is illustrated by generating a hybrid grid about a generic two-element wing. The approach used here extends fictitious surfaces past the trailing edges of both the blades in the direction of the wakes. Then, prisms are generated marching away from both the wing-surface as well as the fictitious surfaces to capture the viscous effects at the wall and the wake region. The grid consists of 9000 boundary nodes of which 5300 are on the surface of the main wing and flap (the rest are on the fictitious surfaces extended into the wakes). A view of the hybrid mesh is shown in the field cut in Figure 25.9. The grid consists of seven prism layers and 53,000 tetrahedra. A completely unstructured mesh in the wake region would require a very large number of tetrahedra.

The prisms in between the main wing and the flap have been *receded* by the procedure described in Section 25.2.1.5 to prevent grid overlapping. Note the grid clustering in the wake and the smooth transition in cell sizes across the domain.

25.3.4 Deforming Hybrid Mesh in 2D

Deformation of a hybrid mesh is now demonstrated via an example of a two-dimensional grid about two circular discs aligned in the tandem direction. This grid has been employed for simulation of vortex-induced vibrations to the two bodies. Figure 25.10a shows the mesh when both cylinders are in their initial position. The thick horizontal and vertical lines are included as a point of reference indicating the

FIGURE 25.9 Field cut of the hybrid grid for the two-element wing. The grid comprises $9K$ boundary nodes (including those on the fictitious surface), seven prism layers and $53K$ tetrahedra. The prisms provide adequate grid clustering in the wake region with fewer cells compared with an all-tetrahedral mesh.

equilibrium position of each cylinder. Figure 25.10b shows the resulting deformed mesh when the two cylinders move away from each other in the transverse direction. Note that the significant displacement of the two cylinders is nicely accommodated by the triangular elements, and connectivity of the mesh is preserved.

25.3.5 Turbomachinery Blade with Tip Clearance

The next case considered is an internal geometry. It is a turbine blade with narrow tip clearance. Figure 25.11 shows two perpendicular field cuts of the hybrid mesh around the blade. The surface was composed of 13,663 triangular faces. The hybrid grid consists of 14 layers of prisms (191,282 prismatic cells), and 415,086 tetrahedral cells. The tetrahedra were able to easily match the prismatic thickness everywhere, including the small gap between the tip of the blade and the shroud. Also, the surface mesh is much finer in the tip region adapting to the features of the geometry. It is important to note that the grid generation scheme was able to mesh an internal geometry as easily as the external geometries presented in the previous sections.

25.3.6 ABB Burner Case

The final case corresponds to flow through a burner, which consists of an annulus diffuser, a swirl producer, and a combustion chamber. This case has been provided by ABB. The geometry has various complexities such as the fuel injection holes, severe cavities, twisted blades that produce the swirl, and vastly different length scales. The geometry has periodic boundaries, and only one burner is being modeled.

Figure 25.12 shows a close-up of the surface triangulation for the swirl producing section. The surface consists of approximately 75,000 triangles. A hybrid mesh of the burner is seen in Figure 25.13, which is a two-dimensional cut along the axis. The view shows that the hybrid grid generator was capable of capturing all the fine features of the geometry, and also clustered points downstream of the swirl producing section. The mesh consists of 415,000 nodes, 521,000 prisms, and 748,000 tetrahedra. A cut

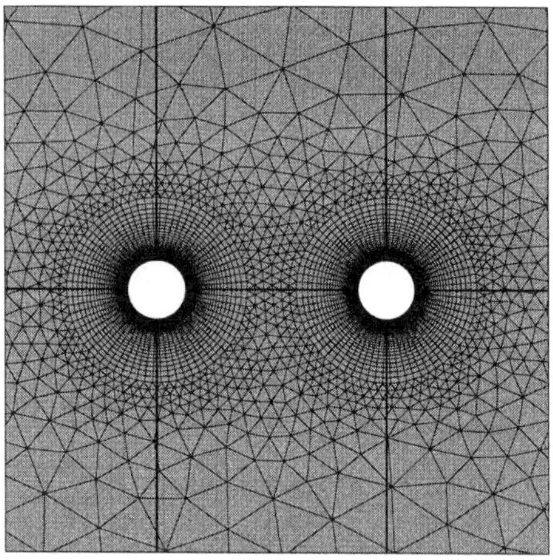

(a)

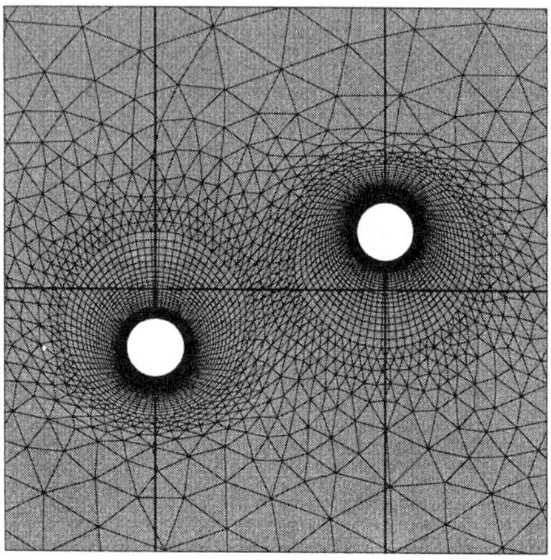

(b)

FIGURE 25.10 Deforming hybrid grids about a tandem cylinder geometry for (a) initial cylinder configuration, (b) cylinders displaced in the transverse direction.

across the swirl producing blades is shown in Figure 25.14. The view shows the hybrid nature of the mesh, and demonstrates the smooth transition in cell sizes even across different element types.

25.4 Research Issues and Summary

Employment of hybrid grids for complex geometries was demonstrated. The prism covered regions of strong flow directionality, such as boundary layers and wakes, while tetrahedra were created elsewhere. The hybrid grid generator consists of two major parts: (1) a special marching method for generation of the prismatic elements, and (2) a combined octree-advancing front technique for generation of the tetrahedra. Narrow gaps and cavities, very disparate length scales, body and flow-field conformity of the mesh, as well as automation were the primary issues that guided the development of the generator.

The use of hybrid grids was demonstrated through complex geometries. The hybrid mesh generator was successful in handling severe cavities and capturing widely varying length scales. Applications included the two main categories of topologies: (1) external and (2) internal.

The marching-vectors procedure to generate the prisms proved to be robust (in avoiding overlapping of prism layers) and efficient. The smoothing operations and the imposition of constraints eliminated

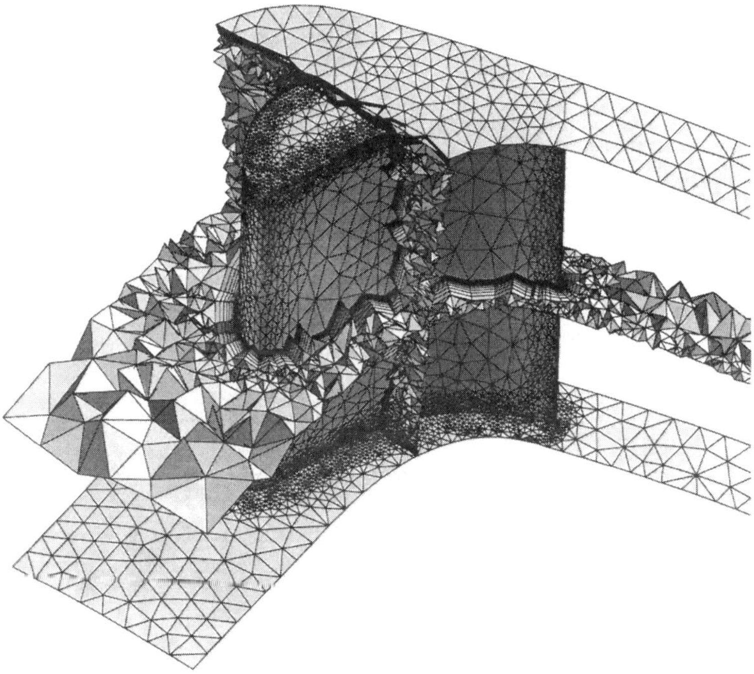

FIGURE 25.11 Surface and volume hybrid mesh for a turbine blade with narrow tip clearance. The volume grid is shown via two field cuts that are on surfaces perpendicular to each other intersecting the blade.

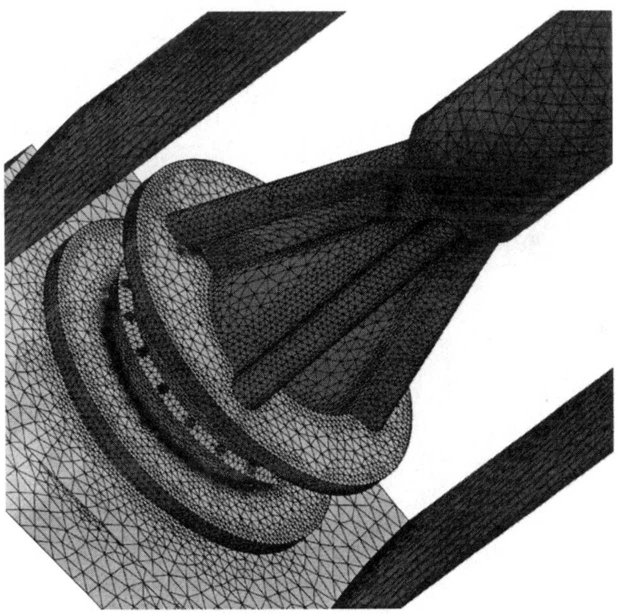

FIGURE 25.12 A close-up view of the swirl-producing section of the ABB Burner geometry. The surface is made up of 75,000 triangles.

surface "ripples" and avoided excessively stretched and skewed meshes. The automatic adjustment of the thickness of the prismatic layer allowed the generation of a single-block, nonoverlapping prismatic mesh even when the surface geometry contained narrow gaps and cavities. The mesh generator allowed for marching along arbitrary parametric surfaces, and was also capable of generating periodic meshes.

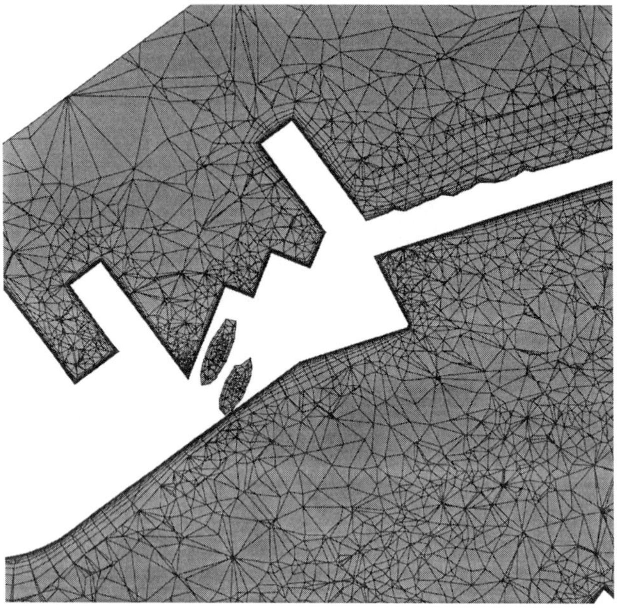

FIGURE 25.13 A close-up view of the hybrid cut along the axis of the burner geometry. The grid generator was capable of automatically handling the small length scales and the severe cavities.

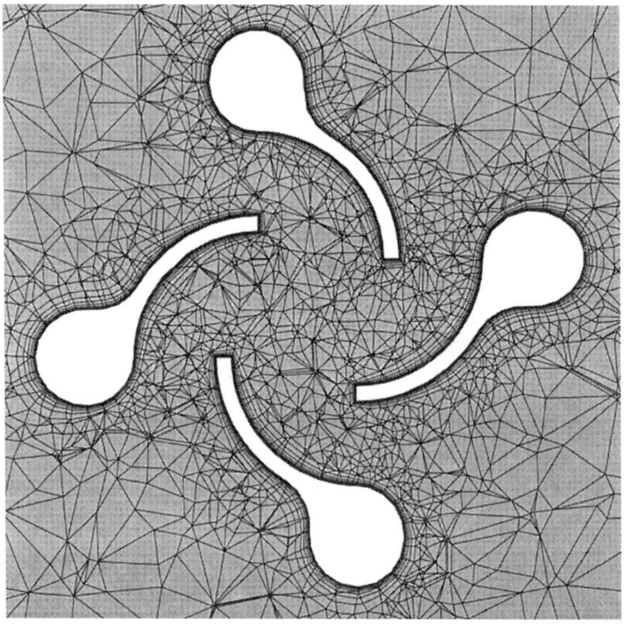

FIGURE 25.14 A cut across the swirl-producing blades of the burner geometry. The hybrid nature of the mesh and the smooth transition in cell sizes are visible.

The octree-advancing front approach provided an automatic method for generating unstructured meshes. The method was effective in generating surface triangulations for different complex geometries including a burner surface. The octree allowed surface triangulations to be generated that captured all of the geometry features. The octree also provided for a smooth variation of grid size over the entire surface mesh.

Anisotropic surface meshes were generated using the octree and minimal user input. These anisotropic meshes resulted in a significant reduction in the number of faces generated. Smooth transition between the different regions of directionality was also accomplished.

Generation of tetrahedra via the advancing front method was also made simpler and more automatic by eliminating the traditional user-defined background mesh for determination of mesh spacing. An automatically generated octree guided the growth of the tetrahedra and enabled a smooth transition of the mesh from the prisms to the tetrahedra in a hybrid mesh. The universality of the octree-advancing front method was demonstrated through its application to different complex geometries. The HSCT aircraft configuration demonstrated that the method is flexible enough to adapt to 200:1 size variations in the local length scale. Local remeshing of the tetrahedral mesh proved very effective in removing areas of abrupt changes in size of the tetrahedra.

Further Information

Additional sources of information on hybrid grids and grid generation, in general, can be found in the proceedings and papers of the following conferences:

- International Conference on Numerical Grid Generation, held every two years
- AIAA Computational Fluid Dynamics Conference, held every two years
- AIAA Aerospace Sciences Meeting, held in Reno, NV every year
- AIAA Applied Aerodynamics Conference, held every year
- International Conference on Finite Elements in Fluids, held every two years
- International Meshing Roundtable, sponsored by Sandia National Labs
- NASA Conference Proceedings on "Unstructured Grid Generation Techniques" (NASA LaRC, CP 10119, Sept. 1993) and "Surface Modeling, Grid Generation, and Related Issues in Computational Fluid Dynamics Solutions" (NASA LeRC, CP 3291, May 1995)

References

1. Thompson, J.F., Warsi, Z.U.A., Mastin, C.W., *Numerical Grid Generation,* North-Holland, New York, 1985.
2. Thompson, J.F. and Weatherill, N.P., Aspects of numerical grid generation: Current Science and Art, *AIAA Paper 93-3539,* 1993.
3. Baker, T.J., Developments and trends in three dimensional mesh generation, *Applied Numerical Mathematics.* 1989, Vol. 5, pp 275–304.
4. Baker, T.J., Mesh generation for the computation of flowfields over complex aerodynamic shapes, *Computers Math. Applic.* 1992, Vol. 24, No. 5/6, pp 103–127.
5. Eiseman, P. R. and Erlebacher, G., Grid Generation for the solution of partial differential equations, NASA CR 178365 and ICASE Report No. 87–57, August 1987.
6. Von Karman Institute (VKI) Lecture series in computational fluid dynamics, *LS 1996-06,* March 25–29, 1996.
7. Kallinderis, Y. and Ward, S., Prismatic grid generation for 3D complex geometries, *Journal of the American Institute of Aeronautics and Astronautics.* October 1993, Vol. 31, No. 10, pp. 1850–1856.
8. Kallinderis, Y., Khawaja, A., McMorris, H., Hybrid prismatic/tetrahedral grid generation for viscous flows around complex geometries, *AIAA Journal.* February 1996, Vol. 34, No. 2, pp 291–298.
9. McMorris, H. and Kallinderis, Y., Octree-advancing front method for generation of unstructured surface and volume meshes, *AIAA Journal.* June 1997, Vol. 35, No. 6, pp 976–984.
10. Nakahashi, K. and Obayashi, S., FDM-FEM approach for viscous flow computations over multiple bodies, *AIAA-87-0605,* 1987.
11. Karman, S. L., SPLITFLOW: A 3D unstructured Cartesian/prismatic grid CFD code for complex geometries, *AIAA-95-0343.* Reno, NV, January 1995.

12. Sharov, D. and Nakahashi, K., Hybrid prismatic/tetrahedral grid generation for viscous flow applications, *AIAA-96-2000, Proc. of the 27th AIAA Fluid Dynamics Conf.* New Orleans, LA, June 1996.
13. Van der Burg, J., Maseland, J., Oskam, B., Development of a fully automated CFD system for three-dimensional flow simulations based on hybrid prismatic-tetrahedral grids, *Proc. of the 5th Int. Conf. on Numerical Grid Generation in Computational Field Simulations.* Mississippi State University, April 1–5, 1996, pp 557–566.
14. Chappell, J., Shaw, J., Leatham, M., The generation of hybrid grids incorporating prismatic regions for viscous flow calculations, *Proc. of the 5th Int. Conf. on Numerical Grid Generation in Computational Field Simulations*, pp 537–546, Mississippi State University, April 1–5, 1996.
15. Noack, R., Steinbrenner, J., Bishop, D., A three-dimensional hybrid grid generation technique with application to bodies in relative motion, *Proc. of the 5th Int. Conf. on Numerical Grid Generation in Computational Field Simulations.* Mississippi State University, April 1–5, 1996, pp 547–556.
16. Kallinderis, Y. and Nakajima, K., Finite element method for incompressible viscous flows with adaptive hybrid grids, *AIAA Journal.* August 1994, Vol. 32, No. 8, pp 1617–1625.
17. Hufford, G.S. and Mitchell, C.R., The generation of hybrid and unstructured grids using curve and area sources, *AIAA-95-0215.* Reno, NV, January 1995.
18. Spragle, G.S., Smith, W.A., Weiss, J. M., Hanging node solution adaption on unstructured grids, *AIAA-95-0216*, Reno, NV, January 1995.
19. Kao, K.H. and Liou, M.S., Direct replacement of arbitrary grid-overlapping by nonstructured grid, *AIAA-95-0346.* Reno, NV, January 1995.
20. Nakahashi, K., FDM-FEM Zonal approach for computations of compressible viscous flows, *Lecture Notes in Physics.* 1986, Springer, Vol. 264, pp 494–498.
21. Weatherill, N.P., Mixed structured–unstructured meshes for aerodynamics flow simulation, *The Aeronautical Journal.* Vol. 94, 134, pp 111–123.
22. Soetrisno, M., Imlay, S.T., Roberts, D.W., A zonal implicit procedure for hybrid structured-unstructured grids, *AIAA-94-0645*, Reno, NV, January 1994.
23. Koomullil, R.P., Soni, B.K., Huang, C.-T., Navier–Stokes Simulation on hybrid grids, *AIAA Paper 96-0768*, Reno, NV, January 1996.
24. Hwang, C.J. and Wu, S.J., Adaptive finite volume approach on mixed quadrilateral-triangular meshes, *AIAA Journal.* January 1993, Vol. 31, No. 1, pp 61–67.
25. Banks, D., Mueller, J.-D., VankeirsBilck, P., An Object oriented approach to hybrid structured/unstructured grid generation, *AIAA Paper 96-0032.* Reno, NV, January 1996.
26. Coirier, W. and Jorgenson, P., A Mixed volume grid approach for the Euler and Navier–Stokes equations, *AIAA Paper 96-0762.* Reno, NV, January 1996.
27. Connell, S.D. and Braaten, M.E., Semistructured mesh generation for 3D Navier–Stokes calculations, *AIAA-95-1679-CP.* San Diego, CA, June 1995.
28. Pirzadeh, S., Viscous unstructured three-dimensional grids by the advancing-layers method, *AIAA Paper 94-0417.* Reno, NV, January 1994.
29. Parthasarathy, V. and Kallinderis, Y., Adaptive prismatic-tetrahedral grid refinement and redistribution for viscous flows, *AIAA Journal.* April 1996, Vol. 34, No. 4, pp 707–716.
30. Minyard, T. and Kallinderis, Y., Octree partitioning of hybrid grids for parallel adaptive viscous flow simulations, *Int. J. for Num. Meth. in Fluids.* January, 1998, Vol. 26, pp 1–22.
31. Lohner, R., Some useful data structures for the generation of unstructured grids, *Communications in Applied Numerical Methods.* 1988, Vol. 4, pp 123–135.
32. Peraire, J., Morgan, K., Peiro, J., Unstructured Finite element mesh generation and adaptive procedures for CFD, in *Application of Mesh Generation to Complex 3D Configurations, AGARD Conference Proceedings* No. 464, 1990, pp 18.1–18.12.
33. Bonet, J. and Peraire, J., An Alternating Digital Tree (ADT) algorithm for 3D geometric searching and intersection problems, *Int. J. for Numerical Methods in Engineering.* 1991, Vol. 31, pp 1–17.

26

Unstructured Grids: Procedures and Applications

Nigel P. Weatherill

26.1 Introduction

The aims of this chapter are to provide some examples of unstructured grids and, moreover, to illustrate the major steps involved in the generation and use of unstructured grids of triangles and tetrahedra. No theory will be presented, since all the basic theory has been introduced in previous chapters.

26.2 Grids Constructed by Delaunay Triangulation — The General Procedure

The Delaunay approach for the construction of unstructured grids is a popular method. It is appropriate, therefore, before discussing real examples, to illustrate the general procedure. Chapters 1 and 16 have discussed the technical aspects of the approach and outlined relevant algorithmic details, so they will not be reproduced here. However, the illustrations presented are based upon the construction of the Delaunay triangulation using the Bowyer [1–6] algorithm.

Consider a circle as shown in Figure 26.1a. It is described as a set of discrete points.

- The first step is to define a convex hull enclosing all the boundary points that describe the geometry. This can be automatically performed given the coordinates of the circle. In this case, four points are used to define the convex hull. A Delaunay triangulation of these four points is performed, and the resulting grid with the geometry is shown in Figure 26.1a The convex hull encloses all the geometry points and is triangulated.

0-8493-2687-7/99/$0.00+$.50

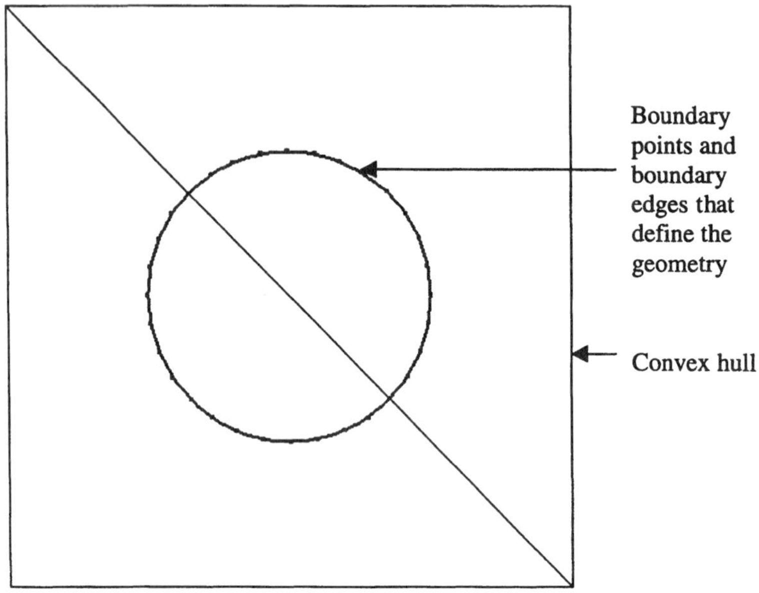

FIGURE 26.1A The convex hull encloses all the geometry points and is triangulated.

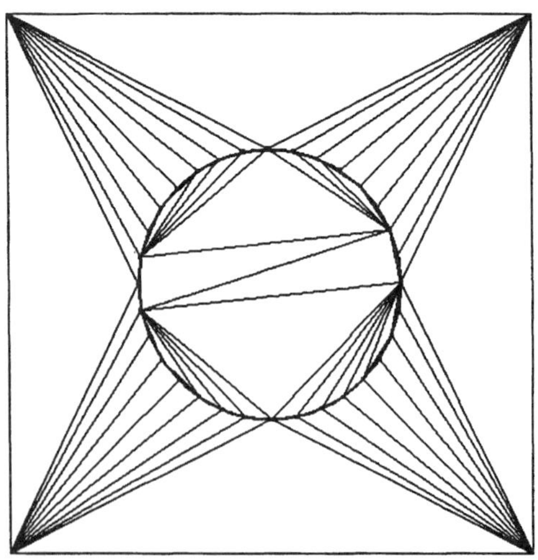

FIGURE 26.1B A Delaunay triangulation of the boundary points is performed.

- Given an initial construction of four points, together with their Delaunay construction, each one of the geometry boundary points is inserted sequentially and connected into the triangulation structure. Figure 26.1b shows the resulting grid after all the boundary points have been inserted.
- To create the grid inside the circle, it is then necessary to systematically refine the triangles inside the circle. There are several methods for performing this task, as described in Chapter 16. However, for this illustration, the insertion strategy involves the addition of points at the centroid of elements until the required point density is achieved (Chapter 1). The grid point density is controlled by the background mesh and any sources that have been specified (see Section 26.3). Points are created by looping over elements within the domain and inserting a point when element refinement

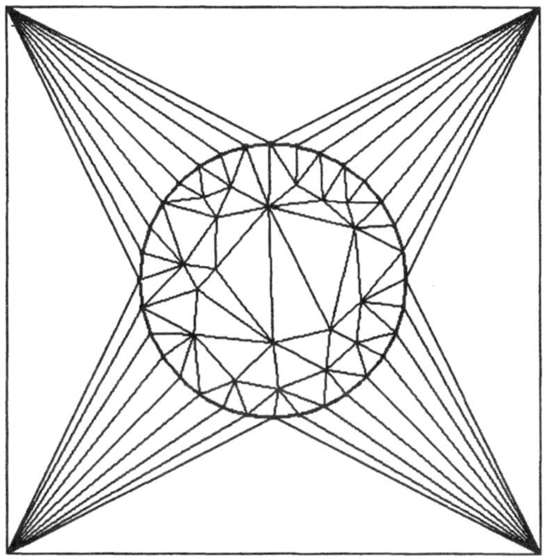

FIGURE 26.1C Points within the domain are inserted in an iterative process until the required point density is obtained. Shown is the grid during the point insertion phase.

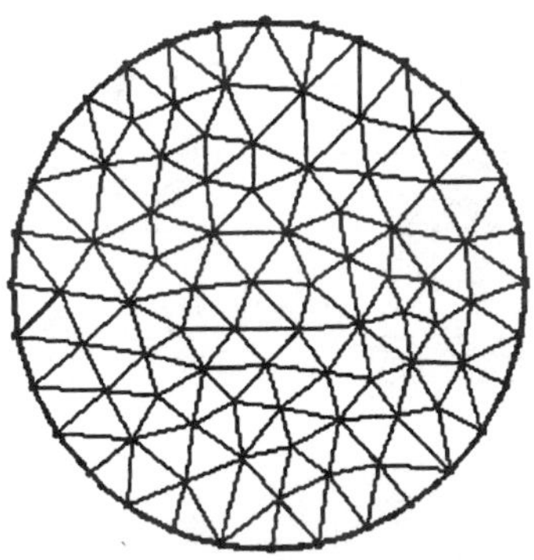

FIGURE 26.1D The final grid after point insertion and the deletion of elements outside the domain of interest.

is required. Points are connected into the triangulation using the Delaunay based algorithm. Figure 26.1c shows the grid after the insertion of some points, although the grid point density criterion throughout the grid has not yet been satisfied. point density criterion throughout the grid has not yet been satisfied.

- Once the grid point density has been achieved, a post-processing step deletes all triangles that are not within the domain, and if appropriate, the grid can then be smoothed using a Laplacian filter. The final grid for this case is shown in Figure 26.1d.

As a further illustration, each stage of the process is illustrated, for a simple geometry, in Figure 26.1e.

The process illustrated here for very simple geometries and small grids highlights the sequence of steps that are applied for the generation of Delaunay grids in both two and three dimensions. These simple geometries do not illustrate a very important requirement in the generation of grids by Delaunay triangulation. It is important that in the final grid the edges of the initial boundary are preserved. This is the so-called *boundary integrity* requirement. Hence, to augment the steps given above, it is necessary to add a final step,

- Ensure that the initial boundary edges are included within the final grid [7].

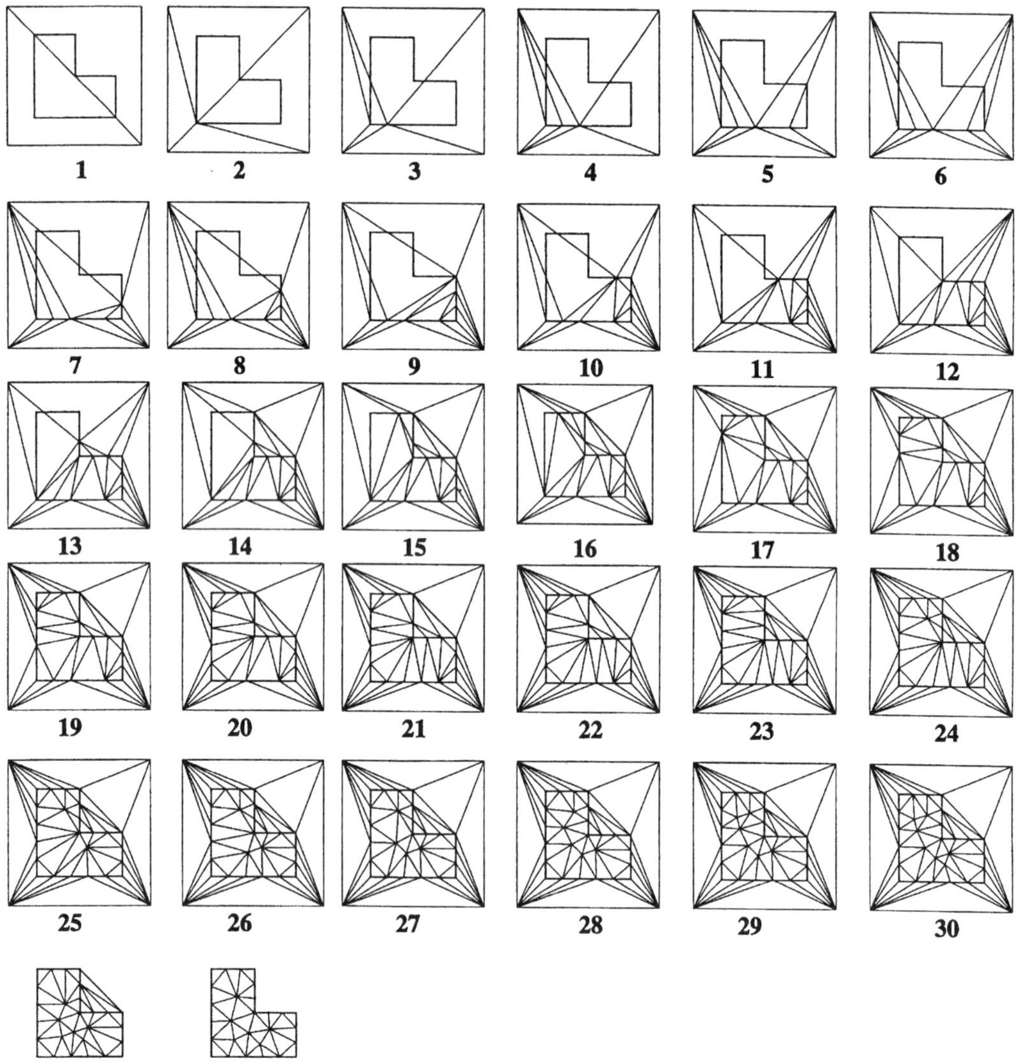

FIGURE 26.1E Each stage of the grid generation phase is shown for a simple geometry.

26.3 Unstructured Grid Control Using a Background Grid and Sources

Figure 26.1a of Section 26.2 shows the points that define the geometry of the circle. However, any mesh generation procedure must provide a suitable mechanism for a user to change the number of points on the boundary of any given geometry — perhaps a coarse discretization is required, or a fine discretization. One of the popular approaches to this problem in the generation of unstructured grids is to use a background grid and sources [Chapters 1 and 17].

Figure 26.2a shows a schematic of a particularly simple background grid. The idea is straightforward.

- Define a mesh that covers the domain Ω to be gridded. The mesh should consist of nodes and have a topologically valid connectivity that defines the elements. In Figure 26.2a the background grid consists of four nodes and two triangles.
- At each node of the grid, a parameter is defined which specifies the point spacing at that position.

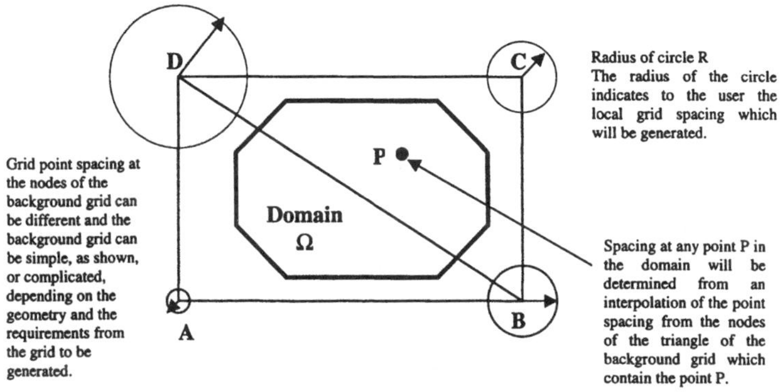

FIGURE 26.2A The background mesh used to control grid point spacing.

- During the grid generation procedure, and included in this is the point discretization of the geometry, the required spacing at any place in the domain is interpolated from the background grid. Delaunay and advancing front methods require such information. Hence, given a position *P* in the domain,
  1. Determine the element, *E*, of the background grid that contains *P*.
  2. Find the nodes $\{n1,n2,n3\}$ of *E*.
  3. Find the point spacing $\{d1,d2,d3\}$ specified at each of the nodes $\{n1,n2,n3\}$.
  4. Using $\{d1,d2,d3\}$, interpolate the spacing at *P*.

This procedure can also be used to generate grids with stretching (Chapter 19 and 20).

This is a very effective way of controlling the element and point density within an unstructured grid. However, it involves the user in generating a suitable grid and specifying the grid point density parameters at each node of the grid. In two dimensions, using graphics user interfaces, this is not too time-consuming and is readily achieved for most geometries. However, in three dimensions it is a nontrivial task. Hence control of a grid by a background grid is usually augmented with the use of sources.

Figure 26.2b shows three basic types of sources. Although there are many variants of the definition of a source now in the literature, the fundamental features of a point source are defined by

- A position, *Q*, within the domain.
- At *Q*, the required grid point spacing is defined, *d*.
- A circle is specified of radius $r1$, within which the user specified grid point density, *d*, is defined.
- A second circle is specified of radius $r2$, where, $r2 > r1$. Within the region defined between the circle radius $r1$ and the circle radius $r2$, the point spacing will decay from *d* to that specified by the background grid.

Hence, for a point source with the structure just defined, the user must specify four parameters. However, this does not involve the intricacies of a mesh connectivity as required with a background grid. In fact, the background grid which accompanies sources is effectively redundant, since uniform spacing everywhere can be the default condition. Hence, no interpolation is required.

The extension to a line source, a triangle source, or even a volume source is straightforward. Figure 26.2b shows, in schematic form, a line and triangle source. These two allow the user to easily specify the required grid point density over regions of the domain.

The concept of a source naturally extends to three dimensions.

26.4 Unstructured Grids of Triangles

The example chosen to illustrate the use of unstructured grids of triangles is the outline of San Francisco Bay. The geometry is defined as a set of discrete points. The boundary can be modeled as a set of NURBS

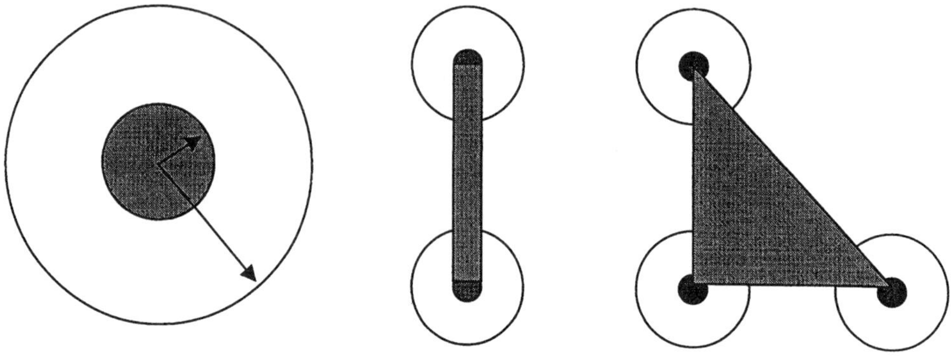

FIGURE 26.2B Point, line, and triangular sources.

FIGURE 26.3A The geometry of San Francisco Bay.

or splines. This then enables an arbitrary point distribution to be generated on the boundary for any given grid density.

- Figure 26.3a shows the geometry as defined by a discrete set of points.
- A background grid is superimposed over the geometry and spacing is defined at the nodes. In Figure 26.3b, the background grid consists of two elements and the specified spacing is shown by the circles around the nodes.
- From the geometrical data, and the background grid, the points which will define the boundary within the grid can be generated. Figure 26.3c shows the point distribution on the boundary.
- Figure 26.3d shows the resulting grid within the domain.
- To illustrate the use of sources, Figure 26.3e shows two line sources that have been designed to construct a grid that will resolve an imaginary deep water channel. The sources that form the line source have different regions of influence, as shown by the circles.

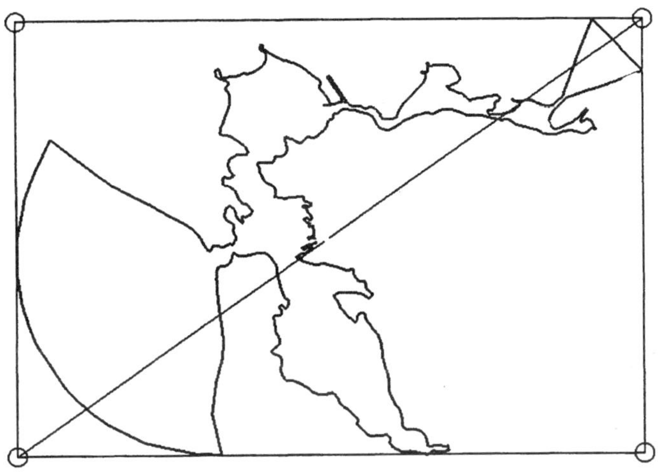

FIGURE 26.3B The geometry of San Francisco Bay, together with the background grid. Note that the circles attached to the nodes of the background grid indicate to the user the spacing specified.

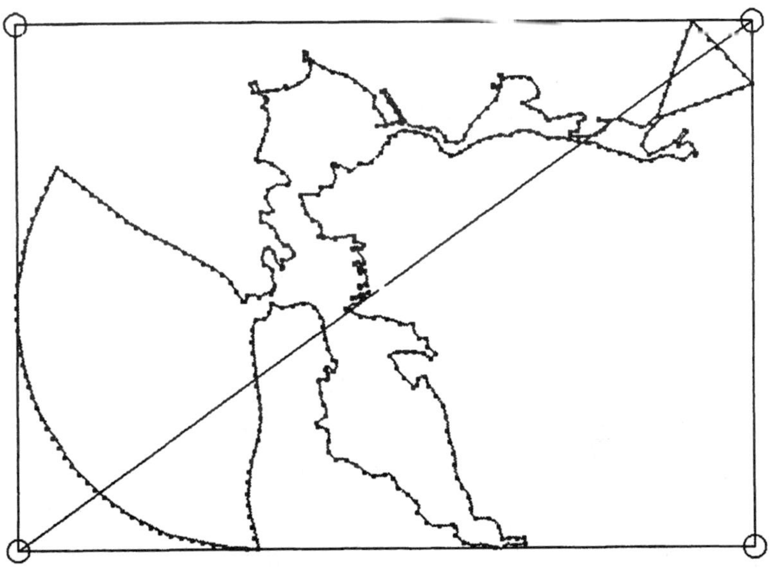

FIGURE 26.3C The boundary grid generated from the point spacing specified on the background grid.

- Figure 26.3f shows the resulting boundary point distribution. The effect of the line sources is apparent. (Compare with the boundary point distribution from the background grid only as shown in Figure 26.3c).
- The resulting grid, controlled by both the background grid and the line sources, is shown in Figure 26.3g.
- Following the generation of a mesh, it is good practice to make an assessment of grid quality [8]. In many cases, such an assessment can be included within a grid generator and only if there is a problem would the user be informed. However, it can also be beneficial to have a stand-alone grid analysis package. The assessment of grid quality in relation to an analysis algorithm is still a topic for much research. However, it is possible to identify geometrical measures of the "goodness" of a grid. Some appropriate measures are shown in Table 26.1. After computation of quality measures they can be presented in the form of histograms, as shown in Figure 26.3h.

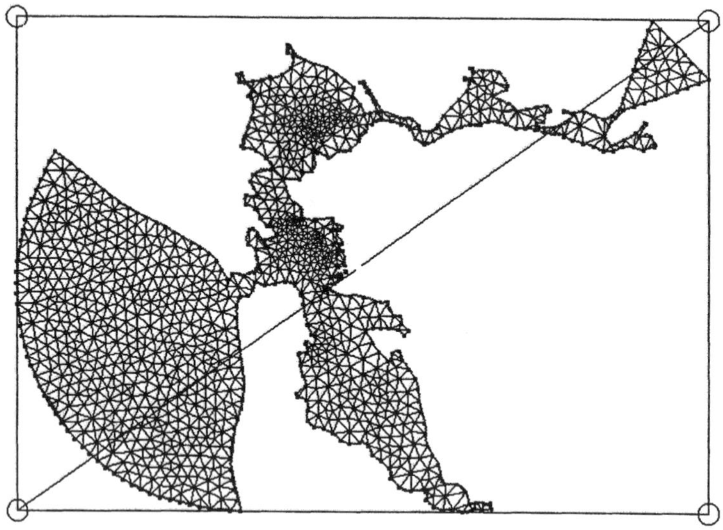

FIGURE 26.3D A grid for the Bay generated with grid control from the background grid.

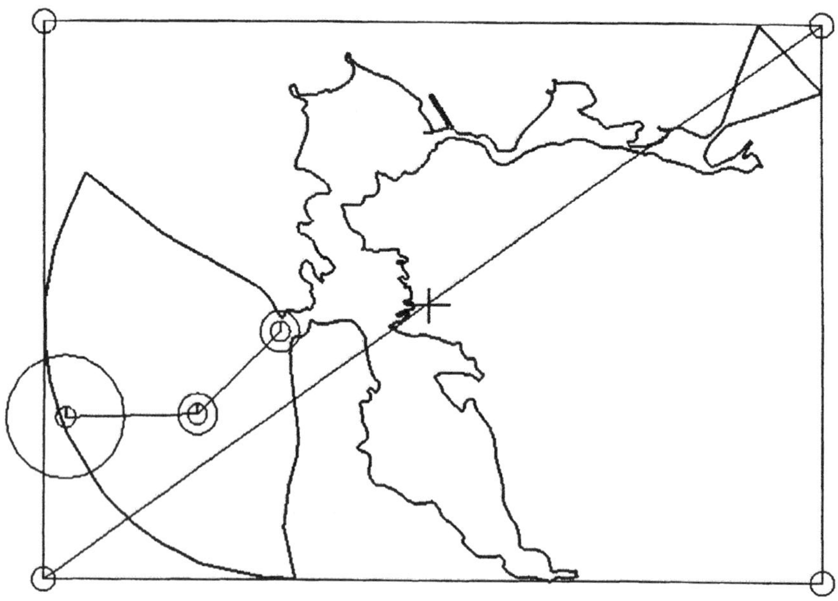

FIGURE 26.3E The geometry of San Francisco Bay with the background grid augmented by two line sources to resolve a deep-water channel.

- It is important to know the location of elements with particular grid quality measures. Figure 26.3i shows the generated grid and the elements that have been highlighted. In practice, such a presentation will involve the user of color.

26.5 Hybrid Grids of Quadrilaterals and Triangles

As a second example of the use of unstructured grids, an approach is presented whereby a hybrid grid is constructed from quadrilaterals and triangles. Shaw (Chapter 23) discusses at length the philosophy behind the use of hybrid grids and presents results for three-dimensional aerospace configurations. The

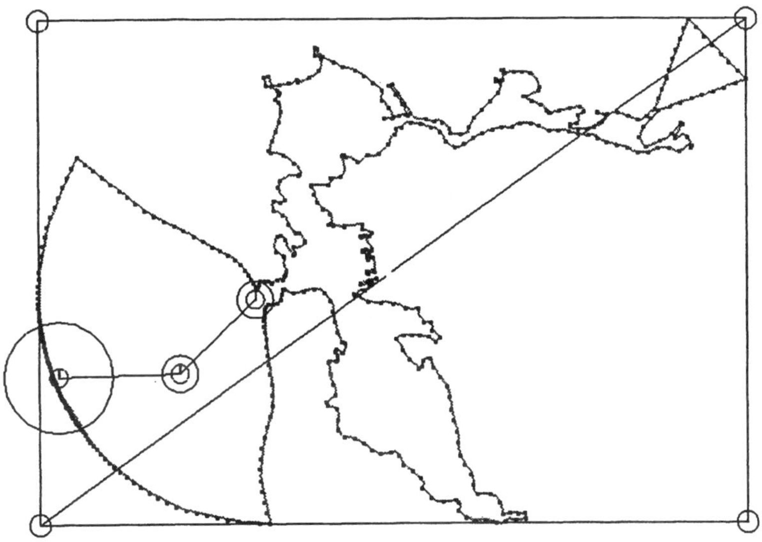

FIGURE 26.3F The boundary grid generated from the point spacing specified on the background grid and the two line sources.

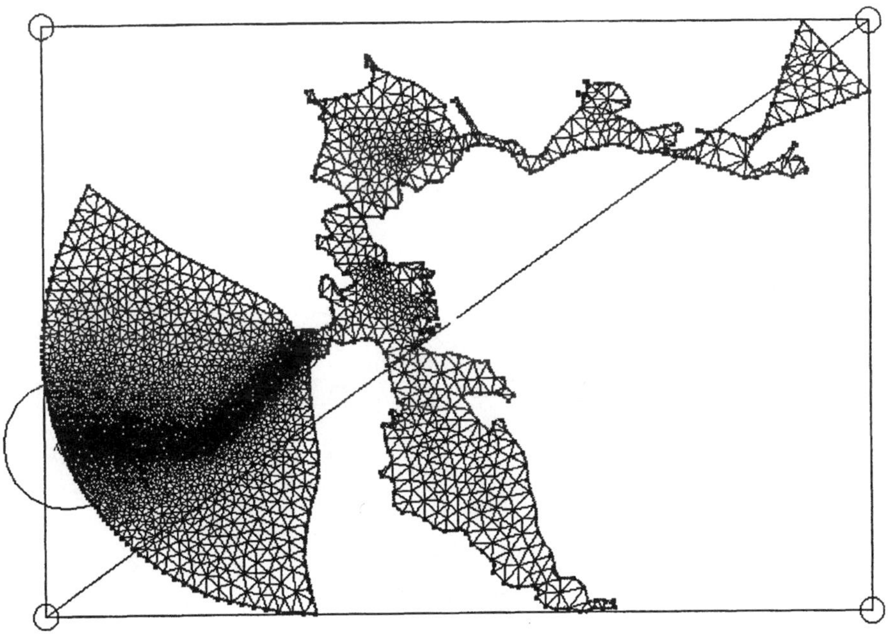

FIGURE 26.3G The grid generated with control from the background grid and the line sources.

example shown here is based upon early work [9,10] and is presented to further elaborate and possibly clarify some of the comments made in Chapter 23. Hybrid grids are also covered in Chapter 25.

Figure 26.4a shows an outline of a four component airfoil system composed of a main airfoil, one leading edge slat and two trailing edge flaps. In the process of generating a hybrid grid,

- The first step is to generate a structured grid around the main component airfoil. Any structured grid technique can be used, but here a conformal mapping grid based upon a Von-Karman–Trefftz transformation is used. Figure 26.4b shows such a grid. The outer boundary, which is not shown, extends about 15 chord lengths away from the airfoil.

TABLE 26.1 Grid Element Quality Parameters

$\beta = \dfrac{\text{Radius of circumscribing sphere}}{\text{Radius of inscribed sphere}}$	$\beta_{\text{equilateral}} = 3.0$
$\sigma = \dfrac{\text{Maximum edge length}}{\text{Radius of inscribed sphere}}$	$\sigma_{\text{equilateral}} = 4.8989$
$\omega = \dfrac{\text{Radius of circumscribed sphere}}{\text{Maximum edge length}}$	$\omega_{\text{equilateral}} = 0.6125$
$\zeta = \dfrac{\text{Maximum edge length}}{\text{Minimum edge length}}$	$\zeta_{\text{equilateral}} = 1.0$
$\alpha = \dfrac{(\text{average element edge length})^3}{\text{Volume}}$	$\alpha_{\text{equilateral}} = 8.479$
$\gamma = \dfrac{(\text{R.M.S edge length})^3}{\text{Volume}}$	$\gamma_{\text{equilateral}} = 8.479$
$K = \dfrac{(\text{Volume})^3}{(\text{Summation of all surface area of triangle faces})^2}$	$K_{\text{equilateral}} = 4.5 \times 10^{-4}$

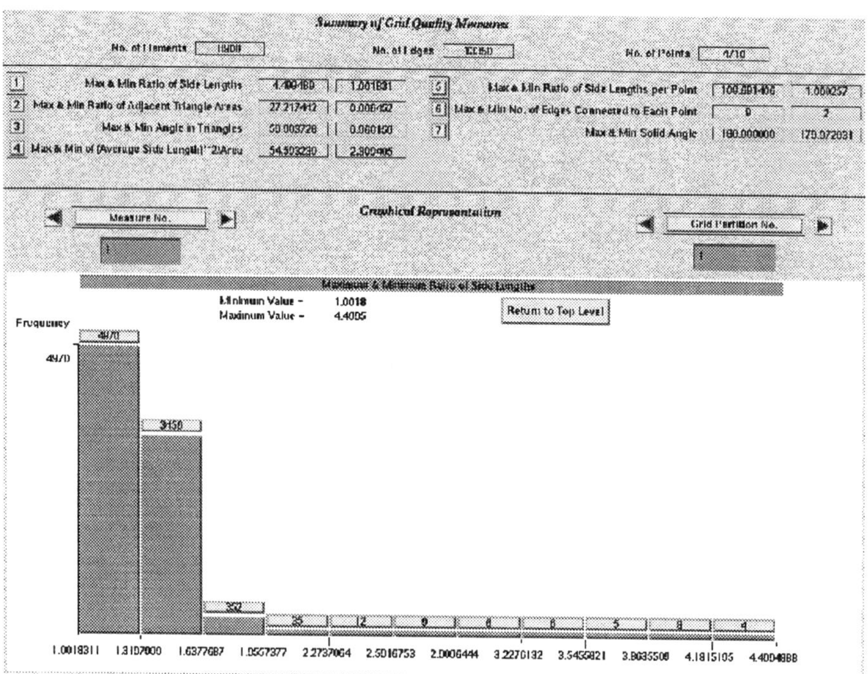

FIGURE 26.3H Typical histogram display of grid quality statistics.

- The next step is to choose one of the flaps or slat components and construct a grid that does not extend too far away from the geometry. In the case illustrated, the second flap is chosen and a structured grid, again generated by a conformal mapping, is produced. Figure 26.4c shows the grid in relation to the main component airfoil.

FIGURE 26.3I Elements with particular grid quality measures can be visualized.

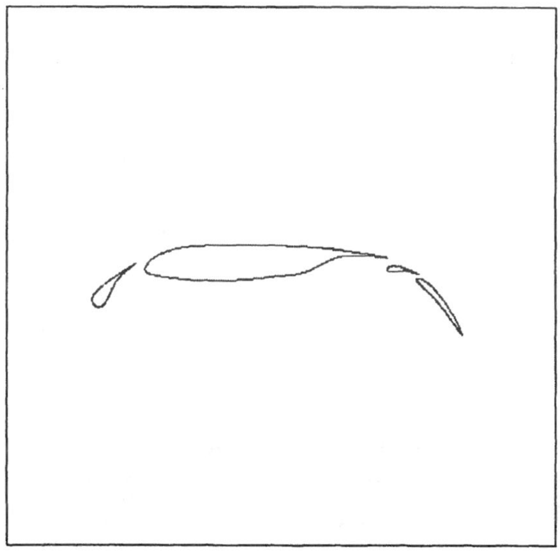

FIGURE 26.4A Geometry of the three component airfoil.

- The grid around the flap is then superimposed over the main component grid, as shown in Figure 26.4d.
- The next step is governed by user input, which specifies how much of the grid around the flap should be preserved. A region of the main component grid is then deleted so that the two overlaid grids do not intersect, as shown in Figure 26.4e.
- In order to connect the two grids it is then necessary to fill the void region by constructing an unstructured grid. This is readily achieved since the boundary points, together with the boundary edge connectivities, can be easily extracted and sent to the Delaunay grid generator [1–6]. An unstructured grid is then generated, as shown in Figure 26.4f.
- A hybrid grid can then be created by connecting together the three generated grids. This is shown in Figure 26.4g.
- To introduce the remaining flap and slat, the process already described is repeated. First, introduce a component grid for the leading edge slat and overlay this over the existing grids, Figure 26.4h.

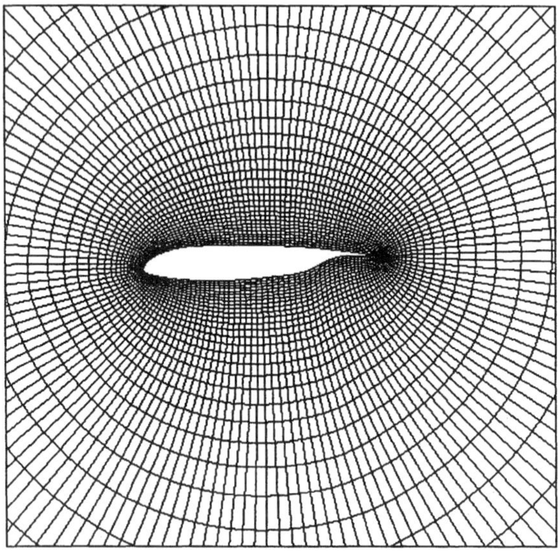

FIGURE 26.4B A structured grid generated from a conformal mapping is constructed around the main component airfoil.

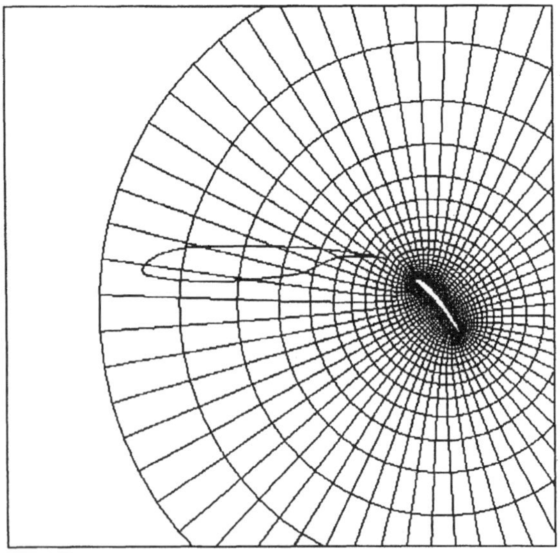

FIGURE 26.4C A structured grid is generated around the flap.

- Preserve a portion of the component grid, determine the empty void region, fill with an unstructured grid and connect all the grids. Figure 26.4i shows the final grid.
- To complete the grid, repeat the process again for the second flap component. The final hybrid grid for the complete four component airfoil is shown in Figure 26.4j.

The quadrilaterals in the final grid could be directly triangulated if a grid of triangles is required. However, this would defeat the objective of generating a hybrid grid. For high Reynolds number viscous flow simulation, it is very easy to modify the structured grid generation so that appropriate point

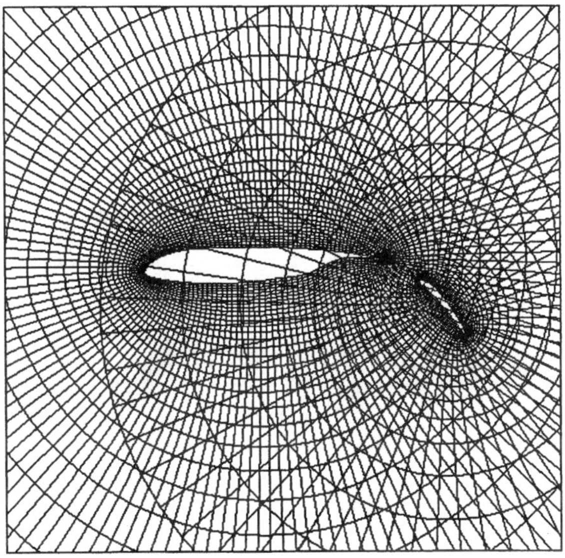

FIGURE 26.4D The structured grid for the flap is superimposed on the grid for the main component.

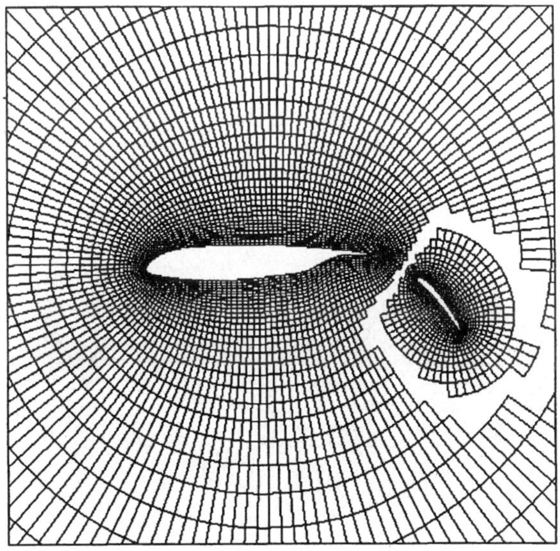

FIGURE 26.4E Overlapping regions of the two grids are deleted leaving two disconnect grids.

clustering in the vicinity of solid boundaries is suitable for capturing boundary layer phenomena. In this way, hybrid grids of the form shown have an important role to play. The major disadvantage of the approach, as illustrated, is that the method is not automatic for general geometries (as defined in the spirit of *automatic* unstructured grid generators), since a structured grid is required and, since this involves a mapping procedure, the method is geometry-specific. However, the approach is potentially powerful in the sense presented by Shaw (Chapter 23). Hybrid grids are suitable for use with finite volume solvers — in particular, an edge-based scheme — since then the fact that different element types are present is not relevant to the solver [9].

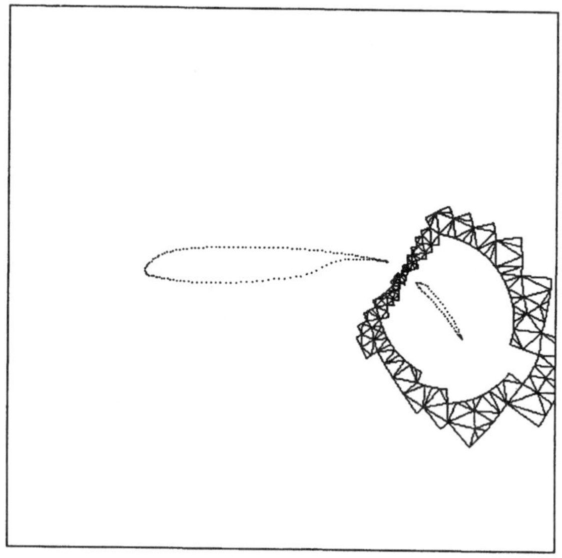

FIGURE 26.4F An unstructured grid is generated within the void domain, thus connecting the two component grids.

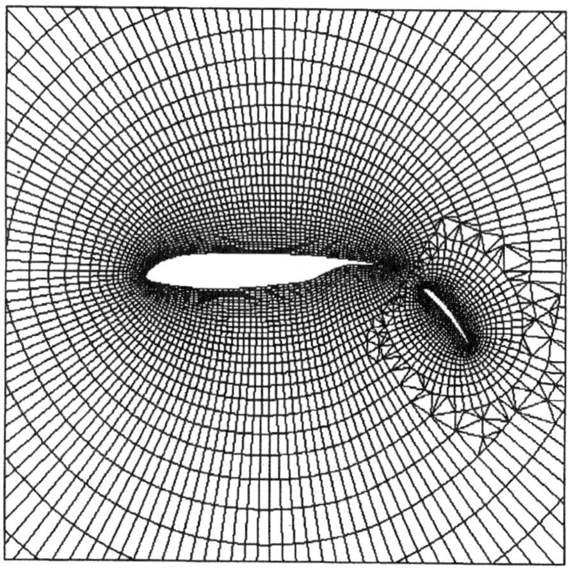

FIGURE 26.4G The two grids are connected by a ribbon of unstructured grid.

26.6 Unstructured Tetrahedral Grids

This section attempts to describe the typical process by which three-dimensional grids of tetrahedra are generated using a Delaunay based approach. Two examples are presented. The first is an aerospace geometry, the Dassault Falcon, consisting of a wing, fuselage, rear-mounted engine, tail and fin, and the second is the geometry of the THRUST Supersonic Car, which broke the world land speed record in October 1997.

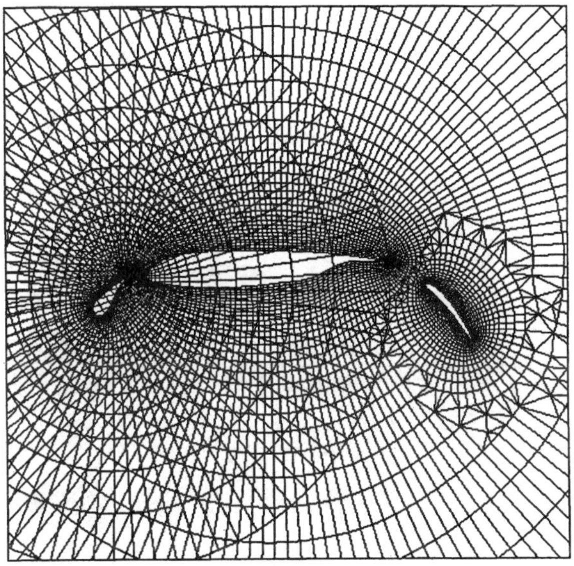

FIGURE 26.4H The third component with a local structured grid is laid over the main component grid.

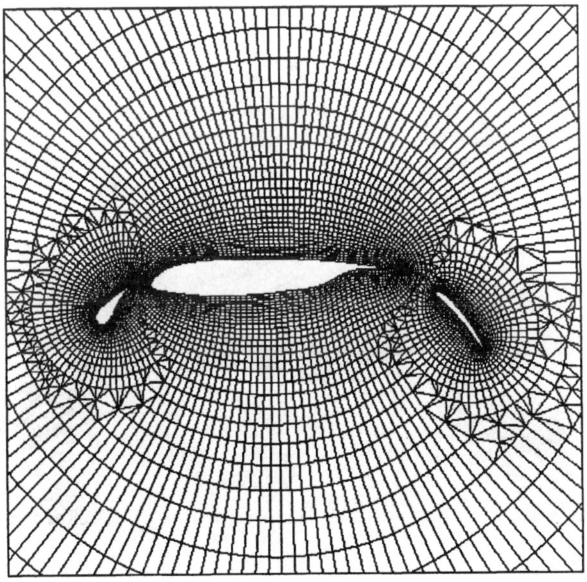

FIGURE 26.4I The component grid is connected to the main grid by a ribbon of unstructured grid.

26.6.1 Dassault Falcon

- Figure 26.5a shows the geometry of the Falcon aircraft. For clarity, only half the aircraft will be considered. The aircraft consists of 12 individual support surfaces.
- It will be assumed that a grid is required around the exterior of the geometry, typical for a flow computation. Hence, the domain must be closed by the addition of bounding surfaces, in this case, an outer hemispherical boundary and a plane of symmetry, as shown in Figure 26.5b.

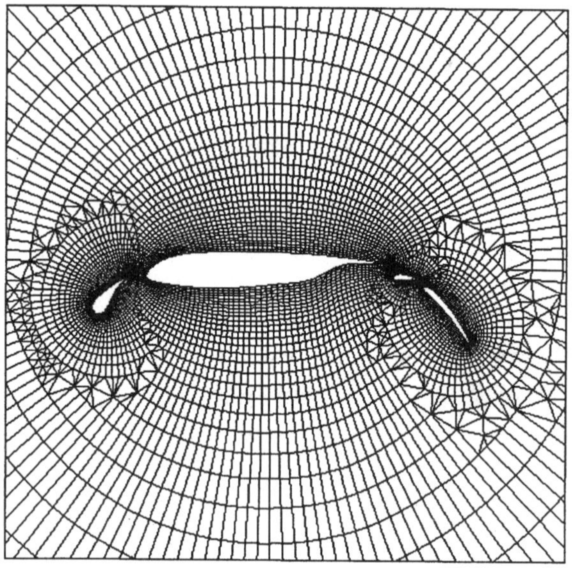

FIGURE 26.4J The second component flap grid is introduced and connected to the main component grid.

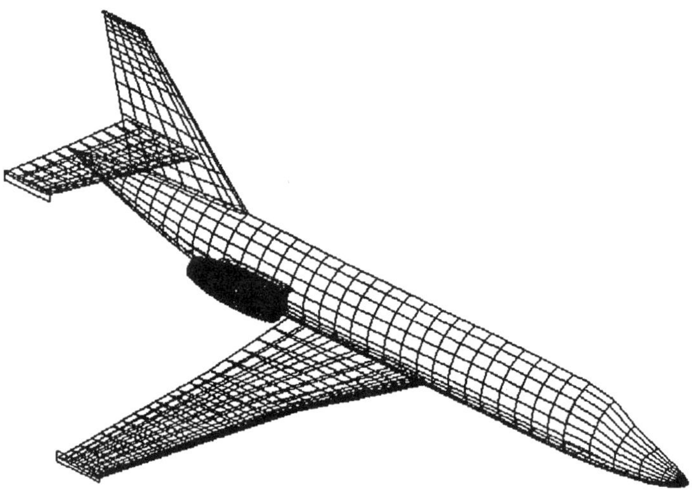

FIGURE 26.5A Shown is the geometry patches. Twelve support surfaces define the shape of the aircraft.

- Given the closed domain, the generation process involves the construction of a grid on the surfaces which define the domain, followed by a tetrahedral grid generated to fill the domain. Before either of these tasks can be performed it is necessary to define the required spacing of elements within the domain. As in the case of the generation of grids in two dimensions, this is performed using a background grid with added sources. For three dimensions it is not particularly beneficial to present a figure which outlines the background grid. Hence, in Figure 26.5c the geometry of the aircraft is shown together with the representation of a line source. This line source, as can be seen, is shown as a thick line along the leading edge of the wing. The two spheres at the end of the line provide the user with an indication of the region of influence of the line source. The definition of a line source by a user is easily performed within a graphics user environment, since points of

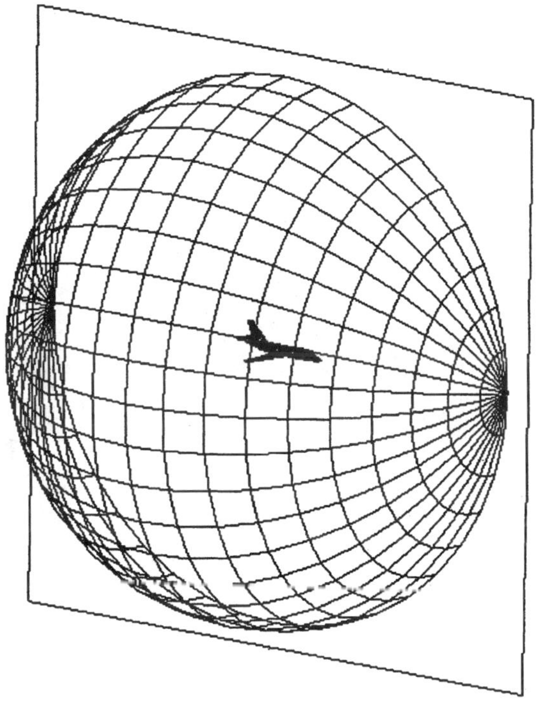

FIGURE 26.5B The region around the aircraft is enclosed by a hemispherical boundary and a plane of symmetry.

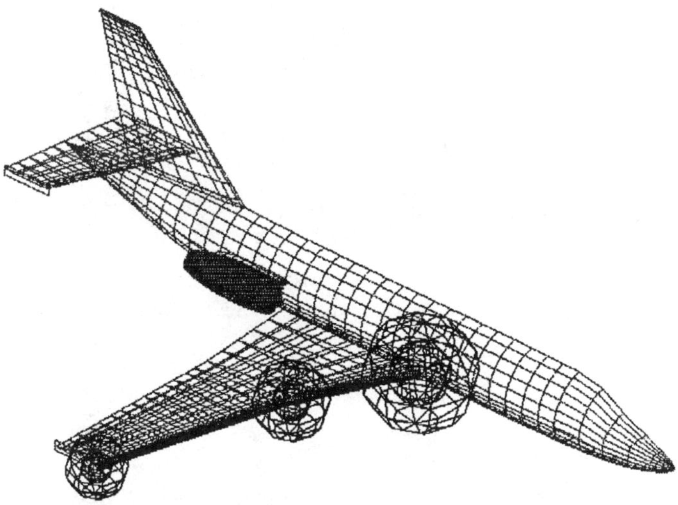

FIGURE 26.5C Sources are used to provide the required grid control. Shown are two line sources along the leading edge of the wind. The outline of the spheres at the ends of the line sources provide the user with an indication of the regions of influence of the sources.

the geometry can be selected and then point sources/line sources created by the push of a button. The line source shown in Figure 26.5c illustrates the concept. However, for a realistic mesh for the Falcon aircraft it is necessary to define many line and point sources, and in such a case it is not effective to show all these in a figure.

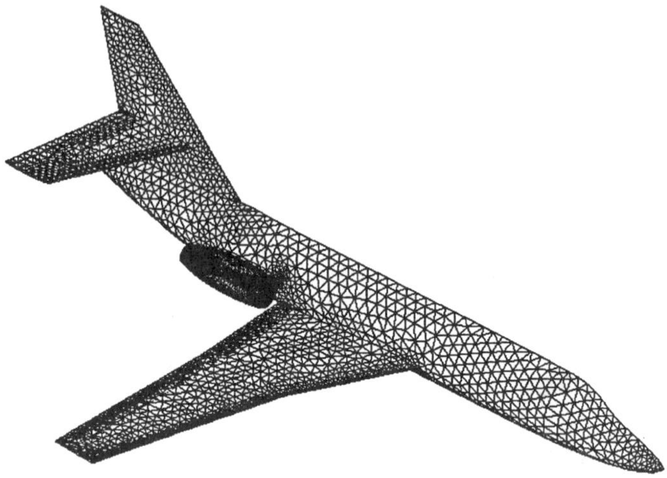

FIGURE 26.5D A coarse surface grid on the aircraft. (Note: The sources used to generate this grid are not the ones shown in Figure 26.5c.)

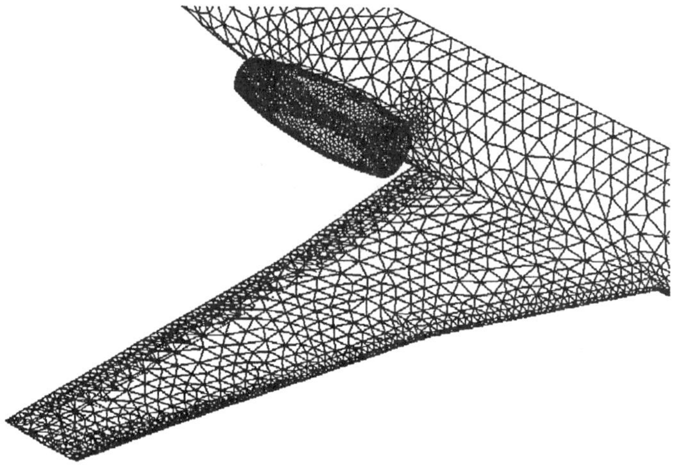

FIGURE 26.5E A close-up view of the surface mesh.

- The surface grid on the aircraft, generated using six point sources and ten line sources, is shown in Figures 26.5d and 26.5e. Surface grid generation is described in detail in Chapters 17 and 19 and in reference [11]. The mesh shown is, for clarity, a coarse mesh, but it does exhibit the required spacing in that the grid has been clustered in regions around leading edges and trailing edges.
- Once the surface grid has been generated on all the boundary surfaces, a volume mesh can be created. It is difficult to view tetrahedra, but Figure 26.5f shows the elements that fall inside a cutting arc. This leads to effective pictures, but arguably these are of little value in assessing grid quality.
- It is necessary to resort to analysis of the grid quality measures to assess the quality of the grid and histograms are a suitable way to project this data, Figure 26.5g.
- If required, elements or nodes within the grid whose associated quality measures are of concern can be viewed, Figure 26.5h.

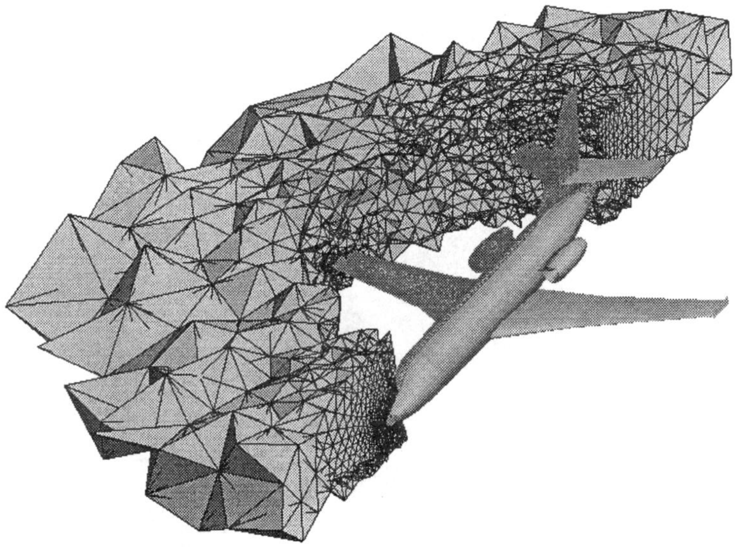

FIGURE 26.5F Sectional cut through the grid of tetrahedra.

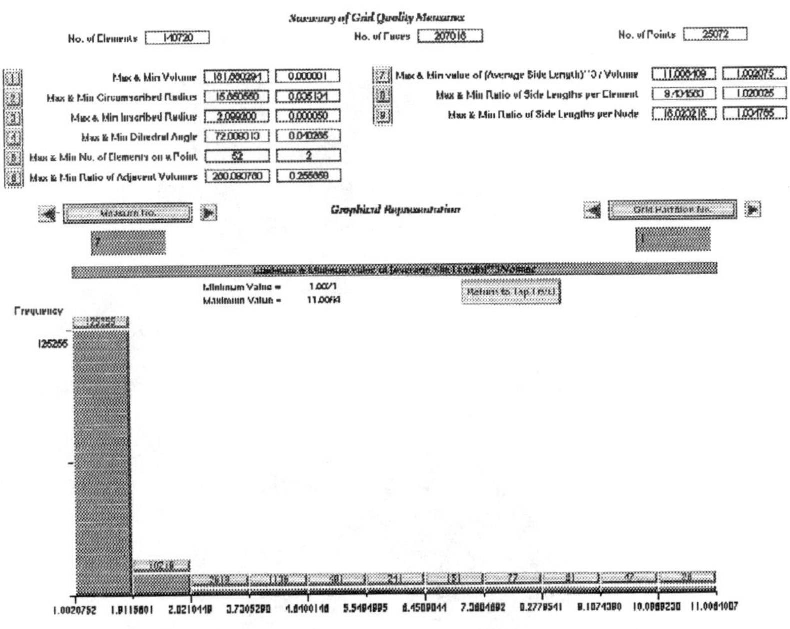

FIGURE 26.5G Histogram of grid quality measures.

- To complete the sequence of figures, the grid generated is suitable for an inviscid flow simulation. Figure 26.5i shows the geometry of the aircraft, streamlines, contours, and sections through the unstructured grid.

26.6.2 THRUST Supersonic Car

- Figure 26.6a shows the geometry of the Thrust car. The car is enclosed within a bounding box, as shown in Figure 26.6b.

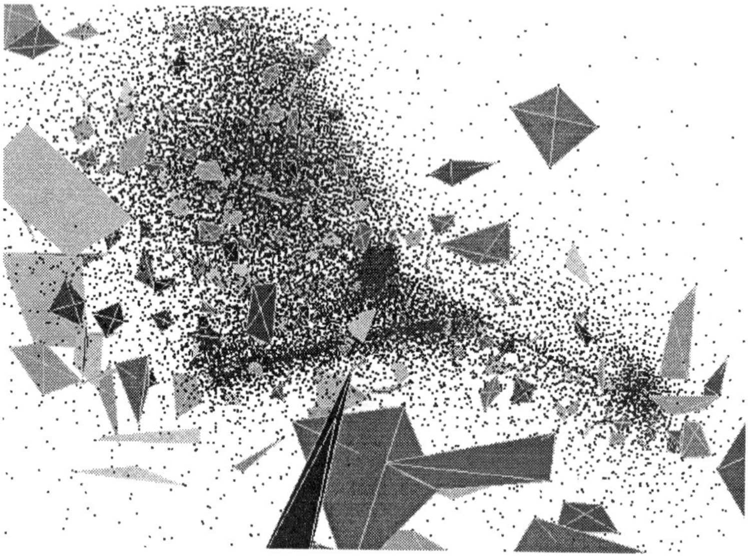

FIGURE 26.5H Elements with specific grid quality measures can be viewed in the mesh.

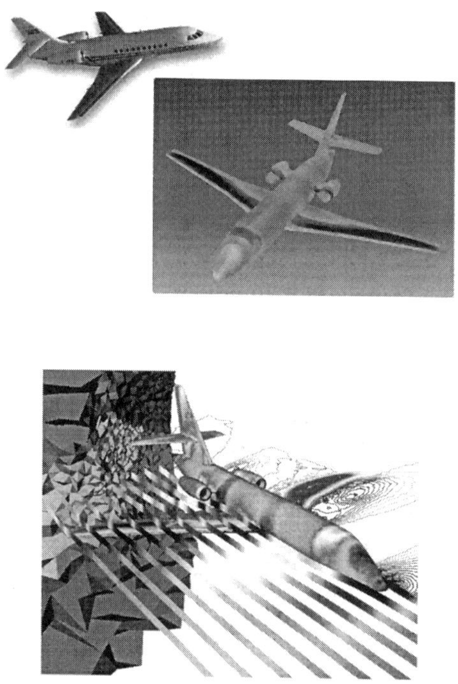

FIGURE 26.5I Flow simulation for the Dassault Falcon.

- Figure 26.6c shows the geometry of the car together with sources to control the grid point density.
- Figure 26.6d shows a grid on the car, plane of symmetry and the ground.
- Figure 26.6e shows a cut through the grid of tetrahedra.
- A typical flow simulation is shown in Figure 26.6f.

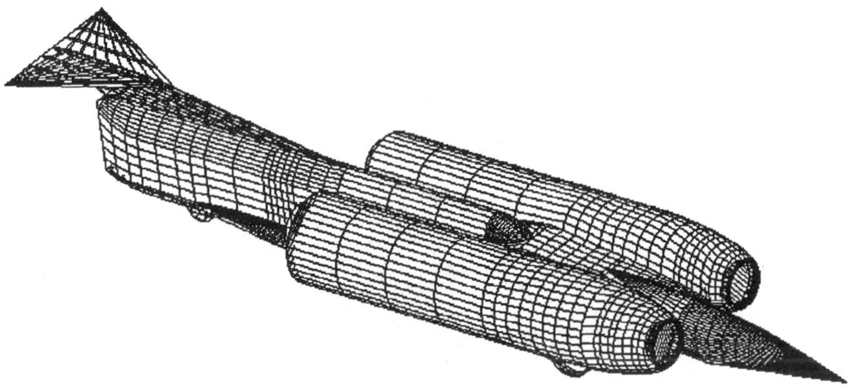

FIGURE 26.6A Geometrical definition of the car.

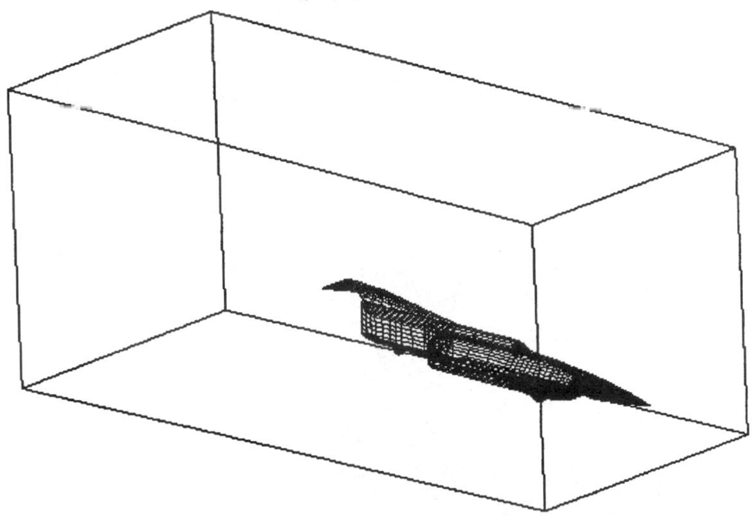

FIGURE 26.6B The car inside a bounding box.

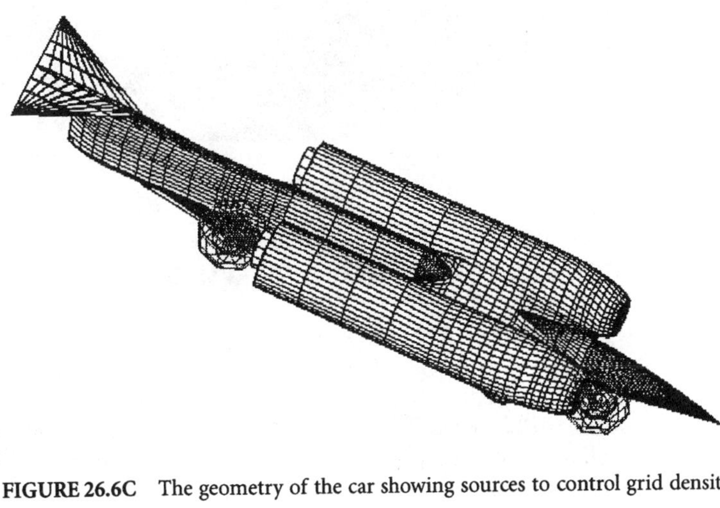

FIGURE 26.6C The geometry of the car showing sources to control grid density.

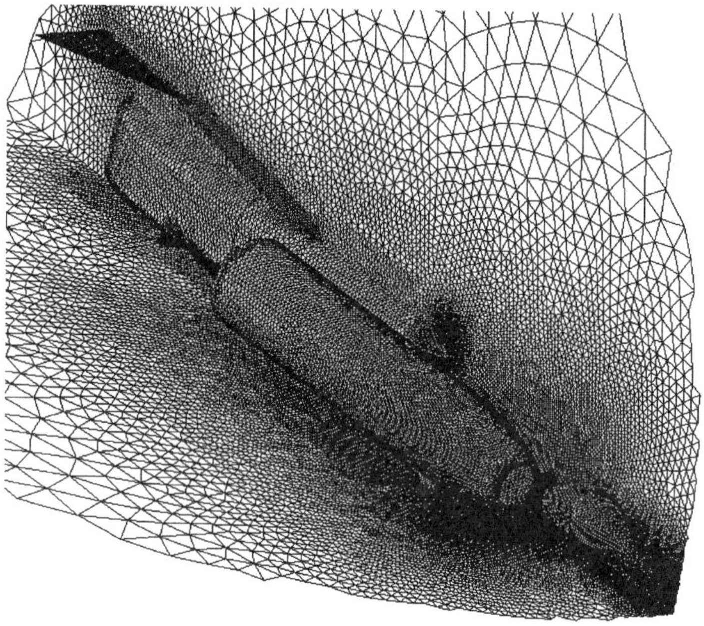

FIGURE 26.6D Surface grid on the car, plane of symmetry and the ground.

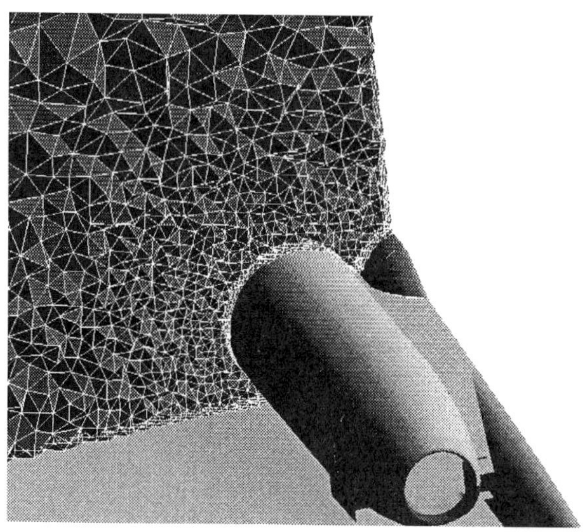

FIGURE 26.6E Cut through the domain of tetrahedra.

26.7 Non-Isotropic Grid Generation for Viscous Flow Simulation

For some applications, the use of regular isotropic elements can lead to very large meshes. A good example of such a case is the simulation of high Reynolds number viscous flows where, to capture boundary layer effects, very small elements are required. It is appropriate, therefore, knowing the physics of boundary layers, to consider a form of *a priori* adaptation to reflect the difference in gradients in flowfield variables across a boundary layer as compared with along a boundary layer in the direction of the flow. If such an approach is followed, then elements with high aspect ratios will be required.

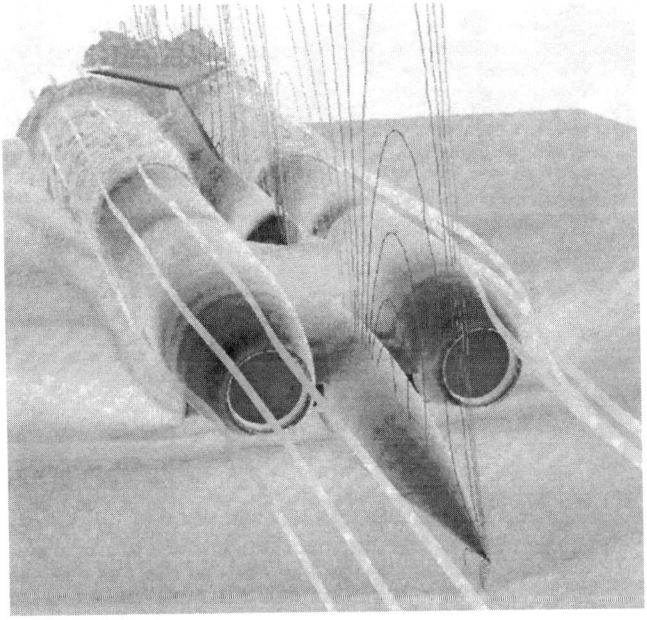

FIGURE 26.6F Flow solution over the car.

The generation of grids that incorporate elements with arbitrary stretching has been the focus of interest for some time. Chapter 20 discusses this issue in some length and presents in detail one approach. An alternative approach is highlighted in Figure 26.7.

- Figure 26.7a shows a grid of quadrilaterals which has been generated using an algebraic approach [14,15]. The approach amounts to growing layers of elements by advancing along lines that are approximately normal to the boundary. These layers of elements are grown until either they self-intersect or reach an aspect ratio of unity.
- After this the domain is filled with regular isotropic elements using the standard Delaunay approach, Figure 26.7c.
- Figures 26.7e show some of the details of this approach within a concave corner.
- The quadrilaterals can be triangulated to provide a grid consisting of triangles, Figures 26.7b, 26.7d, and 26.7f.

This approach is equally applicable in three dimensions where grids of tetrahedra or tetrahedra/prisms can be created.

This method of advancing layers (or advancing normals) is a pragmatic approach and is clearly applicable for solid boundaries. However, it does not take into account other features of viscous flow phenomena, such as wakes. However, it is relatively easy to modify the approach to include a suitable treatment for wakes. The approach adopted is as follows:

- Use an initial mesh to obtain a flow solution.
- From the flow solution, determine the wake lines. Figure 26.8a shows a four-component high lift airfoil system with the computed wake lines.
- Attach the wake lines to the existing geometry and then use the advancing layer approach to construct stretched elements along the geometry boundaries and along the computed wake lines, Figures 26.8b and 26.8c.
- Determine the outer points and edges of the grid generated from the advancing layer stage. These define a boundary which is input data for the Delaunay triangulation phase, Figures 26.8d and 26.8e.

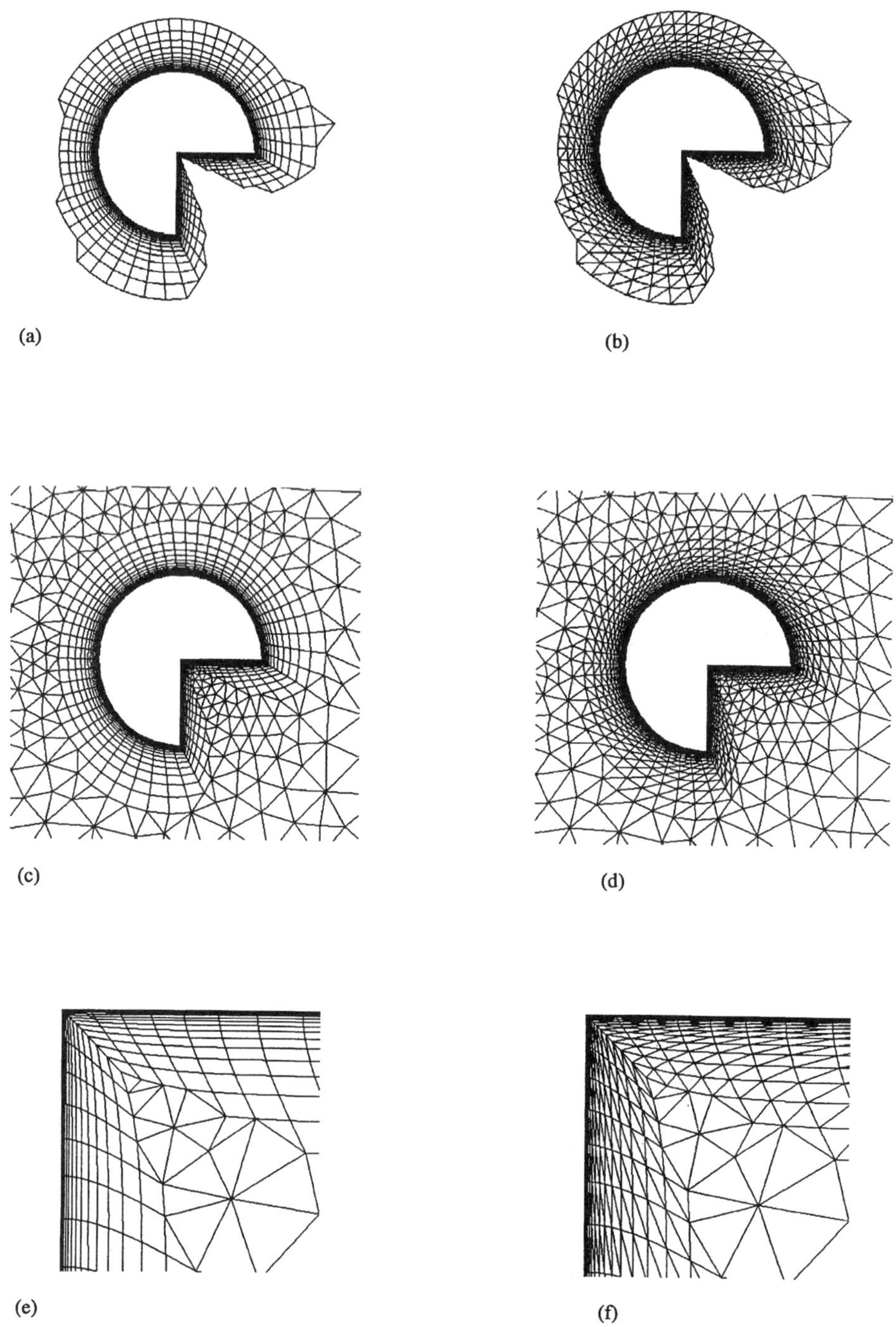

FIGURE 26.7 The advancing layer approach to the generation of stretched elements close to solid boundaries.

- The advancing layers grid and the Delaunay grid are then combined to form the final grid, Figure 26.8f.
- The final grid is suitable for a high Reynolds number viscous flow simulation, Figure 26.8g.

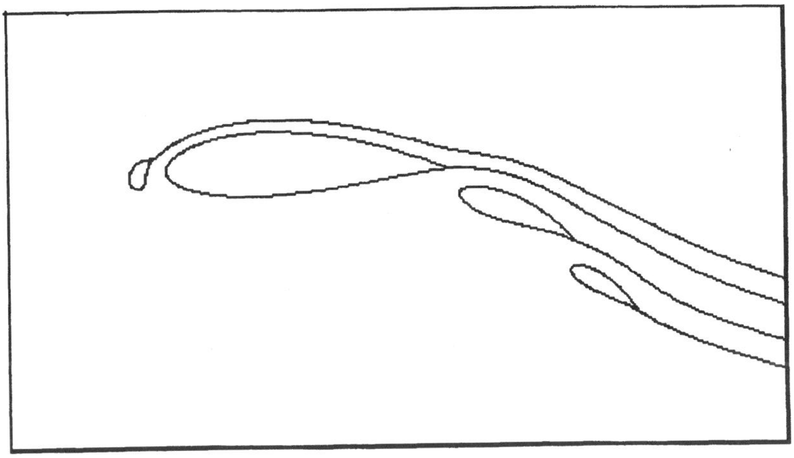

FIGURE 26.8A Geometry of the high-lift airfoils, together with wake lines.

FIGURE 26.8B Highly stretched elements close to the geometrical boundary and the wake lines.

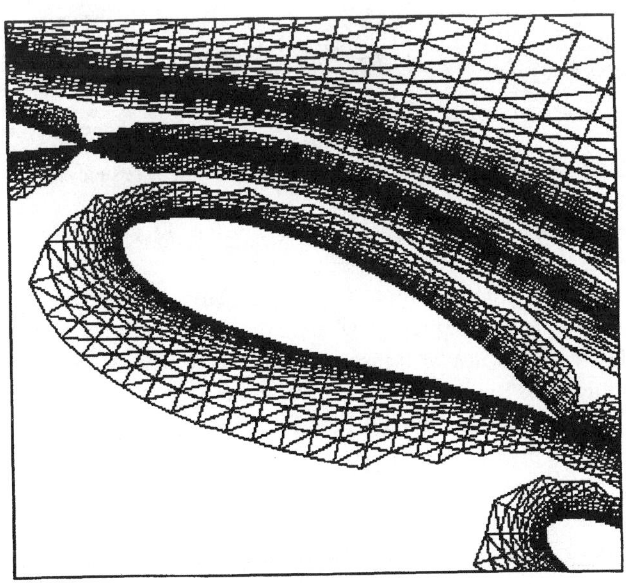

FIGURE 26.8C Close-up view of the stretched elements.

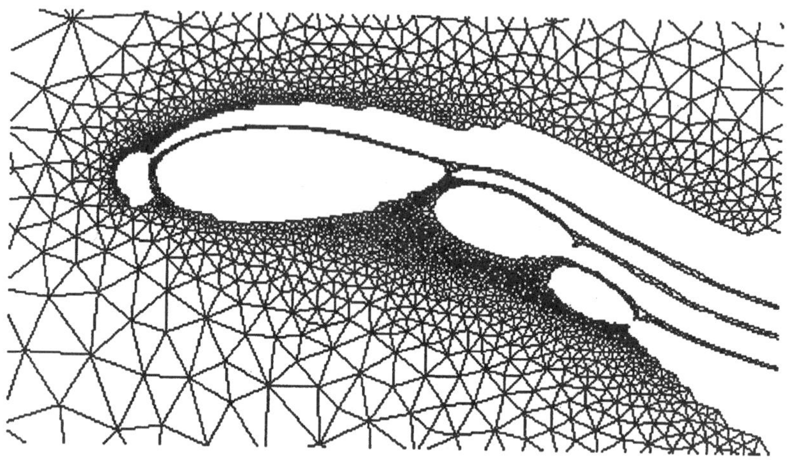

FIGURE 26.8D The isotropic grid generated by Delaunay triangulation.

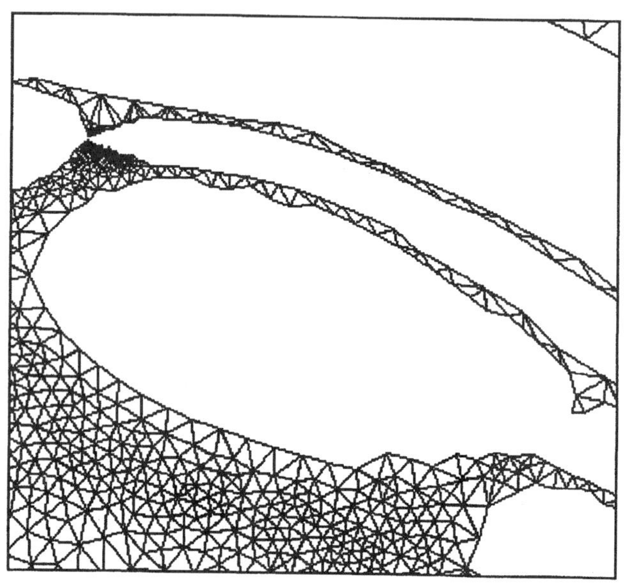

FIGURE 26.8E Close-up view of the unstructured isotropic elements.

26.8 Parallel Unstructured Grid Generation

The introduction of scalable parallel computers is enabling larger problems to be solved in many areas of computational engineering. In computational electromagnetics (CEM), typical simulations employ meshes of five million triangles in two dimensions and many tens of millions of elements in three dimensions. In computational fluid dynamics (CFD), a mesh of at least ten million elements can be required for a high Reynolds number viscous flow simulation over a complete aircraft. As mesh sizes become as large as this, the process of mesh generation on a serial computer can become problematic both in terms of time for generation and memory requirements of computers. Parallel computers afford the potential to relieve this problem.

Chapter 24 discussed in detail many aspects of parallel mesh generation; therefore, here only examples will be given. The approach that will be demonstrated, is based upon geometrical partitioning of the domain [16]. To generate a grid in parallel, the complete domain is divided into a set of smaller

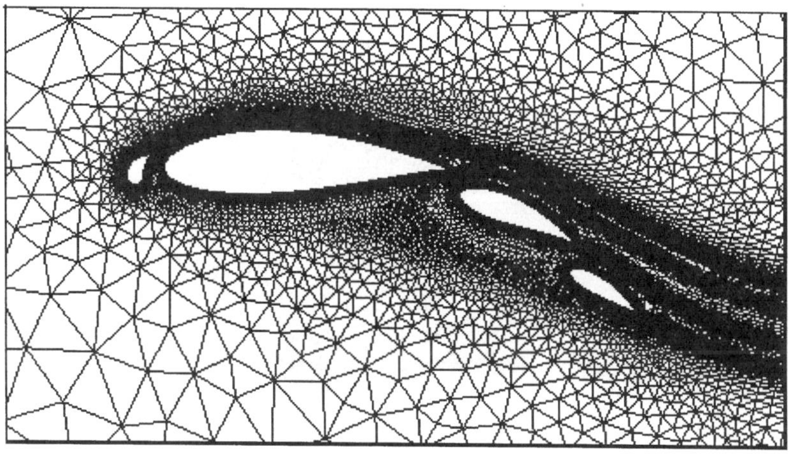

FIGURE 26.8F The final grid.

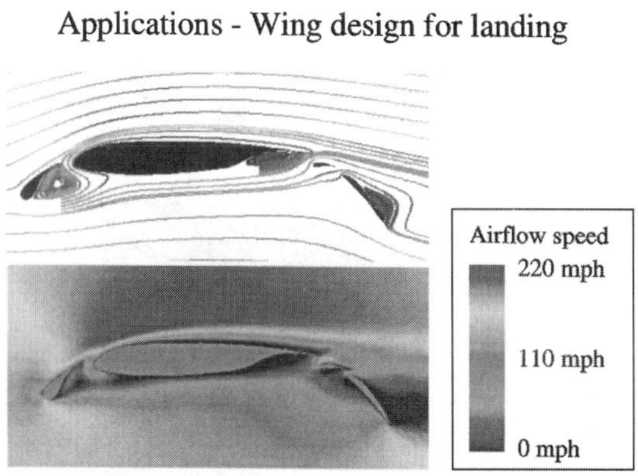

FIGURE 26.8G Flow solution for a high-lift airfoil system.

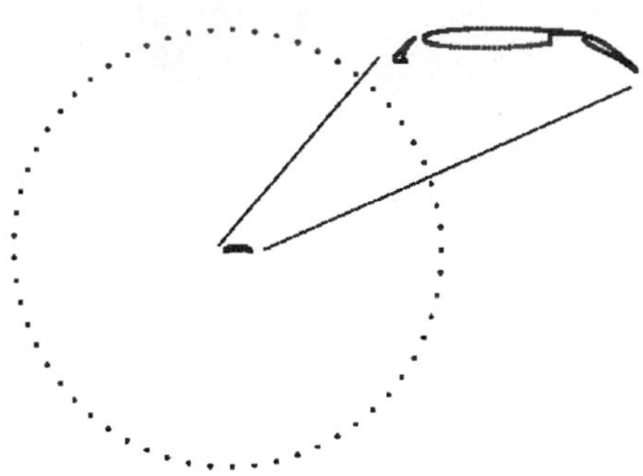

FIGURE 26.9A The inner geometry and the outer boundary is point discretized.

subdomains, and a grid generated in every subdomain independently. A combination of the subdomain grids forms the final grid of the total domain. A manager/worker model is employed, in which the initial work is performed by the manager who then distributes the grid generation tasks to the workers. The manager can recombine all the subdomain grids or, if the grid is particularly large, leave the partitioned grid on disc.

Figure 26.9 shows the general procedure.

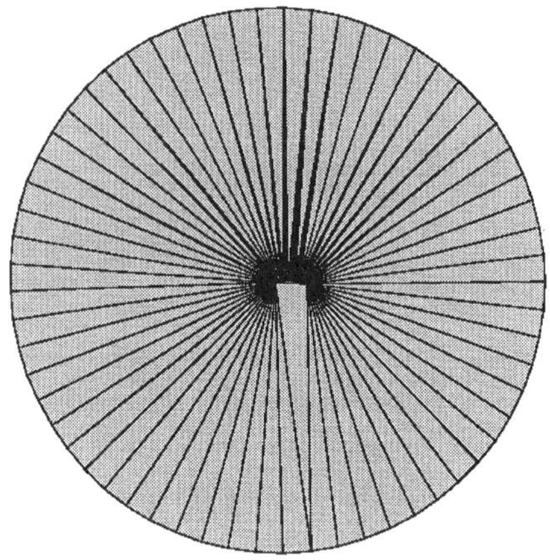

FIGURE 26.9B The initial triangulation formed by triangulating the boundary points.

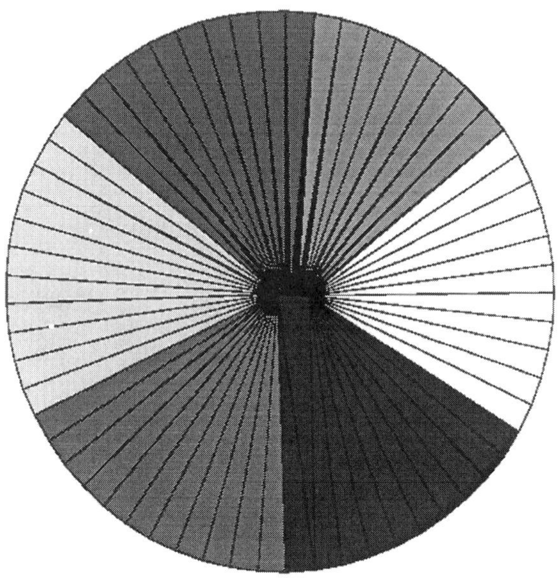

FIGURE 26.9C Domain decomposition.

- The geometry is point discretized, Figure 26.9a.
- The boundary points are connected using a Delaunay algorithm to produce an initial triangulation, as shown in Figure 26.9b.
- A greedy algorithm, with an area criterion, is employed to give a number of equally sized subdomains, Figure 26.9c.
- The interdomain boundaries are discretized leading to a set of independent grid generation tasks, Figure 9d.
- The data for each subdomain grid is distributed to the processors and the grids generated. The distribution of data is performed using the message passing library MPI.

If the number of domains is **N**, and the number of processors is **M**, then static load balances results if **N** = **M**, and dynamic load balancing if **N** > **M**. The parallel procedure is more efficient if dynamic load balancing is employed.

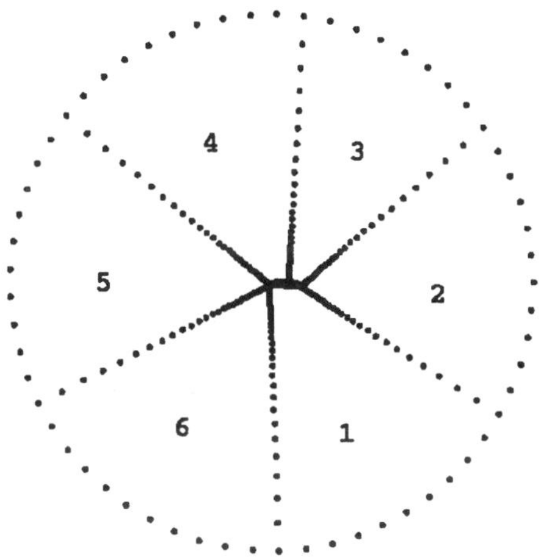

FIGURE 26.9D In this example, six independent grid generation problems are created.

FIGURE 26.10A The initial geometry.

The approach outlined also applies to the generation of grids in three dimensions. The generation of grids on the interdomain boundaries is significantly more difficult [16].

As an illustration of the procedure, Figure 26.10 shows some of the stages in the generation of a grid for a realistic geometry.

- Figure 26.10a shows the initial geometry.
- Figure 26.10b shows the surface grid of triangles.
- Figure 26.10c shows each of the four partitions, first in the form following the initial decomposition, and then after the surface grid has been suitably modified to provide input data for the volume generation.
- Figure 26.10d shows sections cut through the four volume grids.

The procedure outlined is capable of generating very large meshes. As an example, Figure 26.11 shows the profile data of a mesh with almost 50 million tetrahedra. The manipulation of such large meshes becomes very difficult, and the user interaction with a graphics user interface described in the different sections of this chapter is not practical. Therefore, it is necessary to use graphics frameworks based upon parallel computer platforms [17]. Figure 26.12 shows an illustration of the parallel visualization of a large mesh generated by the parallel mesh generator.

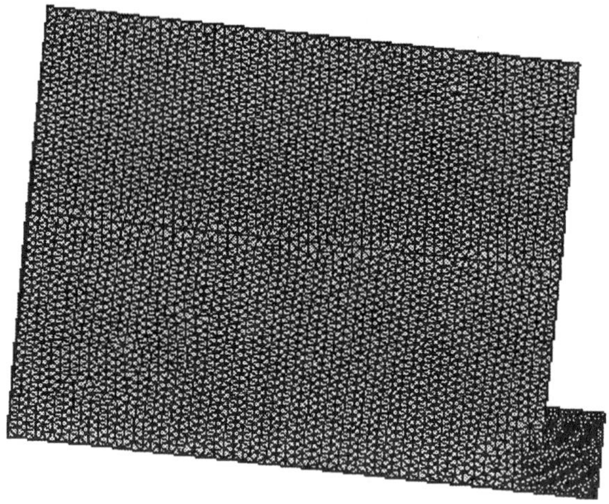

FIGURE 26.10B The surface mesh.

26.9 Summary

In this chapter, an attempt has been made to provide examples of unstructured grids and to indicate the procedures followed in the process of grid generation. In this way, it should augment much of the material presented in the other chapters of this part of the handbook. All the grids have been generated using software developed at Swansea and are snapshots taken of results presented with graphics user interfaces [18,19] (see also the Appendix to this chapter). The literature now provides many impressive examples of grids generated for real-world problems, and the interested reader is directed to proceedings of recent grid generation conferences [20–24] and survey papers [25].

Appendix: Graphics User Interfaces

With the wide-scale availability of high-resolution computer graphics, the process of user-grid interaction has been revolutionized. It is now common practice for grid generation algorithms to be embedded within easy-to-use graphics user interfaces where users can be shown relevant data in a visually meaningful way [18,19]. This technology has reduced both the time taken to generate grids and the training time required for new users to become proficient at generating grids. Images from some typical windows of two graphics user environments for grid generation [18,19] are presented in Figure 26.A1.

Acknowledgment

The author would like to acknowledge Dr. O Hassan, Dr. M. J. Marchant, Mr. R. Said, Mr. E. Turner-Smith, Mr. J. Jones for helping to produce the figures used in the chapter.

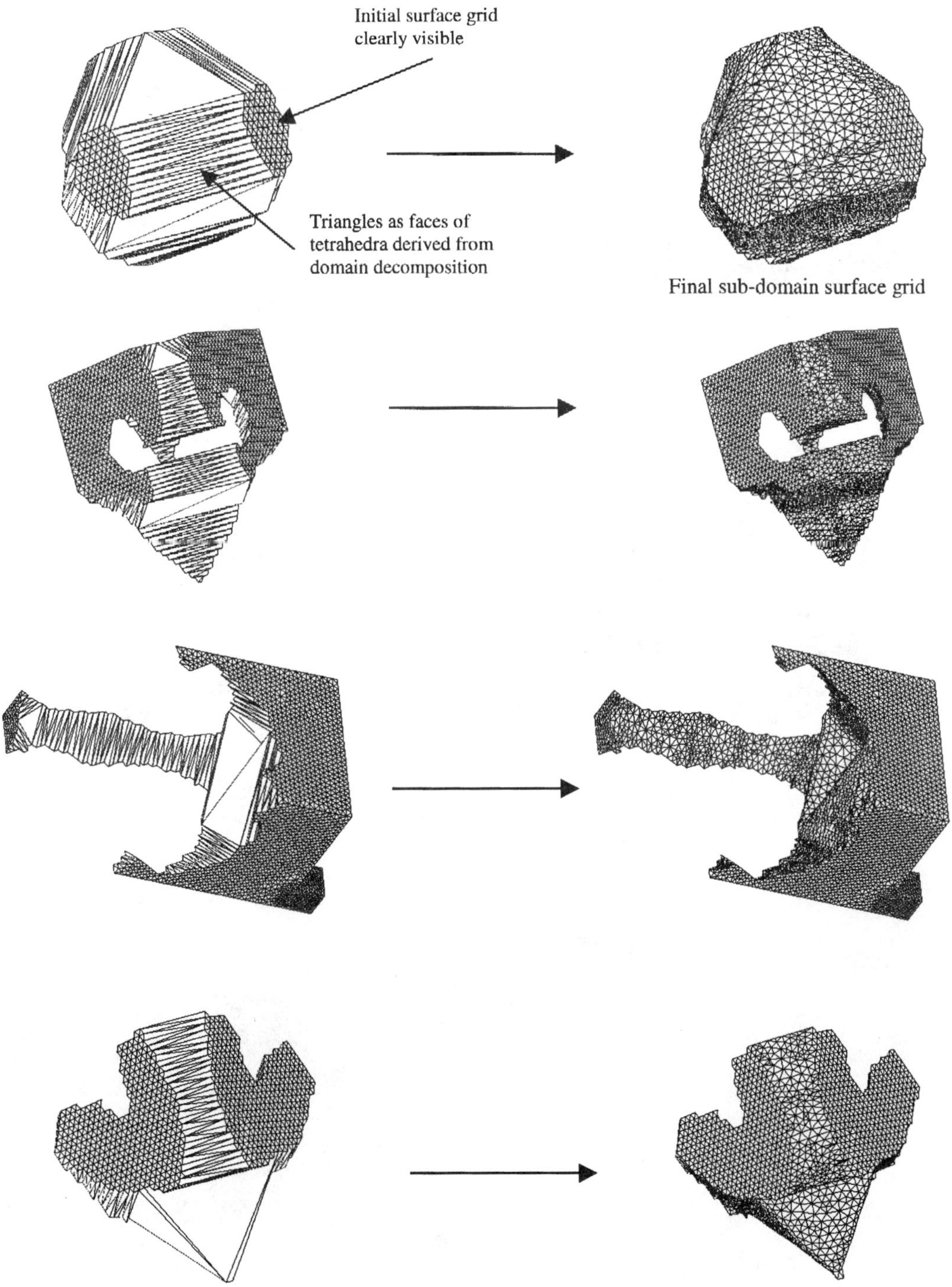

FIGURE 26.10C An example of a domain decomposed into four partitions. Shown is the surface grid, the inter-domain surface triangles (faces of the initial tetrahedra that fill the domain), followed by the final surface grids prior to volume meshing.

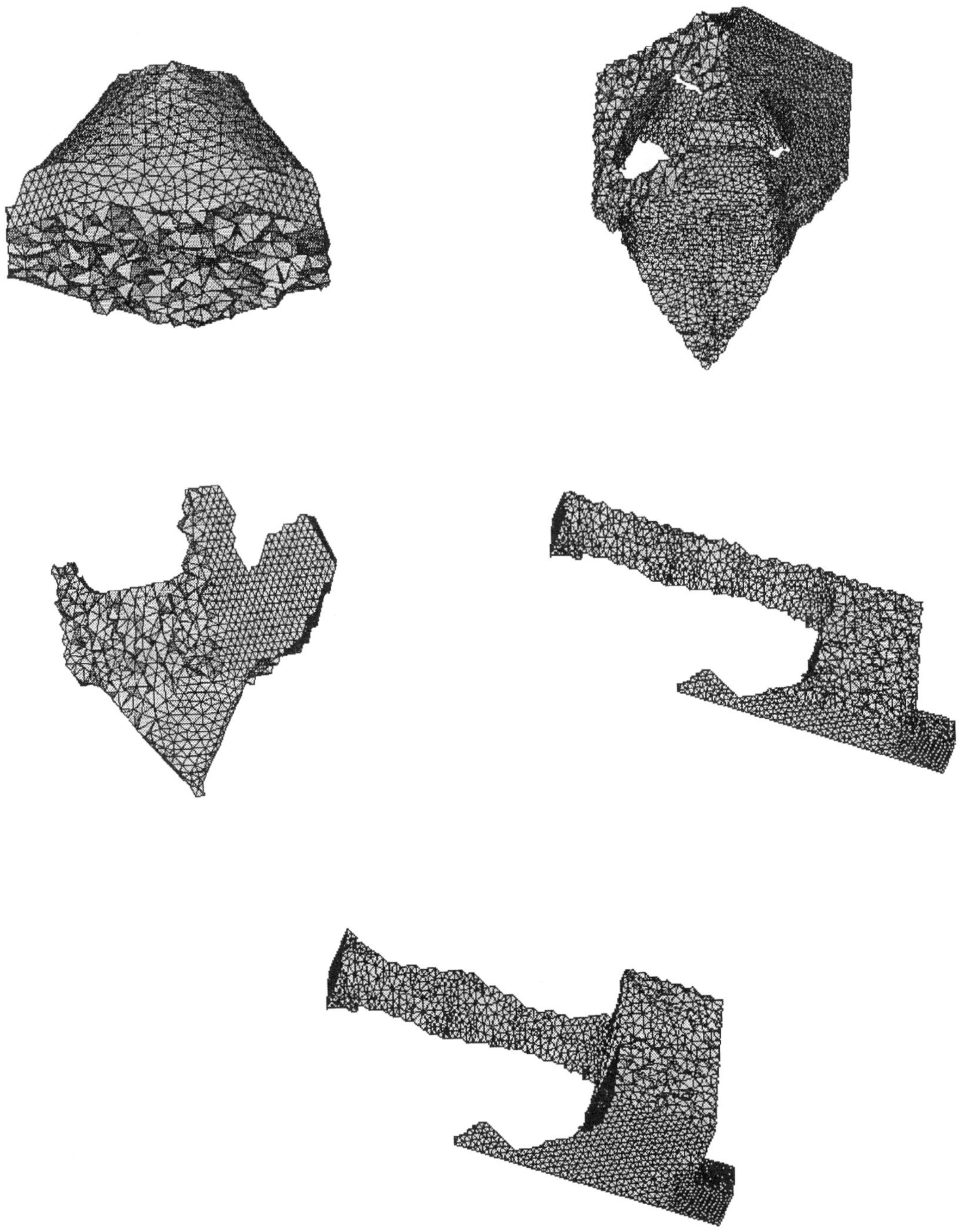

FIGURE 26.10D Sections through the volume grids of the individual partitions.

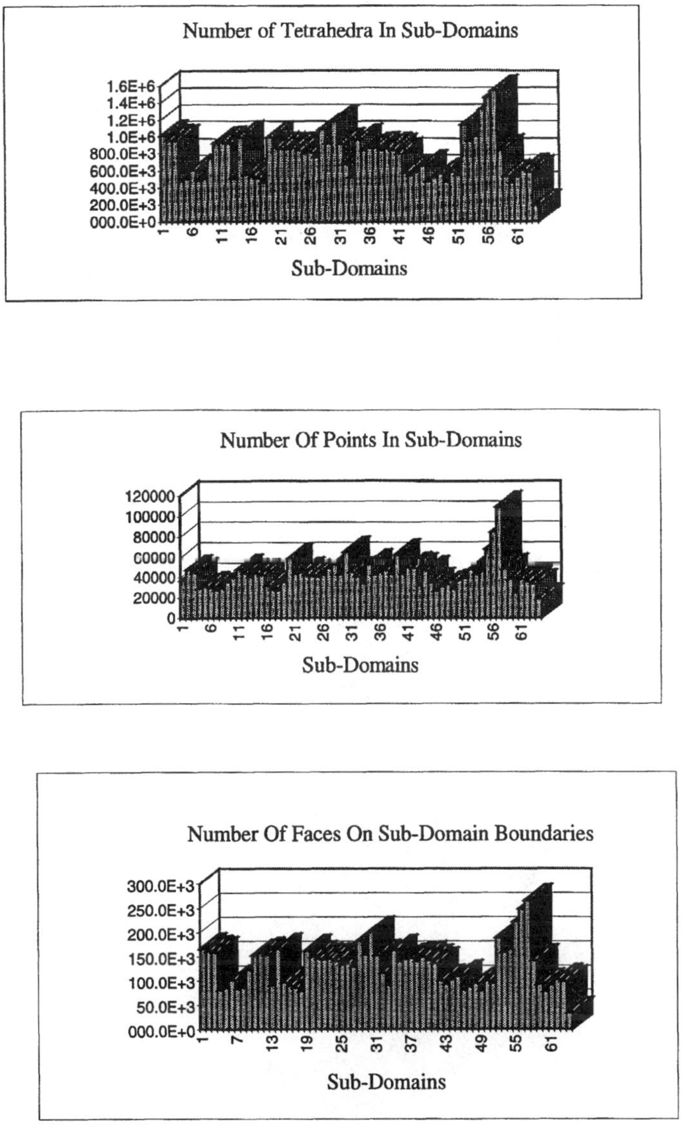

FIGURE 26.11 Details of a mesh of almost 50 million elements generated in parallel.

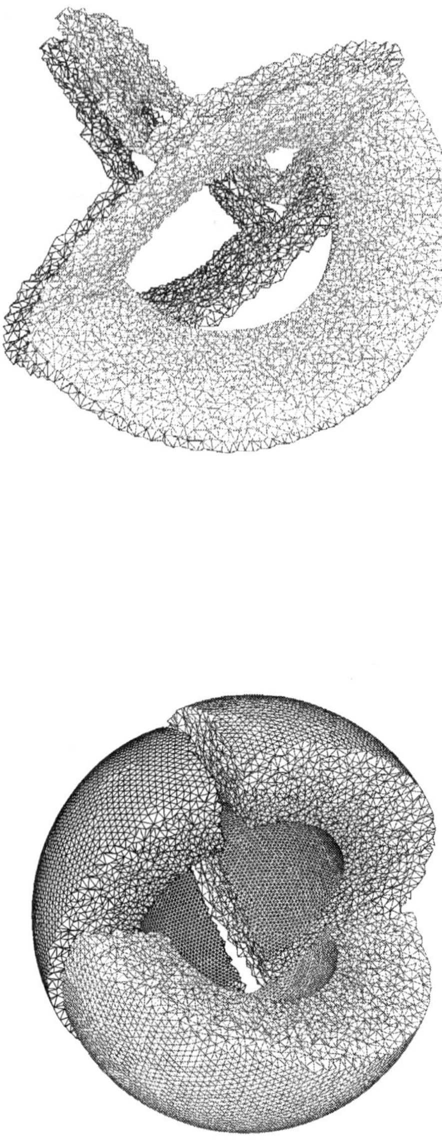

FIGURE 26.12 Sections through a grid computed using parallel visualization.

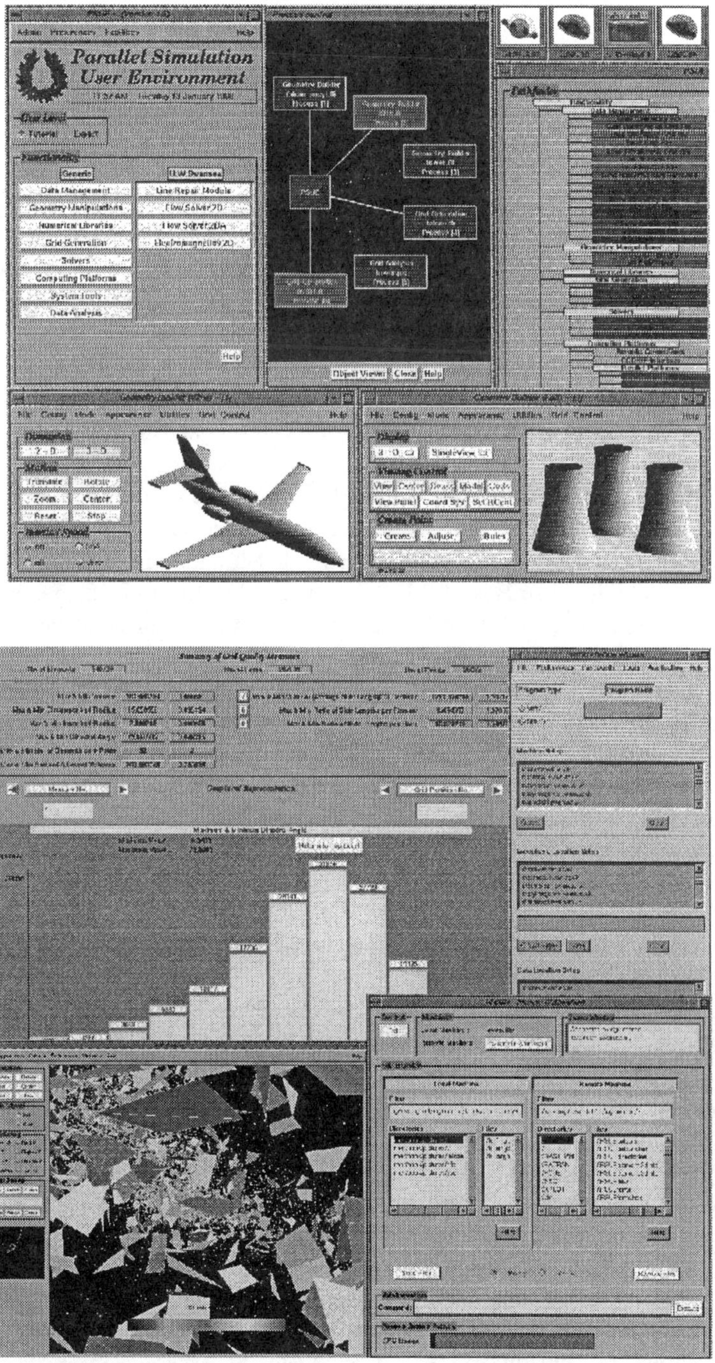

FIGURE 26.A1 Typical Windows environments for more effective interactive grid generation.

References

1. Weatherill, N.P., The generation of unstructured grids using Dirichlet tessellations, Department of Mechanical and Aerospace Engineering, Report No. 1715, Princeton University, 1985.
2. Jameson, A., Baker, T.J., and Weatherill, N.P., Calculation of inviscid transonic flow over a complete aircraft, 24th Aerospace Sciences Meeting, *AIAA Paper 86-0103*. Reno, NV, 1986.
3. Weatherill, N.P., A method for generating irregular computational grids in multiply connected planar domains, *Int. J. for Numerical Methods in Fluids.* 1988, Vol. 8, pp. 181–197.
4. Weatherill, N.P., Delaunay triangulation in computational fluid dynamics, *Computers and Mathematics with Applications.* 1992, Vol. 24, No. 5/6, pp. 129–150.
5. Weatherill, N.P. and Hassan, O., Efficient three-dimensional grid generation using the Delaunay triangulation, *Proc. of the 1st European Computational Fluid Dynamics Conf.* Brussels, Belgium, Hirsch, C., Periaux, J., Kodulla, W., (Eds.), Elsevier, Amsterdam, 1992.
6. Weatherill, N.P. and Hassan, O., Efficient three-dimensional delaunay triangulation with automatic point creation and imposed boundary constraints, *Int. J. for Numerical Methods in Engineering.* 1994, Vol. 37, pp. 2005–2039.
7. Weatherill, N.P., The reconstruction of boundary contours and surfaces in arbitrary unstructured triangular and tetrahedral grids, *Engineering Computations.* 1996, Vol. 3, No. 8, pp. 66–81.
8. Parmley, K.L., Dannenhoffer J.F. III, and Weatherill, N.P., Techniques for the visual evaluation of computational grids, *AIAA Paper 93-3353.* AIAA CFD Meeting Orlando, FL, July 6-9, 1993.
9. Weatherill, N.P., Mixed structured and unstructured meshes for aerodynamic flow simulation, *Aeronautical Journal.* 1990, 94, pp. 111–123.
10. Weatherill, N.P. and Natakusumah, D., The simulation of potential flow around multiple bodies using overlapping connected meshes, *Appl. Math. Comput.*, 1991, 46, pp. 1–21.
11. Peraire, J., Peiro, J., Formaggia, L., Morgan,K., and Zienkiewicz, O.C., Finite element Euler computations in three dimensions, 1988, Vol. 26, pp. 2135–2159.
12. Weatherill, N.P., Mixed structured and unstructured meshes for aerodynamic flow simulation, *Aeronautical Journal.* 1990, 94, pp. 111–123.
13. Weatherill, N.P. and Natakusumah, D., The simulation of potential flow around multiple bodies using overlapping connected meshes, *Appl. Math. Comput.*, 1991, 46, pp. 1–21.
14. Marchant, M.J., Weatherill, N.P., and Hassan, O., FEA.
15. Hassan, O., AIAA.
16. Said, R., Weatherill, N.P., Morgan, K., and Verhoeven, N.A., Distributed Delaunay mesh generation for very large meshes, submitted for publication, January 1998.
17. Jones, J. and Weatherill, N.P., Parallel visualisation, submitted for publication.
18. Marchant, M.J. and Weatherill, N.P., The design of a software tutorial for computational aerodynamics, *Proc. of the Eng. Education Conf., Professional Standards and Quality.* Sheffield, UK, Bramhall, M.D. and Robinson, I.M., (Eds.), SHU Press, 1997.
19. Marchant, M.J., Weatherill, N.P., Turner–Smith, E., Zheng, Y., and Sotirakos, M., A parallel simulation user environment for computational engineering, *Proceedings of the 5th International Conference on Numerical Grid Generation in Computational Field Simulation.* April 1996, Soni, B., Hauser, J., Eiseman, P., Thompson, J.F., (Eds.), MSU Press, 1996.
20. *Proc. of the 1st Int.Conf. on Grid Generation.* Landshut, West Germany, Pineridge Press, UK, 1986.
21. *Proc. of the 2nd Int. Conf. on Grid Generation.* Miami, FL, Pineridge Press, UK, 1988.
22. *Proc. of the 5th Int. Conf. on Grid Generation in Computational Fluid Dynamics and Related Fields.* Starkville, MS, North-Holland, 1991.
23. *Proc. of the 4th Int. Conf. on Grid Generation in Computational Fluid Dynamics and Related Fields.* Swansea, UK., North-Holland, 1994.
24. *Proc. of the 3rd Int. Conf. on Grid Generation in Computational Fluid Dynamics and Related Fields* Barcelona, Spain. North-Holland, 1991.

25. Thompson J.R. and Weatherill, N.P., Aspects of numerical grid generation, AIAA Applied Aerodynamics Meeting, Monterey, CA, August 1993.

III

Surface Definition

Bharat K. Soni

Introduction to Surface Definition

The geometry preparation is the most time-critical and labor-intensive part of the overall grid generation process. Most of the geometrical configurations of interest to practical scientific and engineering problems are designed in the CAD/CAM system as a composition of explicit or implicit analytical entities, semi-analytic parametric-based entities and/or sculptured sets of discrete points. The standard common interface for geometry exchange is IGES (International Graphics Exchange Specification), which is based on the points, curves, and surface definition of geometric entities. There are numerous geometry output formats that require a grid developer to spend a great deal of time manipulating geometrical entities to achieve a useful sculptured geometrical description with appropriate distribution of points. Hence, surface definition associated with all solid geometrical components pertinent to the field region under consideration for grid generation plays a very crucial role in the efficiency and accuracy of the overall grid generation. This part of the handbook is devoted to providing an in-depth description of the mathematics, numerics, technology, and state of the practice of surface definition. In particular, the concentration is placed on the computer-aided geometric design (CAGD) techniques based on the interpolations and approximations involving parametric splines, B-splines and nonuniform rational B-splines (NURBS).

The chapters included in Part III present the mathematical foundations of spline-based geometry definition with pertinent numerics, basic computational and geometry manipulation tools of CAGD and their respective applications in grid generation, and industrial standards for geometry treatments involving practical complex configurations. Basic theory of splines and tools for using splines in engineering work are laid out by Ferguson in Chapter 27. This chapter provides the basic mathematical foundation using a functional approach and discusses the properties and numerical evaluations of general splines. Application of these methodologies in the development of engineering tools is described. The CAGD techniques for curves and surfaces involving widely used deBoor and de Cateljau algorithms are described by Farin in Chapter 28. The discussions also include Bezier and NURBS-based surfaces and their practical applications: surface refinement and reparametrization, approximation of discontinuous surface geometries containing gaps, holes and overlaps, surface–surface intersections are the widely utilized CAGD tools for complex grid generation. The detailed description and development of these tools is provided by Hammann, Razdan, and Jean in Chapter 29. In Chapter 30, the development of grid generation tools based on the NURBS-based surface and volume definition is described. In particular, a step-by-step process to develop NURBS description of widely utilized surface and volume geometrical entities in grid generation is developed.

The development of IGES and NASA–IGES NINO (NURBS-Only) standards with pertinent applications is described by Evans and Miller in Chapter 31. This description also includes the presentation of associated software and documentation for efficiently utilizing these standards.

Recently, the NURBS representation of geometric entities has become the *de facto* standard for geometry description in most of the grid generation systems. Various grid systems presented in Chapter 2 utilize NURBS data structure for geometry and grid generation. The geometry exchange standard, IGES, based on curves and surfaces definition is not suitable for the treatment of trimmed curves that widely appear in industrial CAD geometry design. Therefore, a research concentration has shifted toward using solid modeling-based geometric entities and their utilization in grid generation. Also, a new international standard STEP (Standard for Exchange of Product Data) has been gaining popularity. The standard provides users with the ability to exchange and express useful product information in digital form throughout a product's life cycle. This includes the information needed from conceptual design stage to analysis, manufacturing, and product support and maintenance. However, the utilization of STEP in routine industrial application is still at the research level.

27

Spline Geometry: A Numerical Analysis View

David R. Ferguson

27.1 Background and Introduction

In this Chapter the basic theory of splines and tools for using splines in engineering work are laid out. Mathematical splines, introduced by Schoenberg [18], have become one of the workhorse tools of mathematical modeling and geometry systems. Managing and controlling the wide variety of computer-aided design, manufacturing and engineering (CAD/CAM/CAE) packages and geometry systems in use in a modern engineering company is not only a challenge, but it is also a key to making the enterprise successful. Here we address one aspect of that challenge: the construction, analysis, and management of "geometric" data. By geometric data we include not only basic geometry but also analysis (e.g., pressure, thermal), grids, meshes, kinematic and other data associated with geometry.

To work well, geometry systems, analysis of geometry, conversion from one representation to another, and graphical display of geometry must be based on sound mathematical theories and numerical methods. If the underlying mathematics and numerics are sound, the geometry system will, by and large, perform well, be maintainable, and be adaptable to new needs as they arise. For geometric design, methods

0-8493-2687-7/99/$0.00+$.50

based on polynomial splines using the B-spline representation provide some of the best tools for meeting these goals. Therefore, in this chapter, we concentrate on the basic theory of multivariate, tensor product, polynomial splines and their B-spline representation. The goal is to provide working engineers with the insight and tools needed to use splines effectively in geometric design and related work.

To use splines effectively requires specific knowledge of what constitutes a spline, and familiarity with common methods for working with splines. While most of this Chapter deals with the details of spline theory and application, the remainder of this section discusses the attributes that make a mathematical tool a valuable engineering tool and shows that splines have those attributes.

A mathematical tool or technique is understood and valued if it is simple and familiar, is usable in a number of situations, leads to well-posed problems (i.e., problems in which the solutions are well understood and uniquely defined), produces robust algorithms for computation, and provides techniques for error analysis that practitioners can use to understand how well a problem is being solved and to help manage error. That is, it must be simple, familiar, versatile, and useful. Splines satisfy these criteria. As a natural extension of polynomials, and as a common engineering tool that has been in use for years, they are simple and familiar. Their versatility is shown in their many uses: describing curves and surfaces, data fitting and smoothing, modeling analysis results, and paneling geometry in preparation for analysis are some of the many uses of splines. There is a rich, comprehensive, unifying mathematical theory complete with error analysis to guide practitioners in selecting alternatives and to aid in knowing how well a problem is being solved.

The Chapter is organized into eight sections. Section 27.2 describes the functional or object-oriented approach to splines, the separation of construction from evaluation, and includes extensions to higher dimensions and a discussion of differences with the traditional CAD and CAGD approach. Sections 27.3, 27.4, and 27.5 are the spline theory parts of the chapter. Section 27.3 begins the development of splines as linear vector spaces of mathematical functions. In this section the basic concepts — break-points, knots, degree, order, and continuity — are described. Section 27.4 continues with a discussion of B-splines. Section 27.5 lays out some of the theoretical results on approximation and shows how this theory can be used by the practitioner to manage error and control results.

The remaining sections are devoted to the practical construction and use of splines for representing data and constructing curves, rational curves, and surfaces. In the final section, functional composition is used to address classes of engineering problems where an analysis or geometry depends on another more fundamental geometry as, for example, a mesh depends on the geometry being meshed. Since this is a handbook and not a comprehensive treatment of spline theory, the emphasis in these sections is on matters of interest to the practitioner. Those interested in more exhaustive treatments are advised to consult the books *A Practical Guide to Splines* [3] or *Spline Functions: Basic Theory* [16].

27.2 A Functional Approach to Splines

The premise of this Chapter is that splines are mathematical objects that have a knowable structure, and that structure is useful to understanding and applying splines. Specifically, splines form a finite-dimensional vector space of mathematical functions f mapping an m-dimensional hypercube D into an n-dimensional space. Two crucial ideas follow from this perspective. First, since the spline spaces being considered are finite dimensional, it is possible to determine *a priori* the dimensionality and, hence, the number of conditions needed to specify a spline. Further, it is possible to determine a set of basic splines that can be used to represent other splines. These two aspects provide the framework for the formulation of well-posed problems and robust algorithms for construction and evaluation.

Second, it means that any particular spline can be understood as separate from both its method of construction and its method of evaluation. This has powerful implications. One example might be in highway engineering, where a quadratic spline with one interior knot (see next section) with fixed starting and ending positions and tangents is used to connect two straight sections of highway while avoiding the use of sharp corners. The resulting spline could be used immediately to help design forms by evaluating the spline at a series of way points and also to determine the amount of concrete needed by calculating the arc length of the spline. Moreover, since the spline exists independently of its construction, it may be stored and retrieved later to help determine how well the finished highway met the design goals.

Even more potential benefits come from developing a single evaluator for all tensor product splines. To illustrate the power of a single, standard evaluator, suppose a simulation is built using piecewise linear splines but that later, perhaps years later, the simulation needs to be upgraded by replacing the piecewise linear spline with a smooth, higher-order spline. Such an upgrade might be prohibitively costly if the downstream uses of the original spline — evaluation, calculation of mass properties, etc. — had been based on the assumption that the underlying model was piecewise linear. However, if the evaluating functions are general and only assume a tensor product spline structure, then the cost of the upgrade would be greatly reduced or nonexistent.

The functional approach to splines differs sharply from traditional CAD and CAGD. Traditionally in CAD, the dependence of geometry on the underlying function f is suppressed; the image of f is the sole object of interest, geometry is always planar or spatial, and the preferred development is as generalization of polynomial arcs or patches with an emphasis on parameters (e.g., Bezier points, control points) that can be manipulated interactively to yield various curves and surfaces. By contrast, in the approach undertaken here the underlying function f plays a critical role; properties of f itself become important, geometry is no longer restricted to be planar or spatial, and the emphasis is on defining data requirements that lead to well-posed engineering problems. Having said this, however, it should be pointed out that the end products are often the same. Any Bezier curve or rectangular patched surface can be represented exactly as a spline. Conversely, any spline, as long as it is planar or spatial, can be represented in any of the common CAGD forms. Where the two approaches are incompatible are with higher dimensional objects that have no equivalent representation in CAD and, from the CAGD side, the use of other forms (e.g., triangular patches, radial bases), which are not tensor product based.

27.3 Basics of Spline Theory

In this section the basics of spline theory — degree, order, break-points, knots, continuity, and dimension — are covered. The objective is to describe splines and to determine the conditions required to specify a spline, e.g., the dimension of the spline space. The basic spline is the tensor product spline

$$F : D \subset E^m \rightarrow E^n$$

where $D = [a_1, b_1] \times [a_2, b_2] \times \ldots \times [a_m, b_m]$ is a rectangular parallelapiped. Tensor products are straightforward generalizations of univariate splines, so we begin with simple, univariate splines

$$f : [a,b] \rightarrow E^1.$$

What is a spline? A simple and intuitively pleasing definition is that a spline is a finite sequence of polynomial arcs satisfying certain smoothness conditions at their break-points. The following are four examples.

Example 27.3.1:

$$s_1(t) = |t| = \begin{cases} -t & \text{if } -1 \le t \le 0; \\ t & \text{if } 0 \le t \le 1 \end{cases}.$$

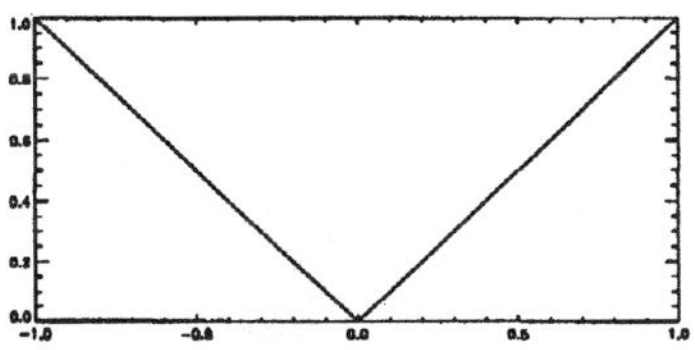

Example 27.3.2:

$$s_2 = \begin{cases} -t^2 & \text{if } -1 \le t < 0; \\ t^2 & \text{if } 0 \le t \le 1 \end{cases}$$

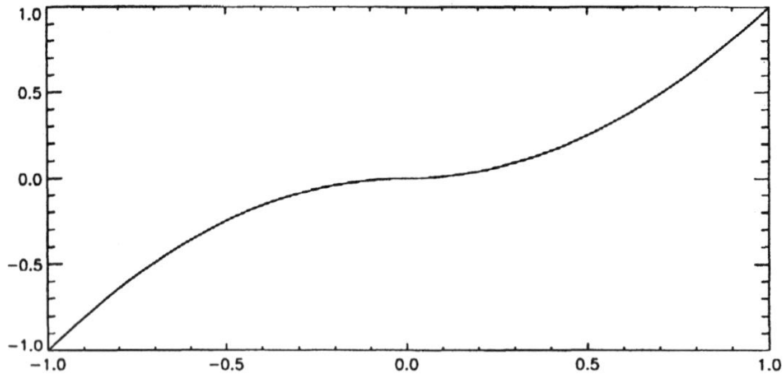

Example 27.3.3:

$$s_3(t) = \begin{cases} t & \text{if } -1 \le t < 0; \\ t^2 & \text{if } 0 \le t \le 1 \end{cases}$$

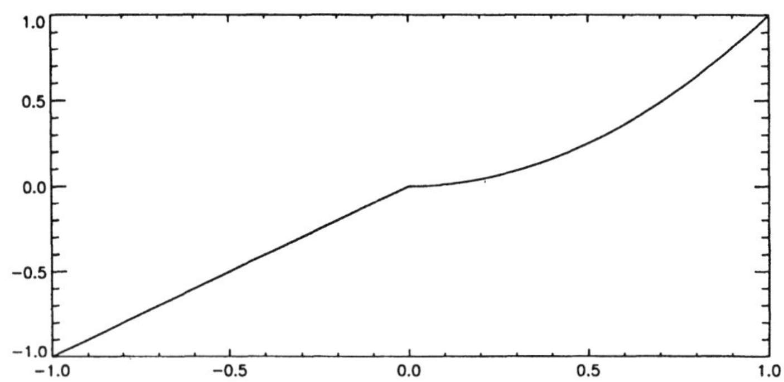

Example 27.3.4:

$$s_4(t) = \begin{cases} t^3 + t^2 & \text{if } -1 \le t < 0; \\ t^2 & \text{if } 0 \le t \le 1 \end{cases}$$

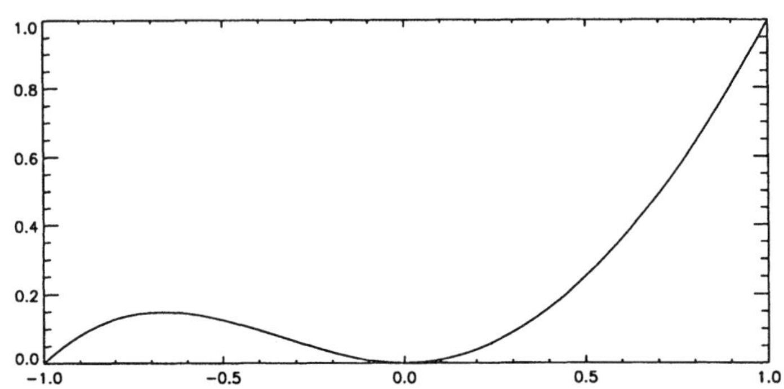

Each example satisfies the working definition, as each is composed of polynomial arcs. In the examples, the degrees are 1, 2, 2, 3 with corresponding orders 2, 3, 3, and 4. Example 27.3 has degree 2 (order 3) and Example 27.4 has degree 3 (order 4) even though, in both cases, there are segments of lower degree. The following is a formal definition.

Definition 27.3.1: Let break-points $a = \xi_0 < \xi_1 < \ldots < \xi_q = b$ and polynomials $p_1, \ldots, p_q$, each of order (degree + 1)* less than or equal to k be given. The function s defined as

$$s(t) = \begin{cases} 0 & \text{if } t < \xi_0; \\ p_1(t) & \text{if } \xi_0 \le t < \xi_1; \\ \cdot & \cdot \\ \cdot & \cdot \\ \cdot & \cdot \\ P_q(t) & \text{if } \xi_{q-1} \le t \le \xi_q; \\ 0 & \text{if } t > \xi_q \end{cases}$$

is a spline function of order k having the indicated break-points.

The spline s is defined for all values of t by using the zero function to extend the definition. This convention provides certain mathematical conveniences in setting up computations. In the examples, –1, 0, and 1 are all break-points.

The next concept is order of continuity. Observe how the examples behave at zero. Direct computation shows that the first is continuous, but there is a continuity break at zero in its derivative. The same holds for the third example. In the second, both the spline and its derivative are continuous, while a break in continuity occurs in the second derivative. In the fourth there is a continuity break in the third derivative. It is important to control the order of continuity or smoothness in order to correctly model phenomena and to assure that the resulting computational models are appropriate for other mathematical operations (e.g., optimization). Knots and multiplicities are used to manage smoothness.

Definition 27.3.2: Let k be the order of the spline s. The break-point ξ is called a knot of order λ if the first break in continuity occurs in the $k - \lambda$ derivative. That is, it is a knot of order λ if

$$s^{(j)}(\xi^-) = s^{(j)}(\xi^+) \text{ for } \quad j = 0, \ldots, k-1-\lambda$$

while

$$s^{(k-\lambda)}(\xi^-) \neq s^{(k-\lambda)}(\xi^+).$$

By this definition, the break-point 0 is a knot of multiplicity 1 in the first, second, and fourth examples and of multiplicity 2 in the third. Because of the convention of extending splines by the zero function, the first and last break-points are knots of multiplicity equal to the order k of the spline.** Thus, the break-points –1 and 1 are knots of multiplicity 2 for the first example, 3 for the next two examples, and 4 for the last example. Note that the number of spans or polynomial pieces of a spline is one less than

*Degree, the greatest exponent, is a classic polynomial exponent. However, order, which is always the degree plus 1, is a more natural parameter when dealing with dimensions of spline spaces and with multiplicities of knots. It would be nice to pick one term and stick with it, but spline theorists use both and the practitioner might as well get used to it. Therefore, in this Chapter we make no effort to exclude one or the other.

**Actually, the multiplicity would be less if the spline or some of its derivatives were zero at the end points but, since that is not the usual case, we assume the knot to be of multiplicity k.

the number of break-points and is not derivable solely from the total number of knots. This fact is sometimes a source of confusion.

There are two commonly accepted ways of representing knots and their multiplicities: either list the knot and its multiplicity explicitly or replicate the knot a number of times corresponding to its multiplicity. Thus, the knots for Example 27.3.3 could be listed as −1, 0, 1 with multiplicities 3, 2, 3 respectively or as −1, −1, −1, 0, 0, 1, 1, 1. Except occasionally, the second representation is preferred.* In either form, the total number of knots of Example 27.3.3 is 8. That is, knots are to be countered with their multiplicities.

Up to this point we have concentrated on individual splines and described the key concepts of order, knots, multiplicities, and smoothness. Now we turn attention to the totality of splines having a specific order and knot set. Understanding properties of a collection of splines is important because these properties are used to represent and construct specific splines that solve particular problems. In particular, it is important to be able to calculate the dimension of the space and to produce a set of basis elements to be used to represent arbitrary splines. Calculating the dimension is the topic of the remainder of the section. Basis elements are covered in the next section.

Let k be the order and m the total number of knots. The dimension of the spline space is m-k. Since the techniques of this Chapter rely on understanding and accepting this formula, it is worthwhile spending time establishing its validity. This can be done by a simple counting argument accounting for the required smoothness of the spline.

Let the break-points be $\xi_0 < \xi_1 < \ldots < \xi_q$ with multiplicities $\lambda_0, \ldots, \lambda_q$. (This is one place where it is convenient to use the distinct knots with multiplicities representation.) Thus, $m = \sum_{j=0}^{q} \lambda_j$ and the number of polynomial pieces is q. Since each polynomial piece can have order at most k, each can be defined by k polynomial coefficients. Thus, there is a total of kq coefficients to be determined, nominally requiring kq equations. However, satisfying the smoothness conditions implied by the multiplicities will reduce the number of required equations as follows. At the knot ξ_i there are $k-\lambda_i$ continuity equations of the form $s^{(j)}(\xi_i^-) = s^{(j)}(\xi_i^+)$ for $j = 0, \ldots, k-1-\lambda_i$. Summing over all the knots gives a total of

$$\sum_{i=0}^{q} (k-\lambda_i) = k(q+1)-m$$

continuity conditions. Subtracting this from the total number of equations required gives

$$kq-(k(q+1)-m) = m-k$$

as the dimension of the spline space.

The following table shows order and knots for the Examples 27.3.1 – 27.3.4, and the dimensions of their associated spline spaces.

The following are three additional properties of splines considered as elements of a function space. First, the derivative of a spline of order k is a spline of order $k-1$ with the same break-points. Any knot of multiplicity k becomes a knot of multiplicity $k-1$, but all others retain their original multiplicity. Second, an antiderivative of a spline of order k is a spline of order $k+1$ with the same break-points. Interior knots have the same multiplicities as before and the endpoints become knots of order $k+1$.

*Preference is a matter of choice. The STEP [17]data exchange standards prefer the first form at this time. The DT _NURBS Library [5] uses the second form. In general, when developing software, simplicity is to be preferred. Because the second form lends itself directly to computation (see Section 27.4) it is our preference.

Example	Order	Knots	Dimension
27.3.1	2	−1, −1, 0, 1, 1	3
27.3.2	3	−1, −1, −1, 0, 1, 1, 1	4
27.3.3	3	−1, −1, −1, 0, 0, 1, 1, 1	5
27.3.4	4	−1, −1, −1, −1, 0, 1, 1, 1, 1	5

Third, any spline of order k can be expressed as a spline of order $k + 1$ with the multiplicity of each knot increased by one.

We close with an observation: not all splines in any particular space necessarily have discontinuities at any particular knot nor do they necessarily have the specified degree. The definition of a spline space as having a particular order (degree) and knot set merely limits the location and multiplicities of knots and the order of the splines in the space; it does not require that the actual order of each spline be equal to the order of the spline space and does not require each of the knots to be active. For example, polynomials of degree $k - 1$ belong to every spline space of order greater than or equal to k even though the polynomials have no discontinuities themselves.

It would be possible to restrict attention only to active knots and full degree, but there are two good reasons for not doing so. First, by allowing inactive knots and less than full degree the spline spaces become closed with respect to taking limits of splines. Second, collections of seemingly disparate splines (e.g., splines of varying degrees, knots, and multiplicities, their derivatives and integrals) can be put together under a common spline framework if inactive knots and less than full degree are accepted. For example, each of the splines of Section 27.3 can be put in the framework of splines of order 4 with the knots −1, −1, −1, −1, 0, 0, 0, 1, 1, 1, 1. That spline space has dimension 7. The triple knot at zero is necessary if s_1 to be included. But for s_4, two of the potential discontinuities are inactive.

27.4 B-Splines

In the last section, examples of splines were given, basic parameters were described, and a formula for the dimension of spline spaces was provided. In this section we look at how splines are represented. We know that there is a certain number of data that are required to define a spline but the actual definition can take many forms. For example, we could simply represent splines, as in the four examples, by sequences of polynomial arcs using the standard power series representation. In this fashion a spline with q segments would be represented by kq coefficients and a corresponding number of constraints as required by the knots and their multiplicities. But this is inherently inefficient and unstable. Inefficient because in most cases it requires the storage of more coefficients than is strictly necessary. Unstable because the power series is inherently unstable [7, 15]. An alternate, more efficient, and more robust scheme uses B-splines as the basic elements of the representation. In this section, B-splines will be examined, again at the practitioner's level. We will use a shortcut definition for B-splines, give examples, provide a formula for evaluation, and list a number of useful properties. We will finish the section by describing a standard representation format for univariate splines. Readers wanting a deeper, more detailed discussion of B-splines are again referred to [3] or [16].

27.4.1 Description and Examples of B-Splines

Start with a set of knots $T = (\tau_1 \leq \ldots \leq \tau_m)$ and assume no knot is repeated more than k times. The dimension of the space of splines of order k with these knots is $m - k$. Now consider the subsets $T_j=(\tau_j, \ldots, \tau_{j+k})$. There are $m - k$ of these subsets, and each one may be used to specify a subspace of splines of order k, the dimension of which is exactly $k + 1 - k = 1$. Thus, for each T_j there is, up to a scaling factor, only one spline in the associated spline subspace. We assert, without proof, that we can choose one spline $B_{j,k}$ from each of the subspaces in such a way that $\sum_{j=1}^{m-k} B_{j,k} \equiv 1$. We will call these functions the B-splines for the knot set T.* The following are four examples of collections of B-splines.**

Example 27.4.1: $T = \{-1, -1, 0, 1, 1\}$ and order = 2. The three B-splines are

$$B_{1,2}(t) = \begin{cases} -t & \text{if } -1 \le t \le 0; \\ 0 & \text{if } 0 < t \le 1. \end{cases}$$

$$B_{2,2}(t) = \begin{cases} t+1 & \text{if } -1 \le t < 0; \\ 1-t & \text{if } 0 \le t \le 1. \end{cases}$$

$$B_{3,2}(t) = \begin{cases} 0 & \text{if } -1 \le t < 0; \\ t & \text{if } 0 \le t \le 1. \end{cases}$$

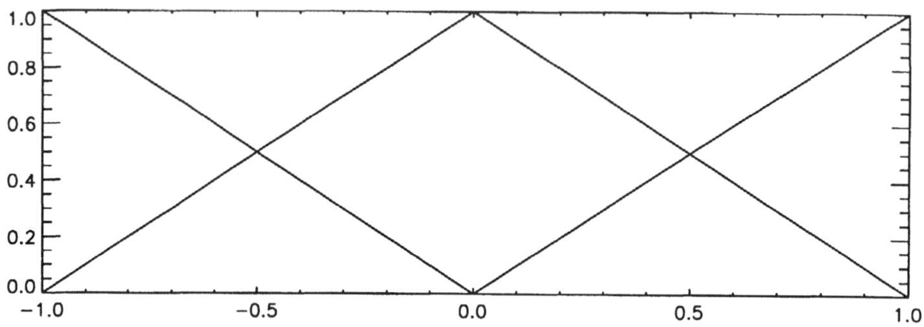

Example 27.4.2: $T = \{-1, -1, -1, 0, 1, 1, 1\}$ and order = 3. Then the four B-splines are

$$B_{1,3}(t) = \begin{cases} t^2 & \text{if } -1 \le t \le 0; \\ 0 & \text{if } 0 < t. \end{cases}$$

$$B_{2,3}(t) = .5\begin{cases} -3t^2 - 2t + 1 & \text{if } -1 < t < 0; \\ (1-t)^2 & \text{if } 0 \le t \le 1. \end{cases}$$

$$B_{3,3}(t) = .5\begin{cases} (t+1)^2 & \text{if } -1 \le t < 0; \\ -3t^2 + 2t + 1 & \text{if } 0 \le t \le 1. \end{cases}$$

$$B_{4,3}(t) = \begin{cases} 0 & \text{if } -1 \le t < 0; \\ t^2 & \text{if } 0 \le t \le 1. \end{cases}$$

*The notation $B_{j,k}$ will be simplified to B_j when there is no danger of confusion about the order. In other cases, if the knots are to be emphasized, we will use $B(t; \tau_j, \ldots, \tau_{j+k})$.

**These B-splines are presented in piecewise polynomial form in order to provide specific examples. It is a useful exercise to verify some properties of B-splines (e.g., positivity, partition of unity, see below). It is also useful to derive the same collections using the normalized divided difference definition [3]. However, neither of these are the standard formulas for B-spline evaluation. The preferred formulas are those of Section 27.4.2.

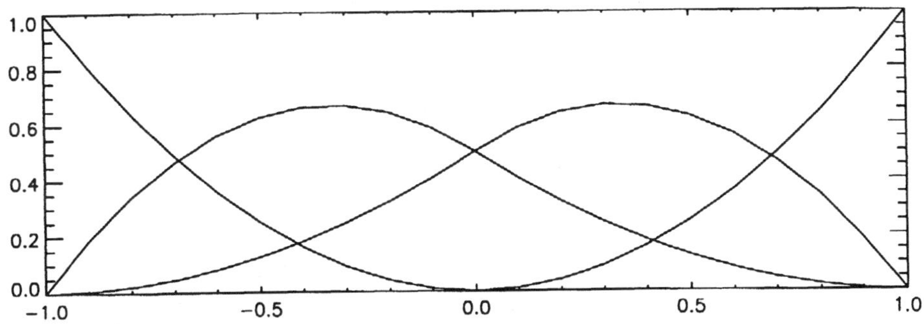

Example 27.4.3: $T = \{-1, -1, -1, 0, 0, 1, 1, 1\}$ and order = 3. Then the five B-splines are

$$M_{1,3}(t) = \begin{cases} t^2 & \text{if } -1 \le t \le 0; \\ 0 & \text{if } 0 < t. \end{cases}$$

$$M_{2,3}(t) = \begin{cases} -2t^2 - 2t & \text{if } -1 \le t < 0; \\ 0 & \text{if } 0 \le t. \end{cases}$$

$$M_{3,3}(t) = \begin{cases} 1 + 2t + t^2 & \text{if } -1 \le t < 0; \\ 1 - 2t + t^2 & \text{if } 0 \le t \le 1. \end{cases}$$

$$M_{4,3}(t) = \begin{cases} 0 & \text{if } -1 \le t < 0; \\ -2t^2 + 2t & \text{if } 0 \le t \le 1. \end{cases}$$

$$M_{5,3}(t) = \begin{cases} 0 & \text{if } -1 \le t < 0; \\ t^2 & \text{if } 0 \le t \le 1. \end{cases}$$

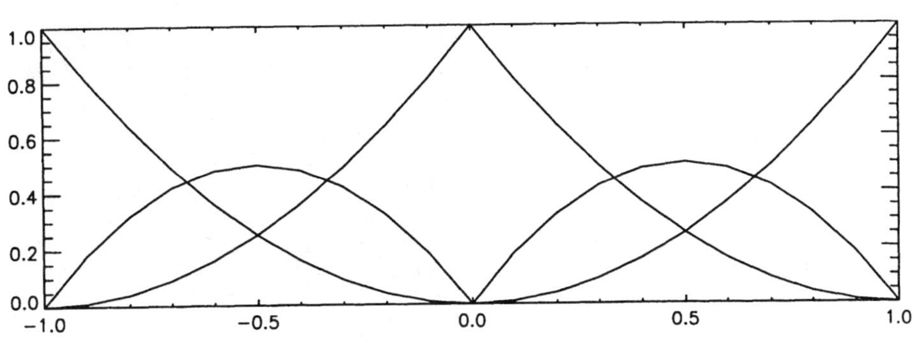

Example 27.4.4: $T = \{-1, -1, -1, -1, 0, 1, 1, 1, 1\}$ and order = 4. Then the five B-splines are

$$B_{1,4}(t) = \begin{cases} -t^3 & \text{if } -1 \le t \le 0; \\ 0 & \text{if } 0 < t. \end{cases}$$

$$B_{2,4}(t) = \left(\tfrac{1}{4}\right)\begin{cases} 7t^3 + 3t^2 - 3t + 1 & \text{if } -1 < t < 0; \\ (1-t)^2 & \text{if } 0 \le t \le 1. \end{cases}$$

$$B_{3,4}(t) = \left(\tfrac{1}{2}\right)\begin{cases} -2t^3 - 3t^2 + 1 & \text{if } -1 \le t < 0; \\ 2t^3 - 3t^2 + 1 & \text{if } 0 \le t \le 1. \end{cases}$$

$$B_{4,4}(t) = \left(\tfrac{1}{4}\right)\begin{cases} (1+t)^3 & \text{if } -1 < t - 0; \\ -7t^3 + 3t^2 + 3t + 1 & \text{if } 0 \le t \le 1. \end{cases}$$

$$B_{5,4}(t) = \begin{cases} 0 & \text{if } -1 \le t < 0; \\ t^3 & \text{if } 0 \le t \le 1. \end{cases}$$

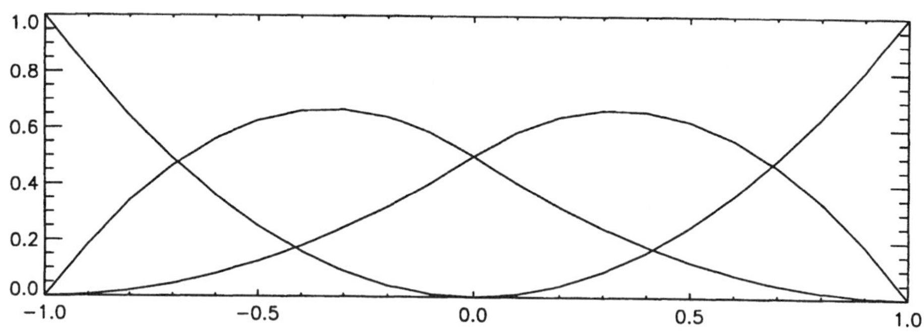

Example 27.4.5: The Bernstein polynomials $\phi_{i,k}$ of order k defined by

$$\phi_{i,k}(t) = \left(k!/i!(k-i)!\right)(1-t)^i t^{k-i}$$

are the B-splines for the knot set consisting of 0 and 1 each with multiplicity k. They are the basis for Bezier curves.

The importance of B-splines lies in the fact that they may be used to represent arbitrary splines. For a spline s there are unique constants $a_1, \ldots, a_{m-k}$ so that

$$s(t) = \sum_{j=1}^{m-k} a_j B_{j,k}(t).$$

For instance, the splines of Examples 27.3.1 – 27.3.4 may be represented by the B-splines of Examples 27.4.1–27.4.4 as

$$s_1 = B_{1,2} + B_{3,2}$$
$$s_2 = -B_{1,3} + B_{4,3}$$
$$s_3 = -M_{1,3} - \left(\tfrac{1}{2}\right)M_{2,3} + M_{5,3}$$
$$s_4 = \left(\tfrac{1}{3}\right)B_{2,4} - \left(\tfrac{1}{3}\right)B_{3,4} + \left(\tfrac{1}{3}\right)B_{4,4} + B_{5,4}$$

Let SC(s) be the number of sign changes of the spline s and SV(c) be the number of sign in the sequence c of B-spline coefficients. The following are some useful properties of B-splines.

Partition of Unity: $$\sum_{j=1}^{m-k} B_{j,k} \equiv 1 \ \ for\ \tau_k \leq t \leq \tau_{m-k+1}$$

Local Support: $$B_{j,k}(t) \equiv 0 \ \ if\ t < \tau_j \ or\ t > \tau_{j+k}$$

Positivity: $$B_{j,k}(t) > 0 \ \ if\ \tau_j < t < \tau_{j+k}$$

Variation Diminishing: $$SC(s) \leq SV(c)$$

Marsden's Identity: $$t^i = \sum_{j=1}^{m-k} a_{i,j} B_{j,k}(t)$$

where

$$a_{ij} = \left(C_i^{k-1}\right)^{-1} \sum_{p_1 < p_2 < \ldots < p_i} \tau_{p_1} \ldots \tau_{p_i}$$

and $C_i^{k-1} = (k-1)!/\ (k-1-i)!i!$ are the binomial coefficients.

These properties have numerical and geometrical significance. The partition of unity idea was used in the shortcut definition of B-splines. The local support property leads to sparse linear equations for solving for B- spline coefficients (see Section 27.6.1) and to methods for local modification of geometry. Positivity leads to robust evaluation algorithms for B-splines.

The variation diminishing property provides a powerful tool for controlling spline curves. For example, if all the coefficients are nonnegative then so is the spline.* The derivative representation and the variation diminishing property are also used to control shape. Control of the sequence of derivative coefficients, $(a_j - a_{j-1})\ /\ (\tau_{j-1+k} - \tau_{j-1})$ is a key to controlling shape. For instance, if there are no sign changes in the sequence then s is monotonic.

27.4.2 Evaluation

The study of splines and B-splines began during World War II with Schoenberg's initial investigation of piecewise polynomials and B-splines [18]. But early methods of evaluating B-splines were so inefficient and inaccurate as to cause one commentator to ask "Will anyone ever use splines for anything useful?" The answer, of course, is yes, but that answer could not be given until the publication of de Boor's 1972 paper "On Calculating with B-spline" [2] in which formulas for the accurate and stable evaluation of B-splines first appeared.

The crucial item for robust evaluation is a formula relating values of a kth-order B-spline to values of a pair of $(k-1)^{st}$ – order B-splines. Let T be a set of m knots and k the order. The formula relating B-splines of one order to those of a lower order is

$$B_{j,k}(t) = \frac{t - \tau_j}{\tau_{j+k-1} - \tau_j} B_{j,k-1}(t) + \frac{\tau_{j+k} - t}{\tau_{j+k} - \tau_{j+1}} B_{j+1,k-1}(t). \tag{27.1}$$

*Warning: The converse is not true. Splines can be nonnegative and still have negative coefficients. For example, the parabola $s(t) = t^2$ is a spline of order 2 with knots –1, –1, –1, 1, 1, 1 and coefficients 1, –1, 1. Yet s is nonnegative.

Notice that this formula means that particular B-splines are evaluated by forming positive combinations of positive quantities thus reducing the danger of errors growing through cancellation effects.

The example $B_{2,3}$ has the knot sequence $(\tau_2, \tau_3, \tau_4, \tau_5) = (-1, -1, 0, 1)$. According to the formula it can be evaluated as

$$B_{2,3}(t;\tau_2,\tau_3,\tau_4,\tau_5) = \frac{(t-\tau_2)}{(\tau_4-\tau_2)} B_{2,2}(t;\tau_2,\tau_3,\tau_4) + \frac{(\tau_5-t)}{(\tau_5-\tau_3)} B_{3,2}(t;\tau_3,\tau_4,\tau_5).$$

Formula 27.1 is used also to obtain a formula for the value of a spline in terms of lower-order B-splines. The formula is

$$s(t) = \sum_{j=1}^{m-k} a_j B_{j,k}(t) = \sum_{j=2}^{m-k} a_j' B_{j,k-1}(t)$$

where the coefficients a_j' are given by

$$a_j' = \frac{(t-\tau_j)a_j + (\tau_{j+k-1}-t)a_{j-1}}{\tau_{j+k-1}-\tau_j}.$$

Using these formulas, algorithms for evaluating both B-splines and splines represented as B-spline series can be developed. The formulas do not require the knots to be simple. Any multiplicity ($\le k$) is allowed. Thus, splines with multiple knots are as easily evaluated as those with simple knots. Most geometry subroutine libraries and geometry modeling systems in use today use these or equivalent formulas for B-spline evaluations. This formula also shows why representing multiplicities as repetitions of knots is often the preferred representation. Simply put, it's the form used in evaluation so it is the form of choice for representation.

27.4.3 Robustness of the B-Spline Representation

B-splines provide a well-conditioned, robust representation for splines. This is an important concept that we illustrate in this section. Again, readers wishing more details or a rigorous development are referred to [3]. We illustrate robustness by examining the effects of error in the coefficients of a spline under various representations. Let knots $\tau_1 \le \tau_2 \le \ldots \le \tau_m$, order k, and the spline

$$s = \sum_{j=1}^{m-k} \alpha_j B_{j,k}$$

be given. If each coefficient α_j is modified by an amount ε_j the result is a second spline

$$s^* = \sum_{j=1}^{m-k} (\alpha_j + \varepsilon_j) B_{j,k}.$$

If each perturbation ε_j is at most ε in size (i.e., $|\varepsilon_j| \le \varepsilon$), then, using the partition of unity property of B-splines, the difference between s and s* may be bounded by

$$|s(t) - s^*(t)| = \left| \sum_{j=1}^{m-k} \varepsilon_j B_{j,k}(t) \right| \le \varepsilon \sum_{j=1}^{m-k} B_{j,k}(t) = \varepsilon.$$

Thus, the difference between the two splines is no bigger than the difference between coefficients.

This is important as it implies that errors in evaluating a B-spline series are no worse than errors in individual coefficients. Thus, the practitioner can, by controlling errors in the coefficients, be assured that errors throughout the spline model are under control. This is not the case for all representations. For example, if a piecewise polynomial representation such as used for the Examples 27.3.1–27.3.4 is used, there is no such guarantee. It is quite possible for the spline error to greatly exceed the errors in the coefficients. The first portion of Example 27.3.4 provides a simple illustration of this. Suppose errors of ε and $-\varepsilon$ are made in the coefficients of t^3 and t^2 respectively. Then, at $t = -1$ the total absolute error is $2|\varepsilon|$, i.e., the error in the function is greater than the errors in any coefficient.

There is another difficulty with errors and the piecewise polynomial representation. Using the B-spline representation guarantees *a priori* that the continuity of the spline is maintained even if coefficients are perturbed. This is not true with the piecewise polynomial representation. Errors in coefficients lead immediately to loss of continuity unless additional care is exercised to assure that the constraints hold.

27.4.4 A Representation Format for Univariate Splines

A compact representation for splines, based on B-splines, is implemented in the Spline Geometry Subprogram Library (DT_NURBS) available from the United States Navy [5]. That representation consists of: the dimension of the parameter space (for univariate splines this is 1), the dimension of the model space (for functions this is 1), the order k of the spline, the number $m-k$ of B-spline coefficients, a parameter *jspan*, the list of knots $\tau_1,\ldots,\tau_m$ and the list of B-spline coefficients $a_1,\ldots,a_{m-k}$. The parameter *jspan* is not part of the mathematical definition of the spline. It is an efficiency enhancement parameter. In the DT_NURBS implementation, it helps the B-spline evaluator rapidly locate knot intervals needed for evaluation thus, enhancing the speed of the basic evaluators. Its value is automatically controlled by software.

A general univariate spline $s(t) = \sum_{j=1}^{m-k} a_j B_{j,k}(t)$ is represented as

$$1, 1, k, m-k, jspan, \tau_1, \ldots, \tau_m, a_1, \ldots a_{m-k}.$$

The planar curve $C(t) = (x(t), y(t))$ where the coordinates $x(t) = \sum_{j=1}^{m-k} x_j B_{j,k}(t)$ and $y(t) = \sum_{j=1}^{m-k} y_j B_{j,k}(t)$ are splines of order k with knots T is represented as

$$1, 2, k, m-k, jspan, \tau_1, \ldots, \tau_m, x_1, \ldots, x_{m-k}, y_1, \ldots, y_{m-k}.$$

Similarly, the space curve $C(t) = (x(t), y(t), z(t))$ is represented as

$$1, 3, k, m-k, jspan, \tau_1, \ldots, \tau_{m,} x_1, \ldots, x_{m-k}, y_1, \ldots, y_{m-k}, z_1, \ldots, z_{m-k}.$$

Example 27.4.6: Consider the curve $C(t) = (t^2, t^3)$ on the interval $[-1,1]$. The order of this spline curve is 4 and the knots are $\{-1,-1,-1,-1,1,1,1,1\}$. Using the Marsden Identity, we calculate the B-spline coefficients of t^2 to be $\{1, \frac{-1}{3}, \frac{-1}{3}, 1\}$ and those for t^3 to be $\{-1,1,-1,1\}$. The DT_NURBS representation of this curve is

$$1, 2, 4, 4, jspan, -1, -1, -1, -1, 1, 1, 1, 1, 1, \frac{-1}{3}, \frac{-1}{3}, 1, -1, 1, -1, 1$$

27.5 Approximation with Splines

We have developed the rudiments of spline theory— splines, spline spaces, and a basis for representation. The mathematical theory actually gives more. It is able to determine how well splines can approximate given functions and data and practitioners can make use of this information in their modeling. How well a spline can approximate a given function or data set is given by the following. Suppose g is given on an interval $[a, b]$ and that g has k continuous derivatives. Let s be a spline of order k that interpolates g at m equidistant points. Then the difference between s and g is bounded by

$$|g(t) - s(t)| \leq Dh^k$$

where $h = (b - a)/(m - 1)$ and the constant D depends only on k and the maximum of $|g^{(k)}|$ but not on the knots. Formulas of this type also apply to best approximation methods such as least squares and minimax approximation. The practitioner can make use of this estimate in two important ways.

The first use of this formula is to estimate the number of knots required to achieve a desired fitting tolerance ε. Suppose splines of order k are being used to approximate a smooth function or to fit a dense set of data that come from a smooth process. Then the error is estimated as approximately Dh^k if the interval between knots is h. This can be solved for the required h if an estimate for D is known. D can be estimated by placing, for some n, $n - 1$ equally spaced knots inside the interval giving a maximum knot step of $h = l/n$ (l is the length of the interval) and then executing the interpolation process and calculating the resulting error. Suppose that error is ε_1. Then, $D \approx \varepsilon_1 n^k$. Thus, to reduce the error to ε, we estimate the number of knots needed to be $m \approx n(\varepsilon_1/\varepsilon)^{1/k}$. Of course, this would have to be tested and possibly refined.

Another, possibly more important application, is to use these estimates to gain a deeper understanding of the data. Suppose that the convergence of the smoothing process is less than expected given our understanding of the smoothness of the data. This means that something is wrong. Probably the process that generated the data is not as smooth as believed. Thus, understanding what the approximating process should produce can lead the user to a more detailed examination and better understanding of the data. The user can use that understanding to improve the fitting process by judiciously inserting knots at specific points.

For example, suppose a curve $(f(t), g(t))$ is given and, for whatever reason, the curve needs to be approximated by a spline. If the curve has a continuously changing tangent then a reasonable practitioner might choose to approximate each coordinate function by a quadratic spline with simple knots, i.e., quadratics that are themselves continuously differentiable. The user would then expect $O(h^3)$ convergence.* However, this might not happen. It is possible for the curve to have a continuous tangent and at the same time for the coordinates to have derivative discontinuities in which case the decrease in error will be much slower than expected. It is possible that convergence could be greatly accelerated by judicious placement of some of the knots including placing double knots.**

This ends the discussion of spline basics — properties, bases, robustness, and approximation power. We now turn attention to algorithms for constructing splices.

27.6 Constructing Spline Functions

In this section we will look at numerical and mathematical operations for defining spline functions. The context is always that the user has selected both an order and a knot set, i.e., has selected a spline space, and now wishes to construct a specific spline function for some purpose. We do not delve, except superficially, into questions of selecting order or knots as these are subjects beyond the scope of this

*The notation $O(h^3)$ is mathematical shorthand meaning the error is proportional to h^3.

**An example is the line segment $s_1(t) = \begin{cases} (t,t) & \text{if } 0 \leq t < 1; \\ (2t-1, 2t-1) & \text{if } 1 \leq t \leq 2 \end{cases}$

Geometrically, this is a line. But the discontinuity in parametric velocity at $t = 1$ makes it difficult to fit with smooth quadratics. Putting a double knot at $t = 1$ will instantly cure this problem.

chapter. We treat both the problem of approximation and of interpolation by splines and begin with the general problem of least-squares approximation.

27.6.1 Least Squares Approximation

Let knots $\tau_1 \le \tau_2 \cdots \le \tau_m$ be given with no multiplicity greater than k. Let a sequence of data pairs $\{(x_i, y_i)\}_{i=0}^{n}$ be given where the abscissae are strictly increasing: $x_i < x_{i+1}$ for each i. (Note: There is no assumption of a relationship between the abscissae and the knots.) The least-squares problem is to choose s so that the objective

$$\sum_{i=1}^{n} \left(s(x_i) - y_i\right)^2$$

is minimized over all choices of splines s of order k having the given knots. Assume that the number of data points is at least equal to the dimension of the spline space, i.e., $n \ge m - k$. Using the B-spline representation, rewrite the minimization problem as

$$\sum_{i=1}^{n} \sum_{j=1}^{m-k} a_j \left(B_j(x_i) - y_i\right)^2 = \text{minimum}.$$

Let

$$A = \begin{pmatrix} B_1(x_1) & B_2(x_1) & \cdots & B_{m-k}(x_1) \\ B_1(x_2) & B_2(x_2) & \cdots & B_{m-k}(x_2) \\ \bullet & \bullet & \cdots & \bullet \\ \bullet & \bullet & \cdots & \bullet \\ \bullet & \bullet & \cdots & \bullet \\ B_1(x_n) & B_2(x_n) & \cdots & B_{m-k}(x_n) \end{pmatrix}$$

The problem then becomes find **a** so that

$$\|A\mathbf{a} - \mathbf{y}\| = \text{minimum}$$

where **a** is the vector of B-spline coefficients $\mathbf{a} = (a_1, \ldots, a_{m-k})$ and **y** is the vector of data values $\mathbf{y} = (y_1, \ldots y_n)$. A is a $n \times (m - k)$ matrix with $n \ge m - k$. If A is of full rank* then there is a unique solution to the problem. The matrix A has full rank if, and only if, there is a subsequence of abscissae $\{x_{i_j}\}_{j=1}^{m-k}$ which satisfies the interlacing conditions [19]

$$\tau_j \le x_{i_j} \le \tau_{j+k}$$

with $\tau_j = x_{i_j}$ only if $\tau_j = \tau_{j+k-1}$ and $x_{i_j} = \tau_{j+k}$ only if $\tau_{j+1} = \tau_{j+k}$.

To illustrate, consider least squares approximation using splines of order 3 and knots $-1, -1, -1, 0, 1, 1, 1$. The interlacing conditions are

$$-1 \le x_1 < 0, -1 < x_2 < 1, -1 < x_3 < 1, 0 < x_4 \le 1.$$

---

*A matrix is a full rank if the number of linearly independent rows (or columns) is equal to the minimum dimension, in this case, $m - k$.

If the interpolation nodes are $-1, -1/2, -1/4, 0$, the interlacing conditions are violated and the resulting matrix

$$\begin{pmatrix} 1 & 0 & 0 & 0 \\ 1/4 & 5/8 & 1/8 & 0 \\ 1/16 & 21/32 & 9/32 & 0 \\ 0 & 1/2 & 1/2 & 0 \end{pmatrix}$$

is not of full rank. On the other hand, if the last node is 1 then the interlacing conditions are satisfied and the matrix

$$\begin{pmatrix} 1 & 0 & 0 & 0 \\ 1/4 & 5/8 & 1/8 & 0 \\ 1/16 & 21/32 & 9/32 & 0 \\ 0 & 1/2 & 1/2 & 1 \end{pmatrix}$$

is of full rank.

27.6.2 Interpolation Methods

We look at two methods of interpolation. The first is a general method where the knots and the interpolation nodes are subject only to the interlacing conditions. The second method is actually a class of methods where the interpolation nodes and knots are closely related, even equal. These methods are (almost) always underdetermined and require a strategy for regularization.

27.6.2.1 General Interpolation

If the number n of data points is equal to the number $m - k$ of coefficients and the interlacing conditions are satisfied, then the solution of the least-squares problems is the solution to the interpolation problem $s(x_i) = y_i$, $i = 1,\ldots, n$. Hence, it is not necessary for interpolation nodes and spline knots to coincide. Often it is preferable if they don't. Even though interpolation can, and often should, take place at points other than the knots, spline interpolation has traditionally begun with interpolation at the knots. We cover that topic next.

27.6.2.2 Underdetermined Problems — Knots at the Data Abscissae

Let an order k and data $\{(x_i, y_i)\}_{i=1}^{n}$ be given with $x_1 < x_2 < \cdots < x_n$. The classic spline interpolation problem is to construct a kth order spline s with break-points $x_2,\ldots, x_{n-1}$ which has $k - 2$ continuous derivatives such that $s(x_i) = y_i$, $i = 1,\ldots, n$. To construct such a function, set x_1 and x_n to be k^{th} order knots and $x_2,\ldots,x_{n-1}$ to be simple (i.e., multiplicity one) knots. Then the spline space has dimension $n + 2(k - 1) - k = n + k - 2$. If the order is 2, then the dimension is equal to the number of equations and the construction of the B-spline coefficients is easily done.

Problems occur if $k > 2$. For example, the cubic spline interpolation problem with $k = 4$ gives dimension $n + 2$ which is 2 more than the number of interpolation conditions. Thus, additional conditions are required. We discuss three (among many) methods for cubic interpolation and one for quadratic interpolation. Each of these methods is easily generalized to higher order splines.

Method 1: Natural Spline Interpolation

In this scheme, equations $s''(x_1) = 0$, $s''(x_n) = 0$ are added giving the linear system $A\mathbf{a} = \mathbf{w}$ to solve for the coefficients $\mathbf{a} = (a_1,\ldots,a_{n+2})$ where $\mathbf{w} = (0, y_1,\ldots,y_n, 0)$ and the matrix is

$$\begin{pmatrix} B_1''(x_1) & B_2''(x_1) & \cdots & B_{n+2}''(x_1) \\ B_1(x_1) & B_2(x_1) & \cdots & B_{n+2}(x_1) \\ B_1(x_2) & B_2(x_2) & \cdots & B_{n+2}(x_2) \\ \bullet & \bullet & \cdots & \bullet \\ \bullet & \bullet & \cdots & \bullet \\ \bullet & \bullet & \cdots & \bullet \\ B_1(x_n) & B_2(x_n) & \cdots & B_{n+2}(x_n) \\ B_1''(x_n) & B_2''(x_n) & \cdots & B_{n+2}''(x_n) \end{pmatrix}$$

The term "natural spline" was coined by I. J. Schoenberg. He called these splines natural because they are equations of (infinitely) thin beams constrained to pass through the points (x_i, y_i). Even though these splines are called "natural" they may not be so natural. That is, there is no *a priori* reason to believe that the data would suggest that the best fitting function should have zero second derivatives at the end points. In fact, the user will see a problem very quickly if natural cubic spline interpolation is used to approximate a function whose second derivatives are not zero at the ends. The rate of convergence will be $O(h^2)$ rather than $O(h^4)$ that might be expected otherwise.* The next method uses a different approach.

Method 2: Complete Spline Interpolation

When derivative information y_1', y_n', is available at the end-points, the complete spline interpolation problem where the conditions $s'(x_1) = y_1'$, $s'(x_n) = y_1'$ are added may be solved. The right-hand side is $\mathbf{w} = (y_1', y_1, \ldots, y_n, y_n')$ and the matrix is

$$\begin{pmatrix} B_1'(x_1) & B_2'(x_1) & \cdots & B_{n+2}'(x_1) \\ B_1(x_1) & B_2(x_1) & \cdots & B_{n+2}(x_1) \\ B_1(x_2) & B_2(x_2) & \cdots & B_{n+2}(x_2) \\ \bullet & \bullet & \cdots & \bullet \\ \bullet & \bullet & \cdots & \bullet \\ \bullet & \bullet & \cdots & \bullet \\ B_1(x_n) & B_2(x_n) & \cdots & B_{n+2}(x_n) \\ B_1'(x_n) & B_2'(x_n) & \cdots & B_{n+2}'(x_n) \end{pmatrix}$$

For this interpolation scheme, the convergence problem observed with the natural spline goes away. Convergence is $O(h^4)$. Both methods discussed so far convert the interpolation at knots into a solvable problem by adding data. The next method removes knots.

Method 3: Not-a-knot Interpolation

In this method, rather than adding equations, variables are removed by deleting x_2 and x_{n-1} from the knot sequence while keeping them as interpolation abscissae. Now, the number of coefficients is exactly n, the right-hand side is $\mathbf{w} = (y_1, \ldots, y_n)$, and the matrix is

---

*The natural cubic spline does solve an interesting and perhaps important variational problem. Among all the possible twice continuously differentiable solutions to the interpolation problem, the natural spline interpolant minimizes the quality $\int_a^b (s'')^2$. Thus, it would appear that natural splines are an attractive starting point for practical interpolation. But in practice, the difficulty mentioned above plus the fact that in applications we are usually concerned with problems about which a lot more is known than simply that the desired model is twice differentiable, make the natural spline less attractive as an all inclusive interpolation tool.

$$\begin{pmatrix} B_1(x_1) & B_2(x_1) & \dots & B_n(x_1) \\ B_1(x_2) & B_2(x_2) & \dots & B_n(x_2) \\ \bullet & \bullet & \dots & \bullet \\ \bullet & \bullet & \dots & \bullet \\ \bullet & \bullet & \dots & \bullet \\ B_1(x_n) & B_2(x_n) & \dots & B_n(x_n) \end{pmatrix}$$

The method is not restricted to removing the second and second-to-last knots. Any two knots may be removed as long as the interlacing conditions hold.

Method 4: Even Degree Interpolation

The methods described above can be easily and naturally modified to handle any odd degree (even order) spline interpolation with knots at the data points. Constructing even degree (odd order) splines presents a slightly different problem. Consider interpolation by quadratic splines. Following the idea that the knots and abscissae must coincide would give knots

$$x_1, x_1, x_1, x_2, \dots, x_{n-1}, x_n, x_n, x_n$$

for a total of $n + 4$ knots and $n + 1$ coefficients but only n equations. We need to select one additional equation. This can be done in a variety of ways, for example, by selecting a condition on the derivative at one of the end points.

There is an elegant alternative. Rather than choosing the points $x_2,\dots,x_{n-1}$ to be the interior knots, select the mid-points $(x_i + x_{i+1})/2$, $i = 1,\dots, n-1$ to be the interior knots. Then, add the equations $s''(x_1) = 0$, $s''(x_n) = 0$ as in the natural cubic spline. As in the case of natural splines, the resulting linear system $A\mathbf{a} = \mathbf{w}$ can be solved for the vector of coefficients. This is an example where interpolation by splines is more conveniently done at nodes other than knots.

27.7 Parametric Curves and Rational Splines

Parametric curves and surfaces are interesting because they provide a convenient method of representing closed curves, e.g., airfoil shapes, without the need to decompose the geometry into a number of domains where it can be represented directly as a function. Rational splines are interesting in that they provide a spline representation for the conic sections so useful in the aerospace and automotive industries. The actual construction of parametric and rational curves usually rely upon repeated application of univariate methods as covered in the previous section or direct construction methods, especially those for conics [6]. In this section we concentrate on what's different when dealing either with parametric curves or rational curves.

27.7.1 Parametric Curves

Given a set of data points $\{(x_i, y_i)\}_{i=1}^{n}$ in the plane, a typical geometric design problem is to construct a curve $C(t) = (x(t),y(t))$ passing through or near the given points. That is, so that $C(t_i) \approx (x_i, y_i)$. Before the problem can be solved, parameter values $\{t_i\}_{i=1}^{n}$ must be known. But these are usually not part of the problem statement and must be constructed from the data.

The construction of parametric values must be done with care. If not, the curves may display unwanted behavior such as wild excesses or self-intersections.* It is generally recognized that a uniform parametrization $t_i = (i-1)/(n-1)$ is prone to giving ill-behaved curves. A more acceptable parametrization is

*Even with a good parametrization the curve may have unwanted characteristics.

that of normalized chord lengths. In this parametrization, the spacing between the parametric values is made proportional to the physical distance between the points themselves. Other alternatives [14] are available.

After the parametrization has been selected, any of the methods based on univariate splines may be used to define a curve by fitting each coordinate in sequence. Thus, to construct a planar curve, construct the two functions $x(t) = \sum_{j=1}^{m-k} x_j B_i(t)$ and $y(t) = \sum_{j=1}^{m-k} y_j B_i(t)$ independently using the parameter values and the x and y data as appropriate. Since the coordinate functions are constructed independently, these methods do not necessarily produce curves that are free from unwanted effects such as self-intersections or wavy shapes. Methods which guarantee that the resulting curves will not have these problems exist. They are, however, highly non-linear and require the simultaneous construction of the components of the curve. As such they are beyond the scope of this chapter ([10, 11]). The B-spline representation of a curve has geometric interest. If the curve is $C(t) = (x(t), y(t))$ where $x(t) = \sum_{j=1}^{m-k} x_j B_i(t)$ and $y(t) = \sum_{j=1}^{m-k} y_j B_i(t)$, let $\mathbf{P}_j = (x_j, y_j)$ and rewrite the curve as

$$C(t)\sum_{j=1}^{m-k} \mathbf{P}_j \mathbf{B}_i(t).$$

The quantities $\mathbf{P}_j$ are called the *control points* of the curve C and the polygon formed by the control points is called the *control polygon.*

The curve matches the control polygon at its first and last points. The tangents to the curve at the first and last points are parallel to the first and last legs of the control polygon. Since, at each t the B-splines are positive and sum to 1, the curve is a convex combination of its control points and is contained in the convex hull of its control polygon (the *convex hull* property). The control polygon provides the basis for much of the interactive curve and surface construction technology such as the classical Bezier methods [8].

Two other interesting properties are: (1) affine transformations (rotation plus translation) of parametric spline curves are accomplished by applying the transformation to each of the control points and (2) if the control polygon is convex, then the curve itself will be convex. (However, convexity of the polygon is not necessary for convexity.)

27.7.2 Rational Splines

A rational spline $R(t)$ is defined as the ratio of two splines:

$$R(t) = \frac{x(t)}{w(t)} = \frac{\sum_{j=1}^{m-k} x_i B_i(t)}{\sum_{j=1}^{m-k} w_i B_i(t)}.$$

A planar rational spline curve is given by

$$C(t) = \left(\frac{x(t)}{w(t)}, \frac{y(t)}{w(t)}\right).$$

The B-spline coefficients $\{w_j\}_{j=1}^{m-k}$ of the denominator are called *weights*. The points $P_i = w_i^{-1}(x_i, y_i)$ are called control points. They, and the corresponding control polygons, have the same properties as the control points and polygons have for planar spline curves.

The interest in rational splines lies in the fact that they provide a representation for conic sections that is compatible with the spline representation. In fact, a conic section starting at Q_0 and ending at Q_2 with initial and final tangents intersecting at a third point Q_1 has the parametrization ([8])

$$r(u) = \frac{w_0 Q_0 (1-u)^2 + 2w_1 Q_1 u(1-u) + w_2 Q_2 u^2}{w_0 (1-u)^2 + 2w_1 u(1-u) + w_2 u^2}$$

which is clearly a quadradic, rational spline and further, the weights (coefficients of the denominator) are non-zero.

27.7.3 Representation of Rational Splines and an Example

The Spline Geometry Subprogram Library (DT_NURBS) [5] represents rational spline functions, curves, and surfaces in homogeneous coordinates which means that the numerators and denominators are stored separately. For example, the function $x(t)/w(t)$ is stored as the parametric curve $(x(t), w(t))$. The fact that it is to be rational is indicated by setting the second element of the spline array set to the negative of the number of dependent variables.* Thus, the rational spline curve

$$C(t) = \left(\frac{x(t)}{w(t)}, \frac{y(t)}{w(t)} \right)$$

is stored as the space curve $(x(t), y(t))$ except with the second element of the array set to –3. The spline evaluator returns the values $x(t)/w(t)$ and $y(t)/w(t)$.

Example 27.7.1: One rational spline parametrization of a circular arc is

$$C(t) = \left(\frac{1-t^2}{1+t^2}, \frac{2t}{1+t^2} \right).$$

It is represented as

$$1,\ -3,3,3,jspan,0,0,0,1,1,1,1,1,0,0,1,2,1,1,2.$$

Storing rational splines this way is a decided departure from standard CAGD practice. In CAGD the convention is to store control points, that is, the B-spline coefficients divided by the weights, in order to make the control polygon the primary data object. This is done to facilitate an interactive approach to curve design. For our purposes, we prefer to store rational splines, curves, and surfaces in B-spline format for two reasons. First, this scheme avoids the continual reconstruction of the actual B-spline coefficients from the weights and control points and, hence, improves computational efficiency.

Second, and more importantly, the B-spline format allows for zero or negative coefficients in the denominator. Example 27.7.1 cannot be stored using the CAGD convention but is easily represented in our convention. Mathematically, using the CAGD representation means that the spline spaces will not be closed with respect to limits which is something we have gone to some pains to require by allowing degenerate degrees and inactive knots. (It is a straightforward matter to construct sequences of rational splines with positive weights which converge to Example 27.7.1. Thus, it is easy to construct rational splines having the CAGD representation which converge to a rational spline that cannot be represented that way.) Using B-spline coefficients directly assures that the resulting spline space is closed with respect to limits.

*This is a temporary state. In the future, it is planned to store the rational information as a specific property of the curve.

28

Computer-Aided Geometric Design

Gerald Farin

28.1 History

CAGD (computer-aided geometric design) dates back to Paris in 1959, when Citroën hired Paul de Faget de Casteljau to develop some mathematical tools. Citroën already had numerically controlled milling machines; but in order to fully utilize them, a link had to be created between the standard blueprints and the milling machines. This link would have to translate the blueprints into formulas that could be evaluated by a program, thus creating the coordinates to drive the milling machine. De Casteljau invented what he called "Courbes à Poles," and what we now know, ironically, as Bézier curves. We will use them throughout this chapter.

Pierre Bézier worked at Rénault, also in Paris, and learned about Citroën's (very secretive) efforts. He was able to create a system with the same functionality himself, and Rénault allowed him to publicize it widely. Thus Bézier curves started to dominate CAGD.

Another development was the introduction of splines — this one being an American contribution. In the late 1950s, J. Ferguson at Boeing developed a package based on interpolating piecewise cubic curves, on C^2 cubic splines, as we would say today. Splines were already known among mathematicians following the discovery of B-splines by I. Schoenberg in 1946. It was most notably C. de Boor who advanced the theory of these curves, based upon practical experience at General Motors.

Based on de Boor's work, Gordon and Risenfeld realized in 1972 that B-splines could be used in much the same way as could Bézier curves. They showed how Bézier curves were just a special case of B-spline curves, thus making possible a unification of systems based on splines (typically American) and those based on Bézier curves (typically French).

One of the most influential American researchers in the field of CAGD was S. Coons, who developed surfaces named after him in the late 1950s. These surfaces have given way to B-spline-based systems now, but another development, also initiated by Coons, has further unified all of CAGD. This is the concept of NURBS, a generalization of piecewise *polynomial* curves to piecewise *rational polynomial* curves. Coons' student K. Vesprille laid down the basic theory of rational B-splines in 1975.

0-8493-2687-7/99/$0.00+$.50

It was quickly realized that they allowed a unified representation of splines and conics. This was important when data were to be transferred between different design systems — spline and Bézier curves were widely used, but conic prevailed in several aircraft design systems, owing to Liming's work.

There have been several instances where CAGD interacted with finite element research, the most notable one being S. Coons' work. Coons patches (including several generalizations) were in use for many years in automotive design. But they also found their way into grid generation for finite elements, where they became known as "transfinite interpolation." Another example is the finite element developed by Clough and Tocher; it was not known in the CAGD community until it was translated into Bernstein-Bézier form.

Today, the main use of CAGD in the context of finite element methods is in grid generation. The geometry of any object is nowadays expressed in the forms of surfaces from some CAD/CAM system, typically using the B-spline or NURBS representation. Grids will have to be created on and around the object. How can we incorporate the CAGD description of the object into the desired grid? In what follows, we will outline the central CAGD techniques to the extent that they will be of use for this problem.

Several books exist on the topic of CAGD, and they should be consulted for more details: Farin [6], Faux and Pratt [8], Hoschek and Lasser [13], Yamaguchi [19]. When we describe results without explicit references, then these texts should be consulted. Another source for up-to-date information is the home page of the journal *CAGD*: http://www.elsevier.nl/locate/comaid.

28.2 Basic Principles

Geometric computation takes place in two- or three-dimensional Euclidean (or affine) space. The objects of the computation are points, denoted by boldface letters: **a, x,** etc. We may obtain points from other points by mapping such as **affine maps.** These are of the form

$$\Phi\mathbf{x} = A\mathbf{x} + \mathbf{v},$$

where A is a square matrix and $\mathbf{v}$ is a translation vector. All affine maps may be thought of as a concatenation of rotations, scalings, shears, and translations.

Affine maps leave **barycentric combinations** unchanged: these are linear combinations where the coefficients sum to one. Thus if

$$\mathbf{x} = \sum_i \alpha_i \mathbf{x}_i; \quad \sum_i \alpha_i = 1,$$

and Φ is an affine map, then also

$$\Phi\mathbf{x} = \sum_i \alpha_i \Phi\mathbf{x}_i.$$

Thus, for example, the midpoint of two points is mapped to the midpoint of the two image points.

Any time we have a relationship between points such as

$$\mathbf{x} = \sum_i \alpha_i \mathbf{x}_i,$$

it is mandatory that the α_i sum to one:* otherwise a simple translation would destroy this relationship. If all α_i are between 0 and 1, then we speak of a **convex combination.** These are known for their inherent numerical stability.

It is possible that the α_i sum to zero; then we have defined a *vector.*

Another basic operation on points is that of **linear interpolation:**

*This is also phrased as "they form a partition of unity."

27.8 Surfaces

In this section definitions of (tensor product) spline surfaces will be given and methods for constructing spline surfaces will be discussed.

27.8.1 Tensor Product Splines

A full generalization of univariate splines would be to consider collections of individual polynomial pieces defined over arbitrary n-dimensional domains. Such a generalization would undoubtedly find application within the engineering disciplines. However, at this time, such spaces are a subject of fundamental mathematical research with many open questions, one of which is the basic question of the dimensionality of the resulting spline spaces. Tensor product splines form an intermediate stage highly useful in their own right but without the theoretical complications. They are direct generalizations of univariate splines curves and, as such, they inherit many of the properties of simple splines. The most general tensor product spline is defined as follows. Let $D = [a_1, b_1] \times [a_2, b_2] \times \ldots \times [a_m, b_m]$ be a rectangular parallelapiped and, for each $[a_i, b_i]$ choose knots $T = \{\tau_{p,i}\}_{p=1}^{P_i}$ and order k_i. Then define

$$F(u_1,\ldots,u_m) = \sum_{i_1=1}^{P_1} \cdots \sum_{i_m=1}^{P_m} f_{i_1,\ldots,i_m} B(u_1, T_{1,i_1}) \ldots B(u_m, T_{m,i_m})$$

where $T_{i,\, i_j} = \{\tau_{i, i_j}, \ldots, \tau_{i,\, i_j+k_i-1}\}$.

Rather than work in such generality however, we will concentrate on the most important case of the bivariate tensor product spline. Let knots $T = \{\tau_p\}_{p=1}^{P}$ and $X = \{\xi_q\}_{q=1}^{Q}$ and orders k_u, k_v be given. The tensor product splines with these knots and orders is the collection of functions

$$f(u,v) = \sum_{p=1}^{P-k_u} \sum_{q=1}^{Q-k_v} f_{ij} B_p(u) C_q(v)$$

where B_p, C_q are the B-splines in u and v, respectively. Thus, f is a function defined on the rectangle $[\tau_1, \tau_P] \times [\xi_1, \xi_Q]$. As in the univariate case, f is taken to be zero outside the rectangle.

There are two types of interpolation and approximation problems that arise with spline surfaces depending on whether the data is given on a rectangular grid or is scattered. Tensor product spline surfaces can be used effectively on both types of problems.

27.8.2 Interpolation and Approximation on a Rectangular Grid

Let a rectangular grid $\{u_i, v_j\}$ and data $\{z_{ij}\}_{(i,j)=(1,1)}^{(n,m)}$ be given where $a = u_1 < \ldots < u_n = b$ and $c = v_1 < \ldots < v_{m=d}$. Choose knots T, X and orders k_u, k_v with $\tau_1 = a$, $\tau_{Pn} = b$, $\xi_1 = c$, $\xi_Q = d$. The approximation problem is to find coefficients for the spline s so that

$$\sum_{i=1}^{n} \sum_{j=1}^{m} \left(z_{ij} - s(u_i, v_j)\right)^2 = \text{minimum}.$$

The same considerations hold here as in the univariate case. Namely, the problem is well posed and a unique solution exists if, and only if, there are subsequences of the u and v points that interface (see Section 27.6) with the knots T and X.

If the knots and the grid points coincide, underdetermined systems similar to those in Section 27.6 arise. The methods of Section 27.6 can be generalized to construct analogous methods for fitting multivariate data.

For these types of problems there are very efficient computational methods that take direct advantage of the rectangular nature of the data and the regular, sparse nature of the resulting matrices.

27.8.3 Interpolation and Approximation of Scattered Data

There is a variety of methods for dealing with scattered data using other mathematical forms such as, for example, radial basis methods. Here we restrict our remarks to the use of tensor product splines for the same problem. The least squares approach is the same as for the rectangularly gridded data. The difference is that the efficient computational methods are no longer available. The resulting matrix will tend to be large and sparse but beyond that will have no easily recognized structure.

There is yet another difference in that there is no equivalent for the interlacing conditions for determining when the matrices are of full rank. This means that the linear algebra needs to be done with some care. It is possible for the systems of equations to be both overdetermined and rank deficient.

27.8.4 Construction of Parametric Spline Surfaces from Rectangular Data

Tensor product parametric spline surfaces are surfaces $S(u,v)=X(u,v), Y(u,v), Z(u,v)$. The coordinate functions X, Y and Z are tensor product polynomial spline surfaces. If data $\{s_{ij} = (x_{ij}, y_{ij}, z_{ij})\}_{(i,j)=(1,1)}^{(n,m)}$ are given, the construction of a surface approximating or interpolating proceeds as in the case of curves, by applying an approximating or interpolating function to each of the coordinates. The problem, as before, is the construction of the parametrization points (u_i, v_j).

27.8.5 Other Methods of Construction of Surfaces

Other methods for the construction of spline surfaces that interpolate a given family or mesh of spline curves include transfinite interpolation [13]. These methods can be generalized to more general situations to include isolated points and disconnected curves [12]. For transfinite interpolation to work, each of the curves in the family needs to have the same order and knots. Thus, transfinite interpolation or blending of curves into surfaces provides another example of where common spline frameworks are needed.

The common spline form for a spline surface is as expected: The first element is the number of independent variables, for surfaces it will be 2.

This ends the discussion of curve and surface construction.We now turn to the final topic, the use of composition of functions to construct piecewise polynomial surfaces.

27.9 Functional Composition

In this last section we discuss functional composition as a valuable tool for building engineering models and handling engineering data. The space of tensor product splines are a valuable engineering tool. However, they have a severe restriction in that they are only applicable to rectangular domains. CAD gets around this restriction by the use of trimmed and joined surfaces. However, this approach appears to be limited to geometric curves and surfaces.

Functional composition has proven to be a useful tool in many application areas. It is useful for building intricate engineering models where the ultimate behavior or performance depends on a number of intermediate phenomena each of which needs to be modeled again as a function of numerous parameters. Aerodynamic performance models are good examples of this use [11]. It is also useful for managing analysis and gridding data that is dependent on a specific geometry or subset of geometry.

Let $f : D \; D \subset R^n \rightarrow R^P$ and $g : E \subset R^m \rightarrow D$ be two tensor product splines. The composition spline h defined by $h(t) = f(g(t))$ is then a mapping $h : E \subset R^m \rightarrow R^n$. If f and g are both univariate splines of respective degrees $l_1 l_2$, then h is a univariate spline of degree $l_1 l_2$. In the general case, similar formulas for degree hold. However, it is not true that composition functions for other than univariate splines have the same type of rectangular knot structure as the general tensor product spline. They have a more general structure.

Piecewise polynomial functions in several variables with nonrectangular knot structures are a valuable engineering tool. However, they are also, at this time, a subject of fundamental mathematical research with many open questions, one of which is the basic question of the dimensionality of the resulting spline spaces. Functional composition, however, gives the practitioner the ability to get some of the benefits of a more general structure while not having to resolve the thorny research issues.

We illustrate some of these ideas with a simple example. Let a surface is given by $f: D \rightarrow R^3$ and suppose part of the surface is submerged in a liquid as illustrated in the Figure below. Now suppose some interaction of the liquid with the surface is to be modeled. This requires restricting attention to the submerged piece of the surface. One way of modeling the piece might be to break-off the piece as a separate surface for the simulation. Functional composition provides another method.

Define a second mapping $g: E \rightarrow D$ where the image of g is simply the pre-image of the wetted surface. The composition map h of g with f provides a convenient model that can be analyzed and simulated. Thus, functional composition provides a convenient method of modeling effects on pieces of geometry.

There are other benefits to this approach. First, suppose the simulation, because of some additional information, is to be altered by changing the placement of the surface in the liquid. This is easily accommodated in the functional composition model by redefining the function g.

Second, suppose that, instead of the placement of the surface changing, the surface itself has changed. The change is transparent in the model because, as the base function f changes, the function g automatically accommodates the change in the wetted surface. This idea is also useful in gridding. By mapping surface grid points to a common parametric domain, the analyst can easily accept minor changes in surface definition without having to regrid because regridding is done directly by functional composition.

Thus the methods of functional composition provide easily updatable models for simulating complicated environments. The parametric domain E could be expanded to include other effects. For example, a time dimension could be added which would allow the wetted surface, or the grid points, or the surface to vary as a function of time. This gives the modeler a convenient method of managing these changes and the data associated with them. For an expansion of these ideas, see [1].

References

1. Ames, R.A. and Ferguson, D.R., Applications to engineering design of the General Geometry, Grid and Analysis (GGA) objectin DT_NURBS, Gridding Conference, Mississippi State University, May 2–3, 1996.
2. De Boor, C., On calculating with B-splines, *J. Approximation Theory,* 1972, 6(1), pp 50–62.
3. De Boor, C., A Practical guide to splines, Springer–Verlag, 1978.
4. Curry, H.B. and Schoenberg, I.J., On polya frequency functions. IV: The fundamental spline functions and their limits, *J. d'Analyse Math.* 1966, 17, pp 71–107.
5. DT_NURBS Spline Geometry Library: dtnet33-199.dt.navy.mil/dtnurbs/doc.htm
6. Farin, G., *Curves and Surfaces for Computer Aided Geometric Design, A Practical Guide, 2nd Edition.* Academic Press, 1990.
7. Farouki, R. and Rajan, V.T., On the numerical condition of polynomials in Bernstein form, *CADG.* 1987, 4, pp 191–216.
8. Faux, I. and Pratt, M., *Computational Geometry for Design and Manufacture.* Ellis Horwood, 1979.
9. Ferguson, D.R., Construction of curves and surfaces using numerical optimization techniques, *CAD.* 1986, Vol. 18, no. 1, pp 15–21.
10. Ferguson, D.R., Frank, P.D., and Jones, A.K., Surface shape control using constrained optimization on the B-spline representation, *CAGD.* 1988, 5, pp 87–103.
11. Ferguson, D.R., Mastro, R.A., and Blakely, R., Modeling and analysis of aerodynamic data, *AIAA* 89-0476, *27th Aerospace Sciences Meeting,* Reno, NV, January 9-12, 1989.
12. Ferguson, D.R. and Grandine, T.A., On the Construction of surfaces interpolating curves: I. A method for handling nonconstant parameter curves, *ACM Transactions on Graphics.* 1990, Vol. 9, No. 2, pp 212–225.

13. Gordon, W., Distributive lattices and the approximation of multivariate functions, Schoenberg, I.E., (Ed.), *Approximation with Special Emphasis on Splines.* 1969, Academic Press, Orlando, FL, pp 223–277.
14. Lee, E.Y., Choosing nodes in parametric curve interpolation, *Computer Aided Design,* 1989, 21(6).
15. Rice, J., On the condition of polynomial and rational forms, *Numer. Math.* 7, pp 426–435.
16. Schumaker, L., *Spline Functions: Basic Theory.* John Wiley & Sons, 1981.
17. STEP (STandard for the Exchange of Product model data) ISO 10303 (Industrial automation systems and integration—Product data representation and exchange), International Organization for Standardization (ISO), Geneva.
18. Schoenberg, I.J., Contributions to the problem of approximation of equidistant data by analytic functions, *Quarterly Applied Math.* 1946, 4, pp 45–99.
19. Schoenberg, I.J. and Whitney, A., On polya frequency functions, III: The positivity of translation determinants with application to the interpolation problem by spline curves, *TAMS,* 1953, 74, pp 246–259.

$$\mathbf{x}(t) = (1-t)\mathbf{a} + t\mathbf{b}. \tag{28.1}$$

Almost all geometric computations may be traced to this simple building block! The above is a computational definition; a geometric one would say that $\mathbf{x}(t)$ is obtained by the affine map Φ that maps [0, 1] to $\bar{a}\bar{b}$. Note that it is not necessary that $t \in [0, 1]$; in those cases, we speak of *extrapolation.*

The bivariate analog of linear interpolation is given as follows: given three points **a, b, c** in $\mathbb{E}^3$, compute points on the plane through them. We think of **a, b, c** as the image of three 2D points **p, q,** $\mathbb{R}^2$. Any point **u** in 2D may be written as $\mathbf{u} = u\mathbf{p} + v\mathbf{q} + w\mathbb{R}^2$ where $u + v + w = 1$. The numbers u, v, w are called **barycentric coordinates** of **u** with respect to **p, q,** $\mathbb{R}^2$. Now the image **x** of **u** will be a point on the plane through **a, b, c** given by

$$\mathbf{x} = u\mathbf{a} + v\mathbf{b} + w\mathbf{c}.$$

The barycentric coordinates of **u** are defined as follows:

$$u = \frac{\text{area}(\mathbf{u},\mathbf{q},\mathbf{r})}{\text{area}(\mathbf{p},\mathbf{q},\mathbf{r})}, \quad v = \frac{\text{area}(\mathbf{p},\mathbf{u},\mathbf{r})}{\text{area}(\mathbf{p},\mathbf{q},\mathbf{r})}, \quad w = \frac{\text{area}(\mathbf{p},\mathbf{q},\mathbf{u})}{\text{area}(\mathbf{p},\mathbf{q},\mathbf{r})}. \tag{28.2}$$

More information on this basic geometry can be found in many texts.*

28.3 Bézier Curves

Any polynomial curve in 2- or 3-space may be expressed as

$$\begin{bmatrix} x(t) \\ y(t) \\ z(t) \end{bmatrix} = \mathbf{x}(t) = \sum_{i=0}^{n} \mathbf{c}_i F_i(t),$$

where the F_i are a set of basis functions for all polynomials of degree n, and the $\mathbf{c}_i$ are the coefficients defining $\mathbf{x}(t)$. The most common choice is to set $F_i(t) = t^i$, i.e., to select the **monomial basis.** "Most common" strictly refers to calculus classes; in numerical and geometric applications, this basis is very unsuitable: the $\mathbf{c}_i$ are almost completely devoid of geometric meaning, and worse, they are extremely sensitive to the slightest round off. The latter observation is due to Farouki and Rajan [7], who demonstrated that a different basis is close to optimal in the sense of numerical stability: this is the **Bernstein basis.** Using it, any curve may be written as

$$\mathbf{x}(t) = \sum_{i=0}^{n} \mathbf{b}_i B_i^n(t), \tag{28.3}$$

where the $B_i^n(t)$, the Bernstein polynomials, are given by

$$B_i^n(t) = \binom{n}{i} t^i (1-t)^{n-i}. \tag{28.4}$$

They are set to zero for $i \notin \{0, \ldots, n\}$. Using Bernstein polynomials, one considers curves over the interval [0, 1], although any other interval could be used equally well. Polynomial curves that are expressed in the Bernstein basis are called **Bézier curves.** Figure 28.1 gives two examples.

---

*For a website, see http://www.eros.cagd.eas.asu.edu/~farin/gbook/gbook_home.html.

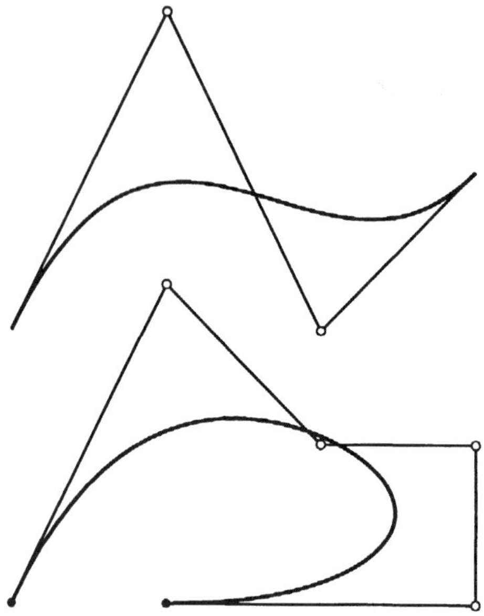

FIGURE 28.1 Bézier curves: top, n = 3, bottom, n = 5.

In order for Eq. 28.3 to be independent of a particular coordinate system, the basis functions must sum to one, i.e., they must form a *partition of unity*. We thus have

$$\sum_{j=0}^{n} B_j^n(t) \equiv 1. \tag{28.5}$$

Bernstein polynomials also satisfy the recursion

$$B_i^n(t) = (1-t)B_i^{n-1}(t) + tB_{i-1}^{n-1}(t). \tag{28.6}$$

It leads directly to the **de Casteljau algorithm** for the evaluation of Bézier curves:

de Casteljau algorithm:
Given: $\mathbf{b}_0, \mathbf{b}_1, \ldots, \mathbf{b}_n \in \mathbb{E}^3$ and $t \in \mathbb{R}$,
set

$$\mathbf{b}_i^r(t) = (1-t)\mathbf{b}_i^{r-1}(t) + t\mathbf{b}_{i+1}^{r-1}(t) \quad \begin{cases} r = 1,\ldots,n \\ i = 0,\ldots,n-r \end{cases} \tag{28.7}$$

and $\mathbf{b}_i^0(t) = \mathbf{b}_i$. Then $\mathbf{b}_0^n(t)$ is the point with parameter value t on the Bézier *curve* $\mathbf{b}^n$.

The polygon **P** formed by $\mathbf{b}_0, \ldots, \mathbf{b}_n$ is called the *Bézier polygon* or *control polygon* of the curve $\mathbf{b}^n$. Similarly, the polygon vertices $\mathbf{b}_i$ are called *control points* or *Bézier points.*

The intermediate coefficients $\mathbf{b}_i^r(t)$ are conveniently written into a triangular array of points, the *de Casteljau scheme.* We give the example of the cubic case:

$$\begin{array}{llll} \mathbf{b}_0 & & & \\ \mathbf{b}_1 & \mathbf{b}_0^1 & & \\ \mathbf{b}_2 & \mathbf{b}_1^1 & \mathbf{b}_0^2 & \\ \mathbf{b}_3 & \mathbf{b}_2^1 & \mathbf{b}_1^2 & \mathbf{b}_0^3. \end{array} \tag{28.8}$$

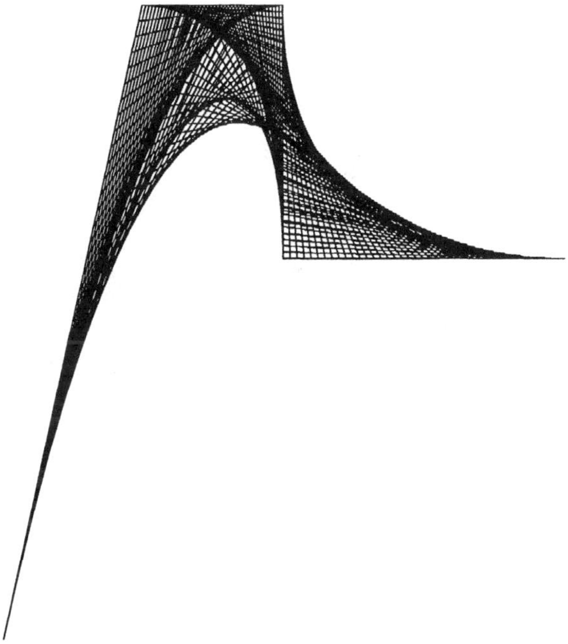

FIGURE 28.2 Several steps of the de Casteljau algorithm.

This triangular array of points seems to suggest the use of a two-dimensional array in writing code for the de Casteljau algorithm. That would be a waste of storage; however, it is sufficient to use the left column only and to overwrite it appropriately.

While the de Casteljau algorithm needs $O(n^2)$ operations for a degree n Bézier curve, its use is still encouraged because of its stability — and if an optimizing compiler is available, it is surprisingly fast!

Figure 28.2 illustrates this important algorithm: A Bézier curve is evaluated at several parameter values, and all intermediate points $\mathbf{b}_i^r$ are connected.

Because of their central role in all of CAGD, we list some of the most important properties of Bézier curves:

Invariance under affine parameter transformations: Algebraically, this property reads

$$\sum_{i=0}^{n} \mathbf{b}_i B_i^n(t) = \sum_{i=0}^{n} \mathbf{b}_i B_i^n\left(\frac{u-a}{b-a}\right). \tag{28.9}$$

It states that we may define a curve over $[a, b]$ as well as over $[0, 1]$.

Convex hull property: Any point on a Bézier curve, as long as its parameter value is between 0 and 1, is in the convex hull of the control polygon. This follows, since for $t \in [0, 1]$, the Bernstein polynomials are nonnegative and they sum to one as shown in Eq. 28.4. This property allows for very cheap interference checks, using the minmax box of the control polygon.

Linear precision: The identity

$$\sum_{j=0}^{n} \frac{j}{n} B_j^n(t) = t, \tag{28.10}$$

has the following application: suppose the polygon vertices $\mathbf{b}_j$ are uniformly distributed on a straight line joining two points $\mathbf{p}$ and $\mathbf{q}$:

$$\mathbf{b}_j = \left(1 - \frac{j}{n}\right)\mathbf{p} + \frac{j}{n}\mathbf{q}; \quad j = 0,\ldots,n.$$

The curve that is generated by this polygon is the straight line between **p** and **q**, i.e., the initial straight line is reproduced.

The **derivative** of a Bézier curve is given by

$$\frac{\mathrm{d}}{\mathrm{d}t}\mathbf{b}^n(t) = n\sum_{j=0}^{n-1}\Delta\mathbf{b}_j B_j^{n-1}(t); \quad \Delta\mathbf{b}_j \in IR^3.$$

here, Δ denotes the forward difference operator $\Delta\mathbf{b}_j = \mathbf{b}_{j+1} - \mathbf{b}_j$. Higher derivatives are given by

$$\frac{\mathrm{d}^r}{\mathrm{d}t^r}\mathbf{b}^n(t) = \frac{n!}{(n-r)!}\sum_{j=0}^{n-r}\Delta^r\mathbf{b}_j B_j^{n-r}(t); \quad \Delta^r\mathbf{b}_j \in IR^3.$$

Bézier curves may be pieced together, thus forming composite curves. Let $\mathbf{b}_0, \ldots, \mathbf{b}_n$ define one curve and $\mathbf{c}_0, \ldots, \mathbf{c}_n$ a second one. Both curves form one continuous curve if $\mathbf{b}_n = \mathbf{c}_0$. They form one smooth curve (no tangent discontinuities) if in addition $\mathbf{b}_{n-1}, \mathbf{c}_0, \mathbf{c}_1$ are collinear. In order to say when they form one C^1 curve, we must define over which intervals they are defined. So let the curve $\mathbf{b}_0, \ldots, \mathbf{b}_n$ be defined over $[a, b]$ and let $\mathbf{c}_0, \ldots, \mathbf{c}_n$ be defined over $[b, c]$. They are C^1 if

$$\mathbf{c}_0 = \frac{(c-b)\mathbf{b}_{n-1} + (b-a)\mathbf{c}_1}{c-a}.$$

If at each level r of the de Casteljau algorithm, we use a different argument t_i instead of t, we obtain a function of n arguments: $\mathbf{b}[t_1, \ldots, t_n]$. This is called the **blossom** of a Bézier curve after L. Ramshaw [17]. It is clear from the definition of a blossom that

$$\mathbf{b}_n^n(t) = \mathbf{b}\left[t^{<n>}\right],$$

where $t^{<n>}$ denotes n–fold repetition of the argument t. One of the most useful properties of blossoms is the following: suppose we wish to redefine our Bézier curve, so that now it is defined over an interval $[a, b]$ instead over $[0, 1]$. It will then be defined by a different control polygon $\mathbf{c}_0, \ldots, \mathbf{c}_n$. The $\mathbf{c}_i$ are simply calculated as

$$\mathbf{c}_i = \mathbf{b}\left[a^{<n-i>}, b^{<i>}\right]; \quad i = 0,\ldots,n. \tag{28.11}$$

If $[a, b] = [0, 1/2]$, then the above is called **subdivision.** It is important for many numerical techniques that need successive control net refinement.

An example is root finding, or, more generally, finding the intersection(s) of a straight line with a curve. A very simple and robust algorithm is the following: find the minmax box that contains the curve. Using the convex hull property, this is done by finding the maximal and minimal coordinate values of the control polygon. Then test if the straight line intersects that box. If not, then there is no intersection. If it does, subdivide the curve at $t = 1/2$ and repeat the above for both halves. The algorithm will terminate if the size of a minmax box is below a given tolerance. While extremely robust, the method is also slow.

We close this section with a collection of formulas.

Power basis $\{t^i\}$ and the Bernstein basis $\{B_i^n\}$ conversion:

$$t^i = \sum_{j=i}^{n} \frac{\binom{j}{i}}{\binom{n}{i}} B_j^n(t)$$

and

$$B_i^n(t) = \sum_{j=i}^{n} (-1)^{j-i} \binom{n}{j}\binom{j}{i} t^j.$$

Recursion:

$$B_i^n(t) = (1-t)B_i^{n-1}(t) + tB_{i-1}^{n-1}(t).$$

Subdivision:

$$B_i^n(ct) = \sum_{j=0}^{n} B_i^j(c) B_j^n(t).$$

Derivative:

$$\frac{\mathrm{d}}{\mathrm{d}t} B_i^n(t) = n\left[B_{i-1}^{n-1}(t) - B_i^{n-1}(t)\right].$$

Integral:

$$\int_0^t B_i^n(x)dx = \frac{1}{n+1} \sum_{j=i+1}^{n+1} B_j^{n+1}(t),$$

$$\int_0^1 B_i^n(x)\mathrm{d}x = \frac{1}{n+1}.$$

Three degree elevation formulas:

$$(1-t)B_i^n(t) = \frac{n+1-i}{n+1} B_i^{n+1}(t),$$

$$tB_i^n(t) = \frac{i+1}{n+1} B_{i+1}^{n+1}(t),$$

$$B_i^n(t) = \frac{n+1-i}{n+1} B_i^{n+1}(t) + \frac{i+1}{n+1} B_{i+1}^{n+1}(t).$$

Product:

$$B_i^m(u) B_j^n(u) = \frac{\binom{m}{i}\binom{n}{j}}{\binom{m+n}{i+j}} B_{i+j}^{m+n}(u).$$

28.4 Cubic Hermite Curves

Suppose one is given two $\mathbf{p}_0, \mathbf{p}_1$ and two tangent vectors $\mathbf{m}_0, \mathbf{m}_1$. The objective is to find a cubic polynomial curve $\mathbf{p}$ that interpolates to these data:

$$\begin{aligned} \mathbf{p}(0) &= \mathbf{p}_0, \\ \dot{\mathbf{p}}(0) &= \mathbf{m}_0, \\ \dot{\mathbf{p}}(1) &= \mathbf{m}_1, \\ \mathbf{p}(1) &= \mathbf{p}_1, \end{aligned}$$

where the dot denotes differentiation.

The interpolant can be written as

$$\mathbf{p}(t) = \mathbf{p}_0 H_0^3(t) + \mathbf{m}_0 H_1^3(t) + \mathbf{m}_1 H_2^3(t) + \mathbf{p}_1 H_3^3(t), \tag{28.12}$$

where

$$\begin{aligned} H_0^3(t) &= B_0^3(t) + B_1^3(t), \\ H_1^3(t) &= \frac{1}{3} B_1^3(t), \\ H_2^3(t) &= -\frac{1}{3} B_2^3(t), \\ H_3^3(t) &= B_2^3(t) + B_3^3(t). \end{aligned} \tag{28.13}$$

The H_i^3 are called cubic Hermite polynomials.

28.5 B-Splines

If continuity higher than C^1 is desired, joining Bézier curves becomes cumbersome, and the B-spline approach is far easier. A B-spline curve consists of several polynomial pieces, or segments, that are connected with a prescribed smoothness. Typically, degree n B-splines have smoothness C^{n-1}. Our development here is similar to that of [6], but is "leaner" because of a change in notation.

We are given a nondecreasing knot sequence $u_0, \ldots, u_K$ and the degree n of a (to be defined) B-spline curve. The curve will be defined by a control polygon

$$\mathbf{P} = \mathbf{d}_0 \mathbf{d}_1 \ldots \mathbf{d}_p,$$

with $p = K - n$. Thus there are as many control points as there are successive n–tuples of knots. Successive knots do not have to be distinct; but no more than n successive knots may coincide. If r successive knots coincide, we speak of a knot of multiplicity r.

Take all spans of n subsequent intervals and map them to the control polygon legs by affine maps:

$$[u_i, u_{i+n}] \rightarrow \mathbf{d}_i \mathbf{d}_{i+1}.$$

This way, each control polygon leg is "engraved" with a span of the knot sequence. Note that, since no knot is allowed to have multiplicity higher than n, none of these spans will be empty.

Now let $\mathcal{L}$ be one of the intervals defined by two successive and different knots. It is part of n spans, and will thus be engraved on n polygon legs. We call the corresponding control polygon $\mathbf{P}^{\mathcal{L}}$ and its control points $\mathbf{d}_i^{\mathcal{L}}$:

$$\mathbf{P}^{\mathrm{L}} = \mathbf{d}_0^{\mathrm{L}}, \ldots, \mathbf{d}_n^{\mathrm{L}}.$$

There is a restriction on $\mathcal{L}$: if it is "too close" to u_0 or to u_K, then there are fewer than n spans $\mathcal{L}_i^n$ containing it — such intervals will not be considered. The admissible intervals will be called *domain intervals.* They are $u_{n-1}, \ldots, u_{K-n+1}$.

Note that we can now write the whole control polygon $\mathbf{P}$ as

$$\mathbf{P} = \mathbf{P}_l \mathbf{P}^{\mathrm{L}} \mathbf{P}_r,$$

with suitably defined left and right subpolygons $\mathbf{P}_l$ and $\mathbf{P}_r$.

There are n spans containing $\mathcal{L}$. We denote them by $\mathcal{L}_i^n$, $i = 1, \ldots, n$. Each of these spans is mapped to a control polygon leg by an affine map Φ_i^n.

$$\Phi_i^n : \mathrm{L}_i^n \rightarrow \mathbf{d}_{i-1}^{\mathrm{L}} \mathbf{d}^{\mathrm{L}} L_i, \quad i = 1, \ldots, n.$$

Let $u \in \mathcal{L}$. Then each of the affine maps Φ_i^n takes u to a point on a control polygon leg, and we define

$$\mathbf{d}_i^1(u) = \Phi_i^n(u); \quad i = 1, \ldots, n. \tag{28.14}$$

We have augmented the knot sequence by one knot u, and we have augmented the control polygon $\mathbf{P}^{\mathcal{L}}$ to a new polygon

$$\mathbf{P}^{\mathrm{L}}[u] = \mathbf{d}_0 \mathbf{d}_1^1(u) \mathbf{d}_2^1(u) \ldots \mathbf{d}_n^1(u) \mathbf{d}_n.$$

We call this process **knot insertion,** after W. Boehm [1].

We utilize affine maps to describe knot insertion. Needless to say that we might have used linear interpolation as well: then we would write

$$\mathbf{d}_i^1(u) = \left(1 - t_i^1\right)\mathbf{d}_{i-1} + t_i^1 \mathbf{d}_i,$$

where t_i^1 is the local parameter in the span $\mathcal{L}_i^n$.

$$t_i^1 = \frac{u - l_i^1}{\left|\mathrm{L}_i^n\right|}.$$

Here, l_i^1 denotes the left endpoint of $\mathcal{L}_i^n$, and $|\mathcal{L}_i^n|$ denotes the length of $\mathcal{L}_i^n$.

The process of knot insertion may be repeated. However, after n steps, the process terminates, resulting in a point on the curve. This process is known as the **de Boor algorithm,** first described in [2]. If u is not already one of the knots, the intermediate de Boor points are found by

$$\mathbf{d}_i^r(u) \Phi_i^{n-r+1}(u); \quad r = 1, \ldots, n; \quad i = r, \ldots, n. \tag{28.15}$$

Here, we have set defined the affine maps $\Phi_i^{n-r+1}(u)$ by

$$\Phi_i^{n-r+1}(u) : \mathrm{L}_i^{n-r+1} \rightarrow \mathbf{d}_{i-1}^r \mathbf{d}_i^r$$

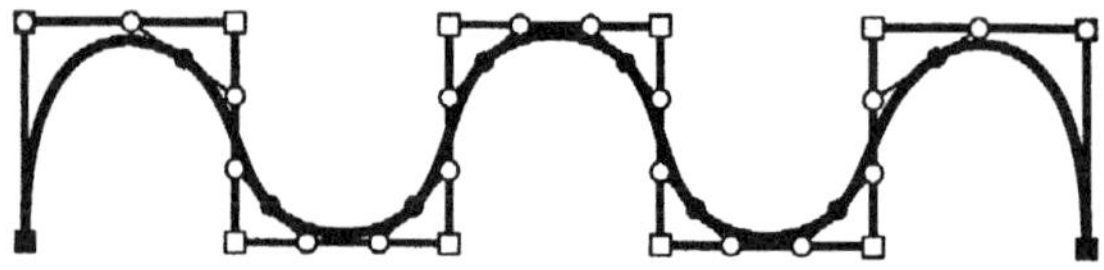

FIGURE 28.3 A cubic B-spline curve, its B-spline polygon (square marks), and the corresponding piecewise cubic Bézier polygon (circular marks).

and $\mathbf{d}_n^n$ is the desired result, i.e., the point on the curve.

Writing the involved affine maps as linear interpolants, we obtain

$$\mathbf{d}_i^r(u) = (1 - t_i^r)\mathbf{d}_{i-1}^{r-1} + t_i^r \mathbf{d}_i^{r-1}, \tag{28.16}$$

where t_i^r is the local parameter in the span $\mathcal{L}_i^{n-r+1}$:

$$t_i^r = \frac{u - l_i^r}{|\mathcal{L}_i^r|}.$$

Here, l_i^r denotes the left endpoint of $\mathcal{L}_i^{n-r+1}$, and $|\mathcal{L}_i^r|$ denotes the length of $\mathcal{L}_i^r$.

If at each level r of the de Boor algorithm, we use a different argument v_r instead of u, we obtain a function of n arguments: $\mathbf{d}^{\mathcal{L}}[v_1, \ldots, v_n]$. This is called the **blossom** of a B-spline curve after L. Ramshaw [17]. It is clear from the definition of a blossom that

$$\mathbf{d}_n^n(u) = \mathbf{d}^{\mathcal{L}}[u^{<n>}].$$

We have already encountered the spans $\mathcal{L}_i^r$. Let us denote the set of $r + 1$ knots in $\mathcal{L}_i^r$ by $\{\mathcal{L}_i^r\}$. The blossom $\mathbf{d}^{\mathcal{L}}$ is well-defined for arguments outside of $\mathcal{L}$. Thus expressions of the form $\mathbf{d}^{\mathcal{L}}[\{\mathcal{L}_i^{n-1}\}]$ formally make sense. In fact, they allow us to write the control points of a B-spline curve as blossom values:

$$\mathbf{d}_i^{\mathcal{L}} = \mathbf{d}^{\mathcal{L}}[\{\mathcal{L}_i^{n-1}\}]; \quad i = 0, \ldots, n. \tag{28.17}$$

The blossom $\mathbf{d}^{\mathcal{L}}$ may also be used to find the Bézier points $\mathbf{b}_i$ of the curve segment corresponding to the segment $\mathcal{L}$. Setting $\mathcal{L} = [u_-, u_+]$, we get

$$\mathbf{b}_i = \mathbf{d}^{\mathcal{L}}[u_-^{<n-i>}, u_+^{<i>}]; \quad i = 0, \ldots, n. \tag{28.18}$$

The simplicity of this formula is striking; in former days, involved papers were written on this conversion problem! That is not to say, however, that Eq. 28.18 is the most efficient way to solve the problem. But it does produce very readable code, which is equally important. A B-spline blossom routine can be obtained via anonymous ftp from enws102.cagd.eas.asu.edu in the directory pub/farin/floppy.

Figure 28.3 gives an example of a cubic B-spline curve.

While our definition of a B-spline curve was recursive, an explicit one also exists. It uses the **B-spline basis functions** $N_i^n(u)$, which are themselves defined recursively:

$$N_i^n(u) = (1 - t_{i+1})N_{i+1}^{n-1}(u) + t_i N_i^{n-1}(u). \tag{28.19}$$

Here, $u \in \mathcal{L}_i^n$ and t_i is the local parameter in $\mathcal{L}_i^n$. The recursion starts with

$$N_i^0(u) = \begin{cases} 1 & \text{if } u_{i-1} \le u < u_i \\ 0 & \text{else} \end{cases}.$$

Using these basis functions, a B-spline curve may be written as

$$\mathbf{d}^n(u) = \sum_{i=0}^{p} \mathbf{d}_i N_i^n(u). \tag{28.20}$$

The **derivative** of a B-spline curve is given by

$$\dot{\mathbf{x}}(u) = \frac{n}{|\mathcal{L}|}\left(\mathbf{d}_n^{n-1}(u) - \mathbf{d}_{n-1}^{n-1}(u)\right).$$

Using blossom notation, this becomes

$$\dot{\mathbf{x}}(u) = n\mathbf{d}^{\mathcal{L}}\left[\vec{1}, u^{<n-1>}\right],$$

where $\vec{1}$ denotes the unit vector on the real line. Written in terms of the B-spline basis, this is

$$\dot{\mathbf{x}}(u) = n\sum_{i=1}^{n} \frac{\Delta \mathbf{d}_{i-1}}{|\mathcal{L}_i^n|} N_i^{n-1}(u),$$

where the $N_i^{n-1}(u)$ are numbered relative to the interval $\mathcal{L}$. Higher derivatives are expressed (and computed!) more easily using just the blossom form:

$$\mathbf{x}^{(r)}(u) = \frac{n!}{(n-r)!}\mathbf{d}^{\mathcal{L}}\left[\vec{1}^r, u^{<n-r>}\right].$$

An implementation remark: the above development only uses knots up to multiplicity n. The most common data format, IGES (initial graphics exchange specification), uses knots of multiplicity $n + 1$ at the domain endpoints. This is not necessary, but it may be important to be aware of. Also, IGES enters multiple knots into the knot sequence as often as their multiplicity implies. It is cleaner programming style to list all knots only once and to keep track of their multiplicities in a separate array. For a particular operation, the knot sequence can then be "expanded."

28.6 Cubic Interpolation and Approximation

An important application of B-spline curves is for interpolation of data points; the cubic case is most often encountered. Here, we are given L data points $\mathbf{x}_i$ and corresponding parameter values u_i. We wish to find a cubic control polygon $\mathbf{d}_0, \ldots, \mathbf{d}_{L+2}$ such that the corresponding B-spline curve passes through the data:

$$\mathbf{x}_i = \sum_{j=0}^{L+1} \mathbf{d}_j N_j^3(u_i); \quad i = 0, \ldots, L.$$

These are $L + 1$ equations for the $L + 3$ unknowns $\mathbf{d}_j$. The common approach is to add two more equations, corresponding to derivative information at the endpoints. The coefficient matrix is obtained by evaluating the B-spline basis functions at the given knots. Since each N_i^3 is nonzero for only three subsequent knots, the matrix is tridiagonal.

If we prescribe the two end derivatives, this amounts to selecting the Bézier points $\mathbf{b}_1$ and $\mathbf{b}_{3L-1}$. We then obtain a linear system of the form

$$\begin{bmatrix} 1 & & & & & \\ \alpha_1 & \beta_1 & \gamma_1 & & & \\ & & \ddots & & & \\ & & & \alpha_{L-1} & \beta_{L-1} & \gamma_{L-1} \\ & & & & & 1 \end{bmatrix} \begin{bmatrix} \mathbf{d}_1 \\ \mathbf{d}_2 \\ \vdots \\ \mathbf{d}_L \\ \mathbf{d}_{L+1} \end{bmatrix} = \begin{bmatrix} \mathbf{r}_0 \\ \mathbf{r}_1 \\ \vdots \\ \mathbf{r}_{L-1} \\ \mathbf{r}_L \end{bmatrix} \tag{28.21}$$

Here we set

$$\mathbf{r}_0 = \mathbf{b}_{1,}$$
$$\mathbf{r}_i = (\Delta_{i-1} + \Delta_i)\mathbf{x}_i,$$
$$\mathbf{r}_L = \mathbf{b}_{3L-1}.$$

The first and last polygon vertices do not cause much of a problem:

$$\mathbf{d}_0 = \mathbf{x}_0, \qquad \mathbf{d}_{L+2} = \mathbf{x}_L.$$

This linear system can be made *symmetric:* we can multiply each equation by a common factor. In particular, we can divide the ith equation through by $\Delta_{i-1}^2 \; \Delta_i^2$. Also, we would have to delete the first and last rows and columns from the system, and update the right-hand side accordingly. The resulting new matrix will now be symmetric; its entries will satisfy $\alpha_{i+1} = \gamma_i$.

If more data points are given than the expected number of spline segments, then **spline approximation** is called for. The most common form is least squares approximation, and it is described now. We are given data points $\mathbf{p}_i$ with $i = 0, \ldots, P$. We wish to find an approximating B-spline curve $\mathbf{p}(u)$ of degree n with L domain knots, i.e., with a knot sequence $u_0, \ldots, u_{L+2n-2}$. We want the curve to be close to the data points in the following sense. Suppose the data point $\mathbf{p}_i$ is associated with a data parameter value w_i*. Then we would like the distance $\|\mathbf{p}_i - \mathbf{p}(w_i)\|$ to be small. Attempting to minimize all such distances then amounts to

$$\text{minimize} \sum_{i=0}^{P} \left\| \mathbf{p}_i - \mathbf{p}(w_i) \right\|^2 \tag{28.22}$$

The squared distances are introduced to simplify our subsequent computations. We shall minimize Eq. 28.22 by finding suitable B-spline control vertices $\mathbf{d}_j$:

$$\text{minimize } f(\mathbf{d}_{0,\ldots,}\mathbf{d}_{L+n-1}) = \sum_{i=0}^{P} \left\| \mathbf{p}_i - \sum_{j=0}^{L+n-1} \mathbf{d}_j N_j^n(w_i) \right\|^2 \tag{28.23}$$

*Note that w_i does not have to be one of the knots.

Thus f is a quadratic form with $L + n$ independent variables $\mathbf{d}_j$. Such functions only have one minimum, and at its location, the partials with respect to the $\mathbf{d}_j$ must vanish: $\partial f/\partial \mathbf{d}_k = \mathbf{0}$*. Thus:

$$\mathbf{0} = \sum_{i=0}^{P}\left[\mathbf{p}_i - \sum_{j=0}^{L+n-1}\mathbf{d}_j N_j^n(w_i)\right]N_k^n(w_i); \quad k = 0,\ldots,L+n-1$$

or

$$\sum_{j=0}^{L+n-1}\mathbf{d}_j\sum_{i=0}^{P}N_j^n(w_i)N_k^n(w_i) = \sum_{i=0}^{P}\mathbf{p}_i N_k^n(w_i); \quad k = 0,\ldots,L+n-1 \tag{28.24}$$

This is a linear system of $L + n$ equations for the unknowns $\mathbf{d}_k$, with a coefficient matrix M whose elements $m_{j,k}$ are given by

$$m_{j,k} = \sum_{i=0}^{P}N_j^n(w_i)N_k^n(w_i); \quad 0 \le j,k \le L+n.$$

These equations are usually called *normal equations*. The symmetric matrix M, although containing many zero entries, is often ill-conditioned — special equation solvers, such as a Cholesky decomposition, should be employed. For more details on the numerical treatment of least squares problems, see [11] or [14].

The matrix M is nonsingular in all "standard" cases. It is obviously singular if the number of data points $P + 1$ is less than the number of domain knots $L + n + 1$. It is also singular if there is a span $[u_{j-1}, u_{j+n}]$ that contains no w_i. In that case, the basis function N_j^n would evaluate to zero for all w_i, resulting in a row of zeroes for M.

We have so far assumed much more than would be available in a practical situation. First, what should the degree n be? In most cases, $n = 3$ is a reasonable choice. The knot sequence poses a more serious problem.

Recall that the data points are typically given without assigned data parameter values w_i. The centripetal parametrization [15] will give reasonable estimates, provided that there is not too much noise in the data. But how many knots u_j shall we use, and what values should they receive? A universal answer to this question does not exist — it will invariably depend on the application at hand. For example, if the data points come from a lesser digitizer, there will be vastly more data points $\mathbf{p}_i$ than knots u_i.

After the curve $\mathbf{p}(u)$ has been computed, we will find that many distance vectors $\mathbf{p}_i - \mathbf{p}(w_i)$ are not perpendicular to $\dot{\mathbf{p}}(w_i)$. This means that the point $\mathbf{p}(w_i)$ on the curve is not the closest point to $\mathbf{p}_i$, and thus $\|\mathbf{p}_i - \mathbf{p}(w_i)\|$ does not measure the distance of $\mathbf{p}_i$ to the curve. This indicates that we could have chosen a better data parameter value w_i corresponding to $\mathbf{p}_i$. We may improve our estimate for w_i by finding the closest point to $\mathbf{p}_i$ on the computed curve and assigning its parameter value $\hat{w}_i$ to $\mathbf{p}_i$. We do this for all i and then recompute the least squares curve with the new $\hat{w}_i$. This process typically converges after three or four iterations. It was named *parameter correction* by J. Hoschek [12].

The new parameter value $\hat{w}_i$ is found using a Newton iteration. We project $\mathbf{p}_i$ onto the tangent at $\mathbf{p}(w_i)$, yielding a point $\mathbf{q}_i$. Then the ratio of the lengths $\|\mathbf{q}_i - \mathbf{p}_i\|/\|\dot{\mathbf{p}}(w_i)\|$ is a measure for the adjustment of w_i. The actual Newton iteration step looks like this:

$$\hat{w}_i = w_i + \left[\mathbf{p}_i - \mathbf{p}(w_i)\right]\frac{\dot{\mathbf{p}}(w_i)}{\|\dot{\mathbf{p}}(w_i)\|}\frac{\Delta u_k}{s_k}. \tag{28.25}$$

*This is shorthand for taking the partials for each of $\mathbf{d}_k$'s components.

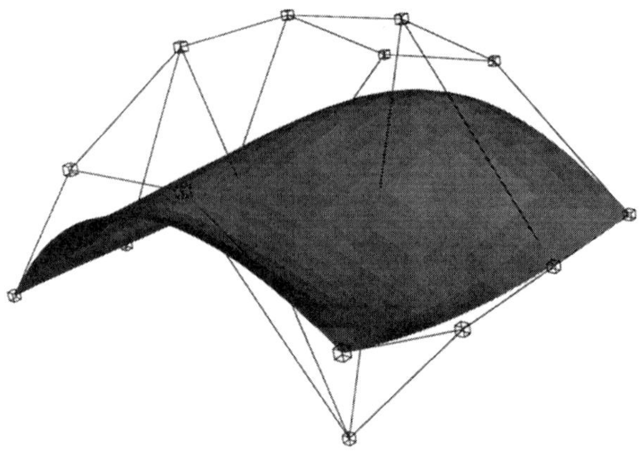

FIGURE 28.4 A rectangular Bézier patch with $m = n - 3$.

In this equation, s_k denotes the arc length of the segment that w_i is in, i.e., $u_k < w_i < u_{k+1}$. This length can safely (and cheaply) be overestimated by the length of the Bézier polygon of the kth segment.*

We finally note that Eq. 28.25 should not be used to compute the point on a curve closest to an arbitrary point $\mathbf{p}_i$. It only works if $\mathbf{p}_i$ is already close to the curve, and if a good estimate w_i is known for the closest point on the curve.

28.7 Bézier Patches

There are two kinds of Bézier patches: rectangular and triangular.

A **rectangular** Bézier **patch** is defined by a rectangular array of control points $\mathbf{b}_{i,j}$; $0 \le i \le m$, $0 \le j \le n$. For the case $m = 3$, $n = 3$, it would look like

$$\begin{matrix} \mathbf{b}_{0,0} & \mathbf{b}_{0,1} & \mathbf{b}_{0,2} & \mathbf{b}_{0,3} \\ \mathbf{b}_{1,0} & \mathbf{b}_{1,1} & \mathbf{b}_{1,2} & \mathbf{b}_{1,3} \\ \mathbf{b}_{2,0} & \mathbf{b}_{2,1} & \mathbf{b}_{2,2} & \mathbf{b}_{2,3} \\ \mathbf{b}_{3,0} & \mathbf{b}_{3,1} & \mathbf{b}_{3,2} & \mathbf{b}_{3,3} \end{matrix}$$

Figure 28.4 gives an example.

A point $\mathbf{b}^{m,n}(u, v)$ on such a surface is given by

$$\mathbf{b}^{m,n}(u,v) = \sum_{i=0}^{m} \sum_{j=0}^{n} \mathbf{b}_{i,j} B_i^m(u) B_j^n(v). \tag{28.26}$$

This surface is a map of the domain $0 \le u, v \le 1$. Its actual degree is $m + n$, since the highest powers in u and v appear in the term $u^m v^n$.

*Hoschek's original development uses $u_{L+n-1} - u_{n-1}$ instead of Δu_k and the length of the total curve instead of s_k. Our formula is cheaper.

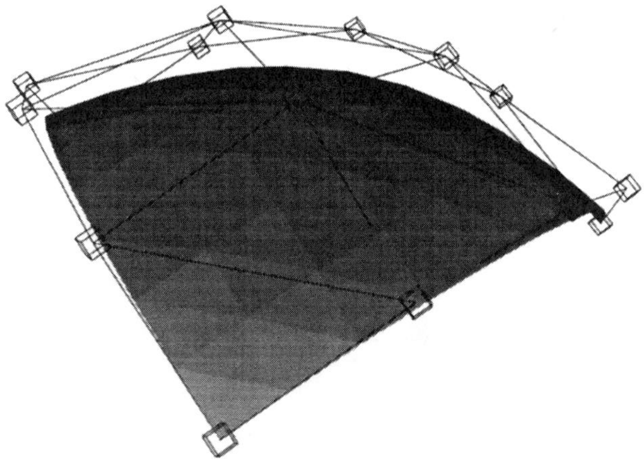

FIGURE 28.5 A triangular Bézier patch with $n = 3$.

The control points of a **triangular patch** are usually given three subscripts, in the example of a quartic patch, this would look like

$$\begin{array}{c} \mathbf{b}_{040} \\ \mathbf{b}_{031}\mathbf{b}_{130} \\ \mathbf{b}_{022}\mathbf{b}_{121}\mathbf{b}_{220} \\ \mathbf{b}_{013}\mathbf{b}_{112}\mathbf{b}_{211}\mathbf{b}_{310} \\ \mathbf{b}_{004}\mathbf{b}_{103}\mathbf{b}_{202}\mathbf{b}_{301}\mathbf{b}_{400} \end{array}$$

Figure 28.5 gives an example.

A point $\mathbf{b}^n(\mathbf{u})$ on the patch is defined by

$$\mathbf{b}^n(\mathbf{u}) = \mathbf{b}_0^n(\mathbf{u}) = \sum_{|\mathbf{j}|=n} \mathbf{b}_{\mathbf{j}} B_{\mathbf{j}}^n(\mathbf{u}) \tag{28.27}$$

where

$$B_{\mathbf{j}}^n(\mathbf{u}) = \frac{n!}{i!j!k!} u^i v^j w^k.$$

Here, $\mathbf{u} = (u, v, w)$ are barycentric coordinates in a domain triangle, implying that $u + v + w = 1$. The actual shape of this domain triangle is immaterial as barycentric coordinates are preserved under affine maps. See Section 28.2 for details.

A de Casteljau algorithm is also defined for this patch type; it is given by

$$\mathbf{b}_{\mathbf{i}}^r(\mathbf{u}) = u\mathbf{b}_{\mathbf{i+e1}}^{r-1}(\mathbf{u}) + v\mathbf{b}_{\mathbf{i+e2}}^{r-1}(\mathbf{u}) + w\mathbf{b}_{\mathbf{i+e3}}^{r-1}(\mathbf{u}), \tag{28.28}$$

where*

$$r = 1, \ldots, n \quad \textit{and} \quad |\mathbf{i}| = n - r$$

*We use the abbreviations **e1** = (1, 0, 0), **e2** = (0, 1, 0), **e3** = (0, 0, 1), and $|\mathbf{i}| = i + j + k$. When we say $|\mathbf{i}| = n$, we mean $i + j + k = n$, always assuming $i, j, k \geq 0$.

And $\mathbf{b}_\mathbf{i}^0(\mathbf{u}) = \mathbf{b}_\mathbf{i}$. Then $\mathbf{b}_\mathbf{0}^n(\mathbf{u})$ is the point on the triangular patch with parameter value $\mathbf{u}$.

For a rectangular patch, the *u*–**partial** is given by

$$\frac{\partial}{\partial u}\mathbf{b}^{m,n}(u,v) = m\sum_{j=0}^{n}\sum_{i=0}^{m-1}\Delta_{1,0}\mathbf{b}_{i,j}B_i^{m-1}(u)B_j^n(v),$$

where $\Delta_{1,0}\mathbf{b}_{i,j} = \mathbf{b}_{i+1,j} - \mathbf{b}_{i,j}$. Higher derivatives are found by repeated application of this formula:

$$\frac{\partial^r}{\partial u^r}\mathbf{b}^{m,n}(u,v) = \frac{m!}{(m-r)!}\sum_{j=0}^{n}\sum_{i=0}^{m-r}\Delta_{1,0}^r\mathbf{b}_{i,j}B_i^{m-r}(u)B_j^n(v),$$

For **derivatives of triangular patches** the notion of partials is not useful; instead, one uses *directional derivatives.* These are taken along a direction $\mathbf{d}$ defined by the difference of two points in barycentric coordinates: $\mathbf{d} = \mathbf{u}_1 - \mathbf{u}_2$. We obtain

$$D_\mathbf{d}\mathbf{b}(\mathbf{u}) = n\sum_{|\mathbf{i}|=n-1}\mathbf{b}_\mathbf{i}^1(\mathbf{d})B_\mathbf{i}^{n-1}(\mathbf{u}),$$

where the $\mathbf{b}_\mathbf{i}^1(\mathbf{d})$ are computed according to Eq. 28.28. Higher derivatives are then given by

$$D_\mathbf{d}^r\mathbf{b}(\mathbf{u}) = \frac{n!}{(n-r)!}\sum_{|\mathbf{i}|=n-r}\mathbf{b}_\mathbf{i}^r(\mathbf{d})B_\mathbf{i}^{n-r}(\mathbf{u}),$$

28.8 Composite Surfaces

As for curves, a single patch is not flexible enough to describe complex shapes. For curves, the B-spline approach works best; it is not very different in the surface case. Just as a rectangular Bézier patch is the tensor product of univariate schemes, a B-spline surface can be written as a tensor product:

$$\mathbf{x}(u,v) = \sum_i\sum_j \mathbf{d}_{ij}N_i^m(u)N_j^n(v), \tag{28.29}$$

where the subscripts run according to the curve case.

An important application is that of *bicubic spline interpolation.* Here, $m = n = 3$, and the problem is as follows: Suppose we have $(K + 1) \times (L + 1)$ data points $\mathbf{x}_{IJ}$ and two knot sequences $u_0, \ldots, u_K$ and $v_0, \ldots, v_L$. For each row of data points, we prescribe two end conditions (e.g., by specifying tangent vectors or Bézier points) and solve the univariate B-spline interpolation problem as described above. As all these interpolation problems use the same tridiagonal coefficient matrix, an $L - U$ decomposition should be performed before the row-by-row loop is entered. We thus produce the elements of an intermediate matrix $\mathbf{D}$.

We now take every column of $\mathbf{D}$ and perform univariate B-spline interpolation on it, again by prescribing end conditions such as clamped end tangents or Bessel tangents. The resulting control points constitute the desired B-spline control net.

The final B-spline control net has two more rows and columns than $\mathbf{X}$.* This is duc to the end conditions; to resolve the apparent discrepancy, we can think of $\mathbf{X}$ as having two additional rows and columns that constitute the end condition data.

*This is inherited from the curve case: there one gets $L + 2$ control points for L data points.

If the data points are not organized in this way, the above approach is not applicable. In case of many points $\mathbf{p}_0, \ldots, \mathbf{p}_L$ without any structure, one can resort to **tensor product least squares approximation.** We assume that we are given knot sequences $u_0, \ldots, u_P$ and $v_0, \ldots, v_Q$.* We also assume that each data point $\mathbf{p}_i$ is assigned a parameter value (s_i, t_i). Then every data point $\mathbf{p}_i$ is associated with a surface point $\mathbf{x}(s_i, t_i)$. Our objective is to minimize all distance squares

$$\sum_{i=0}^{L} \left\| \mathbf{p}_i - \mathbf{x}(s_i, t_i) \right\|^2$$

or

$$\sum_{i=0}^{L} \left\| \mathbf{p}_i - \sum_{c} \sum_{d} \mathbf{d}_{cd} N_c^m(s_i) N_d^n(t_i) \right\|^2 .$$

We use the same approach as we did for curves: interpret the above as a multivariate function of the unknowns $\mathbf{d}_{c,d}$, take partials with respect to each of them, and equate to zero.

This leads to

$$\sum_{i=0}^{L} \mathbf{p}_i \sum_{c} \sum_{d} N_c^m(s_i) N_d^n(t_i) = \sum_{i=0}^{L} \sum_{c} \sum_{d} \mathbf{d}_{cd} \sum_{e} \sum_{f} N_c^m(s_i) N_d^n(t_i) N_e^m(s_i) N_f^n(t_i); c = 0, \ldots, P - m,$$

We have $R = (P - m)(Q - n)$ equations in the same number of unknowns. Yet the structure of a linear system is not clearly discernible. We achieve this by *linearizing:* instead of using two subscripts for counting in our rectangular arrays, we will just use one. Thus $c(\rho)$ and $d(\rho)$ will give the row and column values (c, d) for array element with number ρ and $e(\sigma)$ and $f(\sigma)$ will give the row and column values (e, f) for array element with number σ. We now obtain

$$\sum_{i=0}^{L} \mathbf{p}_i N_{c(\rho)}^m(s_i) N_{d(\rho)}^n(t_i) = \sum_{\rho=0}^{R} \mathbf{d}_\rho \sum_{i=0}^{L} N_{c(\rho)}^m(s_i) N_{d(\rho)}^n(t_i) N_{e(\sigma)}^m(s_i) N_{f(\sigma)}^n(t_i); \sigma = \ldots, R.$$

This has the form of a linear system; the coefficient matrix has elements

$$\alpha_{\rho,\sigma} = \sum_{i=0}^{L} N_{c(\rho)}^m(s_i) N_{d(\rho)}^n(t_i) N_{e(\sigma)}^m(s_i) N_{f(\sigma)}^n(t_i).$$

Figure 28.6 gives an example.

28.9 Rational Curves and Surfaces — NURBS

Some of the most basic curves in this context are the **conic sections.** Comprising the familiar ellipses, parabolas, and hyperbolas, they are all defined by five points in the plane. If the points have coordinates $(x_1, y_1), \ldots, (x_5, y_5)$ the **implicit form** of the conic through them is given by

*In practice, some heuristic is typically necessary to find these.

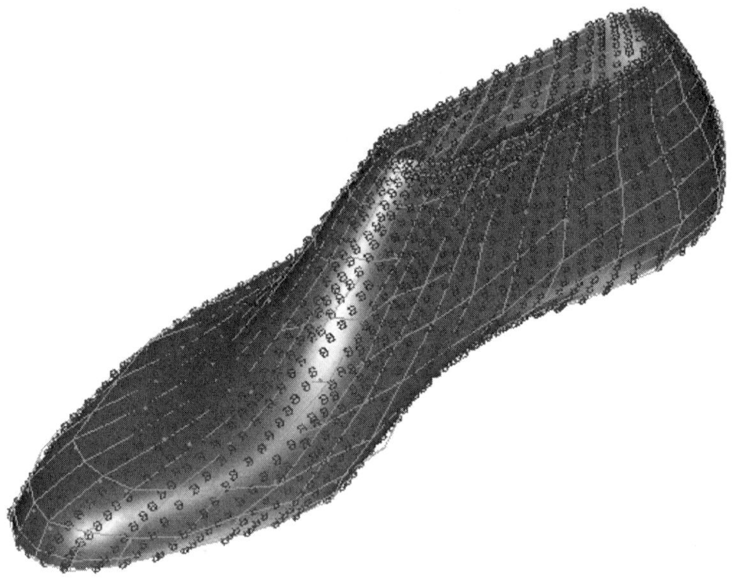

FIGURE 28.6 B-spline approximation: the given data points are marked by square boxes; the control net is superimposed in light gray.

$$f(x,y)=\begin{vmatrix} x^2 & xy & y^2 & x & y & 1 \\ x_1^2 & x_1y_1 & y_1^2 & x_1 & y_1 & 1 \\ x_2^2 & x_2y_2 & y_2^2 & x_2 & y_2 & 1 \\ x_3^2 & x_3y_3 & y_3^2 & x_3 & y_3 & 1 \\ x_4^2 & x_4y_4 & y_4^2 & x_4 & y_4 & 1 \\ x_5^2 & x_5y_5 & y_5^2 & x_5 & y_5 & 1 \end{vmatrix}=0.$$

The implicit form is important when dealing with the IGES data specification. In that data format, a conic is given by its implicit form $f(x, y) = 0$ and two points on it, implying a start and end point $\mathbf{b}_0$ and $\mathbf{b}_2$ of a conic arc. Many applications, however, need the **rational quadratic** form. Now a conic looks like this:

$$\mathbf{x}(t)=\frac{w_0\mathbf{b}_0B_0^2(t)+w_1\mathbf{b}_1B_1^2(t)+w_2\mathbf{b}_2B_2^2(t)}{w_0B_0^2(t)+w_1B_1^2(t)+w_2B_2^2(t)};\quad \mathbf{x}(t),\mathbf{b}_i\in IE^3. \tag{28.30}$$

We call the w_i *weights* and the $\mathbf{b}_i$ the *control polygon.* Without loss of generality, we can assume $w_0 = w_2 = 1$; this is called *standard form* of a conic.

To convert the implicit IGES format to this form, we have to determine $\mathbf{b}_1$ and its weight w_1. First, we find tangents at $\mathbf{b}_0$ and $\mathbf{b}_2$: we know that the gradient of f is a vector that is perpendicular to the conic. The gradient at $\mathbf{b}_0$ is given by f's partials: $\nabla f(\mathbf{b}_0) = [f_x(\mathbf{b}_0), f_y(\mathbf{b}_0)]^{\mathrm{T}}$. The tangent is perpendicular to the gradient and thus has direction $\nabla^{\perp} f(\mathbf{b}_0) = [-f_y(\mathbf{b}_0), f_x(\mathbf{b}_0)]^{\mathrm{T}}$. Thus our tangents are given by

$$\begin{aligned}\mathbf{t}_0(t) &= \mathbf{b}_0+t\nabla^{\perp}f(\mathbf{b}_0) \quad \textbf{and}\\ \mathbf{t}_2(s) &= \mathbf{b}_2+s\nabla^{\perp}f(\mathbf{b}_2).\end{aligned}$$

Their intersection determines $\mathbf{b}_1$. Next, we compute the midpoint $\mathbf{m}$ of $\mathbf{b}_0$ and $\mathbf{b}_2$. Then the line $\overline{\mathbf{mb}_1}$ will intersect our conic in the shoulder point $\mathbf{s}$. This requires the solution of a quadratic equation,* and we find out desired weight w_1 from the relationship

$$w_1 = \text{ratio}(\mathbf{m}, \mathbf{s}, \mathbf{b}_1).$$

If the input is not well defined — imagine $\mathbf{b}_0$ and $\mathbf{b}_2$ being on two different branches of a hyperbola! — then the above quadratic equation may have complex solutions. An error flag would be appropriate here. If the arc between $\mathbf{b}_0$ and $\mathbf{b}_2$ subtends an angle larger than, say, 120°, it should be subdivided. For more details, see [18].

If the control polygon of a conic forms an isosceles triangle, and the weights are given by w_0, w_1, w_2 = 1, $\cos\alpha$, 1, where the angle alpha is given by $\alpha = \angle\, \mathbf{b}_2\mathbf{b}_0\mathbf{b}_1$, then that conic is a circular arc. Note that $\cos\alpha \le 1$.

In general, if the weights of a conic are 1, w, 1, with $w > 0$, then for $w < 1$ the conic is an ellipse, for $w = 1$ it is a parabola, and for $w > 1$ it is a hyperbola.

References for this topic: [3], [4], [5], [9], [16].

A **rational Bézier curve** is defined by

$$\mathbf{x}(t) = \frac{w_0\mathbf{b}_0B_0^n(t) + \cdots + w_n\mathbf{b}_nB_n^n(t)}{w_0B_0^n(t) + \cdots + w_nB_n^n(t)}; \quad \mathbf{x}(t), \mathbf{b}_i \in IE^3. \tag{28.31}$$

The w_i are called *weights;* the $\mathbf{b}_i$ form the control polygon. It is the projection of the 4D control polygon $[w_i\mathbf{b}_i\ w_i]^T$ of a nonrational 4D curve.

If all weights equal one, we obtain the standard nonrational Bézier curve, since the denominator is identically equal to one.** If some w_i are negative, singularities may occur; we will therefore deal only with nonnegative w_i.

For the first **derivative** of a rational Bézier curve, we obtain

$$\dot{\mathbf{x}}(t) = \frac{1}{w(t)}\left[\dot{\mathbf{p}}(t) - \dot{w}(t)\mathbf{x}(t)\right], \tag{28.32}$$

where we have set

$$\mathbf{p}(t) = w(t)\mathbf{x}(t); \quad \mathbf{p}(t), \mathbf{x}(t) \in IE^3.$$

At the endpoint $t = 0$, we find

$$\dot{\mathbf{x}}(0) = \frac{nw_1}{w_0}\Delta\mathbf{b}_0.$$

For higher derivatives, some computation yields

$$\mathbf{x}^{(r)}(t) = \frac{1}{w(t)}\left[\mathbf{p}^{(r)} - \sum_{j=1}^{r}\binom{r}{j}w^{(j)}(t)\mathbf{x}^{(r-j)}(t)\right]. \tag{28.33}$$

---

* The quadratic equation will in general have two solutions. We take the one insider the triangle $\mathbf{b}_0$, $\mathbf{b}_1$, $\mathbf{b}_2$.

**This is also true if the weights are not unity, but are equal to each other — a common factor does not matter.

This is a recursive formula for the rth derivative of a rational Bézier curve. It only involves taking derivatives of polynomial curves.

The first derivative can also be obtained as a byproduct of the de Casteljau algorithm, as described by Floater [10]:

$$\dot{\mathbf{x}}(0) = n\frac{w_0^{n-1}w_1^{n-1}}{\left[w_0^n\right]^2}\left[\mathbf{b}_1^{n-1} - \mathbf{b}_0^{n-1}\right]. \tag{28.34}$$

If B-spline curves are made rational, they are called **NURBS** for nonuniform rational B-splines.* A NURB curve **s** is defined by

$$\mathbf{s}(u) = \frac{\Sigma_{j=0}^{P} w_i \mathbf{d}_i N_i^n(u)}{\Sigma_{j=0}^{P} w_i N_i^n(u)}. \tag{28.35}$$

A rational B-spline curve is given by its knot sequence, its 3D control polygon, and its weight sequence. The control vertices $\mathbf{d}_i$ are the projections of the 4D control vertices $[w_i\mathbf{d}_i \;\; w_i]^{\mathrm{T}}$.

To evaluate a rational B-spline curve at a parameter value u, we may apply the de Boor algorithm to both numerator and denominator of Eq. 28.35 and finally divide through. This corresponds to the evaluation of the 4D nonrational curve with control vertices $[w_i\mathbf{d}_i \;\; w_i]^{\mathrm{T}}$ and to projecting the result into $\mathbb{E}^3$.

Rational surfaces are defined in the same way. The most widely used type, a NURB surface, is defined by

$$\mathbf{s}(u,v) = \frac{\Sigma_i\Sigma_j w_{i,j}\mathbf{d}_{i,j}N_i^m(u)N_j^n(v)}{\Sigma_i\Sigma_j w_{i,j}N_i^m(u)N_j^n(v)}. \tag{28.36}$$

It is one of the most general data formats available. NURB surfaces encompass Bézier patches, both rational and nonrational; quadrics, including spheres; surfaces of revolutions, including tori; and more. As an example, we show how to produce a surface of revolution. These are given by

$$\mathbf{x}(u,v) = \begin{bmatrix} r(v)\cos u \\ r(v)\sin u \\ z(v) \end{bmatrix}.$$

For fixed v, an isoparametric line $v = const.$ traces out a circle of radius $r(v)$, called a *meridian*. Since a circle may be exactly represented by rational quadratic arcs, we may find an exact rational representation of a surface of revolution provided we can represent $r(v)$, $z(v)$ in rational form. Figure 28.7 gives an example.

The most convenient way to define a surface of revolution is to prescribe the (planar) generating curve, or generatrix, given by

$$\mathbf{g}(v) = [r(v), 0, z(v)]^{\mathrm{T}}$$

and by the axis of revolution, in the same plane as **g**. Suppose **g** is given by its control polygon, knot sequence, and weight sequence. We can construct a surface of revolution such that each meridian consists

*A misnomer, since B-splines are defined over a nonuniform knot sequence to begin with.

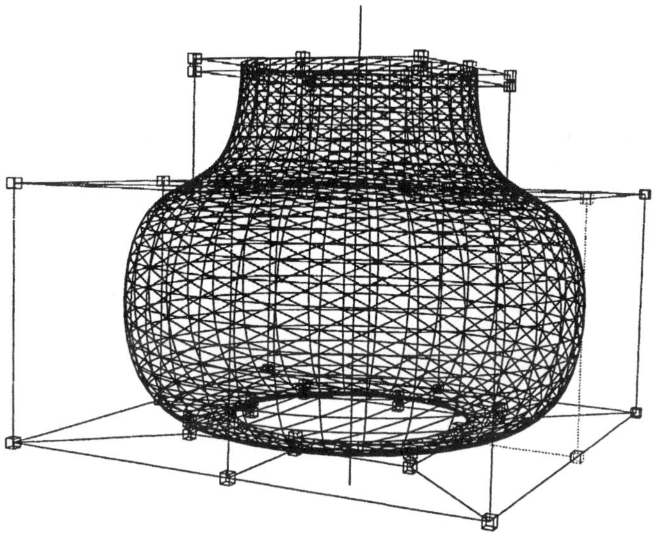

FIGURE 28.7 A NURBS representation of a surface of revolution.

of four rational quadratic arcs. The resulting four control nets then form three concentric squares in the projection into the $z = 0$ plane. The control points at the squares' midpoints are copies of the generatrix control points; their weights are those of the generatrix. The remaining weights, corresponding to the squares' corners, are multiplied by $\cos(45°) = (\sqrt{2})/2$.

Note that although the generatrix can be defined over a knot sequence $\{v_j\}$ with only simple knots, this is not possible for the knots of the meridian circles; we have to use double knots, thereby essentially reducing it to the piecewise Bézier form.

References

1. Boehm, W., Inserting new knots into B-Spline curves, *Computer Aided Design.* 1980, 12, 4, pp. 199–201.
2. de Boor, C., On calculating with B-splines, *J. Approx. Theory.* 1972, 6, 1, pp. 50–62.
3. Farin, G., (Ed.), *NURBS for Curve and Surface Design.* SIAM, Philadelphia, 1991.
4. Farin, G., From conics to NURBS: a tutorial and survey, *IEEE Computer Graphics and Applications.* 1992, 12, 5, pp. 78–86.
5. Farin, G., *NURB Curves and Surfaces.* Peters, A.K., Boston, 1995.
6. Farin, G., *Curves and Surfaces for Computer Aided Geometric Design,* Fourth Edition. Academic Press, 1996.
7. Farouki, R. and Rajan, V., On the numerical condition of polynomials in Bernstein form. *Computer Aided Geometric Design.* 1987, 4, 3, pp. 191–216.
8. Faux, I. and Pratt, M., *Computational Geometry for Design and Manufacture.* Ellis Horwood, 1979.
9. Fiorot, J. and Jeannin, P., *Rational Curves and Surfaces.* Wiley, Chicester, 1992. Translated from the French by Harrison, M.
10. Floater, M., Derivatives of rational Bézier curves. *Computer Aided Geometric Design.* 1993, 10.
11. Hayes, J. and Holladay, J., The least-squares fitting of cubic splines to general data sets. *J. Inst. Maths. Applics.* 1974, 14, pp. 89–103.
12. Hoschek, J., Intrinsic parametrization for approximation, *Computer Aided Geometric Design.* 1988, 5, pp. 27–31.
13. Hoschek, J. and Lasser, D., *Grundlagen der Geometrischen Datenverarbeitung.* Teubner, B.G., Stuttgart, 1989. English translation: *Fundamentals of Computer Aided Geometric Design.* Peters, A.K., 1993.

14. Lawson, C. and Hanson, G., *Solving Least Squares Problems.* SIAM, 1995.
15. Lee, E., Choosing nodes in parametric curve interpolation, *Computer Aided Design.* 21, 6, 1989. Presented at the SIAM Applied Geometry meeting, Albany, NY, 1987.
16. Piegl, L. and Tiller, W., *The Book of NURBS.* Springer Verlag, 1995.
17. Ramshaw, L., Blossoming: a connect-the-dots approach to splines, technical report, Digital Systems Research Center, Palo Alto, CA, 1987.
18. Worsey, A. and Farin, G., Contouring a bivariate quadratic polynomial over a triangle, *Computer Aided Geometric Design.* 1990, 7(1–4), pp. 337–352.
19. Yamaguchi, F., *Curves and Surfaces in Computer Aided Geometric Design.* Springer, 1988.

29

Computer-Aided Geometric Design Techniques for Surface Grid Generation

Bernd Hamann

Brian A. Jean

Anshuman Razdan

29.1 Introduction

This chapter focuses on three computer-aided geometric design (CAGD) techniques that are often needed to "prepare" a complex geometry for grid generation. Standard grid generation methods, as discussed in [George 1991], [Knupp and Steinberg 1993], and [Thompson et al., 1985], assume that parametric surfaces are well parametrized and free of undesired discontinuities.We describe CAGD techniques that are extremely helpful for the preparation of complex geometries for the grid generation process.

0-8493-2687-7/99/$0.00+$.50

29.1.1 Surface Refinement and Reparametrization

One problem that has plagued most grid generation systems is that poorly parametrized surfaces create a poorly distributed grid. This is due to the fact that the grid distributions are performed in the parametric domain and then mapped back to physical space. It is desirable that the parametrization reflect the geometry of the surface in physical space, i.e., the parametrization should mimic the surface in physical space. The more a distribution of points in parametric space resembles the corresponding distribution of points in physical space, the better the parametrization is. Imagine using a uniform parametrization on a sample of points that are not distributed evenly; if one distributes points evenly in parameter space, they are not uniform in physical space. There are ways to achieve a uniform distribution in physical space, but most approaches are based on iterative procedures. To eliminate the need for such iteration procedures, a *chord length* parametrization can be used — it best represents the surface in physical space.

29.1.2 Approximation of Discontinuous Geometries

Grid generation is concerned with discretizing surfaces and surrounding volumes in three-dimensional (3D) space — in this context, it is important that a given geometry is continuous. Before one can generate grids for geometries containing discontinuities, one must approximate the geometry by a set of surface patches which are continuous. The most common problems are gaps between neighbor patches, surfaces with undesired intersections, and overlapping surfaces. We have developed an interactive technique that can be used to approximate a faulty, i.e., discontinuous, geometry by a continuous one. One obtains approximating surface patches by projecting local approximants, *Coons patches*, onto the discontinuous geometry, see [Coons 1974]. Each approximating surface patch is constructed by specifying four boundary curves, computing a Coons patch interpolating the four curves, and projecting the Coons patch onto the given geometry. In the end, one has replaced the entire geometry by a new set of continuous surface patches.

29.1.3 Surface–Surface Intersection

Accurate computation of surface–surface intersection (SSI) curves is essential in many engineering applications including numerical grid generation. SSI curves represent important features that must be captured by the grid. These curves are typically used to trim surfaces; for example, the location where an airplane wing meets the fuselage would be given in terms of the intersection of the wing and the fuselage. Often, geometries defined in terms of standard data exchange formats either do not contain the needed intersection curves, or the curves given in the file may not be in a form that is suitable for the grid generator. A good SSI algorithm must be capable of treating analytic surfaces (e.g., cylinders, cones, etc.), parametric surfaces (e.g., NURBS — nonuniform rational B-splines), surfaces described by discrete data points (e.g., resulting from stereo lithography, Plot3D surface grids, etc.), and combinations of these types. Accuracy must be good enough for packing of grid points allowing high-aspect-ratio cells near the intersection curves, which typically requires that the curves be accurate to at least 10^{-6} units. A good method should be robust and should require a minimum of user input. A user should have to specify only the surfaces to be intersected and the requested tolerance. Use in an interactive environment requires that the method be reasonably fast, i.e., the solution of all but the most demanding problems should take only a few seconds on a state-of-the-art workstation.

29.2 Surface Refinement and Reparametrization — Underlying Principles and Best Practices

The relationship between the parametrization and the control point net of a NURBS surface reflects the relationship between parameter space and range. Each Bézier segment of a curve, for example, may have a normalized local parametrization (t, $0 \le t \le 1.0$). However, there exists a global parametrization that determines the relationship between each segment and the whole curve. The requirements for C^2 continuity between two segments are that the second derivatives at the common break point should match — "from the left and right." The notion of C^r, $r \geq 1$ depends on the interplay between the domain and range configurations. In other words, the first- and the second-order derivatives are dependent on the global parametrization of a NURBS curve. As a rule of thumb, a better curve is obtained if the geometry (range) of the NURBS curve is incorporated into the parametrization. Several parametrization schemes exist, such as *uniform, chord length, centripetal,* and one due to Nielson and Foley [1989]. Each scheme has some favorable aspects; see [Foley 1986, 1987] for a detailed discussion.

In the context of grid generation, the grid can be *smoothed* to correct problems resulting from bad underlying parametrization, but this procedure is rather time-consuming. The other alternative is to reparametrize the surface. The process does not change the geometry in physical space. Without going into the detail, we state that reparametrization is independent of the degree of the rational-polynomial basis functions of NURBS, see [Farin 1995] (see also Chapter 28). The reparametrization will then create a "smooth" parametric domain that will promote high quality grids without jeopardizing accuracy.

The goal then is to *refine* a given, poorly parametrized surface, i.e., to construct a surface that is *chord length* parametrized. A surface $\mathbf{s}(u,v)$ with knot sequence $\{u_0, \ldots, u_{L+2n-2}\}$ and $\{v_0, \ldots, v_{M+2m-2}\}$ is said to be *chord length* parametrized if it has the following properties:

$$\frac{\left\|\mathbf{s}\left(u_{i+1}, v_j\right)-\mathbf{s}\left(u_i, v_j\right)\right\|}{\left\|\mathbf{s}\left(u_{i-1}, v_j\right)-\mathbf{s}\left(u_i, v_j\right)\right\|} \approx \frac{\Delta u_i}{\Delta u_{i-1}} \tag{29.1}$$

and

$$\frac{\left\|\mathbf{s}\left(u_i, v_{j+1}\right)-\mathbf{s}\left(u_i, v_j\right)\right\|}{\left\|\mathbf{s}\left(u_i, v_{j-1}\right)-\mathbf{s}\left(u_i, v_j\right)\right\|} \approx \frac{\Delta v_i}{\Delta v_{i-1}} \tag{29.2}$$

where

$$\Delta u_i = u_{i+1} - u_i, \quad \Delta v_i = v_{i+1} - v_i,$$

and $\| \ \|$ denotes the Euclidean norm, see [Farin 1997].

For interrogation and analysis of a surface, it is desirable that the parametrization and control points reflect the above situation. This is to enable the surface evaluation parameter values to be used as input for subsequent analysis. The question then is, *Can the surface be redefined, within a given tolerance, such that the parametrization is in tune with the geometry of the surface?* In other words, given a poorly parametrized NURBS surface, can one construct a redefined NURBS surface that approximates the given surface within a given tolerance such that it has the properties of a chord length parametrization.

Some of the related research in the areas of reparametrization and curve and surface approximation is reviewed in the following. Previous work related to this research can be categorized in two areas, the first being reparametrization and the second being curve and surface approximation/interpolation methods.

Some work has been done in the area of reparametrization of curves. In [Fuhr and Kallay 1982], a method is described for interpolating a monotone data sequence with a C^1 monotone rational *B-spline* curve of degree 1. If the original curve C and the reparametrization function f are rational *B-splines,* then the reparametrized curve $\tilde{C} = C\,C\,f$ is also a rational *B-spline.* The degree of $\tilde{C}$ is the product of the degrees of C and f. This results in a C^1-continuous spline. We need to achieve C^2 continuity. The algorithm mentioned above ensures that the degree is not raised. This is useful in coming up with a common parametrization for opposite boundary curves on a surface.

In [Crampin et al. 1985] an algorithm is described to transmit a curve by sending discrete points off the original curve, such that the curve can be regenerated at the other end. In order to interpolate a curve effectively, few points should be placed where the radius of curvature is large, but many where it is small.

Yu and Soni [1995] use reparametrization to create grids with different parameter distributions. The reparametrization in the curve case is achieved as follows: Let us consider a NURBS curve with resolution n (number of points), and let

1. $s_1(i)$, $i = 1, \ldots, n$, be the parametric values associated with the desired distribution of the curve in physical space, and
2. $s_2(i)$, $i = 1, \ldots, n$, be the normalized chord length of the curve evaluated at parametric values $s_1(i)$, $i = 1, \ldots, n$.

The $s_2(i)$ values are known, and the $s_1(i)$ values are to be determined such that $\|s_2(i) - s_1(i)\|$ is minimized for all $i = 1, \ldots, n$. This is accomplished by an iterative process. The initial values of $s_1(i)$ are set to be the same as those at which $s_2(i)$ and $s_3(i)$ are evaluated. If the absolute difference $s_2(i) - s_3(i)$ is smaller than a certain tolerance, $s_1(i)$ is set as the desired parametric value. If the difference of $s_2(i) - s_3(i)$ is negative and the absolute value of this difference is greater than the tolerance, $s_1(i)$ should be shifted to a value between $s_1(i-1)$ and $s_1(i)$. The same strategy is applied to the case where $s_2(i) - s_3(i)$ is positive. In this case, the value of $s_1(i)$ should be shifted to a value between $s_1(i)$ and $s_1(i+1)$. The algorithm is further extended to deal with reparametrization of surfaces. Nevertheless, this approach cannot be used directly for the reparametrization of surfaces, it leaves many questions open.

Kim [1993] has attempted to come up with knot placement for NURBS interpolation. He plots the distance between the interpolation points as a monotonically increasing function $f(s)$ over its parametrization. The parametrization is obtained from one of the several methods commonly used. The function can be piecewise linear, piecewise rational–quadratic, or piecewise linear–rational B-spline interpolation. Knot placement is done by dividing the function space into equal number of segments and projecting the division onto the parametric space. This is then used for determining the parametrization.

29.2.1 Approaches to Solving the Problem

Although many different approaches may be applied to solve the problem at hand, the following two are considered:

1. Modify the control parameters of the given surface, with only minimal changes to the surface, i.e., change weights, control points, etc., with the result that the surface exhibits the property of *chord length* parametrization.
2. Construct a new NURBS surface that approximates the given surface, within a given tolerance, such that the surface is chord length parametrized.

29.2.2 Modifying the Existing Surface

There are four design parameters available in the NURBS case that control its behavior (it is assumed that the knot sequence does not have any multiplicities except at the ends); these are degree, control points, weights associated with the control points (we will only deal with the case $w_i > 0$), and parametrization.

Raising (or lowering) the degree does not affect the parametrization and therefore is a non-issue in our case. However, if one were to represent a NURBS curve with its approximation, the approximating polynomial should be at least a cubic. This is due to the fact that cubics are the lowest degree which can represent true space curves.

Modifying the control points modifies the surface itself. It is very difficult to predict the behavior of the surface when its control points are modified. Let us consider the curve case. Moving a control point means affecting (degree – 1) segments on each side. In order to not change the curve itself, the affected neighboring segments would also have to be changed (by moving their control points). This can start a "chain reaction" and convergence might be a problem.

Changing the weights is similar to changing the control points. However, in conjunction with the parametrization, it is possible to keep the curve or surface the same. Thus, it would be a matter of finding the new parametrization (the desired one), and we could possibly change the poorly parametrized curve to a chord length parametrized one without changing the curve itself. Here, the problem is to find the desired knot sequences themselves. This approach, although theoretically appealing, requires as input something that is not known. This first approach, though ideal, does not always result in a convergent solution.

29.2.3 The Surface Approximation Scheme

The second approach is to find another surface as close as possible to the original surface. Let ϵ_{max} be the value which indicates the maximum Euclidean distance the two surfaces are apart from each other. If $\mathbf{s}(u, v)$ is the given surface with knot sequence $\{u_0, \ldots, u_M\}, \{v_0, \ldots, v_N\}$, and $\mathbf{r}(u,v)$ with knot sequence $\{u_0, \ldots, u_K\}, \{v_0, \ldots, v_L\}$ is the approximation to the surface, then we want

$$\max_{u_i, v_j, u_k, v_l} \left\{ \min\left\{ \left\| \mathbf{s}(u_i, v_j) - \mathbf{r}(u_k, v_l) \right\| \right\} \right\} < \varepsilon, \tag{29.3}$$

where $\mathbf{r}(u_k, v_l)$ is the closest point on $\mathbf{r}(u, v)$ for a given point $\mathbf{s}(u_i, v_j)$ on $\mathbf{s}(u, v)$ and ε is the ϵ_{max} bound placed on the *healing* process.

This approach is used in Razdan [1995] to solve the problem at hand. The approximation is based on the assumption that adequate points can be found on the surface, such that when an interpolating surface is passed through them, the resulting surface will be very close to the original surface. The problem then reduces to finding these interpolation points. If, however, the number of points is insufficient, then the error estimation process identifies the point $\mathbf{s}(u_i, v_j)$ on $\mathbf{s}(u, v)$ where the maximum deviation, ϵ_{max}, occurs between $\mathbf{s}$ and $\mathbf{r}$. This information can be used to insert a knot in surface $\mathbf{r}$ such that $\mathbf{r}$ is now forced to interpolate to $\mathbf{s}(u_i, v_j)$.

The construction of the new surface is a two-step process. First, the four boundary curves are determined, then the interior is filled. The reason for tackling the boundary curves first is twofold. The boundary curves provide the spatial bounds to the *filling process*. Second, it works out well to fill using the *outside-in* approach. All computations are based on how well the surface is discretized. We have found that surface evaluation at a density of 10×10 points per knot segment is sufficient.

29.2.4 Boundary Curve Approximation

Approximation of the boundary curves is the first step towards approximating the surface. Each boundary curve is treated individually. The steps for approximating a NURBS curve are:

1. Estimate the number of interpolation points needed.
2. Find interpolation points on the given curve (while keeping such points to a reasonable number).
3. Pass a C^2-continuous interpolating NURBS curve through these points.

The technique for choosing interpolation points uses arc length and curvature distribution characteristics of the given curve. It also uses the *adaptive knot selection* scheme to properly place the knots on the interpolating curve. Details on how this is done can be found in [Razdan 1995]. The underlying principle is to capture as much of the geometric properties of the original curve as possible while trying to keep the number of interpolation points to a minimum. Figure 29.1 is an example of a NURBS curve and its approximation using this technique.

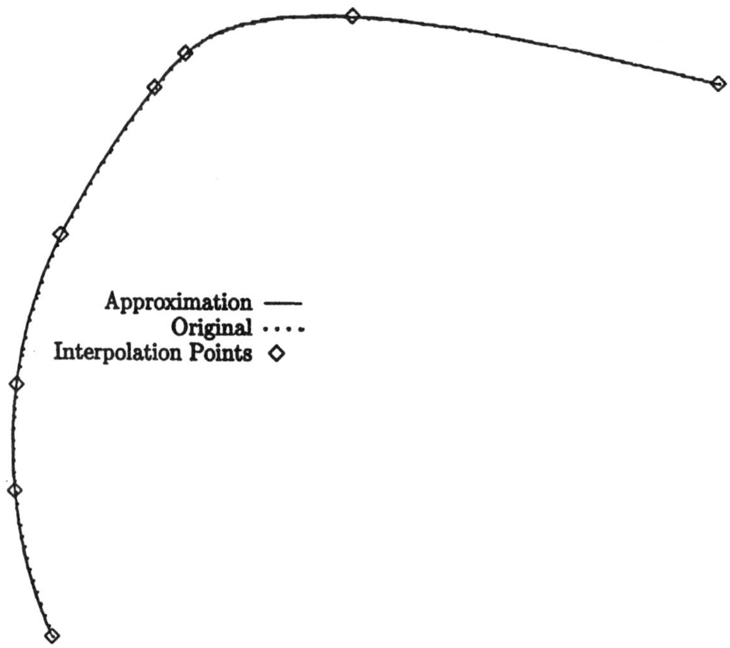

FIGURE 29. 1 NURBS curve interpolation using arc length and curvature.

29.2.5 Finding an Interpolating Surface

The next step in the process is to combine information (parametrization) of opposite boundary curves into one. Although the interpolation points required to describe each of the two boundary curves independently are available, the distribution of these points will not, in general, be satisfactorily represented by a common knot sequence (parametrization). This is due to the fact that the distributions of points in each set depends on the individual curves' curvature and arc length distributions. Choosing a knot sequence of either one of the boundary curves will result in the same initial problem. However, if somehow both curves and the interior surface did have the same distribution of interpolation points, a single parametrization could be used without a problem. But at this time the interior interpolation points are not *fixed.* This is dealt with as follows. First, a *reconciliation* process of the opposite boundary curves is performed to resolve the inequitable interpolation points distribution on the two boundary curves. This ensures that the knot sequences computed after the reconciliation step of the opposite boundary curves will be the same. Two knot sequences result, one for each parametric direction of the approximated surface. Second, the interior interpolation points on the surface are located such that they satisfy the parametrizations in both directions that resulted from the reconciliation process.

We describe briefly the *reconciliation* process between boundary curves. The distance (arc length) between neighboring interpolation points of one boundary curve is computed and tabulated. The same is done for its counterpart, the opposite boundary curve. Next, distances in each set are represented as the fraction of the total arc length of the respective curves. Once the two sets are compiled, they are *reconciled.* For every point in one set, a corresponding point is sought in the second set (and vice-versa) that is the same fraction of distance away from the starting end of the curve it belongs to. If there is no such point within a tolerance, then an auxiliary point is inserted into the set that does not have the point. At the end of this process, both sets have points that are similarly distributed along the length of the original boundary curves. Similarity in distribution means that the ratio of distances between the neighboring points is similar in the two sets. In other words, we have inserted auxiliary knots into the curves so that both curves have the right distribution of knots or points for interpolation. This in turn is nothing but chord length parametrization. This process is applied to the other set of opposite boundary curves.

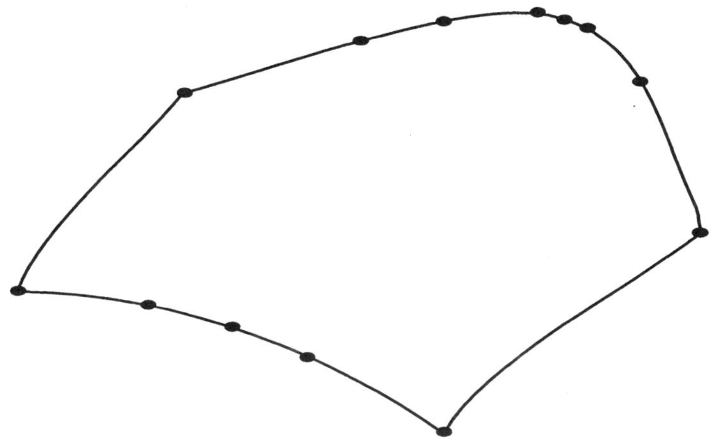

FIGURE 29.2 Boundary interpolation points before reconciliation.

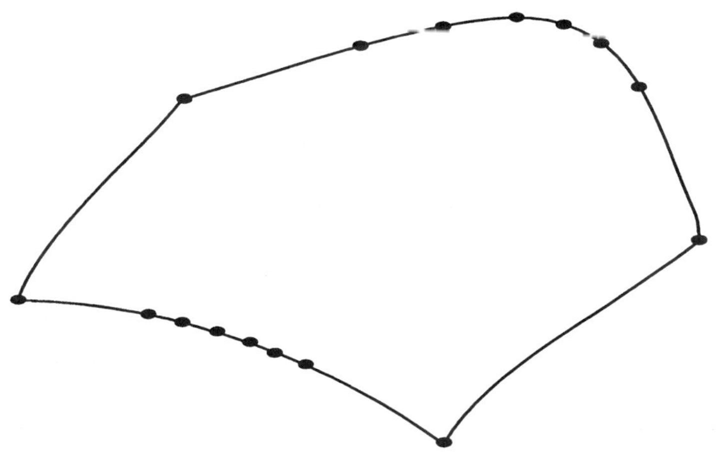

FIGURE 29.3 Boundary interpolation points after reconciliation.

The set of interpolation points is now fixed for all four boundary curves. Figures 29.2 and 29.3 show two sets of interpolation points before and after the reconciliation process. This process of adaptively generating knot sequences based on curvature information and arc length (chord length) is called the *RCA parametrization* (reconciled curvature arc length parametrization).

29.2.6 Finding Interior Interpolation Points

Once the outer framework of the boundary curves is accomplished, the interior of the surface is constructed. This is an iterative process. The parameter values at which the points on the boundary curves will be interpolated are marked on the domain rectangle of the original surface. The corresponding points on the opposite edges are joined with straight lines in the domain rectangle. The intersections of these lines provide (u,v) values in the parameter space of the original surface. This leads to an initial guess for the internal points of the new surface. In Figure 29.4, these points are marked as (1,1), (1,2), (2,1), and (2,2). As is evident from the figure, these points do not reflect the parametrization of the surface. For example, let $u_0 = 0.0$, $u_1 = 0.33$, and $u_2 = 0.66$. In general, point (1,1) will not be half the distance between points (1,0) and (1,2). In an ideal situation the process would stop here. However, for a poorly parametrized surface, this is the starting point for the iterative process.

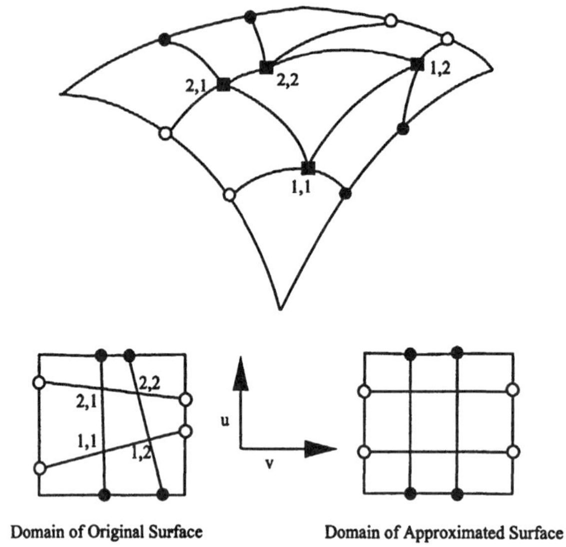

FIGURE 29.4 Surface and its domain rectangle.

We describe an algorithm to find the interior points. As stated above, the first guess of internal interpolation points will probably not satisfy the chosen parametrization. The algorithm iteratively moves each interior point $\mathbf{x}_{i,j}$ a finite distance in the domain and evaluates its relationship (distance) with its immediate four neighbors, $\mathbf{x}_{i-1,j}$, $\mathbf{x}_{i+1,j}$, $\mathbf{x}_{i,j-1}$, and $\mathbf{x}_{i,j+1}$, with respect to the new parametrization. It attempts to find the local minimum for placing this point. The points are always moved in the domain. This is important as it is guaranteed that the corresponding point in the range will always lie on the given surface. The evaluation, whether a particular choice (point location on the surface) is *good* or *bad*, is done based on a *penalty factor.* A high penalty factor means "not good." The penalty factors of all the interior points are computed and sorted in descending order. The point with highest penalty factor is tackled first, since it is most likely to be moved. The algorithm keeps track of points moved in an iteration in a two-dimensional array. In the iterative process, a point is a candidate for relocation if any of its four neighbors have moved since the last iteration. In the case when none of the four neighbors have moved, then local conditions have not changed and repeating the process will not improve the situation. On the other hand, if the local conditions have changed, i.e., one or more of the neighbors has moved since the last iteration, then it is likely that the current point is not the optimum point any more. Thus, it makes sense to apply the algorithm again. The iterative process is terminated when all the points occupy optimum positions. Figure 29.5 shows an example *before* and *after* this procedure is applied.

29.3 Approximation of Discontinuous Geometries — Underlying Principles and Practices

29.3.1 The Algorithm and References

The essential procedure used to approximate a geometry is the construction of a single local approximant. This procedure consists of these steps:

1. Creating four (or selecting four existing) curves as boundary curves for an initial local approximant (Coons patch).
2. Constructing a bilinear Coons patch from the four boundary curves.

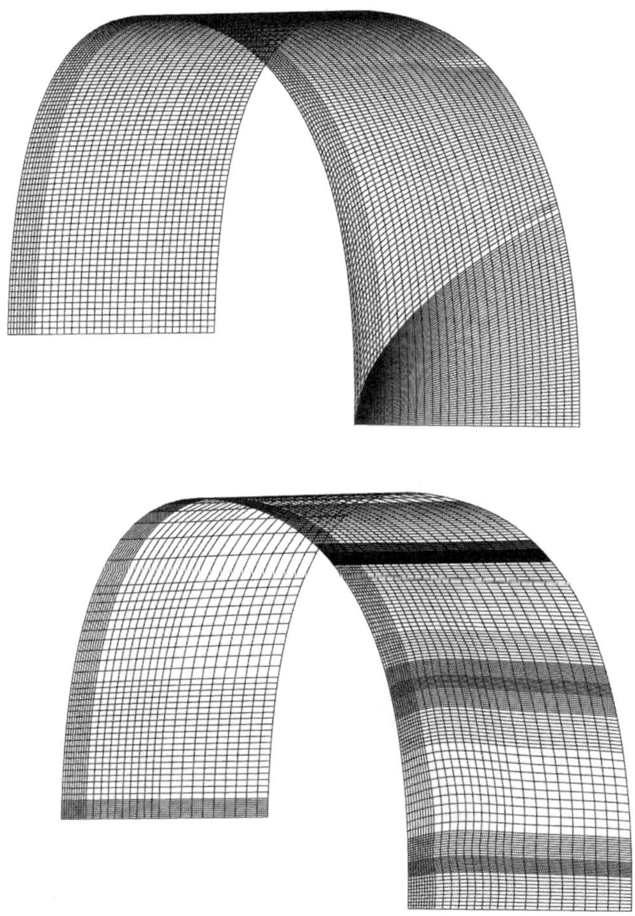

FIGURE 29.5 Surface before and after *healing.*

3. Projecting a curvilinear grid on the Coons patch onto the original geometry.
4. Determining "artificial projections" for those points in the curvilinear mesh that cannot be projected — due to possible gaps in the original surface.
5. Interpolating the points resulting from steps 3 and 4.

One has to perform step 1 interactively, while all other steps can be performed without user interaction. The local surface approximant obtained as the result of this procedure is a bicubic B-spline surface, which is guaranteed to lie within a certain distance of the original surfaces. The distance measure is based on shortest (perpendicular) distances between points on an approximant and the original surfaces. We compute this distance measure only in regions where there is a "clear" correspondence between an approximant and the original surfaces and do not compute it for those parts of an approximant covering a discontinuity.

Once all local approximants are determined and their topology (connectivity) is known, a final step ensures that the overall resulting approximation is continuous by enforcing continuity along shared boundary curves of the local approximants.

The methods that we rely on to approximate a discontinuous geometry are covered in great detail in the literature dealing with CAGD methods for curves and surfaces. References include [Farin 1995, 1997], [Faux and Pratt 1979], [Piegl 1991a, 1991b], and [Piegl and Tiller 1996].

29.3.2 Computing the Initial Coons Patch

The initial local approximant is used to smooth rough data, guide the choices of interpolation points, and serve as a reference for filling in gaps. A user has to specify four continuous curves whose endpoints meet to form a single closed curve — the boundary of a Coons patch. The four boundary curves can span across multiple original surface patches; they can even be above or below the given geometry.

In order to obtain a reasonable surface grid for the Coons patch implied by the four boundary curves, we use a *discrete Coons patch* construction. First, we compute points on the boundary curves distributed uniformly with respect to arc length. We then associate parameter values $(u_{I,0},v_{I,0})$, $(u_{I,N},v_{I,N})$, $(u_{0,J},v_{0,J})$, and $(u_{M,J},v_{M,J})$, $I = 0, \ldots, M$, $J = 0, \ldots, N$, defining the uniformly distributed points on the boundary curves in 3D Euclidean space. The points $\mathbf{x}_{I,J} = (x_{I,J}, y_{I,J}, z_{I,J})$ on the discrete Coons patch are thus defined as

$$\mathbf{x}_{I,J} = \left[\left(1-u_{I,J}\right) u_{I,J}\right]\begin{bmatrix}\mathbf{x}_{0,J}\\ \mathbf{x}_{M,J}\end{bmatrix} + \left[\mathbf{x}_{I,0}\ \mathbf{x}_{I,N}\right]\begin{bmatrix}\left(1-v_{I,J}\right)\\ v_{I,J}\end{bmatrix}$$
$$-\left[\left(1-u_{I,J}\right) u_{I,J}\right]\begin{bmatrix}\mathbf{x}_{0,0} & \mathbf{x}_{0,N}\\ \mathbf{x}_{M,0} & \mathbf{x}_{M,N}\end{bmatrix}\begin{bmatrix}\left(1-v_{I,J}\right)\\ v_{I,J}\end{bmatrix}, \tag{29.4}$$

where $u_{I,J}$, $J = 0, \ldots, N$, varies linearly between $u_{I,0}$ and $u_{I,N}$ and $v_{I,J}$, $I = 0, \ldots, M$, varies linearly between $v_{0,J}$ and $v_{M,J}$, respectively. In general, the points $\mathbf{x}_{I,J}$ do not lie on the given surface patches.

29.3.3 Projecting the Coons Patch onto the Original Surfaces

In order to project each point $\mathbf{x}_{I,J}$ on the Coons patch onto the original surfaces, one must know the approximate surface normal at $\mathbf{x}_{I,J}$, which is used as projection direction. We approximate the unit normal vector at $\mathbf{x}_{I,J}$ by

$$\mathbf{n}_{I,J} \approx \frac{\left(\mathbf{x}_{I+1,J} - \mathbf{x}_{I-1,J}\right)\times\left(\mathbf{x}_{I,J+1} - \mathbf{x}_{I,J-1}\right)}{\left\|\left(\mathbf{x}_{I+1,J} - \mathbf{x}_{I-1,J}\right)\times\left(\mathbf{x}_{I,J+1} - \mathbf{x}_{I,J-1}\right)\right\|}. \tag{29.5}$$

The points $\mathbf{x}_{I,J}$, their associated normal vectors $\mathbf{n}_{I,J}$, and an absolute offset distance d define points on an *upper* and a *lower offset surface* of the initial Coons approximant. The points on the upper offset surface are denoted by $\mathbf{a}_{I,J}$ and the ones on the lower offset surface by $\mathbf{b}_{I,J}$:

$$\mathbf{a}_{I,J} = \mathbf{x}_{I,J} + d\,\mathbf{n}_{I,J} \quad \text{and} \quad \mathbf{b}_{I,J} = \mathbf{x}_{I,J} - d\,\mathbf{n}_{I,J}. \tag{29.6}$$

We relate the offset distance d to the extension of the Coons patch by setting

$$d = \tfrac{1}{8}\left(\left\|\mathbf{x}_{M,0} - \mathbf{x}_{0,0}\right\| + \left\|\mathbf{x}_{M,N} - \mathbf{x}_{M,0}\right\| + \left\|\mathbf{x}_{0,N} - \mathbf{x}_{M,N}\right\| + \left\|\mathbf{x}_{0,0} - \mathbf{x}_{0,N}\right\|\right). \tag{29.7}$$

The choice of d is very important, since it determines the set of original surfaces to be considered for the final local approximation. It is not clear at this point what is the best value for d given an arbitrary geometry. The offset surface construction is shown in Figure 29.6.

It is assumed that the convex set S defined by all the points $\mathbf{a}_{I,J}$ and $\mathbf{b}_{I,J}$ contains the original surfaces that must be considered by the local approximation procedure. The convex hull of S is approximated by computing the 3D bounding box for all the points $\mathbf{a}_{I,J}$ and $\mathbf{b}_{I,J}$. Original surfaces are considered for the local approximation procedure only if they lie partially inside this bounding box. The surfaces lying inside the bounding box are evaluated, using some predefined resolutions, and the resulting point sets

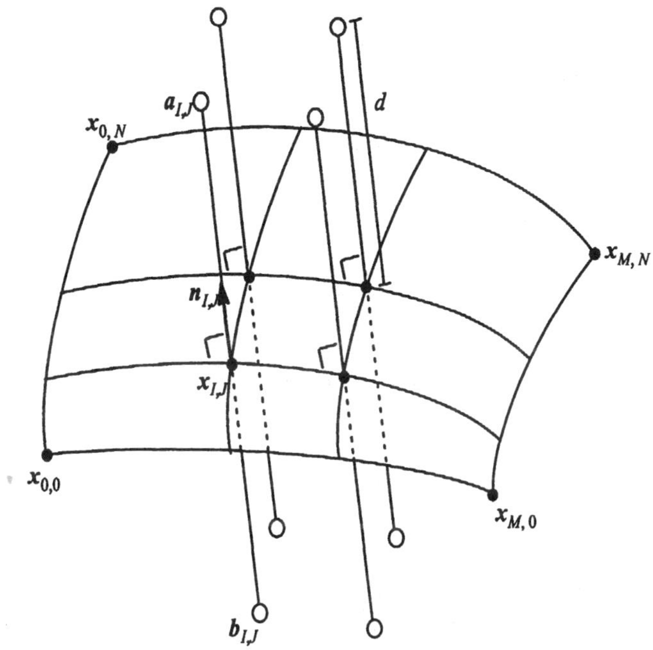

FIGURE 29.6 Local offset surface construction.

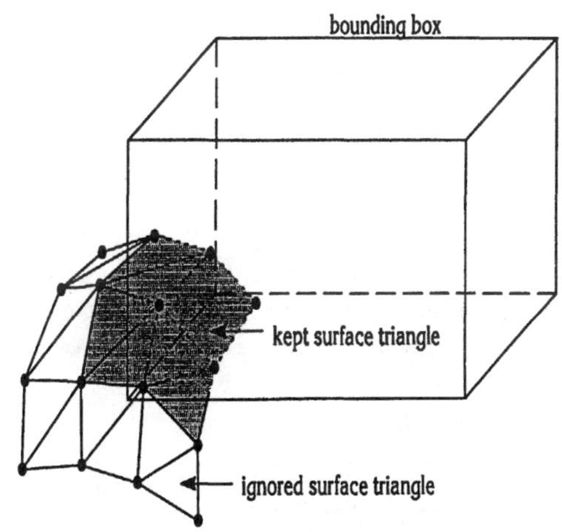

FIGURE 29.7 Clipping original surfaces and surface triangles.

are triangulated. The resulting triangles are then also clipped against the same bounding box — one needs to consider only those triangles lying inside the bounding box when projecting a point $\mathbf{x}_{I,J}$ onto the original surfaces. This is illustrated in Figure 29.7.

Next, each point $\mathbf{x}_{I,J}$ is projected in direction of $\mathbf{n}_{I,J}$ onto the triangles inside the bounding box. The projection of $\mathbf{x}_{I,J}$ must satisfy the condition that it lie between $\mathbf{a}_{I,J}$ and $\mathbf{b}_{I,J}$. In general, it is possible to obtain zero, one, or multiple projections for each point $\mathbf{x}_{I,J}$. If more than one intersection is found, the point closest to the point $\mathbf{x}_{I,J}$ is identified and used in the subsequent steps. If no intersection is found, a bivariate *scattered data approximation* method will be used later to derive "artificial projections."

If the parametric representation of the original surfaces is known the projections, computed as projections onto triangles in a surface triangulation, can be mapped to points that lie exactly on the given surfaces. A projection obtained from intersecting a line segment $\overline{\mathbf{a}_{I,J}\mathbf{b}_{I,J}}$ and a surface triangle can be expressed as a barycentric combination of the vertices of this triangle. Let $\mathbf{p}_{I,J}$, be a projection, and let $\mathbf{p}_1$, $\mathbf{p}_2$, and $\mathbf{p}_3$ be the vertices of the triangle containing the projection. We can express the projection in barycentric form as

$$\mathbf{p}_{I,J} = \bar{u}_1\mathbf{p}_1 + \bar{u}_2\mathbf{p}_2 + \bar{u}_3\mathbf{p}_3 \tag{29.8}$$

and can use the barycentric coordinates in this expression to compute the parameter tuple

$$(\bar{u},\bar{v}) = \bar{u}_1(u_1,v_1) + \bar{u}_2(u_2,v_2) + \bar{u}_3(u_3,v_3) \tag{29.9}$$

where (u_i,v_i) is the parameter tuple of vertex $\mathbf{p}_i$ We can now evaluate the associated original parametric surface $\mathbf{s}(u,v)$ using the parameter tuple $(\bar{u}, \bar{v})$ and replace $\mathbf{p}_{I,J}$ by $\mathbf{s}(\bar{u}, \bar{v})$. We will denote the points obtained by this "map-onto-real-surface step'" by $\mathbf{y}_{I,J}$. (If the parametric representation of the original surfaces is not known, we simply use the intersections with the surface triangulations as final approximation conditions $\mathbf{y}_{I,J}$.)

29.3.4 Computing Additional Approximation Conditions

Due to existing discontinuities, gaps, in the original surfaces the projection strategy might not yield any intersection points for certain line segments $\overline{\mathbf{a}_{I,J}\mathbf{b}_{I,J}}$. We must estimate "artificial projections" to obtain a complete $(M + 1) \times (N + 1)$ array of points required later in the construction of a local B-spline approximant. We use a bivariate scattered data approximation technique, called *Hardy's reciprocal multiquadric method*, see [Franke 1982].

Each intersection point $\mathbf{p}_{I,J}$, obtained by intersecting line segment $\overline{\mathbf{a}_{I,J}\mathbf{b}_{I,J}}$ with the surface triangles, can be written as a linear combination of $\mathbf{a}_{I,J}$ and $\mathbf{b}_{I,J}$, i.e.,

$$\mathbf{p}_{I,J} = \mathbf{p}(t_{I,J}) = (1-t_{I,J})\mathbf{a}_{I,J} + t_{I,J}\mathbf{b}_{I,J}, \quad t_{I,J} \in [0,1]. \tag{29.10}$$

The values $t_{I,J}$ are computed (and stored) when projecting points on the surface triangulation. These values remain unchanged, even if intersection points are mapped onto the real parametric surfaces. Hardy's reciprocal multiquadric method is used to compute a bivariate function $t(u,v)$ that interpolates all parameter values $t_{I,J}$ corresponding to intersection points that have been found. We must consider these conditions:

$$t_{I,J} = t(u_{I,J},v_{I,J}) = \sum_{j\in\{0,\dots,N\}} \sum_{i\in\{0,\dots,M\}} c_{i,j}\left(R + (u_{I,J}-u_{i,j})^2 + (v_{I,J}-v_{i,j})^2\right)^{-\gamma}, \quad I\in\{0,\dots,M\},\ J\in\{0,\dots,N\}, \tag{29.11}$$

where R and γ are fixed parameters and only those values $t_{I,J}$, $u_{I,J}$, $u_{i,j}$, $v_{I,J}$, and $v_{i,j}$ are considered for which an intersection point has been found. Denoting the "mean parameter spacing" in the two parameter directions by $\delta_u = \left(\frac{1}{2}M\right)\sum_{I=0}^{M}(u_{I+1,0}-u_{I,0}) + (u_{I+1,N}-u_{I,N})$ and $\delta_v = \left(\frac{1}{2}N\right)\sum_{J=0}^{N}(v_{0,J+1}-v_{0,J}) + (v_{M,J+1}-v_{M,J})$, we have found that the values $R = 0.5(\delta_u + \delta_v)$ and $\gamma = 0.001$ yield good results. Further investigation is necessary regarding appropriate choices for these parameters.

This global approach, considering *all* projections that have been found, is generally too inefficient. Therefore, we *localize* Hardy's reciprocal method by considering only a relatively small number of found projections to determine an "artificial projection." If there is no intersection between a line segment

$\overline{\mathbf{a}_{I,J}\mathbf{b}_{I,J}}$ and the surface triangulations, we use the K closest found projections. For this purpose, we identify the K found projections whose associated index tuples are closest to the index tuple (I,J). We have found that values for K between five and ten yield good projection estimates. Thus, one has to solve the linear system

$$t_k = \sum_{i=1}^{K} c_i \left(R + (u_k - u_i)^2 + (v_k - v_i)^2 \right)^{-\gamma}, \quad k = 1, \ldots, K, \tag{29.12}$$

where (u_k,v_k) and (u_i,v_i) are parameter values for which projections are known. One must solve such a linear system for each missing projection.

Having determined parameter values $t_{I,J}$ for all line segments $\overline{\mathbf{a}_{I,J}\mathbf{b}_{I,J}}$ for which no projections are nowhere obtain each needed "artificial projection" as the linear combination

$$\mathbf{z}_{I,J} = \left(1 - t_{I,J}\right)\mathbf{a}_{I,J} + t_{I,J}\mathbf{b}_{I,J}. \tag{29.13}$$

The union of all points $\mathbf{y}_{I,J}$ and $\mathbf{z}_{I,J}$ defines an $(M + 1) \times (N + 1)$ curvilinear mesh which we use for the construction of a local B-spline surface approximant.

29.3.5 Constructing a Local Surface Approximant

We use the $(M + 1) \times (N + 1)$ points $\mathbf{y}_{I,J}$ and $\mathbf{z}_{I,J}$ to define a cubic B-spline surface interpolating these points and locally approximating the given surfaces. We denote this B-spline surface as

$$\mathbf{s}(u,v) = \sum_{j=0}^{n} \sum_{i=0}^{m} \mathbf{d}_{i,j} N_i^4(u) N_j^4(v), \tag{29.14}$$

where $\mathbf{d}_{i,j}$ is a B-spline control point, $N_i^4(u)$ and $N_j^4(v)$ are the normalized B-spline basis functions of order four, and $m = 3M$ and $n = 3N$. The (normalized) knot vectors are defined by values ξ_i $(\xi_i < \xi_{i+1})$ and η_j $(\eta_j < \eta_{j+1})$ and have quadruple knots at both ends and triple knots in the interior, i.e.,

$$\begin{aligned} \mathbf{u} &= \left(u_0, u_1, \ldots, u_{m+4}\right) = \left(0,0,0,0,\xi_1,\xi_1,\xi_1,\ldots,\xi_{M-1},\xi_{M-1},\xi_{M-1},1,1,1,1\right) \quad \text{and} \\ \mathbf{v} &= \left(v_0, v_1, \ldots, v_{n+4}\right) = \left(0,0,0,0,\eta_1,\eta_1,\eta_1,\ldots,\eta_{N-1},\eta_{N-1},\eta_{N-1},1,1,1,1\right). \end{aligned} \tag{29.15}$$

For more details regarding B-splines, see, e.g., [Bartels et al. 1987], [Farin 1997], and [Piegl and Tiller 1996]. Here, we are using the indexing scheme used in [Bartels et al. 1987]. We are currently using a uniform knot spacing, i.e., $\xi_i = i/M$ and $\eta_j = j/N$.

Our construction yields local B-spline approximants degenerating to C^1-continuous, piecewise bicubic Bézier surfaces. The control points $\mathbf{d}_{i,j}$ are derived by first using a C^1 cubic interpolation scheme for all rows and columns of points to be interpolated and, second, applying C^1 continuity conditions to obtain the four interior Bézier control points of each bicubic patch constituting a single B-spline approximant. The interior Bézier control points of all bicubic patches are defined by the equations

$$\begin{aligned} \mathbf{d}_{3i\mp1,3j+1} &= \mathbf{d}_{3i\mp1,3j} + \left(\mathbf{d}_{3i,3j+1} - \mathbf{d}_{3i,3j}\right) \quad \text{and} \\ \mathbf{d}_{3i\mp1,3j-1} &= \mathbf{d}_{3i\mp1,3j} + \left(\mathbf{d}_{3i,3j-1} - \mathbf{d}_{3i,3j}\right). \end{aligned} \tag{29.16}$$

29.3.6 Error Estimation

Roughly speaking, the *error* between a locally approximating B-spline surface $\mathbf{s}(u,v)$ and the given geometry is the maximum of the minimum distances between points on the approximant and the original geometry. The existence of discontinuities and overlapping surfaces in the given geometry makes a precise definition of error impossible. We estimate the maximum distance by a discrete error measure. We use the points

$$\begin{aligned}
\mathbf{e}_{I,J} &= \mathbf{s}\left(\left(0.5\left(\xi_I + \xi_{I+1}\right), \eta_J\right)\right), \quad I = 0,\ldots,(M-1),\ J = 0,\ldots,N, \\
\mathbf{f}_{I,J} &= \mathbf{s}\left(\left(\xi_I, 0.5\left(\eta_J + \eta_{J+1}\right)\right)\right), \quad I = 0,\ldots,M,\ J = 0,\ldots,(N-1), \quad \text{and} \\
\mathbf{g}_{I,J} &= \mathbf{s}\left(\left(0.5\left(\xi_I + \xi_{I+1}\right), 0.5\left(\eta_J + \eta_{J+1}\right)\right)\right), \quad I = 0,\ldots,(M-1), J = 0,\ldots,(N-1)
\end{aligned} \tag{29.17}$$

to compute this discrete error measure. An approximating B-spline surface most likely has its greatest deviation from the given geometry in the interior of its constituting bicubic Bézier patches due to the oscillation characteristics of bicubic spline surfaces.

We compute shortest (perpendicular) distances between points on the B-spline approximants and the original surfaces by solving the implied bivariate minimization problem to identify closest points. We do not compute shortest distances for all points $\mathbf{e}_{I,J}$, $\mathbf{f}_{I,J}$, and $\mathbf{g}_{I,J}$; whenever one of these points is associated with a gap in the given geometry we do not compute a closest point for it. If the resulting error estimate is too large for a particular B-spline approximant, the resolution parameters M and N are increased and a new B-spline approximant is computed. In principle, there is no guarantee that this process will converge for arbitrary geometries. Therefore, this iteration is terminated when a maximum resolution is reached. In practice, however, one does not have to worry about this problem as long as the user specifies a "reasonable" set of boundary curves for the initial Coons patch that is projected onto the geometry.

29.3.7 Connecting the Local B-Spline Approximants

Topologically, all B-spline approximants are four-sided entities that can have at most four neighbors. Two B-spline surfaces are neighbors when they share a common boundary curve. Bifurcations (more than two surfaces sharing the same boundary curve) in the set of all B-spline approximants are not allowed. In summary, the B-spline approximants must satisfy these topological conditions:

- Each boundary curve is shared by at most two B-spline approximants.
- A corner point of a B-spline approximant can be common to any number of approximants.
- If a corner point of B-spline approximant is shared by a second approximant then this point is also a corner point of the second approximant (*full-face interface property*).

All B-spline approximants used in the final approximation must be compatible, i.e., they must be bicubic B-spline surfaces, must have the same number of control points, and must have the same knots along common boundary curves. For an arbitrary configuration, this implies that one must use *one* global number of control points in both parameter directions, *one* global knot vector used in both parameter directions, and *one* global order.

However, it is sufficient for practical grid generation purposes to know the connectivity among all B-spline approximants and to know that the *distance* between two neighbor B-spline approximants is smaller than some predefined tolerance. In this context, a distance is implied for two neighbor B-spline approximants by the physically existing gap between the two distinct boundary curves that, topologically, define the common boundary. As long as the gaps between all neighbor B-spline approximants are negligible, then one can use them directly for the generation of a mesh. In the context of mesh generation, we must emphasize that an initial mesh is generated for the set of approximating B-spline surfaces and that this

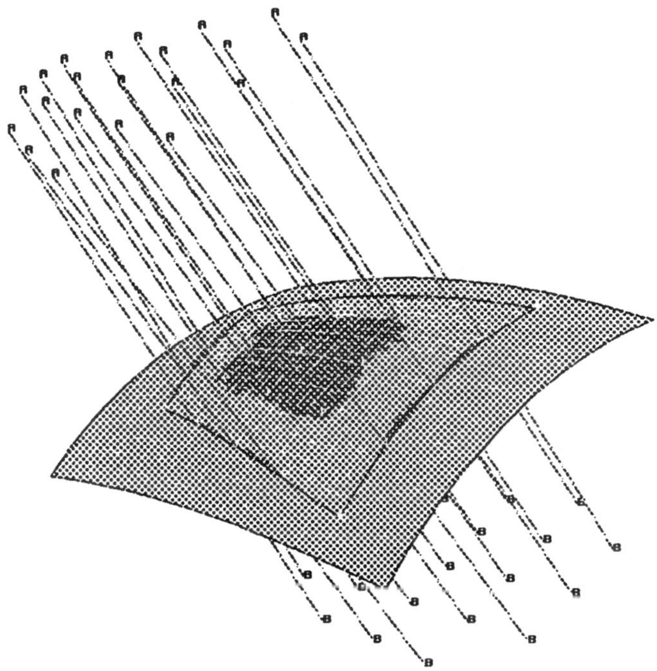

FIGURE 29.8 Approximation of part of single patch.

mesh is finally projected onto the original surfaces — unless an initial mesh point is in a gap region of the given geometry.

The conditions that must be satisfied by the B-spline approximants in order to obtain an overall tangent plane continuous approximation, also called *it gradient-continuous approximation*, are described in [Faux and Pratt 1979]. Essentially, the conditions are coplanarity conditions for certain B-spline control points along shared boundary curves and around shared corner points of B-spline approximants.

This approximation scheme for the "repair" of discontinuous geometries is explained in much more detail in [Hamann 1994] and [Hamann and Jean 1994, 1996].

29.3.8 Examples

Figures 29.8 through 29.11 show single B-spline surfaces approximating various geometries with discontinuities. One can see that the approximating surfaces are lying partially above and partially below the original surfaces. The approximating B-spline surfaces were obtained by specifying combinations of points and curves on the original geometries. Figures 29.8 and 29.9 show the line segments used to obtain sample points. Figures 29.12 and 29.13 show real-world geometries and their approximations (car body and aircraft configuration). Both figures show the original surfaces (top) and their approximations (bottom) consisting of multiple B-spline surfaces.

29.4 Surface–Surface Intersection — Underlying Principles and Best Practices

A good surface–surface intersection (SSI) algorithm should have the following characteristics:

- **Accuracy.** Grid generation for problems with high-gradient regions (such as viscous fluid flow) require a high degree of precision; a good algorithm must yield precise results.
- **Efficiency.** In an interactive environment, all but the most demanding cases should require only a few seconds to solve.

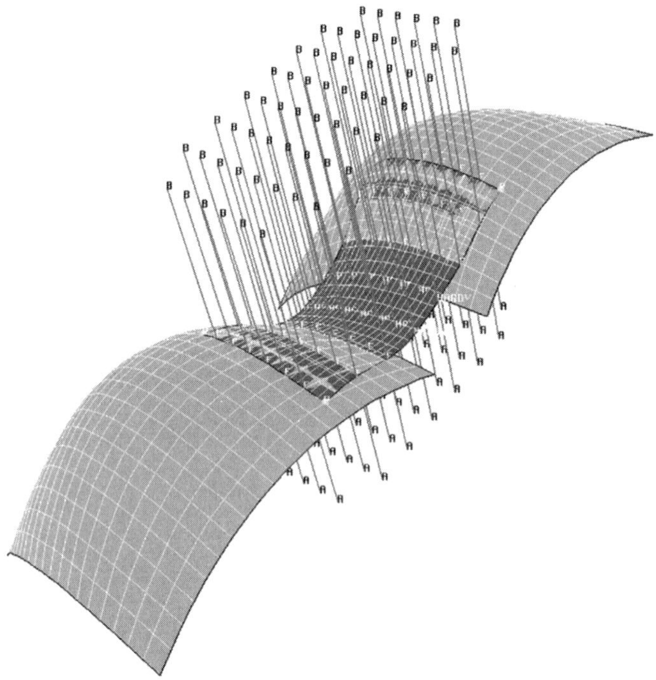

FIGURE 29.9 Approximation of surfaces with gap.

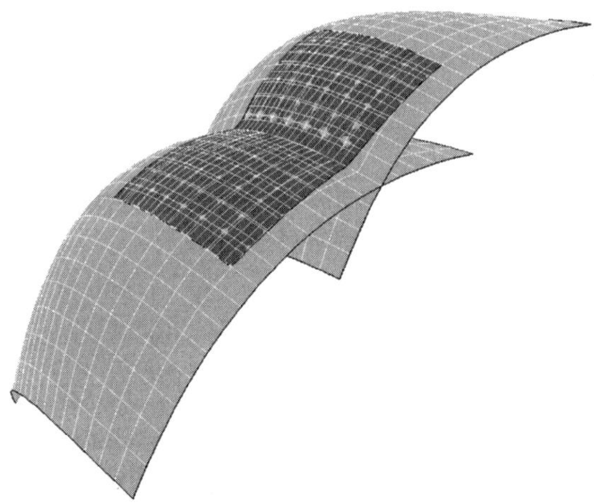

FIGURE 29.10 Approximation of intersecting surfaces.

- **Robustness**. The algorithm should correctly determine multiple intersections among multiple surfaces.
- **Simplicity**. The only action required by the user should be the specification of the surfaces to be intersected and a requested tolerance.

At present, no single algorithm possesses all of these properties. This is due to the fact that an optimal algorithm for a particular intersection problem depends on the type of surfaces involved. For example, the intersection of two planes is a line, while the intersection of two quadrics can be a curve of degree four. The representation of the surfaces must also be considered (i.e., implicit, polyhedral, or parametric). The reader can find a good survey of several types of intersection algorithms in [Hoschek and Lasser 1993].

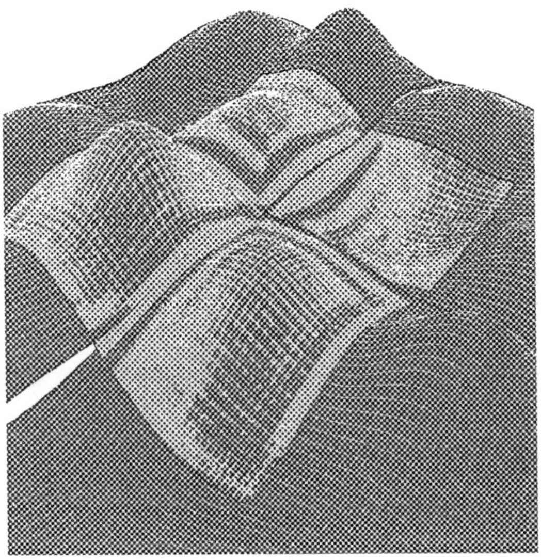

FIGURE 29.11 Approximation of highly oscillating surfaces.

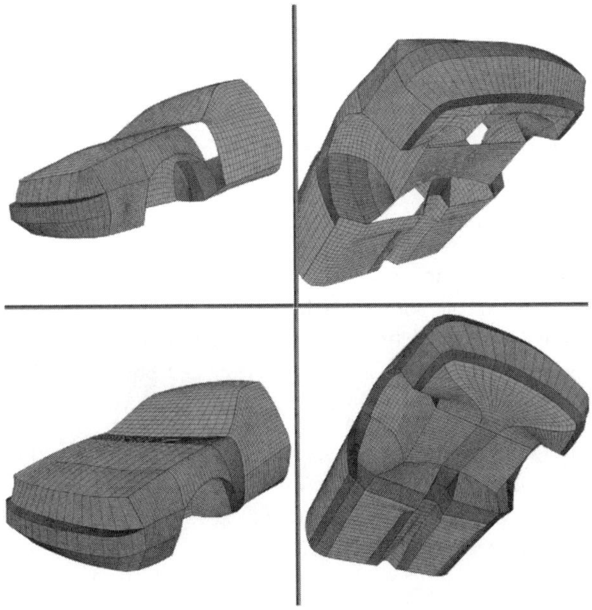

FIGURE 29.12 Approximation of car body.

29.4.1 The Intersection Algorithm.

The SSI algorithm we are describing can operate on surface triangulations, on analytically defined surfaces such as NURBS, or combinations thereof, see [Jean and Hamann 1998]. If the intersection is performed on surface triangulations, then the refinement step described below is skipped, and the piecewise linear curve produced from the intersection of triangles is the end result. If an analytical surface description is known for one or both of the surfaces involved then the refinement step is included. Surface triangulations are frequently encountered in the form of *unstructured surface grids* and are rapidly becoming a standard for data exchange using the StereoLithography format (see [3D Systems, Inc. 1989]). These are the basic steps of the algorithm:

FIGURE 29.13 Approximation of aircraft.

1. Triangulate the surfaces to be intersected.
2. Store each triangulation in a space partitioning tree structure.
3. Intersect the trees to obtain a list of intersecting regions.
4. Intersect the individual triangles within each set of intersecting regions to obtain a collection of line segments.
5. Sort the line segments.
6. Find a starting point for an "intersection loop."
7. Trace the loop storing sample points which are the endpoints of the line segments.
8. If an analytic surface representation is known, refine the sample points.
9. Interpolate the sample points with a spline curve in 3D physical space, 2D parameter space, or both spaces.

The actual intersection algorithm can only operate on two surfaces at a time. When more than two surfaces are to be intersected, the driver calls the intersection operator with successive pairs of surfaces until all possible surface pairs are processed. If the desired result is the intersection of several surfaces, then additional curve-curve intersections may be necessary.

29.4.2 Triangulation

Parametric surfaces are discretized using an adaptive triangulation technique based on recursive subdivision (see [Anderson et al. 1997] and [Samet 1990]). This method triangulates the surface within a specified tolerance without using an excessive number triangles. An example of this method is shown in Figure 29.14. This "adaptive" feature allows the SSI algorithm to more accurately capture important intersection features such as singular points, i.e., points where the normals of the two surfaces are colinear or nearly colinear.

Triangles are stored as a list of vertices and a connectivity table. Each vertex in the triangulation is stored only once in order to reduce memory requirements and to eliminate the possibility of slight edge mismatches due to numerical error. A separate list of associated *uv* parameter values and *uv* connectivity (connectivity in parameter space) is maintained to allow refinement of the calculated intersection curves.

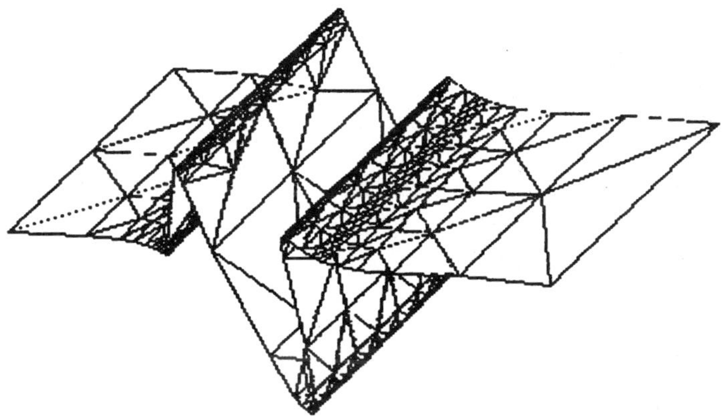

FIGURE 29.14 Adaptive surface triangulation.

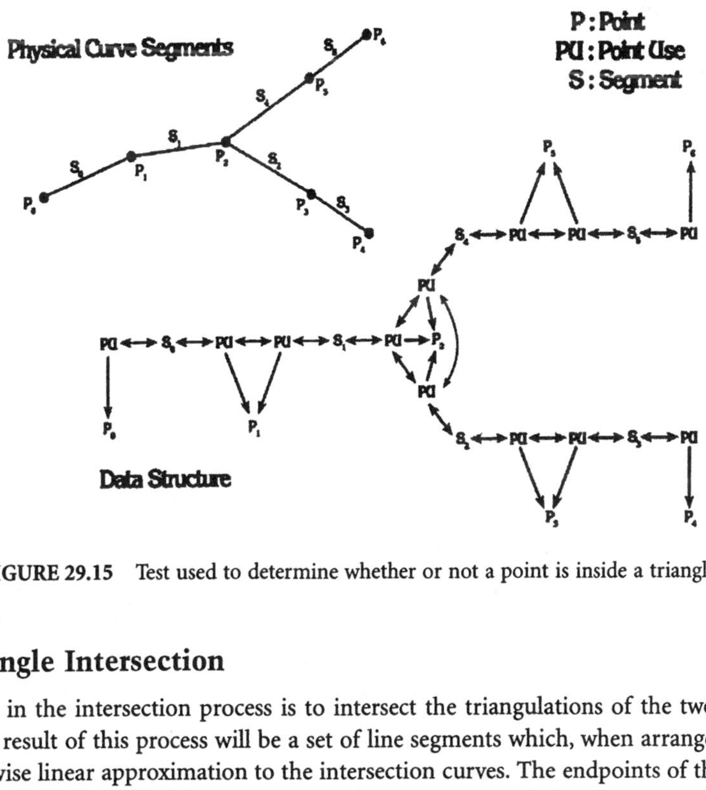

FIGURE 29.15 Test used to determine whether or not a point is inside a triangle.

29.4.3 Triangle Intersection

The first step in the intersection process is to intersect the triangulations of the two surfaces being considered. The result of this process will be a set of line segments which, when arranged properly, will provide a piecewise linear approximation to the intersection curves. The endpoints of the line segments will be used later as an initial guess for the sample points on the true intersection curves. The line segment information is used to determine the topology of the intersection curves and to order the sample points on the curves.

The method intersects two triangles by first performing a bounding box test to see if there is the possibility of the triangles intersecting. If this test is passed, then the edges of the first triangle are intersected with the plane defined by the second triangle and *vice versa.* The points resulting from the edge-plane intersections are then tested to determine if they lie inside the respective triangles. Figure 29.15 illustrates the test that determines whether a point lies inside a triangle. If the area of each of the subtriangles shown is positive, then the point is inside the triangle.

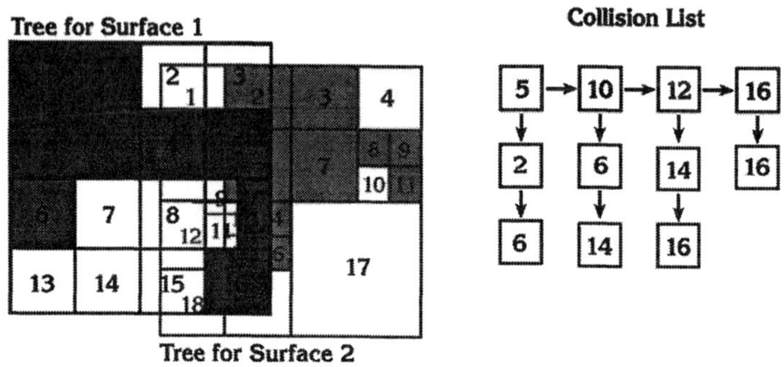

FIGURE 29.16 Intersection of quadtrees and resulting node association list.

The triangle intersection can yield one of three possible results:

1. No points are found that lie within either of the triangles, i.e., the triangles do not intersect.
2. Only one unique point is found, i.e., the triangles intersect only along the edges or at a vertex.
3. Two unique points are found, i.e., the intersection is a line segment.

The SSI method only considers intersections that result in line segments. The first case is ignored for obvious reasons. Case two is ignored because it yields only a point and therefore no information about the topology of the intersection loop.

29.4.4 Intersection Preprocessing Using a Tree Structure

The number of triangles needed to represent a surface may be quite large. The bounding box test discussed above is very fast. However, each triangle of the first surface must be compared with each triangle of the second surface. If steps were not taken to reduce the number of comparisons, this step would dominate the running time of the algorithm. There is a need to efficiently cull triangles that will not be involved in the intersection process. This is achieved by storing the triangles in a tree structure. The tree partitions the space occupied by the triangles and provides quick access to the set of triangles which inhabit a particular region. The tree type we use is a *k-d tree* (see [Samet 1990] and [Bentley 1975]). Given N triangles, the k-d tree will have at most $2N$ nodes with N leaf nodes, each containing exactly one triangle. A node is composed of a bounding box and an integer tag. The bounding box is specified by two points in space and is just large enough to contain the bounding boxes of all its children; the bounding box of a leaf node is just large enough to contain its associated triangle. We use a tag for leaf nodes to identify the triangle that is contained in the leaf.

A separate tree is constructed for each surface. One tree is chosen — it does not matter which one — as the *base tree*, and the remaining tree is referred to as the *target*. The two trees are intersected as follows:

1. Pick a leaf node in the base tree.
2. Intersect base leaf with target tree using recursive bounding box tests.
3. Associate each base leaf with each target leaf that intersects it. If the base leaf does not intersect any target leaf, then the base leaf is not considered.
4. Repeat this procedure for each leaf in the base tree.

The result of the intersection is a set of associations which encompass all possible triangle intersections for the given surfaces. Note that target leaves may be associated with multiple base leaves. However, each base leaf appears only once. This relationship is depicted in Figure 29.16. Note that a two-dimensional quadtree is used to simplify the figure. The k-d tree is a binary tree and can be searched in logarithmic time. Hence, for two surfaces represented by M and N triangles, the tree intersection can be performed in $M\log_2(N)$ time.

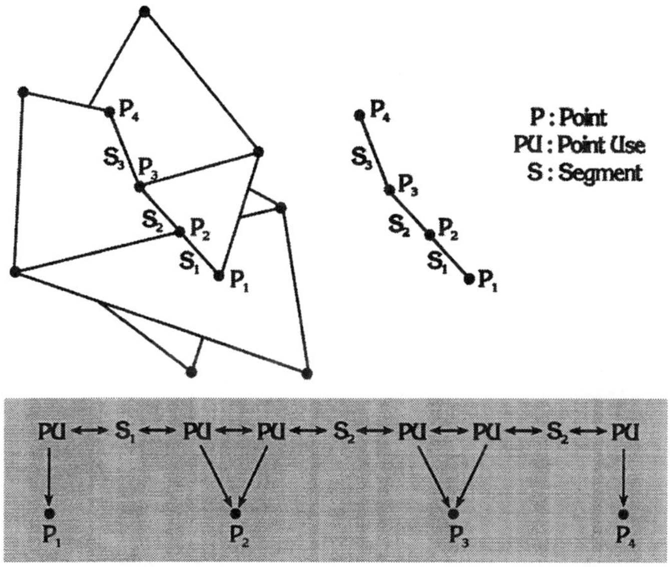

FIGURE 29.17 Data structure used to represent SSI curve topology.

29.4.5 Data Structure, Loop Detection, and Curve Tracing

The points and line segments resulting from the triangle intersection step are stored in a special topology data structure. This data structure provides explicit connectivity between the line segments and allows intersection curves to be easily identified and traced.

The *Point* structure stores the coordinates in physical space, *xyz* space, for the point as well as its associated parameter values for each of the two surfaces. A Point also has an associated circular linked list of *PointUses.* PointUse structures contain connectivity and other topological information about the Point. Each Point in the system is unique. If a new Point is computed having the same coordinates as an existing point, then a Point Use with the appropriate information is added to the list of uses for the existing Point. This list of unique points and uses builds the topology of the intersection curves as the triangles are intersected and does not depend on the order in which the intersections occur.

The PointUse structure contains topological information associated with a Point and a *Segment.* The Segment data structure provides Point connectivity information using PointUses. A Point shared by both Segments has two PointUses. Since both of these PointUses belong to the same Point, they are linked and hence the line segments are linked as well.

The PointUse structure contains a **location** field, which indicates where the PointUse is located on its associated Segment structure. This location is either zero or one indicating the start point or end point of the line segment. The **P** field and **SSISeg_PTR** field are links to the associated Point and Segment structures. So-called **prev** and **next** fields link the PointUse to others (if any) in the circular linked list of PointUses. The **InUse** field is a boolean flag used in the process of tracing the intersection curves.

Detection of individual intersection curves, *loops,* is based on PointUses. In this method, an endpoint of a curve is defined as a point with a number of PointUses not equal to two (closed curves are a special case where all points have two PointUses). If more than two PointUses are present, then the endpoint is a singular point (where three or more curves meet). Intersection curves are automatically broken at a bifurcation point.

Figure 29.17 illustrates this concept. We depict the intersection of four triangles, belonging to two different surfaces, resulting in four intersection points, $\mathbf{p}_1$, $\mathbf{p}_2$, $\mathbf{p}_3$, and $\mathbf{p}_4$. This example yields one loop whose two endpoints are $\mathbf{p}_1$ and $\mathbf{p}_4$. This is a basic outline of the overall curve tracing algorithm:

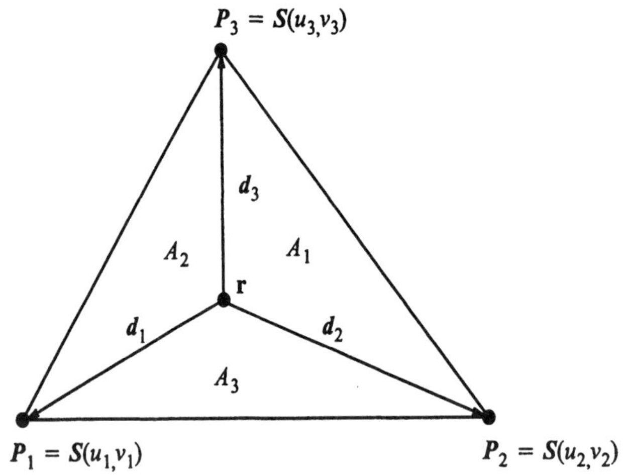

FIGURE 29.18 Stencil of data required to obtain intersection point on exact surface.

1. Find a Point with a number of PointUses ≠ 2 and at least one PointUse with its *InUse* flag set to FALSE. If none can be found, stop.
2. Go to the PointUse with InUse set to FALSE.
3. Set PointUse → InUse = TRUE and add the Point to the ordered list of sample points on the curve. (Remark: The two triangles used to generate the segment are also stored for use in refinement steps.)
4. Go to the Segment associated with the PointUse and set Segment → InUse = TRUE.
5. Go to the opposite PointUse on the Segment.
6. Set PointUse → InUse = TRUE and add its associated Point to the ordered list of sample points on the curve.
7. If the number of PointUses associated with the present Point is two, step to the other PointUse associated with the Point (PointUse → next) and go to step 4; otherwise, continue below.
8. Store the ordered list of sample points for refinement.
9. Repeat.

Should this algorithm terminate and leave certain Segments unused, then one or more of the intersection curves are closed. Closed curves are a special case and are treated separately. Closed curves are found by picking a random starting PointUse from the remaining "unused" PointUses and proceeding with the same basic algorithm. The difference is that the algorithm terminates when the curve is traced back to its starting point.

29.4.6 Refinement

Once all possible curves have been traced, the result is an ordered set of sample points for each intersection curve. In general, these points lie on the triangulation, or, to be more specific, on the piecewise linear surface approximations, but not on the exact analytical surface. The refinement procedure described below maps the points to the surfaces and matches them, within a given tolerance, to the "true" intersection.

The first step in the refinement process is the mapping of the intersection points onto each surface. Each intersection point on the triangulation has references to the triangles containing it. Figure 29.18 shows the stencil of data required to map a point **r** (inside a triangle) to the exact underlying surface. The procedure to do this follows these steps:

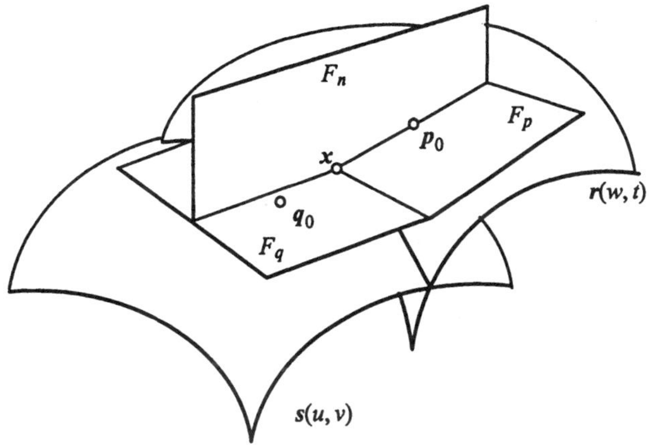

FIGURE 29.19 Point refinement using *auxilliary plane method.*

1. Find vectors $\mathbf{d}_1$, $\mathbf{d}_2$, and $\mathbf{d}_3$ emanating from $\mathbf{r}$ and stopping at the respective triangle vertices $\mathbf{P}_1$, $\mathbf{P}_2$, and $\mathbf{P}_3$.
2. Calculate the sub-triangle areas A_1, A_2, and A_3.
3. Normalize the sub-triangle areas by dividing A_1 by the total triangle area $A_1 + A_2 + A_3$.
4. The normalized sub-triangle areas A_i are the *barycentric coordinates* of $\mathbf{r}$ with respect to the original triangle defined by $\mathbf{P}_1$, $\mathbf{P}_2$, and $\mathbf{P}_3$. Denoting the parameter values of $\mathbf{P}_i$ by (u_i, v_i), we compute $\left(\sum_{i=1}^{3} A_i u_i, \sum_{i=1}^{3} A_i v_i\right)$, which is the parameter value that we use to compute a point on the exact surface replacing $\mathbf{r}$.

The refinement technique used is the *auxiliary plane method* (see Figure 29.19) described in [Hosaka 1992]. The basic steps of this method are:

1. Denote the two "images" of $\mathbf{r}$ on the two underlying parametric surfaces $\mathbf{s}(u,v)$ and $\mathbf{r}(w,t)$ by $\mathbf{q}_0$ and $\mathbf{p}_0$; let $\mathbf{p}_0 = \mathbf{s}(u_0, v_0)$ and $\mathbf{q}_0 = \mathbf{r}(w_0, t_0)$, where u_0, v_0, w_0, and t_0 are the associated parameter values.
2. Calculate the unit normals $\mathbf{n}_p$ and $\mathbf{n}_q$ at $\mathbf{p}_0$ and $\mathbf{q}_0$.
3. Let F_p and F_q be the tangent planes at $\mathbf{p}_0$ and $\mathbf{q}_0$.
4. Calculate the distance values d_p and d_q for the distances between F_p (F_q) and the origin:

$$d_p = \mathbf{n}_p \cdot \mathbf{r}(w_0, t_0), \quad d_p = \mathbf{n}_q \cdot \mathbf{s}(u_0, v_0). \tag{29.18}$$

5. Construct a plane F_n which is orthogonal to both F_p and F_q and passes through $\mathbf{p}_0$. The unit normal $\mathbf{n}_n$ of F_n and its distance from the origin d_n are:

$$\mathbf{n}_n = \frac{\mathbf{n}_p \times \mathbf{n}_q}{\left\|\mathbf{n}_p \times \mathbf{n}_q\right\|}, \tag{29.19}$$

$$d_n = \mathbf{n}_n \cdot \mathbf{r}(w_0, t_0). \tag{29.20}$$

6. Calculate the intersection point $\mathbf{x}$ of the planes F_p, F_q, and F_n as

$$\mathbf{x} = \frac{d_p\left(\mathbf{n}_q \times \mathbf{n}_p\right) + d_q\left(\mathbf{n}_n \times \mathbf{n}_p\right) + d_n\left(\mathbf{n}_p \times \mathbf{n}_q\right)}{\left[\mathbf{n}_p, \mathbf{n}_q, \mathbf{n}_n\right]}, \tag{29.21}$$

where $[\mathbf{v}_1,\mathbf{v}_2,\mathbf{v}_3]$ is the *scalar triple product* $(\mathbf{v}_1 \times \mathbf{v}_2) \cdot \mathbf{v}_3$ of three 3D vectors, see [Hosaka 1992]. (Remark: The point $\mathbf{x}$ is an approximate intersection point and, in general, will lie neither on $\mathbf{s}(u,v)$ nor on $\mathbf{r}(w,t)$.)

7. The point $\mathbf{x}$ must be mapped back to the exact surfaces and new points $\mathbf{p}_0$ and $\mathbf{q}_0$ calculated. We compute the difference vectors $\delta\mathbf{p}_0 = \mathbf{x} - \mathbf{p}_0$ and $\delta\mathbf{q}_0 = \mathbf{x} - \mathbf{q}_0$ and compute the values

$$\acute{\mathbf{r}}_w = \mathbf{r}_w \times \mathbf{n}_p, \quad \acute{\mathbf{r}}_t = \mathbf{r}_t \times \mathbf{n}_p \tag{29.22}$$

and

$$\acute{\mathbf{s}}_u = \mathbf{s}_u \times \mathbf{n}_q, \quad \acute{\mathbf{s}}_v = \mathbf{s}_v \times \mathbf{n}_q, \tag{29.23}$$

where $\mathbf{r}_w$, $\mathbf{r}_t$, $\mathbf{s}_u$, and $\mathbf{s}_v$ are the partial derivative vectors (not normalized) at $\mathbf{p}_0$ and $\mathbf{q}_0$. Considering that for infinitesimally small increments the two equations $\mathbf{r}_w\delta w + \mathbf{r}_t\delta t =$ $\delta\mathbf{p}_0$ and $\mathbf{s}_u\delta u + \mathbf{s}_v\delta v = \delta\mathbf{q}_0$ hold, we can compute the increments for the parameter values as

$$\delta w = \frac{\acute{\mathbf{r}}_t \cdot \delta\mathbf{p}_0}{\acute{\mathbf{r}}_t \cdot \mathbf{r}_w}, \quad \delta t = \frac{\acute{\mathbf{r}}_w \cdot \delta\mathbf{p}_0}{\acute{\mathbf{r}}_w \cdot \mathbf{r}_t}, \tag{29.24}$$

$$\delta u = \frac{\acute{\mathbf{s}}_v \cdot \delta\mathbf{q}_0}{\acute{\mathbf{s}}_v \cdot \mathbf{s}_u} \quad \delta v = \frac{\acute{\mathbf{s}}_u \cdot \delta\mathbf{q}_0}{\acute{\mathbf{s}}_u \cdot s_v}. \tag{29.25}$$

The updated values of $\mathbf{p}_0$ and $\mathbf{q}_0$ are thus given by

$$\mathbf{p}_0 = \mathbf{r}\left(w_0 + \delta w, t_0 + \delta t\right), \quad \mathbf{q}_0 = \mathbf{s}\left(u_0 + \delta u, v_0 + \delta v\right). \tag{29.26}$$

8. Steps 2 through 7 are repeated until $\|\mathbf{p}_0 - \mathbf{q}_0\|$ is within a specified tolerance.

Convergence of this method is very good, even for "poor" initial values of $\mathbf{p}_0$ and $\mathbf{q}_0$. The curve defined by the triangle intersections may or may not meet the requested tolerance. Intersection points can be added or deleted as necessary. Additional intersection points are obtained using the refinement algorithm with starting points based on the known intersection points. The final representation of the curve depends on the requirements of a particular application. Common representations are piecewise linear or cubic curve representations in physical and/or parameter space.

29.5 Research Issues and Summary

In conclusion, we summarize the presented techniques, describe possible improvements, and point out remaining research issues.

29.5.1 Surface Refinement and Reparametrization

We have given techniques for *refining* the parametrization of a NURBS surface. The surface approximation method performs best when the interior surface geometry more or less follows the geometry of the boundary curves. Future work could be directed at the analysis of geodesic curvature distribution of isoparametric curves on the surface and using it as an interrogation tool. Data reduction is another research issue. In some cases, this is achieved as a side effect. Torsion characteristics of the boundary curves are not exploited. Torsion could be incorporated into the scheme in the same way as curvature and arc length to find "key interpolation points" where torsion may be a factor.

29.5.2 Approximation of Discontinuous Geometries

The interactive method for the approximation of discontinuous geometries allows the "replacement" of entire 3D geometries while preserving original boundary curves of given surfaces, if so desired. The method can be used to approximate geometries with gaps, transverse surface overlaps, and undesired surface intersections. The final approximation is a set of bicubic B-spline surfaces determining an overall continuous surface approximation — with negligible gaps between neighbor B-spline surfaces. One should explore whether it is possible to reduce the required user input further, i.e., whether one can construct local approximants automatically without having a user specify the boundary curves of the initial Coons approximants. Currently, the resulting B-spline surfaces approximating the given discontinuous geometry are stored as NURBS surfaces — with all control point weights being one. Choosing control point weights in a more "clever" way might allow the generation of equally good approximants with fewer control points.

29.5.3 Surface–Surface Intersection

The SSI algorithm discussed above is only one of many possible approaches to solving this difficult problem. The advantages of this algorithm are its speed and accuracy and the ability to operate automatically. The method relies on triangles that are locally planar approximations to the underlying surface; therefore, the algorithm can have difficulties when resolving intersection curve topologies near critical points, where critical points occur when the normals of both intersecting surfaces are exactly or nearly collinear. In the region around critical points, the intersection of the surface triangulations may not accurately reflect the intersection of the underlying surfaces, hence causing the algorithm to fail.

Acknowledgments

This work was supported by the National Grid Project consortium and the National Science Foundation under contract EEC-8907070 to Mississippi State University. Special thanks go to all members of the research and development team of the National Grid Project, which was performed at the NSF Engineering Research Center for Computational Field Simulation, Mississippi State University. Part of the work was carried out by the CAGD research group at Arizona State University.

Further Information

The following journals, magazines, and conference proceedings frequently cover topics related to the problems discussed in this chapter: *Computer-Aided Design* (Elsevier), *Computer Aided Geometric Design* (Elsevier), *Journal of Computational Physics* (Academic Press), *Transactions on Graphics* (ACM), *Transactions on Visualization and Computer Graphics* (IEEE), *The Visual Computer* (Springer-Verlag), *Computer Graphics and Applications* (IEEE), *SIGGRAPH* proceedings (ACM), and *Supercomputing* proceedings (ACM/IEEE). In addition, the *SIAM Conference on Geometric Design*, which is organized every other year, is an excellent source of information.

References

1. Anderson, J., Khamayseh, A., and Jean, B.A., Adaptive resolution refinement, Technical Report, Los Alamos National Laboratory, Los Alamos, NM, 1997.

2. Bartels, R.H., Beatty, J.C. and Barsky, B.A., *An Introduction to Splines for Use in Computer Graphics and Geometric Modeling*, Morgan Kaufmann, Los Altos, CA, 1987.
3. Bentley, J., Multidimensional binary search trees used for associative searching, *Communications of the ACM.* 1975, 18, 9.
4. Coons, S.A., Surface patches and B-spline curves, Barnhill, R.E. and Riesenfeld, R.F., (Eds.), *Computer Aided Geometric Design*, Academic Press, San Diego, CA, 1974, pp 1–16.
5. Crampin, M., Guifo R., and Read, G.A., Linear approximation of curves with bounded curvature and a data reduction algorithm, *Computer Aided Design*, 1985, 17,6, pp 257–261.
6. Farin, G., *NURB Curves and Surfaces*, AK Peters, Wellesley, MA, 1995.
7. Farin, G., *Curves And Surfaces For Computer Aided Geometric Design*, 4th Edition, Academic Press, San Diego, CA, 1997.
8. Faux, I.D. and Pratt, M.J., *Computational Geometry for Design and Manufacture*, Ellis Horwood, New York, NY, 1979.
9. Foley, T., Local control of interval tension using weighted splines, *Computer Aided Geometric Design.* 1986, 3, 4, pp 225–230.
10. Foley, T., Interpolation with interval and point tension controls using cubic weighted ν-splines, *ACM Trans. on Math. Software*, 1987, 13,1, pp 68–96.
11. Foley T. and Nielson G.M., A Survey of applications of an affine invariant norm, Lyche, T. and Schumaker, L.L., (Eds.), *Mathematical Methods in Computer Aided Geometric Design*, Academic Press, San Diego, CA, 1989, pp 445–467.
12. Franke, R., Scattered data interpolation: tests of some methods, *Math. Comp.* 1982, 38, pp 181–200.
13. Fuhr, R.D. and Kallay, M., Monotone Linear Rational Spline Interpolation, *Computer Aided Geometric Design*, 1982, 9, pp 313–319.
14. George, P.L., *Automatic Mesh Generation*, Wiley & Sons, New York, 1991.
15. Hamann, B., Construction of B-spline approximations for use in numerical grid generation, *Applied Mathematics and Computation*, 1994, 65, 1–3, pp 295–314.
16. Hamann, B. and Jean, B.A., Interactive techniques for correcting CAD/CAM data, Weatherill, N.P., Eiseman, P.R., Häuser, J., and Thompson, J.F., (Eds.), *Numerical Grid Generation in Computational Fluid Dynamics and Related Fields*, Pineridge Press, Swansea, U.K., 1994, pp 317–328.
17. Hamann, B. and Jean, B.A., Interactive surface correction based on a local approximation scheme, *Computer Aided Geometric Design*, 1996, 13, 4, pp 351–368.
18. Hosaka, M., *Modeling of Curves and Surfaces in CAD/CAM.* Springer–Verlag, New York, 1992.
19. Hoschek, J. and Lasser, D., *Fundamentals of Computer Aided Geometric Design*, AK Peters, Wellesley, MA, 1993.
20. Jean, B.A. and Hamann, B., An efficient surface-surface intersection algorithm based on surface triangulations and space partitioning trees, to appear in *Mathematical Engineering in Industry*, 1998.
21. Kim, T.W., Knot placement for NURB interpolation, M.S. thesis, Department of Computer Science and Engineering, Arizona State University, Tempe, AZ, 1993.
22. Knupp, P.M. and Steinberg, S., *Fundamentals of Grid Generation.* CRC Press, Boca Raton, FL, 1993.
23. Piegl, L.A., On NURBS: A survey, *IEEE Computer Graphics and Applications.* 1991a, 11, 1, pp 55–71.
24. Piegl, L.A., Rational B-spline curves and surfaces for CAD and graphics, Rogers, D.F. and Earnshaw, R.A., (Eds.) *State of the Art in Computer Graphics*, 1991b, Springer-Verlag, New York, pp 225–269.
25. Piegl, L.A. and Tiller, W., *The NURBS Book*, 2nd Edition, Springer-Verlag, New York, 1996.
26. Samet, H., *The Design and Analysis of Spatial Data Structures.* Addison Wesley, New York, 1990.
27. Thompson, J.F., Warsi, Z.U.A., and Mastin, C.W., *Numerical Grid Generation*, North-Holland, New York, 1985.
28. 3D Systems, Inc., *Stereo Lithography Interface Specification.* 1989.
29. Yu, T. and Soni, B.K., Application of NURBS in numerical grid generation, *Computer Aided Design*, 1995, 27, pp 147–157.

30

NURBS in Structured Grid Generation

Tzu-Yi Yu

Bharat K. Soni

30.1 Introduction

The parametric-based *nonuniform rational B-spline* (NURBS) is a widely utilized representation for geometrical entities in CAD/CAM/CAE systems. The convex hull, local support, shape-preserving forms, affine invariance, and variation diminishing properties of NURBS are extremely attractive in engineering design applications. These properties with associated mathematical and numerical development of NURBS, including evaluation algorithms and knot insertion and degree elevation schemes, are described in detail in Chapters 27 and 28 of this handbook. The first commercial product that used the NURBS to represent geometry came from the Structural Dynamics Research Cooperation (SDRC) in 1983. The Boeing company proposed the NURBS as an IGES (initial graphics exchange specification) standard in 1981, and now the NURBS curve and NURBS surface have been adopted as the IGES geometric entities 126 and 128. The IGES format has become the *de facto* standard IO (input/output) for exchanging data between various CAD/CAM and CAE systems. Recently, the IGES entities 126 and 128 have become increasingly popular in grid (mesh) generation, computational field simulations (CFS) and in general computer aided engineering analysis and simulation systems. In view of this popularity, the NASA Surface Modeling and Grid Generation Steering Committee established a NASA-IGES (a subset of the standard IGES) format in 1992, and has further proposed the NINO (NASA-IGES NURBS ONLY) standard. Detailed description of IGES entities and the NASA-IGES and NINO standards are presented in Chapter 31 of this handbook. Most of the geometrical configurations of interest to practical CFS problems are designed in the CAD/CAM systems and are available to an analyst in an IGES format. The geometry

0-8493-2687-7/99/$0.00+$.50

preparation which is considered as the most critical and labor intensive part of CFS involves the discrete–sculptured definitions of all boundaries/surfaces, with a desired point distribution and smoothness and orthogonality criteria, associated with the domain of interest. The algorithms associated with geometry preparation and structured grid generation based on the NURBS are presented in this chapter.

The NURBS-based geometry preparation for addressing complex CFS problems encountered in industrial environments involves

1. Transformation of widely utilized explicitly/implicitly/discretely defined IGES geometric entities into a common data structure involving NURBS
2. Surface reparametrization for poorly defined surfaces and repairing of faulty surfaces (most common faults involve gaps, overlaps, and undesired discontinuity between neighboring surface patches) and pertinent geometric entities
3. Geometrical operations allowing projections, intersections (surface–surface intersections), composition, union, and other related transformations essential for surface grid generation with desired topological criteria
4. Grid point distribution with desired stretching and quality criteria on domain boundaries/surfaces

The algorithms for transforming widely utilized geometric entities into NURBS, composition of curves and surfaces and their respective NURBS definitions, grid point distribution, and surface/volume reparametrization are presented in this chapter. However, the algorithms for surface reparametrization, approximation of faulty surfaces, and surface–surface intersections are described in Chapter 29 of this handbook.

The transfinite interpolation (*TFI*) technique is a widely used algebraic method for structured grid generation (see Chapter 3). The NURBS formulation of *TFI* method [Yu, 1995] for surface and volume grids is described in this chapter with appropriate illustrations. The NURBS control-volume-based three-dimensional grid generation algorithms for ruled, extruded, and composite volume and volumes of revolution are also presented in this chapter. The applications of these algorithms facilitate the NURBS common data structure for geometry preparation and grid generation.

30.2 NURBS Formulation

The parametric representation of a NURBS curve/surface/volume involves control polygons with weights, knot vectors (vectors in higher dimension), and orders (representative of degree of polynomial in each direction).

A NURBS curve of order k is defined [Farin, 1993] as

$$C(t) = \frac{\sum_{i=0}^{n} W_i d_i N_i^k(t)}{\sum_{i=0}^{n} W_i N_i^k(t)} \tag{30.1}$$

where the d_i, $i = 0, \ldots,$ n denote the deBoor control polygon and the W_i are the weights associated with each control point. The $N_i^k(t)$ is the normalized B-spline basis function of order k and is defined over a knot vector $T = T_i$, $i = 0, \ldots, n + k$ by the recurrence relations

$$\begin{aligned} N_i^k(t) &= \frac{(t - T_i) N_i^{k-1}(t)}{T_{i+k-1} - T_i} + \frac{(T_{i+k} - t) N_{i+1}^{k-1}(t)}{T_{i+k} - T_{i+1}} \\ N_i^1(t) &\begin{cases} = 1 & \text{if } T_i \le t < T_{i+1} \\ = 0 & \text{otherwise} \end{cases} \end{aligned} \tag{30.2}$$

Throughout this chapter, it is assumed that the knot vector has the form $T = \{0, \ldots, 0, T_k, \ldots, T_n, 1, \ldots, 1\}$ with the multiplicity k for the knot value 0 and 1 on both ends of the knot vector. If the knot vectors do not match this format, the knot insertion [Yu, 1995] techniques must be used to achieve the multiplicity of k on the ends of the knot vector, and if the end knot values are not 0 and 1, the knot vector must be normalized by the last knot value to match this format. Because shifting and scaling (normalizing) of the knot values does not alter the underlying geometry, the basis function defined in Eq. 30.2 can be normalized appropriately.

The NURBS surface is defined as a tensor product extension of the curve representation in two directions and is formulated as

$$S(s,t) = \frac{\sum_{i=0}^{m}\sum_{j=0}^{n} W_{ij} d_{ij} N_i^{k1}(s) N_j^{k2}(t)}{\sum_{i=0}^{m}\sum_{j=0}^{n} W_{ij} N_i^{k1}(s) N_j^{k2}(t)} \tag{30.3}$$

where d_{ij} denotes the 3D control net, and W_{ij} are the *weights* associated with each control point. The $N_i^{k1}(s)$ and $N_j^{k2}(t)$ denote the normalized B-spline basis functions of *order* $k1$ and $k2$ over the two knot vectors $T_1 = T_i$, $i = 0, \ldots, m + k1$ and $T_2 = T_j$, $j = 0, \ldots, n + k2$ in the I and J directions, respectively. The definition of the B-spline basis functions of the NURBS surface is exactly the same as in Eq. 30.2.

The formula for the 3D NURBS volume is defined analogous to the NURBS surface and is a 3D tensor product form written as

$$V(s,t,u) = \frac{\sum_{i=0}^{m}\sum_{j=0}^{n}\sum_{l=0}^{p} W_{ijl} d_{ijl} N_i^{k1}(s) N_j^{k2}(t) N_l^{k3}(u)}{\sum_{i=0}^{m}\sum_{j=0}^{n}\sum_{l=0}^{p} W_{ijl} N_i^{k1}(s) N_j^{k2}(t) N_l^{k3}(u)} \tag{30.4}$$

The d_{ijl} form the 3D control volume, and the W_{ijl} are *weights* associated with each control point. The $N_i^{k1}(s)$, $N_j^{k2}(t)$, and $N_l^{k3}(u)$ are the normalized B-spline basis functions of *order* $k1$, $k2$, and $k3$ over the three knot vectors $T_1 = T_i$, $i = 0, \ldots, m + k1$, $T_2 = T_j$, $j = 0, \ldots, n + k2$ and $T_3 = T_l$, $l = 0, \ldots, p + k3$ in the I, J, and L directions (i.e., the s, t, u directions), respectively.

30.3 Transforming and Generating Procedures

Transforming procedures for the widely encountered non-NURBS geometric curves and surfaces, including the *TFI* method of surface generation, to NURBS representation are described in this section. To model a NURBS entity, according to Eqs. 30.1–30.4, one should define the control polygons (or control net/volume), weights, knot vectors, and the respective order for the polynomial in each direction.

It is well known that most commonly used geometric entities in engineering design can be analytically transformed to a NURBS representation [Piegl, 1987, 1989, 1991]. However, there are many practical issues not addressed in the transforming procedures published in the open literature. For example, the IGES representation of the implicit conic arc, and important entity, is not contained in those references. The transforming algorithm for a general circular arc (a circular arc with arbitrary starting and ending points) is also missing. The procedure for transforming a surface of revolution into a NURBS is provided only for a 360° revolution, but many grid generation applications require a specified range of angle, such as 60°. Also, several transforming algorithms are never discussed in any of the literature. These include the transfinite interpolation (also known as the *TFI*) for a NURBS surface/volume and the modeling of

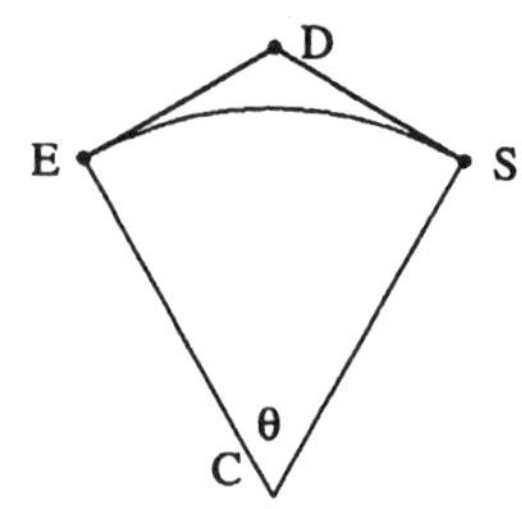

FIGURE 30.1 The basic control triangle for a circular arc.

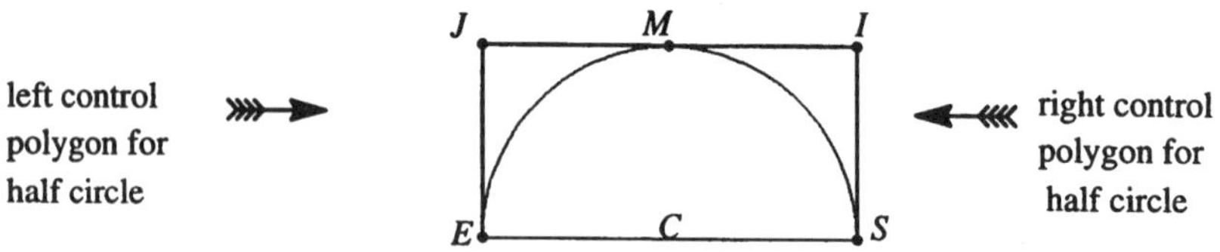

FIGURE 30.2 The NURBS control polygon for a semicircle.

the superellipse as a NURBS curve. The enhancements and generalizations of the transforming procedures were accomplished [Yu, 1995] to meet needs arising from the grid generation process for complex geometries defined in a CAD/CAM system.

These algorithms for transforming various non-NURBS definitions to NURBS representations are described in the following section. (Only the generalized (or enhanced) algorithms are described. The other transforming procedures that can be found in open literature are not repeated here.)

30.3.1 General Circular Arc to NURBS Curve

A circular arc as defined in the IGES standard is represented by a center point, starting point, and ending point within a given constant Z plane. The two endpoints and the center point form an arbitrary sector angle that does not necessarily start from zero. It has been shown that any circular arc with a sector angle less than or equal to 90° can be represented by NURBS [Piegl, 1989, 1991]. The basic control polygon for this NURBS representation is shown in Figure 30.1.

In Figure 30.1 C is the center point, S is the starting point, and E is the ending point. The sector angle SCE (θ) is less than or equal to 90°. The two tangent lines SD and ED intersect at D. The *order* of this control polygon is 3, with the control points S, D, E (hence, the n in Eq. 30.1 will be set to 2) and the *weights* are 1, cos ($\theta/2$), and 1, respectively. The associated knot vector is (0, 0, 0, 1, 1, 1). A circular arc with a sector angle greater than 90° and less than or equal to 180° can be represented by two arcs with one half of the original sector angle. For each of these two sections the previous procedure can be used to evaluate the corresponding control polygon. A 180° circular arc represented by two control polygons is illustrated in Figure 30.2.

From Figure 30.2 the two control polygons can be combined and the common point M can be eliminated. The resulting NURBS information is setting the control polygon to $SIMJE$ (hence, the n is 4), the knot vector to (0, 0, 0, 1/2, 1/2, 1, 1, 1) and the *weights* to (1., cos(θ/n), 1., cos(θ/n), 1.). A similar procedure can be used for circular arcs between 180° and 270° (with n equal to 6) resulting in a final knot vector of (0, 0, 0, 1/3, 1/3, 2/3, 2/3, 1, 1, 1) and *weights* (1., cos(θ/n), 1., cos(θ/n), 1., cos(θ/n), 1., cos(θ/n),1.), and a knot vector of (0, 0, 0, .25, .25, .5, .5, .75, .75, 1, 1, 1) and *weights* (1., cos(θ/n), 1., cos(θ/n), 1., cos(θ/n), 1., cos(θ/n), 1.) for arcs between 270° and 360° for n equal to 8. These four cases are shown in Figure 30.3.

This approach handles all possible circular arcs with no extra computation (such as knot insertion) involved. Furthermore, this algorithm results in good distribution for all cases.

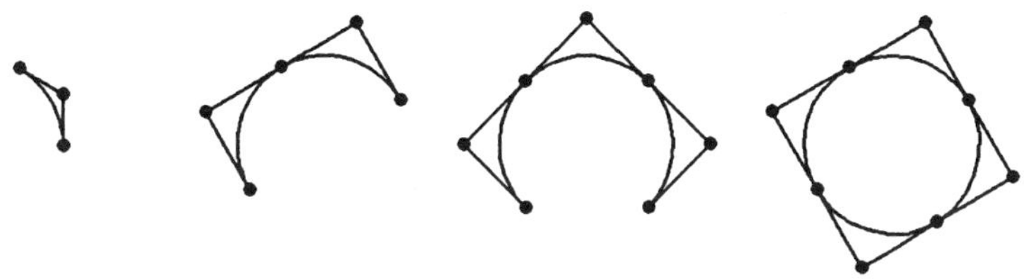

FIGURE 30.3 Arbitrary circular arc with the NURBS control polygons.

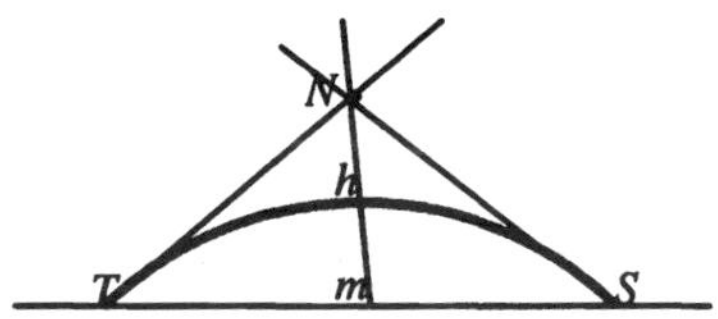

FIGURE 30.4 Basic NURBS control polygon for a conic arc.

30.3.2 Conic Arc to NURBS Curve

The transforming procedure for a conic arc was discussed in [Piegl, 1989, 1991], where the authors described the case of three given control points and changing the weight (conic shape factor) to produce a different family of conic arcs (elliptic, hyperbolic, or parabolic arcs). However, that case is completely different from the one discussed in this section. The conic arc defined in IGES is represented by an implicit form $Ax^2 + Bxy + Cy^2 + Dx + Ey + F = 0$, with starting point S and ending point T supplied (counterclockwise). The transforming procedure for a basic conic arc is illustrated in Figure 30.4. In this figure, m is the middle point of line TS. Since the two endpoints are known, the two slopes of the tangent lines at the end points can be obtained. The equations describing these two tangent lines can be formed and the intersection point N can be determined. This is accomplished as follows: Differentiate the implicit form of the conic equation to obtain $2Ax + By + Bxy' + 2Cyy' + D + Ey' = 0$. Solving the equation yields

$$y' = (2Ax + By + D)/(-2Cy - Bx - E) \tag{30.5}$$

Substitution of the coordinates of the two endpoints S and T into Eq. 30.5 yields the two desired straight lines. The shoulder point h can then be obtained by solving for the intersection of the line Nm and the given implicit equation. The control triangle is then defined by the polygon SNT (hence, the n is 2 for this case) with *weights* of (1, mh/hN, 1). The *order* can be set to 3 and the knot vector is defined in a manner analogous to the circular arc. As long as this basic control triangle can be found, the procedure used for the circular arc with the sector angle greater than 90° can be applied to the conic arc by simply combining the different control triangles together to form the final control polygon and by setting the proper knot vector. The definition of the sector angle θ for the conic arc is only applied to the elliptic arc; for the parabolic or hyperbolic arcs, three control points are sufficient to form the control polygon. Hence, for a parabolic or hyperbolic arc, the knot vector is always (0., 0., 0., 1.0., 1.0., 1.0) with n equal to 2 Figure 30.5 shows different conic arcs represented by the NURBS using this algorithm. From left to right, (I) elliptic arc with equation $2x^2 + 4xy + 5y^2 - 4x - 22y + 7 = 0$, formed by two NURBS control polygons, (II) parabolic arc with equation $4x^2 - 4xy + y^2 - 2x - 14y + 7 = 0$, (III) hyperbolic arc with equation $2x^2 + 4/3xy - 2y^2 - 16 = 0$.

30.3.3 Cubic Parametric Curve to NURBS Curve

The cubic parametric curve defined in IGES format is a sequence of parametric polynomial segments. More precisely, it is composed of N ($N \geq 1$) pieces of cubic parametric segments. For each parametric curve, it is defined as

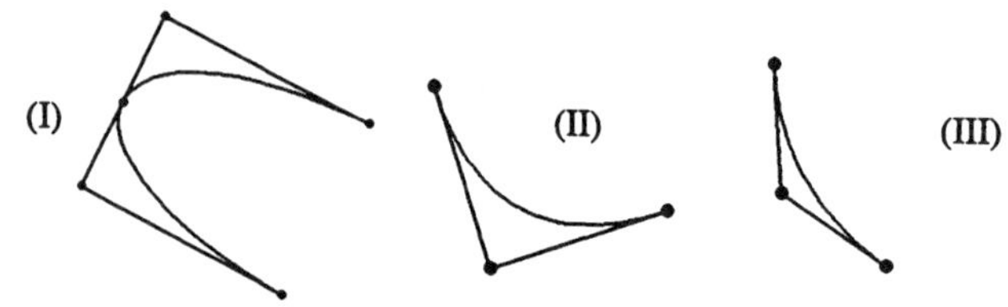

FIGURE 30.5 NURBS control polygon represent different conic arcs.

$$C(u) = a + bt + ct^2 + dt^3 \quad T(i) \le u \le T(i+1) \text{ and } t = u - T(i) \tag{30.6}$$

$T(i)$, $i = 1, \ldots, N + 1$ are the breakpoints. It has been proven [Farin, 1993] that the cubic Bézier curve is a special case of a B-spline curve with knots vector of (0, 0, 0, 0, 1, 1, 1, 1) (no interior knot value). Also, the B-spline curve is a special case of a NURBS curve with all weights equal to 1. The mathematical transformation from a parametric cubic spline curve in IGES definition to NURBS is accomplished as follows.

The matrix form of each simple cubic parametric curve, according to Eq. 30.6, can be expressed as $C(t) = [1\ t\ t^2\ t^3]\ I_{4\times4}\ [a\ b\ c\ d]^T$ where $I_{4\times4}$ is the identity matrix and $[a\ b\ c\ d]^T$ is the transposed matrix containing the coefficients of the cubic curve. The matrix form of the cubic Bézier curve is expressed as $C(t) = [1\ t\ t^2\ t^3]\ B_{4\times4}\ [b_0\ b_1\ b_2\ b_3]^T$. The $B_{4\times4}$ is the cubic Bézier matrix, and $[b_0\ b_1\ b_2\ b_3]^T$ is the transposed matrix containing the Bézier control polygon. The strategy is to first transform the cubic parametric curve to Bézier form, since a Bézier curve can be treated as a special case of a NURBS curve. Each segment of the parametric spline curve is transformed to a Bézier curve by finding the associated Bézier control polygon. This is done by setting the two matrix equations to be equal, as shown in Eq. 30.7:

$$\text{Bezier} = \begin{bmatrix} 1\ t\ t^2\ t^3 \end{bmatrix} \begin{bmatrix} 1 & 0 & 0 & 0 \\ -3 & 3 & 0 & 0 \\ 3 & -6 & 3 & 0 \\ -1 & 3 & -3 & 1 \end{bmatrix} \begin{bmatrix} b_0 \\ b_1 \\ b_2 \\ b_3 \end{bmatrix} = \text{Cubic curve} = \begin{bmatrix} 1\ t\ t^2\ t^3 \end{bmatrix} \begin{bmatrix} 1 & 0 & 0 & 0 \\ 0 & 1 & 0 & 0 \\ 0 & 0 & 1 & 0 \\ 0 & 0 & 0 & 1 \end{bmatrix} \begin{bmatrix} a_i \\ b_i \\ c_i \\ d_i \end{bmatrix} \tag{30.7}$$

and solving the Eq. 30.7 for the Bézier control polygon. Since the cubic parametric spline defined in IGES is composed of N pieces of cubic curves, the range of parametric value t for each piece is not the same as that of the Bézier curve. Hence, a reparameterization of the cubic parametric curve is necessary. For each piece of cubic curve, the coefficients $[a_i\ b_i\ c_i\ d_i]^T$ can be obtained from the given data; therefore, the final equation to solve (for each segment) is

$$\begin{bmatrix} b_0 \\ b_1 \\ b_2 \\ b_3 \end{bmatrix} = \frac{1}{3} \begin{bmatrix} 3 & 0 & 0 & 0 \\ 3 & 1 & 0 & 0 \\ 3 & 2 & 1 & 0 \\ 3 & 3 & 3 & 3 \end{bmatrix} \begin{bmatrix} a_i \\ b_i h \\ c_i h^2 \\ d_i h^3 \end{bmatrix} \tag{30.8}$$

where $h = T(i + 1) - T(i)$ and $T(i)$ is the break value. After all the Bézier control polygons have been obtained, one can join them together and set the multiplicity of joint knot value equal to 3 to form the final B-spline curve. For example, if two cubic Bézier control polygons are obtained, the final knot vector will be set as (0, 0, 0, 0, 0.5, 0.5, 0.5, 1, 1, 1, 1), and the final curve would be C^0 continuous with *order* equal to 4 and all *weights* equal to 1. Figure 30.6 (not applying the knot removal algorithm) demonstrates this approach.

FIGURE 30.6 B-spline control polygon for parametric curve with two segments.

30.3.4 Composite Curve to NURBS Curve

A composite curve in the IGES format is defined as a curve entity consisting of lists of constituent curves. The constituent curve can be any parameterization curve except another composite curve. And this entity is a directed curve, which means the direction of the composite curve is induced by the direction of the constituent curves in the following manner. The start point for the composite curve is the start point of the first curve entity appearing in the defining list, and the terminate point for the composite curve is the terminate point of the last constituent curve appearing in the defining list. Within the defining list itself, the terminate point of each constituent curve entity has the same coordinates as the start point of the succeeding curve entity. It is quite difficult to represent the composite curve precisely without transforming all the constituent curves to the NURBS form. After transforming all curve entities (like straight lines, circular arcs, conic arc, parametric curves and rational B-spline curves), the "NURBS Joining" algorithm for all the constituent NURBS curves is performed to form the NURBS representation for the composite curve. The procedure is illustrated as follows.

Suppose two constituent curves C_1 and C_2 (already transformed to NURBS definition) form a composite curve. Then the first step is to perform the degree of elevation [Piegl, 1991] of the lower-degree curve so that the curves can have the same *order.* The second step is to adjust the knot vector of the second curve C_2 so that the first knot value of the second curve can have the same value as the last knot value of the first curve. Shifting the knot vector will not change the original NURBS curve because the basis function is a "normalized" basis function. The third step is to build up the final knot vector by joining the two knot vectors into one knot vector, and to set that knot value at the joint point to have the multiplicity equal to (*order* –1). For example, if the first knot vector is [0, 0, 0, 1, 1, 1] and the second knot vector is [2, 2, 2, 3, 3, 3], the second knot vector is adjusted by shifting –1 to each value. Thus, the second knot vector becomes [1, 1, 1, 2, 2, 2]. Suppose the *order* of these two curves is 3, then the final knot vector should be [0, 0, 0, 1, 1, 2, 2, 2] (one may notice the interior knot 1 has multiplicity of (*order* –1) = 2). The fourth step is to match the *weights* by timing the ratio of (the last *weight* of the first curve / the first *weight* of the second curve) to all the *weights* of the second curve so that the *weights* at the joint point for the two curves are the same. The last step is to build up the final control polygon and *weights* by throwing the first control point and *weight* of the second curve away and joining the others as one control polygon and one *weights* vector. After these procedures have been applied to all the constituent curves, a composite NURBS curve should be formed. One more procedure that may apply to this final curve is to perform the knot removal to remove the redundant knot vector [Tiller, 1983, 1992]. Figure 30.7 shows this algorithm for transforming the composite curve consisting of, from right to left, one straight line, one circular arc, one straight line, one ellipse arc, and another straight line.

30.3.5 Superellipse to NURBS Curve

A superelliptic arc can be described as Eq. 30.9:

$$\left(\frac{x}{a}\right)^{\eta} + \left(\frac{y}{b}\right)^{\eta} = 1 \tag{30.9}$$

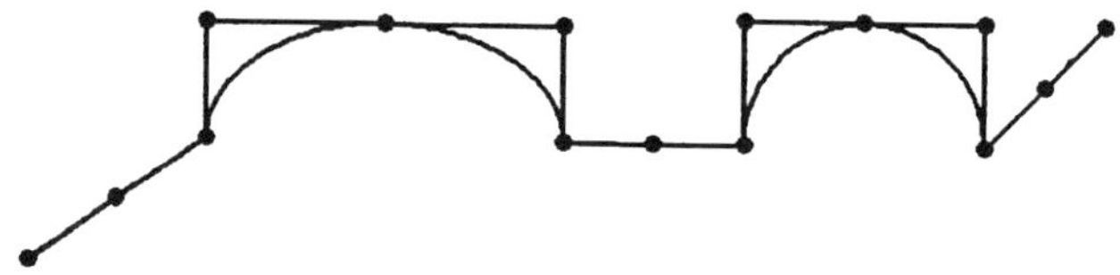

FIGURE 30.7 NURBS control polygon for a composite curve.

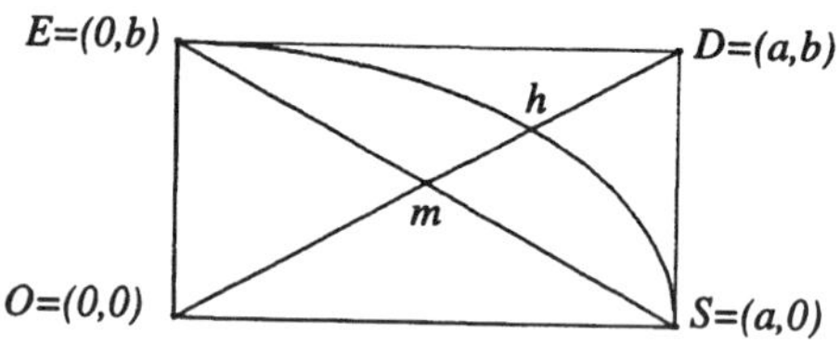

FIGURE 30.8 A superellipse arc with the NURBS control polygon.

where a is the semimajor and b is the semiminor axis of the superellipse. Special cases of Eq. 30.9 include a circle (with $a = b$, and $\eta = 2$), an ellipse (with $a \neq b$, and $y = 2$) and a rectangle (with $a \neq b$, and $\eta = \infty$).

The superellipse is a commonly used geometric description in aerospace design. An example of this is the modeling of a transition duct used for the test of a single-engine nozzle [Reichert et al., 1994]. The transition duct was designed by using a sequence of constant area, superelliptic cross sections according to Eq. 30.9. In this chapter the transforming of this superellipse to a NURBS curve is presented as follows.

This transforming approach is a combination of the circular arc and the conic arc algorithms. Consider a superellipse with semimajor a and semiminor b in the first quadrant, as shown in Figure 30.8. This arc starts at the point $(a, 0)$ and ends at the point $(0, b)$. Two tangent lines intersect at the point (a, b). Similar to the algorithm for the circular arc, these three points can be used as the NURBS control polygon while setting the *order* to be 3 with knot vector (0., 0., 0., 1., 1., 1.). The *weights* at the starting and ending control polygon can be set to 1.0. The only problem remaining is determining the *weight* at the middle point D of the control polygon. This is done similarly to the algorithm of the conic arc. The straight line OD is constructed to intersect with the line SE and the superelliptic arc at the points of m and h. The *weight* at point D is then set as the ratio of (hm/hD).

This approach is self-explanatory. When the exponent of the superellipse η increases, the arc is changing from a circular arc to a rectangular arc, this means that point h is approaching the control point D. Also, the distance of hD is decreasing, and as a result, the *weight* at point D is increased. This situation matches the NURBS theory — a NURBS curve is pulled toward the control point when the *weight* of this control point increases. The mathematical verification can also be done by comparing the h point with the shoulder point evaluation from the NURBS representation. Since the variables (a, b, and η) of the superellipse are all given, the h can be solved from the intersection of the line OD and the arc. On the other hand, after the entire NURBS representation is set up for this superellipse, the shoulder point h can also be evaluated with the parametric value $t = 0.5$. Comparing the locations of these h's, one can find out that the relative deviation is as small as 1.0e – 9. Table 30.1 shows the selected values of the exponent η of the superellipse and the corresponding values of *weights*.

From Table 30.1 one can also notice that in the case of a circular arc or an elliptic arc (when $\eta = 2$), the corresponding *weight* (for the sector angle equal to 90°) is the same as cos(90°/2.0), which has been discussed in circular/elliptic arc section.

30.3.6 Bicubic Parametric Spline Surface to NURBS Surface

The cubic parametric spline surface defined by IGES is composed of M by N cubic patches, as illustrated in Figure 30.9. For each cubic patch, it can be represented as Eq. 30.10:

TABLE 30.1 The Relationship between Exponent η and Weights

η	Weight
2.000000	0.7071067807
2.076143	0.7615055209
2.184741	0.8391550277
2.310944	0.9294727665
2.446475	1.0265482055
2.736506	1.2345144266
2.894152	1.3476587943
3.064489	1.4699782629
3.250206	1.6034070829
3.676614	1.9099667660
3.924127	2.0880154404
4.201364	2.2875017047
4.515468	2.5136151423
4.875638	2.7729511992
5.293192	3.0736854139
5.786112	3.4287875496
6.375087	3.8531827169
7.047038	4.3374610450
7.759080	4.8507150955
8.451551	5.3499221183
9.061041	5.7893464878
9.533431	6.1299460466
9.999865	6.4662654998
10.00000	6.4663630857

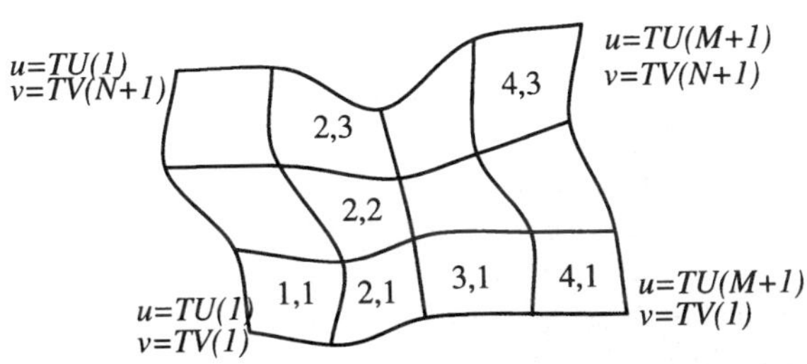

FIGURE 30.9 A 4×4 Bicubic parametric patch.

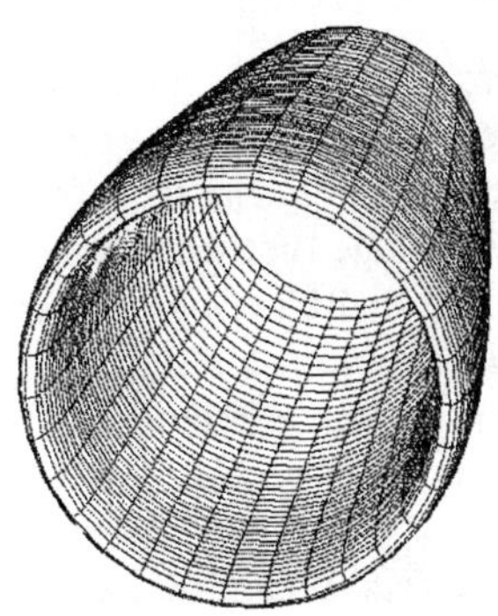

FIGURE 30.10 B-spline surface converted from the bicubic parametric surface.

$$S(u,v) = a + bs + cs^2 + ds^3 + t\left(e + fs + gs^2 + hs^3\right) + t^2\left(k + ls + ms^2 + ns^3\right) + t^3\left(o + ps + qs^2 + rs^3\right) \quad (30.10)$$

Two sets of breakpoint vectors are $TU(i), \ldots, TU(M+1)$ and $TV(i), \ldots, TV(N+1)$, where $TU(i) \leq u \leq TU(i+1)$ $i = 1, \ldots, M$, and $s = u - TU(i)$, and $TV(i) \leq u \leq TV(i+1)$ $i = 1, \ldots, N$ and $t = v - TV(i)$.

The strategy for transforming this entity to a B-spline tensor product surface is similar to the one for the cubic parametric spline. The matrix form for the parametric cubic spline surface, according to Eq. 30.10, can be expressed in a matrix form, as shown in Eq. 30.11.

$$S(u,v) = \begin{bmatrix} 1 & s & s^2 & s^3 \end{bmatrix} \begin{bmatrix} a & e & k & o \\ b & f & l & p \\ c & g & m & q \\ d & h & n & r \end{bmatrix} \begin{bmatrix} 1 \\ t \\ t^2 \\ t^3 \end{bmatrix} \quad (30.11)$$

The matrix form of the Bézier surface with Bézier control points B_{ij} can be expressed as Eq. 30.12.

$$S(u,v) = \begin{bmatrix} 1 & u & u^2 & u^3 \end{bmatrix} \begin{bmatrix} 1 & 0 & 0 & 0 \\ -3 & 3 & 0 & 0 \\ 3 & -6 & 3 & 0 \\ -1 & 3 & -3 & 1 \end{bmatrix} B_{i,j} \begin{bmatrix} 1 & -3 & 3 & -1 \\ 0 & 3 & -6 & 3 \\ 0 & 0 & 3 & -3 \\ 1 & 0 & 0 & 1 \end{bmatrix} \begin{bmatrix} 1 \\ v \\ v^2 \\ v^3 \end{bmatrix} \quad (30.12)$$

The coefficients of this cubic parametric surface are contained in the given data set; therefore, the variables of Eq. 30.11 are all known, and the only unknown for Eq. 30.12 is matrix term of Bézier control points B_{ij}. Hence, the Bézier control points for each bicubic patch are obtained by setting Eq. 30.11 equal to Eq. 30.12 and solving the matrix Eq. 30.13 with reparameterization:

$$\left[B_{ij}\right] = \frac{1}{3} \begin{bmatrix} 3 & 0 & 0 & 0 \\ 3 & 1 & 0 & 0 \\ 3 & 2 & 1 & 0 \\ 3 & 3 & 3 & 3 \end{bmatrix} \begin{bmatrix} a & eh_2 & kh_2^2 & oh_2^3 \\ bh_1 & fh_1h_2 & lh_1h_2^2 & ph_1h_2^3 \\ ch_1^2 & gh_1^2h_2 & mh_1^2h_2^2 & qh_1^2h_2^3 \\ dh_1^3 & hh_1^3h_2 & nh_1^3h_2^2 & rh_1^3h_2^3 \end{bmatrix} \left(\frac{1}{9} \begin{bmatrix} 9 & 9 & 9 & 9 \\ 0 & 3 & 6 & 9 \\ 0 & 0 & 3 & 9 \\ 0 & 0 & 0 & 9 \end{bmatrix} \right) \quad \begin{matrix} h_1 = TU(i+1) - TU(i) \\ h_2 = TV(j+1) - TV(j) \end{matrix} \quad (30.13)$$

After all Bézier control patches $B_{i,j}$ are obtained, one can join each subpatch to form the final B-spline surface by setting the multiplicity of the knot value at the junction to 3 in both directions (I and J). Figure 30.10 shows a nacelle of an engine converted from the bicubic parametric surface.

30.3.7 Surface of Revolution to NURBS Surface

The surface of revolution has been discussed in many places [Piegl 1987, 1991] [Piegl and Tiller, 1989] [Tiller, 1992]. However, the more general algorithm is presented here. IGES defines the surface of revolution as the surface formed by rotating a boundary curve (called generatrix) with respect to a straight line (axis of revolution) from a starting angle (not necessarily zero) to an ending angle. This general algorithm for creating the surface grid by NURBS revolution can be stated as follows.

As a first step, the axis of revolution is transformed (translated or rotated or both) so that it is coincident with the Z axis. It is assumed that the generatrix is defined as a NURBS curve with the control polygon d_0 to d_m, *order* equal to k, and *weights* are w_0 to w_m. Next, for each control point d_i (on generatrix $i = 0, \ldots, m$), the surface control net d_{ij} $j = 0, \ldots, n$ at the ith cross section is constructed according to the starting and ending angle by utilizing the circular arc algorithm. Based on the procedure described in Section 30.3.1, n is determined by the sector angle (equal to the difference between ending and starting angles). For example, if the angle is less than 90°, n is equal to 2, if the angle is in the range of 90°–180°, n is equal

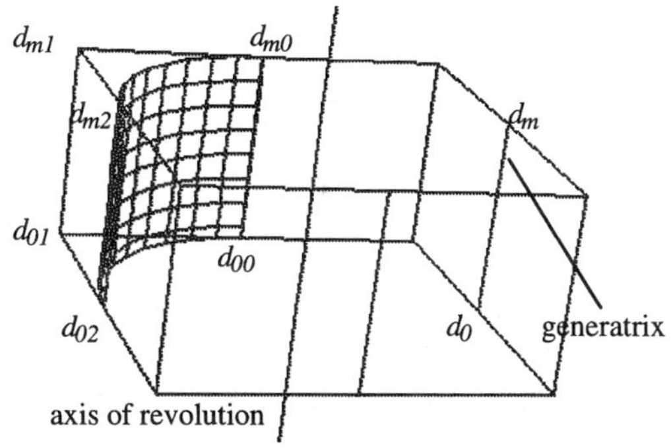

FIGURE 30.11 Illustration of the generalized algorithm for surface of revolution by NURBS.

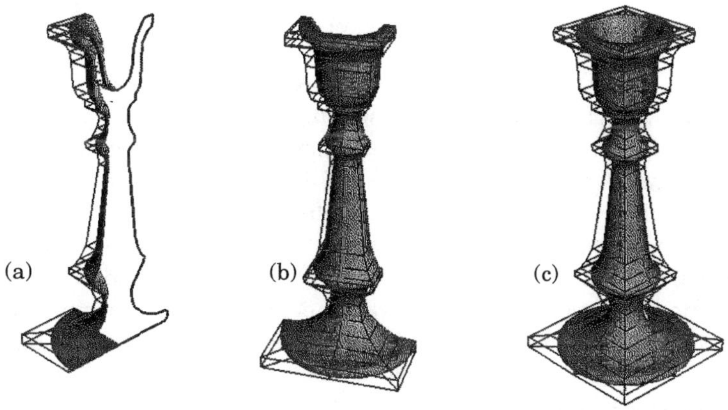

FIGURE 30.12 NURBS surface of revolution. (a) 90° ~ 180°, (b) 90° ~ 270°, (c) full revolution.

to 4, etc. For the section angle θ, the *weights* are set as $W_{ij} = w_i, w_i \cos(\theta/n), w_i, w_i \cos(\theta/n), \ldots$, (repeat $w_i, w_i \cos(\theta/n)$ with total $n + 1$ terms). The knot vector in direction $I(s)$ is the same as the one of the generatrix while the other one in direction $J(t)$ is determined according to the procedure described in Section 30.3.1. The control net and the *weights* are then transferred back to the original coordinates by reversing the translating/rotating operations. Figure 30.11 shows the construction of the associated control polygon at each cross section for the case of n equal to 2 (section angle equal to 90°).

The final NURBS definition for the constructed surface in Figure 30.11 contains d_{ij} $i = 0, \ldots, m$, $j = 0, \ldots, 2$ as the control net. The *order* and knot vector in the direction I are simply those of the generatrix, while the *order* in the J direction will be set as 3 and the knot vector is set as (0, 0, 0, 1, 1, 1). *Weights* are $W_{ij} = (w_i, w_i\cos(90/2), w_i)$ $i = 0, \ldots, m$ for all j. Figure 30.12 illustrates an example for this algorithm. This figure displays the "candle stand" NURBS control nets as well as the revolved surfaces for different starting and ending angles. The left figure also shows the generatrix.

30.3.8 Transfinite Interpolation for NURBS Surface

In many of the numerical grid generation applications, the *H*-type grid can be generated easily by the transfinite interpolation algorithm (*TFI*) [Shih, 1994] [Thompson, 1985] (see Chapter 3). As a matter of fact, the *TFI* algorithm is the most frequently used function for the numerical grid generation package.

TFI, also referred to as *Coons–Gordon patch,* is a bivariate interpolation constructed from the superposition of a set of univariate interpolation schemes by the formation of the Boolean sum projector [Thompson, 1985]. In other words, given a set of boundaries (or isoparametric curves), the *TFI* is a function that constructs the interior surface grid bounded by the given boundaries. The Boolean sum operator for a surface is defined in Eq. 30.14.

$$PS = P_\xi \oplus P_\eta = P_\xi(S) + P_\eta(S) - P_\xi P_\eta(S) \tag{30.14}$$

where $P_\xi(S)$ interpolates the ξ direction of boundaries (the given isoparametric curves) and $P_\eta(S)$ interpolates the η direction of boundaries, while $P_\xi P_\eta(S)$ captures the failures of $P_\xi(S)$ and $P_\eta(S)$. The final surface *PS* bidirectionally interpolates the given curves. There are many functions that can be applied to *TFI*. For example, one can use the linear, quadratic, or even cubic interpolation for function *P* in Eq. 30.14. Taking the linear interpolation for a surface with the resolution *N* by *M* as an example, the Eq. 30.13 can be rewritten as Eq. 30.15:

$$\begin{aligned} R_{ij} = {} & \left(1-s_{ij}\right)R_{ij} + s_{ij}R_{Nj} + \left(1-t_{ij}\right)R_{i1} + t_{ij}R_{iM} - \\ & \left(\left(1-s_{ij}\right)\left(1-t_{ij}\right)R_{11} + \left(1-s_{ij}\right)t_{ij}R_{1M} + s_{ij}\left(1-t_{ij}\right)R_{N1} + s_{ij}t_{ij}R_{NM}\right) \end{aligned} \tag{30.15}$$

Variables R_{ij} in Eq. 30.15 are the control vertices, which need to be determined. For the NURBS case, the R_{ij} could be d_x, d_y, d_z (control points) and w_{ij} (the *weights*).

This *TFI* function is a fundamental tool for generating grids in many grid applications. However, Eqs. 30.14 and 30.15 *cannot* be applied to NURBS *TFI* directly, because when four NURBS curves are given to generate a NURBS *TFI* surface, the interior control points can be created according to Eq. 30.15 (for the bilinear interpolation) by supplying the control vertices of the boundaries without a problem. The problem comes when determining the interior *weights*. The addition and subtraction operations in Eq. 30.15 may lead to the interior weights being negative or zero values. Any negative *weights* will destroy the convex hull property of a NURBS entity, while any zero *weights* will make the control vertices lose their influence. This obstacle can be overcome by the modified NURBS *TFI* [Lin and Hewitt, 1994]. The formula for this is shown in Eq. 30.16:

$$\begin{vmatrix} WP(S) \\ W \end{vmatrix}_{ij} = \begin{vmatrix} W_\xi W_\eta\left(P_\xi(S) + P_\eta(S) - P_\xi P_\eta(S)\right) \\ W_\xi W_\eta \end{vmatrix}_{ij} \tag{30.16}$$

Each term of Eq. 30.16 (for the case of linear interpolation of *P*) is defined as follows: the $P_\xi(S)$ represents a NURBS "ruled surface" with *weights* of W_ξ formed in ξ direction (hence, the *order* is 2, knot vector is [0, 0, 1, 1] in ξ direction). $P_\eta(S)$ represents another NURBS "ruled surface" with *weights* of W_η formed in η direction (*order* is 2, knot vector is [0, 0, 1, 1] in η direction) and the $P_\xi P_\eta(S)$ is a NURBS surface constructed by using the four corner points as the control net with *orders* 2 by 2 in ξ and η directions. This is demonstrated in Figure 30.13.

After creating the intermediate surfaces of $P_\xi(S)$, $P_\eta(S)$, and $P_\xi P_\eta(S)$, one has to perform the "knot insertion" [Piegl, 1991] and "degree elevation" [Piegl, 1991] [Tiller, 1992] algorithms to these three surfaces to ensure all of them have the same *orders* and same knot vectors in both the ξ and η directions. If the NURBS surfaces have the same *orders* and same knot vectors, then the dimension of the control net is the same also. Therefore, the control net of the final NURB *TFI* surface can be obtained by adding the control nets of $P_\xi(S)$, $P_\eta(S)$ and subtracting those of $P_\xi P_\eta(S)$, while the *weights* are determined by multiplying W_ξ and W_η.

Comparing the NURBS *TFI* with the traditional *TFI* shows that the NURBS *TFI* needs more computation because the *weights* need to be handled properly. In addition, the knot insertion and degree elevation algorithms need extra computation. However, this function is fundamental and useful when there is a need to create *H*-type grids. Also, when generating the volume grids for a nozzle, this function is particularly useful to create the inlet and outlet surfaces. Figure 30.14 demonstrates this example.

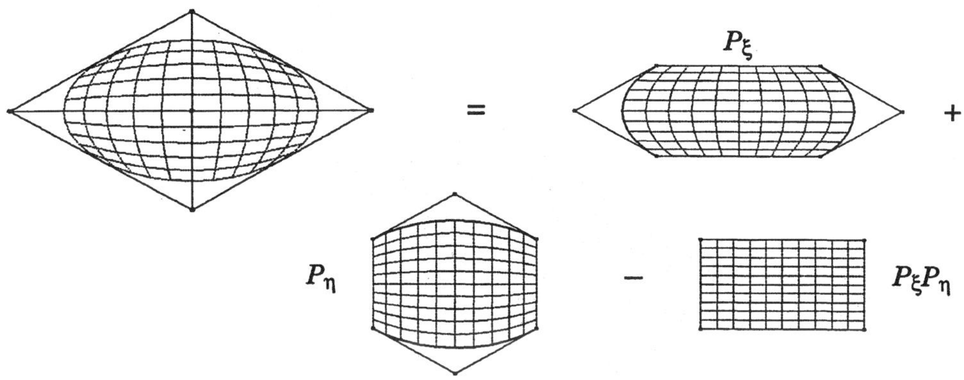

FIGURE 30.13 An illustration of NURBS *TFI* surface.

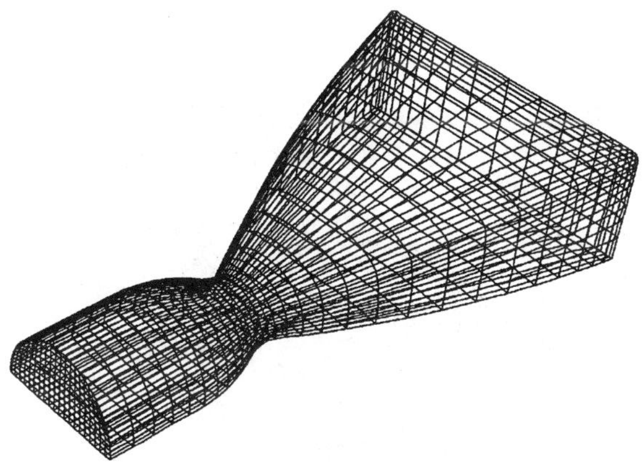

FIGURE 30.14 NURBS *TFI* creates the inlet/outlet surface for a circular–rectangle nozzle.

30.3.9 Cascading Technique for NURBS Surface

As discussed in the previous transforming procedures, the surface of revolution algorithm can be used to model symmetric surfaces. In CFD applications, some of the geometries for analysis are symmetric objects, such as the simulation of the flow passing around a missile. Generally speaking, the surface of revolution algorithm can be used to model a "simple" symmetric surface, but for many CFD applications, the real geometric objects interact with other objects and cannot be modeled by rotating a boundary curve to form a surface of revolution, as shown in Figure 30.15, where a surface blade with the boundary is intersected with a fin. Even though this surface is symmetric, the surface of revolution (SOR) algorithm fails to model it.

This situation also occurs with the "cascade" surface. A cascade surface is usually referred to as the "blade-to-blade" surface in turbomachinery [Shih, 1994]. Even though most of the cascade surfaces are axisymmetric, they cannot be modeled by the NURBS surface revolution algorithm. Also, in the grid generation area, creating the surface grid for the cascade is a challenge. The difficulty of modeling the surface grids for a 3D cascade surface is that when the blade leading edge (or trailing edge) circle radius is too big, such as in a turbine, or if the blade setting angle (the blade angle) is very low, it is hard to generate a well behaved *H*-type grid. Particularly, there is often a grid crossing near the leading edge (or trailing edge) for such a geometry. Traditionally, this kind of surface grid is generated by transforming the 3D surface of the (x, y, z) coordinates to 2D parametric (m', σ) space, griding in the parametric space and then transforming the coordinates back to 3D physical space according to the relation of the (m', σ). Detailed information can be found in [Shih, 1994]. In this section NURBS modeling approach is presented for modeling this type of geometry.

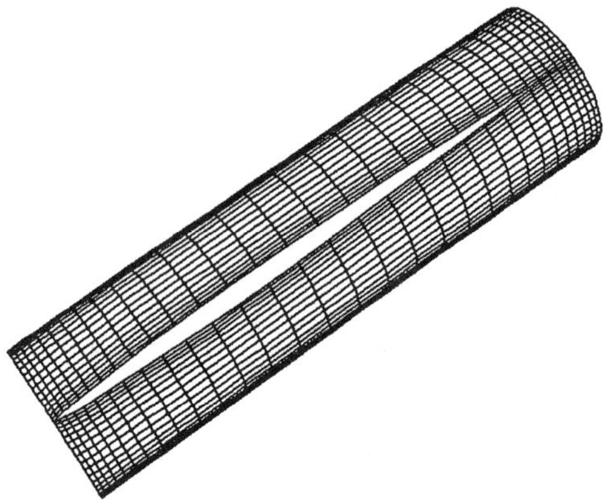

FIGURE 30.15 A symmetric surface blade.

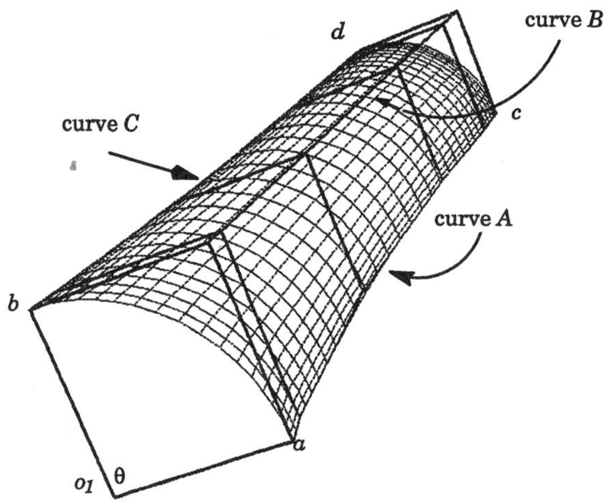

FIGURE 30.16 Illustration of modeling cascade surface by the NURBS control net.

The NURBS algorithm for modeling the cascade surface is described as follows: given a boundary curve of a cascade surface, transform it to a B-spline curve (as curve *A* shown in Figure 30.16) by the interpolation technique. A plane that bisects the surface sector angle (the θ, angle of ao_1b shown in Figure 30.16) is created and then the "mirror" function [Yu, 1995] is used to reflect curve *A* with this plane to create curve *C*. Curve *C* will have the same *order*, knot vector, and number of control points as curve *A*. The next step is to create a straight line on the plane that contains the points o_1, *a*, and *c*. This is done by projecting the control polygons of curve *A* to the plane and setting the *order* and knot vector of the line to be the same as those of curve *A*. After this line is created, the surface of revolution algorithm is performed to rotate this line with respect to the center of o_1o_2 for a total sector angle of θ. A NURBS tabulated cylinder [Yu and Soni, 1995] with sector angle θ will be generated after this step. However, since this surface is not the desired cascade surface, the first and last iso-control polygons (in the axis direction) of this surface must be replaced with the existing B-spline curves *A* and *C*. Because the tabulated cylinder is created by rotating a line that has the same *order* and knot vector as those of curve *A*, it is secure to replace the two control polygons of the surface with *A* and *C* without altering the entire shape of the surface. The control net, with curves *A* and *C* replacing the first and last iso-control polygons, is the final desired NURBS control net. A missile configuration, composed of the surface of revolution and cascade surface, is shown in Figure 30.17 to demonstrate this algorithm.

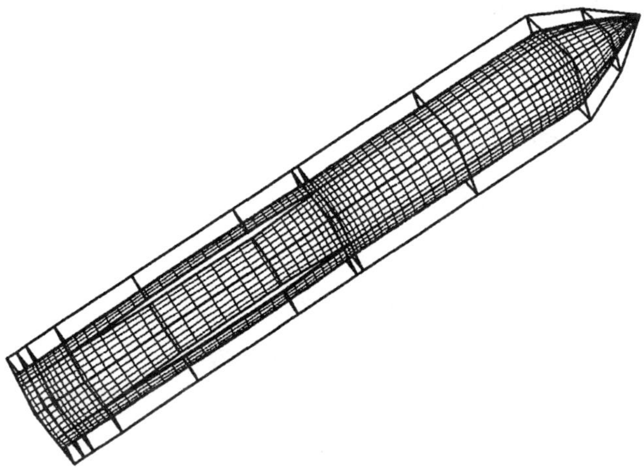

FIGURE 30.17 A missile surface grid modeled by the NURBS control net.

30.4 Grid Redistribution

The NURBS entity (curve, surface, or volume) is presented as a parametric format, and a grid point on a NURBS entity is generated by evaluating the parametric value t (or s, t for surface, s, t, u for the volume). The placement of grid points on the physical geometry of interest with the desired stretching/concentration criteria is of key importance for CFS analysis. This in turn requires the reparametrization of parametric values t (s, t for surface and s, t, u for the volume) such that when NURBS formula is evaluated, the desired distribution criteria are met on the physical geometry. For example, evenly distributed parametric values t may not result in a sequence of evenly distributed grid points of $C(t)$ on the physical NURBS curve. The location of the control polygon, the value of *weights*, and even the knot vector are all possible factors in evaluation of the NURBS entity. Changing any of these factors could result in an unexpected (or undesired) packing of the grid points (lines or surface) in the physical geometry. This situation is referred to as "bad parametrization" and is remedied by performing reparametrization. The problem in this case is calculating the proper parametric values to obtain the desired distribution on the physical NURBS entity without altering the NURBS definition (control polygon, *weights*, and knot vector). The reparametrization algorithm is presented for the three-dimensional NURBS volume entity. The respective algorithms for the one-dimensional (curve) and the two-dimensional (surface) NURBS entities can be easily deduced from the three-dimensional scheme.

30.4.1 Reparametrization Algorithm

Before discussing the algorithms, it is necessary to define several notations. For the NURBS tensor product volume with resolution ni, nj, and nl, the 3D arrays are defined as follows:

1. $(vs_1(i,j,l), vt_1(i,j,l), vu_1(i,j,l))$, $i = 1, \ldots, ni$, $j = 1, \ldots, nj$, and $l = 1, \ldots, nl$ are the parametric distribution volume associated with the desired distribution of the volume in physical space.
2. $(vs_2(i,j,l), vt_2(i,j,l), vu_2(i,j,l))$, $i = 1, \ldots, ni$, $j = 1, \ldots, nj$, and $l = 1, \ldots, nl$ are the normalized chord length of the volume with desired distribution in direction I, J, and L, respectively.
3. $(vs_3(i,j,l), vt_3(i,j,l), vu_3(i,j,l))$, $i = 1, \ldots, ni$, $j = 1, \ldots, nj$, and $l = 1, \ldots, nl$ are the normalized chord length of the volume evaluated at parametric values $(vs_1(i,j,l), vt_1(i,j,l), vu_1(i,j,l))$, $i = 1, \ldots, ni$, $j = 1, \ldots, nj$, and $l = 1, \ldots, nl$.

These variables are explained as follows: If the designer would like to have the final volume grid, say, evenly distributed, then $(vs_2(i,j,l), vt_2(i,j,l))$ would be a 3D array that contains the even distribution, and $(vs_1(i,j,l), vt_1(i,j,l))$ would be the parametric values that are to be determined such that $(vs_3(i,j,l), vt_3(i,j,l))$, the normalized chord length of final volume, would be the same as ($vs_2(i,j,l)$, $vt_2(i,j,l)$) or within certain tolerance. The algorithm for finding the desired parametric values is illustrated by the pseudo-code Algorithm I:

Algorithm I

For *each* parametric value, search the index of I, J, and L such that

$$\begin{aligned}
&\left(vs_2(i,j,l), vt_2(i,j,l), vu_2(i,j,l)\right) \text{ is located within the cells of}\\
&\left(vs_3(I,J,L), vt_3(I,J,L), vu_3(I,J,L)\right)\\
&\left(vs_3(I+1,J,L), vt_3(I+1,J,L), vu_3(I+1,J,L)\right)\\
&\left(vs_3(I,J+1,L), vt_3(I,J+1,L), vu_3(I,J+1,L)\right)\\
&\left(vs_3(I,J,L+1), vt_3(I,J,L+1), vu_3(I,J,L+1)\right)\\
&\left(vs_3(I+1,J+1,L), vt_3(I+1,J+1,L), vu_3(I+1,J+1,L)\right)\\
&\left(vs_3(I+1,J,L+1), vt_3(I+1,J,L+1), vu_3(I+1,J,L+1)\right)\\
&\left(vs_3(I,J+1,L+1), vt_3(I,J+1,L+1), vu_3(I,J+1,L+1)\right)\\
&\left(vs_3(I+1,J+1,L+1), vt_3(I+1,J+1,L+1), vu_3(I+1,J+1,L+1)\right)
\end{aligned}$$

After the (I, J, L) is found, for *each* parametric value, solve the α, β, and γ for Eq. 30.17:

$$\begin{aligned}
&\left(vs_2(i,j,l), vt_2(i,j,l), vu_2(i,j,l)\right) = (1-\alpha)(1-\beta)(1-\gamma)\left(vs_3(I,J,L), vt_3(I,J,L), vu_3(I,J,L)\right)\\
&+\alpha(1-\beta)(1-\gamma)\left(vs_3(I+1,J,L), vt_3(I+1,J,L), vu_3(I+1,J,L)\right)\\
&+(1-\alpha)\beta(1-\gamma)\left(vs_3(I,J+1,L), vt_3(I,J+1,L), vu_3(I,J+1,L)\right)\\
&+(1-\alpha)(1-\beta)\gamma\left(vs_3(I,J,L+1), vt_3(I,J,L+1), vu_3(I,J,L+1)\right)\\
&+\alpha\beta(1-\gamma)\left(vs_3(I+1,J+1,L), vt_3(I+1,J+1,L), vu_3(I+1,J+1,L)\right)\\
&+\alpha(1-\beta)\gamma\left(vs_3(I+1,J,L+1), vt_3(I+1,J,L+1), vu_3(I+1,J,L+1)\right)\\
&+(1-\alpha)\beta\gamma\left(vs_3(I,J+1,L+1), vt_3(I,J+1,L+1), vu_3(I,J+1,L+1)\right)\\
&+\alpha\beta\gamma\left(vs_3(I+1,J+1,L+1), vt_3(I+1,J+1,L+1), vu_3(I+1,J+1,L+1)\right)
\end{aligned} \tag{30.17}$$

After α, β, and γ are obtained, the new parametric values are determined as shown in Eq. 30.18:

$$\begin{aligned}
&\left(vs(i,j,l), vt(i,j,l), vu(i,j,l)\right) = (1-\alpha)(1-\beta)(1-\gamma)\left(vs_1(I,J,L), vt_1(I,J,L), vu_1(I,J,L)\right)\\
&+\alpha(1-\beta)(1-\gamma)\left(vs_1(I+1,J,L), vt_1(I+1,J,L), vu_1(I+1,J,L)\right)\\
&+(1-\alpha)\beta(1-\gamma)\left(vs_1(I,J+1,L), vt_1(I,J+1,L), vu_1(I,J+1,L)\right)\\
&+(1-\alpha)(1-\beta)\gamma\left(vs_1(I,J,L+1), vt_1(I,J,L+1), vu_1(I,J,L+1)\right)\\
&+\alpha\beta(1-\gamma)\left(vs_1(I+1,J+1,L), vt_1(I+1,J+1,L), vu_1(I+1,J+1,L)\right)\\
&+\alpha(1-\beta)\gamma\left(vs_1(I+1,J,L+1), vt_1(I+1,J,L+1), vu_1(I+1,J,L+1)\right)\\
&+(1-\alpha)\beta\gamma\left(vs_1(I,J+1,L+1), vt_1(I,J+1,L+1), vu_1(I,J+1,L+1)\right)\\
&+\alpha\beta\gamma\left(vs_1(I+1,J+1,L+1), vt_1(I+1,J+1,L+1), vu_1(I+1,J+1,L+1)\right)
\end{aligned} \tag{30.18}$$

Finally, replace the $vs_1(i,j,l)$, $vt_1(i,j,l)$, $vu_1(i,j,l)$ with $vs(i,j,l)$, $vt(i,j,l)$, $vu(i,j,l)$.

30.4.2 Singularity Control

It can easily be shown that the reparametrization algorithm presented here will fail if the underlying NURBS curve/surface/volume contains any singularities. Using a NURBS curve as an example, if all the points on this curve collapse to one point, then this NURBS curve is a singular curve. The same definition can be applied to NURBS surface and volume. If any one of the iso-parametric lines on a NURBS surface collapses to one point, then this line is defined as a singular line. For a volume, if any one of the iso-planes on a NURBS volume collapses to a line, then this plane is a singular plane. When these singularity problems occur, the total arc length from calculating it according to the reparameterization algorithms will be zero. Since the normalized arc lengths are obtained by dividing the total arc length by the individual ones, this will lead to the operation of 0/0, which is mathematically undefined. These singularity problems are often encountered in CFS applications requiring structured grids. For example, the surface grid that represents the canopy of an aircraft has a singular line at the nose position: a surface grid that represents the missile has a singular line in the nose position; and the volume grid of a cylinder (or any cylindrical pipe) has a singular plane at the axial direction. In each of these cases, the singularity problems occurred because the control points collapse to one point (for the surface case) or one line (for the volume case). While evaluating the NURBS entities with certain parametric values by utilizing these collapsed control points, the singularity problem arises. Hence, it is necessary to enhance the algorithm to handle this problem.

The strategy for the enhancement of the algorithm is related to the machine accuracy (also called machine precision). The machine accuracy, commonly represented as symbol ε, is defined as the smallest positive real number such that $1 + \varepsilon > 1$. On the *Silicon Graphics Personal Iris,* this number is equal to 10.0^{-16} (double precision). In many numerical simulations, this number is needed to represent the finite precision arithmetic of the computer architecture. For example, the convergence criteria of an iteration scheme is dependent upon the machine accuracy. A variable expected to be *zero* in numerical representation may not be reached due to the finite precision of the computer memory representation. Therefore, in many numerical applications, the exact *zero* is replaced by a value related to ε; for example, if a variable is less than $\sqrt{\varepsilon}$; then that variable is assumed to be equal to *zero.* This concept is also utilized to avoid the singularity problems. Using the NURBS surface with a singular line (say, occurring at grid line index $i = 0$) as an example, the grid line evaluated with the parametric values of $(ss_1(0,j), st_1(0,j), j = 0, \ldots, nj$ will shrink to one singular point due to the control vertices d_{0j} collapsing to one point. However, if one perturbs these parametric values by a small value, perhaps $\sqrt{\varepsilon}$ and then reevaluates the surface, the returned grid line will not be the same as the singular one since these parametric values are no longer exactly *zero.* Instead, it will return a grid line with a small but recognizable total arc length. Even though the total arc length is small, the normalization process will make the values of $(ss_3(0,j), st_3(0,j)\ J = 0, \ldots,$ nj to $0.0 \sim 1.0$ and will avoid the uncertain situation of 0/0.

The associated algorithm applied to a 3D NURBS volume is presented as Algorithm II, and Figure 30.18 shows a 3D NURBS cylindric pipe evaluated with even parametric values. Notice that in its L direction, the surface degenerates to a singular line. The result of reparameterization for this volume is shown in Figure 30.19.

Algorithm II

$$for(k = 0; k < nk; k++)$$
$$for(j = 0; j < nj; j++) \quad vs_1(0,j,k) += \varepsilon;\ vs_1(ni-1,j,k) -= \varepsilon;$$
$$for(k = 0; k < nk; k++)$$
$$for(i = 0; i < ni; i++) \quad vt_1(i,0,k) += \varepsilon;\ ct_1(i,nj-1,k) -= \varepsilon;$$
$$for(j = 0; j < nj; j++)$$
$$for(i = 0; i < ni; i++) \quad vu_1(i,j,0) += \varepsilon;\ vu_1(i,j,nj-1) -= \varepsilon;$$

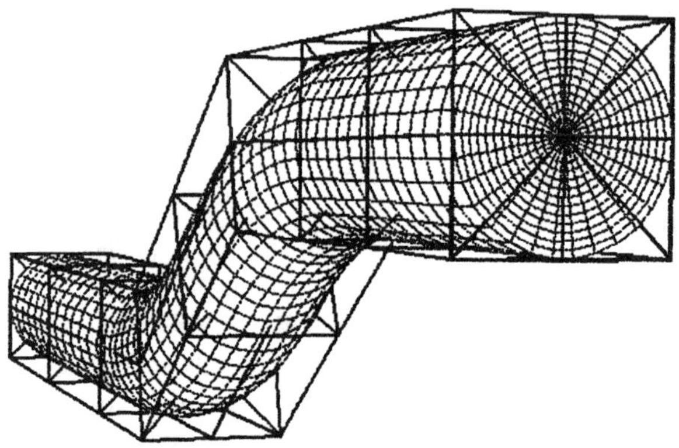

FIGURE 30.18 A NURBS volume grid with a singular plane in flow direction.

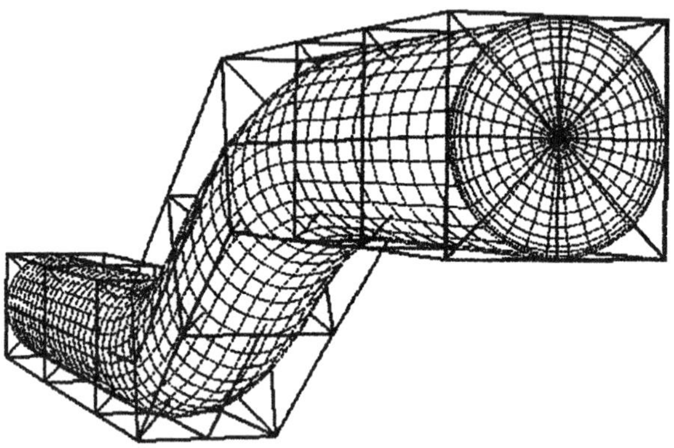

FIGURE 30.19 The reparameterization algorithm for a NURBS volume grid with singularity.

30.5 Volume Grid Generation by NURBS Control Volume

Volume grid generation algorithms have been utilized in many CFD analysis procedures. A widely used technique to algebraically generate a three-dimensional volume grid is to utilize the transfinite interpolation algorithm based on the bounding surface grids. However, the volume generation techniques are seldom applied to CAD/CAM applications. Even though NURBS representation has been widely used in many industry applications, the geometry modeled by the NURBS volume approaches are seldom discussed in the computer-aided geometry design (CAGD) literature. In this chapter, using the NURBS volume to model the geometry for the volume grid is presented. Instead of storing the surface/volume grid points, one can store the associated control polygon (or control net for the surface/volume) with the associated weights to reduce the memory requirement. This is especially useful for volume grid generation. Even though computer memory availability has been dramatically improved, a complicated geometry usually consumes a great deal of computer memory for the volume grid. Storing the NURBS control net to reduce the size of an entire volume grid is demonstrated in the examples of this section.

The ultimate objective is to explore various NURBS control volume options applicable to three-dimensional grid generation. In this section, the development of NURBS ruled volume, NURBS extruded volume, NURBS volume of revolution, NURBS composite volume, and transfinite interpolation (*TFI*) NURBS volume are discussed.

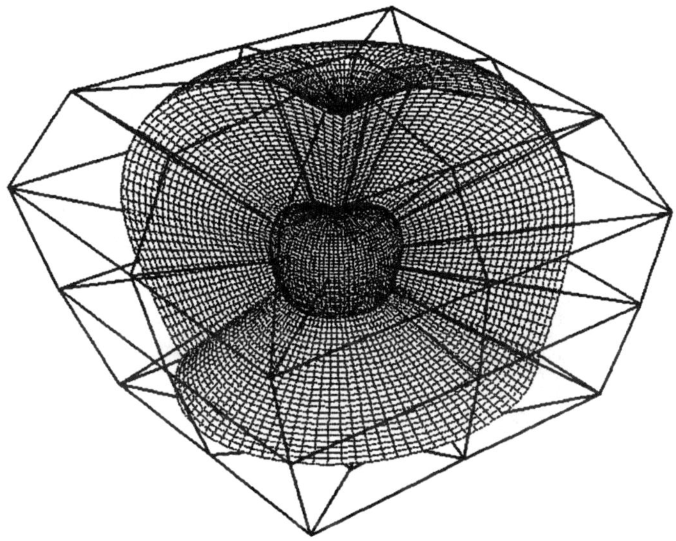

FIGURE 30.20 A volume grid created by NURBS control volume (ruled volume option).

30.5.1 Ruled Volume

The easiest 3D NURBS volume to generate is the ruled NURBS volume. The algorithm is described as follows: Given two NURBS surfaces, the first step to form a ruled volume is to make the knot vectors of the surfaces be in the same range of [(0~l]. Next, considering the *I* direction of both surfaces, the degree raising technique is used to raise the low degree (*order* – 1) of the surface. This procedure will yield a new knot vector and new control net. If the new knot vector differs from the other knot vector, then the knot insertion algorithm is performed to merge them into one final knot vector. Then these steps are applied to the knot vectors in the *J* direction of both surfaces. After this step, the two NURBS surfaces will have the same *orders* and the same knot vectors in both *I* and *J* directions. This means that the resolutions of the control nets of both surfaces will be the same. Finally, connect the corresponding control points together to form the 3D NURBS volume. The orders and knot vectors of the final volume in the *I* and *J* directions will be the same as those of the surfaces after degree elevation and knot insertion, and the *order* in *L* direction will be set as 2 with the knot vector set as (0,0,1,1). Figure 30.20 shows a 3D "apple-like" NURBS volume and its volume grid, while Figure 30.2l shows a missile configuration with the control volume formed by this algorithm.

30.5.2 An Extruded Volume

The generation of a NURBS extruded volume is an extension of the extruded surface definition. The extruded surface is defined as a surface formed by moving a line segment parallel to itself along a curve. In other words, given a NURBS curve, one can generate another curve by extruding the given curve with a distance α along a vector *V*. Similarly, the NURBS extruded volume is defined as follows: Given a NURBS surface with control net d_{ij} and the associate *weights*, knot vectors and *orders*, the new surface $\bar{d}_{ij}$ can be generated by "extruding" the given surface with a distance α along a vector *V*. Mathematically, this new NURBS extruded surface can be described as $\bar{d}_{ij} = d_{ij} + \alpha V$ with the same *orders*, same *weights* and same knot vectors as those of the given surface. After this step the algorithm of "ruled volume" can be applied to these surfaces to form a final NURBS volume. Figure 30.22 shows a 3D NURBS extruded volume.

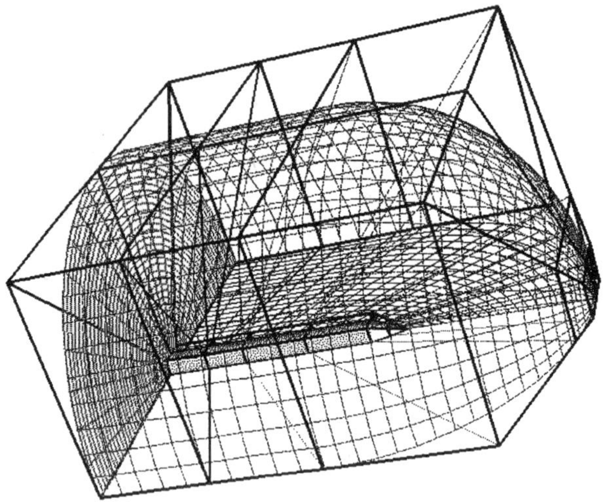

FIGURE 30.21 A missile volume grid modeled by NURBS ruled volume.

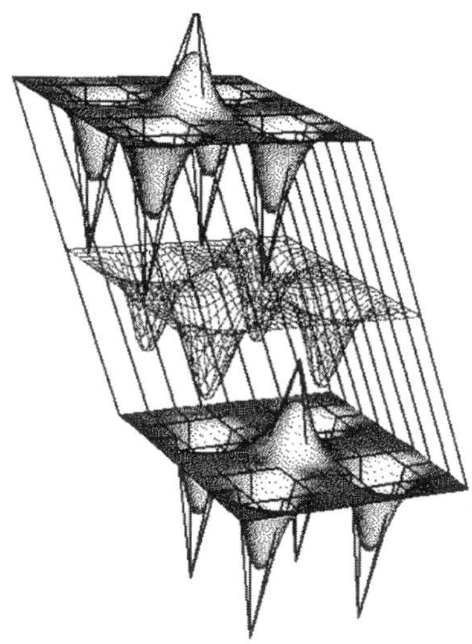

FIGURE 30.22 A volume grid created by the NURBS extruded volume option.

30.5.3 Volume of Revolution

Another commonly used approach for generating volume grids is the "revolution" method. A revolution resulting in a surface is known as a "surface of revolution," while a revolution resulting in a volume is then known as a "volume of revolution." The fact that this modeling technique can be used only for symmetric geometries is not limiting, since many objects in real-world applications, such as turbomachinery configurations, are symmetric. The extension of a volume revolution modeled by NURBS is presented as follows: the definition of surface of revolution is a surface formed by rotating a given curve with respect to an arbitrary straight line from a starting angle to an ending angle. Likewise, the volume of revolution is defined as a volume formed by rotating a given NURBS surface with respect to an arbitrary axis of revolution from any starting angle to an ending angle.

The general algorithm is outlined as follows: the first step is translating/rotating the axis of revolution by a proper transformation matrix so that it is coincident with the Z axis. This transformation matrix

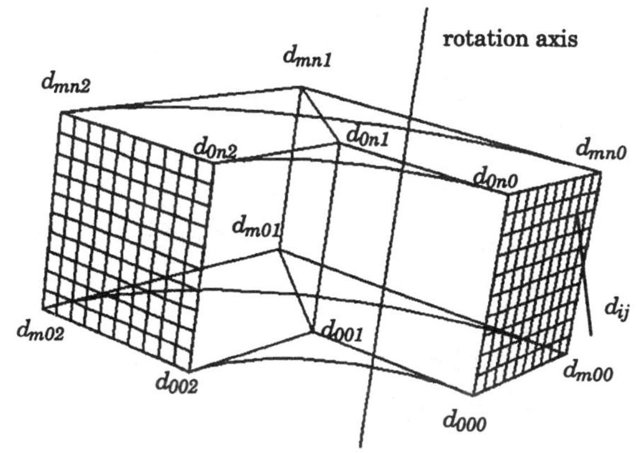

FIGURE 30.23 Illustration of constructing the NURBS volume of revolution.

is also applied to the given NURBS surface so that the entire surface can be kept in the same position as the axis of rotation. It is assumed that the surface is defined (or transformed) as NURBS with the control net d_{ij} order $k1$ and $k2$, *weights* W_{ij} and two knot vectors. The second step is to construct, for each control net d_{ij} (on the generatrix $i = 0, \ldots, m, j = 0, \ldots, n$), the control volume d_{ijl} $l = 0 \ldots p$ at each jth cross section through the starting and ending angle by utilizing the circular arc algorithm. In other words, this approach constructs the NURBS control net at each J-constant plane by revolving the control polygon d_{iJ} with respect to L direction and then "stacks" them together to form a final NURBS volume. Figure 30.23 demonstrates this approach. The general procedure of generating the NURBS circular arc is described in a previous section. The p for the last dimension of control volume is determined by the sector angle θ (equal to the difference between ending and starting angle). For example, if the angle is less than 90°, p is equal to 2. If the angle is in the range of 90° ~ 180°, p is equal to 4; if in the range of 180° ~ 270°, p is 6; if it is greater than 270°, p should be 8. For the sector angle θ, the *weights* are set as (in each J constant plane, $J = 0, .. n$) $W_{iJp} = w_{ip}\, w_{iJ} \cos(\theta/p), w_{ip}\, w_{iJ}, \cos(\theta/p), \ldots$ $i = 0, \ldots, m$ (repeat $w_{ip}\, w_{iJ}\cos(\theta/p)$ with total $p + 1$ terms). The knot vectors in directions of $I(s)$ and $J(t)$ are the same as the ones of the given surface, while the knot vector in direction $L(u)$ is determined according to the circular arc procedure. For example, when p is 2, the associated knot vector is set as (0, 0, 0, 1, 1, 1); for the case of p equal to 4, the knot vector is set as (0, 0, 0, 1/2, 1/2, 1, 1, 1); for the case of p equal to 6, the knot vector is set as (0, 0, 0, 1/3, 1/3, 2/3, 2/3, 1,1,1); and for the case of p equal to 8, the knot vector is (0,0,0, .25, .25, .5, .5, .75, .75,1,1,1). Also, the *orders* in I and J are set to be $k1$ and $k2$ (as the orders of the original surface), while 3 is set as the *order* of L direction.

Because the NURBS has the translate and rotate invariant properties, the inverse transformation matrix can be applied to the control volume (without altering the *weights* and knot vectors) returning the volume to the original coordinates. Figure 30.21 shows a 3D volume grid and its control volume, according to this algorithm. This example was developed by revolving the *TFI* surface from 0° to 180°. Because the NURBS surface *TFI* technique needs four boundary curves to define a surface, this results in an "*H*" type surface grid. Revolving this "*H*" type *TFI* surface creates "*H*" type NURBS control volume and yields the "*H*" volume grid. This topology can be changed by revolving another "0" type NURBS surface to form an "0" type volume grid, as shown in Figure 30.24. Notice that the sizes of this control volume are only $3 \times 3 \times 5$ (for "*H*" type grids) and $9 \times 9 \times 5$ (for "0" type grids), yet the resolution of the entire volume grid can be any number (for this case, $31 \times 31 \times 61$).

30.5.4 Composite Volume

A composite NURBS volume is defined as a volume consisting of lists of constituent volumes. The composing procedure is stated as follows: Suppose two constituent NURBS volumes V_1 and V_2 form a composite volume. Assume that V_1 has control volume $d_1[0{:}m_1, 0{:}n_1, 0{:}l_1]$, weight W_1 $[0{:}m_1, 0{:}n_1,$

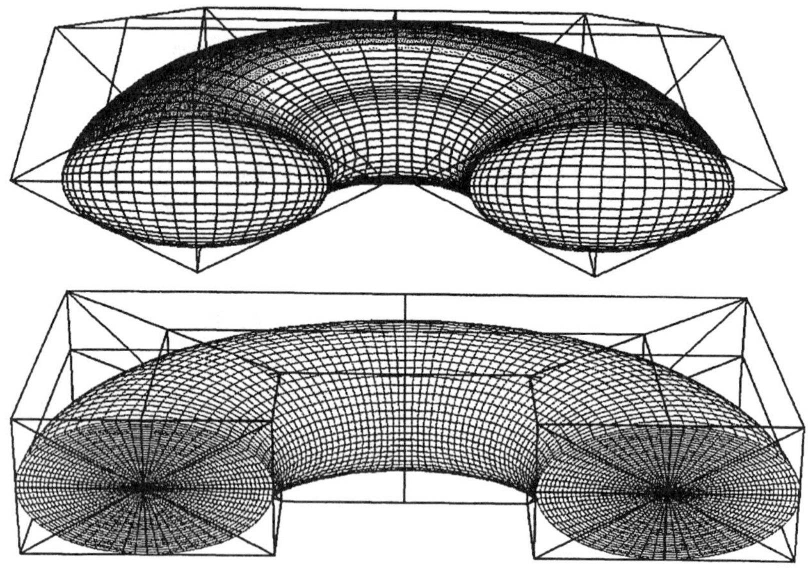

FIGURE 30.24 *H* and *O* type volume grids created by NURBS volume of revolution.

$0{:}l_1]$, three knot vectors $knot\_i)_1$, $knot\_j)_1$, $knot\_l)_1$ and orders $k\_i)_1$, $k\_j)_1$, $k\_l)_1$ while V_2 has control volume $d_2[0{:}m_2, 0{:}n_2, 0{:}l_2]$, weight $W_2[0{:}m_2, 0{:}n_2, 0{:}l_2]$, three knot vectors knot $knot\_i)_2$, $knot\_j)_2$, $knot\_l)_2$ and orders $k\_i)_2$, $k\_j)_2$, $k\_l)_2$. There are many possible combinations of the two volumes joined together. For example, one can join the volumes in the *I* direction with the interface of the *J*, *L* surface, or join in the *L* direction with the interface of the *I*, *J* surface, etc. Even though there are many cases, the procedure is similar. When joining in the *I* direction, for example, the first step is to perform the degree elevation to V_1 and V_2 so that these two volumes can have compatible degrees in the *I*, *J*, and *L* directions. If the two knot vectors in the *J* direction for V_1 and V_2 are not the same, they are merged together by setting the final knot vector as $\{knot\_j)_1 \ / \ knot\_j)_2\}$; then the knot insertion is applied to V_1 and V_2 in the *J* dimension. The same procedure should be applied to the *L* direction if $knot\_l)_1$ and $knot\_l)_2$ are not the same. After this step, V_1 and V_2 will have the same degree in three directions, and the number of control points and knot vectors in the *J* and *L* directions will be the same. The second step is to adjust the knot vector $knot\_i)_2$ so that its first knot value can be the same as the last knot value $knot\_i)_1$. Shifting the knot vector will not change the original NURBS because the basis function is a "normalized" basis function. The third step is to build up the final knot vector by joining the two knot vectors into one knot vector and setting that knot value at the joint point to have the multiplicity equal to (*order* –1). For example, if the knot vector $knot\_i)_1$ is [0., 0., 0., 1., 1., 1.] and the knot vector $knot\_i)_2$ is [2., 2., 2., 3., 3., 3.], the second knot vector is adjusted by shifting –1 to each value. Thus, the $knot\_i)_2$ becomes [1,1,1,2,2,2]. Suppose the final *order* of these two volumes in the *I* direction is 3; then, the final knot vector should be [0., 0., 0., 1., 1., 2., 2., 2.] (notice the interior knot 1 has multiplicity of (*order* –1) = 2). The fourth step is to match the *weights* at the interface surface by multiplying the ratio of $W_1[m_1, j, l] \ / \ W_2[0, j, l]$ to $W_2[i, j, l]$ for $i = 0, \ldots, m_2$, $j = 0, \ldots, n_1$, and $l = 0, \ldots, l1]$. The last step is to construct the final control volume and *weights* by removing the $d_2[0{:}0, 0{:}n_2, 0{:}l_2]$ and $W_2[0{:}0, 0{:}n_2, 0{:}l_2]$ and joining the others as one control volume and *weights*. Figure 30.25 demonstrates this algorithm.

Generally speaking, it is difficult to model a complicated geometry by a single NURBS control volume. However, one can construct the individual control volume and then utilize this composite algorithm to merge for a final volume. Figure 30.26 demonstrates the flexibility and the advantage of this approach. The NURBS control volume is used to model the internal pipes. The griding of the turning portions of the pipe can be constructed by "volume of revolution" without any difficulties. Assembling all the sub-NURBS volumes makes the final single block NURBS control volume.

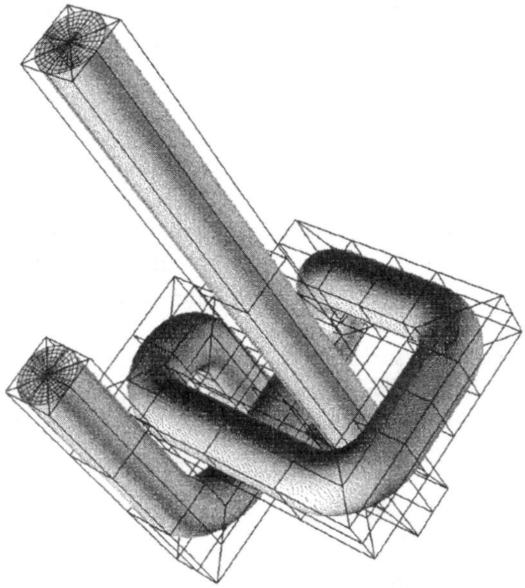

FIGURE 30.25 A volume grid for the turning pipe created by NURBS composite volume.

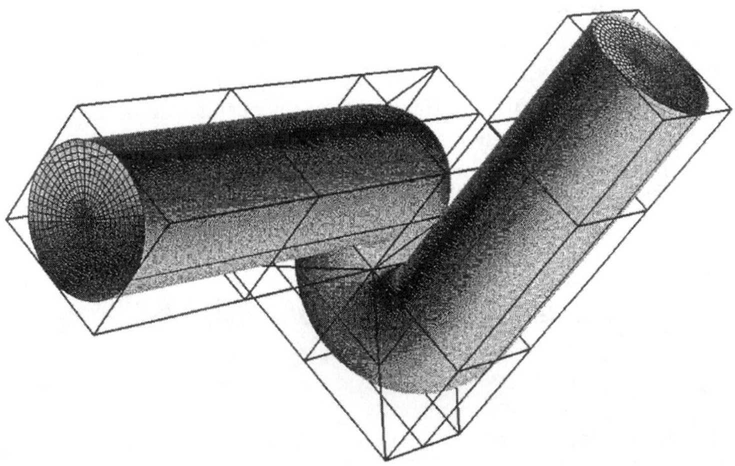

FIGURE 30.26 Volume grid created by NURBS composite volume.

30.5.5 Transfinite Interpolation Volume

Similar to the NURBS *TFI* surface algorithm [Thompson et al., 1985], this approach is frequently used to generate an *H*-type volume grid (see Chapter 3) Instead of providing four NURBS curves for a *TFI* surface, this algorithm requires six NURBS surfaces to generate a NURBS *TFI* volume. This algorithm is the extension from the surface to volume. The *Boolean* sum equation is defined as Eq. 30.19

$$
\begin{aligned}
PV = P_\xi \oplus P_\eta \oplus P_\zeta = \\
P_\xi V + P_\eta V + P_\zeta V - P_\xi P_\eta V - P_\eta P_\zeta V - P_\xi P_\zeta V + P_\xi P_\eta P_\zeta V
\end{aligned}
\tag{30.19}
$$

The *P* could be any interpolation function, such as the linear, quadratic hermit or the cubic interpolation. The traditional definitions of each term in Eq. 30.19 can be found in [Thompson 1985, 1992]. However, as one can find the formulas from the reference, the traditional *TFI* approach [Thompson, 1985], [Soni, 1993] cannot be applied to the process of generating a NURBS *TFI* control volume because the addition and subtraction operations in Eq. 30.19 may lead to zero or negative *weights* in the interior control volume. Any zero *weight* will make the corresponding control point lose its influence and the negative *weights* will create undesirable grids, such as the unbounded grids or crossing grids. Hence, when applying Eq. 30.19 to a NURBS *TFI* volume, it is necessary to redefine the individual terms listed in Eq. 30.19. The procedure is as follows: Suppose the six NURBS surfaces are all predefined, and the surfaces of S_1 and S_2 are used for the ξ direction. S_1 and S_2 have the same orders of *k2*, *k3*, and same number of control points of $n \times L$ (refer to Eq. 30.3). If the orders of these two surfaces do not match, one should perform the degree-raising algorithm to the low degree surface. If the resolutions of the control points of S_1 and S_2 are different, then the knot insertion algorithm should be used to make them the same. The same procedures should be applied to the surface of S_3, S_4 (with the *orders* of *k*1, *k*3 and the resolutions of control net of $m \times L$) and S_5, S_6 (with the *orders* of *k*1, *k*2 and the resolutions of control net of $m \times n$). After this step, each term for a linear NURBS *TFI* volume can be defined as follows: the $P_\xi V$ is a NURBS volume that is created by using the surfaces of S_1 and S_2 with the algorithm of "ruled NURBS volume" (described in the previous section of this chapter). Hence, the *three* orders of $P_\xi V$ are 2, *k*2, and *k*3, while the resolution of the control volume is $2 \times n \times L$. The same procedures should be applied to $P_\eta V$ and $P_\zeta V$. Therefore, the *orders* of $P_\eta V$ are *k*1, 2, *k*3 with the resolution of the control volume of $m \times 2 \times L$, while the *orders* of $P_\zeta V$ are *k*l, *k*2, 2 with the resolution of the control volume of $m \times n \times 2$. $P_\xi P_\eta(V)$ is a NURBS volume that is created by utilizing the boundaries (in ζ direction) of S_1, S_2 and the corner points of S_3, S_4, S_5, S_6. In other words, it has *orders* of 2, 2, *k*3 and the dimension of control volume of 2, 2, *L*. The $P_\eta P_\zeta(V)$ and $P_\xi P_\zeta(V)$ are defined analogously — the *orders* of the $P_\eta P_\zeta(V)$ are *k*1, 2, 2 and the resolution of the control volume is *m*, 2, 2, while the *orders* of the $P_\xi P_\zeta(V)$ are 2, *k2*, *2* and the resolution of the control volume is *2*, *n*, 2. The last term of $P_\xi P_\eta P_\zeta(V)$ is simply a NURBS control volume constructed by all the corner points of the six surfaces. Hence, the *orders* of this volume are 2, 2, 2 and the size of control volume is $2 \times 2 \times 2$. These seven control volumes are illustrated in Figure 30.27.

After these seven intermediate control volumes are created, Eq. 30.20 below should be used for the final linear NURBS TFL This equation will avoid the creation of any undesired interior *weights.*

$$\begin{vmatrix} WP \\ W \end{vmatrix}_{ijk} = \begin{vmatrix} W_\xi W_\eta W_\zeta \left(P_\xi + P_\eta + P_\zeta - P_\xi P_\eta - P_\eta P_\zeta + P_\xi P_\eta P_\zeta \right) \\ W_\xi W_\eta W_\zeta \end{vmatrix}_{ijk} \tag{30.20}$$

In addition to the algorithm of NURBS *TFI* surface, one has to perform the knot insertion and degree elevation algorithms on all seven intermediate control volumes to ensure all of them have the same orders and same knot vectors in all the ξ, η, and ζ directions, respectively. After this step is completed, the sizes of all the control volumes will be the same. Hence, the final control volume for NURBS *TFI* can be obtained by adding the corresponding control points of $P_\xi(V)$, $P_\eta(V)$, $P_\zeta(V)$, $P_\xi P_\eta P_\zeta(V)$ and subtracting those of $P_\xi P_\eta(V)$, $P_\eta P_\zeta(V)$ and $P_\xi P_\zeta(V)$, while the *weights* are determined by multiplication of W_ξ, W_η, and W_ζ. Figure 30.28 shows an *H*-type nozzle generated according to this approach.

30.6 Conclusion and Summary

The geometry modeling techniques used in computer-aided geometric design have been extended and applied to numerical grid generation for CFS simulation. The generalized algorithms that convert the non-NURBS entities to NURBS (or B-spline) representations have been presented. These algorithms can be utilized to bridge the gap between the grid generation and the CAD/CAM systems. The formulation of NURBS has been extended from curves, surfaces to full 3D NURBS control volumes to model the CFS configurations. The development of the redistribution schemes on volume grids with singularity is

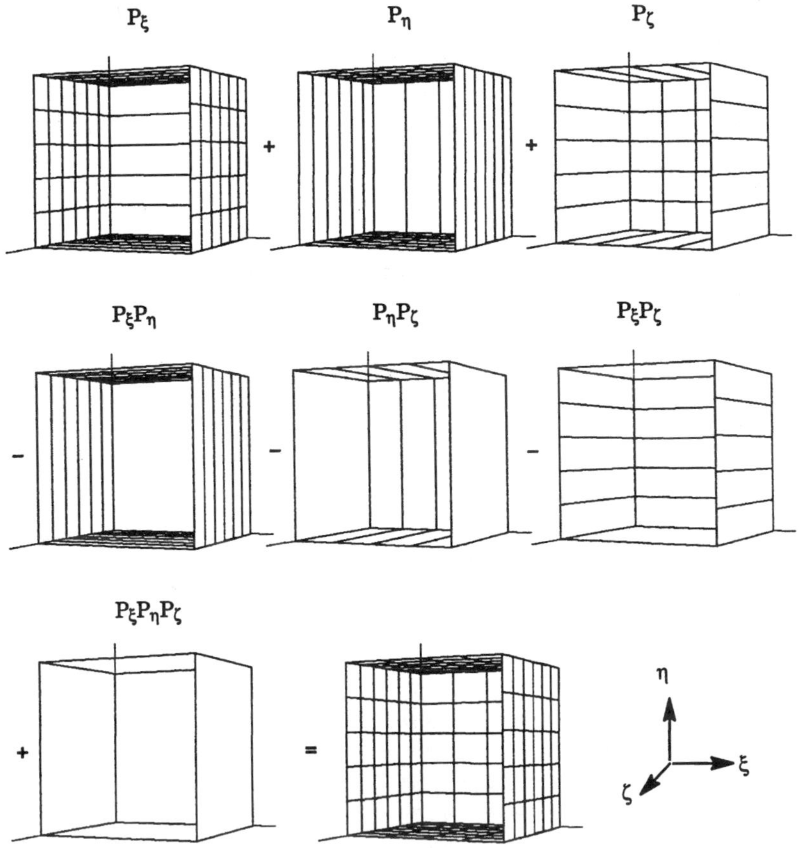

FIGURE 30.27 Illustration of NURBS *TFI* volume.

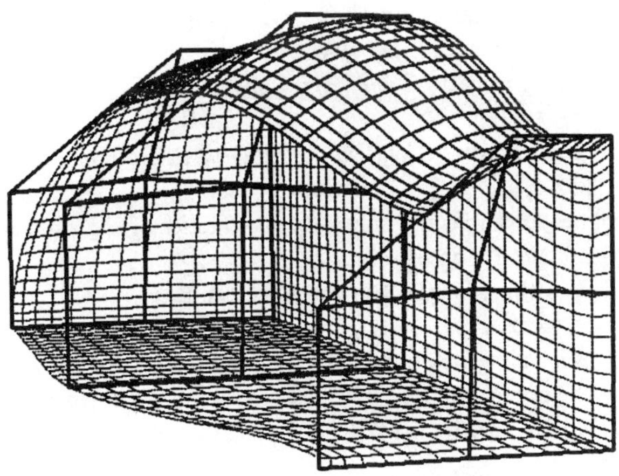

FIGURE 30.28 A volume grid for a nozzle created by NURBS volume *TFI* option.

demonstrated by computational examples. The applications of these reparametrization techniques to precise grid distribution control with accurate geometry fidelity have been demonstrated. In addition, the applications of NURBS to grid generation presented in this chapter have proven the versatility of NURBS in the CFS simulation processes.

Acknowledgment

Part of the contents of this chapter (Figures 30.1–30.6, 30.9, 30.11–12, Eqs. 30.5, 30.7, 30.8, 30.10, 30.11) were reprinted from T. Y. Yu and B. K. Soni, *Computer-Aided Design,* pp 147–157, with kind permission from Elsevier Science Ltd. The algorithms presented in this chapter were developed under a contract from NASA/Marshall Space Flight Center at Mississippi State University.

References

Farin, G.E., *Curves and Surfaces for Computer Aided Geometric Design: A Practical Guide,* 3rd Edition. Academic Press, 1993.

Lin, F. and Hewitt, W.T., Expressing Coons–Gordon surfaces as NURBS, *J. of Computer Aided Design.* Feb.1994, Vol. 26, No. 2, pp 145–155.

Piegl, L., Infinite Control Points – A method for representing surfaces of revolution using boundary data, *IEEE Computer Graphics & Applications.* Mar. 1987, Vol. 7, No. 3, pp 45–55.

Peigl, L., Interactive data interpolation by rational Bézier curves, *IEEE Computer Graphics & Applications.* Apr. 1987, Vol. 7, No. 4, pp 45–58.

Piegl, L., On NURBS: a survey, *IEEE Computer Graphics & Applications.* Jan. 1991, Vol. 11, No. 1, pp 57–71.

Piegl, L., Rational B-Spline curves and surfaces for CAD and graphics, *State of the Art in Computer Graphics Visualization and Modeling.* Rogers, D.F. and Earnshaw, R.A., (Eds.), Springer-Verlag, 1991, pp 225–269.

Piegl, L. and Tiller, W., Curve and surface constructions using rational B-splines, *Computer Aided Design.* 1987, Vol. 19, No. 9, pp 485–498.

Piegl, L. and Tiller, W., A menagerie of rational B-splines, *IEEE Computer Graphics & Applications.* Sept. 1989, Vol. 9, No. 5, pp 48–56.

Reichert, B.A., Hingst, W.R., Okiishi, T.H., Circular-to-rectangular transition duct flow without and with inlet swirl, *J. of Propulsion and Power.* Jan.-Feb. 1994, Vol. 10, No. 1, pp 88–100.

Shih, M.H., Towards a comprehensive computational simulation system for turbomachinery, Ph.D. dissertation, Mississippi State University, May 1994.

Soni, B.K., Grid generation for internal flow configurations, *Computers and Mathematics with Applications.* 1993, Vol. 24, No. 5/6, pp 151–163.

Thompson, J.F., A survey of grid generation techniques in computational fluid dynamics, *AIAA-83-0447,* AIAA 21st Aerospace Sciences Meeting, 1983.

Thompson, J.F., National grid project, *Computing Systems in Engineering.*1992, Vol. 3, No. 1–4, pp 393–399.

Thompson, J.F., Warsi, Z.U.A., Mastin, C.W., *Numerical Grid Generation Foundations and Applications.* North-Holland, 1985.

Tiller, W., Rational B-splines for curve and surface representation, *IEEE Computer Graphics & Applications.* Sep.1983, Vol. 3, No. 10, pp 61–69.

Tiller, W., Knot removal algorithms for NURBS curves and surfaces, *Computer Aided Design.* Aug. 1992, Vol. 24, No. 8, pp 445–453.

Yu, T.Y., CAGD Techniques in grid generation, Ph.D. dissertation, Mississippi State University, Dec. 1995.

Yu, T.Y., and Soni, B.K., The application of NURBS in numerical grid generation, *J. of Computer-Aided Design.* Feb. 1995, Vol. 27, No. 2, pp 147–157.

31

NASA IGES and NASA-IGES NURBS-Only Standard

Austin L. Evans

David P. Miller

31.1 Introduction

31.1.1 Purpose

This chapter is intended to provide background on the NASA Geometry Data Exchange Specification for Computational Fluid Dynamics (NASA-IGES) [RP1338, 1994] and the NURBS-Only subset of NASA-IGES. This will elucidate the logic behind the standard. Documentation in this area will be referenced to provide additional sources of information for future reference. Sample NASA-IGES compatible software will also be discussed. This should facilitate the usage of the NASA-IGES protocol for rapid and accurate data transfer, and should serve to promote the use of an accurate and unified geometry representation method for CFD research.

31.1.2 Scope

This chapter contains an updated synopsis of the NASA-IGES specification along with information on the follow-on activities to the standard. This material has been divided into six sections. The first is this introductory section, which provides some background in this area. Section 31.2 relates the underlying principles and the logic behind the standard and its application. Section 31.3 includes the recommended best practices for use while implementing the NASA-IGES standard. Section 31.4 notes future research issues in this area. The fifth and sixth sections contain the references and bibliography for further information.

0-8493-2687-7/99/$0.00+$.50

31.1.3 Background

The geometry data received by NASA scientists for analysis and modification has been supplied in numerous formats that often require hundreds of hours of manipulation to achieve a format capable of being utilized by analysis software. It has been estimated that this accounts for from 70% to 80% of the analysis cycle time. This modified data set usually has lost a level of accuracy from the original data and often may not maintain the design intent of the original data as developed on the original designer's system. If multidisciplinary analysis is added into the analysis cycle, the problem can increase by one order of magnitude for each discipline. In some cases so much fidelity has been lost between design geometry and the hardware that test data is nearly impossible to relate directly to the analysis.

In the spring of 1991, the NASA Surface Modeling and Grid Generation Steering Committee determined that one of the leading detriments to the grid generation process was the lack of a standard method of transferring complex vehicle geometries between various software systems. A subcommittee for Geometry Exchange Specification composed of technical personnel from the Ames, Langley, and Lewis Research Centers was formed to develop a data exchange format.

Following an analysis of existing and proposed standards, the Subcommittee for Geometry Exchange Specification selected the existing Initial Graphics Exchange Specification (IGES) format [IGES, 1995] as the basis for a NASA standard. In the U.S., IGES is by far the most widely used product data exchange specification. The latest version of the IGES specification (Version 5.3) provides an adequate set of geometric entities to cover the current data transfer needs for computational fluid dynamics (CFD) research. Plans were made to take advantage of the developing STEP standard when moving beyond a CFD-only standard.

A subset of the IGES capability was selected, and a draft NASA Technical Specification was released in September of 1991 entitled "NASA Geometry Data Exchange Specification Utilizing IGES." In the specification, the rational B-spline was chosen as the most stable format to represent all types of geometry and was selected as the primary geometry representation method. In April of 1992, this subset of entities was proposed to the IGES/PDES Organization (IPO) for acceptance as an official IGES application protocol (AP). The IPO did not feel comfortable with restricting geometry entities to a limited subset in an AP. As the restriction on entities was the key to the usability of this specification, the NASA geometry subcommittee chose to proceed with the completion of this document and the development of software to utilize data based on this standard without pursuing official IPO acceptance. Since files conforming to this specification are valid IGES files, there should be minimal impact on industry conversion to utilizing NASA-IGES.

The standard IGES file format is very complex. The IGES documentation is also very large and complex. Utilizing IGES data files requires expert knowledge of the associated format. Even though the NASA-IGES specification contains significantly fewer entities, it still inherits a major portion of the complexity of the IGES file format. It is unreasonable to expect most scientists and CFD software developers to spend the time necessary to understand the file format and to handle the files directly. This IGES file complexity problem has led to the development of the main body of the specification.

It should be noted that the IGES entities allowed under this specification and other related information are contained in summary form in this chapter. Reference in this chapter to "NASA-IGES specification" or "NASA-IGES files" refers to the subset of IGES entities specified in the tables in this chapter and IGES files conforming to that specification.

31.1.4 NASA Support

The NASA-IGES specification has the direct support of the NASA Surface Modeling and Grid Generation Steering Committee representing the NASA Headquarters Office of Aerospace Science and Technology (OAST), three NASA Research Centers — Ames, Langley, and Lewis — and two operational NASA facilities — Johnson Space Center and Marshall Space Flight Center.

These NASA facilities are committed to utilizing this specification for geometry representation for design and analysis of aerospace vehicles utilizing CFD techniques. Several pilot software implementation programs were undertaken at these centers. NASA-IGES compatible software is presented in Section 31.4 of this chapter. In addition, most CAD systems have moved toward being NURBS-based or compatible. This means that they are more or less NASA-IGES compatible.

31.2 Underlying Principles (the CFD Process)

NASA research centers support studies in a variety of scientific areas. Utilizing computer simulation, NASA supports extensive research on analysis of the behavior of complex physical fields. Examples of physical field analysis include computational fluid dynamics (CFD), computational electromagnetics (CEM), heat transfer, and finite element modeling (FEM). Virtually any field that utilizes partial differential equations (PDE) performs some form of field solution calculation. Most of these fields study the effects of a phenomenon around a particular object. The numerical data that provide a mathematical description of that object are called the geometry data or model data. For these computer simulations to be useful in the design process, the geometry data must be passed among many groups rapidly and accurately. Even in the case of pure research, the geometry data must be shared with other groups. For example, in a typical fluid dynamics study, the computational solutions are compared with wind tunnel data. The manufacturer of the wind tunnel model must have an accurate geometry definition from the computational scientist.

The NASA-IGES specification addresses the geometry data transfer and geometry data usage requirements for these complex field simulations. The specific research area most applicable at this time is CFD. The remainder of this section focuses mainly on the CFD process but is generally applicable to other field simulation processes.

31.2.1 The CFD Analysis Process

Research in CFD is accomplished by modeling the fluid as a discrete set of points and computing the velocity and other properties of the fluid at each of these points. The set of points is referred to as a "grid" or "mesh." A general representation of the CFD analysis process and the data transferred is shown in Figure 31.1.

Two forms of geometry definition are utilized by grid generation tools: surface definition and solid definition. In surface definition, the surfaces of the object to be analyzed are required by the grid generation tool. Currently, a majority of the grid generation tools require this surface geometry definition. Most of the grid generation tools utilizing surface definition are used interactively in the grid generation process.

Solid definition is usually required by tools that perform automatic grid generation. Currently, only a few grid generation tools utilize solid geometry definition. The NASA-IGES specification is intended for surface geometry only; future enhancements may include solid definition.

31.2.2 The CFD Design Process

When CFD is used for design, the analysis process must be done a great many times in order to determine an optimal design. If a computer-aided design (CAD) system is available to the analysis group, the geometry is modified on the CAD system, the new data are transferred to the grid generator, and the grid generation process is repeated. This "pipeline" is repeated for each different design modification. Even though most current CAD systems can provide geometry in IGES format today, transferring new geometry for each analysis has to be done via data conversion, since not all grid generation tools currently read IGES data. This conversion is tedious and usually introduces additional errors.

CAD systems are often not available to the analysis group. In these cases, the grid is modified and the rest of the grid generation process is repeated. Current tools perform this geometry modification through changes to the grid representing the surface of the model. Such tools are limited in their capabilities and introduce additional data errors. Once a particular configuration is selected, the grid representing the

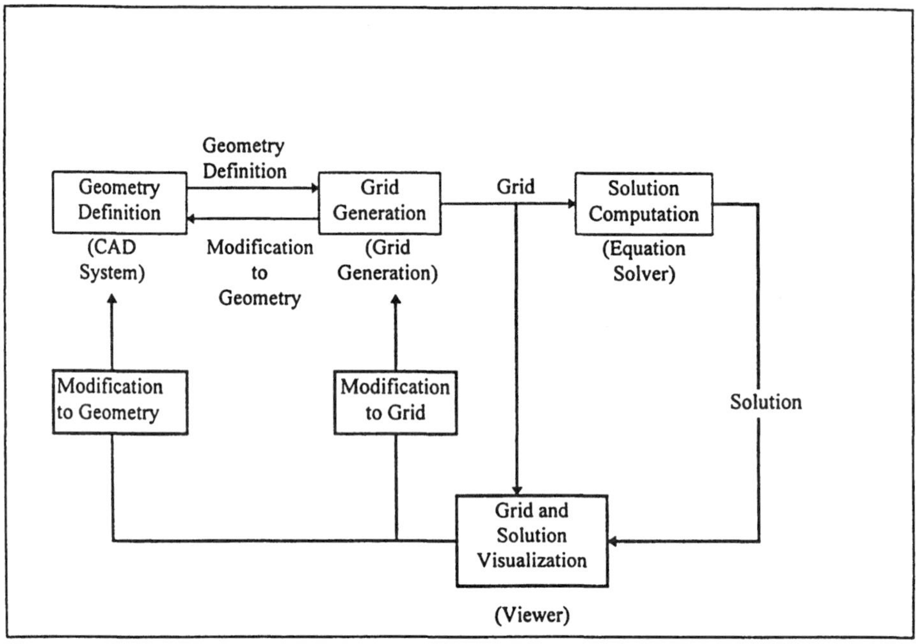

FIGURE 31.1 CFD analysis process. Source: RP1338, NASA Geometry Data Exchange Specification for Computational Fluid Dynamics (NASA-IGES), National Aeronautics and Space Administration, Washington, DC, 1994.

surface of the model is transferred to the CAD system for reintegration with the original model. The methods for this data transfer are not standardized, and the data must be converted into the particular CAD system's format. This entire process is tedious and usually introduces additional data conversion errors.

31.2.3 Problems with Pre-NASA-IGES Methods

The pre-NASA-IGES methods for data transfer and grid generation are very time- and manpower-intensive and often require data approximation during conversion and use.

When the NASA-IGES specification was published, most grid generation software could not utilize the geometry definition generated by CAD systems directly. The geometry had to be massaged into the different *ad hoc* formats required by the different types of grid generation software, and there are as many formats as there are grid generation packages. These formats frequently utilize only discrete point information for representing the geometry and do not retain complete information about the geometry.

Intensive human interaction and extensive manipulation may be required to convert the geometry into the particular format required by a piece of grid generation software. These operations are laborious and error-prone. The geometric information, e.g., surface curvatures, lost during this conversion is either extremely difficult or impossible to recover. Sometimes, the incompleteness of geometric information imposes severe limitations on the capability of the grid generation software.

Consistent utilization of NASA-IGES would dramatically improve these areas. This will be discussed in Section 31.3.

31.2.4 CFD Design Utilizing NASA-IGES

If both the geometry definition system and the grid generator can utilize the NASA-IGES specification data format without any conversion errors, geometry data can be passed back and forth quickly and accurately. A series of design modifications could be generated on a CAD system and transferred to the grid generation software in minutes. The first configuration may require a fair amount of time to perform surface gridding, volume gridding, and solution computation. Successive iterations should be available in very little time if the gridding programs could rapidly regenerate new grids from new geometry data that has similar topology to the previous data. The errors identified in Section 31.2.3 could be eliminated entirely if both the CAD system and the grid generation software operate on the same geometry data.

The NASA-IGES specification is designed to be bidirectional. Software systems should be capable of both reading and writing data in this NASA-IGES format. This will require grid generation programs to read in NASA-IGES data, perform any modifications directly on the NASA-IGES geometry rather than the computational grid, and to write out modified surfaces in NASA-IGES format.

31.2.5 CFD Design Utilizing the Supplied Database Information Format

To facilitate the use of this data transfer method, NASA is developing software for several functions such as reading, writing, and translating NASA-IGES data. (See Section 31.3) Utilizing these programs and their implementation of the abstract database, Standard Data Access Interface, (SDAI) [Evans, 1997], grid generation software can utilize NASA-IGES geometry through their existing in-memory databases without handling NASA-IGES files directly. One possible scheme for such in-memory data access is through a shared memory architecture utilizing the reader–writer software. Alternatively, the user may choose to utilize NASA-IGES data files for transfer, using this document for an understanding of the mathematics behind the geometric entities while using the IGES document to understand the file format.

Since different grid generators may require different internal database formats to satisfy individual needs, a grid generator may need to convert the in-memory database obtained from the reader–writer to its internal in-memory database format. The process for a grid generator to use the reader–writer is expressed in Figure 31.2. All the stages in the process can be done internally and automatically by the grid generator, and the reader-writer will do most of the work of incorporating NASA-IGES files.

31.2.6 General Information on Data Description

The geometry and nongeometry information is described and defined in the NASA-IGES standard [RP1338, 1994]. Tables summarizing this information are included in this chapter. The information is separated into logical units, each of which is called an *object* or an *entity.* The word *entity is* used in this chapter. An entity represents either a complete geometric concept or a complete bit of information. However, in some instances an entity becomes meaningful in the database only after it is attached to another entity.

31.2.6.1 Entity Description Overview

This chapter does not provide all the details necessary to utilize IGES files. The developer of software for reading or writing any IGES files will need to review the IGES document and the RP1338 for a complete description of the file format. This chapter does provide a listing of the entities and restrictions placed on NASA-IGES and NASA-IGES NURBS-Only files. NASA-IGES is a subset of standard IGES, and NASA-IGES NURBS-Only is a subset of NASA-IGES. All NASA-IGES NURBS-Only and NASA-IGES files are valid IGES files.

Throughout this chapter reference is made to NASA-IGES. These comments also pertain to NASA-IGES NURBS-Only files. All comments and restrictions are the same for both types of files except as noted. NASA-IGES files can include seven additional entities not allowed in NASA-IGES NURBS-Only files. These are identified in Sections 31.3.2.

This specification is a subset of the Initial Graphics Exchange Specification (IGES) Version 3 [IGES, 1995], National Institute of Standards and Technology (NIST) number NISTIR 4412. There are no items in this specification that do not adhere to the standard IGES format. There are no IGES entities requiring or utilizing any implement defined types.

There are five classes of IGES entities: (1) curve and surface geometry entities, (2) constructive solid geometry (CSG) entities, (3) B-Rep solid entities, (4) annotation entities, and (5) structure entities.

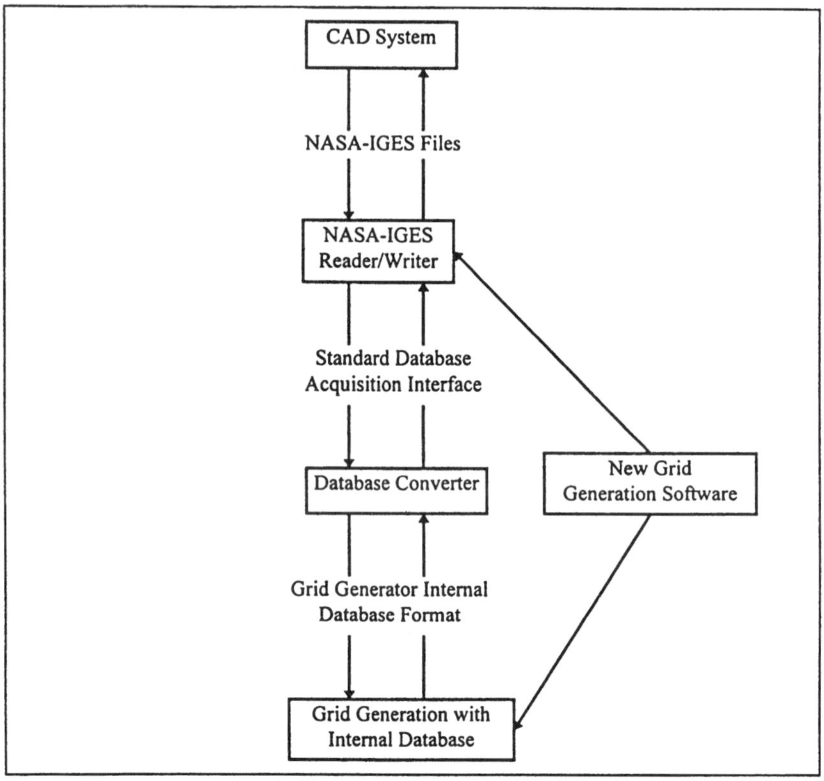

FIGURE 31.2 Grid generation with NASA-IGES file reader–writer. Source: RP 1338, NASA Geometry Data Exchange Specification for Computational Fluid Dynamics (NASA-IGES), National Aeronautics and Space Administration, Washington, DC, 1994.

31.2.6.2 Coordinate System

All of the entities are defined in a local coordinate system that forms the *definition space* of the entity. The local coordinate system is usually the most convenient and stable coordinate system to define the entity. However, the designed model usually resides in a different coordinate system, called the *model space*. The local coordinate system may coincide with the model space coordinate system. If not, one or more coordinate transformation matrices must be used to bring the entity from its definition space position to its final model space position.

A model may be designed at an enlarged or reduced size. To obtain its real-world size, the dimensions of a model as specified in the database must be divided by the factor Model Space Scale (RP 1338, Section 6.2).

A *transformation* matrix *pointer* is associated with every entity. This pointer is either 0, for the identity rotation matrix and zero translation vector, or a transformation matrix entity that will be applied to the entity in the process of bringing the entity to the model space. In fact, a transformation matrix entity contains a transformation matrix pointer. Hence, it is possible to store successive transformations under one transformation matrix pointer. (See RP 1338, Section 5.1 for more information.)

Since the database is hierarchical, i.e., an entity may be a part of another entity, recursively, multiple transformation matrices, following the hierarchy, may be necessary to bring an entity from its definition space to the model space. For example, if entity A is a part of the definition of entity B, entity A will be transformed by the transformation matrix associated with A first and then by that associated with B.

All coordinate systems are right-handed.

31.2.6.3 Common Information

Information common to all entities is not described for each individual entity. This information includes, but is not limited to, color, level, form, and transformation matrix pointer.

Some information common to the entire model and data files is also contained in the database. This includes, but is not limited to, text identifying the model, measuring system units, and data of file creation. This information corresponds to the global section of the IGES files. The common and global information and other database related issues are discussed in Section 6 of the NASA-IGES standard (RP1338, 1994).

31.3 Best Practices

This section is divided into four subsections. The first, Multidisciplinary Data Exchange Standards, relates follow-on activities that have occurred since the NASA-IGES standard was published. The second, Summary of Entity Types and Recommended Usage, describes the contents of the standard and how to use its elements. In the third section, three case studies are presented. The last section, Other NASA-IGES compatible software, lists some additional NASA-IGES-compatible software not mentioned in the case studies.

31.3.1 Multidisciplinary Data Exchange Standards

The charter of the NASA Geometry Exchange Standard Subcommittee was to develop a standard that was focused on CFD. It was intended that the standard work that was done would be expanded to cover other disciplines. The section gives information on that effort.

Turbomachinery characteristics are strongly influenced by a combination of aerodynamic, thermal, and structural effects. The predictions of turbomachinery performance for off-design conditions usually require the inclusion of the thermal and structural displacements to determine blade operating shape. Also, blade, rotor, and casing deflections and tip clearance changes have significant effects on aerodynamic stability and impact to the overall operability of turbomachinery.

The ability to address these multidisciplinary aerodynamic, thermal, and structural effects related to turbomachinery does not require a tightly coupled and fully integrated aeroelastic analysis. The analysis of steady state operating points can be achieved by exchanging boundary condition data between multiple disciplines required for turbomachinery analysis. Stand alone computational fluid dynamics (CFD) and thermal/structural finite element analysis (FEA) codes can be loosely coupled in this manner to obtain turbomachinery analysis results for the steady state coupled effects in a small number of loosely coupled iterations.

An enhanced method for exchanging this boundary condition data between aerodynamics CFD grids and thermal/structural FEA analysis grids has been developed by NASA, DOD Navy, and Boeing. This enhanced method associates the boundary condition data from each analysis discipline's grid with the geometric representation. This method also moves toward a standardized method for the exchange of loosely coupled engineering analysis information. This methodology relies heavily on the use of nonuniform rational B-spline (NURBS) as the technique to represent both the geometry and the boundary conditions information. NURBS mathematics (see Chapter 30) has become the *de facto* standard in the CAD/CAM industry for definition of geometric information (see Chapter 31). Therefore, the multidisciplinary method focuses on associating the engineering data in the NURBS mathematical form. The method also relies on the NASA-IGES subset with the inclusion of IGES user defined-extensions (IGES 5000 entities) for the exchange of both the geometric and boundary data definitions.

The NASA-IGES subset was initially established and adopted by the NASA Centers for the exchange of geometric definitions between CAD/CAM systems and aerodynamic CFD analysis. NASA Lewis and DOD Navy are exploring prototype methods to extend the NASA-IGES geometric subset to multidisciplinary analysis for turbomachinery and for future ISO STEP engineering analysis standardization. The

TABLE 31.1 NASA-IGES Conversion Map

>From Entity Type (,Form)	>To Entity Type (,Form)
Copious Data, Type 106, Form 11	Rational B-Spline Curve Type 126, Form 0, Degree 1, PROP1 1, PROP3 1, PROP4 0
Copious Data, Type 106, Form 12	Rational B-Spline Curve Type 126, Form 0, Degree 1, PROP3 1, PROP4 0
Copious Data, Type 106, Form 13	Rational B-Spline Curve Type 126, Form 0, Degree 1, PROP3 1, PROP4 0, the information about the vectors associated with the points will be lost
Copious Data, Type 106, Form 63	Rational B-Spline Curve Type 126, Form 0, Degree 1, PROP1 1, PROP2 1, PROP3 1, PROP4 0
Parametric Spline Curve, Type 112	Rational B-Spline Curve, Type 126
Parametric Spline Surface, Type 114	Rational B-Spline Surface, Type 128
Ruled Surface, Type 118	Rational B-Spline Surface, Type 128
Surface of Revolution, Type 120	Rational B-Spline Surface, Type 128
Tabulated Cylinder, Type 122	Rational B-Spline Surface, Type 128
Offset Curve, Type 130	Rational B-Spline Curve, Type 126, Circular Arc Type 100 or Line Type 110 on exact conversion.
Offset Surface, Type 140	Rational B-Spline Surface, Type 128
Trimmed Parametric Surface, Type 144	Bounded Surface, Type 143
Definition Levels, Type 406, Form 1	The entity with this property is placed in the first level identified by this Definition Levels entity

Source: RP1338, NASA Geometry Data Exchange Specification for Computational Fluid Dynamics (NASA-IGES), National Aeronautics and Space Administration, Washington, DC, 1994.

TABLE 31.2 NASA-IGES-NURBS-Only Conversion Map

>From Entity Type (,Form)	>To Entity Type (,Form)
Circular Arc, Type 100	Rational B-Spline Curve, Type 126, Form 2, Degree 1,PROP1 1, PROP3 0, PROP4 0
Composite Curve, Type 102	Rational B-Spline Curve, Type 126, Forms 1, 2, 3, 4, or 5 as appropriate, Degree 1, PROP1 1, PROP3 1, PROP4 0
Conic Arc, Type 104	Rational B-Spline Curve, Type 126, Forms 3, 4, or 5 as appropriate, Degree 1, PROP1 1, PROP4 0
Copious Data, Type 106, Form 1	Rational B-Spline Curve, Type 126, Form 0, Degree 1, PROP1 1, PROP3 1, PROP4 0
Copious Data, Type 106, Forms 2 or 3	Rational B-Spline Curve, Type 126, Form 0, Degree 1 PROP3 1, PROP4 0
Line, Type 110	Rational B-Spline Curve, Type 126, Form 1, Degree 1, PROP1 1, PROP3 1, PROP4 0
Singular Subfigure, Instance Type 408	A copy of the geometry using original entities. These entities are then converted as specified in these Conversion Maps

Source: RP1338, NASA Geometry Data Exchange Specification for Computational Fluid Dynamics (NASA-IGES), National Aeronautics and Space Administration, Washington, DC, 1994.

NASA-IGES subset of the overall IGES specification is the basis for the data exchange method and the prototype development. This subset was adopted and supported by the DT_NURBS Spline Subroutine Library [U.S. Navy, 1993]. The development of and extension to the multidisciplinary capabilities were built upon the initial capabilities in the DT_NURBS library.

31.3.2 Summary of Entity Types and Recommended Usage

This section contains summaries and recommended usage of the entity types utilized by the NASA-IGES specification as listed in the several tables included in this chapter. Since NASA-IGES is a limited subset of IGES entities, recommend conversion from non-NASA-IGES entities to NASA-IGES entities has been included in Tables 31.1 and 31.2. The entities in the tables are grouped by function. Table 31.3 contains a summary list, ordered by IGES entity type number, of all the entities allowed in NASA-IGES and

TABLE 31.3 Summary of NASA-IGES Entities

IGES Entity No.	Entity Name	NASA-IGES	NURBS-Only
Entity 0	Null entity	yes	yes
Entity 100	Circular arc	yes	no
Entity 102	Composite curve	yes	no
Entity 104	Conic arc	yes	no
Entity 106	Copious data	yes	no
Entity 110	Line	yes	no
Entity 116	Point	yes	no
Entity 124	Transformation matrix	yes	yes
Entity 126	Rational B-spline curve	yes	yes
Entity 128	Rational B-spline surface	yes	yes
Entity 141	Boundary	yes	yes
Entity 142	Curve on a parametric surface	yes	yes
Entity 143	Bounded surface	yes	yes
Entity 212	General note	yes	yes
Entity 308	Subfigure definition	yes	no
Entity 314	Color definition	yes	yes
Entity 402	Associativity instance	yes	yes
Entity 406	Property, Form 15: name	yes	yes
Entity 408	Singular subfigure instance	yes	no

Source: RP1338, NASA Geometry Data Exchange Specification for Computational Fluid Dynamics (NASA-IGES), National Aeronautics and Space Administration, Washington, DC, 1994.

TABLE 31.4 Geometric Entities Allowed in NASA-IGES NURB-Only Files

IGES Entity No.	Entity Name	Entity Class (see [IGES, 1995])
Entity 124	Transformation matrix	Geometry
Entity 126	Rational B-spline curve	Geometry
Entity 128	Rational B-spline surface	Geometry
Entity 141	Boundary	Geometry
Entity 142	Curve on a parametric surface	Geometry
Entity 143	Bounded surface	Geometry

Source: RP1338, NASA Geometry Data Exchange Specification for Computational Fluid Dynamics (NASA-IGES), National Aeronautics and Space Administration, Washington, DC, 1994.

NASA-IGES NURBS-Only data files. Tables 31.4–31.6 contain summary groupings of the entities by recommended usage. It is desirable to represent all geometric objects utilizing the following entities that are available in NASA-IGES and NASA-IGES NURBS-Only files. Each entity section has three subsections covering the following: (1) Usage: Explaining the general usage and how to use any options. (2) Recommendations: Listing recommended practices, such as explaining any specific usage that is desired but not required, listing any alternate entities that may be preferred over this one, and what application each entity is good for and itemizing exactly for what this entity should be used. (3) Restrictions: Listing specific restrictions such as forms and options that are not allowed. These are additional restrictions to those in IGES Version 5.1. If no restriction are mentioned in this section, then only the restrictions in IGES apply.

Entity 0 : Null Entity

This entity is used to remove an entity from the current file without renumbering the entire file. This entity is a good method for manually removing entities from a specific IGES file without utilizing much of the user's time by not having to reorder and repack an IGES file.

TABLE 31.5 NASA-IGES Entities Not Allowed in NASA-IGES NURBS-Only Files

IGES Entity No.	Entity Name	Entity Class (see [IGES, 1995])
Entity 100	Circular arc	Geometry
Entity 102	Composite curve	Geometry
Entity 104	Conic arc	Geometry
Entity 106	Copious data	Geometry
Entity 110	Line	Geometry
Entity 116	Point	Geometry
Entity 308	Subfigure definition	Structure
Entity 408	Singular subfigure instance	Structure

Source: RP1338, NASA Geometry Data Exchange Specification for Computational Fluid Dynamics (NASA-IGES), National Aeronautics and Space Administration, Washington, DC, 1994.

TABLE 31.6 Nongeometric Entities Allowed in NASA-IGES and NASA-IGES NURBS-Only Files

IGES Entity No.	Entity Name	Entity Class (see [IGES, 1995])
Entity 0	Null entity	Structure
Entity 212	General note	Annotation
Entity 314	Color definition	Structure
Entity 402	Associativity instance	Structure
Entity 406	Property, Form 15: name	Structure

Source: RP1338, NASA Geometry Data Exchange Specification for Computational Fluid Dynamics (NASA-IGES), National Aeronautics and Space Administration, Washington, DC, 1994.

Entity 100: Circular Arc

This entity is used to transfer circular arcs, including full circles. A circular arc should be transferred through this entity, although Entity Type 126 could be used. The receiving system may convert the data to a B-spline format as necessary. This entity is not allowed in NASA-IGES NURBS-Only files.

Entity 102: Composite Curve

This entity is used to transfer a curve composed of several parametrized curves. Note that a composite curve entity is not allowed as a component of a composite curve entity. A connect point entity (not a NASA-IGES entity) or a point entity in a composite curve should be ignored. This does not invalidate the geometry of a composite curve. However, if a parametric spline curve (not a NASA-IGES entity) is in a composite curve, the composite curve should be ignored by a nonrestrictive reader. This entity is not allowed in NASA-IGES NURBS-Only files.

Entity 104: Conic Arc

This entity can be used to represent many types of conic sections. It is recommended that this entity not be used. Conics can be accurately represented by B-splines (Entity Type 126). In order to maintain compatibility with many older systems, this entity is included in this specification. If the sending system knows the conic type, the form of this entity should be set to indicate the type. The entity should be put into its canonical position by the sending system as indicated in Appendix C of IGES V5.3. This entity is not allowed in NASA-IGES NURBS-Only files.

Entity 106: Copious Data

This entity is used to transfer an ordered list of points. This entity with Forms 1 to 3 is recommended for transferring a list of points from a cross-section curve. However, the cross-section curve itself should be transferred, instead of points on the curve, since the curve retains more information that may be useful in the receiving system. In addition to the point coordinate data, a vector is associated with every point in the parameter data section of this entity with Form 3. It is recommended that, if Form 3 is used, this vector be set as the direction vector of the cross-section curve. For other recommended usages of this entity, see Entity 402. Only Forms 1 to 3 are included in this specification. This entity is not allowed in NASA-IGES NURBS-Only files.

Entity 110: Line

The line entity is used to transfer line segments. It is preferred to transfer line segments by this entity rather than by Entity Type 126, since this is a commonly used and more compact representation. This entity is not allowed in NASA-IGES NURBS-Only files.

Entity 116: Point

This entity is used to transfer a point in space. The list of points on a curve or the mesh of points on a surface should be transferred through the appropriate entities, see Entity Type 106 and Entity Type 402. The pointer (PD Index 4) in the parameter data section, which points to the subfigure definition entity specifying the display symbol, will be ignored. The display symbol will be determined by the receiving system. This entity is not allowed in NASA-IGES NURBS-Only files.

Entity 124: Transformation Matrix

The transformation matrix is used to transform an entity from its local coordinate system to its true model space position. A number of entities are required by IGES to be transferred in their canonical definition space. For these entities, a transformation matrix is required to relocate them to their true position. Only Form 0 and Form 1 are included in this chapter. The other forms, for view transformation and finite element modeling, are not included.

Entity 126: Rational B-Spline Curve

This format is used as the primary entity for curve transfer. All the other curve types, excluding lines (Entity Type 110), conics (Entity Type 104), and circular arcs (Entity Type 100), must be converted (maybe with approximation) to this entity for transfer. This is the most flexible format to represent curves and is recommended for transferring all curves. All lines, circular arcs, and conics can be represented by this entity. This entity contains forms that identify each curve type. If the sending system knows the form of the curve, the form of this entity should be set appropriately. All parametric splines can also be represented by this entity. Software for the required conversion to this entity can be obtained from National Institute of Standards and Technology (NIST), U.S. Department of Commerce, Gaithersburg, MD.

Entity 128: Rational B-Spline Surface

This entity is used as the primary entity for surface transfer. All the other surface types must be converted (maybe with approximation) to this entity for transfer. This is the most flexible format to represent surfaces and is recommended for transferring all surfaces. This entity has forms for some analytic surfaces. If the sending system can determine the Form of the surface, the Form of this entity should be set appropriately.

Entity 141: Boundary

This entity should be used with Entity Type 143. It describes one boundary of a bounded surface. There are two types in this entity. Type 0 transfers only model space curves, and the surface may not be parametric. Type 1 transfers both parameter and model space curves, and the surface has to be parametric. Only Type 1 is used in the NASA-IGES specification.

Entity 142: Curve on a Parametric Surface

This entity is used to transfer a curve on a parametric surface when its parameter space curve is important. A curve on a parametric surface may be a curve from the projection of another curve onto the surface, a curve from the intersection of two surfaces, or an isoparametric curve. IGES provides the curve on parametric surface entity for use in either of two ways. It can be used with the trimmed surface entity (Type 144) to form a trimmed surface. Entity 144 is not allowed under this specification, so this use is not allowed. The boundary entity (Type 141) should be used for this purpose. The other use for this entity is to simply represent a curve on a surface. This is the only use allowed for this entity under this specification.

Entity 143: Bounded Surface

The bounded surface entity is used to transfer a bounded surface, a surface whose domain space is relimited (trimmed back) from its original domain. It should be used with Entity Type 141, boundary entity. This entity should be used instead of Entity Type 144, the trimmed parametric surface, for a surface with relimited domain, since Entity Type 144 disallows surfaces with poles or seams, which limits its usage. There are two types in this entity. Type 0 transfers only model space curves, and the surface may not be parametric. Type 1 transfers both parameter and model space curves, and the surface has to be parametric. Only Type 1 is used.

Entity 212: General Note

This entity is used to pass textual information about the geometry. This can include such information as the history of the object, relevant airfoil section numbers, and reference documents. A general note entity can exist separately or can be associated with another entity or entities. This entity is recommended as the entity for transferring relevant nongeometric design information. Form 0, which states that the text strings in the note are not related to each other positionally, is the only form included in this specification. This is also the default form. Font 1, the default font style for the ASCII character set, is the only font included in this specification. This allows the receiving system to use its default font for display.

Entity 308: Subfigure Definition

The subfigure definition and subfigure instance entities allow one copy of the geometry to be placed in many locations in a design without duplicating the geometry. For example, in a turbine engine design, all the turbine blades on the same stage are identical in shape. Only the geometry for one generic blade must be defined by using a subfigure definition entity. All the blades can then be created with the subfigure instance entity. The user should be discriminatory and exercise sound judgment in using this entity. For example, it is a good practice to represent turbine blades with instances, since this reduces file sizes tremendously and makes processing of the files much easier. However, to represent the two wings of an aircraft with an instance may not be wise, since the geometry for the wings is not stored explicitly in an instance; if the user decides to build a CFD grid on the wings, the grid generation software must create the geometry first. The grid generation software will probably not have the capability to create the geometry for an instance. This entity is not allowed in NASA-IGES NURBS-Only files.

Entity 314: Color Definition

Entity 314 is used to define additional colors (there are nine predefined colors in IGES). There are no recommendations on this entity, as its usage is self-evident.

Entity 402: Associativity Instance

This entity is used to group geometry entities into classes. It contains pointers to the grouped entities, called the members of the class. There are 18 predefined forms (classes), of which four (Forms 1, 7, 14, 15) are for grouping. Forms 1 and 7 are for unordered groups, i.e., the entities pointed to by this entity are an unordered set. Forms 14 and 15 are for ordered groups, i.e., there is an order specified for the entities pointed to by this entity; the order is defined by the sequence of the pointers specified within this entity. Unordered groups are frequently used to group surfaces from the same object, hence creating one group per object. Currently, very few CAD systems utilize the ordered group forms. Ordered groups are recommended for grouping a sequence of cross sections and associating them as a surface. In the recommended usage of the ordered forms, it is not required that the curves be from one surface; this information is irrelevant to this entity. The curves could be sliced from numerous surfaces. In this usage, the members of the class will be the cross-section curves. Ordered groups are also recommended to define a mesh of points (either topologically rectangular or nonrectangular) from a surface. In this usage, the members of the class will be the copious data entity (Entity Type 106, Forms 1–3). The same format is recommended to transfer points on a surface sampled along a list of cross-section curves on the surface. Only Forms 1, 7, 14, and 15 are included in this specification.

Entity 406: Property, Form15: Name

This entity is used to associate a name (or brief description) to an entity or a group of entities. All it contains is a text string that is the name. This entity would be appropriate for grouping a portion of the object together, such as a wing, and assigning it a name "wing." Longer comments should be handled through the general note entity.

Entity 408: Singular Subfigure Instance

The singular subfigure instance entity creates one instance of a subfigure, which is defined by a subfigure definition entity. See Section 3.4.15 of RP1338 for more information. See Entity 308 for recommended usage. This entity is not allowed in NASA-IGES NURBS-Only files.

31.3.3 Case Studies:

31.3.3.1 Blade Surface Geometry Modeling

Solid geometry modeling has become increasingly important in designing turbomachinery blading. The blade designs include turbines, pumps, compressors, fans, propellers, etc. In all these applications, the blade design is critical for achieving optimal overall performance. Since the underlying function is to smoothly change the fluid velocity around the blade, it generally consists of pararnetric sculptured surface models. In order to manufacture the blade using contemporary computer-aided manufacturing technology, the blade must be in a standardized portable format. The blade is represented as a nonuniform rational B-spline (NURBS) [Piegl, 1991] surface and is written to a standard NASA IGES (Initial Graphics Exchange Specification) file [NIST, 1990] which is portable to most design, analysis, and manufacturing applications.

A new methodology for interactive design of turbomachinery blades has been developed using methods that provide users with an interface that is intuitive to designers while operating with standardized geometric forms. BladeCAD [Miller, et al., 1996] introduces a new design technique that was motivated by the need to modify blade geometry on general surfaces of revolution, while providing intuitive interaction techniques. The blade is constructed as a three-dimensional space curve that characterizes both the shape of the blade and the stream surfaces. These surfaces and curves are represented as NURBS surfaces and curves with control point specification. Entity 128 is the recommendation for this purpose, see Section 31.3.2. This surface representation is a departure from point specification of blades that designers have used in the past. The point data specification would eventually be incorporated into a CAD system. To accomplish this, the blade point data would be resplined and interpreted by CAD operators who would essentially remodel the blade.

In the development of the surface model, a subset to the IGES file specification was proposed and adopted by the NASA Centers for Geometry Definitions [RP1338, 1994]. The subset was considerably smaller than the full IGES standard [NIST, 1990], which reduces the total number of entities required for an IGES file. The subset was adopted and included in the DT_NURBS library [U.S. Navy, 1997], which is used to generate the surface description of the airfoil. The last Fortran version of this library was released in 1997. Future versions of the library will be in the C++ language. Geometry for typical axial flow fans, propeller, centrifugal compressors, and turbines generated from BladeCAD are shown in Figures 31.3–31.7. The geometry is saved in an IGES standard file for different applications including grid generation or stress analysis.

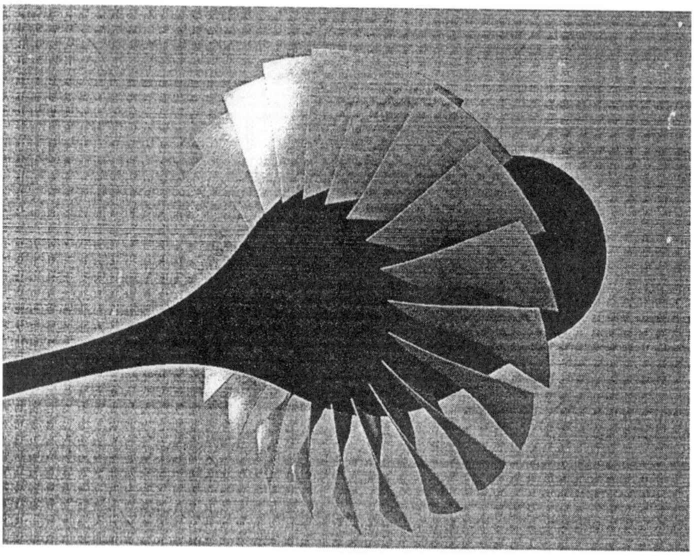

FIGURE 31.3 AST quiet fan blade.

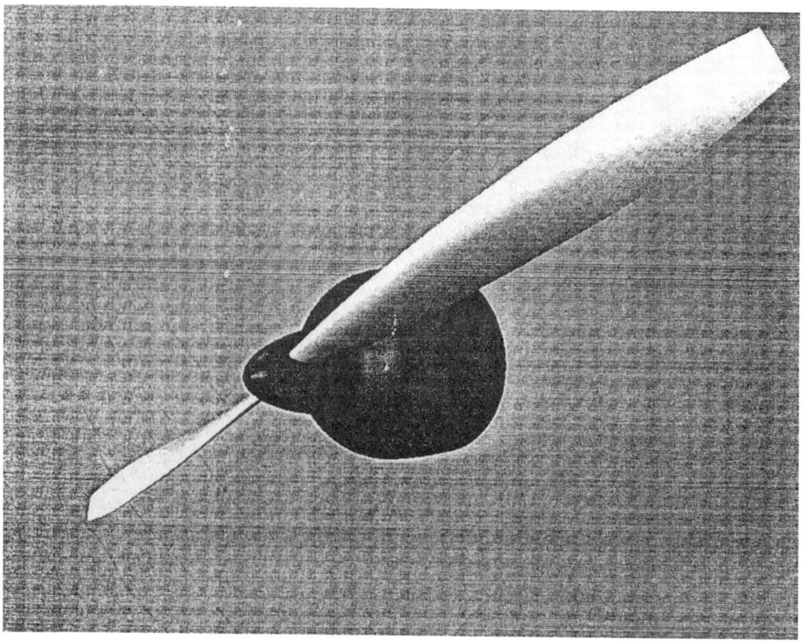

FIGURE 31.4 General aviation propeller.

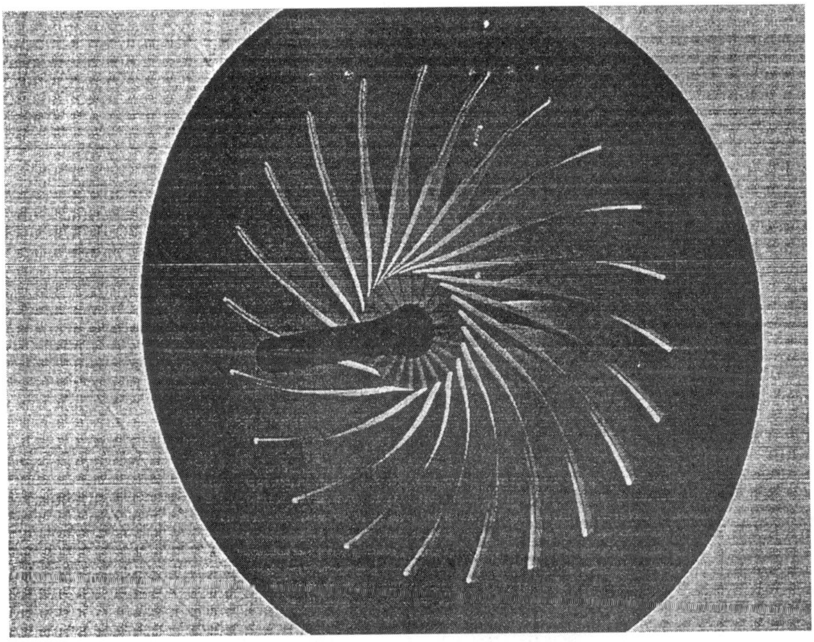

FIGURE 31.5 Centrifugal compressor.

FIGURE 31.6 Axial flow turbine rotor.

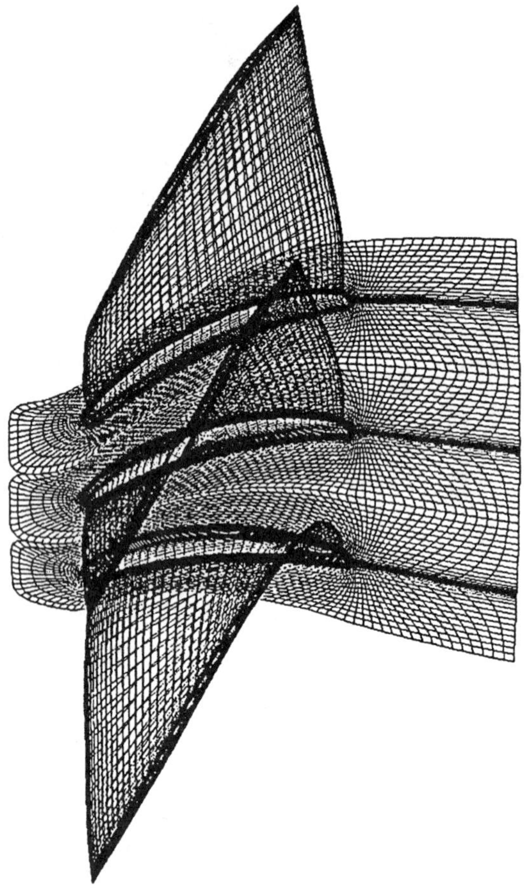

FIGURE 31.7 3D C-grid generated for a transonic fan blade.

31.3.3.2 Computational Fluid Dynamics

Once the blade geometry has been specified, the aerodynamic flow field must be computed to determine the aerodynamic performance associated with the blade design. In order to perform a computational fluid dynamics simulation on the geometry, the fluid domain must first be gridded in order to perform the computation. Since the geometry was written in IGES format, grid generation packages must be able to reconstruct the geometry as specified in the IGES file. Many grid packages can be used to read an IGES formatted geometry. A sampling of some of these code are GRIDGEN [Steinbrenner, 1990], GridPro [Program Development Corporation, 1995], APTGRID [Beach, 1995], TIGER [Shih, 1994], NTIGG [Mokhtar, 1994] and CFD-GEOM [CFDRC 1995]. Since other grid packages have or are adding this capability, a thorough search should be made before choosing a grid package. These types of codes usually take the geometry and subdivide the domain to obtain a grid for the computational codes. Once the grid has been generated, there are a number of flow solvers that can be used to obtain the aerodynamic performance associated with the geometry just constructed. Figure 31.5 shows the grid domain created from a high-speed fan design. In Figure 31.6, the flow solution obtained using RVC3D [Chima 1991], a 3-dimensional Navier–Stokes flow solver, is shown.

After the 3D flow solution has been obtained, the flow field properties are then mapped back to the surface using the DT_NURBS utility library, which uses the original NURBS description of the surface geometry, the discrete grid specification and the flow quantities obtained from the flow solver. This produces an IGES file that has pressures and temperatures mapped to the surface geometry for computing structural analysis of the blade with the blade aerodynamic loads and surface temperatures.

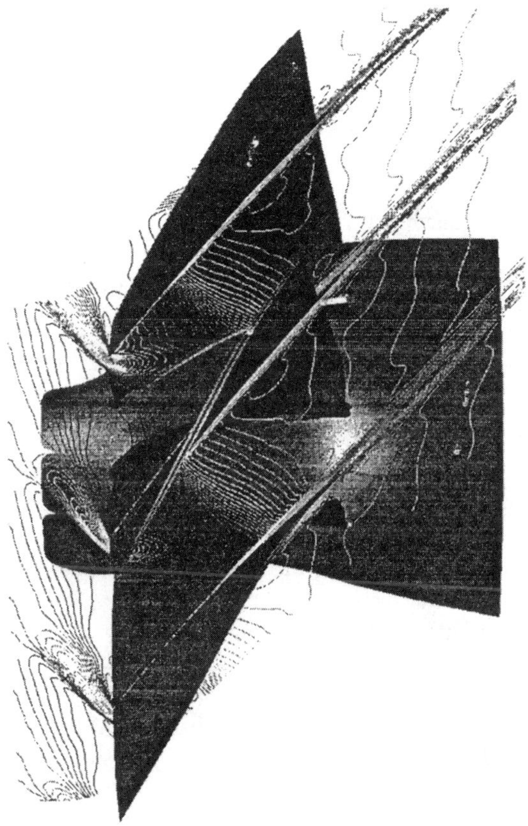

FIGURE 31.8 3D flow solution through a transonic fan blade.

31.3.3.3 Multidisciplinary Geometry, Grid, and Analysis Association

The loosely coupled multidisciplinary methodology relies on the construction of a concept called a "subrange surface." The subrange surface concept is an entity developed in the DT_NURBS library for loose multidisciplinary coupling. It allows scalar or vector component values such as surface pressure or displacement components in an engineer analysis context to be associated with an existing underlying geometric definition in a general way. Subsequent evaluations of the underlying geometric entity will yield interpolated values of the scalar or vector components such as pressure and displacement as an example. The actual interpolation of the boundary condition information is done with NURBS. The B-spline definition for the boundary information may have different order and knot spacing from the underlying geometric definition.

As an example, consider a geometric NURBS surface definition for the blade airfoil. The Cartesian coordinates (x,y,z) on the airfoil surface are defined with a B-spline function f. The function f is defined over the parametric domain u and v.

$$\text{Geometric B-spline-Surface } \{x,y,z\} = f(u,v) \tag{31.1}$$

Furthermore, consider that the analysis results from some aerodynamic CFD grid that produces the surface pressure P and temperature T at discrete points on surface $f(u,v)$.

$$\text{Analysis Grid } x_{ij}, y_{ij}, z_{ij}, P_{ij}, T_{ij} \text{ where } i = 1, nx \text{ and } j = 1, ny$$

The corresponding surface parameters u_{ij} and v_{ij} are either known from the original aerodynamic CFD grid discretization or can be calculated. A second B-spline function g with parameters s and t can be constructed from the value of u, v, P, T so that:

$$\text{Boundary Condition B-spline Function } \{u,v,P,T\} = g(s,t) \tag{31.2}$$

Subsequently the function g in Eq. 31.2 can be evaluated using the parameters s and t. The evaluation of the boundary condition B-spline function g produces values of u, v, P, T. The parametric values u and v obtained from the evaluation of the function g can then be used to evaluate the function f in Eq. 31.1 to produce geometric values of x, y, z. The evaluation of the f and g functions can be composed together to produce the following:

$$x, y, z, P, T = f\{g\{s, t\}\} \tag{31.3}$$

Therefore, the evaluation of the composition of functions using parametric values s and t in Eq. 31.3 would produce x, y, z, P, T. In this example, the geometric and boundary condition data needed for a structural FEA (finite element analysis) grid is generated. The DT_NURBS library has been developed by NASA, DOD Navy, and Boeing to provide this encapsulated functionality for the subrange methods and association of each analysis discipline's geometry, grid, and analysis (GGA) data.

To demonstrate this fundamental discipline couple methodology and technique, NASA Lewis has developed several prototype multidisciplinary coupling tools. The prototype software was used to demonstrate methodology for steady state aeroelastic analysis problems for turbomachinery blading. NASA has developed mapping and interpolation prototypes for pre and post processors aerodynamic CFD analysis APTGRID [Beach 1995] and FEA structural analysis SABER [Thorp, 1995] based on the DT_NURBS library's GGA concept. For this prototype development and the testing of these methods the mapping and interpolation software was incorporated directly into both the CFD and FEA grid generators. This proved to be the most convenient approach, but is not a necessity.

The process was applied to the prediction of the hot running blade shape of the NASA Transonic Rotor 37 stage. The Rotor 37 rotor stage was used because experimental and analytical CFD and FEA data was plentiful for this turbomachinery test case. Experimental data also included measure tip location and displacement at operating speed from NASA Lewis rig testing. Both the pre- and postprocessing tools APTGRID and SABER along with the VSTAGE CFD and NASTRAN FEA analysis codes were used to solve the steady-state aeroelastic problem. In this specific application, two aero/structural iterations were sufficient to achieve a converged solution based on both pressure and displacement criteria to within acceptable accuracy.

The loosely coupled geometry, grid and analysis method has proven to be an accurate and practical approach for loosely coupling aerodynamic CFD to thermal/structural FEA. Further, work is ongoing to enhance and expand this method to larger dimensional problems in terms of geometric complexity and data exchange proportions. Plans are to incorporate this loosely coupled methodology into the ISO 10303 Standard for the Exchange of Product Data (STEP), Part 42, for engineering analysis data exchange.

31.3.4 Other NASA-IGES Compatible Software

Although much NASA-IGES compatible software has been referenced in the above section, some was not covered. This section will deal with three additional codes. This is not meant to be a comprehensive list. The reader is encouraged to search other sources for additional codes. Searching the Web is a good starting point.

It should be noted that while the codes included here are generally free of charge, many have restrictions on their release. The reader will need to check with the point of contact to determine the pertinent restrictions. Since these are free codes, the reader should be cautioned that their quality will vary widely.

The following list of NASA-IGES compatible software contains the name of the code followed by a brief description and a point of contact (POC) for additional information.

- *NASA-IGES Translator (NigesT)* — POC: Jin Chou (415) 424-1202. NigesT is a noninteractive program that reads NASA-IGES CAD data files, which is a subset of the IGES standard, and converts those entities it understands into NURBS. The resulting file is a NASA-IGES NURBS-Only file (NINO). (This is highly recommended by the author. Since most CAD systems output extraneous IGES information that is of no use to a grid generator, NigesT can be used to filter out this information. If you are having trouble generating a grid from a IGES file generated by a NASA-IGES compatible CAD system and the grid generator is NASA-IGES compatible, run the file through the NigesT software.)
- *Portable Extensible Viewer (PEV)* — POC: npss-pev-request@lerc.nasa.gov. PEV is a program designed to read, write, evaluate, display, graphically manipulate, and analyze NURBS data. The NURBS data may be stored in several predefined file formats (including NASA-IGES) or in a file format that can be read in by a user-defined function. The data may be multidisciplinary, including not only geometry information, but pressure and temperature defined over multiple time steps, and various conditions.
- *Standard Data Access Interface (SDAI) for STEP Repositories* — POCs: Jeff Meister (216) 433-6731. Austin L. Evans (216) 433-8313. The C++ SDAI implements classes and methods specified in the standard data access interface for the C++ programming language, ISO/CD 10303-23. The SDAI is a function level interface that provides a standard model and syntax for creating and accessing STEP-based entities contained within a database. Thus, the SDAI enables developers to build applications free of storage method specific function calls.

31.4 Research Issues and Summary

There are many open issues in the area of geometry and data exchange standards due to the fact that not all the standards have been defined nor fully implemented. This is especially true as one moves away from simple surface representation and toward solid models and subrange surface data representation. The state of the art in the solid modeling area is currently at the same level surface modeling was at in 1991 when our team began working on NASA-IGES. There is great difficulty in sharing solid models between various CAD systems. What is a solid model in one system is not read as a solid model when transferred to another system. The use of subrange surfaces to map and exchange data between discipline codes is not yet an accepted practice.

It is currently planned that these issues will be addressed by the STEP standard. The PDES Inc., a business/government consortium, is currently coordinating the U.S. input to the STEP standards. A NASA PDES Working Group has been formed to work with PDES. This group is forging ahead with the work started by the team that wrote the NASA-IGES standard. One of the projects this group is working on is to push for the incorporation of the GGA, subrange, data exchange method into ISO 10303 Part 42. Once this has been accomplished and a consistent solid modeling standard has been implemented, the goal of being able to widely use a common, or master, model across heterogeneous hardware and software systems will be achieved.

Further Information

A good introduction to curves, surfaces and NURBS representation of geometry and data can be found in the following reading list. The first two books are highly recommended.

Farin, G., *Curves and Surfaces for Computer Aided Geometric Design, A Practical Guide*, Third Edition. Academic Press, 1993.

Piegl, L. and Tiller, W., *The NURBS Book, Monographs in Visual Communications.* Springer, 1995.

Bartels, R.H., Beatty, J.C., Barsky, B.A., *An Introduction to Splines for Use in Computer Graphics and Geometric Modeling*, Morgan–Kaufmann, Palo Alto, CA, 1987.

Boehm, W., Farin, G., Kahmann, J., A Survey of Curve and Surface Methods in CAGD, *Computer Aided Geometry Design.* July 1984, Vol. 1, No. 1, pp. 1–60.

de Boor, C., *A Practical Guide to Splines.* Springer Verlag, New York, 1978.

Lee, E.T.Y., Rational quadratic Bézier representation for conics, *Geometric Modeling: Algorithms and New Trends.* Farin, G., (Ed.), SIAM Philadelphia, 1987, pp. 3–19.

Tiller, W. Rational B-splines for curve and surface representation, *CG&A.* Sept. 1983, Vol. 3, No. 10, pp. 61–69.

Piegland, L. and Tiller, W., curve and surface constructions using rational B-splines, *Computer-Aided Design,* Nov. 1987, Vol. 19, No. 9, pp. 485–498.

Piegland, L. and Tiller, W., A menagerie of rational B-spline circles, *IEEE Computer Graphics and Applications.* Sept. 1989, Vol. 9, No. 5, pp. 48–56.

References

American National Standard Institute. Dimensioning and tolerancing, (Y14.5M-1982), 1982.

Beach, T. APTGRID, *User's Guide and Reference Manual,* 1995.

Chima, R.V., Viscous three-dimensional calculations of transonic fan performance, NASA TM103800. Presented at the 77th Symposium of the Propulsion and Energetics Panel CFD Techniques for Propulsion Applications, San Antonio, TX, May 1991.

CFDRC. CFD-GEOM, CFD Research Corporation, 1995.

Evans, A.L., et al. NPSS Software Catalog, Version 1.0, NASA Lewis Research, Cleveland, Ohio, 1997.

Farin, G. *Curves and Surfaces for Computer Aided Geometric Design.* Academic Press, 1988.

Initial Graphics Exchange Specification (IGES), Version 5.3, distributed by National Computer Graphics Association, Administrator, IGES/PDES Organization, 2722 Merrilee Drive, Suite 200, Fairfax, VA, 1995.

Miller, P.L., Oliver, J.H., Miller, D.P., and Tweedt, D.L. BladeCAD: An interactive geometric design tool for turbomachinery blades, *NASA Technical Memorandum 107262,* presented at the 41st Gas Turbine and Aeroengine Congress, Birmingham, UK, June 1996.

Mokhtar, J. and Oliver, J.H. Parametric volume models for interactive three-dimensional grid generation, advances in design automation, 1994, Vol. 1, pp. 435–442.

NIST (National Institute of Standards and Technology), *Initial Graphics Exchange Specification,* Version 5.3. 1990.

Piegl, L. On NURBS: A survey, *IEEE Computer Graphics and Applications,* 1991, Vol. 11, No. 1, pp. 55–71.

Piegl, L. and Tiller, W. *The NURBS Book,* Springer Verlag, Berlin, 1995.

Program Development Corporation, GridPro™/az3000, *User's Guide and Reference Manual,* 1995.

RP1338, NASA Geometry Data Exchange Specification for Computational Fluid Dynamics (NASA-IGES), National Aeronautics and Space Administration, Washington, D.C., 1994.

Shih, A.M. Toward a comprehensive computational simulation system for turbomachinery, Ph.D. thesis, Mississipi State University, 1994.

Steinbrenner, J., et. al. The Gridgen 3D multiple block grid generation system, Final report WRDC-TR-90-3022, 1990.

Thomas, G. *Calculus and Analytic Geometry.* Addison-Wesley, 1960.

U.S. Navy, DT_NURBS Spline Geometry Subprogram Library Users' Manual, Version 3.5, Naval Surface Warfare Center, David Taylor Model Basin, Bethesda, MD, 1997.

IV

Adaptation and Quality

Bharat K. Soni

Introduction to Adaptation and Quality

The accuracy of numerical simulation of a physical field problem depends not only on the formal order of approximation but also on the distribution of grid points in the computational domain. The quality of the grid based on the geometric characteristics influenced by the numerical scheme under consideration and solution characteristics influenced by the field properties being simulated is extremely important in view of improving the accuracy and convergence rate of the simulation process. The usual practice is to evaluate and improve grid quality based on the geometric characteristics and known general physical solution characteristics (for example, number of points needed in the boundary layer in case of viscous fluid simulation), and then perform grid adaptation by coupling the grid generation with the field simulation procedure.

The chapters included in Part IV provide detailed descriptions of grid quality evaluations and grid adaptation procedures. First, the grid quality requirements based on the truncation error analysis associated with the finite difference and finite volume discretization are discussed by Mastin in Chapter 32. The importance of grid stretching with well-behaved aspect ratio, and grid smoothness and near-orthogonality requirements are mathematically developed in this chapter. A systematic mathematical treatment of grid optimization and grid quality improvement is presented by Jacquotte in Chapter 33. The grid quality and optimization is carried out by developing meaningful measures of cell deformation utilizing functional analysis in this chapter. This analysis is further extended to develop error indicators and their utilization in grid adaptation.

There are three basic strategies that may be employed in dynamically adaptive grids coupled with the PDEs of the physical problem. The first approach is to redistribute a fixed number of points. In this approach, points move from regions of relatively small error to regions of large error. While the global order of approximation cannot be increased by such movement of points, it is possible to improve approximation locally. As long as the redistribution of points does not seriously deplete the number of points in other regions, this is a viable approach. The second approach involves local refinement. In this approach, points are added (or removed) locally in a fixed point structure in regions of relatively large error. Here there is, or course, no depletion of points in other regions and therefore no formal increase of error order occurs. However, the computer time and storage increase with refinement and data structures can be difficult. This approach is well suited to unstructured grids. In the last approach, the solution method is changed locally to higher order approximation in regions of relatively large error. This again increases formal global accuracy but involves great complexity of implementation in field simulation software. This approach has not had any significant application in field solvers involving multiple dimensions.

In Chapter 34, the grid quality measures discussed in Chapters 32 and 33 are utilized in the development of the dynamic grid adaptation technique by McRae and Laflin. The grid adaptation procedure is based on the grid redistribution strategy (r–refinement) by improving grid quality on the local solution and is developed for the structured grids. The technique ensures the preservation of the field characteristics.

The grid control and grid adaptation algorithms applicable to unstructured grids using grid refinement and grid movement are discussed by Hassan and Probert in Chapter 35. A detailed description starting from the generation of unstructured grids using the Delaunay triangulation method to the development of error indicators, grid movement, and grid refinement is given in detail with practical demonstrations.

The mathematical analysis of the grid generation naturally leads to variational methods. Khairullina, Sidorov, and Ushakova in Chapter 36 exploit the variational method for optimal grid generation. The basic mathematical foundation of the variational approach is presented and extended to generate adaptive grids using the combination of variational integrals representative of geometric and physical field characteristics.

The mathematical foundation and numerical treatment associated with the dynamically moving grids are described by Zegeling in Chapter 37. The moving grid techniques are critical, especially in the treatment of time-accurate PDEs allowing temporally moving/changing/deforming and adapting geometry/grids. Here a discrete approach of variational methods for mesh optimization and adaptation is employed and discussed for grid adaptation.

32

Truncation Error on Structured Grids

C.Wayne Mastin

32.1 Introduction

A structured grid determines a natural curvilinear coordinate system in the region spanned by the grid. With a curvilinear coordinate system defined, a partial differential equation can be transformed from Cartesian coordinates to curvilinear coordinates using the classical change of variables techniques of applied mathematics. A difference approximation of the differential equation can be obtained from the equation in curvilinear coordinates by forming difference approximations of the derivatives with respect to the curvilinear coordinates (see Chapter 2). An error analysis reveals that the accuracy of the approximation is related to the quality of the grid.

One-dimensional distribution (or stretching) functions are used for distributing grid points along boundary curves of planar regions and surfaces and along edges of three-dimensional regions. Hoffman [1] and Vinokur [5] have analyzed the effect of the grid on truncation error for one-dimensional problems. These results were further developed and extended to two-dimensional problems by Thompson and Mastin [4] and Mastin [3]. Extensions to higher dimensions is straightforward, but lengthy. The problem of accurately and efficiently estimating the truncation error in any dimension remains open. Some progress in that area was made by Lee and Tsuei [2].

The "order" of a difference representation refers to the exponential rate of decrease of the truncation error with the point spacing. On a uniform grid this concerns simply the behavior of the error as the point spacing decreases. With a nonuniform point distribution, there is some ambiguity in the interpretation of order, in that the spacing may be decreased locally either by increasing the number of points in the field or by changing the distribution of a fixed number of points. Both of these could, of course, be done simultaneously, or the points could even be moved randomly, but to be meaningful the order of a difference representation must relate to the error behavior as the point spacing is decreased according to some pattern. This is a moot point with uniform spacing, but two senses of order on a nonuniform grid emerge: the behavior of the error (1) as the number of points in the field is increased while maintaining the same relative point distribution over the field, and (2) as the relative point distribution is changed so as to reduce the spacing locally with a fixed number of points in the field.

0-8493-2687-7/99/$0.00+$.50

On curvilinear coordinate systems the definition of order of a difference representation is integrally tied to point distribution functions. The order is determined by the error behavior as the spacing varies with the points fixed in a certain distribution, either by increasing the number of points or by changing a parameter in the distribution, not simply by consideration of the points used in the difference expression as being unrelated to each other. Actually, global order is meaningful only in the first sense, since as the spacing is reduced locally with a fixed number of points in the field, the spacing somewhere else must certainly increase. This second sense of order on a nonuniform grid then is relevant only locally in regions where the spacing does in fact decrease as the point distribution is changed.

In the following section an illustrative error analysis is given. The general development from which this is taken appears in Thompson and Mastin [4], together with references to related work.

32.2 Order on Nonuniform Spacing

A general one-dimensional point distribution function can be written in the form

$$x(\xi) = q\left(\frac{\xi}{N}\right) \quad 0\xi \le N \tag{32.1}$$

In the following analysis, x will be considered to vary from 0 to 1. (Any other range of x can be constructed simply by multiplying the distribution functions given here by an appropriate constant.) With this form for the distribution function, the effect of increasing the number of points in a discretization of the field can be seen explicitly by defining the values of ξ at the points to be successive integers from 0 to N. In this form, N+1 is then the number of points in the discretization, so that the dependence of the error expressions on the number of points in the field will be displayed explicitly by N. This form removes the confusion that can arise in interpretation of analyses based on a fixed interval ($0 \le \xi \le 1$), where variation of the number of points is represented by variation of the interval $\Delta\xi$. The form of the distribution function, i.e., the relative concentration of points in certain areas while the total number of points in the field is fixed, is varied by changing parameters in the function.

Considering the first derivative in one dimension,

$$f_x = \frac{f_\xi}{x_\xi} \tag{32.2}$$

with a central difference for f_ξ we have the following difference expression (with $\Delta\xi = 1$ as noted above):

$$f_\xi = \frac{1}{2x_\xi}(f_{i+1} - f_{i-1}) + T_1 \tag{32.3}$$

where T_1 is the truncation error. A Taylor series expansion then yields

$$T_1 = -\frac{1}{6}\frac{f_{\xi\xi\xi}}{x_\xi} - \frac{1}{120}\frac{f_{\xi\xi\xi\xi\xi}}{x_\xi}\dots \tag{32.4}$$

Here the metric coefficient, x_ξ, is considered to be evaluated analytically, and hence has no error. (The case of numerical evaluation of the metric coefficients is considered in section 32.3.)

The series in Eq. 32.4 cannot be truncated without further consideration since the ξ-derivatives of f are dependent on the point distribution. Thus if the point distribution is changed, either through the addition of more points or through a change in the form of the distribution function, these derivatives will change. Since the terms of the series do not contain a power of some quantity less than unity, there is no indication that the successive terms become progressively smaller.

It is thus not meaningful to give the truncation error in terms of ξ-derivative of f. Rather, it is necessary to transform these ξ-derivatives to x-derivatives which, of course, are not dependent on the point distribution. The first ξ-derivative follows from Eq. 32.2:

$$f_\xi = x_\xi f_\xi \tag{32.5}$$

Then

$$f_{\xi\xi} = x_{\xi\xi} f_x + x_\xi (f_x)_\xi = x_{\xi\xi} f_x + x_\xi^2 f_{xx} \tag{32.6}$$

and

$$f_{\xi\xi\xi} = x_{\xi\xi\xi} f_x + 3 x_\xi x_{\xi\xi} f_{xx} + x_\xi^3 f_{xxx} \tag{32.7}$$

Each term in $f_{\xi\xi\xi}$ contains three ξ-differentiations. This holds true for all higher derivatives also, so that each term in $f_{\xi\xi\xi\xi\xi}$ will contain five ξ-differentiations, etc.

32.2.1 Order with Fixed Distribution Function

From Eq. 32.1 we have

$$x_\xi = \frac{q'}{N'} \quad x_{\xi\xi} = \frac{q''}{N^2}, \quad x_{\xi\xi\xi} = \frac{q'''}{N^3} \tag{32.8}$$

Therefore, if the number of points in the grid is increased while keeping the same relative point distribution, it is clear that each term in $f_{\xi\xi\xi}$ will be proportional to $1/N^3$, and each term in $f_{\xi\xi\xi\xi\xi}$ will be proportional to $1/N^5$, etc.

It then follows that the series in Eq. 32.4 can be truncated in this case, so that the truncation error is given by the first term, which is, using Eq. 32.7,

$$T_i = -\frac{1}{6}\frac{x_{\xi\xi\xi}}{x_\xi} f_\xi - \frac{1}{2} x_{\xi\xi} f_{xx} - \frac{1}{6} x_\xi^2 f_{xxx} \tag{32.9}$$

The first two terms arise from the nonuniform spacing, while the last term is the familiar term that occurs with uniform spacing as well.

From Eq. 32.9 it is clear that the difference representation Eq. 32.3 is second order regardless of the form of the point distribution function, in the sense that the truncation error goes to zero as $1/N^2$ as the number of points increases. This means that the error will be quartered when the number of points is doubled in the same distribution function. *Thus all difference representations maintain their order on a nonuniform grid with any distribution of points in the formal sense of the truncation error decreasing as the number of points is increased while maintaining the same relative points distribution over the field.*

The critical point here is that the same relative point distribution, i.e., the same distribution function, is used as the number of points in the field is increased. If this is the case, then the error will be decreased by a factor that is a power of the inverse of the number of points in the field as this number is increased. Random addition of points will, however, not maintain order. In a practical vein this means that with twice as many points, the solution will exhibit one fourth of the error (for second-order representations in the transformed plane) when the same point distribution function is used. However, if the number of points is doubled without maintaining the same relative distribution, the error reduction may not be as great as one fourth.

From the standpoint of formal order in this sense, there is no need for concern over the form of the point distribution. However, formal order in this sense relates only to the behavior of the truncation error as the number of points is increased, and the coefficients in the series may become large as the parameters in the distribution are altered to reduce the local spacing with a given number of points in the field. Thus, although the error will be reduced by the same order for all point distributions as the number of points is increased, certain distributions will have smaller error than others with a given number of points in the field, since the coefficients in the series, while independent of the number of points, are dependent on the distribution function.

32.2.2 Order with Fixed Number of Points

An alternate sense of order for point distributions is based on expansion of the truncation error in a series in ascending powers of the spacing, x_ξ, with the number of points in the grid kept fixed and the point distribution changed to decrease the local spacing. From Eq. 32.9, second order requires that

$$x_{\xi\xi\xi} \sim x_\xi^3 \quad \text{and} \quad x_{\xi\xi} \sim x_\xi^2 \tag{32.10}$$

This is a severe restriction that is unlikely to be satisfied. This is understandable, however, since with a fixed number of points the spacing must necessarily increase somewhere when the local spacing is decreased.

The difference between these two approaches to order should be kept clear. The first approach concerns the behavior of the truncation error as the number of points in the field increases with a fixed relative distribution of points. The series there is power series in the inverse of the number of points in the field, and formal order is maintained for all point distributions. The coefficients in the series may, however, become large for some distribution functions as the local spacing decreases for any given number of points. The other approach concerns the behavior of the error as the local spacing decreases with a fixed number of points in the field. This second sense of order is thus more stringent, but the conditions seem to be unattainable.

32.3 Effect of Numerical Metric Coefficients

The above analysis has assumed the use of exact values of x_ξ, the metric coefficient. If the metric coefficient is evaluated numerically, we have, in place of Eq. 32.3, the difference expression

$$f_\xi = \frac{f_{i+1} - f_{i-1}}{x_{i+1} - x_{i-1}} + T_2 \tag{32.11}$$

The Taylor expansion yields

$$T_2 = f_x - \{\Big(f_x(x_{i+1} - x_{i-1}) + \frac{1}{2} f_{xx}[(x_{i+1} - x_i)^2 - (x_{i-1} - x_i)^2]\Big)$$
$$+ \frac{1}{6} f_{xxx}[(x_{i+1} - x_i)^3 - (x_{i-1} - x_i)^3]\}/(x_{i+1} - x_{i-1})$$

or

$$T_2 = -\frac{1}{2} f_{xx}(x_{i+1} - 2x_i + x_{i-1})$$
$$-\frac{1}{6} f_{xxx} \frac{(x_{i+1} - x_i)^3 - (x_{i-1} - x_i)^3}{(x_{i+1} - x_{i-1})} \tag{32.12}$$

The coefficient of f_{xx} here is the difference representation of $x_{\xi\xi}$, while that of f_{xxx} reduces to a difference expression of $x_\xi{}^2$. We thus have T_2 given by the first two terms of the T_1, and the first term of T_1 has been eliminated from the truncation error by evaluating the metric coefficient numerically rather than analytically.

Thus the use of numerical evaluation of the coordinate derivative, rather than exact analytical evaluation, eliminates the f_x term from the truncation error. Since this term is the most troublesome part of the error, being dependent on the derivative being represented, it is clear that *numerical evaluation of the metric coefficients by the same difference representation used for the function whose derivative is being represented is preferable over exact analytical evaluation.* It should be understood that there is no incentive, *per se,* for accuracy in the metric coefficients, since the object is simply to represent a discrete solution accurately, not to represent the solution on some particular coordinate system. The only reason for using any function at all to define the point distribution is to ensure a smooth distribution. There is no reason that the representations of the coordinate derivatives have to be accurate representations of the analytical derivatives of that particular distribution function.

We are thus left with truncation error of the form

$$T = -\frac{1}{2} x_{\xi\xi} f_{xx} - \frac{1}{6} x_\xi^2 f_{xxx} \tag{32.13}$$

when the metric coefficient is evaluated numerically. As noted above, the last term occurs even with uniform spacing. The first term is proportional to the second derivative of the solution and hence represents a numerical diffusion, which is dependent on the rate-of-change of the grid point spacing. This numerical diffusion may even be negative and hence destabilizing. Attention must therefore be paid to the variation of the spacing, and large changes in spacing from point to point cannot be tolerated, else significant truncation error will be introduced.

32.4 Evaluation of Distribution Functions

The above error analysis can be of value in judging the suitability of distribution functions for one-dimensional grid generation. Table 32.1 contains a listing of popular distribution functions along with the ratios

$$L_2 = \frac{x_{\xi\xi}(0)}{x_\xi^2(0)}, \quad L_3 = \frac{x_{\xi\xi\xi}(0)}{x_\xi^3(0)} \tag{32.14}$$

All distribution functions are defined in terms of the normalized computational variable

$$\bar{\xi} = \frac{\xi}{N}$$

Each of these distribution functions can be used to construct a grid on the unit interval $0 \le x \le 1$ with the grid points clustered at the endpoint $x = 0$. The spacing at $x = 0$ decreases with increasing values of the parameter α. Other distribution functions that force clustering at both endpoints and at interior points have been considered by Vinokur [5].

From the values of L_2 and L_3 in Table 32.1, it can be seen that for each distribution function at least one of these values becomes infinite as the grid spacing at $x = 0$ approaches zero. A careful analysis, as in Thompson and Mastin [4], will reveal that some of the distribution functions are better at preserving formal order than others. Figure 32.1 contains plots of the distribution functions $x = 0$ approaches zero. A careful analysis, as in Thompson and Mastin [4], will reveal that some of the distribution functions

TABLE 32.1 Distribution Functions and Error Coefficients at $x = 0$

Function	$x(\bar{\xi})$	L_2	L_3
Exponential	$\frac{\ell^{\alpha\bar{\xi}} - 1}{\ell^{\alpha} - 1}$	$\ell^{\alpha} - 1$	$(\ell^{\alpha} - 1)^2$
Hyperbolic tangent	$1 - \frac{\tanh\ \alpha(1-\bar{\xi})}{\tanh\ \alpha}$	$2\sinh^2\alpha$	$\frac{1}{2}(3\tanh^2\alpha - 1)\sinh^2 2\alpha$
Hyperbolic sine	$\frac{\sinh\ \alpha\bar{\xi}}{\sinh\ \alpha}$	0	$\sinh^2\alpha$
Error Function	$1 - \frac{erf\,\alpha(1-\bar{\xi})}{erf\,\alpha}$	$\sqrt{\pi\alpha e^{\alpha^2} erf\,\alpha}$	$\frac{\pi}{2}(2\alpha^2 - 1)(e^{\alpha^2} erf\,\alpha)^2$
Tangent $(0 \le \alpha \le \frac{\pi}{2})$	$\frac{\tan\ \alpha\bar{\xi}}{\tan\ \alpha}$	0	$2\tan^2\alpha$
Arctangent	$1 - \frac{\tan^{-1}\alpha(1-\bar{\xi})}{\tan^{-1}\alpha}$	$2\alpha\tan^{-1}\alpha$	$2(3\alpha^2 - 1)(\tan^{-1}\alpha)^2$
Sine $(0 \le \alpha \le \frac{\pi}{2})$	$1 - \frac{\sin\alpha(1-\bar{\xi})}{\sin\alpha}$	$\tan^2\alpha$	$-\tan^2\alpha$
Logarithm	$1 - \frac{\ln[1+\alpha(1-\bar{\xi})]}{\ln(1+\alpha)}$	$\ln(1+\alpha)$	$2[\ln(1+\alpha)]^2$
Inverse hyperbolic tangent $(0 \le \alpha \le \frac{\pi}{2})$	$\frac{\tanh^{-1}\alpha\bar{\xi}}{\tanh^{-1}\alpha}$	0	$2(\tanh^{-1}\alpha)^2$
Quadratic $(0 \le \alpha \le 1)$	$(1-\alpha)\bar{\xi} + \alpha\bar{\xi}^2$	$\frac{2\alpha}{(1-\alpha)^2}$	0

are better at preserving formal order than others. Figure 32.1 contains plots of the distribution functions $x = x(\bar{\xi})$ with a value of

$$\frac{dx}{d\xi} = 0.1$$

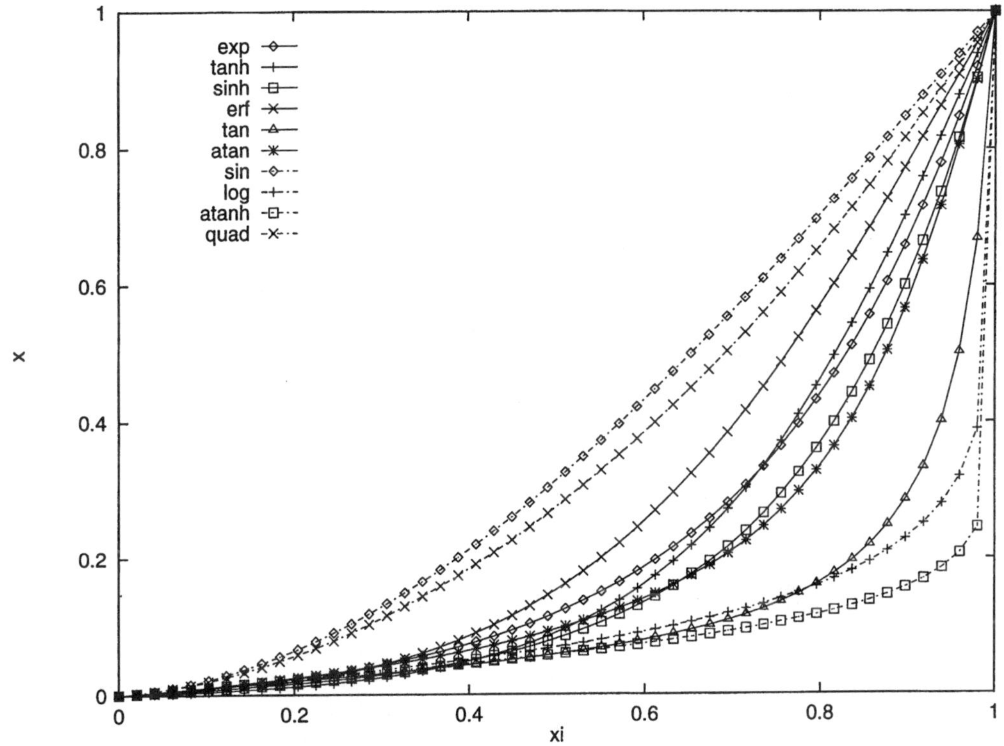

FIGURE 31-01 Distribution functions in the unit interval [0,1].

This would then give a spacing at $x = 0$ of $0.1/N$. The symbols are uniformly spaced in the $\bar{\xi}$ direction. Thus, the distribution of grid points imposed by each function is determined by the x coordinate of each symbol.

The curves plotted in Figure 32.1 reveal properties of some of the distribution functions which would make them unsuitable for use in grid generation. The tangent, logarithm, and inverse hyperbolic tangent functions concentrate nearly all points near $x = 0$ and few points near $x = 1$. The sine and quadratic functions give a more uniform distribution of points on the interval [0,1] at the expense of large variations in grid spacings at $x = 0$. While this may not be important for some problems, it would be a poor choice for solving boundary layer problems. The changes in grid spacings are more apparent in the magnified view of the distribution functions in Figure 32.2.

The change in slope of the sine and quadratic curves are much greater than the other curves which have a more linear behavior near $x = 0$. This behavior is further verified by the expression for L_2 in Table 32.1. Note the asymptotic behavior of L_2 for the sine and quadratic functions as α approaches $\pi/2$ and 1, respectively. This indicates very large changes in grid spacings correspond to small grid spacings at $x = 0$. For this particular grid spacing, the following distribution functions do a good job of distributing the points on the unit interval without excessive variations in grid spacings anywhere on the interval: exponential, hyperbolic tangent, hyperbolic sine, error function, and arctangent. For smaller grid spacings, it was noted by Thompson and Mastin [4] that the arctangent concentrated too many points near $x = 0$. Therefore, based on the observations presented here and the more detailed analysis of error coefficients in Thompson and Mastin [4], the following conclusions can be reached concerning the suitability of the various distribution functions in generating computational grids for solving boundary value problems.

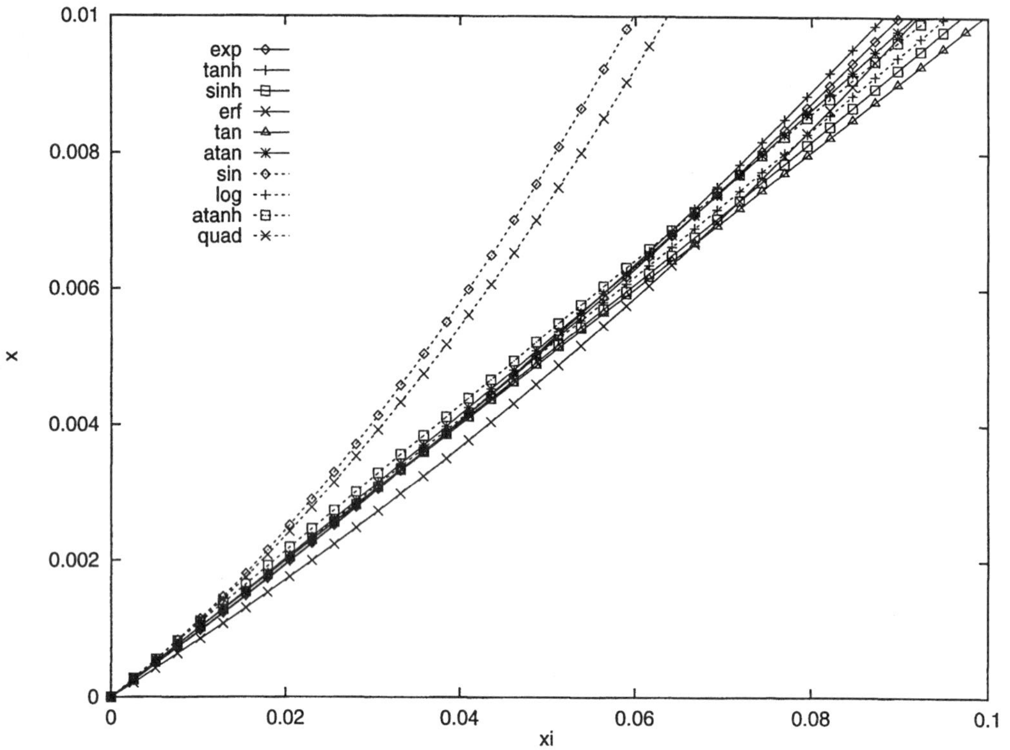

FIGURE 31-02 Distributions functions near x=0.

1. The exponential is not as good as the hyperbolic tangent or the hyperbolic sine. (See Chapter 3 for implementation procedures.)
2. The hyperbolic sine is the best function in the lower part of the boundary layer. Otherwise this function is not as good as the hyperbolic tangent.
3. The error function and the hyperbolic tangent are the best functions outside the boundary layer. Between these two, the hyperbolic tangent is the better inside, while the error function is the better outside. The error function is, however, more difficult to use.
4. The logarithm, sine, tangent, arctangent, inverse hyperbolic tangent, quadratic, and the inverse hyperbolic sine are not suitable.

Although, as has been shown, all distribution functions maintain order in the formal sense with nonuniform spacing as the number of points in the field is increased, these comparisons of particular distribution functions show that considerable error can arise with nonuniform spacing in actual applications. If the spacing doubles from one point to the next we have, approximately, $x_{\xi\xi} = 2x_\xi - x_\xi = x_\xi$ so that the ratio of the first term in Eq. 32.13 to the second is inversely proportional to the spacing x_ξ. Thus for small spacing, such a rate-of-change of spacing would clearly be much too large. Obviously, all of the error terms are of less concern where the solution does not vary greatly. *The important point is that the spacing not be allowed to change too rapidly in high gradient regions such as boundary layers or shocks.*

32.5 Two-Dimensional Forms

The two-dimensional transformation (see Chapter 2) of the first derivative is given by

$$f_x = \left(y_\eta f_\xi - y_\xi f_\eta\right) / \sqrt{g} \tag{32.15}$$

where the Jacobian of the transformation is

$$\sqrt{g} = x_\xi y_\eta - x_\eta y_\xi \tag{32.16}$$

With two-point central difference representations for all derivatives, the leading term of the truncation error is

$$\begin{aligned} T_x &= \frac{1}{2\sqrt{g}}\left(y_\xi x_\eta x_{\eta\eta} - x_\xi y_\eta x_{\xi\xi}\right) f_{\xi xx} + \frac{1}{2\sqrt{g}}\left(y_\xi y_\eta\right)\left(y_{\eta\eta} - y_{\xi\xi}\right) f_{yy} \\ &+ \frac{1}{2\sqrt{g}}\left(y_\xi y_\eta\left(x_{\eta\eta} - x_{\xi\xi}\right) + x_\eta y_\xi y_{\eta\eta} - x_\xi y_\eta y_{\xi\xi}\right) f_{xy} \\ &+ \text{second-order terms in the spacing} \end{aligned} \tag{32.17}$$

where the coordinate derivatives are to be understood here to represent central difference expressions, e.g.,

$$x_\xi = \frac{1}{2}\left(x_{i+1,j} - x_{i-1,j}\right), \qquad x_\eta = \frac{1}{2}\left(x_{i,j+1} - x_{i,j-1}\right)$$

$$x_{\xi\xi} = x_{i+1,j} - 2x_{ij} + x_{i-1,j} \quad x_{\eta\eta} = x_{i,j+1} - 2x_{ij} + x_{i,j-1}$$

These contributions to the truncation error arise from the nonuniform spacing. The familiar terms proportional to a power of the spacing occur in addition to these terms, as has been noted.

Sufficient conditions can now be stated for maintaining the order of the difference representations, with a fixed number of points in each distribution. First, as in the one-dimensional case, the ratios

$$\frac{x_{\xi\xi}}{\left|\mathbf{r}_\xi\right|^2}, \frac{y_{\xi\xi}}{\left|\mathbf{r}_\xi\right|^2}, \frac{x_{\eta\eta}}{\left|\mathbf{r}_\eta\right|^2}, \frac{y_{\eta\eta}}{\left|\mathbf{r}_\eta\right|^2}$$

must be bounded as x_ξ, x_η, y_ξ, y_η approach zero. A second condition must be imposed which limits the rate at which the Jacobian approaches zero. This condition can be met by simply requiring the $\cot\theta$ remain bounded, where ϕ is the angle between the ξ and η coordinate lines. The fact that this bound on the nonorthogonality imposes the correct lower bound on the Jacobian follows from the fact that $|\cot\phi| \le M$ implies

$$g \ge \frac{1}{M^2+1}\left|\mathbf{r}_\xi\right|^2 \cdot \left|\mathbf{r}_\eta\right|^2 \tag{32.18}$$

With these conditions on the ratios of second to first derivatives, and the limit on the nonorthogonality satisfied, the order of the first derivative approximations is maintained in the sense that the contributions to the truncation error arising for the nonuniform spacing will be second-order terms in the grid spacing.

The truncation error terms for second derivatives that are introduced when using a curvilinear coordinate system are very lengthy and involve both second and third derivatives of the function f. However, it can be shown that the same sufficient conditions, together with the condition that

$$\frac{x_{\xi\eta}}{\left|\mathbf{r}_\xi\right| \cdot \left|\mathbf{r}_\eta\right|} \quad \text{and} \quad \frac{y_{\xi\eta}}{\left|\mathbf{r}_\xi\right| \cdot \left|\mathbf{r}_\eta\right|}$$

remain bounded, will insure that the order of the difference representations is maintained.

It was noted above that a limit on the nonorthogonality, imposed by Eq. 32.18, is required for maintaining the order of difference representations. The degree to which nonorthogonality affects truncation error can be stated more precisely, as follows. The truncation error for a first derivative f_x can be written

$$T_x = \left(y_\eta T_\xi - y_\xi T_\eta\right) / \sqrt{g} \tag{32.19}$$

where T_ξ and T_η are the truncation errors for the difference expressions of f_ξ and f_η. Now all coordinate derivatives can be expressed using directions cosines of the angles of inclination, ϕ_ξ and ϕ_η of the ξ and η coordinate lines. After some simplification, the truncation error has the form

$$T_x = \frac{1}{\sin\left(\phi_\eta - \phi_\xi\right)}\left(\sin\phi_\eta \cos\phi_\xi \frac{T_\xi}{x_\xi} - \sin\phi_\xi \cos\phi_\eta \frac{T_\eta}{x_\eta}\right) \tag{32.20}$$

Therefore, the truncation error, in general, varies inversely with the sine of the angle between the coordinate lines. Note that there is also a dependence on the direction of the coordinate lines.

Reasonable departure from orthogonality ($\phi \leq 45°$) is therefore of little concern when the rate-of-change of grid spacing is reasonable. Large departure from orthogonality may be more of a problem at boundaries where one-sided difference expressions are needed. Therefore, grids should probably be made as nearly orthogonal at the boundaries as is practical.

This analysis has been primarily concerned with the effect of the grid on the truncation error. Clearly the higher-order solution derivatives are just as important in analyzing error. The numerical dissipation that arises in the solution of boundary layer problems is a result of variations in both grid spacing and solution gradients. No prescription has been given for measuring truncation error, but the results of this analysis will hopefully give the computational scientist or engineer some insight into how a grid can effect solution error and how the grid might be improved to increase accuracy in the numerical solution.

References

1. Hoffman, J.D., Relationship between the truncation errors of centered finite-difference approximation on uniform and nonuniform meshes, *J. of Computational Physics.* 1982, Vol. 46, pp 469–474.
2. Lee, D. and Tsuei, Y.M., A formula for estimation of truncation errors of convection terms in a curvilinear coordinate system, *J. of Computational Physics.* 1992, Vol. 98, pp 90–100.
3. Mastin, C.W., Error analysis and difference equations on curvilinear coordinate systems, *Large Scale Scientific Computation.* Parter, S.V. (Ed.), Academic Press, Orlando, FL, 1984.
4. Thompson, J.F., and Mastin, C.W., Order of difference expressions in curvilinear coordinate systems, *ASME J. of Fluids Engineering.* 1985, Vol. 107, pp 241–250.
5. Vinokur, M., On one-dimensional stretching functions for finite difference calculations, *J. of Computational Physics.* 1983, Vol. 50, pp 215–234.

33

Grid Optimization Methods for Quality Improvement and Adaptation

Olivier-Pierre Jacquotte

0-8493-2687-7/99/$0.00+$.50

33.1 Introduction

All the mesh generation methods, in particular those presented in this handbook, have been developed over 30 years to construct, as efficiently as possible, **grids with good quality**. By quality, one often means purely geometric quality: the grid should be as regular and orthogonal as possible in order to limit the truncation errors introduced in nonuniform grids or in boundary condition computations on grids that are excessively deformed and skewed (see Chapter 32). One also wants to include the feature for the grid to fit to a physical field in the domain, which means that the points are located in accordance to the characteristics of the solution for the computation of which the grid is used. Unfortunately, grid generators do not necessarily produce the grids that satisfy all users' requirements, and one is often interested in ways to *a posteriori* improve a grid for better quality or to better adapt it to a solution: grid optimization precisely consists in the improvement of an existing grid toward the best one with respect to given criteria resulting from the geometry or the physics of the problem it is constructed for.

To come up with an optimization method, one needs to build a criterion σ, function of a mesh, that will drive the optimization process; besides obvious data concerning the overall domain shape, it is important to introduce in the criterion information on the refinements desired in the mesh. This information is twofold, either related to purely geometric refinements, or to adaptive ones. First, any code user is always able to foresee what type of grid refinement should be obtained, and where to place grid points in the domain before solving the associated governing equations. This can be a finer mesh in areas where the user wants to capture details on the domain geometry, close to some part of the body for instance in external fluid mechanics, or close to discontinuities in the boundary conditions (point forces, mixed free/fixed boundary) in strength of materials. This can also be a coarser mesh in the far field or in parts of the domain where the user *a priori* knowns that the solution will not vary significantly or, conversely, stretched cells (very refined in one direction in comparison to the other ones) where it varies rapidly in areas, such as a boundary layer, for which one has a reasonable idea about the location.

However, when one has solved the problem equations and has gotten a physical solution, the adaptation of the grid will require that σ takes this solution into account; this requires that relevant information is first extracted from it, then transformed into adaptive refinement data and introduced into σ.

In this chapter optimization criteria are introduced in a constructive way: basic functionals σ searching for uniformity and orthogonality of the grids are first described — two classes of functionals will be presented (Sections 33.2 and 33.3); then we show how this basic formulation can be enhanced in order to take desired refinements (Section 33.4) or adaptation (Section 33.5) into account and ways to modify these functionals are presented. Practical reference values that fix for each mesh cell its desired size are used to prescribe refinements: they can be defined either from an initial mesh only for quality improvement, or with the introduction of information from a solution for adaptation.

There are two approaches to construct mesh optimization criteria: a first approach, only applicable to structured grids, is to consider the mesh as a transformation from the unit square or cube (depending on the space dimension) onto the computational domain, rather than a set of cells or points, and to define optimization criteria on this transformation; the book by Knupp and Steinberg [12] thoroughly describes this *continuous* approach. Here we have chosen the *discrete* approach: elementary quality measures are first constructed locally by study of the geometry of the cells (basically its shape); for that, we will use of so-called least square formulation (LSF),* enforcing desired geometric properties in such a weak sense. Then these local contributions are added to obtain a global criterion; continuous description can finally be recovered going back to the transformation from the unit square or cube. This latter more heuristic approach, though less formal, enables the discovery of a class of functionals impossible to put forward otherwise.

Conversely to the discrete approach, which is meaningful only for structured grids, the discrete approach enables the extension of the variational method to unstructured grids: the mesh quality measure

* A least-squares formulation replaces the problem "find x such that $f(x)=0$" by the formulation "find x that minimizes $f(x)^2$."

and optimization criteria presented next have a meaning for unstructured meshes where the notion of grid line does not exist; this is shown in Section 33.7.

33.1.1 Notation and General Framework of the Chapter

In this chapter we mainly consider the optimization of a single structured domain Ω; the way to handle multidomain configurations is described in Section 33.6.3. We consider a three-dimensional structured grid formed by *i*max × *j*max × *k*max nodes $\mathbf{x}_{ijk}$ of coordinates $(x_{ijk}, y_{ijk}, z_{ijk})$; we use

$$\Delta u = 1/(i\text{max}-1);\ \ \Delta v = 1/(j\text{max}-1);\ \ \Delta w = 1/(k\text{max}-1) \tag{33.1}$$

The unit square or cube will be denoted by $\hat{\Omega}$. We also consider the mesh, i.e., a *discrete* set of points, in a *continuous* manner: each index (i,j,k) corresponds to a point

$$\mathbf{u}_{ijk} = (u_i, v_j, w_k) = ((i-1)\Delta u, (j-1)\Delta v, (k-1)\Delta w) \tag{33.2}$$

in $\hat{\Omega}$, and the mesh can be considered the transformation of a uniform mesh of $\hat{\Omega}$ (with cell size $\Delta u \times \Delta v \times \Delta w$) by a piecewise continuous trilinear transformation $\mathbf{x}(\mathbf{u})$ defined by

$$\mathbf{x}\left(\mathbf{u}_{ijk}\right) = \mathbf{x}_{ijk} \tag{33.3}$$

The transformation $\mathbf{x}(\mathbf{u})$ is called the mesh function. In the following, one will often refer to "orthogonal" grids; of course, exact orthogonality of the cells cannot be achieved, so rather than looking for grids with orthogonal cells, one will be looking for a mesh function $\mathbf{x}(\mathbf{u})$ such that the two families of curves obtained by having u and v vary separately are orthogonal.

33.2 Regularity–Orthogonality Formulation

33.2.1 Measure of the Orthogonality

As noted in Chapter 32 and above in this chapter, orthogonality between gridlines is often sought; this quality can be used to build a first optimization criterion where the orthogonality is enforced in a weak sense, in a least-squares formulation. In order to implement the LSF for the orthogonality, we consider the angles between the edges around a node in the mesh; among these angles, orthogonality is desired for four of them in 2D and twelve of them in 3D. Instead of directly considering the values of the angle in a LSF that could be written "minimize $(angle - 90°)^2$,"* the scalar products between the edges forming these angles are considered to define a node function; in 2D this function can be written for node (i, j),

$$\begin{aligned}\sigma_{ij} &= \left[(\mathbf{x}_{ij} - \mathbf{x}_{i+1j})\cdot(\mathbf{x}_{ij} - \mathbf{x}_{ij+1})\right]^2 + \left[(\mathbf{x}_{ij} - \mathbf{x}_{i-1j})\cdot(\mathbf{x}_{ij} - \mathbf{x}_{ij+1})\right]^2 \\ &+ \left[(\mathbf{x}_{ij} - \mathbf{x}_{i+1j})\cdot(\mathbf{x}_{ij} - \mathbf{x}_{ij-1})\right]^2 + \left[(\mathbf{x}_{ij} - \mathbf{x}_{i-1j})\cdot(\mathbf{x}_{ij} - \mathbf{x}_{ij-1})\right]^2\end{aligned} \tag{33.4}$$

When all the angles around the node are 90°, σ_{ij} is 0.

In order to be consistant with finite element formulations used in this chapter, we rewrite this orthogonality measure function considering edges $\mathbf{r}_i^e$, $i = 1.4$ in cell e (numbered in such a way that $\mathbf{r}_i^e$ and $\mathbf{r}_{i+1}^e$ are adjacent), so the expression of the orthogonality measure in a cell becomes

$$\sigma_{ortho}^e = \left[\mathbf{r}_1^e \cdot \mathbf{r}_2^e\right]^2 + \left[\mathbf{r}_2^e \cdot \mathbf{r}_3^e\right]^2 + \left[\mathbf{r}_3^e \cdot \mathbf{r}_4^e\right]^2 + \left[\mathbf{r}_4^e \cdot \mathbf{r}_1^e\right]^2 \tag{33.5}$$

---

*This approach is turned down because of practical difficulties in the optimization due to the nonpolynomial nature of this criterion.

33.2.2 Measure Of The Regularity

A second quality that is often required of a mesh is the so-called regularity property by which uniformity of the node partition in the the domain is looked for; of course, a strict uniformity is not always desirable and refinements are often *a priori* necessary in order to handle geometrical details or physical phenomena (such as boundary layers) for which the location is known before any computation. In Section 33.4 we will present the ways to obtain nonuniform grids, where refinements can be carefully controlled, but here, as a first step, uniformity is looked for.

As mentioned earlier, the mesh quality measure must exhibit terms in a square form in order to come up with a LSF. One must first note that, contrary to the orthogonality that could be defined at the node or cell level, the regularity property is global: it concerns all the cells with respect to one another. However, one will try here to exhibit cell contributions to the global measure.

The construction of the regularity function can more easily be explained and understood by considering a uniform mesh between 0 and 1: $\mathbf{x}_i = (i-1) / (i\max - 1)$. This mesh realizes the minimum of the functional

$$\sigma_{reg1D} = \sum_{i=1,i\max-1} \left|\mathbf{x}_{i+1} - \mathbf{x}_i\right|^2 \tag{33.6}$$

The minimization of each of the term in the summation leads to the meaningless collapse of the cells, but the global minimization has a sense because of the constraints prescribed on the boundaries 0 and 1, imposing that these points stay fixed.

In two and three dimensions, a functional can be defined using similar expressions at the element level; in 2D the elementary regularity functional is

$$\sigma^e_{reg} = \left|\mathbf{r}^e_1\right|^2 + \left|\mathbf{r}^e_2\right|^2 + \left|\mathbf{r}^e_3\right|^2 + \left|\mathbf{r}^e_4\right|^2 \tag{33.7}$$

In 3D, this functional includes twelve terms.

33.2.3 Global Functional

For each cell it is then possible to define a measure of its quality by combining both criteria and obtain

$$\sigma^e = \lambda\sigma^e_{reg} + (1-\lambda)\sigma^e_{ortho} \tag{33.8}$$

where λ is a parameter chosen between 0 and 1 to weigh the contributions. These elementary contributions are summed over the elements and we obtain a global quantity measure σ:

$$\sigma = \sum_{e\in\text{Mesh}} \sigma^e \tag{33.9}$$

For $\lambda = 1$, the pure regularity is enforced; the minimization of this functional leads to an uncollapsed mesh because of boundary conditions that force the nodes to remain on the domain boundary.

The discrete functionals can be rewritten in a continuous form. Noting that the cell edges $\mathbf{r}^e_i$ are derivatives of the mesh function $\mathbf{x}(\mathbf{u})$ with respect u, v, or w depending of their orientation, the regularity and orthogonality functionals, respectively, become

$$\begin{aligned}\sigma_{reg}(\mathbf{x}) &= \int_{\hat\Omega} \left|\nabla_{\mathbf{u}}\mathbf{x}\right|^2 d\mathbf{u} = \int_{\hat\Omega} \left(\left|\mathbf{x}_u\right|^2 + \left|\mathbf{x}_v\right|^2 + \left|\mathbf{x}_w\right|^2\right) d\mathbf{u} \\ &= \int_{\hat\Omega} \left(\left|x_u\right|^2 + \left|y_u\right|^2 + \left|z_u\right|^2 + \left|x_v\right|^2 + \left|y_v\right|^2 + \left|z_v\right|^2 + \left|x_w\right|^2 + \left|y_w\right|^2 + \left|z_w\right|^2\right) d\mathbf{u}\end{aligned} \tag{33.10}$$

and

$$\begin{aligned}\sigma_{ortho}(\mathbf{x}) &= \int_{\hat{\Omega}} \left(|\mathbf{x}_u \cdot \mathbf{x}_v|^2 + |\mathbf{x}_v \cdot \mathbf{x}_w|^2 + |\mathbf{x}_w \cdot \mathbf{x}_u|^2 \right) d\mathbf{u} \\ &= \int_{\hat{\Omega}} \left(|x_u x_v + y_u y_v + z_u z_v|^2 + |x_v x_w + y_v y_w + z_v z_w|^2 + |x_w x_u + y_w y_u + z_w z_u|^2 \right) d\mathbf{u}\end{aligned} \tag{33.11}$$

and the mesh obtained with the method corresponds to the mesh function minimizing $\lambda\sigma_{reg}(\mathbf{x}) + (1 - \lambda)\sigma_{ortho}(\mathbf{x})$. In fact, these compact expressions for σ_{reg} and σ_{ortho} are not exactly the transcription of the functionals introduced above since derivations with respect to u, v, and w, introduce division of the edges by Δu, Δv, and Δw; rigorous expressions will be given in Section 33.4, after introducing nonuniform reference cells.

33.2.4 Origin of the Regularity–Orthogonality Functional

The approach using these regularity and orthogonality functionals was put forward by Brackbill and Saltzman [3]; they used a smoothness integral measuring the roughness of the grid:

$$I_{reg} = \int_{\Omega} \left(|\nabla\xi|^2 + |\nabla\eta|^2 + |\nabla\zeta|^2 \right) d\mathbf{x} \tag{33.12}$$

and the orthogonality measure

$$I_{ortho} = \int_{\Omega} \left(|\nabla\xi \cdot \nabla\eta|^2 + |\nabla\eta \cdot \nabla\zeta|^2 + |\nabla\zeta \cdot \nabla\xi|^2 \right) J^s d\mathbf{x} \tag{33.13}$$

where J, the cell volume, was introduced to emphasize orthogonality more strongly in smaller ($s < 1$) or larger ($s > 1$) cells. The grid was obtained from the Euler equations associated to the minimization of the combination

$$I = \lambda I_{reg} + (1 - \lambda) I_{ortho} \tag{33.14}$$

The refinements in the grids can be obtained by the minimization of another integral that will be introduced in the section devoted to the adaptation.

33.2.5 Discussion of the Regularity–Orthogonality Functional

The regularity–orthogonality functionals introduced by Kennon and Dulikravitch [11], and Carcaillet [6] are the discrete equivalent of the continuous approach put forward by Brackbill and Saltzman [3]: their construction is quite easy to explain and to understand, and mechanical interpretation can be provided for both functionals σ_{reg} and σ_{ortho}. If we first consider a spring between two points $\mathbf{x}_1$ and $\mathbf{x}_2$ at rest (equilibrium position) when the points are in the same position, the potential energy of this system is $\frac{1}{2}\|\mathbf{x}_2 - \mathbf{x}_1\|^2$ (for a unit spring stiffness). If all the grid nodes joined two-by-two by an edge are linked by such a spring, the potential energy of the system is precisely the regularity functional*: the mesh optimized with respect to the regularity criterion is therefore the equilibrium position of the system, which minimizes its potential energy. Similarly, one now considers that the same points are linked by rods, rigid but free to extend, and that rods corresponding to edges where orthogonality is desired (adjacent edges) are linked by torsion springs, at rest when the angle is 90°, with potential

*Up to within a multiplicative factor.

energy* $r_i^2 r_{i+1}^2 cos^2(\mathbf{r}_i, \mathbf{r}_{i+1})$: the potential energy of the system is here the orthogonality functional** and the mesh optimized with respect to the orthogonality criterion is the equilibrium position of this second mechanical system.

A rapid look at the elementary contribution shows that polynomials of degree 2 and 4 of the node coordinates are obtained for σ_{reg} and σ_{ortho}: this indicates that computation of the functional and its gradients are quick to implement and that grid optimization using these criteria can be easily coded. Even though the method provides satisfying results and considerably improves grids in numerous cases [6, 11], results can also be disappointing, even difficult or impossible to obtain in mildly severe cases. This can happen in particular in the neighborhood of nonconvex boundaries where points can move outside of a boundary; problems with convergence of the optimization algorithm may also occur, in which case the solution is either not found, or oscillates in the iterative optimization process between several values that minimize the functional.

These difficulties can be imputed to the lack of mathematical support of the method and to the absence of the properties that ensure the uniqueness of the solution and/or*** the convergence of the optimization algorithms; in particular, a key condition for the well-posedness of the problem is missing — the convexity of the functional to be minimized. The functional presented next overcomes this difficulty. It also tries to avoid the mesh overlapping by embodying a volume control term, that tends to prevent algebraic cell volumes from becoming negative.

33.3 Deformation Formulation

In this section we construct another optimization criterion that overcomes the difficulties evoked above. Here the construction of the criterion is based on principles of contiuum mechanics — the nonlinear elasticity theory — and is characterized among several choices by the appropriate mathematical property — the convexity. We first define a measure of deformations in space, and next we apply this concept to the definition of the deformation of a cell with respect to an "optimum" desired mesh cell. This mesh cell, used as reference to define the deformation measure, is first taken as the unit cube $[0, 1] \times [0, 1] \times [0, 1]$.

33.3.1 Measure of the Cell Deformation

We consider the transformation $\mathbf{x}(\chi)$ which transforms points $\chi = (\xi, \eta, \zeta)$ of the reference unit cube into points $\mathbf{x} = (x, y, z)$ of the mesh cell considered, and we denote as $\mathbf{F}$ the matrix which is the gradient of this transformation ($\mathbf{F} = \nabla\mathbf{x}$). In order to identify a correct measure of the mesh cell deformation, four axioms and properties can be stated:

1. *First-order dependence*: the measure of the deformation in $\mathbf{x}$ depends only on the gradient of the transformation in this point, $\sigma = \sigma(\mathbf{x}, \mathbf{F})$.
2. *Material indifference principle*: the measure depends only on the form of the mesh cell and is independent of the orthonormal basis in which it is evaluated, $\sigma(\mathbf{F}) = \sigma(\mathbf{QF})$ for any orthogonal matrix $\mathbf{Q}$.
3. *Isotropy property*: the measure is independent of the axis system in the reference space, $\sigma(\mathbf{F}) = \sigma(\mathbf{FQ})$ for any orthogonal matrix $\mathbf{Q}$.
4. *Homogeneity property*: the measure is independent of the cell position, $\sigma = \sigma(\mathbf{F})$.

The material indifference axiom and the isotropy property result in the so-called objectivity of the measure: the measure is independent of the orthonormal basis chosen to represent $\mathbf{F}$, $\sigma(\mathbf{F}) = \sigma(\mathbf{QFQ}^T)$ for any orthogonal matrix $\mathbf{Q}$. This property is also consistent with the general physical principle according

* The classical linear definition of the torsion stiffness is $K\theta^2$ where θ is the deviation from 90° $((\mathbf{r}_i, \mathbf{r}_{i+1}) - 90°)$ and does not include the length factors.

** Once again, up to within a multiplicative factor.

***Both are related.

to which any observable quantity of an intrinsic nature must be independent of the orthonormal basis in which it is computed.

From these four properties, it is possible to demonstrate that measure σ of the deformation of the reference unit cube in a current mesh cell depends only on invariants I_1, I_2, and I_3 of the deformation matrix **C**, also called right Cauchy–Green matrix of transformation $\mathbf{x}(\chi)$:

$$\sigma = \sigma(I_1, I_2, I_3) \tag{33.15}$$

where

$$I_1 = trace\ \mathbf{C}, \quad I_2 = trace\ \mathrm{Cof}\ \mathbf{C} \text{ and } I_3 = \det \mathbf{C} \tag{33.16}$$

and

$$\mathbf{C} = \nabla x^{\mathrm{T}} \cdot \nabla x = \mathbf{F}^{\mathrm{T}} \cdot \mathbf{F} \tag{33.17}$$

In Appendix A, we recall the definition of the cofactor matrix, as well as certain properties of the invariants. We also introduce the determinant of $\mathbf{F}(I_3 = J^2)$, which measures the cell volume. In addition, σ is required to depend on the orientation of the cell in physical space: indeed the dependence of σ on I_3 makes it impossible to see if the cell is volume positive or not, in which case overlapping may occur. In order for the cell deformation to effectively see the cell orientation, the dependence of σ on I_3 must then be replaced by a dependence on J, giving

$$\sigma = \sigma(I_1, I_2, J) \tag{33.18}$$

This is also achieved by restricting the material indifference principle and the isotropy property to the *direct* orthogonal matrices **Q**.

33.3.2 Characterization of Functional σ

The second step consists of insuring that the minimization of functional σ is correctly stated and that it gives the least deformed mesh possible in the sense specified above. This leads to setting hypotheses on σ and on its first two derivatives for so-called rigid transformations, which preserve the form of the cell and its orientation. These transformations verify the following equivalent properties:

1. $\mathbf{x}(\chi)$ is a rigid transformation
2. **F** is a direct orthogonal matrix
3. $\mathbf{C} = \mathbf{Id}$ and $det\ \mathbf{F} = +1$
4. $(I_1, I_2, J) = (3, 3, 1)$

The rigid transformations will next be characterized by index 0. In order to ensure the mathematical properties allowing the minimization problem to be well-posed (unique solution) and giving the certainty that the usual minimization algorithms will converge efficiently toward a unique solution, three properties are assumed:

1. A normalization: σ is zero for a rigid transformation $\sigma|_0 = 0$.
2. An equilibrium condition in the mechanical sense: the reference configuration is an unstressed (also called natural) state and therefore σ is stationary for rigid transformations $\mathbf{D}_\mathrm{F}\sigma|_0 = 0$ where $\mathbf{D}_\mathrm{F}$ denotes the derivative of σ with respect to gradient **F** of the transformation.
3. A convexity condition: σ is convex in the neighborhood of the rigid transformations, $\mathbf{D}^2_\mathrm{F}\sigma|_0 = 0$.

These three properties restrict the possible choices for the functional. It can in effect be shown that they require the function of the invariants $\sigma(I_1, I_2, I_3)$ and its derivatives to verify the following in point (3, 3, 1):

$$\sigma = 0 \tag{33.19a}$$

$$\sigma_1 + 2\sigma_2 + \sigma_3 = 0 \tag{33.19b}$$

$$-\sigma_2 - \sigma_3 > 0 \tag{33.19c}$$

$$2(\sigma_2 + \sigma_3) + 3(\sigma_{11} + 4\sigma_{22} + \sigma_{33}) + 6(\sigma_{12} + 2\sigma_{23} + \sigma_{13}) > 0 \tag{33.19d}$$

where index i denotes the derivative of σ with respect to invariant I_i.

33.3.3 Mechanical Interpretation of the Method

The type of functional exhibited has been known for some 20 years to solid mechanics specialists, in particular to those familiar with nonlinear elasticity. Indeed, these functions of the invariants are used for the modelization of the behavior of nonlinear elastic materials such as rubbers; more precisely, they represent the deformation of the material. The mesh generation method proposed can therefore be mechanically interpreted as follows: given a domain (for instance in two dimensions) in which it is desired to construct a mesh, considering a rectangular rubber membrane at rest on which a regular grid has been drawn, the mesh obtained by the method described is the grid obtained when the membrane is stretched so that its boundary coincides with the boundary of the domain. The approach described herein is inspired from continuum mechanics by contrast with that presented in Section 33.2 which tends to assimilate the mesh to a discrete network of springs and torsion bars. This approach also yields a variational formulation that has satisfactory mathematical properties for solving the minimisation problem, contrary to those described in [14] for the continuum case and above for the discrete case.

33.3.4 Cell Deformation and Measure of the Mesh Quality

Before introducing explicit expressions for the function σ of the invariants, we note that up to here we have only defined the measure of a transformation $\mathbf{x}(\chi)$ and obtained σ which is still a continuous function defined pointwise in the unit cube. Indeed, for a hexahedral cell e with nodes $\mathbf{x}^e_{lmn}$, with l, m, n equal to 0 or 1, according to the value of $\chi = (\xi, \eta, \zeta)$ $\mathbf{x}^e(\chi)$ can be written as:

$$\mathbf{x}^e(\chi)\begin{cases}(1-\xi)(1-\eta)(1-\zeta)\mathbf{x}^e_{000} + \xi(1-\eta)(1-\zeta)\mathbf{x}^e_{100} + (1-\xi)\eta(1-\zeta)\mathbf{x}^e_{010} + \xi\eta(1-\zeta)\mathbf{x}^e_{110} \\ \quad +(1-\xi)(1-\eta)\zeta\mathbf{x}^e_{001} + \xi(1-\eta)\zeta\mathbf{x}^e_{101} + (1-\xi)\eta\zeta\mathbf{x}^e_{011} + \xi\eta\zeta\mathbf{x}^e_{111}\end{cases} \tag{33.20}$$

Similarly, **F**, **C** and its invariants are functions of χ, which leads to $\sigma^e(\chi)$ as soon as the function σ of the invariants is known.

For each point χ in the reference unit cube, the function $\sigma^e(\chi)$ gives an expression of the deformation in the neighborhood of the point $\mathbf{x}^e(\chi)$ in the deformed cell. A measure of the *cell deformation* σ^e_{def} is obtained by integration (exactly or using an integration rule) this function over the unit cube. The measure of the *mesh deformation* is finally obtained in a way similar to the regularity–orthogonality formulation by summation of the elementary contributions; this leads to a function of the node coordinates whose form (polynomial or else) depends, of course, on the form of the function σ of the invariants. In the next section we introduce several expressions for this function, for which geometrical interpretation are given.

33.3.5 Choice of the Functionals in Two and Three Dimensions

Several simple functions (polynomials) of the invariants were exhibited in the nonlinear elasticity theory. In particular, several authors proposed the function

$$\sigma(I_1, I_2, J) = C_1(I_1 - I_3 - 2) + C_2(I_2 - 2I_3 - 1) + C_3(J-1)^2 \tag{33.21}$$

where the convexity condition requires that constants C_1, C_2, and C_3 must satisfy

$$3C_3 > 4(C_1 + C_2) > 0 \tag{33.22}$$

As far as the optimization of meshes is concerned, it is interesting to select a few special cases from this family of functions, leading to geometric interpretations.

33.3.5.1 Two-Dimensional Functional

A useful functional in two dimensions can be obtained from the above results by considering plane transformations:

$$x = x(\xi, \eta), \quad y = y(\xi, \eta) \tag{33.23}$$

In this case, 3 × 3 matrices **F** and **C** consist of 2 × 2 blocks, $\tilde{\mathbf{F}}$ and $\tilde{\mathbf{C}}$, made complete by 0, 0, and 1 on the last rows and columns. The invariants of **C** are then related by the equation:

$$I_2 = I_1 + I_3 - 1 \tag{33.24}$$

and the first two terms in Eq. 33.21 are the same. The invariants of **C** verify

$$\tilde{I}_1 = I_1 - 1 \quad \text{and} \quad \tilde{I}_3 = I_3 \tag{33.25}$$

They are used to define the 2D equivalent of the functional

$$\sigma_{2D} = C(\tilde{I}_1 - \tilde{I}_3 - 1) + K(\tilde{J} - 1)^2 \tag{33.26}$$

where $K > C > 0$. This can be written again in the form

$$\sigma_{2D} = C(\tilde{I}_1 - 2\tilde{J}) + (K - C)(\tilde{J} - 1)^2 \tag{33.27}$$

Below, symbols $\tilde{}$ are dropped. This expression includes two terms preceeded by positive constants; the first term, $(I_1 - 2J)$, can be interpreted by considering matrix **F**:

$$\mathbf{F} = \begin{bmatrix} x_\xi & x_\eta \\ y_\xi & y_\eta \end{bmatrix} \tag{33.28}$$

and we have

$$I_1 = x_\xi^2 + y_\xi^2 + x_\eta^2 + y_\eta^2 \quad \text{and} \quad J = x_\xi y_\eta - x_\eta y_\xi \tag{33.29}$$

and therefore

$$I_1 - 2J = (x_\xi - y_\eta)^2 + (x_\eta + y_\xi)^2 \tag{33.30}$$

This term, when minimized, represents a LSF of the Cauchy–Riemann relations:

$$\begin{cases} x_\xi - y_\eta = 0 \\ y_\xi + x_\eta = 0 \end{cases} \tag{33.31}$$

which ensures that the cell is a square. These relations can also be written

$$\mathbf{x}_\xi = \mathbf{x}_\eta \times \mathbf{k} \tag{33.32}$$

where **k** is the unit vector positively orthogonal to the 2D plane. These relations also mean that the 2D transformation $\mathbf{x}(\xi, \eta)$ is conformal.

In the minimization of the second term $(J-1)^2$, one recognizes a LSF of constraint $J = 1$; this contribution to the functional forces the cell to have an area J to remain as close as possible to 1, thus it prevents the cell overlapping, occuring when J becomes negative.

When summation of the elementary contributions is made, the mesh obtained by minimization of the resulting functional corresponds to a mesh function minimizing

$$C\int_{\hat{\Omega}} \left[(x_u - y_v)^2 + (x_v + y_u)^2\right] d\mathbf{u} + (K - C)\int_{\hat{\Omega}} (x_u y_v - x_v y_u - 1)^2 d\mathbf{u} \tag{33.33}$$

For the same reasons as in Section 33.2.3, this expression is not exactly the expression corresponding to the functional described up to here; Section 33.4 will come back to this issue.

33.3.5.2 Surface Functional

Functional σ_{2D} can also be used and interpreted in geometric terms in two dimensions but on a surface. In effect, in this case matrix **F** (where symbols $\tilde{}$ are dropped) is now a 3×2 matrix defined by

$$\mathbf{F} = \begin{bmatrix} x_\xi & x_\eta \\ y_\xi & y_\eta \\ z_\xi & z_\eta \end{bmatrix} \tag{33.34}$$

Matrix **C** is a 2×2 matrix whose first invariant is equal to

$$I_1 = x_\xi^2 + y_\xi^2 + z_\xi^2 + x_\eta^2 + y_\eta^2 + z_\eta^2 \tag{33.35}$$

Previously given definition of J (as the determinant of **F**) no longer applies but, considering the area elements on the surface, can be replaced by

$$J = (\mathbf{x}_\xi \times \mathbf{x}_\eta) \cdot \mathbf{n} = det(\mathbf{x}_\xi, \mathbf{x}_\eta, \mathbf{n}) = \begin{vmatrix} x_\xi & x_\eta & n_1 \\ y_\xi & y_\eta & n_2 \\ z_\xi & z_\eta & n_3 \end{vmatrix} \tag{33.36}$$

where **n** designates the vector normal to the surface. Functional σ_{2D} then represents a LSF of the Cauchy–Riemann equations on a surface, translating the conformity of transformation $\mathbf{x}(\xi, \eta)$ and which can be written

$$\mathbf{x}_\xi = \mathbf{x}_\eta \times \mathbf{n} \tag{33.37}$$

33.3.5.3 Three-Dimensional Functional

In 3D, the following choice

$$C_1 = C_2 \equiv C \quad ; \quad C_3 \equiv 3C + K \tag{33.38}$$

allows certain interesting geometric interpretations. In this case, the general expression given for σ in Section 33.3.3 becomes

$$\sigma_{3D} = C\left(I_1 + I_2 - 6J\right) + K\,(J-1)^2 \tag{33.39}$$

and the first term $(I_1 + I_2 - 6J)$ can be interpreted by recalling that the aim is to minimize the deformation of each mesh cell and therefore to obtain **F** as close as possible to a rigid transformation. In addition to the four properties mentioned above, such a transformation is also characterized by

$$\mathbf{F} = Cof\ \mathbf{F} \quad \text{and} \quad det\ \mathbf{F} = +1 \tag{33.40}$$

A least-squares formulation of this is obtained by minimizing

$$C\|\mathbf{F} - Cof\ \mathbf{F}\|^2 + K|det\ \mathbf{F} - 1|^2 \tag{33.41}$$

It is verified that this expression is effectively equal to σ_{3D}, by choosing

$$\|\mathbf{A}\|^2 = Trace(\mathbf{A}^{\mathrm{T}} \cdot \mathbf{A}) \tag{33.42}$$

as matrix norm, then remarking that

$$\|\mathbf{F} - Cof\ \mathbf{F}\|^2 = I_1 + I_2 - 6J \tag{33.43}$$

In addition to the interpretation given above for $(J-1)^2$, we note here that this term completes the first one by requiring **F** to be a *direct* orthogonal matrix ($det\ \mathbf{F} = +1$).

In addition, the nine equations contained in $\mathbf{F} = Cof\,\mathbf{F}$ and detailed in Appendix A are equivalent to the vector equations:

$$\begin{Bmatrix} \mathbf{x}_\xi = \mathbf{x}_\eta \times \mathbf{x}_\zeta \\ \mathbf{x}_\eta = \mathbf{x}_\zeta \times \mathbf{x}_\xi \\ \mathbf{x}_\zeta = \mathbf{x}_\xi \times \mathbf{x}_\eta \end{Bmatrix} \tag{33.44}$$

These equations can also be interpreted as 3D extensions of the Cauchy–Riemann equations and can be considered for generalizing the conformity concept in three dimensions: they mean that the basis $\{\mathbf{x}_\xi, \mathbf{x}_\eta, \mathbf{x}_\zeta\}$ is orthonormal and that the deformed cell is the unit cube. As before, summation of the elementary contributions leads to a global functional; its continuous correspondant can be written (with the reservations already made in Sections 33.2.3 and 33.3.5.1)

$$C\int_{\hat{\Omega}}\left\{\left\|\mathbf{x}_u-\mathbf{x}_v\times\mathbf{x}_w\right\|^2+\left\|\mathbf{x}_v-\mathbf{x}_w\times\mathbf{x}_u\right\|^2+\left\|\mathbf{x}_w-\mathbf{x}_u\times\mathbf{x}_v\right\|^2\right\}d\mathbf{u}+K\int_{\hat{\Omega}}(\det[\mathbf{x}_u,\mathbf{x}_v,\mathbf{x}_w]-1)^2 d\mathbf{u} \quad (33.45)$$

Remark: In 2D, the equation $I_1 - 2J = 0$ leads to a set of transformations for which J can take any value and the term $(J-1)^2$ must be added for the minimization in order to fix J. However, in 3D, $I_1 + I_2 - 6J = 0$ leads to transformations for which J can only take the values 0 and ± 1, which means that the functional $I_1 + I_2 - 6J$ contains some information about the volume. For this reason, the term $(J-1)^2$ need not be added in 3D, and the minimization of $I_1 + I_2 - 6J$ is sufficient to avoid cell overlapping.

33.4 Handling of an Initial Grid

33.4.1 General Principle

As mentioned in the introduction, it is important to have the possibility to include data in the optimization functionals that will lead to mesh refinements. Besides this reason, a scaling of the functionals appears necessary. Indeed, one can notice that the terms in σ^e_{reg}, σ^e_{ortho}, and σ^e_{def} are not homogeneous in orders: if h is the order of the cell size in an isotropic grid, i.e., a grid where the cell edges have the same order of magnitude in all directions, we have the following orders for the functionals encountered up to here:

$$\sigma^e_{reg}=O(h^2),\quad \sigma^e_{ortho}=O(h^4),\quad I^e_1=O(h^2),\quad I^e_2=O(h^4),\quad J^e=O(h^3) \quad (33.46)$$

and the scaling is necessary in order to be able to meaningfully add these contributions in either of the functionals. These expressions also show that the functionals are not homogeneous in unit: if h is measured in a given unit (in inches, millimeters, or in any unit system), various powers of this unit are added, which is not acceptable. Also, if the desired grid is anisotropic, i.e., differences in length between adjacent cell edges are large, contributions forming each of these elementary measures have different orders of magnitude, so scaling also appears important for the functionals to take all directions into account.

Moreover, even though one can remark that σ_{reg} and σ_{ortho}, taken separately, are consistant and have a meaning (terms of the same scale — lengths to the second or to the fourth — are added), this is not the case for the functionals defined with expressions carelessly adding invariants (as in $I_1 - 2J$, $I_1 + I_2 - 6J$, $(J-1)^2$ ...), and scaling appears more crucial for the deformation criterion.

This scaling will be done by the introduction of *reference lengths* chosen to be the length of cell edges; though the optimization method tends to realize the following compromise: first to obtain ratios between edge lengths as close as possible (in a LSF sense) to the ratios between reference lengths, and second, to ensure the orthogonality between grid lines. Moreover, if reference data are properly scaled, values of the edge lengths themselves will be looked for by the optimization, rather than ratios between them. As a consequence, the regularity and smoothness of these reference data will appear in the mesh optimized with the functionals where they have been introduced; for example, if large discontinuities between reference lengths occur, they will also appear in the optimized grid. Conversely, if one wants a smooth partition of the nodes in the mesh, smooth reference data will have to be given.

These reference edges can be the edges in the mesh initial; if the interior of the initial mesh is not satisfactory, interpolations of edge lengths between the boundaries can be used and smooth values can be obtained after a careful construction limited to the boundaries. However, other contraints may prevent the data from being smooth on the boundaries, leading to irregular reference data propagated by the interpolation from the boundaries to the interior of the domain. In any case (definition directly from the initial mesh, or by interpolation from the boundaries), a Laplace filter can be applied to the unsatisfactory data in order to obtain smoothness of the reference values.

33.4.2 Regularity–Orthogonality Formulation

In 2D the following expressions for the regularity and orthogonality functionals take the reference data a^e_i (length of a reference edge $\mathbf{a}^e_i$) into account:

$$\sigma^e_{reg} = \frac{|\mathbf{r}^e_1|^2}{a^{e2}_1} + \frac{|\mathbf{r}^e_2|^2}{a^{e2}_2} + \frac{|\mathbf{r}^e_3|^2}{a^{e2}_3} + \frac{|\mathbf{r}^e_4|^2}{a^{e2}_4} \tag{33.47}$$

$$\sigma^e_{ortho} = \frac{[\mathbf{r}^e_1 \cdot \mathbf{r}^e_2]^2}{a^{e2}_1 a^{e2}_2} + \frac{[\mathbf{r}^e_2 \cdot \mathbf{r}^e_3]^2}{a^{e2}_2 a^{e2}_3} + \frac{[\mathbf{r}^e_3 \cdot \mathbf{r}^e_4]^2}{a^{e2}_3 a^{e2}_4} + \frac{[\mathbf{r}^e_4 \cdot \mathbf{r}^e_1]^2}{a^{e2}_4 a^{e2}_1} \tag{33.48}$$

Analogously, continuous functionals can also be considered: introducing piecewise constant reference functions $a(\mathbf{u})$, $b(\mathbf{u})$, and $c(\mathbf{u})$ that are the actual lengths that one wants to obtain in the i-, j- and k-directions for the cell of index $\mathbf{u} = (u,v,w)$ in the unit cube, respectively divided by Δu, Δv, and Δw (in order to properly consider derivations with respect to u, v, and w), one obtains

$$\sigma_{reg}(\mathbf{x}) = \int_{\hat\Omega} \left(\frac{|\mathbf{x}_u|^2}{a^2} + \frac{|\mathbf{x}_v|^2}{b^2} + \frac{|\mathbf{x}_w|^2}{c^2} \right) d\mathbf{u} \tag{33.49}$$

and

$$\sigma_{ortho}(\mathbf{x}) = \int_{\hat\Omega} \left(\frac{|\mathbf{x}_u \cdot \mathbf{x}_v|^2}{a^2 b^2} + \frac{|\mathbf{x}_v \cdot \mathbf{x}_w|^2}{b^2 c^2} + \frac{|\mathbf{x}_w \cdot \mathbf{x}_u|^2}{c^2 a^2} \right) d\mathbf{u} \tag{33.50}$$

If the initial mesh defined by its mesh function $\mathbf{x}^0$ is used to construct the reference lengths, the expressions become

$$\sigma_{reg}(\mathbf{x}) = \int_{\hat\Omega} \left(\frac{|\mathbf{x}_u|^2}{|\mathbf{x}^o_u|^2} + \frac{|\mathbf{x}_v|^2}{|\mathbf{x}^o_v|^2} + \frac{|\mathbf{x}_w|^2}{|\mathbf{x}^o_w|^2} \right) d\mathbf{u} \tag{33.51}$$

and

$$\sigma_{ortho}(\mathbf{x}) = \int_{\hat\Omega} \left(\frac{|\mathbf{x}_u \cdot \mathbf{x}_v|^2}{|\mathbf{x}^o_u|^2 |\mathbf{x}^o_v|^2} + \frac{|\mathbf{x}_v \cdot \mathbf{x}_w|^2}{|\mathbf{x}^o_v|^2 |\mathbf{x}^o_w|^2} + \frac{|\mathbf{x}_w \cdot \mathbf{x}_u|^2}{|\mathbf{x}^o_w|^2 |\mathbf{x}^o_u|^2} \right) d\mathbf{u} \tag{33.52}$$

Derivative norms $|\mathbf{x}^0_u|$, $|\mathbf{x}^0_v|$, and $|\mathbf{x}^0_w|$ are averaged in each element in order to have piecewise constant reference functions, and thus polynomial expressions under the integrals. If the initial grid is not smooth, these derivatives are irregular: the Laplace filter can be applied to them and the criteria σ_{reg} and σ_{ortho} are then computed using these smoothed reference data. Before carrying out the optimization, a visualization of these reference lengths can be useful in order to verify their regularity. As an example, Figure 33.1 shows a plot obtained before the optimization of an O-grid about an airfoil with iso-value lines of the longitudinal reference length drawn; their smoothness ensures that the optimized grid will be regular, whatever the initialization is.

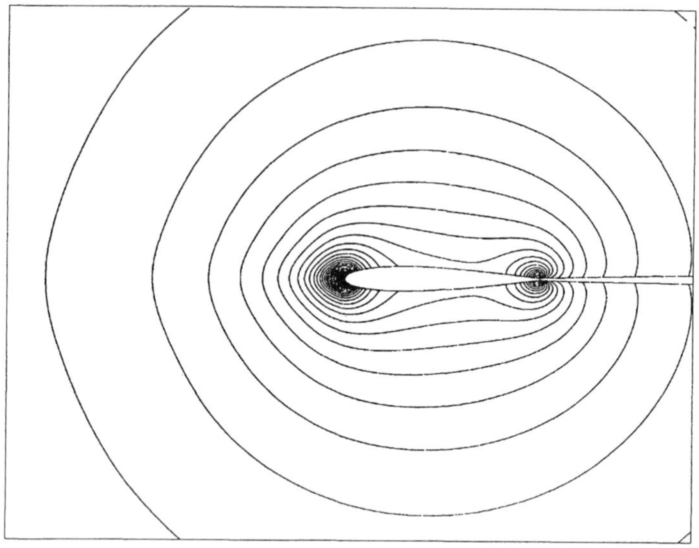

FIGURE 33.1 Visualization of the functional reference parameters before optimization — iso-a contours.

33.4.3 Deformation Formulation

33.4.3.1 Reference Configurations

Until now, the cell deformation measure has been defined with respect to the unit cube. It is, however, possible to generalize this definition and to specify for each mesh cell the dimensions a^e, b^e, and c^e of a reference rectangular parallelepiped with respect to which one desires to compute the cell deformation. In this case the transformation $\mathbf{x}^e(\chi)$ is defined for each cell: it goes from the reference cell $[0,a^e] \times [0,b^e] \times [0,c^e]$ to the current cell and is given by

$$\mathbf{x}^e(\chi) = \begin{cases} \left(1-\frac{\xi}{a^e}\right)\left(1-\frac{\eta}{b^e}\right)\left(1-\frac{\zeta}{c^e}\right)\mathbf{x}^e_{000} + \frac{\xi}{a^e}\left(1-\frac{\eta}{b^e}\right)\left(1-\frac{\zeta}{c^e}\right)\mathbf{x}^e_{100} \\ +\left(1-\frac{\xi}{a^e}\right)\frac{\eta}{b^e}\left(1-\frac{\zeta}{c^e}\right)\mathbf{x}^e_{010} + \frac{\xi}{a^e}\frac{\eta}{b^e}\left(1-\frac{\zeta}{c^e}\right)\mathbf{x}^e_{110} + \left(1-\frac{\xi}{a^e}\right)\left(1-\frac{\eta}{b^e}\right)\frac{\zeta}{c^e}\mathbf{x}^e_{001} \\ +\frac{\xi}{a^e}\left(1-\frac{\eta}{b^e}\right)\frac{\zeta}{c^e}\mathbf{x}^e_{101} + \left(1-\frac{\xi}{a^e}\right)\frac{\eta}{b^e}\frac{\zeta}{c^e}\mathbf{x}^e_{011} + \frac{\xi}{a^e}\frac{\eta}{b^e}\frac{\zeta}{c^e}\mathbf{x}^e_{111} \end{cases} \quad (33.53)$$

As before σ^e is then obtained and the local contribution σ^e is computed by integration over $[0,a^e] \times [0,b^e] \times [0,c^e]$.

As previously mentioned, the minimization of the 2D deformation functional can be interpreted as a LSF of Cauchy–Riemann relations. Introducing $a(u,v)$ and $b(u,v)$ the desired edge lengths (respectively divided by $(\Delta u, \Delta v)$, minimization of I_1–$2J$ and $(J-1)^2$ respectively corresponds to a LSF of

$$\begin{cases} \frac{x_u}{a} - \frac{y_v}{b} = 0 \\ \frac{y_u}{a} + \frac{x_v}{b} = 0 \end{cases} \quad \left(\text{or } \frac{\mathbf{x}_u}{a} = \frac{\mathbf{x}_v}{b} \times \mathbf{k}\right) \quad \text{and} \quad x_u y_v - y_u x_v = ab \quad (33.54)$$

The first set of equations, extention of Cauchy–Riemann relations, means that the cell is only homothetic, but not equal, to the rectangle of sides a and b, and the second equation adjusts the size of the cell so that its volume is ab; combination of both represents an LSF of "the current cell is the rectangle of sides a and b." As another consequence of the linearity of Cauchy–Riemann relations in a and b, it is noteworthy that the relevant value for the reference configuration is the ratio b/a rather than both values a and b, which is equivalent to saying that the current cell, which is looked for when solving the LSF of Cauchy–Rieman relations, is homothetic to a rectangle of side lengths 1 and b/a. The ratio $f = b/a$ that plays an important role here is called the *distortion function.*

Finally, one can note that the extended Cauchy–Riemann relations enable the definition of a distortion function that will have the effect to force the optimized grid to present the refinements found in its initial grid $\mathbf{x}^0$; indeed one can use

$$f = \frac{|\mathbf{x}_v^o|}{|\mathbf{x}_u^o|} \tag{33.55}$$

or use values smoothed by a Laplace filter as before.

33.4.3.2 Toward Conformity and "Exact" Orthogonality in 2D

The minimization of the functionals being an LSF of the relations written just above does not mean that the minimum 0 is necessarily reached and that these relations are necessarily satisfied by the solution; in particular, the mesh may not be orthogonal. For example, if in a rectangular domain of side lengths α and β, one looks for a mesh where all the cells are homothetic to the rectangle of side lengths, a and b constant and independent of ξ and η, Cauchy–Riemann relations can be reached if and only if the following relation is satisfied:

$$\frac{\alpha}{a\ (imax-1)} = \frac{\beta}{b\,(jmax-1)} \tag{33.56}$$

More generally, the existence of such a constraint can be extended for an arbitrary domain: for an arbitrary ratio b/a, Cauchy–Riemann relations cannot be satisfied in general, but there exists a constant μ such that

$$\begin{cases} \mu\dfrac{x_u}{a} - \dfrac{y_v}{b} = 0 \\ \mu\dfrac{y_u}{a} + \dfrac{x_v}{b} = 0 \end{cases} \quad \left(\text{or} \quad \mu\frac{\mathbf{x}_u}{a} = \frac{\mathbf{x}_v}{b} \times \mathbf{k}\right) \tag{33.57}$$

The existence of this so-called conformal module μ is a consequence of the Riemann Mapping Theorem; this parameter depends only on the domain shape and on the distortion function. Going back to the example of the rectangular domain, if the ratio b/a does not satisfy the constraint above, it should be modified and replaced by $\mu b/a$, with μ defined by

$$\mu = \frac{a}{b}\frac{\beta}{\alpha}\frac{imax-1}{jmax-1} \tag{33.58}$$

(If the constraints are satisfied, $\mu = 1$). Another interesting interpretation of the parameter μ can be obtained when looking at the particular case where all cells are first looked for as squares ($b/a = 1$) in a domain with arbitrary shape. In this case μ represents the rectangular aspect of the domain or the domain results from the deformation of the rectangle of side lengths 1 and μ by a conformal transformation. Going back to the mesh obtained, when the number of cells is equal in both directions ($imax = jmax$), all cells are homothetic to the rectangle $1 \times \mu$.

In the general case (arbitrary Ω, a, and b), there is no easy way to obtain μ; however, this parameter satisfies

$$\mu = \tilde{\mu}(\mathbf{x}) \text{ with } \tilde{\mu}^2(\mathbf{x}) = \frac{\int_{\hat{\Omega}} \frac{a}{b} \frac{|\mathbf{x}_v|}{|\mathbf{x}_u|} d\mathbf{u}}{\int_{\hat{\Omega}} \frac{b}{a} \frac{|\mathbf{x}_u|}{|\mathbf{x}_v|} d\mathbf{u}} \tag{33.59}$$

The combination of this relation with the minimization of the conformality functional (σ_{CR}, only based on the Cauchy–Riemann relations $\sigma_{CR}(I_1, J) = I_1 - 2J$), provides a way to construct orthogonal grids in 2D. If we note $\sigma_{CR}(f, .)$ the functional using the distortion function f, a fixed point algorithm enables the adjustment of μ and the computation of the corresponding transformation satisfying the Cauchy–Riemann relations:

1. Choose ε; set $\mu^0 = 1$ and $n \leftarrow 0$
2. $\mathbf{x}^n = ArgMin\ \sigma_{CR}\ (\mu^n f, \mathbf{x})$
3. $\mu^{n+1} = \tilde{\mu}\ (\mathbf{x}^n)$
4. If $|\mu^{n+1} - \mu^n| \geq \varepsilon$, set $n \leftarrow n + 1$ and go to Step 2; otherwise $\mu = \mu^{n+1}$, stop.

Remark 1: This procedure can be possible only if the boundary nodes are allowed to move from their initial position, still remaining along the boundary curve.

Remark 2: Even though orthogonal coordinate system do not seem to exist in three dimensions, one may use an approach similar to the conformal technique just presented to get closer to orthogonality by adjustment of the reference functions; this is presented in Appendix B.

We show in Figures 33.2 to 33.5 how this procedure can be used to construct orthogonal grids that respect certain conditions on the boundaries. The grids are C-grids around an airfoil. Starting from an arbitrary initial grid (Figure 33.2) that is used only to give the overall shape of the domain (here the nodes are uniformly partitioned in arc length on the boundary and the inside grid is constructed by transfinite interpolation from this data on the boundary), the variational method (with $a = b = 1$) combined with the iterative adjustment of μ, constructs an orthogonal grid (thus a mesh function $\mathbf{x}^0(\mathbf{u})$) where all the cells are homothetic to the same rectangle $1 \times \mu$ (Figure 33.3); here μ is close to 2.6. However, the method determines the position of the nodes on the boundaries and it might be interesting (or it is prescribed) to construct the grid with another partition on these boundaries in order to respect special points or desired refinement. The family $\mathbf{x}^0(\mathbf{u})$ found can nevertheless be used.

We suppose that the four boundary sides are paratrametrized by $\mathbf{x}_i(s_i^{max}t)$ $(i = 1,4)$, where s_i^{max} is the length of the boundary arc and t is a parameter varying from 0 to 1, such that $s_i^{max}t$ is the curvilinear arc length. In a first step, we suppose that the respect of the boundary is desired on two *adjacent* sides (sharing a boundary corner), e.g., sides 4 and 1, respectively, corresponding to $u = 0$ and $v = 0$. The prescription of node partition on these sides means that two *distribution functions* $R_4(t)$ and $R_1(t)$ are given and that the desired mesh function $\mathbf{x}(\mathbf{u})$ must verify:

$$\begin{cases} \mathbf{x}(0,v) = \mathbf{x}_4\left(s_4^{max} R_4(v)\right) \\ \mathbf{x}(u,0) = \mathbf{x}_1\left(s_1^{max} R_1(u)\right) \end{cases} \tag{33.60}$$

which is not necessarily the case for $\mathbf{x}^0(\mathbf{u})$; indeed, this mesh function may verify (or at least what it verifies can be written)

$$\begin{cases} \mathbf{x}^0(0,v) = \mathbf{x}_4\left(s_4^{max} R_4^0(v)\right) \\ \mathbf{x}^0(u,0) = \mathbf{x}_1\left(s_1^{max} R_1^0(u)\right) \end{cases} \tag{33.61}$$

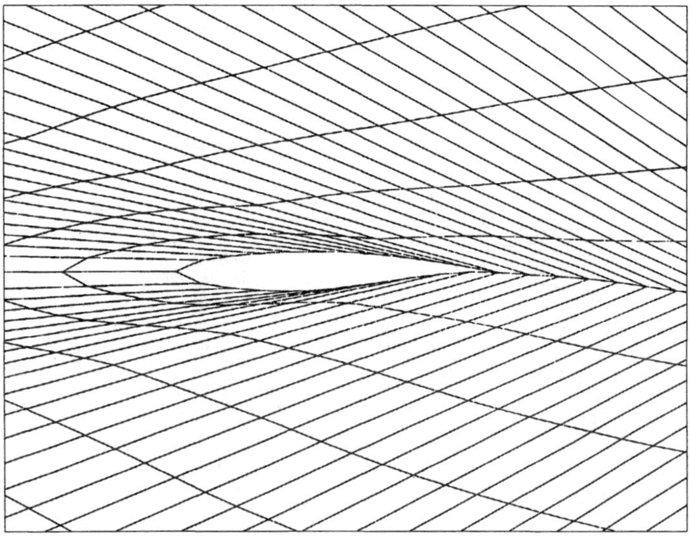

FIGURE 33.2 Initial grid around an airfoil — uniform repartition on the boundaries oand transfinite interpolation.

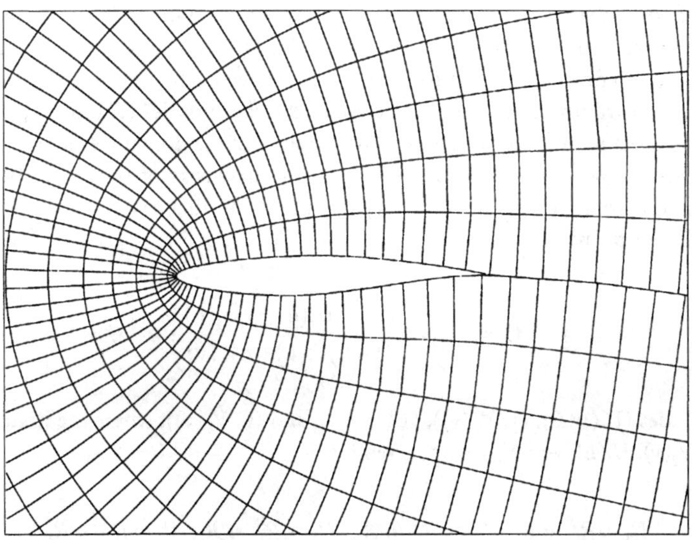

FIGURE 33.3 Conformal grid obtained with $a = b = 1$; $\mu \approx 2.6$.

A property of the distribution functions is that they are *strictly monotonous* from [0, 1] onto [0, 1], so their inverse exists; then one can easily show that the mesh function $\mathbf{x}(\mathbf{u})$,

$$\mathbf{x}(\mathbf{u}) = \mathbf{x}^0\left(R_1^{0-1}\left(R_1(u)\right), \left(R_4^{0-1} R_4(v)\right)\right) \tag{33.62}$$

satisfies the desired refinements on the two boundaries labeled 1 and 4 and is orthogonal. Geometrically, one can say that, starting from the two networks of "parallel" curves obtained by having u and v vary separately in $\mathbf{x}^0(\mathbf{u})$, one constructs two new networks, each of them parallel to the first ones, leading to $\mathbf{x}(\mathbf{u})$; the modification of the spacing between each of the networks is driven on the boundaries by R_1^{0-1} o R_1 and R_4^{0-1} o R_4. Practically, the construction of $\mathbf{x}(\mathbf{u})$ from $\mathbf{x}^0(\mathbf{u})$ amounts to an algebraic transformation

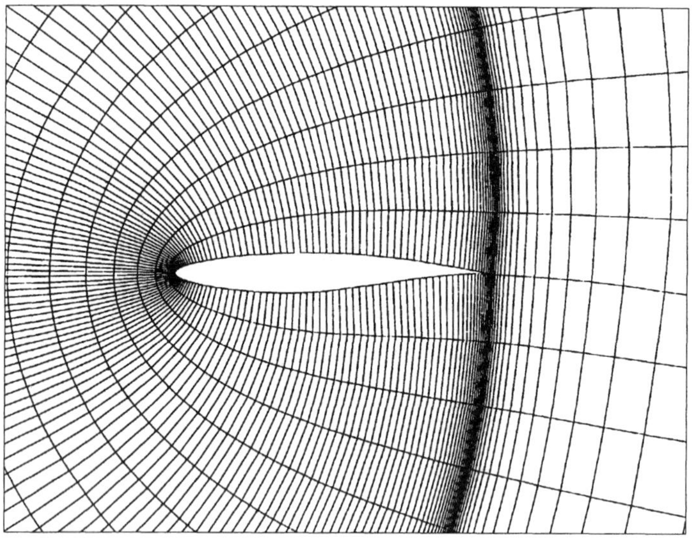

FIGURE 33.4 Orthogonal grid respecting the distribution and the profile and wake.

— it easy to implement and cost-efficient. Figure 33.4 shows the mesh obtained from the previous one (Figure 33.3) prescribing the respect of a node distribution given on the airfoil and the wake so that two nodes (one on each side of the profile) coincide with the leading edge and so that the mesh refines close to this point. Finally, a refinement closed to the profile has been achieved by a last modification using an appropriate node distribution on one of the downstream boundaries (Figure 33.5).

If one wants to respect node distribution on opposite sides of the domain, exact orthogonality cannot be achieved anymore; however, one can use $\mathbf{x}^0(\mathbf{u})$ and the networks of parallel curves it underlies to improve orthogonality, while respecting boundary conditions. In that case, one constructs a grid in the unit square by the interpolation

$$\mathrm{U} = \begin{pmatrix} U \\ V \end{pmatrix} = \begin{pmatrix} (1-v)U_1(u) + vU_3(u) \\ (1-u)V_4(v) + uV_2(v) \end{pmatrix} \tag{33.63}$$

between the four sides $(U_1(u),0)$, $(1, V_2(v))$, $(U_3(u), 1)$, and $(0, V_4(v))$, where the node distributions on the sides, $U_1(u)$, $V_2(v)$, $U_3(u)$, and $V_4(u)$, are given by

$$U_1 = R_1^{0^{-1}} \mathbf{o}\, R_1,\;\; V_2 = R_2^{0^{-1}} \mathbf{o}\, R_2,\;\; U_3 = R_3^{0^{-1}} \mathbf{o}\, R_3,\;\; V_4 = R_4^{0^{-1}} \mathbf{o}\, R_4 \tag{33.64}$$

Then

$$\mathbf{x}(\mathbf{u}) = \mathbf{x}^0(\mathbf{U}) \tag{33.65}$$

satisfies the boundary conditions; however the orthogonality of $\mathbf{x}(\mathbf{u})$ depends on the orthogonality of $\mathrm{U}(\mathbf{u})$. This orthogonality is improved when the conformal grid $\mathbf{x}^0(\mathbf{u})$ is determined using non-constant reference functions a and b, and more astutely functions where the desired distribution on the boundaries have been introduced. The following choice satisfies this prerequisite:

$$\begin{pmatrix} a \\ b \end{pmatrix} = \begin{pmatrix} (1-v)R_1(u)s_1^{\max} + vR_3(u)s_3^{\max} \\ (1-u)R_4(v)s_4^{\max} + uR_2(v)s_2^{\max} \end{pmatrix} \tag{33.66}$$

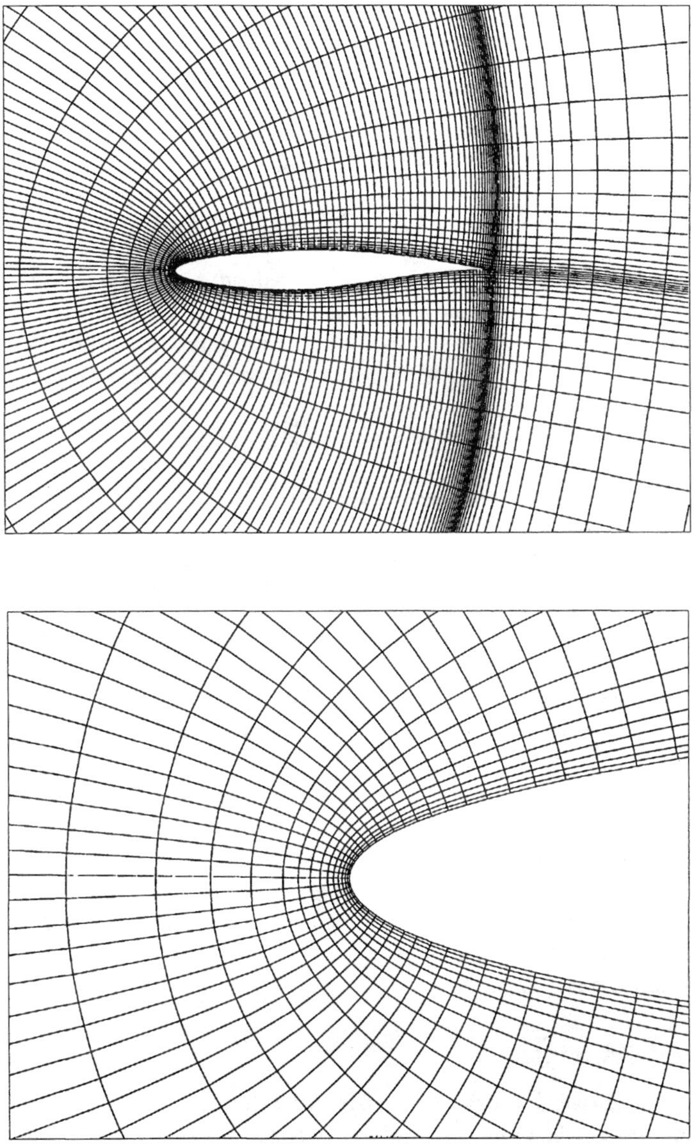

FIGURE 33.5 Orthonal grid respecting the distribution and the profile and wake and refined closed to the profile. (a) View around the profile, (b) close-up on the leading edge.

33.4.4 Summary of the Optimization

So far, we have presented two optimization methods based on the definition of mesh quality measures. These criteria rely on reference data constructed to obtain desired refinements, the method handling then the compromise between respecting these refinements and the overall orthogonality of the grid. It is possible to construct these data using an initial mesh; the corresponding mesh optimisation algorithm writes

1. Construct an initial mesh $\mathbf{M}^0$.
2. Choose an optimization criterion $\sigma(\mathbf{M}^0, .)$.
3. Find $\mathbf{M}_{opt} = \text{ArgMin}\ \sigma(\mathbf{M}^0, \mathbf{M})$.

Smoothness of the node partition in the domain requires the regularity of the reference data. Refer to [10] for details on the implementation of the deformation method and to [10, 9] for an evaluation of the benefits of the optimization in 3D.

33.5 Handling of Adaptation

33.5.1 Introduction

In times where continuous efforts toward efficiency, cost reduction, and budget optimization are made, mesh adaptation has become a major subject of interest and investigation, since the location of grid nodes largely dictates the level of accuracy achievable for a given problem, and since their number directly determines the computational cost. In order to properly master adaptation, three questions need to be answered: where, how, and when to adapt? By "where to adapt," we mean that it is important to know what is the best possible location for the nodes to obtain the optimal accuracy of the result. Once this location is known, one answers the second question by use of an appropriate mesh generation technique capable of taking this location into account and places the nodes accordingly. Finally, the last question concerns the coupling between the flow solver and the adaptation process and asks how coupled the codes should be. For example, what is the optimal level of convergence of the flow solution before any mesh adaptation: full convergence, one order of magnitude before convergence, or a few iterations. How does one have to couple the codes and include the node displacements into the flow equations?

Here we concentrate on the second of these questions (how?) and we indicate how optimization presented above can be modified and used to construct adapted grids. Of course, once a grid has been adapted, it is used for a new computation, so its purely geometric quality must be ensured. Therefore, the process that leads to an adapted grid must improve grid quality, or at least preserve it, if optimization has already been performed.

As described above, the optimization methods tend to construct orthogonal grids with edge lengths as close as possible to given reference lengths; they rely on the construction of criteria defined using these given reference lengths. A first way to obtain adapted grids is to introduce appropriate reference lengths in the criteria. In order to preserve the geometric quality of the grids, smoothness of these adaptation data will of course be required. Another way consists in defining appropriate adaptation criteria. Both ways are described in the next paragraphs.

33.5.2 General Principle

Adaptation is most often driven by an adaptation parameter that represents the ratio between the desired adapted size of the cell and the original one; if this parameter is smaller than 1, the cell size tends to decrease and the mesh is locally refined. By size, lengths, areas, or volumes can be considered, but lengths are the most interesting to use, because this enables *anisotropic* adaptation where a cell can be refined or unrefined in function of the direction.

Both methods presented in the previous sections rely on reference lengths: this implies that anisotropic adaptation can be handled by them. To do so, one introduces for each cell e three adaptation weights ω_a^e, ω_b^e, and ω_c^e associated to each index direction; the adaptated mesh is obtained by optimization of the functional constructed with reference data $(a^e\omega_a^e, b^e\omega_b^e, c^e\omega_c^e)$ for the cell e. Of course the smoothness of the node partition is ensured by the regularity of these new reference data. Once again, a visualization of the modified reference lengths can be useful before carrying out the adaptation. For the example described in Section 33.4.3.1 (an O-grid about an airfoil), a plot similar to that of Figure 33.1 is shown in Figure 33.6 for the field $a\omega_a$, where a is the reference length in the longitudinal direction. Here an initial mesh (Figure 33.7) is used to compute a solution represented by iso-value lines in Figure 33.8, the field used for this representation is the one used for the evaluation of the weight ω_a described below (Section 33.5.3). Figure 33.9 shows the resulting adapted grid.

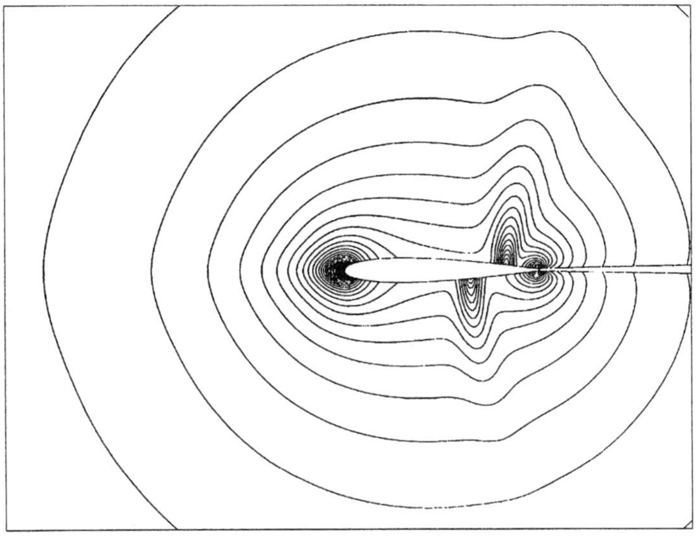

FIGURE 33.6 Visualization of the functional reference parameters before adaptation — iso-$\omega_a a$ contours.

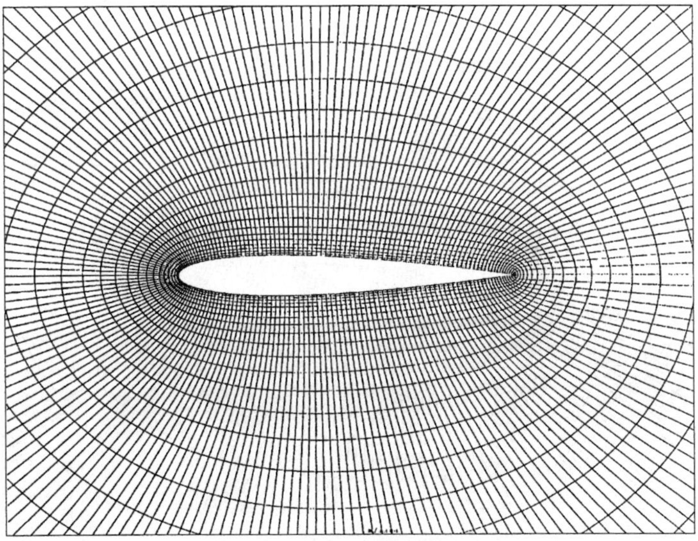

FIGURE 33.7 Initial grid before adaptation.

The adaptation algorithm is as follows:

1. Construct an initial mesh $\mathbf{M}^0$ and compute a solution Φ^0.
2. Choose an adaptive optimization criterion $\sigma(\mathbf{M}^0, \Phi^0, \cdot)$.
3. Find $\mathbf{M}_{adapt} = \text{ArgMin } \sigma(\mathbf{M}^0, \Phi^0, \mathbf{M})$.

It might be interesting to loop on the adaption process; in this case the adaptation loop writes

1. Choose an adaptive optimization criterion $\sigma(\mathbf{M}, \Phi, \cdot)$; set $n \leftarrow 0$.
2. Construct the mesh $\mathbf{M}^n$ and compute a solution Φ^n.
3. Find $\mathbf{M}^{n+1} = \text{ArgMin } \sigma(\mathbf{M}^n, \Phi^n, \mathbf{M})$.
4. If adaption is not satisfactory, set $n \leftarrow n + 1$ and go to Step 2; otherwise, $\mathbf{M}_{adapt} = \mathbf{M}^{n+1}$, stop.

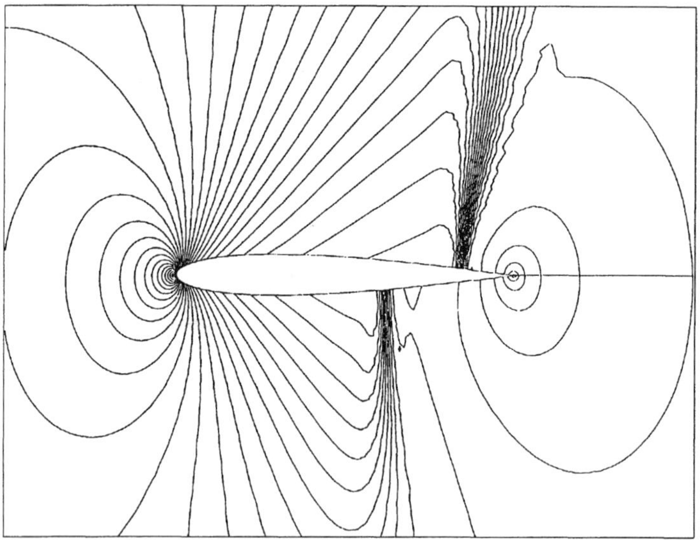

FIGURE 33.8 Solution used to determine the adaptation weights.

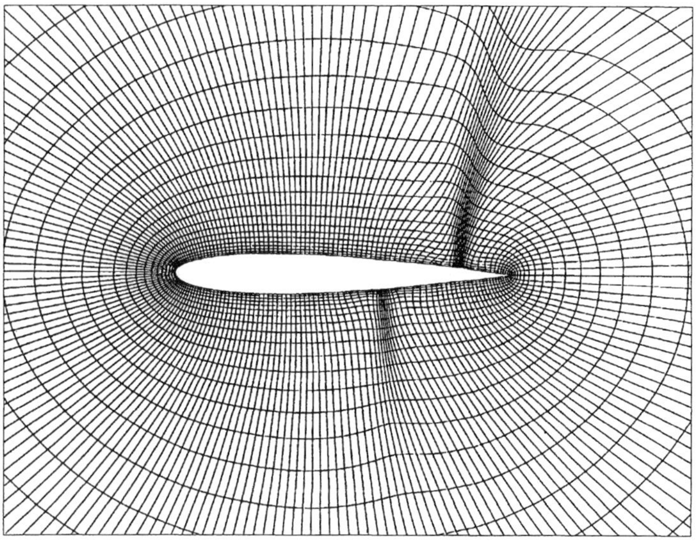

FIGURE 33.9 Adapted grid.

The dependence of σ on $\mathbf{M}^0$ corresponds to the determination of the reference data set (a^e, b^e, c^e) from the initial mesh as described in Section 33.4; similarly, the dependence of σ on Φ^0 corresponds to the determination of the adaptation data set $(\omega_a^e, \omega_b^e, \omega_c^e)$ from the initial solution. Thus, in order to fully use the possibility to perform anisotropic adaptation, ways to extract this three-dimensional field from a solution must be put forward; we indicate here two ways.

33.5.3 Use of Error Indicators

The most commonly used strategy in structured grid adaptation is based on the use of error indicators, rough information about the possible location of the error; they are generally determined as gradients of a variable characteristic of the physical solution and, in that case, the adaptation refines the grid where large variations of this variable occur. This variable can be, for instance, the displacement, the strain or the stress in solid mechanics, the pressure, the Mach number, or the density in fluids; usually, it is up to the user to choose the variable, but programs should be written in such a way that, knowing which variable will drive the adaptation, the user is able to select it easily in the grid generation code, which then performs the adaptation without any new user input.

Within our general framework, the adaptation data are computed according to the following steps.

1. The variable gradient components are computed and normalized between 0 and 1:

$$\theta = \frac{|\nabla\Phi| - |\nabla\Phi|_{min}}{|\nabla\Phi|_{max} - |\nabla\Phi|_{min}} \tag{33.67}$$

   The gradient considered can be

   a. Either the real physical gradient in the cell edge directions ($\nabla_a\Phi = lim_{\Delta a \to 0}(\Phi_{i+1} - \Phi_i)/\Delta a$),
   b. or logical gradient in the mesh line directions, where only differences are considered ($\nabla_i\Phi = \Phi_{i+1} - \Phi_i$); the latter one corresponds to a variation per cell in each index direction and turns out to be more suitable.

2. These normalized gradients are then modified to sharpen the adaptation area: a threshold θ_{th} is chosen between 0 and 1; above (respectively below) this value, the gradient is set to 1 (resp. to 0):

$$\tilde{\theta} = 1 \text{ if } \theta \geq \theta_{th}; \quad \tilde{\theta} = 0 \quad \text{otherwise} \tag{33.68}$$

   Alternatively, one can fix a number N(or percentage N/N_{tot}) of cells; the gradient of the N cells with highest values will be set to 1. This procedure also allows emphasis of weak phenomena.

3. For each component, adaptation parameters are calculated as a linear function of $\tilde{\theta}$: this linear function is determined by the refinement ($\omega_0 < 1$) or enlargement ($\omega_1 > 1$) parameters:

$$\omega_\alpha = \omega_1\left(1 - \tilde{\theta}_\alpha\right) + \omega_0\tilde{\theta}_\alpha \quad \text{for} \quad \alpha = a, b, c \tag{33.69}$$

   Of course ω_0 and ω_1 cannot be chosen independently. The volume of the domain remaining the same before and after adaptation, one has

$$\sum_e a^e\, b^e\, c^e = \sum_e \omega_a^e a^e\, \omega_b^e b^e\, \omega_c^e c^e \tag{33.70}$$

   A practical way to satisfy this volume constraint consists in choosing the ratio ρ between the enlargement and refinement parameters

$$\rho = \omega_1/\omega_0 > 1 \tag{33.71}$$

   and calculate by the following equality deduced from the volume equation and the definition of the adaptation parameters

$$\omega_0^{\,3} = \frac{\sum_e a^e b^e c^e}{\sum_e a^e b^e c^e \left[1 - \tilde{\theta}_a + \rho\tilde{\theta}_a\right]\left[1 - \tilde{\theta}_b + \rho\tilde{\theta}_b\right]\left[1 - \tilde{\theta}_c + \rho\tilde{\theta}_c\right]} \tag{33.72}$$

4. The adaptation reference data ($a^e\omega_a^e$, $b^e\omega_b^e$, $c^e\omega_c^e$) are smoothed to ensure the smoothness of the adaptated mesh.

This methodology requires four choices: the adaptation variable, the type of gradient, the number or percentage of cell in Step 2, and ρ. Here also, these choices should be easy to make in an adaptation module, and any interactive way to select or pick parameter values is of most interest.

33.5.4 Use Of Error Estimators

The estimation of errors started in the 1950s by so-called *a priori* estimates that were derived as soon as algorithms to solve PDEs numerically were developed; they were aimed at proving that these algorithms were robust, in the sense that they led to approximate solutions converging indeed towards the exact solution when one refines the mesh. An *a priori* estimation can be written (with obvious notations)

$$\left\| u - u^h \right\| \leq F(h,u) \tag{33.73}$$

This estimate is a function of the unknown u and cannot be explicitly evaluated. For example, for the classical Laplace operator ($-\Delta u = f$ with Dirichlet boundary conditions), solved with a finite element method (FEM), this estimate is

$$\left| u - u^h \right|_{H^1(\Omega)} \leq C h^s \|u\|_{H^{s+1}(\Omega)} \tag{33.74}$$

where u belongs to $H^{s+1}(\Omega)$ ($0 \leq s \leq k$) and k is the degree of the approximation in the elements ($k \geq 1$).

Conversely, an *a posteriori* estimate involves the numerical solution of the problem, which is known after a computation, and can be written

$$\left\| u - u^h \right\| \leq G\left(h, u^h\right) \tag{33.75}$$

They can only be evaluated once the computation is made and measure the quality of the solution. These kinds of estimates appeared more recently, in the late 1970s, with Babuska's works [1], later extended by Oden [13], Bank [2] and others in the late 1980s [5]. All these researchers are FEM specialists, and therefore these *a posteriori* estimates are always derived for this type of discretization. Because of the generality of this type of discretization, these theories have been applied when the power of the FEM can be most taken advantage of, that is, with unstrutured meshes; of course, it has been so for "classical" unstructured conformal triangular (or tetraedrical) meshes, but also for more peculiar types of mesh, such as the ones introduced by Oden [13].

In order to be useful for mesh adaptation, *a posteriori* error estimates should also be able to give information on the localization of the error; expressions such as the previous one are global and do not give this information. However, η, the second member in the *a posteriori* inequality, can often be written as

$$\eta = \left(\sum_e \eta^e \right)^{1/2} \tag{33.76}$$

with

$$C\eta^e \leq \left\| u - u^h \right|_{\Omega^e} \right\| \leq \eta^e$$

which provides necessary and sufficient information to perform adaptation. Indeed, both inequalities are important: the upper bound ensures that the numerical solution is obtained everywhere within a prescribed tolerance, making sure that refinements (or coarsenings) are sufficient; the lower bound enables the optimization of the adaptation, making sure that refinements are necessary, but not too excessive.

These local estimations can be used to elaborate an adaptation strategy based on the principle of equidistribution of the error [7, 15]: in one dimension a point distribution is set so as to make the product of the spacing in the adapted mesh and the error constant over the points:

$$\eta\,\Delta x_{adapt} = Constant \tag{33.77}$$

This basic equation is modified in the following way to obtain expressions applicable in three dimensions:

1. The initial mesh on which the error is computed is taken into account:

$$\eta\,\frac{\Delta x_{adapt}}{\Delta x_{init}} = Constant \tag{33.78}$$

   which becomes

$$\eta\,\frac{V_{adapt}}{V_{init}} = C_1 \tag{33.79}$$

   in three dimensions.
2. Since the error can be 0 in certain elements, it is modified, and a strictly positive quantity is considered:

$$\tilde{\eta} = 1 + C_2\eta \quad \text{or} \quad \tilde{\eta} = 1/\left(1 + C_2'\eta\right) \tag{33.80}$$

   and the equidistribution is considered for $\tilde{\eta}$ instead of η; this leads to

$$\frac{V_{adapt}}{V_{init}} = \frac{C_1}{\left(1 + C_2\eta\right)} \quad \text{or} \quad \frac{V_{adapt}}{V_{init}} = C_1\,\left(1 + C_2'\eta\right) \tag{33.81}$$

At this point the strategies based on the use of error indicators (Section 33.5.3) and error estimators merge, since we can write

$$\frac{V_{adapt}}{V_{init}} = \omega \tag{33.82}$$

with

$$\omega = \omega_a \omega_b \omega_c \tag{33.83}$$

in the first approach, and

$$\omega = \frac{C_1}{(1 + C_2\eta)} \quad \text{or} \quad \omega = C_1(1 + C_2'\eta) \tag{33.84}$$

for the second. Equivalently, the use of the directional weights ω_α introduced in Section 33.5.3 as a function of a gradient, leading to a volume term $\omega(\nabla\Phi)$ by multiplication of the three contributions, is equivalent to the equirepartition of the error η estimated by

$$\omega(\nabla\Phi) = \frac{C_1}{(1 + C_2\eta)} \quad \text{or} \quad \omega(\nabla\Phi) = C_1(1 + C_2'\eta) \tag{33.85}$$

However, using error indicators (Section 33.5.3), it was possible to identify in the weight three directional contributions. A source of directional information needs to be chosen: a variable is selected (once again, displacement, strain or stress in solid mechanics, pressure, Mach number or density in fluids, but this can be the error η itself) and its normalized gradient (n_a, n_b, n_c) written in a basis linked to the initial cell is used to split the global weight in three by

$$\omega_\alpha = \omega ** n_\alpha^2 \quad \text{for} \quad \alpha = a, b, c \tag{33.86}$$

The weights satisfy

$$\omega_a \omega_b \omega_c = \omega \tag{33.87}$$

and when large variations occur in one direction, this choice of splitting performs the adaptation on the corresponding edge.

The constants C_1 and C_2 (or C_2') are then adjusted in a similar way to what is done using error indicators (Step 3): C_1 and $C_1/(1 + C_2\eta_{max})$ (or $C_1(1 + C_2\ \eta_{max})$ represent volume enlargement and refinement parameters; one of them (or their ratio) can be chosen and the other one is computed by volume consideration. It turns out that the use of C_2' instead of C_2 leads to explicit expressions and in more convenient to select because of this last step.

33.5.5 Formulation Using Volume Integral

Following the introduction of the functionals put forward by Brackbill and Saltzman [3] in Section 33.2.4, we mention that concentration of grid points can be obtained by minimizing the integral I_w defined as

$$I_w = \int_\Omega \eta^2 J\, dx \tag{33.88}$$

where η is a weight function that can be a measure of the error; the minimization of this functional causes the cells to be small where this weight is large and is equivalent to finding a grid realizing the equi-repartition of ηJ, which is the principle used to introduce the error estimations as a driver for the adaptation in the previous paragraph. Once again the integral I_w cannot introduce anisotropy in the adaptation and turns out not to be very useful.

33.6 Optimization Algorithm

33.6.1 General Algorithm

Practically, the different functionals defined above can be written as polynomials of the mesh node coordinates. The degree of the polynomial is different depending on the contribution:

–degree 2 for I_{reg} and 4 for I_{ortho},

–degree 2 for I_{def} in 2D with $\sigma = I_1 - 2J$ and 4 with $\sigma = (J-1)^2$,

–degree 4 for I_{def} in 3D with $\sigma = I_1 + I_2 - 6J$,

but 4 is the maximun degree for any complete functional.

A conjugate gradient algorithm is used for the minimization of the polynomial. It can be written as:

Initialization: Choose X_0(the initial mesh), $\varepsilon, H_0 = 0$

Iteration loop:

1. $G_n = D\sigma(X_n)$
2. $\lambda_n = (G_n - G_{n-1}) \cdot G_n / G_{n-1}^2$
3. $H_n = G_n + \lambda_n H_{n-1}$
4. **Descent Step**: Find $\rho_n = ArgMin\, \sigma(X_n + \rho H_n)$
5. $X_n = X_{n-1} + \rho_n H_n$
6. if $|H_n| > \varepsilon$ go to 1.

This algorithm essentially requires the computation of the functional derivatives with respect to the node coordinates at the current configuration $D\sigma(X_n)$. The gradient represents a set of vectors, each of them associated to a node and are obtained by summation of contributions coming from the differents to which the node belongs:

$$D\sigma|_n = \sum_{e/n \in e} D\sigma^e|_n \tag{33.89}$$

At each iteration the minimization of the polynomial

$$P(\rho) = \sigma(X + \rho H) \tag{33.90}$$

must be performed; however, the solution of this one-dimensional minimization problem is also solution of

$$P'(\rho) = D\sigma(X + \rho H) \cdot H = 0 \tag{33.91}$$

Since σ is a polynomial of degree 4 of the mesh node coordinates, $P(\rho)$ has the same degree and $P'(\rho)$ is of degree 3, leading to 3 roots for the equation $P'(\rho) = 0$.

In the general case, these three roots must be checked to determine which one leads to the minimum of $\sigma(X_n + \rho H_n)$. However, an important and major difference between the variational methods presented in this chapter appears in this optimization algorithm, and shows the benefit of the mathematical background, and more precisely the convexity condition introduced in the definition of the deformation formulation: the polynomial $P'(\rho)$ has three roots, but when the convexity condition is prescribed, out of these three roots, only one is real — the other two are complex conjuguate. The descent step simply is

Descent Step: Find ρ_n solution of $D\sigma(X_n + \rho H_n) \cdot H_n = 0$

This remark shows that the calculation of the functional is in fact never required, which makes the deformation method very efficient and easy to implement.

33.6.2 Handling of Conditions on the Boundary

The conjugate gradient algorithm that is used to minimize the functionals can be interpreted as an iterative calculation of displacements that pull the nodes from their initial positions at one step to their position at the next step. Nodes that are interior to the domain are free to move and are driven by the value of the functional gradient. On part of the boundary whose shape does not matter, one may want to let the nodes entirely free: in that case, the algorithm optimizes the boundary shape and the domain deforms as iterations go (the volume control term prevents from any catastrophic blow up or collapse). One can also fix a set of boundary nodes: this is done by simply fixing to 0 the corresponding displacement vector in the algorithm.

Between these two types of conditions, one can consider an intermediate condition where the nodes are able to move on a given curve of surface (Σ): taking the boundary geometry into account is possible but makes the functional lose its polynomial nature, complicating the exact calculation of the gradient and the descent step. This difficulty is overcome by a *projected* conjugate gradient algorithm that locally linearizes the problem or, equivalently, supposes that the boundary is plane. The descent direction is projected on the curve or surface tangent to the boundary ($T(\Sigma)$) and the node position is projected on the prescribed curve or surface. The descent (3) and updating (5) steps are:

$$\text{3.a. } H_{n-1/2} = G_n + \lambda_n H_{n-1}$$

$$\text{3.b. } H_n = \Pi_{T(\Sigma)}\left(H_{n-\frac{1}{2}}\right)$$

$$\text{5.a. } X_{n-\frac{1}{2}} = X_{n-1} + \rho_n H_n$$

$$\text{5.b. } X_n = \Pi_{\Sigma}\left(X_{n-\frac{1}{2}}\right)$$

At equilibrium (convergence of the algorithm), the gradient is not zero but is orthogonal to the boundary; therefore, the node does not move anymore.

This type of boundary condition turns out to be indispensable, both for mesh quality enhancement and for adaptation: first it improves the orthogonality across a boundary (indeed it has been shown that exact orthogonality can be reached in a continuous approach [8]); second, it allows the capture of phenomena attached to a boundary — shocks in particular. Once again, we refer to [9] for a thorough study of the benefits of this kind of boundary condition in the case of the adaptation to the shocks developing in the flow around the ONERA M6 wing.

33.6.3 Handling of Multidomain Topologies

The multidomain approach consists in decomposing the global computational domain into sub-domains, each meshed with a structured grid. Several decomposition topologies can be adopted: with or without subdomain overlapping, with or without node coincidence at subdomain interfaces. The variationals methods presented here with the discrete approach are able to handle decompositions with node coincidence, presenting the stiffest constaints, and respect this coincidence as iteration go, that is to say, keep the nodes belonging to different subdomains (and thus topologically different) at the same relative position (geometrical identity).

As mentioned in Sections 33.2.3 and 33.3.4, the global functional associated to one subdomain is obtained by summation of elementary contributions. In the multidomain case, summation over each subdomain is then performed:

$$\sigma = \sum_{e \in \text{Mesh}} \sigma^e = \sum_{SubDomain \in \text{Mesh}} \left[\sum_{e \in SubDomain} \sigma^e \right] \tag{33.92}$$

For each node, the assembly process consists in adding the gradients associated to this node in each element. In the multidomain approach, the assembly process can be extended to the case where a node belongs to several subdomains. The gradient component associated to an interface node will be evaluated by summation of the gradients computed in all subdomains to which it belongs:

$$D\sigma|_n = \sum_{SubDomain \in \text{Mesh}} \left[\sum_{\substack{e \in SubDomain \\ \text{such that } n \in e}} D\sigma^e|_n \right] \tag{33.93}$$

The minimum information needed for a general multidomain topology consists, for each node located on an interface, in the number of subdomains to which it belongs, and for each of them, the indices of the node in this subdomain.

Remark: This general assembly process enables the handling of so-called multiple nodes in a unique domain, that are, for example, the nodes on the wake in a C-mesh around a profile. The multidomain algorithm is necessary to treat this kind of topology if one wants to optimize the position of such nodes.

Here also, this type of treatment turns out to be indispensable, both for mesh quality enhancement and for adaptation: it enables the optimization of the mesh inside a multidomain topology; in that case the interface shape is optimized. It also enables a proper adaptation to phenomena that occur close to or across a domain interface.

33.7 Extension to Unstructured Meshes

As mentioned in the introduction, the discrete approach chosen in this chapter enables the extension of the variational methods to unstructured meshes: first, most of the criteria can be defined for any type of element (quads or triangles, hexahedra or tetrahedra), the only criterion being meaningless is the orthogonality; second, the principle of summation of elementary contributions can be used, whatever the mesh topology is, structured or unstructured. So the methodology developed in the previous sections can be applied to any kind of mesh, knowing that it requires that the topology remains fixed, which means that the connexions between the nodes remains the same; however, we recall that one of the most important advantages of unstructured meshes with respect to structured grids is the flexibility provided by their arbitrary topology. Therefore, this optimization and variational strategy based on a fixed topology can be useful when one judges too high the cost implied by modification of the topology, and/or when sufficient improvement can be achieved just by displacement of the nodes.

33.7.1 Regularity Criterion

In a way similar to the construction done in Section 33.2.2, one defines the local regularity measure by summation of the squared edge lengths:

$$\sigma^e_{reg} = \sum_{i \text{ edge of } e} \left|\mathbf{r}^e_i\right|^2 \tag{33.94}$$

By summation of these elementary contributions, one obtains a global measure where each edge is counted as many times as the number of elements to which it belongs ($\eta(i)$):

$$\sigma_{reg} = \sum_{e} \sum_{i \text{ edge of } e} \left|\mathbf{r}^e_i\right|^2 = \sum_{\text{edges "}i\text{"}} \eta(i) \left|\mathbf{r}_i\right|^2 \tag{33.95}$$

with

$$\eta(i) = \sum_{e/i \text{ edge of } e} 1 \tag{33.96}$$

This functional tends to make uniform the node partition inside the domain. In order to obtain refinements, reference lengths a^e_i are introduced:

$$\sigma^e_{reg} = \sum_{i \text{ edge of } e} \frac{\left|\mathbf{r}^e_i\right|^2}{a^{e2}_i} \tag{33.97}$$

These reference lengths could be the length of the edges in the initial mesh; however, the corresponding global functional would then be minimum for the initial mesh. It is preferable to use an elementary reference length a^e

$$\sigma_{reg}^{e} = \frac{1}{a^{e2}} \sum_{i\ edge\ of\ e} \left|\mathbf{r}_i^e\right|^2 \tag{33.98}$$

then

$$\sigma_{reg} = \sum_{e} \frac{1}{a^{e2}} \sum_{i\ edge\ of\ e} \left|\mathbf{r}_i^e\right|^2 = \sum_{edges\ "i"} \tilde{\eta}(i) \left|\mathbf{r}_i\right|^2 \tag{33.99}$$

with

$$\tilde{\eta}(i) = \sum_{e/i\ edge\ of\ e} \frac{1}{a^{e2}} \tag{33.100}$$

This approach has the advantage of automatically averaging the contributions from the elements surrounding each edge through the term $\tilde{\eta}(i)$. Note that one parameter is sufficient to scale the local function. This parameter can be a length, as presented, or an area/volume, from which the length is deduced by a root.

The mechanical interpretation of the method in terms of springs linking the nodes is still valid.

33.7.2 Deformation Formulation

In the unstructured case, the deformation is evaluated with respect to a reference element that is a triangle or a tetrahedron: an equilateral element is chosen and, once again, one parameter — its edge length or a reference volume — is sufficient to scale the local contribution. The same procedure is used to compute the measure of a current cell with respect the equilateral reference element, here $\mathbf{x}^e(\chi)$ is linear. Once again the rubber membrane anology can be given to interpret the method in mechanical terms. This approach has been successfully implemented by Cabello [4].

33.7.3 Adaption

The ingredients put forward for the adaptation of structured grids can be used in the unstructured case through the reference area/volume introduced above: an adaptive optimization is performed using $\omega^e V^e$ as reference cell for the element e where V^e is the volume of the cell in the initial mesh mesh, and ω^e the volumetric adaptation weight as introduced in Section 33.5. Due to the uniqueness of the parameter scaling the elementary contribution, one cannot achieve anisotropy of the adaptation.

33.8 Summary and Research Issues

We have presented a discrete approach of variational methods for mesh optimization and adaptation; this kind of approach enables the geometrical construction of local, then global grid quality measures on which the optimization relies. It also leads to mechanical interpretation of the method. The "finite element" nature of the method shown in the use of summation over elementary contributions, combined with the conjugate gradient algorithm used to obtain the optimal grids, make these methods easy to computationally implement in structured, multidomain, unstructured, or hybrid topologies; this represents a major advantage of these methods in times where it appears that compromises taking advantage of the different kinds of mesh topology must be made.

More specifically, one of the methods presented merges in other continuous formulations and the other one, deduced from principles of mechanics, gives well-known geometrical results back. The first one has been introduced in commercial codes, while the second one has drawn the attention of researchers, attracted by its sound mathematical basis, to improve its computational performance.

These variational methods have all the same limits due to the rigidity of the topology that needs to remain fixed; however, they are one of the bricks necessary in any mesh generation code to improve geometrical property and overall quality of the grids or meshes. Concerning their adaptation capability, they enable the movement of nodes toward regions of interest, once one knowns where these regions are; for this reason, their presence as a necessary brick in any code is once again justified. We have presented ways to locate these interesting areas that are well suited to the node displacement procedure, but research is being done to improve this determination; its results will have direct repercussions on the use of these variational methods.

As mentioned before, another challenge arising as soon as one faces adaptation is the optimization of the adaptation process and the search of the best way to couple the grid generator and the solver: in the iterative or unsteady solution of steady problems, or in the solution of unsteady problems, a new parameter — a (pseudo-)time — must be dealt with, not only for the physical problem of interest, but also for the mesh. Current research tackles this issue.

Finally, note that in both of these subjects of research, answers to the questions labeled above as *where* and *when* to adapt are not specific to the variational methods, but are general and concern all the grid and mesh generation community.

Appendix A

We have the following corresponding expressions for a 3 × 3 matrix **A** and its cofactor matrix *Cof* **A**:

$$\mathbf{A} = \begin{vmatrix} a & d & g \\ b & e & h \\ c & f & i \end{vmatrix} \;;\; Cof\,\mathbf{A} = \begin{vmatrix} ei - fh & hc - ib & bf - ce \\ fg - di & ia - gc & cd - af \\ dh - eg & gb - ha & ae - bd \end{vmatrix}$$

We have the following properties:

$$\mathbf{A} \cdot Cof\ \mathbf{A}^{\mathrm{T}} = Cof\ \mathbf{A}^{\mathrm{T}} \cdot \mathbf{A} = (\det \mathbf{A})\mathbf{Id}$$

and therefore for an invertible matrix **A**,

$$\mathbf{A}^{-1} = Cof\ \mathbf{A}^{T} / \det \mathbf{A}$$

The second invariant I_2 of **A** satisfies

$$I_2 = trace\ Cof\ \mathbf{A} = \left(trace(\mathbf{A}^2) - trace(\mathbf{A})^2\right)/2$$

The matrix can also be written considering its three column vectors (**A**=[**u**, **v**, **w**]). We then have

$$Cof\ \mathbf{A} = [\mathbf{u} \times \mathbf{v}, \mathbf{v} \times \mathbf{w}, \mathbf{w} \times \mathbf{u}]$$

Thus for the gradient of a transformation $\mathbf{x}(\chi)$, we have

$$\mathbf{F} = \nabla\mathbf{x} = \begin{bmatrix} x_\xi & x_\eta & x_\zeta \\ y_\xi & y_\eta & y_\zeta \\ z_\xi & z_\eta & z_\zeta \end{bmatrix} = \left[\mathbf{x}_\xi\,,\ \mathbf{x}_\eta\,,\ \mathbf{x}_\zeta\right]$$

$$Cof\,\mathbf{F} = \begin{bmatrix} y_\eta z_\zeta - z_\eta y_\zeta & y_\zeta z_\xi - z_\zeta y_\xi & y_\xi z_\eta - z_\xi y_\eta \\ z_\eta x_\zeta - x_\eta z_\zeta & z_\zeta x_\xi - x_\zeta z_\xi & z_\xi x_\eta - x_\xi z_\eta \\ x_\eta y_\zeta - y_\eta x_\zeta & x_\zeta y_\xi - y_\zeta x_\xi & x_\xi y_\eta - y_\xi x_\eta \end{bmatrix} = \left[\mathbf{x}_\eta \times \mathbf{x}_\zeta\ ,\ \mathbf{x}_\zeta \times \mathbf{x}_\xi\ ,\ \mathbf{x}_\xi \times \mathbf{x}_\eta\right]$$

and therefore,

$$\mathbf{F} = Cof\,\mathbf{F} \Leftrightarrow \begin{cases} \mathbf{x}_\xi = \mathbf{x}_\eta \times \mathbf{x}_\zeta \\ \mathbf{x}_\eta = \mathbf{x}_\zeta \times \mathbf{x}_\xi \\ \mathbf{x}_\zeta = \mathbf{x}_\xi \times \mathbf{x}_\eta \end{cases}$$

Appendix B

As in two dimensions, it is possible to introduce functions a, b, and c defining the dimensions of the reference parallelepiped (up to within a division by Δu, Δv, and Δw); the minimization of $I_1 + I_2 - 6J$ is then equivalent to a LSF of the generalized and extended Cauchy–Riemann relations:

$$\begin{cases} \dfrac{\mathbf{x}_u}{a} = \dfrac{\mathbf{x}_v}{b} \times \dfrac{\mathbf{x}_w}{c} \\ \dfrac{\mathbf{x}_v}{b} = \dfrac{\mathbf{x}_w}{c} \times \dfrac{\mathbf{x}_u}{a} \\ \dfrac{\mathbf{x}_w}{c} = \dfrac{\mathbf{x}_u}{a} \times \dfrac{\mathbf{x}_v}{b} \end{cases}$$

In a way similar to the 2D conformal approach, we relax this system and introduce three parameters λ, μ, and ν constant in the domain, such that

$$\begin{cases} \dfrac{\mathbf{x}_u}{\lambda a} = \dfrac{\mathbf{x}_v}{\mu b} \times \dfrac{\mathbf{x}_w}{\nu c} \\ \dfrac{\mathbf{x}_v}{\mu b} = \dfrac{\mathbf{x}_w}{\nu c} \times \dfrac{\mathbf{x}_u}{\lambda a} \\ \dfrac{\mathbf{x}_w}{\nu c} = \dfrac{\mathbf{x}_u}{\lambda a} \times \dfrac{\mathbf{x}_v}{\mu b} \end{cases}$$

From these equations it is possible to obtain symetrical expressions for λ, μ, and ν with the form

$$\lambda^4(\mathbf{x}) = \frac{\displaystyle\int_{\hat\Omega} \frac{b}{ac} \frac{|\mathbf{x}_u \times \mathbf{x}_w|}{|\mathbf{x}_v|} d\mathbf{u}}{\displaystyle\int_{\hat\Omega} \frac{ac}{b} \frac{|\mathbf{x}_v|}{|\mathbf{x}_u \times \mathbf{x}_w|} d\mathbf{u}} \cdot \frac{\displaystyle\int_{\hat\Omega} \frac{c}{ab} \frac{|\mathbf{x}_u \times \mathbf{x}_v|}{|\mathbf{x}_w|} d\mathbf{u}}{\displaystyle\int_{\hat\Omega} \frac{ab}{c} \frac{|\mathbf{x}_w|}{|\mathbf{x}_u \times \mathbf{x}_v|} d\mathbf{u}}$$

For constant reference functions ($a = b = c = 1$), these three parameters measure the parallelepiped appearance of the domain. A fixed point algorithm on $\mathbf{x}$, λ, μ, and ν enables the adjustment of the three parameters.

The Cauchy–Riemann relations above provide formulae for a computation of reference functions from a initial mesh function $\mathbf{x}^0$:

$$a^2 = \frac{|\mathbf{x}_u^o \times \mathbf{x}_v^o|}{|\mathbf{x}_v^o|} \cdot \frac{|\mathbf{x}_u^o \times \mathbf{x}_w^o|}{|\mathbf{x}_w^o|}; \quad b^2 = \frac{|\mathbf{x}_v^o \times \mathbf{x}_u^o|}{|\mathbf{x}_u^o|} \cdot \frac{|\mathbf{x}_v^o \times \mathbf{x}_w^o|}{|\mathbf{x}_w^o|}; \quad c^2 = \frac{|\mathbf{x}_w^o \times \mathbf{x}_v^o|}{|\mathbf{x}_v^o|} \cdot \frac{|\mathbf{x}_w^o \times \mathbf{x}_u^o|}{|\mathbf{x}_u^o|}$$

References

1. Babuska, I. and Rheinboldt, W.C., Error estimates for adaptative finite element computations, *SIAM J. Numer, Anal.* 1978, 15, pp 736–754.
2. Bank, R.E., Analysis of a local *a posteriori* error estimate for elliptic equations, *Accuracy Estimates and Adaptivity for Finite Elements.* John Wiley and Sons, NewYork, 1996, pp 119–128.
3. Brackbill, J.U. and Saltman, J.S., Adaptive zoning for singular problems in two dimensions, *J. Comp. Phys.* 1982, 46, pp 342–368.
4. Cabello, J., Löhner, R., and Jacquotte, O.-P., A variational method for the optimization of two- and three-dimensional unstructured meshes, *AIAA Paper* No 92-0450 and *ONERA T.P.* N° 1992–24, 30th Aerospace Sciences Meeting and Exhibit, Reno, NV, Jan. 6–9, 1992.
5. *Calcul d'Erreur a posteriori et Adaptation de Maillage*, Ecole CEA-EDF-INRI, Org. by le Tallec, P. and Perthan, B., Rocquencourt, Sept. 18–21, 1995.
6. Carcaillet, R., Optimization of three-dimensional computational grids and generation of flow adaptive computational grids, *AIAA Paper.* 86-0156, 1986.
7. Eiseman, P.R., Orthogonal Grid Generation, *Numerical Grid Generation.* Thompson, J.F., (Ed.), North Holland, 1982, pp 193–226.
8. Jacquotte, O.-P., A mechanical model for a new mesh generation method in computational fluid dynamics, *Comp. Meth. Appl. Mech. Eng.* 1988, 66, pp 323–338.
9. Jacquotte, O.-P., Coussement, G., and Catherall, D., Evaluation of mesh and solution quality obtained by optimization and adaptation, to appear in *Experimentation, Modelling and Combustion in Flow, Turbulence and Combustion.* Wiley-Interscience, 1997.
10. Jacquotte, O.-P., Desbois, F., Coussement, G., and Gaillet, C., Contribution to the development of a multiblock grid optimisation and adaption code, *Multiblock Grid Generation, Notes on Numerical Fluid Mechanics,* Weatherhill, N.P., Marchant, M.J., King, D.A., (Eds.), Vieweg, 1993, 44.
11. Kennon, S.R. and Dulikravitch, G.S., *A Posteriori* optimization on computational grids, *AIAA Paper* 85-0483 and 85-0486, 1985.
12. Knupp, P. and Steinberg, S., *Fundamentals of Grid Generation,* CRC Press, Boca Raton, FL, 1994.
13. Oden, J.T., Stroboulis, T., and Devloo, D., Adaptative finite element methods for the analysis of inviscid compressible flows: part 1. fast refinement/unrefinement and moving mesh methods for unstructured meshes, *Comp. Meths. Appl. Mech. Eng.* 1986, 59, pp 327–362.
14. Saltzman, J.S. and Brackbill, J.U., Application and generalizations of variational methods for generating adaptive meshes, *Numerical Grid Generation.* Thompson, J.F., (Ed.), North Holland, 1982, pp 865–884.
15. Thompson, J.F. and Kim, H.J., Three-dimensional adaptive grid generation on a composite-block grid, *AIAA Journal.* 1989, 28, p 3.

34

Dynamic Grid Adaption and Grid Quality

D. Scott McRae

Kelly R. Laflin

34.1 Introduction

Many natural physical processes can be described by conservation laws that can be expressed as integral equations. Conservation, in this instance, implies that these equations must account for local changes of dependent quantities, for the effect of fluxes of these quantities across the chosen domain surface, and for any resulting forces, changes in energy levels, etc. An exact evaluation of these integral equations would require complete functional knowledge of the temporal and spatial distribution of the conserved quantities on domain interiors and boundaries. Since such *a priori* physical knowledge of a given problem is unlikely, available information must be used to obtain as complete an approximation to the exact solution as is practical. This statement identifies the two central opposing issues in the process of obtaining a description of an unknown physical process: accuracy versus practicality. To illustrate these issues, consider that the integral statements of those conservation laws are formulated based on consideration of the fluid as a continuum. Since we do not usually know the continuous distribution *a priori*, we could, instead, assign a location and appropriate kinematic and state variables to each molecule in the fluid. The integrals could then be evaluated, including appropriate interactions between the molecules. The problem is that a vanishingly small domain in even low density fluids would immediately overtax the largest available computers (given that we had sufficient knowledge of interactions). The most accurate approach to our problem is immediately tempered by practical considerations.

This leads to a "tool-driven" approach to evaluation of conservation laws that involves choosing discrete domains of the fluid and using statistical averages of the properties and locations of these discrete domains in order that our "tool" (the digital computer) will be able to produce results in a reasonable temporal period. Our task then is to balance the need for the averages to be representational of the fluid in the discrete domains versus the need to limit the number of domains that can be stored and processed in

0-8493-2687-7/99/$0.00+$.50

the computer. Note that the requirements will be similar whether we consider the fluid as a continuum that we divide into discrete domains or as collections of particles existing in discrete domains (such as the direct simulation Monte Carlo method).

The task is then to distribute the discrete domains in which we define our fluid properties such that those associated with each domain are accurately resolved, both spatially and temporally, to the extent permitted by the available resources. The remainder of this chapter will present a process for evaluating how adequately we distribute the domains and will present dynamic solution adaptive mesh procedures that will automatically redistribute cell volume based on solution interpolation and grid quality measures. The discussion will be restricted to body-fitted structured meshes, although the ideas apply locally to unstructured meshes and to Cartesian mesh interfaces with general geometries.

34.2 Problem Statement

Mathematical statements of the above issues can be obtained from considering an integral statement for the conservation of linear momentum in a fluid system defined in a domain with surface S and volume V:

$$\frac{\partial}{\partial t}\int_{V}\overline{U}dV+\int_{S}\underline{F}\cdot\hat{n}dA=0 \tag{34.1}$$

where the quantity to be conserved is $\overline{U} = \overline{U}(x_i, t)$ and the tensor $\underline{\underline{F}} = \overline{F}(\overline{U})$ contains terms that describe surface stresses on V and the flux of $\overline{U}$ across S. A differential statement of the conservation law can be obtained by invoking Gauss' theorem and requiring Eq. 34.1 to be valid for arbitrarily small volumes:

$$\frac{\partial\overline{U}}{\partial t}+\nabla\cdot\underline{F}=0 \tag{34.2}$$

A discrete statement of either Eq. 34.1 or Eq. 34.2 can be obtained by subdividing V and defining values of U either as averages over the smaller subdivided volumes or at nodes. In either case, the discrete form of the equations leads to similar issues of accuracy. Divided differences of the dependent variables occur in either case.

The fundamental issue that results can be illustrated by examining an exact expression for a derivative obtained by a Taylor series expansion between two spatial points located at x_i and x_{i+1}:

$$u=u(x,t)$$

$$u_x\big|_i=\frac{u_{i+1}-u_i}{\Delta x}-\frac{\Delta x}{2!}u_{xx}\big|_i-\frac{\Delta x^2}{3!}u_{xxx}\big|_i-\dots \tag{34.3}$$

where $\Delta x = x_{i+1} - x_i$.

We obtain an approximate form of the first derivative by truncating the higher derivative terms on the RHS of the expression. If $u(x, t)$ is continuous and the approximation is consistent, then this approximate value will approach the exact value of the derivative as $\Delta x \to 0$. Since we cannot in general afford small Δx everywhere, then a reasonable compromise would be to make Δx small only where the derivative terms in the truncation error are large. This example illustrates the most basic fundamental issue, which affects the accuracy of our solution and points to a possible beneficial approach.

However, other issues must be addressed, especially when two-dimensional and three-dimensional solutions are considered. Brackbill and Saltzman [1] developed a fundamental means of optimizing mesh smoothness and orthogonality with the basic property of cell volume distribution to the maximum extent possible through the use of variational calculus. Within the precepts of a structured mesh, this excellent work illustrates the interrelational issues and demonstrates that they are not independent. This approach has been further developed by others as noted in the references cited by Luong, Thompson, and Gatlin

[16]. As an alternative to solving the Euler–Lagrange equations in order to obtain the mesh as done in [1], Luong et al. add cell aspect ratios to the issues considered and develop weight functions based on a generalization of the equidistribution principle by Eiseman [5].

34.3 Theory and Principles

The adaptive grid techniques set forth below require the following conditions for full implementation. Exceptions, qualifications, and current research will be noted as appropriate.

1. Dependent variables are defined at discrete structured grid nodes in the physical domain. If the grid is divided into contiguous blocks, no hanging blocks or "singular" locations must be present. (This may be relaxed in 2D, [9].)
2. The boundaries of the physical domain must be stationary. (Moving boundary research is in progress.)
3. The basic techniques require the existence of a one-to-one transformation to a parametric/computational space. This requirement can be made local rather than global by a more advanced implementation.
4. The solution is always considered to be known relative to an inertial coordinate system. Any changes to the mesh should preserve the solution in the inertial system.
5. Mesh changes will be accomplished by grid node relocation only (i.e., *r*-refinement).

As noted above, our approach to achieving the goal of reducing mesh spacing dynamically, where needed, relies on the concept of the solution being defined relative to a known inertial coordinate system. If we then superimpose an additional coordinate system in motion relative to the inertial system, the vector and scalar dependent variables as referenced to the moving system remain unchanged in the inertial system. This implies that, at any given time, we can transform the solution from one system to the other by a simple interpolation. Since we are considering vector quantities, further explanation is required.

34.3.1 Fundamentals

Consider the 1D conservation law

$$u_t + f_x = 0 \qquad u = u(x,t) \qquad f = f(u) \tag{34.4}$$

subject to a transformation to a "computational" coordinate system

$$\tau = t$$
$$\xi = \xi(x,t)$$

for which, according to the usual requirements, the inverse exists:

$$t = \tau$$
$$x = x(\xi,\tau)$$

Performing the transformation and returning to conservation law form,

$$(x_\xi U)_\tau + (f - x_\tau U)_\xi = 0 \tag{34.5}$$

The quantity x_τ can be interpreted as a "mesh velocity." This interpretation requires that the independent variable x also be allowed to indicate the present location of a grid node position vector in inertial space. The "mesh speed" is a temporal derivative of this position vector.

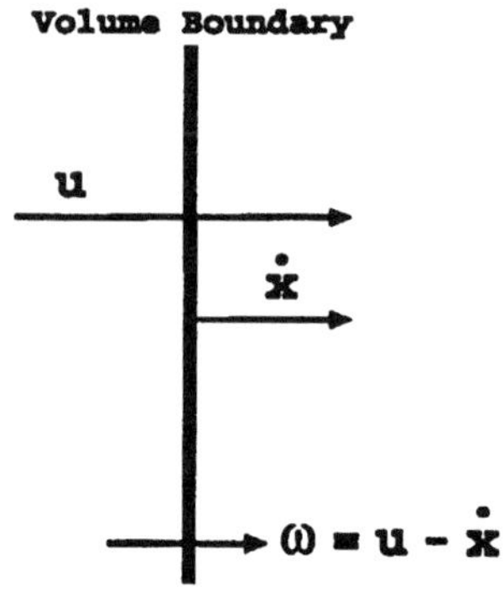

FIGURE 34.1 Moving boundary velocity interface.

If we let $f = cu$, where c is a wave speed, then the second term of Eq. 34.5 is

$$(u[c - x_\tau])_\xi$$

We observe that the quantity x_τ is in reality a correction of the wave speed (or the characteristic slope) for the movement of the mesh. Clearly, if the translating mesh is moving exactly at c, then the solution is stationary relative to this mesh. In case f is a more general flux, x_τ can be considered to be a correction of the flux convective velocity for the relative motion of the translating mesh.

Another perspective is revealed when the equation is discretized. Using backwards Euler as an example,

$$(x_\xi U)_i^{n+1} = (x_\xi U)_i^n - \frac{\Delta\tau}{\Delta\xi}\left[f_i^n - f_{i-1}^n\right] + \frac{\Delta\tau}{\Delta\xi}\left[(x_\tau U)_i^n - (x_\tau U)_{i-1}^n\right] \tag{34.6}$$

If we know the local mesh movement $x_\tau|_i^n$ and $x_\tau|_{i-1}^n$, then numerical approximations can be substituted into the last term:

$$\frac{\Delta\tau}{\Delta\xi}\left[\left(\frac{x_i^{n+1} - x_i^n}{\Delta\tau}\right)U_i - \left(\frac{x_{i-1}^{n+1} - x_{i-1}^n}{\Delta\tau}\right)U_{i-1}\right] \tag{34.7}$$

The time step $\Delta\tau$ cancels and the remaining terms can be considered to be an interpolation (or redistribution) of the solution to the n+1 mesh locations (for the linear wave equation, the $\Delta\tau(cu)_\xi$ term behaves similarly when $c\Delta\tau$ is interpreted properly). This redistribution is easier to relate to a physical process by considering a discrete finite volume integral.

We have chosen to use the fact that "there is only a single solution in inertial space" in an algorithm that seeks to: (1) provide an appropriately resolved mesh at each time step, and (2) preserve the inertial solution (i.e., temporal accuracy). We have included mesh adaptation and solution redistribution in the integration of Eq. 34.5 in two different ways. In the first, Eq. 34.5 is integrated exactly as shown, with the time marching scheme determining the number of algorithm steps. The grid speed, x_τ is found through use of information at the nth time level. When solved as a single, unsplit vector equation, the grid speed serves to modify the convective flux in the locally moving coordinate system (see Figure 34.1). If an explicit solver is used to integrate the equation in this form, mesh movement may have to be restricted in order to maintain stability.

In PDE form, the second split-equation technique proceeds [2] as follows for the transformed conservation law:

$$(x_\xi U)_\tau + f_\xi - (x_\tau U)_\xi = 0 \tag{34.8}$$

First integrate

$$(x_\xi U')_\tau = -f_\xi \tag{34.9}$$

Then, adapt the mesh to improve resolution of the U' solution (note that the grid upon which U' is obtained is fixed in time). Then to obtain the final solution distribution, integrate

$$(x_\xi U)_\tau = (x_\xi U')_\tau + (x_\tau U')_\xi \tag{34.10}$$

Note that the use of U' in the last term introduces, for explicit solvers, a nonlinearity that is still an issue for debate and examination. The integral statement for conservation of a dependent variable U over a domain V can be obtained through application of Leibnitz' rule [2] or through physical arguments. When generalized for changing V, with the exception of any nonlinear effects caused by use of U' rather than U in Eq. 34.10 and the difference in x_τ computation level, both approaches should produce the same mathematical result. However, as will be illustrated below, implementation of the two approaches may be dissimilar.

The second two-step procedure serves to couple the mesh more closely to the solution, thereby ensuring that the mesh upon which the first step (i.e., the flow solver) of the procedure occurs resolves the nth level solution well. It then adapts the mesh to the results of this step and interpolates this solution to the new mesh such that temporal accuracy is preserved. The word "preserved" is appropriate, since the adaptive process only concerns spatial resolution; the task is to conduct this process (specifically, the interpolation to the new grid) such that the temporal accuracy inherent in the first step of the algorithm is carried forward to the new grid.

As noted above, the issues are somewhat more complex for an integral conservation law but the task is exactly the same, i.e., to resolve spatially the solution while preserving time accuracy. The integral statement for conservation of a dependent variable U over a domain V when generalized for changing V becomes

$$\frac{\partial}{\partial t}\int_V U\,dV - \oint_S U\,\dot{\vec{x}} \cdot d\vec{S} + \oint_S \vec{A} \cdot d\vec{S} = 0 \tag{34.11}$$

where

$$U = [\rho, \rho\vec{V}, E_t]^T$$

and

$$\vec{A} = E\hat{i} + F\hat{j} + G\hat{k} \qquad E = E(U), \text{etc.} \qquad \dot{\vec{x}} = \dot{x}\hat{i} + \dot{y}\hat{j} + \dot{z}\hat{k}$$

Note that Eq. 34.6 and Eq. 34.7 in Benson and McRae [2] are oversimplifications of the discretized form of this integral.

The first and last integrals in Eq. 34.11 are the standard forms that we normally encounter. The second integral is the correction to the conserved quantity for the gain/loss due to movement of the cell sides independently of the fluid velocity. An illustration of this movement in both 1D and 3D is given in Figures 34.1 and 34.2. The value of the second integral in Eq. 34.11 for each cell is the sum of the conserved quantity U contained in the volume swept by the cell faces as the grid translates.

The split form of the algorithm can be expressed as follows, using a multistage Runge–Kutta time-stepping algorithm for solution Step (1) where i indicates the ith stage of the multistage Runge–Kutta algorithm:

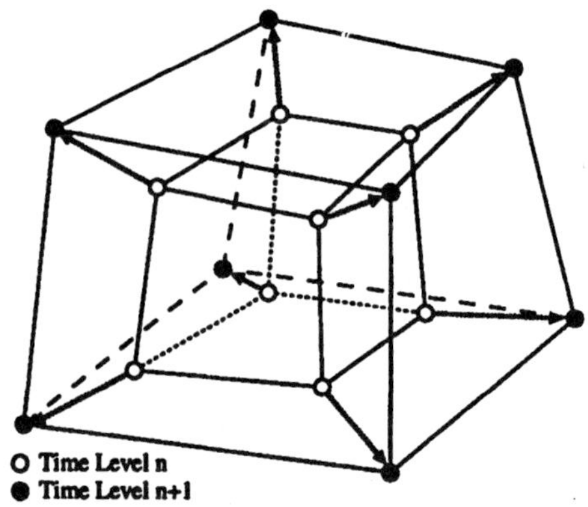

FIGURE 34.2 Cell at time levels n and $n+1$.

Step (1)

$$U^{(i)} = U^{(i-1)} - \alpha^{(i)} \frac{\Delta t}{V^n} \left\{ \Delta_\xi \hat{E}^{(i-1)} + \Delta_\eta \hat{F}^{(i-1)} + \Delta_\zeta \hat{G}^{(i-1)} \right\} \tag{34.12}$$

The mesh is then adapted to the results of this step. The final step is (where (2) indicates the results of Step (1)):

Step (2)

$$(UV)^{n+1} = V^n U^{(2)} + \left\{ \Delta_\xi (U^{(2)} \Delta^n V) + \Delta_\eta (U^{(2)} \Delta^n V) + \Delta_\zeta (U^{(2)} \Delta^n V) \right\} \tag{34.13}$$

In this equation, the term $\Delta^n V$ represents the change in volume between the n and $n+1$ time level (Figure 34.2). As a cautionary note, care must be taken to insure that $\Delta^n V$ includes all of the swept volume as indicated in Figure 34.2. If this final step is carried out with sufficient accuracy, the result will be the solution obtained at Step (1) expressed on a grid that will give very high spatial resolution for the next application of Step (1). *This is the fundamental and only goal of grid adaptation* when applied to an explicit solution technique.

34.3.2 Adaptive Algorithm Implementations (DSAGA, SIERRA)

Within the framework noted previously, the next task is to set forth the versions of the adaptive algorithm and to examine their strengths and weaknesses. The original version of the adaptive algorithm was reported at the Third International Grid Conference in Barcelona [6]. This version was designated DSAGA (dynamic solution adaptive grid algorithm) and was developed for both 2D and 3D. The original adaptive algorithm (DSAGA) can be used with either the split or unsplit form of the conservation law. The only difference occurs in the choice of data upon which to base the adaption decision. Since the manner in which these data are processed such that a criteria for adaption results has been both controversial and widely differing among researchers, we will offer a brief rationale herein for the original approach. Also, DSAGA can be coded and implemented with relative ease. For this reason we will use it as a basis for introducing the new developments that followed.

Although many successful time-varying solutions were obtained, DSAGA has limitations due to the need for weight function tailoring when both strong and weak flow features must be resolved simultaneously and due to problems with stability when large grid movement is necessary to resolve moving solution features.

The two limitations noted above plus others are addressed in the new code SIERRA [13,14,15]. This new code replaces the solution redistribution step of DSAGA but retains the basic structure. Ease of use, stability, and feature resolution are much improved with SIERRA. The details of SIERRA will be presented in a later section.

34.3.2.1 DSAGA

The steps of DSAGA are as follows:

1. Use an available grid generator (preferably elliptical partial differential equation based) to obtain an initial structured, body-fitted grid.
2. Obtain the numerical approximations to the metric derivatives that define numerically a one-to-one transformation to a parametric space. These initial transformation metrics and their approximations remain temporally fixed.
3. A discrete source-term distribution is obtained based on selected parameters and criteria. This step is crucial to successful adaptation.
4. The discrete source term distribution is input to the Poisson solver (in our case Eiseman's "mass weighted algorithm") to find new solution-dependent node locations.
5. The new grid node locations are then used to find a "grid velocity" (finite difference) or input to a finite volume redistribution algorithm (split or unsplit conservation law). In either case, the final step results in a solution at the $n+1$ time step on a grid that has been adapted to the chosen criteria at the n or $n+1$ time level.

Step 1 on the preceding list is standard. We must always define an initial mesh which, in most applications, is body conforming. The cartesian cut-cell meshes are not appropriate as initial meshes for this adaptation method. Step 2 is not always standard, as many modern finite-volume codes compute cell volume and surface areas in physical space, thereby negating the need for a computational or parametric space. However, the use of a coordinate transformation avoids expensive searches which may otherwise be needed to maintain structured connectivity after grid adaptation. The transformation is used in Step 5 and will also be important to the goal of developing "stand alone" versions of the adaptive algorithm.

Step 3 involves first selecting the solution parameters and/or features that require increased resolution. Two obvious candidates are viscous layers and shock waves (note that *any* solution feature can be chosen). Once these features are chosen, then parameters must be selected that vary appropriately at the feature location. For instance, static pressure would not be an appropriate choice on which to base a viscous layer weight function. It is usually necessary to select more than one parameter in order to resolve multiple flow features. Once the parameters are chosen, first or second differences of each parameter are calculated to produce a set of raw weight functions. To respond to the obvious question as to why not divided differences, the problem is that a divided difference may become very large as the mesh spacing decreases.

The usual stated goal for an adaptive mesh algorithm is to promote equal distribution of the approximation error such that the solution is uniformly accurate. This equidistribution concept must, however, remain a goal in most algorithms. Many such algorithms (including the present technique) use an iterative "solution" to a Poisson's type differential equation in order to determine new mesh mode locations. Since the goal is equidistribution of error, it would seem reasonable to base the source term for Poisson's equation on the truncated approximation error terms. Unfortunately, as revealed by a Fourier analysis, these terms are in general oscillatory and change sign depending on local solution behavior. A solution to Poisson's equation depends on both the magnitude and sign of the source term on the RHS. Sign change alone will change locally the mesh obtained through Poisson's type solvers from clustered to declustered in character. This effect will be dominant if the leading approximation error term is second order (i.e., third derivative for convective flux terms). Therefore, the grid solver imposes the requirement that the raw truncation error distribution be processed to create a source term distribution that will give an acceptable solution of Poisson's equation. The first step involves the calculation of a solution (and grid quality, for SIERRA) dependent raw weight function at each mesh node. This, in its simplest form, may be composed of a linear combination of individual first or second differences of the dependent variables. This proceeds by first taking either a first difference, a second difference, or both at each mesh node in the domain. Next, the absolute value operator is applied to all values obtained. A normalization coefficient is then defined by

$$\alpha_k \equiv 1 / \underset{k=1,m}{MAX}(\phi_k).$$

If the final weight function is to include dependence on more than one dependent variable, a biasing coefficient γ_k can then be chosen to determine specific influence of each term in the linear combination. The partially processed weight function at each node is then described by

$$\omega = \sum_k \gamma_k \alpha_k |\phi_k| \tag{34.14}$$

This semi-raw weight function may contain values differing by many orders of magnitude. It also may contain very large spatial gradients which can result in unacceptable skewness or volume shear in the mesh. A procedure to limit the variation of the weight function which adjusts (somewhat) to the current distribution results from obtaining an average value of ω. The minimum value of ω is increased to a percentage of this average. All maxima greater than a chosen multiple of the average are truncated. The resulting distribution is then smoothed to reduce mesh skewness and shear.

After these processing steps, there may remain regions of interest in which the weight function is small compared to the coordinate maximum. If this occurs, the multiple of the average weight used to truncate maximum can be reduced, thereby reducing the maxima relative to the small values. An expansion function is then used to return the weight function to a maximum level appropriate for the degree of adaptation desired. This step results in increasing the magnitude of the small value regions relative to the maximum.

In Step 4 the weight function obtained above is input to a modified Eiseman's mean-value relaxation algorithm. This algorithm begins with a designated stencil of mesh nodes (9 for 2D, or 27 for 3D) and associated weights from Step 3. The algorithm is then applied to locate the center of mass, which is the geometric location at which a body can be replaced by a point with the same total mass. This can be determined for the computational cell in three dimensions by applying the following equation for each coordinate in turn:

$$\xi_{cm_{i,j,k}} = \frac{\sum_{k-1}^{k+1}\sum_{j-1}^{j+1}\sum_{i-1}^{i+1} \omega_{i,j,k}\xi_{i,j,k}}{\sum_{k-1}^{k+1}\sum_{j-1}^{j+1}\sum_{i-1}^{i+1} \omega_{i,j,k}} \tag{34.15}$$

This determines the movement of the mesh node at *i, j, k* to the center of mass for each stencil. This calculation is repeated for every point in the parametric domain except that a reduced stencil is used at boundaries. the mesh nodes are locally redistributed until a movement criteria is satisfied.

The problem of grid-point crossover needs to be addressed. Crossover occurs when the center of mass of the local cell is outside the cell boundaries. There are two cases in which this situation is likely to occur: in the vicinity of concave curved boundaries (which are not present in the parametric space for single block grids and for restricted arrangements of multiple-block grids) or in the interior of the mesh. Adapting the mesh in parametric space reduces greatly the possibility of crossover. At the beginning of each global adaption step, all stencils in parametric space describe rectangular figures, which implies that the center of mass will always be inside the stencil. The mesh in parametric space may become sufficiently distorted for crossover to occur in this case in 3D but this is seldom observed.

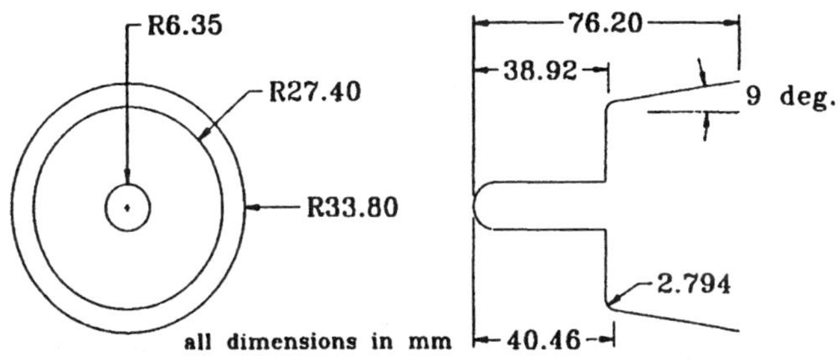

FIGURE 34.3 Sample cell in parametric space after grid relocation.

Prior to application of the mass-weighted algorithm, the relationship between the forward and inverse space metrics is obtained by excluding the time terms in the unsteady mapping. Since the mapping defines a parallelepiped in the parametric space and $\Delta\xi = \Delta\eta = \Delta\zeta = 1$ by definition, the original nodes or grid points have integer values which correspond directly to the i, j, k that are used to reference the arrays, i.e.,

$$int(\xi^{\circ}, \eta^{\circ}, \zeta^{\circ}) = i, j, k \tag{34.16}$$

After application of the center-of-mass algorithm, the mesh node positions have been changed in parametric space and are no longer located at integer values of ξ, η, and ζ. This requires that a mapping to determine the new x, y, and z locations in physical space from the new ξ, η, and ζ positions in parametric space must be obtained. Beginning with the differential dx,

$$dx = \frac{\partial x}{\partial \xi} d\xi + \frac{\partial x}{\partial \eta} d\eta + \frac{\partial x}{\partial \zeta} d\zeta \tag{34.17}$$

This differential can be approximated by finite differences:

$$\Delta x = x_{\xi}\Delta\xi + x_{\eta}\Delta\eta + x_{\zeta}\Delta\zeta \tag{34.18}$$

The differences are chosen to be just the new location of the mesh node, referenced with i, j, k, minus a nearest original position, denoted with the superscript (°). The metric derivatives are also identified with the superscript (°), since the transformation is only determined initially:

$$x_{i,j,k} - x^{\circ} = x_{\xi}^{\circ}(\xi_{i,j,k} - \xi^{\circ}) + x_{\eta}^{\circ}(\eta_{i,j,k} - \eta^{\circ}) + x_{\zeta}^{\circ}(\zeta_{i,j,k} - \zeta^{\circ}) \tag{34.19}$$

If the mesh node at i, j, k is moved to a new position in the parametric space (Figure 34.3), the corresponding new position in physical space must be determined. Truncating the new coordinate locations to integer values identifies the vertex nearest the origin of the reference parallelepiped cell that now contains the mesh nodes:

$$\begin{aligned} l &= int(\xi_{i,j,k}) \\ m &= int(\eta_{i,j,k}) \\ n &= int(\zeta_{i,j,k}) \end{aligned} \tag{34.20}$$

The vertex of the cube of the original parallelepiped that is closest to the new ξ, η, ζ position, shown in Figure 34.3, is given by the nearest integer function:

$$\begin{aligned} ln &= nint(\xi_{i,j,k}) = l + 1 \\ m &= nint(\eta_{i,j,k}) = m + 1 \\ nn &= nint(\zeta_{i,j,k}) = n \end{aligned} \tag{34.21}$$

Recall that the original ξ, η, and ζ were defined to be integers that corresponded directly to the reference coordinates *i*, *j*, *k*, and therefore, the values defined in Eq. 34.20 and Eq. 34.21 correspond directly to the array positions for *x*, *y*, *z* of the original grid point at those respective vertices.

In order to completely define Eq. 34.19 the metrics x_ξ, x_η, and x_ζ, are approximated such that they represent the distance between adjacent nodes in the ξ-, η-, and ζ-directions. The metrics are stored in arrays as forward differences and therefore, for the example cell in Figure 34.3, they are based at the point *l*, *mn*, *nn* for the ξ-direction, *ln*, *m*, *nn* for the η-direction, and *ln*, *mn*, *n* for the ζ-direction. By using the integer value of *ln* in place of ξ° in Eq. 34.19, this will subtract the distance $x^\circ_\xi(\xi - \xi^\circ)$ if $\xi_{i,j,k}$ is closer to the ξ-axis than the nearest original point and add the distance if $\xi_{i,j,k}$ is greater than the nearest original point. The result is similar for η° and ζ°. Therefore, a final expression for the new value of x in the physical space is

$$\begin{aligned} x^n_{i,j,k} &= x^\circ_{ln,mn,nn} + x^\circ_{\xi_{l,mn,nn}}(\xi_{i,j,k} - ln) + x^\circ_{\eta_{ln,m,nn}}(\eta_{i,j,k} - mn) \\ &+ x^\circ_{\zeta_{ln,mn,n}}(\zeta_{i,j,k} - mn) \end{aligned} \tag{34.22}$$

which is simply a Taylor series expansion in three dimensions utilizing the initial grid as a reference grid. Similar equations can be derived for y and z by substituting for x.

The above can be shown to preserve the original boundary shape. Choosing the boundary where $\eta =$ const. = 1, note that a term drops out of Eq. 34.22 leaving

$$x^n_{i.j.k} = x^\circ_{ln,mn,n} + x^\circ_{\xi_{ln,mn,nn}}(\xi_{i,j,k} - ln) + x^\circ_{\zeta_{ln,mn,n}}(\zeta_{i,j,k} - nn) \tag{34.23}$$

Since the new position $\xi_{i,j,k}$, $\eta_{i,j,k}$, $\zeta_{i,j,k}$ is restricted to the plane in parametric space where $\eta_{i,j,k} =$ const., the new position of $x_{i,j,k}$, $y_{i,j,k}$, $z_{i,j,k}$ in the physical space must also be restricted to the boundary surface defined by the mapping.

34.4 Grid Quality

Obtaining a grid that will allow a well-resolved, accurate computational solution is the goal of all mesh generation efforts. However, determining whether you have generated such a grid remains an area for research. In the context of dynamic grid adaptation, we are effectively regenerating the grid as often as each time step as the solution evolves, which means that an initially "good" grid will have to be constantly reevaluated.

Grid quality has been a topic of many previous investigations and discussions. Rather than survey prior work, some observations will be offered based on our own experience. This discussion is intended to focus on the primary issues that must be addressed in order to achieve our stated goal.

The first observation, and most important, is that mesh "quality" cannot be determined with-out considering the function/solution to be resolved by the mesh. This statement underlies all of our adaptive mesh research. An example is provided by considering a shock wave crossing a 2D Cartesian mesh diagonally (i.e., 45° to cell face). If the shock wave is planar, both exact and approximate 1D Riemann solvers can be applied normal to the shockwave with accurate results. However, the Riemann solvers in most formulations are applied to fluxes projected on normals to cell faces, resulting in maximum misalignment with a shock wave at 45° to all cell faces.

Recognition of this inaccuracy has led to research in so called "rotated" or "2D" upwind schemes, which either align a local axis with the largest gradient or seek to solve a 2D Riemann problem. This recognition, along with the discussion at the beginning of the chapter, points to the two main grid quality requirements of a structured adaptive mesh: (1) to reduce grid spacing where derivatives are large in the solver error structure, and (2) to align, to the maximum extent possible, the cell surfaces with large gradients in the flow.

Note that we have made no mention, until now, of the grid attribute usually considered to be a fundamental problem of structured adapted grids: grid cell skewness. In our method, skewness inevitably results if the cell surfaces align with shock waves, for instance. As will be noted in the results section, the benefits of alignment far exceed any possible problems due to cell skewness. In fact, we have found that the resolution of a continuously defined shock wave solution becomes much poorer as the mesh changes from aligned with the shock wave but skewed in one region to near Cartesian but at 45° to the shockwave in another region along the same shockwave. The cell volumes were of similar order in both regions.

The question of cell skewness was addressed by Thompson et al. [22] by evaluating analytically the leading truncation error for central difference representation of a first derivative on a skewed cell with parallel surfaces. Eq. 34.13 in Chapter V of [22] illustrates several points:

$$T_x = -\frac{1}{2}x_{\xi\xi}f_{xx} + \frac{1}{2}\left(\frac{y_\xi}{x_\xi}\right)y_{nn}f_{yy} - \frac{1}{2}\left(\frac{y_\xi}{x_\xi}\right)x_{\xi\xi}f_{xy} \tag{34.24}$$

The first term on the RHS is present in all cases in which x_ξ varies. The second and third terms represent contribution to the truncation error due to skewness for this restricted cell geometry. The ratio (y_ξ/x_ξ) represents the cotangent of the included angle between the x and y coordinates. Analysis of this equation reveals that

1. Skewness has no effect on the solution when the metric derivatives of the transformation are constant or when the solution varies linearly. Note that constant metric derivatives correspond to even mesh spacing.
2. As noted in [10], $y_\xi/x_\xi > 1$ is required for the contributions from skewness to have a larger coefficient than the first term in T_x.

This again supports the conclusion that mesh "quality" should be examined only together with the solution and agrees with the conclusion reached in [22].

The doctoral research [15] of the second author addressed mathematically the question of grid quality, stability, and accuracy of r-refinement adaptation (movement of grid locations rather than subdivision). The results of this research have been included in a solver-independent efficient r-refinement algorithm (SIERRA). Although this algorithm is evolved from DSAGA and uses the basic mass-weighted algorithm for node relocation, important advances have been made in the remainder of the steps.

34.5 SIERRA

34.5.1 Weight Function

A weight function [9] that inherently includes grid geometry as part of an assessment of the solution resolution is given by

$$\omega_i = \left|\int_{\Omega_i}(\phi(r) - \phi_i)dV\right| \tag{34.25}$$

where ϕ_i is the computed piecewise constant representation of the solution scaler function in the *i*th grid cell, as computed by the flow solver on the previous grid. It is assumed that $\phi(r_i) = \phi_i$, where $r_i \in \Omega_i$ and is the position vector of the cell center and Ω_i is the domain of the *i*th cell. Using the mean-value theorem, this weight function can be expressed as

$$\omega_i = V_i \left| (\bar{\phi}_i - \phi_i) \right| \tag{34.26}$$

where

$$\bar{\phi}_i \equiv \frac{1}{V_i} \int_{W_i} f(r) dV \tag{34.27}$$

is the volume-averaged value of $\phi(r)$ over the *i*th grid cell which has volume V_i. Eq. 34.26 shows that the weight function is a measure of how well conservation of the variable is predicted by the piecewise constant representation of the solution $\phi(r)$.

In order to determine how the magnitude of this weight function is influenced by the behavior of the solution and the grid geometry, the solution scalar function $\phi(r)$ is expanded in a Taylor series about r_i. The resulting expression is substituted into Eq. 34.25, and the volume integration is performed. This procedure results in

$$\omega_i = \left| \phi_x I_x + \phi_y I_y + \phi_z I_z + \frac{1}{2} \phi_{xx} I_{xx} + \frac{1}{2} \phi_{yy} I_{yy} + \frac{1}{2} \phi_{zz} I_{zz} \right. \tag{34.28}$$

$$\left. + \phi_{xy} I_{xy} + \phi_{yz} I_{yz} + \phi_{xz} I_{xz} + O(\Delta r^3) V_i \right|$$

where

$$\begin{aligned} I_x &\equiv \int_{\Omega_i} \Delta x_i dV \\ I_{xx} &\equiv \int_{\Omega_i} \Delta x_i^2 dV \\ I_{xy} &\equiv \int_{\Omega_i} \Delta x_i \Delta y_i dV \end{aligned} \tag{34.29}$$

are various moments of inertia of region Ω_i about r_i, and

$$\Delta x_i = (x - x_i) \qquad \Delta y_i = (y - y_i) \qquad \Delta z_i = (z - z_i) \tag{34.30}$$

where x_i, y_i, and z_i are the position coordinates of r_i. The terms I_y, I_z, I_{yy}, I_{zz}, I_{yz}, I_{xz}, ... are defined similarly to I_x, I_{xx}, and I_{xy}.

Eq. 34.28 shows that each term of the weight function is comprised of the product of a derivative of the solution function $\phi(r)$ and a moment of inertia of the grid cell. The derivatives of $\phi(r)$ are evaluated at the r_i and the various moments of inertia, I_x, I_y, I_z, I_{xx}, I_{yy}, I_{zz}, I_{xy}, ..., are defined relative to point r_i. The first moments of inertia multiply the solution gradient, and the second moments of inertia multiply the solution curvature.

If it is assumed that the *r*-refinement adaptation process iteratively adjusts the grid so that the magnitude of the weight function is reduced to a minimum uniform value, then characteristics of the converged adapted grid can be determined by examining Eq. 34.28.

The terms I_x, I_y, and I_z are the first moments of inertia of region Ω_i. They give the relative displacement coordinates of the center of mass of region Ω_i to the support point r_i. These terms can be made zero by repositioning the support point so that it is coincident with the center of mass of Ω_i. Therefore, the terms I_x, I_y, and I_z promote even grid-node spacing but will not discourage grid-cell skewing.

The second moment of inertia term I_{xy} will vanish when the support point r_i is coincident with the center of mass of Ω_i and when Ω_i exhibits x-y symmetry. Similarly, I_{yz} and I_{xz} will vanish when the support point r_i is coincident with the center of mass of Ω_i and when Ω_i exhibits y–z and x–z symmetry, respectively. Note that Eq. 34.28 will result for any orientation of the orthogonal coordinate system with respect to an inertial frame of reference. Therefore, if region Ω_i exhibits symmetry about three orthogonal axes, then I_{xy}, I_{yz}, and I_{xz} will vanish regardless of how the axes are rotated. The terms I_{xy}, I_{yz}, and I_{xz} influence the shape of the grid-cells and promote grid-cell orthogonality.

The terms I_{xx}, I_{yy}, and I_{zz} are second moments of inertia, that only vanish in the limit of zero spacing in the x, y, and z directions, respectively. The magnitude of the terms vary quadratically with grid-cell spacing. These terms effect grid-node clustering.

From the above analysis, the minimum obtainable weight function for a fixed grid-node density is given by

$$\omega_i = \left| \frac{1}{2}\phi_{xx}I_{xx} + \frac{1}{2}\phi_{yy}I_{yy} + \frac{1}{2}\phi_{zz}I_{zz} + O(\Delta r^4)V_i \right| \tag{34.31}$$

which is obtained when the grid is orthogonal and evenly spaced. Further reduction of the magnitude of the weight function can only be achieved through decreases in the I_{xx}, I_{yy}, and I_{zz} terms, i.e., through grid-node clustering.

The relation expressing the minimum weight function for a fixed grid-node density given by Eq. 34.31 was found by considering evenly spaced orthogonal grids. This expression can also be obtained through proper orientation of the grid-cell with respect to the solution field. The dependency of ω_i on the orientation of Ω_i in the solution field is better examined by rewriting Eq. 34.28 in the equivalent form:

$$\omega_i = \left| \phi_x I_x + \phi_y I_y + \phi_z I_z + \int_{\Omega_i} ((\Delta r_i)^T [\Phi](\Delta r_i)) dV + O(\Delta r^4)V_i \right| \tag{34.32}$$

where

$$[\Phi] = \begin{bmatrix} \phi_{xx} & \phi_{xy} & \phi_{xz} \\ \phi_{xy} & \phi_{yy} & \phi_{yz} \\ \phi_{xz} & \phi_{yz} & \phi_{zz} \end{bmatrix} \tag{34.33}$$

and

$$(\Delta r_i) = \begin{pmatrix} \Delta x_i \\ \Delta y_i \\ \Delta z_i \end{pmatrix} = \begin{pmatrix} x - x_i \\ y - y_i \\ z - z_i \end{pmatrix} \tag{34.34}$$

where $(\Delta r_i)^T$ is the transpose of (Δr_i).

The matrix $[\Phi]$ is symmetric and is composed of the second derivatives of the solution field evaluated at r_i. It is analogous to the point stress tensor of fluid [6] and solid mechanics [18]. Because $[\Phi]$ is symmetric, it satisfies certain properties [11], which include the fact that it can be diagonalized to

$$[\Phi'] = \begin{bmatrix} \phi'_{xx} & 0 & 0 \\ 0 & \phi'_{yy} & 0 \\ 0 & 0 & \phi'_{zz} \end{bmatrix} \tag{34.35}$$

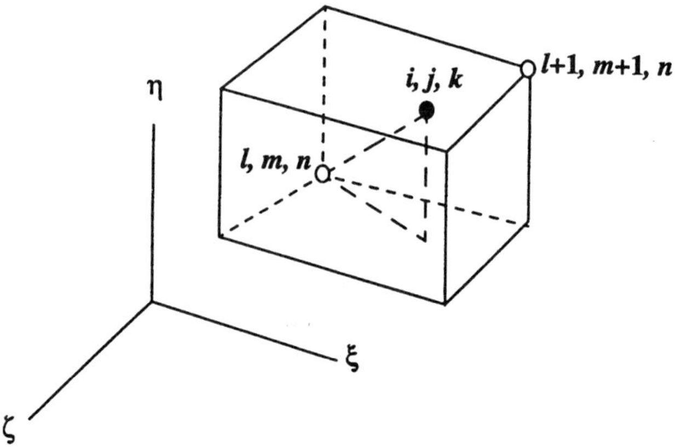

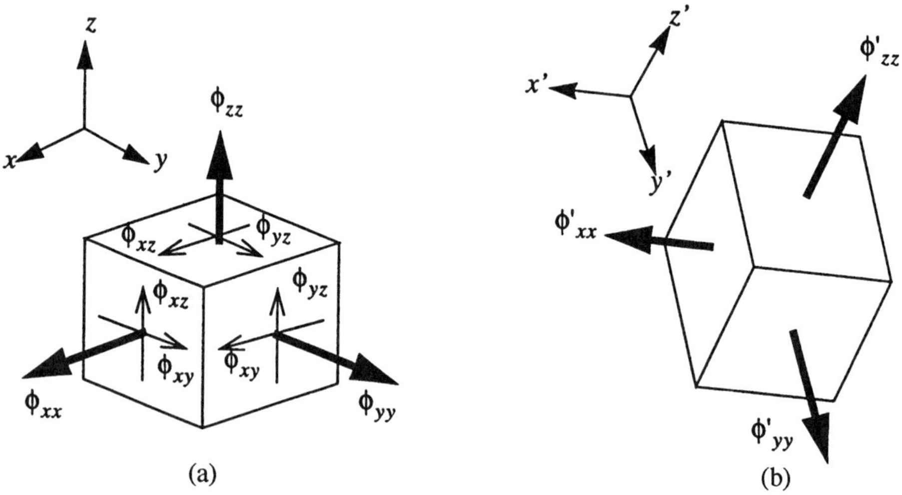

FIGURE 34.4 (a) Reference axes arbitrarily oriented in solution field, (b) Reference axes aligned with principal directions of solution curvature.

by rotating the (x, y, z) reference coordinate system of Figure 34.4a to coincide with the principal directions of the solution curvature, which coincide with the directions of the (x', y', z') coordinate system of Figure 34.4b.

Assuming that the principal directions of the solution curvature are nearly equal throughout region Ω_i, the weight function will be reduced to

$$\omega_i = \left| \phi_{x'} I_{x'} + \phi_{y'} I_{y'} + \phi_{z'} I_{z'} + \frac{1}{2}\phi_{x'x'} I_{x'x'} + \frac{1}{2}\phi_{y'y'} I_{y'y'} + \frac{1}{2}\phi_{z'z'} I_{z'z'} + O(\Delta r^4) V_i \right| \tag{34.36}$$

when the sides of region Ω_i are oriented so that they are normal to and parallel with the (x', y', z') coordinate directions.

If the support points are evenly distributed, then $I_{x'} = I_{y'} = I_{z'} =$ and the resulting weight function is given by

$$\omega_i = \left| \frac{1}{2}\phi_{x'x'} I_{x'x'} + \frac{1}{2}\phi_{y'y'} I_{y'y'} + \frac{1}{2}\phi_{z'z'} I_{z'z'} + O(\Delta r^4) V_i \right| \tag{34.37}$$

where $I_{x'}$, $I_{y'}$, $I_{z'}$, $I_{x'x'}$, $I_{y'y'}$, $I_{z'z'}$, $I_{x'y'}$, ... are moments of inertia of region calculated in the (x', y', z') coordinate system. Because

$$I_1 = \phi_{xx} + \phi_{yy} + \phi_{zz} = \phi_{x'x'} + \phi_{y'y'} + \phi_{z'z'} \tag{34.38}$$

is invariant for a symmetric matrix, Eq. 34.31 and Eq. 34.37 are equivalent.

It is concluded from this analysis that the adapted grid is expected to exhibit both grid-node clustering and grid-node alignment adaptation processes. When cell edges are not aligned normal to the principal directions of solution curvature, the grid cells are expected to exhibit orthogonality.

An efficient discrete approximation of the weight function given by Eq. 34.31 is obtained by transforming the analytic expression of the weight function in physical space (x, y, z) to an equivalent expression in computational space (ξ, η, ζ). This is accomplished by transforming Δx_i, Δy_i, Δz_i, and each of the derivatives of $\phi(r)$ appearing in Eq. 34.31 into equivalent expressions in computational space, using the transformation

$$\begin{aligned} \xi &= \xi(x, y, z) \\ \eta &= \eta(x, y, z) \\ \zeta &= \zeta(x, y, z) \end{aligned} \tag{34.39}$$

Upon performing the transformation and algebraic manipulations, the weight function expressed in computational space reduces to

$$\omega_i = \frac{V_i}{2}\left|(\hat{\nabla}^2\phi)_i + \text{HOT}\right| \tag{34.40}$$

where

$$\hat{\nabla}^2 \equiv \frac{\partial^2}{\partial \xi^2} + \frac{\partial^2}{\partial \eta^2} + \frac{\partial^2}{\partial \zeta^2} \tag{34.41}$$

is the Laplacian operator defined in computational space (ξ, η, ζ) and HOT denotes higher-order terms.

Eq. 34.40 is efficiently approximated by

$$\omega_i = \frac{V_i}{2}\left|(\Delta^2\phi)_i\right| \tag{34.42}$$

where Δ^2 is a discrete approximation of the Laplacian $\hat{\nabla}^2$. The quantity $(\Delta^2\phi)_i$ reduces to an undivided-difference expression, because of the unit spacing of the computational grid.

The discretized weight function given by Eq. 34.42 is expressed in terms of the discrete computed solution variables by the formula

$$\omega_i = \frac{V_i}{2}\left|\sum_{k=1}^{N_k}(\alpha_k\phi_k) - \alpha_i\phi_i\right| \tag{34.43}$$

where

$$\alpha_i = \sum_{k=1}^{N_k}(\alpha_k) \tag{34.44}$$

$N_k = 4$

	1	
1	-4	1
	1	

(a)

$N_k = 8$

1	4	1
4	-20	4
1	4	1

(b)

FIGURE 34.5 (a) Five-point discrete approximation stencil of the Laplacian, (b) Nine-point discrete approximation stencil of the Laplacian.

Here, the number of distinct discrete values of the ith solution vector $(\phi_k \neq \phi_i)$ is used in the discrete approximation $(\Delta^2\phi)_i$, and α_k are constant coefficients of the values ϕ_k that define the discrete approximation. The coefficient of the value ϕ_i is α_i and is dependent on the values α_k Eq. 34.43. Figures 34.5a and 34.5b show the stencils of two discrete approximations of $(\hat{\nabla}^2\phi)_i$ in two dimensions. The boxes represent the discrete values ϕ_k and the numbers in the box give the value of the coefficient α_k associated with ϕ_k. The center box represents the discrete value ϕ_i and contains the value of $-\alpha_i$.

If each of the discrete values $\phi_k = \phi(r_k)$ of Eq. 34.43 are expanded in Taylor series about the r_i, in physical space, then

$$\omega_i = \Big|\phi_x R_x + \phi_y R_y + \phi_z R_z + \frac{1}{2}\phi_{xx}R_{xx} + \frac{1}{2}\phi_{yy}R_{yy} + \frac{1}{2}\phi_{zz}R_{zz} + \phi_{xy}R_{xy} + \phi_{yz}R_{yz} + \phi_{zx}R_{zx} + O(\Delta r^3)\,V_i\Big| \tag{34.45}$$

results, where

$$\begin{aligned} R_x &= \left(\frac{V_i}{2}\right)\sum_{k=1}^{N_k}\alpha_k(x_k - x_i) \approx (V_i)\overline{\Delta x_i} \equiv \int_{\Omega_i}\Delta x_i dV = I_x \\ R_{xx} &= \left(\frac{V_i}{2}\right)\sum_{k=1}^{N_k}\alpha_k(x_k - x_i)^2 \approx (V_i)\overline{\Delta x_i}\,\overline{x_i} \equiv \int_{\Omega_i}\Delta x_i^2 dV = I_{xx} \\ R_{xy} &= \left(\frac{V_i}{2}\right)\sum_{k=1}^{N_k}\alpha_k(x_k - x_i)(y_k - y_i) \approx (V_i)\overline{\Delta x_i}\,\overline{\Delta y_i} \equiv \int_{\Omega_i}\Delta x_i\Delta y_i dV = I_{xy} \end{aligned} \tag{34.46}$$

The relations given by Eq. 34.46 show that the approximate discrete weight function will behave similarly to the analytic weight function, if the terms R_x, R_y, R_z, R_{xx}, R_{yy}, R_{zz}, R_{xy}, R_{yz}, R_{xz}, are close approximations of I_x, I_y, I_z, I_{xx}, I_{yy}, I_{zz}, I_{xy}, I_{yz}, I_{xz}, respectively.

Note that for the stencil given in Figure 34.5.a, the term will go to zero for an evenly spaced skewed cell. However, the term I_{xy} will not be zero unless the grid cell is orthogonal. Therefore, using the stencil of Figure 34.5.b to approximate the Laplacian may result in highly skewed cells. Orthogonality can be enforced by considering the stencil shown in Figure 34.5.b. For this stencil, R_{xy} will go to zero only if the cell is orthogonal.

If the weight function is to be formed from a set of N_l dependent variables, $\phi_i^{(l)}$, then it is defined as

$$\omega_i = \sqrt{\sum_{l=1}^{N_l}(\omega_i^{(l)})^2} = \left\| w_i^{(l)} \right\|_2 \tag{34.47}$$

where

$$\omega_i^{(l)} = \frac{V_i}{2}\left|\Delta^2\phi^{(l)}\right)_i\Big| \tag{34.48}$$

A large range in the magnitude of the variables may occur in the computational domain. Therefore, it may be desirable to scale the weight function by the solution. The weight function can be scaled by using the relation

$$\omega_i = \frac{\sqrt{\sum_{l=1}^{N_l}(\omega_i^{(l)})^2}}{\sqrt{\sum_{l=1}^{N_l}(\phi_i^{(l)})^2 + \mathrm{E}}} \tag{34.49}$$

where the constant $\mathrm{E} > 0$ is a small number that prevents a division by zero if $\phi_i^{(l)} = 0$.

Control over the grid-node density distribution is gained by using the weight function given by Eq. 34.48 or Eq. 34.49 with $\omega_i^{(l)}$ defined as

$$\omega_i^{(l)} = V_i^{1+w_1}\left|\Delta^2\left(\phi^{(l)}\left(\frac{V}{V_i}\right)^{w_2}\right)_i + \omega_{min}\right| \tag{34.50}$$

where ω_{min}, w_1, and w_2 are user specified parameters.

The parameter w_1 controls whether emphasis is placed on small or large volume grid cells. If $w_1 > 0$, then larger cells will be weighted more heavily than smaller cells, relative to the non-modified weight function given by Eq. 34.48 or Eq. 34.49. Similarly, if $w_1 < 0$, then small cells will be weighted more heavily than larger cells, relative to the nonmodified weight function. A consequence of choosing $w_1 > 0$ is that weak solution features, e.g., shock waves, will be less resolved, than when $w_1 = 0$ is specified. A consequence of choosing $w_1 < 0$ is that smooth flow regions may be underresolved.

The parameter $w_2 \geq 0$ allows control over the rate of change of the cell volumes in the grid. Setting $w_2 > 0$ will tend to prevent the evacuation of grid nodes from regions of uniform flow and will promote grid cell orthogonality. Note that if the value of w_2 is such that

$$\left(\frac{V}{V_i}\right)^{w_2} > \phi^{(l)} \tag{34.51}$$

then adaptation to the solution will be lost.

The parameter ω_{min} is the minimum allowable weight function value and is typically set to

$$10 \text{ X machine zero} \leq \omega_{min} \leq 1\times 10^{-2}$$

The upper range of values are specified if it is desired to adapt the grid only to regions associated with prominent errors, as indicated by the weight function.

Because of machine round-off errors, the weight function will contain noise that must be eliminated so that smooth grids can be produced. The noise is eliminated by applying an elliptic smoother to the weight function [14]. Typically, two to five passes of the weight function through the elliptic smoother are sufficient to produce a smooth grid.

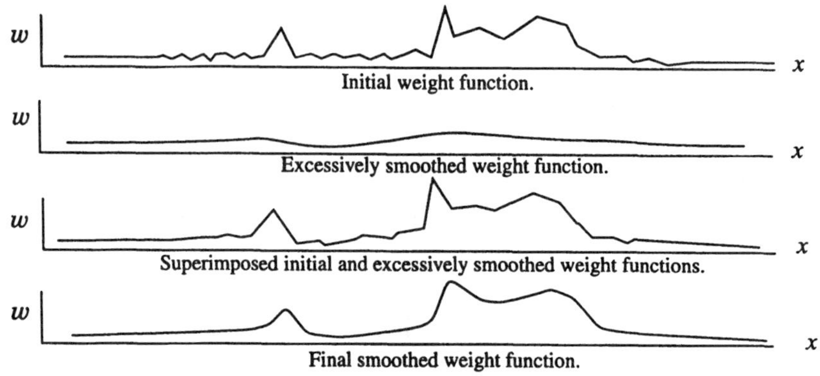

FIGURE 34.6 Weight function excessive smoothing procedure.

The weight function in uniform regions of the flow has a zero value. If an explicit method is used to reposition the grid nodes, then the movement of the grid nodes in these regions will be slight. In order to increase the movement of the grid nodes from nonactive regions of the computational domain to regions of interest, the following procedure is used [15]. The initial weight function values are smoothed excessively using the elliptic smoother. The excessively smoothed weight function values are then superimposed with the initial weight function values and again smoothed to eliminate any noise that might be present. This procedure is depicted in Figure 34.6.

34.5.2 Transformation to Physical Space

The transformation from parametric space to physical space (Eq. 34.18) can also be written as

$$\Delta r_i = r_\xi \Delta\xi + r_\eta \Delta\eta + r_\zeta \Delta\zeta \tag{34.52}$$

where

$$\Delta r_i = r_i^{(\text{new})} - r_i^{(\text{old})}$$

represents the change in the x, y, and z position coordinates of grid-node i in physical space, and

$$\Delta\xi = \xi_i^{(\text{new})} - \xi_i^{(\text{old})} \qquad \Delta\eta = \eta_i^{(\text{new})} - \eta_i^{(\text{old})} \qquad \Delta\zeta = \zeta_i^{(\text{new})} - \zeta_i^{(\text{old})} \tag{34.53}$$

are the grid-node position changes in parametric space. The transformation given by Eq. 34.53 can lead to grid-line crossover if the grid cell is distorted, i.e., if the grid-cell geometry significantly deviates from a parallelogram.

The higher-order transformation

$$\Delta r_i = r_\xi \Delta\xi + r_\eta \Delta\eta + r_\zeta \Delta\zeta + r_{\xi\eta} \Delta\xi\Delta\eta + r_{\eta\zeta} \Delta\eta\Delta\zeta + r_{\xi\zeta} \Delta\xi\Delta\zeta + r_{\xi\eta\zeta} \Delta\xi\Delta\eta\Delta\zeta \tag{34.54}$$

which includes cross-derivative terms, can be used to reduce the occurrence of grid-line crossover.

34.5.3 Grid Adaptation Cut-Off Criteria

The adaptation process is stopped when any one of a number of user specified tolerances is exceeded. For example, the adaptation process will stop if the maximum number of allowed adaptive iterations is exceeded; the maximum grid-node translation distance is below a specified value, e.g., the grid is converged; the standard deviation of the weight function is below a specified value, e.g., the weight function

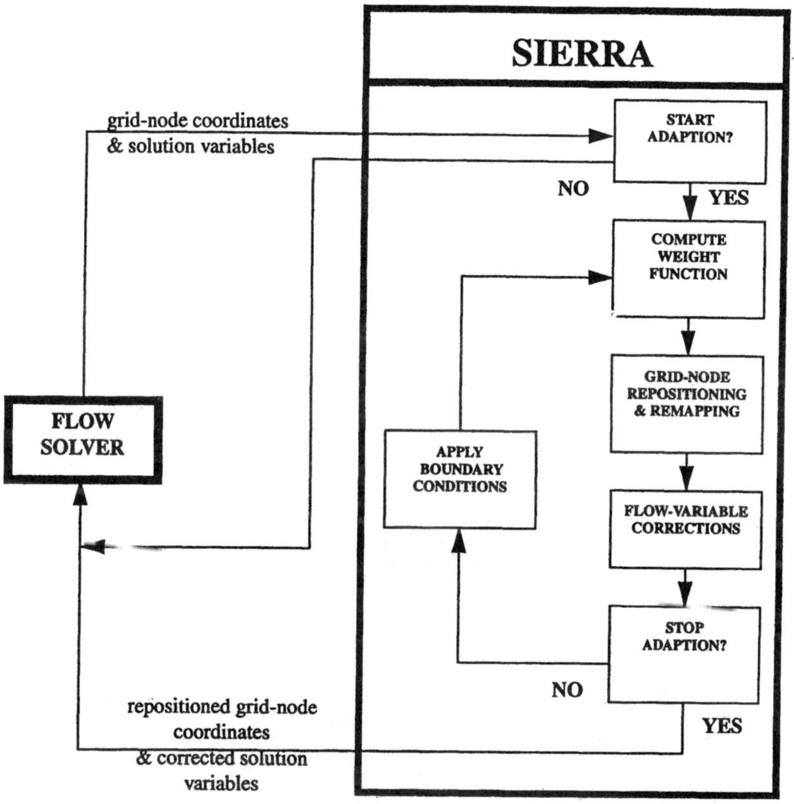

FIGURE 34.7 SIERRA flow chart.

is equally distributed; the maximum value of the weight function is below a specified value, e.g., the solution error measure is small; or the percent change in the global value of any of the solution variables exceeds a specified value, e.g., global conservation is violated.

34.5.4 Interim Steps

An interim step procedure can been added to the solution-variable correction procedure to increase the accuracy of the variable corrections. The interim-step procedure is performed by dividing the time step $\Delta t_g = t_g^{n_g+1} - t_g^{n_g}$ into M smaller interim steps, δt_g, i.e., $\Delta t_g = M\delta t_g$. If the change in position of a grid-node v over the time step Δt_g is given by $\Delta r_v = r_v^{n_g+1} - r_v^{n_g}$, then the change in position of grid-node v over the time step δt_g is $\delta r_v = (\Delta r_v)/M$. Because the grid-node movement over each interim step is a fraction of the total grid-node movement, the magnitude of the cell side-sweep volumes (CSSV) associated with each interim step δt_g is smaller than the magnitudes of the CSSVs associated with the time step Δt_g.

The solution-variable correction $\overline{U}^{n_g+1}$ is obtained by iteratively applying the approximate CIE,

$$\overline{U}^{n_g+\beta_m} = \frac{1}{V^{n_g+\beta_m}}\left((V\overline{U})^{n_g+\beta_{m-1}} + \sum_{p=1}^{N_p}(\overline{V}\,\overline{U})_p\Big|_{n_g+\beta_{m-1}}^{n_g+\beta_m}\right) \tag{34.55}$$

M number of times. The interim step counter is denoted by m, where $m = 1, 2, \ldots, M$. Here, $\beta_m = m/M$ and $\beta_0 = 0$ so that $\overline{U}^{n_g+\beta_0} = \overline{U}^{n_g}$ and $\overline{U}^{n_g+\beta_M} + \overline{U}^{n_g+1}$.

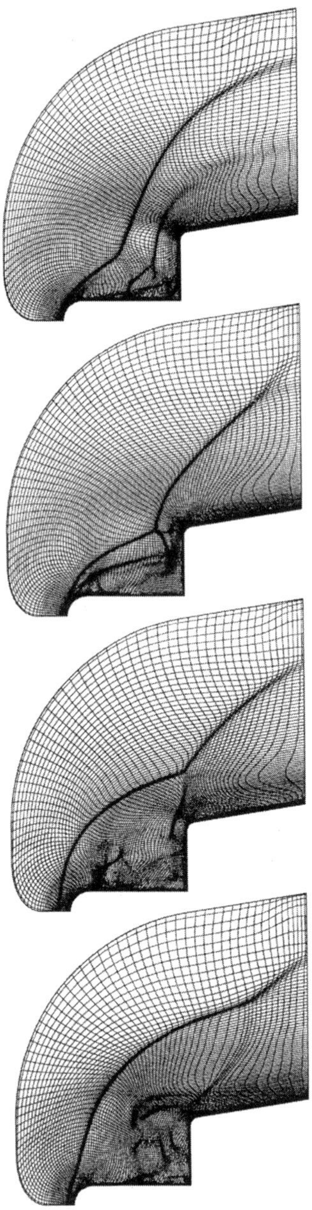

FIGURE 34.8 Spike-tipped body geometry.

In Eq. 34.55, the cell volume at time $t_g^{n_g+\beta_m}$ is $V_g^{n_g+\beta_m}$, the cell volume at time $t_g^{n_g+\beta_{m-1}}$ is $V_g^{n_g+\beta_{m-1}}$, and $V|_{n_g+\beta_{m-1}}^{n_g+\beta_m}$ is the volume swept out by cell side p from time $V_g^{n_g+\beta_{m-1}}$ to time $t_g^{n_g+\beta_m}$. The cell volumes and the CSSVs associated with the interim-step procedure are computed according to formulas presented in [9], using the grid-node locations defined at the appropriate interim step. The position coordinates of grid-node v at time $t_g^{n_g+\beta_m}$ are given by

$$
\begin{aligned}
x_v^{n_g+\beta_m} &= x_v^{n_g+\beta_{m-1}} + \left(x_v^{n_g+1} - x_v^{n_g}\right)/\,M \\
y_v^{n_g+\beta_m} &= y_v^{n_g+\beta_{m-1}} + \left(y_v^{n_g+1} - y_v^{n_g}\right)/\,M \qquad (34.56) \\
z_v^{n_g+\beta_m} &= z_v^{n_g+\beta_{m-1}} + \left(z_v^{n_g+1} - z_v^{n_g}\right)/\,M
\end{aligned}
$$

which assumes that the grid-node moves with a constant velocity over the time step, Δt_g. In general, choosing the value of M so that $\delta r_v < \frac{\Delta L}{8}$, where ΔL is the local dimension of the cell in the direction of the grid-node movement, will produce accurate solution-variable corrections.

34.6 Results

In order to illustrate the operation and effectiveness of DSAGA and SIERRA, we have included selected results. These are chosen in order to illustrate the adaptive techniques rather than to highlight the particular application.

To begin, some observations based on our experience are offered:

1. Alignment of the mesh with physical features in the flow is more important than achieving minimum spacing.
2. If the mesh is aligned with the feature as in 1 (above), skewness does not noticeably degrade the solution.
3. Worst-case resolution of strong features, such as shock waves, occurs when they are diagonal to a low aspect ratio Cartesian-like grid. Note that upwind solvers may contribute to this behavior.

We will indicate locations in these results that support these observations.

The initial goal for DSAGA was to improve accuracy for unsteady flow calculations, with steady-state accuracy improvement as a converged result. Unfortunately, the body of detailed experimental data for unsteady flows is not large. One data set that is frequently used was obtained for supersonic flow over a spike-nosed bluff conical body at supersonic flow conditions for which a self-excited oscillatory flow occurs. Some high-frequency data [4,21] were obtained that we have used for comparison [10]. Figure 34.9 illustrates the shape of the spiked-nosed body. Figure 34.9 contains results at four time steps during the oscillatory cycle. In this case the 100 × 100 grid was mapped such that 100 points lie on the spike and 100 points on the cone [10]. This mapping also resolves the spike-cone junction well, which proved to be crucial for obtaining the correct oscillation frequency. Figure 34.10a gives the Fourier analysis of the pressure signal compared with experiment at a point on the bluff cone face, and the waveform is shown in Figure 34.10b.

The ability of SIERRA to enhance solution quality is demonstrated first by numerical simulations of a laminar viscous supersonic channel flow [15], using both a static evenly spaced fine grid and an r-refined adapted grid. The static grid (121 streamwise by 91 crossflow, evenly spaced nodes) is used as the initial grid for the r-refined grid simulation. A 15 degree compression ramp and a 15 degree expansion corner are used as a shock and expansion wave generator. Volume weight parameters were $w_1 = 1$, $w_2 = 0$, and $\omega_{min} = 1 \times 10^{-6}$. One interim step, a single RK procedure, third-order accurate cell side average flux values, and a conservative limiter were employed by SIERRA.

Figure 34.11 illustrates the channel geometry and shows the SIERRA weight function distribution for a solution obtained on the initial 121 × 91 static grid. This plot is useful for determining where higher resolution would reduce interpolation error. The results of repeating this solution with the mesh adapted by SIERRA are shown in Figure 34.12. Figure 34.13 shows the weight function distribution for this case.

It is apparent that use of SIERRA has resolved the solution to the extent that the density contours approach the detail present in a schlieren photograph. Of particular note is the manner in which the compression waves at viscous layer separations and reattachment coalesce to form shock waves. Also, the flow structure can be analyzed by examining the adapted grid alone.

Figure 34.14 shows details of the vortical structure where the ramp shock wave interacts with the upper viscous layer. The resolution of the impinging shock wave and the alignment with the flow direction reveals three vortex structures with a full saddle point between two of them.

The mesh independence of the adapted result was assessed by repeating the solution on a 533 × 721 evenly spaced static grid. This would place approximately 95 mesh lines in the vortical structure resolved by 17 to 18 lines in the adapted case. Figure 34.15 illustrates the streamlines for the same region

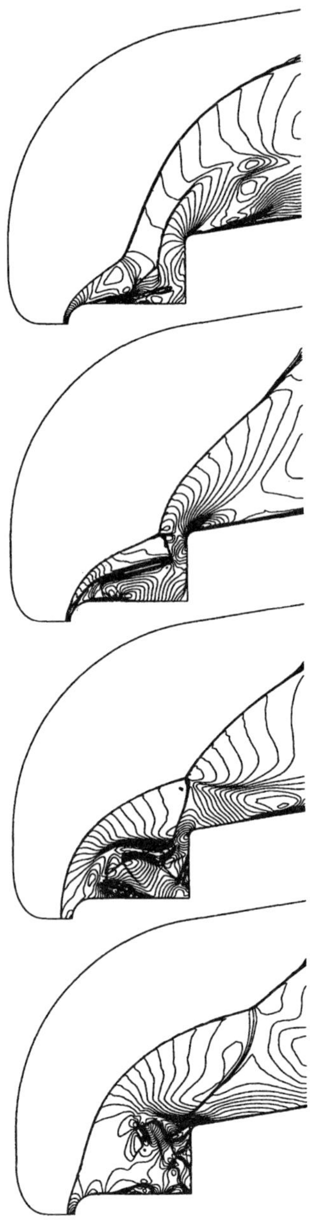

FIGURE 34.9 Adapted grid and Mach contours series during oscillation cycle over spike-tipped body, 100×100 grid.

shown in Figure 34.14. Note that little change has occurred, indicating that the adapted solution may be approaching grid independence for this case with a relatively small total number of nodes.

The adapted grid for this case provides excellent support for statements made in the grid quality section. The following observations are appropriate:

1. The grid lines have been aligned to a great extent with the strong features of the flow.
2. Because of this alignment grid, skewness has been *increased* in the shock transitions rather than decreased. In spite of this, it is obvious that an excellent solution has been obtained, hence our earlier statement that skewness does not degrade the solution appreciably if the mesh is aligned locally with the solution features.

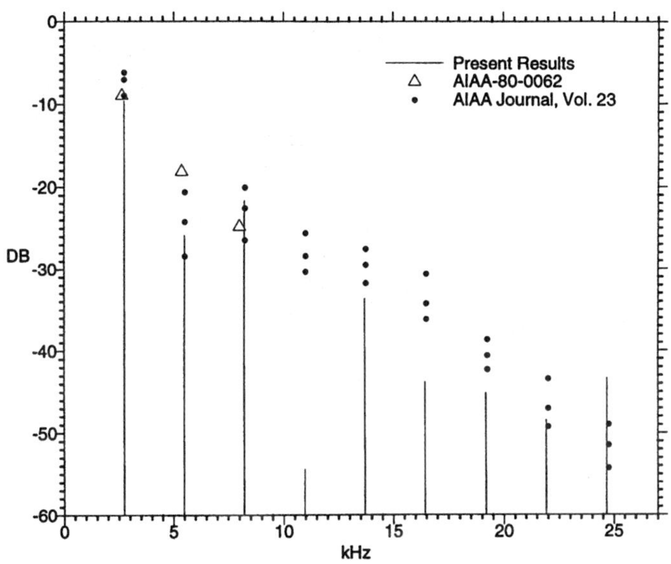

FIGURE 34.10a Comparison of computed spectral data [17,18] with experiment [16], 100 × 100 grid.

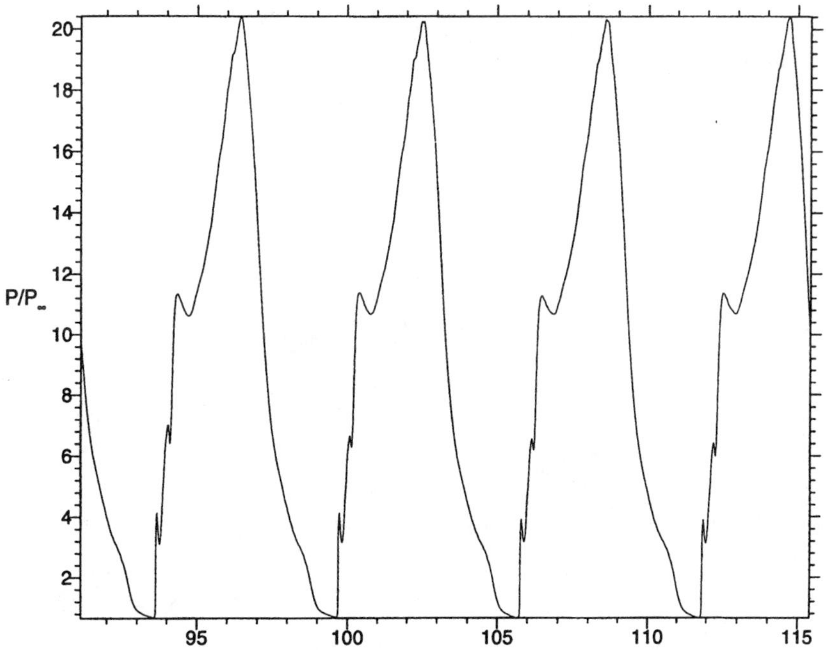

FIGURE 34.10b Computed pressure waveform on bluff face of cone, 100 × 100 grid.

3. Also due to the alignment, this well resolved solution was obtained with relatively large minimum cell volumes. For example, the large vortical structure on the upper surface was resolved by only 17–18 mesh lines in the direction normal to the surface.
4. For steady solutions, mesh cells can be evacuated from constant property regions without solution degradation. (Note that this may not be appropriate for unsteady flows with rapidly translating features.)

FIGURE 34.11 Static grid weight function distribution for 2D viscous laminar supersonic channel flow.

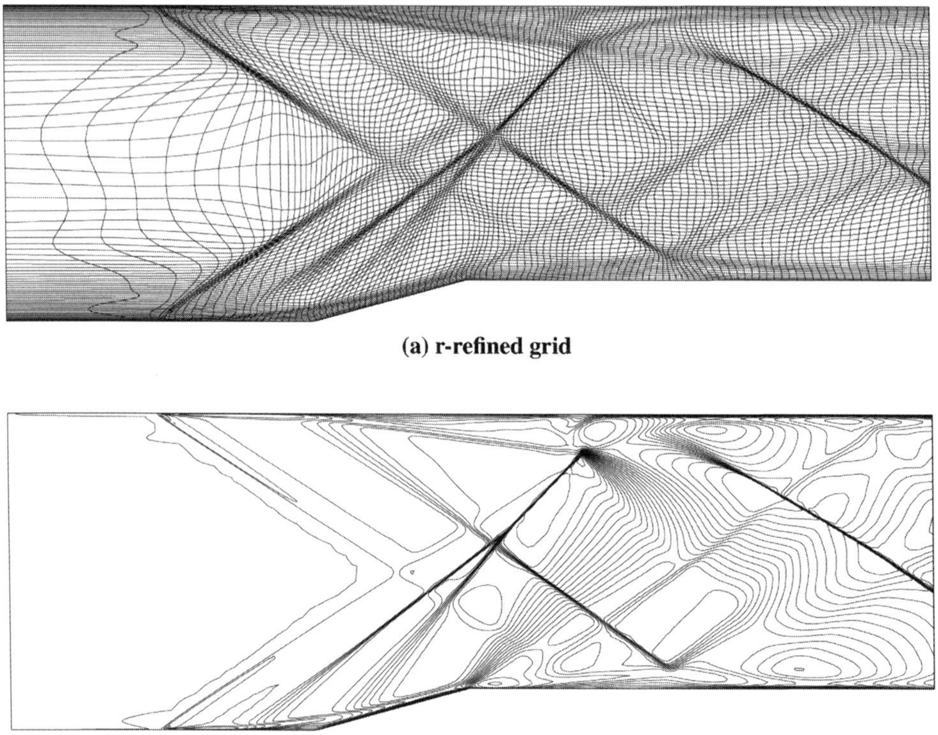

(a) r-refined grid

(b) density contours

FIGURE 34.12 *r*-Refined grid and density contours for 2D viscous laminar supersonic channel flow. Inflow Mach number is 2.0.

The next demonstration of SIERRA will illustrate dynamic adaptation to an impulsively started inviscid flow in the above 2D geometry. The conditions are $M = 1.8$ and 97×31 grid nodes. The developing flow was adapted each time step with $w_1 = 0.50$, $w_2 = 0$, and $\omega_{min} = 1 \times 10^{-6}$. As this solution begins (Figure 34.16), SIERRA moves nearly all of the nodes to the vicinity of the ramp. The initial development of the shock and expansion waves is highly resolved. As these features move into the outer flow, points are redistributed to maintain resolution in the disturbed portion of the domain. The constant property region remains nearly evacuated of nodes.

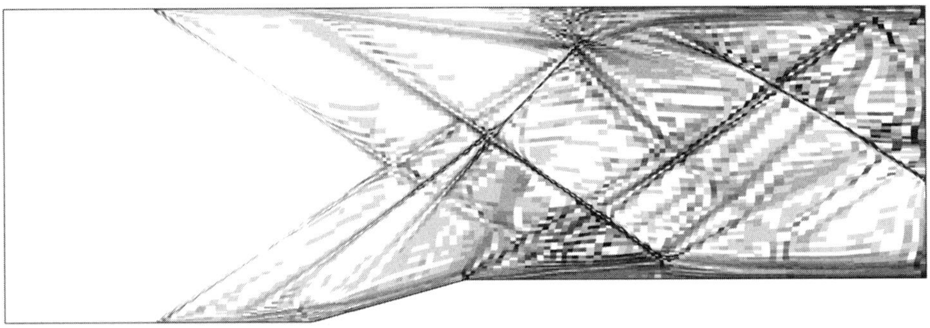

FIGURE 34.13 *r*-Refined grid weight function distribution for 2D viscous laminar supersonic channel flow. Inflow Mach number is 2.0.

It is interesting to note that none of these meshes appear to meet conventional standards of quality. Skewness, high aspect ratio cells, rapid cell volume change, and large line curvature are present in each of the grids shown. Yet examination of the Mach contours for smoothness and resolution reveals that the grids are, in fact, allowing the solver to produce a continuously well- resolved dynamic solution.

34.6.1 Experimental Comparisons

Numerical simulations of two experimental investigations were conducted using SIERRA with CFL3D [12] SIERRA was modified to read in the CFL3D grid and restart files, perform the *r*-refinement adaptation, and rewrite the new grid and redistributed primitive flow variables to the CFL3D grid and restart files. Grid adaptation was performed every tenth time-iteration step of the flow solver, as only steady state simulations were considered. This method of coupling SIERRA with CFL3D is not computationally efficient, but it illustrates how SIERRA can be used completely independently with a flow solver to provide *r*-refinement adaptation capability. No modifications of any kind were made to the CFL3D source code or input file.

For both test problems, SIERRA employs one interim step with the one RK procedure, third-order accurate cell side average value (CSAV) approximations, and the conservative limiter. Grid-node movements were restricted such that the CSSV restrictions given in Section 34.5.4 were satisfied for the 2D and 3D simulation, respectively. The volume weight parameters of the first simulation were $w_1 = 1$, $w_2 = 0$, and $\omega_{\min} = 1 \times 10^{-12}$. The volume weight parameters of the second simulation were $w_1 = 0$, $w_2 = 1 \times 10^{-8}$, and $\omega_{\min} = 1 \times 10^{-15}$.

34.6.1.1 Hypersonic 2D Compression Corner

The first experimental test problem is a Mach 14.1 2D flow over a compression corner that is formed by a wedge intersecting a flat plate at 18°. This test case was experimentally investigated by Holden and Moselle [7] in the Calspan 48-inch Shock Tunnel. The freestream conditions are $M_\infty = 14.1$, $T_\infty = 160°$R, and Reynolds number of $Re = 7.2 \times 10^4$ per foot, so the flow is considered to be laminar. The wall temperature is $T_w = 535°$ R. The wedge begins $x_L = 1.0$ foot from the leading edge of the plate. The results of a previous numerical investigation of this experiment that used CFL3D [19,20] led to the correction of the originally released experimental data. The present numerical results are compared with the corrected experimental data.

As a test of how well a laminar viscous flow could be resolved with very few points, the simulation was first performed with a 49 × 33 evenly spaced initial grid. Results were adequate in all but heat transfer.

The case was then repeated with an initial grid of 101 × 51. Relatively small changes occurred in surface pressure and skin friction but heat transfer is improved. Figures 34.17 and 34.18 show results of the 101 × 51 SIERRA-adapted grid and solution. Previous simulations of this flow used larger numbers of grid cells.

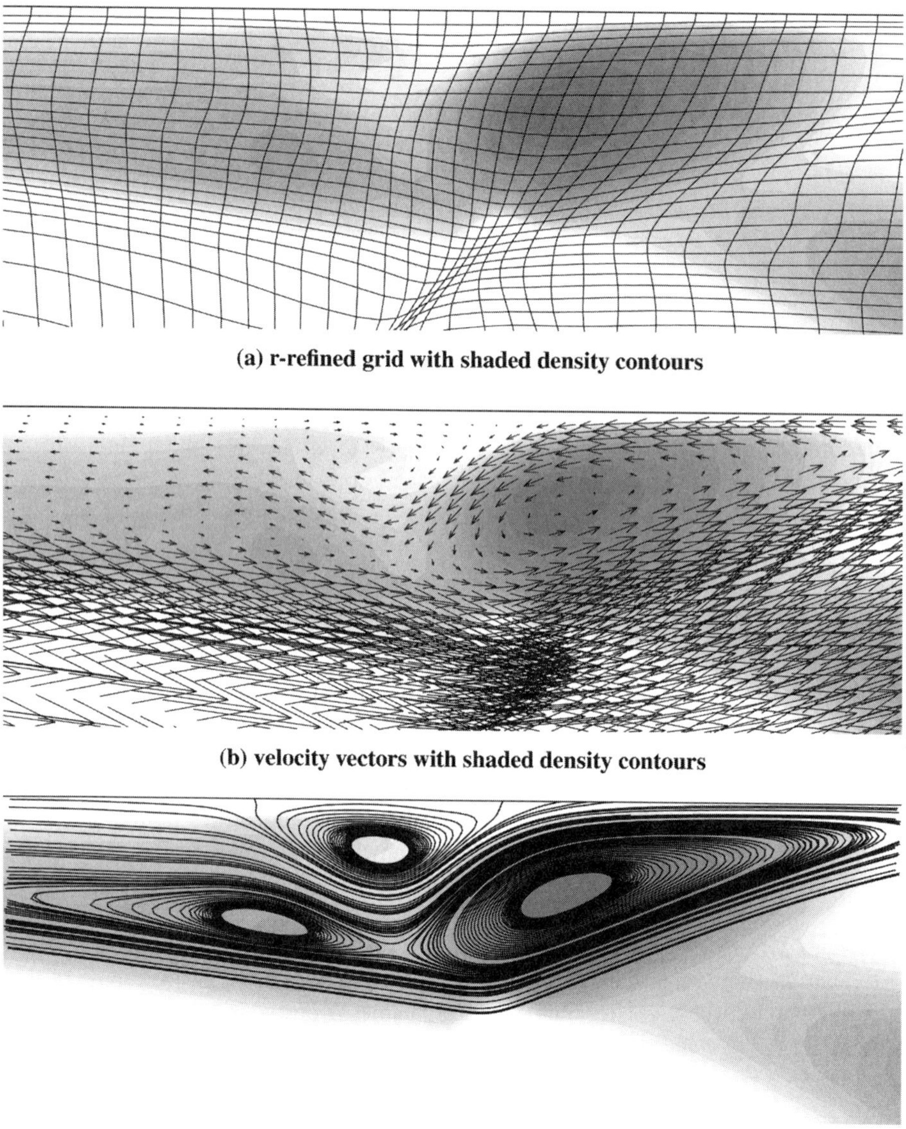

(a) r-refined grid with shaded density contours

(b) velocity vectors with shaded density contours

(c) streamlines with shaded density contours

FIGURE 34.14 Upper-surface shock induced boundary layer separation region predicted by adapted grid computation.

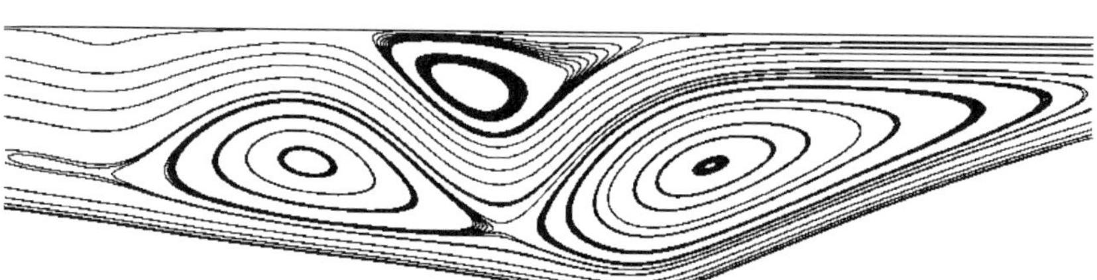

FIGURE 34.15 Streamlines in upper surface separated boundary layer obtained from fine static grid computation.

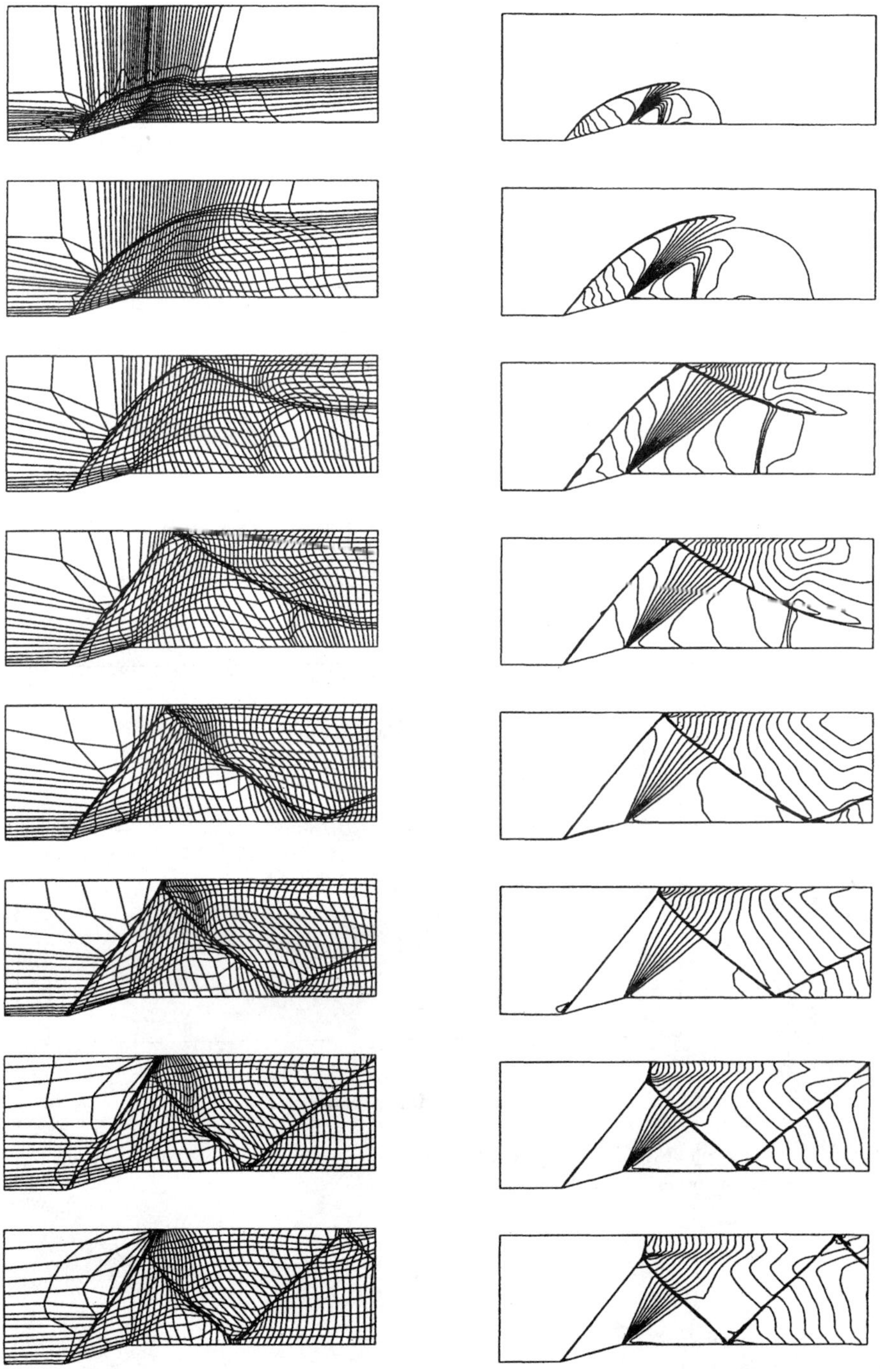

FIGURE 34.16 Developing grid and solution of 2D inviscid supersonic channel flow.

34.6.1.2 Supersonic 3D Symmetric Corner Flow

The final example is the supersonic flow in a 3D symmetric corner formed by the intersection of two 9.48° wedges. The freestream conditions are $M_\infty = 3.0$, $T_\infty = 105°$ K, and the Reynolds number is $Re = 0.39 \times 10^6$ per meter, with a wall temperature of $T_w = 294°$ K. Experimental data were obtained for this

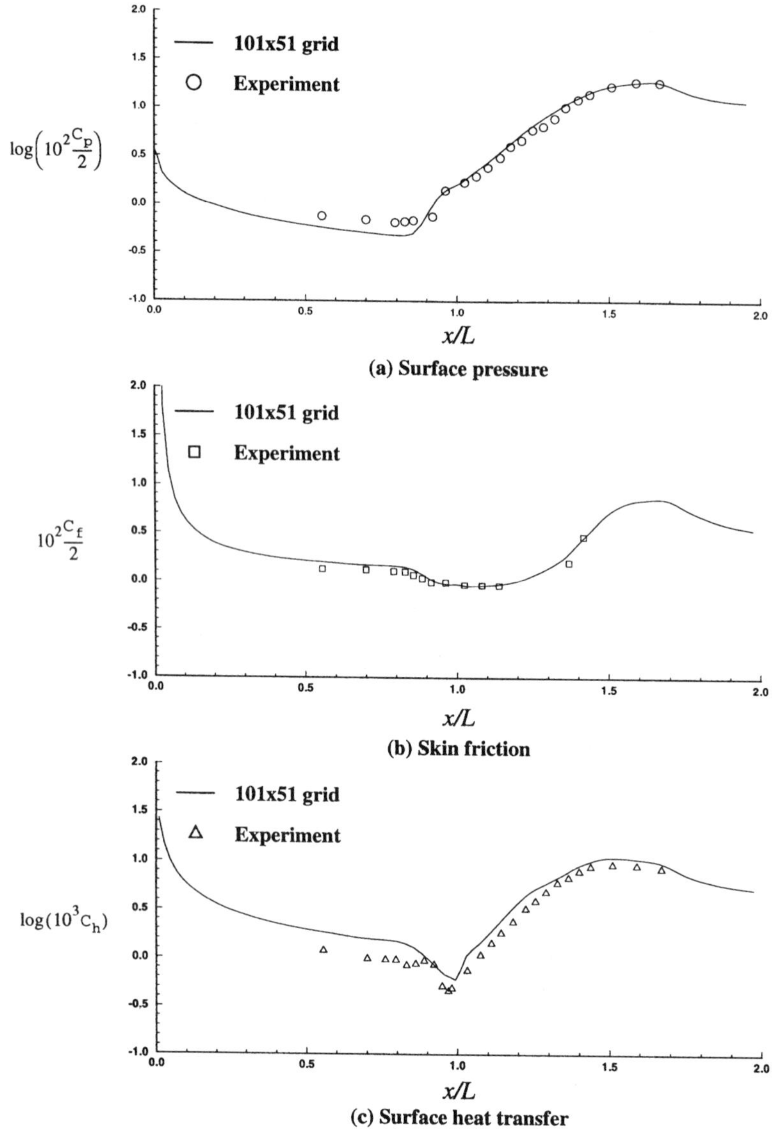

FIGURE 34.17 Comparison of CFL3D *r*-refined 101 × 51 grid computations and experiment, Mach 14.1 flow over an 18-degree compression corner.

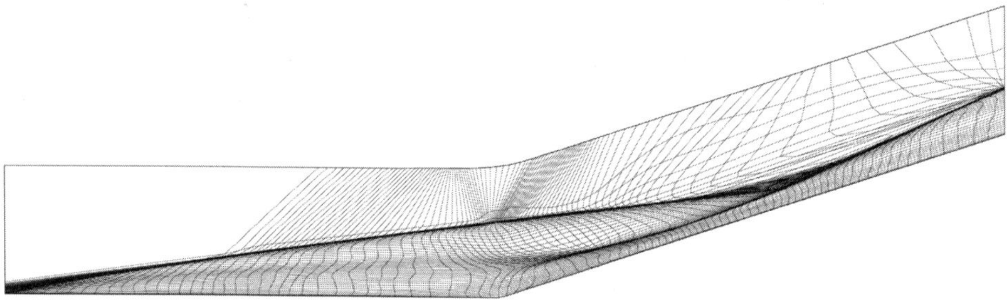

FIGURE 34.18 *r*-refined grid for Mach 14.1 flow over an 18-degree compression corner.

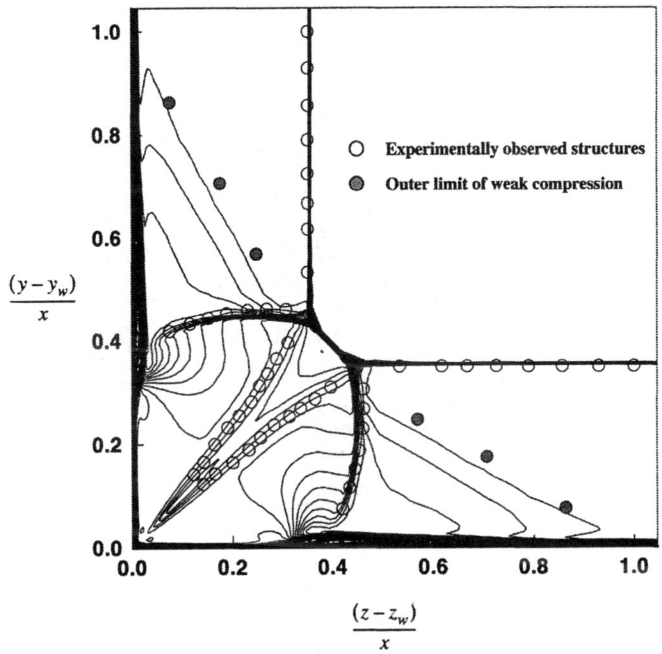

FIGURE 34.19 Computed Mach contours at Re = 0.39×10^6, Mach 3.0 symmetric corner flow.

flow by West and Korkegi [24] The computations were started from freestream conditions and a uniform $57 \times 57 \times 57$ initial grid.

Experimental pitot tube pressure surveys and surface pressure distributions in the crossflow plane were obtained at $Re_x = 3.07 \times 10^6$ so that the flow was considered to be laminar. Computed crossflow plane Mach contours at this Reynolds number are shown in Figure 34.19 and are compared to the experimentally observed flow structure. Embedded internal shocks extend from the oblique corner and wedge shock intersections toward the wedge surface, where the boundary layer is separated. Weak separation induced compression waves from which intersect the embedded internal shocks. Also, curvature of the slip lines that extend from the intersection of the oblique corner shocks and wedge shocks toward the wedge intersection is induced by crossflow expansion. The computed flow structures are highly resolved and are in excellent agreement with the experimental pitot tube pressure survey observations. The agreement with wedge surface data is less adequate, but is better than previous fixed grid results and a fixed grid $57 \times 113 \times 113$ solution with CFL3D. The reduced level of agreement is attributed to the fact that transition is evident just past the location at which data were collected, indicating that the data may have been transitional.

The converged *r*-refined grid for this simulation is shown in Figure 34.20. Adaptation to the shocks and boundary layer are evident in the crossflow plane grid. Adaptation to the regions of weak compression waves can also be seen. Note that large nonorthogonal grid cells remain in the uniform flow regions where the spatial resolution of the flow is not required. The wedge surface grids indicate extensive grid-node clustering near the boundary layer reattachment point, just inside of the embedded internal shocks, and at the intersection of the wedges. Note that grid cells along the wedge surfaces where properties vary linearly exhibit orthogonality and have smoothly varying volumes.

34.7 Summary and Conclusions

Two algorithms, DSAGA and SIERRA, for dynamic *r*-refinement adaption of structured grids have been described and demonstrated. The goal for these algorithms is to improve spatial resolution of numerical solutions to conservation laws while preserving temporal accuracy. This is accomplished by defining a

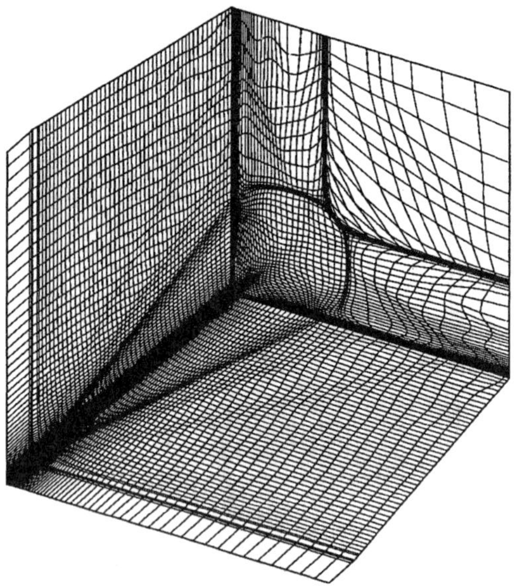

FIGURE 34.20 *r*-Refined grid for Mach 3.0 symmetric corner flow.

grid which moves relative to the original inertially defined mesh. The transformed conservation law is then split into two steps in which a new solution is obtained on the last available initial or adapted grid. A weight function is calculated based on this new solution that is large where additional resolution is needed. This weight function is used in a mass-weighted algorithm to relocate points such that resolution is improved. The solution is then redistributed to these new node locations which becomes the input to the next marching step of the flow solver. Therefore, for each marching step that uses initial data from a previously adapted solution, the solution is well resolved and truncation error will be reduced. Temporal accuracy remains that provided by the solver.

The original algorithm, DSAGA, was used to introduce the details of the parametric space upon which adaption occurs and the simple algorithm that allows transform of the new mesh locations to physical space without searches. The mass-weighted algorithm is also described. Results are shown for dynamic adaption of a self-excited excillatory flow with excellent agreement with experimental data from spectral frequencies of 2.8 KHz to 25 KHz.

The new algorithm, SIERRA, contains important advances over DSAGA. Rather than using specific algorithm truncation error as a weight function criteria, SIERRA is based on a measure of how well the local cell volume and orientation resolves the solution. This solver-independent error criteria uses a determinant of local grid quality to form the weight function used to adapt the mesh. This means that mesh quality is *based on the local solution*, not a set of preconceived standards.

SIERRA also contains an interim step algorithm for improving the accuracy and robustness of the redistribution of the solution to the new adapted grid. Improved techniques are included for ensuring that conservation is preserved when the conserved quantities contained in the swept volumes are calculated.

Results obtained through use of SIERRA were shown for 2D viscous and inviscid flows and 3D viscous flows. A steady viscous laminar solution on a 101 × 51 grid adapted by use of SIERRA was shown to be extremely well resolved when compared with a 533 × 755 fixed grid solution. Density contour plots for this case approach Schlieren photograph resolution. A developing inviscid flow in the same geometry is shown to be extremely well resolved and clean, even though the grid appears to be of poor quality by conventional standards. As a further example, SIERRA was used for uncoupled adaption with the NASA code CFL3D. This interaction involved periodic output of the mesh and solution from CFL3D. The mesh was adapted and the solution redistributed by SIERRA after which the CFL3D restart file was overwritten. Excellent results were obtained as compared with experiment and fixed grid solutions.

The r-refinement algorithms, DSAGA and SIERRA, were shown to greatly improve results on grids with few mesh nodes. Based on this and prior work, we offer the following observations and conclusions for the reader:

1. Grid quality can only be assessed in terms of the local solution variation.
2. Alignment of the grid with strong solution features is at least as important as the reduction of cell volumes at those features.
3. Skewness of the mesh cells causes problems only when inappropriate for the local solution or when some part of the solver is not transformed or projected accurately.
4. Dynamic adaption of both steady and unsteady flows with temporal accuracy preserved was demonstrated.
5. SIERRA, in stand-alone form, can be used to provide single-grid block mesh adaption for any code using a structured body-fitted mesh. Some work remains in the area of complex surface definition for moving surface nodes.

34.8 Research Issues, Current and Future

Presently, various versions of these algorithms are being applied to simulate unstart of hypersonic aircraft inlets and to improve accuracy of environmental air quality models. Dynamic *r*-refinement for 2D unstructured meshes has been implemented and shown to improve mesh characteristics.

A future task is the extension of SIERRA to allow adaption of 3D multiblock grids. We anticipate that this extension may reduce portability, since block numbering and structure, etc. tends to vary between codes and grid generators. This work will be based on a current 2D multiblock version of DSAGA. We also plan to develop further the weight function and redistribution routines presently in SIERRA. Finally, much work remains in the area of geometry definition and in the interaction between solvers, models, and moving grids.

References

1. Brackbill, J. U. and Saltzman, J., An adaptive computation mesh for the solution of singular perturbation problems, *Numerical Grid Generation Techniques*, NASA Conference Publication 2166, pp. 193-196, 1980.
2. Benson, R. A. and McRae, D. S., Time accurate simulation of unsteady flows with a dynamic solution adaptive mesh, Proceedings of the 4th International Conference on Numerical Grid Generation in Computational Fluid Dynamics and Related Fields, Swansea, U.K., April 1994.
3. Benson, R. and McRae, D. S., A solution adaptive mesh algorithm for dynamic/static refinement of two and three dimensional grids, 3rd International Conference on Numerical Grid Generation in Computational Fluid Dynamics and Related Fields, Barcelona, Spain, June 1991.
4. Calarese, W. and Hankey, W. L., Modes of shock-wave oscillations on spike tipped bodies, *AIAA Journal*, Vol. 23, No. 2, pp. 185–192, February 1985.
5. Eiseman, P. R., Adaptive grid generation, *Computer Methods in Applied Mechanics and Engineering*, Vol. 64, No. 1–3, pp. 321–376, October 1987.
6. Hentschel, R. and Hirschel, E. H., Self adaptive flow computations on structured grids, Proceedings of the Second European Computational Fluid Dynamics Conference, pp. 242–249, September 1994.
7. Holden, M. S. and Moselle, J. R., Theoretical and experimental studies of the shock wave-boundary layer interaction on compression surfaces in hypersonic flow, ARL 70-0002, Aerospace Research Laboratories, Wright-Patterson AFB, OH, January 1970.
8. Ilinca, A., Camareo, R., Trepanier, J. Y., and Reggio, M., Error estimator and adaptive moving grids for finite volume schemes, *AIAA J.*, Vol. 33, No. 11, pp. 2058–2065, November 1995.
9. Ingram, C. L., Laflin, K. R., and McRae, D. S., A structured multi-block solution-adaptive mesh algorithm with mesh quality assessment, Proceedings of the ICASE LaRC Workshop on Adaptive Grid Methods, Hampton, VA, Nov. 7–9, 1994.

10. Ingram, C. L. and McRae, D. S., Extension of a dynamic solution - adaptive grid algorithm and sober to general structured multi-block configurations, AIAA 96-0294, AIAA 34th Aerospace Sciences Meeting, Reno, NV, Jan. 1996.
11. Kim, Y.-M. and Gatlin, B., Incompressible viscous flows on adaptive multi-block grids, AIAA Paper 93-0770, January 1993.
12. Krist, S. L., Biedron, R. T., and Rumsey, C. L., CFL3D user's manual (version 5.0), *Aerodynamic and Acoustic Methods Branch*, NASA Langley Research Center, 1996.
13. Laflin, K. R. and McRae, D. S., Solution-dependent grid-quality assessment and enhancement, 5th International Conference on Numerical Grid Generation in Computational Field Simulations, April 1-5, 1996.
14. Laflin, K. and McRae, D. S., Three-dimensional dynamic viscous flow computations using near-optimal grid redistribution algorithm, Proceedings, First AFOSR Conference on Dynamic Motion CFD, Rutgers Univ., New Brunswick, NJ, June 2–5, 1996, pp. 245–268.
15. Laflin, K.R., Solver-independent *r*-refinement adaptation for dynamic numerical simulations, Ph.D. Dissertation, Department of Mechanical and Aerospace Engineering, N.C. State University, Raleigh, NC, 1997.
16. Luong, P. V., Thompson, J. F., and Gatlin, B., Solution-adaptive and quality-enhancing grid generation, *J. Aircraft*, Vol. 30, No. 2, pp. 227-234, 1993.
17. Marchant, M. J. and Weatherill, N. P., Adaptivity techniques for compressible inviscid flows, *Computer Methods in Applied Mechanics and Engineering*, North Holland, 106, pp. 83–106, 1993.
18. Marchant, M. J. and Weatherill, N. P., Adaptivity techniques for compressible inviscid flows, *Computer Methods in Applied Mechanics and Engineering*. 1993, North Holland, 106, pp 83–106.
19. Rudy, D. H., Thomas, J. L., Gnoffo, P. A., and Chakravarthy, S. R., A validation study of four Navier-Stokes codes for high-speed flows, AIAA Paper 89-1838, 1989.
20. Rudy, D. H., Thomas, J. L., Kumar, A., Gnoffo, P. A., and Chakravarthy, S. R., Computation of laminar hypersonic compression-corner flows, *J. Aircraft*, Vol. 29, No. 7, pp. 1108–1113, 1991.
21. Shang, J. S., Hankey, W. L., and Smith, R. E., Flow oscillations of spike-tipped bodies, *AIAA Paper 80-0062*, AIAA 18th Aerospace Sciences Meeting, Pasadena, CA, January 1980.
22. Thompson, J. F., Warsi, Z. U. A., and Mastin, C. W., *Numerical Grid Generation*, North Holland, NY, 1985.
23. Warren, G. P., Anderson, W. K., Thomas, J. L., and Krist, S. L., Grid convergence for adaptive methods, *AIAA Paper 91-1592*, April 1992.
24. West, J. E. and Korkegi, R. H., Supersonic interaction in the corner of intersecting wedges at high reynolds numbers, *AIAA J.*, Vol. 10, No. 5, pp 652–656, May 1972.

NCSU Adaptive Grid Bibliography

Benson, R. and McRae, D. S., A three-dimensional dynamic solution adaptive mesh algorithm, AIAA 90-1566, AIAA 21st Fluid, Plasma Dynamics, and Lasers Conference, Seattle, WA, June 1990.

Benson, R. A. and McRae, D. S., Numerical-simulations using a dynamic solution-adaptive grid algorithm, with applications to unsteady internal flows, AIAA 92-2719, 10th Applied Aerodynamics Conference, Palo Alto, CA, June 1992.

Benson, R. A. and McRae, D. S., Numerical simulations of the unstart phenomena in a supersonic inlet/diffuser, AIAA 93-2239, 29th AIAA/SAE/ASME/ASEE Joint Propulsion Conference, Monterey, CA, June 1993.

Benson, R. A. and McRae, D. S., Unsteady transients in a supersonic inlet subject to freestream perturbations and dynamic attitude changes, AIAA 94-0581, 32nd Aerospace Sciences Meeting, Reno, NV, Jan. 1994a.

Carpenter, J. G. and McRae, D. S., Adaption of unstructured meshes using node movement, 5th International Conference on Numerical Grid Generation in Computational Field Simulations, Mississippi State University, April 1–5, 1996.

Ingram, C. L., McRae, D. S., and Benson, R. S., Time accurate simulation of a self-excited oscillatory supersonic external flow with a multi-block solution adaptive mesh algorithm, AIAA 93-3387, 11th Computational Fluid Dynamics Conference, Orlando, FL, July 1993.

Ingram, C. L. and McRae, D. S., Extension of a dynamic solution–adaptive grid algorithm and sober to general structured multi-block configurations, AIAA 96-0294, AIAA 34th Aerospace Sciences Meeting, Reno, NV, Jan. 1996.

Laflin, Kelly R. and McRae, D. S., Stable, Temporally-accurate computations on highly dynamic moving grids, 5th International Conference on Numerical Grid Generation in Computational Field Simulations, Mississippi State University, April 1-5, 1996a.

Neaves, M. D. and McRae, D. S., Numerical investigation of the unstart phenomenon in an axisymmetric supersonic inlet, *Proceedings, International Symposium on Computational Fluid Dynamics in Aeropropulsion, AD-Vol. 49,* ASME, San Francisco, CA, Nov. 12-17, 1995, pp. 149-156.

Neaves, M. D. and McRae, D. S., Numerical investigation of axisymmetric and three-dimensional supersonic inlet flow dynamics using a solution adaptive mesh, 5th International Conference on Numerical Grid Generation in Computational Field Simulations, Mississippi State University, April 1-5, 1996.

Odman, M. T., Mathur, R., Alapaty, K., Srivastava, R. K., McRae, D. S., and Yamartino, R. J., Nested and adaptive grids for multiscale air quality modeling, *Proceedings of the 1995 Joint Summer Research Conference on Analysis of Multi-Fluid Flows and Interfacial Instabilities,* Bay City, MI, SIAM.

Srivastava, R. K., Odman, M. T., and McRae, D., Governing equations of atmospheric pollutant transport, International Specialty Conference on Acid Rain and Electric Utilities, Air & Waste Management Association, Pittsburgh, PA, 1995.

Srivastava, R. K., McRae, D. S., and Odman, M. T., Application of solution adaptive grid techniques to air quality modeling, 5th International Conference on Numerical Grid Generation in Computational Field Simulations, Mississippi State University, April 1–5, 1996.

35

Grid Control and Adaptation

O. Hassan

E. J. Probert

35.1 Introduction

The recent rapid development of solution algorithms in the field of computational mechanics means that presently it is possible to attempt the numerical solution of a wide range of practical problems. The essential prerequisite to a solution process of this type is the construction of an appropriate mesh to represent the computational domain of interest. A widely used approach [17,20] has been to divide the computational domain into a structured assembly of quadrilateral or hexahedral cells, with the structure in the mesh being apparent from the fact that each interior nodal point is surrounded by exactly the same number of mesh cells (or elements). Generally, such meshes are constructed by mapping the domain of interest into a square or cube and then constructing a regular mesh over the mapped domain. The mapping can be accomplished by the use of conformal techniques or differential equations or algebraic methods. To the analyst, a major advantage arising from the use of a structured mesh is that an appropriate solution method can be selected from among the large number of algorithms that are generally available for implementation on meshes of this type. The major disadvantage of the approach is the fact that it is not always possible to guarantee that an acceptable mesh will be produced following the application of a mapping method to regions of general shape. This difficulty can be alleviated by initially constructing an appropriate subdivision of the computational domain into blocks and then producing a mesh by applying the mapping method to each block separately. This results in a powerful multiblock method of mesh generation [1] that has proved extremely successful in a wide variety of applications. However, for domains of extremely complex shape, the elapsed time required by the general analyst to produce a mesh by this approach can be significant, and the approach can still result in the generation of elements of poor quality.

The alternative approach is to divide the computational domain into an unstructured assembly of computational cells. The notable feature of an unstructured mesh is that the number of cells surrounding a typical interior node of the mesh is not necessarily constant. We will be concentrating our attention upon the use of triangular meshes. The methods normally adopted to generate unstructured triangular

0-8493-2687-7/99/$0.00+$.50

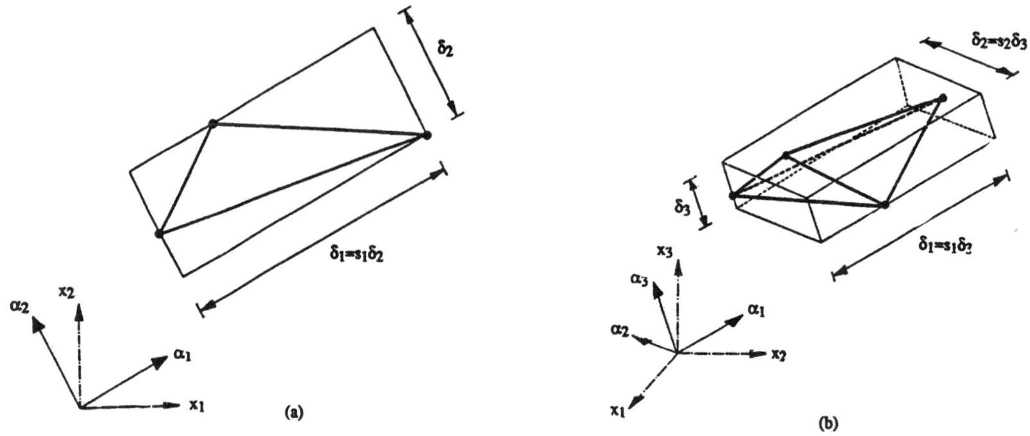

FIGURE 35.1 Characterization of the mesh. (a) In two dimensions (b) in three dimensions.

meshes are based upon either the Delaunay [2] or the advancing front [15] approaches. Discretization methods for the equations of fluid flow that are based upon integral procedures, such as the finite volume or the finite element method, are natural candidates for use with unstructured meshes. The principal advantage of the unstructured approach is that it provides a very powerful tool for discretizing domains of complex shape [5,14], especially if triangles are used in two dimensions and tetrahedra are used in three dimensions. In addition, unstructured mesh methods naturally offer the possibility of incorporating adaptivity [6]. Disadvantages following from adopting the unstructured grid approach are that the number of alternative solution algorithms is currently rather limited and that their computational implementation places large demands on both computer memory and CPU [4]. Further, these algorithms are rather sensitive to the quality of the grid being employed, and so great care has to be taken in the generation process. The improvement of grid quality is a problem of major importance, particularly as grid generation techniques mature, and it is an issue that will be addressed in this chapter.

35.2 Unstructured Mesh Control

35.2.1 Characterization of an Unstructured Mesh

The provision of an adequate mechanism of mesh control is a key ingredient in ensuring the generation of a mesh of the desired form. To achieve this, the user needs to be able to specify, to the mesh generator, the geometrical characteristics of the required mesh. In the approach described here, the geometrical characteristics of a general unstructured mesh of triangular (2D) or tetrahedral (3D) elements are considered to be defined locally in terms of certain *mesh parameters.* For a Delaunay approach (see Chapter 16) the parameter used is element size, δ. In the case of an advancing front approach (see Chapter 17) a set of N mutually orthogonal directions $\underline{\alpha}_i$; i = 1, ... N, and N associated element sizes δ_i; i = 1, ... N (see Figure 35.1) where N (= 2 or 3), is the number of dimensions. Thus, at a certain point, if all N element sizes are equal, the mesh in the vicinity of that point will consist of approximately equilateral elements. To aid the advancing front mesh generation procedure, a transformation **T** that is a function of $\underline{\alpha}_i$ and δ_i is defined. This transformation is represented by a symmetric $N \times N$ matrix and maps the physical space onto a space in which elements, in the neighborhood of the point being considered, will be approximately equilateral with unit average size. This new space is referred to as the normalized space. For a general mesh this transformation will be a function of position. The transformation **T** is the result of superimposing N scaling operations with factors $1/\delta_i$ in each $\underline{\alpha}_i$ direction. Thus

$$\mathbf{T}(\underline{\alpha}_i, \delta_i) = \sum_{i=1}^{N} \frac{1}{\delta_i} \underline{\alpha}_i \otimes \underline{\alpha}_i \tag{35.1}$$

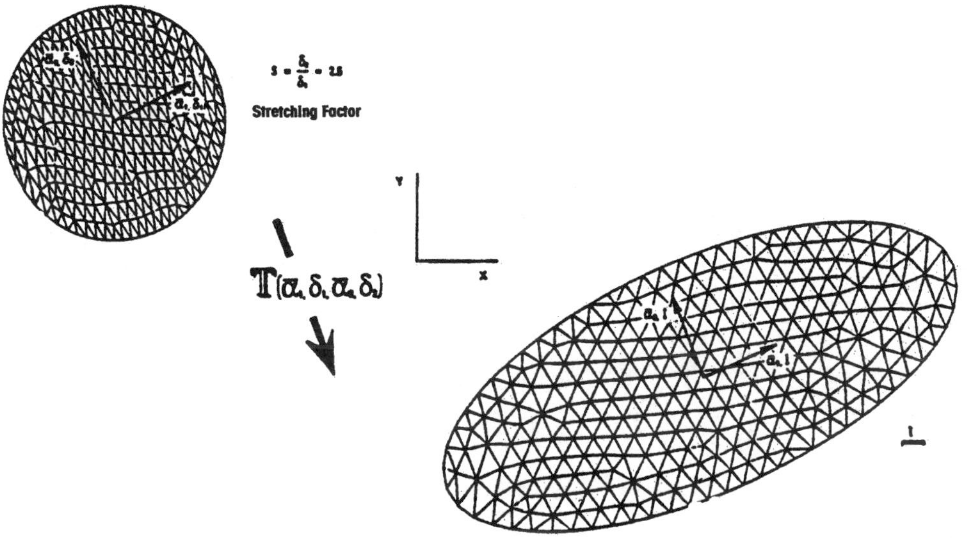

FIGURE 35.2 The effect of transformation **T** for a constant distribution of mesh parameter.

where ⊗ denotes the tensor product of two vectors. The effect of this transformation in two dimensions is illustrated in Figure 35.2 for the case of constant mesh parameters throughout the domain.

35.2.2 Advancing Front Grid Control

The algorithmic procedure for mesh generation by the advancing front method is based upon the method originally proposed in [5] for two dimensions and then extended to three dimensions in [12,13]. The approach has the distinctive feature that elements, i.e., triangles or tetrahedra, and points are generated simultaneously. This enables the generation of elements of variable size and stretching. The mechanism that can be employed to achieve the necessary degree of control over the characteristics of the generated mesh in this context is to define the required spatial distribution of the mesh parameters by means of a background mesh and /or by the use of sources.

35.2.2.1 The Background Mesh

The background mesh is used for interpolation purposes only and is made up of triangles in two dimensions and tetrahedra in three dimensions. Values of $\underline{\alpha}_i$ and δ_i, and hence **T**, are defined at the nodes of the background mesh. At any point within an element of the background grid, the transformation **T** is computed by linearly interpolating its components from the element nodal values. The background mesh employed must cover the region to be discretized (see Figure 35.3). In the generation of an initial mesh for the analysis of a particular problem, the background mesh will usually consist of a small number of elements. The generation of the background mesh can in this case be accomplished without resorting to sophisticated procedures, e.g., a background mesh consisting of a single element can be used to impose the requirement of linear or constant spacing and stretching through the computational domain. The generation process is always carried out in the normalized space. The transformation **T** is repeatedly used to transform regions in the physical space into regions in the normalized space. In this way the process is greatly simplified, as the desired size for a side, triangle, or tetrahedra in this space is always unity. After the element has been generated, the coordinates of the newly created point, if any, are transformed back to the physical space using the inverse transformation. The effect of prescribing a variable mesh spacing and stretching is illustrated in Figure 35.3 for a rectangular domain and using a background grid consisting of two triangular elements.

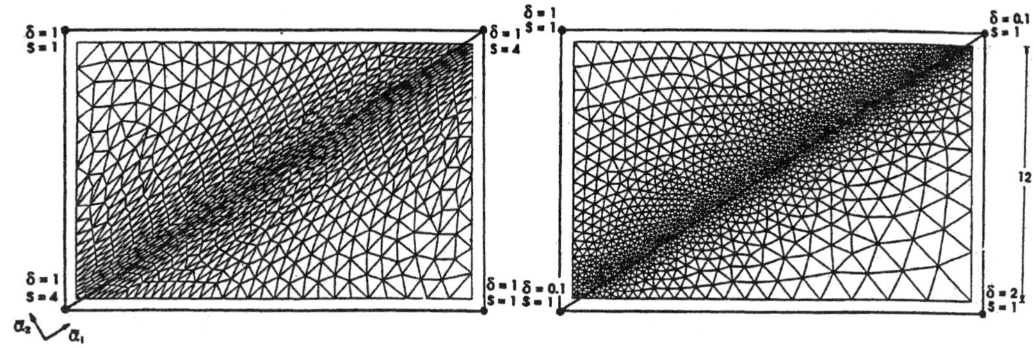

FIGURE 35.3 Variable mesh spacing and stretching for a rectangular domain using a background mesh consisting of two elements.

35.2.2.2 Sources

The requirement of constructing an adequate background grid for complex geometries has proved to be a significant barrier to the successful use of the approach by the inexperienced user. To alleviate this problem, the concept of the use of point, line, and plane sources can be added to the process of defining the variation of the grid parameters over the computational domain. For example [11], with the location of a point source specified, the nodal spacing δ defined by the source at location **x** is determined as

$$\begin{aligned} \delta(x) &= \delta_s & |x_1| &< r_1 \\ \delta(x) &= \delta_s e^{|x_1|\ln\left(\frac{2}{r_2 - r_1}\right)} & |x_1| &\geq r_1 \end{aligned} \tag{35.2}$$

where $|\mathbf{x}_1|$ denotes the distance from **x** to the point source and δ_s, r_1 and r_2 are user-specified constants. Line and plane sources can be constructed in a similar fashion. Point, line, and plane sources defined in this way provide an isotropic distribution in which the element size is specified to be the same in all directions. When combined with the background mesh, the mesh generator will, at a location **x**, consider the required mesh size to be the minimum of the spacing defined at **x** by the background mesh and by all the active sources. To illustrate the simplicity of using sources to aid the mesh generation process, consider the problem of producing an adequate mesh for the simulation of inviscid aerodynamic flow over a wing. It is well known that the mesh employed should be clustered in the vicinity of the leading and trailing edges of the wing, while larger elements can be employed elsewhere. A mesh of this type is readily generated by using a background mesh consisting of one tetrahedral element supplemented by line sources lying along the leading and trailing edges of the wing. Figure 35.4 shows a mesh that has been generated on a wing surface when this approach is followed.

35.2.3 Delaunay Grid Control

The Delaunay grid generation approach is based on a simple geometrical construction. Given a set of points, a tiling is constructed with the property that each point has an associated region closer to that point than to any other point. The boundary of the tile is formed by the perpendicular bisectors of the lines joining each point and its immediate neighbors. If points having a common tile boundary are connected, then a triangulation of the points is obtained. Points for connection by the Delaunay algorithm can be derived in many ways. Two ways that have been used include superposition methods and points generated from an independent technique (e.g., structured grid methods [18]). The former approach gives rise to good-quality grids in the interior of regions, but grid quality can deteriorate where the

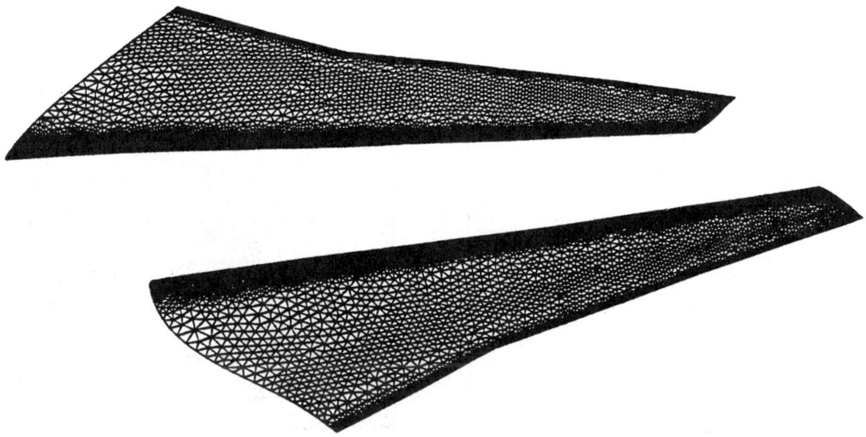

FIGURE 35.4 Discretization of the surface of a wing.

tetrahedra and the connections are constrained by the boundaries. The latter approach is restrictive for general geometries. New methods have been developed which are flexible, easy and efficient to implement, require minimal manual user input, and provide good grid quality.

35.2.3.1 Automatic Point Creation Driven by the Boundary Point Distribution

For grid generation purposes, the boundary of the domain is defined by points and associated connectivities. It will be assumed that the grid points on the surface reflect appropriate variations in surface slope and curvature. Ideally any method which automatically creates points should ensure that the boundary point distribution is extended into the domain in a spatially smooth manner. The method used employs a similar idea to interpolating from a background grid as described in the advancing front method, but here the Delaunay triangulation is used to provide automatically an equivalent background grid whose node spacing is derived from the given boundary point spacing.

Consider, in two dimensions, boundary line segments on which points have been distributed that enclose a domain. It is required to distribute points within the region so as to construct a smooth distribution of points. For each point on the boundary, a typical length scale for the point can be computed as the average of the two lengths of the connected edges. No points should be placed within a distance comparable to the defined length scale, since this would inevitably define a badly formed triangle. Hence, for each point, i, it is appropriate to define a region Γ_i within which no interior point should be placed. In the Delaunay triangulation algorithm, the surface or boundary points are connected together to form an initial triangulation. Points can be placed anywhere within the interior but not inside any of the regions Γ_i already identified. Hence, points are placed at the centroid of each of the formed triangles and then a test is performed to determine if any of the points lie within any Γ_i. If a point lies within Γ_i, it must be rejected; if it does not, then it can be included and connected using the Delaunay triangulation algorithm. Once a point has been inserted, it too must have associated with it a length scale which defines an effective region Γ_i for point exclusion. A newly inserted point takes a length scale from interpolation of the length scales from the nodes that formed the triangle from which it was created. In this way a smooth transition between boundaries of interior points can be ensured. This process of point insertion continues until no point can be added because the union of all Γ_i covers the entire interior domain.

The interpolation of the boundary point distribution function is linear throughout the field. If required, this can be modified to provide a weighting towards the boundaries so as to ensure greater point density in such regions. The implementation of such a procedure involves a scaling of the point distribution of the nodes that form an element on the surface.

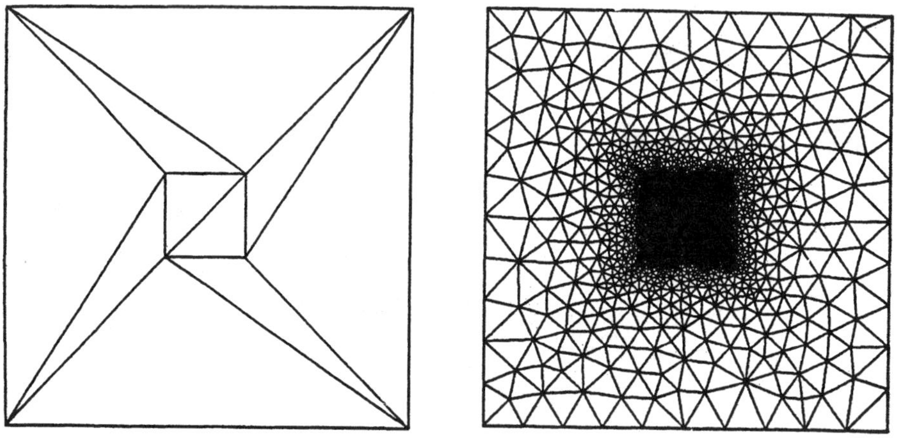

FIGURE 35.5 The effect of the background grid on the control of grid point clustering. (a) The background grid with specified point spacing, (b) the generated grid.

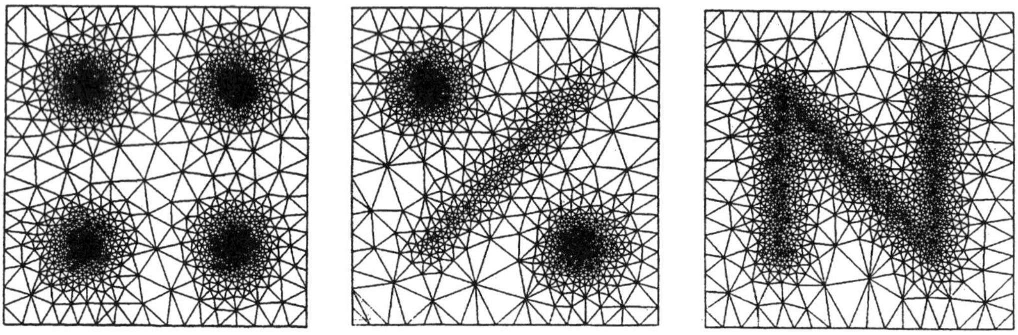

FIGURE 35.6 Effects of point and line sources

35.2.3.2 Automatic Point Creation Controlled by a Background Mesh

Another way to control the point spacing in the domain is to use a background mesh [6,15]. A mesh is overlaid over the domain, and at each node a point spacing is specified. The point distribution function, δ, for a prospective point is obtained from the interpolated spacing from the background mesh. Figure 35.5 shows a grid within a rectangular domain that has used the background grid shown to ensure grid clustering.

35.2.3.3 Automatic Point Creation by the Use of Sources

In some cases the boundary point distribution is not the best distribution to use to construct an efficient grid, while the construction of an adequate background grid mesh for a three-dimensional geometry is a tedious process. However, the use of point and line sources has proved to be a successful technique for the advancing front method, and it also proves effective when implemented with the Delaunay triangulation procedure. The spacing δ at a point is taken to be the minimum of the spacing interpolated from the boundary point distribution and the spacing obtained from all the sources using Eq. 35.2. Examples of the use of the sources approach are shown in Figure 35.6.

35.3 Mesh Quality Enhancement

35.3.1 Mesh Cosmetics

In the case of simple geometries, and for regularly spaced elements, the mesh generation procedure will often prove satisfactory. However, for more complex configurations, or in situations where variation in

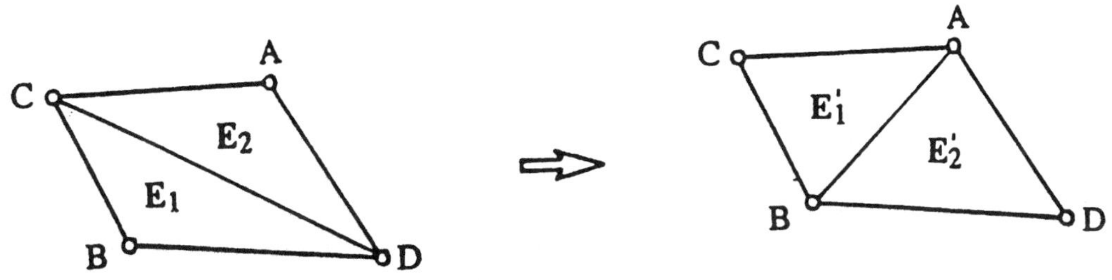

FIGURE 35.7 Diagonal swapping in two dimensions.

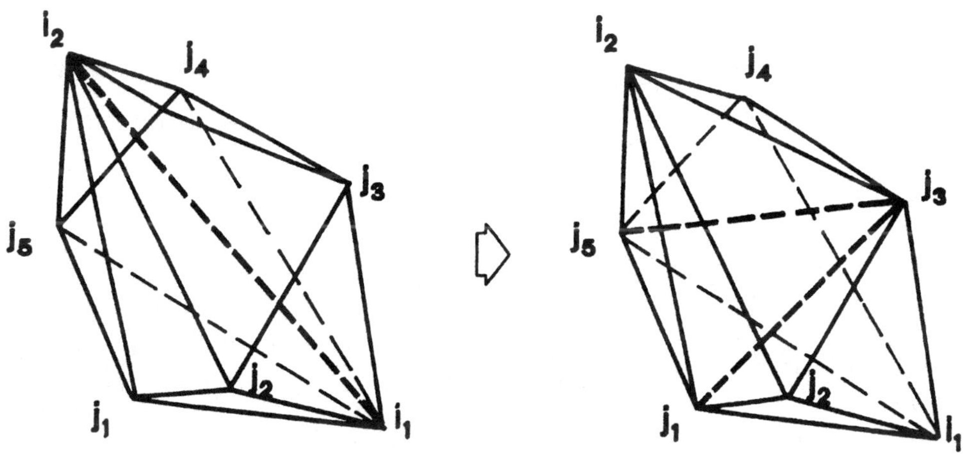

FIGURE 35.8 Skew polygon.

element size is rapid and considerable, deformed elements (i.e., elements with a minimum dihedral angle less than some specified tolerance) may appear. In these situations there are several operations that can be performed to enhance the quality of the mesh that has been generated. Four possible operations are diagonal swapping, element reconnection, element removal, and mesh smoothing. These devices are described below and should be carried out in the following order.

35.3.1.1 Edge Swapping

For a mesh of triangular elements, local diagonal swapping is a straightforward procedure performed on a pair of adjacent elements to improve the regularity of the triangles. This situation is illustrated in Figure 35.7. The connectivity of the existing pair of elements is changed if the minimum angle occurring in the new pair of triangles is greater than the minimum angle in the existing pair. In three dimensions, it is possible, although more difficult, to enhance grid quality through the implementation of an edge swapping procedure. The method can be described algorithmically as follows:

Loop over sides
- If (side i1-i2 is not a boundary side) then
  - 1- list all elements which have i1-i2 as an edge
  - 2- determine the minimum dihedral angle(dh) for the elements in list
  - 3- if (dh) is less than α, then
    - 3.1 form the skew polygon from the nodes of the elements excluding i1 and i2, i.e., j1-j2-j3-j4-j5 (Figure 35.8)
    - 3.2 from the sides of the skew polygon determine the two adjacent sides containing the smallest angle n1-n2-n3

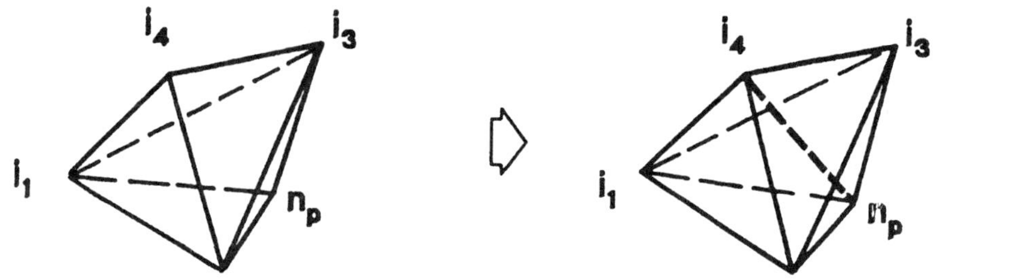

FIGURE 35.9 Nodal reconnection.

3.3 form two tetrahedral elements n1-n2-n3-i1, n1-n2-n3-i2
3.4 check that neither of the two new elements contains a dihedral angle smaller than (dh)
3.5 update the skew polygon
3.6 go to step 3.2
End if
End if
End loop

35.3.1.2 Nodal Reconnection

A search is made over all distorted elements (containing dihedral angle less than α), and their neighbors, and the possibility of creating a new element by reconnecting the connectivities of a distorted element and one of its neighbors is investigated. This procedure results in the creation of three elements out of the original pair of adjacent elements, as illustrated in Figure 35.9. The creation and reconnection is performed if the minimum dihedral angle in the new configuration is greater than that in the existing one. The reconnection procedure will not apply for meshes generated using the Delaunay method, as this situation should not arise. For meshes generated by the advancing front method, a significant improvement in mesh quality results from implementing this technique.

35.3.1.3 Edge Deletion

If badly deformed elements (containing dihedral angle less than β) are still present after the previous two operations have been performed, then an attempt is made to remove these elements from the mesh. This is achieved by collapsing one of the sides of the deformed element so that its nodes coincide. When investigating an element, the decision of which side to collapse is made by considering each side in turn and examining the adjacent elements that would exist if that particular side were to be removed. The chosen configuration is the one with the largest minimum dihedral angle.

35.3.1.4 Spatial Smoothing

The sides of the element in the mesh are replaced by springs of unit stiffness. The force $\underline{F}_{ij}$ exerted by the spring connecting nodes i and j is taken to be:

$$\underline{F}_{ij} = \underline{x}_i - \underline{x}_j \tag{35.3}$$

where $\underline{x}_i$ and $\underline{x}_j$ are the position vectors of nodes i and j, respectively. For badly deformed elements the resulting nodal forces will not be in equilibrium, whereas for regions of well-formed elements the resulting nodal forces will nearly vanish. A relaxation procedure is adopted that moves the nodes until nodal equilibrium of forces is achieved. The new nodal position is accepted provided an improvement in the dihedral angle of the surrounding elements results from the smoothing procedure. A few passes are usually enough to ensure local smoothing of the mesh.

35.3.2 Grid Quality Statistics

It is difficult to display unstructured three-dimensional grids in a way that provides effective information of the grid quality. Planar cuts taken through the unstructured grid provide some information on grid point density, but do not provide any useful information on the quality of the grid connectivity.

For further information on grid quality, statistics of the grid should be computed. Statistics that can be computed include the ratio of the dihedral angle within a tetrahedron to the optimum angle, the ratio of volumes of two adjacent tetrahedra, the ratio of the maximum to minimum side length per element and per point, and the number of elements surrounding a point. For comparison, Figure 35.10 shows grid statistics for two distinct grids that have been generated using the same surface grid: one using the Delaunay approach and the other using the advancing front. The advancing front grid contained 231,507 elements and 42,410 points and the Delaunay grid contained 233,182 elements and 40,442 points. The plots of the number of elements surrounding a point show two distinct maxima; one is centered around the optimum value of 24 elements per point, and a second at approximately 12, which is the optimum number of element connections to a boundary point. The plot of dihedral angle shows that the distribution is centered about an angle just less than the optimum of about 72°. The ratios of volumes of two adjacent elements are also well distributed, indicating smoothly varying element sizes. The smoothness of the grid is also confirmed by the plots of maximum side length to minimum side length both per element and per point. The two grids are comparable in the measures chosen. The improvements in mesh quality that can be obtained by the implementation of the four mesh cosmetic operations described in Section 35.3.1 and applied to the previous mesh are displayed in Table 35.1.

Figure 3.11 shows a further illustration of the mesh quality enhancement that can be achieved by varying the control parameters α and β. Practical experience shows no significant improvement can be gained from adopting a value of α greater than 50°. In addition, the value of β should be restricted to approximately 10° to avoid the removal of an inordinately large number of points, which would adversely affect the mesh resolution.

This mesh cosmetic procedure can prove vital in the case of time-dependent problems, where the solution is advanced at the minimum time step, which is related to the minimum element height. Traditionally this problem is circumvented by using a local time step for the badly distorted elements, hence avoiding the requirement of an excessive number of time steps to perform the simulation. However, the use of local time stepping can cause a deterioration in solution quality. The following computational electromagnetic example demonstrates a reduction in the number of time steps required to perform a calculation and in the number of elements running at local time step. In this example all nodal points connected to elements below a specified minimum height are advanced at local time step. The improvements that can be obtained through application of the mesh cosmetics are clearly shown in Figure 35.12 and Table 35.2.

35.4 Mesh Adaption

35.4.1 Introduction

The procedures described above allow for the computation of an initial approximation to the steady state solution of a given problem. This approximation can generally be improved by adapting the mesh in some manner. Here, we follow the approach of using the computed solution to predict the desired characteristics (i.e., element size and shape) for a new, adapted mesh. The ultimate aim of the adaptation procedure is to predict the characteristics of the optimal mesh. This can be defined as the mesh in which the number of degrees of freedom required to achieve a specified level of accuracy is a minimum. Alternatively, it can be interpreted as the mesh in which a given number of degrees of freedom are distributed in such a manner that the highest possible solution accuracy is achieved. We have made an attempt to develop a heuristic adaptive strategy that uses error estimates based upon concepts from interpolation theory. The possible presence of discontinuities in the solution is taken into account and, in addition, the procedure provides

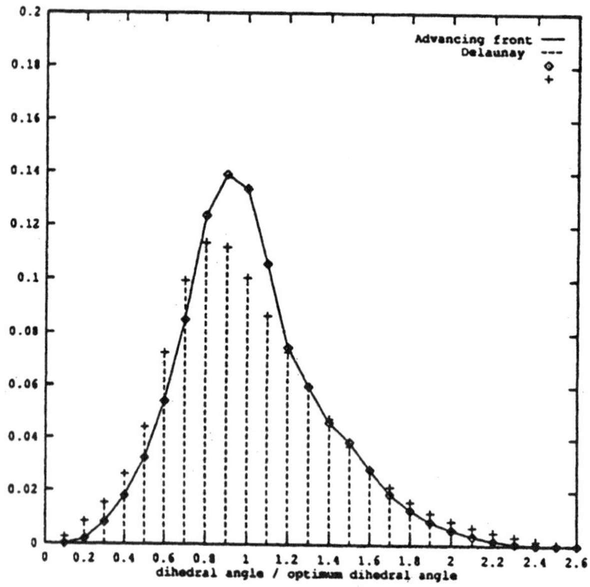

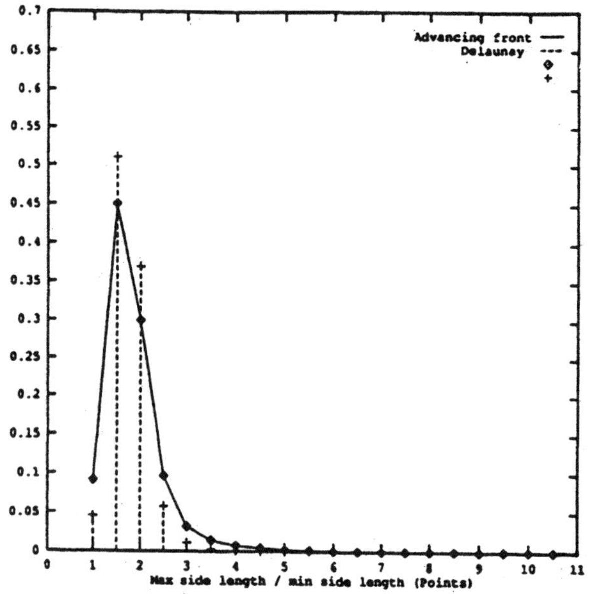

FIGURE 35.10 Grid statistics for grids around an Onera M6 wing.

TABLE 35.1 Improvements in Mesh Quality Around an Onera M6 Wing

Mesh	Number of Elements	Number of Points	Min. Volume	Min. Dihedral	Ratio of Adjacent Volumes
Initial	202,091	35,482	2.2e-05	0.032	2891
$\alpha = 70, \beta = 10$	187,463	35,478	1.1e-05	6.270	17
$\alpha = 50, \beta = 10$	188,270	35,481	1.5e-05	6.2	17

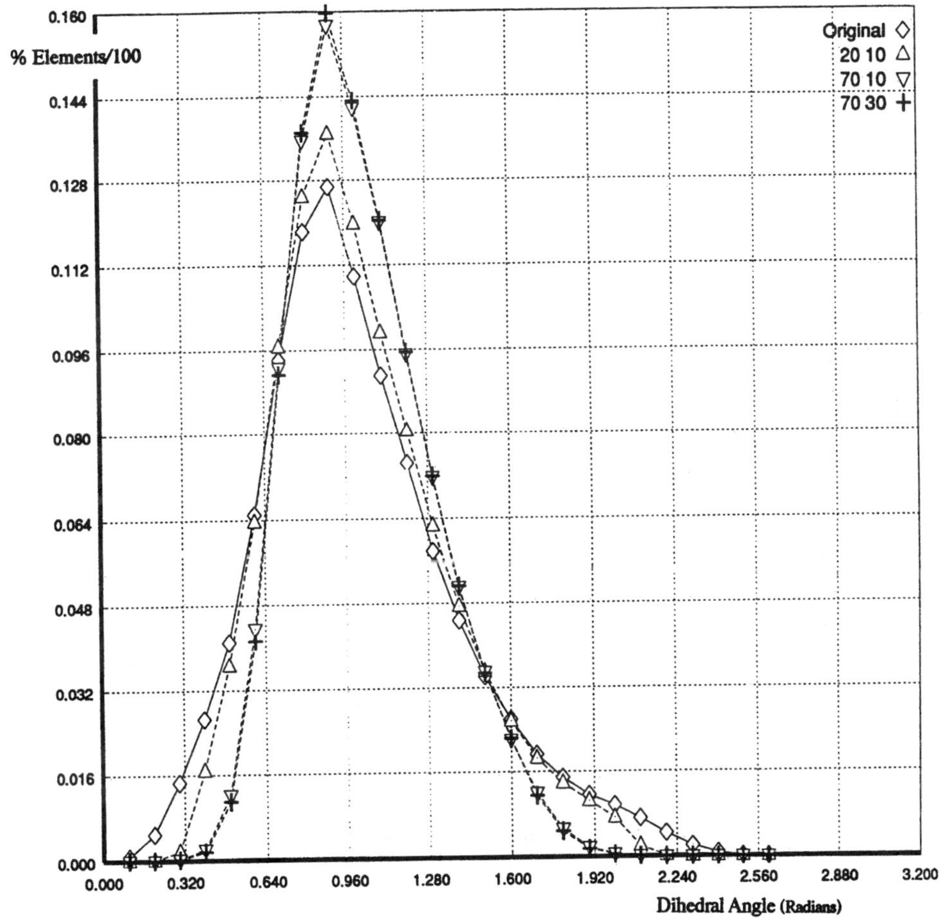

FIGURE 35.11 Mesh quality enhancement by varying the control parameters α and β.

information about any directionality that may be present in the solution. The advantages of using directional error indicators become apparent when we consider the nature of the solutions to be computed involving flows with shocks, contact discontinuities, etc. Such features can be most economically represented on meshes that are stretched in appropriate directions. Although these error estimates have no associated mathematical rigor, considerable success has been achieved with their use in practical situations.

The computed error, estimated from the current solution, is transformed into a spatial distribution of "optimal" mesh spacings that are interpolated using the current mesh. The current mesh is then modified with the objective of meeting these optimal distribution of mesh characteristics as closely as possible. Three alternative procedures will be discussed here for performing the mesh adaption. The resulting mesh is employed to produce a new solution and this procedure can repeated several times until the user is satisfied with the quality of the computed solution.

35.4.2 Error Indicator in 1D

The development of a method for error indication is considerably simplified if we restrict consideration to problems involving a single scalar variable. For this reason, when solving the Euler equations, a key variable is identified and then the mesh adaptation is based on an error analysis for that variable alone. The choice of the best variable to use as a key variable remains an open question, but the Mach number has been adopted for the computations reported in the chapter.

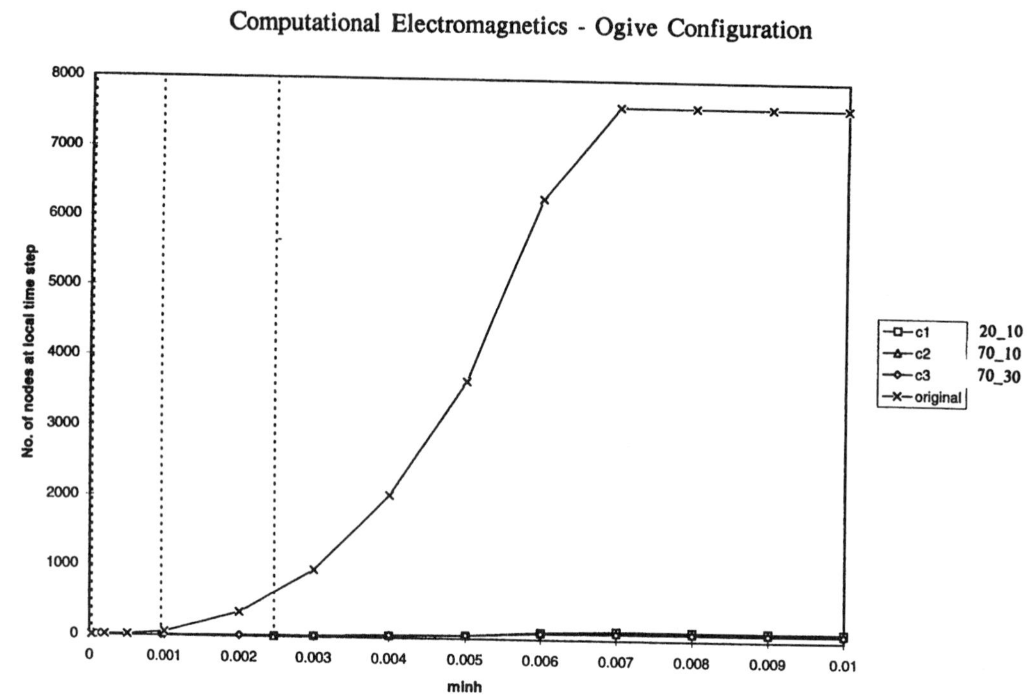

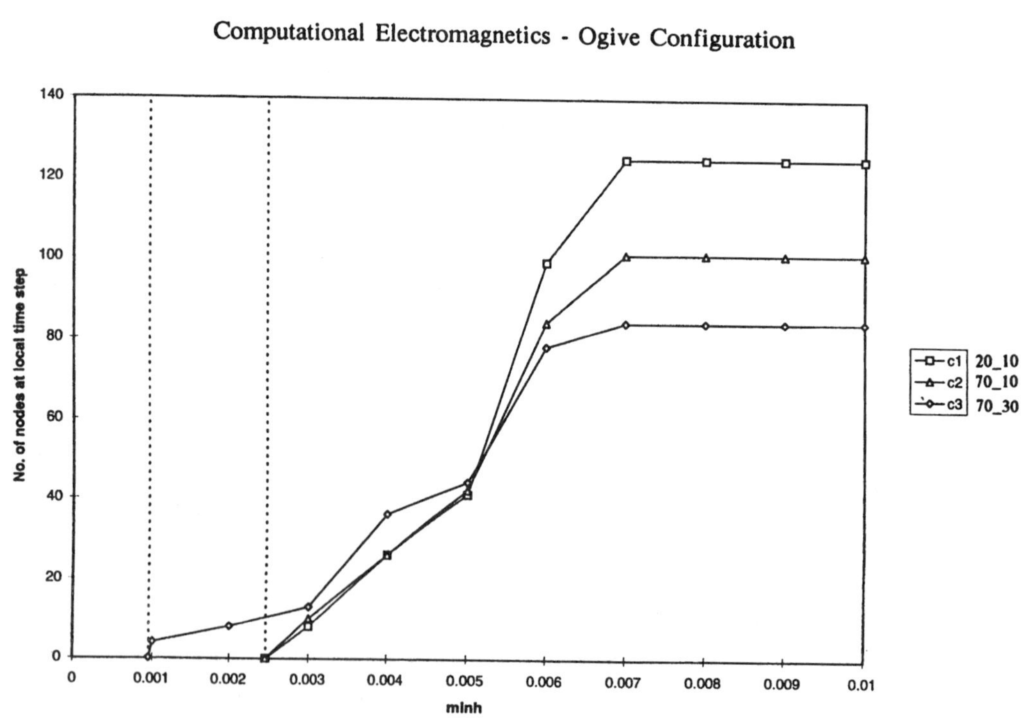

FIGURE 35.12 Improvement in the number of nodes violating the minimum time step.

TABLE 35.2 Computational Electromagnetic Example: Computational Improvements Achieved Through the Implementation of Mesh Cosmetics

	Number of Points at Local Time Step				Number of Time Steps per Cycle			
Min. Height	Initial Mesh	Mesh 1	Mesh 2	Mesh 3	Initial Mesh	Mesh 1	Mesh 2	Mesh 3
3.e-6	0	0	0	0	15135	204	204	1000
0.0001	8	0	0	0	5001	204	204	1000
0.0005	12	0	0	0	5001	204	204	1000
0.001	52	0	0	4	1000	204	204	500
0.005	3670	41	42	44	101	101	101	101
0.01	7624	125	101	84	79	79	79	79

Consider first the one-dimensional situation in which the exact values of the key variable σ are approximated by a piecewise linear function $\hat{\sigma}$. The error E is then defined as

$$E = \sigma(x^1) - \hat{\sigma}(x^1) \tag{35.4}$$

We note here that if the exact solution is a linear function of x^1, then the error will vanish. This is because our approximation has been obtained using piecewise linear finite element shape functions. Moreover, if the exact solution is not linear, but is smooth, then it can be represented, to any order of precision, using polynomial shape functions.

To a first order of approximation, the error E can be evaluated as the difference between a quadratic finite element solution $\tilde{\sigma}$ and the linear computed solution. To obtain a piecewise quadratic approximation, one could obviously solve a new problem using quadratic shape functions. This procedure, however, although possible, is not advisable as it would be even more costly than the original computation. An alternative approach for estimating a quadratic approximation from the linear finite element solution is therefore employed. Assuming that the nodal values of the quadratic and linear approximations coincide, i.e., the nodal values of E are zero, a quadratic solution can be constructed on each element, once the value of the second derivative is known. Thus the variation of the error E within an element e can be expressed as

$$E_e = \frac{1}{2}\zeta(h_e - \zeta)\left.\frac{d^2\tilde{\sigma}}{dx^{12}}\right|_e \tag{35.5}$$

where ζ denotes a local element coordinate and h_e denotes the element length. A procedure for estimating the second derivative of a piecewise linear function is described below.

The root-mean-square value E_e^{RMS} of this error over the element can be computed as

$$E_e^{RMS} = \left\{\int_0^{h_e} \frac{E_e^{\,2}}{h_e} d\zeta\right\}^{1/2} = \frac{1}{\sqrt{120}} h_e^{\,2} \left|\frac{d^2\tilde{\sigma}}{dx^{12}}\right|_e \tag{35.6}$$

where | . | stands for absolute value.

We define the "optimal" mesh, for a given degree of accuracy, as the mesh in which this root- mean-square error is equal over each element. In the present context, this requirement may be regarded as being somewhat arbitrary. However, it has been shown [9] that the requirement of equidistribution of the error leads to optimal results when applied to certain elliptic problems. This requirement is therefore written as

$$h_e^{\,2}\left|\frac{d^2\tilde{\sigma}}{dx^{12}}\right| = C \tag{35.7}$$

where C denotes a positive constant.

Finally, the requirement of Eq. 35.7 suggests that the "optimal" spacing δ on the new adapted mesh should be computed according to

$$\delta^2 \left| \frac{d^2 \tilde{\sigma}}{dx^{1^2}} \right| = C \tag{35.8}$$

The first derivative of the computed solution on a mesh of linear elements will be piecewise constant and discontinuous across elements. Therefore, straightforward differentiation of $\hat{\sigma}$ leads to a second derivative which is zero inside each element and is not defined at the nodes. However, by using a recovery process, based upon a variational or weighted residual statement [21], it is possible to compute nodal values of the second derivatives from element values of the first derivatives of $\hat{\sigma}$. The use of Eq. 35.8 then yields directly a nodal value of the "optimal" spacing for the new mesh.

35.4.3 Extension to Multidimensions

Equation 35.8 can be directly extended to the N-dimensional case by writing the quadratic form

$$\delta_\beta^{\,2} \left(\sum_{i;j=1}^{n} m^{ij} \beta^i \beta^j \right) = C \tag{35.9}$$

where $\underline{\beta}$ is an arbitrary unit vector, δ_β is the spacing along the direction of $\underline{\beta}$, and m^{ij} are the components of a $N \times N$ symmetric matrix of second derivatives:

$$m^{ij} = \frac{\partial^2 \hat{\sigma}}{\partial x^i \partial x^j} \tag{35.10}$$

These derivatives are computed, at each node of the current mesh, by using the N-dimensional equivalent of the procedure presented in the previous section. The meaning of Eq. 35.9 is graphically illustrated in Figure 35.13, which shows how the value of the spacing in the β direction can be obtained as the distance from the origin to the point of intersection of the vector β with the surface of an ellipsoid. The directions and lengths of the axes of the ellipsoid are the principal directions and eigenvalues of the matrix $\boldsymbol{m}$, respectively.

Several alternative procedures exist for modifying an existing mesh in such a way that the requirement expressed by Eq. 35.9 is more closely satisfied. Three such methods will be described here. In the first procedure, called *mesh enrichment*, the nodes of the current mesh are kept fixed but some new nodes/elements are created. In the second procedure, referred to as *mesh movement*, the total number of elements and nodes remains fixed but their position is altered. Finally, in the *adaptive remeshing* algorithm, the mesh adaption is accomplished by completely regenerating a new mesh.

35.4.4 Mesh Enrichment

In order to adapt a mesh using mesh enrichment, a sweep over all the sides in the mesh is made and the "optimal" spacing in the direction of each side is computed according to expression 35.9. For each side, the matrix $\boldsymbol{m}$ is taken to be the average of its value at the two nodes of the side. The enrichment procedure consists of introducing an additional node for each side for which the calculated spacing is less than the length of the side. For interior sides, this additional node is placed at the mid-point of the side, whereas for boundary sides, it is necessary to refer to the boundary definition and to ensure that the new node is placed on the true boundary. When any side is subdivided in this manner, the elements associated with that side will also need to be subdivided in order to preserve the consistency of the final mesh.

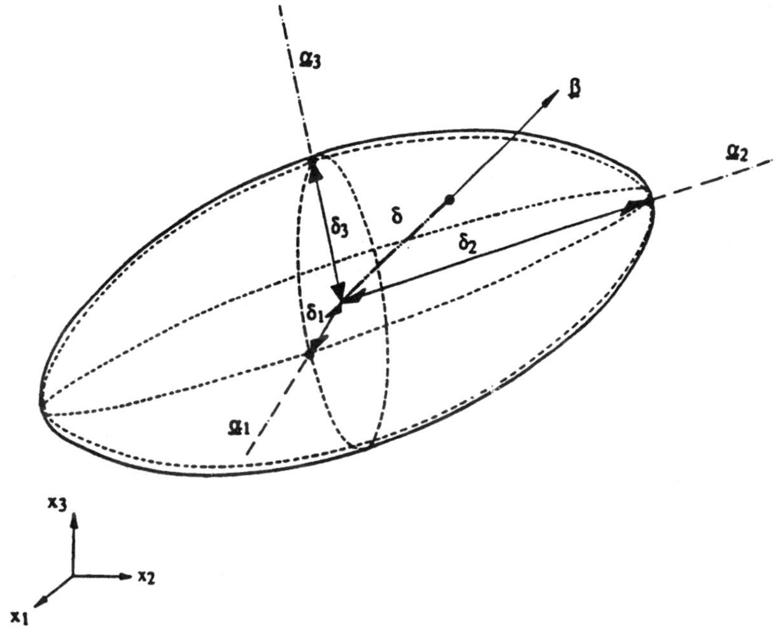

FIGURE 35.13 Determination of the value of the spacing along the β direction.

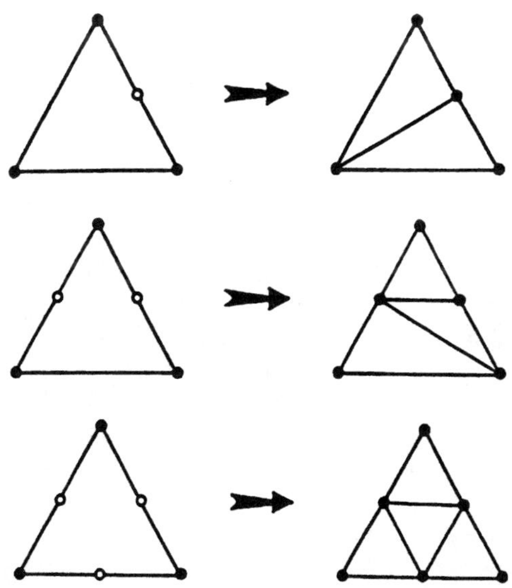

FIGURE 35.14 Mesh enrichment: three possible cases of refinement.

Figure 35.14 illustrates the three possible ways in which this element subdivision might have to be performed in two dimensions. The number of sides to be refined depends on the choice of the constant C in Eq. 35.9. To avoid excessive refinement in the vicinity of discontinuities, a minimum threshold value for the computed spacing can be used. When the mesh enrichment procedure has been completed, the values of the unknowns at the new nodes are linearly interpolated from the original mesh and the solution algorithm is restarted. This procedure has been successfully implemented in two and three dimensions, and several impressive demonstrations of the power of this technique have been made [6,10,13].

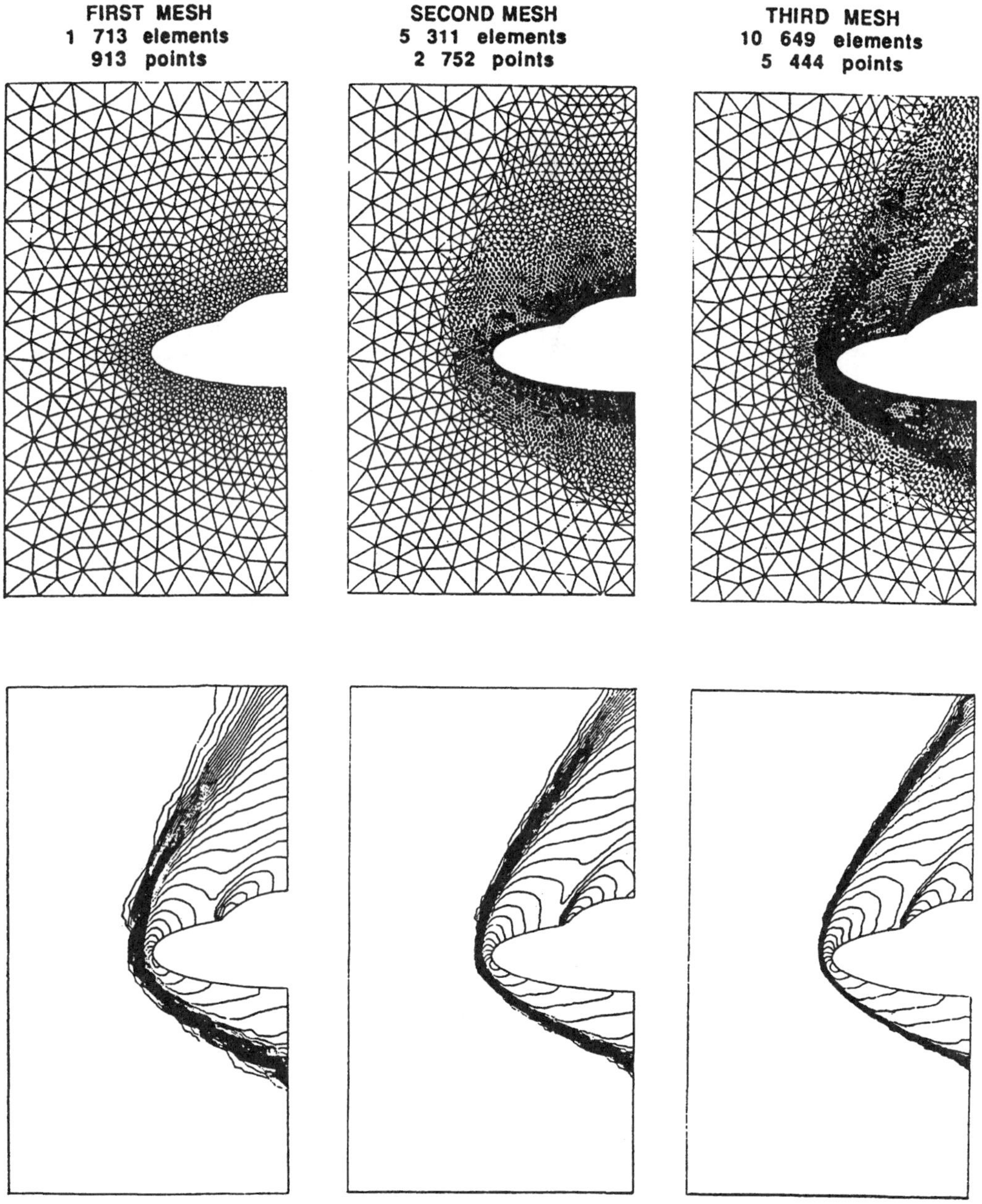

FIGURE 35.15 Supersonic flow past a double ellipse configuration. Sequence of meshes and solutions obtained using adaptive enrichment.

The application of the enrichment procedure in the solution of a two-dimensional example is illustrated in Figure 35.15. The problem solved is a Mach 8.15 flow past a double ellipse configuration at 30° angle of attack. The initial mesh and two adaptively enriched meshes are shown together with the computed Mach number solutions. The application of the enrichment algorithm in three dimensions is shown in Figure 35.16. The inviscid flow past a double ellipsoid is solved. The free stream Mach number is 8.15 at 30°. The starting mesh and the refined mesh are shown together with the corresponding Mach number controus.

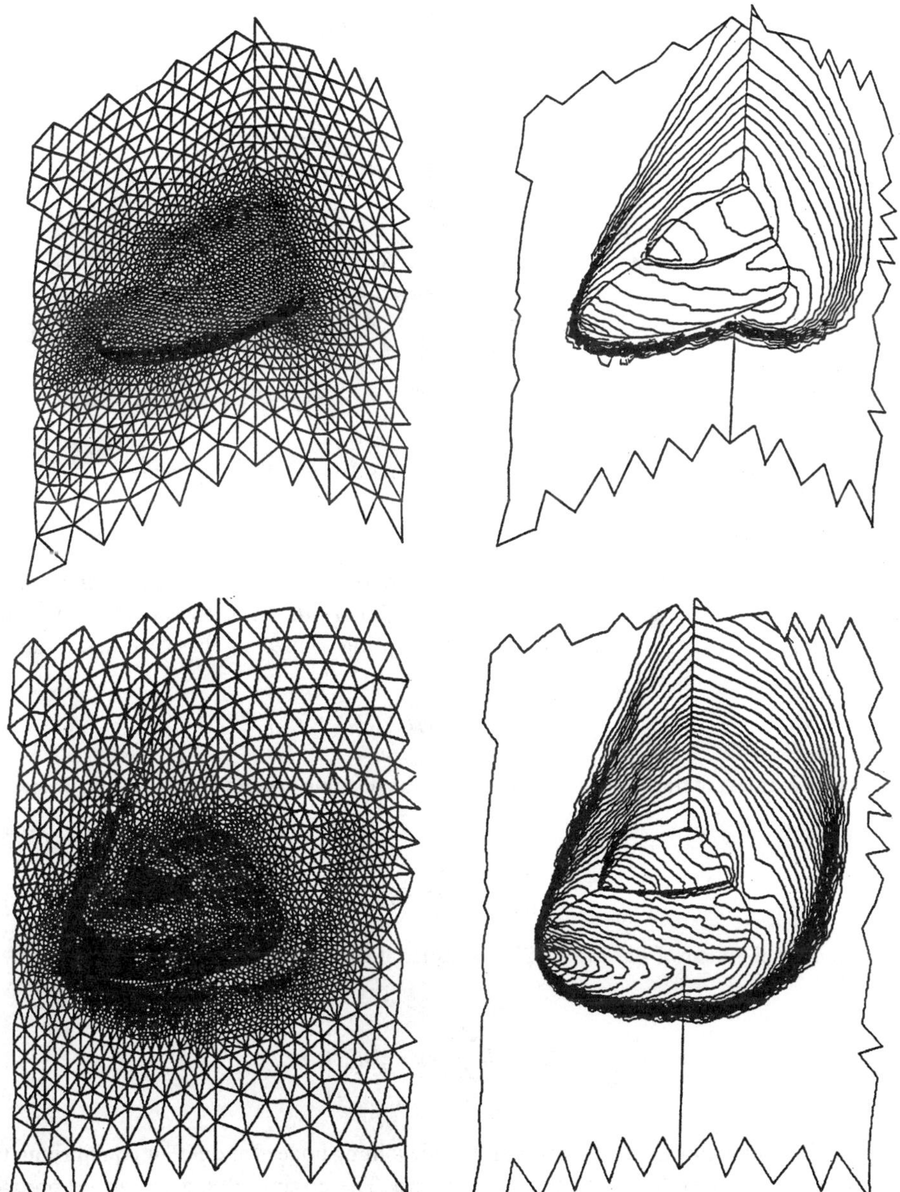

FIGURE 35.16 Supersonic flow past a double ellipsoid configuration. Sequence of meshes and solutions obtained using adaptive enrichment.

It can be observed, from the examples presented, how the quality of the solution is significantly improved by the application of the enrichment procedure. The main drawback of the approach is that the number of elements increases considerably following each application of the procedure. This means that, in the simulation of practical three-dimensional problems, only a small number of such adaptations can be contemplated.

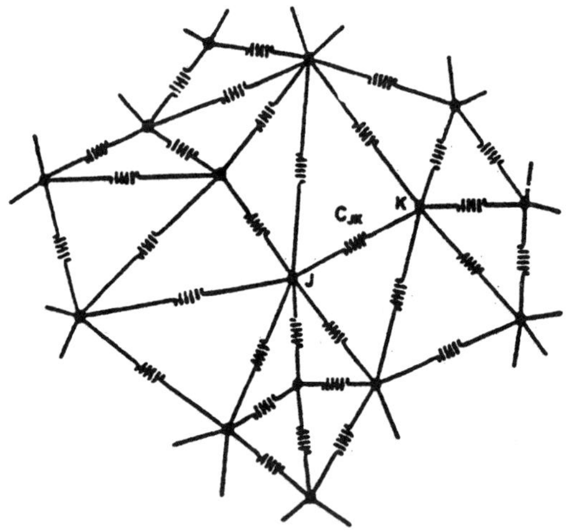

FIGURE 35.17 Mesh movement: element sides are replaced by springs.

35.4.5 Mesh Movement

For the mesh movement alogrithm, the element sides are considered as springs of prescribed stiffness and the nodes are moved until the spring system is in equilibrium. Consider two adjacent nodes J and K as shown in Figure 35.17. The force f_{JK} exerted by the spring connecting these two nodes can be taken to be

$$\underline{f}_{JK} = C_{JK}(\underline{r}_J - \underline{r}_K) \tag{35.11}$$

where C_{JK} is the stiffness of the spring and $\underline{r}_J$ and $\underline{r}_K$ are the position vectors of nodes J and K, respectively. Assuming that

$$h = |\underline{r}_J - \underline{r}_K| \tag{35.12}$$

the adaptation requirement of Eq. 35.11 will be satisfied if the spring stiffnesses are defined as

$$C_{JK} = h\left|\sum_{i;j=1}^{N} m^{ij} n_{JK}{}^{i} n_{JK}{}^{i}\right| \tag{35.13}$$

Here $\underline{n}_{JK}$ is the unit vector in the direction of the side joining nodes J and K. For equilibrium, the sum of spring forces at each node should be equal to zero. The assembled system can be brought into equilibrium by simple iteration. In each iteration, a loop is performed over all the interior nodes and new nodal coordinates are calculated according to the expression

$$\underline{r}_J{}^{NEW} = \frac{\sum_{K=1}^{S_J} C_{JK}\underline{r}_K}{\sum_{K=1}^{S_J} \underline{r}_K} \tag{35.14}$$

where the summation extends over the number of nodes, S_J, which surround node J. Sufficient convergence is normally achieved after three to five passes through this procedure.

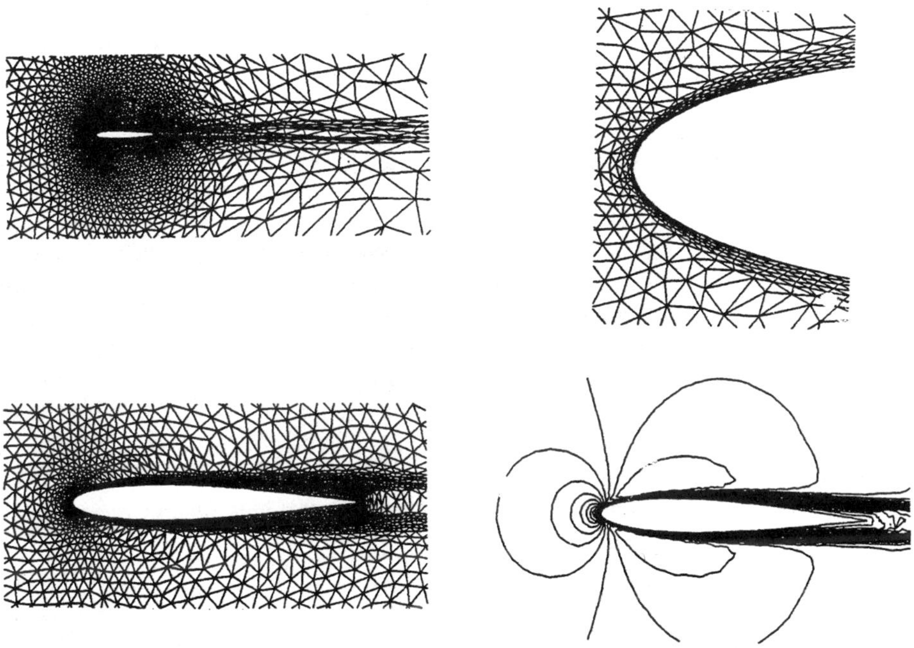

FIGURE 35.18 Example of node movement on an unstructured grid.

This technique will not necessarily produce meshes of better quality, as badly formed elements can appear in regions (such as shocks) in which the spring coefficients C_{JK} vary rapidly over a short distance. To avoid this problem, the definition of the value of C_{JK} given in Eq. 35.13 can be replaced by an expression of the form

$$C_{JK}^{MOD} = 1 + \frac{AC_{JK}}{B + C_{JK}} \tag{35.15}$$

This can be regarded as a blending function definition for the spring stiffnesses, and it has been constructed so as to ensure that, with a suitable choice for the constants A and B, excessively small or excessively large element sizes are avoided. This, in turn, means that meshes of acceptable quality will be produced. More sophisticated procedures for controlling the quality of the mesh during movement can also be devised [11], and mesh movement algorithms have been successfully used in two- and three-dimensional flow simulations on both structured and unstructured meshes [7,11].

The mesh movement algorithm described has been applied to the problem of viscous flow past an aerofoil. Figure 35.18 shows the initial mesh and the final mesh obtained after applying the mesh movement routine every 500 time steps for 9 times. It can be seen that the final mesh inherited all the solution features solutions produced following a series of mesh movement adaption.

In some cases the improvement obtained using this method is minor. This is because the algorithm does not allow for the creation of new nodes, and so the quality of the final solution is very much dependent on the topology of the initial mesh. This is a major drawback of the mesh movement strategy. A possible remedy to this problem is to combine mesh enrichment and mesh movement procedures.

35.4.6 Adaptive Remeshing

The basic idea of the adaptive remeshing technique is to use the computed solution to provide information on the spatial distribution of the mesh parameters. This information will be used by the mesh generator described in Sections 35.1 and 35.2 to generate a completely new adapted mesh for the problem under investigation.

The "optimal" values for the mesh parameters are calculated at each node of the current mesh. The directions $\underline{\alpha}_i$; i = 1, ..., *N*, are taken to be the principal directions of the matrix $\boldsymbol{m}$. The corresponding mesh spacings are computed from the eigenvalues e_i; $i = 1, \ldots, N$, as

$$\delta_i = \sqrt{\frac{C}{e_i}} \qquad \text{for} \quad i = 1, \ldots, N \tag{35.16}$$

The spatial distribution of the mesh parameters is defined when a value is specified for the constant *C*. The total number of elements in the adapted mesh will depend upon the choice of this constant. For smooth regions of the flow, this constant will determine the value of the root-mean-square error in the key variable that we are willing to accept. Therefore this constant should be decreased each time a new mesh adaption is performed. On the other hand, solutions of the Euler equations are known to exhibit discontinuities. At such discontinuities, the root-mean-square error will always remain large, and therefore a different strategy is needed in the vicinity of such features.

In the practical implementation of the present method, two threshold values for the computed spatial distribution of spacing are used: a minimum spacing δ_{min} and a maximum spacing δ_{max}, so that

$$\delta_{min} \leq \delta_i \leq \delta_{max} \qquad \text{for} \quad i = 1, \ldots, N \tag{35.17}$$

The reason for defining the maximum value δ_{max} is to account for the possibility of a vanishing eigenvalue in Eq. 35.16 which would render that expression meaningless. The value of δ_{max} is chosen as the spacing that will be used in the regions where the flow is uniform (the far field, for instance). On the other hand, maximum values of the second derivatives occur near the discontinuities (if any) of the flow where the error indicator will demand that smaller elements are required. By imposing a minimum value δ_{min} for the mesh size, we attempt to avoid an excessive concentration of elements near discontinuities. As the flow algorithm is known to spread discontinuities over a fixed number of elements (i.e., two or three), δ_{min} is therefore set to a value that is considered appropriate to ensure that discontinuities are represented to a required accuracy. This treatment also accounts for the presence of shocks of different strength in which, since the numerical values of the second derivative are different, Eq. 35.16 will assign them different mesh spacings (e.g., larger spacings in the vicinity of weaker shocks).

The total number of elements generated in the new mesh will now depend on the values selected for *C*, δ_{max}, and δ_{min}. However, it turns out that this number is mainly determined by the choice of the constant *C*, which is somewhat arbitrary. The criterion employed here is to select a value that produces a computationally affordable number of elements.

The adaptive remeshing strategy presented in this section is illustrated in Figure 35.19 by showing the various stages during the adaptation process. Figure 35.19a shows the initial mesh employed for the computation of the supersonic flow past a double ellipse configuration. The Mach number contours of the solution obtained on the inital mesh are shown in Figure 35.19b. The flow conditions are a free stream Mach number of 8.15 and an angle of attack of 30°. The application of expression 35.16 to the solution obtained produces the distribution of spacing and stretching displayed in Figures 35.19c and 35.19d respectively. In Figure 35.19d, the contours corresponding to the value of the minimum spacing occuring in any direction is shown, whereas in Figure 35.19c the value and the direction of stretching are displayed in the form of a vector field. The magnitude of the vector represents the amount of stretching, i.e., ratio between maximum and minimum spacings, and the direction of the vector indicates the direction along which the spacing is maximum. In this example, expression 35.17 has been applied to the computed spacings with values of $\delta_{max} = 15$ and $\delta_{min} = 0.9$. Figures 35.19e and 35.19h show various stages during the regeneration process. The completed mesh is shown in Figure 35.19h.

The regeneration process uses the current mesh as the background mesh. Such a background mesh clearly represents accurately the geometry of the computational domain. In this case, the number of elements to be generated, denoted by N_e, can be estimated as follows. Once the values of *C*, δ_{max}, and

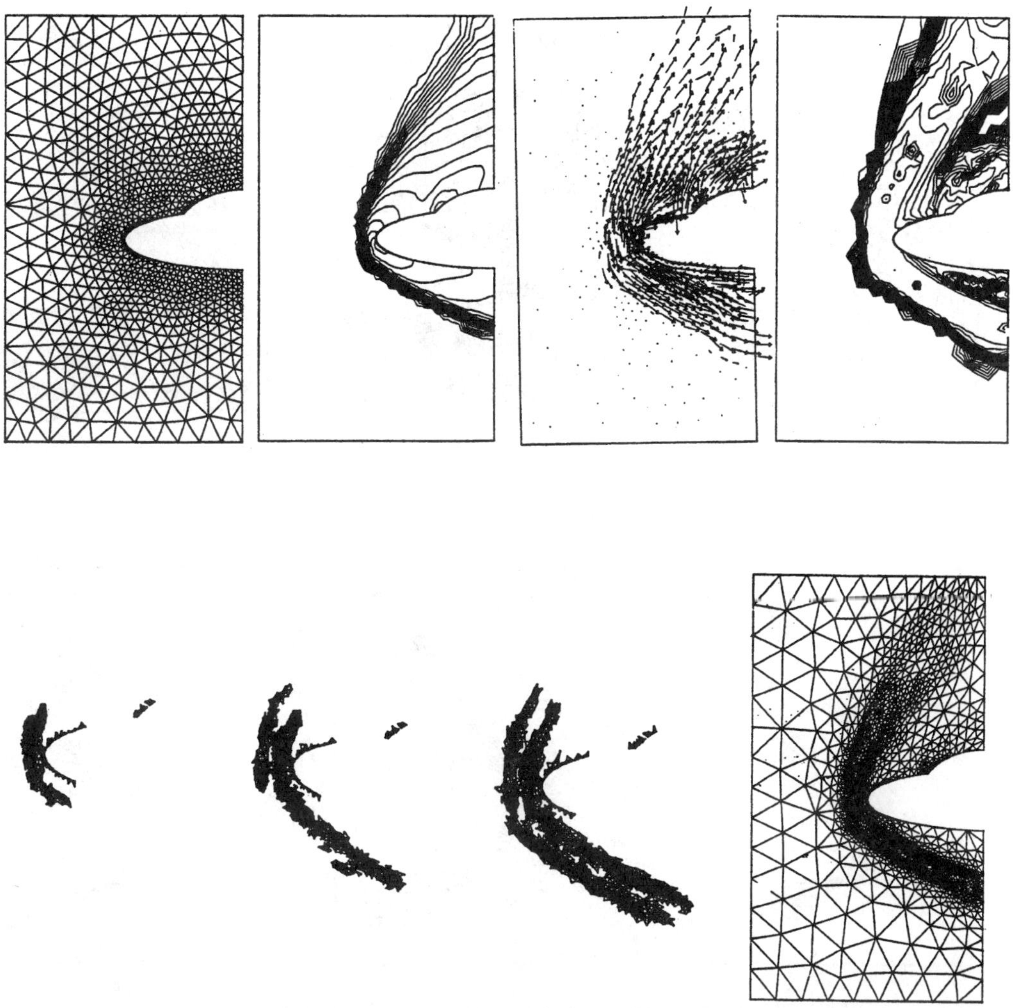

FIGURE 35.19 Illustration of the adaptive remeshing procedure.

δ_{min} have been selected, the spatial distribution of mesh parameters d_i, $\underline{\alpha}_i$; $i = 1, ..., N$ is computed. For each element of the background mesh, the values of the transformation **T** is computed at the centroid. The transformation is applied to the nodes of the element and its volume V_e in the normalized space is computed. The number of elements N_e is assumed to be proportional to the total volume in the unstretched space, i.e.,

$$N_e \approx \chi \sum_{e=1}^{N_b} V_e \tag{35.18}$$

where N_b is the number of elements in the background mesh. The value of χ is calculated as a statistical average of the values obtained for several generated meshes. The calculated value is $\chi \approx 9$. This procedure gives estimates of the value of N_e with an error of less than 20%, which is accurate enough for most practical purposes. If the estimated value of N_e is either too big or too small, then the value of C is reduced or increased and the process repeated until the value of C produces a number of elements which is regarded as being computationally acceptable.

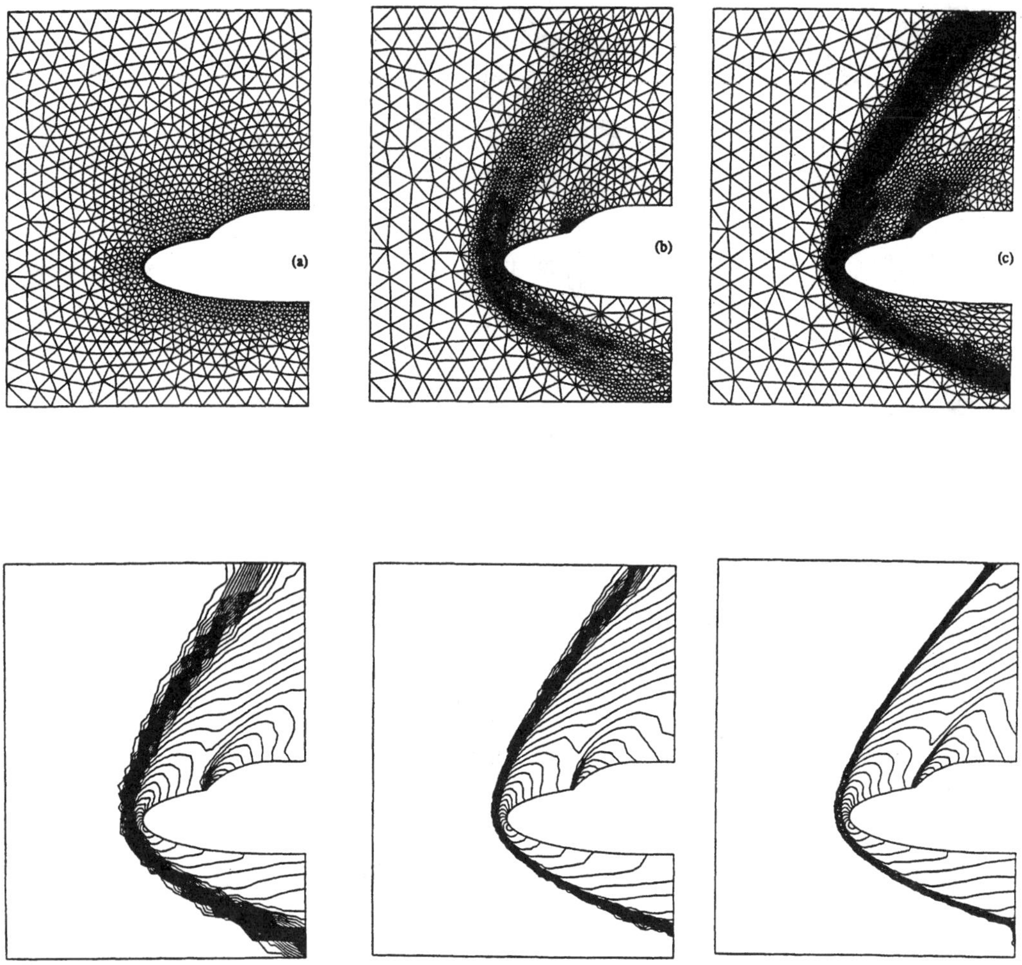

FIGURE 35.20 Adaptive remeshing applied to the supersonic flow past a double ellipse.

TABLE 35.3 Double Ellipse (M_∞ = 8.15, α =300): Mesh Characteristics

Mesh	Elements	Points	δ_{min}
1	2027	1110	4.0
2	3557	1864	0.9
3	6403	3294	0.25

The adaptive remeshing procedure is applied twice to the problem of flow past a double ellipse. The flow conditions are those previously considered for this configuration. The inital and two adapted meshes and the solutions for Mach number are shown in Figure 35.20. The characteristics of the meshes employed are displayed in Table 35.3.

It is observed how the application of the adaptive procedure, when compared to the enrichment strategy, allows for a larger increase in the resolution at the expense of a smaller increase on total number of elements. On the other hand the remeshing procedure does not suffer from the limitations inherent in the mesh movement algorithm.

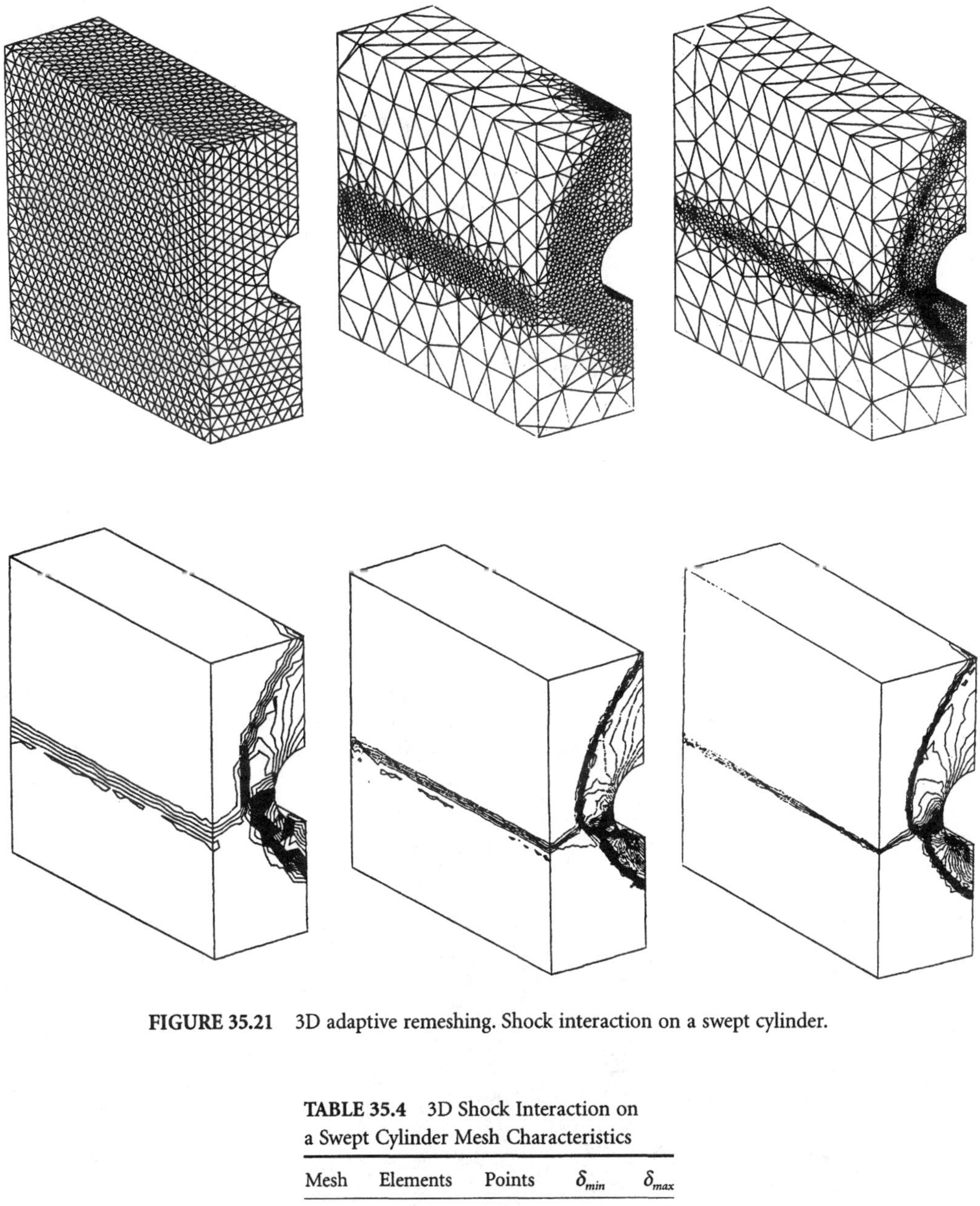

FIGURE 35.21 3D adaptive remeshing. Shock interaction on a swept cylinder.

TABLE 35.4 3D Shock Interaction on a Swept Cylinder Mesh Characteristics

Mesh	Elements	Points	δ_{min}	δ_{max}
1	51 190	10 041	1.0	1.0
2	100 071	18 660	0.5	3.0
3	171 800	31 083	0.18	3.0

The application of this method in three dimensions is demonstrated on the solution of shock interaction on a swept cylinder. The numerical simulation has been carried out for a sweep angle of 15° on a cylinder of diameter D equal to 3 in. and length L equal to 9 in. The undisturbed free stream Mach number is 8.03. The fluid which has been turned by the shock generator enters the computational domain with a Mach number of 5.26. The initial mesh and those obtained after two adaptive remeshings and the density contours distribution are shown in Figure 35.21 The characteristics of the meshes are shown in Table 35.4.

The potential advantages of the adaptive remeshing procedure are clearly illustrated in this three-dimensional example. The final adapted mesh has a resolution of more than five times that of the inital mesh, whereas the total number of degrees of freedom increases by only a factor of 3.4.

35.4.7 Grid Adaptation Using the Delaunay Triangulation with Sources

Here we outline a method that uses the automatic point creation and the ideas outlined for point clustering using sources [19]. The new approach is a combination of *h*-refinement and remeshing and recovers both these procedures for given input parameters. The technique is equally applicable for steady and transient adaptation. The main steps are as follows.

Algorithm I

1. Generate an initial mesh capable of providing an initial solution.
2. Obtain a flow solution.
3. Derive sources.
   a. On the line segments between surfaces.
   b. On surface triangles on the surfaces.
   c. In the field.
4. Generate the adapted surface grid.
5. Generate the adapted field grid.
6. Return to step 2.

Once a flow solution has been obtained the sources are derived by detecting regions in the domain where solution or error activity is high. Several approaches have been implemented, including taking measures of gradients within an element and introducing directional measures of the gradient in the direction of the velocity vector. Typically, density is used as the basis of the error indicator. Once an element has been identified as requiring enrichment, a source is defined with a position inside the element and a strength that is obtained by performing a statistical analysis of the error measure as computed for all elements. A minimum and maximum source strength is set, which controls the degree of enrichment to be provided by the sources.

35.4.7.1 Surface Adaptation

Grid adaptation on the configuration surface is performed as outlined in Algorithm II.

Algorithm II

1. Input the previous surface mesh.
2. Derive the surface sources.
   a. On line segments between surfaces.
   b. On triangles on the surface.
3. Perform adaptation on line segments between surfaces.
   a. Insert points on line segment and connect to surrounding points.
   b. Modify the values of the point distribution function at the surrounding points.
4. Perform adaptation on surface triangles.
   a. Insert a point at the position of the source and connect to form triangles using a "local Delaunay edge swapping" algorithm.
   b. Modify the values of the point distribution function at the nodes that form the element.
5. Perform the automatic point creation with a specified value of concentration factor α_a to generate additional points, connecting the points with a "local Delaunay edge swapping" algorithm.

In the surface triangulation grid adaptation the point connection is performed by a direct connection between the new point and the three points that form the triangle that contains it, followed by several implementations of a "local Delaunay incircle criterion" diagonal swapping routine. This latter approach is used, since a two-dimensional Delaunay algorithm is not applicable on a three-dimensional surface. It is noted that if α_a is large, typically the order of 10^3, then the automatic point creation algorithm will not create any additional points and the surface grid is refined in the standard *h*-refinement manner.

In the generation of adapted surface grids it is necessary to ensure that the added points are placed on the geometrical surface of the configuration. The traditional method is to use the given geometrical definition of the configuration. However, for complex configurations this data can be very extensive, involving very large data sets. For grid adaptation, where it is the aim to couple the grid generation fully within a flow or solution module, the use of such potentially large data files can be problematic. An alternative method for adding points onto surface geometries is explored here. The method adopted is to reconstruct the surface geometry using a transfinite, visually continuous, triangular interpolant [8]. It is viewed that this approach is more efficient and applicable than returning to the original geometrical definition of the surfaces. However, it is relatively easy to provide the necessary calls to the geometry data base if this is desired. The interpolant utilizes outward surface normals, unlike such methods as the Ferguson patch, which uses partial derivatives on boundaries. The resulting reconstructed surface is G^1 in that the surface has a continuously varying outward normal vector. When compared with results obtained using linear interpolation, it is apparent that the use of the G^1 patch to calculate the position of points being inserted reduces the displacement error by a factor in excess of 4, for both the average and maximum displacement values.

35.4.7.2 Field Adaptation

Grid adaptation in the field is performed as follows.

Algorithm III

1. Generate a mesh from the nonadapted surface mesh with a concentration factor α_1. If appropriate, a different concentration factor α can be used from the previous grid or input the previous volume mesh.
2. Input the additional surface points that are included in the adapted surface grid and connect with the Delaunay algorithm.
3. Input the field sources.
   a. Determine the elements that contain the sources.
   b. Insert a point at the position of the source and connect with the Delaunay algorithm.
   c. Modify the values of the point distribution function at the nodes in the element.
4. Perform automatic point creation to generate the adapted field with a concentration factor α_2.

Steps 1 and 2 are straightforward to apply. Step 3 requires a searching process to find the elements that contain the sources. This type of search is similar to the one used in the Delaunay algorithm to find all spheres that contain a point. Hence, in the implementation of Step 3a. the Delaunay algorithm search routine is used with the addition of a routine to determine the element rather than the sphere which contains the source. The important issue in the search is that a tree data structure, which is essential for an efficient implementation of the Delaunay algorithm, is used.

If the parameter α_2 is small, typically in the range 0.8 to 1.4, then points will, in general, be added by the automatic point creation procedure until the point distribution satisfies that which was specified with the sources. If, however, α_2 is large, say the order of 10^3, then after the insertion of a point corresponding to the position of the source, the automatic point creation procedure will not add points. In this way, with the appropriate values of α, the proposed adaptation procedure degenerates to standard *h*-refinement. This was also the case for the surface grid as considered in Section 35.4.7.1. It is clear, therefore, that the method proposed generalizes *h*-refinement so that an arbitrary number of points can be added. Furthermore, since α_1 can be varied it is possible to regenerate a mesh prior to the inclusion of sources so that once features in the flowfield have been detected and sources defined, the initial mesh can be coarsened. Hence, the proposed method has a remeshing capability to ensure that with successive adaptation the number of grid points does not always increase. As with remeshing, the proposed procedure can result in a final adapted mesh having fewer points than the initial mesh.

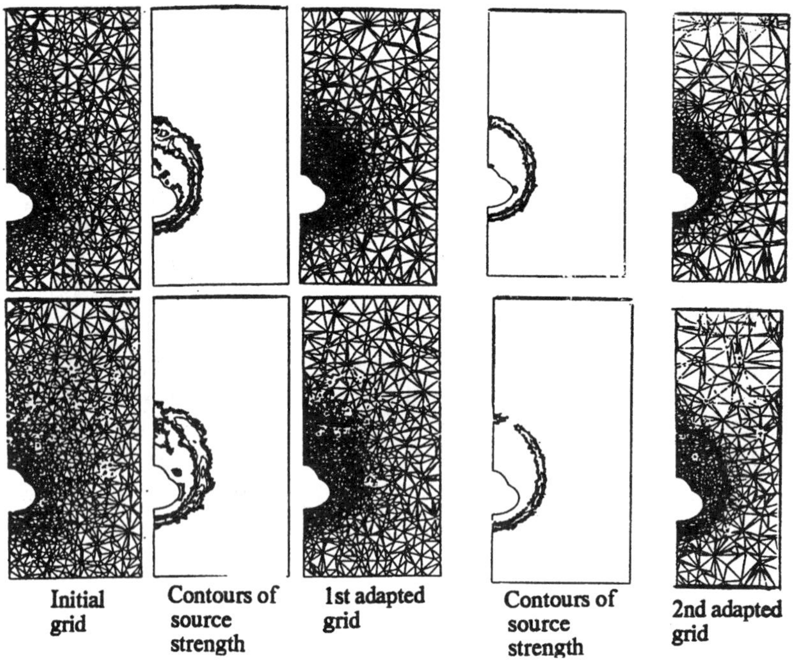

FIGURE 35.22 Hypersonic flow over a double ellipsoid. Meshes and source strength.

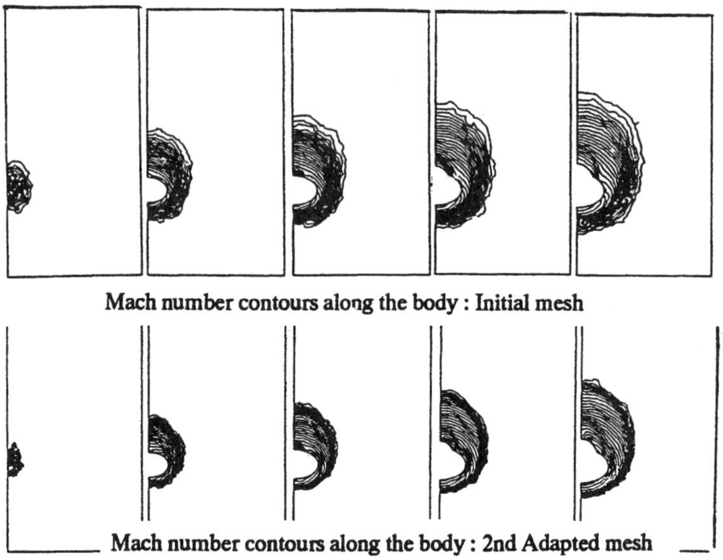

FIGURE 35.23 Hypersonic flow over a double ellipsoid. Mach number contours.

Two examples are now presented of the application of the grid adaptation method described here. The first example is the hypersonic flow over a double ellipsoid. The flow conditions are Mach number of 8.15 and 30° of incidence. Figure 35.22 shows cuts through the initial and the adapted meshes together with the distribution of the source strength. Several views of the flow solutions obtained using this method on the initial and second adapted grids is shown in Figure 35.23.

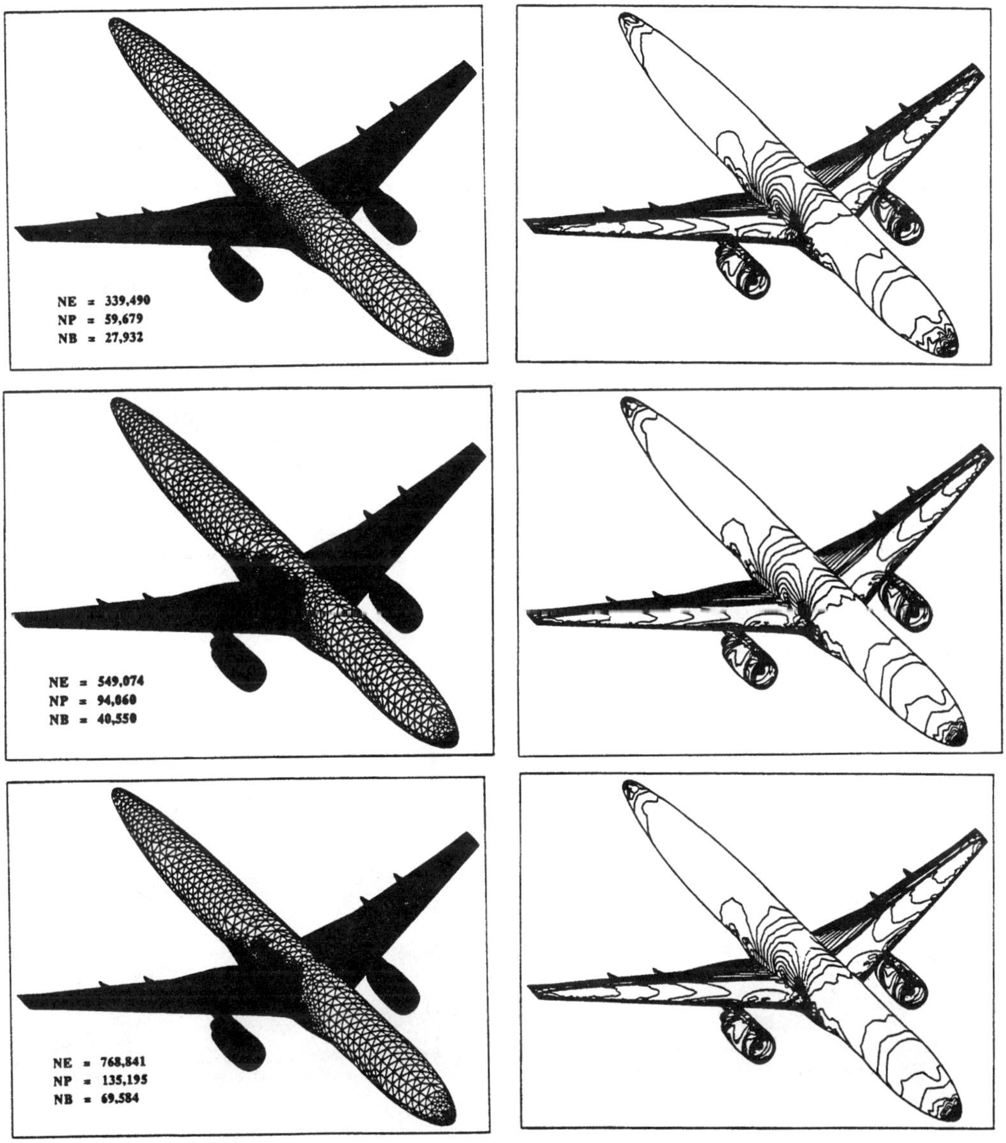

FIGURE 35.24 Adapted B60 configuration. Surface meshes and contours of pressure.

The next example is that of a transport wing-body-pylon-nacelle configuration. Figures 35.24 and 35.25 show the results of grid adaptation of the B60 configuration. The freestream Mach number was 0.801 and the angle of attack 2.738°. For the simulation the engine conditions imposed were a jet pressure ratio of 2.477, an engine mass flow ratio of 2.733 lb/s, and a jet total temperature of 370.04 K.

It is clear from these results the distinct effects of the grid adaptation. The shock wave resolution is greatly improved both on the wing and in the field, and the comparison of the pressure coefficient on the wing with experiment shows an incremental improvement.

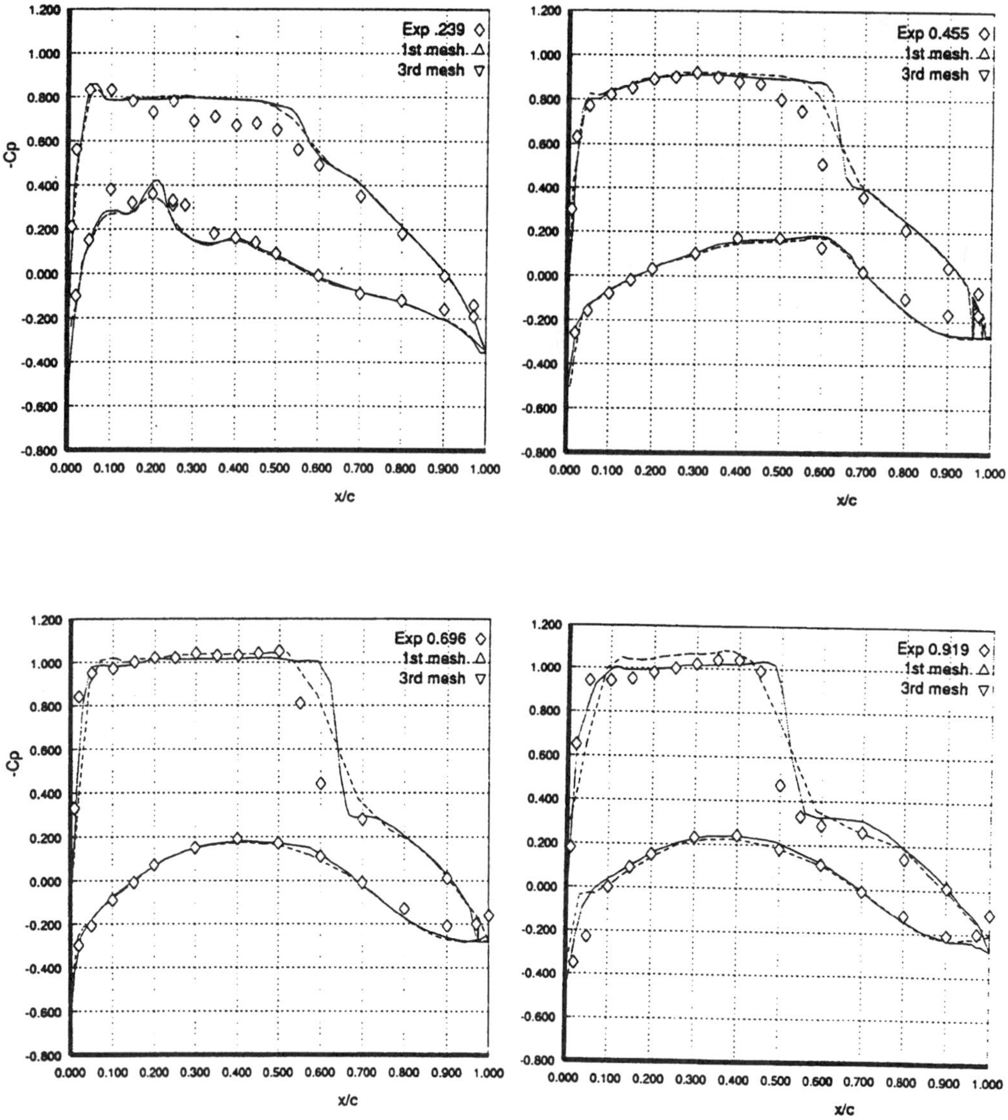

FIGURE 35.25 Adapted B60 configuration. Coefficient of pressure on the wing and nacelle.

References

1. Allwright, S., Multiblock topology specification and grid generation for complete aircraft configurations, *Applications of Mesh Generation to Complex 3-D Configurations,* AGARD Conference Proceedings. 1990, No. 464, 11.1–11.11.
2. Baker, T.J., Unstructured mesh generation by a generalized Delaunay algorithm, *Applications of Mesh Generation to Complex 3-D Configurations,* AGARD Conference Proceedings. 1990, No. 464, 20.1–20.10.
3. Donéa, J. and Giuliani, S., A simple method to generate high-order accurate convection operators for explicit schemes based on linear finite elements, *Int. J. Num. Meth. Fluids 1,* 1981, pp 63–79.

4. Formaggia, L., Peraire, J., Morgan, K., and Peiro, J., Implementation of a 3D explicit Euler solver on a CRAY computer, *Proc. 4th Int. Symposium on Science and Engineering on CRAY Supercomputers*, Minneapolis, 1988, pp 45–65.
5. Jameson, A., Baker, T.J., and Weatherill, N.P., Calculation of inviscid transonic flow over a complete aircraft, *AIAA Paper 86-0102,* 1986.
6. Löhner, R., Morgan, K., and Zienkiewicz, O.C., Adaptive grid refinement for the compressible Euler equations, Babuska, I., et al., (Ed.), *Accuracy Estimates and Adaptive Refinements in Finite Element Computations,* Wiley, 1986, pp 281–297.
7. Nakahashi, K. and Deiwert, G.S., A practical adaptive-grid method for complex fluid flow problems, *Lecture Notes in Physics.* Springer Verlag, 1985, Vol. 218, pp 422–426,
8. Nielson, G.M., The side vertex method for interpolation in triangles, *Journal of Approximation Theory,* 1979, 25, pp 318–336.
9. Oden, J.T., Grid optimisation and adaptive meshes for finite element methods, University of Texas at Austin, Notes, 1983.
10. Palmerio, B., Billey, V., Dervieux, A., and Periaux, J., Self-adaptive Mesh Refinements And Finite Element Methods For Solving the Euler equations, *Numerical Methods for Fluid Dynamics II,* Morton, K.W. and Baines, M.J., (Eds.), 1985, Clarendon Press, Oxford, pp 369–388.
11. Palmerio, B. and Dervieux, B., 2D and 3D Unstructured mesh adaption relying on physical analogy, *Proc. of the Second International Conference on Numerical Grid Generation in Computational Fluid Mechanics,* Miami Beach, FL, 1988.
12. Peraire, J., Morgan, K., and Peiro, J., Unstructured finite element mesh generation and adaptive procedures for CFD, *Applications of Mesh Generation to Complex 3-D Configurations,* AGARD Conference Proceedings, 1990, No. 464, 18.1–18.12.
13. Peraire, J., Morgan, K. Peiro, J., and Zienkiewicz, O.C., An adaptive finite element method for high speed flows, *AIAA Paper 87-0558,* 1987.
14. Peraire, J. Peiro, J., Formaggia, L, Morgan, K., and Zienkiewicz, O.C., Finite element Euler computations in three dimensions, *Int. J. Num. Meth. Eng.* 26, 1988.
15. Peraire, J., Vahdati, M., Morgan, K., and Zienkiewicz, O.C., Adaptive remeshing for compressible flow computations, *J. Comp. Phys.* 1987, 72, pp 449–466.
16. Peiro, J., Peraire, J., and Morgan, K., FELISA System Reference Manual. Part I: Basic Theory, *Technical Report CR/821/94,* University of Wales, Swansea, 1994.
17. Thompson, J.F., Warsi, Z.U.A., and Mastin, C.W., *Numerical Grid Generation — Foundations and Applications.* North-Holland, 1985.
18. Watson, D.F., Computing the *n*-dimensional Delaunay Tessellation with application to Voronoï polytopes, *The Computer Journal.* 1981, 24, pp 167–172.
19. Weatherill, N.P., Hassan, O., Marchant, M.J., and Marcum, D.L., Adaptive inviscid flow solutions for aerospace geometries on efficiently generated unstructured tetrahedral meshes, *AIAA CFD Conference,* July, 1992.
20. Weatherill, N.P., Mesh generation in computational fluid dynamics, *von Karman Institute for Fluid Dynamics,* Lecture Series 1989-04, Brussels, 1989.
21. Zienkiewicz, O.C., and Morgan, K., *Finite Elements and Approximation,* Wiley, 1983.

36

Variational Methods of Construction of Optimal Grids

O.B. Khairullina

A.F. Sidorov

O.V. Ushakova

36.1 Introduction

Although the variational methods of construction of curvilinear grids in complex domains require realization of the solution of rather laborious problems (minimization of functionals for functions of many variables or solution of the appropriate Euler–Ostrogradsky equations (E-O)), nevertheless they give an opportunity to generate grids with good computational properties. As a rule, with the help of the variational approaches structured or block-structured grids in simply connected and multiply connected domains can be generated with distinct grid topology.

The following criteria of grid optimality are mostly used in the solution of the boundary value problems associated with the pertinent partial differential equations.

1. Closeness to uniformity (U). The volumes of the neighboring elementary cells of a grid should be of the same size. Otherwise, it is difficult to build difference approximations of sufficient accuracy for the differential equations. Besides, the conditionality of the systems of difference equations approximator on the constructed grid a system of differential equations is sharply worsened.

0-8493-2687-7/99/$0.00+$.50

2. Closeness to orthogonality (O). The coordinate lines or surfaces of various families in each block should not cross at angles close to 0 or π. Otherwise, again the conditionality of systems of difference equations is worsened.
3. Adaptation (A). The curvilinear grid should follow the properties of a given function (family of functions) or should change in iterative or nonstationary processes in accordance with the solution of boundary value problems.

The concentration of grid lines should take place, in particular, in zones of large gradients, for which adaptive grid is generated.

These criteria, especially (U) and (A), are contradictory. As a rule, they are applied by means of weight parameters determining the values of optimality criteria.

The most widely used is the approach where smooth nondegenerate mapping of some simple domain in the space of parameters (rectangle, parallelepiped, their combinations) onto the given domain in the space of initial variables is searched. A set of functions that define the required mapping should minimize some variational functional with a given boundary or natural conditions. The set of such functionals is rather wide (some examples can be found in Chapter 35.)

In the overwhelming majority of cases, integral variational functionals, formalizing the optimality criteria, contain first partial derivatives of functions realizing the mapping. The E-O equations for them is the system of partial differential equations of the second order, as a rule, of elliptic type. These approaches in the literature have gotten enough attention, and they will be described in this chapter very briefly, by way of review.

The main contents of the chapter are concerned with the presentation of another concept of constructing grids, developing mainly in works of Russian scientists during the past 30 years [25]. The main feature of the approach is associated with the special way of formalization of criterion (U) which gives a nonlinear variational functional containing both first and second partial derivatives of the functions realizing the mapping. This continuous functional arises naturally after the consideration of a discrete functional minimization of the measure of a relative error of a nonuniform grid in comparison with uniform grid. Such formalization leads to a system of E-O equations of the fourth order, hyperbolic in a wide sense. It has enabled consideration of new wider types of boundary conditions, as well as development of effective algorithms and programs of grid generation for the complex domains.

The economic and effective procedures of calculation of grids are connected with the use of iterative processes based on the special nonstationary modification of E-O equations, as well as on the direct geometrical ways of minimization of discrete functionals formalizing all three optimality criteria.

In Section 36.2 of this chapter, a brief review of variational functionals for constructing structured grids is presented. The deduction of discrete functionals formalizing criteria (U), (O), (A) is carried out, and the analysis of their properties in one-dimensional cases is given.

Section 36.3 is devoted to the description of effective algorithms that allow the construction of two-dimensional optimal smooth grids with simple and complex topology in simply connected and multiply connected domains. The description of capabilities of two programs MOPS-2a and LADA for generation of optimal and adaptive grids is given. A new way of automatic generation of an initial approximation of a grid is considered. Examples of grids and results of their testing are shown.

In Section 36.4 a number of applications of geometrically optimal grids to the numerical solution of problems of hydrodynamic and gasdynamic flows in axially symmetric channels involving complex geometries is described. In the construction of fast iterative processes of the solution of these stationary problems, the requirements on grids are very high, since the parameters of flows change in a wide range. Examples of such calculations are given.

In the conclusion of this chapter the capabilities of the approach under development for generation of three-dimensional grids and problems arising here as well as for parallelizing the algorithms for computing optimal grids are briefly described.

36.2 Constructions of the Functionals Formalizing the Optimality Criteria

36.2.1 Analysis of the Functionals (U) and (A) in One-Dimensional Case

We shall consider first the possibilities of representations of the functional (U). Let on the segment $L = [0, M]$ it be required to construct grid nodes x_i ($i = 0, 1, \ldots, N$) with given lengths of boundary intervals A and B at the ends. The grid should be closest in the some metric to the uniform grid. For evaluation of a measure of deviation of grids from uniform grids we shall use two functionals,

$$J_U^{(1)} = \sum_{i=1}^{N-1}\left(\frac{h_{i+1}}{h_i} - 1\right)^2, \tag{36.1}$$

$$J_U^{(2)} = \sum_{i=1}^{N-1}\left(h_{i+1} - h_i\right)^2, \tag{36.2}$$

($h_i = x_i - x_{i-1}$, $i = 1,2, \ldots, N$, $h_1 = A$, $h_N = B$, $x_0 = 0$, $x_N = M$, $M > A + B$), which need to be minimized.

Usually it is more convenient to use the continuous formulation of these problems.

Let $x = x(\xi)$ transform the parametric segment $\xi \in [0, N]$ into the segment L so that $x_i = x(i)$, $i = 0, \ldots, N$. We shall consider

$$h_i \approx y(i), \quad h_{i+1} - h_i \approx y'(i), \quad \frac{h_{i+1} - h_i}{h_i} \approx \frac{y'(i)}{y(i)}, \quad i = 0, \ldots, N-1$$

where $y(\xi) = x_\xi(\xi)$, $\xi \in [0, N]$. Obviously, the relation $\sum_{i=1}^{N} h_i = M$ must be satisfied. Then instead of the discrete functionals 36.1 and 36.2 it is possible to consider the continuous functionals

$$I_U^{(1)} = \int_0^N \frac{x_{\xi\xi}^2}{x_\xi^2}\, d\xi, \tag{36.3}$$

$$I_u^{(2)} = \int_0^N x_{\xi\xi}^2\, d\xi. \tag{36.4}$$

The minimization of the functionals $I_U^{(1)}$ and $I_U^{(2)}$ should be considered under the conditions

$$\int_0^N x_\xi\, d\xi = M, \quad x_\xi(0) = A, \quad x_\xi(N) = B. \tag{36.5}$$

Thus, isoperimetric variational problems arise with analytical solution.

The extremal functions for Eq. 36.3 have one of three possible forms:

$$y(\xi) = a_1 \cos^{-2}\left(a_2\xi + a_3\right), \tag{36.6}$$

$$y(\xi) = b_1 \operatorname{ch}^{-2}(b_2\xi + b_3), \tag{36.7}$$

$$y(\xi) = c_1(\xi + c_2)^{-2}, \tag{36.8}$$

where *ch* is a designation of a hyperbolic cosine. The constants a_k, b_k, c_k are defined from the conditions Eq. 36.5. If in this case the value

$$q = \frac{M}{\sqrt{ABN}}$$

is less than 1, the representation Eq. 36.6 applies, if $q > 1$ – Eq. 36.7, and, finally, if $q = 1$ – Eq. 36.8 applies. The positive solution exists at any N, $A > 0$, $B > 0$, $A + B < M$. The problem can be solved analytically also for the functional Eq. 36.4, but the condition of the positiveness of the solution ($h_k >$ 0) is not always satisfied here. For example, at $A = B$ it is satisfied only under the condition

$$\frac{M}{N} - \frac{1}{3}A > 0.$$

For this reason, hereinafter in constructing the multidimensional functional during the generalization, preference is given to the functional Eq. 36.1 and is analog Eq. 36.3, though in the literature the generalization of the functionals Eqs. 36.2, 36.4, which leads to linear E-0 equations in the parametric spaces, is very frequently used.

It turns out that the grids constructed on the basis of Eqs. 36.6–36.8 [29] have a number of useful properties. Thus in [28] it has been shown that $h_{i+1} - h_i \approx 0(N^{-2})$ at large N and it is possible to approximate more precisely the derivatives of high orders.

In [40, 41] it has been shown that at the expense of choice only of the boundary values $A(\varepsilon, N)$, $B(\varepsilon, N)$ constructed on the basis of such grids, usual difference schemes for the solution of boundary value problems for ordinary equations containing the small parameter ε have the property of uniform convergence on parameter ε at $N \to \infty$. Thus, this construction of the functional in a number of cases allows adaptation of grids to the properties of the boundary value problem solution at the expense only of choice of boundary intervals.

Let us consider now some ways of formalization of criterion (*A*), when the grids should automatically concentrate in the zones of large gradients of a given function $\Phi(x)$ or system of functions $\{\Phi_i(x)\}$.

Let us use as a discrete measure of adaptation

$$J_A = \sum_{i=1}^{N} \left[\Phi(x_i) - \Phi(x_{i-1})\right]^2 h_i^2. \tag{36.9}$$

The functional J_A presents a sum of squares of the areas of rectangles (Figure 36.1), the vertices of which belong to the curve $f = \Phi(x)$. The minimization of J_A with the choice of the nodes x_i results in concentrations of a grid in zones of large gradients of the function Φ.

If $x = x(\xi)$, the continuous counterpart of the functional J_A will be of the form

$$I_A = \int_0^N \Phi_x^2 x_\xi^4 d\xi.$$

Let $\lambda_U \geq 0$ and $\lambda_A \geq 0$ — some constant weight coefficients. The general functional for construction of a grid satisfying criteria (U) and (A) will have the form

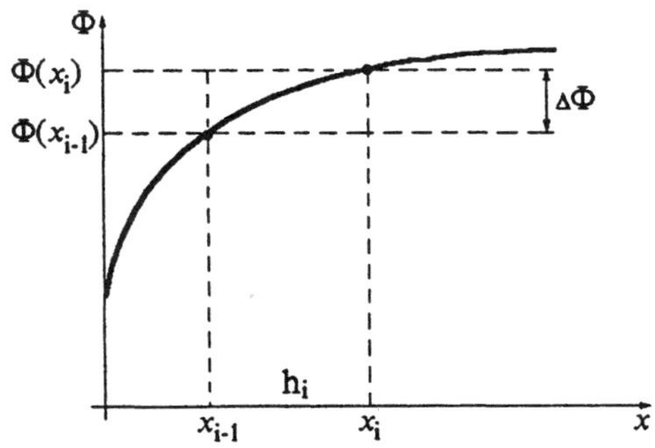

FIGURE 36.1

$$I = \lambda_U \int_0^N \frac{x_{\xi\xi}^2}{x_\xi^2} d\xi + \lambda_A \int_0^N \Phi_x^2 x_\xi^4 d\xi. \tag{36.10}$$

Note that if $\lambda_A \neq 0$, we do not manage to get rid of second derivatives in Eq. 36.10 in the first integral by means of function x_ξ.

Two boundary conditions for the function $x(\xi)$ are obvious:

$$x(0) = 0, \quad x(N) = M. \tag{36.11}$$

At $\lambda_U = 0$ and $\Phi' \neq 0$ from Eqs. 36.10, 36.11 we get the solution in the implicit form:

$$\xi(x) = \frac{N\int_0^x \sqrt[4]{\Phi_x^2(\zeta)}d\zeta}{\int_0^M \sqrt[4]{\Phi_x^2(\zeta)}d\zeta}.$$

The analogs of this solution are used frequently (see [33]) for construction of adaptive grids. Instead of Eq. 36.10) it is possible to use functionals of a more general form ($k = 1, 2$);

$$I_k = \lambda_U \int_0^N \frac{x_{\xi\xi}^2}{x_\xi^2} d\xi + \int_0^N w_k(x(\xi)) x_\xi^4 d\xi,$$

$$w_1(x) = b_0 + \sum_{k=1}^{s} b_k \left(\frac{d^k \Phi(x)}{dx^k}\right)^2, \tag{36.12}$$

$$w_2(x) = c_0 + \sum_{j=1}^{l} c_j \left(\frac{d\Phi_j}{dx}\right)^2 \quad \text{(for a system of functions)}$$

where b_k, c_j – nonnegative weight constants. At $b_k = 0$, $k = 1, \ldots, s$ the minimization of Eq. 36.12 gives the uniform grid $x(\xi) = M\xi/N$.

Besides the conditions Eq. 36.11 for I Eq. 36.10, it is necessary to set two more boundary conditions. These can be, for example, conditions $x_\xi(0) = A$, $x_\xi(N) = B$ (see Eq. 36.5) or natural boundary conditions $x_{\xi\xi}(0) = x_{\xi\xi}(N) = 0$.

The E-O equation for the functional Eq. 36.12 has the form

$$\lambda_U\left(\mu'''\mu^2 - 4\mu''\mu'\mu + 3\mu'^3\right) - \frac{3}{2}w_k'\mu^8 - 6w_k\mu^6\mu' = 0, \quad \mu = x_\xi.$$

Without an analytical solution here we need to use numerical methods, in particular, the method of reaching the steady-state condition during the solution of appropriate boundary value problems [34]. In [35] the theorem of existence and uniqueness of the solution of the boundary value problems for wide classes of functions $w_k(x)$ has been proven.

36.2.2 Construction of Two-Dimensional and Three-Dimensional Functionals (U), (O), (A)

We shall consider, at first, an elementary situation in the two-dimensional case. Let G be a simply connected domain in the plane x, y that is considered as a curvilinear quadrangle $ABCD$ with the given vertices. We shall seek the functions

$$x = x(\xi,\eta), \quad y = y(\xi,\eta) \tag{36.13}$$

mapping at integers N, M a parametric rectangle $P = \{[0, N] \times [0, M]\}$ onto a given domain G. Eq. 36 13 determine at $\xi = i$, $\eta = j$ ($i = 0, \ldots, N, j = 0, \ldots, M$) the equations of coordinate lines in the parametric form, if the Jacobian D of the mapping is nondegenerate.

The variational approach by Brackbill and Saltzman [3], generalizing the Winslow approach [42] for generation of grids, consists of minimization of the functional

$$\iint_P \frac{1}{D}\left(x_\xi^2 + y_\xi^2 + x_\eta^2 + y_\eta^2\right)d\xi d\eta. \tag{36.14}$$

As a rule, it is assumed that the functions x, y on ∂P are given, i.e., the arrangement of nodes on the boundary ∂G is given. The E-O equations for Eq. 36.14 give rise to elliptic generators of grids. Algorithms for construction of such grids are described in Chapter 4.

In [3] to the functional Eq. 36.14 the functionals

$$I_0 = \iint_P \left(x_\xi y_\xi + x_\eta y_\eta\right)^2 d\xi d\eta,$$

$$I_A = \iint_P D^2 W(x,y) d\xi d\eta,$$

responsible for criteria (O) and (A) were also added. Here $W = W(x, y)$ — some positive weight function, dependent on the solution, under which the adaptation of a grid is carried out.

Note that earlier in [39] the variational principles for construction of a moving grid, adapted to the solution of gas dynamics problems were formulated.

In [6] one can find the algorithm for the solution of the variational problem of minimizing the functional

$$\iint_P \left[l\left(x_\xi^2 + y_\xi^2\right) + \frac{1}{l}\left(x_\eta^2 + y_\eta^2\right)\right] d\xi d\eta$$

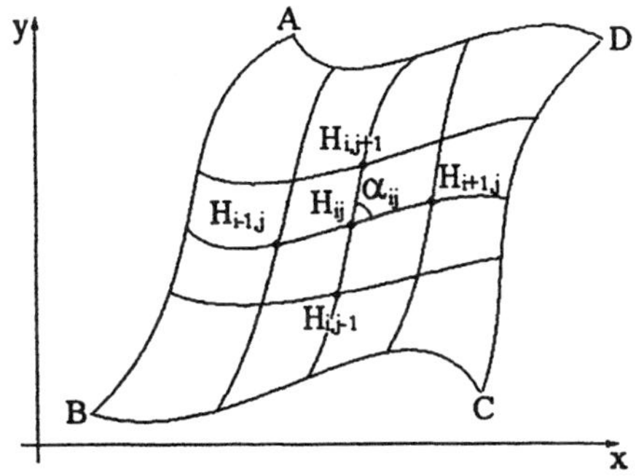

FIGURE 36.2

where I is a parameter. In [2] consideration was given to the functional

$$\iint_P \frac{1}{\sin[\beta(\eta)-\alpha(\xi)]}\left\{\exp[q_1(\eta)-q_2(\xi)]\left(x_\xi^2+y_\xi^2\right)\right. \\ \left.+\exp[q_2(\xi)-q_1(\eta)]\left(x_\eta^2+y_\eta^2\right)-2\cos[\beta(\eta)-\alpha(\xi)]\left(x_\xi x_\eta+y_\xi y_\eta\right)\right\}d\xi d\eta \tag{36.15}$$

where functions $q_1(\eta)$, $q_2(\xi)$, $\alpha(\xi)$, $\beta(\eta)$ have to be found in the process of minimizing the functional Eq. 36.15 on the class of functions $x(\xi, \eta)$, $y(\xi, \eta)$ with given values on the boundary ∂P.

These present construction of continuous functionals, as well as a wide range of other possible representations and other principles in the background of grid generation, are described in detail in [32, 33] and in the recently published survey [20].

Note that very often for variational methods of optimal grid generation, not only continuous functionals but their discrete counterparts are used. Let us introduce some of them.

In [7] for optimization of three-dimensional grids, the sum for all inner nodes of corresponding local measures has been chosen as the measures of uniformity and orthogonality. The local measure of uniformity for each inner node is a sum of squares of lengths of the vectors connecting each node with neighboring nodes, and the local measure of orthogonality is a sum of scalar products of those vectors. In [4] the sum of squares of cell areas has been considered as a measure of uniformity. The base for construction of discrete measures of adaptation is the equidistribution principle formulated in [32].

Let us introduce discrete functionals used in the given approach. Let the grid with nodes H_{ij} be constructed in a curvilinear quadrangle *ABCD*. We shall denote by $r_{i\pm1,j}$, $r_{i,j\pm1}$ the Euclidean distances between nodes H_{ij} and $H_{i\pm,j}$, H_{ij} and $H_{i,j\pm1}$, by $\alpha_{ij}^{(k)}$ angles between lines connecting the node H_{ij} sequentially with the nodes $H_{i+1,j}$, $H_{i,j+1}$, $H_{i-1,j}$, $H_{i,j-1}$, by $\Phi(H_{ij})$ the value in the node H_{ij} of a given function $\Phi(x, y)$ under which the adaptation is carried out and by S_{ij} area of a cell defined by the nodes H_{ij}, $H_{i+1,j}$, $H_{i,j+1}$, $H_{i+1,j+1}$ (Figure 36.2).

The functionals

$$J_U = \sum_{(i,j)\in P_h}\left[\left(r_{i+1,j}-r_{i-1,j}\right)^2\left(\frac{1}{r_{i+1,j}^2}+\frac{1}{r_{i-1,j}^2}\right)+\left(r_{i,j+1}-r_{i,j-1}\right)^2\left(\frac{1}{r_{i,j+1}^2}+\frac{1}{r_{i,j-1}^2}\right)\right], \tag{36.16}$$

$$J_A = \sum_{(i,j)\in P_h} S_{ij}\left[\left|\Phi\left(H_{ij}\right)-\Phi\left(H_{i+1,j}\right)\right|+\left|\Phi\left(H_{ij}\right)-\Phi\left(H_{i,j+1}\right)\right|\right] \tag{36.17}$$

are direct generalizations of the one-dimensional functionals Eq. 36.1 and Eq. 36.9. Minimization of the functional

$$J_O = \sum_{(i,j)\in P_h} \sum_{k=1}^{4} \sin^{-2} \alpha_{ij}^{(k)} \tag{36.18}$$

should ensure the closeness of grids to orthogonality ($\alpha_{ij}^{(k)} \neq 0, \pi$). The general discrete functional has the form

$$J = \lambda_U J_U + \lambda_O J_O + \lambda_A J_A \tag{36.19}$$

where λ_U, λ_O, λ_A are weight coefficients.

Using the functions x, y from Eq. 36.13, mapping the domain G onto a parametric rectangle P, we shall write continuous counterparts of the discrete functionals Eqs. 36.16–36.19 in the form

$$I_U = \iint \left[\frac{1}{g_{11}^2} (g_{11})_{\xi\xi}^2 + \frac{1}{g_{22}^2} (g_{22})_{\eta\eta}^2 \right] d\xi d\eta, \tag{36.20}$$

$$I_O = \iint \left[\frac{g_{11} g_{22}}{D^2} \right] d\xi d\eta, \tag{36.21}$$

$$I_A = \iint W D^2 d\xi d\eta, \quad W = \Phi_x^2 + \Phi_y^2 + \alpha, \quad \alpha = \text{const} > 0, \tag{36.22}$$

$$I = \lambda_U I_U + \lambda_O I_O + \lambda_A I_A. \tag{36.23}$$

$$g_{11} = x_\xi^2 + y_\xi^2, \; g_{22} = x_\eta^2 + y_\eta^2, \; D = \det\{W\} = x_\xi y_\eta - x_\eta y_\xi. \tag{36.24}$$

Similarly, it is possible to construct functionals $J_U(\Omega)$, $J_O(\Omega)$, $J_A(\Omega)$ for generation of grids in a curvilinear quadrangle $G(\Omega)$ on a surface S determined in R^3 by the parametric equations

$$x_i = x_i(\mu_1, \mu_2), \quad i = 1,2,3, \quad (\mu_1, \mu_2) \in \Omega \tag{36.25}$$

(Ω is a limited area in a parametric plane μ_1, μ_2).

The form of functionals Eqs. 36.20–36.22 is retained, if instead of x, y we use functions $\mu_1 = \mu_1(\xi, \eta)$, $\mu_2 = \mu_2(\xi, \eta)$, $(\xi, \eta) \in P$ and for Eq. 36.24 substitute the expressions

$$g_{ii} + \sum_{j,k=1,2} \gamma_{jk} \frac{\partial \mu_j}{\partial p_i} \frac{\partial \mu_k}{\partial p_i}, \quad \gamma_{jk} = \sum_{l=1}^{3} \frac{\partial x_l}{\partial \mu_j} \frac{\partial x_l}{\partial \mu_k}, \quad i = 1,2,$$

$$D = D_1 \cdot \det\left\{ \frac{\partial \mu_j}{\partial p_k} \right\}_{j,k=1,2} \neq 0, \quad D_1 = \gamma_{11}\gamma_{22} - \gamma_{12}^2, \quad p_1 = \xi, \; p_2 = \eta$$

After determining the functions μ_1, μ_2 the relations $x_i(\xi, \eta) = x_i(\mu_1(\xi, \eta), \mu_2(\xi, \eta))$ Eq. 36.25 will define at ξ = const. and η = const. two sets of coordinate lines lying on the surface S.

We shall consider now a three-dimensional domain G representing a three-dimensional curvilinear hexahedron with 8 given vertices, 12 curvilinear edges and 6 curvilinear sides. We shall search for functions

$$x_i = x_i(\xi_1, \xi_2, \xi_3), \quad i = 1,2,3, \tag{36.26}$$

mapping a rectangle parallelepiped $P\{[0, N_1] \times [0, N_2] \times [0, N_3]\}$ onto a given domain G with preservation of the correspondence of vertices, edges, and sides.

The generalization of functionals Eqs. 36.20–36.22 in the three-dimensional case is based on the consideration of the discrete counterparts Eqs. 36.16 and 36.17. The general functional with weight coefficients λ_U, λ_O, λ_A has the form

$$I = \lambda_U \iiint_P \left[\sum_{k=1}^{3} \frac{1}{g_{kk}^2} \left(\frac{\partial g_{kk}}{\partial \xi_k} \right)^2 \right] d\xi_1 d\xi_2 d\xi_3 + \lambda_O \iiint_P \sum_{k=1, ij \neq k}^{3} \left[\frac{1}{g_{kk}} \frac{G_i G_j}{D^2} \right] d\xi_1 d\xi_2 d\xi_3 +$$
$$+ \lambda_A \iiint_P \left[\sum_{k=1}^{3} \left(\frac{\partial \Phi}{\partial x_k} \right)^2 + \alpha \right] D^2 d\xi_1 d\xi_2 d\xi_3, \quad \alpha = \text{const.} > 0. \tag{36.27}$$

At ξ_j = const., j = 1,2,3, the formulas Eq. 36.26 determine the families of coordinate surfaces in the domain G.

36.2.3 Boundary Conditions. The Analysis of Boundary Value Problems in the Two-Dimensional Case

In the algorithm of grid generation, various ways of boundary node arrangement are possible. Most frequently the nodes on the boundary of the domain are considered to be given and fixed. This way is used in generation of block-structured grids, when the domain is cut on subregions and on their common boundaries the nodes should coincide. If grids in separate blocks are calculated independently from each other, the smoothness of grid lines on the interfaces of blocks is broken. The smoothness of grid lines and movement of nodes on lines of block interfaces in correspondence with the considered optimality criteria are achieved by special organization of overlapping of blocks, as realized in the program MOPS-2a.

In construction of adaptive grids it is more natural to determine the boundary nodes in the process of calculation from some requirements on the grid at the boundary, i.e., to consider moving boundary nodes.

In some methods the algorithm of grid generation allows fixed and given slopes of coordinate lines to the boundary of the domain to be considered. In [33] it is remarked that the application of grids very much different from orthogonal ones near boundaries can cause additional difficulties in approximation of boundary conditions during the solution of the problems on such grids. Therefore, frequently grids orthogonal or near-orthogonal or near-orthogonal at the boundary are considered (see Chapter 7).

In the suggested approach, it is possible to realize all boundary conditions listed above. As was already mentioned in the introduction, the summand $\lambda_U I_U$ is leading here. It has the second order of the integrated expression in the functional I. This allows in the variational problem arbitrariness in the choice of unknown functions and their first derivatives on the boundary of the domain. It is possible to fix or to leave free both the location of boundary nodes and the slope of coordinate lines at the boundary.

In the programs MOPS-2a and LADA, the nodes on the boundary of the domain are considered to be given and fixed:

$$x_i\big|_{\partial P} = l_i(\xi), \quad s \in \partial P, \quad i = 1,2 \tag{36.28}$$

($l_i(\xi)$ — given functions of node coordinates on the boundary ∂P).

In addition to Eq. 36.28 it is possible to consider also necessary boundary conditions originating from minimization of the functional Eq. 36.23 on the class of functions satisfying on the boundary ∂P the conditions Eq. 36.28. These are natural boundary conditions for derivatives:

$$V_i \frac{\partial x_j}{\partial \xi_i}\Big|_{\xi_i=0,N_i} = 0, \quad V_i = \frac{1}{g_{ii}^2}\frac{\partial g_{ii}}{\partial \xi_i}\Big|_{\xi_i=0,N_i} \quad j=1,2, \quad i=1,2. \tag{36.29}$$

Other variants of the boundary conditions were considered in [36], where the algorithm with moving boundary nodes and coordinate lines orthogonal to the boundary was described.

Unfortunately, theorems of existence of the solution, uniqueness of it, and the correctness of the posed problems in contrast to the one-dimensional case are at the moment unknown. Only formal reasons (eight functions $l_i(\bar{\xi})$ are given: there is the arbitrariness in eight functions) and the large experience of calculations of grids confirms a hypothesis about the existence of such theorems.

The summand I_U not only determines boundary conditions, but also the type of a system of the E-O equations. They system of E-O equations for functionals Eq. 36.27 in the two-dimensional and three-dimensional cases is too cumbersome. The structure can be presented in the form

$$\sum_{k=1}^{n} \frac{\partial x_k}{\partial \xi_i}\frac{\partial^4 x_k}{\partial \xi_i^4} + L_i(x_1,\ldots,x_n) = 0, \quad i=1,\ldots,n, \; n=2,3 \tag{36.30}$$

where $L_i(x_i, \ldots, x_n)$ — nonlinear forms containing partial derivatives of functions x_k not higher than third order.

Let the equation

$$\Psi(\xi_1,\ldots,\xi_n) = 0$$

be the equation of characteristic variety for the system of Eq. 36.30. From 36.30 it follows that the differential equation for Ψ has the form

$$\Psi_{\xi_1}^4 \cdot \ldots \cdot \Psi_{\xi_n}^4 = 0.$$

Thus, the system of Eq. 36.30 is hyperbolic in a wide sense [19], and the lines or planes ξ_i = const. are characteristics.

If in Eq. 36.27 we put $\lambda_U = \lambda_A = 0$ and consider only the functional responsible for the closeness of grids to orthogonality, then the direct analysis of the system of E-O equations [30] shows that this system is on the second order of a mixed elliptic–hyperbolic type so that the boundary problem with data Eq. 36.28 is incorrectly formulated. Thus, the introduction of the summand with $\lambda_U \neq 0$ plays the important regularizing role.

36.3 Effective Algorithms of Optimal Grid Generation

The variation methods are the most natural for generation of optimal grids. The implementation of effective algorithms, however, involves overcoming of a lot of difficulties. Numerical procedures for grid generation based only on the solution of the E-O equations frequently [17, 37] suffer from several problems:

1. The bulky form of the E-O equations results in large numbers of arithmetical operations.
2. For stability of calculations in the iterative schemes, the small time step should be selected, and that has an effect on the number of iterations required to reach the steady-state condition.

3. The contradictoriness of the requirements included in the basis of a variational method leads to natural difficulties in the choice of control parameters defining the value of one or another criterion of optimality. Variation of the weight coefficients in a wide range can cause instability of the numerical procedure in the solution of the equations [3].

In the approach here, at any positive weight coefficients the type of the E-O system does not vary. However, since at $\lambda_U = 0$, $\lambda_O \neq 0$ the system becomes of a mixed elliptic–hyperbolic type, then for stability of calculations in the solution of the equations the weight coefficients should be selected so that the contribution of summands corresponding to I_O and I_A does not exceed I_U. Otherwise in a discrete solution the problem can turn out to be unstable. The detailed recommendations for choice of weight coefficients in the variational methods based on the solution of the E-O equations, for the example of the Brackbill–Saltzman equations, are given in [17, 33]. Note that numerical solution of the E-O equations is not the only way for implementation of the variational principles. The direct methods of minimization of discrete functionals [7] and [21] can be more effective in generation of grids (see also Chapter 33).

In the approach here, the effective procedures of calculation of grids are realized by special iterative processes that uses a solution of special nonstationary modifications of the E-O equations and direct geometric minimization of discrete functionals (Sections 36.3.1 and 36.3.2).

In Section 36.3.3 an algorithm for two-dimensional optimal adaptive grid generation in simply-connected domains using only direct methods of minimization of functional is described.

36.3.1 Organization of the Iterative Process

The algorithm for optimal curvilinear grids generation was developed according to the requirements for the automatic generation of grids (universality, cost-effectiveness, reliability, minimum of a used information) [25] and optimally criteria of grids (U), (O), when the functional Eq. 36.23

$$I = \lambda_U I_U + \lambda_O I_O \tag{36.31}$$

is minimized at a given arrangement of nodes Eq. 36.28 on the boundary with $\lambda_A = 0$.

For organization of the iterative process, we use the solution of an auxiliary nonstationary system for the E-O Eq. 36.30,

$$\begin{aligned} \alpha_{11}x_t + \alpha_{12}y_t &= x_\xi x_{\xi\xi\xi\xi} + y_\xi y_{\xi\xi\xi\xi} + L_1(x,y), \\ \alpha_{21}x_t + \alpha_{22}y_t &= x_\eta x_{\eta\eta\eta\eta} + y_\eta y_{\eta\eta\eta\eta} + L_2(x,y), \end{aligned} \tag{36.32}$$

where α_{ij} $(i, j = 1, 2)$ are parameters, $x = x(\xi, \eta, t)$, $y = y(\xi, \eta, t)$. If a matrix $A = \{\alpha_{ij}\}$ is taken in the form $A = -W^*$ where W^* is the matrix conjugate to a matrix W Eq. 36.24 [26], in the approximation of "frozen" coefficients the analysis of a short linear system with constant coefficients obtained from Eq. 36.32 shows that the Cauchy problem with periodic initial data is correct.

The set of equations 36.32 at $A = -W^*$ can be used for the calculation of moving grids varying in time, when the form of the domain $G(t)$ varies. Instead of boundary conditions Eq. 36.28 it is then necessary to use nonstationary boundary conditions

$$x\big|_{\partial P} = l_1(\bar{\xi}, t),\ y\big|_{\partial P} = l_2(\bar{\xi}, t),\ \bar{\xi} \in \partial P$$

determining the deformation of the boundary $\partial G(t)$ in time. The functions $l_i(s, t)$ should be defined beforehand or during the solution of a nonstationary system of the differential equations. The parametric domain P remains constant.

Using new unknown functions V_i from Eq. 36.29 and a designation

$$K_1 = g_{11}^2\left(V_{1\xi\xi} - F_1\right),\ K_2 = g_{22}^2\left(V_{2\eta\eta} - F_2\right),$$

(F_1, F_2 are functions dependent on L_1, L_2 Eq. 36.32, we write Eq. 36.32, steady to perturbations, in the form

$$x_t = \frac{1}{D}\left(K_1 y_\eta - K_2 y_\xi\right),\ y_t = \frac{1}{D}\left(K_2 x_\xi - K_1 x_\eta\right), \tag{36.33}$$

The formulated problem is reduced to the problem of search for $x = x(\xi, \eta, t)$, $y = y(\xi, \eta, t)$ defined together with their partial derivatives at each moment of time t in the rectangle $P = \{[0, N] \times [0, M]\}$ satisfying the set of Eq. 36.33, boundary conditions Eq. 36.28 and some initial conditions $x(\xi, \eta, 0) = x^0(\xi, \eta)$, $y(\xi, \eta, 0) = y^0(\xi, \eta)$.

In [30] a sign of the first variation of a functional Eq. 36.31 is investigated. It turns out that functions

$$x^\tau = x^t + x_t^t\tau,\ y^\tau = y^t + y_t^t\tau$$

give to the functional I the value no greater than x^t, y^t;

$$I\left(x^\tau, y^\tau\right) \le I\left(x^t, y^t\right).$$

On the basis of this, the explicit difference iterative scheme for calculation of the coordinates of a grid [26] is developed as

$$x^{n+1}(\xi,\eta) = x^n(\xi,\eta) + \tau Q_1^n,\quad y^{n+1}(\xi,\eta) = y^n(\xi,\eta) + \tau Q_2^n$$

where τ is time step, $x^n(\xi, \eta)$, $y^n(\xi, \eta)$ are the coordinates of the grid node on the nth iteration ($n = 0, 1, \ldots$) at the moment of time $t = n\tau$; Q_1^n, Q_2^n are discrete approximations of right sides of the system Eq. 36.33 in the corresponding point $(\xi, \eta) \in P$. In the calculation of any point of a grid the pattern of nine nearest points (Figure 36.3) is considered. There the problem of the choice of a step τ emerges. Numerous calculations have shown that for organization of movement of all points of a grid on each iteration and in each point, the step should be variable and such that the calculated point should not leave the pattern and self-crossing cells should not arise. It has been found that when the value of the functional $I(\xi, \eta)$ at the point (ξ, η) is large, then the value of τ is small. Movement of all points is ensured if $\tau(\xi, \eta)$ at the point (ξ, η) is selected so that

$$\tau(\xi,\eta)\left|I(\xi,\eta)\right| < B = 0.5\min\left(|d_1|, |d_2|\right)$$

where d_1, d_2 – are diagonals of the quadrangle $PQRT$.

The recalculation of points at each iteration by this method even in the case of a poor initial approximation (with patterns where the angle at point (ξ, η) is small, with nonconvex patterns or stretched along one dimension) provides the movement of all points, but leads to slow stabilization of all nodes.

For the faster stabilization of nodes the iterative process, realized in the program MOPS-2a ([12–15]). has been constructed by the following way. The calculation of optimal grids is ordered on bordering lines (Figure 36.4). On odd iterations the calculation is carried out from the central bordering line up to that near the boundary, and on even from bordering line near boundary up to central. Thus on odd iterations the grid is stretching faster in the direction to the center of the domain, and on even iterations information about geometry of the boundary is transmitted more completely inside of the domain.

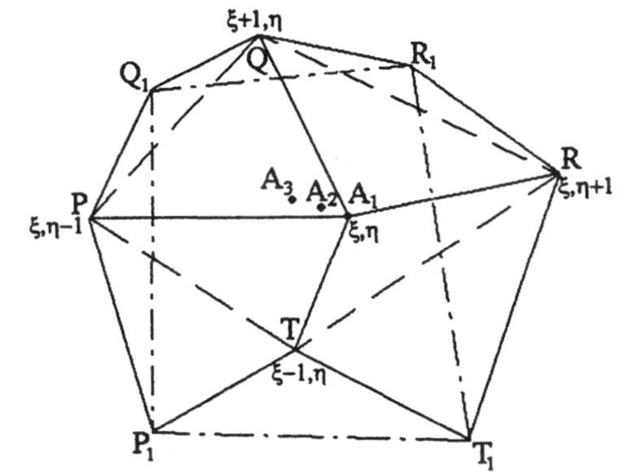

FIGURE 36.3

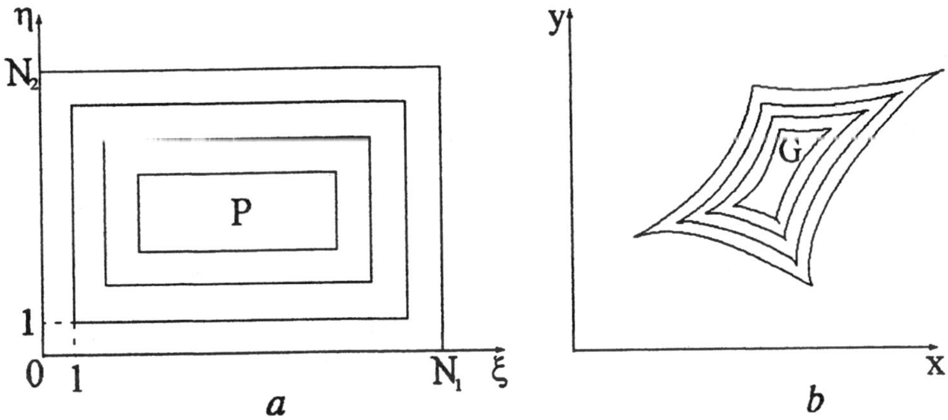

FIGURE 36.4

On each iteration for calculation of coordinates of a point (ξ, η) there are considered three points A_1, A_2, A_3 located uniformly on the segment connecting the point (ξ, η) with the center of gravity of the pattern. In the case, for example, of nonconvex patterns it can turn out that A_2 or A_3 get out of the pattern, and then the coordinates of these points are recalculated with a half step; if recalculated points get out of the pattern again, the movement of a point is organized in the direction of an interior diagonal toward pattern concavity. For each point $A_i (i = 1, 2, 3)$, the coordinates of points $\tilde{A}_i$ are calculated from E-O equations under the explicit difference scheme with a variable step. At six points, values of the functional $I(\xi, \eta)$ are calculated and the minimal value is selected. At a new point (ξ, η) a point corresponding to this value of a functional is selected.

After a given number of iterations l the correction of a grid is carried out, i.e., at iterations, the number of which is multiple of the number l, movement of points is organized not toward the center of gravity of the pattern but toward the point of intersection of diagonals of the quadrangle $P_1 Q_1 R_1 T_1$.

On each iteration a summarized value of a functional I for all calculated points of a grid is computed. The calculation of a grid is considered complete if a relative variation of I on two adjacent iterations is no more than 0.1%. The calculation of a grid can be continued at other values of weight $\lambda = \lambda_U / \lambda_O$ and the number of corrections l.

36.3.2 Multiply Connected Optimal Grids in Two-Dimensional Domains. The Program MOPS-2a

On the basis of the algorithm described in Section 36.3.1, optimal curvilinear block-structured grid in simply connected and multiply connected domains with simple and complex topology are constructed,

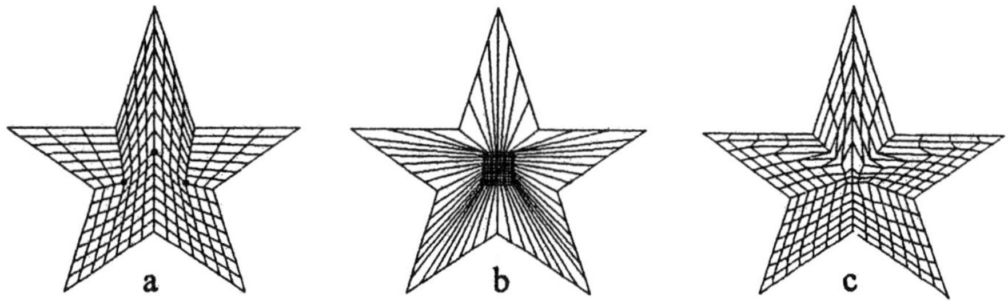

FIGURE 36.5

but the mapping of a given domain G in the plane (x, y) onto a set of rectangles P in a parametrical plane (ξ, η) and inverse mapping can be ambiguous. Such grids contain the elements of basis grids of *O, C, H* type [33]. The grids generated by MOPS-2a are characterized by smoothness of grid lines on the boundaries of block interfaces. To realize that we use the method of overlapping of blocks. The automatic organization of a method allows a reduction and simplification of the volume of input information for calculation of grids.

36.3.2.1 Initial Approximation of Grid

The process of the construction of grids includes some preliminary stages: first of all the choice of topology of grid, which specifies the direction of coordinate lines of a curvilinear grid, i.e., the structure and to a large degree the quality of grids. This process is carried out by the performer of the calculation. In the proposed method the algorithms for dividing the domain into blocks, describing the boundary of blocks, constructing the initial approximation of a grid, and overlapping of blocks are formalized and automated by the program.

At the construction of the initial approximation, the boundary of the domain is represented by a single or several closed curves, each of which is described by a set of specific nodes connected by straight lines or arcs of circles of given radii in a specific direction.

The initial approximation of a grid is automatically generated for different input information:

- For given coordinates of intersection points of typical horizontal and vertical lines that divide blocks into convex or rather close to convex subblocks, the opposite sides are automatically divided into a given number of equal segments. The points of a partition are connected by straight segments (three points in Figure 36.5a) [12].
- If the block is of a star-shaped typed, it is possible to insert in it the corner of some quadrangle with a uniform grid, which is simultaneously a near-boundary bordering line and a fictitious interior boundary (Figure 36.5b) [13].
- For minimal information (specific vertices of blocks and number of points on both sets of coordinate lines) with application of method of *R*-functions (Figure 36.5c) [5].

For construction of grids in multiply connected regions, the domains are divided into blocks — curvilinear quadrangles, the vertices of which belong to the boundary of domain. We shall name the dividing lines as the interior boundaries of the domain. If the domain contains the elements of grids of *H, O, C* types, as slits (*O*–, *C*–grids), and splits (*H*–grids) should coincide with coordinate lines in plane (ξ, η) and be grid lines in plane (x, y). The domain is divided into blocks for the purpose of selection of simply connected subregions from multiply connected, in which structured grids are generated, or with the purpose of selection of subregions with simple configurations, in which for generations of the grid initial approximation the minimum of information is required.

The points of a grid are numbered on horizontal lines and vertical lines; thus, in each block k the grid is determined by a set of coordinates $\{x(\xi, \eta), y(\xi, \eta)\}$ where $\xi = N_{1k}, \ldots, N_{Nk}, \eta = M_{1k}, \ldots, M_{Mk}$. N_{1k}, N_{Nk}, M_{1k}, M_{Mk} should be matched with appropriate N_{1l}, N_{Nl}, M_{1l}, M_{Ml} $(l = 1,2, \ldots)$ of adjacent blocks. The common block grid in the domain is obtained at the expense of the combination of grids in all

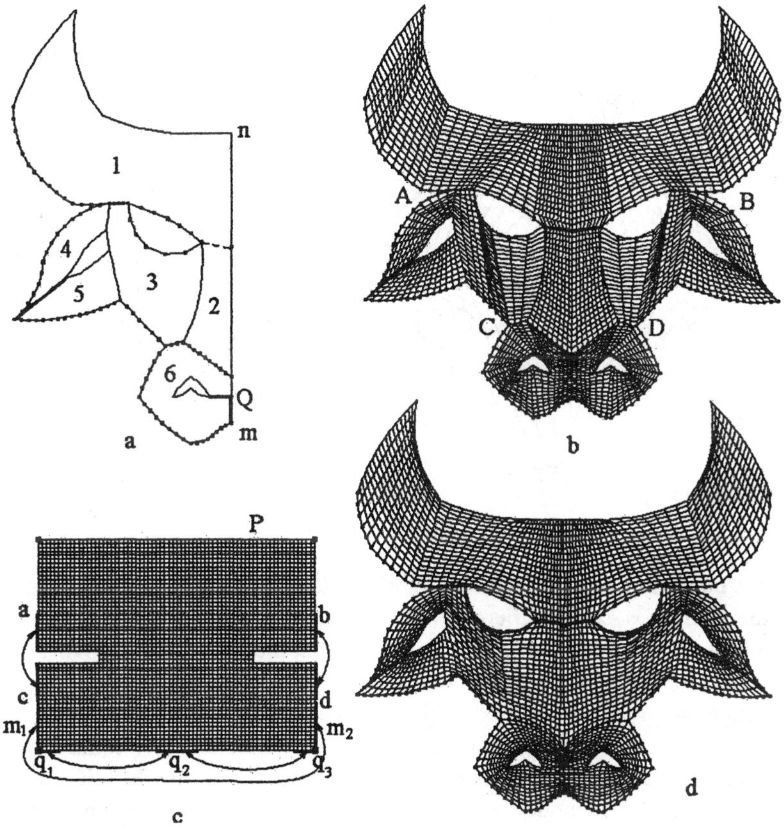

FIGURE 36.6

blocks covering this domain. if the least values N_{1l}, M_{1l} ($l = 1,2, \ldots$) are equal to unity, the greater values N_{Nl}, M_{Ml} ($l = 1,2, \ldots$) define the size ($M \times N$) of block grid of the domain.

The coordinates of grid nodes are stored in a matrix that is filled by a "flag" method. The image of the domain is inscribed in the rectangle of size $M \times N$. If its point does not belong to a specific domain, then a "flag" (a large number) is inserted into the corresponding element of the matrix. Thus, the structure of the matrix is determined by the geometry of the domain P in the parametric plane (Figure 36.6c). There are two columns in the matrix to store the coordinates of the boundaries of a split, if this line is vertical, or two lines, if the line is horizontal (*ab*, *cd*). The cut in the plane (ξ, η) has two images, so that two matrix elements (for example, the elements of the columns q_1m_1, q_3m_2) correspond to each point (the cut Q_m) of a slit in plane (x, y). One point can carry out a few slits (a point Q (Figure 36.6a) and three slits); therefore, more than two matrix elements (q_i, $i = 1, 2, 3$) may correspond to endpoints of the slit. In Figure 36.6c the arrows indicate the correspondence between cuts that are singled out by bold lines.

In Figure 36.6a the given boundary is presented and six blocks are marked, in which a grid of an initial approximation is generated by one of the above described methods. Then it is symmetrically mapped over the axis *mn* (Figure 36.6b). The markers select grid lines on which splits are located.

36.3.2.2 Automatic Overlapping of Subdomains

To construct a block-structured optimal grid we consider each of the blocks as a given simply connected domain. In the blocks the grid is generated by the method with a prescribed node arrangement on the boundary. If in each block the grid is built independently, not connected with coordinates of grids of adjacent adjoining blocks, on the boundaries of block interfaces a smoothness of coordinate lines will be lost. For the solution of the problems, the unknown quantities of which have large gradients in the neighborhood of the boundaries of interfaces, the grid lacking smoothness is considered to be unsuitable.

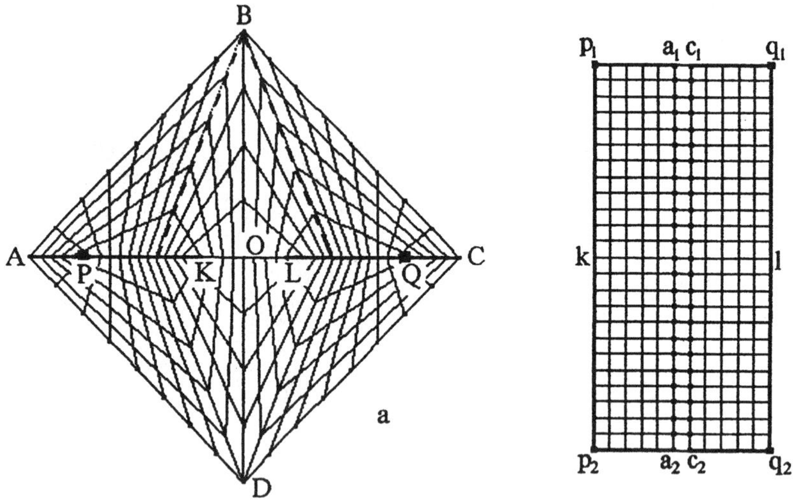

FIGURE 36.7

Let us apply a method of overlapping. Each block, which has as its boundary a part of the interior boundary of the domain, is extended beyond this boundary on one coordinate strip. Thus we take as the boundary of the block the vertical or horizontal line from the adjacent block. When we perform the calculation on each iteration in all blocks successively, we calculate the grid points on the interior boundaries of blocks in the correspondence with the given optimality criteria.

It is rather difficult to realize this method (in a logic sense) for multiply connected domains with complex topology when in the domain there are slits and splits, and on which it is also necessary to provide the movement of grid nodes. In this case we are to analyze a large number of geometric possibilities of block interfaces. The solution of this problem has allowed the volume of input data to be reduced and quality of calculated grids to be improved.

The split is two parts of the boundary (*AB*, *CB* in Figure 36.7a), the points of which have different coordinates in initial plane (x, y) but identical in curvilinear coordinates (ξ, η): the slit has identical coordinates in (x, y) and different in (ξ, η).

The presence of a split is determined by the program in generating the boundary. If the coordinates of two different points of the boundary fall on one element of the matrix and there are more than two such adjacent points on horizontal or vertical lines (and correspondingly matrix elements), this line is a split. After determining these lines are storing the ambiguity of their mappings, the matrix of grid coordinates is extended on the appropriate number of lines, if the splits are horizontals (Figure 36.6c), or on an appropriate number of columns if they are verticals (one column in Figure 36.7b). Coordinates of the boundary of a grid (Figure 36.6a) are enumerated, and the initial approximation of grid (Figure 36.6b, 36.7a) is constructed.

In order to reveal the slits, the parts of the boundary of the domain are automatically analyzed by the coordinates of their endpoints after the initial approximation of grid is constructed in the whole given domain. All slits are numbered by a certain way, and the splits are labeled by a special marker.

In order to organize block overlapping we determine the type of the boundary. A block is topologically equivalent to a rectangle and has four sides. A side may be rigid, when its points belong to the boundary of the domain and the coordinates of these points are specified; it may be movable, when grid points can move during calculation; it may be a slit; it may be rigid or movable, but lying on a coordinate line on which split is located; it may be mixed, when the boundary is a combination of parts of different types. In order to construct a smooth grid the overlapping of blocks is organized through movable sides and slits (Figure 36.8d, shaded strips).

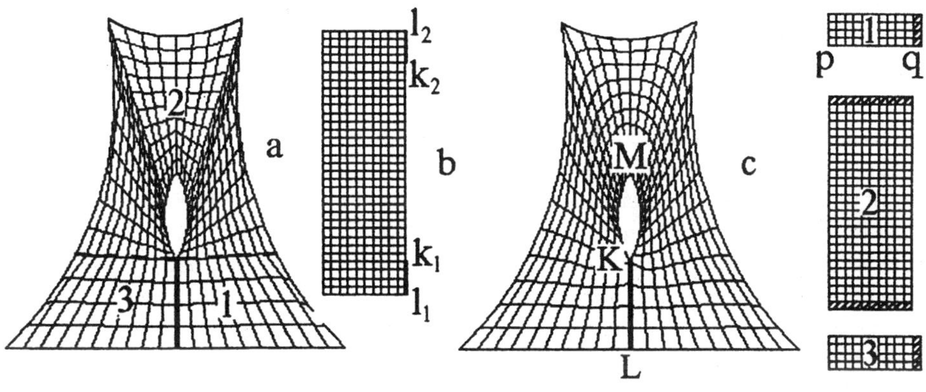

FIGURE 36.8

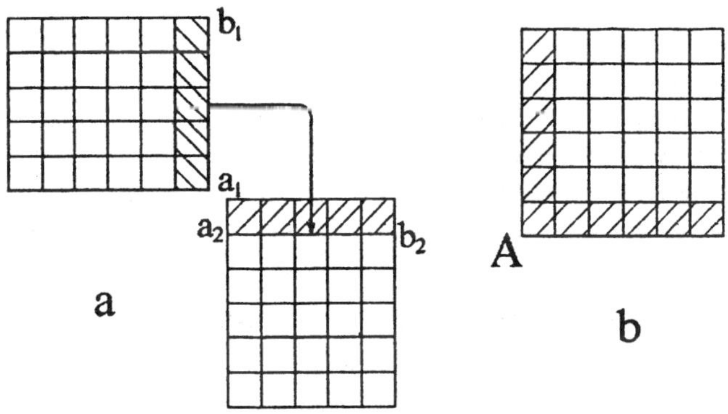

FIGURE 36.9

The analysis of blocks is carried out by the program. If the boundary of the block is rigid, the coordinates of all its points to this moment of time are calculated and are written in the matrix. If the boundary of the block is movable (*KO* (Figure 36.7a)), two of its endpoints are connected by straight lines, and grid points of the boundary of the block are calculated by the method of linear interpolation. On the mixed boundary (*ABO* (Figure 36.7a)), the parts of rigid (*AB*) and movable (*OB*) boundaries are selected and the block is automatically divided into two (Figure 36.7a) or more blocks (Figure 36.8a) so that through the chosen movable boundaries hereinafter to realize overlapping of them (dashed lines are the line of decomposition). All cuts are enumerated. The number of a sit is assigned to the corresponding side. It is common for some two blocks to have sides with the same numbers (p_1a_1, p_2a_2 (Figure 36.7a), l_1k_1, l_2k_2 (Figure 36.8a)).

If a grid is calculated in the block with slits, another block with a slit of the same number is searched for to organize block overlapping. The first block is extended on one coordinate strip beyond the slit and coordinates of points of the slit, and the adjacent grid line from another block are transferred to the strip (Figure 36.9a).

The next step in the analysis of block boundaries is to check for the possibility of the blocks overlapping. For example, if one of the block sides is a slit, its adjacent sides cannot be movable (*pq* in Figure 36.8d); if two adjacent sides are movable, the point at the intersection of coordinate lines bordering these sides should belong to given domain (point *A* in Figure 36.9b).

If block sides belong to one side of a slit, the automatic check of a possibility of organization of overlapping of blocks is carried out similarly, but with the use of working columns (lines) of matrices.

As a result of the above discussions on automatic organization of overlapping blocks, the volume of input information for the calculation of a grid in comparison with hand organization of overlapping was reduced by 4–20 times, depending on the complexity of the configuration of the domain and its topology. So the domain represented on Figure 36.6b, after the analysis of the boundary and description of six blocks for input data, was divided automatically into 42 subregions to organize the block overlapping.

Testing of the algorithm and program MOPS-2a according to criteria from work [22] has shown that for construction of grids closer to uniform, it is necessary to select the weight λ in the range 0.1–0.3 and for grids closer to orthogonal — from 1 up to 10. The optimal numbers of correction are $l = 2, 3$. For calculation of grids on average 4–20 iterations are required. The number of iterations depends on the initial approximation, number of correction l and the weight λ. The quality of grids essentially depends on the choice of its topological image. The computation time for the grid (Figure 36.6) of size 72×54 on PC/486 (40 MHz) (nine iterations) is ≈ 0.5 min.

36.3.3 Algorithm of Two-Dimensional Optimal Adaptive Grid Generation. The Program LADA

This algorithm represents the iterative procedure of minimization of the functional J (Eq. 36.19). The calculation begins with some initial approximation — a non-self-intersecting grid. At each time step the calculations are carried out along bordering lines in the counterclockwise direction moving from the boundary (Figure 36.4). While defining the node (i, j), the other nodes are fixed, and the position of a node is found from the condition of nondegeneracy of a grid and the condition of a minimum of the functional J on a special set of points Ω_1 or Ω_2. During the calculations the coordinates of nodes are replaced by new ones.

36.3.3.1 Set of Points for Minimization of the Functional J

Two sets of points determined by means of special points $C_{\omega+}$, $C_{\omega-}$, $C^*_{\omega-}$ are considered. To construct them we use the equidistribution principle [32] for weight function $\omega = \alpha + \Phi_x^2 + \Phi_y^2$, $\alpha = \text{const.} > 0$ for each point on its own segment. We find the point $C_{\omega+}$, if the cell C_{ij} determined by points $H^n_{i,j-1}$, $H^n_{i,j+1}$, $H^n_{i-1,j}$, $H^n_{i+1,j}$ (n is the number of iteration) is convex. For its determination the equidistribution principle is applied on the segment $[C_i, C_j]$ where the points C_i, C_j are found from the same principle [32] on intervals $[H^n_{i-1,j+1}, H^n_{i-1,j-1}, H^n_{i+1,j-1}, H^n_{i+1,j+1}]$ correspondingly. For a nonconvex cell C_{ij} we find the point $C_{\omega-}$ on its interior diagonal. Similarly, for the point $C^*_{\omega-}$ we shall find for nonconvex cell $C^*_{ij} = \{H^n_{i-1,j+1}, H^n_{i-1,j-1}, H^n_{i-1,j-1}, H^n_{i+1,j+1}\}$. Then we shall construct the set of points H_+, H_-, H^*_-.

$$\left\{ H^k \middle| \overrightarrow{H^n_{ij} H^k} = \overrightarrow{H^n_{ij} C_\omega} \frac{k}{3} \quad k = 0,1,2,3 \right\}$$

where points C_ω coincides with corresponding point $C_{\omega+}$, $C_{\omega-}$, $C^*_{\omega-}$. After this we shall define the sets Ω_1 and Ω_2 by means of Table 36.1.

TABLE 36.1 Sets of Points Ω_1 and Ω_2

Cell C_{ij}	Cell C^*_{ij}	Set Ω_1	Set Ω_2
Convex	Nonconvex	H_+	H_+
	Nonconvex		$H_+ \cup H^*_-$
Nonconvex	Convex	$H_+ \cup H_-$	$H_+ \cup H_-$
	Nonconvex		

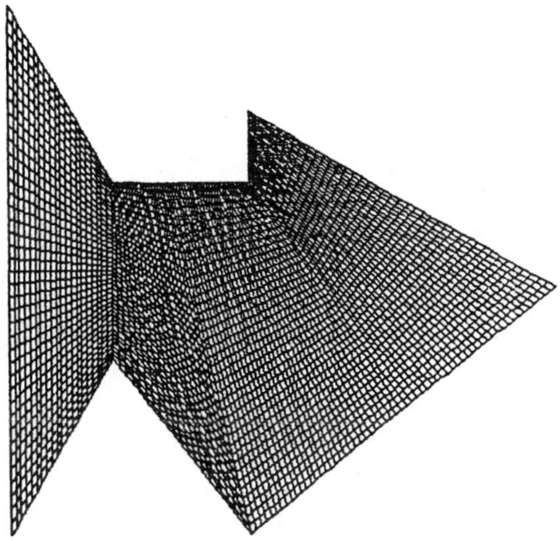

FIGURE 36.10

Note, that if ω = const., the point $C_{\omega+}$ coincides with the center of gravity of a cell C_{ij}, and $C_{\omega-}$, $C^*_{\omega-}$ — with the middles of interior diagonals of cells C_{ij}, C^*_{ij} respectively.

The values of functions $\Phi(x, y)$ at nodes C_i, C_j, $C_{\omega+}$, $C_{\omega-}$, $C^*_{\omega-}$, H^k are calculated by a linear interpolation, and the derivatives Φ_x, Φ_y; according to formulas

$$\Phi_x = \frac{1}{D}\left(\Phi_\xi y_\eta - \Phi_\eta y_\xi\right), \quad \Phi_y = \frac{1}{D}\left(\Phi_\eta x_\xi - \Phi_\xi x_\eta\right),$$

where $\Phi = \Phi(x(\xi, \eta), y(\xi, \eta))$. The derivatives x_ξ, y_ξ, x_η, y_η, Φ_ξ, Φ_η inside the domain are approximated by central differences and on the boundary ∂P by one-sided differences.

36.3.3.2 Organization of Calculations

In the program LADA the following two ways of calculations are utilized:

- *Global search for Minimum.* Sets of points Ω_1 or Ω_2 are calculated. We choose a new node H^{n+1}_{ij} from a selected set of points that give the minimal contribution to the functional J and that with other nodes form noncrossing grid. In this case in the whole domain all optimality criteria are taken into account.
- *Local Search for Minimum.* If cells C_{ij}, C^*_{ij} are convex, we consider $H^{n+1}_{ij} = C_{\omega+}$. If for $C_{\omega+}$ we have self-intersecting cells, and if we have other cases for C_{ij}, C^*_{ij} in the definition of a node H^{n+1}_{ij}, we carry out a global search for the minimum on the selected set of points. In this method in the whole domain only one criterion of adaptation is taken into account. The method of organization of calculations is given in [37].

36.3.3.3 About the Choice of Control Parameters

The methods of the construction of an initial approximation are described in detail in Section 36.3.2. The grid constructed by one of those methods is represented in Figure 36.10. Note that for the algorithm it is important that the initial approximation is not a self-crossing grid. As an initial approximation for generation of adaptive grids, the optimal grids constructed by the given algorithm without criterion of the adaptation ($\lambda_A = 0$) can be used. Such initial approximation is represented in Figure 36.11a.

The constants λ_U, α are selected equal to 1. The parameters λ_O, λ_A are chosen from the requirements on the quality of a grid, estimated with the help of Eqs. 36.16 and 36.17, and according to criteria offered

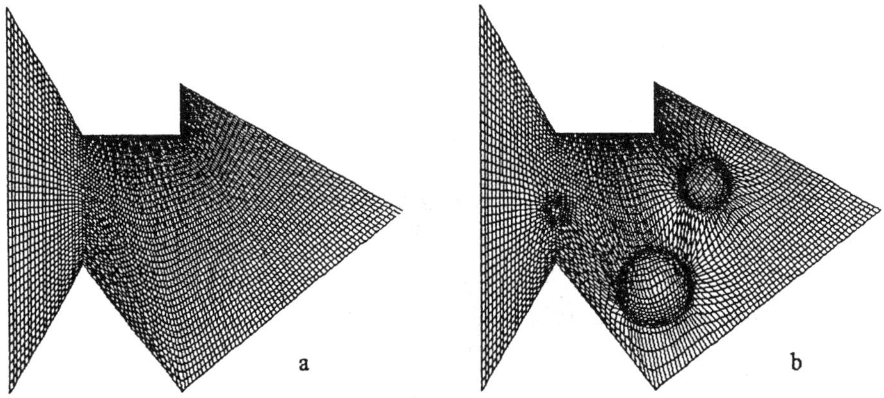

FIGURE 36.11

in [22]. The most frequently used values $\lambda_O = 10^k$, $k = -2, -1, 0, 1, 2$, λ_A depend on values of the function Φ. In the local search for the minimum, λ_A is supposed equal to zero. For the initial approximation in Figure 36.10, $J_U = 49.71$, $J_O = 23925.92$. For the optimal grid in Figure 36.11a, $\lambda_O = 0.05$, $\lambda_A = 0$, $J_U = 11.08$, JO = 22776.7. For an adaptive grid in Figure 36.11b, $\lambda_O = 0.05$, $\lambda_A = 10^6$, $J_U = 236.2801$, $J_O = 22770.05$, $J_A = 2.57$.

The choice of the methods of calculations and the set of points for minimization of the functional Eq. 36.19 is made from the requirements on the quality of a grid and effectiveness of algorithm. Global searching for the minimum and set of points 1 are more effective. An example of a grid for function $z = \Phi(x, y) = \sum_{i=1}^{3} \Phi_i\,(x, y)$ where

$$\Phi_1(x,y) = \exp\left\{-\frac{1}{\varepsilon_1}\left[(x-a_{11})^2 + (y-a_{12})^2\right]\right\},$$

$$\Phi_i(x,y) = \begin{cases} \exp\left[-\dfrac{1}{\varepsilon_i}\left[(x-a_{i1})^2 + (y-a_{i2})^2 - r_i^2\right]\right], & (x-a_{i1})^2 + (y-a_{i2})^2 > r_i^2, \\ -\dfrac{1}{\varepsilon_i}\left[(x-a_{i1})^2 + (y-a_{i2})^2 - r_i^2\right], & (x-a_{i1})^2 + (y-a_{i2})^2 \le r_i^2, \end{cases}$$

$$i = 2,3$$

is demonstrated in Figure 36.11b. Here $\varepsilon_i = 0.001$, $i = 1, 2, 3$, $r_2 = 0.15$, $r_3 = 0.1$, $a_{11} = 0.3$, $a_{12} = 0.7$, $a_{21} = 0.7$, $a_{22} = 0.4$, $a_{31} = 0.9$, $a_{32} = 0.8$.

36.4 Simulation of Rotational Flows of Gas in Channels of Complex Geometries by Means of Optimal Grids

Frequently in technological installations there are axisymmetrical channels of complex configurations in which complicated nonstationary hydro- and gasdynamic flows occur. In constructions of such installations, one of the important points is knowledge both of the structures of the flows and the parameters describing them. For the purpose of reducing field tests, effective numerical methods that permit calculations that can rather quickly and reliably predict parameters of flows are necessary. The development of numerical methods for calculation of gas flows in channels with complicated geometries is connected with large difficulties. These are complex geometries of calculated domains, large range of flow velocities, formation of many rotational zones with closed streamlines caused by interaction of counter streams. As a rule, the calculations described in the publications (for example, [9, 24, 31]) are connected with serious restrictions on geometry of channels or on structure of flows.

The application of optimal smooth block-structured curvilinear grids, described in Section 36.3.2, has appeared as the rather essential factor in solving the problems [1, 10, 11, 16]. Good approximating qualities of used grids and mappings [28, 40, 41] has become the basis of attained results.

So, axisymmetrical simply connected channels of complicated configurations are considered. The surfaces of channels consist of parts of a porous surface through which gas is blown in solid walls, and parts for exit of gas.

For modeling the gas stream in the channel, some simplifications [23] are introduced. We consider that the sizes of boundary layers, increasing along walls, are small in comparison with transversal sizes of channels; boundary layers do not interact with each other; gas that is blown in is homogeneous; and gas flow is stationary. Then for the numerical simulation of gasdynamic processes in channels it is possible to use the model of perfect gas, the flow of which satisfies the Euler equations.

For numerical simulation the Euler equations are written in the stream functions φ- vortex function ω [9] in integral form in curvilinear coordinates (ξ, η):

$$\oint_C \left[\frac{1}{\rho r\Delta}\left(A_1\varphi_\eta - A_3\varphi_\xi\right)d\xi - \frac{1}{\rho r\Delta}\left(A_2\varphi_\xi - A_3\varphi_\eta\right)d\eta\right] = \iint_{G_C} \omega\Delta d\xi d\eta,$$

$$\oint_C \frac{\omega}{r}\left(\varphi_\xi d\xi + \varphi_\eta d\eta\right) = -\oint_C \frac{\rho}{2}\left(V_\xi^2 d\xi + V_\eta^2 d\eta\right),$$

$$\oint_C H\left(\varphi_\xi d\xi + \varphi_\eta d\eta\right) = 0,$$

where

$$A_1 = x_\xi^2 + r_\xi^2,\quad A_2 = x_\eta^2 + r_\eta^2,\quad A_3 = x_\xi x_\eta + r_\xi r_\eta,\quad \Delta = x_\xi r_\eta - x_\eta r_\xi.$$

Velocity vectors V_1, V_2, stream function φ, vortex function ω, enthalpy H, pressure P, and density ρ must satisfy the relations

$$V_1 = -\frac{1}{\rho r\Delta}\left(\varphi_\xi x_\eta - \varphi_\eta x_\xi\right),\quad V_2 = -\frac{1}{\rho r\Delta}\left(\varphi_\xi r_\eta - \varphi_\eta r_\xi\right),$$

$$\omega = \frac{1}{\Delta}\left(V_{1\xi}x_\eta - V_{1\eta}x_\xi + V_{2\xi}r_\eta - V_{2\eta}r_\xi\right),$$

$$P = P_0 - \int_{L(M_0,M)} \left(\frac{\rho}{2}V_\xi^2 + \frac{\omega}{r}\varphi_\xi\right)d\xi - \int_{L(M_0,M)} \left(\frac{\rho}{2}V_\eta^2 + \frac{\omega}{r}\varphi_\eta\right)d\eta,$$

$$\varphi = \varphi_0 + \int_{\xi_0}^{\xi} r(\mu,\eta)\rho(\mu,\eta)\left[V_1(\mu,\eta)r_\xi(\mu,\eta) - V_2(\mu,\eta)x_\xi(\mu,\eta)\right]d\mu$$

$$+\int_{\eta_0}^{\eta} r(\xi,\nu)\rho(\xi,\nu)\left[V_1(\xi,\nu)r_\eta(\xi,\nu) - V_2(\xi,\nu)x_\eta(\xi,\nu)\right]d\nu,$$

where x, r are cylindrical coordinates, G_C is the arbitrary domain with the smooth boundary C from a given domain G, $L(M_0, M)$ being the arbitrary curve, connecting the point $M \in G$ with the point M_0, in which pressure P_0 is given, φ_0 is an arbitrary constant, ξ_0, η_0 is the coordinate of a point of the beginning of going around the boundary at calculation of stream function.

To calculate the subsonic flow we specify at the exit the mass flow of gas, at the entrance parts the density and velocity in the direction of the normal, on solid walls the condition of nonpenetration, and on the axis of a symmetry the condition of symmetry. The boundary conditions should satisfy the relation of mass balance.

The flow in the subsonic region is calculated by a finite difference iterative method, being the modification of the approach [9], in which it is supposed that there are no closed streamlines. In the proposed method, the special approximations of integral equations in the subsonic zone, taking into account the peculiarities of curvilinear grids and also the direction of a stream turns out to be successful. The pressure is calculated by the method of coordinated approximations [8] permitting to avoid the origin of parasitic fluctuations.

To solve the algebraic linear system of equations obtained during the approximation, the matrix of which at formation of closed rotational streams is stiff, we use on each iteration a direct economic method with a regularization essentially taking into account block-diagonal structure of matrices.

The offered method is realized in the programs SOKOL [1, 10, 11, 16]. The following results are obtained:

- The use of optimal curvilinear grids removes restrictions on class of considered configurations of channels.
- The offered method allows calculation of effective both compressible and incompressible streams with numerous rotational zones.
- It is necessary to take into account a compressibility of a medium.
- In channels with a nozzle part, taking account of parameters of a stream in the transonic part allows input data, obtained with some error from experiments, to be corrected.
- The calculations can be carried out for different types of boundary conditions.

Taking into account compressibility of gas and its parameters in the transonic regime, the correct boundary conditions in a series of cases of the domains lead to completely different structure of the flow in channels, namely the formation of closed rotational streams of gas.

Figure 36.12b demonstrates the streamlines of gas flow obtained in calculation of rotational flow of compressible fluid in the model channel, when gas moved on lateral surface *CD* with constant velocity. On the surface *EF* the velocity was set piecewise constant, at end-wall of the channel *AB* — under the cosine law [18]. The density of gas on sides where it is blown in is constant. On the exit *KL* massflow of gas was prescribed from the relation of the mass balance. With this input data three closed vortices have been obtained as the result of calculation. In Figure 36.12 there is a calculated grid that has been cut through one grid line for visualization.

36.5 Conclusion

The iterative algorithms for the calculation of three-dimensional grids can be constructed on the same approaches used in Section 36.3, ideas of a combination of explicit iterative methods of the solution of the system of Eq. 36.32 and direct local minimization of the functional Eq. 36.27. Though we do not have effective automated programs in the three-dimensional case, the first positive experience in this direction was described in [27]. For three-dimensional star-shaped domains (they can also evolve in time), a direct transferring of algorithms, used in MOPS-2a and LADA, is possible. More complicated is the question about dividing the complex three-dimensional domain into star-shaped blocks which now is practically not automated.

At present, the problem of implementing the algorithms for parallel computation of grids of large dimension with number of cells greater than 10^6 (for some problems of continuum mechanics requiring large volume of calculations, simulation could be realized only by utilizing the parallel processors) is

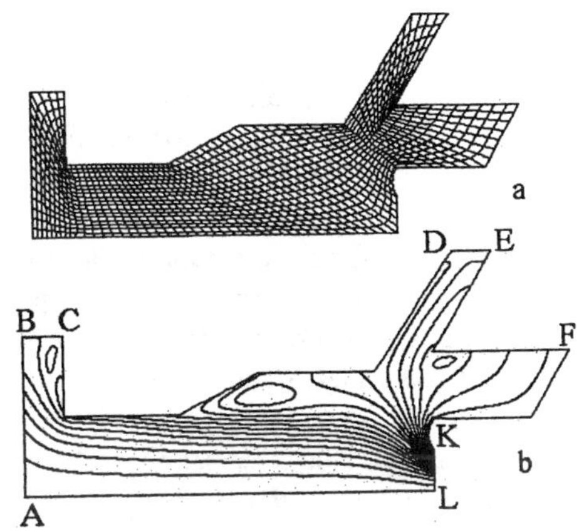

FIGURE 36.12

critical. Such problems include, in particular, the problems of gas dynamics with large deformations that need to be calculated both on moving and on stationary grids.

Algorithms in Section 36.3 describe a few ways of parallelizations. These are parallelizing according to blocks for the computation of block-structured grids; parallelizing explicit iterative processes according to groups of neighboring cells [38]; and use of decomposition methods in the solution of E-O equations by iterative methods.

References

1. Akkmadeev, V.F., Sidorov, A.F., Spiridonov, F.F., Khairullina, O.B., On three methods of numerical modelling of subsonic flows in symmetric channels of complex form, *Modelling in Mechnics. Novosibirsk, CC and ITPM SB AS USSR.* 4(21), 5, (1190), pp. 15–25.
2. Belinskii, P.P., Godunov, S.K., Ivanov, Y.B., Yanenko, I.K., The use of one class of quasiconformal mappings to construct difference grids in regions with curvilinear coordinates, *Zh. Vychisl. Mat. Mat. Fiz..* 1975, 15, pp. 149–1511.
3. Brackbill, J.U. and Saltzman, J.C., Adaptive zoning for singular problems in two dimensions, *J. Comp. Phys.* 1982, 46, 3, pp. 342–368.
4. Deitachmayer, G.S. and Droegemeier, K.K., Application of continuous dynamic grid adaptation techniques to meteorological modelling. part i: basic formulation and accuracy, *Monthly Weather Review,* 1992, 120, 8, pp. 1675–1706.
5. Gasilova, I.A., Algorithm of automatical generation of initial approximation of curvilinear grid for star type domains, *Voprosy Atomnoy Nauki i Tekniki Ser.: Matem. Modelirovanie Fizicheskikh Processov,* 1994, 3, pp. 33–40.
6. Godunov, S.K. and Prokopov, G.P., Calculation of conformal mappings in constructions of difference grids, *Zh. Vychisl. Mat. Mat. Fiz.* 1967, 7, pp. 1031–1059.
7. Kennon, S.R. and Dulikravich, G.S., Generation of computational grids using optimization, *AIAA J.* 1986, 24, 7, pp. 1069–1073.
8. Khahimsyanov, G.S. and Yaushev, I.K., Calculation of pressure in two-dimensional stationary problems of hydrodynamics, *Problem of Dynamics of Viscous Liquid, Novosibirsh,* 1985, pp. 280–284.
9. Khakimzjanov, G.S. and Yaushev, I.K., Iteration method for calculation of two-dimension stable flows of ideal compressed fluid, *Novosibirsk.* (Preprint N. 4-87./ AS USSR, SB, ITPM, 1987).
10. Khairullina, O.B., Calculation of stationary subsonic vortical flows of ideal gas in symmetric channel of complex geometries, *Questions of Atomic Science and Techniques. S. Mathematical Modelling of Physical Processes.* 1990, pp. 32–39.

11. Khairullina, O.B., RDT IMM — Complex of programs for calculation of steady subsonic flows in channels of complex geometries, *Calculation Technologies. 1, 2.* (School-seminar's Works on the Complexes of Mathphysic's Programs) Novosibirsk, 1992, pp. 327–333.
12. Khairullina, O.B., Methods of block optimal grids generation in two-dimentional multi-connected domains, *Questions of Atomic Science and Techniques. S. Mathematical Modelling of Physical Processes,* 1, 1992, pp. 62–66.
13. Khairullina, O.B., Block-regular optimal grids generation, *Questions of Atomic Science and Techniques. S. Mathematical Modelling of Physical Processes,* 1, 1994, pp. 19–25.
14. Khairullina, O.B., Acceleration of iteration process convergence at block-regular grids generation, *Questions of Atomic Science and Techniques. S. Mathematical Modelling of Physical Processes, 1–2,* 1995, pp. 54–59.
15. Khairullina, O.B., Method of constructing block regular optimal grids in two-dimensional multiply connected domains of complex geometries, *Russian Journal of Numerical Analysis and Mathematical Modelling.* 1996, 11, 4, pp. 343–358.
16. Khairullina, O.B., To the calculation of flow of gas in channels of complex configurations, *Prikladnaya Mekhanica i Tecnicheskaya Fizika.* 1996, 37, 2, pp. 103–108.
17. Kreis, R.I., Thames, F.C., Hassan, H.A., Application of a variational method for generating adaptive grids, *AIAA J.* 1986, 24, 3, pp. 404–410.
18. Kulik, F.Ye., Rotational axially symmetric averaged flow and damping of acoustic waves in combustion cameras of rocket engines, *Rocket Engineering and Astronautics,* 1966, 4, 8, pp. 195–197.
19. Kurant, R., Equations with partial derivatives, *M. Mir,* 1963, p 830.
20. Liseikin, V.D., The construction of structured adaptive grids — a review, *Comp. Math. Phys.* 1996, 36, 1, pp. 1–32.
21. Nakahashi, K. and Deiwert, G.S., Three-dimensional adaptive grid method, *AIAA Journal.* 1986, 6, pp. 948–954.
22. Prokopov, G.P., On the organizing the comparison of algorithms and programs of generation of regular two-dimensional difference grids, *Voprosy Atomnoy Nauki i Tekniki. Ser.: Matem. Modelirovanye Fizicheskikh Processov.* 1989, 3, pp. 98–108.
23. Raizberg, V.A., Erokhin, B.T., Samsonov, K.P., Basis of theory of solid propellant jet systems working processes, *M. Mahinostrojenie,* 1972.
24. Serra, R.A., Calculation of Inner gas flows by stabilisation methods, *Rocket Technique and Cosmonautics.* 1972, 10, 5, pp. 55–63.
25. Serezhnikova, T.I., Sidorov, A.F., Ushakova, O.V., On one method of construction of optimal curvilinear grids and its applications, *Soviet Journal of Numerical Analysis and Mathematical Modelling.* 1989, 4, 2, pp. 137–155.
26. Shabashova, T.I., On one economized method of optimal difference grids generation, *Numerical Methods of Continuity Medium Mechanic.* 1983, 14, 5, pp. 139–157.
27. Shabashova, T.I., The construction of optimal curvilinear grids in three-dimensional regions, *Chisl. Metody. Mekhan. Sploshnoi Sredy.* 17, pp. 144–155, Vychisl. Tsentr, ITPM SO Akad. Nauk SSSR, Novosibirsk, 1979.
28. Shirokovskaya, O.S., A remark to A.F. Sidorov's paper: On one algorithm for computing optimal difference grids, *Zh. Bychisl. Mat. Mat. Fiz.* 1969, 9, pp. 468–469.
29. Sidorov, A.F., On one algorithm for computing optimal difference grids, *Proc. Steklov. Math. Institute.* 1966, 24, pp. 147–151.
30. Sidorov, A.F. and Shabashova, T.I., On one method of computation of optimal difference grids for multidimensional domains, *Chisl. Metody Mechaniki Sploshnoy Sredy Novosibirsk,* 1981, 12, pp. 106–123.
31. Tchoi, D., Merkl, Ch.L., Application of stabilisation method to calculate low velocity flows, *Aerodynamic Techniques,* 1986, 7, pp. 29–37.

32. Thomson, J.F., A survey of dynamically adaptive grids in the numerical solution of partial differential equations, *Applied Numerical Mathematics.* 1985, 1, pp. 3–27.
33. Thompson, J.F., Warsi, Z.U.A., Wayne, M.C., *Numerical Grid Generation: Found Applications.* North-Holland, NY, 1985, p 483.
34. Ushakova, O.V., On one iterative scheme of solution of an equation with a small parameter on an adaptive grid, *Analiticheskie Chisl. Metody Issledovania Zadach Mekhaniki Sploshnoy Sredy.* Urals Scientific Center, USSA Acad. Sci., Sverdlovsk, 1987, pp. 119–124.
35. Ushakova, O.V., Theorem of existence and uniquience of the solution of the boundary value problem for generation of one-dimensional optimal adaptive grid, *Modelirovanie v Mekhanike.* Novosibirsk, 1989, 3, 2, pp. 134–141.
36. Ushakova, O.V., Iterational procedure for computing optimal adaptive grids, *Approximate Methods of Investigations of Non-Linear Problems of Continuum Mechanics.* Sverdlovsk, 1992, pp. 58–65.
37. Ushakova, O.V., LADA — Efficient algorithm and program of generation of two-dimensional curvilinear optimal adaptive grids in simply-connected complex geometry domains, *Voprosy Atomnoy Nauki i Tekhniki. Ser.: Matem. Modelirovanye Fizicheskikh Processov.* 1994, 3, pp. 47–56.
38. Ushakova, O.V., Parallel algorithms and program of optimal adaptive grids generation, *Algorithms and Program Tools of Parallel Computations.* Institute of Mathematics and Mechanics, Urals Branch, Russian Academy of Science, Yekaterinburg, 1995, pp. 182–195.
39. Yanenko, N.N., Danaev, N.T., Liseikin, V.D., A variational method of constructing grids, Chisl. Melody Mekhan. Sploshnoi Sredy, Bychisl. Tsentr, ITPM SO Akad. Nauk SSSR, Novosibirsk, 8, 1977, 4, pp. 157–163.
40. Yemel'yanov, K.V., Applying optimal difference grids to problems with singular perturbations, *Comp. Maths Math. Phys.* 1994, 34, 6, pp. 804–814.
41. Yemel'yanov, K.V., On optimal grids and their application to the solution of problems with a singular perturbation, *Russian Journal of Numerical Analysis and Mathematical Modelling.* 1995, 10, 4, pp. 299–310.
42. Winslow, A.M., Numerical solution of quasilinear poisson equation in nonuniform triangle mesh, *J. Comput. Phys.* 1966, 1, 2, pp. 149–172.

37

Moving Grid Techniques

Paul A. Zegeling

37.1 Introduction

Traditional numerical techniques to solve time-dependent partial differential equations (PDEs) integrate on a uniform spatial grid that is kept fixed on the entire time interval. If the solutions have regions of high spatial activity, a standard fixed-grid technique is computationally inefficient, since to afford an accurate numerical approximation, it should contain, in general, a very large number of grid points. The grid on which the PDE is discretized then needs to be locally refined. Moreover, if the regions of high spatial activity are moving in time, like for steep moving fronts in reaction–diffusion or hyperbolic equations, then techniques are needed that also adapt (move) the grid in time.

In the realm of adaptive techniques for time-dependent PDEs, we can roughly distinguish between two classes of methods. The first class, denoted by the term *h*-refinement, consists of the so-called static-regridding methods. For these methods, the grid is adapted only at discrete time levels. The main advantage of this type of technique is their conceptual simplicity and robustness, in the sense that they permit the tracking of a varying number of wave fronts. A drawback, however, is that interpolation must be used to transfer numerical quantities from the old grid to new grids. Also, numerical dispersion, appearing, for instance, when hyperbolic PDEs are numerically approximated, is not fully annihilated with *h*-refinement. Another disadvantage of static-regridding is the fact that it does not produce "smoothing" in the time direction, with the consequence that the time-stepping accuracy therefore will demand small time steps. Examples of this type of methods can be found in Arney et al. [4,5], Berger et al. [8], Trompert et al. [42].

The second class of methods, denoted by the term *r*-refinement (*re*distribute or *re*locate), has the special feature of moving the spatial grid continuously and automatically in the space–time domain while the discretization of the PDE and the moving-grid procedure are intrinsically coupled. Moving-grid techniques use a fixed number of grid points, without need of interpolation and let the grid points dynamically move with the underlying feature of the PDE (wave, pulse, front, ...). Examples of *r*-refinement based methods can be found in Hawken et al. [22], Thompson [41], Zegeling [49] and later on in this chapter. Since the number of grid points is held fixed throughout the course of computation, problems could arise if several steep fronts would act in different regions of the spatial domain. For

0-8493-2687-7/99/$0.00+$.50

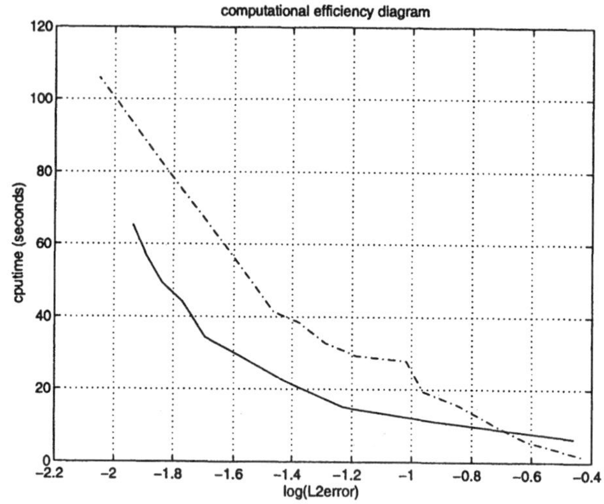

FIGURE 37.1 Computational effort as a function of the L_2-error: fixed (dashed) vs. moving grid (solid).

example, the grid is following one wave front, while a second front arises somewhere else. No "new" grid is created for the new wave front, but rather the "old" one has to adjust itself abruptly to cope with the newly developed front. Another difficulty is of a topological nature, usually referred to as "grid-distortion" or "mesh-tangling." Especially for higher dimensions this may cause problems, since the accuracy of the numerical approximation of the derivatives depends highly on the grid. Therefore, moving-grid techniques often need additional regularization terms to prevent this from happening or to at least slow down the grid degeneration process. Another possibility is to combine static-regridding with moving grid techniques, as is done in *h–r*-refinement methods (see, e.g., Arney et al. [5] or Petzold [36]).

During the last decade, moving grid techniques have been shown to be very useful for solving parabolic and hyperbolic partial differential equations involving fine scale structures such as steep moving moving fronts, emerging steep layers, pulses, and shocks. Using *r*-refinement for these types of PDEs can save up to several factors in terms of numbers of spatial grid points, if the mesh is moved properly, i.e., without distortion and well-adapted to the underlying PDE solution. For a typical situation, Figure 37.1 displays the computational efficiency of moving grids compared to fixed uniform grids, i.e., the relation between computational effort (measured in CPU seconds) and the error in the numerical solution (measured as the L_2-error).

In one space dimension, moving-grid methods have been applied successfully to many different types of PDE systems (see, e.g., Carlson et al. [13], Zegeling et al. [46]). In two space dimensions, however, application of moving-grid methods is far less trivial than in 1D. For instance, there are many possibilities to treat the one-dimensional boundary and to discretize the spatial domain, each having their own difficulties for specific PDEs. Furthermore, in 2D the chances for grid distortion to occur are much greater due to the extra degree of freedom (see Zegeling et al. [47]). In the following sections several moving grid techniques for time-dependent PDEs are discussed.

It should be noted that, in all cases, the method of lines is used, i.e., first the PDE is discretized in the spatial direction yielding a large (stiff) system of initial value ODEs. Then, time-integration of this ODE system, arising from semidiscretizing the PDEs in the discussed examples, is performed by using the integrator of Petzold [35].

37.2 Underlying Principles

Before examining some moving-grid techniques, it is necessary to prepare a time-dependent PDE for the grid movement. This can be done by defining a coordinate transformation from the physical space (a nonuniform grid for the original PDE) to the computational space, where a uniform grid is used.

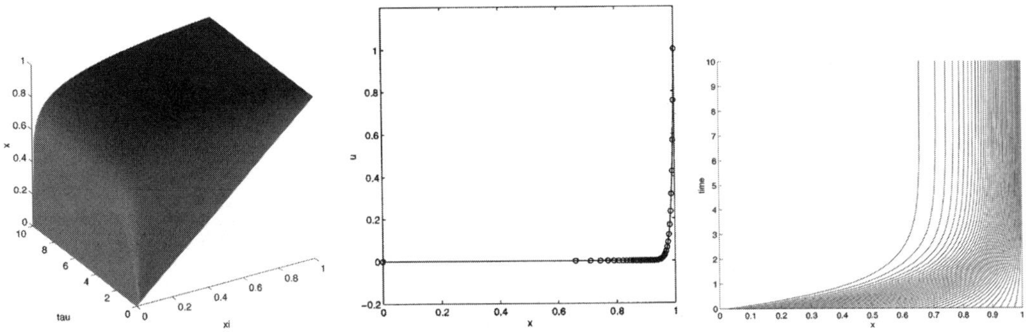

FIGURE 37.2 Transformation (1) (left), solution at $\theta = 10$ (middle), and grid history (right).

37.2.1 Transformation of Variables

Underlying all moving grid methods is a transformation between grids. Let, e.g., in one space dimension, a general time-dependent transformation be given by $x = x(\xi,\theta)$, $t = \theta$, which carries points from the uniform ξ-space into corresponding points in nonuniform x-space. As an example, such a transformation could be given by

$$x(\xi,\theta) = e^{-\theta}\xi + \left(1 - e^{-\theta}\right)\frac{1}{v}\ln\left(1 + \left(e^{v} - 1\right)\xi\right), \quad \text{for } \theta \in [0,10], \xi \in [0,1], v > 0. \tag{37.1}$$

In Figure 37.2 this transformation is displayed for $v = 10$. This transformation and its grid (uniform in ξ direction and therefore stretched in x direction) can be used to follow a PDE solution that ends in a steep boundary layer at $x = 1$ and $t = \theta >> 1$. For example, we could take $u(x,t) = (1 - e^{-t})\,\dfrac{e^{\lambda x} - 1}{e^{\lambda} - 1}$ as a possible PDE solution, with $\lambda = 100$ and $\theta = 10$. Starting with a uniform grid at $t = \theta = 0$, i.e., $x(\xi,0) = \xi$, a moving grid is obtained as shown in the two right plots of Figure 37.2.

Consider now the time-dependent PDE in two space dimensions (the one-dimensional case is obtained by freezing the second space direction),

$$\frac{\partial u}{\partial t} = \delta\Delta u - \beta \cdot \nabla u + S(u,\underline{x},t) \equiv \mathrm{L}(\mathrm{u}), \tag{37.2}$$

for $\underline{x} \in \Omega \subset I\!R^2$, $t > 0$ with given boundary conditions on $\partial\Omega$ and initial condition for $t = 0$. The PDE operator $\mathcal{L}$ contains spatial derivatives of u. We seek for a solution $u(\underline{x},t)$ with $\underline{x} \in \Omega \equiv [0,1]^2$ and $t \in [0,T]$. For general domains Ω, an extra transformation will be needed between the parametric and the physical domain (see Chapter 2).

For the two-dimensional PDE Eq. 37.2 we can define a transformation $x = x(\xi,\eta,\theta)$, $y = y(\xi,\eta,\theta)$, $t = \theta$. Then applying the chain rule for differentiation we get

$$\frac{\partial u}{\partial \theta} = \frac{\partial u}{\partial t} + \frac{\partial u}{\partial x}\frac{\partial x}{\partial \theta} + \frac{\partial u}{\partial y}\frac{\partial y}{\partial \theta}, \tag{37.3}$$

where

$$\frac{\partial u}{\partial x} = 0 + \frac{\partial u}{\partial \xi}\frac{\partial \xi}{\partial x} + \frac{\partial u}{\partial \eta}\frac{\partial \eta}{\partial x}, \quad \text{and} \quad \frac{\partial u}{\partial y} = 0 + \frac{\partial u}{\partial \xi}\frac{\partial \xi}{\partial y} + \frac{\partial u}{\partial \eta}\frac{\partial \eta}{\partial y}.$$

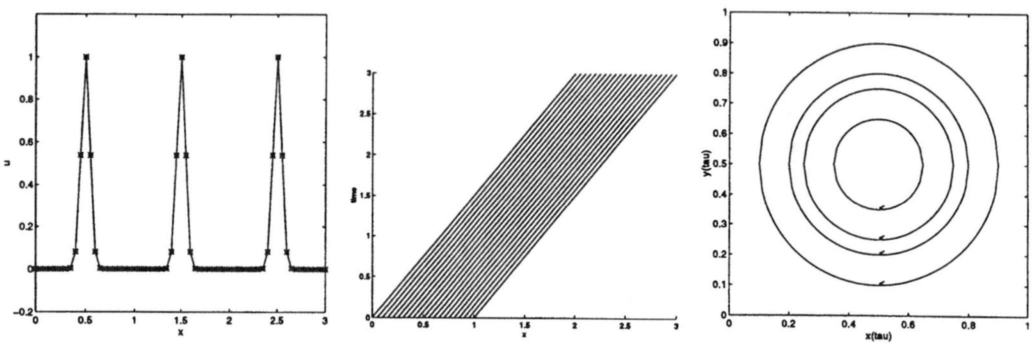

FIGURE 37.3 Using the method of characteristics in 1D (left and middle); right, example of characteristics in 2D that will certainly twist the underlying grid.

Substituting these equations in PDE Eq. 37.2, the Lagrangian form of the PDE is obtained

$$\dot{u} - u_x\dot{x} - u_y\dot{y} = \mathrm{L}(u), \tag{37.4}$$

where the dot stands for $\frac{\partial}{\partial\theta}$, and u_x, u_y for $\frac{\partial u}{\partial x}$ and $\frac{\partial u}{\partial y}$, respectively. Semidiscretizing Eq. 37.4 in the spatial direction, we get a system of ordinary differential equations (ODEs). To complete the system, additional equations (ODEs or PDEs) for the grid movement $\dot{x}$ and $\dot{y}$ are required. This will be presented in the following sections.

37.2.2 The Method of Characteristics (MoC)

One of the "simplest" choices for letting the grid move and implicitly defining the transformation is to make use of the characteristic equations of the PDE. This is, of course, only feasible for a small class of hyperbolic systems. If we consider the transport equation $\frac{\partial u}{\partial t} = -\beta\nabla u + \gamma$, then MoC (see Courant et al. [15]) leads to $\frac{\partial}{\partial\theta}\underline{x} = \beta$ and $\frac{\partial u}{\partial\theta} = \gamma$. Note that if these equations are combined, then we obtain the equivalent equation $\frac{\partial u}{\partial\theta} - \nabla u \cdot \frac{\partial}{\partial\theta}\underline{x} = \beta\nabla u + \gamma$, which is the original PDE but now in the computational domain.

Using moving-grid equations based on MoC, we can produce extremely accurate numerical solutions for this type of PDE. This is shown for $\beta = 1$, $\gamma = 0$ in a 1D situation with 21 grid points in Figure 37.3. In the case of $x := \xi$, $\forall\theta \geq 0$ (a nonmoving uniform grid), numerical solutions would have produced unwanted oscillations and/or severe unnatural damping. The MoC approach is not well-suited for general hyperbolic PDEs; however, a standard counterexample is given by the choice $\beta = u$, $\gamma = 0$ (Burgers' equation), for which the PDE characteristics collide at some point of time and therefore must give colliding grid points. In higher space dimensions this situation will only deteriorate. This feature is also shown in Figure 37.3 (right plot) for the 2D case, where $\beta = \pi(y - \frac{1}{2}, \frac{1}{2} - x)^T$. The characteristic trajectories are now given by circles around $(x,y) = (\frac{1}{2}, \frac{1}{2})$ on which the time-variable θ varies. Using MoC to move the grid would produce a twisted and distorted grid. It should therefore be clear that, in general, MoC is *not* the way to let the grid move, at least without additional remeshing.

37.2.3 Equidistribution

One of the most widely spread concepts to adapt and move a grid in one space dimension is given by the so-called equidistribution principle; cf. De Boor [11], Ren et al. [38].

In this case the coordinate transformation is explicitly given as

$$\int_0^{x(\xi,t)} M(\tilde{x},t)d\tilde{x} = \xi \int_0^1 M(\tilde{x},t)d\tilde{x}, \tag{37.5}$$

where $M > 0$ is a so-called monitor or weight function, usually depending on first- and second-order spatial derivatives of the PDE solution. If we select $N-1$ time-dependent grid points defining the spatial grid,

$$X: 0 = X_0 < \ldots < X_i(t) < X_{i+1}(t) < \ldots < X_N = 1, t > 0,$$

and using a uniform grid in the ξ-direction ($\xi_i = i/N$), Eq. 37.5 can be "discretized" as

$$\int_{x(\xi_{i-1},t)}^{x(\xi_i,t)} Md\tilde{x} = \frac{1}{N}\int_0^1 Md\tilde{x}, \quad \text{for } i = 1,\ldots,N, \tag{37.6}$$

with $x(\xi_i,t) = X_i(t)$. We can also differentiate Eq. 37.5 twice with respect to ξ to obtain the PDE

$$\frac{\partial}{\partial\xi}\left(\frac{\partial x}{\partial\xi}M\right) = 0. \tag{37.7}$$

Using the midpoint rule for evaluating the integrals in Eq. 37.6, we obtain yet another formula that describes equidistribution:

$$\Delta X_{i-1}M_{i-1} = \Delta X_i M_i, \ 1 \le i \le N-1, \tag{37.8}$$

where $M_i \equiv M|_{x=X_{i+1/2}}$ and $\Delta X_i = X_{i+1} - X_i$. This discretized form, which is equivalent to $\Delta X_i\, M_i = const.$, states that the grid should be moved to places where the weight function M dominates. More precisely, the grid cells ΔX_i should be small where M_i is large, and ΔX_i should be large where M_i is small, respectively, since the product of both quantities is constant. In other words, referring to Eq. 37.6, the grid points are redistributed by "distributing the weight function M equally over all subintervals." It is also noted that PDE Eq. 37.7 can be obtained by minimizing the energy integral $I = \int_0^1 Mx_\xi^2\, d\xi$, which can be taken to represent the energy of a system of springs with spring constants M, cf. Thompson [41]. The grid point distribution then would represent the equilibrium state of such a spring system. As an example in 1D the Lagrangian PDE Eq. 37.4 could be combined with the moving grid PDE (cf. Eq. 37.7) $\frac{\partial x}{\partial\theta} = \frac{\partial}{\partial\xi}(\frac{\partial x}{\partial\xi}M)$, where θ is now playing the role of an artificial time-variable. In Figure 37.4 (left and middle) the grid and solution (- -) are shown for this case ($N = 21$) with the arc-length monitor $M = \sqrt{1+u_x^2}$. The exact "solution" $u = \sin^{100}(\pi x)$ is being used. It is clearly seen that the first derivative of u is overemphasized. Some smoothing is therefore needed to provide more regularly distributed grid ratios. This will be worked out in the next subsection.

In two space dimensions there is no straightforward extension of this principle; see, however, Section 37.3.1 and Baines [6], Dwyer et al. [20], Huang et al. [25] for some ways to define equidistribution-like methods in higher dimensions.

37.3 Best Practices

37.3.1 Moving Finite Differences (MFD)

Starting from the equidistribution principle described by Eq. 37.8, it is easy to derive a moving grid technique with a "smooth" behavior in space and time. For this purpose we introduce the point-concentration values $n_i \equiv (\Delta X_i)^{-1}$, $0 \le i \le N$, and the relation Eq. 37.8 is rewritten as

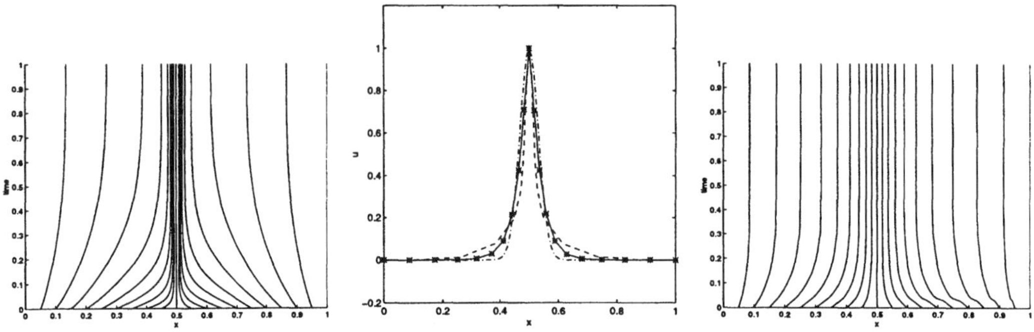

FIGURE 37.4 Left: grid for the equidistribution Eq. 37.8; middle: solution u (with - -), the exact solution (with -.), solution for $\kappa = 2$ (with-*); right: smoothed grid.

$$n_{i-1}/M_{i-1} = n_i/M_i, \quad 1 \le i \le N-1. \tag{37.9}$$

When using Eqs. 37.8 or 37.9 there is little control over the grid movement. For example, it can happen that the grid distance ΔX_i varies extremely rapidly over X (see Figure 37.4; left plot) or that for evolving time the trajectories $X_i(t)$ tend to oscillate. Too large a variation in ΔX_i may be detrimental to spatial accuracy, and temporal grid oscillations are likely to hinder the numerical time-stepping since the grid trajectories are computed automatically by numerical integration. Therefore, two grid-smoothing procedures are added: one for generating a spatially smooth grid and the other for avoiding temporal grid oscillations. This involves a modification of system Eq. 37.9. Instead of Eq. 37.9 the grid motion is now given by the system of ordinary differential equations

$$\left(\tilde{n}_{i-1} + \tau \frac{\mathrm{d}}{\mathrm{dt}} \tilde{n}_{i-1}\right) \Big/ M_{i-1} = \left(\tilde{n}_i + \tau \frac{\mathrm{d}}{\mathrm{dt}} \tilde{n}_i\right) \Big/ M_i, \quad t > 0, \ 1 \le i \le N, \tag{37.10}$$

where $\tilde{n}_i = n_i - \kappa(\kappa + 1)\,(\tilde{n}_{i+1} - 2\tilde{n}_i + \tilde{n}_{i-1})$ with $\kappa \ge 0$. The parameter κ is connected with the spatial grid-smoothing. It can be proved, Verwer et al. [43], that the moving grid defined by Eq. 37.10 satisfies

$$\frac{\kappa}{\kappa+1} \le \frac{\Delta X_{i+1}(t)}{\Delta X_i(t)} \le \frac{\kappa+1}{\kappa} \quad \forall i, \ t \ge 0, \tag{37.11}$$

showing that we have control over the variation in ΔX_i for all points of time. The parameter $\tau \ge 0$ in Eq. 37.10 is connected with the temporal grid-smoothing and serves to act as a delay factor for the grid movement. The introduction of the temporal derivative of the grid X (via $\frac{d}{dt}\ \tilde{n}_i$ in Eq. 37.10 forces the grid to adjust over a time interval of length τ from old to new monitor values, whichprovides a tool for suppressing grid oscillations in time.

Combining system Eq. 37.10 with the 1D semidiscrete form of Eq. 37.4 gives the stiff ODE system

$$A_{mfd}(\eta_1, \tau)\dot{\eta}_1 = G_{mfd}(\eta_1), \tag{37.12}$$

with $\eta_1 \equiv (\ldots, U_i, X_i, \ldots)^T$. A well-known choice for the monitor is $M_i = \sqrt{1 + \alpha \frac{(U_{i+1} - U_{i-1})^2}{(X_{i+1} - X_{i-1})^2}}$, where $\alpha \ge 0$ is an adaptivity parameter. For $\alpha = 1$ we have the arc-length monitor (see Section 37.2.3) which places grid points along uniform arc-length intervals. For $\alpha = 0$ the monitor function $M = 1$, and then Eq. 37.10 yields a uniform grid, while for $\alpha > 1$ the adaptivity increases as the first spatial derivative u_x is

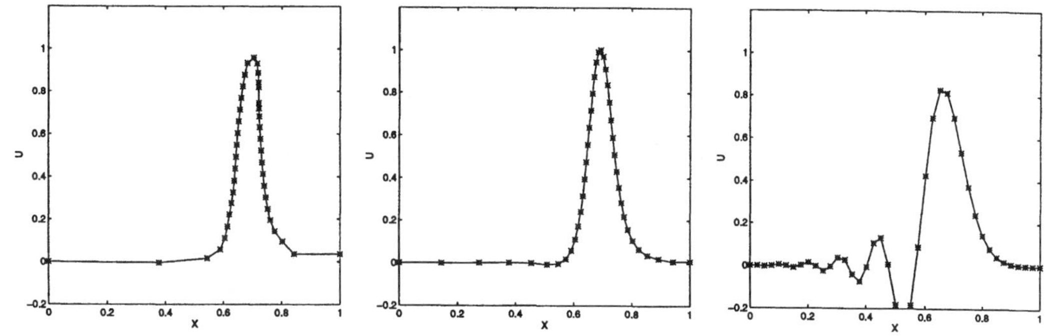

FIGURE 37.5 Numerical solutions with too little spatial smoothing (left; $\kappa = 0.2$), with "standard" spatial smoothing (middle; $\kappa = 2$), and with too much smoothing (right; $\kappa = 100$).

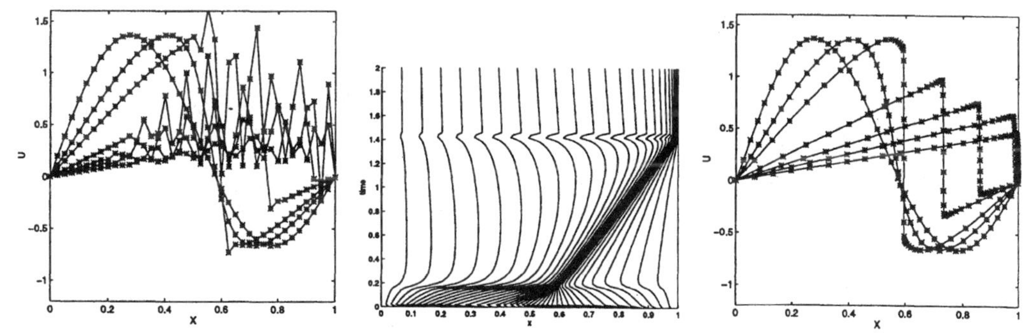

FIGURE 37.6 Numerical solutions of the 1D Burger's Eq. 37.14 with finite differences; left: uniform grid solutions; middle and right: the grid evolution and solution with moving grids.

more emphasized. A "standard" choice for the three method parameters is: $\alpha = 1$, $\kappa = 2$, $\tau = 10^{-3}$ (see Furzeland et al. [21]). In Figure 37.5 the effect of spatial smoothing is depicted at $t = \frac{1}{2}$ when Eq. 37.10 is applied to the scalar advection equation $\frac{\partial u}{\partial t} + \frac{\partial u}{\partial x} = 0$ with the analytical solution $u^*(x,t) = \sin^{50}(\pi(x - t + \frac{3}{10}))$. Note that too little or too much smoothing may give rise to irregular grids (left) and oscillatory solutions (right), whereas "standard" smoothing produces regular grid positioning and solution behavior (middle).

It is interesting to note that Huang et al. have derived a continuous formulation for Eq. 37.10 in terms of the transformation variables ξ and θ. The ODEs in Eq. 37.10 are then semi-discretized versions of "their" PDE,

$$\frac{\partial}{\partial \xi}\left[\frac{\tilde{n} + \tau \dot{\tilde{n}}}{M}\right] = 0, \tag{37.13}$$

where $n \equiv 1/\frac{\partial x}{\partial \xi}$ (the inverse of the Jacobian of the transformation), $\tilde{n} \equiv (I - \frac{\kappa(\kappa+1)}{N}\frac{\partial^2}{\partial \xi^2}$ and $M = \sqrt{1 + \alpha u_x^2}$. Figure 37.6 shows numerical results for this moving-grid method ($N = 41$) when applied to Burger's equation with spatial operator

$$L(u) = \delta \frac{\partial^2 u}{\partial x^2} - u\frac{\partial u}{\partial x}, \tag{37.14}$$

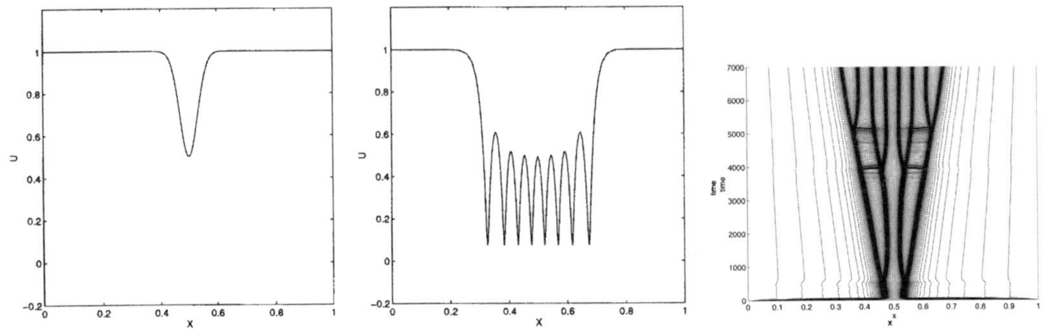

FIGURE 37.7 Moving finite differences for the 1D reaction–diffusion system (Eq. 37.15) at $t = 0$ (left), $t = 7000$ (middle), and the moving grid (right).

and $\delta = 5\ 10^{-4}$, $u|_{t=0} = \frac{1}{2}\sin(\pi x) + \sin(2\pi x)$, $u|_{\partial\Omega} = 0$. In the left plot the well-known "wiggles" are seen for the nonmoving grid case. The moving grid (middle and right plot) follows the sharpening of the solution and moving front satisfactorily.

Figure 37.7 shows further numerical results for this method when applied to a system of reaction–diffusion equations with

$$\begin{aligned} L_1(u,v) &= \Delta u - uv^2 + A(1-u), \\ L_2(u,v) &= 10^{-2}\Delta v + uv^2 - Bv, \end{aligned} \tag{37.15}$$

and constants A and B, an initial steep pulse in the middle of the domain and Dirichlet boundary conditions (see Doelman et al. [19] for more details).

As stated before, in two dimensions no proper mathematical definition for equidistribution exists. However, it is possible to define one-dimensional equidistribution (with smoothing) along coordinate lines in 2D. For example (see also Zegeling [49]), one can define the moving grid by

$$\begin{aligned} \frac{\partial}{\partial \xi}\left[\frac{\tilde{n} + \tau\dot{\tilde{n}}}{M_{(x)}}\right] &= 0, \quad \text{with} \quad n \equiv 1/x_\xi, \\ \frac{\partial}{\partial \eta}\left[\frac{\tilde{m} + \tau\dot{\tilde{m}}}{M_{(y)}}\right] &= 0, \quad \text{with} \quad m \equiv 1/y_\eta, \end{aligned} \tag{37.16}$$

where

$$M_{(x)} \equiv \sqrt{1+\alpha u_x^2}, \quad M_{(y)} \equiv \sqrt{1+\alpha u_y^2},$$

and

$$\tilde{n} \equiv \left(I - \frac{\kappa(\kappa+1)}{N}\frac{\partial^2}{\partial \xi^2}\right)n, \quad \tilde{m} \equiv \left(I - \frac{\kappa(\kappa+1)}{N}\frac{\partial^2}{\partial \eta^2}\right)m.$$

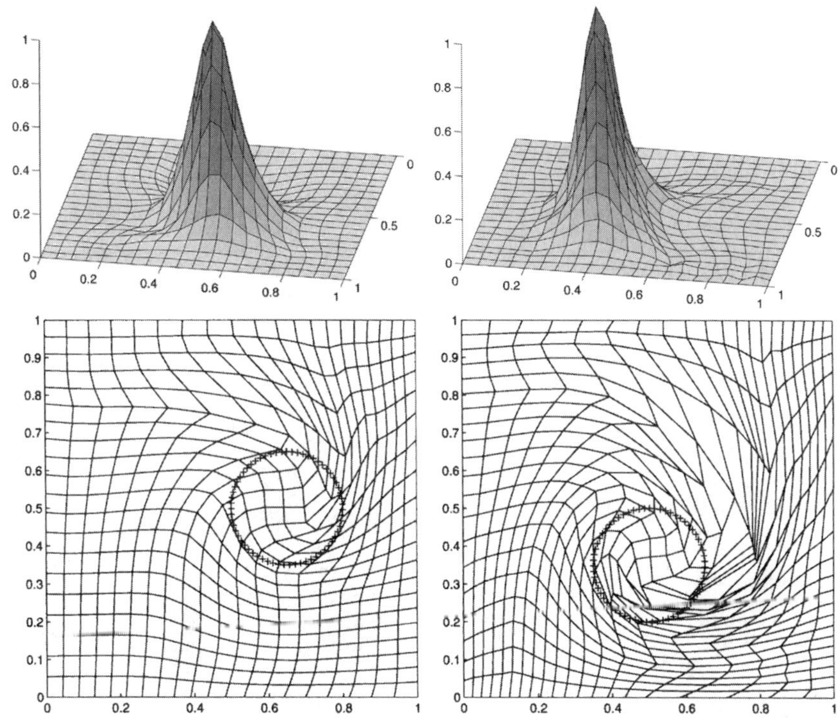

FIGURE 37.8 Moving finite difference results for the 2D advection PDE Eq. 37.18. With + the position of the pulse is depicted.

At the boundary, Neumann conditions for the grid are imposed: $\frac{\partial n}{\partial \xi}|_{x=0} = \frac{\partial n}{\partial \xi}|_{x=1} = \frac{\partial m}{\partial \eta}|_{y=0} = \frac{\partial m}{\partial \eta}|_{y=1} = 0$. Semidiscretizing the PDEs in Eq. 37.16 in the spatial direction with central differences and defining $\eta_2 \equiv (\ldots, U_i, X_i, Y_i, \ldots)^T$, it can be written as:

$$\mathrm{A}_{mfd}(\eta_2, \tau)\dot{\eta}_2 = G_{mfd}(\eta_2). \tag{37.17}$$

Figure 37.8 shows solutions and grids for the hyperbolic PDE with

$$\mathrm{L}(u) = \pi\left(y - \tfrac{1}{2}\right)\frac{\partial u}{\partial x} + \pi\left(\tfrac{1}{2} - x\right)\frac{\partial u}{\partial y}, \tag{37.18}$$

for $u|_{t=0} = e^{-100\left(\left(x-\frac{1}{2}\right)^2 + \left(y - \frac{13}{20}\right)^2\right)}$, $u|_{\partial\Omega} = 0$, and two points of time: $t = \frac{1}{2}$ and $t = 1$. The solution of the PDE is a pulse that rotates without change of shape around the center of the domain. This is a difficult test problem for standard numerical techniques. In the moving grid case almost no numerical diffusion or oscillations appear, in contrast with the nonmoving situation (see also Table 37.1). A second example is a model used in the field of water resources. It is an advection–dispersion equation with a moving front that starts from the left boundary and moves into center of the domain. A practical situation is described by the spatial PDE operator

$$\mathrm{L}(u) = 10^{-3}\frac{\partial^2 u}{\partial x^2} + 10^{-2}\frac{\partial^2 u}{\partial y^2} - \frac{\partial u}{\partial x}, \tag{37.19}$$

TABLE 37.1 Numerical Results for the 2D Advection Model Eq. 37.18 Using MFE, MFD, and Uniform Nonmoving Grids (FFE and FFD)

Method	*Umax* ($t = 0.5$)	*Umin* ($t = 0.5$)	*Umax* ($t = 1.0$)	*Umin* ($t = 1.0$)	Grid	Solution
FFE	0.7863	−0.0011	0.6338	−0.0022	Uniform	Numerically diffused
MFE	1.0027	−0.0040	1.0056	−0.0258	Distorted	Almost exact
FFD	0.8985	−0.0914	0.7784	−0.1637	Uniform	Inaccurate
MFD	0.9430	−0.0106	0.9360	−0.0283	Adaptive	Oscillatory

Note: Maximum and minimum values of the solution should be 1 and 0, respectively.

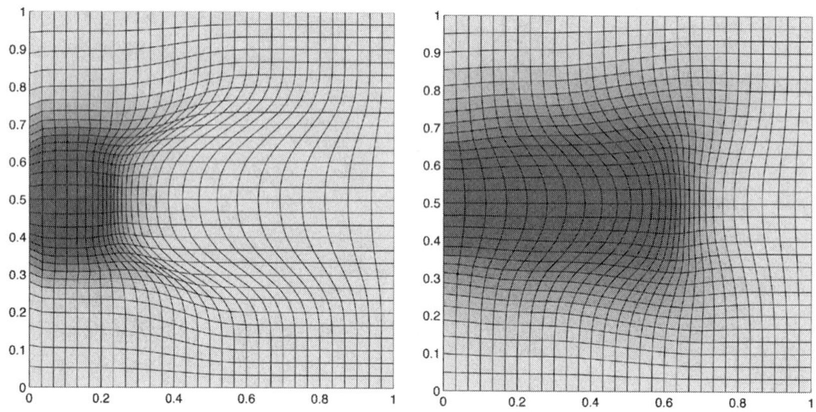

FIGURE 37.9 Moving finite difference results for the 2D advection–dispersion PDE (Eq. 37.19) at $t = 0.06$ (left) and $t = 0.48$ (right).

with initial condition $u|_{t=0} = \frac{1}{4}(1 + \tanh(50(\frac{1}{32} - (y - \frac{1}{2})^2)))(1 + \tanh(50(\frac{1}{32} - x^2)))$, and Neumann boundary conditions, except for that part of the boundary $x = 0$ where the solution is initially maximal (there a Dirichlet condition is imposed). In Figure 37.9 the grids, which are nicely located near the steep front, are displayed for $t = 0.06$ and $t = 0.48$.

37.3.2 Moving Finite Elements (MFE)

A two-dimensional moving grid technique (MFE) based on the minimization of the PDE residual is obtained by approximating the PDE solution u with piecewise-linear finite element basis functions (see Baines [6], Miller et al. [33], Zegeling [48]). There are several ways to describe this method. Here we follow the concept of the transformation between the physical and computational domain:

$$u \approx U = \sum_{j\in J} U_j(\theta)\alpha_j(\xi,\eta), \quad x \approx X = \sum_{j\in J} X_j(\theta)\alpha_j(\xi,\eta), \quad y \approx Y = \sum_{j\in J} Y_j(\theta)\alpha_j(\xi,\eta), \tag{37.20}$$

where α_j are the standard "hat" functions on 2D having a limited support and J stands for the index set of the grid points. Substituting Eq. 37.20 into the time-dependent PDE model gives, in general, a non-zero PDE residual $U_t - \mathcal{L}(U)$. To obtain equations for the grid movement, a minimization procedure ("least squares") is applied with respect to the yet unknown variables $\dot{U}_i, \dot{X}_i, \dot{Y}_i$ of the following quantity:

$$\int_{\Omega_{\xi,\eta}} \left(\dot{U} - U_x\dot{X} - U_y\dot{Y} - \mathrm{L}(U)\right)^2 \mathcal{J}\,d\xi d\eta \quad \forall i \in J. \tag{37.21}$$

Here $\mathcal{J}$ denotes the Jacobian of the transformation. After rewriting Eq. 37.21 in the physical coordinates, we obtain the system

$$\begin{aligned}&\int_\Omega (U_t - \mathrm{L}(U))\alpha_i \mathrm{d}x\mathrm{d}y = 0, \quad \forall i \in J,\\&\int_\Omega (U_t - \mathrm{L}(U))U_x\alpha_i \mathrm{d}x\mathrm{d}y = 0, \quad \forall i \in J,\\&\int_\Omega (U_t - \mathrm{L}(U))U_y\alpha_i \mathrm{d}x\mathrm{d}y = 0, \quad \forall i \in J,\end{aligned} \tag{37.22}$$

Working out the inner products and adding small regularization terms $P_{1,2}$ and $Q_{1,2}$ to keep the finite-element parametrization nondegenerate, yields for $i \in J$,

$$\begin{aligned}&\sum_{l\in J} <\alpha_i,\alpha_l>\dot{U}_l + <\alpha_i,\beta_l>\dot{X}_l + <\alpha_i,\gamma_l>\dot{Y}_l = <\alpha_i, \mathrm{L}_i(U)>\\&\sum_{l\in J} <\beta_i,\alpha_l>\dot{U}_l + <\beta_i,\beta_l>\dot{X}_l + <\beta_i,\gamma_l>\dot{Y}_l + P_1(\varepsilon_1^2) = <\beta_i, \mathrm{L}_i(U)> + Q_1(\varepsilon_2^2)\\&\sum_{l\in J} <\gamma_i,\alpha_l>\dot{U}_l + <\gamma_i,\beta_l>\dot{X}_l + <\gamma_i,\gamma_l>\dot{Y}_l + P_2(\varepsilon_1^2) = <\gamma_i, \mathrm{L}_i(U)> + Q_2(\varepsilon_2^2),\end{aligned}$$

where $\beta_i = -U_x\alpha_i$, $\gamma_i = -U_y\alpha_i$, and $< \bullet, \bullet >$ is the standard L_2-inner product. Using $\eta_2 = (\ldots, U_i, X_i, Y_i, \ldots)^T$ as before, this can be rewritten as

$$\mathrm{A}_{mfe}(\eta_2, \varepsilon_1^2)\dot{\eta}_2 = G_{mfe}(\eta_2, \varepsilon_2^2). \tag{37.23}$$

The small parameters ε_1^2 and ε_2^2 serve to keep the extended mass-matrix $\mathcal{A}_{mfe}$ and the right-hand side G_{mfe} nonsingular, respectively. It is worthwhile to note that the previous derivation can be done in higher space dimensions as well.

The more sophisticated GWMFE (see Carlson et al. [13, 14]) uses an additional gradient-weighting term in the inner products of the form $<w(\nabla U)\bullet, \bullet>$. However, in general, the results shown below hold, for the greater part, also for GWMFE, possibly with some minor modifications.

37.3.2.1 Some Properties of the Moving Grid for MFE

Consider now the PDE Eq. 37.2 in one or two space dimensions. In one space dimension it can be shown, Zegeling et al. [48], that for $J \to \infty$ and $\varepsilon_1^2 = \varepsilon_2^2 = 0$, the grid moves as a perturbed method of characteristics:

$$\frac{\partial x}{\partial \theta} = \beta + \delta\left(2\frac{u_{xxx}}{u_{xx}} - 3\frac{\xi_{xx}}{\xi_x}\right), \tag{37.24}$$

where ξ is the spatial coordinate in the computational domain. Numerical solutions of Eq. 37.23 for Burger's equation Eq. 37.14, clearly indicating property Eq. 37.24, are given in Figure 37.10. From Eq. 37.24 it can be derived that for steady-state situations ($\frac{\partial x}{\partial \theta} = \frac{\partial u}{\partial t} = 0$) an equidistribution-like relation holds for the grid:

$$\frac{\partial x}{\partial \xi}|u_{xx}|^{2/3}|u_x|^{1/3} = \text{const.} \tag{37.25}$$

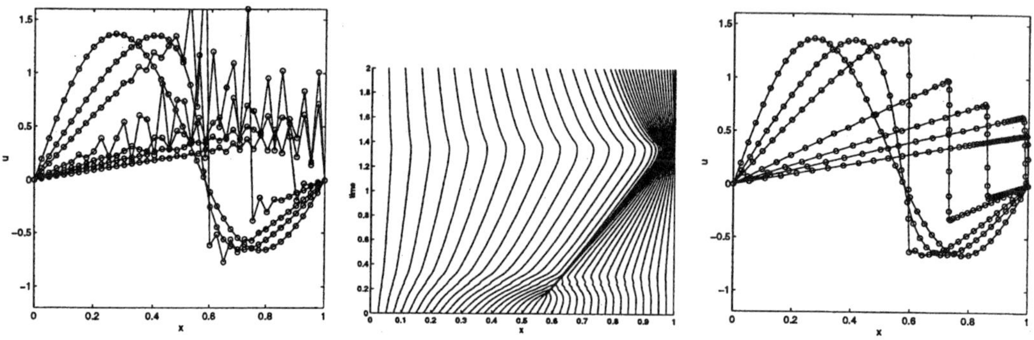

FIGURE 37.10 Numerical solutions of the 1D Burger's Eq. 37.14 with finite elements. Left: (oscillatory) uniform grid solutions; middle and right: the grid evolution and (nonoscillatory) solution with moving grids.

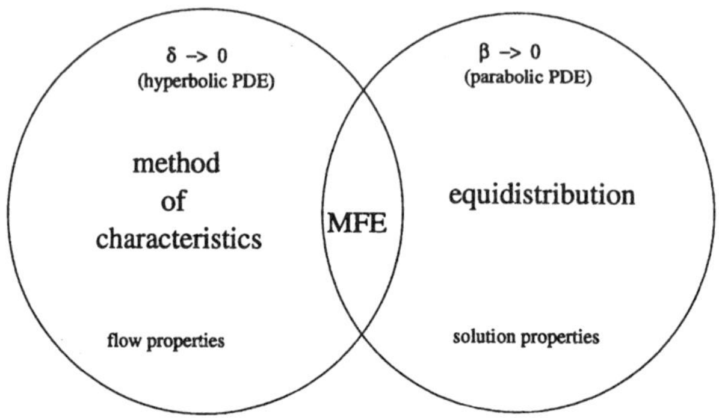

FIGURE 37.11 The moving finite element method has a relation both with equidistribution and with MoC.

In two space dimensions it is known that the grid moves in a similar way.

$$\begin{aligned}\frac{\partial x}{\partial \theta} &= \beta_1 + \delta\phi_1,\\ \frac{\partial y}{\partial \theta} &= \beta_2 + \delta\phi_2.\end{aligned} \tag{37.26}$$

However, an explicit formulation for the perturbation functions ϕ_1 and ϕ_2 has not yet been derived. Numerical experiments suggest that they should depend on first- and second-order spatial derivatives. This behavior "between" equidistribution (Eq. 37.25) and the method of characteristics (Eq. 37.24) is illustrated in Figures 37.11 and 37.12. In Figure 37.11 it is concluded that the grid in the method follows the flow of a hyperbolic PDE, whereas for diffusion dominated PDEs the grids concentrate near regions of high spatial activity (first- and second-order derivatives of the solution). Figure 37.12 confirms this property by letting the diffusion coefficient δ decrease from 1 to 10^{-3} for the PDE with

$$\mathrm{L}(u) = \delta\Delta u + \left(x - \frac{1}{2}\right)\frac{\partial u}{\partial x} - \left(y - \frac{1}{2}\right)\frac{\partial u}{\partial y} + f(x,y,t), \tag{37.27}$$

and $u|_{t=0} = 0$, $u|_{\partial\Omega} = 0$. The source term $f(x, y, t)$ is defined as

$$f(x,y,t) = u_t^* - \delta\Delta u^* - \left(x - \frac{1}{2}\right)\frac{\partial u^*}{\partial x} + \left(y - \frac{1}{2}\right)\frac{\partial u^*}{\partial y},$$

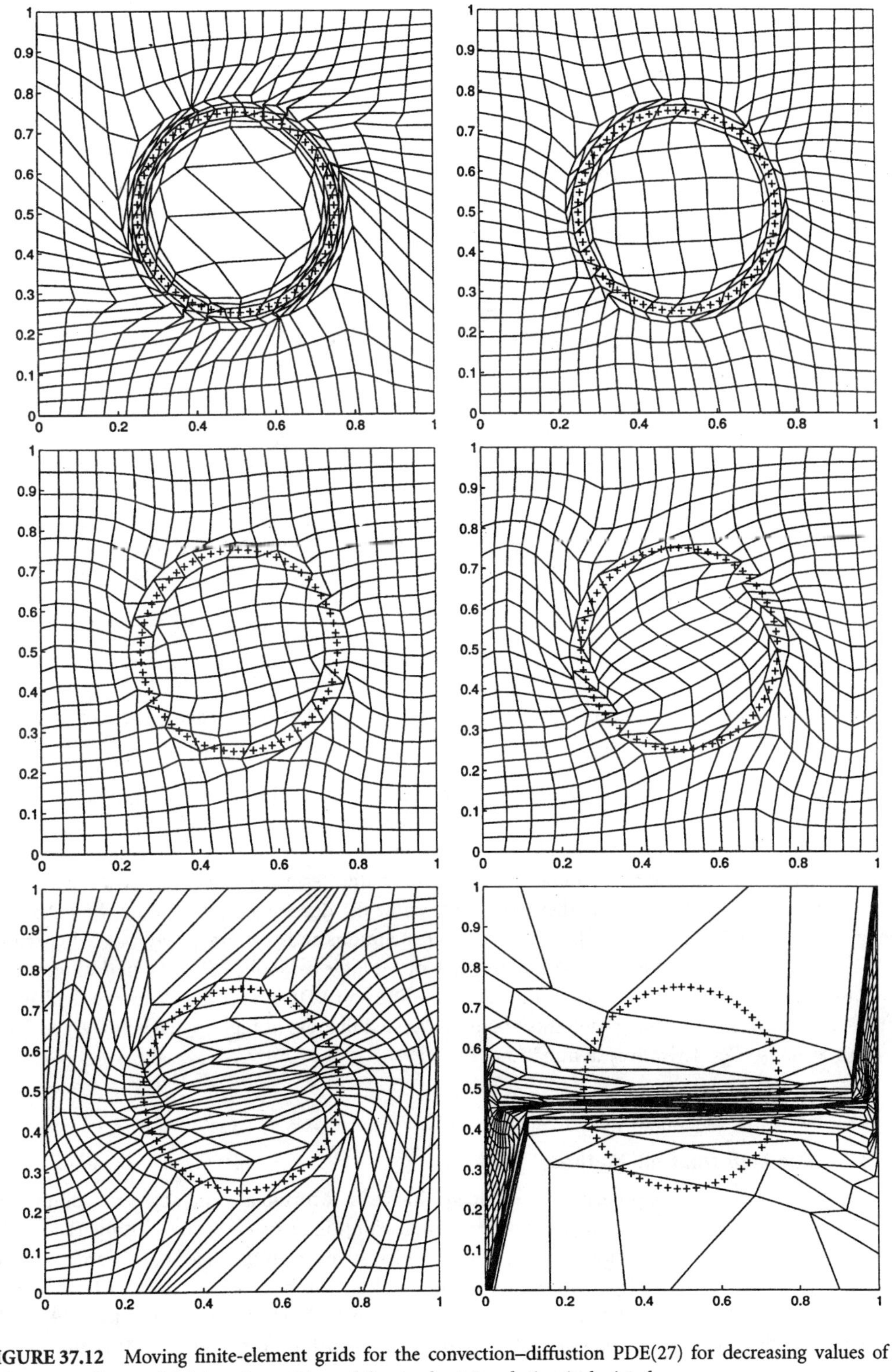

FIGURE 37.12 Moving finite-element grids for the convection–diffustion PDE(27) for decreasing values of the diffusion coefficient δ. With + the position of the steady-state solution is depicted.

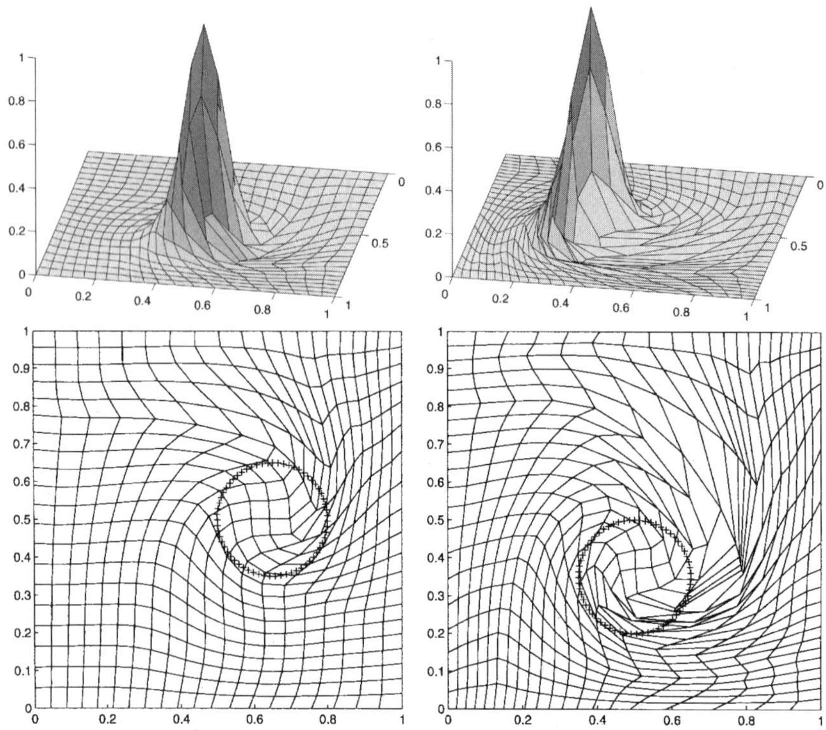

FIGURE 37.13 Moving finite-element results for the 2D advection PDE(18). With + the position of the pulse is depicted.

such that $u^*(x,y,t) = \frac{1}{2}(1 - e^{-t})(1 + \tanh(100\,(\frac{1}{16} - (x - \frac{1}{2})^2 - (y - \frac{1}{2})^2)))$ is the exact solution of the PDE model. This means that in steady-state we always must have the same solution, which is a steep circular "hat" in the middle of the domain (depicted by +'s in the figure). We see that the grid is "equidistributed" for larger values of δ and "distorted," following the first derivative terms, for lower values of the diffusion parameter (i.e., perturbed MoC). Another example to show the dependence of MFE on the PDE characteristics is given in Figure 37.13 and Table 37.1, where solutions and grids are given for the hyperbolic PDE Eq. 37.18. To stress the equidistribution property of MFE for parabolic PDEs, numerical results for MFE when applied to the 2D version of the reaction–diffusion PDE system Eq. 37.15 are depicted in Figure 37.14. For this model the grid points are nicely located in areas of high spatial activity, i.e., where first- and second-order derivatives dominate.

37.3.3 Related Approaches

37.3.3.1 The Deformation Method

Recently, a new moving grid approach was developed which can be formulated in "any" space dimension. In some sense, it can be seen as an extension of the equidistribution principle to higher dimensions. This approach, also denoted by the "deformation method," which stems from the theory of volume elements of a compact Riemannian manifold [30, 31], was first used for given steep functions by Bochev et al. [10], steady-state PDEs by Liao et al. [31], and time-dependent PDEs in 1D by Semper et al. [39]. To be consistent with the previous sections we will describe the ideas behind the method in two dimensions, although it can be done in a more general context.

The movement of the grid in the deformation method is described by the grid PDEs

$$\frac{\partial x}{\partial \theta} = -v_1 / W_l,$$
$$\frac{\partial y}{\partial \theta} = -v_2 / W_l, \tag{37.28}$$

where the vector field $v \equiv (v_1, v_2)^T$ should satisfy

$$\nabla \cdot v = -\frac{\partial W_l}{\partial t}, \quad v\big|_{\partial\Omega} = 0. \tag{37.29}$$

Here W_l is a (scaled) positive weight function, e.g., $W_l = M_l / \int_\Omega M_l d\Omega$, with (unscaled) $M_l = 1 + \alpha_l u^2 + \beta_l \|\nabla u\|^2_2$, such that $\int_\Omega W_l d\Omega = 1$, $\forall t = \theta \geq 0$. It can be shown that from Eqs. 37.28 and 37.29 follows

$$\det(\mathcal{J}) \cdot W_l = 1, \quad \forall t = \theta \geq 0, \tag{37.30}$$

where $\mathcal{J}$ is the Jacobian of the transformation as mentioned in Section 37.2.1.

In one space dimension, Eq. 37.30 reduces to

$$\frac{\partial x}{\partial \xi} W_l = 1, \quad \forall t = \theta \geq 0, \tag{37.31}$$

giving an equidistribution relation which is an integral of PDE Eq. 37.7 with integration constant equal to 1. A consequence of Eq. 37.30 is that the Jacobian of the transformation will always remain non-zero if W_l is positive. In a discretized form this means that the grid cannot distort, since the transformation is "held" nonsingular. For the 1D case a straightforward integration of Eq. 37.29 yields

$$v = -\int_0^x \frac{\partial W_l}{\partial t} d\tilde{x}, \tag{37.32}$$

defining the moving grid equation uniquely. In 2D, however, no unique solution exists for Eq. 37.29, which means that, for example, a least-squares technique has to be used to define the vector field v. On the other hand, it is possible to construct *one* solution that satisfies Eq. 37.29 in two space dimensions:

$$v_1 = \frac{1}{2}\left(-\int_0^x \frac{\partial W_l}{\partial t} d\tilde{x} + h(x) \int_0^1 \frac{\partial W_l}{\partial t} d\tilde{x} + h'(y) \int_0^x \int_0^1 \frac{\partial W_l}{\partial t} d\tilde{y}\, d\tilde{x} \right), \tag{37.33}$$

$$v_2 = \frac{1}{2}\left(-\int_0^y \frac{\partial W_l}{\partial t} d\tilde{y} + h(y) \int_0^1 \frac{\partial W_l}{\partial t} d\tilde{y} + h'(x) \int_0^y \int_0^1 \frac{\partial W_l}{\partial t} d\tilde{x}\, d\tilde{y} \right), \tag{37.34}$$

where $h(\zeta) = \frac{1}{2}(1 + \cos(\zeta))$. In Figure 37.15, deformating grids are shown for a scalar PDE with $\mathcal{L}(u) = -\cos(\pi t) \dfrac{\partial u}{\partial x}$, $u|_{t=0} = \sin^{10}(\pi x)$, $u|_{\partial\Omega} = 0$, and the exact solution $u^*(x,t) = \sin^{10}(\pi\,(x - \sin(\pi t)/\pi))$. The difference in positioning of the grid points can be seen clearly, depending on the choices for the parameters α_l, β_l in M_l. The third parameter γ_l comes from an additional term $\gamma_l\, u^2_{xx}$ in M_l to emphasize second-order derivatives.

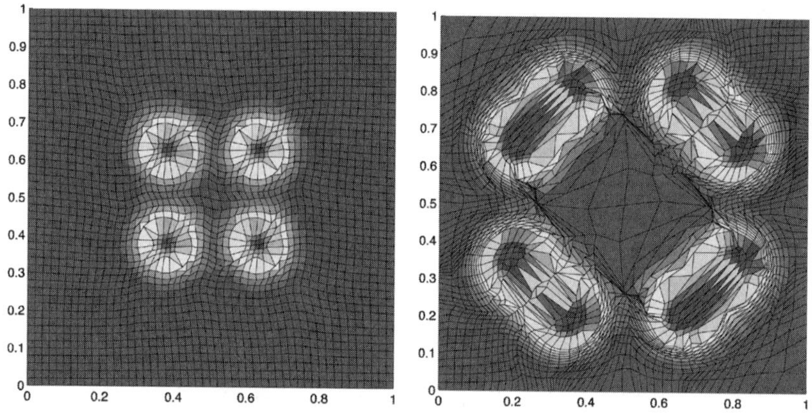

FIGURE 37.14 Moving finte-element results for the 2D reaction-diffusion system (15) at $t = 10$ and $t = 500$.

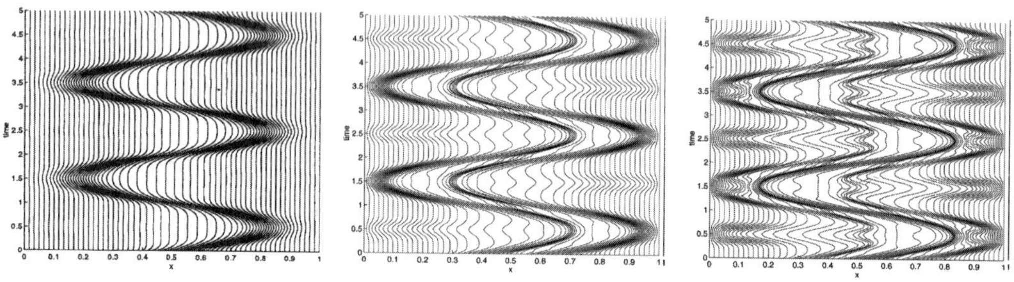

FIGURE 37.15 Grids for the deformation method in 1D; (left), $\alpha_l = \gamma_l = 0$, $\beta_l = 10^{-2}$ (middle) and $\alpha_l = \beta_l = 0$, $\gamma_l = 10^{-4}$ (right).

A second example is given by using the 2D PDE operator $\mathcal{L}(u) = \Delta u + f(x,y,t)$, with $u|_{t=0} = 0$ and $u|_{\partial\Omega} = 0$. The right hand side function is defined as $f(x,y,t) = u_t^* - \Delta u^*$ such that the exact solution of the PDE is $u^*(x,y,t) = (1 - e^{-t})(1 + \sin^{10}(\pi x)\sin^{10}(\pi y))$. Figure 37.16 (two upper plots) shows the grids for two values of α_l at steady state ($t = 10$). The two lower plots give grids for the same model but now for MFD (left) and MFE (right). Note that MFD positions its grid points near high first-order derivatives (as constructed), whereas MFE concentrates its grid at points with high second derivatives (as conjectured by Eq. 37.26). Further numerical experiments should be performed to get a complete picture and to draw final conclusions on the robustness and efficiency of the deformation method.

37.3.3.2 Other Techniques

In this subsection a range of other (important) moving grid techniques will be noted. Each method is only briefly highlighted with references for more detailed information. Note that this list is far from complete. For a more extensive overview, the reader is referred to papers such as Thompson [41] and Hawken et al. [22].

In Huang et al. [24] the idea of so-called moving-mesh PDEs (MMPDEs) is introduced. In fact, Eqs. 37.7 and 37.28, 37.32 can be derived as special cases of this idea. Starting from Eq. 37.7 one can create different kinds of PDEs describing the mesh movement in a continuous setting. A two-dimensional MMPDE is analyzed in Huang et al. [24]. There the grid velocities $\frac{\partial x}{\partial \theta}$ and $\frac{\partial y}{\partial \theta}$ are derived from a heat flow equation, which arises using a mesh adaptation functional that is motivated from the theory of harmonic maps. Both adaptivity and a suitable level of mesh orthogonality can be preserved.

In Arney et al. [3] a moving mesh technique for hyperbolic PDE systems in two space dimensions is described. The mesh movement is based on an algebraic node movement function determined from the

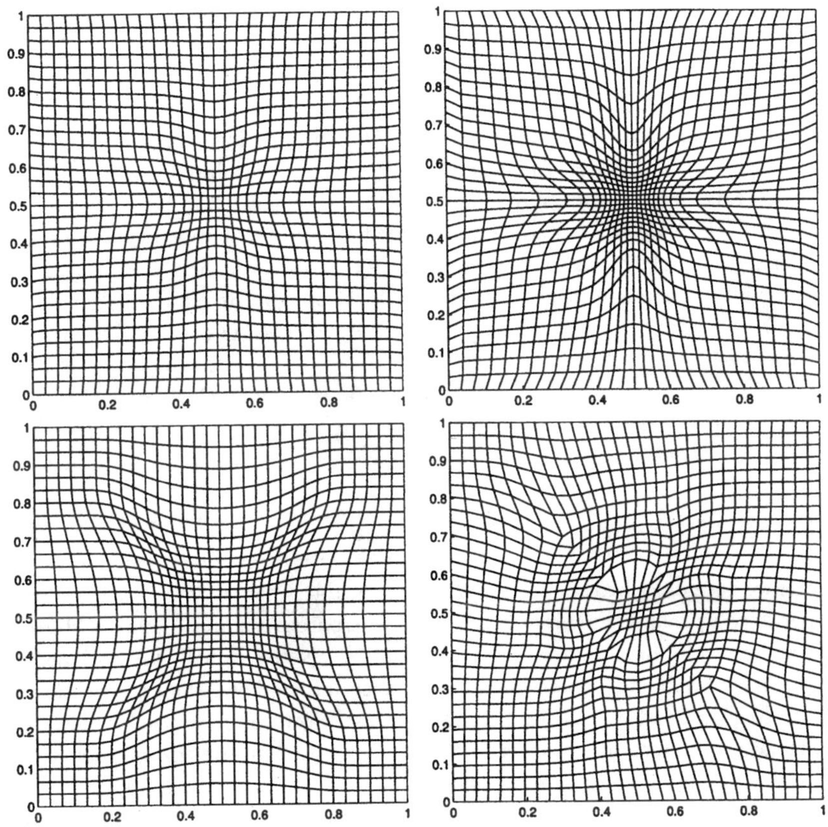

FIGURE 37.16 Moving-grid results for a 2D diffusion PDE. The upper two figures show grids for the deformation method ($\alpha_l = 2$ left and$\alpha_l = 10$ right), the lower two figures show grids for MFD (left) and MFE (right).

geometry and propagation of regions having significant discretization error indicators. Error clusters are moved according to the differential equation $\ddot{\mathbf{r}} + \lambda\dot{\mathbf{r}} = 0$, where **r** is the position vector of the center of an error cluster. Several numerical examples are given there, among others, for the hyperbolic PDE Eq. 37.18 and for the Euler equations for a perfect inviscid fluid. Also an example is given where two pulses rotate in an opposite direction, indicating the need for static rezoning, i.e., *h*-refinement combined with *r*-refinement.

In Rai et al. [37] grid speed equations are given in terms of time-derivatives of the variables ξ in 1D and ξ and η in 2D. Their idea is to relocate the mesh points by attracting other grid points to regions where $|u_\xi|$ is larger than its average value $|u_\xi|_{av}$ and repelling points from regions where $|u_\xi|$ is smaller than $|u_\xi|_{av}$ The attraction is attenuated by an inverse power of the point separation in the transformed domain. The collective attraction of all other points is then made to induce a velocity for each grid point. In Anderson et al. [1,2], the relation of equidistribution with Poisson grid generators and other possible choices for the grid movement are discussed.

In Delillo et al. [17] the grid is moved through an adaptation procedure that is based on a tension spring analogy, with spring constants depending on gradients in the flow of the PDE. This approach is closely related to the ideas of Brackbill et al. [12], Rai et al. [37] and the equidistribution principle.

One of the first moving grid methods stems from Yanenko et al. [44]. They use a variational scheme that allows the grid some movement with the PDE solution and keeping control over the possible grid distortions. Their ideas are based on minimizing a functional that depends on three measures: (preventing) grid distortion, movement with the flow, and refinement whenever the gradients of the solution become large.

Another variational approach is described by Brackbill et al. [12], who obtain an adaptive moving grid from the Euler equations for minimization of $I = \lambda_s I_s + \lambda_v I_v + \lambda_o I_o$, where $I_s = \int_\Omega ((\nabla\xi)^2 + (\nabla\eta)^2)d\Omega$ represents the smoothness of the grid, $I_0 = \int(\nabla\xi \cdot \nabla\eta)^2 d\Omega$ stands for the orthogonality in the grid, and $I_v = \int_\Omega W\mathcal{J}d\Omega$ denotes the weighted volume variation ("adaptivity"). The W and $\mathcal{J}$ are a monitor function, and the Jacobian of the transformation, respectively. Deriving the Euler equations for this variational problem yields a system of elliptic PDEs for the grid variables. In Dietachmayer et al. [18], this variational method is closely followed and applied to PDEs from meteorological models.

In Lee et al. [29] a moving grid is studied that is based on equidistribution of a weight function. Their grid is smoothed by coupling neighboring weight function values to neighboring grid points. In the formulation, the influence of the neighboring values of the weight function is assumed to decay exponentially with the distance from a reference grid point. Partial control over the skewness of the grid is then obtained as well.

Other interesting papers on moving-grid techniques can be found in Coyle et al. [16] (on the stability of the grid selection procedure), in Kuprat [28] (on moving finite elements for surfaces), in Kansa et al. [27] (application to gas dynamic equations), and Smooke et al. [40] (application to chemical reactions).

37.4 Research Issues and Summary

In this Chapter we have described several major moving grid techniques. It is clear that these techniques could be superior compared with their nonmoving counterparts. As a final remark in this context, Table 37.1 displays the results for the 2D advection model Eq. 37.18. Note especially the small percentage errors of MFE and MFD for *Umax* and *Umin*, whereas FFE ("fixed" FE) and FFD show the well-known damping of the peak of the pulse, and oscillations behind the pulse. However, a user should always be aware of the appearance of grid distortion, whatever method is being used for the grid movement.

In one space dimension moving grid techniques are now well established. Both MFD as (GW)MFE (and other techniques as well) have been applied to a large number of PDE models stemming from various application areas. A clear example to illustrate the difference between the residual-minimization based MFE and the equidistribution-based MFD is given in Figure 37.17. The PDE model belonging to this example is the advection-diffusion equation with

$$L(u) = \delta\frac{\partial^2 u}{\partial x^2} - \frac{\partial u}{\partial x}, \tag{37.35}$$

and $\delta = 10^{-3}$, $u|_{t=0} = e^{-20x}$, $u|_{x=0} = 1$, $u|_{x=1} = 0$. The solutions are oscillation-free for both moving grid methods, but the grids obey completely different criteria.

For parabolic models such as for the 2D spatial operator

$$L(u) = \Delta u + \frac{e^{20}}{4}(2-u)e^{-20/u}, \tag{37.36}$$

with $u|_{t=0} = 1 + \sin^{30}(\pi x)\sin^{30}(\pi y)$ and $u|_{\partial\Omega} = 1$, similar equidistribution-type behavior is observed. In Figure 37.18 grids for both methods are displayed for large points of time (steady-state). The difference between the two grids is mainly reflected in the positioning of the grid points near areas of high first- or second-order spatial derivatives.

It must be noted that (GW)MFE and the deformation method can be formulated, in principle, in "any" space dimension. The main research must therefore be focused on efficient moving grid methods in two and three space dimensions. For (GW)MFE one must realize its connection with the method of characteristics for hyperbolic equations, and as a consequence the possibility of grid degeneration.

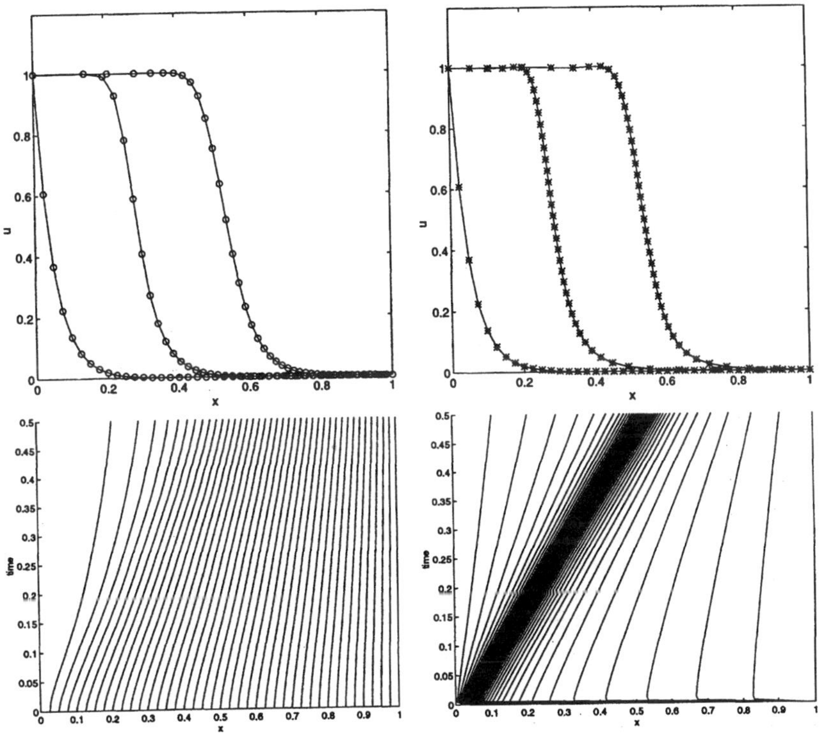

FIGURE 37.17 MFE (left) and MFD (right) results for the 1D advection–diffusion equation (35). Upper two figures show solutions on a moving grid. The lower two figures show the grid movement in time (all runs with $\delta = 10^{-3}$).

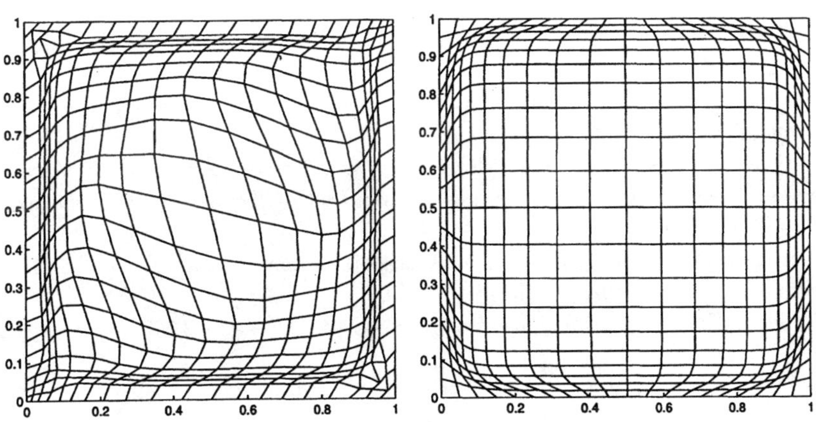

FIGURE 37.18 Steady-state grids for the 2D reaction–diffusion PDE (36); left: MFE, right: MFD.

The MMPDE-approach and the deformation method are new techniques that still have to be examined and tested further. Finally, for general real-life applications, a combination of h- and r-refinement could be beneficial.

Further Information

Papers on moving grid techniques are published in various journals, including the *Journal of Computational Physics, Numerical Methods for PDEs, Applied Numerical Mathematics, SIAM Journal on Scientific*

Computing, *SIAM Journal on Numerical Analysis*, *International Journal for Numerical Methods in Engineering*, and the *International Journal for Numerical Methods in Fluids*.

Proceedings of several conferences and workshops present a number of papers on this subject; for example, *Adaptive Methods for Partial Differential Equations*, SIAM, Philadelphia, 1989, J.E. Flaherty, P.J. Paslow, M.S. Shephard and J.D. Vasilakis, (Eds.), or *Grid Adaptation in Computational PDEs*, as a special issue of *Applied Numerical Mathematics*, 1997.

More detailed are the works of Zegeling [47] for moving finite differences, Carlson et al. [13,14] for moving finite elements, and Thompson [41], Hawken et al. [22] for an overview of moving grid techniques.

Moving grid codes are available at http://www.cwi.nl/gollum/MOVGRD.html and http://www.math.purdue.edu/carlson/. The former is a code (see also Blom et al. [9]) for a general class of time-dependent PDEs using a moving finite difference technique based on equidistribution with smoothing in the spatial and temporal direction. The latter uses a moving finite element technique (see, e.g., Carlson et al. [13,14]) with a gradient-weighted inner product.

References

1. Anderson, D.A., Application of adaptive grids to transient problems, *Adaptive Computational Methods for PDEs*. Babuška, I., Chandra, J., Flaherty, J.E. (Eds.), SIAM, Philadelphia, 1983.
2. Anderson, D.A., Equidistribution Schemes, Poisson generators, and adaptive grids, *Appl. Math. and Comput.* 1987, Vol. 24, pp 211–227.
3. Arney, D.C. and Flaherty, J.E., A Two-dimensional mesh moving technique for time-dependent partial differential equations, *J. Comput. Phys.* 1986, Vol. 67, pp 124–144.
4. Arney, D.C. and Flaherty, J.E., An adaptive local refinement method for time-dependent partial differential equations, *Appl. Numer. Math.* 1989, Vol. 5, pp 257–274.
5. Arney, D.C. and Flaherty, J.E., An adaptive mesh-moving and local refinement method for time-dependent partial differential equations, *Appl. Math. Comp.* 1990, Vol. 5, pp 257–274.
6. Baines, J.J., *Moving Finite Elements*. Clarendon Press, Oxford, 1994.
7. Baines, M.J., Properties of a grid movement algorithm, numerical analysis report 8/95, 1995, University of Reading.
8. Berger, M.J. and Oliger, J., Adaptive mesh refinement for hyperbolic partial differential equations, *J. Compu. Phys.*, 1984, Vol. 53, pp 484–512.
9. Blom, J.G. and Zegeling, P.A., Algorithm 731: A moving-grid interface for systems of one-dimensional time-dependent partial differential equations, *ACM Transactions in Mathematical Software*, 1994, Vol. 20, N3, pp 194–214.
10. Bochev, P., Liao, G., and de la Pena, G., Analysis and computation of adaptive moving grids by deformation, *Numer. Meth. for PDEs*. 1996, Vol. 12, pp 489–506.
11. de Boor, C., Good approximation by splines with variable knots, II, *Springer Lecture Series 363*. Springer-Verlag, NY, 1973.
12. Brackbill, J.U. and Saltzman, J.S., Adaptive zoning for singular problems in two dimensions, *J. Comput. Phys.* 1982, Vol. 46, pp 342–368.
13. Carlson, N. and Miller, K., Design and application of a gradient-weighted moving finite element code, Part I, in 1D, Technical Report 236. 1994, Purdue University.
14. Carlson, N. and Miller, K., Design and application of a gradient-weighted moving finite element code, part II, in 2D, Technical Report 237. 1994, Purdue University.
15. Courant, R. and Hilber, D., *Methods of Mathematical Physics*, Vol 2. Wiley, NY, 1962.
16. Coyle, J.M., Flaherty, J.E., and Ludwig, R., On the stability of mesh equidistribution strategies for time-dependent partial differential equations, *J.Comput. Phys.* 1986, Vol. 62, pp 26–39.
17. DeLillo, T.K. and Jordan, K.E., Some experiments with a dynamic grid technique for fluid flow codes, *Advances in Computer Methods for Partial Differential Equations*. Vichnevetsky, R. and Stepleman, R.S. (Eds.), IMACS, 1987.

18. Dietachmayer, G.S. and Droegemeier, K.K., Application of continuous dynamic grid adaption techniques to meteorological modeling, part I: basic formulation and accuracy, *Monthly Weather Review.* 1992, Vol. 120, N8, pp 1675–1706.
19. Doelman, A., Kaper, T.J., and Zegeling, P.A., Pattern formation in the 1-D Gray–Scott model, *Nonlinearity* Vol. 10, pp 523–563, 1997.
20. Dwyer, H.A., Sanders, B.R., and Raiszadek, F., Ignition and flame propagation studies with adaptive numerical grids, *Combustion and Flame.* 1983, Vol. 52, pp 11–23.
21. Furzeland, R.M., Verwer, J.G., and Zegeling, P.A., A numerical study of three moving grid methods for one-dimensional partial differential equations which are based on the method of lines, *J. Comput. Phys.* 1990, Vol. 89, pp 349–388.
22. Hawken, D.F., Gottlieb, J.J., and Hansen, J.S., Review of some adaptive node-movement techniques in finite-element and finite-difference solutions of partial differential equations, *J. Comput. Phys.* 1991, Vol. 95, pp 254–302.
23. Huang, W. and Russell, R.D., Analysis of moving mesh partial differential equations with spatial smoothing, research report No. 93–17. 1993, Simon Fraser University, Burbaby, B.C.
24. Huang, W., Ren, Y., and Russell, R.D., Moving mesh partial differential equations (mmpdes) based on the equidistribution principle, *SIAM J. Numer. Anal.* 1994, Vol. 31, N3, pp 709–730.
25. Huang, W. and Russell, R.D., Moving mesh strategy based upon a heat flow equation for two dimensional problems, technical report No. 96-04-03, 1996, Dept. of Maths., University of Kansas.
26. Huang, W. and Sloan, D.M., A simple adaptive grid method in two dimensions, *SIAM J. Sci. Comput.* 1994, Vol. 15, pp 776–797.
27. Kansa, E.J., Morgan, D.L., and Morris, L.K., A simplified moving finite difference scheme: application to dense gas dispersion, *SIAM J. Sci. Comput.* 1984, Vol. 5, pp 667–683.
28. Kuprat, A., Adaptive smoothing techniques for 3-D unstructured meshes, *5th International Conference on Numerical Grid Generation in Computational Field Simulation.* Soni, B.K., Thompson, J.F., Haeuser, J., and Eiseman, P. (Eds.), 1996, Starksville, MSU.
29. Lee, D. and Tsuei, Y.M., A modified adaptive grid method for recirculating flows, *Int. J. for Numer. Meth. in Fluids.* 1992, Vol. 14, pp 775–791.
30. Liao, G. and Anderson, D., A new approach to grid generation, *Applic. Anal.* 1992, Vol. 44, pp 285–298.
31. Liao, G. and Su, J., Grid generation via deformation, *Appl. Math. Let.* 1992, Vol. 5, N3.
32. Liu, F., Ji, S., and Liao, G., An adaptive grid method and its application to steady Euler flow calculations, *SIAM J. Sci. Comput.* 1996.
33. Miller, K. and Miller, R.N., Moving finite elements I, *SIAM J. Numer. Anal.* 1981, Vol. 18, pp 1019–1032.
34. Miller, K., Moving finite elements II, *SIAM J. Numer. Anal.* 1981, Vol. 18, pp 1033–1057.
35. Petzold, L.R., A description of DASSL: A Differential/Algebraic System Solver, *IMACs Trans. on Scientific Computation.* Stepleman, R.S. (Ed.), 1983.
36. Petzold, L.R., Observations on an adaptive moving grid method for one-dimensional systems of partial differential equations, *Appl. Num. Math.* 1987, Vol. 3, pp 347–360.
37. Rai, M.M. and Anderson, D.A., Grid evolution in time asymptotic problems, *J. Comput. Phys.* 1981, Vol. 43, pp 327–344.
38. Ren, Y. and Russell, R.D., Moving mesh techniques based upon equidistribution, and their stability, *SIAM J. Sci. Stat. Comp.* 1992, Vol. 13, N6, pp 1265–1286.
39. Semper, W. and Liao, G., A moving grid finite-element method using grid deformation, *Numer. Meth. for PDEs.* 1995, Vol. 11, pp 603–615.
40. Smooke, M.D. and Koszykowski, M.L., Two-dimensional fully adaptive solutions of solid–solid alloying reactions, *J. Comput. Phys.* 1986, Vol. 62, pp 1–25.
41. Thompson, J.F., A survey of dynamically-adaptive grids in the numerical solution of partial differential equations, *Appl. Numer. Maths.* 1985, Vol. 1, pp 3–27.

42. Trompert, R.A. and Verwer, J.G., A Static-regridding method for two-dimensional parabolic partial differential equations, *Appl. Numer. Maths.* 1991, Vol. 8, pp 65–90.
43. Verwer, J.G., Blom, J.G., Furzeland, R.M., and Zegeling, P.A., A moving-grid method for one-dimensional pdes baed on the method of lines, *Adaptive Methods for Partial Differential Equations, SIAM.* Flaherty, J.E., Paslow, P.J., Shephard, M.S., Vasilakis, J.D. (Eds.), Philadelphia, 1989.
44. Yanenko, N.N., Kroshko, E.A., Liseikin, V.V., Fomin, V.M., Shapeev, V.P., and Shitov, Yu A., Methods for the construction of moving grids for problems of fluid dynamics with big deformations, *Lecture Notes in Physics, Springer-Verlag.* 1976, Vol. 59, pp 454–459.
45. Zegeling, P.A., Moving-grid methods for time-dependent parial differential equations, CWI-Tract No. 94, Centre for Mathematics and Comp. Science, Amsterdam, 1993.
46. Zegeling, P.A., Verwer, J.G., and von Eijkeren, J.C.H., Application of a moving-grid method to a class of 1D brine transport problems in porous media, *Int. J. for Numer. Meth. in Fluids.* 1992, Vol. 15, N2, pp 175–191.
47. Zegeling, P.A. and Blom, J.G., A note on the grid movement induced by MFE, *Int. J. for Numer. Meth. in Eng.* 1992, Vol. 35, N3, pp 623–636.
48. Zegeling, P.A., Moving-finite-element solution of time-dependent partial differential equations in two space dimensions, *Comp. Fluid Dyn.* 1993, Vol. 1, pp 135–159.
49. Zegeling, P.A., A Dynamically moving adaptive grid method based on a smoothed equidistribution principle along coordinate lines, *5th International Conference on Numerical Grid Generation in Computational Field Simulation,* Soni, B.K., Thompson, J.F., Haeuser, J., and Eiseman, P. (Eds.), Starksville, MSU, 1996.

Appendix A: Grid Software

Bharat K. Soni, Mississippi State University

A multitude of computer codes to facilitate grid generation strategies involving multiblock structured, overset/chimera, unstructured, hybrid and generalized grids are available in a public domain or commercial world. The information and basic description of these codes was solicited from the grid community at large. This section is devoted to present code descriptions contributed by various sources. In particular, a brief description on the following grid systems is included in this section:

CAGI
CFD-GEOM
CUBIT
FULLMULGRI
GENIE++
GENSY-IGS
GeoMesh 3.4
GRIDGEN
GridTool
GRID2D/3D
GUM-B
HYPGEN
ICEM CFD
IITZeus
LaGriT
MACGS
3DMAGGS
MegaCads
Mesh2d, Mesh3d
MulGri
OVERGRID
PMAG
SiGMA
TGrid 3.0
TIGER
VGM
VGRIDns

0-8493-2687-7/99/$0.00+$.50

Name of Grid Code : CAGI

Outline of Capabilities :

CAGI (Computer Aided Grid Interface) is a software system for preparing surface grids by importing IGES files from the CAD/CAM output or by constructing geometrical entities interactively.

Computing Environment :

CAGI is written in C and FORTRAN 77 utilizing GL and FORMS graphics/GUI libraries and can be executed on most Silicon Graphics workstations.

Code Descriptions :

The CAGI geometry definition and surface grid generation are based on NURBS(Non –Uniform Rational B–Spline). CAGI is developed in a modular fashion using FORMS library for the graphical user interface.The main modules include: (i) Geometry transformer – allows transfer of IGES files to the CAGI and converts the associated geometric entities into standard representation allowing NURBS format; (ii) Geometry manipulation module – contains NURBS tool kit to perform knot insertion,degree elevation,projections,point distribution with desired stretching on curves/surfaces and various transformations useful in preparing surface grids; (iii) Geometry construction – facilitates the generation of widely utilized geometrical entities by interactively supplying parameters to define analytical entities. However, the geometrical entities are automatically converted and stored NURBS format in the database. These modules are linked together with a common graphical interface and graphical visualization module.

Source/Availability :

CAGI is distributed in an executable form to US companies/institutes and can be requested by contacting Bharat K. Soni (bsoni@erc.msstate.edu) at NSF Engineering Research Center, Mississippi State University or Dr. Paul McConnaughey, CFD Branch, NASA/Marshall Space Flight Center(NASA/MSFC), Huntsville, Alabama. CAGI was developed at the NSF Engineering Research Center, Mississippi State University under the sponsorship of NASA/MSFC.

Example Demonstrations :

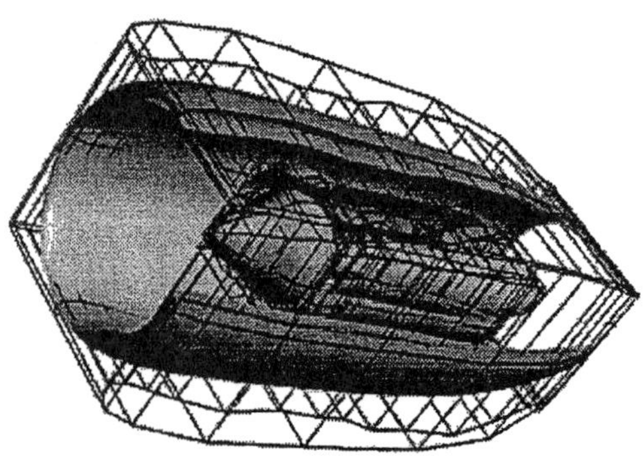

NURBS control net for multiple duct engine

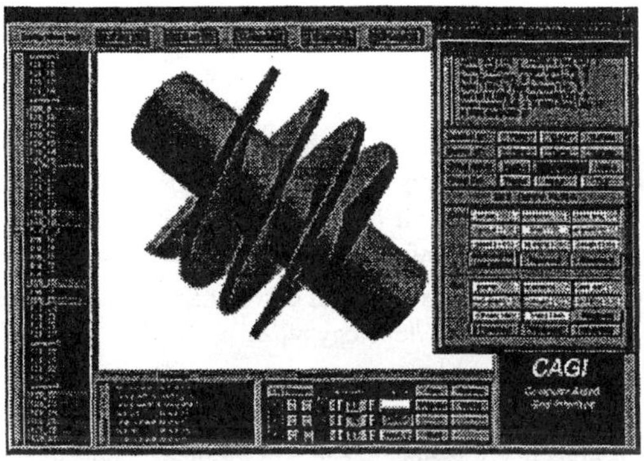

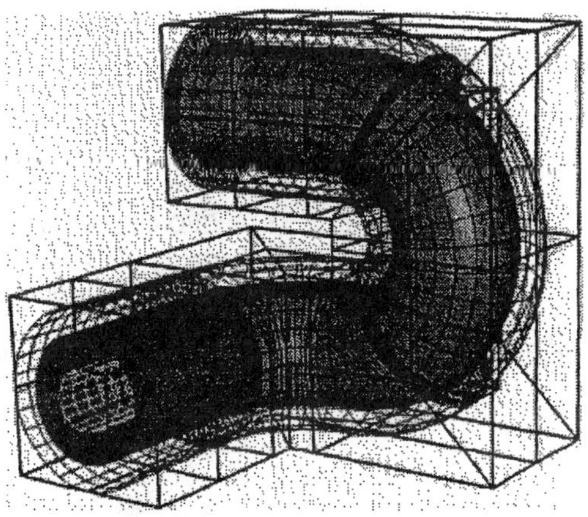

A few configurations using CAGI

References :

[1] http://www.erc.msstate.edu/~bsoni/research .

[2] Yu, T., Soni, B., and Shih, M., "CAGI: Computer Aided Grid Interface", AIAA 95–0243.

[3] Yu, T., and Soni, B., "Application of NURBS in Grid Generation", Journal of Computer Aided Design, Vol.27,No.2, pp. 147–157, 1995.

[4] Yu, T., and Soni, B., "NURBS in Grid Generation", a chapter in the surface generation section of the current CRC Handbook on Numerical Grid Generation.

[5] Yu, T., "CAGD Techniques in Grid Generation", a Ph.D dissertation,Computational Engineering program, Mississippi State University, 1995.

Name of Grid Code CFD-GEOM

Outline of Capabilities

- **Geometry:** CFD-GEOM is based on solid modeling techniques for geometry creation. CFD-GEOM will generate solid models from scratch, import CAD data via IGES format and directly from UniGraphics and Pro-E CAD software, or extract geometry and topology information from external structured surface grids such as Plot3D.
- Automatic "dirty geometry" checking such as point and curve filtering, edge filtering, checking for trimmed surface-set closure, and other tools for geometry/topology repair.
- **Grid:** Variety of grid types supported: multi-block structured, semi-structured prisms, and unstructured (2-D triangles/3-D tetrahedrals) or any combination thereof (hybrid).
- Automatic and instantaneous updating of geometry and grid for model revisions.

Computing Environment

Available on SGI, DEC, IBM, HP, SUN workstations and WIndows NT, Windows-95, Linux PCs

Code Description

The geometry required for grid generation can be created within CFD-GEOM, imported from other CAD packages via the IGES format, or extracted from external structured surface grids. CFD-GEOM generates boundary fitted, structured multi-block and/or unstructured grids, and combinations thereof (hybrids) for arbitrary geometries in two and three dimensions. All geometry and grid entities are connected by a bi-directional relational database allowing manipulations to the geometry or grid entities to be automatically and instantaneously reflected throughout the system.

Grid Generation Process

Grid generation is a two step process, the first step being geometry definition, the second being the actual grid generation itself.

Geometry definition can be achieved either by importing external files or creating internally using CFD-GEOM's geometry modeling capability which uses NURBS technology. Both trimmed and untrimmed representations of geometry is supported by CFD-GEOM.

Structured grid generation is accomplished by a bottom-up approach: edges and faces are defined first coinciding with curves and surfaces on given geometry from which block-grids are readily built. CFD-GEOM supports multi-block structured topology definition using the Trans-Finite Interpolation schemes (TFI) for grid generation. It offers various TFI schemes for generating surface and volume grids including enforced boundary orthogonality and smoothing algorithms. User is given the option of projecting spatial TFI grids to surface sets. Multi-to-one interface matching and grid quality assessment are available.

For unstructured grid generation, the user defines a domain from a watertight trimmed model, for which to create a grid. Surface grid generation is accomplished by an automatic and controllable algorithm which grids each trimmed surface with due regard to appropriate clustering for geometric features. The tetrahedral grid is generated by an automatic Advancing Front Algorithm to fill the volume. Semi structured prismatic grids are generated by simple extrusion and revolution of the triangulation or by advancing a triangulation along the desired face normals. Any of the three methods can be combined in order to form a hybrid grid system.

Examples:

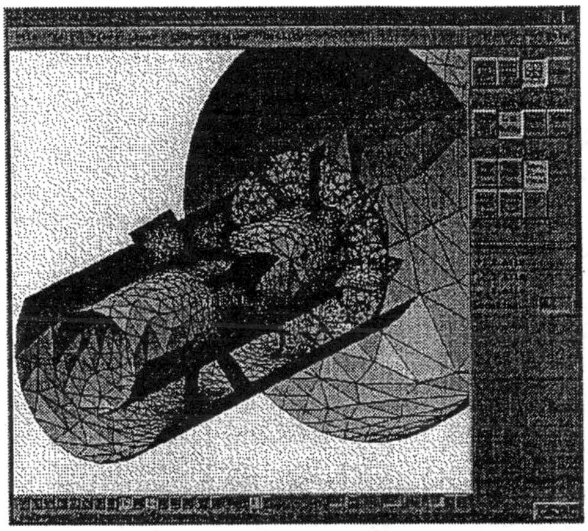

Unstructured grid example showing surface triangulation and slice through volume grid

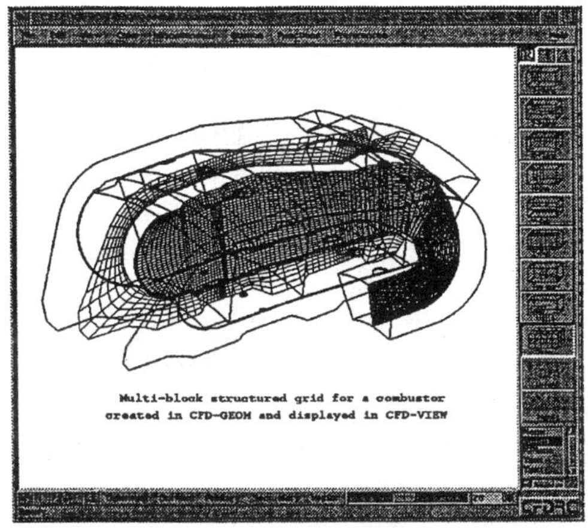

Multiblock structured grid for an aircraft combustor

Source/Availability

CFD Research Corporation, 215 Wynn Drive, Huntsville, AL 35758

Tel: (205) 726-4800
Fax: (205) 726-4806

Key References

1. Hufford G.S., Harrand V.J., Patel B.C. and Mitchell C.R., NASA-CP-3291, May 1995

2. Hufford G.S. and Mitchell C.R., AIAA-95-015, Jan. 1995

3. Coirier W.J. and Powell K.G., AIAA-95-0566, Jan. 1995

4. Hufford G.S., and Harrand V.J., AIAA-96-0053, Jan. 1996

5. Hufford G.S., Mitchell C.R., and Harrand V.J., AIAA-96-0096, June. 1996

6. Whitmire J.B., Dollar T.W., AIAA-98-0219, Jan 1998

Remarks

CFD-GEOM is a comprehensive grid-generation system which provides fast and efficient algorithms for high quality structured, unstructured, and hybrid grid systems for complex geometries. Its tight integration between geometry and grid generation modules, and other uniques features, gives users speed and flexibility in generating grids for various designs, expediting parametric studies.

Name of Grid Code: CUBIT

Outline of Capabilities: A two- and three- dimensional unstructured mesh generation tool for fully automatic hexahedral meshing on complex geometries.

Computing Environment: CUBIT is written in C++ and executes on most UNIX operating systems. It is used daily on Sun Solaris, HP, and Silicon Graphics workstations.
CUBIT uses mouse-driven HOOPS-based graphics in an x-window environment and a command line interface, and can be run in both interactive and batch modes.

Code Description: CUBIT features geometry creation, algebraic command preprocessing, geometry consolidation and decomposition, hexahedral (and some tetrahedral) mesh generation, and high-order finite element types. Geometry can be imported, created, and/or modified using a Virtual Geometry Interface to solid modeling engines such as ACIS and ProE. The interface features a unique merging capability for non-manifold geometry and contiguous meshing. CUBIT can also create simple geometries based on internal primitives. Geometry can be discretized into a finite element mesh using a combination of techniques including paving, mapping, submapping, sweeping, whisker weaving, and mixed element hex/tet plastering. Mesh quality can be enhanced using state-of-the-art smoothers and optimization techniques. Mesh refinement may be achieved via a dicing algorithm. Physical boundary conditions can be applied to the mesh through the geometry to generate analysis input files. CUBIT exports meshes in Exodus II format.

Grid Generation Process: Using a command line interface and an x-windows environment, the modeler imports geometry or creates his own inside CUBIT. The geometry is manipulated via Boolean imprinting and decomposition operations to make it amenable to meshing. Various meshing schemes are assigned to surfaces and volumes and then the model is meshed. The mesh is evaluated via mesh quality measures. Boundary conditions can be assigned and the model exported. The goal of the CUBIT Project is to reduce total time to mesh via full-automation

Example. The geometry in the figure to the right was created via ACIS to create prisms and octahedral solid bodies. Global element sizing and interval settings were specified to ensure consistent mesh density. Simple mapping algorithms were then all that was needed to create the mesh. The journal ile for this mesh was about 10 lines long.

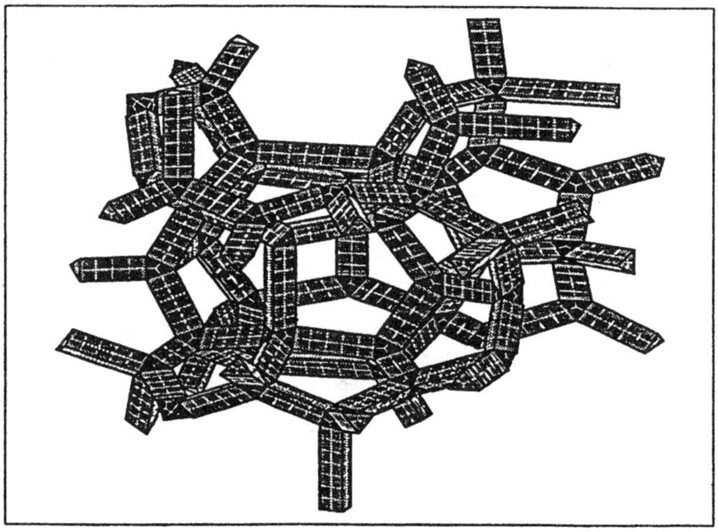

Source/Availability: CUBIT is distributed in statically linked executable form for each supported platform. For information concerning CUBIT liscencing contact

Marilyn K. Smith
Technology Programs Department
Division 1503, MS 0833
Sandia National Laboratories
P.O. Box 5800
Albuquerque, NM 87185-0833
Fax: (505)-844-9297, Email: mksmith@sandia.gov

For further technical information contact

Timothy J. Tautges
Parallel Computing Sciences Dept.
Sandia National Laboratories
M/S 0441, P.O. Box 5800
Albuquerque, NM 87185-0441
Ph: (505)-844-5388
Email: tjtautg@sandia.gov.

Key References:

T.D. Blacker, M.B. Stephenson, "Paving: A new approch to automated quadrilateral mesh genertqaion," Int. J. Numerical Methods, 32:811-847, 1991.

T.J. Tautges, T.D. Blacker, and S.A. Mitchell, "The Whisker Weaving Algorithm. A Connectivity-Based Method for Constructing All-Hexahedral Finite Element Meshes," Int. J. Num. Meth. 1996.

Web Page. http://sass577.endo.sandia.gov/SEACAS/CUBIT/cubit.html

Grid Code: *FullMulgri*
Rosa Maria Spitaleri
Istituto per le Applicazioni del Calcolo-V.le del Policlinico 137, Rome, Italy

FullMulgri is a modular code which couples elliptic grid generation with full multigrid computation [1,2,3,5]. The algorithm adds to the multigrid classical cycles, starting from a given finest grid G^M, a computing process which starts on the coarsest grid G^0, solving or relaxing the restricted problem, and comes back to the finest grid by interpolating and executing MG cycles for each incremented level k, as shown in Figure 1. Since it combines nested iteration with multigrid cycles, MG^{λ}-cycle, (λ=1 $\subset$ V-cycle, λ=2 $\subset$ W-cycle) in a full multigrid MG^{λ}-cycle, we have called this procedure *FullMulgri* [3,4]. It leads to obtain better performances. Being G^0 the coarsest grid the solution or relaxation processes are computationally very cheap. Experimental results are available on how many iterations can be saved by adding full computation to the classical multigrid cycling. Indeed a full multigrid MG^{λ}-cycle is more expensive than a single MG^{λ}-cycle, although the difference in the cost decreases for higher-dimensional grid generation problems. Remembering that usually υ_0 relaxation sweeps are used on the coarsest grid G^0 with $\upsilon_0>1$ we know how taking into account the related cost in evaluating algorithm performance. On the light of performance evaluation results the algorithm seems to be very promising to reduce computational cost [4].
The main module *FullMulgri* controls the code *Mulgri* , with separated grid code description in this book, as its main subroutine. The code, written in FORTRAN 77, is portable in computational environments equipped with the usual libraries for mathematical applications. The user can easily integrate the code with the visualization modules in the last release of *Visgen* [2], running on Silicon Graphics Workstations.
Let $G^0, G^1, \ldots, G^M$ be a sequence of M+1 grids, with decreasing mesh sizes: $h_0, h_1, \ldots, h_M$, $h_k=2h_{k+1}$ for k=0,..., M-1. The code solves the transformed grid generation quasi-linear system

$$\mathbf{LX=0} \quad \mathbf{in}\ \Omega^* \quad , \quad \mathbf{\Lambda X=\Phi} \quad \mathbf{on}\ \partial\Omega^* ,$$

in the Cartesian co-ordinates $\mathbf{X}$ (ξ,η), where $\partial\Omega^*$ is the boundary of the computational domain Ω^*. Let us give the related discrete equations

$$\text{(1)} \qquad \mathbf{L^kX^k=\hat{F}^k} \quad \mathbf{on}\ \mathbf{G^k} \quad , \quad \mathbf{\Lambda^kX^k=\hat{\Phi}^k} \quad \mathbf{on}\ \partial\mathbf{G^k}$$

where

$$\mathbf{\hat{F}^k = L^k(I^k_{k+1}X^{k+1}_a) + I^k_{k+1}(\hat{F}^{k+1} - L^{k+1}X^{k+1}_a)}$$
$$\mathbf{\hat{\Phi}^k = \Lambda^k(I^k_{k+1}X^{k+1}_a) + I^k_{k+1}(\hat{\Phi}^{k+1} - \Lambda^{k+1}X^{k+1}_a)}.$$

$\mathbf{X^{k+1}_a}$ is the current approximation on the finer grid G^{k+1} and I^k_{k+1} an appropriate fine-to-coarse transfer operator (restriction).
Focusing on a couple of grids k+1,k,we can specify the related relaxation and multigrid operators in the following forms:

$$\text{Relax}^{\upsilon_i}(\,\bullet\,;L^{k+1},F^{k+1},\Lambda^{k+1},\hat{\Phi}^{k+1})\ ,\ i=1,2, \qquad MG^{\lambda}(\,\bullet\,,k+1,L^{k+1},F^{k+1},\Lambda^{k+1},\hat{\Phi}^{k+1}).$$

On the coarsest grid G^0 often solution procedures or relaxation sweeps are applied, therefore we denote the related operators

$$\text{Solve}(\,\bullet\,;L^0,F^0)\ , \qquad \text{Relax}^{\upsilon_j}(\,\bullet\,;L^0,F^0,\Lambda^0,\hat{\Phi}^0)\ ,\ j=0,1,2.$$

Let be λ=1 $\subset$ V-cycle. We define the following full multigrid algorithm (Fig 1):
step1 starting from the coarsest level k=0 we compute the approximation $\overline{X}^0$ by

solving or relaxing, an appropriate number of times υ_0, the system $L^0X^0=F^0$, $F^0=0$,

step2 k=k+1; we apply an appropriate interpolation operator $\bar{I}_k^{k+1}$ to the coarser approximation X^k and we obtain the approximation X^{k+1} on the finer grid,

step3 we apply a multigrid V(2,1)-cycle by assuming X^{k+1} as initial approximation in the pre-smoothing phase and we compute the better approximation $\tilde{X}^{k+1}$,

step4 If k=M then the algorithm stops and $\tilde{X}^M$ is the required initial guess. If k<M then we come back to the step 3.

The above algorithm is initial guess generator for the multigrid grid generation algorithm in *Mulgri* [4]. The main computational components of the resulting Full -FAS multigrid algorithm are :

- coarser grid construction (hierarchy of grids) → meshsize doubling (standard coarsening)
- relaxation on the coarsest grid G^0 ($\text{Relax}^{\upsilon_0}(\,\bullet\,;L^0,\hat{F}^0,\Lambda^0,\hat{\Phi}^0)$) → Gauss-Seidel relaxation,
- FMV interpolation ($\bar{I}_k^{k+1}$)→piecewise linear(Fig.1)
- $MG^{\lambda}(\,\bullet\,;L^{k+1},\hat{F}^{k+1},\Lambda^{k+1},\hat{\Phi}^{k+1})$ → $V(2,1)(\,\bullet\,;L^{k+1},\hat{F}^{k+1},\Lambda^{k+1},\hat{\Phi}^{k+1})$
- $FAS^{\lambda}(\,\bullet\,;L^{k+1},\hat{F}^{k+1},\Lambda^{k+1},\hat{\Phi}^{k+1})$ → $V(2,1)(\,\bullet\,;L^{k+1},\hat{F}^{k+1},\Lambda^{k+1},\hat{\Phi}^{k+1})$

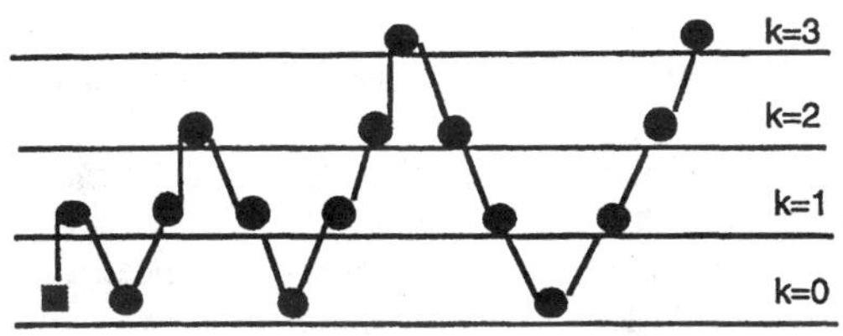

Figure 1. Full Multigrid V-cycle

- relaxation for error smoothing ($\text{Relax}^{\upsilon_1}$, $\text{Relax}^{\upsilon_2}$) → Gauss-Seidel relaxation ,
- relaxation orderings → lexicographic , red-black, four-colours, zebra-line
- restriction (fine-to-coarse-grid transfer operator) (I_{k+1}^k, II_{k+1}^k) → straight injection
- prolongation (coarse-to-fine-grid transfer operator) (I_k^{k+1}) → piecewise linear interpolation

References

[1] A. Brandt. Multi-level Adaptive Solutions to Boundary-Value Problems, *Math. Comp.* 31: 333-355 (1977).

[2] A. Brandt. , Guide to Multigrid Development, in Multigrid Methods,Lect.Notes Math. 960, Springer Verlag,1982

[3] W. Hackbusch, Multigrid Methods and Applications, Springer Verlag, Heidelberg 1985.

[4] R.M. Spitaleri, Full-FAS Multigrid Grid Generation Algorithms, *Proceeding IMACS -WC97* (A.Sydow Ed.) 2:231-236, 1997

[5] J. F. Thompson, Z. U. A. Warsi and C. W. Mastin. Boundary-Fitted Coordinate Systems for Numerical Solutions of Partial Differential Equations - A Review, *J. Comp. Phys.* 47: 1-108 (1982).

Name of Grid Code : GENIE++

Outline of Capabilities :

GENIE++ is a two– and three– dimensional multiblock structured grid generation system for simple to moderately complex configurations. The code can be executed semi–interactively or in a batch mode in computational or geometry domain. The journal file is created automatically containing all functional operations during the interactive mode.

Computing Environment :

GENIE++ is written in C and FORTRAN 77 utilizing GL and OpenGL graphics libraries and can be executed on most UNIX operating systems. A separate version is available which utilizes LINUX and MESA/OpenGL for graphics to execute on PCs.

Code Descriptions :

The grid generation is accomplished by using several techniques either separately or in combination to quickly and economically generate sculptured geometries and grids on/around arbitrary configurations. Initial structured grids are performed using variety of algebraic techniques. Grid clustering is accomplished with either exponential or hyperbolic tangent stretching functions which allows actual/relative desired point distribution. Grid smoothness and quality enhancements can be accomplished by using an elliptic solver with proper forcing functions or B–Spline based algebraic techniques. B–Spline and NURBS based algorithms are employed for surface definition, redistribution and for smoothing boundary/surface/volume grids. The built–in sculptured geometry construction facility, automatic curve/surface generation for interior boundaries/surfaces, NURBS based surface redistribution, weighted transfinite interpolation schemes based on lagrange/hermite interpolators, control function evaluations based on the optimal combination of algebraic grid and hyperbolic equations and on–line graphical visualization of the generation process are the salient features of GENIE++ which result in a significant time savings for a simple/moderately complex geometry grid applications.

Grid Generation Process :

The generation process utilizes a menu–driven text window for user dialogues with the computer and graphics window for process visualization. The user dialogues with the system are carried out allowing command line I/Os in the text window and transformation and other options are allowed in the graphics window. At a given instant the user can be in physical mode constructing sculptured geometrical entities or in the computational mode defining grid points/lines/surfaces/volumes utilizing hierarchies commands and functional options. The notation G(i1–i2,j1–j2,k1–k2) is used to identify point/curve/surface/volume in the computational mode. The construction/generation process consist of execution of boundary/surface/volume definition, grid measures, I/O, utility and multiblock strategy modules.

Source/Availability :

GENIE++ is distributed in an executable form and can be requested by contacting Bharat K. Soni (bsoni@erc.msstate.edu) at NSF Engineering Research Center, Mississippi State University.

Example Demonstrations :

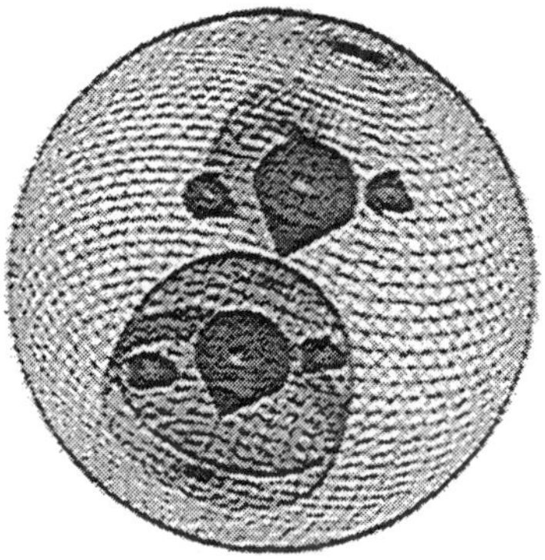

A chimera grid

Full volume grid of a fighter aircraft

References :

[1] http://www.erc.msstate.edu/~bsoni/research .

[2] Soni, B., "Grid Generation for Internal Flow Configurations", Computers and Mathematics with Applications, Vol.24, No. 5/6, pp. 151–163, 1993.

[3] Soni, B., Thompson J.F., Stokes, M., and Shih, M., "GENIE++, EAGLEView, and TIGER : General and Special Purpose Graphically Interactive Grid Systems" AIAA–92–0071.

[4] Soni, B., "Elliptic Grid Generation Systems: Control Functions Revisited", Journal of Applied Mathematics and Computations, 65:1–2, pp 345–354, 1994.

[5] Soni, B., and Thornburg, H., "CAGD Techniques in Structured Grid Generation and Adaptation", Lecture Notes in Physics, Numerical Methods in Fluid Dynamics, Springer, Deshpande, et al(eds.), pp. 166–172,1994.

GENSY- IGS
Rosa Maria Spitaleri
Istituto per le Applicazioni del Calcolo-V.le del Policlinico 137, Rome, Italy

GENSY is an Interactive Graphics System (IGS) which consists of integrable software modules to generate, visualize and evaluate two- and three-dimensional grids. Both the expert and novice users can interactively apply the numerical methods and easily assign and update method parameters. They can also easily require visualization techniques, to exert both visual control over numerical computation and computed grids, and therefore modify and optimize the generation process on the base of the obtained visual information.
The Graphical User Interface (GUI) has been designed by the library FORMS which collects advanced graphics tools of the standard-de-facto Graphical Library (GL) of the Silicon Graphics computing environment. Therefore it can be considered standard itself and portable. Being mixed-type this GUI includes several interaction tools (menus, dialogue boxes, buttons, icons) as Figure 1 shows. Among the GUI attributes we underline the friendliness of its page layout, which is self explaining and fast-to-familiarize [1,2,3].
In dealing with two-dimensional problems, the following sequence describes the main grid generation processes of the system:

- ***multiblock grid generation:***
  - a- subdivision of the physical domain Ω into an appropriate number of pieces, called blocks, with appropriate connections, by defining artificial boundaries in Ω,
  - b- mapping of the physical domain Ω on the transformed domain Ω^* which is an appropriate assembly of $N \geq 1$ computational blocks B_l^*, $\Omega^* = \bigcup_l B_l^*, l = 1,\ldots,N$,
  - c- definition of criteria of solution acceptance,
  - d-***elliptic grid generation*** ,
    - 1-definition of the Poisson system (1) $\Delta\Xi = P(\Xi)$, $\Xi=(\xi,\eta)$ on the current block B_l,
    - 2-activation of sources of attraction by defining the control function $P(\Xi)$ in (1),
    - 3-solving, in the rectangle B_l^*, the transformed system :

$$\alpha x_{\xi\xi} - 2\beta x_{\xi\eta} + \gamma x_{\eta\eta} + J^2(Px_\xi + Qx_\eta) = 0$$
$$\alpha y_{\xi\xi} - 2\beta y_{\xi\eta} + \gamma y_{\eta\eta} + J^2(Py_\xi + Qy_\eta) = 0$$

where

$$\alpha = x_\eta^2 + y_\eta^2, \quad \beta = x_\xi x_\eta + y_\xi y_\eta$$
$$\gamma = x_\xi^2 + y_\xi^2, \quad J = x_\xi y_\eta - x_\eta y_\xi$$

- ***one-*** or ***multi-grid computation***
  - application of a non-linear either onegrid or multigrid full approximation storage (FAS) solution algorithms,
  - application of non-linear relaxation algorithms with appropriate orderings: lexicographic, red-black, four-colors, zebra-line,
  - definition of criteria for anisotropy control,
  - choice of a multigrid cycle (V or W),

4-updating of values on artificial boundaries of the next neighbouring block,

e- iteration over all the blocks to achieve the desired solution on the whole domain according to the connection conditions.

- ***isoparametric grid generation***
  - a- subdivision of the physical domain Ω into an appropriate number of pieces, called regions, with appropriate connections, by defining artificial boundaries in Ω,
  - b- definition of the regions as isoparametric elements with four or eight nodes,
  - c- assignment of row and column numbers for gridding each regions,
  - d- ***generation of quadrilateral or triangular elements***
    - 1- generation of quadrilateral or triangular elements saving element connections,
    - 2- renumbering of elements and element nodes.

To illustrate system application we refer to the well known test problem called 2D double ellipsoid and the 3-block discretization in Figure 1. The step sequence providing the screen layout in this Figure is: —› ***File*** (storage of the problem geometry) —› —› ***Grid*** —› convergence history visualization √ *Elliptic MG Generation* —› *Multigrid Method* —›

Icon —› Dialogue box —› multigrid button —› the finest grid dimensions (16 and 64) and grid number (2) writing—› grid number updating (2 —›3 —›1)[1].

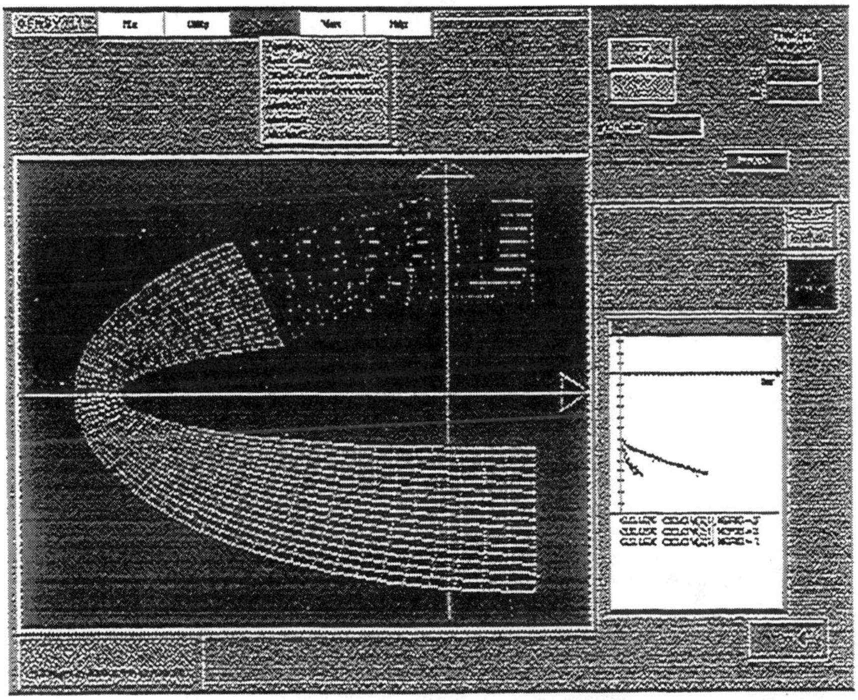

Figure 1. GENSY: Graphical User Interface

The system should be considered always in progress so that co-operations for system development and application are welcome. Multigrid substructure can be also easily transferred to existing grid generation systems or updated to handle others applications.

References

[1] M.Sagona, R.M.Spitaleri, A Guide to GENSY, *Quad.IAC* n. 18, Roma, 1997

[2] R.M. Spitaleri , R.Cossu and A. R. Caldarone, GENSY: A Module for Multiblock Multigrid Grid Generation, in *Numerical Grid Generation in Computational Field Simulations*, (B.K.Soni, J.K.Thompson, J. Hauser and P.Eiseman Eds), Mississipi State University, MS, pp. 459-468, 1996.

[3] R.M.Spitaleri , A.R. Caldarone, Interactive Multiblock Multigrid Grid Generation by GENSY, *Proc. ESM97*, (A. R. Kaylan and A. Lehmann Eds), pp. 180-184, 1997

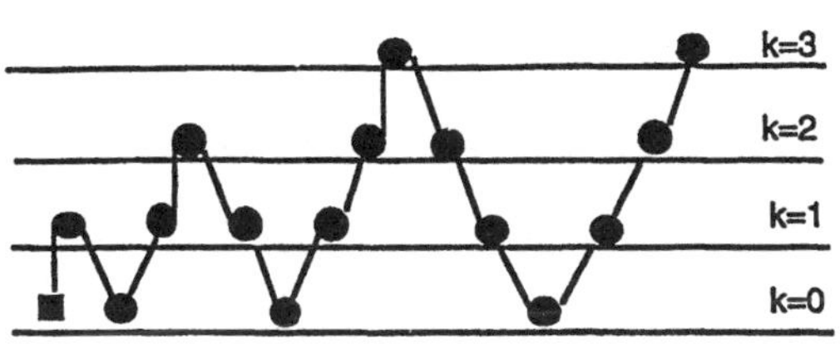

Geometry may be layered to manage large models effectively

Existing geometries may be merged into other geometries

Mesh Generation

Four mesh types: 2-D quadrilaterals, 3-D hexahedra, 2-D triangular meshes, 3-D triangular-surface meshes

Multiple transfinite interpolation options: Eriksson, Hermite-Orthogonal

Multiple smoothing options: Laplace, Poisson, Thomas-Middlecoff

Geometry-independent: generate complicated meshes with little or no predefined geometry

Handle "dirty" geometries without difficulties or additional work

True surface projection: When surface geometry is available, mesh surfaces can be projected directly onto them

Powerful, interactive tools for creating and modifying mesh topology

Point-and-click to assign boundary zones

Unrestricted, interactive assignment of edge node distributions

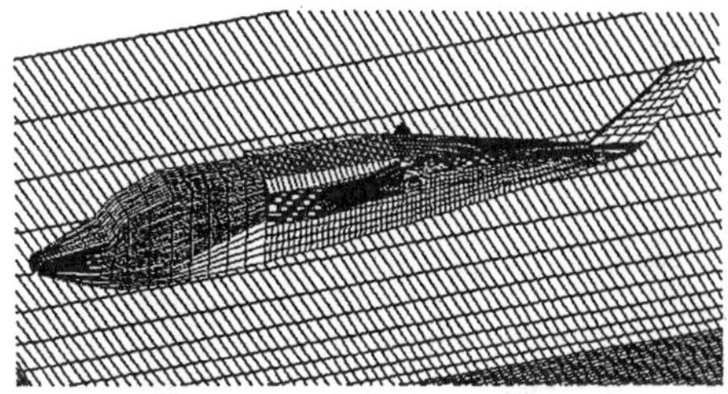

Surfaces of structured mesh around helicopter. Original geometry imported from PATRAN.

Mesh Diagnostics & Editing

Histogram display of aspect ratio, skewness, determinant, Jacobian, angle deviation, adjacent side ratio, warp

Interactive pan/sweep of contours through mesh

Interactive highlighting of mesh elements with diagnostic values in selected ranges

Interactive repositioning of mesh nodes

Supported Hardware

GeoMesh is supported on most common UNIX, Windows 95, and Windows/NT platforms. Please contact Fluent Inc. for details.

Triangular mesh on a low pressure turbine passage

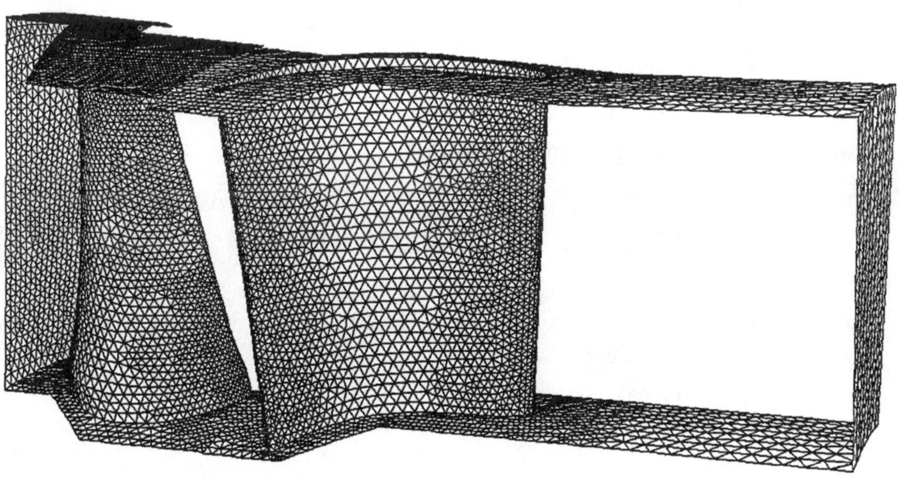

FLUENT INCORPORATED

10 Cavendish Court, Lebanon, NH 03766
tel: 800-445-4454 • fax: 603-643-3967
email: info@fluent.com • www.fluent.com

Name of the Grid Code Gradmesh

Outline of Capabilities The whole code is written in FORTRAN-77, so it can be practically implemented on any computer. The program can generated graded meshes in multiconnected domains, whose boundary is modelled by using B-spline curves. The density of the mesh is defined by a scalar function having its domain the closure of the considered domain. After grid generation and triangulation the Laplacian smoothing is applied. The variety of points insertion criteria ensure good grid quality.

Computing Environment DOS, UNIX systems having standard FORTRAN-77 compiler, The code uses as INPUT data and then produces file OUTPUT containing all necessary information about the mesh.

Code description The code contains 2353 lines of FORTRAN-77 statements. It has got separate procedure of reading and writing DATA. Input DATA must be prepared on file INPUT, thus results are transfered to the file OUTPUT. The following COMMON blocks are used:// input common block:

common/gridin/XA,YA,XB,YB,NLOOP,

D,NSEGLP(20),P(2,1000),EPS

output common block:

common/griout/NINNOD,NBDNOD,PINT(2,20000),PBOUND(2,5000),

1NSEGG(2000), NLOOP1,NTRIAN(3,40000),LTRIAN,NEIGHB(5000)

and scratch common blocks:

common/work/WORK(40000)

common/work/NONDLN(2,10000),LNONDL,IDELAU(2,10000),LDELAN

1 ,NINTP(20000),LINTP,NCAND(900),LCAND,RX(900),ANG(90)

common /work/NSURO(1000),LSURO

common/work/NONDLN(2,10000),LNONDL,IDELAU(2,10000),LDELAN

1 ,NINTP(20000),LINTP,NCAND(90),LCAND,RZ(90),ANG(90)

The whole code cosists of the collection of procedures performing reading data, points generation, triangulation, smoothing and results writing. The program uses a user supplied procedure:

FUNCTION RO(x,y,i)

evaluating the mesh density function.

Grid Generation Process The algorithm of mesh generation and triangulation can be divided into the following steps:

- boundary points generation with a given mesh density function ρ,
- internal points generation according to a given mesh density function ρ,
- triangulation of the domain on the previously obtained grid points,
- Laplacian smoothing of the mesh.

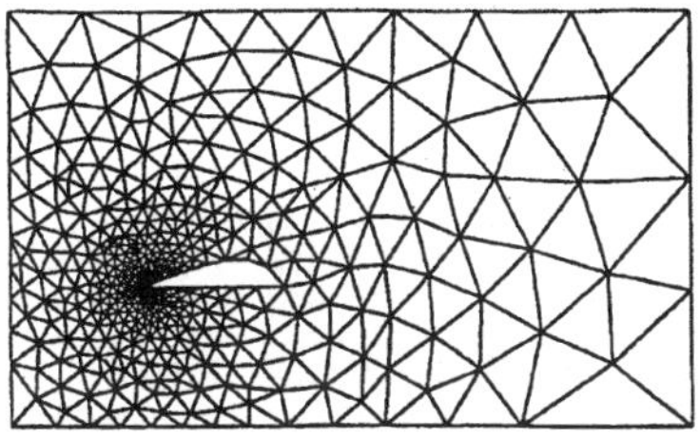

It is important to emphasize that triangulation is performed after points generation.

Example Demonstration Let consider the following INPUT file:

```
6 6 20
-0.5 -0.5 2 -.5 2 1 -0.5 1 0.5 0 0 0
2
4 1 2 3 4
2 5 6
1 0 1 2 1
2 0 2 3 1
3 0 3 4 1
4 0 4 1 1
5 0 5 6 1
6 2 6 5 1
3
0 0 0.4 0.2 0.5 0.0
3
```

The OUTPUT file looks as follows:

```
"the number of nodes"
"coordinates of nodes"
"for every element the number of its nodes and numbers of neighbors along all sides
```

The FUNCTION RO(x,y,i) looks as follows:

```
double precision function ro(x,y,i)
implicit real*8(a-h,o-z)
implicit integer*4 (i-n)
r1=sqrt(x*x+y*y)
ro=0.2*r1+0.01
return
end
```

The obtained mesh is shown in the Fig.1.

Source/Availability The source can be obtained by Email: jkucwaj@usk.pk.edu.pl

Name of Grid Code : Gridgen

Outline of Capabilities :

Multiple block structured grids on imported or internally created geometry.

Computing Environment :

Unix workstations (Silicon Graphics, IBM, Hewlett–Packard, DEC, Sun)

Code Descriptions :

Gridgen is used by engineers and scientists to construct meshes for use with CFD and finite–element analysis software. In use since 1984, it is recognized for its coherent graphical interface and for giving its users fine control of grid spacing, orthogonality, and smoothness, all of which affect the accuracy of analysis results.

Gridgen is a complete preprocessing toolkit that includes geometry modeling, IGES import, grid generation, and direct interfaces to analysis software packages including TEAM, NPARC, VSAERO, INCA, GASP, CFX, FLUENT, FLUENT/UNS, RAMPANT, TASCflow, STAR–CD, Phoenics, PATRAN, and NASTRAN.

Grids are built in Gridgen beginning with curves, the lowest dimensional level of the grid hierarchy and proceeding to surface and volume elements. The geometry model, called a database in Gridgen, serves as the foundation of the grid hierarchy. The engineer constructs curves on top of the database. The curves are in turn used to build the surface and volume components. Because the geometry model and the curves, surfaces, and volumes of the grid are each distinct entities changes to grid or geometry can be propogated forward and backward through the grid system.

Gridgen automates as much of the grid generation minutiae and bookkeeping as possible so that the engineer can concentrate on topology and grid quality.

Source/Availability :

Gridgen is available in executable format only. Gridgen is a fully supported commercial grid generator available from Pointwise, Inc. See http://www.pointwise.com for more information.

Example Demonstrations :

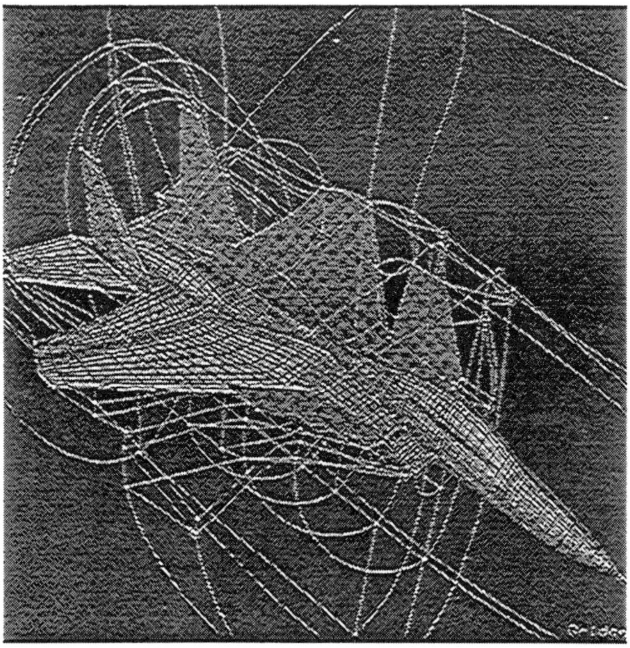

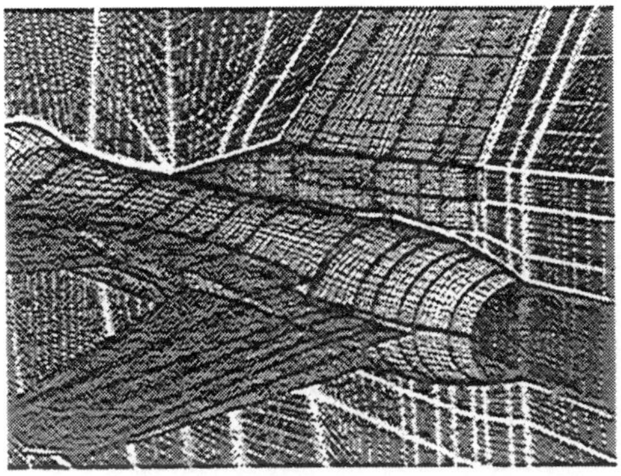

Example 2

The second example is a multiple block grid for Navier–Stokes analysis of engine nozzle performance on an F–16 aircraft. The symmetry plane, the nozzle exhaust and the aircraft surface grid are shown in different shades of grey.

References :

[1] Gridgen User Manual, Version 12; Pointwise, Inc.; Fort Worth, TX; July, 1997.
[2] Steinbrenner, J.P. and Chawner, J.R.; "Gridgen's Synergistic Implementation of CAD and Grid Geometry Modeling," Numerical Grid Generation in Computational Fluid Simulations, ed. by B.K. Soni, et al., Mississippi State University, April 1996.
[3] Steinbrenner, J.P. and Chawner, J.R.; "Automatic Structured Grid Generation Using Gridgen (Some Restrictions Apply)," NASA CP–3291; May, 1995.

GridTool

A Surface Modeling and Grid Generation Tool

(http://geolab.larc.nasa.gov/GridTool/)

GridTool is designed primarily for use with unstructured grid generation systems (e.g. TetrUSS http://ad–www.larc.nasa.gov/tsab/tetruss/). Most grid generation programs represent geometry by a set of structured points which is not consistent with surface representations used in Computer Aided Design (CAD) systems. The purpose of GridTool is to bridge the gap between these two software types. GridTool uses Non–Uniform Rational B–Spline (**NURBS**) surfaces to represent the geometry. These surfaces can be read from an **IGES** file or created internally. Currently, GridTool supports the VGRID (http://www.vigyan.com/vigyan/vgrid.html) and FELISA (http://ab00.larc.nasa.gov/~kbibb/felisa.html) systems, and can be extended easily to support other unstructured grid generation

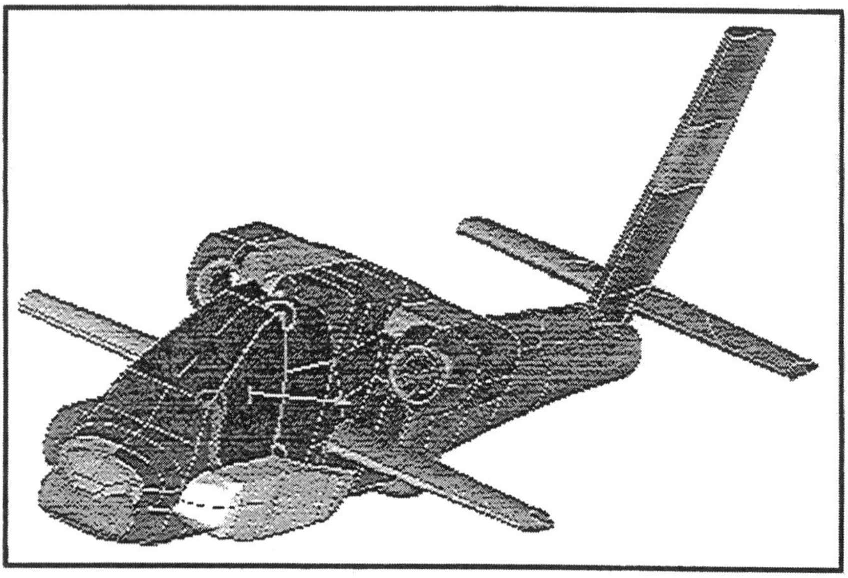

Potential Commercial Uses

- Design and analysis of transportation vehicles such as aircraft, automobiles, trucks,ships, and submarines
- Biomedical technology
- Design of turbomachinery and pumps.
- Analysis of flow through pipes and ducts.
- Design of internal combustion engines.
- Heat transfer analysis and design of heat exchangers.
- Structural analyses using finite element methods.
- Electromagnetic applications.

Benefits

- IGES Compatible
- Non–Uniform Rational B–Spline (NURBS)
- Fast and convenient generation of grids around complex configurations.
- Quick design cycle turn–around.
- Usable by non–experts.
- Minimal training required.
- Applicable to a wide variety of 2– and 3–D configurations.

The Technology

GridTool is an interactive program. It is written entirely in ANSI "C", the interface is based on the FORMS library , and the graphics is based on the GL library. The code has been tested successfully on IRIS workstations running IRIX 5.3 and above. The memory is allocated dynamically, therefore, memory size will depend on the complexity of geometry/grid. GridTool data structure is based on a link–list structure which allows the required memory to expand and contract dynamically according to the user's data size and action. Data structure contains several types of objects such as points, curves, patches, sources and surfaces. GridTool is free and is only available in US. GridTool has on–line documentation available through www. (http://geolab.larc.nasa.gov/GridTool)

Key Words

Computer Aided Design (CAD), NonUniform Rational B–Spline (NURBS), Numerical grid generation, unstructured grids, Computational Fluid Dynamics, Finite Element Analysis

Reference

Samareh, Jamshid, "GridTool: A Surface Modeling and Grid Generation Tool," Proceedings of the Workshop on Surface Modeling, Grid Generation, and Related Issues in CFD Solutions, NASA CP–3291, May 9–11, 1995.

Contact

If your company is interested in this technology, or if you desire additional information, please contact:

Jamshid A. Samareh Mail Stop 159
Langely Research Center (NASA)
Hampton, VA 23681–2199
Phone: (804) 864–5776
Fax: (804) 864–9713
E–mail: j.a.samareh@larc.nasa.gov

World–Wide Web information about this technology is available at URL:

http://geolab.larc.nasa.gov/GridTool/

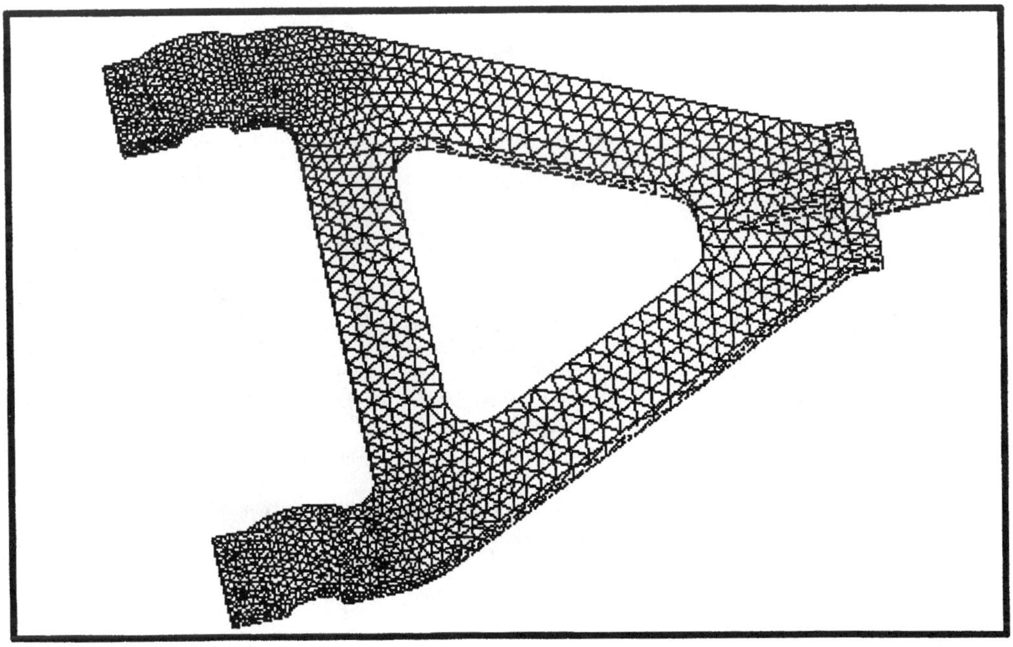

GRID2D/3D – An Algebraic Grid Generator

Tom I-P. Shih
Department of Mechanical Engineering
Carnegie Mellon University

GRID2D/3D is a code for generating single- and multi-block structured grids in two- and three-dimensions by using transfinite interpolation. Capabilities include: (1) ability to generate lines and surfaces that describe the problem geometry or the domain decomposition for multi-block, (2) ability to control grid distribution in an explicit manner, (3) ability to ensure grid orthogonality next to boundaries, (4) ability to smooth discontinuities that propagate into the domain from discontinuities on boundaries, and (5) tools to assist the user in using GRID2D/3D and in visualizing grids generated by it.

The details of the methods used in GRID2D/3D to generate grids are given in References 1, 4, and 5. The essence is as follows. Parametric representation and gridding of surfaces are obtained by using a technique referred to as three-dimensional bi-directional Hermite interpolation. This technique ensures C^1 continuity across patches that make up surfaces. Each patch is described by four edge curves that are constructed by using either cubic or tension splines from user specified discrete points. Parametric representation and gridding of each block, enclosed by six surfaces, is generated by using either the two-, four-, or six-boundary method in which opposite surfaces are connected by Hermite interpolation. Control of grid point distributions is accomplished by stretching functions – both the traditional one-dimensional ones as well as multidimensional ones. Orthogonality of grid lines on boundaries is controlled by coefficients for normalized derivatives used in the Hermite interpolation formulas, a set of consistency conditions, and a linear blending from non-orthogonal boundary intersections to orthogonal grid-line intersections. Boundary discontinuities that propagate into the domain are smooth by repeated application of a Laplacian operator.

GRID2D/3D is written in Fortran 77 and can be run on any computer with a Fortran compiler. The graphics tools for visualizing the grids generated, however, are written for a PC running DOS. References 1 to 3 give a listing of the Fortran programs as well as a user's manual.

Two multi-block grids generated by using GRID2D/3D for a three-dimensional internal coolant passage with baffles and pin fins in a radial turbine blade (see Fig. 1) are shown in Figs. 2 and 3. Figure 2 shows an overlapping grid for the coolant passage in two-dimensions (see Ref. 6). Figure 3 shows a C^1 continuous multi-block grid for the same coolant passage in two- and three-dimensions (see Ref. 7).

To get GRID2D/3D, contact Professor Tom I-P. Shih, Department of Mechanical Engineering, Carnegie Mellon University, Pittsburgh, PA 15213-3890; phone: (412) 268-2503; e-mail: ts2k@andrew.cmu.edu.

Key References:

1. Shih, T. I-P., Bailey, R.T., Nguyen, H.-L., and Roelke, R.J., "GRID2D/3D – A Computer Program for Generating Grid Systems in Complex-Shaped Two- and Three-Dimensional Spatial Domains. Part 1: Theory and Method," NASA Technical Memorandum 102453, March 1990.
2. Bailey, R.T., Shih, T. I-P., Nguyen, H.-L., and Roelke, R.J., "GRID2D/3D – A Computer Program for Generating Grid Systems in Complex-Shaped Two- and Three-Dimensional Spatial Domains. Part 2: User's Manual and Program Listing," NASA Technical Memorandum 102454, April 1990.
3. Steinthorsson, E., Shih, T. I-P., and Roelke, R.J., "GRID3D-v2: An Updated Version of the GRID2D/3D Computer Program for Generating Grid Systems in Complex-Shaped Three-Dimensional Spatial Domains," NASA Technical Memorandum 103766, June 1991.
4. Shih, T. I-P., Bailey, R.T., Nguyen, H.-L., and Roelke, R.J., "Algebraic Grid Generation for Complex Geometries," International Journal for Numerical Methods in Fluid, Vol. 13, 1991, pp. 1-31.
5. Steinthorsson, E., Shih, T. I-P., and Roelke, R.J., "Enhancing Control of Grid Distribution in Algebraic Grid Generation," International Journal for Numerical Methods in Fluid, Vol. 15, 1992, pp. 297-311.

6. Stephens, M.A., Rimlinger, M.J., Shih, T.I-P., and Civinskas, K.C., "Chimera Grids in Computing Flow and Heat Transfer in Blade Cooling," AIAA Journal of Propulsion and Power, Vol. 11, No. 2, 1995, pp. 213-220.
7. Steinthorsson, E., Shih, T.I-P., and Roelke, R.J., "Algebraic Grid Generation for Coolant Passages of Turbine Blades with Serpentine Channels and Pin Fins," AIAA Paper 91-2366, June 1991.

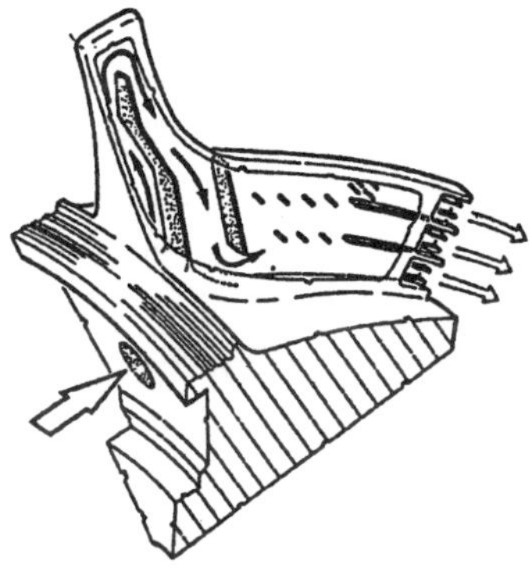

Fig. 1. An internal coolant passage with pin fins and baffles in a radial turbine blade.

Fig. 2. An overlapped multi-block grid (2-D view only).

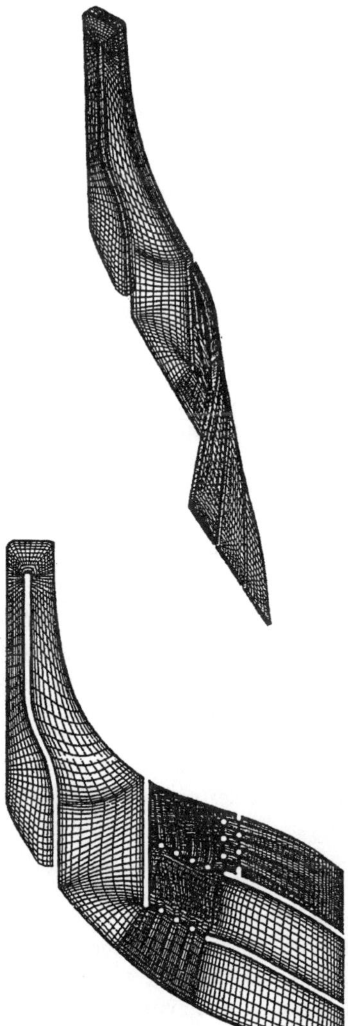

Fig. 3. C^1 continuous multi-block grid (2-D and 3-D views).

GUM-B

*G*eneral *U*nstructured *M*ulti-*B*lock Structured Grid Generator

Main Developer(s): Mike Remotigue (structured, topology, graphics, GUI design, I/O, CAD) and Brian Jean (graphics, geometry/CAD library).
Other Developer(s): Adam Gaither (topology model), Kelly Gaither (graphics library), Gary Newcomb (GuiDe w/library), William Chan (HYPGEN).

Capabilities

o IGES Reader/Writer
o Spline Plot3D/XYZ data into NURBS
o Save State
o Grid Editing (Reactive Topology)
o Option to automatically create faces and blocks
o Automatic connectivity bookkeeping
o User-friendly Graphical User Interface
o Internal CAD package
o Customizable
o Algebraic, elliptic, and hyperbolic grid generation

Computing Environment

Version 2.2 available on SGI Workstation with IRIX 6.2 or higher OS and SUN with Solaris OS and Creator 3D graphics. OpenInventor Graphics is required on all platforms.

Code Description

A Multi-Block structured grid code that allows for the connection of arbitrarily connected (unstructured) blocks. Geometry can be imported via an IGES file or discrete data that is splined to Non-Uniform Rational B-Splines (NURBS) or created from the internal CAD package. A non-manifold solid modeling data structure is used to bookkeep adjacency information, underlying geometry orientation, and grid information. The user validates the domain decomposition before any grid points are distributed on the edges and throughout the domain. Point information is automatically propagated through computationally similar boundaries. Grid points can be output in plot3d format or a hex dump. Boundary condition information can be dumped in a separate file.

Grid Generation Process

Traditional with a twist. The grid is developed upon the normal hierarchy of points, edges, faces, and blocks. However, faces and blocks can be created automatically rather than manual selection. The user gets immediate graphical feedback of point, edge, face, and block creation. These are approximate descriptions. Grid information can be presented on request. The user need not, in most cases, worry about orientation of edges, faces and blocks. The bookkeeping is automatic. Grid information is automatically oriented "correctly" for the given use. Facilities exist to reorient the grid information for output purposes.

Faces and blocks are labeled for recreation. Faces are either labeled as GEOMETRIC, FOUR CURVE, CARPET, SPHERICAL, CYLINDRICAL, REVOLVED, PLANAR. GEOMETRIC is an actual geometric representation, i.e. NURBS. If a grid is generated, it will be created in the parametric space of the geometry. All other faces are generated using arclength transfinite interpolation. The CARPET is a case were the tfi grid is then projected onto an underlying set of geometric surfaces. SPHERICAL, CYLINDRICAL, and PLANAR map the tfi grid to a spherical, cylindrical, or planar reference. REVOLVED rotates the indicated boundary to create the surface grid.

Modifications in geometry and topology can be accomplished through the use of the reactive topology. Points can be relocated and edges and faces can be replaced without having to regenerate the whole domain decomposition. Faces will need to be relabeled and then the grid can be reinitialized rapidly.

Examples

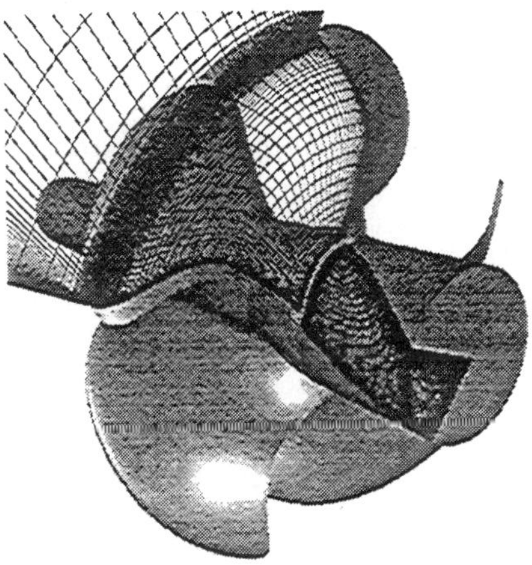
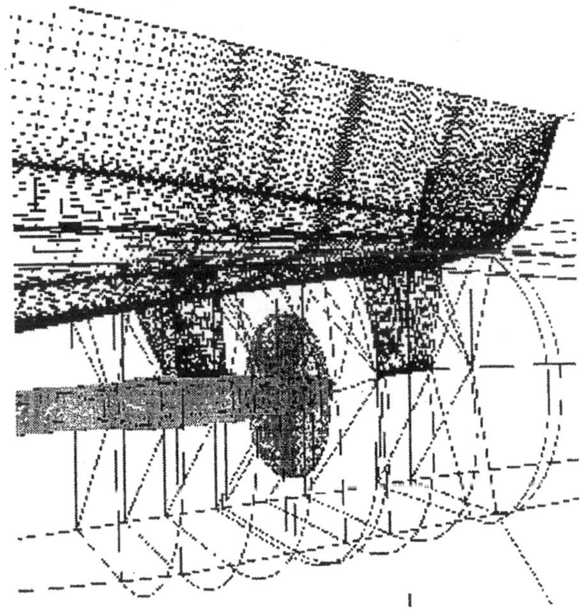

Figure 1: Example of grids produced by GUM-B

Source/Availablity

No source available. Executable availability is currently limited to affiliates of the National Science Foundation/Engineering Research Center at Mississippi State University. Interested parties should contact the director of the ERC.

Key References

M.Y. Jiang, and M. Remotigue. 1998. GUM-B Grid Generation Code and Applications. In preparation for the Proceedings of the 6th International Conference on Numerical Grid Generation in Computational Field Simulation, held at University of Greenwich, Avery Hill Campus, London, England.

A. Gaither, B. Jean., M. Remotigue, and J. Whitmire. 1996. NGP: defining a grid generation paradigm based on NURBS and solid modelling topology. In Proceedings of the 5th International Conference on Numerical Grid Generation in Computational Field Simulation, held at Mississippi State University, April 1-5, 1996, edited by B. Soni, J. Thompson and P. Eisman, 393-402. Mississippi State, MS: NSF Engineering Research Center for Computational Field Simulation.

A. Gaither, K. Gaither, B. Jean, M. Remotigue, J. Whitmire, B. Soni, J. Thompson, J. Dannenhoffer, and N. Weatherill. 1995. The National Grid Project: A system overview. In Proceedings of the Surface Modeling, Grid Generation, and Related Issues in Computational Fluid Dynamic Solutions workshop, by the NASA steering committee on surface modeling and grid generation. Cleveland, Ohio: NASA Lewis Research Center, NASA conference publication 3291, 423-446.

Remarks

HYPGEN is courtesy of William Chan, see HYPGEN for further information and distribution.

Contribution to Grid Codes Chapter for CRC Handbook of Grid Generation

HYPGEN

William M. Chan

Outline of Capabilities

HYPGEN is a field grid generator for structured grids. Given a surface grid (a curve in two dimensions, or a surface in three dimensions), a field grid is generated by solving hyperbolic partial differential equations [Chan and Steger, 1992].

Computing Environment

HYPGEN can be used in batch mode or called from the graphical user interface HGUITK packaged with the code. In batch mode, the code can be run on any machine that supports Fortran 77. The graphical user interface is written in Tcl/Tk and uses PLOT3D [Walatka et al., 1990] as the grid display tool. Any machine with Tcl/Tk can run HGUITK. However, visualization of the resulting grid is restricted to machines on which PLOT3D can be used, e.g. Silicon Graphics workstations. HYPGEN is also hooked up to the OVERGRID unified overset grid generation interface. See the description on OVERGRID in this chapter for further details.

Code Description

A field grid is generated by marching from a user-supplied initial curve (2-D) or initial surface (3-D). Marching distances and grid spacings are specified by the user. Excellent orthogonality and grid clustering characteristics are achieved with the hyperbolic scheme. However, only limited control of the side and outer boundaries is available, e.g. constant planes, periodic, singular axis, and others. The resulting grids are particularly suitable for the overset grid approach to computing flow on complex configurations. The speed of the code is extremely fast: 220000 points per CPU second on the Cray C-90, 46000 points per CPU second on an SGI R10000.

HYPGEN provides two grid quality checks on the computed field grid: a tetrahedral volume decomposition check and a Jacobian computation check as performed by the overset flow solver OVERFLOW [Buning et al., 1997]. If cells with negative Jacobians are present, these are reported to the user. Smoothing parameters may then be adjusted and the code re-run to eliminate the bad cells.

Grid Generation Process

A surface grid file in PLOT3D format is required as input to HYPGEN. In batch mode, an input parameters file has to be supplied also. Under HGUITK mode, the input parameters can be selected by the user via various entry and button widgets, or they can be read in from a file. Input parameters controlled by the user include the boundary conditions to be applied at the grid boundaries, the distance to march out from the initial surface, the normal grid spacing at the initial surface and/or the normal grid spacing at the outer boundary, and several grid smoothing parameters. The code outputs the field grid in PLOT3D format and reports any negative cell Jacobians found.

Example Demonstrations

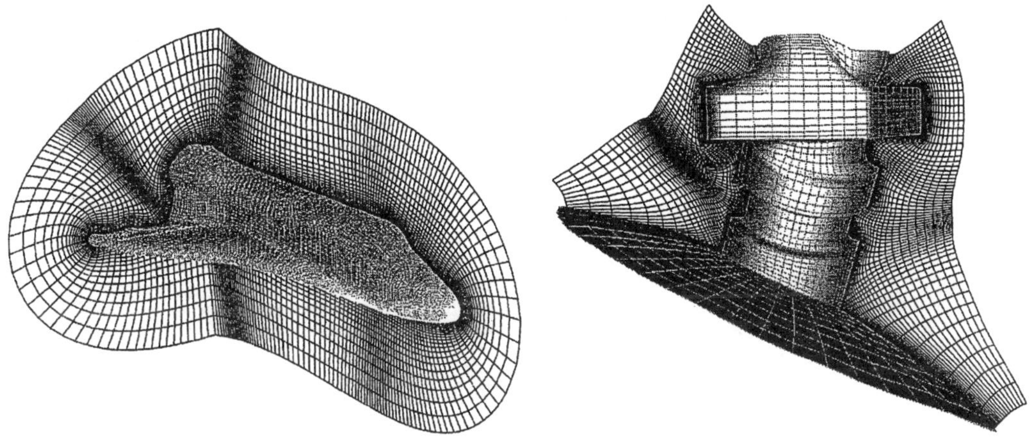

Figure 1: Examples of field grids produced by HYPGEN

Source/Availability

HYPGEN is developed at NASA Ames Research Center. It is available to companies, government laboratories and universities of the U.S. for internal use. The code can be obtained by sending a request letter on company letterhead to the author. The NASA code dissemination process requires the code requestor to sign an agreement form. Code dissemination occurs after the request is approved by NASA management. A user's manual for HYPGEN is available in the form of a NASA TM [Chan, Chiu, and Buning, 1993].

References

Buning, P. G., Jespersen, D. C., Pulliam, T. H., Chan, W. M., Slotnick, J. P., Krist, S. E. and Renze, K. J., "OVERFLOW User's Manual (version 1.7u)," NASA Ames Research Center, 1997.

Chan, W. M. and Steger, J. L., "Enhancements of a Three-Dimensional Hyperbolic Grid Generation Scheme," *Appl. Math. and Comput.* **51**, 181–205, 1992.

Chan, W. M., Chiu, I. T. and Buning, P. G., "User's Manual for the HYPGEN Hyperbolic Grid Generator and the HGUI Graphical User Interface," NASA TM 108791, October, 1993.

Walatka, P. P., Buning, P. G., Pierce, L. and Elson, P. A., "PLOT3D User's Manual," NASA TM 101067, 1990.

Remarks

See chapter on Hyperbolic Grid Generation Systems for further details.

Name of Grid Code	*ICEM CFD*
Outline of Capabilities	*ICEM CFD is software for pre- and post-processing for engineering applications such as computational fluid dynamics and structural analysis. The grid generation tools offer the capability to parametrically create grids from geometry in multiblock structured, unstructured hexahedral, tetrahedral, hybrid grids consisting of hexahedral, tetrahedral, pyramidal and prismatic cells; and cartesian grid formats combined with boundary conditions. Over 60 flow solver and structural analysis translators are provided.*
Computing Environment	*The ICEM CFD is supported on the following workstations: Silicon Graphics, Hewlett Packard, Sun Microsystems, IBM, Digital, Across networks via X-Windows, Intel PC's running Windows/NT 4.0.*
Code Description	*The modules of ICEM CFD are:*

ICEM CFD Direct CAD Interfaces - *This module provides a direct interface between ICEM CFD and CAD systems. Geometry necessary for grid generation can be input into ICEM CFD directly for grid generation without requiring a translation using formats such as IGES. The interface will provide the tools for tagging CAD geometry, assigning mesh sizes and boundary conditions. Geometries parametrically modified in other CAD Systems can be re-meshed with ease in ICEM CFD's environment after the initial topology is set. Users can extend the parametric geometry creation features of any CAD system to computational grid generation for analysis with ICEM CFD.*

ICEM CFD Unput - *This module handles input of geometry for grid generation. The geometry sources can be NURBS geometry from various CAD systems, surface meshes in Nastran and Patran format, Plot3d format and Stereolithography (STL) format. CAD surfaces can be intersected with triangular surface meshes. The creation of material points, editing of meshing parameters, editing or defining of boundary conditions, extraction of curves and points if necessary and segmentation of surfaces/curves are available features.*

ICEM CFD Hexa - *This grid generation tool based on a Global Block topology provides a top-down approach to grid generation. The semi-automated meshing module of ICEM CFD presents rapid generation of multi-block structured or unstructured hexahedral volume meshes. ICEM CFD HEXA represents a new approach to grid generation where the operations most often performed by experts are automated and made available at the touch of a button. Blocks can be interactively adjusted to the underlying CAD geometry. The grid is projected onto the underlying CAD geometry automatically.*

ICEM CFD Tetra - *ICEM CFD Tetra has been tuned to take full advantage of the object oriented unstructured meshing technology. No tedious up-front triangular surface meshing is needed to provide well-balanced start meshes. ICEM CFD Tetra works directly from the CAD surfaces and fills the volume with tetrahedral elements using the Octree approach. Prescribed curves and points can define positions of edges and vertices. A powerful smoothing algorithm provides the element quality. The user simply selects the bounding CAD geometry needed. ICEM CFD Tetra generates the volume mesh and the* *surface mesh on the object surfaces.*

ICEM CFD Prism *- Hybrid tetrahedral grid consisting of layers of prism elements near the boundary surfaces and tetrahedral elements in the interior can be generated with ICEM CFD Prism.*

ICEM CFD Hybrid *– This module allows combining of hexahedral, tetrahedral, and prismatic meshes, If two regions, one with a hexahedral grid and another with a tetrahedral grid, share surfaces, the two volume meshes can be merged into a single mesh. At the boundary, pyramidal cells are created automatically.*

ICEM CFD COMAK *- This module provides parametric geometry modeling for grid generation.*

ICEM CFD Autohexa *– This module is a fully automatic hexahedral mesh generator. It is a simple object-based modeling system utilizing cartesian geometry: rectangular prisms, cylinders and cones, inclined planes, and polygonal prisms.*

ICEM CFD Leo *- The grid post-processing tool of ICEM CFD is the grid postprocessing tool that provides, grid visualization for both unstructured and structured grids, boundary condition editing, global smoothing options available tuned to deliver remarkably improved grid quality and grid smoother.*

ICEM CFD Translators *- The ICEM CFD Translators provides input to more than 60 CFD and Structural Analysis codes.*

Grid Generation Process

In ICEM CFD environment, geometry is either acquired using a provided geometry translator (i.e. IGES, VDAFS, ProSTEP) or a ICEM CFD Direct CAD Interface (CATIA, I-DEAS, ProE, Unigraphics). During geometry acquisition the surfaces, curves and points are tagged and mesh size information can be attached. The user selects the appropriate grid generation module to create the computational grid. The boundary conditions can also be attached to the tagging information. The mesh, boundary conditions and topology information then can be translated into over than 60 flow solvers and structural analysis codes.

Source and Availability

For sales Information please refer to:
ICEM CFD Engineering
Berkeley, CA Office
2855 Telegraph Ave.
Suite 501
Berkeley, CA 94704 U.S.A.
Phone: +1-510-549-1890
FAX: +1-510-549-8523

Key References

Remarks

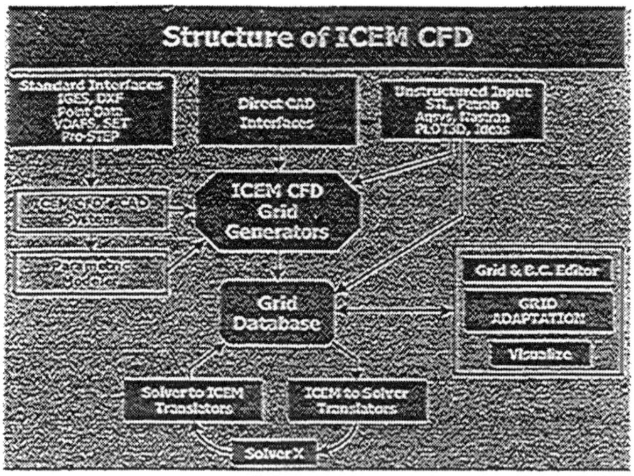
Structure of ICEM CFD
Standard Interfaces
IGES, DXF
Point Data
VDAFS, SET
Pro-STEP
Direct CAD
Interfaces
Unstructured Input
ICEM CFD
Grid
Generators
Grid & B.C. Editor
Grid
Database
GRID
ADAPTATION
Visualize
Solver to ICEM
Translators
ICEM to Solver
Translators
Solver X

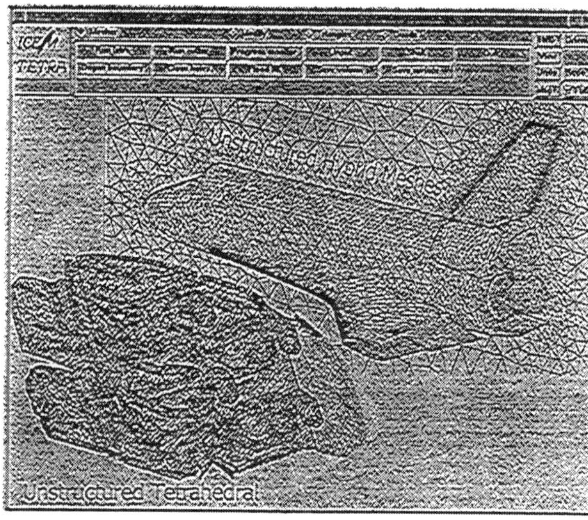
ICEM
TETRA
Unstructured Hybrid Meshes
Unstructured Tetrahedral

ICEM
HEXA
Multiblock Structured Meshes
Unstructured Hexahedral

Title: IITZeus

Name and Affiliation of the Contributor

Dr. G. R. Shevare, Dept. of Aerospace Engg. Indian Institute of Technology, Bombay, India, Fax : (91) 022 5782602, 5783480, email : shevare@aero.iitb.ernet.in

Application

The Aim is to descretise a complex 3-D geometry for a computational method. The emphasis is on automatic surface and semi-automatic volume multi-blocking and their generation. The meshes are for general utility and not necessarily for a numerical method or a specific field of application.

Grid Description

Basic Grid Information : Structured, unstructured, composite grids are generated. Surface structured grids are generated and adapted using variational principal. Volume structured grids are smoothened and adapted by an elliptic solver. For unstructured surface grids both Delaunay and advancing front methods are available. The volume tetrahedrons are made by a Delaunay method.

Size,Quality,Distribution,.. : There are no limitations in terms of size of grid points. It is possible to get a statistics of quality of grids.

Grid Strategy (Multiblock/Chimera/Unstructured/Hybrid,..) : The software not only uses multi-block strategy. It builds the complete geometry as the assembly of blocks. However the meshes in individual blocks can be structured or unstructured. Thus the software is geometric modeller-cum-grid generator for multi-block grid generation.

Generation Method (Algebraic/Elliptic/Delaunay.....) : Structured Grids : Algebraic followed Variational method for surface grid. Algebraic followed by an elliptic solver in volume Unstructured : Delaunay and advancing front in 2D and on surfaces only Delaunay in volume

Geometry Specification : The geometry module supports points, segments, piecewise linear curve, B-spline curve, piecewise bilinear piecewise surface, surface of revolution, B-spline surface, Coon's patch and their intersections. Triangulated surface and panelled surface are available. Volumes are modelled as an assembly of hexahedral blocks. Each block is modelled B-rep of surfaces. Block face can be a piecewise bilinear surface (structured grids) or a triangulated patch (i.e. a triangulated surface with outer boundary as a concatenation of four piecewise linear edges). If a face is not explicitly provided it assumed to be a bilinear Coon's patch. Generated Coon's patch gets discretised to a piecewise bilinear surface or a triagulated patch depending the kind grids required in the block. An edge of a block is piecewise linear curve (discretised curve). If is not provided implicitly of explicitly a straight line between two vertices is assumed.

Reference/Remarks

Developed by IIT Bombay with a support from National Centre for Software Technology, Bombay, Aeronautical Development Agency, Bangalore and Aeronautics Research & Board, New Delhi.

Computing Environment :

Platform : An UNIX machine, **Language** : ANSI C, **Graphics Interface**: X, Motif. The software has its own graphics user interface.

Code Description

Basic modules : Graphics User Interface, Surface Modeller, Automatic Surface Multi-block Module, Semi-automatic Volume Modeller, Surface Structured Grid Generator, Surface Unstructured Grid Generator, 3-D Structured Grid Generator, 3-D Tetrahedron Generator.
Databases: for internal entities, geometric entities, Surface and Volume Blocks, and Tetrahedrons.
Miscellaneous: Mathematical tools.

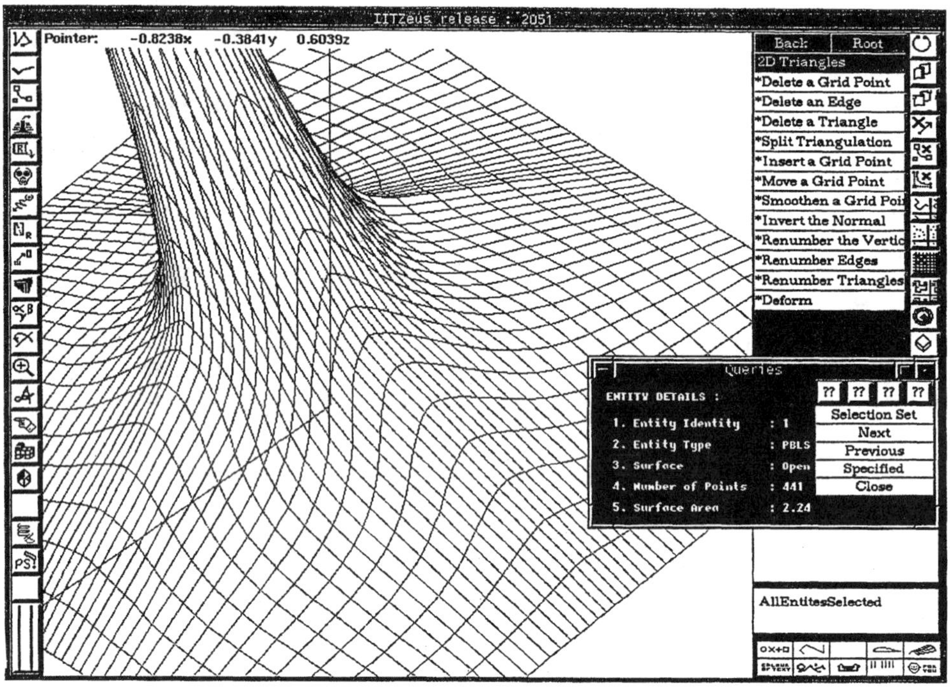

Figure 1: IITZeus screen showing grids adapting to a solution

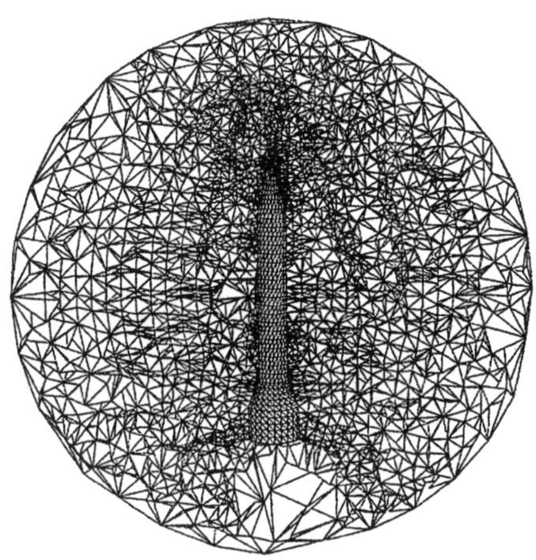

Figure 2: 3-D unstructured grids using Delaunay triangulation

LaGriT -- Los Alamos Grid Toolbox

The LaGriT toolbox provides a full set of grid generation and grid optimization tools for constructing and maintaining unstructured grids for both static and dynamic applications. The grid generation tools respect intricate geometries, allow optional user control over node placement, and provide hybrid grid capabilities. Although LaGriT is primarily a 3D Delaunay tetrahedral code, it supports 2D and 3D hybrid grids as well as homogenous grids. The grid optimization tools allow adjustment of node density using refinement, simple and adaptive smoothing, and reconnection techniques. The LaGriT toolbox methods as well as the LaGriT data structures are designed as objects in order to be user accessible and extensible. This design feature promotes the separation of the physical based simulation from the grid maintenance chores, but provides full access to grid data structures when required. WEB page available at: http://www.t12.lanl.gov/~dcg/

LaGriT is written in a combination of FORTRAN77 and C. It runs on all UNIX workstation platforms. The recommended viewer is GMV (The General Mesh Viewer) available from Los Alamos National Laboratory. LaGriT can also be instructed to produce AVS files.

LaGriT is designed as a toolbox and is currently implemented as a library that can be linked into a driver code. LaGriT commands can be issued from the driver code. In most applications, initial grid generation will be the first step, and subsequently, the driver code will request grid optimization which is especially important in time dependent simulations. LaGriT provides access routines to the grid data structures, thus allowing the driver code to modify the grid as needed. Application areas include semiconductor modeling and fluid flow and transport through geologic strata.

LaGriT uses a modified Watson point insertion method to tetrahedralize the initial point set. Material interfaces are then resolved by adding points on edges that cross material interfaces. LaGriT can also generate brick type hexahedral grids and 2D triangular and quad grids. 2D grids can also be 'extracted' from existing 3D grids.

LaGriT is available only as a node-locked executable for users within the United States. Users must sign a license agreement which is available from dgeorge@lanl.gov.

Khamayseh, Ahmed; Ortega,Frank; Trease, Harold, "Ray Tracing for Point Distribution in Unstructured Grid Generation", LA-UR-95-4470.

Khamayseh, Ahmed; Ortega, Frank; Kuprat, Andrew, "A Robust Point Location algorithm for General Polyhedra", Journal of Computer Aided Geometric Design, LA-UR-95-4465

D.F. Watson, "Computing the n-Dimensional Delaunay Tessalation with Applications to Voronoi Polytopes," The Computer Journal, Vol. 24, 1981, pp.167-172.

D. A. Field, "Implementing Watson's Algorithm in Three Dimensions," Proceedings of the Second Annual ACM Symposium on Computational Geometry, 1986, pp.246-259.

Painter, J.W. and Marshall, J.C., "Three-Dimensional Reconnection and Fluxing Algorithms," Proceedings of the Next Free-Lagrange Conference, Jackson Lake Lodge, Wyoming, June 3-7, 1990, Springer-Verlag Press, Vol. 395, pp. 139-148.

Trease, H.E. "Parallel Nearest Neighbor Calculations," Proceedings of the first Free-Lagrange Conference, Lecture Notes in Physics, Springer-Verlag, Vol. 395, pp. 149-156, 1985.

Randolph E. Bank and R. Kent Smith, "Mesh Smoothing Using A Posteriori Error Estimates", SIAM J. Num. Anal. Vol. 34, Issue 3, pp. 979-997 (1997))

Ahmed Khamayseh and Andrew Kuprat, "Anisotropic Smoothing and Solution Adaption for Unstructured Grids," LA-UR-95-2205, International Journal for Numerical Methods in Engineering, Vol. 39, pp. 3163-3174 (1996).

Kuprat, Andrew, et al. "Moving Adaptive Unstructured 3-D Meshes in Semiconductor Process Modeling Applications", VLSI Design, Vol. 6, 1997, LA-UR-95-4128.

Andrew Kuprat, David Cartwright, J. Tinka Gammel, Denise George, Brian Kendrick, David Kilcrease, Harold Trease, and Robert Walker, "X3D Moving Grid Methods for Semiconductor Applications", LA-UR-97-2891, to appear in VLSI Design.

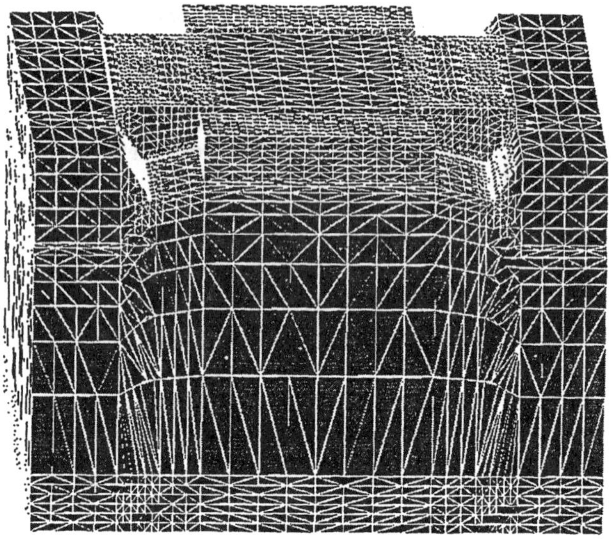

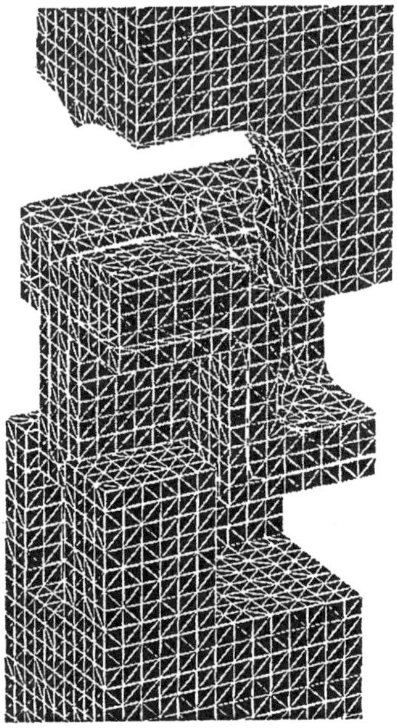

McDonnell Aircraft Computational Grid System (MACGS)

MACGS consists of three modules:

- ZONI3G – Geometry Manipulation, Domain Decomposition, and Surface Grid Generation
- GMAN – Volume Grid Generation, Boundary Condition and Zone Connectivity Specification
- GPRO – File format conversion and zone manipulations

Important System Features

- Capability for multi-zone structured grids, overset grids, unstructured grids, and hybrid grids
- Boundary conditions/zone connectivity interpolation factors stored in grid file
- Common data file structures with the WIND flow solver provided through the NPARC Alliance
- Runs on a variety of computer platforms including Hewlett Packard, Silicon Graphics, IBM, X-terminals

Zonal Interactive Graphics for Grid Generation (ZONI3G) is an interactive graphical program based on the USAF Interactive Graphics for Geometry Generation (I3G) program. It accepts geometry defined by discrete points from a variety of sources. Curves and surfaces can also be constructed within the system, but ZONI3G is not intended to replace the CAD system. ZONI3G can also access analytically defined geometry directly from a Unigraphics CAD (Computer Aided Design) model and can generate points on the geometry or project existing point definitions onto the geometry. Extensive point distribution options along with algebraic and elliptic algorithms are used to generate desirable point distributions on geometry surfaces, and on off-body surfaces which form the boundaries of a structured zone. Alternately, unstructured (triangular) meshes can be generated on any surface with resolution based on boundary point distribution, sources, and clustering parameters. For structured grids, six faces are associated to form a three-dimensional zone. As the zone is defined, adjacent edges are checked for compatibility at common edge points. Completed zones are then output into a common grid file (see below). An interface to the NASA HYPGEN code can be used to generate 3-D zones hyperbolically from a single surface grid.

Grid MANager (GMAN) is an interactive graphical program that can operate in either a command line or batch mode. It has two main functions.

The first function of GMAN is generating the interior grid of a zone. It incorporates several different algebraic interpolation options, elliptic forcing functions, and 1-D redistribution options, all of which can operate on the entire zone or any user-specified subset. Extensive quality checking algorithms allow the user to rapidly identify problem areas and apply corrective action.

The second function of GMAN is boundary condition and zone connectivity specification. MACGS takes the approach that boundary conditions type specifications are primarily functions of the configuration geometry. In GMAN, the user identifies areas of solid wall, symmetry/reflection, inflow/outflow, bleed, jets, inlet, etc. that are associated with the grid. The actual implementation of these types and related parameters is determined by the flow solver at run time. GMAN also determines the connectivity between adjacent zones, which need not be point matched. Interpolation factors defining the physical relationship between boundary points and the cells of an abutting zone face are stored in the grid file.

Overset (Chimera) grid methods are incorporated as part of the boundary conditions specification. The user is given several methods for identifying holes and fringe points. Given the fringe points, 3-D interpolation factors are calculated which identify the location of the fringe point within a cell of an overlapping zone. The ability to define and modify holes and fringes interactively streamlines the process. A suite of quality checks is also provided to rapidly detect boundary condition, coupling, and fringe errors.

Grid PROcessor (GPRO) provides file conversion to and/or from several common formats, such as PLOT3D, EAGLE, NURBS, etc. It also supports manipulation of entire zones. One example is an algorithm to change the spacing normal to the wall and/or change the number of points in the boundary layer region of a zone.

Grid Generation Process

The overall process is fairly basic: acquire geometry, prepare the geometry for gridding, define zone boundaries, generate surface grids on zone boundaries, generate the interior grid, check grid quality and correct flaws, specify boundary conditions and connectivity, and perform boundary condition checks. The order in which these steps are performed is flexible in MACGS. For example, the user can complete a step for all zones before proceeding to the next step or complete any or all steps for one zone before generating the next zone.

Example Demonstrations

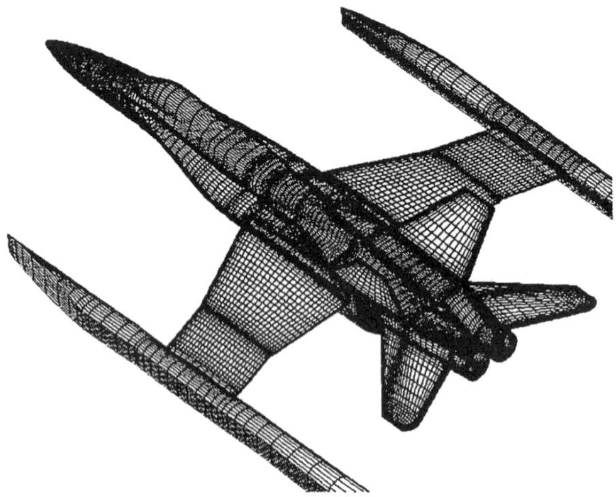

Model with Boom Support (every other grid point shown)

Hybrid Grid for Pilot/Ejection Seat

Source/Availability – MACGS is proprietary software of The Boeing Company. Release of the ZONI3G module is also subject to Air Force restrictions. (An unrestricted replacement for ZONI3G is currently in development.) MACGS or portions thereof have been provided to government laboratories, and other institutions on a case-by-case basis. Inquiries can be directed to Dr. Raymond Cosner, Mailcode S1067126, The Boeing Company, P.O. Box 516, St. Louis, MO 63166.

References

Gatzke, T. D., LaBozzetta, W. F., Finfrock, G. P., Johnson, J. A., and Romer, W. W., "MACGS: A Zonal Grid Generation System for Complex Aero-Propulsion Configurations," AIAA Paper No. 91-2156, June 1991.

Gatzke, T. D., and Weatherill, N. P., "Unstructured Grid Generation Using Interactive Three-Dimensional Boundary and Efficient Three-Dimensional Volume Methods," AIAA Paper No. 93-3452, August 1993.

LaBozzetta, W. F., Gatzke, T. D., Ellison, S., Finfrock, G. P., and Fisher, M. S., "MACGS – Towards The Complete Grid Generation System," AIAA Paper No. 94-1923, June 1994.

Gatzke, T. D., and Melson, T. G., "Generating Grids Directly On CAD Database Surfaces Using A Parametric Evaluator Approach," NASA CP 3291, May 1995.

Remarks

The Boeing "common file" data structures simplify the entire grid generation process. The grid is created in the format used by the WIND flow solver. Boundary conditions are included in the file eliminating the need to keep track of additional files. In-house post-processing tools also use these same files and make use of the boundary condition information. Smart libraries allow the efficient binary data files to be used on multiple platforms without conversion and are self-documenting (i.e., the file can be queried to find out what information is stored). A suite of tools for manipulating common files is also available.

The Three-Dimensional Multi-block Advanced Grid Generation System (3DMAGGS)

Stephen J. Alter*
Lockheed Martin Engineering & Sciences Company, Hampton, Virginia

Summary

The Three-Dimensional Multi-block Advanced Grid Generation System[1] (3DMAGGS) is designed to be a fully 3D computational domain volume discretization tool. The code operates by taking previously defined block faces for each computational side of a structured grid, and generates the interior volume. The code can also make use of plane, spherical, ellipsoidal and cylindrical surfaces with Neumann boundary conditions to develop a volume around a defined wall surface grid. The 3DMAGGS code extends the technology of the Three-Dimensional GRids About anything by Poisson's Equations[2] (3DGRAPE) code through the implementation of:

- Initialization via Trans-Finite Interpolation[3] (TFI) for blocks with all faces discretized.
- Optimal convergence rate determination for the Point Successive Over Relaxation (PSOR) method.
- User specified cell height specifications at any boundary where orthogonality is activated.
- Interior grid controls with Thomas & Middlecoff[4] explicit source terms and hybrid combination with Steger & Sorenson[5] implicit source terms.
- Limiting truncation error for low precision machines.
- Linking to GRIDGEN data sets.

The 3DMAGGS code was designed to be a replacement to the GRIDGEN3D software, by offering more flexible application of source terms through cell height controls as well as the controlling of blocks to be generated and the type of source terms each block will use. As such, the code has a preprocessor, called PREMAGGS, which converts GRIDGEN ASCII grid data into 3DMAGGS inputs. The 3DMAGGS code has built in cell size specification functions, as well as the capability to input the cell sizes from an external file. This external file is usually generated with the PREMAGGS code, where two dimensional TFI and the local arclength cell sizing[6] methods can be used to compute the cell heights.

Advantages Over Other 3D Solvers

For an example of the 3DMAGGS grid generator, an X33 VentureStar volume grid is created. The dimensions of this grid are (253 × 361 × 37) for (I,J,K) limits. As shown in figure 1, the 3DMAGGS code offers the advantage of true restart capabilities, which is evident by the smooth decay of the solver average movement residuals. The 3DMAGGS code simply picks up where the previous run had finished. For this example, 500 iterations were used during each execution. When the restart capability is combined with the time to compute a single point per iteration, at roughly 3.2 micro seconds per iteration per grid point on an

*Senior Aeronautical Engineer

Silicon Graphics Incorporated R10000 with optimized compilation, the time to generate a volume grid is significantly less than other grid generators.

The capabilities of the 3DMAGGS code to generate high quality volume grids is demonstrated in figure 2. Notice that the grid in the concave regions of the X33 vertical fin remains at bay. The 3DMAGGS code tends to preserve the user specified cell heights at a boundary between the beginning and ending iteration, while producing a smooth grid approaching the wall. The differences between the specified and the resultant cell sizes on the wall of the X33 are shown in figure 3.

Finally, the 3DMAGGS code offers grid sequencing, which has been shown to improve grid convergence by nearly 50%[2] as compared to full dimensionality grid generation. Although the effects of grid sequencing are debatable, the method does improve the preconditioning of the source terms used in the fine grid by generating a grid that is close to the final converged solution. During the developmental phase of the 3DMAGGS code, it was discovered that the use of 3DTFI instead of linear point-to-point lines between opposing boundaries for initial grid generation, produces a grid that is more akin to the final grid. The improved match between the initial and final grids can reduce the number of iterations for convergence by 75%.[1] The reduction in iterations can be attributed to the improved preconditioning of the source terms and the reduction in size of corrections used in the Steger and Sorenson source term formulations.

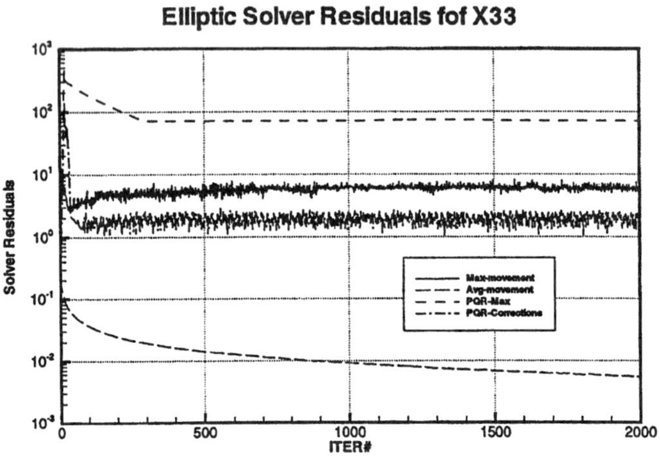

Figure 1: Convergence history of point movement for an X33 VentureStar.

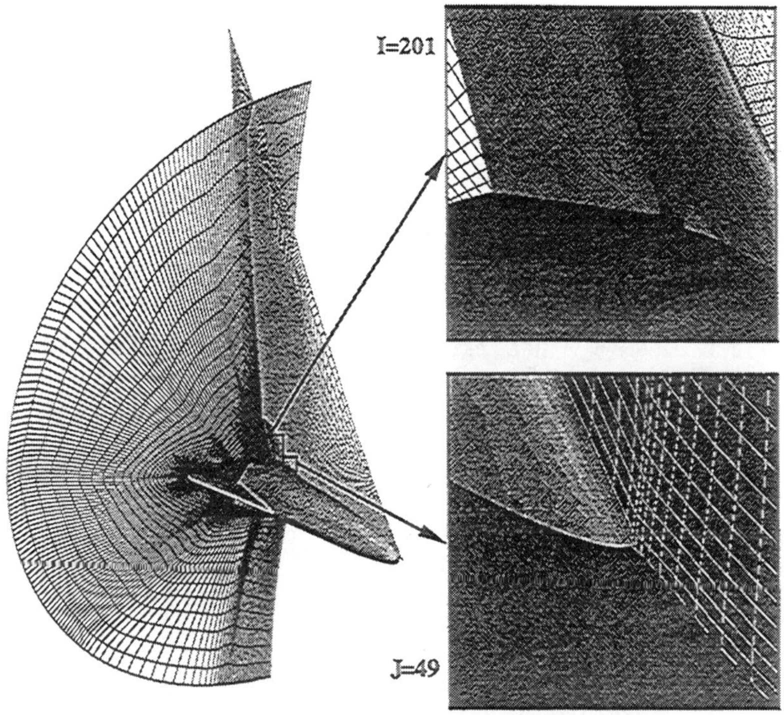

Figure 2: Grid quality in concave regions.

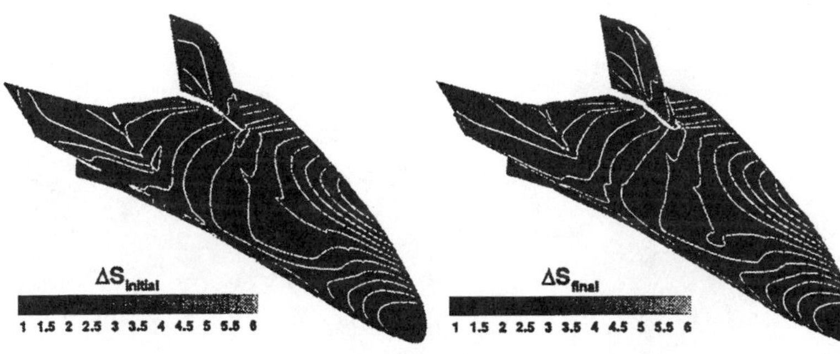

Figure 3: Cell size comparison from specified to final converged grid.

Acknowledgment

This work was performed under NASA Langley Research Center contract NAS1-96014. The author wishes to express his appreciation to Dr. David Schuster for his editorial comments, William Kleb for his contributions to the typesetting of this document and K. James Weilmuenster for permitting this work to be done.

References

[1] Alter, S. J. and Weilmuenster, K. J., "The Three–Dimensional Multi–block Advanced Grid Generation System (3DMAGGS)," NASA TM–108985, April 1993.

[2] Sorenson, R. L., "The 3DGRAPE Book: Theory, Users' Manual, Examples," NASA TM–102224, July 1989.

[3] Soni, B. K., "Two– and Three–Dimensional Grid Generation for Internal Flow Applications of Computational Fluid Dynamics," AIAA Paper 85–1526, 1985.

[4] Thomas, P. D. and Middlecoff, J. F., "Direct Control of the Grid Point Distribution in Meshes Generated by Elliptic Equations," *AIAA Journal*, Vol. 18, No. 6, June 1979, pp. 652–656.

[5] Steger, J. L. and Sorenson, R. L., "Automatic Mesh-Point Clustering Near A Boundary in Grid Generation with Elliptic Partial Differential Equations," Vol. 33, December 1979, pp. 405–410.

[6] Alter, S. J. and Weilmuenster, K. J., "Cell Volume Control at a Surface for Three–Dimensional Grid Generation Packages," *Software Systems for Surface Modeling and Grid Generation*, edited by R. E. Smith, Vol. CP–3143, NASA, 1992, pp. 273–298.

Name of the grid code: MegaCads (**Multi**block-**E**lliptic-**G**rid-Generation-**A**nd-**CAD**-**S**ystem)

Organisation: DLR, Institute of Design Aerodynamics, Lilienthalplatz 7, D-38108 Braunschweig, Germany,

Authors: O. Brodersen, A. Ronzheimer, M. Hepperle, R. Ziegler, T. Kunert, J. Wild, B. Schöning, P. Niederdrenk, C.-C. Rossow and Partners of the german MEGAFLOW project: H. Schwarten (Daimler-Benz Aerospace Airbus), J. Yan (University of Berlin)

Computing Environment: SGI, DEC, HP, SUN, IBM, PC (Linux), NEC-SX4, CRAY

Language: Ansi C

Graphic: X11/Motif, OpenGL/Mesa

Availability: MegaCads is available for the partners of the german MEGAFLOW project

Code Description:

MegaCads is a parametric multiblock structured grid generation system providing interactive (Fig. 1) and batch capabilities and supporting non-overlapping as well as overlapping grids [1]. It enables the user to define block topologies, to distribute points on them, to create surface grids and to fill volume blocks. It supports a parametric design and restart concept using a script language. For the block topology definition basic CAGD techniques and a grid topology data structure are integrated. It is also the purpose that MegaCads serves as a grid generation development platform. Therefore the source code will be distributed to the partners. Software quality management using CASE-Tools is introduced to simplify the maintenance of the system. MegaCads is still under development and is besides the flow solver FLOWer part of the german project MEGAFLOW.

The whole block topology design is performed in a parametric way. Grid spacings are defined as a fraction of a reference length. Trimming of curves and surfaces is performed relatively to their parameter values. Due to this parametric definition of the complete grid generation sequence and because typical reference lengths of a geometry are used (e.g. airfoil chord length), grid generation processes can be reused for different geometries automatically. Currently the non-interactive version (MegaBatch) is linked to FLOWer for the optimization of high-lift configurations.

Each grid generation process, stored in a script file, includes the process name and its parameters. Blocks of processes can be combined to functions which can be used with their appropriate parameters. For multiple reuses of script statements, loops can be introduced. In addition an interactive editing tool for the script, MegaEdit, is available.

MegaCads supports the idea that smooth grids can be obtained when elliptic methods of 1D, 2D and 3D are combined. Therefore different elliptic and biharmonic grid generation techniques using multigrid acceleration are integrated.

Applications:

Figure 2 shows parts of a Navier-Stokes grid for the DLR-F6 aircraft configuration. The grid consists of 55 blocks and 3.3 million points. All components are wrapped into C-grids in streamwise and O-grids in spanwise direction. These blocks are integrated into H-grids resulting in a HCO-topology.
Figure 3 displays the DLR-ALVAST high-lift configuration consisting of a fuselage and a wing with a nose slat (complete wingspan) and an inboard and out board flap. The grid has 53 blocks and 3.5 million points. The slat and the flaps are surrounded by O-grids which expand in spanwise direction to the far field. The O-grids are integrated into H-grids.

References:

[1] O. Brodersen, M. Hepperle, A. Ronzheimer, C.-C. Rossow, B. Schöning : *The Parametric grid generation System MegaCads.* 5th International Conference on Numerical Grid Generation in Computational Field Simulations, ed. B.K. Soni, J.F.Thompson, J.Häuser, P.Eiseman, NSF Engineering Research Center, Mississippi State University, Vol.1, pp. 353-362, 1996

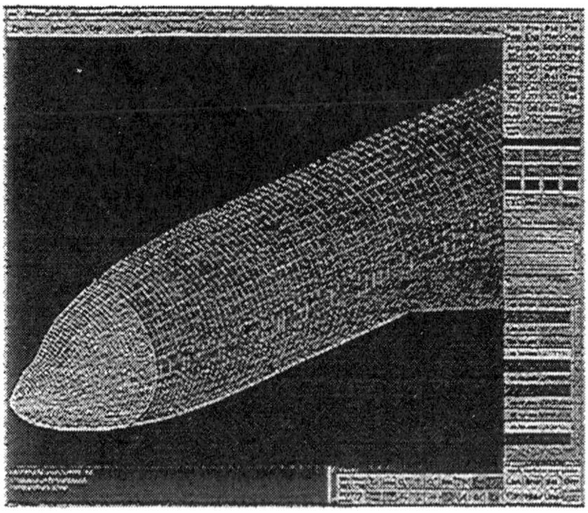

Figure 1: MegaCads user interface

Figure 2: Navier Stokes grid, DLR-F6

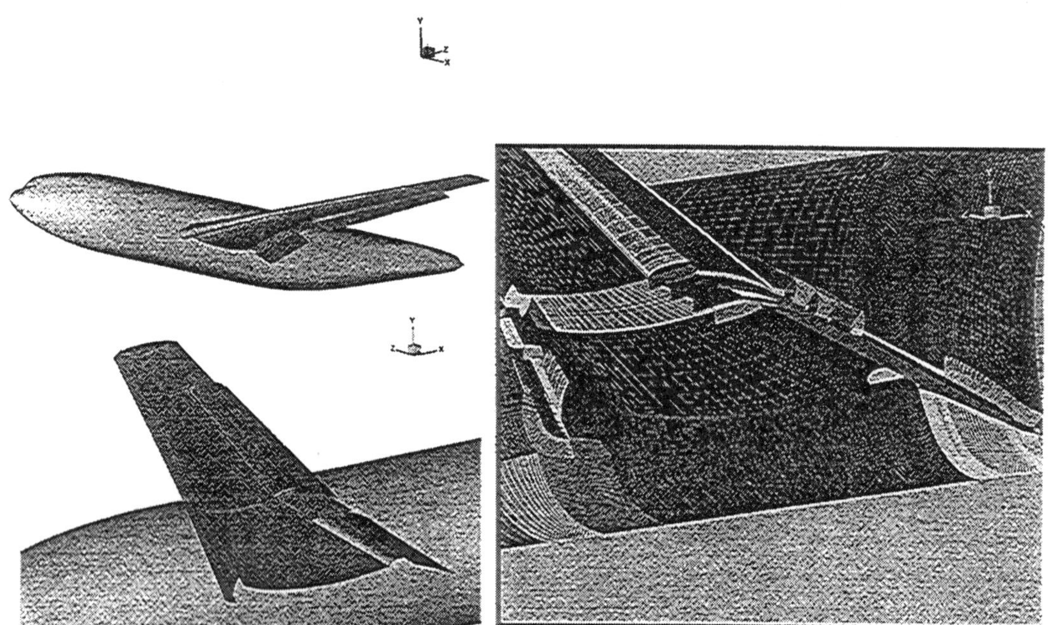

Figure 3: Grid for the DLR-ALVAST high-lift configuration

B. Kaan KARAMETE

Department of Engineering Sciences
Middle East Technical University
Ankara,06531,Turkey

Name of the grid code

Mesh2d,Mesh3d.

Outline of Capabilities

In terms of satisfying the boundary conformity of unstructured grids, which is strickly required for a correct triangulation, the conventional strategies of mesh generation algorithms have some drawbacks. Delaunay Triangulation should be modified to take non-convexity of the geometry into account. Mostly, a seperate algorithm is devised for sweeping-off the undesired elements since the Delaunay method generates undesired elements inevitably. On the other hand, it is too costly to preserve the boundary integrity in moving front strategies since it is required to make geometric intersection checks at the formation of each cell.

In terms of dividing the unstructured domain into equal element blocks for parallel computing purposes which has become one of the top topics in computational field simulations of our age, the usual methods have again some disadvantages. The Delaunay algorithm operates in a break-up fashion. The triangle connectivity is reorganized at the insertion each node into the triangulation. Therefore, the Delaunay method has no moving front. It means that a post seperator algorithm should be devised to divide the domain into blocks. On the other hand, moving front strategy do have a front. However, the front may unfortunately be composed of many interface nodes and zig-zags. Since one of the fronts would be an interface, the communication time between the processors increases with the increase in the number of interface nodes. Since the front propagates in a circular shape into the interior of the domain, the comprimise on the number of interface nodes is an open question. Furthermore, the exact number of triangles to be generated is unknown a-priori. Without knowing the total number of triangles before generating the mesh, it is not possible to calculate the number of elements in each block.

This new method is designed to take care of the inefficiencies of the conventionally used methods mentioned above. The generated triangles and tetrahedrons all have empty circumdisk feature meaning that the grids are of Delaunay property. There is no geometric intersection checks to satisfy the boundary conformity for complex geometries. The natural propagation path of the front can be used as a seperator to divide the domain into blocks having exactly the same number of elements. The method can be regarded to be an efficient combination of Delaunay and moving front strategies. In 3D, the surface triangulation effort of the user is relaxed. The surface triangles start to appear on the surface as a direct consequence of interior

- Assign the nodes at the side of outward normal of GT as candidate nodes.
- Compute circumspheres of trial tetrahedrons composed of GT and candidate nodes.
- Find the normal distances from the GT plane to the center of the circumspheres.
- The node giving the minimum signed distance forms up the tetrahedron.

Source & Availability

The code is open for public requests if it is requested personally from the author by E-Mail. E-Mail address: kaan@rorqual.cc.metu.edu.tr. The code is completely available only for use of academical purposes.

Key References

1. D. H. Mclain,'Two dimensional interpolation from random data',*The Computer Journal*, **19-2**, 178-181 (1976).
2. J. Bonet and J. Peraire, 'An Alternating Digital Tree (ADT) algorithm for geometric searching and intersection problem', *Int. J. Numer. Meth. Eng.*, **31**, 1-17 (1991).
3. N.P. Weatherill and O. Hassan, 'Efficient Three Dimensional Delaunay Triangulation with Automatic Point Creation and Imposed Boundary Constraints', *Int. J. Numer. Meth. Eng.*, **37**, 2005-2039 (1994).
4. R. Löhner et al. 'Parallel Unstructured Grid Generation', *Comp. Meth. Appl. Mech. and Eng.*, **95**, 343-357 (1992).
5. B. Kaan Karamete, 'A Flexible Unstructured Mesh Generation Algorithm Suitable for Block Partitioning',*The Fifth International Conference on Numerical Grid Generation in Computational Field Simulations*, Mississippi State University, 1-5 Apr. 279-289 (1996).
6. B. Kaan Karamete, T. Tokdemir and M. Ger, 'Unstructured Grid Generation and a Simple Triangulation Algorithm for Arbitrary 2D Geometries using Object Oriented Programming', *Int. J. Numer. Meth. Eng.*, Revised and Accepted. (In press).
7. B. Kaan Karamete, H.U. Akay, T. Tokdemir and M. Ger, 'A Simple Unstructured Tetrahedral Mesh Generation Algorithm for Complex Geometries', *Math. and Comp. Modelling J.*, Accepted without revision. (In press).

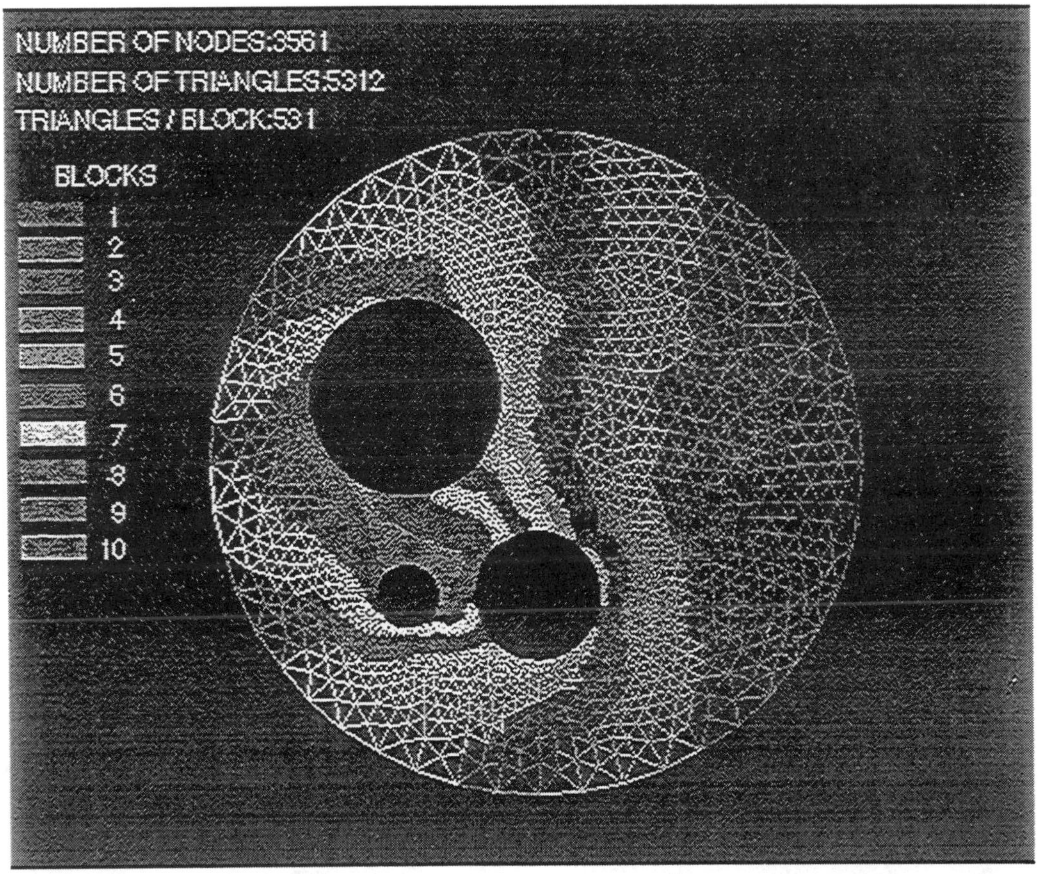
NUMBER OF NODES:3561
NUMBER OF TRIANGLES:5312
TRIANGLES / BLOCK:531
BLOCKS
1
2
3
4
5
6
7
8
9
10

Number of Grid Control Points: 153

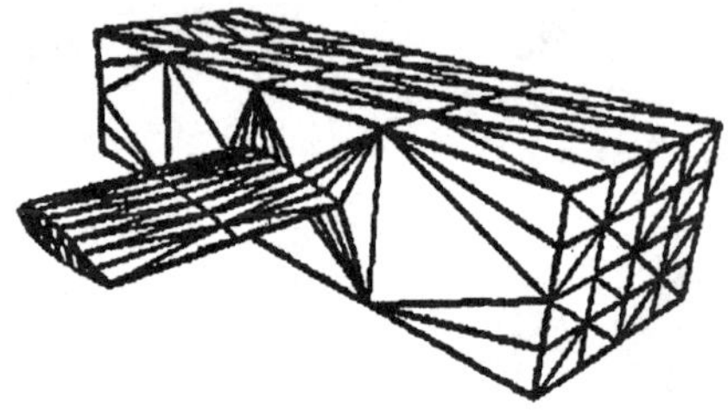

iteration no. 1

Number of Nodes : 153 Number of Tetrahedrons: 352

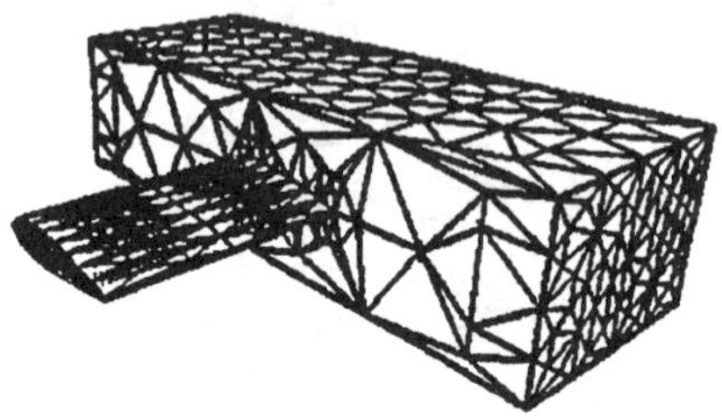

iteration no. 2

Number of Nodes : 1784 Number of Tetrahedrons: 5316

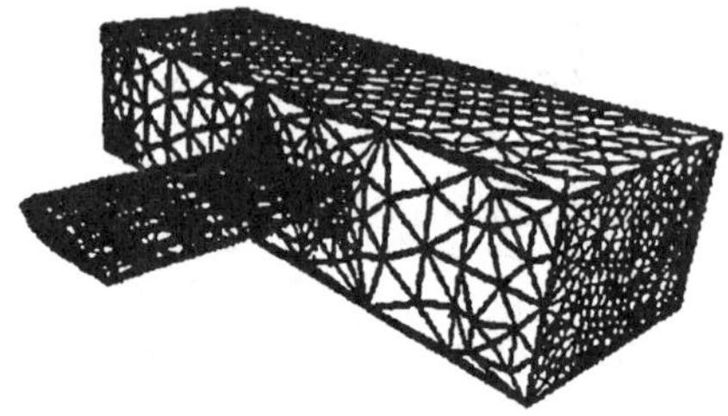

iteration no. 3

Number of Nodes : 16980 Number of Tetrahedrons: 51036

Grid Code: *Mulgri*
Rosa Maria Spitaleri
Istituto per le Applicazioni del Calcolo-V.le del Policlinico 137, Rome, Italy

Mulgri is a modular code for the generation of 2D or 3D boundary-fitted grids (BFGs) on oneblock or multiblock domains [3,4], which couples elliptic generation with multigrid computation for faster applications [1,6].
Let $G^0, G^1, \ldots, G^M$ be a sequence of M+1 grids, with decreasing mesh sizes: $h_0, h_1, \ldots, h_M$, $h_k=2h_{k+1}$ for k=0,..., M-1. The code solves the grid generation quasi-linear system

$$\mathbf{LX}=0 \quad \text{in } \mathbf{B}^* \quad , \quad \Lambda\mathbf{X}=\Phi \quad \text{on } \partial\mathbf{B}^*$$

in the Cartesian co-ordinates $\mathbf{X}$ (ξ,η), where ∂B^* is the boundary of the block B^*. Central differencing provides the discrete equations, along with the associated boundary conditions,

$$L^M X^M = 0 \quad \text{on } G^M \quad , \quad \Lambda^M X^M = \Phi^M \quad \text{on } \partial G^M$$

where ∂G^M is the set of boundary points of the finest grid G^M. The general form of the correction equation of the FAS algorithm and the related boundary conditions are:

$$L^k X^k = \hat{F}^k \quad \text{on } G^k \quad , \quad \Lambda^k X^k = \hat{\Phi}^k \quad \text{on } \partial G^k \tag{1}$$

where

$$\hat{F}^k = L^k(I^k_{k+1}X^{k+1}_a) + I^k_{k+1}(\hat{F}^{k+1} - L^{k+1}X^{k+1}_a)$$
$$\hat{\Phi}^k = \Lambda^k(I^k_{k+1}X^{k+1}_a) + I^k_{k+1}(\hat{\Phi}^{k+1} - \Lambda^{k+1}X^{k+1}_a).$$

X^{k+1}_a is the current approximation on the finer grid G^{k+1} and I^k_{k+1} an appropriate fine-to-coarse transfer operator (restriction).
Indeed the code, which can be used with any number of grids appropriate to the fineness of the grid G^M, has a modular structure, mainly related to the main component of the multigrid computation:

- -relaxation for error smoothing,
- -restriction (fine -to-coarser operator),
- -prolongation (coarse-to-fine operator),
- -multigrid cycling (grid change).

The main module *Mulgri* controls twelve subroutines and auxiliary files for data input and output. The code, written in FORTRAN 77, is portable in computational environments equipped with the usual libraries for mathematical applications. The user can easily integrate the code with the visualization modules in the last release of *Visgen* [2], running on Silicon Graphics Workstations, or interactively use this basic algorithm by the system GENSY.
The code implements a full approximation storage (FAS) algorithm which can be recursively defined on the basis of the FAS *two-grid* method. One iteration step of the (k+1,k) two-grid method for Eqs (1), computing X^{k+1}_{new} from X^{k+1}_{old}, is composed of the following substeps:

1) *Pre-smoothing*: compute $\overline{X}^{k+1}$ by applying υ_1 (≥0) sweeps of a given (non-linear) relaxation method with starting guess X^{k+1}_{old}

$$\overline{X}^{k+1} = \text{Relax}^{\upsilon_1}(X^{k+1}_{old}; L^{k+1}, \hat{F}^{k+1}, \Lambda^{k+1}, \hat{\Phi}^{k+1}).$$

2) *Coarse-grid correction*:

- computation of the residual: $\hat{f}^{k+1} = \hat{F}^{k+1} - L^{k+1}\overline{X}^{k+1}$
- restriction of the residual: $\hat{f}^k = I^k_{k+1}\hat{f}^{k+1}$

- restriction of the current approximation: $\overline{X}^k = \Pi_{k+1}^k \overline{X}^{k+1}$
- computation of the exact solution X^k of: $L^k X^k = \hat{f}^k + L^k \overline{X}^k$
- computation of the correction: $V^k = X^k - \overline{X}^k$
- interpolation of the correction: $V^{k+1} = I_k^{k+1} V^k$
- computation of the correct approximation : $\hat{X}^{k+1} = \overline{X}^{k+1} + V^{k+1}$

3) *Post-smoothing*:: compute X_{new}^{k+1} by applying υ_2 (≥ 0) sweeps of the given relaxation method with starting guess $\hat{X}^{k+1}$

$$X_{new}^{k+1} = \text{Relax}^{\upsilon_2}(\hat{X}^{k+1}; L^{k+1}, \hat{F}^{k+1}, \Lambda^{k+1}, \hat{\Phi}^{k+1}).$$

The non-linear relaxation procedures is applied υ_1 times ($\text{Relax}^{\upsilon_1}$) before and υ_2 times ($\text{Relax}^{\upsilon_2}$) after the coarse-grid correction. Indeed in multigrid schemes Eqs (1) are not solved exactly but still relaxed in multigrid cycles using coarser grids.
In dealing with the main multigrid components, the operators and algorithms we use in the FAS procedure are the following ones:

- relaxation for error smoothing ($\text{Relax}^{\upsilon_1}$, $\text{Relax}^{\upsilon_2}$)→ Gauss-Seidel relaxation ,
- relaxation orderings → lexicographic , red-black, four-colours, zebra-line
- restriction (fine-to-coarse-grid transfer operator) (I_{k+1}^k, Π_{k+1}^k)→ straight injection
- prolongation (coarse-to-fine-grid transfer operator) (I_k^{k+1})
  →piecewise linear interpolation
- coarser grid construction (hierarchy of grids)
  → meshsize doubling (standard coarsening)
- multigrid cycling (grid changing) → V, W cycles.

The code is available for co-operations to advance multigrid grid generation. Anisotropy control can be exerted by application of criteria related to the main term coefficients of the generation system (1)[5]. Results provided by a large experimentation can be acquired by interested users and enlarged with problem solutions dealing with a spectrum of application areas.

References

[1] A. Brandt. Multi-level Adaptive Solutions to Boundary-Value Problems, *Math. Comp.* 31: 333-355 (1977).

[2]R.M. Spitaleri, "VISGEN: A Modular Software for Grid Visualization and Analysis", *Quad. IAC* n 3,1993 (B4)

[3] Spitaleri R.M., Micacchi V." Multiblock Multigrid Grid Generation",*Zeitschrift fuer Angewandte Mathematik und Mechanik--ZAMM*, 6, 1,pp.75-78, 1996.

[4] Spitaleri R.M., Micacchi V., "A Multiblock Multigrid Grid Generation Method for Complex Simulations", *Mathematics and Computers in Simulation* , (to appear).

[5] Spitaleri R.M., Regolo,V., Multiblock Multigrid Grid Generation Methods Overcoming Multigrid Anisotropy, *App. Math. &Comp.*, 84(2-3): 247-276 (1997).

[6] J. F. Thompson, Z. U. A. Warsi and C. W. Mastin. Boundary-Fitted Coordinate Systems for Numerical Solutions of Partial Differential Equations - A Review, *J. Comp. Phys.* 47: 1-108 (1982).

Contribution to Grid Codes Chapter for CRC Handbook of Grid Generation

OVERGRID

William M. Chan

Outline of Capabilities

OVERGRID is a graphical user interface for communicating with grid generation modules tailored for overset grids. Current capabilities include functions for grid editing, grid point redistribution, grid projection, hyperbolic and algebraic surface grid generation, hyperbolic field grid generation and Cartesian box grid generation. A scripting feature allows rapid batch-mode replay of actions performed by the user.

Computing Environment

OVERGRID is written in C, Tcl/Tk, OpenGL and Fortran. It can be used on any machine with X windows, e.g., unix workstations and personal computers.

Code Description

The graphical display (wireframe or smooth shaded) of the input geometry and grids is controlled by the C code with OpenGL library calls. Widgets for toggling display of geometric properties, and widgets for creating and modifying input parameters for the various grid modules are constructed using Tcl/Tk. The grid manipulation and generation modules are called in batch mode via Tcl scripts and the results are automatically loaded into OVERGRID for visualization.

Toggles are available for the display of geometric properties of entities in memory. These include surface normals, tangent vectors, identification numbers and bounding boxes. Grid manipulation functions are available for swapping J/K, K/L, and L/J indices; reversing any of the J, K, or L indices; mirroring about X, Y, or Z=0 planes; scale, translate and rotate operations; extraction of grid subsets; extrapolation of grid planes; concatenation of multiple grids; smoothing of grid subsets; splining, redistribution and projection of grids.

Surface grid generation capabilities include schemes for hyperbolic marching [Chan and Buning, 1995], algebraic marching, and transfinite interpolation [Chan and Meakin, 1997]. Field/volume grid generation capabilities include schemes for hyperbolic marching [Chan and Steger, 1992] and stretched Cartesian box grid generation. The script generation capability automatically records all operations performed by the user into a Tcl script (see usage in the next section).

Grid Generation Process

Currently, OVERGRID reads input geometry or grid files in PLOT3D format [Walatka et. al, 1990]. Work is in progress to allow input of NURBS and trimmed NURBS surfaces as surface geometry descriptions. The surface grid generation step involves identification and extraction of seam curves, and generation of seam surface grids from the seam curves. Seam curves are curves along surface discontinuities, intersection lines between components, and high curvature contours. Hyperbolic or algebraic methods are used to generate surface grids for domains bounded by one or multiple seam curves, respectively. Hyperbolic field grids are then grown from each of the overlapping surface grids. The collection of hyperbolic field grids are then embedded in a Cartesian box grid with a uniform core spacing and stretched outer layers. If a parameter change (such as grid spacing or number of grid points) is required after completion of all grid generation steps, the automatically generated script can be modified and replayed rapidly in batch mode to recreate the entire process.

Source/Availability

OVERGRID is a product of the Rotorcraft Aeromechanics Branch at NASA Ames Research Center and has been supported by the DOD HPC Modernization Program under the CHSSI initiative. It is available to companies, government laboratories and universities of the U.S. for internal use. The code can be requested by sending a letter on company letterhead to the author. Formal release of the code requires fully executed Space Act Agreement documents between the requesting organization and NASA Ames Research Center. Code dissemination occurs after the request is approved by NASA management.

Example Demonstrations

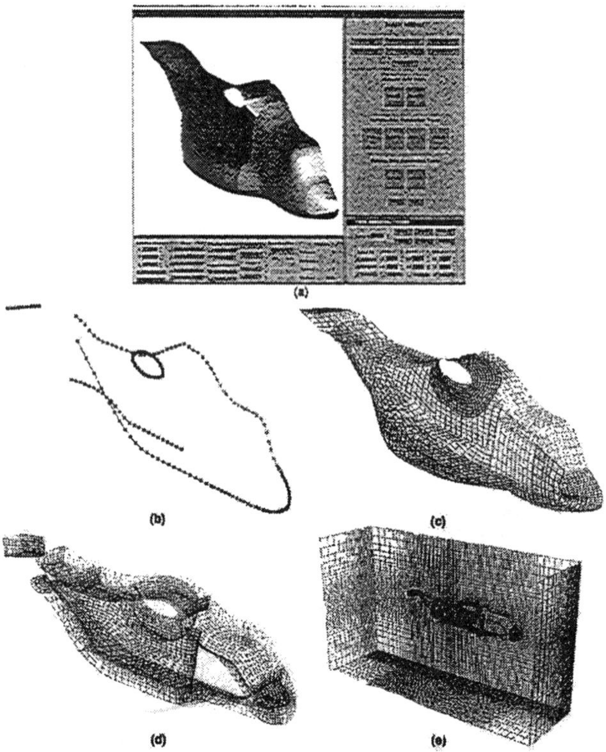

Figure 1: Grid generation process using OVERGRID. (a) OVERGRID windows with surface geometry display. (b) Seam curves. (c) Overlapping hyperbolic and algebraic surface grids. (d) Slices of overlapping hyperbolic volume grids. (e) Stretched Cartesian box grid.

References

Chan, W. M. and Steger, J. L., "Enhancements of a Three-Dimensional Hyperbolic Grid Generation Scheme," *Appl. Math. and Comput.* **51**, 181–205, 1992.

Chan, W. M. and Buning, P. G., "Surface Grid Generation Methods for Overset Grids," *Computers and Fluids*, **24**, No. 5, 509–522, 1995.

Chan, W. M. and Meakin, R. L., "Advances Towards Automatic Surface Domain Decomposition and Grid Generation for Overset Grids," AIAA Paper 97-1979, Proceedings of the AIAA 13th Computational Fluid Dynamics Conference, Snowmass, Colorado, 1997.

Walatka, P. P., Buning, P. G., Pierce, L. and Elson, P. A., "PLOT3D User's Manual," NASA TM 101067, 1990.

Remarks

Work is in progress to expand the capabilities of OVERGRID. The major items being worked on include automatic seam curve detection, the use of NURBS and trimmed NURBS surfaces as input geometry definition, and many others. Further concepts on overset grid generation can be found in the chapters on 'Composite Overset Structured Grids' and 'Hyperbolic Methods for Surface and Field Grid Generation'.

Name of Grid Code : PMAG

Outline of Capabilities :

PMAG (Parallel Multiblock Adaptive Grid) system is design to perfrom grid adaptation on arbitrarily connected/overlapped structured multiblock toloplogies.

Computing Environment :

PMAG is written in C and FORTRAN 77 utilizing MPI as a message passing interface and is executable on unix machines with MPI.

Code Descriptions :

The grid redistribution strategy is employed for grid adaptation in PMAG. The redistribution is accomplished by solving elliptic generation system with appropriate control functions. The control functions are calculated using the boolean sum of contributions from geometric and solution features based weight functions. In computation of weight functions, all dependent variables, and their normalized/scaled derivatives and curvatures are used and are evaluated with minimum user input. These weight functions have been proven to detect flow features of varyiing strengths, including vortices, boundary layers, vortex–vortex and vortex–surface interactions, separated shear layers and multiple shocks of different intensity. The boundary point movement is accomplished by defining boundary surfaces as NURBS. Robust solution interpolation /search scheme based on projection method is accessible as one of the functionality of the system. The grid blocks are computed in individual processes that may be distributed over multiple processors. PMAG can also be used as a multiblock elliptic generator.

Source/Availability :

PMAG source code is distributed and can be requested by contacting Bharat K. Soni (bsoni@erc.msstate.edu) at NSF Engineering Research Center, Mississippi State University .

Example Demonstrations :

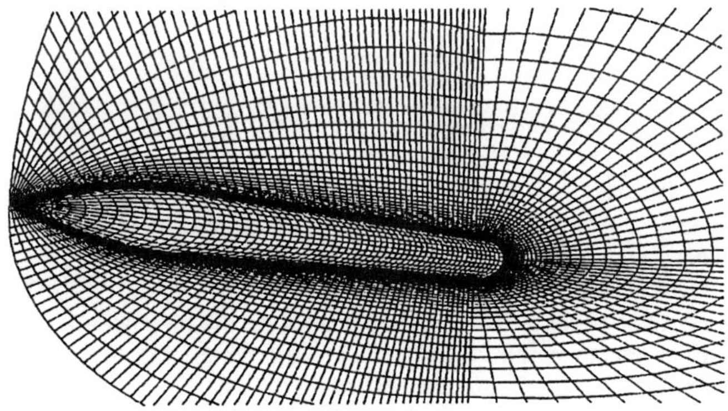

Grid around missile before adaptation

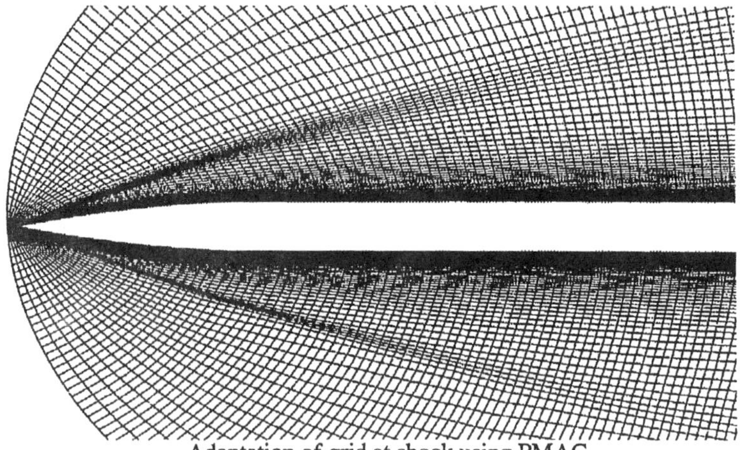

Adaptation of grid at shock using PMAG

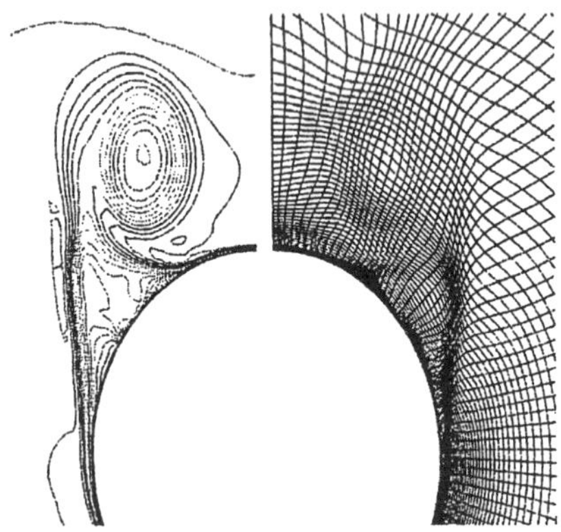

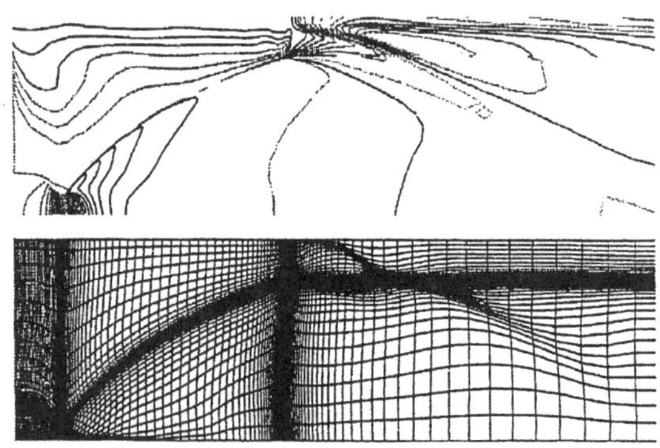

Adaptation of grid at vortex using PMAG

Nozzle solution with adapted grid

References :

[1] http://www.erc.msstate.edu/~bsoni/research .

[2] Apte, M., "Parallel Adaptive Structured Grid Generatorn for Structured Multiblock Domains" ,Masters Thesis, Computational Engineering,Mississippi State University, May 1997.

[3] Soni,B., Thornburg,H., Koomullil,R., Apte, M., and Madhavan, A., "PMAG:Parallel Multiblock Adaptive Grid system, to be published in the proceedings of the 6th International Conference on Numerical Grid Generation in Computational Field Simulations", Cross, Eiseman, Hauser, Thompson and Soni(eds.) , University of Greenwich, London, July 6–9,1998.

[4] Thornburg, H., and Soni, B., "A Structured Grid Based Solution–Adaptive Technique for Complex Separated Flows", Journal of Applied Mathematics and Computation 89:259–273,1998.

SiGMA — 2-D Quasi-Isometric Grid Generator

Gennadii A. Chumakov
Sobolev Institute of Mathematics
Novosibirsk, Russia
`chumakov@math.nsc.ru`

Sergei G. Chumakov
University of Wisconsin
Madison, WI
`chumakov@math.wisc.edu`

Capabilities. The new SiGMA package uses the state-of-art methods in regular 2-D grid generation and is capable of constructing regular quasi-isometric grid of high quality in arbitrary given domain. The method is described in [1]. Moreover, we can obtain a grid that is nearly orthogonal everywhere or a quasi-conformal grid, when the boundary is a smooth closed curve .

Features. • Good control of the cell size as the grid is refined • Parallelogram cells in the corners as the grid is refined • Semi-automatic control of the grid angle over the field • Digital geometry specification • Fixed or "free" boundary points • Semi-automatic adaptation • Supports regular, C and O topologies

Computing environment. The package contains a number of routines written in FORTRAN language. The current version is compiled on PC486DX/66 and requires only basic FORTRAN compiler (e.g., Microsoft FORTRAN, any version).

Grid generation process. SiGMA package uses a new structured grid generation method, which is based on the theory of quasi-isometric mappings and conformal representation of spherical and hyperbolic geometries. The grid in a physical domain $\mathcal{D}$ is generated by a *quasi-isometric* mapping of the unit square $\mathcal{R}$ onto $\mathcal{D}$.

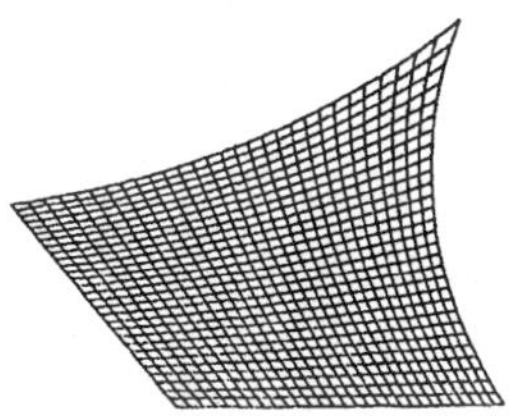

Figure 1. Hyperbolic geometry

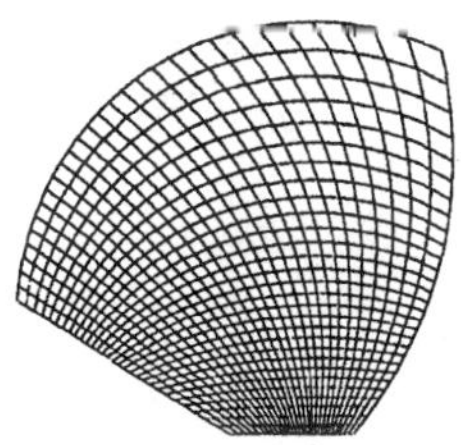

Figure 2. Spherical geometry

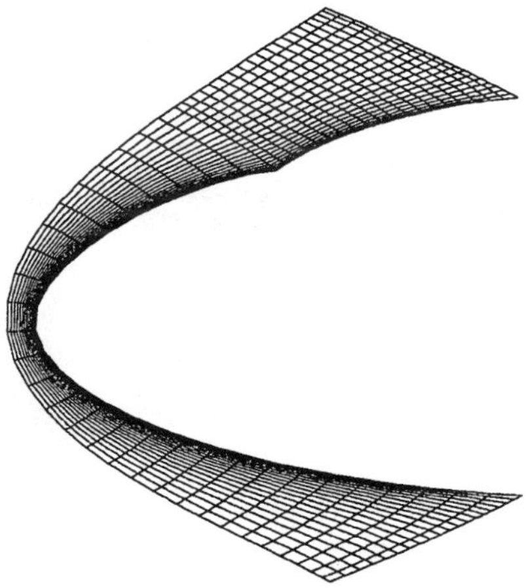

Combination of free and fixed boundary points

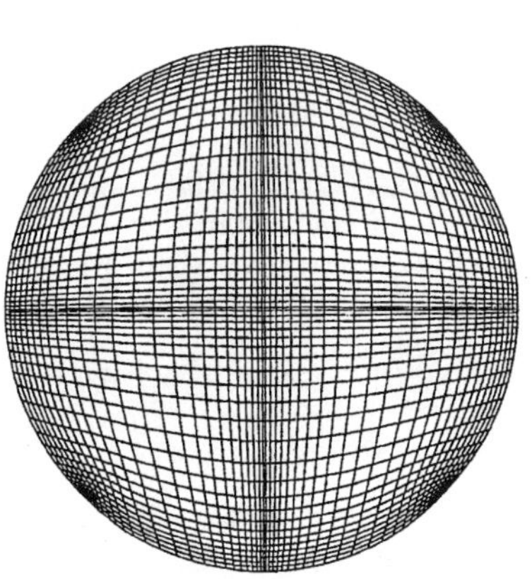

Quasi-conformal grid with adaptation

1. G. Chumakov, S. Chumakov, A Method for the 2-D Quasi-Isometric Regular Grid Generation. *J. Comput. Phys.* (in press)

Rezoning of boundaries (zone subdivision, zone combination)

Creation of boundary zone groups

Step through most-skewed faces

Local smoothing

Triangular/Tetrahedral Volume Meshing

Automatic 2-D triangular or 3-D tetrahedral meshing (Bowyer/Watson Delaunay scheme)

Multiple region mesh generation (multiple fluid/solid zones)

Global mesh refinement based on size and skewness of existing cells

Local face and volume mesh refinement in selected regions

Face swapping and smoothing to reduce skewness

Interactive control over skewness limits, boundary closeness, cell size, and number of cells during mesh generation and refinement

Viscous Boundary Layer Meshing

Automatic growth of triangular (wedge) and rectangular (hexahedral) prisms

Grow normal to selected surfaces, or along specified vectors

Control layer size, growth, and orthogonality

Automatically retriangulates adjacent surfaces

Automatically uses adjacent quadrilateral faces

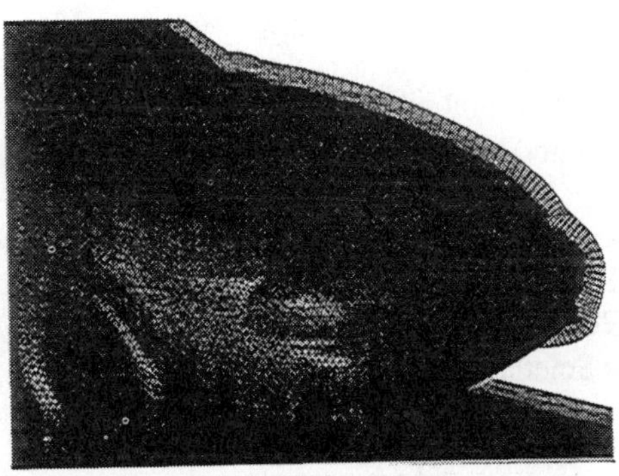

Viscous hybrid mesh on an automotive exterior. Prisms provide an efficient, high-quality mesh around the vehicle surface, and retain the ease of triangular surface meshing.

Pyramid Transitions

Automatically create pyramids on selected quadrilateral faces

Provide conformal transitions between hexahedral and tetrahedral regions

Non-Conformal Interfaces

Two meshes on same surface with unshared nodes

Easily join separately-created mesh regions, or meshes of different element type

Leverages non-conformal interface capabilities in FLUENT/UNS and RAMPANT 4.2

Mesh Diagnostics

Graphical display of unmeshed, free, or multiply-connected boundary faces/edges

Graphical display of faces or cells by skewness, size, aspect ratio, and other quality metrics

Reporting of mesh size (number of elements), minimum/maximum/average face or cell skewness, minimum/maximum/average face or cell size

Histogram display of boundary face size or skewness

Histogram display of cell size or skewness

Quantitative XY-plotting of data

Interface and Graphics

Fully interactive graphical and text-based user interfaces

Journaling and transcripting

Diagnostics and error trapping

Client-server architecture for remote execution

Powerful graphics, mesh visualization, and animation (complete Cortex 3.2 features)

On-screen, mouse-based view manipulation (rotation, translation, magnification)

Light shading, color display of boundary surfaces

Face/cell shrinkage for improved visualization

Graphical probing of data

Extensive hardcopy options

Supported Hardware

TGrid is supported on most common UNIX, Windows 95, and Windows/NT platforms. Please contact Fluent Inc. for details.

Index of Grid Codes

Code Title : TIGER (Turbomachinery Interactive Grid genERation)
Affiliation : Catalpa Research, Inc.
13014 Astalot Drive, Suite C
Huntsville, AL 35803-1802, U. S. A.
TEL: 205-883-2045 / FAX: 205-881-0508
Capability :

- Customized for turbomachinery applications
- Suitable for internal, external, and internal-external flow fields
- Built-in intelligent processes allow very quick turn-around time to obtain a high-quality numerical grid
- Boundary conditions and flow initialization for many popular turbomachinery flow solver

Plateform : SGI IRIX 5.3 or above, 32MB RAM, Other UNIX/LINUX and NT/Win95 versions will be released summer 1998.
Languages : C^{++}/Fortran
Interface : Motif
Grpahics : OpenGL
Grid Generation Process :

- Geometry Data Preparation (Utilities available to convert geometry raw data in various formats)
- Parameter Input (Usually less than 20 simple numbers)
- Ruler line specification (graphically)
- Frame setup/ modification (graphically)
- Grid and boundary conditions output (automatically)

Source : Catalpa Research, Inc.
13014 Astalot Drive, Suite C,
Huntsville, AL 35803-1802, U. S. A.
TEL: 205-883-2045/ FAX: 205-881-0508
Contact : Dr. Alan M. Shih (shih@catalpa.net)
References :

1. Shih, M.H., *"Towards A Comprehensive Computational Simulation System for Turbontachinery',* Dissertation, Mississippi State University, May, 1994.
2. Shih, M.H., Yu, T-Y and Soni, B.K., *"TIGER: A Turbomachinery Grid Generation/ Simulation Sys*tem", AIAA32nd Aerospace Sciences Meeting, AIAA-94-0207, Reno, NV, January 1994.

Pictures :

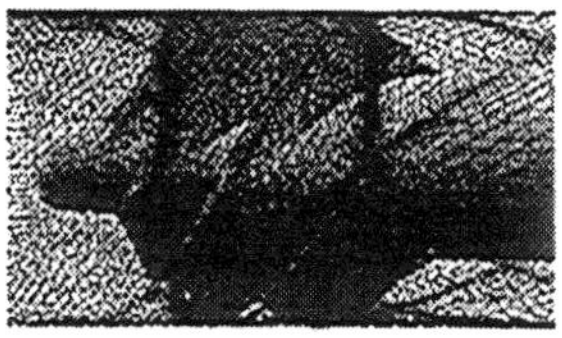

Index of Grid Codes

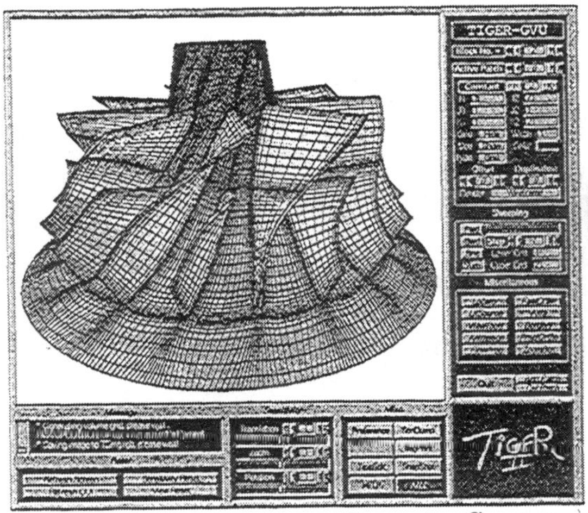

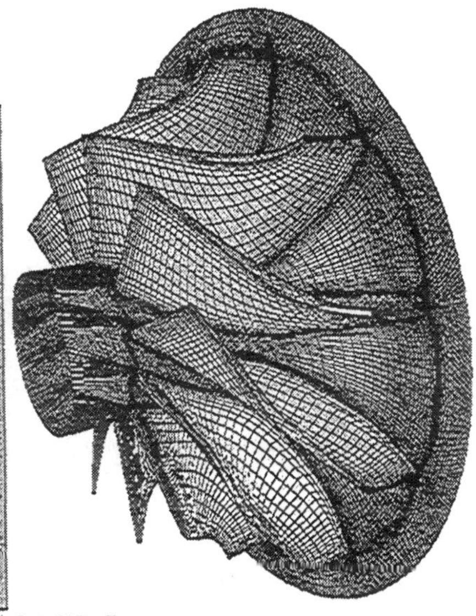

Geometry Courtesy of ACD, Inc.

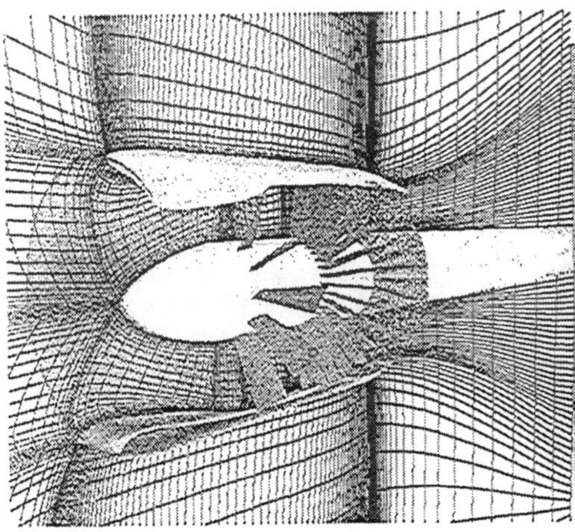

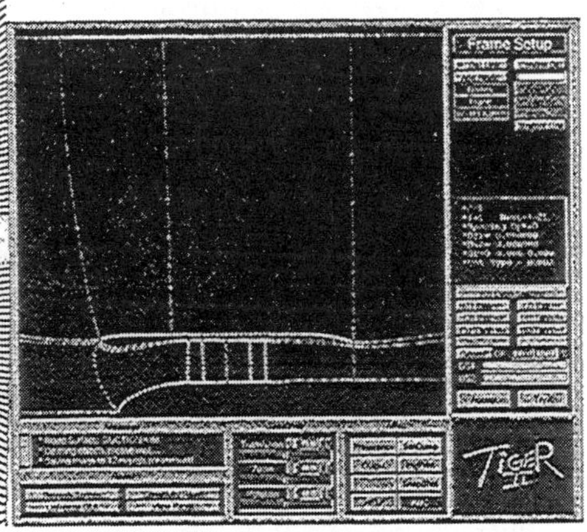

The Volume Grid Manipulator (VGM)

Stephen J. Alter*
Lockheed Martin Engineering & Sciences Company, Hampton, Virginia

Summary

The Volume Grid Manipulator[1] (VGM) is designed to be a tool that augments any grid generation process. It is designed to make a cohesive implementation of Computer Aided Design (CAD), grid generation, and Computational Fluid Dynamic (CFD) codes. The gaps that usually exist between these types of codes consist of data format differences, but can also be attributed to grid quality and grid usability. The VGM tool fills the gaps between these codes by offering numerous data formats for input and output, as well as a user programmable language that contains 12 different commands. Each command performs a very simple manipulation, but when these commands are combined with one another, a variety of complex grid altering techniques results. All techniques afforded by VGM enable the improvement of structured surface and volume grids towards high fidelity CFD grids, including:

- Conversion of inviscid spaced grids to viscous clustered grids.
- Multiple zonal adaption to multiple flow structures to improve the efficient use of available grid points.
- Grid generation from the six known faces of a computational block.
- Smoothing of kinked grids through the use of Trans-Finite Interpolation[2] (TFI), and two new techniques called Parametric Re-mapping (PRM) and Hermite Vector Interpolation (HVI). Smoothing with Laplacian operators on physical grid points and grids mapped to local coordinate systems for improving directional dependencies is also available with the scripting language.
- Extrusion of existing volume grids to develop appended wake domains for CFD simulations.
- Topological changes of existing grids.
- Domain decomposition and reconstruction from multiple block discretizations.

This is a small list of the capabilities of the VGM tool. Each of these alterations can be broken down into a series of the 12 different VGM commands, because each command, though simple in its abilities, has various controlling parameters (i.e., arguments). As examples of some of the above manipulation capabilities, multiple zonal adaption in a volume grid, conversion of volume inviscid to viscous grid, and the smoothing of a surface grid will be shown.

Examples

For the multiple zonal adaption, assuming a supersonic flow field over a simple sphere-cone configuration, the grid will be adapted to the wall and the outer domain bow shock. The bow shock is assumed to lie along the iso-parametric surface shown in figure 1a. The grid is adapted by choosing intermediate planes down the body to define the location of the flow structures, and PRM is used to bridge these distributions down

*Senior Aeronautical Engineer

the body, as shown in figure 1b. The result is a smooth grid that adheres to the physical problem, based on the chosen flow structures.

The conversion of inviscid volume grids to viscous grids is accomplished by clustering a grid in the computational domain using a distribution function such as Vinokur's.[3] By redistributing a grid in the computational domain, the distribution is the same everywhere in the computational domain, but not in the physical domain. In the physical domain, the redistribution becomes a scaling that is uniform at each independent grid line. The side effect is that the grid in the other computational directions that do not have an altered clustering maintains identical characteristics. As shown in figure 2, an inviscid grid on an X33 VentureStar is clustered in the K-direction and the I- and J-direction lines do not change in character. Hence the quality of the original grid is preserved while the clustering needed to resolve boundary layer gradients is obtained. The conversion of inviscid volume grids to viscous grids is very simple, yet offers many advantages to generating viscous grids from the start. First, the development of inviscid grids with a Poisson solver is usually faster than generating viscous grids because the number of points required to do inviscid computations is much smaller. Since most Poisson solvers employ SOR algorithms which tend to require a number of iterations that is linearly proportional to the number of grid points, the fewer the grid points used to develop a volume grid the faster the convergence. The second advantage is that the conversion process does not degrade the grid quality, but usually enhances orthogonality in the direction of the distribution change.

The smoothing of surface and volume grids can be done using a variety of methods. Two of the most powerful are that of PRM and the Laplacian operator. Shown in figure 3 is a surface grid that has a discontinuous lines in the I-direction. This discontinuity can be eliminated with PRM by computing the normalized arclength distribution in the J-direction, smoothing in the I-direction and redistributing the grid based on the new arclength mapping. The PRM method has a dependent based hermitian spline function which is used in this case. The Laplacian operator can also be used in this manner on the original normalized arclength, but the smoothing is done as a sweep through the grid in the I-direction. Any number of sweeps can be done, which usually results in the smoothing of the discontinuous region, also shown in figure 3. The power of PRM and Laplacian type smoothing is that the grid remains on the surface as it is manipulated.

Conclusion

In summary, the VGM code offers numerous techniques to improve the quality of discretized domains towards the generation of high fidelity grids. The tool can improve less than adequate grids in a quick and efficient manner. The language of the VGM code is simple, and provides the combination of commands to do sophisticated and intricate grid alterations with algebraic techniques. Use of the VGM code can significantly reduce the time necessary to develop volume grids for CFD simulations, and improve CFD simulations through grid adaption and topology changes.

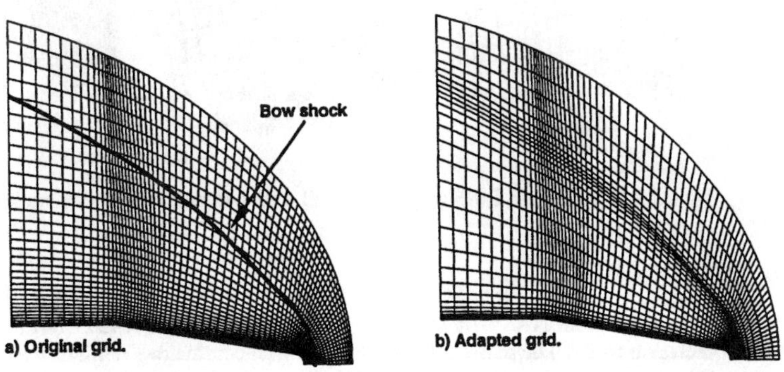

Figure 1: Multiple zonal grid adaption on a simple sphere cone.

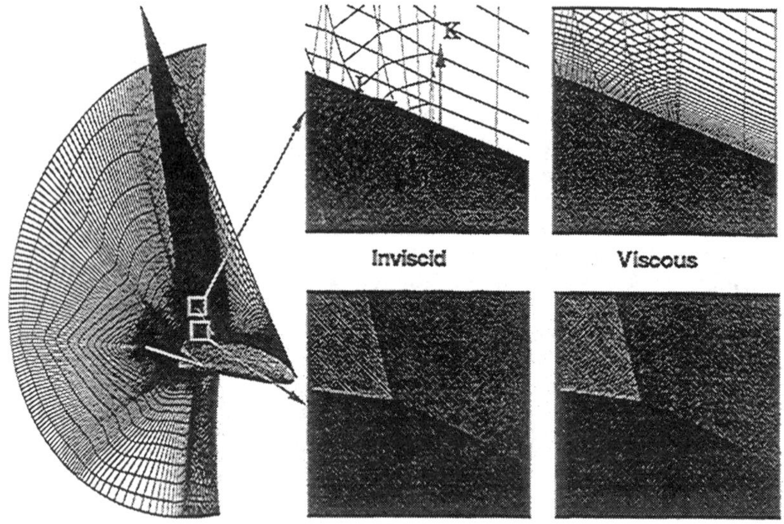

Figure 2: Conversion of inviscid grid to a viscous one.

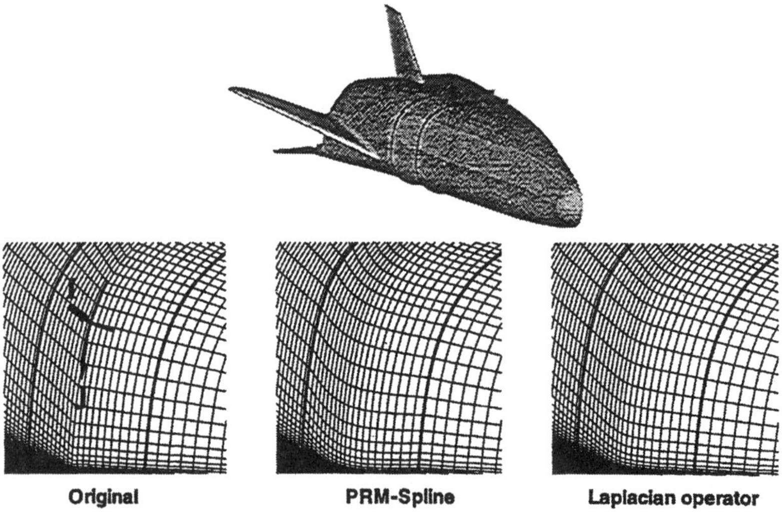

Figure 3: Parametric Re-Mapping (PRM) and Laplacian operator smoothing.

Acknowledgment

This work was performed under NASA Langley Research Center contract NAS1-96014. The author wishes to express his appreciation to Dr. David Schuster for his editorial comments, William Kleb for his contributions to the typesetting of this document and K. James Weilmuenster for permitting this work to be done.

References

[1] Alter, S. J., "The Volume Grid Manipulator (VGM): A Grid Reusability Tool," NASA CR–4772, April 1997.

[2] Soni, B. K., "Two– and Three–Dimensional Grid Generation for Internal Flow Applications of Computational Fluid Dynamics," AIAA Paper 85–1526, 1985.

[3] Vinokur, M., "On One-Dimensional Stretching Functions for Fininte-Difference Calculations," NASA CR–3313, 1993.

Name of Grid Code : VGRIDns

Outline of Capabilities :

1) Convenient generation of tetrahedral grids for the computation of viscous and inviscid flows around complex configurations.
2) Smooth grids by means of the source concept and solving an elliptic PDE on a 'transparent' Cartesian background grid.
3) User control over grid distribution through simple adjustment of source parameters such as spacing, intensity, and direction of source propagation.
4) Graphics capabilities for convenient visualization of surface and volume grids and interactive manipulation of grid distribution and quality.
5) Capable of generating anisotropic stretched grids for improved efficiency of grid clustering.
6) Restart capability through a recurrent local/global renumbering system to reduce computer memory requirement for generating large grids on small workstations.
7) Grid quality improvement by a combination of the Laplacian ("spring") smoothing and the local remeshing.
8) Grid movement capability.
9) User–friendly; extensively used by engineers and researchers from industry, universities, and government organizations.

Computing Environment :

VGRIDns is written in FORTRAN and C languages. It runs on most UNIX–based platforms in the batch mode. The graphic version runs on the Silicon Graphics workstations.

Code Descriptions :

The unstructured grid generation code VGRIDns is based on the Advancing–Front (AFM) and the Advancing–Layers (ALM) methods. The generation of a 'Navier–Stokes' grid is accomplished with VGRIDns in three main steps: (1) generation of triangular surface grid by the AFM and ALM, (2) generation of thin–layered tetrahedral grid in the boundary layer by the ALM, and (3) generation of regular (inviscid) tetrahedral grid outside the boundary layer by the AFM. Although the generation of a grid is divided into separate stages with this approach, the entire process is performed in a single run with automatic transitions from one stage to another. For the generation of an inviscid grid, the Advancing–Layers step is simply skipped in VGRIDns.

Both the Advancing–front and the Advancing–Layers methods are based on the marching techniques. A grid is generated, in these methods, by originating tetrahedral cells from triangulated boundaries and marching the volume mesh into the computational domain. Unlike the conventional AFM, which introduces cells in the field in a totally unstructured manner, the ALM generates layers of thin tetrahedral cells in a more orderly fashion while maintaining many advantageous features of the AFM. The new strategy reduces many grid generation complexities which usually arise from the floating–point operations of extremely small numbers associated with viscous grids.

During the marching process, the grid clustering information is provided by a 'transparent' Cartesian background grid overlaying the entire domain. Included in the background grid are a number of prescribed 'point' and 'line' sources for defining local grid clustering. By solving an elliptic equation, the grid characteristics are smoothly diffused from the sources onto the background grid nodes. The problem is similar to that of the transfer of heat in a conducting medium.

Two main operations are involved in the ALM: 1) computation of surface vectors along which the grid points are distributed and 2) construction of a pattern of compatible cell connectivities within the thin layers. The surface vectors are calculated using a robust iterative algorithm based on an 'equal–angle' criterion. The connectivities among the tetrahedral cells are predetermined by an efficient, all–integer algorithm which eliminates the need for a series of computer–intensive, floating–point calculations as used in the conventional AFM.

Thin layers of tetrahedral cells are formed by inserting new points along the surface vectors and connecting the points according to the predetermined connectivity pattern. The layers continue advancing until some criteria based on the background grid information and/or the approaching fronts stop them on all fronts. At this point, the process automatically switches from the advancing–layers to the advancing–front mode to generate regular grid outside the boundary layer. With a common background grid controlling both methods, the transition from thin layers to the regular grid becomes gradual and continuous. Also the fact that the number of layers vary from one location to another adds to the flexibility of the method and the smoothness of the generated grids.

Another feature of VGRIDns which is of practical importance is its ability to generate multi–directional, anisotropicly stretched grids. With such a capability, fewer points are distributed in the directions of reduced flow gradient with no loss of grid resolution in other essential direction(s). Grid stretching results in at least a factor of three reduction in overall number of cell.

Source/Availability :

VGRIDns is developed by the U.S. government and is in the public domain.

To obtain a copy of the code executable, please contact:

Shahyar Z. Pirzadeh
Mail Stop 499, NASA Langley Research Center,
Hampton, VA 23681–0001 Phone : (757) 864–2245
Fax : (757) 864–8469 E–mail: s.pirzadeh@larc.nasa.gov

Example Demonstrations :

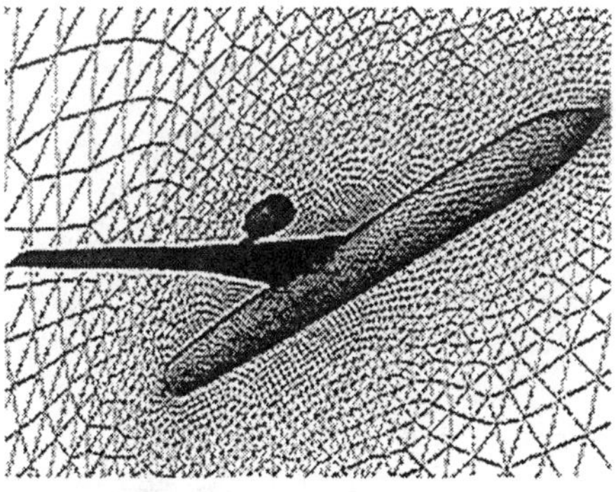

Surface grid and triangulation on the plane of symmetry.

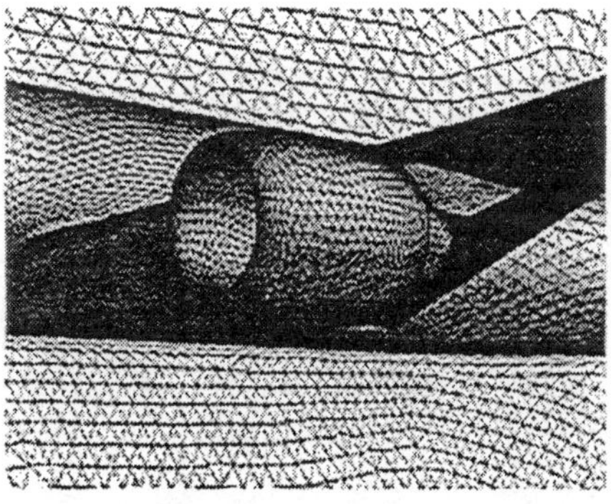

Close–up view of wing/nacelle/pylon showing anisotropic stretched grid at the wing and nacelle leading edges.

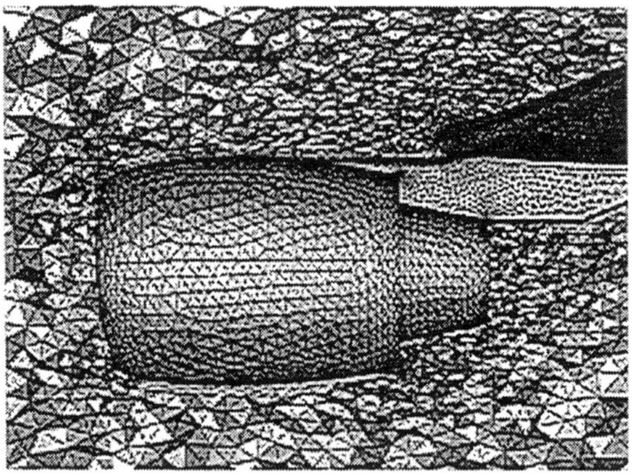

Close–up view of nacelle/pylon surface grid
and a cross section of the volume grid.

References :

[1] Lohner, R. and Parikh, P., "Generation of Three–Dimensional Unstructured Grids by the Advancing Front Method," International Journal of Numerical Methods in Fluids, Vol. 8, 1988, pp. 1135–1149.

[2] Pirzadeh, S., "Structured Background Grids for Generation of Unstructured Grids by Advancing Front Method," AIAA Journal, Vol. 31, No. 2, 1993, pp. 257–265.

[3] Pirzadeh, S., "Three–Dimensional Unstructured Viscous Grids by the Advancing–Layers Methods," AIAA Journal, Vol. 34, No. 1, 1996, pp. 43–49.

[4] Pirzadeh, S., "Progress Toward a User–Oriented Unstructured Viscous Grid Generator," AIAA Paper 96–0031, 1996.

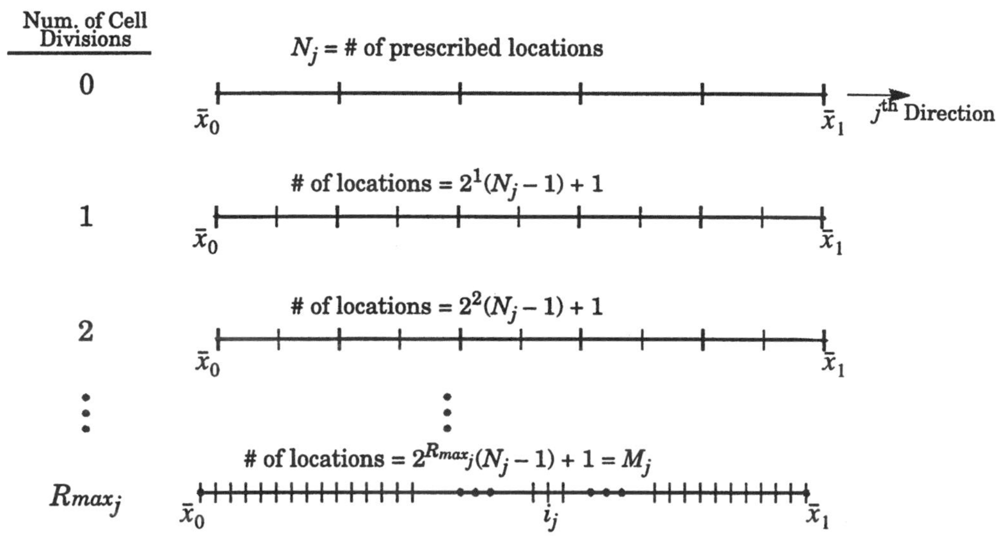

Figure A-1: Specification of integer coordinate locations for a coordinate direction with N_j prescribed locations.

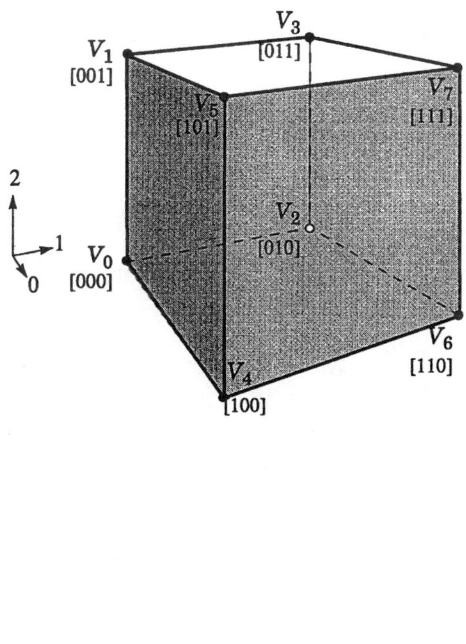

Appendix B : Grid Configurations

Bharat K. Soni, Mississippi State University

Various grid configurations involving structured multiblock grids, unstructured grids, and hybrid/generalized grids are presented with brief description of grid strategy, generation methodology, geometry specification and pictorial views. The information and basic description of these configurations were solicited from the grid community at large. This section is devoted to present grid configurations contributed by various sources. In particular, grid configurations contributed by following developers are included in this section:

DO–328 Euler Mesh

Hans Bleecke and Thomas Streit

2D Quasi–Isometric Grid Around an Airfoil

Gennadii A.Chumakov and Sergei G.Chumakov

Space Shuttle Launch Vehicle Chimera Grid System

Reynaldo J.Gomez III

X–24A Lifting Body Chimera Grid System

James S. Greathouse

CUBIT Continuum Element Mesh: Open Celled Foam Skeleton

David M.Hensinger

Automatic Generation of Unstructured Grids for Viscous Flow Applications Using Directional Refinement

A.R. Jahangirian and L.J. Johnson

0-8493-2687-7/99/$0.00+$.50

Hybrid Prismatic/Tetrahedral Grid for a Burner Geometry
Yannis Kallinderis

Hybrid Prismatic/Tetrahedral Grid for High Speed Civil Transport Aircraft
Yannis Kallinderis

Tests for Two–Dimensional Grid Generation : Suggestions
Khairullina O.B., Sidorov A.F., and Ushakova O.V

Hybrid Grid Generation for CFD Applications
Roy P. Koomullil and Bharat K. Soni

Grid Generation with Mesh Density Functions
Jan Kucwaj

Tetrahedral Viscous Grid Around a Complete F–16 Aircraft Configuration
Shayar Z. Pirzadeh

Orthogonal Grid Generation on Surfaces
R.Ganesh Rajagopalan and V. Baskaran

Impeller with Multiple Splitter Blades
Ming–Hsin Shih

Internal Combustion Engine Port Flow Configuration
Ming–Hsin Shih

Moving Overlapped Grids for Direct Numerical Simulation of Particle–Fluid Interactions
Tom I–P. Shih

DO-328 Euler mesh

Hans Bleecke,Thomas Streit
DLR, Institute of Design Aerodynamics, Braunschweig

Purpose:

Design of a laminar wing for a regional airplane (wing, body and nacelle). The DLR FLOWer code is used to solve the Euler equations and to perform the wing design.

Grid description:

The mesh which is used for the Euler solver has 2,650,532 points and is divided in 36 blocks. For mesh generation the mesh was subdivided in more than 300 blocks. It is a structured mesh. The mesh includes a mixture of different topology types. The configurations surface is sourrounded by a thin layer consisting off O-type mesh blocks. Therefore the used topology is also suited to generate a mesh for a Navier-Stokes application. The required modifications to accomplish this, would be mostly restricted to the innermost layer of blocks.

This mesh was generated with the DOGRID grid code, using elliptical mesh generation.

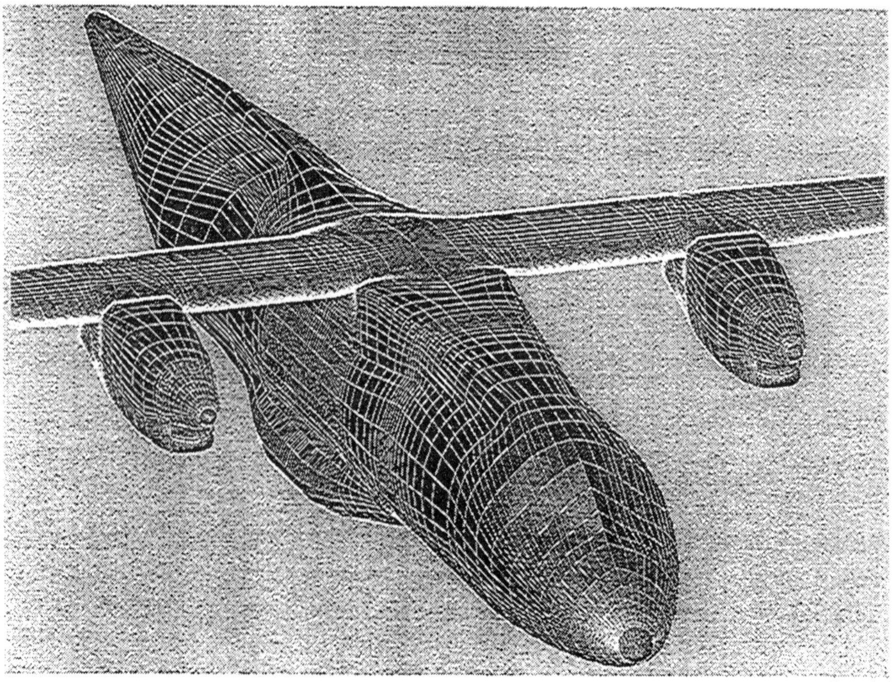

Surface mesh for DO-328 configuration (coarse mesh)

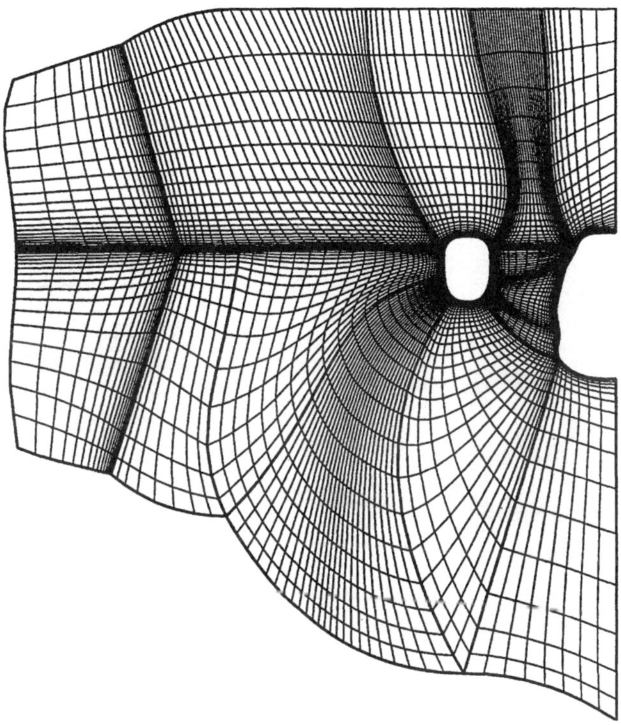

Mesh for a cross section placed near to the wings leading edge. It shows inner blocks and the toplogy used for the lower and upper wingside.
DO-328 Euler mesh

2-D Quasi-Isometric Grid Around an Airfoil

Gennadii A. Chumakov
Sobolev Institute of Mathematics
Novosibirsk, Russia
`chumakov@math.nsc.ru`

Sergei G. Chumakov
University of Wisconsin
Madison, WI
`chumakov@math.wisc.edu`

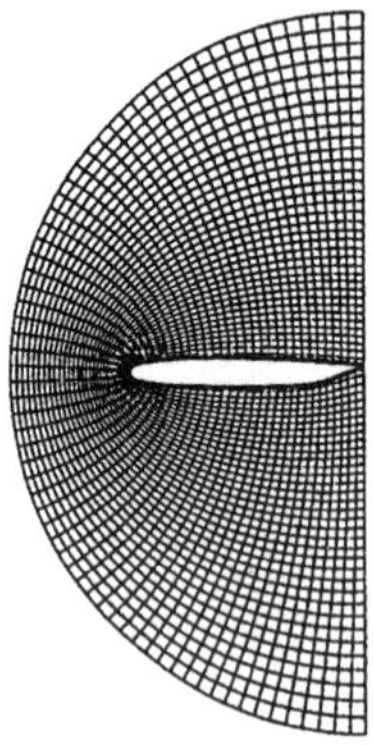

Figure 1.
A quasi-isometric grid around an airfoil.

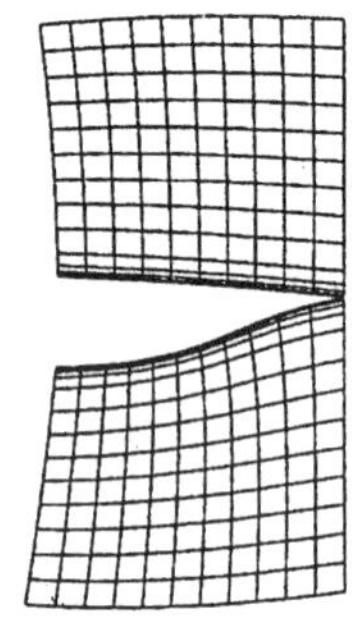

Figure 2.
The grid cells remain parallelograms near corners as the grid is refined.

Grid description

The grid given above is a product of a new structured grid generation method developed in [1-3], and based on the theory of quasi-isometric mappings, conformal representation of spherical and hyperbolic geometries and numerical grid generation [4]. The grid is generated by a *quasi-isometric* mapping of the unit square $\mathcal{R}$ onto the physical domain $\mathcal{D}$; the mapping is the unique solution of the following BVP: given a quasi-isometric correspondence between $\partial\mathcal{R}$ and $\partial\mathcal{D}$, to extend it inside $\mathcal{R}$ as the quasi-isometric solution of the appropriate Beltrami system with coefficients that contain one unknown parameter, which has to be found as well. The method allows more direct control of the grid cells' size and angle as the grid is refined. Among others effects, the grid cells remain parallelograms near the corners of the given quadrangular domain. The illustration to this fact is given in the Figure 2.

We used the grid code SiGMA, a FORTRAN package compiled on a 66MHz PC. It took about one hour of computing time to generate the 70×40 grid of the above configuration.

References

1. S. Chumakov, G. Chumakov: A Method for Generation of 2-D Quasi-isometric Grids. *Abstracts of the Third Mississippi State Conference on Differential Equations & Computational Simulations*, Mississippi State University, May 16-17, 1997.

2. S. K. Godunov, V. M. Gordienko and G. A. Chumakov: Variational Principle for 2-D Regular Quasi-isometric Grid Generation. *Comp. Fluid Dyn.*, Vol. 5, p.99–118. (1995)

3. G. A. Chumakov: Conformal Parameterization of Curvilinear Quadrangles by Means of Geodesic Quadrangles on Surfaces of Constant Positive Curvature. *Siberian Math. J., Vol. 34(1)*, pp. 172–180. (1993)

4. J. F. Thompson, Z. U. A. Warsi, C. W. Mastin: *Numerical Grid Generation*, North-Holland, New York (1985)

Space Shuttle Launch Vehicle Chimera Grid SystemReynaldo

J. Gomez
III/Mail Code EG3/NASA Johnson Space Center

Description:

This grid system was developed to simulate the flow over the complete Space Shuttle Launch Vehicle (SSLV) during ascent. This Chimera/overset grid system consists of 113 zones containing a total of 16,646,854 volume points. The vehicle surface is represented by 264,048 points with a typical cell size of 4" x 4" with higher resolutions in regions with significant changes in curvature. Surface grids were generated using the ICEM–CFD code. The original geometry was received in IGES format. Volume grids were created using the hyperbolic grid generator HYPGEN developed at NASA Ames Research Center. Typical viscous volume grids extend 60 inches outward from the SSLV surfaces. The initial spacing normal to the surface is equal to 0.00016 inches corresponding to a y+ value approximately equal to 1.0 at a distance 10 percent from the External Tank nose. Overall communication is achieved through a set of Cartesian background grids completely enclosing the SSLV. A single farfield grid encloses the Cartesian grids and extends outward approximately 10 Orbiter lengths. Grid communication was established using the PEGSUS code developed at Arnold Engineering Development Center. Time required to generate this grid system was on the order of 2 to 2.5 man–years, including the initial learning curve for most of the grid generation tools used in this effort. The following individuals generated some part of this grid system: Daniel G. Pearce, Scott A. Stanley, Lockheed Engineering and Sciences Co., Reynaldo J. Gomez and Gerald J. Le Beau, NASA Johnson Space Center, William M. Chan and Ing–Tsau Chiu, MCAT Institute.

Surface grids on the aft portion of the Space Shuttle Launch Vehicle

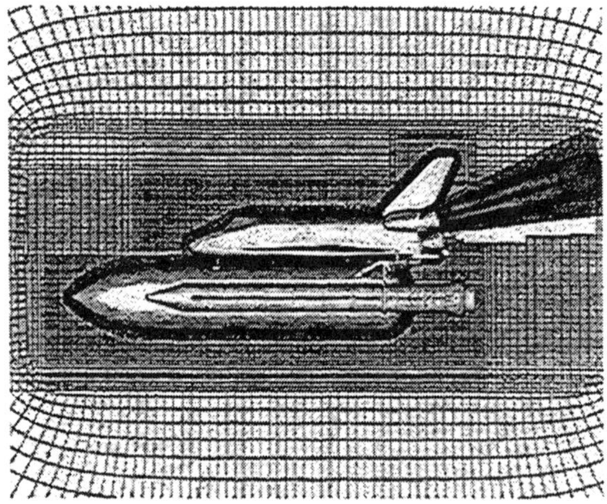

Volume grids in the y = 0 symmetry plane (only grid planes that lie in the y = 0 plane are shown)

References:

Pearce, Daniel G. et al., "Development of a Large Scale Chimera Grid System for the Space Shuttle Launch Vehicle," AIAA–93–0533–CP, January 1993.

"ICEM–CFD User's Guide Version 3.0," Control Data Technical Publications," August 1992.

Chan, W.M. and Steger, J.L., "Enhancements of a Three–Dimensional Hyperbolic Grid Generation Scheme," Appl. Math. and Comput. Vol.51, 181–205, 1992.

Benek, J.A., Donegan, T.L., and Suhs, N.E., "Extended Chimera Grid Embedding Scheme with Application to Viscous Flows," AIAA–87–1126–CP, July 1987.

X–24A Lifting Body Chimera Grid System

James S. Greathouse
Mail Code EG3/NASA Johnson Space Center

Description:

This grid system was developed to simulate the flow over the X–24A lifting body. This geometry was used as a basis for development of X–38 Crew Transfer Vehicle. This Chimera/overset grid system consists of 40 zones containing a total of 4,132,652 volume points. The vehicle surface is represented by 27,040 points with the maximum spacing of approximately 3" and clustered in regions of high curvature and at corners. Most of the surface grids were created using the hyperbolic surface grid generator SURGRD developed at NASA Ames Research Center, with a few generated using GRIDGEN from Pointwise. The original geometry was received as an IGES file of an early X–38 design, and was modified back to the X–24A shape using the National Grid Project code developed at Mississippi State. Some other modifications at the base of the vehicle were made in interest of simplicity. Volume grids were created using the hyperbolic grid generator HYPGEN developed at NASA Ames Research Center. Typical viscous volume grids extend 50 inches outward from the surfaces with an initial normal spacing equal to 0.00015 inches, keeping maximum y+ values in the range of 1.0 to 1.5 in the transonic flow regime. Overall communication is achieved through two Cartesian background grids completely enclosing the X–24 and a single far field grid that encloses the Cartesian grids and extends outward approximately 20 vehicle lengths. Grid communication was established using the PEGSUS code developed at Arnold Engineering Development Center. Time required to generate this grid system was on the order 1 man month which included development of a semi–automated system to change the position of all 8 of the control surfaces.

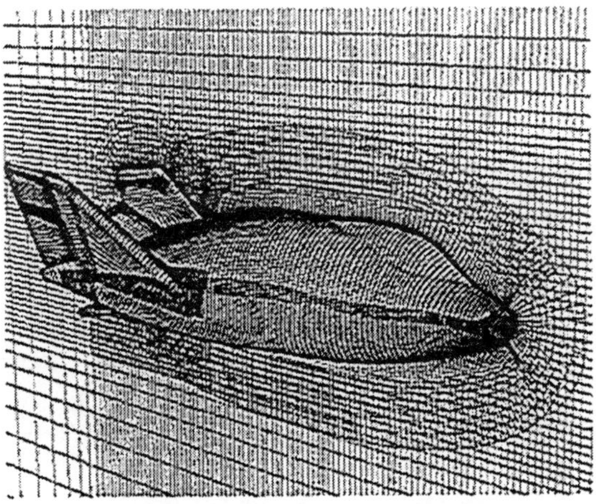

Side view of X–24A grid system showing volume grids on centerline

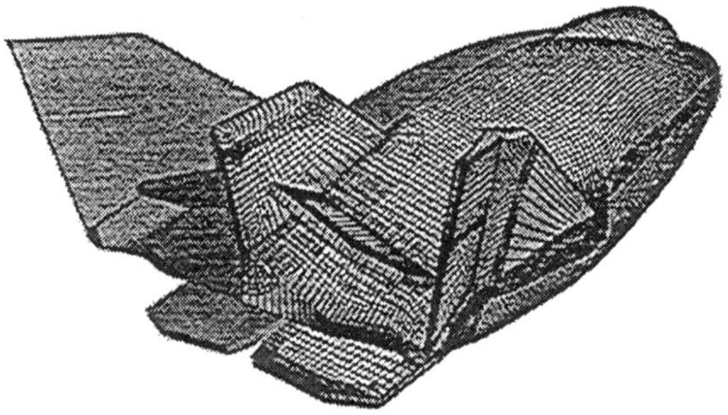

Aft–quarter view of X–24A showing surface grids
and 8 control surfaces

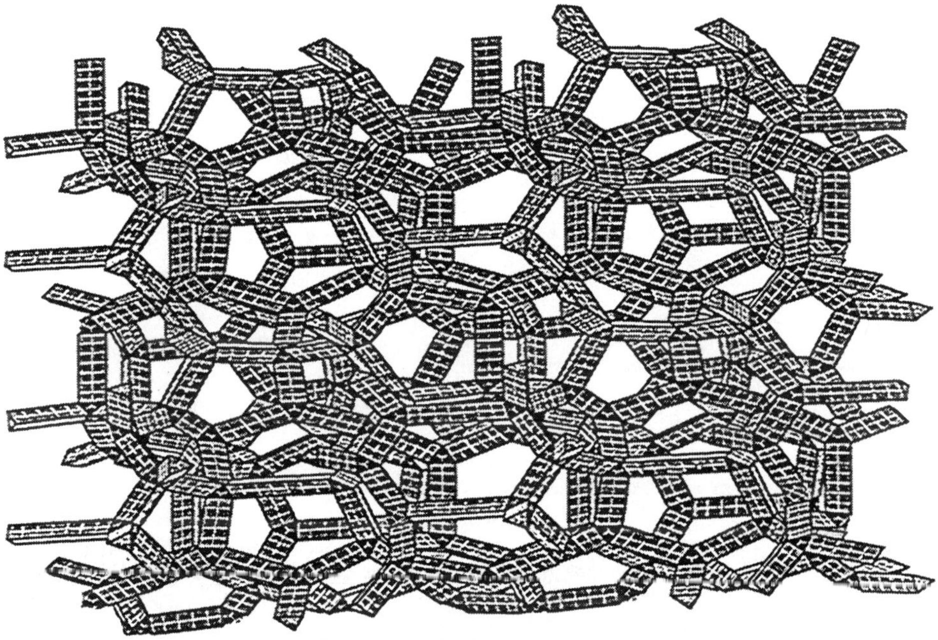

CUBIT Continuum Element Mesh: Open Celled Foam Skeleton

David M. Hensinger, Sandia National Laboratories

Application: Calculation of Structural Properties of Open Celled Foams

Description: The information specifying the geometry of the foam consisted of the vertices of the structure and the connectivity for the triangular prisms and octahedra which are the basic structural components of the open celled foam. Geometry specification was read by an ACIS linked application which created prisms and octahedra solid bodies. A second application decomposed the prisms and the octahedra. The prisms were decomposed into three hexahedral beams. The octahedra were decomposed into 8 tetrahedra. The tetrahedra were decomposed into 4 hexahedra. The geometry prior to meshing consisted entirely of six sided, cube-like, bodies which could be mapped.

The model shown in the figure began as 46 octahedra and 96 prisms. The geometry model meshed in CUBIT contained 3 bodies for every prism and 32 bodies for every octahedra for a total of 1760 bodies. The figure shows the final mesh, composed of four copies of a smaller mesh translated and joined together. Without the capability to join meshes, it would have been impossible to automatically select all prism faces which join up to the face of the octahedra.

Meshing within cubit required approximately two WorkStation hours. The main difficulty was to ensure that all faces of octahedra and ends of prisms had the same mesh density. This was ensured by directly setting the intervals on the octahedras' edges after a global element sizing was specified.

Automatic Generation of Unstructured Grids for Viscous Flow Applications Using Directional Refinement

A.R Jahangirian
L.J Johnston

Department of Mechanical Engineering
UMIST
PO Box 88, Manchester M60 1QD, U.K.

A method to automatically-generate unstructured grids for application to complex two-dimensional turbulent aerodynamic flows is presented.

Techniques have been developed to produce high-quality grids suitable for both turbulent shear-flows and inviscid regions. The procedure is based on a combination of grid enrichment techniques whereby new grid-points and point connectivities are created simultaneously. An efficient data structure is used, so that no search operation is required through the list of edges. The method is used for generating the grid around a realistic SKF 1.1 wing/maneouvre flap section. The initial starting grid contains only 50 vertices including 6 vertices around each element. The final grid consists of 5870 vertices and 11476 cells. Required CPU time including the field grid generation and I/O operations was 5.80 seconds on a HP-735, 125 MHz computer workstation.

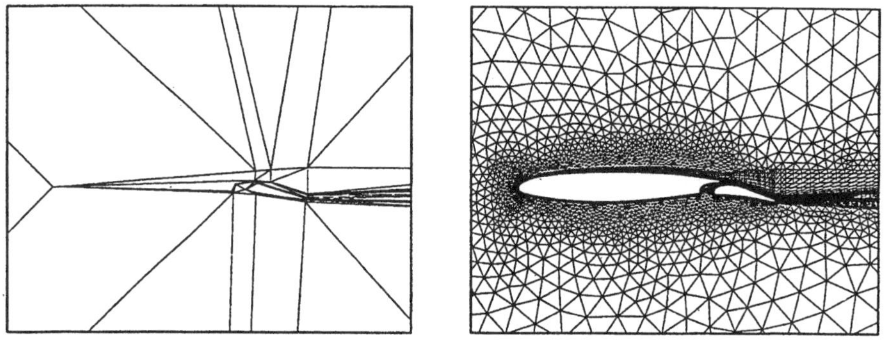

a) Initial starting grid b) Final grid

Reference

A.R Jahangirian and L.J Johnston,' Automatic Generation of Adaptive Unstructured Grids for Viscous Flow Applications', Numerical Grid Generation in Computational Field Simulations, B.K. Soni, J.F. Thompson, J. Hauser, P.R. Eiseman (Eds) , pp. 219-228, 1996

Hybrid Prismatic/Tetrahedral Grid for a Burner Geometry

Yannis Kallinderis *

Dept. of Aerospace Engineering and Engineering Mechanics

The University of Texas, Austin, TX 78712

e-mail: kallind@mail.utexas.edu

website: http://acelab.ae.utexas.edu

1 Application

A hybrid prismatic/tetrahedral grid about a Burner geometry is generated. The case corresponds to internal flow through a burner, which consists of an annulus diffuser, a swirl producer and a combustion chamber. The geometry has been provided by ABB. It has various complexities such as the fuel injection holes, severe cavities, twisted blades which produce the swirl, and vastly different length scales. The geometry has periodic boundaries, and only one burner is being modeled.

2 Grid Description

A trimmed NURBS surface description of the model is used to generate the initial triangulation. The triangles are then marched away from the body surface via an algebraic scheme to produce prismatic layers. The final prismatic layer is used as the starting point for the advancing frong tetrahedra generation scheme. Figure 1 shows a close-up of the surface triangulation for the swirl producing section. The surface consists of approximately 75,000 triangles. A hybrid mesh of the burner is seen in Figure 2, which is a two-dimensional cut along the axis. The view shows that the hybrid grid generator was capable of capturing all the fine features of the geometry, and also clustered points downstream of the swirl producing section. The mesh consists of 415,000 nodes, 521,000 prisms, and 748,000 tetrahedra. A cut across the swirl producing blades is shown in Figure 3. The view shows the hybrid nature of the mesh, and demonstrates the smooth transition in cell sizes even

*Associate Professor

across different element types. The entire grid generation process (from surface mesh generation to final hybrid grid) took on the order of 10 hours of CPU time on an IBM RS/6000 workstation. Details of the grid generation method can be found in [1, 2, 3].

References

[1] von Karman Institute (VKI) Lecture Series in Computational Fluid Dynamics, LS 1996-06, March 25-29, 1996.

[2] Y. Kallinderis, A. Khawaja and H. McMorris, "Hybrid Prismatic/Tetrahedral Grid Generation for Viscous Flows Around Complex Geometries," *AIAA Journal*, Vol. 34, No. 2, pp. 291-298, February 1996.

[3] McMorris, H. and Kallinderis, Y., " Octree-Advancing Front Method for Generation of Unstructured Surface and Volume Meshes", *AIAA Journal,* Vol. 35, No.6, pp. 976-984, June 1997.

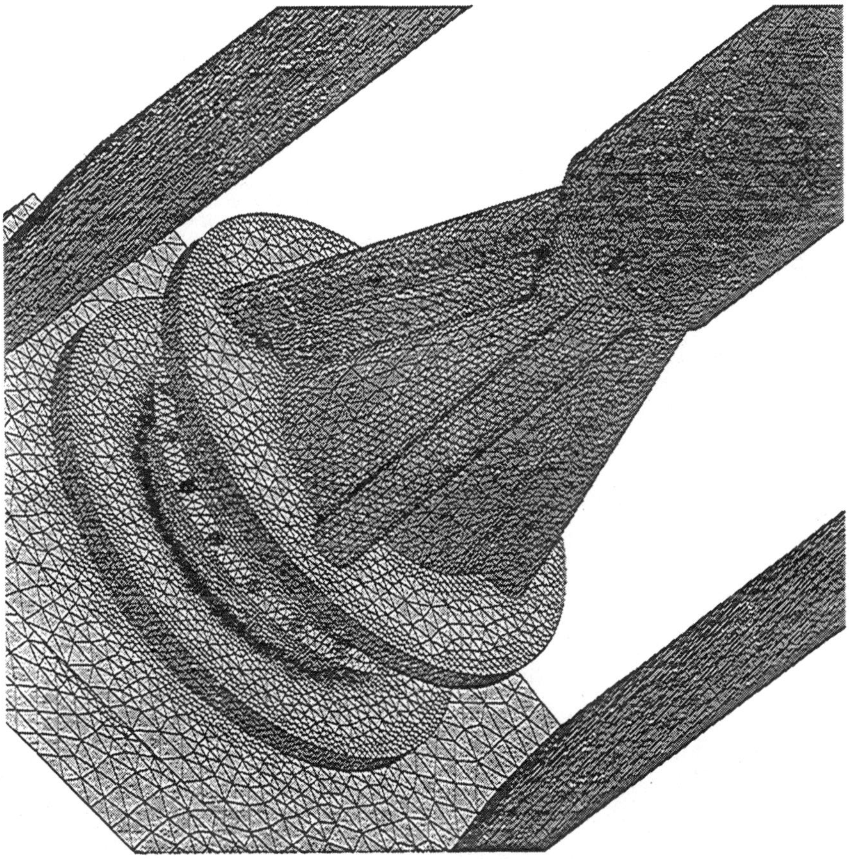

A close-up view of the swirl producing section of the ABB Burner geometry. The surface is made up of 75,000 triangles.

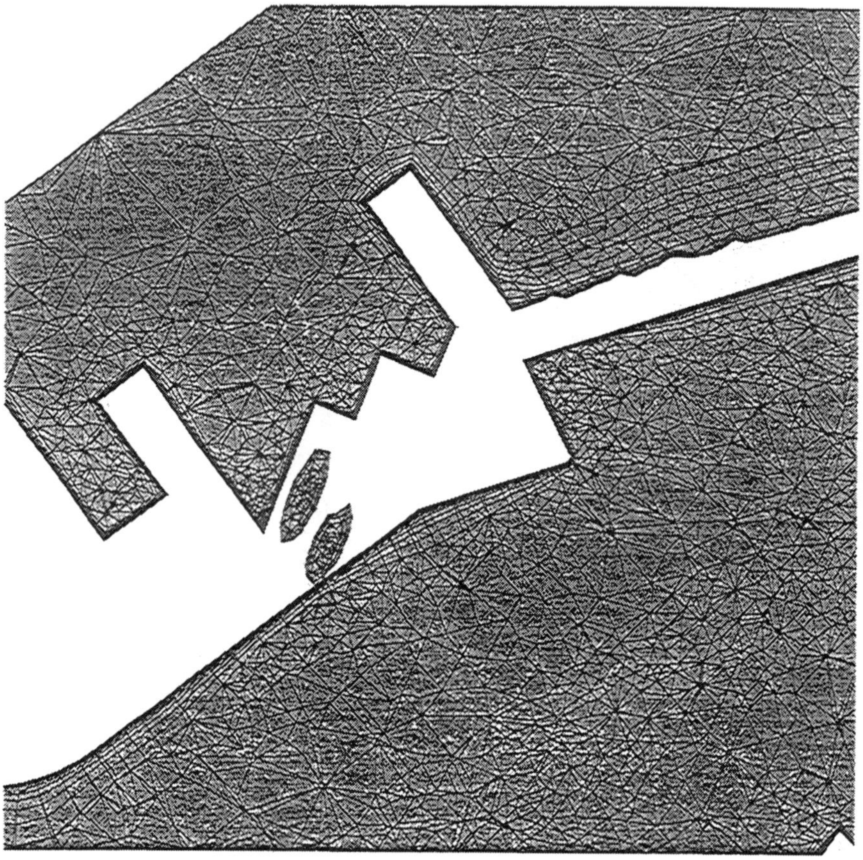

Figure A close-up view of the hybrid cut along the axis of the burner geometry. The grid generator was capable of automatically handling the small length scales and the severe cavities.

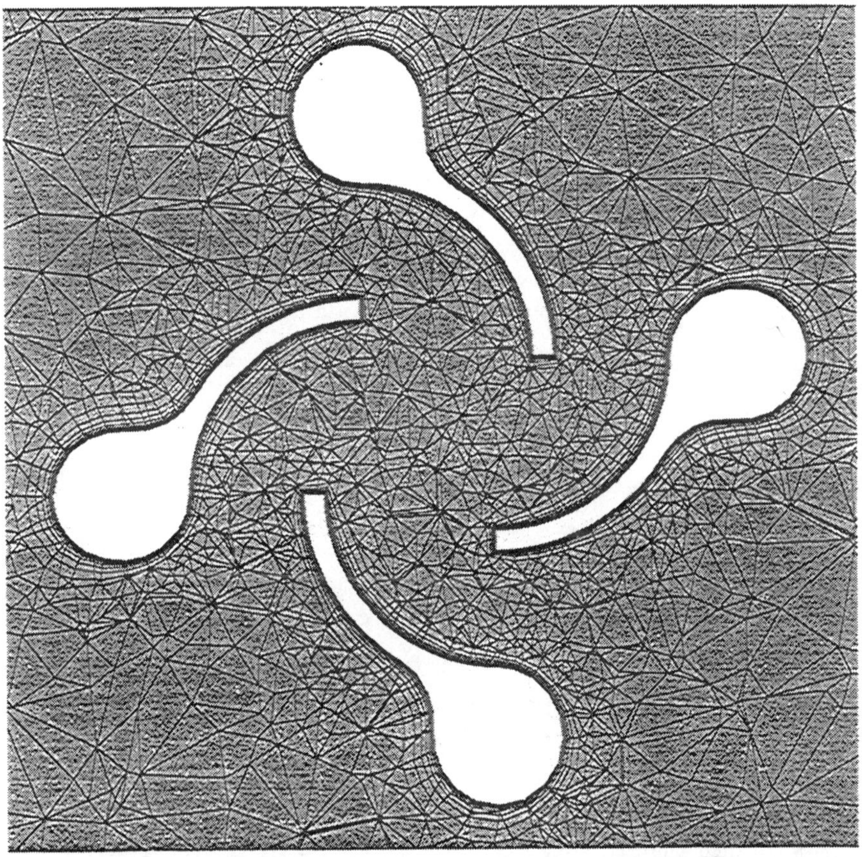

Figure A cut across the swirl producing blades of the burner geometry. The hybrid nature of the mesh and the smooth transition in cell sizes are visible.

Hybrid Prismatic/Tetrahedral Grid for High Speed Civil Transport Aircraft

Yannis Kallinderis *

Dept. of Aerospace Engineering and Engineering Mechanics

The University of Texas, Austin, TX 78712

e-mail: kallind@mail.utexas.edu

website: http://acelab.ae.utexas.edu

1 Application

A hybrid prismatic/tetrahedral grid about a High Speed Civil Transport (HSCT) aircraft is generated. The HSCT aircraft is a next generation aircraft being designed to travel at supersonic speeds. It has a double-delta wing configuration emerging from the nose. The cavity between the engine and the wing presents a challenge in grid generation.

2 Grid Description

A trimmed NURBS surface description of the model is used to generate the initial triangulation. The triangles are then marched away from the body surface via an algebraic scheme to produce prismatic layers. The final prismatic layer is used as the starting point for the advancing frong tetrahedra generation scheme. A view of the grid is shown in Figure 1. The grid is shown on two surfaces that are perpendicular to each other. The first is the symmetry plane with the quadrilateral faces corresponding to the prisms and the triangular faces corresponding to the tetrahedra. The second surface is a field cut intersecting the fuselage and engine. A field cut is a cut through the discretized grid showing all the cells that intersect the plane cut, thus emphasizing the 3-D nature of the grid. Figure 2 illustrates the widely varying length scales of the hybrid grid. The field cut shows portion of the fuselage, the wing, as well as the engine. The varying thickness of the prismatic layer

*Associate Professor

is dictated not only by the thickness of the boundary layer, but also by the size of the cavity between the engine and the wing. The tetrahedral part of the mesh is very dense in the cavity area in order to match the sizes of the local prisms and becomes isotropic away from the cavity.

The grid consists of approximately 100,000 prisms and 500,000 tetrahedra, and a total of 160,000 nodes. The entire grid generation process (from surface mesh generation to final hybrid grid) took on the order of 7 hours of CPU time on an IBM RS/6000 workstation. Details of the grid generation method can be found in [1, 2, 3].

References

[1] von Karman Institute (VKI) Lecture Series in Computational Fluid Dynamics, LS 1996-06, March 25-29, 1996.

[2] Y. Kallinderis, A. Khawaja and H. McMorris, "Hybrid Prismatic/Tetrahedral Grid Generation for Viscous Flows Around Complex Geometries," *AIAA Journal*, Vol. 34, No. 2, pp. 291-298, February 1996.

[3] McMorris, H. and Kallinderis, Y., " Octree-Advancing Front Method for Generation of Unstructured Surface and Volume Meshes", *AIAA Journal*, Vol. 35, No.6, pp. 976-984, June 1997.

Figure 1: View of the hybrid mesh around the HSCT aircraft with engines on two different planes that are perpendicular to each other. The first plane is that of the symmetry while the second is a field cut intersecting the fuselage and engine.

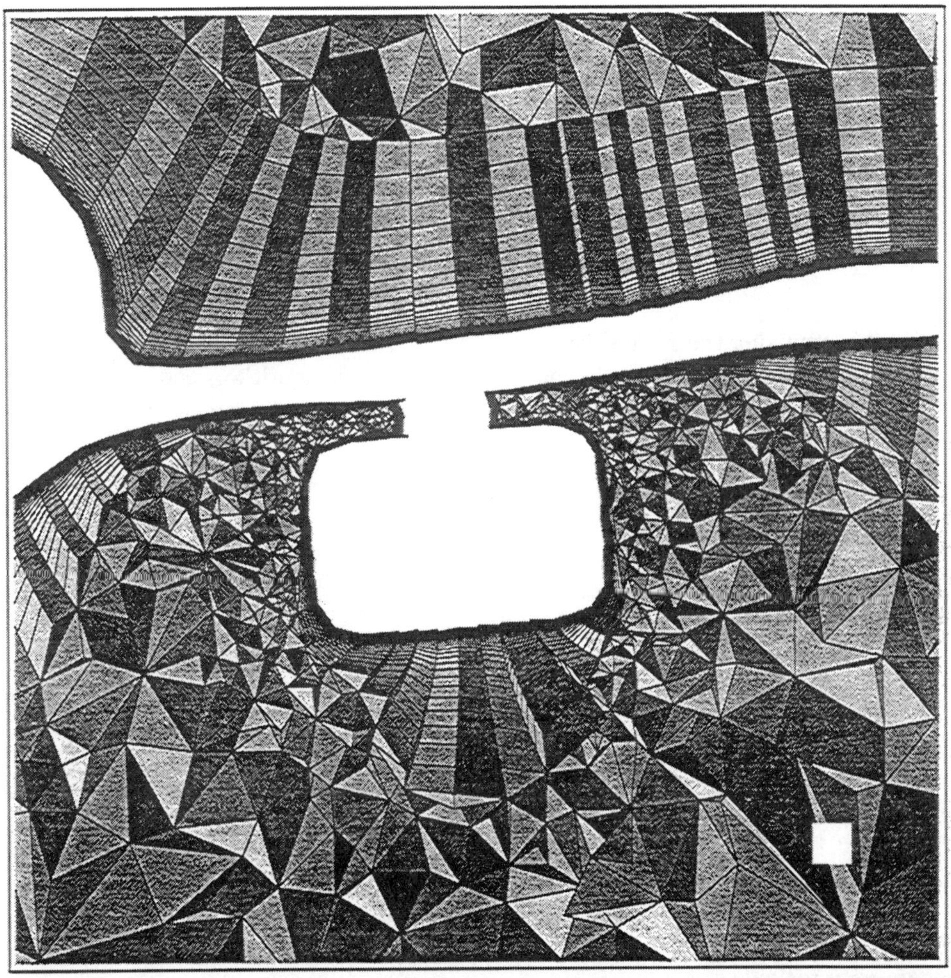

Figure 2: Close up of the hybrid grid for the HSCT aircraft around the engine cavity. The tetrahedral mesh is very dense here compared to other regions so as to match the thin local prism cell sizes.

TESTS FOR TWO-DIMENSIONAL GRID GENERATION

Suggestions

Khairullina O.B., Sidorov A.F. and Ushakova O.V.
Institute of Mathematics and Mechanics, UB RAS, Ekaterinburg

While organizing the *V* conference on grid generation we had an attempt to conduct a competition with the purpose of comparison of various algorithms and programs of grid generation.

For testing two-dimensional techniques the following tests have been suggested by G.P.Prokopov:

Class 1. *For simply–connected domains. To generate two-dimensional regular (structured) grid with a given number of cells.*

Case 1.1 *Domain - a five-edged star.*

Case 1.2 *Domain: from the square* $0 \leq x \leq 1, 0 \leq y \leq 1$ *the square* $0.5 - a < x < 0.5$, $0 < y < a$ *has been cut. A left boundary of the curvilinear quadrange –* $x = 0$, *the right one* $x = 0$. *Values of* a: $a = 0.3$, $a = 0.4$, $0.5 - \varepsilon$.

Class 2. *To generate a block-structured grid in simply–connected and multiply–connected domains.*

Case 2.1 *Domain: in a circle of radius* $R = 1$ *three squares with the side* $a = 0.4$ *have been cut.*

Case 2.2 *In case of 2.1 one square is substituted by a circle surrounded it with radius* $R = 0.2\sqrt{2}$, *the second one – by equilateral triangle with the side* $a = 0.4$, *third – by a square with the side* $a = 0.1$.

As a main criterion of quality of a grid - criterion of uniformity has been suggested: relation of minimum and maximum of grid cell areas.

In our programs *MOPS-2* and *LADA* on the basis of [1] six various criteria of quality of two-dimensional grid are calculated. We shall present them. We consider, that the grid nodes are connected by straight lines. The cell vertices are numbered from 1 to 4 in the counterclockwise direction. (fig. 1)

We shall calculate the areas σ_i, $i = 1, 2, 3, 4$ of triangles formed by diagonals of quadrangle according the formulas

$$2\sigma_i = (x_i - x_{i-1})(y_{i+1} - y_i) - (x_{i+1} - x_i)(y_i - y_{i-1}),$$

as well as length of quadrangle's sides

$$l_i = \sqrt{(x_{i+1} - x_i)^2 + (y_{i+1} - y_i)^2}$$

(we consider here, that $(x, y)_0 = (x, y)_4$, $(x, y)_5 = (x, y)_1$).

In the programs the following criteria are calculated.

Cases 2.1 - 2.2

a) Cut figures are asymmetric.

b) To generate a structured grid with ambiguous mappins of boundaries of the cut figures (O, C, H – grids).

We consider more difficult and interesting in addition to the **cases 2.1-2.2** such tests:

1) To generate a block-structured grid in the multiply-connected domain - fig. 5.

2) To generate block-structured grid, using different topologies in "petals and stem of the flower" (fig. 6), thus the matrix of coordinates of grid should be maximally filled .

Tests 2,3,4 from [1] are also very good . We shall sypply their description from [1].

Test 2. *The domain is a prescribed part between sections $x = x_l$ and $x = x_r$ of a "pipe of variable section" represented on fig. 7. Its top and bottom boundaries are prescribed accordingly by the formulas*

$$y_t(x) = d_0 + d_1 cos 2R_0 x;$$

$$y_b(x) = d_2 cos 3R_0 x.$$

It is convenient to assume as scaling size $R_0 = \pi$. . Parameters d_0, d_1, d_2 are free.

For calculation of extreme points of pipe width $z(x) = y_t(x) - y_b(x)$ it is convenient to choose $d_1 = d/2$, $d_2 = d/3$. Then the equation for defining such points $z'(x) = 0$ has the form $sin 2\pi x - sin 3\pi x = 0$. Its roots at the segment $0 \le x \le 1$ are following:

$$x_1 = 0; \quad x_2 = 1/5; \quad x_3 = 3/5; \quad x_4 = 1$$

and further symmetrically relatively to the points $x = 0$ to the left and $x = 1$ to the right. Corresponding values of width of a pipe are:

$$z_1 = d_0 + d/6; \quad z_2 = d_0 + 5d/6 cos 36^o;$$

$$z_3 = d_0 - 5d/6 cos 18^o; \quad z_4 = d_0 + 5d/6.$$

For this test parameters can be choosen by this way:

$$d_0 = 0.6, d = -0.6, d_1 = -0.3, d_2 = -0.2.$$

Methodical purpose of this test is getting qualitative characteristic of grid lines behaviour near the convexities of a boundary. By choozing parameters x_l and x_r various situations with extention and narrowing of a pipe are received. It is most natural to choose as the law of arranging of grid nodes on the boundary the law of uniform arranging of nodes with respect to the arch length. Correspondence between top and bottom boundary nodes can be maintained by sequence of coordinates $\{x_j\}$. It is appropriate to consider the modification of the test 2, domain of which is represented on fig. 8. In it each half-wave sinusoid is replaced by an arch of a circle, basing on the same segments in direction of coordinate x, as on fig. 7. The arch can be conveniently characterized by the value Θ which is equal to half of its central angle (see fig.9).

$$\delta_2 = (R+Q)\frac{H}{\sqrt{H^2+(R+Q)^2}} - R;$$

$$\delta_3 = \sqrt{H^2+(L-Q-R)^2} - 2R.$$

For excepting self-crossing of a contour of the domain it is necessary the fulfilment of the following restrictions:

$$H > H_{cr} = R;$$

$$Q > Q_{cr} = R\left[\frac{1}{\sqrt{1-(R/H)^2}} - 1\right];$$

$$L > L_{cr} = Q + R + \sqrt{4R^2 - H^2}, \text{ if } H < 2R.$$

The configuration of the domain becomes more and more difficult for grid generating when even one of parameters H, Q, L close to critical. Parameters can be choosen by the values:

$$H = 1, R = 0.6, Q = 0.4, L = 2.$$

The vertices of the quadranle can be chosen in different ways. The simplest among them is shown on the fig.11.

As to adaptive grids, a question concerning the quality criteria remains opened. As one of the most natural tests next one can be suggested: which of grids gives the best approximation of the first derivative under the standard formulas. It is possible to offer different kinds of functions, on which to check up the grid node concentration in appropriate places of the domain. For example

$$z = \begin{cases} \exp\left(-\left((x-a)^2+(y-b)^2-r^2\right)/\varepsilon\right), & (x-a)^2+(y-b)^2 > r^2, \\ -\left(\left((x-a)^2+(y-b)^2-r^2\right)/\varepsilon\right), & (x-a)^2+(y-b)^2 \geq r^2, \end{cases}$$

$$z = \exp\left(-\left((x-a)^2+(y-b)^2\right)/\varepsilon\right).$$

Here $a, b, r, \varepsilon = \text{const}$.

Besides criteria of quality of grids at evaluation of efficiency of algorithms and programs we also think it is necessary to take into account following:

1. Volume of the input information;

2. Time of grid calculation;

3. Economy of storage of information and coordinates of nodes.

REFERENCES

1. G.P.Prokopov, On the organizing the comparison of algorithms and programs of generation of regular two–dimensional difference grids, Voprosy Atomnoy Nauki i Tekhniki. Ser.: Matem. Modelirovanye Fizicheskikh Processov, 1989, No.3, pp.98-108.

2. S.K. Godunov, A.V.Zabrodin and others, Numerical solving multy – dimensional problems of gas dynamics. Moscow. 1976. 400 p.

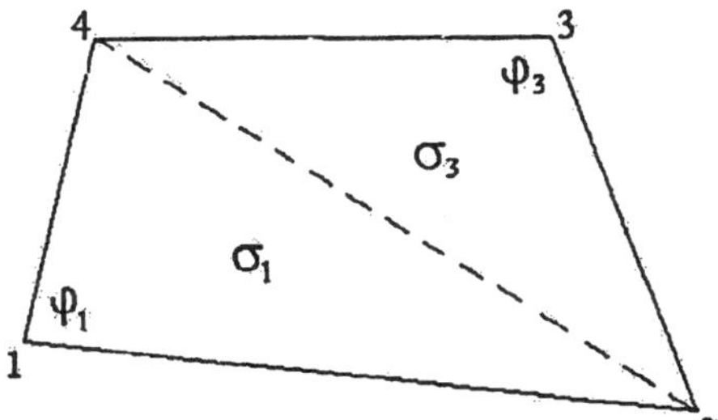
4
3
φ_3
σ_3
σ_1
φ_1
1
2

1

3

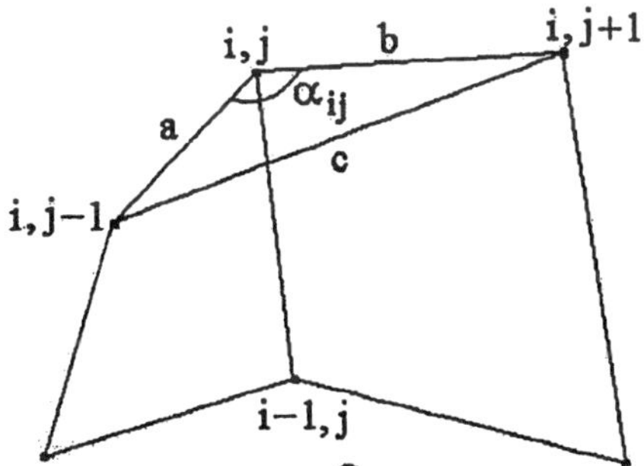
i, j
b
i, j+1
α_{ij}
a
c
i, j−1
i−1, j

2

4

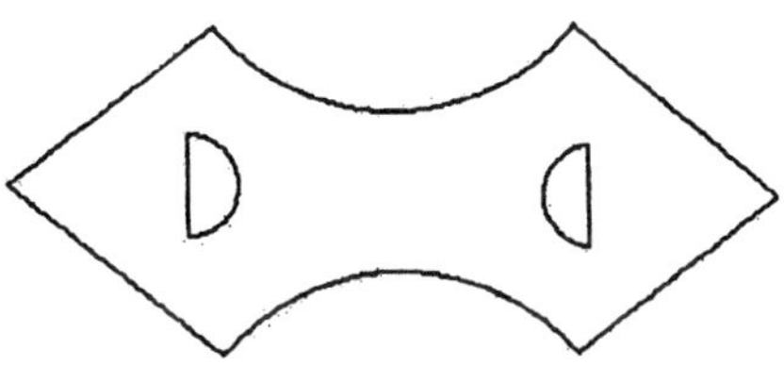

5

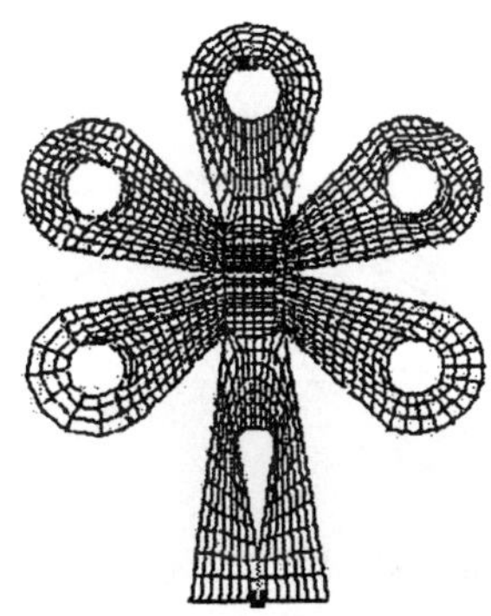

6

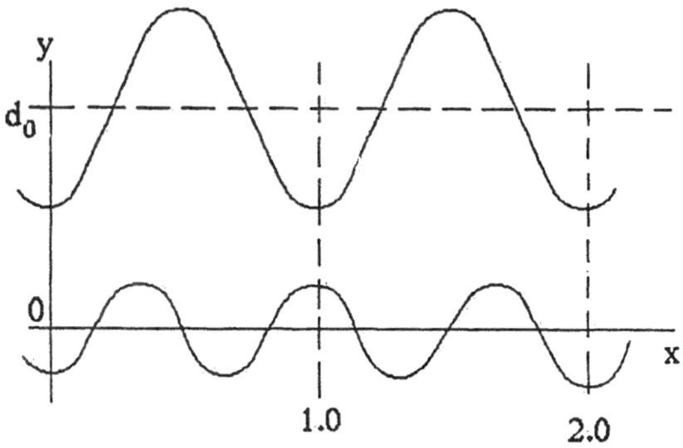
y
d_0
0
x
1.0
2.0

7

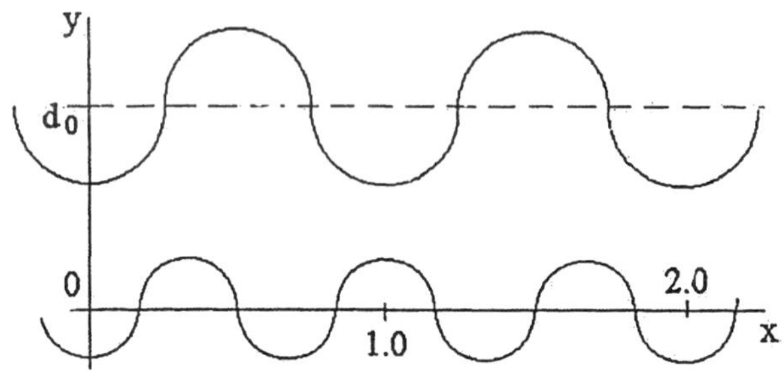
y
d_0
0
1.0
2.0
x

8

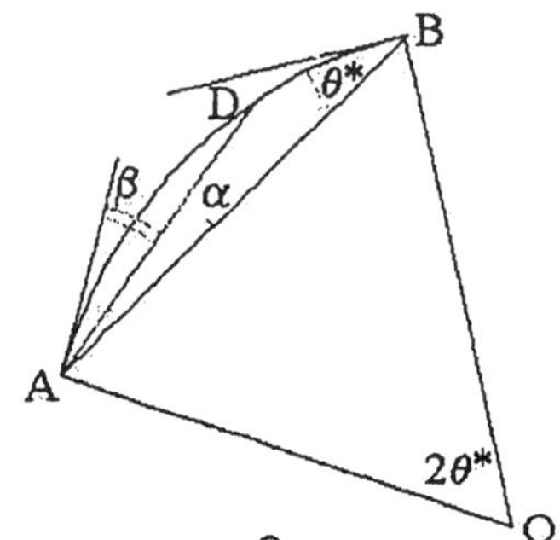
B
D
θ*
β
α
A
2θ*
O

9

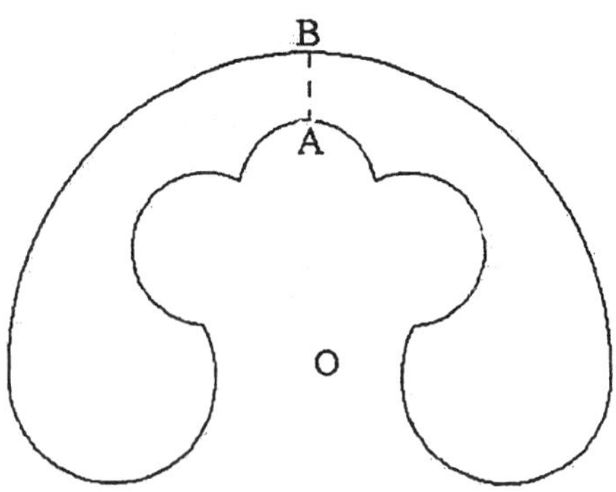
B
A
O

10

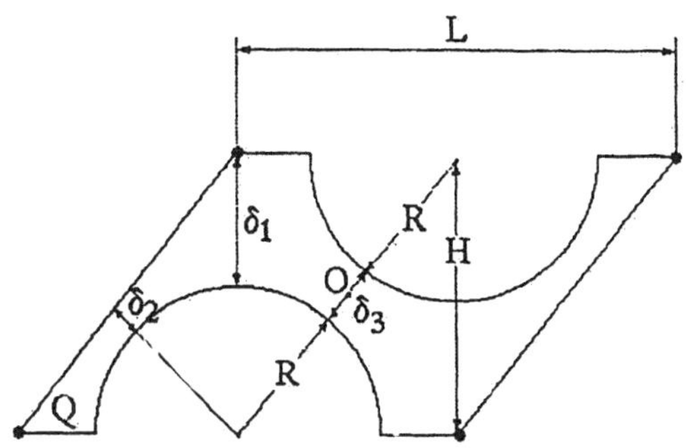
L
δ_1
R
H
O
δ_2
δ_3
R
Q

11

Hybrid Grid Generation for CFD Applications

Roy P. Koomullil and Bharat K. Soni.
NSF Engineering Research Center
Mississippi State University

Hybrid grids combines the capability of unstructured grids in handling complex geometries and capability of structured grids in resolving viscous boundary layers. Structured grid is kept in the vicinity of bodies and the rest of the domain is filled with unstructured grids. Structured grid near the body is generated using advancing layer method utilizing the surface normals (e.g. Figure. 1). Complex bodies are decomposed in to simple geometric entities before generating structured grids. The structured grids from different entities are trimmed by checking the overlapping regions and aspect ratio of the cells. Unstructured grid in the void between the trimmed grids is generated using Delaunay triangulation. Different steps involved in hybrid grid generation are demonstrated in Figure. 2. Examples of hybrid grid generated using this approach are shown in Figures. 3 & 4. An example of a generalized grid with different number of sided polygons together with the Mach number contours for a freestream Mach number of 0.3 is shown in Figure. 5.

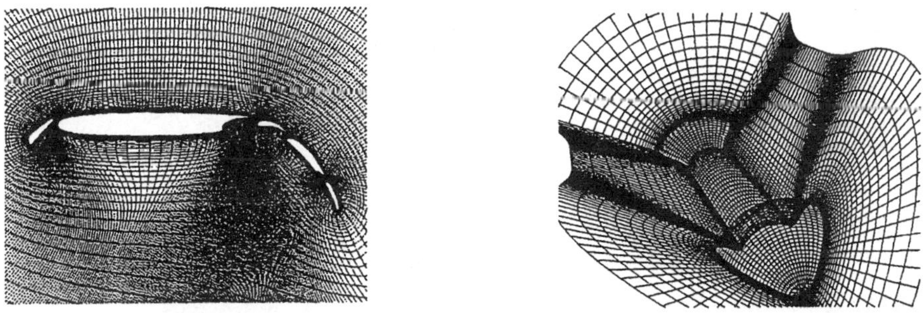

(a) Single Block Structured Grid for 5 Element Airfoils

(b) Structured Grid For an Axisymmetric Body

Figure. 1 Structured Grids Using Advancing Layer Method.

(a) Structured Grid Around Three Components

(b) Trimmed Structured Grid

(c) Resulting Hybrid Grid

Figure. 2 Steps involved in Hybrid Grid Generation.

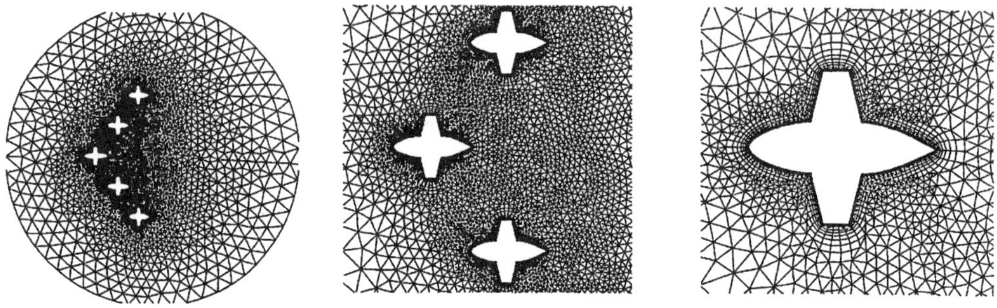

Figure. 3 Different Views of Hybrid Grid for Arbitrary Bodies.

(a) Grid Around 9 Cylinders (b) Grid for 4 Element Airfoil

Figure. 4 Examples of Hybrid Grids.

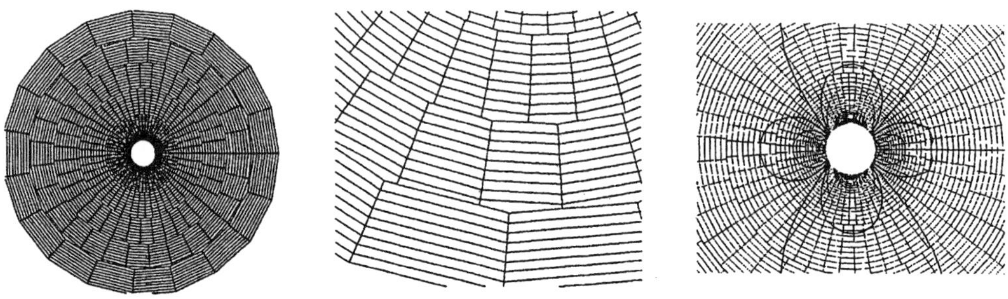

(a) Overall Grid (b) Interface Between Layers (c) Mach Number Contours

Figure. 5 Example of Generalized Element Grid.

Key References

[1] Huang, Chi–Ti *"Hybrid Grid Generation System,"* Master Thesis, Department of Aerospace Engineering, Mississippi State University. August 1996.

[2] Koomullil, R. P., and Soni, B. K., *"Generalized Grid Techniques in Computational Field Simulation"*, Accepted to present at the 6th International Conference on Numerical Grid Generation in Computational Field Simulation, University of Greenwich, Avery Hill Campus, London, UK, July 6–9, 1998.

[3] Koomullil, R. P., *"Flow Simulation System for Generalized Static and Dynamic Grids,"* Ph.D. Dissertation, Department of Aerospace Engineering, Mississippi State University, May 1997.

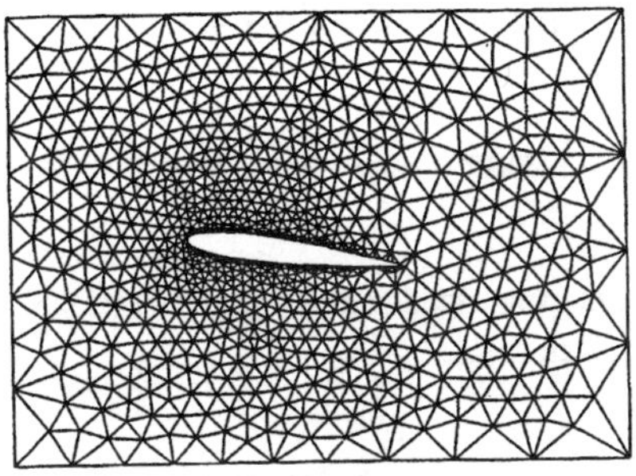

Figure 1: NACA0012 airfoil, 1384 triangles and 736 nodes.

GLOSSARY OF GRID CONFIGURATION

Jan Kucwaj

The Technical University of Kraków
Section of Applied Mathematics
University Computer Centre
ul. Warszawska 24, 31-155 Kraków
E-mail: jkucwaj@usk.pk.edu.pl

Title Grid Generation with Mesh Density Functions.

Name and affiliation of the contributor Dr. Jan Kucwaj

The Technical University of Kraków
Section of Applied Mathematics
University Computer Centre
ul. Warszawska 24, 31-155 Kraków
E-mail: jkucwaj@usk.pk.edu.pl

Application The grid generation code is used in finite element application, especially in fluid mechanics. Because of using B-spline curves to boundary representation the code may be applied to practically arbitrary geometries. The user supplied function defining the size of the mesh gives the possibility of generation of graded meshes.

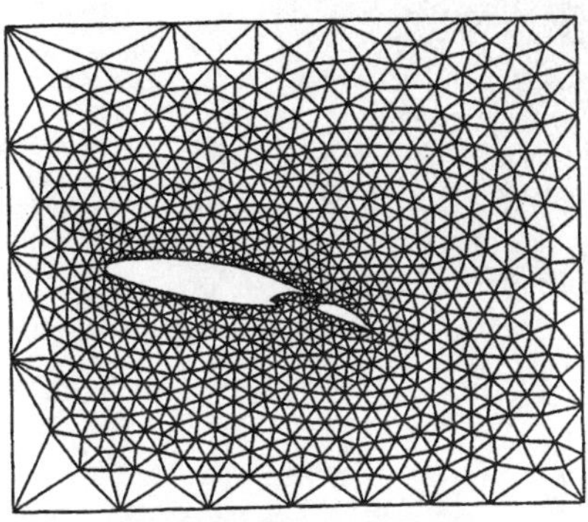

Figure 2: NACA0066 with flap, 1740 triangles and 939 nodes.

Grid description This is an unstructured grid generation code. The size of the mesh defines a positive function function, taking into account this function the points are appriopriately distribu-ted by using advancing front technique.

The applied method uses two runs of advancing front technique implementation, first for grid insertion, and second for tringulation. The triangulation is performed by Delaunay advancing front method [1, 2]. The method is a generalization of Delaunay triangulation to non-convex multiconnected domains.

Geometry is defined by specification of external loop and internal closed curves surrounding the domain. Anyone of them may cosist of straight line segment, arch of circle or B-spline curve. The whole code is written in FORTRAN-77. It needs INPUT file and produces OUTPUT file giving all the DATA about generated mesh. The execution time depends upon number of generated triangles.

References

[1] J. Kucwaj: The Modelling and Triangulation of geometrically complicated plane domains appearing in Fluid Mechanics, Journal of Theoretical and Applied Mechanics,No2, vol. 35, 1997, pp. 369-392.

[2] S. H. Lo: Delaunay Triangulation of non-convex planar domains, Int. J. numer. Meth. in Engn., 28, 2695-2707, 1989.

Tetrahedral Viscous Grid Around a Complete F–16 Aircraft Configuration

Shahyar Z. Pirzadeh
NASA Langley Research Center

Application:

The objective of this work is to accurately predict the aerodynamic loads on a generic finned store held in external carriage on a complex F–16 aircraft configuration.

Grid Description:

The geometry, defined in the IGES format, consists of a wing, fuselage with a flow–through inlet, horizontal and vertical tails, an external tank mounted on the inner wing pylon, and a generic finned store held underneath the outer wing pylon with a small gap in between. The geometry was modeled with stings at the aft ends of the fuselage and the store as in the wind tunnel experiment. The generated tetrahedral unstructured grid contains 1,428,779 cells, a total of 255,959 grid points, 31,297 boundary points, and 62,586 surface triangles. An exponential distribution of grid points normal to the no–slip surfaces has resulted in 6 layers of cells in the boundary layer with a Y+ of 30 at the wall. Such a point distribution is normally adequate for solving he Navier–Stokes equations with the sublayer region of the boundary layer modeled with a wall function. The grid was generated using the grid utility program GridTool and the unstructured grid generation code VGRIDns [1]. The strategy used in VGRIDns for generating viscous grids is based on a combination of the Advancing–Front and the Advancing–Layers methods. The labor time required for generating the present grid, starting from the CAD surface definition to the post-processing of the final volume grid, has been about 4 days. VGRIDns generates approximately 1450 cells per second with the Advancing–Front and about 3900 cells per second with the Advancing–Layers method using a Silicon Graphics Octane workstation with a R10000 processor.

Unstructured–grid, Navier–Stokes solution of a F–16 configuration:

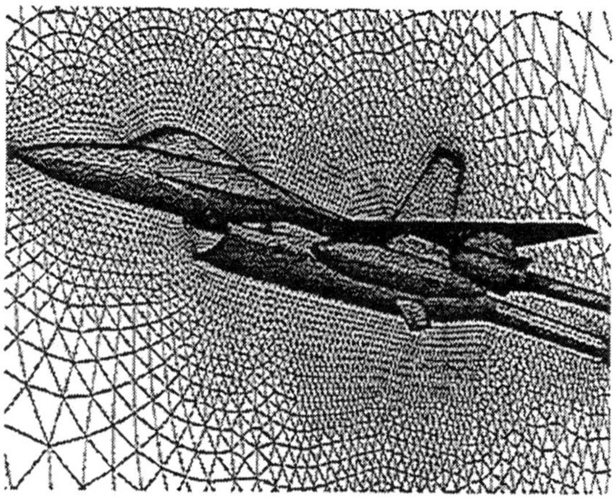

Surface grid and triangulation on the plane of symmetry.

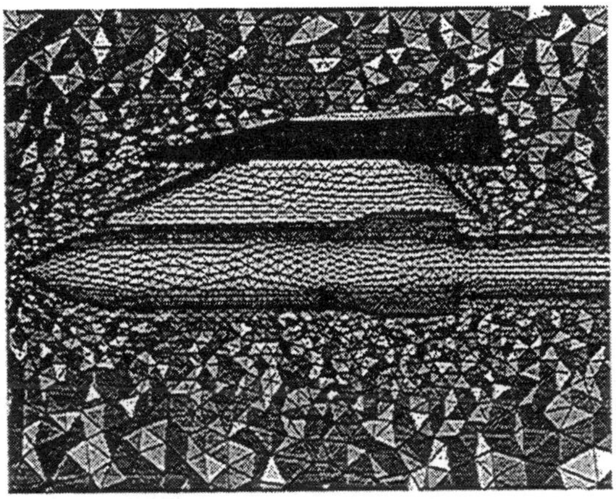

Surface grid along with a cross section of the volume grid around the store

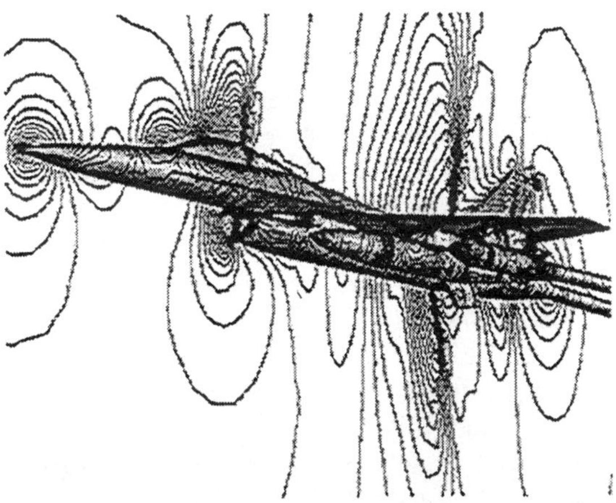

Surface pressure contours generated with the Navier–Stokes solver USM3Dns [2]

Remarks: The F–16 geometry was provided by the Naval Air Systems Command, Patuxent River MD.

Reference:

1) Pirzadeh, S., "Three–Dimensional Unstructured Viscous Grids by the Advancing–Layers Methods," AIAA Journal, Vol. 34, No. 1, 1996, pp. 43–49.

2) Frink, N.T., "Assessment of an Unstructured–Grid Method for Predicting 3–D Turbulent Viscous Flows," AIAA Paper 96–0292, 1996.

Orthogonal Grid Generation on Surfaces

R. Ganesh Rajagopalan
Computational Fluid Dynamics Center
Iowa State University
Ames, IA 50011

V. Baskaran
Sukra Helitek Inc

Motivation: Surface grid generation is an important part of the necessary steps to be taken for solving flow over complex three dimensional bodies. Orthogonality of the grid is a measure of the quality of the grid and a method based on the surface Beltramian is used to numerically generate orthogonal surface grids in the illustration presented.

Description: The Beltramian, a second order partial differential equation, yields a system of two coupled elliptic surface grid generating equations when discretized. The metric tensor of the generating equations are constrained to yield orthogonal grids on the surface. As an example, one fourth of the surface of a right circular cylinder of radius 20 units is considered as the surface on which the grids are to be generated. A (40× 40) grid on the surface is initially generated using transfinite interpolation. The grids are uniformly distributed on the east and west boundaries. On the south boundary the distance between successive grid points are increased by a constant ratio of 1.05. Similarly the distance between the successive grid points are decreased by a constant ratio of 1.05 on the north boundary. With this distribution as the Dirichlet boundary condition, the discretized elliptic equations are solved, with under-relaxation, to yield the final grid displayed. The generated grid is nearly 100% orthogonal.

Reference: Rajagopalan, R. G., Baskaran, V., "Elliptic Generation of Orthogonal Grid in Arbitrary Surfaces," International Conference on Numerical Grid Generation in Computational Fluid Dynamics and Related Fields, April 1996, Mississippi, USA.

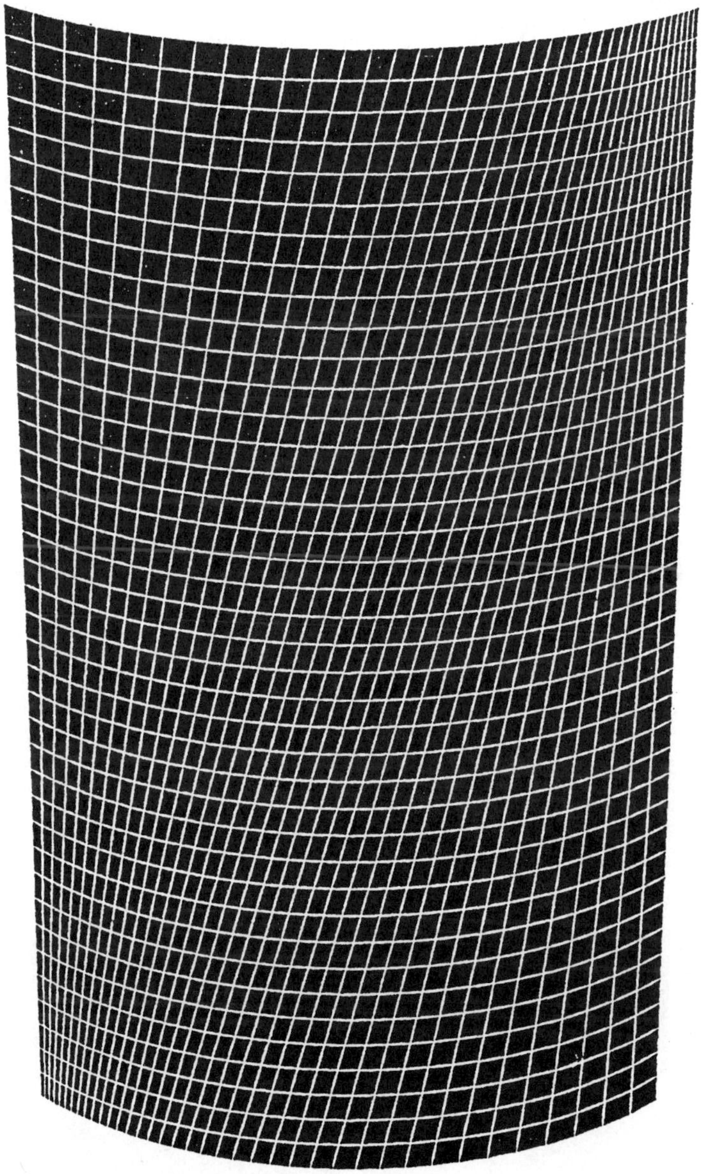

Figure 1: Initial Grid on the Right Circular Cylinder

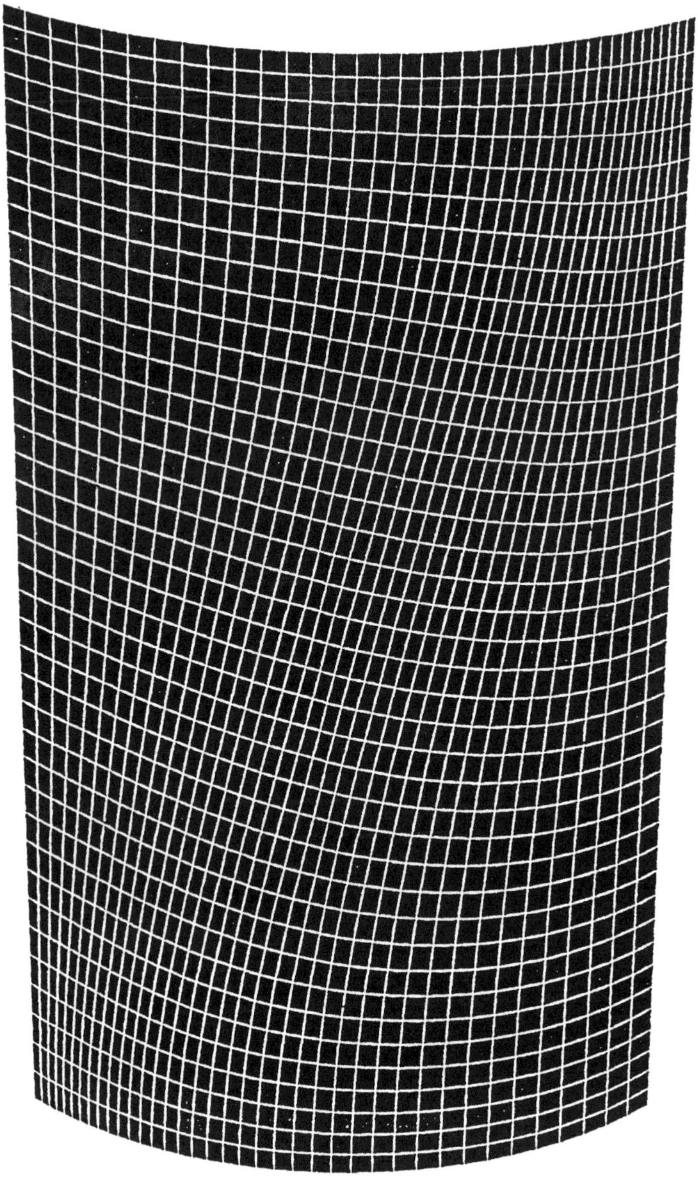

Figure 2: Orthogonal Grid on the Right Circular Cylinder

Title : Impeller with Multiple Splitter Blades

Name : Dr. Ming-Hsin Shih

Affiliation : Catalpa Research, Inc.
13014 Astalot Drive, Suite C
Huntsville, AL 35803-1802, U. S. A.
TEL: 205-883-2045 / FAX: 205-881-0508

Application : Future re-usable launch systems will require higher thrust-to-weight (T/W) ratio engines than currently available. one of the key technologies that will enable significant improvements in T/W ratio is advanced unshrouded impeller technology. However, the numerical grid generation about a general turbomachinery configuration is a tedious and time consuming task. This is especially true for an impeller configuration with multiple splitter blades. In order to facilitate the generation process and reduce man-hours required to produce a turbomachinery grid, customized procedures and methodologies are needed.

Description : Numerical grid for this type of configuration can be generated with ease using TIGER (Turbomachinery Interactive Grid genERation) code developed at Catalpa Research, Inc.. It is a customized grid generation system that allows a user to generate turbomachinery grid for internal, external, or internal-external configurations. It allows both algebraic and elliptic grid generation approach. Its special surface smoothing algorithm provide a robust way to generate high-quality surface grid. A user can create a typical turbomachinery grid in a matter of hours, or even minutes, instead of days or even weeks..

Remarks : For more information, please refer to **http://www.catalpa.net**

Reference :

1. Shih, M.H., *"Towards A Comprehensive Computational Simulation System for Turbontachinery',* Dissertation, Mississippi State University, May, 1994.
2. Shih, M.H., Yu, T-Y and Soni, B.K.,*"TIGER: A Turbomachinery Grid Generation/ Simulation System*", AIAA32nd Aerospace Sciences Meeting, AIAA-94-0207, Reno, NV, January 1994.

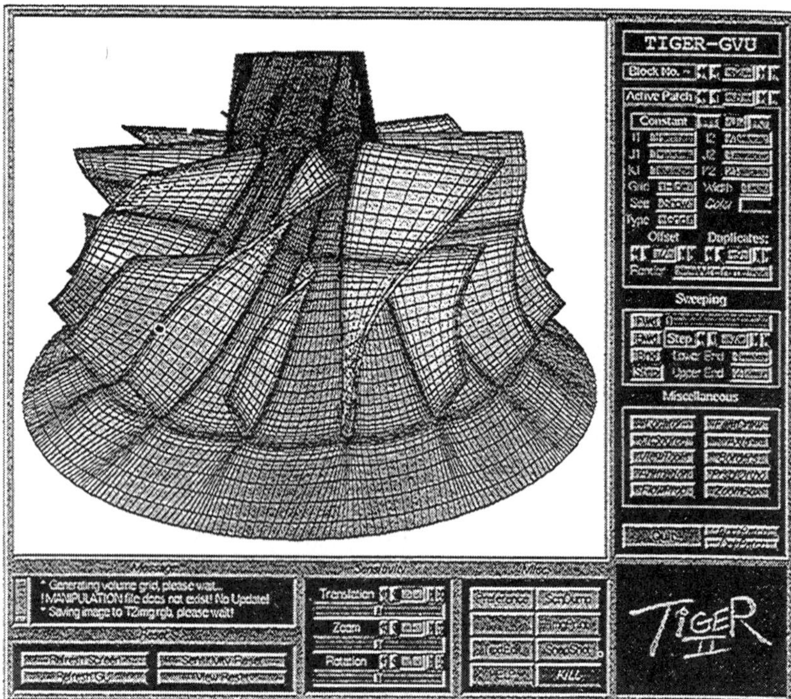

(Geometry Courtesy of ACD, Inc.)

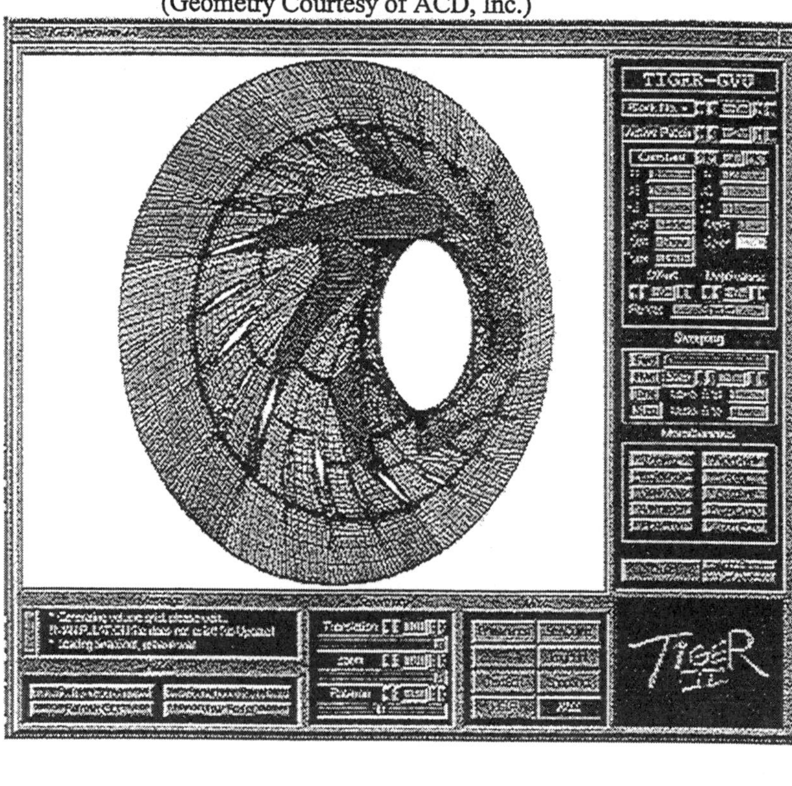

Title : Internal Combustion Engine Port Flow Configuration

Name : Dr. Ming-Hsin Shih

Affiliation : Engineering Sciences, Inc.
1900 Golf Road, Suite D
Huntsville, AL 35802
U. S. A.

Application : The flow field involving the port-valve configuration affects the breathing capacity of the internal combustion (IC) engine and thus its performance. It governs the flame propagation rate in spark-ignition engines and is directly related to the operating and emission characteristics of an IC engine. The objective of this ongoing work is to evaluate, validate and improve computational fluid dynamics (CFD) prediction methods for this important automotive application.

Description : This grid was created using PF_GRID developed at Engineering Sciences, Inc.. It is a customized automatic grid generation tool that decomposes the entire domain into 5 grid blocks using both Weighted Transfinite Interpolation (Algebraic) and Poisson Grid System (Elliptic). It maintains the structured grid topology for better computation efficiency. The geometry was specified in a short input file. With PF_GRID, a user can create a grid for a design in a matter of minutes.

Remarks : For more information, please refer to **http://www.esi-al.com/gallery/pf/**

Reference :

1. Chen, Y.S. and Shih, M. H., "CFD APPLICATIONS TO INTERNAL COMBUSTION ENGINE PORT-FLOW DESIGNS," FEDSM97-3024, 1997 ASME Fluids Engineering Division Summer Meeting, June 22 - 26, 1997, Vancouver, B. C., Canada.

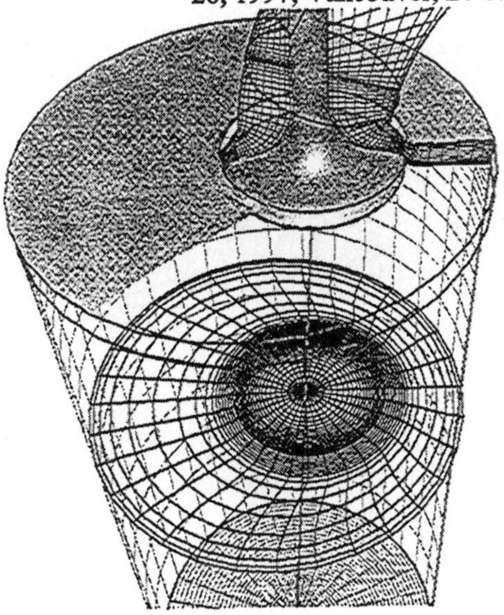

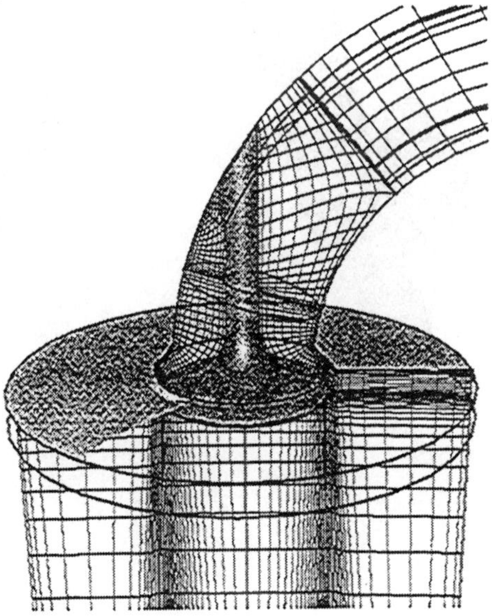

Moving Overlapped Grids for Direct Numerical Simulation of Particle-Fluid Interactions

Tom I-P. Shih
Department of Mechanical Engineering
Carnegie Mellon University

Many flows of practical interest involve particles (liquid droplets or solid particles) dispersed in a fluid (gas or liquid). Examples include sprays, particle-laden flows, and slurries. For these flows, it is important to understand the interactions that take place among the particles and between the particles and the fluid. Direct numerical simulation of such flows – in which the flow around each moving particle is resolved – will greatly increase our understanding of the relevant physics. The use of moving overlapping grids enables and greatly simplifies such a simulation. With this approach, an O-O grid is attached to and moves with each particle through an H-H grid for the flow domain. Figure 1 shows typical O-O grids generated about spherical particles. Figure 2 is a close-up view of the grid in the overlapped region where an O-O grid overlaps and cuts a hole in the H-H grid. The reader is referred to Refs. 1 to 3 for details.

Key References:

1. Dasgupta, A., Shih, T.I-P., Kundu, K.P., and Deur, J.M., "Numerical Simulation of Flow past an Array of Moving Spherical Particles," AIAA Paper 94-3283, June 1994.
2. Dasgupta, A. and Shih, T.I-P., "Oscillating Flow past Stationary Particles," AIAA Paper 95-0492, Jan. 1995. Accepted for publication by the AIAA Journal of Propulsion and Power.
3. Shih, T.I-P., "Direct Simulation of Particle-Fluid Interactions," invited paper, ASME Fluids Engineering Division Summer Meeting, July 1996.

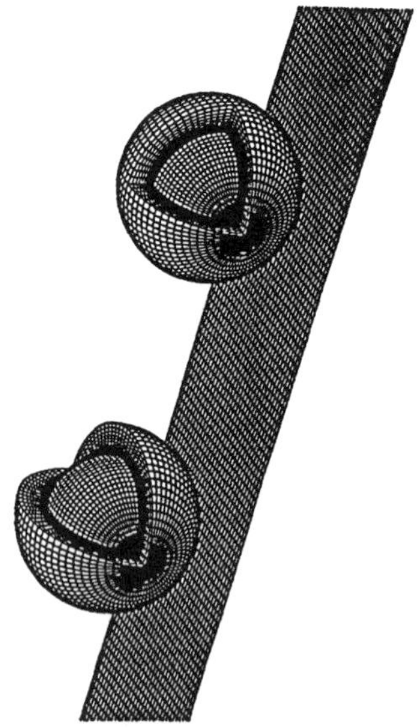

Fig. 1. Moving overlapping grids – one O-O grid for for each particle.

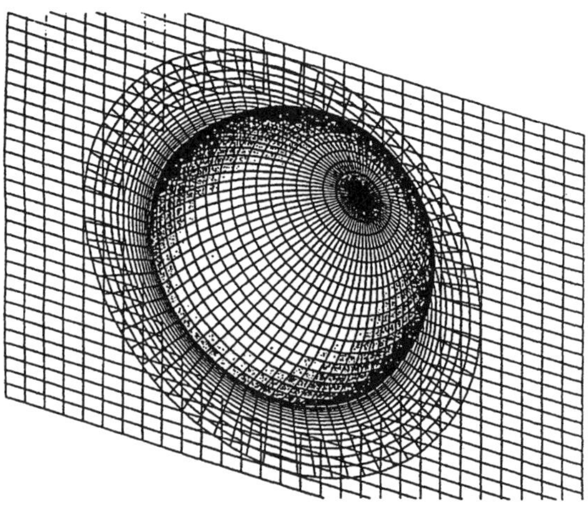

Fig. 2. Close-up view in overlapped region.

Overlapped/Patched Grids for Flows with Many Holes

Tom I-P. Shih
Department of Mechanical Engineering
Carnegie Mellon University

Many flows of practical importance involve holes in which the flow through each hole must be resolved. Examples include control of shock-wave/boundary-layer interactions by bleed and film cooling of turbine blades. For these flow geometries, the use of overlapped/patched structured grids greatly simplifies the generation of high-quality grids. This is because the flow domain, made complicated by the presence of holes, can be arbitrarily partitioned with overlaps before gridding (see Refs. 1 and 2). Figures 1 and 2 show grids generated for flow domains with holes. In Fig. 1, the grids for holes overlapped other grids for the rest of the flow domain (Ref. 3). Note that each hole is made up of an O-H grid to resolve the cylindrical geometry of the hole and an H-H grid to remove the centerline singularity associated with O-H grids. This grid was generated by using algebraic grid generation (Ref. 4). In Fig. 2, grids for holes are patched to other grids (Ref. 5). Figure 2 illustrates how the number of grid points needed to resolve flows issuing from holes can be reduced through grid embedding.

Key References:

1. Shih, T.I-P., Rimlinger, M.J., and Chyu, W.J., "Three-Dimensional Shock-Wave/Boundary-Layer Interactions with Bleed," AIAA Journal, Vol. 31, No. 10, 1993, pp. 1819-1826.
2. Stephens, M.A., Rimlinger, M.J., Shih, T.I-P., and Civinskas, K.C., "Chimera Grids in Computing Flow and Heat Transfer in Blade Cooling," AIAA Journal of Propulsion and Power, Vol. 11, No. 2, 1995, pp. 213-220.
3. Rimlinger, M.J., Shih, T.I-P., and Chyu, W.J., "Three-Dimensional Shock-Wave/Boundary-Layer Interactions with Bleed through Rows of Holes," AIAA Journal of Propulsion and Power, Vol. 12, No. 2, 1996, pp. 217-224.
4. Chyu, W.J., Rimlinger, M.J., and Shih, T.I-P., "A Procedure for Automating CFD Simulations of an Inlet-Bleed Problem," NASA Workshop on Surface Modelling, Grid Generation, and Related Issues in CFD Solutions," NASA CP-3291, NASA - Lewis Research Center, May 1995, pp. 731-749.
5. Lin, Y.-L., Stephens, M.A., and Shih, T.I-P., "Computation of Leading-Edge Film Cooling with Injection through Rows of Compound-Angled Holes," ASME Paper 97-GT-298, June 1997.

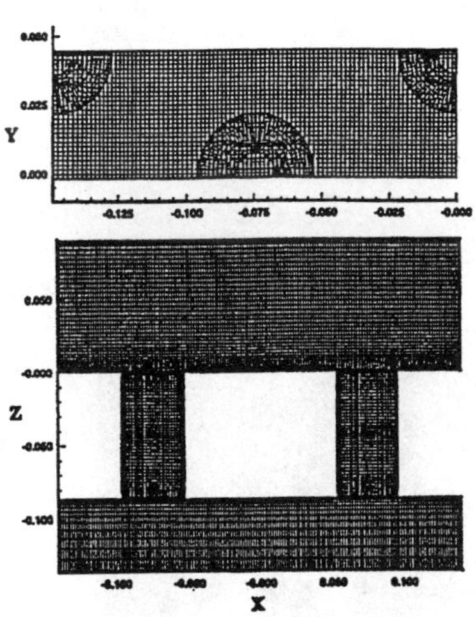

Fig. 1. Overlapped grids with holes.

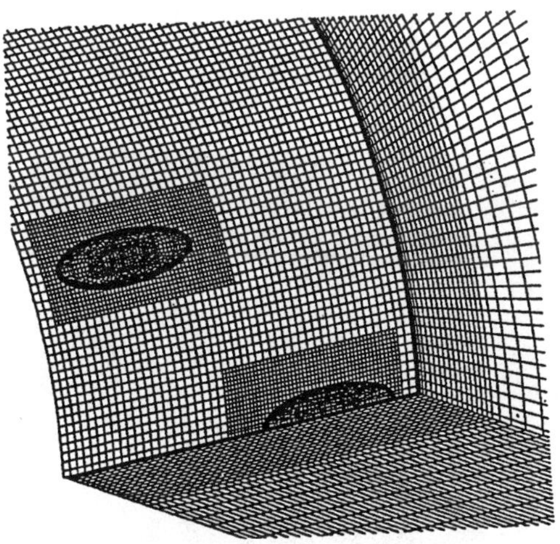

Fig. 2. Overlapped/patched grids with grid embedding.

Keiji Yano† and Shoichiro Nakamura††
†Department of Aerospace Engineering, Applied Mechanics, and Aviation
††Department of Mechanical Engineering
The Ohio State University, Columbus, OH, 43210, USA

Application: A grid system which consists of sixteen subdomains has been generated for analyzing the flow in an entire centrifugal heart pump using the parallel computer T3D [1].

Grid Description: A geometrical model for the IVAS centrifugal heart pump includes the inlet eye, rotating blades, a volute, and a diffuser. The domain decomposition partitions its complex geometry into a number of simple geometrical subdomains, in which a structured algebraic grid is generated. The grids for the inlet eye and blade regions are on the rotating frame, while grids for the volute and the diffuser are on the non-rotating frame. The edges of the rotating grids slide over the edges of the non-rotating grids. A linear interpolation is used to pass variables from the rotating subdomain grid to the non-rotating subdomain grid, or vice versa. When the two adjacent subdomains do not slide from each other, the grid points are fully aligned in two directions between the two adjacent subdomain grids so that no interpolation is necessary. The 2-1/2D (quasi three dimensions) unsteady incompressible Navier-Stokes equations have been solved on the grid system using parallel computing.

Reference: [1] K.Yano, S.Nakamura, Proc. 15th IMACS (1997).

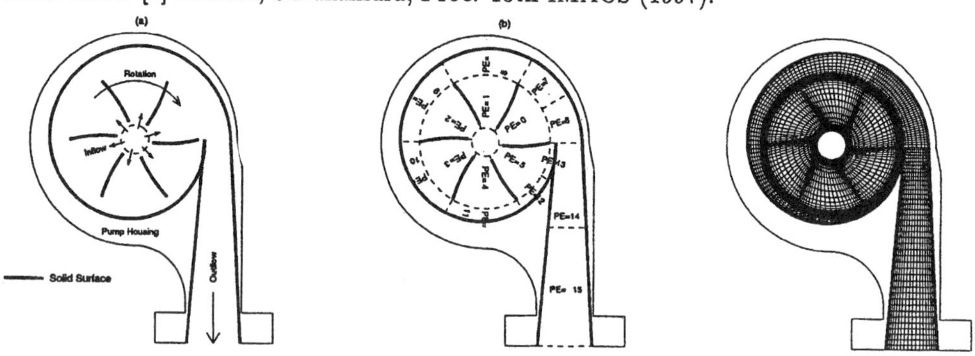

Figure 1: Domain decomposition of the IVAS pump. (a) Flow domain. (b) Subdomains, PE0∼PE5: a rotating frame, PE6∼PE15: a non-rotating frame. (c) 2D grids.

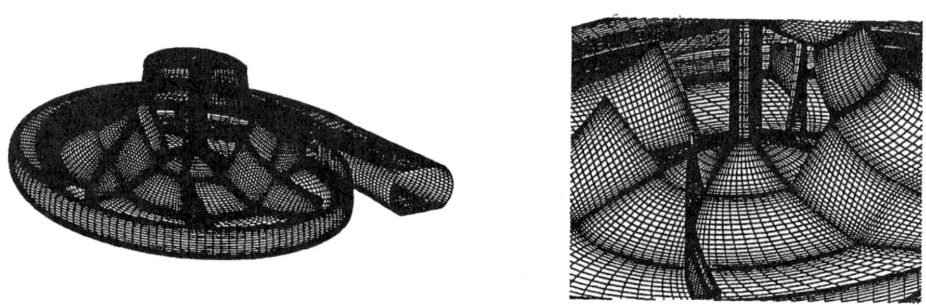

Figure 2: 3D grid for the IVAS pump.

A Structured Grid for Simulation of V-22 Aircraft Forward-Flight Flowfield

Tsze C. Tai
U.S. Naval Surface Warfare Center, Carderock Division
Bethesda, MD 20817

Objective

To simulate the flowfield about a complete V-22 tiltrotor aircraft configuration at forward flight by using multi-zone thin-layer Navier-Stokes method.

Grid Description

A curvilinear, body-conforming structure grid is generated based on the surface grid shown in Fig. 1. The cylindrical grid topology, which is basically made up of an H-O mixed type with H-type in the longitudinal plane and O-type in the crossflow plane, is adopted. The outer cylindrical surface is set at about 9.5 chord lengths from the aircraft centerline. The advantage of this grid topology is that it permits exacting patching of the grid between grid zones in a multi-zone solution domain. Unlike the Chemira scheme, the present approach avoids grid overlapping and solution interpolation at the interface. Thus not only it is faster in flow solving but also yields more accurate results than a Chemira grid.

Fig. 1 - Surface grid for V-22 aircraft in forward-flight mode

The V-22 geometry is intrinsically complex because of the nacelle mounted at the wing tip and the sponson protrusion at the bottom of the fuselage. To overcome the difficulty, a two-step grid generation procedure is employed. The first step performs the basic grid generation, and the second step is devoted to grid refinement. A total of 25 blocks were used in generating the basic grid that covers half of the aircraft (half of the solution domain). The basic grid is then refined if it contains intersecting grid lines. The partial grid containing intersecting grid lines is then taken out for regeneration. A six-sided hole in the basic grid is opened and the corresponding six faces of the hole are used as new boundaries for the elliptic generator. The overall grid is obtained by simply reflecting the existing grid about the y = 0 symmetric plane. The full-span model is needed for aircraft flying at yaw. Figure 2 shows a cross section of the grid that comprises the wing, nacelle, fuselage, and sponson. The final grid for wing-nacelle-fuselage configuration has a total number of 159 x 57 x 209 grid points. With the tail sections added, this number has increased to 179 x 57 x 245.

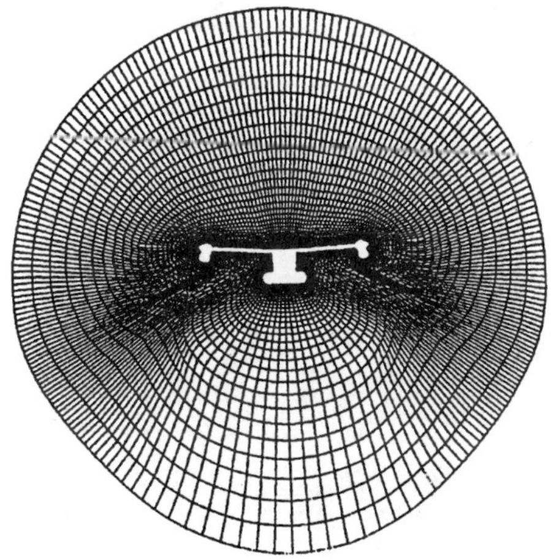

Fig. 2 - A cross section of the final volume grid

The NASA Ames 3DGRAPE code[1] was used for all basic and refinement grid generations. The resulting grid has been applied and briefly described in our previous work.[2,3]

References

[1] Sorenson, R.L., "The 3DGRAPE Book: Theory, Users' Manual, Examples," NASA TM 102224, July 1989.

[2] Tai, T.C., "Simulation and Analysis of V-22 Tiltrotor Aircraft Forward-Flight Flowfield," *Journal of Aircraft*, Vol. 33, No. 2, March-April 1996.

[3] Tai, T.C., "Simulation and Analysis of V-22 Aircraft Maneuvering Aerodynamics, AIAA Paper 96-2477, June 1996.

Index

A

D

E

V

W

X

Z